De Gruyter Textbook

Bhattacharyya · Distributions

Pulin Kumar Bhattacharyya

Distributions

Generalized Functions with Applications in Sobolev Spaces

De Gruyter

Mathematics Subject Classification 2010: 46FXX , 46F10, 46F12, 35E05, 46E35, 46XX, 46-01, 35-01, 35J40.

ISBN: 978-3-11-026927-7
e-ISBN: 978-3-11-026929-1

Library of Congress Cataloging-in-Publication Data

Bhattacharyya, Pulin K.
Distributions : generalized functions with applications in Sobolev spaces / by Pulin K. Bhattacharyya.
p. cm. – (De Gruyter textbook)
Includes bibliographical references and index.
ISBN 978-3-11-026927-7 (hardcover : alk. paper) –
ISBN 978-3-11-026929-1 (e-book) 1. Theory of distributions (Functional analysis)–Textbooks. 2. Sobolev spaces–Textbooks. I. Title.
QA324.B46 2012
515′.782–dc23
2011042975

Bibliographic information published by the Deutsche Nationalbibliothek

The Deutsche Nationalbibliothek lists this publication in the Deutsche Nationalbibliografie; detailed bibliographic data are available in the internet at http://dnb.dnb.de.

Typesetting: Da-TeX Gerd Blumenstein, Leipzig, www.da-tex.de
Printing and binding: Hubert & Co. GmbH & Co. KG, Göttingen
∞ Printed on acid-free paper
Printed in Germany

www.degruyter.com

To my daughter
Marina (*Tonushree*)

Preface

The term *distribution*[1] was introduced by the celebrated French mathematician Laurent Schwartz in his *Theory of Distributions* (*Théorie des distributions*), which was developed in the late 1940s to denote new mathematical objects such as the popularly (though incorrectly) named Dirac delta function $\delta(x)$ and its derivatives $\delta^{(k)}(x)$, and idealized concepts such as the density of mass/charge at a point, the magnitude of instantaneous force applied at a point, the excitation caused by an instantaneous source of heat placed at a point, etc. The definitions of these new mathematical objects, which do not have any point-values and cannot be represented by the usual functions having point-values, were intended to represent some kind of physical distribution or spread of mass, charge, force, etc. over an interval on $\mathbb{R}$, an area in $\mathbb{R}^2$, a volume in $\mathbb{R}^3$, etc. *Theory of Distributions* provides rigorous mathematical foundations for these new mathematical objects and also generalizes in some sense the notion of functions from classical analysis which have point-values. Hence, (Schwartz) distributions are also called *generalized functions*—specifically by the Russian school of Gelfand–Schilov–Vilenkin–Graev (see [1]) and also by many other non-Russian mathematicians—though Courant [2] prefers to call distributions *ideal functions*. In our treatment, although distributions, generalized functions and ideal functions are all synonyms, we find the term *distributions* more appropriate and more exact from the physical point of view, and will therefore use it in the remainder of the present book (including the title).

Owing to the nice properties of distributions, *Theory of Distributions* found early favour with physicists, and has had a profound influence on the development of topological vector spaces, nuclear spaces, etc., becoming an integral part of modern functional analysis. But its impact on mathematical physics is the most profound. Consequently, we have considered Sobolev spaces as the most important application in general, being essential tools for boundary value problems of elliptic partial differential equations. There are several very good books on functional analysis (for example, Rudin [3] and Yoshida [4]), on partial differential equations (for example, Hörmander [5]) and on mathematical physics (for example, Vladimirov [6]), which contain a chapter or two or more on distributions, but the prohibitively brief and concise treatment of the topics, possibly combined with the necessarily high mathematical level of their presentation, make them not easily understandable for applied scientists. The solitary exception to this is [7] by Laurent Schwartz himself (the 1966 translation of the original, *Mathématiques pour les sciences physiques*), which contains an excep-

[1] Distributions must *not* be confused with *probability* or *statistical* distributions, since these are completely different objects (see also the last part of Section 1.11).

tionally well-written chapter on distributions giving almost all the basic results. It does not, however, contain vector-valued distributions, and Sobolev spaces are not even mentioned, possibly due to historical reasons. Almost all good books on distributions/generalized functions were published over four decades ago, were written by mathematicians for mathematicians and were primarily addressed at pure mathematicians to present distributions as new mathematical objects. The best of these is probably the monograph [8], *Théorie des distributions* by Laurent Schwartz himself, which has only ever been published in French, the other one being the monumental work of Gelfand–Schilov–Vilenkin–Graev in five volumes, *Generalized Functions* [1], which contains diverse applications of distributions to different branches of higher mathematics. Two additional important books are *Topological Vector Spaces and Distributions* by J. Horvath [9] and *Topological Vector Spaces, Distributions and Kernels* by F. Trèves [10], which give further developments of the theory of distributions, dealing particularly with linear topological vector space aspects, and are therefore suitable for researchers in the theory of distributions. *Linear Partial Differential Equations* by L. Hörmander [5] gives almost all the basic results on distributions. This is probably the most elegant, concise presentation of the results of the theory of distributions, but also probably in the most difficult style for an applied scientist.

Generalized Functions and Partial Differential Equations by A. Friedman [11] is specially oriented to the study of partial differential equations in the distributional sense and a highly specialized book. There are also many interesting books which discuss the theory of distributions but which are not available in English, and thus will not be discussed here.

Although *Sobolev Spaces* by R. A. Adams [12] is possibly one of the best reference books on Sobolev spaces, specifically for imbedding results, the books of Lions [13], [14]; Lions and Magenes [15]; Nečas [16]; Grisvard [17], [18], [19]; etc. contain more interesting results on Sobolev spaces for application to boundary value problems of partial differential equations.

The treatment of the topics in all of these books is far beyond the reach of the average reader belonging to the large community of applied mathematicians, physicists and engineers. But a book giving a rigorous treatment of distributions and their applications in a simple style and form such that the proofs and results are understandable to the applied community of readers is very much in demand. To our knowledge, such a book dedicated solely to distributions and their application primarily to Sobolev spaces is conspicuous by its absence. Hence, the rationale for writing the present book is to fill this gap, and the scope of the book has been increased by including some additional topics and innumerable examples of different applications in order to widen the readership circle. This book therefore differs from all the good books mentioned above in the style, form and content of the presentation of the theory of distributions, and is addressed in principle to the large community of applied mathematicians, engineers, physicists, etc. In general, it follows the principles of presentation of concepts with proper motivations, the gradual development of concepts with suitable examples and counterexamples and, finally, identifying and listing all the important properties and results so that an applied scientist or engineer can easily apply these results.

- For easy reading, Appendices A–D have been added.
- Innumerable examples are given with all the intermediate steps and explanations which are not usually given in the advanced treatises.
- Proofs also include all the intermediate steps and necessary explanations to make them easily understandable.
- Practical applications are given, such as the physical interpretation of the duality principle, discussions on physical versus mathematical distributions and the application of convolution of distributions to the R-L-C circuit in electrical engineering and in the heat flow problem in a rod.
- Distributional derivatives of discontinuous piecewise smooth functions of several variables have been dealt with in all details together with their application in the construction of *finite element spaces*, a new concept which will be extremely useful in understanding the mathematical foundation of finite element methods for boundary value problems.
- Different methods of construction of elementary solutions of linear differential operators with constant coefficients have been given with a lot of details, which will be useful in boundary integral methods and boundary element methods.
- Convolution matrices, determinants and the convolution system of equations etc. in Schwartz's convolution algebra $\mathcal{A}$ [7] have been dealt with in detail.
- Unusually for a book on distributions, systematic treatment has been given to Fourier transforms of tempered distributions and their applications to Sobolev spaces $H^s(\mathbb{R}^n)$ of arbitrary order $s \in \mathbb{R}$; the nice properties of Sobolev spaces on $\mathbb{R}^n$ and also the problems of extension of these properties to domains $\Omega \neq \mathbb{R}^n$; compactness results in Sobolev spaces; Sobolev's imbedding results; Sobolev spaces on manifolds Γ, which are boundaries of a domain Ω in $\mathbb{R}^n$; trace theorems for Sobolev spaces on $\mathbb{R}^n$, $\mathbb{R}^n_+$, C^m-regular domains Ω and polygonal domains in $\mathbb{R}^2$; etc.
- Almost all the basic results of the theory of distributions are contained in this book. It can therefore be read as an introduction to advanced treatises on distributions such as, for example, *Théorie des distributions* by Laurent Schwartz.
- Finally, the present book is written *by* an applied scientist, meant *for* the applied community and will serve as a reference-cum-text book *of* that same applied community.

The present book grew out of a course taught in the Department of Mathematics, Indian Institute of Technology, Delhi, which was tailored to the needs of the applied community of mathematicians, engineers, physicists, etc. who were interested in studying the problems of mathematical physics in general and their approximate solutions on computer in particular. Although this book contains almost all the topics which will be essential for the study of Sobolev spaces and their application to elliptic boundary value problems and their finite element approximations, many additional topics of interest have been included, along with many interesting examples, for specific applied disciplines and engineering: elementary solutions, derivatives of discontinuous functions of several variables, delta-convergent sequences of functions, Fourier series of distributions, convolution system of equations, etc. Moreover, the topics have been presented in such a manner that the reader may concentrate on topics of his or her interest, omitting others. While teaching engineers and others, the author found that even mathematically alert students of applied disciplines found extreme difficulty with

1. brief presentations, without sufficient explanation and motivation;
2. omission of the intermediate steps of involved computations in some problems and of justifications, however trivial these might be;
3. mathematical notations, which can instil fear or distaste when not judiciously chosen.

The author has therefore addressed all these problems with sufficient care and due respect so that readers from applied disciplines and engineering should find that the theory of distributions and their applications are within their reach and understandable for subsequent successful and active application in the study of boundary value problems and their approximate solutions on computers. The book can be used either as a reference book or as a text book for different courses as shown separately later.

Of all the books on distributions in English, French, German and Russian to which I have had access, the monographs *Théorie des distributions* and *Mathematics for Physical Sciences* helped me most in understanding various aspects of the theory of distributions, and their profound influence on me is reflected throughout the present book. For this I express my deepest sense of gratitude and indebtedness to the celebrated author of these monographs, Professor Laurent Schwartz (1915–2002).

With very respectful sentiments of gratitude and indebtedness and *très bons souvenirs de l'époque de sa grandeur*, I recall the distinguished French mathematician Professor Jacques-Louis Lions (1928–2001), who kindly gave me all help, encouragement and opportunity to do research in shell analysis during 1976–77 in the wonderful research environment created by him in the Institut National de Recherche en Informatique et en Automatique (INRIA), France (called IRIA, France at that time). During this stay of mine at INRIA I was exposed to the great achievements of the new French school of research in applied mathematics developed by Professor Lions, which, in fact, inspired me and influenced my research and academic life in a definitive manner. For this I owe a lot to Professor Jacques-Louis Lions and express my grateful indebtedness to him.

In addition to my feelings of gratitude and indebtedness to Prof. L. Schwartz mentioned above, I would like to acknowledge further that almost all the basic results, concepts, theorems, etc. on distributions presented in this book belong uniquely to Professor L. Schwartz, although his name is not specifically mentioned in each case in the book.

In spite of my best efforts to make the manuscript free from mistakes, some may still remain, having defied all rigid checks and correction operations. For this, sincere regret and apology are expressed by me. I further request sympathetic readers to send their criticisms of the book and suggestions for improvement to me at `pulinkb@yahoo.com` or `pulinkbh@rediffmail.com`, which will be thankfully acknowledged in future editions.

Finally, all my efforts will only be fruitful if the readers are benefited and find the book readable and interesting. *Bon courage* to all readers!

New Delhi, September 2011 P. K. Bhattacharyya

Contents

How to use this book in courses

This book can be used either as a reference book or as a text book for many specialized advanced courses, as shown below.

Reference book

As the book contains almost all the basic results on distributions (generalized functions) and Sobolev spaces, and also many other applications, it can be used as a reference book by applied mathematicians, functional analysts, physicists, engineers and also by Ph.D. scholars and postdoctoral fellows in computational mathematics, mechanics and engineering disciplines.

Text book

It can also be used as a text book for self study or for different courses at the M.S./ M.Sc., M.Phil., M.Tech. and Ph.D. levels. For example:

Theory of Distributions or Generalized Functions

Syllabus for a two-semester course

Chapters 1–9 (omitting some sections if necessary, particularly those containing specialized applications).

Introduction to Theory of Distributions or Generalized Functions

Syllabus for a one-semester course

Chapter 1, Sections 1.1–1.3, 1.5, 1.8, 1.11; Chapter 2, Sections 2.1, 2.2, 2.5, 2.8, 2.9, 2.12; Chapter 3, Sections 3.1.1, 3.1.2, 3.3; Chapter 4, Sections 4.1, 4.2; Chapter 5, Sections 5.1–5.7; Chapter 6, Sections 6.3–6.7; Chapter 7, Sections 7.2, 7.6, 7.7; Chapter 8, Sections 8.1–8.7.

Sobolev Spaces with Distributions

Syllabus for a two-semester course

Chapter 1, Sections 1.2–1.4, 1.8; Chapter 2, Sections 2.1–2.3, 2.5, 2.8, 2.9, 2.12, 2.13; Chapter 3, Sections 3.1.1–3.1.3, 3.4; Chapter 4, Sections 4.2, 4.3; Chapter 5, Sections 5.1–5.3, 5.5, 5.6; Chapter 6, Sections 6.2–6.4, 6.8; Chapter 7, Sections 7.2, 7.6, 7.7; Chapter 8, Sections 8.2–8.6, 8.9–8.14; Chapter 9, Sections 9.4–9.6; Appendix D: Boundary Geometry.

Sobolev Spaces

Syllabus for a one-semester course

Chapter 1, Sections 1.1–1.3, 1.5, 1.8; Chapter 2, Sections 2.1, 2.2, 2.5, 2.13; Chapter 3, Section 3.4; Chapter 4, Sections 4.2, 4.3; Chapter 5, Sections 5.2, 5.5; Chapter 6, Sections 6.3, 6.4, 6.8; Chapter 7, Sections 7.2, 7.6; Chapter 8, Sections 8.2, 8.3, 8.5, 8.6, 8.9–8.14; Appendix D: Boundary Geometry. Some sections and some special results may be omitted without disturbing the sequence of topics.

Fourier Series, Fourier and Laplace Transforms

Syllabus for a half-semester course

Chapter 1, Sections 1.2, 1.3, 1.8, 1.10; Chapter 2, Sections 2.1–2.3, 2.5, 2.9, 2.11; Chapter 5, Sections 5.5, 5.6; Chapter 6, Sections 6.2, 6.3, 6.7; Chapter 7, Sections 7.1–7.3, 7.6, 7.7; Chapter 8, Sections 8.1–8.8.

Dirac Distributions and their Properties

Syllabus for a half-semester course

Chapter 1, Sections 1.1–1.3, 1.6–1.8, 1.10, 1.11; Chapter 2, Sections 2.1–2.3, 2.5, 2.6, 2.10, 2.11; Chapter 3, Sections 3.1–3.3; Chapter 5, Sections 5.5–5.7; Chapter 6, Sections 6.3, 6.4, 6.7; Chapter 7, Sections 7.2, 7.7; Chapter 8, Sections 8.2, 8.8.

Elementary Solutions for Boundary Integral Equations

Syllabus for a half-semester course

Chapter 1, Sections 1.1–1.3, 1.5, 1.8; Chapter 2, Sections 2.1–2.3, 2.5, 2.14; Chapter 3, Sections 3.2, 3.3; Chapter 5, Sections 5.5–5.7; Chapter 6, Sections 6.3, 6.4, 6.9; Chapter 7, Section 7.2; Chapter 8, Sections 8.2–8.7.

Additional topics may be included or some sections may be deleted in any of these courses.

Other courses are possible, for example:

- Distributions and Differential Equations;
- Introduction to Vector-Valued Distributions for Evolution Equations of Parabolic and Hyperbolic Types.

A few chapters of this book have been class tested in a short course on distributions entitled *Selected Topics* for M.Tech. students and Ph.D. research scholars in the Electrical Engineering Dept., I.I.T., Delhi, in 2006–2008. The responses from the participants of the course were quite encouraging for the author of the book.

Acknowledgment

The author expresses his heartiest thanks to Prof. Neela Nataraj of I.I.T., Mumbai, Prof. Kallol Ghosh of Jadavpur University, Calcutta and Prof. B. N. Mandal of I.S.I., Calcutta for their assistance in the preparation of the manuscript in LaTeX.

Dr. Subhashish Roy Chaudhury of Jadavpur University and Mr. Hariharan, Mr. Anoop Nair, Mr. Girish, Mr. Himanshu Tyagi, Tarun Vir Singh, Mr. Sarat and Mr. Srikamal of I.I.T., Delhi helped in typing the manuscript in LaTeX, for which the author expresses grateful thanks to all of them.

For almost all the figures, graphs and tables, which were primarily drawn in the U.S. by Mr. Ashish Das, a former student of I.I.T., Delhi, and his energetic wife, Anindita, the author expresses special thanks to the couple.

Mr. Anoop Nair, Mr. Girish, Mr. Srikamal, Mr. Srihari, Mr. Sarat and Mr. Mohit Garg took the responsibility of preparing the final version of the manuscript in the VLSI Laboratory, I.I.T., Delhi, for which the author thanks them again, along with their supervisors, Prof. G. S. Visweswaran and Prof. Jayadeva, and Prof. B. Bhaumik for all their help. The author thanks Prof. S. C. Dutta Roy for his written comments on the application of convolution to R-L-C circuit analysis in Chapter 6.

The author thanks his former students Prof. S. Balasundaram and Dr. S. Gopalsamy for taking an interest in the publication of the book. In particular, the author is grateful to Dr. Gopalsamy for taking the trouble to read and correct a few chapters.

With feelings of gratitude, the author recalls the encouragement of many colleagues from I.I.T., Delhi; J.N.U., New Delhi; Jadavpur University, Calcutta, where the author taught in recent years. In particular, Prof. Suresh Chandra, Prof. B. R. Handa, Dr. W. Shukla, Mr. A. Nagabhusanam of I.I.T., Delhi, Prof. Karmeshu of J.N.U., New Delhi and Prof. A. K. Pani of I.I.T., Mumbai are thanked for their interest in the publication of the book.

The author expresses grateful thanks to Prof. Olivier Pironneau of Pierre and Marie Curie University, Paris, Prof. Michel Bernadou of Leonard de Vinci University, La Defense, France and Prof. Maurice A. Jaswon of City University, London for their active interest in the book.

The author thanks the Dept. of Electrical Engineering, I.I.T., Delhi, for providing the facilities for putting some chapters of the present book to class test at a highest level departmental course, and also thanks Dept. of Mathematics, I.I.T., Delhi, for giving the opportunity to start the teaching of distributions about three decades ago. Prof. Balasundaram took up the extensively laborious job of preparing the errata of all Chapters and Appendices of the book in typed form. Without his sincere and encouragement I could not have completed the whole task of correction of the galley-proof by this time. For all these, I express my most grateful thanks to Prof. S. Balasundaram.

Finally, the author thanks De Gruyter for accepting my book for publication in their text book series and also their editorial and publishing division and in particular, Madame Friederike Dittberner, Madame Anja Möbius, Madame Ulrike Swientek and Mr. Christoph von Friedeburg for their excellent professional cooperation and gracious help on all occasions.

Notation

This section summarizes the notation used within this book. Where page numbers are given, these are either the page of the first occurrence of the notation or the page of its definition in Appendices A–D.

Logical symbols

$\exists x$	there exists x
$\forall x$	for every x
$\Longrightarrow$	Logical implication: $(P) \overset{\text{implies}}{\Longrightarrow} (Q)$
$\not\Longrightarrow$	Logical non-implication: $(P) \overset{\text{does not imply}}{\not\Longrightarrow} (Q)$

Set notations

$a \in A$	a is an element of the set A
$\emptyset$	empty set
$A \subset B$	A is a subset of the set B
$A \subset\subset B$	A is a *compact* subset of B
$A^{\complement} \equiv \complement A$	complement of the set A
$B \setminus A$	complement of A in B
$A \cup B$	the union of the sets A and B
$A \cap B$	the intersection of the sets A and B
$A \times B$	Cartesian product of the sets A and B
A^n	$\underbrace{A \times A \times \cdots \times A}_{n \text{ times}}$
$A \pm B$	(p. 303)

Number systems

$\mathbb{N}$	$\{1, 2, 3, \ldots, n, \ldots\}$; the set of all natural numbers
$\mathbb{N}_0$	$\mathbb{N} \cup \{0\} = \{0, 1, 2, \ldots\}$
$\mathbb{Z}$	$\{0, \pm 1, \pm 2 \ldots, \pm n, \ldots\}$
$\mathbb{Q}$	the set of all rational numbers
$\mathbb{R}$	the set of all real numbers $\equiv]-\infty, \infty[$
$\mathbb{R}_+$	$]0, \infty[$
$\mathbb{R}_-$	$]-\infty, 0[$
$\mathbb{R}^n_+$, $\mathbb{R}^n_-$	(p. 553)
$\mathbb{C}$	the set of all complex numbers (p. 5)
$\mathbb{F}$	number field $\mathbb{R}$ or $\mathbb{C}$

$\mathbb{R}^n$	$\underbrace{\mathbb{R}\times\mathbb{R}\times\cdots\times\mathbb{R}}_{n \text{ times}}$
$\mathbb{C}^n$	$\underbrace{\mathbb{C}\times\mathbb{C}\times\cdots\times\mathbb{C}}_{n \text{ times}}$

Multi-index notations

$\boldsymbol{\alpha}$	$(\alpha_1, \alpha_2, \dots, \alpha_n)$, $\alpha_i \in \mathbb{N}_0$ (p. 5)
$\lvert\boldsymbol{\alpha}\rvert$	$\alpha_1 + \alpha_2 + \cdots + \alpha_n$, $\alpha_i \in \mathbb{N}_0$ (p. 5)
$\mathbf{x}$	$(x_1, x_2, \dots, x_n) \in \mathbb{R}^n$ (p. 5)
$\boldsymbol{\xi}$	$(\xi_1, \xi_2, \dots, \xi_n) \in \mathbb{R}^n$
$\mathbf{x}^{\boldsymbol{\alpha}}$	$x_1^{\alpha_1} x_2^{\alpha_2} \cdots x_n^{\alpha_n}$
$\boldsymbol{\xi}^{\boldsymbol{\alpha}}$	$\xi_1^{\alpha_1} \xi_2^{\alpha_2} \cdots \xi_n^{\alpha_n}$
$f(\mathbf{x})$	$f(x_1, x_2, \dots, x_n)$
$\partial^{\boldsymbol{\alpha}} \equiv D^{\boldsymbol{\alpha}} f$	$\frac{\partial^{\lvert\boldsymbol{\alpha}\rvert} f}{\partial x_1^{\alpha_1} \partial x_2^{\alpha_2} \dots \partial x_n^{\alpha_n}}$
$\partial_i^{\alpha_i} f \equiv D_i^{\alpha_i} f$	$\frac{\partial^{\alpha_i} f}{\partial x_i^{\alpha_i}}$ (p. 5)
$\boldsymbol{\alpha} \le \boldsymbol{\beta}$	$\boldsymbol{\alpha} \le \boldsymbol{\beta} \iff \alpha_i \le \beta_i$ for $1 \le i \le n$ (p. 5)
$\boldsymbol{\alpha}!$	$\alpha_1! \alpha_2! \cdots \alpha_n!$
$\boldsymbol{\alpha} - \boldsymbol{\beta}$	$(\alpha_1 - \beta_1, \alpha_2 - \beta_2, \dots, \alpha_n - \beta_n)$ with $\alpha_i - \beta_i \ge 0$ (p. 5)

Notations used for properties in $\mathbb{R}^n$

$\mathbf{x}$	$(x_1, \dots, x_n) \in \mathbb{R}^n$, an ordered n-tuple of real numbers x_i, $1 \le i \le n$
$\mathbf{x} + \mathbf{y}$	$(x_1 + y_1, \dots, x_n + y_n) \in \mathbb{R}^n$
$\alpha\mathbf{x}$	$(\alpha x_1, \dots, \alpha x_n) \in \mathbb{R}^n$
$\mathbf{0}$	$(0, 0, \dots, 0) \in \mathbb{R}^n$
Ω	open subset of $\mathbb{R}^n$
$\overline{\Omega}$	closure of Ω in $\mathbb{R}^n$
$d_2(\mathbf{x}, \mathbf{y}) = d(\mathbf{x}, \mathbf{y})$	the usual (Euclidean) metric in $\mathbb{R}^n$
$d_1(\mathbf{x}, \mathbf{y})$	$\sum_{i=1}^n \lvert x_i - y_i \rvert$
$d_\infty(\mathbf{x}, \mathbf{y})$	$\max_{1\le i\le n}\{\lvert x_i - y_i\rvert\}$: other equivalent metrics in $\mathbb{R}^n$
$\lVert\mathbf{x}\rVert = \lVert\mathbf{x}\rVert_2$	the usual (Euclidean) norm in $\mathbb{R}^n$ with $\lVert\mathbf{x}\rVert^2 = x_1^2 + x_2^2 + \cdots + x_n^2$
$\lVert\mathbf{x}\rVert_1$	$\sum_{i=1}^n \lvert x_i \rvert$
$\lVert\mathbf{x}\rVert_\infty$	$\max_{1\le i\le n}\{\lvert x_i\rvert\}$: other equivalent norms in $\mathbb{R}^n$
$B(\mathbf{0}; \varepsilon)$	open ball, $\{\mathbf{x}; \mathbf{x} \in \mathbb{R}^n, \lVert\mathbf{x}\rVert < \varepsilon, \varepsilon > 0\}$
$\overline{B}(\mathbf{0}; \varepsilon)$	closed ball: closure of $B(\mathbf{0}; \varepsilon)$ in $\mathbb{R}^n$ (p. 9)
$S(\mathbf{0}; \varepsilon)$	sphere, $\{\mathbf{x}; \mathbf{x} \in \mathbb{R}^n, \lVert\mathbf{x}\rVert = \varepsilon\}$ in $\mathbb{R}^n$
$\langle\mathbf{x}, \mathbf{y}\rangle_{\mathbb{R}^n}$	inner product of $\mathbf{x}, \mathbf{y} \in \mathbb{R}^n = \sum_{i=1}^n x_i y_i$

$\|\mathbf{x}\|$	$\langle \mathbf{x}, \mathbf{x} \rangle_{\mathbb{R}^n}^{1/2}$
$\mathbf{x} \perp \mathbf{y}$	$\mathbf{x} \perp \mathbf{y} \Longleftrightarrow \mathbf{x} \in \mathbb{R}^n$ is orthogonal to $\mathbf{y} \in \mathbb{R}^n$

Mappings

$f : X \longrightarrow Y$	mapping f from set X into set Y
$f : \Omega \subset \mathbb{R}^n \longrightarrow \mathbb{R}$	real-valued functions f from Ω into $\mathbb{R}$
$f : \Omega \subset \mathbb{R}^n \longrightarrow \mathbb{C}$	complex-valued function f from Ω into $\mathbb{C}$
$J : V \longrightarrow \mathbb{R}$	functional J from vector space V into $\mathbb{R}$
$A : V \longrightarrow W$	operator A from vector space V into vector space W
$L : V \longrightarrow \mathbb{R}$	linear functional from V into $\mathbb{R}$
$\hookrightarrow: X \longrightarrow Y$	continuous imbedding operator
$\hookrightarrow\hookrightarrow: X \longrightarrow Y$	compact imbedding operator

Notations for usual derivatives in the point-wise sense

$[\frac{dH}{dx}(x)]$, $[\frac{dH}{dx}](x)$, $H'(x)$	(p. 107)
$[\frac{\partial f}{\partial x_i}(\mathbf{x})]$, $[\frac{\partial f}{\partial x_i}](\mathbf{x})$	partial derivative of f with respect to x_i at the point $\mathbf{x}$ in the usual point-wise sense (pp. 119, 212)
J_0, J_k, J_{kl}	jump of f, $[\frac{\partial f}{\partial x_k}(\mathbf{x})]$,$[\frac{\partial^2 f}{\partial x_k \partial x_l}(\mathbf{x})]$ across Γ_0 (p. 213)
$[\Delta f](\mathbf{x})$	Laplacian of f in the usual point-wise sense (p. 223)

Notations for distributional derivatives

$\frac{\partial T}{\partial x_i}$, $\partial^{\boldsymbol{\alpha}} T$	$\frac{\partial^{\|\alpha\|} T}{\partial x_1^{\alpha_1} \ldots \partial x_n^{\alpha_n}}$ (pp. 100–101)
$\partial^{\boldsymbol{\alpha}} f$	$\frac{\partial^{\|\alpha\|} f}{\partial x_1^{\alpha_1} \ldots \partial x_n^{\alpha_n}}$ (p. 102)
$\frac{dH}{dx}$	(p. 105)
$\Delta T_f \equiv \Delta f$	Laplacian of f in the distributional sense (p. 223)

Multiple integrals

$\int_\Omega f(\mathbf{x}) d\mathbf{x}$	$\int_\Omega \cdots \int f(x_1, \ldots, x_n)\, dx_1 \ldots dx_n$ with $d\mathbf{x} = dx_1 dx_2 \ldots dx_n$ (p. 21)

Linear spaces used for distributions on $\boldsymbol{\Omega} \subset \mathbb{R}^n$

$\overline{\Omega}$	closure of Ω in $\mathbb{R}^n$, for Ω with boundary Γ: $\overline{\Omega} \equiv \Omega \cup \Gamma$
$\mathcal{F}(\Omega)$	space of real-valued (resp. complex-valued) functions (p. 738)
$C^k(\Omega), C^\infty(\Omega)$	(p. 7)
$C_0^0(\Omega)$	$C_0(\Omega)$ (p. 7)
$C_0^m(\Omega)$	(p. 13)
$C_0^\infty(\Omega)$	(p. 7)

$C^m(\overline{\Omega})$	(p. 20)
$C^{k,\lambda}(\overline{\Omega})$	(p. 754)
$L^p(\Omega), 1 \leq p \leq \infty$	(p. 788)
$L^1(\Omega)$	(p. 780)
$L^2(\Omega)$	(p. 797)
$L^1_{\text{loc}}(\Omega)$	(p. 16)
$\mathcal{D}(\Omega)$	$C_0^\infty(\Omega)$ (p. 9)
$\mathcal{D}^m(\Omega)$	(p. 13)
$\mathcal{D}_K(\Omega)$	(p. 13)
$\mathcal{D}^*(\Omega)$	algebraic dual space of $\mathcal{D}(\Omega)$ (p. 25)
$\mathcal{D}'(\Omega)$	space of distributions on Ω = algebraic and *topological* dual of $\mathcal{D}(\Omega)$ (p. 26)
$\mathcal{D}''(\Omega)$	$(\mathcal{D}'(\Omega))'$ = the second dual space of $\mathcal{D}(\Omega)$ (p. 263)
$\mathcal{E}(\mathbb{R}^n)$	$C^\infty(\mathbb{R}^n)$ (p. 287)
$\mathcal{E}'(\mathbb{R}^n)$	dual space of $\mathcal{E}(\mathbb{R}^n)$ (p. 288)

Linear spaces used for distributions on circle Γ

$\mathcal{D}(\Gamma) = C^\infty(\Gamma)$	test space of C^∞-functions on Γ (p. 77)
$\mathcal{D}'(\Gamma)$	space of distributions on Γ = dual space of $\mathcal{D}(\Gamma)$ (p. 78)
$\mathcal{D}_T(\mathbb{R})$	space of periodic functions on $\mathbb{R}$ with period $T > 0$ (p. 78)
$\mathcal{D}'_T(\mathbb{R})$	dual space of $\mathcal{D}_T(\mathbb{R})$ (p. 78)

Linear spaces used in tempered distributions on $\mathbb{R}^n$

$S(\mathbb{R}^n)$	(p. 407)
$S'(\mathbb{R}^n)$	(p. 426)
$\theta_M(\mathbb{R}^n)$	(p. 456)
$\theta'_C(\mathbb{R}^n)$	(p. 457)
$S(\Gamma)$	$S(\mathbb{R}^{n-1})$ (p. 527)
$\mathcal{D}'^+$	$\mathcal{D}'^+(\mathbb{R})$ (p. 492)

Linear spaces used in vector-valued functions/distributions

$]0, T[$	time interval with $T > 0$
$C^0(\overline{\Omega} \times [0, T])$	(p. 712)
$C^0([0, T]; V)$	(p. 715)
$C^k([0, T]; V)$	(p. 715)
$C^k(]0, T[; V)$	
$C^\infty(]0, T[; V)$	
$\mathcal{D}(]0, T[; V)$	$C_0^\infty(]0, T[; V)$ (p. 715)

$L^2(\Omega \times]0,T[)$	$(0 < T < +\infty)$ (p. 714)
$L^p(0,T;V)$	$L^p(]0,T[;V)$ (p. 716)
$\mathcal{D}'(]0,T[;V)$	(p. 718)
$L^1_{\mathrm{loc}}(0,T;V)$	(p. 720)
$L^2(0,T;V)$	(p. 722)
$\mathbb{E}(0,T;V,W)$	(p. 725)
$\mathbb{W}_1(0,T;V)$	(p. 725)
$\mathbb{W}_2(0,T;V)$	(p. 728)

Sobolev spaces with Hilbert space structure on $\Omega \subset \mathbb{R}^n$ or on $\mathbb{R}^n$

$H^m(\Omega)$	(p. 178)
$H^1(\Omega)$	(p. 180)
$H^2(\Omega)$	(p. 181)
$H^m_0(\Omega)$	(p. 187)
$H^1_0(\Omega)$, $H^2_0(\Omega)$	(p. 187)
$H^{-m}(\Omega)$	(p. 191)
$H^s(\overline{\Omega})$	(p. 546)
$H^s(\Omega)$	(p. 574)
$H^\sigma(\Omega)$	(p. 574)
$H^s_0(\Omega)$	(p. 579)
$H^{-s}(\Omega)$	(p. 579)
$H^s_{00}(\Omega)$	(p. 585)
$X^s(\Omega)$	(p. 587)
$(H^s_{00}(\Omega))'$	(p. 591)
$H^s(\mathbb{R}^n)$	(p. 502)
$H^m(\mathbb{R}^n)$	(p. 507)
$\mathcal{H}^m(\mathbb{R}^m)$	(p. 507)
$H^\sigma(\mathbb{R}^n)$, $H^s(\mathbb{R}^n)$	$(s = [s] + \sigma, 0 < \sigma < 1)$ (p. 509)
$H^{s-\frac{1}{2}}(\mathbb{R}^{n-1})$	(p. 528)
$H^{s-j-\frac{1}{2}}(\mathbb{R}^{n-1})$	(p. 530)

Sobolev spaces with Banach space structure on $\Omega \subset \mathbb{R}^n$

$W^{m,p}(\Omega)$, $1 \le p < \infty$	(p. 196)
$W^{m,\infty}(\Omega)$	(p. 196)
$W^{m,p}_0(\Omega)$	(p. 200)
$W^{-m,q}(\Omega)$	(p. 203)
$W^{s,p}(\Omega)$	(p. 580)
$W^{-s,q}(\Omega)$	(p. 583)
$W^{s,p}(\bar{\Omega})$	(p. 584)
$W^{s,p}_{00}(\Omega)$	(p. 591)

$X_s^p(\Omega)$	(p. 592)
$(W_{00}^{s,p}(\Omega))'$	(p. 593)

Sobolev spaces with Hilbert space structure on boundary Γ and $\Gamma_0 \subset \Gamma$

$H^s(\Gamma)$	(p. 637)
$L^2(\Gamma)$	(p. 638)
$H^{-s}(\Gamma)$	(p. 641)
$L^p(\Gamma)$	(p. 635)
$H^s(\Gamma_0)$	(p. 642)
$H_0^s(\Gamma_0)$	(p. 643)
$H_{00}^{1/2}(\Gamma_0)$	(p. 643)
$H_{00}^{3/2}(\Gamma_0)$	(p. 643)
$H_0^s(\Gamma_i)$	(p. 651)
$H^{-s}(\Gamma_i)$	(p. 651)
$(H_{00}^s(\Gamma_i))'$	(p. 651)

Sobolev spaces with Banach space structure on boundary Γ and $\Gamma_0 \subset \Gamma$

$W^{s,p}(\Gamma)$	(p. 636)
$W^{-s,q}(\Gamma)$	(p. 643)
$W^{s,p}(\Gamma_0)$	(p. 642)
$W_{00}^{s,p}(\Gamma_0)$	(p. 642)
$W_0^{s,p}(\Gamma_0)$	(p. 642)
$(W_0^{s,p}(\Gamma_0))'$	(p. 643)

Notations for duality pairing

$\langle\cdot,\cdot\rangle_{V\times V'}$	duality pairing between V and V' (p. 715)
$\langle\cdot,\cdot\rangle_{\mathcal{D}'(\Omega)\times\mathcal{D}(\Omega)}$	duality pairing between $\mathcal{D}'(\Omega)$ and $\mathcal{D}(\Omega)$ (p. 26)
$\langle\cdot,\cdot\rangle_{\mathcal{D}'(\mathbb{R}^n)\times\mathcal{D}(\mathbb{R}^n)}$	duality pairing between $\mathcal{D}'(\mathbb{R}^n)$ and $\mathcal{D}(\mathbb{R}^n)$ (p. 34)
$T(\phi)$, $\langle T,\phi\rangle$, (T,ϕ)	value of distribution $T \in \mathcal{D}'(\Omega)$ at a test function ϕ (p. 26)
$\langle\cdot,\cdot\rangle_{S'(\mathbb{R}^n)\times S(\mathbb{R}^n)}$	duality pairing between $S'(\mathbb{R}^n)$ and $S(\mathbb{R}^n)$ (p. 436)

Notations for inner products

$\langle\cdot,\cdot\rangle_V$ in V	
$\langle\langle\cdot,\cdot\rangle\rangle_W$ in W	
$\langle\cdot,\cdot\rangle_V$	(p. 765)
$\langle\cdot,\cdot\rangle_{L^2(\Omega)}$	$\langle\cdot,\cdot\rangle_{0,\Omega}$ (p. 210)
$\langle\cdot,\cdot\rangle_{L^2(T)}$	$\langle\cdot,\cdot\rangle_T$ (p. 158)
$\langle\cdot,\cdot\rangle_{H^m(\Omega)}$	$\langle\cdot,\cdot\rangle_{m,\Omega}$ (p. 179)
$\langle\cdot,\cdot\rangle_{H^s(\Omega)}$	$\langle\cdot,\cdot\rangle_{s,\Omega}$ (p. 575)

$\langle\cdot,\cdot\rangle_{H^s(\mathbb{R}^n)}$	$\langle\cdot,\cdot\rangle_{s,\mathbb{R}^n}$ (p. 502)
$\langle\cdot,\cdot\rangle_{H^s_{00}(\Omega)}$	$\langle\cdot,\cdot\rangle_{00,s,\Omega}$ (p. 585)
$\langle\cdot,\cdot\rangle_{H^1(\Omega)}$	$\langle\cdot,\cdot\rangle_{1,\Omega}$ (p. 181)
$[\cdot,\cdot]_{0,\Omega}$	(p. 703)
$\langle\!\langle\cdot,\cdot\rangle\!\rangle_{0,\Omega}$	$\langle\!\langle(\cdot),(\cdot)\rangle\!\rangle_{0,\Omega}$ (p. 703)
$\langle\cdot,\cdot\rangle_{H^s(\Gamma)}$	$\langle\cdot,\cdot\rangle_{s,\Gamma}$ (p. 650)
$\langle\cdot,\cdot\rangle_{H(\mathrm{div};\Omega)}$	(p. 687)
$\langle\cdot,\cdot\rangle_{H(\Delta;\Omega)}$	$\langle\cdot,\cdot\rangle_{0,\Delta,\Omega}$ (p. 693)
$\langle[\cdot],[\cdot]\rangle_{H^m(\Omega)/M}$	(p. 192)

Notations for semi-norms $|\cdot|_V$ in V

$\lvert\cdot\rvert_{H^m(\Omega)}$	$\lvert\cdot\rvert_{m,\Omega}$ (p. 179)
$\lvert\cdot\rvert_{W^{m,p}(\Omega)}$	$\lvert\cdot\rvert_{m,p,\Omega}$ (p. 196)
$\lvert[\cdot]\rvert_{W^{m,p}(\Omega)/P_{m-1}}$	(p. 206)
$\lvert\cdot\rvert_{C^k(\overline{\Omega})}$	(p. 753)
$\lvert\cdot\rvert_{C^{k,\lambda}(\overline{\Omega})}$, $\mathbf{\lvert}\cdot\mathbf{\rvert}_{C^{k,\lambda}(\overline{\Omega})}$	(p. 754)
$p(\cdot)$	semi-norm (p. 748)
$q_{\boldsymbol{\alpha},\boldsymbol{\beta}}(\cdot)$	semi-norm in $S(\mathbb{R}^n)$ (p. 408)
$q_{l,m}(\cdot)$	semi-norm in $S(\mathbb{R}^n)$ (p. 408)
$q^*_{\boldsymbol{\alpha},\boldsymbol{\beta}}(\cdot)$	semi-norm in $S(\mathbb{R}^n)$ (p. 408)
$q^*_{l,m}(\cdot)$	semi-norm in $S(\mathbb{R}^n)$ (p. 408)
$p_{\boldsymbol{\alpha}}$	semi-norm (p. 14)
$p_{K,m(K)}$, $\tilde{p}_{K,m(K)}$	semi-norm (p. 27)

Notations for norms $\|\cdot\|_V$ in V, $|\!|\!|\cdot|\!|\!|_W$ in W

$\|\cdot\|_{C^k(\overline{\Omega})}$, $|\!|\!|\cdot|\!|\!|_{C^k(\overline{\Omega})}$	(p. 752)
$\|\cdot\|_{C^{0,\lambda}(\overline{\Omega})}$	(p. 754)
$\|\cdot\|_{L^p(\Omega)}$	(p. 749)
$\|\cdot\|_{L^2(\Omega)}$	(p. 766)
$\|\cdot\|_{L^2(T)}$	$\|\cdot\|_T$ (p. 159)
$\|\cdot\|_{\mathcal{L}(V;W)}$	(p. 760)
$\|\cdot\|_{\mathcal{L}(V;\mathbb{R})}$	(p. 760)
$\|\cdot\|_{H^m(\Omega)}$	$\|\cdot\|_{m,\Omega}$ (p. 179)
$\|\cdot\|_{H^1(\Omega)}$	$\|\cdot\|_{1,\Omega}$ (p. 181)
$\|\cdot\|_{H^{-m}(\Omega)}$	$\|\cdot\|_{-m,\Omega}$ (p. 269)
$\|\cdot\|_{H^s(\Omega)}$	$\|\cdot\|_{s,\Omega}$ (p. 575)
$\|\cdot\|_{H^s_{00}}$	$\|\cdot\|_{00,s,\Omega}$ (p. 585)
$\|\cdot\|_{H^s(\mathbb{R}^n)}$	$\|\cdot\|_{s,(\mathbb{R}^n)}$ (p. 528)
$\|\cdot\|_{H^s(\overline{\Omega})}$	(p. 546)
$\|\cdot\|_{X^s(\Omega)}$	(p. 587)

$\|\cdot\|_{m,p,\Omega}$	(p. 196)
$\|\cdot\|_{m,2,\Omega}$	(p. 200)
$\|\|u\|\|_{m,2,\Omega}$	(p. 200)
$\|\cdot\|_{L^p(\mathrm{div};\Omega)}$	(p. 686)
$\|\cdot\|_{H(\Delta;\Omega)}$	(p. 693)
$\|\cdot\|_{H^2(\Lambda,\Omega)}$	$\|.\|_{2,\Lambda,\Omega}$ (p. 702)
$\|\|\cdot\|\|_{0,\Omega}$	(p. 703)
$\|\cdot\|_{\mathbb{E}(0,T;V,W)}$	(p. 725)
$\|\cdot\|_{\mathbb{W}_1(0,T;V)}$	(p. 726)
$\|\cdot\|_{\mathbb{W}_2(0,T;V)}$	(p. 728)
$\|\cdot\|_V$	(p. 748)
$\|\cdot\|_\infty$	(p. 749)
$\|[\cdot]\|_{H^m(\Omega)/M}$	(p. 191)
$\|\|[\cdot]\|\|_{H^m(\Omega)/P_{m-1}}$	(p. 193)
$\|[\cdot]\|_{W^{m,p}(\Omega)/M}$	(p. 203)

Notations for tensor product

$f \otimes g$	tensor product of functions f and g (p. 298)
$1_{\mathbf{x}} \otimes g(\mathbf{y}),\ f(\mathbf{x}) \otimes 1_{\mathbf{y}}$	(p. 299)
$T_{\mathbf{x}} \otimes S_{\mathbf{y}}$	$T_{\mathbf{x}} \equiv T(\mathbf{x}),\ S_{\mathbf{y}} \equiv S(\mathbf{y})$ (p. 301)
$\langle T_{\mathbf{x}} \otimes S_{\mathbf{y}}, \phi(\mathbf{x}) \cdot \psi(\mathbf{y})\rangle$	(p. 301)
$\langle T_{\mathbf{x}} \otimes S_{\mathbf{y}}, \chi(\mathbf{x}, \mathbf{y})\rangle$	(p. 301)
$\mathrm{supp}(T_{\mathbf{x}} \otimes T_{\mathbf{y}})$	(p. 301)

Notations for convolutions

$f * g$	convolution of functions f and g (p. 304)
$T * \phi$	convolution of distribution T and test function ϕ (p. 315)
$\rho_\varepsilon * f$	regularization of f by convolution with ρ_ε (p. 308)
$\delta * f$	convolution of Dirac distribution δ and f (p. 322)
$T * S$	convolution of two distributions T and S (p. 317)
$\delta * T$	convolution of Dirac distribution δ and T (p. 322)
$\mathrm{supp}(T * S)$	support of $T * S$ (p. 321)
$S_1(s) * S_2(s)$	convolution of distributions S_1, S_2 on circle Γ (p. 336)
$\check{T}$	(p. 316)
$\tau_{\mathbf{a}} f,\ \tau_{\mathbf{a}} T$	(pp. 323, 72)

Convolution algebra $\mathcal{A}$

$\mathcal{A}$	convolution algebra (p. 367)
$[A]$	$(A_{ij})_{1\le i,j\le n}$ with $A_{ij} \in \mathcal{A}$

$[A] * [B]$	convolution matrix product in $\mathcal{A}$ (p. 371)
$\Delta^*(A)$	convolution determinant of $[A]$ (p. 372)
$[E] = [A^{*-1}]$	convolution inverse of $[A]$ (p. 373)

Fourier transform $\mathcal{F}$ and co-transform $\overline{\mathcal{F}}$

$\mathcal{F}, \overline{\mathcal{F}}$	(p. 383)
$\hat{f}$	$\mathcal{F} f$ (p. 383)
$\check{f}, (\check{\mathcal{F}} f)$	(p. 389)
$\hat{T}$	$\mathcal{F} T$: Fourier transform of tempered distributions $T \in S'(\mathbb{R}^n)$ (p. 435)

Notations for notions of convergences

$\to$	strong convergence (p. 763)
$\rightharpoonup$	weak convergence (p. 764)
$\rightharpoonup^*$	weak-$*$ convergence (p. 764)

Laplace transform

$\mathcal{L}$	(p. 492)
$\mathcal{L}^{-1}$	inverse Laplace transform (p. 497)

Notations for trace operators ,, ,j, ,$_j^*$ and their right-hand inverses

γ_0	trace operator (p. 528)
γ_j	trace operator (p. 527)
$\boldsymbol{\nu}$	trace operator (p. 708)
γ_j	$\gamma_j : H^s(\mathbb{R}^n) \longmapsto H^{s-j-\frac{1}{2}}(\mathbb{R}^{n-1})$ (p. 530)
γ	$\gamma : H^s(\mathbb{R}^n) \longrightarrow \prod_{j=0}^m H^{s-j-\frac{1}{2}}(\mathbb{R}^{n-1})$ (p. 545)
γ_j^*	(p. 663)
γ^*	trace operator (p. 666)
γ_j^*	$\gamma_j^* : H^m(\mathbb{R}^n_+) \longrightarrow H^{m-j-1/2}(\mathbb{R}^{n-1})$ (p. 654)
γ^*	$\gamma^* : H^m(\mathbb{R}^n_+) \longrightarrow \prod_{j=0}^{m-1} H^{m-j-1/2}(\mathbb{R}^{n-1})$ (p. 654)
γ_j^*	$\gamma_j^* : H^m(\Omega) \longrightarrow H^{m-j-1/2}(\Gamma)$ (p. 666)
γ^*	$\gamma^* : H^m(\Omega) \longrightarrow \prod_{j=0}^{m-1} H^{m-j-1/2}(\Gamma)$ (p. 666)
$H_0^m(\Omega)$	$\mathrm{Ker}(\gamma^*)$ (p. 670)
γ_j	$\gamma_j : W^{s,p}(\mathbb{R}^n) \longrightarrow W^{s-p-1/p,p}(\mathbb{R}^{n-1})$ (p. 670)
γ	$\gamma : W^{s,p}(\mathbb{R}^n) \longrightarrow \prod_{j=0}^k W^{s-j-1/p,p}(\mathbb{R}^{n-1})$ (p. 671)
γ_{lj}	$\gamma_{lj} : H^s(\Omega) \longrightarrow \prod_{k=0}^l H^{s-k-1/2}(\Gamma_j)$ (p. 672)
γ	$\gamma : \mathbf{v} \in L^p(\mathrm{div}; \Omega) \longmapsto \gamma\mathbf{v} \in W^{-1/p,p}(\Gamma)$ (p. 687)

$\boldsymbol{\nu}$	$\boldsymbol{\nu} : H^2(\Lambda, \Omega) \longrightarrow H^{-3/2}(\Gamma) \times H^{-1/2}(\Gamma)$ (p. 709)
ρ_j	right-hand inverse $\rho_j : H^{s-j-1/2}(\mathbb{R}^{n-1}) \longrightarrow H^s(\mathbb{R}^n)$ (p. 544)
$\boldsymbol{\rho}$	$\boldsymbol{\rho} : \prod_{j=0}^{m-1} H^{m-j-1/2}(\mathbb{R}^{n-1}) \longrightarrow H^m(\mathbb{R}^n_+)$ (p. 654)
ρ	$\rho : \prod_{j=0}^{m-1} H^{m-j-1/2}(\Gamma) \longrightarrow H^m(\Omega)$ (p. 666)
ρ	$\rho : \prod_{j=0}^{k} W^{s-j-1/p,p}(\mathbb{R}^{n-1}) \longrightarrow W^{s,p}(\mathbb{R}^n)$ (p. 671)
ρ	$\rho : W^{-1/p,p}(\Gamma) \longrightarrow L^p(\text{div}; \Omega)$ (p. 687)

General notations used in the book

$\Omega \subset \mathbb{R}^n$	an open subset of $\mathbb{R}^n$
$\Gamma \subset \mathbb{R}^n$	boundary of $\Omega \subset \mathbb{R}^n$ with $\overline{\Omega} = \Omega \cup \Gamma$
$\lvert \cdot \rvert_V$	semi-norm in V, $p(\cdot)$ in V (p. 748)
$\lVert \cdot \rVert_V$	norm in V (p. 748)
$\lvert\lvert\lvert \cdot \rvert\rvert\rvert_V$	(p. 703)
$\lVert \cdot \rVert_{s,\Omega}$	norm in $H^s(\Omega)$ (p. 575)
$\lVert \cdot \rVert_{00,s,\Omega}$	norm in $H^s_{00}(\Omega)$ (p. 585)
$\lVert \cdot \rVert_{s,p,\Omega}$	norm in $W^{s,p}(\Omega)$ (p. 581)
$\lVert \cdot \rVert_{00,s,p,\Omega}$	norm in $W^{s,p}_{00}(\Omega)$ (p. 591)
$d(\cdot,\cdot)$	metric/distance function in normed linear space X: $d(x, y) = \lVert x - y \rVert_X$ (p. 732)
$\langle \cdot,\cdot \rangle_V$	inner product in V (p. 765)
$\langle\langle \cdot,\cdot \rangle\rangle_V$	(p. 703)
$\langle \cdot,\cdot \rangle_{V' \times V}$	duality pairing between V' and V (p. 715)
δ_{ij}	Kronecker delta (p. 2)
$\delta = \delta_0 = \delta(x)$	Dirac delta function with concentration at $0 \in \mathbb{R}$ (p. 1)
$\delta_a = \delta(x - a)$	Dirac delta function with concentration at $a \in \mathbb{R}$ (p. 1)
$\delta_{\mathbf{a}} = \delta(\mathbf{x} - \mathbf{a})$	Dirac delta function with concentration at $\mathbf{a} \in \mathbb{R}^n$ (p. 5)
$\delta_{\mathbb{S}}$	Dirac delta function with concentration on surface $\mathbb{S} \subset \mathbb{R}^n$ (p. 36)
δ_Γ	Dirac delta function with concentration on $\Gamma \subset \mathbb{R}^n$ (p. 90)
$\text{supp}(\phi)$	support of a continuous function ϕ (p. 6)
$\text{supp}(\phi) \subset\subset \Omega$	$\text{supp}(\phi)$ compact in Ω (p. 6)
$H(x)$	Heaviside function on $\mathbb{R}$ (p. 3)
$\ln(x \pm i0)$	(p. 41)

$x^\lambda, x_+^\lambda, x_-^\lambda$	(p. 49)
Pf	Finite part; $\mathrm{Pf}(\frac{1}{x^k}) \equiv$ Pseudo-function (finite part of) $\mathrm{Pf}(\frac{1}{x^k})$ (p. 42)
c.p.v.	Cauchy principal value $\frac{1}{x}$ (p. 39)
$J_{\mathbf{F}}(\mathbf{x})$	Jacobian of $\mathbf{F} : \mathbb{R}^n \longrightarrow \mathbb{R}^n$ at $\mathbf{x}$ (p. 68)
$\mathbf{F}T$	image of T under $\mathbf{F}$ (p. 69)
ρ_ε	regularizing function (p. 308)
ρ_m	regularizing sequence (p. 307)
$\phi(\mathbf{x})$	test function in $\mathcal{D}(\Omega)$ (p. 5)
$\widetilde{\Phi}_T(\mathbf{x})$	periodic function in $\mathcal{D}_T(\mathbb{R})$ with period $T > 0$ (p. 78)
$\Phi(s)$	test function in $\mathcal{D}(\Gamma)$ (p. 77)
$\frac{\partial u}{\partial n_A}, \frac{\partial u}{\partial n_{A^*}}$	conormal derivatives of u with respect to A, A^* (p. 689)
T, S, etc.	generic distributions on Ω (p. 25)
T, S, etc.	generic tempered distributions (p. 424)
T	period of periodic distributions S on $\mathbb{R}$ (p. 77)
T_f	(regular) distribution in $\mathcal{D}'(\Omega)$ defined by $f \in L^1_{\mathrm{loc}}(\Omega)$ (p. 29)
$T \in \mathcal{E}'(\mathbb{R}^n)$	distribution $T \in \mathcal{D}'(\mathbb{R}^n)$ with compact support in $\mathbb{R}^n$ (p. 288)
$T(\phi)$	duality pairing between $T \in \mathcal{D}'(\Omega)$ and $\phi \in \mathcal{D}(\Omega)$ (p. 26)
$\langle T, \phi \rangle_{\mathcal{D}'(\Omega)\times\mathcal{D}(\Omega)}$	$T(\phi)$ (p. 26)
(T, ϕ)	$T(\phi)$ (p. 26)
$\tau_{\mathbf{b}}$	translation operator (p. 72)
$\tau_{\mathbf{b}}T, \tau_{\mathbf{b}}\delta$	(p. 72)
$T \downarrow_{\Omega_0}$	restriction to Ω_0 of T (p. 280)
$X \hookrightarrow Y$	continuous imbedding of X into Y,
$X \hookrightarrow\hookrightarrow Y$	compact imbedding from X into Y (p. 761)

Notations used in the analysis of R-L-C circuit

$\delta(t)$	unit excitation impulse (p. 377)
$E(t)$	impulse response (p. 377)
R	resistance (p. 377)
L	inductance (p. 377)
C	capacitance (p. 377)
$Z(t)$	impedance (p. 378)
$A(t)$	admittance (p. 378)

Notations used for traces in $H^2(\Lambda, \Omega)$ for plate bending problems

Λ	plate bending operator (p. 699)
Φ	tensor-valued functions (p. 700)
$\Phi = (\phi_{ij})$	bending moment tensor field (p. 700)
M_n	normal moment (p. 700)
M_{nt}	twisting moment (p. 700)
Q_n	vertical shear (p. 700)
K_n	Kirchhoff force (p. 700)

Chapter 1

Schwartz distributions

1.1 Introduction: Dirac's delta function $\delta(x)$ and its properties

Although the theory of distributions of Laurent Schwartz [8] has diverse applications in various branches of mathematics, we will be primarily concerned with its application in the theory of Sobolev spaces and elliptic partial differential equations. In fact, distributions are also called *generalized functions* [1], since the theory of distributions in some senses generalizes the notion of function in classical analysis by including not only the usual functions f with point-values (i.e. functions f having point-values $f(\mathbf{x})$ at the points $\mathbf{x}$) but also *new* mathematical objects. These include the Dirac distribution δ (popularly, but incorrectly, called the delta function $\delta(\mathbf{x})$) and its derivatives $\delta^{(k)}$, and also idealized concepts such as the density of a material point, the density of a point charge or dipole moment, the spatial volume density of simple or double layers in electrostatics, the magnitude of an instantaneous force applied at a point, etc. which frequently arise in physics and mechanics. But according to Courant and Hilbert [2, p. 766], 'the term "Ideal functions" seems much indicative of the true role of this concept ... This role is indeed that of functions, almost as the role of real numbers is that of ordinary numbers'. It is further stated in [2, p. 767], '"Distributions" are most appropriately introduced as ideal elements in function spaces'. Moreover, the theory of distributions provides rigorous mathematical foundations for all the new mathematical objects mentioned above. Indeed, in the late 1920s, Dirac[1] introduced the so-called delta function $\delta(\mathbf{x})$ which violated existing mathematical principles, having the following properties [21], [22], [23], [24]:

- $\delta(x)$ is defined, and not only continuous but equal to zero, for all $x \neq 0$, i.e.

$$\delta(x) = \begin{cases} 0 & \forall x \neq 0 \\ \infty & \text{for } x = 0 \end{cases}$$

such that $\delta(-x) = \delta(x)\ \forall x \in \mathbb{R}$, with

$$\int_{-\infty}^{\infty} \delta(-x)dx = \int_{-\infty}^{\infty} \delta(x)dx = 1. \tag{1.1.1}$$

[1] Paul Dirac was a Nobel Laureate in Physics and one of the founders of quantum mechanics.

Then, by shifting the origin through $a \in \mathbb{R}$,

$$\delta(x-a) = \delta(a-x) = \begin{cases} 0 & \forall x \neq a \\ \infty & \text{for } x = a \end{cases}$$

with $\int_{-\infty}^{\infty} \delta(x-a)dx = \int_{-\infty}^{\infty} \delta(a-x)dx = 1$.

• The 'most important property' of $\delta(x)$ [21, p. 58] is given by: $\forall$ functions f continuous on $\mathbb{R} \equiv]-\infty, \infty[$,

$$\int_{-\infty}^{\infty} f(x)\delta(x-a)dx = f(a);^{2} \tag{1.1.2}$$

$$\int_{-\infty}^{\infty} f(x)\delta(x-a)dx = \int_{-\infty}^{\infty} f(a)\delta(x-a)dx = f(a); \tag{1.1.3}$$

$$\int_{-\infty}^{\infty} f(x)\delta(a-x)dx = \int_{-\infty}^{\infty} f(a)\delta(a-x)dx = f(a).$$

Moreover, for $f(x) = 1\ \forall x$ and for $a = 0$, $\delta(x-a) = \delta(x)$ and (1.1.2) reduces to (1.1.1).

• δ is not only continuous but also *infinitely differentiable*, such that, for every k-times differentiable function f on $\mathbb{R}$, $\int_{-\infty}^{\infty} f(t)\delta^{(k)}(x-t)dt = f^{(k)}(x)\ \forall k \in \mathbb{N}$,

[2] An intuitive (though incorrect) derivation of formula (1.1.2): Consider the system $(f_i)_{i=-n}^{n}$ with $f_i \in \mathbb{R}$ and the (Kronecker) delta ($\delta = (\delta_{ij})_{-n \le i,j \le n}, \delta_{ij} = 0$ for $i \neq j$ and $\delta_{ij} = 1$ for $i = j$). For fixed j, $-n \le j \le n$, $\sum_{i=-n}^{n} f_i \delta_{ij} = f_j$.

Consider a real-valued function f defined on discrete numbers $i = -n, -(n-1), \ldots, 0, 1, \ldots, n$ with $f(i) = f_i \in \mathbb{R}\ \forall i = -n, -(n-1), \ldots 0, 1, \ldots, n$, and set $\delta(i-j) = \delta_{ij}$ for $-n \le i, j \le n$. Then

$$\sum_{i=-n}^{n} f_i \delta_{ij} = \sum_{i=-n}^{n} f(i)\delta(i-j) = f(j) \quad \forall \text{ fixed } j,\ -n \le j \le n, \tag{$*$}$$

which holds for *discrete* variables i and j. Now we may intuitively try to extend ($*$), which is a correct definition, to the continuous case of a real variable $x \in \mathbb{R} =]-\infty, \infty[$, and for any fixed real number $a \in \mathbb{R}$, by replacing:

- '$i \in \{0, \pm 1, \ldots, \pm n\}$' by '$x \in]-\infty, \infty[$'; '$j = 0, \pm 1, \ldots, \pm n$' by '$a \in]-\infty, \infty[$';
- '$f(i)$' by '$f(x)$' $\in \mathbb{R}$, i.e. f is defined for all $x \in \mathbb{R}$ and may be continuous at a;
- '$\delta(i-j)$' by '$\delta(x-a)$', with $\delta(x-a) = 0\ \forall x \neq a$, $a \in \mathbb{R}$, and $\delta(x-a) = 1$ for $x = a$;
- '$\sum_{-n}^{n}(\cdot)$' by '$\int_{-\infty}^{\infty}(\cdot)dx$'; and
- $$\text{'}\sum_{-n}^{n} f(i)\delta(i-j) = f(j)\text{' by '}\int_{-\infty}^{\infty} f(x)\delta(x-a)dx = f(a)\text{'}. \tag{$**$}$$

The *intuitive* definition ($**$) is *mathematically incorrect* (although ($*$) is correct), since the value of the integrand $f(x)\delta(x-a) = 0\ \forall x \neq a$, and hence the value of the integral $\int_{-\infty}^{\infty} f(x)\delta(x-a)dx = 0$, not $f(a)\ \forall$ fixed $a \in \mathbb{R}$ as claimed in ($**$) (see Proposition 1.3.2). The name 'delta function' probably follows from this intuitive approach. See Section 1.11 for more interesting details.

$\forall x \in \mathbb{R}$,

$$\Longrightarrow \quad \int_{-\infty}^{\infty} f(t)\delta^{(k)}(a-t)dt = f^{(k)}(a) \quad \forall a \in \mathbb{R},\ \forall k \in \mathbb{N}, \tag{1.1.4}$$

where $(\cdot)^{(k)} = \frac{d^k(\cdot)}{dx^k}$.

- Another interesting property listed by Dirac [21, p. 61] is $x\delta(x) = 0$. Then, for constants A and B, $A = B \Longrightarrow \frac{A}{x} = \frac{B}{x} + C\delta(x)$ with unknown constant C, i.e. $\frac{A}{x} \neq \frac{B}{x}$ in general.

- For the Heaviside function $H(x) = \begin{cases} 1 \text{ for } x > 0 \\ 0 \text{ for } x < 0 \end{cases}$,

$$\frac{dH}{dx}(x) = \delta(x). \tag{1.1.5}$$

- Finally, Dirac listed the 'remarkable formula' used in the quantum theory of collision processes:

$$\frac{d}{dx} \ln x = 1/x - \imath\pi\delta(x) \quad (\imath = \sqrt{-1}). \tag{1.1.6}$$

The properties of the so-called delta function $\delta(x)$ in (1.1.1) are contradictory, and hence such a function can *not* exist (see also Proposition 1.3.2 and equations (1.11.6) and (1.11.7)):

1. If $\delta(x) = 0\ \forall x \neq 0$ and $\delta(0) = \infty$ (as defined in (1.1.1)), then its Lebesgue/Riemann integral is zero, i.e. $\int_{-\infty}^{\infty} \delta(x)dx = 0$, but, according to (1.1.1),

$$\int_{-\infty}^{\infty} \delta(x)dx = 1. \tag{1.1.7}$$

2. $\delta(x) = 0\ \forall x \neq 0$ and $\delta(0) = \infty$

$$\Longrightarrow \quad \forall \alpha > 0, \quad \alpha\delta(x) = \begin{cases} 0 & \forall x \neq 0 \\ \infty & \text{for } x = 0 \end{cases}$$

$$\Longrightarrow \quad \text{its integral} \int_{-\infty}^{\infty} \alpha\delta(x)dx = 1, \tag{1.1.8}$$

but, from (1.1.1),

$$\forall \alpha > 0, \quad \int_{-\infty}^{\infty} \alpha\delta(x)dx = \alpha \int_{-\infty}^{\infty} \delta(x)dx = \alpha \cdot 1 = \alpha. \tag{1.1.9}$$

In other words, (1.1.7), (1.1.8) and (1.1.9) are contradictory results!

Hence, mathematicians immediately pointed out that from the point of view of existing 'rigorous' mathematics, all this is nonsense, but Dirac believed '...advancement in physics is to be associated with a continuous modification and generalization of the axioms at the base of mathematics rather than with a logical development of any one mathematical scheme on a fixed foundation' ([22] as quoted by [6]), and applied formulae (1.1.1)–(1.1.6) in quantum mechanics with great success to obtain physically meaningful and important results. Consequently, the physics community and the applied mathematicians accepted (1.1.1)–(1.1.6) as 'correct'.

It was perfectly clear to Dirac himself that $\delta(x)$ was *not* a function in the classical sense of the term. In fact, he stated,

> $\delta(x)$ is not a function of x according to the usual mathematical definition of a function, which requires a function to have a definite value for each point in its domain, but is something more general, which we may call an "improper function" to show up its difference from a function defined by the usual definition. Thus, $\delta(x)$ is not a quantity which can be generally used in mathematical analysis like an ordinary function. [21, p. 58].

What was actually important for him was to consider $\delta(x)$ and $\delta^{(k)}(x)$ as 'symbolic devices', i.e. how $\delta(x)$ and $\delta^{(k)}(x)$ act as operators on continuous functions and k-times differentiable functions f according to the formulae (1.1.1) and (1.1.4) respectively.

In fact, it took almost three decades to discover the mathematical foundations of a correct formulation of the definition and properties of Dirac's delta function, and it turned out that Dirac's brilliant intuitive results had been right in all cases (1.1.1)–(1.1.6), if:

1. the defining integrals in (1.1.1) and (1.1.2) are considered meaningless and dispensed with;
2. δ and $\delta^{(k)}$ (respectively $\delta_{\mathbf{a}}$ and $\delta_{\mathbf{a}}^{(k)}$) are considered to be continuous linear functionals (i.e. not functions with point-values) on a suitable test space of functions. (Then, the defining meaningless integrals in (1.1.1)–(1.1.2) can be replaced by the 'duality pairings' between the test spaces and their duals).
3. the usual point-wise concept of the derivative of a function is replaced by a new notion of derivatives, in which (1.1.5) and (1.1.6) are to be understood.

All of these will be explained eventually in this chapter and the subsequent chapters.

In order to obtain a very large space of continuous linear functionals, which were called *distributions* by Laurent Schwartz[3] and contain, in particular, new mathemati-

[3]The history of the discovery of the theory of distributions by Laurent Schwartz is similar to that of the 'search for iron which led to the discovery of gold'! As in the discovery of X-rays by Röntgen in physics, formula (1.1.2) did not motivate Schwartz to the discovery of distributions. In fact, an altogether different problem of approximations associated with polyharmonic functions posed by Choquet and

cal objects δ and its derivatives $\delta^{(k)}$ (resp $\delta_{\mathbf{a}}$ and $\delta_{\mathbf{a}}^{(k)}$), Schwartz introduced the smallest space of 'sufficiently good' functions ϕ (i.e. the test space of functions) on which these continuous linear functionals or distributions will be defined, i.e. $\forall$ test functions ϕ,

$$\delta(\phi) = \langle \delta, \phi \rangle = \phi(0); \quad \delta^{(k)}(\phi) = \langle \delta^{(k)}, \phi \rangle = (-1)^k \phi^{(k)}(0) \tag{1.1.10}$$

(resp. $\delta_{\mathbf{a}}(\phi) = \langle \delta_{\mathbf{a}}, \phi \rangle = \phi(\mathbf{a})$; $\delta_{\mathbf{a}}^{(k)}(\phi) = \langle \delta_{\mathbf{a}}^{(k)}, \phi \rangle = (-1)^k \phi^{(k)}(\mathbf{a})$), which will be the definitions instead of (1.1.1), (1.1.2) and (1.1.4) (see also (6.3.22)–(6.3.24) and Section 1.11).

Hence, first of all, we will define this 'smallest' space of 'sufficiently good' functions, i.e. the common set of functions on which all continuous linear functionals are defined.

Multi-index notations We first define the multi-index and other notations to be used here:

- $\mathbb{N} = \{1, 2, \ldots, n, \ldots\}$; $\mathbb{N}_0 = \mathbb{N} \cup \{0\} = \{0, 1, 2, \ldots, n, \ldots\}$; $\mathbb{Z} = \{0, \pm 1, \pm 2, \ldots, \pm n, \ldots\}$; $\mathbb{R}$ = the set of all real numbers; $\mathbb{C}$ = the set of all complex numbers;
- $\boldsymbol{\alpha} = (\alpha_1, \alpha_2, \ldots, \alpha_n)$ with $\alpha_i \in \mathbb{N}_0$, i.e. $\alpha_i \geq 0$ for $1 \leq i \leq n$; $|\boldsymbol{\alpha}| = \alpha_1 + \cdots + \alpha_n \geq 0$, i.e. $|\boldsymbol{\alpha}| \in \mathbb{N}_0$; $|\boldsymbol{\alpha}| = 0 \iff \alpha_i = 0$ for $1 \leq i \leq n$;
- $\mathbb{R}^n = \underbrace{\mathbb{R} \times \mathbb{R} \times \cdots \times \mathbb{R}}_{n \text{ times}}$ = the linear space of all ordered n-tuples of real numbers;
- $$\mathbf{x} = (x_1, x_2, \ldots, x_n) \in \mathbb{R}^n; \quad \mathbf{x}^{\boldsymbol{\alpha}} = x_1^{\alpha_1} x_2^{\alpha_2} \cdots x_n^{\alpha_n}; \tag{1.1.11}$$
- For derivatives with respect to variables $x_1, x_2, \ldots, x_n$, $n \geq 2$, $f(\mathbf{x}) = f(x_1, x_2, \ldots, x_n)$, we use the notations:
  - $$\frac{\partial f}{\partial x_i} = \partial_i(f); \quad \frac{\partial^{\alpha_i} f}{\partial x_i^{\alpha_i}} = \frac{\partial}{\partial x_i}\left(\frac{\partial}{\partial x_i}\left(\cdots\underbrace{\left(\frac{\partial f}{\partial x_i}\right)\cdots}_{\alpha_i \text{ times}}\right)\right) = \partial_i^{\alpha_i} f;$$
  - $$\frac{\partial^{|\boldsymbol{\alpha}|} f}{\partial x_1^{\alpha_1} \partial x_2^{\alpha_2} \ldots \partial x_n^{\alpha_n}} = \partial_1^{\alpha_1} \ldots \partial_n^{\alpha_n} f = \partial^{(\alpha_1, \ldots, \alpha_n)} f = \partial^{\boldsymbol{\alpha}} f; \tag{1.1.12}$$
  - $\partial^{\boldsymbol{\alpha}} f \equiv f$ for $\boldsymbol{\alpha} = \mathbf{0}$, i.e. $\alpha_1 = \alpha_2 = \cdots = \alpha_n = 0$;
  - for $n = 1$ and $f(x)$, we use the usual notations: $\frac{df}{dx}$, $\frac{d^\alpha f}{dx^\alpha}$ $\forall \alpha \in \mathbb{N}$;
- other equivalent notations for derivatives of f ($n \geq 2$, $f(x_1, x_2, \ldots, x_n)$) are:

Deny was solved by Schwartz for higher dimensions $n > 2$, leading him to the theory of distributions for which he was awarded the Fields Medal (the highest honour for original mathematical creativity and discovery, mathematics's Nobel prize).

– $D_i f \equiv \partial_i f;\ D_i^{\alpha_i} f = \partial_i^{\alpha_i} f;\ D^{\boldsymbol{\alpha}} f = \partial^{\boldsymbol{\alpha}} f;$ etc. $\forall \boldsymbol{\alpha} = (\alpha_1, \dots, \alpha_n);$ (1.1.13)

– for $\boldsymbol{\alpha} = \mathbf{0}$, i.e. $\alpha_1 = \alpha_2 = \cdots = \alpha_n = 0$, $D^{\boldsymbol{\alpha}} f \equiv f$;
– for $n = 1$ and $f(x)$, $Df = \frac{df}{dx}$, $D^\alpha f = \frac{d^\alpha f}{dx^\alpha}$ $\forall \alpha \in \mathbb{N}$; for $\alpha = 0$, $D^\alpha f \equiv f$;
– $\boldsymbol{\alpha} = (\alpha_1, \dots, \alpha_n)$, $\boldsymbol{\beta} = (\beta_1, \dots, \beta_n)$, $\boldsymbol{\alpha} \le \boldsymbol{\beta} \iff \alpha_i \le \beta_i\ \forall 1 \le i \le n$;
– $\boldsymbol{\alpha}! = \alpha_1!\alpha_2! \cdots \alpha_n!$.

1.2 Test space $\mathcal{D}(\Omega)$ of Schwartz

1.2.1 Support of a continuous function

Let $\Omega \subset \mathbb{R}^n$ be an open subset of $\mathbb{R}^n$ and $C^0(\Omega)$ be the linear space of functions ϕ continuous on Ω. Then the *support* of ϕ, denoted by supp(ϕ), is the closure of the set $\{\mathbf{x} : \mathbf{x} \in \Omega, \phi(\mathbf{x}) \neq 0\}$ in $\mathbb{R}^n$, i.e.

$$\operatorname{supp}(\phi) = \overline{\{\mathbf{x} : \mathbf{x} \in \Omega, \phi(\mathbf{x}) \neq 0\}} \quad \text{in } \mathbb{R}^n. \tag{1.2.1}$$

For example,

1. for $\Omega =]0, \pi[\subset \mathbb{R}$, $\phi_1(x) = \sin x\ \forall x \in]0, \pi[\implies \operatorname{supp}(\phi_1) = \overline{]0, \pi[} = [0, \pi]$ in $\mathbb{R} \implies \operatorname{supp}(\phi_1) = \overline{\Omega}$ with $\Omega \subset \operatorname{supp}(\phi_1)$;
2. for $\Omega = \mathbb{R}$, $\phi_2(x) = e^x\ \forall x \in \mathbb{R} \implies \operatorname{supp}(\phi_2) = \overline{\mathbb{R}} \equiv \mathbb{R}$ ($\mathbb{R}$ is also closed!);
3. for $\Omega =]-\pi, \pi[$, $\phi_3(x) = \cos x\ \forall\ |x| < \pi/2$ and $\phi_3(x) = 0$ otherwise on $\Omega \implies \operatorname{supp}(\phi_3) = \overline{]-\pi/2, \pi/2[} = [-\pi/2, \pi/2] \subset \Omega \subset \mathbb{R}$;
4. for $\Omega = B(\mathbf{0}; 2) = \{(x_1, x_2) : (x_1, x_2) \in \mathbb{R}^2, (x_1^2 + x_2^2)^{1/2} < 2\}$,
   $\phi_4(x_1, x_2) = +\sqrt{1 - x_1^2 - x_2^2}\ \forall (x_1, x_2) \in B(\mathbf{0}; 1)$ and $\phi_4(x_1, x_2) = 0$ otherwise $\implies \operatorname{supp}(\phi_4) \subset B(\mathbf{0}; 2)$.

Hence, according to our definition (1.2.1), supp(ϕ) may be a proper subset of Ω, or Ω may be a proper subset of supp(ϕ), or $\operatorname{supp}(\phi) = \Omega = \mathbb{R}^n$, $\mathbb{R}^n$ being both open and closed.

We note the following:

- supp(ϕ) is a closed set.
- For $\Omega = \mathbb{R}^n$, let $\operatorname{supp}(\phi) \neq \mathbb{R}^n$. Then, supp($\phi$) is the smallest closed subset of $\mathbb{R}^n$ outside which ϕ vanishes. Conversely,

$$\mathbb{R}^n \setminus \operatorname{supp}(\phi) = \text{the complement of supp}(\phi) \text{ in } \mathbb{R}^n \text{ is the largest open subset of } \mathbb{R}^n \text{ in which } \phi \text{ vanishes.} \tag{1.2.1a}$$

- Let $\text{supp}(\phi) \subset \Omega \subset \mathbb{R}^n$, and $\text{supp}(\phi)$ be a bounded subset of Ω. Then $\text{supp}(\phi)$ is a compact (i.e. closed and bounded) subset of Ω, which we denote by the notation:

$$\text{supp}(\phi) \subset\subset \Omega \quad (\text{i.e. } \subset\subset \iff \text{compact subset}). \tag{1.2.2}$$

Let $C^k(\Omega), C_0^k(\Omega), C^\infty(\Omega)$, $\forall k \in \mathbb{N}_0$, be the spaces defined by:

$$\begin{aligned} C^k(\Omega) &= \{\phi : \partial^\alpha \phi \in C^0(\Omega)\ \forall |\boldsymbol{\alpha}| \le k\}; \\ C_0^k(\Omega) &= \{\phi : \phi \in C^k(\Omega), \text{supp}(\phi) \subset\subset \Omega\} \\ &= C^k(\Omega) \cap C_0(\Omega). \end{aligned} \tag{1.2.3}$$

For $k = 0$, $C_0(\Omega) \equiv C_0^0(\Omega)$,

$$C_0^0(\Omega) = \{\phi : \phi \in C^0(\Omega), \text{supp}(\phi) \subset\subset \Omega\}. \tag{1.2.4}$$

For example, ϕ_3 in (3) above belongs to $C_0(]-\pi,\pi[)$, but $\phi_3' \notin C_0^1(]-\pi,\pi[)$, since $\phi' \notin C^1(]-\pi,\pi[)$ ($\phi'(x) = -\sin x$ on $]-\pi/2, \pi/2[$ and $\phi'(x) = 0$ otherwise $\Longrightarrow$ $\phi'(\frac{\pi}{2}^+) = 0, \phi'(\frac{\pi}{2}^-) = 1$).

$$\begin{aligned} C^\infty(\Omega) = \{\phi : \phi \in C^k(\Omega)\ \forall k \in \mathbb{N}_0\} = \bigcap_{k=0}^{\infty} C^k(\Omega) \\ = \text{the set of all continuous functions whose partial derivatives} \\ \text{of all orders with respect to } \textit{all} \text{ the variables are continuous in } \Omega. \end{aligned}$$

Then,

$$C_0^\infty(\Omega) = \{\phi : \phi \in C^\infty(\Omega), \text{supp}(\phi) \subset\subset \Omega\} = C^\infty(\Omega) \cap C_0(\Omega). \tag{1.2.5}$$

As an example of a function $\phi \in C_0^\infty(\Omega)$, we consider:

Example 1.2.1. Let $\Omega = \mathbb{R} =]-\infty, \infty[$, and let ϕ be a function constructed from Cauchy's infinitely differentiable function, which assumes the value $\exp(-1/t^2)$ for $t > 0$ and the value zero for $t \le 0$, by:

$$\phi(x) = \begin{cases} e^{(-\frac{1}{1-x^2})} & \text{for } |x| < 1 \\ 0 & \text{for } |x| \ge 1 \end{cases} \tag{1.2.6}$$

Then, $\phi \in C_0^\infty(\mathbb{R})$ with $\text{supp}(\phi) = [-1, 1] \subset\subset \mathbb{R}$ (Figure 1.1).

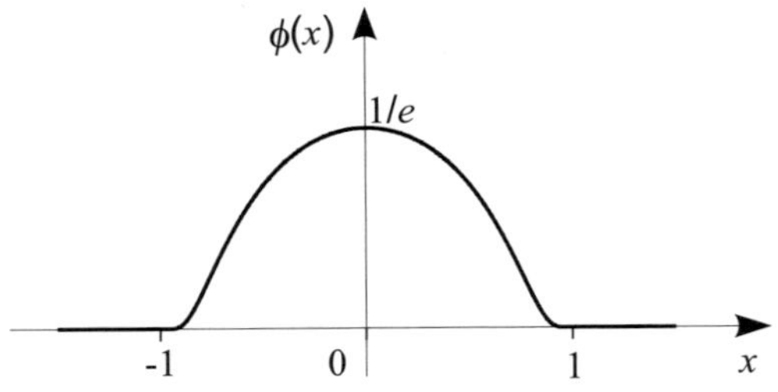

Figure 1.1 Graph of $\phi(x)$ with $\text{supp}(\phi) = [-1, 1]$

Proof. $\phi{\downarrow}_{]-1,1[}(x) = \exp(\frac{-1}{(1-x^2)})$ is continuous and infinitely differentiable for $|x| < 1$, and $\phi = 0$ is continuous and infinitely differentiable for $|x| > 1 \Longrightarrow \phi \in C^\infty(\mathbb{R} \setminus \{-1, 1\})$; i.e. ϕ is continuous and infinitely differentiable on $\mathbb{R} \setminus \{-1, +1\}$. It remains to show that ϕ is, in fact, continuous and infinitely differentiable at ± 1 also – then ϕ will belong to $C_0^\infty(\mathbb{R})$, with $\text{supp}(\phi) = [-1, 1]$. Since ϕ is a symmetric function in x, it is sufficient to show the continuity and infinite differentiability of ϕ at $x = +1$ only.

Continuity of ϕ at $x = +1$: $\lim_{x\to 1^-} \phi(x) = \lim_{x\to 1^-}(e^{\frac{-1}{(1-x^2)}}) = 0$ and

$$\lim_{x\to 1^+} \phi(x) = \lim_{x\to 1^+} 0 = 0 \quad \Longrightarrow \quad \lim_{x\to 1^-} \phi(x) = \lim_{x\to 1^+} \phi(x) = \phi(1) = 0$$
$$\Longrightarrow \quad \phi \text{ is continuous at } x = 1.$$

Infinite differentiability of ϕ at $x = +1$: Set $u(x) = 1 - x^2$ so that $\phi(x) = e^{-1/u}$ for $|x| < 1$. Then, $\frac{du}{dx} = -2x$, $\frac{d}{dx}(-\frac{1}{u}) = -2xu^{-2}$, $\phi'(x) = -2xu^{-2}e^{-1/u} \Longrightarrow \forall |x| < 1$, $\phi'(x) = p_1(x)u^{-2}e^{-1/u}$, where $p_1(x) = -2x$ is a polynomial in x with the subscript '1' in $p_1(x)$ corresponding to the order of the derivative in $\phi'(x)$. Then, $\forall |x| < 1$, $\phi''(x) = [(p_1(x))'u^2 + (4x)p_1(x)u + (p_1(x)^2]u^{-4}e^{-1/u} = p_2(x)u^{-4}e^{-1/u}$, where $p_2(x)$ is a polynomial in x with the subscript '2' corresponding to the order of $\phi''(x)$. Thus, by induction, we get, $\forall k \in \mathbb{N}$, $\phi^{(k)}(x) = p_k(x)u^{-2k}e^{-1/u}$, $\forall |x| < 1$, where $p_k(x)$ is a polynomial in x with the subscript k corresponding to the order of $\phi^{(k)}(x)$. Hence, $\forall k \in \mathbb{N}$, $\lim_{x\to 1^-} \phi^{(k)}(x) = \lim_{x\to 1^-}[p_k(x)u^{-2k}e^{-1/u}] = p_k(1)\cdot 0 = 0$ (by L'Hospital's rule, $e^{-1/u}\cdot u^{-2k} \to 0$ as $x \to 1^-$). Again, $\forall |x| > 1$, $\phi(x) = 0 \Longrightarrow \forall |x| > 1$, $\forall k \in \mathbb{N}$, $\phi^{(k)}(x) = 0 \Longrightarrow \lim_{x\longrightarrow 1^+} \phi^{(k)}(x) = 0\ \forall k \in \mathbb{N}$. Thus, $\phi^{(k)}(x)$ is defined for $0 < |x - 1| < r$ with $r > 0$, and $\lim_{x\to 1} \phi^{(k)}(x) = \lim_{x\to 1^-} \phi^{(k)}(x) = \lim_{x\to 1^+} \phi^{(k)}(x) = 0\ \forall k \in \mathbb{N}$.

Hence, for $k = 1$, $\phi'(1) = \lim_{x\to 1} \frac{\phi(x)-\phi(1)}{x-1} = \lim_{x\to 1} \phi'(1 + \theta_1(x - 1)) = 0$ ($\theta_1 \in]0, 1[$). Thus, $\phi'(1)$ is well defined and $\phi'(1) = 0$. Similarly, $\phi''(1) = \lim_{x\to 1} \frac{\phi'(x)-\phi'(1)}{x-1} = \lim_{x\to 1} \phi''(1 + \theta_2(x - 1)) = 0(\theta_2 \in]0, 1[)$. Thus, in general, if it is known that for some $k \in \mathbb{N}, \phi^{(k)}(1) = 0$, then

$$\begin{aligned}\phi^{(k+1)}(1) &= \lim_{x\to 1} \frac{\phi^{(k)}(x) - \phi^{(k)}(1)}{x-1} \\ &= \lim_{x\to 1} \phi^{(k+1)}(1 + \theta_k(x-1)) = 0 \quad (0 < \theta_k < 1).\end{aligned}$$

Hence, we have $\phi(1) = \phi'(1) = \cdots = \phi^{(k)}(1) = \cdots = 0$. As explained earlier, by virtue of the symmetry of ϕ, $\phi(-1) = \phi'(-1) = \cdots = \phi^{(k)}(-1) = \cdots = 0$, i.e. $\phi \in C_0^\infty(\mathbb{R})$ with $\text{supp}(\phi) = [-1, -1] \subset\subset \mathbb{R}$, with $\phi^{(k)}(x) = 0 \forall k \in \mathbb{N}_0, \forall x$ outside $[-1, 1]$. □

In general, we can define, $\forall \varepsilon > 0$:

$$\phi_\varepsilon(x) = \begin{cases} e^{(-\frac{\varepsilon^2}{\varepsilon^2 - x^2})} & \text{for } |x| < \varepsilon \\ 0 & \text{for } |x| \geq \varepsilon. \end{cases} \tag{1.2.6a}$$

Then, $\phi_\varepsilon \in C_0^\infty(\mathbb{R})$ with $\text{supp}(\phi_\varepsilon) = [-\varepsilon, \varepsilon] \subset\subset \mathbb{R}$.
$\phi_\varepsilon(x)$ and its derivatives $\phi_\varepsilon^{(k)}(x) \forall k \in \mathbb{N}$ vanish outside $[-\varepsilon, \varepsilon]$.

Example 1.2.2. For $n \geq 2$, let $\Omega = \mathbb{R}^n$, $\|\mathbf{x}\|^2 = x_1^2 + x_2^2 + \cdots + x_n^2$, $\forall \varepsilon > 0$, $\phi_\varepsilon : \mathbb{R}^n \longrightarrow \mathbb{R}$ be defined by:

$$\phi_\varepsilon(\mathbf{x}) = \begin{cases} e^{(-\frac{\varepsilon^2}{\varepsilon^2 - \|\mathbf{x}\|^2})} & \text{for } \|\mathbf{x}\| < \varepsilon \\ 0 & \text{for } \|\mathbf{x}\| \geq \varepsilon. \end{cases} \tag{1.2.6b}$$

Then, ϕ_ε belongs to $C_0^\infty(\mathbb{R}^n)$ with $\text{supp}(\phi_\varepsilon) = \overline{B(\mathbf{0}; \varepsilon)} = \{\mathbf{x} : \mathbf{x} \in \mathbb{R}^n, \|\mathbf{x}\| \leq \varepsilon\} \subset\subset \mathbb{R}^n$.
$\phi_\varepsilon(\mathbf{x}) = 0, \partial^{\boldsymbol{\alpha}} \phi_\varepsilon(\mathbf{x}) = 0 \quad \forall |\boldsymbol{\alpha}| \in \mathbb{N}$ and $\forall \mathbf{x}$ outside $B(\mathbf{0}; \varepsilon)$.

1.2.2 Space $\mathcal{D}(\Omega)$

Definition 1.2.1. Let $\Omega \subset \mathbb{R}^n$ be an open subset of $\mathbb{R}^n$. Then the space $C_0^\infty(\Omega)$ (1.2.5) is called the *test space* (or, equivalently, the *space of test functions*) and denoted by $\mathcal{D}(\Omega)$ (using the notation of Schwartz [8]); i.e. $\mathcal{D}(\Omega) = C_0^\infty(\Omega) = \{\phi : \phi \in C^\infty(\Omega), \text{supp}(\phi) \subset\subset \Omega\}$ if it is equipped with the following notion of *convergence*:
A sequence $(\phi_n)_{n=1}^\infty$ in $\mathcal{D}(\Omega)$ is said to converge to $\phi \in \mathcal{D}(\Omega)$ if and only if:

I. $\exists$ a compact subset $K \subset\subset \Omega$ such that $\text{supp}(\phi_n - \phi) \subset K \ \forall n \in \mathbb{N}$ (i.e. the supports of all $(\phi_n - \phi)$ are contained in the same compact set K ($\forall n \in \mathbb{N}$)).

II. $$\partial^{\boldsymbol{\alpha}} \phi_n \longrightarrow \partial^{\boldsymbol{\alpha}} \phi \text{ \emph{uniformly} in } \Omega \text{ as } n \to \infty \ \forall \text{ fixed } \boldsymbol{\alpha} \text{ with } |\boldsymbol{\alpha}| \in \mathbb{N}_0 \tag{1.2.7}$$

(i.e. the derivatives of any given order $|\boldsymbol{\alpha}| \in \mathbb{N}_0$ of ϕ_n converge *uniformly* to the corresponding derivatives of ϕ of the same order $|\boldsymbol{\alpha}|$ as $n \to \infty$, and for $\boldsymbol{\alpha} = \mathbf{0}$, with $|\boldsymbol{\alpha}| = 0, \partial^{\boldsymbol{\alpha}} \varphi_n = \phi_n, \partial^{\boldsymbol{\alpha}} \phi = \phi, \phi_n \to \phi$ *uniformly* as $n \to \infty$ (see also Remark 1.2.1)).

Examples of convergence in $\mathcal{D}(\Omega)$

For the sake of simplicity, we will consider the case $\Omega = \mathbb{R}$. Let $\phi \in \mathcal{D}(\mathbb{R})$ be defined by (1.2.6) with $\text{supp}(\phi) = [-1, 1]$, which is used to define the sequences in the following examples.

Example 1.2.3. Let $(\phi_n)_{n=1}^{\infty}$ be a sequence in $\mathcal{D}(\mathbb{R})$ defined by: $\phi_n(x) = \frac{1}{n}\phi(x)$ $\forall x \in \mathbb{R}, \forall n \in \mathbb{N}$. Hence, $\text{supp}(\phi_n) = \text{supp}(\phi) = [-1, 1] \subset\subset \mathbb{R} \forall n \in \mathbb{R}$, and we can choose $K = [-1, 1] \subset\subset \mathbb{R}$ such that $\text{supp}(\phi_n) \subset K \forall n \in \mathbb{N}$ and $\frac{d^\alpha \phi_n}{dx^\alpha} \to 0$ (null function) *uniformly* in $\mathbb{R}$ as $n \to \infty \forall \alpha \geq 0 \Longrightarrow \phi_n \to 0$ in $\mathcal{D}(\mathbb{R})$ as $n \to \infty$ (Figure 1.2).

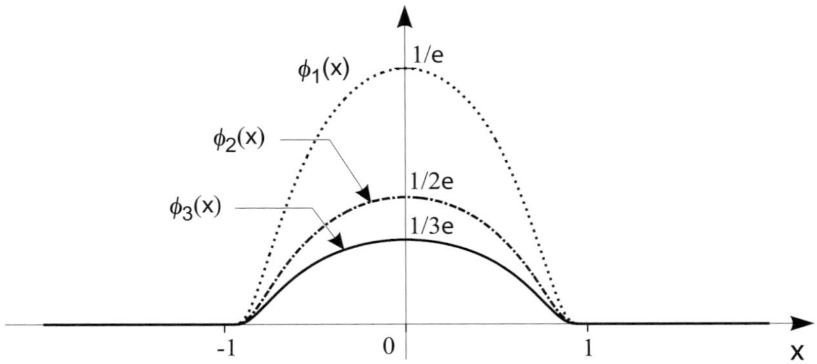

Figure 1.2 Graphs of $\phi_n(x) = \frac{1}{n}\phi(x)$ with $\text{supp}(\phi_n) = [-1, 1], n = 1, 2, 3, \ldots$

Counterexample 1.2.4. $\forall n \in \mathbb{N}$, let ϕ_n be defined by:

$$\phi_n(x) = \frac{1}{n}\phi(\frac{x}{n}) = \begin{cases} \frac{1}{n}\exp\left(\frac{-n^2}{n^2-x^2}\right) & \text{for } |x| < n \\ 0 & \text{for } |x| \geq n \end{cases}$$

Then $(\phi_n)_{n=1}^{\infty}$ is a sequence in $\mathcal{D}(\mathbb{R})$ with $\text{supp}(\phi_n) = [-n, n] \subset\subset \mathbb{R} \forall$ fixed $n \in \mathbb{N}$.

But $\text{supp}(\phi_n) = [-n, n] \to]-\infty, \infty[$ as $n \to \infty \Longrightarrow$ there does *not* exist any compact subset $K \subset\subset \mathbb{R}$ containing $\text{supp}(\phi_n)$ for *all* $n \in \mathbb{N}$.

Hence, $(\phi_n)_{n=1}^{\infty}$ does *not* converge to the null function 0 in $\mathcal{D}(\mathbb{R})$, although $\frac{d^\alpha \phi_n}{dx^\alpha} \to 0$ *uniformly* in $\mathbb{R}$ as $n \to \infty$ $\forall \alpha \geq 0$ (Figure 1.3).

Example 1.2.5. For fixed non-zero $a \in \mathbb{R}$ and $\forall n \in \mathbb{N}$, let ϕ_n be defined by:

$$\phi_n(x) = \frac{1}{n}\phi\left(\frac{x}{a}\right) = \begin{cases} \frac{1}{n}\exp\left(\frac{-a^2}{a^2-x^2}\right) & \text{for } |x| < a \\ 0 & \text{for } |x| \geq a \end{cases}$$

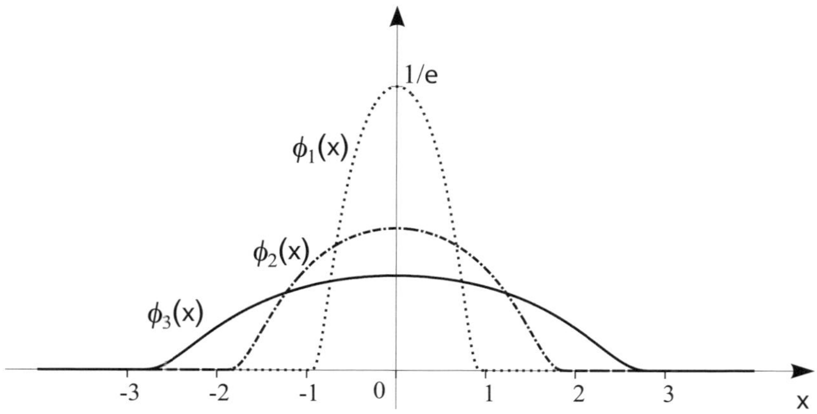

Figure 1.3 Graphs of $\phi_n(x) = \frac{1}{n}\phi(\frac{x}{n})$ with $\text{supp}(\phi_n) = [-n, n], n = 1, 2, 3, \ldots$

$\Longrightarrow (\phi_n)_{n=1}^{\infty}$ is a sequence in $\mathcal{D}(\mathbb{R})$ with $\text{supp}(\phi_n) = [-a, a] \subset\subset \mathbb{R}\, \forall n \in \mathbb{N}$.

Hence, we can choose $K = [-a, a] \subset\subset \mathbb{R}$ satisfying $\text{supp}(\phi_n) \subset K \quad \forall n \in \mathbb{N}$ and $\frac{d^\alpha \phi_n}{dx^\alpha} \to 0$ *uniformly* in $\mathbb{R}$ as $n \to \infty\ \forall \alpha \geq 0 \Longrightarrow \phi_n \to 0$ in $\mathcal{D}(\mathbb{R})$ as $n \to \infty$.

Counterexample 1.2.6. $\forall n \in \mathbb{N}$, let ϕ_n be defined by:

$$\phi_n(x) = \frac{1}{n}\phi(x - n) = \begin{cases} \frac{1}{n}\exp\left(\frac{-1}{1-(x-n)^2}\right) & \text{for } |x - n| < 1 \\ 0 & \text{for } |x - n| \geq 1 \end{cases}$$

$\Longrightarrow \phi_n \in \mathcal{D}(\mathbb{R})$ with $\text{supp}(\phi_n) = \{x : x \in \mathbb{R}, |x - n| \leq 1\} = [n - 1, n + 1]\ \forall$ fixed $n \in \mathbb{N}$.

But there does *not* exist any compact set $K \subset\subset \mathbb{R}$ such that $\text{supp}(\phi_n) \subset K\ \forall n \in \mathbb{N}$.

Hence, $(\phi_n)_{n=1}^{\infty}$ does *not* converge to the null function 0 in $\mathcal{D}(\mathbb{R})$, although $\frac{d^\alpha \phi_n}{dx^\alpha} \to 0$ *uniformly* in $\mathbb{R}$ as $n \to \infty\ \forall \alpha \geq 0$ (Figure 1.4).

Counterexample 1.2.7. $\forall n \in \mathbb{N}$, let ϕ_n be defined by:

$$\phi_n = \frac{1}{n}\phi(nx) = \begin{cases} \frac{1}{n}\exp\left(\frac{-1}{1-(nx)^2}\right) \text{ for } |nx| < 1 (\text{ i.e. for } |x| < \frac{1}{n}) \\ 0 \text{ for } |x| \geq \frac{1}{n} \end{cases}$$

$\Longrightarrow \phi_n \in \mathcal{D}(\mathbb{R})$ with $\text{supp}(\phi_n) = \overline{B}(0; \frac{1}{n})\ \forall n \in \mathbb{N}$, i.e. ϕ_n has support shrinking to 0 as $n \to \infty$. Hence, condition I in Definition 1.2.1, $\text{supp}(\phi_n) \subset [-1, 1]$, holds $\forall n \in \mathbb{N}$, but for $k \geq 2$, $\frac{d^k \phi_n}{dx^k}$ does *not* converge to 0 on any neighbourhood of 0 as $n \to \infty$, i.e. condition II in (1.2.7) does *not* hold.

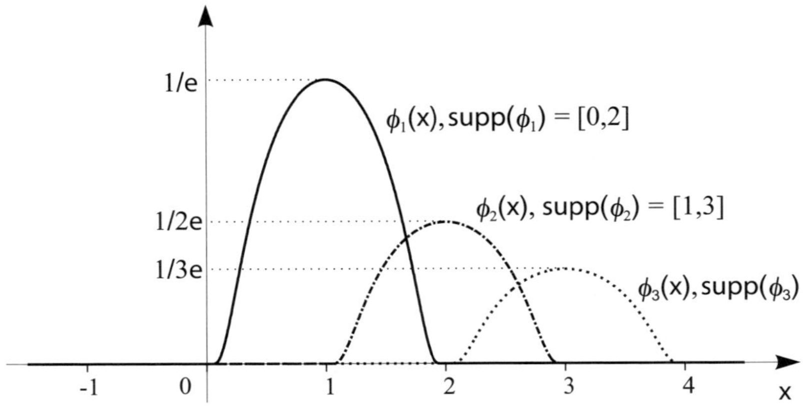

Figure 1.4 Graphs of $\phi_n(x) = \frac{1}{n}\phi(x-n)$ with $\text{supp}(\phi_n) = [n-1, n+1], n = 1, 2, 3, \ldots$

Remark 1.2.1.

- Since the uniform convergence of derivatives of *all* orders are involved in this notion of convergence, it is a notion of convergence of *infinite order*.
- The derivatives of each order *taken separately* converge uniformly, i.e. derivatives of all orders need not *simultaneously* converge uniformly.
- The elements $\phi \in \mathcal{D}(\Omega)$ are called *test functions*.
- $\phi = 0 \in \mathcal{D}(\Omega)$, since the null function $0 \in C^\infty(\Omega)$, $\text{supp}(0) = \emptyset$, the empty set, which is a *compact subset* of Ω.
- $\phi \in \mathcal{D}(\Omega) \Longrightarrow \exists$ a compact subset $K \subset\subset \Omega$ such that $\phi(\mathbf{x}) = 0, \partial^{\boldsymbol{\alpha}}\phi(\mathbf{x}) = 0$ $\forall \mathbf{x} \in \Omega \setminus K$ (i.e. $\mathbf{x}$ lying outside K) $\forall |\boldsymbol{\alpha}| \in \mathbb{N}$.
- Ω is a *bounded* domain with *boundary* $\Gamma, \phi \in \mathcal{D}(\Omega)$. Hence:

  I. $\phi(\mathbf{x}) = 0,\ \partial^{\boldsymbol{\alpha}}\phi(\mathbf{x}) = 0\ \forall \mathbf{x} \in \Gamma,\ \forall |\boldsymbol{\alpha}| \in \mathbb{N}$.

  II. For *sufficiently small* $\varepsilon > 0$, $\exists$ an ε-boundary layer of Γ in Ω such that $\phi(\mathbf{x}) = 0,\ \partial^{\boldsymbol{\alpha}}\phi(\mathbf{x}) = 0\ \forall |\boldsymbol{\alpha}| \in \mathbb{N}$, $\forall \mathbf{x}$ belonging to the ε-boundary layer of Γ in Ω.

  For *sufficiently small* $\varepsilon > 0$, define

$$\Omega_\varepsilon = \{\mathbf{x} : \mathbf{x} \in \Omega, d(\mathbf{x}, \Gamma) = \inf_{\mathbf{y} \in \Gamma} d(\mathbf{x}, \mathbf{y}) > \varepsilon\} \subset \Omega. \tag{1.2.8}$$

  Then $\Omega \setminus \Omega_\varepsilon$ is called an *ε-boundary layer of* Γ in Ω (Figure 1.5).

For more details, see Theorems A.0.5.1 and A.0.5.3, Appendix A.

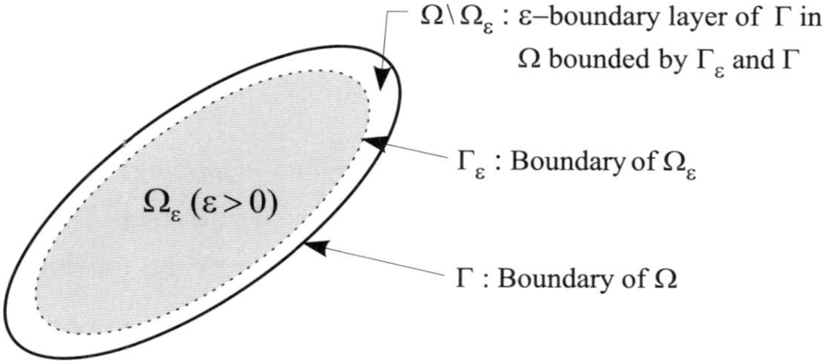

Figure 1.5 ε-boundary layer of Γ in Ω

1.2.3 Space $\mathcal{D}^m(\Omega)$

Definition 1.2.2. $\forall m \in \mathbb{N}_0$, the space $C_0^m(\Omega)$ defined by (1.2.3),

$$C_0^m(\Omega) = \{\phi : \partial^{\boldsymbol{\alpha}} \phi \in C^0(\Omega) \forall |\boldsymbol{\alpha}| \leq m, \operatorname{supp}(\phi) \subset\subset \Omega\},$$

will be denoted by $\mathcal{D}^m(\Omega) \equiv C_0^m(\Omega)$ if it is equipped with the following notion of convergence: a sequence $(\phi_n)_{n=1}^{\infty}$ in $\mathcal{D}^m(\Omega)$ converges to $\phi \in \mathcal{D}^m(\Omega)$ if and only if:

I. $\exists$ a compact subset $K \subset\subset \Omega$ of Ω such that

$$\operatorname{supp}(\phi_n - \phi) \subset K \quad \forall n \in \mathbb{N}; \tag{1.2.9}$$

II. $\partial^{\boldsymbol{\alpha}} \phi_n \to \partial^{\boldsymbol{\alpha}} \phi$ *uniformly* in Ω as $n \to \infty$ $\forall |\boldsymbol{\alpha}| \leq m$.

Convergence of a sequence in $\mathcal{D}(\Omega) \Longrightarrow$ its convergence in $\mathcal{D}^m(\Omega)$

For example, the sequence (ϕ_n) in Examples 1.2.3 and 1.2.5, which converges in $\mathcal{D}(\mathbb{R})$, also converges in $\mathcal{D}^m(\mathbb{R})$, whereas the sequences in Examples 1.2.4 and 1.2.6 do *not* converge in both $\mathcal{D}(\mathbb{R})$ and $\mathcal{D}^m(\mathbb{R})$, since condition I for convergence is the same in $\mathcal{D}(\mathbb{R})$ and $\mathcal{D}^m(\mathbb{R})$ $\forall m \in \mathbb{N}_0$.

$$\mathcal{D}(\mathbb{R}) \hookrightarrow \mathcal{D}^m(\mathbb{R}) \quad \forall m \in \mathbb{N}_0 \text{with continuous imbedding} \tag{1.2.10}$$

$$\mathcal{D}^m(\Omega) \equiv \mathcal{D}(\Omega) \quad \Longleftrightarrow \quad m = \infty. \tag{1.2.11}$$

1.2.4 Space $\mathcal{D}_K(\Omega)$

Definition 1.2.3. $\forall$ fixed *compact* subsets $K \subset\subset \Omega$ of $\Omega \subset \mathbb{R}^n$, $\mathcal{D}_K(\Omega)$ is defined by the set:

$$\mathcal{D}_K(\Omega) = \{\phi : \phi \in \mathcal{D}(\Omega), \operatorname{supp}(\phi) \subset K\}. \tag{1.2.12}$$

Then, $\mathcal{D}(\Omega) = \bigcup_{K \subset\subset \Omega} \mathcal{D}_K(\Omega)$.

- $\forall$ multi-index $\boldsymbol{\alpha}$ with $|\boldsymbol{\alpha}| \in \mathbb{N}_0$, semi-norms $p_{\boldsymbol{\alpha}}$ on $\mathcal{D}_K(\Omega)$ are defined by:

$$p_{\boldsymbol{\alpha}}(\phi) = \max_{\mathbf{x}\in K} |\partial^{\boldsymbol{\alpha}}\phi(\mathbf{x})| \quad \forall \phi \in \mathcal{D}_K(\Omega). \tag{1.2.13}$$

- The family $\{p_{\boldsymbol{\alpha}}\}$ of all these semi-norms defines a *topology* on $\mathcal{D}_K(\Omega)$.
- $\mathcal{D}(\Omega)$ is the *strict inductive limit* of the spaces $\mathcal{D}_K(\Omega)$ (see [8], [9], [10], [11]).
- The topology of $\mathcal{D}(\Omega)$ is defined with the help of the topology of $\mathcal{D}_K(\Omega)$ (see [8], [9], [25]).

Now we will collect the important properties of $\mathcal{D}(\Omega)$.

1.2.5 Properties of $\mathcal{D}(\Omega)$

Property 1: $\mathcal{D}(\Omega)$ is a *linear* space: $\phi_1, \phi_2 \in \mathcal{D}(\Omega)$, $\alpha_1, \alpha_2 \in \mathbb{R} \Longrightarrow \alpha_1\phi_1 + \alpha_2\phi_2 \in \mathcal{D}(\Omega)$, $\phi = 0 \in \mathcal{D}(\Omega)$.

Property 2: $\mathcal{D}(\Omega)$ is *not* normable. $\mathcal{D}(\Omega)$ is a locally convex topological vector space. (1.2.14)

Property 3: Topology on $\mathcal{D}(\Omega)$: For this, it is necessary to introduce the notion of *inductive limit topology*, for which we refer to Schwartz [8, p. 66], Horvath [9, p. 132], and Trèves [25]. But in our treatment and for the applications in this book, it is sufficient to know the concept of convergence in $\mathcal{D}(\Omega)$ from (1.2.7), which is, in fact, called the *pseudo-topology of Schwartz* in $\mathcal{D}(\Omega)$ by Lions [13], Lions–Magenes [15], Nečas [16], etc.[4]

Property 4: $\phi_1, \phi_2 \in \mathcal{D}(\Omega) \Longrightarrow \phi_1\phi_2 \in \mathcal{D}(\Omega)$ (1.2.15)
with $\operatorname{supp}(\phi_1\phi_2) = \operatorname{supp}(\phi_1) \cap \operatorname{supp}(\phi_2)$;

$\phi \in \mathcal{D}(\Omega), \psi \in C^\infty(\Omega) \Longrightarrow \psi\phi \in \mathcal{D}(\Omega)$ (1.2.16)
with $\operatorname{supp}(\psi\phi) = \operatorname{supp}(\psi) \cap \operatorname{supp}(\phi)$.

Property 5:

$$\phi \in \mathcal{D}(\Omega) \Longrightarrow \partial^\alpha\phi \in \mathcal{D}(\Omega) \quad \forall |\alpha| \in \mathbb{N}. \tag{1.2.17}$$

Property 6: $\phi_n \to \phi$ in $\mathcal{D}(\Omega)$ in the sense of (1.2.7) $\Longrightarrow \partial^{\boldsymbol{\alpha}}\phi_n \to \partial^{\boldsymbol{\alpha}}\phi$ in $\mathcal{D}(\Omega)$ $\forall |\boldsymbol{\alpha}| \in \mathbb{N}$. (1.2.18)

[4] Moreover, Trèves also writes in [10], 'In teaching it [sophisticated functional analysis of topological vector spaces] can easily be, and most of the time is, bypassed. Most of the basic tenets of the theory can be stated and proved using solely [sequences] of test functions or distributions. The great success and usefulness of distribution theory lies in its simplicity and in the easy, automatic nature of operations. Many accused it of being shallow… With the easy part taken care of, analysts could push further and take care of the finer and more difficult points.'

Property 7: For sufficiently small $d > 0$, the *d-neighbourhood* of a compact set $K \subset\subset \Omega$ is the set K_d of points in Ω whose distance from K is $\le d$, i.e.

$$K_d = \{\mathbf{x} : \mathbf{x} \in \Omega, d(\mathbf{x}; K) = \inf_{\mathbf{y}\in K} d(\mathbf{x}, \mathbf{y}) \le d\}. \tag{1.2.19}$$

Then, $K_d \subset\subset \Omega$ is a compact set containing $K \subset K_d$ (Figure 1.6). For more details on the properties of compact sets in $\mathbb{R}^n$, see Section A.0.5, Appendix A.

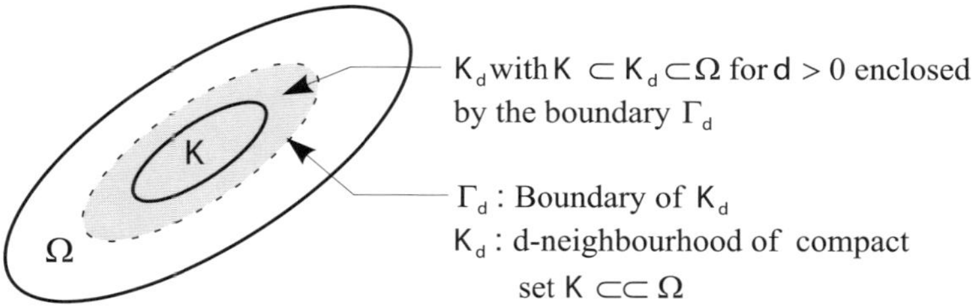

Figure 1.6 d-neighbourhood of K_d in Ω with $K \subset K_d \subset \Omega$ for $d > 0$

Theorem 1.2.1 (Approximation Theorem). *Let* $f \in C_0(\Omega) \equiv C_0^0(\Omega)$ (1.2.4) *be a continuous function with compact support* $K = \text{supp}(f) \subset\subset \Omega$, *and* $K_d \subset\subset \Omega$ *be a* d*-neighbourhood of* K *with* $d > 0$ *as defined by* (1.2.19). *Then, for any* $\varepsilon > 0, \exists$ *a function* $\phi \in \mathcal{D}(\Omega)$ *with* $\text{supp}(\phi) \subset K_d$ *such that* $\sup_{\mathbf{x}\in\Omega}|f(\mathbf{x}) - \phi(\mathbf{x})| \le \varepsilon$, *i.e.* $\forall f \in C_0(\Omega), \exists\phi \in \mathcal{D}(\Omega)$ arbitrarily close *to* f. (1.2.20)

In other words, $\mathcal{D}(\Omega)$ *is* dense *in* $C_0(\Omega)$.

Proof. Let $f \in C_0(\Omega)$ with compact $\text{supp}(f) = K \subset\subset \Omega$, and let $\tilde{f}$ be its null extension to $\mathbb{R}^n$, i.e. $\tilde{f}(\mathbf{x}) = f(\mathbf{x})\ \forall \mathbf{x} \in \Omega$ and $\tilde{f}(\mathbf{x}) = 0\ \forall \mathbf{x} \in \mathbb{R}^n \setminus \Omega$, $\text{supp}(\tilde{f}) = \text{supp}(f) = K$, $\tilde{f} \in C_0(\mathbb{R}^n)$.

Using *regularizing functions* ρ_δ (see Definition 6.2.1) with the properties (6.2.17)–(6.2.19): $\forall \delta > 0$, $\rho_\delta(\mathbf{x}) = \frac{1}{k}\rho(\frac{\mathbf{x}}{\delta})$, where $\rho(\frac{\mathbf{x}}{\delta}) = \exp(\frac{-\delta^2}{\delta^2 - \|\mathbf{x}\|^2})$ for $\|\mathbf{x}\| < \delta$ and $\rho(\frac{\mathbf{x}}{\delta}) = 0$ for $\|\mathbf{x}\| \ge \delta$, $0 \le \rho_\delta(\mathbf{x}) \le 1$, $\text{supp}(\rho_\delta) = \bar{B}(\mathbf{0}; \delta) = \{\mathbf{x} : \|\mathbf{x}\| \le \delta\}$; $\int_{\mathbb{R}^n} \rho_\delta(\mathbf{x})d\mathbf{x} = 1$, $\rho_\delta \in \mathcal{D}(\mathbb{R}^n)$, $k = \int_{\mathbb{R}^n} \rho(\frac{\mathbf{x}}{\delta})d\mathbf{x}$, define $\psi_\delta(\mathbf{x}) = \int_{\mathbb{R}^n} \tilde{f}(\mathbf{x} - \boldsymbol{\xi})\rho_\delta(\boldsymbol{\xi})d\boldsymbol{\xi} = \int_{\mathbb{R}^n} \tilde{f}(\boldsymbol{\xi})\rho_\delta(\mathbf{x} - \boldsymbol{\xi})d\boldsymbol{\xi}$ (by change of variables) $= \int_K f(\boldsymbol{\xi})\rho_\delta(\mathbf{x} - \boldsymbol{\xi})d\boldsymbol{\xi}\ \forall \mathbf{x} \in \mathbb{R}^n$, since $\text{supp}(\tilde{f}) = \text{supp}(f) = K$.

Then $\psi_\delta \in C_0^\infty(\mathbb{R}^n)$ with $\text{supp}(\psi_\delta) \subset K_d$, if we choose $0 < \delta \le d$ (see the proof of Lemma 6.2.1 for all the details); since $\rho_\delta \in C_0^\infty(\mathbb{R}^n)$ with $\text{supp}(\rho_\delta) = \bar{B}(\mathbf{0}, \delta)$, ψ_δ can be differentiated indefinitely with respect to variables $x_1, x_2, \dots, x_n$, which can be carried out under the integral sign $\int$, i.e.

- $\forall |\boldsymbol{\alpha}| \in \mathbb{N}$, $\partial^{\boldsymbol{\alpha}}\psi_\delta(\mathbf{x}) = \int_K f(\boldsymbol{\xi})\partial_{\mathbf{x}}^{\boldsymbol{\alpha}}[\rho_\delta(\mathbf{x} - \boldsymbol{\xi})]d\boldsymbol{\xi} \Longrightarrow \psi_\delta \in C^\infty(\mathbb{R}^n)$;
- $\rho_\delta(\mathbf{x} - \boldsymbol{\xi}) = 0$ for $\|\mathbf{x} - \boldsymbol{\xi}\| \ge \delta$ for all $\boldsymbol{\xi} \in K \Longrightarrow \psi_\delta(\mathbf{x}) = \int_K f(\boldsymbol{\xi})\rho_\delta(\mathbf{x} - \boldsymbol{\xi})d\boldsymbol{\xi} = 0$ for $\mathbf{x} \notin K_\delta$ with $K_\delta = \{\mathbf{x} : d(\mathbf{x}; K) = \inf_{\boldsymbol{\zeta}\in K} \|\mathbf{x} - \boldsymbol{\zeta}\| \le \delta\}$. But

$\mathbf{x} \notin K_\delta \Longrightarrow$ for all $\boldsymbol{\xi} \in K$, $\|\mathbf{x}-\boldsymbol{\xi}\| \geq \inf_{\boldsymbol{\zeta}\in K} \|\mathbf{x}-\boldsymbol{\zeta}\| > \delta$. Hence $\mathrm{supp}(\psi_\delta)$ will be contained in $K_d \subset\subset \Omega$, i.e. $\mathrm{supp}(\psi_\delta) \subset K_d \subset\subset \Omega$, *if we choose* $\delta \leq d$. (1.2.20a)

Since $\int_{\mathbb{R}^n} \rho_\delta(\boldsymbol{\xi})d\boldsymbol{\xi} = 1$,

$$\tilde{f}(\mathbf{x}) - \psi_\delta(\mathbf{x}) = \tilde{f}(\mathbf{x}) \int_{\mathbb{R}^n} \rho_\delta(\boldsymbol{\xi})d\boldsymbol{\xi} - \int_{\mathbb{R}^n} \tilde{f}(\mathbf{x}-\boldsymbol{\xi})\rho_\delta(\boldsymbol{\xi})d\boldsymbol{\xi}$$
$$= \int_{\mathbb{R}^n} [\tilde{f}(\mathbf{x}) - \tilde{f}(\mathbf{x}-\boldsymbol{\xi})]\rho_\delta(\boldsymbol{\xi})d\boldsymbol{\xi} = \int_{\|\boldsymbol{\xi}\|\leq\delta} [\tilde{f}(\mathbf{x}) - \tilde{f}(\mathbf{x}-\boldsymbol{\xi})]\rho_\delta(\boldsymbol{\xi})d\boldsymbol{\xi} \tag{1.2.20b}$$

(since $\rho_\delta(\boldsymbol{\xi}) = 0$ for $\|\boldsymbol{\xi}\| \geq \delta$).

But $\tilde{f}$ is continuous in $\mathbb{R}^n$ and $\mathrm{supp}(\tilde{f}) = \mathrm{supp}(f) = K$ is *compact*. Hence, $\tilde{f}$ is *uniformly continuous* and consequently, $\forall\varepsilon > 0, \exists\eta > 0$ such that $|\tilde{f}(\mathbf{x})-\tilde{f}(\mathbf{x}-\boldsymbol{\xi})| \leq \varepsilon\ \forall\|\boldsymbol{\xi}\| \leq \eta$.

Then $|\tilde{f}(\mathbf{x}) - \tilde{f}(\mathbf{x}-\boldsymbol{\xi})| \leq \varepsilon\ \forall\|\boldsymbol{\xi}\| \leq \delta \leq \eta$ (if we choose $\delta \leq \eta$) $\Longrightarrow$ from (1.2.20b),

$$|\tilde{f}(\mathbf{x}) - \psi_\delta(\mathbf{x})| \leq \int_{\|\boldsymbol{\xi}\|\leq\delta} |\tilde{f}(\mathbf{x}) - \tilde{f}(\mathbf{x}-\boldsymbol{\xi})|\rho_\delta(\boldsymbol{\xi})d\boldsymbol{\xi} \leq \varepsilon \int_{\mathbb{R}^n} \rho_\delta(\boldsymbol{\xi})d\boldsymbol{\xi} = \varepsilon\cdot 1 = \varepsilon \tag{1.2.20c}$$

for $0 < \delta \leq \eta$.

Define $\phi = \psi_\delta\downarrow_\Omega$ with $\psi_\delta \in C_0^\infty(\mathbb{R}^n)$, $\mathrm{supp}(\psi_\delta) \subset K_d \subset\subset \Omega$, $0 < \delta \leq d$ and $\delta \leq \eta$ (by virtue of (1.2.20a) and (1.2.20c)). Then, $\phi \in C_0^\infty(\Omega)$ with $\mathrm{supp}(\phi) = \mathrm{supp}(\psi_\delta) \subset K_d \subset\subset \Omega$ and, $\forall\mathbf{x} \in \Omega$, $|f(\mathbf{x}) - \phi(\mathbf{x})| = |\tilde{f}(\mathbf{x}) - \psi_\delta(\mathbf{x})| \leq \varepsilon$ for $0 < \delta \leq d$ and $\delta \leq \eta$. Thus, for $0 < \delta \leq d$ and $\delta \leq \eta$, $\exists\phi \in \mathcal{D}(\Omega)$ with $\mathrm{supp}(\phi) \subset K_d$ such that $\sup_{\mathbf{x}\in\Omega} |f(\mathbf{x}) - \phi(\mathbf{x})| \leq \varepsilon$. □

Theorem 1.2.2. *Let $\Omega \subset \mathbb{R}^n$ be an open subset of $\mathbb{R}^n$ and $L^1_{\mathrm{loc}}(\Omega)$ be the space of* locally integrable functions *on Ω, i.e.*

$$L^1_{\mathrm{loc}}(\Omega) = \{f : \forall K \subset\subset \Omega \int_{\mathring{K}} |f(\mathbf{x})|\, d\mathbf{x} < +\infty, \mathring{K} = int(K)\}. \tag{1.2.21}$$

If $u \in L^1_{\mathrm{loc}}(\Omega)$ and $\int_\Omega u\psi d\mathbf{x} = 0 \forall\psi \in C_0(\Omega)$, then $u(\mathbf{x}) = 0$ almost everywhere (a.e. – see Definition B.1.3.1 in Appendix B) on Ω. (1.2.21a)

Proof. Following the interesting proof in [26, pp. 61–63], we give the proof in two steps (see also [12, p. 60]).

Step 1. Let Ω be a *bounded* open subset of $\mathbb{R}^n$ and $u \in L^1(\Omega)$. In Appendix B, the density of $C_0(\Omega)$ in $L^1(\Omega)$ is given by Theorem B.3.3.4. Hence, for any given $\varepsilon > 0, \exists\psi \in C_0(\Omega)$ such that $\|u - \psi\|_{L^1(\Omega)} \leq \varepsilon$. (1.2.21b)

Then, using (1.2.21a) and (1.2.21b) and Hölder's inequality for $(u - \psi) \in L^1(\Omega)$ and $\phi \in C_0(\Omega) \subset L^\infty(\Omega)$, we get

$$\left|\int_\Omega \psi\phi d\mathbf{x} - \int_\Omega u\phi d\mathbf{x}\right| = \left|\int_\Omega (\psi - u)\phi d\mathbf{x}\right| \leq \|\psi - u\|_{L^1(\Omega)} \cdot \|\phi\|_{L^\infty(\Omega)}$$

$$\Longrightarrow \quad \left|\int_\Omega \psi\phi dx - 0\right| = \left|\int_\Omega \psi\phi d\mathbf{x}\right| \leq \varepsilon\|\phi\|_{L^\infty(\Omega)} \forall\phi \in C_0(\Omega). \tag{1.2.21c}$$

Now we will construct a function $\phi_0 \in C_0(\Omega)$ with $-1 \leq \phi_0(\mathbf{x}) \leq 1 \ \forall \mathbf{x} \in \Omega$ and some additional properties to be explained now.

For *fixed* $\varepsilon > 0$ satisfying (1.2.21b), define subsets A and B of Ω by:

$$A = \{\mathbf{x} : \mathbf{x} \in \Omega, \psi(\mathbf{x}) \geq \varepsilon\}, \quad B = \{\mathbf{x} : \mathbf{x} \in \Omega, \psi(\mathbf{x}) \leq -\varepsilon\}. \tag{1.2.21d}$$

Then A and B are *closed* subsets of a bounded set $\Omega \subset \mathbb{R}^n$ (see the proof of Proposition A.1.3.2 in Appendix A) $\Longrightarrow$ A and B are *compact* subsets of Ω, i.e. $A \subset\subset \Omega, B \subset\subset \Omega$. Again, $\mathbf{x} \in A \Longrightarrow \psi(\mathbf{x}) \geq \varepsilon \Longrightarrow \mathbf{x} \notin B$ and vice versa. Hence, A and B are *disjoint, closed* sets, i.e. $A \cap B = \emptyset$. Moreover, $A, B \subset K = \operatorname{supp}(\psi) \subset\subset \Omega$. Using Proposition A.1.3.2 in Appendix A, we can define $\phi_0 \in C_0(\Omega)$ such that $\phi_0(\mathbf{x}) = +1$ for $\mathbf{x} \in A$, $\phi_0(\mathbf{x}) = -1$ for $\mathbf{x} \in B$ and $-1 \leq \phi_0(\mathbf{x}) \leq 1$ $\forall \mathbf{x} \in \Omega$ with $\|\phi_0\|_{L^\infty(\Omega)} = \sup_{\mathbf{x}\in\Omega} |\phi_0(\mathbf{x})| = 1.$ (1.2.21e)

Set $E = A \cup B$. Then, from (1.2.21d) and (1.2.21e), for fixed $\varepsilon > 0$, with (1.2.21b), $|\psi(\mathbf{x})| \geq \varepsilon \forall \mathbf{x} \in E$, $|\psi(\mathbf{x})| < \varepsilon \forall \mathbf{x} \in \Omega \setminus E$, $|\psi(\mathbf{x})| = \psi(\mathbf{x})\phi_0(\mathbf{x}) \ \forall \mathbf{x} \in E$ and $|\psi(\mathbf{x})\phi_0(\mathbf{x})| \leq |\psi(\mathbf{x})| \ \forall \mathbf{x} \in \Omega \setminus E$.

$$\begin{aligned}\int_E |\psi(\mathbf{x})| d\mathbf{x} &= \int_E \psi(\mathbf{x})\phi_0(\mathbf{x}) d\mathbf{x} = \int_\Omega \psi(\mathbf{x})\phi_0(\mathbf{x}) d\mathbf{x} - \int_{\Omega\setminus E} \psi(\mathbf{x})\phi_0(\mathbf{x}) d\mathbf{x} \\ &\leq \left|\int_\Omega \psi(\mathbf{x})\phi_0(\mathbf{x}) d\mathbf{x}\right| + \left|\int_{\Omega\setminus E} \psi(\mathbf{x})\phi_0(\mathbf{x}) d\mathbf{x}\right| \\ &\leq \left|\int_\Omega \psi(\mathbf{x})\phi_0(\mathbf{x}) d\mathbf{x}\right| + \int_{\Omega\setminus E} \left|\psi(\mathbf{x})\phi_0(\mathbf{x})\right| d\mathbf{x}.\end{aligned} \tag{1.2.21f}$$

But $|\int_\Omega \psi(\mathbf{x})\phi_0(\mathbf{x}) d\mathbf{x}| \leq \varepsilon\|\phi_0\|_{L^\infty(\Omega)} = \varepsilon$ (using first (1.2.21c) with $\phi = \phi_0$ and then (1.2.21e)). Then $\int_E |\psi(\mathbf{x})| d\mathbf{x} \leq \varepsilon + \int_{\Omega\setminus E} |\psi(\mathbf{x})\phi_0(\mathbf{x})| d\mathbf{x} \leq \varepsilon + \int_{\Omega\setminus E} |\psi(\mathbf{x})| d\mathbf{x}$ (using (1.2.21e) and (1.2.21f)). Hence,

$$\begin{aligned}\int_\Omega |\psi(\mathbf{x})| d\mathbf{x} &= \int_E |\psi(\mathbf{x})| d\mathbf{x} + \int_{\Omega\setminus E} |\psi(\mathbf{x})| d\mathbf{x} \\ &\leq \left(\varepsilon + \int_{\Omega\setminus E} |\psi(\mathbf{x})| d\mathbf{x}\right) + \int_{\Omega\setminus E} |\psi(\mathbf{x})| d\mathbf{x} \\ &= \varepsilon + 2\int_{\Omega\setminus E} |\psi(\mathbf{x})| d\mathbf{x} \leq \varepsilon + 2\varepsilon \int_{\Omega\setminus E} d\mathbf{x} \leq \varepsilon[1 + 2\mu(\Omega)],\end{aligned}$$

since $|\psi(\mathbf{x})| < \varepsilon \forall x \in \Omega \setminus E$ and $\text{vol}(\Omega \setminus E) \le \text{vol}(\Omega) = \mu(\Omega)$ (μ being the n-dimensional Lebesgue volume measure) $\Longrightarrow \|\psi\|_{L^1(\Omega)} \le \varepsilon[1 + 2\mu(\Omega)]$. Finally,

$$\begin{aligned}\|u\|_{L^1(\Omega)} = \|(u-\psi)+\psi\|_{L^1(\Omega)} &\le \|u-\psi\|_{L^1(\Omega)} + \|\psi\|_{L^1(\Omega)} \\ &\le \varepsilon + \varepsilon[1+2\mu(\Omega)] = \varepsilon[2+2\mu(\Omega)]\end{aligned}$$

$\forall$ fixed $\varepsilon > 0$ satisfying (1.2.21b) $\Longrightarrow \|u\|_{L^1(\Omega)} \le \lim_{\varepsilon \to 0^+} [\varepsilon(2 + 2\mu(\Omega)] = 0$, i.e. $\|u\|_{L^1(\Omega)} = 0 \Longrightarrow u(\mathbf{x}) = 0$ almost everywhere on Ω.

Step 2. Now we will consider the general case of $\Omega \subset \mathbb{R}^n$ and $u \in L^1_{\text{loc}}(\Omega)$ (i.e. u may or may *not* belong to $L^1(\Omega)$). Let $\Omega \subset \mathbb{R}^n$ be an open subset of $\mathbb{R}^n$. Set $\Omega = \bigcup_m \Omega_m$ with $m \in \mathbb{N}$, where Ω_m is a *bounded* subset of Ω and $\overline{\Omega}_m$ is a compact subset of Ω $\forall m \in \mathbb{N}$, i.e. $\Omega_m \subset \overline{\Omega}_m \subset\subset \Omega$ for $m \in \mathbb{N}$. In fact, for $m \in \mathbb{N}$, Ω_m can be constructed, *for example*, as follows: let Ω^c be the complement of Ω in $\mathbb{R}^n$, i.e. $\Omega^c = \{\mathbf{x} : \mathbf{x} \in \mathbb{R}^n, \mathbf{x} \notin \Omega\}$.

Then, for $m \in \mathbb{N}$, define $\Omega_m = \{\mathbf{x} : \mathbf{x} \in \Omega, d(\mathbf{x}; \Omega^c) > \frac{1}{m+1}$ and $\|\mathbf{x}\|_{\mathbb{R}^n} < m\}$.

To fix our ideas, consider the simple, i.e. one-dimensional, case.

For $\Omega =]0, \infty[$, $\Omega^c =]-\infty, 0]$, and $\Omega_1 =]\frac{1}{2}, 1[$, $\Omega_2 =]\frac{1}{3}, 2[, \ldots, \Omega_m =]\frac{1}{m+1}, m[, \ldots; \overline{\Omega}_1 = [\frac{1}{2}, 1], \overline{\Omega}_2 = [\frac{1}{3}, 2], \ldots, \overline{\Omega}_m = [\frac{1}{m+1}, m], \ldots$ being the corresponding compact subsets of $]0, \infty[$. Then, $]0, \infty[= \bigcup_{m \in \mathbb{N}} \Omega_m$. Let $u \in L^1_{\text{loc}}(\Omega)$. Then, $u{\downarrow_{\Omega_m}} \in L^1(\Omega_m)$ $\forall m \in \mathbb{N}$.

Define $C_0(\Omega_m) = \{\phi : \phi \in C_0(\Omega), \text{supp}(\phi) \subset\subset \Omega_m\}$. Hence, $\forall m \in \mathbb{N}$, $\phi \in C_0(\Omega_m) \Longrightarrow \phi \in C_0(\Omega)$ and, $\forall$ fixed m,

$$\int_{\Omega_m} (u{\downarrow_{\Omega_m}})\phi dx = \int_{\Omega} u\phi dx = 0 \quad \forall \phi \in C_0(\Omega_m). \tag{1.2.21g}$$

Since, $\forall$ fixed m, Ω_m is a *bounded* domain, $u{\downarrow_{\Omega_m}} \in L^1(\Omega_m)$ satisfying (1.2.21g), from Step 1, $u{\downarrow_{\Omega_m}} = 0$ almost everywhere on Ω_m. For fixed m, define $E_m = \{\mathbf{x} : \mathbf{x} \in \Omega_m, u{\downarrow_{\Omega m}}(\mathbf{x}) \neq 0\}$.

Then $u{\downarrow_{\Omega_m}} = 0$ a.e. on $\Omega_m \Longrightarrow$ the n-dimensional (Lebesgue) volume measure $\mu(E_m) = 0$ $\forall m \in \mathbb{N}$. Set $E_0 = \bigcup_m E_m$. Then $E_0 = \{\mathbf{x} : \mathbf{x} \in \Omega, u(\mathbf{x}) \neq 0\}$ with $\mu(E_0) = 0$, since $0 \le \mu(E_0) \le \sum_m \mu(E_m) = 0$. Hence, $u(\mathbf{x}) = 0$ for $\mathbf{x} \in \Omega \setminus E_0$ with $\mu(E_0) = 0 \Longrightarrow u(\mathbf{x}) = 0$ for almost all $\mathbf{x} \in \Omega$. □

Corollary 1.2.1. *Let $\Omega \subset \mathbb{R}^n$ be an open subset of $\mathbb{R}^n$ and $u \in L^p(\Omega)$, $1 \le p \le \infty$, such that $\int_\Omega u(\mathbf{x})\phi d\mathbf{x} = 0$ $\forall \phi \in C_0(\Omega)$. Then $u(\mathbf{x}) = 0$ almost everywhere on Ω.*

Proof. Let $u \in L^p(\Omega)$, $1 \le p \le \infty$. Then, $\forall$ bounded $\Omega_0 \subset \Omega$ with $\overline{\Omega}_0 \subset\subset \Omega$, $u{\downarrow_{\Omega_0}} \in L^p(\Omega_0) \hookrightarrow L^1(\Omega_0)$, $1 \le p \le \infty$, by the imbedding result for L^p-spaces on every bounded domain Ω_0 ((B.4.1.12) in Appendix B). Hence, $u \in L^1_{\text{loc}}(\Omega)$ and $\int_\Omega u\phi d\mathbf{x} = 0 \forall \phi \in C_0(\Omega)$. Then, by Theorem 1.2.2, $u(\mathbf{x}) = 0$ almost everywhere on Ω. □

Theorem 1.2.3. *Let $\Omega \subset \mathbb{R}^n$ be an open (bounded or unbounded) subset of $\mathbb{R}^n$. Then $C_0(\Omega)$ is a* dense *subspace of $L^p(\Omega)$ for $1 \le p < \infty$.*

Proof. For $p = 1$, see the proof of Theorem B.3.3.4 in Appendix B. Hence, it remains to prove the result for $1 < p < \infty$. The mapping $v \in L^p(\Omega) \longmapsto \langle u, v\rangle_{(L^p(\Omega))' \times L^p(\Omega)} = \int_\Omega uv d\mathbf{x}$ is a *bounded, linear* functional on $L^p(\Omega)$ for $u \in (L^p(\Omega))' \equiv L^q(\Omega)$ with $1 < q = \frac{p}{p-1} < \infty, 1 < p < \infty$, since, by virtue of Hölder's inequality, $|\langle u, v\rangle| = |\int_\Omega uv d\mathbf{x}| \le \|u\|_{L^q)(\Omega)} \cdot \|v\|_{L^p(\Omega)}\ \forall v \in L^p(\Omega), 1 < p < \infty$. As a consequence of the Hahn–Banach theorem, $C_0(\Omega) \subset L^p(\Omega), 1 < p < \infty$, will be a *dense* subspace of $L^p(\Omega)$ if the *vanishing* of this bounded, linear functional on $C_0(\Omega) \subset L^p(\Omega), 1 < p < \infty$, implies that it is a *null functional* on $L^p(\Omega)$, i.e. if $\langle u, \phi\rangle = \int_\Omega u\phi d\mathbf{x} = 0\ \forall \phi \in C_0(\Omega)$, then $u = 0$ in $L^q(\Omega) \equiv (L^p(\Omega))'$.

Hence, we assume that $\langle u, \phi\rangle_{(L^p(\Omega)' \times L^p(\Omega)} = \int_\Omega u\phi d\mathbf{x} = 0\ \forall \phi \in C_0(\Omega) \subset L^p(\Omega)$, $1 < p < \infty$. We are to prove that $u = 0$ in $(L^p(\Omega))' = L^q(\Omega)$ with $1 < q = \frac{p}{p-1} < \infty$. In fact, $u \in L^q(\Omega) \equiv (L^p(\Omega))'$ with $q = \frac{p}{p-1}, 1 < p < \infty$. Then, $\forall$ bounded $\Omega_0 \subset \Omega$ with $\overline{\Omega}_0 \subset\subset \Omega$, $u\downarrow_{\Omega_0} \in L^q(\Omega_0) \hookrightarrow L^1(\Omega_0)$ by the imbedding result for bounded domain $\Omega_0 \Longrightarrow u \in L^1_{\text{loc}}(\Omega)$ and $\langle u, \phi\rangle = \int_\Omega u\phi d\mathbf{x} = 0\ \forall \phi \in C_0(\Omega) \Longrightarrow u = 0$ a.e. on Ω by Theorem 1.2.2 $\Longrightarrow u = 0$ in $L^q(\Omega) \equiv (L^p(\Omega))'$, $1 < p < \infty$. Hence, $C_0(\Omega)$ is a dense subspace of $L^p(\Omega), 1 < p < \infty$, and the result follows. □

Remark 1.2.1A. For $p = \infty$, $C_0(\Omega)$ is *not* a dense subspace of $L^\infty(\Omega)$.

Property 8:

Theorem 1.2.3A. *Let $\Omega \subset \mathbb{R}^n$ be an open subset of $\mathbb{R}^n$ and $L^1_{\text{loc}}(\Omega)$ be the space of* locally *integrable functions on Ω.*

If $u \in L^1_{\text{loc}}(\Omega)$ and $\int_\Omega u(\mathbf{x})\phi(\mathbf{x}) = 0\ \forall \phi \in \mathcal{D}(\Omega)$, then $u(\mathbf{x}) = 0$ almost everywhere on Ω. (1.2.22)

Proof. Let $u \in L^1_{\text{loc}}(\Omega)$ satisfy (1.2.22). Let $\psi \in C_0(\Omega)$ be any continuous function with compact $\text{supp}(\psi) = K \subset\subset \Omega$. Then we show that $\int_\Omega u(\mathbf{x})\psi(\mathbf{x})d\mathbf{x} = 0$. By the Approximation Theorem 1.2.1 in Property 7, for any $\varepsilon > 0$ and $d > 0, \exists \phi \in \mathcal{D}(\Omega)$ with $\text{supp}(\phi) \subset K_d, \text{supp}(\psi) = K \subset K_d \subset \Omega$, such that $\sup_{\mathbf{x}\in\Omega} |\psi(\mathbf{x}) - \phi(\mathbf{x})| \le \varepsilon$. Since $\psi(\mathbf{x}) - \phi(\mathbf{x}) = 0\ \forall \mathbf{x} \in \Omega \setminus K_d$,

$$\left|\int_\Omega u(\mathbf{x})(\psi(\mathbf{x}) - \phi(\mathbf{x}))d\mathbf{x}\right| \le \sup_{\mathbf{x}\in\Omega} |\psi(\mathbf{x}) - \phi(\mathbf{x})| \int_{K_d} |u(\mathbf{x})|d\mathbf{x} \le \varepsilon \int_{K_d} |u(\mathbf{x})|d\mathbf{x}. \tag{1.2.23}$$

But $\int_\Omega u(\mathbf{x})\phi(\mathbf{x})d\mathbf{x} = 0 \Longrightarrow |\int_\Omega u(\mathbf{x})\psi(\mathbf{x})d\mathbf{x}| \le \varepsilon \int_{K_d} |u(\mathbf{x})|d\mathbf{x} \to 0$ as $\varepsilon \to 0$ with fixed u and $d > 0$. Hence, $\int_\Omega u(\mathbf{x})\psi(\mathbf{x})d\mathbf{x} = 0 \quad \forall \psi \in C_0(\Omega)$. Now, applying Theorem 1.2.2, we get the result (1.2.22). □

Property 9:

Corollary 1.2.2. *For* $1 \leq p \leq \infty, u \in L^p(\Omega)$,

$$\int_\Omega u(\mathbf{x})\phi(\mathbf{x})dx = 0 \ \forall \phi \in \mathcal{D}(\Omega) \quad \Longrightarrow \quad u = 0 \text{ in } L^p(\Omega) \text{ (i.e. } u(\mathbf{x}) = 0 \text{ a.e. on } \Omega). \tag{1.2.24}$$

Proof. From the proof of Corollary 1.2.1, $L^p(\Omega) \subset L^1_{\text{loc}}(\Omega) \Longrightarrow$ the result by Theorem 1.2.3A. □

Property 10: Density Results

- $\mathcal{D}(\Omega)$ is *dense* in $\mathcal{D}^m(\Omega)$ (see Definition 1.2.2) $\forall m \in \mathbb{N}_0$ with $\mathcal{D}(\Omega) \hookrightarrow \mathcal{D}^m(\Omega)$, the imbedding being a *dense, continuous* one. (1.2.25)

Theorem 1.2.3B. $\mathcal{D}(\Omega)$ *is* dense *in* $L^p(\Omega)$ *for* $1 \leq p < \infty$. (1.2.26)
In particular, $\mathcal{D}(\Omega)$ *is* dense *in* $L^2(\Omega)$. (1.2.27)

Proof. The proof is given in Chapter 6 using *convolutions*. See the proof of Theorem 6.8.3. □

- $\mathcal{D}(\Omega)$ is *not* dense in $L^\infty(\Omega)$, *nor* in $C^\infty(\overline{\Omega})$, *nor* in $C^m(\overline{\Omega})$ $\forall m \in \mathbb{N}_0$.

Here, we show that $\mathcal{D}(\Omega)$ is not dense in $C^m(\overline{\Omega})$ $\forall m \in \mathbb{N}_0$, where $C^m(\overline{\Omega}) = \{\phi : \partial^{\boldsymbol{\alpha}}\phi$ is bounded and uniformly continuous in Ω $\forall |\boldsymbol{\alpha}| \leq m\}$ is a *Banach space* for the norm $\|\cdot\|_{C^m(\overline{\Omega})}$ or $|||\cdot|||_{C^m(\overline{\Omega})}$ defined by:

$$\|\phi\|_{C^m(\overline{\Omega})} = \|\phi\|_{m,\infty,\Omega} = \sum_{|\boldsymbol{\alpha}|=0}^{m} \sup_{\mathbf{x}\in\Omega} |\partial^{\boldsymbol{\alpha}}\phi(\mathbf{x})|, \quad \text{or}$$
$$|||\phi|||_{C^m(\overline{\Omega})} = \max_{0\leq|\boldsymbol{\alpha}|\leq m} \sup_{\mathbf{x}\in\Omega} |\partial^{\boldsymbol{\alpha}}\phi(\mathbf{x})|, \tag{1.2.28}$$

both the norms in (1.2.28) being *equivalent* (see Definition A.4.1.3 and (A.4.1.7) (equivalent norms) in Appendix A).

Proof. Let $(\phi_n)_{n=1}^\infty$ be a sequence in $\mathcal{D}(\Omega)$ with $\text{supp}(\phi_n) = K_n \subset\subset \Omega$ $\forall n \in \mathbb{N}$. Let $u \in C^m(\overline{\Omega})$ such that

$$\begin{aligned} \|u - \phi_n\|_{C^m(\overline{\Omega})} &= \sum_{|\boldsymbol{\alpha}|=0}^{m} \sup_{\mathbf{x}\in\Omega} |\partial^{\boldsymbol{\alpha}}(u - \phi_n)(\mathbf{x})| \\ &= \sum_{|\boldsymbol{\alpha}|=0}^{m} \max\left\{ \sup_{\mathbf{x}\in\Omega-K_n} |\partial^{\boldsymbol{\alpha}}u(\mathbf{x})|, \sup_{\mathbf{x}\in K_n} |\partial^{\boldsymbol{\alpha}}(u - \phi_n)(\mathbf{x})| \right\}. \end{aligned} \tag{1.2.29}$$

For $\|u - \phi_n\|_{C^m(\overline{\Omega})} < \varepsilon$ with $\varepsilon > 0$, it is necessary that

$$\sum_{|\boldsymbol{\alpha}|=0}^{m} \sup_{\mathbf{x}\in\Omega-K_n} |\partial^{\boldsymbol{\alpha}} u(\mathbf{x})| < \varepsilon, \tag{1.2.30}$$

which must hold $\forall \varepsilon > 0$ and $\forall n > n_0(\varepsilon), n_0(\varepsilon) \in \mathbb{N}$. But this will be possible only if $u \in C^m(\overline{\Omega})$ satisfies some additional conditions. In other words, $\forall u \in C^m(\overline{\Omega})$, (1.2.30) will *not* hold and, consequently, there does *not* exist a sequence (ϕ_n) in $\mathcal{D}(\Omega)$ such that $\phi_n \to u$ in $C^m(\overline{\Omega})$ *in general*. Hence $\mathcal{D}(\Omega)$ is *not* dense in $C^m(\overline{\Omega})$ $\forall m \in \mathbb{N}_0$. □

Identification of additional conditions such that (1.2.30) holds

1. If Ω is a *bounded* domain with a sufficiently smooth boundary Γ, then the condition (1.2.30) implies that u and all derivatives $\partial^{\boldsymbol{\alpha}} u$ of order $|\boldsymbol{\alpha}| \leq m$ must vanish on the boundary Γ. (1.2.31)
2. If $\Omega = \mathbb{R}^n$, then (1.2.30) implies that $u(\mathbf{x})$ and $\partial^{\boldsymbol{\alpha}} u(\mathbf{x})$ must vanish as $\|\mathbf{x}\| \to \infty$ $\forall |\boldsymbol{\alpha}| \leq m$. (1.2.32)

If $u \in C^m(\overline{\Omega})$ possess either additional property 1 or 2, then there will exist a sequence $(\phi_n)_{n=1}^{\infty}$ in $\mathcal{D}(\Omega)$ such that $\phi_n \to u$ in $C^m(\overline{\Omega})$ as $n \to \infty$ (see also [27, p. 70]). (1.2.33)

Property 11: Poincaré–Friedrichs inequality

Theorem 1.2.4. *Let $\phi \in \mathcal{D}(\mathbb{R}^n)$. Then, the following properties hold:*

I. $\forall i = 1, 2, \ldots, n$,

$$\int_{\mathbb{R}^n} |\phi(\mathbf{x})|^2 d\mathbf{x} = -2 \int_{\mathbb{R}^n} x_i \phi(\mathbf{x}) \frac{\partial \phi}{\partial x_i}(\mathbf{x}) d\mathbf{x}. \tag{1.2.34}$$

II. *For* bounded *open subset $\Omega \subset \mathbb{R}^n$, $\exists$ a constant $C > 0$ such that, $\forall \phi \in \mathcal{D}(\mathbb{R}^n)$ with* $\operatorname{supp}(\phi) \subset \Omega$,

$$\int_{\Omega} |\phi(\mathbf{x})|^2 d\mathbf{x} \leq C \int_{\Omega} |\nabla \phi|^2 d\mathbf{x} = C \sum_{i=1}^{n} \int_{\Omega} \left| \frac{\partial \phi}{\partial x_i}(\mathbf{x}) \right|^2 d\mathbf{x}, \tag{1.2.35}$$

i.e. $\forall \phi \in \mathcal{D}(\Omega)$, (1.2.35) *holds.* (1.2.36)

Proof.

I. $\int_{\mathbb{R}^n} x_i\phi(\mathbf{x})\frac{\partial\phi}{\partial x_i}(\mathbf{x})d\mathbf{x} = \int_{\mathbb{R}}\cdots\int_{\mathbb{R}}\cdots(\int_{\mathbb{R}} x_i\phi(\mathbf{x})\frac{\partial\phi}{\partial x_i}dx_i)dx_1\dots dx_{i-1}dx_{i+1}\dots dx_n$ (since $\phi\in\mathcal{D}(\mathbb{R}^n)\implies\operatorname{supp}(\phi)$ is a compact subset of $\mathbb{R}^n$ and $x_i\phi(\mathbf{x})\frac{\partial\phi}{\partial x_i}(\mathbf{x})\in L^1(\mathbb{R}^n)$, we can apply Fubini's Theorem 7.1.2C on the interchange of the order of integration).

Since the variables $x_1,\dots,x_{i-1},x_{i+1},\dots,x_n$ are treated as parameters in the integral involving the variables x_i, we can apply an integration by parts to get

$$\int_{\mathbb{R}}\underbrace{x_i\phi(\mathbf{x})}_{u}\underbrace{\frac{\partial\phi}{\partial x_i}dx_i}_{dv} = (x_i\phi(\mathbf{x})\phi(\mathbf{x}))\Big|_{x_i=-\infty}^{x_i=+\infty} - \int_{\mathbb{R}}\frac{\partial}{\partial x_i}(x_i\phi(\mathbf{x}))\phi(\mathbf{x})dx_i.$$

Since $\phi\in\mathcal{D}(\mathbb{R}^n)\implies\phi(\mathbf{x})=0$ for $x_i=\pm\infty$, we have

$$\begin{aligned}\int_{\mathbb{R}} x_i\phi(\mathbf{x})\frac{\partial\phi}{\partial x_i}(\mathbf{x})dx_i &= -\int_{\mathbb{R}}\frac{\partial}{\partial x_i}(x_i\phi(\mathbf{x}))\phi(\mathbf{x})dx_i\\ &= -\Big(\int_{\mathbb{R}}\Big[(\phi(\mathbf{x}))^2 + x_i\frac{\partial\phi}{\partial x_i}(\mathbf{x})\phi(\mathbf{x})\Big]dx_i\Big)\end{aligned}$$

$\implies \int_{\mathbb{R}}|\phi(\mathbf{x})|^2dx_i = -2\int_{\mathbb{R}} x_i\phi(\mathbf{x})\frac{\partial\phi}{\partial x_i}(\mathbf{x})dx_i$.

Integrating with respect to the other variables $x_1,x_2,\dots,x_{i-1},x_{i+1},\dots,x_n$, we get

$$\begin{aligned}&\underbrace{\int_{\mathbb{R}}\cdots\int_{\mathbb{R}}\cdots\int_{\mathbb{R}}}_{n\text{ times}}|\phi(x_1,x_2,\dots x_n)|^2dx_1dx_2\dots dx_n\\ &\qquad= -2\underbrace{\int_{\mathbb{R}}\cdots\int_{\mathbb{R}}\cdots\int_{\mathbb{R}}}_{n\text{ times}} x_i\phi(\mathbf{x})\frac{\partial\phi}{\partial x_i}(\mathbf{x})dx_1dx_2\dots dx_n\end{aligned}$$

$\implies \int_{\mathbb{R}^n}|\phi(\mathbf{x})|^2d\mathbf{x} = -2\int_{\mathbb{R}^n} x_i\phi(\mathbf{x})\frac{\partial\phi}{\partial x_i}(\mathbf{x})d\mathbf{x}$.

II. Since $\mathcal{D}(\Omega)\subset L^2(\Omega)$, $\forall\phi\in\mathcal{D}(\mathbb{R}^n)$ with $\operatorname{supp}(\phi)\subset\subset\Omega$, we can apply the Cauchy–Schwarz inequality (see (B.4.1.7)–(B.4.1.8) with $p=q=2$ in Appendix B):

$$\begin{aligned}\int_{\Omega}|\phi(\mathbf{x})|^2d\mathbf{x} &= \Big|-2\int_{\Omega} x_i\phi(\mathbf{x})\frac{\partial\phi}{\partial x_i}(\mathbf{x})d\mathbf{x}\Big| = 2\Big|\Big\langle x_i\phi,\frac{\partial\phi}{\partial x_i}\Big\rangle_{L^2(\Omega)}\Big|\\ &\le 2\|x_i\phi\|_{L^2(\Omega)}\Big\|\frac{\partial\phi}{\partial x_i}\Big\|_{L^2(\Omega)} = 2\Big(\int_{\Omega}(x_i)^2|\phi(\mathbf{x})|^2d\mathbf{x}\Big)^{\frac{1}{2}}\Big(\int_{\Omega}\Big|\frac{\partial\phi}{\partial x_i}\Big|^2d\mathbf{x}\Big)^{\frac{1}{2}}\\ &\le 2C_1\Big(\int_{\Omega}|\phi(\mathbf{x})|^2d\mathbf{x}\Big)^{\frac{1}{2}}\Big(\int_{\Omega}\Big|\frac{\partial\phi}{\partial x_i}\Big|^2d\mathbf{x}\Big)^{\frac{1}{2}},\end{aligned}$$

since Ω is bounded and $\mathbf{x} \in \Omega$, $\exists C_1(\Omega) > 0$ such that $|x_i| \leq C_1$.

$$\Longrightarrow \quad \int_\Omega |\phi(\mathbf{x})|^2 d\mathbf{x} \leq 2C_1 \Big(\int_\Omega |\phi(\mathbf{x})|^2 d\mathbf{x} \Big)^{\frac{1}{2}} \Big(\int_\Omega \Big| \frac{\partial \phi}{\partial x_i} \Big|^2 d\mathbf{x} \Big)^{\frac{1}{2}}.$$

For $\phi = 0$, the inequality is trivially satisfied. Otherwise, dividing both sides by $(\int_\Omega |\phi(\mathbf{x})|^2 d\mathbf{x})^{\frac{1}{2}}$, we get

$$\Big(\int_\Omega |\phi(\mathbf{x})|^2 d\mathbf{x} \Big)^{\frac{1}{2}} \leq 2C_1 \Big(\int_\Omega \Big| \frac{\partial \phi}{\partial x_i} \Big|^2 d\mathbf{x} \Big)^{\frac{1}{2}}$$

$$\Longrightarrow \quad \int_\Omega |\phi(\mathbf{x})|^2 d\mathbf{x} \leq 4C_1^2 \int_\Omega \Big| \frac{\partial \phi}{\partial x_i} \Big|^2 d\mathbf{x} \leq 4C_1^2 \sum_{i=1}^{n} \int_\Omega \Big| \frac{\partial \phi}{\partial x_i} \Big|^2 d\mathbf{x}$$

$\Longrightarrow$ the inequality (1.2.35) holds with $C = 4C_1^2 > 0$. □

Property 12: For the solution of many problems later, we will need the following result:

Proposition 1.2.1. *Let $\phi \in \mathcal{D}(\mathbb{R})$ such that* $\operatorname{supp}(\phi) \subset [-A, A] \subset \mathbb{R}$ *with $A > 0$. For fixed $n \in \mathbb{N}_0$, let $\psi : \mathbb{R} \longrightarrow \mathbb{R}$ be a function defined by:*

$$\psi(x) = \begin{cases} \frac{1}{x^{n+1}} [\phi(x) - \sum_{k=0}^{n} \frac{x^k}{k!} \phi^{(k)}(0)] & \text{for } x \neq 0 \\ \frac{\phi^{(n+1)}(0)}{(n+1)!} & \text{for } x = 0. \end{cases} \tag{1.2.37}$$

Then,

I. *ψ is continuous on $\mathbb{R}$ and $\exists C > 0$, independent of x, such that*

$$\sup_{|x| \leq A} |\psi(x)| \leq C \sup_{|x| \leq A} |\phi^{(n+1)}(x)|; \tag{1.2.38}$$

II.
$$\phi(x) = \phi(0) + x\phi'(0) + \frac{x^2}{2!}\phi''(0) + \cdots + \frac{x^n}{n!}\phi^{(n)}(0) + x^{n+1}\psi(x). \tag{1.2.39}$$

Proof.

I. $\phi \in \mathcal{D}(\mathbb{R}) \equiv C_0^\infty(\mathbb{R})$, we can write Taylor's formula for ϕ with the remainder R_n in integral form:

$$\phi(x) = \phi(0) + \frac{x}{1!}\phi'(0) + \cdots + \frac{x^n}{n!}\phi^{(n)}(0) + \frac{x^{n+1}}{n!} \int_0^1 (1-t)^n \phi^{(n+1)}(tx) dt$$

$$\Longrightarrow \phi(x) - \sum_{k=0}^{n} \frac{x^k}{k!} \phi^{(k)}(0) = \frac{x^{n+1}}{n!} \int_0^1 (1-t)^n \phi^{(n+1)}(tx) dt.$$

$\Longrightarrow$ For $x \neq 0$, $\psi(x) = \frac{1}{x^{n+1}}[\phi(x) - \sum_{k=0}^{n} \frac{x^k}{k!}\phi^k(0)] = \frac{1}{n!}\int_0^1 (1-t)^n \phi^{(n+1)}(tx)dt$, which is continuous for all $x \neq 0$. (1.2.40)

Then,

$$\begin{aligned}\lim_{x\to 0} \psi(x) &= \lim_{x\to 0} \frac{1}{n!}\int_0^1 (1-t)^n \phi^{(n+1)}(tx)dt \\ &= \frac{1}{n!}\int_0^1 (1-t)^n [\lim_{x\to 0} \phi^{(n+1)}(tx)]dt \\ &= \frac{1}{n!}\int_0^1 (1-t)^n \phi^{(n+1)}(0)dt \\ &= \frac{\phi^{(n+1)}(0)}{n!} \cdot \frac{-(1-t)^{n+1}}{n+1}\bigg|_0^1 = \frac{\phi^{(n+1)}(0)}{(n+1)!} = \psi(0)\end{aligned}$$

(by definition (1.2.37)).

$\Longrightarrow$ ψ is continuous at $x = 0$. Hence, together with (1.2.40), ψ is continuous on $\mathbb{R}$.

- $-A \le x \neq 0 \le A$: from (1.2.40),

$$|\psi(x)| \le \frac{1}{n!} \sup_{\substack{|tx|\le A \\ 0\le t\le 1}} |\phi^{(n+1)}(tx)| \int_0^1 (1-t)^n dt$$

$$\Longrightarrow \quad |\psi(x)| \le \frac{1}{n!} \sup_{|x|\le A} |\phi^{(n+1)}(x)| \cdot \frac{1}{n+1} = \frac{1}{(n+1)!} \sup_{|x|\le A} |\phi^{(n+1)}(x)|$$

for $0 < |x| \le A$.

- $x = 0$: from (1.2.37),

$$|\psi(0)| = \frac{|\phi^{(n+1)}(0)|}{(n+1)!} \le \frac{1}{(n+1)!} \sup_{|x|\le A} |\phi^{(n+1)}(x)|.$$

Hence, $\forall x \in [-A, A]$, $|\psi(x)| \le C \sup_{|x|\le A} |\phi^{(n+1)}(x)|$ with $C = \frac{1}{(n+1)!} > 0 \Longrightarrow \sup_{|x|\le A} |\psi(x)| \le C \sup_{|x|\le A} |\phi^{(n+1)}(x)|$.

II. In particular, for $n = 1$,

$$\phi(x) = \phi(0) + x\psi(x), \tag{1.2.41}$$

for $n = 2$,

$$\phi(x) = \phi(0) + x\phi'(0) + x^2\psi(x) \tag{1.2.42}$$

etc. □

Remark 1.2.2 (Test space $\mathcal{D}(\Omega)$ of complex-valued functions). We have considered the test space $\mathcal{D}(\Omega)$ consisting of only real-valued test functions $\phi \in C_0^\infty(\Omega)$ since, except in Fourier transform and Fourier analysis, we do not require and will not use complex vector spaces (of complex-valued functions) *in general*.

But all the important results and properties of $\mathcal{D}(\Omega)$ stated above for the real case can be extended to the complex case with minor, if any, modifications, i.e. to the test space $\mathcal{D}(\Omega)$ consisting of complex-valued test functions $\phi \in C_0^\infty(\Omega)$ with

$$\begin{gathered}\phi(\mathbf{x}) = \psi(\mathbf{x}) + i\chi(\mathbf{x}), \quad \psi(\mathbf{x}) = \mathrm{Re}(\phi(\mathbf{x})), \quad \chi(\mathbf{x}) = \mathrm{Im}(\phi(\mathbf{x})) \quad \forall \mathbf{x} \in \Omega, \\ \mathrm{supp}(\psi) \subset\subset \Omega, \quad \mathrm{supp}(\chi) \subset\subset \Omega \quad \text{and} \quad \psi \in C_0^\infty(\Omega), \chi \in C_0^\infty(\Omega), \end{gathered} \tag{1.2.43}$$

assuming that $\mathcal{D}(\Omega)$ is a complex-vector space, which is obviously closed under the addition operation and for multiplication by complex numbers $\alpha \in \mathbb{C}$.

Convergence in a complex test space $\mathcal{D}(\Omega)$ is similar to that in (1.2.7): $(\phi_n) \to \phi$ in $\mathcal{D}(\Omega)$ if and only if:

I. $\exists$ a compact subset $K \subset\subset \Omega$ such that $\mathrm{supp}(\phi_n - \phi) \subset K \quad \forall n \in \mathbb{N}$;

II. $\partial^{\boldsymbol{\alpha}} \phi_n \to \partial^{\boldsymbol{\alpha}} \phi$ uniformly in Ω as $n \to \infty$, $\forall \boldsymbol{\alpha}$. (1.2.44)

Thus, the extension of $\mathcal{D}(\Omega)$ to complex functions is a straightforward affair.

1.3 Space $\mathcal{D}'(\Omega)$ of (Schwartz) distributions

1.3.1 Algebraic dual space $\mathcal{D}^\star(\Omega)$

Let $\mathcal{D}(\Omega)$ be the space of real-valued test functions ϕ on Ω (see Definition 1.2.1), and $T : \phi \in \mathcal{D}(\Omega) \longmapsto T(\phi) \in \mathbb{R}$ be a *linear* functional on $\mathcal{D}(\Omega)$, i.e.

$$\forall \alpha_i \in \mathbb{R},\ \forall \phi_i \in \mathcal{D}(\Omega)\ (i = 1, 2), \quad T(\alpha_1\phi_1 + \alpha_2\phi_2) = \alpha_1 T(\phi_1) + \alpha_2 T(\phi_2), \tag{1.3.1}$$

other notations equivalent to $T(\phi)$ being $\langle T, \phi\rangle$ or (T, ϕ) or $[T, \phi]$. For example,

$$\langle T, \alpha_1\phi_1 + \alpha_2\phi_2\rangle = \alpha_1\langle T, \phi_1\rangle + \alpha_2\langle T, \phi_2\rangle \quad \forall \alpha_i \in \mathbb{R},\ \forall \phi_i \in \mathcal{D}(\Omega),\ i = 1, 2.$$

As convenient, we will use interchangeably any one of the different notations indicating the same object $T(\phi)$:

$$T(\phi) = \langle T, \phi\rangle = (T, \phi) = [T, \phi] \quad \forall \phi \in \mathcal{D}(\Omega). \tag{1.3.2}$$

Then $\mathcal{D}^\star(\Omega) = \{T : T \text{ is a linear functional on } \mathcal{D}(\Omega)\}$ (1.3.3)
is the linear space of all linear functionals T defined on $\mathcal{D}(\Omega)$, called the *algebraic dual space of* $\mathcal{D}(\Omega)$.

Continuity of linear functional $T \in \mathcal{D}^\star(\Omega)$ on $\mathcal{D}(\Omega)$

Definition 1.3.1. A linear functional $T \in \mathcal{D}^\star(\Omega)$ defined on $\mathcal{D}(\Omega)$ is said to be *continuous* on $\mathcal{D}(\Omega)$ if and only if

$$\phi_n \to \phi \text{ in } \mathcal{D}(\Omega) \text{ in the sense of (1.2.7)} \quad \Longrightarrow \quad T(\phi_n) \to T(\phi) \text{ in } \mathbb{R} \text{ as } n \to \infty. \tag{1.3.4}$$

1.3.2 Distributions and the space $\mathcal{D}'(\Omega)$ of distributions on Ω

Definition 1.3.2. A continuous linear functional $T \in \mathcal{D}^\star(\Omega)$ is called a *distribution on* $\Omega \subset \mathbb{R}^n$. Then $T(\phi) \in \mathbb{R}$ is the value of the distribution T at $\phi \in \mathcal{D}(\Omega)$, other equivalent notations used for $T(\phi)$ being those in (1.3.2), i.e. $T(\phi) = \langle T, \phi\rangle = (T, \phi) = [T, \phi]\ \forall \phi \in \mathcal{D}(\Omega)$.

Space $\mathcal{D}'(\Omega)$ of distributions on Ω The set of all distributions on Ω, i.e. of all continuous linear functionals defined on $\mathcal{D}(\Omega)$, forms a vector space denoted by $\mathcal{D}'(\Omega)$ if we define the sum $T_1 + T_2 \in \mathcal{D}'(\Omega)$ of distributions $T_1, T_2 \in \mathcal{D}'(\Omega)$, the product $\alpha T \in \mathcal{D}'(\Omega)$ of $\alpha \in \mathbb{R}$ and $T \in \mathcal{D}'(\Omega)$ and the null distribution $0 \in \mathcal{D}'(\Omega)$ by:

Sum $T_1 + T_2 \in \mathcal{D}'(\Omega)$: $(T_1 + T_2)(\phi) = T_1(\phi) + T_2(\phi),\ \forall \phi \in \mathcal{D}(\Omega)$;

Product $\alpha T \in \mathcal{D}'(\Omega)$: $(\alpha T)(\phi) = \alpha \cdot T(\phi)\ \forall \phi \in \mathcal{D}(\Omega),\ \forall \alpha \in \mathbb{R}$; (1.3.5)

Null distribution $0 \in \mathcal{D}'(\Omega)$: $T = 0$ in $\mathcal{D}'(\Omega) \Longleftrightarrow T(\phi) = 0\ \forall \phi \in \mathcal{D}(\Omega)$.

Definition 1.3.3. The linear space $\mathcal{D}'(\Omega)$ of all distributions T on Ω (i.e. continuous linear functionals on $\mathcal{D}(\Omega)$) is called the *space of (Schwartz) distributions on* Ω, with

$$T(\phi) = \langle T, \phi\rangle_{\mathcal{D}'(\Omega)\times\mathcal{D}(\Omega)} = (T, \phi)_{\mathcal{D}'(\Omega)\times\mathcal{D}(\Omega)} = [T, \phi]_{\mathcal{D}'(\Omega)\times\mathcal{D}(\Omega)} \quad \forall \phi \in \mathcal{D}(\Omega),$$

where $\langle \cdot, \cdot \rangle$ or $(\cdot, \cdot)$ or $[\cdot, \cdot]$ denotes the *duality pairing* between $\mathcal{D}'(\Omega)$ and $\mathcal{D}(\Omega)$, i.e. any one of these notations can be used interchangeably to mean the same thing.

Remark 1.3.1. $\mathcal{D}^\star(\Omega)$ in (1.3.3) is the set of all linear functionals on $\mathcal{D}(\Omega)$, which may be or may not be continuous on $\mathcal{D}(\Omega)$, whereas $\mathcal{D}'(\Omega)$ is the set of all linear functionals of $\mathcal{D}^\star(\Omega)$, which *are* continuous on $\mathcal{D}(\Omega)$. Hence, $\mathcal{D}'(\Omega) \subset \mathcal{D}^\star(\Omega)$ is a subspace of $\mathcal{D}^\star(\Omega)$.

There are currently no known examples of linear functionals discontinuous on $\mathcal{D}(\Omega)$, and there is very little chance of ever encountering such a discontinuous linear functional in practical applications. But the existence of linear functionals discontinuous on $\mathcal{D}(\Omega)$ can be established mathematically with the help of the Axiom of Choice [7], [28].

Physical interpretation of duality The abstract concepts of *duality* and *duality pairing* can be interpreted physically with the help of problems from mechanics (or optimization). Consider a conservative mechanical system (for example, a deformable elastic body). We can associate with this system a vector space V of admissible displacement fields v. Then there exists another vector space F such that there is a *duality* between the vector space V of admissible displacements v and the vector space F, the elements of which are the admissible force fields f acting on the system. In fact, an element $f \in F$ of the *dual* F associates with each element $v \in V$ a *scalar* $\langle f, v\rangle \in \mathbb{R}$ called the *work* done by the force $f \in F$ for the displacement $v \in V$. Thus, the rôle of $f \in F$ of the *dual* F is to associate a *scalar* $\langle f, v\rangle \in \mathbb{R}$ to each $v \in V$.

1.3.3 Characterization, order and extension of a distribution

Characterization of a distribution $T \in \mathcal{D}'(\Omega)$

Proposition 1.3.1. *A linear functional T on $\mathcal{D}(\Omega)$ is a* distribution *on Ω if and only if any one of the following two equivalent properties hold:*

I. *$\phi_n \to 0$ in $\mathcal{D}(\Omega)$ in the sense of* (1.2.7) $\Longrightarrow T(\phi_n) = \langle T, \phi_n\rangle \to 0$ *in $\mathbb{R}$ as $n \to \infty$ (i.e. T is continuous on $\mathcal{D}(\Omega)$ and hence $T \in \mathcal{D}'(\Omega)$).*

II. *$\forall$ compact subsets $K \subset\subset \Omega$ of Ω, $\exists$ an integer $m(K) \geq 0$ and a constant $C_K > 0$ such that, $\forall \phi \in \mathcal{D}(\Omega)$ with* $\operatorname{supp}(\phi) \subset K$ *(i.e. $\forall \phi \in \mathcal{D}_K(\Omega)$),*

$$|T(\phi)| = |\langle T, \phi\rangle| \leq C_K \max_{|\boldsymbol{\alpha}| \leq m(K)} \max_{\mathbf{x} \in K} |\partial^{\boldsymbol{\alpha}} \phi(\mathbf{x})| = C_K(p_{K,m(K)}(\phi)) \quad (1.3.6)$$

(or, equivalently, $|T(\phi)| \leq \tilde{C}_K \sum_{|\boldsymbol{\alpha}| \leq m(K)} \max_{\mathbf{x} \in K} |\partial^{\boldsymbol{\alpha}} \phi(\mathbf{x})| = \tilde{C}_K(\tilde{p}_{K,m(K)}(\phi))$), where $p_{K,m(K)}(\phi) = \max_{|\boldsymbol{\alpha}| \leq m(K)} \max_{\mathbf{x} \in K} |\partial^{\boldsymbol{\alpha}} \phi(\mathbf{x})|$ (or, equivalently, $\tilde{p}_{K,m(K)}(\phi) = \sum_{|\boldsymbol{\alpha}| \leq m(K)} \max_{\mathbf{x} \in K} |\partial^{\boldsymbol{\alpha}} \phi(\mathbf{x})|$, $|\boldsymbol{\alpha}| = \alpha_1 + \alpha_2 + \cdots + \alpha_n$).

Proof.

I. follows from Definition 1.3.2 of a distribution $T \in \mathcal{D}'(\Omega)$.

II. Assume that T is a linear functional on $\mathcal{D}(\Omega)$, which satisfies (1.3.6). We are to show that T is a distribution on Ω. Let (ϕ_n) be a sequence in $\mathcal{D}(\Omega)$ such that $\phi_n \to 0$ in $\mathcal{D}(\Omega)$. Then $\exists$ a compact subset $K \subset\subset \Omega$ such that $\operatorname{supp}(\phi_n) \subset K$ $\forall n \in \mathbb{N}$, and $\partial^{\boldsymbol{\alpha}} \phi_n \to 0$ *uniformly* $\forall |\boldsymbol{\alpha}|$ as $n \to \infty \Longrightarrow \forall |\boldsymbol{\alpha}| \in \mathbb{N}_0$, $\max_{\mathbf{x} \in \Omega} |\partial^{\boldsymbol{\alpha}} \phi_n(\mathbf{x})| = \max_{\mathbf{x} \in K} |\partial^{\boldsymbol{\alpha}} \phi_n(\mathbf{x})| \to 0$ as $n \to \infty$.

Hence, $|\langle T, \phi_n\rangle| \leq C_K \sum_{|\boldsymbol{\alpha}| \leq m} \max_{\mathbf{x} \in K} |\partial^{\boldsymbol{\alpha}} \phi_n(\mathbf{x})| \to 0$ as $n \to \infty$, i.e. II $\Longrightarrow$ I.

Conversely, let $T \in \mathcal{D}'(\Omega)$ be a distribution. Suppose that (1.3.6) does *not* hold for T, i.e. $\exists$ a compact set $K \subset\subset \Omega$ such that (1.3.6) does *not* hold for *all* m and for *all* $C > 0$. Hence, for this compact set K, (1.3.6) also does *not* hold for $C = m = n$, i.e. $\exists \phi_n \in \mathcal{D}_K(\Omega)$ such that $|\langle T, \phi_n\rangle| > n \sum_{|\boldsymbol{\alpha}|\le n} \max_{\mathbf{x}\in K} |\partial^{\boldsymbol{\alpha}} \phi_n(\mathbf{x})| \forall n \in \mathbb{N}$. Since $\langle T, \phi_n\rangle \neq 0$, we can define $\psi_n = \frac{\phi_n}{|\langle T,\phi_n\rangle|}$ such that $|\langle T, \psi_n\rangle| = |\langle T, \frac{\phi_n}{|\langle T,\phi_n\rangle|}\rangle| = \frac{1}{|\langle T,\phi_n\rangle|}|\langle T, \phi_n\rangle| = 1$ $\forall n \in \mathbb{N}$.

Then,

$$
\begin{aligned}
1 = |\langle T, \psi_n\rangle| &> \frac{n}{|\langle T, \phi_n\rangle|} \sum_{|\boldsymbol{\alpha}|\le n} \max_{\mathbf{x}\in K} |\partial^{\boldsymbol{\alpha}} \phi_n(\mathbf{x})| \\
&= n \sum_{|\boldsymbol{\alpha}|\le n} \max_{\mathbf{x}\in K} |\partial^{\boldsymbol{\alpha}} \psi_n(\mathbf{x})| \quad \forall n \in \mathbb{N}
\end{aligned}
$$

$\Longrightarrow \sum_{|\boldsymbol{\alpha}|\le n} \max_{\mathbf{x}\in K} |\partial^{\boldsymbol{\alpha}} \psi_n(\mathbf{x})| < \frac{1}{n} \forall n \in \mathbb{N}$

$\Longrightarrow \max_{\mathbf{x}\in K} |\partial^{\boldsymbol{\alpha}} \psi_n(\mathbf{x})| < \frac{1}{n} \forall n \in \mathbb{N}, \forall |\boldsymbol{\alpha}| \le n$, with $\operatorname{supp}(\psi_n) \subset \operatorname{supp}(\phi_n) \subset K \Longrightarrow \forall \boldsymbol{\alpha}$, the sequence $(\partial^{\boldsymbol{\alpha}} \psi_n)$ tends to 0 uniformly in Ω as $n \to \infty \Longrightarrow \psi_n \to 0$ in $\mathcal{D}(\Omega)$ as $n \to \infty$. But $|\langle T, \psi_n\rangle| = 1 \ \forall n \in \mathbb{N} \Longrightarrow \langle T, \psi_n\rangle$ does *not* tend to 0 as $n \to \infty$. Hence, T is *not* a distribution, which contradicts the hypothesis that $T \in \mathcal{D}'(\Omega)$. Therefore, our original assumption that $T \in \mathcal{D}'(\Omega)$ does not satisfy (1.3.6) is wrong, i.e. $T \in \mathcal{D}'(\Omega)$ must satisfy (1.3.6). In other words, we have proved I $\Longrightarrow$ II. □

Order of a distribution $T \in \mathcal{D}'(\Omega)$

Definition 1.3.4. A distribution $T \in \mathcal{D}'(\Omega)$ is said to be of *order* $m_0 \ge 0$ if and only if, for any compact subset $K \subset\subset \Omega$ of Ω, $m_0 \ge 0$ is the smallest integer with $m(K) \le m_0$ for which $T(\phi) = \langle T, \phi\rangle$ satisfies the inequality (1.3.6). That is, $\forall \phi \in \mathcal{D}(\Omega)$ with $\operatorname{supp}(\phi) \subset K$,

$$|T(\phi)| = |\langle T, \phi\rangle| \le C_K \max_{|\boldsymbol{\alpha}|\le m_0} \max_{\mathbf{x}\in K} |\partial^{\boldsymbol{\alpha}} \phi(\mathbf{x})| = C_K(p_{K,m_0}(\phi)).$$

Measures

Definition 1.3.5. Distributions of order zero are called *measures*.

Remark 1.3.2. The measures defined here are in bijective correspondence with measures defined on $\mathbb{R}^n$, although we are *not* interested in pursuing this any further.

Equality of distributions in $\mathcal{D}'(\Omega)$

$T_1, T_2 \in \mathcal{D}'(\Omega)$ are *equal* $\Longleftrightarrow T_1 - T_2 = 0$ in $\mathcal{D}'(\Omega) \Longleftrightarrow \langle T_1, \phi\rangle = \langle T_2, \phi\rangle$ $\forall \phi \in \mathcal{D}(\Omega)$. (1.3.7)

Without proof,[5] we state the extension results for distributions of order $m \ge 0$.

[5] The proof is based on the density of $\mathcal{D}(\Omega)$ in $\mathcal{D}^m(\Omega) \forall m \in \mathbb{N}_0$ (i.e. $\mathcal{D}(\Omega) \hookrightarrow \mathcal{D}^m(\Omega)$ with dense,

Extension of a distribution of order $m \geq 0$

Theorem 1.3.1. *A distribution $T \in \mathcal{D}'(\Omega)$ of order $m \geq 0$ can be extended to a* unique, continuous, linear *functional $\overline{T}$ on $\mathcal{D}^m(\Omega)$, i.e. $\overline{T} \in \mathcal{D}'^m(\Omega) \equiv (\mathcal{D}^m(\Omega))'$ with $\langle \overline{T}, \phi \rangle = \langle T, \phi \rangle$ $\forall \phi \in \mathcal{D}(\Omega)$ and, $\forall K \subset\subset \Omega$, $\exists C_K > 0$ such that, $\forall \phi \in \mathcal{D}^m(\Omega)$ with* supp$(\phi) \subset K$,

$$|\langle \overline{T}, \phi \rangle| \leq C_K \max_{|\boldsymbol{\alpha}| \leq m} \max_{\mathbf{x} \in K} |\partial^{\boldsymbol{\alpha}} \phi(\mathbf{x})| = C_K(p_{K,m}(\phi)). \tag{1.3.8}$$

Corollary 1.3.1. *If $T \in \mathcal{D}'(\Omega)$ is of order* 0, *then T can be extended to a* unique, continuous, linear *functional $\overline{T}$ on $\mathcal{D}^0(\Omega) \equiv C_0^0(\Omega) \equiv C_0(\Omega)$ with $\langle \overline{T}, \phi \rangle = \langle T, \phi \rangle$ $\forall \phi \in \mathcal{D}(\Omega)$ and, $\forall K \subset\subset \Omega$, $\exists C_K > 0$ such that, $\forall \phi \in C_0(\Omega)$ with* supp$(\phi) \subset K$,

$$|\overline{T}(\phi)| \leq C_K \max_{\mathbf{x} \in K} |\phi(\mathbf{x})|. \tag{1.3.9}$$

We agree to denote the extended functional by the same notation T. Hence, from now on, T will also denote the *extended continuous linear functional* on $\mathcal{D}^m(\Omega)$.

1.3.4 Examples of distributions

Example 1.3.1 (Usual functions). For open subsets $\Omega \subset \mathbb{R}^n$, every locally integrable (summable) function $f \in L^1_{\text{loc}}(\Omega)$ (see (1.2.21) for definition) defines a distribution $T_f \in \mathcal{D}'(\Omega)$ of order 0 by:

$$T_f(\phi) = \langle T, \phi \rangle = \int_\Omega f(\mathbf{x})\phi(\mathbf{x})d\mathbf{x} \quad \forall \phi \in \mathcal{D}(\Omega). \tag{1.3.10}$$

In fact,

1. $\forall \phi \in \mathcal{D}(\Omega)$, supp$(\phi) = K \subset\subset \Omega$ with $\phi(\mathbf{x}) = 0$ $\forall \mathbf{x}$ lying outside K;
2. $f \in L^1_{\text{loc}}(\Omega)$ is integrable on this support K, i.e. $\int_K |f(\mathbf{x})|d\mathbf{x} < +\infty$;
3. ϕ is continuous, and hence $f\phi$ is also integrable on K, i.e.

$$\left|\int_\Omega f(\mathbf{x})\phi(\mathbf{x})d\mathbf{x}\right| = \left|\int_K f(\mathbf{x})\phi(\mathbf{x})d\mathbf{x}\right| \leq \int_K |f(\mathbf{x})||\phi(\mathbf{x})|d\mathbf{x} \tag{1.3.11}$$
$$\leq \max_{\mathbf{x} \in K} |\phi(\mathbf{x})| \int_K |f(\mathbf{x})|d\mathbf{x} < +\infty$$

$$\Longrightarrow T_f(\phi) = \langle T_f, \phi \rangle \in \mathbb{R} \ \forall \phi \in \mathcal{D}(\Omega) \Longrightarrow T_f \text{ is a functional on } \mathcal{D}(\Omega).$$

---

continuous imbedding (1.2.25)) and on the application of Corollary A.7.3.7 of the well-known Hahn–Banach Theorem A.7.2.1 (Section A.7, Appendix A) for the continuous, linear extension of continuous, linear functionals on $\mathcal{D}(\Omega)$ to $\mathcal{D}^m(\Omega)$ (see also [9], [25], for example).

T_f is a linear functional on $\mathcal{D}(\Omega)$ $\forall \alpha_1, \alpha_2 \in \mathbb{R}, \forall \phi_1, \phi_2 \in \mathcal{D}(\Omega)$,

$$\begin{aligned}\langle T_f, \alpha_1\phi_1 + \alpha_2\phi_2\rangle &= \int_\Omega f(\mathbf{x})(\alpha_1\phi_1(\mathbf{x}) + \alpha_2\phi_2(\mathbf{x}))d\mathbf{x} \\ &= \alpha_1 \int_\Omega f(\mathbf{x})\phi_1(\mathbf{x})d\mathbf{x} + \alpha_2 \int_\Omega f(\mathbf{x})\phi_2(\mathbf{x})d\mathbf{x} \\ &= \alpha_1\langle T_f, \phi_1\rangle + \alpha_2\langle T_f, \phi_2\rangle.\end{aligned}$$

From (1.3.11), $\forall$ compact subsets $K \subset\subset \Omega$ of Ω, $\forall \phi \in \mathcal{D}(\Omega)$ with $\operatorname{supp}(\phi) \subset K$,

$$|T_f(\phi)| \le \max_{\mathbf{x}\in K} |\phi(\mathbf{x})| \int_K |f(\mathbf{x})|d\mathbf{x} = C_K \max_{\mathbf{x}\in K} |\phi(\mathbf{x})|, \qquad C_K = \int_K |f(\mathbf{x})|d\mathbf{x} < +\infty.$$

$\Longrightarrow$ the inequality (1.3.6) holds with $C_K > 0$ and $m(K) = 0$ for all $K \subset\subset \Omega$

$\Longrightarrow$ by Proposition 1.3.1, T_f is a distribution on Ω, i.e. $T_f \in \mathcal{D}'(\Omega)$ and, by Definition 1.3.4, $T_f \in \mathcal{D}'(\Omega)$ is a distribution of order $m_0 = 0$.

In particular, for $\Omega = \mathbb{R}^n$, $f \in L^1_{\text{loc}}(\mathbb{R}^n)$ defines a distribution $T_f \in \mathcal{D}'(\mathbb{R}^n)$ of order 0 by

$$T_f(\phi) = \langle T_f, \phi\rangle = \int_{\mathbb{R}^n} f(\mathbf{x})\phi(\mathbf{x})d\mathbf{x} \quad \forall \phi \in \mathcal{D}(\mathbb{R}^n). \tag{1.3.12}$$

Theorem 1.3.2. *Let $f, g \in L^1_{\text{loc}}(\Omega)$ be any two locally summable (integrable) functions on $\Omega \subset \mathbb{R}^n$, and $T_f, T_g \in \mathcal{D}'(\Omega)$ be the distributions defined by f and g respectively, i.e. $\langle T_f, \phi\rangle = \int_\Omega f(\mathbf{x})\phi(\mathbf{x})d\mathbf{x}$ and $\langle T_g, \phi\rangle = \int_\Omega g(\mathbf{x})\phi(\mathbf{x})d\mathbf{x}$ $\forall \phi \in \mathcal{D}(\Omega)$. Then,*

$$T_f = T_g \text{ in } \mathcal{D}'(\Omega) \quad \Longleftrightarrow \quad f = g \text{ in } L^1_{\text{loc}}(\Omega). \tag{1.3.13}$$

Proof. Let $f = g$ in $L^1_{\text{loc}}(\Omega)$. Then $f(\mathbf{x}) = g(\mathbf{x})$ a.e. on Ω

$\Longrightarrow \langle T_f, \phi\rangle = \int_\Omega f(\mathbf{x})\phi(\mathbf{x})d\mathbf{x} = \int_\Omega g(\mathbf{x})\phi(\mathbf{x})d\mathbf{x} = \langle T_g, \phi\rangle\ \forall \phi \in \mathcal{D}(\Omega)$

$\Longrightarrow T_f = T_g \in \mathcal{D}'(\Omega)$ (by (1.3.7)).

Let $T_f = T_g$ in $\mathcal{D}'(\Omega)$. Then, from (1.3.7), $\langle T_f, \phi\rangle = \langle T_g, \phi\rangle\ \forall \phi \in \mathcal{D}(\Omega)$

$\Longrightarrow \int_\Omega f(\mathbf{x})\phi(\mathbf{x})d\mathbf{x} = \int_\Omega g(\mathbf{x})\phi(\mathbf{x})d\mathbf{x}\ \forall \phi \in \mathcal{D}(\Omega)$

$\Longrightarrow \int_\Omega h(\mathbf{x})\phi(\mathbf{x})d\mathbf{x} = 0\ \forall \phi \in \mathcal{D}(\Omega)$, where $h(\mathbf{x}) = f(\mathbf{x}) - g(\mathbf{x})$ for almost all $\mathbf{x} \in \Omega$, and $f - g = h \in L^1_{\text{loc}}(\Omega)$. Then, by Theorem 1.2.3A, $h = 0$ in $L^1_{\text{loc}}(\Omega)$, i.e. $f = g$ in $L^1_{\text{loc}}(\Omega)$. □

Important consequences of Theorem 1.3.2

- Distinct functions in $L^1_{\text{loc}}(\Omega)$ define distinct distributions in $\mathcal{D}'(\Omega)$ with $\Omega \subset \mathbb{R}^n$, i.e.

$$f \neq g \text{ in } L^1_{\text{loc}}(\Omega) \quad \Longrightarrow \quad T_f \neq T_g \text{ in } \mathcal{D}'(\Omega). \tag{1.3.14}$$

- As a consequence of (1.3.14), there will be no chance of confusion if we identify a locally summable function $f \in L^1_{\text{loc}}(\Omega)$ with the distribution $T_f \in \mathcal{D}'(\Omega)$ defined by it. Hence, we *will* make this identification, and the same notation f will be used instead of T_f to denote the distribution it defines: $f \in L^1_{\text{loc}}(\Omega) \Longrightarrow f \in \mathcal{D}'(\Omega) \Longrightarrow L^1_{\text{loc}}(\Omega) \subset \mathcal{D}'(\Omega)$ such that

$$\langle f, \phi\rangle = \langle T_f, \phi\rangle = \int_\Omega f(\mathbf{x})\phi(\mathbf{x})d\mathbf{x} \forall \phi \in \mathcal{D}(\Omega). \tag{1.3.15}$$

- In particular, $\langle T, \phi\rangle = \int_\Omega \phi(\mathbf{x})d\mathbf{x}\ \forall\phi \in \mathcal{D}(\Omega)$ defines a distribution T which will be identified with $f = 1$, i.e. $\langle 1, \phi\rangle = \int_\Omega \phi(\mathbf{x})d\mathbf{x}\ \forall\phi \in \mathcal{D}(\Omega)$. (1.3.16)
- All continuous functions (which include k-time differentiable or Hölder continuous or Lipschitz continuous functions) on $\overline{\Omega}$ and L^p-functions $(1 \le p \le \infty)$ on Ω (see Appendix B for details) are locally summable on Ω, and hence define distributions, i.e. $\mathcal{D}(\Omega)$, $C^0(\overline{\Omega})$, $C^k(\overline{\Omega})$, $C^{k,\lambda}(\overline{\Omega})$, $L^p(\Omega)$, $1 \le p \le \infty$, are *subspaces* of $L^1_{\text{loc}}(\Omega)$ (for $C^k(\overline{\Omega})$, $C^{k,\lambda}(\overline{\Omega})$, $0 \le \lambda \le 1$, see Appendix A, Section A.4 for details, and for $L^p(\Omega)$, $1 \le p \le \infty$, see Appendix B, Section B.4).

$$\Longrightarrow \quad \mathcal{D}(\Omega), C^0(\overline{\Omega}), C^k(\overline{\Omega}), C^{k,\lambda}(\overline{\Omega}), L^p(\Omega) \subset \mathcal{D}'(\Omega). \tag{1.3.17}$$

In other words, *the usual functions on Ω are distributions*. For example, for $n = 1, \Omega =]a,b[, f(x) = \sin x$ or $\cos x$ or e^x or x^m or $1/\sqrt{|x|}$ etc. are all elements of $L^1_{\text{loc}}(]a,b[)$ and, hence, elements of $\mathcal{D}'(]a,b[)$ in the following sense: $\forall\phi \in \mathcal{D}(]a,b[)$,

$$\langle \sin x, \phi\rangle = \int_a^b \sin x\phi(x)dx; \quad \langle e^x, \phi\rangle = \int_a^b e^x\phi(x)dx;$$
$$\langle \cos x, \phi\rangle = \int_a^b \cos x\phi(x)dx; \quad \langle x^m, \phi\rangle = \int_a^b x^m\phi(x)dx;$$
$$\left\langle \frac{1}{\sqrt{|x|}}, \phi\right\rangle = \int_a^b \frac{1}{\sqrt{|x|}}\phi(x)dx; \tag{1.3.18}$$

etc., where a, b may have arbitrary values including $-\infty$ to $+\infty$ respectively, since $\sin x\phi(x)$, $e^x\phi(x)$, $\cos x\phi(x)$, $\frac{1}{\sqrt{|x|}}\phi(x)$, etc. vanish outside $[a_0, b_0] \subset\subset]-\infty, \infty[$ for $\text{supp}(\phi) \subset [a_0, b_0]$, and the corresponding integrals in (1.3.18) exist.

Similarly, for $n = 2$, $\Omega =]a,b[\times]c,d[\subset \mathbb{R}^2$, $f(x_1, x_2) = \sin x_1 \cos x_2$ or $e^{x_1}\sin x_2$ or $x_1^m e^{x_2}$ etc. are all elements of $L^1_{\text{loc}}(]a,b[\times]c,d[)$ and, hence, elements of $\mathcal{D}'(]a,b[\times]c,d[)$ in the following sense: $\forall\phi\in\mathcal{D}(]a,b[\times]c,d[)$,

$$\langle \sin x_1 \cos x_2, \phi\rangle = \int_a^b\int_c^d \sin x_1 \cos x_2 \phi(x_1,x_2)dx_1dx_2;$$

$$\langle e^{x_1} \sin x_2, \phi\rangle = \int_a^b\int_c^d e^{x_1} \sin x_2 \phi(x_1,x_2)dx_1dx_2;$$

$$\langle x_1^m e^{x_2}, \phi\rangle = \int_a^b\int_c^d x_1^m e^{x_2} \phi(x_1,x_2)dx_1dx_2; \tag{1.3.19}$$

where a,b (resp. c,d) may have arbitrary values including $-\infty$ and $+\infty$ respectively.

Similarly, the definitions as elements of $\mathcal{D}'(\Omega)$ for other elementary functions f of several variables belonging to $L^1_{\text{loc}}(\Omega)$ can be written.

- $f(x) = \frac{1}{x} \notin L^1_{\text{loc}}(\mathbb{R})$, i.e. is not *locally* integrable on $\mathbb{R}$ (see Example 1.3.7), and hence does *not* define a distribution on $\mathbb{R}$.
- $f(x) = \frac{1}{x}$ for $x > 0$ (i.e. on $\mathbb{R}^+ =]0,\infty[$) is both continuous and *locally* summable on $\mathbb{R}^+$, but *not* summable on $\mathbb{R}^+$, i.e. $\frac{1}{x}\in L^1_{\text{loc}}(\mathbb{R}^+) \Longrightarrow \frac{1}{x}$ defines a distribution on $]0,\infty[$ by

$$\left\langle \frac{1}{x},\phi\right\rangle = \int_0^\infty \frac{1}{x}\phi(x)dx \quad \forall\phi\in\mathcal{D}(]0,\infty[), \tag{1.3.20}$$

  since $\phi(0) = 0\ \forall\phi\in\mathcal{D}(]0,\infty[)$.
- $f(x) = \frac{1}{x}$ for $x < 0$ (i.e. on $\mathbb{R}^- =]-\infty,0[$) is both continuous and locally summable on $\mathbb{R}^-$, i.e. $\frac{1}{x}\in L^1_{\text{loc}}(\mathbb{R}^-) \Longrightarrow \frac{1}{x}$ defines a distribution on $]-\infty,0[$ by:

$$\left\langle \frac{1}{x},\phi\right\rangle = \int_{-\infty}^0 \frac{1}{x}\phi(x)dx \quad \forall\phi\in\mathcal{D}(]-\infty,0[), \tag{1.3.21}$$

  since $\phi(0) = 0\ \forall\phi\in\mathcal{D}(]-\infty,0[)$.
- $f(x) = \frac{1}{x}$ for $x\in\mathbb{R}\setminus\{0\}$ is locally integrable on $\mathbb{R}\setminus\{0\} = \mathbb{R}^-\cup\mathbb{R}^+$ and defines a distribution on $\mathbb{R}\setminus\{0\}$ by:

$$\left\langle \frac{1}{x},\phi\right\rangle = \int_{-\infty}^\infty \frac{1}{x}\phi(x)dx \quad \forall\phi\in\mathcal{D}(\mathbb{R}\setminus\{0\}). \tag{1.3.22}$$

- For Cauchy principal value $\int_{-\infty}^{+\infty}\frac{1}{x}\phi(x)dx\ \forall\phi\in\mathcal{D}(\mathbb{R})$, see Example 1.3.8.
- $\ln|x|$ is locally integrable on $\mathbb{R}$ (see Example 1.4.1), and hence defines a distribution by

$$\langle \ln|x|,\phi\rangle = \int_{-\infty}^\infty ln|x|\phi(x)dx \quad \forall\phi\in\mathcal{D}(]-\infty,\infty[).$$

- In fact, every test function $\phi \in \mathcal{D}(\Omega) \subset L^1_{\text{loc}}(\Omega)$ defines a distribution $T_\phi = \phi \in \mathcal{D}'(\Omega)$ in the sense:

$$\langle \phi, \psi \rangle = \int_\Omega \phi(\mathbf{x})\psi(\mathbf{x})d\mathbf{x} \quad \forall \psi \in \mathcal{D}(\Omega). \tag{1.3.23}$$

Hence, $\mathcal{D}(\Omega) \subset \mathcal{D}'(\Omega)$. (1.3.24)

- But $\int_\Omega u(\mathbf{x})\phi(\mathbf{x})d\mathbf{x}$ may *not* exist for $u \in \mathbb{M}(\Omega)$ and $\phi \in \mathcal{D}(\Omega)$, $\mathbb{M}(\Omega)$ being the space of Lebesgue measurable functions on Ω (see Appendix B), since $\mathbb{M}(\Omega)$ is *not* a subspace of $L^1_{\text{loc}}(\Omega)$ (i.e. $\mathbb{M}(\Omega) \not\subset L^1_{\text{loc}}(\Omega)$). (1.3.25)

Regular distributions on $\Omega \subset \mathbb{R}^n$

Definition 1.3.6. A distribution $T \in \mathcal{D}'(\Omega)$ is called a *regular distribution* if and only if T can be identified with a locally (integrable) summable function $f \in L^1_{\text{loc}}(\Omega)$ such that (1.3.15) holds. In other words, for a regular distribution $T \in \mathcal{D}'(\Omega)$, $\exists$ a unique $f \in L^1_{\text{loc}}(\Omega)$ such that the integral representation of T in (1.3.15) holds, i.e.

$$\langle T, \phi \rangle = \langle f, \phi \rangle = \int_\Omega f(\mathbf{x})\phi(\mathbf{x})d\mathbf{x} \quad \forall \phi \in \mathcal{D}(\Omega). \tag{1.3.26}$$

Functions of $\mathcal{D}(\Omega)$, $C^0(\overline{\Omega})$, $C^k(\overline{\Omega})$, $C^{k,\lambda}(\overline{\Omega})$, $L^p(\Omega)$, $1 \le p \le \infty$ (see Appendices A and B), which are subspaces of $L^1_{\text{loc}}(\Omega)$ (see (1.3.17)) define regular distributions of $\mathcal{D}'(\Omega)$. (1.3.18)–(1.3.23) represent some examples of regular distributions.

Singular distributions on $\Omega \subset \mathbb{R}^n$

Definition 1.3.7. A distribution $T \in \mathcal{D}'(\Omega)$ is called a *singular distribution* if it is *not* a regular distribution. Hence, for singular distributions $T \in \mathcal{D}'(\Omega)$, there does *not* exist any locally (integrable) summable function $f \in L^1_{\text{loc}}(\Omega)$ with which T can be identified.

In other words, for a singular distribution $T \in \mathcal{D}'(\Omega)$, an integral representation of T with the help of a function $f \in L^1_{\text{loc}}(\Omega)$ in the form (1.3.26) is *not* possible.

Examples of singular distributions

Example 1.3.2.

- The Dirac distribution at the origin $\mathbf{0} \in \mathbb{R}^n$ (i.e. mass/charge/force concentrated at the origin), denoted by δ, or equivalently by $\delta_{\mathbf{0}}$ or by $\delta(\mathbf{x})$, is defined by:

$$\langle \delta, \phi \rangle = \langle \delta_{\mathbf{0}}, \phi \rangle = \langle \delta(\mathbf{x}), \phi \rangle = \phi(\mathbf{0}) \quad \forall \phi \in \mathcal{D}(\mathbb{R}^n), \tag{1.3.27}$$

where $T = \delta = \delta_{\mathbf{0}} = \delta(\mathbf{x}) \in \mathcal{D}'(\mathbb{R}^n)$ (which is proved below).

- The Dirac distribution at the point $\mathbf{a} \in \mathbb{R}^n$ (i.e. mass/charge/force concentrated at the point $\mathbf{a}$), denoted by $\delta_{\mathbf{a}}$ or $\delta(\mathbf{x} - \mathbf{a})$, is defined by:

$$\langle \delta_{\mathbf{a}}, \phi \rangle = \langle \delta(\mathbf{x} - \mathbf{a}), \phi \rangle = \phi(\mathbf{a}) \quad \forall \phi \in \mathcal{D}(\mathbb{R}^n), \tag{1.3.28}$$

where $T = \delta_{\mathbf{a}} = \delta(\mathbf{x} - \mathbf{a}) \in \mathcal{D}'(\mathbb{R}^n)$, from which, for $\mathbf{a} = \mathbf{0} \in \mathbb{R}^n$, we get $\delta_{\mathbf{0}} = \delta = \delta(\mathbf{x})$ and (1.3.27). Hence it is sufficient to show that $\delta_{\mathbf{a}}$ is a distribution of $\mathcal{D}'(\mathbb{R}^n)$.

- $\delta_{\mathbf{a}}$ is a linear functional on $\mathcal{D}(\mathbb{R}^n)$ $\forall$ fixed $\mathbf{a} \in \mathbb{R}^n$: using (1.3.28), $\forall \phi_1, \phi_2 \in \mathcal{D}(\mathbb{R}^n)$,

$$\langle \delta_{\mathbf{a}}, \phi_1 + \phi_2 \rangle = (\phi_1 + \phi_2)(\mathbf{a}) = \phi_1(\mathbf{a}) + \phi_2(\mathbf{a}) = \langle \delta_{\mathbf{a}}, \phi_1 \rangle + \langle \delta_{\mathbf{a}}, \phi_2 \rangle;$$
$$\langle \delta_{\mathbf{a}}, \alpha\phi \rangle = (\alpha\phi)(\mathbf{a}) = \alpha\phi(\mathbf{a}) = \alpha \langle \delta_{\mathbf{a}}, \phi \rangle \quad \forall \alpha \in \mathbb{R},\ \forall \phi \in \mathcal{D}(\mathbb{R}^n).$$

- $|\langle \delta_{\mathbf{a}}, \phi \rangle| = |\phi(\mathbf{a})| \le \max_{\mathbf{x} \in K} |\phi(\mathbf{x})|$ $\forall$ fixed compact subsets $K \subset\subset \mathbb{R}^n$, $\forall \phi \in \mathcal{D}(\mathbb{R}^n)$ with $\operatorname{supp}(\phi) \subset K$, i.e. $\forall \phi \in \mathcal{D}_K(\mathbb{R}^n) \implies$ the inequality (1.3.6) holds with $C_K = 1$, $m(K) = m_0 = 0$ $\forall$ compact subsets $K \subset\subset \mathbb{R}^n \implies$ by Proposition 1.3.1 and Definition 1.3.4, $\forall$ fixed $\mathbf{a} \in \mathbb{R}^n$, $\delta_{\mathbf{a}}$ defined by (1.3.28) is a distribution of order 0 in $\mathcal{D}'(\mathbb{R}^n)$. Hence, the Dirac distribution $\delta_{\mathbf{a}} \in \mathcal{D}'(\mathbb{R}^n)$ defined by (1.3.28) is a *measure* by Definition 1.3.5, and is also called the *Dirac measure*.

Remark 1.3.3.

- Equation (1.3.28) is often written in the (incorrect) form:

$$\int_{\mathbb{R}^n} \delta(\mathbf{x} - \mathbf{a})\phi(\mathbf{x})d\mathbf{x} = \phi(\mathbf{a}) \quad \forall \phi \in \mathcal{D}(\mathbb{R}^n), \tag{1.3.29}$$

where the integral is meaningless, and hence must be understood as the duality pairing $\langle \delta_{\mathbf{a}}, \phi \rangle_{\mathcal{D}'(\mathbb{R}^n) \times \mathcal{D}(\mathbb{R}^n)}$ between $\delta_{\mathbf{a}} \in \mathcal{D}'(\mathbb{R}^n)$ and $\phi \in \mathcal{D}(\mathbb{R}^n)$. But we will try to avoid this, and write in the correct form (1.3.28) unless stated otherwise.

- The Dirac distribution (measure) $\delta_{\mathbf{a}}$ is usually called the *delta function* $\delta(\mathbf{x}-\mathbf{a})$ (see (1.1.2)) with mass/charge/load etc. concentrated at the point $\mathbf{a} \in \mathbb{R}^n$. Hence, the precise mathematical definition of the delta function $\delta(\mathbf{x}-\mathbf{a})$ is given by (1.3.28) and does *not* suffer from the contradictory properties in (1.1.7)–(1.1.9), since $\delta(\mathbf{x}-\mathbf{a})$ is the distribution $\delta_{\mathbf{a}}$ and *not* a function, and the meaningless integral representation in (1.1.2) is replaced by the correct representation (1.3.28) with the duality pairing $\langle \delta_{\mathbf{a}}, \phi \rangle_{\mathcal{D}'(\mathbb{R}^n)\times\mathcal{D}(\mathbb{R}^n)}$ (see also Section 1.11).
- In fact, we will prove that (1.3.28) can *not* be written in the integral form (1.3.29) to justify that $\delta_{\mathbf{a}} \in \mathcal{D}'(\mathbb{R}^n)$ is a singular distribution.

Proposition 1.3.2. *There does not exist any locally integrable function $f \in L^1_{\text{loc}}(\mathbb{R}^n)$ such that*

$$\langle \delta_{\mathbf{a}}, \phi \rangle = \int_{\mathbb{R}^n} f(\mathbf{x})\phi(\mathbf{x})d\mathbf{x} = \phi(\mathbf{a}) \quad \forall \phi \in \mathcal{D}(\mathbb{R}^n). \tag{1.3.30}$$

(In other words, the Dirac distribution $\delta_{\mathbf{a}} = \delta(\mathbf{x}-\mathbf{a})$ in (1.3.28) *can not be represented by a locally integrable function $f \in L^1_{\text{loc}}(\mathbb{R}^n)$ and is thus a* singular distribution.*)*

Proof. Without loss of generality, and for the sake of simplicity, we consider $\mathbf{a} = \mathbf{0} \in \mathbb{R}^n$, i.e. $\delta = \delta_{\mathbf{0}} = \delta(\mathbf{x})$. $\forall \varepsilon > 0$, let $\phi_\varepsilon \in \mathcal{D}(\mathbb{R}^n)$ be a test function defined by:

$$\phi_\varepsilon(\mathbf{x}) = \begin{cases} e^{-\frac{\varepsilon^2}{\varepsilon^2-\|\mathbf{x}\|^2}} & \text{for } \mathbf{x} \in B(\mathbf{0};\varepsilon) = \{\mathbf{x} : \|\mathbf{x}\| = (x_1^2+\cdots+x_n^2)^{1/2} < \varepsilon\} \\ 0 & \text{for } \|\mathbf{x}\| \geq \varepsilon \end{cases}$$

with $\operatorname{supp}(\phi_\varepsilon) = \bar{B}(\mathbf{0};\varepsilon) = \{\mathbf{x} : \|\mathbf{x}\| \leq \varepsilon\}$, $\phi_\varepsilon(\mathbf{x}) \leq e^{-1}$ $\forall \mathbf{x} \in \mathbb{R}^n$.

Then $\langle \delta, \phi_\varepsilon \rangle = \phi_\varepsilon(\mathbf{0}) = \frac{1}{e}$ $\forall \varepsilon > 0 \Longrightarrow \lim_{\varepsilon\to 0}\langle \delta, \phi_\varepsilon \rangle = \lim_{\varepsilon\to 0}\phi_\varepsilon(\mathbf{0}) = \frac{1}{e}$.

Suppose that the *contrary* holds, i.e. $\exists f \in L^1_{\text{loc}}(\mathbb{R}^n)$ such that δ can be represented by f: $\langle \delta, \phi \rangle = \langle f, \phi \rangle = \int_{\mathbb{R}^n} f(\mathbf{x})\phi(\mathbf{x})d\mathbf{x} = \phi(\mathbf{0})$ $\forall \phi \in \mathcal{D}(\mathbb{R}^n)$. Then we have

$$\langle f, \phi_\varepsilon \rangle = \langle \delta, \phi_\varepsilon \rangle = \phi_\varepsilon(\mathbf{0}) = \frac{1}{e} \quad \Longrightarrow \quad \lim_{\varepsilon\to 0}\langle f, \phi_\varepsilon \rangle = \frac{1}{e}. \tag{1.3.31}$$

Since $f \in L^1_{\text{loc}}(\mathbb{R}^n)$, $\phi_\varepsilon \in \mathcal{D}(\mathbb{R}^n)$ $\forall \varepsilon > 0$, we have

$$\langle f, \phi_\varepsilon \rangle = \int_{\mathbb{R}^n} f(\mathbf{x})\phi_\varepsilon(\mathbf{x})d\mathbf{x} = \int_{B(\mathbf{0};\varepsilon)} f(\mathbf{x})\phi_\varepsilon(\mathbf{x})d\mathbf{x},$$

because $\phi(\mathbf{x}) = 0$ for all $\mathbf{x}$ outside $B(\mathbf{0};\varepsilon)$,

$$\begin{aligned} \Longrightarrow \quad |\langle f, \phi_\varepsilon \rangle| &= \left| \int_{B(\mathbf{0};\varepsilon)} f(\mathbf{x})\phi_\varepsilon(\mathbf{x})d\mathbf{x} \right| \leq \sup_{\mathbf{x}\in B(\mathbf{0};\varepsilon)} \phi_\varepsilon(\mathbf{x}) \int_{B(\mathbf{0};\varepsilon)} |f(\mathbf{x})|d\mathbf{x} \\ &= \frac{1}{e}\int_{B(\mathbf{0},\varepsilon)} |f(\mathbf{x})|d\mathbf{x} < +\infty, \end{aligned}$$

since $f \in L^1_{\text{loc}}(\Omega)$.

But Lebesgue n-dimensional volume measure $\mu(B(\mathbf{0};\varepsilon)) \to 0$ as $\varepsilon \to 0$,

$$\Longrightarrow \quad \int_{B(\mathbf{0};\varepsilon)} |f(\mathbf{x})| d\mathbf{x} \to 0 \quad \text{as } \varepsilon \to 0$$

$$\Longrightarrow \quad \frac{1}{e} \int_{B(\mathbf{0};\varepsilon)} |f(\mathbf{x})| d\mathbf{x} \to 0 \quad \text{as } \varepsilon \to 0.$$

Hence, $\lim_{\varepsilon \to 0} \langle f, \phi_\varepsilon \rangle = 0$, which contradicts (1.3.31), i.e. our assumption is wrong and the result follows.

Thus, the Dirac distribution $\delta_{\mathbf{a}}$ can not be written in the form (1.3.30) and is a singular distribution. □

Remark 1.3.4. For more details, see Section 1.11.

Example 1.3.3. For an open subset $\Omega \subset \mathbb{R}^n$, functional T defined by: $\forall$ fixed $\boldsymbol{\alpha}$ with $|\boldsymbol{\alpha}| = m$ and fixed point $\mathbf{a} \in \Omega$,

$$T(\phi) = \langle T, \phi \rangle = \partial^{\boldsymbol{\alpha}} \phi(\mathbf{a}) \quad \forall \phi \in \mathcal{D}(\Omega) \tag{1.3.32}$$

is a distribution of order m in $\mathcal{D}'(\Omega)$.

T defined by (1.3.32) is a linear functional: $\forall \phi_1, \phi_2 \in \mathcal{D}(\Omega)$,

$$\langle T, \phi_1 + \phi_2 \rangle = \partial^{\boldsymbol{\alpha}}(\phi_1 + \phi_2)(\mathbf{a}) = \partial^{\boldsymbol{\alpha}}\phi_1(\mathbf{a}) + \partial^{\boldsymbol{\alpha}}\phi_2(\mathbf{a}) = \langle T, \phi_1 \rangle + \langle T, \phi_2 \rangle$$
$$\langle T, \lambda\phi \rangle = \partial^{\boldsymbol{\alpha}}(\lambda\phi)(\mathbf{a}) = \lambda \partial^{\boldsymbol{\alpha}}\phi(\mathbf{a}) = \lambda \langle T, \phi \rangle \quad \forall \lambda \in \mathbb{R},\ \forall \phi \in \mathcal{D}(\Omega).$$

T satisfies the inequality (1.3.6) with $C_K = 1$ and $m(K) = m$ for all compact subsets $K \subset\subset \Omega$: $\forall \phi \in \mathcal{D}(\Omega)$ with $\operatorname{supp}(\phi) \subset K$,

$$|T(\phi)| = |\langle T, \phi \rangle| = |\partial^{\boldsymbol{\alpha}}\phi(\mathbf{a})| \le \max_{|\boldsymbol{\alpha}|=m} \max_{\mathbf{x} \in K} |\partial^{\boldsymbol{\alpha}}\phi(\mathbf{x})|$$

$\Longrightarrow$ by Proposition 1.3.1 and Definition 1.3.4, T is a distribution of order m on Ω, i.e. $T \in \mathcal{D}'(\Omega)$. Hence, for $m \ge 1$, T is *not* a measure, although it is still a singular distribution.

In particular, for $\Omega = \mathbb{R}^n$ and $m = 0$, we get $T = \delta_{\mathbf{a}} \in \mathcal{D}'(\mathbb{R}^n)$ defined by (1.3.28), i.e. $m = 0, T = \delta_{\mathbf{a}}$ is a measure.

Example 1.3.4 (Dirac distribution $\delta_{\mathbb{S}}$ on a surface $\mathbb{S}$, see Section 1.11 for interesting details).

• Let $x_1 = 0$ be the equation of a hyperplane $\mathbb{S}$ in $\mathbb{R}^n$. Then $\delta_{\mathbb{S}} \in \mathcal{D}'(\mathbb{R}^n)$ is a Dirac distribution (measure) concentrated on $\mathbb{S}$ (i.e. with mass/charge/force etc. concentrated on the hyperplane $\mathbb{S} \subset \mathbb{R}^n$): $\forall \phi \in \mathcal{D}(\mathbb{R}^n)$ with $\operatorname{supp}(\phi) \subset K \subset\subset \mathbb{R}^n$,

$$\langle \delta_{\mathbb{S}}, \phi \rangle = \int_{\mathbb{R}^{n-1}} \phi(0, x_2, \ldots, x_n) dx_2 dx_3 \ldots dx_n. \tag{1.3.33}$$

Since $\delta_{\mathbb{S}}$ is a linear functional on $\mathcal{D}(\mathbb{R}^n)$ and, $\forall \phi \in \mathcal{D}(\mathbb{R}^n)$ with $\text{supp}(\phi) \subset K$,

$$
\begin{aligned}
|\langle \delta_{\mathbb{S}}, \phi \rangle| &= \left| \int_{\mathbb{R}^{n-1}} \phi(0, x_2, \dots, x_n) dx_2 \dots dx_n \right| \\
&= \left| \int_{K \cap \mathbb{R}^{n-1}} \phi(0, x_2, \dots, x_n) dx_2 \dots dx_n \right| \leq C_K \max_{\mathbf{x} \in K} |\phi(\mathbf{x})|
\end{aligned}
$$

with $C_K = \int_{K \cap \mathbb{S}} dx_2 \dots dx_n \leq \int_K dx_1 dx_2 \dots dx_n =$ n-dimensional volume measure $\mu(K) < +\infty$ ($\text{supp}(\phi) \subset K \subset\subset \mathbb{R}^n$)

$\Longrightarrow |\langle \delta_{\mathbb{S}}, \phi \rangle|$ satisfies the inequality (1.3.6) with $C_K < +\infty$ and $m(K) = m_0 = 0$ for *all* compact subsets K with $\text{supp}(\phi) \subset K$

$\Longrightarrow$ by Proposition 1.3.1 and Definition 1.3.4, $\delta_{\mathbb{S}}$ defined by (1.3.33) is a distribution of order 0 in $\mathcal{D}'(\mathbb{R}^n)$ and, hence, a measure by Definition 1.3.5.

• Let $f(x_1, x_2, \dots, x_n) = 0$ be the equation of the smooth (i.e. infinitely differentiable) hypersurface $\mathbb{S} \subset \mathbb{R}^n$ such that $\frac{\partial f}{\partial x_j} \neq 0$ for $1 \leq j \leq n$ (i.e. $\nabla f \neq \mathbf{0}$). *Leray form* ω [1, Vol.1, p. 348] for the hypersurface $\mathbb{S}$ is the exterior differential form such that

$$
df \wedge \omega = dx_1 \wedge dx_2 \wedge \dots \wedge dx_n = dx_1 dx_2 \dots dx_n.
$$

Then on $\mathbb{S}$ with $\frac{\partial f}{\partial x_j} \neq 0$, $1 \leq j \leq n$, we have

$$
\begin{aligned}
\omega &= \frac{dx_2 \wedge dx_3 \wedge \dots \wedge dx_n}{\partial f / \partial x_1} = -\frac{dx_1 \wedge dx_3 \wedge \dots \wedge dx_n}{\partial f / \partial x_2} \\
&= (-1)^{j-1} \frac{dx_1 \wedge \dots \wedge dx_{j-1} \wedge dx_{j+1} \wedge \dots \wedge dx_n}{\partial f / \partial x_j} \qquad (1 \leq j \leq n),
\end{aligned}
$$

and Dirac distribution $\delta_{\mathbb{S}} \in \mathcal{D}'(\mathbb{R}^n)$ is defined by: for fixed j with $1 \leq j \leq n$,

$$
\langle \delta_{\mathbb{S}}, \phi \rangle = \int_{\mathbb{S}} \phi \omega = (-1)^{j-1} \int_{\mathbb{S}} \frac{\phi(\mathbf{x})}{\partial f / \partial x_j} dx_1 dx_2 \dots dx_{j-1} dx_{j+1} \dots dx_n. \quad (1.3.34)
$$

The integral in (1.3.34) is taken over the hypersurface $\mathbb{S} : f = 0$, which is why $\delta_{\mathbb{S}}$ is said to be *concentrated* on hypersurface $\mathbb{S} : f(\mathbf{x}) = 0$.

Since $\delta_{\mathbb{S}}$ is a linear functional on $\mathcal{D}(\mathbb{R}^n)$ and $|\langle \delta_{\mathbb{S}}, \phi \rangle| \leq C_K \max_{\mathbf{x} \in K} |\phi(\mathbf{x})|$ satisfies the inequality (1.3.6) with $C_K = \int_{K \cap \mathbb{S}} \omega < +\infty$ and $m(K) = m_0 = 0$ for all compact subsets K of $\mathbb{R}^n$, with $\text{supp}(\phi) \subset K$, $\delta_{\mathbb{S}}$ defined by (1.3.34) is also a distribution of order 0 by Proposition 1.3.1 and Definition 1.3.4, and, hence, a measure by Definition 1.3.5. For more details on (1.3.34), see [1, Vol.1, pp. 210–211, 348–349]).

Dirac distributions $\delta_{\mathbb{S}} \in \mathcal{D}'(\mathbb{R}^n)$ defined by (1.3.33) and (1.3.34) are singular distributions on $\mathbb{R}^n$, since $\langle \delta_{\mathbb{S}}, \phi \rangle \forall \phi \in D(\mathbb{R}^n)$ can not be represented by an n-dimensional volume integral over $\mathbb{R}^n$.

Example 1.3.5 (Magnetic dipole moment layer on smooth surface $\mathbb{S}$; see Section 1.11 for interesting details). Let $\mathbb{S}$ be a smooth, bounded surface in $\mathbb{R}^3$, dS the surface area measure on $\mathbb{S}$, $\frac{\partial \phi}{\partial n}$ the derivative of ϕ in the direction of the unit normal $\hat{\mathbf{n}}$ to $\mathbb{S}$ and ρ be a continuous function on $\mathbb{S}$ defining the surface moment density on $\mathbb{S}$. Then,

$$\langle T, \phi \rangle = \int_{\mathbb{S}} \rho(\mathbf{x}) \frac{\partial \phi}{\partial n} dS \quad \forall \phi \in \mathcal{D}(\mathbb{R}^3) \tag{1.3.35}$$

defines a distribution T on $\mathbb{R}^3$, which is the *normally oriented magnetic dipole distribution* over $\mathbb{S}$ with surface moment density ρ, since T is a linear functional on $\mathcal{D}(\mathbb{R}^3)$ and

$$|\langle T, \phi \rangle| = \left| \int_{\mathbb{S}} \rho(\mathbf{x}) \frac{\partial \phi}{\partial n}(\mathbf{x}) dS \right| \leq \max_{\mathbf{x} \in K} \left| \frac{\partial \phi}{\partial n}(\mathbf{x}) \right| \int_{\mathbb{S} \cap \mathbb{K}} |\rho(\mathbf{x})| dS \ \forall \phi \in \mathcal{D}(\mathbb{R}^3)$$

with $\mathrm{supp}(\phi) \subset K$.

Since $|\frac{\partial \phi}{\partial n}(\mathbf{x})| = |\langle \nabla \phi(\mathbf{x}), \ \hat{\mathbf{n}} \rangle_{\mathbb{R}^3}| \leq \sqrt{3} \ \max_{1 \leq i \leq 3} \ \max_{\mathbf{x} \in K} \ |\partial_i \phi(\mathbf{x})|$, $|\langle T, \phi \rangle| \leq \max_{|\boldsymbol{\alpha}| \leq 1} \max_{\mathbf{x} \in K} |\partial^{\boldsymbol{\alpha}} \phi(\mathbf{x})| C_K$ with $C_K = \sqrt{3} \int_{\mathbb{S} \cap \mathbb{K}} |\rho(\mathbf{x})| dS < +\infty$ with $\mathrm{supp}(\phi) \subset K$, and for all compact subsets $K \subset\subset \mathbb{R}^3$,

$\Longrightarrow$ by Proposition 1.3.1 and Definition 1.3.4, T is a distribution of order $m_0 = 1$.

Remark 1.3.5. T defined by (1.3.35) is a singular distribution (see Section 1.11 for explanations). This distribution must *not* be confused with the regular distribution defined by *volume* density f, which is identified with the function f itself, i.e. $\langle f, \phi \rangle = \langle T, \phi \rangle = \int_{\mathbb{R}^3} f(\mathbf{x}) \phi(\mathbf{x}) \ \forall \phi \in \mathcal{D}(\mathbb{R}^3)$.

We will also meet with distributions of the following form in the definition of derivatives of distributions later.

Example 1.3.6. For open subset $\Omega \subset \mathbb{R}^n$ of $\mathbb{R}^n$, let $f \in L^1_{\mathrm{loc}}(\Omega)$ be a fixed locally integrable (summable) function on Ω. Then the functional T defined by: $\forall$ fixed $\boldsymbol{\alpha}$ with $|\boldsymbol{\alpha}| = m \in \mathbb{N}$, $\forall \phi \in \mathcal{D}(\Omega)$,

$$T(\phi) = \langle T, \phi \rangle = \int_{\Omega} f(\mathbf{x}) \partial^{\boldsymbol{\alpha}} \phi(\mathbf{x}) d\mathbf{x} = \langle f, \partial^{\boldsymbol{\alpha}} \phi \rangle \tag{1.3.36}$$

is also a distribution of order m in $\mathcal{D}'(\Omega)$, since $\forall \phi \in \mathcal{D}(\Omega)$ with $\mathrm{supp}(\phi) \subset K$,

$$|T(\phi)| = |\langle T, \phi \rangle| = \left| \int_{\Omega} f(\mathbf{x}) \partial^{\boldsymbol{\alpha}} \phi(\mathbf{x}) d\mathbf{x} \right|$$

$$\leq \max_{|\boldsymbol{\alpha}|=m} \max_{\mathbf{x} \in K} |\partial^{\boldsymbol{\alpha}} \phi(\mathbf{x})| \int_K |f(\mathbf{x})| d\mathbf{x} = C_K \max_{|\boldsymbol{\alpha}|=m} \max_{\mathbf{x} \in K} |\partial^{\boldsymbol{\alpha}} \phi(\mathbf{x})|,$$

with $C_K = \int_K |f(\mathbf{x})|d\mathbf{x} < +\infty$, $m(K) = m$ for all compact subsets $K \subset\subset \Omega$ of Ω. In particular, for $\Omega = \mathbb{R}^n$ and fixed $f \in L^1_{\text{loc}}(\mathbb{R}^n)$, $T \in \mathcal{D}'(\mathbb{R}^n)$ defined by: $\forall$ fixed $\boldsymbol{\alpha}$ with $|\boldsymbol{\alpha}| = m$,

$$T(\phi) = \langle T, \phi\rangle = \int_\Omega f(\mathbf{x})\partial^{\boldsymbol{\alpha}}\phi(\mathbf{x})d\mathbf{x} \quad \forall\phi \in \mathcal{D}(\mathbb{R}^n) \tag{1.3.37}$$

is also a distribution of order m in $\mathcal{D}'(\mathbb{R}^n)$.

Example 1.3.7. $f(x) = \frac{1}{x}$ is *not* locally summable on $\mathbb{R}$. Hence, $\frac{1}{x}$ does not define a distribution on $\mathbb{R}$.

In fact, for $a > 0$, $\int_{-a}^0 \frac{1}{|x|}dx = \infty$, $\int_0^a \frac{1}{x}dx = \infty$, and hence, $\forall$ compact intervals $[-a, a]$ with $a > 0$, $\int_{-a}^a \frac{1}{|x|}dx = \infty$; $\frac{1}{|x|}$ is not integrable in the neighbourhood of $x = 0$, and consequently $\frac{1}{x}$ is not locally integrable (summable) on $\mathbb{R} =]-\infty, \infty[$. Hence, $\int_{-\infty}^\infty \frac{1}{x}\phi(x)dx$ is *not* defined $\forall\phi \in \mathcal{D}(\mathbb{R})$. But $\frac{1}{x}$ is locally summable on $\mathbb{R} \setminus \{0\}$ and defines a distribution on $\mathbb{R} \setminus \{0\}$ (see (1.3.22)).

Example 1.3.8 (Cauchy principal value).

• c.p.v. $\int_{-\infty}^\infty \frac{1}{x}\phi(x)dx$ defines a distribution T on $\mathbb{R}$ by: $\langle T, \phi\rangle =$ c.p.v. $\int_{-\infty}^\infty \frac{1}{x}\phi(x)\, dx$ $\forall\phi \in \mathcal{D}(\mathbb{R})$, since c.p.v. $\int_{-\infty}^\infty \frac{1}{x}\phi(x)dx =$ c.p.v. $\int_{-a}^a \frac{1}{x}\phi(x)dx = \lim_{\varepsilon\to 0+}[\int_{-a}^{-\varepsilon} \frac{1}{x}\phi(x)dx + \int_\varepsilon^a \frac{1}{x}\phi(x)dx]$ exists $\forall$ compact intervals $[-a, a]$ with $\text{supp}(\phi) \subset]-a, a[$, $a > 0$.

• $$\langle T, \phi\rangle = \text{c.p.v.} \int_{-\infty}^\infty \frac{1}{x}\phi(x)dx = \lim_{\varepsilon\to 0+} \int_{|x|\geq\varepsilon} \frac{1}{x}\phi(x)dx \quad \forall\phi \in \mathcal{D}(\mathbb{R}) \tag{1.3.38}$$

defines a distribution on $\mathbb{R}$.

For this we are to show that T is a continuous linear functional on $\mathcal{D}(\mathbb{R})$. By virtue of the properties of integral and limit, $\langle T, \alpha_1\phi_1 + \alpha_2\phi_2\rangle = \alpha_1 \lim_{\varepsilon\to 0+} \int_{|x|\geq\varepsilon} \frac{1}{x}\phi_1(x)\, dx + \alpha_2 \lim_{\varepsilon\to 0+} \int_{|x|\geq\varepsilon} \frac{1}{x}\phi_2(x)dx = \alpha_1\langle T, \phi_1\rangle + \alpha_2\langle T, \phi_2\rangle \quad \forall\phi_1, \phi_2 \in \mathcal{D}(\mathbb{R})$.

• T is continuous on $\mathcal{D}(\mathbb{R}) \iff \langle T, \phi_n\rangle =$ c.p.v. $\int_{-\infty}^\infty \frac{1}{x}\phi_n(x)dx \to 0$ in $\mathbb{R}$ as $\phi_n \to 0$ in $\mathcal{D}(\mathbb{R})$. Let $]-a, a[$ be any bounded interval such that

$$\text{supp}(\phi_n) \subset]-a, a[\quad \forall n \in \mathbb{N},\ a > 0. \tag{1.3.39}$$

Hence,

$$\begin{aligned} \text{c.p.v.} \int_{-\infty}^\infty \frac{1}{x}\phi_n(x)dx &= \text{c.p.v.} \int_{-a}^a \frac{1}{x}\phi_n(x)dx \\ &= \text{c.p.v.} \int_{-a}^a \frac{\phi_n(0)}{x}dx + \text{c.p.v.} \int_{-a}^a \frac{\phi_n(x) - \phi_n(0)}{x}dx. \end{aligned} \tag{1.3.40}$$

But

$$\text{c.p.v.} \int_{-a}^{a} \frac{\phi_n(0)}{x} dx = \phi_n(0) \left(\text{c.p.v.} \int_{-a}^{a} \frac{1}{x} dx \right) = \phi_n(0) \cdot 0 = 0, \tag{1.3.41}$$

since $\frac{1}{x}$ is an odd function, $(\int_{-a}^{-\varepsilon} \frac{1}{x} dx + \int_{\varepsilon}^{a} \frac{1}{x} dx) = 0 \; \forall \varepsilon > 0$ and c.p.v. $\int_{-a}^{a} \frac{1}{x} dx = \lim_{\varepsilon \to 0^+} (\int_{-a}^{-\varepsilon} \frac{1}{x} dx + \int_{\varepsilon}^{a} \frac{1}{x} dx) dx = 0$.

Since $\lim_{x \to 0} \frac{\phi_n(x) - \phi_n(0)}{x} = \phi_n'(0)$, the integrals $\int_{-a}^{0} \frac{\phi_n(x) - \phi_n(0)}{x} dx$ and $\int_{0}^{a} \frac{\phi_n(x) - \phi_n(0)}{x} dx$ exist (are finite) and, consequently,

$$\text{c.p.v.} \int_{-a}^{a} \frac{\phi_n(x) - \phi_n(0)}{x} dx = \int_{-a}^{a} \frac{\phi_n(x) - \phi_n(0)}{x} dx. \tag{1.3.42}$$

Moreover, by the Mean Value Theorem,

$$\begin{aligned} &|\phi_n(x) - \phi_n(0)| \le |x| \max_{x \in [-a,a]} |\phi_n'(x)| \\ \Longrightarrow \quad & \left| \frac{\phi_n(x) - \phi_n(0)}{x} \right| \le \max_{x \in [-a,a]} |\phi_n'(x)| \\ \Longrightarrow \quad & \left| \int_{-a}^{a} \frac{\phi_n(x) - \phi_n(0)}{x} dx \right| \le \max_{x \in [-a,a]} |\phi_n'(x)| \int_{-a}^{a} dx = 2a \max_{x \in [-a,a]} |\phi_n'(x)| \end{aligned} \tag{1.3.43}$$

Using (1.3.41)–(1.3.43) in (1.3.40), we get

$$\left| \text{c.p.v.} \int_{-\infty}^{\infty} \frac{\phi_n(x)}{x} dx \right| \le 2a \max_{x \in [-a,a]} |\phi_n'(x)| \to 0 \quad \text{in } \mathbb{R} \text{ as } n \to \infty, \tag{1.3.44}$$

since $\operatorname{supp}(\phi_n) \subset \,]-a, a[\; \forall n \in \mathbb{N}$, $\phi_n \to 0$ in $\mathcal{D}(\mathbb{R}) \Longrightarrow \phi_n'(x) \to 0$ uniformly in $\mathbb{R}$ as $n \to \infty \Longrightarrow \max_{x \in [-a,a]} |\phi_n'(x)| \to 0$ as $n \to \infty$.

Thus, $\langle T, \phi_n \rangle = \text{c.p.v.} \int_{-\infty}^{\infty} \frac{\phi_n(x)}{x} dx \to 0$ as $n \to \infty \Longrightarrow T$ is continuous on $\mathcal{D}(\mathbb{R}) \Longrightarrow T$ is a distribution, which we agree to denote by c.p.v. $\frac{1}{x}$ such that

$$\left\langle \text{c.p.v.} \frac{1}{x}, \phi \right\rangle = \text{c.p.v.} \int_{-\infty}^{\infty} \frac{1}{x} \phi(x) dx \quad \forall \phi \in \mathcal{D}(\mathbb{R}), \tag{1.3.45}$$

i.e. c.p.v. $\frac{1}{x} \in \mathcal{D}'(\mathbb{R})$. From (1.3.44), c.p.v. $\frac{1}{x}$ is a distribution of order 1.

1.3.5 Distribution defined on test space $\mathcal{D}(\Omega)$ of complex-valued functions

Up to now we have considered distributions T defined on $\mathcal{D}(\Omega)$ consisting of real-valued test functions ϕ such that $T(\phi) \in \mathbb{R}$. But in *some* situations we are to consider test space $\mathcal{D}(\Omega)$ consisting of *complex-valued* test functions $\phi \in C_0^\infty(\Omega)$ satisfying

(1.2.43) and (1.2.44); the complexification of $\mathcal{D}(\Omega)$ is a quite straightforward affair (see Remark 1.2.2), since almost all results stated for the real case in this section can be extended to the complex case with *minor* or *no* modifications at all. For example, $\mathcal{D}(\Omega)$ is now a complex vector space, which is obviously closed under multiplication by complex numbers $\alpha \in \mathbb{C}$. Hence, for the complex test space $\mathcal{D}(\Omega)$ consisting of complex-valued functions ϕ of real variables $x_1, \dots, x_n$ in $\Omega \subset \mathbb{R}^n$, satisfying (1.2.43) and (1.2.44), T is a distribution if and only if

- $T : \phi \in \mathcal{D}(\Omega) \longmapsto T(\phi) = \langle T, \phi \rangle \in \mathbb{C}$ is linear on $\mathcal{D}(\Omega)$: (1.3.46)

$$T(\phi_1 + \phi_2) = T(\phi_1) + T(\phi_2) \quad \forall \phi_1, \phi_2 \in \mathcal{D}(\Omega);$$
$$T(\alpha\phi) = \alpha T(\phi) \quad \forall \alpha \in \mathbb{C},\ \forall \phi \in \mathcal{D}(\Omega). \tag{1.3.47}$$

- T is continuous on $\mathcal{D}(\Omega)$: $\phi_n \to 0$ in $\mathcal{D}(\Omega)$ in the sense of (1.2.44) (together with (1.2.43)) $\Longrightarrow T(\phi_n) \to 0$ in $\mathbb{C}$ as $n \to \infty$. (1.3.48)

$\mathcal{D}'(\Omega) = \{T : T$ is a continuous linear functional on $\mathcal{D}(\Omega)$ satisfying (1.3.46)–(1.3.48)$\}$ is the space of distributions defined on complex test space $\mathcal{D}(\Omega)$. (1.3.49)

All the basic results of this section for the real case will also hold for the complex case with the necessary modifications satisfying (1.2.43) and (1.2.44), (1.3.46)–(1.3.49) and without any additional theoretical complications, since the variables are real and the role of $i = \sqrt{-1}$ is that of a parameter in all operations of differentiation and integration.

Hence, the complexification of $\mathcal{D}'(\Omega)$ is also a straightforward affair and consequently we will ignore the complexification aspects of later problems.

1.4 Some more examples of interesting distributions

Example 1.4.1. Function $\ln|x|$ defines a regular distribution on $\mathbb{R}$.

Solution. Function $\ln|x|$ $(x \neq 0)$ is integrable in the neighbourhood of 0 (see Example 2.3.6 for an alternate proof) since $\forall \varepsilon < 1$, $|x|^{\varepsilon}\big|\ln|x|\big| \to 0$ as $|x| \to 0$ and consequently, $\big|\ln|x|\big| \le \frac{1}{|x|^{\varepsilon}}$ $\forall x \neq 0$ in some neighbourhood $]-a, a[$ of 0 with $a > 0$, $\int_{-a}^{a} \big|\ln|x|\big| dx \le \int_{-a}^{a} \frac{1}{x|^{\varepsilon}} dx < \infty$ for $\varepsilon < 1 \Longrightarrow \ln|x|$ is locally integrable on $\mathbb{R}$, i.e. $\ln|x| \in L^1_{\text{loc}}(\mathbb{R})$ and defines a regular distribution on $\mathbb{R}$ by

$$\langle \ln|x|, \phi \rangle = \int_{-\infty}^{\infty} (\ln|x|)\phi(x)dx \quad \forall \phi \in \mathcal{D}(\mathbb{R}). \tag{1.4.1}$$

Example 1.4.2. Functions $\ln(x + i0)$ and $\ln(x - i0)$ define regular distributions on $\mathbb{R}$.

Solution. Functions $\ln(x \pm i0)$ are defined by :

$$\ln(x + i0) = \lim_{y \to 0+} \ln(x + iy); \quad \ln(x - i0) = \lim_{y \to 0+} \ln(x - iy). \tag{1.4.2}$$

In fact, for fixed $y > 0$ in the upper half-plane of $z = x + iy$,

$$\ln z = \ln|z| + i \arg z = \ln\sqrt{x^2+y^2} + i \arctan\left(\frac{y}{x}\right) \quad \text{is analytic.} \tag{1.4.3}$$

But $\lim_{y\to 0+} \ln\sqrt{x^2+y^2} = \lim_{y\to 0+} \frac{1}{2}\ln(x^2+y^2) = \frac{1}{2}\ln x^2 = \ln|x|$, and

$$\lim_{y\to 0+} i \arctan\left(\frac{y}{x}\right) = \begin{Bmatrix} i\pi & \text{for } x < 0 \\ 0 & \text{for } x > 0 \end{Bmatrix} = i\pi H(-x), \tag{1.4.4}$$

where the Heaviside function $H(-x) = 1$ for $x < 0$, $H(-x) = 0$ for $x > 0$.

Hence,

$$\lim_{y\to 0+} \ln(x+iy) = \begin{cases} \ln|x| + i\pi & \text{for } x < 0 \\ \ln|x| & \text{for } x > 0 \end{cases}$$

$$\Longrightarrow \quad \ln(x+i0) = \lim_{y\to 0+} \ln(x+iy) = \ln|x| + i\pi H(-x). \tag{1.4.5}$$

Similarly, for fixed $y > 0$, $\ln(x-iy) = \frac{1}{2}\ln(x^2+y^2) - i\arctan(\frac{y}{x})$

$$\Longrightarrow \ln(x-i0) = \lim_{y\to 0+} \ln(x-iy) = \begin{cases} \ln|x| - i\pi & \text{for } x < 0 \\ \ln|x| & \text{for } x > 0 \end{cases} \tag{1.4.6}$$

$$= \ln|x| - i\pi H(-x).$$

Since $\ln|x| \in L^1_{\text{loc}}(\mathbb{R})$ (see Example 1.4.1) and $H(-x) \in L^1_{\text{loc}}(\mathbb{R})$, $\ln(x \pm i0) \in L^1_{\text{loc}}(\mathbb{R})$ and define regular distributions by: $\forall \phi \in \mathcal{D}(\mathbb{R})$,

$$\begin{aligned} \langle \ln(x+i0), \phi\rangle &= \int_{\mathbb{R}} \ln(x+i0)\phi(x)dx \\ &= \int_{-\infty}^{\infty} \ln|x|\phi(x)dx + i\pi\int_{-\infty}^{0}\phi(x)dx; \end{aligned} \tag{1.4.7}$$

$$\begin{aligned} \langle \ln(x-i0), \phi\rangle &= \int_{\mathbb{R}} \ln(x-i0)\phi(x)dx \\ &= \int_{-\infty}^{\infty} \ln|x|\phi(x)dx - i\pi\int_{-\infty}^{0}\phi(x)dx. \end{aligned} \tag{1.4.8}$$

Distributions defined by pseudo-functions (or finite part) Pf $\frac{1}{x^k}$, Pf $\frac{H(\pm x)}{x^k}$

From Examples 1.3.7 and 1.3.8, we know that $\frac{1}{x} \notin L^1_{\text{loc}}(\mathbb{R})$ and hence does not define a distribution on $\mathbb{R}$, but c.p.v. $\int_{-\infty}^{\infty} \frac{\phi(x)}{x}dx$ is well defined and defines a distribution on $\mathbb{R}$ denoted by c.p.v. $\frac{1}{x}$. For $k \geq 2$, $\forall \phi \in \mathcal{D}(\mathbb{R})$, $\lim_{\varepsilon \to 0+} \int_{|x| \geq \varepsilon} \frac{\phi(x)}{x^k} dx = \pm\infty$, but $\forall \varepsilon > 0$, if we can write

$$\int_{|x| \geq \varepsilon} \frac{\phi(x)}{x^k} dx = I(\varepsilon) + F(\varepsilon), \tag{1.4.9}$$

where $I(\varepsilon)$ is called the *infinite part* of the divergent integral, since $I(\varepsilon)$ tends to $\pm\infty$ as $\varepsilon \to 0^+$, and $F(\varepsilon)$ is called the *finite part* (*partie finie* (Pf) in French), since $F(\varepsilon)$ tends to a finite limit F as $\varepsilon \to 0^+$. Then,

$$\lim_{\varepsilon \to 0+} \left[\int_{|x| \geq \varepsilon} \frac{\phi(x)}{x^k} dx - I(\varepsilon) \right] = \lim_{\varepsilon \to 0+} F(\varepsilon) = F \in \mathbb{R}, \quad \text{and} \tag{1.4.10}$$

$$F = \text{Pf} \int_{-\infty}^{\infty} \frac{\phi(x)}{x^k} dx \quad \forall \phi \in \mathcal{D}(\mathbb{R}) \tag{1.4.11}$$

is called the *finite part* (Pf) of the divergent integral $\int_{-\infty}^{\infty} \frac{\phi(x)}{x^k} dx$, $k \geq 2$. This concept of the separation of the finite part from a divergent integral is due to Hadamard [29, p. 38].

Then, *pseudo-function* Pf $\frac{1}{x^k}$ $(k \geq 2)$, i.e. (1.4.11) defines a distribution on $\mathbb{R}$ by: $\forall \phi \in \mathcal{D}(\mathbb{R})$ with $\text{supp}(\phi) \subset [-A, A]$, $A > 0$,

$$\begin{aligned} \left\langle \text{Pf} \frac{1}{x^k}, \phi \right\rangle &= \text{Pf} \int_{-\infty}^{\infty} \frac{\phi(x)}{x^k} dx = \lim_{\varepsilon \to 0+} \left[\int_{|x| \geq \varepsilon} \frac{\phi(x)}{x^k} dx - I(\varepsilon) \right] \\ &= \lim_{\varepsilon \to 0+} \left[\int_{-A}^{-\varepsilon} \frac{\phi(x)}{x^k} dx + \int_{\varepsilon}^{A} \frac{\phi(x)}{x^k} dx - I(\varepsilon) \right], \end{aligned} \tag{1.4.12}$$

where $I(\varepsilon) \to \pm\infty$ as $\varepsilon \to 0^+$ is the infinite part of the divergent integral.

The expression in the square bracket on the r.h.s. of (1.4.12), which is obtained by subtracting the infinite part $I(\varepsilon)$, has finite limit $F \in \mathbb{R}$ as $\varepsilon \to 0^+$, i.e. we are considering the finite part (*partie finie*) of the divergent integral denoted by Pf $\int_{-\infty}^{\infty} \frac{\phi(x)}{x^k} dx$.

Determination of $I(\varepsilon)$ Let $\phi \in \mathcal{D}(\mathbb{R})$ with $\text{supp}(\phi) = K \subset [-A, A]$, $A > 0$. Then, by Proposition 1.2.1, we have $\phi(x) = \phi(0) + \frac{x}{1!}\phi'(0) + \cdots + \frac{x^{k-1}}{(k-1)!}\phi^{(k-1)}(0) + x^k \psi(x)$, where

$$\psi \in C^0(\mathbb{R}) \quad \text{with} \sup_{|x| \leq A} |\psi(x)| \leq C \sup_{|x| \in K} |\phi^{(k)}(x)|, \quad C > 0. \tag{1.4.13}$$

Hence $\forall \varepsilon > 0$, for $k \geq 2$,

$$\begin{aligned} \int_{\varepsilon \leq |x| \leq A} \frac{\phi(x)}{x^k} dx &= \phi(0) \int_{\varepsilon \leq |x| \leq A} \frac{1}{x^k} dx + \phi'(0) \int_{\varepsilon \leq |x| \leq A} \frac{dx}{x^{k-1}} + \cdots \\ &+ \frac{\phi^{(k-1)}(0)}{(k-1)!} \int_{\varepsilon \leq |x| \leq A} \frac{1}{x} dx + \int_{\varepsilon \leq |x| \leq A} \psi(x) dx. \end{aligned} \tag{1.4.14}$$

But $\frac{1}{x}$ is an odd function on $\varepsilon \leq |x| \leq A \Longrightarrow \int_{\varepsilon\leq|x|\leq A} \frac{dx}{x} = 0.$ (1.4.15)
Using (1.4.14) and (1.4.15),

$$\int_{\varepsilon\leq|x|\leq A} \frac{\phi(x)}{x^k}dx = \sum_{j=1}^{k-1} \frac{\phi^{(j-1)}(0)}{(j-1)!} \int_{\varepsilon\leq|x|\leq A} \frac{1}{x^{k-(j-1)}}dx + \int_{\varepsilon\leq|x|\leq A} \psi(x)dx$$

$$= \sum_{j=1}^{k-1} \frac{\phi^{(j-1)}(0)}{(j-1)!}\left[\left(\frac{x^{-(k-j+1)+1}}{-(k-j+1)+1}\right)\Big|_{-A}^{-\varepsilon} + \left(\frac{x^{-(k-j+1)+1}}{-(k-j+1)+1}\right)\Big|_{\varepsilon}^{A}\right]$$

$$+ \int_{\varepsilon\leq|x|\leq A} \psi(x)dx$$

$$= \left[\sum_{j=1}^{k-1} \frac{\phi^{(j-1)}(0)}{(j-1)!} \cdot \frac{1}{j-k} \cdot \frac{(-1)^{j-k}-1}{\varepsilon^{k-j}}\right]$$

$$+ \left[\sum_{j=1}^{k-1} \frac{\phi^{(j-1)}(0)}{(j-1)!} \cdot \frac{1}{j-k} \cdot \frac{1-(-1)^{j-k}}{A^{k-j}}\right] + \int_{\varepsilon\leq|x|\leq A} \psi(x)dx.$$

We set

$$\sum_{j=1}^{k-1} C_j\phi^{(j-1)}(0) = \sum_{j=1}^{k-1}\left(\frac{1-(-1)^{j-k}}{(j-1)!(j-k)A^{k-j}}\right)\phi^{(j-1)}(0) \quad \text{and}$$

$$I(\varepsilon) = \sum_{j=1}^{k-1} \frac{\phi^{(j-1)}(0)}{(j-k)(j-1)!} \cdot \frac{(-1)^{j-k}-1}{\varepsilon^{k-j}}, \tag{1.4.16}$$

where constants C_j do not depend on ε, but depend on A, $I(\varepsilon) \to$ infinity ($\pm\infty$) as $\varepsilon \to 0^+$. Then $\int_{\varepsilon\leq|x|\leq A} \frac{\phi(x)}{x^k}dx = I(\varepsilon) + \sum_{j=1}^{k-1} C_j\phi^{(j-1)}(0) + \int_{\varepsilon\leq|x|\leq A} \psi(x)dx$

$$\Longrightarrow \langle \mathrm{Pf}\,\frac{1}{x^k}, \phi\rangle = \mathrm{Pf}\int_{-\infty}^{\infty} \frac{\phi(x)}{x^k}dx = \lim_{\varepsilon\to 0+}\left[\int_{\varepsilon\leq|x|\leq A} \frac{\phi(x)}{x^k}dx - I(\varepsilon)\right]$$

$$= \sum_{j=1}^{k-1} C_j\phi^{(j-1)}(0) + \int_{|x|\leq A} \psi(x)dx,$$

the r.h.s. being a finite number (since $\psi \in C^0(\mathbb{R})$),

$$\Longrightarrow \quad \left|\left\langle \mathrm{Pf}\frac{1}{x^k}, \phi\right\rangle\right| \leq \sum_{j=1}^{k-1} |C_j| \sup_{x\in K} |\phi^{(j-1)}(x)| + (\sup_{|x|\leq A} |\psi(x)|)2A$$

$$\leq \sum_{j=1}^{k-1} |C_j| \sup_{x\in K} |\phi^{(j-1)}(x)| + C \sup_{|x|\in A} |\phi^{(k)}(x)| \quad \text{(using(1.4.13))}$$

$$\leq C^* \sum_{j=0}^{k} \sup_{x\in K} |\phi^{(j)}(x)|, C^* \geq \max\{|C_1|, \ldots, |C_{k-1}|; C\} > 0$$

$\Longrightarrow \mathrm{Pf}\,\frac{1}{x^k}$ defines a distribution of order k. Thus,

$$\left\langle \mathrm{Pf}\frac{1}{x^k}, \phi\right\rangle = \lim_{\varepsilon\to 0+} \left[\int_{-\infty}^{-\varepsilon} \frac{\phi(x)}{x^k} dx + \int_{\varepsilon}^{\infty} \frac{\phi(x)}{x^k} dx - I(\varepsilon)\right] \tag{1.4.17}$$

with

$$I(\varepsilon) = \sum_{j=1}^{k-1} \frac{(-1)^{j-k} - 1}{\varepsilon^{k-j}(j-1)!(j-k)} \phi^{(j-1)}(0) \tag{1.4.18}$$

defines a distribution of order k on $\mathbb{R}$.

Remark 1.4.1. Gelfand and Schilov [1] call this procedure *regularization* of divergent integrals of functions with algebraic singularities and give an exhaustive analysis of this involved problem. For all details, we refer to [1].

Example 1.4.3. $\forall \phi \in \mathcal{D}(\mathbb{R})$,

$$\left\langle \mathrm{Pf}\frac{1}{x^2}, \phi\right\rangle = \mathrm{Pf}\int_{-\infty}^{\infty} \frac{\phi(x)}{x^2} dx = \lim_{\varepsilon\to 0+} \left[\int_{|x|\geq\varepsilon} \frac{\phi(x)}{x^2} dx - 2\frac{\phi(0)}{\varepsilon}\right] \tag{1.4.19}$$

defines a distribution on $\mathbb{R}$.

Proof. From (1.4.17) and (1.4.18), for $k = 2$, $j = 1$, $I(\varepsilon) = \frac{(-1)-1}{\varepsilon 0!(-1)}\phi(0) = \frac{2\phi(0)}{\varepsilon}$

$$\Longrightarrow \quad \left\langle \mathrm{Pf}\frac{1}{x^2}, \phi\right\rangle = \mathrm{Pf}\int_{-\infty}^{\infty} \frac{\phi(x)}{x^2} dx$$

$$= \lim_{\varepsilon\to 0+} \left[\int_{-\infty}^{-\varepsilon} \frac{\phi(x)}{x^2} dx + \int_{\varepsilon}^{\infty} \frac{\phi(x)}{x^2} dx - 2\frac{\phi(0)}{\varepsilon}\right]$$

$\forall \phi \in \mathcal{D}(\mathbb{R})$ defines a distribution of order 2 on $\mathbb{R}$. □

Example 1.4.4. $\forall \phi \in \mathcal{D}(\mathbb{R})$,

$$\left\langle \mathrm{Pf}\, \frac{H(x)}{x^k}, \phi \right\rangle = \lim_{\varepsilon \to 0+} \left[\int_\varepsilon^\infty \frac{\phi(x)}{x^k} dx - I(\varepsilon) \right] \tag{1.4.20}$$

with

$$I(\varepsilon) = -\sum_{j=1}^{k-1} \frac{\phi^{(j-1)}(0)}{(j-k)\varepsilon^{k-j}} \cdot \frac{1}{(j-1)!} - \frac{\phi^{(k-1)}(0)}{(k-1)!} \ln \varepsilon, \tag{1.4.21}$$

defines a distribution of order k, H being the Heaviside function: $H(x) = 1$ for $x > 0$, $H(x) = 0$ for $x < 0$.

Proof. From (1.4.13) and (1.4.14), we get: $\forall \phi \in \mathcal{D}(\mathbb{R})$ with $\mathrm{supp}(\phi) = K \subset [-A, A]$, $A > 0$,

$$\begin{aligned}
\int_\varepsilon^A \frac{\phi(x)}{x^k} dx &= \sum_{j=1}^{k-1} \frac{\phi^{(j-1)}(0)}{(j-1)!} \int_\varepsilon^A \frac{dx}{x^{k-(j-1)}} + \frac{\phi^{(k-1)}(0)}{(k-1)!} \int_\varepsilon^A \frac{dx}{x} + \int_\varepsilon^A \psi(x) dx \\
&= \left[\sum_{j=1}^{k-1} \frac{\phi^{(j-1)}(0)}{(j-1)!} \frac{-1}{(j-k)\varepsilon^{k-j}} - \frac{\phi^{(k-1)}(0)}{(k-1)!} \ln \varepsilon \right] + \sum_{j=1}^{k} C_j \phi^{(j-1)}(0) \\
&\quad + \int_\varepsilon^A \psi(x) dx,
\end{aligned}$$

where $I(\varepsilon) = -\sum_{j=1}^{k-1} \frac{\phi^{(j-1)}(0)}{(j-1)!(j-k)\varepsilon^{k-j}} - \frac{\phi^{(k-1)}(0)}{(k-1)!} \ln \varepsilon$ is the infinite part; coefficients C_j do not depend on ε, but depend on A.

$$\begin{aligned}
\left\langle \mathrm{Pf}\, \frac{H(x)}{x^k}, \phi \right\rangle &= \lim_{\varepsilon \to 0+} \left[\int_\varepsilon^A \frac{\phi(x)}{x^k} dx - I(\varepsilon) \right] \\
&= \sum_{j=1}^{k} C_j \phi^{(j-1)}(0) + \lim_{\varepsilon \to 0+} \int_\varepsilon^A \psi(x) dx,
\end{aligned}$$

which is a finite number, and $|\langle \mathrm{Pf}\, \frac{H(x)}{x^k}, \phi \rangle| \le \tilde{C} \sum_{j=0}^{k} \sup_{x \in K} |\phi^{(j)}(x)|$ with $\tilde{C} = \tilde{C}(A) > 0$,

$\Longrightarrow$ Pf $\frac{H(x)}{x^k}$ is a distribution of order k. In particular, for $k = 2,\ j = 1$,

$$I(\varepsilon) = -\frac{\phi(0)}{(-1)0!\varepsilon} - \frac{\phi'(0)}{1!}\ln\varepsilon = \frac{\phi(0)}{\varepsilon} - \phi'(0)\ln\varepsilon, \tag{1.4.22}$$

$$\left\langle \operatorname{Pf}\frac{H(x)}{x^2}, \phi\right\rangle = \operatorname{Pf}\int_0^\infty \frac{\phi(x)}{x^2}dx = \lim_{\varepsilon\to 0+}\left[\int_\varepsilon^\infty \frac{\phi(x)}{x^2}dx - \frac{\phi(0)}{\varepsilon} + \phi'(0)\ln\varepsilon\right], \tag{1.4.23}$$

$$\left\langle \operatorname{Pf}\frac{H(-x)}{x^2}, \phi\right\rangle = \operatorname{Pf}\int_{-\infty}^0 \frac{\phi(x)}{x^2}dx = \lim_{\varepsilon\to 0+}\left[\int_{-\infty}^{-\varepsilon} \frac{\phi(x)}{x^2}dx - \frac{\phi(0)}{\varepsilon} - \phi'(0)\ln\varepsilon\right] \tag{1.4.24}$$

(with $I(\varepsilon) = \frac{\phi(0)}{\varepsilon} + \phi'(0)\ln\varepsilon$ in (1.4.24)) are distributions on $\mathbb{R}$. □

Example 1.4.5.

1. $$\operatorname{Pf}\frac{1}{x} = \text{c.p.v.}\,\frac{1}{x} \quad \text{in } \mathcal{D}'(\mathbb{R}); \tag{1.4.25}$$

2. $$\left\langle \operatorname{Pf}\frac{H(x)}{x}, \phi\right\rangle = \operatorname{Pf}\int_0^\infty \frac{\phi(x)}{x}dx = \lim_{\varepsilon\to 0+}\left[\int_\varepsilon^\infty \frac{\phi(x)}{x}dx + \phi(0)\ln\varepsilon\right]; \tag{1.4.26}$$

3. $$\left\langle \operatorname{Pf}\frac{H(-x)}{x}, \phi\right\rangle = \operatorname{Pf}\int_{-\infty}^0 \frac{\phi(x)}{x}dx = \lim_{\varepsilon\to 0+}\left[\int_{-\infty}^{-\varepsilon} \frac{\phi(x)}{x}dx - \phi(0)\ln\varepsilon\right]. \tag{1.4.27}$$

Proof.

1. $\langle \operatorname{Pf}\frac{1}{x}, \phi\rangle = \operatorname{Pf}\int_{-\infty}^\infty \frac{\phi(x)}{x}dx = \lim_{\varepsilon\to 0+}\int_{|x|\geq\varepsilon}\frac{\phi(x)}{x}dx = \langle \text{c.p.v.}\,\frac{1}{x}, \phi\rangle\ \forall\phi \in \mathcal{D}(\mathbb{R}) \quad \Longrightarrow \operatorname{Pf}\frac{1}{x} = \text{c.p.v.}\,\frac{1}{x}$ in $\mathcal{D}'(\mathbb{R})$.

2. For $\phi \in \mathcal{D}(\mathbb{R})$ with $\operatorname{supp}(\phi) = K \subset [-A, A]$, $A > 0$, $\phi(x) = \phi(0) + x\psi(x)$ with $\psi \in C^0(\mathbb{R})$, $\sup_{|x|\leq A}|\psi(x)| \leq C\sup_{x\in K}|\phi^{(1)}(x)|$,

$$\begin{aligned}\int_\varepsilon^\infty \frac{\phi(x)}{x}dx = \int_\varepsilon^A \frac{\phi(x)}{x}dx &= \phi(0)\int_\varepsilon^A \frac{1}{x}dx + \int_\varepsilon^A \psi(x)dx \\ &= \phi(0)[\ln A - \ln\varepsilon] + \int_\varepsilon^A \psi(x)dx \\ &= \underbrace{-\phi(0)\ln\varepsilon}_{I(\varepsilon)} + \phi(0)\ln A + \int_\varepsilon^A \psi(x)dx\end{aligned}$$

$\Longrightarrow \quad \lim_{\varepsilon\to 0+}[\int_\varepsilon^A \frac{\phi(x)}{x}dx - I(\varepsilon)] = \phi(0)\ln A + \int_0^A \psi(x)dx$, which is a finite number, with $I(\varepsilon) = -\phi(0)\ln\varepsilon$.

$$\Longrightarrow \quad \left\langle \mathrm{Pf}\,\frac{H(x)}{x}, \phi \right\rangle = \mathrm{Pf}\int_0^\infty \frac{\phi(x)}{x}dx = \lim_{\varepsilon\to 0+}\left[\int_\varepsilon^\infty \frac{\phi(x)}{x}dx + \phi(0)\ln\varepsilon\right]$$
$$= \phi(0)\ln A + \int_0^A \psi(x)dx = \text{a finite number}$$

$$\Longrightarrow \quad \left\langle \mathrm{Pf}\,\frac{H(x)}{x}, \phi \right\rangle = \lim_{\varepsilon\to 0+}\left[\int_\varepsilon^\infty \frac{\phi(x)}{x}dx + \phi(0)\ln\varepsilon\right] \quad \forall\phi\in\mathcal{D}(\mathbb{R})$$

defines a distribution on $\mathbb{R}$, since

$$\left|\langle \mathrm{Pf}\,\frac{H(x)}{x}, \phi\rangle\right| \le |\phi(0)|\cdot|\ln A| + CA\sup_{x\in K}|\phi^{(1)}(x)|$$
$$\le C_1^*\sum_{j=0}^{1}\sup_{x\in K}|\phi^{(j)}(x)|,$$

with $C_1^* = C_1^*(A) > 0$.

3. For $\phi\in\mathcal{D}(\mathbb{R})$ as defined in the proof of (2), we have, $\forall\varepsilon>0$,

$$\int_{-\infty}^{-\varepsilon}\frac{\phi(x)}{x}dx = \phi(0)\int_{-A}^{-\varepsilon}\frac{1}{x}dx + \int_{-A}^{-\varepsilon}\psi(x)dx$$
$$= \phi(0)[\ln\varepsilon - \ln A] + \int_{-A}^{-\varepsilon}\psi(x)dx$$
$$= \phi(0)\ln\varepsilon - \phi(0)\ln A + \int_{-A}^{-\varepsilon}\psi(x)dx$$

$$\Longrightarrow \quad \left\langle \mathrm{Pf}\,\frac{H(-x)}{x}, \phi \right\rangle = \mathrm{Pf}\int_{-\infty}^0 \frac{\phi(x)}{x}dx = \lim_{\varepsilon\to 0+}\left[\int_{-\infty}^{-\varepsilon} \frac{\phi(x)}{x}dx - \phi(0)\ln\varepsilon\right]$$
$$= -\phi(0)\ln A + \int_{-A}^0 \psi(x)dx = \text{a finite number}$$

$$\Longrightarrow \quad \left\langle \mathrm{Pf}\,\frac{H(-x)}{x}, \phi \right\rangle = \lim_{\varepsilon\to 0+}\left[\int_{-\infty}^{-\varepsilon} \frac{\phi(x)}{x}dx - \phi(0)\ln\varepsilon\right] \quad \forall\phi\in\mathcal{D}(\mathbb{R})$$

defines a distribution on $\mathbb{R}$, since

$$\left|\left\langle \mathrm{Pf}\,\frac{H(-x)}{x}, \phi\right\rangle\right| \le |\phi(0)|\,|\ln A| + CA\sup_{x\in K}|\phi^{(1)}(x)|$$
$$\le C_2^*\sum_{j=0}^{1}\sup_{x\in K}|\phi^{(j)}(x)|,$$

with $C_2^* = C_2^*(A) > 0$. □

Remark 1.4.2. From (1.4.26) and (1.4.27), the result (1.4.25) follows.

Distributions defined by $x^\lambda, x_+^\lambda, x_-^\lambda$

Function x^λ Let $\lambda = a + ib \in \mathbb{C}$ be a complex number with $a, b \in \mathbb{R}$. Then, for real $x \neq 0$, x^λ is defined by:

$$x^\lambda = e^{\lambda \ln x} = x^a e^{ib \ln x} \quad \text{for } x > 0 \tag{1.4.28}$$

$$x^\lambda = [(-1)(-x)]^\lambda = e^{i\pi\lambda}(-x)^\lambda = |x|^a . e^{-b\pi + i(b \ln |x| + a\pi)} \quad \text{for } x < 0 \tag{1.4.29}$$

$$\Longrightarrow |x^\lambda| \leq |x|^a \ \forall x \neq 0, \forall \lambda = a + ib$$

$\Longrightarrow x^\lambda \in L^1_{\text{loc}}(\mathbb{R})$ if $|x|^a$ is locally integrable on $\mathbb{R}$. But $|x|^a$ is locally integrable on $\mathbb{R}$ for $a = \text{Re}(\lambda) > -1$:

$$\int_{-A}^{A} |x^\lambda| dx \leq \int_{-A}^{A} |x|^a dx < +\infty \quad \forall a = \text{Re}(\lambda) > -1, \ \forall A > 0. \tag{1.4.30}$$

Distribution x^λ For λ with $\text{Re}(\lambda) > -1$, $x^\lambda \in L^1_{\text{loc}}(\mathbb{R})$ and defines a regular distribution $T_{x^\lambda} = x^\lambda \in \mathcal{D}'(\mathbb{R})$ by:

$$\langle T_{x^\lambda}, \phi \rangle = \langle x^\lambda, \phi \rangle = \int_{-\infty}^{\infty} x^\lambda \phi(x) dx \quad \forall \phi \in \mathcal{D}(\mathbb{R}). \tag{1.4.31}$$

For $\lambda = -k$, $k \in \mathbb{N}$, $\text{Pf}\,\frac{1}{x^k} \in \mathcal{D}'(\mathbb{R})$ is defined by (1.4.17). (1.4.32)

Functions x_+^λ, x_-^λ For $\lambda \in \mathbb{C}$ with $\text{Re}(\lambda) > -1$, functions x_+^λ, x_-^λ are defined by:

$$x_+^\lambda = \begin{cases} x^\lambda & \text{for } x > 0 \\ 0 & \text{for } x \leq 0 \end{cases}; \quad x_-^\lambda = \begin{cases} 0 & \text{for } x \geq 0 \\ |x|^\lambda & \text{for } x < 0 \end{cases}. \tag{1.4.33}$$

Distributions x_+^λ, x_-^λ For $\text{Re}(\lambda) > -1$, functions $x_+^\lambda, x_-^\lambda \in L^1_{\text{loc}}(\mathbb{R})$ are locally integrable (see (1.4.30)) and hence define regular distributions on $\mathbb{R}$ by:

$$\langle x_+^\lambda, \phi \rangle = \int_0^{\infty} x^\lambda \phi(x) dx; \tag{1.4.34}$$

$$\langle x_-^\lambda, \phi \rangle = \int_{-\infty}^{0} |x|^\lambda \phi(x) dx. \tag{1.4.35}$$

Alternative forms of (1.4.34) and (1.4.35) for $\mathrm{Re}(\lambda) > -2$ and $\lambda \neq -1$ Since for $\mathrm{Re}(\lambda) > -1$, $\int_0^1 x^\lambda = \frac{x^{\lambda+1}}{\lambda+1}\big|_0^1 = \frac{1}{\lambda+1} \in \mathbb{C}, \quad \forall\phi \in \mathcal{D}(\mathbb{R})$,

$$\begin{aligned}\langle x_+^\lambda, \phi\rangle &= \int_0^\infty x^\lambda \phi(x)dx \\ &= \int_0^1 x^\lambda[\phi(x) - \phi(0)]dx + \phi(0)\int_0^1 x^\lambda dx + \int_1^\infty x^\lambda\phi(x)dx \\ &= \int_0^1 x^\lambda[\phi(x) - \phi(0)]dx + \int_1^\infty x^\lambda\phi(x)dx + \frac{\phi(0)}{\lambda+1}.\end{aligned} \tag{1.4.36}$$

Since $\check{\phi}(x) = \phi(-x)\ \forall x, \forall\phi \in \mathcal{D}(\mathbb{R})$, for $\mathrm{Re}(\lambda) > -1$,

$$\begin{aligned}\langle x_-^\lambda, \phi\rangle = \langle x_+^\lambda, \check{\phi}\rangle &= \int_0^\infty |x|^\lambda \phi(-x)dx \\ &= \int_0^1 x^\lambda[\phi(-x) - \phi(0)]dx \\ &\quad + \int_1^\infty x^\lambda\phi(-x)dx + \frac{\phi(0)}{\lambda+1} \quad \forall\phi \in \mathcal{D}(\mathbb{R}).\end{aligned} \tag{1.4.37}$$

• Distributions x_+^λ and x_-^λ defined by (1.4.36) and (1.4.37) respectively hold for $\mathrm{Re}(\lambda) > -2$, and $\lambda \neq -1$. Consider the r.h.s. of (1.4.36). In fact, by Proposition 1.2.1, $\phi(x) = \phi(0) + x\psi(x)$ with $\psi \in C^0(\mathbb{R}) \Longrightarrow$ the first integral $\int_0^1 x^\lambda[\phi(x) - \phi(0)]dx = \int_1^\infty x^{\lambda+1}\psi(x)dx$ is well defined for $\mathrm{Re}(\lambda+1) > -1 \Longrightarrow \mathrm{Re}(\lambda) > -2$.

• The second integral $\int_1^\infty x^\lambda\phi(x)dx$ is well defined $\forall\lambda$, since $x \geq 1$.

• The third expression $\frac{\phi(0)}{(\lambda+1)}$ is defined for $\lambda \neq -1$. Hence, the expression on the r.h.s. of (1.4.36) is well defined for $\mathrm{Re}(\lambda) > -2$ and $\lambda \neq -1$.

Similarly, we can prove that the r.h.s. of (1.4.37) is well defined for $\mathrm{Re}(\lambda) > -2$ and $\lambda \neq -1$.

• Moreover, in a similar way we can show that for $\mathrm{Re}(\lambda) > -n$, $\lambda \neq -1, -2, \ldots, -(n-1)$,

$$\begin{aligned}\langle x_+^\lambda, \phi\rangle &= \int_0^\infty |x|^\lambda\phi(x)dx \\ &= \int_0^1 x^\lambda\left[\phi(x) - \sum_{j=1}^n \frac{x^{j-1}\phi^{(j-1)}(0)}{(j-1)!}\right]dx \\ &\quad + \int_1^\infty |x|^\lambda\phi(x)dx + \sum_{k=1}^n \frac{\phi^{(k-1)}(0)}{(\lambda+k)(k-1)!}.\end{aligned} \tag{1.4.38}$$

Remark 1.4.3. Formulae (1.4.36) and (1.4.38) are *not* contradictory ones. (1.4.38) is an *analytic* extension of (1.4.36) to the left of $\lambda = -1$ with $\lambda \neq -2, -3, \dots, -(n-1)$ (see [1] for all details).

1.5 Multiplication of distributions by C^∞-functions

For arbitrary distributions $S \in \mathcal{D}'(\Omega)$ and $T \in \mathcal{D}'(\Omega)$, the product TS or ST is *not* defined in general. (1.5.1)

For example, for $n = 1, \Omega = \mathbb{R}$, $f(x) = \frac{1}{\sqrt{|x|}} (x \neq 0) \in L^1_{\text{loc}}(\mathbb{R})$ and defines a regular distribution $T_f = f$ on $\mathbb{R}$ since, $\forall$ compact $[a,b] \subset \mathbb{R}$ containing 0,

$$\int_a^b \frac{1}{\sqrt{|x|}} dx = \lim_{\varepsilon_1 \to 0+} \int_a^{-\varepsilon_1} \frac{1}{\sqrt{|x|}} dx + \lim_{\varepsilon_2 \to 0+} \int_{\varepsilon_2}^b \frac{1}{\sqrt{|x|}} dx < +\infty.$$

But $f^2 = f.f = \frac{1}{|x|} \notin L^1_{\text{loc}}(\mathbb{R})$ (see Example 1.3.7) and hence f^2 does *not* define a distribution, i.e. T_{f^2} is not a distribution. But in particular cases, the product ST of distributions $S, T \in \mathcal{D}'(\Omega)$, may have a meaning, when for an 'irregular' $T \in \mathcal{D}'(\Omega)$, $S \in \mathcal{D}'(\Omega)$ is 'highly regular'. For example, let $T \in \mathcal{D}'(\Omega)$ be an arbitrary distribution and $S \in \mathcal{D}'(\Omega)$ be defined by an infinitely differentiable function $\alpha \in C^\infty(\Omega)$ (i.e. $\alpha \in C^\infty(\Omega) \Longrightarrow \alpha \in L^1_{\text{loc}}(\Omega) \Longrightarrow \alpha$ defines a regular distribution $S \in \mathcal{D}'(\Omega)$ by $S(\phi) = \int_\Omega \alpha\phi d\mathbf{x} \quad \forall \phi \in \mathcal{D}(\Omega)$). Then, αT is defined and a distribution. Hence, we have:

Definition 1.5.1. Let $T \in \mathcal{D}'(\Omega), \alpha \in C^\infty(\Omega)$. Then $\alpha T \in \mathcal{D}'(\Omega)$ and is defined by:

$$\langle \alpha T, \phi \rangle = \langle T, \alpha\phi \rangle \quad \forall \phi \in \mathcal{D}(\Omega), \tag{1.5.2}$$

with $\alpha\phi \in \mathcal{D}(\Omega)$.

In particular,

$$f \in L^1_{\text{loc}}(\Omega),\ g \in C^\infty(\Omega) \quad \Longrightarrow \quad fg \in L^1_{\text{loc}}(\Omega) \quad \Longrightarrow \quad T_{fg} \in \mathcal{D}'(\Omega). \tag{1.5.3}$$

Justification of Definition 1.5.1

$\alpha\phi \in C_0^\infty(\Omega)$ with $\text{supp}(\alpha\phi) = \text{supp}(\alpha) \cap \text{supp}(\phi) \subset \text{supp}(\phi) \subset\subset \Omega$, $\phi_n \to 0$ in $\mathcal{D}(\Omega) \Longrightarrow \alpha\phi_n \to 0$ in $\mathcal{D}(\Omega)$, which is established by applying Leibniz' theorem to successive differentiations of $(\alpha\phi)$.

αT is a continuous linear functional on $\mathcal{D}(\Omega)$: $\phi_n \to 0$ in $\mathcal{D}(\Omega) \Longrightarrow \alpha\phi_n \to 0$ in $\mathcal{D}(\Omega) \Longrightarrow \langle \alpha T, \phi_n \rangle = \langle T, \alpha\phi_n \rangle \to 0$ in $\mathbb{R}$, since $T \in \mathcal{D}'(\Omega)$.

Example 1.5.1. $\forall \alpha \in C^{\infty}(\mathbb{R}^n)$ and for Dirac distribution $\delta \in \mathcal{D}'(\mathbb{R}^n)$, the product $\alpha\delta$ is defined by:

$$\alpha\delta = \alpha(\mathbf{0})\delta.$$

In fact,

$$\langle \alpha\delta, \phi \rangle = \langle \delta, \alpha\phi \rangle = \alpha(\mathbf{0})\phi(\mathbf{0}) = \alpha(\mathbf{0})\langle \delta, \phi \rangle = \langle \alpha(\mathbf{0})\delta, \phi \rangle \quad \forall \phi \in \mathcal{D}(\mathbb{R}^n)$$

$\Longrightarrow \alpha\delta = \alpha(\mathbf{0})\delta$ in $\mathcal{D}'(\mathbb{R}^n)$, i.e. any product in which δ occurs, is *proportional* to δ.

In particular, for $n = 1$, $\alpha(x) = x \ \forall x \in \mathbb{R}$, $x\delta = 0 \in \mathcal{D}'(\mathbb{R})$. (1.5.4)

For $\alpha(x) = p(x) = a_0 + a_1 x + \cdots + a_n x^n$,

$$p(x)\delta = p(0)\delta = a_0\delta \in \mathcal{D}'(\mathbb{R}). \tag{1.5.5}$$

Example 1.5.2. $\forall \alpha \in C^{\infty}(\mathbb{R}^n)$ and for Dirac distribution $\delta_{\mathbf{a}} \in \mathcal{D}'(\mathbb{R}^n)$ defined by $\langle \delta_{\mathbf{a}}, \phi \rangle = \phi(\mathbf{a}) \ \forall \phi \in \mathcal{D}(\mathbb{R}^n)$, the product $\alpha\delta_{\mathbf{a}}$ is given by $\alpha\delta_{\mathbf{a}} = \alpha(\mathbf{a})\delta_{\mathbf{a}}$.

Indeed,

$$\langle \alpha\delta_{\mathbf{a}}, \phi \rangle = \langle \delta_{\mathbf{a}}, \alpha\phi \rangle = \alpha(\mathbf{a})\phi(\mathbf{a}) = \alpha(\mathbf{a})\langle \delta_{\mathbf{a}}, \phi \rangle = \langle \alpha(\mathbf{a})\delta_{\mathbf{a}}, \phi \rangle \quad \forall \phi \in \mathcal{D}(\mathbb{R}^n) \tag{1.5.6}$$

$\Longrightarrow \alpha\delta_{\mathbf{a}} = \alpha(\mathbf{a})\delta_{\mathbf{a}}$ in $\mathcal{D}'(\mathbb{R}^n)$.

In particular, for $n = 1$, $\alpha(x) = x \ \forall x \in \mathbb{R}$,

$$x\delta_a = a\delta_a \tag{1.5.7}$$

$$p(x)\delta_a = p(a)\delta_a; \tag{1.5.8}$$

$$e^{bx}\delta_a = e^{ab}\delta_a; \tag{1.5.9}$$

$$(x-a)\delta_a = 0. \tag{1.5.10}$$

Example 1.5.3. Show that the following equalities hold in $\mathcal{D}'(\mathbb{R})$:

$$\begin{aligned} &1. \quad x\left(\text{c.p.v.}\,\frac{1}{x}\right) = 1;\\ &2. \quad x\left(\text{Pf}\,\frac{H(x)}{x}\right) = H;\\ &3. \quad x\left(\text{Pf}\,\frac{1}{x}\right) = x(\text{c.p.v.}\,\frac{1}{x}) = 1. \end{aligned} \tag{1.5.11}$$

Solution 1. $\phi \in \mathcal{D}(\mathbb{R}) \Longrightarrow \int_{-\infty}^{\infty} \phi(x)dx$ exists $\Longrightarrow$ c.p.v. $\int_{-\infty}^{\infty} \phi(x)dx = \int_{-\infty}^{\infty} \phi(x)dx$.

1. $\forall \phi \in \mathcal{D}(\mathbb{R})$,

$$\left\langle x\left(\text{c.p.v.}\,\frac{1}{x}\right), \phi\right\rangle = \left\langle \text{c.p.v.}\,\frac{1}{x}, x\phi\right\rangle = \text{c.p.v.}\int_{-\infty}^{\infty}\frac{1}{x}(x\phi(x))dx$$
$$= \int_{-\infty}^{\infty}\phi(x)dx = \langle 1, \phi\rangle \Longrightarrow x\left(\text{c.p.v.}\,\frac{1}{x}\right) = 1 \quad \text{in } \mathcal{D}'(\mathbb{R}).$$

2. $\forall \phi \in \mathcal{D}(\mathbb{R})$,

$$\left\langle x\left(\text{Pf}\,\frac{H(x)}{x}\right), \phi\right\rangle = \left\langle \text{Pf}\,\frac{H(x)}{x}, x\phi\right\rangle = \text{Pf}\int_{0}^{\infty}\frac{1}{x}(x\phi)dx = \int_{0}^{\infty}\phi(x)dx$$
$$= \int_{-\infty}^{\infty} H(x)\phi(x)dx = \langle H, \phi\rangle \Longrightarrow x\left(\text{Pf}\,\frac{H(x)}{x}\right) = H \quad \text{in } \mathcal{D}'(\mathbb{R}).$$

3. The result follows from (1) and Example 1.4.5(1).

Example 1.5.4. Calculate $x \cdot x_+^\lambda$ and $x \cdot x_-^\lambda$ for $\text{Re}(\lambda) > -2$ and $\lambda \neq -1$.
Solution. Using (1.4.36), for $\text{Re}(\lambda) > -2$ and $\lambda \neq -1$,

$$\begin{aligned}
\langle x \cdot x_+^\lambda, \phi\rangle &= \langle x_+^\lambda, x\phi\rangle \\
&= \int_0^1 x^\lambda[x\phi(x) - (x\phi)(0)]dx + \int_1^\infty x^\lambda x\phi(x)dx + \frac{(x\phi)(0)}{\lambda+1} \\
&= \int_0^1 x^{\lambda+1}\phi(x) + \int_1^\infty x^{\lambda+1}\phi dx = \int_0^\infty x^{\lambda+1}\phi(x)dx \quad \forall \phi \in \mathcal{D}(\mathbb{R}),
\end{aligned}$$
$$\Longrightarrow \quad x \cdot x_+^\lambda = x_+^{\lambda+1} \quad \text{in } \mathcal{D}'(\mathbb{R}), \quad \text{since } (x\phi)(0) = 0 \text{ (using (1.4.33)).} \tag{1.5.12}$$
$$\begin{aligned}
\langle x \cdot x_-^\lambda, \phi\rangle &= \langle x_-^\lambda, x\phi\rangle = \langle x_+^\lambda, (x\phi)^\vee\rangle = \langle x_+^\lambda, -x\check{\phi}\rangle \quad \text{(by (1.4.37))} \\
&= -\int_0^\infty x^\lambda x\phi(-x)dx = -\int_0^\infty x^{\lambda+1}\phi(-x)dx = -\langle x_+^{\lambda+1}, \check{\phi}\rangle \\
&= -\langle x_-^{\lambda+1}, \phi\rangle = \langle -x_-^{\lambda+1}, \phi\rangle \quad \forall \phi \in \mathcal{D}(\mathbb{R})
\end{aligned}$$
$$\Longrightarrow \quad x \cdot x_-^\lambda = -x_-^{\lambda+1} \quad \text{in } \mathcal{D}'(\mathbb{R}). \tag{1.5.13}$$

Multiplication of several distributions

The product of several distributions is well defined if all of the distributions, *except at most one of them*, are C^∞-functions and the product is associative. Otherwise, the product of several distributions is *not* defined in general. In particular, even if the product of several distributions is defined, it may not be associative.

$$\alpha, \beta \in C^\infty(\Omega),\ T \in \mathcal{D}'(\Omega) \quad \Longrightarrow \quad \alpha(\beta T) = (\alpha\beta)T \in \mathcal{D}'(\Omega). \tag{1.5.14}$$

In fact,

$$\begin{aligned}\langle \alpha(\beta T), \phi\rangle &= \langle \beta T, \alpha\phi\rangle = \langle T, \beta\alpha\phi\rangle = \langle T, \alpha\beta\phi\rangle \\ &= \langle \alpha T, \beta\phi\rangle = \langle \beta\alpha T, \phi\rangle = \langle (\alpha\beta)T, \phi\rangle \quad \forall \phi \in \mathcal{D}(\Omega). \qquad (1.5.15)\end{aligned}$$

Counterexample 1.5.5. Let c.p.v. $\frac{1}{x} \in \mathcal{D}'(\mathbb{R})$ be the distribution defined in Example 1.3.8: $\langle \text{c.p.v.} \frac{1}{x}, \phi\rangle = \text{c.p.v.} \int_{-\infty}^{\infty} \frac{\phi(x)}{x} dx = \lim_{\varepsilon\to 0^+} \int_{|x|\geq\varepsilon} \frac{\phi(x)}{x} dx \ \forall \phi \in \mathcal{D}(\mathbb{R})$. Then the product of Dirac distribution δ and c.p.v. $\frac{1}{x}$ can not be defined.

Proof. Set $T = \delta$, $S = \text{c.p.v.} \frac{1}{x}$. Suppose that TS, ST are defined with $TS = ST$. Then, for $\alpha \in C^\infty(\mathbb{R})$, we would have $\alpha(TS) = (\alpha T)S = T(\alpha S)$. Choose $\alpha(x) = x \ \forall x \in \mathbb{R}$. Then, we would have $x(\delta \, \text{c.p.v.} \frac{1}{x}) = (x\delta) \, \text{c.p.v.} \frac{1}{x} = 0 \cdot \text{c.p.v.} \frac{1}{x} = 0 \in \mathcal{D}'(\mathbb{R})$, $x(\delta \, \text{c.p.v.} \frac{1}{x}) = x(\text{c.p.v.} \frac{1}{x}\delta) = (x \, \text{c.p.v.} \frac{1}{x}) \cdot \delta = 1 \cdot \delta = \delta \in \mathcal{D}'(\mathbb{R})$, ($x \, \text{c.p.v.} \frac{1}{x} = 1 \in \mathcal{D}'(\mathbb{R})$ from Example 1.5.3(1)), which is impossible.

Hence, the product of δ and c.p.v. $\frac{1}{x}$ can not be defined. □

Remark 1.5.1. δ and c.p.v. $\frac{1}{x}$ are two distributions, x is a C^∞-function and $x\delta = 0 \in \mathcal{D}'(\mathbb{R})$, $x(\text{c.p.v.} \frac{1}{x}) = 1 \in \mathcal{D}'(\mathbb{R})$ are well defined.

1.6 Problem of division of distributions

Case of single variable ($n = 1$) Let $\Omega \subset \mathbb{R}$ be an open subset of $\mathbb{R}$, in which a given function f does not vanish, i.e. $f(x) \neq 0 \ \forall x \in \Omega$. Let $S \in \mathcal{D}'(\Omega)$ be a given distribution and $f \in C^\infty(\Omega)$. Then, $\exists$ one and only one distribution $T \in \mathcal{D}'(\Omega)$ such that $fT = S$ in $\mathcal{D}'(\Omega)$, since $\frac{1}{f} \in C^\infty(\Omega)$ and, multiplying both sides by $\frac{1}{f}$, we get $\frac{1}{f}(fT) = \frac{1}{f}S \iff T = S\frac{1}{f} \in \mathcal{D}'(\Omega)$, i.e. for $f \in C^\infty(\Omega)$ with $f(x) \neq 0$ for any $x \in \Omega$, the division of S by f is the multiplication of S by $\frac{1}{f} \in C^\infty(\Omega)$.

In particular, for $S = 0$ in $\mathcal{D}'(\Omega)$ and $f \in C^\infty(\Omega)$ with $f(x) \neq 0 \ \forall x \in \Omega$,

$$fT = 0 \text{ in } \mathcal{D}'(\Omega) \quad \Longrightarrow \quad T = 0 \text{ in } \mathcal{D}'(\Omega). \qquad (1.6.1)$$

Example 1.6.1. For $f(x) = 1+x^2 \in C^\infty(\mathbb{R})$, $(1+x^2)T = 0$ in $\mathcal{D}'(\mathbb{R}) \Longrightarrow T = 0$ in $\mathcal{D}'(\mathbb{R})$.

But if $f(x) = 0$ for some $x \in \Omega$, the situation is completely different. For example, if for $f \in C^\infty(\Omega)$, $f(x) = 0$ for some $x \in \Omega \subset \mathbb{R}$, then $fT = 0$ in $\mathcal{D}'(\Omega)$ does *not* imply $T = 0$ in $\mathcal{D}'(\Omega)$. In fact, we have:

Proposition 1.6.1. $\forall$ *real* $a \in \mathbb{R}$,

$$(x-a)T = 0 \textit{ in } \mathcal{D}'(\mathbb{R}) \quad \iff \quad T = C\delta_a, \qquad (1.6.2)$$

where δ_a is the Dirac distribution with mass/charge/force etc. concentrated at a, $C \in \mathbb{R}$ being an arbitrary constant.

In particular, for $a = 0$,

$$xT = 0 \text{ in } \mathcal{D}'(\mathbb{R}) \iff T = C\delta_0 = C\delta. \tag{1.6.3}$$

Proof. $T = C\delta_a$

$$\begin{aligned}\implies \quad \langle (x-a)T, \phi\rangle &= \langle T, (x-a)\phi\rangle = \langle C\delta_a, (x-a)\phi\rangle = C\langle \delta_a, (x-a)\phi\rangle \\ &= C[(x-a)\phi(x)]_{x=a} = C\cdot[0\cdot\phi(a)] = 0 \quad \forall\phi \in \mathcal{D}(\mathbb{R}).\end{aligned}$$

Converse: Now we show that $(x-a)T = 0 \implies T = C\delta_a$.

In fact, $\langle T, (x-a)\phi\rangle = 0 \quad \forall\phi \in \mathcal{D}(\mathbb{R}) \implies T$ vanishes on all $\psi = (x-a)\phi \in \mathcal{D}(\mathbb{R})$, i.e. on *all* $\psi \in \mathcal{D}(\mathbb{R})$ which vanish at $x = a$. Let $\theta \in \mathcal{D}(\mathbb{R})$ be a given fixed function such that $\theta(a) = 1$. Then every $\chi \in \mathcal{D}(\mathbb{R})$ is of the form: $\chi = \lambda\theta + \psi$ with $\psi = (x-a)\phi \in \mathcal{D}(\mathbb{R})$, $\phi \in \mathcal{D}(\mathbb{R})$ such that $\lambda = \chi(a), \psi(a) = 0$. Hence, $T(\chi) = T(\lambda\theta + \psi) = \lambda T(\theta) + T(\psi) = C\chi(a) + \langle T, (x-a)\phi\rangle$, where $\lambda = \chi(a), C = T(\theta), \langle T, (x-a)\phi\rangle = \langle (x-a)T, \phi\rangle = 0$ (by hypothesis)

$\implies T(\chi) = C\chi(a) = C\langle \delta_a, \chi\rangle = \langle C\delta_a, \chi\rangle \quad \forall\chi \in \mathcal{D}(\mathbb{R})$

$\implies T = C\delta_a, C$ being an arbitrary constant. □

Remark 1.6.1. Let $T \in \mathcal{D}'(\mathbb{R} - \{a\})$ be a distribution on the open set $\mathbb{R} - \{a\}$; then $\frac{1}{(x-a)} \in C^\infty(\mathbb{R} - \{a\})$ does not vanish on the open set $\mathbb{R} - \{a\}$. Then $(x-a)T = S \in \mathcal{D}'(\mathbb{R} - \{a\})$ has the solution

$$T = \frac{1}{x-a} S \in \mathcal{D}'(\mathbb{R} - \{a\}). \tag{1.6.4}$$

Example 1.6.2. Find the general solution (i.e. all the solutions) of the following equations for T in $\in \mathcal{D}'(\mathbb{R})$: for $a, b \in \mathbb{R}, a \neq b$,

1. $(x-a)T = \delta_b$;
2. $(x-a)(x-b)T = 0$;
3. $(1+x^2)(1-x^2)T = 0$;
4. $p(x)T = 0$, where $p(x) = (x-a_1)(x-a_2)\dots(x-a_n)$ with $a_1 < a_2 < \cdots < a_n$. (1.6.5)

Solution. 1. For $(x-a)T = \delta_b$, where δ_b is the Dirac distribution with force/mass/charge etc. concentrated at $b \in \mathbb{R}$, $T = T_h + T_p$ is the general solution of $(x-a)T = \delta_b$, where T_p is a particular solution of $(x-a)T = \delta_b$, i.e. $(x-a)T_p = \delta_b$, and T_h is the general solution of the corresponding homogeneous equation $(x-a)T = 0$, which is obtained by putting the right-hand side equal to $0 \in \mathcal{D}'(\mathbb{R})$, since $(x-a)T = \delta_b \implies (x-a)(T_p + T_h) = (x-a)T_p + (x-a)T_h = \delta_b + (x-a)T_h = \delta_b \implies (x-a)T_h = 0 \implies T_h$ satisfies the corresponding homogeneous equation.

Then, by Proposition 1.6.1, $T_h = C_1\delta_a$, $C_1 \in \mathbb{R}$ being an arbitrary constant.

Construction of T_p: $T_p = \frac{\delta_b}{b-a}$ with $a \neq b$. Indeed, $\langle (x-a)T_p, \phi\rangle = \langle T_p, (x-a)\phi\rangle = \langle \frac{\delta_b}{b-a}, (x-a)\phi\rangle = \frac{1}{b-a}\langle \delta_b, (x-a)\phi\rangle = \frac{1}{b-a}[(b-a)\phi(b)] = \phi(b) = \langle \delta_b, \phi\rangle \quad \forall \phi \in \mathcal{D}(\mathbb{R}) \Longrightarrow (x-a)T_p = \delta_b$ with $T_p = \frac{\delta_b}{b-a}$. Hence, the general solution is

$$T = C_1\delta_a + \frac{\delta_b}{b-a}, \quad C_1 \in \mathbb{R} \text{ being an arbitrary constant}. \tag{1.6.6}$$

Remark 1.6.2. Case $a = b$ can not be discussed here, since derivatives of Dirac distributions will be involved, which will be introduced in the next chapter.

2. $(x-a)(x-b)T = 0 \in \mathcal{D}'(\mathbb{R})(b \neq a)$. Set $S = (x-b)T \in \mathcal{D}'(\mathbb{R})$. Then $(x-a)S = 0$ in $\mathcal{D}'(\mathbb{R}) \Longrightarrow S = C_1\delta_a$ by Proposition 1.6.1, $C_1 \in \mathbb{R}$ being an arbitrary constant $\Longrightarrow (x-b)T = C_1\delta_a$ in $\mathcal{D}'(\mathbb{R})$. From the solution (1.6.6) of (1), $T = C_1\frac{\delta_a}{a-b} + C_2\delta_b$ in $\mathcal{D}'(\mathbb{R})$ is the general solution, which can be rewritten as

$$T = d_1\delta_a + d_2\delta_b, \quad d_1 \text{ and } d_2 \text{ being arbitrary real constants.} \tag{1.6.7}$$

3. $(1+x^2)(1-x^2)T = 0$ in $\mathcal{D}'(\mathbb{R})$. Since $(1+x^2)^{-1} \in C^\infty(\mathbb{R})$ does not vanish on $\mathbb{R}$, $(1+x^2)^{-1}(1+x^2)(1-x^2)T = 0$ in $\mathcal{D}'(\mathbb{R}) \Longrightarrow (1-x^2)T = 0 \Longrightarrow (x+1)(x-1)T = 0$. Then from the solution (1.6.7) of (2) with $a = 1, b = -1$,

$$T = d_1\delta_1 + d_2\delta_{-1} \quad \text{is the required general solution,} \tag{1.6.8}$$

where δ_1 (resp. δ_{-1}) is the Dirac distribution with concentration at $x = 1$ (resp. -1), and d_1 and d_2 are arbitrary real constants.

4. $p(x)T = 0$ in $\mathcal{D}'(\mathbb{R}) \Longrightarrow (x-a_1)(x-a_2)\dots(x-a_n)T = 0$, where $a_1 < a_2 < \dots < a_n$.

Set $S_1 = (x-a_2)\cdots(x-a_n)T$ in $\mathcal{D}'(\mathbb{R})$. Then $(x-a_1)S_1 = 0 \Longrightarrow S_1 = C_1\delta_{a_1}$ by Proposition 1.6.1, C_1 being an arbitrary constant.

Then $(x-a_2)\dots(x-a_n)T = C_1\delta_{a_1} \in \mathcal{D}'(\mathbb{R})$. Set $S_2 = (x-a_3)\dots(x-a_n)T \in \mathcal{D}'(\mathbb{R})$. Then $(x-a_2)S_2 = C_1\delta_{a_1} \Longrightarrow S_2 = C_2\delta_{a_2} + C_1\frac{\delta_{a_1}}{a_1-a_2}$ (by solution (1.6.6) of (1) with $a_2 = a, a_1 = b) \Longrightarrow (x-a_3)\cdots(x-a_n)T = C_2\delta_{a_2} + C_1\frac{\delta_{a_1}}{a_1-a_2}$.

Then $(x-a_4)\dots(x-a_n)T = C_3\delta_{a_3} + C_2\frac{\delta_{a_2}}{a_2-a_3} + C_1\frac{\delta_{a_1}}{(a_1-a_2)(a_1-a_3)}$.

Continuing in this way, finally we get

$$T = \sum_{k=1}^{n} d_k\delta_{a_k} \quad \text{with } d_k = \frac{C_k}{(a_k-a_{k+1})\cdots(a_k-a_n)}, \; 1 \le k \le n-1, \; d_n = C_n, \tag{1.6.9}$$

where δ_{a_k} is the Dirac distribution with concentration at a_k, $1 \le k \le n$.

1.7 Even, odd and positive distributions

Even and odd distributions

$\forall \phi \in \mathcal{D}(\mathbb{R}), \check{\phi}(x) = \phi(-x)\ \forall x \in \mathbb{R}, \check{\phi} \in \mathcal{D}(\mathbb{R})$.

Definition 1.7.1. A distribution $T \in \mathcal{D}'(\mathbb{R})$ is called

even if and only if

$$T(\check{\phi}) = T(\phi), \quad \text{i.e. } T(\phi(-x)) = T(\phi(x)) \quad \forall \phi \in \mathcal{D}(\mathbb{R}); \tag{1.7.1}$$

odd if and only if

$$T(\check{\phi}) = -T(\phi), \quad \text{i.e. } T(\phi(-x)) = -T(\phi(x)) \quad \forall \phi \in \mathcal{D}(\mathbb{R}). \tag{1.7.2}$$

Examples of even distributions are:

1. f is even in $L^1_{\text{loc}}(\mathbb{R}) \iff f(x) = f(-x)$ for almost all $x \in \mathbb{R}$. Then

$$T_f(\check{\phi}) = \langle f, \check{\phi}\rangle = \int_{\mathbb{R}} f(x)\phi(-x)dx = \int_{\mathbb{R}} f(-x)\phi(x)dx \quad \text{(by change of variable)}$$
$$= \int_{\mathbb{R}} f(x)\phi(x)dx = T_f(\phi) \quad \forall \phi \in \mathcal{D}(\mathbb{R}) \quad (\text{since } f(x) = f(-x) \text{ a.e. on } \mathbb{R})$$

$\Longrightarrow T_f$ defined by even $f \in L^1_{\text{loc}}(\mathbb{R})$ is an even distribution. (1.7.3)

2. $$\langle \delta^{(2m)}, \check{\phi}\rangle = (-1)^{2m}\langle \delta, (\check{\phi})^{(2m)}\rangle = (\check{\phi})^{(2m)}(0) = (\phi)^{(2m)}(0)$$
$$= \langle \delta, \phi^{(2m)}\rangle = (-1)^{2m}\langle \delta^{(2m)}, \phi\rangle = \langle \delta^{(2m)}, \phi\rangle \quad \forall \phi \in \mathcal{D}(\mathbb{R})$$

(For derivatives of δ, see (2.3.8) in Chapter 2)

$\Longrightarrow \delta^{(2m)}$ (i.e. even-order derivatives of Dirac distribution δ) are even distributions for all $m \in \mathbb{N}_0$. (1.7.4)

3. $|x|^\lambda = x_+^\lambda + x_-^\lambda$, with x_+^λ, x_-^λ defined by (1.4.35) and (1.4.36) respectively, is an even distribution. (1.7.5)

Examples of odd distributions are:

4. f is *odd* in $L^1_{\text{loc}}(\mathbb{R}) \iff f(x) = -f(-x)$ for almost all $x \in \mathbb{R}$. Then

$$T_f(\check{\phi}) = \int_{\mathbb{R}} f(x)\phi(-x)dx = \int_{\mathbb{R}} f(-x)\phi(x)dx \quad \text{(by change of variables)}$$
$$= -\int_{\mathbb{R}} f(x)\phi(x)dx = -T_f(\phi) \quad \forall \phi \in \mathcal{D}(\mathbb{R})$$

$\Longrightarrow T_f$ defined by odd $f \in L^1_{\text{loc}}(\mathbb{R})$ is an odd distribution. (1.7.6)

5. $\langle \delta^{(2m+1)}, \check{\phi} \rangle = (-1)^{(2m+1)} \langle \delta, (\check{\phi})^{(2m+1)} \rangle = \cdots = (-1)^{(2m+1)} \langle \delta^{(2m+1)}, \phi \rangle$
$= -\langle \delta^{(2m+1)}, \phi \rangle \quad \forall \phi \in \mathcal{D}(\mathbb{R})$ (see (2.3.8) in Chapter 2)

$\Longrightarrow \delta^{(2m+1)}$ (i.e. odd-order derivatives of Dirac distribution δ) are odd distributions for all $m \in \mathbb{N}_0$. (1.7.7)

6. $|x|^\lambda \operatorname{sgn} x = x_+^\lambda - x_-^\lambda$, with x_+^λ, x_-^λ defined by (1.4.35) and (1.4.36) respectively, is an odd distribution. (1.7.8)

Every distribution $T \in \mathcal{D}'(\mathbb{R})$ can be expressed as a sum of an even distribution $T_E \in \mathcal{D}'(\mathbb{R})$ and an odd distribution $T_0 \in \mathcal{D}'(\mathbb{R})$, i.e.

$$T \in \mathcal{D}'(\mathbb{R}) \quad \Longrightarrow \quad T = T_E + T_0, \tag{1.7.9}$$

the decomposition being a unique one.

Proof. Define T_E and T_0 by: $T_E(\phi) = \frac{1}{2}[T(\phi) + T(\check{\phi})]$,

$$T_0(\phi) = \frac{1}{2}[T(\phi) - T(\check{\phi})] \quad \forall \phi \in \mathcal{D}(\mathbb{R}). \tag{1.7.10}$$

Then $T_E(\check{\phi}) = \frac{1}{2}[T(\check{\phi}) + T(\check{\check{\phi}})] = \frac{1}{2}[T(\check{\phi}) + T(\phi)] = T_E(\phi) \Longrightarrow T_E$ is even.

$$T_0(\check{\phi}) = \frac{1}{2}[T(\check{\phi}) - T(\check{\check{\phi}})] = \frac{1}{2}[T(\check{\phi}) - T(\phi)] = -\frac{1}{2}[T(\phi) - T(\check{\phi})] = -T_0(\phi)$$

$\Longrightarrow T_0$ is odd.

$$(T_E + T_0)(\phi) = T_E(\phi) + T_0(\phi) = T(\phi) \quad \forall \phi \in \mathcal{D}(\mathbb{R})$$

$\Longrightarrow T_E + T_0 = T$ in $\mathcal{D}'(\mathbb{R})$.

Uniqueness: Let $T'_E, T'_0 \in \mathcal{D}'(\mathbb{R})$ such that $T'_E + T'_0 = T_E + T_0 = T$.
Then, $T - T'_E = T'_0$ is odd

$$\begin{aligned}
&\Longrightarrow \quad \langle T - T'_E, \phi \rangle = -\langle T - T'_E, \check{\phi} \rangle = -T(\check{\phi}) + T'_E(\check{\phi}) \\
&\Longrightarrow \quad T(\phi) - T'_E(\phi) = -T(\check{\phi}) + T'_E(\check{\phi}) \\
&\Longrightarrow \quad 2T'_E(\check{\phi}) = T(\phi) + T(\check{\phi}) \\
&\Longrightarrow \quad T'_E(\phi) = T'_E(\check{\phi}) = \frac{1}{2}[T(\phi) + T(\check{\phi})] = T_E(\phi) \quad \forall \phi \in \mathcal{D}(\mathbb{R}) \\
&\Longrightarrow \quad T'_E = T_E \quad \text{in } \mathcal{D}'(\mathbb{R}).
\end{aligned}$$

Similarly, $T'_0 = T_0$ in $\mathcal{D}'(\mathbb{R})$. □

Positive distributions

Definition 1.7.2. A distribution $T \in \mathcal{D}'(\Omega)$ with $\Omega \subset \mathbb{R}^n$ is called *positive*, i.e. $T \geq 0$, if and only if $T(\phi) = \langle T, \phi \rangle \geq 0 \ \forall \phi \in \mathcal{D}(\mathbb{R})$ with $\phi \geq 0$. (1.7.11)

$T_1 \geq T_2$ in $\mathcal{D}'(\mathbb{R}) \iff T_1 - T_2 \geq 0 \iff T_1(\phi) \geq T_2(\phi) \ \forall \phi \in \mathcal{D}(\mathbb{R})$ with $\phi \geq 0$. (1.7.12)

1.8 Convergence of sequences of distributions in $\mathcal{D}'(\Omega)$

Definition 1.8.1. Let $(T_n)_{n=1}^{\infty}$ be a sequence of distributions in $\mathcal{D}'(\Omega)$ with $\Omega \subset \mathbb{R}^n$. Then, if $\exists T \in \mathcal{D}'(\Omega)$ such that

$$\langle T_n, \phi \rangle \to \langle T, \phi \rangle \text{ in } \mathbb{R} \text{ (resp. } \mathbb{C}) \ \forall \phi \in \mathcal{D}(\Omega) \quad \text{as } n \to \infty, \tag{1.8.1}$$

the sequence $(T_n)_{n=1}^{\infty}$ is said to *converge* to $T \in \mathcal{D}'(\Omega)$, and we write $T_n \to T$ in $\mathcal{D}'(\Omega)$ as $n \to \infty$. Then T is called the *limit* of the sequence (T_n) in $\mathcal{D}'(\Omega)$:

$$\lim_{n\to\infty} T_n = T \quad \text{in } \mathcal{D}'(\Omega). \tag{1.8.2}$$

Proposition 1.8.1. *Let $(T_n)_{n=1}^{\infty}$ be a sequence of distributions in $\mathcal{D}'(\Omega)$ such that $\lim_{n\to\infty}\langle T_n, \phi \rangle$ exists in $\mathbb{R}$ (resp. $\mathbb{C}$) $\forall \phi \in \mathcal{D}(\Omega)$. Then the sequence $(T_n)_{n=1}^{\infty}$ has a limit in $\mathcal{D}'(\Omega)$, i.e. $\exists$ a unique $T \in \mathcal{D}'(\Omega)$ such that*

$$\langle T_n, \phi \rangle \to \langle T, \phi \rangle \text{ in } \mathbb{R} \text{ (resp. } \mathbb{C}) \ \forall \phi \in \mathcal{D}(\Omega) \quad \text{as } n \to \infty. \tag{1.8.3}$$

Remark 1.8.1. As a consequence of Proposition 1.8.1, for the convergence of a sequence (T_n) in $\mathcal{D}'(\Omega)$ it is *not* necessary to know its limit explicitly, i.e. it is sufficient to show that $\lim_{n\to\infty}\langle T_n, \phi \rangle$ exists in $\mathbb{R}$ (resp. $\mathbb{C}$) $\forall \phi \in \mathcal{D}(\Omega)$.

Examples of convergence of sequences of distributions

Example 1.8.1. $T_n = \sin nx$ in $\mathcal{D}'(\Omega)$ $\forall n \in \mathbb{N} \Longrightarrow \lim_{n\to\infty} \sin nx = 0$ in $\mathcal{D}'(\mathbb{R})$, i.e.

$$\lim_{n\to\infty}\langle T_n, \phi \rangle = \lim_{n\to\infty}\langle \sin nx, \phi \rangle = \lim_{n\to\infty}\int_{-\infty}^{\infty} \sin nx\phi(x)dx = 0 \quad \forall \phi \in \mathcal{D}(\mathbb{R}). \tag{1.8.4}$$

In fact, $\forall \phi \in \mathcal{D}(\mathbb{R}), \exists A > 0$ such that $\operatorname{supp}(\phi) \subset [-A, A]$. Then, integrating by parts, we get $\int_{-\infty}^{\infty} \sin nx\phi(x)dx = \int_{-A}^{A} \frac{\cos nx}{n}\phi'(x)dx$, since $\phi(\pm A) = 0$.

$$\text{But } \left|\int_{-A}^{A} \frac{\cos nx}{n}\phi'(x)dx\right| \leq \frac{1}{n}\max_{x\in[-A,A]}|\phi'(x)|\int_{-A}^{A} dx$$
$$= \frac{2A}{n}\max_{x\in[-A,A]}|\phi'(x)| \to 0 \quad \text{as } n \to \infty$$

$\Longrightarrow \lim_{n\to\infty} \int_{-A}^{A} \frac{\cos nx}{n}\phi'(x)dx = 0 \quad \forall\phi \in \mathcal{D}(\mathbb{R})$ with $\mathrm{supp}(\phi) \subset [-A, A]$.

Hence, $\lim_{n\to\infty}\langle T_n, \phi\rangle = \lim_{n\to\infty}\langle \sin nx, \phi\rangle = 0 \ \forall\phi \in \mathcal{D}(\mathbb{R})$.

Then, by Proposition 1.8.1, $\exists T = 0 \in \mathcal{D}'(\mathbb{R})$ such that $\lim_{n\to\infty}\langle \sin nx, \phi\rangle = \langle 0, \phi\rangle \ \forall\phi \in \mathcal{D}(\mathbb{R}) \Longrightarrow \lim_{n\to\infty} \sin nx = 0$ in $\mathcal{D}'(\mathbb{R})$.

Example 1.8.2. $T_n = \frac{\sin nx}{x}$ in $\mathcal{D}'(\mathbb{R})$ $\forall n \in \mathbb{N} \Longrightarrow \lim_{n\to\infty} \frac{\sin nx}{x} = \pi\delta$ in $\mathcal{D}'(\mathbb{R})$, i.e., $\forall\phi \in \mathcal{D}(\mathbb{R})$,

$$\lim_{n\to\infty}\langle T_n, \phi\rangle = \lim_{n\to\infty}\left\langle \frac{\sin nx}{x}, \phi\right\rangle = \lim_{n\to\infty}\int_{-\infty}^{\infty} \frac{\sin nx}{x}\phi(x)dx = \pi\phi(0) = \langle \pi\delta, \phi\rangle. \tag{1.8.5}$$

Indeed, $\forall\phi \in \mathcal{D}(\mathbb{R}), \exists A > 0$ such that $\mathrm{supp}(\phi) \subset [-A, A]$, and

$$\begin{aligned}\int_{-\infty}^{\infty} \frac{\sin nx}{x}\phi(x)dx &= \int_{-A}^{A} \frac{\sin nx}{x}\phi(x)dx \\ &= \int_{-A}^{A} \sin nx \frac{\phi(x) - \phi(0)}{x}dx + \phi(0)\int_{-A}^{A} \frac{\sin nx}{x}dx.\end{aligned} \tag{1.8.6}$$

From Proposition 1.2.1, $\phi(x) = \phi(0) + x\psi(x)$ with $\psi \in C^0(\mathbb{R})$,

$$\Longrightarrow \quad \int_{-A}^{A} \sin nx \frac{\phi(x) - \phi(0)}{x}dx = \int_{-A}^{A} \sin nx\psi(x)dx \to 0 \quad \text{as } n \to \infty,$$

since $\psi \in C^0([-A, A]) \subset L^1(]-A, A[)$ by the Riemann–Lebesgue theorem[6] (see also (2.11.7i)),

$$\lim_{n\to\infty}\int_{-A}^{A} \sin nx\psi(x)dx = 0. \tag{1.8.7}$$

[6]**Riemann–Lebesgue Theorem:** For compact $[a, b] \subset \mathbb{R}$, let $f \in L^1(]a, b[)$. Then,

I. $\lim_{\lambda\to\infty} \int_a^b f(x)\sin\lambda x dx = 0$;

II. $\lim_{\lambda\to\infty} \int_a^b f(x) cos\lambda x dx = 0$.

Proof. We give the proof of (I) and (II) for $f \in C^1([a, b]) \subset L^1(]a, b[)$. Hence, for $f \in C^1([a, b])$, $\int_a^b f(x)\sin\lambda x dx = -f(x)\frac{\cos\lambda x}{\lambda}|_a^b + \frac{1}{\lambda}\int_a^b f'(x)\cos\lambda x dx \Longrightarrow |\int_a^b f(x)\sin\lambda x dx| \le \frac{1}{\lambda}[|f(b)| + |f(a)| + \int_a^b |f'(x)|dx] \to 0$ as $\lambda \to \infty \Longrightarrow \lim_{\lambda\to\infty} \int_a^b f(x)\sin\lambda x dx = 0$.

The linear functional $l(f) = \lim_{\lambda\to\infty} \int_a^b f(x)\sin\lambda x dx \ \forall f \in L^1(]a, b[)$ is continuous on $L^1(]a, b[)$, i.e. $|l(f)| \le \lim_{\lambda\to\infty} \int_a^b |f(x)|dx \Longrightarrow |l(f)| \le \|f\|_{L^1(]a,b[)} \ \forall f \in L^1(]a, b[)$. The continuous linear functional l vanishes on $C^1([a, b])$, which is *dense* in $L^1(]a, b[)$. Hence, the result follows as a consequence of Hahn–Banach Theorem. □

Putting $x = \frac{\xi}{n}$ with $dx = \frac{d\xi}{n}$, we get $\int_{-A}^{A} \frac{\sin nx}{x} dx = \int_{-nA}^{nA} \frac{\sin \xi}{\xi} d\xi \to \int_{-\infty}^{\infty} \frac{\sin \xi}{\xi} d\xi = \pi$ as $n \to \infty$, i.e.

$$\lim_{n\to\infty} \int_{-A}^{A} \frac{\sin nx}{x} dx = \pi. \tag{1.8.8}$$

Finally, from (1.8.6)–(1.8.8), $\lim_{n\to\infty} \langle \frac{\sin nx}{x}, \phi \rangle = 0 + \phi(0)\pi = \langle \pi\delta, \phi \rangle \ \forall \phi \in \mathcal{D}(\mathbb{R}) \Longrightarrow \lim_{n\to\infty} \frac{\sin nx}{x} = \pi\delta$ in $\mathcal{D}'(\mathbb{R})$.

Example 1.8.3. $T_n = U_n$ with $U_n(x) = \int_{-n}^{n} e^{ixy} dy \ \forall x \in \mathbb{R}, \forall n \in \mathbb{N}$

$$\Longrightarrow \quad \lim_{n\to\infty} \int_{-n}^{n} e^{ixy} dy = 2\pi\delta \text{ in } \mathcal{D}'(\mathbb{R}) \text{ with } \langle \delta, \phi \rangle = \phi(0) \quad \forall \phi \in \mathcal{D}(\mathbb{R}), \tag{1.8.9}$$

i.e.

$$\lim_{n\to\infty} \langle T_n, \phi \rangle = \lim_{n\to\infty} \left\langle \int_{-n}^{n} e^{ixy} dy, \phi \right\rangle = \langle 2\pi\delta, \phi \rangle = 2\pi\phi(0). \tag{1.8.10}$$

In fact, $\int_{-n}^{n} e^{ixy} dy = \frac{e^{ixy}}{ix} \Big|_{y=-n}^{y=n} = \frac{e^{ixn} - e^{-ixn}}{ix} = 2 \cdot \frac{\sin nx}{x}$

$$\Longrightarrow \quad \lim_{n\to\infty} \left\langle \int_{-n}^{n} e^{ixy} dy, \phi \right\rangle = \lim_{n\to\infty} \int_{-n}^{n} 2\frac{\sin nx}{x}\phi(x)dx = \lim_{n\to\infty} \left\langle 2\frac{\sin nx}{x}, \phi \right\rangle$$
$$= 2\pi\phi(0) = \langle 2\pi\delta, \phi \rangle \quad \text{(by (1.8.5))} \quad \forall \phi \in \mathcal{D}(\mathbb{R}).$$

$$\Longrightarrow \quad \lim_{n\to\infty} \int_{-n}^{n} e^{ixy} dy = 2\pi\delta \quad \text{in } \mathcal{D}'(\mathbb{R}).$$

Example 1.8.4. For $\lambda > 0$, let $T_\lambda \in \mathcal{D}'(\mathbb{R})$ be defined by:

$$\langle T_\lambda, \phi \rangle = \int_{-\frac{\pi}{2}}^{\frac{\pi}{2}} \frac{\cos \lambda x}{x} [\phi(x) - \phi(0)] dx \quad \forall \phi \in \mathcal{D}(\mathbb{R}). \tag{1.8.11}$$

Find

1. $\lim_{\lambda\to\infty} T_\lambda$;
2. $\lim_{\lambda\to 0^+} T_\lambda$ in $\mathcal{D}'(\mathbb{R})$.

Solution.

1. By Proposition 1.2.1, $\phi(x) = \phi(0) + x\psi(x)$ with $\psi \in C^0(\mathbb{R})$. Then $\phi(x) - \phi(0) = x\psi(x) \ \forall x \Longrightarrow \langle T_\lambda, \phi \rangle = \int_{\frac{-\pi}{2}}^{\frac{\pi}{2}} \cos \lambda x \psi(x) dx$, where ψ is continuous on $[\frac{-\pi}{2}, \frac{\pi}{2}]$ and hence belongs to $L^1(]\frac{-\pi}{2}, \frac{\pi}{2}[)$. Consequently, by the Riemann–Lebesgue theorem (see footnote p. 60),

$$\lim_{\lambda\to\infty} \int_{\frac{-\pi}{2}}^{\frac{\pi}{2}} \cos \lambda x \psi(x) dx = 0. \tag{1.8.12}$$

Hence, $\lim_{\lambda\to\infty} \langle T_\lambda, \phi \rangle = 0 \ \forall \phi \in \mathcal{D}(\mathbb{R}) \Longrightarrow \lim_{\lambda\to\infty} T_\lambda = 0$ in $\mathcal{D}'(\mathbb{R})$.

2. $\lambda \to 0^+ \implies \cos\lambda x \to 1 \implies \int_{-\frac{\pi}{2}}^{\frac{\pi}{2}} \cos\lambda x \psi(x)dx \to \int_{\frac{-\pi}{2}}^{\frac{\pi}{2}} \psi(x)dx$ by Lebesgue's dominated convergence theorem (see Theorem B.3.2.2 in Appendix B), since $|\cos\lambda x\psi(x)| \le |\psi(x)| \in L^1(]\frac{-\pi}{2}, \frac{\pi}{2}[)$.
Hence, $\lim_{\lambda\to 0^+}\langle T_\lambda, \phi\rangle = \int_{\frac{-\pi}{2}}^{\frac{\pi}{2}} \frac{1}{x}[\phi(x) - \phi(0)]dx\ \forall\phi \in \mathcal{D}(\mathbb{R})$.
Let $T \in \mathcal{D}'(\mathbb{R})$ be defined by:

$$\langle T, \phi\rangle = \int_{-\frac{\pi}{2}}^{\frac{\pi}{2}} \frac{1}{x}[\phi(x) - \phi(0)]dx \quad \forall\phi \in \mathcal{D}(\mathbb{R}). \tag{1.8.13}$$

Then $\lim_{\lambda\to 0^+} T_\lambda = T$ in $\mathcal{D}'(\mathbb{R})$ with T defined by (1.8.13).

Example 1.8.5. $\forall\varepsilon > 0$, let f_ε be defined by $f_\varepsilon(x) = \frac{\varepsilon}{2}|x|^{\varepsilon-1}\ \forall x \in \mathbb{R}$. Then show that

$$\lim_{\varepsilon\to 0^+} f_\varepsilon = \delta \quad \text{in } \mathcal{D}'(\mathbb{R}). \tag{1.8.14}$$

Proof. Since $|x|^\lambda$ is locally integrable on $\mathbb{R}$ for $\lambda = \varepsilon - 1 > -1\quad \forall\varepsilon > 0$ (see (1.4.30)), $|x|^{\varepsilon-1}$ is locally integrable on $\mathbb{R}$ and defines a distribution on $\mathbb{R}$, and consequently, $f_\varepsilon \in \mathcal{D}'(\mathbb{R})\ \forall\varepsilon > 0$ with $\langle f_\varepsilon, \phi\rangle = \frac{\varepsilon}{2}\int_{\mathbb{R}} |x|^{\varepsilon-1}\phi(x)dx$.

For $\phi \in \mathcal{D}(\mathbb{R})$ with $\text{supp}(\phi) = K \subset [-A, A]$, $A > 0$,

$$\begin{aligned}\langle f_\varepsilon, \phi\rangle &= \frac{\varepsilon}{2}\int_{-A}^{0} (-x)^{\varepsilon-1}\phi(x)dx + \frac{\varepsilon}{2}\int_0^A x^{\varepsilon-1}\phi(x)dx \\ &= \frac{\varepsilon}{2}\int_0^A x^{\varepsilon-1}\phi(-x)dx + \frac{\varepsilon}{2}\int_0^A x^{\varepsilon-1}\phi(x)dx \\ &= \frac{\varepsilon}{2}\int_0^A x^{\varepsilon-1}[\phi(x) + \phi(-x)]dx.\end{aligned}$$

From Proposition 1.2.1, $\phi(\pm x) = \phi(0) \pm x\psi(\pm x)$ with $\sup_{|x|\le A}|\psi(x)| \le C\sup_{|x|\le A}|\phi'(x)| \implies \phi(x) + \phi(-x) = 2\phi(0) + x[\psi(x) - \psi(-x)]$

$$\implies \quad \langle f_\varepsilon, \phi\rangle = \frac{\varepsilon}{2}2\phi(0)\int_0^A x^{\varepsilon-1}dx + \frac{\varepsilon}{2}\int_0^A x^\varepsilon[\psi(x) - \psi(-x)]dx. \tag{1.8.15}$$

Then, for $0 < \varepsilon < 1$, $\varepsilon\int_0^A x^{\varepsilon-1}dx = \varepsilon\frac{x^\varepsilon}{\varepsilon}|_0^A = A^\varepsilon$, which tends to 1 as $\varepsilon \to 0^+$,

$$\implies \quad \lim_{\varepsilon\to 0^+}\left[\phi(0)\left(\varepsilon\int_0^A x^{\varepsilon-1}dx\right)\right] = \phi(0)\cdot\lim_{\varepsilon\to 0^+} A^\varepsilon = \phi(0). \tag{1.8.16}$$

Again, for $0 < \varepsilon < 1$, $\frac{\varepsilon}{2}|\int_0^A x^\varepsilon[\psi(x) - \psi(-x)]dx| \le \frac{\varepsilon}{2}\int_0^A x^\varepsilon|[\psi(x) - \psi(-x)]|dx \le \frac{\varepsilon}{2}\cdot 2\cdot C\sup_{|x|\le A}|\phi'(x)|\cdot\int_0^A x^\varepsilon dx \le C\sup_{|x|\le A}|\phi'(x)|\cdot\frac{A^{1+\varepsilon}}{1+\varepsilon}\cdot\varepsilon \to 0$ as $\varepsilon \to 0^+$

$$\implies \quad \lim_{\varepsilon\to 0^+}\frac{\varepsilon}{2}\int_0^A x^\varepsilon[\psi(x) - \psi(-x)]dx = 0. \tag{1.8.17}$$

Then, from (1.8.15)–(1.8.17), $\lim_{\varepsilon\to 0^+}\langle f_\varepsilon,\phi\rangle = \phi(0)+0 = \phi(0) = \langle\delta,\phi\rangle\,\forall\phi\in\mathcal{D}(\mathbb{R}) \Longrightarrow \lim_{\varepsilon\to 0^+} f_\varepsilon = \delta$ in $\mathcal{D}'(\mathbb{R})$. □

Example 1.8.6.

1. Let (f_n) be a sequence of functions on $\mathbb{R}$ defined by:

$$f_n(x) = \begin{cases} n^2 & \text{for } |x| < \frac{1}{n} \\ 0 & \text{for } |x| \geq \frac{1}{n} \end{cases} \qquad \forall n \in \mathbb{N}. \tag{1.8.18}$$

   Show that

   (a) $(f_n(x))$ converges to 0 in $\mathbb{R}$ $\forall x \neq 0$;

   (b) (f_n) does *not* converge in $\mathcal{D}'(\mathbb{R})$.

2. Let (g_n) be a sequence defined by:

$$g_n(x) = \begin{cases} \frac{n}{2} & \text{for } |x| < \frac{1}{n} \\ 0 & \text{for } |x| \geq \frac{1}{n} \end{cases} \qquad \forall n \in \mathbb{N}. \tag{1.8.19}$$

   Show that

   (a) (g_n) converges to δ in $\mathcal{D}'(\mathbb{R})$;

   (b) $(g_n(x))$ converges to 0 for $x \neq 0$.

Proof.

1. (a) $\forall x \neq 0$, $f_n(x) \to 0$ in $\mathbb{R}$ as $n \to \infty$.
   In fact, $\forall$ fixed $x_0 \neq 0$, $\exists n_0 \in \mathbb{N}$ such that $\forall n \geq n_0$, $\frac{1}{n_0} < |x_0| \Longrightarrow f_n(x_0) = 0\ \forall n \geq n_0 \Longrightarrow \forall x_0 \neq 0$, $f_n(x_0) \to 0$ in $\mathbb{R}$ as $n \to \infty$.
   (b) $\langle f_n,\phi\rangle = \int_{-\frac{1}{n}}^{\frac{1}{n}} n^2\phi(x)dx = n^2\int_{-\frac{1}{n}}^{\frac{1}{n}}\phi(x)dx$.
   $\exists\phi\in\mathcal{D}(\mathbb{R})$ such that $\phi(x) = 1\ \forall|x| \leq 1$. Then, for such a $\phi\in\mathcal{D}(\mathbb{R})$ with $\phi(x) = 1\forall|x|\leq 1$, $\langle f_n,\phi\rangle = n^2\int_{-\frac{1}{n}}^{\frac{1}{n}} 1dx = n^2.\frac{2}{n} = 2n \to \infty$ as $n\to\infty$, i.e. $(\langle f_n,\phi\rangle)$ does *not* converge in $\mathbb{R}$, and hence (f_n) does *not* converge in $\mathcal{D}'(\mathbb{R})$.

2. (a) $\langle g_n,\phi\rangle = \int_{-\frac{1}{n}}^{\frac{1}{n}} \frac{n}{2}\phi(x)dx = \frac{n}{2}\cdot\frac{2}{n}\phi(\xi_n) = \phi(\xi_n)$ with $|\xi_n| \leq \frac{1}{n}$.
   But $\xi_n \to 0$ as $n\to\infty$, and $\xi_n \to 0 \Longrightarrow \phi(\xi_n)\to\phi(0)$, since ϕ is continuous. Hence,

$$\lim_{n\to\infty}\langle g_n,\phi\rangle = \phi(0) = \langle\delta,\phi\rangle\ \forall\phi\in\mathcal{D}(\mathbb{R}) \quad\Longrightarrow\quad \lim_{n\to\infty} g_n = \delta \text{ in } \mathcal{D}'(\mathbb{R}).$$

   (b) For $x \neq 0$, $g_n(x) \to 0$ in $\mathbb{R}$ as $n \to \infty$ (see proof of 1(a)). □

Example 1.8.7. Let $(\delta_n)_{n=0}^{\infty}$ be a sequence of Dirac distributions δ_n with unit mass/charge/force etc. concentrated at $x = n \in \mathbb{N}_0$, $\langle \delta_n, \phi \rangle = \phi(n)$. Show that for an arbitrary system $(a_n)_{n=0}^{\infty}$ of real numbers a_n, the sequence $(a_n\delta_n)_{n=0}^{\infty}$ of distributions in $\mathcal{D}'(\mathbb{R})$ converges to $0 \in \mathcal{D}'(\mathbb{R})$.

Proof. Set $T_n = a_n\delta_n \in \mathcal{D}'(\mathbb{R})\ \forall n \in \mathbb{N}$. Then (T_n) converges in $\mathcal{D}'(\mathbb{R})$ iff $(\langle T_n, \phi\rangle)$ with $\langle T_n, \phi\rangle = \langle a_n\delta_n, \phi\rangle = a_n\langle\delta_n, \phi\rangle = a_n\phi(n)$ converges in $\mathbb{R}\ \forall\phi \in \mathcal{D}(\mathbb{R})$.

$\forall$ fixed $\phi \in \mathcal{D}(\mathbb{R})$, $\operatorname{supp}(\phi)$ is a compact subset of $\mathbb{R} \Longrightarrow \exists n_0 \in \mathbb{N}$ such that $\operatorname{supp}(\phi) \subset [-n_0, n_0] \Longrightarrow \forall n \geq n_0, \phi(n) = 0$

$$\Longrightarrow \quad \langle T_n, \phi\rangle = a_n\phi(n) = 0\ \forall n \geq n_0 \quad \Longrightarrow \quad \lim_{n\to\infty}\langle T_n, \phi\rangle = 0 \text{ in } \mathbb{R}\ \forall\phi \in \mathcal{D}(\mathbb{R})$$

$$\Longrightarrow \quad \lim_{n\to\infty}\langle T_n, \phi\rangle = \langle 0, \phi\rangle\ \forall\phi \in \mathcal{D}(\mathbb{R}) \quad \Longrightarrow \quad T_n = a_n\delta_n \to 0 \in \mathcal{D}'(\mathbb{R})$$

for arbitrary choice of $a_n \in \mathbb{R}$. □

Example 1.8.8. Prove that $\lim_{y\to 0+} \ln(x \pm iy) = \ln(x \pm i0)$ in $\mathcal{D}'(\mathbb{R})$, where

$$\ln(x \pm iy) = \ln\sqrt{x^2 + y^2} \pm i \arctan\left(\frac{y}{x}\right) \quad \text{for } y > 0, \tag{1.8.20}$$

$$\ln(x \pm i0) = \lim_{y\to 0+} \ln(x \pm iy) = \ln|x| \pm i\pi H(-x) \quad \text{with}$$

$$\ln|x| \pm i\pi H(-x) = \begin{cases} \ln|x| & \text{for } x > 0 \\ \ln|x| \pm i\pi & \text{for } x < 0. \end{cases} \tag{1.8.21}$$

Proof. $\forall$ fixed $y > 0$, $\ln(x \pm iy)$ is locally integrable on $\mathbb{R} \Longrightarrow \ln(x \pm iy) \in \mathcal{D}'(\mathbb{R})$ $\forall$ fixed $y > 0$, with

$$\langle \ln(x \pm iy), \phi\rangle = \int_{\mathbb{R}} \ln(x \pm iy)\phi(x)dx \quad \forall\phi \in \mathcal{D}(\mathbb{R}). \tag{1.8.22}$$

Again, $\ln(x \pm i0) \in \mathcal{D}'(\mathbb{R})$ is defined by (1.4.7) and (1.4.8): $\forall\phi \in \mathcal{D}(\mathbb{R})$,

$$\langle \ln(x \pm i0), \phi\rangle = \int_{\mathbb{R}} \ln|x|\phi(x)dx \pm i\pi \int_{-\infty}^{0} \phi(x)dx. \tag{1.8.23}$$

We will show that $\lim_{y\to 0+}\langle \ln(x + iy), \phi\rangle = \langle \ln(x + i0), \phi\rangle\ \forall\phi \in \mathcal{D}(\mathbb{R})$

$$\begin{aligned} \Longrightarrow \quad \lim_{y\to 0+}\int_{\mathbb{R}} \ln(x + iy)\phi(x)dx &= \int_{\mathbb{R}} \ln(x + i0)\phi(x)dx \\ &= \int_{\mathbb{R}} \lim_{y\to 0+} \ln(x + iy)\phi(x)dx, \end{aligned} \tag{1.8.24}$$

which we will prove using Lebesgue's Theorem B.3.2.2 on dominated convergence in Appendix B.

In fact, $\lim_{y\to 0+}[\ln|(x+iy)|\phi(x)] = \ln|x|\phi(x)$ for almost all $x \in \mathbb{R}$.
For $|x+iy| > 1$ and $0 < y < 1$,

$$\big|\ln|x+iy|\big| = \ln|x+iy| = \ln\sqrt{x^2+y^2} = \frac{1}{2}\ln(x^2+y^2) \le \frac{1}{2}\ln(x^2+1),$$

and for $|x+iy| < 1$, and $y > 0$,

$$\big|\ln|x+iy|\big| = \ln\left(\frac{1}{|x+iy|}\right) \le \ln\left(\frac{1}{|x|}\right) = \big|\ln|x|\big|.$$

Hence, $\big|\ln|x+iy|\phi(x)\big| \le [\frac{1}{2}\ln(x^2+1) + \big|\ln|x|\big|]|\phi(x)|$ for $0 < y < 1$ and for almost all $x \in \mathbb{R}$, where $[\frac{1}{2}\ln(x^2+1) + \big|\ln|x|\big|]|\phi(x)|$ is integrable on $\mathbb{R}$ $\forall\phi \in \mathcal{D}(\mathbb{R})$ (see Example 1.4.1).

Hence, by Lebesgue's dominated convergence theorem, $\forall\phi \in \mathcal{D}(\mathbb{R})$,

$$\lim_{y\to 0+}\int_{\mathbb{R}} \ln|(x+iy)|\phi(x)dx = \int_{\mathbb{R}} (\lim_{y\to 0+} \ln|(x+iy)|)\phi(x)dx = \int_{\mathbb{R}} \ln|x|\phi(x)dx. \tag{1.8.25}$$

For $x > 0$, $\lim_{y\to 0+}[\arctan\frac{y}{x}] = 0$, and for $x < 0$, $\lim_{y\to 0+}[\arctan\frac{y}{x}] = \pi$, i.e. $\lim_{y\to 0+}[\arctan\frac{y}{x}]\phi(x) = \pi H(-x)\phi(x)$ for almost all $x \in \mathbb{R}$ and $\forall\phi \in \mathcal{D}(\mathbb{R})$.

Moreover, for $y > 0$, $|\arctan\frac{y}{x}| \le \pi \implies |\arctan\frac{y}{x}\phi(x)| \le \pi|\phi(x)|$ for almost all $x \in \mathbb{R}$, $\forall y > 0$. Hence, again by Lebesgue's dominated convergence theorem, $\forall\phi \in \mathcal{D}(\mathbb{R})$,

$$\begin{aligned}\lim_{y\to 0+}\int_{\mathbb{R}}\left(\arctan\frac{y}{x}\right)\phi(x)dx &= \int_{\mathbb{R}} \lim_{y\to 0+}\left[\left(\arctan\frac{y}{x}\right)\phi(x)\right]dx \\ &= \int_{\mathbb{R}} \pi H(-x)\phi(x)dx. \end{aligned} \tag{1.8.26}$$

Finally, from (1.8.24)–(1.8.26), we get, $\forall\phi \in \mathcal{D}(\mathbb{R})$,

$$\begin{aligned}\lim_{y\to 0+}\int_{\mathbb{R}} \ln(x+iy)\phi(x)dx &= \int_{\mathbb{R}} \ln|x|\phi(x)dx + i\int_{\mathbb{R}} \pi H(-x)\phi(x)dx \\ &= \int_{\mathbb{R}} [\ln|x| + i\pi H(-x)]\phi(x)dx = \int_{\mathbb{R}} \ln(x+i0)\phi(x)dx\end{aligned}$$

$$\implies \quad \lim_{y\to 0+} \ln(x+iy) = \ln(x+i0) \quad \text{in } \mathcal{D}'(\mathbb{R}).$$

Similarly, we can show that $\lim_{y\to 0+}\ln(x-iy) = \ln(x-i0)$ in $\mathcal{D}'(\mathbb{R})$. □

Example 1.8.9. Show that $\lim_{y\to+\infty} e^{ixy}(\text{c.p.v. }\frac{1}{x}) = i\pi\delta$ in $\mathcal{D}'(\mathbb{R})$.

Proof. $\forall$ fixed $y > 0$, $e^{ixy}(\text{c.p.v.}\ \frac{1}{x}) \in \mathcal{D}'(\mathbb{R})$ is defined by: $\forall \phi \in \mathcal{D}(\mathbb{R})$,

$$\left\langle e^{ixy}\left(\text{c.p.v.}\ \frac{1}{x}\right), \phi\right\rangle = \left\langle \text{c.p.v.}\ \frac{1}{x}, e^{ixy}\phi\right\rangle = \lim_{\varepsilon\to 0+}\int_{\varepsilon\le|x|}\frac{e^{ixy}\phi(x)}{x}dx. \quad (1.8.27)$$

Thus, for $\phi \in \mathcal{D}(\mathbb{R})$ with $\text{supp}(\phi) = K \subset [-A, A]$, $A > 0$, $\forall$ fixed $y > 0$,

$$\begin{aligned} I(\varepsilon) = \int_{\varepsilon\le|x|\le A}\frac{e^{ixy}\phi(x)}{x}dx &= \int_{-A}^{-\varepsilon}\frac{e^{ixy}\phi(x)}{x}dx + \int_{\varepsilon}^{A}\frac{e^{ixy}\phi(x)}{x}dx \\ &= -\int_{\varepsilon}^{A}\frac{e^{-ixy}\phi(-x)}{x}dx + \int_{\varepsilon}^{A}\frac{e^{ixy}\phi(x)}{x}dx. \end{aligned} \quad (1.8.28)$$

From Proposition 1.2.1, $\phi(\pm x) = \phi(0) \pm x\psi(\pm x)$ with $\psi \in C^0(\mathbb{R})$.
$\Longrightarrow$ For fixed $y > 0$,

$$\begin{aligned} I(\varepsilon) &= \int_{\varepsilon}^{A}\frac{(e^{ixy} - e^{-ixy})\phi(0) + x[e^{ixy}\psi(x) + e^{-ixy}\psi(-x)]}{x}dx \\ &= \underbrace{2i\phi(0)\int_{\varepsilon}^{A}\frac{\sin xy}{x}dx}_{I_1(\varepsilon)} + \underbrace{\int_{\varepsilon}^{A}[e^{ixy}\psi(x) + e^{-ixy}\psi(-x)]dx}_{I_2(\varepsilon)} = I_1(\varepsilon) + I_2(\varepsilon) \end{aligned}$$

For *fixed* $y > 0$,

$$\begin{aligned} \lim_{\varepsilon\to 0+} I_1(\varepsilon) &= 2i\phi(0)\lim_{\varepsilon\to 0+}\int_{\varepsilon}^{A}\frac{\sin xy}{x}dx \\ &= 2i\phi(0)\lim_{\varepsilon\to 0+}\int_{y\varepsilon}^{yA}\frac{\sin s}{s}ds = 2i\phi(0)\int_{0}^{yA}\frac{\sin s}{s}ds, \end{aligned} \quad (1.8.29)$$

since $\frac{\sin s}{s}$ has a removable discontinuity at $s = 0$, i.e. $\frac{\sin s}{s}$ is continuous at $s = 0$.
Similarly, for fixed $y > 0$,

$$\lim_{\varepsilon\to 0+} I_2(\varepsilon) = \int_{0}^{A}[e^{ixy}\psi(x) + e^{-ixy}\psi(-x)]dx, \quad (1.8.30)$$

since the integrand is continuous at $x = 0$. Then, from (1.8.29) and (1.8.30), $\langle e^{ixy}(\text{c.p.v.}\ \frac{1}{x}), \phi\rangle = 2i\phi(0)\int_0^{yA}\frac{\sin s}{s}ds + \int_0^A[e^{ixy}\psi(x) + e^{-ixy}\psi(-x)]dx$. But $\lim_{y\to\infty}\int_0^{yA}\frac{\sin s}{s}ds = \int_0^\infty\frac{\sin s}{s}ds = \frac{\pi}{2}$ and $\lim_{y\to\infty}\int_0^A e^{ixy}\psi(x)dx = 0$ by the Riemann–Lebesgue theorem. Hence, $\lim_{y\to\infty}\langle e^{ixy}(\text{c.p.v.}\ \frac{1}{x}), \phi\rangle = 2i\phi(0).\frac{\pi}{2} + 0 = i\pi\phi(0) = \langle i\pi\delta, \phi\rangle$ $\forall\phi \in \mathcal{D}(\mathbb{R}) \Longrightarrow \lim_{y\to\infty} e^{ixy}(\text{c.p.v.}\ \frac{1}{x}) = i\pi\delta$ in $\mathcal{D}'(\mathbb{R})$. $\square$

1.9 Convergence of series of distributions in $\mathcal{D}'(\Omega)$

Definition 1.9.1. Let $\sum_{n=1}^{\infty} T_n$ be a series of distributions $T_n \in \mathcal{D}'(\Omega) \quad \forall n \in \mathbb{N}$. Let $S_N = \sum_{n=1}^{N} T_n \forall N \in \mathbb{N}$. Then, if the sequence $(S_N)_{N=1}^{\infty}$ converges in $\mathcal{D}'(\Omega)$, the series $\sum_{n=1}^{\infty} T_n$ is said to *converge* in $\mathcal{D}'(\Omega)$ or, equivalently, the series $\sum_{n=1}^{\infty} T_n$ converges in $\mathcal{D}'(\Omega)$ if and only if the number series $\sum_{n=1}^{\infty} \langle T_n, \phi\rangle$ converges in $\mathbb{R}$ (resp. $\mathbb{C}$) $\forall \phi \in \mathcal{D}(\Omega)$.

Then the series $\sum_{n=1}^{\infty} T_n$ is called *summable* in $\mathcal{D}'(\Omega)$. (1.9.1)

- $\sum_{n=1}^{\infty} T_n$ converges in $\mathcal{D}'(\Omega) \iff \exists$ a unique $T \in \mathcal{D}'(\Omega)$ such that $T = \sum_{n=1}^{\infty} T_n$ with

$$\langle T, \phi\rangle = \sum_{n=1}^{\infty} \langle T_n, \phi\rangle \quad \forall \phi \in \mathcal{D}(\Omega). \tag{1.9.2}$$

- $\sum_{i \in I} T_i$ converges in $\mathcal{D}'(\Omega) \iff \sum_{i \in I} \langle T_i, \phi\rangle$ converges in $\mathbb{R}$ (resp. $\mathbb{C}$) $\forall \phi \in \mathcal{D}(\Omega)$. (1.9.3)

For more interesting properties of series of distributions, see Chapter 2.

Example 1.9.1. For $a > 0$, let $\sum_{n=0}^{\infty} a^n \delta_n$ and $\sum_{n=-\infty}^{n=+\infty} a^n \delta_n$ be two given series in $\mathcal{D}'(\mathbb{R})$, δ_n being Dirac distributions (with unit mass/charge/force etc.) concentrated at $x = \pm n \in \mathbb{Z}$. Show that both the series of distributions converge in $\mathcal{D}'(\mathbb{R})$.

Proof. Set $S_N = \sum_{n=0}^{N} a^n \delta_n \ \forall N \in \mathbb{N}$ with $a > 0$. Then (S_N) is a sequence of distributions in $\mathcal{D}'(\mathbb{R})$ with $\langle S_N, \phi\rangle = \langle \sum_{n=0}^{N} a^n \delta_n, \phi\rangle = \sum_{n=0}^{N} a^n \langle \delta_n, \phi\rangle = \sum_{n=0}^{N} a^n \phi(n) \ \forall \phi \in \mathcal{D}(\mathbb{R})$, $\forall N \in \mathbb{N}$. Now, $\forall$ fixed $\phi \in \mathcal{D}(\mathbb{R})$, $\exists n_0 \in \mathbb{N}$ such that $\operatorname{supp}(\phi) \subset [-n_0, n_0] \implies \forall n > n_0$, $\phi(n) = 0 \implies \lim_{N\to\infty} \langle S_N, \phi\rangle = \sum_{n=0}^{n_0} a^n \phi(n) \in \mathbb{R} \implies (S_N)$ converges to a distribution $S \in \mathcal{D}'(\mathbb{R})$ as $N \to \infty$ $\forall a > 0$, and $\sum_{n=0}^{\infty} a^n \delta_n$ converges with $S = \sum_{n=0}^{\infty} a^n \delta_n \ \forall a > 0$. Similarly, setting $T_N = \sum_{n=-N}^{n=+N} a^n \delta_n$ and proceeding in the same way, we can show that $\lim_{N\to\infty} \langle T_N, \phi\rangle = \sum_{n=-n_0}^{n=n_0} a^n \phi(n) \in \mathbb{R} \ \forall$ fixed $\phi \in \mathcal{D}(\mathbb{R})$ with $\operatorname{supp}(\phi) \subset [-n_0, n_0] \implies (T_N)$ converges to a distribution $T \in \mathcal{D}'(\mathbb{R})$ $\forall a > 0$ as $N \to \infty \implies \sum_{n=-\infty}^{n=+\infty} a^n \delta_n$ converges with $T = \sum_{n=-\infty}^{n=+\infty} a^n \delta_n \ \forall a > 0$. □

Remark 1.9.1. More interesting problems will be discussed in Chapter 2.

1.10 Images of distributions due to change of variables, homogeneous, invariant, spherically symmetric, constant distributions

Let $\mathbf{F} : \mathbb{R}^n \to \mathbb{R}^n$ be an invertible, infinitely differentiable mapping from $\mathbb{R}^n$ onto itself defined by:

$$\forall \mathbf{x} \in \mathbb{R}^n, \quad \boldsymbol{\xi} = \mathbf{F}(\mathbf{x}) \in \mathbb{R}^n, \tag{1.10.1}$$

which defines the bijective correspondence between the variables $x_1, x_2, \dots, x_n$ and the variables $\xi_1, \xi_2, \dots, \xi_n$, i.e. $\xi_i = \xi_i(x_1, x_2, \dots, x_n) = f_i(x_1, x_2, \dots, x_n)$, $1 \le i \le n$, with

$$\mathbf{x} = (x_1, x_2, \dots, x_n), \quad \boldsymbol{\xi} = (\xi_1, \xi_2, \dots, \xi_n), \quad \mathbf{F} = (f_1, f_2, \dots, f_n),$$
$$f_i \in C^\infty(\mathbb{R}^n) \quad \forall i. \tag{1.10.2}$$

Let $\mathbf{F}^{-1} : \mathbb{R}^n \to \mathbb{R}^n$ also be an infinitely differentiable mapping from $\mathbb{R}^n$ onto itself defined by: $\forall \boldsymbol{\xi} \in \mathbb{R}^n$, $\mathbf{x} = \mathbf{F}^{-1}(\boldsymbol{\xi}) \in \mathbb{R}^n$, i.e. $x_i = x_i(\xi_1, \dots, \xi_n) = g_i(\xi_1, \dots, \xi_n), 1 \le i \le n, \mathbf{F}^{-1} = (g_1, g_2, \dots g_n)$, $g_i \in C^\infty(\mathbb{R}^n)\ \forall i$.

In other words, $\mathbf{F}$ is a C^∞-diffeomorphism from $\mathbb{R}^n$ onto itself and

$$J_{\mathbf{F}}(\mathbf{x}) = \det\left(\left(\frac{\partial \xi_i}{\partial x_j}(\mathbf{x})\right)_{1\le i,j\le n}\right) = \det\left(\left(\frac{\partial f_i}{\partial x_j}(\mathbf{x})\right)_{1\le i,j\le n}\right) \tag{1.10.3}$$

is the Jacobian of $\mathbf{F}$ such that $J_{\mathbf{F}}(\mathbf{x}) \neq 0$, and the Jacobian of $\mathbf{F}^{-1}$ is

$$J_{\mathbf{F}^{-1}}(\boldsymbol{\xi}) = \det\left(\left(\frac{\partial x_i}{\partial \xi_j}(\boldsymbol{\xi})\right)_{1\le i,j\le n}\right) = \frac{1}{J_{\mathbf{F}}} \neq 0. \tag{1.10.4}$$

Let $f(\mathbf{x})$ be a function *locally integrable* on $\mathbb{R}^n$ in variables $x_1, x_2, \dots, x_n$ defining the corresponding (regular) distribution in $\mathcal{D}'(\mathbb{R}^n)$ by:

$$\langle f(\mathbf{x}), \phi(\mathbf{x})\rangle = \int_{\mathbb{R}^n} f(\mathbf{x})\phi(\mathbf{x})d\mathbf{x} \quad \forall \phi \in \mathcal{D}(\mathbb{R}^n). \tag{1.10.5}$$

Now, under change of variables, $\boldsymbol{\xi} = \mathbf{F}(\mathbf{x})$, $\mathbf{x} = \mathbf{F}^{-1}(\boldsymbol{\xi})$ defined in (1.10.1)–(1.10.4), we have

$$\int_{\mathbb{R}^n} f(\mathbf{x})\phi(\mathbf{x})d\mathbf{x} = \int_{\mathbb{R}^n} f(\mathbf{F}^{-1}(\boldsymbol{\xi}))\phi(\mathbf{F}^{-1}(\xi))|J_{\mathbf{F}^{-1}}(\boldsymbol{\xi})|d\boldsymbol{\xi}, \tag{1.10.6}$$

which we can rewrite in distribution notation as follows:

$$\langle f(\mathbf{x}), \phi(\mathbf{x})\rangle = \langle g(\boldsymbol{\xi}), \psi(\boldsymbol{\xi})\rangle \quad \forall \phi \in \mathcal{D}(\mathbb{R}^n), \tag{1.10.7}$$

where $\psi(\boldsymbol{\xi}) = \phi \circ \mathbf{F}^{-1}(\boldsymbol{\xi})|J_{\mathbf{F}^{-1}(\boldsymbol{\xi})}|$. Then, (1.10.7) defines the function g:

$$g(\boldsymbol{\xi}) = (\mathbf{F}f)(\boldsymbol{\xi}) = f(\mathbf{F}^{-1}(\boldsymbol{\xi})), \tag{1.10.8}$$

which corresponds to the function $f(\mathbf{x})$ in new variables $\xi_1, \xi_2, \dots, \xi_n$, i.e. g is the *image* of f under $\mathbf{F}$.

Now, we will extend this definition (1.10.7)–(1.10.8) to the case of an arbitrary distribution $T \in \mathcal{D}'(\mathbb{R}^n)$. Let $S, T \in \mathcal{D}'(\mathbb{R}^n)$ be two distributions on $\mathbb{R}^n$ in variables $\xi_1, \xi_2, \dots, \xi_n$ and $x_1, x_2, \dots, x_n$ respectively. By abuse of notations, we will write $T(\mathbf{x})$ (resp. $S(\boldsymbol{\xi})$) to denote that T (resp. S) is associated with the variables $x_1, \dots, x_n$ (resp. $\xi_1, \dots, \xi_n$). $T(\mathbf{x})$ (resp. $S(\boldsymbol{\xi})$) must *not* be understood as the value of T (resp. S) at the point $\mathbf{x}$ (resp. $\boldsymbol{\xi}$), since distributions cannot have point-values. An alternative notation could have been $T_{\mathbf{x}}$ (resp. $S_{\boldsymbol{\xi}}$) to indicate the corresponding variables. (1.10.9)

Definition 1.10.1. Let $T(\mathbf{x}) \in \mathcal{D}'(\mathbb{R}^n)$ be a distribution on $\mathbb{R}^n$ in variables $x_1, x_2, \dots, x_n$, and $\mathbf{F}$ be the invertible C^∞-diffeomorphism on $\mathbb{R}^n$ defined in (1.10.1)–(1.10.4). Then the corresponding distribution $S(\boldsymbol{\xi}) = (\mathbf{F}T)(\boldsymbol{\xi})$ in the new variables $\xi_1, \dots, \xi_n$ is defined by the continuous linear functional on $\mathcal{D}(\mathbb{R}^n)$: $\forall \psi, \phi \in \mathcal{D}(\mathbb{R}^n)$,

$$\begin{aligned}\langle S(\boldsymbol{\xi}), \psi(\boldsymbol{\xi})\rangle &= \langle (\mathbf{F}T)(\boldsymbol{\xi}), \psi(\boldsymbol{\xi})\rangle = \langle T(\mathbf{F}^{-1}(\boldsymbol{\xi})), \psi(\boldsymbol{\xi})\rangle \\ &= \langle T(\mathbf{x}), \psi(\mathbf{F}\,(\mathbf{x}))|J_{\mathbf{F}}(\mathbf{x})|\rangle = \langle T(\mathbf{x}), \phi(\mathbf{x})\rangle \end{aligned} \tag{1.10.10}$$

with

$$(\mathbf{F}T)(\boldsymbol{\xi}) = (T \circ \mathbf{F}^{-1})(\boldsymbol{\xi}) = T(\mathbf{F}^{-1}(\boldsymbol{\xi})), \tag{1.10.11}$$

$$\phi(\mathbf{x}) = \psi(\mathbf{F}(\mathbf{x}))|J_{\mathbf{F}}(\mathbf{x})|, \; \psi(\boldsymbol{\xi}) = \phi(\mathbf{F}^{-1}(\boldsymbol{\xi}))|J_{\mathbf{F}^{-1}}(\boldsymbol{\xi})|. \tag{1.10.12}$$

Hence, $S = \mathbf{F}T$ is called the *image* of T under $\mathbf{F}$.

The mapping $\phi \in \mathcal{D}(\mathbb{R}^n) \mapsto \phi \circ \mathbf{F}^{-1}|J_{\mathbf{F}^{-1}}| = \psi \in \mathcal{D}(\mathbb{R}^n)$ (resp. $\psi \in \mathcal{D}(\mathbb{R}^n) \mapsto \psi \circ \mathbf{F}|J_{\mathbf{F}}| = \phi \in \mathcal{D}(\mathbb{R}^n)$) is linear and continuous from $\mathcal{D}(\mathbb{R}^n)$ into $\mathcal{D}(\mathbb{R}^n)$. Hence, $\mathbf{F}T = T \circ \mathbf{F}^{-1}$ defined by (1.10.11) is linear and continuous on $\mathcal{D}(\mathbb{R}^n)$ and, consequently, a distribution on $\mathbb{R}^n$.

Example 1.10.1. Under the change of variable $\boldsymbol{\xi} = \mathbf{F}(\mathbf{x})$ in (1.10.1)–(1.10.4), Dirac distribution $\delta = \delta_{\mathbf{0}} = \delta(\mathbf{x})$ with mass/charge/force etc. concentrated at $\mathbf{0}$ is transformed into

$$\mathbf{F}\delta = |J_{\mathbf{F}}(\mathbf{0})|\delta_{\mathbf{F}(\mathbf{0})}, \tag{1.10.13}$$

where $\delta_{\mathbf{F}(\mathbf{0})} = \delta(\boldsymbol{\xi} - \mathbf{F}(\mathbf{0}))$ is the Dirac distribution with mass/charge/force etc. concentrated at $\mathbf{F}(\mathbf{0})$, $|J_{\mathbf{F}}(\mathbf{0})| = |J_{\mathbf{F}}(\mathbf{x})|_{\mathbf{x}=\mathbf{0}}$ = the absolute value of the Jacobian of $\mathbf{F}$ at $\mathbf{x} = \mathbf{0}$.

In fact,

$$\begin{aligned}\langle \mathbf{F}\delta, \psi\rangle &= \langle \delta(\mathbf{F}^{-1}(\boldsymbol{\xi}), \psi(\boldsymbol{\xi})\rangle = \langle \delta(\mathbf{x}), \psi(\mathbf{F}(\mathbf{x}))|J_F(\mathbf{x})|\rangle = \psi(\mathbf{F}(\mathbf{0}))|J_F(\mathbf{0})| \\ &= \langle \delta_{\mathbf{F}(\mathbf{0})}, \psi(\boldsymbol{\xi})\rangle|J_F(\mathbf{0})| = \langle |J_F(\mathbf{0})|\delta_{\mathbf{F}(\mathbf{0})}, \psi\rangle \quad \forall \psi \in \mathcal{D}(\mathbb{R}^n) \\ \Longrightarrow \quad & \mathbf{F}\delta = |J_F(\mathbf{0})|\delta_{\mathbf{F}(\mathbf{0})}.\end{aligned}$$

Change of variables defined by $F(x, y) = x^2 + y^2$, where F is not invertible

Let $F : \mathbb{R}^2 \to \mathbb{R}$ be defined by

$$F(x, y) = x^2 + y^2 = \xi \in \mathbb{R} \quad \forall (x, y) \in \mathbb{R}^2. \tag{1.10.14}$$

$\forall T \in \mathcal{D}'(\mathbb{R}^2)$, let $FT \in \mathcal{D}'(\mathbb{R})$ be the image of T under F defined by

$$\langle FT, \phi\rangle = \langle T, \phi \circ F\rangle = \langle T, \phi(F(x, y))\rangle \quad \forall \phi \in \mathcal{D}(\mathbb{R}). \tag{1.10.15}$$

Example 1.10.2. Under the change of variables defined by (1.10.14), let $FT \in \mathcal{D}'(\mathbb{R})$ be the image of $T \in \mathcal{D}'(\mathbb{R}^2)$ defined by (1.10.15). Then find

1. $F\delta_{(a,b)}$;
2. FH;
3. $F1$;

where $\delta_{(a,b)} \in \mathcal{D}'(\mathbb{R}^2)$ is the Dirac distribution with mass/charge/force etc. concentrated at $(a, b) \in \mathbb{R}^2$:

$$\langle \delta_{(a,b)}, \psi(x, y)\rangle = \psi(a, b) \quad \forall \psi \in \mathcal{D}(\mathbb{R}^2); \tag{1.10.16}$$

$H = H(x, y) = 1$ for $x > 0$ and $y > 0$ and $H(x, y) = 0$ otherwise in $\mathbb{R}^2$; $1 = 1(x, y)\ \forall (x, y) \in \mathbb{R}^2$. (1.10.17)

Solution.

1. From (1.10.15),

$$\begin{aligned}\langle F\delta_{(a,b)}, \phi(\xi)\rangle &= \langle \delta_{(a,b)}, \phi \circ F(x, y)\rangle = \langle \delta_{(a,b)}, \phi(F(x, y))\rangle \\ &= \langle \delta_{(a,b)}, \phi(x^2 + y^2)\rangle = \phi(a^2 + b^2) = \langle \delta_{a^2+b^2}, \phi(\xi)\rangle \\ \Longrightarrow \quad & F\delta_{(a,b)} = \delta_{a^2+b^2} \text{ in } \mathcal{D}'(\mathbb{R}), \quad \forall \phi \in \mathcal{D}(\mathbb{R}).\end{aligned} \tag{1.10.18}$$

2. $\langle FH, \phi\rangle = \langle H(x, y), \phi \circ F(x, y)\rangle = \langle H(x, y), \phi(F(x, y))\rangle$

$$= \int_0^\infty \int_0^\infty \phi(x^2 + y^2)dxdy = \int_0^\infty \int_0^{\frac{\pi}{2}} \phi(r^2)rdrd\theta$$

with change of variables:

$$x = r\cos\theta, \quad y = r\sin\theta, \quad J = r, \quad x^2 + y^2 = r^2, \tag{1.10.19}$$

$$\Longrightarrow \quad \langle FH, \phi\rangle = \int_0^\infty \phi(t)\frac{dt}{2}\int_0^{\frac{\pi}{2}} d\theta = \frac{\pi}{4}\int_0^\infty \phi(t)dt \quad (\text{setting } t = r^2)$$

$$= \frac{\pi}{4}\int_{-\infty}^{\infty} H(x)\phi(x)dx = \langle \frac{\pi}{4}H(x), \phi\rangle \quad \forall \phi \in \mathcal{D}(\mathbb{R})$$

$\Longrightarrow FH = \frac{\pi}{4}H(x)$ in $\mathcal{D}'(\mathbb{R})(H(x)$ is the Heaviside function).

3. $$\langle F1, \phi\rangle = \langle 1, \phi \circ F(x, y)\rangle = \int_{-\infty}^{\infty}\int_{-\infty}^{\infty} 1 \cdot \phi(x^2 + y^2)dxdy$$

$$= \int_0^\infty \int_0^{2\pi} \phi(r^2)rdrd\theta = 2\pi\int_0^\infty \phi(t)\frac{dt}{2} = \pi\int_{-\infty}^{\infty} H(x)\phi(x)dx$$

$$\Longrightarrow \langle F1, \phi\rangle = \langle \pi H(x), \phi\rangle \ \forall \phi \in \mathcal{D}(\mathbb{R}) \Longrightarrow F1 = \pi H(x) \text{ in } \mathcal{D}'(\mathbb{R}). \tag{1.10.20}$$

Invertible affine transformation of variables

In particular, let $\mathbf{F} : \mathbb{R}^n \to \mathbb{R}^n$ be an invertible affine mapping satisfying (1.10.1)–(1.10.4) such that

$$\mathbf{F}(\mathbf{x}) = A\mathbf{x} + \mathbf{b} = \boldsymbol{\xi} \quad \forall \mathbf{x} \in \mathbb{R}^n, \tag{1.10.21}$$

where $A = (a_{ij})_{1\le i,j\le n}$ is a *non-singular* square matrix of order n, and $\mathbf{b} \in \mathbb{R}^n$ is a *fixed* vector.

Then,

$$\mathbf{F}^{-1}(\boldsymbol{\xi}) = A^{-1}(\boldsymbol{\xi} - \mathbf{b}) = \mathbf{x} \quad \forall \boldsymbol{\xi} \in \mathbb{R}^n,$$

$$J_{\mathbf{F}} = |A| \neq 0, \quad J_{\mathbf{F}^{-1}} = \frac{1}{|A^{-1}|}, \tag{1.10.22}$$

and $\mathbf{F}$ is a C^∞*-diffeomorphism* from $\mathbb{R}^n$ onto itself. Hence, for $T(\mathbf{x}) \in \mathcal{D}'(\mathbb{R}^n)$, the image $S = \mathbf{F}T$ of T under $\mathbf{F}$ is defined by: $S(\boldsymbol{\xi}) = \mathbf{F}T(\boldsymbol{\xi}) = T(\mathbf{F}^{-1}(\boldsymbol{\xi})) = T(A^{-1}(\boldsymbol{\xi} - \mathbf{b}))$ such that, $\forall \phi, \psi \in \mathcal{D}(\mathbb{R}^n)$ related by (1.10.12),

$$\langle S(\boldsymbol{\xi}), \psi(\boldsymbol{\xi})\rangle = \langle \mathbf{F}T(\boldsymbol{\xi}), \psi(\boldsymbol{\xi})\rangle = \langle T(\mathbf{A}^{-1}(\boldsymbol{\xi} - \boldsymbol{b})), \psi(\boldsymbol{\xi})\rangle$$

$$= \langle T(\mathbf{x}), \psi(A\mathbf{x} + \mathbf{b})|\det(A)|\rangle = \langle T(\mathbf{x}), \phi(\mathbf{x})\rangle \tag{1.10.23}$$

with

$$\phi(\mathbf{x}) = \psi \circ \mathbf{F}(\mathbf{x})|J_{\mathbf{F}}| = \psi(A\mathbf{x} + \mathbf{b})|\det(A)|$$

$$\psi(\boldsymbol{\xi}) = \phi \circ \mathbf{F}^{-1}(\boldsymbol{\xi})|J_{\mathbf{F}^{-1}}| = \phi(A^{-1}(\boldsymbol{\xi} - \mathbf{b})) \cdot \frac{1}{|\det(A)|}. \tag{1.10.24}$$

Example 1.10.3. Let $\mathbf{F}(\mathbf{x}) = A\mathbf{x} = \mathbf{y} \in \mathbb{R}^n$ with $\mathbf{b} = \mathbf{0}$, $\mathbf{x} \in \mathbb{R}^n$, $A = (a_{ij})_{1 \le i,j \le n}$, $\det(A) \neq 0$ and $f \in L^1(\mathbb{R}^n)$. Then the image $\mathbf{F}f = Af$ of f under $\mathbf{F}$ is defined by: $\forall \psi \in \mathcal{D}(\mathbb{R}^n)$,

$$\begin{aligned}\langle Af(\mathbf{y}), \psi(\mathbf{y})\rangle &= \langle AT_f(\mathbf{y}), \psi(\mathbf{y})\rangle = \langle T_f(A^{-1}(\mathbf{y}), \psi(\mathbf{y})\rangle \\ &= \int_{\mathbb{R}^n} f(A^{-1}(\mathbf{y}))\psi(\mathbf{y})d\mathbf{y} = \int_{\mathbb{R}^n} f(\mathbf{x})\psi(A\mathbf{x})|\det(A)|d\mathbf{x} \\ &= \langle f(\mathbf{x}), \phi(\mathbf{x})\rangle \quad \text{with } \phi(\mathbf{x}) = \psi(A\mathbf{x})|\det(A)|. \end{aligned} \tag{1.10.25}$$

Important examples are *translation*, *rotation*, *reflection* at the origin, etc., which can be retrieved as particular cases of the affine transformation in (1.10.21)–(1.10.24), as shown below.

Translation $\mathbf{F}(\mathbf{x}) = \mathbf{x} + \mathbf{b} = \boldsymbol{\xi}$ with $A = I$ (identity matrix), $\mathbf{b} \neq \mathbf{0}$, $\mathbf{F}^{-1}(\boldsymbol{\xi}) = \boldsymbol{\xi} - \mathbf{b} = \mathbf{x}$ with $A^{-1} = I$, $|J_{\mathbf{F}}| = |I| = |J_{\mathbf{F}^{-1}}| = 1$ in (1.10.21)–(1.10.24).

Then $S = \mathbf{F}T = T$ with $S(\boldsymbol{\xi}) = (\mathbf{F}T)(\boldsymbol{\xi}) = T(\mathbf{F}^{-1}(\boldsymbol{\xi})) = T(\boldsymbol{\xi} - \mathbf{b})$ such that $\forall \phi, \psi \in \mathcal{D}(\mathbb{R}^n)$ satisfying (1.10.24):

$$\begin{aligned}\langle \mathbf{F}T, \psi\rangle &= \langle (\mathbf{F}T)(\boldsymbol{\xi}), \psi(\boldsymbol{\xi})\rangle = \langle T(\boldsymbol{\xi} - \mathbf{b}), \psi(\boldsymbol{\xi})\rangle \\ &= |J_{\mathbf{F}}|\langle T(\mathbf{x}), \psi(\mathbf{x} + \mathbf{b})\rangle = \langle T(\mathbf{x}), \psi(\mathbf{x} + \mathbf{b})\rangle = \langle T(\mathbf{x}), \phi(\mathbf{x})\rangle. \end{aligned} \tag{1.10.26}$$

Example 1.10.4. $\forall \mathbf{b} \in \mathbb{R}^n$, $\forall T \in \mathcal{D}'(\mathbb{R}^n)$, $\tau_{\mathbf{b}}T = \mathbf{F}T$ is called the *translated distribution* of T by the vector $\mathbf{b}$, and is defined by: $(\tau_{\mathbf{b}}T)(\boldsymbol{\xi}) = T(\boldsymbol{\xi} - \mathbf{b})$ $\forall \boldsymbol{\xi} \in \mathbb{R}^n$ such that

$$\langle \tau_{\mathbf{b}}T, \psi\rangle = \langle T(\mathbf{x}), \psi(\mathbf{x} + \mathbf{b})\rangle \quad \forall \psi \in \mathcal{D}(\mathbb{R}^n). \tag{1.10.27}$$

Example 1.10.5. For Dirac distribution $\delta = \delta(\mathbf{x})$ (with mass/charge/force etc.) concentrated at $\mathbf{x} = \mathbf{0} \in \mathbb{R}^n$, $\tau_{\mathbf{b}}\delta = \delta_{\mathbf{b}} = \delta(\boldsymbol{\xi} - \mathbf{b})$, which is the Dirac distribution with (mass/charge/force etc.) concentrated at $\boldsymbol{\xi} = \mathbf{b}$. In fact, from (1.10.27),

$$\langle \tau_{\mathbf{b}}\delta, \psi\rangle = \langle \delta(\mathbf{x}), \psi(\mathbf{x} + \mathbf{b})\rangle = \psi(\mathbf{b}) = \langle \delta_{\mathbf{b}}, \psi\rangle \quad \forall \psi \in \mathcal{D}(\mathbb{R}^n)$$

$\Longrightarrow \tau_{\mathbf{b}}\delta = \delta_{\mathbf{b}}$ in $\mathcal{D}'(\mathbb{R}^n)$.

Remark 1.10.1. Translation is often written as $(\tau_{\mathbf{b}}T)(\mathbf{x}) = T(\mathbf{x} - \mathbf{b})$ instead of introducing another variable $\boldsymbol{\xi}$.

Example 1.10.6. For $T = T_f = f \in L^1_{\text{loc}}(\mathbb{R}^n)$ and the translation defined by (1.10.26),

$$\int_{\mathbb{R}^n} f(\boldsymbol{\xi} - \mathbf{b})\psi(\boldsymbol{\xi})d\boldsymbol{\xi} = \int_{\mathbb{R}^n} f(\mathbf{x})\psi(\mathbf{x} + \mathbf{b})d\mathbf{x} \quad \forall \psi \in \mathcal{D}(\mathbb{R}^n). \tag{1.10.28}$$

Rotation $\mathbf{F}(\mathbf{x}) = A\mathbf{x} = \boldsymbol{\xi}, \mathbf{F}^{-1}(\boldsymbol{\xi}) = A^{-1}\boldsymbol{\xi} = A^t\boldsymbol{\xi} = \mathbf{x}$ with $\mathbf{b} = \mathbf{0}$, where A is an *orthogonal* matrix with

$$A^{-1} = A^t; \quad |J_{\mathbf{F}}| = |A| = |A^{-1}| = 1; \tag{1.10.29}$$

(using (1.10.24))

$$\phi(\mathbf{x}) = \psi(A\mathbf{x}), \quad \psi(\boldsymbol{\xi}) = \phi(A^t\boldsymbol{\xi}) \quad \forall\phi, \psi \in \mathcal{D}(\mathbb{R}^n). \tag{1.10.30}$$

Then the image $\mathbf{F}T$ of $T \in \mathcal{D}'(\mathbb{R}^n)$ under rotation is defined by: $(\mathbf{F}T)(\boldsymbol{\xi}) = T(A^t\boldsymbol{\xi})$ such that $\forall\phi, \psi \in \mathcal{D}(\mathbb{R}^n)$ satisfying (1.10.30),

$$\langle \mathbf{F}T, \psi\rangle = \langle T(A^t\boldsymbol{\xi}), \psi(\boldsymbol{\xi})\rangle = \langle T(\mathbf{x}), \psi(A\mathbf{x})\rangle |\det(A)| = \langle T(\mathbf{x}), \phi(\mathbf{x})\rangle. \tag{1.10.31}$$

Example 1.10.7. For $T = T_f = f \in L^1_{\text{loc}}(\mathbb{R}^n)$ and the rotation defined by (1.10.30),

$$\int_{\mathbb{R}^n} f(A^t\boldsymbol{\xi})\psi(\boldsymbol{\xi})d\boldsymbol{\xi} = \int_{\mathbb{R}^n} f(\mathbf{x})\psi(A\mathbf{x})d\mathbf{x} = \int_{\mathbb{R}^n} f(\mathbf{x})\phi(\mathbf{x})d\mathbf{x}. \tag{1.10.32}$$

Example 1.10.8. For Dirac distribution $\delta = \delta_{\mathbf{0}} = \delta(\mathbf{x})$ with mass/charge/force etc. concentrated at $\mathbf{x} = \mathbf{0}$, $\mathbf{F}\delta = \delta$ under rotation in $\mathbb{R}^n$ (see *invariance* later). $\forall\psi \in \mathcal{D}(\mathbb{R}^n)$,

$$\begin{aligned}\langle \mathbf{F}\delta, \psi\rangle &= \langle \delta(A^t\boldsymbol{\xi}), \psi(\boldsymbol{\xi})\rangle = \langle \delta(\mathbf{x}), \psi(A\mathbf{x})\rangle = \psi(A\mathbf{0}) \quad \text{(by definition of } \delta)\\ &= \psi(\mathbf{0}) = \langle \delta(\boldsymbol{\xi}), \psi(\boldsymbol{\xi})\rangle = \langle \delta, \psi\rangle \quad \Longleftrightarrow \quad \mathbf{F}\delta = \delta.\end{aligned}$$

Reflection at the origin $\mathbf{F}(\mathbf{x}) = -\mathbf{x} = \boldsymbol{\xi}, \mathbf{F}^{-1}(\boldsymbol{\xi}) = -\boldsymbol{\xi} = \mathbf{x}$ with $\mathbf{b} = \mathbf{0}$,

$$A = A^{-1} = -I, \quad |J_{\mathbf{F}}| = |J_{\mathbf{F}^{-1}}| = |\det(A)| = 1, \tag{1.10.33}$$

(from (1.10.24))

$$\phi(\mathbf{x}) = \psi(-\mathbf{x}), \quad \psi(\boldsymbol{\xi}) = \phi(-\boldsymbol{\xi}) \quad \forall\phi, \psi \in \mathcal{D}(\mathbb{R}^n). \tag{1.10.34}$$

Then the image $\mathbf{F}T$ of $T \in \mathcal{D}'(\mathbb{R}^n)$ under reflection at the origin is defined by $(\mathbf{F}T)(\boldsymbol{\xi}) = T(\mathbf{F}^{-1}(\boldsymbol{\xi}) = T(-\boldsymbol{\xi})$ such that $\forall\phi, \psi \in \mathcal{D}(\mathbb{R}^n)$ satisfying (1.10.34),

$$\begin{aligned}\langle \mathbf{F}T, \psi\rangle &= \langle T(-\boldsymbol{\xi}), \psi(\boldsymbol{\xi})\rangle = \langle T(\mathbf{x}), \psi(-\mathbf{x})\rangle |\det(A)|\\ &= \langle T(\mathbf{x}), \psi(-\mathbf{x})\rangle = \langle T(\mathbf{x}), \phi(\mathbf{x})\rangle.\end{aligned} \tag{1.10.35}$$

Example 1.10.9. For Dirac distribution $\delta = \delta_{\mathbf{0}} = \delta(\mathbf{x})$, $(\mathbf{F}\delta)(\boldsymbol{\xi}) = \delta(-\mathbf{x}) = \delta(\mathbf{x})$. In fact

$$\begin{aligned}\langle \mathbf{F}\delta, \psi\rangle &= \langle \delta(-\boldsymbol{\xi}), \psi(\boldsymbol{\xi})\rangle = \langle \delta(\mathbf{x}), \psi(-\mathbf{x})\rangle |\det(A)| = \langle \delta(\mathbf{x}), \psi(-\mathbf{x})\rangle\\ &= \psi(\mathbf{0}) = \langle \delta, \psi(\boldsymbol{\xi})\rangle = \langle \delta, \psi\rangle \quad \forall\psi \in \mathcal{D}(\mathbb{R}^n)\end{aligned}$$

$\Longleftrightarrow$ $\mathbf{F}\delta = \delta$, i.e. $\delta(-\boldsymbol{\xi}) = \delta(\boldsymbol{\xi})$, which can be written as $\delta(-\mathbf{x}) = \delta(\mathbf{x})$ (see (1.1.1)).

Example 1.10.10. For $T = T_f = f \in L^1_{\text{loc}}(\mathbb{R}^n)$ and reflection at the origin, $\forall\phi,\ \psi \in \mathcal{D}(\mathbb{R}^n)$ satisfying (1.10.34), $\int_{\mathbb{R}^n} f(-\boldsymbol{\xi})\psi(\boldsymbol{\xi})d\boldsymbol{\xi} = \int_{\mathbb{R}^n} f(\mathbf{x})\psi(-\mathbf{x})d\mathbf{x} = \int_{\mathbb{R}^n} f(\mathbf{x})\phi(\mathbf{x})d\mathbf{x}$.

Homothetic transformation $\mathbf{F}(\mathbf{x}) = \alpha\mathbf{x} = \boldsymbol{\xi}$ with $\alpha > 0, \mathbf{b} = \mathbf{0}$;

$$\mathbf{F}^{-1}(\boldsymbol{\xi}) = \frac{1}{\alpha}\boldsymbol{\xi} = \mathbf{x}; \quad A = \alpha I, \quad A^{-1} = \frac{1}{\alpha}I,$$
$$J_{\mathbf{F}} = |A| = \alpha^n, \quad |A^{-1}| = \frac{1}{\alpha^n}. \tag{1.10.36}$$

From (1.10.24),

$$\phi(\mathbf{x}) = \psi(\alpha\mathbf{x}) \cdot |A| = \alpha^n\psi(\alpha\mathbf{x}); \quad \psi(\boldsymbol{\xi}) = \phi\left(\frac{1}{\alpha}\boldsymbol{\xi}\right) \cdot \frac{1}{\alpha^n}. \tag{1.10.37}$$

Then the image $\mathbf{F}T$ of T under homothetic transformation with $\alpha > 0$ is given by: $(\mathbf{F}T)(\boldsymbol{\xi}) = T(\frac{1}{\alpha}\boldsymbol{\xi})$ such that $\forall\phi,\ \psi \in \mathcal{D}(\mathbb{R}^n)$ satisfying (1.10.37),

$$\begin{aligned}\langle \mathbf{F}T, \psi\rangle &= \left\langle T\left(\frac{1}{\alpha}\boldsymbol{\xi}\right), \psi(\boldsymbol{\xi})\right\rangle = \langle T(\mathbf{x}), \psi(\mathbf{F}(\mathbf{x}))\rangle |J_{\mathbf{F}}(\mathbf{x})| = \alpha^n \langle T(\mathbf{x}), \psi(\alpha\mathbf{x})\rangle \\ &= \langle T(\mathbf{x}), \phi(\mathbf{x})\rangle = \langle T, \phi\rangle.\end{aligned} \tag{1.10.38}$$

Homogeneous distributions A distribution $T \in \mathcal{D}'(\mathbb{R}^n)$ is called *homogeneous of degree* $d \in \mathbb{R}$ if and only if, $\forall\lambda > 0$, $T(\lambda\mathbf{x}) = \lambda^d T(\mathbf{x})$ in $\mathcal{D}'(\mathbb{R}^n)$ such that

$$\langle T(\lambda\mathbf{x}), \phi(\mathbf{x})\rangle = \lambda^d \langle T(\mathbf{x}), \phi(\mathbf{x})\rangle \quad \forall\phi \in \mathcal{D}(\mathbb{R}^n). \tag{1.10.39}$$

Example 1.10.11. For homogeneous T of degree d defined by: $T = T_f = f \in L^1_{\text{loc}}(\mathbb{R}^n)$, (1.10.39) implies that $\int_{\mathbb{R}^n} f(\lambda\mathbf{x})\phi(\mathbf{x})d\mathbf{x} = \lambda^d \int_{\mathbb{R}^n} f(\mathbf{x})\phi(\mathbf{x})d\mathbf{x}\ \forall\phi \in \mathcal{D}(\mathbb{R}^n) \Longrightarrow \int_{\mathbb{R}^n}[f(\lambda\mathbf{x}) - \lambda^d f(\mathbf{x})]\phi(\mathbf{x})d\mathbf{x} = 0\ \forall\phi \in \mathcal{D}(\mathbb{R}^n) \Longrightarrow f(\lambda\mathbf{x}) - \lambda^d f(\mathbf{x}) = 0$ a.e. in $\mathbb{R}^n$ (by Theorem 1.2.3A) $\Longrightarrow f(\lambda\mathbf{x}) = \lambda^d f(\mathbf{x})$ a.e. in $\mathbb{R}^n \Longrightarrow f$ is a homogeneous function of degree d in $L^1_{\text{loc}}(\mathbb{R}^n)$.

Equation (1.10.39) can be rewritten as follows: Using (1.10.23) and (1.10.24),

$$\begin{gathered}\mathbf{F}(\mathbf{x}) = \lambda\mathbf{x} = \boldsymbol{\xi}, \quad \mathbf{F}^{-1}(\boldsymbol{\xi}) = \frac{1}{\lambda}\boldsymbol{\xi} = \mathbf{x}, \quad \lambda > 0, \quad A = \lambda I, \quad A^{-1} = \frac{1}{\lambda}I, \\ J_{\mathbf{F}} = |A| = \lambda^n, \quad |A|^{-1} = \lambda^{-n}, \quad \phi(\mathbf{x}) = \psi(\lambda\mathbf{x}) \cdot \lambda^n, \quad \psi(\boldsymbol{\xi}) = \phi\left(\frac{1}{\lambda}\boldsymbol{\xi}\right)\lambda^{-n}.\end{gathered} \tag{1.10.40}$$

Then $(\mathbf{F}T)(\boldsymbol{\xi}) = T(\mathbf{F}^{-1}(\boldsymbol{\xi})) = T(\frac{1}{\lambda}\boldsymbol{\xi})$ such that, $\forall\phi, \psi \in \mathcal{D}(\mathbb{R}^n)$ satisfying (1.10.40),

$$\langle \mathbf{F}T(\boldsymbol{\xi}), \psi(\boldsymbol{\xi})\rangle = \left\langle T\left(\frac{1}{\lambda}\boldsymbol{\xi}\right), \psi(\boldsymbol{\xi})\right\rangle = \langle T(\mathbf{x}), \psi(\lambda\mathbf{x})\rangle\lambda^n$$
$$\Longrightarrow \left\langle \left(\frac{1}{\lambda}\right)^d T(\boldsymbol{\xi}), \psi(\boldsymbol{\xi})\right\rangle = \lambda^{-d}\langle T(\boldsymbol{\xi}), \psi(\boldsymbol{\xi})\rangle = \lambda^n\langle T(\mathbf{x}), \psi(\lambda\mathbf{x})\rangle.$$

Alternative definition of homogeneous distributions From the previous steps we have: $\langle T(\frac{1}{\lambda}\boldsymbol{\xi}), \psi(\boldsymbol{\xi})\rangle = \lambda^{-d}\langle T(\boldsymbol{\xi}), \psi(\boldsymbol{\xi})\rangle = \lambda^n\langle T(\mathbf{x}), \psi_\lambda(\mathbf{x})\rangle$ with $\psi_\lambda(\mathbf{x}) = \psi(\lambda\mathbf{x}) \Longrightarrow \langle T, \psi_\lambda\rangle = \lambda^{-(d+n)}\langle T, \psi\rangle \quad \forall\psi \in \mathcal{D}(\mathbb{R}^n)$. Hence, an alternative equivalent definition of homogeneous distributions is as follows: A distribution $T \in \mathcal{D}'(\mathbb{R}^n)$ is called *homogeneous of degree 'd'* $\in \mathbb{R}$ if and only if

$$\langle T, \phi_\lambda\rangle = \lambda^{-(d+n)}\langle T, \phi\rangle \quad \forall\phi \in \mathcal{D}(\mathbb{R}^n). \tag{1.10.41}$$

Invariance A distribution $T \in \mathcal{D}'(\mathbb{R}^n)$ is called *invariant* under transformation of variables $\mathbf{F}(\mathbf{x}) = \boldsymbol{\xi}$ defined in (1.10.21)–(1.10.24) if and only if $\mathbf{F}T = T$ with $(\mathbf{F}T)(\boldsymbol{\xi}) = T(\mathbf{F}^{-1}(\boldsymbol{\xi}))$, i.e.

$$\langle \mathbf{F}T, \psi\rangle = \langle T(\mathbf{F}^{-1}(\boldsymbol{\xi})), \psi(\boldsymbol{\xi})\rangle = \langle T, \psi\rangle \quad \forall\psi \in \mathcal{D}(\mathbb{R}^n). \tag{1.10.42}$$

Example 1.10.12. Dirac distribution $\delta = \delta_{\mathbf{0}} = \delta(\mathbf{x})$ is invariant under rotation in $\mathbb{R}^n$, i.e. $\mathbf{F}\delta = \delta$ (see Example 1.10.8).

Spherically symmetric distribution A distribution $T \in \mathcal{D}'(\mathbb{R}^n)$ is called *spherically symmetric* if and only if it is invariant with respect to all rotations. (1.10.43)

Important examples of spherically symmetric distributions are Dirac δ distributions and distributions defined by functions

$$f = f(r) \in L^1_{\text{loc}}(\mathbb{R}^n) \quad \text{with } r^2 = \sum_{i=1}^n x_i^2. \tag{1.10.44}$$

1.10.1 Periodic distributions

Periodic functions on $\mathbb{R}$ A function $\tilde{f} : \mathbb{R} \longrightarrow \mathbb{C}$ is called *periodic with period* $p \in \mathbb{R}$ if and only if $\tilde{f}(x+p) = \tilde{f}(x)\forall x \in \mathbb{R}$, which can be rewritten as $\tau_{-p}\tilde{f} = \tilde{f}$, where $(\tau_{-p}\tilde{f})(x) = \tilde{f}(x-(-p)) = \tilde{f}(x+p)\forall x \in \mathbb{R}$.

We will use $\tilde{f}$ rather than f to denote that it is a periodic function.

$\forall a \in \mathbb{R}$, $]a, a+p[$ is called a *period interval* of periodic $\tilde{f}$ with period p.
$\tilde{f} \in C^k(\mathbb{R})$ is periodic on $\mathbb{R} \Longrightarrow \tilde{f}^{(l)}$ is also periodic on $\mathbb{R}$ for $l \leq k$.
$\tilde{f} \in L^1_{\text{loc}}(\mathbb{R})$ is periodic on $\mathbb{R}$ with period p

$$\Longrightarrow \quad \int_a^{a+p} \tilde{f}(x)dx = \int_b^{b+p} \tilde{f}(x)dx. \tag{1.10.45a}$$

In fact, for $b \neq a$,

$$\int_b^{b+p} \tilde{f}(x)dx = \int_a^{a+p} \tilde{f}(x)dx + \underbrace{\int_{a+p}^{b+p} \tilde{f}(x)dx - \int_a^{b} \tilde{f}(x)dx}_{I}.$$

By change of variable, $x = \eta + p$, $dx = d\eta$,

$$\int_{a+p}^{b+p} \tilde{f}(x)dx = \int_a^{b} \tilde{f}(\eta + p)d\eta = \int_a^{b} \tilde{f}(\eta)d\eta = \int_a^{b} \tilde{f}(x)dx \quad \Longrightarrow \quad I = 0.$$

Periodic functions on $\mathbb{R}^n$ A function $\tilde{f} : \mathbb{R}^n \longrightarrow \mathbb{C}$ is called *periodic* with period $\mathbf{p} \in \mathbb{R}^n$ if and only if $\tilde{f}(\mathbf{x}+\mathbf{p}) = \tilde{f}(\mathbf{x})\ \forall \mathbf{x} \in \mathbb{R}^n$, which can be written as $\tau_{-\mathbf{p}}\tilde{f} = \tilde{f}$, where $(\tau_{-\mathbf{p}}\tilde{f})(\mathbf{x}) = \tilde{f}(\mathbf{x}+\mathbf{p})\ \forall \mathbf{x} \in \mathbb{R}^n$.

Periodic distributions Let $\tilde{f} \in L^1_{\text{loc}}(\mathbb{R}^n)$ be a *periodic* function with period $\mathbf{p} \in \mathbb{R}^n$. Then $\tilde{f} \in \mathcal{D}'(\mathbb{R}^n)$ and $\tau_{\mathbf{p}}\tilde{f} \in \mathcal{D}'(\mathbb{R}^n)$ is a *regular* distribution on $\mathbb{R}^n$ defined by:

$$\begin{aligned}
\langle(\tau_{\mathbf{p}}\tilde{f})(\mathbf{x}), \phi(\mathbf{x})\rangle_{\mathcal{D}'(\mathbb{R}^n)\times\mathcal{D}(\mathbb{R}^n)} &= \langle\tilde{f}(\mathbf{x}-\mathbf{p}), \phi(\mathbf{x})\rangle_{\mathcal{D}'(\mathbb{R}^n)\times\mathcal{D}(\mathbb{R}^n)} \\
&= \int_{\mathbb{R}^n} \tilde{f}(\mathbf{x}-\mathbf{p})\phi(\mathbf{x})d\mathbf{x} = \int_{\mathbb{R}^n} \tilde{f}(\boldsymbol{\eta})\phi(\boldsymbol{\eta}+\mathbf{p})d\boldsymbol{\eta} \\
&= \langle\tilde{f}(\mathbf{x}), \phi(\mathbf{x}+\mathbf{p})\rangle_{\mathcal{D}'(\mathbb{R}^n)\times\mathcal{D}(\mathbb{R}^n)} \quad \forall\phi \in \mathcal{D}(\mathbb{R}^n).
\end{aligned}$$

For periodic $\tilde{f} \in L^1_{\text{loc}}(\mathbb{R}^n)$ with period $\mathbf{p}$, we have $\tau_{\mathbf{p}}\tilde{f} = \tilde{f}$ and $\langle(\tau_{\mathbf{p}}\tilde{f})(\mathbf{x}), \phi(\mathbf{x})\rangle_{\mathcal{D}'(\mathbb{R}^n)\times\mathcal{D}(\mathbb{R}^n)} = \langle\tilde{f}(\mathbf{x}), \phi(\mathbf{x}+\mathbf{p})\rangle_{\mathcal{D}'(\mathbb{R}^n)\times\mathcal{D}(\mathbb{R}^n)}\ \forall\phi \in \mathcal{D}(\mathbb{R}^n)$ or, equivalently, $\langle(\tau_{\mathbf{p}}\tilde{f})(\mathbf{x}), \phi(\mathbf{x})\rangle_{\mathcal{D}'(\mathbb{R}^n)\times\mathcal{D}(\mathbb{R}^n)} = \langle\tilde{f}(\mathbf{x}), \phi(\mathbf{x})\rangle\forall\phi \in \mathcal{D}(\mathbb{R}^n)$, i.e. $\tau_{\mathbf{p}}\tilde{f} = \tilde{f} \in \mathcal{D}'(\mathbb{R}^n)$, any one of which may be used to define periodic distributions on $\mathbb{R}^n$ with period $\mathbf{p}$. Hence:

Definition 1.10.1A. A distribution $T \in \mathcal{D}'(\mathbb{R}^n)$ is called *periodic* with *period* $\mathbf{p} \in \mathbb{R}^n$ if and only if $\tau_{\mathbf{p}}T = T \in \mathcal{D}'(\mathbb{R}^n)$ or, equivalently, $\langle T(\mathbf{x}), \phi(\mathbf{x}+\mathbf{p})\rangle = \langle T(\mathbf{x}), \phi(\mathbf{x})\rangle$, or

$$\langle(\tau_{\mathbf{p}}T)(\mathbf{x}), \phi(\mathbf{x})\rangle = \langle T(\mathbf{x}), (\tau_{-\mathbf{p}}\phi)(\mathbf{x})\rangle = \langle T(\mathbf{x}), \phi(\mathbf{x})\rangle \quad \forall\phi \in \mathcal{D}(\mathbb{R}^n). \tag{1.10.45b}$$

(By abuse of notation, we have written $T(\mathbf{x})$ to show that the variable involved is $\mathbf{x}$, although $T(\mathbf{x})$ is *not* the value of T at $\mathbf{x}$ as distributions T cannot have point values.)

In other words, a distribution $T \in \mathcal{D}'(\mathbb{R}^n)$ *invariant under translations through the vector* $\mathbf{p} = \mathbf{b}$ is called *periodic* with periodic $\mathbf{p}$.

For $n = 1$, $T \in \mathcal{D}'(\mathbb{R})$ is periodic with period p

$$\Longleftrightarrow \quad \langle \tau_p T, \phi\rangle = \langle T, \tau_{-p}\phi\rangle = \langle T, \phi(x+p)\rangle = \langle T, \phi\rangle \forall \phi \in \mathcal{D}(\mathbb{R}) \quad \text{(see (1.10.27))}.$$

For example, for $p = 1$, $T \in \mathcal{D}'(\mathbb{R})$ is periodic with period 1 if and only if

$$\langle \tau_1 T, \phi\rangle = \langle T, \tau_{-1}\phi\rangle = \langle T, \phi(x+1)\rangle = \langle T, \phi\rangle \forall \phi \in \mathcal{D}(\mathbb{R}). \tag{1.10.45c}$$

Constant distribution A distribution $T \in \mathcal{D}'(\mathbb{R}^n)$ is *constant* if and only if T has *every* $\mathbf{p} \in \mathbb{R}^n$ as a period. Alternatively, a distribution T *invariant* with respect to all translations is called *constant*. (1.10.46)

Other equivalent notations for period are T (time period), L (length period), etc. Henceforth, we will use T instead of $p \in \mathbb{R}$ as the period of functions and distributions on $\mathbb{R}$, and then T will *not* be a distribution, for which we will use S or any other convenient notation.

Since periodic functions and distributions on $\mathbb{R}$ are very important in applications and can be given Fourier series representations, we will give further interesting details here.

Alternative definition of periodic distributions on $\mathbb{R}$ First of all, we define a completely new test space $\mathcal{D}(\Gamma)$ on circle Γ, and its dual $\mathcal{D}'(\Gamma)$ as the space of distributions on Γ, which have a one-to-one correspondence with the space of periodic distributions on $\mathbb{R}$ with period $T =$ circumference of Γ; this will be essential later for Fourier series of periodic distributions.

Let $\Gamma = \Gamma(\mathbf{0}; r)$ be a circle in the xy-plane with circumference $2\pi r$ and centre at the origin $\mathbf{0}$. Let $A = (r, 0)$ be the point of intersection of Γ and the x-axis, from which the arc length coordinate s of any point $P \in \Gamma$ is measured in the *anti-clockwise* direction as *positive* sense of orientation of Γ. For $r = 1$, $s =$ angle θ in radians.

For functions $\Phi : \Gamma \longrightarrow \mathbb{C}$ on circle Γ with $\Phi = \Phi(s)$ (we have used capital Φ instead of $\phi \in \mathcal{D}(\mathbb{R})$), we can define derivatives: $\frac{d^k\Phi}{ds^k}$, integrals: $\int_\Gamma \Phi(s)ds$, etc. in the usual way. Then $C^\infty(\Gamma)$ denotes the linear space of all $\Phi = \Phi(s)$ on Γ such that $\Phi \in C^k(\Gamma)$ $\forall k \in \mathbb{N}$.

Definition 1.10.2. The test space

$$\mathcal{D}(\Gamma) = \{\Phi : \Phi : \Gamma \longrightarrow \mathbb{C}, \Phi \in C^\infty(\Gamma)\} \tag{1.10.47}$$

is the linear space of all infinitely differentiable complex-valued functions $\Phi = \Phi(s)$ defined on Γ.

Convergence in $\mathcal{D}(\Gamma)$ A sequence $(\Phi_n)_{n=1}^{\infty}$ in $\mathcal{D}(\Gamma)$ is said to converge to $\Phi \in \mathcal{D}(\Gamma)$ if and only if $\frac{d^\alpha \Phi_k}{ds^\alpha} \to \frac{d^\alpha \Phi}{ds^\alpha}$ *uniformly* on Γ $\forall \alpha \in \mathbb{N}_0$ as $n \to \infty$.

Continuity on $\mathcal{D}(\Gamma)$ A linear functional S on $\mathcal{D}(\Gamma)$ is called *continuous* on $\mathcal{D}(\Gamma)$ if and only if $\Phi_k \to \Phi$ in $\mathcal{D}(\Gamma) \Longrightarrow \langle S, \Phi_k \rangle \to \langle S, \Phi \rangle$ in $\mathbb{C}$ as $k \to \infty$.

Since Γ is bounded, $\mathcal{D}(\Gamma)$ is less complicated than $\mathcal{D}(\mathbb{R})$. For example, $\Phi = 1$ belongs to $\mathcal{D}(\Gamma)$, but it does not belong to $\mathcal{D}(\mathbb{R})$.

Definition 1.10.3. $\mathcal{D}'(\Gamma)$ is the linear space of all continuous linear functionals on $\mathcal{D}(\Gamma)$ and is called the *space of distributions on* Γ. Then $S \in \mathcal{D}'(\Gamma) \Longrightarrow \langle S, \Phi \rangle_{\mathcal{D}'(\Gamma) \times \mathcal{D}(\Gamma)}$ is the duality pairing between $\mathcal{D}'(\Gamma)$ and $\mathcal{D}(\Gamma)$.

We will show that the distributions on Γ are intimately related to periodic distributions on $\mathbb{R}$ with period $T = 2\pi r$. For this we need

$$\begin{aligned} \mathcal{D}_T(\mathbb{R}) &= \{\tilde{\Phi} : \tilde{\Phi} : \mathbb{R} \longrightarrow \mathbb{C},\ \tilde{\Phi}(x+T) = \tilde{\Phi}(x)\ \forall x \in \mathbb{R}, \tilde{\Phi} \in C^\infty(\mathbb{R})\} \\ &= C_T^\infty(\mathbb{R}), \end{aligned} \tag{1.10.48}$$

which is the linear space of infinitely differentiable, complex-valued periodic functions $\tilde{\Phi} = \tilde{\Phi}(x)$ on $\mathbb{R}$ with period $T = 2\pi r$, and $\mathcal{D}'_T(\mathbb{R})$ is the linear space of continuous, linear functionals on $\mathcal{D}_T(\mathbb{R})$, which can be identified with the space of periodic distributions on $\mathbb{R}$ with period T. (1.10.48a)

Isomorphism between two sets $\mathcal{D}(\Gamma)$ and $\mathcal{D}_T(\mathbb{R})$ To each $\Phi \in \mathcal{D}(\Gamma)$ with $\Phi = \Phi(s)$, we associate a unique *periodic* $\tilde{\Phi} \in \mathcal{D}_T(\mathbb{R})$ such that

$$\tilde{\Phi}(x) = \Phi(P), \tag{1.10.49}$$

where $P \in \Gamma$ has arc length coordinate $s = x$.

Conversely, to each *periodic* $\tilde{\Phi} \in \mathcal{D}_T(\mathbb{R})$ we associate a *unique* $\Phi \in \mathcal{D}(\Gamma)$ by

$$\Phi(P) = \tilde{\Phi}(x), \tag{1.10.50}$$

where x is *one* of the arc length distances of P from $A(s = 0)$, any two of which differ by kT with $k \in \mathbb{Z}$. Then the correspondence $\Phi \in \mathcal{D}(\Gamma) \longleftrightarrow \tilde{\Phi} \in \mathcal{D}_T(\mathbb{R})$ is one-to-one from $\mathcal{D}(\Gamma)$ onto $\mathcal{D}_T(\mathbb{R})$, i.e. defines an *isomorphism* between these two sets.

$\mathcal{D}_T(\mathbb{R})$ is a subspace of $C^\infty(\mathbb{R})$ (i.e. of $\mathcal{E}(\mathbb{R})$; see Definition 5.7.1, Chapter 5).

$\tilde{\Phi} \in \mathcal{D}_T(\mathbb{R}) \Longrightarrow \tilde{\Phi}$ does *not* belong to $\mathcal{D}(\mathbb{R})$. $\psi \in \mathcal{D}(\mathbb{R}) \Longrightarrow \psi$ does *not* belong to $\mathcal{D}_T(\mathbb{R})$.

Construction of test functions $\Phi \in \mathcal{D}(\Gamma)$ and $\tilde{\Phi} \in \mathcal{D}_T(\mathbb{R})$ in terms of $\phi \in \mathcal{D}(\mathbb{R})$ Let $\phi \in \mathcal{D}(\mathbb{R})$. Then ϕ does not belong to $\mathcal{D}_T(\mathbb{R})$. But if we consider the sum of all translated or shifted functions $\tau_{-lT}\phi$, i.e. $\sum_{l=-\infty}^{\infty}(\tau_{-lT}\phi)(x) = \sum_{l=-\infty}^{\infty}\phi(x+lT)$, which will always converge, since $\text{supp}(\phi) \subset\subset \mathbb{R}$ will imply a finite summation $\sum$. Let this sum be denoted by

$$\tilde{\Phi}(x) = \sum_{l=-\infty}^{\infty} \phi(x + lT) \tag{1.10.51}$$

such that $\tilde{\Phi}(x + T) = \sum_{l=-\infty}^{\infty}\phi(x + (l + 1)T) = \tilde{\Phi}(x)\ \forall x \in \mathbb{R}$, i.e. $\tilde{\Phi}$ is a periodic function on $\mathbb{R}$ with period T, and $\tilde{\Phi} \in C^\infty(\mathbb{R})$ (since $\phi \in C^\infty(\mathbb{R})$). Hence, $\tilde{\Phi} \in \mathcal{D}_T(\mathbb{R})$. To $\tilde{\Phi} \in \mathcal{D}_T(\mathbb{R})$ we associate the unique function $\Phi = \Phi(s)$ on Γ by

$$\Phi(P) = \tilde{\Phi}(x) \quad \forall x \in \mathbb{R}, \tag{1.10.52}$$

where x is one of the arc length coordinates of P (see (1.10.50)) such that

$$\Phi(s) = \tilde{\Phi}(x) = \tilde{\Phi}(x + kT) = \sum_{l=-\infty}^{\infty} \phi(x + lT), \tag{1.10.53}$$

with $k \in \mathbb{Z}$ and $\Phi \in C^\infty(\Gamma)$, i.e. $\Phi \in \mathcal{D}(\Gamma)$.

Example 1.10.13. For $\varepsilon > 0$, let $\phi_\varepsilon \in \mathcal{D}(\mathbb{R})$ be defined as in (1.2.6a): $\phi_\varepsilon(x) = \exp(-\frac{\varepsilon^2}{\varepsilon^2-|x|^2})$ for $|x| < \varepsilon$ and $= 0$ for $|x| \geq \varepsilon$. Choosing $\varepsilon < T/2$, we get $\tilde{\Phi}_\varepsilon \in \mathcal{D}_T(\mathbb{R})$ defined by: $\tilde{\Phi}_\varepsilon(x) = \sum_{l=-\infty}^{\infty}\phi_\varepsilon(x + lT)$. See Figure 1.7 for an illustration.

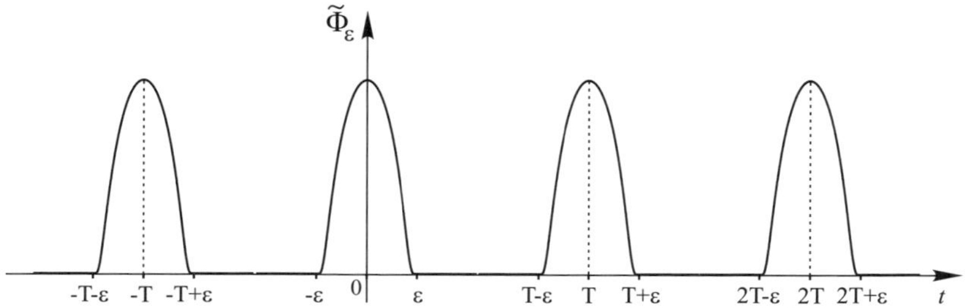

Figure 1.7 Periodic test function $\tilde{\Phi}_\varepsilon \in \mathcal{D}_T(\mathbb{R})$

One-to-one correspondence between distributions on Γ and periodic distributions on $\mathbb{R}$ with period T Let $f \in L^1_{\text{loc}}(\Gamma)$ be a locally summable function on Γ defining a regular distribution on Γ by: $\forall \Phi \in \mathcal{D}(\Gamma)$,

$$\langle f, \Phi \rangle_{\mathcal{D}'(\Gamma)\times\mathcal{D}(\Gamma)} = \int_\Gamma f(s)\Phi(s)ds, \tag{1.10.54}$$

where Φ is associated with the periodic $\tilde{\Phi}$ on $\mathbb{R}$ with period T, and defined by (1.10.52). Let $\tilde{f} \in L^1_{\text{loc}}(\mathbb{R})$ be the associated (with $f \in L^1_{\text{loc}}(\Gamma)$) periodic locally summable function such that $\tilde{f}(x) = f(P)$ a.e., where $P \in \Gamma$ has the arc length coordinate $s = x$, $\tilde{f}(x + kT) = \tilde{f}(x)$ a.e. $\forall k \in \mathbb{Z}$. Then $\tilde{f} \in \mathcal{D}'(\mathbb{R})$ defines a regular distribution in $\mathcal{D}'(\mathbb{R})$ by:

$$\langle \tilde{f}(x), \phi(x) \rangle_{\mathcal{D}'(\mathbb{R}) \times \mathcal{D}(\mathbb{R})} = \int_{\mathbb{R}} \tilde{f}(x)\phi(x)dx \quad \forall \phi \in \mathcal{D}(\mathbb{R}). \tag{1.10.55}$$

Now we will show that the integrals (1.10.54) and (1.10.55) are equal.

$$\begin{aligned}
\int_{-\infty}^{\infty} \tilde{f}(x)\phi(x)dx &= \sum_{l=-\infty}^{\infty} \int_{lT}^{(l+1)T} \tilde{f}(x)\phi(x)dx \\
&= \sum_{l=-\infty}^{\infty} \int_0^T \tilde{f}(\eta + lT)\phi(\eta + lT)d\eta \\
&\qquad\qquad \text{(by change of variables: } x = \eta + lT) \\
&= \int_0^T \tilde{f}(x)\Big(\sum_{l=-\infty}^{\infty} \phi(x + lT)\Big)dx \\
&= \int_0^T \tilde{f}(x)\tilde{\Phi}(x)dx = \int_{\Gamma} f(s)\Phi(s)ds
\end{aligned}$$

($\tilde{f}$ is periodic, $\tilde{\Phi}$ is defined by (1.10.51) and $f(s) = \tilde{f}(x)$, $\Phi(s) = \tilde{\Phi}(x)$ (1.10.53), $ds = dx$). We write

$$\int_0^T \tilde{f}(x)\tilde{\Phi}(x)dx = \langle \tilde{f}(x), \tilde{\Phi}(x) \rangle_{\mathcal{D}'_T(\mathbb{R}) \times \mathcal{D}_T(\mathbb{R})} \quad \text{with } \tilde{\Phi} \in \mathcal{D}_T(\mathbb{R}). \tag{1.10.56}$$

Hence,

$$\langle \tilde{f}, \phi \rangle_{\mathcal{D}'(\mathbb{R}) \times \mathcal{D}(\mathbb{R})} = \langle f, \Phi \rangle_{\mathcal{D}'(\Gamma) \times \mathcal{D}(\Gamma)} = \langle \tilde{f}, \tilde{\Phi} \rangle_{\mathcal{D}'_T(\mathbb{R}) \times \mathcal{D}_T(\mathbb{R})} \quad \forall \phi \in \mathcal{D}(\mathbb{R}). \tag{1.10.57}$$

(1.10.56) shows the relation between the periodic distribution on $\mathbb{R}$ defined by periodic function $\tilde{f} \in L^1_{\text{loc}}(\mathbb{R})$ with period T and the distribution on Γ defined by the associated function $f \in L^1_{\text{loc}}(\Gamma)$ on Γ. Finally, relation (1.10.56) suggests an alternative consistent definition of *periodic distributions* $\tilde{S}$ on $\mathbb{R}$ with period T, which are associated in one-to-one correspondence with the distributions $S \in \mathcal{D}'(\Gamma)$ on Γ, by:

$$\langle \tilde{S}(x), \phi(x) \rangle_{\mathcal{D}'(\mathbb{R}) \times \mathcal{D}(\mathbb{R})} = \langle S(s), \Phi(s) \rangle_{\mathcal{D}'(\Gamma) \times \mathcal{D}(\Gamma)} = \langle \tilde{S}(x), \tilde{\Phi}(x) \rangle_{\mathcal{D}'_T(\mathbb{R}) \times \mathcal{D}_T(\mathbb{R})} \tag{1.10.58}$$

$\forall\phi \in \mathcal{D}(\mathbb{R})$ and $\Phi \in \mathcal{D}(\Gamma)$ with ϕ and $\tilde{\Phi}$ related by (1.10.51), and with $\tilde{\Phi}$ and Φ related by (1.10.53), since

$$\begin{aligned}\langle \tilde{S}, \phi(x)\rangle_{\mathcal{D}'(\mathbb{R})\times\mathcal{D}(\mathbb{R})} &= \langle \tilde{S}, \phi(x+T)\rangle_{\mathcal{D}'(\mathbb{R})\times\mathcal{D}(\mathbb{R})}\\ &= \langle \tilde{S}(x), \tilde{\Phi}(x)\rangle_{\mathcal{D}'_T(\mathbb{R})\times\mathcal{D}_T(\mathbb{R})} = \langle S(s), \Phi(s)\rangle_{\mathcal{D}'(\Gamma)\times\mathcal{D}(\Gamma)}\end{aligned}$$

(if $\phi(x)$ is replaced by the shifted value $\phi(x+T)$, $\tilde{\Phi}$, and consequently Φ, do *not* change).

Example 1.10.14. Let $S = \delta_0 = \delta(s) \in \mathcal{D}'(\Gamma)$ be the Dirac distribution corresponding to unit $(+1)$ mass/charge/force etc. concentrated at $s = 0$. Then the corresponding *periodic* distribution $\tilde{S} = \tilde{\delta}$ on $\mathbb{R}$ with period T, i.e. *periodic* Dirac distributions $\tilde{\delta}$ of unit $(+1)$ mass/charge/force etc. concentrated at points $x = lT \in \mathbb{R}$ for $l \in \mathbb{Z}$, is defined by: $\forall\phi \in \mathcal{D}(\mathbb{R})$,

$$\begin{aligned}\langle \tilde{\delta}(x), \phi(x)\rangle_{\mathcal{D}'(\mathbb{R})\times\mathcal{D}(\mathbb{R})} &= \langle \delta(s), \Phi(s)\rangle_{\mathcal{D}'(\Gamma)\times\mathcal{D}(\Gamma)}\\ &= \Phi(0) = \tilde{\Phi}(0) = \sum_{l=-\infty}^{\infty} \phi(0 + lT)\\ &= \sum_{l=-\infty}^{\infty} \phi(lT) \quad \Big(\sum_{l} \text{ is a finite summation}\Big)\\ &= \sum_{l=-\infty}^{\infty} \langle \delta_{lT}, \phi(x)\rangle_{\mathcal{D}'(\mathbb{R})\times\mathcal{D}(\mathbb{R})}\\ &= \Big\langle \sum_{l=-\infty}^{\infty} \delta_{lT}, \phi(x)\Big\rangle_{\mathcal{D}'(\mathbb{R})\times\mathcal{D}(\mathbb{R})},\end{aligned}$$

since $\langle \delta_{lT}, \phi\rangle = \phi(lT)$ with

$$\delta_{lT} = \delta(x - lT) \quad \Longrightarrow \quad \tilde{\delta} = \sum_{l=-\infty}^{\infty} \delta_{lT} = \sum_{l=-\infty}^{\infty} \delta(x - lT), \tag{1.10.59}$$

where $\tilde{\delta} \in \mathcal{D}'(\mathbb{R})$ is a periodic Dirac distribution on $\mathbb{R}$ with period T, $\delta_{lT} = \delta(x - lT)$ is a Dirac distribution in $\mathcal{D}'(\mathbb{R})$ with concentration at $x = lT$.

Alternatively,

$$\langle \tilde{\delta}(x), \tilde{\Phi}(x)\rangle = \langle \tilde{\delta}, \tilde{\Phi}\rangle = \tilde{\Phi}(0) = \cdots = \Big\langle \sum_{l=-\infty}^{\infty} \delta_{lT}, \phi\Big\rangle_{\mathcal{D}'(\mathbb{R})\times\mathcal{D}(\mathbb{R})}.$$

But

$$\langle \tilde{\delta}, \phi\rangle = \langle \tilde{\delta}, \tilde{\Phi}\rangle = \Big\langle \sum_{l=-\infty}^{\infty} \delta_{lT}, \phi\Big\rangle \; \forall\phi \in \mathcal{D}(\mathbb{R}) \quad \Longleftrightarrow \quad \tilde{\delta} = \sum_{l=-\infty}^{\infty} \delta_{lT} \text{ in } \mathcal{D}'(\mathbb{R}).$$

Remark 1.10.2. From (1.10.57) and (1.10.58), we find that periodic distributions $\tilde{S} \in \mathcal{D}'(\mathbb{R})$ on $\mathbb{R}$ with period T also satisfy the following relation: $\forall \phi \in \mathcal{D}(\mathbb{R})$ with

$$\tilde{\Phi}(x) = \sum_{l=-\infty}^{\infty} \phi(x + lT) \in \mathcal{D}_T(\mathbb{R}),$$

$$\langle \tilde{S}(x), \phi(x) \rangle_{\mathcal{D}'(\mathbb{R}) \times \mathcal{D}(\mathbb{R})} = \langle \tilde{S}(x), \tilde{\Phi}(x) \rangle_{\mathcal{D}'_T(\mathbb{R}) \times \mathcal{D}_T(\mathbb{R})}. \tag{1.10.60}$$

Denoting *the set of all periodic distributions* $\tilde{S}$ *on* $\mathbb{R}$ *with period* T by $\mathcal{D}'_T(\mathbb{R})$, we write formally:

$$\langle \tilde{S}(x), \tilde{\phi}(x) \rangle = \langle \tilde{S}(x), \tilde{\phi}(x) \rangle_{\mathcal{D}'_T(\mathbb{R}) \times \mathcal{D}_T(\mathbb{R})}. \tag{1.10.61}$$

For more details, see Remark 6.7.4 in Section 6.7, Chapter 6.

Example 1.10.15. Let $(p_n)_{n=1}^{\infty}$ be defined by: $p_n(t) = n$ for $0 < t < 1/n$ and $p_n(t) = 0$ for $t < 0$ and $t > 1/n$. Let $\tilde{P}_n$ be defined by: $\tilde{P}_n(t) = \sum_{l=-\infty}^{\infty} p_n(t+lT)$ $\forall t \in \mathbb{R}$, $T > 0$ being a number with $n > 2/T$. Show that

1. $p_n \to \delta$ in $\mathcal{D}'(\mathbb{R})$;
2. $\tilde{P}_n$ defines a periodic distribution on $\mathbb{R}$ with period T $\forall n \in \mathbb{N}$ such that $\lim_{n\to\infty} \tilde{P}_n = \tilde{\delta} = \sum_{l=-\infty}^{\infty} \delta_{lT}$ in $\mathcal{D}'(\mathbb{R})$, where $\tilde{\delta}$ is a periodic Dirac distribution on $\mathbb{R}$, $\delta_{lT} = \delta(x - lT)$ is Dirac distribution of unit (+1) mass/charge/force etc. concentrated at the points $x = lT$ with $l \in \mathbb{Z}$.

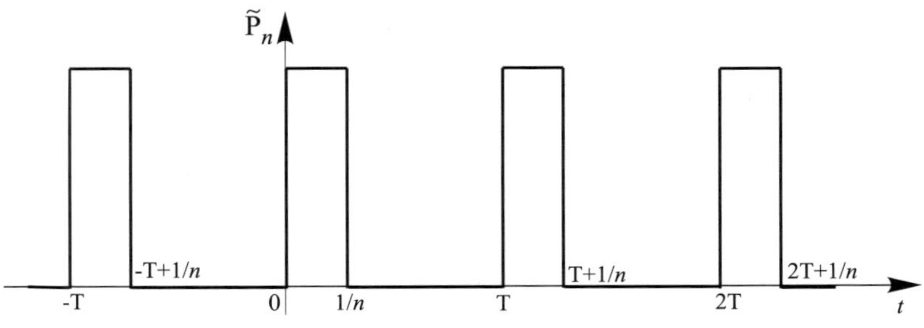

Figure 1.8 Periodic rectangular pulse function $\tilde{P}_n$.

Solution.

1.
$$\lim_{n\to\infty} \langle p_n, \phi \rangle = \lim_{n\to\infty} \int_0^{\frac{1}{n}} n\phi(t)dt = \lim_{n\to\infty} \left(n \int_0^{\frac{1}{n}} [\phi(0) + (\phi(t) - \phi(0))]dt \right)$$
$$= \lim_{n\to\infty} \left(n \cdot \phi(0) \cdot \frac{1}{n} + n \int_0^{\frac{1}{n}} t \cdot \phi'(\tau)dt \right) = \phi(0),$$

since

$$\left| n \int_0^{\frac{1}{n}} t \cdot \phi'(\tau) dt \right| \leq \max_{t\in\mathbb{R}} |\phi'(t)| n \int_0^{\frac{1}{n}} t dt$$
$$= \max_{t\in\mathbb{R}} |\phi'(t)| \cdot n \cdot \frac{1}{2.n^2} \to 0 \quad \text{as } n \to \infty$$

$\Longrightarrow \lim_{n\to\infty} \langle p_n, \phi \rangle = \phi(0) = \langle \delta, \phi \rangle \; \forall \phi \in \mathcal{D}(\mathbb{R})$

$\Longrightarrow \lim_{n\to\infty} p_n = \delta \in \mathcal{D}'(\mathbb{R})$.

2. Since $\operatorname{supp}(p_n) = [0, \frac{1}{n}]$, $\sum_{l=-\infty}^{\infty} p_n(t + lT)$ contains a finite number of terms $\forall t \in \mathbb{R}$. $\tilde{P}_n(t+T) = \sum_{l=-\infty}^{\infty} p_n(t + (l+1)T) = \tilde{P}_n(t) \; \forall t \in \mathbb{R} \Longrightarrow$ $\tilde{P}_n$ is a periodic function on $\mathbb{R}$ with period T, i.e. a *periodic extension* to $\mathbb{R}$ of *rectangular pulse function* p_n (see Figure 1.8).

   Hence, $\tilde{P}_n \in L^1_{\text{loc}}(\mathbb{R})$ is a periodic function on $\mathbb{R}$ with period T and defines a periodic distribution in $\mathcal{D}'(\mathbb{R})$ with period T by:

$$\langle \tilde{P}_n, \phi(t) \rangle_{\mathcal{D}'(\mathbb{R})\times\mathcal{D}(\mathbb{R})} = \int_{-\infty}^{\infty} \tilde{P}_n(t)\phi(t)dt = \sum_{l=-\infty}^{\infty} \int_{lT}^{(l+1)T} \tilde{P}_n(t)\phi(t)dt$$
$$= \sum_{l=-\infty}^{\infty} \int_0^T \tilde{P}_n(\tau + lT)\phi(\tau + lT)d\tau$$
$$= \sum_{l=-\infty}^{\infty} \int_0^T \tilde{P}_n(\tau)\phi(\tau + lT)d\tau$$
$$= \sum_{l=-\infty}^{\infty} \int_0^{\frac{1}{n}} n\phi(t + lT)dt \quad \forall \phi \in \mathcal{D}(\mathbb{R})$$

$$\Longrightarrow \quad \lim_{n\to\infty} \langle \tilde{P}_n(t), \phi(t) \rangle_{\mathcal{D}'(\mathbb{R})\times\mathcal{D}(\mathbb{R})}$$
$$= \lim_{n\to\infty} \left[\sum_{l=-\infty}^{\infty} n \int_0^{\frac{1}{n}} \phi(t + lT)dt \right] \quad \text{(summation is finite)}$$
$$= \sum_{l=-\infty}^{\infty} \lim_{n\to\infty} \left[n \int_0^{\frac{1}{n}} \phi(lT)dt + n \int_0^{\frac{1}{n}} (\phi(t + lT) - \phi(lT))dt \right]$$
$$\text{(interchange of sum and limit is admissible)}$$
$$= \sum_{l=-\infty}^{\infty} \lim_{n\to\infty} \left[\phi(lT) \cdot n \cdot \frac{1}{n} + n \int_0^{\frac{1}{n}} t\phi'(\tau^*)dt \right] = \sum_{l=-\infty}^{\infty} \phi(lT),$$

since $\tau^* \in]lT, t+lT[$, $|n\int_0^{\frac{1}{n}} t\phi'(\tau^*)dt| \le \max_{t\in\mathbb{R}} |\phi'(t)| \cdot n \cdot \frac{1}{2n^2} \to 0$ as $n \to \infty$

$$\Longrightarrow \quad \lim_{n\to\infty} n \int_0^{\frac{1}{n}} t\phi'(\tau^*)dt = 0$$

$$\Longrightarrow \quad \lim_{n\to\infty} \langle \tilde{P}_n, \phi(t)\rangle = \sum_{l=-\infty}^{\infty} \phi(lT) = \sum_{l=-\infty}^{\infty} \langle \delta_{lT}, \phi(t)\rangle$$

$$= \left\langle \sum_{l=-\infty}^{\infty} \delta_{lT}, \phi \right\rangle \quad \forall \phi \in \mathcal{D}(\mathbb{R}) \quad [\delta_{lT} = \delta(x - lT)]$$

$$\Longleftrightarrow \quad \lim_{n\to\infty} \tilde{P}_n = \sum_{l=-\infty}^{\infty} \delta_{lT} = \tilde{\delta} \quad \text{in } \mathcal{D}'(\mathbb{R}).$$

(For convergence of series of distributions, see Definition 1.9.1 and Example 1.9.1.)

$\sum_{l=-\infty}^{\infty} \delta_{lT} = \tilde{\delta} \in \mathcal{D}'(\mathbb{R})$ is a periodic distribution on $\mathbb{R}$ with periodic T: from (1.10.45b),

$$\langle \tau_T \tilde{\delta}, \phi \rangle_{\mathcal{D}'(\mathbb{R})\times\mathcal{D}(\mathbb{R})} = \langle \tilde{\delta}, \tau_{-T}\phi \rangle = \langle \tilde{\delta}, \phi(x+T)\rangle = \left\langle \sum_{l=-\infty}^{\infty} \delta_{lT}, \phi(x+T) \right\rangle$$

$$= \sum_{l=-\infty}^{\infty} \langle \delta_{lT}, \phi(x+T)\rangle = \sum_{l=-\infty}^{\infty} \phi(lT+T) = \sum_{l'=-\infty}^{\infty} \phi(l'T) \quad (l' = l+1)$$

$$= \sum_{l=-\infty}^{\infty} \phi(lT) = \sum_{l=-\infty}^{\infty} \langle \delta_{lT}, \phi(t)\rangle = \left\langle \sum_{l=-\infty}^{\infty} \delta_{lT}, \phi(t) \right\rangle$$

$$= \langle \tilde{\delta}, \phi(t) \rangle \quad \forall \phi \in \mathcal{D}(\mathbb{R})$$

$\Longrightarrow \tau_T\tilde{\delta} = \tilde{\delta}$ in $\mathcal{D}'(\mathbb{R}) \Longrightarrow \tilde{\delta} = \sum_{l=-\infty}^{\infty} \delta_{lT} = \sum_{l=-\infty}^{\infty} \delta(x - lT)$ is a periodic distribution on $\mathbb{R}$ with period T.

1.11 Physical distributions versus mathematical distributions

1.11.1 Physical interpretation of mathematical distributions

A mathematical distribution $T \in \mathcal{D}'(\Omega)$ on $\Omega \subset \mathbb{R}^n$ can be interpreted as the physical distribution or *spread* of mass, force, electric charges, magnetic dipole moments, etc. in Ω. For example, a mathematical distribution $T = T_f \in \mathcal{D}'(\Omega)$ corresponding to a locally integrable function f on Ω can be interpreted as the *physical distribution* of mass/force/charge etc. with density f over a volume Ω. Again, Dirac distribution δ (resp. $\delta_{\mathbf{a}}$) can be interpreted physically as the mass/force/charge etc. with intensity $+1$ concentrated at $\mathbf{x} = \mathbf{0} \in \Omega$ (resp. $\mathbf{x} = \mathbf{a} \in \Omega$). Hence, regular distributions $T = T_f \in \mathcal{D}'(\Omega)$ correspond to physical distributions of mass/force/charge etc. with density $f \in L^1_{\text{loc}}(\Omega)$, whereas the singular Dirac distribution $T = \delta$ (resp. $\delta_{\mathbf{a}}$) corresponds physically to point concentration of mass/force/charge etc. at $\mathbf{0} \in \Omega$ (resp. $\mathbf{a} \in \Omega$) with intensity $+1$.

In physical applications, we will meet with expression $\langle T_f, \phi\rangle$ or $\langle \delta_{\mathbf{a}}, \phi\rangle$, where $\phi \notin \mathcal{D}(\Omega)$ (see 5.6 in Chapter 5 for details). T_f (resp. $\delta_{\mathbf{a}}$) can be given extension to a space of functions ϕ larger than $\mathcal{D}(\Omega)$ (see, for example, Theorem 1.3.1 for a *unique*, continuous extension of T). $T = T_f$ defined by $\langle T_f, \phi\rangle = \int_\Omega f\phi d\mathbf{x}$ can be extended to a larger space of functions ϕ such that $f\phi$ is integrable on Ω, whereas $T = \delta_{\mathbf{a}}$ with $\langle \delta_{\mathbf{a}}, \phi\rangle = \phi(\mathbf{a})$ can be extended to an extremely large space of functions ϕ continuous at $\mathbf{a} \in \Omega$ (see also Theorem 1.3.1 for a *unique* extension). From these examples, we find that different distributions T_f and $\delta_{\mathbf{a}}$ have been extended to different spaces of functions ϕ which cannot be interchanged, i.e. $\langle \delta_{\mathbf{a}}, \phi\rangle$ (resp. $\langle T_f, \phi\rangle$) is *not* defined in general on the space of functions ϕ such that $f\phi$ is integrable on Ω (resp. such that ϕ is continuous at $\mathbf{a} \in \Omega$). Then the question arises: What is the role of $\mathcal{D}(\Omega)$? $\mathcal{D}(\Omega)$ is the common set of functions ϕ on which all distributions on Ω are defined. From these discussions, we find that $\forall T \in \mathcal{D}'(\Omega)$, $\langle T, 1\rangle$, if it is defined, gives the total value. For example,

- for $T = T_f = f \in L^1(\Omega)$,

$$\langle T, 1\rangle = \langle T_f, 1\rangle = \int_\Omega f(\mathbf{x}) d\mathbf{x}, \tag{1.11.1}$$

- for $T = \delta_{\mathbf{a}} \in \mathcal{D}'(\Omega)$,

$$\langle T, 1\rangle = \langle \delta, 1\rangle = +1, \tag{1.11.2}$$

gives the total value of mass/force/charge etc. (distributed) in Ω (see also Section 5.6 for justifications).

1.11.2 Load intensity

The notion of mathematical distributions leads us to the fact that in reality we cannot measure the value of a physical entity *at a point*, but we can only measure mean or average values of the *distribution* or *spread* of a physical entity over sufficiently small neighbourhoods of the given point. Then, considering the system of these mean/average values of the (physical) distribution on the corresponding system of neighbourhoods of the point, if we can find the 'limit' of this system in an appropriate sense as the system of neighbourhoods shrinks to the given point, then this limit will be called the value of the physical entity at the given point. As a simple example, we consider the case of a point load of intensity (magnitude) $+1$ acting at the point $\mathbf{0} \in \mathbb{R}^2(\Omega = \mathbb{R}^2)$. In order to determine the load intensity at $\mathbf{0} \in \mathbb{R}^2$, we distribute *uniformly* (other non-uniform distributions are also possible) this unit load on a circular disc B_ε of radius $\varepsilon > 0$ and centre at $\mathbf{0} \in \mathbb{R}^2$, i.e. $B_\varepsilon = \{\mathbf{x} : \|\mathbf{x}\| = (x_1^2 + x_2^2)^{\frac{1}{2}} \leq \varepsilon\}$. Then, the mean/average load intensity F_ε over B_ε is given, $\forall \varepsilon > 0$, by:

$$F_\varepsilon(\mathbf{x}) = \begin{cases} \frac{1}{\pi\varepsilon^2} & \text{for } \|\mathbf{x}\| = (x_1^2 + x_2^2)^{\frac{1}{2}} \leq \varepsilon \\ 0 & \text{for } \|\mathbf{x}\| > \varepsilon. \end{cases} \tag{1.11.3}$$

Then, $\{F_\varepsilon\}_{\varepsilon>0}$ denotes the system of mean/average load intensities on the system $\{B_\varepsilon\}_{\varepsilon>0}$ of neighbourhoods of $\mathbf{0} \in \mathbb{R}^2$ with the following basic property:

$$\forall \varepsilon > 0, \quad \int_{\mathbb{R}^2} F_\varepsilon(\mathbf{x}) dx_1 dx_2 = \int_{B_\varepsilon} \frac{1}{\pi\varepsilon^2} dx_1 dx_2 = \frac{1}{\pi\varepsilon^2} \cdot \pi\varepsilon^2 = 1, \tag{1.11.4}$$

i.e. the integral of the mean intensity over the whole space $\mathbb{R}^2$ gives the total load of $+1$ for every choice of uniform distributions over B_ε with $\varepsilon > 0$. Then,

$$\lim_{\varepsilon \to 0+} \int_{\mathbb{R}^2} F_\varepsilon(\mathbf{x}) dx_1 dx_2 = 1. \tag{1.11.5}$$

Now, we are interested in the load intensity at $\mathbf{x} = \mathbf{0}$, i.e. for $\varepsilon \to 0^+$. Hence, the first *intuitive* approach suggests to take the 'pointwise limit' of $\{F_\varepsilon\}$ in (1.11.3) as $\varepsilon \to 0^+$, i.e.

$$\lim_{\varepsilon \to 0+} F_\varepsilon(\mathbf{x}) = \begin{cases} +\infty & \text{for } \mathbf{x} = \mathbf{0} \\ 0 & \text{for } \mathbf{x} \neq \mathbf{0}, \end{cases}$$

which is precisely what was done by Dirac to define the delta function $\delta(\mathbf{x})$ in (1.1.1):

$$\lim_{\varepsilon \to 0+} F_\varepsilon(\mathbf{x}) = \delta(\mathbf{x}) = \begin{cases} +\infty & \text{for } \mathbf{x} = \mathbf{0} \\ 0 & \text{for } \mathbf{x} \neq \mathbf{0} \end{cases}. \tag{1.11.6}$$

Then,

$$\int_{\mathbb{R}^2} \lim_{\varepsilon \to 0+} F_\varepsilon(\mathbf{x}) dx_1 dx_2 = \int_{\mathbb{R}^2} \delta(\mathbf{x}) dx_1 dx_2 = \int_{\mathbb{R}^2 - \{\mathbf{0}\}} 0 dx_1 dx_2 = 0, \tag{1.11.7}$$

$\Longrightarrow$ the integral of the delta function $\delta(\mathbf{x}) = \lim_{\varepsilon \to 0+} F_\varepsilon(\mathbf{x})$ gives the total value of the load equal to 0, i.e. this 'pointwise limit' $\delta(\mathbf{x})$ does *not* restore the total load $+1$. Thus, the delta function $\delta(\mathbf{x})$ defined by (1.11.6) can *not* be accepted as the desired load intensity and the 'pointwise limit' of the system $\{F_\varepsilon\}_{\varepsilon>0}$ of mean load intensities defined by (1.11.3) is not acceptable. Hence, we are to renounce this and consider the limit of the system $\{F_\varepsilon\}_{\varepsilon>0}$ in a weaker sense, for example in the sense of convergence in $\mathcal{D}'(\mathbb{R}^2)$ (1.8.1): $\forall\phi\in\mathcal{D}(\mathbb{R}^2)$,

$$\lim_{\varepsilon\to 0^+}\langle F_\varepsilon,\phi\rangle = \lim_{\varepsilon\to 0^+}\int_{\mathbb{R}^2} F_\varepsilon(\mathbf{x})\phi(\mathbf{x})d\mathbf{x} = \lim_{\varepsilon\to 0^+}\int_{B_\varepsilon} F_\varepsilon(\mathbf{x})\phi(\mathbf{x})d\mathbf{x} = \phi(\mathbf{0}). \tag{1.11.8}$$

In fact, for $\phi(\mathbf{x}) = \phi(\mathbf{0}) + [\phi(\mathbf{x}) - \phi(\mathbf{0})]$,

$$\begin{aligned}
\int_{B_\varepsilon} F_\varepsilon\phi d\mathbf{x} &= \frac{\phi(\mathbf{0})}{\pi\varepsilon^2}\int_{B_\varepsilon} d\mathbf{x} + \frac{1}{\pi\varepsilon^2}\int_{B_\varepsilon}[\phi(\mathbf{x}) - \phi(\mathbf{0})]d\mathbf{x}\\
&= \phi(\mathbf{0}) + \frac{1}{\pi\varepsilon^2}\int_{B_\varepsilon}[\phi(\mathbf{x}) - \phi(\mathbf{0})]d\mathbf{x}\\
\Longrightarrow\quad \lim_{\varepsilon\to 0^+}\int_{B_\varepsilon} F_\varepsilon\phi d\mathbf{x} &= \phi(\mathbf{0}) + \lim_{\varepsilon\to 0^+}\frac{1}{\pi\varepsilon^2}\int_{B_\varepsilon}[\phi(\mathbf{x}) - \phi(\mathbf{0})]d\mathbf{x}\\
&= \phi(\mathbf{0}) \quad \forall\phi\in\mathcal{D}(\mathbb{R}^2),
\end{aligned}$$

since

$$\begin{aligned}
\left|\frac{1}{\pi\varepsilon^2}\int_{B_\varepsilon}[\phi(\mathbf{x}) - \phi(\mathbf{0})]d\mathbf{x}\right| &\le \frac{1}{\pi\varepsilon^2}\max_{\mathbf{x}\in B_\varepsilon}|\phi(\mathbf{x}) - \phi(\mathbf{0})|\cdot\pi\varepsilon^2\\
&= \max_{\mathbf{x}\in B_\varepsilon}|\phi(\mathbf{x}) - \phi(\mathbf{0})| \to 0 \quad \text{as } \varepsilon\to 0^+
\end{aligned}$$

$$\Longrightarrow\quad \lim_{\varepsilon\to 0^+}\frac{1}{\pi\varepsilon^2}\int_{B_\varepsilon}[\phi(\mathbf{x}) - \phi(\mathbf{0})]d\mathbf{x} = 0 \quad \forall\phi\in\mathcal{D}(\mathbb{R}^2).$$

Now, using (1.11.8) and the definition of Dirac distribution $\delta\in\mathcal{D}'(\mathbb{R}^2)$, $\langle\delta,\phi\rangle = \phi(\mathbf{0})$, we have: $\forall\phi\in\mathcal{D}(\mathbb{R}^2)$,

$$\lim_{\varepsilon\to 0^+}\langle F_\varepsilon,\phi\rangle = \lim_{\varepsilon\to 0^+}\int_{B_\varepsilon} F_\varepsilon(\mathbf{x})\phi(\mathbf{x})d\mathbf{x} = \phi(\mathbf{0}) = \langle\delta,\phi\rangle$$

(i.e. $\lim_{\varepsilon\to 0^+}\int_{\mathbb{R}^2} F_\varepsilon(\mathbf{x})\phi(\mathbf{x})d\mathbf{x} \neq \int_{\mathbb{R}^2}\lim_{\varepsilon\to 0^+} F_\varepsilon(\mathbf{x})\phi(\mathbf{x})d\mathbf{x} = 0$)

$\Longrightarrow F_\varepsilon \to \delta$ in $\mathcal{D}'(\mathbb{R}^2)$ as $\varepsilon\to 0^+$ (see (1.8.1)).

In other words, the system $\{F_\varepsilon\}$ of mean load intensities is not considered as a system of *functions*, but as a system of regular *distributions* (continuous linear functionals) in $\mathcal{D}'(\mathbb{R}^2)$, whose limit is the (singular) Dirac distribution $\delta\in\mathcal{D}'(\mathbb{R}^2)$, which is *not* a function.

Thus, the point load of intensity $+1$ concentrated at the origin $\mathbf{0}\in\mathbb{R}^2$ will be given by Dirac distribution $\delta\in\mathcal{D}'(\mathbb{R}^2)$ (*not* by the delta function in (1.11.6)), if δ restores the total load $+1$. In fact, from (1.11.2), the total load is given by $\langle\delta,1\rangle = +1$, since $\langle\delta,\phi\rangle = \phi(\mathbf{0})$ is well defined for $\phi\in C^\infty(\mathbb{R}^2)$ with arbitrary support (i.e. $\phi\notin\mathcal{D}(\mathbb{R}^2)$) (see Section 5.6 for details) and consequently, for $\phi = 1\in C^\infty(\mathbb{R}^2)$, $\langle\delta,1\rangle = 1(\mathbf{0}) = 1$.

1.11.3 Electrical charge distribution

We consider another interesting example of physical distribution met with in electrostatics to find the corresponding mathematical distribution. Consider the electrical charge distribution on the real line corresponding to the dipole with electric/magnetic moment $+1$ placed at the origin $0 \in \mathbb{R}$. Two (concentrated) charges $\frac{1}{\varepsilon}$ at $x = \varepsilon$ and $\frac{-1}{\varepsilon}$ at $x = 0$ yield the same electric moment $+1$ with total charges equal to 0 $\forall \varepsilon > 0$.

Hence, the density of charges, which *approximately* corresponds to this dipole of electric/magnetic moment $+1$ at $x = 0$, is given by:

$$T_\varepsilon = \frac{1}{\varepsilon}\delta_\varepsilon - \frac{1}{\varepsilon}\delta \quad \forall \varepsilon > 0, \tag{1.11.9}$$

where δ_ε (resp. $\delta_0 = \delta$) is the Dirac distribution with unit charge concentrated at $x = \varepsilon$ (resp. $x = 0$). Then the dipole is defined as the limit of the system T_ε when $\varepsilon \to 0^+$. But $\forall \varepsilon > 0$, the system T_ε corresponds to the mathematical distribution defined by: $\forall \varepsilon > 0^+$, $\forall \phi \in \mathcal{D}(\mathbb{R})$,

$$\langle T_\varepsilon, \phi\rangle = \left\langle \frac{1}{\varepsilon}\delta_\varepsilon - \frac{1}{\varepsilon}\delta, \phi \right\rangle = \frac{1}{\varepsilon}[\langle \delta_\varepsilon, \phi\rangle - \langle \delta, \phi\rangle] = \frac{\phi(\varepsilon) - \phi(0)}{\varepsilon}$$

$$\implies \quad \lim_{\varepsilon\to 0^+} \langle T_\varepsilon, \phi\rangle = \lim_{\varepsilon\to 0^+} \frac{\phi(\varepsilon) - \phi(0)}{\varepsilon} = \phi'(0) \quad \forall \phi \in \mathcal{D}(\mathbb{R}),$$

which suggests a mathematical definition of *dipole* as the distribution $T \in \mathcal{D}'(\mathbb{R})$:

$$\langle T, \phi\rangle = \phi'(0) \quad \forall \phi \in \mathcal{D}(\mathbb{R}). \tag{1.11.10}$$

In this definition of dipole as the mathematical distribution $T \in \mathcal{D}'(\mathbb{R})$, we do *not* need to construct an approximating system T_ε and then take the limit as $\varepsilon \to 0^+$ (see also Remark 2.3.3).

Remark 1.11.1. $T \in \mathcal{D}'(\mathbb{R})$ defining the dipole of moment $+1$ at $x = 0$ will be related to Dirac distribution δ, and this relationship will give the density of charge distribution corresponding to the dipole of moment $+1$ at $x = 0$. But for this we are to use the results of Chapter 2 on the derivative $\delta' = \frac{d\delta}{dx}$ of Dirac distribution δ given in (2.3.8). Instead of deferring the details we give them here, which the reader may read after going through the derivatives of Dirac distribution δ in the next chapter. Using (2.3.8), from (1.11.10),

$$\langle T, \phi\rangle = \phi'(0) = \langle \delta, \phi'\rangle = -\langle \delta', \phi\rangle = \langle -\delta', \phi\rangle \quad \forall \phi \in \mathcal{D}(\mathbb{R})$$

$$\implies \quad T = -\delta' = -\frac{d\delta}{dx} \quad \text{in } \mathcal{D}'(\mathbb{R}). \tag{1.11.11}$$

Thus, the required density of charge distribution corresponding to the dipole of moment $+1$ is $-\delta'$. Now, we are to check whether this density of charge distribution gives the total charge equal to 0 and the total moment equal to 1:

$$\text{Total charge} = \langle T, 1\rangle = \langle -\delta', 1\rangle = \left\langle \delta, \frac{d1}{dx}\right\rangle = \langle \delta, 0\rangle = 0; \tag{1.11.12}$$

$$\text{Total moment} = \langle -\delta', x\rangle = \left\langle \delta, \frac{dx}{dx}\right\rangle = \langle \delta, 1\rangle = 1. \tag{1.11.13}$$

(For details, see Section 5.6 and the definition of δ' in (2.3.8).)

Now, let us compute the density of charges corresponding to the dipole of moment $+1$ placed at the origin $\mathbf{0} \in \mathbb{R}^3$ and oriented in the given direction of unit vector $\hat{\boldsymbol{\tau}} = (\tau_1, \tau_2, \tau_3) \in \mathbb{R}^3$ with $\|\hat{\boldsymbol{\tau}}\| = 1$.

The density of charges corresponding to this dipole of moment $+1$ oriented in the direction of $\hat{\boldsymbol{\tau}}$ is *approximately* given by:

$$T_\varepsilon = \frac{1}{\varepsilon}\delta_{\varepsilon\hat{\boldsymbol{\tau}}} - \frac{1}{\varepsilon}\delta \quad \forall \varepsilon > 0, \tag{1.11.14}$$

where $\delta_{\varepsilon\hat{\boldsymbol{\tau}}}$ (resp. δ) is the Dirac distribution of charges with unit charge concentrated at $\varepsilon\hat{\boldsymbol{\tau}} \in \mathbb{R}^3$ (resp. $\mathbf{0} \in \mathbb{R}^3$).

Then the dipole is the limit of the system T_ε, when $\varepsilon \to 0^+$. But $\forall \varepsilon > 0$, system T_ε corresponds to the *mathematical distribution* defined by: $\forall \phi \in \mathcal{D}(\mathbb{R}^3)$,

$$\langle T_\varepsilon, \phi\rangle = \left\langle \frac{1}{\varepsilon}\delta_{\varepsilon\hat{\boldsymbol{\tau}}} - \frac{1}{\varepsilon}\delta, \phi\right\rangle = \frac{1}{\varepsilon}[\langle \delta_{\varepsilon\hat{\boldsymbol{\tau}}}, \phi\rangle - \langle \delta, \phi\rangle] = \frac{\phi(\varepsilon\hat{\boldsymbol{\tau}}) - \phi(\mathbf{0})}{\varepsilon} \quad \forall \varepsilon > 0 \tag{1.11.15}$$

$$\implies \lim_{\varepsilon\to 0^+} \langle T_\varepsilon, \phi\rangle = \lim_{\varepsilon\to 0^+} \frac{\phi(\varepsilon\hat{\boldsymbol{\tau}}) - \phi(\mathbf{0})}{\varepsilon} = \frac{\partial\phi}{\partial\tau}(\mathbf{0}) = \langle \nabla\phi(\mathbf{0}), \hat{\boldsymbol{\tau}}\rangle_{\mathbb{R}^3} \quad \forall \phi \in \mathcal{D}(\mathbb{R}^3),$$

which suggests that this dipole of moment $+1$ and oriented in the direction of $\hat{\boldsymbol{\tau}}$ should be defined mathematically as the distribution T *directly* by:

$$\langle T, \phi\rangle = \frac{\partial\phi}{\partial\tau}(\mathbf{0}) = \langle \nabla\phi(\mathbf{0}), \hat{\boldsymbol{\tau}}\rangle_{\mathbb{R}^3} \quad \forall \phi \in \mathcal{D}(\mathbb{R}^3), \tag{1.11.16}$$

where $\langle\cdot,\cdot\rangle_{\mathbb{R}^3}$ denotes the inner product of vectors in $\mathbb{R}^3$,

$$\nabla\phi(\cdot) = \frac{\partial\phi}{\partial x_1}(\cdot)\hat{\mathbf{i}}_1 + \frac{\partial\phi}{\partial x_2}(\cdot)\hat{\mathbf{i}}_2 + \frac{\partial\phi}{\partial x_3}(\cdot)\hat{\mathbf{i}}_3, \quad \langle \hat{\mathbf{i}}_k, \hat{\mathbf{i}}_l\rangle = \delta_{kl}, \quad 1 \le k, l \le 3.$$

This definition does not require the construction of the system $\{T_\varepsilon\}_{\varepsilon>0}$ and taking its limit as $\varepsilon \to 0^+$.

Remark 1.11.2. Following Remark 1.11.1, we can write

$$\langle T, \phi\rangle = \frac{\partial \phi}{\partial \tau}(\mathbf{0}) = \left\langle \delta, \frac{\partial \phi}{\partial \tau}\right\rangle = -\left\langle \frac{\partial \delta}{\partial \tau}, \phi\right\rangle = \left\langle -\frac{\partial \delta}{\partial \tau}, \phi\right\rangle \quad \forall \phi \in \mathcal{D}(\mathbb{R}^3)$$

$$\Longrightarrow \quad T = -\frac{\partial \delta}{\partial \tau} = \langle -\nabla\delta, \hat{\boldsymbol{\tau}}\rangle \quad \text{in } \mathcal{D}'(\mathbb{R}^3) \tag{1.11.17}$$

is the required density of charge distribution corresponding to the dipole of moment $+1$ placed at $x = 0$ and oriented in the direction of $\hat{\boldsymbol{\tau}}$.

Total charge of the dipole is 0:

$$\left\langle -\frac{\partial \delta}{\partial \tau}, 1\right\rangle = \left\langle \delta, \frac{\partial}{\partial \tau} 1\right\rangle = \langle \delta, \langle \nabla 1, \hat{\boldsymbol{\tau}}\rangle_{\mathbb{R}^3}\rangle = \langle \delta, 0\rangle = 0. \tag{1.11.18}$$

Total moment due to this density of charge distribution is $+1$:

$$\begin{aligned}\left\langle -\frac{\partial \delta}{\partial \tau}, \langle \mathbf{x}, \hat{\boldsymbol{\tau}}\rangle_{\mathbb{R}^3}\right\rangle &= \left\langle \delta, \frac{\partial}{\partial \tau}(\langle \mathbf{x}, \hat{\boldsymbol{\tau}}\rangle_{\mathbb{R}^3})\right\rangle = \langle \delta, \langle \nabla(x_1\tau_1 + x_2\tau_2 + x_3\tau_3), \hat{\boldsymbol{\tau}}\rangle_{\mathbb{R}^3}\rangle \\ &= \langle \delta, \langle \hat{\boldsymbol{\tau}}, \hat{\boldsymbol{\tau}}\rangle_{\mathbb{R}^3}\rangle = \langle \delta, \|\hat{\boldsymbol{\tau}}\|^2\rangle = \langle \delta, 1\rangle = 1. \end{aligned} \tag{1.11.19}$$

1.11.4 Simple layer and double layer distributions

In electrostatics we meet with new types of physical distributions of charges called *simple* (or *single*) *layer* and *double layer* distributions on surfaces, which we will describe now and show the corresponding mathematical distribution $T \in \mathcal{D}'(\mathbb{R}^3)$.

Simple layer distributions Let Γ be a (piecewise) smooth, orientable (two-sided) surface (a Möbius strip is *not* admissible) in $\mathbb{R}^3$ and $\rho = \rho(\mathbf{x})$ be a continuous function on Γ defining the *surface density* of continuous electric charge distribution on surface Γ.

A generalization of the *discrete* Dirac distribution $\delta_{\mathbf{a}}$ with charges of intensity $+1$ concentrated at point $\mathbf{a} \in \Gamma$ gives Dirac distribution δ_Γ corresponding to charges concentrated on $\Gamma \subset \mathbb{R}^3$. Then $\rho\delta_\Gamma \in \mathcal{D}'(\mathbb{R}^3)$ is the volume or spatial density of distribution on $\mathbb{R}^3$ corresponding to charges concentrated on the surface Γ with continuous surface density ρ. But corresponding to the *physical distribution* of charges over Γ with surface density ρ, the mathematical distribution $T \in \mathcal{D}'(\mathbb{R}^3)$ is defined by:

$$\langle T, \phi\rangle = \int_\Gamma \rho(\mathbf{x})\phi(\mathbf{x})dS \quad \forall \phi \in \mathcal{D}(\mathbb{R}^3) \tag{1.11.20}$$

(dS being the surface area measure).

This distribution $T \in \mathcal{D}'(\mathbb{R}^3)$ must *not* be confused with the distribution $T_f \in \mathcal{D}'(\mathbb{R}^3)$ defined by the volume density function f, where T_f is identified with the function f itself. But here $T \in \mathcal{D}'(\mathbb{R}^3)$ represents the volume/spatial density of charges in $\mathbb{R}^3$ corresponding to charges distributed on Γ with surface density ρ. In fact, we will show that $T = \rho\delta_\Gamma \in \mathcal{D}'(\mathbb{R}^3)$ with T defined by (1.11.20) by means of generalization of point Dirac distributions $\delta_{\mathbf{a}}$ as follows:

• Γ is bounded in $\mathbb{R}^3$: although this assumption is not necessary, for the sake of simplicity in presentation of all details, we assume that Γ is a bounded, smooth surface subdivided into N distinct subsets $S_1, S_2, \ldots, S_N$ such that $\Gamma = \bigcup_{k=1}^N S_k$ with

$$\begin{aligned} &\mathring{S}_k = \{\mathbf{x} : \mathbf{x} \in S_k, \mathbf{x} \notin \partial S_k, \partial S_k \text{ is the boundary curve of } S_k\}; \\ &\mathring{S}_j \cap \mathring{S}_k = \emptyset, \quad 1 \le j \ne k \le N; \end{aligned} \tag{1.11.21}$$

• $\triangle S_k =$ surface area measure μ_Γ of $S_k = \mu_\Gamma(S_k) > 0$, $1 \le k \le$ N;

• for $\mathbf{x}_k \in \mathring{S}_k$, $1 \le k \le N$, $\delta_{\mathbf{x}_k}$ is the Dirac distribution with charges concentrated at $\mathbf{x}_k$. Then $\rho(\mathbf{x}_k)\triangle S_k$ approximates the total charges on S_k and $\rho(\mathbf{x}_k)\triangle S_k\delta_{\mathbf{x}_k}$ is the Dirac distribution corresponding to approximate total charges on S_k concentrated at the point $\mathbf{x}_k$ $\forall k = 1, 2, \ldots, N$. Then, $\forall N \in \mathbb{N}$, the sum

$$\sum_{k=1}^{N} \rho(\mathbf{x}_k)\triangle S_k\delta_{\mathbf{x}_k} \in \mathcal{D}'(\mathbb{R}^3) \tag{1.11.22}$$

represents the approximate total charges on Γ corresponding to *discrete* distributions of charges on $S_1, S_2, \ldots, S_N$ with intensities $\rho(\mathbf{x}_1)\triangle S_1, \rho(\mathbf{x}_2)\triangle S_2, \ldots, \rho(\mathbf{x}_N)\triangle S_N$ concentrated at the points $\mathbf{x}_1, \mathbf{x}_2, \ldots, \mathbf{x}_N$ respectively.

• The volume/spatial density $\rho\delta_\Gamma \in \mathcal{D}'(\mathbb{R}^3)$ of charges corresponding to the continuous distribution of charges concentrated on surface Γ with surface density ρ is given by the limit of this sum $\sum_{k=1}^N \rho(\mathbf{x}_k)\triangle S_k\delta_{\mathbf{x}_k}$ in $\mathcal{D}'(\mathbb{R}^3)$, when $N \to \infty$ such that $\max_{1\le k\le N}\{\triangle S_k\} \to 0^+$, i.e.

$$\rho\delta_\Gamma = \lim_{N\to\infty} \sum_{k=1}^{N} \rho(\mathbf{x}_k)\triangle S_k\delta_{\mathbf{x}_k} \quad \text{in } \mathcal{D}'(\mathbb{R}^3) \quad \text{as} \quad \max_{1\le k\le N}\{\triangle S_k\} \to 0^+ \tag{1.11.23}$$

$$\begin{aligned} \implies \quad \langle \rho\delta_\Gamma, \phi\rangle &= \lim_{N\to\infty} \left\langle \sum_{k=1}^{N} \rho(\mathbf{x}_k)\triangle S_k\delta_{\mathbf{x}_k}, \phi \right\rangle \quad \forall\phi \in \mathcal{D}(\mathbb{R}^3) \\ &= \lim_{N\to\infty} \left(\sum_{k=1}^{N} \rho(\mathbf{x}_k)\triangle S_k\langle\delta_{\mathbf{x}_k}, \phi\rangle \right) \\ &= \lim_{\substack{N\to\infty \\ \max_K\{\triangle S_K\}\to 0^+}} \sum_{k=1}^{N} \rho(\mathbf{x}_k)\phi(\mathbf{x}_k)\triangle S_k \\ &= \int_\Gamma \rho(\mathbf{x})\phi(\mathbf{x})dS \quad \forall\phi \in \mathcal{D}(\mathbb{R}^3), \end{aligned}$$

since $\rho\phi$ is continuous on Γ and the limit exists by definition of surface integrals on Γ,

$$\implies \quad \langle \rho\delta_\Gamma, \phi\rangle = \int_\Gamma \rho(\mathbf{x})\phi(\mathbf{x})dS \quad \forall \phi \in \mathcal{D}(\mathbb{R}^3). \tag{1.11.24}$$

Then, from (1.11.20) and (1.11.24) we have

$$T = \rho\delta_\Gamma \quad \text{in } \mathcal{D}'(\mathbb{R}^3), \tag{1.11.25}$$

which is called the *simple layer* distribution on the surface Γ. To repeat again, it is the volume density of charges concentrated on the surface Γ with *surface density* ρ. Then $\rho\delta_\Gamma$ must restore the total charges $\int_\Gamma \rho(\mathbf{x})dS$. In fact, $\langle \rho\delta_\Gamma, 1\rangle = \int_\Gamma \rho(\mathbf{x})dS =$ total charges on Γ, since $\langle \rho\delta_\Gamma, \phi\rangle$ is well defined for $\phi \in C^\infty(\mathbb{R}^3)$ with *arbitrary* support (see Section 5.6) and hence, for $\phi = 1 \in C^\infty(\mathbb{R}^3)$, giving the result.

Double layer distributions Let Γ be a smooth, orientable (two-sided) surface such that the unit normal $\hat{\mathbf{n}}$ to Γ is defined at each point on Γ. Let $\nu = \nu(\mathbf{x})$ denote the continuous surface moment density of the normally oriented dipole moment distribution on surface Γ. Then a generalization of the discrete distribution of normally oriented dipole $-\frac{\partial}{\partial n}(\delta_\mathbf{a})$ of unit moment concentrated at the discrete points $\mathbf{a} \in \Gamma$ gives a distribution $-\frac{\partial}{\partial n}(\delta_\Gamma)$ corresponding to normally oriented dipoles of moment intensity $+1$ distributed on the surface Γ. Hence, $-\frac{\partial}{\partial n}(\nu\delta_\Gamma) \in \mathcal{D}'(\mathbb{R}^3)$ is the volume density of distribution corresponding to the distribution of normally oriented dipoles over the surface Γ with surface moment density ν.

But corresponding to the *physical distribution* of normally oriented dipoles over Γ with surface moment density ν, the mathematical distribution $T \in \mathcal{D}'(\mathbb{R}^3)$ is defined by:

$$\langle T, \phi\rangle = \int_\Gamma \nu(\mathbf{x})\frac{\partial\phi}{\partial n}(\mathbf{x})dS \quad \forall \phi \in \mathcal{D}(\mathbb{R}^3) \tag{1.11.26}$$

(dS is the surface area measure).

Now, we will show that $T = -\frac{\partial}{\partial n}(\nu\delta_\Gamma)$ in $\mathcal{D}'(\mathbb{R}^3)$. For this we are to use the results of Chapter 2 on the derivative $\frac{\partial}{\partial n}(\delta_{\mathbf{x}_k})$ of Dirac distribution $\delta_{\mathbf{x}_k} \in \mathcal{D}'(\mathbb{R}^3)$ in (2.3.8).

Instead of deferring the details, we give them here, which the reader may read after going through derivatives in Chapter 2.

- Γ is bounded in $\mathbb{R}^3$: although this assumption is not necessary, for the sake of simplicity in presentation of details we assume that Γ is a bounded, smooth surface subdivided into N distinct subsets $S_1, S_2, \dots, S_N$ such that

$$\Gamma = \bigcup_{k=1}^N S_k \quad \text{with } \overset{\circ}{S}_k = \{\mathbf{x} : \mathbf{x} \in S_k, \mathbf{x} \notin \text{boundary } \partial S_k \text{ of } S_k\}, \quad 1 \le k \le N; \tag{1.11.27}$$

- $\mathring{S}_j \cap \mathring{S}_k = \emptyset$ for $1 \le j \ne k \le N$;
- $\triangle S_k$ = surface area measure of $S_k = \mu_\Gamma(S_k) > 0, 1 \le k \le N$;
- $\hat{\mathbf{n}}_k$ is the unit normal to $\mathring{S}_k$ at $\mathbf{x}_k \in \mathring{S}_k$, $1 \le k \le N$;
- For $1 \le k \le N$, $\nu(\mathbf{x}_k)\triangle S_k$ approximates the total surface moment acting on S_k and $-\frac{\partial}{\partial n_k}(\nu(\mathbf{x}_k)\triangle S_k \delta_{\mathbf{x}_k})$ with $\mathbf{x}_k \in \mathring{S}_k$ is approximately the normally oriented dipole of moment $\nu(\mathbf{x}_k)\triangle S_k$ concentrated at the point $\mathbf{x}_k \in \mathring{S}_k$.
- Then, $\forall N \in \mathbb{N}$, the sum $\sum_{k=1}^N -\frac{\partial}{\partial n_k}[(\nu(\mathbf{x}_k)\triangle S_k \delta_{\mathbf{x}_k})] \in \mathcal{D}'(\mathbb{R}^3)$ approximates the volume density $-\frac{\partial}{\partial n}(\nu\delta_\Gamma) \in \mathcal{D}'(\mathbb{R}^3)$ of the normally oriented dipole distribution on Γ with surface moment density ν, and the (volume) density $-\frac{\partial}{\partial n}(\nu\delta_\Gamma)$ is the limit of this sum $\sum_{k=1}^N -\frac{\partial}{\partial n_k}[(\nu(\mathbf{x}_k)\triangle S_k \delta_{\mathbf{x}_k})]$ in $\mathcal{D}'(\mathbb{R}^3)$ when $N \to \infty$ such that $\max_{1\le k\le N}\{\triangle S_k\} \to 0^+$, i.e.

$$-\frac{\partial}{\partial n}(\nu\delta_\Gamma) = \lim_{N\to\infty}\left[\sum_{k=1}^N -\frac{\partial}{\partial n_k}(\nu(\mathbf{x}_k)\triangle S_k \delta_{\mathbf{x}_k})\right] \quad \text{in } \mathcal{D}'(\mathbb{R}^3) \tag{1.11.28}$$

$$\begin{aligned}
\implies \quad \left\langle -\frac{\partial}{\partial n}(\nu\delta_\Gamma), \phi\right\rangle &= \lim_{N\to\infty}\left\langle \sum_{k=1}^N -\frac{\partial}{\partial n_k}[\nu(\mathbf{x}_k)\triangle S_k \delta_{\mathbf{x}_k}], \phi\right\rangle \quad \forall \phi \in \mathcal{D}(\mathbb{R}^3)\\
&= \lim_{N\to\infty}\sum_{k=1}^N \left\langle -\frac{\partial}{\partial n_k}[\nu(\mathbf{x}_k)\triangle S_k \delta_{\mathbf{x}_k}], \phi\right\rangle\\
&= \lim_{N\to\infty}\sum_{k=1}^N \left\langle \nu(\mathbf{x}_k)\triangle S_k \delta_{\mathbf{x}_k}, \frac{\partial \phi}{\partial n_k}\right\rangle \quad \text{(see (2.3.9))}\\
&= \lim_{N\to\infty}\sum_{k=1}^N \nu(\mathbf{x}_k)\triangle S_k\left\langle \delta_{\mathbf{x}_k}, \frac{\partial \phi}{\partial n_k}\right\rangle \quad \text{with } \max_{1\le k\le N}\{\triangle S_k\} \to 0^+\\
&= \lim_{N\to\infty}\sum_{k=1}^N \nu(\mathbf{x}_k)\frac{\partial \phi}{\partial n_k}(\mathbf{x}_k)\triangle S_k = \int_\Gamma \nu(\mathbf{x})\frac{\partial \phi}{\partial n}(\mathbf{x})dS,
\end{aligned}$$

since $\nu\frac{\partial\phi}{\partial n}$ is continuous on Γ, and consequently the limit exists and the limit is the surface integral $\int_\Gamma \nu\frac{\partial\phi}{\partial n}dS$ by its definition, i.e.

$$\left\langle -\frac{\partial}{\partial n}(\nu\delta_\Gamma), \phi\right\rangle = \int_\Gamma \nu\frac{\partial \phi}{\partial n}dS \quad \forall \phi \in \mathcal{D}(\mathbb{R}^3). \tag{1.11.29}$$

Then, from (1.11.26) and (1.11.29), we get $\langle T, \phi\rangle = \langle -\frac{\partial}{\partial n}(\nu\delta_\Gamma), \phi\rangle \ \forall \phi \in \mathcal{D}(\mathbb{R}^3)$,

$$\implies \quad T = -\frac{\partial}{\partial n}(\nu\delta_\Gamma) \in \mathcal{D}'(\mathbb{R}^3) \tag{1.11.30}$$

is called the *double layer distribution* of a normally oriented dipole over the surface Γ with continuous surface moment density ν.

1.11.5 Relation with probability distribution [7]

A natural interesting question arises in our mind: is there any relation with probability distributions? Instead of entering this new domain of calculus of probability, which is outside the scope of the present treatment of topics based on the *principle of determinism*, we would like to indicate here only when a distribution $T \in \mathcal{D}'(\mathbb{R})$ is related to probability distribution at the *most elementary level* of the calculus of probabilities.

Let $f \in L^1(]-\infty, \infty[)$ be an integrable function on $]-\infty, \infty[$ with

$$f(x) \geq 0, \int_{-\infty}^{\infty} f(x)dx = 1 \tag{1.11.31}$$

such that $f(x)$ defines the probability distribution on $(]-\infty, \infty[)$, and the probability $P \in [0, 1]$ of the variable x lying on $(]a, b[)$ is given by:

$$P(a < x < b) = \int_a^b f(x)dx.$$

Hence,

$$P(-\infty < x < \infty) = \int_{-\infty}^{\infty} f(x)dx = 1 = \langle f, 1\rangle = \langle T_f, 1\rangle. \tag{1.11.32}$$

A distribution $T \in \mathcal{D}'(\mathbb{R})$ defines a probability distribution if and only if

1. T is a positive distribution (see Definition 1.7.2), i.e.

$$T \geq 0 \quad (\Longleftrightarrow \langle T, \phi\rangle \geq 0 \; \forall \phi \in \mathcal{D}(\mathbb{R}) \text{ with } \phi \geq 0); \tag{1.11.33}$$

2. $\langle T, 1\rangle$ is well defined in *some sense*, for example,

$$T \geq 0, \quad \langle T, 1\rangle = \sup_{0\leq\phi\leq 1} \langle T, \phi\rangle \quad \text{with } \phi \in \mathcal{D}(\mathbb{R})) \text{ and } \langle T, 1\rangle = 1 \tag{1.11.34}$$

(see (1.11.1)) [8].

We complete our discussion with the following examples:

(a) $T = \delta_a = \delta(x - a) \in \mathcal{D}'(\mathbb{R})$. Then, $\delta_a \geq 0$, since $\langle \delta_a, \phi\rangle = \phi(a) \geq 0$ $\forall \phi \in \mathcal{D}(\mathbb{R})$ with $\phi \geq 0$. $\langle \delta_a, 1\rangle = 1$ (see (1.11.1)). Hence, δ_a defines a probability distribution according to which the variable x can have only one value a with probability $P = 1$, i.e. the value $x = a$ is *certain*.

(b) $T = \frac{1}{4}\delta_a + \frac{3}{4}\delta_b \in \mathcal{D}'(\mathbb{R})$ with $a \neq b$. Then $\frac{1}{4}\delta_a + \frac{3}{4}\delta_b \geq 0$, since

$$\left\langle \frac{1}{4}\delta_a + \frac{3}{4}\delta_b, \phi \right\rangle = \frac{1}{4}\langle \delta_a, \phi\rangle + \frac{3}{4}\langle \delta_b, \phi\rangle = \frac{1}{4}\phi(a) + \frac{3}{4}\phi(b) \geq 0$$
$$\forall \phi \in \mathcal{D}(\mathbb{R}) \text{ with } \phi \geq 0.$$

$$\left\langle \frac{1}{4}\delta_a + \frac{3}{4}\delta_b, 1 \right\rangle = \frac{1}{4}\langle \delta_a, 1\rangle + \frac{3}{4}\langle \delta_b, 1\rangle = \frac{1}{4}\cdot 1 + \frac{3}{4}\cdot 1 = 1 \quad \text{(see (1.11.1))}.$$

Hence, $\frac{1}{4}\delta_a + \frac{3}{4}\delta_b$ defines a probability distribution, according to which x can take only two values $x = a$ and $x = b$ with the probability $\frac{1}{4}$ and probability $\frac{3}{4}$ respectively.

Remark 1.11.3. Laurent Schwartz [7] is of the opinion that we can *not* logically prove the analogy between physical and mathematical distributions in general. Then the natural question arises: *What are these mathematical distributions, after all?*

The answer, according to him [7], is that *mathematical distributions constitute the mathematically rigorous definitions of physical distributions.*

Chapter 2

Differentiation of distributions and application of distributional derivatives

2.1 Introduction: an integral definition of derivatives of C^1-functions

The notion of the derivatives of distributions is perhaps the most useful and important one in applications, especially in the study of Sobolev spaces and partial differential equations. This is due to the fact that the derivative $\frac{\partial T}{\partial x_i}$ of a distribution T on $\Omega \subset \mathbb{R}^n$ with respect to the variable x_i, $1 \le i \le n$, is defined in such a way that if the distribution T is a function f with continuous partial derivatives on $\overline{\Omega}$, $\overline{\Omega}$ being the closure of Ω in $\mathbb{R}^n$ (i.e. $f \in C^1(\overline{\Omega})$, $T = T_f = f \in \mathcal{D}'(\Omega)$ by (1.3.15) $\Longrightarrow \langle f, \phi\rangle = \langle T_f, \phi\rangle = \int_\Omega f(\mathbf{x})\phi(\mathbf{x})d\mathbf{x}\ \forall \phi \in \mathcal{D}(\Omega)$), then we can retrieve the partial derivative $\frac{\partial f}{\partial x_i}$ of the function f in the usual pointwise sense:

$$\frac{\partial f}{\partial x_i}(\mathbf{x}) = \lim_{\Delta x_i \to 0} \frac{f(x_1, \ldots, x_i + \Delta x_i, \ldots, x_n) - f(x_1, \ldots, x_n)}{\Delta x_i}$$

$$\forall \mathbf{x} = (x_1, x_2, \ldots, x_n) \in \Omega, \quad (2.1.1)$$

and importantly, unlike functions, distributions T on $\Omega \subset \mathbb{R}^n$ are *infinitely differentiable* on Ω and can be differentiated with respect to several variables in an arbitrary order of differentiation with respect to the variables. (For mixed partial derivatives of functions, such a result holds if Schwarz's theorem on mixed partial derivatives of functions holds.) In other words, if a function $f \in C^1(\overline{\Omega})$ (i.e. has bounded and uniformly continuous partial derivatives $\frac{\partial f}{\partial x_i}$, $1 \le i \le n$ on Ω), then the derivative $\frac{\partial T_f}{\partial x_i}$ of the distribution T_f associated with $f \in C^1(\overline{\Omega})$ is the distribution $T_{\frac{\partial f}{\partial x_i}}$ associated with $\frac{\partial f}{\partial x_i} \in C^0(\overline{\Omega})$, i.e. $\forall \phi \in \mathcal{D}(\Omega)$,

$$\left\langle \frac{\partial T_f}{\partial x_i}, \phi \right\rangle = \langle T_{\frac{\partial f}{\partial x_i}}, \phi\rangle = \int_\Omega \frac{\partial f}{\partial x_i}\phi d\mathbf{x} = \left\langle \frac{\partial f}{\partial x_i}, \phi \right\rangle, \quad (2.1.2)$$

$\Longrightarrow$ the distribution $\frac{\partial T_f}{\partial x_i}$ on Ω is the function $\frac{\partial f}{\partial x_i} \in C^0(\overline{\Omega})$ on Ω in the usual pointwise sense (2.1.1), $C^1(\overline{\Omega}) = \{u :$ function u and its partial derivatives $\frac{\partial u}{\partial x_i}$, $1 \le i \le n$, are *bounded* and *uniformly continuous* on Ω, which have unique *continuous* extension to $\overline{\Omega}\}$.

An integral definition of derivatives of functions of $C^1(\overline{\Omega})$

Now we show that for any function $f \in C^1(\overline{\Omega})$ and its derivative $\frac{\partial f}{\partial x_i} \in C^0(\overline{\Omega})$, $(1 \leq i \leq n)$, the following relation holds: $\forall \phi \in \mathcal{D}(\Omega)$,

$$\int_\Omega \frac{\partial f}{\partial x_i}\phi(\mathbf{x})d\mathbf{x} = -\int_\Omega f(\mathbf{x})\frac{\partial \phi}{\partial x_i}(\mathbf{x})d\mathbf{x}, \quad d\mathbf{x} = dx_1 dx_2 \ldots dx_n. \tag{2.1.3}$$

One dimensional case, $n = 1$ First of all, we consider the simplest case of functions of a single variable x: $n = 1, \Omega = \mathbb{R} =]-\infty, \infty[$ with $\overline{\Omega} = \mathbb{R} =]-\infty, \infty[$, $f \in C^1(]-\infty, \infty[)$, $\frac{df}{dx} \in C^0(]-\infty, \infty[)$. Then, by integrating by parts, $\forall \phi \in \mathcal{D}(]-\infty, \infty[)$,

$$\int_{-\infty}^{+\infty} \frac{df}{dx}(x)\phi(x)dx = [f(x)\phi(x)]_{-\infty}^{+\infty} - \int_{-\infty}^{+\infty} f(x)\frac{d\phi}{dx}(x)dx. \tag{2.1.4}$$

But $\phi \in \mathcal{D}(]-\infty, \infty[) \implies \exists$ a bounded, closed interval $[a, b] \subset \mathbb{R}$ such that $\text{supp}(\phi) \subset [a, b] \implies \phi(x) = 0 \quad \forall x$ lying outside $[a, b] \implies [f(x)\phi(x)]_{-\infty}^{+\infty} = 0$.

Hence, $\int_\mathbb{R} \frac{df}{dx}(x)\phi(x)dx = -\int_\mathbb{R} f(x)\frac{d\phi}{dx}(x)dx \quad \forall \phi \in \mathcal{D}(\mathbb{R})$.

Similarly, for $n = 1$, $\Omega =]a, b[\subset \mathbb{R}$ with $-\infty < a < b < +\infty$, $\overline{\Omega} = [a, b]$, $f \in C^1([a, b])$, $\frac{df}{dx} \in C^0([a, b])$, we have

$$\int_a^b \frac{df}{dx}(x)\phi(x)dx = -\int_a^b f(x)\frac{d\phi}{dx}(x)dx \quad \forall \phi \in \mathcal{D}(]a, b[). \tag{2.1.5}$$

The converse result also holds. Suppose that $f \in C^1([a, b])$. If $\exists g \in C^0([a, b])$ such that

$$\int_a^b g(x)\phi(x)dx = -\int_a^b f(x)\frac{d\phi}{dx}(x)dx \quad \forall \phi \in \mathcal{D}(]a, b[), \tag{2.1.6}$$

then $\frac{df}{dx} = g$ in $C^0([a, b])$, i.e. g is the *usual* derivative of f with respect to x on $[a, b]$.

Proof. Since $f \in C^1([a, b])$, $\frac{df}{dx} \in C^0([a, b])$.Then, from (2.1.5), we have

$$-\int_a^b f(x)\frac{d\phi}{dx}(x)dx = \int_a^b \frac{df}{dx}(x)\phi(x)dx = \int_a^b g(x)\phi(x)dx \quad \forall \phi \in \mathcal{D}(]a, b[)$$

$\implies \int_a^b (\frac{df}{dx} - g)\phi dx = 0 \ \forall \phi \in \mathcal{D}(]a, b[)$. But $\frac{df}{dx} - g \in C^0([a, b]) \subset L^1_{\text{loc}}(]a, b[)$. Hence, by Theorem 1.2.3A, $\frac{df}{dx} - g = 0 \implies \frac{df}{dx} = g$ in $C^0([a, b])$. □

Conclusion For functions $f \in C^1([a, b])$, the derivative $g = \frac{df}{dx}$ of f can be defined by the integral relation (2.1.5) instead of the definition:

$$\frac{df}{dx}(x) = \lim_{\Delta x \to 0} \frac{f(x + \Delta x) - f(x)}{\Delta x} \quad \forall x \in]a, b[,$$

both the results being equivalent.

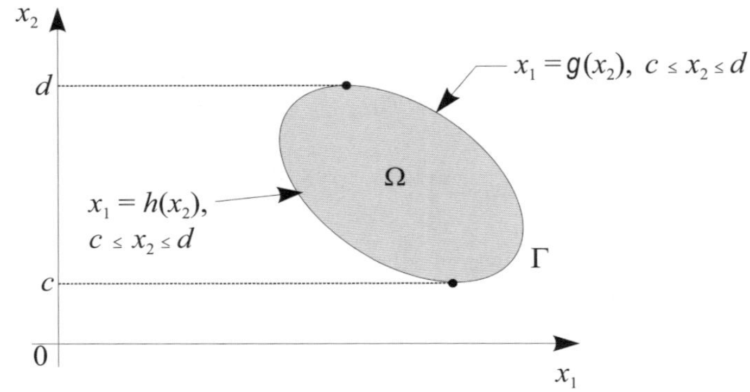

Figure 2.1 Domain Ω enclosed by $x_1 = g(x_2), x_1 = h(x_2), c \le x_2 \le d$, such that a line parallel to the x_1-axis intersects Γ at no more than two points

Two-dimensional case, $n = 2$ Now we consider $\Omega \subset \mathbb{R}^2$ with boundary Γ such that $\Omega = \{(x_1, x_2) : \forall$ fixed $x_2 \in \mathbb{R}$ with $c < x_2 < d, h(x_2) < x_1 < g(x_2)\}$, i.e. Ω is enclosed by two curves $x_1 = g(x_2)$ and $x_1 = h(x_2)$ as shown in Figure 2.1 such that any line parallel to the x_1-axis intersects Γ at no more than two points. (This assumption has been made for the sake of simplicity in presentation. For domains Ω not satisfying this assumption, the proof is slightly modified. For example, Ω may be subdivided into subdomains $\{\Omega_i\}$, in each of which this assumption holds.) Suppose that $f \in C^1(\overline{\Omega})$, $\frac{\partial f}{\partial x_i} \in C^0(\overline{\Omega})$. We prove (2.1.3) for $i = 1$. Transforming the double integral over Ω into iterative (definite) integrals, we get

$$\int_\Omega \frac{\partial f}{\partial x_1}(x_1, x_2)\phi(x_1, x_2)dx_1 dx_2 = \int_c^d dx_2 \int_{x_1=h(x_2)}^{x_1=g(x_2)} \frac{\partial f}{\partial x_1}(x_1, x_2)\phi(x_1, x_2)dx_1,$$

in which x_2 is held fixed on $]c, d[$. Hence, we can integrate by parts the right-hand side integral with respect to x_1: $\forall \phi \in \mathcal{D}(\Omega)$,

$$\int_{x_1=h(x_2)}^{x_1=g(x_2)} \frac{\partial f}{\partial x_1}(x_1, x_2)\phi(x_1, x_2)dx_1 = [f(x_1, x_2)\phi(x_1, x_2)]_{x_1=h(x_2)}^{x_1=g(x_2)} - \int_{x_1=h(x_2)}^{x_1=g(x_2)} f(x_1, x_2)\frac{\partial \phi}{\partial x_1}(x_1, x_2)dx_1,$$

since $\frac{\partial f}{\partial x_1}\phi = \frac{\partial (f\phi)}{\partial x_1} - f\frac{\partial \phi}{\partial x_1}$.

But $\phi \in \mathcal{D}(\Omega) \Longrightarrow \phi\downarrow_\Gamma = 0 \Longrightarrow \phi(x_1, x_2)|_{x_1=g(x_2)} = \phi(x_1, x_2)|_{x_1=h(x_2)} = 0$ $\Longrightarrow [f(x_1, x_2)\phi(x_1, x_2)]_{x_1=h(x_2)}^{x_1=g(x_2)} = 0$. Hence, $\forall \phi \in \mathcal{D}(\Omega)$,

$$\int_{x_1=h(x_2)}^{x_1=g(x_2)} \frac{\partial f}{\partial x_1}(x_1, x_2)\phi(x_1, x_2)dx_1 = -\int_{x_1=h(x_2)}^{x_1=g(x_2)} f(x_1, x_2)\frac{\partial \phi}{\partial x_1}(x_1, x_2)dx_1$$

and

$$\int_\Omega \frac{\partial f}{\partial x_1}(x_1, x_2)\phi(x_1, x_2)dx_1dx_2 = -\int_c^d dx_2 \int_{x_1=h(x_2)}^{x_1=g(x_2)} f(x_1, x_2)\frac{\partial \phi}{\partial x_1}(x_1, x_2)dx_1$$
$$= -\int_\Omega f(x_1, x_2)\frac{\partial \phi}{\partial x_1}(x_1, x_2)dx_1dx_2.$$

Hence, $\forall f \in C^1(\overline{\Omega})$ with $\frac{\partial f}{\partial x_1} \in C^0(\overline{\Omega})$, the relation (2.1.3) holds for $i = 1$:

$$\int_\Omega \frac{\partial f}{\partial x_1}\phi d\mathbf{x} = -\int_\Omega f\frac{\partial \phi}{\partial x_1}d\mathbf{x} \quad \forall \phi \in \mathcal{D}(\Omega) \quad (d\mathbf{x} = dx_1dx_2).$$

Similarly, we can prove (2.1.3) for $i = 2$:

$$\int_\Omega \frac{\partial f}{\partial x_2}\phi d\mathbf{x} = -\int_\Omega f\frac{\partial \phi}{\partial x_2}d\mathbf{x} \quad \forall \phi \in \mathcal{D}(\Omega).$$

***n*-dimensional case** Thus, we can prove that for $f \in C^1(\overline{\Omega})$, $\frac{\partial f}{\partial x_i} \in C^0(\overline{\Omega})$ $\Omega \subset \mathbb{R}^n$, the relation (2.1.3) holds: for $1 \le i \le n$,

$$\int_\Omega \frac{\partial f}{\partial x_i}\phi d\mathbf{x} = -\int_\Omega f\frac{\partial \phi}{\partial x_i}d\mathbf{x} \quad \forall \phi \in \mathcal{D}(\Omega) \quad (d\mathbf{x} = dx_1dx_2\ldots dx_n).$$

The converse result also holds. Suppose that $f \in C^1(\overline{\Omega})$ with bounded $\Omega \subset \mathbb{R}^n$. If $\exists g_i \in C^0(\overline{\Omega})$, $1 \le i \le n$, such that

$$\int_\Omega g_i(\mathbf{x})\phi(\mathbf{x})d\mathbf{x} = -\int_\Omega f(\mathbf{x})\frac{\partial \phi}{\partial x_i}(\mathbf{x})d\mathbf{x} \quad \forall \phi \in \mathcal{D}(\Omega), \tag{2.1.7}$$

then $\frac{\partial f}{\partial x_i} = g_i$ in $C^0(\overline{\Omega})$, $1 \le i \le n$, i.e. g_i is the *usual* partial derivative of f with respect to x_i in Ω.

Proof. Since $f \in C^1(\overline{\Omega})$, $\frac{\partial f}{\partial x_i} \in C^0(\overline{\Omega})$ for $1 \le i \le n$. Then, from (2.1.3), we have: for $1 \le i \le n$,

$$-\int_\Omega f(\mathbf{x})\frac{\partial \phi}{\partial x_i}(\mathbf{x})d\mathbf{x} = \int_\Omega \frac{\partial f}{\partial x_i}(\mathbf{x})\phi(\mathbf{x})d\mathbf{x} = \int_\Omega g_i(\mathbf{x})\phi(\mathbf{x})d\mathbf{x} \quad \forall \phi \in \mathcal{D}(\Omega)$$

$\Longrightarrow \int_\Omega (\frac{\partial f}{\partial x_i} - g_i)\phi d\mathbf{x} = 0$ $\forall \phi \in \mathcal{D}(\Omega) \Longrightarrow \frac{\partial f}{\partial x_i} - g_i = 0$ in $C^0(\overline{\Omega}) \subset L^1_{\text{loc}}(\Omega)$ by Theorem 1.2.3A $\Longrightarrow \frac{\partial f}{\partial x_i} = g_i$ in $C^0(\overline{\Omega})$, i.e. g_i is the usual derivative of f with respect to x_i in $\overline{\Omega}$. □

Conclusion For functions f of n variables with $f \in C^1(\overline{\Omega})$, $\Omega \subset \mathbb{R}^n$, partial derivatives $g_i = \frac{\partial f}{\partial x_i}$ on Ω can be defined by the integral relation in (2.1.3) instead of the usual definition (2.1.1):

$$\begin{aligned} g_i(\mathbf{x}) &= \frac{\partial f}{\partial x_i}(\mathbf{x}) \\ &= \lim_{\Delta x_i \to 0} \frac{f(x_1, x_2, \ldots, x_i + \Delta x_i, \ldots, x_n) - f(x_1, x_2, \ldots, x_n)}{\Delta x_i} \quad \forall \mathbf{x} \in \Omega, \end{aligned}$$

the results being the same.

Remark 2.1.1. It is of general interest that Sobolev [30] extended the integral definition (2.1.3) to (discontinuous and unbounded) functions $f \in L^2(\Omega)$ as follows: suppose that $f \in L^2(\Omega)$ with $\Omega \subset \mathbb{R}^n$. If $\exists g_i \in L^2(\Omega)$, $1 \le i \le n$, such that

$$\int_\Omega g_i \phi d\mathbf{x} = -\int_\Omega f \frac{\partial \phi}{\partial x_i} d\mathbf{x} \quad \forall \phi \in C_0^\infty(\Omega) = \mathcal{D}(\Omega), \tag{2.1.8}$$

then $g_i = \frac{\partial f}{\partial x_i} \in L^2(\Omega)$ is called the *generalized derivative* of $f \in L^2(\Omega)$ with respect to the variable x_i. (Obviously, g_i is not the partial derivative of f in the usual pointwise sense (2.1.1), which, in fact, does not exist in general in Ω for $f \in L^2(\Omega)$.)

2.2 Derivatives of distributions

From (1.3.17), $C^1(\overline{\Omega}), C^0(\overline{\Omega}) \subset \mathcal{D}'(\Omega)$, we can identify $f \in C^1(\overline{\Omega})$ with $T_f \in \mathcal{D}'(\Omega)$ and $\frac{\partial f}{\partial x_i} \in C^0(\overline{\Omega})$ with $T_{\frac{\partial f}{\partial x_i}} \in \mathcal{D}'(\Omega)$. Hence, we set $f = T_f \in \mathcal{D}'(\Omega)$,

$$\Longrightarrow \quad \langle f, \phi \rangle = \langle T_f, \phi \rangle = \int_\Omega f(\mathbf{x})\phi(\mathbf{x}) d\mathbf{x}; \quad \frac{\partial f}{\partial x_i} = T_{\frac{\partial f}{\partial x_i}} \in \mathcal{D}'(\Omega)$$

$$\Longrightarrow \quad \langle T_{\frac{\partial f}{\partial x_i}}, \phi \rangle = \left\langle \frac{\partial f}{\partial x_i}, \phi \right\rangle = \int_\Omega \frac{\partial f}{\partial x_i}(\mathbf{x})\phi(\mathbf{x}) d\mathbf{x} \forall \phi \in \mathcal{D}(\Omega), \tag{2.2.1}$$

where $\langle \cdot, \cdot \rangle$ denotes the duality pairing $\langle \cdot, \cdot \rangle_{\mathcal{D}'(\Omega) \times \mathcal{D}(\Omega)}$ between $\mathcal{D}'(\Omega)$ and $\mathcal{D}(\Omega)$.

Consequently, using (2.2.1), we can rewrite (2.1.3) in the following form:

$$\left\langle \frac{\partial f}{\partial x_i}, \phi \right\rangle = -\left\langle f, \frac{\partial \phi}{\partial x_i} \right\rangle \quad \forall \phi \in \mathcal{D}(\Omega). \tag{2.2.2}$$

Hence, this relation (2.2.2) between function $f \in C^1(\overline{\Omega})$ and its derivative $\frac{\partial f}{\partial x_i} \in C^0(\overline{\Omega})$ will be preserved if we can define the derivative $\frac{\partial T}{\partial x_i}$ of a distribution $T \in \mathcal{D}'(\Omega)$ on $\Omega \subset \mathbb{R}^n$ by the equation:

$$\left\langle \frac{\partial T}{\partial x_i}, \phi \right\rangle = -\left\langle T, \frac{\partial \phi}{\partial x_i} \right\rangle \quad \forall \phi \in \mathcal{D}(\Omega). \tag{2.2.3}$$

In order to prove that this definition (2.2.3) of $\frac{\partial T}{\partial x_i}$ is meaningful, we are to show that $S = \frac{\partial T}{\partial x_i}$ defined by (2.2.3), $\langle S, \phi\rangle = -\langle T, \frac{\partial \phi}{\partial x_i}\rangle$ $\forall \phi \in \mathcal{D}(\Omega)$, is a *continuous, linear* functional on $\mathcal{D}(\Omega)$.

Linearity of S:

$$\begin{aligned}\langle S, \alpha_1\phi_1 + \alpha_2\phi_2\rangle &= -\left\langle T, \frac{\partial}{\partial x_i}(\alpha_1\phi_1 + \alpha_2\phi_2)\right\rangle = -\alpha_1\left\langle T, \frac{\partial \phi_1}{\partial x_i}\right\rangle - \alpha_2\left\langle T, \frac{\partial \phi_2}{\partial x_i}\right\rangle \\ &= \alpha_1\langle S, \phi_1\rangle + \alpha_2\langle S, \phi_2\rangle \quad \forall \phi_1, \phi_2 \in \mathcal{D}(\Omega).\end{aligned}$$

Continuity of S:

$\phi_n \to \phi$ in $\mathcal{D}(\Omega) \Longrightarrow \frac{\partial \phi_n}{\partial x_i} \to \frac{\partial \phi}{\partial x_i}$ in $\mathcal{D}(\Omega)$ as $n \to \infty$. But $T \in \mathcal{D}'(\Omega)$ is continuous on $\mathcal{D}(\Omega)$. Hence, $\frac{\partial \phi_n}{\partial x_i} \to \frac{\partial \phi}{\partial x_i}$ in $\mathcal{D}(\Omega) \Longrightarrow \langle T, \frac{\partial \phi_n}{\partial x_i}\rangle \to \langle T, \frac{\partial \phi}{\partial x_i}\rangle$ in $\mathbb{R}$ as $n \to \infty \Longrightarrow \langle T, \frac{\partial \phi_n}{\partial x_i}\rangle = -\langle S, \phi_n\rangle \to \langle T, \frac{\partial \phi}{\partial x_i}\rangle = -\langle S, \phi\rangle$ in $\mathbb{R}$ as $n \to \infty$. Thus, $\phi_n \to \phi$ in $\mathcal{D}(\Omega) \Longrightarrow \langle S, \phi_n\rangle \to \langle S, \phi\rangle$ in $\mathbb{R}$ as $n \to \infty$.

$\Longrightarrow$ S is a continuous, linear functional on $\mathcal{D}(\Omega) \Longrightarrow S \in \mathcal{D}'(\Omega)$.

Hence, $S = \frac{\partial T}{\partial x_i} \in \mathcal{D}'(\Omega)$ defined by (2.2.3) is a distribution on Ω.

Second-order derivatives

$\frac{\partial^2 T}{\partial x_i \partial x_j}$ of distributions $T \in \mathcal{D}'(\Omega)$ are defined using (2.2.3): $\forall \phi \in \mathcal{D}(\Omega)$, $1 \le i \le n$,

$$\left\langle \frac{\partial^2 T}{\partial x_i \partial x_j}, \phi\right\rangle = \left\langle \frac{\partial}{\partial x_i}\left(\frac{\partial T}{\partial x_j}\right), \phi\right\rangle = -\left\langle \frac{\partial T}{\partial x_j}, \frac{\partial \phi}{\partial x_i}\right\rangle = \left\langle T, \frac{\partial^2 \phi}{\partial x_j \partial x_i}\right\rangle; \tag{2.2.4}$$

$$\left\langle \frac{\partial^2 T}{\partial x_j \partial x_i}, \phi\right\rangle = \left\langle \frac{\partial}{\partial x_j}\left(\frac{\partial T}{\partial x_i}\right), \phi\right\rangle = -\left\langle \frac{\partial T}{\partial x_i}, \frac{\partial \phi}{\partial x_j}\right\rangle = \left\langle T, \frac{\partial^2 \phi}{\partial x_i \partial x_j}\right\rangle. \tag{2.2.5}$$

But $\phi \in \mathcal{D}(\Omega) \Longrightarrow \frac{\partial^2 \phi}{\partial x_i \partial x_j}, \frac{\partial^2 \phi}{\partial x_j \partial x_i}$ are continuous in $\Omega \Longrightarrow \frac{\partial^2 \phi}{\partial x_i \partial x_j} = \frac{\partial^2 \phi}{\partial x_j \partial x_i}$ by Schwarz's theorem. Hence, $\forall \phi \in \mathcal{D}(\Omega)$, $\langle T, \frac{\partial^2 \phi}{\partial x_j \partial x_i}\rangle = \langle T, \frac{\partial^2 \phi}{\partial x_i \partial x_j}\rangle \Longrightarrow \langle \frac{\partial^2 T}{\partial x_i \partial x_j}, \phi\rangle = \langle \frac{\partial^2 T}{\partial x_j \partial x_i}, \phi\rangle$ $\forall \phi \in \mathcal{D}(\Omega) \Longleftrightarrow \frac{\partial^2 T}{\partial x_i \partial x_j} = \frac{\partial^2 T}{\partial x_j \partial x_i}$ in $\mathcal{D}'(\Omega)$, i.e. *any change of order of differentiation of a distribution T is permissible.*

2.2.1 Higher-order derivatives of distributions *T*

For $\boldsymbol{\alpha} = (\alpha_1, \alpha_2, \dots, \alpha_n)$ with integers $\alpha_i \ge 0$ for $i = 1, 2, \dots, n$, $|\boldsymbol{\alpha}| = \alpha_1 + \alpha_2 + \dots + \alpha_n \in \mathbb{N}$,

$$\partial^{\boldsymbol{\alpha}} = \frac{\partial^{\alpha_1}}{\partial x_1^{\alpha_1}}\left(\frac{\partial^{\alpha_2}}{\partial x_2^{\alpha_2}}\left(\cdots\left(\frac{\partial^{\alpha_n}}{\partial x_n^{\alpha_n}}\right)\cdots\right)\right) = \frac{\partial^{\alpha_1+\alpha_2+\dots+\alpha_n}}{\partial x_1^{\alpha_1}\partial x_2^{\alpha_2}\dots\partial x_n^{\alpha_n}} = \frac{\partial^{|\boldsymbol{\alpha}|}}{\partial x_1^{\alpha_1}\partial x_2^{\alpha_2}\dots\partial x_n^{\alpha_n}}$$

denotes the *partial differential operator of order* $|\boldsymbol{\alpha}| \in \mathbb{N}$. For example,

$$\partial^{(1,2,3)} = \frac{\partial^6}{\partial x_1 \partial x_2^2 \partial x_3^3}; \quad \partial^{(4,0)} = \frac{\partial^4}{\partial x_1^4}; \quad \partial^{(0,4)} = \frac{\partial^4}{\partial x_2^4}; \quad \partial^{(2,2)} = \frac{\partial^4}{\partial x_1^2 \partial x_2^2}; \quad \text{etc.}$$

Then the derivative $\partial^{\boldsymbol{\alpha}} T = \frac{\partial^{|\boldsymbol{\alpha}|} T}{\partial x_1^{\alpha_1} \partial x_2^{\alpha_2} \dots \partial x_n^{\alpha_n}}$ (of order $|\boldsymbol{\alpha}| \in \mathbb{N}$) of the distribution $T \in \mathcal{D}'(\Omega)$ on Ω is defined by: $\langle \partial^{\boldsymbol{\alpha}} T, \phi \rangle = (-1)^{|\boldsymbol{\alpha}|} \langle T, \partial^{\boldsymbol{\alpha}} \phi \rangle \ \forall \phi \in \mathcal{D}(\Omega)$ or, equivalently, $\langle \frac{\partial^{|\boldsymbol{\alpha}|} T}{\partial x_1^{\alpha_1} \partial x_2^{\alpha_2} \dots \partial x_n^{\alpha_n}}, \phi \rangle = (-1)^{|\boldsymbol{\alpha}|} \langle T, \frac{\partial^{|\boldsymbol{\alpha}|} \phi}{\partial x_1^{\alpha_1} \partial x_2^{\alpha_2} \dots \partial x_n^{\alpha_n}} \rangle$.

Finally, we state the result:

Theorem 2.2.1. *Every distribution $T \in \mathcal{D}'(\Omega)$ is* infinitely differentiable, *i.e. has successive derivatives of* all *orders that are distributions of $\mathcal{D}'(\Omega)$, and* the order of differentiation may be interchanged in an arbitrary manner. *The derivatives $\partial^{\boldsymbol{\alpha}} T$ of T are defined by: $\forall |\boldsymbol{\alpha}| \in \mathbb{N}, \forall T \in \mathcal{D}'(\Omega)$,*

$$\langle \partial^{\boldsymbol{\alpha}} T, \phi \rangle = (-1)^{|\boldsymbol{\alpha}|} \langle T, \partial^{\boldsymbol{\alpha}} \phi \rangle \quad \forall \phi \in \mathcal{D}(\Omega), \tag{2.2.6}$$

where $\partial^{\boldsymbol{\alpha}} T = \frac{\partial^{|\boldsymbol{\alpha}|} T}{\partial x_1^{\alpha_1} \partial x_2^{\alpha_2} \dots \partial x_n^{\alpha_n}}$, $\partial^{\boldsymbol{\alpha}} \phi = \frac{\partial^{|\boldsymbol{\alpha}|} \phi}{\partial x_1^{\alpha_1} \partial x_2^{\alpha_2} \dots \partial x_n^{\alpha_n}}$, $|\boldsymbol{\alpha}| = \alpha_1 + \alpha_2 + \dots + \alpha_n$.

2.3 Derivatives of functions in the sense of distribution

Let $f \in L^1_{\text{loc}}(\Omega) \subset \mathcal{D}'(\Omega)$ (see (1.3.15)) be a locally summable (integrable) function on Ω (which, in particular, may belong to $C^0(\overline{\Omega})$ or $L^2(\Omega)$ or $L^p(\Omega)$, $1 \le p \le \infty$) with $f = T_f \in \mathcal{D}'(\Omega)$.

Definition 2.3.1. $\forall |\boldsymbol{\alpha}| \in \mathbb{N}$, the derivative $\partial^{\boldsymbol{\alpha}} f = \frac{\partial^{|\boldsymbol{\alpha}|} f}{\partial x_1^{\alpha_1} \partial x_2^{\alpha_2} \dots \partial x_n^{\alpha_n}}$ defined by (2.2.6), i.e.

$$\begin{aligned}\langle \partial^{\boldsymbol{\alpha}} f, \phi \rangle &= \langle \partial^{\boldsymbol{\alpha}} T_f, \phi \rangle = (-1)^{|\boldsymbol{\alpha}|} \langle T_f, \partial^{\boldsymbol{\alpha}} \phi \rangle = (-1)^{|\boldsymbol{\alpha}|} \langle f, \partial^{\boldsymbol{\alpha}} \phi \rangle \\ &= (-1)^{|\boldsymbol{\alpha}|} \int_\Omega f(\mathbf{x}) \frac{\partial^{|\boldsymbol{\alpha}|} \phi(\mathbf{x})}{\partial x_1^{\alpha_1} \partial x_2^{\alpha_2} \dots \partial x_n^{\alpha_n}} d\mathbf{x} \quad \forall \phi \in \mathcal{D}(\Omega)\end{aligned} \tag{2.3.1}$$

is called the *derivative* of the function $f \in L^1_{\text{loc}}(\Omega)$ of *order* $|\boldsymbol{\alpha}| \in \mathbb{N}$ with respect to $x_1, x_2, \dots, x_n$ *in the sense of distribution*, or is equivalently called the *distributional derivative of f* of order $|\boldsymbol{\alpha}|$ with respect to $x_1, x_2, \dots, x_n$.

Some explanations on alternative notations used in (2.3.1) are in order.

Since $f \in L^1_{\text{loc}}(\Omega)$, we set $f = T_f \in \mathcal{D}'(\Omega)$ and, hence, we write: $\langle T_f, \partial^{\boldsymbol{\alpha}} \phi \rangle = \langle f, \partial^{\boldsymbol{\alpha}} \phi \rangle = \int_\Omega f(\mathbf{x}) \partial^{\boldsymbol{\alpha}} \phi(\mathbf{x}) d\mathbf{x}$ on the right-hand side of (2.3.1).

But $f = T_f \in \mathcal{D}'(\Omega) \Longrightarrow \partial^{\boldsymbol{\alpha}} f = \partial^{\boldsymbol{\alpha}} T_f \in \mathcal{D}'(\Omega) \quad \forall |\boldsymbol{\alpha}| \in \mathbb{N}$ and we have written $\langle \partial^{\boldsymbol{\alpha}} f, \phi \rangle = \langle \partial^{\boldsymbol{\alpha}} T_f, \phi \rangle$, which can *not* be defined by an integral *in general*, and consequently we have *not* written an integral representation of $\langle \partial^{\boldsymbol{\alpha}} T_f, \phi \rangle$. In other words, for $f \in L^1_{\text{loc}}(\Omega)$, $\partial^{\boldsymbol{\alpha}} f \in \mathcal{D}'(\Omega)$ may be a singular distribution, and then $\langle \partial^{\boldsymbol{\alpha}} f, \phi \rangle = \langle \partial^{\boldsymbol{\alpha}} T_f, \phi \rangle$ cannot be represented by an integral on Ω and, hence, we have *not* used any integral representation of $\langle \partial^{\boldsymbol{\alpha}} f, \phi \rangle$ on the left-hand side of (2.3.1).

Remark 2.3.1.

• The derivatives of regular distributions may *not* be regular distributions, i.e. may be *singular* distributions. (2.3.2)

• Every continuous function $f \in C^0(\overline{\Omega})$, or even locally summable function $f \in L^1_{\text{loc}}(\Omega)$, can be *successively* differentiated in the distribution sense, i.e. using the formula (2.3.1): $\forall \phi \in \mathcal{D}(\Omega)$,

$$\begin{aligned}\langle \partial^{\boldsymbol{\alpha}} f, \phi\rangle_{\mathcal{D}'(\Omega)\times\mathcal{D}(\Omega)} &= \langle \partial^{\boldsymbol{\alpha}} T_f, \phi\rangle_{\mathcal{D}'(\Omega)\times\mathcal{D}(\Omega)} = (-1)^{|\boldsymbol{\alpha}|}\langle T_f, \partial^{\boldsymbol{\alpha}}\phi\rangle \\ &= (-1)^{|\boldsymbol{\alpha}|}\int_\Omega f(\mathbf{x})\frac{\partial^{|\boldsymbol{\alpha}|}\phi}{\partial x_1^{\alpha_1}\partial x_2^{\alpha_2}\dots\partial x_n^{\alpha_n}}d\mathbf{x}.\end{aligned}$$

Hence, f is a function defining a regular distribution T_f, but $\partial^{\boldsymbol{\alpha}} f \in \mathcal{D}'(\Omega)$ may *not* be a function, i.e. may be a *singular* distribution in $\mathcal{D}'(\Omega)$ (see Example 2.3.2).

• The situation can be understood better by considering the fact that any algebraic equation, say, for example, $ax^2 + bx + c = 0$ with real a, b, c, has *complex* roots (i.e. may *not* have *real* roots). Similarly, any locally summable function has successive derivatives of *all orders* which are *distributions* (i.e. may *not* be functions).

• Suppose that the distributional derivative $\partial^{\boldsymbol{\alpha}} f$ of function $f \in L^1_{\text{loc}}(\Omega)$ is *also a function* belonging to $L^1_{\text{loc}}(\Omega)$, i.e. $f \in L^1_{\text{loc}}(\Omega) \subset \mathcal{D}'(\Omega)$ with $\partial^{\boldsymbol{\alpha}} f \in L^1_{\text{loc}}(\Omega) \Longrightarrow \partial^{\boldsymbol{\alpha}} f = T_{\partial^{\boldsymbol{\alpha}} f} = \partial^{\boldsymbol{\alpha}} T_f \in \mathcal{D}'(\Omega)$

$$\Longrightarrow \quad \langle \partial^{\boldsymbol{\alpha}} f, \phi\rangle = \langle \partial^{\boldsymbol{\alpha}} T_f, \phi\rangle = \langle T_{\partial^{\boldsymbol{\alpha}} f}, \phi\rangle = \int_\Omega \partial^{\boldsymbol{\alpha}} f(\mathbf{x})\phi(\mathbf{x})d\mathbf{x} \quad \forall \phi \in \mathcal{D}(\Omega). \tag{2.3.3}$$

Then, for $f \in L^1_{\text{loc}}(\Omega)$ with $\partial^{\boldsymbol{\alpha}} f \in L^1_{\text{loc}}(\Omega) \subset \mathcal{D}'(\Omega)$, using (2.3.1) and (2.3.3), the equation $\langle \partial^{\boldsymbol{\alpha}} f, \phi\rangle_{\mathcal{D}'(\Omega)\times\mathcal{D}(\Omega)} = (-1)^{|\boldsymbol{\alpha}|}\langle f, \partial^{\boldsymbol{\alpha}}\phi\rangle_{\mathcal{D}'(\Omega)\times\mathcal{D}(\Omega)}$ $\forall \phi \in \mathcal{D}(\Omega)$ can be rewritten in the following integral form: $\forall \phi \in \mathcal{D}(\Omega)$,

$$\int_\Omega \frac{\partial^{|\boldsymbol{\alpha}|} f(\mathbf{x})}{\partial x_1^{\alpha_1}\partial x_2^{\alpha_2}\dots\partial x_n^{\alpha_n}}\phi(\mathbf{x})d\mathbf{x} = (-1)^{|\boldsymbol{\alpha}|}\int_\Omega f(\mathbf{x})\frac{\partial^{|\boldsymbol{\alpha}|}\phi(\mathbf{x})}{\partial x_1^{\alpha_1}\partial x_2^{\alpha_2}\dots\partial x_n^{\alpha_n}}d\mathbf{x}. \tag{2.3.4}$$

Moreover, we have:

Proposition 2.3.1. *Let $f \in L^1_{\text{loc}}(\Omega)$. Then, if $\exists$ a function $g_{\boldsymbol{\alpha}} \in L^1_{\text{loc}}(\Omega)$ such that $\forall \phi \in \mathcal{D}(\Omega)$,*

$$\int_\Omega g_{\boldsymbol{\alpha}}(\mathbf{x})\phi(\mathbf{x})d\mathbf{x} = (-1)^{|\boldsymbol{\alpha}|}\int_\Omega f(\mathbf{x})\frac{\partial^{|\boldsymbol{\alpha}|}\phi(\mathbf{x})}{\partial x_1^{\alpha_1}\partial x_2^{\alpha_2}\dots\partial x_n^{\alpha_n}}d\mathbf{x}, \tag{2.3.5}$$

then $g_{\boldsymbol{\alpha}} = \partial^{\boldsymbol{\alpha}} f = \frac{\partial^{|\boldsymbol{\alpha}|} f}{\partial x_1^{\alpha_1}\partial x_2^{\alpha_2}\dots\partial x_n^{\alpha_n}} \in L^1_{\text{loc}}(\Omega)$ in the sense of distribution.

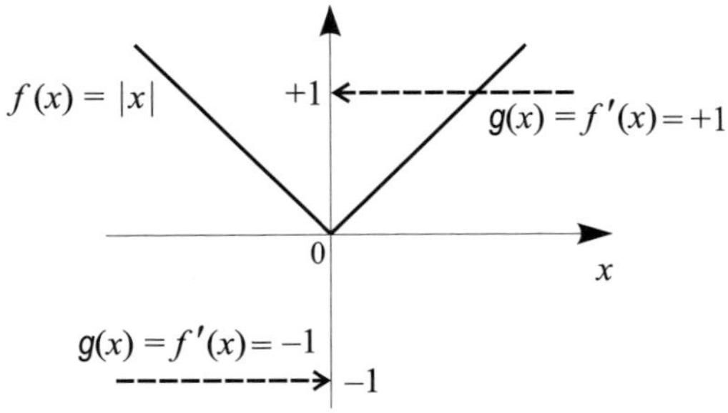

Figure 2.2 Distributional derivative of $|x|$

Proof. $\forall \phi \in \mathcal{D}(\Omega)$

$$(-1)^{|\boldsymbol{\alpha}|} \int_\Omega f(\mathbf{x}) \frac{\partial^{|\boldsymbol{\alpha}|}\phi(\mathbf{x})}{\partial x_1^{\alpha_1} \partial x_2^{\alpha_2} \dots \partial x_n^{\alpha_n}} d\mathbf{x} = (-1)^{|\boldsymbol{\alpha}|} \langle f, \partial^{\boldsymbol{\alpha}} \phi \rangle_{\mathcal{D}'(\Omega) \times \mathcal{D}(\Omega)}$$

$$= \langle \partial^{\boldsymbol{\alpha}} f, \phi \rangle_{\mathcal{D}'(\Omega) \times \mathcal{D}(\Omega)} = \int_\Omega g_{\boldsymbol{\alpha}} \phi d\mathbf{x} = \langle g_{\boldsymbol{\alpha}}, \phi \rangle_{\mathcal{D}'(\Omega) \times \mathcal{D}(\Omega)}$$

$\iff \partial^{\boldsymbol{\alpha}} f = g_{\boldsymbol{\alpha}}$ in $\mathcal{D}'(\Omega)$. But $g_{\boldsymbol{\alpha}} \in L^1_{\text{loc}}(\Omega) \subset \mathcal{D}'(\Omega)$ defines the unique distribution with $g_{\boldsymbol{\alpha}} = \partial^{\boldsymbol{\alpha}} f \in L^1_{\text{loc}}(\Omega)$, $\partial^{\boldsymbol{\alpha}} f$ being in the sense of distribution. □

Examples

Example 2.3.1. For $n = 1$, $\Omega \subset \mathbb{R} =]-\infty, \infty[$,

$$f(x) = |x| = \begin{cases} x & \text{for } x \geq 0 \\ -x & \text{for } x < 0. \end{cases}$$

Then, $\forall$ compact subsets $[a, b] \subset\subset \mathbb{R}$ with $a < b$ and $0 \in]a, b[$,

$$\int_a^0 -x dx + \int_0^b x dx = \frac{b^2 + a^2}{2} < +\infty$$

$\implies f \in L^1_{\text{loc}}(\mathbb{R})$. f is *not* differentiable in the usual pointwise sense at $x = 0$ (see Figure 2.2),

$\implies f$ is *not* differentiable on $\mathbb{R}$ in the usual pointwise sense. But $f \in L^1_{\text{loc}}(\mathbb{R})$

$\Longrightarrow \frac{df}{dx} \in \mathcal{D}'(\mathbb{R})$ is defined in the sense of distribution by, $\forall \phi \in \mathcal{D}(\mathbb{R})$,

$$\begin{aligned}\left\langle \frac{df}{dx}, \phi \right\rangle_{\mathcal{D}'(\mathbb{R})\times\mathcal{D}(\mathbb{R})} &= -\left\langle f, \frac{d\phi}{dx} \right\rangle_{\mathcal{D}'(\mathbb{R})\times\mathcal{D}(\mathbb{R})} = -\int_{-\infty}^{\infty} f \frac{d\phi}{dx} dx \\ &= \int_{-\infty}^{0^-} x \frac{d\phi}{dx} dx - \int_0^{\infty} x \frac{d\phi}{dx} dx \\ &= [x\phi(x)]_{-\infty}^{0^-} - \int_{-\infty}^{0^-} 1 \cdot \phi(x) dx - [x\phi(x)]_{0+}^{\infty} \\ &\quad + \int_0^{\infty} 1 \cdot \phi(x) dx \quad \forall \phi \in \mathcal{D}(\mathbb{R}).\end{aligned}$$

But $\phi \in \mathcal{D}(]-\infty, \infty[)$
$\Longrightarrow \operatorname{supp}(\phi) \subset\subset]-\infty, \infty[$
$\Longrightarrow \exists$ a compact interval $[-a, a]$ with $a > 0$ outside which $\phi(x) = 0$
$\Longrightarrow [x\phi(x)]_{-\infty}^{0^-} = 0 - 0 = 0$, $[x\phi(x)]_0^{\infty} = 0 - 0 = 0$.
Hence,

$$\begin{aligned}\left\langle \frac{df}{dx}, \phi \right\rangle_{\mathcal{D}'(\mathbb{R})\times\mathcal{D}(\mathbb{R})} &= -\int_{-\infty}^{0^-} 1 \cdot \phi(x) dx + \int_0^{\infty} 1 \cdot \phi(x) dx \\ &= \left(\int_{-\infty}^{0^-} (-1) \cdot \phi(x) dx + \int_0^{\infty} (+1) \cdot \phi(x) dx \right) \\ &= \int_{-\infty}^{\infty} g(x)\phi(x) dx \quad \forall \phi \in \mathcal{D}(\mathbb{R}),\end{aligned}$$

where $g(x) = 1$ for $x > 0$ and $= -1$ for $x < 0$ and belongs to $L^1_{\text{loc}}(\mathbb{R})$, since $\forall$ compact interval $[a, b] \subset\subset \mathbb{R}$ with $a < b$, $\int_a^b |g(x)| dx = (b - a) < +\infty$.

$\Longrightarrow \langle \frac{df}{dx}, \phi \rangle = \langle g, \phi \rangle$ with $g \in L^1_{\text{loc}}(\mathbb{R})$ $\forall \phi \in \mathcal{D}(\mathbb{R}) \Longrightarrow \frac{df}{dx} = g \in L^1_{\text{loc}}(\mathbb{R})$ in the sense of distribution by Proposition 2.3.1. Hence, both f and its derivative $\frac{df}{dx} = g$ in the sense of distribution are functions (locally summable on $\mathbb{R}$). In fact, $f'(x) = g(x) = 1$ for $x > 0$ and $= -1$ for $x < 0$ is also in the usual pointwise sense (Figure 2.1).

Example 2.3.2. Let $\Omega \equiv \mathbb{R} \equiv]-\infty, \infty[$ and $H(x)$ be the Heaviside function defined by (1.1.5): $H(x) = 1$ for $x > 0$ and $H(x) = 0$ for $x < 0$. Then, $\forall$ compact interval $[-a, a] \subset\subset \mathbb{R}$ with $a > 0$,

$$\int_{-a}^{a} H(x) dx = \int_{-a}^{0} 0 dx + \int_0^a 1 dx = a < +\infty$$

$$\Longrightarrow \quad H \in L^1_{\text{loc}}(\mathbb{R}) \quad \Longrightarrow \quad \frac{dH}{dx} \in \mathcal{D}'(\mathbb{R}),$$

which is no longer a function and, in fact, $\frac{dH}{dx} \in \mathcal{D}'(\mathbb{R})$ is the *singular Dirac distribution* $\delta = \delta_0$. We prove this now: $\forall \phi \in \mathcal{D}(\mathbb{R})$,

$$\left\langle \frac{dH}{dx}, \phi \right\rangle = -\left\langle H, \frac{d\phi}{dx} \right\rangle = -\int_{-\infty}^{\infty} H(x)\frac{d\phi}{dx}dx = -\int_{-\infty}^{0} 0 \cdot \frac{d\phi}{dx}dx - \int_{0}^{\infty} 1 \cdot \frac{d\phi}{dx}dx$$
$$= -\int_{0}^{\infty} \frac{d\phi}{dx}dx = -[\phi(x)]_0^{\infty} = \phi(0),$$

since $\phi \in \mathcal{D}(\mathbb{R}) \Longrightarrow \text{supp}(\phi) \subset\subset]-\infty, \infty[\Longrightarrow \exists$ a compact interval $[-a, a]$ with $a > 0$ outside which $\phi(x) = 0 \Longrightarrow \phi(x) = 0$ for $x = \infty \Longrightarrow \langle \frac{dH}{dx}, \phi\rangle = \phi(0) = \langle \delta, \phi\rangle \quad \forall \phi \in \mathcal{D}(\mathbb{R})$ (by the definition of Dirac distribution in (1.3.27)) $\Longleftrightarrow \frac{dH}{dx} = \delta \in \mathcal{D}'(\mathbb{R})$ in the sense of distribution $\Longrightarrow \frac{dH}{dx} \notin L^1_{\text{loc}}(\mathbb{R})$ by Proposition 1.3.2, i.e. there does *not* exist any function $g \in L^1_{\text{loc}}(\mathbb{R})$ such that $\langle \frac{dH}{dx}, \phi\rangle = \int_{-\infty}^{\infty} g(x)\phi(x)dx = \phi(0) \ \forall \phi \in \mathcal{D}(\mathbb{R})$. Hence, $H \in L^1_{\text{loc}}(\mathbb{R})$ is the regular distribution $T_H = H \in \mathcal{D}'(\mathbb{R})$, but its derivative $\frac{dT_H}{dx} = \frac{dH}{dx} = \delta \in \mathcal{D}'(\mathbb{R})$ in the sense of distribution is a singular distribution on $\mathbb{R}$.

Remark 2.3.2. For the Heaviside function H in (1.1.5), which is discontinuous with jump $J_0 = H(0^+) - H(0^-) = 1 - 0 = 1$ at $x = 0$, we have *two* different distributions corresponding to *two* different notions of derivatives of $H(x)$.

1. The derivative $\frac{dH}{dx} = \frac{dT_H}{dx} \in \mathcal{D}'(\mathbb{R})$ of $H = T_H \in \mathcal{D}'(\mathbb{R})$ *in the sense of distribution*, i.e. $\frac{dH}{dx} = \frac{dT_H}{dx} = \delta \in \mathcal{D}'(\mathbb{R})$ in the sense of distribution. (We cannot write $\frac{dH}{dx}(x)$ or $\delta(x)$, since these are distributions and not functions, and consequently have no point values.)

2. The distribution $T_{\frac{dH}{dx}(x)}$ defined by the usual ordinary derivative $\frac{dH}{dx}(x)$:

$$\frac{dH}{dx}(x) = \lim_{\Delta x \to 0} \frac{H(x + \Delta x) - H(x)}{\Delta x} \quad \forall x \neq 0$$

(in the *pointwise sense*), i.e. $\frac{dH}{dx}(x) = 0 \ \forall x \neq 0 \Longrightarrow T_{\frac{dH}{dx}(x)} = 0 \in \mathcal{D}'(\mathbb{R})$, since $\forall \phi \in \mathcal{D}(\mathbb{R})$

$$\langle T_{\frac{dH}{dx}(x)}, \phi\rangle = \int_{-\infty}^{\infty} \frac{dH}{dx}(x)\phi(x)dx = \int_{-\infty}^{0-} 0 \cdot \phi dx + \int_{0+}^{\infty} 0 \cdot \phi dx = 0$$

$\Longleftrightarrow T_{\frac{dH}{dx}(x)} = 0 \in \mathcal{D}'(\mathbb{R})$.

Here, by abuse of notation, we have used $\frac{dH}{dx}(x)$ to indicate that it is the *usual ordinary derivative* of H at x in the *pointwise sense*, whereas $\frac{dH}{dx} \in \mathcal{D}'(\mathbb{R})$ is to be understood *in the sense of distribution*.

An equivalent notation for the distribution $T_{\frac{dH}{dx}(x)}$ is $[\frac{dH}{dx}(x)] \in \mathcal{D}'(\mathbb{R})$ such that

$$T_{\frac{dH}{dx}(x)} = \left[\frac{dH}{dx}(x)\right] = 0 \in \mathcal{D}'(\mathbb{R}) \quad \text{and}$$
$$\frac{dT_H}{dx} = \frac{dH}{dx} = \left[\frac{dH}{dx}(x)\right] + J_0\delta \quad \text{in } \mathcal{D}'(\mathbb{R}) \tag{2.3.6}$$

with $J_0 = 1$ (see Chapter 3: Theorem 3.1.1).

Then

$$\frac{dH}{dx} = \delta \neq 0 = \left[\frac{dH}{dx}(x)\right] \quad \text{in } \mathcal{D}'(\mathbb{R}), \tag{2.3.7}$$

i.e. the distributional derivative $\frac{dH}{dx} = \frac{dT_H}{dx} \in \mathcal{D}'(\mathbb{R})$ of the discontinuous Heaviside function H on $\mathbb{R}$ is not equal to the distribution $[\frac{dH}{dx}(x)] = T_{\frac{dH}{dx}(x)} = 0$ defined by the *usual* ordinary derivative $\frac{dH}{dx}(x)$ in the pointwise sense.

$[\frac{dH}{dx}(x)] \in \mathcal{D}'(\mathbb{R})$ corresponds to the null function $0 \in L^1_{\text{loc}}(\mathbb{R})$, whereas $\frac{dH}{dx} \in \mathcal{D}'(\mathbb{R})$ is the Dirac distribution δ, which is *not* a locally summable function on $\mathbb{R}$.

In (2.3.6), the effect of the discontinuity of H with a finite jump $J_0 = 1$ appears in the distributional derivative in the form of a point mass or force or charge as the case may be (see Section 3.1 in Chapter 3 for more details).

Dirac's result (1.1.5), $\frac{dH}{dx} = \delta$, is correct and is to be understood in the sense of distributional derivative and not in the usual pointwise sense of the derivative.

In electrical engineering $H(x)$ is called the *unit step function* and Dirac distribution δ is called the *impulse function*.

Example 2.3.3. Now we find the second-order distributional derivative $\frac{d^2 f}{dx^2}$ of the function $f(x) = |x|$ in Example 2.3.1, where we have shown that $\frac{df}{dx} = g(x) = 1$ for $x > 0$ and $\frac{df}{dx} = g(x) = -1$ for $x < 0$ is a function in $L^1_{\text{loc}}(\mathbb{R})$. Here, we will show that $\frac{d^2 f}{dx^2} = \frac{dg}{dx}$ in the sense of distribution is *not* a function but is a *singular* distribution on $\mathbb{R}$, i.e. $\frac{d^2 f}{dx^2}$ does not belong to $L^1_{\text{loc}}(\mathbb{R})$. In fact, $\forall \phi \in \mathcal{D}(\Omega)$,

$$\begin{aligned}
\left\langle \frac{d^2 f}{dx^2}, \phi \right\rangle &= \left\langle \frac{d}{dx}\left(\frac{df}{dx}\right), \phi \right\rangle = \left\langle \frac{dg}{dx}, \phi \right\rangle = -\left\langle g, \frac{d\phi}{dx} \right\rangle \\
&= -\int_{-\infty}^{\infty} g(x)\frac{d\phi}{dx}(x)dx \quad (\text{since } g \in L^1_{\text{loc}}(\mathbb{R})) \\
&= -\int_{-\infty}^{0-} (-1)\frac{d\phi}{dx}dx - \int_{0+}^{\infty} 1 \cdot \frac{d\phi}{dx}dx \\
&= \phi(x)|_{-\infty}^{0-} - \phi(x)|_{0+}^{\infty} = \phi(0) + \phi(0) = 2\phi(0) = 2\langle \delta, \phi \rangle
\end{aligned}$$

(since $\phi(x) = 0$ for $x \to \pm\infty$ and $\phi(0) = \lim_{x\to 0^-} \phi(x) = \lim_{x\to 0^+} \phi(x)$)
$\implies \langle \frac{d^2 f}{dx^2}, \phi\rangle = \langle 2\delta, \phi\rangle \ \forall\phi \in \mathcal{D}(\mathbb{R}) \iff \frac{d^2 f}{dx^2} = 2\delta \in \mathcal{D}'(\mathbb{R})$ is a singular distribution.

Hence, the first-order distributional derivative $\frac{df}{dx} \in L^1_{\text{loc}}(\mathbb{R})$ is a regular distribution, but the second-order distributional derivative $\frac{d^2 f}{dx^2} \in \mathcal{D}'(\mathbb{R})$ is a singular distribution on $\mathbb{R}$, i.e. not a function (locally summable) on $\mathbb{R}$.

Derivatives of Dirac distributions

Example 2.3.4. Let $\delta \in \mathcal{D}'(\mathbb{R})$ be the Dirac distribution on $\mathbb{R}$ defined by $\langle\delta, \phi\rangle = \phi(0)\ \forall\phi \in \mathcal{D}(\mathbb{R})$. Then the derivatives of Dirac distribution δ on $\mathbb{R}$ are defined by: $\forall k \in \mathbb{N}, \forall\phi \in \mathcal{D}(\mathbb{R})$,

$$\langle\delta^{(k)}, \phi\rangle = (-1)^k\left\langle\delta, \frac{d^k\phi}{dx^k}\right\rangle = (-1)^k\phi^{(k)}(0). \tag{2.3.8}$$

In particular, for $k = 1$, $\langle\delta^{(1)}, \phi\rangle = -\phi^{(1)}(0) = -\frac{d\phi}{dx}(0)$; for $k = 2$, $\langle\delta^{(2)}, \phi\rangle = (-1)^2\phi^{(2)}(0) = \frac{d^2\phi}{dx^2}(0)$, and so on.

Remark 2.3.3. $\delta^{(1)} = \delta'$ is the dipole of moment -1 at the origin 0.

Example 2.3.5. Now we find the second-order distributional derivative $\frac{d^2H}{dx^2}$ of the Heaviside function H of Example 2.3.2, $\frac{dH}{dx} = \delta \in \mathcal{D}'(\mathbb{R})$. Hence, $\forall\phi \in \mathcal{D}(\mathbb{R})$,

$$\left\langle\frac{d^2H}{dx^2}, \phi\right\rangle = \left\langle\frac{d}{dx}\left(\frac{dH}{dx}\right), \phi\right\rangle = -\left\langle\frac{dH}{dx}, \frac{d\phi}{dx}\right\rangle = -\left\langle\delta, \frac{d\phi}{dx}\right\rangle = \langle\delta', \phi\rangle$$

$$\iff \quad \frac{d^2H}{dx^2} = \delta' \in \mathcal{D}'(\mathbb{R}), \quad \text{i.e.} \ \frac{d^2H}{dx^2} \notin L^1_{\text{loc}}(\mathbb{R}), \tag{2.3.9}$$

where δ' is the derivative of the Dirac distribution and the dipole of moment -1 at origin 0 and is defined by: $\langle\delta', \phi\rangle = -\langle\delta, \phi'\rangle = -\phi'(0)$ (see Example 2.3.4).

Similarly,

$$\begin{aligned}\left\langle\frac{d^kH}{dx^k}, \phi\right\rangle &= \left\langle\frac{d}{dx}\left(\frac{d^{k-1}H}{dx^{k-1}}\right), \phi\right\rangle = -\left\langle\frac{d^{k-1}H}{dx^{k-1}}, \frac{d\phi}{dx}\right\rangle \\ &= \cdots = (-1)^{k-1}\left\langle\frac{dH}{dx}, \frac{d^{k-1}\phi}{dx^k}\right\rangle \\ &= (-1)^{k-1}\langle\delta, \phi^{(k-1)}\rangle = \langle\delta^{(k-1)}, \phi\rangle \quad \forall\phi \in \mathcal{D}(\mathbb{R})\end{aligned}$$

$$\iff \quad \frac{d^kH}{dx^k} = \delta^{(k-1)} \in \mathcal{D}'(\mathbb{R}) \quad (\text{i.e.} \ \frac{d^kH}{dx^k} \notin L^1_{\text{loc}}(\mathbb{R})\ \forall k \in \mathbb{N}). \tag{2.3.10}$$

Example 2.3.6. Since $\ln|x|$ ($x \neq 0$) is a locally summable function in the neighbourhood of $x = 0$, $\ln|x| \in L^1_{\text{loc}}(]-\infty, \infty[)$ and defines a *regular* distribution on $\mathbb{R}$.

In fact, $\int \ln|x|dx = x\ln|x| - x + C$ for $x \neq 0$,

$$\lim_{x\to 0} x\ln|x| = \lim_{x\to 0}\frac{\ln|x|}{\frac{1}{x}} = \lim_{x\to 0}\frac{\frac{1}{x}}{\frac{-1}{x^2}} = \lim_{x\to 0}(-x) = 0, \tag{2.3.11}$$

and

$$\int \big|\ln|x|\big|dx = \begin{cases} \int \ln|x|dx & \text{for } \ln|x| \geq 0 \\ -\int \ln|x|dx & \text{for } \ln|x| < 0. \end{cases}$$

$\Longrightarrow \forall$ compact interval $[a,b] \subset\subset \mathbb{R}$, $\int_a^b |\ln|x||dx < +\infty \Longrightarrow \ln|x| \in L^1_{\text{loc}}(]-\infty, \infty[)$. Thus, $\ln|x| \in \mathcal{D}'(\mathbb{R})$ defines a regular distribution on $\mathbb{R} \Longrightarrow$ the distributional derivative $\frac{d}{dx}ln|x|$ of $\ln|x|$ is a distribution on $\mathbb{R}$ defined by:

$$\begin{aligned}\left\langle \frac{d}{dx}\ln|x|, \phi\right\rangle &= -\langle \ln|x|, \phi'\rangle = -\int_{-\infty}^{\infty} \ln|x|\phi'(x)dx \quad \forall \phi \in \mathcal{D}(\mathbb{R}) \\ &= -\int_{-\infty}^{0} \ln|x|\phi'(x)dx - \int_0^{\infty} \ln|x|\phi'(x)dx,\end{aligned} \tag{2.3.12}$$

$$\begin{aligned}\int_0^{\infty} \ln|x|\phi'(x)dx &= \lim_{\varepsilon\to 0+}\int_{\varepsilon}^{\infty} (\ln|x|)\phi'(x)dx \\ &= \lim_{\varepsilon\to 0+}\left[(\ln|x|)\phi(x)|_{\varepsilon}^{\infty} - \int_{\varepsilon}^{\infty}\frac{1}{x}\phi(x)dx\right] \\ &= \lim_{\varepsilon\to 0+}\left[-(\ln\varepsilon)\phi(\varepsilon) - \int_{\varepsilon}^{\infty}\frac{1}{x}\phi(x)dx\right] \quad \forall \phi \in \mathcal{D}(\mathbb{R}).\end{aligned} \tag{2.3.13}$$

Similarly, we have, $\forall \varepsilon > 0$, $\int_{-\infty}^{-\varepsilon} \ln|x|\phi'(x)dx = (\ln\varepsilon)\phi(-\varepsilon) - \int_{-\infty}^{-\varepsilon}\frac{1}{x}\phi(x)dx$ $\forall \phi \in \mathcal{D}(\mathbb{R})$

$$\Longrightarrow \quad \forall \phi \in \mathcal{D}(\mathbb{R}), \quad \int_{-\infty}^{0} \ln|x|\phi'(x)dx = \lim_{\varepsilon\to 0+}\left[(\ln\varepsilon)\phi(-\varepsilon) - \int_{-\infty}^{-\varepsilon}\frac{1}{x}\phi(x)dx\right]. \tag{2.3.14}$$

Then, from (2.3.12)–(2.3.14), we get, $\forall \phi \in \mathcal{D}(\mathbb{R})$,

$$\begin{aligned}\langle \frac{d}{dx}\ln|x|, \phi\rangle = &\lim_{\varepsilon\to 0+}\left[\int_{-\infty}^{-\varepsilon}\frac{1}{x}\phi(x)dx + \int_{\varepsilon}^{\infty}\frac{1}{x}\phi(x)dx\right] \\ &+ \lim_{\varepsilon\to 0+}[(\ln\varepsilon)\phi(\varepsilon) - (\ln\varepsilon)\phi(-\varepsilon)].\end{aligned}$$

But

$$|(\ln\varepsilon)\phi(\varepsilon)-(\ln\varepsilon)\phi(-\varepsilon)| = |(\ln\varepsilon)(\phi(\varepsilon)-\phi(-\varepsilon))| \le |\ln\varepsilon|\cdot|\phi(\varepsilon)-\phi(-\varepsilon)|$$
$$\le |\ln\varepsilon|\cdot 2\varepsilon\cdot \max_{x\in\operatorname{supp}(\phi)}|\phi'(x)| = 2\max_{x\in\operatorname{supp}(\phi)}|\phi'(x)|\cdot|\varepsilon\ln\varepsilon| \to 0 \quad \text{as } \varepsilon\to 0^+$$

(using (2.3.11)).

Hence,

$$\left\langle \frac{d}{dx}\ln|x|,\phi\right\rangle = \text{c.p.v.}\int_{-\infty}^{\infty}\frac{1}{x}\phi(x)dx$$
$$= \left\langle \text{c.p.v.}\,\frac{1}{x},\phi(x)\right\rangle \quad \forall\phi\in\mathcal{D}(\mathbb{R}) \quad \text{(by (1.3.45))}$$

$\Longleftrightarrow \frac{d}{dx}ln|x| = \text{c.p.v.}\,\frac{1}{x}$ in $\mathcal{D}'(\mathbb{R})$, i.e. the distributional derivative $\frac{d}{dx}\ln|x| = \text{c.p.v.}\,\frac{1}{x}\in\mathcal{D}'(\mathbb{R})$ is *not* a regular distribution.

Example 2.3.7 (Derivative of $\ln x\,(x\neq 0)$ in the sense of distribution). Now we are in a position to show that Dirac's remarkable formula for the derivative $\frac{d}{dx}\ln x$ of $\ln x\,(x\neq 0)$ in (1.1.6) is correct. Here, we allow x to be negative, and consequently we are to consider a logarithmic function of a *complex variable* z. From the theory of complex variables, for $z = x+iy$,

$$\ln z = \ln(x+iy) = \ln(x^2+y^2)^{\frac{1}{2}} + i\arctan\left(\frac{y}{x}\right) \tag{2.3.15}$$

is analytic in the *upper half-plane* $y>0$. Then, $\lim_{y\to 0^+}\ln(x+iy) = \ln x + i0$ with the properties:

- the first term $\frac{1}{2}\ln(x^2+y^2)$ on the right-hand side of (2.3.15) converges, decreasing monotonically to $\ln|x|$, i.e. $\lim_{y\to 0^+}\frac{1}{2}\ln(x^2+y^2) = \ln|x|$.
- the second term $i\arctan(\frac{y}{x})$ has its *absolute* value *bounded* by π, and as $y\to 0^+$, it converges to:

$$\tilde{H}(x) = i\pi H(-x) = \begin{cases} i\pi & \text{for } x<0 \\ 0 & \text{for } x>0, \end{cases}$$

i.e.

$$\lim_{y\to 0^+} i\arctan\left(\frac{y}{x}\right) = \tilde{H}(x) = i\pi H(-x), \tag{2.3.16}$$

where the Heaviside function $H(x)$ is defined by (1.1.5) and

$$H(-x) = \begin{cases} 1 & \text{for } x<0 \\ 0 & \text{for } x>0. \end{cases}$$

Hence, we write

$$\ln(x + i0) = \lim_{y\to 0+} \ln(x + iy) = \ln|x| + i\pi H(-x). \tag{2.3.17}$$

Then,

$$\begin{aligned}\left\langle \frac{d}{dx} ln(x+i0), \phi \right\rangle &= \left\langle \frac{d}{dx} \ln|x| + i\pi H(-x)], \phi \right\rangle \quad \forall \phi \in \mathcal{D}(\mathbb{R}) \\ &= \left\langle \frac{d}{dx} \ln|x|, \phi \right\rangle + i\pi \left\langle \frac{d}{dx} H(-x), \phi \right\rangle.\end{aligned}$$

But $\frac{d}{dx} \ln|x| = \text{c.p.v.} \frac{1}{x}$, with

$$\left\langle \text{c.p.v.} \frac{1}{x}, \phi \right\rangle = \text{c.p.v.} \int_{-\infty}^{\infty} \frac{\phi(x)}{x} dx = \lim_{\varepsilon \to 0+} \int_{|x|\ge\varepsilon} \frac{\phi(x)}{x} dx$$

(see Example 2.3.6) and, $\forall \phi \in \mathcal{D}(\mathbb{R})$,

$$\begin{aligned}\left\langle \frac{d}{dx} H(-x), \phi \right\rangle &= -\langle H(-x), \phi' \rangle = -\int_{-\infty}^{\infty} H(-x)\phi'(x) dx \\ &= -\int_{-\infty}^{0} 1 \cdot \phi'(x) dx - \int_{0}^{\infty} 0 \cdot \phi'(x) \\ &= -[\phi(x)]_{x=-\infty}^{x=0} - 0 = -\phi(0) = -\langle \delta, \phi \rangle = \langle -\delta, \phi \rangle \end{aligned} \tag{2.3.18}$$

(by the definition of Dirac distribution δ in (1.3.27)).

Hence,

$$\begin{aligned}\left\langle \frac{d}{dx} \ln(x+i0), \phi \right\rangle &= \left\langle \text{c.p.v.} \frac{1}{x}, \phi \right\rangle + i\pi \langle -\delta, \phi \rangle \\ &= \left\langle \text{c.p.v.} \frac{1}{x} - i\pi\delta, \phi \right\rangle \quad \forall \phi \in \mathcal{D}(\mathbb{R})\end{aligned}$$

$$\iff \quad \frac{d}{dx} \ln(x+i0) = \text{c.p.v.} \frac{1}{x} - i\pi\delta \quad \text{in } \mathcal{D}'(\mathbb{R}). \tag{2.3.19}$$

Thus, writing $\ln x = \ln(x + i0)$, we get Dirac's formula for $\frac{d}{dx} \ln x$ in (1.1.6), in which $\frac{1}{x}$ is to be understood as the c.p.v. $\frac{1}{x}$ (see (1.3.38)).

Example 2.3.8. Prove the following results:

1. $$\frac{d}{dx} \ln(x \pm iy) = \frac{1}{x \pm iy} \quad \forall \text{ fixed } y > 0 \text{ in } \mathcal{D}'(\mathbb{R}); \tag{2.3.20}$$

2. $$\frac{d}{dx} \ln(x - i0) = \text{c.p.v.} \left(\frac{1}{x}\right) + i\pi\delta \quad \text{in } \mathcal{D}'(\mathbb{R}). \tag{2.3.21}$$

Proof.

1. $\forall$ fixed $y > 0$, $\ln(x \pm iy) = \frac{1}{2}\ln(x^2 + y^2) \pm i \arctan(\frac{y}{x})$ (see (1.4.3)). Then, from the definition of the distributional derivative, $\forall \phi \in \mathcal{D}(\mathbb{R})$,

$$\left\langle \frac{d}{dx}\ln(x + iy), \phi \right\rangle = -\langle \ln(x + iy), \phi' \rangle = -\int_{-\infty}^{\infty} \ln(x + iy)\phi'(x)dx$$
$$= -[\ln(x + iy)\phi(x)]_{x=-\infty}^{x=+\infty} + \int_{-\infty}^{\infty} \frac{1}{x + iy}\phi(x)dx$$
$$= \int_{-\infty}^{\infty} \frac{1}{x + iy}\phi(x)dx,$$

which is obtained by integrating by parts with: $u = \frac{1}{2}\ln(x^2+y^2)+i \arctan(\frac{y}{x})$, $dv = \phi'(x)dx$,

$$du = \left[\frac{x}{x^2 + y^2} - i\frac{y}{x^2(1 + (\frac{y}{x})^2)}\right]dx = \frac{x - iy}{x^2 + y^2}dx = \frac{1}{x + iy}dx,$$

$v = \phi(x)$ and applying $\phi(x) = 0$ for $x = \pm\infty$.
$\Longrightarrow \frac{d}{dx}\ln(x + iy) = \frac{1}{x+iy}$ in $\mathcal{D}'(\mathbb{R})$.
Similarly,

$$\frac{d}{dx}\ln(x - iy) = \frac{d}{dx}\left[\frac{1}{2}\ln(x^2 + y^2) - i \arctan\left(\frac{y}{x}\right)\right] = \frac{1}{x - iy} \quad \text{in } \mathcal{D}'(\mathbb{R}).$$

2. As in the case of $\ln(x + i0)$ defined by (2.3.17), we have

$$\ln(x - i0) = \lim_{y\to 0+} \ln(x - iy) = \ln|x| - i\pi H(-x), \tag{2.3.22}$$

with $H(-x) = 1$ for $x < 0$ and $H(-x) = 0$ for $x > 0$. Hence,

$$\left\langle \frac{d}{dx}\ln(x - i0), \phi \right\rangle = \left\langle \frac{d}{dx}[\ln|x| - i\pi H(-x)], \phi \right\rangle$$
$$= \left\langle \text{c.p.v.}\left(\frac{1}{x}\right) - i\pi(-\delta), \phi \right\rangle \quad \forall\phi \in \mathcal{D}(\mathbb{R})$$

(see (2.3.18) and Example 2.3.6)
$\Longrightarrow \frac{d}{dx}\ln(x - i0) = \text{c.p.v.}(\frac{1}{x}) + i\pi\delta$ in $\mathcal{D}'(\mathbb{R})$. □

Derivative of an unbounded and discontinuous function in $\mathbb{R}^2$

Example 2.3.9. Let

$$u = u(x_1, x_2) = \begin{cases} \ln|\ln r| & \text{for } 0 < r = (x_1^2 + x_2^2)^{\frac{1}{2}} < \frac{1}{e} \\ 0 & \text{for } \frac{1}{e} \le r < \infty. \end{cases} \tag{2.3.23}$$

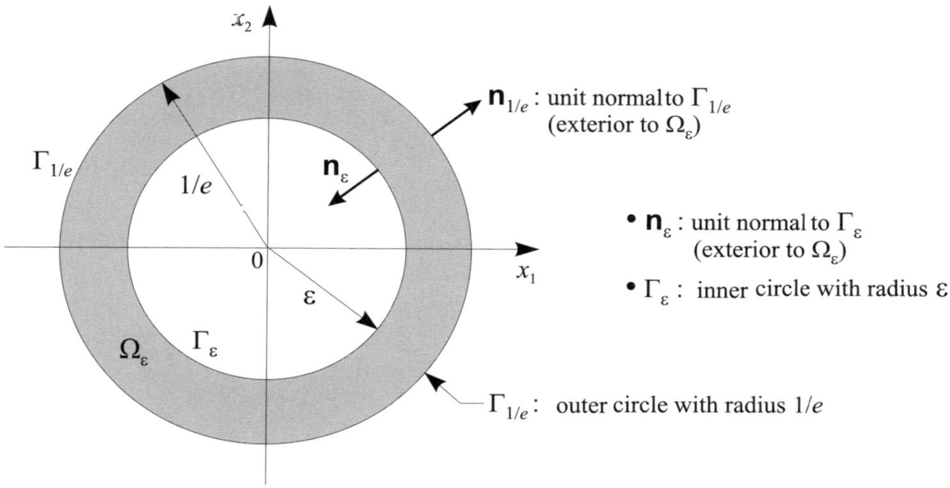

Figure 2.3 Annular domain Ω_ε enclosed by concentric circles $\Gamma_{\frac{1}{e}}$ and Γ_ε with radii $\frac{1}{e}$ and ε respectively

Then,

1. $u \in L^2(\mathbb{R}^2)$;
2. $\frac{\partial u}{\partial x_i} \in L^2(\mathbb{R}^2)$, $i = 1, 2$, where $\frac{\partial u}{\partial x_i}$ is in the distributional sense. (2.3.24)

Proof.

1. Since $|u(x_1, x_2)| \to \infty$ as $r \to 0$, u is unbounded and discontinuous in $\mathbb{R}^2$. But $u \in L^2(\mathbb{R}^2)$. In fact,

$$\int_{\mathbb{R}^2} u(x_1, x_2)^2 dx_1 dx_2 = \int_{0<r<\frac{1}{e}} (\ln|\ln r|)^2 dx_1 dx_2 = \int_0^{\frac{1}{e}} \int_0^{2\pi} (\ln|\ln r|)^2 r dr d\theta$$

$$= 2\pi \int_0^{\frac{1}{e}} (\ln|\ln r|)^2 r dr. \tag{2.3.25}$$

For $0 < r < \frac{1}{e}$, $|\ln(r)| = -\ln(r) = \ln(\frac{1}{r})$, since for $r = \frac{1}{e}, ln(\frac{1}{e}) = -ln(e) = -1$ and $\ln(r)$ monotonically increases from $-\infty$ to -1 as r increases from 0 to $\frac{1}{e}$. Hence, $0 < r < \frac{1}{e} \Longrightarrow e < \frac{1}{r} < \infty \Longrightarrow 1 < \ln(\frac{1}{r}) < \frac{1}{r} \Longrightarrow \ln|\ln r| = \ln(\ln\frac{1}{r}) < \ln\frac{1}{r} = -\ln r \Longrightarrow (\ln|\ln r|)^2 < (\ln r)^2$ for $0 < r < \frac{1}{e}$ (see Figure 2.3)

$$\Longrightarrow \quad \int_0^{\frac{1}{e}} (\ln|\ln r|)^2 r dr \leq \int_0^{\frac{1}{e}} (\ln r)^2 r dr = I_1. \tag{2.3.26}$$

Set $u = (\ln r)^2$, $du = 2(\ln r)\frac{1}{r}dr$, $v = \int r dr = \frac{r^2}{2}$. Hence,

$$I_1 = \frac{r^2(\ln r)^2}{2}\Big|_{r\to 0+}^{r=\frac{1}{e}} - \int_0^{\frac{1}{e}} \frac{r^2}{2}\cdot\frac{2\ln r}{r}dr$$

$$= \frac{1}{2e^2} - \underbrace{\lim_{r\to 0+}\frac{r^2(\ln r)^2}{2}}_{I_2} - \underbrace{\int_0^{\frac{1}{e}} r\ln r dr}_{I_3} = \frac{1}{2e^2} - I_2 - I_3, \qquad (2.3.27)$$

$$I_3 = \frac{r^2\ln r}{2}\Big|_0^{\frac{1}{e}} - \int_0^{\frac{1}{e}} \frac{r^2}{2}\frac{1}{r}dr = -\frac{1}{2e^2} - \underbrace{\lim_{r\to 0+}\frac{r^2\ln r}{2}}_{I_4} - \frac{1}{4e^2},$$

$$I_2 = \lim_{r\to 0+}\frac{(\ln r)^2}{2r^{-2}} = \lim_{r\to 0+}\frac{2\ln r}{2r\cdot(-2)r^{-3}} = \lim_{r\to 0+}\frac{\ln r}{-2r^{-2}}$$

$$= \lim_{r\to 0+}\frac{1}{r(-2)\cdot(-2)r^{-3}} = \lim_{r\to 0+}\frac{r^2}{4} = 0,$$

$$I_4 = \lim_{r\to 0+}\frac{\ln r}{2r^{-2}} = \lim_{r\to 0+}\frac{1}{r\cdot 2\cdot(-2)r^{-3}} = \lim_{r\to 0+}\frac{-r^2}{4} = 0.$$

Hence, $I_3 = -\frac{1}{2e^2} - 0 - \frac{1}{4e^2} = \frac{-3}{4e^2}$
$\Longrightarrow$ from (2.3.27), $I_1 = \frac{1}{2e^2} - 0 - \frac{-3}{4e^2}$
$\Longrightarrow I_1 = \frac{5}{4e^2}$
$\Longrightarrow$ from (2.3.26), $\int_0^{\frac{1}{e}} (\ln|\ln r|)^2 r dr \le \frac{5}{4e^2}$
$\Longrightarrow$ from (2.3.25),

$$\int_{\mathbb{R}^2} |u(x_1,x_2)|^2 dx_1 dx_2 = 2\pi\int_0^{\frac{1}{e}} (\ln|\ln r|)^2 r dr \le \frac{2\pi\cdot 5}{4e^2} < +\infty$$

$\Longrightarrow u \in L^2(\mathbb{R}^2)$.

2. Let $[\frac{\partial u}{\partial x_i}(\mathbf{x})]$ denote the usual partial derivative of u with respect to x_i ($i = 1,2$) in the pointwise sense. For $0 < r < \frac{1}{e}$,

$$\left[\frac{\partial u}{\partial x_i}(\mathbf{x})\right] = \left[\frac{d}{dr}\ln|\ln r|\right]\left[\frac{\partial r}{\partial x_i}\right] = -\frac{1}{\ln r}\left[\frac{d}{dr}(-\ln r)\right]\cdot\frac{x_i}{r}$$

$$= \frac{x_i}{r^2\ln r}, \qquad i = 1,2,$$

since $r = (x_1^2 + x_2^2)^{\frac{1}{2}} \Longrightarrow [\frac{\partial r}{\partial x_i}] = \frac{x_i}{r}$. For $\frac{1}{e} < r < \infty$, $[\frac{\partial u}{\partial x_i}(\mathbf{x})] = 0$.

Thus,

$$\left[\frac{\partial u}{\partial x_i}(\mathbf{x})\right] = \begin{cases} \frac{x_i}{r^2 \ln r} & \text{for } 0 < r < \frac{1}{e} \\ 0 & \text{for } r > \frac{1}{e} \end{cases} \tag{2.3.28}$$

and belongs to $L^2(\mathbb{R}^2)$, i.e.

$$\left[\frac{\partial u}{\partial x_i}(\mathbf{x})\right] \in L^2(\mathbb{R}^2), \quad i = 1, 2. \tag{2.3.29}$$

In fact,

$$\begin{aligned} \int_{\mathbb{R}^2} \left[\frac{\partial u}{\partial x_i}(\mathbf{x})\right]^2 dx_1 dx_2 &= \int_{0<r<\frac{1}{e}} \frac{x_i^2}{r^4(\ln r)^2} dx_1 dx_2 \\ &\leq \int_0^{\frac{1}{e}} \int_0^{2\pi} \frac{r^2}{r^4(\ln r)^2} r dr d\theta = 2\pi \int_0^{\frac{1}{e}} \frac{dr}{r(\ln r)^2} \\ &= 2\pi \int_0^{\frac{1}{e}} \frac{d(\ln r)}{(\ln r)^2} = 2\pi \left[\frac{-1}{\ln r}\right]_0^{\frac{1}{e}} = 2\pi \cdot \frac{-1}{\ln(\frac{1}{e})} = 2\pi < +\infty. \end{aligned}$$

Let $\frac{\partial u}{\partial x_i}$ $(i = 1, 2)$ be the derivative of u with respect to x_i in the distributional sense in $\mathbb{R}^2$.

Then, $\forall \phi \in \mathcal{D}(\mathbb{R}^2)$,

$$\begin{aligned} &\left\langle \frac{\partial u}{\partial x_i}, \phi \right\rangle = -\left\langle u, \frac{\partial \phi}{\partial x_i} \right\rangle = -\int_{\mathbb{R}^2} u(\mathbf{x}) \frac{\partial \phi}{\partial x_i}(\mathbf{x}) d\mathbf{x} = -\int_{0<r<\frac{1}{e}} u(\mathbf{x}) \frac{\partial \phi}{\partial x_i}(\mathbf{x}) d\mathbf{x} \\ &= -\int_{0<r<\frac{1}{e}} \left[\frac{\partial}{\partial x_i}(u\phi)\right] d\mathbf{x} + \int_{0<r<\frac{1}{e}} \left[\frac{\partial u}{\partial x_i}(\mathbf{x})\right] \phi(\mathbf{x}) d\mathbf{x} \\ &= \underbrace{-\lim_{\varepsilon \to 0+} \int_{\Omega_3 : \varepsilon \leq r \leq \frac{1}{e}} \left[\frac{\partial}{\partial x_i}(u\phi)\right] d\mathbf{x}}_{I_5} + \underbrace{\int_{\mathbb{R}^2} \left[\frac{\partial u}{\partial x_i}(\mathbf{x})\right] \phi(\mathbf{x}) d\mathbf{x}}_{I_6} = -I_5 + I_6, \end{aligned} \tag{2.3.30}$$

since $[\frac{\partial u}{\partial x_i}(\mathbf{x})] = 0$ for $r > \frac{1}{e}$.

Applying Green's theorem on the closed annular circular domain $\Omega_\varepsilon = \{\mathbf{x} : \mathbf{x} = (x_1, x_2), \varepsilon \leq r = (x_1^2 + x_2^2)^{\frac{1}{2}} \leq \frac{1}{e}\}$ enclosed by concentric circles Γ_ε and $\Gamma_{\frac{1}{e}}$ with radii ε and $\frac{1}{e}$ respectively, we get

$$\begin{aligned} I_5 &= \lim_{\varepsilon \to 0+} \left[\int_{\Gamma_\varepsilon : r = \varepsilon} u(\mathbf{x})\phi(\mathbf{x}) n_i(\Gamma_\varepsilon) ds + \int_{\Gamma_{\frac{1}{e}}} u(\mathbf{x})\phi(\mathbf{x}) n_i(\Gamma_{\frac{1}{e}}) ds\right] \\ &= \lim_{\varepsilon \to 0+} \int_{\Gamma_\varepsilon : r = \varepsilon} u(\mathbf{x})\phi(\mathbf{x}) n_i(\Gamma_\varepsilon) ds, \end{aligned} \tag{2.3.31}$$

where $n_i(\Gamma_\varepsilon)$ (resp. $n_i(\Gamma_{\frac{1}{e}})$) is the ith component of unit vector $\hat{\mathbf{n}}$ normal to Γ_ε (resp. $\Gamma_{\frac{1}{e}}$), $u = 0$ on $\Gamma_{\frac{1}{e}}$ by definition (2.3.23), and consequently the line integral along $\Gamma_{\frac{1}{e}}$ vanishes.

$$\left|\int_{\Gamma_\varepsilon : r=\varepsilon} u(\mathbf{x})\phi(\mathbf{x}) n_i(\Gamma_\varepsilon) ds\right| = \left|\int_{\Gamma_\varepsilon : r=\varepsilon} \ln|\ln\varepsilon| \phi(\mathbf{x}) n_i(\Gamma_\varepsilon) ds\right|$$
$$\leq \max|\phi(\mathbf{x})| \, \big|\ln|\ln\varepsilon|\big| 2\pi\varepsilon \to 0 \quad \text{as } \varepsilon \to 0^+,$$

since $\forall \varepsilon > 0$, $\ln|\ln\varepsilon| < |\ln\varepsilon| \implies \varepsilon\ln|\ln\varepsilon| < \varepsilon|\ln\varepsilon| \to 0$ as $\varepsilon \to 0^+$ ($\lim_{\varepsilon\to 0^+} \varepsilon|\ln\varepsilon| = \lim_{\varepsilon\to 0^+} \frac{|\ln\varepsilon|}{\frac{1}{\varepsilon}} = \lim_{\varepsilon\to 0^+} \frac{-\frac{1}{\varepsilon}}{-\frac{1}{\varepsilon^2}} = 0$). Hence, from (2.3.31), $I_5 = 0$, and from (2.3.30), $\forall \phi \in \mathcal{D}(\mathbb{R}^2)$,

$$\left\langle \frac{\partial u}{\partial x_i}, \phi \right\rangle = \int_{\mathbb{R}^2} \left[\frac{\partial u}{\partial x_i}(\mathbf{x})\right] \phi(\mathbf{x}) d\mathbf{x} \quad \Longrightarrow \quad \frac{\partial u}{\partial x_i} = \left[\frac{\partial u}{\partial x_i}(\mathbf{x})\right] \in \mathcal{D}'(\mathbb{R}^2)$$

with $[\frac{\partial u}{\partial x_i}(\mathbf{x})]$ defined by (2.3.28) for $i = 1, 2$. But $[\frac{\partial u}{\partial x_i}(\mathbf{x})] \in L^2(\mathbb{R}^2) \Longrightarrow \frac{\partial u}{\partial x_i} \in L^2(\mathbb{R}^2)$ for $i = 1, 2$.

□

Example 2.3.10. Show that in $\mathcal{D}'(\mathbb{R})$,

1. $$\frac{d}{dx}\left(\text{c.p.v.}\left(\frac{1}{x}\right)\right) = -\text{Pf}\left(\frac{1}{x^2}\right); \tag{2.3.32}$$

2. $$\frac{d}{dx}\left[\text{Pf}\left(\frac{H(-x)}{x}\right)\right] = -\text{Pf}\left(\frac{H(-x)}{x^2}\right) + \delta'; \tag{2.3.33}$$

3. $$\frac{d}{dx}\left[\text{Pf}\left(\frac{H(x)}{x}\right)\right] = -\text{Pf}\left(\frac{H(x)}{x^2}\right) - \delta'. \tag{2.3.34}$$

Proof.

1. By definition of distributional derivatives, $\forall \phi \in \mathcal{D}(\mathbb{R})$,

$$\left\langle \frac{d}{dx}\left[\text{c.p.v.}\left(\frac{1}{x}\right)\right], \phi \right\rangle = -\left\langle \text{c.p.v.}\left(\frac{1}{x}\right), \phi' \right\rangle$$
$$= -\lim_{\varepsilon\to 0^+} \int_{|x|\geq\varepsilon} \frac{\phi'(x)}{x} dx = -\lim_{\varepsilon\to 0^+} I(\varepsilon), \tag{2.3.35}$$

where $I(\varepsilon) = \int_{|x|\geq\varepsilon} \frac{\phi'(x)}{x} dx = \int_{|x|\geq\varepsilon} \frac{\phi(x)}{x^2} dx - \frac{\phi(\varepsilon)}{\varepsilon} - \frac{\phi(-\varepsilon)}{\varepsilon}$, which is obtained by integrating by parts: $u = \frac{1}{x}$, $dv = \phi'(x)dx$ and using $\int_{|x|\geq\varepsilon} \cdots =$

$\int_{-\infty}^{-\varepsilon} \dots + \int_{\varepsilon}^{\infty} \dots$ and applying $\phi(x) = 0$ for $x \pm \infty$ ($\int_{|x|\geq\varepsilon} \frac{\phi'(x)}{x} dx = \frac{\phi(x)}{x}\Big|_{-\infty}^{-\varepsilon} + \int_{-\infty}^{-\varepsilon} \frac{\phi(x)}{x^2} dx + \frac{\phi(x)}{x}\Big|_{\varepsilon}^{\infty} + \int_{\varepsilon}^{\infty} \frac{\phi(x)}{x^2} dx$).

But $\phi(x) = \phi(0) + x\phi'(0) + x^2\psi(x)$ with $\psi \in C^0(\mathbb{R})$ by Proposition 1.2.1

$\Longrightarrow \phi(\pm\varepsilon) = \phi(0) \pm \varepsilon\phi'(0) + \varepsilon^2\psi(\pm\varepsilon)$

$\Longrightarrow \frac{\phi(\varepsilon)}{\varepsilon} + \frac{\phi(-\varepsilon)}{\varepsilon} = 2\frac{\phi(0)}{\varepsilon} + \varepsilon[\psi(\varepsilon) + \psi(-\varepsilon)]$

$\Longrightarrow I(\varepsilon) = \int_{|x|\geq\varepsilon} \frac{\phi(x)}{x^2} dx - 2\frac{\phi(0)}{\varepsilon} - \varepsilon[\psi(\varepsilon) + \psi(-\varepsilon)]$.

But $\varepsilon[\psi(\varepsilon) + \psi(-\varepsilon)] \to 0$ as $\varepsilon \to 0^+$, since $\psi \in C^0(\mathbb{R})$

$\Longrightarrow \psi(\pm\varepsilon) \to \psi(0)$ as $\varepsilon \to 0^+$.

$\Longrightarrow \lim_{\varepsilon\to 0^+} I(\varepsilon) = \lim_{\varepsilon\to 0^+}[\int_{|x|\geq\varepsilon} \frac{\phi(x)}{x^2} dx - 2\frac{\phi(0)}{\varepsilon}] = \mathrm{Pf} \int_{-\infty}^{\infty} \frac{\phi(x)}{x^2} dx$ (see (1.4.19)).

Hence, from (2.3.35), $\langle \frac{d}{dx}[\text{c.p.v.}(\frac{1}{x})], \phi\rangle = -\langle \mathrm{Pf}(\frac{1}{x^2}), \phi\rangle = \langle -\mathrm{Pf}(\frac{1}{x^2}), \phi\rangle$ $\forall \phi \in \mathcal{D}(\mathbb{R})$

$\Longrightarrow \frac{d}{dx}[\text{c.p.v.}(\frac{1}{x})] = -\mathrm{Pf}(\frac{1}{x^2})$ in $\mathcal{D}'(\mathbb{R})$.

2.

$$\left\langle \frac{d}{dx}\left[\mathrm{Pf}\left(\frac{H(-x)}{x}\right)\right], \phi\right\rangle = -\left\langle \mathrm{Pf}\left(\frac{H(-x)}{x}\right), \phi'(x)\right\rangle = -\,\mathrm{Pf}\int_{-\infty}^{0} \frac{\phi'(x)}{x} dx$$
$$= -\lim_{\varepsilon\to 0^+} \underbrace{\left[\int_{-\infty}^{-\varepsilon} \frac{\phi'(x)}{x} dx - \phi'(0)\ln\varepsilon\right]}_{I_1(\varepsilon)} \quad \text{(using (1.4.27))}$$
$$= -\lim_{\varepsilon\to 0^+} I_1(\varepsilon) \quad \forall \phi \in \mathcal{D}(\mathbb{R}).$$

Integrating by parts: $u = \frac{1}{x}$, $dv = \phi'(x)dx$, and applying $\phi(x) = 0$ for $x = -\infty$, we get, $\forall \varepsilon > 0$,

$$I_1(\varepsilon) = \int_{-\infty}^{-\varepsilon} \frac{\phi'(x)}{x} dx - \phi'(0)\ln\varepsilon = \frac{\phi(x)}{x}\Big|_{-\infty}^{-\varepsilon} + \int_{-\infty}^{-\varepsilon} \frac{\phi(x)}{x^2} dx - \phi'(0)\ln\varepsilon$$
$$= -\frac{\phi(-\varepsilon)}{\varepsilon} + \int_{-\infty}^{-\varepsilon} \frac{\phi(x)}{x^2} dx - \phi'(0)\ln\varepsilon$$
$$= \int_{-\infty}^{-\varepsilon} \frac{\phi(x)}{x^2} dx - \phi'(0)\ln\varepsilon - \frac{\phi(-\varepsilon)}{\varepsilon} \quad \forall \phi \in \mathcal{D}(\mathbb{R}).$$

But $\phi(x) = \phi(0) + x\phi'(0) + x^2\psi(x)$ with $\psi \in C^0(\mathbb{R}) \Longrightarrow \frac{\phi(-\varepsilon)}{\varepsilon} = \frac{\phi(0)}{\varepsilon} - \phi'(0) + \varepsilon\psi(-\varepsilon)$ with $\varepsilon\psi(-\varepsilon) \to 0$ as $\varepsilon \to 0^+$, since $\psi \in C^0(\mathbb{R})$.

Hence,

$$\lim_{\varepsilon\to 0+} I_1(\varepsilon) = \lim_{\varepsilon\to 0+}\left[\int_{-\infty}^{-\varepsilon}\frac{\phi(x)}{x^2}dx - \frac{\phi(0)}{\varepsilon} + \phi'(0) - \varepsilon\psi(-\varepsilon) - \phi'(0)\ln\varepsilon\right]$$

$$= \lim_{\varepsilon\to 0+}\left[\int_{-\infty}^{-\varepsilon}\frac{\phi(x)}{x^2}dx - \frac{\phi(0)}{\varepsilon} - \phi'(0)\ln\varepsilon\right] + \phi'(0) - \lim_{\varepsilon\to 0}\varepsilon\psi(-\varepsilon)$$

$$= \left\langle \operatorname{Pf}\left(\frac{H(-x)}{x^2}\right), \phi\right\rangle + \phi'(0) = \left\langle \operatorname{Pf}\left(\frac{H(-x)}{x^2}\right), \phi\right\rangle + \langle\delta, \phi'\rangle$$

(using (1.4.24) and $\langle\delta,\phi'\rangle = \phi'(0) = \langle -\delta', \phi\rangle$) $\forall\phi\in\mathcal{D}(\mathbb{R})$, and

$$\left\langle \frac{d}{dx}\left[\operatorname{Pf}\left(\frac{H(-x)}{x}\right)\right], \phi\right\rangle = -\left[\left\langle \operatorname{Pf}\left(\frac{H(-x)}{x^2}\right), \phi\right\rangle - \langle\delta', \phi\rangle\right]$$

$$= \left\langle -\operatorname{Pf}\left(\frac{H(-x)}{x^2}\right) + \delta', \phi\right\rangle \quad \forall\phi\in\mathcal{D}(\mathbb{R})$$

$$\Longrightarrow \quad \frac{d}{dx}\left[\operatorname{Pf}\left(\frac{H(-x)}{x}\right)\right] = -\operatorname{Pf}\left(\frac{H(-x)}{x^2}\right) + \delta' \quad \text{in } \mathcal{D}'(\mathbb{R}).$$

3.

$$\left\langle \frac{d}{dx}\left[\operatorname{Pf}\left(\frac{H(x)}{x}\right)\right], \phi\right\rangle = -\left\langle \operatorname{Pf}\left(\frac{H(x)}{x}\right), \frac{d\phi}{dx}\right\rangle = -\operatorname{Pf}\int_0^{\infty}\frac{\phi'(x)}{x}dx$$

$$= -\lim_{\varepsilon\to 0+}\left[\int_{\varepsilon}^{\infty}\frac{\phi'(x)}{x}dx + \phi'(0)\ln\varepsilon\right] \quad \text{(using (1.4.26))}$$

$$= -\lim_{\varepsilon\to 0+} I_2(\varepsilon).$$

Following the steps of part (2) above, we have: $\forall\phi\in\mathcal{D}(\mathbb{R})$,

$$\lim_{\varepsilon\to 0+} I_2(\varepsilon) = \lim_{\varepsilon\to 0+}\left[\int_{\varepsilon}^{\infty}\frac{\phi(x)}{x^2}dx - \frac{\phi(0)}{\varepsilon} - \phi'(0) - \varepsilon\psi(\varepsilon) + \phi'(0)\ln\varepsilon\right]$$

$$= \lim_{\varepsilon\to 0+}\left[\int_{\varepsilon}^{\infty}\frac{\phi(x)}{x^2}dx - \frac{\phi(0)}{\varepsilon} + \phi'(0)\ln\varepsilon\right] - \phi'(0)$$

(since $\varepsilon\psi(\varepsilon)\to 0$ as $\varepsilon\to 0$)

$$= \left\langle \operatorname{Pf}\left(\frac{H(x)}{x^2}\right), \phi\right\rangle - \phi'(0 = \left\langle \operatorname{Pf}\left(\frac{H(x)}{x^2}\right), \phi\right\rangle - \langle\delta, \phi'\rangle$$

(using (1.4.23)).

Hence,

$$\left\langle \frac{d}{dx}\left[\operatorname{Pf}\left(\frac{H(x)}{x}\right)\right], \phi\right\rangle = \left\langle -\operatorname{Pf}\left(\frac{H(x)}{x^2}\right) - \delta', \phi\right\rangle \quad \forall\phi\in\mathcal{D}(\mathbb{R})$$

$\Longrightarrow$ the result (2.3.34) (using $\langle\delta,\phi'\rangle = -\langle\delta',\phi\rangle$). □

2.4 Conditions under which the two notions of derivatives of functions coincide

A natural question arises: when will the distributional and the usual derivatives of a function coincide? The following theorem (for $n = 2$) answers this interesting question:

Theorem 2.4.1.

I. *Let* $f \in L^1_{\text{loc}}(\mathbb{R}^n)$ *such that* f *is* absolutely continuous[1] *in* x_1 *for almost all* $\hat{\mathbf{x}} = (x_2, \ldots, x_n)$ *(i.e. on almost all parallels to the* x_1 *axis) and its partial derivative* $[\frac{\partial f}{\partial x_1}(\mathbf{x})]$ *in the usual pointwise sense is equal to a* locally summable *function* $g_1 \in L^1_{\text{loc}}(\mathbb{R}^n)$ *a.e. in* $\mathbb{R}^n$. *Then the* distributional derivative $\frac{\partial f}{\partial x_1} \in \mathcal{D}'(\mathbb{R}^n)$ *and the usual partial derivative* $[\frac{\partial f}{\partial x_1}(\mathbf{x})]$ *will coincide, i.e.* $\frac{\partial f}{\partial x_1} = [\frac{\partial f}{\partial x_1}(\mathbf{x})] = g_1$ *a.e. in* $\mathbb{R}^n$.

II. *Let* $f \in L^1_{\text{loc}}(\mathbb{R}^n)$, *and* $\frac{\partial f}{\partial x_1} = g_1 \in L^1_{\text{loc}}(\mathbb{R}^n)$ *be the* distributional derivative *of* f. *Then* f *(after a possible change of its values on a set of points with measure zero) is* absolutely continuous *in* x_1 *on all parallels to the* x_1*-axis and has the* usual partial derivative $[\frac{\partial f}{\partial x_1}(\mathbf{x})] = g_1(\mathbf{x})$ *a.e. in* $\mathbb{R}^n$.

Proof.

I. Let $f \in L^1_{\text{loc}}(\mathbb{R}^n)$ be absolutely continuous in x_1 on almost all parallels to the x_1-axis such that $[\frac{\partial f}{\partial x_1}(\mathbf{x})] = g_1(\mathbf{x})$ a.e. in $\mathbb{R}^n$, $g_1 \in L^1_{\text{loc}}(\mathbb{R}^n)$. $\forall \phi \in \mathcal{D}(\mathbb{R}^n)$, $f\frac{\partial \phi}{\partial x_1} \in L^1(\mathbb{R}^n)$, since $\text{supp}(\phi) \subset\subset \mathbb{R}^n$, $f \in L^1_{\text{loc}}(\mathbb{R}^n)$ and Fubini's theorem can be applied for $f\frac{\partial \phi}{\partial x_1} \in L^1(\mathbb{R}^n)$ and also for $g_1\phi$. Thus, $\forall \phi \in \mathcal{D}(\mathbb{R}^n)$,

$$\left\langle \frac{\partial f}{\partial x_1}, \phi \right\rangle = -\left\langle f, \frac{\partial \phi}{\partial x_1} \right\rangle = -\int_{\mathbb{R}^n} f(\mathbf{x})\frac{\partial \phi}{\partial x_1}(\mathbf{x})d\mathbf{x}$$

$$= \int_{\mathbb{R}} dx_2 \cdots \int_{\mathbb{R}} dx_n \Big(-\int_{\mathbb{R}} f(x_1, \hat{\mathbf{x}})\frac{\partial \phi}{\partial x_1}(x_1, \hat{\mathbf{x}})dx_1 \Big) \quad \text{(by Fubini's theorem)}$$

$$= \int_{\mathbb{R}} dx_2 \cdots \int_{\mathbb{R}} dx_n \Big(-\int_{-\infty}^{\infty} \Big(\Big[\frac{\partial (f\phi)}{\partial x_1}(x_1, \hat{\mathbf{x}})\Big] - \Big[\frac{\partial f}{\partial x_1}(x_1, \hat{\mathbf{x}})\Big]\phi(x_1, \hat{\mathbf{x}}) \Big) dx_1 \Big)$$

$$= \int_{\mathbb{R}} dx_2 \cdots \int_{\mathbb{R}} dx_n \Big((f\phi)(x_1, \hat{\mathbf{x}})\Big|_{x_1=-\infty}^{x_1=+\infty} + \int_{-\infty}^{\infty} g_1(x_1, \hat{\mathbf{x}})\phi(x_1, \hat{\mathbf{x}})dx_1 \Big)$$

---

[1] For $n = 1$, a function f is said to be *absolutely continuous* on $]a,b[$ if and only if it can be represented in the form $f(x) = A + \int_C^x g_1(\xi)d\xi$ for $x \in]a,b[$, where $g_1 \in L^1(]a,b[)$, $C \in]a,b[$ is any point, $A = f(C)$, since $\int_C^C g_1(\xi)d\xi = 0$. Then f is continuous on $]a,b[$, since for $x, x+h \in]a,b[$, $|f(x+h) - f(x)| = |\int_x^{x+h} g_1(\xi)d\xi| \to 0$ as $h \to 0$ and the usual derivative $[\frac{df}{dx}(x)] = g_1(x)$ a.e. on $]a,b[$. At the points $x \in]a,b[$, which form a set with measure 0, where the equality $[\frac{df}{dx}(x)] = g_1(x)$ does *not* hold, f can have a derivative different from $g_1(x)$ or f has no derivative at all.

$$= \int_{\mathbb{R}} dx_2 \cdots \int_{\mathbb{R}} dx_n \int_{\mathbb{R}} g_1(\mathbf{x})\phi(\mathbf{x})dx_1$$

$$\text{(since } \phi(x_1, \hat{\mathbf{x}}) = 0 \text{ for } x_1 = \pm\infty \text{ and } [\frac{\partial f}{\partial x_1}(\mathbf{x})] = g_1(\mathbf{x}) \text{ a.e.)}$$

$$= \int_{R^n} g_1(\mathbf{x})\phi(\mathbf{x})d\mathbf{x} \text{ (by Fubini's Theorem 7.1.2C, since } g_1\phi \in L^1(\mathbb{R}^n))$$

$$= \langle g_1, \phi\rangle \quad \forall \phi \in \mathcal{D}(\mathbb{R}^n) \quad \Longrightarrow \quad \frac{\partial f}{\partial x_1} = g_1 \text{ in } \mathcal{D}'(\mathbb{R}^n).$$

II. Since $g_1 \in L^1_{\text{loc}}(\mathbb{R}^n)$, we can define f_1 by: $f_1(\mathbf{x}) = \int_{a_1}^{x_1} g_1(\xi_1, \hat{\mathbf{x}})d\xi_1$ ($\hat{\mathbf{x}} = (x_2, \ldots, x_n)$) with the properties: f_1 is defined a.e. in $\mathbb{R}^n$, absolutely continuous in x_1 on almost all parallels to the x_1-axis, integrable on all compact sets, i.e. $f_1 \in L^1_{\text{loc}}(\mathbb{R}^n)$ and, finally, $[\frac{\partial f_1}{\partial x_1}(\mathbf{x})] = g_1(\mathbf{x})$ a.e. in $\mathbb{R}^n$ (by virtue of the absolute continuity of f_1). Then, from (1) above, the distributional derivative $\frac{\partial f_1}{\partial x_1} = g_1$ in $\mathcal{D}'(\mathbb{R}^n)$. By hypothesis, $f \in L^1_{\text{loc}}(\mathbb{R}^n)$ with $\frac{\partial f}{\partial x_1} = g_1 \in L^1_{\text{loc}}(\mathbb{R}^n)$. This suggests we write $f = f_1 + Z$ in $\mathcal{D}'(\mathbb{R}^n)$, where Z is a distribution independent of x_1, i.e. the distributional derivative $\frac{\partial Z}{\partial x_1} = 0$. But f and f_1 are functions. Hence, Z is also a function and $\frac{\partial Z}{\partial x_1} = 0 \Longrightarrow$ as a function, Z is independent of x_1 a.e. in $\mathbb{R}^n$. Thus, to complete the proof, we are to show that Z is a function independent of x_1 in the *usual sense*, i.e. $[\frac{\partial Z}{\partial x_1}(\mathbf{x})] = 0$ *everywhere* in $\mathbb{R}^n$. As a function, Z is independent of x_1 a.e. in $\mathbb{R}^n$. Hence, by changing the values of Z on a set of points with measure zero (where Z is *not* independent of x_1), we can make Z independent of x_1 *everywhere* in $\mathbb{R}^n$ in the usual sense. Then Z becomes absolutely continuous in x_1 on all parallels to the x_1-axis. Hence, its usual partial derivative $[\frac{\partial Z}{\partial x_1}(\mathbf{x})] = 0$ *everywhere* in $\mathbb{R}^n$ and the proof is complete.

□

Remark 2.4.1. Let the conditions of (2) in Theorem 2.4.1 hold. In particular, if f and its distributional derivative g_1 are *continuous* functions in $\Omega \subset \mathbb{R}^n$, then $[\frac{\partial f}{\partial x_1}(\mathbf{x})] = g_1(\mathbf{x})$ *everywhere* in Ω (see Schwartz [8]).

Example 2.4.1. Let $f(\mathbf{x}) = \frac{1}{r^\lambda}$ with $r = r(\mathbf{x}) = (x_1^2 + x_2^2 + \cdots + x_n^2)^{1/2} > 0$ and $0 < \lambda < n - 1$. Then, the *distributional derivative* $\frac{\partial}{\partial x_1}(\frac{1}{r^\lambda}) =$ its *usual partial derivative* $[\frac{\partial}{\partial x_1}(\frac{1}{r^\lambda})(\mathbf{x})] = -\frac{\lambda}{r^{\lambda+1}}\frac{\partial r}{\partial x_1} = -\frac{\lambda x_1}{r^{\lambda+2}}$ for $0 < \lambda < n-1$, since $f \in L^1_{\text{loc}}(\mathbb{R}^n)$, $f = \frac{1}{r^\lambda}$ is absolutely continuous in x_1 on almost all parallels to the x_1-axis: $f(x_1, \hat{\mathbf{x}}) = \int_{a_1}^{x_1} -\frac{\lambda\xi_1}{r^{\lambda+2}(\xi_1, \hat{\mathbf{x}})}d\xi_1 = \int_{a_1}^{x_1} \frac{\partial}{\partial\xi_1}(\frac{1}{r^\lambda(\xi_1, \hat{\mathbf{x}})})d\xi_1 = \frac{1}{r^\lambda} - h(\hat{\mathbf{x}})$ for some h; $[\frac{\partial f}{\partial x_1}(\mathbf{x})] = -\frac{\lambda x_1}{r^{\lambda+2}} = g_1(\mathbf{x})$ a.e. with $g_1 \in L^1_{\text{loc}}(\mathbb{R}^n)$ for $0 < \lambda < n - 1$.

In fact, $f \in L^1_{\text{loc}}(\mathbb{R}^n)$ if $\int_{\|\mathbf{x}\|\le 1} \frac{1}{r^\lambda} d\mathbf{x} = S_n \int_0^1 r^{n-\lambda-1} dr < +\infty$ ($S_n = \frac{2(\sqrt{\pi})^n}{\Gamma(n/2)}$ being the surface area of the unit sphere in $\mathbb{R}^n$). Hence, for $n-\lambda-1 > -1$, i.e. for $0<\lambda<n$, $f \in L^1_{\text{loc}}(\mathbb{R}^n)$. $g_1 \in L^1_{\text{loc}}(\mathbb{R}^n)$, if $\int_{\|\mathbf{x}\|\le 1} |-\frac{\lambda x_1}{r^{\lambda+2}}| d\mathbf{x} = \int_{\|\mathbf{x}\|\le 1} \frac{\lambda |x_1|}{r^{\lambda+2}} d\mathbf{x} \le \lambda \int_{\|\mathbf{x}\|\le 1} \frac{1}{r^{\lambda+1}} d\mathbf{x} = \lambda S_n \int_0^1 r^{n-(\lambda+1)-1} dr < +\infty$. Hence, for $n-\lambda-1-1 > -1$, i.e. for $0<\lambda<n-1$, $g_1 \in L^1_{\text{loc}}(\mathbb{R}^n)$. Hence, *both* f and g_1 will belong to $L^1_{\text{loc}}(\mathbb{R}^n)$, if $0<\lambda<n-1$. Then the result follows from Theorem 2.4.1.

2.5 Derivative of product αT with $T \in \mathcal{D}'(\Omega)$ and $\alpha \in C^\infty(\Omega)$

Theorem 2.5.1. *Let $\alpha \in C^\infty(\Omega)$, $T \in \mathcal{D}'(\Omega)$ and $\alpha T \in \mathcal{D}'(\Omega)$ be defined by (1.5.2): $\langle \alpha T, \phi\rangle = \langle T, \alpha\phi\rangle \quad \forall \phi \in \mathcal{D}(\Omega)$. Then the usual formula of the* product rule *of the derivative holds:*

$$\frac{\partial}{\partial x_i}(\alpha T) = \frac{\partial \alpha}{\partial x_i} T + \alpha \frac{\partial T}{\partial x_i} \quad \text{in } \mathcal{D}'(\Omega). \tag{2.5.1}$$

Proof. $\forall \phi \in \mathcal{D}(\Omega)$,

$$\begin{aligned}
\left\langle \frac{\partial}{\partial x_i}(\alpha T), \phi \right\rangle &= -\left\langle \alpha T, \frac{\partial \phi}{\partial x_i} \right\rangle = -\left\langle T, \alpha \frac{\partial \phi}{\partial x_i} \right\rangle = -\left\langle T, \frac{\partial}{\partial x_i}(\alpha\phi) - \frac{\partial \alpha}{\partial x_i}\phi \right\rangle \\
&= -\left\langle T, \frac{\partial}{\partial x_i}(\alpha\phi) \right\rangle + \left\langle T, \frac{\partial \alpha}{\partial x_i}\phi \right\rangle = \left\langle \frac{\partial T}{\partial x_i}, \alpha\phi \right\rangle + \left\langle \frac{\partial \alpha}{\partial x_i} T, \phi \right\rangle \\
&= \left\langle \alpha \frac{\partial T}{\partial x_i} + \frac{\partial \alpha}{\partial x_i} T, \phi \right\rangle
\end{aligned}$$

$\Longrightarrow \frac{\partial}{\partial x_i}(\alpha T) = \alpha \frac{\partial T}{\partial x_i} + \frac{\partial \alpha}{\partial x_i} T$ in $\mathcal{D}'(\Omega)$.

Similarly, $\frac{\partial^2}{\partial x_i^2}(\alpha T) = \frac{\partial^2 \alpha}{\partial x_i^2} T + 2 \frac{\partial \alpha}{\partial x_i}\frac{\partial T}{\partial x_i} + \alpha \frac{\partial^2 T}{\partial x_i^2}$. □

Leibniz's formula for successive differentiation of the product fT holds:

Theorem 2.5.1A. *For $f \in C^\infty(\Omega)$, $T \in \mathcal{D}'(\Omega)$ and $\partial^{\boldsymbol{\alpha}} = \frac{\partial^{|\boldsymbol{\alpha}|}}{\partial x_1^{\alpha_1} \dots \partial x_n^{\alpha_n}}$, $\boldsymbol{\alpha} = (\alpha_1, \alpha_2, \dots, \alpha_n) \in \mathbb{N}_0^n$, $\boldsymbol{\beta} = (\beta_1, \beta_2, \dots, \beta_n) \in \mathbb{N}_0^n$, $\alpha_i - \beta_i \ge 0 \ \forall i = 1, 2, \dots, n$,*

$$\begin{aligned}
\partial^{\boldsymbol{\alpha}}(fT) &= \sum_{\boldsymbol{\beta} \le \boldsymbol{\alpha}} \binom{\boldsymbol{\alpha}}{\boldsymbol{\beta}} (\partial^{\boldsymbol{\beta}} f)(\partial^{\boldsymbol{\alpha}-\boldsymbol{\beta}} T) \\
&= \sum_{\beta_1 \le \alpha_1} \cdots \sum_{\beta_n \le \alpha_n} \binom{\alpha_1}{\beta_1} \cdots \binom{\alpha_n}{\beta_n} (\partial^{\boldsymbol{\beta}} f)(\partial^{\boldsymbol{\alpha}-\boldsymbol{\beta}} T),
\end{aligned} \tag{2.5.2}$$

where $\binom{\boldsymbol{\alpha}}{\boldsymbol{\beta}} = \binom{\alpha_1}{\beta_1}\binom{\alpha_2}{\beta_2} \cdots \binom{\alpha_n}{\beta_n}$ *with* $\binom{\alpha_i}{\beta_i} = \frac{\alpha_i!}{\beta_i!(\alpha_i-\beta_i)!}$, $1 \le i \le n$.

In particular, for $n = 2$, $\boldsymbol{\alpha} = (\alpha_1, \alpha_2)$, $\boldsymbol{\beta} = (\beta_1, \beta_2)$, $f \in C^\infty(\Omega)$,

$$\partial^{\boldsymbol{\alpha}}(fT) = \sum_{\beta_1 \le \alpha_1} \sum_{\beta_2 \le \alpha_2} \binom{\alpha_1}{\beta_1} \binom{\alpha_2}{\beta_2} (\partial^{\boldsymbol{\beta}} f)(\partial^{\boldsymbol{\alpha}-\boldsymbol{\beta}} T).$$

Proof. The result is proved by the method of induction and Theorem 2.5.1. □

Example 2.5.1. For $n = 1$, $\delta \in \mathcal{D}'(\mathbb{R})$, $(\alpha\delta)' = \alpha\delta' + \alpha'\delta = \alpha(0)\delta'$ with $\alpha \in C^\infty(\mathbb{R})$,

$$\alpha\delta' = \alpha(0)\delta' - \alpha'(0)\delta \quad \text{and} \quad \alpha'\delta = \alpha'(0)\delta \quad \text{in } \mathcal{D}'(\mathbb{R}). \tag{2.5.3}$$

In fact, from (2.5.1), $(\alpha\delta)' = \alpha\delta' + \alpha'\delta$ in $\mathcal{D}'(\mathbb{R})$ with

$$\begin{aligned}\langle \alpha\delta', \phi\rangle &= \langle \delta', \alpha\phi\rangle = -\langle \delta, (\alpha\phi)'\rangle = -(\alpha\phi)'(0) = -\alpha(0)\phi'(0) - \alpha'(0)\phi(0)\\ &= -\alpha(0)\langle \delta, \phi'\rangle - \alpha'(0)\langle \delta, \phi\rangle = \alpha(0)\langle \delta', \phi\rangle - \alpha'(0)\langle \delta, \phi\rangle\\ &= \langle \alpha(0)\delta' - \alpha'(0)\delta, \phi\rangle \quad \forall \phi \in \mathcal{D}(\mathbb{R})\end{aligned}$$

$\Longrightarrow \alpha\delta' = \alpha(0)\delta' - \alpha'(0)\delta$ in $\mathcal{D}'(\mathbb{R})$,

$$\langle \alpha'\delta, \phi\rangle = \langle \delta, \alpha'\phi\rangle = \alpha'(0)\phi(0) = \alpha'(0)\langle \delta, \phi\rangle = \langle \alpha'(0)\delta, \phi\rangle \quad \forall \phi \in \mathcal{D}(\mathbb{R})$$

$\Longrightarrow \alpha'\delta = \alpha'(0)\delta$ in $\mathcal{D}'(\mathbb{R})$. Combining these, we have $(\alpha\delta)' = \alpha(0)\delta'$.

In particular, for $\alpha(x) = x^k$ $\forall x \in \mathbb{R}$, $k \in \mathbb{N}$,

$$\alpha(0) = 0, \quad \alpha^{(l)}(0) = \begin{cases} k! & \text{for } l = k\\ 0 & \text{for } l \neq k.\end{cases}$$

For $k = 1$,

$$x\delta' = 0 \cdot \delta' - 1 \cdot \delta = -\delta, \quad x^k\delta' = 0 \quad \forall k \ge 2. \tag{2.5.4}$$

Example 2.5.2. For $n = 1$, $x\delta^{(k)} = -k\delta^{(k-1)}$ $\forall k \ge 2$.

Indeed, $\forall k \ge 2$,

$$\begin{aligned}\langle x\delta^{(k)}, \phi\rangle &= \langle \delta^{(k)}, x\phi\rangle = (-1)^k\langle \delta, (x\phi)^{(k)}\rangle = (-1)^k(x\phi)^{(k)}(0)\\ &= (-1)^k[(x\phi^{(k)})(0) + k\phi^{(k-1)}(0)] = (-1)^k[0 + k\phi^{(k-1)}(0)]\\ &= (-1)^k k\langle \delta, \phi^{(k-1)}\rangle = (-1)^k \cdot (-1)^{k-1}k\langle \delta^{(k-1)}, \phi\rangle = \langle -k\delta^{(k-1)}, \phi\rangle\end{aligned}$$

$\Longrightarrow x\delta^{(k)} = -k\delta^{(k-1)}$ in $\mathcal{D}'(\mathbb{R})$.

Example 2.5.3. $\forall m \in \mathbb{N}$,

$$x^m \delta^{(k)} = \begin{cases} 0 & \text{for } k = 0, 1, \dots, m-1 \\ (-1)^m m! \binom{k}{m} \delta^{(k-m)} & \text{for } k \geq m. \end{cases} \tag{2.5.5}$$

Proof. $\forall \phi \in \mathcal{D}(\mathbb{R})$,

$$\begin{aligned} \langle x^m \delta^{(k)}, \phi \rangle &= \langle \delta^{(k)}, x^m \phi \rangle = (-1)^k \langle \delta, (x^m \phi)^{(k)} \rangle = (-1)^k (x^m \phi)^{(k)}|_{x=0} \\ &= (-1)^k \cdot \Big(\sum_{0 \leq j \leq k} \binom{k}{j} (x^m)^{(j)} \phi^{(k-j)} \Big)\Big|_{x=0}. \end{aligned}$$

For $k = 0, 1, \dots, m-1$ with $0 \leq j \leq k < m$, $(x^m)^{(j)}|_{x=0} = 0$

$$\implies \quad (-1)^k \Big[\sum_{0 \leq j \leq k} \binom{k}{j} (x^m)^{(j)} \phi^{(k-j)} \Big]\Big|_{x=0} = 0.$$

For $k = m, m+1, \dots,$ with $0 \leq j \leq k$, $(x^m)^{(j)}|_{x=0} = (x^m)^{(m)}|_{x=0} = m!$ for $j = m$, and $(x^m)^{(j)}|_{x=0} = 0$ for $j \neq m \implies$ for $k = m, m+1, \dots,$ with $0 \leq j \leq k$,

$$\Big((-1)^k \sum_{0 \leq j \leq k} \binom{k}{j} (x^m)^{(j)} \phi^{(k-j)} \Big)\Big|_{x=0} = (-1)^k \binom{k}{m} m! \phi^{(k-m)}(0) \quad \forall \phi \in \mathcal{D}(\mathbb{R}).$$

Hence,

$$\langle x^m \delta^{(k)}, \phi \rangle = \begin{cases} 0 \quad \text{for } k = 0, 1, \dots, m-1; \\ (-1)^k \binom{k}{m} m! \phi^{(k-m)}(0) = (-1)^m \binom{k}{m} m! \langle \delta^{(k-m)}, \phi \rangle \quad \text{for } k \geq m \end{cases}$$

since $\phi^{(k-m)}(0) = \langle \delta, \phi^{(k-m)}(x) \rangle = (-1)^{k-m} \langle \delta^{(k-m)}, \phi \rangle \ \forall \phi \in \mathcal{D}(\mathbb{R})$

$$\implies \quad x^m \delta^{(k)} = \begin{cases} 0 & \text{for } k = 0, 1, \dots, m-1 \\ (-1)^m \binom{k}{m} m! \delta^{(k-m)} & \text{for } k \geq m. \end{cases}$$ □

Example 2.5.4. Let $\frac{du}{dx} = \operatorname{Pf}(\frac{H(x)}{x^2}) + \delta'$ in $\mathcal{D}'(\mathbb{R})$, where Pf denotes the pseudo-function defined by the finite part of the (divergent) integral: $\operatorname{Pf} \int_{-\infty}^{\infty} \frac{H(x)}{x^2} \phi(x) dx = \operatorname{Pf} \int_0^\infty \frac{\phi(x)}{x^2} dx$, and δ' is the derivative of Dirac distribution δ. Find xu' in $\mathcal{D}'(\mathbb{R})$.

Solution.

$$\begin{aligned}
\langle xu', \phi\rangle &= \langle u', x\phi\rangle = \left\langle \operatorname{Pf}\left(\frac{H(x)}{x^2}\right) + \delta', x\phi\right\rangle \quad \forall \phi \in \mathcal{D}(\mathbb{R}) \\
&= \left\langle \operatorname{Pf}\left(\frac{H(x)}{x^2}\right), x\phi\right\rangle + \langle \delta', x\phi\rangle = \operatorname{Pf}\int_0^\infty \frac{x(\phi(x))}{x^2}dx - \langle \delta, (x\phi)'\rangle \\
&= \operatorname{Pf}\int_0^\infty \frac{\phi(x)}{x}dx - (x\phi'(x) + \phi(x))|_{x=0} \\
&= \operatorname{Pf}\int_0^\infty \frac{\phi(x)}{x}dx - \phi(0) = \operatorname{Pf}\int_0^\infty \frac{\phi(x)}{x}dx - \langle \delta, \phi\rangle \quad \forall \phi \in \mathcal{D}(\mathbb{R})
\end{aligned}$$

Hence, $\langle xu', \phi\rangle = \operatorname{Pf}\int_0^\infty \frac{\phi(x)}{x}dx - \langle \delta, \phi\rangle = \langle \operatorname{Pf}(\frac{H(x)}{x}) - \delta, \phi\rangle \ \forall \phi \in \mathcal{D}(\mathbb{R}) \implies xu' = \operatorname{Pf}(\frac{H(x)}{x}) - \delta$ in $\mathcal{D}'(\mathbb{R})$.

Derivative of image $\mathbf{F}T$ of $T \in \mathcal{D}'(\mathbb{R}^n)$ under F

Let $\mathbf{F}T \in \mathcal{D}'(\mathbb{R}^n)$ be the image of $T \in \mathcal{D}'(\mathbb{R}^n)$ under transformation of variables $\mathbf{F}(x) = \boldsymbol{\xi} \in \mathbb{R}^n$ defined in (1.10.1)–(1.10.4) (resp. (1.10.21)–(1.10.24)). Then the partial derivative $\frac{\partial}{\partial \xi_j}(\mathbf{F}T)(\boldsymbol{\xi}) = \frac{\partial}{\partial \xi_j}[T(\mathbf{F}^{-1}\boldsymbol{\xi}))]$ is given by the *usual chain rule of partial derivatives of composite functions*,

$$\frac{\partial}{\partial \xi_j}(\mathbf{F}T) = \mathbf{F}\left(\sum_{i=1}^n \frac{\partial x_i}{\partial \xi_j}\cdot\frac{\partial T}{\partial x_i}(x(\boldsymbol{\xi}))\right) = \mathbf{F}\left(\frac{\partial x_i}{\partial \xi_j}\cdot\frac{\partial T}{\partial x_i}(x(\boldsymbol{\xi}))\right). \tag{2.5.6}$$

(In (2.5.6), Einstein's summation convention with respect to twice repeated index i has been followed, and will continue to be followed). In fact, $\forall \psi \in \mathcal{D}'(\mathbb{R}^n), 1 \le j \le n$,

$$\begin{aligned}
\left\langle \frac{\partial}{\partial \xi_j}(\mathbf{F}T)(\boldsymbol{\xi}), \psi(\boldsymbol{\xi})\right\rangle &= -\left\langle (\mathbf{F}T)(\boldsymbol{\xi}), \frac{\partial \psi}{\partial \xi_j}(\boldsymbol{\xi})\right\rangle = -\left\langle T(\mathbf{F}^{-1}(\boldsymbol{\xi})), \frac{\partial \psi}{\partial \xi_j}(\boldsymbol{\xi})\right\rangle \\
&= -\left\langle T(\mathbf{x}), \frac{\partial \psi}{\partial \xi_j}(\mathbf{F}(\mathbf{x}))|J_{\mathbf{F}}|\right\rangle,
\end{aligned} \tag{2.5.7}$$

$$\begin{aligned}
\left\langle \mathbf{F}\left(\frac{\partial x_i}{\partial \xi_j}\cdot\frac{\partial T}{\partial x_i}\right), \psi\right\rangle &= \left\langle \frac{\partial x_i}{\partial \xi_j}\cdot\frac{\partial T}{\partial x_i}(\mathbf{F}^{-1}(\boldsymbol{\xi})), \psi(\boldsymbol{\xi})\right\rangle = \left\langle \frac{\partial x_i}{\partial \xi_j}\cdot\frac{\partial T}{\partial x_i}(\mathbf{x}), \psi(\mathbf{F}(\mathbf{x}))|J_{\mathbf{F}}(\mathbf{x})|\right\rangle \\
&= \left\langle \frac{\partial T}{\partial x_i}(\mathbf{x}), \frac{\partial x_i}{\partial \xi_j}(\psi(\mathbf{F}(\mathbf{x}))|J_{\mathbf{F}}(\mathbf{x})|)\right\rangle \\
&= -\left\langle T, \frac{\partial}{\partial x_i}\left[\frac{\partial x_i}{\partial \xi_j}(\psi(\mathbf{F}(\mathbf{x}))|J_{\mathbf{F}}(\mathbf{x})|)\right]\right\rangle \quad \forall \psi \in \mathcal{D}(\mathbb{R}^n),
\end{aligned}$$

which must be equal to the right-hand side of (2.5.7) (the proof, involving the differentiation of determinants and the properties of $J_{\mathbf{F}}$, is omitted here).

2.6 Problem of division of distribution revisited

In Section 1.6, Chapter 1, we introduced the problem of division of distributions defined on $\mathbb{R}$, and considered only those examples in which derivatives of Dirac distributions δ (resp. δ_a) were not involved. Here we will discuss those problems, whose solutions will involve derivatives of Dirac distribution δ (resp. δ_a) on $\mathbb{R}$.

Example 2.6.1. Find the general solution of the following equations for T in $\mathcal{D}'(\mathbb{R})$:

$$\begin{array}{llll} 1. & (x-a)T = \delta_a; & 5. & (x-a)^m T = 0; \\ 2. & (x-a)T = \delta'_b; & 6. & (x-a)^m T = \delta_b^{(k)}; \\ 3. & (x-a)^2 T = 0; & 7. & P(x)T = 0, \\ 4. & (1+x^2)(1-x^2)^2 T = 0; & & \end{array} \tag{2.6.1}$$

where $P(x) = \prod_{i=1}^{k}(x-a_i)^{m_i} Q(x)$ is a polynomial of degree n with $a_i \neq a_j$ for $i \neq j, m_i \in \mathbb{N}$, $Q(x) \neq 0 \; \forall x \in \mathbb{R}$, and δ_a (resp. δ_b) is the Dirac distribution concentrated at $a \in \mathbb{R}$ (resp. $b \in \mathbb{R}$).

Solution.

1. $(x-a)T = \delta_a$ has a particular solution $T_p = -\delta'_a \in \mathcal{D}'(\mathbb{R})$.

   In fact,

$$\begin{aligned} \langle (x-a)T_p, \phi\rangle &= \langle T_p, (x-a)\phi\rangle = \langle -\delta'_a, (x-a)\phi\rangle = \langle \delta_a, ((x-a)\phi)'\rangle \\ &= [(x-a)\phi'(x) + \phi(x)]_{x=a} = \phi(a) = \langle \delta_a, \phi\rangle \end{aligned}$$

   $\Longrightarrow (x-a)T_p = \delta_a$ in $\mathcal{D}'(\mathbb{R})$ with $T_p = -\delta'_a$.

   $(x-a)T = 0$ has the general solution $T_h = C_1\delta_a$ by Proposition 1.6.1. Hence,

$$T = T_p + T_h = -\delta'_a + C_1\delta_a \quad \text{with } C_1 \in \mathbb{R} \tag{2.6.2}$$

   is the general solution of $(x-a)T = \delta_a$.

2. $(x-a)T = \delta'_b$, where $\delta'_b = \frac{d}{dx}\delta_b$, has the general solution $T = T_p + T_h$, $(x-a)T_p = \delta'_b$ and $T_h = C_1\delta_a$ with arbitrary constant $C_1 \in \mathbb{R}$ being the general solution of $(x-a)T = 0$ by Proposition 1.6.1.

   Construction of particular solution T_p: $(x-a)T_p = \delta'_b$

   Case $b \neq a$: $T_p = \frac{\delta_b}{(b-a)^2} + \frac{\delta'_b}{(b-a)}$

Indeed,

$$\begin{aligned}\langle (x-a)T_p, \phi\rangle &= \langle T_p, (x-a)\phi\rangle = \left\langle \frac{\delta_b}{(b-a)^2} + \frac{\delta_b'}{(b-a)}, (x-a)\phi \right\rangle \\ &= \frac{1}{(b-a)^2}\langle \delta_b, (x-a)\phi\rangle - \frac{1}{(b-a)}\langle \delta_b, ((x-a)\phi)'\rangle \\ &= \frac{1}{(b-a)^2}(b-a)\phi(b) - \frac{1}{(b-a)}[(x-a)\phi'(x) + \phi(x)]_{x=b} \\ &= \frac{\phi(b)}{(b-a)} - \frac{1}{(b-a)}[(b-a)\phi'(b) + \phi(b)] \\ &= -\phi'(b) = \langle \delta_b', \phi\rangle \quad \forall \phi \in \mathcal{D}(\mathbb{R})\end{aligned}$$

$\Longrightarrow (x-a)T_p = \delta_b'$ with $T_p = \frac{\delta_b}{(b-a)^2} + \frac{\delta_b'}{(b-a)}$ in $\mathcal{D}'(\mathbb{R})$.
Hence, for $b \neq a$ the general solution of Equation (2) in (2.6.1) is

$$T = \frac{\delta_b}{(b-a)^2} + \frac{\delta_b'}{(b-a)} + C_1\delta_a, \tag{2.6.3}$$

$C_1 \in \mathbb{R}$ being an arbitrary constant.
Case $b = a$: $T_p = -\frac{1}{2}\delta_a''$
In fact,

$$\begin{aligned}\langle (x-a)T_p, \phi\rangle &= \left\langle -\frac{1}{2}\delta_a'', (x-a)\phi \right\rangle = -\frac{1}{2}\langle \delta_a, ((x-a)\phi)''\rangle \\ &= -\frac{1}{2}\langle \delta_a, (x-a)\phi''(x) + 2\phi'(x)\rangle \\ &= -\frac{1}{2}\cdot 2\phi'(a) = -\phi'(a) = \langle \delta_a', \phi\rangle \qquad \forall \phi \in \mathcal{D}(\mathbb{R})\end{aligned}$$

$\Longrightarrow (x-a)T_p = \delta_a'$ with $T_p = -\frac{1}{2}\delta_a''$. Hence, for $b = a$, the general solution of (2) is

$$T = -\frac{1}{2}\delta_a'' + C_1\delta_a, \tag{2.6.4}$$

$C_1 \in \mathbb{R}$ being arbitrary constant.

3. $(x-a)^2 T = 0$. Set $S = (x-a)T$. Then,

$$\begin{aligned}(x-a)S = 0 \quad &\Longrightarrow \quad S = C_1\delta_a \quad \text{by Proposition 1.6.1} \\ &\Longrightarrow \quad (x-a)T = C_1\delta_a \\ &\Longrightarrow \quad T = -C_1\delta_a' + C_2\delta_a = d_1\delta_a' + C_2\delta_a \end{aligned} \tag{2.6.5}$$

is the general solution of $(x-a)^2 T = 0$, d_1 and C_2 being arbitrary real constants, which is obtained from the solution (2.6.2) of Equation (1).

4. $(1+x^2)(1-x^2)^2T = 0$ in $\mathcal{D}'(\mathbb{R})$.
   Since $(1+x^2) \in C^\infty(\mathbb{R})$ does *not* vanish on $\mathbb{R}$,

$$(1+x^2)^{-1}(1+x^2)(1-x^2)^2T = (1-x^2)^2T = 0$$

$\Longrightarrow (x-1)^2(x+1)^2T = 0$ in $\mathcal{D}'(\mathbb{R})$. Set $S = (x+1)^2T \in \mathcal{D}'(\mathbb{R})$. Then $(x-1)^2S = 0$ and, from the solution (2.6.5) of Equation (3) in (2.6.1) with $a = 1$, we get: $S = d_1\delta_1' + C_2\delta_1$, d_1 and C_2 being arbitrary constants.

Hence, $(x+1)^2T = d_1\delta_1' + C_2\delta_1$ has the general solution $T = T_p + T_h$, where T_p is any particular solution with $(x+1)^2T_p = d_1\delta_1' + C_2\delta_1$ and T_h is the general solution of the corresponding homogeneous equation $(x+1)^2T = 0$, which is obtained from Equation (3) in (2.6.1) with $a = -1$: $T_h = C_3\delta_{-1}' + C_4\delta_{-1}$, C_3, C_4 being arbitrary constants.

Construction of T_p satisfying $(x+1)^2T_p = d_1\delta_1' + C\delta_1$: We seek T_p in the form $T_p = \alpha_1\delta_1 + \alpha_2\delta_1'$, where α_1, α_2 are constants, which are determined by putting T_p in the equation:

$$\begin{aligned}
& & \langle (x+1)^2T_p, \phi\rangle &= \langle d_1\delta_1' + C_2\delta_1, \phi\rangle \quad \forall \phi \in \mathcal{D}(\mathbb{R}) \\
\Longrightarrow & & \langle T_p, (x+1)^2\phi\rangle &= \langle \alpha_1\delta_1 + \alpha_2\delta_1', (x+1)^2\phi\rangle \\
& & &= d_1\langle \delta_1', \phi\rangle + C_2\langle \delta_1, \phi\rangle \\
\Longrightarrow & & \alpha_1\langle \delta_1, (x+1)^2\phi\rangle + \alpha_2\langle \delta_1', (x+1)^2\phi\rangle &= -d_1\langle \delta_1, \phi'\rangle + C_2\phi(1) \\
\Longrightarrow & & \alpha_1(1+1)^2\phi(1) - \alpha_2\langle \delta_1, ((x+1)^2\phi)'\rangle &= -d_1\phi'(1) + C_2\phi(1).
\end{aligned}$$

But

$$\begin{aligned}
&4\alpha_1\phi(1) - \alpha_2[(x+1)^2\phi'(x) + 2(x+1)\phi(x)]_{x=1} \\
&\quad = 4\alpha_1\phi(1) - 4\alpha_2[\phi'(1) + \phi(1)] \\
&\Longrightarrow \quad (4\alpha_1 - 4\alpha_2)\phi(1) - 4\alpha_2\phi'(1) = -d_1\phi'(1) + C_2\phi(1) \quad \forall \phi \in \mathcal{D}(\mathbb{R}).
\end{aligned}$$

Equating the coefficients of $\phi(1)$ and $\phi'(1)$, we get $4\alpha_1 - 4\alpha_2 = C_2$, $-4\alpha_2 = -d_1 \Longrightarrow \alpha_2 = \frac{1}{4}d_1$ and $\alpha_1 = \frac{1}{4}(C_2 + d_1)$. Hence, $T_p = \frac{1}{4}(C_2 + d_1)\delta_1 + \frac{1}{4}d_1\delta_1'$.

Finally,

$$T = T_p + T_h = \frac{1}{4}(C_2 + d_1)\delta_1 + \frac{1}{4}d_1\delta_1' + C_3\delta_{-1}' + C_4\delta_{-1} \qquad (2.6.6)$$

is the required general solution of Equation (4), d_1, C_2, C_3, C_4 being arbitrary constants.

5. $(x-a)^m T = 0$. Then $T = \sum_{k=0}^{m-1} \alpha_k \delta_a^{(k)}$ with $\delta_a^{(0)} = \delta_a$, $\delta_a^{(k)} = \frac{d^k \delta_a}{dx^k}$ is the general solution of $(x-a)^m T = 0 \; \forall m \in \mathbb{N}$.

   In fact, it holds for $m = 1$ by Proposition 1.6.1 with $\delta_a^{(0)} = \delta_a, \alpha_0 = C_1$. We prove the result by the method of induction. Suppose that it holds for m. Now we will show that it holds for $m+1$. Indeed,

$$(x-a)^{m+1} T = 0 \implies (x-a)^m \underbrace{(x-a)T}_{U} = 0$$

$$\implies (x-a)^m U = 0 \implies U = \sum_{k=0}^{m-1} \alpha_k \delta_a^{(k)},$$

since it holds for m by assumption. Hence,

$$(x-a)T = U = \sum_{k=0}^{m-1} \alpha_k \delta_a^{(k)}. \tag{2.6.7}$$

$T = T_p + T_h$ is the general solution of (2.6.7), where T_p is a particular solution: $(x-a)T_p = \sum_{k=0}^{m-1} \alpha_k \delta_a^{(k)}$, $T_h = C_1 \delta_a$ is the general solution of $(x-a)T = 0$, C_1 being an arbitrary real constant.

Construction of T_p: First of all, we construct a particular solution T_p of $(x-a)T = \delta_a^{(k)}$, $0 \le k \le m-1$, which we seek in the form $T_p = \alpha_{k+1} \delta_a^{(k+1)}$, where α_{k+1} is an unknown real constant. Then,

$$\begin{aligned}
\langle (x-a)T_p, \phi \rangle &= \langle T_p, (x-a)\phi \rangle = \langle \alpha_{k+1} \delta_a^{(k+1)}, (x-a)\phi \rangle \\
&= \langle \delta_a^{(k)}, \phi \rangle \quad \forall \phi \in \mathcal{D}(\mathbb{R}) \\
\implies \quad & \alpha_{k+1} (-1)^{k+1} \langle \delta_a, ((x-a)\phi)^{(k+1)} \rangle = (-1)^k \langle \delta_a, \phi^{(k)} \rangle \\
\implies \quad & (-1)\alpha_{k+1} [(x-a)\phi^{(k+1)}(x) + (k+1)\phi^{(k)}(x)]_{x=a} = \phi^{(k)}(a) \\
\implies \quad & -\alpha_{k+1} [(k+1)\phi^{(k)}(a)] = \phi^{(k)}(a) \\
\implies \quad & \alpha_{k+1} = -\frac{1}{k+1} \implies T_p = -\frac{1}{k+1} \delta_a^{(k+1)}
\end{aligned}$$

is a particular solution of $(x-a)T = \delta_a^{(k)}$. Hence, a particular solution T_p of $(x-a)T = \sum_{k=0}^{m-1} \alpha_k \delta_a^{(k)}$ is given by:

$$T_p = \sum_{k=0}^{m-1} \alpha_k \cdot \left(-\frac{1}{k+1}\right) \delta_a^{(k+1)} = \sum_{k=0}^{m-1} \beta_{k+1} \delta_a^{(k+1)} \quad \text{with } \beta_{k+1} = -\frac{\alpha_k}{(k+1)}.$$

Then the general solution of $(x-a)^{(m+1)}T = 0$ is given by:

$$T = T_p + T_h = \sum_{k=0}^{m-1} \beta_{k+1}\delta_a^{(k+1)} + C_1\delta_a = \sum_{k=0}^{m} d_k\delta_a^{(k)},$$

d_ks being real constants with $d_0 = c_1, d_k = \beta_k, 1 \le k \le m$.

Thus, by the method of induction, we have proved that $(x-a)^m T = 0$ has the general solution

$$T = \sum_{k=0}^{m-1} \alpha_k\delta_a^{(k)}, \tag{2.6.8}$$

α_ks being arbitrary real constants.

6. The general solution $T \in \mathcal{D}'(\mathbb{R})$ of $(x-a)^m T = \delta_b^{(k)}$ is given by $T = T_h + T_p$, where

$$T_h = \sum_{k=0}^{m-1} \alpha_k\delta_a^{(k)} \tag{2.6.9}$$

with arbitrary constants α_k is the general solution of $(x-a)^m T = 0$ given by (2.6.8) and T_p is any particular solution of $(x-a)^m T = \delta_b^{(k)}$.

Construction of a Particular Solution T_p: $(x-a)^m T_p = \delta_b^{(k)}$ (2.6.10)

Case $b \ne a$: $T_p = C_0\delta_b + C_1\delta_b' + \cdots + C_k\delta_b^{(k)}$, (2.6.11)

where coefficient $C_k = \frac{1}{(b-a)^m}$, and the remaining coefficients $C_{k-1}, C_{k-2}\ldots,$ C_0 are sequentially determined in the given order by putting (2.6.11) in the equation (2.6.10). Equating the coefficients of $\phi^{(j)}(b)$ on both sides of the equality, C_j's are determined, coefficients of $\phi^{(j)}(b)$ on the right-hand side of the equality being zero $\forall j = 0, 1, \ldots, k-1$. For example,

- for $k = 0$,

$$T_p = C_0\delta_b \quad \text{with } C_0 = \frac{1}{(b-a)^m}. \tag{2.6.12}$$

In fact,

$$\begin{aligned}\langle (x-a)^m(C_0\delta_b), \phi\rangle &= C_0\langle \delta_b, (x-a)^m\phi\rangle = C_0(b-a)^m\phi(b)] \\ &= \frac{1}{(b-a)^m}[(b-a)^m\phi(b)] = \phi(b) = \langle \delta_b, \phi\rangle \quad \forall \phi \in \mathcal{D}(\mathbb{R})\end{aligned}$$

$\Longrightarrow (x-a)^m T_p = \delta_b$ with $T_p = C_0\delta_b$, $C_0 = \frac{1}{(b-a)^m}$;

- for $k = 1$,

$$T_p = C_0\delta_b + C_1\delta_b' \quad \text{with } C_1 = \frac{1}{(b-a)^m},\ C_0 = \frac{m}{(b-a)^{m+1}}. \tag{2.6.13}$$

Indeed, $\langle (x-a)^m[C_0\delta_b + C_1\delta_b'], \phi\rangle = \langle \delta_b', \phi\rangle = -\phi'(b)\ \forall \phi \in \mathcal{D}(\mathbb{R})$
$\Longrightarrow C_0\langle \delta_b, (x-a)^m\phi\rangle - C_1\langle \delta_b, ((x-a)^m\phi)'\rangle = -\phi'(b)$
$\Longrightarrow C_0(b-a)^m\phi(b) - C_1[(b-a)^m\phi'(b) + m(b-a)^{m-1}\phi(b)] = -\phi'(b)$.
Equating the coefficients of $\phi^{(j)}(b)$ on both sides of the equality, we have
Coefficient of $\phi'(b)$: $-C_1(b-a)^m = -1 \Longrightarrow C_1 = \frac{1}{(b-a)^m}$
Coefficient of $\phi(b)$: $C_0(b-a)^m - C_1 m(b-a)^{m-1} = 0 \Longrightarrow C_0 = \frac{C_1 m(b-a)^{m-1}}{(b-a)^m} = \frac{m}{(b-a)^{m+1}}$;

- for $k = 2$,

$$T_p = C_0\delta_b + C_1\delta_b' + C_2\delta_b'' \quad \text{with } C_2 = \frac{1}{(b-a)^m}. \tag{2.6.14}$$

In fact, $\langle (x-a)^m[C_0\delta_b + C_1\delta_b' + C_2\delta_b''], \phi\rangle = \langle \delta_b'', \phi\rangle = \phi''(b)$
$\Longrightarrow C_0\langle \delta_b, (x-a)^m\phi\rangle - C_1\langle \delta_b, ((x-a)^m\phi)'\rangle + C_2\langle \delta_b, ((x-a)^m\phi)''\rangle = \phi''(b)$
$\Longrightarrow C_0[(b-a)^m\phi(b)] - C_1[(b-a)^m\phi'(b) + m(b-a)^{m-1}\phi(b)] + C_2[(b-a)^m\phi''(b) + 2m(b-a)^{m-1}\phi'(b) + m(m-1)(b-a)^{m-2}\phi(b)] = \phi''(b)$.
Equating the coefficients of $\phi^{(j)}(b), 0 \le j \le 2$, we get:
Coefficient of $\phi''(b)$: $C_2(b-a)^m = 1 \Longrightarrow C_2 = \frac{1}{(b-a)^m}$
Coefficient of $\phi'(b)$: $-C_1(b-a)^m + C_2 2m(b-a)^{m-1} = 0 \Longrightarrow C_1 = \frac{C_2 2m}{b-a} = \frac{2m}{(b-a)^{m+1}}$
Coefficient of $\phi(b)$: $C_0(b-a)^m - C_1 m(b-a)^{m-1} + C_2 m(m-1)(b-a)^{m-2} = 0$

$$\Longrightarrow \quad C_0 = \frac{1}{(b-a)^2}[C_1 m(b-a) - C_2(m^2-m)] = \frac{m(m+1)}{(b-a)^{m+2}}. \tag{2.6.15}$$

Similarly, coefficients for other values of k are determined sequentially.

Case $b = a$: $(x-a)^m T = \delta_a^{(k)}$ has a particular solution

$$T_p = (-1)^m \frac{\delta_a^{(k+m)}}{m!\binom{k+m}{m}}, \tag{2.6.16}$$

since $(x-a)^m\delta_a^{(k+m)} = (-1)^m m!\binom{k+m}{m}\delta_a^{(k)}$ (see Example 2.5.3 with $a = 0$).

7. For $P(x) = (x - a_1)^{m_1}(x - a_2)^{m_2} \dots (x - a_k)^{m_k} Q(x)$, where $a_1 < a_2 < \dots < a_k$ are k distinct real zeros of $P(x)$ with multiplicities $m_1, m_2, \dots, m_k$ respectively, $Q(x)$ is a polynomial of degree $(n - \sum_{i=1}^{k} m_i) \geq 0$ such that $Q(x) \neq 0 \; \forall x \in \mathbb{R}$, $(x-a_1)^{m_1} \dots (x-a_k)^{m_k} Q(x)T = 0$. Since $Q \in C^\infty(\mathbb{R})$ does *not* vanish on $\mathbb{R}$, multiplying both sides by $\frac{1}{Q(x)}$, we get:

$$(x - a_1)^{m_1}(x - a_2)^{m_2} \cdots (x - a_k)^{m_k} T = 0 \quad \text{in } \mathcal{D}'(\mathbb{R}). \tag{2.6.17}$$

Set $S_1 = (x - a_2)^{m_2}(x - a_3)^{m_3} \cdots (x - a_k)^{m_k} T$. Then $(x - a_1)^{m_1} S_1 = 0$ $\implies S_1 = \sum_{l_1=0}^{m_1-1} \alpha_{l_1} \delta_{a_1}^{(l_1)}$ (using solution (2.6.7) of Equation (5) in (2.6.1) with $m_1 = m$).

Hence, $(x - a_2)^{m_2}(x - a_3)^{m_3} \cdots (x - a_k)^{m_k} T = S_1 = \sum_{l_1=0}^{m_1-1} \alpha_{l_1} \delta_{a_1}^{(l_1)}$.

Now setting $S_2 = (x - a_3)^{m_3} \cdots (x - a_k)^{m_k} T$, we get $(x - a_2)^{m_2} S_2 = \sum_{l_1=0}^{m_1-1} \alpha_{l_1} \delta_{a_1}^{(l_1)}$

$\implies S_2 = \sum_{l_2=0}^{m_2-1} \alpha_{2l_2} \delta_{a_2}^{(l_2)} + \sum_{l_1=0}^{m_1-1} \beta_{2l_1} \delta_{a_1}^{(l_1)}$ (using (2.6.7) and (2.6.12) with $b = a_2 \neq a_1$). Continuing this procedure, each time using (2.6.7) and (2.6.12), we finally get the general solution of Equation (7) in (2.6.1) in the following form:

$$T = \sum_{l_1=0}^{m_1-1} C_{1l_1} \delta_{a_1}^{(l_1)} + \sum_{l_2=0}^{m_2-1} C_{2l_2} \delta_{a_2}^{(l_2)} + \dots + \sum_{l_k=0}^{m_k-1} C_{kl_k} \delta_{a_k}^{(l_k)}, \tag{2.6.18}$$

where C_{jl_j}, $1 \leq j \leq k, 0 \leq l_j \leq m_j$, are real constants.

Problem of division: case of several variables $n \geq 2$

The problem of division for the case of several variables is quite complex. Hörmander [31] has solved the problem of division by polynomials in several variables, and Łojasiewicz [8, p. 126] has solved the more general problem of division of distributions by analytic functions. These problems are beyond the scope of this book.

2.7 Primitives of a distribution and differential equations

Definition 2.7.1. Let $S \in \mathcal{D}'(\mathbb{R})$ be a given distribution. Then a distribution $T \in \mathcal{D}'(\mathbb{R})$ is called a *primitive* of S if and only if

$$\frac{dT}{dx} = S \quad \text{in } \mathcal{D}'(\mathbb{R}). \tag{2.7.1}$$

Theorem 2.7.1.

I. *Every distribution $S \in \mathcal{D}'(\mathbb{R})$ of single variable $x \in \mathbb{R}$ has an* infinite number *of primitives $T \in \mathcal{D}'(\mathbb{R})$.*

II. *For every pair of primitives $T_1, T_2 \in \mathcal{D}'(\mathbb{R})$ of S with $\frac{dT_i}{dx} = S \in \mathcal{D}'(\mathbb{R})$ $(i = 1, 2)$, their difference $T_1 - T_2$ is a constant function C, i.e. $\forall \phi \in \mathcal{D}(\mathbb{R})$,*

$$\langle T_1 - T_2, \phi \rangle = \langle C, \phi \rangle = \int_{-\infty}^{\infty} C\phi(x)dx. \tag{2.7.2}$$

III. *The equation $\frac{dT}{dx} = 0$ in $\mathcal{D}'(\mathbb{R})$ has the general solution $T = C_1$ in $\mathcal{D}'(\mathbb{R})$, i.e.*

$$\langle T, \phi \rangle = \langle C_1, \phi \rangle = \int_{-\infty}^{\infty} C_1\phi(x)dx \quad \forall \phi \in \mathcal{D}(\mathbb{R}), \tag{2.7.3}$$

where C_1 is the constant function (as in the case of ordinary differential equations for functions).

IV. *For $S \in \mathcal{D}'(\mathbb{R})$, the general solution of $\frac{dT}{dx} = S$ is given by:*

$$T = T_0 + C_1, \tag{2.7.4}$$

where $T_0 \in \mathcal{D}'(\mathbb{R})$ is any particular solution *of $\frac{dT}{dx} = S, C_1$ is the constant function and the general solution of the corresponding homogeneous equation $\frac{dT}{dx} = 0$.*

Proof.

I. In order that $T' = S$, it is necessary and sufficient that $\forall \chi \in \mathcal{D}(\mathbb{R})$ with $\chi = \frac{d\psi}{dx}$ for some $\psi \in \mathcal{D}(\mathbb{R})$,

$$T(\chi) = T\left(\frac{d\psi}{dx}\right) = -S(\psi). \tag{2.7.5}$$

But such a $\chi \in \mathcal{D}(\mathbb{R})$ with $\chi = \frac{d\psi}{dx}$, $\psi \in \mathcal{D}(\mathbb{R})$, will exist if and only if

$$\int_{-\infty}^{\infty} \chi(t)dt = 0. \tag{2.7.6}$$

In fact, if for $\chi \in \mathcal{D}(\mathbb{R})$, $\exists \psi \in \mathcal{D}(\mathbb{R})$ such that $\chi = \psi'$, then (2.7.6) holds:

$$\int_{-\infty}^{\infty} \chi(t)dt = \int_{-\infty}^{\infty} \psi'(t)dt = \psi(t)|_{-\infty}^{\infty} = 0,$$

since $\psi \in \mathcal{D}(\mathbb{R})$ vanishes at $\pm\infty$.

Conversely, suppose that (2.7.6) holds. Set $\psi(x) = \int_{-\infty}^{x} \chi(t)dt$. Then, $\psi'(x) = \chi(x) \forall x \in]-\infty, \infty[$ and $\psi \in \mathcal{D}(\mathbb{R})$, since $\chi \in \mathcal{D}(\mathbb{R})$. Define

$$H = \left\{\chi : \chi \in \mathcal{D}(\mathbb{R}), \int_{-\infty}^{\infty} \chi(t)dt = 0\right\} \subset \mathcal{D}(\mathbb{R}). \tag{2.7.7}$$

Then H is a proper subspace of $\mathcal{D}(\mathbb{R})$. T is a linear functional on $\mathcal{D}(\mathbb{R})$ and defined on $H \subset \mathcal{D}(\mathbb{R})$. T will be completely known if its value $T(\theta_0)$ at $\theta_0 \in \mathcal{D}(\mathbb{R}) \setminus H$ is known. Let $\phi \in \mathcal{D}(\mathbb{R})$ be any element. Let $\theta_0 \in \mathcal{D}(\mathbb{R})$ be a fixed element such that $\theta_0 \notin H$, i.e. $\int_{-\infty}^{\infty} \theta_0(x)dx \neq 0$. Now, we choose $\lambda \in \mathbb{R}$ such that $(\phi - \lambda\theta_0) \in H$, i.e. $\int_{-\infty}^{\infty}(\phi - \lambda\theta_0)dx = 0 \Longrightarrow \lambda = \frac{\int_{-\infty}^{\infty} \phi(x)dx}{\int_{-\infty}^{\infty} \theta_0(x)dx}$. For the sake of simplicity, choose θ_0 such that

$$\int_{-\infty}^{\infty} \theta_0(x)dx = 1.$$

Then

$$\lambda = \int_{-\infty}^{\infty} \phi(x)dx. \tag{2.7.8}$$

For such a λ, in (2.7.8), $\forall \phi \in \mathcal{D}(\mathbb{R})$, we have $\phi - \lambda\theta_0 = \chi = \frac{d\psi}{dx} \in H$ with some $\psi \in \mathcal{D}(\mathbb{R})$, i.e. $\phi - \lambda\theta_0$ is the derivative of some other function $\psi \in \mathcal{D}(\mathbb{R})$. Hence, $\forall \phi \in \mathcal{D}(\mathbb{R})$, $\phi = \lambda\theta_0 + \chi$; $\lambda = \int_{-\infty}^{+\infty} \phi(x)dx$; $\chi = \phi - \lambda\theta_0 \in H$ and $\Longrightarrow T(\phi - \lambda\theta_0) = T(\chi) = T(\frac{d\psi}{dx}) = -S(\psi)$ (by (2.7.5)).

$$\Longrightarrow \quad T(\phi) = T(\lambda\theta_0) - S(\psi) = \lambda T(\theta_0) - S(\psi). \tag{2.7.9}$$

Conversely, if T is the linear functional defined on H (2.7.7) by (2.7.5), then T is continuous on $H \cap \mathcal{D}_K(\mathbb{R})$ $\forall$ fixed $K \subset\subset \mathbb{R}$. In fact, if $\chi \in H \cap \mathcal{D}_K(\mathbb{R})$ converges to 0 in $\mathcal{D}_K(\mathbb{R})$, then its primitive $\psi \to 0$ in $\mathcal{D}_{K_0}(\mathbb{R})$ with $K \subset K_0$. Consequently, $S(\psi) \to 0$, since S is a distribution by hypothesis. For every choice of $\gamma = T(\theta_0)$, (2.7.9) defines a linear functional T on $\mathcal{D}(\mathbb{R})$, which is continuous on $\mathcal{D}_K(\mathbb{R})$, $\forall K \subset\subset \mathbb{R}$. Hence, T is continuous on $\mathcal{D}(\mathbb{R})$. In fact, if $\phi \to 0$ in $\mathcal{D}_K(\mathbb{R})$ $\forall$ fixed $K \subset\subset \mathbb{R}$, $\lambda = \int_{-\infty}^{+\infty} \phi(x)dx \to 0$. Then, $\phi - \lambda\theta_0 = \chi \to 0$ in $H \cap \mathcal{D}_L(\mathbb{R})$ with $L = K \cup \operatorname{supp}(\theta_0)$. Consequently, $S(\psi) \to 0$ (as shown earlier) and $\lambda T(\theta_0) \to 0 \Longrightarrow T(\phi) = \lambda T(\theta_0) - S(\psi) \to 0$. So, $T \in \mathcal{D}'(\mathbb{R})$ and satisfies (2.7.5), and is thus a primitive of $S \in \mathcal{D}'(\mathbb{R})$. Hence, there exist an *infinite number* of primitives T of S (as in the case of functions) satisfying (2.7.9). But in this family of primitives in (2.7.9), T can be uniquely fixed by its value $T(\theta_0) = C \in \mathbb{R}$ with $\theta_0 \in \mathcal{D}(\mathbb{R})$, $\theta_0 \notin H$ (otherwise, θ_0 can be chosen arbitrarily). Let T_C denote the distribution defined by the constant function C. Using (2.7.8), equation (2.7.9) can be rewritten as:

$$T(\phi) = T_C(\phi) - S(\psi), \tag{2.7.10}$$

since $T_C(\phi) = \int_{-\infty}^{\infty} C\phi(x)dx = \int_{-\infty}^{\infty} T(\theta_0)\phi(x)dx = T(\theta_0)\int_{-\infty}^{\infty} \phi(x)dx = \lambda T(\theta_0)$ (by (2.7.8)).

II. Let $T_1, T_2 \in \mathcal{D}'(\mathbb{R})$ be any two primitives of $S \in \mathcal{D}'(\mathbb{R})$. Then, from (2.7.9),

$$T_1(\phi) = \lambda T_1(\theta_0) - S(\psi), T_2(\phi) = \lambda T_2(\theta_0) - S(\psi)$$
$$\implies \quad (T_1 - T_2)(\phi) = \lambda(T_1 - T_2)(\theta_0).$$

Set $(T_1 - T_2)(\theta_0) = C$.
Then,

$$(T_1 - T_2)(\phi) = \lambda C = \int_{-\infty}^{\infty} C\phi(x)dx = T_C(\phi) \quad \forall \phi \in \mathcal{D}(\mathbb{R})$$

$\implies T_1 - T_2 = T_C$ is the distribution defined by the constant function C, i.e. they differ by a constant function.

III. $\frac{dT}{dx} = 0$ in $\mathcal{D}'(\mathbb{R}) \implies T(\phi) = \lambda T(\theta_0)$ (using (2.7.9) with $S = 0$) $\implies T(\phi) = C\lambda$ with $C = T(\theta_0)$ and $\lambda = \int_{-\infty}^{\infty} \phi(x)dx$. Then $T(\phi) = C\int_{-\infty}^{\infty} \phi(x)dx = \langle C, \phi\rangle \; \forall \phi \in \mathcal{D}(\mathbb{R}) \implies T = C$ is the general solution in $\mathcal{D}'(\mathbb{R})$, C being the constant function, i.e. the classical or usual solution is also the only solution in the distributional sense.

IV. Let $T_0 \in \mathcal{D}'(\mathbb{R})$ be a particular solution of $\frac{dT}{dx} = S$, i.e. $\frac{dT_0}{dx} = S$. Then, its general solution is $T = T_0 + W$ with $W \in \mathcal{D}'(\mathbb{R})$, which is the new unknown distribution, such that

$$\frac{dT}{dx} = \frac{d}{dx}(T_0 + W) = S \quad \implies \quad \frac{dT_0}{dx} + \frac{dW}{dx} = S$$
$$\implies \quad S + \frac{dW}{dx} = S \quad \implies \quad \frac{dW}{dx} = 0$$

$\implies W \in \mathcal{D}'(\mathbb{R})$ is the general solution of the corresponding homogeneous equation $\frac{dT}{dx} = 0$ in $\mathcal{D}'(\mathbb{R})$. But $W = C$ is the general solution of $\frac{dT}{dx} = 0$, which is defined by (2.7.3). Hence, $T = T_0 + C$ is the general solution of the given non-homogeneous equation $\frac{dT}{dx} = S$.

□

Remark 2.7.1.

1. In calculus, in order to choose a particular primitive of a function, it is sufficient to fix its value at a particular point x_0. Here, for choosing a particular primitive $T \in \mathcal{D}'(\mathbb{R})$ of a distribution, we fix the value of T for a particular function $\theta_0 \in \mathcal{D}(\mathbb{R}), \theta_0 \notin H$, with H defined by (2.7.7).
2. Let $\Omega \subset \mathbb{R}$ be an open subset of $\mathbb{R}$. Then, from (2.7.3) of Theorem 2.7.1, a distribution T with $\frac{dT}{dx} = 0$ on Ω is equal to a *constant function in every connected subset of* Ω, but *not* necessarily equal to the same constant on Ω. For example, $\frac{dH}{dx} = 0$ in $\mathcal{D}'(\mathbb{R} \setminus \{0\})$ with $\Omega = \mathbb{R} \setminus \{0\}$, but $H(x) = 1$ in $]0, \infty[$ and $= 0$ in $]-\infty, 0[$ i.e. two different constant functions on two different connected subsets.

Example 2.7.1.

1. Find a primitive of $u = \text{Pf}(\frac{H(x)}{x}) \in \mathcal{D}'(\mathbb{R})$.
2. Find all the primitives of u.

Solution.

1. Let $v \in \mathcal{D}'(\mathbb{R})$ be a primitive of $u = \text{Pf}(\frac{H(x)}{x}) \in \mathcal{D}'(\mathbb{R})$, which is defined by:

$$\left\langle \text{Pf}\left(\frac{H(x)}{x}\right), \phi \right\rangle = \lim_{\varepsilon \to 0^+} \left[\int_\varepsilon^\infty \frac{\phi(x)}{x} dx + \phi(0) \ln \varepsilon \right] \quad \forall \phi \in \mathcal{D}(\mathbb{R}).$$

Then, by definition, $\frac{dv}{dx} = u$ in $\mathcal{D}'(\mathbb{R}) \Longrightarrow \langle v', \phi \rangle = \langle u, \phi \rangle \Longrightarrow -\langle v, \phi' \rangle = \langle u, \phi \rangle \ \forall \phi \in \mathcal{D}(\mathbb{R}) \Longrightarrow \langle v, \psi \rangle = -\langle u, \phi \rangle \ \forall \psi \in \mathcal{D}(\mathbb{R})$, with $\psi = \phi'$

$$\begin{aligned} &= -\left\langle \text{Pf}\left(\frac{H(x)}{x}\right), \phi \right\rangle = -\,\text{Pf} \int_0^\infty \frac{\phi(x)}{x} dx \\ &= -\lim_{\varepsilon \to 0^+} \left[\int_\varepsilon^\infty \frac{\phi(x)}{x} dx + \phi(0) \ln \varepsilon \right]. \end{aligned}$$

Integrating by parts, we get

$$\begin{aligned} \int_\varepsilon^\infty \frac{\phi(x)}{x} dx &= \phi(x) \ln x \Big|_\varepsilon^\infty - \int_\varepsilon^\infty \ln x \phi'(x) dx \\ &= -\phi(\varepsilon) \ln \varepsilon - \int_\varepsilon^\infty \ln x \phi'(x) dx \end{aligned}$$

$$\Longrightarrow \quad \langle v, \psi \rangle = \lim_{\varepsilon \to 0^+} \left[\int_\varepsilon^\infty \ln x \phi'(x) dx + \phi(\varepsilon) \ln \varepsilon - \phi(0) \ln \varepsilon \right]$$

$\forall \psi \in \mathcal{D}(\mathbb{R})$ with $\psi = \phi'$.

But from Proposition 1.2.1, $\phi(\varepsilon) = \phi(0) + \varepsilon \chi(x)$ with $\chi \in C^0(\mathbb{R}) \Longrightarrow \phi(\varepsilon) \ln \varepsilon = \phi(0) \ln \varepsilon + \varepsilon \chi(\varepsilon) \ln \varepsilon \Longrightarrow \phi(\varepsilon) \ln \varepsilon - \phi(0) \ln \varepsilon = \varepsilon \chi(\varepsilon) \ln \varepsilon \to 0$ as $\varepsilon \to 0^+ \Longrightarrow \lim_{\varepsilon \to 0^+} [\phi(\varepsilon) \ln \varepsilon - \phi(0) \ln \varepsilon] = 0$.

Hence,

$$\begin{aligned} \langle v, \psi \rangle &= \int_0^\infty lnx \phi'(x) dx = \int_0^\infty \ln x \psi(x) dx = \int_{-\infty}^\infty H(x) \ln x \psi(x) dx \\ &= \langle H(x) \ln x, \psi \rangle \quad \forall \psi \in \mathcal{D}(\mathbb{R}) \text{ with } \psi = \phi' \end{aligned}$$

$\Longrightarrow v = H(x) \ln x$ in $\mathcal{D}'(\mathbb{R})$ is a primitive of $u = \text{Pf}(\frac{H(x)}{x})$.

2. By Theorem 2.7.1, the primitives of u are given by $v = H(x) \ln x + C_1$, C_1 being an arbitrary constant function, since the primitives differ by the constant function C_1 defined by (2.7.2).

Proposition 2.7.1. *The general solution of the equation* $\frac{d^m T}{dx^m} = 0$ *in* $\mathcal{D}'(\mathbb{R})$ *(i.e. mth order primitive* $T \in \mathcal{D}'(\mathbb{R})$ *of* $0 \in \mathcal{D}'(\mathbb{R})$*) is a polynomial of degree* $\leq m-1$*, i.e.*

$$T = \alpha_{m-1}x^{m-1} + \alpha_{m-2}x^{m-2} + \cdots + \alpha_1 x + \alpha_0, \tag{2.7.11}$$

α_i*s being arbitrary constants.*

Proof. Let $T \in \mathcal{D}'(\mathbb{R})$ with $\frac{d^m T}{dx^m} = 0$ in $\mathcal{D}'(\mathbb{R})$. Set $S_1 = \frac{d^{m-1}T}{dx^{m-1}} \in \mathcal{D}'(\mathbb{R})$.

Then $\frac{d^m T}{dx^m} = \frac{dS_1}{dx} = 0 \implies S_1 = C_1$ by Theorem 2.7.1, C_1 being the constant function defined by (2.7.3) $\implies \frac{d^{m-1}T}{dx^{m-1}} = C_1 \implies \frac{d}{dx}(\frac{d^{m-2}T}{dx^{m-2}}) = C_1 \implies \frac{d^{m-2}T}{dx^{m-2}} = C_1 x + C_2$, since $\frac{d^{m-2}T}{dx^{m-2}}$ is the primitive of the constant function C_1, which is obviously a polynomial of degree ≤ 1, C_1 and C_2 being arbitrary constants $\implies \frac{d}{dx}(\frac{d^{m-3}T}{dx^{m-3}}) = C_1 x + C_2 \implies \frac{d^{m-3}T}{dx^{m-3}}$ is the primitive of polynomial $C_1 x + C_2$ of degree ≤ 1, its primitive being a polynomial $C_1\frac{x^2}{2} + C_2 x + C_3$ of degree ≤ 2, i.e. $\frac{d^{m-3}T}{dx^{m-3}} = C_1\frac{x^2}{2} + C_2 x + C_3$, C_1, C_2, C_3 being arbitrary constants.

Continuing this procedure, we get:

$$\frac{dT}{dx} = C_1\frac{x^{m-2}}{(m-2)!} + C_2\frac{x^{m-3}}{(m-3)!} + \cdots + C_{m-1},$$

and finally we get

$$\begin{aligned} T &= C_1\frac{x^{m-1}}{(m-1)!} + C_2\frac{x^{m-2}}{(m-2)!} + \cdots + C_{m-1}x + C_m \\ &= \alpha_{m-1}x^{m-1} + \alpha_{m-2}x^{m-2} + \cdots + \alpha_1 x + \alpha_0, \end{aligned}$$

where $(C_i)_{i=1}^m$ (resp. $(\alpha_i)_{i=0}^{m-1}$) are arbitrary constants with $\alpha_{m-1} = \frac{C_1}{(m-1)!}, \alpha_{m-2} = \frac{C_2}{(m-2)!}, \ldots, \alpha_1 = C_{m-1}, \alpha_0 = C_m$. □

Theorem 2.7.2. *Consider the nth-order ordinary linear differential equation with variable coefficients:*

$$a_0(x)T^{(n)} + a_1(x)T^{(n-1)} + \cdots + a_n(x)T = b(x) \quad \textit{for } x \in \mathbb{R}, \tag{2.7.12}$$

where $a_0(x) \neq 0\ \forall x \in \mathbb{R}$ *and* $a_i, b \in C^\infty(\mathbb{R})\ \forall i = 0, 1, \ldots, n$*. Then, if* $T \in \mathcal{D}'(\mathbb{R})$ *is a solution of this equation* (2.7.12)*,* T *is a* C^∞*-function and satisfies the equation* (2.7.12) *in the usual or classical sense.*

Proof. For the sake of simplicity, we give the proof for order $n = 1$. Then,

$$a_0(x)\frac{dT}{dx} + a_1(x)T = b(x) \quad \text{in } \mathcal{D}'(\mathbb{R}). \tag{2.7.13}$$

Since $a_0(x) \neq 0 \ \forall x \in \mathbb{R}$, this first-order linear equation is equivalent to the non-homogeneous linear equation

$$\frac{dT}{dx} + p(x)T = q(x) \quad \text{in } \mathcal{D}'(\mathbb{R}), \tag{2.7.14}$$

where $p(x) = \frac{a_1(x)}{a_0(x)}$, $q(x) = \frac{b(x)}{a_0(x)}$ are C^∞-functions on $\mathbb{R}$, i.e. both (2.7.13) and (2.7.14) have the same solution set. Let $T \in \mathcal{D}'(\mathbb{R})$ be a distribution solution of (2.7.14). We are to show that T is a C^∞-function on $\mathbb{R}$ and coincides with the usual solution of the first-order equation for an unknown function. In fact, let $A(x)$ be the primitive of the C^∞-function $p(x)$, i.e. $\frac{dA}{dx}(x) = p(x)$. Then set $S = e^{A(x)}T \in \mathcal{D}'(\mathbb{R})$:

$$\begin{aligned}
\Longrightarrow \quad \frac{dS}{dx} &= \frac{d}{dx}[e^{A(x)}T] = e^{A(x)}\frac{dT}{dx} + Te^{A(x)} \cdot \frac{dA}{dx}(x) \\
&= e^{A(x)}\left[\frac{dT}{dx} + p(x)T\right] = e^{A(x)}q(x),
\end{aligned}$$

since T is a solution of (2.7.14).

Hence, $S \in \mathcal{D}'(\mathbb{R})$ is a primitive of $e^{A(x)}q(x)$. But $e^{A(x)}q(x)$ is a function in $C^\infty(\mathbb{R})$,

$$\begin{aligned}
&\Longrightarrow \quad S' = \frac{dS}{dx} \text{ is a function belonging to } C^\infty(\mathbb{R}) \\
&\Longrightarrow \quad S \text{ is a function of } C^\infty(\mathbb{R}), \text{ i.e. } T \in C^\infty(\mathbb{R}).
\end{aligned} \tag{2.7.15}$$

Now we will show that $T \in C^\infty(\mathbb{R})$ is the usual solution. In fact, $u_h = Ce^{-A(x)}$ is the general solution of the corresponding *homogeneous* equation: $\frac{dT}{dx} + p(x)T = 0$ in the *usual sense*, $C \in \mathbb{R}$ being the arbitrary constant. Then, by Lagrange's method of variation of parameters, we have $\frac{dC}{dx} = e^{A(x)}q(x)$, from which a particular solution u_0 of (2.7.14) in the *usual sense* is given by $u_0 = e^{-A(x)} \int q(x)e^{A(x)}dx \in C^\infty(\mathbb{R})$. Hence $u = u_0 + u_h = u_0 + Ce^{-A(x)} \in C^\infty(\mathbb{R})$ is the general solution of (2.7.14) in the *usual sense*. Now we will show that $T = u_0 + Ce^{-A(x)} \in C^\infty(\mathbb{R})$ is also the solution in the *distributional sense*. In fact, we can set $T = u_0 + e^{-A(x)}S \in \mathcal{D}'(\mathbb{R})$, where a particular solution of (2.7.14) $u_0 \in C^\infty(\mathbb{R})$ is known, i.e. $\frac{du_0}{dx} + p(x)u_0 = q$ and $S \in \mathcal{D}'(\mathbb{R})$ is a new unknown distribution, such that T satisfies (2.7.14). Then,

$$\begin{aligned}
q &= \frac{dT}{dx} + p(x)T = \frac{d}{dx}[u_0 + e^{-A(x)}S] + p(x)[u_0 + e^{-A(x)}S] \\
&= \left(\frac{du_0}{dx} + p(x)u_0\right) + e^{-A(x)}\frac{dS}{dx} - e^{-A(x)} \cdot p(x)S + p(x)e^{-A(x)}S \\
&= q + e^{-A(x)}\frac{dS}{dx}
\end{aligned}$$

in $\mathcal{D}'(\mathbb{R})$, since $\frac{dA}{dx}(x) = p(x) \implies q = q + e^{-A(x)}\frac{dS}{dx} \implies e^{-A(x)}\frac{dS}{dx} = 0$ in $\mathcal{D}'(\mathbb{R}) \implies \frac{dS}{dx} = 0$ in $\mathcal{D}'(\mathbb{R})$, since $e^{-A(x)} \neq 0\ \forall x \in \mathbb{R} \implies S = C$ by Theorem 2.7.1, $C \in \mathbb{R}$ being an arbitrary constant. Hence, $T = u_0 + Ce^{-A(x)} \in C^\infty(\mathbb{R})$ is also a distributional solution of (2.7.14). □

Remark 2.7.2. The condition that $a_0(x) \neq 0\ \forall x \in \mathbb{R}$ in (2.7.13) is essential, since if $a_0(x) = 0$ for some $x \in \mathbb{R}$, the situation will be completely different, i.e. there might be distributional solutions in addition to classical solutions as a function, or some classical solutions may not be admissible as distributional solutions and will then disappear. We give some standard examples to explain these phenomena.

Example 2.7.2. Find all the solutions of $x\frac{dT}{dx} = 0$ in $\mathcal{D}'(\mathbb{R})$. (2.7.16)

Solution. Here, $a_0(x) = x$ vanishes at $x = 0$. Let $T \in \mathcal{D}'(\mathbb{R})$ be a solution of this equation. Define $S = \frac{dT}{dx} \in \mathcal{D}'(\mathbb{R})$. Then, $xS = 0$ in $\mathcal{D}'(\mathbb{R})$ has the general solution $S = C_1\delta$, $C_1 \in \mathbb{R}$ being an arbitrary constant, δ being the Dirac distribution concentrated at $x = 0$ by Proposition 1.6.1. Hence, $\frac{dT}{dx} = C_1\delta$ has the general solution $T = T_p + T_h$, where T_p is any particular solution with $\frac{dT_p}{dx} = C_1\delta$ and T_h is general solution of the *homogeneous* equation $\frac{dT}{dx} = 0$. For the Heaviside function $H(x) = 1$ for $x > 0$ and $H(x) = 0$ for $x < 0$, $\frac{dH}{dx} = \delta$ (see Example 2.3.2) $\implies T_p = C_1 H$, and $T_h = C_2$ by Theorem 2.7.1 and is defined by (2.7.3).

Hence, $T = C_1 H + C_2$ is the general solution in $\mathcal{D}'(\mathbb{R})$ in the distributional sense; $T_1 = H$ and $T_2 = 1$ are *two* linearly independent solutions in $\mathcal{D}'(\mathbb{R})$!

Example 2.7.3. Consider the equation $2x^3\frac{dT}{dx} + T = 0$ in $\mathcal{D}'(\mathbb{R})$. (2.7.17)

It has only the trivial solution $T = 0 \in \mathcal{D}'(\mathbb{R})$ in the distributional sense. For $x \neq 0$, the usual (i.e. classical) solution of the first-order equation $\frac{dT}{dx} + \frac{1}{2x^3}T = 0$ is given by:

$$T = Ce^{\frac{1}{4x^2}} \quad \text{with } C \neq 0 \text{ or } C = 0. \tag{2.7.18}$$

For $x \neq 0$, i.e. on $\mathbb{R} \setminus \{0\}$, the distributional solution and the usual solution of the equation must coincide with $C \neq 0$ or $C = 0$. But the first choice $C \neq 0$ is *impossible*, since there does not exist any distribution $T \in \mathcal{D}'(\mathbb{R})$ such that $\langle T, \phi\rangle = \int_{-\infty}^{\infty} e^{\frac{1}{4x^2}}\phi(x)dx\ \forall\phi \in \mathcal{D}(\mathbb{R} \setminus \{0\})$ (see Example 5.3.1 in Chapter 5).

Hence, the usual (classical) solutions of the form $Ce^{\frac{1}{4x^2}}$ with $C \neq 0$ will *not* be a distributional solution in $\mathcal{D}'(\mathbb{R})$, and $T = 0 \in \mathcal{D}'(\mathbb{R})$ is the *only* distributional solution, which is also the *usual* trivial solution $C = 0$. In other words, all (non-trivial) classical solutions $Ce^{\frac{1}{4x^2}}$ with $C \neq 0$ will disappear!

Example 2.7.4. Find the general solution of

1. $xy' + y = \delta$ in $\mathcal{D}'(\mathbb{R})$; (2.7.19)

2. $xy' + y = \mathrm{Pf}(\frac{H(x)}{x})$ in $\mathcal{D}'(\mathbb{R})$.

Solution.

1. $xy' + y = (xy)'$. Set $S = xy \in \mathcal{D}'(\mathbb{R})$.

   Then $\frac{dS}{dx} = \delta$ has the general solution $S = S_p + S_h$, where S_p is any particular solution such that $\frac{dS_p}{dx} = \delta$, i.e. $S_p = H(x)$, H being the Heaviside function; $S_h = C_1$ is the general solution of $\frac{dS}{dx} = 0$ by Theorem 2.7.1, C_1 being an arbitrary constant function defined by (2.7.3).

   Hence, $S = xy = H(x) + C_1$ with $y \in \mathcal{D}'(\mathbb{R})$ has the general solution given by $y = y_p + y_h$, where $y_p \in \mathcal{D}'(\mathbb{R})$ is a particular solution satisfying $xy_p = H(x)+C_1$ and $y_h \in \mathcal{D}'(\mathbb{R})$ is the general solution of $xy = 0$ in $\mathcal{D}'(\mathbb{R})$, i.e. $y_h = C_2\delta$, by Proposition 1.6.1.

   Construction of $y_p \in \mathcal{D}'(\mathbb{R})$: We can seek y_p in the form $y_p = \mathrm{Pf}(\frac{H(x)}{x}) + C_1(\text{c.p.v.}(\frac{1}{x}))$, since

$$\begin{aligned}\langle xy_p, \phi\rangle &= \left\langle \mathrm{Pf}\left(\frac{H(x)}{x}\right) + C_1\left(\text{c.p.v.}\left(\frac{1}{x}\right)\right), x\phi\right\rangle \\ &= \left\langle \mathrm{Pf}\left(\frac{H(x)}{x}\right), x\phi\right\rangle + C_1\left\langle \text{c.p.v.}\left(\frac{1}{x}\right), x\phi\right\rangle = \langle H(x), \phi\rangle + \langle C_1, \phi\rangle \\ &= \langle H(x) + C_1, \phi\rangle\end{aligned}$$

   $\Longrightarrow xy_p = H(x) + C_1$ with $y_p = \mathrm{Pf}(\frac{H(x)}{x}) + C_1(\text{c.p.v.}(\frac{1}{x})) \in \mathcal{D}'(\mathbb{R})$.

   Then the general solution is $y = \mathrm{Pf}(\frac{H(x)}{x}) + C_1(\text{c.p.v.}(\frac{1}{x})) + C_2\delta$, C_1, C_2 being arbitrary constants.

2. $xy' + y = (xy)' = \mathrm{Pf}(\frac{H(x)}{x})$ in $\mathcal{D}'(\mathbb{R}) \Longrightarrow xy \in \mathcal{D}'(\mathbb{R})$ is a primitive of $\mathrm{Pf}(\frac{H(x)}{x}) \in \mathcal{D}'(\mathbb{R}) \Longrightarrow xy = H(x)\ln x + C \in \mathcal{D}'(\mathbb{R})$ by Example 2.7.1, C being an arbitrary constant function defined by (2.7.2). Then the general solution $y \in \mathcal{D}'(\mathbb{R})$ of $xy = H(x)\ln x + C$ in $\mathcal{D}'(\mathbb{R})$ is given by $y = y_h + y_p$, where $y_h = C_1\delta \in \mathcal{D}'(\mathbb{R})$ is the general solution of the *homogeneous* equation $xy = 0$ in $\mathcal{D}'(\mathbb{R})$ by Proposition 1.6.1, y_p is any particular solution of $xy = H(x)\ln x + C$ in $\mathcal{D}'(\mathbb{R})$, for example, $y_p = \mathrm{Pf}(\frac{H(x)}{x}\ln x) + C(\text{c.p.v.}(\frac{1}{x})) \in$

$\mathcal{D}'(\mathbb{R})$. In fact,

$$\begin{aligned}\langle xy_p, \phi\rangle &= \left\langle x\left[\operatorname{Pf}\left(\frac{H(x)}{x}\ln x\right) + C\left(\text{c.p.v.}\left(\frac{1}{x}\right)\right)\right], \phi\right\rangle \quad \forall\phi \in \mathcal{D}(\mathbb{R})\\ &= \left\langle \operatorname{Pf}\left(\frac{H(x)}{x}\ln x\right) + C\left(\text{c.p.v.}\left(\frac{1}{x}\right)\right)\right], x\phi\right\rangle\\ &= \operatorname{Pf}\int_0^\infty \frac{\ln x}{x}\cdot x\phi dx + C\left(\text{c.p.v.}\int_{-\infty}^{\infty}\frac{x\phi(x)}{x}dx\right)\\ &= \int_0^\infty \ln x\phi(x)dx + \int_{-\infty}^{\infty} C\phi(x)dx = \int_{-\infty}^{\infty}(H(x)\ln x + C)\phi(x)dx\\ &= \langle H(x)\ln x + C, \phi\rangle\end{aligned}$$

$\Longrightarrow xy_p = H(x)\ln x + C$ in $\mathcal{D}'(\mathbb{R})$. Hence, $y = C_1\delta + \operatorname{Pf}(\frac{H(x)}{x}\ln x) + C(\text{c.p.v.}(\frac{1}{x}))$ is the general solution of Equation (2) in $\mathcal{D}'(\mathbb{R})$.

Example 2.7.5. Find all the solutions of $x^k\frac{d^mT}{dx^m} = 0$ in $\mathcal{D}'(\mathbb{R})$ for $k, m \in \mathbb{N}$. (2.7.20)

Solution. Set $S = \frac{d^mT}{dx^m} = T^{(m)} \in \mathcal{D}'(\mathbb{R})$. Then, $x^kS = 0$ in $\mathcal{D}'(\mathbb{R}) \Longrightarrow S = \sum_{l=0}^{k-1} C_l\delta^{(l)}$ (see Example 2.6.1, Equation (5) with $a = 0, m = k$), where $\delta^{(l)}$ is the lth order derivative of the Dirac distribution δ, $C_0, C_1, \ldots, C_{k-1}$ being arbitrary real constants. Hence,

$$\frac{d^mT}{dx^m} = \sum_{l=0}^{k-1} C_l\delta^{(l)} \quad \text{has the general solution } T = T_h + T_p, \tag{2.7.21}$$

where $T_h = \alpha_{m-1}x^{m-1} + \alpha_{m-2}x^{m-2} + \cdots + \alpha_0$ is the general solution of the *homogeneous* equation $\frac{d^mT}{dx^m} = 0$ by Proposition 2.7.1, T_p is any particular solution of $T^{(m)} = \sum_{l=0}^{k-1} C_l\delta^{(l)}$.

Construction of a particular solution T_p: Since T_p satisfies

$$T^{(m)} = C_0\delta + C_1\delta^{(1)} + \cdots + C_{k-1}\delta^{(k-1)}, \tag{2.7.22}$$

T_p will be in terms of Heaviside function H and possibly also Dirac distributions. In fact, $\frac{dH}{dx} = \delta$, $(x^lH)^{(l)} = d_lH$ with $d_l = l!$, which we prove below using Leibniz's formula (see Example 2.5.3).

$$\begin{aligned}(x^lH)^{(l)} &= (x^l)^{(l)}H + \sum_{p=0}^{l-1}\binom{l}{p}(x^l)^{(p)}H^{(l-p)}\\ &= l!H + \sum_{p=0}^{l-1}\binom{l}{p}(l(l-1)\ldots(l-p+1)x^{l-p})\delta^{(l-p-1)}\end{aligned}$$

with $\binom{l}{p}[l(l-1)\cdots(l-p+1)x^{l-p}] = x^l$ for $p = 0$. But $x^{l-p}\delta^{(l-p-1)} = 0$ for $0 \le p \le l-1$, since $(l-p) > (l-p-1)$ (see (2.5.5)). Hence, $\sum_{p=0}^{l-1}\binom{l}{p}l(l-1)\cdots(l-p+1)x^{l-p}\delta^{(l-p-1)} = 0 \Longrightarrow (x^l H)^{(l)} = l!H = d_l H$ with $d_l = l!$ $\forall l \in \mathbb{N} \Longrightarrow (x^l H)^{(l+1)} = d_l H^{(1)} = d_l\delta \Longrightarrow (x^l H)^{(l+2)} = d_l\delta^{(1)} \Longrightarrow \cdots \Longrightarrow (x^l H)^{(l+s)} = d_l\delta^{(s-1)}$ $\forall s \in \mathbb{N}$, for fixed $l \in \mathbb{N}$.

Case $m < k-1$: Hence, $(x^{(m-1)}H)^{(m)} = d_{m-1}\delta, (x^{m-2}H)^{(m)} = d_{m-2}\delta^{(1)}, \ldots, (x^0 H)^{(m)} = d_0\delta^{(m-1)}$ with $d_l = l!, 0 \le l \le m-1$.

Therefore, from (2.7.22),

$$T_p = \left\{\left[\frac{C_0}{d_{m-1}}(x^{m-1}H) + \frac{C_1}{d_{m-2}}(x^{m-2}H) + \cdots + \frac{C_{m-1}}{d_0}H\right] + (C_m\delta + C_{m+1}\delta^{(1)} + \cdots + C_{k-1}\delta^{(k-1-m)})\right\} \quad \text{in } \mathcal{D}'(\mathbb{R})$$

$$\Longrightarrow \quad T_p = [\beta_1(x^{m-1}H) + \beta_2(x^{m-2}H) + \cdots + \beta_m H] + (\beta_{m+1}\delta + \beta_{m+2}\delta^{(1)} + \cdots + \beta_k\delta^{(k-1-m)}) \tag{2.7.23}$$

is a particular solution, $\{\beta_i\}_{i=1}^k$ being arbitrary real constants.

Then, for $m < k-1$, the general solution is $T = T_h + T_p$, where $T_h = \sum_{k=0}^{m-1}\alpha_k x^k$ and T_p are given by (2.7.21) and (2.7.23) respectively.

Case $m \ge k-1$: T_p is given by the first k terms in the expression in square brackets $[\ldots]$ in (2.7.23), i.e.

$$T_p = \beta_1(x^{m-1}H) + \beta_2(x^{m-2}H) + \cdots + \beta_k(x^{m-k}H) \tag{2.7.24}$$

is a particular solution, $\{\beta_i\}_{i=1}^k$ being arbitrary real constants. Then, for $m \ge k-1$, the general solution is $T = T_h + T_p$, where T_h and T_p are given by (2.7.21) and (2.7.24) respectively.

2.8 Properties of distributions whose distributional derivatives are known

Let $T \in \mathcal{D}'(\Omega)$ be a distribution on Ω. Then, from the properties of derivatives of T, we can ascertain the property of T itself, i.e. whether T is a function f, or T is independent of a variable x_i, or T is a continuously differentiable function, etc. Here, we state the final results:

- $\forall i = 1, 2, \ldots, n$, if $\frac{\partial T}{\partial x_i} = g_i \in C^0(\Omega)$ is a *continuous function* on $\Omega \Longrightarrow$ $\exists f \in C^1(\Omega)$ such that $T = f$ in $\mathcal{D}'(\Omega)$ with $\frac{\partial T}{\partial x_i} = \frac{\partial f}{\partial x_i}$ (in the *distributional sense*), and

$$\left[\frac{\partial f}{\partial x_i}(\mathbf{x})\right] = \frac{\partial f}{\partial x_i} = g_i, \quad 1 \le i \le n, \tag{2.8.1}$$

$[\frac{\partial f}{\partial x_i}(\mathbf{x})]$ being the *usual* partial derivative of f in the *pointwise sense.*

- $$\frac{\partial T}{\partial x_i} = 0 \text{ in } \mathcal{D}'(\mathbb{R}^n) \iff T \text{ is } \textit{independent} \text{ of the variable } x_i. \qquad (2.8.2)$$

Theorem 2.8.1. *Let $S_1, S_2, \ldots, S_k \in \mathcal{D}'(\mathbb{R}^n)$ be given distributions on $\mathbb{R}^n$ and $T \in \mathcal{D}'(\mathbb{R}^n)$. Then the system of k equations for T : $\frac{\partial T}{\partial x_1} = S_1, \frac{\partial T}{\partial x_2} = S_2, \ldots, \frac{\partial T}{\partial x_k} = S_k$ will be consistent if and only if*

$$\frac{\partial S_i}{\partial x_j} - \frac{\partial S_j}{\partial x_i} = 0 \quad \text{in } \mathcal{D}'(\mathbb{R}^n), \quad 1 \le i \ne j \le k. \qquad (2.8.3)$$

Example 2.8.1. Consider the partial differential equation $\frac{\partial^2 T}{\partial x_1 \partial x_2} = 0$ in $\mathcal{D}'(\mathbb{R}^2)$. Set $S_i = \frac{\partial T}{\partial x_i}$ $(i = 1, 2)$. Then, $\frac{\partial S_1}{\partial x_2} - \frac{\partial S_2}{\partial x_1} = 0$ in $\mathcal{D}'(\mathbb{R}^2)$. Hence, $\frac{\partial T}{\partial x_1} = S_1$, $\frac{\partial T}{\partial x_2} = S_2$ is a consistent system. Moreover, $\frac{\partial^2 T}{\partial x_1 \partial x_2} = 0 \Longrightarrow \frac{\partial S_1}{\partial x_2} = 0$ and $\frac{\partial S_2}{\partial x_1} = 0$.

$\frac{\partial S_1}{\partial x_2} = 0 \Longrightarrow S_1 = S_1(x_1)$ is independent of x_2 by (2.8.2), $\frac{\partial S_2}{\partial x_1} = 0 \Longrightarrow S_2 = S_2(x_2)$ is independent of x_1 by (2.8.2), where, by abuse of notations, we have written $S_1(x_1)$ (resp. $S_2(x_2)$) in order to stress the fact that it is independent of the variable x_2 (resp. x_1). Then the general solution of $\frac{\partial^2 T}{\partial x_1 \partial x_2} = 0$ is given by: $T = S_1 + S_2 \in \mathcal{D}'(\mathbb{R}^2)$.

Proposition 2.8.1 ([8]). *Let $T \in \mathcal{D}'(\mathbb{R}^n)$ such that $\partial^{\boldsymbol{\alpha}} T = \frac{\partial^{|\boldsymbol{\alpha}|} T}{\partial x_1^{\alpha_1} \ldots \partial x_n^{\alpha_n}} = 0$ in $\mathcal{D}'(\mathbb{R}^n)$ $\forall |\boldsymbol{\alpha}| = m$ (i.e. all partial derivatives of order m of the distribution T vanish). Then T is a polynomial of degree $\le m - 1$ in n variables $x_1, x_2, \ldots, x_n$.*

2.9 Continuity of differential operator $\partial^{\boldsymbol{\alpha}} : \mathcal{D}'(\Omega) \longrightarrow \mathcal{D}'(\Omega)$

Theorem 2.9.1. *Differentiation is a* linear, continuous *operation from $\mathcal{D}'(\Omega)$ into $\mathcal{D}'(\Omega)$, i.e. $\forall |\boldsymbol{\alpha}| \in \mathbb{N}$,*

$$T_n \to T \text{ in } \mathcal{D}'(\Omega) \quad \Longrightarrow \quad \partial^{\boldsymbol{\alpha}} T_n \to \partial^{\boldsymbol{\alpha}} T \text{ in } \mathcal{D}'(\Omega). \qquad (2.9.1)$$

Proof. By definition (2.2.5), $\partial^{\boldsymbol{\alpha}} T \in \mathcal{D}'(\Omega)$ $\forall T \in \mathcal{D}'(\Omega)$.
$\partial^{\boldsymbol{\alpha}}$ is *linear*:

$$\begin{aligned}
\langle \partial^{\boldsymbol{\alpha}}(\lambda_1 T_1 + \lambda_2 T_2), \phi \rangle &= (-1)^{|\boldsymbol{\alpha}|} \langle (\lambda_1 T_1 + \lambda_2 T_2), \partial^{\boldsymbol{\alpha}} \phi \rangle \\
&= \lambda_1 (-1)^{|\boldsymbol{\alpha}|} \langle T_1, \partial^{\boldsymbol{\alpha}} \phi \rangle + \lambda_2 (-1)^{|\boldsymbol{\alpha}|} \langle T_2, \partial^{\boldsymbol{\alpha}} \phi \rangle \\
&= \lambda_1 \langle \partial^{\boldsymbol{\alpha}} T_1, \phi \rangle + \lambda_2 \langle \partial^{\boldsymbol{\alpha}} T_2, \phi \rangle = \langle \lambda_1 \partial^{\boldsymbol{\alpha}} T_1 + \lambda_2 \partial^{\boldsymbol{\alpha}} T_2, \phi \rangle
\end{aligned}$$

$\forall \phi \in \mathcal{D}(\Omega)$, $\forall \lambda_i \in \mathbb{R}$ (resp. $\mathbb{C}$), $\forall T_i \in \mathcal{D}'(\Omega)$,

$\Longrightarrow \partial^{\alpha}(\lambda_1 T_1 + \lambda_2 T_2) = \lambda_1 \partial^{\alpha} T_1 + \lambda_2 \partial^{\alpha} T_2$ in $\mathcal{D}'(\Omega)$.
$\partial^{\alpha} : \mathcal{D}'(\Omega) \longrightarrow \mathcal{D}'(\Omega)$ is *continuous*: $\forall |\alpha| \in \mathbb{N}$, $\langle \partial^{\alpha} T_n, \phi \rangle = (-1)^{|\alpha|} \langle T_n, \partial^{\alpha} \phi \rangle$ $\forall \phi \in \mathcal{D}(\Omega)$. But $T_n \to T$ in $\mathcal{D}'(\Omega) \Longrightarrow \langle T_n, \phi \rangle \to \langle T, \phi \rangle$ $\forall \phi \in \mathcal{D}(\Omega) \Longrightarrow$ $\langle T_n, \partial^{\alpha} \phi \rangle \to \langle T, \partial^{\alpha} \phi \rangle$ $\forall \phi \in \mathcal{D}(\Omega)$ as $n \to \infty$ (since $\phi \in \mathcal{D}(\Omega) \Longrightarrow \partial^{\alpha} \phi \in \mathcal{D}(\Omega)$) $\forall |\alpha| \in \mathbb{N}) \Longrightarrow (-1)^{|\alpha|} \langle T_n, \partial^{\alpha} \phi \rangle \to (-1)^{|\alpha|} \langle T, \partial^{\alpha} \phi \rangle = \langle \partial^{\alpha} T, \phi \rangle$ $\forall \phi \in \mathcal{D}(\Omega)$.

Hence, $T_n \to T$ in $\mathcal{D}'(\Omega) \Longrightarrow \langle \partial^{\alpha} T_n, \phi \rangle = (-1)^{|\alpha|} \langle T_n, \partial^{\alpha} \phi \rangle \to (-1)^{|\alpha|}$ $\langle T, \partial^{\alpha} \phi \rangle = \langle \partial^{\alpha} T, \phi \rangle$ $\forall \phi \in \mathcal{D}(\Omega) \Longrightarrow \partial^{\alpha} : \mathcal{D}'(\Omega) \to \mathcal{D}'(\Omega)$ is continuous. □

Example 2.9.1. Let $T_n = u_n$ with $u_n(x) = \int_{-n}^{n} y e^{ixy} dy$ $\forall n \in \mathbb{N}$, $\forall x \in \mathbb{R}$, be a sequence in $\mathcal{D}'(\mathbb{R})$. Then, show that

$$\lim_{n \to \infty} \int_{-n}^{n} y e^{ixy} dy = -i 2\pi \delta' \quad \text{in } \mathcal{D}'(\mathbb{R}). \tag{2.9.2}$$

Proof. $\forall n \in \mathbb{N}$,

$$\int_{-n}^{n} y e^{ixy} dy = \int_{-n}^{n} \frac{1}{i} \frac{d}{dx}(e^{ixy}) dy = \frac{1}{i} \frac{d}{dx} \int_{-n}^{n} e^{ixy} dy.$$

From (1.8.10), $\langle \int_{-n}^{n} e^{ixy} dy, \phi \rangle \to \langle 2\pi\delta, \phi \rangle$ as $n \to \infty$ $\forall \phi \in \mathcal{D}(\mathbb{R})$, i.e. $\int_{-n}^{n} e^{ixy} dy$ $\to 2\pi\delta$ in $\mathcal{D}'(\mathbb{R}) \Longrightarrow \langle \frac{d}{dx} \int_{-n}^{n} e^{ixy} dy, \phi \rangle \to \langle \frac{d}{dx}(2\pi\delta), \phi \rangle$ as $n \to \infty$ $\forall \phi \in \mathcal{D}(\mathbb{R})$, since $\frac{d}{dx} : \mathcal{D}'(\mathbb{R}) \to \mathcal{D}'(\mathbb{R})$ is continuous on $\mathcal{D}'(\mathbb{R})$ by Theorem 2.9.1.

Hence, $\lim_{n \to \infty} (\frac{d}{dx} \int_{-n}^{n} e^{ixy} dy) = \frac{d}{dx}(2\pi\delta) = 2\pi\delta'$ in $\mathcal{D}'(\mathbb{R})$

$$\begin{aligned} \Longrightarrow \quad \lim_{n \to \infty} \int_{-n}^{n} y e^{ixy} dy &= \frac{1}{i} \lim_{n \to \infty} \frac{d}{dx} \int_{-n}^{n} e^{ixy} dy \\ &= \frac{1}{i} \cdot 2\pi\delta' = -2i\pi\delta' \quad \text{in } \mathcal{D}'(\mathbb{R}). \end{aligned}$$ □

New distributions $\frac{1}{x+i0}$, $\frac{1}{x-i0}$ and related results

Now, applying Theorem 2.9.1, we can define interesting new distributions $\frac{1}{(x \pm i0)}$ and establish the related results in the following example.

Example 2.9.2. 1. Prove that

$$\frac{1}{x + i0} = \lim_{y \to 0+} \frac{1}{x + iy} = \text{c.p.v.} \left(\frac{1}{x} \right) - i\pi\delta \quad \text{in } \mathcal{D}'(\mathbb{R}); \tag{2.9.3}$$

$$\frac{1}{x - i0} = \lim_{y \to 0+} \frac{1}{x - iy} = \text{c.p.v.} \left(\frac{1}{x} \right) + i\pi\delta \quad \text{in } \mathcal{D}'(\mathbb{R}). \tag{2.9.4}$$

2. Show that $\forall \varepsilon > 0$, $\frac{\varepsilon}{(x^2+\varepsilon^2)}$ defines a distribution in $\mathcal{D}'(\mathbb{R})$ and

$$\lim_{\varepsilon\to 0+} \frac{\varepsilon}{\pi(x^2+\varepsilon^2)} = \delta \quad \text{in } \mathcal{D}'(\mathbb{R}); \tag{2.9.5}$$

$$\lim_{\varepsilon\to 0+} \frac{x}{x^2+\varepsilon^2} = \text{c.p.v.}\left(\frac{1}{x}\right) \quad \text{in } \mathcal{D}'(\mathbb{R}). \tag{2.9.6}$$

3. Show that

$$\lim_{h\to 0+} \lim_{a\to 0+} \frac{a}{2h}\left[\frac{1}{(x-h)^2+a^2} - \frac{1}{(x+h)^2+a^2}\right] = -\pi\delta' \quad \text{in } \mathcal{D}'(\mathbb{R}). \tag{2.9.7}$$

4. Prove that

$$\lim_{t\to\infty} \frac{e^{ixt}}{x-i0} = (i2\pi)\delta \quad \text{in } \mathcal{D}'(\mathbb{R}). \tag{2.9.8}$$

Proof.

1. We have shown in Example 1.4.2 that $\lim_{y\to 0+} \ln(x+iy) = \ln(x+i0)$ in $\mathcal{D}'(\mathbb{R})$, i.e.

$$\lim_{y\to 0+} \int_{-\infty}^{\infty} \ln(x+iy)\phi(x)dx = \int_{-\infty}^{\infty} [\ln|x| + i\pi H(-x)]\phi(x)dx$$
$$= \int_{-\infty}^{\infty} \ln(x+i0)\phi(x)dx \quad \forall\phi \in \mathcal{D}(\mathbb{R}),$$

where $\ln(x+i0) = \ln|x| + i\pi H(-x)$ $\forall x \neq 0$. But from (2.3.19), $\frac{d}{dx}\ln(x+i0) = \text{c.p.v.}(\frac{1}{x}) - i\pi\delta$ in $\mathcal{D}'(\mathbb{R})$. Hence, $\ln(x+iy) \to \ln(x+i0)$ in $\mathcal{D}'(\mathbb{R})$ as $y \to 0^+ \implies \frac{d}{dx}\ln(x+iy) \longrightarrow \frac{d}{dx}ln(x+i0)$ in $\mathcal{D}'(\mathbb{R})$ as $y \to 0^+$ by Theorem 2.9.1.
From (2.3.20), $\frac{d}{dx}ln(x+iy) = \frac{1}{x+iy}$ $\forall$ fixed $y > 0$ in $\mathcal{D}'(\mathbb{R})$

$$\implies \quad \lim_{y\to 0+} \frac{1}{x+iy} = \lim_{y\to 0+} \frac{d}{dx}\ln(x+iy)$$
$$= \frac{d}{dx}\ln(x+i0) = \text{c.p.v.}\left(\frac{1}{x}\right) - i\pi\delta \quad \text{in } \mathcal{D}'(\mathbb{R}).$$

Then we write $\frac{1}{x+i0} = \lim_{y\to 0+} \frac{1}{x+iy} = \text{c.p.v.}(\frac{1}{x}) - i\pi\delta$ in $\mathcal{D}'(\mathbb{R})$, i.e.

$$\int_{-\infty}^{\infty} \frac{1}{x+i0}\phi(x)dx = \lim_{y\to 0+} \int_{-\infty}^{\infty} \frac{\phi(x)}{x+iy}dx$$
$$= \text{c.p.v.} \int_{-\infty}^{\infty} \frac{\phi(x)}{x}dx - \langle i\pi\delta, \phi\rangle$$
$$= \text{c.p.v.} \int_{-\infty}^{\infty} \frac{\phi(x)}{x}dx - i\pi\phi(0) \quad \forall\phi \in \mathcal{D}(\mathbb{R}).$$

Similarly,

$$\frac{1}{x - i0} = \lim_{y \to 0+} \frac{1}{x - iy} = \lim_{y \to 0+} \frac{d}{dx} \ln(x - iy)$$
$$= \frac{d}{dx} \ln(x - i0) = \text{c.p.v.} \left(\frac{1}{x}\right) + i\pi\delta \quad \text{in } \mathcal{D}'(\mathbb{R}),$$

i.e.

$$\int_{-\infty}^{\infty} \frac{1}{x - i0} \phi(x) dx = \lim_{y \to 0+} \int_{-\infty}^{\infty} \frac{\phi(x)}{x - iy} dx$$
$$= \text{c.p.v.} \int_{-\infty}^{\infty} \frac{\phi(x)}{x} dx + i\pi\phi(0) \quad \forall \phi \in \mathcal{D}(\mathbb{R}).$$

2. $\forall \varepsilon > 0$, function $\frac{\varepsilon}{(x^2+\varepsilon^2)}$ is locally summable on $\mathbb{R}$ and, consequently, defines a distribution in $\mathcal{D}'(\mathbb{R})$ by: $\langle \frac{\varepsilon}{(x^2+\varepsilon^2)}, \phi \rangle = \int_{-\infty}^{\infty} \frac{\varepsilon}{(x^2+\varepsilon^2)} \phi(x) dx \; \forall \phi \in \mathcal{D}(\mathbb{R})$. But $\frac{\varepsilon}{(x^2+\varepsilon^2)} = \frac{1}{2i} [\frac{1}{x - i\varepsilon} - \frac{1}{x + i\varepsilon}]$

$$\Longrightarrow \quad \lim_{\varepsilon \to 0+} \int_{-\infty}^{\infty} \frac{\varepsilon}{(x^2 + \varepsilon^2)} \phi(x) dx$$
$$= \lim_{\varepsilon \to 0+} \frac{1}{2i} \left[\int_{-\infty}^{\infty} \frac{\phi(x)}{x - i\varepsilon} dx - \int_{-\infty}^{\infty} \frac{\phi(x)}{x + i\varepsilon} dx \right]$$
$$= \frac{1}{2i} \left[\int_{-\infty}^{\infty} \frac{\phi(x)}{x - i0} dx - \int_{-\infty}^{\infty} \frac{\phi(x)}{x + i0} dx \right]$$
$$= \frac{1}{2i} \left[\text{c.p.v.} \int_{-\infty}^{\infty} \frac{\phi(x)}{x} dx + i\pi\phi(0) - \text{c.p.v.} \int_{-\infty}^{\infty} \frac{\phi(x)}{x} dx + i\pi\phi(0) \right]$$
$$= \frac{1}{2i} 2i\pi\phi(0) = \langle \pi\delta, \phi \rangle \quad \forall \phi \in \mathcal{D}(\mathbb{R})$$

$\Longrightarrow \lim_{\varepsilon \to 0+} \frac{\varepsilon}{(x^2+\varepsilon^2)} = \pi\delta$ in $\mathcal{D}'(\mathbb{R}) \Longrightarrow \lim_{\varepsilon \to 0+} \frac{\varepsilon}{\pi(x^2+\varepsilon^2)} = \delta$ in $\mathcal{D}'(\mathbb{R})$.
From (2.9.3) and (2.9.4),

$$\text{c.p.v.} \left(\frac{1}{x}\right) = \frac{1}{2} \left[\frac{1}{x + i0} + \frac{1}{x - i0} \right] = \frac{1}{2} \lim_{\varepsilon \to 0+} \left[\frac{1}{x + i\varepsilon} + \frac{1}{x - i\varepsilon} \right]$$
$$= \lim_{\varepsilon \to 0+} \frac{x}{(x^2 + \varepsilon^2)} \quad \text{in } \mathcal{D}'(\mathbb{R})$$

i.e.

$$\text{c.p.v.} \int_{-\infty}^{\infty} \frac{\phi(x)}{x} dx = \lim_{\varepsilon \to 0+} \int_{-\infty}^{\infty} \frac{x}{(x^2 + \varepsilon^2)} \phi(x) dx \quad \forall \phi \in \mathcal{D}(\mathbb{R}).$$

3. Setting $y = x - h$, $z = x + h$ $\forall$ fixed $h > 0$, using (2.9.5) with $a = \varepsilon > 0$, we have

$$\lim_{a\to 0+} \frac{a}{(x-h)^2 + a^2} = \lim_{a\to 0+} \frac{a}{y^2 + a^2} = \pi\delta(y)$$
$$= \pi\delta(x-h) = \pi\delta_h \quad \text{in } \mathcal{D}'(\mathbb{R}),$$
$$\lim_{a\to 0+} \frac{a}{(x+h)^2 + a^2} = \lim_{a\to 0+} \frac{a}{z^2 + a^2} = \pi\delta(z)$$
$$= \pi\delta(x+h) = \pi\delta_{-h} \quad \text{in } \mathcal{D}'(\mathbb{R}).$$

Hence,

$$\lim_{h\to 0+} \lim_{a\to 0+} \frac{a}{2h}\left[\frac{1}{(x-h)^2 + a^2} - \frac{1}{(x+h)^2 + a^2}\right]$$
$$= \lim_{h\to 0+} \frac{\pi(\delta_h - \delta_{-h})}{2h} = \pi \lim_{h\to 0+} \frac{\delta_h - \delta_{-h}}{2h} = -\pi\delta' \quad \text{in } \mathcal{D}'(\mathbb{R}).$$

In fact,

$$\lim_{h\to 0+} \frac{1}{2h}\langle \delta_h - \delta_{-h}, \phi\rangle = \lim_{h\to 0+} \frac{1}{2h}[\phi(h) - \phi(-h)]$$
$$= \phi'(0) = \langle \delta, \phi'\rangle = -\langle \delta', \phi\rangle \quad \forall \phi \in \mathcal{D}(\mathbb{R})$$

$\Longrightarrow \lim_{h\to 0+} \frac{\delta_h - \delta_{-h}}{2h} = -\delta'$ in $\mathcal{D}'(\mathbb{R})$.

4. From (2.9.4), $\frac{1}{x-i0} = \text{c.p.v.}(\frac{1}{x}) + i\pi\delta$ in $\mathcal{D}'(\mathbb{R}) \Longrightarrow \frac{e^{ixt}}{x-i0} = e^{ixt}\,\text{c.p.v.}(\frac{1}{x}) + e^{ixt}\cdot i\pi\delta$ in $\mathcal{D}'(\mathbb{R})$.

We are to show that

$$\lim_{t\to\infty}\left(e^{ixt}\,\text{c.p.v.}\left(\frac{1}{x}\right)\right) + \lim_{t\to\infty}(e^{ixt}\cdot i\pi\delta) = i2\pi\delta \quad \text{in } \mathcal{D}'(\mathbb{R}), \tag{2.9.9}$$

$$\langle e^{ixt} i\pi\delta, \phi\rangle = i\pi\langle\delta, e^{ixt}\phi\rangle = i\pi\phi(0) = i\pi\langle\delta,\phi\rangle = \langle i\pi\delta, \phi\rangle, \tag{2.9.10}$$

$\forall \phi \in \mathcal{D}(\mathbb{R})$, $\forall$ fixed $t > 0$

$$\Longrightarrow \quad \lim_{t\to\infty}\langle e^{ixt} i\pi\delta, \phi\rangle = \langle i\pi\delta, \phi\rangle \quad \text{in } \mathbb{C}$$
$$\Longrightarrow \quad \lim_{t\to\infty} e^{ixt} i\pi\delta = i\pi\delta \quad \text{in } \mathcal{D}'(\mathbb{R}). \tag{2.9.11}$$

$\langle e^{ixt}\,\text{c.p.v.}(\frac{1}{x}), \phi\rangle = \langle \text{c.p.v.}(\frac{1}{x}), e^{ixt}\phi\rangle = \lim_{\varepsilon\to 0+}\int_{|x|\geq\varepsilon} \frac{e^{ixt}\phi(x)}{x}dx.$
For $\phi \in \mathcal{D}(\mathbb{R})$ with $\text{supp}(\phi) \subset [-A, A]$, $A > 0$ and for fixed $t > 0$,

$$\left\langle e^{ixt}\,\text{c.p.v.}\left(\frac{1}{x}\right), \phi\right\rangle = \lim_{\varepsilon\to 0+}\left[\int_{-A}^{-\varepsilon} \frac{e^{ixt}\phi(x)}{x}dx + \int_{\varepsilon}^{A} \frac{e^{ixt}\phi(x)}{x}dx\right]$$
$$= \lim_{\varepsilon\to 0+}\left[\int_{\varepsilon}^{A} \frac{e^{ixt}\phi(x)}{x}dx - \int_{\varepsilon}^{A} \frac{e^{-ixt}\phi(-x)}{x}dx\right].$$

But for $\phi \in \mathcal{D}(\mathbb{R})$ with $\operatorname{supp}(\phi) \subset [-A, A]$, $A > 0$, $\phi(\pm x) = \phi(0) \pm x\psi(\pm x)$ with $\psi \in C^0(\mathbb{R})$,

$$\left\langle e^{ixt}\,\text{c.p.v.}\left(\frac{1}{x}\right), \phi\right\rangle$$

$$= \lim_{\varepsilon\to 0+}\Bigg(\underbrace{\phi(0)\int_\varepsilon^A \frac{e^{ixt} - e^{-ixt}}{x}dx}_{I_1(\varepsilon)} + \underbrace{\int_\varepsilon^A [e^{ixt}\psi(x) + e^{-ixt}\psi(-x)]dx}_{I_2(\varepsilon)}\Bigg)$$

$$= \lim_{\varepsilon\to 0+}(I_1(\varepsilon) + I_2(\varepsilon)) \quad \text{for fixed } t > 0. \tag{2.9.12}$$

For fixed $t > 0$,

$$I_1(\varepsilon) = \phi(0)\int_\varepsilon^A \frac{2i\sin xt}{x}dx = 2i\phi(0)\int_\varepsilon^A \frac{\sin xt}{xt}d(xt)$$

$$= 2i\phi(0)\int_{\varepsilon t}^{At}\frac{\sin y}{y}dy \quad (\text{setting } y = xt).$$

But $\frac{\sin y}{y}$ has a *removable* discontinuity at $y = 0$. Hence, for fixed $t > 0$,

$$\lim_{\varepsilon\to 0+}\int_{\varepsilon t}^{At}\frac{\sin y}{y}dy = \int_0^{At}\frac{\sin y}{y}dy$$

$$\Longrightarrow \quad \lim_{\varepsilon\to 0+} I_1(\varepsilon) = 2i\phi(0)\int_0^{At}\frac{\sin y}{y}dy. \tag{2.9.13}$$

Now,

$$\lim_{\varepsilon\to 0+} I_2(\varepsilon) = \int_0^A [e^{ixt}\psi(x) + e^{-ixt}\psi(-x)]dx. \tag{2.9.14}$$

Then, from (2.9.12)–(2.9.14),

$$\lim_{t\to\infty}\left\langle e^{ixt}\,\text{c.p.v.}\left(\frac{1}{x}\right), \phi\right\rangle = 2i\phi(0)\int_0^\infty \frac{\sin y}{y}dy$$

$$+ \lim_{t\to\infty}\int_0^A [e^{ixt}\psi(x) + e^{-ixt}\psi(-x)]dx$$

$$= 2i\phi(0)\frac{\pi}{2} + 0 = i\pi\phi(0) = \langle i\pi\delta, \phi\rangle \quad \forall \phi \in \mathcal{D}(\mathbb{R}),$$

since $\lim_{t\to\infty}\int_0^{At}\frac{\sin y}{y}dy = \int_0^\infty \frac{\sin y}{y}dy = \frac{\pi}{2}$ and $\lim_{t\to\infty}\int_0^A\{\sin xt, \cos xt\}$ $\psi(\pm x)dx = 0$ by the Riemann–Lebesgue Theorem (2.11.7i) [32] $\Longrightarrow \lim_{t\to\infty}$ $\int_0^A e^{\pm ixt}\psi(\pm x)dx = 0$. Hence,

$$\lim_{t\to\infty} e^{ixt}\,\text{c.p.v.}\left(\frac{1}{x}\right) = i\pi\delta \quad \text{in } \mathcal{D}'(\mathbb{R}). \tag{2.9.15}$$

Finally, from (2.9.9), (2.9.11) and (2.9.15), the result (2.9.8) follows:

$$\lim_{t\to\infty} \frac{e^{ixt}}{x-i0} = \lim_{t\to\infty}\left[e^{ixt}\,\text{c.p.v.}\left(\frac{1}{x}\right)\right] + \lim_{t\to\infty}(e^{ixt} i\pi\delta)$$
$$= i\pi\delta + i\pi\delta = 2i\pi\delta \quad \text{in } \mathcal{D}'(\mathbb{R}). \qquad \square$$

Convergence of sequences of derivatives of functions of $L^1_{\text{loc}}(\Omega)$

Proposition 2.9.1. *Let (f_n) be a sequence of functions $f_n \in L^1_{\text{loc}}(\Omega)$ $\forall n \in \mathbb{N}$, $\Omega \subset \mathbb{R}^n$. Then, if*

1. *$f_n(\mathbf{x}) \to f(\mathbf{x})$ a.e. on Ω in the ordinary sense, and*
2. *$\exists M > 0$ such that $|f_n(\mathbf{x})| \le M$ $\forall n \in \mathbb{N}$ a.e. on Ω* (2.9.16)

(or $\exists g \ge 0$ such that $g \in L^1_{\text{loc}}(\Omega)$ and $|f_n(\mathbf{x})| \le g(\mathbf{x})$ $\forall n \in \mathbb{N}$ a.e. on Ω), $\forall$ multi-index $\boldsymbol{\alpha}$, $\partial^{\boldsymbol{\alpha}} f_n \to \partial^{\boldsymbol{\alpha}} f$ in $\mathcal{D}'(\Omega)$ as $n \to \infty$, where derivatives are in the distributional sense.

In particular, for $\boldsymbol{\alpha} = 0$,

$$\partial^{\boldsymbol{\alpha}} f_n = f_n \text{ and } \partial^{\boldsymbol{\alpha}} f = f, \quad \text{i.e. } f_n \to f \text{ in } \mathcal{D}'(\mathbb{R}). \tag{2.9.17}$$

Proof. $f_n \in L^1_{\text{loc}}(\Omega) \implies \langle T_{f_n}, \phi\rangle = \langle f_n, \phi\rangle = \int_\Omega f_n\phi d\mathbf{x}$ $\forall \phi \in \mathcal{D}(\Omega)$ $\forall n \in \mathbb{N}$. By virtue of conditions (1) and (2), we can apply Lebesgue's Theorem B.3.2.3 (Appendix B) on the convergence of a sequence of integrals and we have, for $\phi \in \mathcal{D}(\Omega)$ with $\text{supp}(\phi) = K$,

$$\int_\Omega f_n\phi d\mathbf{x} = \int_K f_n\phi d\mathbf{x} \to \int_K f\phi d\mathbf{x} = \int_\Omega f\phi d\mathbf{x}$$

$\implies \langle f_n, \phi\rangle \to \langle f, \phi\rangle$ as $n \to \infty$ $\forall \phi \in \mathcal{D}(\Omega) \implies f_n \to f$ in $\mathcal{D}'(\Omega)$ as $n \to \infty$
$\implies \partial^{\boldsymbol{\alpha}} f_n \to \partial^{\boldsymbol{\alpha}} f$ in $\mathcal{D}'(\Omega)$, by Theorem 2.9.1. $\square$

In particular, we have

Proposition 2.9.2. *Let (f_n) be a sequence in $L^1_{\text{loc}}(\Omega)$ such that $f_n \to f \in L^1_{\text{loc}}(\Omega)$* uniformly *on every bounded subset of Ω. Then $\partial^{\boldsymbol{\alpha}} f_n \to \partial^{\boldsymbol{\alpha}} f$ in $\mathcal{D}'(\Omega)$ as $n \to \infty$ $\forall |\boldsymbol{\alpha}| \in \mathbb{N}_0$.*

Proof. For $\phi \in \mathcal{D}(\Omega)$ with $\text{supp}(\phi) = K \subset\subset \Omega$,

$$\langle f_n, \phi\rangle = \int_\Omega f_n\phi d\mathbf{x} = \int_K f_n\phi d\mathbf{x} \to \int_K f\phi d\mathbf{x},$$

since

$$\left|\int_K f\phi d\mathbf{x} - \int_K f_n\phi d\mathbf{x}\right| = \left|\int_K (f - f_n)\phi d\mathbf{x}\right|$$
$$\leq \max_{\mathbf{x}\in K} |\phi(\mathbf{x})| \int_K |f(\mathbf{x}) - f_n(\mathbf{x})| dx \to 0 \quad \text{as } n \to \infty$$

$\Longrightarrow \langle f_n, \phi\rangle \to \langle f, \phi\rangle$ as $n \to \infty$ $\forall \phi \in \mathcal{D}(\Omega) \Longrightarrow f_n \to f$ in $\mathcal{D}'(\Omega)$ as $n \to \infty$
$\Longrightarrow \partial^{\boldsymbol{\alpha}} f_n \to \partial^{\boldsymbol{\alpha}} f$ in $\mathcal{D}'(\Omega)$ as $n \to \infty$ $\forall |\boldsymbol{\alpha}| \in \mathbb{N}$, by Theorem 2.9.1. □

2.10 Delta-convergent sequences of functions in $\mathcal{D}'(\mathbb{R}^n)$

Definition 2.10.1. Let (f_j) be a sequence of functions on $\mathbb{R}^n$. Then, if $f_j \to \delta$ in $\mathcal{D}'(\mathbb{R}^n)$ as $j \to \infty$, δ being the Dirac distribution with concentration at $\mathbf{0}$, i.e. if

$$\lim_{j\to\infty} \langle f_j, \phi\rangle = \lim_{j\to\infty} \int_{\mathbb{R}^n} f_j(\mathbf{x})\phi(\mathbf{x}) d\mathbf{x} = \phi(\mathbf{0}) = \langle \delta, \phi\rangle \quad \forall \phi \in \mathcal{D}(\mathbb{R}^n), \tag{2.10.1}$$

(f_j) is called a *delta-convergent sequence of functions* in $\mathcal{D}'(\mathbb{R}^n)$.

A characterization of delta-convergent sequences of functions in $\mathcal{D}'(\mathbb{R}^n)$

Theorem 2.10.1. *Let (f_j) be a sequence of functions on $\mathbb{R}^n$ satisfying:*

1. $f_j(\mathbf{x}) \geq 0$ *for* $\|\mathbf{x}\|_{\mathbb{R}^n} = (x_1^2 + \cdots + x_n^2)^{\frac{1}{2}} \leq k$, $k > 0$*;*
2. $f_j \to 0$ *uniformly as* $j \to \infty$ *on every set* $\{\mathbf{x} : 0 < a \leq \|\mathbf{x}\| \leq \frac{1}{a} < +\infty\}$;
3. $\int_{\|\mathbf{x}\|\leq a} f_j(\mathbf{x}) d\mathbf{x} \to +1$ *as* $j \to \infty$ $\forall a > 0$.

Then $f_j \to \delta$ in $\mathcal{D}'(\mathbb{R}^n)$ as $j \to \infty$ (i.e. (2.10.1) *holds).*

Remark 2.10.1. The property (1), $f_j(\mathbf{x}) \geq 0$ for $\|\mathbf{x}\| \leq k$, is *not* necessary and may be replaced by the following more general condition:

1′. For a suitably chosen $k > 0$, $\exists M > 0$, independent of j, such that

$$\int_{\|\mathbf{x}\|\leq k} |f_j(\mathbf{x})| d\mathbf{x} \leq M \,\forall j \in \mathbb{N}. \tag{2.10.2}$$

Proof of Theorem 2.10.1.

$$\langle f_j, \phi\rangle = \int_{\mathbb{R}^n} f_j(\mathbf{x})\phi(\mathbf{x}) d\mathbf{x}$$
$$= \underbrace{\int_{\|x\|\leq a} f_j(\mathbf{x})\phi(\mathbf{0}) d\mathbf{x}}_{I_1(j)} + \underbrace{\int_{\|\mathbf{x}\|\leq a} f_j(\mathbf{x})[\phi(\mathbf{x}) - \phi(\mathbf{0})] d\mathbf{x}}_{I_2(j)} + \underbrace{\int_{\|\mathbf{x}\|>a} f_j(\mathbf{x})\phi(\mathbf{x}) d\mathbf{x}}_{I_3(j)}$$
$$= I_1(j) + I_2(j) + I_3(j). \tag{2.10.3}$$

Estimate for $|I_2(j)|$: $|\phi(\mathbf{x})-\phi(\mathbf{0})| = |\langle \nabla\phi(\boldsymbol{\xi}), \mathbf{x}\rangle| \leq \|\mathbf{x}\|\sqrt{n}M$ with $M \geq \max_{1\leq i\leq n} \max_{\mathbf{x}\in\mathbb{R}^n} |\frac{\partial\phi}{\partial x_i}(\mathbf{x})|$, $\nabla\phi(\boldsymbol{\xi}) = (\frac{\partial\phi}{\partial x_1}(\boldsymbol{\xi}), \frac{\partial\phi}{\partial x_2}(\boldsymbol{\xi}), \ldots, \frac{\partial\phi}{\partial x_n}(\boldsymbol{\xi}))$
$\Longrightarrow |I_2(j)| \leq M\sqrt{n}a \int_{\|\mathbf{x}\|\leq a} f_j(\mathbf{x})d\mathbf{x}$.

Since properties (2) and (3) hold for any sufficiently small $a > 0$, we can choose $a \leq k$. Thus, for $a \leq k$, $\int_{\|\mathbf{x}\|\leq a} f_j(\mathbf{x})d\mathbf{x} \leq \int_{\|\mathbf{x}\|\leq k} f_j(\mathbf{x})d\mathbf{x} \leq C\, \forall j \in \mathbb{N} \Longrightarrow |I_2(j)| \leq M\sqrt{n}a \int_{\|\mathbf{x}\|\leq a} f_j(\mathbf{x})d\mathbf{x} \leq CM\sqrt{n}a$.

$$\text{Hence, choose } a \leq \frac{\varepsilon}{3CM\sqrt{n}} \text{ such that } |I_2(j)| \leq \frac{\varepsilon}{3}. \tag{2.10.4}$$

Estimate for $|I_3(j)|$: Now we are to make a *final* choice of $a > 0$ satisfying three conditions: $0 < a \leq k$, $a \leq \frac{\varepsilon}{3CM\sqrt{n}}$ and $\operatorname{supp}(\phi) \subset [-\frac{1}{a}, \frac{1}{a}]$ for sufficiently small a.

For such a choice of a, $f_j \to 0$ uniformly as $j \to \infty$ on $\{\mathbf{x} : a \leq \|\mathbf{x}\| \leq \frac{1}{a}\}$

$$\Longrightarrow \quad I_3(j) \to 0 \text{ as } j \to \infty \quad \Longrightarrow \quad \exists j_1 \text{ such that } |I_3(j)| \leq \frac{\varepsilon}{3}\ \forall j \geq j_1. \tag{2.10.5}$$

Estimate for $|I_1(j) - \phi(\mathbf{0})|$: Since $\int_{\|\mathbf{x}\|\leq a} f_j(\mathbf{x})d\mathbf{x} \to 1$ as $j \to \infty$, $I_1(j) \to \phi(\mathbf{0})$ as $j \to \infty \Longrightarrow \exists j_2$ such that $|I_1(j) - \phi(\mathbf{0})| \leq \frac{\varepsilon}{3} \forall j \geq j_2$. Now, choosing $j_0 = \max\{j_1, j_2\}$,

$$\begin{aligned}|\langle f_j, \phi\rangle - \phi(\mathbf{0})| &= |I_1(j) + I_2(j) + I_3(j) - \phi(\mathbf{0})|\\ &\leq |I_1(j) - \phi(\mathbf{0})| + |I_2(j)| + |I_3(j)| \leq \frac{\varepsilon}{3} + \frac{\varepsilon}{3} + \frac{\varepsilon}{3} = \varepsilon \quad \forall j \geq j_0.\end{aligned}$$

Hence, $\lim_{j\to\infty}\langle f_j, \phi\rangle = \phi(\mathbf{0}) = \langle\delta, \phi\rangle\ \forall\phi \in \mathcal{D}(\mathbb{R}^n) \Longrightarrow f_j \to \delta$ in $\mathcal{D}'(\mathbb{R}^n)$ as $j \to \infty$. □

Example 2.10.1. $\forall j \in \mathbb{N}$, $f_j(\mathbf{x}) = \frac{1}{\varepsilon^n} e^{\frac{-\pi r^2}{\varepsilon^2}}$ with $\varepsilon = \frac{1}{j}$, $r^2 = \sum_{i=1}^n x_i^2$. (2.10.6)

$\forall j \in \mathbb{N}$, $f_j(\mathbf{x}) \geq 0$ for all $\mathbf{x} \in \mathbb{R}^n \Longrightarrow$ property (1) holds $\forall k > 0$.

For $\|\mathbf{x}\| = r \geq a$, $f_j(\mathbf{x}) \leq \frac{1}{\varepsilon^n} e^{\frac{-\pi a^2}{\varepsilon^2}} \to 0$ as $\varepsilon \to 0^+ \Longrightarrow f_j \to 0$ uniformly as $j \to \infty$ for $\|\mathbf{x}\| \geq a \Longrightarrow$ property (2) holds $\forall a > 0$.

Changing variables: $x_i = \varepsilon\xi_i$, $1 \leq i \leq n$, with Jacobian $J = \varepsilon^n$, $\forall a > 0$, $\varepsilon = \frac{1}{j}$,

$$\begin{aligned}\int_{\|\mathbf{x}\|\leq a} f_j(\mathbf{x})d\mathbf{x} &= \int_{\|\boldsymbol{\xi}\|\leq ja} f_j(\varepsilon\boldsymbol{\xi})\varepsilon^n d\boldsymbol{\xi} = \int_{\|\boldsymbol{\xi}\|\leq ja} (\frac{1}{\varepsilon^n} e^{-\pi\varepsilon^2\frac{r^2}{\varepsilon^2}})\varepsilon^n d\boldsymbol{\xi}\\ &= \int_{\|\boldsymbol{\xi}\|\leq ja} e^{-\pi r^2} d\boldsymbol{\xi} \to \int_{\mathbb{R}^n} e^{-\pi r^2} d\boldsymbol{\xi} = 1\end{aligned}$$

as $j \to \infty$ ($r^2 = \xi_1^2 + \cdots + \xi_n^2$) since $\int_{-\infty}^{\infty} e^{-\xi_k^2} d\xi_k = \sqrt{\pi}$, $1 \le k \le n$,

$$\int_{\mathbb{R}^n} e^{-\pi r^2} d\boldsymbol{\xi} = \underbrace{\int_{-\infty}^{\infty} \cdots \int_{-\infty}^{\infty}}_{n \text{ times}} e^{-\pi(\xi_1^2+\cdots+\xi_n^2)} d\xi_1 \ldots d\xi_n$$

$$= \int_{-\infty}^{\infty} e^{-\pi\xi_1^2} d\xi_1 \int_{-\infty}^{\infty} e^{-\pi\xi_2^2} d\xi_2 \cdots \int_{-\infty}^{\infty} e^{-\pi\xi_n^2} d\xi_n$$

$$= \int_{-\infty}^{\infty} e^{-\eta_1^2} \frac{d\eta_1}{\sqrt{\pi}} \int_{-\infty}^{\infty} e^{-\eta_2^2} \frac{d\eta_2}{\sqrt{\pi}} \cdots \int_{-\infty}^{\infty} e^{-\eta_n^2} \frac{d\eta_n}{\sqrt{\pi}} \quad \left(\text{setting } \xi_k = \frac{\eta_k}{\sqrt{\pi}}, 1 \le k \le n\right)$$

$$= \frac{\sqrt{\pi}}{\sqrt{\pi}} \cdot \frac{\sqrt{\pi}}{\sqrt{\pi}} \cdots \frac{\sqrt{\pi}}{\sqrt{\pi}} = 1.$$

$\Longrightarrow$ property (3) holds $\forall a > 0$. Hence, by Theorem 2.10.1, $f_j \to \delta$ in $\mathcal{D}'(\mathbb{R}^n)$ as $j \to \infty$.

Example 2.10.2. $\forall j \in \mathbb{N}$, $f_j(\mathbf{x}) = \frac{1}{\varepsilon^n(\sqrt{2\pi})^n} e^{\frac{-r^2}{(2\varepsilon^2)}}$ with $\varepsilon = \frac{1}{j}$, $r^2 = \sum_{i=1}^n x_i^2$. Following the steps of the proof of Example 2.10.1, we can show that all three properties (1)–(3) hold:

$|f_j(\mathbf{x})| \ge 0$ $\forall \mathbf{x} \in \mathbb{R}^n$, $\forall j \in \mathbb{N}$;

$|f_j(\mathbf{x})| \le \frac{1}{\varepsilon^n(\sqrt{2\pi})^n} e^{\frac{-a^2}{(2\varepsilon^2)}}$ for $\|\mathbf{x}\| \ge a$ and $\frac{1}{\varepsilon^n(\sqrt{2\pi})^n} e^{\frac{-a^2}{(2\varepsilon^2)}} \to 0$ as $\varepsilon \to 0^+$;

$\forall a > 0$,

$$\int_{\|\mathbf{x}\| \le a} f_j(\mathbf{x}) d\mathbf{x} = \int_{\|\boldsymbol{\xi}\| \le ja} \frac{1}{(\sqrt{2\pi})^n} e^{\frac{-r^2}{2}} d\boldsymbol{\xi} \to \int_{\mathbb{R}^n} \frac{1}{(\sqrt{2\pi})^n} e^{\frac{-r^2}{2}} d\boldsymbol{\xi} = 1$$

as $j \to \infty$, since

$$\int_{\mathbb{R}^n} e^{\frac{-r^2}{2}} d\boldsymbol{\xi} = \int_{-\infty}^{\infty} e^{\frac{-\xi_1^2}{2}} d\xi_1 \int_{-\infty}^{\infty} e^{\frac{-\xi_2^2}{2}} d\xi_2 \cdots \int_{-\infty}^{\infty} e^{\frac{-\xi_n^2}{2}} d\xi_n$$

$$= \int_{-\infty}^{\infty} e^{-\eta_1^2} \sqrt{2} d\eta_1 \int_{-\infty}^{\infty} e^{-\eta_2^2} \sqrt{2} d\eta_2 \cdots \int_{-\infty}^{\infty} e^{-\eta_n^2} \sqrt{2} d\eta_n$$

$$= (\sqrt{2\pi})(\sqrt{2\pi}) \cdots (\sqrt{2\pi}) = (\sqrt{2\pi})^n.$$

Hence, by Theorem 2.10.1, $f_j \to \delta$ in $\mathcal{D}'(\mathbb{R}^n)$ as $j \to \infty$.

Example 2.10.3. $\forall j \in \mathbb{N}$,

$$f_j(\mathbf{x}) = \begin{cases} \frac{n}{\varepsilon^n S_n} & \text{for } \|\mathbf{x}\| < \varepsilon = \frac{1}{j} \\ 0 & \text{for } \|\mathbf{x}\| \ge \varepsilon = \frac{1}{j}, \end{cases} \tag{2.10.7}$$

where S_n = surface area of an n-dimensional unit sphere $= \frac{2(\pi^{\frac{n}{2}})}{\Gamma(\frac{n}{2})}$ with $S_1 = 2$,

$$S_2 = 2\pi, \quad S_3 = 4\pi, \quad S_4 = 2\pi^2 \quad \text{etc.,}$$

$$\Gamma\left(\frac{1}{2}\right) = \sqrt{\pi}, \quad \Gamma(1) = 1, \quad \Gamma\left(\frac{3}{2}\right) = \frac{1}{2}\Gamma\left(\frac{1}{2}\right) = \frac{1}{2}\sqrt{\pi}. \tag{2.10.8}$$

$f_j(\mathbf{x}) \geq 0 \ \forall \mathbf{x} \in \mathbb{R}^n, \forall j \in \mathbb{N} \Longrightarrow$ property (1) holds for all $k > 0$;
$f_j(\mathbf{x}) = 0$ for $\|\mathbf{x}\| \geq a$ with $a > \frac{1}{j} = \varepsilon \Longrightarrow f_j \to 0$ uniformly in every set

$$\left\{\mathbf{x} : 0 < \frac{1}{j} < a \leq \|\mathbf{x}\| \leq \frac{1}{a} < \infty\right\} \quad \text{as } j \to \infty$$

$\Longrightarrow$ property (2) holds;

$$\int_{\|\mathbf{x}\| \leq a} f_j(\mathbf{x}) d\mathbf{x} = \frac{n}{\varepsilon^n S_n} \int_{\|\mathbf{x}\| \leq \frac{1}{j} \text{ for } a > \frac{1}{j}} d\mathbf{x} = \frac{n}{\varepsilon^n S_n} V_\varepsilon = 1$$

for all sufficiently large $j \in \mathbb{N}$, where V_ε is the volume of the n-dimensional ball $B(\mathbf{0}; \varepsilon)$ with radius $\varepsilon = \frac{1}{j}$, since $V_\varepsilon = \frac{\varepsilon^n S_n}{n}$.

Hence, $\int_{\|\mathbf{x}\| \leq a} f_j(\mathbf{x}) d\mathbf{x} \to 1$ as $j \to \infty$ and property (3) holds.

Consequently, by Theorem 2.10.1, $f_j \to \delta$ in $\mathcal{D}'(\mathbb{R}^n)$ as $j \to \infty$.

Alternative characterization of a delta-convergent sequence of functions

Let (f_j) be a sequence of functions on $\mathbb{R}$ satisfying the following two properties:

1′. For any $M > 0$ with $|a| \leq M$, $|b| \leq M$, $\exists K = K(M) > 0$, independent of a, b, j, such that $|\int_a^b f_j(x) dx| \leq K \ \forall j \in \mathbb{N}$;

2′. $\forall$ fixed $a \neq 0, b \neq 0$,

$$\lim_{j \to \infty} \int_a^b f_j(x) dx = \begin{cases} 1 & \text{for } a < 0 < b \\ 0 & \text{for } a < b < 0 \text{ and } 0 < a < b. \end{cases} \tag{2.10.9}$$

Then (f_j) is a delta-convergent sequence in $\mathcal{D}'(\mathbb{R})$, i.e. $f_j \to \delta$ in $\mathcal{D}'(\mathbb{R})$ as $j \to \infty$ (see [1] for more details).

Example 2.10.4. Let

$$f_t(x) = \frac{1}{2(\sqrt{\pi t})} e^{\frac{-x^2}{4t}} \ (t > 0). \tag{2.10.10}$$

In fact, $f_t(x) > 0 \ \forall x \in \mathbb{R}, \forall t > 0$ and $\int_a^b f_t(x) dx \leq \frac{1}{2\sqrt{\pi t}} \int_{-\infty}^{\infty} e^{\frac{-x^2}{4t}} dx = 1$ (since $\int_{-\infty}^{\infty} e^{-\xi^2} d\xi = \sqrt{\pi}$; setting $x = 2\sqrt{t}\xi, dx = 2\sqrt{t} d\xi, t > 0, \frac{1}{2\sqrt{(\pi t)}} \int_{-\infty}^{\infty} e^{-\xi^2} 2\sqrt{t} d\xi = \frac{1}{\sqrt{\pi}} \int_{-\infty}^{\infty} e^{-\xi^2} d\xi = \frac{\sqrt{\pi}}{\sqrt{\pi}} = 1$). $\Longrightarrow$ property (1′) holds with $K = 1$.

For $a < 0 < b$,

$$\lim_{t\to 0+}\int_a^b f_t(x)dx = \lim_{t\to 0+}\frac{1}{2\sqrt{\pi t}}\int_a^b e^{\frac{-x^2}{(4t)}}dx$$
$$= \lim_{t\to 0+}\frac{1}{2\sqrt{\pi t}}\int_{\frac{b}{\sqrt{t}}}^{\frac{b}{\sqrt{t}}} e^{-\xi^2}2\sqrt{t}d\xi \quad (\text{setting } x = 2\sqrt{t}\xi)$$
$$= \frac{1}{\sqrt{\pi}}\int_{-\infty}^{\infty} e^{-\xi^2}d\xi = \frac{1\sqrt{\pi}}{\sqrt{\pi}} = 1$$

$\Longrightarrow$ property (2′) holds for $a < 0 < b$.

For $0 < a < b$, $x > a \Longrightarrow \frac{x}{a} > 1$,

$$\frac{1}{2\sqrt{\pi t}}\int_a^b e^{\frac{-x^2}{4t}}dx \le \frac{1}{2\sqrt{\pi t}}\int_a^\infty e^{\frac{-x^2}{4t}}dx \le \frac{1}{2\sqrt{\pi t}}\int_a^\infty e^{\frac{-x^2}{4t}}\frac{x}{2t}\frac{2t}{a}dx$$
$$= \frac{2t}{a2\sqrt{\pi t}}[-e^{\frac{-x^2}{4t}}]_a^\infty = \frac{\sqrt{t}}{a\sqrt{\pi}}e^{\frac{-a^2}{4t}} \to 0 \quad \text{as } t \to 0^+$$

$\Longrightarrow$ for $0 < a < b$, $\lim_{t\to 0}\frac{1}{2\sqrt{\pi t}}\int_a^b e^{\frac{-x^2}{4t}}dx = 0$.

Similarly, for $a < b < 0$, $\lim_{t\to 0}\frac{1}{2\sqrt{\pi t}}\int_a^b e^{\frac{-x^2}{4t}}dx = 0$.

Thus, property (2′) holds for $a < 0 < b$, $0 < a < b$, $a < b < 0$.

Hence, (f_t) is a delta-convergent sequence in $\mathcal{D}'(\mathbb{R})$, and $\frac{1}{2\sqrt{\pi t}}e^{(\frac{-x^2}{4t})} \to \delta$ in $\mathcal{D}'(\mathbb{R})$ as $t \to 0^+$.

But there are independent proofs of delta-convergent sequences in $\mathcal{D}'(\mathbb{R})$.

Example 2.10.5. Consider the sequence $f_\varepsilon = \frac{\varepsilon}{\pi(x^2+\varepsilon^2)}$ with $\varepsilon = \frac{1}{j}$. (2.10.11)

Then $f_\varepsilon \to \delta$ in $\mathcal{D}'(\mathbb{R})$ as $\varepsilon \to 0^+$ (see (2.9.5) in Example 2.9.2 for an independent proof).

Example 2.10.6. Let $f_n = \frac{1}{\pi}\cdot\frac{\sin nx}{x}$. (2.10.12)

Then $f_n \to \delta$ in $\mathcal{D}'(\mathbb{R})$ as $n \to \infty$ (see Example 1.8.2 for an independent proof).

Example 2.10.7. Let $f_t(x) = \frac{1}{i2\pi}\frac{e^{ixt}}{(x-i0)}$ for $(t > 0)$. (2.10.13)

Then, $f_t \to \delta$ in $\mathcal{D}'(\mathbb{R})$ as $t \to \infty$ (see the proof of (2.9.8) in Example 2.9.2).

Example 2.10.8. Let $U_n(x) = \frac{1}{2\pi}\int_{-n}^n e^{ixy}dy\ \forall n \in \mathbb{N}$. (2.10.14)

Then, $U_n \to \delta$ in $\mathcal{D}'(\mathbb{R})$ as $n \to \infty$.

In fact,

$$\int_{-n}^{n} e^{ixy}\,dy = \frac{e^{ixy}}{ix}\bigg|_{y=-n}^{y=n} = \frac{e^{inx} - e^{-inx}}{ix} = 2\frac{\sin nx}{x}$$

$$\Longrightarrow \quad \lim_{n\to\infty}\left\langle \frac{1}{2\pi}\int_{-n}^{n} e^{ixy}\,dy, \phi \right\rangle = \lim_{n\to\infty}\left\langle \frac{2}{2\pi}\frac{\sin nx}{x}, \phi \right\rangle = \lim_{n\to\infty}\left\langle \frac{1}{\pi}\frac{\sin nx}{x}, \phi \right\rangle$$
$$= \langle \delta, \phi\rangle \quad \forall \phi \in \mathcal{D}(\mathbb{R}) \quad \text{(see Example 2.10.6)}$$

$\Longrightarrow \frac{1}{2\pi}\int_{-n}^{n} e^{ixy}\,dy \to \delta$ in $\mathcal{D}'(\mathbb{R})$ as $n \to \infty$.

2.11 Term-by-term differentiation of series of distributions

Theorem 2.11.1. *Let $\sum_{n=1}^{\infty} T_n$ be a convergent series of distributions in $\mathcal{D}'(\Omega)$ with its sum $T = \sum_{n=1}^{\infty} T_n, T \in \mathcal{D}'(\Omega)$. Then the series $\sum_{n=1}^{\infty} T_n$ can be differentiated* indefinitely term by term under the summation sign, *i.e.* $\forall$ *multi-index* $\boldsymbol{\alpha} = (\alpha_1, \alpha_2, \ldots, \alpha_n)$,

$$\partial^{\boldsymbol{\alpha}} T = \sum_{n=1}^{\infty} \partial^{\boldsymbol{\alpha}} T_n \quad \textit{with } \partial^{\boldsymbol{\alpha}} = \frac{\partial^{|\boldsymbol{\alpha}|}}{\partial x_1^{\alpha_1} \ldots \partial x_n^{\alpha_n}}. \tag{2.11.1}$$

Proof. Set $S_N = \sum_{n=1}^{N} T_n \in \mathcal{D}'(\Omega)$. Since $\sum_{n=1}^{\infty} T_n$ is convergent in $\mathcal{D}'(\Omega)$ with $T = \sum_{n=1}^{\infty} T_n, T \in \mathcal{D}'(\Omega)$, $S_N \to T$ in $\mathcal{D}'(\Omega)$ as $N \to \infty \Longrightarrow \partial^{\boldsymbol{\alpha}} S_N \to \partial^{\boldsymbol{\alpha}} T$ in $\mathcal{D}'(\Omega)$ as $N \to \infty$ by Theorem 2.9.1. $\Longrightarrow \partial^{\boldsymbol{\alpha}} S_N = \sum_{n=1}^{N} \partial^{\boldsymbol{\alpha}} T_n \to \sum_{n=1}^{\infty} \partial^{\boldsymbol{\alpha}} T_n = \partial^{\boldsymbol{\alpha}} T$ in $\mathcal{D}'(\Omega)$ as $N \to \infty$. □

Remark 2.11.1. For a convergent series of functions (*not* treated as distributions), term-by-term differentiation (in the *usual pointwise sense*) of the series under the summation sign cannot be done unless some additional conditions are fulfilled. Hence, this is a nice property of distributions, which is extremely useful in applications.

Convergence of trigonometric series in $\mathcal{D}'(\mathbb{R})$

Theorem 2.11.2. *Let $\sum_{k=-\infty}^{\infty} C_k e^{i2\pi kx}$ be a trigonometric series in complex form. If the complex coefficients C_k satisfy the condition that*

$$\exists \text{ a constant } M > 0 \text{ and an integer } p \in \mathbb{N}_0 \text{ such that } |C_k| \le M|k|^p \ \forall k \neq 0, \tag{2.11.2}$$

then the following hold:

I. *The series $\sum_{k=-\infty}^{\infty} C_k e^{i2\pi kx}$ converges in $\mathcal{D}'(\mathbb{R})$ with its sum denoted by*

$$T = \sum_{k=-\infty}^{\infty} C_k e^{i2\pi kx}; \tag{2.11.3}$$

II. $$\frac{dT}{dx} = \sum_{k=-\infty}^{\infty} \frac{d}{dx}[C_k e^{i2\pi kx}] = \sum_{k=-\infty}^{\infty} (i2\pi k) C_k e^{i2\pi kx}; \tag{2.11.4}$$

III. $$\tau_1 T = T, \quad i.e.\ \langle \tau_1 T, \phi \rangle = \langle T, \phi \rangle \quad \forall \phi \in \mathcal{D}(\mathbb{R}). \tag{2.11.5}$$

Proof.

I. For *fixed* $p \in \mathbb{N}_0$ satisfying (2.11.2), consider an auxiliary series with the term C_0 omitted: $\sum_{k=-\infty, k\neq 0}^{\infty} \frac{C_k}{(i2\pi k)^{p+2}} e^{i2\pi kx}$. Let $S_N = \sum_{k=-N, k\neq 0}^{N} \frac{C_k}{(i2\pi k)^{p+2}} e^{i2\pi kx}$ denote the sequence of the partial sums of the first $2N$ terms of this auxiliary series. Then, $\forall k \neq 0$,

$$\left| \frac{C_k}{(i2\pi k)^{p+2}} \right| = \frac{|C_k|}{(2\pi)^{p+2}|k|^{p+2}} \le \left(\frac{M}{(2\pi)^{p+2}} \right) \cdot \frac{1}{|k|^2} \quad \text{and} \quad \sum_{k\neq 0} \frac{1}{k^2} < +\infty.$$

Hence, by Weierstrass's M-test, the auxiliary series converges uniformly (and absolutely). Consequently, (S_N) converges uniformly to a continuous function f on $\mathbb{R} \implies (S_N)$ converges to $T_f = f$ in $\mathcal{D}'(\mathbb{R})$ as $N \to \infty \implies \frac{d^{p+2}}{dx^{p+2}} S_N \to \frac{d^{p+2}}{dx^{p+2}} T_f = T_f^{(p+2)}$ in $\mathcal{D}'(\mathbb{R})$ as $N \to \infty$ by Theorem 2.9.1 with $\alpha = p + 2$,

$$\implies \lim_{N\to\infty} \left[\frac{d^{p+2}}{dx^{p+2}} \sum_{\substack{k=-N \\ k\neq 0}}^{N} \frac{C_k}{(i2\pi k)^{p+2}} e^{i2\pi kx} \right]$$

$$= \lim_{N\to\infty} \left(\sum_{\substack{k=-N \\ k\neq 0}}^{N} C_k e^{i2\pi kx} \right) = \sum_{\substack{k=-\infty \\ k\neq 0}}^{\infty} C_k e^{i2\pi kx} = T_f^{(p+2)} \quad \text{in } \mathcal{D}'(\mathbb{R})$$

$\implies \sum_{k=-\infty}^{\infty} C_k e^{i2\pi kx} = C_0 + T_f^{(p+2)} = T$ in $\mathcal{D}'(\mathbb{R})$ with

$$\begin{aligned} \langle T, \phi \rangle &= \langle C_0 + T_f^{(p+2)}, \phi \rangle = \langle C_0, \phi \rangle + \langle T_f^{(p+2)}, \phi \rangle \\ &= \int_{\mathbb{R}} C_0 \phi dx + (-1)^{p+2} \langle T_f, \phi^{(p+2)} \rangle \\ &= \int_{\mathbb{R}} [C_0 \phi + (-1)^{p+2} f \phi^{(p+2)}] dx \quad \forall \phi \in \mathcal{D}(\mathbb{R}). \end{aligned}$$

II. Set $T_N = \sum_{k=-N}^{N} C_k e^{i2\pi kx}$, with $T_N \to T = \sum_{k=-\infty}^{\infty} C_k e^{i2\pi kx}$ in $\mathcal{D}'(\mathbb{R})$ as $N \to \infty$. Hence,

$$\frac{dT_N}{dx} = \sum_{k=-N}^{N} (i2\pi k) C_k e^{i2\pi kx} \to \frac{dT}{dx} = \sum_{k=-\infty}^{\infty} (i2\pi k) C_k e^{i2\pi kx}$$

in $\mathcal{D}'(\mathbb{R})$ as $N \to \infty$ by Theorem 2.9.1.

III. T is a periodic distribution with period 1, i.e.

$$\begin{aligned}\langle \tau_1 T, \phi\rangle &= \langle T, \tau_{-1}\phi\rangle \quad \text{(see (1.10.45b))}\\ &= \langle T, \phi(x+1)\rangle = \langle T, \phi\rangle \quad \forall \phi \in \mathcal{D}(\mathbb{R}).\end{aligned}$$

In fact,

$$\begin{aligned}\langle T, \phi(x+1)\rangle &= \lim_{N\to\infty} \langle T_N, \phi(x+1)\rangle\\ &= \lim_{N\to\infty} \sum_{k=-N}^{N} C_k \int_{\mathbb{R}} e^{i2\pi kx}\phi(x+1)dx\\ &= \lim_{N\to\infty} \sum_{k=-N}^{N} C_k \int_{\mathbb{R}} e^{i2\pi k\xi}\phi(\xi)d\xi\\ &= \lim_{N\to\infty} \langle T_N, \phi\rangle = \langle T, \phi\rangle \quad \forall \phi \in \mathcal{D}(\mathbb{R})\end{aligned}$$

$$\Longrightarrow \quad \tau_1 T = T \text{ in} \mathcal{D}'(\mathbb{R}). \tag{2.11.6}$$

□

Example 2.11.1.

1. Show that series $\sum_{k=-\infty}^{\infty} e^{i2\pi kx}$ converges in $\mathcal{D}'(\mathbb{R})$.
2. Let $S = \sum_{k=-\infty}^{\infty} e^{i2\pi kx}$. Show that $(1-e^{i2\pi x})S = 0$ and $S = C\sum_{k=-\infty}^{\infty}\delta_k$, C being a constant.

Proof.

1. $\sum_{k=-\infty}^{\infty} e^{i2\pi kx}$ converges in $\mathcal{D}'(\mathbb{R})$ by Theorem 2.11.2, since $C_k = 1 \le |k|^p$ $\forall$ fixed $p \in \mathbb{N}_0$, $\forall k \neq 0$.
2. Let $S = \sum_{k=-\infty}^{\infty} e^{i2\pi kx}$ in $\mathcal{D}'(\mathbb{R})$. Then

$$\begin{aligned}(1-e^{i2\pi x})S &= (1-e^{i2\pi x}) \lim_{N\to\infty} \sum_{k=-N}^{N} e^{i2\pi kx}\\ &= \lim_{N\to\infty}\left[\sum_{k=-N}^{N} e^{i2\pi kx} - \sum_{k=-N+1}^{N+1} e^{i2\pi kx}\right]\\ &= \lim_{N\to\infty} [e^{-i2\pi Nx} - e^{i2\pi(N+1)x}] = 0 \quad \text{in } \mathcal{D}'(\mathbb{R}),\end{aligned}$$

since, $\forall\phi \in \mathcal{D}(\mathbb{R})$,

$$\begin{aligned}
&\langle e^{-i2\pi Nx} - e^{i2\pi(N+1)x}, \phi\rangle \\
&= \int_{\mathbb{R}} e^{-i2\pi Nx}\phi(x)dx - \int_{\mathbb{R}} e^{-i2\pi(N+1)x}\phi(x)dx \\
&= -\frac{1}{i2\pi N}\int_{\mathbb{R}} e^{-i2\pi Nx}\phi'(x)dx - \frac{1}{i2\pi(N+1)}\int_{\mathbb{R}} e^{i2\pi(N+1)x}\phi'(x)dx
\end{aligned}$$

$\Longrightarrow |\langle e^{-i2\pi Nx} - e^{i2\pi(N+1)x}, \phi\rangle| \leq \frac{1}{2\pi\cdot N}\int_{\mathbb{R}} |\phi'(x)|dx \to 0$ as $N \to \infty$ $\forall\phi \in \mathcal{D}(\mathbb{R})$.

Hence, $(1 - e^{i2\pi x})S = 0$ in $\mathcal{D}'(\mathbb{R})$ with $(1 - e^{i2\pi x}) = 0$ for $x = 0, \pm1, \pm2, \dots, \pm N, \dots,$

and, $\forall N \in \mathbb{N}$, $\forall x \in]-N, N[$, we can write

$$\begin{aligned}
e^{i2\pi x} - 1 = \alpha(x)[(x - (-N+1))(x - (-N+2))\cdots \\
\cdot x \cdot (x-1)\cdots(x-(N-1))]
\end{aligned}$$

with $\alpha(x) \neq 0$.

Then $\alpha(x)(x-(-N+1))\cdots(x-(N-1))S = 0$ in $\mathcal{D}'(]-N, N[)$ with $\alpha(x) \neq 0$ on $]-N, N[$ (see Chapter 5 for more details on restrictions of distributions)

$\Longrightarrow (x-(-N+1))\cdots(x-(N-1))S = 0$ in $\mathcal{D}'(]-N, N[)$ (see Sections 1.6 and 2.6)

$\Longrightarrow S = \sum_{k=-N+1}^{N-1} C_k\delta_k$ in $\mathcal{D}'(]-N, N[)$ (see (4) in Example 1.6.1).

Hence,

$$\langle S, \phi\rangle = \left\langle \sum_{k=-N+1}^{N-1} C_k\delta_k, \phi\right\rangle = \sum_{k=-N+1}^{N-1} C_k\phi(k) \quad \forall\phi \in \mathcal{D}(]-N, N[),\ \forall N \in \mathbb{N}$$

$\Longrightarrow S = \sum_{k=-\infty}^{\infty} C_k\delta_k$ in $\mathcal{D}'(\mathbb{R})$. Thus, $S = \sum_{k=-\infty}^{\infty} e^{i2\pi kx} = \sum_{k=-\infty}^{\infty} C_k\delta_k$ in $\mathcal{D}'(\mathbb{R})$. But

$$\langle \tau_1 S, \phi\rangle = \langle S, \tau_{-1}\phi\rangle = \langle S, \phi(x+1)\rangle = \langle S, \phi\rangle \quad \forall\phi \in \mathcal{D}(\mathbb{R}), \tag{2.11.7}$$

since

$$\begin{aligned}\langle S, \phi(x+1)\rangle &= \lim_{N\to\infty}\left\langle \sum_{k=-N}^{N} e^{i2\pi kx}, \phi(x+1)\right\rangle \\ &= \lim_{N\to\infty}\sum_{k=-N}^{N}\int_{\mathbb{R}} e^{i2\pi kx}\phi(x+1)dx \\ &= \lim_{N\to\infty}\sum_{k=-N}^{N}\int_{\mathbb{R}} e^{i2\pi k\xi}\phi(\xi)d\xi \quad (x=\xi-1,\ e^{-i2\pi k}=1) \\ &= \lim_{N\to\infty}\langle S_N,\phi\rangle = \langle S,\phi\rangle \quad \forall\phi\in\mathcal{D}(\mathbb{R})\ \ (S_N = \sum_{k=-N}^{N} e^{i2\pi k\xi}).\end{aligned}$$

Hence, for fixed $k \in \mathbb{Z}$, for $\phi \in \mathcal{D}(]k-\frac{1}{2}, k+\frac{1}{2}[)$ with $\phi(k) = 1$, using (2.11.7), we have

$$\begin{aligned}\langle \tau_1 S,\phi\rangle = \langle S,\tau_{-1}\phi\rangle &= \left\langle \sum_{k=-\infty}^{\infty} C_k\delta_k, \phi(x+1)\right\rangle = C_{k-1}\phi(k) = C_{k-1} \\ &= \langle S,\phi\rangle = \left\langle \sum_{k=-\infty}^{\infty} C_k\delta_k, \phi\right\rangle = C_k\phi(k) = C_k\end{aligned}$$

$\Longrightarrow C_k = C_{k-1} = C\ \forall k \in \mathbb{Z}$
$\Longrightarrow S = C\sum_{k=-\infty}^{\infty}\delta_k.$ □

Fourier series of periodic functions and their convergence in the distributional sense in $\mathcal{D}'(\mathbb{R})$

For *periodic* functions and distributions with period $T > 0$, we refer to Section 1.10, Chapter 1.

- $L^2(T)$: Let $L^2(T)$ be a (complex) *Hilbert space* of (equivalence classes of) complex-valued functions $f : \mathbb{R} \longrightarrow \mathbb{C}$ which are *periodic* on $\mathbb{R}$ with period $T > 0$ and equipped with *inner product* $\langle\cdot,\cdot\rangle_T = \frac{1}{T}\langle\cdot,\cdot\rangle_{L^2(]a,a+T[)}$ defined by

$$\langle f, g\rangle_T = \frac{1}{T}\langle f, g\rangle_{L^2(]a,a+T[)} = \frac{1}{T}\int_a^{a+T} f(x)\overline{g(x)}dx \tag{2.11.7a}$$

(see also Table B.3, Appendix B);

• and *norm*:

$$\|f\|_T^2 = \langle f, f\rangle_T = \frac{1}{T}\langle f, f\rangle_{L^2(]a,a+T[)} = \frac{1}{T}\int_a^{a+T} |f(x)|^2 dx,$$

where $]a, a+T[$ is called a *period interval* on $\mathbb{R}$ $\forall a \in \mathbb{R}$ (see (1.10.45a)).

• *Orthogonality* in $L^2(]a, a+T[)$ $\iff$ *orthogonality* in $L^2(T)$, since $\langle f, g\rangle_{L^2(]a,a+T[)} = 0 \iff \langle f, g\rangle_T = 0$.

• *Orthonormal Systems in* $L^2(T)$: $(e^{in\omega x})_{n=-\infty}^{\infty}$ is an *orthonormal system* in $L^2(T)$, since $e^{in\omega x}$ is periodic with period T and $\omega = \frac{2\pi}{T}$, i.e. $e^{in\omega(a+T)} = e^{in\omega a}$ $\forall a \in \mathbb{R}$ and

$$\begin{aligned}\langle e^{im\omega x}, e^{in\omega x}\rangle_T &= \frac{1}{T}\int_a^{a+T} e^{im\omega x}\cdot e^{-in\omega x}dx \\ &= \frac{1}{T}\int_a^{a+T} e^{i(m-n)\omega x}dx = \delta_{mn} \qquad (2.11.7b)\\ &= \begin{cases}1 & \text{for } m = n \\ 0 & \text{for } m \neq n\end{cases} \quad m, n \in \mathbb{Z}.\end{aligned}$$

• $(e^{in\omega x})_{n=-\infty}^{\infty}$ is a *complete* (or *total* in French) orthonormal system in $L^2(T)$ (see also Section A.12, Appendix A), since

$$\langle f, e^{in\omega x}\rangle_T = \frac{1}{T}\int_a^{a+T} f(x)e^{-in\omega x}dx = 0\ \forall n \in \mathbb{Z} \quad \Longrightarrow \quad f = 0 \text{ in } L^2(T).$$

• $L^2(T) \hookrightarrow \mathcal{D}'(\mathbb{R})$, i.e.

$$\begin{aligned} f \in L^2(T) \quad &\Longrightarrow \quad T_f = f \in \mathcal{D}'(\mathbb{R}) \quad \text{and} \\ f_n \to f \text{ in } L^2(T) \quad &\Longrightarrow \quad f_n \to f \text{ in } \mathcal{D}'(\mathbb{R}), \end{aligned} \qquad (2.11.7c)$$

since $\langle f_n, \phi\rangle \to \langle f, \phi\rangle$ $\forall \phi \in \mathcal{D}(\mathbb{R})$ as $n \to \infty$.

Definition 2.11.1. A trigonometric series

$$\sum_{n=-\infty}^{\infty} \langle f, e^{in\omega x}\rangle_T e^{in\omega x} = \sum_{n=-\infty}^{\infty} c_n(f)e^{in\omega x} \qquad (2.11.7d)$$

is called a *Fourier series* in complex form of the periodic function $f \in L^2(T)$ with respect to the orthonormal system $(e^{in\omega x})_{n=-\infty}^{\infty}$ in $L^2(T)$, and

$$c_n(f) = \langle f, e^{in\omega x}\rangle_T = \frac{1}{T}\int_a^{a+T} f(x)e^{-in\omega x}dx \quad \forall n \in \mathbb{Z} \qquad (2.11.7e)$$

is called the *Fourier coefficient* of $f \in L^2(T)$.

For $a = 0$ (resp. $a = -\pi$), $T = 2\pi$, $\omega = 1$, $]a, a+T[\, = \,]0, 2\pi[$ (resp. $]-\pi, \pi[$), we get the standard Fourier series $\sum_{n=-\infty}^{\infty} c_n(f)e^{inx}$ with Fourier coefficients

$$c_n(f) = \frac{1}{2\pi}\int_0^{2\pi} f(x)e^{-inx}dx \quad \text{(resp. } \frac{1}{2\pi}\int_{-\pi}^{\pi} f(x)e^{-inx}dx) \quad \forall n \in \mathbb{Z}. \tag{2.11.7f}$$

Important properties of Fourier coefficients

1. $$\sum_{k=-\infty}^{\infty} |c_k(f)|^2 \le \|f\|_T^2 = \frac{1}{T}\int_a^{a+T} |f(x)|^2 dx \quad \forall f \in L^2(T)$$

   (called *Bessel's inequality*), since, $\forall n \in \mathbb{N}$,

$$\begin{aligned} 0 &\le \left\| f - \sum_{k=-n}^{n} c_k(f)e^{ik\omega x} \right\|_T^2 \\ &= \left\langle f - \sum_{k=-n}^{n} c_k(f)e^{ik\omega x}, f - \sum_{k=-n}^{n} c_k(f)e^{ik\omega x} \right\rangle_T \\ &= \cdots = \|f\|_T^2 - \sum_{k=-n}^{n} |c_k(f)|^2 \end{aligned} \tag{2.11.7g}$$

   (using the orthonormality of $(e^{ik\omega x})$ and properties of the complex inner product).

2. It is well known that for the *complete* orthonormal system $(e^{ik\omega x})_{k=-\infty}^{\infty}$, Bessel's inequality becomes a strict equality, called *Parseval's relation*, which holds $\forall f \in L^2(T)$, i.e.

$$\sum_{k=-\infty}^{\infty} |c_k(f)|^2 = \|f\|_T^2 \quad \forall f \in L^2(T), \tag{2.11.7h}$$

   and the orthonormal system $(e^{ik\omega x})_{k=-\infty}^{\infty}$ is called *closed* in $L^2(T)$.

3. $c_k(f) \to 0$ as $|k| \to \infty$, since the general term $|c_k(f)|^2$ of the *convergent* series $\sum_{k=-\infty}^{\infty} |c_k(f)|^2$ must tend to 0 as $|k| \to \infty$.

4. For periodic function (resp. $L^1_{\text{loc}}(\mathbb{R})$) $f \in L^1(\mathbb{R})$ on $\mathbb{R}$ with period T, Fourier coefficients

$$c_k(f) = \frac{1}{T}\int_a^{a+T} f(x)e^{-ik\omega x}dx \tag{2.11.7i}$$

   are *well defined* for $k \in \mathbb{Z}$ and $c_k(f) \to 0$ as $|k| \to \infty$ by the Riemann–Lebesgue Theorem.

5. Fourier coefficients $c_k(f)$ remain unaltered if the values $f(x)$ are changed on a set of points with measure 0.

Convergence in $L^2(T)$

Theorem 2.11.3. *The Fourier series of $f \in L^2(T)$ converges to f in $L^2(T)$ in the following sense:*

$$\lim_{n\to\infty} \left\| f - \sum_{k=-n}^{n} c_k(f)e^{ik\omega x} \right\|_T^2$$

$$= \lim_{n\to\infty} \frac{1}{T} \int_a^{a+T} \left| f(x) - \sum_{k=-n}^{n} c_k(f)e^{ik\omega x} \right|^2 dx = 0$$

i.e. $f = \sum_{k=-\infty}^{\infty} c_k(f)e^{ik\omega x}$ *in* $L^2(T)$.

Proof. Since $(e^{ik\omega x})_{k=-\infty}^{\infty}$ is a *closed* orthonormal system in $L^2(T)$, by (2.11.7h),

$$\sum_{k=-\infty}^{\infty} |c_k(f)|^2 = \|f\|_T^2.$$

Set $S_n = \sum_{k=-n}^{n} c_k(f)e^{ik\omega x}$. Then, from (2.11.7g),

$$\|f - S_n\|_T^2 = \|f\|_T^2 - \sum_{k=-n}^{n} |c_k(f)|^2 \quad \forall n \in \mathbb{N}.$$

Hence,

$$\lim_{n\to\infty} \|f - S_n\|_T^2 = \lim_{n\to\infty} \left[\|f\|_T^2 - \sum_{k=-n}^{n} |c_k(f)|^2 \right]$$

$$= \|f\|_T^2 - \sum_{k=-\infty}^{\infty} |c_k(f)|^2 = 0$$

$\Longrightarrow f = \lim_{n\to\infty} S_n = \sum_{k=-\infty}^{\infty} c_k(f)e^{ik\omega x}$ in $L^2(T)$. □

Remark 2.11.1A. For *periodic* $f \in L^1(\mathbb{R})$ with period $T > 0$, Fourier series (2.11.7d) of f does *not* converge in general. In fact, Kolmogorov gave an example of a periodic $f_0 \in L^1(\mathbb{R})$ whose Fourier series is divergent everywhere on $\mathbb{R}$ [30] (see also Property (4) of Fourier coefficients above). The convergence of Fourier series of $f \in L^2(T)$ in Theorem 2.11.3 is in the *mean square sense* on any period interval $]a, a + T[$ on $\mathbb{R}$, i.e. the convergence in $L^2(T)$ is neither *uniform* nor *pointwise* in general. But we need *uniform* convergence of Fourier series (as a *sufficient* condition) so that its sum $\sum_{k=-\infty}^{\infty} c_k(f)e^{ik\omega x}$ is a *continuous* function. Moreover, for *term-by-term* differentiation (resp. integration) of Fourier series, we require the *uniform* convergence of the resultant series (obtained by *term-by-term* differentiation (resp. integration)) so that its sum is *continuous*. Now, we will state some *sufficient* conditions for *uniform* convergence. For this, we define

- $C_T^0(\mathbb{R}) = \{f : f$ is a periodic complex-valued function on $\mathbb{R}$ with period T and *continuous* on $\mathbb{R}\} \subset L^2(T)$;
- $C_T^m(\mathbb{R}) = \{f : f$ and its derivatives $f^{(1)}, \ldots, f^{(m)} \in C_T^0(\mathbb{R})\}\ \forall m \in \mathbb{N}$.

Theorem 2.11.4.

I. *For $f \in C_T^2(\mathbb{R})$, the Fourier series $\sum_{k=-\infty}^{\infty} c_k(f)e^{ik\omega x}$ of f converges* uniformly *and* absolutely *to f.*

II. *For $f \in C_T^m(\mathbb{R})$ with $m > 2$, the Fourier series $\sum_{k=-\infty}^{\infty} c_k(f)e^{ik\omega x}$ of f can be differentiated l times* term by term, *and the resultant series converges* uniformly *and* absolutely *such that, $\forall x \in \mathbb{R}$,*

$$f^{(l)}(x) = \sum_{\substack{k=-\infty \\ k\neq 0}}^{\infty} c_k(f)\frac{d^l}{dx^l}(e^{ik\omega x})$$
$$= \sum_{\substack{k=-\infty \\ k\neq 0}}^{\infty} c_k(f)(ik\omega)^l e^{ik\omega x} = \sum_{\substack{k=-\infty \\ k\neq 0}}^{\infty} c_k(f^{(l)})e^{ik\omega x}$$

with $c_k(f)\cdot(ik\omega)^l = c_k(f^{(l)})\ \forall k \in \mathbb{Z}$, i.e. the resultant series is the Fourier series of $f^{(l)}\ \forall l = 1, \ldots, m-2$.

Proof.

I. Let $f \in C_T^2(\mathbb{R}) \subset L^2(T)$. Then, integrating by parts and using the periodicity of $f(x)e^{-ik\omega x}$, we have, $\forall k \neq 0$,

$$c_k(f) = \frac{1}{T}\int_a^{a+T} f(x)e^{-ik\omega x}dx$$
$$= \frac{1\cdot e^{-ik\omega x}}{T\cdot(-ik\omega)}f(x)\Big|_a^{a+T} + \frac{1}{ik\omega}\cdot\frac{1}{T}\int_a^{a+T} f'(x)e^{-ik\omega x}dx$$
$$= \frac{1}{ik\omega}c_k(f').$$

Again integrating by parts and using the periodicity of $f'(x)e^{-ik\omega x}$, we get, $\forall k \neq 0$,

$$c_k(f) = \left(\frac{1}{ik\omega}\right)^2\frac{1}{T}\int_a^{a+T} f''(x)e^{-ik\omega x}dx = \frac{1}{(ik\omega)^2}c_k(f'').$$

For $k = 0$, $c_0(f) = \frac{1}{T}\int_a^{a+T} f(x)dx$

$$\Longrightarrow \quad |c_0(f)| \leq \frac{1}{T}\int_a^{a+T} |f(x)|dx \leq max_{x\in[a,a+T]}|f(x)| = M_0.$$

Then, $\forall k \neq 0$,

$$|c_k(f)| \leq \frac{1}{k^2\omega^2}\frac{1}{T}\int_a^{a+T} |f''(x)|\cdot|e^{-ik\omega x}|dx \leq \frac{M_1}{k^2}$$

with $M_1 = \frac{1}{\omega^2}\max_{x\in[a,a+T]}|f''(x)| \cdot \frac{1}{T}\int_a^{a+T} dx = \frac{1}{\omega^2}\max_{x\in[a,a+T]}$ $|f''(x)|$

$$\Longrightarrow \quad \sum_{k=-\infty}^{\infty} |c_k(f)e^{ik\omega x}| \leq M_0 + \sum_{\substack{k=-\infty\\k\neq 0}}^{\infty} \frac{M_1}{k^2}$$

$$\leq M_2\Bigg(1 + \sum_{\substack{k=-\infty\\k\neq 0}}^{\infty} \frac{1}{k^2}\Bigg),$$

$M_2 = \max\{M_0, M_1\}$, the *majoring series* $M_2(1 + \sum_{\substack{k=-\infty\\k\neq 0}}^{\infty} \frac{1}{k^2})$ being a *convergent* series of *positive* numbers. Hence, by Weierstrass's M-test with this majoring series, Fourier series $\sum_{k=-\infty}^{\infty} c_k(f)e^{ik\omega x}$ converges *uniformly* and *absolutely* on $\mathbb{R}$. Hence, its sum $g(x) = \sum_{k=-\infty}^{\infty} c_k(f)e^{ik\omega x}$ is a *continuous* function on $\mathbb{R}$. Moreover, the uniform convergence of $\sum_{k=-\infty}^{\infty} c_k e^{ik\omega x}$ to g implies its *mean square convergence* in $L^2(T)$, i.e. $g \in L^2(T)$ in the sense of Theorem 2.11.3. But $f \in C_T^2(\mathbb{R}) \subset L^2(T) \Longrightarrow$ by Theorem 2.11.3, $f = \lim_{n\to\infty}\sum_{k=-n}^{n} c_k(f)e^{ik\omega x}$ in the *mean square sense* in $L^2(T)$. Hence, by virtue of the uniqueness of the limit, $f = g$.

II. For $f \in C_T^m(\mathbb{R}) \subset L^2(T)$, we can repeatedly integrate by parts l times with $l \leq m$ and using the periodicity of $f(x)e^{-ik\omega x}$ and its derivatives of order $\leq m$ to get, for $k \neq 0$, $c_k(f) = \frac{1}{(ik\omega)^l}c_k(f^{(l)})$ for $l \leq m$. Then, for $l = m$ and $k \neq 0$, $|c_k(f)| \leq \frac{1}{|ik\omega|^m}|c_k(f^{(m)})| \leq \frac{M_*}{|k|^m|\omega|^m}$ with $|c_k(f^{(m)})| \leq M_* = \max_{a\leq x\leq a+T}|f^{(m)}(x)|$. But for $1 \leq l \leq m-2$,

$$|c_k(f^{(l)})| = |(ik\omega)^l c_k(f)|$$
$$= |k|^l|\omega|^l|c_k(f)| \leq \frac{M_*|k|^l|\omega|^l}{|k|^m|\omega|^m} \leq \frac{\widetilde{M}}{|k|^{m-l}} \leq \frac{\widetilde{M}}{k^2},$$

since $\widetilde{M} = \frac{M_*}{|\omega|^{m-l}}$ and $m - l \geq 2 \Longrightarrow |k|^{m-l} \geq k^2 \Longrightarrow \frac{1}{|k|^{m-l}} \leq \frac{1}{k^2}$.

$\forall l$ with $1 \le l \le m-2$,

$$\sum_{\substack{k=-\infty \\ k\neq 0}}^{\infty} |c_k(f)(ik\omega)^l e^{ik\omega x}| \le \sum |c_k(f^{(l)})| \le \sum_{\substack{k=-\infty \\ k\neq 0}}^{\infty} \frac{\widetilde{M}}{k^2}.$$

Again, by Weierstrass's M-test with convergent majoring series $\sum_{\substack{k=-\infty \\ k\neq 0}}^{\infty} \frac{\widetilde{M}}{k^2}$, the series $\sum_{\substack{k=-\infty \\ k\neq 0}}^{\infty} c_k(f)(ik\omega l)^l e^{ik\omega x}$ converges *uniformly* and *absolutely* with sum $g_l(x)$ as a continuous function on $\mathbb{R}$ for $1 \le l \le m-2$. Now, we are to prove that $g_l(x) = \frac{d^l}{dx^l} f(x)$, $1 \le l \le m-2$. For this it is sufficient to prove this for $l = 1$, since the result can be similarly proved for other values of $l \le m-2$, and we need the following lemma. □

Lemma 2.11.1 ([33, p. 229]). *Let* $(\frac{df_n}{dx})_{n=1}^{\infty}$ *with* $\frac{df_n}{dx}$ *continuous on* $]a,b[$ $\forall n$ *converge uniformly to* g *on* $]a,b[$. *If* $\exists x_0 \in]a,b[$ *such that* $\lim_{n\to\infty} f_n(x_0) = f(x_0)$, *then* $\frac{d}{dx}[\lim_{n\to\infty} f_n] = \lim_{n\to\infty}[\frac{d}{dx} f_n] = g$ *on* $]a,b[$.

Proof. Set $S_n = \sum_{k=-n}^{n} c_k(f) e^{ik\omega x} \forall n \in \mathbb{N}$. Then, $\frac{dS_n}{dx} = \sum_{\substack{k=-n \\ k\neq 0}}^{n} [c_k(f)\, ik\omega] e^{ik\omega x}$ $\forall n \in \mathbb{N}$ with $S_n \to f$ *uniformly* and $\frac{dS_n}{dx} = S_n' \to g_1$ *uniformly* on $\mathbb{R}$ as $n \to \infty$. Hence, the sequences $(S_n')_{n=1}^{\infty}$ and $(S_n)_{n=1}^{\infty}$, which converge *uniformly*, satisfy all the hypotheses of Lemma 2.11.1 and we get $g_1(x) = \lim_{n\to\infty} S_n'(x) = \frac{d}{dx}(\lim_{n\to\infty} S_n) = \frac{d}{dx} f(x)$ for $x \in \mathbb{R}$. □

Remark 2.11.2. For weaker sufficient conditions for pointwise (resp. uniform) convergence of Fourier series, see, for example, [32], [33].

Fourier series in sines and cosines

- $(1; \cos n\omega x; \sin n\omega x)_{n=1}^{\infty}$ is an *orthogonal* system but *not* an *orthonormal* system in $L^2(T)$, the corresponding orthonormal system being $(1; \frac{\cos n\omega x}{\sqrt{1/2}}; \frac{\sin n\omega x}{\sqrt{1/2}})_{n=1}^{\infty}$, since $\forall n \in \mathbb{N}$, $\|\cos n\omega x\|_T = \|\sin n\omega x\|_T = \sqrt{1/2}$.
- $(1; \cos n\omega x; \sin n\omega x)_{n=1}^{\infty}$ is a *complete orthogonal system* in $L^2(T)$, i.e.

$$\langle f, 1\rangle_T = 0, \quad \langle f, \cos n\omega x\rangle_T = 0, \quad \langle f, \sin n\omega x\rangle_T = 0 \quad \forall n \in \mathbb{N} \tag{2.11.7j}$$

$\Longrightarrow f = 0$ in $L^2(T)$.

Definition 2.11.2. The trigonometric series

$$\frac{a_0}{2} + \sum_{n=1}^{\infty}(a_n \cos n\omega x + b_n \sin n\omega x) \tag{2.11.7k}$$

is called the *Fourier series in sines and cosines* of periodic $f \in L^2(T)$ with respect to the orthogonal system $(1; \cos n\omega x; \sin n\omega x)_{n=1}^{\infty}$ in $L^2(T)$, with the *Fourier coefficients* $(a_0/2; a_n; b_n)_{n=1}^{\infty}$ of f defined by:

- $$a_0(f) = 2\langle f, 1\rangle_T = \frac{2}{T}\int_a^{a+T} f(x)dx; \qquad (2.11.7l)$$

- $\forall n \in \mathbb{N}$,

$$a_n(f) = 2\langle f, \cos n\omega x\rangle_T = \frac{2}{T}\int_a^{a+T} f(x)\cos n\omega x dx; \qquad (2.11.7m)$$

$$b_n(f) = 2\langle f, \sin n\omega x\rangle_T = \frac{2}{T}\int_a^{a+T} f(x)\sin n\omega x dx.$$

Bessel's equality (or Parseval's relation) $\forall f \in L^2(T)$, Fourier coefficients satisfy Bessel's equality:

$$\frac{|a_0|^2}{4} + \sum_{k=1}^{\infty} \frac{|a_k|^2 + |b_k|^2}{2} = \|f\|_T^2 = \frac{1}{T}\int_a^{a+T} |f(x)|^2 dx.$$

Hence, $(1; \cos n\omega x; \sin n\omega x)_{n=1}^{\infty}$ is a *closed orthogonal system* in $L^2(T)$.

Convergence of Fourier series in $L^2(T)$

Theorem 2.11.5. *Fourier series* (2.11.7k) *of $f \in L^2(T)$ converges to f in $L^2(T)$ in the mean square sense:*

$$\lim_{n\to\infty} \left\| f - \left[\frac{a_0}{2} + \sum_{k=1}^{n} (a_k \cos k\omega x + b_k \sin k\omega x)\right]\right\|_T^2 = 0.$$

Proof. The proof is exactly similar to that of Theorem 2.11.3, since $(1; \cos n\omega x; \sin n\omega x)_{n=1}^{\infty}$ is a *closed orthogonal system* in $L^2(T)$. □

Uniform convergence of Fourier series

Theorem 2.11.6.

I. *For $f \in C_T^2(\mathbb{R})$, Fourier series* (2.11.7k) *converges* uniformly *and* absolutely *on $\mathbb{R}$.*

II. *For $f \in C_T^m(\mathbb{R})$ with $m > 2$, Fourier series* (2.11.7k) *can be differentiated l times* term by term *and the resultant series converges* uniformly *and* absolutely, *i.e.* $\forall x \in \mathbb{R}$,

$$f^{(l)}(x) = \sum_{k=1}^{\infty}\left[a_k(f)\cdot\frac{d^l}{dx^l}(\cos k\omega x) + b_k(f)\cdot\frac{d^l}{dx^l}(\sin k\omega x)\right]$$
$$= \sum_{k=1}^{\infty}(a_k(f^{(l)})\cdot\cos k\omega x + b_k(f^{(l)})\cdot\sin k\omega x)$$

for $l = 1,\dots,m-2$.

Remark 2.11.3. For weaker sufficient conditions for pointwise (resp. uniformly) convergence of (2.11.7k), see, for example, [32], [33].

Particular Cases For two important particular cases, $a = -\pi$ (resp. $a = 0$), $T = 2\pi$, $\omega = \frac{2\pi}{T} = 1$, $]a, a+T[\, = \,]-\pi,\pi[$ (resp. $]0, 2\pi[$), the *standard forms* of Fourier series are obtained:

- $\sum_{n=-\infty}^{\infty} c_n(f)e^{inx}$ with $c_n(f) = \frac{1}{2\pi}\int_a^{a+2\pi} f(x)e^{-inx}dx$, $a = 0$ (resp. $a = -\pi$) $\forall n \in \mathbb{Z}$;
- $\frac{a_0}{2} + \sum_{n=1}^{\infty}(a_n\cos nx + b_n\sin nx)$ with $\frac{a_0(f)}{2} = \frac{1}{2\pi}\int_a^{a+2\pi} f(x)dx$,

$$a_n(f) = \frac{1}{\pi}\int_a^{a+2\pi} f(x)\cos nx dx,$$
$$b_n(f) = \frac{1}{\pi}\int_a^{a+2\pi} f(x)\sin nx dx \quad \forall n \in \mathbb{N},\ a = 0 \text{ (resp. } a = -\pi)$$

and instead of $L^2(2\pi)$, we may sometimes use the space $L^2(]-\pi,\pi[)$ (resp. $L^2(]0,2\pi[)$) equipped with the usual inner product:

$$\langle f, g\rangle_{L^2(]-\pi,\pi[)} = \int_{-\pi}^{\pi} f(x)\overline{g(x)}dx,$$
$$\langle f, g\rangle_{L^2(]0,2\pi[)} = \int_0^{2\pi} f(x)\overline{g(x)}dx.$$

In fact, $f \in L^2(]a, a+T[)$ can be given periodic extension to $\mathbb{R}$ with period $T = 2\pi$ such that $f \in L^2(2\pi)$ equipped with the inner product $\langle\cdot,\cdot\rangle_T$ in (2.11.7a). (2.11.7n)

Remark 2.11.4.

- For term-by-term *integration* of Fourier series, see the following Example 2.11.2. For Fourier series of *periodic distributions*, see Section 6.7, Chapter 6.

- $\sum_{k=1}^{\infty} \frac{\sin kx}{k}$, which is the Fourier series of the *sawtooth* function in Example 2.11.2, converges at the points of continuity of f, but $\sum_{k=1}^{\infty} \cos kx = \sum_{k=1}^{\infty} \frac{d}{dx}(\frac{\sin kx}{k})$ *diverges* everywhere in the usual *classical sense,* whereas this series $\sum_{k=1}^{\infty} \cos kx$ *converges* in the *distributional sense* in $\mathcal{D}'(\mathbb{R})$ (see Example 2.11.2), the sum being a *distribution* on $\mathbb{R}$.

Examples of Fourier series and their differentiation in $\mathcal{D}'(\mathbb{R})$

See also (6.7.18)–(6.7.44).

Example 2.11.2. Let $f(x) = \frac{(\pi - x)}{2}$ for $0 < x < 2\pi$ be a periodic function (the *sawtooth* function) with period 2π. Prove that

1. $$\sum_{k=1}^{\infty} \cos kx = -\frac{1}{2} + \pi \sum_{k=-\infty}^{\infty} \delta_{2k\pi} \text{ in } \mathcal{D}'(\mathbb{R}) \text{ with } \delta_{2k\pi} = \delta(x - 2k\pi); \tag{2.11.8}$$

2. $$\sum_{k=-\infty}^{\infty} e^{ikx} = 2\pi \sum_{k=-\infty}^{\infty} \delta_{2k\pi} \text{ in } \mathcal{D}'(\mathbb{R}); \tag{2.11.9}$$

3. $$\sum_{k=-\infty}^{\infty} (ik)^l e^{ikx} = 2\pi \sum_{k=-\infty}^{\infty} \delta_{2k\pi}^{(l)} \text{ in } \mathcal{D}'(\mathbb{R}), \tag{2.11.10}$$

where $\delta_{2k\pi}^{(l)} = \frac{d^l}{dx^l}\delta_{2k\pi}$ is the lth-order derivative of Dirac distribution $\delta_{2k\pi} \in \mathcal{D}'(\mathbb{R})$ with mass/charge/force etc. concentrated at points $x = 2k\pi$, $k \in \mathbb{Z}$.

Proof.

1. f is a periodic function with period $T = 2\pi$ on $\mathbb{R}$ and $f \in L^2(]0, 2\pi[)$. Then, by virtue of (2.11.71), $f \in L^2(2\pi)$ with $T = 2\pi$. Hence, by Theorem 2.11.5, $f(x)$ has Fourier series representation $f(x) = \sum_{k=1}^{\infty} \frac{\sin kx}{k}$, since

$$a_0 = \frac{1}{\pi}\int_0^{2\pi} \frac{\pi - x}{2} dx = \frac{1}{2\pi}(\pi x - \frac{1}{2}x^2)\Big|_0^{2\pi} = 0,$$

$$\begin{aligned} a_k &= \frac{1}{\pi}\int_0^{2\pi} \frac{\pi - x}{2} \cos kx dx \\ &= \frac{1}{2\pi}(\pi - x)\frac{\sin kx}{k}\Big|_0^{2\pi} + \frac{1}{2k\pi}\int_0^{2\pi} \sin kx dx = 0 \quad \forall k \in \mathbb{N}, \end{aligned}$$

$$\begin{aligned} b_k &= \frac{1}{\pi}\int_0^{2\pi} \frac{\pi - x}{2} \sin kx dx \\ &= -\frac{1}{2\pi}(\pi - x)\frac{\cos kx}{k}\Big|_0^{2\pi} - \frac{1}{2k\pi}\int_0^{2\pi} \cos kx dx = \frac{1}{k} \quad \forall k \in \mathbb{N} \end{aligned}$$

such that $S_n = \sum_{k=1}^{n} \frac{\sin kx}{k} \to f$ in $L^2(2\pi) \hookrightarrow \mathcal{D}'(\mathbb{R})$ as $n \to \infty$

$$\Longrightarrow \quad S_n \to f \text{ in } \mathcal{D}'(\mathbb{R}) \text{ as } n \to \infty \quad \text{(see (2.13.1) later).} \tag{2.11.11}$$

Consider the auxiliary series $\sum_{k=1}^{\infty} -\frac{\sin kx}{k^3}$, which is obtained by *formally* integrating each term twice. Since $|-\frac{\sin kx}{k^3}| \le \frac{1}{k^2}\ \forall k \in \mathbb{N}$, $\sum_{k=1}^{\infty} -\frac{\sin kx}{k^3}$ converges uniformly and absolutely to a continuous function $F(x)$ on $\mathbb{R}$ by Weierstrass's M-test and $F(x+2\pi) = F(x)$, i.e. F is periodic on $\mathbb{R}$ with period 2π. Hence, $F_n = \sum_{k=1}^{n} -\frac{\sin kx}{k^3} \to F$ uniformly on every compact K as $n \to \infty \Longrightarrow F_n \to F$ in $\mathcal{D}'(\mathbb{R})$ as $n \to \infty$ by Proposition 2.9.2, since $F_n \in L^1_{\text{loc}}(\mathbb{R})$, $F \in L^1_{\text{loc}}(\mathbb{R}) \Longrightarrow \frac{d^2}{dx^2} \sum_{k=1}^{n} -\frac{\sin kx}{k^3} = \sum_{k=1}^{n} \frac{\sin kx}{k} = S_n \to \frac{d^2F}{dx^2} = F''$ in $\mathcal{D}'(\mathbb{R})$ as $n \to \infty$ by Theorem 2.9.1. By virtue of (2.11.11), $f = F'' \in \mathcal{D}'(\mathbb{R})$. Hence, $\sum_{k=1}^{\infty} \frac{\sin kx}{k} = f$ in $\mathcal{D}'(\mathbb{R}) \Longrightarrow \frac{d}{dx} \sum_{k=1}^{\infty} \frac{\sin kx}{k} = \sum_{k=1}^{\infty} \cos kx = \frac{df}{dx}$ in $\mathcal{D}'(\mathbb{R})$ by Theorem 2.11.1 on term-by-term differentiation of series of distributions, where $\frac{df}{dx} \in \mathcal{D}'(\mathbb{R})$ is the distributional derivative of f on $\mathbb{R}$. Since f is piecewise continuous on $\mathbb{R}$ with points of discontinuity at $x = 2\pi k\ \forall k = 0, \pm 1, \pm 2, \ldots$, i.e. $\forall k \in \mathbb{Z}$, where f has finite jump $J_{2k\pi} = [f(x^+) - f(x^-)]_{x=2k\pi} = \frac{\pi}{2} - (-\frac{\pi}{2}) = \pi\ \forall k \in \mathbb{Z}$ and the usual derivatives $[\frac{df}{dx}(x)] = \frac{d}{dx}(\frac{\pi - x}{2}) = \frac{-1}{2}$, the distributional derivative $\frac{df}{dx} \in \mathcal{D}'(\mathbb{R})$ is given by:

$$\begin{aligned} \frac{df}{dx} &= \left[\frac{df}{dx}(x)\right] + \sum_{k=-\infty}^{\infty} J_{2k\pi}\delta_{2k\pi} \quad \text{(see Chapter 3, Theorem 3.1.1)} \\ &= -\frac{1}{2} + \pi \sum_{k=-\infty}^{\infty} \delta_{2k\pi} \quad \text{in } \mathcal{D}'(\mathbb{R}). \end{aligned}$$

Hence, $\sum_{k=1}^{\infty} \cos kx = -\frac{1}{2} + \pi \sum_{k=-\infty}^{\infty} \delta_{2k\pi}$ in$\mathcal{D}'(\mathbb{R})$. (The series $\sum_{k=1}^{\infty} \cos kx$ does *not* converge in the usual classical sense, but it converges in the *distributional sense*).

2. $e^{\pm ikx} = \cos kx \pm i \sin kx$

$$\begin{aligned} \Longrightarrow \qquad \sum_{k=1}^{N} \cos kx &= \frac{1}{2} \sum_{k=1}^{N} (e^{+ikx} + e^{-ikx}) \\ &= \frac{1}{2} \sum_{\substack{k=-N \\ k \ne 0}}^{N} e^{ikx} = \frac{1}{2} \sum_{k=-N}^{N} e^{ikx} - \frac{1}{2} \end{aligned}$$

$$\Longrightarrow \quad \frac{1}{2}\sum_{k=-\infty}^{\infty} e^{ikx} = \lim_{N\to\infty}\left[\frac{1}{2}\sum_{k=-N}^{N} e^{ikx}\right]$$

$$= \sum_{k=1}^{\infty}\cos kx + \frac{1}{2} = \pi \sum_{k=-\infty}^{\infty} \delta_{2k\pi} \quad \text{in } \mathcal{D}'(\mathbb{R}) \quad \text{(using (2.11.8))}$$

$$\Longrightarrow \quad \sum_{k=-\infty}^{\infty} e^{ikx} = 2\pi \sum_{k=-\infty}^{\infty} \delta_{2k\pi} \quad \text{in } \mathcal{D}'(\mathbb{R}).$$

3. By Theorem 2.11.1,

$$\frac{d^l}{dx^l}\left[\sum_{k=-\infty}^{\infty} e^{ikx}\right] = 2\pi\frac{d^l}{dx^l}\sum_{k=-\infty}^{\infty}\delta_{2k\pi} = 2\pi\sum_{k=-\infty}^{\infty}\delta_{2k\pi}^{(l)} \quad \text{in } \mathcal{D}'(\mathbb{R})$$

$$\Longrightarrow \quad \sum_{k=-\infty}^{\infty}\frac{d^l}{dx^l}(e^{ikx}) = \sum_{k=-\infty}^{\infty}(ik)^l e^{ikx} = 2\pi\sum_{k=-\infty}^{\infty}\delta_{2k\pi}^{(l)} \quad \text{in } \mathcal{D}'(\mathbb{R}). \qquad \square$$

Example 2.11.3. Let $S = \sum_{k=-\infty}^{\infty} e^{i2\pi kx} = C\sum_{k=-\infty}^{\infty}\delta_k$ as proved in Example 2.11.1. Consider the series $\sum_{k=-\infty}^{\infty}\frac{1}{1+k^2}e^{i2\pi kx}$ on $\mathbb{R}$, with its sum $f(x) = \sum_{k=-\infty}^{\infty}\frac{1}{1+k^2}e^{i2\pi kx}$.

1. Show that $\frac{d^2 f}{dx^2} - 4\pi^2 f = -4\pi^2 S$ in $\mathcal{D}'(\mathbb{R})$. (2.11.12)
2. (a) Show that the distribution solutions of this equation for f in $\mathcal{D}'(]0,1[)$ (resp. in $\mathcal{D}'(]-1,0[)$) are C^∞-functions.

   (b) Assuming the periodicity of f with a period of 1 (i.e. $f(0) = f(1)$),

   $$f(x) = C_1[e^{-2\pi x}] + C_2 e^{2\pi x} \quad \text{in } \mathcal{D}'(]0,1[) \text{ (resp. in } \mathcal{D}'(]-1,0[)) \tag{2.11.13}$$

   with $C_1 = d_1 e^{\pi}$, $C_2 = d_2 e^{-\pi}$ (resp. $C_1 = d_1 e^{-\pi}$, $C_2 = d_2 e^{\pi}$), d_i being constants, show that f is continuous on $[-1,1]$.
3. Show that for continuous periodic f (i.e. $f(0) = f(1) = f(-1)$), f'' in the sense of distribution on $]-1,1[$ is given by: $f'' = [f''(x)] + 4\pi d(e^{-\pi} - e^{\pi})\delta_0$ and

   $$S = \left(\frac{e^{\pi} - e^{-\pi}}{\pi}d\right)\sum_{k=-\infty}^{\infty}\delta_k \quad \text{with } C = \frac{e^{\pi}-e^{-\pi}}{\pi}d, \delta_0 = \delta. \tag{2.11.14}$$

4. Considering the *two* representations (representation by the series and by (2.11.13)) of $f(x)$ on $]0,1[$, and computing $\int_0^1 f(x)dx$, show that constant

   $$d = \frac{\pi}{e^{\pi} - e^{-\pi}}, \quad \text{i.e. } S = \sum_{k=-\infty}^{\infty}\delta_k = \sum_{k=-\infty}^{\infty} e^{i2\pi kx} \text{ in } \mathcal{D}'(\mathbb{R}). \tag{2.11.15}$$

Proof.

1. $\forall k \in \mathbb{Z}$ with $k \neq 0$, $\frac{1}{1+k^2} < \frac{1}{k^2} \Longrightarrow \sum_{k=-\infty}^{\infty} \frac{e^{i2\pi kx}}{1+k^2}$ converges *uniformly* and absolutely to a continuous function $f(x) = \sum_{k=-\infty}^{\infty} \frac{1}{1+k^2} e^{i2\pi kx}$ on $\mathbb{R} \Longrightarrow \sum_{k=-\infty}^{\infty} \frac{1}{1+k^2} e^{i2\pi kx}$ converges to $T_f = f$ in $\mathcal{D}'(\mathbb{R})$ with

$$\langle T_f, \phi \rangle = \langle f, \phi \rangle = \sum_{k=-\infty}^{\infty} \frac{1}{1+k^2} \int_{\mathbb{R}} e^{i2\pi kx} \phi(x) dx \quad \forall \phi \in \mathcal{D}(\mathbb{R})$$

by Theorem 2.11.2,

$$\Longrightarrow \quad \frac{d^m f}{dx^m} = \sum_{\substack{k=-\infty \\ k \neq 0}}^{\infty} \frac{(i2\pi k)^m}{1+k^2} e^{i2\pi kx} \quad \text{in } \mathcal{D}'(\mathbb{R}) \text{ by Theorem 2.11.1}$$

$$\Longrightarrow \quad \frac{d^2 f}{dx^2} = \sum_{k=-\infty}^{\infty} \frac{-4\pi^2(1+k^2) + 4\pi^2}{1+k^2} e^{i2\pi kx} = 4\pi^2 f - 4\pi^2 S$$

$$\Longrightarrow \quad \frac{d^2 f}{dx^2} - 4\pi^2 f = -4\pi^2 S \quad \text{in } \mathcal{D}'(\mathbb{R}).$$

2a. Since $S = \sum_{k=-\infty}^{\infty} e^{i2\pi kx} = C \sum_{k=-\infty}^{\infty} \delta_k$, we have $\langle S, \phi \rangle = 0$ for $\phi \in \mathcal{D}(]0,1[)$, i.e. $\forall \phi \in \mathcal{D}(\mathbb{R})$ with $\text{supp}(\phi) \subset]0,1[$ ($\langle C\delta_k, \phi \rangle = C\phi(k)$ with $\phi(k) = 0 \ \forall \phi \in \mathcal{D}(]0,1[)$ and $\forall k \in \mathbb{Z}$). Hence, $\langle \frac{d^2 f}{dx^2} - 4\pi^2 f, \phi \rangle = 0$ $\forall \phi \in \mathcal{D}(]0,1[)$

$$\Longrightarrow \quad \frac{d^2 f}{dx^2} - 4\pi^2 f = 0 \quad \text{in } \mathcal{D}'(]0,1[). \tag{2.11.16}$$

Similarly,

$$\frac{d^2 f}{dx^2} - 4\pi^2 f = 0 \quad \text{in } \mathcal{D}'(]-1,0[). \tag{2.11.17}$$

The distributional solutions of (2.11.16) (resp. (2.11.17)) are C^∞-functions and given by the usual solutions by Theorem 2.7.2: $f(x) = C_1 e^{-2\pi x} + C_2 e^{2\pi x}$ in $\mathcal{D}'(]0,1[)$ (resp. $\mathcal{D}'(]-1,0[)$), C_1 and C_2 being arbitrary constants. Setting $C_1 = d_1 e^{\pi}$ (resp. $d_1 e^{-\pi}$), $C_2 = d_2 e^{-\pi}$ (resp. $d_2 e^{\pi}$), we get $f(x) = d_1 e^{-2\pi(x-\frac{1}{2})} + d_2 e^{2\pi(x-\frac{1}{2})}$ on $]0,1[$ (resp. $f(x) = d_1 e^{-2\pi(x+\frac{1}{2})} + d_2 e^{2\pi(x+\frac{1}{2})}$ on $]-1,0[$).

2b. Then, using the periodicity of f with period of 1:

$$\begin{aligned}
\langle \tau_1 f, \phi \rangle &= \langle f, \tau_{-1}\phi \rangle = \langle f, \phi(x+1) \rangle \\
&= \lim_{N\to\infty} \left\langle \sum_{k=-N}^{N} \frac{1}{1+k^2} e^{i2\pi kx}, \phi(x+1) \right\rangle \\
&= \lim_{N\to\infty} \left(\sum_{k=-N}^{N} \frac{1}{1+k^2} \int_{\mathbb{R}} e^{i2\pi kx} \phi(x+1) dx \right) \\
&= \lim_{N\to\infty} \left(\sum_{k=-N}^{N} \frac{1}{1+k^2} \int_{\mathbb{R}} e^{i2\pi k\xi} \phi(\xi) d\xi \right) \\
&= \langle f, \phi \rangle \quad \forall \phi \text{ in } \mathcal{D}(\mathbb{R})
\end{aligned}$$

(setting $\xi = x + 1$, $e^{i2\pi k(\xi-1)} = e^{i2\pi k\xi}$), we have $f(0) = f(1)$ (resp. $f(0) = f(-1)$) $\Longrightarrow d_1 = d_2 = d \Longrightarrow f = d(e^{-2\pi(x-\frac{1}{2})} + e^{2\pi(x-\frac{1}{2})})$ in $]0,1[$ (resp. $f = d(e^{-2\pi(x+\frac{1}{2})} + e^{2\pi(x+\frac{1}{2})})$ in $]-1,0[$), such that $f(0^+) = f(0^-) = d(e^{\pi} + e^{-\pi}) = f(0) = f(-1) = f(1)$. Thus, f is continuous on $[-1,1] \subset \mathbb{R}$ and periodic with period 1.

3. Then,

$$f'(x) = d[-2\pi e^{-2\pi(x-\frac{1}{2})} + 2\pi e^{2\pi(x-\frac{1}{2})}] = g_1(x) \quad \text{for } 0 < x < 1;$$

$$f'(x) = d[-2\pi e^{-2\pi(x+\frac{1}{2})} + 2\pi e^{2\pi(x+\frac{1}{2})}] = g_2(x) \quad \text{for } -1 < x < 0.$$

$J_1(0) =$ jump of the first derivative $f'(x)$ at 0

$$\begin{aligned}
&= f'(0^+) - f'(0^-) = g_1(0^+) - g_2(0^-) \\
&= d[e^{\pi}(-2\pi - 2\pi) + e^{-\pi}(2\pi + 2\pi)] = 4\pi d(e^{-\pi} - e^{\pi}).
\end{aligned}$$

$[f''(x)] =$ the usual second-order ordinary derivative $= 4\pi^2 d[e^{-2\pi(x-\frac{1}{2})} + e^{2\pi(x-\frac{1}{2})}] = 4\pi^2 f$ on $]0,1[$ and $4\pi^2 d[e^{-2\pi(x+\frac{1}{2})} + e^{2\pi(x+\frac{1}{2})}] = 4\pi^2 f$ on $]-1,0[\Longrightarrow [f''(x)] - 4\pi^2 f = 0$ on $]-1,0[\cup]0,1[$.

The second-order distributional derivative f'' in $\mathcal{D}'(]-1,1[)$ is defined, $\forall \phi \in \mathcal{D}(]-1,1[)$, by

$$\begin{aligned}
\langle f'', \phi \rangle &= -\langle f', \phi' \rangle = -\int_1^{0^-} f'\phi' dx - \int_{0^+}^{1} f'\phi' dx \\
&= -f'(x)\phi(x)|_{-1}^{0^-} + \int_{-1}^{0^-} [f''(x)]\phi dx - f'(x)\phi(x)\Big|_{0^+}^{1} + \int_{0^+}^{1} [f''(x)]\phi dx
\end{aligned}$$

$$= \int_{-1}^{1} [f''(x)]\phi dx - f'(0^-)\phi(0) + f'(0^+)\phi(0)$$
$$= \langle [f''(x)], \phi\rangle + [f'(0^+) - f'(0^-)]\langle \delta_0, \phi\rangle$$
$$= \langle [f''(x)] + J_1(0)\delta_0, \phi\rangle$$

with $\delta = \delta_0 \Longrightarrow f'' = [f''(x)] + J_1(0)\delta_0 = [f''(x)] + 4\pi d(e^{\pi} + e^{-\pi})\delta_0$ in $\mathcal{D}'(]-1,1[)$ (see also Chapter 3, Theorem 3.1.1)

$$\begin{aligned}\Longrightarrow \quad f'' - 4\pi^2 f &= ([f''(x)] - 4\pi^2 f) + 4\pi d(e^{-\pi} - e^{\pi})\delta_0 \\ &= 0 + 4\pi d(e^{-\pi} - e^{\pi})\delta_0 = 4\pi d(e^{-\pi} - e^{\pi})\delta_0.\end{aligned}$$

But $f'' - 4\pi^2 f = -4\pi^2 S = -4\pi^2 C \sum_{k=-\infty}^{\infty} \delta_k = -4\pi^2 C\delta_0$ in $\mathcal{D}'(]-1,1[)$ (see also Section 5.3, Chapter 5) $\Longrightarrow -4\pi^2 C\delta_0 = 4\pi d(e^{-\pi} - e^{\pi})\delta_0 \Longrightarrow C = -\frac{e^{-\pi}-e^{\pi}}{\pi} d = \frac{e^{\pi}-e^{-\pi}}{\pi} d$.

Hence, $S = (\frac{e^{\pi}-e^{-\pi}}{\pi} d)\sum_{k=-\infty}^{\infty} \delta_k$.

4. Since the defining series for f is *uniformly* convergent on $\mathbb{R}$, term-by-term integration of the series is possible and we get

$$\begin{aligned}\int_0^1 f(x)dx &= \sum_{k=-\infty}^{\infty} \frac{1}{1+k^2}\int_0^1 e^{i2\pi kx} dx \\ &= 1\int_0^1 dx + \sum_{\substack{k=-\infty \\ k\neq 0}}^{\infty} \frac{1}{1+k^2} \frac{e^{i2\pi kx}}{i2\pi k}\bigg|_0^1 = 1 + 0 = 1\end{aligned}$$

$$\begin{aligned}\int_0^1 f(x)dx &= d\int_0^1 (e^{-2\pi(x-\frac{1}{2})} + e^{2\pi(x-\frac{1}{2})})dx \\ &= d\left[\frac{e^{-2\pi(x-\frac{1}{2})}}{-2\pi} + \frac{e^{2\pi(x-\frac{1}{2})}}{2\pi}\right]_0^1 = d\left(\frac{e^{\pi} - e^{-\pi}}{\pi}\right)\end{aligned}$$

$\Longrightarrow 1 = d(\frac{e^{\pi}-e^{-\pi}}{\pi}) \Longrightarrow d = \frac{\pi}{e^{\pi}-e^{-\pi}}$.

Finally, $C = \frac{e^{\pi}-e^{-\pi}}{\pi} d = (\frac{e^{\pi}-e^{-\pi}}{\pi}) \cdot \frac{\pi}{e^{\pi}-e^{-\pi}} = 1 \Longrightarrow S = \sum_{k=-\infty}^{\infty} \delta_k = \sum_{k=-\infty}^{\infty} e^{i2\pi kx}$ in $\mathcal{D}'(\mathbb{R})$.

□

2.12 Convergence of sequences of $C^k(\overline{\Omega})$ (resp. $C^{k,\lambda}(\overline{\Omega})$) in $\mathcal{D}'(\Omega)$

Proposition 2.12.1. *Let (u_n) be a* convergent *sequence in Banach space $C^k(\overline{\Omega})$ (resp. $C^{k,\lambda}(\overline{\Omega})$, $0 < \lambda < 1$) with $k \in \mathbb{N}_0$ (see Appendix* A, *Sections* A.4.1 *and* A.5.2*) such that $u_n \longrightarrow u \in C^k(\overline{\Omega})$ (resp. $C^{k,\lambda}(\overline{\Omega})$). Then, $\forall$ multi-index $\boldsymbol{\alpha}$ with $|\boldsymbol{\alpha}| \in \mathbb{N}_0$,*

$$\partial^{\boldsymbol{\alpha}} u_n \to \partial^{\boldsymbol{\alpha}} u \text{ in } \mathcal{D}'(\Omega) \quad \text{as } n \to \infty. \tag{2.12.1}$$

In particular, for $\boldsymbol{\alpha} = \mathbf{0}$,

$$\partial^{\boldsymbol{\alpha}} u_n = u_n,\ \partial^{\boldsymbol{\alpha}} u = u, \quad \text{i.e. } u_n \to u \text{ in } \mathcal{D}'(\Omega). \tag{2.12.2}$$

Proof. $u_n \to u$ in $C^k(\overline{\Omega}) \iff \|u - u_n\|_{C^k(\overline{\Omega})} = \max_{0\le|\boldsymbol{\beta}|\le k} \sup_{\mathbf{x}\in\Omega} |\partial^{\boldsymbol{\beta}} u(\mathbf{x}) - \partial^{\boldsymbol{\beta}} u_n(\mathbf{x})|$. Hence, $u_n \to u$ in $C^k(\overline{\Omega}) \Longrightarrow \partial^{\boldsymbol{\beta}} u_n \longrightarrow \partial^{\boldsymbol{\beta}} u$ *uniformly* in every compact subset of Ω as $n \to \infty$, $\forall |\boldsymbol{\beta}| \le k$ and $C^k(\overline{\Omega}) \subset L^1_{\text{loc}}(\Omega)$ $\forall k \in \mathbb{N}_0 \Longrightarrow u_n \to u$ in $\mathcal{D}'(\Omega)$ as $n \to \infty$ by Proposition 2.9.2 $\Longrightarrow \partial^{\boldsymbol{\alpha}} u_n \to \partial^{\boldsymbol{\alpha}} u$ in $\mathcal{D}'(\Omega)$ as $n \to \infty$ $\forall |\boldsymbol{\alpha}| \in \mathbb{N}$ by Theorem 2.9.1.

Similarly, for the space $C^{k,\lambda}(\overline{\Omega})$ of Hölder continuous functions with $k \in \mathbb{N}_0$, index $\lambda \in]0,1[$ (see Appendix A, Definition A.5.3.1), the result can be proved. □

Imbedding results $C^k(\overline{\Omega}) \hookrightarrow \mathcal{D}'(\Omega)$ (resp. $C^{k,\lambda}(\overline{\Omega})) \hookrightarrow \mathcal{D}'(\Omega)$, $0 < \lambda < 1$) $\forall k \in \mathbb{N}_0$, the imbedding operator $\hookrightarrow$ being a continuous one, i.e. $u \in C^k(\overline{\Omega})$ (resp. $u \in C^{k,\lambda}(\overline{\Omega})) \Longrightarrow T_u = u \in \mathcal{D}'(\Omega)$ and

$$u_n \to u \text{ in } C^k(\overline{\Omega}) \text{ (resp. } C^{k,\lambda}(\overline{\Omega})) \quad \Longrightarrow \quad u_n \to u \text{ in } \mathcal{D}'(\Omega). \tag{2.12.3}$$

2.13 Convergence of sequences of $L^p(\Omega)$, $1 \le p \le \infty$, in $\mathcal{D}'(\Omega)$

Theorem 2.13.1. *Let (u_n) be a sequence in $L^p(\Omega)$, $1 \le p \le \infty$, which converges to $u \in L^p(\Omega)$* strongly *or* weakly *or in* weak-$*$ sense *as $n \to \infty$ (see Appendix* B, *Section* B.4*). Then, $\forall$ multi-index $\boldsymbol{\alpha}$ with $|\boldsymbol{\alpha}| \in \mathbb{N}_0$, $\partial^{\boldsymbol{\alpha}} u_n \to \partial^{\boldsymbol{\alpha}} u$ in $\mathcal{D}'(\Omega)$ as $n \to \infty$, i.e.*

$$\langle \partial^{\boldsymbol{\alpha}} u_n, \phi \rangle \to \langle \partial^{\boldsymbol{\alpha}} u, \phi \rangle \text{ in } \mathbb{R} \text{ (resp. } \mathbb{C}) \quad \text{as } n \to \infty\ \forall |\boldsymbol{\alpha}| \in \mathbb{N}_0. \tag{2.13.1}$$

Proof. Strong convergence in $L^p(\Omega)$, $1 \le p \le \infty$: $u_n \xrightarrow{\text{strongly}} u$ in $L^p(\Omega)$ as $n \to \infty \iff \|u - u_n\|_{L^p(\Omega)} \to 0$ as $n \to \infty$. Then, $\forall n \in \mathbb{N}$,

$$\langle T_{u_n}, \phi\rangle = \langle u_n, \phi\rangle = \int_\Omega u_n \phi d\mathbf{x}, \langle T_u, \phi\rangle = \langle u, \phi\rangle = \int_\Omega u\phi d\mathbf{x} \quad \forall \phi \in \mathcal{D}(\Omega) \tag{2.13.2}$$

$$\begin{aligned} \implies \quad |\langle u, \phi\rangle - \langle u_n, \phi\rangle| &= \left|\int_\Omega (u - u_n)\phi d\mathbf{x}\right| \\ &\le \|u - u_n\|_{L^p(\Omega)} \|\phi\|_{L^q(\Omega)} \to 0 \quad \text{as } n \to \infty \end{aligned}$$

(by the Hölder inequality with $\frac{1}{p} + \frac{1}{q} = 1$, $1 \le p, q \le \infty$)
$\implies \langle u_n, \phi\rangle \to \langle u, \phi\rangle$ as $n \to \infty$ $\forall \phi \in \mathcal{D}(\Omega)$
$\implies u_n \to u$ in $\mathcal{D}'(\Omega)$ as $n \to \infty \implies \partial^{\boldsymbol{\alpha}} u_n \to \partial^{\boldsymbol{\alpha}} u$ in $\mathcal{D}'(\Omega)$ by Theorem 2.9.1 $\forall |\boldsymbol{\alpha}| \in \mathbb{N}$.

Weak convergence in $L^p(\Omega)$, $1 \le p < \infty$, $\frac{1}{p} + \frac{1}{q} = 1$, $1 < q \le \infty$: $u_n \xrightarrow{\text{weakly}} u$ in $L^p(\Omega)$, $1 \le p < \infty$

$$\iff \quad \int_\Omega u_n v d\mathbf{x} \to \int_\Omega u v d\mathbf{x} \quad \forall v \in L^q(\Omega) \tag{2.13.3}$$

$$\iff \quad \left|\int_\Omega (u - u_n) v d\mathbf{x}\right| \to 0 \quad \text{as } n \to \infty \ \forall v \in L^q(\Omega),\ 1 < q \le \infty \tag{2.13.4}$$

$\implies |\langle u, \phi\rangle - \langle u_n, \phi\rangle| = |\int_\Omega (u - u_n)\phi d\mathbf{x}| \to 0$ as $n \to \infty$ $\forall \phi \in \mathcal{D}(\Omega) \subset L^q(\Omega)$
$\implies u_n \to u$ in $\mathcal{D}'(\Omega)$ as $n \to \infty \implies \partial^{\boldsymbol{\alpha}} u_n \to \partial^{\boldsymbol{\alpha}} u$ in $\mathcal{D}'(\Omega)$ by Theorem 2.9.1 $\forall |\boldsymbol{\alpha}| \in \mathbb{N}$.

Weak-$*$ convergence in $L^\infty(\Omega) \equiv (L^1(\Omega))'$: For $1 < p < \infty$,

$$\textit{weak-}* \textit{ convergence and weak convergence coincide;} \tag{2.13.5}$$

$u_n \rightharpoonup^* u$ in the weak-$*$ sense in $L^\infty(\Omega)$

$$\begin{aligned} &\iff \quad \int_\Omega u_n v d\mathbf{x} \to \int_\Omega u v d\mathbf{x} \quad \forall v \in L^1(\Omega) \\ &\iff \quad \left|\int_\Omega (u - u_n) v d\mathbf{x}\right| \to 0 \quad \text{as } n \to \infty \ \forall v \in L^1(\Omega). \end{aligned} \tag{2.13.6}$$

Hence, $|\langle u, \phi\rangle - \langle u_n, \phi\rangle| = |\int_\Omega (u - u_n)\phi d\mathbf{x}| \to 0$ as $n \to \infty$ $\forall \phi \in \mathcal{D}(\Omega)$, since $\mathcal{D}(\Omega) \subset L^1(\Omega) \implies u_n \to u$ in $\mathcal{D}'(\Omega) \implies \partial^{\boldsymbol{\alpha}} u_n \to \partial^{\boldsymbol{\alpha}} u$ in $\mathcal{D}'(\Omega)$ $\forall |\boldsymbol{\alpha}| \in \mathbb{N}$ by Theorem 2.9.1. □

Remark 2.13.1. $\forall n \in \mathbb{N}$, $u_n \in L^p(\Omega)$, $u \in L^p(\Omega)$, $1 \le p \le \infty \implies u_n$ and u define *regular* distributions in $\mathcal{D}'(\Omega)$. But $\partial^{\boldsymbol{\alpha}} u_n, \partial^{\boldsymbol{\alpha}} u \in \mathcal{D}'(\Omega)$ are distributions and do *not* belong to $L^1_{\text{loc}}(\Omega)$ in general. Hence, $\langle \partial^{\boldsymbol{\alpha}} u_n, \phi\rangle$, $\langle \partial^{\boldsymbol{\alpha}} u, \phi\rangle$ cannot be defined by integrals in general.

Example 2.13.1. Let

$$u_n(x) = \begin{cases} 0 & \text{for } x < 0 \\ nx & \text{for } 0 < x < \frac{1}{n} \\ 1 & \text{for } x > \frac{1}{n} \end{cases}$$

be a sequence in $L^1(]-1,1[)$.

Then $u_n \to H$ in $L^1(]-1,1[)$ as $n \to \infty$, $H(x) = 1$ for $0 < x < 1$, $H(x) = 0$ for $-1 < x < 0$ being the step function (Heaviside function). By Theorem 2.9.1, $\frac{d}{dx}u_n \to \frac{dH}{dx} = \delta \in \mathcal{D}'(]-1,1[)$ as $n \to \infty$, since

$$\begin{aligned}\left\langle \frac{dH}{dx}, \phi \right\rangle &= -\int_{-1}^{1} H(x)\phi' dx = -\int_{0}^{1} \phi'(x)dx \\ &= -\phi(x)|_0^1 = \phi(0) = \langle \delta, \phi \rangle \quad \forall \phi \in \mathcal{D}(]-1,1[)\end{aligned}$$

$\Longrightarrow \frac{dH}{dx} = \delta$ in $\mathcal{D}'(]-1,1[)$ is a *singular* distribution and does not belong to $L^1(]-1,1[)$. Moreover, $\frac{d^k u_n}{dx^k} \notin L^1_{\text{loc}}(]-1,1[)$ $\forall k \geq 2$, $\forall n \in \mathbb{N}$, i.e. $\frac{d^k u_n}{dx^k} \in \mathcal{D}'(]-1,1[)$ is also a *singular* distribution $\forall k \geq 2$, $\forall n \in \mathbb{N}$.

Imbedding results $L^p(\Omega) \hookrightarrow \mathcal{D}'(\Omega)$, $1 \leq p \leq \infty$, the imbedding $\hookrightarrow$ being a *continuous* one, i.e. $u \in L^p(\Omega) \Longrightarrow T_u = u \in \mathcal{D}'(\Omega)$ and $u_n \xrightarrow{\text{strongly}} u$ or $u_n \xrightarrow{\text{weakly}} u$ or $u_n \rightharpoonup^* u$ (in the weak-$*$ sense) in $L^p(\Omega)$

$$\Longrightarrow \quad u_n \to u \text{ in } \mathcal{D}'(\Omega) \quad \text{as } n \to \infty. \tag{2.13.7}$$

2.14 Transpose (or formal adjoint) of a linear partial differential operator

Let A, a linear partial differential operator of order m with coefficients $a_{\boldsymbol{\alpha}} = a_{(\alpha_1,\alpha_2,\ldots,\alpha_n)} \in C^\infty(\mathbb{R}^n)$ $\forall$ multi-index $\boldsymbol{\alpha}$, be defined by:

$$A = \sum_{0\leq|\boldsymbol{\alpha}|\leq m} a_{\boldsymbol{\alpha}} \partial^{\boldsymbol{\alpha}} = \sum_{0\leq|\boldsymbol{\alpha}|\leq m} a_{(\alpha_1,\alpha_2,\ldots,\alpha_n)} \frac{\partial^{|\boldsymbol{\alpha}|}}{\partial x_1^{\alpha_1} \partial x_2^{\alpha_2} \ldots \partial x_n^{\alpha_n}}. \tag{2.14.1}$$

Then, $\forall T \in \mathcal{D}'(\Omega)$, AT is defined with $AT \in \mathcal{D}'(\Omega)$ such that

$$\begin{aligned}\langle AT, \phi\rangle &= \Big\langle \sum_{0\le|\boldsymbol{\alpha}|\le m} a_{\boldsymbol{\alpha}}\partial^{\boldsymbol{\alpha}} T, \phi\Big\rangle = \sum_{0\le|\boldsymbol{\alpha}|\le m} \Big\langle a_{\boldsymbol{\alpha}}\partial^{\boldsymbol{\alpha}} T, \phi\Big\rangle \\ &= \sum_{0\le|\boldsymbol{\alpha}|\le m} \Big\langle \partial^{\boldsymbol{\alpha}} T, a_{\boldsymbol{\alpha}}\phi\Big\rangle = \sum_{0\le|\boldsymbol{\alpha}|\le m} (-1)^{|\boldsymbol{\alpha}|}\Big\langle T, \partial^{\boldsymbol{\alpha}}(a_{\boldsymbol{\alpha}}\phi)\Big\rangle \\ &= \Big\langle T, \sum_{0\le|\boldsymbol{\alpha}|\le m} (-1)^{|\boldsymbol{\alpha}|}\partial^{\boldsymbol{\alpha}}(a_{\boldsymbol{\alpha}}\phi)\Big\rangle \quad \forall\phi \in \mathcal{D}(\Omega).\end{aligned}$$

Definition 2.14.1. The operator A' defined by, $\forall\phi \in \mathcal{D}'(\Omega)$,

$$A'\phi = \sum_{0\le|\boldsymbol{\alpha}|\le m} (-1)^{|\boldsymbol{\alpha}|}\partial^{\boldsymbol{\alpha}}(a_{\boldsymbol{\alpha}}\phi) = \sum_{0\le|\boldsymbol{\alpha}|\le m} (-1)^{|\boldsymbol{\alpha}|}\frac{\partial^{|\boldsymbol{\alpha}|}(a_{\boldsymbol{\alpha}}\phi)}{\partial x_1^{\alpha_1}\partial x_2^{\alpha_2}\dots\partial x_n^{\alpha_n}} \tag{2.14.2}$$

such that $\langle AT, \phi\rangle = \langle T, A'\phi\rangle$ $\forall T \in \mathcal{D}'(\Omega)$, $\forall\phi \in \mathcal{D}(\Omega)$, is called the *transpose* or *formal adjoint* of the operator A defined by (2.14.1).

Then $\langle A'T, \phi\rangle = \langle T, A\phi\rangle$ $\forall T \in \mathcal{D}'(\Omega)$, $\forall\phi \in \mathcal{D}(\Omega)$. In fact, $\forall T \in \mathcal{D}'(\Omega)$,

$$\begin{aligned}\langle A'T, \phi\rangle &= \Big\langle \sum_{0\le|\boldsymbol{\alpha}|\le m} (-1)^{|\boldsymbol{\alpha}|}\partial^{\boldsymbol{\alpha}}(a_{\boldsymbol{\alpha}}T), \phi\Big\rangle = \sum_{0\le|\boldsymbol{\alpha}|\le m} \langle(-1)^{|\boldsymbol{\alpha}|}\partial^{\boldsymbol{\alpha}}(a_{\boldsymbol{\alpha}}T), \phi\rangle \\ &= \sum_{0\le|\boldsymbol{\alpha}|\le m} \langle a_{\boldsymbol{\alpha}}T, \partial^{\boldsymbol{\alpha}}\phi\rangle = \sum_{0\le|\boldsymbol{\alpha}|\le m} \langle T, a_{\boldsymbol{\alpha}}\partial^{\boldsymbol{\alpha}}\phi\rangle \\ &= \langle T, \sum_{0\le|\boldsymbol{\alpha}|\le m} a_{\boldsymbol{\alpha}}\partial^{\boldsymbol{\alpha}}\phi\rangle = \langle T, A\phi\rangle \ \forall T \in \mathcal{D}'(\Omega), \quad \forall\phi \in \mathcal{D}(\Omega).\end{aligned}$$

$$A' \text{ is the } \textit{transpose} \text{ of } A \quad \Longrightarrow \quad A'' = (A')' = A. \tag{2.14.3}$$

Indeed, $\forall T \in \mathcal{D}'(\Omega)$, $\langle A''T, \phi\rangle = \langle (A')'T, \phi\rangle = \langle T, A'\phi\rangle = \langle AT, \phi\rangle$ $\forall\phi \in \mathcal{D}(\Omega)$
$\Longrightarrow \forall T \in \mathcal{D}'(\Omega)$, $A''T = AT$ in $\mathcal{D}'(\Omega) \Longrightarrow A'' = A$.

Example 2.14.1. $A = \Delta = \frac{\partial^2}{\partial x_1^2} + \frac{\partial^2}{\partial x_2^2} + \dots + \frac{\partial^2}{\partial x_n^2}$ is the *Laplace operator* in n variables

$$\Longrightarrow \quad A' = (-1)^2\frac{\partial^2}{\partial x_1^2} + (-1)^2\frac{\partial^2}{\partial x_2^2} + \dots + (-1)^2\frac{\partial^2}{\partial x_n^2} = \Delta = A.$$

Then, for the Laplace operator Δ, we have $\langle \Delta T, \phi\rangle = \langle T, \Delta\phi\rangle$ $\forall T \in \mathcal{D}'(\Omega)$, $\forall\phi \in \mathcal{D}(\Omega)$.

Example 2.14.2. Let A be the second-order linear partial differential operator with variable coefficients defined, $\forall T \in \mathcal{D}'(\Omega)$, by:

$$AT = -\sum_{i=1}^{n}\sum_{j=1}^{n} \frac{\partial}{\partial x_j}(a_{ij}\frac{\partial T}{\partial x_i}) + \sum_{i=1}^{n} a_i \frac{\partial T}{\partial x_i} + a_0 T,$$

where $a_{ij}, a_i, a_0 \in C^\infty(\mathbb{R}^n)$.

Then the transpose or formal adjoint A' of A is defined, $\forall \phi \in \mathcal{D}(\Omega)$, by:

$$A'\phi = -\sum_{i=1}^{n}\sum_{j=1}^{n} \frac{\partial}{\partial x_j}(a_{ji}\frac{\partial \phi}{\partial x_i}) - \sum_{i=1}^{n} \frac{\partial}{\partial x_i}(a_i\phi) + a_0\phi.$$

Proof.

$$\begin{aligned}
\langle AT, \phi\rangle &= \left\langle -\sum_{i=1}^{n}\sum_{j=1}^{n} \frac{\partial}{\partial x_j}\left(a_{ij}\frac{\partial T}{\partial x_i}\right) + \sum_{i=1}^{n} a_i\frac{\partial T}{\partial x_i} + a_0 T, \phi\right\rangle \\
&= -\sum_{i=1}^{n}\sum_{j=1}^{n}\left\langle \frac{\partial}{\partial x_j}\left(a_{ij}\frac{\partial T}{\partial x_i}\right), \phi\right\rangle + \sum_{i=1}^{n}\left\langle a_i\frac{\partial T}{\partial x_i}, \phi\right\rangle + \langle a_0 T, \phi\rangle \\
&= -\sum_{i=1}^{n}\sum_{j=1}^{n} -\left\langle a_{ij}\frac{\partial T}{\partial x_i}, \frac{\partial \phi}{\partial x_j}\right\rangle - \sum_{i=1}^{n}\left\langle T, \frac{\partial}{\partial x_i}(a_i\phi)\right\rangle + \langle T, a_0\phi\rangle \\
&= -\sum_{i=1}^{n}\sum_{j=1}^{n}\left\langle T, \frac{\partial}{\partial x_i}\left(a_{ij}\frac{\partial \phi}{\partial x_j}\right)\right\rangle - \left\langle T, \sum_{i=1}^{n}\frac{\partial}{\partial x_i}(a_i\phi)\right\rangle + \langle T, a_0\phi\rangle \\
&= \left\langle T, -\sum_{i=1}^{n}\sum_{j=1}^{n}\frac{\partial}{\partial x_i}\left(a_{ij}\frac{\partial \phi}{\partial x_j}\right) - \sum_{i=1}^{n}\frac{\partial}{\partial x_i}(a_i\phi) + a_0\phi\right\rangle \\
&= \left\langle T, -\sum_{i=1}^{n}\sum_{j=1}^{n}\frac{\partial}{\partial x_j}\left(a_{ji}\frac{\partial \phi}{\partial x_i}\right) - \sum_{i=1}^{n}\frac{\partial}{\partial x_i}(a_i\phi) + a_0\phi\right\rangle \\
&= \langle T, A'\phi\rangle \quad \forall \phi \in \mathcal{D}(\Omega).
\end{aligned}$$

□

2.15 Applications: Sobolev spaces $H^m(\Omega)$, $W^{m,p}(\Omega)$

2.15.1 Sobolev Spaces

We will now show the most important application of distributions in the definition of Sobolev spaces, which are the basic tools in the study of boundary value problems of partial differential equations on $\Omega \subset \mathbb{R}^n$. Sobolev spaces may be Hilbert spaces, usually denoted by $H^m(\Omega)$ for $m \in \mathbb{N}$ or $H^s(\Omega)$ for $s \in \mathbb{R}$, or Banach spaces denoted by $W^{m,p}(\Omega)$, $1 \leq p \leq \infty$, $\forall m \in \mathbb{N}$, or $W^{s,p}(\Omega)$ for $s \in \mathbb{R}$, $1 \leq p \leq \infty$, such that for $p = 2$, $H^m(\Omega) \equiv W^{m,2}(\Omega)$, $H^s(\Omega) \equiv W^{s,2}(\Omega)$ $\forall m \in \mathbb{N}$, $\forall s \in \mathbb{R}$ respectively. Among the family of all these Sobolev spaces, the spaces $H^m(\Omega)$ of integral order $m \in \mathbb{N}$ and their subspaces are the most important ones owing to their nice Hilbert structure. Hence, we will begin by studying their elementary defining properties.

2.15.2 Space $H^m(\Omega)$

Definition 2.15.1. Let Ω be an open subset of $\mathbb{R}^n$. Then $H^m(\Omega)$ is the set of all (*equivalence classes* $[u]$ of) real-valued functions $u \in L^2(\Omega)$ whose derivatives $\partial^{\boldsymbol{\alpha}} u$ in the *distributional sense* (2.3.1) also belong to $L^2(\Omega)\forall$ multi-index $\boldsymbol{\alpha}$ with $|\boldsymbol{\alpha}| = \alpha_1 + \alpha_2 + \cdots + \alpha_n \leq m$, i.e.

$$H^m(\Omega) = \{u : u \in L^2(\Omega), \partial^{\boldsymbol{\alpha}} u \in L^2(\Omega) \quad \forall |\boldsymbol{\alpha}| \leq m\}, \tag{2.15.1}$$

where the distributional derivatives $\partial^{\boldsymbol{\alpha}} u$ are defined by (2.3.1):

$$\int_\Omega \partial^{\boldsymbol{\alpha}} u(\mathbf{x})\phi(\mathbf{x})d\mathbf{x} = (-1)^{|\boldsymbol{\alpha}|} \int_\Omega u(\mathbf{x})\partial^{\boldsymbol{\alpha}}\phi(\mathbf{x})d\mathbf{x} \quad \forall \phi \in \mathcal{D}(\Omega). \tag{2.15.2}$$

Then the following properties hold for functions of $H^m(\Omega)$:

$$u = v \text{ in } H^m(\Omega) \iff u(\mathbf{x}) = v(\mathbf{x}),\ \partial^{\boldsymbol{\alpha}} u(\mathbf{x}) = \partial^{\boldsymbol{\alpha}} v(\mathbf{x}) \quad \forall |\boldsymbol{\alpha}| \leq m \text{ a.e. on } \Omega; \tag{2.15.3}$$

$$u = 0 \text{ in } H^m(\Omega) \iff u(\mathbf{x}) = 0,\ \partial^{\boldsymbol{\alpha}} u(\mathbf{x}) = 0 \quad \forall |\boldsymbol{\alpha}| \leq m \text{ a.e. on } \Omega; \tag{2.15.4}$$

$$u, v = H^m(\Omega) \iff (u+v)(\mathbf{x}) = u(\mathbf{x}) + v(\mathbf{x}) \quad \text{a.e. on } \Omega; \tag{2.15.5}$$

$$(\lambda u)(\mathbf{x}) = \lambda u(\mathbf{x}) \quad \text{a.e. on } \Omega\ \forall \lambda \in \mathbb{R}. \tag{2.15.6}$$

Consequently, $H^m(\Omega)$ is a *linear space*, whose elements $u \in H^m(\Omega)$ are distributions on Ω, (2.15.7)
i.e.

$$\langle u, \phi \rangle = \int_\Omega u\phi d\mathbf{x}, \quad \langle \partial^{\boldsymbol{\alpha}} u, \phi \rangle = \int_\Omega (\partial^{\boldsymbol{\alpha}} u)\phi d\mathbf{x} \quad \forall \phi \in \mathcal{D}(\Omega),\ \forall |\boldsymbol{\alpha}| \leq m. \tag{2.15.8}$$

Proposition 2.15.1. *$H^m(\Omega)$ equipped with the inner product $\langle\cdot,\cdot\rangle_{H^m(\Omega)} = \langle\cdot,\cdot\rangle_{m,\Omega}$ defined, $\forall u, v \in H^m(\Omega)$, by*

$$\begin{aligned}\langle u, v\rangle_{m,\Omega} &= \langle u, v\rangle_{L^2(\Omega)} + \sum_{1\le|\boldsymbol{\alpha}|\le m} < \partial^{\boldsymbol{\alpha}} u, \partial^{\boldsymbol{\alpha}} v >_{L^2(\Omega)} \\ &= \int_\Omega u(\mathbf{x})v(\mathbf{x})d\mathbf{x} + \sum_{1\le|\boldsymbol{\alpha}|\le m} \int_\Omega \partial^{\boldsymbol{\alpha}} u(\mathbf{x})\partial^{\boldsymbol{\alpha}} v(\mathbf{x})d\mathbf{x} \\ &= \sum_{0\le|\boldsymbol{\alpha}|\le m} \int_\Omega \partial^{\boldsymbol{\alpha}} u(\mathbf{x})\partial^{\boldsymbol{\alpha}} v(\mathbf{x})d\mathbf{x}, \end{aligned} \tag{2.15.9}$$

where $\partial^{\boldsymbol{\alpha}}(\cdot) = (\cdot)$ for $|\boldsymbol{\alpha}| = 0$, is an inner product space *(see Appendix* A, *Section* A.12, *and also Table* B.3 *in Appendix* B*).*

Then the corresponding norm $\|\cdot\|_{m,\Omega}$ *and* semi-norm $|\cdot|_{m,\Omega}$ *are given, $\forall u \in H^m(\Omega)$, by:*

$$\begin{aligned}\|u\|_{m,\Omega} = \langle u, u\rangle_{m,\Omega}^{\frac{1}{2}} &= \Big(\|u\|^2_{L^2(\Omega)} + \sum_{1\le|\boldsymbol{\alpha}|\le m} \|\partial^{\boldsymbol{\alpha}} u\|^2_{L^2(\Omega)}\Big)^{\frac{1}{2}} \\ &= \Big(\int_\Omega |u(\mathbf{x})|^2 d\mathbf{x} + \sum_{1\le|\boldsymbol{\alpha}|\le m} \int_\Omega |\partial^{\boldsymbol{\alpha}} u(\mathbf{x})|^2 d\mathbf{x}\Big)^{\frac{1}{2}} \\ &= \Big(\sum_{0\le|\boldsymbol{\alpha}|\le m} \int_\Omega |\partial^{\boldsymbol{\alpha}} u(\mathbf{x})|^2 d\mathbf{x}\Big)^{\frac{1}{2}}; \end{aligned} \tag{2.15.10}$$

$$|u|_{m,\Omega} = \Big(\sum_{|\boldsymbol{\alpha}|=m} \|\partial^{\boldsymbol{\alpha}} u\|^2_{L^2(\Omega)}\Big)^{\frac{1}{2}} = \Big(\sum_{|\boldsymbol{\alpha}|=m} \int_\Omega |\partial^{\boldsymbol{\alpha}} u(\mathbf{x})|^2 d\mathbf{x}\Big)^{\frac{1}{2}}. \tag{2.15.11}$$

Theorem 2.15.1. *$\forall m \in \mathbb{N}$, $H^m(\Omega)$ equipped with the inner product $\langle\cdot,\cdot\rangle_{m,\Omega}$ defined in* (2.15.9) *is a* Hilbert space.

Proof. It is sufficient to show that every Cauchy sequence in $H^m(\Omega)$ converges to an element in $H^m(\Omega)$. Let (u_k) be a Cauchy sequence in $H^m(\Omega)$, i.e.

$$\begin{aligned}&\|u_k - u_l\|^2_{m,\Omega} = \|u_k - u_l\|^2_{L^2(\Omega)} + \sum_{1\le|\boldsymbol{\alpha}|\le m} \|\partial^{\boldsymbol{\alpha}} u_k - \partial^{\boldsymbol{\alpha}} u_l\|^2_{L^2(\Omega)} \to 0 \\ &\quad \text{as } k, l \to \infty \\ &\iff \|u_k - u_l\|_{L^2(\Omega)} \to 0 \text{ and } \|\partial^{\boldsymbol{\alpha}} u_k - \partial^{\boldsymbol{\alpha}} u_l\|_{L^2(\Omega)} \to 0 \\ &\quad \text{for } 1 \le |\boldsymbol{\alpha}| \le m \text{ as } k, l \to \infty\end{aligned}$$

$\Longrightarrow (u_k)$ and $(\partial^{\boldsymbol{\alpha}} u_k)$ are Cauchy sequences in $L^2(\Omega)$ $\forall \boldsymbol{\alpha}$ with $1 \le |\boldsymbol{\alpha}| \le m$. But $L^2(\Omega)$ is a Hilbert space, i.e. a complete space (see Appendix B, Theorem B.4.1.2). Hence, $\exists u \in L^2(\Omega)$ and $w_{\boldsymbol{\alpha}} \in L^2(\Omega)$, $\forall \boldsymbol{\alpha}$ with $1 \le |\boldsymbol{\alpha}| \le m$, such that

$$u_k \to u \text{ in } L^2(\Omega) \text{ and } \partial^{\boldsymbol{\alpha}} u_k \to w_{\boldsymbol{\alpha}} \text{ in } L^2(\Omega) \text{ as } k \to \infty. \tag{2.15.12}$$

Since $L^2(\Omega) \hookrightarrow \mathcal{D}'(\Omega)$, the imbedding being a continuous one (see (2.13.7)), $u_k \to u$ in $L^2(\Omega) \Longrightarrow u_k \to u$ in $\mathcal{D}'(\Omega)$ by Theorem 2.13.1, i.e. $\langle u_k, \phi\rangle \to \langle u, \phi\rangle$ $\forall \phi \in \mathcal{D}(\Omega)$; $\partial^{\boldsymbol{\alpha}} u_k \to w_{\boldsymbol{\alpha}}$ in $L^2(\Omega) \Longrightarrow \partial^{\boldsymbol{\alpha}} u_k \to w_{\boldsymbol{\alpha}}$ in $\mathcal{D}'(\Omega)$ by Theorem 2.13.1, i.e.

$$\langle \partial^{\boldsymbol{\alpha}} u_k, \phi\rangle \to \langle w_{\boldsymbol{\alpha}}, \phi\rangle \quad \forall \phi \in \mathcal{D}(\Omega),\ \forall \boldsymbol{\alpha} \text{ with } 1 \le |\boldsymbol{\alpha}| \le m, \text{ as } k \to \infty. \tag{2.15.13}$$

But $\partial^{\boldsymbol{\alpha}} : \mathcal{D}'(\Omega) \longrightarrow \mathcal{D}'(\Omega)$ is continuous by Theorem 2.9.1. Hence, $u_k \to u$ in $\mathcal{D}'(\Omega) \Longrightarrow \partial^{\boldsymbol{\alpha}} u_k \to \partial^{\boldsymbol{\alpha}} u$ in $\mathcal{D}'(\Omega)$ $\forall \boldsymbol{\alpha}$ with $1 \le |\boldsymbol{\alpha}| \le m$, i.e.

$$\langle \partial^{\boldsymbol{\alpha}} u_k, \phi\rangle \to \langle \partial^{\boldsymbol{\alpha}} u, \phi\rangle \quad \forall \phi \in \mathcal{D}(\Omega). \tag{2.15.14}$$

Since the limit is unique, from (2.15.13) and (2.15.14), we have $\partial^{\boldsymbol{\alpha}} u = w_{\boldsymbol{\alpha}} \in \mathcal{D}'(\Omega)$ $\forall \boldsymbol{\alpha}$ with $1 \le |\boldsymbol{\alpha}| \le m$. Therefore, $u \in L^2(\Omega)$, $\partial^{\boldsymbol{\alpha}} u = w_{\boldsymbol{\alpha}} \in L^2(\Omega)$ (by (2.15.12)) $\forall \boldsymbol{\alpha}$ with $1 \le |\boldsymbol{\alpha}| \le m$. Hence, $u \in H^m(\Omega)$ and

$$\|u - u_k\|_{m,\Omega}^2 = \|u - u_k\|_{L^2(\Omega)}^2 + \sum_{1 \le |\boldsymbol{\alpha}| \le m} \|\partial^{\boldsymbol{\alpha}} u - \partial^{\boldsymbol{\alpha}} u_k\|_{L^2(\Omega)}^2 \to 0 \quad \text{as } k \to \infty.$$

Thus, Cauchy sequence (u_k) in $H^m(\Omega)$ converges to $u \in H^m(\Omega)$. Hence, $H^m(\Omega)$ is a complete space, i.e. a Hilbert space. □

Examples.

- $L^2(\Omega)$: We let $m = 0$ such that $H^0(\Omega) = L^2(\Omega)$ with

$$\langle u, v\rangle_{0,\Omega} = \langle u, v\rangle_{L^2(\Omega)} = \int_\Omega u(\mathbf{x})v(\mathbf{x})d\mathbf{x},$$

$$\|u\|_{0,\Omega} = \langle u, u\rangle_{0,\Omega}^{\frac{1}{2}} = \left(\int_\Omega |u(\mathbf{x})|^2 d\mathbf{x}\right)^{\frac{1}{2}}. \tag{2.15.15}$$

• $H^1(\Omega)$: For $m = 1, \Omega \subset \mathbb{R}^n$, $H^1(\Omega)$ is equipped with inner product $\langle \cdot, \cdot \rangle_{1,\Omega}$, norm $\|\cdot\|_{1,\Omega}$ and semi-norm $|\cdot|_{1,\Omega}$ defined by:

$$
\begin{aligned}
\langle u, v\rangle_{1,\Omega} &= \langle u, v\rangle_{0,\Omega} + \sum_{i=1}^{n} \left\langle \frac{\partial u}{\partial x_i}, \frac{\partial v}{\partial x_i} \right\rangle_{0,\Omega} \\
&= \int_\Omega \left(uv + \frac{\partial u}{\partial x_1}\frac{\partial v}{\partial x_1} + \cdots + \frac{\partial u}{\partial x_n}\frac{\partial v}{\partial x_n} \right) d\mathbf{x}; \qquad (2.15.16)\\
\|u\|_{1,\Omega} &= \langle u, u\rangle_{1,\Omega}^{\frac{1}{2}} = \left[\langle u, u\rangle_{0,\Omega} + \sum_{i=1}^{n} \left\langle \frac{\partial u}{\partial x_i}, \frac{\partial u}{\partial x_i} \right\rangle_{0,\Omega} \right]^{\frac{1}{2}} \\
&= \left[\int_\Omega \left(|u(\mathbf{x})|^2 + \left| \frac{\partial u}{\partial x_1}(\mathbf{x}) \right|^2 + \cdots + \left| \frac{\partial u}{\partial x_n}(\mathbf{x}) \right|^2 \right) d\mathbf{x} \right]^{\frac{1}{2}}; \qquad (2.15.17)\\
|u|_{1,\Omega} &= \left[\sum_{i=1}^{n} \left\langle \frac{\partial u}{\partial x_i}, \frac{\partial u}{\partial x_i} \right\rangle_{0,\Omega} \right]^{\frac{1}{2}} = \left[\sum_{i=1}^{n} \int_\Omega \left| \frac{\partial u}{\partial x_i}(\mathbf{x}) \right|^2 d\mathbf{x} \right]^{\frac{1}{2}}. \qquad (2.15.18)
\end{aligned}
$$

• $H^2(\Omega)$: For $m = 2$, $n = 2$, $\Omega \subset \mathbb{R}^2$, $H^2(\Omega)$ is equipped with inner product $\langle \cdot, \cdot \rangle_{2,\Omega}$, norm $\|\cdot\|_{2,\Omega}$ and semi-norm $|\cdot|_{2,\Omega}$ defined, $\forall u, v \in H^2(\Omega)$, by:

$$
\begin{aligned}
\langle u, v\rangle_{2,\Omega} &= \langle u, v\rangle_{1,\Omega} + \sum_{1 \le i,j \le 2} \left\langle \frac{\partial^2 u}{\partial x_i \partial x_j}, \frac{\partial^2 v}{\partial x_i \partial x_j} \right\rangle_{0,\Omega} \\
&= \int_\Omega \left(uv + \frac{\partial u}{\partial x_1}\frac{\partial v}{\partial x_1} + \frac{\partial u}{\partial x_2}\frac{\partial v}{\partial x_2} + \frac{\partial^2 u}{\partial x_1^2}\frac{\partial^2 v}{\partial x_1^2} \right. \\
&\qquad \left. + 2\frac{\partial^2 u}{\partial x_1 \partial x_2}\frac{\partial^2 v}{\partial x_1 \partial x_2} + \frac{\partial^2 u}{\partial x_2^2}\frac{\partial^2 v}{\partial x_2^2} \right) dx_1 dx_2; \qquad (2.15.19)\\
\|u\|_{2,\Omega} &= \langle u, u\rangle_{2,\Omega}^{\frac{1}{2}} \\
&= \left(\int_\Omega \left[|u|^2 + \left| \frac{\partial u}{\partial x_1} \right|^2 + \left| \frac{\partial u}{\partial x_2} \right|^2 + \left| \frac{\partial^2 u}{\partial x_1^2} \right|^2 \right. \right. \\
&\qquad \left. \left. + 2\left| \frac{\partial^2 u}{\partial x_1 \partial x_2} \right|^2 + \left| \frac{\partial^2 u}{\partial x_2^2} \right|^2 \right] dx_1 dx_2 \right)^{\frac{1}{2}}; \qquad (2.15.20)\\
|u|_{2,\Omega} &= \left(\int_\Omega \left[\left| \frac{\partial^2 u}{\partial x_1^2} \right|^2 + 2\left| \frac{\partial^2 u}{\partial x_1 \partial x_2} \right|^2 + \left| \frac{\partial^2 u}{\partial x_2^2} \right|^2 \right] dx_1 dx_2 \right)^{\frac{1}{2}}. \qquad (2.15.21)
\end{aligned}
$$

2.15.3 Examples of functions belonging to or not belonging to $H^m(\Omega)$

Case of single variable ($n = 1$)

Example 2.15.1. For $\Omega =]-1, 1[$, consider $u(x) = |x| \forall x \in]-1, 1[$. Show that

1. $u \in H^1(]-1, 1[)$, but
2. $u \notin H^2(]-1, 1[)$.

Proof.

1. $\int_{-1}^{1} |u(x)|^2 dx = \int_{-1}^{1} x^2 dx = \frac{x^3}{3}\Big|_{-1}^{1} = \frac{1}{3} + \frac{1}{3} = \frac{2}{3} < +\infty \Longrightarrow u \in L^2(]-1, 1[)$. From Example 2.3.1, the distributional derivative
$$\frac{du}{dx} = g(x) = \begin{cases} 1 & \text{for } 0 < x < 1 \\ -1 & \text{for } -1 < x < 0 \end{cases}$$
and
$$\int_{-1}^{1} |g(x)|^2 dx = \int_{-1}^{0} 1 dx + \int_{0}^{1} 1 dx = 2 < +\infty$$
$\Longrightarrow g \in L^2(]-1, 1[) \Longrightarrow \frac{du}{dx} \in L^2(]-1, 1[)$. Thus, $u, \frac{du}{dx} \in L^2(]-1, 1[) \Longrightarrow u \in H^1(]-1, 1[)$.

2. From Example 2.3.3, the second-order distributional derivative $\frac{d^2u}{dx^2} = 2\delta \notin L^1_{\text{loc}}(]-1, 1[)$ by Proposition 1.3.2, δ being the Dirac distribution (with mass/charge/force etc.) concentrated at 0. Hence, $\frac{d^2u}{dx^2} \notin L^2(]-1, 1[)$, since $L^2(]-1, 1[) \subset L^1_{\text{loc}}(]-1, 1[)$. Thus $u \notin H^2(]-1, 1[)$.

□

Case of two variables ($n = 2$)

Example 2.15.2. For $\Omega = \mathbb{R}^2$, consider function u defined in (2.3.23) in Example 2.3.9:
$$u(x_1, x_2) = \begin{cases} \ln|\ln r| & \text{for } 0 < r = (x_1^2 + x_2^2)^{\frac{1}{2}} < \frac{1}{e} \\ 0 & \text{for } \frac{1}{e} \le r < \infty. \end{cases}$$

Show that this unbounded and discontinuous function in $\mathbb{R}^2$ belongs to $H^1(\mathbb{R}^2)$.

Proof. From (2.3.24), $u \in L^2(\mathbb{R}^2)$ and $\frac{\partial u}{\partial x_1}, \frac{\partial u}{\partial x_2} \in L^2(\mathbb{R}^2)$. Hence, $u \in H^1(\mathbb{R}^2)$. □

Example 2.15.3. For $\Omega =]0,1[\times]0,1[\subset \mathbb{R}^2$ with $\overline{\Omega} = [0,1]\times[0,1]$ (see Figure 2.4), consider the discontinuous, piecewise polynomial function u defined by:

$$u(x_1,x_2) = \begin{cases} 1 + 4x_1 - 2x_2 & \text{in } \mathring{T}_1 = \{(x_1,x_2) : 0 < x_1 < x_2 < 1\} \\ -2 + 4x_1 + 4x_2 & \text{in } \mathring{T}_2 = \{(x_1,x_2) : 0 < x_2 < x_1 < 1\} \end{cases}$$

with $\mathring{T}_i = \mathrm{int}(T_i)$, $T_1 \cup T_2 = \overline{\Omega}$, $T_1 \cap T_2 = \Gamma_0$, the diagonal of $\overline{\Omega}$ joining $(0,0)$ and $(1,1)$. Then $u \notin H^1(\Omega)$. For the proof and more details, we refer to Example 3.1.2 in Chapter 3.

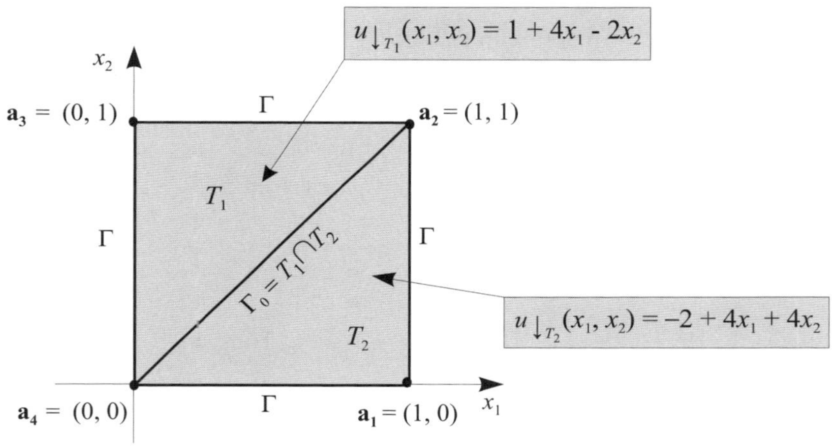

Figure 2.4 Piecewise polynomial function u on $\overline{\Omega} = T_1 \cup T_2$ with discontinuity across Γ_0

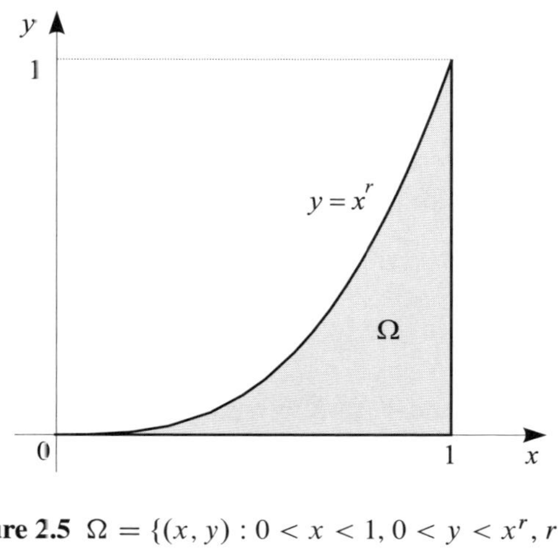

Figure 2.5 $\Omega = \{(x,y) : 0 < x < 1, 0 < y < x^r, r > 0\}$

Example 2.15.4. Let $\Omega \subset \mathbb{R}^2$ be defined by $\Omega = \{(x,y) : 0 < x < 1, 0 < y < x^r, r > 0\}$ (see Figure 2.5), and $u(x,y) = x^\alpha \ \forall (x,y) \in \Omega$. Show that $u \in H^1(\Omega)$ if $2\alpha + r > 1$. Can α be negative such that $u \in H^1(\Omega)$?

Solution.

$$\int_\Omega |u(x,y)|^2 dxdy = \int_0^1 \left(\int_{y=0}^{y=x^r} x^{2\alpha} dy \right) dx = \int_0^1 x^{2\alpha} \left(\int_{y=0}^{y=x^r} dy \right) dx$$
$$= \int_0^1 x^{2\alpha+r} dx < +\infty \quad \text{if } 2\alpha + r > -1$$

$\Longrightarrow u \in L^2(\Omega)$ if $2\alpha + r > -1$.

For $x > 0$, u is a C^∞-function in Ω with $\frac{\partial u}{\partial x} = \alpha x^{\alpha-1}$, $\frac{\partial u}{\partial y} = 0$ in Ω, since the usual partial derivatives and distributional derivatives of u will coincide in Ω. Hence, $\frac{\partial u}{\partial y} \in L^2(\Omega) \ \forall \alpha, r$, and

$$\int_\Omega \left| \frac{\partial u}{\partial x} \right|^2 dxdy = \int_0^1 \left(\int_{y=0}^{y=x^r} \alpha^2 x^{2\alpha-2} dy \right) dx$$
$$= \alpha^2 \int_0^1 x^{2\alpha-2+r} dx < +\infty \quad \text{if } 2\alpha - 2 + r > -1$$

or $2\alpha + r > 1$. Hence, $\frac{\partial u}{\partial x} \in L^2(\Omega)$ for $2\alpha + r > 1$. Thus, $u, \frac{\partial u}{\partial x}, \frac{\partial u}{\partial y} \in L^2(\Omega)$ for $2\alpha + r > 1$,

$\Longrightarrow u \in H^1(\Omega)$ for $2\alpha + r > 1$.

α *can* be negative, if r is sufficiently large such that $2\alpha + r > 1$.

2.15.4 Separability of $H^m(\Omega)$

Proposition 2.15.2. *$\forall m \in \mathbb{N}$, $H^m(\Omega)$ is* separable, *i.e. $\exists$ a countably dense subset of $H^m(\Omega)$.*

The proof of Proposition 2.15.2 depends on the following well-known results:

1. The product of separable spaces is also separable. (For example, since $L^2(\Omega)$ is separable (see (B.4.3.5), Appendix B),

$$\Longrightarrow \quad (L^2(\Omega))^k = \underbrace{L^2(\Omega) \times L^2(\Omega) \times \cdots \times L^2(\Omega)}_{k \text{ times}}$$

is also separable.) (2.15.21a)

2. If X is a Hilbert space and $Y \subset X$ is a *closed* subspace of X, then the separability of X implies the separability of Y. (2.15.21b)

In fact, let $(x_k)_{k\in\mathbb{N}}$ be a dense sequence (hence countable) in X, and $P_Y : X \longrightarrow Y$ be the *projection* operator. Define $y_k = P_Y x_k \ \forall k \in \mathbb{N}$. Then, $(y_k)_{k\in\mathbb{N}}$ is a dense sequence in Y, from which its separability follows.

3. Let X, Y be Hilbert spaces and $\mathcal{J} : X \longrightarrow Z \subset Y$ be an *isometric isomorphism* from X onto $\mathcal{J}X = Z \subset Y$. If Y is separable, then $\mathcal{J}^{-1}Z = X$ is also separable. (2.15.21c)

Proof of Proposition 2.15.2. Set $N = N(n,m)$ defined by:

$$N(n,m) = \sum_{0\le|\boldsymbol{\alpha}|\le m} \text{Card}\left\{(\alpha_1,\alpha_2,\dots,\alpha_n) : 0 \le |\boldsymbol{\alpha}| = \sum_{i=1}^{n} \alpha_i \le m,\ \alpha_i \in \mathbb{N}_0\right\}.$$
$$= N.$$

Define the product space

$$(L^2(\Omega))^N = \underbrace{L^2(\Omega) \times L^2(\Omega) \times \cdots \times L^2(\Omega)}_{N \text{ times}}.$$

By (2.15.21a), $(L^2(\Omega))^N$ is a *separable* product space. $\forall u \in H^m(\Omega)$, define $(\partial^{\boldsymbol{\alpha}} u)_{0\le|\boldsymbol{\alpha}|\le m}$ by:

$$(\partial^{\boldsymbol{\alpha}} u)_{0\le|\boldsymbol{\alpha}|\le m} = \left(u; \left(\frac{\partial u}{\partial x_i}\right)_{1\le i\le n}; \left(\frac{\partial^2 u}{\partial x_i \partial x_j}\right)_{1\le i,j\le n}; \dots; \left(\frac{\partial^m u}{\partial x_1^{\alpha_1}\dots\partial x_n^{\alpha_n}}\right)_{|\boldsymbol{\alpha}|=m}\right).$$

Then $(\partial^{\boldsymbol{\alpha}} u)_{0\le|\boldsymbol{\alpha}|\le m} \in (L^2(\Omega))^N$ $\forall u \in H^m(\Omega)$. Let $\mathcal{J} : H^m(\Omega) \longrightarrow (L^2(\Omega))^N$ be defined, $\forall u \in H^m(\Omega)$, by $\mathcal{J}u = (\partial^{\boldsymbol{\alpha}} u)_{0\le|\boldsymbol{\alpha}|\le m} \in (L^2(\Omega))^N$ with

$$\|\mathcal{J}u\|_{(L^2(\Omega))^N} = \left(\|u\|^2_{L^2(\Omega)} + \sum_{1\le|\boldsymbol{\alpha}|\le m} \|\partial^{\boldsymbol{\alpha}} u\|^2_{L^2(\Omega)}\right)^{1/2} = \|u\|_{m,\Omega}.$$

Hence, $\mathcal{J}$ is an *isometry* from $H^m(\Omega)$ *onto* $\mathbb{W} = \mathcal{J}(H^m(\Omega))$ with $\mathbb{W} \subsetneq (L^2(\Omega))^N$. Moreover, $\mathcal{J}$ is linear, continuous and injective from $H^m(\Omega)$ *onto* $\mathbb{W}$, and $\mathcal{J}^{-1}$ is also linear and continuous from $\mathbb{W}$ *onto* $H^m(\Omega)$, i.e. $\mathcal{J}$ is an *isometric isomorphism* from $H^m(\Omega)$ *onto* $\mathbb{W}$, which is a *closed* subspace of the separable product space $(L^2(\Omega))^N$. In fact, let $(\mathbf{u}_k)_{k\in\mathbb{N}}$ with $\mathbf{u}_k = (\partial^{\boldsymbol{\alpha}} u_k)_{0\le|\boldsymbol{\alpha}|\le m} = \mathcal{J}u_k$ $\forall k \in \mathbb{N}$ be a Cauchy sequence in $\mathbb{W} \equiv \mathcal{J}(H^m(\Omega))$. Then $\|\mathbf{u}_k - \mathbf{u}_l\|_{(L^2(\Omega))^N} = \|\mathcal{J}u_k - \mathcal{J}u_l\|_{(L^2(\Omega))^N} = \|u_k - u_l\|_{H^m(\Omega)} \to 0$ as $k, l \to \infty$. Hence, $(u_k)_{k\in\mathbb{N}}$ is a Cauchy sequence in $H^m(\Omega) \Longrightarrow \exists u \in H^m(\Omega)$ such that $u_k \to u$ in $H^m(\Omega)$ as $k \to \infty \Longrightarrow \mathcal{J}u_k \to \mathcal{J}u$ in $(L^2(\Omega))^N$ with $\mathcal{J}u \in \mathcal{J}(H^m(\Omega)) \equiv \mathbb{W} \Longrightarrow \mathbf{u}_k \to \mathbf{u} \in \mathbb{W}$ with $\mathbf{u} = \mathcal{J}u = (\partial^{\boldsymbol{\alpha}} u)_{0\le|\boldsymbol{\alpha}|\le m}$. Hence, $\mathbb{W}$ is also *separable* by (2.15.21b). Then $H^m(\Omega) = \mathcal{J}^{-1}(\mathbb{W})$ is also separable by (2.15.21c). □

Reflexivity of $H^m(\Omega)$ (i.e. $H^m(\Omega) \equiv (H^m(\Omega))''$) follows from the fact that *every* Hilbert space is reflexive (see also the proof of reflexivity of $W^{m,p}(\Omega)$ for $1 < p < \infty$ in Theorem 2.15.4, from which the result is obtained for $p = 2$).

2.15.5 Generalized Poincaré inequality in $H^m(\Omega)$

Theorem 2.15.1A. *Let $\Omega \subset \mathbb{R}^n$ be a bounded domain with a sufficiently smooth boundary Γ (for example, a Lipschitz continuous boundary (see Appendix D)). Then, $\forall u \in H^m(\Omega)$, $\exists C > 0$ such that $\|u\|^2_{m,\Omega} \le C[|u|^2_{m,\Omega} + \sum_{|\alpha|\le m-1} |\int_\Omega \partial^\alpha u(\mathbf{x})d\mathbf{x}|^2]$.*

Proof. Suppose that the contrary holds, i.e. the inequality does *not* hold for any constant C. In other words, we can find a sequence $(u_k)_{k=1}^\infty$ in $H^m(\Omega)$ with $\|u_k\|_{m,\Omega} = 1$ such that

$$1 = \|u_k\|^2_{m,\Omega} > k\left(|u_k|^2_{m,\Omega} + \sum_{|\alpha|\le m-1}\left|\int_\Omega \partial^\alpha u_k(\mathbf{x})d\mathbf{x}\right|^2\right)$$

$$\Longrightarrow \quad \left(|u_k|^2_{m,\Omega} + \sum_{|\alpha|\le m-1}\left|\int_\Omega \partial^\alpha u_k(\mathbf{x})d\mathbf{x}\right|^2\right) < \frac{1}{k} \ \forall k \in \mathbb{N}$$

$$\Longrightarrow \quad |u_k|^2_{m,\Omega} \to 0 \text{ and } \left|\int_\Omega \partial^\alpha u_k(\mathbf{x})d\mathbf{x}\right|^2 \to 0 \ \forall |\alpha| \le m-1 \text{ as } k \to \infty. \tag{2.15.21d}$$

Hence, $|u_k|^2_{m,\Omega} \to 0 \Longrightarrow \partial^\alpha u_k \to 0$ in $L^2(\Omega)$ $\forall|\alpha| = m$ as $k \to \infty$ and $\int_\Omega \partial^\alpha u_k(\mathbf{x})d\mathbf{x} \to 0$ as $k \to \infty$ $\forall|\alpha| \le m-1$. But $(u_k)_{k=1}^\infty$ is a *bounded* sequence with $\|u_k\|_{m,\Omega} = 1$ $\forall k \in \mathbb{N}$ in Hilbert space $H^m(\Omega)$, which is *reflexive*. Hence, we can extract a *weakly* convergent subsequence $(u_{k_l})_{l=1}^\infty$ from the sequence $(u_k)_{k=1}^\infty$ in $H^m(\Omega)$. Let $u \in H^m(\Omega)$ such that $u_{k_l} \rightharpoonup u$ *weakly* in $H^m(\Omega)$ as $l \to \infty$. But by the Rellich–Kondraschov Theorem 8.11.4, $H^m(\Omega) \hookrightarrow\hookrightarrow H^{m-1}(\Omega)$, i.e. $H^m(\Omega)$ is *compactly imbedded*[2] (see Section 8.11, Chapter 8 for more details) in $H^{m-1}(\Omega)$. Hence, $u_{k_l} \rightharpoonup u$ *weakly* in $H^m(\Omega) \Longrightarrow u_{k_l} \to u$ *strongly* in $H^{m-1}(\Omega)$, i.e. $\|u - u_{k_l}\|_{m-1,\Omega} \to 0$ as $l \to \infty \Longrightarrow u_{k_l} \to u$ in $L^2(\Omega)$ and $\partial^\alpha u_{k_l} \to \partial^\alpha u$ in $L^2(\Omega)$ for $1 \le |\alpha| \le m-1$. But $u_{k_l} \to u$ in $L^2(\Omega) \Longrightarrow u_{k_l} \to u$ in $\mathcal{D}'(\Omega)$ (by (2.13.1))

$$\Longrightarrow \partial^\alpha u_{k_l} \to \partial^\alpha u \text{ in } \mathcal{D}'(\Omega) \text{ for } 1 \le |\alpha| \le m \text{ (by Theorem 2.9.1).} \tag{2.15.21e}$$

From (2.15.21d), $|u_k|_{m,\Omega} \to 0 \Longrightarrow \partial^\alpha u_k \to 0$ in $L^2(\Omega)$ $\forall|\alpha| = m$ as $k \to \infty \Longrightarrow \partial^\alpha u_{k_l} \to 0$ in $L^2(\Omega)$ $\forall|\alpha| = m$ as $l \to \infty$

$$\Longrightarrow \partial^\alpha u_{k_l} \to 0 \text{ in } \mathcal{D}'(\Omega) \ \forall|\alpha| = m \text{ as } l \to \infty. \tag{2.15.21f}$$

Then, from (2.15.21e) and (2.15.21f), $\partial^\alpha u = 0$ in $L^2(\Omega)\forall|\alpha| = m$ (by virtue of the uniqueness of the limit). Thus, $\partial^\alpha u_{k_l} \to \partial^\alpha u$ in $L^2(\Omega)$ for $0 \le |\alpha| \le m$

[2] Let X and Y be Banach spaces and $A : X \longrightarrow Y$ be a linear operator from X into Y. Then A is called *compact* from X into Y if and only if $x_n \rightharpoonup x$ *weakly* in X *implies* $Ax_n \to Ax$ *strongly* in Y as $n \to \infty$ (see Appendix A, Definition A.16.1.1). For $X \hookrightarrow Y$, the imbedding operator $\hookrightarrow: X \longrightarrow\ \hookrightarrow X \subset Y$ is called *compact* from X into Y if and only if $x_n \rightharpoonup x$ *weakly* in X *implies* $x_n \to x$ in Y as $n \to \infty$, since $x_n \in X \longmapsto\ \hookrightarrow x_n = x_n \in Y, x \in X \longmapsto\ \hookrightarrow x = x \in Y$. Then, for $X \hookrightarrow Y$, the imbedding operator $\hookrightarrow$ is compact from X into Y, X is called compactly imbedded in Y and we write $X \hookrightarrow\hookrightarrow Y$.

with $\partial^{\boldsymbol{\alpha}} u = 0\ \forall|\boldsymbol{\alpha}| = m \Longrightarrow u_{k_l} \to u$ in $H^m(\Omega)$ as $l \to \infty$ with $\partial^{\boldsymbol{\alpha}} u = 0$ in $L^2(\Omega)\ \forall|\boldsymbol{\alpha}| = m$. Then $\partial^{\boldsymbol{\alpha}} u = 0$ in $\mathcal{D}'(\Omega)\ \forall|\boldsymbol{\alpha}| = m$ and Ω is a connected set $\Longrightarrow u \in P_{m-1}$, i.e. u is a polynomial of degree $\leq m-1$(by Proposition 2.8.1). But, from (2.15.21d), $\forall|\boldsymbol{\alpha}| \leq m-1$, $\lim_{l\to\infty}\int_\Omega \partial^{\boldsymbol{\alpha}} u_{k_l} d\mathbf{x} = 0$ and, $\forall|\boldsymbol{\alpha}| \leq m-1$,

$$\left|\int_\Omega \partial^{\boldsymbol{\alpha}} u_{k_l} d\mathbf{x} - \int_\Omega \partial^{\boldsymbol{\alpha}} u d\mathbf{x}\right| = \left|\int_\Omega (\partial^{\boldsymbol{\alpha}} u_{k_l} - \partial^{\boldsymbol{\alpha}} u) d\mathbf{x}\right|$$
$$\leq \|1\|_{L^2(\Omega)} \|\partial^{\boldsymbol{\alpha}} u_{k_l} - \partial^{\boldsymbol{\alpha}} u\|_{L^2(\Omega)} \to 0 \quad \text{as } l \to \infty.$$

Hence, $\lim_{l\to\infty}\int_\Omega \partial^{\boldsymbol{\alpha}} u_{k_l} d\mathbf{x} = \int_\Omega \partial^{\boldsymbol{\alpha}} u d\mathbf{x} = 0$ for $0 \leq |\boldsymbol{\alpha}| \leq m-1$. Then $u \in P_{m-1}$ with $\int_\Omega \partial^{\boldsymbol{\alpha}} u d\mathbf{x} = 0$ for $0 \leq |\boldsymbol{\alpha}| \leq m-1 \Longrightarrow u = 0$. In fact, $\forall|\boldsymbol{\alpha}| = m-1$, $\partial^{\boldsymbol{\alpha}} u =$ constant $a_{\boldsymbol{\alpha}}$ with $\int_\Omega a_{\boldsymbol{\alpha}} d\mathbf{x} = 0 \Longrightarrow a_{\boldsymbol{\alpha}} = 0\ \forall|\boldsymbol{\alpha}| = m-1$

$\Longrightarrow u$ is a polynomial of degree $\leq m-2$ with $\int_\Omega \partial^{\boldsymbol{\alpha}} u d\mathbf{x} = 0\ \forall|\boldsymbol{\alpha}| = m-2 \Longrightarrow u$ is a polynomial of degree $\leq m-3$ with $\int_\Omega \partial^{\boldsymbol{\alpha}} u d\mathbf{x} = 0\ \forall|\boldsymbol{\alpha}| = m-3 \Longrightarrow \cdots \Longrightarrow$ for $|\boldsymbol{\alpha}| = 0$, $u = a_0$ with $\int_\Omega a_0 d\mathbf{x} = 0 \Longrightarrow a_0 = 0$.

Thus, $u_k \to u$ in $H^m(\Omega)$ with $u = 0$ in $H^m(\Omega)$ and we meet with a contradiction, since $0 = \|u\|_{m,\Omega} = \lim_{l\to\infty}\|u_{k_l}\|_{m,\Omega} = \lim_{k\to\infty}\|u_k\|_{m,\Omega} = 1$. Hence, our original assumption is wrong and the inequality holds. □

2.15.6 Space $H_0^m(\Omega)$

Definition 2.15.2. Let $\Omega \subset \mathbb{R}^n$ be an open subset of $\mathbb{R}^n$. Then, $\forall m \in \mathbb{N}$, $H_0^m(\Omega)$ is the *closure* of $\mathcal{D}(\Omega) \equiv C_0^\infty(\Omega)$ in the norm $\|\cdot\|_{m,\Omega}$ of $H^m(\Omega)$, i.e.

$$H_0^m(\Omega) = \overline{\mathcal{D}(\Omega)} = \overline{C_0^\infty(\Omega)} \quad \text{in } H^m(\Omega). \tag{2.15.22}$$

In other words, $\mathcal{D}(\Omega) \equiv C_0^\infty(\Omega)$ is *dense* in $H_0^m(\Omega)\ \forall m \in \mathbb{N}$.

Alternative characterization of $H_0^m(\Omega)$

For domains Ω with sufficiently smooth boundary Γ (see Appendix D), there is an alternative characterization of $H_0^m(\Omega)$ with the help of trace theorems (see, for example, Theorem 8.9.11 for $\Omega = \mathbb{R}^n$ with $\Gamma = \mathbb{R}^{n-1}$), for which we refer to Lions [13], Lions and Magenes [15], Nečas [16], Grisvard [17], [19], etc. For example, for circular, elliptic, polygonal domains $\Omega \subset \mathbb{R}^2$ with circular, elliptic, polygonal boundaries $\Gamma \subset \mathbb{R}^2$ respectively,

$$\begin{aligned} H_0^1(\Omega) &= \{u : u \in H^1(\Omega), u{\downarrow_\Gamma} = 0\} \equiv \overline{\mathcal{D}(\Omega)} \quad \text{in } H^1(\Omega); \\ H_0^2(\Omega) &= \left\{u : u \in H^2(\Omega), u{\downarrow_\Gamma} = 0, \frac{\partial u}{\partial n}{\downarrow_\Gamma} = 0\right\} \equiv \overline{\mathcal{D}(\Omega)} \quad \text{in } H^2(\Omega). \end{aligned} \tag{2.15.23}$$

Theorem 2.15.2. $\forall m \in \mathbb{N}$, $H_0^m(\Omega)$ *defined by* (2.15.22) *and equipped with inner product* $\langle\cdot,\cdot\rangle_{m,\Omega}$ *is also a* separable Hilbert space.

Proof. By Definition 2.15.2, $H_0^m(\Omega)$ is a closed subspace of $H^m(\Omega)$. But $H^m(\Omega)$ is a Hilbert space by Theorem 2.15.1, and every closed subspace of Hilbert space $H^m(\Omega)$ equipped with the inner product $\langle \cdot, \cdot \rangle_{m,\Omega}$ is also a Hilbert space. Its separability follows from (2.15.21a), since $H_0^m(\Omega)$ is a *closed* subspace of the separable Hilbert space $H^m(\Omega)$.

Owing to its Hilbert space structure, $H_0^m(\Omega)$ is reflexive. □

Orthogonal complement of $H_0^m(\Omega)$ in $H^m(\Omega)$

Proposition 2.15.3. *The* orthogonal complement *of $H_0^m(\Omega)$ in $H^m(\Omega)$ is the linear space of all $u \in H^m(\Omega)$ which satisfy the following equation:*

$$\sum_{0 \le |\boldsymbol{\alpha}| \le m} (-1)^{|\boldsymbol{\alpha}|} \partial^{2\boldsymbol{\alpha}} u = 0 \quad \text{in } \mathcal{D}'(\Omega), \tag{2.15.23a}$$

where partial derivatives $\partial^{2\boldsymbol{\alpha}} u$ are in the distributional sense.

Proof. Let $u \in H^m(\Omega)$. Then u belongs to the orthogonal complement of $H_0^m(\Omega)$ if and only if $\langle u, v \rangle_{m,\Omega} = 0 \ \forall v \in H_0^m(\Omega) \iff \langle u, \phi \rangle_{m,\Omega} = 0 \ \forall \phi \in \mathcal{D}(\Omega)$, since the inner product $\langle u, \cdot \rangle_{m,\Omega}$ is continuous on $H_0^m(\Omega)$ and $\mathcal{D}(\Omega)$ is dense in $H_0^m(\Omega)$. Thus, for $u \in H^m(\Omega)$ belonging to the orthogonal complement of $H_0^m(\Omega)$, $\forall \phi \in \mathcal{D}(\Omega)$,

$$0 = \langle u, \phi \rangle_{m,\Omega} = \sum_{0 \le |\boldsymbol{\alpha}| \le m} \langle \partial^{\boldsymbol{\alpha}} u, \partial^{\boldsymbol{\alpha}} \phi \rangle_{m,\Omega} = \sum_{0 \le |\boldsymbol{\alpha}| \le m} \int_\Omega \partial^{\boldsymbol{\alpha}} u \partial^{\boldsymbol{\alpha}} \phi \, d\mathbf{x}$$

$$= \sum_{0 \le |\boldsymbol{\alpha}| \le m} (-1)^{|\boldsymbol{\alpha}|} \langle \partial^{2\boldsymbol{\alpha}} u, \phi \rangle_{\mathcal{D}'(\Omega) \times \mathcal{D}(\Omega)} = \left\langle \sum_{0 \le |\boldsymbol{\alpha}| \le m} (-1)^{|\boldsymbol{\alpha}|} \partial^{2\boldsymbol{\alpha}} u, \phi \right\rangle_{\mathcal{D}'(\Omega) \times \mathcal{D}(\Omega)}$$

$\iff \sum_{0 \le |\boldsymbol{\alpha}| \le m} (-1)^{|\boldsymbol{\alpha}|} \partial^{2\boldsymbol{\alpha}} u = 0$ in $\mathcal{D}'(\Omega)$. □

Corollary 2.15.1. *$\mathcal{D}(\Omega)$ is* dense *in $L^2(\Omega) \equiv H^0(\Omega)$.*

Proof. $m = 0 \Longrightarrow \alpha = 0 \Longrightarrow \partial^{2\alpha} u = u$. Hence, $u \in H^0(\Omega) = L^2(\Omega)$ belongs to the orthogonal complement of $H_0^0(\Omega) = \overline{\mathcal{D}(\Omega)}$ in $H^0(\Omega) \iff \langle u, \phi \rangle_{0,\Omega} = 0$ $\forall \phi \in \mathcal{D}(\Omega) \iff \sum_{|\boldsymbol{\alpha}|=0} (-1)^{|\boldsymbol{\alpha}|} \partial^{2\boldsymbol{\alpha}} u = u = 0$ in $\mathcal{D}'(\Omega)$ by Proposition 2.15.3; $\Longrightarrow u = 0$ in $L^2(\Omega) \Longrightarrow \mathcal{D}(\Omega)$ is dense in $L^2(\Omega) \equiv H^0(\Omega)$. □

Norm equivalence in $H_0^m(\Omega)$

Theorem 2.15.3. *The mapping*

$$u \in H_0^m(\Omega) \longmapsto |u|_{m,\Omega} = \Big(\sum_{|\boldsymbol{\alpha}| = m} \|\partial^{\boldsymbol{\alpha}} u\|_{0,\Omega}^2 \Big)^{1/2} = \Big(\sum_{|\boldsymbol{\alpha}| = m} \int_\Omega |\partial^{\boldsymbol{\alpha}} u(\mathbf{x})|^2 d\mathbf{x} \Big)^{1/2}$$

defines a norm *in* $H_0^m(\Omega)$ equivalent *to the original norm* $\|\cdot\|_{m,\Omega}$ *induced by* $H^m(\Omega)$ $\forall m \in \mathbb{N}$, *i.e.* $\exists C_1, C_2 > 0$ *such that*

$$C_1\|u\|_{m,\Omega} \leq |u|_{m,\Omega} \leq C_2\|u\|_{m,\Omega} \quad \forall u \in H_0^m(\Omega). \tag{2.15.23b}$$

$(H_0^m(\Omega); |\cdot|_{m,\Omega})$ *equipped with the new norm defined by* $|\cdot|_{m,\Omega}$ *is a* Hilbert space $\forall m \in \mathbb{N}$. (2.15.23c)

In particular, for $m = 1$, $(H_0^1(\Omega); |\cdot|_{1,\Omega})$ equipped with the norm defined by $|\cdot|_{1,\Omega}$ (see Example 2.15.5 later) is a *Hilbert space*. (2.15.23d)

For $m = 2$, $(H_0^2(\Omega); |\cdot|_{2,\Omega})$ equipped with the norm defined by $|\cdot|_{2,\Omega}$ (see Example 2.15.5 later) is a *Hilbert space*. (2.15.23e)

Proof of Theorem 2.15.3. *The semi-norm* $|\cdot|_{m,\Omega}$ *is a norm in* $H_0^m(\Omega)$: First of all, we temporarily assume that (2.15.23b) holds. Then, for $u \in H_0^m(\Omega)$ satisfying (2.15.23b), $|u|_{m,\Omega} = 0 \Longrightarrow \|u\|_{m,\Omega} \leq \frac{1}{C_1}|u|_{m,\Omega} = 0 \Longrightarrow \|u\|_{m,\Omega} = 0 \Longrightarrow u = 0$ in $H_0^m(\Omega)$. Hence, the semi-norm $|\cdot|_{m,\Omega}$ is a *norm* in $H_0^m(\Omega)$.

Now we prove (2.15.23b) for $m = 1, 2$, which will be met with in applications.

Case $m = 1$: Let $u \in H_0^1(\Omega)$. Then, $|u|_{1,\Omega} \leq \|u\|_{1,\Omega}$ with $C_2 = 1$. (2.15.23f)

It remains to prove that $C_1\|u\|_{1,\Omega} \leq |u|_{1,\Omega}$ with $C_1 > 0$. For this we will apply the density of $\mathcal{D}(\Omega)$ in $H_0^1(\Omega)$ and the result (1.2.36) in Section 1.2. In fact, $\exists$ a sequence $(\phi_k)_{k=1}^{\infty}$ in $\mathcal{D}(\Omega)$ such that $\|u - \phi_k\|_{1,\Omega} \to 0$ as $k \to \infty$. Then $\|\phi_k\|_{1,\Omega} \leq \|u - \phi_k\|_{1,\Omega} + \|u\|_{1,\Omega} \to \|u\|_{1,\Omega}$ as $k \to \infty$, i.e. $\lim_{k\to\infty}\|\phi_k\|_{1,\Omega} = \|u\|_{1,\Omega}$; and $|\phi_k|_{1,\Omega} \leq |u - \phi_k|_{1,\Omega} + |u|_{1,\Omega} \leq \|u - \phi_k\|_{1,\Omega} + |u|_{1,\Omega} \to |u|_{1,\Omega}$ as $k \to \infty$ (since $\|u - \phi_k\|_{1,\Omega} \to 0$ as $k \to \infty$) $\Longrightarrow \lim_{k\to\infty}|\phi_k|_{1,\Omega} = |u|_{1,\Omega}$. (2.15.23g)

From (1.2.35) and (1.2.36), $\|\phi_k\|_{0,\Omega}^2 \leq C\|\nabla\phi_k\|_{0,\Omega}^2 = C|\phi_k|_{1,\Omega}^2 \forall k \in \mathbb{N}$. Hence, $\|\phi_k\|_{1,\Omega}^2 = |\phi_k|_{1,\Omega}^2 + \|\phi_k\|_{0,\Omega}^2 \leq (1+C)|\phi_k|_{1,\Omega}^2 \Longrightarrow (\frac{1}{1+C})\|\phi_k\|_{1,\Omega}^2 \leq |\phi_k|_{1,\Omega}^2 \Longrightarrow C_1\|\phi_k\|_{1,\Omega} \leq |\phi_k|_{1,\Omega}$ with $C_1 = \sqrt{1/(1+C)} > 0$. Thus, we have proved: $\forall k \in \mathbb{N}$, $\exists C_1 > 0$ such that $C_1\|\phi_k\|_{1,\Omega} \leq |\phi_k|_{1,\Omega} \Longrightarrow C_1 \lim_{k\to\infty}\|\phi_k\|_{1,\Omega} \leq \lim_{k\to\infty}|\phi_k|_{1,\Omega}$. Then, using (2.15.23g) and (2.15.23f) with $C_2 = 1$,

$$C_1\|u\|_{1,\Omega} \leq |u|_{1,\Omega} \leq C_2\|u\|_{1,\Omega}. \tag{2.15.23h}$$

Case $m = 2$: Let $u \in H_0^2(\Omega)$. Then $|u|_{2,\Omega} \leq \|u\|_{2,\Omega}$ with $C_2 = 1$. Again applying the density of $\mathcal{D}(\Omega)$ in $H_0^2(\Omega)$, (1.2.35), (1.2.36) and (2.15.23h), we get $\tilde{C}_1\|u\|_{2,\Omega} \leq |u|_{2,\Omega}$. In fact, $\exists$ a sequence $(\psi_k)_{k=1}^{\infty}$ in $\mathcal{D}(\Omega)$ such that $\|u - \psi_k\|_{2,\Omega} \to 0$ as $k \to \infty$. Hence, $\lim_{k\to\infty}\|\psi_k\|_{2,\Omega} = \|u\|_{2,\Omega}$ and $\lim_{k\to\infty}|\psi_k|_{2,\Omega} = |u|_{2,\Omega}$ (see the steps for the case $m = 1$). Moreover, $\|u - \psi_k\|_{1,\Omega}^2 \leq \|u - \psi_k\|_{2,\Omega}^2 \to 0$ as $k \to \infty \Longrightarrow \psi_k \to u$ in $H^1(\Omega) \Longrightarrow u \in H_0^1(\Omega)$ (by virtue of the density of $\mathcal{D}(\Omega)$ in $H_0^1(\Omega)$).Then, $\forall k \in \mathbb{N}$, $\forall i = 1, 2, \ldots, n$, $\frac{\partial\psi_k}{\partial x_i} \in L^2(\Omega)$, $\frac{\partial}{\partial x_j}(\frac{\partial\psi_k}{\partial x_i}) \in L^2(\Omega)$

$\forall i, j = 1, 2, \ldots, n$

$$\Longrightarrow \quad \left\| \frac{\partial \psi_k}{\partial x_i} \right\|_{0,\Omega}^2 \le C \sum_{j=1}^{n} \left\| \frac{\partial^2 \psi_k}{\partial x_i \partial x_j} \right\|_{0,\Omega}^2$$

$$\Longrightarrow \quad |\psi_k|_{1,\Omega}^2 = \sum_{i=1}^{n} \left\| \frac{\partial \psi_k}{\partial x_i} \right\|_{0,\Omega}^2 \le C \sum_{i=1}^{n} \sum_{j=1}^{n} \left\| \frac{\partial^2 \psi_k}{\partial x_i \partial x_j} \right\|_{0,\Omega}^2 = C |\psi_k|_{2,\Omega}^2. \tag{2.15.23i}$$

Hence, $\forall k \in \mathbb{N}$,

$$\begin{aligned} \|\psi_k\|_{2,\Omega}^2 &= |\psi_k|_{2,\Omega}^2 + \|\psi_k\|_{1,\Omega}^2 \le |\psi_k|_{2,\Omega}^2 + \frac{1}{C_1^2} |\psi_k|_{1,\Omega}^2 \quad \text{(using (2.15.23h))} \\ &\le |\psi_k|_{2,\Omega}^2 + \frac{C}{C_1^2} |\psi_k|_{2,\Omega}^2 \quad \text{(by (2.15.23i))} \\ &= \left(1 + \frac{C}{C_1^2}\right) |\psi_k|_{2,\Omega}^2 \end{aligned}$$

$\Longrightarrow \tilde{C}_1 \|\psi_k\|_{2,\Omega} \le |\psi_k|_{2,\Omega}$ with $\tilde{C}_1 = \sqrt{1/(1 + \frac{C}{C_1^2})} > 0 \ \forall k \in \mathbb{N}$. Hence, $\tilde{C}_1 \lim_{k\to\infty} \|\psi_k\|_{2,\Omega} \le \lim_{k\to\infty} |\psi_k|_{2,\Omega} \Longrightarrow \tilde{C}_1 \|u\|_{2,\Omega} \le |u|_{2,\Omega}$

$\Longrightarrow \forall u \in H_0^2(\Omega)$, $\tilde{C}_1 \|u\|_{2,\Omega} \le |u|_{2,\Omega} \le \tilde{C}_2 \|u\|_{2,\Omega}$ with $\tilde{C}_2 = C_2 = 1$, $\tilde{C}_1 > 0$. (2.15.23j)

Similarly, by the method of induction, the general case of $m \in \mathbb{N}$ can be easily proved.

$(H_0^m(\Omega); |\cdot|_{m,\Omega})$ *is a Hilbert space*: From Theorem 2.15.2, $H_0^m(\Omega)$ is a Hilbert space equipped with the *original* norm $\|\cdot\|_{m,\Omega}$. Now, we will show that $H_0^m(\Omega)$ equipped with the *new* norm defined by the semi-norm $|\cdot|_{m,\Omega}$, i.e. $(H_0^m(\Omega); |\cdot|_{m,\Omega})$, is a Hilbert space. For this we consider a Cauchy sequence $(u_k)_{k=1}^{\infty}$ in $(H_0^m(\Omega); |\cdot|_{m,\Omega})$, i.e. $|u_k - u_l|_{m,\Omega} \to 0$ as $k, l \to \infty$. Using norm equivalence (2.15.23b), $\|u_k - u_l\|_{m,\Omega} \le \frac{1}{C_1} |u_k - u_l|_{l,\Omega} \to 0 \Longrightarrow (u_k)_{k=1}^{\infty}$ is also a Cauchy sequence in Hilbert space $(H_0^m(\Omega); \|\cdot\|_{m,\Omega})$ equipped with the original norm $\|\cdot\|_{m,\Omega} \Longrightarrow \exists u \in H_0^m(\Omega)$ such that $\|u - u_k\|_{m,\Omega} \to 0$ as $k \to \infty$, i.e. $\lim_{k\to\infty} u_k = u$ in $(H_0^m(\Omega); \|\cdot\|_{m,\Omega})$. But $|u - u_k|_{m,\Omega} \le \|u - u_k\|_{m,\Omega} \to 0$ as $k \to \infty \Longrightarrow \lim_{k\to\infty} u_k = u$ in $(H_0^m(\Omega); |\cdot|_{m,\Omega})$. Thus, every Cauchy sequence $(u_k)_{k=1}^{\infty}$ converges to an element $u \in (H_0^m(\Omega); |\cdot|_{m,\Omega})$, i.e. $(H_0^m(\Omega); |\cdot|_{m,\Omega})$ is a Hilbert space.

$(H_0^m(\Omega); |\cdot|_{m,\Omega})$ is a Hilbert space and hence reflexive.

The separability follows from the separability of $H^m(\Omega)$, since every closed subspace of a separable Hilbert space is separable. □

Example 2.15.5.

1. For $m = 1$, $|u|_{1,\Omega} = \int_\Omega (|\frac{\partial u}{\partial x_1}|^2 + \cdots + |\frac{\partial u}{\partial x_n}|^2)^{1/2} d\mathbf{x}$ is a *norm* in $H_0^1(\Omega)$ *equivalent* to the original norm $\|u\|_{1,\Omega}$ given in (2.15.17).
2. For $m = 2$, $n = 2$, $|u|_{2,\Omega} = (\int_\Omega [|\frac{\partial^2 u}{\partial x_1^2}|^2 + 2|\frac{\partial^2 u}{\partial x_1 \partial x_2}|^2 + |\frac{\partial^2 u}{\partial x_2^2}|^2] dx_1 dx_2)^{1/2}$ is a *norm* in $H_0^2(\Omega)$ *equivalent* to the original norm $\|u\|_{2,\Omega}$ in (2.15.20).

2.15.7 Space $H^{-m}(\Omega)$

$H^{-m}(\Omega)$ is defined as the *dual* of $H_0^m(\Omega)$, i.e. $H^{-m}(\Omega) \equiv (H_0^m(\Omega))'$ (see Section 4.3, Chapter 4 for full details and other results).

2.15.8 Quotient space $H^m(\Omega)/M$

Definition 2.15.2A. Let M be a *closed* subspace of $H^m(\Omega)$. Then the *quotient space* $H^m(\Omega)/M$ (i.e. the quotient of $H^m(\Omega)$ by M) is the linear space of equivalence classes $[u]$ of functions $u \in H^m(\Omega)$ satisfying the property:

$$u, v \in [u] \iff u - v \in M \iff u = v \pmod{M},$$

i.e. $[u] = u + M$, $u + M$ being the *coset* of u relative to M. $H^m(\Omega)/M$ is equipped with the usual *quotient norm* $\|[u]\|_{H^m(\Omega)/M}$ defined, $\forall [u] \in H^m(\Omega)/M$, by:

$$\|[u]\|_{H^m(\Omega)/M} = \inf_{u \in [u]} \|u\|_{H^m(\Omega)} = \inf_{w \in M} \|u + w\|_{H^m(\Omega)}. \qquad (2.15.24)$$

- $\phi : H^m(\Omega) \longrightarrow H^m(\Omega)/M$ is the linear mapping called *canonical surjection* from $H^m(\Omega)$ *onto* $H^m(\Omega)/M$ and defined, $\forall u \in H^m(\Omega)$, by:
  $\phi u = [u] = u + M \in H^m(\Omega)/M$, $u + M$ being the coset of u relative to M; (2.15.24a)
  and $\forall [u] = u + M \in H^m(\Omega)/M$ with $u \in H^m(\Omega)$, $\phi(u) = [u]$.
- *Linearity of ϕ:* $\phi(u + v) = (u + v) + M = (u + M) + (v + M) = \phi u + \phi v \forall u, v \in H^m(\Omega)$,

$$\phi(\alpha u) = (\alpha u) + M = \alpha(u + M) = \alpha \phi u \quad \forall \alpha \in \mathbb{R},\ \forall u \in H^m(\Omega); \qquad (2.15.24b)$$

- *Kernel of ϕ:* Ker(ϕ)$= M$, i.e. $\forall u \in M$, $\phi u = [0] \in H^m(\Omega)/M$; (2.15.24c)
- *Continuity of ϕ:*

$$\begin{aligned}\|\phi u\|_{H^m(\Omega)/M} &= \|[u]\|_{H^m(\Omega)/M} \\ &= \inf_{u \in [u]} \|u\|_{H^m(\Omega)} \le \|u\|_{H^m(\Omega)} \quad \forall u \in H^m(\Omega) \qquad (2.15.24d)\end{aligned}$$

  with $\phi u = [u] \in H^m(\Omega)/M$.

Lemma 2.15.1. *Let $\phi : H^m(\Omega) \to H^m(\Omega)/M$ be the* canonical surjection *defined above, and $M^\perp$ be the orthogonal complement of M such that $H^m(\Omega) = M \oplus M^\perp$. Then the restriction ρ of ϕ to $M^\perp : \rho = \phi\downarrow_{M^\perp} : M^\perp \longrightarrow H^m(\Omega)/M$ is a linear* bijection *from $M^\perp$* onto *$H^m(\Omega)/M$. Consequently, $\rho^{-1} : H^m(\Omega)/M \longrightarrow M^\perp$ is* also *a linear bijection.*

Proof. Let $u \in H^m(\Omega)$ with $u = v+w$, $v \in M$, $w \in M^\perp$. The linearity of ρ follows from that of ϕ. Then, $\phi u = \phi v + \phi w = \phi w \in H^m(\Omega)/M$, since $\mathrm{Ker}(\phi) = M$ (by (2.15.24c)) and $\phi v = [0]\ \forall v \in M$. Hence, ϕ maps $M^\perp$ *onto* $H^m(\Omega)/M$, i.e. $\rho(M^\perp) = H^m(\Omega)/M$. Now, we are to show that ρ is *one-to-one*. Let $w_1, w_2 \in M^\perp$ such that $\rho w_1 = \phi w_1 \in H^m(\Omega)/M$ and $\rho w_2 = \phi w_2 \in H^m(\Omega)/M$. Then $\rho w_1 = \rho w_2 \Longrightarrow \phi w_1 - \phi w_2 = [0] \Longrightarrow \phi(w_1 - w_2) = [0] \Longrightarrow w_1 - w_2 \in \mathrm{Ker}(\phi) \Longrightarrow w_1 - w_2 \in M$. But $w_1 - w_2 \in M^\perp \Longrightarrow w_1 - w_2 \in M \cap M^\perp = \{\mathbf{0}\} \Longrightarrow w_1 - w_2 = \mathbf{0} \Longrightarrow w_1 = w_2$. Thus, ρ is *one-to-one* and the result follows. □

Two *important* quotient spaces are $H^m(\Omega)/H_0^m(\Omega)$ and $H^m(\Omega)/P_{m-1}$, where $M = H_0^m(\Omega)$ and P_{m-1} respectively, P_{m-1} is the (closed) subspace of polynomials of degree $\le m-1$ in n variables $x_1, x_2, \ldots, x_n$ defined on Ω. Since $H^m(\Omega)$ is a Banach space and M is a *closed* subspace of $H^m(\Omega)$, we have the classical result:

Proposition 2.15.4A. *The quotient space $H^m(\Omega)/M$ equipped with the quotient norm $\|[\cdot]\|_{H^m(\Omega)/M}$ in* (2.15.24) *is a* Banach space.

In fact, $H^m(\Omega)/M$ is a Hilbert space:

Proposition 2.15.4B. *$H^m(\Omega)/M$ is a* Hilbert space *for the norm $\|[\cdot]\|_{H^m(\Omega)/M}$ in* (2.15.24).

Proof. Since $H^m(\Omega)/M$ is a Banach space by Proposition 2.15.4A, it is sufficient to define an inner product $\langle[\cdot],[\cdot]\rangle_{H^m(\Omega)/M}$ in $H^m(\Omega)/M$ such that the corresponding norm satisfies (2.15.24). Since M is a *closed* subspace of Hilbert space $H^m(\Omega)$, we can write $H^m(\Omega) = M \oplus M^\perp$ (i.e. the direct sum of M and $M^\perp$), where $M^\perp$ is the *orthogonal complement* of M in $H^m(\Omega)$. Then, $\forall u \in H^m(\Omega)$, $u = u_M + u_{M^\perp}$ with $u_M \in M$, $u_{M^\perp} \in M^\perp$ such that $\langle u_M, u_{M^\perp}\rangle_{m,\Omega} = 0$, the decomposition being a unique one. $\forall [u] \in H^m(\Omega)/M$, $\exists$ *precisely* one element $u_{M^\perp} \in [u]$ with $u_{M^\perp} \in M^\perp$, since the mapping $[u] \in H^m(\Omega)/M \longmapsto u_{M^\perp} \in M^\perp$ is a *bijection* from $H^m(\Omega)/M$ *onto* $M^\perp$ by Lemma 2.15.1. Hence, we can define an *inner product* in $H^m(\Omega)/M$ by: $\forall [u], [v] \in H^m(\Omega)/M$ with

$$u = u_M + u_{M^\perp} \in [u], \quad v = v_M + v_{M^\perp} \in [v],$$
$$\langle [u],[v]\rangle_{H^m(\Omega)/M} = \langle u_{M^\perp}, v_{M^\perp}\rangle_{H^m(\Omega)}. \tag{2.15.24e}$$

Then

$$\|[u]\|^2_{H^m(\Omega)/M} = \|u_{M^\perp}\|^2_{m,\Omega} \quad \forall [u] \in H^m(\Omega)/M, \tag{2.15.24f}$$

with $u = u_M + u_{M-} \in [u]$.

Now, we are to show that this norm (2.15.24f) satisfies (2.15.24). In fact, from (2.15.24), we have

$$\begin{aligned}\|[u]\|^2_{H^m(\Omega)/M} &= [\inf_{u\in[u]}(\|u\|_{m,\Omega})]^2 = \inf_{u\in[u]}\|u\|^2_{m,\Omega} \\ &= \inf_{u\in[u]}(\|u_M\|^2_{m,\Omega} + \|u_{M^\perp}\|^2_{m,\Omega}) = \|u_{M^\perp}\|^2_{m,\Omega},\end{aligned}$$

since the mapping $[u] \longleftrightarrow u_{M^\perp}$ is *one-to-one* and the infimum is realized for $u_M = 0$ and $u_{M^\perp} \in M^\perp$. Hence, Banach space $H^m(\Omega)/M$ equipped with the corresponding inner product $\langle[\cdot],[\cdot]\rangle_{H^m(\Omega)/M}$ in (2.15.24e) is a Hilbert space. □

2.15.9 Quotient space $H^m(\Omega)/P_{m-1}$

Theorem 2.15.3A. *Let* $\Omega \subset \mathbb{R}^n$ *be a* bounded *domain with* Lipschitz continuous *boundary* Γ *(see Appendix* D*). Then* $\exists C_1, C_2 > 0$ *such that,* $\forall [u] \in H^m(\Omega)/P_{m-1}$ *with*

$$u \in H^m(\Omega), \quad \phi u = [u], \quad C_1\|[u]\|_{H^m(\Omega)/P_{m-1}} \le |u|_{m,\Omega} \le C_2\|[u]\|_{H^m(\Omega)/P_{m-1}}. \tag{2.15.24g}$$

Another norm $|||[u]|||_{H^m(\Omega)/P_{m-1}}$ equivalent to *the original quotient norm* $\|[u]\|_{H^m(\Omega)/P_{m-1}}$ *is given by:*

$$\begin{aligned}|||[u]|||_{H^m(\Omega)/P_{m-1}} &= \langle[u],[u]\rangle^{1/2}_{H^m(\Omega)/P_{m-1}} = \left(\sum_{|\boldsymbol{\alpha}|=m}\int_\Omega |\partial^{\boldsymbol{\alpha}} u(\mathbf{x})|^2 d\mathbf{x}\right)^{1/2} \\ &= |u|_{m,\Omega}.\end{aligned} \tag{2.15.24h}$$

$H^m(\Omega)/P_{m-1}$ *is a* Hilbert space *with inner product* $\langle[\cdot],[\cdot]\rangle_{H^m(\Omega)/P_{m-1}}$ *defined by:*

$$\langle[u],[v]\rangle_{H^m(\Omega)/P_{m-1}} = \sum_{|\boldsymbol{\alpha}|=m}\int_\Omega \partial^{\boldsymbol{\alpha}} u \partial^{\boldsymbol{\alpha}} v d\mathbf{x}. \tag{2.15.24i}$$

Proof. See the proof of Theorem 2.15.7 with $p = 2$. □

2.15.10 Other equivalent norms in $H^m(\Omega)$

Theorem 2.15.3B. *Let* $\Omega \subset \mathbb{R}^n$ *be a* bounded *domain with Lipschitz continuous boundary* Γ *(see Appendix D). Then we have the following results:*

I. Equivalent norm in $H^m(\Omega)$: $\forall u \in H^m(\Omega)$, $\exists C_1, C_2 > 0$ *such that*

$$C_1\|u\|_{m,\Omega} \le \Big(\|u\|^2_{L^2(\Omega)} + \sum_{|\alpha|=m} \|\partial^{\alpha} u\|^2_{L^2(\Omega)}\Big)^{1/2} \le C_2\|u\|_{m,\Omega}; \tag{2.15.25a}$$

II. Equivalent norm in $H^1(\Omega)$: *For* $\Gamma_0 \subset \Gamma$ *with measure* $\mu(\Gamma_0) > 0$, $\forall u \in H^1(\Omega)$, $\exists C_1, C_2 > 0$ *such that*

$$C_1\|u\|_{1,\Omega} \le \Big(\int_{\Gamma_0} |u(\mathbf{x})|^2 dS + \sum_{i=1}^{n} \int_{\Omega} \Big|\frac{\partial u}{\partial x_i}\Big|^2 d\mathbf{x}\Big)^{1/2} \le C_2\|u\|_{1,\Omega}, \tag{2.15.25b}$$

which is also called Friedrichs' inequality *(Nečas [16]);*

III. Equivalent norm in $H^2(\Omega)$: $\forall u \in H^2(\Omega)$, $\exists C_1, C_2 > 0$ *such that*

$$C_1\|u\|_{2,\Omega} \le \Big(\int_{\Gamma} |u(\mathbf{x})|^2 dS + \sum_{|\alpha|=2} \int_{\Omega} |\partial^{\alpha} u|^2 d\mathbf{x}\Big)^{1/2} \le C_2\|u\|_{2,\Omega}. \tag{2.15.25c}$$

(The same constants $C_1, C_2 > 0$ *have been used to denote* different *values in the different inequalities* (2.15.25a)–(2.15.25c).*)*

An interesting counterexample has been given in [16]:

Example 2.15.6. If $\Omega \subset \mathbb{R}^3$ is an open ellipsoid with boundary Γ in $\mathbb{R}^3$ and $u \in H^3(\Omega)$, an inequality analogous to (2.15.25c) for $H^3(\Omega)$ with ellipsoidal domain Ω does *not* hold, since the left-hand side inequality becomes:

$$C_1\|u\|_{3,\Omega} \le \Big(\int_{\Gamma} |u(\mathbf{x})|^2 dS + \sum_{|\alpha|=3} \int_{\Omega} |\partial^{\alpha} u|^2 d\mathbf{x}\Big)^{1/2}, \tag{2.15.25d}$$

which does *not* hold in this case. In fact, the right-hand side of inequality (2.15.25d) is *not* a norm in $H^3(\Omega)$ for ellipsoidal Ω. For $u \in H^3(\Omega)$ with $u(x_1, x_2, x_3) = p_2(x_1, x_2, x_3) = \alpha_1^2 x_1^2 + \alpha_2^2 x_2^2 + \alpha_3^2 x_3^2 - 1 = 0$ (equation of an ellipsoidal surface Γ) with $\alpha_1, \alpha_2, \alpha_3 > 0$ $\forall (x_1, x_2, x_3) \in \Gamma$, p_2 being a polynomial of degree 2 in three variables x_1, x_2, x_3, both the integrals in (2.15.25d) vanish and $u = p_2 \neq 0$ on Ω. But if Ω is *not* an open ellipsoid and has a Lipschitz continuous boundary Γ, inequality (2.15.25d) holds in $H^3(\Omega)$.

Proof of Theorem 2.15.3B. We give the scheme of the proof, for example, for (I), as follows: first of all, it is shown that $W_1 = H^m(\Omega)$ equipped with the new norm defined by the right-hand side expression relative to the first inequality in (2.15.25a) is *complete*. Let $W_2 = H^m(\Omega)$ equipped with the original norm $\|\cdot\|_{m,\Omega}$. Hence, W_1 and W_2 are Banach spaces. Then $\mathcal{I} : W_1 \longrightarrow W_2$ is a bijective linear, continuous mapping from W_1 onto W_2 and as a consequence (Corollary A.8.1.1, Appendix A) of the Open Mapping Theorem A.8.1.3, $\mathcal{I}^{-1} : W_2 \longrightarrow W_1$ is also continuous, i.e. $\mathcal{I}$ is an isomorphism from W_1 onto W_2 and the result follows. □

2.15.11 Density results

- $\mathcal{D}(\Omega)$ is *dense* in $H_0^m(\Omega)$ by Definition 2.15.2. (2.15.26a)
- $\mathcal{D}(\mathbb{R}^n)$ is *dense* in $H^m(\mathbb{R}^n)$ $\forall m \in \mathbb{N}$ (see Theorem 8.9.6). (2.15.26b)
- For $m_2 < m_1$, $H_0^{m_1}(\Omega)$ is *dense* in $H_0^{m_2}(\Omega)$(by virtue of (2.15.26a) and imbedding (2.15.27b)). (2.15.26c)
- $\mathcal{D}(\Omega)$ is *dense* in $L^2(\Omega)$ (see Corollary 2.15.1 and Theorem 6.8.3). But $\mathcal{D}(\Omega)$ is *not* dense in $H^m(\Omega)$ for $\Omega \neq \mathbb{R}^n$ with measure $\mu(\mathbb{R}^n \backslash \Omega) > 0$, $\forall m \in \mathbb{N}$. See Proposition 2.15.7 for the proof with $p = 2$. See also (2.15.27f). (2.15.26d)
- $\mathcal{D}(\overline{\Omega}) = \{\psi : \exists \phi \in \mathcal{D}(\mathbb{R}^n)$ such that $\psi = \phi{\downarrow_\Omega}\}$ (see Definition 8.10.2) (2.15.26e)

  is *dense* in $H^m(\Omega)$ for any Ω with the *m-extension property* (see Theorems 8.10.1 and 8.10.2 in Section 8.10, Chapter 8, which together prove the result).
- For *arbitrary* domain Ω, $\mathcal{D}(\overline{\Omega}) \cap H^m(\Omega)$ (resp. $C^m(\Omega) \cap H^m(\Omega)$) is *dense* in $H^m(\Omega)$ (see Theorem 2.15.8). (2.15.26f)
- For other density results, see Chapters 6 and 8.

2.15.12 Algebraic inclusions ($\subset$) and imbedding ($\hookrightarrow$) results

For m_1, $m_2 \in \mathbb{N}$ with $m_1 > m_2$, the following results hold:

1. $H^{m_1}(\Omega) \hookrightarrow H^{m_2}(\Omega)$, i.e. $H^{m_1}(\Omega) \subset H^{m_2}(\Omega)$ (*algebraic inclusion*) and $\exists C > 0$ such that $\|u\|_{m_2,\Omega} \leq C\|u\|_{m_1,\Omega}$ $\forall u \in H^{m_1}(\Omega)$ (*continuity* of $\hookrightarrow$), which follow from Definition 2.15.1 and (2.15.10) with $C = 1$; (2.15.27a)
2. $H_0^{m_1}(\Omega) \hookrightarrow H_0^{m_2}(\Omega)$, i.e. $H_0^{m_1}(\Omega) \subset H_0^{m_2}(\Omega)$ and $\|u\|_{m_2,\Omega} \leq C\|u\|_{m_1,\Omega}$ with $C = 1$, $\forall u \in H_0^{m_1}(\Omega)$; (2.15.27b)
3. $\mathcal{D}(\Omega) \subset H_0^m(\Omega) \subset H^m(\Omega) \subset \mathcal{D}'(\Omega)$ $\forall m \in \mathbb{N}$; (2.15.27c)
4. for $\Omega = \mathbb{R}^n$, $H_0^m(\mathbb{R}^n) \equiv H^m(\mathbb{R}^n)$ (see Theorem 8.9.6); (2.15.27d)
5. for $\Omega \subsetneq \mathbb{R}^n$ with measure $\mu(\mathbb{R}^n \setminus \Omega) > 0$, $H_0^m(\Omega) \subsetneq H^m(\Omega)$. (2.15.27e)

But we have the counterexample, for $n \geq 2$ and $\Omega = \mathbb{R}^n \setminus \{\mathbf{0}\}$ with $\mu(\mathbb{R}^n \setminus \Omega) = \mu(\{\mathbf{0}\}) = 0$, $H_0^1(\Omega) = H^1(\Omega)$ (see Brezis [26]). (2.15.27f)

For more imbedding results, see Section 4.3, Section 8.9 (Theorems 8.9.4 and 8.9.5, (8.9.33), Proposition 8.9.2), Section 8.10 (Proposition 8.10.3), Sections 8.11 and 8.12.

For *compact imbedding* ($\hookrightarrow\hookrightarrow$) results, see Sections 8.11 and 8.12, Chapter 8.

Space $H^s(\Omega)$ for arbitrary $s \in \mathbb{R}$ For $\Omega = \mathbb{R}^n$, $H^s(\mathbb{R}^n)$ with $s > 0$ defined with the help of Fourier transforms of tempered distributions, and its dual $H^{-s}(\mathbb{R}^n)$, are studied in Section 8.9, Chapter 8.

For $\Omega \subsetneq \mathbb{R}^n$, $H^s(\Omega)$ with $s > 0$ and their closed subspaces $H_0^s(\Omega)$, $H_{00}^s(\Omega)$, and the dual spaces $H^{-s}(\Omega)$, $(H_{00}^s(\Omega))'$, are defined in Sections 8.10 and 8.11, Chapter 8.

2.15.13 Space $W^{m,p}(\Omega)$ with $m \in \mathbb{N}, 1 \leq p \leq \infty$

Definition 2.15.3. Let $\Omega \subset \mathbb{R}^n$ be an open subset of $\mathbb{R}^n$. Then, $\forall m \in \mathbb{N}$ and $1 \leq p \leq \infty$, $W^{m,p}(\Omega)$ is the set of all (*equivalence classes* $[u]$ of) real-valued functions $u \in L^p(\Omega)$ whose distributional derivatives $\partial^{\boldsymbol{\alpha}} u \in L^p(\Omega) \forall |\boldsymbol{\alpha}| \leq m$, i.e.

- for $1 \leq p < \infty$,

$$W^{m,p}(\Omega) = \{u : u \in L^p(\Omega), \partial^{\boldsymbol{\alpha}} u \in L^p(\Omega) \ \forall |\boldsymbol{\alpha}| \leq m\}, \tag{2.15.28}$$

- for $p = \infty$,

$$W^{m,\infty}(\Omega) = \{u : u \in L^\infty(\Omega), \partial^{\boldsymbol{\alpha}} u \in L^\infty(\Omega) \ \forall |\boldsymbol{\alpha}| \leq m\}.$$

Then $W^{m,p}(\Omega), 1 \leq p \leq \infty$, is a *linear space*.

Proposition 2.15.5. $\forall m \in \mathbb{N}$, $1 \leq p \leq \infty$, $W^{m,p}(\Omega)$ *is a* normed linear space *equipped with the norm* $\|\cdot\|_{m,p,\Omega}$ *and the semi-norm* $|\cdot|_{m,p,\Omega}$ *defined,* $\forall u \in W^{m,p}(\Omega)$*, by:*

- *for* $1 \leq p < \infty$,

$$\begin{aligned} \|u\|_{m,p,\Omega} &= \left(\|u\|_{L^p(\Omega)}^p + \sum_{1 \leq |\boldsymbol{\alpha}| \leq m} \|\partial^{\boldsymbol{\alpha}} u\|_{L^p(\Omega)}^p \right)^{\frac{1}{p}} \\ &= \left[\int_\Omega |u|^p d\mathbf{x} + \sum_{1 \leq |\boldsymbol{\alpha}| \leq m} \int_\Omega |\partial^{\boldsymbol{\alpha}} u|^p d\mathbf{x} \right]^{\frac{1}{p}}, \end{aligned} \tag{2.15.29}$$

- *for* $p = \infty$,

$$\|u\|_{m,\infty,\Omega} = \|u\|_{L^\infty(\Omega)} + \sum_{1\le|\boldsymbol{\alpha}|\le m} \|\partial^{\boldsymbol{\alpha}} u\|_{L^\infty(\Omega)}$$

$$= \operatorname*{ess\,sup}_{\mathbf{x}\in\Omega} |u(\mathbf{x})| + \sum_{1\le|\boldsymbol{\alpha}|\le m} (\operatorname*{ess\,sup}_{\mathbf{x}\in\Omega} |\partial^{\boldsymbol{\alpha}} u(\mathbf{x})|) \quad \textit{or} \tag{2.15.30}$$

$$\|u\|_{m,\infty,\Omega} = \max_{0\le|\boldsymbol{\alpha}|\le m} \{\operatorname*{ess\,sup}_{\mathbf{x}\in\Omega} |\partial^{\boldsymbol{\alpha}} u(\mathbf{x})|\}. \tag{2.15.31}$$

The norms (2.15.30) *and* (2.15.31) *are* equivalent *in* $W^{m,\infty}(\Omega)$. *For other equivalent norms in* $W^{m,p}(\Omega)$, *see* (2.15.36a) *and* (2.15.36b) *later.* ess sup *is the* essential supremum *(see Appendix B, Definition B.2.1.2).*

- *for* $1 \le p < \infty$,

$$|u|_{m,p,\Omega} = \Big(\sum_{|\boldsymbol{\alpha}|=m} \|\partial^{\boldsymbol{\alpha}} u\|^p_{L^p(\Omega)}\Big)^{\frac{1}{p}} = \Big(\sum_{|\boldsymbol{\alpha}|=m} \int_\Omega |\partial^{\boldsymbol{\alpha}} u|^p d\mathbf{x}\Big)^{\frac{1}{p}}; \tag{2.15.32}$$

- *for* $p = \infty$,

$$|u|_{m,\infty,\Omega} = \sum_{|\boldsymbol{\alpha}|=m} \|\partial^{\boldsymbol{\alpha}} u\|_{L^\infty(\Omega)} = \sum_{|\boldsymbol{\alpha}|=m} \operatorname*{ess\,sup}_{\mathbf{x}\in\Omega} |\partial^{\boldsymbol{\alpha}} u(\mathbf{x})| \quad \textit{or}$$

$$|u|_{m,\infty,\Omega} = \max_{|\boldsymbol{\alpha}|=m} \{\operatorname*{ess\,sup}_{\mathbf{x}\in\Omega} |\partial^{\boldsymbol{\alpha}} u(\mathbf{x})|\}. \tag{2.15.33}$$

Theorem 2.15.4. $\forall m \in \mathbb{N}$, $1 \le p \le \infty$, $W^{m,p}(\Omega)$ *equipped with the norm* $\|\cdot\|_{m,p,\Omega}$ *defined in* (2.15.29)–(2.15.31) *is a* Banach space.

I. *For* $1 \le p < \infty$, $W^{m,p}(\Omega)$ *is* separable.

II. *For* $1 < p < \infty$, $W^{m,p}(\Omega)$ *is* reflexive.

Proof. Since $L^p(\Omega)$ is a Banach space for $1 \le p \le \infty$, the proof is exactly similar to that of Theorem 2.15.1 if $L^2(\Omega)$ is replaced by $L^p(\Omega)$ and 'Hilbert space' by 'Banach space'.

I. The proof is exactly similar to that given for the separability of $H^m(\Omega) \equiv W^{m,2}(\Omega)$ in Proposition 2.15.2 if we replace $L^2(\Omega)$ by $L^p(\Omega)$ with $1 \le p < \infty$ everywhere; $\mathbb{W}$ by $\mathbb{W}_p \subset (L^p(\Omega))^N$; $H^m(\Omega)$ by $W^{m,p}(\Omega)$ with $1 \le p < \infty$; and use the separability of $L^p(\Omega)$ for $1 \le p < \infty$; the consequent separability of $(L^p(\Omega))^N$ for $1 \le p < \infty$; the separability of the *closed* subspace $\mathbb{W}_p \equiv \mathcal{J}[W^{m,p}(\Omega)]$ of $(L^p(\Omega))^N$; and, finally, the separability of $\mathcal{J}^{-1}(\mathbb{W}_p) \equiv W^{m,p}(\Omega)$, $\mathcal{J}$ being the *isometric isomorphism* (see Appendix A) from $W^{m,p}(\Omega)$ onto $\mathbb{W}_p \subset (L^p(\Omega))^N$, $1 \le p < \infty$: $\|\mathcal{J}u\|_{(L^p(\Omega))^N} = \|u\|_{m,p,\Omega}$ $\forall u \in W^{m,p}(\Omega)$.

II. For the proof of the reflexivity of $W^{m,p}(\Omega)$, $1 < p < \infty$, we prepare the following results:

(a) Since $L^q(\Omega) = (L^p(\Omega))'$ for $1 < p < \infty$, $\frac{1}{p} + \frac{1}{q} = 1$,

$$[(L^p(\Omega))^N]' = [\underbrace{L^p(\Omega) \times \cdots \times L^p(\Omega)}_{N \text{ times}}]' = [\underbrace{L^q(\Omega) \times \cdots \times L^q(\Omega)}_{N \text{ times}}]$$
$$= [(L^q(\Omega))^N].$$

Then, $\forall l \in [(L^p(\Omega))^N]'$, $\exists \mathbf{v} = (v_i)_{1 \le i \le N} \in (L^q(\Omega))^N$ such that $\forall \mathbf{u} = (u_i)_{1 \le i \le N} \in [L^p(\Omega)]^N$, we have

$$l(\mathbf{u}) = \sum_{i=1}^{N} \langle u_i, v_i \rangle_{L^p(\Omega) \times L^q(\Omega)} = \sum_{i=1}^{N} \int_\Omega u_i(\mathbf{x}) v_i(\mathbf{x}) d\mathbf{x}. \quad (2.15.33a)$$

(b) *Continuous linear functionals on* $W^{m,p}(\Omega)$, $1 < p < \infty$: Let $L \in (W^{m,p}(\Omega))'$ be a continuous linear functional on $W^{m,p}(\Omega)$ with $|L(u)| \le \|L\|_{(W^{m,p}(\Omega))'} \|u\|_{m,p,\Omega}$ $\forall u \in W^{m,p}(\Omega)$. Define an *isometric isomorphism* (see Appendix A) $\mathcal{J} : W^{m,p}(\Omega) \to \mathbb{W}_p \subset (L^p(\Omega))^N$ (see details in the proof of Proposition 2.15.2 for the separability of $H^m(\Omega) \equiv W^{m,2}(\Omega)$) from $W^{m,p}(\Omega)$ *onto* $\mathbb{W}_p$ such that $\mathcal{J}u = (\partial^{\boldsymbol{\alpha}} u)_{0 \le |\boldsymbol{\alpha}| \le m} \in \mathbb{W}_p \subset (L^p(\Omega))^N$ $(N = \sum_{0 \le |\boldsymbol{\alpha}| \le m} 1 = \text{Card}\{(\alpha_1, \ldots, \alpha_n) : 0 \le |\boldsymbol{\alpha}| \le m\})$, $\|\mathcal{J}u\|_{(L^p(\Omega))^N} = \|(\partial^{\boldsymbol{\alpha}} u)_{0 \le |\boldsymbol{\alpha}| \le m}\|_{(L^p(\Omega))^N} = \|u\|_{m,p,\Omega}$ $\forall u \in W^{m,p}(\Omega)$, $\mathbb{W}_p$ being a *closed* subspace of $(L^p(\Omega))^N$. Then we can define a linear functional L_0 on $\mathbb{W}_p$ by: $L_0(\mathcal{J}u) = L(u)$ $\forall u \in W^{m,p}(\Omega)$ with $\|L_0\|_{\mathbb{W}_{\mathbb{P}}'} = \sup_{\|\mathcal{J}u\|_{(L^p(\Omega))^N} \le 1} |L_0(\mathcal{J}u)| = \sup_{\|u\|_{m,p,\Omega} \le 1} |L(u)| = \|L\|_{(W^{m,p}(\Omega))'}$. Hence, L_0 is a *continuous*, linear functional on the *closed* subspace $\mathbb{W}_p$ of $(L^p(\Omega))^N$, and by Corollary A.7.3.1 of the Hahn–Banach Theorem A.7.2.1 (see Appendix A), L_0 can be given a norm-preserving extension to $(L^p(\Omega))^N$ such that $\tilde{L}_0 \in [(L^p(\Omega))^N]'$ and $\tilde{L}_0(\mathcal{J}u) = L_0(\mathcal{J}u) = L(u)$ $\forall u \in W^{m,p}(\Omega)$ with $\|\tilde{L}_0\|_{[(L^p(\Omega))^N]'} = \|L_0\|_{\mathbb{W}_{\mathbb{P}}'}$. Then $\exists \mathbf{v} = (v_{\boldsymbol{\alpha}})_{0 \le |\boldsymbol{\alpha}| \le m} \in (L^q(\Omega))^N$ such that $\forall \mathbf{u} = (u_{\boldsymbol{\alpha}})_{0 \le |\boldsymbol{\alpha}| \le m} \in (L^p(\Omega))^N$, $\tilde{L}_0(\mathbf{u}) = \sum_{0 \le |\boldsymbol{\alpha}| \le m} \langle u_{\boldsymbol{\alpha}}, v_{\boldsymbol{\alpha}} \rangle_{L^p(\Omega) \times L^q(\Omega)}$. Hence, $\forall u \in W^{m,p}(\Omega)$,

$$L(u) = L_0(\mathcal{J}u) = \tilde{L}_0(\mathcal{J}u) = \tilde{L}_0((\partial^{\boldsymbol{\alpha}} u)_{0 \le |\boldsymbol{\alpha}| \le m})$$
$$= \sum_{0 \le |\boldsymbol{\alpha}| \le m} \langle \partial^{\boldsymbol{\alpha}} u, v_{\boldsymbol{\alpha}} \rangle_{L^p(\Omega) \times L^q(\Omega)}$$
$$\Longrightarrow \quad L(u) = \sum_{0 \le |\boldsymbol{\alpha}| \le m} \int_\Omega \partial^{\boldsymbol{\alpha}} u(\mathbf{x}) v_{\boldsymbol{\alpha}}(\mathbf{x}) d\mathbf{x} \quad \forall u \in W^{m,p}(\Omega).$$
(2.15.34)

Let $(u_k)_{k=1}^{\infty}$ be a bounded sequence in $W^{m,p}(\Omega)$, i.e. $\exists C > 0$ such that $\|u\|_{m,p,\Omega} \leq C\ \forall k \in \mathbb{N}$. For the proof of reflexivity of Banach space $W^{m,p}(\Omega)$, by the Eberlein–Schmulyan Theorem A.11.1.2 in Appendix A (see [4]), for example, it is necessary and sufficient to show that we can extract a *subsequence* $(u_{k_l})_{l=1}^{\infty}$ of the sequence $(u_k)_{k=1}^{\infty}$ such that $(u_{k_l})_{l=1}^{\infty}$ converges *weakly* in $W^{m,p}(\Omega)$. $\forall|\boldsymbol{\alpha}| \leq m$, $\forall k \in \mathbb{N}$, $\partial^{\boldsymbol{\alpha}} u_k \in L^p(\Omega)$ with $\|\partial^{\boldsymbol{\alpha}} u_k\|_{L^p(\Omega)} \leq \|u_k\|_{m,p,\Omega} \leq C \implies \forall|\boldsymbol{\alpha}| \leq m$, $(\partial^{\boldsymbol{\alpha}} u_k)_{k=1}^{\infty}$ is a bounded sequence in the *reflexive* Banach space $L^p(\Omega)$ for $1 < p < \infty$. Hence, we can extract a subsequence $(u_{k_l})_{l=1}^{\infty}$ from $(u_k)_{k=1}^{\infty}$ such that $u_{k_l} \rightharpoonup u$ *weakly* in $L^p(\Omega)$ and $\partial^{\boldsymbol{\alpha}} u_{k_l} \rightharpoonup w_{\boldsymbol{\alpha}}$ *weakly* in $L^p(\Omega)$ for $1 \leq |\boldsymbol{\alpha}| \leq m$. But $u_{k_l} \rightharpoonup u$ in $L^p(\Omega) \implies u_{k_l} \to u$ in $\mathcal{D}'(\Omega)$ (by (2.13.3) and (2.13.4)) $\implies \partial^{\boldsymbol{\alpha}} u_{k_l} \to \partial^{\boldsymbol{\alpha}} u$ in $\mathcal{D}'(\Omega)$ (by Theorem 2.9.1). Again, $\partial^{\boldsymbol{\alpha}} u_{k_l} \rightharpoonup w_{\boldsymbol{\alpha}}$ in $L^p(\Omega) \implies \partial^{\boldsymbol{\alpha}} u_{k_l} \to w_{\boldsymbol{\alpha}}$ in $\mathcal{D}'(\Omega)$ $\forall$ fixed $\boldsymbol{\alpha}$ with $1 \leq |\boldsymbol{\alpha}| \leq m$. But the limit is unique. Hence, $w_{\boldsymbol{\alpha}} = \partial^{\boldsymbol{\alpha}} u\ \forall|\boldsymbol{\alpha}| \leq m$. Thus, $\partial^{\boldsymbol{\alpha}} u_{k_l} \rightharpoonup \partial^{\boldsymbol{\alpha}} u$ in $L^p(\Omega)$ $\forall|\boldsymbol{\alpha}| \leq m \implies \int_\Omega \partial^{\boldsymbol{\alpha}} u_{k_l}(\mathbf{x}) v_{\boldsymbol{\alpha}}(\mathbf{x}) d\mathbf{x} \to \int_\Omega \partial^{\boldsymbol{\alpha}} u(\mathbf{x}) v_{\boldsymbol{\alpha}}(\mathbf{x}) d\mathbf{x}\ \forall v_{\boldsymbol{\alpha}} \in L^q(\Omega)$ with $|\boldsymbol{\alpha}| \leq m$, $1 < p, q < \infty$. Hence, from (2.15.34), $\forall L \in (W^{m,p}(\Omega))'$, $L(u_{k_l}) = \sum_{0 \leq |\boldsymbol{\alpha}| \leq m} \int_\Omega \partial^{\boldsymbol{\alpha}} u_{k_l}(\mathbf{x}) v_{\boldsymbol{\alpha}}(\mathbf{x}) d\mathbf{x} \to \sum_{0 \leq |\boldsymbol{\alpha}| \leq m} \int_\Omega \partial^{\boldsymbol{\alpha}} u(\mathbf{x}) v_{\boldsymbol{\alpha}}(\mathbf{x}) d\mathbf{x} = L(u)$.
Thus, $u_{k_l} \rightharpoonup u$ weakly in $W^{m,p}(\Omega)$, and the reflexivity of $W^{m,p}(\Omega)$ follows.

□

In particular, for $p = 2$, $W^{m,2}(\Omega) \equiv H^m(\Omega)$ equipped with inner product $\langle \cdot, \cdot \rangle_{m,\Omega}$ is a *Hilbert space.* (2.15.35a)

For $p \neq 2$, $W^{m,p}(\Omega)$ cannot be equipped with an inner product, and hence is *not* a Hilbert space. (2.15.35b)

We let $m = 0$ such that $W^{0,p}(\Omega) \equiv L^p(\Omega)$ equipped with the norm $\|u\|_{L^p(\Omega)} = \|u\|_{0,p,\Omega}$.

$1 \leq p < \infty$,

$$\|u\|_{L^p(\Omega)} = \|u\|_{0,p,\Omega} = \left(\int_\Omega |u(\mathbf{x})|^p d\mathbf{x} \right)^{\frac{1}{p}}. \tag{2.15.35c}$$

$p = \infty$,

$$\|u\|_{L^\infty(\Omega)} = \|u\|_{0,\infty,\Omega} = \operatorname*{ess\,sup}_{\mathbf{x} \in \Omega} |u(\mathbf{x})|. \tag{2.15.35d}$$

Norm equivalence in $W^{m,p}(\Omega)$, $1 \leq p < \infty$ In many situations, *equivalent* norms in $W^{m,p}(\Omega)$ are quite useful. For bounded $\Omega \subset \mathbb{R}^n$, $m \in \mathbb{N}$, $1 \leq p < \infty$, the following norms $|||\cdot|||_{m,p,\Omega}$ and $\mathbf{|} \cdot \mathbf{|}_{m,p,\Omega}$ defined, $\forall u \in W^{m,p}(\Omega)$, by:

1.
$$|||u|||_{m,p,\Omega} = |||u|||_{W^{m,p}(\Omega)} = (\|u\|_{L^p(\Omega)}^p + |u|_{m,p,\Omega}^p)^{1/p}$$
$$= \left(\|u\|_{L^p(\Omega)}^p + \sum_{|\boldsymbol{\alpha}|=m} \|\partial^{\boldsymbol{\alpha}} u\|_{L^p(\Omega)}^p \right)^{1/p}; \tag{2.15.36a}$$

2. $$|\!\!|\cdot|\!\!|_{m,p,\Omega} = \|u\|_{L^p(\Omega)} + |u|_{m,p,\Omega} = \|u\|_{L^p(\Omega)} + \sum_{|\boldsymbol{\alpha}|=m} \|\partial^{\boldsymbol{\alpha}} u\|_{L^p(\Omega)}, \tag{2.15.36b}$$

are *equivalent* to the original norm $\|u\|_{m,p,\Omega}$ given by (2.15.29), $|\cdot|_{m,p,\Omega}$ being the *semi-norm* in $W^{m,p}(\Omega)$ defined by (2.15.32), i.e. $\exists C_1, C_2 > 0$ such that $C_1 \|u\|_{m,p,\Omega} \le |\!|\!|u|\!|\!|_{m,p,\Omega}$ (resp. $|\!\!|\cdot|\!\!|_{m,p,\Omega}) \le C_2\|u\|_{m,p,\Omega}$ $\forall u \in W^{m,p}(\Omega)$ (it is understood that constants C_1 and C_2 have *different* values).

For $p = 2$, the second equivalent norm $|\!\!|\cdot|\!\!|_{m,2,\Omega}$ in (2.15.36b) is *not* a Hilbert norm, i.e. $(W^{m,2}(\Omega); |\!\!|\cdot|\!\!|_{m,2,\Omega})$ is a Banach space, but *not* a Hilbert space, whereas the *first* equivalent norm $|\!|\!|u|\!|\!|_{m,2,\Omega}$ in (2.15.36a) *is* a Hilbert norm and the corresponding Banach space $(W^{m,2}(\Omega); |\!|\!|u|\!|\!|_{m,2,\Omega})$ is a Hilbert space with the inner product $\langle\langle\cdot,\cdot\rangle\rangle_{m,\Omega}$ such that

$$|\!|\!|u|\!|\!|_{m,2,\Omega} = \langle\langle u, u\rangle\rangle_{m,\Omega}^{1/2} = \left(\langle u, u\rangle_{L^2(\Omega)} + \sum_{|\boldsymbol{\alpha}|=m} \langle \partial^{\boldsymbol{\alpha}} u, \partial^{\boldsymbol{\alpha}} u\rangle_{L^2(\Omega)}\right)^{1/2}. \tag{2.15.36c}$$

Example 2.15.7. For $m = 2$, $1 \le p < \infty$, $W^{2,p}(\Omega)$ can be equipped with

$$|\!|\!|u|\!|\!|_{2,p,\Omega} = \left(\int_\Omega |u(\mathbf{x})|^p d\mathbf{x} + \sum_{1\le i,j\le n} \int_\Omega \left|\frac{\partial^2 u(\mathbf{x})}{\partial x_i \partial x_j}\right|^p d\mathbf{x}\right)^{1/p}; \tag{2.15.36d}$$

$$|\!\!|u|\!\!|_{2,p,\Omega} = \left(\int_\Omega |u(\mathbf{x})|^p d\mathbf{x}\right)^{1/p} + \sum_{1\le i,j\le n} \left(\int_\Omega \left|\frac{\partial^2 u(\mathbf{x})}{\partial x_i \partial x_j}\right|^p d\mathbf{x}\right)^{1/p}, \tag{2.15.36e}$$

which are *equivalent* to the original norm:

$$\|u\|_{2,p,\Omega} = \left(\int_\Omega |u(\mathbf{x})|^p d\mathbf{x} + \sum_{i=1}^{n} \int_\Omega \left|\frac{\partial u}{\partial x_i}\right|^p + \sum_{1\le i,j\le n} \int_\Omega \left|\frac{\partial^2 u}{\partial x_i \partial x_j}\right|^p d\mathbf{x}\right)^{1/p}. \tag{2.15.36f}$$

For $p = 2$, $|\!|\!|\cdot|\!|\!|_{2,2,\Omega}$ is a *Hilbert norm* in $H^2(\Omega) \equiv W^{2,2}(\Omega)$ (2.15.36g) given in (2.15.36c) with $\langle\langle u, v\rangle\rangle_{m,\Omega} = \langle u, v\rangle_{L^2(\Omega)} + \sum_{|\alpha|=m} \langle \partial^{\boldsymbol{\alpha}} u, \partial^{\boldsymbol{\alpha}} v\rangle_{L^2(\Omega)}$, whereas $|\!\!|\cdot|\!\!|_{2,2,\Omega}$ is *not* a Hilbert norm.

When there is *no* chance of confusion, we will use the same notation $\|\cdot\|_{m,p,\Omega}$ to denote any *equivalent* norm.

2.15.14 Space $W_0^{m,p}(\Omega)$, $1 \le p < \infty$

Definition 2.15.4. Let $\Omega \subset \mathbb{R}^n$ be an open subset of $\mathbb{R}^n$. Then, $\forall m \in \mathbb{N}$, $1 \le p < \infty$, $W_0^{m,p}(\Omega)$ is the *closure* of $\mathcal{D}(\Omega) \equiv C_0^\infty(\Omega)$ in the norm $\|\cdot\|_{m,p,\Omega}$ of $W^{m,p}(\Omega)$,

i.e.

$$W_0^{m,p}(\Omega) = \overline{\mathcal{D}(\Omega)} = \overline{C_0^\infty(\Omega)} \quad \text{in } W^{m,p}(\Omega),\ 1 \le p < \infty. \tag{2.15.37}$$

In other words, $\mathcal{D}(\Omega) = C_0^\infty(\Omega)$ is *dense* in $W_0^{m,p}(\Omega)$ for $1 \le p < \infty$, $\forall m \in \mathbb{N}_0$.

In particular, for $p = 2$, $W_0^{m,2}(\Omega) = H_0^m(\Omega)$. (2.15.38)

Alternative characterization of $W_0^{m,p}(\Omega)$, $m \in \mathbb{N}$, $1 < p < \infty$

As in the case of $H_0^m(\Omega)$ (see the examples in (2.15.23)), for domains with sufficiently smooth boundary Γ, there is an alternative characterization of $W_0^{m,p}(\Omega)$ with the help of trace theorems (see, for example, Theorem 8.9.11 for $p = 2, \Omega = \mathbb{R}^n, \Gamma = \mathbb{R}^{n-1}$), for which we refer to Nečas [16], Grisvard [17], [18], [19], etc. For example, for $m = 1, p = 2, W_0^{1,2}(\Omega) = H_0^1(\Omega)$ and $m = 2, p = 2, W_0^{2,2}(\Omega) = H_0^2(\Omega)$ are characterized by (2.15.23).

Theorem 2.15.4A. *For $m \in \mathbb{N}$ and $1 \le p < \infty$, $W_0^{m,p}(\Omega)$ is a* separable Banach space. *For $1 < p < \infty$, $W_0^{m,p}(\Omega)$ is* reflexive.

Proof. The proof is similar to that of Theorem 2.15.2 by virtue of (2.15.37), and Theorem 2.15.4. □

Null extension of functions of $W_0^{m,p}(\Omega)$, $1 \le p < \infty$

Theorem 2.15.5. *Let $\Omega \subset \mathbb{R}^n$ be an open set in $\mathbb{R}^n$ and $u \in W_0^{m,p}(\Omega)$ with $1 \le p < \infty$, and $\widetilde{u}$ be its null extension*

$$\widetilde{u}(\mathbf{x}) = \begin{cases} u(\mathbf{x}) & \text{for } \mathbf{x} \in \Omega \\ 0 & \text{for } \mathbf{x} \in \mathbb{R}^n \setminus \Omega. \end{cases}$$

Then

I. $\forall |\boldsymbol{\alpha}| \le m$, $\partial^{\boldsymbol{\alpha}} \widetilde{u} = \widetilde{(\partial^{\boldsymbol{\alpha}} u)}$ *in the distributional sense in $\mathbb{R}^n$, where*

$$\widetilde{(\partial^{\boldsymbol{\alpha}} u)}(\mathbf{x}) = \begin{cases} \partial^{\boldsymbol{\alpha}} u(\mathbf{x}) & \text{for } \mathbf{x} \in \Omega \\ 0 & \text{for } \mathbf{x} \in \mathbb{R}^n \setminus \Omega; \end{cases}$$

II. $\tilde{u} \in W^{m,p}(\mathbb{R}^n)$ *with* $\|\tilde{u}\|_{m,p,\mathbb{R}^n} = \|u\|_{m,p,\Omega}$ (isometry).

In particular, for $p = 2$, $u \in H_0^m(\Omega) \Longrightarrow \tilde{u} \in H^m(\mathbb{R}^n)$ with $\|\tilde{u}\|_{m,\mathbb{R}^n} = \|u\|_{m,\Omega}$.

Proof. By definition, $\mathcal{D}(\Omega)$ is dense in $W_0^{m,p}(\Omega)$ for $1 \le p < \infty$ and $\forall m \in \mathbb{N}$. Hence, $\exists$ a sequence (ϕ_k) in $\mathcal{D}(\Omega)$ such that $\phi_k \to u$ in $W_0^{m,p}(\Omega) \Longrightarrow \partial^{\boldsymbol{\alpha}} \phi_k \to \partial^{\boldsymbol{\alpha}} u$ in $L^p(\Omega)$ $\forall |\boldsymbol{\alpha}| \le m$, $1 \le p < \infty$ as $k \to \infty$. But $\partial^{\boldsymbol{\alpha}} u \in L^p(\Omega) \Longrightarrow \widetilde{\partial^{\boldsymbol{\alpha}} u} \in L^p(\mathbb{R}^n)$ with $\|\partial^{\boldsymbol{\alpha}} u\|_{L^p(\Omega)} = \|\widetilde{(\partial^{\boldsymbol{\alpha}} u)}\|_{L^p(\mathbb{R}^n)}$.

I. From the definition of distributional derivatives in (2.3.1): $\forall\psi \in \mathcal{D}(\mathbb{R}^n)$, $\forall|\boldsymbol{\alpha}| \le m$,

$$\begin{aligned}
\langle \partial^{\boldsymbol{\alpha}}\tilde{u}, \psi\rangle &= (-1)^{|\boldsymbol{\alpha}|}\langle \tilde{u}, \partial^{\boldsymbol{\alpha}}\psi\rangle = (-1)^{|\boldsymbol{\alpha}|}\int_{\mathbb{R}^n} \tilde{u}(\mathbf{x})\partial^{\boldsymbol{\alpha}}\psi(\mathbf{x})d\mathbf{x} \\
&= (-1)^{|\boldsymbol{\alpha}|}\int_{\Omega} u(\mathbf{x})\partial^{\boldsymbol{\alpha}}\psi(\mathbf{x})d\mathbf{x} = (-1)^{|\boldsymbol{\alpha}|}\lim_{k\to\infty}\int_{\Omega}\phi_k(\mathbf{x})\partial^{\boldsymbol{\alpha}}\psi(\mathbf{x})d\mathbf{x} \\
&\quad (\text{since } \phi_k \xrightarrow{\text{strongly}} u \text{ in } L^p(\Omega) \Longrightarrow \phi_k \rightharpoonup u \textit{ weakly in } L^p(\Omega) \\
&\quad \text{with } \partial^{\boldsymbol{\alpha}}\psi \in L^q(\Omega), \frac{1}{p}+\frac{1}{q}=1) \\
&= \lim_{k\to\infty}(-1)^{|\boldsymbol{\alpha}|}\int_{\Omega}\phi_k(\mathbf{x})\partial^{\boldsymbol{\alpha}}\psi(\mathbf{x})d\mathbf{x} \\
&= \lim_{k\to\infty}\int_{\Omega}\partial^{\boldsymbol{\alpha}}\phi_k(\mathbf{x})\psi(\mathbf{x})d\mathbf{x} \\
&= \int_{\Omega}\partial^{\boldsymbol{\alpha}}u(\mathbf{x})\psi(\mathbf{x})d\mathbf{x} \\
&\quad (\text{since } \partial^{\boldsymbol{\alpha}}\phi_k \xrightarrow{\text{strongly}} \partial^{\boldsymbol{\alpha}}u \text{ in } L^p(\Omega) \Longrightarrow \partial^{\boldsymbol{\alpha}}\phi_k \rightharpoonup \partial^{\boldsymbol{\alpha}}u \textit{ weakly in } L^p(\Omega)) \\
&= \int_{\mathbb{R}^n}\widetilde{\partial^{\boldsymbol{\alpha}}u}(\mathbf{x})\psi(\mathbf{x})d\mathbf{x} \\
&= \langle \widetilde{\partial^{\boldsymbol{\alpha}}u}, \psi\rangle \quad \forall\psi \in \mathcal{D}(\mathbb{R}^n)
\end{aligned}$$

$\Longleftrightarrow \partial^{\boldsymbol{\alpha}}\tilde{u} = (\widetilde{\partial^{\boldsymbol{\alpha}}u})$ in $\mathcal{D}'(\mathbb{R}^n)$ with $(\widetilde{\partial^{\boldsymbol{\alpha}}u}) \in L^p(\mathbb{R}^n)$ $\forall|\boldsymbol{\alpha}| \le m$.

II.
$$\begin{aligned}
\|\tilde{u}\|^p_{m,p,\mathbb{R}^n} &= \sum_{0\le|\boldsymbol{\alpha}|\le m}\int_{\mathbb{R}^n}|\partial^{\boldsymbol{\alpha}}\tilde{u}|^p d\mathbf{x} = \sum_{0\le|\boldsymbol{\alpha}|\le m}\int_{\mathbb{R}^n}|\widetilde{\partial^{\boldsymbol{\alpha}}u}|^p d\mathbf{x} \quad (\text{by (I)}) \\
&= \sum_{0\le|\boldsymbol{\alpha}|\le m}\int_{\Omega}|\partial^{\boldsymbol{\alpha}}u|^p d\mathbf{x} = \|u\|^p_{m,p,\Omega}
\end{aligned}$$

$\Longrightarrow \|\tilde{u}\|_{m,p,\mathbb{R}^n} = \|u\|_{m,p,\Omega}$.

□

Proposition 2.15.6. *Let Ω be a C^1-regular bounded domain with boundary Γ (see Definition* 8.10.4 *and also Appendix D). Let $u \in W^{1,p}(\Omega)$, $1 < p < \infty$, such that $u\downarrow_\Gamma = 0$. Then $u \in W_0^{1,p}(\Omega)$.*

Norm equivalence in $W_0^{m,p}(\Omega)$, $1 \le p < \infty$

The norm equivalence in $W_0^{m,p}(\Omega)$ follows from the results given by:

Proposition 2.15.7. *Let $\Omega \subset \mathbb{R}^n$ be a bounded domain. Then $\exists\, C = C(\Omega, p) > 0$ such that $\forall u \in W_0^{1,p}(\Omega)$, $1 \le p < \infty$,*

1.
$$\|u\|_{W^{1,p}(\Omega)} \le C\,\|\nabla u\|_{(L^p(\Omega))^n} = C\left[\sum_{i=1}^{n}\left\|\frac{\partial u}{\partial x_i}\right\|_{L^p(\Omega)}^p\right]^{1/p} = C\,|u|_{1,p,\Omega}. \tag{2.15.39a}$$

In general, $\forall m \in \mathbb{N}$, $\forall u \in W_0^{m,p}(\Omega)$, $1 \le p < \infty$, $\exists C = C(\Omega, m, p) > 0$ such that:

2.
$$\|u\|_{L^p(\Omega)} \le C\,|u|_{m,p,\Omega} = C\Big(\sum_{|\alpha|=m}\|\partial^\alpha u\|_{L^p(\Omega)}^p\Big)^{\frac{1}{p}}. \tag{2.15.39b}$$

3. *The semi-norm $|u|_{m,p,\Omega}$ is a norm in $W_0^{m,p}(\Omega)$* equivalent *to the original norm $\|\cdot\|_{m,p,\Omega}$ defined by* (2.15.29). (2.15.39c)
4. *$(W_0^{m,p}(\Omega); |\cdot|_{m,p,\Omega})$ is a separable Banach space and reflexive for* $1 < p < \infty$. (2.15.39d)

Proof. The proof is similar to that of Theorem 2.15.3, if we replace $L^2(\Omega)$ by $L^p(\Omega)$ and $H_0^m(\Omega)$ by $W_0^{m,p}(\Omega)$, $1 \le p < \infty$, $m \in \mathbb{N}$, and use the density of $\mathcal{D}(\Omega)$ in $W_0^{m,p}(\Omega)$ along with (2.15.36a). □

2.15.15 Space $W^{-m,q}(\Omega)$

$W^{-m,q}(\Omega)$ with $m \in \mathbb{N}$, $1 < q \le \infty$, $1 \le p < \infty$, $\frac{1}{p} + \frac{1}{q} = 1$, is defined as the *dual* of $W_0^{m,p}(\Omega)$, i.e. $\quad W^{-m,q}(\Omega) = (W_0^{m,p}(\Omega))'$ (2.15.40)
(see Section 4.3 for more details and other results).

2.15.16 Quotient space $W^{m,p}(\Omega)/M$ for $m \in \mathbb{N}$, $1 \le p < \infty$

Definition 2.15.5. Let M be a *closed* subspace of Banach space $W^{m,p}(\Omega)$ for $m \in \mathbb{N}$, $1 \le p < \infty$. Then, the *quotient space* $W^{m,p}(\Omega)/M$ (i.e. the quotient of $W^{m,p}(\Omega)$ by M) is a *Banach space* of equivalence classes $[u]$ of functions $u \in W^{m,p}(\Omega)$ satisfying the property: $u, v \in [u] \iff u - v \in M \iff u = v$ (mod M)), i.e.$[u] = u + M$, $u + M$ being the *coset* of u relative to M and equipped with the usual *quotient norm* $\|[\cdot]\|_{W^{m,p}(\Omega)/M}$ defined, $\forall [u] \in W^{m,p}(\Omega)/M$, by:

$$\|[u]\|_{W^{m,p}(\Omega)/M} = \inf_{u\in[u]}\|u\|_{m,p,\Omega} = \inf_{w\in M}\|u + w\|_{m,p,\Omega}. \tag{2.15.41}$$

For canonical surjection $\phi : W^{m,p}(\Omega) \to W^{m,p}(\Omega)/M$, (2.15.24a)–(2.15.24d) hold with $H^m(\Omega)$ replaced by $W^{m,p}(\Omega)$.

For $p=2$, $W^{m,2}(\Omega)/M \equiv H^m(\Omega)/M$ is a *Hilbert space* (see Proposition 2.15.4B).

The most important quotient space is $W^{m,p}(\Omega)/P_{m-1}$ with $M = P_{m-1}$, P_{m-1} being the N-dimensional (hence, *closed*) subspace of polynomials of degree $\leq m-1$ in n variables $x_1, \ldots, x_n$ defined on Ω, since this is used in many problems and specifically in error estimates for finite element approximations (see Bernadou [34], Ciarlet [35]); the other one being the case $M = W_0^{m,p}(\Omega)$, which will *not* be dealt with here. For $p = 2$, see Proposition 2.15.4B.

Let P_{m-1} be the linear space of polynomials of degree $\leq m-1$ in n variables $x_1, \ldots, x_n$ defined on Ω with $\dim(P_{m-1}) = \binom{n+(m-1)}{m-1} = N < +\infty$. Hence, P_{m-1} is a *closed* subspace of $W^{m,p}(\Omega)$ for $m \in \mathbb{N}$, $1 \leq p < \infty$. Then we have:

Lemma 2.15.2. *Let* $\Omega \subset \mathbb{R}^n$ *be a* bounded *domain with Lipschitz continuous boundary* Γ. *Then* $\exists$ *continuous linear functionals* $\{l_i\}_{i=1}^N$ *on* $W^{m,p}(\Omega)$ *such that* $\forall q \in P_{m-1}$,

$$\sum_{i=1}^N |\langle l_i, q\rangle|^p = 0 \iff q = 0 \text{ in } P_{m-1}. \tag{2.15.42}$$

Proof. Let $\{q_i\}_{i=1}^N$ be a basis in P_{m-1}. Then $\exists$ a unique dual basis $\{l_i\}_{i=1}^N$ in P'_{m-1} such that

$$\langle l_i, q_j\rangle = \delta_{ij}, \quad i \leq i, j \leq N. \tag{2.15.43}$$

Since P_{m-1} is an N-dimensional *closed* subspace of $W^{m,p}(\Omega)$, in which all norms are equivalent, l_i's are *continuous* on P_{m-1} in the norm $\|\cdot\|_{m,p,\Omega}$ of $W^{m,p}(\Omega)$. Hence, by Corollary A.7.3.1 of the Hahn–Banach Theorem A.7.2.1 in Appendix A, each l_i, $i \leq i \leq N$, can be extended to a *continuous*, linear functional on $W^{m,p}(\Omega)$ (each extended continuous functional will still be denoted by l_i, $i \leq i \leq N$) such that (2.15.42) holds.

In fact, for $q \in P_{m-1}$, $\sum_{i=1}^N |\langle l_i, q\rangle|^p = 0 \iff \langle l_i, q\rangle = 0$, $1 \leq i \leq N$, for $q \in P_{m-1}$. Hence, it is sufficient to show that $\langle l_i, q\rangle = 0$, $1 \leq i \leq N \iff q = 0$ in P_{m-1}. For $q \in P_{m-1}$, $q = \sum_{j=1}^N \alpha_j q_j$. Then, for $1 \leq i \leq N$, $\langle l_i, q\rangle = \sum_{j=1}^N \alpha_j l_i(q_j) = \sum_{j=1}^N \alpha_j \delta_{ij} = \alpha_i$ (by (2.15.43)) and $\langle l_i, q\rangle = 0 \iff \alpha_i = 0$ for $1 \leq i \leq N$. Hence, $q = 0$ in P_{m-1}.

$(l_i)_{i=1}^N$ can be constructed in different ways. For example, for bounded Ω, for $\boldsymbol{\alpha} = \boldsymbol{\alpha}(i)$, $0 \leq |\boldsymbol{\alpha}| \leq m-1$, $\langle l_{\boldsymbol{\alpha}}, q\rangle = \int_\Omega \partial^{\boldsymbol{\alpha}} q(\mathbf{x}) d\mathbf{x}$, or $\langle l_{\boldsymbol{\alpha}}, q\rangle = \int_\Omega \mathbf{x}^{\boldsymbol{\alpha}} q(\mathbf{x}) d\mathbf{x}$, etc. □

Theorem 2.15.6. *Let* $\Omega \subset \mathbb{R}^n$ *be a* bounded *domain with Lipschitz continuous boundary* Γ *(see Definition* D.2.3.1, *Appendix* D*) and* $\{l_i\}_{i=1}^N$ *be the* continuous *linear functionals on* $W^{m,p}(\Omega)$, $m \in \mathbb{N}$, $1 \leq p < \infty$ *(for the sake of simplicity,* $p = \infty$ *is* not *considered) satisfying* (2.15.42) *and* (2.15.43). *Then* $\exists C_1, C_2 > 0$ *such that*

$$C_1 \|u\|_{m,p,\Omega} \leq \left(|u|^p_{m,p,\Omega} + \sum_{i=1}^N |\langle l_i, u\rangle|^p \right)^{1/p} \leq C_2 \|u\|_{m,p,\Omega}. \tag{2.15.44}$$

Proof. The right-hand side inequality holds: Since $l_i \in (W^{m,p}(\Omega))'$, $\exists \tilde{C}_i > 0$ such that $|\langle l_i, u\rangle| \leq \tilde{C}_i \|u\|_{m,p,\Omega}$ $\forall u \in W^{m,p}(\Omega)$. Hence, $\forall u \in W^{m,p}(\Omega)$, $\sum_{i=1}^{N} |\langle l_i, u\rangle|^p \leq \tilde{C} \|u\|^p_{m,p,\Omega}$ with $\tilde{C} \geq \sum_{i=1}^{N} \tilde{C}_i^p > 0$ and $|u|_{m,p,\Omega} \leq \|u\|_{m,p,\Omega}$. Then,

$$\left(|u|^p_{m,p,\Omega} + \sum_{i=1}^{N} |\langle l_i, u\rangle|^p \right)^{1/p} \leq C_2 \|u\|_{m,p,\Omega} \tag{2.15.45}$$

with $C_2 \geq (1 + \tilde{C})^{1/p} > 0$.

The left-hand side inequality holds: We give an indirect proof. Suppose that the left-hand side inequality does *not* hold for any $C_1 > 0$. Then, it does not hold for $C_1 = \frac{1}{k}$ with $k \in \mathbb{N}$, i.e. $\exists u_k \in W^{m,p}(\Omega)$ such that after its normalization $\|u_k\|_{m,p,\Omega} = 1$ $\forall k \in \mathbb{N}$,

$$\frac{1}{k}\|u_k\|_{m,p,\Omega} = \frac{1}{k} > \left(|u_k|^p_{m,p,\Omega} + \sum_{i=1}^{N} |\langle l_i, u_k\rangle|^p \right)^{1/p} \quad \forall k \in \mathbb{N}.$$

Then $(|u_k|^p_{m,p,\Omega} + \sum_{i=1}^{N} |\langle l_i, u_k\rangle|^p) \to 0$ as $k \to \infty \Longrightarrow |u_k|^p_{m,p,\Omega} \to 0$ and $\sum_{i=1}^{N} |\langle l_i, u_k\rangle|^p \to 0$ as $k \to \infty \Longrightarrow \lim_{k\to\infty} \partial^{\boldsymbol{\alpha}} u_k = 0$ in $L^p(\Omega)$ $\forall |\alpha| = m$ and $\lim_{k\to\infty} \langle l_i, u_k\rangle = 0,\ 1 \leq i \leq N$. (2.15.46)

Since $W^{m,p}(\Omega) \hookrightarrow\hookrightarrow W^{m-1,p}(\Omega)$, i.e. the imbedding $\hookrightarrow\hookrightarrow$ is compact from $W^{m,p}(\Omega)$ to $W^{m-1,p}(\Omega)$ by the Rellich–Kondraschov Theorem 8.11.4 in Section 8.11, Chapter 8, we can extract a subsequence $(u_{k_l})_{l\in\mathbb{N}}$ of the *bounded* sequence $(u_k)_{k\in\mathbb{N}}$ in $W^{m,p}(\Omega)$ such that $(u_{k_l})_{l\in\mathbb{N}}$ converges *strongly* in $W^{m-1,p}(\Omega)$, i.e. $\exists u \in W^{m-1,p}(\Omega)$ such that

$$\lim_{l\to\infty} \|u - u_{k_l}\|_{m-1,p,\Omega} = 0. \tag{2.15.47}$$

From (2.15.46), $\lim_{k\to\infty} \partial^{\boldsymbol{\alpha}} u_k = 0$ in $L^p(\Omega)$ $\forall |\boldsymbol{\alpha}| = m \Longrightarrow \lim_{l\to\infty} \partial^{\boldsymbol{\alpha}} u_{k_l} = 0$ in $L^p(\Omega)$ $\forall |\boldsymbol{\alpha}| = m$. Then, from (2.15.47), $\partial^{\boldsymbol{\alpha}} u_{k_l} \to \partial^{\boldsymbol{\alpha}} u$ in $L^p(\Omega)$ $\forall |\boldsymbol{\alpha}| \leq m - 1$ and $\partial^{\boldsymbol{\alpha}} u_{k_l} \to 0$ in $L^p(\Omega)$ $\forall |\boldsymbol{\alpha}| = m \Longrightarrow \partial^{\boldsymbol{\alpha}} u_{k_l} \to \partial^{\boldsymbol{\alpha}} u$ in $L^p(\Omega)$ $\forall |\boldsymbol{\alpha}| \leq m$ with $\partial^{\boldsymbol{\alpha}} u = 0$ $\forall |\boldsymbol{\alpha}| = m$

$$\Longrightarrow \quad \lim_{l\to\infty} u_{k_l} = u \quad \text{in } W^{m,p}(\Omega) \text{ with } \partial^{\boldsymbol{\alpha}} u = 0\ \forall |\boldsymbol{\alpha}| = m. \tag{2.15.48}$$

Since Ω is a connected set and $\partial^{\boldsymbol{\alpha}} u = 0$ $\forall |\boldsymbol{\alpha}| = m$, u is a polynomial of degree $\leq m - 1$ by Proposition 2.8.1, i.e. $u \in P_{m-1}$, $P_{m-1} \subset W^{m,p}(\Omega)$ being a closed subspace of $W^{m,p}(\Omega)$. $u_{k_l} \to u$ in $W^{m,p}(\Omega)$ with $u \in P_{m-1} \Longrightarrow u_{k_l} \rightharpoonup u$ weakly in $W^{m,p}(\Omega)$ with $u \in P_{m-1} \Longrightarrow \forall i = 1, 2, \ldots, N$,

$$\langle l_i, u_{k_l}\rangle \to \langle l_i, u\rangle \quad \Longrightarrow \quad \sum_{i=1}^{N} |\langle l_i, u_{k_l}\rangle|^p \to \sum_{i=1}^{N} |\langle l_i, u\rangle|^p \quad \text{as } l \to \infty.$$

But, from (2.15.46), $\sum_{i=1}^{N} |\langle l_i, u_{k_l}\rangle|^p \to 0$ as $l \to \infty$ and the limit is unique. Hence, $\sum_{i=1}^{N} |\langle l_i, u\rangle|^p = 0$ for $u \in P_{m-1} \iff u = 0$ in $W^{m,p}(\Omega)$ by (2.15.42). Thus we have a contradiction, since $0 = \|u\|_{m,p,\Omega} = \lim_{l\to\infty} \|u_{k_l}\|_{m,p,\Omega} = \lim_{k\to\infty} \|u_k\|_{m,p,\Omega} = 1$. Hence, our original assumption is wrong and the result follows. □

Norm equivalence in quotient space $W^{m,p}(\Omega)/P_{m-1}$

For $W^{m,p}(\Omega)/P_{m-1}$ equipped with the norm $\|[\cdot]\|_{W^{m,p}(\Omega)/P_{m-1}}$, the mapping $[u] \in W^{m,p}(\Omega)/P_{m-1} \longmapsto |[u]|_{W^{m,p}(\Omega)/P_{m-1}} = |u|_{m,p,\Omega}$ with $u \in [u]$ is a priori a *semi-norm* $|[u]|_{W^{m,p}(\Omega)/P_{m-1}}$. In fact, $\forall q \in P_{m-1}$, $\partial^{\boldsymbol{\alpha}} q = 0 \ \forall |\boldsymbol{\alpha}| = m \implies \forall u \in W^{m,p}(\Omega)$,

$$\begin{aligned} & \partial^{\boldsymbol{\alpha}}(u+q) = \partial^{\boldsymbol{\alpha}} u \quad \forall q \in P_{m-1} \ \forall |\boldsymbol{\alpha}| = m \\ \implies\ & \|\partial^{\boldsymbol{\alpha}}(u+q)\|_{L^p(\Omega)} = \|\partial^{\boldsymbol{\alpha}} u\|_{L^p(\Omega)} \quad \forall |\boldsymbol{\alpha}| = m \\ \implies\ & |u+q|_{m,p,\Omega} = |u|_{m,p,\Omega} \quad \forall u \in [u] \end{aligned}$$

with $[u] \in W^{m,p}(\Omega)/P_{m-1}$, $\forall q \in P_{m-1}$. Then,

$$\begin{aligned} \|[u]\|_{W^{m,p}(\Omega)/P_{m-1}} &= \inf_{q\in P_{m-1}} \|u+q\|_{m,p,\Omega} \\ &= \inf_{q\in P_{m-1}} \Big[|u|^p_{m,p,\Omega} + \sum_{1\le|\boldsymbol{\alpha}|\le m-1} \|\partial^{\boldsymbol{\alpha}}(u+q)\|^p_{L^p(\Omega)} \Big]^{1/p} \\ &= |u|_{m,p,\Omega} + \inf_{q\in P_{m-1}} \Big[\sum_{1\le|\boldsymbol{\alpha}|\le m-1} \|\partial^{\boldsymbol{\alpha}}(u+q)\|^p_{L^p(\Omega)} \Big]^{1/p}. \end{aligned} \tag{2.15.49}$$

This suggests the definition of the semi-norm $|[\cdot]|_{W^{m,p}(\Omega)/P_{m-1}}$ should be:

$$|[u]|_{W^{m,p}(\Omega)/P_{m-1}} = |u|_{m,p,\Omega} \quad \forall [u] \in W^{m,p}(\Omega)/P_{m-1} \text{ with } u \in [u]. \tag{2.15.50}$$

Now we show that the semi-norm $|[\cdot]|_{W^{m,p}(\Omega)/P_{m-1}}$ is a *norm* in $W^{m,p}(\Omega)/P_{m-1}$ *equivalent* to the original norm $\|[\cdot]\|_{W^{m,p}(\Omega)/P_{m-1}}$.

Theorem 2.15.7. *Let $\Omega \subset \mathbb{R}^n$ be a bounded domain with Lipschitz continuous boundary Γ (see Definition* D.2.3.1, *Appendix* D*). Then $\exists C_1, C_2 > 0$ such that, $\forall [u] \in W^{m,p}(\Omega)/P_{m-1}$, $m \in \mathbb{N}$, $1 \le p < \infty$, with $u \in [u]$,*

$$C_1 \|[u]\|_{W^{m,p}(\Omega)/P_{m-1}} \le |u|_{m,p,\Omega} = |[u]|_{W^{m,p}(\Omega)/P_{m-1}} \le C_2 \|[u]\|_{W^{m,p}(\Omega)/P_{m-1}}. \tag{2.15.51}$$

For $p = 2$, $W^{m,2}(\Omega)/P_{m-1} \equiv H^m(\Omega)/P_{m-1}$ is a *Hilbert space* equipped with the inner product $\langle[\cdot],[\cdot]\rangle_{H^m(\Omega)/P_{m-1}}$: $\forall[u],[v] \in H^m(\Omega)/P_{m-1}$ with $u \in [u]$, $v \in [v]$,

$$\langle[u],[v]\rangle_{H^m(\Omega)/P_{m-1}} = \sum_{|\boldsymbol{\alpha}|=m} \langle\partial^{\boldsymbol{\alpha}} u, \partial^{\boldsymbol{\alpha}} v\rangle_{L^2(\Omega)} = \sum_{|\boldsymbol{\alpha}|=m} \int_\Omega \partial^{\boldsymbol{\alpha}} u(\mathbf{x}) \partial^{\boldsymbol{\alpha}} v(\mathbf{x}) d\mathbf{x}. \tag{2.15.52}$$

Proof of Theorem 2.15.7. $\forall u \in W^{m,p}(\Omega)$, the inequality (2.15.44) holds with $l_i \in (W^{m,p}(\Omega))'$, $\langle l_i, q_j\rangle = \delta_{ij}$, $1 \le i,j \le N = \dim(P_{m-1})$, $(q_i)_{i=1}^N$ being a basis in P_{m-1}. But $\forall u \in W^{m,p}(\Omega)$, $\exists q_u \in P_{m-1}$ such that $\langle l_i, u + q_u\rangle = 0$ for $1 \le i \le N$. (2.15.53)

In fact, we define $q_u = -\sum_{j=1}^N \langle l_j, u\rangle q_j \in P_{m-1}$. Then, for $1 \le i \le N$,

$$\langle l_i, q_u\rangle = -\sum_{j=1}^N \langle l_j, u\rangle\langle l_i, q_j\rangle = -\sum_{j=1}^N \langle l_j, u\rangle\delta_{ij} = -\langle l_i, u\rangle$$

$\Longrightarrow \langle l_i, q_u\rangle + \langle l_i, u\rangle = 0$ for $1 \le i \le N$

$\Longrightarrow \langle l_i, u + q_u\rangle = 0\ \forall i = 1,2,\dots,N$. Then, for $q_u \in P_{m-1}$ satisfying (2.15.53), $\langle l_i, u + q_u\rangle = 0$, we get from (2.15.44): $\exists C_1 > 0$ such that $\forall[u] \in W^{m,p}(\Omega)/P_{m-1}$ with $u \in [u]$,

$$\begin{aligned} C_1\|[u]\|_{W^{m,p}(\Omega)/P_{m-1}} &= C_1(\inf_{w\in P_{m-1}} \|u + w\|_{m,p,\Omega}) \le C_1\|u + q_u\|_{m,p,\Omega} \\ &\le (|u + q_u|^p_{m,p,\Omega} + 0)^{1/p} = |u + q_u|_{m,p,\Omega} = |u|_{m,p,\Omega}, \end{aligned}$$

since for $q_u \in P_{m-1}$, $|u + q_u|_{m,p,\Omega} = |u|_{m,p,\Omega}$. Thus, we have proved the left-hand side inequality:

$$\begin{aligned} &C_1\|[u]\|_{W^{m,p}(\Omega)/P_{m-1}} \le |u|_{m,p,\Omega} = |[u]|_{W^{m,p}(\Omega)/P_{m-1}} \quad \text{(by (2.15.50))} \\ &\le \|[u]\|_{W^{m,p}(\Omega)/P_{m-1}} \quad \forall[u] \in W^{m,p}(\Omega)/P_{m-1} \text{ with } u \in [u]; \end{aligned}$$

the right-hand side inequality holds with $C_2 = 1 > 0$ and the result follows.

Since the semi-norm $|[u]|_{W^{m,p}(\Omega)/P_{m-1}} = |u|_{W^{m,p}(\Omega)}$ is a norm in $W^{m,p}(\Omega)/P_{m-1}$ *equivalent to* the original norm $\|[u]\|_{W^{m,p}(\Omega)/P_{m-1}}$, for $p = 2$, the result follows immediately. □

2.15.17 Density results

- $\mathcal{D}(\Omega)$ is *dense* in $W_0^{m,p}(\Omega)$ for $1 \le p < \infty$, $m \in \mathbb{N}$ and *arbitrary* Ω (2.15.54a)

  by Definition 2.15.4.

- $\mathcal{D}(\mathbb{R}^n)$ is *dense* in $W^{m,p}(\mathbb{R}^n)$ for $1 \le p < \infty$, $m \in \mathbb{N}_0$ by Theorem 6.8.9. (2.15.54b)

- For *fixed* p with $1 \le p < \infty$ and $m_2 < m_1$, $W_0^{m_1,p}(\Omega)$ is *dense* in $W_0^{m_2,p}(\Omega)$. (2.15.54c)

- $\mathcal{D}(\overline{\Omega})$ defined by (2.15.26e) (see also Definition 8.10.2) is *dense* in $W^{m,p}(\Omega)$ for Ω with the *m-extension property* (for example, Ω with a Lipschitz continuous boundary, or Ω a C^m-regular domain (see Definition 8.10.4 and Appendix D)), $1 \le p < \infty$, $m \in \mathbb{N}$ (see also (8.10.98d) Section 8.10). (2.15.54d)

Theorem 2.15.8 (Meyers and Serrin (see Adams [12, p. 52])). *For* arbitrary *domain* $\Omega \subset \mathbb{R}^n$, $\mathcal{D}(\overline{\Omega}) \cap W^{m,p}(\Omega)$ *(resp.* $C^{m,p}(\Omega) \cap W^{m,p}(\Omega)$*),* $1 \le p < \infty$, $m \in \mathbb{N}$, *is* dense *in* $W^{m,p}(\Omega)$, *i.e.* $\forall u \in W^{m,p}(\Omega)$, $\exists$ *a sequence* $(u_k)_{k\in\mathbb{N}}$ *in* $\mathcal{D}(\overline{\Omega}) \cap W^{m,p}(\Omega)$ *(resp.* $C^{m,p}(\Omega) \cap W^{m,p}(\Omega)$*) such that* $u_k \to u$ *in* $W^{m,p}(\Omega)$ *as* $k \to \infty$.

2.15.18 A non-density result

Proposition 2.15.8. *For* $\Omega \neq \mathbb{R}^n$ *with measure* $\mu(\mathbb{R}^n \setminus \Omega) > 0$, $\mathcal{D}(\Omega)$ *is* not *dense in* $W^{m,p}(\Omega)$ $\forall m \in \mathbb{N}, 1 \le p < \infty$. *(Precisely speaking, for* Ω *with* $\complement\Omega = \mathbb{R}^n \setminus \Omega$ *not* $(m; q)$*-polar,* $\frac{1}{p} + \frac{1}{q} = 1$ *(see Lions* [13]*).)*

In particular, for $p = 2$, $\mathcal{D}(\Omega)$ *is* not *dense in* $H^m(\Omega)$. (2.15.54e)

Proof. Suppose that the contrary holds, i.e. $\mathcal{D}(\Omega)$ is *dense* in $W^{m,p}(\Omega)$. Consider the continuous, linear mapping $\phi \in \mathcal{D}(\Omega) \longmapsto \tilde{\phi} \in \mathcal{D}(\mathbb{R}^n)$, $\tilde{\phi}$ being the null extension to $\mathbb{R}^n$ of ϕ as defined in Theorem 2.15.5 with $\|\tilde{\phi}\|_{m,p,\mathbb{R}^n} = \|\phi\|_{m,p,\Omega}$. Since, by our assumption, $\mathcal{D}(\Omega)$ is dense in $W^{m,p}(\Omega)$, this linear continuous mapping can be extended by continuity to a *continuous* linear mapping $u \in W^{m,p}(\Omega) \longmapsto \tilde{u} \in W^{m,p}(\mathbb{R}^n)$, $\tilde{u}$ being the null extension to $\mathbb{R}^n$ of u, from $W^{m,p}(\Omega)$ into $W^{m,p}(\mathbb{R}^n)$ with $\|\tilde{u}\|_{m,p,\mathbb{R}^n} = \|u\|_{m,p,\Omega}$, which is impossible if we choose, for example, $\Omega = B(\mathbf{0}; 1) =$ unit ball in $\mathbb{R}^n$ with boundary Γ and $u = 1$ in Ω. Then $\mu(\mathbb{R}^n \setminus \Omega) = \mu(\mathbb{R}^n \setminus B(\mathbf{0}; 1)) > 0$ and $u = 1 \in W^{m,p}(\Omega) \forall m \in \mathbb{N}, 1 \le p < \infty$, but its null extension $\tilde{u} \notin W^{m,p}(\mathbb{R}^n)$. In fact, the distributional derivative $\frac{\partial \tilde{u}}{\partial x_k} \in \mathcal{D}'(\mathbb{R}^n)$ of discontinuous $\tilde{u}$ is given by an expression containing the Dirac distribution δ_Γ concentrated on the boundary Γ. (For $n = 2$, see (3.1.29) in Chapter 3 with $[\frac{\partial \tilde{u}}{\partial x_k}(\mathbf{x})] = 0$, $\Gamma_0 = \Gamma$, jump $J_0 = 0 - 1 = -1$, $\Theta_k = \mathbf{n} \cdot \mathbf{i}_k$.) Hence, $\frac{\partial \tilde{u}}{\partial x_k} \notin L^1_{\text{loc}}(\mathbb{R}^n) \Longrightarrow \frac{\partial \tilde{u}}{\partial x_k} \notin L^p(\mathbb{R}^n)$ $\Longrightarrow \tilde{u} \notin W^{1,p}(\mathbb{R}^n) \Longrightarrow \tilde{u} \notin W^{m,p}(\mathbb{R}^n)$, which contradicts that $\tilde{u} \in W^{m,p}(\mathbb{R}^n)$ $\forall m \in \mathbb{N}, 1 \le p < \infty$. Hence, our original assumption that $\mathcal{D}(\Omega)$ is dense in $W^{m,p}(\Omega)$ is *wrong* and the result follows. □

- For *bounded* domain $\Omega \subset \mathbb{R}^n$ with a *sufficiently smooth* (for example, C^m-regular or Lipschitz continuous (see Definition D.2.3.1 in Appendix D)) boundary, $\mathcal{D}(\Omega)$ is *not* dense in $W^{m,p}(\Omega)$, $1 \le p \le \infty$, $m \in \mathbb{N}$. (2.15.54f)

- For *arbitrary* $\Omega \subset \mathbb{R}^n$, $\mathcal{D}(\Omega)$ is *not* dense in $W^{m,\infty}(\Omega)$ for any $m \in \mathbb{N}_0$. (2.15.54g)
- For *arbitrary* domain $\Omega \subset \mathbb{R}^n$, $\mathcal{D}(\overline{\Omega})$ is *not* dense in $W^{m,p}(\Omega)$, $1 \leq p < \infty$, $m \in \mathbb{N}$ (see also Theorem 2.15.8). (2.15.54h)

  For example, for $\Omega \subset \mathbb{R}^2$ in the example in Remark 8.10.2, which does not possess the m-extension property for $m = 1$, $\mathcal{D}(\overline{\Omega})$ is not dense in $W^{1,p}(\Omega)$, $1 \leq p < \infty$. In fact, for $\Omega = \{(x_1, x_2) : 0 < |x_1| < 1, 0 < x_2 < 1\}$ and $u \in W^{1,2}(\Omega) \equiv H^1(\Omega)$ defined by $u(x_1, x_2) = 1$ for $0 < x_1 < 1$ and $= 0$ for $-1 < x_1 < 0$, $x_2 \in]0, 1[$, (see (8.10.26a)), $\nexists$ any sequence $(u_k)_{k\in\mathbb{N}}$ in $\mathcal{D}(\overline{\Omega})$ such that $u_k \to u$ in $W^{1,2}(\Omega)$ as $k \to \infty$.

For other density results, see Section 8.10, Chapter 8.

2.15.19 Algebraic inclusion $\subset$ and imbedding ($\hookrightarrow$) results

1. For *fixed* p with $1 \leq p \leq \infty$ and $m_1 > m_2$, $W^{m_1,p}(\Omega) \hookrightarrow W^{m_2,p}(\Omega)$, i.e. $W^{m_1,p}(\Omega) \subset W^{m_2,p}(\Omega)$ (*algebraic inclusion*) and $\exists C > 0$ such that $\|u\|_{m_2,p,\Omega} \leq C\|u\|_{m_1,p,\Omega}$ $\forall u \in W^{m_1,p}(\Omega)$ (continuity of $\hookrightarrow$), which follows from Definition 2.15.3 and (2.15.29)–(2.15.31). with $C = 1$; (2.15.55a)
2. For *fixed* p with $1 \leq p < \infty$ and $m_1 > m_2$, $W_0^{m_1,p}(\Omega) \hookrightarrow W_0^{m_2,p}(\Omega)$, i.e. $W_0^{m_1,p}(\Omega) \subset W_0^{m_2,p}(\Omega)$ and $\exists C > 0$ such that $\|u\|_{m_2,p,\Omega} \leq C\|u\|_{m_1,p,\Omega}$ $\forall u \in W_0^{m_1,p}(\Omega)$. (2.15.55b)
3. $\mathcal{D}(\Omega) \hookrightarrow W_0^{m,p}(\Omega) \hookrightarrow W_{m,p}(\Omega) \hookrightarrow \mathcal{D}'(\Omega)$ $\forall m \in \mathbb{N}$, $1 \leq p < \infty$. (2.15.55c)
4. For $\Omega = \mathbb{R}^n$, $W_0^{m,p}(\mathbb{R}^n) \equiv W^{m,p}(\mathbb{R}^n)$ (see Theorem 8.9.6). (2.15.55d)
5. For $\Omega \subsetneq \mathbb{R}^n$ with measure $\mu(\mathbb{R}^n \setminus \Omega) > 0$, $W_0^{m,p}(\Omega) \subsetneq W_{m,p}(\Omega)$, $1 \leq p < \infty$, $m \in \mathbb{N}$ (see counterexample in (2.15.27f)). (2.15.55e)

For more imbedding ($\hookrightarrow$) results, see Sections 4.3, 8.10 and 8.12.
For compact imbedding ($\hookrightarrow\hookrightarrow$) results, see Sections 8.11 and 8.12, Chapter 8.

2.15.20 Space $W^{s,p}(\Omega)$ for arbitrary $s \in \mathbb{R}$

For $W^{s,p}(\Omega)$ with $s > 0$, $1 \leq p < \infty$, and its closed subspaces $W_0^{s,p}(\Omega)$, $W_{00}^{s,p}(\Omega)$ for $s > 0$, $1 < p < \infty$, and their dual spaces $W^{-s,q}(\Omega)$, $(W_{00}^{s,p}(\Omega))'$, see Sections 8.10–8.12, Chapter 8.

Remark 2.15.1. Since we will be primarily concerned with real-valued functions, we have not considered Sobolev spaces $H^m(\Omega)$ (resp. $W^{m,p}(\Omega), 1 \leq p \leq \infty$) of (equivalence classes $[u]$ of) *complex-valued* functions $u \in L^2(\Omega)$, $\partial^{\boldsymbol{\alpha}} u \in L^2(\Omega)$ $\forall |\boldsymbol{\alpha}| \leq m$ (resp. $u \in L^p(\Omega)$, $\partial^{\boldsymbol{\alpha}} u \in L^p(\Omega)$). Complexification of $H^m(\Omega)$ (resp. $W^{m,p}(\Omega), 1 \leq p \leq \infty$) is a straightforward procedure. For example, for $H^m(\Omega) =$

$\{u : u(\mathbf{x})$ is a complex-valued function of real variables $\mathbf{x}$, $u \in L^2(\Omega)$, $\partial^{\boldsymbol{\alpha}} u \in L^2(\Omega)$ $\forall |\boldsymbol{\alpha}| \le m\}$, (2.15.56a)
where $L^2(\Omega)$ is a (complex) Hilbert space with *complex inner product* $\langle \cdot, \cdot \rangle_{L^2(\Omega)}$ such that

$$\langle u, v \rangle_{L^2(\Omega)} = \int_\Omega u(\mathbf{x})\overline{v(\mathbf{x})} d\mathbf{x} \quad \forall u, v \in L^2(\Omega). \tag{2.15.56b}$$

Then, (2.15.2)–(2.15.6) hold, and $u \in H^m(\Omega) \Longrightarrow \lambda u \in H^m(\Omega) \forall \lambda \in \mathbb{C}$. Hence, $H^m(\Omega)$ is a complex vector space.

$H^m(\Omega)$ equipped with complex inner product $\langle \cdot, \cdot \rangle_{m,\Omega}$ defined, $\forall u, v \in H^m(\Omega)$, by:

$$\begin{aligned} \langle u, v \rangle_{m,\Omega} &= \langle u, v \rangle_{L^2(\Omega)} + \sum_{1 \le |\boldsymbol{\alpha}| \le m} \langle \partial^{\boldsymbol{\alpha}} u, \partial v \rangle_{L^2(\Omega)} \\ &= \int_\Omega u(\mathbf{x})\overline{v(\mathbf{x})} d\mathbf{x} + \sum_{1 \le |\boldsymbol{\alpha}| \le m} \int_\Omega \partial^\alpha u(\mathbf{x}) \overline{\partial^\alpha v(\mathbf{x})} d\mathbf{x}, \end{aligned} \tag{2.15.56c}$$

where $\overline{v(\mathbf{x})}$, $\overline{\partial^{\boldsymbol{\alpha}} v(\mathbf{x})}$ are the complex conjugates of $v(\mathbf{x})$ and $\partial^{\boldsymbol{\alpha}} v(\mathbf{x})$ respectively, i.e.

$$u(\mathbf{x}) = u_1(\mathbf{x}) + i u_2(\mathbf{x}) \quad \text{with } u_1(\mathbf{x}) = \text{Re}[u(\mathbf{x})], u_2(\mathbf{x}) = \text{Im}[u(\mathbf{x})], \tag{2.15.57a}$$

$$v(\mathbf{x}) = v_1(\mathbf{x}) + i v_2(\mathbf{x}) \quad \text{with } v_1(\mathbf{x}) = \text{Re}[v(\mathbf{x})], v_2(\mathbf{x}) = \text{Im}[v(\mathbf{x})] \tag{2.15.57b}$$

$$\Longrightarrow \quad \overline{v(\mathbf{x})} = \overline{v_1(\mathbf{x}) + i v_2(\mathbf{x})} = v_1(\mathbf{x}) - i v_2(\mathbf{x}) \quad \text{a.e. on } \Omega, \tag{2.15.57c}$$

$\overline{\partial^{\boldsymbol{\alpha}} v(\mathbf{x})} = \partial^{\boldsymbol{\alpha}} v_1(\mathbf{x}) - i \partial^{\boldsymbol{\alpha}} v_2(\mathbf{x})$ a.e. on Ω, is a (complex) Hilbert space. (2.15.57d)

All other formulae and results remain unchanged or undergo minor changes owing to (2.15.56a)–(2.15.57d).

Hence, these minor changes (2.15.56a)–(2.15.57d) due to complexification will be introduced in later chapters, if necessary, without any accompanying explanation.

Chapter 3

Derivatives of piecewise smooth functions, Green's formula, elementary solutions, applications to Sobolev spaces

3.1 Distributional derivatives of piecewise smooth functions

Motivation Piecewise smooth functions in general, and piecewise polynomials in particular, are the workhorses needed for efficient and elegant solution of interpolation and approximation problems. In many methods of approximation (for example, the finite element method for elliptic boundary value problems [34], [35], [36]), construction of finite element subspaces of Sobolev spaces $H^1(\Omega)$, $H^2(\Omega)$, etc. with the help of piecewise polynomials is essential. For this, the distributional derivatives of these piecewise smooth functions must belong to $L^2(\Omega)$, which suggests studying their differentiation in detail. (For the construction of subspaces of $H^m(\Omega)$, $m = 1, 2$, see Section 3.4 at the end of the chapter).

3.1.1 Case of single variable ($n = 1$)

Let $\{x_i\}_{i=1}^n$ be a set of n points on $]a, b[$, $a < x_1 < x_2 < \cdots < b$. Let f be a *piecewise smooth* function having *continuous* ordinary derivatives $f'(x), f''(x), \ldots, f^{(m)}(x)$ in the *usual pointwise sense* everywhere on $]a,b[$ except at the points $\{x_i\}_{i=1}^n$, where $f(x), f'(x), f''(x), \ldots, f^{(m)}(x)$ have discontinuities of the *first* kind, i.e. the right-hand side and left-hand side limits of $f(x), f'(x), \ldots, f^{(m)}(x)$ at $\{x_i\}_{i=1}^n$ exist: for $0 \leq k \leq m$,

$$\begin{aligned} f^{(k)}(x_i^+) &= \lim_{x \to x_i^+} f^{(k)}(x), \\ f^{(k)}(x_i^-) &= \lim_{x \to x_i^-} f^{(k)}(x), \\ f^{(0)}(x) &= f(x), \end{aligned} \tag{3.1.1}$$

such that $f^{(k)}(x)$ has *finite* jumps $J_i^{(k)}$ at x_i:

$$J_i^{(k)} = f^{(k)}(x_i^+) - f^{(k)}(x_i^-); \quad J_i^{(0)} = J_i. \tag{3.1.2}$$

The kth-order derivative of f in the *sense of distribution* will be denoted by $\frac{d^k f}{dx^k}$ or $f^{(k)} \in \mathcal{D}'(]a,b[)$, and the distributions defined by the *usual ordinary derivatives* $\frac{d^k f}{dx^k}(x)$ or $f^{(k)}(x)$ in the *pointwise sense* will be denoted by $[\frac{d^k f}{dx^k}(x)]$ or $[f^{(k)}(x)] \in \mathcal{D}'(]a,b[)$ for $1 \le k \le m$, i.e. $\frac{d^k f}{dx^k} = \frac{d^k T_f}{dx^k} \in \mathcal{D}'(]a,b[)$ with

$$\left\langle \frac{d^k f}{dx^k}, \phi \right\rangle = \left\langle \frac{d^k T_f}{dx^k}, \phi \right\rangle = (-1)^k \langle T_f, \phi^{(k)} \rangle = (-1)^k \langle f, \phi^{(k)} \rangle$$
$$= (-1)^k \int_a^b f(x) \frac{d^k \phi}{dx^k}(x) dx \quad \forall \phi \in \mathcal{D}(]a,b[), \tag{3.1.3}$$

and $[\frac{d^k f}{dx^k}(x)] = T_{[\frac{d^k f}{dx^k}(x)]} \in \mathcal{D}'(]a,b[)$ with

$$\left\langle \left[\frac{d^k f}{dx^k}(x)\right], \phi \right\rangle = \left\langle T_{[\frac{d^k f}{dx^k}(x)]}, \phi \right\rangle = \int_a^b \left[\frac{d^k f}{dx^k}(x)\right] \phi(x) dx \quad \forall \phi \in \mathcal{D}(]a,b[). \tag{3.1.4}$$

Note that for $1 \le k \le m$, $[\frac{d^k f}{dx^k}(x)] \in \mathcal{D}'(]a,b[)$ is a *regular* distribution, whereas $\frac{d^k f}{dx^k} \in \mathcal{D}'(]a,b[)$ is *not* a regular distribution *in general*. In fact, they are related by:

Theorem 3.1.1. *Let f be a piecewise smooth function having derivatives $f'(x)$, $f''(x), \ldots, f^{(m)}(x)$ (in the usual pointwise sense) which are continuous everywhere on $]a,b[$ except at the points $\{x_i\}_{i=1}^n$ of discontinuity (of the* first *kind) of $f^{(k)}(x)$, $0 \le k \le m$, where conditions* (3.1.1) *and* (3.1.2) *hold (see Figure 3.1). Let $\frac{d^k f}{dx^k}$ be the kth-order* distributional *derivative of f on $]a,b[$, and $[\frac{d^k f}{dx^k}(x)]$ be the distribution defined by the* ordinary derivative *$\frac{d^k f}{dx^k}(x)$ in the* usual pointwise sense *in* (3.1.3) *and* (3.1.4) *respectively.*

Then $\frac{d^k f}{dx^k}$ and $[\frac{d^k f}{dx^k}(x)]$ in $\mathcal{D}'(]a,b[)$ are related by:

$$k = 1: \qquad \frac{df}{dx} = \left[\frac{df}{dx}(x)\right] + \sum_{i=1}^n J_i \delta_{x_i};$$
$$k = 2: \qquad \frac{d^2 f}{dx^2} = \left[\frac{d^2 f}{dx^2}(x)\right] + \sum_{i=1}^n J_i \delta_{x_i}^{(1)} + \sum_{i=1}^n J_i^{(1)} \delta_{x_i}; \tag{3.1.5}$$
$$\vdots \qquad \vdots$$

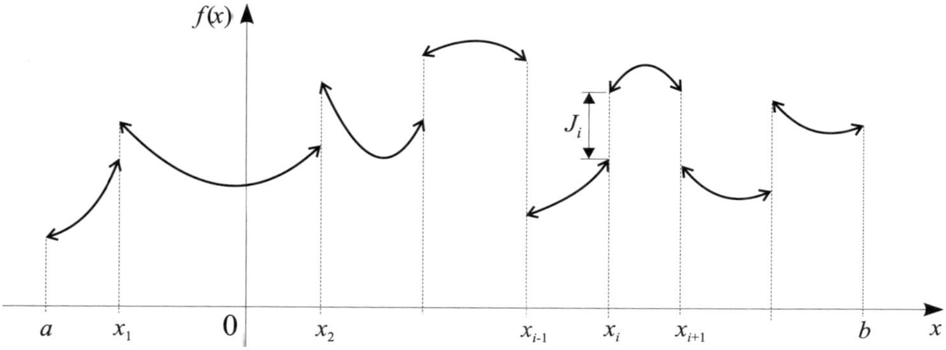

Figure 3.1 Piecewise smooth function f with finite jumps J_i at points of discontinuity of the first kind at $x_1 < x_2 < \cdots < x_n$

$$\frac{d^k f}{dx^k} = \left[\frac{d^k f}{dx^k}(x)\right] + \sum_{i=1}^{n} J_i \delta_{x_i}^{(k-1)} + \sum_{i=1}^{n} J_i^{(1)} \delta_{x_i}^{(k-2)} + \cdots$$
$$+ \sum_{i=1}^{n} J_i^{(k-1)} \delta_{x_i}, \quad 1 \le k \le m,$$

where $\delta_{x_i} \in \mathcal{D}'(]a,b[)$ *is the* singular *Dirac distribution with mass/charge/force etc., as the case may be, concentrated at* x_i *and defined by* $\langle \delta_{x_i}, \phi\rangle = \phi(x_i)\ \forall \phi \in \mathcal{D}(]a,b[)$, $1 \le i \le n$ *(see* (1.3.28)*);* $\delta_{x_i}^{(k)} \in \mathcal{D}'(]a,b[)$ *is the* k*th-order derivative of* δ_{x_i} *defined by* $\langle \delta_{x_i}^{(k)}, \phi\rangle = (-1)^k \langle \delta_{x_i}, \phi^{(k)}\rangle = (-1)^k \phi^{(k)}(x_i)\ \forall \phi \in \mathcal{D}(]a,b[)$ *(see* (2.3.8)*).*

Proof. We give the proof for $k = 1$, since for $k \ge 2$ the proof is similar. From the definition (3.1.3), we have:

$$\left\langle \frac{df}{dx}, \phi \right\rangle = -\left\langle f, \frac{d\phi}{dx} \right\rangle = -\int_a^b f(x)\frac{d\phi}{dx}(x)dx \quad \forall \phi \in \mathcal{D}(]a,b[), \tag{3.1.6}$$

since f is piecewise continuous on $]a,b[$ with finite jumps $J_i = f(x_i^+) - f(x_i^-)$ at the points x_i of discontinuity, $1 \le i \le n$, and consequently $f \in L^1_{\text{loc}}(]a,b[)$.

$$\int_a^b f(x)\phi'(x)dx = \int_a^{x_1^-}(\ldots) + \cdots + \int_{x_{i-1}^+}^{x_i^-}(\ldots) + \int_{x_i^+}^{x_{i+1}^-}(\ldots) + \cdots + \int_{x_n^+}^{b}(\ldots). \tag{3.1.7}$$

Integrating by parts and applying the properties of $\phi \in \mathcal{D}(]a,b[)$ ($\phi(x_i) = \phi(x_i^+) = \phi(x_i^-)$ by virtue of the continuity of ϕ at x_i, $\phi(a) = \phi(b) = 0$), we have

$$\int_a^{x_1^-} f(x)\phi'(x)dx = f(x_1^-)\phi(x_1) - \int_a^{x_1^-} f'(x)\phi(x)dx,$$

and for $(1 \le i \le n-1)$,

$$\int_{x_i^+}^{x_{i+1}^-} f(x)\phi'(x)dx = f(x_{i+1}^-)\phi(x_{i+1}) - f(x_i^+)\phi(x_i) - \int_{x_i^+}^{x_{i+1}^-} f'(x)\phi(x)dx,$$

$$\int_{x_n^+}^{b} f(x)\phi'(x)dx = -f(x_n^+)\phi(x_n) - \int_{x_n^+}^{b} f'(x)\phi(x)dx \quad (\phi(b)=0). \quad (3.1.8)$$

$f'(x)$ is piecewise continuous on $]a,b[$ with finite jumps $J_i^{(1)} = f'(x_i^+) - f'(x_i^-)$ at the points x_i of the discontinuity of $f'(x)$, $1 \le i \le n$ (see (3.1.2)) $\Longrightarrow f'(x) \in L^1_{\text{loc}}(]a,b[)$. Hence, we can combine the integrals in (3.1.8) into a single integral on $]a,b[$ and write (3.1.7) as follows: $\forall \phi \in \mathcal{D}(]a,b[)$,

$$\int_a^b f(x)\frac{d\phi}{dx}(x)dx = -\int_a^b f'(x)\phi(x)dx - \sum_{i=1}^{n}[f(x_i^+) - f(x_i^-)]\phi(x_i).$$

But from the definition of the distribution $[\frac{df}{dx}(x)] \in \mathcal{D}'(]a,b[)$ in (3.1.4), we have

$$\left\langle \left[\frac{df}{dx}(x)\right], \phi \right\rangle = \int_a^b f'(x)\phi(x)dx \quad \forall \phi \in \mathcal{D}(]a,b[).$$

Hence,

$$\int_a^b f(x)\frac{d\phi}{dx}dx = -\left\langle \left[\frac{df}{dx}(x)\right], \phi \right\rangle - \sum_{i=1}^{n} J_i \langle \delta_{x_i}, \phi \rangle = -\left\langle \left[\frac{df}{dx}(x)\right] + \sum_{i=1}^{n} J_i \delta_{x_i}, \phi \right\rangle.$$

Finally, from (3.1.6), we get:

$$\left\langle \frac{df}{dx}, \phi \right\rangle = \left\langle \left[\frac{df}{dx}(x)\right] + \sum_{i=1}^{n} J_i \delta_{x_i}, \phi \right\rangle \quad \forall \phi \in \mathcal{D}(]a,b[)$$

$$\Longleftrightarrow \quad \frac{df}{dx} = \left[\frac{df}{dx}(x)\right] + \sum_{i=1}^{n} J_i \delta_{x_i} \quad \text{in } \mathcal{D}'(]a,b[). \qquad \square$$

Remark 3.1.1.

- The effect of the discontinuity of f at x_i, $1 \le i \le n$, appears in the form of a point mass/charge/force, as the case may be, concentrated at x_i in the derivative of f in the *distributional sense*.
- $[\frac{df}{dx}(x)] \in \mathcal{D}'(]a,b[)$ is a *regular* distribution defined by the ordinary derivative $f'(x)$ in the usual pointwise sense, whereas the distributional derivative $\frac{df}{dx} \in \mathcal{D}'(]a,b[)$ of f is a *singular* distribution, i.e. the distributional derivative and the usual derivative in the pointwise sense do *not* coincide here!

Example 3.1.1. Consider the sawtooth function $f(x) = \frac{1}{2} - \frac{x}{2\pi}$ for $x \in]0, 2\pi[$ such that f is 2π-periodic on $]-\infty, \infty[$, i.e. $f(x \pm 2\pi) = f(x)$. Then f has discontinuities at $x = 0, \pm 2\pi, \pm 4\pi, \dots$ with jump $J_k = \frac{1}{2} - (-\frac{1}{2}) = 1$ at $x_k = \pm 2k\pi$ with $k = 0, 1, \dots, \infty$ (see Figure 3.2).

Hence, $f' = [\frac{df}{dx}(x)] + \sum_{k=0}^{\infty} 1 \cdot \delta_{\pm 2k\pi} = -\frac{1}{2\pi} + \sum_{k=-\infty}^{\infty} \delta_{2k\pi}$, where $\delta_{2k\pi} = \delta(x - 2k\pi)$ is Dirac distribution with unit mass/charge/force concentrated at $x = 2k\pi$, $k = 0, \pm 1, \pm 2, \dots, \pm\infty$. In Example 2.11.2, it is shown that $\sum_{k=-\infty}^{\infty} \delta_{2k\pi}$ converges in $\mathcal{D}'(]-\infty, \infty[)$. Moreover, the series $\sum_{k=-\infty}^{\infty} \delta_{2k\pi}$ is a *periodic Dirac distribution* on $\mathbb{R}$ with period 2π (see Section 1.10).

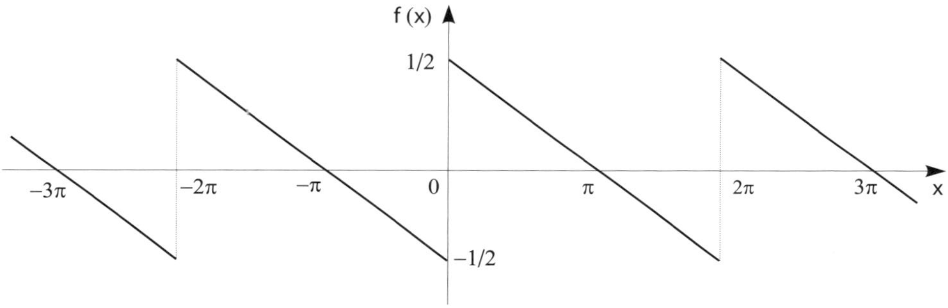

Figure 3.2 Sawtooth function $f(x) = \frac{1}{2} - \frac{x}{2\pi}, x \in]0, 2\pi[$, with periodic extension $f(x \pm 2k\pi) = f(x)$ and jumps $J_k = 1$ at $x_k = \pm 2k\pi, k = 0, 1, 2, \dots$

3.1.2 Case of two variables ($n = 2$)

Let $\Omega \subset \mathbb{R}^2$ be a bounded domain in $\mathbb{R}^2$ with a piecewise smooth boundary Γ such that $\overline{\Omega} = \Omega \cup \Gamma$ and the unit vector $\hat{\mathbf{n}}$ normal to Γ and exterior to Ω is defined almost everywhere on Γ. The (positive) direction of the unit vector $\hat{\boldsymbol{\tau}}$ tangent to Γ is obtained by rotating the exterior unit normal $\hat{\mathbf{n}}$ through $\frac{\pi}{2}$ in the anticlockwise direction (see Figure 3.3), with $\hat{\mathbf{n}} = \cos(\hat{\mathbf{n}}, \hat{\mathbf{i}}_1)\hat{\mathbf{i}}_1 + \cos(\hat{\mathbf{n}}, \hat{\mathbf{i}}_2)\hat{\mathbf{i}}_2$, where $(\hat{\mathbf{n}}, \hat{\mathbf{i}}_k)$ is the angle between $\hat{\mathbf{n}}$ and $\hat{\mathbf{i}}_k$ measured in the anticlockwise direction, $\hat{\mathbf{i}}_k$ being the unit vector in the positive direction of x_k-axis. Measuring angles in the anticlockwise direction, we get $(\hat{\boldsymbol{\tau}}, \hat{\mathbf{i}}_2) = (\hat{\mathbf{n}}, \hat{\mathbf{i}}_1)$,

$$(\hat{\boldsymbol{\tau}}, \hat{\mathbf{i}}_1) = \pi + (\hat{\mathbf{n}}, \hat{\mathbf{i}}_2) \implies \cos(\hat{\boldsymbol{\tau}}, \hat{\mathbf{i}}_1) = -\cos(\hat{\mathbf{n}}, \hat{\mathbf{i}}_2), \cos(\hat{\boldsymbol{\tau}}, \hat{\mathbf{i}}_2) = \cos(\hat{\mathbf{n}}, \hat{\mathbf{i}}_1), \tag{3.1.9}$$

$$dx_1 = \cos(\hat{\boldsymbol{\tau}}, \hat{\mathbf{i}}_1)ds = -\cos(\hat{\mathbf{n}}, \hat{\mathbf{i}}_2)ds, \quad dx_2 = \cos(\hat{\boldsymbol{\tau}}, \hat{\mathbf{i}}_2)ds = \cos(\hat{\mathbf{n}}, \hat{\mathbf{i}}_1)ds, \tag{3.1.10}$$

where ds is the arc length measure.

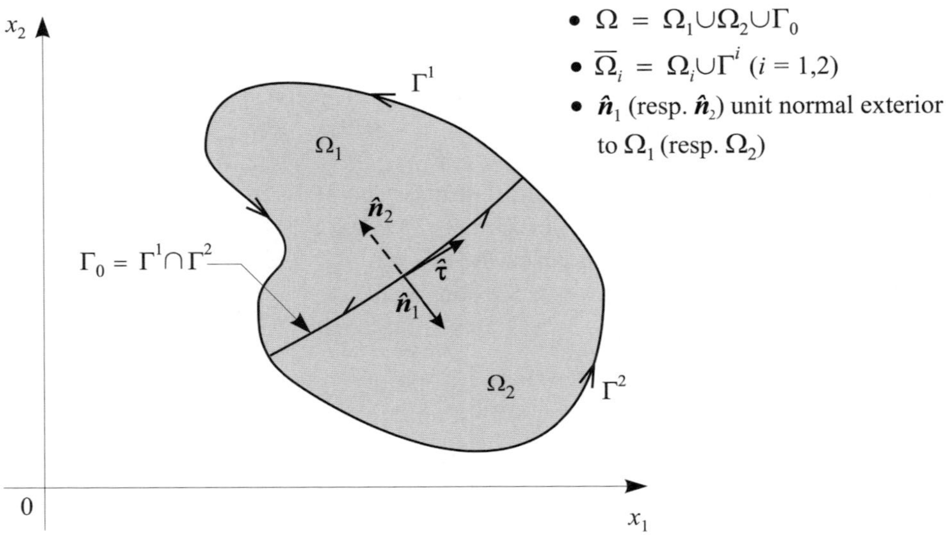

Figure 3.3 Piecewise smooth curve $\overline{\Gamma}_0$ subdividing domain Ω into two subdomains Ω_1 and Ω_2 with closed boundaries Γ^1 and Γ^2 respectively

Let $\Gamma_0 \subset \Omega$ be a piecewise smooth curve which subdivides Ω into two subdomains Ω_1 and Ω_2 with boundaries Γ^1 and Γ^2 respectively such that:

- $\Omega = \Omega_1 \cup \Omega_2 \cup \Gamma_0$, $\overline{\Omega} = \overline{\Omega}_1 \cup \overline{\Omega}_2$, $\overline{\Omega}_i = \Omega_i \cup \Gamma^i$ (see Figure 3.3);

- $\Gamma^i = \Gamma_i \cup \Gamma_0$ $(i = 1, 2)$, $\Gamma^1 \cap \Gamma^2 = \Gamma_0$,

- $\Gamma = \Gamma_1 \cup \Gamma_2$, $\text{meas}(\Gamma_1 \cap \Gamma_2) = 0.$ (3.1.11)

Let f, $[\frac{\partial f}{\partial x_k}(\mathbf{x})]$, $[\frac{\partial^2 f}{\partial x_k \partial x_l}(\mathbf{x})] \in C^0$ $(\overline{\Omega} \setminus \Gamma_0)$ with $\overline{\Omega} \setminus \Gamma_0 = \overline{\Omega} - \Gamma_0$ be bounded and continuous functions in the complement of Γ_0 in $\overline{\Omega}$ (i.e. discontinuous with finite jumps across Γ_0), $[\frac{\partial f}{\partial x_k}(\mathbf{x})]$ and $[\frac{\partial^2 f}{\partial x_k \partial x_l}(\mathbf{x})]$ being partial derivatives of f *in the usual pointwise sense*: for example, for $k = 1$,

$$\left[\frac{\partial f}{\partial x_1}(\mathbf{x})\right] = \lim_{\Delta x_1 \to 0} \frac{f(x_1 + \Delta x_1, x_2) - f(x_1, x_2)}{\Delta x_1}.$$

Define

$$f_i = f \downarrow_{\overline{\Omega}_i - \Gamma_0} \text{ with } f_i(\mathbf{x}) = f(\mathbf{x}) \ \forall \mathbf{x} \in \overline{\Omega}_i - \Gamma_0 \tag{3.1.12}$$

such that f_i and its partial derivatives $[\frac{\partial f_i}{\partial x_k}(\mathbf{x})]$, $[\frac{\partial^2 f_i}{\partial x_k \partial x_l}(\mathbf{x})]$ (in the usual pointwise sense) can be *continuously extended* to Γ_0 as follows: $\forall \boldsymbol{\xi} \in \Gamma_0$,

$$f_1(\boldsymbol{\xi}) = \lim_{\mathbf{x}\to\boldsymbol{\xi},\mathbf{x}\in\Omega_1} f(\mathbf{x}) = f(\boldsymbol{\xi}^-); \quad f_2(\boldsymbol{\xi}) = \lim_{\mathbf{x}\to\boldsymbol{\xi},\mathbf{x}\in\Omega_2} f(\mathbf{x}) = f(\boldsymbol{\xi}^+); \tag{3.1.13}$$

$$\left[\frac{\partial f_1}{\partial x_k}(\boldsymbol{\xi})\right] = \lim_{\mathbf{x}\to\boldsymbol{\xi},\mathbf{x}\in\Omega_1}\left[\frac{\partial f}{\partial x_k}(\mathbf{x})\right] = \left[\frac{\partial f}{\partial x_k}(\boldsymbol{\xi}^-)\right]; \tag{3.1.14}$$

$$\left[\frac{\partial f_2}{\partial x_k}(\boldsymbol{\xi})\right] = \lim_{\mathbf{x}\to\boldsymbol{\xi},\mathbf{x}\in\Omega_2}\left[\frac{\partial f}{\partial x_k}(\mathbf{x})\right] = \left[\frac{\partial f}{\partial x_k}(\boldsymbol{\xi}^+)\right];$$

$$\left[\frac{\partial^2 f_1}{\partial x_k \partial x_l}(\boldsymbol{\xi})\right] = \lim_{\mathbf{x}\to\boldsymbol{\xi},\mathbf{x}\in\Omega_1}\left[\frac{\partial^2 f}{\partial x_k \partial x_l}(\mathbf{x})\right] = \left[\frac{\partial^2 f}{\partial x_k \partial x_l}(\boldsymbol{\xi}^-)\right];$$

$$\left[\frac{\partial^2 f_2}{\partial x_k \partial x_l}(\boldsymbol{\xi})\right] = \lim_{\mathbf{x}\to\boldsymbol{\xi},\mathbf{x}\in\Omega_2}\left[\frac{\partial^2 f}{\partial x_k \partial x_l}(\mathbf{x})\right] = \left[\frac{\partial^2 f}{\partial x_k \partial x_l}(\boldsymbol{\xi}^+)\right]. \tag{3.1.15}$$

Then the jumps J_0, J_k $(k = 1,2)$, $J_{kl} = J_{lk}$ $(k,l = 1,2)$ of f, $[\frac{\partial f}{\partial x_k}(\mathbf{x})]$ and $[\frac{\partial^2 f}{\partial x_k \partial x_l}(\mathbf{x})]$ across Γ_0 respectively are defined, $\forall \mathbf{x} \in \Gamma_0$, by:

$$J_0 = f(\mathbf{x}^+) - f(\mathbf{x}^-) = f_2(\mathbf{x}) - f_1(\mathbf{x}); \tag{3.1.16}$$

$$J_k = \left[\frac{\partial f}{\partial x_k}(\mathbf{x}^+)\right] - \left[\frac{\partial f}{\partial x_k}(\mathbf{x}^-)\right] = \left[\frac{\partial f_2}{\partial x_k}(\mathbf{x})\right] - \left[\frac{\partial f_1}{\partial x_k}(\mathbf{x})\right]; \tag{3.1.17}$$

$$J_{kl} = \left[\frac{\partial^2 f}{\partial x_k \partial x_l}(\mathbf{x}^+)\right] - \left[\frac{\partial^2 f}{\partial x_k \partial x_l}(x^-)\right] = \left[\frac{\partial^2 f_2}{\partial x_k \partial x_l}(\mathbf{x})\right] - \left[\frac{\partial^2 f_1}{\partial x_k \partial x_l}(\mathbf{x})\right]; \tag{3.1.18}$$

such that J_0, J_k, J_{kl} are functions *continuous* on Γ_0.

$f, [\frac{\partial f}{\partial x_k}(\mathbf{x})], [\frac{\partial^2 f}{\partial x_k \partial x_l}(\mathbf{x})] \in C^0(\overline{\Omega} \setminus \Gamma_0)$ satisfying (3.1.12)–(3.1.15) will belong to $L^1_{\text{loc}}(\Omega)$ (since two-dimensional (area) measure $\mu(\Gamma_0) = 0$) and will define regular distributions $T_f, T_{[\frac{\partial f}{\partial x_k}(\mathbf{x})]}, T_{[\frac{\partial^2 f}{\partial x_k \partial x_l}(\mathbf{x})]} \in \mathcal{D}'(\Omega)$ respectively, $\forall \phi \in \mathcal{D}(\Omega)$, by:

$$\langle T_f, \phi \rangle = \langle f, \phi \rangle = \int_\Omega f(\mathbf{x})\phi(\mathbf{x})dx_1 dx_2; \tag{3.1.19}$$

$$\langle T_{[\frac{\partial f}{\partial x_k}(\mathbf{x})]}, \phi \rangle = \left\langle \left[\frac{\partial f}{\partial x_k}(\mathbf{x})\right], \phi \right\rangle = \int_\Omega \frac{\partial f}{\partial x_k}(\mathbf{x})\phi(\mathbf{x})dx_1 dx_2; \tag{3.1.20}$$

$$\left\langle T_{\left[\frac{\partial^2 f}{\partial x_k \partial x_l}(\mathbf{x})\right]}, \phi \right\rangle = \left\langle \left[\frac{\partial^2 f}{\partial x_k \partial x_l}(\mathbf{x})\right], \phi \right\rangle = \int_\Omega \frac{\partial^2 f}{\partial x_k \partial x_l}(\mathbf{x})\phi(\mathbf{x})dx_1 dx_2. \tag{3.1.21}$$

Let $\frac{\partial T_f}{\partial x_k} = \frac{\partial f}{\partial x_k} \in \mathcal{D}'(\Omega)$ and $\frac{\partial^2 T_f}{\partial x_k \partial x_l} = \frac{\partial^2 f}{\partial x_k \partial x_l} \in \mathcal{D}'(\Omega)$ denote the first- and second-order *distributional derivatives* of f defined, $\forall \phi \in \mathcal{D}(\Omega)$, by:

$$\left\langle \frac{\partial T_f}{\partial x_k}, \phi \right\rangle = \left\langle \frac{\partial f}{\partial x_k}, \phi \right\rangle = -\left\langle f, \frac{\partial \phi}{\partial x_k} \right\rangle = -\int_\Omega f \frac{\partial \phi}{\partial x_k} dx_1 dx_2; \tag{3.1.22}$$

$$\left\langle \frac{\partial^2 T_f}{\partial x_k \partial x_l}, \phi \right\rangle = \left\langle \frac{\partial^2 f}{\partial x_k \partial x_l}, \phi \right\rangle = \left\langle f, \frac{\partial^2 \phi}{\partial x_k \partial x_l} \right\rangle = \int_\Omega f \frac{\partial^2 \phi}{\partial x_k \partial x_l} dx_1 dx_2. \tag{3.1.23}$$

Since f and its partial derivatives $[\frac{\partial f}{\partial x_k}(\mathbf{x})]$, $[\frac{\partial^2 f}{\partial x_k \partial x_l}(\mathbf{x})]$ in the usual pointwise sense are discontinuous across Γ_0 with finite jumps J_0, J_k, J_{kl} defined by(3.1.16), (3.1.17) and (3.1.18) respectively, the *distributional* derivatives $\frac{\partial T_f}{\partial x_k} = \frac{\partial f}{\partial x_k}, \frac{\partial^2 T_f}{\partial x_k \partial x_l} = \frac{\partial^2 f}{\partial x_k \partial x_l}$ defined by (3.1.22) and (3.1.23) will *not* be equal to the regular distributions $T_{[\frac{\partial f}{\partial x_k}(\mathbf{x})]} = [\frac{\partial f}{\partial x_k}(\mathbf{x})] \in \mathcal{D}'(\Omega)$, $T_{[\frac{\partial^2 f}{\partial x_k \partial x_l}(\mathbf{x})]} = [\frac{\partial^2 f}{\partial x_l \partial x_k}(\mathbf{x})] \in \mathcal{D}'(\Omega)$ defined by (3.1.20) and (3.1.21) respectively as in the case of a single variable (see Theorem 3.1.1).

The relations between the distributional derivatives $\frac{\partial T_f}{\partial x_k} = \frac{\partial f}{\partial x_k}, \frac{\partial^2 T_f}{\partial x_k \partial x_l} = \frac{\partial^2 f}{\partial x_k \partial x_l}$ and the partial derivatives $[\frac{\partial f}{\partial x_k}(\mathbf{x})]$ and $[\frac{\partial^2 f}{\partial x_k \partial x_l}(\mathbf{x})]$ in the usual pointwise sense will now be established, but for this we need Green's Theorem from calculus.

Theorem 3.1.2 (Green's Theorem [32]). *Let $D \subset \mathbb{R}^2$ be a domain with a piecewise smooth boundary ∂D and $\overline{D} = D \cup \partial D$ (see Figure 3.4). Let P, Q, $[\frac{\partial P}{\partial x_2}(\mathbf{x})]$ and $[\frac{\partial Q}{\partial x_1}(\mathbf{x})] \in C^0(\overline{D})$. Then*

$$\begin{aligned} \int_D \left[\frac{\partial Q}{\partial x_1}(\mathbf{x}) - \frac{\partial P}{\partial x_2}(\mathbf{x})\right] dx_1 dx_2 &= \oint_{\partial D} P dx_1 + Q dx_2 \\ &= \oint_{\partial D} \left[-P \cos(\hat{\mathbf{n}}, \hat{\mathbf{i}}_2) + Q \cos(\hat{\mathbf{n}}, \hat{\mathbf{i}}_1)\right] ds, \end{aligned} \tag{3.1.24}$$

where $dx_1 = -\cos(\hat{\mathbf{n}}, \hat{\mathbf{i}}_2) ds$, $dx_2 = \cos(\hat{\mathbf{n}}, \hat{\mathbf{i}}_1) ds$, $\hat{\mathbf{n}}$ being the unit normal exterior to D, and the line integral over ∂D is to be understood in the anticlockwise direction so that the domain D remains on the left-hand side while moving along ∂D.

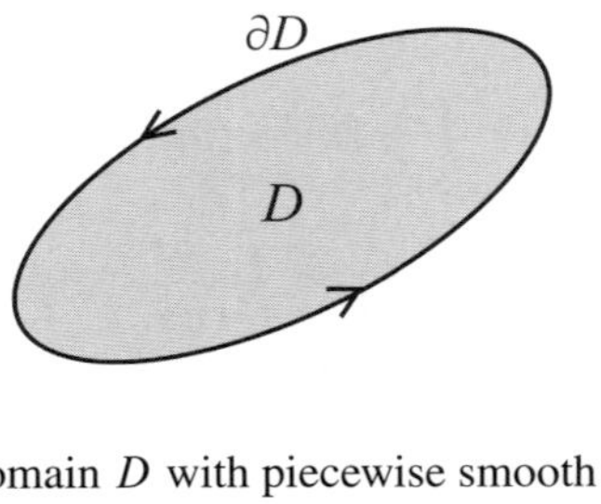

Figure 3.4 Domain D with piecewise smooth boundary ∂D

Theorem 3.1.3. *Let $\Omega \subset \mathbb{R}^2$ be a domain with a piecewise smooth boundary Γ and $\Gamma_0 \subset \Omega$ be a piecewise smooth curve subdividing Ω into Ω_1 and Ω_2 with boundaries Γ^1 and Γ^2 respectively such that* (3.1.9)–(3.1.11) *hold. Let $f, [\frac{\partial f}{\partial x_k}(\mathbf{x})], [\frac{\partial^2 f}{\partial x_k \partial x_l}(\mathbf{x})] \in C^0(\overline{\Omega} \setminus \Gamma_0)$ satisfy* (3.1.12)–(3.1.15) *(consequently* (3.1.19)–(3.1.21)*). Then the distributional derivatives $\frac{\partial f}{\partial x_k} = \tilde{f}$ and $\frac{\partial^2 f}{\partial x_k \partial x_l} = \frac{\partial^2 T_f}{\partial x_k \partial x_l} \in \mathcal{D}'(\Omega)$ of f and the partial derivatives $[\frac{\partial f}{\partial x_k}(\mathbf{x})], [\frac{\partial^2 f}{\partial x_k \partial x_l}(\mathbf{x})]$ of f in the usual pointwise sense are related, for $1 \le k, l \le 2$, by:*

$$\frac{\partial T_f}{\partial x_k} = T_{[\frac{\partial f}{\partial x_k}(\mathbf{x})]} + J_0 \cos\theta_k \delta_{\Gamma_0} \text{ in } \mathcal{D}'(\Omega), \tag{3.1.25}$$

$$\frac{\partial^2 T_f}{\partial x_k \partial x_l} = T_{[\frac{\partial^2 f}{\partial x_k \partial x_l}(\mathbf{x})]} + \frac{\partial}{\partial x_l}(J_0 \cos\theta_k \delta_{\Gamma_0}) + J_k \cos\theta_l \delta_{\Gamma_0} \quad \text{in } \mathcal{D}'(\Omega), \tag{3.1.26}$$

which can be written equivalently *as follows: for $1 \le k, l \le 2$,*

$$\frac{\partial f}{\partial x_k} = \left[\frac{\partial f}{\partial x_k}(\mathbf{x})\right] + J_0 \cos\theta_k \delta_{\Gamma_0} \text{ in } \mathcal{D}'(\Omega), \tag{3.1.27}$$

$$\frac{\partial^2 f}{\partial x_k \partial x_l} = \left[\frac{\partial^2 f}{\partial x_k \partial x_l}(\mathbf{x})\right] + \frac{\partial}{\partial x_l}(J_0 \cos\theta_k \delta_{\Gamma_0}) + J_k \cos\theta_l \delta_{\Gamma_0} \quad \text{in } \mathcal{D}'(\Omega), \tag{3.1.28}$$

where $T_{[\frac{\partial f}{\partial x_k}(\mathbf{x})]}, T_{[\frac{\partial^2 f}{\partial x_k \partial x_l}(\mathbf{x})]} \in \mathcal{D}'(\Omega)$ are regular distributions defined by (3.1.21) *and* (3.1.22) *respectively; the jumps J_0, J_k are continuous functions of $\mathbf{x} \in \Gamma_0$ and defined by* (3.1.16) *and* (3.1.17) *respectively; $\theta_k = (\hat{\mathbf{n}}, \hat{\mathbf{i}}_k)$, $\hat{\mathbf{n}}$ being the unit normal to Γ_0 and exterior to Ω_1 (see Figure* 3.3*);*

$$\langle J_0 \cos\theta_k \delta_{\Gamma_0}, \phi \rangle = \int_{\Gamma_0} J_0 \cos\theta_k \phi \, ds \quad \forall \phi \in \mathcal{D}(\Omega), \tag{3.1.29}$$

$$\langle J_k \cos\theta_l \delta_{\Gamma_0}, \phi \rangle = \int_{\Gamma_0} J_k \cos\theta_l \phi \, ds \quad \forall \phi \in \mathcal{D}(\Omega) \tag{3.1.30}$$

define Dirac distributions corresponding to mass or charge placed on Γ_0 with a linear density $J_0 \cos\theta_k$ and $J_k \cos\theta_l$ respectively. In particular, for $l = k$, we get the important formula:

$$\frac{\partial^2 T_f}{\partial x_k^2} = T_{[\frac{\partial^2 f}{\partial x_k^2}(\mathbf{x})]} + \frac{\partial}{\partial x_k}(J_0 \cos\theta_k \delta_{\Gamma_0}) + J_k \cos\theta_k \delta_{\Gamma_0} \quad \text{in } \mathcal{D}'(\Omega) \tag{3.1.31}$$

or, equivalently,

$$\frac{\partial^2 f}{\partial x_k^2} = \left[\frac{\partial^2 f}{\partial x_k^2}(\mathbf{x})\right] + \frac{\partial}{\partial x_k}(J_0 \cos\theta_k \delta_{\Gamma_0}) + J_k \cos\theta_k \delta_{\Gamma_0} \quad \text{in } \mathcal{D}'(\Omega). \tag{3.1.32}$$

Proof. The distributional derivative $\frac{\partial T_f}{\partial x_k} = \frac{\partial f}{\partial x_k} \in \mathcal{D}'(\Omega)$ of f discontinuous on Γ_0 is given, $\forall \phi \in \mathcal{D}(\Omega)$, by:

$$\begin{aligned}
\left\langle \frac{\partial T_f}{\partial x_k}, \phi \right\rangle &= -\left\langle T_f, \frac{\partial \phi}{\partial x_k} \right\rangle = -\left\langle f, \frac{\partial \phi}{\partial x_k} \right\rangle = -\int_{\Omega} f \frac{\partial \phi}{\partial x_k} dx_1 dx_2 \\
&= -\int_{\Omega_1} f_1 \frac{\partial \phi}{\partial x_k} dx_1 dx_2 - \int_{\Omega_2} f_2 \frac{\partial \phi}{\partial x_k} dx_1 dx_2 \\
&= -\sum_{i=1}^{2} \left(\int_{\Omega_i} f_i \frac{\partial \phi}{\partial x_k} dx_1 dx_2 \right) \\
&= -\sum_{i=1}^{2} \int_{\Omega_i} \left\{ \left[\frac{\partial (f_i \phi)}{\partial x_k} (\mathbf{x}) \right] - \left[\frac{\partial f_i}{\partial x_k} (\mathbf{x}) \right] \phi \right\} dx_1 dx_2 \\
&= \sum_{i=1}^{2} \int_{\Omega_i} \left[\frac{\partial f_i}{\partial x_k} (\mathbf{x}) \right] \phi \, dx_1 dx_2 - \sum_{i=1}^{2} \int_{\Omega_i} \left[\frac{\partial (f_i \phi)}{\partial x_k} (\mathbf{x}) \right] dx_1 dx_2, \quad (3.1.33)
\end{aligned}$$

where $f_i = f \downarrow_{\overline{\Omega_i} - \Gamma_0}$ $(i = 1, 2)$ satisfies (3.1.12)–(3.1.18), and $[\frac{\partial f_i}{\partial x_k}(\mathbf{x})] \in C^0(\overline{\Omega_i})$ with $\overline{\Omega_i} = \Omega_i \cup \Gamma^i$, $\Gamma^i = \Gamma_i \cup \Gamma_0$ (see Figure 3.3). Now we will transform the second double integral over Ω_i in (3.1.33) into a line integral along the boundary Γ^i of Ω_i in the anticlockwise direction $(i = 1, 2)$ using (3.1.24). For this, set $\Omega_i = D$, $\Gamma^i = \Gamma_i \cup \Gamma_0 = \partial D$ with $i = 1, 2$, $P = 0$, $Q = f_i \phi$ with $\phi \in \mathcal{D}(\Omega)$, $\phi \downarrow_\Gamma = 0$, and $k = 1$, and we have, from (3.1.24): $\int_{\Omega_i} [\frac{\partial}{\partial x_1}(f_i \phi)] dx_1 dx_2 = \int_{\Gamma^i} f_i \phi dx_2$ such that

$$\begin{aligned}
\int_{\Gamma^1} f_1 \phi dx_2 &= \int_{\Gamma_1} f_1 \phi dx_2 + \int_{\Gamma_0} f_1 \phi dx_2 = \int_{\Gamma_0} f_1 \phi \cos(\hat{\mathbf{n}}, \hat{\mathbf{i}}_1) ds, \\
\int_{\Gamma^2} f_2 \phi dx_2 &= \int_{\Gamma_2} f_2 \phi dx_2 + \int_{\Gamma_0^-} f_2 \phi dx_2 = -\int_{\Gamma_0} f_2 \phi \cos(\hat{\mathbf{n}}, \hat{\mathbf{i}}_1) ds, \quad (3.1.34)
\end{aligned}$$

where $\phi \downarrow_{\Gamma_i} = 0 \Longrightarrow \int_{\Gamma_i} f_i \phi dx_2 = 0$, the curve Γ_0 traversed in the opposite direction is denoted by Γ_0^-, $\hat{\mathbf{n}}$ is the unit normal to Γ_0, the exterior to Ω_1 (see Figure 3.3), $dx_2 = \cos(\hat{\mathbf{n}}, \hat{\mathbf{i}}_1) ds$ (see (3.1.10)). Hence, from (3.1.34),

$$\begin{aligned}
-\sum_{i=1}^{2} \int_{\Omega_i} \left[\frac{\partial}{\partial x_1} (f_i \phi(\mathbf{x})) \right] dx_1 dx_2 &= \int_{\Gamma_0} [f_2(\mathbf{x}) - f_1(\mathbf{x})] \phi(\mathbf{x}) \cos(\hat{\mathbf{n}}, \hat{\mathbf{i}}_1) ds \\
&= \int_{\Gamma_0} J_0 \phi \cos(\hat{\mathbf{n}}, \hat{\mathbf{i}}_1) ds \quad \text{(by (3.1.16))}, \quad (3.1.35)
\end{aligned}$$

where $J_0 = J_0(\mathbf{x}) = f_2(\mathbf{x}) - f_1(\mathbf{x})$ $\forall \mathbf{x} \in \Gamma_0$ is defined by (3.1.16).

For $Q = 0$, $P = f_i\phi$ with $\phi \in \mathcal{D}(\Omega)$ and $k = 2$, we have, from (3.1.24), $-\int_{\Omega_i}[\frac{\partial}{\partial x_2}(f_i\phi)(\mathbf{x})]dx_1dx_2 = \int_{\Gamma^i} f_i\phi dx_1$ such that

$$\int_{\Gamma^1} f_1\phi dx_1 = \int_{\Gamma_1} f_1\phi dx_1 + \int_{\Gamma_0} f_1\phi dx_1 = -\int_{\Gamma_0} f_1\phi \cos(\hat{\mathbf{n}}, \hat{\mathbf{i}}_2)ds, \quad (3.1.36)$$

$$\int_{\Gamma^2} f_2\phi dx_1 = \int_{\Gamma_2} f_2\phi dx_1 + \int_{\Gamma_0^-} f_2\phi dx_1 = \int_{\Gamma_0} f_2\phi \cos(\hat{\mathbf{n}}, \hat{\mathbf{i}}_2)ds,$$

where $dx_1 = -\cos(\hat{\mathbf{n}}, \hat{\mathbf{i}}_2)ds$ (see (3.1.10)), $\phi\downarrow_\Gamma = 0 \Longrightarrow \int_{\Gamma_i} f_i\phi dx_1 = 0$; $\int_{\Gamma_0^-}(\cdot)dx_1 = -\int_{\Gamma_0}(\cdot)dx_1$. Hence, from (3.1.36),

$$-\sum_{i=1}^{2}\int_{\Omega_i}\left[\frac{\partial}{\partial x_2}(f_i\phi)(\mathbf{x})\right]dx_1dx_2 = \int_{\Gamma_0}\left[f_2(\mathbf{x}) - f_1(\mathbf{x})\right]\phi(x)\cos(\hat{\mathbf{n}}, \hat{\mathbf{i}}_2)ds$$
$$= \int_{\Gamma_0} J_0\phi\cos(\hat{\mathbf{n}}, \hat{\mathbf{i}}_2)ds \quad \text{(by (3.1.16))}. \qquad (3.1.37)$$

Thus, from (3.1.33), (3.1.35) and (3.1.37), we have, for $k = 1, 2$,

$$-\sum_{i=1}^{2}\int_{\Omega_i}\left[\frac{\partial}{\partial x_k}(f_i\phi)(\mathbf{x})\right]dx_1dx_2 = \int_{\Gamma_0} J_0\phi\cos(\hat{\mathbf{n}}, \hat{\mathbf{i}}_k)ds; \qquad (3.1.38)$$

$$\begin{aligned}\left\langle\frac{\partial T_f}{\partial x_k}, \phi\right\rangle &= \int_{\Omega_1}\left[\frac{\partial f_1}{\partial x_k}(\mathbf{x})\right]\phi dx_1dx_2 + \int_{\Omega_2}\left[\frac{\partial f_2}{\partial x_k}(\mathbf{x})\right]\phi dx_1dx_2 + \int_{\Gamma_0} J_0\phi\cos\theta_k ds\\ &= \int_{\Omega}\left[\frac{\partial f}{\partial x_k}(\mathbf{x})\right]\phi dx_1dx_2 + \int_{\Gamma_0} J_0\phi\cos\theta_k ds\\ &= \langle T_{[\frac{\partial f}{\partial x_k}(\mathbf{x})]}, \phi\rangle + \langle J_0\cos\theta_k\delta_{\Gamma_0}, \phi\rangle\\ &= \langle T_{[\frac{\partial f}{\partial x_k}(\mathbf{x})]} + J_0\cos\theta_k\delta_{\Gamma_0}, \phi\rangle \quad \forall\phi \in \mathcal{D}(\Omega)\end{aligned}$$

$\Longleftrightarrow \frac{\partial T_f}{\partial x_k} = T_{[\frac{\partial f}{\partial x_k}(\mathbf{x})]} + J_0\cos\theta_k\delta_{\Gamma_0}$ in $\mathcal{D}'(\Omega)$, which can be written equivalently in the form (3.1.27).

Now we prove (3.1.26)/(3.1.28). Since the partial derivative $[\frac{\partial^2 f}{\partial x_k\partial x_l}(\mathbf{x})]$ in the usual pointwise sense is continuous in $\overline{\Omega} - \Gamma_0$, the order of differentiation can be interchanged, i.e. $[\frac{\partial^2 f}{\partial x_k \partial x_l}(\mathbf{x})] = [\frac{\partial^2 f}{\partial x_l \partial x_k}(\mathbf{x})]\ \forall \mathbf{x} \in \overline{\Omega} - \Gamma_0$. Then,

$$\begin{aligned}
\left\langle \frac{\partial^2 T_f}{\partial x_k \partial x_l}, \phi \right\rangle &= \left\langle \frac{\partial}{\partial x_l}\left(\frac{\partial T_f}{\partial x_k}\right), \phi \right\rangle = -\left\langle \frac{\partial T_f}{\partial x_k}, \frac{\partial \phi}{\partial x_l} \right\rangle \\
&= -\left\langle T_{[\frac{\partial f}{\partial x_k}(\mathbf{x})]} + J_0 \cos\theta_k \delta_{\Gamma_0}, \frac{\partial \phi}{\partial x_l} \right\rangle \quad \text{(using (3.1.25))} \\
&= -\left\langle T_{[\frac{\partial f}{\partial x_k}(\mathbf{x})]}, \frac{\partial \phi}{\partial x_l} \right\rangle - \left\langle J_0 \cos\theta_k \delta_{\Gamma_0}, \frac{\partial \phi}{\partial x_l} \right\rangle \\
&= -\int_\Omega \left[\frac{\partial f}{\partial x_k}(\mathbf{x})\right] \frac{\partial \phi}{\partial x_l} dx_1 dx_2 + \left\langle \frac{\partial}{\partial x_l}(J_0 \cos\theta_k \delta_{\Gamma_0}), \phi \right\rangle \\
&\qquad \forall \phi \in \mathcal{D}(\Omega). \qquad (3.1.39)
\end{aligned}$$

But

$$\begin{aligned}
-\int_\Omega \left[\frac{\partial f}{\partial x_k}(\mathbf{x})\right] \frac{\partial \phi}{\partial x_l} dx_1 dx_2 &= -\int_\Omega \left\{ \frac{\partial}{\partial x_l}\left[\frac{\partial f}{\partial x_k}(\mathbf{x})\phi\right] - \left[\frac{\partial^2 f}{\partial x_k \partial x_l}(\mathbf{x})\right]\phi \right\} dx_1 dx_2 \\
= \int_\Omega \left[\frac{\partial^2 f}{\partial x_k \partial x_l}(\mathbf{x})\right]\phi dx_1 dx_2 &- \int_\Omega \frac{\partial}{\partial x_l}\left[\frac{\partial f}{\partial x_k}(\mathbf{x})\phi\right] dx_1 dx_2 \quad \forall \phi \in \mathcal{D}(\Omega).
\end{aligned}$$
(3.1.40)

Replacing 'f' by '$[\frac{\partial f}{\partial x_k}(\mathbf{x})]$', '$f_i$' by '$[\frac{\partial f_i}{\partial x_k}(\mathbf{x})]$', '$x_k$' by '$x_l$' and '$\theta_k$' by '$\theta_l$' in (3.1.35)–(3.1.38), and using the definition of J_k in (3.1.17), we get:

$$\begin{aligned}
-\int_\Omega \frac{\partial}{\partial x_l}\left[\frac{\partial f}{\partial x_k}(\mathbf{x})\phi(\mathbf{x})\right] dx_1 dx_2 &= \int_{\Gamma_0} \left(\left[\frac{\partial f_2}{\partial x_k}(\mathbf{x})\right] - \left[\frac{\partial f_1}{\partial x_k}(\mathbf{x})\right]\right)\phi \cos\theta_l ds \\
&= \int_{\Gamma_0} J_k \cos\theta_l \phi ds = \langle J_k \cos\theta_l \delta_{\Gamma_0}, \phi\rangle \quad \forall \phi \in \mathcal{D}(\Omega),
\end{aligned}$$
(3.1.41)

where $J_k = J_k(\mathbf{x}) = [\frac{\partial f_2}{\partial x_k}(\mathbf{x})] - [\frac{\partial f_1}{\partial x_k}(\mathbf{x})]\ \forall \mathbf{x} \in \Gamma_0$. Then, from (3.1.39)–(3.1.41), we get:

$$\begin{aligned}
\left\langle \frac{\partial^2 T_f}{\partial x_k \partial x_l}, \phi \right\rangle &= \int_\Omega \left[\frac{\partial^2 f}{\partial x_k \partial x_l}(\mathbf{x})\right]\phi dx_1 dx_2 + \left\langle \frac{\partial}{\partial x_l}(J_0 \cos\theta_k \delta_{\Gamma_0}), \phi \right\rangle \\
&\quad + \langle J_k \cos\theta_l \delta_{\Gamma_0}, \phi\rangle \\
= \langle T_{[\frac{\partial^2 f}{\partial x_k \partial x_l}(\mathbf{x})]}, \phi\rangle &+ \left\langle \frac{\partial}{\partial x_l}(J_0 \cos\theta_k \delta_{\Gamma_0}), \phi \right\rangle + \langle J_k \cos\theta_l \delta_{\Gamma_0}, \phi\rangle \quad \forall \phi \in \mathcal{D}(\Omega) \\
\Longrightarrow \quad \frac{\partial^2 T_f}{\partial x_k \partial x_l} &= T_{[\frac{\partial^2 f}{\partial x_k \partial x_l}(\mathbf{x})]} + \frac{\partial}{\partial x_l}(J_0 \cos\theta_k \delta_{\Gamma_0}) + J_k \cos\theta_l \delta_{\Gamma_0} \quad \text{in } \mathcal{D}'(\Omega),
\end{aligned}$$

which can be written equivalently in the form (3.1.28). □

Corollary 3.1.1. *Under the assumptions of Theorem* 3.1.3, *for* f, $[\frac{\partial^2 f}{\partial x_k^2}(x)]$ *belonging to* $C^0(\overline{\Omega} \setminus \Gamma_0)$ *and satisfying* (3.1.12)–(3.1.21),

$$\Delta T_f = T_{[\Delta f(\mathbf{x})]} + J_n \delta_{\Gamma_0} + \frac{\partial}{\partial n}(J_0 \delta_{\Gamma_0}) \quad \text{in } \mathcal{D}'(\Omega), \tag{3.1.42}$$

or, equivalently,

$$\Delta f = [\Delta f(\mathbf{x})] + J_n \delta_{\Gamma_0} + \frac{\partial}{\partial n}(J_0 \delta_{\Gamma_0}) \quad \text{in } \mathcal{D}'(\Omega), \tag{3.1.43}$$

where $\Delta T_f = \Delta f = \frac{\partial^2 f}{\partial x_1^2} + \frac{\partial^2 f}{\partial x_2^2} \in \mathcal{D}'(\Omega)$ *is the Laplacian of (discontinuous)* f *in the sense of distribution;* $[\Delta f(\mathbf{x})] = [\frac{\partial^2 f}{\partial x_1^2}(\mathbf{x})] + [\frac{\partial^2 f}{\partial x_2^2}(\mathbf{x})] \in \mathcal{D}'(\Omega)$ *is the Laplacian of* f *in* $\Omega \setminus \Gamma_0$ *in the usual pointwise sense,* $[\frac{\partial^2 f}{\partial x_k^2}(\mathbf{x})]$ *being the second order partial derivative of* f *with respect to* x_k *in the usual pointwise sense;*

$$J_n = \left[\frac{\partial f}{\partial n}(\mathbf{x}^+)\right] - \left[\frac{\partial f}{\partial n}(\mathbf{x}^-)\right] = \left[\frac{\partial f_2}{\partial n}(\mathbf{x})\right] - \left[\frac{\partial f_1}{\partial n}(\mathbf{x})\right] \quad \forall \mathbf{x} \in \Gamma_0 \tag{3.1.44}$$

is the jump of the normal derivative of f *in the direction of* $\hat{\mathbf{n}}$ *in crossing* Γ_0, $\hat{\mathbf{n}}$ *being the unit normal exterior to* Ω_1 *(see Figure* 3.3*), such that the distribution* $J_n \delta_{\Gamma_0} \in \mathcal{D}'(\Omega)$ *is defined by:*

$$\langle J_n \delta_{\Gamma_0}, \phi \rangle = \int_{\Gamma_0} J_n \phi \, ds \quad \forall \phi \in \mathcal{D}(\Omega); \tag{3.1.45}$$

$\frac{\partial}{\partial n}(J_0 \delta_{\Gamma_0}) \in \mathcal{D}'(\Omega)$ *is the distribution defined,* $\forall \phi \in \mathcal{D}(\Omega)$, *by:*

$$\left\langle \frac{\partial}{\partial n}(J_0 \delta_{\Gamma_0}), \phi \right\rangle = \left\langle -J_0 \delta_{\Gamma_0}, \frac{\partial \phi}{\partial n} \right\rangle = -\int_{\Gamma_0} J_0 \frac{\partial \phi}{\partial n} ds. \tag{3.1.46}$$

Remark 3.1.2. The distribution $\langle T, \phi \rangle = -\int_{\Gamma_0} J_0 \frac{\partial \phi}{\partial n} ds \ \forall \phi \in \mathcal{D}(\Omega)$ in (3.1.46) is formed by *dipoles* directed along the normal $\hat{\mathbf{n}}$ with a linear moment density $-J_0$ along Γ_0. Obviously, this distribution can also be represented by $\frac{\partial}{\partial n}(J_0 \delta_{\Gamma_0})$ as in (3.1.46) (for details, see Section 1.11 in Chapter 1).

Proof of Corollary 3.1.1. From (3.1.31)–(3.1.32), we have, for $k = 1, 2$,

$$\frac{\partial^2 T_f}{\partial x_k^2} = T_{[\frac{\partial^2 f}{\partial x_k^2}(\mathbf{x})]} + \frac{\partial}{\partial x_k}(J_0 \cos\theta_k \delta_{\Gamma_0}) + J_k \cos\theta_k \delta_{\Gamma_0} \quad \text{in } \mathcal{D}'(\Omega) \quad \text{or}$$

$$\frac{\partial^2 f}{\partial x_k^2} = \left[\frac{\partial^2 f}{\partial x_k^2}(x)\right] + \frac{\partial}{\partial x_k}(J_0 \cos\theta_k \delta_{\Gamma_0}) + J_k \cos\theta_k \delta_{\Gamma_0} \quad \text{in } \mathcal{D}'(\Omega).$$

Hence,

$$\Delta T_f = \sum_{k=1}^{2} \frac{\partial^2 T_f}{\partial x_k^2} \tag{3.1.47}$$

$$= \sum_{k=1}^{2} T_{[\frac{\partial^2 f}{\partial x_k^2}(\mathbf{x})]} + \sum_{k=1}^{2} \frac{\partial}{\partial x_k}(J_0 \cos\theta_k \delta_{\Gamma_0}) + \sum_{k=1}^{2} J_k \cos\theta_k \delta_{\Gamma_0} \quad \text{in } \mathcal{D}'(\Omega).$$

$$\left\langle \sum_{k=1}^{2} T_{[\frac{\partial^2 f}{\partial x_k^2}(\mathbf{x})]}, \phi \right\rangle = \langle T_{[\frac{\partial^2 f}{\partial x_1^2}(\mathbf{x})]} + T_{[\frac{\partial^2 f}{\partial x_2^2}(\mathbf{x})]}, \phi\rangle \quad \forall \phi \in \mathcal{D}(\Omega)$$

$$= \int_\Omega \left[\frac{\partial^2 f}{\partial x_1^2}(\mathbf{x})\right] \phi \, dx_1 dx_2 + \int_\Omega \left[\frac{\partial^2 f}{\partial x_2^2}(\mathbf{x})\right] \phi \, dx_1 dx_2$$

$$= \int_\Omega [\Delta f(\mathbf{x})] \phi \, dx_1 dx_2 \quad \forall \phi \in \mathcal{D}(\Omega)$$

$$\Longrightarrow \quad \sum_{k=1}^{2} T_{[\frac{\partial^2 f}{\partial x_k^2}(\mathbf{x})]} = T_{[\Delta f(\mathbf{x})]} \quad \text{in } \mathcal{D}'(\Omega). \tag{3.1.48}$$

$$\sum_{k=1}^{2} J_k \cos\theta_k = \sum_{k=1}^{2} \left(\left\{ \left[\frac{\partial f}{\partial x_k}(\mathbf{x}^+)\right] - \left[\frac{\partial f}{\partial x_k}(\mathbf{x}^-)\right] \right\} \cos\theta_k \right)$$

$$= \sum_{k=1}^{2} \left[\frac{\partial f}{\partial x_k}(\mathbf{x}^+)\right] \cos\theta_k - \sum_{k=1}^{2} \left[\frac{\partial f}{\partial x_k}(\mathbf{x}^-)\right] \cos\theta_k$$

$$= \frac{\partial f}{\partial n}(\mathbf{x}^+) - \frac{\partial f}{\partial n}(\mathbf{x}^-) = J_n$$

$$\Longrightarrow \quad \sum_{k=1}^{2} J_k \cos\theta_k \delta_{\Gamma_0} = J_n \delta_{\Gamma_0} \quad \text{in } \mathcal{D}'(\Omega) \tag{3.1.49}$$

is the distribution defined by $\langle J_n \delta_{\Gamma_0}, \phi\rangle = \int_{\Gamma_0} J_n \phi \, ds \ \forall \phi \in \mathcal{D}(\Omega)$.

$$\left\langle \sum_{k=1}^{2} \frac{\partial}{\partial x_k}(J_0 \cos\theta_k \delta_{\Gamma_0}), \phi \right\rangle = -\sum_{k=1}^{2} \left\langle J_0 \cos\theta_k \delta_{\Gamma_0}, \frac{\partial \phi}{\partial x_k} \right\rangle$$

$$= -\sum_{k=1}^{2} \int_{\Gamma_0} J_0 \cos\theta_k \frac{\partial \phi}{\partial x_k} ds$$

$$= -\int_{\Gamma_0} J_0 \left(\sum_{k=1}^{2} \frac{\partial \phi}{\partial x_k} \cos\theta_k \right) ds = -\int_{\Gamma_0} J_0 \frac{\partial \phi}{\partial n} ds$$

$$= -\left\langle J_0 \delta_{\Gamma_0}, \frac{\partial \phi}{\partial n} \right\rangle = \left\langle \frac{\partial}{\partial n}(J_0 \delta_{\Gamma_0}), \phi \right\rangle \quad \forall \phi \in \mathcal{D}(\Omega)$$

$$\Longrightarrow \quad \sum_{k=1}^{2} \frac{\partial}{\partial x_k}(J_0 \cos\theta_k \delta_{\Gamma_0}) = \frac{\partial}{\partial n}(J_0 \delta_{\Gamma_0}) \quad \text{in } \mathcal{D}'(\Omega). \tag{3.1.50}$$

Then, combining (3.1.47)–(3.1.50), we get the result (3.1.42)/(3.1.43). □

Remark 3.1.3.

- The jump J_n of the normal derivative $\frac{\partial f}{\partial n}$ in crossing Γ_0 is *independent* of the choice of the normal $\hat{\mathbf{n}}$ to Γ_0. If the direction is reversed, then the sign of J_k changes and the sign of the normal derivative itself also changes, so that the jump J_n of the normal derivative ultimately remains unchanged.
- The formula (3.1.50) does *not* depend on the choice of the normal $\hat{\mathbf{n}}$. If the direction of the normal is reversed, the signs of *both* J_0 and $\frac{\partial \phi}{\partial n}$ change.
- The essential point is to use the *same sense* in the evaluation of the jump J_k and the choice of the normal $\hat{\mathbf{n}}$ to Γ_0.

Remark 3.1.4. Let $\hat{\mathbf{n}}_1$ and $\hat{\mathbf{n}}_2$ be unit normals to Γ_0 exterior to Ω_1 and Ω_2 respectively such that $\hat{\mathbf{n}}_1 \cdot \hat{\mathbf{n}}_2 = -1$ (see Figure 3.3). Then, *without introducing a sense in crossing* Γ_0, the expressions for J_n in (3.1.49) and $\frac{\partial}{\partial n}(J_0 \delta_{\Gamma_0})$ in (3.1.50) can be replaced respectively by:

$$J_n = \left[\frac{\partial f}{\partial n_1}(\mathbf{x}) \right] + \left[\frac{\partial f}{\partial n_2}(\mathbf{x}) \right] \quad \text{(see (3.1.44))}, \tag{3.1.51}$$

and

$$\frac{\partial}{\partial n}(J_0 \delta_{\Gamma_0}) = \frac{\partial}{\partial n_1}(f_1 \delta_{\Gamma_0}) + \frac{\partial}{\partial n_2}(f_2 \delta_{\Gamma_0}). \tag{3.1.52}$$

Then the formula (3.1.42)/(3.1.43) can be rewritten equivalently as:

$$\Delta T_f = \Delta f = [\Delta f(\mathbf{x})] + \left(\left[\frac{\partial f}{\partial n_1}(\mathbf{x}) \right] + \left[\frac{\partial f}{\partial n_2}(\mathbf{x}) \right] \right) \delta_{\Gamma_0}$$
$$+ \frac{\partial}{\partial n_1}(f_1 \delta_{\Gamma_0}) + \frac{\partial}{\partial n_2}(f_2 \delta_{\Gamma_0}) \quad \text{in } \mathcal{D}'(\Omega). \tag{3.1.53}$$

Example 3.1.2. Let $\Omega =]0,1[\times]0,1[\subset \mathbb{R}^2$ be the *open unit square* with the square boundary Γ such that $\Omega = \overset{\circ}{T}_1 \cup \overset{\circ}{T}_2 \cup \overset{\circ}{\Gamma}_0$, where $\overset{\circ}{T}_i = \text{int}(T_i)$, T_1 and T_2 being two closed triangles, $\overset{\circ}{\Gamma}_0 = \Gamma_0 - \{\mathbf{a}_2, \mathbf{a}_4\}$:

$$T_1 = \{(x_1,x_2) : 0 \le x_1 \le x_2 \le 1\} \quad \text{with vertices } \mathbf{a}_2(1,1), \mathbf{a}_3(0,1), \mathbf{a}_4(0,0);$$
$$T_2 = \{(x_1,x_2) : 0 \le x_2 \le x_1 \le 1\} \quad \text{with vertices } \mathbf{a}_1(1,0), \mathbf{a}_2(1,1), \mathbf{a}_4(0,0);$$
$$\Gamma_0 = T_1 \cap T_2 = \{(x_1,x_2) : 0 \le x_1 = x_2 \le 1\} = [\mathbf{a}_4, \mathbf{a}_2] = \text{join of } \mathbf{a}_4 \text{ and } \mathbf{a}_2;$$
$$\overline{\Omega} = \Omega \cup \Gamma = T_1 \cup T_2 \quad \text{(see Figure 3.5).}$$

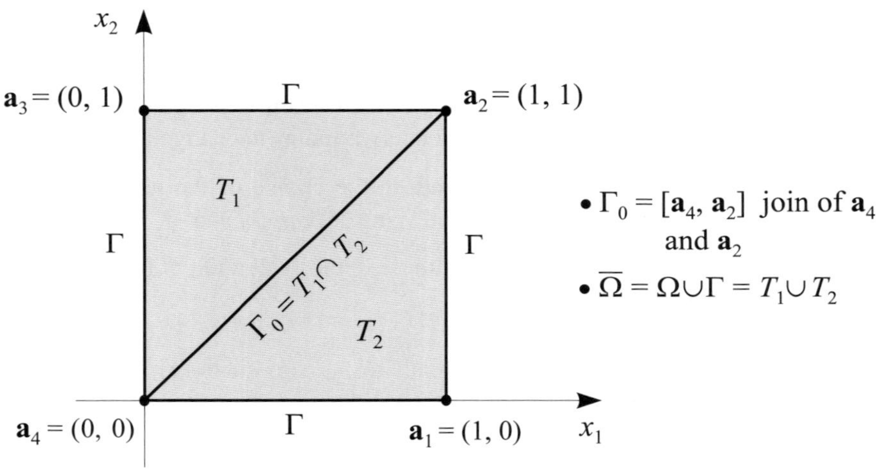

Figure 3.5 Piecewise polynomial function $u = u(x_1,x_2)$ on $\overline{\Omega}$ with discontinuity across Γ_0 (see also Figure 2.3)

Let f be a function discontinuous across Γ_0 defined by:

$$f(x_1,x_2) = \begin{cases} p(x_1,x_2) = 1 + 4x_1 - 2x_2 & \text{in } \overset{\circ}{T}_1 \\ q(x_1,x_2) = -2 + 4x_1 + 4x_2 & \text{in } \overset{\circ}{T}_2 \end{cases} \tag{3.1.54}$$

such that p and q are given *natural polynomial extensions* to T_1 and T_2.

Let

$$\left[\frac{\partial f}{\partial x_i}(\mathbf{x})\right] = \begin{cases} \frac{\partial p}{\partial x_i}(\mathbf{x}) & \text{in } \overset{\circ}{T}_1 \\ \frac{\partial q}{\partial x_i}(\mathbf{x}) & \text{in } \overset{\circ}{T}_2 \end{cases} \quad (1 \le i \le 2) \tag{3.1.55}$$

be the usual partial derivative of f with respect to x_i, and $\frac{\partial f}{\partial x_i}$ be the distributional derivative of f with respect to x_i. Then,

1. $f \in L^2(\Omega)$; (3.1.56)

2. $[\frac{\partial f}{\partial x_i}(\mathbf{x})] \in L^2(\Omega)$; (3.1.57)

3. $\frac{\partial f}{\partial x_i} \notin L^2(\Omega)$. (3.1.58)

Solution.

1.
$$\int_\Omega |f(x_1,x_2)|^2 dx_1dx_2 = \int_{\overset{\circ}{T}_1} |p(x_1,x_2)|^2 dx_1dx_2 + \int_{\overset{\circ}{T}_2} |q(x_1,x_2)|^2 dx_1dx_2 < +\infty$$

$$\Longrightarrow f \in L^2(\Omega).$$

2.
$$\left[\frac{\partial f}{\partial x_1}(\mathbf{x})\right] = \begin{cases} \frac{\partial}{\partial x_1}(1+4x_1-2x_2) = 4 & \text{in } \overset{\circ}{T}_1 \\ \frac{\partial}{\partial x_1}(-2+4x_1+4x_2) = 4 & \text{in } \overset{\circ}{T}_2 \end{cases}$$

$$\Longrightarrow \left[\frac{\partial f}{\partial x_1}(\mathbf{x})\right] = 4 \quad \text{in } \Omega.$$

$$\Longrightarrow \int_\Omega \left[\frac{\partial f}{\partial x_1}(\mathbf{x})\right]^2 dx_1dx_2 = 4^2 \int_\Omega dx_1dx_2 = 16 < +\infty$$

$$\Longrightarrow \left[\frac{\partial f}{\partial x_1}(\mathbf{x})\right] \in L^2(\Omega).$$

$$\left[\frac{\partial f}{\partial x_2}(\mathbf{x})\right] = \begin{cases} -2 & \text{in } \overset{\circ}{T}_1 \\ 4 & \text{in } \overset{\circ}{T}_2 \end{cases}$$

$$\Longrightarrow \int_\Omega \left[\frac{\partial f}{\partial x_2}(\mathbf{x})\right]^2 dx_1dx_2 = \int_{\overset{\circ}{T}_1} 4dx_1dx_2 + \int_{\overset{\circ}{T}_2} 16dx_1dx_2 = 4\cdot\frac{1}{2} + 16\cdot\frac{1}{2} < +\infty$$

$$\Longrightarrow \left[\frac{\partial f}{\partial x_2}(\mathbf{x})\right] \in L^2(\Omega).$$

Hence, $[\frac{\partial f}{\partial x_k}(\mathbf{x})] \in L^2(\Omega)$, $k = 1, 2$.

3. $f \in L^2(\Omega) \Longrightarrow f \in L^1_{\text{loc}}(\Omega) \subset \mathcal{D}'(\Omega)$ with $f = T_f \in \mathcal{D}'(\Omega)$.
Then the distributional derivative $\frac{\partial f}{\partial x_k} = \frac{\partial}{\partial x_k} T_f$ is defined, $\forall \phi \in \mathcal{D}(\Omega)$, by:

$$\left\langle \frac{\partial f}{\partial x_k}, \phi \right\rangle_{\mathcal{D}'(\Omega)\times\mathcal{D}(\Omega)} = -\left\langle f, \frac{\partial \phi}{\partial x_k} \right\rangle = -\int_\Omega f \frac{\partial \phi}{\partial x_k} dx_1 dx_2; \tag{3.1.59}$$

$$\left\langle \frac{\partial f}{\partial x_1}, \phi \right\rangle_{\mathcal{D}'(\Omega)\times\mathcal{D}(\Omega)} = -\int_\Omega f \frac{\partial \phi}{\partial x_1} dx_1 dx_2$$
$$= -\sum_{i=1}^{2} \int_{\mathring{T}_i} \left[\frac{\partial}{\partial x_1}(f\phi)(\mathbf{x}) \right] dx_1 dx_2 + \sum_{i=1}^{2} \int_{\mathring{T}_i} \left[\frac{\partial f}{\partial x_1}(\mathbf{x}) \right] \phi \, dx_1 dx_2. \tag{3.1.60}$$

Let ∂T_i be the triangular boundary of $\mathring{T}_i$ such that $T_i = \mathring{T}_i \cup \partial T_i$ $(i = 1, 2)$, and $\Gamma_0 \subset \partial T_i$, $(\partial T_i - \Gamma_0) \subset \Gamma$. Then, applying Green's Theorem 3.1.2, we get

$$\int_{\mathring{T}_1} \left[\frac{\partial}{\partial x_1}(f\phi)(\mathbf{x}) \right] dx_1 dx_2 = \int_{\mathring{T}_1} \left[\frac{\partial}{\partial x_1}(p\phi)(\mathbf{x}) \right] dx_1 dx_2$$
$$= \int_{\partial T_1} p\phi \, dx_2 = \int_{\partial T_1 - \Gamma_0} p\phi \, dx_2 + \int_{\Gamma_0} p\phi \, dx_2$$
$$= \int_{\Gamma_0} p\phi \, dx_2 \quad \forall \phi \in \mathcal{D}(\Omega),$$

since $\phi \in \mathcal{D}(\Omega) \Longrightarrow \phi\downarrow_\Gamma = 0 \Longrightarrow \phi\downarrow_{\partial T_1 - \Gamma_0} = 0$. Hence,

$$\int_{\mathring{T}_1} \left[\frac{\partial}{\partial x_1}(p\phi)(\mathbf{x}) \right] dx_1 dx_2 = \int_{\Gamma_0} p\phi \, dx_2 = \int_{\Gamma_0} p\phi \cos(\hat{\mathbf{n}}, \hat{\mathbf{i}}_1) ds,$$

where $\hat{\mathbf{n}} = \cos(\hat{\mathbf{n}}, \hat{\mathbf{i}}_1)\hat{\mathbf{i}}_1 + \cos(\hat{\mathbf{n}}, \hat{\mathbf{i}}_2)\hat{\mathbf{i}}_2$ is the unit normal exterior to Γ_0. Again, applying Green's Theorem 3.1.2, we get

$$\int_{\mathring{T}_2} \left[\frac{\partial}{\partial x_1}(f\phi)(\mathbf{x}) \right] dx_1 dx_2 = \int_{\mathring{T}_2} \left[\frac{\partial}{\partial x_1}(q\phi)(\mathbf{x}) \right] dx_1 dx_2$$
$$= \int_{\Gamma_0^-} q\phi \, dx_2 = -\int_{\Gamma_0} q\phi \, dx_2 = -\int_{\Gamma_0} q\phi \cos(\hat{\mathbf{n}}, \hat{\mathbf{i}}_1) ds,$$

where the orientation of Γ_0^- is from $\mathbf{a}_2$ to $\mathbf{a}_4$, i.e. opposite to that of Γ_0. Hence

$$-\sum_{i=1}^{2} \int_{\mathring{T}_i} \left[\frac{\partial}{\partial x_1}(f\phi)(\mathbf{x}) \right] dx_1 dx_2 = \int_{\Gamma_0} (q - p)\phi \cos(\hat{\mathbf{n}}, \hat{\mathbf{i}}_1) ds$$
$$= \int_{\Gamma_0} J\phi \cos(\hat{\mathbf{n}}, \hat{\mathbf{i}}_1) ds, \tag{3.1.61}$$

where J = jump of f across Γ_0

$$= (-2 + 4x_1 + 4x_2) - (1 + 4x_1 - 2x_2) = -3 + 6x_2 \quad \forall (x_1, x_2) \in \Gamma_0. \tag{3.1.62}$$

Then, from (3.1.60), (3.1.61) and (3.1.62) we get

$$\begin{aligned}
\left\langle \frac{\partial f}{\partial x_1}, \phi \right\rangle_{\mathcal{D}'(\Omega)\times\mathcal{D}(\Omega)} &= \sum_{i=1}^{2} \int_{\mathring{T}_i} \left[\frac{\partial f}{\partial x_1}(\mathbf{x}) \right] \phi\, dx_1 dx_2 + \int_{\Gamma_0} J\phi \cos(\hat{\mathbf{n}}, \hat{\mathbf{i}}_1) ds \\
&= \int_{\Omega} \left[\frac{\partial f}{\partial x_1}(\mathbf{x}) \right] \phi\, dx_1 dx_2 + \langle J \cos(\hat{\mathbf{n}}, \hat{\mathbf{i}}_1)\delta_{\Gamma_0}, \phi \rangle \\
&= \left\langle \left[\frac{\partial f}{\partial x_1}(\mathbf{x}) \right], \phi \right\rangle + \langle J \cos(\hat{\mathbf{n}}, \hat{\mathbf{i}}_1)\delta_{\Gamma_0}, \phi \rangle \\
&= \left\langle \left[\frac{\partial f}{\partial x_1}(\mathbf{x}) \right] + J \cos(\hat{\mathbf{n}}, \hat{\mathbf{i}}_1)\delta_{\Gamma_0}, \phi \right\rangle \quad \forall \phi \in \mathcal{D}(\Omega),
\end{aligned}$$

where $\langle \delta_{\Gamma_0}, \phi \rangle = \int_{\Gamma_0} \phi(x_1, x_2) ds\ \forall \phi \in \mathcal{D}(\Omega)$ is the Dirac distribution with mass/charge/force etc. concentrated along Γ_0; $\cos(\hat{\mathbf{n}}, \hat{\mathbf{i}}_1) = 1/2$; $J(\mathbf{x}) = -3 + 6x_2$ (by (3.1.50));

$$= \langle \delta_{\Gamma_0}, J \cos(\hat{\mathbf{n}}, \hat{\mathbf{i}}_1)\phi \rangle = \int_{\Gamma_0} J \cos(\hat{\mathbf{n}}, \hat{\mathbf{i}}_1)\phi\, ds \quad \forall \phi \in \mathcal{D}(\Omega)$$

$\Longrightarrow \frac{\partial f}{\partial x_1} = [\frac{\partial f}{\partial x_1}(\mathbf{x})] + J \cos(\hat{\mathbf{n}}, \hat{\mathbf{i}}_1)\delta_{\Gamma_0} \in \mathcal{D}'(\Omega)$ (see (3.1.27)). $[\frac{\partial f}{\partial x_1}(\mathbf{x})] \in L^2(\Omega)$, but $\delta_{\Gamma_0} \notin L^1_{\text{loc}}(\Omega) \Longrightarrow \delta_{\Gamma_0} \notin L^2(\Omega) \Longrightarrow J \cos(\hat{\mathbf{n}}, \hat{\mathbf{i}}_1)\delta_{\Gamma_0} \notin L^2(\Omega)$, ($J \neq 0$, $\cos(\hat{\mathbf{n}}, \hat{\mathbf{i}}_1) = \frac{1}{2}$). Consequently, $[\frac{\partial f}{\partial x_1}(\mathbf{x})] + J \cos(\hat{\mathbf{n}}, \hat{\mathbf{i}}_1)\delta_{\Gamma_0} \notin L^2(\Omega)$. Hence, $\frac{\partial f}{\partial x_1} \notin L^2(\Omega)$.

For $k = 2$, similarly applying Green's Theorem 3.1.2, we have

$$\begin{aligned}
-\int_{\mathring{T}_1} \left[\frac{\partial}{\partial x_2}(f\phi)(\mathbf{x}) \right] dx_1 dx_2 &= \int_{\Gamma_0} p\phi\, dx_1 = -\int_{\Gamma_0} p\phi \cos(\hat{\mathbf{n}}, \hat{\mathbf{i}}_2) ds, \\
-\int_{\mathring{T}_2} \left[\frac{\partial}{\partial x_2}(f\phi)(\mathbf{x}) \right] dx_1 dx_2 &= \int_{\Gamma_0^-} q\phi\, dx_1 = -\int_{\Gamma_0} q\phi\, dx_1 \\
&= \int_{\Gamma_0} q\phi \cos(\hat{\mathbf{n}}, \hat{\mathbf{i}}_2) ds
\end{aligned}$$

$$\begin{aligned}
\Longrightarrow -\sum_{i=1}^{2} \int_{\mathring{T}_i} \left[\frac{\partial}{\partial x_2}(f\phi)(\mathbf{x}) \right] dx_1 dx_2 &= \int_{\Gamma_0} (q - p)\phi \cos(\hat{\mathbf{n}}, \hat{\mathbf{i}}_2) ds \\
&= \int_{\Gamma_0} J\phi \cos(\hat{\mathbf{n}}, \hat{\mathbf{i}}_2) ds. \qquad (3.1.63)
\end{aligned}$$

Following the steps of the proof for $k = 1$, finally, from (3.1.63), we have

$$\left\langle \frac{\partial f}{\partial x_2}, \phi \right\rangle_{\mathcal{D}'(\Omega)\times\mathcal{D}(\Omega)} = \int_\Omega \left[\frac{\partial f}{\partial x_2}(\mathbf{x})\right] dx_1 dx_2 + \int_{\Gamma_0} J\phi \cos(\hat{\mathbf{n}}, \hat{\mathbf{i}}_2) ds$$

$$= \left\langle \left[\frac{\partial f}{\partial x_2}(\mathbf{x})\right] + J \cos(\hat{\mathbf{n}}, \hat{\mathbf{i}}_2)\delta_{\Gamma_0}, \phi \right\rangle \quad \forall \phi \in \mathcal{D}(\Omega) \quad \text{(see (3.1.27))}$$

$\Longrightarrow \frac{\partial f}{\partial x_2} = [\frac{\partial f}{\partial x_2}(\mathbf{x})] + J \cos(\hat{\mathbf{n}}, \hat{\mathbf{i}}_2)\delta_{\Gamma_0} \notin L^2(\Omega)$ (see the steps for the case $k = 1$). Hence, the distributional derivative $\frac{\partial f}{\partial x_k} \notin L^2(\Omega)$ for $k = 1, 2$.

3.1.3 Case of three variables ($n = 3$)

Let $\Omega \subset \mathbb{R}^3$ be a domain in $\mathbb{R}^3$ bounded by an orientable piecewise smooth boundary surface Γ such that the unit normal $\hat{\mathbf{n}} = \cos(\hat{\mathbf{n}}, \hat{\mathbf{i}}_1)\hat{\mathbf{i}}_1 + \cos(\hat{\mathbf{n}}, \hat{\mathbf{i}}_2)\hat{\mathbf{i}}_2 + \cos(\hat{\mathbf{n}}, \hat{\mathbf{i}}_3)\hat{\mathbf{i}}_3$ to Γ and exterior to Ω is defined almost everywhere on Γ and $\Gamma_0 \subset \Omega$ be an orientable piecewise smooth surface subdividing Ω into subdomains Ω_1 and Ω_2 with boundary surfaces $\Gamma^1 = \Gamma_1 \cup \Gamma_0$ and $\Gamma^2 = \Gamma_2 \cup \Gamma_0$ such that conditions analogous to (3.1.11) for $n = 2$ also hold for $n = 3$:

$$\Omega = \Omega_1 \cup \Omega_2 \cup \Gamma_0, \quad \overline{\Omega} = \overline{\Omega}_1 \cup \overline{\Omega}_2, \quad \overline{\Omega}_i = \Omega_i \cup \Gamma^i, \quad \Gamma^i = \Gamma_i \cup \Gamma_0,$$
$$\Gamma^1 \cap \Gamma^2 = \Gamma_0, \quad \Gamma = \Gamma_1 \cup \Gamma_2, \text{ (surface area)} \quad \text{meas}(\Gamma_1 \cap \Gamma_2) = 0. \tag{3.1.64}$$

Let functions $f, [\frac{\partial f}{\partial x_k}(\mathbf{x})], [\frac{\partial^2 f}{\partial x_k \partial x_l}(\mathbf{x})] \in C^0(\overline{\Omega} \setminus \Gamma_0)$ with finite jumps J_0, J_k, J_{kl} across Γ_0 respectively, as defined by formulae (3.1.16), (3.1.17) and (3.1.18) respectively such that $f_i = f\downarrow_{\overline{\Omega}-\Gamma_0}$, $[\frac{\partial f_i}{\partial x_k}(\mathbf{x})]$, $[\frac{\partial f_i}{\partial x_k \partial x_l}(\mathbf{x})]$, $i = 1, 2$, can be *continuously* extended to Γ_0 using (3.1.12)–(3.1.15) with $\mathbf{x} = (x_1, x_2, x_3) \in \Omega_i$ (resp. $\boldsymbol{\xi} = (\xi_1, \xi_2, \xi_3) \in \Gamma_0$).

Then the distributional derivatives $\frac{\partial f}{\partial x_k} = \frac{\partial T_f}{\partial x_k} \in \mathcal{D}'(\Omega)$ and $\frac{\partial^2 f}{\partial x_k \partial x_l} = \frac{\partial^2 T_f}{\partial x_k \partial x_l} \in \mathcal{D}'(\Omega)$, and the partial derivatives $[\frac{\partial f}{\partial x_k}(\mathbf{x})] = T_{[\frac{\partial f}{\partial x_k}(\mathbf{x})]} \in \mathcal{D}'(\Omega)$, $[\frac{\partial^2 f}{\partial x_k \partial x_l}(\mathbf{x})] = T_{[\frac{\partial^2 f}{\partial x_k \partial x_l}(\mathbf{x})]} \in \mathcal{D}'(\Omega)$ in the usual pointwise sense, are related by the same types of formulae as in (3.1.25) and (3.1.26) for $1 \le k, l \le 3$, i.e.:

Theorem 3.1.4. *For functions $f, [\frac{\partial f}{\partial x_k}(\mathbf{x})], [\frac{\partial^2 f}{\partial x_k \partial x_l}(\mathbf{x})] \in C^0(\overline{\Omega} \setminus \Gamma_0)$ with finite jumps J_0, J_k, J_{kl} across Γ_0 respectively, the following formulae hold for $1 \le k, l \le 3$:*

$$\frac{\partial T_f}{\partial x_k} = T_{[\frac{\partial f}{\partial x_k}(\mathbf{x})]} + J_0 \cos\theta_k \delta_{\Gamma_0} \quad \textit{in } \mathcal{D}'(\Omega), \tag{3.1.65}$$

$$\frac{\partial^2 T_f}{\partial x_k \partial x_l} = T_{[\frac{\partial^2 f}{\partial x_k \partial x_l}(\mathbf{x})]} + \frac{\partial}{\partial x_l}(J_0 \cos\theta_k \delta_{\Gamma_0}) + J_k \cos\theta_l \delta_{\Gamma_0} \quad \textit{in } \mathcal{D}'(\Omega), \tag{3.1.66}$$

which can be written equivalently *as:*

$$\frac{\partial f}{\partial x_k} = \left[\frac{\partial f}{\partial x_k}(\mathbf{x})\right] + J_0 \cos\theta_k \delta_{\Gamma_0} \quad \text{in } \mathcal{D}'(\Omega), \tag{3.1.67}$$

$$\frac{\partial^2 f}{\partial x_k \partial x_l} = \left[\frac{\partial^2 f}{\partial x_k \partial x_l}(\mathbf{x})\right] + \frac{\partial}{\partial x_l}(J_0 \cos\theta_k \delta_{\Gamma_0}) + J_k cos\theta_l \delta_{\Gamma_0} \quad \text{in } \mathcal{D}'(\Omega), \tag{3.1.68}$$

where $T_{[\frac{\partial f}{\partial x_k}(\mathbf{x})]}$ *and* $T_{[\frac{\partial^2 f}{\partial x_k \partial x_l}(\mathbf{x})]} \in \mathcal{D}'(\Omega)$ *are regular distributions defined by* (3.1.20) *and* (3.1.21) *with* $\Omega \subset \mathbb{R}^3$, $\mathbf{x} = (x_1, x_2, x_3)$ *and '*$dx_1 dx_2$*' replaced by '*$dx_1 dx_2 dx_3$*';* $\theta_k = (\hat{\mathbf{n}}, \hat{\mathbf{i}}_k)$*;* $\forall \mathbf{x} \in \Gamma_0$,

$$\begin{aligned} J_0 &= f(\mathbf{x}^+) - f(\mathbf{x}^-) = f_2(\mathbf{x}) - f_1(\mathbf{x}), \\ J_k &= \left[\frac{\partial f}{\partial x_k}(\mathbf{x}^+)\right] - \left[\frac{\partial f}{\partial x_k}(\mathbf{x}^-)\right] \\ &= \frac{\partial f_2}{\partial x_k}(\mathbf{x}) - \frac{\partial f_1}{\partial x_k}(\mathbf{x}) \quad (\textit{see also } (3.1.16)\text{–}(3.1.17)), \end{aligned} \tag{3.1.69}$$

$$\langle J_0 \cos\theta_k \delta_{\Gamma_0}, \phi\rangle = \int_{\Gamma_0} J_0 \cos\theta_k \phi \, dA \quad \forall \phi \in \mathcal{D}(\Omega), \tag{3.1.70}$$

$$\langle J_k \cos\theta_l \delta_{\Gamma_0}, \phi\rangle = \int_{\Gamma_0} J_k \cos\theta_l \phi \, dA \quad \forall \phi \in \mathcal{D}(\Omega), \tag{3.1.71}$$

define Dirac distributions corresponding to mass or charge or load placed on Γ_0 *with a surface area density* $J_0 \cos\theta_k$ *in* (3.1.70) *and a surface area density* $J_k \cos\theta_l$ *in* (3.1.71)*,* dA *being surface area measure in* (3.1.70) *and* (3.1.71)*. In particular, for* $l = k$ *in* (3.1.66)*, we get, for* $1 \le k \le 3$,

$$\frac{\partial^2 T_f}{\partial x_k^2} = T_{[\frac{\partial^2 f}{\partial x_k^2}(\mathbf{x})]} + \frac{\partial}{\partial x_k}(J_0 \cos\theta_k \delta_{\Gamma_0}) + J_k \cos\theta_k \delta_{\Gamma_0} \quad \text{in } \mathcal{D}'(\Omega) \quad \textit{or}$$

$$\frac{\partial^2 f}{\partial x_k^2} = \left[\frac{\partial^2 f}{\partial x_k^2}(\mathbf{x})\right] + \frac{\partial}{\partial x_k}(J_0 \cos\theta_k \delta_{\Gamma_0}) + J_k \cos\theta_k \delta_{\Gamma_0} \quad \text{in } \mathcal{D}'(\Omega). \tag{3.1.72}$$

Consequently,

$$\begin{aligned} \Delta T_f &= T_{[\Delta f(\mathbf{x})]} + \sum_{k=1}^{3}(J_k \cos\theta_k \delta_{\Gamma_0}) + \sum_{k=1}^{3} \frac{\partial}{\partial x_k}(J_k \cos\theta_k \delta_{\Gamma_0}) \\ &= T_{[\Delta f(\mathbf{x})]} + J_n \delta_{\Gamma_0} + \frac{\partial}{\partial n}(J_0 \delta_{\Gamma_0}) \quad \text{in } \mathcal{D}'(\Omega) \end{aligned} \tag{3.1.73}$$

(using a formula for $n = 3$ analogous to (3.1.49) for $n = 2$), or

$$\Delta f = [\Delta f(\mathbf{x})] + J_n \delta_{\Gamma_0} + \frac{\partial}{\partial n}(J_0 \delta_{\Gamma_0}) \quad \textit{in } \mathcal{D}'(\Omega) \tag{3.1.74}$$

with

$$\sum_{k=1}^{3} J_k \cos\theta_k \delta_{\Gamma_0} = J_n \delta_{\Gamma_0}, \quad \sum_{k=1}^{3} \frac{\partial}{\partial x_k}(J_0 \cos\theta_k \delta_{\Gamma_0}) = \frac{\partial}{\partial n}(J_0 \delta_{\Gamma_0}) \tag{3.1.75}$$

(see (3.1.49) and (3.1.50) for $n = 2$), where $\Delta T_f = \Delta f = \sum_{k=1}^{3} \frac{\partial^2 f}{\partial x_k^2}$ is the Laplacian of (discontinuous) f in the distributional sense in $\Omega \subset \mathbb{R}^3$, and $[\Delta f(\mathbf{x})] = [\frac{\partial^2 f(\mathbf{x})}{\partial x_1^2}] + [\frac{\partial^2 f(\mathbf{x})}{\partial x_2^2}] + [\frac{\partial^2 f(\mathbf{x})}{\partial x_3^2}]$ is the Laplacian of f in the usual pointwise sense in $\Omega - \Gamma_0$.

Proof.

$$\begin{aligned}\left\langle \frac{\partial T_f}{\partial x_k}, \phi \right\rangle &= -\left\langle T_f, \frac{\partial \phi}{\partial x_k} \right\rangle = -\left\langle f, \frac{\partial \phi}{\partial x_k} \right\rangle = -\int_\Omega f \frac{\partial \phi}{\partial x_k} dx_1 dx_2 dx_3 \\ &= -\int_{\Omega_1} f_1 \frac{\partial \phi}{\partial x_k} dx_1 dx_2 dx_3 - \int_{\Omega} f_2 \frac{\partial \phi}{\partial x_k} dx_1 dx_2 dx_3 \\ &= -\sum_{i=1}^{2} \int_{\Omega_i} \frac{\partial}{\partial x_k}(f_i \phi)(\mathbf{x}) dx_1 dx_2 dx_3 + \sum_{i=1}^{2} \int_{\Omega_i} \left[\frac{\partial f_i}{\partial x_k}(\mathbf{x})\right] \phi dx_1 dx_2 dx_3.\end{aligned} \tag{3.1.76}$$

We need the Divergence Theorem of Gauss–Ostrogradski:

Theorem 3.1.5 ([32]). *Let $D \subset \mathbb{R}^3$ be a bounded (volume) domain in $\mathbb{R}^3$ with piecewise smooth orientable boundary surface ∂D and $\overline{D} = D \cup \partial D$. Then, for $P_k = P_k(x_1, x_2, x_3)$, $1 \le k \le 3$, with $P_k, \frac{\partial P_k}{\partial x_k} \in C^0(\overline{D})$,*

$$\int_D \left(\frac{\partial P_1}{\partial x_1} + \frac{\partial P_2}{\partial x_2} + \frac{\partial P_3}{\partial x_3} \right) dx_1 dx_2 dx_3 = \int_{\partial D} (P_1 \cos\theta_1 + P_2 \cos\theta_2 + P_3 \cos\theta_3) dA, \tag{3.1.77}$$

where $\cos\theta_k = \cos(\hat{\mathbf{n}}, \hat{\mathbf{i}}_k) = \hat{\mathbf{n}} \cdot \hat{\mathbf{i}}_k$, $1 \le k \le 3$.

Now we continue with the proof.

We transform the triple integral $\int_{\Omega_i} \frac{\partial}{\partial x_k}(f_i, \phi) dx_1 dx_2 dx_3$ in (3.1.76) into a surface integral over $\Gamma^i = \Gamma_i \cup \Gamma_0$, $i = 1, 2$, for $k = 1, 2, 3$, as in the case of two variables in steps (3.1.34)–(3.1.37) and get the result. In fact, for this we set $\Omega_i = D$, $\Gamma^i = \partial D$ in (3.1.77). Then, for *fixed* $k = 1, 2, 3$, $P_k = f_i \phi \in C^0(\overline{\Omega}_i)$,

$\frac{\partial P_k}{\partial x_k} = \frac{\partial}{\partial x_k}(f_i\phi) \in C^0(\overline{\Omega}_i), \overline{\Omega}_i = \Omega_i \cap \Gamma^i$ with $\phi \in \mathcal{D}(\Omega), \phi\downarrow_\Gamma = 0$ and $P_j = 0$ for $1 \le j \ne k \le 3$, from (3.1.77) we get, for $i = 1, 2$:

$$
\begin{aligned}
-\int_{\Omega_i}\left[\frac{\partial}{\partial x_k}(f_i\phi)(\mathbf{x})\right]dx_1dx_2dx_3 &= -\int_{\Gamma^i=\Gamma_i\cup\Gamma_0}(f_i\phi)\cos(\hat{\mathbf{n}}_i,\hat{\mathbf{i}}_k)dA \\
&= -\int_{\Gamma_0}(f_i\phi)\cos(\hat{\mathbf{n}}_i,\hat{\mathbf{i}}_k)dA \quad (\text{since } \phi\downarrow_{\Gamma_i} = 0),
\end{aligned}
\tag{3.1.78}
$$

where $\hat{\mathbf{n}}_i$ is the unit normal exterior to Ω_i with $\hat{\mathbf{n}}_2 = -\hat{\mathbf{n}}_1$, $\hat{\mathbf{n}}_1 = \hat{\mathbf{n}}$. Hence, from (3.1.76) and (3.1.78),

$$
\begin{aligned}
\left\langle\frac{\partial T_f}{\partial x_k},\phi\right\rangle &= \sum_{i=1}^{2}\int_{\Omega_i}\left[\frac{\partial f_i}{\partial x_k}(\mathbf{x})\right]\phi dx_1dx_2dx_3 \\
&\quad -\left[\int_{\Gamma_0}f_1\phi\cos(\hat{\mathbf{n}}_1,\hat{\mathbf{i}}_k)dA - \int_{\Gamma_0}f_2\phi\cos(\hat{\mathbf{n}}_2,\hat{\mathbf{i}}_k)dA\right] \\
&= \int_{\Omega}\left[\frac{\partial f}{\partial x_k}(\mathbf{x})\right]\phi dx_1dx_2dx_3 \\
&\quad + \int_{\Gamma_0}[f_2(\mathbf{x}) - f_1(\mathbf{x})]\phi(\mathbf{x})\cos(\hat{\mathbf{n}}_1,\hat{\mathbf{i}}_k)dA \quad (\hat{\mathbf{n}}_2 = -\hat{\mathbf{n}}_1) \\
&= \langle T_{[\frac{\partial f}{\partial x_k}(\mathbf{x})]},\phi\rangle + \int_{\Gamma_0}J_0\cos(\hat{\mathbf{n}},\hat{\mathbf{i}}_k)\phi dA \quad \forall\phi\in\mathcal{D}(\Omega) \\
&\qquad (\hat{\mathbf{n}} = \hat{\mathbf{n}}_1, \text{ see (3.1.69) for } J_0)
\end{aligned}
$$

$$
\Longrightarrow \quad \frac{\partial T_f}{\partial x_k} = T_{[\frac{\partial f}{\partial x_k}(\mathbf{x})]} + J_0\cos\theta_k\delta_{\Gamma_0} \quad \text{in } \mathcal{D}'(\Omega) \quad (\theta_k = \cos(\hat{\mathbf{n}},\hat{\mathbf{i}}_k)). \tag{3.1.79}
$$

$$
\begin{aligned}
\left\langle\frac{\partial^2 T_f}{\partial x_k\partial x_l},\phi\right\rangle &= \left\langle\frac{\partial}{\partial x_l}\left(\frac{\partial T_f}{\partial x_k}\right),\phi\right\rangle = -\left\langle\frac{\partial T_f}{\partial x_k},\frac{\partial\phi}{\partial x_l}\right\rangle \\
&= -\left\langle T_{[\frac{\partial f}{\partial x_k}(\mathbf{x})]} + J_0\cos\theta_k\delta_{\Gamma_0},\frac{\partial\phi}{\partial x_l}\right\rangle \\
&= -\left\langle T_{[\frac{\partial f}{\partial x_k}(\mathbf{x})]},\frac{\partial\phi}{\partial x_l}\right\rangle - \left\langle J_0\cos\theta_k\delta_{\Gamma_0},\frac{\partial\phi}{\partial x_l}\right\rangle \\
&= -\int_{\Omega}\left[\frac{\partial f}{\partial x_k}(\mathbf{x})\right]\frac{\partial\phi}{\partial x_l}dx_1dx_2dx_3 \\
&\quad + \left\langle\frac{\partial}{\partial x_l}(J_0\cos\theta_k\delta_{\Gamma_0}),\phi\right\rangle \quad \forall\phi\in\mathcal{D}(\Omega).
\end{aligned}
\tag{3.1.80}
$$

But

$$-\int_{\Omega}\left[\frac{\partial f}{\partial x_k}(\mathbf{x})\right]\frac{\partial \phi}{\partial x_l}dx_1dx_2dx_3 = \int_{\Omega}\left[\frac{\partial^2 f}{\partial x_k\partial x_l}(\mathbf{x})\phi\right]dx_1dx_2dx_3 - \int_{\Omega}\frac{\partial}{\partial x_l}\left(\left[\frac{\partial f}{\partial x_k}(\mathbf{x})\right]\phi\right)dx_1dx_2dx_3. \quad (3.1.81)$$

Now, we will transform the last volume integral in (3.1.81) into a surface integral.

Since $\int_{\Omega}(\dots) = \sum_{i=1}^{2}\int_{\Omega_i}(\dots)$, applying the results of (3.1.78) and (3.1.79) with 'f_i' replaced by '$[\frac{\partial f_i}{\partial x_k}(\mathbf{x})]$' and subscript '$k$' by '$l$', and using $J_k = [\frac{\partial f_2}{\partial x_k}(\mathbf{x})] - [\frac{\partial f_1}{\partial x_k}(\mathbf{x})]$ on Γ_0, we get $\forall\phi \in \mathcal{D}(\Omega)$ $(\phi\downarrow_{\Gamma_i} = 0)$:

$$\begin{aligned}
-\int_{\Omega}\frac{\partial}{\partial x_l}\left(\left[\frac{\partial f}{\partial x_k}(\mathbf{x})\right]\phi\right)d\mathbf{x} &= -\sum_{i=1}^{2}\int_{\Omega_i}\frac{\partial}{\partial x_l}\left(\left[\frac{\partial f_i}{\partial x_k}(\mathbf{x})\right]\phi\right)d\mathbf{x}\\
&= -\sum_{i=1}^{2}\left(\int_{\Gamma^i=\Gamma_i\cup\Gamma_0}\left[\frac{\partial f_i}{\partial x_k}(\mathbf{x})\right]\phi\cos(\hat{\mathbf{n}}_i,\hat{\mathbf{i}}_l)dA\right)\\
&= -\int_{\Gamma_0}\left(\left[\frac{\partial f_1}{\partial x_k}(\mathbf{x})\right]\cos(\hat{\mathbf{n}}_1,\hat{\mathbf{i}}_l) + \left[\frac{\partial f_2}{\partial x_k}(\mathbf{x})\right]\cos(\hat{\mathbf{n}}_2,\hat{\mathbf{i}}_l)\right)\phi dA\\
&= \int_{\Gamma_0}\left(\left[\frac{\partial f_2}{\partial x_k}(\mathbf{x})\right] - \left[\frac{\partial f_1}{\partial x_k}(\mathbf{x})\right]\right)\phi\cos(\hat{\mathbf{n}}_1,\hat{\mathbf{i}}_l)dA\\
&= \int_{\Gamma_0} J_k\cos\theta_l\phi dA = \langle J_k\cos\theta_l\delta_{\Gamma_0},\phi\rangle \quad \left(\int_{\Gamma_i}(\dots)\phi dA = 0\right), \qquad (3.1.82)
\end{aligned}$$

where $\hat{\mathbf{n}} = \hat{\mathbf{n}}_1$, $\hat{\mathbf{n}}_2 = -\hat{\mathbf{n}}_1 = -\hat{\mathbf{n}}$ have been used. Combining (3.1.80)–(3.1.82), we get the result, $\forall\phi \in \mathcal{D}(\Omega)$:

$$\begin{aligned}
\left\langle\frac{\partial^2 T_f}{\partial x_k\partial x_l},\phi\right\rangle &= \int_{\Omega}\left[\frac{\partial^2 f}{\partial x_k\partial x_l}(\mathbf{x})\right]\phi dx_1dx_2dx_3 + \left\langle\frac{\partial}{\partial x_l}(J_0\cos\theta_k\delta_{\Gamma_0}),\phi\right\rangle\\
&\quad + \langle J_k\cos\theta_l\delta_{\Gamma_0},\phi\rangle\\
&= \left\langle T_{[\frac{\partial^2 f}{\partial x_k\partial x_l}(\mathbf{x})]} + \frac{\partial}{\partial x_l}(J_0\cos\theta_k\delta_{\Gamma_0}) + J_k\cos\theta_l\delta_{\Gamma_0},\phi\right\rangle
\end{aligned}$$

$$\iff \quad \frac{\partial^2 T_f}{\partial x_k\partial x_l} = T_{[\frac{\partial^2 f}{\partial x_k\partial x_l}(\mathbf{x})]} + \frac{\partial}{\partial x_l}(J_0\cos\theta_k\delta_{\Gamma_0}) + J_k\cos\theta_l\delta_{\Gamma_0} \quad \text{in } \mathcal{D}'(\Omega) \text{ or}$$

$$\frac{\partial^2 f}{\partial x_k\partial x_l} = \left[\frac{\partial^2 f}{\partial x_k\partial x_l}(\mathbf{x})\right] + \frac{\partial}{\partial x_l}(J_0\cos\theta_k\delta_{\Gamma_0}) + J_k\cos\theta_l\delta_{\Gamma_0} \quad \text{in } \mathcal{D}'(\Omega).$$

Following the proof of Corollary 3.1.1 and Remarks 3.1.3 and 3.1.4, with necessary modifications for $n = 3$, we get the result (3.1.73)/(3.1.74). □

Remark 3.1.5. Since Remarks 3.1.3 and 3.1.4 can be extended to the case $n = 3$ with the necessary modifications, formulae (3.1.51) and (3.1.52) hold with the necessary modifications for $n = 3$, and formula (3.1.74) can be rewritten equivalently, for $\hat{\mathbf{n}}_1 . \hat{\mathbf{n}}_2 = -1$, as:

$$\begin{aligned}\Delta f &= [\Delta f(\mathbf{x})] + \left(\left[\frac{\partial f}{\partial n_1}(\mathbf{x})\right] + \left[\frac{\partial f}{\partial n_2}(\mathbf{x})\right]\right)\delta_{\Gamma_0} \\ &\quad + \frac{\partial}{\partial n_1}(f_1\delta_{\Gamma_0}) + \frac{\partial}{\partial n_2}(f_2\delta_{\Gamma_0}) \quad \text{in } \mathcal{D}'(\Omega). \end{aligned} \tag{3.1.83}$$

3.2 Unbounded domain $\Omega \subset \mathbb{R}^n$, Green's formula

Let $f, [\frac{\partial f}{\partial x_k}(\mathbf{x})], [\frac{\partial^2 f}{\partial x_k \partial x_l}(\mathbf{x})]$ be continuous in the complement $\mathbb{R}^n \setminus \Gamma_0$ $(n = 2, 3)$ of Γ_0 with finite jumps J_0, J_k, J_{kl} defined by (3.1.16),(3.1.17) and (3.1.18) respectively for the case $n = 2$ (similar formulae for the case $n = 3$ are obtained with $\mathbf{x} = (x_1, x_2, x_3)$, $\boldsymbol{\xi} = (\xi_1, \xi_2, \xi_3)$), Γ_0 being a piecewise smooth curve in $\mathbb{R}^2$ (resp. piecewise smooth surface in $\mathbb{R}^3$). Then Theorem 3.1.3, Corollary 3.1.1 and Remarks 3.1.2, 3.1.4 for $n = 2$ (resp. Theorem 3.1.4, Remark 3.1.5 for $n = 3$) will hold, and we have

$$\begin{aligned}\Delta T_f &= T_{[\Delta f(\mathbf{x})]} + J_n\delta_{\Gamma_0} + \frac{\partial}{\partial n}(J_0\delta_{\Gamma_0}) \\ &= T_{[\Delta f(\mathbf{x})]} + \left(\left[\frac{\partial f}{\partial n_1}(\mathbf{x})\right] + \left[\frac{\partial f}{\partial n_2}(\mathbf{x})\right]\right)\delta_{\Gamma_0} \\ &\quad + \left(\frac{\partial}{\partial n_1}(f_1\delta_{\Gamma_0}) + \frac{\partial}{\partial n_2}(f_2\delta_{\Gamma_0})\right) \quad \text{in } \mathcal{D}'(\mathbb{R}^n), \end{aligned} \tag{3.2.1}$$

$$\begin{aligned}\Delta f &= [\Delta f(\mathbf{x})] + J_n\delta_{\Gamma_0} + \frac{\partial}{\partial n}(J_0\delta_{\Gamma_0}) \\ &= [\Delta f(\mathbf{x})] + \left(\left[\frac{\partial f}{\partial n_1}(\mathbf{x})\right] + \left[\frac{\partial f}{\partial n_2}(\mathbf{x})\right]\right)\delta_{\Gamma_0} \\ &\quad + \frac{\partial}{\partial n_1}(f_1\delta_{\Gamma_0}) + \frac{\partial}{\partial n_2}(f_2\delta_{\Gamma_0}) \quad \text{in } \mathcal{D}'(\mathbb{R}^n), \end{aligned} \tag{3.2.2}$$

where the distributions $J_n\delta_{\Gamma_0}$, $\frac{\partial}{\partial n}(J_0\delta_{\Gamma_0})$ are defined by (3.1.75) for $n = 3$ (resp. (3.1.49) and (3.1.50) for $n = 2$); for $\hat{\mathbf{n}}_1 \cdot \hat{\mathbf{n}}_2 = -1$, using (3.1.53) for $n = 2$ and (3.1.83) for $n = 3$, the alternative expression in (3.2.1)/(3.2.2) is obtained.

Green's formula Let $\Omega \subset \mathbb{R}^n$ $(n = 2, 3)$ be a domain in $\mathbb{R}^n$ with an orientable smooth boundary Γ such that the unit vector $\hat{\mathbf{n}}$ normal to Γ and exterior to Ω is defined *everywhere* on Γ. Let $f \in C^2(\Omega) \cap C^1(\overline{\Omega})$ such that $f(\mathbf{x}) = 0$ $\forall \mathbf{x} \in \mathbb{R}^n \setminus \overline{\Omega}$, i.e. f

is defined on $\mathbb{R}^n$ such that f and its normal derivative (in the pointwise sense) $[\frac{\partial f}{\partial n}(\mathbf{x})]$ have finite jumps J_0 and J_n across Γ defined, $\forall \mathbf{x} \in \Gamma$, by:

$$J_0 = f(\mathbf{x}^+) - f(\mathbf{x}) = 0 - f(\mathbf{x}) = -f(\mathbf{x}),$$
$$J_n = \left[\frac{\partial f}{\partial n}(\mathbf{x}^+)\right] - \left[\frac{\partial f}{\partial n}(\mathbf{x})\right] = 0 - \left[\frac{\partial f}{\partial n}(\mathbf{x})\right] = -\left[\frac{\partial f}{\partial n}(\mathbf{x})\right]. \tag{3.2.3}$$

Then we have:

Theorem 3.2.1 (Green's formula). $\forall \phi \in \mathcal{D}(\mathbb{R}^n)$,

$$\int_\Omega (f\Delta\phi - \phi[\Delta f(\mathbf{x})])d\mathbf{x} = \int_\Gamma \left(f\frac{\partial \phi}{\partial n} - \phi\left[\frac{\partial f}{\partial n}(\mathbf{x})\right]\right)dS$$
$$= \left\langle \frac{\partial}{\partial n}(-f\delta_\Gamma), \phi\right\rangle + \left\langle -\left[\frac{\partial f}{\partial n}(\mathbf{x})\right]\delta_\Gamma, \phi\right\rangle, \tag{3.2.4}$$

where, $\forall \boldsymbol{\xi} \in \Gamma$, $f(\boldsymbol{\xi}) = \lim_{\mathbf{x}\longrightarrow\boldsymbol{\xi},\mathbf{x}\in\Omega} f(\mathbf{x})$, $[\frac{\partial f}{\partial n}(\boldsymbol{\xi})] = \lim_{\mathbf{x}\longrightarrow\boldsymbol{\xi},\mathbf{x}\in\Omega}[\frac{\partial f}{\partial n}(\mathbf{x})]$ *is the normal derivative of* f *from within* Ω *in the usual pointwise sense;* $[\Delta f(\mathbf{x})] = \sum_{k=1}^n [\frac{\partial^2 f}{\partial x_k^2}(\mathbf{x})]$ *is the Laplacian of* f *in the usual pointwise sense;* $d\mathbf{x} = dx_1dx_2$ *and* $dS = ds$ *for* $n = 2$ *(resp.* $dS = dA$ *and* $d\mathbf{x} = dx_1dx_2dx_3$ *for* $n = 3$*).*

Proof. From (3.1.43) for $n = 2$ (resp. (3.1.74) for $n = 3$) with $\Gamma = \Gamma_0$, $J_0 = -f(\mathbf{x})$, $J_n = -[\frac{\partial f}{\partial n}(\mathbf{x})]$ (3.2.3), we have, $\forall \phi \in \mathcal{D}(\mathbb{R}^n)$,

$$\langle \Delta f, \phi\rangle = \langle [\Delta f(\mathbf{x})], \phi\rangle + \left\langle -\frac{\partial f}{\partial n}(\mathbf{x})\delta_\Gamma, \phi\right\rangle + \left\langle \frac{\partial}{\partial n}(-f\delta_\Gamma), \phi\right\rangle$$
$$= \int_\Omega \phi[\Delta f(\mathbf{x})]d\mathbf{x} - \int_\Gamma \left\{\left[\frac{\partial f}{\partial n}(\mathbf{x})\right]\phi - f\frac{\partial \phi}{\partial n}\right\}dS$$

($dS = ds$ for $n = 2$, $dS = dA$ for $n = 3$), since $f(\mathbf{x}) = 0$ outside $\overline{\Omega}$. But $\forall \phi \in \mathcal{D}(\mathbb{R}^n)$,

$$\langle \Delta f, \phi\rangle = \langle f, \Delta\phi\rangle = \int_\Omega f\Delta\phi d\mathbf{x},$$

since $f \in L^1_{\text{loc}}(\mathbb{R}^n)$ and $f(\mathbf{x}) = 0$ outside $\overline{\Omega}$.

Hence,

$$\int_\Omega f\Delta\phi d\mathbf{x} = \int_\Omega \phi[\Delta f(\mathbf{x})]d\mathbf{x} + \int_\Gamma \left(f\frac{\partial \phi}{\partial n} - \phi\left[\frac{\partial f}{\partial n}(\mathbf{x})\right]\right)dS \quad \forall \phi \in \mathcal{D}(\mathbb{R}^n)$$
$$\Longrightarrow \quad \int_\Omega (f\Delta\phi - \phi[\Delta f(\mathbf{x})])d\mathbf{x} = \int_\Gamma \left(f\frac{\partial \phi}{\partial n} - \phi\left[\frac{\partial f}{\partial n}(\mathbf{x})\right]\right)dS$$
$$= \left\langle \frac{\partial}{\partial n}(-f\delta_\Gamma), \phi\right\rangle + \left\langle -\left[\frac{\partial f}{\partial n}(\mathbf{x})\right]\delta_\Gamma, \phi\right\rangle,$$

where

$$\left\langle \frac{\partial}{\partial n}(-f\delta_\Gamma), \phi \right\rangle = \left\langle f\delta_\Gamma, \frac{\partial \phi}{\partial n} \right\rangle = \int_\Gamma f \frac{\partial \phi}{\partial n} dS;$$

$$\left\langle -\left[\frac{\partial f}{\partial n}(\mathbf{x})\right]\delta_\Gamma, \phi \right\rangle = -\int_\Gamma \phi \left[\frac{\partial f}{\partial n}(\mathbf{x})\right] dS. \qquad \square$$

Proposition 3.2.1. *$\forall \varepsilon > 0$, let $\Omega_\varepsilon = \{\mathbf{x} : \mathbf{x} \in \mathbb{R}^n, \|\mathbf{x}\| > \varepsilon\}$ and $\Gamma_\varepsilon = \{\mathbf{x} : \|\mathbf{x}\| = \varepsilon\}$ be its boundary with radius ε. Then, $\forall f \in C^\infty(\mathbb{R}^n \setminus \{\mathbf{0}\})$, $\forall \phi \in \mathcal{D}(\mathbb{R}^n)$,*

$$\int_{\Omega_\varepsilon} \{f(\mathbf{x})\Delta\phi(\mathbf{x}) - [\Delta f(\mathbf{x})]\phi(\mathbf{x})\} d\mathbf{x} = \int_{\Gamma_\varepsilon} \left[\phi(\mathbf{x})\frac{\partial f}{\partial r}(\mathbf{x}) - f(\mathbf{x})\frac{\partial \phi}{\partial r}(\mathbf{x})\right] dS, \tag{3.2.5}$$

where $[\Delta f(\mathbf{x})]$ is the Laplacian of f in the usual pointwise sense; $\frac{\partial}{\partial r} = -\frac{\partial}{\partial n_\varepsilon}$, $\frac{\partial}{\partial r}(\cdot)$ being the derivative in the radial direction $-\hat{\mathbf{n}}_\varepsilon$, $\hat{\mathbf{n}}_\varepsilon$ being the exterior unit normal to Γ_ε; $d\mathbf{x} = dx_1dx_2dx_3$ and $dS = dA$ (surface area measure) for $n = 3$ (resp. $d\mathbf{x} = dx_1dx_2$ and $dS = ds$ (arc length measure) for $n = 2$).

Proof. For $\phi \in \mathcal{D}(\mathbb{R}^n)$, $\exists R > 0$ such that $\mathrm{supp}(\phi) \subset \{\mathbf{x} : \|\mathbf{x}\| < R\}$.

Define $\Omega_{\varepsilon,R} = \{\mathbf{x} : \varepsilon < \|\mathbf{x}\| < R\}$ enclosed by Γ_ε and Γ_R with radius ε and R respectively. Since $f, \phi \in C^\infty(\overline{\Omega}_{\varepsilon,R})$, the classical Green's formula of calculus [32] holds:

$$\begin{aligned}\int_{\Omega_{\varepsilon,R}} \{f(\mathbf{x})\Delta\phi(\mathbf{x}) - [\Delta f(\mathbf{x})]\phi(\mathbf{x})\} d\mathbf{x} &= \int_{\Gamma_\varepsilon} \left[f(\mathbf{x})\frac{\partial \phi}{\partial n_\varepsilon}(\mathbf{x}) - \phi(\mathbf{x})\frac{\partial f}{\partial n_\varepsilon}(\mathbf{x})\right] dS \\ &+ \int_{\Gamma_R} \left[f(\mathbf{x})\frac{\partial \phi}{\partial n_R}(\mathbf{x}) - \phi(\mathbf{x})\frac{\partial f}{\partial n_R}(\mathbf{x})\right] dS,\end{aligned} \tag{3.2.6}$$

where $\hat{\mathbf{n}}_R$ is the unit exterior normal to Γ_R.

Since $\phi\downarrow_{\Gamma_R} = \frac{\partial \phi}{\partial n_R}\downarrow_{\Gamma_R} = 0$, $\int_{\Omega_\varepsilon}\{\dots\}d\mathbf{x} = \int_{\Omega_{\varepsilon,R}}\{\dots\}d\mathbf{x}$ and $\int_{\Gamma_R}(\dots) = 0$, and $\frac{\partial}{\partial n_\varepsilon}(\cdot) = -\frac{\partial}{\partial r}(\cdot)$ in (3.2.6), the result (3.2.5) follows from (3.2.6). $\square$

Example 3.2.1. For $r = \|\mathbf{x}\| = (x_1^2 + x_2^2 + \dots + x_n^2)^{1/2}$ and $\phi \in \mathcal{D}(\mathbb{R}^n)$, prove that

1. for $n = 2$,

$$\int_{\Omega_\varepsilon:\|\mathbf{x}\|>\varepsilon} \ln r \Delta\phi dx_1dx_2 = \int_{\Gamma_\varepsilon:\|\mathbf{x}\|=\varepsilon} \left[\phi(\mathbf{x})\frac{1}{\varepsilon} - \ln\varepsilon\frac{\partial \phi}{\partial r}\right] ds; \tag{3.2.7}$$

2. for $n = 3$,

$$\int_{\Omega_\varepsilon:\|\mathbf{x}\|>\varepsilon} \frac{1}{r}\Delta\phi dx_1dx_2dx_3 = \int_{\Gamma_\varepsilon:\|\mathbf{x}\|=\varepsilon} \left[\phi(\mathbf{x})\frac{-1}{\varepsilon^2} - \frac{1}{\varepsilon}\frac{\partial \phi}{\partial r}\right] dS. \tag{3.2.8}$$

Solution.

1. $\frac{\partial}{\partial x_i}(\ln r) = \frac{1}{r}\cdot\frac{\partial r}{\partial x_i} = \frac{x_i}{r^2}$; $\frac{\partial^2}{\partial x_i^2}(\ln r) = \frac{\partial}{\partial x_i}(\frac{x_i}{r^2}) = \frac{1}{r^2} - \frac{2x_i^2}{r^4}$; $\Longrightarrow [\Delta \ln r(\mathbf{x})] = \sum_{i=1}^{2}\frac{\partial^2}{\partial x_i^2}\ln r = \frac{2}{r^2} - \frac{2r^2}{r^4} = 0\ \forall \mathbf{x} \neq \mathbf{0}$. Since $\ln r \in C^\infty(\mathbb{R}^2 \setminus \{\mathbf{0}\})$ and $[\Delta \ln r(\mathbf{x})] = 0\ \forall \mathbf{x}\neq\mathbf{0}$, from (3.2.5),
$$\int_{\Omega_\varepsilon} \ln r\Delta\phi dx_1dx_2 = \int_{\Gamma_\varepsilon}\left[\phi(\mathbf{x})\frac{\partial \ln r}{\partial r}(\mathbf{x}) - \ln r\frac{\partial\phi}{\partial r}(\mathbf{x})\right]ds$$
$$= \int_{\Gamma_\varepsilon}\left[\phi(\mathbf{x})\frac{1}{\varepsilon} - \ln\varepsilon\frac{\partial\phi}{\partial r}(\mathbf{x})\right]ds,$$
since $\frac{\partial}{\partial r}\ln r\downarrow_{\Gamma_\varepsilon} = \frac{1}{r}\downarrow_{\Gamma_\varepsilon} = \frac{1}{\varepsilon}$ and $\ln r\downarrow_{\Gamma_\varepsilon} = \ln\varepsilon$.
2. For $n=3$, $\frac{\partial}{\partial x_i}(\frac{1}{r}) = -\frac{1}{r^2}\cdot\frac{\partial r}{\partial x_i} = -\frac{x_i}{r^3}$, $\frac{\partial^2}{\partial x_i^2}(\frac{1}{r}) = \frac{\partial}{\partial x_i}(-\frac{x_i}{r^3}) = -\frac{1}{r^3} + 3\cdot\frac{x_i}{r^4}\cdot\frac{x_i}{r}$ $\Longrightarrow [\Delta(\frac{1}{r})(\mathbf{x})] = \sum_{i=1}^{3}\frac{\partial^2}{\partial x_i^2}(\frac{1}{r}) = -\frac{3}{r^3} + 3\frac{r^2}{r^5} = 0\ \forall\mathbf{x}\neq\mathbf{0}$.
Since $\frac{1}{r}\in C^\infty(\mathbb{R}^3\setminus\{\mathbf{0}\})$ and $[\Delta(\frac{1}{r})(\mathbf{x})] = 0\ \forall\mathbf{x}\neq\mathbf{0}$, we get the result (3.2.8) from (3.2.5) by writing $\frac{\partial}{\partial r}(\frac{1}{r})\downarrow_{\Gamma_\varepsilon} = -\frac{1}{r^2}\downarrow_{\Gamma_\varepsilon} = -\frac{1}{\varepsilon^2}$ and $\frac{1}{r}\downarrow_{\Gamma_\varepsilon} = \frac{1}{\varepsilon}$.

3.3 Elementary solutions

Laplacian $\boldsymbol{\Delta(\frac{1}{r^{n-2}})}$ of $\boldsymbol{\frac{1}{r^{n-2}}}$ in the distribution sense ($n \geq 3$)

$\boldsymbol{\frac{1}{r^{n-2}}}$ is a harmonic function in $\mathbb{R}^n - \{\mathbf{0}\}$ for $n \neq 2$ Every constant function $f(\mathbf{x}) = C\ \forall\mathbf{x}\in\mathbb{R}^n$ is harmonic everywhere in $\mathbb{R}^n$, since $[\Delta f(\mathbf{x})] = 0\ \forall\mathbf{x}\in\mathbb{R}^n$. Now we have to show that a non-constant function $f = f(r)$ with
$$r = \|\mathbf{x}\| = \left(\sum_{i=1}^{n} x_i^2\right)^{\frac{1}{2}} \quad\text{and}\quad f(r) = \begin{cases} C_1\frac{1}{r^{n-2}} + C_2 & \text{for } n\neq 2\ (r\neq 0),\\ C_1\ln\frac{1}{r} + C_2 & \text{for } n = 2\ (r\neq 0),\end{cases}$$
where C_1, and C_2 are constants, is harmonic in $\mathbb{R}^n - \{\mathbf{0}\}$, i.e. $[\Delta f(\mathbf{x})] = 0\ \forall\mathbf{x}\in(\mathbb{R}^n - \{\mathbf{0}\})$.

Since, for $r\neq 0$,
$$\left[\frac{\partial f}{\partial x_i}(\mathbf{x})\right] = \frac{df}{dr}\cdot\frac{\partial r}{\partial x_i} = \frac{df}{dr}\cdot\frac{x_i}{r},$$
$$\left[\frac{d^2 f}{\partial x_i^2}(\mathbf{x})\right] = \frac{\partial}{\partial x_i}\left(\frac{df}{dr}\frac{x_i}{r}\right) = \frac{d^2 f}{dr^2}\left(\frac{x_i^2}{r^2}\right) + \frac{df}{dr}\left(\frac{1}{r} - \frac{x_i^2}{r^3}\right), \tag{3.3.1}$$
$$[\Delta f(\mathbf{x})] = \sum_{i=1}^{n}\left[\frac{\partial^2 f}{\partial x_i^2}(\mathbf{x})\right] = \sum_{i=1}^{n}\left[\frac{d^2 f}{dr^2}\left(\frac{x_i^2}{r^2}\right) + \frac{df}{dr}\left(\frac{1}{r} - \frac{x_i^2}{r^3}\right)\right]$$
$$= \left[\frac{d^2 f}{dr^2}(r)\right] + \frac{n-1}{r}\left[\frac{df}{dr}(r)\right]$$

for $r \neq 0$. Hence, f is a harmonic function of r for $r \neq 0 \Longrightarrow [\Delta f(\mathbf{x})] = 0$

$$\Longrightarrow \quad f \text{ must satisfy } \left[\frac{d^2 f}{dr^2}(r) + \frac{n-1}{r}\frac{df}{dr}(r)\right] = 0. \tag{3.3.2}$$

Setting $g = \frac{df}{dr}$, we get, from (3.3.2), $\frac{dg}{dr} + \frac{n-1}{r} g = 0$, which has the general solution $g = C_1 \frac{1}{r^{n-1}} \Longrightarrow f = C_1 \int \frac{1}{r^{n-1}} dr + C_2$

$$\Longrightarrow \quad f = \begin{cases} C_1 \frac{1}{r^{n-2}} + C_2 \text{ with constants } C_1, C_2 & \text{for } n \neq 2 \ (r \neq 0), \\ C_1 \ln \frac{1}{r} + C_2 \text{ with constants } C_1, C_2 & \text{for } n = 2 \ (r \neq 0) \end{cases}$$

is the general solution of (3.3.2), since $\ln r$ and $\ln \frac{1}{r} = -\ln r$ are *linearly dependent* solutions of (3.3.2) for $n = 2$.

Properties of $\frac{1}{r^{n-2}}$ for $n \geq 3$ Function $\frac{1}{r^{n-2}}$ is

- a C^∞-function in $\mathbb{R}^n - \{\mathbf{0}\}$ with a singularity at the origin $r = 0$;
- a harmonic function:

$$\left[\Delta\left(\frac{1}{r^{n-2}}\right)(\mathbf{x})\right] = 0 \quad \forall \mathbf{x} \in \mathbb{R}^n - \{\mathbf{0}\}; \tag{3.3.3}$$

- a locally summable function in $\mathbb{R}^n$, i.e. $\frac{1}{r^{n-2}} \in L^1_{\text{loc}}(\mathbb{R}^n)$, since $n - 2 < n$;

$\Longrightarrow \int_{\|\mathbf{x}\| \leq R} \frac{1}{r^{n-2}} d\mathbf{x} < +\infty \ \forall R > 0$. As $\frac{1}{r^{n-2}} \in L^1_{\text{loc}}(\mathbb{R}^n)$, it defines a regular distribution $T_{\frac{1}{r^{n-2}}} \in \mathcal{D}'(\mathbb{R}^n)$. But the Laplacian $[\Delta(\frac{1}{r^{n-2}})(\mathbf{x})] = 0$ in the pointwise sense $\forall \mathbf{x} \in (\mathbb{R}^n - \{\mathbf{0}\})$, and $\mathbf{x} = \mathbf{0}$ is a singularity of $\frac{1}{r^{n-2}}$ as a function. Hence, we expect that $\Delta T_{\frac{1}{r^{n-2}}} \in \mathcal{D}'(\mathbb{R}^n)$ will be a singular distribution, which will involve the Dirac distribution concentrated at the origin $\mathbf{x} = \mathbf{0}$.

For the sake of simplicity, we establish the result for the most important case of $n = 3$, i.e. for $\Delta T_{\frac{1}{r}}$: $\forall \phi \in \mathcal{D}(\mathbb{R}^3)$,

$$\langle \Delta T_{\frac{1}{r}}, \phi \rangle = \langle T_{\frac{1}{r}}, \Delta\phi \rangle = \left\langle \frac{1}{r}, \Delta\phi \right\rangle = \int_{\mathbb{R}^3} \frac{1}{r} \Delta\phi \, d\mathbf{x} = \lim_{\varepsilon \to 0+} \int_{\|\mathbf{x}\| > \varepsilon} \frac{1}{r} \Delta\phi \, d\mathbf{x}. \tag{3.3.4}$$

Define an auxiliary function f_ε by:

$$f_\varepsilon(r) = \begin{cases} 0 & \text{for } r < \varepsilon; \\ \frac{1}{r} & \text{for } r > \varepsilon \end{cases} \tag{3.3.5}$$

with $\varepsilon > 0$.

Then $f_\varepsilon \in L^1_{\mathrm{loc}}(\mathbb{R}^3)$ is discontinuous on the spherical surface S_ε with radius $r = \varepsilon$, normal $\hat{\mathbf{n}}_\varepsilon = \hat{\mathbf{r}}$, $\hat{\mathbf{r}}$ being the unit vector in the direction of $\mathbf{r} = r \cdot \hat{\mathbf{r}}$ such that $\frac{\partial}{\partial n_\varepsilon} = \frac{\partial}{\partial r}$. Hence, the jumps J_0 and $J_{n_\varepsilon} = J_r$ of f_ε and $\frac{\partial f_\varepsilon}{\partial n_\varepsilon} = \frac{\partial f_\varepsilon}{\partial r}$ respectively in crossing S_ε are given by:

$$\begin{aligned} J_0 &= f_\varepsilon(\varepsilon^+) - f_\varepsilon(\varepsilon^-) = \frac{1}{r} - 0 = \frac{1}{r}, \\ J_r &= \frac{\partial f_\varepsilon}{\partial r}(\varepsilon^+) - \frac{\partial f_\varepsilon}{\partial r}(\varepsilon^-) = \frac{\partial}{\partial r}\left(\frac{1}{r}\right) - 0 = -\frac{1}{r^2}. \end{aligned} \tag{3.3.6}$$

f_ε is harmonic in $\mathbb{R}^3 \setminus S_\varepsilon \Longrightarrow [\Delta f_\varepsilon(\mathbf{x})] = 0 \ \forall \mathbf{x} \in \mathbb{R}^3 \setminus S_\varepsilon$

$$\Longrightarrow \quad \int_{\mathbb{R}^3} [\Delta f_\varepsilon(\mathbf{x})]\phi d\mathbf{x} = 0 \quad \forall \phi \in \mathcal{D}(\mathbb{R}^3). \tag{3.3.7}$$

$f_\varepsilon \in L^1_{\mathrm{loc}}(\mathbb{R}^3) \Longrightarrow \langle T_{f_\varepsilon}, \Delta\phi\rangle = \langle \Delta T_{f_\varepsilon}, \phi\rangle \ \forall \phi \in \mathcal{D}(\mathbb{R}^3)$.

$$\begin{aligned} \langle T_{f_\varepsilon}, \Delta\phi\rangle &= \langle f_\varepsilon, \Delta\phi\rangle = \int_{\mathbb{R}^3} f_\varepsilon \Delta\phi d\mathbf{x} = \int_{\|\mathbf{x}\|>\varepsilon} \frac{1}{r}\Delta\phi d\mathbf{x} \\ \Longrightarrow \quad & \lim_{\varepsilon\to 0+} \int_{\|\mathbf{x}\|=r>\varepsilon} \frac{1}{r}\Delta\phi d\mathbf{x} = \lim_{\varepsilon\to 0+} \langle T_{f_\varepsilon}, \Delta\phi\rangle = \lim_{\varepsilon\to 0+} \langle \Delta T_{f_\varepsilon}, \phi\rangle \\ &= \lim_{\varepsilon\to 0+} \left[\int_{\mathbb{R}^3} [\Delta f_\varepsilon(\mathbf{x})]\phi d\mathbf{x} + \left\langle \frac{\partial}{\partial n_\varepsilon}(J_0 \delta_{S_\varepsilon}), \phi \right\rangle + \langle J_{n_\varepsilon} \delta_{S_\varepsilon}, \phi\rangle \right] \end{aligned}$$

(using (3.2.1)/(3.2.2) with $S_\varepsilon = \Gamma_0$, $\hat{\mathbf{n}}_\varepsilon = \hat{\mathbf{n}}$). Applying (3.3.4)–(3.3.7),

$$\begin{aligned} \langle \Delta T_{\frac{1}{r}}, \phi\rangle &= \lim_{\varepsilon\to 0+} \left[\left\langle \frac{\partial}{\partial r}(\frac{1}{r}\delta_{S_\varepsilon}), \phi \right\rangle + \langle J_r \delta_{S_\varepsilon}, \phi\rangle \right] \\ &= \lim_{\varepsilon\to 0+} \left[-\left\langle \frac{1}{r}\delta_{S_\varepsilon}, \frac{\partial\phi}{\partial r} \right\rangle + \langle -\frac{1}{r^2}\delta_{S_\varepsilon}, \phi \rangle \right] \quad \forall \phi \in \mathcal{D}(\mathbb{R}^3) \\ &= \lim_{\varepsilon\to 0+} \left(-\int_{S_\varepsilon} \frac{1}{\varepsilon}\frac{\partial\phi}{\partial r} dS \right) + \lim_{\varepsilon\to 0+} \left(\int_{S_\varepsilon} -\frac{1}{\varepsilon^2}\phi dS \right), \end{aligned} \tag{3.3.8}$$

since $r = \varepsilon$ on S_ε.

Now we will show that, $\forall\phi \in \mathcal{D}(\mathbb{R}^3)$, the first and second surface integrals in (3.3.8) tend to 0 and $-4\pi\phi(\mathbf{0})$ respectively as $\varepsilon \to 0^+$. $\forall\phi \in \mathcal{D}(\mathbb{R}^3)$,

$$\left|-\int_{S_\varepsilon} \frac{1}{\varepsilon}\frac{\partial\phi}{\partial r}dS\right| \leq \frac{1}{\varepsilon}\max_{\mathbf{x}\in S_\varepsilon}\left|\frac{\partial\phi}{\partial r}(\mathbf{x})\right| \int_{S_\varepsilon} dS = \frac{1}{\varepsilon}\max_{\mathbf{x}\in S_\varepsilon}\left|\frac{\partial\phi}{\partial r}(\mathbf{x})\right| \cdot 4\pi\varepsilon^2$$

$$\leq \max_{\mathbf{x}\in S_\varepsilon}\left|\frac{\partial\phi}{\partial r}(\mathbf{x})\right| \cdot 4\pi\varepsilon \to 0 \text{ as } \varepsilon \to 0^+$$

$$\Longrightarrow \quad \lim_{\varepsilon\to 0^+}\left(-\int_{S_\varepsilon} \frac{1}{\varepsilon}\frac{\partial\phi}{\partial r}dS\right) = 0; \tag{3.3.9}$$

$$\int_{S_\varepsilon} -\frac{1}{\varepsilon^2}\phi dS = -\int_{S_\varepsilon} \frac{1}{\varepsilon^2}\phi(\mathbf{0})dS + \int_{S_\varepsilon} -\frac{1}{\varepsilon^2}(\phi(\mathbf{x}) - \phi(\mathbf{0}))dS. \tag{3.3.10}$$

But

$$-\int_{S_\varepsilon} \frac{1}{\varepsilon^2}\phi(\mathbf{0})dS = -\frac{\phi(\mathbf{0})}{\varepsilon^2}\int_{S_\varepsilon} dS = -\frac{\phi(\mathbf{0})}{\varepsilon^2}4\pi\varepsilon^2 = -4\pi\phi(\mathbf{0}). \tag{3.3.11}$$

Using the mean-value theorem and Schwarz's inequality (see (B.4.3.1) in Appendix B), for $\mathbf{x} \in S_\varepsilon$ with $\|\mathbf{x}\| = \varepsilon$,

$$|\phi(\mathbf{x}) - \phi(\mathbf{0})| = \left|\sum_{i=1}^{3} x_i \frac{\partial\phi}{\partial x_i}(\boldsymbol{\xi})\right| = |\langle \mathbf{x}, \nabla\phi(\boldsymbol{\xi})\rangle|$$

$$\leq \|\mathbf{x}\|\|\nabla\phi(\boldsymbol{\xi})\| \leq \varepsilon\sqrt{3}\max_{1\leq i\leq 3}\max_{\mathbf{x}\in\operatorname{supp}(\phi)}\left|\frac{\partial\phi}{\partial x_i}(\mathbf{x})\right| \quad \forall\phi \in \mathcal{D}(\mathbb{R}^3),$$

since

$$\|\nabla\phi(\boldsymbol{\xi})\|^2 = \sum_{i=1}^{3}\left|\frac{\partial\phi}{\partial x_i}(\boldsymbol{\xi})\right|^2 \leq 3\left(\max_{1\leq i\leq 3}\left|\frac{\partial\phi}{\partial x_i}(\boldsymbol{\xi})\right|\right)^2$$

$$\leq 3\left(\max_{1\leq i\leq 3}\max_{\mathbf{x}\in\operatorname{supp}(\phi)}\left|\frac{\partial\phi}{\partial x_i}(\mathbf{x})\right|\right)^2.$$

Hence,

$$\left|\int_{S_\varepsilon} -\frac{1}{\varepsilon^2}(\phi(\mathbf{x}) - \phi(\mathbf{0}))dS\right| \leq \frac{1}{\varepsilon^2}\varepsilon\sqrt{3}\max_{1\leq i\leq 3}\left(\max_{\mathbf{x}\in\operatorname{supp}(\phi)}\left|\frac{\partial\phi}{\partial x_i}(\mathbf{x})\right|\right)\int_{S_\varepsilon} dS$$

$$= \sqrt{3}\left(\max_{1\leq i\leq 3}\max_{\mathbf{x}\in\operatorname{supp}(\phi)}\left|\frac{\partial\phi}{\partial x_i}(\mathbf{x})\right|\right)\frac{1}{\varepsilon}\cdot 4\pi\varepsilon^2 \to 0 \quad \text{as } \varepsilon \to 0^+ \ \forall\phi \in \mathcal{D}(\mathbb{R}^3)$$

$$\Longrightarrow \quad \lim_{\varepsilon\to 0^+}\int_{S_\varepsilon} -\frac{1}{\varepsilon^2}(\phi(\mathbf{x}) - \phi(\mathbf{0}))dS = 0.$$

Then, from (3.3.10) and (3.3.11),

$$\lim_{\varepsilon\to 0+}\int_{S_\varepsilon} -\frac{1}{\varepsilon^2}\phi dS = -4\pi\phi(\mathbf{0}). \tag{3.3.12}$$

Finally, from (3.3.8), (3.3.9) and (3.3.12), we get, $\forall\phi\in\mathcal{D}(\mathbb{R}^3)$:

$$\langle \Delta T_{\frac{1}{r}}, \phi\rangle = -4\pi\phi(\mathbf{0}) = -4\pi\langle\delta,\phi\rangle = \langle -4\pi\delta,\phi\rangle$$

$\Longrightarrow \Delta T_{\frac{1}{r}} = -4\pi\delta$, $\delta\in\mathcal{D}'(\mathbb{R}^3)$ being the Dirac distribution concentrated at the singular point $\mathbf{x}=\mathbf{0}\in\mathbb{R}^3$. Similarly, $\Delta(T_{\frac{1}{r^{n-2}}})$ can be evaluated $\forall n\neq 2$.

For $n = 2$, Laplacian of $\ln r$ in the distributional sense in $\mathbb{R}^2$ $\ln r$ is locally integrable in $\mathbb{R}^2$, i.e. $\ln r\in L^1_{\text{loc}}(\mathbb{R}^2)$: for this, it is sufficient to show that $\int_{\|\mathbf{x}\|\le 1}|\ln r|\,d\mathbf{x}<+\infty$, since $\int_{1\le\|\mathbf{x}\|\le R}|\ln r|d\mathbf{x}<+\infty\ \forall R>1$.

$$\begin{aligned}\int_{\|\mathbf{x}\|\le 1}|\ln r|d\mathbf{x} &= \int_0^1\int_0^{2\pi}|\ln r|rdrd\theta = -\int_0^{2\pi}d\theta\int_0^1 r\ln rdr\\ &= -2\pi\int_0^1 r\ln rdr = -2\pi\left[\frac{r^2}{2}\ln r\Big|_0^1 - \int_0^1\frac{r^2}{2r}dr\right]\\ &= -2\pi\left(\frac{-r^2}{4}\Big|_0^1\right) = \frac{\pi}{2}<+\infty.\end{aligned}$$

Hence, $\ln r\in\mathcal{D}'(\mathbb{R}^2)$, and the Laplacian of $\ln r$ in the distributional sense is defined by, $\forall\phi\in\mathcal{D}(\mathbb{R}^2)$, $\langle\Delta\ln r,\phi\rangle = \langle\ln r,\Delta\phi(\mathbf{x})\rangle = \int_{\mathbb{R}^2}(\ln r)\Delta\phi(\mathbf{x})d\mathbf{x}$. $\ln r\in L^1_{\text{loc}}(\mathbb{R}^2)$, but its distributional derivatives of order two are *not* locally integrable. Hence, we cannot apply integration by parts. Since $\ln r\in L^1_{\text{loc}}(\mathbb{R}^2)$, we can apply Lebesgue's Theorem and write

$$\langle\Delta\ln r,\phi\rangle = \int_{\mathbb{R}^2}(\ln r)\Delta\phi(\mathbf{x})d\mathbf{x} = \lim_{\varepsilon\to 0+}\int_{\Omega_\varepsilon:\|\mathbf{x}\|>\varepsilon}(\ln r)\Delta\phi(\mathbf{x})d\mathbf{x} = \lim_{\varepsilon\to 0+}I_1(\varepsilon),$$

where $\Omega_\varepsilon = \{\mathbf{x}:\mathbf{x}\in\mathbb{R}^2, \|\mathbf{x}\|>\varepsilon\}$ with boundary $\Gamma_\varepsilon = \{\mathbf{x}:\|\mathbf{x}\|=\varepsilon\}$, and, from (3.2.7),

$$\begin{aligned}I_1(\varepsilon) &= \int_{\Gamma_\varepsilon:\|\mathbf{x}\|=\varepsilon}\left[\phi(\mathbf{x})\cdot\frac{1}{\varepsilon} - \ln\varepsilon\cdot\frac{\partial\phi}{\partial r}\right]ds\\ &= \int_{\Gamma_\varepsilon}\frac{\phi(\mathbf{x})}{\varepsilon}ds - \int_{\Gamma_\varepsilon}\ln\varepsilon\frac{\partial\phi}{\partial r}ds = I_2(\varepsilon)-I_3(\varepsilon)\end{aligned}$$

$\Longrightarrow \lim_{\varepsilon\to 0+}I_1(\varepsilon) = \lim_{\varepsilon\to 0+}I_2(\varepsilon) - \lim_{\varepsilon\to 0+}I_3(\varepsilon)$.

For finding the right-hand side limits, Γ_ε is represented by $x_1 = \varepsilon\cos\theta$, $x_2 = \varepsilon\sin\theta$.

Then $I_2(\varepsilon) = \int_0^{2\pi} \frac{\tilde{\phi}(\varepsilon,\theta)}{\varepsilon}\varepsilon d\theta = \int_0^{2\pi} \tilde{\phi}(\varepsilon,\theta)d\theta$ with $\tilde{\phi}(\varepsilon,\theta) = \phi(\varepsilon\cos\theta, \varepsilon\sin\theta)$.

But

$$\begin{aligned}\lim_{\varepsilon\to 0+} I_2(\varepsilon) &= \lim_{\varepsilon\to 0+}\int_0^{2\pi} \tilde{\phi}(\varepsilon,\theta)d\theta = \int_0^{2\pi} \lim_{\varepsilon\to 0+} \tilde{\phi}(\varepsilon,\theta)d\theta \\ &= \int_0^{2\pi} \tilde{\phi}(0,\theta)d\theta = \int_0^{2\pi} \phi(0,0)d\theta = \phi(\mathbf{0})\int_0^{2\pi} d\theta = 2\pi\phi(\mathbf{0}),\end{aligned}$$

since $\tilde{\phi}(0,\theta) = \phi(0\cdot\cos\theta, 0\cdot\sin\theta) = \phi(0,0) = \phi(\mathbf{0})$.

Again,

$$\left|\frac{\partial\phi}{\partial r}\right| = |\nabla\phi\cdot(-\hat{\mathbf{n}}_\varepsilon)| \le \|\nabla\phi\|\,\|-\hat{\mathbf{n}}_\varepsilon\| \le \sqrt{2}\max_{1\le i\le 2}\max_{\operatorname{supp}(\phi)}\left|\frac{\partial\phi}{\partial x_i}(\mathbf{x})\right| = C_1.$$

Then

$$\begin{aligned}|I_3(\varepsilon)| &= \left|\int_{\Gamma_\varepsilon} \ln\varepsilon\frac{\partial\phi}{\partial r}ds\right| = \left|\int_0^{2\pi} (\ln\varepsilon)\varepsilon\frac{\partial\phi}{\partial r}d\theta\right| \\ &\le |\varepsilon\ln\varepsilon|C_1\int_0^{2\pi} d\theta = 2\pi C_1|\varepsilon\ln\varepsilon| \to 0 \quad \text{as } \varepsilon\to 0^+\end{aligned}$$

$\Longrightarrow \lim_{\varepsilon\to 0+} I_3(\varepsilon) = 0$. Hence,

$$\begin{aligned}&\lim_{\varepsilon\to 0+} I_1(\varepsilon) = \lim_{\varepsilon\to 0+} I_2(\varepsilon) - \lim_{\varepsilon\to 0+} I_3(\varepsilon) = 2\pi\phi(\mathbf{0}) \\ &\Longrightarrow \quad \langle\Delta\ln r, \phi\rangle = 2\pi\phi(\mathbf{0}) = \langle 2\pi\delta, \phi\rangle \quad \forall\phi\in\mathcal{D}(\mathbb{R}^2) \\ &\Longrightarrow \quad \Delta\ln r = 2\pi\delta \quad \text{in } \mathcal{D}'(\mathbb{R}^2).\end{aligned} \tag{3.3.13}$$

The Laplacian

$$\Delta\ln\left(\frac{1}{r}\right) = -\Delta\ln r = -2\pi\delta \quad \text{in } \mathcal{D}'(\mathbb{R}^2). \tag{3.3.14}$$

Theorem 3.3.1. *For $n \neq 2$,*

$$\Delta T_{\frac{1}{r^{n-2}}} = \Delta\left(\frac{1}{r^{n-2}}\right) = -(n-2)S_n\delta \quad \text{in } \mathcal{D}'(\mathbb{R}^n), \tag{3.3.15}$$

where

$$S_n = \frac{2(\pi^{\frac{n}{2}})}{\Gamma(\frac{n}{2})} = \text{surface area of the unit } n\text{-sphere } \left(\sum_{i=1}^n x_i^2 = 1\right) \tag{3.3.16}$$

with $\Gamma(\frac{1}{2}) = \sqrt{\pi}$, $\Gamma(\frac{n}{2}+1) = (\frac{n}{2})\Gamma(\frac{n}{2})$*;* δ *is the (n-dimensional) Dirac distribution concentrated at* $\mathbf{0} \in \mathbb{R}^n$*.*

For $n = 2$*,*

$$\Delta T_{\ln \frac{1}{r}} = \Delta\left(\ln \frac{1}{r}\right) = -2\pi\delta \quad \textit{in } \mathcal{D}'(\mathbb{R}^2). \tag{3.3.17}$$

For $n = 1$*,*

$$\Delta(T_{|x|}) = \Delta(|x|) = 2\delta \quad \textit{in } \mathcal{D}'(\mathbb{R}). \tag{3.3.18}$$

For $n = 3$*,*

$$\Delta T_{\frac{1}{r}} = \Delta\left(\frac{1}{r}\right) = -4\pi\delta \quad \textit{in } \mathcal{D}'(\mathbb{R}^3). \tag{3.3.19}$$

Elementary (fundamental) solution of Laplace operator

Definition 3.3.1. A function $E \in C^\infty(\mathbb{R}^n \setminus \{\mathbf{0}\})$ which is locally integrable on $\mathbb{R}^n$ (i.e. $E \in \mathcal{D}'(\mathbb{R}^n)$ is a distribution on $\mathbb{R}^n$) is called an *elementary* or *fundamental solution* of the Laplace operator $\Delta = \sum_{i=1}^n \frac{\partial^2}{\partial x_1^2}$ if it satisfies the equation:

$$\Delta E = \delta \quad \text{in } \mathcal{D}'(\mathbb{R}^n). \tag{3.3.20}$$

Elementary solution E of the Laplace operator is *not* unique. (3.3.21)

In fact, any function E_0 harmonic in $\mathbb{R}^n$ (i.e. $\Delta E_0 = 0$ in $\mathbb{R}^n$), which is a C^∞-function in $\mathbb{R}^n$, is also an elementary solution of Δ. Thus, $E + E_0$ is also an elementary solution of Δ, i.e. $\Delta(E + E_0) = \Delta E + \Delta E_0 = \delta + 0 = \delta$ in $\mathcal{D}'(\mathbb{R}^n)$.

Theorem 3.3.2.

$$E_n = -\frac{1}{(n-2)S_n} \cdot \frac{1}{r^{n-2}} \quad \textit{for } n \neq 2,\ r = (x_1^2 + \cdots + x_n^2)^{\frac{1}{2}} \tag{3.3.22}$$

and

$$E_2 = -\frac{1}{2\pi} \ln \frac{1}{r} \quad \textit{for } n = 2,\ r = (x_1^2 + x_2^2)^{\frac{1}{2}} \tag{3.3.23}$$

are elementary solutions of the Laplace operator Δ*, where* $S_n = \frac{2(\pi^{\frac{n}{2}})}{\Gamma(\frac{n}{2})}$*.*

Proof. $E_n \in C^\infty(\mathbb{R}^n \setminus \{\mathbf{0}\})$ and $E_n \in L^1_{\text{loc}}(\mathbb{R}^n)$, since for $n \neq 2$, $\frac{1}{r^{n-2}} \in C^\infty(\mathbb{R}^n \setminus \{\mathbf{0}\})$ and $\frac{1}{r^{n-2}} \in L^1_{\text{loc}}(\mathbb{R}^n)$, and $\ln r \in C^\infty(\mathbb{R}^2 \setminus \{\mathbf{0}\})$, $\ln r \in L^1_{\text{loc}}(\mathbb{R}^2)$ for $n = 2$. Consequently, $E_n \in \mathcal{D}'(\mathbb{R}^n)$ and, from Theorem 3.3.1, $\Delta E_n = -\frac{1}{(n-2)S_n}\Delta(\frac{1}{r^{n-2}}) = -\frac{(-(n-2)S_n\delta)}{(n-2)S_n} = \delta$ in $\mathcal{D}'(\mathbb{R}^n)$ for $n \neq 2$ and $\Delta E_2 = -\frac{1}{2\pi}\Delta(\ln \frac{1}{r}) = -\frac{-2\pi\delta}{2\pi} = \delta$ in $\mathcal{D}'(\mathbb{R}^2)$ for $n = 2$. □

Elementary solution of ordinary differential operator Let L_n be the nth-order linear ordinary differential operator defined by:

$$L_n \equiv \frac{d^n}{dt^n} + a_1(t)\frac{d^{n-1}}{dt^{n-1}} + \cdots + a_n(t), \tag{3.3.24}$$

where $a_k \in C^\infty(]-\infty,\infty[)$ $\forall k = 1, 2, \ldots, n$.

Definition 3.3.2. A distribution $E \in \mathcal{D}'(]-\infty,\infty[)$ is called an *elementary* or *fundamental solution* of L_n on $]-\infty,\infty[$ if and only if

$$L_n E = \delta \quad \text{in } \mathcal{D}'(]-\infty,\infty[). \tag{3.3.25}$$

Elementary solution $E \in \mathcal{D}'(]-\infty,\infty[)$ of (3.3.24) is *not* unique. (3.3.26)

Construction of elementary solution $E \in \mathcal{D}'(]-\infty,\infty[)$ of (3.3.24) Let $y = y(t)$ with $y \in C^\infty([0,\infty[)$ be the unique solution of the Cauchy problem $L_n y = 0$,

$$y(0) = y'(0) = \cdots = y^{(n-2)}(0) = 0, \quad y^{(n-1)}(0) = 1. \tag{3.3.27}$$

Let $H = H(t)$ be the Heaviside function. Set

$$E(t) = H(t)y(t) = \begin{cases} y(t) & \text{for } t > 0; \\ 0 & \text{for } t < 0. \end{cases} \tag{3.3.28}$$

Then $E \in \mathcal{D}'(]-\infty,\infty[)$ is an elementary solution of L_n, i.e. $L_n E = \delta$. In fact, $\forall \phi \in \mathcal{D}(]-\infty,\infty[)$,

$$\begin{aligned}\langle E', \phi\rangle &= -\left\langle E, \frac{d\phi}{dt}\right\rangle = -\int_{-\infty}^{\infty} H(t)y(t)\frac{d\phi}{dt}dt = -\int_{0+}^{\infty} y(t)\phi'(t)dt \\ &= -y(t)\phi(t)|_{0+}^{\infty} + \int_{0+}^{\infty} y'(t)\phi(t)dt = \int_{-\infty}^{\infty} H(t)y'(t)\phi(t)dt,\end{aligned}$$

since $y(0^+) = y(0) = 0$, $\phi(t) = 0$ for $t = \infty \Longrightarrow E'(t) = H(t)y'(t)$ $\forall t \neq 0$. Similarly,

$$E^{(k)}(t) = H(t)y^{(k)}(t) \quad \forall k = 2, 3, \ldots, n-1. \tag{3.3.29}$$

Then

$$\begin{aligned}\langle E^{(n)}, \phi\rangle &= -\left\langle E^{(n-1)}, \frac{d\phi}{dt}\right\rangle = -\int_{0+}^{\infty} y^{(n-1)}(t)\phi'(t)dt \\ &= -y^{(n-1)}(t)\phi(t)|_{0+}^{\infty} + \int_{-\infty}^{\infty} H(t)y^{(n)}(t)\phi(t)dt \\ &= y^{(n-1)}(0)\phi(0) + \langle H(t)y^{(n)}, \phi\rangle = \langle\delta, \phi\rangle + \langle H(t)y^{(n)}, \phi\rangle, \\ &= \langle \delta + H(t)y^{(n)}, \phi\rangle \quad \forall \phi \in \mathcal{D}(\mathbb{R})\end{aligned}$$

since $y^{(n-1)}(0^+) = y^{(n-1)}(0) = 1$, $\langle \delta, \phi \rangle = \phi(0)$,

$$\Longrightarrow \quad E^{(n)} = \delta + H(t)y^{(n)} \quad \text{in } \mathcal{D}'(\mathbb{R}). \tag{3.3.30}$$

Then, from (3.3.29) and (3.3.30),

$$\begin{aligned} L_n E &= E^{(n)} + a_1 E^{(n-1)} + \cdots + a_n E \\ &= \delta + H(t)y^{(n)} + a_1 H(t) y^{(n-1)} + \cdots + a_n H(t) y \\ &= \delta + H(t)[L_n y] = \delta \quad \text{in } \mathcal{D}'(\mathbb{R}), \end{aligned}$$

since, by definition of $y(t)$, $L_n y = 0$.

$\Longrightarrow L_n E = \delta$ in $\mathcal{D}'(]-\infty, \infty[)$, i.e. $E = H(t)y(t)$ is an elementary solution of (3.3.24).

In particular,

- for $n = 1$, $a_1 = a$, $y = e^{-at}$ is the solution of $\frac{dy}{dt} + ay = 0$ with $y(0) = 1$, and $E_1(t) = H(t)e^{-at}$ is an elementary solution of $\frac{d}{dt} + a$; (3.3.31)
- for $n = 2$, $y = \frac{\sin at}{a}$ is the solution of $\frac{d^2 y}{dt^2} + a^2 y = 0$ with $y(0) = 0$, $y'(0) = 1$, and $E_2(t) = H(t)\frac{\sin at}{a}$ is an elementary solution of $\frac{d^2}{dt^2} + a^2$. (3.3.32)

Elementary solution of $\Delta + k^2$, $k > 0$

Example 3.3.1. Show that

$$E_\lambda(\mathbf{x}) = \lambda\left(\frac{e^{ikr}}{-4\pi r}\right) + (1-\lambda)\frac{e^{-ikr}}{-4\pi r}, \tag{3.3.33}$$

with $\lambda \in \mathbb{R}$, $r = (x_1^2 + x_2^2 + x_3^2)^{1/2}$, is an elementary solution of $\Delta + k^2$, $\Delta = \sum_{j=1}^{3} \frac{\partial^2}{\partial x_j^2}$, $k > 0$, $E_0(\mathbf{x}) = -\frac{e^{-ikr}}{4\pi r}$ being the only elementary solution satisfying the Sommerfeld condition [2, p. 315].

Proof. $\frac{e^{\pm ikr}}{r} = \frac{\cos kr}{r} \pm i \frac{\sin kr}{r}$ is a C^∞-function in $\mathbb{R}^3 \setminus \{\mathbf{0}\}$, since $\frac{\cos kr}{r} \in C^\infty(\mathbb{R}^3 \setminus \{\mathbf{0}\})$ and $\frac{\sin kr}{r} \in C^\infty(\mathbb{R}^3)$, $r = 0$ being a *removable* singular point. Hence $[(\Delta + k^2)(\frac{e^{\pm ikr}}{r})](\mathbf{x})$ in the usual pointwise sense can be found as follows: $\forall \mathbf{x} \in \mathbb{R}^3 \setminus \{\mathbf{0}\}$,

$$\left[\frac{\partial}{\partial x_j}\left(\frac{1}{r}\right)\right](\mathbf{x}) = -\frac{x_j}{r^3}; \quad \left[\frac{\partial}{\partial x_j} e^{ikr}\right](\mathbf{x}) = \frac{ikx_j}{r} \cdot e^{ikr};$$

$$\left[\frac{\partial^2}{\partial x_j^2} e^{ikr}\right](\mathbf{x}) = \left(\frac{ik}{r} - \frac{ikx_j^2}{r^3} - \frac{k^2 x_j^2}{r^2}\right) e^{ikr}; \quad [\Delta(e^{ikr})](\mathbf{x}) = \left(\frac{2ik}{r} - k^2\right) e^{ikr};$$

$$\left[\Delta\left(\frac{e^{ikr}}{r}\right)\right](\mathbf{x}) = e^{ikr}\left[\Delta\left(\frac{1}{r}\right)\right](\mathbf{x}) + 2\sum_{j=1}^{3}\left[\frac{\partial}{\partial x_j}\left(\frac{1}{r}\right)\right](\mathbf{x}) \cdot \left[\frac{\partial}{\partial x_j}(e^{ikr})\right](\mathbf{x})$$
$$+ \frac{1}{r}[\Delta e^{ikr}](\mathbf{x})$$
$$= e^{ikr}\left[0 + 2\sum_{j=1}^{3}\frac{-ikx_j^2}{r^4} + \frac{1}{r}\left(\frac{2ik}{r} - k^2\right)\right] = -k^2\frac{e^{ikr}}{r} \tag{3.3.34}$$

$$\Longrightarrow \quad \left[(\Delta + k^2)\left(\frac{e^{ikr}}{r}\right)\right](\mathbf{x}) = 0 \quad \forall \mathbf{x} \in \mathbb{R}^3 \setminus \{\mathbf{0}\} \tag{3.3.35}$$

$$\Longrightarrow \quad \left[(\Delta + k^2)\frac{\cos kr}{r}\right](\mathbf{x}) = 0, \quad \left[(\Delta + k^2)\frac{\sin kr}{r}\right](\mathbf{x}) = 0 \quad \forall \mathbf{x} \in \mathbb{R}^3 \setminus \{\mathbf{0}\},$$

but

$$\frac{\sin kr}{r} \in C^\infty(\mathbb{R}^3) \quad \Longrightarrow \quad \left[(\Delta + k^2)\frac{\sin kr}{r}\right](\mathbf{x}) = 0 \quad \forall \mathbf{x} \in \mathbb{R}^3.$$

$\Longrightarrow (\Delta + k^2)\frac{\sin kr}{r} = 0$ in $\mathcal{D}'(\mathbb{R}^3)$ in the distributional sense. (3.3.36)
$\forall R > 0$,

$$\int_{\|\mathbf{x}\| \le R}\left|\frac{\cos kr}{r}\right| d\mathbf{x} = \lim_{\varepsilon \to 0+}\int_{\varepsilon \le \|\mathbf{x}\| \le R}\left|\frac{\cos kr}{r}\right| d\mathbf{x} \le \lim_{\varepsilon \to 0+}\int_{\varepsilon \le \|\mathbf{x}\| \le R}\frac{1}{r} d\mathbf{x}$$
$$= \lim_{\varepsilon \to 0+}\int_{\varepsilon}^{R}\int_{0}^{\pi}\int_{0}^{2\pi}\frac{1}{r}r^2 \sin\theta\, dr\, d\theta\, d\phi = 4\pi \cdot \frac{R^2}{2} < +\infty$$

$\Longrightarrow \frac{\cos kr}{r} \in L^1_{\text{loc}}(\mathbb{R}^3)$. Now we will find $(\Delta + k^2)\frac{\cos kr}{r}$ in $\mathcal{D}'(\mathbb{R}^3)$ $\forall \phi \in \mathcal{D}(\mathbb{R}^3)$:

$$\left\langle(\Delta + k^2)\frac{\cos kr}{r}, \phi\right\rangle = \left\langle\Delta\left(\frac{\cos kr}{r}\right), \phi\right\rangle + \left\langle k^2\frac{\cos kr}{r}, \phi\right\rangle$$
$$= \left\langle\frac{\cos kr}{r}, \Delta\phi\right\rangle + \left\langle\frac{\cos kr}{r}, k^2\phi\right\rangle = \left\langle\frac{\cos kr}{r}, (\Delta + k^2)\phi\right\rangle$$
$$= \lim_{\varepsilon \to 0+}\underbrace{\int_{\|\mathbf{x}\| \ge \varepsilon}\frac{\cos kr}{r}(\Delta + k^2)\phi\, d\mathbf{x}}_{I_1(\varepsilon)} = \lim_{\varepsilon \to 0+} I_1(\varepsilon). \tag{3.3.37}$$

Since $\frac{\cos kr}{r} \in C^\infty(\mathbb{R}^3 \setminus \{\mathbf{0}\})$, using (3.2.5) we have

$$\begin{aligned}
&\int_{\|\mathbf{x}\|\geq\varepsilon} \left\{\frac{\cos kr}{r}(\Delta+k^2)\phi - \left[(\Delta+k^2)\left(\frac{\cos kr}{r}\right)\right](\mathbf{x})\phi(\mathbf{x})\right\}d\mathbf{x} \\
&= \int_{\|\mathbf{x}\|\geq\varepsilon} \left\{\frac{\cos kr}{r}\Delta\phi - \left[\Delta\left(\frac{\cos kr}{r}\right)\right](\mathbf{x})\phi(\mathbf{x})\right\}d\mathbf{x} \\
&= \int_{\Gamma_\varepsilon:\|\mathbf{x}\|=\varepsilon} \left[\phi(\mathbf{x})\frac{\partial}{\partial r}\left(\frac{\cos kr}{r}\right) - \frac{\cos kr}{r}\frac{\partial\phi}{\partial r}\right]dS, \quad \left(\frac{\partial(\cdot)}{\partial r} = -\frac{\partial}{\partial n_\varepsilon}(\cdot)\right),
\end{aligned}$$

and using (3.3.35): $\int_{\|\mathbf{x}\|\geq\varepsilon}[(\Delta+k^2)\frac{\cos kr}{r}](\mathbf{x})\phi(\mathbf{x})d\mathbf{x} = 0$. Hence, $\forall\varepsilon > 0$, from (3.3.37),

$$\begin{aligned}
I_1(\varepsilon) &= \int_{\|\mathbf{x}\|\geq\varepsilon} \frac{\cos kr}{r}(\Delta+k^2)\phi d\mathbf{x} \\
&= \int_{\Gamma_\varepsilon:\|\mathbf{x}\|=\varepsilon} \left\{\phi(\mathbf{x})\left(\frac{-kr\sin kr - \cos kr}{r^2}\right)_{r=\varepsilon} - \left(\frac{\cos kr}{r}\right)_{r=\varepsilon}\frac{\partial\phi}{\partial r}\right\}\varepsilon^2 d\omega \\
&= \underbrace{-k\varepsilon\sin k\varepsilon\int_{\Gamma_\varepsilon:\|\mathbf{x}\|=\varepsilon}\phi(\mathbf{x})d\omega}_{I_2(\varepsilon)} \underbrace{-\cos k\varepsilon\int_{\Gamma_\varepsilon:\|\mathbf{x}\|=\varepsilon}\phi(\mathbf{x})d\omega}_{I_3(\varepsilon)} \underbrace{-\varepsilon\cos k\varepsilon\int_{\Gamma_\varepsilon:\|\mathbf{x}\|=\varepsilon}\frac{\partial\phi}{\partial r}d\omega}_{I_4(\varepsilon)} \\
&= -I_2(\varepsilon) - I_3(\varepsilon) - I_4(\varepsilon),
\end{aligned}$$

where $dS = \varepsilon^2 d\omega$, $d\omega$ being the surface area measure on the *unit sphere*.

$$\left|\frac{\partial\phi}{\partial r}\right| = \left|\nabla\phi\cdot(-\hat{\mathbf{n}}_\varepsilon)\right| \leq \|\nabla\phi(\mathbf{x})\|\|\hat{\mathbf{n}}_\varepsilon\| \leq \sqrt{3}\left[\max_{1\leq i\leq 3}\max_{\mathbf{x}\in\mathrm{supp}(\phi)}\left|\frac{\partial\phi}{\partial x_i}(\mathbf{x})\right|\right] = C;$$

$$\int_{\|\mathbf{x}\|=1} d\omega = \int_0^\pi \sin\theta d\theta\int_0^{2\pi} d\phi = (-\cos\theta|_0^\pi)2\pi = 4\pi;$$

$$\begin{aligned}
&|I_4(\varepsilon)| \leq \varepsilon|\cos k\varepsilon|\left(C\int_{\|\mathbf{x}\|=1} d\omega\right) \leq (4\pi C\varepsilon) \to 0 \quad \text{as } \varepsilon\to 0^+ \\
&\qquad\qquad \Longrightarrow \quad \lim_{\varepsilon\to 0^+} I_4(\varepsilon) = 0; \\
&|I_2(\varepsilon)| \leq k\varepsilon|\sin k\varepsilon|\left(\max_{\mathbf{x}\in\mathrm{supp}(\phi)}|\phi(\mathbf{x})|\int_{\|\mathbf{x}\|=1} d\omega\right) \leq (4\pi k\max|\phi(\mathbf{x})|)\varepsilon \to 0 \\
&\qquad\qquad \Longrightarrow \quad \lim_{\varepsilon\to 0^+} I_2(\varepsilon) = 0;
\end{aligned}$$

$$I_3(\varepsilon) = \cos k\varepsilon\int_0^\pi\int_0^{2\pi}\phi(\varepsilon\sin\theta\cos\phi, \varepsilon\sin\theta\sin\phi, \varepsilon\cos\theta)\sin\theta d\theta d\phi$$

with $x_1 = \varepsilon\sin\theta\cos\phi$, $x_2 = \varepsilon\sin\theta\sin\phi$, $x_3 = \varepsilon\cos\theta$, $0 \leq \theta \leq \pi$, $0 \leq \phi \leq 2\pi$.

But $\lim_{\varepsilon\to 0+} \phi(\varepsilon\sin\theta\cos\phi, \varepsilon\sin\theta\sin\phi, \varepsilon\cos\theta) = \phi(0,0,0) = \phi(\mathbf{0})$, and $\lim_{\varepsilon\to 0+}\cos k\varepsilon = 1$,

$$\begin{aligned}\Longrightarrow \quad \lim_{\varepsilon\to 0+} I_3(\varepsilon) &= \lim_{\varepsilon\to 0+}\cos k\varepsilon \cdot \lim_{\varepsilon\to 0+}\int_0^{\pi}\int_0^{2\pi}\phi(\cdot,\cdot,\cdot)\sin\theta d\theta d\phi \\ &= 1\cdot \int_0^{\pi}\int_0^{2\pi}\lim_{\varepsilon\to 0+}\phi(\cdot,\cdot,\cdot)\sin\theta d\theta d\phi \\ &= \phi(\mathbf{0})\int_0^{\pi}\int_0^{2\pi}\sin\theta d\theta d\phi = 4\pi\phi(\mathbf{0}).\end{aligned}$$

Hence,

$$\begin{aligned}\lim_{\varepsilon\to 0+} I_1(\varepsilon) &= -\lim_{\varepsilon\to 0+} I_2(\varepsilon) - \lim_{\varepsilon\to 0+} I_3(\varepsilon) - \lim_{\varepsilon\to 0+} I_4(\varepsilon) \\ &= -4\pi\phi(\mathbf{0}) \quad \forall\phi\in\mathcal{D}(\mathbb{R}^3).\end{aligned}$$

From (3.3.37), $\langle(\Delta+k^2)\frac{\cos kr}{r},\phi\rangle = -4\pi\phi(\mathbf{0}) = \langle -4\pi\delta,\phi\rangle \Longrightarrow (\Delta+k^2)\frac{\cos kr}{r} = -4\pi\delta$ in $\mathcal{D}'(\mathbb{R}^3) \Longrightarrow (\Delta+k^2)\frac{e^{ikr}}{r} = -4\pi\delta$ in $\mathcal{D}'(\mathbb{R}^3)$, since $(\Delta+k^2)\frac{\sin kr}{r} = 0$ in $\mathcal{D}'(\mathbb{R}^3)$ (by (3.3.36)). Hence, $E = -\frac{e^{ikr}}{4\pi r}$ is an elementary solution of $\Delta+k^2$ in $\mathcal{D}'(\mathbb{R}^3)$. Replacing '$i$' by '$-i$' in (3.3.34) and (3.3.35) and following the subsequent steps with the necessary modifications, we can show that $\bar{E} = -\frac{e^{-ikr}}{4\pi r}$ is also an elementary solution of $\Delta+k^2$. Hence, $\forall\lambda\in\mathbb{R}$, $E_\lambda = \lambda(\frac{e^{ikr}}{-4\pi r}) + (1-\lambda)\frac{e^{-ikr}}{-4\pi r}$ is also an elementary solution of $\Delta+k^2$.

Obviously, $\frac{-\cos kr}{4\pi r}$ is an elementary solution of $\Delta+k^2$ in real form. (3.3.38) □

Construction of elementary solution with the help of pseudofunctions Pf(r^λ)
Since a detailed analysis of Pf(r^λ) for $\lambda\in\mathbb{C}$ and $r = (x_1^2+\cdots+x_n^2)^{\frac{1}{2}}$ in the distributional sense with proofs and justifications is prohibitively long and involved, and, moreover, such an analysis will *not* be required later for other topics, we have decided to state only the *final* results, without any proof, for which we refer to [1] and [8]. But we have shown how these interesting results can be used to construct elementary solutions of operators encountered in real applications. For this, we begin with the case $n = 1$, i.e. with the distribution defined by the pseudo-function (finite part) $\mathrm{Pf}(x_+^\lambda) = \mathrm{Pf}(x^\lambda)_{x>0}$ for $\lambda\in\mathbb{C}$ [8]. $\forall\phi\in\mathcal{D}(\mathbb{R})$,

$$\begin{aligned}\langle \mathrm{Pf}(x^\lambda)_{x>0},\phi\rangle &= \mathrm{Pf}\int_0^\infty x^\lambda\phi(x)dx = \lim_{\varepsilon\to 0+}\left[\int_\varepsilon^\infty x^\lambda\phi(x)dx - I(\varepsilon)\right] \\ &= \lim_{\varepsilon\to 0-}\left[\int_\varepsilon^\infty x^\lambda\phi(x)dx + \sum_{j=0}^{k=k(\lambda)}\frac{\phi^{(j)}(0)}{j!}\frac{\varepsilon^{\lambda+j+1}}{\lambda+j+1}\right], \quad (3.3.39)\end{aligned}$$

where $I(\varepsilon)$ is the *infinite part* of the divergent integral (see also (1.4.12) and (1.4.18));

$$\operatorname{Pf}(x^\lambda)_{x>0} = (x^\lambda)_{x>0} \quad \text{for } \operatorname{Re}(\lambda) > -1; \tag{3.3.40}$$

$k = k(\lambda)$ in (3.3.39) depends on the value of λ, for $\lambda + j + 1 = 0$ (i.e. for $\lambda = -m$ with $m \in \mathbb{N}$ and $-m + j + 1 = 0$), the term $\frac{\varepsilon^0}{0}$ must be replaced by $\ln \varepsilon$ (see (1.4.22)). (3.3.41)

Distribution defined by a pseudofunction (or finite part) $\operatorname{Pf}(r^\lambda)$ Now we will extend the definition in (3.3.39)–(3.3.41) to $\operatorname{Pf}(r^\lambda)$ with $r = r(\mathbf{x}) = (x_1^2 + x_2^2 + \cdots + x_n^2)^{1/2}$ $\forall \mathbf{x} = (x_1, x_2, \ldots, x_n) \in \mathbb{R}^n$, $\lambda \in \mathbb{C}$, as follows. Let $r = \varepsilon$ denote an n-dimensional sphere with radius $\varepsilon > 0$. Then, for $\lambda \in \mathbb{C}$, the distribution $\operatorname{Pf}(r^\lambda)$ is defined similarly by [8] (see also [1]): $\forall \phi \in \mathcal{D}(\mathbb{R}^n)$,

$$\begin{aligned}\langle \operatorname{Pf}(r^\lambda), \phi\rangle &= \operatorname{Pf}\int_{\mathbb{R}^n} r^\lambda \phi(\mathbf{x})d\mathbf{x} = \lim_{\varepsilon\to 0+}\left[\int_{r\ge\varepsilon} r^\lambda\phi(\mathbf{x})d\mathbf{x} - I(\varepsilon)\right] \\ &= \lim_{\varepsilon\to 0+}\left[\int_{r\ge\varepsilon} r^\lambda\phi(\mathbf{x})d\mathbf{x} + \sum_k H_k\Delta^k\phi(\mathbf{0})\frac{\varepsilon^{\lambda+n+2k}}{\lambda+n+2k}\right],\end{aligned} \tag{3.3.42}$$

where

$$H_k = \frac{(\sqrt{\pi})^n}{2^{2k-1}k!\Gamma(\frac{n}{2}+k)} \quad \text{with } H_0 = \frac{2(\sqrt{\pi})^n}{\Gamma(n/2)}; \tag{3.3.43}$$

$$\Delta = \frac{\partial^2}{\partial x_1^2} + \frac{\partial^2}{\partial x_2^2} + \cdots + \frac{\partial^2}{\partial x_n^2};$$

$$\Delta\phi(\mathbf{0}) = \langle\delta, \Delta\phi(\mathbf{x})\rangle = \langle\Delta\delta, \phi\rangle \quad \forall\phi \in \mathcal{D}(\mathbb{R}^n);$$

the integer k depends on λ; for $\lambda + n + 2k = 0$ (i.e. for negative integer $\lambda = -m$ with $m \in \mathbb{N}$ and $-m + n + 2k = 0$), the term $\frac{\varepsilon^0}{0}$ must be replaced by $\ln\varepsilon$. (3.3.44)

For the deduction of the 'exotic' expressions in square brackets [...] in (3.3.42) and of H_k in (3.3.43), the *analyticity* of $F(\lambda) = \langle \operatorname{Pf}(r^\lambda), \phi\rangle$, except for non-positive, even integral values of $\lambda + n$, and its Laurent series expansion have been used to find the residue at simple poles, along with other artifices (see [1], [11, p. 293]).

For $\operatorname{Re}(\lambda) > -n$, $\operatorname{Pf}(r^\lambda) = r^\lambda$, i.e., $\forall\phi \in \mathcal{D}(\mathbb{R}^n)$,

$$\langle \operatorname{Pf}(r^\lambda), \phi\rangle = \langle r^\lambda, \phi\rangle = \int_{\mathbb{R}^n} r^\lambda\phi(\mathbf{x})d\mathbf{x}. \tag{3.3.45}$$

Laplacian $\Delta[\operatorname{Pf}(r^m)]$ for integral values of λ [8] For $\lambda = m \in \mathbb{Z}$ with $m + n = -2p + 2$, $p \in \mathbb{N}_0$ (i.e. $p \ge 0$, $m + n - 2 = -2p$),

$$\Delta[\operatorname{Pf}(r^m)] = m(m+n-2)\operatorname{Pf}(r^{m-2}) + \frac{(2-n-4p)(\sqrt{\pi})^n}{2^{2p-1}p!\Gamma(\frac{n}{2}+p)}\Delta^p\delta, \tag{3.3.46}$$

where δ is the Dirac distribution with concentration at $\mathbf{0}$.

For $m + n \neq$ an even integer ≤ 2,

$$\Delta[\mathrm{Pf}(r^m)] = m(m+n-2)\,\mathrm{Pf}(r^{m-2}) \quad [8]. \tag{3.3.47}$$

The most important particular case of (3.3.45)/(3.3.46) is $m + n = 2$, i.e. $p = 0$ and $m = -(n-2) > -n$ with $n \neq 2$. Then, from (3.3.45) and (3.3.46), for $n \neq 2$, we have

$$\mathrm{Pf}\left(\frac{1}{r^{n-2}}\right) = \frac{1}{r^{n-2}} \quad \text{(i.e. the symbol Pf is useless), and}$$

$$\Delta\left(\frac{1}{r^{n-2}}\right) = \frac{-(n-2)2(\sqrt{\pi})^n}{\Gamma(\frac{n}{2})}\delta = -(n-2)H_0\delta = -(n-2)S_n\delta, \tag{3.3.48}$$

where $H_0 = S_n = 2(\sqrt{\pi})^n/\Gamma(n/2)$ = surface area of the n-dimensional unit sphere (see (3.3.43) and Theorem 3.3.1 for $n \neq 2$).

Case $n = 2$: $(n-2)H_0 = 0$ and we are to replace $\frac{1}{r^{n-2}}$ by $\ln\frac{1}{r}$, i.e., for $n = 2$, $\ln\frac{1}{r}$ plays the role of $\frac{1}{r^{n-2}}$.

$$\Delta\left(\ln\frac{1}{r}\right) = -2\pi\delta \quad \text{for } n = 2 \tag{3.3.49}$$

(see the independent proof of (3.3.13)–(3.3.14)).

Iterated Laplacian $\Delta^k[\mathrm{Pf}(r^m)]$, $\Delta^k[r^{2k-n}\ln r]$ Using (3.3.45)–(3.3.48), the iterated Laplacian of the distribution $\mathrm{Pf}(r^m)$ can be found. Here, we state the following important final results, which will be of use in the construction of an elementary solution of iterated Laplace operator Δ^k. We consider the following two cases:

1. For $k \in \mathbb{N}$, $2k - n > -n$, from (3.3.48), $\mathrm{Pf}(r^{2k-n}) = r^{2k-n}$. Then $\Delta^k(r^{2k-n})$ is given in [8, p. 47] (see also [1] for more details): for $2k - n < 0$ or $2k - n > 0$ and *odd* n,

$$\Delta^k(r^{2k-n}) = (2k-n)(2k-2-n)\cdots(4-n)(2-n)2^{k-1}(k-1)!H_0\delta, \tag{3.3.50}$$

   where H_0 is given by (3.3.43).

2. For $2k - n \geq 0$ and *even* n, we give the formula for $\Delta^k[r^{2k-n}\ln r]$ (valid for this case only):

$$\Delta^k[r^{2k-n}\ln r] = [(2k-n)(2k-2-n)\cdots(4-n)(2-n)]2^{k-1}(k-1)!H_0\delta, \tag{3.3.51}$$

   where the factor 0 of the expression in square brackets [...] in (3.3.51) must be *omitted*. (From (3.3.51), $\Delta\ln r = -\Delta\ln\frac{1}{r} = 2\pi\delta$ in (3.3.49) can be retrieved with the choice of $k = 1$ and $n = 2$). For example, for $k = 2$, $n = 2$, $2k - n = 2$, we are to omit the factor $2 - n = 2 - 2 = 0$ in square brackets $[\ldots]$ in (3.3.51) and we get, from (3.3.51):

$$\Delta^2(r^2 \ln r) = [(4-2)]2^{2-1}(2-1)!H_0\delta = 4H_0\delta \quad \text{with } H_0 = 2\pi. \tag{3.3.52}$$

Elementary solution E of iterated Laplace operator Δ^k From (3.3.50) and (3.3.51), we get an elementary solution E of Δ^k, i.e. $\Delta^k E = \delta$ in $\mathcal{D}'(\mathbb{R}^n)$, as follows:

1. For $2k - n < 0$ or for $2k - n \geq 0$ *and odd* n, $\exists$ a constant $B_{k,n}$ such that

$$\Delta^k(B_{k,n} r^{2k-n}) = \delta, \tag{3.3.53}$$

   i.e. $E(\mathbf{x}) = B_{k,n} r^{2k-n}$ is an elementary solution of Δ^k for $2k - n < 0$, or for $2k - n \geq 0$ *and odd* n, $B_{k,n}$ is determined from (3.3.50).

2. For $2k - n \geq 0$ *and even* n, $\exists$ a constant $A_{k,n}$ such that

$$\Delta^k[A_{k,n} r^{2k-n} \ln r] = \delta, \tag{3.3.54}$$

   i.e. $E(\mathbf{x}) = A_{k,n} r^{2k-n} \ln r$ is an elementary solution of Δ^k for $2k - n \geq 0$ *and even* n, $A_{k,n}$ is determined from (3.3.51).

Combining (3.3.53) and (3.3.54) for *arbitrary* $k \in \mathbb{N}$ and n, $\exists$ constants $A_{k,n}$ and $B_{k,n}$, *one* of these two constants being equal to 0 in *each* case, such that

$$\Delta^k[r^{2k-n}(A_{k,n} \ln r + B_{k,n})] = \delta, \tag{3.3.55}$$

from which (3.3.53) (resp. (3.3.54)) is obtained for $A_{k,n} = 0$ (resp. $B_{k,n} = 0$). For details, we refer to [1], [8]. Hence,

$$E(\mathbf{x}) = \begin{cases} A_{k,n} r^{2k-n} \ln r, & \text{if } 2k - n \geq 0 \textit{ and } n \text{ is } \textit{even}; \\ B_{k,n} r^{2k-n} & \text{if } 2k - n < 0, \text{ or if } 2k - n \geq 0 \textit{ and } n \text{ is } \textit{odd} \end{cases} \tag{3.3.56}$$

is an *elementary solution* of Δ^k, from which we can retrieve an elementary solution of the Laplace operator Δ with $k = 1$, i.e.

- for $k = 1, n \neq 2$, $E(\mathbf{x}) = B_{1,n} \cdot \frac{1}{r^{n-2}}$ with $B_{1,n} = -\frac{1}{(n-2)S_n}$ (since for $n \geq 3$, $2k - n = 2 - n < 0$, and for $n = 1$, $2k - n = 2 - 1 > 0$, n being an odd number),
- for $k = 1, n = 2$, $E(\mathbf{x}) = A_{1,2} \ln r$ with $A_{1,2} = \frac{1}{2\pi}$ (since $2k - n = 2 - 2 = 0$ and n is even),

are elementary solutions of the Laplace operator Δ given in Theorem 3.3.2.

Elementary solution of the operator $(1 - \frac{\Delta}{4\pi^2})^k$ Here we will state, as usual, the final results given in [8], which depend heavily on the use of results of classical Bessel function theory for the evaluation of the involved integrals obtained by the introduction of n-dimensional spherical coordinates (see, for example, [7, p. 334]). Following [8, p. 47], we define the distributions $L_m \in \mathcal{D}'(\mathbb{R}^n)$ by:

$$L_m = \frac{2(\sqrt{\pi})^m}{\Gamma(m/2)} \operatorname{Pf}[r^{\frac{m-n}{2}} K_{\frac{n-m}{2}}(2\pi r)] \quad \text{for } m \neq -2k \text{ with } k \in \mathbb{N}_0; \tag{3.3.57}$$

$$L_{-2k} = \left(1 - \frac{\Delta}{4\pi^2}\right)^k \delta \quad \text{for } m = -2k \text{ with } k \in \mathbb{N}_0; \tag{3.3.58}$$

$$L_0 = \delta, \tag{3.3.59}$$

where K_ν is the classical function from the theory of Bessel functions defined, $\forall \nu \in \mathbb{R}$, by:

$$K_\nu(x) = \pi \frac{[I_{-\nu}(x) - I_\nu(x)]}{2 \sin \pi \nu}, \quad \text{with } I_\nu(x) = \left(\frac{x}{2}\right)^\nu \sum_{k=0}^{\infty} \frac{(\frac{x}{2})^{2k}}{k!\Gamma(\nu + k + 1)}, \tag{3.3.60}$$

which converges exponentially to 0 at infinity. For L_m defined in (3.3.57)–(3.3.59), the following recurrent relation holds [8, p. 47]: for $m \neq -2k$ with $k \in \mathbb{N}_0$,

$$\left(1 - \frac{\Delta}{4\pi^2}\right)^2 L_m = \left(1 - \frac{\Delta}{4\pi^2}\right) L_{m-2} = L_{m-4}, \quad \ldots \tag{3.3.61}$$

Hence,

$$\left(1 - \frac{\Delta}{4\pi^2}\right)^k L_m = L_{m-2k} \quad (m \neq -2k \text{ with } k \in \mathbb{N}_0). \tag{3.3.62}$$

In particular, for $m = 2k$, $(1 - \frac{\Delta}{4\pi^2})^k L_{2k} = L_0 = \delta$ (by (3.3.59) and (3.3.57))

$$\Longrightarrow \quad \text{From (3.3.57)}, \quad L_{2k} = \frac{2\pi^k}{(k-1)!} \operatorname{Pf}[r^{k-\frac{n}{2}} K_{\frac{n}{2}-k}(2\pi r)] \tag{3.3.63}$$

is an elementary solution of $(1 - \frac{\Delta}{4\pi^2})^k$.

Elementary solution E of the biharmonic operator Δ^2

Example 3.3.2. Find an elementary solution E of the biharmonic operator Δ^2.

Solution. Case $n = 1$: $\Delta^2 = \frac{d^4}{dx^4}$, $r = |x|\ \forall x \in \mathbb{R}$, $k = 2 \Longrightarrow 2k - n = 4 - 1 > 0$, but n is *odd*. From (3.3.50) and (3.3.53),

$$E(x) = B_{2,1} r^{4-1} = \frac{1}{12}|x|^3, \tag{3.3.64}$$

since from (3.3.43), $H_0 = \frac{2\sqrt{\pi}}{\Gamma(1/2)} = 2$, and from (3.3.50), $B_{2,1} = \frac{1}{(4-1)(2-1)21!2} = \frac{1}{12}$.

In fact,

$$\langle \Delta^2 E, \phi\rangle = \langle E, \Delta^2\phi\rangle = \int_{-\infty}^{\infty} \frac{1}{12}|x|^3 \frac{d^4\phi}{dx^4} dx$$
$$= \frac{1}{12}\Big[\int_{-\infty}^{0} (-x)^3\phi^{(4)}(x)dx + \int_{0}^{\infty} x^3\phi^{(4)}(x)dx\Big] \quad \forall\phi \in \mathcal{D}(\mathbb{R}).$$

But

$$\int_{-\infty}^{0} -x^3\phi^{(4)}(x)dx = -x^3\phi^{(3)}(x)|_{\infty}^{0} + 3\int_{-\infty}^{0} x^2\phi^{(3)}(x)dx$$
$$= 3x^2\phi^2(x)|_{-\infty}^{0} - 6\int_{-\infty}^{0} x\phi^{(2)}(x)dx$$
$$= -6x\phi^{(1)}(x)|_{-\infty}^{0} + 6\int_{-\infty}^{0} \phi^{(1)}(x)dx = 6\phi(x)|_{-\infty}^{0} = 6\phi(0),$$

since $x^k\phi^{(j)}(x) = 0$ for $x = 0, -\infty$ ($\phi \in \mathcal{D}(\mathbb{R}) \Longrightarrow \phi^{(k)}(\pm\infty) = 0\ \forall k$).

Similarly, $\int_0^\infty x^3\phi^{(4)}(x)dx = 6\phi(0)$.

Hence, $\langle \Delta^2 E, \phi\rangle = \frac{1}{12}[6\phi(0) + 6\phi(0)] = \phi(0) = \langle\delta, \phi\rangle\ \forall\phi \in \mathcal{D}(\mathbb{R})$

$\Longrightarrow \Delta^2 E = \frac{d^4}{dx^4}E = \delta$ with $E = \frac{1}{12}|x|^3\ \forall x \in \mathbb{R}$.

Case $n = 2$:

$$\Delta^2 = \Delta\Delta = \frac{\partial^4}{\partial x_1^4} + 2\frac{\partial^4}{\partial x_1^2\partial x_2^2} + \frac{\partial^4}{\partial x_4^2}, \quad r = (x_1^2 + x_2^2)^{1/2}, \quad k = 2$$

$\Longrightarrow 2k - n = 4 - 2 > 0$ and n is *even*. Hence, from (3.3.51) and (3.3.54), an elementary solution E of $\Delta\Delta$ is given by [37]:

$$E(\mathbf{x}) = A_{2,2}r^{2k-n}\ln r = \frac{1}{8\pi}r^2\ln r, \tag{3.3.65}$$

since from (3.3.43), $H_0 = \frac{2(\sqrt{\pi})^2}{\Gamma(\frac{2}{2})} = 2\pi$, and from (3.3.52), $A_{2,2} = \frac{1}{8\pi}$.

In fact, $r^2 \in C^\infty(\mathbb{R}^2)$, $\ln r \in L^1_{\text{loc}}(\mathbb{R}^2)$ with its distributional derivative $\frac{\partial}{\partial x_k}\ln r =$ the usual partial derivative $[\frac{\partial}{\partial x_k}\ln r(\mathbf{x})] = \frac{x_k}{r^2} \in L^1_{\text{loc}}(\mathbb{R}^2)$. Hence, applying Leibniz's formula (2.5.2) with $f = r^2$,

$$\Delta(r^2\ln r) = \sum_{k=1}^{2}\frac{\partial^2}{\partial x_k^2}(r^2\ln r) = \ln r\Delta r^2 + 2\sum_{k=1}^{2}\frac{\partial r^2}{\partial x_k}\cdot\frac{\partial\ln r}{\partial x_k} + r^2\Delta\ln r$$
$$= 4\ln r + 2\sum_{k=1}^{2} 2x_k\cdot\frac{x_k}{r^2} + r^2 2\pi\delta = 4\ln r + 4 + 0 \quad \text{in } \mathcal{D}'(\mathbb{R}^2),$$

since $\Delta \ln r = 2\pi\delta$ (by (3.3.14)).

$$\langle r^2 2\pi\delta, \phi\rangle = 2\pi\langle\delta, r^2\phi\rangle = 2\pi(r^2\phi)(\mathbf{0}) = 2\pi\cdot 0 = 0 \quad \forall\phi \in \mathcal{D}(\mathbb{R}^2)$$

$\Longrightarrow r^2 2\pi\delta = 0$ in $\mathcal{D}'(\mathbb{R}^2)$.

Then,

$$\begin{aligned}\Delta\Delta E &= \frac{1}{8\pi}\Delta(\Delta(r^2\ln r)) = \frac{1}{8\pi}\Delta(4\ln r + 4)\\ &= \frac{1}{8\pi}\cdot 4\cdot 2\pi\delta + 0 = \delta \quad \text{in } \mathcal{D}'(\mathbb{R}^2).\end{aligned}$$

Case $n = 3$:

$$\Delta^2 = \frac{\partial^4}{\partial x_1^4} + \frac{\partial^4}{\partial x_2^4} + \frac{\partial^4}{\partial x_3^4} + 2\frac{\partial^4}{\partial x_1^2\partial x_2^2} + 2\frac{\partial^4}{\partial x_2^2\partial x_3^2} + 2\frac{\partial^4}{\partial x_3^2\partial x_1^2},$$

$r = (x_1^2 + x_2^2 + x_3^2)^{1/2}$, $k = 2 \Longrightarrow 2k - n = 4 - 3 > 0$, but n is *odd*. Hence, from (3.3.50) and (3.3.53), an elementary solution E of Δ^2 is given by:

$$E(\mathbf{x}) = B_{2,3}r = -\frac{1}{8\pi}r, \tag{3.3.66}$$

since from (3.3.43), $H_0 = \frac{2(\sqrt{\pi})^3}{\Gamma(3/2)} = 4\pi$, and from (3.3.50), $B_{2,3} = \frac{1}{(4-3)(2-3)2!4\pi}$ $= -\frac{1}{8\pi}$. In fact, $\Delta\Delta E = \Delta[\Delta(-\frac{1}{8\pi}r)] = -\frac{1}{8\pi}\Delta[\Delta r]$. For $r = (x_1^2 + x_2^2 + x_3^2)^{\frac{1}{2}}$, $\frac{1}{r} \in L^1_{\text{loc}}(\mathbb{R}^3)$ with its distributional derivative $\frac{\partial}{\partial x_k}(\frac{1}{r})$ = the usual partial derivative $[\frac{\partial}{\partial x_k}(\frac{1}{r})](\mathbf{x}) = -\frac{x_k}{r^3}$, $\Delta r^2 = 6$ in $\mathcal{D}'(\mathbb{R}^3)$. Since $r^2 \in C^\infty(\mathbb{R}^3)$, $r = r^2\cdot\frac{1}{r} \in L^1_{\text{loc}}(\mathbb{R})$, applying Leibniz's formula (2.5.2), we have

$$\begin{aligned}\Delta r &= \Delta\Big(r^2\cdot\frac{1}{r}\Big) = \frac{1}{r}\Delta r^2 + 2\sum_{k=1}^{3}\frac{\partial r^2}{\partial x_k}\cdot\frac{\partial}{\partial x_k}\Big(\frac{1}{r}\Big) + r^2\Delta\frac{1}{r}\\ &= \frac{6}{r} + 2\sum_{k=1}^{3} 2x_k\cdot\frac{-1}{r^2}\cdot\frac{x_k}{r} + r^2(-4\pi\delta) = \frac{6}{r} - \frac{4}{r} - 0 = \frac{2}{r} \quad \text{in } \mathcal{D}'(\mathbb{R}^2),\end{aligned}$$

since $\Delta(\frac{1}{r}) = -4\pi\delta$ in $\mathcal{D}'(\mathbb{R}^3)$. Thus, $\Delta\Delta r = \Delta(\frac{2}{r}) = 2(-4\pi\delta) = -8\pi\delta$ in $\mathcal{D}'(\mathbb{R}^3)$. Hence,

$$\Delta\Delta E = \Delta(\Delta\frac{-1}{8\pi}r) = -\frac{1}{8\pi}\Delta\Delta r = \delta \quad \text{in } \mathcal{D}'(\mathbb{R}^3).$$

An elementary solution E of $\Delta^3 = (\frac{\partial^2}{\partial x_1^2} + \frac{\partial^2}{\partial x_2^2})^3$ in $\mathcal{D}'(\mathbb{R}^2)$ is given by (3.3.51) and (3.3.54):

$$E = A_{3,2}r^4\ln r = \frac{1}{128\pi}r^4\ln r, \tag{3.3.67}$$

since $k = 3, n = 2 \Longrightarrow 2k - n = 4$ and n is even, $H_0 = 2\pi$, and

$$A_{3,2} = \frac{1}{[(6-2)(4-2)]2^{3-1}(3-1)!2\pi} = \frac{1}{128\pi}.$$

In fact, $r^4 \in C^\infty(\mathbb{R}^2)$, $\ln r \in L^1_{\text{loc}}(\mathbb{R}^2)$ with its distributional derivative $\frac{\partial}{\partial x_k} \ln r =$ the usual partial derivative $[\frac{\partial}{\partial x_k} \ln r(\mathbf{x})] = \frac{x_k}{r^2} \in L^1_{\text{loc}}(\mathbb{R}^2)$. Then, applying Leibniz's formula (2.5.2), we have

$$\Delta^3(r^4 \ln r) = \Delta^2(\Delta(r^4 \ln r)) = \Delta^2\bigg(\ln r \Delta r^4 + 2\sum_{k=1}^{2} \frac{\partial r^4}{\partial x_k} \cdot \frac{\partial \ln r}{\partial x_k} + r^4 \Delta \ln r\bigg)$$

$$= \Delta^2(16r^2 \ln r + 2 \cdot 4 \cdot r^2 + r^4 2\pi\delta) = 16\Delta^2(r^2 \ln r) + 2 \cdot 4\Delta^2(r^2) + \Delta^2(0)$$

$$= 16 \cdot 8\pi\delta + 0 + 0 = 128\pi\delta \quad \text{in } \mathcal{D}'(\mathbb{R}^2),$$

since $\Delta^2(r^2 \ln r) = 8\pi\delta$ in $\mathcal{D}'(\mathbb{R}^2)$ (by (3.3.65)); $\Delta^2(r^2) = 0$ in $\mathcal{D}'(\mathbb{R}^2)$;

$$\langle r^4 2\pi\delta, \phi\rangle = 2\pi\langle \delta, r^4\phi\rangle = 2\pi(r^4\phi)(\mathbf{0}) = 0 \quad \forall \phi \in \mathcal{D}(\mathbb{R}^2)$$

$\Longrightarrow r^4 2\pi\delta = 0$ in $\mathcal{D}'(\mathbb{R}^2)$.

Hence,

$$\Delta^3 E = \Delta^3\bigg(\frac{1}{128\pi} r^4 \ln r\bigg) = \frac{1}{128\pi}\Delta^3(r^4 \ln r) = \frac{1}{128\pi} \cdot 128\pi\delta = \delta$$

in $\mathcal{D}'(\mathbb{R}^2)$.

An elementary solution E of $\Delta^4 = (\frac{\partial^2}{\partial x_1^2} + \frac{\partial^2}{\partial x_2^2})^4$ in $\mathcal{D}'(\mathbb{R}^2)$, which arises in elastic cylindrical shell analysis, is given by (3.3.51) and (3.3.54):

$$E = A_{4,2} r^6 \ln r = \frac{1}{4608\pi} r^6 \ln r, \tag{3.3.68}$$

since $k = 4, n = 2 \Longrightarrow 2k - n = 6 > 0$ and n is *even*, $H_0 = 2\pi$, and

$$A_{4,2} = \frac{1}{(8-2)(6-2)(4-2)2^{4-1}(4-1)!2\pi} = \frac{1}{4608\pi}.$$

Indeed,

$$\Delta^4[r^6 \ln r] = \Delta^3(\Delta(r^6 \ln r)) = \Delta^3\bigg(\ln r \Delta(r^6) + 2\sum_{k=1}^{2} \frac{\partial r^6}{\partial x_k} \cdot \frac{\partial \ln r}{\partial x_k} + r^6 \Delta \ln r\bigg)$$

$$= \Delta^3(36r^4 \ln r + 2 \cdot 6 \cdot r^4 + r^6 2\pi\delta) = 36\Delta^3(r^4 \ln r) + 12\Delta^3(r^4) + \Delta(0)$$

$$= 36 \cdot 128\pi\delta + 0 + 0 = 4608\pi\delta,$$

since $\Delta^3(r^4 \ln r) = 128\pi\delta$ in $\mathcal{D}'(\mathbb{R}^2)$ (by (3.3.67)); $\Delta^3(r^4) = 0$; $\langle r^6 2\pi\delta, \phi\rangle = 2\pi\langle \delta, r^6\phi\rangle = 0 \ \forall \phi \in \mathcal{D}(\mathbb{R}^2) \Longrightarrow r^6 2\pi\delta = 0$ in $\mathcal{D}'(\mathbb{R}^2)$. Hence, $\Delta^4 E = \Delta(\frac{1}{4608\pi} r^6 \ln r) = \delta$ in $\mathcal{D}'(\mathbb{R}^2)$.

Remark 3.3.1. We will come back to this topic in a more useful setting in Chapter 8 after the introduction of Fourier transform of tempered distributions.

3.4 Applications

Construction of finite-dimensional subspaces of Sobolev space $H^m(\Omega)$

In Section 2.15, Chapter 2, we introduced Sobolev spaces $H^m(\Omega)$, $W^{m,p}(\Omega)$, etc., and stated their elementary defining properties for arbitrary domain $\Omega \subset \mathbb{R}^n$. Now we will construct *finite-dimensional subspaces* of $H^m(\Omega)$, when $\Omega \subset \mathbb{R}^2$ is a *polygonal domain* in $\mathbb{R}^2$ for $m = 1, 2$. These subspaces of $H^m(\Omega)$ are used as the finite element spaces in the finite element method of approximation of the solution of elliptic boundary value problems in Ω.

Case of two variables ($n = 2$) Let $\Omega \subset \mathbb{R}^2$ be a bounded polygonal domain with $\overline{\Omega} = \Omega \cup \Gamma$, Γ being the polygonal boundary of Ω.

Triangulation τ_h of the closed polygonal domain $\overline{\Omega}$ [34], [35], [36] Let τ_h denote a triangulation (i.e. subdivision) of $\overline{\Omega}$ into *closed* triangles $T_1, T_2, \dots, T_N$ (see Figure 3.6) such that:

- $\tau_h = \{T_1, T_2, \dots, T_N\}$, $h = \max_{1 \le i \le N}\{\text{dia}(T_i)\}$, $\text{dia}(T_i) = \sup_{\mathbf{x},\mathbf{y} \in T_i} d(\mathbf{x}, \mathbf{y})$;
- $\overline{\Omega} = \bigcup_{i=1}^N T_i$, $\mathring{T}_i = \text{int}(T_i)$, ∂T_i = boundary of T_i with $T_i = \mathring{T}_i \cup \partial T_i$ $\forall i$; (3.4.1)
- $\mathring{T}_i \cap \mathring{T}_j = \emptyset$ for $1 \le i \ne j \le N$;
- For $i \ne j$, $T_i \cap T_j = \emptyset$ (*empty set*) or a *common vertex* or a *common side*. (3.4.2)

Remark 3.4.1.

- *Triangulation* τ_h and closed *triangles* T_i, $1 \le i \le N$, should be understood in a *generalized sense*, i.e. for $n = 2$, the T_i's may be *quadrilaterals*, and for some closed domains $\overline{\Omega}$, the T_i's may be *rectangles* or a combination of *rectangles* and *triangles* or *quadrilaterals* etc. satisfying (3.4.1) and (3.4.2). In the following discussions, $T \in \tau_h$ will imply closed triangles in the usual sense.
- For arbitrary $\Omega \subset \mathbb{R}^2$ with *sufficiently smooth* curved boundary Γ, we will consider a *polygonal approximation* Γ_h to Γ and denote the polygonal domain inside Γ_h by Ω_h with $\overline{\Omega}_h = \Omega_h \cup \Gamma_h$, such that $\overline{\Omega}_h$ is an approximation to $\overline{\Omega}$. Then, closed polygonal domain $\overline{\Omega}_h$ will be triangulated instead of $\overline{\Omega}$ as shown above.

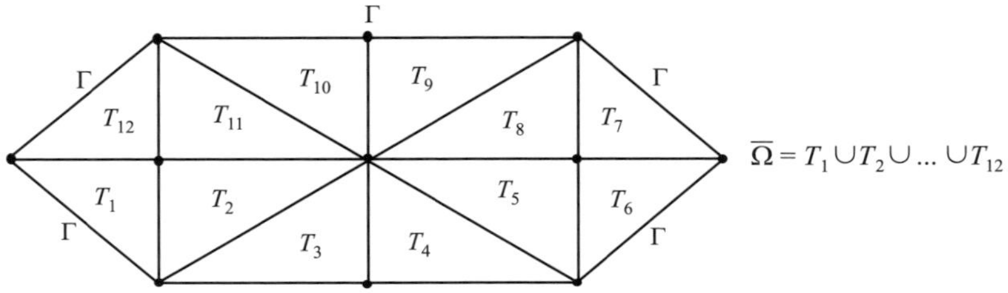

(a) Admissible triangulation τ_h of $\overline{\Omega} = \Omega \cup \Gamma$ with $\tau_h = \{T_i\}$, $i = 1, \ldots, 12$.

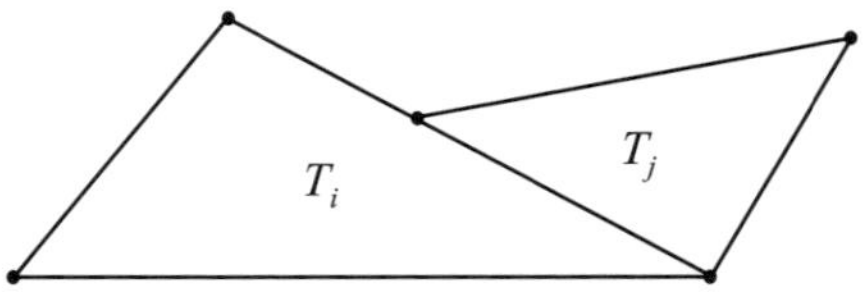

(b) Inadmissible triangles T_i and T_j, since $T_i \cap T_j$ is not a side of T_i.

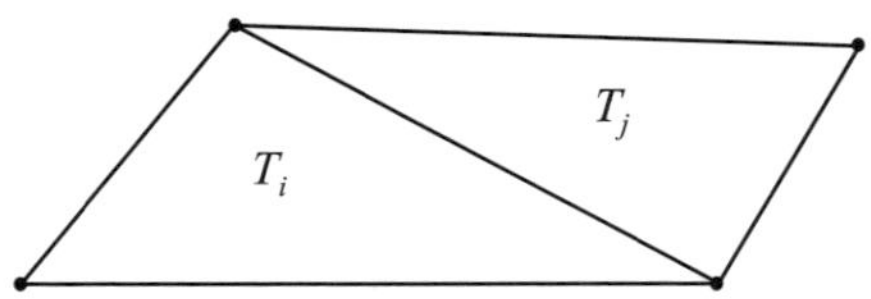

(c) Admissible triangles T_i and T_j, since $T_i \cap T_j$ is a common side of T_i and T_j.

Figure 3.6 Triangulation τ_h of $\overline{\Omega}$

- This definition of triangulation τ_h can be extended to the triangulation of a closed polyhedral domain $\overline{\Omega} \subset \mathbb{R}^3$ into closed tetrahedrons, for example, T_i, $1 \le i \le N$, simply by replacing (3.4.2) by: For $n = 3$, $1 \le i \ne j \le N$, $T_i \cap T_j = \emptyset$ (*empty set*) or a *common vertex* or a *common edge* or a *common face*.
- For $n = 2$, we have $\bigcup_{i=1}^{N} \partial T_i$ = set of sides of triangles of $\tau_h = \bigcup_{1 \le i \ne j \le N} (\partial T_i \cap \partial T_j) \cup \Gamma$.
- $\partial T_i \cap \partial T_j = L_k$ $(1 \le k \le 3)$ = inter-triangular side common to T_i and T_j $(i \ne j)$.
- $P_m(A)$ = linear space of polynomials of degree $\le m$ in two variables x_1, x_2 defined in $A \subset \overline{\Omega} \subset \mathbb{R}^2 = \{p : p$ is a polynomial of degree $\le m$ in x_1, x_2 in $A \subset \overline{\Omega}\}$.
- $P_m(A) = \text{Span}\{1; x_1, x_2; x_1^2, x_1 x_2, x_2^2; \ldots; x_1^m, x_1^{m-1} x_2, \ldots, x_2^m\}$ with $\dim(P_m(A)) = 1 + 2 + 3 + \cdots + (m+1) = (m+1)(m+2)/2$. For example, $P_1(A) = \text{Span}\{1; x_1, x_2\}$, $\dim(P_1(A)) = 3$, $\dim P_2(A) = 6$, etc.

Finite-dimensional vector space X_h To each triangulation τ_h of $\overline{\Omega}$ defined by (3.4.1) and (3.4.2), we associate a set X_h of real-valued functions v_h, whose restriction to each triangle $T \in \tau_h$ is a polynomial of degree $\le m$ in two variables x_1, x_2,

i.e.

$$X_h = \{v_h : v_h \text{ is a real-valued function with } v_h{\downarrow}_T \in P_m(T)\ \forall T \in \tau_h, m \in \mathbb{N}_0\}. \tag{3.4.3}$$

Properties of functions of X_h

- $v_h \in X_h$ is a piecewise polynomial of degree $\leq m$ in each $T \in \tau_h$; $\forall$ fixed $T \in \tau_h$ with $\overset{\circ}{T} = \text{int}(T)$, $v_h{\downarrow}_{\overset{\circ}{T}} \in P_m(\overset{\circ}{T})$ has a unique, natural polynomial extension v_T to T of $v_h{\downarrow}_{\overset{\circ}{T}}$, and we set $v_T = v_h{\downarrow}_T \in P_m(T)$. (3.4.4)
- $v_h \in X_h$ is discontinuous, in general, across inter-triangular sides of the triangulation τ_h.
- $v_{T_i} \neq v_{T_j}$ for $i \neq j$, i.e. in different triangles, polynomials v_T are different. (3.4.5)
- If $v_{T_i} = v_{T_j}$ on their common side $T_i \cap T_j$ $(i \neq j)$, then v_h is *continuous* across the common side $T_i \cap T_j$. (3.4.6)

Proposition 3.4.1. *X_h defined by* (3.4.4) *is a finite-dimensional subspace of $L^2(\Omega)$, i.e.*

$$X_h \subset L^2(\Omega). \tag{3.4.7}$$

Proof. Since $P_m(T)$ is a vector space $\forall T \in \tau_h, \forall m \in \mathbb{N}_0, X_h$ is obviously a vector space. By virtue of property (3.4.6), $\dim(X_h) = \sum_{T\in\tau_h} \dim(P_m(T)) = \text{Card}\{\tau_h\} \cdot \dim(P_m(T)) = N \cdot (m+1)(m+2)/2 < +\infty$, since $\text{Card}\{\tau_h\} = N$ and $\dim(P_m(T)) = (m+1)(m+2)/2 \Longrightarrow X_h$ is a finite-dimensional vector space.

Let $v_h \in X_h$. Then

$$\begin{aligned}\int_\Omega |v_h(\mathbf{x})|^2 dx_1 dx_2 &= \sum_{T\in\tau_h} \int_T |v_T(\mathbf{x})|^2 dx_1 dx_2 \\ &\leq \max_{T\in\tau_h}(\max_{\mathbf{x}\in T} |v_T(\mathbf{x})|^2) \cdot (\text{area measure of } \Omega) < +\infty\end{aligned}$$

$\Longrightarrow v_h \in L^2(\Omega)$. Hence, $v_h \in X_h \Longrightarrow v_h \in L^2(\Omega) \Longrightarrow X_h \subset L^2(\Omega)$.

But $X_h \not\subset H^1(\Omega)$ in general. (3.4.8)

To justify this, consider Example 3.1.2, in which $\Omega =]0,1[\times]0,1[$, $\tau_h = \{T_1, T_2\}$ with $N = 2, \overline{\Omega} = T_1 \cup T_2$, T_1 with vertices $\mathbf{a}_2, \mathbf{a}_3, \mathbf{a}_4$; T_2 with vertices $\mathbf{a}_1, \mathbf{a}_2, \mathbf{a}_4$; $\Gamma_0 = T \cap T_2 = [\mathbf{a}_4, \mathbf{a}_2] =$ join of $\mathbf{a}_4$ and $\mathbf{a}_2$. Set

$$v_h = f = \begin{cases} p(x_1,x_2) = 1 + 4x_1 - 2x_2 & \text{in } \overset{\circ}{T}_1 \\ q(x_1,x_2) = -2 + 4x_1 + 4x_2 & \text{in } \overset{\circ}{T}_2 \end{cases}$$

such that p and q are given natural polynomial extensions to T_1 and T_2 respectively:

$$v_{T_1} = v_h{\downarrow}_{T_1} = 1 + 4x_1 - 2x_2 \in P_1(T_1) \quad \forall (x_1, x_2) \in T_1,$$
$$v_{T_2} = v_h{\downarrow}_{T_2} = -2 + 4x_1 + 4x_2 \in P_1(T_2) \quad \forall (x_1, x_2) \in T_2,$$

v_h being discontinuous across inter-triangular side $\Gamma_0 = [\mathbf{a}_4, \mathbf{a}_2]$ with jump $J = (q - p)(x_1, x_2) = -3 + 6x_2$ across Γ_0.

Then $v_h \in X_h$ with $v_h{\downarrow}_T \in P_1(T)\ \forall T \in \tau_h$. From (3.1.56), $v_h \in L^2(\Omega)$, but from (3.1.58), the *distributional derivative* $\frac{\partial v_h}{\partial x_i} = \frac{\partial f}{\partial x_i} \notin L^2(\Omega)$. Consequently, $v_h \notin H^1(\Omega)$, although $v_h \in X_h \subset L^2(\Omega)$, i.e. $X_h \not\subset H^1(\Omega)$. □

But we have the following important results.

Theorem 3.4.1. *Let $V_h \subset X_h$ be a vector space defined by:*

$$V_h = \{v_h : v_h \in X_h, v_h \in C^0(\overline{\Omega})\}. \tag{3.4.9}$$

Then $V_h \subset H^1(\Omega)$ is a finite-dimensional subspace of $H^1(\Omega)$.

Proof. Let $v_h \in V_h$. Then $v_h \in C^0(\overline{\Omega}) \subset L^2(\Omega) \Longrightarrow v_h \in L^2(\Omega)$. We are to prove that the distributional derivative $\frac{\partial v_h}{\partial x_i} \in L^2(\Omega)$ $(i = 1, 2)$ in order that $v_h \in H^1(\Omega)$. Set $v_T = v_h{\downarrow}_T \in P_m(T)\ \forall T \in \tau_h$. Then, in each $\mathring{T} = \operatorname{int}(T)$ of $T \in \tau_h$, the distributional derivative $\frac{\partial v_T}{\partial x_i}$ and the usual partial derivative $[\frac{\partial v_T}{\partial x_i}(\mathbf{x})]$ in the pointwise sense coincide in $\mathring{T}$, both being polynomials of degree $\leq m-1$. Define a new function w_i such that $w_i{\downarrow}_{\mathring{T}} = [\frac{\partial v_T}{\partial x_i}(\mathbf{x})] \in P_{m-1}(\mathring{T})$ $(i = 1, 2)$ in $\mathring{T}\ \forall T \in \tau_h$. Then w_i is defined a.e. on Ω and $\int_\Omega |w_i(\mathbf{x})|^2 dx_1 dx_2 = \sum_{T \in \tau_h} \int_{\mathring{T}} (w_i{\downarrow}_{\mathring{T}}(\mathbf{x}))^2 dx_1 dx_2 = \sum_{T \in \tau_h} \int_{\mathring{T}} [\frac{\partial v_T}{\partial x_i}(\mathbf{x})]^2 dx_1 dx_2 < +\infty \Longrightarrow w_i \in L^2(\Omega)$. Let $w_i{\downarrow}_T \in P_{m-1}(T)$ be the natural polynomial extension to $T = \mathring{T} \cup \partial T$ of $w_i{\downarrow}_{\mathring{T}}$, ∂T being the boundary of T (resp. $\mathring{T}$) $\forall T \in \tau_h$. Now we will show that the distributional derivative $\frac{\partial v_h}{\partial x_i} = w_i$ in $\mathcal{D}'(\Omega)$, i.e. $\frac{\partial v_h}{\partial x_i} = w_i \in L^2(\Omega)$, $i = 1, 2$. In fact, $\forall \phi \in \mathcal{D}(\Omega)$,

$$\begin{aligned}
\langle w_i, \phi \rangle &= \int_\Omega w_i(\mathbf{x})\phi(\mathbf{x}) dx_1 dx_2 = \sum_{T \in \tau_h} \int_{\mathring{T}} w_{iT}(\mathbf{x})\phi(\mathbf{x}) dx_1 dx_2 \\
&= \sum_{T \in \tau_h} \int_{\mathring{T}} \left[\frac{\partial v_T}{\partial x_i}(\mathbf{x})\right] \phi(\mathbf{x}) dx_1 dx_2 \\
&= \sum_{T \in \tau_h} \left[\int_{\mathring{T}} \left[\frac{\partial}{\partial x_i}(v_T \phi)(\mathbf{x})\right] dx_1 dx_2 - \int_{\mathring{T}} v_T \frac{\partial \phi}{\partial x_i} dx_1 dx_2 \right].
\end{aligned} \tag{3.4.10}$$

Applying Green's Theorem 3.1.2 to the first double integral over each $\mathring{T}$ with boundary $\partial T = L_1 \cup L_2 \cup L_3$, $\{L_k\}_{k=1}^3$ being the three sides of $T = \mathring{T} \cup \partial T$, we have

$$\int_{\mathring{T}} \left[\frac{\partial}{\partial x_i} (v_T \phi) \right] dx_1 dx_2 = \int_{\partial T = \bigcup_{k=1}^3 L_k} v_T \phi \cos(\hat{\mathbf{n}}_T, x_i) ds,$$

where $\hat{\mathbf{n}}_T$ is the unit vector normal to ∂T, $\hat{\mathbf{n}}_T \cdot \hat{\mathbf{i}}_k = \cos(\hat{\mathbf{n}}_T, x_k)$, and ds is the element arc length measured along ∂T. Then

$$\sum_{T \in \tau_h} \int_{\mathring{T}} \left[\frac{\partial}{\partial x_i} (v_T \phi) \right] dx_1 dx_2 = \sum_{T \in \tau_h} \int_{\partial T} v_T \phi \cos(\hat{\mathbf{n}}_T, x_i) ds$$

$$= \sum_{L_j \subset \partial T, L_j \subset \Omega} \int_{L_j} v_T \phi \cos(\hat{\mathbf{n}}_T, x_i) ds + \int_{\Gamma} v_h \phi \cos(\hat{\mathbf{n}}_T, x_i) ds, \qquad (3.4.11)$$

where, $\forall$ inter-triangular sides L_j, $\int_{L_j} (\ldots) ds + \int_{L_j^-} (\ldots) ds = 0$, since v_h and ϕ are continuous across each L_j and the line integrals along L_j and L_j^- have opposite senses of orientation, and consequently

$$\sum_{L_j \subset \partial T, L_j \subset \Omega} \int_{L_j} v_T \phi \cos(\hat{\mathbf{n}}_T, x_i) ds = 0; \qquad (3.4.12)$$

Γ is obtained as the union of the boundary sides $L_j \subset \Gamma$ of all boundary triangles, at least one side of which is a part of Γ, and

$$\int_{\Gamma} v_h \phi \cos(\hat{\mathbf{n}}_T, x_i) ds = 0, \qquad (3.4.13)$$

since $\phi \in \mathcal{D}(\Omega) \Longrightarrow \phi\downarrow_\Gamma = 0$. Then, from (3.4.10)–(3.4.13), we have

$$\langle w_i, \phi \rangle = \sum_{T \in \tau_h} - \int_{\mathring{T}} v_T \frac{\partial \phi}{\partial x_i} dx_1 dx_2 = - \int_{\Omega} v_h \frac{\partial \phi}{\partial x_i} dx_1 dx_2$$

$$= - \left\langle v_h, \frac{\partial \phi}{\partial x_i} \right\rangle = \left\langle \frac{\partial v_h}{\partial x_i}, \phi \right\rangle \quad \forall \phi \in \mathcal{D}(\Omega)$$

$\Longrightarrow w_i = \frac{\partial v_h}{\partial x_i}$ in $\mathcal{D}'(\Omega)$, but $w_i \in L^2(\Omega)$
$\Longrightarrow \frac{\partial v_h}{\partial x_i} \in L^2(\Omega)$ $(i = 1, 2)$.

Hence, $v_h \in L^2(\Omega)$, $\frac{\partial v_h}{\partial x_i} \in L^2(\Omega)$, $i = 1, 2 \Longrightarrow v_h \in H^1(\Omega)$. Thus $V_h \subset H^1(\Omega)$. But $V_h \subset X_h$, X_h being a finite-dimensional vector space $\Longrightarrow V_h$ is also a finite-dimensional vector space. Hence, V_h is a finite-dimensional subspace of $H^1(\Omega)$. □

Remark 3.4.2. For the inclusion $V_h \subset H^1(\Omega)$ to hold, the minimal value of m to be used in defining $P_m(T)$ in (3.4.3) and (3.4.9) is 1, i.e.

$$V_h = \{v_h : v_h \in C^0(\overline{\Omega}), v_h{\downarrow}_T \in P_m(T) \quad \forall T \in \tau_h, m \geq 1\} \subset H^1(\Omega), \quad (3.4.14)$$

since for $m = 0$, $v_h{\downarrow}_T \in P_0(T) \Longrightarrow v_h{\downarrow}_T = C \neq 0$, and in different triangles $T_i \in \tau_h$, different constants C, i.e. $v_h{\downarrow}_{T_i} = C_i \neq 0$, $v_h{\downarrow}_{T_k} = C_k \neq 0$ with $C_i \neq C_k$ in general for

$$i \neq k \quad \Longrightarrow \quad v_h \notin C^0(\overline{\Omega}). \quad (3.4.15)$$

Theorem 3.4.2. *Let τ_h be a triangulation of $\overline{\Omega}$ defined by* (3.4.1) *and* (3.4.2) *and $V_h \subset X_h$ be defined by:*

$$\begin{aligned} V_h &= \{v_h : v_h \in X_h, v_h \in C^1(\overline{\Omega})\} \\ &= \{v_h : v_h \in C^1(\overline{\Omega}), v_h{\downarrow}_T \in P_m(T)\ \forall T \in \tau_h\}. \end{aligned} \quad (3.4.16)$$

Then $V_h \subset H^2(\Omega)$.

Proof. The proof is exactly similar to that of Theorem 3.4.1, and is left as an exercise for the reader. □

Remark 3.4.3.

- The minimal value of m to be used in $P_m(T)$ in defining V_h in (3.4.16) is 5 (see [35]), i.e. (3.4.16) is to be replaced by:

$$V_h = \{v_h : v_h \in C^1(\overline{\Omega}), v_h{\downarrow}_T \in P_m(T)\ \forall T \in \tau_h, m \geq 5\} \subset H^2(\Omega). \quad (3.4.17)$$

- For $V_h \subset H^k(\Omega)$, $V_h \subset X_h$ is defined by:

$$V_h = \{v_h : v_h \in C^{k-1}(\overline{\Omega}), v_h{\downarrow}_T \in P_m(T)\ \forall T \in \tau_h, m \geq 4(k-1)+1\}, \quad (3.4.18)$$

  from which (3.4.14) and (3.4.17) can be retrieved as particular cases for $k = 1, 2$ respectively.

Chapter 4

Additional properties of $\mathcal{D}'(\Omega)$

4.1 Reflexivity of $\mathcal{D}(\Omega)$ and density of $\mathcal{D}(\Omega)$ in $\mathcal{D}'(\Omega)$

Reflexivity of $\mathcal{D}(\Omega)$

First of all, we will show that $\mathcal{D}(\Omega)$ can be identified with a subspace of $\mathcal{D}''(\Omega)$, which is the (topological) dual space of $\mathcal{D}'(\Omega)$, i.e. $\mathcal{D}''(\Omega) \equiv (\mathcal{D}'(\Omega))'$.
To every $\phi \in \mathcal{D}(\Omega)$, we can associate a *linear functional* L_ϕ on $\mathcal{D}'(\Omega)$ defined by

$$L_\phi(T) = T(\phi) = \langle T, \phi \rangle \quad \forall T \in \mathcal{D}'(\Omega), \tag{4.1.1}$$

since

$$\begin{aligned} L_\phi(\alpha_1 T_1 + \alpha_2 T_2) &= \langle \alpha_1 T_1 + \alpha_2 T_2, \phi \rangle = \alpha_1 \langle T_1, \phi \rangle + \alpha_2 \langle T_2, \phi \rangle \\ &= \alpha_1 L_\phi(T_1) + \alpha_2 L_\phi(T_2) \quad \forall T_i \in \mathcal{D}'(\Omega),\ \forall \alpha_i \in \mathbb{R}. \end{aligned}$$

$\forall \phi \in \mathcal{D}(\Omega)$, L_ϕ defined by (4.1.1) is a *continuous* linear functional on $\mathcal{D}'(\Omega)$, i.e. $L_\phi \in \mathcal{D}''(\Omega) \equiv (\mathcal{D}'(\Omega))'$, which is the second (topological) dual space of $\mathcal{D}(\Omega)$, since

$$T_n \to 0 \text{ in } \mathcal{D}'(\Omega) \quad \Longrightarrow \quad L_\phi(T_n) = \langle T_n, \phi \rangle \to 0 \text{ in } \mathbb{R} \text{ as } n \to \infty\ \forall \phi \in \mathcal{D}(\Omega). \tag{4.1.2}$$

The mapping $\phi \in \mathcal{D}(\Omega) \longmapsto L_\phi \in \mathcal{D}''(\Omega)$ is *one-to-one*. (4.1.3)
In fact,

$$\begin{aligned} L_{\phi_1} = L_{\phi_2} \in \mathcal{D}''(\Omega) \quad &\Longrightarrow \quad L_{\phi_1}(T) = L_{\phi_2}(T) \quad \forall T \in \mathcal{D}'(\Omega) \\ &\Longrightarrow \quad \langle T, \phi_1 \rangle = \langle T, \phi_2 \rangle \quad \forall T \in \mathcal{D}'(\Omega) \\ &\Longrightarrow \quad \langle T, \phi_1 - \phi_2 \rangle = 0 \quad \forall T \in \mathcal{D}'(\Omega) \\ &\Longrightarrow \quad \phi_1 - \phi_2 = 0 \quad \Longrightarrow \quad \phi_1 = \phi_2 \quad \text{in } \mathcal{D}(\Omega). \end{aligned} \tag{4.1.4}$$

Conversely, suppose that $\exists \psi = \phi_1 - \phi_2 \neq 0$ in $\mathcal{D}(\Omega)$ such that $\langle T, \psi \rangle = 0\ \forall T \in \mathcal{D}'(\Omega)$. Then $\exists$ at least one $\mathbf{x}_0 \in \Omega$ such that $\psi(\mathbf{x}_0) \neq 0$. Choose $T = \delta_{\mathbf{x}_0} = \delta(\mathbf{x} - \mathbf{x}_0) \in \mathcal{D}'(\Omega)$, $\delta_{\mathbf{x}_0}$ being the Dirac distribution concentrated at $\mathbf{x}_0$. Then $\langle T, \psi \rangle = \langle \delta_{\mathbf{x}_0}, \psi \rangle = \psi(\mathbf{x}_0) \neq 0$, which contradicts the fact that $\langle T, \psi \rangle = 0\ \forall T \in \mathcal{D}'(\Omega)$.
Hence, $\mathcal{D}(\Omega)$ can be identified with a subspace of $\mathcal{D}''(\Omega)$, i.e. $\mathcal{D}(\Omega) \subset \mathcal{D}''(\Omega)$. (4.1.5)

It remains to show that the mapping $\phi \in \mathcal{D}(\Omega) \longmapsto L_\phi \in \mathcal{D}''(\Omega)$ is *surjective* from $\mathcal{D}(\Omega)$ onto $\mathcal{D}''(\Omega)$. Then the mapping will be *bijective* from $\mathcal{D}(\Omega)$ onto $\mathcal{D}''(\Omega)$. But the proof of the surjectivity of the mapping from $\mathcal{D}(\Omega)$ onto $\mathcal{D}''(\Omega)$ is quite involved, since we are to show that $\mathcal{D}(\Omega)$ is a Montel space, every Montel space being a reflexive one [8], [9], [25]. We have decided *not* to overburden the reader with these, which can be found in [8], [9], [25], and will not be required in our treatment later. Without proof, we agree to accept that the mapping $\phi \in \mathcal{D}(\Omega) \longmapsto L_\phi \in \mathcal{D}''(\Omega)$ is *onto*. Then, from (4.1.3) and (4.1.5), this mapping is bijective, and, from (4.1.2), it is continuous.

Hence, $\mathcal{D}(\Omega)$ can be identified with $\mathcal{D}''(\Omega)$, i.e. $\forall L \in \mathcal{D}''(\Omega)$,

$$\exists \text{ a unique } \phi \in \mathcal{D}(\Omega) \text{ such that } L(T) = \langle T, \phi\rangle \ \forall T \in \mathcal{D}'(\Omega). \tag{4.1.6}$$

In other words, we have agreed to accept.

Theorem 4.1.1. *$\mathcal{D}(\Omega)$ is* reflexive, *i.e.* $\mathcal{D}(\Omega) \equiv \mathcal{D}''(\Omega)$.

Density of $\mathcal{D}(\Omega)$ in $\mathcal{D}'(\Omega)$

Theorem 4.1.2. *$\mathcal{D}(\Omega)$ is a* dense *subspace of $\mathcal{D}'(\Omega)$.*

Proof. Here we give a direct proof. (For an alternative proof based on convolution, see Theorem 6.5.1.)

$\mathcal{D}(\Omega) \hookrightarrow \mathcal{D}'(\Omega)$ (i.e. $\hookrightarrow: \mathcal{D}(\Omega) \longrightarrow \mathcal{D}'(\Omega)$ is an imbedding operator). To every $\phi \in \mathcal{D}(\Omega) \subset L^1_{\text{loc}}(\Omega)$, we can associate a *unique* distribution $T_\phi \in \mathcal{D}'(\Omega)$ defined by:

$$\langle \phi, \psi\rangle_{\mathcal{D}'(\Omega)\times\mathcal{D}(\Omega)} = \langle T_\phi, \psi\rangle = \int_\Omega \phi(\mathbf{x})\psi(\mathbf{x})d\mathbf{x} \quad \forall\, \psi \in \mathcal{D}(\Omega). \tag{4.1.7}$$

$\hookrightarrow: \mathcal{D}(\Omega) \longrightarrow \mathcal{D}'(\Omega)$ is a (linear, continuous) imbedding operator from $\mathcal{D}(\Omega)$ onto the subspace $(\hookrightarrow\mathcal{D}(\Omega)) \subset \mathcal{D}'(\Omega)$: $\forall\phi \in \mathcal{D}(\Omega)$, $\hookrightarrow \phi = T_\phi \in \mathcal{D}'(\Omega)$, with T_ϕ defined by (4.1.7).

$\hookrightarrow$ *is linear:* $\hookrightarrow(\alpha_1\phi_1 + \alpha_1\phi_2) = T_{\alpha_1\phi_1+\alpha_2\phi_2}$, with

$$\begin{aligned}\langle T_{\alpha_1\phi_1+\alpha_2\phi_2}, \psi\rangle &= \int_\Omega (\alpha_1\phi_1 + \alpha_2\phi_2)\psi d\mathbf{x} = \alpha_1 T_{\phi_1}(\psi) + \alpha_2 T_{\phi_2}(\psi)\\ &= \langle \alpha_1 T_{\phi_1} + \alpha_2 T_{\phi_2}, \psi\rangle \quad \forall\psi \in \mathcal{D}(\Omega)\end{aligned}$$

$\Longrightarrow T_{\alpha_1\phi_1+\alpha_2\phi_2} = \alpha_1 T_{\phi_1} + \alpha_2 T_{\phi_2} \Longrightarrow \hookrightarrow(\alpha_1\phi_1 + \alpha_2\phi_2) = \alpha_1{\hookrightarrow}\phi_1 + \alpha_2{\hookrightarrow}\phi_2$.

$\hookrightarrow$ *is one-to-one:* $\hookrightarrow\phi_1 = \hookrightarrow\phi_2$ in $\mathcal{D}'(\Omega) \Longrightarrow T_{\phi_1} = T_{\phi_2}$ in $\mathcal{D}'(\Omega)$

$\Longrightarrow \int_\Omega \phi_1\psi d\mathbf{x} = \int_\Omega \phi_2\psi d\mathbf{x} \Longrightarrow \int_\Omega [\phi_1(\mathbf{x}) - \phi_2(\mathbf{x})]\psi(\mathbf{x})d\mathbf{x} = 0 \ \forall\psi \in \mathcal{D}(\Omega)$.

$\Longrightarrow \phi_1 - \phi_2 = 0$ in $\mathcal{D}(\Omega)$ by Corollary 1.2.1, since $\phi_1 - \phi_2 \in \mathcal{D}(\Omega) \subset L^p(\Omega)$, $1 \le p < \infty \Longrightarrow \phi_1 = \phi_2$ in $\mathcal{D}(\Omega)$.

$\hookrightarrow$ *is continuous:* $\phi_n \to 0$ in $\mathcal{D}(\Omega) \Longrightarrow \exists$ a compact subset $K \subset\subset \Omega$ with $\text{supp}(\phi_n) \subset K\ \forall n \in \mathbb{N}$ such that $\max_{\mathbf{x}\in K} |\phi_n(\mathbf{x})| \to 0$ as $n \to \infty$

$\Longrightarrow (\hookrightarrow\phi_n) = T_{\phi_n} \longrightarrow \hookrightarrow 0 = 0$ in $\mathcal{D}'(\Omega)$. In fact, for $\text{supp}(\phi_n) \subset K \subset\subset \Omega$ $\forall n \in \mathbb{N}$,

$$|\langle \hookrightarrow\phi_n, \psi\rangle - 0| = |\langle T_{\phi_n}, \psi\rangle| = \left|\int_\Omega \phi_n(\mathbf{x})\psi(\mathbf{x})d\mathbf{x}\right| = \left|\int_K \phi_n(\mathbf{x})\psi(\mathbf{x})d\mathbf{x}\right|$$

$$\le \max_{\mathbf{x}\in K\subset\subset\Omega} |\phi_n(\mathbf{x})| \int_K |\psi(\mathbf{x})|d\mathbf{x} \to 0 \int_K \psi(\mathbf{x})d\mathbf{x} = 0 \quad \text{as } n \to \infty.$$

Therefore, $\hookrightarrow: \mathcal{D}(\Omega) \longrightarrow \mathcal{D}'(\Omega)$ is continuous. Hence, $\mathcal{D}(\Omega) \hookrightarrow \mathcal{D}'(\Omega)$, $\hookrightarrow$ being a continuous injection, so $\mathcal{D}(\Omega)$ can be identified with a subspace of $\mathcal{D}'(\Omega)$.

In order to prove that $\mathcal{D}(\Omega)$ is a *dense* subspace of $\mathcal{D}'(\Omega)$, it is sufficient to show that a continuous linear functional $L \in \mathcal{D}''(\Omega) = (\mathcal{D}'(\Omega))'$, which vanishes on $\mathcal{D}(\Omega)$, is a null functional in $\mathcal{D}''(\Omega)$, i.e. $L(\phi) = \langle L, \phi\rangle = 0\ \forall\phi \in \mathcal{D}(\Omega) \Longrightarrow L = 0$ in $\mathcal{D}''(\Omega)$. (This is a well-known result, which follows from the Hahn–Banach Theorem A.7.2.1 (see Corollary A.7.3.6, Appendix A)). Since $\mathcal{D}(\Omega)$ is reflexive by Theorem 4.1.1, $\forall L$ on $\mathcal{D}'(\Omega)\ \exists\psi \in \mathcal{D}(\Omega)$ such that $\langle L, T\rangle = \langle T, \psi\rangle\ \forall T \in \mathcal{D}'(\Omega)$. But $\langle L, \phi\rangle = \langle L, T_\phi\rangle = \langle T_\phi, \psi\rangle = 0\ \forall\phi \in \mathcal{D}(\Omega)$. We are to show that $L = 0$. In particular,

$$\langle L, \psi\rangle = \langle L, T_\psi\rangle = \langle T_\psi, \psi\rangle = \int_\Omega \psi(\mathbf{x})\psi(\mathbf{x})d\mathbf{x} = \int_\Omega |\psi(\mathbf{x})|^2 d\mathbf{x} = 0$$

$\Longrightarrow \psi = 0$ in $\mathcal{D}(\Omega) \Longrightarrow \langle L, T\rangle = \langle T, 0\rangle = 0\ \forall T \in \mathcal{D}'(\Omega) \Longrightarrow L = 0$ in $\mathcal{D}''(\Omega)$. Hence, $\mathcal{D}(\Omega)$ is dense in $\mathcal{D}'(\Omega)$. □

4.2 Continuous imbedding of dual spaces of Banach spaces in $\mathcal{D}'(\Omega)$

Spaces of distributions

We have already shown in Section 1.3, Chapter 1, that all important function spaces are algebraically contained in the space $L_{\text{loc}}(\Omega)$ of locally integrable functions on Ω and, hence, in $\mathcal{D}'(\Omega)$. Now, we will discuss the *continuous* imbedding of general Banach spaces in $\mathcal{D}'(\Omega)$.

Let V be a Banach space. Then

$$\mathcal{D}(\Omega) \hookrightarrow V \quad \Longrightarrow \quad \begin{cases} \text{I. } \mathcal{D}(\Omega) \subset V \textit{ algebraically}, \\ \text{II. } \phi_n \to \phi \text{ in } \mathcal{D}(\Omega) \implies \phi_n \to \phi \text{ in } V, \end{cases} \tag{4.2.1}$$

(to be read as '$\mathcal{D}(\Omega)$ *is continuously imbedded in* V');

$$V \hookrightarrow \mathcal{D}'(\Omega) \quad \Longrightarrow \quad \begin{cases} \text{I. } V \subset \mathcal{D}'(\Omega) \textit{ algebraically}, \\ \text{II. } v_n \to v \text{ in } V \implies v_n \to v \text{ in } \mathcal{D}'(\Omega), \end{cases} \tag{4.2.2}$$

(to be read as '*V is continuously imbedded in $\mathcal{D}'(\Omega)$*').

Definition 4.2.1. A Banach space V is called a *subspace of distributions* if and only if $V \hookrightarrow \mathcal{D}'(\Omega)$ in the sense of (4.2.2).

Banach spaces $C^k(\overline{\Omega})$, $C^{k,\lambda}(\overline{\Omega})$ $(0 < \lambda < 1)$, $k \in \mathbb{N}_0$, $L^p(\Omega)$, $1 \le p \le \infty$, are spaces of distributions, since (4.2.2) holds with

$$V = C^k(\overline{\Omega}), C^{k,\lambda}(\overline{\Omega}), L^p(\Omega). \tag{4.2.3}$$

But the most important class of spaces of distributions are *Sobolev spaces*,which are either Hilbert spaces or Banach spaces (see Chapter 2). We now collect some important results which will be very useful in the study of duals of Sobolev spaces to be introduced in the following section.

Imbedding of dual spaces of Banach spaces in $\mathcal{D}'(\Omega)$

Let V and W be Banach spaces such that $V \hookrightarrow W$ *and* V is *dense* in W,

$$\text{the imbedding } \hookrightarrow \text{ being a continuous injection from } V \text{ into } W. \tag{4.2.4}$$

Let

$$V' = \{v : v \text{ is a continuous linear functional on } V\} \quad \text{and}$$
$$W' = \{w : w \text{ is a continuous linear functional on } W\}$$

be the (topological) dual spaces of V and W respectively. Then V' and W' are also Banach spaces by Theorem A.8.1.2, and

$$W' \hookrightarrow V' \text{ implying } \begin{cases} \text{I. } W' \subset V' \textit{ algebraically}, \\ \text{II. } w_n \to w \text{ in } W' \implies w_n \to w \text{ in } V', \end{cases} \tag{4.2.5}$$

i.e. the imbedding $\hookrightarrow$ is *continuous* from W' into V' with the following identification: $w \in W'$ is identified with $w{\downarrow}_V \in V'$ such that

$$\langle w{\downarrow}_V, v\rangle_{V' \times V} = \langle w, v\rangle_{W' \times W} \quad \forall w \in W',\ \forall v \in V \hookrightarrow W \text{ satisfying (4.2.4).} \tag{4.2.6}$$

Justification of (4.2.6) This identification in (4.2.6) is possible as a consequence of the density of V in W in (4.2.4).

Set $\bar{w} = w{\downarrow}_V \in V'$. Then, $w = \bar{w}{\uparrow}_W \in W'$ is the *unique continuous, linear extension* to W of $\bar{w} \in V$, since V is dense in W, and by the Hahn–Banach Theorem A.7.2.1 (see Corollary A.7.3.7), this *unique* extension is obtained.

But this result will still be meaningful with V replaced by $\mathcal{D}(\Omega)$, which is *no longer* a Banach space, W being a Banach space, i.e.

$$\begin{aligned} &\mathcal{D}(\Omega) \hookrightarrow W \textit{ and } \mathcal{D}(\Omega) \text{ is a dense subspace of } W \\ \Longrightarrow \quad &W' \hookrightarrow \mathcal{D}'(\Omega), \text{ i.e. } W' \text{ is a } \textit{subspace of distributions}. \end{aligned} \tag{4.2.7}$$

Example 4.2.1. For $1 \leq p < \infty$, $L^p(\Omega) = W$ is a Banach space such that $\mathcal{D}(\Omega) \hookrightarrow L^p(\Omega)$ and $\mathcal{D}(\Omega)$ is dense in $L^p(\Omega)$ (see (1.2.25)). Hence, $(L^p(\Omega))' \equiv L^q(\Omega) \hookrightarrow \mathcal{D}'(\Omega)$ with $1 < q \leq \infty$, $\frac{1}{p} + \frac{1}{q} = 1$, i.e. $(L^p(\Omega))' \equiv L^q(\Omega)$ is a subspace of distributions for $1 \leq p < \infty$, $1 < q \leq \infty$.

Remark 4.2.1. If $\mathcal{D}(\Omega) \hookrightarrow W$ in the sense of (4.2.1), but $\mathcal{D}(\Omega)$ is *not* a dense subspace of W, then W' will *not* be a subspace of distributions and, hence, W' cannot be identified with a subspace of $\mathcal{D}'(\Omega)$. In other words, $W' \not\subset \mathcal{D}'(\Omega)$. (4.2.8)

Example 4.2.2.

1. For $p = \infty$, $\mathcal{D}(\Omega) \hookrightarrow L^\infty(\Omega)$, but $\mathcal{D}(\Omega)$ is *not* dense in $L^\infty(\Omega)$.

   Hence,

   $$(L^\infty(\Omega))' \not\subset \mathcal{D}'(\Omega), \tag{4.2.9}$$

   i.e. $(L^\infty(\Omega))'$ cannot be identified with a subspace of $\mathcal{D}'(\Omega)$. $(L^\infty(\Omega))'$ is *not* a subspace of distributions.

2. $\mathcal{D}(\Omega) \hookrightarrow C^{k,\lambda}(\overline{\Omega})$, $\forall k \in \mathbb{N}_0$, $0 < \lambda < 1$, but $\mathcal{D}(\Omega)$ is *not* dense in $C^{k,\lambda}(\overline{\Omega})$ (see Definition A.5.3.1 in Appendix A) (for $\lambda = 0$, $C^{k,0}(\overline{\Omega}) \equiv C^k(\overline{\Omega})$ $\forall k \in \mathbb{N}_0$). Hence,

   $$(C^{k,\lambda}(\overline{\Omega}))' \not\subset \mathcal{D}'(\Omega), \tag{4.2.10}$$

   i.e. $(C^{k,\lambda}(\overline{\Omega}))'$, $0 < \lambda < 1$, cannot be identified with a subspace of $\mathcal{D}'(\Omega)$ and is *not* a subspace of distributions $\forall k \in \mathbb{N}_0$.

There will be important examples in the family of Sobolev spaces.

Case of triple spaces

Let V and W be Banach spaces with their (topological) dual spaces V' and W' (which are also Banach spaces). If the following continuous imbeddings hold:

$$\mathcal{D}(\Omega) \hookrightarrow V \hookrightarrow W \text{ with } \mathcal{D}(\Omega) \textit{ dense} \text{ in } V \text{ and } V \textit{ dense} \text{ in } W,$$
$$\text{then } W' \hookrightarrow V' \hookrightarrow \mathcal{D}'(\Omega). \tag{4.2.11}$$

That is, W' and V' are subspaces of distributions, and W' is also a subspace of V' with continuous imbedding from W' into V'.

An important application of (4.2.11) will be for the family of Sobolev spaces in Section 4.3.

Multiplication by a function If, for any *fixed* function ϕ on Ω, the mapping $v \in V \longmapsto \phi v \in W$ is *linear* and *continuous* from V into W with $\mathcal{D}(\Omega) \hookrightarrow V \hookrightarrow W$ satisfying (4.2.11), then the mapping $w \in W' \longmapsto \phi w \in V'$ is *linear* and *continuous* from W' into V' with $W' \hookrightarrow V' \hookrightarrow \mathcal{D}'(\Omega)$ *and*

$$\langle \phi w, v\rangle_{V'\times V} = \langle w, \phi v\rangle_{W'\times W} \quad \forall w \in W',\ \forall v \in V'. \tag{4.2.12}$$

Differentiation If, for any multi-index $\boldsymbol{\alpha}$, the mapping $\partial^{\boldsymbol{\alpha}} : v \in V \longmapsto \partial^{\boldsymbol{\alpha}} v \in W$ is *linear* and *continuous* from V into W with $\mathcal{D}(\Omega) \hookrightarrow V \hookrightarrow W$ satisfying (4.2.11), then the mapping $\partial^{\boldsymbol{\alpha}} : w \in W' \longmapsto \partial^{\boldsymbol{\alpha}} w \in V'$ defined by:

$$\langle \partial^{\boldsymbol{\alpha}} w, v\rangle_{V'\times V} = (-1)^{|\boldsymbol{\alpha}|}\langle w, \partial^{\boldsymbol{\alpha}} v\rangle_{W'\times W} \quad \forall w \in W',\ \forall v \in V \tag{4.2.13}$$

is *linear* and *continuous* from W' into V', with $W' \hookrightarrow V' \hookrightarrow \mathcal{D}'(\Omega)$.

Restriction and null extension Let $\Omega_1 \subset \Omega_2 \subset \mathbb{R}^n$ and $\tilde{v}$ be the *null extension* to Ω_2 of a function v defined on Ω_1, i.e.

$$\tilde{v}(\mathbf{x}) = v(\mathbf{x}) \text{ for } \mathbf{x} \in \Omega_1 \quad \text{and} \quad \tilde{v}(\mathbf{x}) = 0 \text{ for } \mathbf{x} \in \Omega_2 \setminus \Omega_1. \tag{4.2.14}$$

Let $V(\Omega_1), W(\Omega_2)$ be Banach spaces and $\mathcal{D}(\Omega_1), \mathcal{D}(\Omega_2)$ test spaces such that $\mathcal{D}(\Omega_1) \hookrightarrow V(\Omega_1)$, $\mathcal{D}(\Omega_2) \hookrightarrow W(\Omega_2)$ with $\mathcal{D}(\Omega_1)$ dense in $V(\Omega_1)$ and $\mathcal{D}(\Omega_2)$ dense in $W(\Omega_2)$. Then, if the null extension mapping $v \longmapsto \tilde{v}$ is continuous from $V(\Omega_1)$ into $W(\Omega_2)$, the restriction mapping $u \longmapsto u{\downarrow_{\Omega_1}}$ is *continuous* from $W'(\Omega_2)$ into $V'(\Omega_1)$ with

$$\langle u{\downarrow_{\Omega_1}}, v\rangle_{V'(\Omega_1)\times V(\Omega_1)} = \langle u, \tilde{v}\rangle_{W'(\Omega_2)\times W(\Omega_2)} \quad \forall u \in W'(\Omega_2),\ \forall v \in V'(\Omega_1). \tag{4.2.15}$$

4.3 Applications: Sobolev spaces $H^{-m}(\Omega)$, $W^{-m,q}(\Omega)$

Space $H^{-m}(\Omega)$, $m \in \mathbb{N}$

Definition 4.3.1. Let $\Omega \subset \mathbb{R}^n$ be an open subset of $\mathbb{R}^n$. Let $H_0^m(\Omega) = \overline{\mathcal{D}(\Omega)}$ in the norm $\|\cdot\|_{m,\Omega}$ of $H^m(\Omega)$ $\forall m \in \mathbb{N}$ (Definition 2.15.2). Then the (topological) dual $(H_0^m(\Omega))'$ of $H_0^m(\Omega)$ is the linear space of all continuous linear functionals L defined on $H_0^m(\Omega)$ and denoted by $H^{-m}(\Omega)$, i.e.

$$H^{-m}(\Omega) \equiv (H_0^m(\Omega))' = \{L : L \text{ is a continuous linear functional on } H_0^m(\Omega)\}. \tag{4.3.1}$$

Theorem 4.3.1. *$\forall m \in \mathbb{N}$, $H^{-m}(\Omega)$ equipped with the usual dual norm $\|\cdot\|_{-m,\Omega}$ defined, $\forall L \in H^{-m}(\Omega)$, by*

$$\|L\|_{-m,\Omega} = \sup_{u \in H^m(\Omega)-\{0\}} \frac{|L(u)|}{\|u\|_{m,\Omega}} = \sup\{|L(u)| : u \in H_0^m(\Omega), \|u\|_{m,\Omega} \leq 1\} \tag{4.3.2}$$

is a Banach Space.

$H^{-m}(\Omega)$ equipped with inner product $\langle\cdot,\cdot\rangle_{-m,\Omega}$ induced by $\langle\cdot,\cdot\rangle_{m,\Omega}$ (with the help of the Riesz Representation Theorem A.13.1.1 *and Proposition* A.13.1.2 *in Appendix* A *on Hilbert space $H_0^m(\Omega)$ equipped with inner product $\langle\cdot,\cdot\rangle_{m,\Omega}$) is a* Hilbert space *$\forall m \in \mathbb{N}$, i.e. $\forall L_1, L_2 \in H^{-m}(\Omega)$,*

$$\langle L_1, L_2\rangle_{-m,\Omega} = \langle u_{L_1}, u_{L_2}\rangle_{m,\Omega}, \tag{4.3.3}$$

where $u_{L_1}, u_{L_2} \in H_0^m(\Omega)$ are Riesz representers of L_1 and L_2 respectively.

Proof. The dual space $H^{-m}(\Omega)$ of a Banach space $H_0^m(\Omega)$ equipped with the norm $\|\cdot\|_{-m,\Omega}$ in (4.3.2) is a Banach space (Theorem A.8.1.2 and Section A.13 in Appendix A). Banach space $H^{-m}(\Omega)$ equipped with inner product $\langle\cdot,\cdot\rangle_{-m,\Omega}$ becomes a Hilbert space.

Imbedding results $\mathcal{D}(\Omega) \hookrightarrow H_0^m(\Omega)$ in the sense of (4.2.1) and $\mathcal{D}(\Omega)$ is dense in $H_0^m(\Omega)$ by definition of $H_0^m(\Omega)$

$$\Longrightarrow \quad H^{-m}(\Omega) \equiv (H_0^m(\Omega))' \hookrightarrow \mathcal{D}'(\Omega) \text{ in the sense of (4.2.5).} \tag{4.3.4}$$

$H^{-m}(\Omega)$ can be identified with a subspace of $\mathcal{D}'(\Omega)$, and hence

$$H^{-m}(\Omega) \text{ is a subspace of distributions on } \Omega. \tag{4.3.5}$$

In fact, for a distribution $T \in \mathcal{D}'(\Omega)$ to be in $H^{-m}(\Omega)$, it is necessary and sufficient that T is continuous on $\mathcal{D}(\Omega)$ equipped with the norm $\|\cdot\|_{m,\Omega}$ induced by $H_0^m(\Omega)$. (4.3.6)

□

Remark 4.3.1.

- For $\Omega \neq \mathbb{R}^n$, $\mathcal{D}(\Omega) \hookrightarrow H^m(\Omega)$, but $\mathcal{D}(\Omega)$ is *not* dense in $H^m(\Omega)$ ($H_0^m(\Omega) \neq H^m(\Omega)$). Hence, $(H^m(\Omega))' \not\subset \mathcal{D}'(\Omega)$ by Remark 4.2.1, i.e.

$$(H^m(\Omega))' \text{ is } \textit{not} \text{ a subspace of distributions.} \tag{4.3.7}$$

- For $\Omega = \mathbb{R}^n$, $\overline{\mathcal{D}(\mathbb{R}^n)} \equiv H_0^m(\mathbb{R}^n) \equiv H^m(\mathbb{R}^n)$ $\forall m \in \mathbb{N}$. Then $\mathcal{D}(\mathbb{R}^n) \hookrightarrow H^m(\mathbb{R}^n) \equiv H_0^m(\mathbb{R}^n)$ and $\mathcal{D}(\mathbb{R}^n)$ is dense in $H^m(\mathbb{R}^n)$. Hence, by (4.2.5), $H^{-m}(\mathbb{R}^n) \equiv (H^m(\mathbb{R}^n))' \hookrightarrow \mathcal{D}'(\mathbb{R}^n)$ in the sense of (4.2.2), i.e.

$$(H^m(\mathbb{R}^n))' \equiv H^{-m}(\mathbb{R}^n) \text{ is a subspace of distributions.} \tag{4.3.8}$$

Hilbert triple For a *Hilbert triple* $H_0^m(\Omega), L^2(\Omega), H^{-m}(\Omega)$, we are allowed to make only one identification of a Hilbert space with its dual, i.e. $L^2(\Omega) \equiv (L^2(\Omega))'$ by the Riesz representation theorem, and no more identification is possible. In other words, we cannot identify $H_0^m(\Omega)$ with its dual $H^{-m}(\Omega)$. Then we will have:

$$\mathcal{D}(\Omega) \hookrightarrow H_0^m(\Omega) \hookrightarrow \underbrace{L^2(\Omega) \equiv (L^2(\Omega))'}_{\text{Pivot space}} \hookrightarrow H^{-m}(\Omega) \hookrightarrow \mathcal{D}'(\Omega). \tag{4.3.9}$$

Isometric Isomorphism

Theorem 4.3.2. *The mapping* $\sum_{0\le|\boldsymbol{\alpha}|\le m}(-1)^{|\boldsymbol{\alpha}|}\partial^{2\boldsymbol{\alpha}} : H_0^m(\Omega) \longrightarrow H^{-m}(\Omega)$ *is an* isometric isomorphism *(i.e.* canonical isomorphism*) from* $H_0^m(\Omega)$ *onto* $H^{-m}(\Omega)$.

Proof. Let $v \in H_0^m(\Omega) \hookrightarrow \mathcal{D}'(\Omega)$. Then, $\forall$ multi-index $\boldsymbol{\alpha}$, $\partial^{2\boldsymbol{\alpha}} v \in \mathcal{D}'(\Omega)$.

Set $T = \sum_{0\le|\boldsymbol{\alpha}|\le m}(-1)^{|\boldsymbol{\alpha}|}\partial^{2\boldsymbol{\alpha}} v \in \mathcal{D}'(\Omega)$. We will show that T is an element of $H^{-m}(\Omega)$.

In fact, $\forall \phi \in \mathcal{D}(\Omega)$,

$$\begin{aligned}\langle T, \phi\rangle &= \Big\langle \sum_{0\le|\boldsymbol{\alpha}|\le m}(-1)^{|\boldsymbol{\alpha}|}\partial^{2\boldsymbol{\alpha}} v, \phi\Big\rangle = \sum_{0\le|\boldsymbol{\alpha}|\le m}\langle \partial^{\boldsymbol{\alpha}} v, \partial^{\boldsymbol{\alpha}}\phi\rangle \\ &= \sum_{0\le|\boldsymbol{\alpha}|\le m}\int_\Omega \partial^{\boldsymbol{\alpha}} v \partial^{\boldsymbol{\alpha}}\phi d\mathbf{x} = \langle v, \phi\rangle_{m,\Omega},\end{aligned}$$

where $\langle\cdot,\cdot\rangle_{m,\Omega}$ denotes the inner product (2.15.9) in $H^m(\Omega)$

$$\Longrightarrow \quad |\langle T, \phi\rangle| = |\langle v, \phi\rangle_{m,\Omega}| \le \|v\|_{m,\Omega}\|\phi\|_{m,\Omega} \quad \forall \phi \in \mathcal{D}(\Omega).$$

Hence, T is a linear functional, which is continuous on $\mathcal{D}(\Omega) \subset H_0^m(\Omega)$ in the norm $\|\cdot\|_{m,\Omega}$ induced by $H_0^m(\Omega)$. But $\mathcal{D}(\Omega)$ is *dense* in $H_0^m(\Omega)$ in the norm

$\|\cdot\|_{m,\Omega}$. Hence, by the Hahn–Banach Theorem A.7.2.1 (Appendix A), T can be extended to a unique, continuous, linear functional L on $H_0^m(\Omega)$ such that

$$L(\phi) = \langle T, \phi \rangle = \langle v, \phi \rangle_{m,\Omega} \quad \forall \phi \in \mathcal{D}(\Omega),$$
$$L(u) = \langle v, u \rangle_{m,\Omega} \quad \forall u \in H_0^m(\Omega),$$

with $\|T\|_{H^{-m}(\Omega)} = \|L\|_{H^{-m}(\Omega)} = \|L\|_{-m,\Omega} = \|v\|_{m,\Omega}$ for $v \in H_0^m(\Omega)$.

Hence,

$$\|L\|_{-m,\Omega} = \Big\| \sum_{0\le|\boldsymbol{\alpha}|\le m} (-1)^{|\boldsymbol{\alpha}|}\partial^{2\boldsymbol{\alpha}} v \Big\|_{-m,\Omega} = \|v\|_{m,\Omega} \quad \forall v \in H_0^m(\Omega)$$
$$\Longrightarrow \quad \sum_{0\le|\boldsymbol{\alpha}|\le m} (-1)^{|\boldsymbol{\alpha}|}\partial^{2\boldsymbol{\alpha}} \text{ is an } \textit{isometry} \text{ from } H_0^m(\Omega) \text{ into } H^{-m}(\Omega)$$
$$\Longrightarrow \quad \sum_{0\le|\boldsymbol{\alpha}|\le m} (-1)^{|\boldsymbol{\alpha}|}\partial^{2\boldsymbol{\alpha}} \text{ is an } \textit{injection} \text{ from } H_0^m(\Omega) \text{ into } H^{-m}(\Omega).$$

Now, it remains to show that the mapping $\sum_{0\le|\boldsymbol{\alpha}|\le m}(-1)^{|\boldsymbol{\alpha}|}\partial^{2\boldsymbol{\alpha}} : H_0^m(\Omega) \longrightarrow H^{-m}(\Omega)$ is indeed a surjection. Since $L \in H^{-m}(\Omega) \equiv (H_0^m(\Omega))'$ and $H_0^m(\Omega)$ is a Hilbert space with inner product $\langle \cdot,\cdot \rangle_{m,\Omega}$, by the Riesz Representation Theorem A.13.1.1 in Appendix A, $\exists$ a unique $v \in H_0^m(\Omega)$ such that $L(u) = \langle u, v \rangle_{m,\Omega}$ $\forall u \in H_0^m(\Omega)$ with $\|L\|_{-m,\Omega} = \|v\|_{m,\Omega}$.

Again, by definition, $T = L{\downarrow}_{\mathcal{D}(\Omega)} \Longrightarrow \forall \phi \in \mathcal{D}(\Omega)$,

$$L(\phi) = \langle T, \phi \rangle = \langle v, \phi \rangle_{m,\Omega} = \sum_{0\le|\boldsymbol{\alpha}|\le m} \langle \partial^{\boldsymbol{\alpha}} v, \partial^{\boldsymbol{\alpha}} \phi \rangle_{0,\Omega}$$
$$= \sum_{0\le|\boldsymbol{\alpha}|\le m} \langle \partial^{\boldsymbol{\alpha}} v, \partial^{\boldsymbol{\alpha}} \phi \rangle_{\mathcal{D}'(\Omega)\times\mathcal{D}(\Omega)} = \sum_{0\le|\boldsymbol{\alpha}|\le m} (-1)^{|\boldsymbol{\alpha}|} \langle \partial^{2\boldsymbol{\alpha}} v, \phi \rangle_{\mathcal{D}'(\Omega)\times\mathcal{D}(\Omega)}$$
$$\Longrightarrow \quad \langle T, \phi \rangle_{\mathcal{D}'(\Omega)\times\mathcal{D}(\Omega)} = \Big\langle \sum_{0\le|\boldsymbol{\alpha}|\le m} (-1)^{|\boldsymbol{\alpha}|}\partial^{2\boldsymbol{\alpha}} v, \phi \Big\rangle_{\mathcal{D}'(\Omega)\times\mathcal{D}(\Omega)} \quad \forall \phi \in \mathcal{D}(\Omega)$$
$$\Longrightarrow \quad T = \sum_{0\le|\boldsymbol{\alpha}|\le m} (-1)^{|\boldsymbol{\alpha}|}\partial^{2\boldsymbol{\alpha}} v \quad \text{in } \mathcal{D}'(\Omega), \tag{4.3.10}$$

i.e. $\forall L \in H^{-m}(\Omega)$, $\exists v \in H_0^m(\Omega)$ such that (4.3.10) holds.

Hence, $\sum_{0\le|\boldsymbol{\alpha}|\le m}(-1)^{|\boldsymbol{\alpha}|}\partial^{2\boldsymbol{\alpha}}$ is a surjection from $H_0^m(\Omega)$ onto $H^{-m}(\Omega)$. □

Density of $\mathcal{D}(\Omega)$ in $H^{-m}(\Omega)$

Corollary 4.3.1. *$\mathcal{D}(\Omega)$ is dense in Hilbert space $H^{-m}(\Omega)$.*

Proof. From Theorem 4.3.2, the mapping $\sum_{0\le|\boldsymbol{\alpha}|\le m}(-1)^{|\boldsymbol{\alpha}|}\partial^{2\boldsymbol{\alpha}} : H_0^m(\Omega) \longrightarrow H^{-m}(\Omega)$ is an isometric isomorphism from $H_0^m(\Omega)$ onto $H^{-m}(\Omega)$, i.e. a canonical surjective isometry from $H_0^m(\Omega)$ onto $H^{-m}(\Omega)$, which maps *dense* subspaces of $H_0^m(\Omega)$ onto *dense* subspaces of $H^{-m}(\Omega)$. But $\mathcal{D}(\Omega)$ is a dense subspace of $H_0^m(\Omega)$, which is mapped onto *itself* (i.e. onto $\mathcal{D}(\Omega)$) in $H^{-m}(\Omega)$ under the canonical surjective isometry $\sum_{0\le|\boldsymbol{\alpha}|\le m}(-1)^{|\boldsymbol{\alpha}|}\partial^{2\boldsymbol{\alpha}}$. Therefore, the range space $\mathcal{D}(\Omega) \subset H^{-m}(\Omega)$ is a dense subspace of $H^{-m}(\Omega)$. □

Structure of the elements of $H^{-m}(\Omega)$

Theorem 4.3.3. *A distribution $T \in \mathcal{D}'(\Omega)$ on Ω belongs to $H^{-m}(\Omega)$ (i.e. T can be extended to an element of $H^{-m}(\Omega)$) if and only if, for multi-index $\boldsymbol{\alpha}$ with $|\boldsymbol{\alpha}| \le m$, $\exists$ (equivalence classes of) functions $f_{\boldsymbol{\alpha}} \in L^2(\Omega)$ such that*

$$T = \sum_{0\le|\boldsymbol{\alpha}|\le m} \partial^{\boldsymbol{\alpha}} f_{\boldsymbol{\alpha}}, \tag{4.3.11}$$

where derivatives $\partial^{\boldsymbol{\alpha}} f_{\boldsymbol{\alpha}}$ are in the distribution sense.

Proof. Let $T = \sum_{0\le|\boldsymbol{\alpha}|\le m} \partial^{\boldsymbol{\alpha}} f_{\boldsymbol{\alpha}}$ with $f_{\boldsymbol{\alpha}} \in L^2(\Omega)$. Then $T \in \mathcal{D}'(\Omega)$ and, $\forall \phi \in \mathcal{D}(\Omega)$,

$$\begin{aligned}\langle T, \phi\rangle &= \Big\langle \sum_{0\le|\boldsymbol{\alpha}|\le m} \partial^{\boldsymbol{\alpha}} f_{\boldsymbol{\alpha}}, \phi\Big\rangle = \sum_{0\le|\boldsymbol{\alpha}|\le m} (-1)^{|\boldsymbol{\alpha}|}\langle f_{\boldsymbol{\alpha}}, \partial^{\boldsymbol{\alpha}}\phi\rangle \\ &= \sum_{0\le|\boldsymbol{\alpha}|\le m} (-1)^{|\boldsymbol{\alpha}|}\int_\Omega f_{\boldsymbol{\alpha}}\partial^{\boldsymbol{\alpha}}\phi\, d\mathbf{x} = \sum_{0\le|\boldsymbol{\alpha}|\le m} (-1)^{|\boldsymbol{\alpha}|}\langle f_{\boldsymbol{\alpha}}, \partial^{\boldsymbol{\alpha}}\phi\rangle_{0,\Omega}\end{aligned}$$

(where $\langle\cdot,\cdot\rangle_{0,\Omega}$ denotes the inner product in $L^2(\Omega)$)

$$\begin{aligned}\Longrightarrow \quad |\langle T, \phi\rangle| &\le \sum_{0\le|\boldsymbol{\alpha}|\le m} |\langle f_{\boldsymbol{\alpha}}, \partial^{\boldsymbol{\alpha}}\phi\rangle_{0,\Omega}| \le \sum_{0\le|\boldsymbol{\alpha}|\le m} \|f_{\boldsymbol{\alpha}}\|_{0,\Omega}\|\partial^{\boldsymbol{\alpha}}\phi\|_{0,\Omega} \\ &\le \Big(\sum_{0\le|\boldsymbol{\alpha}|\le m} \|f_{\boldsymbol{\alpha}}\|_{0,\Omega}\Big)\|\phi\|_{m,\Omega}\end{aligned}$$

(since $\|\partial^{\boldsymbol{\alpha}}\phi\|_{0,\Omega} \le \|\phi\|_{m,\Omega}$ for $|\boldsymbol{\alpha}| \le m$), where $\|\cdot\|_{0,\Omega}$ (resp. $\|\cdot\|_{m,\Omega}$) denotes the norm in $L^2(\Omega)$ (resp. $H^m(\Omega)$),

$\Longrightarrow T$ is a continuous, linear functional on $\mathcal{D}(\Omega) \subset H_0^m(\Omega)$ in the norm $\|\cdot\|_{m,\Omega}$ of $H_0^m(\Omega) \subset H^m(\Omega)$. Hence, $T \in H^{-m}(\Omega) \equiv (H_0^m(\Omega))'$ by (4.3.6).

Conversely, let $T \in H^{-m}(\Omega)$. Then, by Theorem 4.3.2 on the isometric isomorphism of $\sum_{0\le|\alpha|\le m}(-1)^{|\alpha|}\partial^{2\alpha}$ from $H_0^m(\Omega)$ onto $H^{-m}(\Omega)$, $\exists v \in H_0^m(\Omega)$ such that

$$T = \sum_{0\le|\alpha|\le m} (-1)^{|\alpha|}\partial^{2\alpha} v.$$

$\forall$ multi-index α with $|\alpha| \le m$, set $f_\alpha = (-1)^{|\alpha|}\partial^\alpha v$. Then $f_\alpha \in L^2(\Omega)$ $\forall|\alpha| \le m$, since $v \in H_0^m(\Omega) \Longrightarrow \partial^\alpha v \in L^2(\Omega)$ $\forall|\alpha| \le m$.

Hence,

$$T = \sum_{0\le|\alpha|\le m} \partial^\alpha[(-1)^{|\alpha|}\partial^\alpha v] = \sum_{0\le|\alpha|\le m} \partial^\alpha f_\alpha. \qquad \square$$

Example 4.3.1. Show that the Dirac distribution $\delta = \delta_0$ with concentration at $0 \in \mathbb{R}$ belongs to $H^{-1}(]-1,1[)$.

Proof. From Example 2.3.2, $\delta = \frac{dH}{dx} \in \mathcal{D}'(]-1,1[)$, where $H(x) = 1$ for $0 < x < 1$ and $H(x) = 0$ for $-1 < x < 0$. But $\int_{-1}^{1}(H(x))^2 dx = \int_{-1}^{0} 0 dx + \int_0^1 1 dx = 1 \Longrightarrow H \in L^2(]-1,1[)$. Hence, $\delta = \frac{dH}{dx}$ with $H \in L^2(]-1,1[) \Longrightarrow$ by Theorem 4.3.3, $\delta \in H^{-1}(]-1,1[)$. $\square$

4.3.1 Space $W^{-m,q}(\Omega)$, $1 < q \le \infty$, $m \in \mathbb{N}$

Definition 4.3.2. Let $\Omega \subset \mathbb{R}^n$ be an open subset of $\mathbb{R}^n$ and $W_0^{m,p}(\Omega) = \overline{\mathcal{D}(\Omega)}$ in the norm $\|\cdot\|_{m,p,\Omega}$ of $W^{m,p}(\Omega)$ $\forall m \in \mathbb{N}$, $1 \le p < \infty$ (Definition 2.15.4). Then the (topological) dual $(W_0^{m,p}(\Omega))'$ of $W_0^{m,p}(\Omega)$ is the linear space of all continuous, linear functionals L defined on $W_0^{m,p}(\Omega)$ and denoted by $W^{-m,q}(\Omega)$ with $1 < q \le \infty$, $\frac{1}{p} + \frac{1}{q} = 1$; i.e. $\forall m \in \mathbb{N}$, $1 \le p < \infty$, $1 < q \le \infty$ with $\frac{1}{p} + \frac{1}{q} = 1$,

$$W^{-m,q}(\Omega) \equiv (W_0^{m,p}(\Omega))' = \{L : L \text{ is a continuous linear functional on } W_0^{m,p}(\Omega)\}. \tag{4.3.12}$$

Theorem 4.3.4. *$\forall m \in \mathbb{N}$, $1 < q \le \infty$, $W^{-m,q}(\Omega)$ equipped with the usual dual norm $\|\cdot\|_{-m,q,\Omega}$ defined by: $\forall L \in W^{-m,q}(\Omega)$, for $1 \le p < \infty$, $1 < q \le \infty$, with $\frac{1}{p} + \frac{1}{q} = 1$,*

$$\begin{aligned}\|L\|_{-m,q,\Omega} &= \sup_{u\in W_0^{m,p}(\Omega)-\{0\}} \frac{|L(u)|}{\|u\|_{m,p,\Omega}} \\ &= \sup\{|L(u)| : u \in W_0^{m,p}(\Omega),\ \|u\|_{m,p,\Omega} \le 1\},\end{aligned} \tag{4.3.13}$$

is a Banach space.

Proof. $\forall m \in \mathbb{N}$, $1 \le p < \infty$, $W_0^{m,p}(\Omega)$ is a Banach space with norm $\|\cdot\|_{m,p,\Omega}$. Hence, its (topological) dual space $(W_0^{m,p}(\Omega))' \equiv W^{-m,q}(\Omega)$ defined by (4.3.12) and equipped with the norm $\|\cdot\|_{-m,q,\Omega}$ defined by (4.3.13) is also a Banach space for $1 < q \le \infty$, $\frac{1}{p} + \frac{1}{q} = 1$. □

Imbedding results

- $\mathcal{D}(\Omega) \hookrightarrow W_0^{m,p}(\Omega)$ in the sense of (4.2.1) *and* $\mathcal{D}(\Omega)$ is dense in $W_0^{m,p}(\Omega)$ $\forall m \in \mathbb{N}$, $1 \le p < \infty \Longrightarrow W^{-m,q}(\Omega) \equiv (W_0^{m,p}(\Omega))' \hookrightarrow \mathcal{D}'(\Omega)$ in the sense of (4.2.2) by (4.2.5), $1 < q \le \infty$, $\frac{1}{p} + \frac{1}{q} = 1$. (4.3.14)
- $W^{-m,q}(\Omega)$, $m \in \mathbb{N}$, $1 < q \le \infty$, can be identified with a subspace of $\mathcal{D}'(\Omega)$, and hence $W^{-m,q}(\Omega)$ is a subspace of distributions on Ω. (4.3.15)
- For a distribution $T \in \mathcal{D}(\Omega)$ to be in $W^{-m,q}(\Omega)$, it is *necessary and sufficient* that T is *continuous* on $\mathcal{D}(\Omega) \subset W_0^{m,p}(\Omega)$ equipped with the norm $\|\cdot\|_{m,p,\Omega}$ induced by $W^{m,p}(\Omega)$, $m \in \mathbb{N}$, $1 \le p < \infty$, $\frac{1}{p} + \frac{1}{q} = 1$. (4.3.16)

Remark 4.3.2. For $\Omega \ne \mathbb{R}^n$, $m \in \mathbb{N}$ and $1 \le p < \infty$, $\mathcal{D}(\Omega)$ is *not* dense in $W^{m,p}(\Omega)$. Hence, by Remark 4.2.1,

$$(W^{m,p}(\Omega))' \not\subset \mathcal{D}'(\Omega), \tag{4.3.17}$$

i.e. $(W^{m,p}(\Omega))'$ is *not* a subspace of distributions.

- For $p = \infty$ and any $\Omega \subset \mathbb{R}^n$, $\mathcal{D}(\Omega)$ is not dense in $W^{m,\infty}(\Omega)$, $m \in \mathbb{N}$. Hence, $(W^{m,\infty}(\Omega))' \not\subset \mathcal{D}'(\Omega)$ by Remark 4.2.1, i.e. $\forall m \in \mathbb{N}$ and $\forall \Omega \subset \mathbb{R}^n$,

$$(W^{m,\infty}(\Omega))' \text{ is } \textit{not} \text{ a subspace of distributions.} \tag{4.3.18}$$

- For $\Omega = \mathbb{R}^n$, $\mathcal{D}(\mathbb{R}^n)$ is dense in $W^{m,p}(\mathbb{R}^n) \equiv W_0^{m,p}(\mathbb{R}^n)$ for $m \in \mathbb{N}$, $1 \le p < \infty$.
  Hence, $(W^{m,p}(\mathbb{R}^n))' \equiv W^{-m,q}(\mathbb{R}^n) \hookrightarrow \mathcal{D}'(\mathbb{R}^n)$, i.e. for $1 \le p < \infty$,

$$(W^{m,p}(\mathbb{R}^n))' \equiv W^{-m,q}(\mathbb{R}^n) \text{ is a subspace of distributions on } \mathbb{R}^n, \tag{4.3.19}$$

$(\frac{1}{p} + \frac{1}{q} = 1)$.

Structure of elements of $W^{-m,q}(\Omega)$ We have similar results as those for $H^{-m}(\Omega)$ in Theorem 4.3.3.

Theorem 4.3.5. *A distribution $T \in \mathcal{D}'(\Omega)$ on Ω belongs to $W^{-m,q}(\Omega)$, $m \in \mathbb{N}$, $1 < q \le \infty$ (i.e. T can be extended to an element of $W^{-m,q}(\Omega)$) if and only if, for* *(some) multi-index $\boldsymbol{\alpha}$ with $|\boldsymbol{\alpha}| \leq m$, $\exists$ (equivalence classes of) functions $f_{\boldsymbol{\alpha}} \in L^q(\Omega)$, $1 < q \leq \infty$, such that*

$$T = \sum_{0 \leq |\boldsymbol{\alpha}| \leq m} \partial^{\boldsymbol{\alpha}} f_{\boldsymbol{\alpha}}, \tag{4.3.20}$$

where derivatives $\partial^{\boldsymbol{\alpha}} f_{\boldsymbol{\alpha}}$ are in the distributional sense.

Proof. The proof is similar to that for Theorem 4.3.3. □

Example 4.3.2. Show that function $\frac{1}{x} \in W^{-1,q}(]0,1[)$ with $1 < q < \infty$.

Proof. $\ln x \in L^1_{\text{loc}}(]0,1[)$ (see (Example 1.4.1)) $\Longrightarrow \ln x \in \mathcal{D}'(]0,1[) \Longrightarrow (\ln x)' = \frac{1}{x} \in \mathcal{D}'(]0,1[)$. Hence, we can write $\frac{1}{x} = \frac{d}{dx}(\ln x)$ with $\ln x \in L^q(]0,1[)$ for $1 \leq q < \infty$ (see (2.3.11)) $\Longrightarrow$ by Theorem 4.3.5, $\frac{1}{x} \in W^{-1,q}(]0,1[)$ for $1 < q < \infty$. □

Example 4.3.3. For $mp > n$ (i.e. $m - \frac{n}{p} > 0$) with $m \in \mathbb{N}$, $1 \leq p < \infty$, Dirac distribution $\delta \in W^{-m,q}(\mathbb{R}^n)$ with $\frac{1}{p} + \frac{1}{q} = 1$.

Solution. $\mathcal{D}(\mathbb{R}^n)$ is dense in $W^{m,p}(\mathbb{R}^n)$, i.e. $W_0^{m,p}(\mathbb{R}^n) \equiv W^{m,p}(\mathbb{R}^n)$ for $m \in \mathbb{N}$, $1 \leq p < \infty$, by Theorem 6.8.9 (see also (2.15.54b)). Moreover, for $mp > n$, $W^{m,p}(\mathbb{R}^n) \hookrightarrow C^{\mathbf{0}}(\mathbb{R}^n)$ by Sobolev's imbedding results (8.12.20g) (see Theorem 8.12.2 in Section 8.12). Hence, $\forall \phi \in \mathcal{D}(\mathbb{R}^n)$, $|\langle \delta, \phi \rangle| = |\phi(\mathbf{0})| \leq \|\phi\|_{C^0(\mathbb{R}^n)} \leq C\|\phi\|_{W^{m,p}(\mathbb{R}^n)} \Longrightarrow$ Dirac distribution $\delta \in \mathcal{D}'(\mathbb{R}^n)$ with mass/charge/force etc. concentrated at $\mathbf{0}$ is a continuous, linear functional on $\mathcal{D}(\mathbb{R}^n)$ in the norm of $W^{m,p}(\mathbb{R}^n)$. But $\mathcal{D}(\mathbb{R}^n)$ is dense in $W^{m,p}(\mathbb{R}^n)$. Hence, by Corollary A.7.3.7 of the Hahn–Banach Theorem A.7.2.1 in Appendix A, δ can be extended to a *unique* continuous, linear functional on $W^{m,p}(\mathbb{R}^n)$ such that this extended unique, continuous, linear functional, which will still be denoted by δ, will belong to $W^{-m,q}(\mathbb{R}^n) \equiv (W^{m,p}(\mathbb{R}^n))'$, i.e. $\delta \in W^{-m,q}(\mathbb{R}^n)$ with

$$|\langle \delta, u \rangle|_{W^{-m,q}(\mathbb{R}^n) \times W^{m,p}(\mathbb{R}^n)} \leq C\|u\|_{W^{m,p}(\mathbb{R}^n)} \quad \forall u \in W^{m,p}(\mathbb{R}^n). \tag{4.3.20a}$$

Restriction and null extension in Sobolev spaces Let $\Omega_1 \subset \Omega_2 \subset \mathbb{R}^n$. For $1 < p < \infty$, the *restriction* mapping $v \longmapsto v\downarrow_{\Omega_1}$ is *continuous* (obviously, *linear*) from $W^{m,p}(\Omega_2)$ into $W^{m,p}(\Omega_1)$, with

$$\|v\downarrow_{\Omega_1}\|_{m,p,\Omega_1} \leq \|v\|_{m,p,\Omega_2} \tag{4.3.21}$$

(which follows from the definition and holds also for $p = 1$ and ∞).

For $V(\Omega_1) = W_0^{m,p}(\Omega_1)$, $W(\Omega_2) = W_0^{m,p}(\Omega_2)$, $1 < p < \infty$, $m \in \mathbb{N}$, satisfying (4.2.15) and (4.2.11), the *null extension* mapping $v \longmapsto \tilde{v}$ is *continuous* from $W_0^{m,p}(\Omega_1)$ into $W_0^{m,p}(\Omega_2)$ (see Theorem 2.15.5 for an analogous proof). Then, by virtue of (4.2.15), the restriction mapping $u \longmapsto u\downarrow_{\Omega_1}$ is *continuous* from

$(W_0^{m,p}(\Omega_2))'$ into $(W_0^{m,p}(\Omega_1))'$, i.e. from $W^{-m,q}(\Omega_2)$ into $W^{-m,q}(\Omega_1)$ with $1 < p < \infty$, $\frac{1}{p} + \frac{1}{q} = 1$, $\forall u \in W^{-m,q}(\Omega_2)$,

$$\langle u{\downarrow_{\Omega_1}}, v\rangle_{W^{-m,q}(\Omega_1)\times W_0^{m,p}(\Omega_1)} = \langle u, \tilde{v}\rangle_{W^{-m,q}(\Omega_2)\times W_0^{m,p}(\Omega_2)} \quad \forall v \in W_0^{m,p}(\Omega_1). \tag{4.3.22}$$

Duals of closed subspaces and quotient spaces [3]

Let M be a *closed* subspace of Banach space V, the norm in M being that induced by V, i.e. $\|\cdot\|_M = \|\cdot\|_V$, such that V/M is the *quotient space* of V by M, which is also a Banach space equipped with the *quotient norm* $\|[\cdot]\|_{V/M}$ defined in (2.15.24) (resp. (2.15.41)) for $V = H^m(\Omega)$ (resp. $V = W^{m,p}(\Omega)$).

Let M' and V' be the duals of M and V respectively. Since M and V are Banach spaces, M' and V' are also Banach spaces equipped with the *dual norms* $\|\cdot\|_{M'}$ and $\|\cdot\|_{V'}$ respectively.

Annihilator M^0 of M (see Rudin [3])

Definition 4.3.3. The *annihilator* M° of M is the closed subspace of V' defined by:

$$M^0 = \{l : l \in V',\ \langle l, u\rangle = 0\ \forall u \in M\} \subset V', \tag{4.3.23}$$

which is also a Banach space equipped with the norm $\|\cdot\|_{V'}$ induced by V'.

Hence, V'/M^0 is the quotient space of V' by M^0, which is also a Banach space equipped with the corresponding quotient norm $\|[\cdot]\|_{V'/M^0}$ defined by [3]:

$$\|[\tilde{l}]\|_{V'/M^0} = \inf_{\tilde{l}\in[\tilde{l}]} \|\tilde{l}\|_{V'} = \inf_{m_0\in M^0} \|\tilde{l} + m_0\| \le \|\tilde{l}\|_{V'}. \tag{4.3.24}$$

Then the dual M' of the *closed* subspace M of V and the dual $(V/M)'$ of the *quotient space* V/M are intimately related to the *annihilator* M^0. In fact, M' is *identified* with the quotient space V'/M^0 and $(V/M)'$ with M^0 by writing:

$$M' = V'/M^0 \quad \text{and} \quad (V/M)' = M^0, \tag{4.3.25}$$

since these are *isometrically isomorphic*.

Since M is a *closed* subspace of Banach space V, by Corollary A.7.3.1 of the Hahn–Banach Theorem A.7.2.1 in Appendix A,

$$\forall l \in M',\ \exists \tilde{l} \in V' \text{ such that } \tilde{l}(u) = l(u)\ \forall u \in M,\ \|\tilde{l}\|_{V'} = \|l\|_{M'}. \tag{4.3.26}$$

Then, we have the following result on isometric isomorphisms:

Theorem 4.3.6.

I. *Let* $J : M' \longrightarrow V'/M^0$ *be the mapping defined by:*

$$\forall l \in M', \quad Jl = [\tilde{l}] = \tilde{l} + M^0 \in V'/M^0, \tag{4.3.27}$$

with $\tilde{l} \in V'$ *satisfying* (4.3.26). *Then* J *is an* isometric isomorphism *from* M' onto V'/M^0.

II. *Let* $\phi : V \longrightarrow V/M$ *be the* canonical surjection *from* V *onto* V/M *with the same properties given by (2.15.24a)–(2.15.24d), and* $\tilde{J} : (V/M)' \longrightarrow M^0$ *be the mapping defined by:* $\forall \Lambda \in (V/M)'$, $(\tilde{J}\Lambda)u = (\Lambda\phi)u = \Lambda(\phi u)$ $\forall u \in V$ *such that*

$$(\tilde{J}\Lambda)u = (\Lambda\phi)u = 0 \ \forall u \in M, \quad \textit{i.e. } \tilde{J}\Lambda \in M^0. \tag{4.3.28}$$

Then $\tilde{J}$ *is an* isometric isomorphism *from* $(V/M)'$ onto M^0.

Proof.

I. J is *well defined*: We are to show that (4.3.27) holds for *any* choice of extension $\tilde{l}$ of l, since $\tilde{l}$ is not unique. Let $\tilde{l}$, $\tilde{l}_1 \in V'$ be Hahn–Banach extensions of $l \in M'$. Then $(\tilde{l} - \tilde{l}_1)(u) = l(u) - l(u) = 0$ $\forall u \in M$. Hence, $\tilde{l} - \tilde{l}_1 = m \in M^0 \Longrightarrow \tilde{l} = \tilde{l}_1 + m$ with $m \in M^0 \Longrightarrow \tilde{l} = \tilde{l}_1 \pmod{M^0} \Longrightarrow \tilde{l} + M^0 = \tilde{l}_1 + M^0 = [\tilde{l}] \in V'/M^0$, i.e. $\forall$ choices of Hahn–Banach extension $\tilde{l} \in V'$, $\tilde{l} + M^0 = [\tilde{l}] \in V'/M^0$ and J is well defined by (4.3.27).

J is *linear*: $J(\alpha_1 l_1 + \alpha_2 l_2) = (\alpha_1 \tilde{l}_1 + \alpha_2 \tilde{l}_2) + M^0 = \alpha_1(\tilde{l}_1 + M^0) + \alpha_2(\tilde{l}_2 + M^0) = \alpha_1 J l_1 + \alpha_2 J l_2$ $\forall \alpha_i \in \mathbb{R}$.

J is *onto*: $\forall \tilde{l} \in V'$, $\exists l \in M'$ such that $l = \tilde{l}\downarrow_M$. Then, $\forall [\tilde{l}] \in V'/M^0$, $\exists l \in M'$ such that $Jl = \tilde{l} + M^0 = [\tilde{l}] \in V'/M^0$.

J is *one-to-one* and *continuous*: If J is an isometry, then J is *one-to-one* and *continuous*, since $\|Jl\|_{V'/M^0} = \|l\|_{M'} = 0 \Longrightarrow l = 0 \Longrightarrow J$ is *one-to-one* and *continuous* from M' onto V'/M^0. Hence, the proof will be complete if we can show that J is an *isometry*, since J will be a *continuous, linear* bijection from M' onto V'/M^0 and J^{-1} will also be linear and its continuity will follow from Corollary A.8.1.1 of the Open Mapping Theorem A.8.1.3 in Appendix A.

J is an *isometry*: For any *fixed* $l \in M'$, let $\tilde{l} \in V'$ be an extension to V' of l. Then, $l(u) = \tilde{l}(u)$ $\forall u \in M \subset V$. Hence,

$$\begin{aligned} \|l\|_{M'} &= \sup_{u \in M} \frac{|l(u)|}{\|u\|_V} = \sup_{u \in M} \frac{|\tilde{l}(u)|}{\|u\|_V} \\ &\le \sup_{u \in V} \frac{|\tilde{l}(u)|}{\|u\|_V} = \|\tilde{l}\|_{V'} \quad \Longrightarrow \quad \|l\|_{M'} \le \|\tilde{l}\|_{V'}. \end{aligned} \tag{4.3.29}$$

But $\forall m_0 \in M^0$, $\tilde{l} + m_0 \in V'$ is also an extension of $l \in M'$ to V, since $(\tilde{l} + m_0)(u) = \tilde{l}(u) = l(u)\ \forall u \in M$. Hence, from (4.3.29), $\|l\|_{M'} \le \|\tilde{l} + m_0\|_{V'}$ $\forall m_0 \in M^0$,

$$\Longrightarrow \|l\|_{M'} \le \inf_{m_0 \in M^0} \|\tilde{l} + m_0\|_{V'} = \|\tilde{l} + M^0\|_{V'/M^0} = \|Jl\|_{V'/M^0} \le \|\tilde{l}\|_{V'} \tag{4.3.30}$$

by (4.3.24).

But $\forall l \in M'$, $\exists$ a Hahn–Banach extension $\tilde{l} \in V'$ of l such that (4.3.26) holds. Then, using (4.3.30) and (4.3.26), we get the result:

$$\|l\|_{M'} = \|\tilde{l} + M^0\|_{V'/M^0} = \|Jl\|_{V'/M^0} = \|\tilde{l}\|_{V'},$$

i.e. J defined by (4.3.27) is an *isometric isomorphism* from M' onto V'/M^0.

II. $\tilde{J}$ is *well defined*: Let $\Lambda \in (V/M)'$ and $u \in V$ with $\phi u \in V/M$, ϕ being the *canonical surjection* from V onto V/M. Then the mapping $u \in V \longmapsto \Lambda\phi u$ is well defined and a *continuous*, *linear* functional on V, which vanishes on $\text{Ker}(\phi) = M$. Hence, $\Lambda\phi \in M^0$ and $\tilde{J}\Lambda = \Lambda\phi \in M^0\ \forall \Lambda \in (V/M)'$, and $\tilde{J}$ is well defined.

$\tilde{J}$ is *linear*: $\Lambda_i \in (V/M)' \Longrightarrow \alpha_1\Lambda_1 + \alpha_2\Lambda_2 \in (V/M)'\ \forall \alpha_1, \alpha_2 \in \mathbb{R}$

$$\Longrightarrow \quad \tilde{J}(\alpha_1\Lambda_1 + \alpha_2\Lambda_2) = (\alpha_1\Lambda_1 + \alpha_2\Lambda_2)\phi$$
$$= \alpha_1\Lambda_1\phi + \alpha_2\Lambda_2\phi = \alpha_1\tilde{J}\Lambda_1 + \alpha_2\tilde{J}\Lambda_2.$$

$\tilde{J}$ is *onto*: Let $m_0 \in M^0$ be any *fixed* element, with $\mathcal{N}_0 = \text{Ker}(m_0)$. Then $M \subset \mathcal{N}_0 \subset V$, since $m_0(u) = 0\ \forall u \in M$. Hence, $\exists$ a linear functional L on V/M such that $L\phi = m_0 \in M^0$ with the null space $\mathcal{N}(L)$ of $L = \{\phi u : u \in \mathcal{N}_0\} = \phi(\mathcal{N}_0)$. But ϕ is the *continuous* (see (2.15.24d)) canonical surjection from V onto V/M. Hence, $\phi(\mathcal{N}_0)$ with $\mathcal{N}_0 =$ null space of $m_0 \in M^0$ is a *closed* subspace of V/M. In fact, $\mathcal{N}_0$ is a closed subspace of V. Then $u_k \to u$ in $\mathcal{N}_0 \Longrightarrow \phi u_k \to \phi u$ in $\phi(\mathcal{N}_0)$ by virtue of the *continuity* of quotient mapping ϕ. Hence, $\phi(\mathcal{N}_0) = \mathcal{N}(L)$ is a *closed* subspace of V/M, L being a *non-null linear* functional on V/M. Then, by a well-known theorem (see Rudin [3, p. 14]), L is *continuous* on V/M, i.e. $L \in (V/M)'$. Hence, for any fixed $m_0 \in M^0$, $\exists L \in (V/M)'$ such that $\tilde{J}L = L\phi = m_0 \in M^0$.

$\tilde{J}$ is an *isometry*: Let $L \in (V/M)'$ be any *fixed* element. Let $[u] \in V/M$ with $\|[u]\|_{V/M} = 1$. Then $\exists v_0 \in V$ with $\|v_0\|_V = 1 + \varepsilon\ \forall \varepsilon > 0$ such that $\|[u]\|_{V/M} \le \|v_0\|_V$ (by (2.15.24d)) with $\phi v_0 = [u] \in V/M$. Hence,

$$|\langle L, [u]\rangle| = |L\phi v_0| = |(\tilde{J}L)v_0| \le \|\tilde{J}L\|_{V'}\|v_0\|_V \le (1+\varepsilon)\|\tilde{J}L\|_{V'} \quad \forall \varepsilon > 0.$$

Then,

$$\|L\|_{(V/M)'} = \sup_{\|[u]\|_{V/M}=1} |\langle L, [u]\rangle| \leq \|\tilde{J}L\|_{V'} = \|\tilde{J}L\|_{M^0}. \tag{4.3.31}$$

But $\|\phi u\|_{V/M} \leq \|u\|_V \ \forall u \in V$ (see (2.15.24d)). Then

$$|(\tilde{J}L)(u)| = |L(\phi u)| \leq \|L\|_{(V/M)'}\|\phi u\|_{V/M} \leq \|L\|_{(V/M)'}\|u\|_V \quad \forall u \in V$$
$$\Longrightarrow \quad \|\tilde{J}L\|_{M^0} = \|\tilde{J}L\|_{V'} \leq \|L\|_{(V/M)'}. \tag{4.3.32}$$

From (4.3.31) and (4.3.32), the *isometry* of $\tilde{J}$ follows: $\|\tilde{J}L\|_{M^0} = \|L\|_{(V/M)'}$ $\forall L \in (V/M)'$, from which it also follows immediately that $\tilde{J}$ is *continuous* and *one-to-one* (the proof is similar to that of the injectivity and the continuity of J given earlier), i.e. $\tilde{J}$ is a *continuous, linear bijection* from $(V/M)'$ onto M^0. Consequently, $\tilde{J}^{-1} : M^0 \longrightarrow (V/M)'$ is also linear and its continuity follows from Corollary A.8.1.1 of the Open Mapping Theorem A.8.1.3 in Appendix A. Hence, $\tilde{J}$ is an *isometric isomorphism* from $(V/M)'$ onto M^0. □

Chapter 5

Local properties, restrictions, unification principle, space $\mathcal{E}'(\mathbb{R}^n)$ of distributions with compact support

5.1 Null distribution in an open set

Although a distribution $T \in \mathcal{D}'(\Omega)$ with $\Omega \subset \mathbb{R}^n$ does not have point values, i.e. a value at a given point $\mathbf{x} \in \Omega$, we can say that T is *zero* in an open subset $\Omega_0 \subset \Omega$.

Definition 5.1.1. A distribution $T \in \mathcal{D}'(\Omega)$ is a *zero* or *null* distribution in an open set $\Omega_0 \subset \Omega$ if and only if

$$T(\phi) = 0 \quad \forall \phi \in \mathcal{D}(\Omega) \text{ with } \operatorname{supp}(\phi) \subset \Omega_0 \text{ (i.e. } \forall \phi \in \mathcal{D}(\Omega_0)). \tag{5.1.1}$$

A distribution $T \in \mathcal{D}'(\Omega)$ is a *null distribution in a neighbourhood of a point* $\mathbf{x}_0 \in \Omega$ if and only if T is zero in an open set U containing the point $\mathbf{x}_0$, i.e. $T = 0$ in a neighbourhood U of $\mathbf{x}_0 \in \Omega$

$$\iff \quad \langle T, \phi \rangle = 0 \; \forall \phi \in \mathcal{D}(\Omega) \text{ with } \operatorname{supp}(\phi) \subset U, \tag{5.1.2}$$

(i.e. $\forall \phi \in \mathcal{D}(U)$).

$T = 0$ in $\mathcal{D}'(\Omega) \iff \forall \mathbf{x}_0 \in \Omega$, $\exists$ an open set $U \subset \Omega$ with $\mathbf{x}_0 \in U$ such that $T = 0$ in U, i.e. $\forall$ open sets $U \subset \Omega$,

$$\langle T, \phi \rangle = 0 \quad \forall \phi \in \mathcal{D}(\Omega) \text{ with } \operatorname{supp}(\phi) \subset U \text{ (i.e. } \forall \phi \in \mathcal{D}(U)). \tag{5.1.3}$$

5.2 Equality of distributions in an open set

Definition 5.2.1. Two distributions $T_1, T_2 \in \mathcal{D}'(\Omega)$ are said to be *equal* in an open set $\Omega_0 \subset \Omega$ if and only if $T_1 - T_2 = 0$ in Ω_0, i.e.

$$\langle T_1, \phi \rangle = \langle T_2, \phi \rangle \quad \forall \phi \in \mathcal{D}(\Omega) \text{ with } \operatorname{supp}(\phi) \subset \Omega_0 \text{ (i.e. } \forall \phi \in \mathcal{D}(\Omega_0)). \tag{5.2.1}$$

5.3 Restriction of a distribution to an open set

Definition 5.3.1. Let $\Omega_0 \subset \Omega$ be an open subset of Ω and $T \in \mathcal{D}'(\Omega)$ be a distribution on Ω. Then the *restriction to* Ω_0 *of the distribution* T is the distribution

$T\downarrow_{\Omega_0} = T_0 \in \mathcal{D}'(\Omega_0)$ on Ω_0 defined by:

$$\langle T_0, \phi\rangle = \langle T\downarrow_{\Omega_0}, \phi\rangle = \langle T, \phi\rangle \ \forall \phi \in \mathcal{D}(\Omega) \text{ with } \operatorname{supp}(\phi) \subset \Omega_0 \text{ (i.e. } \forall \phi \in \mathcal{D}(\Omega_0)). \tag{5.3.1}$$

A distribution $T \in \mathcal{D}'(\Omega)$ with $\Omega \subset \mathbb{R}^n$ can *not* necessarily be extended to a distribution on $\mathbb{R}^n$. In fact, we have:

Example 5.3.1. For $e^{1/x^2} \in \mathcal{D}'(\mathbb{R} \setminus \{0\})$, there does *not* exist any distribution $T \in \mathcal{D}'(\mathbb{R})$ on $\mathbb{R}$ such that $T\downarrow_{\mathbb{R}\setminus\{0\}} = e^{1/x^2} \in \mathcal{D}'(\mathbb{R} \setminus \{0\})$, i.e.

$$\langle T, \phi\rangle = \langle e^{1/x^2}, \phi\rangle = \int_{-\infty}^{\infty} e^{1/x^2}\phi(x)dx \quad \forall \phi \in \mathcal{D}(\mathbb{R} \setminus \{0\}). \tag{5.3.2}$$

In other words, the distribution $e^{1/x^2} \in \mathcal{D}'(\mathbb{R} \setminus \{0\})$ on $\mathbb{R} \setminus \{0\}$ can *not* be extended to a distribution $T \in \mathcal{D}'(\mathbb{R})$ on $\mathbb{R}$.

Proof. The scheme of the proof is as follows: we choose a sequence (ϕ_n) in $\mathcal{D}(\mathbb{R})$ such that $\phi_n \to 0$ in $\mathcal{D}(\mathbb{R})$. Then we show that $\langle T, \phi_n\rangle = \int_{-\infty}^{\infty} e^{1/x^2}\phi_n(x)dx \to \infty$ as $n \to \infty$, establishing that T is *not* a distribution on $\mathbb{R}$.

Let $\phi \in \mathcal{D}(\mathbb{R})$ such that $\operatorname{supp}(\phi) \subset]1, 2[$, $0 \le \phi(x) \le 1$ $\forall x \in \mathbb{R}$ and $\phi(x) = 1$ for $a \le x \le b$ with $1 < a < b < 2$.

Define $\phi_n = e^{-n}\phi(nx)$ $\forall n \in \mathbb{N}$. Then $\phi_n \in \mathcal{D}(\mathbb{R})$ with $\operatorname{supp}(\phi_n) \subset \{x : 1 < nx < 2\} = \{x : \frac{1}{n} < x < \frac{2}{n}\}$ $\forall n \in \mathbb{N}$. Hence, $\forall n \in \mathbb{N}$, $\operatorname{supp}(\phi_n) \subset [0, 2]$, $[0, 2]$ being a compact subset of $\mathbb{R}$. Moreover, $\forall m \in \mathbb{N}_0$, $\phi_n^{(m)}(x) = e^{-n}n^m\phi^{(m)}(nx)$, and $\sup_{x\in\mathbb{R}} |\phi_n^{(m)}(x)| \le (n^m \cdot e^{-n} \cdot \sup_{1\le y\le 2} |\phi^{(m)}(y)|) \to 0$ as $n \to \infty$, since $\forall m \in \mathbb{N}_0$, $(n^m \cdot e^{-n}) \to 0$ as $n \to \infty \implies \phi_n \to 0$ in $\mathcal{D}(\mathbb{R})$ as $n \to \infty$.

Suppose that the contrary holds, i.e. $\exists T \in \mathcal{D}'(\mathbb{R})$ which *extends* e^{1/x^2} to $\mathbb{R}$ satisfying (5.3.2). In other words, $\exists T \in \mathcal{D}'(\mathbb{R})$ which coincides with e^{1/x^2} on $\mathbb{R} \setminus \{0\}$. Then,

$$\phi_n \to 0 \text{ in } \mathcal{D}(\mathbb{R}) \quad \Longrightarrow \quad \langle T, \phi_n\rangle \to 0 \text{ as } n \to \infty. \tag{5.3.3}$$

But $\operatorname{supp}(\phi_n) \subset]1/n, 2/n[\subset \mathbb{R} \setminus \{0\}$ $\forall n \in \mathbb{N}$. Since $\phi_n(x) \ge 0$ and $\phi(nx) = 1$ for $1 < a < nx < b < 2$, i.e. for $a/n < x < b/n$, $\phi_n(x) = e^{-n}\phi(nx) = e^{-n}$ for $1/n < a/n < x < b/n < 2/n$. Hence,

$$\begin{aligned}\langle T, \phi_n\rangle &= \int_{1/n}^{2/n} e^{1/x^2}\phi_n(x)dx \\ &\ge \int_{a/n}^{b/n} e^{1/x^2}\phi_n(x)dx = \int_{a/n}^{b/n} e^{1/x^2} \cdot e^{-n}dx\end{aligned}$$

(since the integrand is positive) $\Longrightarrow \langle T, \phi_n \rangle \geq e^{-n} \int_{a/n}^{b/n} e^{1/x^2} dx$. But for $0 < x \leq b/n$, $\frac{n^2}{b^2} \leq \frac{1}{x^2} \Longrightarrow e^{n^2/b^2} \leq e^{1/x^2}$ for $\frac{a}{n} \leq x \leq \frac{b}{n} \Longrightarrow \langle T, \phi_n \rangle \geq e^{-n} \cdot e^{n^2/b^2} \int_{a/n}^{b/n} dx = e^{n(\frac{n-b^2}{b^2})} \cdot \frac{b-a}{n}$. For $n \geq 2b^2, n - b^2 \geq b^2 \Longrightarrow n(\frac{n-b^2}{b^2}) \geq n \Longrightarrow e^{\frac{n(n-b^2)}{b^2}} \geq e^n$.

Hence, for $n \geq 2b^2$, $\langle T, \phi_n \rangle \geq e^n \cdot \frac{b-a}{n} \to \infty$ as $n \to \infty$, which contradicts (5.3.3) that $\langle T, \phi_n \rangle \to 0$ as $n \to \infty$. Hence, our assumption that T extends e^{1/x^2} is wrong and the result follows. □

Restriction mappings Now we can state some important results on restriction mappings, which will be used in the study of Sobolev spaces and their duals.

Let Ω_1, Ω_2 be open subsets of $\mathbb{R}^n$ such that $\Omega_1 \subset \Omega_2$, and $V_1(\Omega_1)$ and $V_2(\Omega_2)$ be Banach spaces of functions on Ω_1 and Ω_2 respectively such that:

1. $\mathcal{D}(\Omega_1) \hookrightarrow V_1(\Omega_1)$ and $\mathcal{D}(\Omega_1)$ is *dense* in $V_1(\Omega_1)$;
2. $\mathcal{D}(\Omega_2) \hookrightarrow V_2(\Omega_2)$ and $\mathcal{D}(\Omega_2)$ is *dense* in $V_2(\Omega_2)$. (5.3.4)

Then, by (4.2.4)–(4.2.5),

$$V_1'(\Omega_1) \hookrightarrow \mathcal{D}'(\Omega_1), \quad V_2'(\Omega_2) \hookrightarrow \mathcal{D}'(\Omega_2), \tag{5.3.5}$$

i.e. $V_1'(\Omega_1)$ and $V_2'(\Omega_2)$ are subspaces of distributions.

Restriction operator from $V_2'(\Omega_2)$ into $V_1'(\Omega_1)$ For the definition of the restriction operator from $V_2'(\Omega_2)$ to $V_1'(\Omega_1)$, we need to define the *null extension of functions*.

Definition 5.3.2. Let v be a function on $\Omega_1 \subset \Omega_2 \subset \mathbb{R}^n$. Then $\tilde{v}$ defined by

$$\tilde{v}(\mathbf{x}) = \begin{cases} v(\mathbf{x}) & \text{for } \mathbf{x} \in \Omega_1 \\ 0 & \text{for } \mathbf{x} \in \Omega_2,\ \mathbf{x} \notin \Omega_1 \text{ (i.e. for } \mathbf{x} \in \Omega_2 \setminus \Omega_1) \end{cases} \tag{5.3.6}$$

is called the *null extension* of v to Ω_2 or *extension of* v *by* 0 to Ω_2.

For $v \in V_1(\Omega_1)$, $\tilde{v}$ may *not* belong to $V_2(\Omega_2)$. Hence, we make the assumption that for $v \in V_1(\Omega_1)$, let $\tilde{v} \in V_2(\Omega_2)$.

Now, by virtue of (5.3.5) and Definition 5.3.1 of the restriction of a distribution to an open subset, we can define the restriction operator $\rho : V_2'(\Omega_2) \to V_1'(\Omega_1)$ by: $\forall w \in V_2'(\Omega_2)$,

$$\rho w = w\downarrow_{\Omega_1} \in V_1'(\Omega_1), \tag{5.3.7}$$

with

$$\langle \rho w, v \rangle_{V_1'(\Omega_1) \times V_1(\Omega_1)} = \langle w\downarrow_{\Omega_1}, v \rangle_{V_1'(\Omega_1) \times V_1(\Omega_1)} = \langle w, \tilde{v} \rangle_{V_2'(\Omega_2) \times V_2(\Omega_2)} \tag{5.3.8}$$

$\forall w \in V_2'(\Omega_2)$, $\forall v \in V_1(\Omega_1)$ with $\tilde{v}$ = null extension of v to Ω_2 such that $\tilde{v} \in V_2(\Omega_2)$.

Continuity of restriction operator $\rho : V_2'(\Omega_2) \to V_1'(\Omega_1)$ If the null extension mapping $v \in V_1(\Omega_1) \longmapsto \tilde{v} \in V_2(\Omega_2)$ is *linear* and *continuous* from $V_1(\Omega_1)$ into $V_2(\Omega_2)$, i.e. $\exists C > 0$ such that

$$\|\tilde{v}\|_{V_2(\Omega_2)} \leq C\|v\|_{V_1(\Omega_1)} \quad \forall v \in V_1(\Omega_1), \tag{5.3.9}$$

then the restriction operator ρ defined by (5.3.7)–(5.3.8) is *linear* and *continuous* from $V_2'(\Omega_2)$ into $V_1'(\Omega_1)$. (5.3.10)

In fact,

$$\begin{aligned} |\langle \rho w, v\rangle| = |\langle w, \tilde{v}\rangle| &\leq \|w\|_{V_2'(\Omega_2)}\|\tilde{v}\|_{V_2(\Omega_2)} \\ &\leq C\|w\|_{V_2'(\Omega_2)}\|v\|_{V_1(\Omega_1)} \quad \text{(by (5.3.9))} \\ \Longrightarrow \quad \|\rho w\|_{V_1'(\Omega_1)} &\leq C\|w\|_{V_2'(\Omega_2)} \quad \forall w \in V_2'(\Omega_2). \end{aligned} \tag{5.3.11}$$

5.4 Unification principle

From the *local* knowledge of a distribution on a family of open sets, the following theorem on the *Unification Principle* (called *principe du recollement des morceaux* in French, and also known as the *Principle of Localization*) allows us to have a *global* knowledge of the distribution on the union of these open sets.

Theorem 5.4.1 (Unification Principle Theorem of Schwartz). *Let $\Omega \subset \mathbb{R}^n$ be an open subset of $\mathbb{R}^n$, $\{\Omega_i\}_{i\in I}$ be a* finite *or* infinite *family of open sets in $\mathbb{R}^n$ such that $\Omega = \bigcup_{i\in I}\Omega_i$ is an open set, and $\{T_i\}_{i\in I}$ be a family of distributions $T_i \in \mathcal{D}'(\Omega_i)$ on Ω_i $\forall i \in I$, $I \subset \mathbb{N}$ being a set of indices. Suppose that $\forall i, j \in I$ $(i \neq j)$, if Ω_i, Ω_j have nonempty intersection $\Omega_i \cap \Omega_j \neq \emptyset$ and $T_i = T_j$ on $\Omega_i \cap \Omega_j$, then $\exists$ a* unique *distribution $T \in \mathcal{D}'(\Omega)$ on Ω such that $T\downarrow_{\Omega_j} = T_j$ on each Ω_j, $j \in I$, i.e. $\forall j \in I$,*

$$\langle T_j, \phi\rangle = \langle T, \phi\rangle \quad \forall \phi \in \mathcal{D}(\Omega) \textit{ with } \operatorname{supp}(\phi) \subset \Omega_j \ (\textit{i.e. } \forall \phi \in \mathcal{D}(\Omega_j)). \tag{5.4.1}$$

Proof. Assume that $I = \mathbb{N}$ and $T_i \in \mathcal{D}'(\Omega_i)$ is a distribution on Ω_i $\forall i \in \mathbb{N}$. Let $\{\Theta_i\}_{i=1}^{\infty}$ be a *partition* of unity subordinate to the family of open sets $\{\Omega_i\}_{i=1}^{\infty}$ of Ω with $\bigcup_{i\in\mathbb{N}}\Omega_i = \Omega$, i.e. $\Theta_i \in \mathcal{D}(\Omega_i)$ $\forall i \in \mathbb{N}$, $\sum_{i=1}^{\infty}\Theta_i(\mathbf{x}) = 1$ $\forall \mathbf{x} \in \Omega$ (see Appendix C). Let $\phi \in \mathcal{D}(\Omega)$ with compact support, i.e. $\operatorname{supp}(\phi) \subset\subset \Omega$. Then define $\phi_i = \Theta_i\phi$ with $\phi_i \in \mathcal{D}(\Omega_i)$, $\operatorname{supp}(\phi_i) \subset\subset \Omega_i$ $\forall i \in \mathbb{N}$ such that $\phi(\mathbf{x}) = \sum_{i=1}^{\infty}\phi_i(\mathbf{x}) = \sum_{i=1}^{\infty}\Theta_i(\mathbf{x})\phi(\mathbf{x})$ $\forall \mathbf{x} \in \Omega$. But $\operatorname{supp}(\phi)$ is *compact* in Ω. Hence, $\operatorname{supp}(\phi)$ will intersect a *finite number* of the (compact) supports of Θ_i such that the series $\sum_i \Theta_i\phi$ will contain only a *finite number* of terms, i.e. $\phi = \sum_{i=1}^{\infty} \phi_i = \sum_{k=1}^{N} \phi_{i_k} = \sum_{k=1}^{N} \Theta_{i_k}\phi$. Since $\Theta_i\phi = \phi_i \in \mathcal{D}(\Omega_i)$, $\langle T_i, \phi_i\rangle$ is well defined $\forall i \in \mathbb{N}$.

But $\phi_i \in \mathcal{D}(\Omega_i) \Longrightarrow \phi_i \in \mathcal{D}(\Omega) \Longrightarrow \langle T, \phi_i\rangle$ is also well defined. Hence, $\forall\phi \in \mathcal{D}(\Omega_i)$, we set $\langle T, \phi_i\rangle = \langle T_i, \phi_i\rangle = \langle T_i, \Theta_i\phi\rangle$ with $\phi \in \mathcal{D}(\Omega)$. Then we define

$$\langle T, \phi\rangle = \Big\langle T, \sum_i \Theta_i\phi\Big\rangle = \sum_i \langle T, \Theta_i\phi\rangle = \sum_i \langle T_i, \Theta_i\phi\rangle \quad \forall\phi \in \mathcal{D}(\Omega), \tag{5.4.2}$$

which uniquely defines $T \in \mathcal{D}'(\Omega)$ in terms of $T_i \in \mathcal{D}'(\Omega_i)$. □

Existence of $T \in \mathcal{D}'(\Omega)$ In the proof given above, we assumed the existence of $T \in \mathcal{D}'(\Omega)$. Now, we establish its existence. For this we define T by the right-hand side of (5.4.2): $\langle T, \phi\rangle = \sum_i \langle T_i, \Theta_i\phi\rangle$ $\forall\phi \in \mathcal{D}(\Omega)$. Then T is a linear functional on $\mathcal{D}(\Omega)$ by virtue of the linearity of $T_i \in \mathcal{D}'(\Omega_i)$ $\forall i \in \mathbb{N}$. Now, for the continuity of T on $\mathcal{D}(\Omega)$, by Proposition 1.3.1, it is sufficient to show that $\forall$ compact $K \subset\subset \Omega$, T is continuous in $\mathcal{D}_K(\Omega)$, i.e. $\forall\phi \in \mathcal{D}(\Omega)$ with $\operatorname{supp}(\phi) \subset K$, the series $\sum_{i=1}^{\infty} \langle T_i, \Theta_i\phi\rangle$ consists only of the finite number of terms for which the support of Θ_i intersects K, i.e. $\forall\phi \in \mathcal{D}_K(\Omega)$, $\langle T, \phi\rangle = \sum_{k=1}^{N_0} \langle T_{l_k}, \Theta_{l_k}\phi\rangle$, where each $T_{l_k} \in \mathcal{D}'(\Omega_{l_k})$ is *continuous*. Hence, it is easily shown that $T \in \mathcal{D}'_K(\Omega)$ $\forall K \subset\subset \Omega$. Therefore, T is continuous and belongs to $\mathcal{D}'(\Omega)$.

$T\downarrow_{\Omega_j} = T_j$ To complete the proof, we are to show that $T\downarrow_{\Omega_j} = T_j$. Let $\phi \in \mathcal{D}(\Omega)$ with $\operatorname{supp}(\phi) \subset\subset \Omega_j$, i.e. $\phi \in \mathcal{D}(\Omega_j)$, and $\phi_i = \Theta_i\phi$ with $i \neq j$. Then $\phi_i \in \mathcal{D}(\Omega_i)$ with $\operatorname{supp}(\phi_i) = \operatorname{supp}(\Theta_i) \cap \operatorname{supp}(\phi) \subset (\Omega_i \cap \Omega_j)$, for $\phi_i = \Theta_i\phi \neq 0$ with $\operatorname{supp}(\phi_i) \neq \emptyset$ and $\operatorname{supp}(\phi_i) \subset \operatorname{supp}(\phi_j)$. Hence, $\Omega_i \cap \Omega_j \neq \emptyset \Longrightarrow T_i = T_j$ in $\Omega_i \cap \Omega_j$ (by hypothesis) $\Longrightarrow \langle T_j, \Theta_i\phi\rangle = \langle T_i, \Theta_i\phi\rangle$ $\forall\phi \in \mathcal{D}(\Omega_j)$. Then, $\forall\phi = \sum_i \Theta_i\phi \in \mathcal{D}(\Omega_j)$ (finite number of terms in the summation $\sum$),

$$\langle T_j, \phi\rangle = \Big\langle T_j, \sum_i \Theta_i\phi\Big\rangle = \sum_i \langle T_j, \Theta_i\phi\rangle = \sum_i \langle T_i, \Theta_i\phi\rangle = \langle T, \phi\rangle$$

$\Longrightarrow T\downarrow_{\Omega_j} = T_j$.

Consequences of the unification principle

- A distribution $T = 0$ in a family of open sets $\{\Omega_i\}_{i \in I_0}$ with $I_0 \subset I$

$$\Longrightarrow \quad T = 0 \text{ on the their union } \bigcup_{i \in I_0} \Omega_i \text{ (see also (5.1.3)).} \tag{5.4.3}$$

- The union of all open sets Ω_i in which $T = 0$ is the *largest* open set in which $T = 0$.

5.5 Support of a distribution

Definition 5.5.1. The *support* of a distribution T on Ω is the *smallest closed subset* of Ω outside which $T = 0$ or, equivalently, the *support* of a distribution T on Ω is the *complement of the largest open set* in which $T = 0$.

$$\operatorname{supp}(T) = A \subset \Omega \quad \Longleftrightarrow \quad \langle T, \phi \rangle = 0 \ \forall \phi \in \mathcal{D}(\Omega) \text{ with } \operatorname{supp}(\phi) \subset A^{\complement} \tag{5.5.1}$$

($A^{\complement}$ = complement of A in Ω).

A point $\mathbf{x}_0 \in \operatorname{supp}(T) \Longleftrightarrow T \neq 0$ in an open set U containing $\mathbf{x}_0$ (i.e. in a neighbourhood of $\mathbf{x}_0$). (5.5.2)

$$\operatorname{supp}(T) \cap \operatorname{supp}(\phi) = \emptyset \quad \Longrightarrow \quad \langle T, \phi \rangle = 0. \tag{5.5.3}$$

Examples of supports of distributions

- f is a continuous function with $\operatorname{supp}(f) = K \subset \Omega$ and $T_f \in \mathcal{D}'(\Omega)$ is the distribution defined by f

$$\Longrightarrow \quad \operatorname{supp}(T_f) = \operatorname{supp}(f) = K. \tag{5.5.4}$$

- The support of Dirac distribution $\delta_{\mathbf{a}}$, $\langle \delta_{\mathbf{a}}, \phi \rangle = \phi(\mathbf{a}) \ \forall \phi \in \mathcal{D}(\mathbb{R}^n)$, is $\{\mathbf{a}\}$, $\mathbf{a}$ being the point at which the mass/force/charge etc. is concentrated:

$$\operatorname{supp}(\delta_{\mathbf{a}}) = \{\mathbf{a}\}, \tag{5.5.5}$$

since $\langle \delta_{\mathbf{a}}, \phi \rangle = 0 \ \forall \phi \in \mathcal{D}(\mathbb{R}^n)$ with $\operatorname{supp}(\phi) \subset (\mathbb{R}^n \setminus \{\mathbf{a}\})$ (i.e. $\forall \phi \in \mathcal{D}(\mathbb{R}^n \setminus \{\mathbf{a}\})$), $\mathbb{R}^n \setminus \{\mathbf{a}\}$ being the *largest* open set in which $\delta_{\mathbf{a}} = 0$.

- For Dirac distribution $\delta_{\mathbb{S}}$ with mass/charge/force concentrated on the hypersurface $\mathbb{S} \subset \mathbb{R}^n$ defined by the equation $x_n = 0$,

$$\langle \delta_{\mathbb{S}}, \phi \rangle = \int_{\mathbb{R}^{n-1}} \phi(x_1, x_2 \ldots, x_{n-1}, 0) dx_1, dx_2 \ldots dx_{n-1} \quad \forall \phi \in \mathcal{D}(\mathbb{R}^n),$$

the support of $\delta_{\mathbb{S}}$ is the surface $\mathbb{S} \subset \mathbb{R}^n$:

$$\operatorname{supp}(\delta_{\mathbb{S}}) = \mathbb{S}, \tag{5.5.6}$$

since $\langle \delta_{\mathbb{S}}, \phi \rangle = 0 \ \forall \phi \in \mathcal{D}(\mathbb{R}^n)$ with $\operatorname{supp}(\phi) \subset (\mathbb{R}^n \setminus \mathbb{S})$, $(\mathbb{R}^n \setminus \mathbb{S})$ being the largest open set in which $\delta_{\mathbb{S}} = 0$.

- The support of c.p.v.$(\frac{1}{x})$ is $]-\infty, \infty[\ = \mathbb{R}$, since c.p.v. $\frac{1}{x}$ is the locally integrable function $\frac{1}{x}$ in the open set $\mathbb{R} \setminus \{0\}$, (i.e. $1/x \in L^1_{\mathrm{loc}}(\mathbb{R} \setminus \{0\})$) and the closure of this set is $\mathbb{R}$. (5.5.7)

5.6 Distributions with compact support

Definition 5.6.1. A distribution $T \in \mathcal{D}'(\mathbb{R}^n)$ is said to have *compact support* K in $\mathbb{R}^n$ if and only if $K \subset\subset \mathbb{R}^n$ is a compact subset of $\mathbb{R}^n$ and $K = \text{supp}(T)$.

A distribution T has compact support $K \subset\subset \mathbb{R}^n$ if and only if

$$\langle T, \phi \rangle = 0 \quad \forall \phi \in \mathcal{D}(\mathbb{R}^n) \text{ with } \text{supp}(\phi) \cap K = \emptyset. \tag{5.6.1}$$

Definition of $\langle T, \phi \rangle$ for $\phi \in C^\infty(\mathbb{R}^n)$ with arbitrary support and $T \in \mathcal{D}'(\mathbb{R}^n)$ with compact support Let $T \in \mathcal{D}'(\mathbb{R}^n)$ be a distribution with compact support $K_0 \subset\subset \mathbb{R}^n$, and $\phi \in C^\infty(\mathbb{R}^n)$ be an infinitely differentiable function with *arbitrary support*, i.e. ϕ *may not* belong to $\mathcal{D}(\mathbb{R}^n)$. Let α be a function of $\mathcal{D}(\mathbb{R}^n)$ such that $\alpha(\mathbf{x}) = 1\ \forall \mathbf{x} \in U$ with $K_0 \subset U, U$ being an open set containing K_0 (i.e. U is a neighbourhood of K_0). Then $\alpha\phi \in \mathcal{D}(\mathbb{R}^n)$ with $\text{supp}(\alpha\phi) \subset\subset \mathbb{R}^n$ compact in $\mathbb{R}^n$ and $\langle T, \alpha\phi \rangle$ is well defined, since $T \in \mathcal{D}'(\mathbb{R}^n)$. Now we show that the value $\langle T, \alpha\phi \rangle$ is independent of the choice of the function α. Let $\alpha, \beta \in \mathcal{D}(\mathbb{R}^n)$ be any two functions such that $\alpha(\mathbf{x}) = \beta(\mathbf{x}) = 1\ \forall \mathbf{x} \in U$ with $K_0 \subset U$. Then $(\alpha - \beta)\phi = 0$ in $K_0 \Longrightarrow \text{supp}((\alpha - \beta)\phi)$ is contained in the *complement* $K_0^{\complement}$ of K_0

$$\begin{aligned}
&\Longrightarrow \quad \text{supp}(T) \cap \text{supp}((\alpha - \beta)\phi) = \emptyset \quad \Longrightarrow \quad \langle T, (\alpha - \beta)\phi \rangle = 0 \\
&\Longrightarrow \quad \langle T, \alpha\phi - \beta\phi \rangle = \langle T, \alpha\phi \rangle - \langle T, \beta\phi \rangle = 0 \\
&\Longrightarrow \quad \langle T, \alpha\phi \rangle = \langle T, \beta\phi \rangle \quad \forall \phi \in C^\infty(\mathbb{R}^n).
\end{aligned} \tag{5.6.2}$$

$\langle T, \alpha\phi \rangle$ *coincides with* $\langle T, \phi \rangle$ if $\phi \in C^\infty(\mathbb{R}^n)$ and $T \in \mathcal{D}'(\mathbb{R}^n)$ has compact support $K_0 \subset\subset \mathbb{R}^n$ and $\alpha \in \mathcal{D}(\mathbb{R}^n)$ with $\alpha(\mathbf{x}) = 1\ \forall \mathbf{x} \in U, K_0 \subset U$. Hence, we can set

$$\langle T, \phi \rangle = \langle T, \alpha\phi \rangle \quad \text{for all } \phi \in C^\infty(\mathbb{R}^n). \tag{5.6.3}$$

Thus, for a distribution $T \in \mathcal{D}'(\mathbb{R}^n)$ with compact support, $\langle T, \phi \rangle$ is well defined when $\phi \in C^\infty(\mathbb{R}^n)$ is an infinitely differentiable function with *arbitrary support* in $\mathbb{R}^n$.

Important properties of distribution T with compact support Let T be a distribution with compact support $K_0 \subset\subset \mathbb{R}^n$. Then the following properties hold:

- $\langle T, \phi \rangle$ is well defined $\forall \phi \in C^\infty(\mathbb{R}^n)$ with *arbitrary support*, and

$$\langle T, \phi \rangle = \langle T, \alpha\phi \rangle, \tag{5.6.4}$$

  where $\alpha \in \mathcal{D}(\mathbb{R}^n)$, $\alpha(\mathbf{x}) = 1$ in a neighbourhood U of K_0 with $K_0 \subset U$.

- $\langle T, \phi \rangle$ depends only on the values of ϕ in the neighbourhood U of $K_0 \subset U$. (5.6.5)

In particular, if $\phi(\mathbf{x}) = 1 \ \forall \mathbf{x} \in \mathbb{R}^n$, $\phi \in C^\infty(\mathbb{R}^n)$ with $\text{supp}(\phi) = \mathbb{R}^n$ (i.e. $\phi = 1 \notin \mathcal{D}(\mathbb{R}^n)$),

- $\langle T, 1\rangle$ is well defined and usually called the *total mass/charge/force* etc. or the *integral* of T. (5.6.6)

5.7 Space $\mathcal{E}'(\mathbb{R}^n)$ of distributions with compact support

5.7.1 Space $\mathcal{E}(\mathbb{R}^n)$

Definition 5.7.1. $\mathcal{E}(\mathbb{R}^n) \equiv C^\infty(\mathbb{R}^n)$ is the linear space of infinitely differentiable functions having arbitrary support in $\mathbb{R}^n$, equipped with the following notion of *convergence*: a sequence $(\phi_m)_{m=1}^\infty$ converges to 0 in $\mathcal{E}(\mathbb{R}^n)$ if and only if, $\forall\alpha$ with $|\alpha| \in \mathbb{N}_0$,

$$\partial^\alpha \phi_m \to 0 \textit{ uniformly} \text{ on every } \textit{compact} \text{ set as } m \to \infty. \tag{5.7.1}$$

$$\mathcal{D}(\mathbb{R}^n) \subset \mathcal{E}(\mathbb{R}^n),\ \phi_m \to 0 \text{ in } \mathcal{D}(\mathbb{R}^n) \quad \Longrightarrow \quad \phi_m \to 0 \text{ in } \mathcal{E}(\mathbb{R}^n). \tag{5.7.2}$$

But $\phi_m \to 0$ in $\mathcal{E}(\mathbb{R}^n)$ does *not* imply $\phi_m \to 0$ in $\mathcal{D}(\mathbb{R}^n)$.

Definition 5.7.2. A linear functional $L : \mathcal{E}(\mathbb{R}^n) \longrightarrow \mathbb{R}$ on $\mathcal{E}(\mathbb{R}^n)$ is *continuous* on $\mathcal{E}(\mathbb{R}^n)$ if and only if

$$\phi_m \to 0 \text{ in } \mathcal{E}(\mathbb{R}^n) \quad \Longrightarrow \quad L(\phi_m) \to 0 \text{ in } \mathbb{R} \text{ as } m \to \infty. \tag{5.7.3}$$

- A distribution $T \in \mathcal{D}'(\mathbb{R}^n)$ with compact support $K \subset\subset \mathbb{R}^n$ is a continuous, linear functional L on $\mathcal{E}(\mathbb{R}^n)$.

In fact,

$$L(\phi) = \langle T, \phi\rangle = \langle T, \alpha\phi\rangle \quad \forall\phi \in \mathcal{E}(\mathbb{R}^n), \tag{5.7.4}$$

with $\alpha \in \mathcal{D}(\mathbb{R}^n)$, $\alpha(\mathbf{x}) = 1 \ \forall\mathbf{x} \in U$, and $\text{supp}(T) \subset U$ defines a *continuous*, *linear* functional on $\mathcal{E}(\mathbb{R}^n)$.

- Conversely, a *continuous*, *linear* functional L on $\mathcal{E}(\mathbb{R}^n)$ defines a distribution $T \in \mathcal{D}'(\mathbb{R}^n)$ with compact support. (5.7.5)

Indeed, $\mathcal{D}(\mathbb{R}^n) \subset \mathcal{E}(\mathbb{R}^n)$, $\phi_m \to 0$ in $\mathcal{D}(\mathbb{R}^n) \Longrightarrow \phi_m \to 0$ in $\mathcal{E}(\mathbb{R}^n)$.

Hence, L is a linear functional on $\mathcal{E}(\mathbb{R}^n) \Longrightarrow L$ is a linear functional on $\mathcal{D}(\mathbb{R}^n)$ and $\phi_m \to 0$ in $\mathcal{D}(\mathbb{R}^n) \Longrightarrow L(\phi_m) \to 0$ in $\mathbb{R}$ as $m \to \infty$

$\Longrightarrow L$ is a distribution in $\mathcal{D}'(\mathbb{R}^n)$ with $L(\phi) = \langle T, \phi\rangle \ \forall\phi \in \mathcal{D}(\mathbb{R}^n)$.

• $\text{supp}(T)$ is a *compact* subset of $\mathbb{R}^n$.

Suppose that the contrary holds, i.e. $\text{supp}(T)$ is unbounded in $\mathbb{R}^n$. Then it is possible to choose a sequence (ϕ_m) in $\mathcal{D}(\mathbb{R}^n)$ such that $\phi_m = 0$ in $\{\mathbf{x} : \|\mathbf{x}\| < m\}$ $\forall m \in \mathbb{N}$, i.e. $\exists$ a sequence (ϕ_m) with $\phi_m \in \mathcal{D}(\mathbb{R}^n)$ and $\text{supp}(\phi_m) \subset\subset \{\mathbf{x} : \|\mathbf{x}\| \geq m\}$ such that $\text{supp}(T) \cap \text{supp}(\phi_m) \neq \emptyset$ (since $\text{supp}(T) \cap \text{supp}(\phi_m) = \emptyset \Longrightarrow \langle T, \phi_m\rangle = 0$) and $\langle T, \phi_m\rangle = 1$ $\forall m \in \mathbb{N}$. Consequently, $\partial^{\boldsymbol{\alpha}}\phi_m \to 0$ *uniformly* on every compact set as $m \to \infty$, $\forall |\boldsymbol{\alpha}| \in \mathbb{N}_0$. Thus, $\phi_m \to 0$ in $\mathcal{E}(\mathbb{R}^n)$ as $m \to \infty \Longrightarrow L(\phi_m) = \langle T, \phi_m\rangle \to 0$ as $m \to \infty$, since L is continuous on $\mathcal{E}(\mathbb{R}^n)$ by hypothesis. But $L(\phi_m) = \langle T, \phi_m\rangle = 1$ $\forall m \in \mathbb{N} \Longrightarrow L(\phi_m) \to 1$ as $m \to \infty$. Thus, we meet with a contradiction and our assumption that $\text{supp}(T)$ is not bounded is wrong. Hence, $\text{supp}(T)$ is a compact subset of $\mathbb{R}^n$.

Now, we state this result as:

Theorem 5.7.1. *Every distribution $T \in \mathcal{D}'(\mathbb{R}^n)$ with compact support can be extended to a unique, continuous, linear functional L on $\mathcal{E}(\mathbb{R}^n)$, i.e.*

$$L(\phi) = \langle T, \phi\rangle \quad \forall \phi \in \mathcal{D}(\mathbb{R}^n) \subset \mathcal{E}(\mathbb{R}^n). \tag{5.7.6}$$

Consequently, L can be *identified* with $T \in \mathcal{D}'(\mathbb{R}^n)$ having compact support.

5.7.2 Space $\mathcal{E}'(\mathbb{R}^n)$

Definition 5.7.3. The distributions $T \in \mathcal{D}'(\mathbb{R}^n)$ with compact support form a vector space denoted by $\mathcal{E}'(\mathbb{R}^n)$, which is the *dual space* of $\mathcal{E}(\mathbb{R}^n) \equiv C^\infty(\mathbb{R}^n)$ by virtue of Theorem 5.7.1.

Now we collect the important properties of $\mathcal{E}'(\mathbb{R}^n)$:

Property 1 $T \in \mathcal{E}'(\mathbb{R}^n)$ if and only if T is a distribution with compact support in $\mathbb{R}^n$, i.e. $T \in \mathcal{D}'(\mathbb{R}^n)$ and $\text{supp}(T) \subset\subset \mathbb{R}^n$.

$$\mathcal{E}'(\mathbb{R}^n) \subset \mathcal{D}'(\mathbb{R}^n). \tag{5.7.7}$$

Property 2 $T \in \mathcal{E}'(\mathbb{R}^n) \Longrightarrow T(\phi)$ is well defined for every $\phi \in C^\infty(\mathbb{R}^n) \equiv \mathcal{E}(\mathbb{R}^n)$ with *arbitrary support*, and is defined by (5.6.4). (5.7.8)

In particular, if $\phi(\mathbf{x}) = 1$ $\forall \mathbf{x} \in \mathbb{R}^n$, $\phi \in C^\infty(\mathbb{R}^n) \equiv \mathcal{E}(\mathbb{R}^n)$ with $\text{supp}(\phi) = \mathbb{R}^n$ (i.e. $\phi = 1 \notin \mathcal{D}(\mathbb{R}^n)$). Hence, $\forall T \in \mathcal{E}'(\mathbb{R}^n)$, $\langle T, 1\rangle$ is well defined and usually called the *integral of T* or total mass/charge/force etc. (see (5.6.6) also). (5.7.9)

For a continuous function $f \in C_0(\mathbb{R}^n)$ with *compact support* in $\mathbb{R}^n$, $T = T_f \in \mathcal{E}'(\mathbb{R}^n)$ (since $T_f \in \mathcal{D}'(\mathbb{R}^n)$ and $\text{supp}(T_f) = \text{supp}(f) \subset\subset \mathbb{R}^n$),

$$\langle T_f, 1\rangle = \langle f, 1\rangle = \int_{\mathbb{R}^n} f(\mathbf{x})d\mathbf{x} = \int_{\text{supp}(f)} f(\mathbf{x})d\mathbf{x} \tag{5.7.10}$$

denotes the *total mass/force/charge etc.* corresponding to the volume density distribution of mass/force/charge etc. defined by $f \in C_0(\mathbb{R}^n)$.

For $T = \delta_{\mathbf{a}} \in \mathcal{E}'(\mathbb{R}^n)$ with compact support $\mathrm{supp}(\delta_{\mathbf{a}}) = \{\mathbf{a}\}$ (the Dirac distribution $\delta_{\mathbf{a}}$ with mass/force/charge etc. concentrated at $\{\mathbf{a}\}$), $\langle \delta_{\mathbf{a}}, 1\rangle = +1$ denotes the *total mass/charge/force* etc. (5.7.11)

$T \in \mathcal{E}'(\mathbb{R}^n)$, $\langle T, r^2\rangle$ with $r^2 = \sum_{i=1}^n x_i^2$ denotes the *total moment of inertia*: (5.7.12)

$$\langle T_f, r^2\rangle = \langle f, r^2\rangle = \int_{\mathbb{R}^n} f(\mathbf{x}) r^2 d\mathbf{x} \quad \text{for } f \in C_0(\mathbb{R}^n) \tag{5.7.13}$$

is the total moment of inertia. For $T = \delta_{\mathbf{a}} \in \mathcal{E}'(\mathbb{R}^n)$ with $\mathrm{supp}(\delta_{\mathbf{a}}) = \{\mathbf{a}\}$,

$\langle \delta_{\mathbf{a}}, r^2\rangle = \|\mathbf{a}\|^2 = \sum_{i=1}^n a_i^2$ is the total moment of inertia. (5.7.14)

For a continuous function $f \in C_0(\mathbb{R}^3)$ with compact support, $T = T_f \in \mathcal{E}'(\mathbb{R}^3)$ with $\mathrm{supp}(T_f) = \mathrm{supp}(f) \subset\subset \mathbb{R}^3$, $\langle T_f, \frac{1}{\|\mathbf{x}-\mathbf{b}\|}\rangle$ with $\frac{1}{\|\mathbf{x}-\mathbf{b}\|} \in C^\infty(\mathbb{R}^3 \setminus \{\mathbf{b}\})$ defines *Newtonian potential* at $\mathbf{b} \in \mathbb{R}^3$ with

$$\left\langle T_f, \frac{1}{\|\mathbf{x}-\mathbf{b}\|}\right\rangle = \int_{\mathbb{R}^3} \frac{f(\mathbf{x})}{[(x_1-b_1)^2 + (x_2-b_2)^2 + (x_3-b_3)^2]^{1/2}} d\mathbf{x}, \tag{5.7.15}$$

where $\|\mathbf{x}-\mathbf{b}\| = [(x_1-b_1)^2 + (x_2-b_2)^2 + (x_3-b_3)^2]^{1/2}$.

For $T = \delta_{\mathbf{a}} \in \mathcal{E}'(\mathbb{R}^3)$ with $\mathrm{supp}(\delta_{\mathbf{a}}) = \{\mathbf{a}\}$ and $\mathbf{b} \neq \mathbf{a}$,

$$\left\langle \delta_{\mathbf{a}}, \frac{1}{\|\mathbf{x}-\mathbf{b}\|}\right\rangle = \frac{1}{\|\mathbf{a}-\mathbf{b}\|} = \frac{1}{[\sum_{i=1}^3 (a_i-b_i)^2]^{1/2}}. \tag{5.7.16}$$

Property 3 Every distribution $T \in \mathcal{E}'(\mathbb{R}^n)$ is of *finite order* $m_0 \in \mathbb{N}_0$.

Property 4 *Local structures of distributions with compact support* [8, p. 91].

$$T \in \mathcal{E}'(\mathbb{R}^n) \quad \Longrightarrow \quad T = \sum_{|\boldsymbol{\alpha}| \le m_0} \partial^{\boldsymbol{\alpha}} f_{\boldsymbol{\alpha}} \tag{5.7.17}$$

(the representation being a *non-unique* one), where $f_{\boldsymbol{\alpha}} \in C_0(\mathbb{R}^n)$ with $|\boldsymbol{\alpha}| \le m_0$ with $\mathrm{supp}(f_{\boldsymbol{\alpha}}) \subset U$, $\mathrm{supp}(T) \subset U$, U being an arbitrary neighbourhood of $\mathrm{supp}(T)$, m_0 being the order of T.

Remark 5.7.1. $\forall T \in \mathcal{D}'(\mathbb{R}^n)$, $\exists f_{\boldsymbol{\alpha}} \in C_0(\mathbb{R}^n)$ such that $T = \sum_{\boldsymbol{\alpha}} \partial^{\boldsymbol{\alpha}} f_{\boldsymbol{\alpha}}$ in $\mathcal{D}'(\mathbb{R}^n)$ in the sense that $\forall K \subset\subset \mathbb{R}^n$, $\forall \phi \in \mathcal{D}(\mathbb{R}^n)$ with $\mathrm{supp}(\phi) \subset K$ (i.e. $\forall \phi \in \mathcal{D}_K(\mathbb{R}^n)$), $\exists m_0 = m_0(K) \in \mathbb{N}$ such that

$$\langle T, \phi\rangle = \left\langle \sum_{|\boldsymbol{\alpha}| \le m_0} \partial^{\boldsymbol{\alpha}} f_{\boldsymbol{\alpha}}, \phi\right\rangle = \sum_{|\boldsymbol{\alpha}| \le m_0} (-1)^{|\boldsymbol{\alpha}|} \int_{\mathbb{R}^n} f_{\boldsymbol{\alpha}} \partial^{\boldsymbol{\alpha}} \phi d\mathbf{x}, \tag{5.7.18}$$

the proof of which involves a partition of unity and convolution to be introduced in Chapter 6.

We then have the following interesting result.

Example 5.7.1. Show that Dirac distribution $\delta \in \mathcal{E}'(\mathbb{R})$ on $\mathbb{R}$ can *not* be equal to the derivative of certain order $m \in \mathbb{N}$ of a *single* continuous function $f \in C_0(\mathbb{R})$ with compact support in $\mathbb{R}$, i.e. we cannot write $\frac{d^m f}{dx^m} = \delta \in \mathcal{D}'(\mathbb{R})$ for a single $f \in C_0(\mathbb{R})$, $m \in \mathbb{N}$.

Proof. Suppose that the contrary holds, i.e. $\exists f \in C_0(\mathbb{R})$ with compact support in $\mathbb{R}$ such that $\frac{d^m f}{dx^m} = \delta$ in $\mathcal{D}'(\mathbb{R})$ for some $m \in \mathbb{N}$. Since $\delta\downarrow_{\mathbb{R}\setminus\{0\}} = 0 \in \mathcal{D}'(\mathbb{R}\setminus\{0\})$, f must satisfy $\frac{d^m f}{dx^m} = 0$ in $\mathbb{R}\setminus\{0\}$, whose general solution is a polynomial of degree $\leq m-1$ (see Proposition 2.7.1 and Remark 2.7.1): $f(x) = \alpha_0 x^{m-1} + \alpha_1 x^{m-2} + \cdots + \alpha_{m-1}$. But $f \in C_0(\mathbb{R}) \Longrightarrow f$ must have compact support

$$\Longrightarrow \quad f = 0 \text{ in } \mathbb{R}\setminus\{0\}, \text{ i.e. } \alpha_0 = \alpha_1 = \cdots = \alpha_{m-1} = 0, \tag{5.7.19}$$

since every *non-null* polynomial has support $\mathbb{R}$. Hence, $f = 0$ on $\mathbb{R}\setminus\{0\}$. But f is continuous on $\mathbb{R} \Longrightarrow f = 0$ on $\mathbb{R} \Longrightarrow \frac{d^m f}{dx^m} = 0$ in $\mathcal{D}'(\mathbb{R})$, which contradicts the hypothesis that $\frac{d^m f}{dx^m} = \delta \neq 0$ in $\mathcal{D}'(\mathbb{R})$. Thus, our assumption is wrong and the result follows. □

Independent proof of (5.7.19). $\frac{f(x)}{x^{m-1}} = \alpha_0 + \frac{\alpha_1}{x} + \cdots + \frac{\alpha_{m-1}}{x^{m-1}}$ in $\mathbb{R}\setminus\{0\}$
$\Longrightarrow \lim_{|x|\to\infty} \frac{f(x)}{x^{m-1}} = 0$, since f has compact support in $\mathbb{R} \Longrightarrow$ for sufficiently large $|x|$, $f(x) = 0$.

But $\lim_{|x|\to\infty}(\alpha_0 + \frac{\alpha_1}{x} + \cdots + \frac{\alpha_{m-1}}{x^{m-1}}) = \alpha_0 = \lim_{|x|\to\infty} \frac{f(x)}{x^{m-1}} = 0$. Thus, $\alpha_0 = 0$.

Hence, $f(x) = \alpha_1 x^{m-2} + \cdots + \alpha_{m-1}$ in $\mathbb{R}\setminus\{0\}$.

Similarly, we can show sequentially that $\alpha_1 = \alpha_2 = \cdots = \alpha_{m-1} = 0$. □

Property 5 $\forall T \in \mathcal{E}'(\mathbb{R}^n) \subset \mathcal{D}'(\mathbb{R}^n)$, $\forall$ multi-index $\boldsymbol{\alpha}$, $\partial^{\boldsymbol{\alpha}} T \in \mathcal{E}'(\mathbb{R}^n)$ is defined by:

$$\langle \partial^{\boldsymbol{\alpha}} T, \phi\rangle = (-1)^{|\boldsymbol{\alpha}|}\langle T, \partial^{\boldsymbol{\alpha}}\phi\rangle \quad \forall \phi \in C^\infty(\mathbb{R}^n) = \mathcal{E}(\mathbb{R}^n). \tag{5.7.20}$$

Example 5.7.2. For $T = \delta_a$ and $\phi(x) = 1\ \forall x \in \mathbb{R}$,

$$\langle \delta_a', 1\rangle = -\langle \delta_a, \frac{d1}{dx}\rangle = -\langle \delta_a, 0\rangle = 0. \tag{5.7.21}$$

Example 5.7.3. For $T = \partial^{\boldsymbol{\alpha}} S \in \mathcal{E}'(\mathbb{R}^n)$, $\langle T, 1\rangle = 0$. In fact,

$$\langle T, 1\rangle = \langle \partial^{\boldsymbol{\alpha}} S, 1\rangle = (-1)^{|\boldsymbol{\alpha}|}\langle S, \partial^{\boldsymbol{\alpha}} 1\rangle = (-1)^{|\boldsymbol{\alpha}|}\langle S, 0\rangle = 0. \tag{5.7.22}$$

Property 6 Let $T \in \mathcal{E}'(\mathbb{R}^n) \subset \mathcal{D}'(\mathbb{R}^n)$ and $f \in C^\infty(\mathbb{R}^n)$ such that $f(\mathbf{x}) = 0$ $\forall \mathbf{x} \in \operatorname{supp}(T)$. Then, fT does *not* vanish in $\mathcal{D}'(\mathbb{R}^n)$ *in general.*

Example 5.7.4. Let $T = \delta' \in \mathcal{E}'(\mathbb{R}) \subset \mathcal{D}'(\mathbb{R})$ with $\operatorname{supp}(\delta') = \{0\}$ and $f(x) = x$ $\forall x \in \mathbb{R}$. Then $f \in C^\infty(\mathbb{R})$ with $f(x) = 0$ for $x \in \operatorname{supp}(\delta') = \{0\}$. But $\forall \phi \in \mathcal{D}(\mathbb{R})$,

$$\begin{aligned}\langle f\delta', \phi\rangle &= \langle x\delta', \phi\rangle = \langle \delta', x\phi\rangle = -\langle \delta, (x\phi)'\rangle \\ &= -[x\phi'(x) + \phi(x)]_{x=0} = -\phi(0) = \langle -\delta, \phi\rangle \\ &\implies \quad x\delta' = -\delta \neq 0 \text{ in } \mathcal{D}'(\mathbb{R}).\end{aligned}$$

Example 5.7.5. Show that the following hold for $T \in \mathcal{E}'(\mathbb{R}) \subset \mathcal{D}'(\mathbb{R})$, $\phi \in \mathcal{D}(\mathbb{R})$:

1. $\phi T = 0$ in $\mathcal{D}'(\mathbb{R}) \implies \langle T, \phi\rangle = 0$;
2. $\langle T, \phi\rangle = 0$ does *not* imply $\phi T = 0$ in $\mathcal{D}'(\mathbb{R})$.

Proof.

1. Let $\phi \in \mathcal{D}(\mathbb{R})$. Choose $\psi \in \mathcal{D}(\mathbb{R})$ such that $\psi(x) = 1$ for $x \in \operatorname{supp}(\phi)$. Then $\psi\phi(x) = \phi(x)$ $\forall x \in \mathbb{R}$, and $0 = \langle \phi T, \psi\rangle = \langle T, \phi\psi\rangle = \langle T, \phi\rangle$ $\forall \phi \in \mathcal{D}(\mathbb{R})$.
2. For $n = 1$, $T = \delta' \in \mathcal{E}'(\mathbb{R})$, choose $\phi \in \mathcal{D}(\mathbb{R})$ with $\phi = 1$ in U, $\{0\} = \operatorname{supp}(\delta') \subset U$. Then $\phi(0) = 1$ and $\phi'(0) = 0$. Hence, $\langle T, \phi\rangle = \langle \delta', \phi\rangle = -\langle \delta, \phi'\rangle = -\phi'(0) = 0$. Now we show that $\phi\delta' \neq 0$ in $\mathcal{D}'(\mathbb{R})$. For this, choose $\psi \in \mathcal{D}(\mathbb{R})$ with $\psi'(0) \neq 0$:

$$\begin{aligned}\langle \phi T, \psi\rangle &= \langle \phi\delta', \psi\rangle = \langle \delta', \phi\psi\rangle = -\langle \delta, (\phi\psi)'\rangle \\ &= -[\phi'(x)\psi(x) + \phi(x)\psi'(x)]_{x=0} \\ &= -0 \cdot \psi(0) - \phi(0)\psi'(0) = -\psi'(0) \neq 0 \\ &\implies \quad \phi\delta' \neq 0 \text{ in } \mathcal{D}'(\mathbb{R}).\end{aligned}$$

□

Property 7 For $T \in \mathcal{E}'(\mathbb{R}^n)$ with $\operatorname{supp}(T) = K \subset\subset \mathbb{R}^n$, $T(\phi) \neq 0$ *in general* for functions $\phi \in \mathcal{D}(\mathbb{R}^n)$ with $\phi(\mathbf{x}) = 0$ $\forall \mathbf{x} \in K$.

Example 5.7.6. Let $T = \delta' \in \mathcal{E}'(\mathbb{R})$ with $K = \operatorname{supp}(\delta') = \{0\}$. Let $\psi \in \mathcal{D}(\mathbb{R})$ with $\psi(0) \neq 0$. Define $\phi = x\psi \in \mathcal{D}(\mathbb{R})$ with $\phi(0) = 0$, i.e. $\phi(x) = 0$ on $\operatorname{supp}(\delta')$.

But $\langle \delta', \phi\rangle = \langle \delta', x\psi\rangle = -\langle \delta, (x\psi)'\rangle = -[x\psi'(x) + \psi(x)]_{x=0} = -\psi(0) \neq 0$.

But we have:

Proposition 5.7.1. *Let $T \in \mathcal{E}'(\mathbb{R}^n)$ with* $\operatorname{supp}(T) = K \subset\subset \mathbb{R}^n$, *$m_0$ being the order of T (by Property 3). If $\phi \in \mathcal{D}(\mathbb{R}^n)$ such that $\phi(\mathbf{x}) = \partial^{\boldsymbol{\alpha}}\phi(\mathbf{x}) = 0$ $\forall \mathbf{x} \in K$, $\forall |\boldsymbol{\alpha}| \leq m_0$, then $T(\phi) = 0$.*

Example 5.7.7. Let $T = \partial^{\boldsymbol{\alpha}}\delta \in \mathcal{E}'(\mathbb{R}^n) \subset \mathcal{D}'(\mathbb{R}^n)$ be a distribution of order $|\boldsymbol{\alpha}| = m_0$ with $\operatorname{supp}(\partial^{\boldsymbol{\alpha}}\delta) = \{\mathbf{0}\}$. Then, $\forall\phi \in \mathcal{D}(\mathbb{R}^n)$ with $\phi(\mathbf{0}) = \partial^{\boldsymbol{\alpha}}\phi(\mathbf{0}) = 0\ \forall|\boldsymbol{\alpha}| \le m_0$, we have

$$\langle \partial^{\boldsymbol{\alpha}}\delta, \phi\rangle = (-1)^{|\boldsymbol{\alpha}|}\langle \delta, \partial^{\boldsymbol{\alpha}}\phi\rangle = (-1)^{|\boldsymbol{\alpha}|}\langle \delta, \partial^{\boldsymbol{\alpha}}\phi\rangle = (-1)^{|\boldsymbol{\alpha}|}\partial^{\boldsymbol{\alpha}}\phi(\mathbf{0}) = 0.$$

Property 8

Theorem 5.7.2. *A distribution $T \in \mathcal{E}'(\mathbb{R}^n)$ with $\operatorname{supp}(T) = \{\mathbf{0}\}$ is a finite, linear combination of Dirac distribution δ and its derivatives $\partial^{\boldsymbol{\alpha}}\delta$ (with mass/charge/force etc. concentrated at $\mathbf{0}$).*

Proof. Let $T \in \mathcal{E}'(\mathbb{R}^n)$ with $\operatorname{supp}(T) = \{\mathbf{0}\}$. Then, by Property 3, T is of finite order $m_0 \in \mathbb{N}_0$. $\forall\phi \in \mathcal{D}(\mathbb{R}^n)$. Using Maclaurin's Theorem, we have

$$\psi(\mathbf{x}) = \phi(\mathbf{x}) - \phi(\mathbf{0}) - \sum_k x_k \frac{\partial\phi}{\partial x_k}(\mathbf{0}) - \cdots - \sum_{|\boldsymbol{\alpha}|=m_0} \frac{1}{\boldsymbol{\alpha}!}\mathbf{x}^{\boldsymbol{\alpha}}\partial^{\boldsymbol{\alpha}}\phi(\mathbf{0})$$

$$\Longrightarrow \quad \psi(\mathbf{0}) = \partial^{\boldsymbol{\beta}}\psi(\mathbf{0}) = 0 \quad \forall|\boldsymbol{\beta}| \le m_0$$

$$\Longrightarrow \quad T(\psi) = 0 \quad \text{by Proposition 5.7.1}$$

$$\Longrightarrow \quad T(\psi) = T(\phi) - T\left[\phi(\mathbf{0}) + \sum_{k=1}^{n} x_k \frac{\partial\phi}{\partial x_k}(\mathbf{0}) + \cdots + \sum_{|\boldsymbol{\alpha}|=m_0} \frac{1}{\boldsymbol{\alpha}!}\mathbf{x}^{\boldsymbol{\alpha}}\partial^{\boldsymbol{\alpha}}\phi(\mathbf{0})\right] = 0$$

$$\Longrightarrow \quad T(\phi) = \phi(\mathbf{0})\langle T, 1\rangle + \sum_{k=1}^{n} \frac{\partial\phi}{\partial x_k}(\mathbf{0})\langle T, x_k\rangle + \cdots + \sum_{|\boldsymbol{\alpha}|=m_0} \partial^{\boldsymbol{\alpha}}\phi(\mathbf{0})\left\langle T, \frac{\mathbf{x}^{\boldsymbol{\alpha}}}{\boldsymbol{\alpha}!}\right\rangle,$$

where $1, x_k, \dots, \frac{\mathbf{x}^{\boldsymbol{\alpha}}}{\boldsymbol{\alpha}!} \in C^\infty(\mathbb{R}^n)$ and hence $\langle T, 1\rangle$, $\langle T, x_k\rangle, \dots, \langle T, \frac{\mathbf{x}^{\boldsymbol{\alpha}}}{\boldsymbol{\alpha}!}\rangle$ are well defined for $T \in \mathcal{E}'(\mathbb{R}^n)$.

(Then, $b_0 = \langle T, 1\rangle$, $b_\alpha = \langle T, \frac{\mathbf{x}^\alpha}{\alpha!}\rangle \quad \forall|\alpha| \le m_0$)

$$\Longrightarrow \quad \forall\phi \in \mathcal{D}(\mathbb{R}^n), \quad T(\phi) = b_0\phi(\mathbf{0}) + \sum_{|\boldsymbol{\alpha}|=1} b_{\boldsymbol{\alpha}}\partial^{\boldsymbol{\alpha}}\phi(\mathbf{0}) + \cdots + \sum_{|\boldsymbol{\alpha}|=m_0} b_{\boldsymbol{\alpha}}\partial^{\boldsymbol{\alpha}}\phi(\mathbf{0})$$

$$= \sum_{|\boldsymbol{\alpha}|=0}^{m_0} b_{\boldsymbol{\alpha}}\partial^{\boldsymbol{\alpha}}\phi(\mathbf{0}) = \sum_{|\boldsymbol{\alpha}|=0}^{m_0} b_{\boldsymbol{\alpha}}\langle \delta, \partial^{\boldsymbol{\alpha}}\phi\rangle$$

$$= \sum_{|\boldsymbol{\alpha}|=0}^{m_0} (-1)^{|\boldsymbol{\alpha}|} b_\alpha\langle \partial^{\boldsymbol{\alpha}}\delta, \phi\rangle$$

$$\Longrightarrow \quad T = \sum_{|\boldsymbol{\alpha}|=0}^{m_0} a_{\boldsymbol{\alpha}}\partial^{\boldsymbol{\alpha}}\delta \text{ in } \mathcal{D}'(\mathbb{R}^n) \text{ with } a_{\boldsymbol{\alpha}} = (-1)^{|\boldsymbol{\alpha}|}b_{\boldsymbol{\alpha}}\ \forall|\boldsymbol{\alpha}| \le m_0. \tag{5.7.23}$$

□

Example 5.7.8. Show that $T = \sum_{|\boldsymbol{\alpha}|\le m_0} a_{\boldsymbol{\alpha}}\partial^{\boldsymbol{\alpha}}\delta = 0$ in $\mathcal{E}'(\mathbb{R}^n)$ with $a_{\boldsymbol{\alpha}} \in \mathbb{R}\ \forall|\boldsymbol{\alpha}| \le m_0$ if and only if $a_{\boldsymbol{\alpha}} = 0\ \forall|\boldsymbol{\alpha}| \le m_0$.

Proof. For multi-index $\boldsymbol{\beta}_0$ with $|\boldsymbol{\beta}_0| \le m_0$, let $\mathbf{x}^{\boldsymbol{\beta}_0} \in C^\infty(\mathbb{R}^n)$. Hence,

$$\begin{aligned}\langle T, \mathbf{x}^{\boldsymbol{\beta}_0}\rangle &= \Big\langle \sum_{|\boldsymbol{\alpha}|\le m_0} a_{\boldsymbol{\alpha}}\partial^{\boldsymbol{\alpha}}\delta, \mathbf{x}^{\boldsymbol{\beta}_0}\Big\rangle = \sum_{|\boldsymbol{\alpha}|\le m_0} a_{\boldsymbol{\alpha}}\langle\partial^{\boldsymbol{\alpha}}\delta, \mathbf{x}^{\boldsymbol{\beta}_0}\rangle \\ &= \sum_{|\boldsymbol{\alpha}|\le m_0} a_{\boldsymbol{\alpha}}(-1)^{|\boldsymbol{\alpha}|}\langle\delta, \partial^{\boldsymbol{\alpha}}(\mathbf{x}^{\boldsymbol{\beta}_0})\rangle = \sum_{|\boldsymbol{\alpha}|\le m_0} a_{\boldsymbol{\alpha}}(-1)^{|\boldsymbol{\alpha}|}[\partial^{\boldsymbol{\alpha}}(\mathbf{x}^{\boldsymbol{\beta}_0})](\mathbf{0}), \\ \langle T, \mathbf{x}^{\boldsymbol{\beta}_0}\rangle &= \langle 0, \mathbf{x}^{\boldsymbol{\beta}_0}\rangle = 0.\end{aligned}$$

But

$$[\partial^{\boldsymbol{\alpha}}(\mathbf{x}^{\boldsymbol{\beta}_0})](\mathbf{0}) = \begin{cases} 0 & \text{for } \boldsymbol{\alpha} \ne \boldsymbol{\beta}_0 \\ \boldsymbol{\beta}_0! & \text{for } \boldsymbol{\alpha} = \boldsymbol{\beta}_0 \end{cases}$$

$$\Longrightarrow \langle T, \mathbf{x}^{\boldsymbol{\beta}_0}\rangle = a_{\boldsymbol{\beta}_0}(-1)^{|\boldsymbol{\beta}_0|}\boldsymbol{\beta}_0! = 0 \Longleftrightarrow a_{\boldsymbol{\beta}_0} = 0\ \forall|\boldsymbol{\beta}_0| \le m_0. \qquad \square$$

Example 5.7.9. Find all the distributions $T \in \mathcal{E}'(\mathbb{R}^n)$ with $\operatorname{supp}(T) = \{\mathbf{0}\}$ which are invariant under the transformations $\mathbf{F}_i : (x_1, \dots, x_i, \dots, x_n) \longmapsto (x_1, \dots, -x_i, \dots, x_n)$ from $\mathbb{R}^n$ into $\mathbb{R}^n$, $1 \le i \le n$ (using (1.10.21)–(1.10.23)).

Solution. $\mathbf{F}_i : \mathbb{R}^n \longrightarrow \mathbb{R}^n$ is defined by the diagonal matrix A_i of order n: $A_i = \lceil 1, 1, \dots, -1, \dots, 1 \rfloor$ with $a_{ii} = -1$, $a_{jj} = +1$ for $j \ne i$, $a_{jk} = 0$ for $1 \le j \ne k \le n$, $|\det(A_i)| = 1$, $A_i^{-1} = A_i\ \forall i = 1, 2, \dots, n$, $A_i\mathbf{x} = \boldsymbol{\xi}$ with $\xi_j = x_j$ for $1 \le j \ne i \le n$, $\xi_i = -x_i$. But $T \in \mathcal{E}'(\mathbb{R}^n)$ with $\operatorname{supp}(T) = \{\mathbf{0}\} \Longrightarrow T = \sum_{|\boldsymbol{\alpha}|\le m_0} a_{\boldsymbol{\alpha}}\partial^{\boldsymbol{\alpha}}\delta$ by Theorem 5.7.1, m_0 being the order of $T \in \mathcal{E}'(\mathbb{R}^n)$. Now, for invariance (1.10.42), we are to show that $\mathbf{F}_i T = T = \sum_{|\boldsymbol{\alpha}|\le m_0} a_{\boldsymbol{\alpha}}\partial^{\boldsymbol{\alpha}}\delta$, $1 \le i \le n$, where $\mathbf{F}_i T = T(\mathbf{F}_i^{-1}) = T(A_i^{-1})$ (see (1.10.21)–(1.10.23)),

$$\begin{aligned}\langle \mathbf{F}_i T, \psi\rangle &= \langle T(A_i^{-1}\boldsymbol{\xi}), \psi(\boldsymbol{\xi})\rangle = \langle T(\mathbf{x}), \psi(A_i\mathbf{x})\rangle|\det(A_i)|, \ (|\det(A_i)| = 1) \\ &= \Big\langle \sum_{|\boldsymbol{\alpha}|\le m_0} a_{\boldsymbol{\alpha}}\partial^{\boldsymbol{\alpha}}\delta, \psi(A_i\mathbf{x})\Big\rangle = \sum_{|\boldsymbol{\alpha}|\le m_0} (-1)^{|\boldsymbol{\alpha}|}a_{\boldsymbol{\alpha}}\langle\delta, \partial^{\boldsymbol{\alpha}}(\psi(A_i\mathbf{x}))\rangle \\ &= \sum_{|\boldsymbol{\alpha}|\le m_0} (-1)^{|\boldsymbol{\alpha}|}a_{\boldsymbol{\alpha}}[\partial^{\boldsymbol{\alpha}}\psi(A_i\mathbf{x})](\mathbf{0}) \quad \forall\psi \in \mathcal{D}(\mathbb{R}^n).\end{aligned}$$

But

$$\partial^{\boldsymbol{\alpha}}(\psi(A_i\mathbf{x}))(\mathbf{0}) = (-1)^{\alpha_i}\partial^{\boldsymbol{\alpha}}\psi(\mathbf{0}) = (-1)^{\alpha_i}\langle\delta, \partial^{\boldsymbol{\alpha}}\psi\rangle = (-1)^{\alpha_i}(-1)^{|\boldsymbol{\alpha}|}\langle\partial^{\boldsymbol{\alpha}}\delta, \psi\rangle.$$

Hence,

$$\langle \mathbf{F}_i T, \psi \rangle = \sum_{|\boldsymbol{\alpha}| \leq m_0} a_{\boldsymbol{\alpha}} (-1)^{|\boldsymbol{\alpha}|} (-1)^{\alpha_i} (-1)^{|\boldsymbol{\alpha}|} \langle \partial^{\boldsymbol{\alpha}} \delta, \psi \rangle$$
$$= \sum_{|\boldsymbol{\alpha}| \leq m_0} a_{\boldsymbol{\alpha}} (-1)^{\alpha_i} \langle \partial^{\boldsymbol{\alpha}} \delta, \psi \rangle \quad \forall \psi \in \mathcal{D}(\mathbb{R}^n)$$

$\implies \mathbf{F}_i T = \sum_{|\boldsymbol{\alpha}| \leq m_0} a_{\boldsymbol{\alpha}} (-1)^{\alpha_i} \partial^{\boldsymbol{\alpha}} \delta$.
Since T is invariant under $\mathbf{F}_i$, $\sum_{|\boldsymbol{\alpha}| \leq m_0} a_{\boldsymbol{\alpha}} (-1)^{\alpha_i} \partial^{\boldsymbol{\alpha}} \delta = \sum_{|\boldsymbol{\alpha}| \leq m_0} a_{\boldsymbol{\alpha}} \partial^{\boldsymbol{\alpha}} \delta$
$\implies \sum_{|\boldsymbol{\alpha}| \leq m_0} [(-1)^{\alpha_i} - 1] a_{\boldsymbol{\alpha}} \partial^{\boldsymbol{\alpha}} \delta = 0$ in $\mathcal{E}'(\mathbb{R}^n)$, $1 \leq i \leq n$
$\implies [(-1)^{\alpha_i} - 1] a_{\boldsymbol{\alpha}} = 0 \; \forall |\boldsymbol{\alpha}| \leq m_0$ (from Example 5.7.9), $1 \leq i \leq n$.
$\implies \forall \boldsymbol{\alpha}$ with $a_{\boldsymbol{\alpha}} \neq 0$, $[(-1)^{\alpha_i} - 1] = 0$, $1 \leq i \leq n \implies \alpha_i$ is even $\forall i = 1, 2, \ldots, n$
$\implies T = \sum_{2|\boldsymbol{\beta}| \leq m_0} a_{2\beta_1, 2\beta_2, \ldots, 2\beta_n} \partial_1^{2\beta_1} \partial_2^{2\beta_2} \ldots \partial_n^{2\beta_n} \delta$. (5.7.24)
Hence, every distribution of the form (5.7.24) will be invariant under the given transformation $\mathbf{F}_i$ defined above.

Property 9 Convergence in $\mathcal{E}'(\mathbb{R}^n)$.

Convergence of sequences of distributions with compact support in $\mathcal{E}'(\mathbb{R}^n)$

Definition 5.7.4. A sequence (T_k) in $\mathcal{E}'(\mathbb{R}^n)$ is said to converge to $T \in \mathcal{E}'(\mathbb{R}^n)$ if and only if

$$\lim_{k \to \infty} \langle T_k, \phi \rangle = \langle T, \phi \rangle \quad \forall \phi \in \mathcal{E}(\mathbb{R}^n) = C^\infty(\mathbb{R}^n). \tag{5.7.25}$$

Convergence of series of distributions with compact support in $\mathcal{E}'(\mathbb{R}^n)$

Definition 5.7.5. Let $\sum_{k=1}^\infty a_k T_k$ with $a_k \in \mathbb{R}$ (resp. $\mathbb{C}$) and $T_k \in \mathcal{E}'(\mathbb{R}^n)$ $\forall k \in \mathbb{N}$. Then the series $\sum_{k=1}^\infty a_k T_k$ is said to converge in $\mathcal{E}'(\mathbb{R}^n)$ if and only if the sequence (S_N) of the partial sums of the first N terms of the series converges to $S \in \mathcal{E}'(\mathbb{R}^n)$, i.e. $S_N = \sum_{k=1}^N a_k T_k$ $\forall N \in \mathbb{N}$ and

$$\lim_{N \to \infty} \langle S_N, \phi \rangle = \langle S, \phi \rangle \quad \forall \phi \in \mathcal{E}(\mathbb{R}^n) \equiv C^\infty(\mathbb{R}^n); \tag{5.7.26}$$

$S \in \mathcal{E}'(\mathbb{R}^n)$ is called the *sum* of the series $\sum_{k=1}^\infty a_k T_k$, and we write

$$S = \sum_{k=1}^\infty a_k T_k. \tag{5.7.27}$$

Example 5.7.10. Find a sequence (T_n) of distributions $T_n \in \mathcal{E}'(\mathbb{R})$ with $\operatorname{supp}(T_n) = \{0\}$ $\forall n \in \mathbb{N}$ such that the sequence (S_n) defined by $\langle S_n, \phi\rangle = \langle T_n, \phi\rangle - \sum_{j=1}^{n} \phi(\frac{1}{j})$ $\forall \phi \in \mathcal{D}(\mathbb{R})$ converges in $\mathcal{D}'(\mathbb{R})$, using $\phi\left(\frac{1}{j}\right) = \phi(0) + \frac{1}{j}\phi'(0) + \frac{1}{j^2}\psi\left(\frac{1}{j}\right)$ by Proposition 1.2.1 for ϕ with $\operatorname{supp}(\phi) = [-A, A]$, $A > 0$.

Solution. By Proposition 1.2.1, $\forall \phi \in \mathcal{D}(\mathbb{R})$ with $\operatorname{supp}(\phi) = K \subset [-A, A]$, $A > 0$, we have $\phi(x) = \phi(0) + x\phi'(0) + \cdots + \frac{x^{m-1}}{(m-1)!}\phi^{(m-1)}(0) + x^m\psi(x)$, with $\psi \in C^0(\mathbb{R})$ and $\sup|\psi(x)| \le C \sup|\phi^{(m)}(x)|$. $\forall$ choices of $m \ge 2$, we will get a sequence (T_n).

Case $m = 2$: $\phi(\frac{1}{j}) = \phi(0) + \frac{1}{j}\phi'(0) + \frac{1}{j^2}\psi(\frac{1}{j})$, and

$$\begin{aligned}\langle S_n, \phi\rangle &= \langle T_n, \phi\rangle - \sum_{j=1}^{n}\left[\phi(0) + \frac{1}{j}\phi'(0) + \frac{1}{j^2}\psi\left(\frac{1}{j}\right)\right] \\ &= \langle T_n, \phi\rangle - n\phi(0) - \left(\sum_{j=1}^{n}\frac{1}{j}\right)\phi'(0) - \sum_{j=1}^{n}\frac{1}{j^2}\psi\left(\frac{1}{j}\right). \end{aligned} \tag{5.7.28}$$

But $\frac{1}{j^2}|\psi(\frac{1}{j})| \le \frac{1}{j^2}\sup|\psi(x)| \le \frac{C}{j^2}\sup|\phi''(x)| \le \frac{M}{j^2}$, (where M is independent of j and n). $|\sum_{j=1}^{n}\frac{1}{j^2}\psi(\frac{1}{j})| \le M\sum_{j=1}^{n}\frac{1}{j^2} \to$ a finite limit in $\mathbb{R}$ as $n \to \infty$.

Hence, if we set

$$\langle T_n, \phi\rangle = n\phi(0) + \left(\sum_{j=1}^{n}\frac{1}{j}\right)\phi'(0) \quad \forall n \in \mathbb{N}, \tag{5.7.29}$$

$|\langle S_n, \phi\rangle| \le M \sum\limits_{j=1}^{n} \frac{1}{j^2} \to$ a finite limit as $n \to \infty$ implies.

$\langle S_n, \phi\rangle$ converges in $\mathbb{R}$ as $n \to \infty$ $\forall \phi \in \mathcal{D}(\mathbb{R})$.

$\Longrightarrow$ (S_n) converges in $\mathcal{D}'(\mathbb{R})$ for the sequence (T_n) in $\mathcal{D}'(\mathbb{R})$ with

$$\begin{aligned}\langle T_n, \phi\rangle &= n\phi(0) + \left(\sum_{j=1}^{n}\frac{1}{j}\right)\phi'(0) = \langle n\delta, \phi\rangle + \left(\sum_{j=1}^{n}\frac{1}{j}\right)\langle \delta, \phi'\rangle \\ &= \left\langle n\delta - \left(\sum_{j=1}^{n}\frac{1}{j}\right)\delta', \phi\right\rangle \quad \forall \phi \in \mathcal{D}(\mathbb{R})\end{aligned}$$

$\Longrightarrow$ for $m = 2$, $T_n = n\delta - (\sum_{j=1}^{n}\frac{1}{j})\delta'$ in $\mathcal{D}'(\mathbb{R})$ with $\operatorname{supp}(T_n) = \{0\} \forall n \in \mathbb{N} \Longrightarrow T_n \in \mathcal{E}'(\mathbb{R})$ $\forall n \in \mathbb{N}$.

For $m \ge 3$, T_n can be found in a similar way.

Example 5.7.11. Show that the series $\sum_{n=0}^{\infty} a^n\delta_n$ with $a > 0$ and $\langle \delta_n, \phi\rangle = \phi(n)$ $\forall \phi \in \mathcal{E}(\mathbb{R}) \equiv C^\infty(\mathbb{R})$ $\forall n \in \mathbb{N}$ (which converges in $\mathcal{D}'(\mathbb{R})$ for arbitrary $a > 0$ (see Example 1.9.1)), does *not* converge in $\mathcal{E}'(\mathbb{R})$ for any $a > 0$.

Proof. Set $S_N = \sum_{n=0}^{N} a^n \delta_n$. Since $\delta_n \in \mathcal{E}'(\mathbb{R})\ \forall n \in \mathbb{N}_0$, $S_N \in \mathcal{E}'(\mathbb{R})\ \forall N \in \mathbb{N}$. Then, $\forall \phi \in C^\infty(\mathbb{R}) \equiv \mathcal{E}(\mathbb{R})$, $\langle S_N, \phi\rangle$ is well defined and

$$\begin{aligned}\langle S_N, \phi\rangle &= \left\langle \sum_{n=0}^{N} a^n \delta_n, \phi\right\rangle \\ &= \sum_{n=0}^{N} a^n \langle \delta_n, \phi\rangle = \sum_{n=0}^{N} a^n \phi(n) \quad \forall N \in \mathbb{N},\ \forall \phi \in C^\infty(\mathbb{R}).\end{aligned}$$

For $a \geq 1$, let $\phi(x) = 1\ \forall x \in \mathbb{R}$. Then $\phi \in C^\infty(\mathbb{R})$ and $\langle S_N, 1\rangle = \sum_{n=0}^{N} a^n \to \infty$ as $N \to \infty$, since $a \geq 1$. Hence, the sequence (S_N) does *not* converge in $\mathcal{E}'(\mathbb{R})$ for $a \geq 1 \Longrightarrow \sum_{n=0}^{\infty} a^n \delta_n$ does *not* converge in $\mathcal{E}'(\mathbb{R})$ for $a \geq 1$.

For $0 < a < 1$, let $\phi(x) = (\frac{1}{a})^x = e^{\ln[(\frac{1}{a})^x]} = e^{x \ln(\frac{1}{a})}\ \forall x \in \mathbb{R}$. Then $\phi \in C^\infty(\mathbb{R})$ and $\langle S_N, \phi\rangle = \sum_{n=0}^{N} a^n \phi(n) = \sum_{n=0}^{N} a^n (\frac{1}{a})^n = 1 + \cdots + 1 = N + 1 \to \infty$ as $N \to \infty$.

$\Longrightarrow$ the sequence (S_N) does *not* converge in $\mathcal{E}'(\mathbb{R})$ for $0 < a < 1$.

$\Longrightarrow \sum_{n=0}^{\infty} a^n \delta_n$ does *not* converge in $\mathcal{E}'(\mathbb{R})$ for $0 < a < 1$. Thus, we have proved the result that $\sum_{n=0}^{\infty} a^n \delta_n$ does *not* converge in $\mathcal{E}'(\mathbb{R})$ for $a > 0$. □

5.8 Definition of $\langle T, \phi\rangle$ for $\phi \in C^\infty(\mathbb{R}^n)$ and $T \in \mathcal{D}'(\mathbb{R}^n)$ with non-compact support

The procedure of definition of $\langle T, \phi\rangle$ for distributions T with compact support and $\phi \in C^\infty(\mathbb{R}^n)$ with arbitrary support can be extended to distributions T with *non-compact* support in the following situation.

For a distribution $T \in \mathcal{D}'(\mathbb{R}^n)$ with non-compact support and $\phi \in C^\infty(\mathbb{R}^n)$ with arbitrary support, if the *intersection* of the supports of T and ϕ is *compact*, i.e. if $K_0 = \operatorname{supp}(T) \cap \operatorname{supp}(\phi)$ is a compact subset of $\mathbb{R}^n$, then $\langle T, \phi\rangle$ can always be defined. (5.8.1)

Let $\alpha \in \mathcal{D}(\mathbb{R}^n)$ be a function with $\alpha(\mathbf{x}) = 1\ \forall \mathbf{x} \in$ a *compact* neighbourhood U of K_0, i.e. $K_0 \subset U$. Then $\alpha\phi \in \mathcal{D}(\mathbb{R}^n)$ and we set

$$\langle T, \phi\rangle = \langle T, \alpha\phi\rangle \quad \forall \phi \in C^\infty(\mathbb{R}^n), \tag{5.8.2}$$

the right-hand side of which does *not* depend on the choice of α.

Let α, β be two functions of $\mathcal{D}(\mathbb{R}^n)$ such that $\alpha(\mathbf{x}) = \beta(\mathbf{x}) = 1\ \forall \mathbf{x} \in$ a compact neighbourhood U of K_0, $K_0 \subset U$. Then $(\alpha - \beta)\phi \in \mathcal{D}(\mathbb{R}^n)$ with $\operatorname{supp}((\alpha - \beta)\phi)$ contained in $\operatorname{supp}(\phi) \cap \operatorname{supp}(\alpha - \beta)$. But $\alpha(\mathbf{x}) - \beta(\mathbf{x}) = 0\ \forall \mathbf{x} \in K_0 \Longrightarrow \operatorname{supp}(\alpha - \beta) \subset K_0^{\complement}$ = complement of K_0.

Hence, $\operatorname{supp}((\alpha - \beta)\phi)$ is contained in $\operatorname{supp}(\phi)$ *and also* in $K_0^{\complement}$, i.e. outside K_0.

$\mathbf{x} \in \operatorname{supp}(\phi)$, $\mathbf{x} \in K_0 \implies \mathbf{x} \in \operatorname{supp}(T)$. But $\mathbf{x} \in \operatorname{supp}(\phi)$, $\mathbf{x} \notin K_0 \implies \mathbf{x} \notin \operatorname{supp}(T) \implies \operatorname{supp}((\alpha - \beta)\phi)$ is outside the support of T
$\implies \langle T, (\alpha - \beta)\phi\rangle = 0$, since $\operatorname{supp}(T) \cap \operatorname{supp}((\alpha - \beta)\phi) = \emptyset$
$\implies \langle T, \alpha\phi\rangle = \langle T, \beta\phi\rangle$. Thus, $\forall \phi \in C^\infty(\mathbb{R}^n)$ with compact intersection $\operatorname{supp}(T) \cap \operatorname{supp}(\phi)$,

$$\langle T, \phi\rangle = \langle T, \alpha\phi\rangle. \tag{5.8.3}$$

Chapter 6
Convolution of distributions

6.1 Tensor product

Let $\Omega_{\mathbf{x}} \subset \mathbb{R}^n$, $\Omega_{\mathbf{y}} \subset \mathbb{R}^m$ be open subsets of $\mathbb{R}^n$ and $\mathbb{R}^m$ whose generic points are $\mathbf{x} = (x_1, x_2, \ldots, x_n) \in \Omega_{\mathbf{x}}$, $\mathbf{y} = (y_1, y_2, \ldots, y_m) \in \Omega_{\mathbf{y}}$. Then,

$$\Omega_{\mathbf{x}} \times \Omega_{\mathbf{y}} = \{(\mathbf{x}, \mathbf{y}) : \mathbf{x} \in \Omega_{\mathbf{x}}, \mathbf{y} \in \Omega_{\mathbf{y}}\} \subset \mathbb{R}^n \times \mathbb{R}^m = \mathbb{R}^{n+m} \tag{6.1.1}$$

is an open subset of $\mathbb{R}^n \times \mathbb{R}^m$ with $(\mathbf{x}, \mathbf{y}) = (x_1, x_2, \ldots, x_n; y_1, y_2, \ldots, y_m)$.

By abuse of notations, functions $\mathbf{x} \in \Omega_{\mathbf{x}} \longmapsto f(\mathbf{x})$, $\mathbf{y} \in \Omega_{\mathbf{y}} \longmapsto g(\mathbf{y})$, and $(\mathbf{x}, \mathbf{y}) \in \Omega_{\mathbf{x}} \times \Omega_{\mathbf{y}} \longmapsto h(\mathbf{x}, \mathbf{y})$ will be denoted by $f(\mathbf{x}), g(\mathbf{y})$ and $h(\mathbf{x}, \mathbf{y})$ respectively.

Let $\mathcal{D}(\Omega_{\mathbf{x}})$, $\mathcal{D}(\Omega_{\mathbf{y}})$ and $\mathcal{D}(\Omega_{\mathbf{x}} \times \Omega_{\mathbf{y}})$ be the test spaces of infinitely differentiable real- or complex-valued functions with compact support in $\mathbb{R}^n$, $\mathbb{R}^m$ and $\mathbb{R}^n \times \mathbb{R}^m$ respectively, and let their dual spaces $\mathcal{D}'(\Omega_{\mathbf{x}})$, $\mathcal{D}'(\Omega_{\mathbf{y}})$ and $\mathcal{D}'(\Omega_{\mathbf{x}} \times \Omega_{\mathbf{y}})$ be the spaces of distributions $T_{\mathbf{x}}$, $S_{\mathbf{y}}$ and $W_{\mathbf{x},\mathbf{y}}$ (or equivalently $T(\mathbf{x})$, $S(\mathbf{y})$, $W(\mathbf{x}, \mathbf{y})$) on $\Omega_{\mathbf{x}}$, $\Omega_{\mathbf{y}}$ and $\Omega_{\mathbf{x}} \times \Omega_{\mathbf{y}}$ respectively.

Remark 6.1.1. $T_{\mathbf{x}}$ or $T(\mathbf{x})$, $S_{\mathbf{y}}$ or $S(\mathbf{y})$, $W_{\mathbf{x},\mathbf{y}}$ or $W(\mathbf{x}, \mathbf{y})$ do *not* indicate the values of T, S, W at $\mathbf{x}$, $\mathbf{y}$ and $(\mathbf{x}, \mathbf{y})$ respectively, since a distribution cannot have a value at a point.

The notation $T_{\mathbf{x}}$ or $T(\mathbf{x})$, $S_{\mathbf{y}}$ or $S(\mathbf{y})$, $W_{\mathbf{x},\mathbf{y}}$ or $W(\mathbf{x}, \mathbf{y})$ is used to indicate that the generic points of $\Omega_{\mathbf{x}}$, $\Omega_{\mathbf{y}}$, $\Omega_{\mathbf{x}} \times \Omega_{\mathbf{y}}$ are denoted by $\mathbf{x}$, $\mathbf{y}$, $(\mathbf{x}, \mathbf{y})$ respectively.

Tensor product of functions

Definition 6.1.1. Let $f(\mathbf{x})$, $g(\mathbf{y})$ be functions on $\Omega_{\mathbf{x}}$ and $\Omega_{\mathbf{y}}$ respectively. Then the *tensor product* $f(\mathbf{x}) \otimes g(\mathbf{y})$ of $f(\mathbf{x})$ and $g(\mathbf{y})$ is the function $h(\mathbf{x}, \mathbf{y})$ on $\Omega_{\mathbf{x}} \times \Omega_{\mathbf{y}}$ defined by:

$$h(\mathbf{x}, \mathbf{y}) = f(\mathbf{x}) \otimes g(\mathbf{y}) = f(\mathbf{x}) \cdot g(\mathbf{y}) \quad \forall (\mathbf{x}, \mathbf{y}) \in \Omega_x \times \Omega_y. \tag{6.1.2}$$

Example 6.1.1.

- For $m = n = 1$, $\Omega_x = \Omega_y = \mathbb{R}$, $\Omega_x \times \Omega_y = \mathbb{R} \times \mathbb{R} = \mathbb{R}^2$.
  1. $f(x) = \sin mx$, $g(y) = \cos ny$, $h(x, y) = \sin mx \otimes \cos ny = \sin mx . \cos ny$ $\forall (x, y) \in \mathbb{R}^2$;
  2. $f(x) = e^{i2\pi x}$, $g(y) = e^{i2\pi y}$, $h(x, y) = e^{i2\pi x} \otimes e^{i2\pi y} = e^{i2\pi x} \cdot e^{i2\pi y} = e^{i2\pi(x+y)}$ $\forall (x, y) \in \mathbb{R}^2$.

- For $m = 2, n = 1, \Omega_{\mathbf{x}} = \mathbb{R}^2, \Omega_{\mathbf{y}} = \mathbb{R}, f(x_1, x_2) = \sin mx_1 \cos nx_2, g(y) = e^{-y^2}, h(x_1, x_2; y) = (\sin mx_1 \cos nx_2) \otimes e^{-y^2} = e^{-y^2} \sin mx_1 \cos nx_2$ $\forall (x_1, x_2; y) \in \mathbb{R}^2 \times \mathbb{R}$.

Let $1_{\mathbf{x}} : \mathbf{x} \longmapsto 1_{\mathbf{x}}(\mathbf{x}) = 1 \ \forall \mathbf{x} \in \mathbb{R}^n$, and $1_{\mathbf{y}} : \mathbf{y} \longmapsto 1_{\mathbf{y}}(\mathbf{y}) = 1 \ \forall \mathbf{y} \in \mathbb{R}^n$. (6.1.3)
A function g (resp. f) is said to be *independent of* $\mathbf{x}$ (resp. $\mathbf{y}$) if it is of the form:

$$\forall (\mathbf{x}, \mathbf{y}) \in \Omega_{\mathbf{x}} \times \Omega_{\mathbf{y}}, \quad g(\mathbf{y}) = 1_{\mathbf{x}} \otimes g(\mathbf{y}) \quad (\text{resp. } f(\mathbf{x}) = f(\mathbf{x}) \otimes 1_{\mathbf{y}}). \tag{6.1.4}$$

Generalized integrals dependent on a parameter

The classical theorems of calculus on the continuity and differentiability of integrals dependent on a parameter (under the integral sign $\int$) has been extended to distributions by [8]; since it will be used later to define the tensor product of distributions.

Let $\phi(\mathbf{x}; \boldsymbol{\lambda})$ with $\boldsymbol{\lambda} = (\lambda_1, \lambda_2, \ldots, \lambda_m) \in \boldsymbol{\Lambda} \subset \mathbb{R}^m$ be a function not only of $\mathbf{x} = (x_1, x_2, \ldots, x_n) \in \Omega_{\mathbf{x}}$, but also of parameter $\boldsymbol{\lambda}$ such that for every *fixed* value $\boldsymbol{\lambda} \in \boldsymbol{\Lambda}$, $\phi(\mathbf{x}; \boldsymbol{\lambda})$ can be considered to be a function of $\mathbf{x}$ only, and $\phi(\mathbf{x}; \boldsymbol{\lambda}) \in \mathcal{D}(\Omega_{\mathbf{x}})$ $\forall$ *fixed* $\boldsymbol{\lambda} \in \boldsymbol{\Lambda}$.

Let $T_{\mathbf{x}} \in \mathcal{D}'(\Omega_{\mathbf{x}})$ be a fixed distribution on $\Omega_{\mathbf{x}}$, in which $\mathbf{x}$ is a generic point. Then, $\forall$ fixed $\boldsymbol{\lambda} \in \boldsymbol{\Lambda}$ with $\phi(\mathbf{x}; \boldsymbol{\lambda}) \in \mathcal{D}(\Omega_{\mathbf{x}})$,

$$I(\boldsymbol{\lambda}) = T_{\mathbf{x}}[\phi(\mathbf{x}; \boldsymbol{\lambda})] = \langle T_{\mathbf{x}}, \phi(\mathbf{x}; \boldsymbol{\lambda}) \rangle, \tag{6.1.5}$$

well defined for each value of $\boldsymbol{\lambda} \in \boldsymbol{\Lambda}$, will be called a *generalized integral* dependent on the parameter $\boldsymbol{\lambda}$. In fact, if $T_{\mathbf{x}} = f(\mathbf{x}) \in L^1_{\text{loc}}(\Omega_{\mathbf{x}})$, then we can rewrite (6.1.5) in integral form: $\forall$ fixed $\boldsymbol{\lambda} \in \boldsymbol{\Lambda}$ with $\phi(\mathbf{x}; \boldsymbol{\lambda}) \in \mathcal{D}(\Omega_{\mathbf{x}})$,

$$I(\boldsymbol{\lambda}) = T_{\mathbf{x}}[\phi(\mathbf{x}; \boldsymbol{\lambda})] = \int_{\Omega_{\mathbf{x}}} f(\mathbf{x}) \phi(\mathbf{x}; \boldsymbol{\lambda}) d\mathbf{x}. \tag{6.1.6}$$

Otherwise (i.e. if $T_{\mathbf{x}} \in \mathcal{D}'(\Omega_{\mathbf{x}})$ is a *singular distribution*), we can *not* write $I(\boldsymbol{\lambda})$ using the integral sign $\int$ as in (6.1.6). Hence, we have called $I(\boldsymbol{\lambda})$ in (6.1.5) the *generalized integral*.

We agree to accept the result (see Schwartz [8]):

Lemma 6.1.1. *$I(\boldsymbol{\lambda})$ defined by* (6.1.5) *is continuous in $\boldsymbol{\lambda}$ and infinitely differentiable (in the* usual pointwise sense*) with respect to $\lambda_i, 1 \le i \le m$, i.e. $\forall$ multi-index $\boldsymbol{\alpha} = (\alpha_1, \alpha_2, \ldots, \alpha_m)$ and* $\partial^{\boldsymbol{\alpha}}_{\boldsymbol{\lambda}} = \frac{\partial^{|\boldsymbol{\alpha}|}}{\partial^{\alpha_1}_{\lambda_1} \partial^{\alpha_2}_{\lambda_2} \ldots \partial^{\alpha_m}_{\lambda_m}}$,

$$\begin{aligned} \partial^{\boldsymbol{\alpha}}_{\boldsymbol{\lambda}} I(\boldsymbol{\lambda}) &= \partial^{\boldsymbol{\alpha}}_{\boldsymbol{\lambda}} T_{\mathbf{x}}[\phi(\mathbf{x}; \boldsymbol{\lambda})] = T_{\mathbf{x}}[\partial^{\boldsymbol{\alpha}}_{\boldsymbol{\lambda}} \phi(\mathbf{x}; \boldsymbol{\lambda})] = T_{\mathbf{x}}\left[\frac{\partial^{|\boldsymbol{\alpha}|} \phi(\mathbf{x}; \boldsymbol{\lambda})}{\partial^{\alpha_1}_{\lambda_1} \partial^{\alpha_2}_{\lambda_2} \cdots \partial^{\alpha_m}_{\lambda_m}}\right] \\ &= \left\langle T_{\mathbf{x}}, \frac{\partial^{|\boldsymbol{\alpha}|} \phi(\mathbf{x}; \boldsymbol{\lambda})}{\partial^{\alpha_1}_{\lambda_1} \partial^{\alpha_2}_{\lambda_2} \cdots \partial^{\alpha_m}_{\lambda_m}} \right\rangle \quad \forall \boldsymbol{\lambda} \in \boldsymbol{\Lambda} \textit{ with } \phi(\mathbf{x}; \boldsymbol{\lambda}) \in \mathcal{D}(\Omega_{\mathbf{x}}). \end{aligned} \tag{6.1.7}$$

Corollary 6.1.1. *Let $\phi(\mathbf{x},\mathbf{y})$ be a function of $\mathbf{x} \in \Omega_\mathbf{x}$ which depends on parameter $\mathbf{y} \in \Omega_\mathbf{y}$ and is infinitely differentiable with respect to variables $\mathbf{x},\mathbf{y}$ in $\Omega_\mathbf{x} \times \Omega_\mathbf{y}$ such that, for fixed $\mathbf{y} \in \Omega_\mathbf{y}$, $\phi(\mathbf{x},\mathbf{y}) \in \mathcal{D}(\Omega_\mathbf{x})$. Then the generalized integral*

$$I(\mathbf{y}) = T_\mathbf{x}[\phi(\mathbf{x},\mathbf{y})] = \langle T_\mathbf{x}, \phi(\mathbf{x},\mathbf{y})\rangle, \tag{6.1.8}$$

well defined for each value of $\mathbf{y} \in \Omega_\mathbf{y}$, is an infinitely differentiable function of $\mathbf{y}$ in $\Omega_\mathbf{y}$. Moreover, if $\phi(\mathbf{x},\mathbf{y}) \in \mathcal{D}(\Omega_\mathbf{x} \times \Omega_\mathbf{y})$ with $\operatorname{supp}(\phi(\mathbf{x},\mathbf{y})) \subset\subset \Omega_\mathbf{y}$ *for fixed $\mathbf{x} \in \Omega_\mathbf{x}$, then* $\operatorname{supp}(I(\mathbf{y})) \subset\subset \Omega_\mathbf{y}$, *$I(\mathbf{y}) \in \mathcal{D}(\Omega_\mathbf{y})$ and*

$$S_\mathbf{y}(I(\mathbf{y})) = \langle S_\mathbf{y}, I(\mathbf{y})\rangle = \langle S_\mathbf{y}, \langle T_\mathbf{x}, \phi(\mathbf{x},\mathbf{y})\rangle\rangle. \tag{6.1.9}$$

Linear functional defined by tensor product $f \otimes g$

Let $f(\mathbf{x}) \in L^1(\Omega_\mathbf{x})$, $g(\mathbf{y}) \in L^1(\Omega_\mathbf{y})$ and $\chi(\mathbf{x},\mathbf{y}) = \phi(\mathbf{x})\psi(\mathbf{y})$, with $\phi(\mathbf{x}) \in \mathcal{D}(\Omega_\mathbf{x})$, $\psi(\mathbf{y}) \in \mathcal{D}(\Omega_\mathbf{y})$. Then define a linear functional $L_{f\otimes g}$ by: $\forall \chi(\mathbf{x},\mathbf{y}) = \phi(\mathbf{x})\psi(\mathbf{y})$ with $\phi(\mathbf{x}) \in \mathcal{D}(\Omega_\mathbf{x})$, $\psi(\mathbf{y}) \in \mathcal{D}(\Omega_\mathbf{y})$,

$$\begin{aligned} L_{f\otimes g}(\chi) &= \langle f(\mathbf{x}) \otimes g(\mathbf{y}), \phi(\mathbf{x})\psi(\mathbf{y})\rangle = \int_{\Omega_\mathbf{x}\times\Omega_\mathbf{y}} f(\mathbf{x})g(\mathbf{y})\phi(\mathbf{x})\psi(\mathbf{y})d\mathbf{x}d\mathbf{y} \\ &= \int_{\Omega_x} f(\mathbf{x})\phi(\mathbf{x})d\mathbf{x} \int_{\Omega_y} g(\mathbf{y})\psi(\mathbf{y})d\mathbf{y} = L_f(\phi)\cdot L_g(\psi) = \langle f,\phi\rangle \cdot \langle g,\psi\rangle \end{aligned}$$

$$\Longrightarrow \quad \langle f(\mathbf{x}) \otimes g(\mathbf{y}), \phi(\mathbf{x})\psi(\mathbf{y})\rangle = \langle f,\phi\rangle \cdot \langle g,\psi\rangle \quad \forall \phi \in \mathcal{D}(\Omega_\mathbf{x}),\ \forall \psi \in \mathcal{D}(\Omega_\mathbf{y}). \tag{6.1.10}$$

If $\chi(\mathbf{x},\mathbf{y}) \in \mathcal{D}(\Omega_x \times \Omega_y)$ is *not* of the tensor product form (i.e. $\chi(\mathbf{x},\mathbf{y}) \neq \phi(\mathbf{x})\psi(\mathbf{y})$), then applying Fubini's Theorem (see Theorem 7.1.2C), we have, $\forall \chi \in \mathcal{D}(\Omega_\mathbf{x} \times \Omega_\mathbf{y})$:

$$\begin{aligned} \langle f(\mathbf{x}) \otimes g(\mathbf{y}), \chi(\mathbf{x},\mathbf{y})\rangle &= \int_{\Omega_\mathbf{x}\times\Omega_\mathbf{y}} f(\mathbf{x})g(\mathbf{y})\chi(\mathbf{x},\mathbf{y})d\mathbf{x}d\mathbf{y} \\ &= \begin{cases} \int_{\Omega_\mathbf{x}} f(\mathbf{x})(\int_{\Omega_\mathbf{y}} g(\mathbf{y})\chi(\mathbf{x},\mathbf{y})d\mathbf{y})d\mathbf{x} = \langle f(\mathbf{x}), \langle g(\mathbf{y}), \chi(\mathbf{x},\mathbf{y})\rangle\rangle \\ \int_{\Omega_\mathbf{y}} g(\mathbf{y})(\int_{\Omega_\mathbf{x}} f(\mathbf{x})\chi(\mathbf{x},\mathbf{y})d\mathbf{x})d\mathbf{y} = \langle g(\mathbf{y}), \langle f(\mathbf{x}), \chi(\mathbf{x},\mathbf{y})\rangle\rangle \end{cases} \end{aligned}$$

$\Longrightarrow$ for $f \in L^1(\Omega_\mathbf{x})$, $g \in L^1(\Omega_\mathbf{y})$ and $\forall \chi \in \mathcal{D}(\Omega_\mathbf{x} \times \Omega_\mathbf{y})$,

$$\langle f(\mathbf{x}) \otimes g(\mathbf{y}), \chi(\mathbf{x},\mathbf{y})\rangle = \langle f(\mathbf{x}), \langle g(\mathbf{y}), \chi(\mathbf{x},\mathbf{y})\rangle\rangle \tag{6.1.11}$$

$$= \langle g(\mathbf{y}), \langle f(\mathbf{x}), \chi(\mathbf{x},\mathbf{y})\rangle\rangle. \tag{6.1.12}$$

Tensor product of distributions

Following formulae (5.1.10)–(6.1.12), Lemma 6.1.1 and Corollary 6.1.1, we can define the tensor product of distributions as follows:

Definition 6.1.2. Let $T_{\mathbf{x}} \in \mathcal{D}'(\Omega_{\mathbf{x}})$, $S_{\mathbf{y}} \in \mathcal{D}'(\Omega_{\mathbf{y}})$ be any two distributions on $\Omega_{\mathbf{x}}$ and $\Omega_{\mathbf{y}}$ respectively. Then the *tensor product* $T_{\mathbf{x}} \otimes S_{\mathbf{y}}$ of two distributions $T_{\mathbf{x}}$ and $S_{\mathbf{y}}$ is the unique distribution $W_{\mathbf{x},\mathbf{y}} \in \mathcal{D}'(\Omega_{\mathbf{x}} \times \Omega_{\mathbf{y}})$ defined by:

$$\langle W_{\mathbf{x},\mathbf{y}}, \phi(\mathbf{x})\psi(\mathbf{y})\rangle = \langle T_{\mathbf{x}} \otimes S_{\mathbf{y}}, \phi(\mathbf{x})\psi(\mathbf{y})\rangle = \langle T_{\mathbf{x}}, \phi(\mathbf{x})\rangle \cdot \langle S_{\mathbf{y}}, \psi(\mathbf{y})\rangle \qquad (6.1.13)$$

$\forall \chi(\mathbf{x},\mathbf{y}) = \phi(\mathbf{x})\psi(\mathbf{y}) \in \mathcal{D}(\Omega_x) \times \mathcal{D}(\Omega_y)$;

$$\langle W_{\mathbf{x},y}, \chi(\mathbf{x},\mathbf{y})\rangle = \langle T_{\mathbf{x}} \otimes S_{\mathbf{y}}, \chi(\mathbf{x},\mathbf{y})\rangle = \langle S_{\mathbf{y}}, \langle T_{\mathbf{x}}, \chi(\mathbf{x},\mathbf{y})\rangle\rangle = \langle T_{\mathbf{x}}, \langle S_{\mathbf{y}}, \chi(\mathbf{x},\mathbf{y})\rangle\rangle \qquad (6.1.14)$$

$\forall \chi(\mathbf{x},\mathbf{y}) \in \mathcal{D}(\Omega_{\mathbf{x}} \times \Omega_{\mathbf{y}})$ *not* in the tensor product form.

For every $\chi(\mathbf{x},\mathbf{y}) = \phi(\mathbf{x})\psi(\mathbf{y})$ with $\phi(\mathbf{x}) \in \mathcal{D}(\Omega_{\mathbf{x}})$, $\psi(\mathbf{y}) \in \mathcal{D}(\Omega_{\mathbf{y}})$, we get (6.1.13) from (6.1.14), i.e. $\forall \phi(\mathbf{x})\psi(\mathbf{y}) \in \mathcal{D}(\Omega_{\mathbf{x}}) \times \mathcal{D}(\Omega_{\mathbf{y}}) \subset \mathcal{D}(\Omega_{\mathbf{x}} \times \Omega_{\mathbf{y}})$,

$$\langle T_{\mathbf{x}}, \langle S_{\mathbf{y}}, \phi(\mathbf{x})\psi(\mathbf{y})\rangle\rangle = \langle S_{\mathbf{y}}, \langle T_{\mathbf{x}}, \phi(\mathbf{x})\psi(\mathbf{y})\rangle\rangle = \langle T_{\mathbf{x}}, \phi(\mathbf{x})\rangle \cdot \langle S_{\mathbf{y}}, \psi(\mathbf{y})\rangle. \qquad (6.1.15)$$

Support of tensor product $T_{\mathbf{x}} \otimes S_{\mathbf{y}}$

$$\operatorname{supp}(T_{\mathbf{x}} \otimes T_{\mathbf{y}}) = \operatorname{supp}(T_{\mathbf{x}}) \times \operatorname{supp}(S_{\mathbf{y}}) = \{(\mathbf{x},\mathbf{y}) : \mathbf{x} \in \operatorname{supp}(T_{\mathbf{x}}), \mathbf{y} \in \operatorname{supp}(S_{\mathbf{y}})\}. \qquad (6.1.16)$$

Example 6.1.2. Let $\Omega_{\mathbf{x}} = \mathbb{R}^l$, $\Omega_{\mathbf{y}} = \mathbb{R}^m$, $T_{\mathbf{x}} = \delta(\mathbf{x}) \in \mathcal{D}'(\mathbb{R}^n)$ and $S_{\mathbf{y}} = \delta(\mathbf{y}) \in \mathcal{D}'(\mathbb{R}^m)$. Then their tensor product $\delta(\mathbf{x}) \otimes \delta(\mathbf{y})$ is defined, $\forall \phi(\mathbf{x},\mathbf{y}) \in \mathcal{D}(\mathbb{R}^l \times \mathbb{R}^m)$, by:

$$\langle \delta(\mathbf{x}) \otimes \delta(\mathbf{y}), \phi(\mathbf{x},\mathbf{y})\rangle = \langle \delta(\mathbf{x}), \langle \delta(\mathbf{y}), \phi(\mathbf{x},\mathbf{y})\rangle\rangle = \langle \delta(\mathbf{x}), \phi(\mathbf{x},\mathbf{0})\rangle = \phi(\mathbf{0},\mathbf{0}),$$

i.e. $\delta(\mathbf{x}) \otimes \delta(\mathbf{y}) = \delta_{\mathbf{x},\mathbf{y}}$ with $(\mathbf{x},\mathbf{y}) \in \mathbb{R}^l \times \mathbb{R}^m$,

$$\operatorname{supp}(\delta(\mathbf{x}) \otimes \delta(\mathbf{y})) = (\mathbf{0},\mathbf{0}) \in \mathbb{R}^l \times \mathbb{R}^m. \qquad (6.1.17)$$

A distribution is said to be *independent* of $\mathbf{x}$ (resp. $\mathbf{y}$) if it is of the form $1_{\mathbf{x}} \otimes S_{\mathbf{y}}$ (resp. $T_{\mathbf{x}} \otimes 1_{\mathbf{y}}$), where $1_{\mathbf{x}}$ (resp. $1_{\mathbf{y}}$) is defined by (6.1.3). This definition is suggested by (6.1.4). Then, $\forall\phi \in \mathcal{D}(\Omega_{\mathbf{x}} \times \Omega_{\mathbf{y}})$,

$$\langle 1_{\mathbf{x}} \otimes S_{\mathbf{y}}, \phi\rangle = \langle 1_{\mathbf{x}}, \langle S_{\mathbf{y}}, \phi(\mathbf{x}, \mathbf{y})\rangle\rangle = \int_{\Omega_{\mathbf{x}}} 1 \cdot \langle S_{\mathbf{y}}, \phi(\mathbf{x}, \mathbf{y})\rangle d\mathbf{x} = \int_{\Omega_{\mathbf{x}}} \langle S_{\mathbf{y}}, \phi(\mathbf{x}, \mathbf{y})\rangle d\mathbf{x}$$

$$= \langle S_{\mathbf{y}}, \langle 1_{\mathbf{x}}, \phi(\mathbf{x}, \mathbf{y})\rangle\rangle = \langle S_{\mathbf{y}}, \int_{\Omega_{\mathbf{x}}} \phi(\mathbf{x}, \mathbf{y}) d\mathbf{x}\rangle; \tag{6.1.18}$$

$$\langle T_{\mathbf{x}} \otimes 1_{\mathbf{y}}, \phi\rangle = \langle T_{\mathbf{x}}, \langle 1_{\mathbf{y}}, \phi(\mathbf{x}, \mathbf{y})\rangle\rangle = \langle T_{\mathbf{x}}, \int_{\Omega_{\mathbf{y}}} \phi(\mathbf{x}, \mathbf{y}) d\mathbf{y}\rangle$$

$$= \langle 1_{\mathbf{y}}, \langle T_{\mathbf{x}}, \phi(\mathbf{x}, \mathbf{y})\rangle\rangle = \int_{\Omega_{\mathbf{y}}} \langle T_{\mathbf{x}}, \phi(\mathbf{x}, \mathbf{y})\rangle d\mathbf{y}, \tag{6.1.19}$$

since $1_{\mathbf{x}}(\mathbf{x}) = 1\ \forall \mathbf{x} \in \Omega_{\mathbf{x}}$, $1_{\mathbf{y}}(\mathbf{y}) = 1\ \forall \mathbf{y} \in \Omega_{\mathbf{y}}$.

Derivatives of tensor product

Let $\partial_{\mathbf{x}}^{\boldsymbol{\alpha}} = \frac{\partial^{|\boldsymbol{\alpha}|}}{\partial_{x_1}^{\alpha_1}\partial_{x_2}^{\alpha_2}\dots\partial_{x_n}^{\alpha_n}}$, $\partial_{\mathbf{y}}^{\boldsymbol{\beta}} = \frac{\partial^{|\boldsymbol{\beta}|}}{\partial_{y_1}^{\beta_1}\partial_{y_2}^{\beta_2}\dots\partial_{y_m}^{\beta_m}}$. Then

$$\partial_{\mathbf{x}}^{\boldsymbol{\alpha}}\partial_{\mathbf{y}}^{\boldsymbol{\beta}}[T_{\mathbf{x}} \otimes S_{\mathbf{y}}] = \partial_{\mathbf{x}}^{\boldsymbol{\alpha}} T_{\mathbf{x}} \otimes \partial_{\mathbf{y}}^{\boldsymbol{\beta}} S_{\mathbf{y}}. \tag{6.1.20}$$

In fact,

$$\begin{aligned}
\langle \partial_{\mathbf{x}}^{\boldsymbol{\alpha}} T_{\mathbf{x}} \otimes \partial_{\mathbf{y}}^{\boldsymbol{\beta}} S_{\mathbf{y}}, \phi\rangle &= \langle \partial_{\mathbf{x}}^{\boldsymbol{\alpha}} T_{\mathbf{x}}, \langle \partial_{\mathbf{y}}^{\boldsymbol{\beta}} S_{\mathbf{y}}, \phi\rangle\rangle = \langle \partial_{\mathbf{x}}^{\boldsymbol{\alpha}} T_{\mathbf{x}}, (-1)^{|\boldsymbol{\beta}|}\langle S_{\mathbf{y}}, \partial_{\mathbf{y}}^{\boldsymbol{\beta}}\phi(\mathbf{x}, \mathbf{y})\rangle\rangle \\
&= (-1)^{|\boldsymbol{\beta}|}\langle \partial_{\mathbf{x}}^{\boldsymbol{\alpha}} T_{\mathbf{x}}, \langle S_{\mathbf{y}}, \partial_{\mathbf{y}}^{\boldsymbol{\beta}}\phi(\mathbf{x}, \mathbf{y})\rangle\rangle \\
&= (-1)^{|\boldsymbol{\beta}|}\langle S_{\mathbf{y}}, \langle \partial_{\mathbf{x}}^{\boldsymbol{\alpha}} T_{\mathbf{x}}, \partial_{\mathbf{y}}^{\boldsymbol{\beta}}\phi(\mathbf{x}, \mathbf{y})\rangle\rangle \\
&= (-1)^{|\boldsymbol{\beta}|}(-1)^{|\boldsymbol{\alpha}|}\langle S_{\mathbf{y}}, \langle T_{\mathbf{x}}, \partial_{\mathbf{x}}^{\boldsymbol{\alpha}}\partial_{\mathbf{y}}^{\boldsymbol{\beta}}\phi(\mathbf{x}, \mathbf{y})\rangle\rangle \\
&= (-1)^{|\boldsymbol{\beta}|}(-1)^{|\boldsymbol{\alpha}|}\langle T_{\mathbf{x}} \otimes S_{\mathbf{y}}, \partial_{\mathbf{x}}^{\boldsymbol{\alpha}}\partial_{\mathbf{y}}^{\boldsymbol{\beta}}\phi(\mathbf{x}, \mathbf{y})\rangle \quad \text{(by definition of } T_{\mathbf{x}} \otimes S_{\mathbf{y}}) \\
&= \langle \partial_{\mathbf{x}}^{\boldsymbol{\alpha}}\partial_{\mathbf{y}}^{\boldsymbol{\beta}}(T_{\mathbf{x}} \otimes S_{\mathbf{y}}), \phi(\mathbf{x}, \mathbf{y})\rangle \quad \forall\phi \in \mathcal{D}(\Omega_{\mathbf{x}} \times \Omega_{\mathbf{y}}),
\end{aligned}$$

since $T_{\mathbf{x}} \otimes S_{\mathbf{y}}$ is a distribution.

Tensor product of several distributions

Let $\Omega_{\mathbf{x}} \subset \mathbb{R}^l$, $\Omega_{\mathbf{y}} \subset \mathbb{R}^m$, $\Omega_{\mathbf{z}} \subset \mathbb{R}^n$ and $T_{\mathbf{x}} \in \mathcal{D}'(\Omega_{\mathbf{x}})$, $S_{\mathbf{y}} \in \mathcal{D}'(\Omega_{\mathbf{y}})$, $R_{\mathbf{z}} \in \mathcal{D}'(\Omega_{\mathbf{z}})$.

Then the *tensor product* $T_{\mathbf{x}} \otimes S_{\mathbf{y}} \otimes R_{\mathbf{z}}$ of the three distributions $T_{\mathbf{x}}, S_{\mathbf{y}}, R_{\mathbf{z}}$ is a distribution on $\Omega_{\mathbf{x}} \times \Omega_{\mathbf{y}} \times \Omega_{\mathbf{z}}$ defined, $\forall\phi(\mathbf{x}) \in \mathcal{D}(\Omega_{\mathbf{x}})$, $\psi(\mathbf{y}) \in \mathcal{D}(\Omega_{\mathbf{y}})$, $\chi(\mathbf{z}) \in$

$\mathcal{D}(\Omega_{\mathbf{z}})$, by:

$$\langle T_{\mathbf{x}} \otimes S_{\mathbf{y}} \otimes R_{\mathbf{z}}, \phi(\mathbf{x})\psi(\mathbf{y})\chi(\mathbf{z})\rangle = \langle T_{\mathbf{x}}, \phi(\mathbf{x})\rangle\langle S_{\mathbf{y}}, \psi(\mathbf{y})\rangle \cdot \langle R_{\mathbf{z}}, \chi(\mathbf{z})\rangle; \tag{6.1.21}$$

$$\begin{aligned}\langle T_{\mathbf{x}} \otimes S_{\mathbf{y}} \otimes R_{\mathbf{z}}, \phi(\mathbf{x},\mathbf{y},\mathbf{z})\rangle &= \langle T_{\mathbf{x}}, \langle S_{\mathbf{y}} \otimes R_{\mathbf{z}}, \phi(\mathbf{x},\mathbf{y},\mathbf{z})\rangle\rangle \quad (6.1.22)\\ &\qquad \forall \phi(\mathbf{x},\mathbf{y},\mathbf{z}) \in \mathcal{D}(\Omega_{\mathbf{x}} \times \Omega_{\mathbf{y}} \times \Omega_{\mathbf{z}})\\ &= \langle T_{\mathbf{x}} \otimes S_{\mathbf{y}}, \langle R_{\mathbf{z}}, \phi(\mathbf{x},\mathbf{y},\mathbf{z})\rangle\rangle. \quad (6.1.23)\end{aligned}$$

Example 6.1.3. Let $H(x_i)$ be the Heaviside function in variable $x_i, 1 \le i \le 3$, defined by $H(x_i) = 1$ for $x_i > 0$ and $H(x_i) = 0$ for $x_i < 0$.

Then $H(x_1, x_2, x_3) = H(x_1) \otimes H(x_2) \otimes H(x_3)$ is a Heaviside function in 3 variables, which equals 1 in the octant $x_1 > 0, x_2 > 0, x_3 > 0$ and equals 0 otherwise.

Now $\frac{dH(x_i)}{dx_i} = \delta = \delta_{x_i}$, $1 \le i \le 3$ (x_i is a *generic point* of $\mathbb{R}$). Hence,

$$\begin{aligned}\frac{\partial^3 H}{\partial x_1 \partial x_2 \partial x_3} &= \frac{\partial^3}{\partial x_1 \partial x_2 \partial x_3}(H(x_1) \otimes H(x_2) \otimes H(x_3))\\ &= \frac{dH(x_1)}{dx_1} \otimes \frac{dH(x_2)}{dx_2} \otimes \frac{dH(x_3)}{dx_3} = \delta_{x_1} \otimes \delta_{x_2} \otimes \delta_{x_3} = \delta_{x_1,x_2,x_3}.\end{aligned} \tag{6.1.24}$$

6.2 Convolution of functions

Notations Let $A, B \subset \mathbb{R}^n$ be any two non empty sets in $\mathbb{R}^n$. Then

$$A \pm B = \{\mathbf{z} : \mathbf{z} \in \mathbb{R}^n \quad \text{such that } \mathbf{z} = \mathbf{x} \pm \mathbf{y} \text{ with } \mathbf{x} \in A, \mathbf{y} \in B\}. \tag{6.2.1}$$

Some properties of $A + B$ are summarized here:

- If A *or* B is *open*, then $A + B$ is *open*. (6.2.2)
- If A *and* B are *compact*, then $A + B$ is *compact*. (6.2.3)
- If one of the two sets is *closed* and the other one is *compact*, then $A + B$ is *closed*. (6.2.4)

Proof of (6.2.4). Assume that A is *compact* and B is *closed* in $\mathbb{R}^n$. Let $(\mathbf{x}_m)_{m=1}^{\infty}$ be a Cauchy sequence in $A + B$ with $\mathbf{x}_m = \mathbf{a}_m + \mathbf{b}_m$, $\mathbf{a}_m \in A$, $\mathbf{b}_m \in B$ $\forall m \in \mathbb{N}$. Then $(\mathbf{x}_m)$ is a Cauchy sequence in $\mathbb{R}^n$, which is complete. Hence, $\exists \mathbf{x} \in \mathbb{R}^n$ such that $\mathbf{x}_m \longrightarrow \mathbf{x}$ in $\mathbb{R}^n$ as $m \to \infty$. We are to show that $\mathbf{x} \in A + B$ with $\mathbf{x} = \mathbf{a} + \mathbf{b}$, $\mathbf{a} \in A$, $\mathbf{b} \in B$. Since A is *compact* in $\mathbb{R}^n$ and $(\mathbf{a}_m)_{m=1}^{\infty}$ is a *bounded* sequence in A, $\exists$ a subsequence $(\mathbf{a}_{m_k})_{k=1}^{\infty}$ of $(\mathbf{a}_m)_{m=1}^{\infty}$ in A such that $\mathbf{a}_{m_k} \longrightarrow \mathbf{a} \in A$ as $k \to \infty$. Then $\mathbf{x}_{m_k} = \mathbf{a}_{m_k} + \mathbf{b}_{m_k}$ $\forall k \in \mathbb{N}$, with $\mathbf{x}_{m_k} \longrightarrow \mathbf{x}$ in $\mathbb{R}^n$, $\mathbf{a}_{m_k} \longrightarrow \mathbf{a}$ in A as $k \to \infty$ and $\mathbf{b}_{m_k} = \mathbf{x}_{m_k} - \mathbf{a}_{m_k} \in B$ $\forall k \in \mathbb{N}$. Hence, $\lim_{k\to\infty} \mathbf{b}_{m_k} = \lim_{k\to\infty}(\mathbf{x}_{m_k} - \mathbf{a}_{m_k}) = \mathbf{x} - \mathbf{a} \in \mathbb{R}^n$ with $\mathbf{x} \in \mathbb{R}^n$, $\mathbf{a} \in A$. But B is *closed* in $\mathbb{R}^n$ and $(\mathbf{b}_{m_k})_{k=1}^{\infty}$ converges in

$\mathbb{R}^n \Longrightarrow \exists \mathbf{b} \in B$ such that $\lim_{k\to\infty} \mathbf{b}_{m_k} = \mathbf{b} = \mathbf{x}-\mathbf{a} \in B$ by virtue of the uniqueness of the limit. Hence, $\mathbf{b} = \mathbf{x} - \mathbf{a} \Longrightarrow \mathbf{x} = \mathbf{a} + \mathbf{b}$ with $\mathbf{a} \in A$, $\mathbf{b} \in B \Longrightarrow \mathbf{x} \in A + B$. Thus, $A + B$ is *closed*. □

Convolution of functions integrable on $\mathbb{R}^n$

Let $f, g \in L^1(\mathbb{R}^n)$ be functions integrable on $\mathbb{R}^n$. Then the *tensor product* $f(\boldsymbol{\xi}) \otimes g(\boldsymbol{\eta})$ is integrable on $\mathbb{R}^n \times \mathbb{R}^n \equiv \mathbb{R}^{2n}$, since

$$\begin{aligned}\int_{\mathbb{R}^n}\int_{\mathbb{R}^n} |f(\boldsymbol{\xi}) \otimes g(\boldsymbol{\eta})| d\boldsymbol{\xi}\, d\boldsymbol{\eta} &= \int_{\mathbb{R}^n}\int_{\mathbb{R}^n} |f(\boldsymbol{\xi}) g(\boldsymbol{\eta})| d\boldsymbol{\xi}\, d\boldsymbol{\eta} \\ &= \int_{\mathbb{R}^n} |f(\boldsymbol{\xi})| d\boldsymbol{\xi} \int_{\mathbb{R}^n} |g(\boldsymbol{\eta})| d\boldsymbol{\eta} < +\infty.\end{aligned}$$

By change of variables $\mathbf{x} = \boldsymbol{\xi} + \boldsymbol{\eta}$, $\mathbf{t} = \boldsymbol{\eta}$ with their Jacobian $= 1$ and $d\boldsymbol{\xi}\, d\boldsymbol{\eta} = d\mathbf{x} d\mathbf{t}$, we get

$$\int_{\mathbb{R}^n}\int_{\mathbb{R}^n} f(\boldsymbol{\xi}) g(\boldsymbol{\eta}) d\boldsymbol{\xi}\, d\boldsymbol{\eta} = \int_{\mathbb{R}^n}\int_{\mathbb{R}^n} f(\mathbf{x}-\mathbf{t}) g(\mathbf{t}) d\mathbf{x} d\mathbf{t}.$$

Then, by Fubini's Theorem, the function $h(\mathbf{x}) = \int_{\mathbb{R}^n} f(\mathbf{x}-\mathbf{t}) g(\mathbf{t}) d\mathbf{t}$ is defined a.e. in $\mathbb{R}^n$ and $h \in L^1(\mathbb{R}^n)$

$$\Longrightarrow \quad h = f * g \in L^1(\mathbb{R}^n) \quad \forall f, g \in L^1(\mathbb{R}^n). \tag{6.2.5}$$

In general, we have the following results (see Schwartz [8, p. 151] and also Vladimirov [6, pp. 60–61]):

Theorem 6.2.1. *If $f \in L^p(\mathbb{R}^n)$, $g \in L^q(\mathbb{R}^n)$ with $1 \le p, q \le \infty$ and $\frac{1}{p} + \frac{1}{q} \ge 1$, then*

I. $h(\mathbf{x}) = (f * g)(\mathbf{x}) = \int_{\mathbb{R}^n} f(\mathbf{x}-\boldsymbol{\xi}) g(\boldsymbol{\xi}) d\boldsymbol{\xi}$ is well defined for almost all $\mathbf{x} \in \mathbb{R}^n$ such that:

II. $h = f * g \in L^r(\mathbb{R}^n)$ with $\frac{1}{r} = \frac{1}{p} + \frac{1}{q} - 1$, $1 \le r \le \infty$; (6.2.6)

III. $\|h\|_{L^r(\mathbb{R}^n)} = \|f * g\|_{L^r(\mathbb{R}^n)} \le \|f\|_{L^p(\mathbb{R}^n)} \|g\|_{L^q(\mathbb{R}^n)}$, (6.2.7)

i.e.

$$\left(\int_{\mathbb{R}^n} |h(\mathbf{x})|^r d\mathbf{x}\right)^{1/r} \le \left(\int_{\mathbb{R}^n} |f(\mathbf{x})|^p d\mathbf{x}\right)^{1/p} \left(\int_{\mathbb{R}^n} |g(\mathbf{x})|^q d\mathbf{x}\right)^{1/q}. \tag{6.2.8}$$

Proposition 6.2.1. *Let $f \in L^1(\mathbb{R}^n)$ and $g \in L^p(\mathbb{R}^n)$ with $1 \le p \le \infty$. Then $h(\mathbf{x}) = (f * g)(\mathbf{x}) = \int_{\mathbb{R}^n} f(\mathbf{x}-\boldsymbol{\xi}) g(\boldsymbol{\xi}) d\boldsymbol{\xi}$ is well defined for almost all $\mathbf{x} \in \mathbb{R}^n$ such that $h = f * g \in L^p(\mathbb{R}^n)$, with*

$$\|h\|_{L^p(\mathbb{R}^n)} = \|f * g\|_{L^p(\mathbb{R}^n)} \le \|f\|_{L^1(\mathbb{R}^n)} \|g\|_{L^p(\mathbb{R}^n)}. \tag{6.2.9}$$

Proof. Replacing 'q' by 'p' and 'p' by '1' in Theorem 6.2.1, we get $\frac{1}{p} + \frac{1}{q} = \frac{1}{1} + \frac{1}{p} \geq 1$ and $\frac{1}{r} = \frac{1}{1} + \frac{1}{p} - 1 = \frac{1}{p}$, i.e. $r = p$. Then, by Theorem 6.2.1, the results follow. □

Remark 6.2.1. For $f \geq 0$, $g \geq 0$ and $p = q = r = 1$, the inequality (6.2.8) becomes an equality:

$$\|h\|_{L^1(\mathbb{R}^n)} = \|f * g\|_{L^1(\mathbb{R}^n)} = \|f\|_{L^1(\mathbb{R}^n)}\|g\|_{L^1(\mathbb{R}^n)}. \tag{6.2.10}$$

Proposition 6.2.2. *If $f, g \in L^2(\mathbb{R}^n)$, then $h = f * g \in L^\infty(\mathbb{R}^n)$.* (6.2.11)

Proof. $p = q = 2 \Longrightarrow r = \infty$ and

$$\|h\|_{L^\infty(\mathbb{R}^n)} = \operatorname*{ess\,sup}_{\mathbf{x}\in\mathbb{R}^n} |h(\mathbf{x})| \leq \|f\|_{L^2(\mathbb{R}^n)}\|g\|_{L^2(\mathbb{R}^n)}.$$

For $r = \infty$, h exists *everywhere* and is *continuous* on $\mathbb{R}^n$. Moreover, for $r = \infty$, $h(\mathbf{x}) \to 0$ for $\|\mathbf{x}\| \to \infty$ except when $p = \infty$ or $q = \infty$. (6.2.11a) □

Theorem 6.2.2. *Let $f \in L^1(\mathbb{R}^n)$ and $g \in L^p(\mathbb{R}^n)$ with $1 \leq p \leq \infty$. Then*

$$\operatorname{supp}(f * g) \subset \operatorname{supp}(f) + \operatorname{supp}(g). \tag{6.2.12}$$

Proof. By Proposition 6.2.1, $(f * g)(\mathbf{x})$ is well defined and $(f * g)(\mathbf{x}) = \int_{\mathbb{R}^n} f(\mathbf{x} - \boldsymbol{\xi})g(\boldsymbol{\xi})d\boldsymbol{\xi}$. But $f(\mathbf{x} - \boldsymbol{\xi}) = 0$ for $\mathbf{x} - \boldsymbol{\xi} = \mathbf{y} \notin \operatorname{supp}(f) \Longrightarrow f(\mathbf{x} - \boldsymbol{\xi}) = 0$ for $\boldsymbol{\xi} = \mathbf{x} - \mathbf{y} \notin \mathbf{x} - \operatorname{supp}(f)$, where $\mathbf{x} - \operatorname{supp}(f) = \{\mathbf{z} : \mathbf{z} = \mathbf{x} - \mathbf{y} \text{ with } \mathbf{y} \in \operatorname{supp}(f)\}$. Hence,

$$(f * g)(\mathbf{x}) = \int_{(\mathbf{x}-\operatorname{supp}(f))\cap\operatorname{supp}(g)} f(\mathbf{x} - \boldsymbol{\xi})g(\boldsymbol{\xi})d\boldsymbol{\xi}.$$

But $\mathbf{x} \in (\operatorname{supp}(f) + \operatorname{supp}(g)) \Longrightarrow \mathbf{x} = \mathbf{y} + \boldsymbol{\xi}$ with $\mathbf{y} \in \operatorname{supp}(f)$, $\boldsymbol{\xi} \in \operatorname{supp}(g) \Longrightarrow \mathbf{x} - \mathbf{y} = \boldsymbol{\xi} \in \operatorname{supp}(g)$ with $\mathbf{y} \in \operatorname{supp}(f) \Longrightarrow (\mathbf{x} - \operatorname{supp}(f)) \cap \operatorname{supp}(g) \neq \emptyset$. Hence, $\mathbf{x} \notin (\operatorname{supp}(f) + \operatorname{supp}(g)) \Longrightarrow (\mathbf{x} - \operatorname{supp}(f)) \cap \operatorname{supp}(g) = \emptyset \Longrightarrow (f * g)(\mathbf{x}) = \int_{(\mathbf{x}-\operatorname{supp}(f))\cap\operatorname{supp}(g)} f(\mathbf{x} - \boldsymbol{\xi})g(\boldsymbol{\xi})d\boldsymbol{\xi} = 0$ for almost all $\mathbf{x} \notin (\operatorname{supp}(f) + \operatorname{supp}(g))$

$$\begin{aligned}
&\Longrightarrow \quad (f * g)(\mathbf{x}) = 0 \quad \text{a.e. in the complement of } (\operatorname{supp}(f) + \operatorname{supp}(g)) \\
&\Longrightarrow \quad (f * g)(\mathbf{x}) = 0 \quad \text{a.e. in the interior of } [(\operatorname{supp}(f) + \operatorname{supp}(g))^{\complement}] \\
&\Longrightarrow \quad \operatorname{supp}(f * g) \subset \overline{\operatorname{supp}(f) + \operatorname{supp}(g)} = \operatorname{supp}(f) + \operatorname{supp}(g).
\end{aligned}$$
□

Remark 6.2.2.

- If $\operatorname{supp}(f)$ and $\operatorname{supp}(g)$ are *both compact*, then $\operatorname{supp}(f) + \operatorname{supp}(g)$ is *compact* by (6.2.3), and $\operatorname{supp}(f * g)$ is also *compact*. (6.2.13)
- If only one of the supports, i.e. $\operatorname{supp}(f)$ *or* $\operatorname{supp}(g)$ is *compact*, then $\operatorname{supp}(f * g)$ is *not* compact in general. (6.2.14)

Regularization with the help of convolution

Regularizing functions Let $\phi \in \mathcal{C}_0^\infty(\mathbb{R}^n)$ (see Example 1.2.1 and Figure 1.1 for $n = 1$) be defined by:

$$\phi(\mathbf{x}) = \begin{cases} \exp(-1/(1 - \|\mathbf{x}\|^2)) & \text{for } \|\mathbf{x}\| < 1 \\ 0 & \text{for } \|\mathbf{x}\| \geq 1 \end{cases} \tag{6.2.15}$$

with the properties $\phi(\mathbf{x}) \geq 0$, $\operatorname{supp}(\phi) = \bar{B}(\mathbf{0}; 1)$, $\int_{\mathbb{R}^n} \phi(\mathbf{x}) d\mathbf{x} > 0$.

Define $\rho(\mathbf{x}) = \alpha\phi(\mathbf{x})$ with $\alpha > 0$ such that

$$\int_{\mathbb{R}^n} \rho(\mathbf{x}) d\mathbf{x} = 1, \quad \text{i.e. } \alpha = 1\Big/\int_{\mathbb{R}^n} \phi(\mathbf{x}) d\mathbf{x}. \tag{6.2.16}$$

Definition 6.2.1. For $\varepsilon > 0$, the functions

$$\rho_\varepsilon(\mathbf{x}) = \varepsilon^{-n} \rho(\mathbf{x}/\varepsilon) \tag{6.2.17}$$

with $\rho(\mathbf{x}/\varepsilon) = \alpha\phi(\mathbf{x}/\varepsilon)$ are called *regularizing functions* or *mollifiers*, with the following properties $\forall \varepsilon > 0$:

$$\rho_\varepsilon \in C_0^\infty(\mathbb{R}^n); \quad \operatorname{supp}(\rho_\varepsilon) = \text{closed ball } \bar{B}(\mathbf{0}; \varepsilon); \quad \int_{\mathbb{R}^n} \rho_\varepsilon(\mathbf{x}) d\mathbf{x} = 1. \tag{6.2.18}$$

In fact,

$$\int_{\mathbb{R}^n} \rho_\varepsilon(\mathbf{x}) d\mathbf{x} = \int_{\bar{B}(\mathbf{0};\varepsilon)} \varepsilon^{-n} \rho(\mathbf{x}/\varepsilon) d\mathbf{x} = \varepsilon^{-n} \int_{\bar{B}(\mathbf{0};1)} \rho(\mathbf{y}) \varepsilon^n d\mathbf{y} = 1, \tag{6.2.19}$$

which is obtained by the change of variables $\mathbf{x} = \varepsilon\mathbf{y}$ with $|J| = \varepsilon^n$.

An alternative notation used for a regularizing function is J_ε.

Definition 6.2.2. $\forall \varepsilon > 0$, the *convolution* $\rho_\varepsilon * u$ defined by

$$(\rho_\varepsilon * u)(\mathbf{x}) = \int_{\mathbb{R}^n} \rho_\varepsilon(\mathbf{x} - \boldsymbol{\xi}) u(\boldsymbol{\xi}) d\boldsymbol{\xi} \tag{6.2.20}$$

for functions u, for which the integral over $\mathbb{R}^n$ on the right-hand side of (6.2.20) is well defined, is called a *regularization* or *mollification* of u with the help of regularizing functions ρ_ε with $\varepsilon > 0$.

Remark 6.2.3.

- The importance of $\rho_\varepsilon * u$ is due to the fact that that $\rho_\varepsilon * u$ behaves much like u, but $\rho_\varepsilon * u$ is *extremely smooth* or *regular* – hence the name *regularization* or *mollification*. This fact will be established in a series of theorems below.

- In [30, p. 329], $\rho_\varepsilon * u$ is called *Sobolev's regularization* of u with ρ_ε, $\varepsilon > 0$. But [26, p. 71] states that the technique of regularization by convolution was introduced by Leray and Friedrichs. Finally, according to [10, p. 1075], in 1926 Norbert Wiener used regularization, i.e. convolution with compactly supported C^∞-functions, to approximate continuous functions f by smooth ones. Choosing $\varepsilon = 1/m$ in the right-hand side expression of (6.2.17), we get

$$\left(\frac{1}{m}\right)^{-n} \rho(m\mathbf{x}) = m^n \rho(m\mathbf{x}) \quad \forall m \in \mathbb{N}.$$

We set

$$\rho_m(\mathbf{x}) = m^n \rho(m\mathbf{x}) = \alpha m^n \phi(m\mathbf{x}), \tag{6.2.21}$$

where ϕ and α are defined by (6.2.15) and (6.2.16) respectively.

Definition 6.2.3. A sequence $(\rho_m)_{m\in\mathbb{N}}$ with ρ_m defined by (6.2.21) and having the properties, $\forall m \in \mathbb{N}$,

$$\rho_m(\mathbf{x}) \geq 0,\ \rho_m \in C_0^\infty(\mathbb{R}^n); \quad \operatorname{supp}(\rho_m) = \bar{B}(\mathbf{0}; 1/m); \quad \int_{\mathbb{R}^n} \rho_m(\mathbf{x})d\mathbf{x} = 1, \tag{6.2.22}$$

is called the *sequence of regularizing functions* or *sequence of mollifiers* and, $\forall m \in \mathbb{N}$,

$$(\rho_m * u)(\mathbf{x}) = \int_{\mathbb{R}^n} \rho_m(\mathbf{x} - \boldsymbol{\xi})u(\boldsymbol{\xi})d\boldsymbol{\xi} \tag{6.2.23}$$

is called the *regularization* of u, for which the integral in (6.2.23) is well defined, with the help of the sequence (ρ_m). *By abuse of terminology*, $(\rho_\varepsilon)_{\varepsilon>0}$ is also called a *sequence* of regularizing functions. (6.2.24)

Elementary properties of convolution

By change of variables, $\forall \varepsilon > 0$,

$$(\rho_\varepsilon * u)(\mathbf{x}) = \int_{\mathbb{R}^n} \rho_\varepsilon(\mathbf{x} - \boldsymbol{\xi})u(\boldsymbol{\xi})d\boldsymbol{\xi} = \int_{\mathbb{R}^n} \rho_\varepsilon(\boldsymbol{\xi})u(\mathbf{x} - \boldsymbol{\xi})d\boldsymbol{\xi} = (u * \rho_\varepsilon)(\mathbf{x}), \tag{6.2.25}$$

$$\alpha_1 \rho_\varepsilon * u_1 + \alpha_2 \rho_\varepsilon * u_2 = \rho_\varepsilon * (\alpha_1 u_1 + \alpha_2 u_2). \tag{6.2.26}$$

In (6.2.25) and (6.2.26) it is assumed that the defining integral of the convolution is well defined.

Proposition 6.2.3. *Let $u \in C^0(\mathbb{R}^n)$. Then $\rho_\varepsilon * u \to u$ uniformly on every compact subset K of $\mathbb{R}^n$ as $\varepsilon \to 0^+$.*

Proof. Let $K \subset\subset \mathbb{R}^n$ be any fixed compact subset of $\mathbb{R}^n$. Since $u \in C^0(\mathbb{R}^n)$, $\forall \eta > 0$, $\exists \delta = \delta(K, \eta) > 0$ such that

$$|u(\mathbf{x} - \boldsymbol{\xi}) - u(\mathbf{x})| < \eta \quad \forall \mathbf{x} \in K, \ \forall \boldsymbol{\xi} \tag{6.2.27}$$

with $\|\mathbf{x} - (\mathbf{x} - \boldsymbol{\xi})\| = \|\boldsymbol{\xi}\| < \delta$, i.e. $\forall \boldsymbol{\xi} \in B(\mathbf{0}; \delta)$. But

$$\begin{aligned}
(\rho_\varepsilon * u)(\mathbf{x}) - u(\mathbf{x}) &= (u * \rho_\varepsilon)(\mathbf{x}) - u(\mathbf{x}) \\
&= \int_{\mathbb{R}^n} u(\mathbf{x} - \boldsymbol{\xi})\rho_\varepsilon(\boldsymbol{\xi})d\boldsymbol{\xi} - u(\mathbf{x})\int_{\mathbb{R}^n} \rho_\varepsilon(\boldsymbol{\xi})d\boldsymbol{\xi} \quad (\text{by (6.2.18)}, \textstyle\int_{\mathbb{R}^n} \rho_\varepsilon(\boldsymbol{\xi})d\boldsymbol{\xi} = 1) \\
&= \int_{\mathbb{R}^n} [u(\mathbf{x} - \boldsymbol{\xi}) - u(\mathbf{x})]\rho_\varepsilon(\boldsymbol{\xi})d\boldsymbol{\xi} = \int_{B(\mathbf{0};\varepsilon)} [u(\mathbf{x} - \boldsymbol{\xi}) - u(\mathbf{x})]\rho_\varepsilon(\boldsymbol{\xi})d\boldsymbol{\xi}.
\end{aligned}$$

Hence, for any fixed compact $K \subset\subset \mathbb{R}^n$, $\forall \eta > 0$, $\exists \delta = \delta(K, \eta) > 0$ such that $\forall \mathbf{x} \in K$,

$$\begin{aligned}
|(\rho_\varepsilon * u)(\mathbf{x}) - u(\mathbf{x})| &\le \int_{B(\mathbf{0};\varepsilon)} |u(\mathbf{x} - \boldsymbol{\xi}) - u(\mathbf{x})|\rho_\varepsilon(\boldsymbol{\xi})d\boldsymbol{\xi}| \\
&\le \eta \int_{B(\mathbf{0};\varepsilon)} \rho_\varepsilon(\boldsymbol{\xi})d\boldsymbol{\xi} = \eta \quad \forall \varepsilon < \delta \quad (\text{by (6.2.27), (6.2.18)})
\end{aligned}$$

$\Longrightarrow \rho_\varepsilon * u \to u$ uniformly on $K \subset\subset \mathbb{R}^n$ as $\varepsilon \to 0^+$. But K is any compact of subset of $\mathbb{R}^n$. Hence, the result holds for every compact subset of $\mathbb{R}^n$. □

Lemma 6.2.1. *Let $f \in C_0(\mathbb{R}^n) \equiv \mathcal{D}^0(\mathbb{R}^n)$ be a continuous function with compact support in $\mathbb{R}^n$. Let functions f_ε be defined, $\forall \varepsilon > 0$, by:*

$$f_\varepsilon(\mathbf{x}) = (\rho_\varepsilon * f)(\mathbf{x}) = \int_{\mathbb{R}^n} \rho_\varepsilon(\mathbf{x} - \boldsymbol{\xi}) f(\boldsymbol{\xi})d\boldsymbol{\xi}. \tag{6.2.28}$$

Then

I. $f_\varepsilon \in C_0^\infty(\mathbb{R}^n) \equiv \mathcal{D}(\mathbb{R}^n)$;

II. $\lim_{\varepsilon\to 0^+} f_\varepsilon = \lim_{\varepsilon\to 0^+}(\rho_\varepsilon * f) = f$ *in* $\mathcal{D}^0(\mathbb{R}^n) \equiv C_0(\mathbb{R}^n)$, *i.e.* $C_0^\infty(\mathbb{R}^n)$ *is* dense *in* $C_0(\mathbb{R}^n)$. (6.2.29)

Proof.

I. *Compact support of f_ε $\forall \varepsilon > 0$:* Let $f \in C_0(\mathbb{R}^n)$. Then, $\exists$ a compact set $K \subset\subset \mathbb{R}^n$ such that $\text{supp}(f) = K$. Since $\rho_\varepsilon, f \in L^1(\mathbb{R}^n)$, using (6.2.12),

$$\text{supp}(f_\varepsilon) = \text{supp}(\rho_\varepsilon * f) \subset \text{supp}(\rho_\varepsilon) + \text{supp}(f) = \bar{B}(\mathbf{0}; \varepsilon) + K = K_\varepsilon \tag{6.2.30}$$

$\forall$ fixed $\varepsilon > 0$, $K_\varepsilon \subset\subset \mathbb{R}^n$ being a compact subset of $\mathbb{R}^n$ by (6.2.13).
Hence, $\forall \varepsilon < \varepsilon_0$ with $\varepsilon_0 > 0$, $\exists$ a *fixed compact* set $K_0 \subset\subset \mathbb{R}^n$ such that

$$\mathrm{supp}(f_\varepsilon) \subset K_0. \tag{6.2.31}$$

Continuity and infinite differentiability of f_ε $\forall \varepsilon > 0$: The *continuity* of f_ε (resp. the *infinite differentiability* of f_ε) follows from the classical theorems on integrals depending on parameter $\mathbf{x} \in K_\varepsilon$, since in the integral $\int_{\mathbb{R}^n} \rho_\varepsilon(\mathbf{x} - \boldsymbol{\xi}) f(\boldsymbol{\xi}) d\boldsymbol{\xi} = \int_{K=\mathrm{supp}(f)} \rho_\varepsilon(\mathbf{x}-\boldsymbol{\xi}) f(\boldsymbol{\xi}) d\boldsymbol{\xi}$ on a compact set K (precisely speaking on $(\mathbf{x} - \mathrm{supp}(\rho_\varepsilon)) \cap K$), the integrand function $F(\mathbf{x}-\boldsymbol{\xi}) = \rho_\varepsilon(\mathbf{x}-\boldsymbol{\xi}) f(\boldsymbol{\xi})$ is continuously dependent on parameter $\mathbf{x}$ (resp. infinitely differentiable with respect to the variables $x_1, x_2, \ldots, x_n$) on the compact set K_ε $\forall$ fixed $\varepsilon > 0$. Hence, $f_\varepsilon \in C^\infty(\mathbb{R}^n)$ and $\mathrm{supp}(f_\varepsilon) \subset K_\varepsilon \subset\subset \mathbb{R}^n$ $\forall$ fixed $\varepsilon > 0$, i.e. $f_\varepsilon \in C_0^\infty(\mathbb{R}^n)$. (6.2.32)

II. *Uniform convergence of $f_\varepsilon \to f$ in $\mathcal{D}^0(\mathbb{R}^n) \equiv C_0(\mathbb{R}^n)$*: Since $f \in C_0(\mathbb{R}^n) \Longrightarrow f \in C^0(\mathbb{R}^n)$, by Proposition 6.2.3, $f_\varepsilon = \rho_\varepsilon * f \to f$ *uniformly* as $\varepsilon \to 0^+$ on every compact subset of $\mathbb{R}^n$. Hence, $\forall \varepsilon < \varepsilon_0$ with $\varepsilon_0 > 0$, $\exists$ a *fixed compact* set K_0 such that $\mathrm{supp}(f_\varepsilon) \subset K_0$ *and* $f_\varepsilon \to f$ uniformly as $\varepsilon \to 0^+$, i.e. $f_\varepsilon \to f$ in $\mathcal{D}^0(\mathbb{R}^n) \equiv C_0(\mathbb{R}^n)$ as $\varepsilon \to 0^+$. Thus, $C_0^\infty(\mathbb{R}^n)$ is *dense* in $C_0(\mathbb{R}^n)$.

□

Corollary 6.2.1. *$\forall m \in \mathbb{N}$, let $f \in C_0^m(\mathbb{R}^n) \equiv \mathcal{D}^m(\mathbb{R}^n)$ with compact support in $\mathbb{R}^n$. Let $f_\varepsilon = \rho_\varepsilon * f$ be defined by* (6.2.28). *Then*

I. *$f_\varepsilon \in C_0^\infty(\mathbb{R}^n)$ $\forall$ fixed $\varepsilon > 0$;*

II. *$\forall$ multi-index $\boldsymbol{\alpha} = (\alpha_1, \alpha_2, \ldots, \alpha_n)$ with $0 \le |\boldsymbol{\alpha}| \le m$, $\partial_{\mathbf{x}}^{\boldsymbol{\alpha}} f_\varepsilon(\mathbf{x}) = \partial_{\mathbf{x}}^{\boldsymbol{\alpha}}(\rho_\varepsilon * f)(\mathbf{x}) \to \partial_{\mathbf{x}}^{\boldsymbol{\alpha}} f(\mathbf{x})$ in $\mathcal{D}^m(\mathbb{R}^n)$ as $\varepsilon \to 0^+$ (derivatives being in the usual pointwise sense), i.e. $\mathcal{D}(\mathbb{R}^n)$ is* dense *in $\mathcal{D}^m(\mathbb{R}^n)$.* (6.2.33)

Proof.

I. $f \in C_0^m(\mathbb{R}^n) \Longrightarrow f \in C_0(\mathbb{R}^n) \Longrightarrow f_\varepsilon \in C_0^\infty(\mathbb{R}^n)$ $\forall$ fixed $\varepsilon > 0$ by Lemma 6.2.1.

II. By change of variables $\mathbf{z} = \mathbf{x} - \boldsymbol{\xi}$ with $|J| = 1$, we have $f_\varepsilon(\mathbf{x}) = \int_{\mathbb{R}^n} \rho_\varepsilon(\mathbf{z}) f(\mathbf{x} - \mathbf{z}) d\mathbf{z}$. Then, $\forall$ multi-index $\boldsymbol{\alpha}$ with $|\boldsymbol{\alpha}| \le m$ and $\partial_{\mathbf{x}}^{\boldsymbol{\alpha}} = \frac{\partial^{|\boldsymbol{\alpha}|}}{\partial x_1^{\alpha_1} \ldots \partial x_n^{\alpha_n}}$,

$$\partial_{\mathbf{x}}^{\boldsymbol{\alpha}} f_\varepsilon(\mathbf{x}) = \partial_{\mathbf{x}}^{\boldsymbol{\alpha}} \int_{\mathbb{R}^n} \rho_\varepsilon(\mathbf{z}) f(\mathbf{x} - \mathbf{z}) d\mathbf{z} = \int_{\mathbb{R}^n} \rho_\varepsilon(z) \partial_{\mathbf{x}}^{\boldsymbol{\alpha}} f(\mathbf{x} - \mathbf{z}) d\mathbf{z},$$

i.e. the derivative $\partial_{\mathbf{x}}^{\boldsymbol{\alpha}} f_\varepsilon(\mathbf{x})$ can be obtained by differentiating in the *usual pointwise sense* under the integral sign $\int$. But $f \in C_0^m(\mathbb{R}^n) \Longrightarrow \partial_{\mathbf{x}}^{\boldsymbol{\alpha}} f \in C_0(\mathbb{R}^n)$ $\forall |\boldsymbol{\alpha}| \le m$. Then, by Lemma 6.2.1 the result is obtained, i.e. $\forall \varepsilon > \varepsilon_0$ with

$\varepsilon_0 > 0$, $\exists$ a *fixed compact set* $K_0 \subset\subset \mathbb{R}^n$ such that $\mathrm{supp}(f_\varepsilon) \subset K_0$, and, $\forall \boldsymbol{\alpha}$ with $0 \le |\boldsymbol{\alpha}| \le m$, $\partial_{\mathbf{x}}^{\boldsymbol{\alpha}} f_\varepsilon \to \partial_{\mathbf{x}}^{\boldsymbol{\alpha}} f$ *uniformly* as $\varepsilon \to 0^+$, i.e. $f_\varepsilon \to f$ in $\mathcal{D}^m(\mathbb{R}^n) \equiv C_0^m(\mathbb{R}^n)$ as $\varepsilon \to 0^+$. Hence, $C_0^\infty(\mathbb{R}^n) \equiv \mathcal{D}(\mathbb{R}^n)$ is *dense* in $\mathcal{D}^m(\mathbb{R}^n)$. □

Some important results of regularization in L^p-spaces

Proposition 6.2.4. *Let $\phi \in C_0(\mathbb{R}^n)$. Then $\rho_\varepsilon * \phi \in L^p(\mathbb{R}^n)$ and*

$$\rho_\varepsilon * \phi \to \phi \text{ in } L^p(\mathbb{R}^n) \quad \text{as } \varepsilon \to 0^+ \quad \text{for } 1 \le p < \infty, \tag{6.2.34}$$

i.e. $C_0(\mathbb{R}^n)$ is dense in $L^p(\mathbb{R}^n)$, $1 \le p < \infty$.

Proof. Let $\phi \in C_0(\mathbb{R}^n) \subset L^p(\mathbb{R}^n)$, $1 \le p < \infty$, with compact $\mathrm{supp}(\phi) = K \subset\subset \mathbb{R}^n$. By Theorem 6.2.2, $\mathrm{supp}(\rho_\varepsilon * \phi) \subset \bar{B}(\mathbf{0};\varepsilon) + K = K_\varepsilon$, K_ε being a compact set by (6.2.13) $\forall$ fixed $\varepsilon > 0$. Hence, $\exists$ a *fixed compact* set K_0 with $K, K_\varepsilon \subset K_0$ $\forall \varepsilon < \varepsilon_0$ with some $\varepsilon_0 > 0$. Then $(\rho_\varepsilon * \phi)(\mathbf{x}) = 0$, $\phi(\mathbf{x}) = 0$ $\forall \mathbf{x} \notin K_0$. By Lemma 6.2.1 and Proposition 6.2.3, $\rho_\varepsilon * \phi \in C_0^\infty(\mathbb{R}^n) \subset L^p(\mathbb{R}^n)$ for $1 \le p < \infty$ and $(\rho_\varepsilon * \phi)(\mathbf{x}) \to \phi(\mathbf{x})$ *uniformly* on compact set K_0, i.e. $\forall \eta > 0$, $\exists \delta = \delta(K_0, \eta) > 0$ such that $\forall \mathbf{x} \in K_0$, $|(\rho_\varepsilon * \phi)(\mathbf{x}) - \phi(\mathbf{x})| \le \frac{\eta}{[\mu(K_0)]^{1/p}}$ $\forall \varepsilon < \delta$. Then

$$\begin{aligned}
\|\rho_\varepsilon * \phi - \phi\|_{L^p(\mathbb{R}^n)}^p &= \int_{\mathbb{R}^n} |(\rho_\varepsilon * \phi)(\mathbf{x}) - \phi(\mathbf{x})|^p \, d\mathbf{x} \\
&= \int_{K_0} |(\rho_\varepsilon * \phi)(\mathbf{x}) - \phi(\mathbf{x})|^p \, d\mathbf{x} \\
&\le \frac{\eta^p}{\mu(K_0)} \mu(K_0) = \eta^p \quad \forall \varepsilon < \varepsilon_0^* = \min\{\varepsilon_0, \delta\}
\end{aligned}$$

$\Longrightarrow \forall \eta > 0$, $\exists \varepsilon_0^* > 0$ such that $\|\rho_\varepsilon * \phi - \phi\|_{L^p(\mathbb{R}^n)} \le \eta$ $\forall \varepsilon < \varepsilon_0^*$, i.e. $\lim_{\varepsilon \to 0^+} (\rho_\varepsilon * \phi) = \phi$ in $L^p(\mathbb{R}^n)$, $1 \le p < \infty$. □

Density result

Proposition 6.2.5. *$C_0^\infty(\mathbb{R}^n) \equiv \mathcal{D}(\mathbb{R}^n)$ is dense in $L^p(\mathbb{R}^n)$, $1 \le p < \infty$.*

Proof. Let $u \in L^p(\mathbb{R}^n)$, $1 \le p < \infty$. Let $\eta > 0$. Then, by virtue of the density of $C_0(\mathbb{R}^n)$ in $L^p(\mathbb{R}^n)$, $1 \le p < \infty$, $\exists \phi \in C_0(\mathbb{R}^n)$ with compact $\mathrm{supp}(\phi) = K \subset\subset \mathbb{R}^n$ such that

$$\|u - \phi\|_{L^p(\mathbb{R}^n)} < \eta/2. \tag{6.2.35}$$

By Lemma 6.2.1, $\rho_\varepsilon * \phi \in C_0^\infty(\mathbb{R}^n)$ $\forall$ fixed $\varepsilon > 0$, and by Proposition 6.2.4, $\rho_\varepsilon * \phi \to \phi$ in $L^p(\mathbb{R}^n)$, $1 \le p < \infty$, as $\varepsilon \to 0^+$, i.e. $\forall \eta > 0$, $\exists \varepsilon_0 > 0$ such that

$$\|\rho_\varepsilon * \phi - \phi\|_{L^p(\mathbb{R}^n)} < \eta/2 \quad \forall \varepsilon < \varepsilon_0. \tag{6.2.36}$$

Then, combining (6.2.35) and (6.2.36), we have: $\forall \eta > 0$, $\exists \varepsilon_0 > 0$ such that

$$\begin{aligned}\|u - \rho_\varepsilon * \phi\|_{L^p(\mathbb{R}^n)} &\le \|u - \phi\|_{L^p(\mathbb{R}^n)} + \|\phi - \rho_\varepsilon * \phi\|_{L^p(\mathbb{R}^n)} \\ &< \eta/2 + \eta/2 = \eta \quad \forall \varepsilon < \varepsilon_0,\end{aligned}$$

i.e. for $u \in L^p(\mathbb{R}^n)$, $1 \le p < \infty$, $\forall \eta > 0$, $\exists (\rho_\varepsilon * \phi) \in C_0^\infty(\mathbb{R}^n)$ with $\varepsilon < \varepsilon_0$ such that $\|u - \rho_\varepsilon * \phi\|_{L^p(\mathbb{R}^n)} < \eta$. Thus, $C_0^\infty(\mathbb{R}^n)$ is dense in $L^p(\mathbb{R}^n)$, $1 \le p < \infty$. □

Theorem 6.2.3. *Let $\rho_\varepsilon \in C_0^\infty(\mathbb{R}^n)$ be defined by* (6.2.17) $\forall \varepsilon > 0$. *Then:*

I. *If $u \in L^1_{\text{loc}}(\mathbb{R}^n)$, then*

$$\rho_\varepsilon * u \in C^\infty(\mathbb{R}^n) \quad \forall \varepsilon > 0. \tag{6.2.37}$$

Moreover, if $u \in L^1_{Loc}(\mathbb{R}^n)$ has compact *support, i.e.* $\operatorname{supp}(u) = K \subset\subset \mathbb{R}^n$ *is a compact subset of $\mathbb{R}^n$,*

$$\rho_\varepsilon * u \in C_0^\infty(\mathbb{R}^n) \quad \forall \varepsilon > 0. \tag{6.2.38}$$

II. *If $u \in L^p(\mathbb{R}^n)$ with $1 \le p < \infty$, then*

$$\rho_\varepsilon * u \in L^p(\mathbb{R}^n), \quad \|\rho_\varepsilon * u\|_{L^p(\mathbb{R}^n)} \le \|u\|_{L^p(\mathbb{R}^n)} \quad \forall \varepsilon > 0; \tag{6.2.39}$$

$$\lim_{\varepsilon \to 0+} \|u - \rho_\varepsilon * u\|_{L^p(\mathbb{R}^n)} = 0. \tag{6.2.40}$$

Proof.

I. $\forall$ *fixed* $\mathbf{x} \in \mathbb{R}^n$, the function $\boldsymbol{\xi} \in \mathbb{R}^n \longmapsto \rho_\varepsilon(\mathbf{x} - \boldsymbol{\xi})u(\boldsymbol{\xi})$ is *integrable* on $\mathbb{R}^n$, since $\rho_\varepsilon \in C_0^\infty(\mathbb{R}^n)$ has compact support in $\mathbb{R}^n$ and $u \in L^1_{\text{loc}}(\mathbb{R}^n)$ is integrable on compact subsets of $\mathbb{R}^n$. Hence, $(\rho_\varepsilon * u)(\mathbf{x})$ is well defined $\forall \mathbf{x} \in \mathbb{R}^n$, $\forall \varepsilon > 0$.

Continuity of $\rho_\varepsilon * u$: Let $\mathbf{x} \in \mathbb{R}^n$ be any fixed point and $(\mathbf{x}_m)$ be a sequence in $\mathbb{R}^n$ such that $\mathbf{x}_m \to \mathbf{x}$ in $\mathbb{R}^n$ as $m \to \infty$. Then, $\forall$ fixed $\boldsymbol{\xi} \in \mathbb{R}^n$, $\rho_\varepsilon(\mathbf{x}_m - \boldsymbol{\xi}) \to \rho_\varepsilon(\mathbf{x} - \boldsymbol{\xi})$ as $m \to \infty$ since $\rho_\varepsilon(\mathbf{x} - \boldsymbol{\xi})$ is continuous in $\mathbf{x}$. Now we will show that $(\rho_\varepsilon * u)(\mathbf{x}_m) \to (\rho_\varepsilon * u)(\mathbf{x})$ $\forall \varepsilon > 0$ with the help of Lebesgue's Dominated Convergence Theorem B.3.2.2 (Appendix B), as follows.

$\forall \varepsilon > 0$ define $f_m(\boldsymbol{\xi}) = \rho_\varepsilon(\mathbf{x}_m - \boldsymbol{\xi})u(\boldsymbol{\xi})$ $\forall m \in \mathbb{N}$ and $f(\boldsymbol{\xi}) = \rho_\varepsilon(\mathbf{x} - \boldsymbol{\xi})u(\boldsymbol{\xi})$ such that $f_m(\boldsymbol{\xi}) = [\rho_\varepsilon(\mathbf{x}_m - \boldsymbol{\xi})]u(\boldsymbol{\xi}) \to f(\boldsymbol{\xi})$ as $m \to \infty$ a.e. in $\mathbb{R}^n$. Moreover, $\mathbf{x}_m \to \mathbf{x}$ as $m \to \infty \Longrightarrow \exists M > 0$ such that $\|\mathbf{x}_m\| \le M$ $\forall m \in \mathbb{N}$ (since $(\mathbf{x}_m)$ is then a bounded sequence in $\mathbb{R}^n$).

$$(\rho_\varepsilon * u)(\mathbf{x}_m) = \int_{(\mathbf{x}_m - \operatorname{supp}(\rho_\varepsilon)) \cap \operatorname{supp}(u)} \rho_\varepsilon(\mathbf{x}_m - \boldsymbol{\xi})u(\boldsymbol{\xi})d\boldsymbol{\xi} \quad \forall \text{fixed } \varepsilon > 0.$$

But $\boldsymbol{\xi} \in \mathbf{x}_m - \operatorname{supp}(\rho_\varepsilon) \Longrightarrow \|\boldsymbol{\xi}\| \le \|\mathbf{x}_m\| + \|\mathbf{z}\| \le M + \varepsilon = M_\varepsilon$ with $\mathbf{z} \in \operatorname{supp}(\rho_\varepsilon) \Longrightarrow \mathbf{x}_m - \operatorname{supp}(\rho_\varepsilon)$ is a bounded set $\forall$ fixed $\varepsilon > 0$. Let K_0 be a fixed compact set such that $\mathbf{x}_m - \text{supp}(\rho_\varepsilon) \subset K_0\ \forall m \in \mathbb{N}$. Then $\boldsymbol{\xi} \notin K_0 \Longrightarrow \boldsymbol{\xi} \notin \mathbf{x}_m - \text{supp}(\rho_\varepsilon) \Longrightarrow \rho_\varepsilon(\mathbf{x}_m - \boldsymbol{\xi}) = 0$ for $\boldsymbol{\xi} \notin K_0$. Hence, $\forall m \in \mathbb{N}$,

$$|f_m(\boldsymbol{\xi})| = |\rho_\varepsilon(\mathbf{x}_m - \boldsymbol{\xi})\chi_{K_0}(\boldsymbol{\xi})u(\boldsymbol{\xi})| \leq \sup_{\mathbf{z}\in\mathbb{R}^n} \rho_\varepsilon(\mathbf{z})\chi_{K_0}(\boldsymbol{\xi})|u(\boldsymbol{\xi})|$$
$$= \|\rho_\varepsilon\|_{L^\infty(\mathbb{R}^n)}\chi_{K_0}(\boldsymbol{\xi})|u(\boldsymbol{\xi})| = g(\boldsymbol{\xi}) \quad \text{a.e. in } \mathbb{R}^n,$$

where

$$\chi_{K_0}(\boldsymbol{\xi}) = \begin{cases} 1 & \text{for } \boldsymbol{\xi} \in K_0 \\ 0 & \text{for } \boldsymbol{\xi} \notin K_0 \end{cases}$$

is the *characteristic function* of K_0 and $g(\boldsymbol{\xi}) \geq 0$ is the majoring integrable function in $\mathbb{R}^n$:

$$\int_{\mathbb{R}^n} g(\boldsymbol{\xi})d\boldsymbol{\xi} = \|\rho_\varepsilon\|_{L^\infty(\mathbb{R}^n)} \int_{K_0} |u(\boldsymbol{\xi})|d\boldsymbol{\xi} < +\infty,$$

since $u \in L^1_{\text{loc}}(\mathbb{R}^n)$.
Hence, $f_m(\boldsymbol{\xi}) \to f(\boldsymbol{\xi})$ a.e. in $\mathbb{R}^n$ as $m \to \infty$, and $\forall m \in \mathbb{N}$, $|f_m(\boldsymbol{\xi})| \leq g(\boldsymbol{\xi})$ a.e. in $\mathbb{R}^n$ with $g \in L^1(\mathbb{R}^n)$. Then, by Lebesgue's Dominated Convergence Theorem, $\forall$ fixed $\varepsilon > 0$,

$$(\rho_\varepsilon * u)(\mathbf{x}_m) = \int_{\mathbb{R}^n} \rho_\varepsilon(\mathbf{x}_m - \boldsymbol{\xi})u(\boldsymbol{\xi})d\boldsymbol{\xi} \to \int_{\mathbb{R}^n} \rho_\varepsilon(\mathbf{x} - \boldsymbol{\xi})u(\boldsymbol{\xi})d\boldsymbol{\xi}$$
$$= (\rho_\varepsilon * u)(\mathbf{x}) \quad \text{as } m \to \infty.$$

Thus, $\mathbf{x}_m \to \mathbf{x} \Longrightarrow (\rho_\varepsilon * u)(\mathbf{x}_m) \to (\rho_\varepsilon * u)(\mathbf{x})$ as $m \to \infty$. Hence, $\rho_\varepsilon * u \in C^0(\mathbb{R}^n)\ \forall$ fixed $\varepsilon > 0$.
Infinite differentiability of $(\rho_\varepsilon * u)$: Following the proof of Lemma 6.2.1, the differentiation can be carried out under the integral sign, and, $\forall |\boldsymbol{\alpha}| \in \mathbb{N}$, $\partial^{\boldsymbol{\alpha}}_{\mathbf{x}}(\rho_\varepsilon * u)(\mathbf{x}) = \int_{K_0} \partial^{\boldsymbol{\alpha}}_{\mathbf{x}}[\rho_\varepsilon(\mathbf{x} - \boldsymbol{\xi})]u(\boldsymbol{\xi})d\boldsymbol{\xi}$ exists $\forall \mathbf{x} \in \mathbb{R}^n$. Hence,

$$\rho_\varepsilon * u \in C^\infty(\mathbb{R}^n) \quad \forall \text{ fixed } \varepsilon > 0. \tag{6.2.41}$$

Compact support of $(\rho_\varepsilon * u)$ *for* u *with compact support*: Let $u \in L^1_{\text{loc}}(\mathbb{R}^n)$ such that $\text{supp}(u) = K \subset\subset \mathbb{R}^n$. Then $u \in L^1(\mathbb{R}^n)$ with $\text{supp}(u) = K$ and, by Theorem 6.2.2, $\text{supp}(\rho_\varepsilon * u) \subset \bar{B}(\mathbf{0}; \varepsilon) + K = K_\varepsilon$, K_ε being a compact set by (6.2.13) $\forall$ fixed $\varepsilon > 0$. Hence, from (6.2.41) $\rho_\varepsilon * u \in C^\infty(\mathbb{R}^n)$ with $\text{supp}(\rho_\varepsilon * u) \subset\subset \mathbb{R}^n\ \forall$ fixed $\varepsilon > 0 \Longrightarrow \rho_\varepsilon * u \in C^\infty_0(\mathbb{R}^n)\ \forall$ fixed $\varepsilon > 0$.

II. $\rho_\varepsilon \in C^\infty_0(\mathbb{R}^n)\ \forall \varepsilon > 0 \Longrightarrow \rho_\varepsilon \in L^1(\mathbb{R}^n)\ \forall \varepsilon > 0$. Then, by Proposition 6.2.1, $\forall u \in L^p(\mathbb{R}^n)$, $1 \leq p < \infty$, $\rho_\varepsilon * u \in L^p(\mathbb{R}^n)\ \forall \varepsilon > 0$ and

$$\|\rho_\varepsilon * u\|_{L^p(\mathbb{R}^n)} \leq \|\rho_\varepsilon\|_{L^1(\mathbb{R}^n)}\|u\|_{L^p(\mathbb{R}^n)} = 1 \cdot \|u\|_{L^p(\mathbb{R}^n)} = \|u\|_{L^p(\mathbb{R}^n)}. \tag{6.2.42}$$

In fact, we can prove (6.2.42) without using Proposition 6.2.1 as follows:
Case $1 < p < \infty$: For $1 < p < \infty$ with $\frac{1}{p} + \frac{1}{q} = 1$, using Hölder's inequality:

$$\begin{aligned}
|(\rho_\varepsilon * u)(\mathbf{x})| &\le \left| \int_{\mathbb{R}^n} [\rho_\varepsilon(\mathbf{x}-\boldsymbol{\xi})]^{1/q}([\rho_\varepsilon(\mathbf{x}-\boldsymbol{\xi})]^{1/p} u(\boldsymbol{\xi}))d\boldsymbol{\xi} \right| \\
&\le \left(\int_{\mathbb{R}^n} \rho_\varepsilon(\mathbf{x}-\boldsymbol{\xi})d\boldsymbol{\xi} \right)^{1/q} \left(\int_{\mathbb{R}^n} \rho_\varepsilon(\mathbf{x}-\boldsymbol{\xi})|u(\boldsymbol{\xi})|^p d\boldsymbol{\xi} \right)^{1/p}
\end{aligned}$$

$$\begin{aligned}
\Longrightarrow \quad & \int_{\mathbb{R}^n} |(\rho_\varepsilon * u)(\mathbf{x})|^p d\mathbf{x} \\
&\le \int_{\mathbb{R}^n} \left(\int_{\mathbb{R}^n} \rho_\varepsilon(\mathbf{x}-\boldsymbol{\xi})|u(\boldsymbol{\xi})|^p d\boldsymbol{\xi} \right) d\mathbf{x} \quad (\text{since } \textstyle\int_{\mathbb{R}^n} \rho_\varepsilon(\mathbf{x}-\boldsymbol{\xi})d\boldsymbol{\xi} = 1) \\
&= \int_{\mathbb{R}^n} \rho_\varepsilon(\mathbf{x}-\boldsymbol{\xi})d\mathbf{x} \int_{\mathbb{R}^n} |u(\boldsymbol{\xi})|^p d\boldsymbol{\xi} \quad (\text{by Fubini's Theorem 7.1.2C}) \\
&= 1 \cdot \|u\|^p_{L^p(\mathbb{R}^n)} < +\infty.
\end{aligned}$$

Case $p = 1$:

$$\begin{aligned}
\int_{\mathbb{R}^n} |(\rho_\varepsilon * u)(\mathbf{x})| d\mathbf{x} &\le \int_{\mathbb{R}^n} \int_{\mathbb{R}^n} \rho_\varepsilon(\mathbf{x}-\boldsymbol{\xi})|u(\boldsymbol{\xi})| d\boldsymbol{\xi} d\mathbf{x} \\
&= \int_{\mathbb{R}^n} \int_{\mathbb{R}^n} (\rho_\varepsilon(\mathbf{x}-\boldsymbol{\xi})d\mathbf{x})|u(\boldsymbol{\xi})| d\boldsymbol{\xi} \quad (\text{by Fubini's Theorem 7.1.2C}) \\
&= \|u\|_{L^1(\mathbb{R}^n)}.
\end{aligned}$$

Now we will prove (6.2.40). Let $\eta > 0$. Then, by virtue of the density of $C_0^\infty(\mathbb{R}^n)$ in $L^p(\mathbb{R}^n)$, $1 \le p < \infty$ (see Proposition 6.2.5), $\exists \phi \in C_0^\infty(\mathbb{R}^n)$ such that

$$\|u - \phi\|_{L^p(\mathbb{R}^n)} \le \eta/3. \tag{6.2.43}$$

Using (6.2.42),

$$\|\rho_\varepsilon * u - \rho_\varepsilon * \phi\|_{L^p(\mathbb{R}^n)} = \|\rho_\varepsilon * (u - \phi)\|_{L^p(\mathbb{R}^n)} \le \|u - \phi\|_{L^p(\mathbb{R}^n)} \le \eta/3. \tag{6.2.44}$$

Moreover, $\phi \in C_0^\infty(\mathbb{R}^n)$, $\rho_\varepsilon \in C_0^\infty(\mathbb{R}^n) \Longrightarrow \rho_\varepsilon * \phi \to \phi$ *uniformly* on *every compact* subset of $\mathbb{R}^n$ by Proposition 6.2.4, and, by Theorem 6.2.2, $\operatorname{supp}(\rho_\varepsilon * \phi) \subset \bar{B}(\mathbf{0}; \varepsilon) + \operatorname{supp}(\phi) \subset K_\varepsilon$, K_ε being a compact set by (6.2.13). Hence, for all sufficiently small $\varepsilon < \varepsilon_0$ with $\varepsilon_0 > 0$, $\exists$ a fixed compact set K_0 such that $K_\varepsilon \subset K_0$. Then $\rho_\varepsilon * \phi \to \phi$ *uniformly* on K_0 as $\varepsilon \to 0^+ \Longrightarrow \forall \mathbf{x} \in K_0$, $\forall \eta > 0, \exists \delta = \delta(K_0, \eta) > 0$ such that

$$|(\rho_\varepsilon * \phi)(\mathbf{x}) - \phi(\mathbf{x}| \le \frac{\eta}{3[\mu(K_0)]^{1/p}} \quad \forall \varepsilon < \varepsilon_0^*$$

with $\varepsilon_0^* = \min\{\varepsilon_0, \delta\} > 0$, where $\mu(K_0) = n$-dimensional Lebesgue volume measure of K_0. Then, $\forall \eta > 0$, $\exists \varepsilon_0^* > 0$ such that, $\forall \varepsilon < \varepsilon_0^*$,

$$\begin{aligned}\|\rho_\varepsilon * \phi - \phi\|^p_{L^p(\mathbb{R}^n)} &= \int_{K_0} |(\rho_\varepsilon * \phi)(\mathbf{x}) - \phi(\mathbf{x})|^p d\mathbf{x} \\ &\leq \frac{\eta^p}{3^p[\mu(K_0)]}\mu(K_0) = \left(\frac{\eta}{3}\right)^p. \end{aligned} \tag{6.2.45}$$

Hence, from (6.2.43)–(6.2.45), $\forall \eta > 0$, $\exists \varepsilon_0^* > 0$ such that

$$\begin{aligned}\|\rho_\varepsilon * u - u\|_{L^p(\mathbb{R}^n)} &\leq \|\rho_\varepsilon * u - \rho_\varepsilon * \phi\|_{L^p(\mathbb{R}^n)} \\ &\quad + \|\rho_\varepsilon * \phi - \phi\|_{L^p(\mathbb{R}^n)} + \|\phi - u\|_{L^p(\mathbb{R}^n)} \\ &\leq \eta/3 + \eta/3 + \eta/3 = \eta \quad \forall \varepsilon < \varepsilon_0^* \end{aligned}$$

$\Longrightarrow \lim_{\varepsilon\to 0^+} \|\rho_\varepsilon * u - u\|_{L^p(\mathbb{R}^n)} = 0.$ □

Construction of cut-off functions $\phi \in C_0^\infty(\Omega)$ with $\phi(\mathbf{x}) = 1$ $\forall \mathbf{x} \in K \subset\subset \Omega$

In many situations we are to replace a given continuous function by another one with *compact* support such that the two functions are identical on a *large compact* set. This is achieved by multiplying the given function by a suitable *cut-off function* (see also other forms of cut-off function in (6.8.55), (7.5.3)). For example, for a given function $\psi \in C(\Omega)$, *continuous* on Ω with *non-compact* support in Ω, and a given *compact* set $K \subset\subset \Omega$, it is required to construct a continuous function $\phi_0 \in C_0(\Omega)$ with *compact* support in Ω such that $\phi_0(\mathbf{x}) = \psi(\mathbf{x})$ $\forall \mathbf{x} \in K \subset\subset \Omega$. Let $\phi \in C_0^\infty(\Omega) = \mathcal{D}(\Omega)$ be a test function with $\phi(\mathbf{x}) = 1$ $\forall \mathbf{x} \in K_0$ with $K \subset K_0 \subset \operatorname{supp}(\phi) \subset\subset \Omega$ and $0 \leq \phi(\mathbf{x}) \leq 1$ $\forall \mathbf{x} \in \Omega$. Then ϕ is the required cut-off function, since $\phi_0 = \phi\psi$ has the desired properties: $\phi_0 \in C_0(\Omega)$, $\phi_0(\mathbf{x}) = \phi(\mathbf{x})\psi(\mathbf{x}) = 1 \cdot \psi(\mathbf{x}) = \psi(\mathbf{x})$ $\forall \mathbf{x} \in K \subset\subset \Omega$, $\operatorname{supp}(\phi_0) = \operatorname{supp}(\phi) \cap \operatorname{supp}(\psi) \subset \operatorname{supp}(\phi) \subset\subset \Omega$.[1] The existence of such a cut-off function $\phi \in \mathcal{D}(\Omega)$ is given by:

Theorem 6.2.4. *Let $\Omega \subset \mathbb{R}^n$ be an open subset of $\mathbb{R}^n$ and $K \subset\subset \Omega$ be a compact subset of Ω. Then, $\exists \phi \in C_0^\infty(\Omega)$ with $0 \leq \phi(\mathbf{x}) \leq 1$ $\forall \mathbf{x} \in \Omega$ such that $\phi = 1$ in a compact neighbourhood $K_\varepsilon = \{\mathbf{y} : \mathbf{y} \in \Omega,\ d(\mathbf{y}, K) = \inf_{\mathbf{x}\in K} \|\mathbf{x}-\mathbf{y}\| \leq \varepsilon\}$ of K in Ω for a* sufficiently small $\varepsilon > 0$ *(see* (1.2.19) *and Figure 1.6). Consequently, $\phi(\mathbf{x}) = 1$ $\forall \mathbf{x} \in K$ with $K \subset K_\varepsilon$.*

[1] See Section 5.6, Chapter 5, where such a cut-off function $\alpha \in \mathcal{D}(\mathbb{R}^n)$, with $\alpha(\mathbf{x}) = 1$ $\forall \mathbf{x} \in K_0$, K_0 being a (compact) neighbourhood of $\operatorname{supp}(T) = K \subset\subset \mathbb{R}^n$, and $0 \leq \alpha(\mathbf{x}) \leq 1$ $\forall \mathbf{x} \in \mathbb{R}^n$, has been used to define $\langle T, \phi\rangle = \langle T, \alpha\phi\rangle$ for $\phi \in \mathcal{E}(\mathbb{R}^n) \equiv C^\infty(\mathbb{R}^n)$ with *non-compact* support and distribution $T \in \mathcal{E}'(\mathbb{R}^n)$ with *compact* support.

Proof. By *sufficiently small* $\varepsilon > 0$, we mean that $\|\mathbf{x} - \mathbf{y}\| \geq 4\varepsilon$ for $\mathbf{x} \in K$ and $\mathbf{y} \in \Omega^{\complement}$ = complement of Ω in $\mathbb{R}^n$. For sufficiently small $\varepsilon > 0$, define compact neighbourhoods $K_\varepsilon, K_{2\varepsilon}, K_{3\varepsilon}$ of K in Ω by $K_\delta = \{\mathbf{y} : \mathbf{y} \in \Omega, d(\mathbf{y}, K) = \inf_{\mathbf{x} \in K} \|\mathbf{x} - \mathbf{y}\| \leq \delta, \delta > 0\}$ with $K \subset K_\delta$ and $\delta = \varepsilon, 2\varepsilon, 3\varepsilon$ (see (1.2.19) and Figure 1.6). (6.2.46)

Let $\chi_{K_{2\varepsilon}}$ be the characteristic function of $K_{2\varepsilon}$, i.e. $\chi_{K_{2\varepsilon}}(\mathbf{x}) = 1$ for $\mathbf{x} \in K_{2\varepsilon}$ and $= 0$ for $\mathbf{x} \in \mathbb{R}^n \setminus K_{2\varepsilon}$. For $\varepsilon > 0$ satisfying (6.2.46), let ρ_ε be the regularizing functions defined in (5.2.18)–(6.2.19). Then, define $\phi = \chi_{K_{2\varepsilon}} * \rho_\varepsilon \in C^\infty(\mathbb{R}^n)$ by Theorem 6.2.3, since $\chi_{K_{2\varepsilon}} \in L^2(\mathbb{R}^n)$, and $\operatorname{supp}(\phi) \subset \operatorname{supp}(\chi_{K_{2\varepsilon}}) + \operatorname{supp}(\rho_\varepsilon) = K_{2\varepsilon} + \bar{B}(\mathbf{0}; \varepsilon) \implies \operatorname{supp}(\phi) \subset K_{3\varepsilon} \subset\subset \Omega$ (by Theorem 6.2.2) $\implies \phi \in C_0^\infty(\Omega)$ with $\phi(\mathbf{x}) = 1$ in K_ε, $K \subset K_\varepsilon$. In fact, $1 - \phi = (1 - \chi_{K_{2\varepsilon}}) * \rho_\varepsilon$, since $1 * \rho_\varepsilon = 1$. Then $\operatorname{supp}(1 - \phi) \subset \operatorname{supp}(1 - \chi_{K_{2\varepsilon}}) + \bar{B}(\mathbf{0}; \varepsilon) = \overline{K_{2\varepsilon}^{\complement}} + \bar{B}(\mathbf{0}; \varepsilon) \implies 1 - \phi(\mathbf{x}) = 0$ for $\mathbf{x} \in K_\varepsilon \implies \phi(\mathbf{x}) = 1$ for $\mathbf{x} \in K_\varepsilon$ with $K \subset K_\varepsilon$. □

6.3 Convolution of two distributions

Convolution $T * \phi$ of a distribution T and a test function ϕ

Let $f \in L^1(\mathbb{R}^n)$ and $\phi \in \mathcal{D}(\mathbb{R}^n) \subset L^1(\mathbb{R}^n)$. Then $f * \phi$ is well defined by:

$$\forall \phi \in \mathcal{D}(\mathbb{R}^n), (f * \phi)(\mathbf{x}) = \int_{\mathbb{R}^n} f(\mathbf{x} - \boldsymbol{\xi})\phi(\boldsymbol{\xi})d\boldsymbol{\xi} = \int_{\mathbb{R}^n} f(\boldsymbol{\xi})\phi(\mathbf{x} - \boldsymbol{\xi})d\boldsymbol{\xi}$$
$$= \int_{\mathbb{R}^n} f(\boldsymbol{\xi})(\tau_{\mathbf{x}}\check{\phi})(\boldsymbol{\xi}) = \langle f, \tau_{\mathbf{x}}\check{\phi}\rangle = \langle T_f, \tau_{\mathbf{x}}\check{\phi}\rangle, \quad (6.3.1)$$

where $\check{\phi}, \tau_{\mathbf{x}}\check{\phi}$ are defined by:

$$\check{\phi}(\mathbf{x}) = \phi(-\mathbf{x}) \quad \forall \mathbf{x} \in \mathbb{R}^n,$$
$$(\tau_{\mathbf{x}}\phi)(\boldsymbol{\xi}) = \phi(\boldsymbol{\xi} - \mathbf{x}) \quad \forall \boldsymbol{\xi} \in \mathbb{R}^n, \ \forall \text{ fixed } \mathbf{x} \in \mathbb{R}^n,$$
$$(\tau_{\mathbf{x}}\check{\phi})(\boldsymbol{\xi}) = \check{\phi}(\boldsymbol{\xi} - \mathbf{x}) = \phi(-(\boldsymbol{\xi} - \mathbf{x})) = \phi(\mathbf{x} - \boldsymbol{\xi});$$

$T_f \in \mathcal{D}'(\mathbb{R}^n)$ is the distribution defined by $f \in L^1(\mathbb{R}^n)$.

Equation (6.3.1) suggests the definition of the convolution $T * \phi$ of an arbitrary distribution $T \in \mathcal{D}'(\mathbb{R}^n)$ and $\phi \in \mathcal{D}(\mathbb{R}^n)$ by:

Definition 6.3.1A. The convolution $T * \phi$ of $T \in \mathcal{D}'(\mathbb{R}^n)$ and $\phi \in \mathcal{D}(\mathbb{R}^n)$ is defined by:

$$(T * \phi)(\mathbf{x}) = \langle T, \tau_{\mathbf{x}}\check{\phi}\rangle = \langle T_{\boldsymbol{\xi}}, \check{\phi}(\boldsymbol{\xi} - \mathbf{x})\rangle = \langle T_{\boldsymbol{\xi}}, \phi(\mathbf{x} - \boldsymbol{\xi})\rangle. \quad (6.3.2)$$

Remark 6.3.1. $T * \phi$ is a C^∞-function on $\mathbb{R}^n$ (see Lemma 6.1.1, Corollary 6.1.1 and also Section 6.4 later). In particular, $(T * \phi)(\mathbf{0}) = \langle T, \check{\phi}\rangle$.

Let $T = T_f$ and $\check{T} = T_{\check{f}}$ be the regular distributions in $\mathcal{D}'(\mathbb{R}^n)$ defined by f and $\check{f}$ respectively. Then

$$\begin{aligned}\langle \check{T}, \phi\rangle = \langle \check{f}, \phi\rangle &= \int_{\mathbb{R}^n} f(-\mathbf{x})\phi(\mathbf{x})d\mathbf{x} = \int_{\mathbb{R}^n} f(\mathbf{x})\phi(-\mathbf{x})d\mathbf{x}\\ &= \langle f, \check{\phi}\rangle = \langle T, \check{\phi}\rangle \quad \forall \phi \in \mathcal{D}(\mathbb{R}^n)\end{aligned}$$

$\Longrightarrow \langle \check{T}, \phi\rangle = \langle T, \check{\phi}\rangle\ \forall \phi \in \mathcal{D}(\mathbb{R}^n)$, which suggests the following definition:

Definition 6.3.1B. For every distribution $T \in \mathcal{D}'(\mathbb{R}^n)$, the distribution $\check{T}$ is defined by:

$$\langle \check{T}, \phi\rangle = \langle T, \check{\phi}\rangle \quad \forall \phi \in \mathcal{D}(\mathbb{R}^n). \tag{6.3.3}$$

Since the mapping $\phi \in \mathcal{D}(\mathbb{R}^n) \longmapsto \langle \check{T}, \phi\rangle$ is continuous on $\mathcal{D}(\mathbb{R}^n)$, $\check{T} \in \mathcal{D}'(\mathbb{R}^n)$ $\forall T \in \mathcal{D}'(\mathbb{R}^n)$. Hence,

$$\langle \check{T}, \check{\phi}\rangle = \langle T, \check{\check{\phi}}\rangle = \langle T, \phi\rangle \quad \forall \phi \in D(\mathbb{R}^n). \tag{6.3.4}$$

$$\check{T} * \phi \text{ is a } C^\infty\text{-function } \forall \phi \in \mathcal{D}(\mathbb{R}^n) \text{ and } \check{T} * \phi \notin \mathcal{D}(\mathbb{R}^n). \tag{6.3.5}$$

Case when both distributions are functions Let $f, g \in L^1(\mathbb{R}^n)$ such that $T = T_f$ and $S = S_g$ are regular distributions defined by f and g respectively and $\check{S} = S_{\check{g}}$. Then $T * S$ is defined and a regular distribution defined by $f * g$. In fact, $\forall \phi \in \mathcal{D}(\mathbb{R}^n)$,

$$\langle T * S, \phi\rangle = \langle f * g, \phi\rangle = \int_{\mathbb{R}^n}\int_{\mathbb{R}^n} f(\mathbf{x}-\mathbf{t})g(\mathbf{t})\phi(\mathbf{x})d\mathbf{t}d\mathbf{x} \tag{6.3.6}$$

$$= \int_{\mathbb{R}^n}\int_{\mathbb{R}^n} f(\boldsymbol{\xi})g(-\boldsymbol{\eta})\phi(\boldsymbol{\xi}-\boldsymbol{\eta})d\boldsymbol{\xi}d\boldsymbol{\eta} \tag{6.3.7}$$

(by change of variables: $\boldsymbol{\xi} = \mathbf{x}-\mathbf{t}$, $\boldsymbol{\eta} = -\mathbf{t}$ with $|J| = 1$)

$$\begin{aligned}&= \int_{\mathbb{R}^n} f(\boldsymbol{\xi})\Big(\int_{\mathbb{R}^n} \check{g}(\boldsymbol{\eta})\phi(\boldsymbol{\xi}-\boldsymbol{\eta})d\boldsymbol{\eta}\Big)d\boldsymbol{\xi}\\ &= \int_{\mathbb{R}^n} f(\boldsymbol{\xi})(\check{g} * \phi)(\boldsymbol{\xi})d\boldsymbol{\xi}\\ &= \langle f, \check{g} * \phi\rangle = \langle T_f, S_{\check{g}} * \phi\rangle = \langle T, \check{S} * \phi\rangle\end{aligned}$$

$$\Longrightarrow \quad \langle T * S, \phi\rangle = \langle T, \check{S} * \phi\rangle \quad \forall \phi \in \mathcal{D}(\mathbb{R}^n). \tag{6.3.8}$$

Since $\check{S} * \phi$ is a C^∞-function and does *not* belong to $\mathcal{D}(\mathbb{R}^n)$, $\langle T, \check{S} * \phi\rangle$ is well defined $\forall T \in \mathcal{E}'(\mathbb{R}^n) \subset \mathcal{D}'(\mathbb{R}^n)$ i.e. $\forall$ distributions T with *compact support*. Thus, (6.3.6) can be accepted as the definition of the convolution $T * S$, when $T \in \mathcal{E}'(\mathbb{R}^n) \subset \mathcal{D}'(\mathbb{R}^n)$ is a distribution with *compact* support and $S \in \mathcal{D}'(\mathbb{R}^n)$ is an *arbitrary* distribution. Then, $T * S$ is a distribution, since the mapping $\phi \in \mathcal{D}(\mathbb{R}^n) \longmapsto \langle T, \check{S} * \phi\rangle$ is continuous, and we have:

Definition 6.3.2. The convolution $T * S$ of $T \in \mathcal{E}'(\mathbb{R}^n) \subset \mathcal{D}'(\mathbb{R}^n)$ and $S \in \mathcal{D}'(\mathbb{R}^n)$ is a distribution defined by:

$$\langle T * S, \phi\rangle = \langle T, \check{S} * \phi\rangle \quad \forall \phi \in \mathcal{D}(\mathbb{R}^n). \tag{6.3.9}$$

Alternative definition of the convolution $T * S$ in terms of the tensor product $T \otimes S$

Again, we consider $f, g \in L^1(\mathbb{R}^n)$ with $T = T_f$ and $S = T_g$. Then $h = f * g \in L^1(\mathbb{R}^n)$ is a distribution defined, $\forall \phi \in \mathcal{D}(\mathbb{R}^n)$, by:

$$\langle T_h, \phi\rangle = \langle h, \phi\rangle = \langle f * g, \phi\rangle = \int_{\mathbb{R}^n} \Big(\int_{\mathbb{R}^n} f(\mathbf{x}-\mathbf{t}) g(\mathbf{t}) d\mathbf{t} \Big) \phi(\mathbf{x}) d\mathbf{x}.$$

By changing the variables, $\boldsymbol{\xi} = \mathbf{x} - \mathbf{t}$, $\boldsymbol{\eta} = \mathbf{t}$ with $|J| = 1$, we have, $\forall \phi \in \mathcal{D}(\mathbb{R}^n)$,

$$\begin{aligned} \langle f * g, \phi\rangle &= \int_{\mathbb{R}^n} f(\xi) \Big(\int_{\mathbb{R}^n} g(\eta)\phi(\xi+\eta) d\eta \Big) d\xi \\ &= \langle f(\xi), \langle g(\eta), \phi(\xi+\eta)\rangle\rangle = \langle f(\xi) \otimes g(\eta), \phi(\xi+\eta)\rangle \end{aligned} \tag{6.3.10}$$

$$\begin{aligned} \Longrightarrow \quad \langle T_f * T_g, \phi\rangle &= \langle T_{f*g}, \phi\rangle = \langle f * g, \phi\rangle \\ &= \langle f(\xi) \otimes g(\eta), \phi(\xi+\eta)\rangle \quad \forall \phi \in \mathcal{D}(\mathbb{R}^n). \end{aligned}$$

Thus, $\forall f, g, \in L^1(\mathbb{R}^n)$, the *convolution* $T_f * T_g = T_{f*g} = f * g$ is a *distribution* in $\mathcal{D}'(\mathbb{R}^n)$ defined by (5.3.10) in terms of the *tensor product* $f(\boldsymbol{\xi}) \otimes g(\boldsymbol{\eta})$. Hence, it suggests the possibility of defining the convolution $T * S$, if it exists, of two distributions $T \in \mathcal{D}'(\mathbb{R}^n)$, $S \in \mathcal{D}'(\mathbb{R}^n)$ by: $\forall \phi \in \mathcal{D}(\mathbb{R}^n)$, $\langle T * S, \phi\rangle = \langle T_{\xi} \otimes S_{\eta}, \phi(\xi+\eta)\rangle$ or, equivalently,

$$\langle T * S, \phi\rangle = \langle T(\xi) \otimes S(\eta), \phi(\xi+\eta)\rangle. \tag{6.3.11}$$

Remark 6.3.2. An immediate remark is in order: for $T \in \mathcal{E}'(\mathbb{R}^n)$ and $S \in \mathcal{D}'(\mathbb{R}^n)$, the two definitions (6.3.9) and(6.3.11) coincide. In fact, using the definitions of tensor product $T_{\xi} \otimes S_{\eta}$, $\check{S}$ and $\check{\phi}(\xi - \eta) = \check{\phi}_{\xi}(-\eta) = \phi_{\xi}(\eta) = \phi(\xi+\eta)$ $\forall$ fixed $\boldsymbol{\xi} \in \mathbb{R}^n$, we get

$$\begin{aligned} \langle T * S, \phi\rangle &= \langle T_{\xi} \otimes S_{\eta}, \phi(\xi+\eta)\rangle = \langle T_{\xi}, \langle S_{\eta}, \check{\phi}(\xi+\eta)\rangle\rangle \\ &= \langle T_{\xi}, (S * \check{\phi}(\xi))\rangle = \langle T, \check{S} * \phi\rangle \quad \forall \phi \in \mathcal{D}(\mathbb{R}^n). \end{aligned} \tag{6.3.12}$$

For the moment we ignore this most important case of $T \in \mathcal{E}'(\mathbb{R}^n) \subset \mathcal{D}'(\mathbb{R}^n)$ with compact support and $S \in \mathcal{D}'(\mathbb{R}^n)$; we will start with the *probable* definition (6.3.9), which can be accepted as the definition of the convolution $T * S$ of two distributions T and S *if* the right-hand side of (6.3.9) is well defined for every $\phi(\mathbf{x}) \in \mathcal{D}(\mathbb{R}^n)$.

Since $T_{\boldsymbol{\xi}} \otimes S_{\boldsymbol{\eta}} \in \mathcal{D}'(\mathbb{R}^n \times \mathbb{R}^n)$ is well defined as a distribution for *arbitrary* T *and* S, the right-hand side of (6.3.9) will make sense if $\phi(\boldsymbol{\xi} + \boldsymbol{\eta}) \in \mathcal{D}(\mathbb{R}^{2n})$, for $\phi(\mathbf{x}) \in \mathcal{D}(\mathbb{R}^n)$.

But for $\phi(\mathbf{x}) \in \mathcal{D}(\mathbb{R}^n)$, $\phi(\boldsymbol{\xi} + \boldsymbol{\eta}) \in C^\infty(\mathbb{R}^n \times \mathbb{R}^n)$ (as the function $\phi(\mathbf{x}(\boldsymbol{\xi}, \boldsymbol{\eta})) = \phi(\boldsymbol{\xi} + \boldsymbol{\eta})$ with $\mathbf{x} = \boldsymbol{\xi} + \boldsymbol{\eta}$ of $\boldsymbol{\xi}$ and $\boldsymbol{\eta}$), $\operatorname{supp}\phi(\boldsymbol{\xi} + \boldsymbol{\eta})$ is *not* compact in $\mathbb{R}^{2n}$, and consequently $\phi(\boldsymbol{\xi} + \boldsymbol{\eta}) \notin \mathcal{D}(\mathbb{R}^{2n})$.

For example, for $n = 1$, $\phi(x) \in \mathcal{D}(\mathbb{R})$ is defined by:

$$\phi(x) = \begin{cases} e^{-\frac{1}{1-x^2}} & \text{for } |x| \le 1 \\ 0 & \text{for } |x| > 1 \end{cases}$$

with $K = \operatorname{supp}(\phi) = [-1, 1]$.

Set $x = \xi + \eta$, with $\xi, \eta \in \mathbb{R}$. Then $\operatorname{supp}(\phi(\xi + \eta)) = \{(\xi, \eta) : (\xi, \eta) \in \mathbb{R}^2 \text{ with } \xi + \eta \in [-1, 1]\}$, which is the infinite strip $-1 \le \xi + \eta \le +1$ parallel to $\xi + \eta = 0$ (see Figure 6.1) and *not* bounded in $\mathbb{R}^2$, and hence *not* compact in $\mathbb{R}^2$.

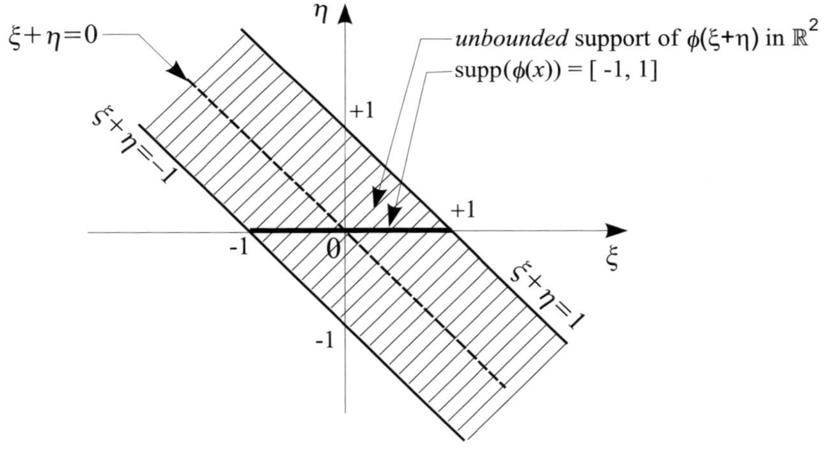

Figure 6.1 Infinite strip bounded by $-1 \le \xi + \eta \le +1$, parallel to $\xi + \eta = 0$, defining the *unbounded* support $\phi(\xi + \eta)$ in $\mathbb{R}^2$

Thus, we arrive at the important conclusion:

For *arbitrary* distributions $T, S \in \mathcal{D}'(\mathbb{R}^n)$, the right-hand side of (6.3.9) will *not* make sense, since $\phi(\boldsymbol{\xi} + \boldsymbol{\eta}) \notin \mathcal{D}(\mathbb{R}^{2n})$ for $\phi(\mathbf{x}) \in \mathcal{D}(\mathbb{R}^n)$, i.e. for arbitrary distributions $T, S \in \mathcal{D}'(\mathbb{R}^n)$, $T * S$ is *not* defined. Now we identify the situations in which the right-hand side of (6.3.9) *will* have a meaning. Let $A = \operatorname{supp}(T_{\boldsymbol{\xi}})$, $B = \operatorname{supp}(S_{\boldsymbol{\eta}})$.

Then, from (6.1.16),

$$\operatorname{supp}(T_{\boldsymbol{\xi}} \otimes S_{\boldsymbol{\eta}}) = A \times B \subset \mathbb{R}^n \times \mathbb{R}^n \quad \text{with } A \times B = \{(\boldsymbol{\xi}, \boldsymbol{\eta}) : \boldsymbol{\xi} \in A, \boldsymbol{\eta} \in B\}.$$

We state the result as follows:

Theorem 6.3.1. *If the set*

$$\begin{aligned} &\operatorname{supp}(T_{\boldsymbol{\xi}} \otimes S_{\boldsymbol{\eta}}) \cap \operatorname{supp}(\phi(\boldsymbol{\xi} + \boldsymbol{\eta})) \\ &= \{(\boldsymbol{\xi}, \boldsymbol{\eta}) : \boldsymbol{\xi} \in A, \boldsymbol{\eta} \in B, \boldsymbol{\xi} + \boldsymbol{\eta} \in \operatorname{supp}(\phi), \phi \in \mathcal{D}(\mathbb{R}^n)\} \end{aligned} \tag{6.3.13}$$

is bounded *in $\mathbb{R}^n \times \mathbb{R}^n$ (i.e. for $\boldsymbol{\xi} \in A$, $\boldsymbol{\eta} \in B$, $\boldsymbol{\xi} + \boldsymbol{\eta}$ will only remain bounded if both $\boldsymbol{\xi}$ and $\boldsymbol{\eta}$ remain bounded), the right-hand side of* (6.3.9) *is well defined for $\phi(\mathbf{x}) \in \mathcal{D}(\mathbb{R}^n)$, and the convolution $T * S = S * T$ is a distribution defined by* (6.3.9).

Remark 6.3.3. For $n = 1$, the condition relating the supports of distributions T and S in Theorem 6.3.1 can be replaced by the following: *if both the distributions T and S on $\mathbb{R}$ have supports bounded from the left*, i.e. *their supports are contained in* $]a, \infty[$, or *bounded from the right*, i.e. their supports are contained in $]-\infty, b[$, then $T * S$ exists.

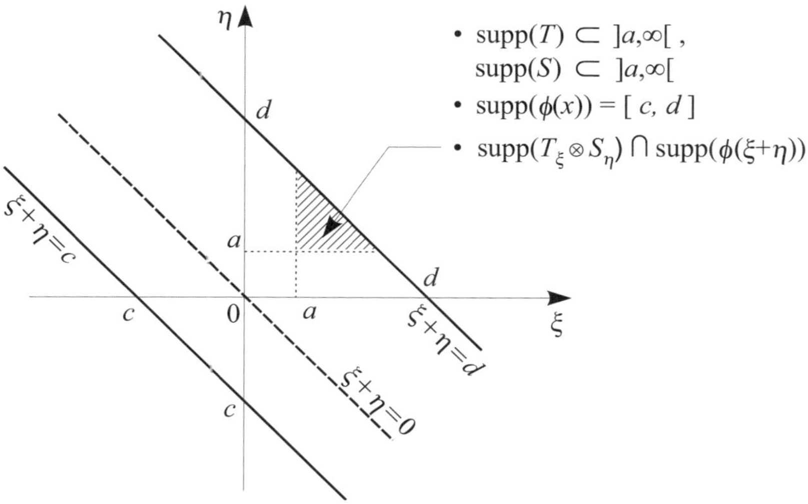

Figure 6.2 Boundedness of $\operatorname{supp}(T_\xi \otimes S_\eta) \cap \operatorname{supp}(\phi(\xi + \eta))$ in $\mathbb{R}^2$ with T and S having supports bounded from the left, i.e. contained in $]a, \infty[$

In fact, suppose that T and S have supports bounded from the left, i.e. contained in $]a, +\infty[$.

Hence, $\xi \geq a, \eta \geq a$. Then, for $\text{supp}(\phi(x)) = [c, d]$, $\text{supp}(T_\xi \otimes S_\eta) \cap \text{supp}(\phi(\xi + \eta))$ is bounded in $\mathbb{R} \times \mathbb{R}$ (see Figure 6.2). Hence, $T * S$ exists.

Now we will study the important case in which the existence of the convolution depends on the property of compactness of the support of T or (and) S. (We have already given the definition of $T * S$ in (6.3.7) (resp. (6.3.9)) when T has compact support, and shown that both the definitions coincide).

Case I. At least one of the distributions T or S has compact support Let $A = \text{supp}(T)$ and $B = \text{supp}(S)$ be the supports of T and S such that at least one of them, say A, is *compact*. Let $\phi \in \mathcal{D}(\mathbb{R}^n)$ with $\text{supp}(\phi) = K$ compact in $\mathbb{R}^n$. Then the intersection $I = \text{supp}(T_\xi \otimes S_\eta) \cap (\phi(\xi + \eta))$ is a bounded subset of $\mathbb{R}^n \times \mathbb{R}^n = \mathbb{R}^{2n}$. In fact, $\text{supp}(T_\xi \otimes S_\eta) = A \times B$ and $\text{supp}(\phi(\xi + \eta)) = \{(\xi, \eta) : \xi, \eta \in \mathbb{R}^n$ with $\xi + \eta \in K\}$.

Then $I = \{(\xi, \eta) : \xi \in A, \eta \in B, \xi + \eta \in K\}$. But $\xi \in A$, $\xi + \eta \in K \Longrightarrow \eta = (\xi + \eta) - \xi \in (K - A) \Longrightarrow I \subset A \times (K - A)$, which is a compact subset of $\mathbb{R}^{2n}$ $\Longrightarrow I$ is bounded in $\mathbb{R}^{2n}$.

Now we introduce a function $\alpha(\xi) \in \mathcal{D}(\mathbb{R}^n)$ such that $\alpha(\xi) = 1$ $\forall \xi \in U$ with $A \subset U$, U being a neighbourhood of the compact support A of T_ξ, and $\text{supp}(\alpha(\xi))$ is compact. Then the function $\alpha(\xi)\phi(\xi + \eta)$ has compact support in $\mathbb{R}^{2n}$, and $\alpha(\xi)\phi(\xi + \eta) = \phi(\xi + \eta)$ $\forall (\xi, \eta) \in V$ with $A \times B \subset V$, V being a neighbourhood of $\text{supp}(T_\xi \otimes S_\eta)$. Then $\langle T * S, \phi \rangle$ is well defined as a linear functional $\forall \phi \in \mathcal{D}(\mathbb{R}^n)$:

$$\langle T * S, \phi \rangle = \langle T_\xi \otimes S_\eta, \phi(\xi + \eta) \rangle = \langle T_\xi \otimes S_\eta, \alpha(\xi)\phi(\xi + \eta) \rangle \quad \forall \phi \in \mathcal{D}(\mathbb{R}^n).$$

Continuity of $T * S$ on $\mathcal{D}(\mathbb{R}^n)$ Let (ϕ_m) be a sequence in $\mathcal{D}(\mathbb{R}^n)$ such that $\forall m \in \mathbb{N}$, $\text{supp}(\phi_m) \subset K_0$, K_0 being a *fixed compact* subset of $\mathbb{R}^n$ and $\phi_m \to 0$ in $\mathcal{D}(\mathbb{R}^n)$ as $m \to \infty$. Then, $\forall m \in \mathbb{N}$, $\alpha(\xi)\phi_m(\xi + \eta)$ has support contained in a fixed compact subset of $\mathbb{R}^n \times \mathbb{R}^n$ and converges, along with all derivatives, uniformly to 0, i.e. $\alpha(\xi)\phi_m(\xi + \eta) \to 0$ in $\mathcal{D}(\mathbb{R}^n \times \mathbb{R}^n)$ as $m \to \infty$. But $T_\xi \otimes S_\eta \in \mathcal{D}'(\mathbb{R}^n \times \mathbb{R}^n)$ and, consequently, $\langle T_\xi \otimes S_\eta, \alpha(\xi)\phi_m((\xi + \eta)) \rangle \to 0$ as $m \to \infty$. Hence, $\phi_m \to 0$ in $\mathcal{D}(\mathbb{R}^n) \Longrightarrow \langle T * S, \phi_m \rangle = \langle T_\xi \otimes S_\eta, \phi_m(\xi + \eta) \rangle = \langle T_\xi \otimes S_\eta, \alpha(\xi)\phi_m(\xi + \eta) \rangle \to 0$ in $\mathbb{R}$ as $m \to \infty \Longrightarrow$ the linear functional $T * S$ is continuous on $\mathcal{D}(\mathbb{R}^n)$ and, consequently, a distribution on $\mathbb{R}^n$. Hence, we have:

Theorem 6.3.2. *Let T and S be arbitrary distributions on $\mathbb{R}^n$ such that* at least one *of the two has compact support. Then the convolution of T and S, denoted by $T * S$ or $S * T$, is well defined as a distribution on $\mathbb{R}^n$ such that*

$$\langle T * S, \phi \rangle = \langle T_\xi \otimes S_\eta, \phi((\xi + \eta)) \rangle \quad \forall \phi \in \mathcal{D}(\mathbb{R}^n). \tag{6.3.14}$$

Moreover, using the definition of the tensor product of two distributions (see (6.1.14)*), we can write:* $\forall \phi \in \mathcal{D}(\mathbb{R}^n)$,

$$\begin{aligned}\langle T * S, \phi\rangle &= \langle T_\xi \otimes S_\eta, \phi((\xi + \eta))\rangle \\ &= \langle T_\xi, \langle S_\eta, \phi((\xi + \eta))\rangle\rangle = \langle S_\eta, \langle T_\xi, \phi((\xi + \eta))\rangle\rangle. \end{aligned} \tag{6.3.15}$$

In fact, for $T \in \mathcal{D}'(\mathbb{R}^n)$ with compact support A and $\phi \in \mathcal{D}(\mathbb{R}^n)$, $\langle T_\xi, \phi((\xi + \eta))\rangle = h(\eta)$ is a well defined function of η which is infinitely differentiable on $\mathbb{R}^n$ (see Remark 6.2.1) and has compact support in $\mathbb{R}^n$, i.e. $h \in \mathcal{D}(\mathbb{R}^n)$. Then, for $S_\eta \in \mathcal{D}'(\mathbb{R}^n)$ with *arbitrary support*, $\langle S_\eta, h(\eta)\rangle$ is well defined $\forall h(\eta) \in \mathcal{D}(\mathbb{R}^n)$. Thus,

$$\langle S_\eta, h(\eta)\rangle = \langle S_\eta, \langle T_\xi, \phi(\xi + \eta)\rangle\rangle. \tag{6.3.16}$$

Again, for $S_\eta \in \mathcal{D}'(\mathbb{R}^n)$ with *arbitrary support*, $g(\xi) = \langle S_\eta, \phi(\xi + \eta)\rangle$ is a well defined function of ξ which is infinitely differentiable on $\mathbb{R}^n$ and may have *arbitrary support* in $\mathbb{R}^n$ (see Remark 6.2.1), i.e. $g \in \mathcal{E}(\mathbb{R}^n)$. But $T_\xi \in \mathcal{E}'(\mathbb{R}^n)$, i.e. $T \in \mathcal{D}'(\mathbb{R}^n)$ is a distribution with compact support A. Hence, $\langle T_\xi, g(\xi)\rangle$ is well defined. Thus,

$$\langle T_\xi, g(\xi)\rangle = \langle T_\xi, \langle S_\eta, \phi(\xi + \eta)\rangle\rangle \tag{6.3.17}$$

is well defined. Then (6.3.16) and (6.3.17) must give the same result such that (6.3.15) holds, i.e. $\langle T * S, \phi\rangle = \langle T_\xi, \langle S_\eta, \phi(\xi + \eta)\rangle\rangle = \langle S_\eta, \langle T_\xi, \phi(\xi + \eta)\rangle\rangle \ \forall \phi \in \mathcal{D}(\mathbb{R}^n)$.

Support of the convolution of two distributions

Theorem 6.3.3. *Let T and S be two distributions with $A = \operatorname{supp}(T)$ and $B = \operatorname{supp}(S)$ such that either T or S has compact support, i.e. either A or B is* compact *in $\mathbb{R}^n$. Then*

$$\operatorname{supp}(T * S) \subset A + B. \tag{6.3.18}$$

In particular, if A *and* B are compact, then $T * S$ has compact support, i.e.

$$T \in \mathcal{E}'(\mathbb{R}^n), S \in \mathcal{E}'(\mathbb{R}^n) \quad \Longrightarrow \quad T * S \in \mathcal{E}'(\mathbb{R}^n). \tag{6.3.19}$$

For $n = 1$,

$$\operatorname{supp}(T) \subset]a, \infty[, \quad \operatorname{supp}(S) \subset]b, \infty[, \quad \operatorname{supp}(T * S) \subset]a + b, \infty[. \tag{6.3.20}$$

Proof of Theorem 6.3.3. Since either A or B is compact and A and B are closed sets, $A + B$ is closed (see (6.2.4)). Let $\Omega = (A + B)^{\complement} =$ the complement of the closed set $A + B$. Then Ω is open and we are to show that $\langle T * S, \phi\rangle = 0 \ \forall \phi \in \mathcal{D}(\mathbb{R}^n)$ with $\operatorname{supp}(\phi) \subset \Omega$. For such a ϕ, the support of $\phi(\xi + \eta)$ in $\mathbb{R}^n \times \mathbb{R}^n$ is contained in the open set defined by $\xi + \eta \in \Omega$ (i.e. $\xi + \eta \notin (A + B)$). But $\operatorname{supp}(T_\xi \otimes$

$S_{\eta}) = A \times B = \{(\boldsymbol{\xi}, \boldsymbol{\eta}) : \boldsymbol{\xi} \in A, \boldsymbol{\eta} \in B\} \subset \mathbb{R}^n \times \mathbb{R}^n$. Hence, $(\boldsymbol{\xi}, \boldsymbol{\eta}) \in A \times B \Longrightarrow \boldsymbol{\xi} + \boldsymbol{\eta} \in A + B \Longrightarrow \text{supp}(T_{\boldsymbol{\xi}} \otimes S_{\boldsymbol{\eta}}) = A \times B \subset A + B$. Consequently, $\text{supp}(\phi(\boldsymbol{\xi} + \boldsymbol{\eta})) \cap \text{supp}(T_{\boldsymbol{\xi}} \otimes S_{\boldsymbol{\eta}}) = \emptyset \Longrightarrow \langle T_{\boldsymbol{\xi}} \otimes S_{\boldsymbol{\eta}}, \phi(\boldsymbol{\xi} + \boldsymbol{\eta})\rangle = 0 \ \forall \phi \in \mathcal{D}(\mathbb{R}^n)$ with $\text{supp}(\phi) \subset \Omega \Longrightarrow \langle T * S, \phi\rangle = \langle T_{\boldsymbol{\xi}} \otimes S_{\boldsymbol{\eta}}, \phi(\boldsymbol{\xi} + \boldsymbol{\eta})\rangle = 0 \ \forall \phi \in \mathcal{D}(\mathbb{R}^n)$ with $\text{supp}(\phi) \subset \Omega \Longrightarrow \text{supp}(T * S) \subset \Omega^C = A + B$. □

In particular, if A and B are compact, then $A + B$ is compact (6.2.3). Hence, $\text{supp}(T * S)$ is a closed subset of the compact set $A + B$, i.e. $\text{supp}(T * S)$ is compact.

Example 6.3.1.

1. Let $\delta = \delta_{(\mathbf{0})}$ be the Dirac distribution with mass/force/charge concentrated at $\mathbf{0} \in \mathbb{R}^n$ such that $\langle \delta, \phi\rangle = \phi(\mathbf{0}) \ \forall \phi \in \mathcal{D}(\mathbb{R}^n)$. Hence, δ is a distribution with compact support $\{\mathbf{0}\}$, and $\delta * T$ exists $\forall T \in \mathcal{D}'(\mathbb{R}^n)$ by Theorem 6.3.2 and is given by $\delta * T = T \ \forall T \in \mathcal{D}'(\mathbb{R}^n)$, since $\langle \delta * T, \phi\rangle = \langle \delta_{\boldsymbol{\xi}} \otimes T_{\boldsymbol{\eta}}, \phi(\boldsymbol{\xi} + \boldsymbol{\eta})\rangle = \langle T_{\boldsymbol{\eta}}, \langle \delta_{\boldsymbol{\xi}}, \phi(\boldsymbol{\xi} + \boldsymbol{\eta})\rangle\rangle = \langle T_{\boldsymbol{\eta}}, \phi(\mathbf{0} + \boldsymbol{\eta})\rangle = \langle T_{\boldsymbol{\eta}}, \phi(\boldsymbol{\eta})\rangle = \langle T, \phi\rangle \ \forall \phi \in \mathcal{D}(\mathbb{R}^n)$
$\Longrightarrow \delta * T = T$, i.e. *in convolution, δ is the unity.* (6.3.21)

Remark 6.3.4. In physics and mechanics, for $T = T_f$, $\delta * T_f = \delta * f$ is usually written incorrectly in the form:

$$\int_{\mathbb{R}^n} \delta(\mathbf{x} - \boldsymbol{\xi}) f(\boldsymbol{\xi}) d\boldsymbol{\xi} = \int_{\mathbb{R}^n} f(\mathbf{x} - \boldsymbol{\xi}) \delta(\boldsymbol{\xi}) d\boldsymbol{\xi} = f(\mathbf{x}), \tag{6.3.21a}$$

which is, in fact, the definition of δ function given by Dirac, although the integral sign has *no* meaning.

2. $\delta' * T = T' \ \forall \ T \in \mathcal{D}'(\mathbb{R}^n)$. (6.3.22)
In fact, $\langle \delta' * T, \phi\rangle = \langle \delta'_{\boldsymbol{\xi}} \otimes T_{\boldsymbol{\eta}}, \phi(\boldsymbol{\xi} + \boldsymbol{\eta})\rangle = \langle T_{\boldsymbol{\eta}}, \langle \delta'_{\boldsymbol{\xi}}, \phi(\boldsymbol{\xi} + \boldsymbol{\eta})\rangle\rangle = \langle T_{\boldsymbol{\eta}}, -\phi'(\boldsymbol{\eta})\rangle = -\langle T, \phi'\rangle = \langle T', \phi\rangle \ \forall \phi \in \mathcal{D}(\mathbb{R}^n) \Longrightarrow \delta' * T = T'$ in $\mathcal{D}'(\mathbb{R}^n)$.
In particular, $\delta' = \frac{\partial \delta}{\partial x_k} \Longrightarrow \frac{\partial \delta}{\partial x_k} * T = \frac{\partial T}{\partial x_k} \ \forall T \in \mathcal{D}'(\mathbb{R}^n)$.
In physics and mechanics, for $T = T_f$, $\frac{\partial}{\partial x_k}(\delta * T_f) = \frac{\partial \delta}{\partial x_k} * T_f = \frac{\partial \delta}{\partial x_k} * f = \frac{\partial f}{\partial x_k}$ is usually written in the form:
$$\int_{\mathbb{R}^n} \frac{\partial \delta}{\partial x_k}(\mathbf{x} - \boldsymbol{\xi}) f(\boldsymbol{\xi}) d\boldsymbol{\xi} = \frac{\partial f}{\partial x_k}, \tag{6.3.23}$$
although the integral sign has *no* meaning here. Equation (6.3.23) can be *formally* obtained from the (incorrect) formula (6.3.21a) by ‘differentiating under the integral sign’!

3. $\forall$ multi-index $\boldsymbol{\alpha} = (\alpha_1, \alpha_2 \ldots, \alpha_n)$ with $\partial^{\boldsymbol{\alpha}} = \frac{\partial^{|\alpha|}}{\partial_{x_1}^{\alpha_1} \ldots \partial_{x_n}^{\alpha_n}}$, $\partial^{\boldsymbol{\alpha}} \delta * T = \partial^{\boldsymbol{\alpha}} T$ $\forall T \in \mathcal{D}'(\mathbb{R}^n)$, since $\langle \partial^{\boldsymbol{\alpha}} \delta * T, \phi\rangle = \langle \partial^{\boldsymbol{\alpha}} \delta_{\boldsymbol{\xi}} \otimes T_{\boldsymbol{\eta}}, \phi(\boldsymbol{\xi} + \boldsymbol{\eta})\rangle = \langle T_{\boldsymbol{\eta}}, \langle \partial^{\boldsymbol{\alpha}} \delta_{\boldsymbol{\xi}}, \phi(\boldsymbol{\xi} + \boldsymbol{\eta})\rangle\rangle = \langle T_{\boldsymbol{\eta}}, (-1)^{|\alpha|} \partial^{\boldsymbol{\alpha}} \phi(\boldsymbol{\eta})\rangle = (-1)^{|\alpha|} \langle T, \partial^{\boldsymbol{\alpha}} \phi\rangle = \langle \partial^{\boldsymbol{\alpha}} T, \phi\rangle \ \forall \phi \in \mathcal{D}(\mathbb{R}^n)$
$$\Longrightarrow \quad \partial^{\boldsymbol{\alpha}} \delta * T = \partial^{\boldsymbol{\alpha}} T \quad \text{in } \mathcal{D}'(\mathbb{R}^n). \tag{6.3.24}$$

Translation operator $\tau_{\mathbf{a}}$ For $\mathbf{a} \in \mathbb{R}^n$, let $\tau_{\mathbf{a}}$ be the translation operator defined, for any function f on $\mathbb{R}^n$, by:

$$(\tau_{\mathbf{a}} f)(\mathbf{x}) = f(\mathbf{x} - \mathbf{a}). \tag{6.3.25}$$

Then

$$(\tau_{\xi} \check{f})(\eta) = \check{f}(\eta - \xi) = f(-(\eta - \xi)) = f(\xi - \eta). \tag{6.3.26}$$

Let $f \in L^1(\mathbb{R}^n)$. Then $\check{f}$ and $\tau_{\mathbf{a}} f$ belong to $L^1(\mathbb{R}^n)$ such that $\check{T} = T_{\check{f}}$ and $\tau_{\mathbf{a}} T = T_{\tau_{\mathbf{a}} f}$ are distributions defined by $\check{f}$ and $\tau_{\mathbf{a}} f$ respectively:

$$\langle \tau_{\mathbf{a}} T, \phi \rangle = \langle T_{\tau_{\mathbf{a}} f}, \phi \rangle = \int_{\mathbb{R}^n} (\tau_{\mathbf{a}} f)(\mathbf{x}) \phi(\mathbf{x}) d\mathbf{x} = \int_{\mathbb{R}^n} f(\mathbf{x} - \mathbf{a}) \phi(\mathbf{x}) d\mathbf{x}.$$

Changing variables defined by $\xi = \mathbf{x} - \mathbf{a}$ with the absolute value of the Jacobian of the transformation equal to 1, we get

$$\begin{aligned} \int_{\mathbb{R}^n} f(\mathbf{x} - \mathbf{a}) \phi(\mathbf{x}) d\mathbf{x} &= \int_{\mathbb{R}^n} f(\xi) \phi(\xi + \mathbf{a}) d\xi \\ &= \langle T_f, \phi(\xi + \mathbf{a}) \rangle = \langle T_{\xi}, \phi(\xi + \mathbf{a}) \rangle = \langle T_{\mathbf{x}}, \phi(\mathbf{x} + \mathbf{a}) \rangle. \end{aligned}$$

Hence,

$$\langle \tau_{\mathbf{a}} T, \phi \rangle = \langle T_{\mathbf{x}}, \phi(\mathbf{x} + \mathbf{a}) \rangle \quad \forall \phi \in \mathcal{D}(\mathbb{R}^n), \tag{6.3.27}$$

which suggests we define the *translated* or *shifted* distribution $\tau_{\mathbf{a}} T$ for arbitrary $T \in \mathcal{D}'(\mathbb{R}^n)$ by: $\forall T \in \mathcal{D}'(\mathbb{R}^n)$,

$$\langle \tau_{\mathbf{a}} T, \phi \rangle = \langle T_{\mathbf{x}}, \phi(\mathbf{x} + \mathbf{a}) \rangle \quad \forall \phi \in \mathcal{D}(\mathbb{R}^n). \tag{6.3.28}$$

Since $(\tau_{\mathbf{a}} f)(\mathbf{x}) = f(\mathbf{x} - \mathbf{a})$, $\tau_{\mathbf{a}} T_{\mathbf{x}}$ is denoted by $T_{\mathbf{x}-\mathbf{a}}$, or $T(\mathbf{x} - \mathbf{a})$ (the latter is more frequently used in physics and mechanics). For example, for $T_{\mathbf{x}} = \delta(\mathbf{x})$ (the Dirac distribution associated with variable $\mathbf{x}$ with charge/mass/force concentrated at $\mathbf{x} = \mathbf{0}$), $\tau_{\mathbf{a}} \delta_{\mathbf{x}} = \delta_{\mathbf{x}-\mathbf{a}}$ or $\delta(\mathbf{x} - \mathbf{a})$ is the *translated* Dirac distribution associated with $\mathbf{x}$ with charge/force/mass concentrated at $\mathbf{a}$.

Remark 6.3.5. In $T(\mathbf{x} - \mathbf{a})$ (resp. $\delta(\mathbf{x} - \mathbf{a})$), no point values of T (resp. δ) at $\mathbf{x} - \mathbf{a}$ are to be understood, since $T(\mathbf{x} - \mathbf{a})$ (resp. $\delta(\mathbf{x} - \mathbf{a})$) is a distribution, not a function. Hence, we have written $T_{\mathbf{x}-\mathbf{a}}$, but $\phi(\mathbf{x} + \mathbf{a})$ is the *point value* of the function ϕ at the point $\mathbf{x} + \mathbf{a}$.

The mapping $\phi \in \mathcal{D}(\mathbb{R}^n) \longmapsto \langle \tau_{\mathbf{a}} T, \phi \rangle$ is continuous. Hence, $T \in \mathcal{D}'(\mathbb{R}^n) \Longrightarrow \tau_{\mathbf{a}} T \in \mathcal{D}'(\mathbb{R}^n)$ is a distribution.

Example 6.3.2. Let $\delta_{\mathbf{a}}$ be the Dirac distribution with mass/charge/force concentrated at $\mathbf{a} \in \mathbb{R}^n$. Then

$$\delta_{\mathbf{a}} * T = \tau_{\mathbf{a}} T \quad \forall T \in \mathcal{D}'(\mathbb{R}^n), \tag{6.3.29}$$

where $\tau_{\mathbf{a}} T$ is defined by (6.3.28), since $\langle \delta_{\mathbf{a}} * T, \phi \rangle = \langle \delta_{\mathbf{a},\boldsymbol{\xi}} \otimes T_{\boldsymbol{\eta}}, \phi(\boldsymbol{\xi} + \boldsymbol{\eta}) \rangle = \langle T_{\boldsymbol{\eta}}, \langle \delta_{\mathbf{a},\boldsymbol{\xi}}, \phi(\boldsymbol{\xi} + \boldsymbol{\eta}) \rangle \rangle = \langle T_\eta, \phi(\mathbf{a} + \boldsymbol{\eta}) \rangle = \langle \tau_{\mathbf{a}} T, \phi \rangle \ \forall \phi \in \mathcal{D}(\mathbb{R}^n) \Longrightarrow \delta_{\mathbf{a}} * T = \tau_{\mathbf{a}} T$.

Remark 6.3.6. Again, in physics and mechanics, for $T = T_f, \delta_{\mathbf{a}} * T_f = \delta_{\mathbf{a}} * f = \tau_{\mathbf{a}} f$ is usually written in the form:

$$\begin{aligned} \int_{\mathbb{R}^n} \delta_{\mathbf{a}}(\boldsymbol{\xi}) f(\mathbf{x} - \boldsymbol{\xi}) d\boldsymbol{\xi} &= \int_{\mathbb{R}^n} \delta(\boldsymbol{\xi} - \mathbf{a}) f(\mathbf{x} - \boldsymbol{\xi}) d\boldsymbol{\xi} \\ &= \int_{\mathbb{R}^n} \delta((\mathbf{x} - \mathbf{a}) - \boldsymbol{\xi}) f(\boldsymbol{\xi}) d\boldsymbol{\xi} = f(\mathbf{x} - \mathbf{a}) = \tau_{\mathbf{a}} f(\mathbf{x}), \end{aligned} \tag{6.3.30}$$

which, in fact, follows from the definition of $\delta(\mathbf{x} - \mathbf{a})$ given by Dirac.

4. $\delta_{\mathbf{a}} * \delta_{\mathbf{b}} = \delta_{\mathbf{a}+\mathbf{b}}$, since $T = \delta_{\mathbf{b}} \Longrightarrow \delta_{\mathbf{a}} * \delta_{\mathbf{b}} = \tau_{\mathbf{a}} \delta_{\mathbf{b}}$

$$\begin{aligned} &\Longrightarrow \quad \langle \tau_{\mathbf{a}} \delta_{\mathbf{b}}, \phi \rangle = \langle \delta_{\mathbf{b}}, \phi(\mathbf{x} + \mathbf{a}) \rangle = \phi(\mathbf{b} + \mathbf{a}) = \langle \delta_{\mathbf{a}+\mathbf{b}}, \phi \rangle \quad \forall \phi \in \mathcal{D}(\mathbb{R}^n) \\ &\Longrightarrow \quad \tau_{\mathbf{a}} \delta_{\mathbf{b}} = \delta_{\mathbf{a}} * \delta_{\mathbf{b}} = \delta_{\mathbf{a}+\mathbf{b}}, \end{aligned} \tag{6.3.31}$$

where $\delta_{\mathbf{a}+\mathbf{b}}$ is the Dirac distribution with mass/force/charge concentrated at the point $\mathbf{a} + \mathbf{b} \in \mathbb{R}^n$.

Case II. For $\boldsymbol{\xi} \in A = \operatorname{supp}(T)$, $\eta \in B = \operatorname{supp}(S)$, $\boldsymbol{\xi} + \eta$ remains bounded only if both $\boldsymbol{\xi}$ and η remain bounded Case I is a particular situation of Case II: A is compact $\Longrightarrow \boldsymbol{\xi}$ is bounded. If $\boldsymbol{\xi} + \boldsymbol{\eta}$ is bounded, then $\boldsymbol{\eta} = (\boldsymbol{\xi} + \boldsymbol{\eta}) - \boldsymbol{\xi}$ is also bounded $\Longrightarrow$ Case II holds.

For $f, g \in L^1_{\text{loc}}(\mathbb{R}^n)$, $T_f * S_g = f * g$ is not defined *in general*, since $\operatorname{supp}(\phi(\boldsymbol{\xi} + \boldsymbol{\eta}))$ is *not* bounded in $\mathbb{R}^n \times \mathbb{R}^n$ and the double integral on $\mathbb{R}^n \times \mathbb{R}^n$ in (6.3.8), $\iint_{\mathbb{R}^n \times \mathbb{R}^n} f(\boldsymbol{\xi}) g(\boldsymbol{\eta}) \phi(\boldsymbol{\xi} + \boldsymbol{\eta}) d\boldsymbol{\xi} d\boldsymbol{\eta}$ has *no* meaning. But we have:

Theorem 6.3.4. *Let T_f and S_g be the distributions defined by functions $f, g \in L^1_{\text{loc}}(\mathbb{R}^n)$. Let $A = \operatorname{supp}(T_f) = \operatorname{supp}(f)$ and $B = \operatorname{supp}(S_g) = \operatorname{supp}(g)$ satisfy the following condition:*

$$\begin{aligned} &\operatorname{supp}(f(\boldsymbol{\xi}) \otimes g(\boldsymbol{\eta})) \cap \operatorname{supp}(\phi(\boldsymbol{\xi} + \boldsymbol{\eta})) \\ &= \{(\boldsymbol{\xi}, \boldsymbol{\eta}) : \boldsymbol{\xi} \in A, \boldsymbol{\eta} \in B, \boldsymbol{\eta} + \boldsymbol{\xi} \in K, K = \operatorname{supp}(\phi(\mathbf{x}))\} \end{aligned}$$

is bounded *in* $\mathbb{R}^n \times \mathbb{R}^n$. *Then the convolution* $T_f * S_g$ *is well defined by* (6.3.9) *and a distribution* W_{f*g} *defined by* $f * g = h \in L^1_{\text{loc}}(\mathbb{R}^n)$*:* $\forall \phi \in \mathcal{D}(\mathbb{R}^n)$,

$$\langle T_f * S_g, \phi \rangle = \langle W_{f*g}, \phi \rangle = \langle f * g, \phi \rangle = \int_{\mathbb{R}^n} h(\mathbf{x})\phi(\mathbf{x})d\mathbf{x},$$

where

$$h(\mathbf{x}) = (f * g)(\mathbf{x}) = \int_{\mathbb{R}^n} f(\mathbf{x}-\mathbf{t})g(\mathbf{t})d\mathbf{t} = \int_{\mathbb{R}^n} f(\mathbf{t})g(\mathbf{x}-\mathbf{t})d\mathbf{t} \quad \text{a.e. in } \mathbb{R}^n. \tag{6.3.32}$$

Proof. From definition (6.3.9), $\forall \phi \in \mathcal{D}(\mathbb{R}^n)$,

$$\langle T_f * T_g, \phi \rangle = \langle f(\xi) \otimes g(\eta), \phi(\eta + \xi) \rangle = \iint_{\mathbb{R}^n \times \mathbb{R}^n} f(\xi)g(\eta)\phi(\xi + \eta)d\xi d\eta, \tag{6.3.33}$$

since the right-hand side double integral on $\mathbb{R}^n \times \mathbb{R}^n$ exists by virtue of the assumption on A and B. In fact, f and $g \in L^1_{\text{loc}}(\mathbb{R}^n) \Longrightarrow f(\xi)g(\eta)\phi(\xi+\eta)$ is locally summable on $\mathbb{R}^n \times \mathbb{R}^n$, but $\operatorname{supp}(f(\xi)g(\eta)\phi(\xi+\eta)) = \{(\xi,\eta) : \xi \in A, \eta \in B, \eta+\xi \in K, K = \operatorname{supp}(\phi(\mathbf{x}))\}$ is a bounded subset of $\mathbb{R}^n \times \mathbb{R}^n$. Hence, $f(\xi)g(\eta)\phi(\xi+\eta)$ is integrable on $\mathbb{R}^n \times \mathbb{R}^n$, i.e. $f(\xi)g(\eta)\phi(\xi+\eta) \in L^1(\mathbb{R}^n)$. Then, by Fubini's Theorem 7.1.2C, we can interchange the order of integration. In order to show (6.3.32), we are to change the variables: $\mathbf{x} = \xi + \eta$, $\mathbf{t} = \eta$, the Jacobian of this transformation being $+1$, i.e. from (6.3.33),

$$\langle T_f * T_g, \phi \rangle = \iint_{\mathbb{R}^n \times \mathbb{R}^n} f(\xi - \mathbf{t})g(\mathbf{t})\phi(\mathbf{x})d\mathbf{x}d\mathbf{t} = \int_{\mathbb{R}^n} d\mathbf{x} \int_{\mathbb{R}^n} \phi(\mathbf{x})f(\mathbf{x}-\mathbf{t})g(\mathbf{t})d\mathbf{t};$$

since $f(\mathbf{x}-\mathbf{t})g(\mathbf{t})\phi(\mathbf{x}) \in L^1(\mathbb{R}^n \times \mathbb{R}^n)$, by Fubini's Theorem 7.1.2C, $\int_{\mathbb{R}^n} f(\xi - \mathbf{t})g(\mathbf{t})\phi(\mathbf{x})d\mathbf{t}$ exists for almost all values of $\mathbf{x} \in \mathbb{R}^n$. Hence, the integral $\int_{\mathbb{R}^n} f(\mathbf{x} - \mathbf{t})g(\mathbf{t})d\mathbf{t}$ has a meaning for almost all values of $\mathbf{x}$, and $h(\mathbf{x}) = \int_{\mathbb{R}^n} f(\mathbf{x}-\mathbf{t})g(\mathbf{t})d\mathbf{t} = (f * g)(\mathbf{x})$ is a function well defined for almost all values of $\mathbf{x} \in \mathbb{R}^n$. Again, by Fubini's Theorem 7.1.2C, $\int_{\mathbb{R}^n} \phi(\mathbf{x})h(\mathbf{x})d\mathbf{x}$ exists and $\phi(\mathbf{x})h(\mathbf{x}) \in L^1(\mathbb{R}^n) \Longrightarrow h(\mathbf{x})\phi(\mathbf{x}) \in L^1_{\text{loc}}(\mathbb{R}^n)$ with $\phi \in \mathcal{D}(\mathbb{R}^n) \Longrightarrow h \in L^1_{\text{loc}}(\mathbb{R}^n)$.

Hence,

$$\begin{aligned}\langle T_f * S_g, \phi \rangle = \int_{\mathbb{R}^n} h(\mathbf{x})\phi(\mathbf{x})d\mathbf{x} &= \int_{\mathbb{R}^n} (f * g)(\mathbf{x})\phi(\mathbf{x})d\mathbf{x} \\ &= \langle W_{f*g}, \phi \rangle \quad \forall \phi \in \mathcal{D}(\mathbb{R}^n)\end{aligned}$$

$\Longrightarrow T_f * S_g = W_{f*g}$ with $f * g = h$. Substituting $\xi = \mathbf{t}$, $\xi + \eta = \mathbf{x}$ in (6.3.33) and proceeding similarly, we get $h(\mathbf{x}) = \int_{\mathbb{R}^n} f(\mathbf{t})g(\mathbf{x}-\mathbf{t})d\mathbf{t}$. □

Remark 6.3.7. For functions f and g, their convolution $f * g$ may exist, even if their supports do not satisfy the condition in Theorem 6.3.4. For example,

- $f, g \in L^1(\mathbb{R}^n)$, $f * g$ always exists and $f * g \in L^1(\mathbb{R}^n)$ with $\|f * g\|_{L^1(\mathbb{R}^n)} \leq \|f\|_{L^1(\mathbb{R}^n)} \|g\|_{L^1(\mathbb{R}^n)}$ (see (6.2.5)).
- For $f \in L^1(\mathbb{R}^n)$ and g *bounded* in $\mathbb{R}^n$, $f * g$ exists and is a *continuous* function *bounded* on $\mathbb{R}^n$ with the properties: $h(\mathbf{x}) = (f * g)(\mathbf{x})$ is defined $\forall \mathbf{x} \in \mathbb{R}^n$;

$$|h(\mathbf{x})| \leq \left(\int_{\mathbb{R}^n} |f(\mathbf{t})| d\mathbf{t} \right) \sup_{\mathbf{t} \in \mathbb{R}^n} |g(\mathbf{t})|$$
$$\Longrightarrow \quad \|h\|_{L^\infty(\mathbb{R}^n)} \leq \|f\|_{L^1(\mathbb{R}^n)} \|g\|_{L^\infty(\mathbb{R}^n)}. \tag{6.3.34}$$

In particular, for $f, g \in C(\mathbb{R})$ with $\mathrm{supp}(f)$ and $\mathrm{supp}(g)$ contained in $]0, \infty[$, i.e. $f(x) = 0$, $g(x) = 0$ for $x \leq 0$,

$$h(x) = (f * g)(x) = \int_{-\infty}^{\infty} f(x-t) g(t) dt = \int_0^{\infty} f(x-t) g(t) dt$$
$$= \int_0^x f(x-t) g(t) dt + \int_x^{\infty} f(x-t) g(t) dt.$$

- For $x \leq 0$, $x - t \leq 0 \ \forall t \geq 0 \Longrightarrow f(x-t) = 0 \Longrightarrow \int_0^x$ and $\int_x^\infty$ vanish $\Longrightarrow h(x) = 0$ for $x \leq 0$.
- For $x \geq 0$, $x - t \leq 0 \ \forall t \geq x \Longrightarrow f(x-t) = 0 \Longrightarrow \int_x^\infty$ vanishes $\Longrightarrow h(x) = \int_0^x f(x-t) g(t) dt \ \forall x \geq 0$.

Thus,

$$h(x) = f * g(x) = \begin{cases} 0 & \text{for } x \leq 0 \\ \int_0^x f(x-t) g(t) dt & \text{for } x \geq 0. \end{cases} \tag{6.3.35}$$

Example 6.3.3. For any complex number $a \in \mathbb{C}$, let f and g be the functions defined by:

$$f(x) = H(x) e^{ax} \frac{x^{\alpha-1}}{\Gamma(\alpha)}, \quad g(x) = H(x) e^{ax} \frac{x^{\beta-1}}{\Gamma(\beta)},$$

where $\alpha > 0$, $\beta > 0$ and $H(x)$ denotes the Heaviside function. Then

$$H(x) e^{ax} \frac{x^{\alpha-1}}{\Gamma(\alpha)} * H(x) e^{ax} \frac{x^{\beta-1}}{\Gamma(\beta)} = H(x) e^{ax} \frac{x^{\alpha+\beta-1}}{\Gamma(\alpha+\beta)}.$$

Solution. f and g are continuous for $x \leq 0$, since $f(x) = 0$, $g(x) = 0 \ \forall x \leq 0$, and f and g are also continuous for $x > 0$. Hence, f and g are continuous on $\mathbb{R}$.

Then $f * g$ is given by (6.3.35): $\forall x \le 0$, $f * g(x) = 0$, $\forall x \ge 0$,

$$\begin{aligned} f * g(x) &= \int_{\mathrm{C}}^{x} \frac{(x-t)^{\alpha-1}}{\Gamma(\alpha)} \cdot \frac{t^{\beta-1}}{\Gamma(\beta)} e^{a(x-t)} \cdot e^{at} dt \\ &= \frac{e^{cx} \cdot x^{\alpha-1} \cdot x^{\beta-1} \cdot x}{\Gamma(\alpha)\Gamma(\beta)} \int_0^1 (1-\xi)^{\alpha-1} \xi^{\beta-1} d\xi \\ &\qquad \text{(by change of variables: } t = x\xi, dt = xd\xi) \\ &= \frac{e^{ax} x^{\alpha+\beta-1}}{\Gamma(\alpha)\Gamma(\beta)} \cdot \frac{\Gamma(\alpha)\Gamma(\beta)}{\Gamma(\alpha+\beta)} = \frac{e^{ax} \cdot x^{\alpha+\beta-1}}{\Gamma(\alpha+\beta)} \end{aligned}$$

(since the Beta function $B(\alpha,\beta) = \int_0^1 (1-\xi)^{\alpha-1}\xi^{\beta-1} d\xi = \frac{\Gamma(\alpha)\Gamma(\beta)}{\Gamma(\alpha+\beta)}$).

Hence,

$$f * g(x) = H(x) e^{ax} \frac{x^{\alpha+\beta-1}}{\Gamma(\alpha+\beta)} = \begin{cases} 0 & \text{for } x \le 0 \\ e^{ax} \frac{x^{\alpha+\beta-1}}{\Gamma(\alpha+\beta)} & \text{for } x \ge 0. \end{cases}$$

6.4 Regularization of distributions by convolution

Let $T \in \mathcal{D}'(\mathbb{R}^n)$ be a distribution and $\rho \in C^\infty(\mathbb{R}^n)$ be a function infinitely differentiable in the usual pointwise sense such that either T has *compact support*, i.e. $T \in \mathcal{E}'(\mathbb{R}^n)$, or ρ has *compact support*, i.e. $\rho \in \mathcal{D}(\mathbb{R}^n)$. Then, for fixed $\mathbf{x}$, $\rho(\mathbf{x}-\mathbf{t})$ is a function of $\mathbf{t}$ only and, as a function of $\mathbf{t}$, $\rho(\mathbf{x}-\mathbf{t}) \in C^\infty(\mathbb{R}^n)$ is infinitely differentiable. For $T \in \mathcal{E}'(\mathbb{R}^n) \subset \mathcal{D}'(\mathbb{R}^n)$ with compact support, and for $\rho(\mathbf{x}-\mathbf{t}) \in \mathcal{E}(\mathbb{R}^n) = C^\infty(\mathbb{R}^n)$ $\forall$ fixed $\mathbf{x} \in \mathbb{R}^n$, $h(\mathbf{x}) = \langle T_{\mathbf{t}}, \rho(\mathbf{x}-\mathbf{t})\rangle$ is a well-defined function of $\mathbf{x}$. Similarly, for $T \in \mathcal{D}'(\mathbb{R}^n)$ and for $\rho(\mathbf{x}-\mathbf{t}) \in \mathcal{D}(\mathbb{R}^n)$ for fixed $\mathbf{x} \in \mathbb{R}^n$, $h(\mathbf{x}) = \langle T_t, \rho(\mathbf{x}-\mathbf{t})\rangle$ is a well-defined function of $\mathbf{x}$.

In *both* the situations, $h(\mathbf{x}) = \langle T_t, \rho(\mathbf{x}-\mathbf{t})\rangle$ has the following properties:

- h is infinitely differentiable in the usual pointwise sense;
- $\forall$ multi-index $\boldsymbol{\alpha} = (\alpha_1, \alpha_2, \dots, \alpha_n)$ with $\partial_{\mathbf{x}}^{\boldsymbol{\alpha}} = \frac{\partial^{|\alpha|}}{\partial_{x_1}^{\alpha_1}\partial_{x_2}^{\alpha_2}\dots\partial_{x_n}^{\alpha_n}}$, $\partial_{\mathbf{x}}^{\boldsymbol{\alpha}} h(\mathbf{x}) = \partial_{\mathbf{x}}^{\boldsymbol{\alpha}} \langle T_{\mathbf{t}}, \rho(\mathbf{x}-\mathbf{t})\rangle = \langle T_{\mathbf{t}}, \partial_{\mathbf{x}}^{\boldsymbol{\alpha}} \rho(\mathbf{x}-\mathbf{t})\rangle$ (see Remark 6.2.1);
- h defines a distribution on $\mathbb{R}^n$ which coincides with $T * \rho$, $T * \rho$ being a distribution since either T or ρ has compact support, and ρ defines a regular distribution T_ρ in $\mathcal{D}'(\mathbb{R}^n)$.

In fact, $\forall \phi \in \mathcal{D}(\mathbb{R}^n)$,

$$\begin{aligned} \langle T * \rho, \phi\rangle &= \langle T_\xi \otimes \rho_\eta, \phi(\boldsymbol{\xi}+\boldsymbol{\eta})\rangle = < T_{\boldsymbol{\xi}}, \langle \rho(\boldsymbol{\eta}), \phi(\boldsymbol{\xi}+\boldsymbol{\eta})\rangle > \\ &= \langle T_{\boldsymbol{\xi}}, \int_{\mathbb{R}^n} \rho(\boldsymbol{\eta})\phi(\boldsymbol{\xi}+\boldsymbol{\eta}) d\boldsymbol{\eta}\rangle = \langle T_{\boldsymbol{\xi}}, \int_{\mathbb{R}^n} \rho(\mathbf{x}-\boldsymbol{\xi})\phi(\mathbf{x}) d\mathbf{x}\rangle \end{aligned}$$

(putting $\mathbf{x} = \boldsymbol{\xi} + \boldsymbol{\eta}$, $|J| = 1$ and $d\mathbf{x} = d\boldsymbol{\eta}$)

$$= \langle T_{\boldsymbol{\xi}}, \langle \phi(\mathbf{x}), \rho(\mathbf{x} - \boldsymbol{\xi}) \rangle \rangle = \langle T_{\boldsymbol{\xi}} \otimes \phi(\mathbf{x}), \rho(\mathbf{x} - \boldsymbol{\xi}) \rangle$$
$$= \langle \phi(\mathbf{x}), \langle T_{\boldsymbol{\xi}}, \rho(\mathbf{x} - \boldsymbol{\xi}) \rangle \rangle = \langle \phi(\mathbf{x}), h(\mathbf{x}) \rangle = \langle h, \phi \rangle$$

$\Longrightarrow T * \rho = h = \langle T_{\boldsymbol{\xi}}, \rho(\mathbf{x} - \boldsymbol{\xi}) \rangle \in C^{\infty}(\mathbb{R}^n)$ (but the support of h is not compact *in general* and hence $h \notin \mathcal{D}(\mathbb{R}^n)$ in general). We state this result in the following form:

Theorem 6.4.1. *The convolution $T * \rho$ of a distribution T and an infinitely differentiable function ρ, at least one of which has compact support (i.e. either $T \in \mathcal{E}'(\mathbb{R}^n)$, $\rho \in \mathcal{E}(\mathbb{R}^n)$ or $T \in \mathcal{D}'(\mathbb{R}^n)$, $\rho \in \mathcal{D}(\mathbb{R}^n)$) is defined by:*

$$T * \rho(\mathbf{x}) = \langle T_{\boldsymbol{\xi}}, \rho(\mathbf{x} - \boldsymbol{\xi}) \rangle \in C^{\infty}(\mathbb{R}^n), \tag{6.4.1}$$

and the infinitely differentiable (in the usual pointwise sense) function $T * \rho$ is called regularization *or* mollification *of the distribution T by ρ with*

$$\partial_{\mathbf{x}}^{\boldsymbol{\alpha}}(T * \rho)(\mathbf{x}) = T * \partial_{\mathbf{x}}^{\boldsymbol{\alpha}} \rho(\mathbf{x}), \tag{6.4.2}$$

where $\partial_{\mathbf{x}}^{\boldsymbol{\alpha}} \rho = \frac{\partial^{|\alpha|} \rho}{\partial_{x_1}^{\alpha_1} \ldots \partial_{x_n}^{\alpha_n}}$ is in the usual pointwise sense. Then, ρ is called the regulator *or* mollifier.

Example 6.4.1.

1. For $\rho(\mathbf{x}) = 1 \; \forall \mathbf{x} \in \mathbb{R}^n$, $T * 1 = \langle T, 1 \rangle =$ constant. (6.4.3)
   - $T = \delta$, $\delta * 1 = \langle \delta, 1 \rangle = 1 =$ constant. (6.4.4)
   - $T = T_f$, $T_f * 1 = \langle T_f, 1 \rangle = \int_{\mathbb{R}^n} f(\mathbf{x}) d\mathbf{x} =$ constant. (6.4.5)
2. For polynomial $p \in P_m$ of degree $\leq m$ and $T \in \mathcal{E}'(\mathbb{R}^n)$,

   $$T * p = \langle T_t, p(\mathbf{x} - \mathbf{t}) \rangle \tag{6.4.6}$$

   is a polynomial of degree $\leq m$. For the sake of simplicity, we consider the case of polynomial $p \in P_m$ in a single variable. Then,

   $$(T * p)(x) = \langle T_t, p(x - t) \rangle = \langle T_t, p((-t) + x) \rangle$$
   $$= \left\langle T_t, \sum_{k=0}^{m} \frac{x^k}{k!} p^{(k)}(-t) \right\rangle = \sum_{k=0}^{m} \frac{x^k}{k!} \langle T_t, p^{(k)}(-t) \rangle = \sum_{k=0}^{m} \frac{\alpha_k}{k!} x^k \in P_m, \tag{6.4.7}$$

   where $\alpha_k = \langle T_t, p^{(k)}(-t) \rangle \in \mathbb{R} \; \forall k = 0, 1, 2 \ldots m$, $p^{(k)}(-t) = \frac{d^k p}{dt^k}(-t)$ $\forall k = 0, 1, 2 \ldots m$.

Continuity of the convolution operation

We agree to accept the following result without proof (see Schwartz [8, p. 170]).

Theorem 6.4.2. *Let (T_k) and (S_k) be any two sequences of distributions in $\mathcal{D}'(\mathbb{R}^n)$ such that $T_k \to T$ and $S_k \to S$ in $\mathcal{D}'(\mathbb{R}^n)$ as $k \to \infty$. Let $A, B \subset \mathbb{R}^n$ be* fixed *closed sets in $\mathbb{R}^n$ with* $\text{supp}(T_k) \subset A$, $\text{supp}(S_k) \subset B$ *$\forall k \in \mathbb{N}$ such that for $\boldsymbol{\xi} \in A$, $\boldsymbol{\eta} \in B$, $\boldsymbol{\xi} + \boldsymbol{\eta}$ remains* bounded *only if both $\boldsymbol{\xi}$ and $\boldsymbol{\eta}$ remain* bounded, *i.e. $T_k * S_k$ is defined $\forall k \in \mathbb{N}$ and $T * S$ is also defined. Then, $T_k * S_k \to T * S$ in $\mathcal{D}'(\mathbb{R}^n)$ as $k \to \infty$.*

6.5 Approximation of distributions by C^∞-functions

Proposition 6.5.1. *Every distribution $T \in \mathcal{D}'(\mathbb{R}^n)$ is the limit of C^∞-functions (i.e. functions belonging to $C^\infty(\mathbb{R}^n)$) in $\mathcal{D}'(\mathbb{R}^n)$. In other words,*

$$\mathcal{E}(\mathbb{R}^n) \equiv C^\infty(\mathbb{R}^n) \text{ is dense in } \mathcal{D}'(\mathbb{R}^n). \tag{6.5.1}$$

Proof. $\forall \varepsilon > 0$, let ρ_ε be defined by (6.2.15).

Then $\rho_\varepsilon \in C_0^\infty(\mathbb{R}^n)$ with $\text{supp}(\rho_\varepsilon) = \overline{B}(0;\varepsilon) = \{\mathbf{x} : \|\mathbf{x}\| \le \varepsilon\} \subset\subset \mathbb{R}^n$.

$\lim_{\varepsilon\to 0+} \int_{\mathbb{R}^n} \rho_\varepsilon(\mathbf{x})\phi(\mathbf{x})d\mathbf{x} = \phi(\mathbf{0})$ $\forall \phi \in \mathcal{D}(\mathbb{R}^n)$. In fact,

$$\begin{aligned}
\int_{\mathbb{R}^n} \rho_\varepsilon(\mathbf{x})\phi(\mathbf{x})d\mathbf{x} - \phi(\mathbf{0}) &= \int_{\mathbb{R}^n} [\rho_\varepsilon(\mathbf{x})\phi(\mathbf{x}) - \phi(\mathbf{0})\rho_\varepsilon(\mathbf{x})]d\mathbf{x} \\
&= \int_{\mathbb{R}^n} \rho_\varepsilon(\mathbf{x})(\phi(\mathbf{x}) - \phi(\mathbf{0}))d\mathbf{x} \\
&= \int_{\overline{B}(0;\varepsilon)} \rho_\varepsilon(\mathbf{x})(\phi(\mathbf{x}) - \phi(\mathbf{0}))d\mathbf{x}
\end{aligned}$$

$$\begin{aligned}
\implies \quad \left| \int_{\mathbb{R}^n} \rho_\varepsilon(\mathbf{x})\phi(\mathbf{x})d\mathbf{x} - \phi(\mathbf{0}) \right| &\le \max_{\|\mathbf{x}\|\le\varepsilon} |\phi(\mathbf{x}) - \phi(\mathbf{0})| \int_{\overline{B}(0,\varepsilon)} \rho_\varepsilon(\mathbf{x})d\mathbf{x} \\
&= \max_{\|\mathbf{x}\|\le\varepsilon} |\phi(\mathbf{x}) - \phi(\mathbf{0})|,
\end{aligned}$$

since $\int_{\overline{B}(0,\varepsilon)} \rho_\varepsilon(\mathbf{x})d\mathbf{x} = 1$. Since $\phi \in \mathcal{D}(\mathbb{R}^n)$ is continuous in $\mathbb{R}^n$, $\forall \eta > 0$, $\exists \varepsilon_0 = \varepsilon_0(\eta) > 0$ such that $|\phi(\mathbf{x}) - \phi(\mathbf{0})| < \eta$ $\forall \|\mathbf{x}\| \le \varepsilon$ with $\varepsilon < \varepsilon_0$. Hence, $\forall \eta > 0$, $\exists \varepsilon_0 = \varepsilon_0(\eta) > 0$ such that

$$\left| \int_{\mathbb{R}^n} \rho_\varepsilon(\mathbf{x})\phi(\mathbf{x})d\mathbf{x} - \phi(\mathbf{0}) \right| \le \eta \quad \forall \varepsilon < \varepsilon_0,$$

i.e.

$$\lim_{\varepsilon\to 0+} \int_{\mathbb{R}^n} \rho_\varepsilon(\mathbf{x})\phi(\mathbf{x})d\mathbf{x} = \phi(\mathbf{0}). \tag{6.5.2}$$

$\rho_\varepsilon \to \delta$ in $\mathcal{D}'(\mathbb{R}^n)$ as $\varepsilon \to 0^+$: $\forall \phi \in \mathcal{D}(\mathbb{R}^n)$,

$$\lim_{\varepsilon \to 0^+} \langle \rho_\varepsilon, \phi \rangle = \lim_{\varepsilon \to 0^+} \int_{\mathbb{R}^n} \rho_\varepsilon(\mathbf{x})\phi(\mathbf{x})d\mathbf{x} = \phi(0) = \langle \delta, \phi \rangle$$
$$\Longrightarrow \quad \lim_{\varepsilon \to 0^+} \rho_\varepsilon = \delta \in \mathcal{D}'(\mathbb{R}^n). \tag{6.5.3}$$

$T * \rho_\varepsilon \to T$ as $\varepsilon \to 0^+$: Since $\rho_\varepsilon \in \mathcal{D}(\mathbb{R}^n)$ $\forall \varepsilon > 0$, $T * \rho_\varepsilon$ is a C^∞-function $\forall$ distributions $T \in \mathcal{D}(\mathbb{R}^n)$ by Theorem 6.4.1. But $\rho_\varepsilon \to \delta$ in $\mathcal{D}'(\mathbb{R}^n)$ as $\varepsilon \to 0^+ \Longrightarrow T * \rho_\varepsilon \to T * \delta = T$ in $\mathcal{D}'(\mathbb{R}^n)$ as $\varepsilon \to 0^+$ by Theorem 6.4.2, i.e. $\lim_{\varepsilon \to 0^+} \langle T * \rho_\varepsilon, \phi \rangle = \langle T * \delta, \phi \rangle = \langle T, \phi \rangle$ $\forall \phi \in \mathcal{D}(\mathbb{R}^n)$. Thus, $\forall T \in \mathcal{D}'(\mathbb{R}^n)$, $\exists$ a sequence of C^∞-functions $T * \rho_\varepsilon \in C^\infty(\mathbb{R}^n)$ such that

$$\lim_{\varepsilon \to 0^+} (T * \rho_\varepsilon) = T \in \mathcal{D}'(\mathbb{R}^n). \tag{6.5.4}$$

□

Density of $\mathcal{D}(\mathbb{R}^n)$ in $\mathcal{D}'(\mathbb{R}^n)$

Theorem 6.5.1. $\mathcal{D}(\mathbb{R}^n)$ *is* dense *in* $\mathcal{D}'(\mathbb{R}^n)$.

Proof. $\mathcal{D}(\mathbb{R}^n)$ is dense in $\mathcal{E}(\mathbb{R}^n)$ which is dense in $\mathcal{D}'(\mathbb{R}^n)$ by (6.5.1), and $\mathcal{D}(\mathbb{R}^n) \hookrightarrow \mathcal{E}(\mathbb{R}^n) \hookrightarrow \mathcal{D}'(\mathbb{R}^n)$, the imbeddings being continuous ones. Hence, $\mathcal{D}(\mathbb{R}^n)$ is dense in $\mathcal{D}'(\mathbb{R}^n)$. For an alternative proof, see Theorem 4.1.2. □

We agree to accept the following result without proof.

Proposition 6.5.2. *Let $\delta \in \mathcal{D}'(\mathbb{R}^n)$ be the Dirac distribution (measure) with concentration at $\mathbf{0} \in \mathbb{R}^n$. Then $\exists$ a sequence (p_k) of polynomials p_k such that $p_k \to \delta = \delta_{\mathbf{0}}$ in $\mathcal{D}'(\mathbb{R}^n)$ as $k \to \infty$, i.e.*

$$\langle p_k, \phi \rangle \to \langle \delta, \phi \rangle = \phi(\mathbf{0}) \quad \text{as } k \to \infty \ \forall \phi \in \mathcal{D}(\mathbb{R}^n). \tag{6.5.5}$$

Weierstrass's Approximation Theorem Let $T \in \mathcal{E}'(\mathbb{R}^n) \subset \mathcal{D}'(\mathbb{R}^n)$ be a distribution with compact support. Let (p_k) be a sequence of polynomials such that $p_k \to \delta$ in $\mathcal{D}'(\mathbb{R}^n)$, i.e. $\lim_{k \to \infty} \langle p_k, \phi \rangle = \langle \delta, \phi \rangle = \phi(\mathbf{0})$. Then, analogously to Weierstrass's theorem on the approximation of continuous functions on compact sets by polynomials, we have:

Theorem 6.5.2. $\forall T \in \mathcal{E}'(\mathbb{R}^n)$ *with compact support in $\mathbb{R}^n$, $\exists$ a sequence (p_k) of polynomials p_k with $p_k \to \delta$ (the Dirac distribution) in $\mathcal{D}'(\mathbb{R}^n)$ as $k \to \infty$ such that*

$$\lim_{k \to \infty} (T * p_k) = T \quad \text{in } \mathcal{D}'(\mathbb{R}^n). \tag{6.5.6}$$

Proof. From Proposition 6.5.2, $\exists$ a sequence (p_k) of polynomials such that $p_k \to \delta$ in $\mathcal{D}'(\mathbb{R}^n)$ as $k \to \infty$. $T * p_k$ is well defined $\forall T \in \mathcal{E}'(\mathbb{R}^n)$, and $\forall$ polynomials p_k with $k \in \mathbb{N}$ (see (6.4.6)–(6.4.7)). Moreover, $T * \delta = T$ is defined $\forall T$. From Theorem 6.4.2, $T * p_k \to T * \delta = T$ as $k \to \infty$. □

6.6 Convolution of several distributions

Let T, S, R in $\mathcal{D}'(\mathbb{R}^n)$ be distributions such that, $\forall \phi \in \mathcal{D}(\mathbb{R}^n)$, $\operatorname{supp}(T_\xi \otimes S_\eta \otimes R_\zeta) \cap \operatorname{supp} \phi(\xi + \eta + \zeta)$ is a bounded set in $\mathbb{R}^n \times \mathbb{R}^n \times \mathbb{R}^n = \mathbb{R}^{3n}$. Then the convolution $T * S * R$ of T, S and R is well defined and a distribution defined, $\forall \phi \in \mathcal{D}(\mathbb{R}^n)$, by:

$$\langle T * S * R, \phi \rangle = \langle T_\xi \otimes S_\eta \otimes R_\zeta, \phi(\xi + \eta + \zeta) \rangle. \tag{6.6.1}$$

Remark 6.6.1.

- Even if $T * S * R$ is not defined, $(T * S) * R$ and $T * (S * R)$ *may* be defined and may *not* be equal. For example, for the Heaviside function H with its derivative $\frac{dH}{dx} = \delta$ (in the sense of distribution), $1 * \delta' * H$ does *not* exist, but $(1 * \delta') * H$ and $1 * (\delta' * H)$ both exist, ($\delta' = \frac{d\delta}{dx}$). In fact,
$$(1 * \delta') * H = (\delta' * 1) * H = 0 * H = 0,$$
$$1 * (\delta' * H) = 1 * H' = 1 * \delta = \delta * 1 = 1,$$
i.e. $(1 * \delta') * H$ and $1 * (\delta' * H)$ exist, but are *not* equal.
- $T * S * R$ defined by (6.6.1) exists $\Longrightarrow$ it is associative, i.e.
$$T * S * R = T * (S * R) = (T * S) * R. \tag{6.6.2}$$

The following result is of importance in applications:

Theorem 6.6.1. *If at least* two *of the three distributions T, S, R have compact support in $\mathbb{R}^n$, then the convolution $T * S * R$ exists and is* associative *and* commutative.

Moreover, for $n = 1$, if T, S and R have supports bounded from the left, i.e. contained in $]a, \infty[$ (resp. bounded from the right, i.e. contained in $]-\infty, b[$), then $T * S * R$ exists and is associative and commutative.

Example 6.6.1. Let $H(x)$ be the Heaviside function with $H(x) = 1$ for $x \geq 0$ and $H(x) = 0$ for $x < 0$, and let $\Gamma(x)$ be the Gamma function. Then, for $\alpha, \beta > 0$ and $\lambda \in \mathbb{C}$, prove that

1. $$\left[H(x)e^{\lambda x} \cdot \frac{x^{\alpha-1}}{\Gamma(\alpha)}\right] * \left[H(x)e^{\lambda x} \cdot \frac{x^{\beta-1}}{\Gamma(\beta)}\right] = \left[H(x)e^{\lambda x} \frac{x^{\alpha+\beta-1}}{\Gamma(\alpha+\beta)}\right]. \tag{6.6.3}$$

   In particular, for $\lambda = 0$, $\alpha, \beta > 0$, $H_\alpha(x) = \frac{H(x)x^{\alpha-1}}{\Gamma(\alpha)}$, $H_\beta(x) = \frac{H(x)x^{\beta-1}}{\Gamma(\beta)}$,
$$(H_\alpha * H_\beta)(x) = H_{\alpha+\beta}(x).$$

2.
$$\underbrace{H(x)e^{\lambda x} * H(x)e^{\lambda x} * \cdots * H(x)e^{\lambda x}}_{n \text{ distributions}} = H(x)e^{\lambda x}\frac{x^{n-1}}{(n-1)!}.$$

Solution.

1.
$$\left[H(x)e^{\lambda x}\cdot\frac{x^{\alpha-1}}{\Gamma(\alpha)}\right] * \left[H(x)e^{\lambda x}\cdot\frac{x^{\beta-1}}{\Gamma(\beta)}\right]$$
$$= \int_{-\infty}^{\infty} H(x-\xi)e^{\lambda(x-\xi)}\frac{(x-\xi)^{\alpha-1}}{\Gamma(\alpha)}\frac{H(\xi)e^{\lambda\xi}\xi^{\beta-1}}{\Gamma(\beta)}d\xi$$
$$= \begin{cases} 0 & \text{for } x < 0 \\ \underbrace{\int_0^x \frac{(x-\xi)^{\alpha-1}\xi^{\beta-1}e^{\lambda(x-\xi)}e^{\lambda\xi}}{\Gamma(\alpha)\Gamma(\beta)}d\xi}_{I} & \text{for } x \geq 0, \end{cases} \tag{6.6.4}$$

since $H(\xi) = 0$ for $\xi < 0$, $H(x-\xi) = 0$ for $x < \xi < 0$, $H(\xi) = 1$ for $\xi \geq 0$, $H(x-\xi) = 1$ for $0 \leq \xi \leq x$. By change of the variable, $\xi = xt$ with $d\xi = x dt$, we get the right-hand side integral I in (6.6.4) as follows:

$$I = \frac{x^{\alpha+\beta-1}e^{\lambda x}}{\Gamma(\alpha)\Gamma(\beta)}\int_0^1 (1-t)^{\alpha-1}t^{\beta-1}dt = \frac{e^{\lambda x}\cdot x^{\alpha+\beta-1}}{\Gamma(\alpha)\Gamma(\beta)}B(\alpha,\beta),$$

where the *Beta function* $B(\alpha,\beta) = \frac{\Gamma(\alpha)\Gamma(\beta)}{\Gamma(\alpha+\beta)}$.

$$\Longrightarrow \quad I = \frac{e^{\lambda x}x^{\lambda+\beta-1}}{\Gamma(\alpha)\Gamma(\beta)}\cdot\frac{\Gamma(\alpha)\Gamma(\beta)}{\Gamma(\alpha+\beta)} = \frac{e^{\lambda x}x^{\lambda+\beta-1}}{\Gamma(\alpha+\beta)} \quad \text{for } x \geq 0. \tag{6.6.5}$$

Then, from (6.6.4) and (6.6.5), we get

$$\left[\frac{H(x)e^{\lambda x}x^{\alpha-1}}{\Gamma(\alpha)}\right] * \left[\frac{H(x)e^{\lambda x}x^{\beta-1}}{\Gamma(\beta)}\right] = \begin{cases} 0 & \text{for } x < 0 \\ \frac{e^{\lambda x}x^{\lambda+\beta-1}}{\Gamma(\alpha+\beta)} & \text{for } x \geq 0 \end{cases}$$
$$= H(x)e^{\lambda x}\cdot\frac{x^{\alpha+\beta-1}}{\Gamma(\alpha+\beta)}.$$

2. For $\alpha = \beta = 1$, $\Gamma(1) = 1$ and $\Gamma(n) = (n-1)!$,

$$[H(x)e^{\lambda x}] * [H(x)e^{\lambda x}] = H(x)e^{\lambda x}\frac{x^{1+1-1}}{\Gamma(2)} = H(x)e^{\lambda x}\cdot\frac{x}{1!};$$

$$[H(x)e^{\lambda x}] * [H(x)e^{\lambda x}] * [H(x)e^{\lambda x}] = H(x)e^{\lambda x}\frac{x^{2+1-1}}{\Gamma(2+1)} = H(x)e^{\lambda x}\cdot\frac{x^2}{2!};$$

$$\underbrace{[H(x)e^{\lambda x}] * [H(x)e^{\lambda x}] * \cdots * [H(x)e^{\lambda x}]}_{n \text{ terms}} = H(x)e^{\lambda x}\frac{x^{n-1}}{(n-1)!}.$$

Remark 6.6.2. Convolutions are used in defining derivatives of *fractional* order. For example, for $f \in \mathcal{D}'(\mathbb{R})$ with $\operatorname{supp}(f) \subset [0, \infty[$, i.e. $f \in \mathcal{D}'^{+}$ (see Sections 6.9 and Section 8.8 for more detail) and $D = \frac{d}{dx}$, we define $D^{1/2} f = D(H_{1/2} * f)$, where $H_{1/2}(x)$ is obtained from (6.6.3) with $\alpha = 1/2$, $\lambda = 0$, i.e. $H_{1/2}(x) = \frac{H(x)x^{1/2-1}}{\Gamma(1/2)} = \frac{H(x)}{\sqrt{\pi}\cdot\sqrt{x}}$, $(\Gamma(1/2) = \sqrt{\pi})$. Then, for $f \in \mathcal{D}'(\mathbb{R})$ with $\operatorname{supp}(f) \subset [0, \infty[$,

$$
\begin{aligned}
(H_{1/2} * f) &= \int_{-\infty}^{\infty} H_{1/2}(x-\xi) f(\xi) d\xi = \int_{0}^{\infty} H_{1/2}(x-\xi) f(\xi) d\xi \\
&= \int_{0}^{\infty} \frac{H(x-\xi) f(\xi)}{\sqrt{\pi}\sqrt{x-\xi}} d\xi = \int_{0}^{x} \frac{f(\xi)}{\sqrt{\pi}\sqrt{x-\xi}} d\xi,
\end{aligned}
$$

since

$$
H(x-\xi) = \begin{cases} 0 & \text{for } x-\xi < 0, \text{ i.e. for } \xi > x \geq 0, \\ 1 & \text{for } x-\xi \geq 0, \text{ i.e. for } x \geq \xi \geq 0. \end{cases}
$$

Hence, $D^{1/2} f = D(H_{1/2} * f) = \frac{d}{dx} \int_0^x \frac{f(\xi)}{\sqrt{\pi}\sqrt{x-\xi}} d\xi$ (see Vladimirov [6, pp. 88–89] for more detail).

6.7 Derivatives of convolution, convolution of distributions on a circle Γ and their Fourier series representations on Γ

Since the Dirac distribution δ and its derivatives have compact support in $\mathbb{R}^n$, by Theorem 6.6.1 $\partial^{\boldsymbol{\alpha}} \delta * T * S$ will be well defined and given by (6.6.1) ∀ multi-index $\boldsymbol{\alpha}$, $\partial^{\boldsymbol{\alpha}} = \frac{\partial^{|\alpha|}}{\partial_{x_1}^{\alpha_1} \cdots \partial_{x_n}^{\alpha_n}}$, if either T or S has compact support, and it will be associative and commutative (6.6.2).

Proposition 6.7.1. *Let* $T \in \mathcal{E}'(\mathbb{R}^n) \subset \mathcal{D}'(\mathbb{R}^n)$, $S \in \mathcal{D}'(\mathbb{R}^n)$. *Then* $T * S \in \mathcal{D}'(\mathbb{R}^n)$, *and the* distributional derivatives *of* $T * S$ *are given by:*

$$
\frac{\partial}{\partial x_k}(T * S) = \frac{\partial T}{\partial x_k} * S = T * \frac{\partial S}{\partial x_k}; \tag{6.7.1}
$$

$$
\frac{\partial^2}{\partial x_i \partial x_j}(T * S) = \frac{\partial^2 T}{\partial x_i \partial x_j} * S = T * \frac{\partial^2 S}{\partial x_i \partial x_j} \tag{6.7.2}
$$

$$
= \frac{\partial T}{\partial x_i} * \frac{\partial S}{\partial x_j} = \frac{\partial T}{\partial x_j} * \frac{\partial S}{\partial x_i}. \tag{6.7.3}
$$

In general, ∀ *multi-index* $\boldsymbol{\alpha} = (\alpha_1, \alpha_2, \ldots, \alpha_n)$,

$$
\partial^{\boldsymbol{\alpha}}(T * S) = \partial^{\boldsymbol{\alpha}} T * S = T * \partial^{\boldsymbol{\alpha}} S. \tag{6.7.4}
$$

Proof. From (6.3.21)–(6.3.24), we know that $\delta * S = S$, $\frac{\partial \delta}{\partial x_k} * S = \frac{\partial S}{\partial x_k}$, $\partial^{\boldsymbol{\alpha}} \delta * S = \partial^{\boldsymbol{\alpha}} S$ $\forall$ distributions $S \in \mathcal{D}'(\mathbb{R}^n)$. Now, applying the associative and commutative properties, we have:

$$\begin{aligned}\frac{\partial}{\partial x_k}(T * S) &= \frac{\partial \delta}{\partial x_k} * (T * S) = \left(\frac{\partial \delta}{\partial x_k} * T\right) * S = \frac{\partial T}{\partial x_k} * S \\ &= \frac{\partial \delta}{\partial x_k} * (S * T) = \left(\frac{\partial \delta}{\partial x_k} * S\right) * T = \frac{\partial S}{\partial x_k} * T = T * \frac{\partial S}{\partial x_k}.\end{aligned}$$

Applying this formula, we get

$$\begin{aligned}\frac{\partial^2}{\partial x_i \partial x_j}(T * S) &= \frac{\partial}{\partial x_i}\left[\frac{\partial}{\partial x_j}(T * S)\right] = \frac{\partial}{\partial x_i}\left(\frac{\partial T}{\partial x_j} * S\right) \\ &= \frac{\partial}{\partial x_i}\left(\frac{\partial T}{\partial x_j}\right) * S = \frac{\partial T}{\partial x_j} * \frac{\partial S}{\partial x_i} = \frac{\partial^2 T}{\partial x_i \partial x_j} * S; \\ \frac{\partial^2}{\partial x_i \partial x_j}(T * S) &= \frac{\partial}{\partial x_i}\left[\frac{\partial}{\partial x_j}(T * S)\right] = \frac{\partial}{\partial x_i}\left(T * \frac{\partial S}{\partial x_j}\right) \\ &= T * \frac{\partial}{\partial x_i}\left(\frac{\partial S}{\partial x_j}\right) = \frac{\partial T}{\partial x_i} * \frac{\partial S}{\partial x_j} = T * \frac{\partial^2 S}{\partial x_i \partial x_j}.\end{aligned}$$

Similarly, (6.7.4) can be proved. □

Remark 6.7.1.

- For arbitrary distributions T and S, $T * S$ is *not* defined. Even if $T * S$ is not defined, $\partial^{\boldsymbol{\alpha}} T * S$ and $T * \partial^{\boldsymbol{\alpha}} S$ *may* be defined and may *not* be equal. For example, for the Heaviside function H with $\frac{dH}{dx} = \delta$, $\frac{dH}{dx} * 1 = \delta * 1 = 1$, but $H * \frac{d1}{dx} = H * 0 = 0$, i.e. $\frac{dH}{dx} * 1 \neq H * \frac{d1}{dx}$, since $H * 1$ is not defined, and consequently (6.7.1) does *not* hold. Thus, from the existence of $\partial^{\boldsymbol{\alpha}} f * g$ and $f * \partial^{\boldsymbol{\alpha}} g$, we cannot conclude that their equality in formula (6.7.4) holds.
- The formulae (6.7.1)–(6.7.4) do *not* hold for functions *in general*, if the derivatives of f are in the usual pointwise sense. For example, for $f \in C_0(\mathbb{R})$ with *compact support* and Heaviside function H with $\frac{dH}{dx} = \delta$ in the distributional sense,

$$\frac{d}{dx}(H * f) = \frac{dH}{dx} * f = \delta * f = f.$$

But if we consider the derivative of H in the usual pointwise sense, $[\frac{dH}{dx}](x) = 0$ $\forall x \neq 0$, then

$$\left[\frac{dH}{dx}\right] * f = \int_{-\infty}^{\infty} f(t)\left[\frac{dH}{dx}\right](x - t)dt = 0,$$

i.e. $\frac{d}{dx}(H * f) \neq [\frac{dH}{dx}] * f$. Hence, in the formula for functions, the derivatives are *in the sense of distributions*.

Example 6.7.1. Let

$$P(\xi) = \xi^m - a_1\xi^{m-1} + \cdots + a_{m-1}\xi + a_m = \prod_{i=1}^{m}(\xi - \lambda_i) \tag{6.7.5}$$

be a polynomial of degree m with real roots $\lambda_1, \lambda_2, \ldots \lambda_m$ such that the λ_is may or may not be distinct numbers, and the corresponding ordinary differential operator $P(D)$ be defined by

$$P(D) \equiv \frac{d^m}{dx^m} + a_1\frac{d^{m-1}}{dx^{m-1}} + \cdots + a_m = \left(\frac{d}{dx} - \lambda_1\right)\cdots\left(\frac{d}{dx} - \lambda_m\right) \tag{6.7.6}$$

with $D = \frac{d}{dx}$.

Then, show that

$$P(D)\delta = \left(\frac{d\delta}{dx} - \lambda_1\delta\right) * \left(\frac{d\delta}{dx} - \lambda_2\delta\right) * \cdots * \left(\frac{d\delta}{dx} - \lambda_m\delta\right), \tag{6.7.7}$$

δ being the Dirac distribution.

Solution. Using (6.3.21),

$$\begin{aligned}
P(D)\delta &= P(D)\delta * \delta = \left[\left(\frac{d}{dx} - \lambda_1\right)\left(\frac{d}{dx} - \lambda_2\right)\cdots\left(\frac{d}{dx} - \lambda_m\right)\right]\delta * \delta \\
&= \left(\frac{d}{dx} - \lambda_1\right)\left[\left(\frac{d}{dx} - \lambda_2\right)\cdots\left(\frac{d}{dx} - \lambda_m\right)\right]\delta * \delta \\
&= \left[\left(\frac{d}{dx} - \lambda_2\right)\cdots\left(\frac{d}{dx} - \lambda_m\right)\right]\delta * \left(\frac{d}{dx} - \lambda_1\right)\delta \quad \text{(by (6.7.4))} \\
&= \left(\frac{d}{dx} - \lambda_2\right)\left[\left(\frac{d}{dx} - \lambda_3\right)\cdots\left(\frac{d}{dx} - \lambda_m\right)\right]\delta * \left(\frac{d\delta}{dx} - \lambda_1\delta\right) * \delta \quad \text{(by (6.3.21))} \\
&= \left[\left(\frac{d}{dx} - \lambda_3\right)\cdots\left(\frac{d}{dx} - \lambda_m\right)\right]\delta * \left(\frac{d\delta}{dx} - \lambda_1\delta\right) * \left(\frac{d}{dx} - \lambda_2\right)\delta = \cdots \\
&= \left(\frac{d}{dx} - \lambda_m\right)\delta * \left(\frac{d\delta}{dx} - \lambda_1\delta\right) * \left(\frac{d\delta}{dx} - \lambda_2\delta\right) * \cdots * \left(\frac{d\delta}{dx} - \lambda_{m-1}\delta\right) \\
&= \left(\frac{d\delta}{dx} - \lambda_1\delta\right) * \left(\frac{d\delta}{dx} - \lambda_2\delta\right) * \cdots * \left(\frac{d\delta}{dx} - \lambda_m\delta\right).
\end{aligned}$$

In particular, for $\lambda_1 = \lambda_2 = \cdots = \lambda_m = \lambda$, $P(D) = (\frac{d}{dx} - \lambda)^m$ and

$$P(D)\delta = \left(\frac{d\delta}{dx} - \lambda\delta\right) * \cdots * \left(\frac{d}{dx} - \lambda\delta\right) = \left(\frac{d\delta}{dx} - \lambda\delta\right)^{*m}. \tag{6.7.8}$$

Convolution of distributions of $\mathcal{D}'(\Gamma)$ defined on a circle Γ

Consider the test space $\mathcal{D}(\Gamma)$ (see Definition 1.10.2) of infinitely differentiable functions on a circle $\Gamma = \Gamma(\mathbf{0}; r)$, and the dual $\mathcal{D}'(\Gamma)$ of distributions on the circle Γ (see Definition 1.10.3). Since arcs of different lengths on Γ can be *added*, we can define the convolution of distributions on Γ. For this we begin with functions on Γ. Let $f, g \in L^1_{\text{loc}}(\Gamma)$. Then their *convolution* $f * g$ on Γ is defined by: $\forall$ arc length s,

$$h(s) = f * g(s) = \int_\Gamma f(s-\xi)g(\xi)d\xi = \int_\Gamma f(\xi)g(s-\xi)d\xi. \tag{6.7.9}$$

Similarly, for *any pair* of distributions $S_1, S_2 \in \mathcal{D}'(\Gamma)$, their convolution $S_1 * S_2 \in \mathcal{D}'(\Gamma)$ is well defined, since $S_1, S_2 \in \mathcal{D}'(\Gamma)$ have *bounded* support, and is given, $\forall \phi \in \mathcal{D}(\Gamma)$, by:

$$\langle S_1 * S_2, \phi\rangle_{\mathcal{D}'(\Gamma)\times\mathcal{D}(\Gamma)} = \langle S_1(\xi) \otimes S_2(\eta), \phi(\xi+\eta)\rangle = \langle S_1(\xi), \langle S_2(\eta), \phi(\xi+\eta)\rangle\rangle. \tag{6.7.10}$$

$S_1 * S_2 = S_2 * S_1$ in $\mathcal{D}'(\Gamma)$ $\forall S_1, S_2 \in \mathcal{D}'(\Gamma)$.

Example 6.7.2. For $\delta(s) \in \mathcal{D}'(\Gamma)$ and $S \in \mathcal{D}'(\Gamma)$, $\delta(s) * S = S$ in $\mathcal{D}'(\Gamma)$, since

$$\begin{aligned}\langle \delta * S, \phi\rangle = \langle S * \delta, \phi\rangle &= \langle S(\xi), \langle \delta(\eta), \phi(\xi+\eta)\rangle\rangle \\ &= \langle S(\xi), \phi(\xi+0)\rangle = \langle S, \phi\rangle \quad \forall \phi \in \mathcal{D}(\Gamma) \qquad (6.7.11)\\ \iff \quad & \delta * S = S \in \mathcal{D}'(\Gamma) \quad \text{(see also (6.3.21))}. \qquad (6.7.12)\end{aligned}$$

For $\delta'(s) \in \mathcal{D}'(\Gamma)$ and $S \in \mathcal{D}'(\Gamma)$, $\delta' * S = \delta * S' = S'$ in $\mathcal{D}'(\Gamma)$, where $\delta'(s) = \frac{d\delta}{ds}$ (see also (6.3.22)). In fact,

$$\begin{aligned}\langle \delta' * S, \phi\rangle &= \langle S(\xi), \langle \delta'(\eta), \phi(\xi+\eta)\rangle\rangle = \left\langle S(\xi), (-1)\left\langle \delta(\eta), \frac{\partial\phi}{\partial\eta}(\xi+\eta)\right\rangle\right\rangle \\ &= -\langle S(\xi), \phi'(\xi+0)\rangle = \langle S', \phi\rangle \quad \forall\phi \in \mathcal{D}(\Gamma) \\ \iff \quad & \delta' * S = S' = \delta * S' \quad \text{in } \mathcal{D}'(\Gamma). \qquad (6.7.13)\end{aligned}$$

$S_1, S_2 \in \mathcal{D}'(\Gamma)$

$$\implies \quad \frac{d(S_1 * S_2)}{ds} = \frac{dS_1}{ds} * S_2 = S_1 * \frac{dS_2}{ds} \quad \text{in } \mathcal{D}'(\Gamma). \tag{6.7.14}$$

Indeed, from (6.7.12)–(6.7.13), $\delta * S_i = S_i$, $\delta' * S_i = S_i'$ in $\mathcal{D}'(\Gamma)$. Then, applying the commutative and associative properties, we have

$$\begin{aligned}\frac{d(S_1 * S_2)}{ds} &= \frac{d\delta}{ds} * (S_1 * S_2) = \left(\frac{d\delta}{ds} * S_1\right) * S_2 = \frac{dS_1}{ds} * S_2 \\ &= S_1' * S_2 = S_1 * S_2'.\end{aligned}$$

The last equality in (6.7.14) is similarly established.

Remark 6.7.2. In (6.7.9), $f * g$ has been defined for $f, g \in L^1_{\text{loc}}(\Gamma)$. Let $\tilde{f}, \tilde{g}$ be the associated *periodic* functions on $\mathbb{R}$ with period $T > 0$ defined in (1.10.55). Then, in general, $\tilde{f} * \tilde{g}$ is *not* defined, since $\int_{-\infty}^{\infty} f(x-\xi)g(\xi)d\xi$ does not exist for arbitrary periodic $\tilde{f}, \tilde{g} \in L^1_{\text{loc}}(\mathbb{R})$ owing to the fact that $\text{supp}(\tilde{f})$ and $\text{supp}(\tilde{g})$ are *not* bounded on $\mathbb{R}$. Hence, for periodic $\tilde{f}, \tilde{g}$ on $\mathbb{R}$ with period T, we define

$$h(x) = \int_a^{a+T} \tilde{f}(x-\xi)\tilde{g}(\xi)d\xi = \int_a^{a+T} \tilde{f}(\xi)\tilde{g}(x-\xi)d\xi. \qquad (6.7.15)$$

Fourier series of distributions in $\mathcal{D}'(\Gamma)$

Let $f \in L^1_{\text{loc}}(\Gamma)$ be a locally summable function on Γ. Thus, from Property 4 of *Fourier coefficients* in (2.11.7e), we can define $c_k(f)$ by: for $T = 2\pi r$,

$$c_k(f) = \frac{1}{T}\int_\Gamma f(s)e^{-ik\omega s}ds = \frac{1}{T}\langle f, e^{-ik\omega s}\rangle_{\mathcal{D}'(\Gamma)\times\mathcal{D}(\Gamma)} \quad \forall k \in \mathbb{Z}, \qquad (6.7.16)$$

which suggests we define *Fourier coefficients of distributions* $S \in \mathcal{D}'(\Gamma)$ by:

$$c_k(S) = \frac{1}{T}\langle S, e^{-ik\omega s}\rangle_{\mathcal{D}'(\Gamma)\times\mathcal{D}(\Gamma)} \quad \forall k \in \mathbb{Z}. \qquad (6.7.17)$$

Fourier series of distributions on Γ

The trigonometric series $\sum_{k=-\infty}^{\infty} c_k(S)e^{ik\omega s}$ with

$$c_k(S) = \frac{1}{T}\langle S, e^{-ik\omega s}\rangle_{\mathcal{D}'(\Gamma)\times\mathcal{D}(\Gamma)} \qquad (6.7.18)$$

is called the *Fourier series* of $S \in \mathcal{D}'(\Gamma)$.

It is interesting to note that if a trigonometric series $\sum_{k=-\infty}^{\infty} d_k e^{ik\omega s}$ converges to a distribution $S \in \mathcal{D}'(\Gamma)$, then $\sum_{k=-\infty}^{\infty} d_k e^{ik\omega s}$ is the Fourier series of $S \in \mathcal{D}'(\Gamma)$ with $d_k = c_k(S)\ \forall k \in \mathbb{Z}$. (6.7.19)

In fact, $\sum_{k=-\infty}^{\infty} d_k e^{ik\omega s}$ converges to $S \in \mathcal{D}'(\Gamma)$

$$\iff \left\langle \sum_{k=-n}^{n} d_k e^{ik\omega s}, \phi \right\rangle_{\mathcal{D}'(\Gamma)\times\mathcal{D}(\Gamma)} \to \langle S, \phi\rangle_{\mathcal{D}'(\Gamma)\times\mathcal{D}(\Gamma)}$$

in $\mathbb{C}\ \forall\phi \in \mathcal{D}(\Gamma)$ as $n \to \infty$. Since

$$\langle e^{ik\omega s}, e^{-il\omega s}\rangle = \int_\Gamma e^{ik\omega s}e^{-il\omega s}ds = \int_0^T e^{ik\omega x}e^{-il\omega x}dx = T\delta_{kl} \quad \forall k, l \in \mathbb{Z}$$

(by (2.11.7d)), for $\phi = e^{-il\omega s} \in \mathcal{D}(\Gamma)$,

$$\left\langle \sum_{k=-n}^{n} d_k e^{ik\omega s}, e^{-il\omega s} \right\rangle = \sum_{k=-n}^{n} d_k T\delta_{kl} = Td_l, \quad -n \le l \le n\ \forall n \in \mathbb{N}$$

$\Longrightarrow \forall l \in \mathbb{Z},\ Td_l = \langle S, e^{-il\omega s}\rangle = Tc_l(S) \Longrightarrow d_l = c_l(S)\ \forall l \in \mathbb{Z} \Longrightarrow \sum_{k=-\infty}^{\infty} d_k e^{ik\omega s} = \sum_{k=-\infty}^{\infty} c_k(S)e^{ik\omega s}$ with $d_k = c_k(S)\ \forall k \in \mathbb{Z}$.

Example 6.7.3. Let $\delta \in \mathcal{D}'(\Gamma)$ be the Dirac distribution with unit mass/charge/force concentrated at $s = 0$ on Γ. Then the *Fourier coefficients* $c_k(\delta)$ of $\delta \in \mathcal{D}'(\Gamma)$ are given by $c_k(\delta) = \frac{1}{T}\langle \delta, e^{-ik\omega s}\rangle = \frac{1}{T}\cdot e^{-ik\omega s}|_{s=0} = \frac{1}{T}\ \forall k \in \mathbb{Z}$, and the *Fourier series* of $\delta \in \mathcal{D}'(\Gamma)$ is given by:

$$\sum_{k=-\infty}^{\infty} c_k(\delta)e^{ik\omega s} = \sum_{k=-\infty}^{\infty} \frac{1}{T}e^{ik\omega s}. \tag{6.7.20}$$

Theorem 6.7.1. *The Fourier series* $\sum_{k=-\infty}^{\infty} \frac{1}{T}e^{ik\omega s}$ *of* $\delta \in \mathcal{D}'(\Gamma)$ *in* (6.7.20) *converges to* $\delta \in \mathcal{D}'(\Gamma)$*:*

$$\delta = \sum_{k=-\infty}^{\infty} \frac{1}{T}e^{ik\omega s} \in \mathcal{D}'(\Gamma). \tag{6.7.21}$$

Proof. For the convergence of Fourier series (6.7.20), we can apply Theorem 2.11.2 on the convergence of trigonometric series. Since $c_k(\delta) = \frac{1}{T} \le |k|^p$ for some $p \in \mathbb{N}$ $\forall |k| \in \mathbb{N}$, Fourier series (6.7.20) converges in $\mathcal{D}'(\Gamma)$. Let the distribution $S \in \mathcal{D}'(\Gamma)$ be its sum, i.e. $S = \sum_{k=-\infty}^{\infty} \frac{1}{T}e^{ik\omega s}$ in $\mathcal{D}'(\Gamma)$. Multiplying term by term by $e^{i\omega s}$, we get $e^{i\omega s}S = \sum_{k=-\infty}^{\infty} \frac{1}{T}e^{i(k+1)\omega s} = \sum_{k'=-\infty}^{\infty} \frac{1}{T}e^{ik'\omega s} = S \Longrightarrow (e^{i\omega s} - 1)S = 0$ in $\mathcal{D}'(\Gamma) \Longrightarrow \alpha(s)S = 0$ with $\alpha(s) = (e^{i\omega s} - 1)$.

$\alpha \in C^\infty(\Gamma)$ vanishes only at $s = 0$, i.e. $\alpha(0) = 0$, but $\alpha'(0) = i\omega e^{ik\omega\cdot 0} = i\omega \neq 0$. Hence, $\alpha(s)$ has a *simple zero* only at $s = 0$ and, by Proposition 1.6.1, $\alpha(s)S = 0$ in $\mathcal{D}'(\Gamma) \Longrightarrow S = C\delta$, where δ is a Dirac distribution concentrated at $s = 0$ and C is a constant. Since $S = \sum_{k=-\infty}^{\infty} \frac{1}{T}e^{ik\omega s}$ in $\mathcal{D}'(\Gamma)$ with $\frac{1}{T}$ = the Fourier coefficient of $S = c_k(S) = c_k(C\delta) = \frac{1}{T}\langle C\delta, e^{-ik\omega s}\rangle = \frac{1}{T}\cdot C\cdot\langle \delta, e^{-k\omega s}\rangle = \frac{1}{T}\cdot C\cdot 1$. Hence, $\frac{1}{T} = \frac{1}{T}\cdot C \Longleftrightarrow C = 1$. Thus, $S = \delta$ in $\mathcal{D}'(\Gamma)$. □

Alternative method. $S = C\delta$ in $\mathcal{D}'(\Gamma) \Longrightarrow C\delta = \sum_{k=-\infty}^{\infty} \frac{1}{T}e^{ik\omega s}$

$\Longrightarrow \langle C\delta, \phi(s)\rangle = \langle \sum_{k=-\infty}^{\infty} \frac{1}{T}e^{ik\omega s}, \phi(s)\rangle = \sum_{k=-\infty}^{\infty} \frac{1}{T}\langle e^{ik\omega s}, \phi\rangle$

$\Longrightarrow C\phi(0) = \sum_{k=-\infty}^{\infty} \frac{1}{T}\langle e^{ik\omega s}, \phi\rangle\ \forall \phi \in \mathcal{D}(\Gamma)$. In particular, for $\phi = 1 \in \mathcal{D}(\Gamma)$, $\phi(0) = 1$ and

$$\langle e^{ik\omega s}, 1\rangle = \int_\Gamma e^{ik\omega s}ds = \begin{cases} \operatorname{meas}(\Gamma) = 2\pi r = T & \text{for } k = 0, \\ 0 & \text{for } k \neq 0,\ k \in \mathbb{Z}. \end{cases}$$

Hence, for $\phi = 1$, $\phi(0) = 1$ and $C\phi(0) = C = \frac{1}{T}\cdot T + 0 = 1$. □

Remark 6.7.3. Since $e^{ik\omega s} \in \mathcal{D}(\Gamma)$, the associated periodic function $e^{ik\omega x}$ on $\mathbb{R}$ with period T belongs to $\mathcal{D}_T(\mathbb{R})$ such that $e^{ik\omega s} = e^{ik\omega x} = e^{ik\omega(x+lT)}$, and $\sum_{k=-\infty}^{\infty} \frac{1}{T} e^{ik\omega x}$ converges to the periodic distribution $\tilde{\delta} \in \mathcal{D}'(\mathbb{R})$ on $\mathbb{R}$ with period T (see also Examples 2.11.2 and 2.11.3). In fact, using (1.10.57),

$$\begin{aligned}
&\left\langle \sum_{k=-n}^{n} \frac{1}{T} e^{ik\omega x}, \phi(x) \right\rangle_{\mathcal{D}'(\mathbb{R})\times\mathcal{D}(\mathbb{R})} = \left\langle \sum_{k=-n}^{n} \frac{1}{T} e^{ik\omega s}, \Phi(s) \right\rangle_{\mathcal{D}'(\Gamma)\times\mathcal{D}(\Gamma)} \\
&\to \langle \delta, \Phi(s) \rangle_{\mathcal{D}'(\Gamma)\times\mathcal{D}(\Gamma)} \quad (\text{as } n \to \infty) \\
&\quad = \Phi(0) = \tilde{\Phi}(0) \quad (\text{see (1.10.53)}) \\
&\quad = \sum_{l=-\infty}^{\infty} \phi(x + lT)|_{x=0} \quad (\text{see (1.10.59)}) \\
&\quad = \sum_{l=-\infty}^{\infty} \phi(lT) = \sum_{l=-\infty}^{\infty} \langle \delta(x - lT), \phi(x) \rangle_{\mathcal{D}'(\mathbb{R})\times\mathcal{D}(\mathbb{R})} \\
&\quad = \left\langle \sum_{l=-\infty}^{\infty} \delta(x - lT), \phi(x) \right\rangle \quad \forall \phi \in \mathcal{D}(\mathbb{R}) \text{ as } n \to \infty \\
&\Longrightarrow \quad \lim_{n\to\infty} \sum_{k=-n}^{n} \frac{1}{T} e^{ik\omega x} = \sum_{l=-\infty}^{\infty} \delta(x - lT) \quad \text{in } \mathcal{D}'(\mathbb{R}) \\
&\Longrightarrow \quad \sum_{k=-\infty}^{\infty} \frac{1}{T} e^{ik\omega x} = \sum_{l=-\infty}^{\infty} \delta(x - lT) = \tilde{\delta}.
\end{aligned} \tag{6.7.22}$$

Hence,

$$\sum_{k=-\infty}^{\infty} \frac{1}{T} e^{ik\omega x} = \sum_{l=-\infty}^{\infty} \delta(x - lT) = \tilde{\delta}, \tag{6.7.23}$$

where $\tilde{\delta}$ is the periodic Dirac distribution with concentrated unit mass/charge/force at the points $x = lT$ on the real line $\forall l \in \mathbb{Z}$. (For another alternative proof see Theorem 6.7.5, and see also the proof of (2.11.9) in Example 2.11.2.)

Convergence of Fourier series (6.7.18) of distributions on Γ In Theorem 6.7.1, we have already proved the convergence of the Fourier series (6.7.20) of Dirac distribution δ. Now, with the help of convolutions of distributions of $\mathcal{D}'(\Gamma)$, the properties of which are similar to those of convolutions of distributions on $\mathbb{R}$, we will prove the convergence of the Fourier series.

Theorem 6.7.2. *Let $\sum_{k=-\infty}^{\infty} c_k(S)e^{ik\omega s}$ be the Fourier series of a distribution $S \in \mathcal{D}'(\Gamma)$ on Γ. Then the series converges to $S \in \mathcal{D}'(\Gamma)$, i.e.*

$$S = \sum_{k=-\infty}^{\infty} c_k(S)e^{ik\omega s}. \tag{6.7.24}$$

Proof. Consider Fourier series (6.7.21), $\delta = \sum_{k=-\infty}^{\infty} \frac{1}{T}e^{ik\omega s}$. By virtue of the continuity of convolution (Theorem 6.4.2), which also holds for distributions on Γ, term-by-term convolution of Fourier series (6.7.21) with any $S \in \mathcal{D}'(\Gamma)$ can be performed, and we get, from (6.7.12):

$$S = S * \delta = S * \sum_{k=-\infty}^{\infty} \frac{1}{T}e^{ik\omega s} = \sum_{k=-\infty}^{\infty} \frac{1}{T}(S * e^{ik\omega s}), \tag{6.7.25}$$

which converges in $\mathcal{D}'(\Gamma)$. We will show that the series in (6.7.24) and (6.7.25) are the same. Since $e^{ik\omega s}$ is a C^∞-function on Γ and Theorem 6.4.1 is applicable for distributions $S \in \mathcal{D}'(\Gamma)$, i.e. the convolution product of a distribution on Γ with a C^∞-function on Γ is a C^∞-function, we have

$$S * e^{ik\omega s} = \langle S(\xi), e^{ik\omega(s-\xi)}\rangle = \langle S(\xi), e^{-ik\omega\xi}e^{ik\omega s}\rangle \tag{6.7.26}$$

$$= e^{ik\omega s}\langle S(\xi), e^{-ik\omega\xi}\rangle = e^{ik\omega s}Tc_k(S). \tag{6.7.27}$$

Then, from (6.7.24)–(6.7.26), we have

$$\sum_{k=-\infty}^{\infty} \frac{1}{T}Tc_k(S)e^{ik\omega s} = \sum_{k=-\infty}^{\infty} c_k(S)e^{ik\omega s} = S \in \mathcal{D}'(\Gamma). \qquad \square$$

Term-by-term differentiation of Fourier series (6.7.18)

Theorem 6.7.3. *Fourier series* (6.7.18) *of the distribution $S \in \mathcal{D}'(\Gamma)$ can be differentiated* term by term, *and the sum of the resultant series obtained by m-times term-by-term differentiation of the series is the* Fourier series *of*

$$S^{(m)} = \frac{d^m S}{ds^m} \ \forall m \in \mathbb{N}, \quad \text{i.e. } S^{(m)} = \sum_{k=-\infty}^{\infty} c_k(S)(ik\omega)^m e^{ik\omega s}. \tag{6.7.28}$$

Proof. By Theorem 6.7.2, $S = \sum_{k=-\infty}^{\infty} c_k(S)e^{ik\omega s}$ in $\mathcal{D}'(\Gamma)$, i.e.

$$S_n = \sum_{k=-n}^{n} c_k(S)e^{ik\omega s} \to S = \sum_{k=-\infty}^{\infty} c_k(S)e^{ik\omega s} \quad \text{in } \mathcal{D}'(\Gamma) \quad \text{as } n \to \infty.$$

But $\forall m \in \mathbb{N}$, $\frac{d^m}{ds^m} : \mathcal{D}'(\Gamma) \longrightarrow \mathcal{D}'(\Gamma)$ is *continuous* (see Theorem 2.9.1, which also holds in $\mathcal{D}'(\Gamma)$). Hence, $S_n \to S$ in $\mathcal{D}'(\Gamma)$ as $n \to \infty \Longrightarrow \frac{d^m}{ds^m} S_n \to \frac{d^m}{ds^m} S = S^{(m)}$ in $\mathcal{D}'(\Gamma)$ $\forall m \in \mathbb{N} \Longrightarrow \sum_{k=-n}^{n} c_k(S)(ik\omega)^m e^{ik\omega s} \to S^{(m)}$ in $\mathcal{D}'(\Gamma)$ as $n \to \infty \Longrightarrow S^{(m)} = \sum_{k=-\infty}^{\infty} c_k(S)(ik\omega)^m e^{ik\omega s}$ in $\mathcal{D}'(\Gamma)$ $\forall m \in \mathbb{N}$. The right-hand side series is a trigonometric series, which converges to the distribution $S^{(m)} \in \mathcal{D}'(\Gamma) \Longrightarrow$ the trigonometric series is the *Fourier series* of $S^{(m)}$ (see (6.7.19)), i.e. Fourier coefficients $c_k(S^{(m)}) = c_k(S)(ik\omega)^m$ $\forall k \in \mathbb{Z}$, $\forall m \in \mathbb{N}$. □

Fourier series of convolutions of distributions in $\mathcal{D}'(\Gamma)$

Let $S_1, S_2 \in \mathcal{D}'(\Gamma)$. Then $S_1 * S_2 \in \mathcal{D}'(\Gamma)$ is defined by (6.7.10) and the *Fourier coefficients* $c_k(S_1 * S_2)$ of $S_1 * S_2 \in \mathcal{D}'(\Gamma)$ are defined, $\forall k \in \mathbb{Z}$, by:

$$
\begin{aligned}
c_k(S_1 * S_2) &= \frac{1}{T}\langle S_1 * S_2, e^{-ik\omega s}\rangle = \frac{1}{T}\langle S_1(\xi) \otimes S_2(\eta), e^{-ik\omega(\xi+\eta)}\rangle \quad \text{(by (6.7.10))}\\
&= \frac{1}{T}\langle S_1(\xi), \langle S_2(\eta), e^{-ik\omega\eta} \cdot e^{-ik\omega\xi}\rangle\rangle = \frac{1}{T}\langle S_1(\xi), e^{-ik\omega\xi}\rangle\langle S_2(\eta), e^{-ik\omega\eta}\rangle\\
&= c_k(S_1) \cdot T \cdot \left(\frac{1}{T}\langle S_2(\eta), e^{-ik\omega\eta}\rangle\right) = T c_k(S_1) c_k(S_2).
\end{aligned} \tag{6.7.29}
$$

The *Fourier series* of $S_1 * S_2 \in \mathcal{D}'(\Gamma)$ is defined by:

$$
\sum_{k=-\infty}^{\infty} c_k(S_1 * S_2) e^{ik\omega s} = T \sum_{k=-\infty}^{\infty} c_k(S_1) \cdot c_k(S_2) e^{ik\omega s}. \tag{6.7.30}
$$

Theorem 6.7.4. *Fourier series* (6.7.30) *of $S_1 * S_2 \in \mathcal{D}'(\Gamma)$ converges to $S_1 * S_2 \in \mathcal{D}'(\Gamma)$, i.e. $S_1 * S_2 = T \sum_{k=-\infty}^{\infty} c_k(S_1) \cdot c_k(S_2) e^{ik\omega s}$.*

Proof. Let $S_1, S_2 \in \mathcal{D}'(\Gamma)$ with $S_1 * S_2 \in \mathcal{D}'(\Gamma)$ and $S_i = \sum_{k=-\infty}^{\infty} c_k(S_i) e^{ik\omega s}$ in $\mathcal{D}'(\Gamma)$ $(i = 1, 2)$. Let $S_1^n, S_2^n \in \mathcal{D}'(\Gamma)$ be defined by $S_1^n = \sum_{k=-n}^{n} c_k(S_1) e^{ik\omega s}$, $S_2^n = \sum_{k=-n}^{n} c_k(S_2) e^{ik\omega s}$ such that $S_1^n * S_2^n \to S_1 * S_2$ in $\mathcal{D}'(\Gamma)$ as $n \to \infty$ by virtue of the continuity of the convolution operation (see Theorem 6.4.2, which also holds in $\mathcal{D}'(\Gamma)$). In fact,

$$
\begin{aligned}
S_1^n * S_2^n &= \sum_{k=-n}^{n} \sum_{l=-n}^{n} c_k(S_1) c_l(S_2) e^{ik\omega s} * e^{il\omega s}\\
&= \sum_{k=-n}^{n} \sum_{l=-n}^{n} c_k(S_1) c_l(S_2) \int_\Gamma e^{ik\omega(s-\xi)} e^{il\omega\xi} d\xi \quad \text{(by (6.7.9))}\\
&= \sum_{k=-n}^{n} \sum_{l=-n}^{n} c_k(S_1) c_l(S_2) \cdot e^{ik\omega s} \cdot \int_\Gamma e^{-ik\omega\xi} e^{il\omega\xi} d\xi
\end{aligned}
$$

$$= \sum_{k=-n}^{n} \sum_{l=-n}^{n} c_k(S_1)c_l(S_2) \cdot e^{ik\omega s} \cdot T\delta_{kl} \quad \text{(using (2.11.7b))}$$

$$= \sum_{k=-n}^{n} Tc_k(S_1)c_k(S_2)e^{ik\omega s} \to S_1 * S_2 = \sum_{k=-\infty}^{\infty} c_k(S_1 * S_2)e^{ik\omega s}, \tag{6.7.31}$$

which is the Fourier series of $S_1 * S_2 \in \mathcal{D}'(\Gamma)$ with Fourier coefficients $c_k(S_1 * S_2) = Tc_k(S_1)c_k(S_2)$. □

Example 6.7.4. With the help of Fourier series (6.7.18) of distributions $S \in \mathcal{D}'(\Gamma)$, and using Theorem 6.7.3 on term-by-term differentiation of Fourier series, prove the following:

1. $\delta * S = S$;
2. $\delta' * S = S'$ in $\mathcal{D}'(\Gamma)$.

Solution.

1. From (6.7.21), $\delta = \sum_{k=-\infty}^{\infty} \frac{1}{T} e^{ik\omega s}$. By Theorem 6.7.2, $\forall S \in \mathcal{D}'(\Gamma)$, $S = \sum_{k=-\infty}^{\infty} c_k(S)e^{ik\omega s}$. Then, from Theorem 6.7.4,

$$\delta * S = \sum_{k=-\infty}^{\infty} c_k(\delta * S)e^{ik\omega s} = \sum_{k=-\infty}^{\infty} T \cdot c_k(\delta) \cdot c_k(S)e^{ik\omega s}$$
$$= \sum_{k=-\infty}^{\infty} T \cdot \frac{1}{T} \cdot c_k(S)e^{ik\omega s} = \sum_{k=-\infty}^{\infty} c_k(S)e^{ik\omega s} = S.$$

2. By Theorem 6.7.3, $\delta' = \frac{d\delta}{ds} = \sum_{k=-\infty}^{\infty} \frac{1}{T}(ik\omega)e^{ik\omega s}$ with $c_k(\delta') = \frac{(ik\omega)}{T}$. For $S \in \mathcal{D}'(\Gamma)$ with $S = \sum_{k=-\infty}^{\infty} c_k(S)e^{ik\omega s}$, by Theorem 6.7.3 on term-by-term differentiation, $S' = \frac{dS}{ds} = \sum_{k=-\infty}^{\infty}[c_k(S)(ik\omega)]e^{ik\omega s}$. By Theorem 6.7.4,

$$\delta' * S = \sum_{k=-\infty}^{\infty} T \cdot c_k(\delta') \cdot c_k(S)e^{ik\omega s}$$
$$= \sum_{k=-\infty}^{\infty} T \cdot \frac{ik\omega}{T} . c_k(S)e^{ik\omega s} = \sum_{k=-\infty}^{\infty} [c_k(S)(ik\omega)]e^{ik\omega s} = S'.$$

Remark 6.7.4. We have followed Schwartz [7] to study Fourier series of distributions in $\mathcal{D}'(\Gamma)$, which are in *one-to-one correspondence* with *periodic distributions* on $\mathbb{R}$ with period T = circumference of Γ. Since $\mathcal{D}_T(\mathbb{R}) \not\subseteq \mathcal{D}(\mathbb{R})$ and $\mathcal{D}(\mathbb{R}) \not\subseteq \mathcal{D}_T(\mathbb{R})$, i.e. $\mathcal{D}_T(\mathbb{R})$ is not related to $\mathcal{D}(\mathbb{R})$ (see (1.10.48)), we did *not* consider $\mathcal{D}'_T(\mathbb{R})$, although $\mathcal{D}'_T(\mathbb{R})$ and $\mathcal{D}'(\mathbb{R})$ are related by (1.10.57) (see (6.7.32) below), $\mathcal{D}'_T(\mathbb{R})$ being the dual of $\mathcal{D}_T(\mathbb{R})$. But an alternative treatment of Fourier series of periodic distributions on $\mathbb{R}$ with period T is also interesting and can be given as follows (see [30]).

For periodic $\tilde{f} \in L^1_{\text{loc}}(\mathbb{R})$ with period T, we have shown in (1.10.54)–(1.10.57): $\forall \phi \in \mathcal{D}(\mathbb{R})$,

$$\begin{aligned}\langle \tilde{f}, \tilde{\Phi}\rangle_{\mathcal{D}'_T(\mathbb{R})\times\mathcal{D}_T(\mathbb{R})} &= \int_0^T \tilde{f}(x)\tilde{\Phi}(x)dx = \int_{\mathbb{R}} \tilde{f}(x)\phi(x)dx = \langle \tilde{f}, \phi\rangle_{\mathcal{D}'(\mathbb{R})\times\mathcal{D}(\mathbb{R})}\\ &= \langle f(s), \Phi(s)\rangle_{\mathcal{D}'(\Gamma)\times\mathcal{D}(\Gamma)} = \int_\Gamma f(s)\Phi(s)ds, \qquad (6.7.32)\end{aligned}$$

where $\tilde{\Phi}(x) = \sum_{l=-\infty}^{\infty} \phi(x+lT)$ (finite summation $\sum$) with $\tilde{\Phi} \in \mathcal{D}_T(\mathbb{R})$ and $f(s)$ (resp. $\Phi(s)$) is a function on Γ associated with the periodic $\tilde{f}(x)$ (resp. $\tilde{\Phi}(x)$) on $\mathbb{R}$. Then, for periodic $\tilde{f_1}, \tilde{f_2} \in L^1_{\text{loc}}(\mathbb{R})$ with period T, from (6.7.32), we have

$$\begin{aligned}&\langle \tilde{f_1}, \tilde{\Phi}\rangle_{\mathcal{D}'_T(\mathbb{R})\times\mathcal{D}_T(\mathbb{R})} = \langle \tilde{f_2}, \tilde{\Phi}\rangle_{\mathcal{D}'_T(\mathbb{R})\times\mathcal{D}_T(\mathbb{R})} \quad \forall\tilde{\Phi} \in \mathcal{D}'_T(\mathbb{R})\\ \Longrightarrow\quad &\langle \tilde{f_1}, \phi\rangle_{\mathcal{D}'(\mathbb{R})\times\mathcal{D}(\mathbb{R})} = \langle \tilde{f_2}, \phi\rangle_{\mathcal{D}'(\mathbb{R})\times\mathcal{D}(\mathbb{R})}\end{aligned}$$

$\Longrightarrow \tilde{f_1} = \tilde{f_2}$ a.e. in $\mathbb{R}$. Hence, distinct periodic functions of $L^1_{\text{loc}}(\mathbb{R})$ with period T define distinct linear functionals in $\mathcal{D}'_T(\mathbb{R})$. We need the following results to introduce the notion of convergence in $\mathcal{D}_T(\mathbb{R})$ and the continuity of linear functionals in $\mathcal{D}'_T(\mathbb{R})$.

Lemma 6.7.1. *Let $\tilde{\Phi} \in \mathcal{D}_T(\mathbb{R})$ such that its Fourier series is given by $\sum_{k=-\infty}^{\infty} c_k(\tilde{\Phi})$ $e^{ik\omega x}$, with $c_k(\tilde{\Phi}) = \frac{1}{T}\int_0^T \tilde{\Phi}(x)e^{-ik\omega x}dx\ \forall k \in \mathbb{Z}$. Then,*

I. $S_n(\tilde{\Phi}) = \sum_{k=-n}^{n} c_k(\tilde{\Phi})e^{ik\omega x} \to \tilde{\Phi}$ uniformly *on* $\mathbb{R}$*;*

II. $\forall m \in \mathbb{N}$, $\frac{d^m}{dx^m}[S_n(\tilde{\Phi})(x)] \to \tilde{\Phi}^{(m)}(x) = \sum_{k=-\infty}^{\infty}(ik\omega)^m c_k(\tilde{\Phi})e^{ik\omega x}$ uniformly *on* $\mathbb{R}$ *as* $n \to \infty$. (6.7.33)

Proof. The results follow from Theorem 2.11.4, since $\mathcal{D}_T(\mathbb{R}) = C_T^\infty(\mathbb{R})$. □

Now, using the results of Lemma 6.7.1, we can introduce the notion of convergence in $\mathcal{D}_T(\mathbb{R})$ and $\mathcal{D}'_T(\mathbb{R})$ as follows:

Convergence in $\mathcal{D}_T(\mathbb{R})$ Let $(\tilde{\Phi}_n)$ be a sequence in $\mathcal{D}_T(\mathbb{R})$. Then

$$\begin{aligned}&\tilde{\Phi}_n \to \tilde{\Phi} \text{ in } \mathcal{D}_T(\mathbb{R})\\ \Longleftrightarrow\quad &\begin{cases}\text{I. } \tilde{\Phi}_n \to \tilde{\Phi} \textit{ uniformly} \text{ as } n \to \infty;\\ \text{II. } \tilde{\Phi}_n^{(m)} \to \tilde{\Phi}^{(m)} \textit{ uniformly} \text{ as } n \to \infty\ \forall m \in \mathbb{N}.\end{cases} \qquad (6.7.34)\end{aligned}$$

Convergence in $\mathcal{D}'_T(\mathbb{R})$ Let $(\tilde{S}_n)$ be a sequence in $\mathcal{D}'_T(\mathbb{R})$. Then, $\forall \tilde{\Phi} \in \mathcal{D}_T(\mathbb{R})$,

$$\begin{aligned}&\tilde{S}_n \to \tilde{S} \text{ in } \mathcal{D}'_T(\mathbb{R})\\ &\iff \quad \langle \tilde{S}_n, \tilde{\Phi}\rangle_{\mathcal{D}'_T(\mathbb{R})\times\mathcal{D}_T(\mathbb{R})} \to \langle \tilde{S}, \tilde{\Phi}\rangle_{\mathcal{D}'_T(\mathbb{R})\times\mathcal{D}_T(\mathbb{R})} \text{ as } n \to \infty. \end{aligned} \tag{6.7.35}$$

Convergence of $\tilde{S}_n$ in $\mathcal{D}'_T(\mathbb{R}) \Longrightarrow$ convergence of $\tilde{S}_n$ in $\mathcal{D}'(\mathbb{R})$ i.e.

$$\tilde{S}_n \to \tilde{S} \text{ in } \mathcal{D}'_T(\mathbb{R}) \quad \Longrightarrow \quad \tilde{S}_n \to \tilde{S} \text{ in } \mathcal{D}'(\mathbb{R}) \text{ as } n \to \infty. \tag{6.7.36}$$

In fact, $\forall \phi \in \mathcal{D}(\mathbb{R})$, $\langle \tilde{S} - \tilde{S}_n, \phi\rangle_{\mathcal{D}'(\mathbb{R})\times\mathcal{D}(\mathbb{R})} = \langle \tilde{S} - \tilde{S}_n, \tilde{\Phi}\rangle_{\mathcal{D}'_T(\mathbb{R})\times\mathcal{D}_T(\mathbb{R})} \to 0$ for $\tilde{\Phi} \in \mathcal{D}'_T(\mathbb{R})$, since $\forall \phi \in \mathcal{D}(\mathbb{R})$, $\tilde{\Phi} \in \mathcal{D}_T(\mathbb{R})$ is defined by:

$$\tilde{\Phi}(x) = \sum_{l=-\infty}^{\infty} \phi(x + lT) \quad \text{(finite summation } \textstyle\sum\text{)}. \tag{6.7.37}$$

Continuity of functionals on $\mathcal{D}_T(\mathbb{R})$ $\tilde{S} \in \mathcal{D}'_T(\mathbb{R})$ is continuous on $\mathcal{D}_T(\mathbb{R})$ if and only if $\tilde{\Phi}_n \to \tilde{\Phi}$ in $\mathcal{D}_T(\mathbb{R})$ in the sense of (6.7.34),

$$\Longrightarrow \quad \langle \tilde{S}, \tilde{\Phi}_n\rangle_{\mathcal{D}'_T(\mathbb{R})\times\mathcal{D}_T(\mathbb{R})} \to \langle \tilde{S}, \tilde{\Phi}\rangle_{\mathcal{D}'_T(\mathbb{R})\times\mathcal{D}_T(\mathbb{R})} \text{ as } n \to \infty. \tag{6.7.38}$$

Definition 6.7.1. $\mathcal{D}'_T(\mathbb{R})$ is the linear space of *continuous* (in the sense of (6.7.38)) linear functionals defined on $\mathcal{D}_T(\mathbb{R})$.

$$\tilde{S} \in \mathcal{D}'_T(\mathbb{R}) \quad \iff \quad \tilde{S} \text{ is a } \textit{periodic distribution} \text{ on } \mathbb{R} \text{ with period } T. \tag{6.7.39}$$

Fourier series of periodic distributions on $\mathbb{R}$

Definition 6.7.2. A trigonometric series

$$\sum_{k=-\infty}^{\infty} c_k(\tilde{S})e^{ik\omega x} \tag{6.7.40}$$

is called the *Fourier series* of the *periodic distribution* $\tilde{S} \in \mathcal{D}'_T(\mathbb{R}) \subset \mathcal{D}'(\mathbb{R})$ on $\mathbb{R}$ with period T, $c_k(\tilde{S})$ being the *Fourier coefficients* of $\tilde{S} \in \mathcal{D}'_T(\mathbb{R}) \subset \mathcal{D}'(\mathbb{R})$ defined, $\forall k \in \mathbb{Z}$, by:

$$c_k(\tilde{S}) = \frac{1}{T}\langle \tilde{S}, e^{-ik\omega x}\rangle_{\mathcal{D}'_T(\mathbb{R})\times\mathcal{D}_T(\mathbb{R})}, \tag{6.7.41}$$

where $e^{ik\omega x} \in \mathcal{D}_T(\mathbb{R})$ $\forall k \in \mathbb{Z}$.

Example 6.7.5. Let $\tilde{\delta}$ be the periodic Dirac distribution (see (1.10.59)) on $\mathbb{R}$ with period T. Then, for $\tilde{\delta} \in \mathcal{D}'_T(\mathbb{R}) \subset \mathcal{D}'(\mathbb{R})$, the Fourier coefficients $c_k(\tilde{\delta})$ are given by:

$$c_k(\tilde{\delta}) = \frac{1}{T}\langle \tilde{\delta}, e^{-ik\omega x}\rangle = \frac{1}{T}e^{-ik\omega x}|_{x=0} = \frac{1}{T} \quad \forall k \in \mathbb{Z}, \tag{6.7.42}$$

and the Fourier series of the periodic Dirac distribution $\tilde{\delta}$ is given by $\sum_{k=-\infty}^{\infty} \frac{1}{T} e^{ik\omega x}$ (see (6.7.23)).

Convergence and term-by-term differentiation of Fourier series (6.7.39)

Theorem 6.7.5. *The Fourier series* $\sum_{k=-\infty}^{\infty} \frac{1}{T} e^{ik\omega x}$ *of* $\tilde{\delta}$ *in* (6.7.42) *converges in* $\mathcal{D}'(\mathbb{R})$ *to* $\tilde{\delta} \in \mathcal{D}'(\mathbb{R})$*:*

$$\sum_{k=-\infty}^{\infty} \frac{1}{T} e^{ik\omega x} = \sum_{k=-\infty}^{\infty} \delta(x - lT) = \tilde{\delta} \quad \text{in } \mathcal{D}'(\mathbb{R}), \tag{6.7.43}$$

where $\delta(x - lT)$ *is the Dirac distribution with unit mass/force/charge concentrated at* $x = lT$, $l \in \mathbb{Z}$; $\tilde{\delta} = \sum_{k=-\infty}^{\infty} \delta(x - lT)$ *is the periodic Dirac distribution (see* (1.10.59)*).*

Proof. For the particular case $T = 2\pi, \omega = \frac{2\pi}{T} = 1$, see the proof of (2.11.9) in Example 2.11.2. We give here an alternative independent proof. Consider $\sum_{k=-n}^{n} \frac{e^{ik\omega x}}{T}$ in $\mathcal{D}'(\mathbb{R})$. Then, $\forall \phi \in \mathcal{D}(\mathbb{R})$,

$$\begin{aligned}
\left\langle \sum_{k=-n}^{n} \frac{e^{ik\omega x}}{T}, \phi \right\rangle_{\mathcal{D}'(\mathbb{R})\times\mathcal{D}(\mathbb{R})} &= \sum_{k=-n}^{n} \frac{1}{T} \int_{-\infty}^{\infty} e^{ik\omega x}\phi(x)dx \\
&= \sum_{k=-n}^{n} \frac{1}{T} \sum_{l=-\infty}^{\infty} \int_{lT}^{(l+1)T} e^{ik\omega x}\phi(x)dx \\
&= \sum_{k=-n}^{n} \frac{1}{T} \sum_{l=-\infty}^{\infty} \int_{0}^{T} e^{ik\omega\xi}\phi(\xi + lT)d\xi \quad \text{(by change of variables } x = \xi + lT) \\
&= \sum_{k=-n}^{n} \frac{1}{T} \int_{0}^{T} e^{ik\omega x} \sum_{l=-\infty}^{\infty} (\phi(x + lT))dx \quad \text{(finite summation with respect to } l) \\
&= \sum_{k=-n}^{n} \frac{1}{T} \int_{0}^{T} e^{ik\omega x}\tilde{\Phi}(x)dx \quad \text{(by (1.10.51))} \\
&= \sum_{k=-n}^{n} \frac{1}{T} \int_{0}^{T} \tilde{\Phi}(x)e^{-ik\omega x}dx
\end{aligned}$$

(both sides of the last equality contain the same terms), where $\tilde{\Phi} \in \mathcal{D}_T(\mathbb{R})$ is a C^∞-function which is periodic on $\mathbb{R}$ with period T such that its uniformly and absolutely convergent Fourier series is given by Theorem 2.11.4/Lemma 6.7.1:

$$\tilde{\Phi} = \sum_{k=-\infty}^{\infty} c_k(\tilde{\Phi}) e^{ik\omega x} = \sum_{k=-\infty}^{\infty} \langle \tilde{\Phi}, e^{ik\omega x} \rangle_T e^{ik\omega x} \quad \text{(see (2.11.7d))}$$

$$= \sum_{k=-\infty}^{\infty} \left(\frac{1}{T} \int_0^T \tilde{\Phi}(x) e^{-ik\omega x} dx \right) e^{ik\omega x},$$

i.e. $\tilde{\Phi}(x) = \lim_{n\to\infty} S_n(\tilde{\Phi})(x)$ with $S_n(\tilde{\Phi})(x) = \sum_{k=-n}^{n} (\frac{1}{T} \int_0^T \tilde{\Phi}(x) e^{-ik\omega x} dx)$ $e^{ik\omega x}$ and

$$S_n(\tilde{\Phi})(0) = \sum_{k=-n}^{n} \left(\frac{1}{T} \int_0^T \tilde{\Phi}(x) e^{-ik\omega x} dx \right) e^{ik\omega\cdot 0}$$

$$= \sum_{k=-n}^{n} \left(\frac{1}{T} \int_0^T \tilde{\Phi}(x) e^{-ik\omega x} dx \right)$$

such that

$$\lim_{n\to\infty} S_n(\tilde{\Phi})(0) = \sum_{k=-\infty}^{\infty} \left(\frac{1}{T} \int_0^T \tilde{\Phi}(x) e^{-ik\omega x} dx \right) = \tilde{\Phi}(0).$$

Hence,

$$\left\langle \sum_{k=-n}^{n} \frac{e^{ik\omega x}}{T}, \phi \right\rangle_{\mathcal{D}'(\mathbb{R})\times\mathcal{D}(\mathbb{R})} = S_n(\tilde{\Phi})(0) \to \tilde{\Phi}(0) \quad \text{as } n \to \infty$$

$$\Longrightarrow \quad \lim_{n\to\infty} \left\langle \sum_{k=-n}^{n} \frac{e^{ik\omega x}}{T}, \phi \right\rangle = \lim_{n\to\infty} S_n(\tilde{\Phi})(0)$$

$$= \tilde{\Phi}(0) = \sum_{l=-\infty}^{\infty} \phi(0 + lT) = \sum_{l=-\infty}^{\infty} \phi(lT)$$

$$\Longrightarrow \quad \left\langle \sum_{k=-\infty}^{\infty} \frac{e^{ik\omega x}}{T}, \phi \right\rangle_{\mathcal{D}'(\mathbb{R})\times\mathcal{D}(\mathbb{R})} = \sum_{l=-\infty}^{\infty} \phi(lT)$$

$$= \left\langle \sum_{l=-\infty}^{\infty} \delta(x - lT), \phi \right\rangle_{\mathcal{D}'(\mathbb{R})\times\mathcal{D}(\mathbb{R})} \qquad \forall \phi \in \mathcal{D}(\mathbb{R})$$

$$\Longrightarrow \quad \sum_{k=-\infty}^{\infty} \frac{e^{ik\omega x}}{T} = \sum_{l=-\infty}^{\infty} \delta(x - lT) = \tilde{\delta} \quad \text{in } \mathcal{D}'(\mathbb{R}),$$

which is the periodic Dirac distribution $\tilde{\delta}$ with period T (see (1.10.59). □

Theorem 6.7.6.

I. *The Fourier series* $\sum_{k=-\infty}^{\infty} c_k(\tilde{S})e^{ik\omega x}$ *of the periodic distribution* $\tilde{S} \in \mathcal{D}'_T(\mathbb{R}) \subset \mathcal{D}'(\mathbb{R})$ *converges in* $\mathcal{D}'(\mathbb{R})$ *to* $\tilde{S}$.

II. $\forall m \in \mathbb{N}$,

$$\tilde{S}^{(m)} = \frac{d^m \tilde{S}}{dx^m} = \sum_{l=-\infty}^{\infty} c_k(\tilde{S})(ik\omega)^m e^{ik\omega x}, \tag{6.7.44}$$

i.e. m-times term-by-term differentiation of the Fourier series of $\tilde{S}$ *is admissible* $\forall m \in \mathbb{N}$.

Proof.

I. Let $\tilde{S} \in \mathcal{D}'_T(\mathbb{R})$ be a periodic distribution in $\mathcal{D}'(\mathbb{R})$ with period T. We are to show that $\lim_{n\to\infty}\langle\sum_{k=-n}^{n} c_k(\tilde{S})e^{ik\omega x}, \phi\rangle = \langle\tilde{S}, \phi\rangle$ $\forall \phi \in \mathcal{D}(\mathbb{R})$. Set $\tilde{S}_n = S_n(\tilde{S}) = \sum_{k=-n}^{n} c_k(\tilde{S})e^{ik\omega x}$ $\forall n \in \mathbb{N}$. Then $\tilde{S}_n \in L^1_{\text{loc}}(\mathbb{R})$ and is periodic on $\mathbb{R}$ with period T. Hence, from (6.7.32) we have, $\forall \phi \in \mathcal{D}(\mathbb{R})$,

$$\begin{aligned}
\left\langle \tilde{S}_n, \phi \right\rangle_{\mathcal{D}'(\mathbb{R})\times\mathcal{D}(\mathbb{R})} &= \left\langle \tilde{S}_n, \tilde{\Phi} \right\rangle_{\mathcal{D}'_T(\mathbb{R})\times\mathcal{D}_T(\mathbb{R})} = \int_0^T S_n(\tilde{S})(x)\tilde{\Phi}(x)dx \\
&= \int_0^T \sum_{k=-n}^{n} c_k(\tilde{S})e^{ik\omega x}\tilde{\Phi}(x)dx = \sum_{k=-n}^{n} c_k(\tilde{S}) \int_0^T \tilde{\Phi}(x)e^{ik\omega x}dx \\
&= \sum_{k=-n}^{n} \frac{1}{T}\langle \tilde{S}, e^{-ik\omega x}\rangle_{\mathcal{D}'_T(\mathbb{R})\times\mathcal{D}_T(\mathbb{R})}\left(\int_0^T \tilde{\Phi}(x)e^{ik\omega x}dx\right) \quad \text{(using (6.7.41))} \\
&= \sum_{k=-n}^{n} \left\langle \tilde{S}, \left(\frac{1}{T}\int_0^T \tilde{\Phi}(x)e^{ik\omega x}dx\right)e^{-ik\omega x}\right\rangle_{\mathcal{D}'_T(\mathbb{R})\times\mathcal{D}_T(\mathbb{R})} \\
&= \left\langle \tilde{S}, \sum_{k=-n}^{n} \left(\frac{1}{T}\int_0^T \tilde{\Phi}(x)e^{ik\omega x}dx\right)e^{-ik\omega x}\right\rangle_{\mathcal{D}'_T(\mathbb{R})\times\mathcal{D}_T(\mathbb{R})} \\
&= \left\langle \tilde{S}, \sum_{k=-n}^{n} c_{-k}(\tilde{\Phi})e^{-ik\omega x}\right\rangle_{\mathcal{D}'_T(\mathbb{R})\times\mathcal{D}_T(\mathbb{R})} = \langle \tilde{S}, S_n(\tilde{\Phi})\rangle_{\mathcal{D}'_T(\mathbb{R})\times\mathcal{D}_T(\mathbb{R})},
\end{aligned}$$

where $\forall n \in \mathbb{N}$, $S_n(\tilde{\Phi}) = \sum_{k=-n}^{n} c_{-k}(\tilde{\Phi})e^{-ik\omega x} = \sum_{k=-n}^{n} c_k(\tilde{\Phi})e^{ik\omega x} \in \mathcal{D}_T(\mathbb{R})$ is the nth partial sum of the Fourier series $\sum_{k=-\infty}^{\infty} c_k(\tilde{\Phi})e^{ik\omega x}$ of $\tilde{\Phi} \in \mathcal{D}_T(\mathbb{R})$. Hence,

$$\begin{aligned}
\langle \tilde{S}_n, \phi\rangle_{\mathcal{D}'(\mathbb{R})\times\mathcal{D}(\mathbb{R})} &= \langle \tilde{S}_n, \tilde{\Phi}\rangle_{\mathcal{D}'_T(\mathbb{R})\times\mathcal{D}_T(\mathbb{R})} \\
&= \langle \tilde{S}, S_n(\tilde{\Phi})\rangle_{\mathcal{D}'_T(\mathbb{R})\times\mathcal{D}_T(\mathbb{R})} \quad \forall \phi \in \mathcal{D}(\mathbb{R}).
\end{aligned} \tag{6.7.45}$$

By Lemma 6.7.1, $S_n(\tilde{\Phi}) \to \tilde{\Phi}$ *uniformly* as $n \to \infty$ and $S_n^{(m)}(\tilde{\Phi}) \to \tilde{\Phi}^{(m)}$ *uniformly* $\forall m \in \mathbb{N}$ as $n \to \infty$. Hence, $S_n(\tilde{\Phi}) \to \tilde{\Phi}$ in $\mathcal{D}_T(\mathbb{R})$ as $n \to \infty$ by (6.7.34) $\Longrightarrow \langle \tilde{S}, S_n(\tilde{\Phi})\rangle_{\mathcal{D}'_T(\mathbb{R})\times\mathcal{D}_T(\mathbb{R})} \to \langle \tilde{S}, \tilde{\Phi}\rangle_{\mathcal{D}'_T(\mathbb{R})\times\mathcal{D}_T(\mathbb{R})}$ as $n \to \infty$ by (6.7.35). From (6.7.32),

$$\langle \tilde{S}, \tilde{\Phi}\rangle_{\mathcal{D}'_T(\mathbb{R})\times\mathcal{D}_T(\mathbb{R})} = \langle \tilde{S}, \phi\rangle_{\mathcal{D}'(\mathbb{R})\times\mathcal{D}(\mathbb{R})}. \tag{6.7.46}$$

Hence, from (6.7.45) and (6.7.46), $\forall \phi \in \mathcal{D}(\mathbb{R})$,

$$\begin{aligned}\lim_{n\to\infty} \langle \tilde{S}_n, \phi\rangle_{\mathcal{D}'(\mathbb{R})\times\mathcal{D}(\mathbb{R})} &= \lim_{n\to\infty} \langle \tilde{S}, S_n(\tilde{\Phi})\rangle_{\mathcal{D}'_T(\mathbb{R})\times\mathcal{D}_T(\mathbb{R})} \\ &= \langle \tilde{S}, \tilde{\Phi}\rangle_{\mathcal{D}'_T(\mathbb{R})\times\mathcal{D}_T(\mathbb{R})} = \langle \tilde{S}, \phi\rangle_{\mathcal{D}'(\mathbb{R})\times\mathcal{D}(\mathbb{R})}\end{aligned}$$

$\Longleftrightarrow \lim_{n\to\infty} \tilde{S}_n = \sum_{k=-\infty}^{\infty} c_k(\tilde{S})e^{ik\omega x} = \tilde{S}$ in $\mathcal{D}'(\mathbb{R})$.

II. Since $\tilde{S} = \sum_{k=-\infty}^{\infty} c_k(\tilde{S})e^{ik\omega x}$ in $\mathcal{D}'(\mathbb{R})$, $\tilde{S}^{(m)} \in \mathcal{D}'(\mathbb{R})\forall m \in \mathbb{N}$. In fact, $\forall m \in \mathbb{N}$, $\frac{d^m}{dx^m} : \mathcal{D}'(\mathbb{R}) \to \mathcal{D}'(\mathbb{R})$ is continuous on $\mathcal{D}'(\mathbb{R})$. Hence, $\tilde{S}_n = \sum_{k=-n}^{n} c_k(\tilde{S})e^{ik\omega x} \to \tilde{S}$ in $\mathcal{D}'(\mathbb{R})$ (proved above in (I)) $\Longrightarrow \frac{d^m \tilde{S}_n}{dx^m} = \sum_{k=-n}^{n} c_k(\tilde{S})(ik\omega)^m e^{ik\omega x} \to \frac{d^m \tilde{S}}{dx^m}$ in $\mathcal{D}'(\mathbb{R})$ as $n \to \infty$ with $\tilde{S}^{(m)} = \sum_{k=-n}^{n} c_k(\tilde{S})(ik\omega)^m e^{ik\omega x} \in \mathcal{D}'(\mathbb{R})$, which is the Fourier series of $\tilde{S}^{(m)}$. In fact, from (6.7.41), $\forall m \in \mathbb{N}$,

$$\begin{aligned}c_k(\tilde{S}^{(m)}) &= \frac{1}{T}\langle \tilde{S}^{(m)}, e^{-ik\omega x}\rangle_{\mathcal{D}'_T(\mathbb{R})\times\mathcal{D}_T(\mathbb{R})} \\ &= \frac{1}{T}(-1)^m \left\langle \tilde{S}, \frac{d^m(e^{-ik\omega x})}{dx^m}\right\rangle_{\mathcal{D}'_T(\mathbb{R})\times\mathcal{D}_T(\mathbb{R})} \\ &= \frac{1}{T}(-1)^m \langle \tilde{S}, (-1)^m (ik\omega)^m e^{-ik\omega x}\rangle_{\mathcal{D}'_T(\mathbb{R})\times\mathcal{D}_T(\mathbb{R})} \\ &= \frac{1}{T}(ik\omega)^m \langle \tilde{S}, e^{-ik\omega x}\rangle_{\mathcal{D}'_T(\mathbb{R})\times\mathcal{D}_T(\mathbb{R})} = (ik\omega)^m c_k(\tilde{S}) \quad \forall k \in \mathbb{Z},\end{aligned}$$

since $c_k(\tilde{S}) = \frac{1}{T}\langle \tilde{S}, e^{-ik\omega x}\rangle_{\mathcal{D}'_T(\mathbb{R})\times\mathcal{D}_T(\mathbb{R})}$. □

Example 6.7.6.

1. $\tilde{\delta}(x-a)$ is represented by the *complex* Fourier series $\tilde{\delta}(x-a) = \sum_{k=-\infty}^{\infty} \frac{1}{T} e^{ik\omega(x-a)}$ with Fourier coefficients $c_k(\tilde{\delta}(x-a)) = \frac{1}{T}\langle \tilde{\delta}(x-a), e^{-ik\omega(x-a)}\rangle = \frac{1}{T}e^{-ik\omega(x-a)}|_{x=a} = \frac{1}{T}e^{-ik\omega\cdot 0} = \frac{1}{T}$ $\forall k \in \mathbb{Z}$, *and* by the *real* Fourier series $\tilde{\delta}(x-a) = \frac{1}{T} + \frac{2}{T}\sum_{k=1}^{\infty} \cos(k\omega(x-a))$.
2. $\forall m \in \mathbb{N}$, for the derivatives $\frac{d^m}{dx^m}\tilde{\delta}(x-a) = \tilde{\delta}^{(m)}(x-a)$, we have:
   - $\tilde{\delta}^{(m)}(x-a) = \sum_{k=-\infty}^{\infty} \frac{1}{T}(ik\omega)^m e^{ik\omega(x-a)}$ with $c_k(\tilde{\delta}^{(m)}(x-a)) = \frac{1}{T}(ik\omega)^m$;

- $\tilde{\delta}^{(2m)}(x-a) = \sum_{k=1}^{\infty}(-1)^m \frac{2}{T}(k\omega)^{2m}\cos(k\omega(x-a))$;
- $\tilde{\delta}^{(2m-1)}(x-a) = \sum_{k=1}^{\infty}(-1)^m \frac{2}{T}(k\omega)^{2m}\sin(k\omega(x-a))$;

where $\tilde{\delta}(x-a)$ is the periodic Dirac distribution with unit mass/force/charge etc. concentrated at $x = a + lT$ with $l \in \mathbb{Z}$.

Remark 6.7.5. In Example 6.7.6, the real Fourier series is obtained by writing the nth partial sum:

$$S_n = \sum_{k=-n}^{n} c_k(\tilde{\delta}(x-a))e^{ik\omega(x-a)} = \sum_{k=-n}^{n} \frac{1}{T}e^{ik\omega(x-a)}$$
$$= \frac{1}{T} + \sum_{k=1}^{n} \frac{1}{T}[e^{ik\omega(x-a)} + e^{-ik\omega(x-a)}] = \frac{1}{T} + \sum_{k=1}^{n} \frac{2}{T}\cos(k\omega(x-a))$$
$$\to \tilde{\delta}(x-a) = \frac{1}{T} + \sum_{k=1}^{\infty} \frac{2}{T}\cos(k\omega(x-a)) \quad \text{in } \mathcal{D}'(\mathbb{R}) \text{ as } n \to \infty.$$

The other results are obtained by m or $2m$ or $2m-1$ times, as the case may be, term-by-term differentiation of the Fourier series.

6.8 Applications

Application to partial differential equations

Let $P(\partial) = \sum_{|\boldsymbol{\alpha}|\le m} a_{\boldsymbol{\alpha}}\partial^{\boldsymbol{\alpha}} = \sum_{|\boldsymbol{\alpha}|\le m} a_{(\alpha_1,\dots,\alpha_n)} \frac{\partial^{|\boldsymbol{\alpha}|}}{\partial_{x_1}^{\alpha_1}\partial_{x_2}^{\alpha_2}\cdots\partial_{x_n}^{\alpha_n}}$ be a partial differential operator with constant coefficients $a_{\boldsymbol{\alpha}} \in \mathbb{R}$ $\forall$ multi-index $\boldsymbol{\alpha} = (\alpha_1, \alpha_2, \dots, \alpha_n)$. Then

$$P(\partial)(\delta * T) = P(\partial)\delta * T = P(\partial)T = \sum_{|\boldsymbol{\alpha}|\le m} a_{\boldsymbol{\alpha}}\partial^{\boldsymbol{\alpha}}T \quad \forall T \in \mathcal{D}'(\mathbb{R}^n). \tag{6.8.1}$$

Proof.

$$P(\partial)(\delta * T) = \Big(\sum_{|\boldsymbol{\alpha}|\le m} a_{\boldsymbol{\alpha}}\partial^{\boldsymbol{\alpha}}\Big)(\delta * T) = \sum_{|\boldsymbol{\alpha}|\le m} a_{\boldsymbol{\alpha}}(\partial^{\boldsymbol{\alpha}}(\delta * T))$$
$$= \sum_{|\boldsymbol{\alpha}|\le m} a_{\boldsymbol{\alpha}}(\partial^{\boldsymbol{\alpha}}T * \delta) \quad \text{(by (6.7.4))}$$
$$= \sum_{|\boldsymbol{\alpha}|\le m} a_{\boldsymbol{\alpha}}\,\partial^{\boldsymbol{\alpha}}T = P(\partial)T.$$

□

Example 6.8.1.

1. $\Delta = \sum_{i=1}^{n} \frac{\partial^2}{\partial x_i^2} \Longrightarrow \Delta(\delta * T) = \Delta T * \delta = \Delta T$;
2. $\Delta\Delta = \frac{\partial^4}{\partial x_1^4} + 2\frac{\partial^4}{\partial x_1^2 \partial x_2^2} + \frac{\partial^4}{\partial x_2^4} \Longrightarrow \Delta\Delta(\delta * T) = \Delta\Delta T * \delta = \Delta\Delta T$; (6.8.2)
3. $\Box = \frac{\partial^2}{\partial t^2} - \Delta \Longrightarrow \Box(\delta * T) = \Box T * \delta = \Box T$. (6.8.3)
4. Consider the partial differential equation $P(\partial)$. For a given distribution $S \in \mathcal{E}'(\mathbb{R}^n)$ with compact support in $\mathbb{R}^n$, find $T \in \mathcal{D}'(\mathbb{R}^n)$ such that $P(\partial)T = S$ in $\mathcal{D}'(\mathbb{R}^n)$. (6.8.4)

Theorem 6.8.1 (Malgrange and Ehrenpreis [38], [5]). *Let $S \in \mathcal{E}'(\mathbb{R}^n) \subset \mathcal{D}'(\mathbb{R}^n)$ be a distribution with compact support in $\mathbb{R}^n$, and $E \in \mathcal{D}'(\mathbb{R}^n)$ be an elementary solution of $P(\partial)$, i.e. $P(\partial)E = \delta$ in $\mathcal{D}'(\mathbb{R}^n)$. Then $T = S * E$ is a solution of the equation* (6.8.4).

Proof. Since $S \in \mathcal{E}'(\mathbb{R}^n)$ is a distribution with compact support in $\mathbb{R}^n$, $S * E$ is well defined. Then, using (6.7.4), $P(\partial)(S * E) = S * P(\partial)E = S * \delta = \delta * S = S$. □

Example 6.8.2. For given $f \in \mathcal{E}'(\mathbb{R}^n)$ (i.e. a distribution with compact support in $\mathbb{R}^n$), $n = 2$ or 3, find $u \in \mathcal{D}'(\mathbb{R}^n)$ such that

1. $-\Delta u = f$ in $\mathcal{D}'(\mathbb{R}^n)$;
2. for $n = 2$, $\Delta\Delta u = f$ in $\mathcal{D}'(\mathbb{R}^2)$.

Solution.

1. From Theorem 3.3.2, an elementary solution $E \in \mathcal{D}'(\mathbb{R}^n)$ of $-\Delta$, for $r = \|\mathbf{x}\|_{\mathbb{R}^n} = \sum_{i=1}^{n} x_i^2$, is given by $E = \frac{1}{4\pi r}$ for $n = 3$; $E = \frac{1}{2\pi} \ln \frac{1}{r}$ for $n = 2$. Hence, by Theorem 6.8.1, $u = f * E = E * f$ with
$$u(\mathbf{x}) = \begin{cases} \frac{1}{4\pi} \int_{\mathbb{R}^3} \frac{f(\boldsymbol{\xi})}{\|\mathbf{x}-\boldsymbol{\xi}\|_{\mathbb{R}^3}} d\boldsymbol{\xi} & \text{for } n = 3 \\ \frac{1}{2\pi} \int_{\mathbb{R}^2} f(\boldsymbol{\xi}) \ln \frac{1}{\|\mathbf{x}-\boldsymbol{\xi}\|_{\mathbb{R}^2}} d\boldsymbol{\xi} & \text{for } n = 2 \end{cases} \tag{6.8.5}$$
is a solution of $-\Delta u = f$.
2. For $n = 2$, from (3.3.65), $E = \frac{1}{8\pi} r^2 \ln r \in \mathcal{D}'(\mathbb{R}^2)$ with $r = \|\mathbf{x}\| = (x_1^2 + x_2^2)^{\frac{1}{2}}$ is an elementary solution of $\Delta\Delta$. Hence, from Theorem 6.8.1,
$$\begin{aligned} u = f * E = E * f &= \int_{\mathbb{R}^2} E(\mathbf{x} - \boldsymbol{\xi}) f(\boldsymbol{\xi}) d\boldsymbol{\xi} \\ &= \frac{1}{8\pi} \int_{\mathbb{R}^2} \|\mathbf{x} - \boldsymbol{\xi}\|^2 \ln \|\mathbf{x} - \boldsymbol{\xi}\| f(\boldsymbol{\xi}) d\boldsymbol{\xi} \end{aligned} \tag{6.8.6}$$
is a solution of $\Delta\Delta u = f$ in $\mathcal{D}'(\mathbb{R}^2)$.

Application to potentials

Let $\mu = \mu(\mathbf{x})$ be the density of a continuous distribution of mass with compact support in $\mathbb{R}^n$ ($n = 3$ or 2) and E be the elementary solution of $-\Delta$, i.e. $-\Delta E = \delta$. Then the *Newtonian volume potential* $U = U_\mu$ of this mass distribution with density μ is defined by:

$$U_\mu = \int_{\mathbb{R}^3} \frac{\mu(\boldsymbol{\xi})}{4\pi \|\mathbf{x} - \boldsymbol{\xi}\|} d\boldsymbol{\xi} = E * \mu \quad \text{for } n = 3, \tag{6.8.7}$$

which is well defined, since $\operatorname{supp}(\mu) \subset\subset \mathbb{R}^3$.

The *logarithmic potential* $U = U_\mu$ in $\mathbb{R}^2$ of the mass distribution with density $\mu = \mu(\mathbf{x})$ is defined by:

$$U_\mu = \int_{\mathbb{R}^2} \frac{1}{2\pi} \ln \frac{1}{\|\mathbf{x} - \boldsymbol{\xi}\|} \mu(\boldsymbol{\xi}) d\boldsymbol{\xi} = E * \mu \quad \text{for } n = 2. \tag{6.8.8}$$

This suggests that we define the potential of a distribution $T \in \mathcal{D}'(\mathbb{R}^n)$ as follows:

Potential of a distribution with compact support

Definition 6.8.1. Let $T \in \mathcal{E}'(\mathbb{R}^n)$ be a distribution with compact support in $\mathbb{R}^n$. Then its *potential* $U_T \in \mathcal{D}'(\mathbb{R}^n)$ is defined by

$$U_T = E * T, \tag{6.8.9}$$

where E is an elementary solution of $-\Delta$, i.e. $-\Delta E = \delta$ in $\mathcal{D}'(\mathbb{R}^n)$ (see Theorem 3.3.2).

Hence, (6.8.9) is well defined $\forall T \in \mathcal{E}'(\mathbb{R}^n)$, and

$$\Delta U_T = \Delta(E * T) = \Delta E * T = -\delta * T = -T, \tag{6.8.10}$$

$$U_{\Delta T} = E * \Delta T = \Delta(E * T) = \Delta E * T = -\delta * T = -T. \tag{6.8.11}$$

Example 6.8.3. For $f \in \mathcal{E}'(\mathbb{R}^n)$, $\Delta u = -f$ has the solution

$$u = E * f = \begin{cases} \frac{1}{4\pi} \int_{\mathbb{R}^3} \frac{f(\boldsymbol{\xi})}{\|\mathbf{x}-\boldsymbol{\xi}\|_{\mathbb{R}^3}} d\boldsymbol{\xi} & \text{for } n = 3, \\ \frac{1}{2\pi} \int_{\mathbb{R}^2} f(\boldsymbol{\xi}) \ln \frac{1}{\|\mathbf{x}-\boldsymbol{\xi}\|_{\mathbb{R}^2}} d\boldsymbol{\xi} & \text{for } n = 2. \end{cases}$$

Potential of simple layer surface charge distribution Let Γ be a compact, orientable, smooth surface in $\mathbb{R}^3$, $\rho = \rho(\mathbf{x})$ be a continuous function on Γ defining the *surface density* of electric charge distribution with compact support on Γ and

$T = \rho\delta_\Gamma \in \mathcal{D}'(\mathbb{R}^3)$ be the corresponding *volume density of distribution* on $\mathbb{R}^3$ defined by (1.11.24)–(1.11.25): $\forall\phi \in \mathcal{D}(\mathbb{R}^3)$,

$$\langle T, \phi\rangle = \langle \rho\delta_\Gamma, \phi\rangle = \int_\Gamma \rho(\mathbf{x})\phi(\mathbf{x})dS, \tag{6.8.12}$$

with $\operatorname{supp}(\rho\delta_\Gamma) = \Gamma \subset\subset \mathbb{R}^3$ (for more details, see Section 1.11). Then, from (6.8.9), the potential $U_0 = U_{\rho\delta_\Gamma}$ of the simple layer charge distribution with surface density ρ is defined by:

$$U_0 = E * \rho\delta_\Gamma = \int_\Gamma E(\mathbf{x}-\boldsymbol{\xi})\rho(\boldsymbol{\xi})dS(\boldsymbol{\xi}) = \frac{1}{4\pi}\int_\Gamma \frac{\rho(\boldsymbol{\xi})}{\|\mathbf{x}-\boldsymbol{\xi}\|}dS(\boldsymbol{\xi}). \tag{6.8.13}$$

Justification of (6.8.13): $\forall\phi \in \mathcal{D}(\mathbb{R}^3)$,

$$\begin{aligned}\langle U_0, \phi\rangle &= \langle E * \rho\delta_\Gamma, \phi\rangle = \langle \rho\delta_\Gamma(\boldsymbol{\xi}), \langle E(\mathbf{x}), \phi(\mathbf{x}+\boldsymbol{\xi})\rangle\rangle \\ &= \left\langle \rho\delta_\Gamma(\boldsymbol{\xi}), \frac{1}{4\pi}\int_{\mathbb{R}^3}\frac{\phi(\mathbf{x}+\boldsymbol{\xi})}{\|\mathbf{x}\|}d\mathbf{x}\right\rangle = \frac{1}{4\pi}\int_\Gamma \rho(\boldsymbol{\xi})\left(\int_{\mathbb{R}^3}\frac{\phi(\mathbf{x}+\boldsymbol{\xi})}{\|\mathbf{x}\|}d\mathbf{x}\right)dS(\boldsymbol{\xi}).\end{aligned}$$

Changing variables: $\mathbf{x} = \mathbf{y} - \mathbf{t}$, $\mathbf{t} = \boldsymbol{\xi}$ with $|J| = 1$, we have

$$\langle U_0, \phi\rangle = \frac{1}{4\pi}\int_\Gamma \rho(\boldsymbol{\xi})\left(\int_{\mathbb{R}^3}\frac{\phi(\mathbf{y})}{\|\mathbf{y}-\boldsymbol{\xi}\|}d\mathbf{y}\right)dS(\boldsymbol{\xi}).$$

Since $\int_\Gamma |\rho(\boldsymbol{\xi})| \int_{\mathbb{R}^3} \frac{|\phi(\mathbf{y})|}{\|\mathbf{y}-\boldsymbol{\xi}\|}d\mathbf{y}dS(\boldsymbol{\xi})$ exists, we can change the order of integration by Fubini's Theorem. Then, setting $\mathbf{x} = \mathbf{y}$, we get, $\forall\phi \in \mathcal{D}(\mathbb{R}^3)$,

$$\begin{aligned}\langle U_0, \phi\rangle &= \frac{1}{4\pi}\int_{\mathbb{R}^3}\phi(\mathbf{x})\left(\int_\Gamma \frac{\rho(\boldsymbol{\xi})}{\|\mathbf{x}-\boldsymbol{\xi}\|}dS(\boldsymbol{\xi})\right)d\mathbf{x} = \left\langle \frac{1}{4\pi}\int_\Gamma \frac{\rho(\boldsymbol{\xi})}{\|\mathbf{x}-\boldsymbol{\xi}\|}dS(\boldsymbol{\xi}), \phi\right\rangle \\ \Longrightarrow \quad U_0 &= \frac{1}{4\pi}\int_\Gamma \frac{\rho(\boldsymbol{\xi})}{\|\mathbf{x}-\boldsymbol{\xi}\|}dS(\boldsymbol{\xi}) \quad \text{in } \mathcal{D}'(\mathbb{R}^3).\end{aligned}$$

Potential of double layer surface distribution of dipoles Let Γ be a compact, orientable, smooth surface such that the unit normal $\hat{\mathbf{n}}$ is defined at each point of Γ, and $T = -\frac{\partial}{\partial n}(\nu\delta_\Gamma) \in \mathcal{D}'(\mathbb{R}^3)$ be the *volume density* of the double layer distribution of a normally oriented dipole distribution on Γ with *continuous surface moment density* $\nu = \nu(\mathbf{x})$, which is defined by (1.11.29)–(1.11.30) (for details, see Section 1.11): $\forall\phi \in \mathcal{D}(\mathbb{R}^3)$,

$$\langle T, \phi\rangle = \left\langle -\frac{\partial}{\partial n}(\nu\delta_\Gamma), \phi\right\rangle = \int_\Gamma \nu\frac{\partial\phi}{\partial n}dS, \tag{6.8.14}$$

with $\operatorname{supp}(-\frac{\partial}{\partial n}(\nu\delta_\Gamma)) \subset \Gamma \subset\subset \mathbb{R}^3$. Then, from (6.8.9), the *potential* $U_1 = U_{-\frac{\partial}{\partial n}(\nu\delta_\Gamma)}$ of the double layer surface distribution of normally oriented dipoles with surface moment density ν is given by:

$$U_1 = E * \left(-\frac{\partial}{\partial n}(\nu\delta_\Gamma)\right) = \frac{1}{4\pi}\frac{1}{\|\mathbf{x}\|} * \frac{\partial}{\partial n}(\nu\delta_\Gamma)$$
$$= \frac{1}{4\pi}\int_\Gamma \nu(\boldsymbol{\xi})\frac{\partial}{\partial n(\boldsymbol{\xi})}\left[\frac{1}{\|\mathbf{x}-\boldsymbol{\xi}\|}\right]dS(\boldsymbol{\xi}). \qquad (6.8.15)$$

Justification of (6.8.15): $\forall\phi \in \mathcal{D}(\mathbb{R}^3)$,

$$\langle U_1,\phi\rangle = \left\langle E * \left(-\frac{\partial}{\partial n}(\nu\delta_\Gamma)\right),\phi\right\rangle = \frac{-1}{4\pi}\left\langle\frac{1}{\|\mathbf{x}\|} * \frac{\partial}{\partial n}(\nu\delta_\Gamma),\phi\right\rangle$$
$$= \frac{-1}{4\pi}\left\langle\frac{\partial}{\partial n(\boldsymbol{\xi})}(\nu\delta_\Gamma)(\boldsymbol{\xi}),\left\langle\frac{1}{\|\mathbf{x}\|},\phi(\mathbf{x}+\boldsymbol{\xi})\right\rangle\right\rangle$$
$$= \frac{-1}{4\pi}\left\langle\frac{\partial}{\partial n(\boldsymbol{\xi})}(\nu\delta_\Gamma)(\boldsymbol{\xi}),\int_{\mathbb{R}^3}\frac{\phi(\mathbf{x}+\boldsymbol{\xi})}{\|\mathbf{x}\|}d\mathbf{x}\right\rangle.$$

Changing variables, $\mathbf{x} = \mathbf{y}-\mathbf{t}$, $\mathbf{t} = \boldsymbol{\xi}$ with $|J| = 1$, we have:

$$\langle U_1,\phi\rangle = \frac{-1}{4\pi}\left\langle\frac{\partial}{\partial n(\boldsymbol{\xi})}(\nu\delta_\Gamma)(\boldsymbol{\xi}),\int_{\mathbb{R}^3}\frac{\phi(\mathbf{y})}{\|\mathbf{y}-\boldsymbol{\xi}\|}d\mathbf{y}\right\rangle$$
$$= \frac{1}{4\pi}\left\langle(\nu\delta_\Gamma)(\boldsymbol{\xi}),\frac{\partial}{\partial n(\boldsymbol{\xi})}\int_{\mathbb{R}^3}\frac{\phi(\mathbf{y})}{\|\mathbf{y}-\boldsymbol{\xi}\|}d\mathbf{y}\right\rangle$$
$$= \frac{1}{4\pi}\int_\Gamma \nu(\boldsymbol{\xi})\left(\int_{\mathbb{R}^3}\frac{\partial}{\partial n(\boldsymbol{\xi})}\left(\frac{1}{\|\mathbf{y}-\boldsymbol{\xi}\|}\right)\phi(\mathbf{y})d\mathbf{y}\right)dS(\boldsymbol{\xi}).$$

Since the iterated integral $\int_\Gamma |\nu(\boldsymbol{\xi})|(\int_{\mathbb{R}^3}|\phi(\mathbf{y})|\,|\frac{\partial}{\partial n(\boldsymbol{\xi})}(\frac{1}{\|\mathbf{y}-\boldsymbol{\xi}\|})|d\mathbf{y})dS(\boldsymbol{\xi})$ exists, we can change the order of integration by Fubini's Theorem (Theorem 7.1.2C). Then, setting $\mathbf{x} = \mathbf{y}$, we get, $\forall\phi \in \mathcal{D}(\mathbb{R}^3)$,

$$\langle U_1,\phi\rangle = \frac{1}{4\pi}\int_{\mathbb{R}^3}\phi(\mathbf{x})\left(\int_\Gamma \nu(\boldsymbol{\xi})\frac{\partial}{\partial n(\boldsymbol{\xi})}\left(\frac{1}{\|\mathbf{x}-\boldsymbol{\xi}\|}\right)dS(\boldsymbol{\xi})\right)d\mathbf{x}$$
$$= \left\langle\frac{1}{4\pi}\int_\Gamma \nu(\boldsymbol{\xi})\frac{\partial}{\partial n(\boldsymbol{\xi})}\left[\frac{1}{\|\mathbf{x}-\boldsymbol{\xi}\|}\right]dS(\boldsymbol{\xi}),\phi\right\rangle \quad \forall\phi \in \mathcal{D}(\mathbb{R}^3)$$
$$\Longrightarrow \quad U_1 = \frac{1}{4\pi}\int_\Gamma \nu(\boldsymbol{\xi})\frac{\partial}{\partial n(\boldsymbol{\xi})}\left[\frac{1}{\|\mathbf{x}-\boldsymbol{\xi}\|}\right]dS(\boldsymbol{\xi}) \quad \text{in } \mathcal{D}'(\mathbb{R}^3).$$

Remark 6.8.1. Many authors define the potentials in (6.8.7), (6.8.13) and (6.8.15) by omitting the factor $\frac{1}{4\pi}$ (resp. (6.8.8) by omitting the factor $\frac{1}{2\pi}$). But following Schwartz [7], we have accepted (6.8.9) as the definition of the potential of a distribution.

Application of regularization in Sobolev spaces $H^m(\Omega)$ (see (2.15.1))

In Section 6.2 of this chapter, we stated important results for regularization in $L^p(\mathbb{R}^n)$. Now we will state, first of all, analogous results for regularizations in $L^p(\Omega)$.

Let $\Omega \subset \mathbb{R}^n$ be an open subset of $\mathbb{R}^n$ and $u \in L^1_{\text{loc}}(\Omega)$ be a function *locally integrable* on Ω and $\tilde{u}$ be its *null extension* to $\mathbb{R}^n$:

$$\tilde{u}(\mathbf{x}) = \begin{cases} u(\mathbf{x}) & \text{for } \mathbf{x} \in \Omega \\ 0 & \text{for } \mathbf{x} \in \mathbb{R}^n \setminus \Omega. \end{cases} \tag{6.8.16}$$

$\forall \varepsilon > 0$, let $\rho_\varepsilon * \tilde{u}$ be defined by (6.2.20) for $u \in L^1_{\text{loc}}(\Omega)$:

$$(\rho_\varepsilon * \tilde{u})(\mathbf{x}) = \int_{\mathbb{R}^n} \rho_\varepsilon(\mathbf{x} - \boldsymbol{\xi})\tilde{u}(\boldsymbol{\xi})d\boldsymbol{\xi} = \int_\Omega \rho_\varepsilon(\mathbf{x} - \boldsymbol{\xi})u(\boldsymbol{\xi})d\boldsymbol{\xi}. \tag{6.8.17}$$

Since $u \in L^1_{\text{loc}}(\Omega)$, the integral in (6.8.17) exists for *all* $\mathbf{x} \in \mathbb{R}^n$, i.e. $(\rho_\varepsilon * \tilde{u})(\mathbf{x})$ is well defined $\forall \mathbf{x} \in \mathbb{R}^n$. In fact, $u \in L^1_{\text{loc}}(\Omega) \Longrightarrow \tilde{u} \in L^1_{\text{loc}}(\mathbb{R}^n)$

$$\Longrightarrow \quad \rho_\varepsilon * \tilde{u} \in C^\infty(\mathbb{R}^n) \quad \text{by Theorem 6.2.3.} \tag{6.8.18}$$

Now we collect the important results in the following theorems. For this we define the function u_ε by: $\forall \varepsilon > 0$, $\forall u \in L^1_{\text{loc}}(\Omega)$,

$$u_\varepsilon = (\rho_\varepsilon * \tilde{u})\downarrow_\Omega, \tag{6.8.19}$$

with $u_\varepsilon \in C^\infty(\Omega)$ such that $\forall \mathbf{x} \in \Omega$, $u_\varepsilon(\mathbf{x})$ is defined by:

$$u_\varepsilon(\mathbf{x}) = \int_{\mathbb{R}^n} \rho_\varepsilon(\mathbf{x} - \boldsymbol{\xi})\tilde{u}(\boldsymbol{\xi})d\boldsymbol{\xi} = \int_\Omega \rho_\varepsilon(\mathbf{x} - \boldsymbol{\xi})u(\boldsymbol{\xi})d\boldsymbol{\xi}. \tag{6.8.20}$$

Theorem 6.8.2. *Let $u \in L^p(\Omega)$ with $1 \le p < \infty$. Then,*

I. $\rho_\varepsilon * \tilde{u} \in C^\infty(\mathbb{R}^n)$, $u_\varepsilon = (\rho_\varepsilon * \tilde{u})\downarrow_\Omega \in C^\infty(\Omega)$ $\forall \varepsilon > 0$; (6.8.21)

II. $u_\varepsilon = (\rho_\varepsilon * \tilde{u})\downarrow_\Omega \in L^p(\Omega)$ *with* $\|u_\varepsilon\|_{L^p(\Omega)} \le \|u\|_{L^p(\Omega)}$ $\forall \varepsilon > 0$; (6.8.22)
$u_\varepsilon = (\rho_\varepsilon * \tilde{u})\downarrow_\Omega \to u$ *in* $L^p(\Omega)$ *as* $\varepsilon \to 0^+$. (6.8.23)

III. *Moreover, if $u \in L^p(\Omega)$, $1 \le p < \infty$, has support contained in a compact set $K \subset\subset \Omega$, and if $d(K, \partial\Omega)$ = distance between K and $\partial\Omega > \varepsilon$, $\partial\Omega$ being the boundary of Ω, then $u_\varepsilon \in C_0^\infty(\Omega)$.* (6.8.24)

IV. *If $u \in L^p(\Omega)$, $1 \le p < \infty$, is continuous at a point $\mathbf{x} \in \Omega$, then $u_\varepsilon(\mathbf{x}) = (\rho_\varepsilon * \tilde{u})\downarrow_\Omega(\mathbf{x}) \to u(\mathbf{x})$ as $\varepsilon \to 0^+$, the convergence being uniform on every compact set of points at which u is continuous.*

Proof.

I. Since $L^p(\Omega) \subset L^1_{\text{loc}}(\Omega)$ for $1 \le p < \infty$ and $\tilde{u} \in L^1_{\text{loc}}(\mathbb{R}^n)$, from (6.2.34), $\rho_\varepsilon * \tilde{u} \in C^\infty(\mathbb{R}^n)$. Then $u_\varepsilon = (\rho_\varepsilon * \tilde{u})\downarrow_\Omega \in C^\infty(\Omega)$ $\forall \varepsilon > 0$.

II. For $u \in L^p(\Omega)$, $1 \le p < \infty$, $\tilde{u} \in L^p(\mathbb{R}^n)$. Then, by Theorem 6.2.3, $(\rho_\varepsilon * \tilde{u}) \in L^p(\mathbb{R}^n)$ with $\|\rho_\varepsilon * \tilde{u}\|_{L^p(\mathbb{R}^n)} \le \|\tilde{u}\|_{L^p(\mathbb{R}^n)}$ and $\rho_\varepsilon * \tilde{u} \to \tilde{u}$ in $L^p(\mathbb{R}^n)$ as $\varepsilon \to 0^+$. (6.8.25)

Hence, $u_\varepsilon = (\rho_\varepsilon * \tilde{u})\downarrow_\Omega \in L^p(\Omega)$, $1 \le p < \infty$, and, by (6.8.25),

$$\begin{aligned}\|u_\varepsilon\|^p_{L^p(\Omega)} &= \int_\Omega |(\rho_\varepsilon * \tilde{u})\downarrow_\Omega(\mathbf{x})|^p d\mathbf{x} \\ &\le \int_{\mathbb{R}^n} |(\rho_\varepsilon * \tilde{u})(\mathbf{x})|^p d\mathbf{x} = \|\rho_\varepsilon * \tilde{u}\|^p_{L^p(\mathbb{R}^n)} \le \|\tilde{u}\|^p_{L^p(\mathbb{R}^n)} \\ \Longrightarrow \quad \|u_\varepsilon\|_{L^p(\Omega)} &\le \|\tilde{u}\|_{L^p(\mathbb{R}^n)} = \|u\|_{L^p(\Omega)} \quad \forall \varepsilon > 0,\ 1 \le p < \infty.\end{aligned}$$

Then,

$$\begin{aligned}\|u_\varepsilon - u\|^p_{L^p(\Omega)} &\le \int_{\mathbb{R}^n} |(\rho_\varepsilon * \tilde{u})(\mathbf{x}) - \tilde{u}(\mathbf{x})|^p d\mathbf{x} = \|\rho_\varepsilon * \tilde{u} - \tilde{u}\|^p_{L^p(\mathbb{R}^n)} \\ \Longrightarrow \quad \|u_\varepsilon - u\|_{L^p(\Omega)} &\le \|\rho_\varepsilon * \tilde{u} - \tilde{u}\|_{L^p(\mathbb{R}^n)} \to 0 \quad \text{as } \varepsilon \to 0^+ \text{ (by (6.8.25))}.\end{aligned}$$

III. For $u \in L^p(\Omega)$, $\rho_\varepsilon * \tilde{u} \in C^\infty(\mathbb{R}^n)$ and $u_\varepsilon = (\rho_\varepsilon * \tilde{u})\downarrow_\Omega \in C^\infty(\Omega)$ $\forall \varepsilon > 0$. From Theorem 6.2.2, $\operatorname{supp}(\rho_\varepsilon * \tilde{u}) \subset \bar{B}(\mathbf{0};\varepsilon) + \operatorname{supp}(u) \subset \bar{B}(\mathbf{0};\varepsilon) + K = K_\varepsilon$, K_ε being a compact subset of $\mathbb{R}^n$ by (6.2.3). But $d(K, \partial\Omega) > \varepsilon \Longrightarrow K_\varepsilon \subset \Omega$ is a *compact* subset of Ω for all *sufficiently small* $\varepsilon > 0$ satisfying $d(K, \partial\Omega) > \varepsilon$. Hence, $\forall \varepsilon > 0$ with $d(K,\partial\Omega) > \varepsilon$, $\operatorname{supp}(u_\varepsilon) \subset \operatorname{supp}(\rho_\varepsilon * \tilde{u}) \subset K_\varepsilon \subset\subset \Omega$. Thus, $u_\varepsilon \in C_0^\infty(\Omega)$ $\forall \varepsilon > 0$ with $d(K, \partial\Omega) > \varepsilon$. (6.8.26)

IV. Let $\mathbf{x} \in \Omega$ be a point of *continuity* of $u \in L^p(\Omega)$, $1 \le p < \infty$. Then $\tilde{u}$ is also continuous at the point $\mathbf{x}$, and $u(\mathbf{x}) = u(\mathbf{x}) \int_{\mathbb{R}^n} \rho_\varepsilon(\mathbf{x} - \boldsymbol{\xi}) d\boldsymbol{\xi} = \int_{\mathbb{R}^n} \rho_\varepsilon(\mathbf{x} - \boldsymbol{\xi}) u(\mathbf{x}) d\boldsymbol{\xi}$, since $\int_{\mathbb{R}^n} \rho_\varepsilon(\mathbf{x} - \boldsymbol{\xi}) d\boldsymbol{\xi} = \int_{\mathbb{R}^n} \rho_\varepsilon(\mathbf{y}) d\mathbf{y} = 1$ $\forall \varepsilon > 0$

$$\begin{aligned}\Longrightarrow \quad |u_\varepsilon(\mathbf{x}) - u(\mathbf{x})| = |(\rho_\varepsilon * \tilde{u})\downarrow_\Omega(\mathbf{x}) - u(\mathbf{x})| &\le \int_{\mathbb{R}^n} \rho_\varepsilon(\mathbf{x} - \boldsymbol{\xi}) |\tilde{u}(\boldsymbol{\xi}) - \tilde{u}(\mathbf{x})| d\boldsymbol{\xi} \\ &= \int_{\|\mathbf{x}-\boldsymbol{\xi}\| \le \varepsilon} \rho_\varepsilon(\mathbf{x} - \boldsymbol{\xi}) |\tilde{u}(\boldsymbol{\xi}) - \tilde{u}(\mathbf{x})| d\boldsymbol{\xi} \\ &\le \sup_{\|\mathbf{x}-\boldsymbol{\xi}\| \le \varepsilon} |\tilde{u}(\boldsymbol{\xi}) - \tilde{u}(\mathbf{x})| \int_{\mathbb{R}^n} \rho_\varepsilon(\mathbf{x} - \boldsymbol{\xi}) d\boldsymbol{\xi} \\ &= \sup_{\|\mathbf{x}-\boldsymbol{\xi}\| \le \varepsilon} |\tilde{u}(\boldsymbol{\xi}) - \tilde{u}(\mathbf{x})| \to 0 \qquad (6.8.27)\end{aligned}$$

as $\varepsilon \to 0^+$ (since $\int_{\mathbb{R}^n} \rho_\varepsilon(\mathbf{x} - \boldsymbol{\xi}) d\boldsymbol{\xi} = \int_{\mathbb{R}^n} \rho_\varepsilon(\mathbf{y}) d\mathbf{y} = 1$ and $\tilde{u}$ is continuous at $\mathbf{x} \in \Omega$), the convergence being uniform on all *compact* subsets of points of continuity of u in Ω (see Proposition 6.2.4 for more detail). □

Density result

Theorem 6.8.3. *Let $\Omega \subset \mathbb{R}^n$ be any open subset of $\mathbb{R}^n$. Then $C_0^\infty(\Omega) \equiv \mathcal{D}(\Omega)$ is* dense *in $L^p(\Omega)$, $1 \le p < \infty$.*

Proof. Let $u \in L^p(\Omega)$, $1 \le p < \infty$. Let $\eta > 0$. Then, by virtue of the density of $C_0(\Omega)$ in $L^p(\Omega)$, $1 \le p < \infty$ (see Appendix B, Theorem B.3.3.4 and Theorem 1.2.3 of Chapter 1), $\exists \phi \in C_0(\Omega)$ with compact $\operatorname{supp}(\phi) = K \subset\subset \Omega$ such that $\|u - \phi\|_{L^p(\Omega)} < \frac{\eta}{2}$. (6.8.28)

Let $\tilde{\phi}$ be the null extension to $\mathbb{R}^n$ of $\phi \in C_0(\Omega)$: $\tilde{\phi}(\mathbf{x}) = \phi(\mathbf{x})$ for $\mathbf{x} \in \Omega$ and $\tilde{\phi}(\mathbf{x}) = 0$ for $\mathbf{x} \in \mathbb{R}^n \setminus \Omega$ with $\operatorname{supp}(\tilde{\phi}) = \operatorname{supp}(\phi) = K \subset\subset \Omega$. Then $\tilde{\phi} \in C_0(\mathbb{R}^n) \subset L^p(\mathbb{R}^n)$ for $1 \le p < \infty$. Hence, by Proposition 6.2.5,

$$\|\rho_\varepsilon * \tilde{\phi} - \tilde{\phi}\|_{L^p(\mathbb{R}^n)} \to 0 \quad \text{as } \varepsilon \to 0^+. \tag{6.8.29}$$

Moreover, by Lemma 6.2.1, $\exists \varepsilon_0 > 0$ such that $\rho_\varepsilon * \tilde{\phi} \in C_0^\infty(\mathbb{R}^n)$ with $\operatorname{supp}(\rho_\varepsilon * \tilde{\phi}) \subset K_0 \subset\subset \Omega$ $\forall \varepsilon < \varepsilon_0$, K_0 being a fixed *compact* set in Ω. Define $u_\varepsilon = (\rho_\varepsilon * \tilde{\phi})\downarrow_\Omega$ $\forall \varepsilon > 0$. Then $u_\varepsilon \in C_0^\infty(\Omega)$ with $\operatorname{supp}(u_\varepsilon) \subset \operatorname{supp}(\rho_\varepsilon * \tilde{\phi}_\varepsilon) \subset K_0 \subset\subset \Omega$ $\forall \varepsilon < \varepsilon_0$. (6.8.30)

Then,

$$\begin{aligned}\|u_\varepsilon - \phi\|^p_{L^p(\Omega)} &= \int_\Omega |(\rho_\varepsilon * \tilde{\phi})\downarrow_\Omega(\mathbf{x}) - \phi(\mathbf{x})|^p d\mathbf{x} \\ &\le \int_{\mathbb{R}^n} |(\rho_\varepsilon * \tilde{\phi})(\mathbf{x}) - \tilde{\phi}(\mathbf{x})|^p d\mathbf{x} = \|\rho_\varepsilon * \tilde{\phi} - \tilde{\phi}\|^p_{L^p(\mathbb{R}^n)} \to 0\end{aligned}$$

as $\varepsilon \to 0^+$ (by (6.8.29)) $\Longrightarrow \forall \eta > 0, \exists \varepsilon_1 > 0$ such that

$$\|u_\varepsilon - \phi\|_{L^p(\mathbb{R}^n)} < \frac{\eta}{2} \quad \forall \varepsilon < \varepsilon_1. \tag{6.8.31}$$

Hence, combining (6.8.28) and (6.8.31), we get: $\forall \eta > 0$, $\exists \varepsilon_0^* = \min\{\varepsilon_0, \varepsilon_1\} > 0$ such that, for $1 \le p < \infty$,

$$\|u - u_\varepsilon\|_{L^p(\Omega)} \le \|u - \phi\|_{L^p(\Omega)} + \|\phi - u_\varepsilon\|_{L^p(\Omega)} < \frac{\eta}{2} + \frac{\eta}{2} = \eta \quad \forall \varepsilon < \varepsilon_0^*,$$

i.e. for $u \in L^p(\Omega)$, $1 \le p < \infty$, $\forall \eta > 0$, $\exists u_\varepsilon \in C_0^\infty(\Omega)$ with $\varepsilon < \varepsilon_0^*$ such that $\|u - u_\varepsilon\|_{L^p(\Omega)} < \eta$. Thus, $C_0^\infty(\Omega) \equiv \mathcal{D}(\Omega)$ is dense in $L^p(\Omega)$, $1 \le p < \infty$. □

Now we can deal with the results in Sobolev spaces $H^m(\Omega)$ defined by (2.15.1)–(2.15.3). Let $u \in H^m(\Omega)$ with $m \in \mathbb{N}$. Then its distributional derivatives $\partial^{\boldsymbol{\alpha}} u \in L^2(\Omega)$ $\forall |\boldsymbol{\alpha}| \le m$ are defined by (2.15.2). Let $\tilde{u}$ and $\widetilde{\partial^{\boldsymbol{\alpha}} u}$ be the null extensions to $\mathbb{R}^n$ of u and $\partial^{\boldsymbol{\alpha}} u$ $\forall |\boldsymbol{\alpha}| \le m$, respectively:

$$\tilde{u}(\mathbf{x}) = \begin{cases} u(\mathbf{x}) & \text{for } \mathbf{x} \in \Omega \\ 0 & \text{for } \mathbf{x} \in \mathbb{R}^n \setminus \Omega \end{cases} \quad \text{and} \quad \widetilde{\partial^{\boldsymbol{\alpha}} u}(\mathbf{x}) = \begin{cases} (\partial^{\boldsymbol{\alpha}} u)(\mathbf{x}) & \text{for } \mathbf{x} \in \Omega \\ 0 & \text{for } \mathbf{x} \in \mathbb{R}^n \setminus \Omega \end{cases} \tag{6.8.32}$$

Then $\tilde{u} \in L^2(\mathbb{R}^n)$ with $\|\tilde{u}\|_{L^2(\mathbb{R}^n)} = \|u\|_{L^2(\Omega)}$ and, $\forall|\boldsymbol{\alpha}| \le m$, $\widetilde{\partial^{\boldsymbol{\alpha}} u} \in L^2(\mathbb{R}^n)$, although

$$\tilde{u} \notin H^m(\mathbb{R}^n) \text{ and } \partial^{\boldsymbol{\alpha}}\tilde{u} \notin L^2(\mathbb{R}^n) \text{ for } 1 \le |\boldsymbol{\alpha}| \le m \text{ in general.} \tag{6.8.33}$$

For example, for $n = 1$ and $\Omega =]0,1[$, $u(x) = 1\ \forall x \in]0,1[$, $\tilde{u} \notin H^1(\mathbb{R})$, since

$$\begin{aligned}\left\langle \frac{d\tilde{u}}{dx}, \phi \right\rangle_{\mathcal{D}'(\mathbb{R})\times\mathcal{D}(\mathbb{R})} &= -\left\langle \tilde{u}, \frac{d\phi}{dx} \right\rangle = -\int_0^1 1 \cdot \frac{d\phi}{dx}dx = -\phi(x)|_{x=0}^{x=1} = -\phi(1) + \phi(0) \\ &= -\langle \delta_1, \phi\rangle + \langle \delta_0, \phi\rangle = \langle -\delta_0 + \delta_1, \phi\rangle \quad \forall \phi \in \mathcal{D}(\mathbb{R}),\end{aligned}$$

where δ_0, δ_1 are Dirac distributions with concentration of mass/force/charge etc. at 0 and 1 respectively

$$\implies \quad \frac{d\tilde{u}}{dx} = -\delta_0 + \delta_1 \notin L^2(\mathbb{R}) \quad \implies \quad \tilde{u} \notin H^1(\mathbb{R}). \tag{6.8.34}$$

By Theorem 6.2.3, $\forall \varepsilon > 0$, $(\rho_\varepsilon * \tilde{u}) \in L^2(\mathbb{R}^n)$ and $(\rho_\varepsilon * \widetilde{\partial^{\boldsymbol{\alpha}} u}) \in L^2(\mathbb{R}^n)$ $\forall|\boldsymbol{\alpha}| \le m$, and consequently $(\rho_\varepsilon * \tilde{u})\downarrow_\Omega \in L^2(\Omega)$, $(\rho_\varepsilon * \widetilde{\partial^{\boldsymbol{\alpha}} u})\downarrow_\Omega \in L^2(\Omega)$ $\forall|\boldsymbol{\alpha}| \le m$. Then we have the following result:

Theorem 6.8.4. *Let $u \in H^m(\Omega)$. Then, for $\mathbf{x} \in \Omega$ with $d(\mathbf{x}, \partial\Omega) > \varepsilon$, $\partial\Omega$ being the boundary of Ω,*

$$\partial^{\boldsymbol{\alpha}}[(\rho_\varepsilon * \tilde{u})\downarrow_\Omega](\mathbf{x}) = (\rho_\varepsilon * \widetilde{\partial^{\boldsymbol{\alpha}} u})\downarrow_\Omega(\mathbf{x}) \quad \forall|\boldsymbol{\alpha}| \le m. \tag{6.8.35}$$

Proof. $\forall \mathbf{x} \in \Omega$, $(\rho_\varepsilon * \tilde{u})\downarrow_\Omega(\mathbf{x}) = \int_{\mathbb{R}^n} \rho_\varepsilon(\mathbf{x} - \boldsymbol{\xi})\tilde{u}(\boldsymbol{\xi})d\boldsymbol{\xi} = \int_\Omega \rho_\varepsilon(\mathbf{x} - \boldsymbol{\xi})u(\boldsymbol{\xi})d\boldsymbol{\xi}$ and

$$(\rho_\varepsilon * \widetilde{\partial^{\boldsymbol{\alpha}} u})\downarrow_\Omega(\mathbf{x}) = \int_{\mathbb{R}^n} \rho_\varepsilon(\mathbf{x} - \boldsymbol{\xi})\widetilde{\partial^{\boldsymbol{\alpha}} u}(\boldsymbol{\xi})d\boldsymbol{\xi} = \int_\Omega \rho_\varepsilon(\mathbf{x} - \boldsymbol{\xi})(\partial^{\boldsymbol{\alpha}} u)(\boldsymbol{\xi})d\boldsymbol{\xi} \quad \forall|\boldsymbol{\alpha}| \le m. \tag{6.8.36}$$

Then, from Theorem 6.8.2, $(\rho_\varepsilon * \tilde{u})\downarrow_\Omega \in C^\infty(\Omega)$. Hence, $\forall|\boldsymbol{\alpha}| \le m$, $\partial^{\boldsymbol{\alpha}}[(\rho_\varepsilon * \tilde{u})\downarrow_\Omega](\mathbf{x})$ is well defined $\forall \mathbf{x} \in \Omega$, $\forall \varepsilon > 0$ and can be obtained by differentiating (in the *usual pointwise sense*) under the integral sign: $\forall|\boldsymbol{\alpha}| \le m$,

$$\begin{aligned}\partial^{\boldsymbol{\alpha}}[(\rho_\varepsilon * \tilde{u})\downarrow_\Omega](\mathbf{x}) &= \partial^{\boldsymbol{\alpha}}_{\mathbf{x}} \int_\Omega \rho_\varepsilon(\mathbf{x} - \boldsymbol{\xi})u(\boldsymbol{\xi})d\boldsymbol{\xi} = \int_\Omega \partial^{\boldsymbol{\alpha}}_{\mathbf{x}}[\rho_\varepsilon(\mathbf{x} - \boldsymbol{\xi})]u(\boldsymbol{\xi})d\boldsymbol{\xi} \\ &= \int_\Omega \partial^{\boldsymbol{\alpha}}_{\boldsymbol{\xi}}[\rho_\varepsilon(\mathbf{x} - \boldsymbol{\xi})](-1)^{|\boldsymbol{\alpha}|} \cdot u(\boldsymbol{\xi})d\boldsymbol{\xi} \\ &= (-1)^{|\boldsymbol{\alpha}|} \int_\Omega u(\boldsymbol{\xi})\partial^{\boldsymbol{\alpha}}_{\boldsymbol{\xi}}[\rho_\varepsilon(\mathbf{x} - \boldsymbol{\xi})]d\boldsymbol{\xi}.\end{aligned} \tag{6.8.37}$$

For *fixed* $\mathbf{x} \in \Omega$, set $\phi_\varepsilon(\boldsymbol{\xi}) = \rho_\varepsilon(\mathbf{x} - \boldsymbol{\xi})$ $\forall \boldsymbol{\xi} \in \Omega$, $\forall \varepsilon > 0$. Then, for *fixed* $\mathbf{x} \in \Omega$, $d(\mathbf{x}, \partial\Omega) > \varepsilon \implies \phi_\varepsilon \in C_0^\infty(\Omega) \equiv \mathcal{D}(\Omega)$. In fact, for *fixed* $\mathbf{x} \in \Omega$, the function $\boldsymbol{\xi} \in$

$\mathbb{R}^n \longmapsto \rho_\varepsilon(\mathbf{x}-\boldsymbol{\xi})$ belongs to $C_0^\infty(\mathbb{R}^n)$ by definition of ρ_ε. Then, for *all sufficiently small* $\varepsilon > 0$, $\phi_\varepsilon \in C^\infty(\Omega)$ with $\phi_\varepsilon(\boldsymbol{\xi}) = 0\ \forall \boldsymbol{\xi} \in \Omega$ with $\|\mathbf{x}-\boldsymbol{\xi}\| > \varepsilon$. Thus, for $\boldsymbol{\xi} \in \partial\Omega$ and $\mathbf{x} \in \Omega$, $\|\mathbf{x}-\boldsymbol{\xi}\| \geq \inf_{\boldsymbol{\eta}\in\partial\Omega} \|\mathbf{x}-\boldsymbol{\eta}\| = d(\mathbf{x}, \partial\Omega) > \varepsilon \Longrightarrow \rho_\varepsilon(\mathbf{x}-\boldsymbol{\xi}) = 0$ $\forall \mathbf{x} \in \Omega$ with $d(\mathbf{x}, \partial\Omega) > \varepsilon \Longrightarrow \text{supp}(\phi_\varepsilon) \subset\subset \Omega$ is a compact subset of Ω. Hence, $\phi_\varepsilon \in C_0^\infty(\Omega) \equiv \mathcal{D}(\Omega)$ for *all* sufficiently small $\varepsilon > 0$ with $d(\mathbf{x}, \partial\Omega) > \varepsilon$. Then, from (6.8.37), $\forall |\boldsymbol{\alpha}| \leq m$,

$$\begin{aligned}\partial^{\boldsymbol{\alpha}}[(\rho_\varepsilon * \tilde{u})\downarrow_\Omega](\mathbf{x}) &= (-1)^{|\boldsymbol{\alpha}|}\langle u, \partial_{\boldsymbol{\xi}}^{\boldsymbol{\alpha}}\phi_\varepsilon\rangle_{\mathcal{D}'(\Omega)\times\mathcal{D}(\Omega)} = \langle \partial_{\boldsymbol{\xi}}^{\boldsymbol{\alpha}} u, \phi_\varepsilon\rangle_{\mathcal{D}'(\Omega)\times\mathcal{D}(\Omega)} \\ &= \int_\Omega \partial^{\boldsymbol{\alpha}} u(\boldsymbol{\xi})\phi_\varepsilon(\boldsymbol{\xi})d\boldsymbol{\xi} = \int_{\mathbb{R}^n} \rho_\varepsilon(\mathbf{x}-\boldsymbol{\xi})\widetilde{\partial^{\boldsymbol{\alpha}} u}(\boldsymbol{\xi})d\boldsymbol{\xi} \qquad (6.8.38)\end{aligned}$$

for $\mathbf{x} \in \Omega$ with $d(\mathbf{x}, \partial\Omega) > \varepsilon$, since for $u \in H^m(\Omega)$, its distributional derivatives $\partial^{\boldsymbol{\alpha}} u \in L^2(\Omega)\ \forall |\boldsymbol{\alpha}| \leq m$. From (6.8.36)–(6.8.38), the result follows: $\partial^{\boldsymbol{\alpha}}[(\rho_\varepsilon * \tilde{u})\downarrow_\Omega](\mathbf{x}) = (\rho_\varepsilon * \widetilde{\partial^{\boldsymbol{\alpha}} u})\downarrow_\Omega(\mathbf{x})$ for $\mathbf{x} \in \Omega$ with $d(\mathbf{x}, \partial\Omega) > \varepsilon$, i.e. for *all sufficiently small* $\varepsilon > 0$. □

Theorem 6.8.5. *Let* $\Omega, \Omega_1 \subset \mathbb{R}^n$ *be open subsets of* $\mathbb{R}^n$ *such that* $\overline{\Omega}_1 \subset\subset \Omega$ *is a compact subset of* Ω *(i.e.* Ω_1 *is relatively compact in* Ω*). Let* $u \in H^m(\Omega)$ *and* $\tilde{u}$ *and* $\widetilde{\partial^{\boldsymbol{\alpha}} u}$ *be the null extensions to* $\mathbb{R}^n$ *of* u *and* $\partial^{\boldsymbol{\alpha}} u$ $(\forall |\boldsymbol{\alpha}| \leq m)$ *respectively, as defined by* (6.8.32). *If* $u_\varepsilon = (\rho_\varepsilon * \tilde{u})\downarrow_\Omega\ \forall \varepsilon > 0$, *then*

$$u_\varepsilon\downarrow_{\Omega_1} \to u\downarrow_{\Omega_1} \text{ in } H^m(\Omega_1) \text{ as } \varepsilon \to 0^+. \qquad (6.8.39)$$

Proof. Let $u \in H^m(\Omega)$. Then $\tilde{u}, \widetilde{\partial^{\boldsymbol{\alpha}} u} \in L^2(\mathbb{R}^n)\ \forall |\boldsymbol{\alpha}| \leq m$. So, $\rho_\varepsilon * \tilde{u} \in L^2(\mathbb{R}^n) \Longrightarrow u_\varepsilon = (\rho_\varepsilon * \tilde{u})\downarrow_\Omega \in L^2(\Omega)$

$$\Longrightarrow \quad u_\varepsilon\downarrow_{\Omega_1} \in L^2(\Omega_1)\ \forall \varepsilon > 0 \text{ and } u_\varepsilon\downarrow_{\Omega_1} \to u\downarrow_{\Omega_1} \text{ in } L^2(\Omega_1) \text{ as } \varepsilon \to 0^+ \qquad (6.8.40)$$

by Theorem 6.8.2 with $p = 2$. Again, from Theorem 6.8.4, for all sufficiently small $\varepsilon > 0$ with $d(\Omega_1, \partial\Omega) > \varepsilon$, $\forall |\boldsymbol{\alpha}| \leq m$, $(\partial^{\boldsymbol{\alpha}} u_\varepsilon)(\mathbf{x}) = (\rho_\varepsilon * \widetilde{\partial^{\boldsymbol{\alpha}} u})(\mathbf{x})\ \forall \mathbf{x} \in \Omega_1$

$$\Longrightarrow \quad (\rho_\varepsilon * \widetilde{\partial^{\boldsymbol{\alpha}} u})\downarrow_{\Omega_1} = (\partial^{\boldsymbol{\alpha}} u_\varepsilon)\downarrow_{\Omega_1} = \partial^{\boldsymbol{\alpha}}(u_\varepsilon\downarrow_{\Omega_1}), \qquad (6.8.41)$$

since $u_\varepsilon \in C^\infty(\Omega) \cap L^2(\Omega)$ by (6.8.21) and (6.8.22). But

$$u \in H^m(\Omega) \quad \Longrightarrow \quad (\partial^{\boldsymbol{\alpha}} u)\downarrow_{\Omega_1} = \partial^{\boldsymbol{\alpha}}(u\downarrow_{\Omega_1}) \quad \forall |\boldsymbol{\alpha}| \leq m. \qquad (6.8.42)$$

In fact, $\int_\Omega (\partial^{\boldsymbol{\alpha}} u)\phi d\mathbf{x} = (-1)^{|\boldsymbol{\alpha}|} \int_\Omega u\partial^{\boldsymbol{\alpha}}\phi\ \forall \phi \in \mathcal{D}(\Omega)$. Hence, $\forall |\boldsymbol{\alpha}| \leq m$,

$$\begin{aligned}\int_{\Omega_1} (\partial^{\boldsymbol{\alpha}} u)\downarrow_{\Omega_1}\phi d\mathbf{x} &= (-1)^{|\boldsymbol{\alpha}|}\int_{\Omega_1} (u\downarrow_{\Omega_1})\partial^{\boldsymbol{\alpha}}\phi d\mathbf{x} \quad \forall \phi \in \mathcal{D}(\Omega_1) \\ &= \langle \partial^{\boldsymbol{\alpha}}(u\downarrow_{\Omega_1}), \phi\rangle_{\mathcal{D}'(\Omega_1)\times\mathcal{D}(\Omega_1)} \quad \forall \phi \in \mathcal{D}(\Omega_1)\end{aligned}$$

$\Longrightarrow (\partial^{\boldsymbol{\alpha}} u)\downarrow_{\Omega_1} = \partial^{\boldsymbol{\alpha}}(u\downarrow_{\Omega_1})$ in $\mathcal{D}'(\Omega_1)$, but $(\partial^{\boldsymbol{\alpha}} u)\downarrow_{\Omega_1} \in L^2(\Omega_1)\ \forall |\boldsymbol{\alpha}| \leq m$.

Hence, $\forall|\boldsymbol{\alpha}| \leq m$, $\partial^{\boldsymbol{\alpha}}(u{\downarrow_{\Omega_1}}) = (\partial^{\boldsymbol{\alpha}}u){\downarrow_{\Omega_1}}$ in $L^2(\Omega_1)$. Again, applying Theorem 6.8.2, $\forall|\boldsymbol{\alpha}| \leq m$,

$$(\rho_\varepsilon * \widetilde{\partial^{\boldsymbol{\alpha}}u}){\downarrow_{\Omega_1}} \to (\partial^{\boldsymbol{\alpha}}u){\downarrow_{\Omega_1}} \quad \text{in } L^2(\Omega_1) \text{ as } \varepsilon \to 0^+, \tag{6.8.43}$$

since

$$\int_{\Omega_1} |(\rho_\varepsilon * \widetilde{\partial^{\boldsymbol{\alpha}}u}){\downarrow_{\Omega_1}}(\mathbf{x}) - (\partial^{\boldsymbol{\alpha}}u){\downarrow_{\Omega_1}}(\mathbf{x})|^2 d\mathbf{x}$$
$$\leq \int_\Omega |(\rho_\varepsilon * \widetilde{\partial^{\boldsymbol{\alpha}}u}){\downarrow_\Omega}(\mathbf{x}) - \partial^{\boldsymbol{\alpha}}u(\mathbf{x})|^2 d\mathbf{x} \to 0$$

as $\varepsilon \to 0^+$. Hence, from (6.8.41),(6.8.42) and (6.8.43), $\forall|\boldsymbol{\alpha}| \leq m$,

$$(\rho_\varepsilon * \widetilde{\partial^{\boldsymbol{\alpha}}u}){\downarrow_{\Omega_1}} = \partial^{\boldsymbol{\alpha}}(u_\varepsilon{\downarrow_{\Omega_1}}) \to (\partial^{\boldsymbol{\alpha}}u){\downarrow_{\Omega_1}} = \partial^{\boldsymbol{\alpha}}(u{\downarrow_{\Omega_1}}) \quad \text{in } L^2(\Omega_1) \text{ as } \varepsilon \to 0^+. \tag{6.8.44}$$

Finally, from (6.8.40) and (6.8.44), we have $u_\varepsilon{\downarrow_{\Omega_1}} \to u{\downarrow_{\Omega_1}}$ in $L^2(\Omega_1)$ and $\partial^{\boldsymbol{\alpha}}(u_\varepsilon{\downarrow_{\Omega_1}}) \to \partial^{\boldsymbol{\alpha}}(u{\downarrow_{\Omega_1}})$ in $L^2(\Omega_1)$ as $\varepsilon \to 0^+$ $\forall|\boldsymbol{\alpha}| \leq m$
$\Longrightarrow u_\varepsilon{\downarrow_{\Omega_1}} \to u{\downarrow_{\Omega_1}}$ in $H^m(\Omega_1)$ as $\varepsilon \to 0^+$. □

The following discussion needs two auxiliary spaces, $C^{*m}(\Omega)$ and $H^{*m}(\Omega)$, which will be introduced now.

Space $C^{*m}(\Omega)$

Definition 6.8.2. Let $\Omega \subset \mathbb{R}^n$ be any open subset of $\mathbb{R}^n$, and $C^m(\Omega)$ be the space of m times continuously differentiable (in the usual sense) functions on Ω. Then $C^{*m}(\Omega)$ is the subspace of $C^m(\Omega)$ defined by $C^{*m}(\Omega) = \{u : u \in C^m(\Omega)$ such that, $\forall|\boldsymbol{\alpha}| \leq m,\ \partial^{\boldsymbol{\alpha}}u \in L^2(\Omega)\}$, which is equipped with the inner product $\langle\cdot,\cdot\rangle_{m,\Omega}$ and the norm $\|\cdot\|_{m,\Omega}$ defined by:

$$\langle u, v\rangle_{C^{*m}(\Omega)} = \langle u, v\rangle_{m,\Omega} = \left(\sum_{0\leq|\boldsymbol{\alpha}|\leq m} \int_\Omega \partial^{\boldsymbol{\alpha}}u\,\partial^{\boldsymbol{\alpha}}v\,d\mathbf{x}\right) \quad \forall u, v \in C^{*m}(\Omega); \tag{6.8.45}$$

$$\|u\|_{C^{*m}(\Omega)} = \|u\|_{m,\Omega} = \langle u, u\rangle^{1/2} = \left(\sum_{0\leq|\boldsymbol{\alpha}|\leq m} \int_\Omega |\partial^{\boldsymbol{\alpha}}u|^2 d\mathbf{x}\right)^{1/2} \quad \forall u \in C^{*m}(\Omega).$$

For *arbitrary* $\Omega \subset \mathbb{R}^n$, $C^{*m}(\Omega) \subset C^m(\Omega)$, $C^{*m}(\Omega) \subset L^2(\Omega)$. (6.8.46)

Space $H^{*m}(\Omega)$

Definition 6.8.3. $H^{*m}(\Omega) \equiv \overline{C^{*m}(\Omega)}$ in the norm $\|\cdot\|_{m,\Omega}$ defined by (6.8.45), i.e. $H^{*m}(\Omega)$ is the *completion* of $C^{*m}(\Omega)$ in the norm $\|\cdot\|_{m,\Omega}$.

Following Agmon [39, p. 2], we give another useful characterization as follows: $u \in H^{*m}(\Omega) \iff u \in L^2(\Omega)$ *and* $\exists$ a sequence $(u_k)_{k=1}^{\infty}$ in $C^{*m}(\Omega)$ such that $u_k \to u$ in $L^2(\Omega)$ and $(\partial^{\boldsymbol{\alpha}} u_k)_{k=1}^{\infty}$ is a Cauchy sequence in $L^2(\Omega)$ $\forall|\boldsymbol{\alpha}| \le m$ as $k \to \infty$. (6.8.47)

Now we state some important results without proof (see, for example, Agmon [39, pp. 2–4]).

Theorem 6.8.6.

I. $H^{*m}(\Omega)$ *equipped with the inner product* $\langle\cdot,\cdot\rangle_{m,\Omega}$ *in* (6.8.45) *is a* Hilbert space.

II. $H^{*m}(\Omega) \subset H^m(\Omega)$, *where* $H^m(\Omega)$ *is defined by* (2.15.1)–(2.15.3).

III. *For* Ω *satisfying the* segment property,[2] $H^{*m}(\Omega) \equiv H^m(\Omega)$.

Theorem 6.8.7 (Leibniz Rule). *Let* $u \in H^m(\Omega)$ *and* $v \in C^m(\Omega)$ *such that* v *and its usual partial derivatives (in the pointwise sense)* $\partial^{\boldsymbol{\alpha}} v$ *of order* $|\boldsymbol{\alpha}| \le m$ *are bounded in* Ω. *Then,*

I. $uv \in H^m(\Omega)$;

II. $\partial^{\boldsymbol{\alpha}}(uv) = \sum_{\boldsymbol{\beta}\le\boldsymbol{\alpha}} \binom{\boldsymbol{\alpha}}{\boldsymbol{\beta}} \partial^{\boldsymbol{\beta}} u \partial^{\boldsymbol{\alpha}-\boldsymbol{\beta}} v$ $\forall|\boldsymbol{\alpha}| \le m$, *where* $\boldsymbol{\beta} \le \boldsymbol{\alpha} \iff \beta_i \le \alpha_i$ $\forall i = 1, 2, \ldots, n$ *and* $\binom{\boldsymbol{\alpha}}{\boldsymbol{\beta}} = \binom{\alpha_1}{\beta_1}\binom{\alpha_2}{\beta_2}\cdots\binom{\alpha_n}{\beta_n}$.

Proof. First we will consider the case of $u \in H^{*m}(\Omega) \subset H^m(\Omega)$.

Let $u \in H^{*m}(\Omega)$. Then, $\exists$ a sequence (u_k) in $C^{*m}(\Omega)$ such that

$$u_k \to u \quad \text{and} \quad \partial^{\boldsymbol{\alpha}} u_k \to u_{\boldsymbol{\alpha}} \quad \text{in } L^2(\Omega)\ \forall|\boldsymbol{\alpha}| \le m. \tag{6.8.48}$$

But v and $\partial^{\boldsymbol{\alpha}} v$ are continuous and bounded on Ω $\forall|\boldsymbol{\alpha}| \le m$. Hence, $\forall|\boldsymbol{\alpha}| \le m$,

$$\partial^{\boldsymbol{\alpha}}(u_k v) = \sum_{\boldsymbol{\beta}\le\boldsymbol{\alpha}} \binom{\boldsymbol{\alpha}}{\boldsymbol{\beta}} \partial^{\boldsymbol{\beta}} u_k \partial^{\boldsymbol{\alpha}-\boldsymbol{\beta}} v \to \sum_{\boldsymbol{\beta}\le\boldsymbol{\alpha}} \binom{\boldsymbol{\alpha}}{\boldsymbol{\beta}} \partial^{\boldsymbol{\beta}} u \partial^{\boldsymbol{\alpha}-\boldsymbol{\beta}} v \quad \text{in } L^2(\Omega) \text{ as } k \to \infty. \tag{6.8.49}$$

So the sequence (w_k) with $w_k = u_k v$ $\forall k \in \mathbb{N}$ is a Cauchy sequence in $C^{*m}(\Omega)$. Hence, $\exists w \in H^{*m}(\Omega)$ such that $w_k = u_k v \to w = uv$ and

$$\partial^{\boldsymbol{\alpha}} w_k = \partial^{\boldsymbol{\alpha}}(u_k v) \to w_{\boldsymbol{\alpha}} \quad \text{in } L^2(\Omega) \text{ as } k \to \infty\ \forall|\boldsymbol{\alpha}| \le m. \tag{6.8.50}$$

---

[2]Roughly speaking, Ω has the *segment property* if at each point $\mathbf{x}$ on the boundary Γ of Ω, there exists a *linear segment* L with origin at $\mathbf{x}$ such that $L \setminus \{\mathbf{x}\} \subset \Omega$. For more details, see Appendix D.
Ω has the segment property $\implies$ Ω is *not* locally a *two-sided* domain (see Appendix D).

But $L^2(\Omega) \subset \mathcal{D}'(\Omega) \Longrightarrow w_k = u_k v \to w = uv$ in $\mathcal{D}'(\Omega)$ and

$$\partial^{\boldsymbol{\alpha}}(u_k v) \to w_{\boldsymbol{\alpha}} \text{ in } \mathcal{D}'(\Omega) \text{ as } k \to \infty. \tag{6.8.51}$$

Since $\partial^{\boldsymbol{\alpha}} : \mathcal{D}'(\Omega) \to \mathcal{D}'(\Omega)$ is continuous, $u_k v \to uv$ in $\mathcal{D}'(\Omega) \Longrightarrow \partial^{\boldsymbol{\alpha}}(u_k v) \to \partial^{\boldsymbol{\alpha}}(uv)$ in $\mathcal{D}'(\Omega)$ as $k \to \infty$. From the uniqueness of the limit, $\partial^{\boldsymbol{\alpha}}(uv) = w_{\boldsymbol{\alpha}} \in L^2(\Omega)$ $\forall|\boldsymbol{\alpha}| \leq m$. Hence,

$$w = uv \in L^2(\Omega),\ \partial^{\boldsymbol{\alpha}} w = \partial^{\boldsymbol{\alpha}}(uv) \in L^2(\Omega)\ \forall|\boldsymbol{\alpha}| \leq m \quad \Longrightarrow \quad w = uv \in H^m(\Omega). \tag{6.8.52}$$

Then, for $u \in H^{*m}(\Omega)$, $\forall|\boldsymbol{\alpha}| \leq m$, $\partial^{\boldsymbol{\alpha}}(uv) = \partial^{\boldsymbol{\alpha}} w = \lim_{k\to\boldsymbol{\alpha}} \partial^{\boldsymbol{\alpha}}(u_k v) = \sum_{\boldsymbol{\beta}\leq\boldsymbol{\alpha}} \binom{\boldsymbol{\alpha}}{\boldsymbol{\beta}} \partial^{\boldsymbol{\beta}} u \partial^{\boldsymbol{\alpha}-\boldsymbol{\beta}} v$ in $L^2(\Omega)$ (by (6.8.52) and (6.8.49)).

Now assume that $u \in H^m(\Omega)$. Then, $\forall$ *relatively compact* $\Omega_1 \subset\subset \Omega$ (i.e. $\overline{\Omega}_1 \subset\subset \Omega$), by Theorem 6.8.5, $u_\varepsilon\downarrow_{\Omega_1} \to u\downarrow_{\Omega_1}$ in $H^m(\Omega_1)$ as $\varepsilon \to 0^+$, where $u_\varepsilon = (\rho_\varepsilon * \tilde{u})\downarrow_\Omega \in C^\infty(\Omega)$. Hence, $u_\varepsilon\downarrow_{\Omega_1} \in C^{*m}(\Omega_1) \to u\downarrow_{\Omega_1}$ in $H^m(\Omega_1) \Longrightarrow u\downarrow_{\Omega_1} \in H^{*m}(\Omega_1)$, since $u_\varepsilon\downarrow_{\Omega_1} \to u\downarrow_{\Omega_1}$ in $L^2(\Omega_1)$, and $\forall|\boldsymbol{\alpha}| \leq m$, $\partial^{\boldsymbol{\alpha}}(u_\varepsilon\downarrow_{\Omega_1}) = (\partial^{\boldsymbol{\alpha}} u_\varepsilon)\downarrow_{\Omega_1} \to \partial^{\boldsymbol{\alpha}}(u\downarrow_{\Omega_1}) = (\partial^{\boldsymbol{\alpha}} u)\downarrow_{\Omega_1}$ in $L^2(\Omega_1)$. Then, using the first part of this proof, $u\downarrow_{\Omega_1} \in H^{*m}(\Omega_1) \Longrightarrow (uv)\downarrow_{\Omega_1} = u\downarrow_{\Omega_1} \cdot v\downarrow_{\Omega_1} \in H^m(\Omega_1)$ and $\partial^{\boldsymbol{\alpha}}[(uv)\downarrow_{\Omega_1}] = \sum_{\boldsymbol{\beta}\leq\boldsymbol{\alpha}} \binom{\boldsymbol{\alpha}}{\boldsymbol{\beta}} \partial^{\boldsymbol{\beta}}(u\downarrow_{\Omega_1})\partial^{\boldsymbol{\alpha}-\boldsymbol{\beta}}(v\downarrow_{\Omega_1})$. But $\partial^{\boldsymbol{\beta}}(u\downarrow_{\Omega_1}) = (\partial^{\boldsymbol{\beta}} u)\downarrow_{\Omega_1}$ and $\partial^{\boldsymbol{\alpha}}[(uv)\downarrow_{\Omega_1}] = [\partial^{\boldsymbol{\alpha}}(uv)]\downarrow_{\Omega_1}$ and $v \in C^m(\Omega)$

$$\Longrightarrow \quad \partial^{\boldsymbol{\alpha}}(uv) = \sum_{\boldsymbol{\beta}\leq\boldsymbol{\alpha}} \binom{\boldsymbol{\alpha}}{\boldsymbol{\beta}} \partial^{\boldsymbol{\beta}} u \partial^{\boldsymbol{\alpha}-\boldsymbol{\beta}} v \quad \text{in } \Omega_1, \tag{6.8.53}$$

which holds $\forall\Omega_1$ with $\overline{\Omega}_1 \subset\subset \Omega$. But the right-hand side of (6.8.53) belongs to $L^2(\Omega)$, since $\partial^{\boldsymbol{\alpha}-\boldsymbol{\beta}} v$ is bounded in Ω, and $\partial^{\boldsymbol{\beta}} u \in L^2(\Omega)$ $\forall|\boldsymbol{\beta}| \leq m$. Thus, $\forall\overline{\Omega}_1 \subset\subset \Omega$, $uv \in H^m(\Omega_1)$, i.e. $uv \in H^m_{\text{loc}}(\Omega) = \{w : w\downarrow_{\overline{\Omega}_1} \in H^m(\Omega_1)\ \forall\overline{\Omega}_1 \subset\subset \Omega\}$. Since the relation (6.8.53) holds $\forall\Omega_1$ with $\overline{\Omega}_1 \subset\subset \Omega$, it will also hold in Ω, i.e. $\forall|\boldsymbol{\alpha}| \leq m$

$$\partial^{\boldsymbol{\alpha}}(uv) = \sum_{\boldsymbol{\beta}\leq\boldsymbol{\alpha}} \binom{\boldsymbol{\alpha}}{\boldsymbol{\beta}} \partial^{\boldsymbol{\beta}} u \partial^{\boldsymbol{\alpha}-\boldsymbol{\beta}} v \quad \text{in } \Omega. \tag{6.8.54}$$

Hence, $\partial^{\boldsymbol{\alpha}}(uv) \in L^2(\Omega)$ $\forall|\boldsymbol{\alpha}| \leq m$ and $uv \in H^m(\Omega)$ with (6.8.54) (see also Agmon for more details [39, p. 10]). □

Density result

Theorem 6.8.8. *The set of functions $u \in H^m(\Omega)$ with compact support in $\overline{\Omega}$ is* dense *in $H^m(\Omega)$.*

Proof. Let $u \in H^m(\Omega)$ with $\operatorname{supp}(u) \subset\subset \overline{\Omega}$. Let $\psi \in \mathcal{D}(\mathbb{R}^n)$ be a cut-off function with the properties: $0 \leq \psi(\mathbf{x}) \leq 1\ \forall \mathbf{x} \in \mathbb{R}^n$ and

$$\psi(\mathbf{x}) = \begin{cases} 1 & \text{for } \|\mathbf{x}\| \leq 1 \\ 0 & \text{for } \|\mathbf{x}\| \geq 2, \end{cases} \tag{6.8.55}$$

as shown in Figure 6.3 for $n = 1$.

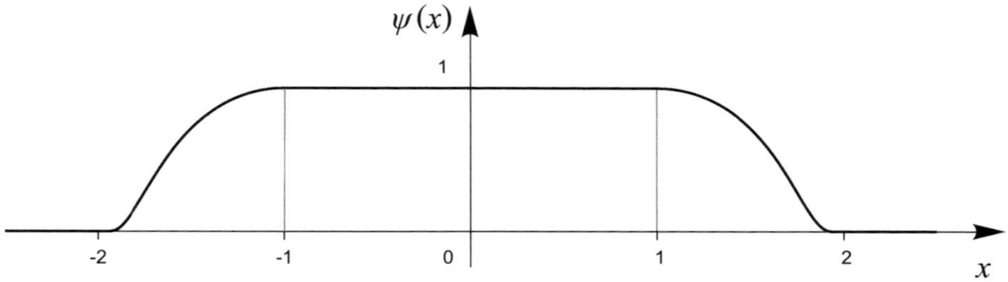

Figure 6.3 Cut-off function ψ on $\mathbb{R}$

Define a sequence $(u_k)_{k=1}^{\infty}$ with $u_k(\mathbf{x}) = \psi(\frac{\mathbf{x}}{k})u(\mathbf{x})\ \forall k \in \mathbb{N}$. Then, by Leibniz's rule, $u_k \in H^m(\Omega)$ and u_k has compact support in $\overline{\Omega}\ \forall$ fixed $k \in \mathbb{N}$. But $\forall |\boldsymbol{\alpha}| \leq m$,

$$\begin{aligned}
\partial^{\boldsymbol{\alpha}} u_k &= \sum_{\boldsymbol{\beta} \leq \boldsymbol{\alpha}} \binom{\boldsymbol{\alpha}}{\boldsymbol{\beta}} \partial^{\boldsymbol{\beta}} \left[\psi\left(\frac{\mathbf{x}}{k}\right) \right] \partial^{\boldsymbol{\alpha}-\boldsymbol{\beta}} u \\
&= \sum_{\substack{\boldsymbol{\beta} \leq \boldsymbol{\alpha} \\ \boldsymbol{\beta} \neq \mathbf{0}}} \binom{\boldsymbol{\alpha}}{\boldsymbol{\beta}} \frac{1}{k^{|\boldsymbol{\beta}|}} (\partial^{\boldsymbol{\beta}} \psi)\left(\frac{\mathbf{x}}{k}\right) \partial^{\boldsymbol{\alpha}-\boldsymbol{\beta}} u + \psi\left(\frac{\mathbf{x}}{k}\right) \partial^{\boldsymbol{\alpha}} u
\end{aligned}$$

$$\Longrightarrow \quad \partial^{\boldsymbol{\alpha}} u_k - \partial^{\boldsymbol{\alpha}} u = \sum_{\substack{\boldsymbol{\beta} \leq \boldsymbol{\alpha} \\ \boldsymbol{\beta} \neq \mathbf{0}}} \binom{\boldsymbol{\alpha}}{\boldsymbol{\beta}} \frac{1}{k^{|\boldsymbol{\beta}|}} (\partial^{\boldsymbol{\beta}} \psi)\left(\frac{\mathbf{x}}{k}\right) \partial^{\boldsymbol{\alpha}-\boldsymbol{\beta}} u + \left[\psi\left(\frac{\mathbf{x}}{k}\right) - 1 \right] \partial^{\boldsymbol{\alpha}} u.$$

Applying the triangular inequality, $|\boldsymbol{\alpha}| \leq m$,

$$\|\partial^{\boldsymbol{\alpha}} u_k - \partial^{\boldsymbol{\alpha}} u\|_{L^2(\Omega)} \leq \Bigg(\sum_{\substack{\boldsymbol{\beta} \leq \boldsymbol{\alpha} \\ \boldsymbol{\beta} \neq \mathbf{0}}} \binom{\boldsymbol{\alpha}}{\boldsymbol{\beta}} \frac{1}{k} \max_{\mathbf{x} \in \mathbb{R}^n} |\partial^{\boldsymbol{\beta}} \psi(\mathbf{x})| \|\partial^{\boldsymbol{\alpha}-\boldsymbol{\beta}} u\|_{L^2(\Omega)}$$

$$+ \left[\int_{\|\mathbf{x}\|>k} |\partial^{\boldsymbol{\alpha}} u(\mathbf{x})|^2 d\mathbf{x} \right]^{1/2} \Bigg) \to 0 \quad \text{as } k \to \infty, \tag{6.8.56}$$

i.e. the right-hand side tends to 0 as $k \to \infty$, since $\psi(\frac{\mathbf{x}}{k}) = 1$ for $\|\mathbf{x}\| \leq k \implies \psi(\frac{\mathbf{x}}{k}) - 1 = 0$ for $\|\mathbf{x}\| \leq k$ and $|\psi(\frac{\mathbf{x}}{k}) - 1| \leq 1$ for $\|\mathbf{x}\| > k$, $\frac{1}{k^{|\boldsymbol{\beta}|}} \leq \frac{1}{k}$ $\forall |\boldsymbol{\beta}|$, $\forall k \in \mathbb{N}$, $\operatorname{supp}(u) \subset\subset \overline{\Omega} \implies$ for all *sufficiently* large $k \in \mathbb{N}$, $\partial^{\boldsymbol{\alpha}} u(\mathbf{x}) = 0$ for $\|\mathbf{x}\| > k$ $\forall |\boldsymbol{\alpha}| \leq m \implies \int_{\|\mathbf{x}\|>k} |\partial^{\boldsymbol{\alpha}} u(\mathbf{x})|^2 d\mathbf{x} = 0$ for all *sufficiently* large $k \in \mathbb{N}$. Hence, $u_k \to u$ in $H^m(\Omega)$ as $k \to \infty$, i.e. $\forall u \in H^m(\Omega)$, $\exists u_k = \psi_k u \in H^m(\Omega)$ with $\operatorname{supp}(u_k) \subset\subset \overline{\Omega}$, such that $u_k \to u$ in $L^2(\Omega)$ and $\partial^{\boldsymbol{\alpha}} u_k \to \partial^{\boldsymbol{\alpha}} u$ in $L^2(\Omega)$ $\forall |\boldsymbol{\alpha}| \leq m \implies u_k \to u$ in $H^m(\Omega)$ as $k \to \infty$. □

Theorem 6.8.9. *$\mathcal{D}(\mathbb{R}^n)$ is* dense *in $W^{m,p}(\mathbb{R}^n)$ $\forall m \in \mathbb{N}$, $1 \leq p < \infty$.*

Proof. The subspace $W_0 = \{u : u \in W^{m,p}(\mathbb{R}^n), \operatorname{supp}(u) \subset\subset \mathbb{R}^n\}$ *is dense in* $W^{m,p}(\mathbb{R}^n)$ $\forall m \in \mathbb{N}$, $1 \leq p < \infty$: Let $u \in W^{m,p}(\mathbb{R}^n)$, and $\psi \in \mathcal{D}(\mathbb{R}^n)$ be the cut-off function in (6.8.55). Define $u_k(\mathbf{x}) = \psi(\mathbf{x}/k)u(\mathbf{x})$ $\forall k \in \mathbb{N}$. Then $(u_k)_{k \in \mathbb{N}}$ is a sequence in W_0, $\operatorname{supp}(u_k) \subset\subset \mathbb{R}^n$ $\forall k \in \mathbb{N}$. Following the steps of the proof of Theorem 6.8.8, replacing 'Ω' by '$\mathbb{R}^n$', '$L^2(\Omega)$' by '$L^p(\mathbb{R}^n)$', $1 \leq p < \infty$, in (6.8.56), and introducing necessary minor modifications, we get $u_k \to u$ in $W^{m,p}(\mathbb{R}^n)$ with $m \in \mathbb{N}$, $1 \leq p < \infty$, i.e. $\forall u \in W^{m,p}(\mathbb{R}^n)$, $\exists (u_k)_{k \in \mathbb{N}}$ in W_0 such that $u_k \to u$ in $W^{m,p}(\mathbb{R}^n)$. Hence, W_0 is *dense* in $W^{m,p}(\mathbb{R}^n)$ with $m \in \mathbb{N}$, $1 \leq p < \infty$.
$\mathcal{D}(\mathbb{R}^n)$ is dense in W_0: Let $v \in W_0$, i.e. $v \in W^{m,p}(\mathbb{R}^n)$ with $\operatorname{supp}(v) \subset\subset \mathbb{R}^n$, and $(\rho_k)_{k \in \mathbb{N}}$ be a sequence of regularizing functions as defined in (6.2.2) with $\operatorname{supp}(\rho_k) = \bar{B}(\mathbf{0}; \frac{1}{k})$ $\forall k \in \mathbb{N}$. Define $\phi_k = \rho_k * v$ with $v \in W_0$ $\forall k \in \mathbb{N}$. But $v \in W^{m,p}(\mathbb{R}^n) \implies v \in L^1_{\text{loc}}(\mathbb{R}^n)$ and $v \in W_0 \implies \operatorname{supp}(v)$ is *compact* in $\mathbb{R}^n$. Then $\phi_k \in \mathcal{D}(\mathbb{R}^n)$ $\forall k \in \mathbb{N}$ by (6.2.38). $v \in W_0 \implies v \in L^p(\mathbb{R}^n)$, $\partial^{\boldsymbol{\alpha}} v \in L^p(\mathbb{R}^n)$ for $1 \leq |\boldsymbol{\alpha}| \leq m \implies \phi_k = \rho_k * v \to v$ in $L^p(\mathbb{R}^n)$ and $\partial^{\boldsymbol{\alpha}} \phi_k = \partial^{\boldsymbol{\alpha}}(\rho_k * v) = \rho_k * \partial^{\boldsymbol{\alpha}} v \to \partial^{\boldsymbol{\alpha}} v$ in $L^p(\mathbb{R}^n)$ for $1 \leq |\boldsymbol{\alpha}| \leq m$ as $k \to \infty$ (by (6.2.40) with $\varepsilon = \frac{1}{k} \to 0$); the second equality $\partial^{\boldsymbol{\alpha}}(\rho_k * v) = \rho_k * \partial^{\boldsymbol{\alpha}} v$ follows from (6.7.4). Hence, for $v \in W_0$, $\exists (\phi_k)_{k \in \mathbb{N}}$ in $\mathcal{D}(\mathbb{R}^n)$ such that $\|v - \phi_k\|^p_{m,p,\mathbb{R}^n} = [\|v - \phi_k\|^p_{L^p(\mathbb{R}^n)} + \sum_{1 \leq |\boldsymbol{\alpha}| \leq m} \|\partial^{\boldsymbol{\alpha}} v - \partial^{\boldsymbol{\alpha}} \phi_k\|^p_{L^p(\mathbb{R}^n)}] \to 0$ as $k \to \infty \implies \mathcal{D}(\mathbb{R}^n)$ is *dense* in W_0, and W_0 is *dense* in $W^{m,p}(\mathbb{R}^n)$ (proved earlier). Hence, $\mathcal{D}(\mathbb{R}^n)$ is *dense* in $W^{m,p}(\mathbb{R}^n)$, since $\forall u \in W^{m,p}(\mathbb{R}^n)$, $\forall \varepsilon > 0$, $\exists v_\varepsilon \in W_0$ and $\exists \phi_\varepsilon \in \mathcal{D}(\mathbb{R}^n)$ such that $\|u - v_\varepsilon\|_{m,p,\mathbb{R}^n} < \varepsilon/2$ and $\|v_\varepsilon - \phi_\varepsilon\|_{m,p,\mathbb{R}^n} < \varepsilon/2 \implies \|u - \phi_\varepsilon\|_{m,p,\mathbb{R}^n} < \varepsilon$. □

6.9 Convolution equations (see also Section 8.7, Chapter 8)

A *convolution equation* is an equation of the form

$$A * T = B, \tag{6.9.1}$$

where A and B are *given* distributions, A being the *coefficient* distribution and B being the right-hand-side distribution, and T is the *unknown* distribution. Since the convolution of any two distributions is *not necessarily* defined, we make an assumption that

$$A \in \mathcal{E}'(\mathbb{R}^n), \tag{6.9.2}$$

(i.e. A is a distribution with *compact* support in $\mathbb{R}^n$) such that $A * T$ is well defined for arbitrary $T \in \mathcal{D}'(\mathbb{R}^n)$ by Theorem 6.3.2.

Homogeneous (convolution) equation

For $B = 0$,

$$A * T = 0 \text{ is called } \textit{homogeneous}. \tag{6.9.3}$$

Particular cases of convolution equations are:

1. **Linear partial differential equations with constant coefficients**
   For $A = \sum_{|\boldsymbol{\alpha}|\le m} a_{\boldsymbol{\alpha}} \partial^{\boldsymbol{\alpha}} \delta \in \mathcal{E}'(\mathbb{R}^n)$ with $\operatorname{supp}(A) \subset\subset \mathbb{R}^n$, $a_{\boldsymbol{\alpha}} \in \mathbb{R}$, $\partial^{\boldsymbol{\alpha}} = \frac{\partial^{|\boldsymbol{\alpha}|}}{\partial x_1^{\alpha_1} \cdots \partial x_n^{\alpha_n}}$ $\forall$ multi-index $\boldsymbol{\alpha} = (\alpha_1, \alpha_2, \ldots, \alpha_n)$,

   $$A * T = B \Longrightarrow \sum_{|\boldsymbol{\alpha}|\le m} a_{\boldsymbol{\alpha}} \partial^{\boldsymbol{\alpha}} \delta * T = \sum_{|\boldsymbol{\alpha}|\le m} a_{\boldsymbol{\alpha}} \partial^{\boldsymbol{\alpha}} T = B, \tag{6.9.4}$$

   which is a partial differential equation with constant coefficients $a_{\boldsymbol{\alpha}}$ for the unknown distribution T.

2. **Linear difference equations with constant coefficients**
   For $A = \sum_{i=1}^{N} a_{\mathbf{h}_i} \delta_{\mathbf{h}_i} = \sum_{i=1}^{N} a_{\mathbf{h}_i} \delta(\mathbf{x} - \mathbf{h}_i)$ with $\operatorname{supp}(A) = \{\mathbf{h}_1, \mathbf{h}_2, \ldots, \mathbf{h}_N\} \subset\subset \mathbb{R}^n$, $a_{\mathbf{h}_i} \in \mathbb{R}$,

   $$A * T = B \quad \Longrightarrow \quad \sum_{i=1}^{N} a_{\mathbf{h}_i} \delta_{\mathbf{h}_i} * T = \sum_{i=1}^{N} a_{\mathbf{h}_i} \tau_{\mathbf{h}_i} T = B, \tag{6.9.5}$$

   where $\tau_{\mathbf{h}_i} T = T(\mathbf{x} - \mathbf{h}_i)$ by (6.3.28).

3. **Volterra's integral equations of the first and second kinds**

(a) *First kind:* For $x \geq 0$,

$$\int_0^x K(x-t)f(t)dt = g(x); \tag{6.9.6}$$

(b) *Second kind:* For $x \geq 0$,

$$f(x) + \int_0^x K(x-t)f(t)dt = g(x); \tag{6.9.7}$$

where the *kernel* K and the right-hand-side function g are *given* and *locally summable* on $[0, \infty[$, and f is the *unknown* function. We extend f, g, K by 0 for $x < 0$. Then, setting

- $A = K(x)$ (resp. $A = \delta + K(x)$) with $K \in L^1_{\text{loc}}(\mathbb{R})$, $\text{supp}(K) \subset [0, \infty[$;
- $T = f(x)$ with $\text{supp}(f) \subset [0, \infty[$; (6.9.8)
- $B = g(x)$ with $g \in L^1_{\text{loc}}(\mathbb{R})$, $\text{supp}(g) \subset [0, \infty[$;

we get Volterra's integral equations in *convolution form*:

(a′) *First kind:* $(A * f)(x) = (K * f)(x) = \int_0^x K(x-t)f(t)dt = g(x);$ (6.9.9)

(b′) *Second kind:* $(A * f)(x) = ((\delta + K(x)) * f)(x) = g(x)$

$$\Longrightarrow \quad \delta * f + K * f = g \quad \Longrightarrow \quad f(x) + \int_0^x K(x-t)f(t)dt = g(x). \tag{6.9.10}$$

Remark 6.9.1. There are Volterra's equations, which are *not* convolution equations.

For example, $\int_0^x K(x,\xi)f(\xi)d\xi = g(x)$ is Volterra's equation of the *first kind*, where the kernel $K(x,\xi)$ is a function of *two* variables, with $x \geq 0$, $0 \leq \xi \leq x$.

4. **Integro-differential equations**

These can be obtained by combining diverse types of linear differential and integral equations. See (6.10.3) for an example.

Systems of convolution equations

A system of m convolution equations for m *unknown distributions* $T_1, T_2, \ldots, T_m$ is given, for $1 \leq i \leq m$, by:

$$A_{i1} * T_1 + A_{i2} * T_2 + \cdots + A_{im} * T_m = B_i, \tag{6.9.11}$$

where $A_{ij} \in \mathcal{E}'(\mathbb{R}^n)$, $1 \leq i, j \leq m$, are the *given* m^2 *coefficient distributions* such that $A_{ij} * T_j$ is well defined $\forall i, j = 1, 2, \ldots, m$, and $B_1, B_2, \ldots, B_m$ are the *given* right-hand-side distributions.

Elementary solution (see also Section 3.3 and Section 8.7)

Definition 6.9.1. A distribution $E \in \mathcal{D}'(\mathbb{R}^n)$ satisfying the convolution equation (6.9.1) with $B = \delta$, i.e. $A * E = \delta$, is called an *elementary* or *fundamental solution* of the convolution equation. Then E is denoted by A^{*-1} or simply by A^{-1}, i.e.

$$A * E = A * A^{*-1} = A * A^{-1} = A^{-1} * A = \delta. \tag{6.9.12}$$

If A^{-1} (i.e. E) exists, the coefficient distribution A is called *invertible*.

Non-uniqueness of A^{-1} In general, elementary solution $E \in \mathcal{D}'(\mathbb{R}^n)$ is *not* unique. Hence, A^{-1} is *not* unique in general. In fact, E is determined up to an additive distribution $E_0 \in \mathcal{D}'(\mathbb{R}^n)$ satisfying $A * E_0 = 0$ (i.e. the corresponding homogeneous equation).

Indeed,

$$A * (E + E_0) = A * E + A * E_0 = \delta + 0 = \delta. \tag{6.9.13}$$

Non-invertibility of A $\exists$ distributions A which are *not* invertible, i.e. for which A^{-1} does not exist. For example, let $A = \phi \in \mathcal{D}(\mathbb{R}^n)$. Then $A \in \mathcal{D}'(\mathbb{R}^n)$ and $A * T = \phi * T = T * \phi = \psi$ is well defined $\forall T \in \mathcal{D}'(\mathbb{R}^n)$. But ψ is a C^∞-function by Definition 6.3.1A and Remark 6.3.1 (see also Theorem 6.4.1). Hence, for $A = \phi$, $A * T$ can *never* be Dirac distribution δ, i.e. $A = \phi \in \mathcal{D}(\mathbb{R}^n) \subset \mathcal{D}'(\mathbb{R}^n)$ is *not* invertible.

Example 6.9.1. For $n = 1$, $A = \frac{d\delta}{dx} = \delta' \in \mathcal{E}'(\mathbb{R})$ has inverse $A^{-1} \in \mathcal{D}'(\mathbb{R})$, but $A^{-1} \notin \mathcal{E}'(\mathbb{R})$. In fact, $A * E = \delta' * E = \delta * \frac{dE}{dx} = \frac{dE}{dx} = \delta$ (applying first (6.3.24) and then (6.3.21)) $\Longrightarrow E = H(x) \Longrightarrow A^{-1} = H(x)$ (the Heaviside function) with $\operatorname{supp}(A^{-1}) = [0, \infty[$, which is *not* compact in $\mathbb{R} \Longrightarrow A^{-1} \notin \mathcal{E}'(\mathbb{R})$.

Example 6.9.2. For $n = 3$, $A = \Delta\delta \in \mathcal{E}'(\mathbb{R}^3)$ with $\Delta = \frac{\partial^2}{\partial x_1^2} + \frac{\partial^2}{\partial x_2^2} + \frac{\partial^2}{\partial x_3^2}$ (the Laplace operator), $A * E = \Delta\delta * E = \delta * \Delta E = \Delta E = \delta$ (applying first (6.3.24) and then (6.3.21)) $\Longrightarrow A^{-1} = E = -\frac{1}{4\pi r} \in \mathcal{D}'(\mathbb{R}^3)$ by Theorem 3.3.2. But $\operatorname{supp}(\frac{-1}{4\pi r}) \equiv \mathbb{R}^3 \Longrightarrow A^{-1} \notin \mathcal{E}'(\mathbb{R}^3)$, although $A \in \mathcal{E}'(\mathbb{R}^3)$.

Example 6.9.3. Every polynomial in derivatives is invertible, i.e. corresponding to polynomial $P(\boldsymbol{\xi}) = \sum_{|\boldsymbol{\alpha}| \le m} a_{\boldsymbol{\alpha}} \boldsymbol{\xi}^{\boldsymbol{\alpha}}$, $A = P(\partial)\delta$ with $P(\partial) = \sum_{|\boldsymbol{\alpha}| \le m} a_{\boldsymbol{\alpha}} \partial^{\boldsymbol{\alpha}}$, $\delta \in \mathcal{E}'(\mathbb{R}^n)$, is invertible. In fact, $P(\partial)\delta * E = \delta \Longrightarrow P(\partial)E = \delta$ has a solution by Theorem 6.8.1.

Methods for constructing elementary solutions can be found in Chapter 8, Section 8.7 and Chapter 3, Section 3.3.

Convolution algebra $\mathcal{A}$

The discussions on convolution equations and their solutions, elementary solutions and the inverses $A^{*-1} = A^{-1}$ suggest the study of convolution equations and their solutions in a special subspace $\mathcal{A}$ of $\mathcal{D}'(\mathbb{R}^n)$, called a *convolution algebra*, which we will now define, following [7].

Definition 6.9.2. A subspace $\mathcal{A} \subset \mathcal{D}'(\mathbb{R}^n)$, i.e. $T_1, T_2 \in \mathcal{A} \Longrightarrow T_1, T_2 \in \mathcal{D}'(\mathbb{R}^n)$ and $\alpha_1 T_1 + \alpha_2 T_2 \in \mathcal{A}\ \forall \alpha_1, \alpha_2 \in \mathbb{C}$, is called a *convolution algebra* if and only if

- $\delta \in \mathcal{A}$ (i.e. the Dirac distribution δ is an element of $\mathcal{A}$); (6.9.14a)
- $T_1, T_2, \dots, T_m \in \mathcal{A} \Longrightarrow T_1 * T_2 * \dots * T_m$ is defined and belongs to $\mathcal{A}$ with

$$T_i * T_j = T_j * T_i \tag{6.9.14b}$$

in $\mathcal{A}$, $1 \le i, j \le m$;

$$T_i * T_j * T_k = T_i * (T_j * T_k) = (T_i * T_j) * T_k \tag{6.9.14c}$$

in $\mathcal{A}$, $1 \le i, j, k \le m$, and so on.

Examples of convolution algebras are:

1. $\mathcal{A} = \mathcal{E}'(\mathbb{R}^n)$ (see Chapter 5, Section 5.6); (6.9.15a)
2. $\mathcal{A} = \mathcal{D}'^{+} = \{T \in \mathcal{D}'(\mathbb{R}) : \operatorname{supp}(T) \subset [0, \infty[\}$ (see Definition 8.8.2); (6.9.15b)
3. $\mathcal{A} = \mathcal{D}'(\Gamma)$ = the space of distributions on a circle Γ (see Section 1.10 and Section 6.7). (6.9.15c)

$\mathcal{D}'(\mathbb{R}^n)$ is *not* a convolution algebra, since (6.9.15b)–(6.9.15c) do *not* hold in general.

Consider the convolution equation (6.9.1) in $\mathcal{A}$, i.e. for *given* $A, B \in \mathcal{A}$, find $T \in \mathcal{A}$ such that

$$A * T = B. \tag{6.9.16}$$

Theorem 6.9.1. *For given $A \in \mathcal{A}$ and arbitrary $B \in \mathcal{A}$, the convolution equation (6.9.16) in $\mathcal{A}$ has at least one solution $T \in \mathcal{A}$ if and only if A has an inverse $A^{*-1} = A^{-1} \in \mathcal{A}$, i.e. $A * E = A * A^{-1} = A^{-1} * A = \delta \in \mathcal{A}$. Then $A^{-1} \in \mathcal{A}$ is unique, and the* unique *solution $T \in \mathcal{A}$ of (6.9.16) is given by $T = A^{-1} * B$.*

Proof. Existence of $A^{-1} \in \mathcal{A}$: Suppose that there exists *at least* one solution $T \in \mathcal{A}$ of (6.9.16) for *arbitrary* $B \in \mathcal{A}$. Then, for $B = \delta \in \mathcal{A}$, $\exists$ a solution $T = E \in \mathcal{A}$ such that $A * E = \delta \Longrightarrow A$ has an inverse $A^{-1} \in \mathcal{A}$.

Uniqueness of $A^{-1} \in \mathcal{A}$: Suppose that $A * E = \delta$ and $A * E_1 = \delta$. Then $A * (E - E_1) = 0 \Longrightarrow A * E_0 = 0$ with $E_0 = E - E_1 \in \mathcal{A}$. But $E_0 = \delta * E_0 =$

$(A^{-1} * A) * E_0 = A^{-1} * (A * E_0) = A^{-1} * 0 = 0 \Longrightarrow E = E_1 \Longrightarrow A^{-1} = E \in \mathcal{A}$ is unique.

Conversely, suppose that $A^{-1} \in \mathcal{A}$ exists with $A * A^{-1} = A^{-1} * A = \delta \in \mathcal{A}$. Then, by taking the convolution of both sides of (6.9.16) with $A^{-1} \in \mathcal{A}$, we get $A^{-1} * A * T = A^{-1} * B$ in $\mathcal{A} \Longrightarrow \delta * T = A^{-1} * B \Longrightarrow T = A^{-1} * B \in \mathcal{A}$.

Again, $T = A^{-1} * B \Longrightarrow A * T = A * A^{-1} * B = \delta * B = B \Longrightarrow T = A^{-1} * B$ is the solution of (6.9.16). Hence, if, for $A \in \mathcal{A}$, $A^{-1} \in \mathcal{A}$ exists, $A * T = B \Longleftrightarrow T = A^{-1} * B \in \mathcal{A}$, i.e. equation (6.9.16) has a unique solution $T = A^{-1} * B \in \mathcal{A}$. □

For solving convolution equation (6.9.16) in $\mathcal{A}$, the essential problem is to find an elementary solution $E \in \mathcal{A}$. (6.9.17)

Application Let $E(x) = H(x)y(x)$ be an elementary solution of the differential operator $P(D) \equiv L_m$ derived in (3.3.25)–(3.3.30), H being the Heaviside function and y being the unique solution of the Cauchy problem $P(D)y = 0$ with the initial conditions $y^{(k)}(0) = 0$, $0 \leq k \leq m-2$, $y^{(m-1)}(0) = 1$.

Then the solution $z = z(x)$ of the corresponding *non-homogeneous* equation $P(D)z = f$ and satisfying the initial conditions $z^{(k)}(0) = z_{k,0}$, $0 \leq k \leq m-1$, where f is a given *sufficiently regular* function, $\{z_{k0}\}_{k=0}^{m-1}$ are *arbitrary* given real numbers, is given by:

$$Hz = Hy * \left[Hf + \sum_{k=0}^{m-1} \gamma_k \delta^{(k)} \right],$$

with

$$\gamma_k = z_{m-k-1,0} + a_1 z_{m-k-2,0} + \cdots + a_{m-k-1} z_{00}, \qquad \delta^{(k)} = \frac{d^k \delta}{dx^k},$$

or, for $x \geq 0$,

$$z(x) = \int_0^x y(x-\xi) f(\xi) d\xi + \sum_{k=0}^{m-1} \gamma_k y^{(k)}. \tag{6.9.18}$$

Proof. $E = Hy \in \mathcal{D}'(\mathbb{R})$ is an elementary solution of $P(D) \equiv L_m$ in (3.3.25)–(3.3.30) with $\operatorname{supp}(E) = \operatorname{supp}(Hy) \subset [0, \infty[\Longrightarrow E = Hy \in \mathcal{D}'^{+}$ and $P(D)E = \delta \Longrightarrow P(D)\delta * E = \delta \Longrightarrow [P(D)\delta]^{-1} = E = Hy \in \mathcal{D}'^{+}$. Moreover, following the steps of the proof of (3.3.29) and using initial values $z_{k,0} = z^{(k)}(0)$, we get

$$\begin{aligned}
(Hz)^{(1)} &= Hz^{(1)} + \delta z(0) = Hz^{(1)} + \delta z_{0,0}; \\
(Hz)^{(2)} &= Hz^{(2)} + \delta z_{1,0} + \delta^{(1)} z_{0,0}; \\
&\ \vdots \\
(Hz)^{(k)} &= Hz^{(k)} + \delta z_{k-1,0} + \delta^{(1)} z_{k-2,0} + \cdots + \delta^{(k-1)} z_{0,0}, \quad 1 \leq k \leq m.
\end{aligned}$$

Then

$$\begin{aligned}P(D)(Hz) = {} & H[z^{(m)} + a_1 z^{(m-1)} + \cdots + a_m z] \\ & + \delta[z_{m-1,0} + a_1 z_{m-2,0} + \cdots + a_{m-1} z_{0,0}] \\ & + \cdots + \delta^{(k)}[z_{m-k-1,0} + a_1 z_{m-k-2,0} + \cdots + a_{m-k-1} z_{0,0}] \\ & + \cdots + \delta^{(m-1)}[z_{0,0}]\end{aligned}$$

with $a_0 = 1$,

$$\begin{aligned}\implies \quad P(D)(Hz) = {} & H \underbrace{[P(D)z]}_{f} \\ & + \sum_{k=0}^{m-1} \underbrace{[z_{m-k-1,0} + a_1 z_{m-k-2,0} + \cdots + a_{m-k-1} z_{0,0}]}_{\gamma_k} \delta^{(k)},\end{aligned}$$

with $\delta^{(0)} = \delta = Hf + \sum_{k-0}^{m-1} \gamma_k \delta^{(k)}$ with $\gamma_k \in \mathbb{R}$. Hence, in the notations of Theorem 6.9.1, the convolution algebra $\mathcal{A} = \mathcal{D}'^{+}$, $A = P(D)\delta \in \mathcal{D}'^{+}$, $A^{-1} = [P(D)\delta]^{-1} = E = Hy \in \mathcal{D}'^{+}$, $T = Hz$, $B = Hf + \sum_{k=0}^{m-1} \gamma_k \delta^{(k)} \in \mathcal{D}'^{+}$. Then, by Theorem 6.9.1, $T = Hz = A^{-1} * B = Hy * [Hf + \sum_{k=0}^{m-1} \gamma_k \delta^{(k)}] \in \mathcal{D}'^{+}$ (see Proposition 8.8.2) $\implies Hz = Hy * Hf + \sum_{k=0}^{m-1} \gamma_k Hy * \delta^{(k)}$. Then, for $x \geq 0$, $z(x) = \int_0^x y(x-\xi) f(\xi) d\xi + \sum_{k=0}^{m-1} \gamma_k y^{(k)}$, since $H(\xi) = 0$ for $\xi < 0$, $H(x-\xi) = 0$ for $\xi > x \geq 0$

$$\begin{aligned}\implies \quad Hy * Hf & = \int_{-\infty}^{\infty} H(x-\xi) \cdot y(x-\xi) H(\xi) f(\xi) d\xi \\ & = \int_0^{\infty} H(x-\xi) y(x-\xi) f(\xi) d\xi = \int_0^x y(x-\xi) f(\xi) d\xi,\end{aligned}$$

and $Hy * \delta^{(k)} = \delta^{(k)} * Hy = \delta * (Hy)^{(k)} = Hy^{(k)} = y^{(k)}$ for $x \geq 0$ and for $0 \leq k \leq m-1$ (by (3.3.29), with $y^{(0)} = y$, ($y^{(k)}(0) = 0$ for $0 \leq k \leq m-2$ by virtue of the initial conditions). □

Remark 6.9.2. Consider Example 6.9.2 with $\mathcal{A} \equiv \mathcal{E}'(\mathbb{R}^3)$, $A = \Delta\delta \in \mathcal{A}$. But $A^{-1} = -\frac{1}{4\pi r} \notin \mathcal{A}$. Hence, Theorem 6.9.1 is *not* applicable and we cannot deduce $T = A^{-1} * B$ from $A * T = B$. In fact, $A^{-1} * A * T$ is not meaningful if T does not have *compact* support (see Theorem 6.6.1). Now we consider the following particular cases:

Case I: supp(T) is *compact*. For $A = \Delta\delta$, if T has compact support, then $B = A * T$ has compact support (see (6.3.19)). Now, if T has compact support and satisfies $A * T = B$, then $T = A^{-1} * B$, which is well defined. But Theorem 6.9.1 is not applicable (since $A^{-1} \notin \mathcal{A}$), and consequently we cannot prove that the solution

is unique. In fact, it is *not* unique, since the corresponding homogeneous equation $\Delta\delta * T = \Delta T = 0$ has an *infinite* number of solutions, for example *harmonic functions* with *non-compact* support.
Case II: supp(B) is *compact*. Then $T = A^{-1} * B$ is well defined, from which the equation $A * T = B$ can be deduced. Hence, $\Delta\delta * T = \Delta T = B$ has a particular solution $T = -\frac{1}{4\pi r} * B$, where B has compact support. Then the general solution is given by $T = -\frac{1}{4\pi r} * B +$ harmonic distributions.
Case III: supp(B) is *not* compact. In this case, we can *not* prove the existence of a solution by this method.

Remark 6.9.3. If A^{-1} does *not* exist, then for $B = \delta$, $A * T = \delta$ has *no* solution. But for $B \neq \delta$, there may exist only one or more than one solution (see Schwartz [7, p. 125]).

Proposition 6.9.1. *Let $\mathcal{A}$ be a convolution algebra. If $A_1, A_2 \in \mathcal{A}$ have inverses $A_1^{-1}, A_2^{-1} \in \mathcal{A}$, then $(A_1 * A_2)^{-1} \in \mathcal{A}$ and $(A_1 * A_2)^{-1} = A_1^{-1} * A_2^{-1} = A_2^{-1} * A_1^{-1}$.*

Proof. $A_1, A_2 \in \mathcal{A}$ have inverses in $\mathcal{A} \Longrightarrow A_1 * A_1^{-1} = \delta$, $A_2 * A_2^{-1} = \delta$. Then $(A_1 * A_2) * (A_1^{-1} * A_2^{-1}) = (A_1 * A_1^{-1}) * (A_2 * A_2^{-1}) = \delta * \delta = \delta \in \mathcal{A} \Longrightarrow$ $(A_1 * A_2)^{-1} = A_1^{-1} * A_2^{-1} = A_2^{-1} * A_1^{-1}$. □

Example 6.9.4. Let $P(D)$ be the differential operator defined by (6.7.6) and $P(D)\delta$ be defined by (6.7.7). Then $[P(D)\delta]^{-1} = H(x)e^{\lambda_1 x} * H(x)e^{\lambda_2 x} * \cdots * H(x)e^{\lambda_m x}$, where H is the Heaviside function.

In particular, for $\lambda_1 = \lambda_2 = \cdots = \lambda_m = \lambda$, $[P(D)\delta]^{-1} = H(x)e^{\lambda x} \cdot \frac{x^{m-1}}{(m-1)!}$.

Proof. $\delta \in \mathcal{D}'(\mathbb{R})$ and supp$(\delta) \subset [0, \infty[\ \Longrightarrow \delta \in \mathcal{D}'^{+}$, which is a convolution algebra $\mathcal{A}$ (see (6.9.15b) $\Longrightarrow P(D)\delta \in \mathcal{D}'^{+}$. But, from (6.7.7),

$$P(D)\delta = \left(\frac{d\delta}{dx} - \lambda_1\delta\right) * \left(\frac{d\delta}{dx} - \lambda_2\delta\right) * \cdots * \left(\frac{d\delta}{dx} - \lambda_m\delta\right) \in \mathcal{D}'^{+},$$

each $(\frac{d\delta}{dx} - \lambda_k\delta)$ being an element of $\mathcal{D}'^{+}$ (see also Section 8.8). From (3.3.24)–(3.3.30), $E(x) = H(x)y(x) \in \mathcal{D}'(\mathbb{R})$, where H is the Heaviside function and y is a C^{∞}-function which is the unique solution of Cauchy problem $P(D)y = 0$ with $y(0) = y'(0) = \cdots = y^{(m-2)}(0) = 0$, $y^{(m-1)}(0) = 1$, is an elementary solution of $P(D)$, i.e.

$$P(D)E = \delta \quad \Longrightarrow \quad P(D)\delta * E = \delta \quad \Longrightarrow \quad [P(D)\delta]^{-1} = E = Hy \in \mathcal{D}'(\mathbb{R}),$$

with supp$([P(D)\delta]^{-1}) =$ supp$(Hy) \subset [0, \infty[\ \Longrightarrow [P(D)\delta]^{-1} \in \mathcal{D}'^{+}$. Moreover, from (3.3.31), the elementary solution of $(\frac{d}{dx} - \lambda_k)$ is $H(x)e^{\lambda_k x}$, i.e. $(\frac{d}{dx} - \lambda_k)E = \delta \Longrightarrow (\frac{d\delta}{dx} - \lambda_k\delta) * E = (\frac{d}{dx} - \lambda_k)E = \delta \Longrightarrow (\frac{d\delta}{dx} - \lambda_k\delta)^{-1} = H(x)e^{\lambda_k x} \in \mathcal{D}'^{+}$.

Now, using (6.7.7) and repeatedly applying Proposition 6.9.1, we have

$$\begin{aligned}[P(D)\delta]^{-1} &= \left[\left(\frac{d\delta}{dx} - \lambda_1\delta\right) * \left(\frac{d\delta}{dx} - \lambda_2\delta\right) * \cdots * \left(\frac{d\delta}{dx} - \lambda_m\delta\right)\right]^{-1} \\ &= \left(\frac{d\delta}{dx} - \lambda_1\delta\right)^{-1} * \left(\frac{d\delta}{dx} - \lambda_2\delta\right)^{-1} * \cdots * \left(\frac{d\delta}{dx} - \lambda_m\delta\right)^{-1} \\ &= H(x)e^{\lambda_1 x} * H(x)e^{\lambda_2 x} * \cdots * H(x)e^{\lambda_m x}.\end{aligned}$$

In particular, for $\lambda_1 = \lambda_2 = \cdots = \lambda_m = \lambda$, from (6.6.3) and (6.7.8), we have

$$\begin{aligned}[P(D)\delta]^{-1} &= \left(\frac{d\delta}{dx} - \lambda\delta\right)^{-*m} \\ &= \underbrace{\left(\frac{d\delta}{dx} - \lambda\delta\right)^{-1} * \left(\frac{d\delta}{dx} - \lambda\delta\right)^{-1} * \cdots * \left(\frac{d\delta}{dx} - \lambda\delta\right)^{-1}}_{m \text{ distributions}} \\ &= H(x)e^{\lambda x}\frac{x^{m-1}}{(m-1)!}.\end{aligned}$$

□

Systems of convolution equations in $\mathcal{A}$

For the system of convolution equations with $A_{ij} \in \mathcal{A}$, $B_i \in \mathcal{A}$, $1 \le i, j \le m$, it will be convenient to use matrix calculus and determinants of distributions in $\mathcal{A}$, with necessary modifications as follows.

Convolution matrix product $[A] * [B]$ in $\mathcal{A}$ Let $[A] = (A_{ij})_{1\le i,j\le m}$, $[B] = (B_{ij})_{1\le i,j\le m}$, with $A_{ij}, B_{ij} \in \mathcal{A}$ such that $A_{ik} * B_{kj} \in \mathcal{A}$ $\forall i, j, k = 1, 2, \ldots, n$ (by virtue of (6.9.15b)). Then we define $[A] * [B]$ as the matrix $[C] = (C_{ij})_{1\le i,j\le m} = [A] * [B]$ with

$$C_{ij} = \sum_{k=1}^{m} A_{ik} * B_{kj} \in \mathcal{A} \quad \forall i, j = 1, 2, \ldots, m, \tag{6.9.19}$$

where the multiplication operation '$\cdot$' in the usual matrix multiplication formula is replaced by convolution operation '$*$' [7].

$$[A] * [B] \neq [B] * [A] \quad \text{in general.} \tag{6.9.20}$$

For example, for $m = 2$, $[A] = \left[\begin{smallmatrix} A_{11} & A_{12} \\ A_{21} & A_{22} \end{smallmatrix}\right]$, $[B] = \left[\begin{smallmatrix} B_{11} & B_{12} \\ B_{21} & B_{22} \end{smallmatrix}\right]$,

$$[C] = [A] * [B] = \begin{bmatrix} C_{11} & C_{12} \\ C_{21} & C_{22} \end{bmatrix}$$
$$= \begin{bmatrix} A_{11} * B_{11} + A_{12} * B_{21} & A_{11} * B_{12} + A_{12} * B_{22} \\ A_{21} * B_{11} + A_{22} * B_{21} & A_{21} * B_{12} + A_{22} * B_{22} \end{bmatrix}, \quad (6.9.21)$$

where $A_{ij}, B_{ij}, A_{ik} * B_{kj} \in \mathcal{A}$, $i, j, k = 1, 2$.

Convolution determinant $\boldsymbol{\Delta^*(A)}$ of $\boldsymbol{[A]}$ in $\boldsymbol{\mathcal{A}}$ The *convolution determinant* $\det^*(A) = \Delta^*(A)$ of $[A] = (A_{ij})_{1 \le i,j \le m}$ with $A_{ij} \in \mathcal{A}$ is a distribution in $\mathcal{A}$ which is obtained by opening the determinant $\Delta^*(A)$ in the usual way and replacing multiplication '$\cdot$' by convolution '$*$', so that $\Delta^*(A) \in \mathcal{A}$ by (6.9.15b). For example, for $[A]$ in (6.9.21),

$$\Delta^*(A) = \begin{vmatrix} A_{11} & A_{12} \\ A_{21} & A_{22} \end{vmatrix}^* = A_{11} * A_{22} - A_{12} * A_{21} \in \mathcal{A}, \quad (6.9.22)$$

where multiplication '$\cdot$' is replaced by convolution '$*$'. Then

$$[A] * [B] = [C] \implies \Delta^*(A) * \Delta^*(B) = \Delta^*(C) \in \mathcal{A}. \quad (6.9.23)$$

For the example in (6.9.21), using (6.9.15b)–(6.9.15c), we have

$$\begin{aligned} \Delta^*(A) * \Delta^*(B) &= (A_{11} * A_{22} - A_{12} * A_{21}) * (B_{11} * B_{22} - B_{12} * B_{21}) \\ &= \underbrace{A_{11} * A_{22} * B_{11} * B_{22}}_{(1)} - \underbrace{A_{11} * A_{22} * B_{12} * B_{21}}_{(2)} \\ &\quad - \underbrace{A_{12} * A_{21} * B_{11} * B_{22}}_{(3)} + \underbrace{A_{12} * A_{21} * B_{12} * B_{21}}_{(4)} \\ &= (1) - (2) - (3) + (4) \end{aligned}$$

$$\Delta^*([A] * [B]) = \Delta^*\left(\begin{bmatrix} A_{11} * B_{11} + A_{12} * B_{21} & A_{11} * B_{12} + A_{12} * B_{22} \\ A_{21} * B_{11} + A_{22} * B_{21} & A_{21} * B_{12} + A_{22} * B_{22} \end{bmatrix} \right)$$
$$= (1) - (2) - (3) + (4) + (5) - (5) + (6) - (6) = (1) - (2) - (3) + (4),$$

where $(5) = A_{11} * B_{11} * A_{21} * B_{12}$, $(6) = A_{12} * B_{21} * A_{22} * B_{22}$.
$\implies \Delta^*(A) * \Delta^*(B) = \Delta^*([A] * [B])$ in $\mathcal{A}$.

Convolution cofactors $\boldsymbol{C_{ij}}$ of $\boldsymbol{A_{ij}}$ in $\boldsymbol{\Delta^*(A)}$ These are defined in the usual way with multiplication '$\cdot$' replaced by convolution '$*$', i.e. convolution cofactor C_{ij} of A_{ij} in $\Delta^*(A)$ is given by:

$$C_{ij} = (-1)^{i+j} \Delta^*([A(i \mid j)]) \in \mathcal{A}, \quad (6.9.24)$$

where $[A(i\,|\,j)]$ is the matrix of order $m-1$ obtained from $[A]$ by deleting its ith row and jth column, $\Delta^*([A(i\,|\,j)])$ being its convolution determinant.

For example, for $m = 3$, $[A] = (A_{ij})_{1\le i,j,\le 3}$,

$$C_{23} = \text{convolution cofactor of } A_{23} = (-1)^{2+3} \begin{vmatrix} A_{11} & A_{12} \\ A_{31} & A_{32} \end{vmatrix}^*$$
$$= -(A_{11} * A_{32} - A_{12} * A_{31}) \in \mathcal{A}.$$

Inverse $\Delta^{*-1}(A) \in \mathcal{A}$ of the determinant $\Delta^*(A)$ in $\mathcal{A}$ In convolution algebra $\mathcal{A}$, Dirac distribution $\delta \in \mathcal{A}$ plays the rôle of *unity* (1), since $\delta * T = T * \delta = T$ $\forall T \in \mathcal{A}$ (see (6.3.21)). Hence,

$$\Delta^*(A) \in \mathcal{A} \text{ has an } \textit{inverse } \Delta^{*-1}(A) \in \mathcal{A}$$
$$\Longleftrightarrow \quad \Delta^*(A) * \Delta^{*-1}(A) = \delta. \tag{6.9.25}$$

Convolution inverse $[A^{*-1}]$ of $[A]$ in $\mathcal{A}$ In the convolution matrix product (6.9.19), the diagonal matrix

$$[\delta I] = (\delta\delta_{ij})_{1\le i,j\le m} = \underbrace{\lceil \delta, \delta, \dots, \delta \rfloor}_{m \text{ diagonal elements}}$$

plays the rôle of the identity matrix $[I]$ in usual matrix multiplication, Kronecker delta, $\delta_{ij} = 1$ for $i = j$ and 0 for $i \ne j$, diagonal elements δ being the Dirac distribution. For example, for $m = 2$, $[\delta I] = \left[\begin{smallmatrix} \delta & 0 \\ 0 & \delta \end{smallmatrix}\right] = \lceil \delta, \delta \rfloor$.

$$\Delta^*([\delta I]) = \underbrace{\delta * \delta * \cdots * \delta}_{(n-1) \text{ convolutions}} = \delta \quad \text{(by (6.3.21))}. \tag{6.9.25a}$$

For $[A] = (A_{ij})_{1\le i,j\le m}$ with $A_{ij} \in \mathcal{A}$, if there exists $[E] = (E_{ij})_{1\le i,j\le m}$ with $E_{ij} \in \mathcal{A}$ such that $[A] * [E] = [\delta I]$, then $[E] = [A^{*-1}]$ is a *convolution inverse* of $[A]$.

Definition 6.9.3. A matrix $[E] = (E_{ij})_{1\le i,j\le m}$ with distributions $E_{ij} \in \mathcal{A}$ is called an *elementary solution* of the convolution matrix equation $[A] * [T] = [B]$ in $\mathcal{A}$ if $[A] * [E] = [\delta I]$. Then $[E]$ is called a *convolution inverse* of $[A]$ and will be denoted by $[A^{*-1}]$, i.e. $[A] * [E] = [A] * [A^{*-1}] = [\delta I]$. (6.9.26)

If $\Delta^{*-1}(A) \in \mathcal{A}$ exists, then $[E] = [A^{*-1}]$ exists and is given by $[E] = \Delta^{*-1}(A) * [C]^t$, with $[C] = (C_{ij})_{1\le i,j\le m}$, C_{ij} = cofactor of A_{ij} in $\Delta^*(A)$ defined by (6.9.26)

$$\Longrightarrow \quad E_{ij} = \Delta^{*-1}(A) * C_{ji}, \quad 1 \le i, j \le m. \tag{6.9.27}$$

For example, for $[A]$ with $\Delta^*(A), \Delta^{*-1}(A) \in \mathcal{A}$ such that $\Delta^*(A) * \Delta^{*-1}(A) = \delta$. Then $[C]^t = \left[\begin{smallmatrix} A_{22} & -A_{12} \\ -A_{21} & A_{11} \end{smallmatrix}\right]$ and $[E] = (E_{ij})_{1\le i,j\le 2} = [A^{*-1}]$ is given by $[E] = \Delta^{*-1}(A) * \left[\begin{smallmatrix} A_{22} & -A_{12} \\ -A_{21} & A_{11} \end{smallmatrix}\right]$ such that $E_{ij} = \Delta^{*-1}(A) * C_{ji}$, $1 \le i, j \le 2$, i.e.

$$\begin{aligned} E_{11} &= \Delta^{*-1}(A) * A_{22}, & E_{12} &= \Delta^{*-1}(A) * (-A_{12}), \\ E_{21} &= \Delta^{*-1}(A) * (-A_{21}), & E_{22} &= \Delta^{*-1}(A) * A_{11}. \end{aligned} \tag{6.9.28}$$

Then

$$\begin{aligned} [A] * [E] &= \begin{bmatrix} A_{11} & A_{12} \\ A_{21} & A_{22} \end{bmatrix} * \begin{bmatrix} \Delta^{*-1}(A) * A_{22} & -\Delta^{*-1}(A) * A_{12} \\ -\Delta^{*-1}(A) * A_{21} & \Delta^{*-1}(A) * A_{11} \end{bmatrix} \\ &= \begin{bmatrix} \Delta^{*-1}(A) * (A_{11} * A_{22} - A_{12} * A_{21}) & \Delta^{*-1}(A) * (-A_{12} * A_{11} + A_{11} * A_{12}) \\ \Delta^{*-1}(A) * (A_{21} * A_{22} - A_{22} * A_{21}) & \Delta^{*-1}(A) * (-A_{21} * A_{12} + A_{22} * A_{11}) \end{bmatrix} \\ &= \begin{bmatrix} \Delta^{*-1}(A) * \Delta^*(A) & 0 \\ 0 & \Delta^{*-1}(A) * \Delta^*(A) \end{bmatrix} = \begin{bmatrix} \delta & 0 \\ 0 & \delta \end{bmatrix} = [\delta I] \end{aligned}$$

$\Longrightarrow [E] = [A^{*-1}]$ is defined by (6.9.27).

System of convolution equations in $\mathcal{A}$ Now we consider the system (6.9.11) in $\mathcal{A}$. For given $A_{ij} \in \mathcal{A}$ and the right-hand side $B_i \in \mathcal{A}$, $1 \le i, j \le m$, find $T_j \in \mathcal{A}$, $1 \le j \le m$ such that $\sum_{j=1}^m A_{ij} * T_j = B_i$, $1 \le i \le m$ or, equivalently,

$$[A]_{m\times m} * [T]_{m\times 1} = [B]_{m\times 1}, \tag{6.9.29}$$

where $[A]_{m\times m} = (A_{ij})_{1\le i,j\le m}$,

$$[T]_{m\times 1} = (T_1, T_2, \dots, T_m)^t = [B]_{m\times 1} = (B_1, B_2, \dots, B_m)^t.$$

Theorem 6.9.2. *For* arbitrary $B_i \in \mathcal{A}$, $1 \le i \le m$, *the system* (6.9.29) *has* at least *one solution* $[T]_{m\times 1} = (T_1, T_2, \dots, T_m)^t$, *with* $T_i \in \mathcal{A}$, *if and only if the convolution determinant* $\Delta^*(A) \in \mathcal{A}$ *has an inverse* $\Delta^{*-1}(A) \in \mathcal{A}$, *i.e.* $\Delta^*(A) * \Delta^{*-1}(A) = \delta$.

Then, the inverse $\Delta^{*-1}(A) \in \mathcal{A}$ *is* unique *and the system* (6.9.29) *has a* unique *solution* $[T]_{m\times 1}$ *for arbitrary* $[B]_{m\times 1}$ *defined by:*

$$[T]_{m\times 1} = [E]_{m\times m} * [B]_{m\times 1}, \tag{6.9.30}$$

$[E]_{m\times m} = [A^{*-1}]_{m\times m}$ *being the convolution inverse of* $[A]_{m\times m}$ *satisfying* (6.9.26) *and given by* (6.9.27).

Note: In the following, the sizes $m \times m$, $m \times 1$ of square matrices and column vectors respectively will *not* be shown.

Proof. Assume that for *arbitrary* $B_i \in \mathcal{A}$, $1 \le i \le m$, the system (6.9.29) has a solution. Then, for $[B] = [\delta I_k] = (0, \dots, 0, \delta, 0, \dots, 0)^t$, let $[T] = [E_k]$, $1 \le k \le m$, be a solution of the system (6.9.29), i.e. $[A] * [E_k] = [\delta I_k]\ \forall k = 1, 2, \dots, m$. Hence, $[A]*[E] = [\delta I]$ with $[E] = [E_1, E_2, \dots, E_m]$, i.e. elementary solution $[E] = [A^{*-1}]$. Then $[E]$ is defined by (6.9.27). Moreover, using (6.9.23),

$$\Delta^*(A) * \Delta^*(E) = \Delta^*(\delta I) = \underbrace{\delta * \delta * \cdots * \delta}_{(n-1)\text{ convolutions}} = \delta \in \mathcal{A}$$

$\Longrightarrow \Delta^{*-1}(A) = \Delta^*(E) \in \mathcal{A}$ exists.
Uniqueness of $\Delta^{*-1}(A)$: Let $\tilde{\Delta}^{*-1}(A)$ be another inverse such that $\Delta^*(A) * \tilde{\Delta}^{*-1}(A) = \delta$. Then, $\Delta^*(A) * [\Delta^{*-1}(A) - \tilde{\Delta}^{*-1}(A)] = 0 \in \mathcal{A}$. Set $\Delta_0^* = \Delta^{*-1}(A) - \tilde{\Delta}^{*-1}(A) \in \mathcal{A}$ with $\Delta^*(A) * \Delta_0^* = 0$. But

$$\Delta_0^* = \Delta_0^* * \delta = \delta * \Delta_0^* = \Delta^{*-1}(A) * \Delta^*(A) * \Delta_0^* = \Delta^{*-1}(A) * 0 = 0.$$

Conversely, suppose that $\Delta^{*-1}(A) \in \mathcal{A}$ exists. Define $[E]$ by (6.9.27): $[E] = \Delta^{*-1}(A)*[C]^t$. Then $[A]*[E] = [E]*[A] = [\delta I]$ by (6.9.26), and $[T] = [E]*[B]$ is a solution of (6.9.29) for arbitrary $[B]$ with $B_i \in \mathcal{A}$. In fact, $[A]*[T] = [A]*[E]*[B] = [\delta I] * [B] = [B]$, i.e. $[T] = [E] * [B]$ is a solution of (6.9.29).

In other words, there exists a solution $[T] = [E] * [B]$ of (6.9.29). (6.9.31)

Now suppose that (6.9.29) has a solution, i.e. $\exists T$ such that (6.9.29) holds. Then, taking convolution of both sides with $[E]$, we have $[E] * [B] = [E] * [A] * [T] = [\delta I] * [T] = [T] \Longrightarrow [T] = [E] * [B]$, i.e. every solution of (6.9.29) is given by (6.9.30). (6.9.32)

Hence, from (6.9.31) and (6.9.32), the *uniqueness* of the solution (6.9.30) follows, i.e. if $\Delta^{*-1}(A) \in \mathcal{A}$ exists, $[A] * [T] = [B] \Longleftrightarrow [T] = [E] * [B]$. □

6.10 Application of convolutions in electrical circuit analysis and heat flow problems

6.10.1 Electric circuit analysis problem [7]

Let us consider an *R-L-C* electrical circuit (see Figure 6.4), i.e. a circuit consisting of *resistor* R, *inductor* L and *capacitor* C and a source of *electromotive force* (EMF) $e(t)$ (see Example 6.10.1 for more details) which is switched on at $t = t_0$ such that a current $i(t)$ begins to flow in this circuit for $t \ge t_0$. Then the EMF $e(t)$ defines the *excitation* of the circuit for $t \ge t_0$, and the current $i(t)$ flowing in the circuit is the *response* of the circuit corresponding to this excitation. *Both* $e(t)$ and $i(t)$ are *zero* for $t < 0$.

Excitations $e(t)$ and their responses $i(t)$ may not just be *functions* on $\mathbb{R}$ with $e(t) = 0$, $i(t) = 0$ for $t < 0$, but also *distributions* in $\mathcal{D}'^+$ (see (6.9.15b) and Chapter 8, Section 8.8) with their supports in $[0, \infty[$. For example, $e(t)$ may be a *transient*

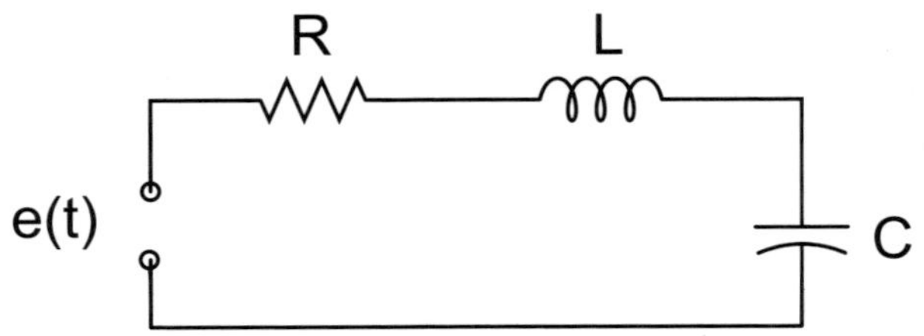

Figure 6.4 *R-L-C* electrical circuit

instantaneous excitation defined by the Dirac distribution δ, i.e. $e(t) = \delta \in \mathcal{D}'^{+}$, which is *not* a function. Thus, distributions in $\mathcal{D}'^{+}$ are used to define excitations of the circuit in practical situations. Then the question arises as to whether distributions are necessary to also define the *current* in the circuit. The answer is an affirmative one. In fact, current is the derivative (in the *distributional sense*) of the quantity of electricity in the circuit. Suppose that a *single constant charge* q is passed through an ammeter at $t = \tau$. Then the quantity of electricity passed equals

$$qH(t-\tau) = \begin{cases} q & \text{for } t \geq \tau \\ 0 & \text{for } t < \tau, \end{cases}$$

H being the Heaviside function, and the current i is the *distributional* derivative (*not* the usual derivative in the pointwise sense), i.e. $i = \frac{d}{dx}[qH(t-\tau)] = q\delta(t-\tau)$, which is a distribution in $\mathcal{D}'^{+}$ and not a function of t. Hence, both excitations and responses of a circuit may be distributions in $\mathcal{D}'^{+}$, rather than functions. Thus, in general, to each *excitation distribution* $e \in \mathcal{D}'^{+}$ there corresponds a *response distribution* $i \in \mathcal{D}'^{+}$. This defines the *excitation–response* relationship in the circuit and, mathematically, an *operator* with the following properties (see Schwartz [7, pp. 134–135]):

1. *Linearity (the 'principle of superposition').* If i_k is the response to the excitation e_k $(k = 1, 2)$, then $\alpha_1 i_1 + \alpha_2 i_2$ is the response to $\alpha_1 e_1 + \alpha_2 e_2$ $\forall \alpha_k$.

2. *Translational invariance (the 'principle of invariance' in time).* If $i(t)$ is the response to $e(t)$, then $i(t-\tau)$ is the response to $e(t-\tau)$, i.e. both undergo the *same* shift τ in time.

3. *For $t < 0$, a null excitation yields a null response.* $e(t) = 0$ for $t < 0 \Longrightarrow i(t) = 0$ for $t < 0$. But $e(t) = 0$ for $t > \tau \geq 0 \not\Longrightarrow i(t) = 0\ \forall t > \tau$, since due to self-induction and capacitance, the response will *not* vanish immediately after the disappearance of excitation $e(t)$ at $t = \tau$. Moreover, by virtue of Property 2, $e(t-\tau) = 0\ \forall t < \tau \Longrightarrow i(t-\tau) = 0\ \forall t < \tau$.

4. *Continuity.* Let $(e_k)_{k=1}^{\infty}$ be a sequence of excitations in $\mathcal{D}'^{+}$ and $(i_k)_{k=1}^{\infty}$ be the sequence of corresponding responses in $\mathcal{D}'^{+}$. Then, if $e_k \to e$ in $\mathcal{D}'^{+}$ as $k \to \infty$, then $i_k \to i$ in $\mathcal{D}'^{+}$, where i is the response to e (for $\mathcal{D}'^{+}$, see Section 8.8, Chapter 8).

Unit impulse and impulse response

The excitation $e(t) = \delta(t) \in \mathcal{D}'^{+}$, where $\delta = \delta(t)$ is the Dirac distribution (*not* a function) representing an instantaneous excitation at $t = 0$, is called the *unit impulse*, with $\langle \delta, 1 \rangle = +1$ (see (1.11.2) and Sections 1.1 and 1.11, Chapter 1). Then the response $i(t) = E(t)$ of the circuit due to the unit impulse excitation $\delta(t)$ is called the *impulse response*. The impulse response $E(t)$ is a function of t with $E(t) = 0$ for $t < 0$ and $E(t) \neq 0$ for $0 \leq t < \tau$, ($\tau > 0$ being a small number), even when $e(t) = 0\ \forall t > 0$ by virtue of Property 3.

If a response function or distribution i corresponds to an excitation function or distribution e satisfying Properties 1–4, and E is the impulse response, then the response i is given by:

$$i = E * e. \tag{6.10.1}$$

In fact, for $e = \delta$, $i = E * \delta = E$ is the impulse response. (6.10.1a)

Remark 6.10.1. Instead of the EMF e, the current i in the circuit can be taken as the excitation. Then e is the response to i and the role of the impulse response is replaced by that of the *impedance* 'Z' [7] (see Example 6.10.1 later) of the circuit such that the response e is given by:

$$e = Z * i. \tag{6.10.2}$$

Example 6.10.1. We consider the R-L-C circuit consisting of three elements:

1. *Resistor* R, which resists the flow of current $i = i(t)$ in the circuit resulting in a *voltage drop* $e_R = Ri$, with current $i(t)$ in amperes and resistance R in ohms;

2. *Inductor* L, which opposes a change in the current i, having an inertia effect in electricity similar to that of mass in mechanics, causing a *voltage drop* $e_L = L\frac{di}{dt}$, i.e. the voltage drop e_L is proportional to the *instantaneous* time rate of change of current $i(t)$, L being the constant of proportionality called *inductance*, measured in henrys;

3. *Capacitor* C, which stores energy causing a *voltage drop* e_C proportional to the *instantaneous* charge $Q(t)$ (in coulombs) on the capacitor, i.e. $e_C = Q/C$, where C is the *capacitance*, measured in farads. Then $\frac{dQ}{dt} = i(t) \Longrightarrow Q(t) = \int_{t_0}^{t} i(\tau)d\tau \Longrightarrow e_C = \frac{1}{C}\int_{t_0}^{t} i(\tau)d\tau$.

By Kirchhoff's second law, the algebraic sum of all the instantaneous voltage drops around any closed loop is zero, and the total voltage drop on a closed loop (or circuit) is equal to the sum of the voltage drops in the rest of the loop. Hence, for the R-L-C circuit with resistor R, inductor L and capacitor C, we have the total voltage drop $e(t) = e_R + e_L + e_C = Ri + L\frac{di}{dt} + \frac{1}{C}\int_0^t i(\tau)d\tau$. Then we have the *integro-differential* equation for the excitation $i(t)$: for $t \geq 0$,

$$Ri + L\frac{di}{dt} + \frac{1}{C}\int_0^t i(\tau)d\tau = e(t), \tag{6.10.3}$$

which can be rewritten as:

$$e(t) = (Z * i)(t) \quad \text{(see (6.10.2))}, \tag{6.10.4}$$

where $e(t)$ is the response to the excitation $i(t)$, with $i(t) = 0$ and $e(t) = 0$ for $t < 0$;

$$Z = R\delta + L\delta' + \frac{1}{C}H \in \mathcal{D}'^{+} \tag{6.10.5}$$

is a *distribution* called the *impedance* [7] of the R-L-C circuit.

$\delta = \delta(t)$ is the unit impulse, $\delta' = \frac{d\delta}{dt}$, $H = H(t)$ is the Heaviside function with $H(t) = 1$ for $t \geq 0$ and $= 0$ for $t > 0$. In fact,

$$\begin{aligned}
(Z * i)(t) &= (R\delta + L\delta' + \frac{1}{C}H) * i = R(\delta * i) + L(\delta' * i) + \frac{1}{C}H * i \\
&= Ri + L(\delta * \frac{di}{dt}) + \frac{1}{C}\int_{-\infty}^{\infty} H(\tau)i(t-\tau)d\tau \\
&= Ri + L\frac{di}{dt} + \frac{1}{C}\int_{-\infty}^{\infty} H(t-\tau)i(\tau)d\tau = Ri + L\frac{di}{dt} + \frac{1}{C}\int_0^t i(\tau)d\tau,
\end{aligned}$$

since by (6.3.21) $\delta * i = \delta$, and $\delta' * i = \frac{di}{dt}$ by (6.3.22), $i(t) = 0$ for $t < 0$ and

$$\begin{aligned}
\int_{-\infty}^{\infty} H(t-\tau)i(\tau)d\tau &= \int_0^{\infty} H(t-\tau)i(\tau)d\tau \\
&= \int_0^t H(t-\tau)i(\tau)d\tau + \int_t^{\infty} H(t-\tau)i(\tau)d\tau \\
&= \int_0^t 1.i(\tau)d\tau + \int_t^{\infty} 0.i(\tau)d\tau = \int_0^t i(\tau)d\tau
\end{aligned}$$

($H(t-\tau) = 0$ for $t < \tau < \infty$). Then $Z \in \mathcal{D}'^{+}$ gives the response e in terms of the excitation i by convolution (6.10.4).

$A \in \mathcal{D}'^{+}$, called the *admittance* [7], is the inverse of Z in the *convolution algebra* $\mathcal{D}'^{+}$ (see (6.9.26)), i.e. $A = Z^{-1}$ with $A * Z = \delta$. (6.10.6)

Then formula (6.10.5) corresponds to equation (6.10.4) with $e = \delta$ and $i = A$, i.e. $\delta = Z * A = A * Z \Longrightarrow A = Z^{-1}$.

Alternative relation Differentiating (see (6.7.1)–(6.7.3)) both sides of (6.10.6) and (6.10.5) with respect to t, we get:

$$\begin{aligned}\frac{d}{dt}(A * Z) = \delta' \quad &\Longrightarrow \quad \frac{d}{dt}(Z * A) = \frac{dZ}{dt} * A = \delta' \\ &\Longrightarrow \quad \left(L\delta'' + R\delta' + \frac{1}{C}\delta\right) * A = \delta' \\ &\Longrightarrow \quad L\frac{d^2A}{dt^2} + R\frac{dA}{dt} + \frac{1}{C}A = \delta',\end{aligned}$$

since $\delta'' * A = \frac{d^2A}{dt^2}$, $\delta' * A = \frac{dA}{dt}$, $\delta * A = A$, $\frac{dH}{dt} = \delta$ (for more details, see Schwartz [7, p. 137]).

Physical interpretation of (6.10.1) Let $e(t) = \delta(t)$ be the unit impulse excitation defined by Dirac's δ function, i.e. *heuristically speaking*, the excitation $e(t)$ will be very large, of the order $\frac{1}{\varepsilon}$, for a very, very short period, $0 \le t \le \varepsilon$ (see also page 2, Footnote 2 and Section 1.11), and $e(t) = 0$ for $t < 0$. Then the impulse response is $i(t) = E(t)$ with $E(t) = 0$ for $t < 0$ by Property 3. By virtue of Property 2, the unit impulse excitation $\delta(t-\tau)$ will generate the impulse response

$$i(t) = E(t-\tau). \tag{6.10.7}$$

We will take recourse to the fact that any excitation $e(t)$ is composed of impulse *point excitations* $e(\tau)\delta(t-\tau)$, as any material body is composed of *point masses* (see also Section 1.11) or any current is composed of a stream of charged particles, etc. (see Schwartz [7, p. 135]). Then, any excitation $e(t)$ can be expressed *heuristically* as the linear combination in the *integral form* of *impulsive excitations* $e(\tau)\delta(t-\tau)$ such that

$$e(t) = \int_{\mathbb{R}} e(\tau)\delta(t-\tau)d\tau = \int_0^t e(\tau)\delta(t-\tau)d\tau \tag{6.10.8}$$

$\Longrightarrow e = e * \delta$ (see Example 6.3.1).

In (6.10.8) the integral has *no* mathematical meaning, it has a *heuristic* interpretation only! Then, by virtue of Property 1 and (6.10.7), the response $i(t)$ is, again heuristically, the corresponding linear combination in the *integral form* of the *impulse responses* $E(t-\tau)$ such that

$$i(t) = \int_{\mathbb{R}} e(t)E(t-\tau)d\tau = \int_0^t e(\tau)E(t-\tau)d\tau \tag{6.10.9}$$

$$\Longrightarrow \quad i = e * E \tag{6.10.10}$$

is the formula giving the excitation–response relation, i.e. the impulse response E gives the response i to any excitation e by (6.10.1)/(6.10.10).

6.10.2 Excitations and responses defined by several functions or distributions [7]

The problem of excitations and responses defined by *several* functions or distributions leads to a system of convolution equations (6.9.11). As an example, we consider an electrical network involving several EMFs as excitations and unknown currents as responses.

For $n = 3$, let $(e_k(t))_{k=1}^3$ be the functions or distributions defining EMFs which together represent the *excitation* of a network. Let the corresponding *response* of the network be defined by the functions or distributions $(i_k(t))_{k=1}^3$, which are the unknown currents in the network. Then, instead of (6.10.10), the excitation–response relation will now be defined by a system of three convolution equations (see (6.9.11)/(6.9.29)) for the unknown currents $i_1(t)$, $i_2(t)$ and $i_3(t)$:

$$\begin{aligned} i_1 &= E_{11} * e_1 + E_{12} * e_2 + E_{13} * e_3, \\ i_2 &= E_{21} * e_1 + E_{22} * e_2 + E_{23} * e_3, \\ i_3 &= E_{31} * e_1 + E_{32} * e_2 + E_{33} * e_3, \end{aligned} \tag{6.10.11}$$

which can be rewritten in matrix form:

$$[\mathbf{i}] = [E] * [\mathbf{e}] \tag{6.10.12}$$

with $[E] = (E_{ij})_{1 \le i,j \le 3}$, $[\mathbf{i}] = (i_1, i_2, i_3)^t$, $[\mathbf{e}] = (e_1, e_2, e_3)^t$.

Then

$$[E] * [\delta I] = [E] \tag{6.10.13}$$

is the *impulse response matrix* (see also (6.10.1a) and (6.9.26)), $[I]$ is the identity matrix,

$$[\delta I] = \begin{bmatrix} \delta & 0 & 0 \\ 0 & \delta & 0 \\ 0 & 0 & \delta \end{bmatrix}$$

is a diagonal matrix of order 3 with Dirac distributions δ as diagonal elements, all other elements being zero. For example, the response to the excitation $(e_1, e_2, e_3)^t = (0, \delta, 0)^t$ is $(E_{12}, E_{22}, E_{32})^t$. Then the element E_{jk} of the impulse response matrix $[E]$ is the value of the current i_j for $e_k = \delta$, $e_l = 0$ for $1 \le l \ne k \le 3$, i.e. E_{32} is the value of i_3 for $e_1 = 0$, $e_2 = \delta$, $e_3 = 0$, or E_{13} is the value of i_1 for $e_1 = 0$, $e_2 = 0$, $e_3 = \delta$.

Application of convolution in the problem of heat flow in a rod [7]

Case of several excitations causing a single response Consider a one-dimensional heat-flow problem in a rod AB of length l with end points A and B defined by $x = 0$ and $x = l$ respectively. In the rod, heat is transmitted only by conduction along the axis of the rod from $x = 0$ to $x = l$, i.e. there is *no* heat exchange by radiation or convection with the exterior of the rod. For this, the rod is assumed to be completely insulated everywhere except at the ends A and B, where the sources of heat are placed. A time-varying quantity of heat is being transmitted by these sources and received by the rod without any heat loss from the initial instant $t = 0$.

Let $q_1(t)$ (resp. $q_2(t)$) be the quantity of heat transmitted by the heat source at the end A (resp. B) and received by the rod per unit time at the instant t. Then the total amount of heat received during $[0, t]$ from the heat source at A (resp. B) is given by:

$$\int_0^t q_1(\tau)d\tau \quad (\text{resp.} \int_0^t q_2(\tau)d\tau). \tag{6.10.14}$$

The temperature of the rod will change due to the flow of heat along the rod by conduction, since there will be *no* loss of heat due to radiation or convection. Let $u = u(x,t)$ denote the temperature at the point $x \in [0,l]$ and at the instant $t \geq 0$. It is assumed that the temperature in the rod is zero for $t < 0$, i.e. $u(x,t) = 0$ for $t < 0$, $\forall x \in [0,l]$. Hence, $u(x,t)$ will denote the change in temperature, and it is required to find $u(x,t)$ in the rod for $t \geq 0$ and $x \in [0,l]$.

Excitation by an instantaneous source of heat and the resultant impulse response Let $q_1(t) = \delta(t) \in \mathcal{D}'^{+}$ (resp. $q_2(t) = \delta(t) \in \mathcal{D}'^{+}$) denote the *instantaneous excitation heat source* placed at A (resp. B) at $t = 0$. Such a heat source at A or B will also be called the *unit (excitation) impulse* $\delta(t)$, with $\langle\delta, 1\rangle = 1$. Due to the unit excitation impulse $\delta(t)$, there will be a change in temperature of the rod, which will be called the *impulse response* and denoted by $E_1(x,t)$ (resp. $E_2(x,t)$) corresponding to $q_1(t) = \delta(t)$ at A (resp. $q_2(t) = \delta(t)$ at B), since the change in temperature will also depend on x. For *fixed* $x \in [0,l]$, $E_1(x,t)$ (resp. $E_2(x,t)$) is a function of t, with $E_i(x,t) = 0\ \forall t < 0$, but $E_i(x,t) \neq 0$ for $0 \leq t < \tau$, τ being a very small number, even when $q_i(t) = 0\ \forall t > 0$, $(i = 1,2)$ (Properties 1–4 of the R-L-C circuit analysis also hold in this case). Moreover, the unit excitation impulse $\delta(t - \tau)$ will yield the impulse response $E_i(x, t - \tau)$ by Property 2. Let $q_1(t)$ (resp. $q_2(t)$) denote an *arbitrary* excitation heat source at A (resp. B) which is *not* necessarily a unit excitation impulse. Then the response u_1 (resp. u_2) due to q_1 at A (resp. q_2 at B) is given by:

$$u_1 = E_1 * q_1 \quad (\text{resp. } u_2 = E_1 * q_2). \tag{6.10.15}$$

In fact, for $q_i(t) = \delta(t)$, $u_i = E_i * \delta = E_i$ $(i = 1,2)$. Hence, the *total* response u due to two arbitrary excitation heat sources q_1 at A and q_2 at B is given by the algebraic sum $u_1 + u_2$ of the responses u_1, u_2 in (6.10.15), i.e.

$$\begin{aligned} u(x,t) &= E_1(x,\cdot) * q_1(\cdot) + E_2(x,\cdot) * q_2(\cdot) \\ &= \int_0^t E_1(x,t-\tau)e_1(\tau)d\tau + \int_0^t E_2(x,t-\tau)e_2(\tau)d\tau. \end{aligned}$$

Here we meet with the case of two excitations causing one response.

Chapter 7

Fourier transforms of functions of $L^1(\mathbb{R}^n)$ and $S(\mathbb{R}^n)$

7.1 Fourier transforms of integrable functions in $L^1(\mathbb{R}^n)$

Let $\mathbf{x} = (x_1, x_2, \ldots, x_n)$ and $\boldsymbol{\xi} = (\xi_1, \xi_2, \ldots, \xi_n) \in \mathbb{R}^n$ be any two points in $\mathbb{R}^n$ with their inner product $\langle \mathbf{x}, \boldsymbol{\xi} \rangle = \mathbf{x} \cdot \boldsymbol{\xi} = x_1\xi_1 + x_2\xi_2 + \cdots + x_n\xi_n$. Then, for functions $f \in L^1(\mathbb{R}^n)$, i.e.

$$\int_{\mathbb{R}^n} |f(\mathbf{x})| d\mathbf{x} = \int_{\mathbb{R}^n} |f(x_1, x_2, \ldots, x_n)| dx_1 dx_2 \ldots dx_n < +\infty, \tag{7.1.1}$$

the *Fourier transform* $\hat{f} = \mathcal{F}[f]$ of f is defined by:

$$\hat{f}(\boldsymbol{\xi}) = \mathcal{F}[f(\mathbf{x})] = \int_{\mathbb{R}^n} f(\mathbf{x}) e^{-i2\pi\langle \mathbf{x}, \boldsymbol{\xi} \rangle} d\mathbf{x}. \tag{7.1.2}$$

Similarly, we can define the *Fourier co-transform* (also called *Fourier transform*) $\bar{\mathcal{F}}[f]$ of $f \in L^1(\mathbb{R}^n)$ by:

$$\bar{\mathcal{F}}[f](\boldsymbol{\xi}) = \int_{\mathbb{R}^n} f(\mathbf{x}) e^{i2\pi\langle \mathbf{x}, \boldsymbol{\xi} \rangle} d\mathbf{x}. \tag{7.1.3}$$

In fact, for $f \in L^1(\mathbb{R}^n)$, $|\int_{\mathbb{R}^n} f(\mathbf{x}) e^{\mp i2\pi\langle \mathbf{x}, \boldsymbol{\xi} \rangle} d\mathbf{x}| \leq \int_{\mathbb{R}^n} |f(\mathbf{x})| d\mathbf{x} < +\infty$, since $|e^{\mp i2\pi\langle \mathbf{x}, \boldsymbol{\xi} \rangle}| = |e^{\mp i2\pi x_1\xi_1}| \cdot |e^{\mp i2\pi x_2\xi_2}| \cdots |e^{\mp i2\pi x_n\xi_n}| = 1$.

Hence, both definitions (7.1.2) and (7.1.3) are well defined for $f \in L^1(\mathbb{R}^n)$ and we have:

Theorem 7.1.1. *Every function $f \in L^1(\mathbb{R}^n)$ has a Fourier transform $\hat{f} = \mathcal{F} f$ and Fourier co-transform $\bar{\mathcal{F}} \hat{f}$ defined by* (7.1.2) *and* (7.1.3) *respectively.*

- $\hat{f} : \mathbb{R}^n \longrightarrow \mathbb{C}$ *is a* complex-valued *function of n real variables* $\xi_1, \xi_2, \ldots, \xi_n$ *with*

$$\boldsymbol{\xi} = (\xi_1, \xi_2, \ldots, \xi_n) \in \mathbb{R}^n, \quad \hat{f}(\boldsymbol{\xi}) \in \mathbb{C}. \tag{7.1.4}$$

- $$\hat{f}(\mathbf{0}) = \int_{\mathbb{R}^n} f(\mathbf{x}) d\mathbf{x}. \tag{7.1.5}$$

- $\hat{f}$ *is* bounded *in* $\mathbb{R}^n$ *with* $\|\widehat{f}\|_\infty = \|\hat{f}\|_{L^\infty(\mathbb{R}^n)} \leq \|f\|_{L^1(\mathbb{R}^n)}$. (7.1.6)

In fact, $\forall \boldsymbol{\xi} \in \mathbb{R}^n$,

$$|\hat{f}(\boldsymbol{\xi})| = \left| \int_{\mathbb{R}^n} f(\mathbf{x}) e^{-i2\pi\langle \mathbf{x}, \boldsymbol{\xi}\rangle} d\mathbf{x} \right| \le \int_{\mathbb{R}^n} |f(\mathbf{x})| d\mathbf{x} = \|f\|_{L^1(\mathbb{R}^n)}.$$

Hence,

$$\|\hat{f}\|_\infty = \|\hat{f}\|_{L^\infty(\mathbb{R}^n)} = \sup_{\boldsymbol{\xi}\in\mathbb{R}^n} |\hat{f}(\boldsymbol{\xi})| \le \|f\|_{L^1(\mathbb{R}^n)}. \tag{7.1.7}$$

Remark 7.1.1. We will show later that for certain classes of functions f (for example, for functions $f \in S(\mathbb{R}^n) \subset L^1(\mathbb{R}^n)$) (see Definition 7.2.3), the Fourier cotransform $\bar{\mathcal{F}} = \mathcal{F}^{-1}$ (i.e. the inverse of $\mathcal{F}$) and $\mathcal{F} = (\bar{\mathcal{F}})^{-1}$ (i.e. the inverse of $\bar{\mathcal{F}}$) are such that $\bar{\mathcal{F}}\mathcal{F} f = \mathcal{F}\bar{\mathcal{F}} f = f \; \forall f \in S(\mathbb{R}^n)$ (see Theorem 7.7.1).

$\bar{\mathcal{F}}(\mathcal{F} f) = \bar{\mathcal{F}}\hat{f}$ is *not* defined $\forall f \in L^1(\mathbb{R}^n)$, since $\hat{f} \notin L^1(\mathbb{R}^n)$ for $f \in L^1(\mathbb{R}^n)$ in general (see Example 7.1.1). In particular, for $n = 1$ and $f \in L^1(]-\infty, \infty[)$,

$$\hat{f}(\xi) = \mathcal{F}[f(x)] = \int_{-\infty}^{\infty} f(x) e^{-i2\pi x\xi} dx, \tag{7.1.8}$$

$$\bar{\mathcal{F}}[f(x)] = \int_{-\infty}^{\infty} f(x) e^{i2\pi x\xi} dx. \tag{7.1.9}$$

Example 7.1.1. Let f be defined by:

$$f(x) = \begin{cases} 1 & \text{for } |x| < 1 \\ 0 & \text{otherwise.} \end{cases}$$

Then $\int_{-\infty}^{\infty} |f(x)| dx = \int_{-1}^{1} 1 dx = 2 < +\infty \Longrightarrow f \in L^1(]-\infty, \infty[)$ and its Fourier transform

$$\hat{f}(\xi) = \mathcal{F}[f(x)] = \int_{-1}^{1} 1 \cdot e^{-i2\pi x\xi} dx = \left(\frac{1}{-i2\pi\xi} e^{-i2\pi x\xi} \right) \Bigg|_{x=-1}^{x=+1}$$

$$= \frac{1}{\pi\xi} \frac{e^{i2\pi\xi} - e^{-i2\pi\xi}}{2i} = \frac{\sin(2\pi\xi)}{\pi\xi} \quad (\xi \ne 0)$$

and, for $\xi = 0$,

$$\hat{f}(0) = \int_{-\infty}^{\infty} f(x) dx = \int_{-1}^{1} 1 dx = 2 = \lim_{\xi\to 0} \hat{f}(\xi) = \lim_{\xi\to 0} \frac{2\sin(2\pi\xi)}{2\pi\xi}$$

$\Longrightarrow \hat{f}$ is continuous at $\xi = 0$.

But $\hat{f}(\xi) = \frac{sin(2\pi\xi)}{\pi\xi}$ is continuous for all ξ. Hence, $\hat{f}$ is continuous on $]-\infty, \infty[$. But $\hat{f}$ is *not* summable on $\mathbb{R} =]\infty, \infty[$, since

$$\int_{-\infty}^{\infty} |\hat{f}(\xi)| = \int_{-\infty}^{\infty} \left| \frac{\sin 2\pi\xi}{\pi\xi} \right| d\xi = 2 \int_{-\infty}^{\infty} \left| \frac{\sin(2\pi\xi)}{2\pi\xi} \right| d\xi \to \infty$$

$\Longrightarrow \hat{f} \notin L^1(]-\infty,\infty[) \Longrightarrow$ Fourier co-transform $\bar{\mathcal{F}}\hat{f}$ of $\hat{f}$ (resp. Fourier transform $\mathcal{F}\hat{f}$ of $\hat{f}$) does *not* exist, i.e. $\bar{\mathcal{F}}\mathcal{F}f$ (resp. $\mathcal{F}\bar{\mathcal{F}}f$) of this function f is *not* defined.

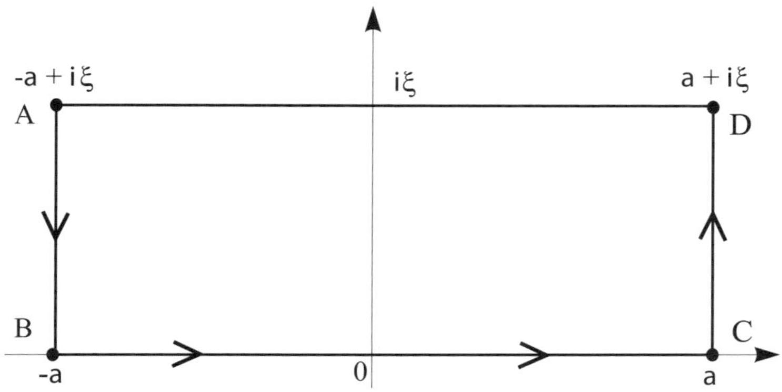

Figure 7.1 Path of integration along ABCD from A $= -a + i\xi$ to D $= a + i\xi$

Example 7.1.2 (Fourier transform $\hat{f}$ is the function f itself, i.e. $\hat{f} = f$). For $f(x) = e^{-\pi x^2}$ $\forall x \in \mathbb{R}$, $\mathcal{F}f = f$, i.e. $\hat{f}(\xi) = f(\xi) = e^{-\pi\xi^2}$.

Proof. Here, we give a proof based on the theory of analytic functions of a complex variable. For an alternative proof, see page 421 later in this chapter. $\int_{-\infty}^{\infty} e^{-\pi x^2} dx = 1$ (the *Gauss integral*) $\Longrightarrow e^{-\pi x^2} \in L^1(]-\infty,\infty[) \Longrightarrow$ its Fourier transform $\hat{f}$ exists and is given by:

$$\hat{f}(\xi) = \int_{-\infty}^{\infty} e^{-\pi x^2} \cdot e^{-i2\pi x\xi} dx = \int_{-\infty}^{\infty} e^{-\pi\xi^2} \cdot e^{-\pi[(x^2-\xi^2)+i2x\xi]} dx$$
$$= e^{-\pi\xi^2} \int_{-\infty}^{\infty} e^{-\pi(x+i\xi)^2} dx.$$

For the evaluation of the last integral for fixed ξ, we introduce complex variable $z = x + iy$ such that for fixed $y = \xi$, $z = x + i\xi$ with $dz = dx$ and $\hat{f}(\xi) = e^{-\pi\xi^2}[\lim_{a\to\infty} \int_{-a+i\xi}^{a+i\xi} e^{-\pi z^2} dz]$.

Since $e^{-\pi z^2}$ is analytic everywhere, the contour integration along fixed ξ from $-a + i\xi$ to $a + i\xi$ with $a > 0$ (see Figure 7.1) can be replaced by:

$$\lim_{a\to\infty} \int_{-a+i\xi}^{a+i\xi} e^{-\pi z^2} dz = \lim_{a\to\infty} \left[\int_{-a+i\xi}^{-a} e^{-\pi z^2} dz + \int_{-a}^{a} e^{-\pi x^2} dx \right.$$
$$\left. + \int_{a}^{a+i\xi} e^{-\pi z^2} dz \right]$$

$$= \lim_{a\to\infty} \int_{-a+i\xi}^{-a} e^{-\pi z^2} dz + \int_{-\infty}^{\infty} e^{-\pi x^2} dx + \lim_{a\to\infty} \int_{a}^{a+i\xi} e^{-\pi z^2} dz,$$

where the first and third integrals vanish.

For example, for the third integral,

$$\left| \int_a^{a+i\xi} e^{-\pi z^2} dz \right| = \left| \int_0^{\xi} e^{-\pi(a+iy)^2} i\,dy \right| \le \left| i \int_0^{\xi} e^{-\pi(a^2-y^2)-i2\pi a y} dy \right|$$
$$\le (e^{-\pi a^2} e^{\pi\xi^2} |\xi|) \to 0 \quad \text{as } a \to \infty$$

for fixed ξ (ξ may be positive or negative, although ξ is positive in Figure 7.1).

Hence, $\lim_{a\to\infty} \int_{-a+i\xi}^{a+i\xi} e^{-\pi z^2} dz = \int_{-\infty}^{\infty} e^{-\pi x^2} dx = 1$, and $\hat{f}(\xi) = e^{-\pi\xi^2} \cdot 1 = e^{-\pi\xi^2} = f(\xi)\ \forall \xi \in \mathbb{R} \Longrightarrow \hat{f} = f$. □

Counterexample 7.1.3. For $n = 1$, $f(x) = 1\ \forall x \in \mathbb{R}$, Fourier transform $\hat{f} = \mathcal{F} f$ (resp. co-transform $\bar{\mathcal{F}} f$) does *not* exist, since $\int_{-\infty}^{\infty} 1 \cdot dx = \infty \Longrightarrow 1 \notin L^1(]-\infty, \infty[)$.

Similarly, $f(x) = x^n$ or e^x or $\sin x$ etc. does *not* possess a Fourier transform $\hat{f} = \mathcal{F} f$ (resp. co-transform $\hat{g} = \bar{\mathcal{F}} f$), since $f \notin L^1(]-\infty, \infty[)$. In other words, many elementary functions do not possess a Fourier transform and co-transform as defined by (7.1.2) and (7.1.3) respectively. This suggests that we should develop *new definitions* of Fourier transform to overcome such annoyances; these will be introduced in Chapter 8.

Counterexample 7.1.4. For $n = 1$, consider $f_1(x) = \frac{1}{\sqrt{1+x^2}}$, $f_2(x) = (f_1(x))^2 = \frac{1}{1+x^2}$, $\forall x \in]-\infty, \infty[$.

$$\int_{-\infty}^{\infty} \frac{dx}{\sqrt{1+x^2}} = \int_{\frac{-\pi}{2}}^{\frac{\pi}{2}} \sec\theta d\theta = \infty \quad (\text{setting } x = \tan\theta)$$

$\Longrightarrow \frac{1}{\sqrt{1+x^2}} \notin L^1(\mathbb{R})$.

$$\int_{-\infty}^{\infty} \frac{1}{1+x^2} dx = \int_{\frac{-\pi}{2}}^{\frac{\pi}{2}} d\theta = \pi \quad (\text{setting } x = \tan\theta)$$

$\Longrightarrow \frac{1}{1+x^2} \in L^1(\mathbb{R})$.

Hence,

$$\frac{1}{\sqrt{1+x^2}} \notin L^1(\mathbb{R}), \quad \text{but } \frac{1}{\sqrt{1+x^2}} \in L^2(\mathbb{R}). \tag{7.1.10}$$

Therefore, the Fourier transform of $\frac{1}{\sqrt{1+x^2}}$ does *not* exist, whereas $f_2(x) = \frac{1}{1+x^2}$ *has* a Fourier transform. (See Remark 8.3.1 for a definition of the Fourier transform of L^2-functions.)

Alternative definitions of Fourier transforms

Although we have accepted and followed Schwartz's definition of Fourier transform $\hat{f}(\boldsymbol{\xi}) = \mathcal{F}[f(\mathbf{x})]$ of $f \in L^1(\mathbb{R}^n)$ in (7.1.2), *various* forms of definition of Fourier transforms are found in mathematical literature.

For example, for $f \in L^1(\mathbb{R}^n)$, the Fourier transform of f can also be defined by any one of the following alternative formulae:

I.
$$\hat{g}_1(\boldsymbol{\xi}) = \frac{1}{(\sqrt{2\pi})^n}\int_{\mathbb{R}^n} f(\mathbf{x})e^{-i\langle \mathbf{x},\boldsymbol{\xi}\rangle}d\mathbf{x}, \quad \text{or} \tag{7.1.11}$$
$$\hat{g}_2(\boldsymbol{\xi}) = \frac{1}{(\sqrt{2\pi})^n}\int_{\mathbb{R}^n} f(\mathbf{x})e^{i\langle \mathbf{x},\boldsymbol{\xi}\rangle}d\mathbf{x};$$

II.
$$\hat{g}_3(\boldsymbol{\xi}) = \int_{\mathbb{R}^n} f(\mathbf{x})e^{-i\langle \mathbf{x},\boldsymbol{\xi}\rangle}d\mathbf{x}, \quad \text{or} \tag{7.1.12}$$
$$\hat{g}_4(\boldsymbol{\xi}) = \int_{\mathbb{R}^n} f(\mathbf{x})e^{i\langle \mathbf{x},\boldsymbol{\xi}\rangle}d\mathbf{x}, \quad \text{or}$$
$$\hat{g}_5(\boldsymbol{\xi}) = \frac{1}{(2\pi)^n}\int_{\mathbb{R}^n} f(\mathbf{x})e^{-i\langle \mathbf{x},\boldsymbol{\xi}\rangle}d\mathbf{x}, \quad \text{etc.}$$

Other choices of definition are also possible.

Relations between $\hat{f}$ and $\hat{g}_i$ Let $\hat{f} = \mathcal{F}f$ be defined by (7.1.2). Then the Fourier transforms $\hat{g}_i$ $(i = 1, 2, 3, 4, 5)$ are related to $\hat{f}$ by the following formulae:

$$\hat{g}_1(\boldsymbol{\xi}) = \frac{1}{(\sqrt{2\pi})^n}\hat{f}\left(\frac{1}{2\pi}\boldsymbol{\xi}\right); \quad \hat{g}_2(\boldsymbol{\xi}) = \frac{1}{(\sqrt{2\pi})^n}\hat{f}\left(\frac{-1}{2\pi}\boldsymbol{\xi}\right); \quad \hat{g}_3(\boldsymbol{\xi}) = \hat{f}\left(\frac{1}{2\pi}\boldsymbol{\xi}\right);$$
$$\hat{g}_4(\boldsymbol{\xi}) = \hat{f}\left(\frac{-1}{2\pi}\boldsymbol{\xi}\right); \quad \hat{g}_5(\boldsymbol{\xi}) = \frac{1}{(2\pi)^n}\hat{f}\left(\frac{\boldsymbol{\xi}}{2\pi}\right) = \frac{1}{\sqrt{2\pi})^n}\hat{g}_1(\boldsymbol{\xi}). \tag{7.1.13}$$

For example,

$$\hat{f}\left(\frac{\pm\boldsymbol{\xi}}{2\pi}\right) = \int_{\mathbb{R}^n} f(\mathbf{x})e^{\mp i\langle \mathbf{x},\boldsymbol{\xi}\rangle}d\mathbf{x}$$
$$\Longrightarrow \quad \frac{1}{(\sqrt{2n})^n}\hat{f}\left(\frac{1}{2\pi}\boldsymbol{\xi}\right) = \frac{1}{(\sqrt{2\pi})^n}\int_{\mathbb{R}^n} f(\mathbf{x})e^{-i\langle \mathbf{x},\boldsymbol{\xi}\rangle}d\mathbf{x} = \hat{g}_1(\boldsymbol{\xi})$$
$$\frac{1}{(\sqrt{2n})^n}\hat{f}\left(\frac{-1}{2\pi}\boldsymbol{\xi}\right) = \frac{1}{(\sqrt{2n})^n}\int_{\mathbb{R}^n} f(\mathbf{x})e^{i\langle \mathbf{x},\boldsymbol{\xi}\rangle}d\mathbf{x} = \hat{g}_2(\boldsymbol{\xi}).$$

Remark 7.1.2.

- The multiplying factor $\frac{1}{(\sqrt{2\pi})^n}$ is introduced for the convenience of some computations in applications such as in multi-dimensional Fourier series, and the associated definition is followed in many applied books.
- We have accepted Schwartz's definition for the sake of elegance in many of the results and formulae to be obtained later. For example, the Fourier transform of Dirac distribution δ will be 1 and the Fourier co-transform of 1 will be δ as tempered distributions (see Chapter 8).

Properties of Fourier transform $\mathcal{F}$ and co-transform $\bar{\mathcal{F}}$

All the properties of Fourier transform $\mathcal{F}$ will also hold for the co-transform $\bar{\mathcal{F}}$, since by replacing '$-i$' by 'i' in the proofs for the case of $\mathcal{F}$ defined by (7.1.2), the corresponding results for the case of the co-transform $\bar{\mathcal{F}}$ defined by (7.1.3) can be obtained. Hence, we will state and prove the results *only* for Fourier transform $\mathcal{F}$.

Property 1 Fourier transform $\mathcal{F}$ is linear.

$$\mathcal{F}[\alpha_1 f_1 + \alpha_2 f_2] = \alpha_1 \mathcal{F} f_1 + \alpha_2 \mathcal{F} f_2 \quad \forall f_1, f_2 \in L^1(\mathbb{R}^n),\ \forall \alpha_1, \alpha_2 \in \mathbb{R}. \tag{7.1.14}$$

In fact,

$$\begin{aligned}\mathcal{F}[\alpha_1 f_1 + \alpha_2 f_2](\boldsymbol{\xi}) &= \int_{\mathbb{R}^n} (\alpha_1 f_1 + \alpha_2 f_2)(\mathbf{x}) e^{-i2\pi\langle \mathbf{x}, \boldsymbol{\xi}\rangle} d\mathbf{x} \\ &= \alpha_1 \int_{\mathbb{R}^n} f_1(\mathbf{x}) e^{-i2\pi\langle \mathbf{x}, \boldsymbol{\xi}\rangle} d\mathbf{x} + \alpha_2 \int_{\mathbb{R}^n} f_2(\mathbf{x}) e^{-i2\pi\langle \mathbf{x}, \boldsymbol{\xi}\rangle} d\mathbf{x} \\ &= (\alpha_1 \mathcal{F} f_1 + \alpha_2 \mathcal{F} f_2)(\boldsymbol{\xi}) \quad \forall \boldsymbol{\xi} \in \mathbb{R}^n,\ \forall \alpha_1, \alpha_2 \in \mathbb{R}.\end{aligned}$$

Property 2 Let $\{f_i\}_{i=1}^n$ be a set of n functions such that $f_i = f_i(x_i)\ \forall x_i \in \mathbb{R}$, $1 \le i \le n$, $f_i \in L^1(\mathbb{R})$ and $\hat{f}_i(\xi_i) = (\mathcal{F} f_i)(\xi_i) = \int_{-\infty}^{\infty} f_i(x_i) e^{-i2\pi x_i \xi_i} dx_i$.

Then, for $f = \bigotimes_{i=1}^n f_i = f_1 \otimes f_2 \otimes \cdots \otimes f_n$ defined by

$$f(\mathbf{x}) = f(x_1, x_2, \ldots, x_n) = f_1(x_1) f_2(x_2) \ldots f_n(x_n) \quad \forall \mathbf{x} = (x_1, x_2, \ldots, x_n),$$

$$\mathcal{F}(f)(\boldsymbol{\xi}) = \int_{\mathbb{R}^n} f(\mathbf{x}) e^{-i2\pi\langle \mathbf{x}, \boldsymbol{\xi}\rangle} d\mathbf{x}$$

$$= \int_{\mathbb{R}^n = \mathbb{R}\times\mathbb{R}\times\cdots\times\mathbb{R}} f_1(x_1) f_2(x_2) \cdots f_n(x_n) e^{-i2\pi(x_1\xi_1 + x_2\xi_2 + \cdots + x_n\xi_n)} dx_1 dx_2 \ldots dx_n$$

$$= \int_{\mathbb{R}} f_1(x_1)e^{-i2\pi x_1\xi_1}dx_1 \int_{\mathbb{R}} f_2(x_2)e^{-i2\pi x_2\xi_2}dx_2 \cdots \int_{\mathbb{R}} f_n(x_n)e^{-i2\pi x_n\xi_n}dx_n$$
$$= (\mathcal{F} f_1)(\xi_1)\cdot(\mathcal{F} f_2)(\xi_2)\cdots(\mathcal{F} f_n)(\xi_n) \quad \forall \boldsymbol{\xi} = (\xi_1,\xi_2,\ldots\xi_n)$$
$$\Longrightarrow \quad \mathcal{F} f = (\mathcal{F} f_1)\otimes(\mathcal{F} f_2)\otimes\cdots\otimes(\mathcal{F} f_n). \tag{7.1.15}$$

Example 7.1.5. $f(\mathbf{x}) = f(x_1,x_2,\ldots,x_n) = e^{-\pi\|\mathbf{x}\|^2} = e^{-\pi x_1^2-\pi x_2^2-\cdots-\pi x_n^2}$
$\Longrightarrow f(\mathbf{x}) = f_1(x_1)\cdot f_2(x_2)\cdots f_n(x_n)$ with $f_i(x_i) = e^{-\pi x_i^2}$ $\forall x_i \in \mathbb{R}$, $\hat{f_i} = \mathcal{F} f_i = f_i$ (see Example 7.1.2)

$$\Longrightarrow \quad \hat{f}(\boldsymbol{\xi}) = (\mathcal{F} f)(\boldsymbol{\xi}) = (\mathcal{F} f_1)(\xi_1)\cdot(\mathcal{F} f_2)(\xi_2)\cdots(\mathcal{F} f_n)(\xi_n)$$
$$= \hat{f}_1(\xi_1)\hat{f}_2(\xi_2)\cdots\hat{f}_n(\xi_n) = e^{-\pi\xi_1^2}\cdot e^{-\pi\xi_2^2}\cdots e^{-\pi\xi_n^2}$$
$$= e^{-\pi(\xi_1^2+\xi_2^2+\cdots+\xi_n^n)} = e^{-\pi\|\boldsymbol{\xi}\|^2} = f(\boldsymbol{\xi}) \quad \forall \boldsymbol{\xi}\in\mathbb{R}^n$$

$\Longrightarrow \mathcal{F} f = (\mathcal{F} f_1)\otimes(\mathcal{F} f_2)\otimes\ldots(\mathcal{F} f_n) = f$.
Similarly, for $f(\mathbf{x}) = e^{-\pi\|\mathbf{x}\|^2}$,

$$\bar{\mathcal{F}} f = f, \tag{7.1.16}$$

since $\bar{\mathcal{F}} f(\xi_i) = \mathcal{F} f(-\xi_i) = e^{-\pi\xi_i^2}$ for $1 \le i \le n$
$\Longrightarrow \bar{\mathcal{F}} f(\boldsymbol{\xi}) = \mathcal{F} f(-\boldsymbol{\xi}) = e^{-\pi\|\boldsymbol{\xi}\|^2} = f(\boldsymbol{\xi})\ \forall\ \boldsymbol{\xi}\in\mathbb{R}^n$.

Property 3 Let $\check{f}$ be the function defined by $\check{f}(\mathbf{x}) = f(-\mathbf{x})$ for $\mathbf{x}\in\mathbb{R}^n$. Then,

I. $\bar{\mathcal{F}}(\check{f}) = \mathcal{F} f$;
II. $\mathcal{F}(\check{f}) = \bar{\mathcal{F}} f$;
III. $\bar{\mathcal{F}}(\overline{f}) = \overline{\mathcal{F} f}$ for complex-valued f;
IV. $(\mathcal{F} f)^\vee = \bar{\mathcal{F}} f$, where $(\mathcal{F} f)^\vee(\boldsymbol{\xi}) = (\mathcal{F} f)(-\boldsymbol{\xi})$;
V. $(\bar{\mathcal{F}} f)^\vee = \mathcal{F} f$. (7.1.17)

Proof.

I. From the definition (7.1.3) of $\bar{\mathcal{F}}$, we have

$$(\bar{\mathcal{F}}(\check{f}))(\boldsymbol{\xi}) = \int_{\mathbb{R}^n} \check{f}(\mathbf{x})e^{+i2\pi\langle\mathbf{x},\boldsymbol{\xi}\rangle}d\mathbf{x} = \int_{\mathbb{R}^n} f(-\mathbf{x})e^{+i2\pi\langle\mathbf{x},\boldsymbol{\xi}\rangle}d\mathbf{x}.$$

By change of variables $\mathbf{x} = -\mathbf{y}$ with Jacobian J, $|J| = 1$, we have $\forall\boldsymbol{\xi}\in\mathbb{R}^n$,

$$(\bar{\mathcal{F}}(\check{f}))(\boldsymbol{\xi}) = \int_{\mathbb{R}^n} f(\mathbf{y})e^{+i2\pi\langle-\mathbf{y},\boldsymbol{\xi}\rangle}|J|d\mathbf{y} = \int_{\mathbb{R}^n} f(\mathbf{y})e^{-i2\pi\langle\mathbf{y},\boldsymbol{\xi}\rangle}d\mathbf{y} = (\mathcal{F} f)(\boldsymbol{\xi})$$

$$\Longrightarrow \bar{\mathcal{F}}(\check{f}) = \mathcal{F} f.$$

II. $\mathcal{F}(\check{f}) = \bar{\mathcal{F}}(\check{f})^\vee = \bar{\mathcal{F}} f$, since $(\check{f})^\vee(\mathbf{x}) = \check{f}(-\mathbf{x}) = f(\mathbf{x})$ for $\mathbf{x} \in \mathbb{R}^n$.

III. For complex-valued $f(\mathbf{x})$,

$$\bar{\mathcal{F}}(\overline{f}) = \int_{\mathbb{R}^n} \overline{f(\mathbf{x})} e^{+i2\pi\langle \mathbf{x},\boldsymbol{\xi}\rangle} d\mathbf{x} = \int_{\mathbb{R}^n} \overline{f(\mathbf{x}) e^{-i2\pi\langle \mathbf{x},\boldsymbol{\xi}\rangle} d\mathbf{x}}$$
$$= \overline{\int_{\mathbb{R}^n} f(\mathbf{x}) e^{-i2\pi\langle \mathbf{x},\boldsymbol{\xi}\rangle} d\mathbf{x}} = \overline{\mathcal{F} f}.$$

IV.
$$(\mathcal{F} f)^\vee(\boldsymbol{\xi}) = (\mathcal{F} f)(-\boldsymbol{\xi}) = \int_{\mathbb{R}^n} f(\mathbf{x}) e^{-i2\pi\langle \mathbf{x},-\boldsymbol{\xi}\rangle} d\mathbf{x}$$
$$= \int_{\mathbb{R}^n} f(\mathbf{x}) e^{i2\pi\langle \mathbf{x},\boldsymbol{\xi}\rangle} d\mathbf{x} = \bar{\mathcal{F}} f.$$

V. $(\bar{\mathcal{F}} f)^\vee = ((\mathcal{F} f)^\vee)^\vee = \mathcal{F} f$. □

Property 4 Let $A : \mathbb{R}^n \longrightarrow \mathbb{R}^n$ be an invertible linear mapping defined by a square matrix of order n, and $f \cdot A : \mathbb{R}^n \longrightarrow \mathbb{R}$ be defined, $\forall f \in L^1(\mathbb{R}^n)$, by $(f \circ A)(\mathbf{x}) = f(A\mathbf{x})$ a.e. in $\mathbb{R}^n$. Then

$$\mathcal{F}[f \circ A](\boldsymbol{\xi}) = \frac{1}{|\det A|}(\mathcal{F} f)(A^{-t}\boldsymbol{\xi}),$$

i.e.

$$\mathcal{F}[f \circ A] = \frac{1}{|\det A|}[(\mathcal{F} f) \circ A^{-t}], \tag{7.1.18}$$

where $A^{-t} = (A^{-1})^t$ is the transpose of A^{-1}.

Proof. Set $\mathbf{y} = A\mathbf{x}$. Then, $\mathbf{x} = A^{-1}\mathbf{y}$ with Jacobian $J = \det(A^{-1}) = \frac{1}{\det(A)}$ for this change of variables, and $\langle \mathbf{x}, \boldsymbol{\xi}\rangle = \langle A^{-1}\mathbf{y}, \boldsymbol{\xi}\rangle = \langle \mathbf{y}, A^{-t}\boldsymbol{\xi}\rangle$. Hence,

$$\mathcal{F}[f \circ A](\boldsymbol{\xi}) = \int_{\mathbb{R}^n} (f \circ A)(\mathbf{x}) e^{-i2\pi\langle \mathbf{x},\boldsymbol{\xi}\rangle} d\mathbf{x} = \int_{\mathbb{R}^n} f(A\mathbf{x}) e^{-i2\pi\langle \mathbf{x},\boldsymbol{\xi}\rangle} d\mathbf{x}$$
$$= \int_{\mathbb{R}^n} f(\mathbf{y}) e^{-i2\pi\langle \mathbf{y},A^{-t}\boldsymbol{\xi}\rangle} |J| d\mathbf{y}$$
$$= \frac{1}{|\det(A)|}\int_{\mathbb{R}^n} f(\mathbf{y}) e^{-i2\pi\langle \mathbf{y},A^{-t}\boldsymbol{\xi}\rangle} d\mathbf{y}$$
$$= \frac{1}{|\det A|}(\mathcal{F} f)(A^{-t}\boldsymbol{\xi}) \quad \forall \boldsymbol{\xi} \in \mathbb{R}^n$$
$$\Longrightarrow \quad \mathcal{F}[f \circ A] = \frac{1}{|\det A|}(\mathcal{F} f \circ A^{-t}).$$
□

Particular cases

(a) $\mathbf{y} = k\mathbf{x}$ with $k \in \mathbb{R}, k \neq 0$

$\Longrightarrow A = k\ I$, I being the identity matrix of order $n \Longrightarrow A^{-1} = \frac{1}{k}I$ and $\det(A) = k^n$. Then

$$\mathcal{F}[f(k\mathbf{x})](\boldsymbol{\xi}) = \frac{1}{|k|^n}(\mathcal{F} f)\Big(\frac{\boldsymbol{\xi}}{k}\Big) = \frac{1}{|k|^n}(\mathcal{F} f)\Big(\frac{\xi_1}{k}, \frac{\xi_2}{k}, \ldots, \frac{\xi_n}{k}\Big). \quad (7.1.19)$$

For $n = 1$, f is *even* (resp. *odd*) $\Longrightarrow \hat{f}$ is *even* (resp. *odd*).
Indeed, f is even $\Longrightarrow f(x) = f(-x)$ for $x \in \mathbb{R} \Longrightarrow \hat{f}(\xi) = \mathcal{F}[f(x)] = \mathcal{F}[f(-x)] = \hat{f}(-\xi)$ with $k = -1$ (from (7.1.19)) $\Longrightarrow \hat{f}(\xi) = \hat{f}(-\xi) \Longrightarrow \hat{f}$ is even.

(b) $\mathbf{y} = S\mathbf{x}$, where $A = S$ is the *rotation matrix* of order n.
Then $SS^t = I$, $\det(S) = 1$, $S^{-1} = S^t \Longrightarrow S^{-t} = S$.
Hence, $\mathcal{F}[(f \circ S)(\mathbf{x})](\boldsymbol{\xi}) = (\mathcal{F} f)(S\boldsymbol{\xi})$.

(c) If f is a radial function, then its Fourier transform will also be a radial function.
For $r = (x_1^2 + x_2^2 + \cdots + x_n^2)^{\frac{1}{2}}$, let $f(\mathbf{x}) = f(x_1, x_2, \ldots, x_n)$ be a radial function Φ of r only: $f(x_1, x_2, \ldots, x_n) = \Phi(r)$. Then its Fourier transform $(\mathcal{F} f)\boldsymbol{\xi} = \hat{f}(\boldsymbol{\xi})$ will be a radial function Ψ of $\rho = (\xi_1^2 + \xi_2^2 + \cdots + \xi_n^2)^{\frac{1}{2}}$ only, i.e.

$$(\mathcal{F} f)(\boldsymbol{\xi}) = \Psi(\rho). \quad (7.1.20)$$

Proof. It is sufficient to show that $\hat{f}(\boldsymbol{\xi})$ is *invariant* under rotation about the origin. Let S be such a rotation matrix. Then, $\forall \mathbf{x} \in \mathbb{R}^n$ with $\|\mathbf{x}\| = r$, $\|S\mathbf{x}\|^2 = \langle S\mathbf{x}, S\mathbf{x}\rangle = \langle \mathbf{x}, S^tS\ \mathbf{x}\rangle = \langle \mathbf{x}, \mathbf{x}\rangle = \|\mathbf{x}\|^2 = r^2$ (since $S^tS = I$). f is a radial function of $\mathbf{x} \Longrightarrow f(\mathbf{x}) = f(S\mathbf{x})$, since $f(\mathbf{x})$ is invariant under the rotation about the origin defined by S. Hence, $f(S\mathbf{x}) = f(\mathbf{x}) = \Phi(r)$. Now we will show that $\hat{f}(\boldsymbol{\xi})$ is *also* invariant under the rotation about the origin defined by S, i.e. $\hat{f}(S\boldsymbol{\xi}) = \hat{f}(\boldsymbol{\xi})$. Set $\eta = S\boldsymbol{\xi}$, i.e. the point $\boldsymbol{\xi}$ is carried to the point η by the rotation defined by S. Hence, $\hat{f}(\eta) = \hat{f}(S\boldsymbol{\xi}) = \int_{\mathbb{R}^n} f(\mathbf{x})e^{-i2\pi\langle \mathbf{x}, S\boldsymbol{\xi}\rangle} d\mathbf{x}$.

But the inner product $\langle\cdot,\cdot\rangle$ is invariant under S^{-1}, and $\langle \mathbf{x}, S\boldsymbol{\xi}\rangle = \langle SS^{-1}\mathbf{x}, S\boldsymbol{\xi}\rangle = \langle S^{-1}\mathbf{x}, S^tS\boldsymbol{\xi}\rangle = \langle S^{-1}\mathbf{x}, \boldsymbol{\xi}\rangle$. Hence, $\hat{f}(S\boldsymbol{\xi}) = \int_{\mathbb{R}^n} f(\mathbf{x})e^{-i2\pi\langle S^{-1}\mathbf{x}, \boldsymbol{\xi}\rangle} d\mathbf{x}$.

By change of variables $\mathbf{y} = S^{-1}\mathbf{x}$ such that $\mathbf{x} = S\mathbf{y}$ with Jacobian $J = \det(S) = 1$, $\hat{f}(S\boldsymbol{\xi}) = \int_{\mathbb{R}^n} f(S\mathbf{y})e^{-i2\pi\langle \mathbf{y}, \boldsymbol{\xi}\rangle} d\mathbf{y} = \int_{\mathbb{R}^n} f(\mathbf{y})e^{-i2\pi\langle \mathbf{y}, \boldsymbol{\xi}\rangle} d\mathbf{y} = \hat{f}(\boldsymbol{\xi})$, since f is invariant under rotation about the origin, i.e. $f(\mathbf{y}) = f(S\mathbf{y})$.

So, $\hat{f}(S\boldsymbol{\xi}) = \hat{f}(\boldsymbol{\xi}) \Longrightarrow \hat{f}$ is invariant under rotation about the origin. Hence, we can write $\hat{f}(\boldsymbol{\xi}) = \Psi(\rho)$ with $\rho = (\xi_1^2 + \xi_2^2 + \cdots + \xi_n^2)^{\frac{1}{2}}$. We refer to [7] for finding the explicit formula of $\Psi(\rho)$ as the Fourier transform of $\Phi(r)$ in terms of Bessel functions. □

Property 5 Let $(\tau_{\mathbf{a}} f)(\mathbf{x})$ be the translation of $f(\mathbf{x})$ by $\mathbf{a} \in \mathbb{R}^n$,

$$\tau_{\mathbf{a}} f(\mathbf{x}) = f(\mathbf{x} - \mathbf{a}), \tag{7.1.21}$$

and $e^{\pm i2\pi\langle \mathbf{a}, \cdot \rangle}$ be denoted by $\chi_{\pm \mathbf{a}}(\cdot)$, such that

$$\chi_{\pm\mathbf{a}}(\mathbf{x}) = e^{\pm i2\pi\langle \mathbf{a},\mathbf{x}\rangle}, \quad \chi_{\pm\mathbf{a}}(\boldsymbol{\xi}) = e^{\pm i2\pi\langle \mathbf{a},\boldsymbol{\xi}\rangle}. \tag{7.1.22}$$

Then

I. $\widehat{\tau_{\mathbf{a}} f} = \chi_{-\mathbf{a}} \hat{f}$;

II. $\widehat{\chi_{\mathbf{a}} f} = \tau_{\mathbf{a}} \hat{f}$. (7.1.23)

Proof.

I. $\widehat{\tau_{\mathbf{a}} f}(\boldsymbol{\xi}) = \int_{\mathbb{R}^n} (\tau_{\mathbf{a}} f)(\mathbf{x}) e^{-i2\pi\langle \mathbf{x},\boldsymbol{\xi}\rangle} d\mathbf{x} = \int_{\mathbb{R}^n} f(\mathbf{x}-\mathbf{a}) e^{-i2\pi\langle \mathbf{x},\boldsymbol{\xi}\rangle} d\mathbf{x}$.
Set $\mathbf{y} = \mathbf{x} - \mathbf{a}$. Then $\mathbf{x} = \mathbf{y} + \mathbf{a}$ with Jacobian $J = |I| = 1$, and

$$\begin{aligned}\int_{\mathbb{R}^n} f(\mathbf{x}-\mathbf{a}) e^{-i2\pi\langle \mathbf{x},\boldsymbol{\xi}\rangle} d\mathbf{x} &= \int_{\mathbb{R}^n} f(\mathbf{y}) e^{-i2\pi\langle \mathbf{y}+\mathbf{a},\boldsymbol{\xi}\rangle} |J| d\mathbf{y} \\ &= \int_{\mathbb{R}^n} f(\mathbf{y}) e^{-i2\pi[\langle \mathbf{y},\boldsymbol{\xi}\rangle + \langle \mathbf{a},\boldsymbol{\xi}\rangle]} d\mathbf{y} \\ &= e^{-i2\pi\langle \mathbf{a},\boldsymbol{\xi}\rangle} \int_{\mathbb{R}^n} f(\mathbf{y}) e^{-i2\pi\langle \mathbf{y},\boldsymbol{\xi}\rangle} d\mathbf{y} = \chi_{-\mathbf{a}}(\boldsymbol{\xi}) \hat{f}(\boldsymbol{\xi})\end{aligned}$$

$$\Longrightarrow \widehat{\tau_{\mathbf{a}} f}(\boldsymbol{\xi}) = \chi_{-\mathbf{a}}(\boldsymbol{\xi}) \hat{f}(\boldsymbol{\xi}) \ \forall \boldsymbol{\xi} \in \mathbb{R}^n \Longrightarrow \widehat{\tau_{\mathbf{a}} f} = \chi_{-\mathbf{a}} \hat{f}.$$

II.
$$\begin{aligned}\widehat{\chi_{\mathbf{a}} f}(\boldsymbol{\xi}) &= \int_{\mathbb{R}^n} (\chi_{\mathbf{a}} f)(\mathbf{x}) e^{-i2\pi\langle \mathbf{x},\boldsymbol{\xi}\rangle} d\mathbf{x} = \int_{\mathbb{R}^n} e^{i2\pi\langle \mathbf{a},\mathbf{x}\rangle} f(\mathbf{x}) e^{-i2\pi\langle \mathbf{x},\boldsymbol{\xi}\rangle} d\mathbf{x} \\ &= \int_{\mathbb{R}^n} f(\mathbf{x}) e^{-i2\pi\langle \mathbf{x},\boldsymbol{\xi}-\mathbf{a}\rangle} d\mathbf{x} = \hat{f}(\boldsymbol{\xi}-\mathbf{a}) = \tau_{\mathbf{a}} \hat{f}(\boldsymbol{\xi}) \quad \forall \boldsymbol{\xi} \in \mathbb{R}^n\end{aligned}$$

$$\Longrightarrow \widehat{\chi_{\mathbf{a}} f} = \tau_{\mathbf{a}} \hat{f}.$$ □

In order to study the deeper properties of Fourier transforms of functions $f \in L^1(\mathbb{R}^n)$, which are continuous or differentiable, we need:

Some auxiliary results on the continuity and differentiability of integrals

Definition 7.1.1A. A function $f = f(x, y)$ with $x, y \in \mathbb{R}$ is called *separately* (or *partially*) *continuous* with respect to *each* variable x and y if and only if f is continuous with respect to *one* variable when the other variable is fixed.

The separate (or partial) continuity of f does not imply its continuity, but the continuity of f in (x, y) implies its separate continuity with respect to x and y.

For example, f defined by:

$$f(x, y) = \begin{cases} \frac{xy}{x^2+y^2} & \text{for } (x, y) \neq (0, 0) \\ 0 & \text{for } (x, y) = (0, 0) \end{cases}$$

is *separately* continuous at $(0, 0)$, but f is *not* continuous at $(0, 0)$.

In fact, for *fixed* y,

$$f_y(\cdot) : x \longmapsto f_y(x) = f(x, y) = \begin{cases} \frac{xy}{x^2+y^2} & \text{for } x \neq 0 \\ 0 & \text{for } x = 0 \end{cases}$$

$\Longrightarrow \lim_{x\to 0} f_y(x) = 0 = f_y(0) \Longrightarrow f_y(\cdot)$ is continuous at $x = 0$ for *fixed* y.

Similarly, $f_x(\cdot)$ is continuous at $y = 0$ for *fixed* x. Hence, f is *separately* continuous at $(0, 0)$. But $f(x, mx) = \frac{m}{1+m^2} \neq 0$ for $m \neq 0$

$\Longrightarrow \lim_{x\to 0\,(m\neq 0)} f(x, mx) = \frac{m}{1+m^2} \neq f(0, 0) \Longrightarrow f$ is *not* continuous at $(0, 0)$.

Definition 7.1.1A holds when $f = f(\mathbf{x}, \mathbf{y})$ with $\mathbf{x} \in \mathbb{R}^n, \mathbf{y} \in \mathbb{R}^m$.

Define function $F : I \subset \mathbb{R} \longrightarrow \mathbb{R}$ by $F(x) = \int_a^b f(x, t)dt\ \forall x \in I \subset \mathbb{R}$ such that the integral exists $\forall x \in I$, the interval $]a, b[\subset \mathbb{R}$ (resp. $I \subset \mathbb{R}$) being a *finite* or *infinite* one. The continuity, differentiability and integrability of F will be dependent on the analogous properties of f, which will be stated now. We agree to accept the following theorems without proof.

Continuity of the integral $F(x) = \int_a^b f(x, t)dt$

Theorem 7.1.2A (Lebesgue). *Let $f = f(x, t)$ be separately continuous with respect to x at $x = x_0 \in I \subset \mathbb{R}$ for almost all $t \in]a, b[$, $]a, b[\subset \mathbb{R}$ being an arbitrary finite or infinite interval of integration. If $g \geq 0$ is a positively valued function of t such that $|f(x, t)| \leq g(t)$ for almost all $t \in]a, b[$ and $\int_a^b g(t)dt < +\infty$, then $F(x) = \int_a^b f(x, t)dt$ is continuous at $x = x_0 \in I$.*

Differentiability of the integral $F(x) = \int_a^b f(x, t)dt$

Theorem 7.1.2B (Lebesgue). *Let $I = [\alpha, \beta] \subset \mathbb{R}$ be a finite interval with $\alpha \leq x \leq \beta$. Let $\frac{\partial f}{\partial x} = f'_x(x, t)$ be the partial derivative of f for almost all $t \in]a, b[$ such that $f'_x(x, t)$ is separately continuous with respect to x for almost all $t \in]a, b[$ and $|f'_x(x, t)| \leq g(t)$ for almost all $t \in]a, b[$, $g(t) \geq 0$ with $\int_a^b g(t)dt < +\infty$. Then, if*

$F(x) = \int_a^b f(x,t)dt$ exists for a particular value of $x = x_0$, the following hold:

I. *$F(x)$ exists for any $x \in [\alpha, \beta]$;*

II. *F is continuous and differentiable and the derivative $F'(x)$ is obtained by differentiating under the integral sign $\int$, i.e. $F'(x) = \int_a^b f'_x(x,t)dt$.*

Integrability of the integral $F(x) = \int_a^b f(x,t)dt$

Theorem 7.1.2C (Fubini). *Let f be summable on rectangle $]\alpha,\beta[\times]a,b[$, i.e. $\iint_{]\alpha,\beta[\times]a,b[} |f(x,t)|dxdt < +\infty$. Then*

I. *F is summable on $]\alpha,\beta[$, i.e. $\int_\alpha^\beta |F(x)|dx < +\infty$;*

II.
$$\int_\alpha^\beta F(x)dx = \int_\alpha^\beta dx \int_a^b f(x,t)dt = \int_a^b dt \int_\alpha^\beta f(x,t)dx$$
$$= \iint_{]\alpha,\beta[\times]a,b[} f(x,t)dxdt \tag{7.1.23a}$$

(i.e. interchange of the order of integration in the iterated integral is possible).

Instead of giving the proof of Fubini's Theorem 7.1.2C (see, for example, [20], [30], [33] for proof), we give some interesting counterexamples to highlight the different situations which may arise. For example, if $f(x,y)$ is *not* summable on $]\alpha,\beta[\times]a,b[$, i.e. $\int_{]\alpha,\beta[\times]a,b[} |f(x,y)|dxdy = \infty$, it may happen that *one* of the iterated integrals in (7.1.23a), $\int_\alpha^\beta dx \int_a^b f(x,y)dy$ or $\int_a^b dy \int_\alpha^\beta f(x,y)dx$, may exist and the other one may *not* exist, or *both* the iterated integrals may exist with *equal* or *unequal* values, i.e. (7.1.23a) does *not* hold in all these cases.

Example 7.1.6. Let $\Omega =]0,1[\times]0,1[\subset \mathbb{R}^2$ and $f(x,y) = \frac{x^2-y^2}{(x^2+y^2)^2}$ $\forall (x,y) \in \Omega$. Then

1. $\int_\Omega \frac{|x^2-y^2|}{(x^2+y^2)^2} dxdy$ does *not* exist.

2. The *iterated integrals* have *different values*, i.e.

$$I_1 = \int_0^1 \left(\int_0^1 \frac{x^2-y^2}{(x^2+y^2)^2} dx \right) dy = -\pi/4;$$
$$I_2 = \int_0^1 \left(\int_0^1 \frac{x^2-y^2}{(x^2+y^2)^2} dy \right) dx = \pi/4.$$

Proof.

1. Consider the first quadrant Q_1 of the unit circle or ball $B(\mathbf{0};1)$ with radius 1 such that $Q_1 \subset \Omega$, and introduce polar coordinates with $x = r\cos\theta$, $y = r\sin\theta$, $|J| = r, 0 < r < 1, 0 < \theta < \pi/2$. Then

$$\int_\Omega \frac{|x^2-y^2|}{(x^2+y^2)^2}dxdy > \int_{Q_1} \frac{|x^2-y^2|}{(x^2+y^2)^2}dxdy = \int_0^1\int_0^{\pi/2} \frac{r^2|\cos\theta|}{r^4} rdrd\theta$$
$$= \int_0^1 \frac{1}{r}\left(\int_0^{\pi/4}\cos 2\theta d\theta + \int_{\pi/4}^{\pi/2}(-\cos 2\theta)d\theta\right)dr = \int_0^1 \frac{dr}{r} = \infty$$

$\Longrightarrow [(x^2-y^2)/(x^2+y^2)^2]$ is *not* summable/integrable on Ω.

2. $I_1 = \int_0^1 I_1(y)dy$, with

$$I_1(y) = \int_0^1 \frac{x^2-y^2}{(x^2+y^2)^2}dx = \int_0^1 \frac{x^2+y^2-2y^2}{(x^2+y^2)^2}dx$$
$$= \int_0^1 \frac{dx}{x^2+y^2} - 2y^2\int_0^1 \frac{dx}{(x^2+y^2)^2}.$$

Set $u = \frac{1}{x^2+y^2}$ with $du = \frac{-2x}{(x^2+y^2)^2}dx$, $v = \int dx = x$. Then

$$\int \frac{dx}{x^2+y^2} = \frac{x}{x^2+y^2} + \int \frac{2x^2}{(x^2+y^2)^2}dx$$
$$= \frac{x}{x^2+y^2} + 2\int \frac{dx}{x^2+y^2} - 2y^2\int \frac{dx}{(x^2+y^2)^2}$$
$$\Longrightarrow \quad \int \frac{dx}{x^2+y^2} - 2y^2\int \frac{dx}{(x^2+y^2)^2} = -\frac{x}{x^2+y^2}$$
$$\Longrightarrow \quad I_1(y) = -\frac{x}{x^2+y^2}\bigg|_0^1 = -\frac{1}{1+y^2}$$
$$\Longrightarrow \quad I_1 = \int_0^1 -\frac{1}{1+y^2}dy = -\tan^{-1}(y)|_0^1 = -\pi/4.$$

Similarly, $I_2 = \int_0^1 I_2(x)$ with

$$I_2(x) = \int_0^1 \frac{x^2-y^2}{(x^2+y^2)^2}dy = -\int_0^1 \frac{y^2-x^2}{(y^2+x^2)^2}dy = -(-\pi/4) = \pi/4,$$

which is obtained by replacing 'x' by 'y' and vice versa. □

Example 7.1.7. Let $\Omega =]-1,1[\times]-1,1[\subset \mathbb{R}^2$ be a square domain and $f(x,y) = \frac{xy}{(x^2+y^2)^2}$ $\forall (x,y) \in \Omega$. Then

1. $\int_\Omega |f(x,y)|dxdy$ does *not* exist;
2. but the corresponding iterated integrals exist and are *equal*, i.e.

$$\int_{-1}^{1}\left(\int_{y=-1}^{y=1} f(x,y)dx\right)dy = \int_{-1}^{1}\left(\int_{x=-1}^{x=1} f(x,y)dy\right)dx = 0.$$

Proof.

1. Consider the unit circle or ball $B(\mathbf{0},1) \subset \Omega$. Introducing polar coordinates, $x = r\cos\theta$, $y = r\sin\theta$, $|J| = r$, we get

$$\int_\Omega |f(x,y)|dxdy > \int_{B(\mathbf{0},1)} |f(x,y)|dxdy = \int_0^1\int_0^{2\pi} \frac{r^2|\cos\theta\sin\theta|}{r^4} r dr d\theta$$
$$= \int_0^1 \frac{1}{r}\left(\int_0^{2\pi} \frac{|\sin 2\theta|}{2} d\theta\right)dr = 2\int_0^1 \frac{1}{r}dr = \infty,$$

since $\int_0^{2\pi} |\sin 2\theta| d\theta = \int_0^{\pi/2} + \int_{\pi/2}^{\pi} + \int_{\pi}^{3\pi/2} + \int_{3\pi/2}^{2\pi} = 1+1+1+1 = 4$,
$\Longrightarrow \int_\Omega |f(x,y)|dxdy = \infty$.
Hence, $f(x,y)$ is *not* summable/integrable on Ω in the sense of Lebesgue.

2. $$\int_{-1}^{1}\left(\int_{x=-1}^{x=1} \frac{xy}{(x^2+y^2)^2}dx\right)dy = \int_{-1}^{1}\frac{y}{2}\left(\int_{x=-1}^{x=1}\frac{2x}{(x^2+y^2)^2}dx\right)dy$$
$$= \frac{1}{2}\int_{-1}^{1} y\cdot\left(\frac{-1}{x^2+y^2}\right)\Bigg|_{x=-1}^{x=1} dy = \frac{1}{2}\int_{-1}^{1} y\cdot 0 = 0.$$

Similarly, $\int_{-1}^{1}(\int_{y=-1}^{y=1} \frac{xy}{(x^2+y^2)^2}dy)dx = 0$.

□

Example 7.1.8. Let $\Omega =]0,\infty[\times]0,\infty[\subset \mathbb{R}^2$ and $f(x,y) = e^{-xy}\sin x$ $\forall (x,y) \in \Omega$. Then the *double integral* $\int_\Omega |f(x,y)|dxdy$ does *not* exist, but the corresponding *iterated integrals* exist and are *equal*.

Proof. $\int_\Omega |f(x,y)|dxdy$ does *not* exist, since if $\int_\Omega |f(x,y)|dxdy < +\infty$, then by Fubini's Theorem 7.1.2C the corresponding iterated integrals in (7.1.23a) exist and are *equal*, i.e.

$$\int_0^\infty\left(\int_{x=0}^{x=\infty} |f(x,y)|dx\right)dy = \int_0^\infty\left(\int_{y=0}^{y=\infty} |f(x,y)|dy\right)dx < +\infty.$$

In fact,

$$\int_0^\infty \left(\int_{y=0}^{y\to\infty} |\sin x| e^{-xy}\, dy \right) dx = \int_0^\infty |\sin x| \left(\int_{y=0}^{y\to\infty} e^{-xy}\, dy \right) dx$$
$$= \int_0^\infty \frac{|\sin x|}{x} dx = \infty.$$

But

$$\int_0^\infty \frac{|\sin x|}{x} dx = \sum_{n=0}^{\infty} \int_{n\pi}^{(n+1)\pi} \frac{|\sin x|}{x} dx = \sum_{n=0}^{\infty} \int_0^\pi \frac{\sin t}{n\pi + t} dt,$$

which is obtained by the change of variable $x = n\pi + t$. Since $(n\pi + t) \le (n+1)\pi$ $\forall t \in [0, \pi]$, we have, $\forall n \in N_0$:

$$\int_0^\pi \frac{\sin t}{n\pi + t} dt \ge \int_0^\pi \frac{\sin t}{(n+1)\pi} dt = \frac{1}{(n+1)\pi} (-\cos t)\Big|_0^\pi = \frac{2}{(n+1)\pi}$$

$$\Longrightarrow \int_0^\infty \frac{|\sin x|}{x} dx = \sum_{n=0}^{\infty} \int_0^\pi \frac{\sin t}{n\pi + t} dt \ge \frac{2}{\pi} \sum_{n=0}^{\infty} \frac{1}{n+1} = \infty,$$

since $1 + \frac{1}{2} + \cdots + \frac{1}{n} + \cdots = \infty$. Hence, $\int_0^\infty \frac{|\sin x|}{x} dx = \infty$. So the iterated integral $\int_0^\infty (\int_{y=0}^{y\to\infty} |\sin x| e^{-xy} dy) dx$ does *not* exist. Consequently, the double integral $\int_0^\infty \int_0^\infty |\sin x e^{-xy}| dx dy$ does *not* exist by Fubini's Theorem 7.1.2C. Hence, the double integral $\int_0^\infty \int_0^\infty \sin x e^{-xy} dx dy$ does not exist in the sense of Lebesgue. But as an improper Riemann integral, we have

$$\int_0^\infty \sin x \left(\int_{y=0}^{y=\infty} e^{-xy}\, dy \right) dx = \int_0^\infty \sin x \left(\frac{e^{-xy}}{-x} \right)\Big|_{y=0}^{y=\infty} dx = \int_0^\infty \frac{\sin x}{x} dx = \pi/2.$$

$$\int_0^\infty \left(\int_{x=0}^{x=\infty} e^{-xy} \sin x\, dx \right) dy = \int_0^\infty \frac{y^2}{1+y^2} \left[-\frac{\sin x e^{-xy}}{y} - \frac{\cos x e^{-xy}}{y^2} \right]\Bigg|_{x=0}^{x=\infty} dy$$
$$= \int_0^\infty \frac{dy}{1+y^2} = \tan^{-1}(y)\Big|_0^\infty = \pi/2.$$

□

Now we state the additional properties of Fourier transform $\hat{f} = \mathcal{F} f$ of $f \in L^1(\mathbb{R}^n)$.

Property 6 $\hat{f}$ is *continuous* in $\mathbb{R}^n$, i.e. $\hat{f} \in C^0(\mathbb{R}^n)$. (7.1.24)

Proof. $f(\mathbf{x}) e^{-i2\pi\langle \mathbf{x}, \boldsymbol{\xi} \rangle}$ is *separately* continuous with respect to $\boldsymbol{\xi}$ for almost all $\mathbf{x} \in \mathbb{R}^n$ and $|f(\mathbf{x}) e^{-i2\pi\langle \mathbf{x}, \boldsymbol{\xi} \rangle}| = |f(\mathbf{x})|$. Choose $g(\mathbf{x}) = |f(\mathbf{x})|$ with $\int_{\mathbb{R}^n} g(\mathbf{x}) d\mathbf{x} = \int_{\mathbb{R}^n} |f(\mathbf{x})| d\mathbf{x} < +\infty$. Hence, by Theorem 7.1.2A, which also holds for several variables, $\hat{f}(\boldsymbol{\xi}) = \int_{\mathbb{R}^n} f(\mathbf{x}) e^{-i2\pi\langle \mathbf{x}, \boldsymbol{\xi} \rangle} d\mathbf{x}$ is a continuous function of $\boldsymbol{\xi} \in \mathbb{R}^n$, i.e. $\hat{f}$ is continuous in $\mathbb{R}^n$ and $\hat{f} \in C^0(\mathbb{R}^n)$. □

Property 7 $\hat{f}$ is *not* summable on $\mathbb{R}^n$ *in general*, i.e. $\hat{f} \notin L^1(\mathbb{R}^n)$ in general (see Example 7.1.1). Hence, $\bar{\mathcal{F}}\hat{f}$ is *not* defined for $f \in L^1(\mathbb{R}^n)$ in general.

Property 8

Proposition 7.1.1. *Let $f, g \in L^1(\mathbb{R}^n)$ with*

$$\hat{f}(\boldsymbol{\xi}) = \int_{\mathbb{R}^n} f(\mathbf{x})e^{-i2\pi\langle \mathbf{x},\boldsymbol{\xi}\rangle}d\mathbf{x}, \quad \hat{g}(\mathbf{x}) = \int_{\mathbb{R}^n} g(\boldsymbol{\xi})e^{-i2\pi\langle \boldsymbol{\xi},\mathbf{x}\rangle}d\boldsymbol{\xi}. \tag{7.1.25}$$

Then

$$\int_{\mathbb{R}^n} \hat{f}(\boldsymbol{\xi})g(\boldsymbol{\xi})d\boldsymbol{\xi} = \int_{\mathbb{R}^n} f(\mathbf{x})\hat{g}(\mathbf{x})d\mathbf{x}; \tag{7.1.26}$$

$$\int_{\mathbb{R}^n} (\bar{\mathcal{F}}f)(\boldsymbol{\xi})g(\boldsymbol{\xi})d\boldsymbol{\xi} = \int_{\mathbb{R}^n} f(\mathbf{x})(\bar{\mathcal{F}}g)(\mathbf{x})d\mathbf{x}. \tag{7.1.27}$$

Proof. From Theorem 7.1.1, $\hat{f}, \hat{g} \in L^\infty(\mathbb{R}^n)$ with

$$|\hat{f}(\boldsymbol{\xi})| \le \|f\|_{L^1(\mathbb{R}^n)}\ \forall \boldsymbol{\xi} \in \mathbb{R}^n, \quad |\hat{g}(\mathbf{x})| \le \|g\|_{L^1(\mathbb{R}^n)}\ \forall \mathbf{x} \in \mathbb{R}^n.$$

Since $f \in L^1(\mathbb{R}^n)$ and $\hat{g} \in L^\infty(\mathbb{R}^n)$, $\hat{f} \in L^\infty(\mathbb{R}^n)$ and $g \in L^1(\mathbb{R}^n)$, by Hölder's inequality (see Theorem B.4.1.1 and Table B.1, Appendix B), $f\hat{g} \in L^1(\mathbb{R}^n)$, $\hat{f}g \in L^1(\mathbb{R}^n)$. Again, for $f, g \in L^1(\mathbb{R}^n)$, their tensor product $f(\mathbf{x}) \otimes g(\boldsymbol{\xi})$ is integrable on $\mathbb{R}^n \times \mathbb{R}^n$, i.e.

$$\int_{\mathbb{R}^n\times\mathbb{R}^n} |f(\mathbf{x})g(\boldsymbol{\xi})|d\mathbf{x}d\boldsymbol{\xi} = \int_{\mathbb{R}^n} |f(\mathbf{x})|d\mathbf{x}\int_{\mathbb{R}^n} |g(\boldsymbol{\xi})|d\boldsymbol{\xi} < +\infty.$$

Consequently, $|f(\mathbf{x})g(\boldsymbol{\xi})e^{-i2\pi\langle \mathbf{x},\boldsymbol{\xi}\rangle}| = |f(\mathbf{x})g(\boldsymbol{\xi})|$ is integrable on $\mathbb{R}^n \times \mathbb{R}^n$. Hence, we can apply Fubini's Theorem 7.1.2C and get

$$\begin{aligned}\iint_{\mathbb{R}^n\times\mathbb{R}^n} f(\mathbf{x})g(\boldsymbol{\xi})e^{-i2\pi\langle \mathbf{x},\boldsymbol{\xi}\rangle}d\mathbf{x}d\boldsymbol{\xi} &= \int_{\mathbb{R}^n} f(\mathbf{x})\left(\int_{\mathbb{R}^n} g(\boldsymbol{\xi})e^{-i2\pi\langle \mathbf{x},\boldsymbol{\xi}\rangle}d\boldsymbol{\xi}\right)d\mathbf{x}\\ &= \int_{\mathbb{R}^n} g(\boldsymbol{\xi})\left(\int_{\mathbb{R}^n} f(\mathbf{x})e^{-i2\pi\langle \mathbf{x},\boldsymbol{\xi}\rangle}d\mathbf{x}\right)d\boldsymbol{\xi}.\end{aligned} \tag{7.1.28}$$

But $\langle \mathbf{x}, \boldsymbol{\xi}\rangle = \langle \boldsymbol{\xi}, \mathbf{x}\rangle$,

$$\int_{\mathbb{R}^n} g(\boldsymbol{\xi})e^{-i2\pi\langle \boldsymbol{\xi},\mathbf{x}\rangle}d\boldsymbol{\xi} = \hat{g}(\mathbf{x}) \quad \text{and}$$

$$\int_{\mathbb{R}^n} f(\mathbf{x})e^{-i2\pi\langle \mathbf{x},\boldsymbol{\xi}\rangle}d\mathbf{x} = \hat{f}(\boldsymbol{\xi}). \tag{7.1.29}$$

Then, from (7.1.28) and (7.1.29), we get the result: $\int_{\mathbb{R}^n} f(\mathbf{x})\hat{g}(\mathbf{x})d\mathbf{x} = \int_{\mathbb{R}^n} g(\boldsymbol{\xi})\hat{f}(\boldsymbol{\xi})d\boldsymbol{\xi}$.

Similarly, (7.1.27) is proved. □

Property 9

Corollary 7.1.1 (Riesz formula). $\forall f, g \in L^1(\mathbb{R}^n)$,

$$\int_{\mathbb{R}^n} \hat{f}(\boldsymbol{\xi}) g(\boldsymbol{\xi}) e^{i2\pi\langle \mathbf{a}, \boldsymbol{\xi}\rangle} d\boldsymbol{\xi} = \int_{\mathbb{R}^n} f(\mathbf{a}+\mathbf{x}) \hat{g}(\mathbf{x}) d\mathbf{x}, \tag{7.1.30}$$

where $\hat{f}(\boldsymbol{\xi})$ and $\hat{g}(\mathbf{x})$ are defined by (7.1.25).

$$\int_{\mathbb{R}^n} (\bar{\mathcal{F}} f)(\boldsymbol{\xi}) g(\boldsymbol{\xi}) e^{-i2\pi\langle \mathbf{a}, \boldsymbol{\xi}\rangle} d\boldsymbol{\xi} = \int_{\mathbb{R}^n} f(\mathbf{a}+\mathbf{x}) \bar{\mathcal{F}} g(\mathbf{x}) d\mathbf{x}. \tag{7.1.31}$$

Proof. $(\widehat{\tau_{-\mathbf{a}} f})(\boldsymbol{\xi}) = \int_{\mathbb{R}^n} (\tau_{-\mathbf{a}} f)(\mathbf{x}) e^{-i2\pi\langle \mathbf{x}, \boldsymbol{\xi}\rangle} d\mathbf{x} = \int_{\mathbb{R}^n} f(\mathbf{x}+\mathbf{a}) e^{-i2\pi\langle \mathbf{x}, \boldsymbol{\xi}\rangle} d\mathbf{x}$.

By the change of variables $\mathbf{y} = \mathbf{x} + \mathbf{a}$, we have $\mathbf{x} = \mathbf{y} - \mathbf{a}$ with Jacobian $J = 1$. Then

$$\begin{aligned}(\widehat{\tau_{-\mathbf{a}} f})(\boldsymbol{\xi}) &= \int_{\mathbb{R}^n} f(\mathbf{y}) e^{-i2\pi\langle \mathbf{y}-\mathbf{a}, \boldsymbol{\xi}\rangle} |J| d\mathbf{y} \\ &= \left(\int_{\mathbb{R}^n} f(\mathbf{y}) e^{-i2\pi\langle \mathbf{y}, \boldsymbol{\xi}\rangle} d\mathbf{y}\right) e^{+i2\pi\langle \mathbf{a}, \boldsymbol{\xi}\rangle} = \hat{f}(\boldsymbol{\xi}) e^{i2\pi\langle \mathbf{a}, \boldsymbol{\xi}\rangle}.\end{aligned}$$

Hence,

$$\begin{aligned}\int_{\mathbb{R}^n} \hat{f}(\boldsymbol{\xi}) g(\boldsymbol{\xi}) e^{i2\pi\langle \mathbf{a}, \boldsymbol{\xi}\rangle} d\boldsymbol{\xi} &= \int_{\mathbb{R}^n} (\widehat{\tau_{-\mathbf{a}} f})(\boldsymbol{\xi}) g(\boldsymbol{\xi}) d\boldsymbol{\xi} \\ &= \int_{\mathbb{R}^n} \left(\int_{\mathbb{R}^n} f(\mathbf{x}+\mathbf{a}) e^{-i2\pi\langle \mathbf{x}, \boldsymbol{\xi}\rangle} d\mathbf{x}\right) g(\boldsymbol{\xi}) d\boldsymbol{\xi} \\ &= \int_{\mathbb{R}^n} f(\mathbf{x}+\mathbf{a}) \left(\int_{\mathbb{R}^n} g(\boldsymbol{\xi}) e^{-i2\pi\langle \boldsymbol{\xi}, \mathbf{x}\rangle} d\boldsymbol{\xi}\right) d\mathbf{x} \\ &= \int_{\mathbb{R}^n} f(\mathbf{x}+\mathbf{a}) \hat{g}(\mathbf{x}) d\mathbf{x},\end{aligned}$$

since $f(\mathbf{x}+\mathbf{a}) g(\boldsymbol{\xi}) \in L^1(\mathbb{R}^n \times \mathbb{R}^n)$, and consequently the order of integration can be interchanged by Fubini's Theorem 7.1.2C.

Similarly, (7.1.31) can be proved. □

Property 10

Theorem 7.1.3. *Let $f \in L^1(\mathbb{R}^n)$ and $\hat{f} = \mathcal{F} f$ be its Fourier transform.*

I. *If $\mathbf{x}^{\boldsymbol{\alpha}} f \in L^1(\mathbb{R}^n) \forall$ multi-index $\boldsymbol{\alpha} = (\alpha_1, \alpha_2, \ldots, \alpha_n)$ with $|\boldsymbol{\alpha}| = \alpha_1 + \alpha_2 + \cdots + \alpha_n \le m$, $\mathbf{x}^{\boldsymbol{\alpha}} = x_1^{\alpha_1} x_2^{\alpha_2} \cdots x_n^{\alpha_n}$, then*

$$\hat{f} \in C^m(\mathbb{R}^n) \quad \text{and} \quad \partial_{\boldsymbol{\xi}}^{\boldsymbol{\alpha}} \hat{f} = \frac{\partial^{|\boldsymbol{\alpha}|} \hat{f}}{\partial \xi_1^{\alpha_1} \partial \xi_2^{\alpha_2} \cdots \partial \xi_n^{\alpha_n}} = \mathcal{F}[(-i2\pi\mathbf{x})^{\boldsymbol{\alpha}} f], \tag{7.1.32}$$

where $(-i2\pi\mathbf{x})^{\boldsymbol{\alpha}} = (-i2\pi x_1)^{\alpha_1} (-i2\pi x_2)^{\alpha_2} \cdots (-i2\pi x_n)^{\alpha_n}$.

II. *If* $f \in C^m(\mathbb{R}^n)$ *and* $\partial_{\mathbf{x}}^{\boldsymbol{\alpha}} f = \frac{\partial^{|\boldsymbol{\alpha}|} f}{\partial x_1^{\alpha_1} \partial x_2^{\alpha_2} \cdots \partial x_n^{\alpha_n}} \in L^1(\mathbb{R}^n)$ $\forall$ *multi-index* $\boldsymbol{\alpha} = (\alpha_1, \alpha_2, \ldots, \alpha_n)$ *with* $|\boldsymbol{\alpha}| \le m$*, then*

$$\boldsymbol{\xi}^{\boldsymbol{\alpha}} \hat{f} \text{ is bounded in } \mathbb{R}^n \text{ (i.e. } \boldsymbol{\xi}^{\boldsymbol{\alpha}} \hat{f} \in L^\infty(\mathbb{R}^n)) \text{ and } (i2\pi\boldsymbol{\xi})^{\boldsymbol{\alpha}} \hat{f} = \mathcal{F}[\partial_{\mathbf{x}}^{\boldsymbol{\alpha}} f], \tag{7.1.33}$$

where $(i2\pi\boldsymbol{\xi})^{\boldsymbol{\alpha}} = (i2\pi\xi_1)^{\alpha_1} (i2\pi\xi_2)^{\alpha_2} \cdots (i2\pi\xi_n)^{\alpha_n}$.

Proof.

I. It is sufficient to prove the result for $\boldsymbol{\alpha} = (1, 0, \ldots, 0)$ with $\alpha_1 = 1$, i.e. if $x_1 f(\mathbf{x}) \in L^1(\mathbb{R}^n)$ ($\int_{\mathbb{R}^n} |x_1 f(\mathbf{x})| d\mathbf{x} < +\infty$), then

(i) $\hat{f}(\boldsymbol{\xi})$ is differentiable with respect to ξ_1 in the usual pointwise sense;

(ii) $\frac{\partial \hat{f}}{\partial \xi_1}(\boldsymbol{\xi}) = \int_{\mathbb{R}^n} (-i2\pi x_1) f(\mathbf{x}) e^{-i2\pi\langle \mathbf{x}, \boldsymbol{\xi}\rangle} d\mathbf{x}$.

For this, we set $F(\mathbf{x}, \boldsymbol{\xi}) = f(\mathbf{x}) e^{-i2\pi\langle \mathbf{x}, \boldsymbol{\xi}\rangle}$ such that $\hat{f}(\boldsymbol{\xi}) = \int_{\mathbb{R}^n} F(\mathbf{x}, \boldsymbol{\xi}) d\mathbf{x}$ $\forall \boldsymbol{\xi} \in \mathbb{R}^n$.

Since $\frac{\partial}{\partial \xi_1}\langle \mathbf{x}, \boldsymbol{\xi}\rangle = \frac{\partial}{\partial \xi_1}(x_1\xi_1 + x_2\xi_2 + \cdots + x_n\xi_n) = x_1$,

$$\frac{\partial F(\mathbf{x}, \boldsymbol{\xi})}{\partial \xi_1} = f(\mathbf{x}) e^{-i2\pi\langle \mathbf{x}, \boldsymbol{\xi}\rangle} \frac{\partial}{\partial \xi_1}[-i2\pi\langle \mathbf{x}, \boldsymbol{\xi}\rangle] = (-i2\pi x_1) f(\mathbf{x}) e^{-i2\pi\langle \mathbf{x}, \boldsymbol{\xi}\rangle},$$

which is separately continuous with respect to $\boldsymbol{\xi}$ in $\mathbb{R}^n$ for almost all $\mathbf{x} \in \mathbb{R}^n$, $|\frac{\partial F}{\partial \xi_1}(\mathbf{x}, \boldsymbol{\xi})| = 2\pi |x_1 f(\mathbf{x})|$ for almost all $\mathbf{x} \in \mathbb{R}^n$ and $\int_{\mathbb{R}^n} 2\pi |x_1 f(\mathbf{x})| d\mathbf{x} < +\infty$; (Lebesgue's) Theorem 7.1.2B on differentiability under the integral sign $\int$ can be applied, i.e. $\hat{f}$ is differentiable with respect to ξ_1 and

$$\frac{\partial \hat{f}}{\partial \xi_1} = \int_{\mathbb{R}^n} \frac{\partial}{\partial \xi_1} F(\mathbf{x}, \boldsymbol{\xi}) d\mathbf{x} = \int_{\mathbb{R}^n} (-i2\pi x_1) f(\mathbf{x}) e^{-i2\pi\langle \mathbf{x}, \boldsymbol{\xi}\rangle} d\mathbf{x}.$$

II. It is sufficient to show that if f and $\frac{\partial f}{\partial x_1}$ belong to $L^1(\mathbb{R}^n) \cap C^0(\mathbb{R}^n)$, then

$$i2\pi\xi_1 \hat{f}(\boldsymbol{\xi}) = \int_{\mathbb{R}^n} \frac{\partial f}{\partial x_1}(\mathbf{x}) e^{-i2\pi\langle \mathbf{x}, \boldsymbol{\xi}\rangle} d\mathbf{x} = \widehat{\frac{\partial f}{\partial x_1}}. \tag{7.1.34}$$

For this, we consider the difference of the two terms in (7.1.34):

$$\begin{aligned}
&\int_{\mathbb{R}^n} \frac{\partial f}{\partial x_1}(\mathbf{x}) e^{-i2\pi\langle \mathbf{x}, \boldsymbol{\xi}\rangle} d\mathbf{x} - i2\pi\xi_1 \int_{\mathbb{R}^n} f(\mathbf{x}) e^{-i2\pi\langle \mathbf{x}, \boldsymbol{\xi}\rangle} d\mathbf{x} \\
&= \int_{\mathbb{R}^n} \left[\frac{\partial f}{\partial x_1}(\mathbf{x}) e^{-i2\pi\langle \mathbf{x}, \boldsymbol{\xi}\rangle} - i2\pi\xi_1 e^{-i2\pi\langle \mathbf{x}, \boldsymbol{\xi}\rangle} f(\mathbf{x})\right] d\mathbf{x} \\
&= \int_{\mathbb{R}^{n-1}} \left(\int_{-\infty}^{\infty} \frac{\partial}{\partial x_1}[f(\mathbf{x}) e^{-i2\pi\langle \mathbf{x}, \boldsymbol{\xi}\rangle}] dx_1\right) dx_2 \ldots dx_n \\
&= \int_{\mathbb{R}^{n-1}} [f(x_1, x_2, \ldots, x_n) e^{-i2\pi\langle \mathbf{x}, \boldsymbol{\xi}\rangle}]_{x_1=-\infty}^{x_1=+\infty} dx_2 \ldots dx_n.
\end{aligned} \tag{7.1.35}$$

For the proof of (7.1.34), we must show that (7.1.35) vanishes. For this, it is sufficient to show that for f and f' (considered as functions of a *single* variable) belonging to $L^1(\mathbb{R}) \cap C^0(\mathbb{R})$, $f(+\infty) = 0$ and $f(-\infty) = 0$. In fact,

- $f' \in L^1(\mathbb{R}) \Longrightarrow \lim_{a\to\pm\infty} \int_0^a f'(t)dt$ exists;
- $f' \in C(\mathbb{R}) \Longrightarrow f(a) - f(0) = \int_0^a f'(t)dt \ \forall a \in \mathbb{R}$.

Hence, for $f' \in L^1(\mathbb{R}) \cap C(\mathbb{R})$,

$$\lim_{a\to\pm\infty} f(a) = \lim_{a\to\pm\infty} [f(0) + \int_0^a f'(t)dt] = f(0) + \lim_{a\to\pm\infty} \int_0^a f'(t)dt$$

$\Longrightarrow f(\pm\infty)$ exists.

But $f \in L^1(\mathbb{R}) \Longrightarrow f(\pm\infty) = 0$.

(Suppose that the contrary holds, i.e. $\lim_{x\to\infty} f(x) = L \neq 0$. Then $\exists M > 0$ such that $|f(x)| > |\frac{L}{2}| \ \forall x > M \Longrightarrow \int_M^\infty |f(x)|dx \geq \int_M^\infty |\frac{L}{2}|dx$, but $\int_M^\infty |\frac{L}{2}|dx = \infty$ for $L \neq 0 \Longrightarrow \int_M^\infty |f(x)|dx = \infty$, which contradicts that f is summable on $]-\infty, \infty[$: $\int_{-\infty}^\infty |f(x)|dx < +\infty$. Hence, our assumption is wrong, i.e. $\lim_{x\to\infty} f(x) = L = 0$. Similarly, we can prove that $\lim_{x\to-\infty} f(x) = 0$.)

□

Property 11 *The Riemann–Lebesgue Property.* Let $f \in L^1(\mathbb{R}^n)$ and $\hat{f}(\boldsymbol{\xi}) = (\mathcal{F} f)(\boldsymbol{\xi})$ be its Fourier transform. Then $\hat{f}(\boldsymbol{\xi}) \to 0$ as $\|\boldsymbol{\xi}\| \to \infty$. (7.1.36)

Proof. Let $\phi \in \mathcal{D}(\mathbb{R}^n) \subset L^1(\mathbb{R}^n)$, $\mathcal{D}(\mathbb{R}^n)$ being a dense subspace of $L^1(\mathbb{R}^n)$. Then, from (7.1.33) with $\alpha_i = 1$, $1 \leq i \leq n$, we have

$$i2\pi\xi_i\hat{\phi}(\boldsymbol{\xi}) = \int_{\mathbb{R}^n} \frac{\partial\phi}{\partial x_i}(\mathbf{x})e^{-i2\pi\langle\mathbf{x},\boldsymbol{\xi}\rangle}d\mathbf{x} = \widehat{\frac{\partial\phi}{\partial x_i}}(\boldsymbol{\xi}).$$

Hence, for $\xi_i \neq 0$,

$$|\hat{\phi}(\boldsymbol{\xi})| = \frac{1}{2\pi|\xi_i|}\left|\widehat{\frac{\partial\phi}{\partial x_i}}(\boldsymbol{\xi})\right|.$$

From (7.1.6),

$$\left\|\widehat{\frac{\partial\phi}{\partial x_i}}(\boldsymbol{\xi})\right\|_\infty = \sup_{\boldsymbol{\xi}\in\mathbb{R}^n}\left|\widehat{\frac{\partial\phi}{\partial x_i}}(\boldsymbol{\xi})\right| \leq \left\|\frac{\partial\phi}{\partial x_i}\right\|_{L^1(\mathbb{R}^n)}$$

$\Longrightarrow$ for $\xi_i \neq 0$, $|\hat{\phi}(\boldsymbol{\xi})| \leq \frac{1}{2\pi|\xi_i|}\|\frac{\partial\phi}{\partial x_i}\|_{L^1(\mathbb{R}^n)}$. But $\|\boldsymbol{\xi}\| \to \infty \Longrightarrow$ at least one of the $|\xi_i| \to \infty$

$\Longrightarrow$ at least one of $\frac{1}{|\xi_i|} \to 0 \Longrightarrow |\hat{\phi}(\boldsymbol{\xi})| \le \frac{1}{2\pi|\xi_i|}\|\frac{\partial\phi}{\partial x_i}\|_{L^1(\mathbb{R}^n)} \to 0$ as $\|\boldsymbol{\xi}\| \to \infty$. Thus,

$$|\hat{\phi}(\boldsymbol{\xi})| \to 0 \quad \text{as } \|\boldsymbol{\xi}\| \to \infty \quad \text{for } \phi \in \mathcal{D}(\mathbb{R}^n). \tag{7.1.37}$$

Then the result $|\hat{f}(\boldsymbol{\xi})| \to 0$ as $\|\boldsymbol{\xi}\| \to \infty\ \forall f \in L^1(\mathbb{R}^n)$ will follow from the density of $\mathcal{D}(\mathbb{R}^n)$ in $L^1(\mathbb{R}^n)$. In fact, let $f \in L^1(\mathbb{R}^n)$. Then $\exists$ a sequence (ϕ_k) in $\mathcal{D}(\mathbb{R}^n)$ such that $\phi_k \to f$ in $L^1(\mathbb{R}^n)$ as $k \to \infty$, i.e. $\|f - \phi_k\|_{L^1(\mathbb{R}^n)} \to 0$ as $k \to \infty$. But $\phi_k \in \mathcal{D}(\mathbb{R}^n)$, $f \in L^1(\mathbb{R}^n) \Longrightarrow (\hat{f} - \hat{\phi}_k) \in C^0(\mathbb{R}^n)\ \forall k \in \mathbb{N}$ (by (7.1.24) of Property 6) with $\|\hat{f} - \hat{\phi}_k\|_\infty = \sup_{\boldsymbol{\xi}\in\mathbb{R}^n} |\hat{f}(\boldsymbol{\xi}) - \hat{\phi}_k(\boldsymbol{\xi})| \le \|f - \phi_k\|_{L^1(\mathbb{R}^n)} \to 0$ as $k \to \infty \Longrightarrow (\hat{\phi}_k)$ converges uniformly to $\hat{f} \in C^0(\mathbb{R}^n)$ in $\mathbb{R}^n \Longrightarrow \forall\varepsilon > 0$, $\exists n_0 = n_0(\varepsilon) \in \mathbb{N}$ such that

$$|\hat{f}(\boldsymbol{\xi}) - \hat{\phi}_k(\boldsymbol{\xi})| \le \|\hat{f} - \hat{\phi}_k\|_\infty < \frac{\varepsilon}{2} \quad \forall k \ge n_0,\ \forall \boldsymbol{\xi} \in \mathbb{R}^n. \tag{7.1.38}$$

From (7.1.38), $\forall k \in \mathbb{N}$, $|\hat{\phi}_k(\boldsymbol{\xi})| \to 0$ as $\|\boldsymbol{\xi}\| \to \infty$. We fix $k = n_0$ such that (7.1.38) holds with $k = n_0$. Then, $\forall\varepsilon > 0$, $\exists M > 0$ such that

$$|\hat{\phi}_{n_0}(\boldsymbol{\xi})| < \varepsilon/2 \quad \text{for } \|\boldsymbol{\xi}\| \ge M. \tag{7.1.39}$$

Hence, for $k = n_0$ satisfying (7.1.39), we have, from (7.1.38) and (7.1.39), $\forall\varepsilon > 0$, $\exists M > 0$ such that $|\hat{f}(\boldsymbol{\xi})| \le |\hat{f}(\boldsymbol{\xi}) - \hat{\phi}_{n_0}(\boldsymbol{\xi})| + |\hat{\phi}_{n_0}(\boldsymbol{\xi})| < \varepsilon/2 + \varepsilon/2 = \varepsilon$ $\forall\|\boldsymbol{\xi}\| \ge M$

$$\Longrightarrow \quad \lim_{\|\boldsymbol{\xi}\|\to\infty} \hat{f}(\boldsymbol{\xi}) = 0, \quad \text{i.e. } \hat{f}(\boldsymbol{\xi}) \to 0 \text{ as } \|\boldsymbol{\xi}\| \to \infty. \qquad \square$$

Example 7.1.9. Let $H(x)e^{-ax}\frac{x^{\alpha-1}}{\Gamma(\alpha)} \in L^1(\mathbb{R})$ with *unbounded* support. Prove that

1. $\mathcal{F}[H(x)e^{-ax}\frac{x^{\alpha-1}}{\Gamma(\alpha)}] = \frac{1}{(a+i2\pi\xi)^\alpha} = \hat{f}(\xi)$ for $a > 0, \alpha > 0$;
2. $\mathcal{F}[H(-x)\frac{e^{-a|x|}|x|^{\alpha-1}}{\Gamma(\alpha)}] = \frac{1}{a-i2\pi\xi)^\alpha}$ for $x < 0$;
3. $\mathcal{F}[e^{-a|x|}] = \frac{2a}{a^2+4\pi^2\xi^2}$.

Proof.

1. $$\mathcal{F}[f](\xi)\hat{f}(\xi) = \int_{-\infty}^{\infty}\left(H(x)e^{-ax}\frac{x^{\alpha-1}}{\Gamma(\alpha)}\right)e^{-i2\pi x\xi}dx$$
$$= \int_0^\infty e^{-(a+i2\pi\xi)x}\cdot\frac{x^{\alpha-1}}{\Gamma(\alpha)}dx = \lim_{\substack{\varepsilon\to0^+\\R\to\infty}}\int_\varepsilon^R e^{-(a+i2\pi\xi)x}\cdot\frac{x^{\alpha-1}}{\Gamma(\alpha)}dx.$$

We will follow [7] to evaluate this improper integral with the help of complex integration along a half-line in the complex plane. For this, we introduce the complex variable z defined by $z = (a + i2\pi\xi)x \in \mathbb{C}$, from which $dx = \frac{dz}{a+i2\pi\xi}$, such that z varies from 0 to ∞ (i.e. $(a + i2\pi\xi)\infty$) along the half-line $]0, (a + i2\pi\xi)\infty[$ in the complex plane of z as x varies from 0 to ∞ along the real axis. Then,

$$\hat{f}(\xi) = \int_{]0,(a+i2\pi\xi)\infty[} e^{-z} \cdot \frac{(\frac{x}{a+i2\pi\xi})^{\alpha-1}}{\Gamma(\alpha)} \cdot \frac{dz}{a+i2\pi\xi}$$
$$= \frac{1}{(a+i2\pi\xi)^{\alpha}\Gamma(\alpha)} \int_{]0,(a+i2\pi\xi)\infty[} e^{-z} \cdot z^{\alpha-1} dz.$$

Since $z^{\alpha-1}$ is a multiple-valued function, we will choose the *branch* of $z^{\alpha-1}$ for $\mathrm{Re}(z) > 0$ such that $e^{-z}z^{\alpha-1}$ is analytic (or holomorphic) for $\mathrm{Re}(z) \in]0, \infty[$. Then, by virtue of the analyticity of $e^{-z}z^{\alpha-1}$ in the right-hand side half-plane of z, the complex integral along $]0, (a + i2\pi\xi)\infty[$ is evaluated by:

$$\int_{]0,(a+i2\pi\xi)\infty[} e^{-z} \cdot z^{\alpha-1} dz = \lim_{\substack{\varepsilon\to 0^+ \\ R\to\infty}} \int_{(a+i2\pi\xi)\varepsilon}^{(a+i2\pi\xi)R} e^{-z} \cdot z^{\alpha-1} dz$$
$$= \lim_{\substack{\varepsilon\to 0^+ \\ R\to\infty}} I(\varepsilon; R).$$

For the evaluation of complex integral $I(\varepsilon; R)$ along the complex half-line $]0, (a + i2\pi\xi)\infty[$ from the point $(a + i2\pi\xi)\varepsilon$ to $(a + i2\pi\xi)R$, we can *replace* this path of integration along the complex half-line from $(a + i2\pi\xi)\varepsilon$ to $(a+i2\pi\xi)R$ by *any suitable path* in the right-hand side half-plane of z in which the integrand is *analytic*. Hence we choose a contour with 4 vertices:

$$A : (a + i2\pi\xi)\varepsilon; \quad B : (\varepsilon, 0); \quad C : (R, 0); \quad D : (a + i2\pi\xi)R,$$

such that $\widehat{AB}$ is a curve joining A and B, BC is a segment of the real axis joining B and C and $\widehat{CD}$ is a curve joining C and D.

$$I(\varepsilon; R) = \int_{[AD]} e^{-z} \cdot z^{\alpha-1} dz$$
$$= \int_{\widehat{AB}} e^{-z} \cdot z^{\alpha-1} dz + \int_{[BC]} e^{-x} \cdot x^{\alpha-1} dx + \int_{\widehat{CD}} e^{-z} \cdot z^{\alpha-1} dz.$$

Then

$$\int_{]0,(a+i2\pi\xi)\infty[} e^{-z} \cdot z^{\alpha-1} dz = \lim_{\substack{\varepsilon\to 0^+ \\ R\to\infty}} I(\varepsilon; R) = \lim_{\substack{\varepsilon\to 0^+ \\ R\to\infty}} \int_{[BC]} e^{-x} \cdot x^{\alpha-1} dx$$
$$= \lim_{\substack{\varepsilon\to 0^+ \\ R\to\infty}} \int_{\varepsilon}^{R} e^{-x} \cdot x^{\alpha-1} dx = \int_0^{\infty} e^{-x} \cdot x^{\alpha-1} dx = \Gamma(\alpha),$$

since

$$\lim_{\varepsilon\to 0+}\underbrace{\int_{\widehat{AB}} e^{-z}\cdot z^{\alpha-1}dz}_{I_1(\varepsilon)}=0,\qquad \lim_{R\to\infty}\underbrace{\int_{\widehat{CD}} e^{-z}\cdot z^{\alpha-1}dz}_{I_2(R)}=0,$$

with

$$\widehat{AB}\colon\begin{cases} x=a\varepsilon \text{ with } y \text{ from } 2\pi\xi\varepsilon \text{ to } 0,\\ y=0 \quad\text{with } x \text{ from } a\varepsilon \text{ to } \varepsilon;\end{cases}\qquad \widehat{CD}\colon\begin{cases} y=0 \quad\text{with } x \text{ from } R \text{ to } aR,\\ x=aR \text{ with } y \text{ from } 0 \text{ to } 2\pi\xi R;\end{cases}$$

$$\begin{aligned} I_1(\varepsilon)&=\int_{\widehat{AB}} e^{-z}\cdot z^{\alpha-1}dz\\ &=\underbrace{\int_{2\pi\xi\varepsilon}^{0} e^{-a\varepsilon-iy}(a\varepsilon+iy)^{\alpha-1}i\,dy}_{I_{1,1}(\varepsilon)}+\underbrace{\int_{a\varepsilon}^{\varepsilon} e^{-x}x^{\alpha-1}dx}_{I_{1,2}(\varepsilon)},\\ I_2(R)&=\int_{\widehat{CD}} e^{-z}\cdot z^{\alpha-1}dz\\ &=\underbrace{\int_{R}^{aR} e^{-x}x^{\alpha-1}dx}_{I_{2,1}(R)}+\underbrace{\int_{0}^{2\pi\xi R} e^{-aR-iy}(aR+iy)^{\alpha-1}i\,dy}_{I_{2,2}(R)}.\end{aligned}$$

We consider the case $0<a<1$ with $a\varepsilon<\varepsilon$, $aR<R$. For $a>1$, the same proof will hold with modifications owing to the sign change.

$\lim_{\varepsilon\to 0+} I_1(\varepsilon)=0$:

$$\begin{aligned} |I_{1,1}(\varepsilon)|&\le e^{-a\varepsilon}\int_0^{2\pi\xi\varepsilon}|a\varepsilon+iy|^{\alpha-1}dy\\ &\le e^{-a\varepsilon}(\sqrt{a^2+4\pi^2\xi^2})^{\alpha-1}\cdot\varepsilon^{\alpha-1}\cdot 2\pi\xi\varepsilon\\ &=2\pi\xi(\sqrt{a^2+4\pi^2\xi^2})^{\alpha-1}\cdot e^{-a\varepsilon}\cdot\varepsilon^{\alpha}\to 0\end{aligned}$$

as $\varepsilon\to 0$, since $e^{-a\varepsilon}\to e^0=1$, $\varepsilon^\alpha\to 0$ as $\varepsilon\to 0$ $\forall\alpha>0$. For $a=1$, $I_{1,2}=0$. Hence, for $a\neq 1$ with $a\varepsilon<\varepsilon$ (similarly, for $a\varepsilon>\varepsilon$) $I_{1,2}(\varepsilon)\to 0$. Finally, $|I_1(\varepsilon)|=|\int_{\widehat{AB}} e^{-z}\cdot z^{\alpha-1}dz|\le(|I_{1,1}(\varepsilon)|+|I_{1,2}(\varepsilon)|)\to 0$ as $\varepsilon\to 0^+$.

$\lim_{R\to\infty} I_2(R)=0$: $|I_{2,1}(R)|=\int_{aR}^{R} e^{-x}x^{\alpha-1}dx=e^{-\xi(R)}\cdot\frac{x^\alpha}{\alpha}\big|_{aR}^{R}$ (by the generalized Mean Value Theorem) $\to 0$ as $R\to\infty$, since $\xi(R)\to\infty$ as $R\to\infty$, $e^{-\xi(R)}R^\alpha$, $e^{-\xi(R)}(aR)^\alpha\to 0$ as $R\to\infty$ (by L'Hospital's rule)

$\forall a > 0$.

$$|I_{2,2}(R)| \le e^{-aR} \int_0^{2\pi\xi R} (\sqrt{a^2R^2 + 4\pi^2\xi^2R^2})^{\alpha-1} dy$$
$$\le 2\pi\xi(\sqrt{a^2 + 4\pi^2\xi^2})^{\alpha-1} \cdot e^{-aR}R^{\alpha} \to 0$$

as $R \to \infty$ $\forall \alpha > 0$ (using L'Hospital's rule).

Hence, $|I_2(R)| = |\int_{\overline{CD}} e^{-z} \cdot z^{\alpha-1} dz| \le (|I_{2,1}(R)| + |I_{2,2}(R)|) \to 0$ as $R \to \infty$.

Thus, $\int_{]0,(a+i2\pi\xi)\infty[} e^{-z} \cdot z^{\alpha-1} dz = \lim_{\varepsilon\to 0^+, R\to\infty} I(\varepsilon; R) = \Gamma(\alpha)$.

Therefore, $\hat{f}(\xi) = \frac{\Gamma(\alpha)}{(a+i2\pi\xi)^{\alpha}\Gamma(\alpha)} = \frac{1}{(a+i2\pi\xi)^{\alpha}}$.

2. For $x < 0$, $-x = |x|$ and $\mathcal{F}[f(-x)] = \hat{f}(-\xi)$, we have for $x < 0$,

$$\mathcal{F}\left[H(-x)e^{-a(-x)}\frac{(-x)^{\alpha-1}}{\Gamma(\alpha)}\right] = \mathcal{F}\left[H(-x)e^{-a|x|}\frac{|x|^{\alpha-1}}{\Gamma(\alpha)}\right]$$
$$= \frac{1}{(a + i2\pi(-\xi))^{\alpha}} = \frac{1}{(a - i2\pi\xi)^{\alpha}}.$$

3. Adding (1) and (2), we have $\mathcal{F}[e^{-a|x|}\frac{|x|^{\alpha-1}}{\Gamma(\alpha)}] = \frac{1}{(a+i2\pi\xi)^{\alpha}} + \frac{1}{a-i2\pi\xi)^{\alpha}}$. Finally, for $\alpha = 1$, we have $\Gamma(1) = 1$ and $|x|^{\alpha-1} = 1$ and $\mathcal{F}[e^{-a|x|}] = \frac{2a}{a^2+4\pi^2\xi^2}$. □

7.2 Space $S(\mathbb{R}^n)$ of infinitely differentiable functions with rapid decay at infinity

Infinitely differentiable functions with rapid decay at infinity

For the sake of simplicity we consider the case of a single variable ($n = 1$).

Definition 7.2.1. An infinitely differentiable complex-valued function $\phi \in C^{\infty}(]-\infty, \infty[)$ of the real variable x, and its derivatives $\phi^{(l)}(x)$ of all orders $l \in \mathbb{N}$, are said to *decrease* or *decay rapidly at infinity* if, $\forall l \in \mathbb{N}_0$, $|\phi^{(l)}(x)|$ with $\phi^{(0)}(x) = \phi(x)$ decreases *more rapidly* than any power of $\frac{1}{|x|}$ as $|x| \to \infty$, i.e. $\forall k, l \in \mathbb{N}_0$,

$$\frac{|\phi^{(l)}(x)|}{1/|x|^k} = |x^k\phi^{(l)}(x)| \to 0 \quad \text{as } |x| \to \infty,$$

or, in other words,

$$\lim_{|x|\to\infty} |x^k\phi^{(l)}(x)| = 0 \quad \forall k, l \in \mathbb{N}_0. \tag{7.2.1}$$

For example, $\phi(x) = e^{-x^2} \in C^\infty(]-\infty, \infty[)$ and its derivatives of all orders decrease faster than any power of $\frac{1}{|x|}$, since

$$\lim_{|x|\to\infty} \left| x^k \frac{d^l}{dx^l}(e^{-x^2}) \right| = 0 \quad \forall k, l \in \mathbb{N}_0 \tag{7.2.2}$$

by L'Hospital's rule.

Functions $\phi \in \mathcal{D}(]-\infty, \infty[)$ and their derivatives of all orders have compact support in $\mathbb{R}$. Consequently, ϕ and all its derivatives $\phi^{(l)}(x)$ of order $l \in \mathbb{N}$ are a fortiori functions with rapid decay at infinity. (7.2.3)

In fact, $\forall k, l \in \mathbb{N}_0$, $x^k \phi^{(l)}(x) = 0 \ \forall x$ lying outside the compact support of ϕ.

Hence, $\lim_{|x|\to\infty} |x^k \phi^{(l)}(x)| = 0 \ \forall k, l \in \mathbb{N}_0$.

Now we state the following lemma which will be used frequently.

Lemma 7.2.1 ([40]). *$\forall$ fixed $k \in \mathbb{N}$, $\exists$ a strictly positive constant $C > 0$, dependent on k, n, such that the following inequalities hold: $\forall$ multi-index $\boldsymbol{\beta} = (\beta_1, \beta_2, \ldots, \beta_n)$ with $|\boldsymbol{\beta}| = \beta_1 + \beta_2 + \cdots + \beta_n$, $\forall \boldsymbol{\xi} = (\xi_1, \xi_2, \ldots, \xi_n) \in \mathbb{C}^n$ with $\|\boldsymbol{\xi}\| = (|\xi_1|^2 + |\xi_2|^2 + \cdots + |\xi_n|^2)^{\frac{1}{2}}$,*

$$C(1 + \|\boldsymbol{\xi}\|^2)^k \le \sup_{|\boldsymbol{\beta}|\le k} |\boldsymbol{\xi}^{\boldsymbol{\beta}}|^2 \le \sup_{|\boldsymbol{\beta}|\le 2k} |\boldsymbol{\xi}^{\boldsymbol{\beta}}| \le (1 + \|\boldsymbol{\xi}\|^2)^k, \tag{7.2.4}$$

where the second inequality is actually an equality.

Proof. $\boldsymbol{\xi}^{\boldsymbol{\beta}} = \boldsymbol{\xi}_1^{\beta_1} \boldsymbol{\xi}_2^{\beta_2} \ldots \boldsymbol{\xi}_n^{\beta_n} \Longrightarrow |\boldsymbol{\xi}^{\boldsymbol{\beta}}|^2 = |\xi_1{}^{\beta_1} \xi_2{}^{\beta_2} \ldots \xi_n{}^{\beta_n}|^2 = |\boldsymbol{\xi}^{2\boldsymbol{\beta}}|$ with $2\boldsymbol{\beta} = (2\beta_1, 2\beta_2, \ldots, 2\beta_n)$

$$\Longrightarrow \quad \sup_{|\boldsymbol{\beta}|\le k} |\boldsymbol{\xi}^{\boldsymbol{\beta}}|^2 = \sup_{|\boldsymbol{\beta}|\le k} |\boldsymbol{\xi}^{2\boldsymbol{\beta}}| = \sup_{|\boldsymbol{\beta}|\le 2k} |\boldsymbol{\xi}^{\boldsymbol{\beta}}|. \tag{7.2.5}$$

But

$$\begin{aligned} |\boldsymbol{\xi}^{2\boldsymbol{\beta}}| &= |\xi_1|^{2\beta_1} |\xi_2|^{2\beta_2} \cdots |\xi_n|^{2\beta_n} \\ &\le \|\boldsymbol{\xi}\|^{2\beta_1} \|\boldsymbol{\xi}\|^{2\beta_2} \cdots \|\boldsymbol{\xi}\|^{2\beta_n} = \|\boldsymbol{\xi}\|^{2(\beta_1+\beta_2+\cdots+\beta_n)} = \|\boldsymbol{\xi}\|^{2|\boldsymbol{\beta}|}. \end{aligned}$$

Hence, for $|\boldsymbol{\beta}| \le 2k$, $|\boldsymbol{\xi}^{2\boldsymbol{\beta}}| \le \|\boldsymbol{\xi}\|^{4k} \le (1 + \|\boldsymbol{\xi}\|^2)^{2k}$
$\Longrightarrow$ For $|\boldsymbol{\beta}| \le 2k$, $|\boldsymbol{\xi}^{\boldsymbol{\beta}}|^2 = |\boldsymbol{\xi}^{2\boldsymbol{\beta}}| \le (1 + \|\boldsymbol{\xi}\|^2)^{2k} \Longrightarrow |\boldsymbol{\xi}^{\boldsymbol{\beta}}| \le (1 + \|\boldsymbol{\xi}\|^2)^k$
$\Longrightarrow \sup_{|\boldsymbol{\beta}|\le 2k} |\boldsymbol{\xi}^{\boldsymbol{\beta}}| \le (1 + \|\boldsymbol{\xi}\|^2)^k$

$$\Longrightarrow \quad \sup_{|\boldsymbol{\beta}|\le k} |\boldsymbol{\xi}^{\boldsymbol{\beta}}|^2 = \sup_{|\boldsymbol{\beta}|\le 2k} |\boldsymbol{\xi}^{\boldsymbol{\beta}}| \le (1 + \|\boldsymbol{\xi}\|^2)^k \quad \text{(using (7.2.5))}. \tag{7.2.6}$$

Now, we prove the first inequality using the multinomial of Newton. Introducing the multi-index notation $\boldsymbol{\beta}! = \beta_1!\beta_2!\dots\beta_n!$, we have

$$\begin{aligned}(1+\|\boldsymbol{\xi}\|^2)^k &= (1+|\xi_1|^2+|\xi_2|^2+\dots+|\xi_n|^2)^k \\ &= \sum_{|\boldsymbol{\beta}|\le k}\frac{k!}{\boldsymbol{\beta}!(k-|\boldsymbol{\beta}|)!}|\xi_1|^{2\beta_1}|\xi_2|^{2\beta_2}\dots|\xi_n|^{2\beta_n} = \sum_{|\boldsymbol{\beta}|\le k}\frac{k!}{\boldsymbol{\beta}!(k-|\boldsymbol{\beta}|)!}|\boldsymbol{\xi}^{2\boldsymbol{\beta}}| \\ &\le C_0(k,n)\sup_{|\boldsymbol{\beta}|\le k}|\boldsymbol{\xi}^{2\boldsymbol{\beta}}| = C_0(k,n)\sup_{|\boldsymbol{\beta}|\le k}|\boldsymbol{\xi}^{\boldsymbol{\beta}}|^2,\end{aligned}$$

with $C_0(k,n) = \sum_{|\boldsymbol{\beta}|\le k}\frac{k!}{\boldsymbol{\beta}!(k-|\boldsymbol{\beta}|)!} > 0$

$$\Longrightarrow \quad C(1+\|\boldsymbol{\xi}\|^2)^k \le \sup_{|\boldsymbol{\beta}|\le k}|\boldsymbol{\xi}^{\boldsymbol{\beta}}|^2 \tag{7.2.7}$$

with $C = \frac{1}{C_0} > 0$. Hence, from (7.2.5)–(7.2.7), we get the result (7.2.4). □

Definition 7.2.2. An infinitely differentiable complex-valued function $\phi \in C^\infty(\mathbb{R}^n)$ of n real variables $x_1, x_2, \dots, x_n \in \mathbb{R}$ and its partial derivatives $\partial^{\boldsymbol{\beta}}\phi(\mathbf{x}) = \frac{\partial^{|\boldsymbol{\beta}|}}{\partial_{x_1}^{\beta_1}\partial_{x_2}^{\beta_2}\dots\partial_{x_n}^{\beta_n}}\phi(\mathbf{x})$ of all orders $|\boldsymbol{\beta}| \in \mathbb{N}$ are said to *decrease* or *decay rapidly at infinity* if $\partial^{\boldsymbol{\beta}}\phi(\mathbf{x})$ of all orders $|\boldsymbol{\beta}| \in \mathbb{N}_0$ (with $\partial^{(\mathbf{0})}\phi(\mathbf{x}) = \phi(\mathbf{x})$) decrease *more rapidly* than $\frac{1}{|\mathbf{x}^{\boldsymbol{\alpha}}|}$ $\forall$ multi-index $\boldsymbol{\alpha}$ with $\mathbf{x}^{\boldsymbol{\alpha}} = x_1^{\alpha_1}x_2^{\alpha_2}\dots x_n^{\alpha_n}$ as $\|\mathbf{x}\| \to \infty$; i.e. if, $\forall \boldsymbol{\alpha}, \boldsymbol{\beta}$,

$$\lim_{\|\mathbf{x}\|\to\infty}\frac{|\partial^{\boldsymbol{\beta}}\phi(\mathbf{x})|}{1/|\mathbf{x}^{\boldsymbol{\alpha}}|} = \lim_{\|\mathbf{x}\|\to\infty}|\mathbf{x}^{\boldsymbol{\alpha}}\partial^{\boldsymbol{\beta}}\phi(\mathbf{x})| = 0 \tag{7.2.8}$$

or, equivalently, $\forall k \in \mathbb{N}_0$, $\forall|\boldsymbol{\beta}| \in \mathbb{N}_0$, $\lim_{\|\mathbf{x}\|\to\infty}|(1+\|\mathbf{x}\|^2)^k\partial^{\boldsymbol{\beta}}\phi(\mathbf{x})| = 0$, which follows from (7.2.8) and Lemma 7.2.1.

For example, for $\|\mathbf{x}\|^2 = {x_1}^2 + {x_2}^2 + \dots + {x_n}^2$, $\phi(\mathbf{x}) = e^{-\|\mathbf{x}\|^2} \in C^\infty(\mathbb{R}^n)$ and the partial derivatives $\partial^{\boldsymbol{\beta}}\phi(\mathbf{x})$ of all orders $|\boldsymbol{\beta}| \in \mathbb{N}$ are functions with rapid decay at infinity, since

$$\lim_{\|\mathbf{x}\|\to\infty}|\mathbf{x}^{\boldsymbol{\alpha}}\partial^{\boldsymbol{\beta}}[e^{-\|\mathbf{x}\|^2}]| = 0 \quad \forall\boldsymbol{\alpha},\boldsymbol{\beta} \in \mathbb{N}_0^n \quad \text{(see (7.2.2))}. \tag{7.2.9}$$

Functions $\phi \in \mathcal{D}(\mathbb{R}^n)$ and their partial derivatives $\partial^{\boldsymbol{\beta}}\phi(\mathbf{x})$ of all orders $|\boldsymbol{\beta}| \in \mathbb{N}$ are a fortiori functions with rapid decay at infinity (see (7.2.3)).

7.2.1 Space $S(\mathbb{R}^n)$

Definition 7.2.3. $S(\mathbb{R}^n)$ is the linear space of all complex-valued functions $\phi \in C^\infty(\mathbb{R}^n)$ which, together with all partial derivatives $\partial^{\boldsymbol{\beta}}\phi(\mathbf{x})$, *decay rapidly at infinity*,

i.e.

$$S(\mathbb{R}^n) = \{\phi : \phi \in C^\infty(\mathbb{R}^n); \text{for all multi-index } \boldsymbol{\alpha}, \boldsymbol{\beta}, \lim_{\|\mathbf{x}\|\to\infty} |\mathbf{x}^{\boldsymbol{\alpha}} \partial^{\boldsymbol{\beta}} \phi(\mathbf{x})| = 0\}, \tag{7.2.10}$$

or, equivalently,

$$S(\mathbb{R}^n) = \{\phi : \phi \in C^\infty(\mathbb{R}^n); \forall k \in \mathbb{N}_0, \forall |\boldsymbol{\beta}| \in \mathbb{N}_0, \lim_{\|\mathbf{x}\|\to\infty} |(1+\|\mathbf{x}\|^2)^k \partial^{\boldsymbol{\beta}} \phi(\mathbf{x})| = 0\}, \tag{7.2.11}$$

which follows from (7.2.10) by virtue of Lemma 7.2.1.

Important properties of functions of $S(\mathbb{R}^n)$

Proposition 7.2.1. *Let $\phi \in S(\mathbb{R}^n)$. Then the following properties hold:*

I. *For all multi-index $\boldsymbol{\alpha}, \boldsymbol{\beta}$,*

   i. $\mathbf{x}^{\boldsymbol{\alpha}} \partial^{\boldsymbol{\beta}} \phi(\mathbf{x})$,

   ii. $\partial^{\boldsymbol{\beta}} (\mathbf{x}^{\boldsymbol{\alpha}} \phi(\mathbf{x}))$,

   iii. $p(\mathbf{x}) \partial^{\boldsymbol{\beta}} \phi(\mathbf{x})$ *for any polynomial $p(\mathbf{x})$ in n variables,*

   are all bounded *in $\mathbb{R}^n$. Consequently,*

   - $\forall \boldsymbol{\alpha}, \boldsymbol{\beta}$,

$$q_{\boldsymbol{\alpha},\boldsymbol{\beta}}(\phi) = \sup_{\mathbf{x}\in\mathbb{R}^n} |\mathbf{x}^{\boldsymbol{\alpha}} \partial^{\boldsymbol{\beta}} \phi(\mathbf{x})| < +\infty; \tag{7.2.12}$$

   - $\forall l, m \in \mathbb{N}_0$,

$$q_{l,m}(\phi) = \sup_{\substack{|\boldsymbol{\alpha}|\le l \\ |\boldsymbol{\beta}|\le m}} q_{\boldsymbol{\alpha},\boldsymbol{\beta}}(\phi) = \sup_{\substack{|\boldsymbol{\alpha}|\le l \\ |\boldsymbol{\beta}|\le m}} \sup_{\mathbf{x}\in\mathbb{R}^n} |\mathbf{x}^{\boldsymbol{\alpha}} \partial^{\boldsymbol{\beta}} \phi(\mathbf{x})| < +\infty; \tag{7.2.13}$$

   - $\forall l \in \mathbb{N}_0, \forall \boldsymbol{\beta}$,

$$(1 + \|\mathbf{x}\|^2)^l \partial^{\boldsymbol{\beta}} \phi(\mathbf{x}) \textit{ is } \text{bounded} \textit{ in } \mathbb{R}^n; \tag{7.2.14}$$

   - $\forall l \in \mathbb{N}_0, \forall \boldsymbol{\beta}$,

$$q^*_{l,\boldsymbol{\beta}}(\phi) = \sup_{\mathbf{x}\in\mathbb{R}^n} |(1 + \|\mathbf{x}\|^2)^l \partial^{\boldsymbol{\beta}} \phi(\mathbf{x})| < +\infty; \tag{7.2.15}$$

   - $\forall l, m \in \mathbb{N}_0$,

$$q^*_{l,m}(\phi) = \sup_{|\boldsymbol{\beta}|\le m} q^*_{l,\boldsymbol{\beta}}(\phi) = \sup_{|\boldsymbol{\beta}|\le m} \sup_{\mathbf{x}\in\mathbb{R}^n} |(1 + \|\mathbf{x}\|^2)^l \partial^{\boldsymbol{\beta}} \phi(\mathbf{x})| < +\infty; \tag{7.2.16}$$

II. • $\forall \boldsymbol{\alpha}, \boldsymbol{\beta}$, $\mathbf{x}^{\boldsymbol{\alpha}} \partial^{\boldsymbol{\beta}} \phi$, $\partial^{\boldsymbol{\beta}}(\mathbf{x}^{\boldsymbol{\alpha}} \phi) \in S(\mathbb{R}^n)$; (7.2.17)

• $\forall l \in \mathbb{N}_0$, $\forall \boldsymbol{\beta}$, $(1 + \|\mathbf{x}\|^2)^l \partial^{\boldsymbol{\beta}} \phi \in S(\mathbb{R}^n)$, (7.2.18)

i.e. $\phi \in S(\mathbb{R}^n) \Longrightarrow$

i. *all derivatives* $\partial^{\boldsymbol{\alpha}} \phi \in S(\mathbb{R}^n)$ $\forall \boldsymbol{\alpha}$;

ii. *the product* $\mathbf{x}^{\boldsymbol{\alpha}} \phi \in S(\mathbb{R}^n)$ $\forall \mathbf{x}^{\boldsymbol{\alpha}}$;

iii. *the product* $(1 + \|\mathbf{x}\|^2)^l \phi \in S(\mathbb{R}^n)$ $\forall l \in \mathbb{N}_0$.

Remark 7.2.1. For *arbitrary* $f \in C^\infty(\mathbb{R}^n)$ and $\phi \in S(\mathbb{R}^n)$, the product $f\phi \notin S(\mathbb{R}^n)$. For example, for $n = 1$, $f(x) = e^{x^2} \in C^\infty(\mathbb{R})$, $\phi(x) = e^{-x^2} \in S(\mathbb{R})$, the product $f\phi = 1 \notin S(\mathbb{R})$. But for $\phi \in S(\mathbb{R}^n)$ and a function $f \in C^\infty(\mathbb{R}^n)$ which has, along with all its derivatives, *polynomial* or *slow* growth at infinity: $\forall$ multi-index $\boldsymbol{\alpha}$, $\exists$ a constant $C > 0$ and an integer (see also (3)) $l = l(\boldsymbol{\alpha}) \in \mathbb{N}_0$ such that $|\partial^{\boldsymbol{\alpha}} f(\mathbf{x})| \le C(1 + \|\mathbf{x}\|^2)^l$ $\forall \mathbf{x} \in \mathbb{R}^n$, the product $f\phi \in S(\mathbb{R}^n)$. (7.2.19)

For example, for $n = 1$, $(ix)^k$ with $k \in \mathbb{N}$ is a C^∞-function of polynomial growth (by Lemma 7.2.1).

Proof of Proposition 7.2.1.

I. It is sufficient to show the boundedness of $\mathbf{x}^{\boldsymbol{\alpha}} \partial^{\boldsymbol{\beta}} \phi(\mathbf{x})$ in $\mathbb{R}^n$, because other cases can be shown similarly with minor modifications. In fact, since $\phi \in S(\mathbb{R}^n)$, by the definition of $S(\mathbb{R}^n)$ in (7.2.10), $\forall$ fixed $\boldsymbol{\alpha}, \boldsymbol{\beta}$, $\lim_{\|\mathbf{x}\| \to \infty} |\mathbf{x}^{\boldsymbol{\alpha}} \partial^{\boldsymbol{\beta}} \phi(\mathbf{x})| = 0 \Longrightarrow \forall \varepsilon \in]0, 1[$, $\exists R > 0$ such that $|\mathbf{x}^{\boldsymbol{\alpha}} \partial^{\boldsymbol{\beta}} \phi(\mathbf{x})| < \frac{\varepsilon}{2}$ $\forall \|\mathbf{x}\| > R$. But $\mathbf{x}^{\boldsymbol{\alpha}} \partial^{\boldsymbol{\beta}} \phi(\mathbf{x})$ is continuous in the compact set $K = \{\mathbf{x} : \mathbf{x} \in \mathbb{R}^n, \|\mathbf{x}\| \le R\} \Longrightarrow \exists M > 0$ such that $|\mathbf{x}^{\boldsymbol{\alpha}} \partial^{\boldsymbol{\beta}} \phi(\mathbf{x})| \le M$ $\forall \mathbf{x} \in K$. Then, $\forall$ fixed $\boldsymbol{\alpha}, \boldsymbol{\beta}$, $\exists M_0 > 0$ such that $\forall \mathbf{x} \in \mathbb{R}^n$, $|\mathbf{x}^{\boldsymbol{\alpha}} \partial^{\boldsymbol{\beta}} \phi(\mathbf{x})| \le M_0 = \max\{\frac{\varepsilon}{2}, M\} \Longrightarrow \forall$ fixed $\boldsymbol{\alpha}, \boldsymbol{\beta}$, $\mathbf{x}^{\boldsymbol{\alpha}} \partial^{\boldsymbol{\beta}} \phi(\mathbf{x})$ is bounded in $\mathbb{R}^n$.

Then the results (7.2.12) and (7.2.13) follow from the property of the existence of the supremum of bounded functions in $\mathbb{R}^n$. Applying Lemma 7.2.1 and using the boundedness of $\mathbf{x}^{\boldsymbol{\alpha}} \partial^{\boldsymbol{\beta}} \phi(\mathbf{x})$ and the property of supremum, the boundedness of $(1 + \|\mathbf{x}\|^2)^l \partial^{\boldsymbol{\beta}} \phi(\mathbf{x})$ in (7.2.14), (7.2.15) and (7.2.16) is proved.

II. Using Leibniz's theorem on the derivatives of products of functions:

$$\partial^{\boldsymbol{\beta}}(uv) = \sum_{\boldsymbol{\gamma} \le \boldsymbol{\beta}} \frac{\boldsymbol{\beta}!}{(\boldsymbol{\beta} - \boldsymbol{\gamma})! \boldsymbol{\gamma}!} \partial^{\boldsymbol{\beta} - \boldsymbol{\gamma}} u \partial^{\boldsymbol{\gamma}} v, \tag{7.2.20}$$

where $\boldsymbol{\beta}! = \beta_1! \beta_2! \ldots, \beta_n!$, $\boldsymbol{\beta} - \boldsymbol{\gamma} = (\beta_1 - \gamma_1, \beta_2 - \gamma_2, \ldots, \beta_n - \gamma_n)$ with $\beta_i - \gamma_i \ge 0$, $(\boldsymbol{\beta} - \boldsymbol{\gamma})! = (\beta_1 - \gamma_1)!(\beta_2 - \gamma_2)! \cdots (\beta_n - \gamma_n)!$, Lemma 7.2.1 and the definition of $S(\mathbb{R}^n)$ in (7.2.10) (resp. (7.2.11)), the result (7.2.17) (resp. (7.2.18)) is proved. □

Alternative definition of $S(\mathbb{R}^n)$

As a consequence of Properties I and II in (7.2.12)–(7.2.18), instead of Definition 7.2.3 ((7.2.10)/(7.2.11)), $S(\mathbb{R}^n)$ can *equivalently* be defined by:

Definition 7.2.4.

$$S(\mathbb{R}^n) = \{\phi : \phi \in C^\infty(\mathbb{R}^n); \forall k, m \in \mathbb{N}_0, \sup_{|\boldsymbol{\beta}|\le m} \sup_{\mathbf{x}\in\mathbb{R}^n} |(1 + \|\mathbf{x}\|^2)^k \partial^{\boldsymbol{\beta}} \phi(\mathbf{x})| = q^*_{k,m}(\phi) < +\infty\} \quad (7.2.21)$$

or, equivalently,

$$S(\mathbb{R}^n) = \{\phi : \phi \in C^\infty(\mathbb{R}^n); \forall \boldsymbol{\alpha}, \boldsymbol{\beta}, q_{\boldsymbol{\alpha},\boldsymbol{\beta}}(\phi) = \sup_{\mathbf{x}\in\mathbb{R}^n} |\mathbf{x}^{\boldsymbol{\alpha}} \partial^{\boldsymbol{\beta}} \phi(\mathbf{x})| < +\infty\}.$$

Proposition 7.2.2. *Let $\phi \in S(\mathbb{R}^n)$. Then, $\forall \boldsymbol{\alpha}, \boldsymbol{\beta}$,*

I. • ϕ, $\mathbf{x}^{\boldsymbol{\alpha}} \partial^{\boldsymbol{\beta}} \phi$, $\partial^{\boldsymbol{\beta}}(\mathbf{x}^{\boldsymbol{\alpha}} \phi(\mathbf{x})) \in L^1(\mathbb{R}^n)$; (7.2.22)

• $(1 + \|\mathbf{x}\|^2)^l \partial^{\boldsymbol{\beta}} \phi \in L^1(\mathbb{R}^n)$ $\forall l \in \mathbb{N}_0$, (7.2.23)

i.e. $S(\mathbb{R}^n) \subset L^1(\mathbb{R}^n)$; (7.2.24)

II. $\sup_{\mathbf{x}\in\mathbb{R}^n} |\phi(\mathbf{x})| \le \sup_{|\boldsymbol{\beta}|\le n} \|\partial^{\boldsymbol{\beta}} \phi\|_{L^1(\mathbb{R}^n)}$. (7.2.25)

Proof.

I. It is sufficient to prove that $\phi \in S(\mathbb{R}^n) \implies \phi \in L^1(\mathbb{R}^n)$ and, $\forall \boldsymbol{\alpha}, \boldsymbol{\beta}$, $\mathbf{x}^{\boldsymbol{\alpha}} \partial^{\boldsymbol{\beta}} \phi(\mathbf{x}) \in L^1(\mathbb{R}^n)$. From (7.2.15), $\sup_{\mathbf{x}\in\mathbb{R}^n} |(1 + \|\mathbf{x}\|^2)^n \phi(\mathbf{x})| = q^*_{n,\mathbf{0}}(\phi) < +\infty$. Then $\phi(\mathbf{x}) = (1 + \|\mathbf{x}\|^2)^n \phi(\mathbf{x}) \frac{1}{(1+\|\mathbf{x}\|^2)^n}$.
But $1 + \|\mathbf{x}\|^2 = 1 + x_1^2 + \cdots + x_n^2 \ge 1 + x_i^2$ $\forall i = 1, 2, \ldots, n$
$\implies \frac{1}{(1+\|\mathbf{x}\|^2)^n} \le \frac{1}{1+x_1^2} \frac{1}{1+x_2^2} \cdots \frac{1}{1+x_n^2}$
$\implies \int_{\mathbb{R}^n} \frac{d\mathbf{x}}{(1+\|\mathbf{x}\|^2)^n} \le \int_{-\infty}^{\infty} \frac{dx_1}{1+x_1^2} \int_{-\infty}^{\infty} \frac{dx_2}{1+x_2^2} \cdots \int_{-\infty}^{\infty} \frac{dx_n}{1+x_n^2} = \pi\pi\cdots\pi = \pi^n$,
since, setting $x_i = \tan\theta$, $\int_{-\infty}^{\infty} \frac{dx_i}{1+x_i^2} = \int_{-\frac{\pi}{2}}^{\frac{\pi}{2}} d\theta = \pi$.

Hence,

$$\int_{\mathbb{R}^n} |\phi(\mathbf{x})| d\mathbf{x} = \int_{\mathbb{R}^n} |(1 + \|\mathbf{x}\|^2)^n \phi(\mathbf{x})| \frac{1}{(1 + \|\mathbf{x}\|^2)^n} d\mathbf{x}$$
$$\le \sup_{\mathbf{x}\in\mathbb{R}^n} |(1+\|\mathbf{x}\|^2)^n \phi(\mathbf{x})| \int_{\mathbb{R}^n} \frac{d\mathbf{x}}{(1 + \|\mathbf{x}\|^2)^n} \le q^*_{n,\mathbf{0}}(\phi)\pi^n < +\infty$$
$$\implies \phi \in L^1(\mathbb{R}^n) \implies S(\mathbb{R}^n) \subset L^1(\mathbb{R}^n). \quad (7.2.26)$$

Now, we will show that $\mathbf{x}^{\boldsymbol{\alpha}} \partial^{\boldsymbol{\beta}} \phi(\mathbf{x}) \in L^1(\mathbb{R}^n)$ $\forall \boldsymbol{\alpha}, \boldsymbol{\beta}$.

From Lemma 7.2.1, $\sup_{|\boldsymbol{\alpha}|\le 2k} |\mathbf{x}^{\boldsymbol{\alpha}}| \le (1+\|\mathbf{x}\|^2)^k$. Hence, $\forall |\boldsymbol{\alpha}| \le l$, $\forall |\boldsymbol{\beta}| \le m$ with $l, m \in \mathbb{N}_0$,

$$|\mathbf{x}^{\boldsymbol{\alpha}} \partial^{\boldsymbol{\beta}} \phi(\mathbf{x})| \le (1 + \|\mathbf{x}\|^2)^{l/2} |\partial^{\boldsymbol{\beta}} \phi(\mathbf{x})| \le |(1 + \|\mathbf{x}\|^2)^{n+l/2} \partial^{\boldsymbol{\beta}} \phi(\mathbf{x})| \frac{1}{(1 + \|\mathbf{x}\|^2)^n}$$

$$\Longrightarrow \int_{\mathbb{R}^n} |\mathbf{x}^{\boldsymbol{\alpha}} \partial^{\boldsymbol{\beta}} \phi(\mathbf{x})| d\mathbf{x} \le \sup_{|\boldsymbol{\beta}|\le m} \sup_{\mathbf{x}\in\mathbb{R}^n} |(1+\|\mathbf{x}\|^2)^{n+l/2} \partial^{\boldsymbol{\beta}} \phi(\mathbf{x})| \int_{\mathbb{R}^n} \frac{d\mathbf{x}}{(1 + \|\mathbf{x}\|^2)^n}$$
$$\le q^*_{n+l/2,m}(\phi)\pi^n < +\infty \quad \text{(using (7.2.16) and (7.2.26))} \tag{7.2.27}$$

$\Longrightarrow \mathbf{x}^{\boldsymbol{\alpha}} \partial^{\boldsymbol{\beta}} \phi \in L^1(\mathbb{R}^n)$ $\forall \boldsymbol{\alpha}, \boldsymbol{\beta}$, since it holds $\forall l, m \in \mathbb{N}_0$.
Now, using Leibniz's theorem (7.2.20) and the integrability of $\mathbf{x}^{\boldsymbol{\alpha}} \partial^{\boldsymbol{\beta}} \phi(\mathbf{x})$ $\forall \boldsymbol{\alpha}, \boldsymbol{\beta}$, we find that $\partial^{\boldsymbol{\beta}} (\mathbf{x}^{\boldsymbol{\alpha}} \phi(\mathbf{x})) \in L^1(\mathbb{R}^n)$ $\forall \boldsymbol{\alpha}, \boldsymbol{\beta}$.

II. $\phi \in S(\mathbb{R}^n) \Longrightarrow \partial^{\boldsymbol{\beta}} \phi \in L^1(\mathbb{R}^n)$ $\forall \boldsymbol{\beta}$ (by (7.2.22) with $\boldsymbol{\alpha} = \mathbf{0}$)
$\Longrightarrow \int_{\mathbb{R}^n} |\frac{\partial^n \phi(\mathbf{x})}{\partial x_1 \partial x_2 \cdots \partial x_n}| d\mathbf{x} < +\infty$ with $\beta = (1, 1, \ldots, 1)$, $|\boldsymbol{\beta}| = n$.
Then, for $\mathbf{x} = (x_1, x_2, \ldots, x_n)$, $\mathbf{a} = (a_1, a_2, \ldots, a_n) \in \mathbb{R}^n$, we have

$$\phi(\mathbf{x}) - \phi(\mathbf{a}) = \int_{a_1}^{x_1} dx_1 \int_{a_2}^{x_2} dx_2 \cdots \int_{a_n}^{x_n} \frac{\partial}{\partial x_1}\left(\frac{\partial}{\partial x_2}\left(\cdots\left(\frac{\partial \phi}{\partial x_n}\right)\cdots\right)\right) dx_n,$$

where $\lim_{a_i \to -\infty, 1\le i\le n} \phi(\mathbf{a}) = 0$, since $\phi \in S(\mathbb{R}^n)$,

$$\Longrightarrow \quad \phi(\mathbf{x}) = \int_{-\infty}^{x_1} \int_{-\infty}^{x_2} \cdots \int_{-\infty}^{x_n} \frac{\partial^n \phi}{\partial x_1 \partial x_2 \cdots \partial x_n} d\mathbf{x}$$

$$\Longrightarrow \quad \sup_{\mathbf{x}\in\mathbb{R}^n} |\phi(\mathbf{x})| \le \int_{\mathbb{R}^n} \left| \frac{\partial^n \phi}{\partial x_1 \partial x_2 \ldots \partial x_n} \right| d\mathbf{x} = \left\| \frac{\partial^n \phi}{\partial x_1 \partial x_2 \ldots \partial x_n} \right\|_{L^1(\mathbb{R}^n)}$$
$$\le \sup_{|\boldsymbol{\beta}|\le n} \|\partial^{\boldsymbol{\beta}} \phi\|_{L^1(\mathbb{R}^n)}.$$

□

Topology of $S(\mathbb{R}^n)$

For all multi-index $\boldsymbol{\alpha}, \boldsymbol{\beta}$, define the *semi-norm* $q_{\boldsymbol{\alpha},\boldsymbol{\beta}}$ in $S(\mathbb{R}^n)$ by (7.2.12):

$$q_{\boldsymbol{\alpha},\boldsymbol{\beta}}(\phi) = \sup_{\mathbf{x}\in\mathbb{R}^n} |\mathbf{x}^{\boldsymbol{\alpha}} \partial^{\boldsymbol{\beta}} \phi(\mathbf{x})| \quad \forall \phi \in S(\mathbb{R}^n). \tag{7.2.28}$$

Similarly, $\forall l, m \in \mathbb{N}_0$, define the *semi-norm* $q^*_{l,m}$ in $S(\mathbb{R}^n)$ by (7.2.16):

$$q^*_{l,m}(\phi) = \sup_{|\boldsymbol{\beta}|\le m} \sup_{\mathbf{x}\in\mathbb{R}^n} |(1 + \|\mathbf{x}\|^2)^l \partial^{\boldsymbol{\beta}} \phi(\mathbf{x})| \quad \forall \phi \in S(\mathbb{R}^n). \tag{7.2.29}$$

Then the *topology* of $S(\mathbb{R}^n)$ is defined by the family $\{q_{\boldsymbol{\alpha},\boldsymbol{\beta}}\}_{\boldsymbol{\alpha},\boldsymbol{\beta}}$ (resp. $\{q^*_{l,m}\}_{l,m}$) of all semi-norms $q_{\boldsymbol{\alpha},\boldsymbol{\beta}}$ (resp. $q^*_{l,m}$).

We do not like to overburden the reader with further details, for which we refer to [4], [8], [9], [25], [3], etc., since in our treatment and applications all we will need is the notion of convergence in $S(\mathbb{R}^n)$, which we introduce now.

Notion of convergence in $S(\mathbb{R}^n)$

Definition 7.2.5. A sequence (ϕ_k) in $S(\mathbb{R}^n)$ is said to converge to $0 \in S(\mathbb{R}^n)$ as $k \to \infty$ if and only if, $\forall \boldsymbol{\alpha}, \boldsymbol{\beta}$, the sequence $(\mathbf{x}^{\boldsymbol{\alpha}} \partial^{\boldsymbol{\beta}} \phi_k(\mathbf{x}))$ (resp. $\forall l, m \in \mathbb{N}_0$ with $|\boldsymbol{\beta}| \le m$, the sequence $((1+\|\mathbf{x}\|^2)^l \partial^{\boldsymbol{\beta}} \phi_k(\mathbf{x}))$ converges *uniformly* to 0 in $\mathbb{C}$ as $k \to \infty$. Hence, we define $\phi_k \to 0$ in $S(\mathbb{R}^n) \Longrightarrow \forall \boldsymbol{\alpha}, \boldsymbol{\beta}$, $\sup_{\mathbf{x}\in\mathbb{R}^n} |\mathbf{x}^{\boldsymbol{\alpha}} \partial^{\boldsymbol{\beta}} \phi_k(\mathbf{x})| = q_{\boldsymbol{\alpha},\boldsymbol{\beta}}(\phi_k) \to 0$ as $k \to \infty$ or, equivalently, $\phi_k \to 0$ in $S(\mathbb{R}^n) \Longrightarrow \forall l, m \in \mathbb{N}_0$,

$$\sup_{|\boldsymbol{\beta}|\le m} \sup_{\mathbf{x}\in\mathbb{R}^n} |(1+\|\mathbf{x}\|^2)^l \partial^{\boldsymbol{\beta}} \phi_k(\mathbf{x})| = q^*_{l,m}(\phi_k) \to 0 \quad \text{as } k \to \infty. \tag{7.2.30}$$

Consequently, we have two equivalent definitions of convergence:

$$\phi_k \to \phi \text{ in } S(\mathbb{R}^n) \quad \Longrightarrow \quad \forall \boldsymbol{\alpha}, \boldsymbol{\beta},\ q_{\boldsymbol{\alpha},\boldsymbol{\beta}}(\phi - \phi_k) \to 0 \text{ as } k \to \infty; \tag{7.2.31}$$

$$\phi_k \to \phi \text{ in } S(\mathbb{R}^n) \quad \Longrightarrow \quad \forall l, m \in \mathbb{N}_0,\ q^*_{l,m}(\phi - \phi_k) \to 0 \text{ as } k \to \infty. \tag{7.2.32}$$

7.3 Continuity of linear mapping from $S(\mathbb{R}^n)$ into $S(\mathbb{R}^n)$

Definition 7.3.1. A linear mapping $A : S(\mathbb{R}^n) \longrightarrow S(\mathbb{R}^n)$ is called *continuous* from $S(\mathbb{R}^n)$ into itself if and only if for semi-norms $q_{\boldsymbol{\alpha},\boldsymbol{\beta}}$ with arbitrary multi-index $\boldsymbol{\alpha}, \boldsymbol{\beta}$, $\exists$ a positive constant $C > 0$, independent of ϕ, and a semi-norm $q_{\boldsymbol{\gamma},\boldsymbol{\delta}}$ with multi-index $\boldsymbol{\gamma}$ and $\boldsymbol{\delta}$ such that, $\forall \boldsymbol{\alpha}, \boldsymbol{\beta}$,

$$q_{\boldsymbol{\alpha},\boldsymbol{\beta}}(A\phi) \le C q_{\boldsymbol{\gamma},\boldsymbol{\delta}}(\phi) \quad \forall \phi \in S(\mathbb{R}^n), \tag{7.3.1}$$

i.e. $\forall \boldsymbol{\alpha}, \boldsymbol{\beta}$, $\sup_{\mathbf{x}\in\mathbb{R}^n} |\mathbf{x}^{\boldsymbol{\alpha}} \partial^{\boldsymbol{\beta}} (A\phi)(\mathbf{x})| \le C \sup_{\mathbf{x}\in\mathbb{R}^n} |\mathbf{x}^{\boldsymbol{\gamma}} \partial^{\boldsymbol{\delta}} \phi(\mathbf{x})|$ $\forall \phi \in S(\mathbb{R}^n)$, or, equivalently, if and only if for all semi-norms $q^*_{l,\boldsymbol{\beta}}$ (i.e. for all $l \in \mathbb{N}_0$ and all multi-index $\boldsymbol{\beta}$), $\exists$ a positive constant $C_1 > 0$, independent of ϕ, and a semi-norm $q^*_{m,\boldsymbol{\delta}}$ with $m \in \mathbb{N}_0$, multi-index $\boldsymbol{\delta}$ such that, $\forall l \in \mathbb{N}_0$, $\forall \boldsymbol{\beta}$,

$$q^*_{l,\boldsymbol{\beta}}(A\phi) \le C_1 q^*_{m,\boldsymbol{\delta}}(\phi) \quad \forall \phi \in S(\mathbb{R}^n), \tag{7.3.2}$$

i.e. $\forall l \in \mathbb{N}_0$, $\forall \boldsymbol{\beta}$, $\sup_{\mathbf{x}\in\mathbb{R}^n} |(1+\|\mathbf{x}\|^2)^l \partial^{\boldsymbol{\beta}} (A\phi)(\mathbf{x})| \le C_1 \sup_{\mathbf{x}\in\mathbb{R}^n} |(1+\|\mathbf{x}\|^2)^m \partial^{\boldsymbol{\delta}} \phi(\mathbf{x})|$ $\forall \phi \in S(\mathbb{R}^n)$.

Consequences If $A : S(\mathbb{R}^n) \longrightarrow S(\mathbb{R}^n)$ is a *continuous* linear mapping, then

$$\phi_k \to \phi \text{ in } S(\mathbb{R}^n) \quad \Longrightarrow \quad A\phi_k \to A\phi \text{ in } S(\mathbb{R}^n) \text{ as } k \to \infty, \tag{7.3.3}$$

$$\phi_k \to 0 \text{ in } S(\mathbb{R}^n) \quad \Longrightarrow \quad A\phi_k \to 0 \text{ in } S(\mathbb{R}^n) \text{ as } k \to \infty. \tag{7.3.4}$$

Applications

$\forall$ multi-index $\boldsymbol{\gamma}, \partial^{\boldsymbol{\gamma}}$: $S(\mathbb{R}^n) \ni \phi \longmapsto \partial^{\boldsymbol{\gamma}}\phi = \frac{\partial^{|\boldsymbol{\gamma}|}\phi}{\partial x_1{}^{\gamma_1}\partial x_2{}^{\gamma_2}\ldots\partial x_n{}^{\gamma_n}} \in S(\mathbb{R}^n)$ is a *continuous linear* operator from $S(\mathbb{R}^n)$ into $S(\mathbb{R}^n)$. (7.3.5)

Proof. From the linearity of the differential operator, the linearity of $\partial^{\boldsymbol{\gamma}}$ follows. It remains to show that (7.3.1) holds. In fact, $\forall \boldsymbol{\alpha}, \boldsymbol{\beta}$,

$$q_{\boldsymbol{\alpha},\boldsymbol{\beta}}(\partial^{\boldsymbol{\gamma}}\phi) = \sup_{\mathbf{x}\in\mathbb{R}^n} |\mathbf{x}^{\boldsymbol{\alpha}}\partial^{\boldsymbol{\beta}}(\partial^{\boldsymbol{\gamma}}\phi(\mathbf{x}))| = \sup_{\mathbf{x}\in\mathbb{R}^n} |\mathbf{x}^{\boldsymbol{\alpha}}\partial^{\boldsymbol{\beta}+\boldsymbol{\gamma}}\phi(\mathbf{x})| = q_{\boldsymbol{\alpha},\boldsymbol{\delta}}(\phi),$$

with $C = 1$, $\boldsymbol{\delta} = \boldsymbol{\beta} + \boldsymbol{\gamma}$ and $\forall\phi \in S(\mathbb{R}^n)$.

Hence, $\partial^{\boldsymbol{\gamma}} : S(\mathbb{R}^n) \longrightarrow S(\mathbb{R}^n)$ is continuous.

Alternatively, $\phi_k \to 0$ in $S(\mathbb{R}^n) \Longrightarrow q_{\boldsymbol{\alpha},\boldsymbol{\beta}}(\phi_k) \to 0\ \forall\boldsymbol{\alpha},\boldsymbol{\beta}$ as $k \to \infty$ by (7.2.31) $\Longrightarrow q_{\boldsymbol{\alpha},\boldsymbol{\beta}}(\partial^{\boldsymbol{\gamma}}\phi_k) = q_{\boldsymbol{\alpha},\boldsymbol{\delta}}(\phi_k) \to 0$ as $k \to \infty \forall\boldsymbol{\alpha},\boldsymbol{\beta} \Longrightarrow \partial^{\boldsymbol{\gamma}}\phi_k \to 0$ in $S(\mathbb{R}^n)$

$\Longrightarrow \partial^{\boldsymbol{\gamma}} : S(\mathbb{R}^n) \longrightarrow S(\mathbb{R}^n)$ is continuous. □

The multiplication of $\phi \in S(\mathbb{R}^n)$ by an *arbitrary monomial* $\mathbf{x}^{\boldsymbol{\gamma}}$ (resp. by *arbitrary polynomial* $(1 + \|\mathbf{x}\|^2)^l$ with $l \in \mathbb{N}$), $\phi \in S(\mathbb{R}^n) \longmapsto \mathbf{x}^{\boldsymbol{\gamma}}\phi \in S(\mathbb{R}^n)$ (resp. $(1 + \|\mathbf{x}\|^2)^l\phi \in S(\mathbb{R}^n)$) defines a *continuous* linear mapping from $S(\mathbb{R}^n)$ into $S(\mathbb{R}^n)$.

Proof. The linearity is obvious. For continuity we are to show that (7.3.1) holds. In fact, using Leibniz's theorem $\partial^{\boldsymbol{\beta}}(\mathbf{x}^{\boldsymbol{\gamma}}\phi) = \sum_{\boldsymbol{\delta}\le\boldsymbol{\beta}} \frac{\boldsymbol{\beta}!}{(\boldsymbol{\beta}-\boldsymbol{\delta})!\boldsymbol{\delta}!}\partial^{\boldsymbol{\beta}-\boldsymbol{\delta}}(\mathbf{x}^{\boldsymbol{\gamma}})\partial^{\boldsymbol{\delta}}\phi$, we find that $\exists$ a constant $C > 0$, independent of ϕ, such that $\forall l, m \in \mathbb{N}_0$ with $|\boldsymbol{\alpha}| \le l, |\boldsymbol{\beta}| \le m$, $q_{l,m}(\mathbf{x}^{\boldsymbol{\gamma}}\phi) \le C q_{l+|\boldsymbol{\gamma}|,m}(\phi)\ \forall\phi \in S(\mathbb{R}^n)$, where $q_{l,m}(\mathbf{x}^{\boldsymbol{\gamma}}\phi) = \sup_{|\boldsymbol{\alpha}|\le l,|\boldsymbol{\beta}|\le m} |\mathbf{x}^{\boldsymbol{\alpha}}\partial^{\boldsymbol{\beta}}(\mathbf{x}^{\boldsymbol{\gamma}}\phi)|$, $q_{l+|\boldsymbol{\gamma}|,m}(\phi) = \sup_{|\boldsymbol{\lambda}|\le l+|\boldsymbol{\gamma}|,|\boldsymbol{\beta}|\le m} |\mathbf{x}^{\boldsymbol{\lambda}}\partial^{\boldsymbol{\beta}}\phi|$ (multi-index $\boldsymbol{\lambda} = (\lambda_1, \lambda_2, \ldots, \lambda_n)$)

$\Longrightarrow$ the mapping $\phi \in S(\mathbb{R}^n) \longmapsto \mathbf{x}^{\boldsymbol{\gamma}}\phi \in S(\mathbb{R}^n)$ is continuous by (7.3.1). Then, applying Lemma 7.2.1, the continuity of the linear mapping $\phi \in S(\mathbb{R}^n) \longmapsto (1 + \|\mathbf{x}\|^2)^l\phi \in S(\mathbb{R}^n)$ can be proved by (7.3.2). □

7.4 Imbedding results

Proposition 7.4.1.

I. $S(\mathbb{R}^n) \hookrightarrow L^p(\mathbb{R}^n)$, $1 \le p \le \infty$, *with* continuous *imbedding* $\hookrightarrow$, *i.e.* $\exists$ *a constant* $C > 0$, *independent of* ϕ, *such that*

$$\|\hookrightarrow\phi\|_{L^p(\mathbb{R}^n)} = \|\phi\|_{L^p(\mathbb{R}^n)} \le C q^*_{n,0}(\phi) \quad \forall\phi \in S(\mathbb{R}^n). \tag{7.4.1}$$

II. $\mathcal{D}(\mathbb{R}^n) \hookrightarrow S(\mathbb{R}^n)$ *with* continuous *imbedding* $\hookrightarrow$, (7.4.2)
where $\mathcal{D}(\mathbb{R}^n)$ *is the space of complex-valued test functions (see* (1.3.46)–(1.3.49)*).*

Proof.

I. *Case* $p = \infty$: The result follows from: $\|\phi\|_{L^\infty(\mathbb{R}^n)} = \sup_{\mathbf{x}\in\mathbb{R}^n} |\phi(\mathbf{x})| = q^*_{0,0}(\phi)$.
Case $p = 1$: The result follows from (7.2.26) with $\|\phi\|_{L^1(\mathbb{R}^n)} \le \pi^n q^*_{n,0}(\phi)$.
Case $p = 2$: Following the proof of Proposition 7.2.2, we have, $\forall\phi \in S(\mathbb{R}^n)$,

$$|\phi(\mathbf{x})| \le (\sup_{\mathbf{x}\in\mathbb{R}^n} |(1+\|\mathbf{x}\|^2)^n\phi(\mathbf{x})|)\frac{1}{(1+\|\mathbf{x}\|^2)^n} \le q^*_{n,0}(\phi)\frac{1}{(1+\|\mathbf{x}\|^2)^n}$$

$$\Longrightarrow \int_{\mathbb{R}^n} |\phi(\mathbf{x})|^2 d\mathbf{x} \le [q*_{n,0}(\phi)]^2 \int_{\mathbb{R}^n} \frac{1}{(1+\|\mathbf{x}\|^2)^{2n}} d\mathbf{x}$$

$$\le \pi^n [q^*_{n,0}(\phi)]^2 < +\infty,$$

since $(1+\|\mathbf{x}\|^2)^{2n} \ge (1+\|\mathbf{x}\|^2)^n \ge (1+x_1^2)(1+x_2^2)\cdots(1+x_n^2)\ \forall\mathbf{x}\in\mathbb{R}^n$ $\Longrightarrow \int_{\mathbb{R}^n} \frac{1}{(1+\|\mathbf{x}\|^2)^{2n}} d\mathbf{x} \le \int_{\mathbb{R}^n} [\frac{1}{(1+\|\mathbf{x}\|^2)^n}] d\mathbf{x} \le \pi^n$ (see proof of Proposition 7.2.2).
Hence, $\phi \in L^2(\mathbb{R}^n) \Longrightarrow S(\mathbb{R}^n) \subset L^2(\mathbb{R}^n)$ with

$$\|\boldsymbol{\phi}\|_{L^2(\mathbb{R}^n)} = \left(\int_{\mathbb{R}^n} |\phi(\mathbf{x})|^2 d\mathbf{x}\right)^{\frac{1}{2}} \le C q^*_{n,0}(\phi) \quad \forall\phi \in S(\mathbb{R}^n),$$

with $C = \pi^{\frac{n}{2}} > 0$.
Cases $p \in]2,\infty[$, $p \in]1,2[$: These can be proved as in the case of $p = 2$ with $C = \pi^{\frac{n}{p}}$.

II. Let K be any compact subset of $\mathbb{R}^n$ and $\mathcal{D}_K(\mathbb{R}^n)$ be defined by

$$\mathcal{D}_K(\mathbb{R}^n) = \{\phi : \phi \in \mathcal{D}(\mathbb{R}^n), \operatorname{supp}(\phi) \subset K\},$$

with semi-norm $p_{K,m}$ defined, $\forall m \in \mathbb{N}_0$, by

$$p_{K,m}(\phi) = \sup_{|\boldsymbol{\beta}|\le m} \sup_{\mathbf{x}\in K} |\partial^{\boldsymbol{\beta}}\phi(\mathbf{x})| \quad \forall\phi \in \mathcal{D}_K(\mathbb{R}^n).$$

Then it is sufficient to show the continuity of the imbedding $\mathcal{D}_K(\mathbb{R}^n) \hookrightarrow S(\mathbb{R}^n)$ $\forall$ compact $K \subset\subset \mathbb{R}^n$, from which the continuous imbedding (7.4.2) will follow.
In fact, $\forall\phi \in \mathcal{D}_K(\mathbb{R}^n)$, $\forall l, m \in \mathbb{N}_0$,

$$q_{l,m}(\phi) = \sup_{|\boldsymbol{\alpha}|\le l, |\boldsymbol{\beta}|\le m} \sup_{\mathbf{x}\in\mathbb{R}^n} |\mathbf{x}^{\boldsymbol{\alpha}}\partial^{\boldsymbol{\beta}}\phi(\mathbf{x})| = \sup_{|\boldsymbol{\alpha}|\le l, |\boldsymbol{\beta}|\le m} \sup_{\mathbf{x}\in K} |\mathbf{x}^{\boldsymbol{\alpha}}\partial^{\boldsymbol{\beta}}\phi(\mathbf{x})|$$

(since $\phi(\mathbf{x}) = 0$ outside K)

$$\le (\sup_{|\boldsymbol{\alpha}|\le l} \sup_{\mathbf{x}\in K} |\mathbf{x}^{\boldsymbol{\alpha}}|) \sup_{|\boldsymbol{\beta}|\le m} \sup_{\mathbf{x}\in K} |\partial^{\boldsymbol{\beta}}\phi(\mathbf{x})| < +\infty$$

$\Longrightarrow \forall l, m \in \mathbb{N}_0$, $q_{l,m}(\phi) \le C_{K,l}\, p_{K,m}(\phi)$ $\forall \phi \in \mathcal{D}_K(\mathbb{R}^n)$ $\forall K \subset\subset \mathbb{R}^n$ with $C_{K,l} = \sup_{|\boldsymbol{\alpha}|\le l} \sup_{\mathbf{x}\in K} |\mathbf{x}^{\boldsymbol{\alpha}}|$

$\Longrightarrow \forall$ compact $K \subset\subset \mathbb{R}^n$, the imbedding $\mathcal{D}_K(\mathbb{R}^n) \hookrightarrow S(\mathbb{R}^n)$ is continuous,

$\Longrightarrow$ the imbedding $\mathcal{D}(\mathbb{R}^n) \hookrightarrow S(\mathbb{R}^n)$ is continuous. □

7.5 Density results

Proposition 7.5.1.

I. $\mathcal{D}(\mathbb{R}^n)$ *is a* dense *subspace of* $S(\mathbb{R}^n)$; (7.5.1)

II. $S(\mathbb{R}^n)$ *is a* dense *subspace of* $L^2(\mathbb{R}^n)$. (7.5.2)

Proof.

I. We will prove the density of $\mathcal{D}(\mathbb{R}^n)$ in $S(\mathbb{R}^n)$ with the help of cut-off functions. Let $\psi \in \mathcal{D}(\mathbb{R}^n)$ such that $\psi(\mathbf{x}) = 1$ for $\|\mathbf{x}\| \le 1$. Now, $\forall k \in \mathbb{N}$, define $\psi_k(\mathbf{x}) = \psi(\frac{\mathbf{x}}{k})$ $\forall \mathbf{x} \in \mathbb{R}^n$. Then (ψ_k) is a sequence of cut-off functions with the following properties:

- $\psi_k \in \mathcal{D}(\mathbb{R}^n)$ $\forall k \in \mathbb{N}$;
- $\psi_k(\mathbf{x}) = \psi(\frac{\mathbf{x}}{k}) = 1$ for $\|\mathbf{x}\| \le k$ $\forall k \in \mathbb{N}$; (7.5.3)
- $\psi_k(\mathbf{x}) - 1 = 0$ for $\|\mathbf{x}\| \le k$ $\forall k \in \mathbb{N}$;
- $$\sup_{k\in\mathbb{N}} \sup_{\mathbf{x}\in\mathbb{R}^n} |\partial^{\boldsymbol{\beta}} \psi_k(\mathbf{x})| = \sup_{k\in\mathbb{N}} \sup_{\mathbf{x}\in\mathbb{R}^n} \left| \partial^{\boldsymbol{\beta}} \psi\left(\frac{\mathbf{x}}{k}\right) \right| \le \sup_{\mathbf{x}\in\mathbb{R}^n} |\partial^{\boldsymbol{\beta}} \psi(\mathbf{x})| \quad \forall \boldsymbol{\beta}. \tag{7.5.4}$$

Let $\phi \in S(\mathbb{R}^n)$. Define $\phi_k = \phi\psi_k$ $\forall k \in \mathbb{N}$. Hence, $\forall k \in \mathbb{N}$, $\phi_k \in \mathcal{D}(\mathbb{R}^n)$ (since $\phi \in C^\infty(\mathbb{R}^n)$, $\psi_k \in \mathcal{D}(\mathbb{R}^n)$ $\forall k \in \mathbb{N}$). Then the sequence (ϕ_k) converges to ϕ in $S(\mathbb{R}^n)$, for which we are to show that the sequence of semi-norms $q_{\boldsymbol{\alpha},\boldsymbol{\beta}}(\phi - \phi_k) = q_{\boldsymbol{\alpha},\boldsymbol{\beta}}(\chi_k)$ (with $\chi_k = \phi_k - \phi = \phi(\psi_k - 1)$) tends to 0 as $k \to \infty$ $\forall \boldsymbol{\alpha}, \boldsymbol{\beta}$ (see (7.2.31)). In fact, $\partial^{\boldsymbol{\beta}} \chi_k = \sum_{\boldsymbol{\gamma}\le\boldsymbol{\beta}} \frac{\boldsymbol{\beta}!}{(\boldsymbol{\beta}-\boldsymbol{\gamma})!\boldsymbol{\gamma}!} \partial^{\boldsymbol{\beta}-\boldsymbol{\gamma}}(\psi_k - 1)\partial^{\boldsymbol{\gamma}}\phi$ with $\partial^{\boldsymbol{\beta}-\boldsymbol{\gamma}}(\psi_k - 1) = 0$ for $\|\mathbf{x}\| \le k$ (by virtue of (7.5.3)).
Hence, $\forall \boldsymbol{\alpha}, \boldsymbol{\beta}$, $|\mathbf{x}^{\alpha}\partial^{\beta}\chi_k(\mathbf{x})| \le \sum_{\boldsymbol{\gamma}\le\boldsymbol{\beta}} \frac{\boldsymbol{\beta}!}{(\boldsymbol{\beta}-\boldsymbol{\gamma})!\boldsymbol{\gamma}!} |\partial^{\boldsymbol{\beta}-\boldsymbol{\gamma}}(\psi_k(\mathbf{x}) - 1)||\mathbf{x}^{\boldsymbol{\alpha}}\partial^{\boldsymbol{\gamma}}\phi(\mathbf{x})|$

$$\begin{aligned}
\Longrightarrow \quad & \sup_{\mathbf{x}\in\mathbb{R}^n} |\mathbf{x}^{\boldsymbol{\alpha}}\partial^{\boldsymbol{\beta}}\chi_k(\mathbf{x})| \\
& \le \sum_{\boldsymbol{\gamma}\le\boldsymbol{\beta}} \frac{\boldsymbol{\beta}!}{(\boldsymbol{\beta}-\boldsymbol{\gamma})!\boldsymbol{\gamma}!} \sup_{\|\mathbf{x}\|>k} |\partial^{\boldsymbol{\beta}-\boldsymbol{\gamma}}(\psi_k(\mathbf{x}) - 1)| \sup_{\|\mathbf{x}\|>k} |\mathbf{x}^{\boldsymbol{\alpha}}\partial^{\boldsymbol{\gamma}}\phi(\mathbf{x})| \\
& \le \left(\sum_{\boldsymbol{\gamma}\le\boldsymbol{\beta}} \frac{\boldsymbol{\beta}!}{(\boldsymbol{\beta}-\boldsymbol{\gamma})!\boldsymbol{\gamma}!} \sup_{\|\mathbf{x}\|>k} |\partial^{\boldsymbol{\beta}-\boldsymbol{\gamma}}(\psi_k(\mathbf{x}) - 1)| \right) \sup_{\boldsymbol{\gamma}\le\boldsymbol{\beta}} \sup_{\|\mathbf{x}\|>k} |\mathbf{x}^{\boldsymbol{\alpha}}\partial^{\boldsymbol{\gamma}}\phi(\mathbf{x})|.
\end{aligned} \tag{7.5.5}$$

But $\forall k \in \mathbb{N}$,

$$\begin{aligned}\sup_{\|\mathbf{x}\|>k} |\partial^{\boldsymbol{\beta}-\boldsymbol{\gamma}}(\psi_k(\mathbf{x})-1)| &\le \sup_{k\in\mathbb{N}}\sup_{\mathbf{x}\in\mathbb{R}^n} |\partial^{\boldsymbol{\beta}-\boldsymbol{\gamma}}\psi_k(\mathbf{x})|+1 \\ &\le \sup_{\mathbf{x}\in\mathbb{R}^n} |\partial^{\boldsymbol{\beta}-\boldsymbol{\gamma}}\psi(\mathbf{x})|+1 \quad \forall \boldsymbol{\gamma} \le \boldsymbol{\beta} \quad \text{(by (7.5.4))}\end{aligned}$$

$\Longrightarrow \forall k \in \mathbb{N}$,

$$\sup_{\boldsymbol{\gamma}\le\boldsymbol{\beta}}\sup_{\|\mathbf{x}\|>k} |\partial^{\boldsymbol{\beta}-\boldsymbol{\gamma}}(\psi(\mathbf{x})-1)| \le \sup_{\boldsymbol{\gamma}\le\boldsymbol{\beta}}\sup_{\mathbf{x}\in\mathbb{R}^n} |\partial^{\boldsymbol{\beta}-\boldsymbol{\gamma}}\psi(\mathbf{x})|+1 = M_{\boldsymbol{\beta}}+1, \tag{7.5.6}$$

since for $\psi \in \mathcal{D}(\mathbb{R}^n)$, $\exists M_{\boldsymbol{\beta}} > 0$ such that

$$\sup_{\boldsymbol{\gamma}\le\boldsymbol{\beta}}\sup_{\mathbf{x}\in\mathbb{R}^n} |\partial^{\boldsymbol{\beta}-\boldsymbol{\gamma}}\psi(\mathbf{x})| \le M_{\boldsymbol{\beta}}. \tag{7.5.7}$$

Hence, from (7.5.5)–(7.5.7), we have, $\forall k \in \mathbb{N}$, $\forall \boldsymbol{\alpha}, \boldsymbol{\beta}$,

$$\begin{aligned} q_{\boldsymbol{\alpha},\boldsymbol{\beta}}(\chi_k) &= \sup_{\mathbf{x}\in\mathbb{R}^n} |\mathbf{x}^{\boldsymbol{\alpha}}\partial^{\boldsymbol{\beta}}\chi_k(\mathbf{x})| \\ &\le \Big(\sum_{\boldsymbol{\gamma}\le\boldsymbol{\beta}} \frac{\boldsymbol{\beta}!}{(\boldsymbol{\beta}-\boldsymbol{\gamma})!\boldsymbol{\gamma}!}\Big) \sup_{\boldsymbol{\gamma}\le\boldsymbol{\beta}}\sup_{\|\mathbf{x}\|>k} |\partial^{\boldsymbol{\beta}-\boldsymbol{\gamma}}(\psi(\mathbf{x})-1)| \cdot \sup_{\boldsymbol{\gamma}\le\boldsymbol{\beta}}\sup_{\|\mathbf{x}\|>k} |\mathbf{x}^{\boldsymbol{\alpha}}\partial^{\boldsymbol{\gamma}}\phi(\mathbf{x})| \\ &\le \Big(\sum_{\boldsymbol{\gamma}\le\boldsymbol{\beta}} \frac{\boldsymbol{\beta}!}{(\boldsymbol{\beta}-\boldsymbol{\gamma})!\boldsymbol{\gamma}!}\Big)(M_{\boldsymbol{\beta}}+1) \sup_{\boldsymbol{\gamma}\le\boldsymbol{\beta}}\sup_{\|\mathbf{x}\|>k} |\mathbf{x}^{\boldsymbol{\alpha}}\partial^{\boldsymbol{\gamma}}\phi(\mathbf{x})|. \end{aligned}$$

Set $C(\boldsymbol{\beta}) = \sum_{\boldsymbol{\gamma}\le\boldsymbol{\beta}} \frac{\boldsymbol{\beta}!}{(\boldsymbol{\beta}-\boldsymbol{\gamma})!\boldsymbol{\gamma}!}(M_{\boldsymbol{\beta}}+1) > 0$.
Then $\forall \boldsymbol{\alpha}, \boldsymbol{\beta}$, $q_{\boldsymbol{\alpha},\boldsymbol{\beta}}(\chi_k) \le C(\boldsymbol{\beta}) \sup_{\boldsymbol{\gamma}\le\boldsymbol{\beta}} \sup_{\|\mathbf{x}\|>k} |\mathbf{x}^{\boldsymbol{\alpha}}\partial^{\boldsymbol{\gamma}}\phi(\mathbf{x})| \to 0$ as $k \to \infty$, since $\phi \in S(\mathbb{R}^n) \Longrightarrow \forall \boldsymbol{\alpha}, \boldsymbol{\gamma}$, $\mathbf{x}^{\boldsymbol{\alpha}}\partial^{\boldsymbol{\gamma}}\phi(\mathbf{x}) \in S(\mathbb{R}^n)$ by Proposition 7.2.1 $\Longrightarrow$ $\forall \boldsymbol{\alpha}, \boldsymbol{\gamma}$, $\mathbf{x}^{\boldsymbol{\alpha}}\partial^{\boldsymbol{\gamma}}\phi(\mathbf{x}) \to 0$ as $\|\mathbf{x}\| \to \infty$ (by (7.2.10)) $\Longrightarrow \sup_{\|\mathbf{x}\|>k} |\mathbf{x}^{\boldsymbol{\alpha}}\partial^{\boldsymbol{\gamma}}\phi(\mathbf{x})| \to 0$ as $k \to \infty$ $\forall \boldsymbol{\alpha}, \boldsymbol{\gamma} \Longrightarrow \forall \boldsymbol{\alpha}, \boldsymbol{\beta}$, $q_{\boldsymbol{\alpha},\boldsymbol{\beta}}(\chi_k) \to 0$ as $k \to \infty$. Hence, $\mathcal{D}(\mathbb{R}^n)$ is dense in $S(\mathbb{R}^n)$, since $\phi_k \to \phi$ in $S(\mathbb{R}^n)$ as $k \to \infty$.

II. Let $f \in L^2(\mathbb{R}^n)$. Since $\mathcal{D}(\mathbb{R}^n)$ is dense in $L^2(\mathbb{R}^n)$, $\exists$ a sequence (ϕ_k) in $\mathcal{D}(\mathbb{R}^n)$ such that $\|f-\phi_k\|_{L^2(\mathbb{R}^n)} \to 0$ as $k \to \infty$. (7.5.8)

But $\mathcal{D}(\mathbb{R}^n) \hookrightarrow S(\mathbb{R}^n)$ by (7.4.2) $\Longrightarrow \phi_k \in S(\mathbb{R}^n)$ $\forall k \in \mathbb{N}$ such that (7.5.8) holds. Hence, for $f \in L^2(\mathbb{R}^n)$, $\exists$ a sequence (ϕ_k) in $S(\mathbb{R}^n)$ such that $\phi_k \to f$ as $k \to \infty$. Thus, $S(\mathbb{R}^n)$ is dense in $L^2(\mathbb{R}^n)$. □

7.6 Fourier transform of functions of $S(\mathbb{R}^n)$

By virtue of the imbedding result $S(\mathbb{R}^n) \hookrightarrow L^1(\mathbb{R}^n)$ (7.4.1), *all* the Properties 1–11 of Fourier transform $\hat{f} = \mathcal{F} f$ (resp. co-transform $\overline{\mathcal{F}} f$) stated in Theorems 7.1.1–7.1.3, in Proposition 7.1.1 and Corollary 7.1.1 for functions $f \in L^1(\mathbb{R}^n)$ will *also* hold for functions $\phi \in S(\mathbb{R}^n)$.

For $f \in L^1(\mathbb{R}^n)$, its Fourier transform $\hat{f} \notin L^1(\mathbb{R}^n)$ *in general* (see Example 7.1.1), but for every $\phi \in S(\mathbb{R}^n)$ with its Fourier transform $\hat{\phi} = \mathcal{F}\phi \in S(\mathbb{R}^n) \hookrightarrow L^1(\mathbb{R}^n)$:

Theorem 7.6.1.

I. $\forall \phi \in S(\mathbb{R}^n)$, *Fourier transform* $\hat{\phi} = \mathcal{F}\phi \in S(\mathbb{R}^n)$ *and co-transform* $\overline{\mathcal{F}}\phi \in S(\mathbb{R}^n)$ *are defined by* (7.1.2) *and* (7.1.3):

$$\hat{\phi}(\boldsymbol{\xi}) = (\mathcal{F}\phi)(\boldsymbol{\xi}) = \int_{\mathbb{R}^n} \phi(\mathbf{x}) e^{-i2\pi\langle \mathbf{x}, \boldsymbol{\xi}\rangle} d\mathbf{x}, \tag{7.6.1}$$

$$(\overline{\mathcal{F}}\phi)(\boldsymbol{\xi}) = \int_{\mathbb{R}^n} \phi(\mathbf{x}) e^{i2\pi\langle \mathbf{x}, \boldsymbol{\xi}\rangle} d\mathbf{x}. \tag{7.6.2}$$

II. $\mathcal{F} : S(\mathbb{R}^n) \longrightarrow S(\mathbb{R}^n)$ *(resp.* $\overline{\mathcal{F}} : S(\mathbb{R}^n) \longrightarrow S(\mathbb{R}^n)$*) is* continuous *from* $S(\mathbb{R}^n)$ *into* $S(\mathbb{R}^n)$.

Proof. We give the proof for $\mathcal{F}$. Then, replacing 'i' by '$-i$' in the proof, the results for $\overline{\mathcal{F}}$ are obtained.

I. $\phi \in S(\mathbb{R}^n) \Longrightarrow \hat{\phi} = \mathcal{F}\phi \in C^\infty(\mathbb{R}^n)$: Let $\phi \in S(\mathbb{R}^n)$. Then, by Proposition 7.2.2, $\mathbf{x}^{\boldsymbol{\beta}}\phi \in L^1(\mathbb{R}^n)$ $\forall$ multi-index $\boldsymbol{\beta}$ with $|\boldsymbol{\beta}| \in \mathbb{N}_0$, and, from Theorem 7.1.3, $\hat{\phi} = \mathcal{F}\phi \in C^{|\boldsymbol{\beta}|}(\mathbb{R}^n)$ $\forall |\boldsymbol{\beta}| \in \mathbb{N}_0$, i.e. $\hat{\phi} \in C^\infty(\mathbb{R}^n)$ *and* $\partial_{\boldsymbol{\xi}}^{\boldsymbol{\beta}}\hat{\phi} = \frac{\partial^{|\boldsymbol{\beta}|}\hat{\phi}}{\partial\xi_1^{\beta_1}\ldots\partial\xi_n^{\beta_n}} = \mathcal{F}[(-i2\pi\mathbf{x})^{\boldsymbol{\beta}}\phi]$ $\forall |\boldsymbol{\beta}| \in \mathbb{N}_0$, where $(-i2\pi\mathbf{x})^{\boldsymbol{\beta}} = (-i2\pi x_1)^{\beta_1}(-i2\pi x_2)^{\beta_2}\cdots(-i2\pi x_n)^{\beta_n} = (-i2\pi)^{|\boldsymbol{\beta}|}\mathbf{x}^{\boldsymbol{\beta}}$ with $|\boldsymbol{\beta}| = \beta_1 + \beta_2 + \cdots + \beta_n$.

$\hat{\phi} \in S(\mathbb{R}^n)$: For this it remains to show that $q_{\boldsymbol{\alpha},\boldsymbol{\beta}}(\hat{\phi}) < +\infty$ for all multi- index $\boldsymbol{\alpha}, \boldsymbol{\beta}$ (by Definition 7.2.4). Applying Proposition 7.2.2, $\mathbf{x}^{\boldsymbol{\alpha}}\partial_{\mathbf{x}}^{\boldsymbol{\beta}}\phi(\mathbf{x})$ and $\partial_{\mathbf{x}}^{\boldsymbol{\alpha}}(\mathbf{x}^{\boldsymbol{\beta}}\phi(\mathbf{x})) \in L^1(\mathbb{R}^n)$ $\forall \boldsymbol{\alpha}, \boldsymbol{\beta}$ with $\|\mathbf{x}^{\boldsymbol{\alpha}}\partial_{\mathbf{x}}^{\boldsymbol{\beta}}\phi\|_{L^1(\mathbb{R}^n)} < +\infty$, $\|\partial_{\mathbf{x}}^{\boldsymbol{\alpha}}(\mathbf{x}^{\boldsymbol{\beta}}\phi)\|_{L^1(\mathbb{R}^n)} < +\infty$. Then, from Theorem 7.1.3, we have, $\forall \boldsymbol{\alpha}, \boldsymbol{\beta}$,

$$(i2\pi\boldsymbol{\xi})^{\boldsymbol{\alpha}}\partial_{\boldsymbol{\xi}}^{\boldsymbol{\beta}}\hat{\phi}(\boldsymbol{\xi}) = (i2\pi\boldsymbol{\xi})^{\boldsymbol{\alpha}}\mathcal{F}[(-i2\pi\mathbf{x})^{\boldsymbol{\beta}}\phi(x)] = \mathcal{F}[\partial_{\mathbf{x}}^{\boldsymbol{\alpha}}\{(-i2\pi\mathbf{x})^{\boldsymbol{\beta}}\phi\}].$$

Applying (7.1.7):

$$\begin{aligned}
\forall \boldsymbol{\alpha}, \boldsymbol{\beta},\ \|(i2\pi\boldsymbol{\xi})^{\boldsymbol{\alpha}}\partial_{\boldsymbol{\xi}}^{\boldsymbol{\beta}}\hat{\phi}\|_\infty &= \|\mathcal{F}[\partial_{\mathbf{x}}^{\boldsymbol{\alpha}}\{(-i2\pi\mathbf{x})^{\boldsymbol{\beta}}\phi\}]\|_{L^\infty(\mathbb{R}^n)} \\
&\le \|\partial_{\mathbf{x}}^{\boldsymbol{\alpha}}[(-i2\pi\mathbf{x})^{\boldsymbol{\beta}}\phi]\|_{L^1(\mathbb{R}^n)} \\
\Longrightarrow \quad \forall \boldsymbol{\alpha}, \boldsymbol{\beta},\ |i2\pi|^{|\boldsymbol{\alpha}|}\|\boldsymbol{\xi}^{\boldsymbol{\alpha}}\partial_{\boldsymbol{\xi}}^{\boldsymbol{\beta}}\hat{\phi}\|_\infty &= (2\pi)^{|\boldsymbol{\alpha}|}\|\boldsymbol{\xi}^{\boldsymbol{\alpha}}\partial_{\boldsymbol{\xi}}^{\boldsymbol{\beta}}\hat{\phi}\|_{L^\infty(\mathbb{R}^n)} \\
&\le |i2\pi|^{|\boldsymbol{\beta}|}\|\partial_{\mathbf{x}}^{\boldsymbol{\alpha}}(\mathbf{x}^{\boldsymbol{\beta}}\phi)\|_{L^1(\mathbb{R}^n)} \\
\Longrightarrow \quad \forall \boldsymbol{\alpha}, \boldsymbol{\beta},\ q_{\boldsymbol{\alpha},\boldsymbol{\beta}}(\hat{\phi}) &= \sup_{\boldsymbol{\xi}\in\mathbb{R}^n}(|\boldsymbol{\xi}^{\boldsymbol{\alpha}}\partial_{\boldsymbol{\xi}}^{\boldsymbol{\beta}}\hat{\phi}(\xi)|) = \|\boldsymbol{\xi}^{\boldsymbol{\alpha}}\partial_{\boldsymbol{\xi}}^{\boldsymbol{\beta}}\hat{\phi}(\boldsymbol{\xi})\|_{L^\infty(\mathbb{R}^n)} \\
&\le (2\pi)^{|\boldsymbol{\beta}|-|\boldsymbol{\alpha}|}\|\partial_{\mathbf{x}}^{\boldsymbol{\alpha}}(\mathbf{x}^{\boldsymbol{\beta}}\phi)\|_{L^1(\mathbb{R}^n)} < +\infty \qquad (7.6.3)\\
&\Longrightarrow \quad \hat{\phi} \in S(\mathbb{R}^n).
\end{aligned}$$

Hence, $\phi \in S(\mathbb{R}^n) \Longrightarrow \hat{\phi} = \mathcal{F}\phi \in S(\mathbb{R}^n)$.

II. $\phi_k \to 0$ in $S(\mathbb{R}^n) \Longrightarrow \partial_{\mathbf{x}}^{\boldsymbol{\alpha}}(\mathbf{x}^{\boldsymbol{\beta}}\phi_k) \to 0$ in $S(\mathbb{R}^n)$ $\forall \boldsymbol{\alpha}, \boldsymbol{\beta}$, since $\partial_{\mathbf{x}}^{\boldsymbol{\alpha}} : S(\mathbb{R}^n) \longrightarrow S(\mathbb{R}^n)$ and $\phi_k \in S(\mathbb{R}^n) \longmapsto \mathbf{x}^{\boldsymbol{\beta}}\phi_k \in S(\mathbb{R}^n)$ are continuous and hence their composition $\phi_k \in S(\mathbb{R}^n) \longmapsto \partial_{\mathbf{x}}^{\boldsymbol{\alpha}}(\mathbf{x}^{\boldsymbol{\beta}}\phi_k) \in S(\mathbb{R}^n)$ is continuous from $S(\mathbb{R}^n)$ into $S(\mathbb{R}^n)$, i.e. $\phi_k \to 0$ in $S(\mathbb{R}^n) \Longrightarrow \partial_{\mathbf{x}}^{\boldsymbol{\alpha}}(\mathbf{x}^{\boldsymbol{\beta}}\phi_k) \to 0$ in $S(\mathbb{R}^n)$ $\forall \boldsymbol{\alpha}, \boldsymbol{\beta}$. As a consequence of the imbedding result $S(\mathbb{R}^n) \hookrightarrow L^1(\mathbb{R}^n)$ (Proposition 7.4.1), $\phi_k \to 0$ in $S(\mathbb{R}^n) \Longrightarrow \partial_{\mathbf{x}}^{\boldsymbol{\alpha}}(\mathbf{x}^{\boldsymbol{\beta}}\phi_k) \to 0$ in $L^1(\mathbb{R}^n)$ $\forall \boldsymbol{\alpha}, \boldsymbol{\beta}$,

$$\Longrightarrow \quad \forall \boldsymbol{\alpha}, \boldsymbol{\beta},\ \|\partial_{\mathbf{x}}^{\boldsymbol{\alpha}}(\mathbf{x}^{\boldsymbol{\beta}}\phi_k)\|_{L^1(\mathbb{R}^n)} \to 0 \text{ as } k \to \infty. \tag{7.6.4}$$

But from (7.6.3) and (7.6.4), $\forall \boldsymbol{\alpha}, \boldsymbol{\beta}$,

$$q_{\boldsymbol{\alpha},\boldsymbol{\beta}}(\mathcal{F}\phi_k) \le (2\pi)^{|\boldsymbol{\beta}|-|\boldsymbol{\alpha}|}\|\partial_{\mathbf{x}}^{\boldsymbol{\alpha}}(\mathbf{x}^{\boldsymbol{\beta}}\phi_k)\|_{L^1(\mathbb{R}^n)} \to 0$$

as $k \to \infty \Longrightarrow \mathcal{F}\phi_k = \hat{\phi}_k \to 0$ in $S(\mathbb{R}^n)$ as $k \to \infty$, i.e. $\phi_k \to 0$ in $S(\mathbb{R}^n) \Longrightarrow \mathcal{F}\phi_k \to 0$ in $S(\mathbb{R}^n) \Longrightarrow \mathcal{F} : S(\mathbb{R}^n) \longrightarrow S(\mathbb{R}^n)$ is continuous from $S(\mathbb{R}^n)$ into $S(\mathbb{R}^n)$.

□

7.7 Fourier inversion theorem in $S(\mathbb{R}^n)$

For $f \in L^1(\mathbb{R}^n)$, $\hat{f} = \mathcal{F}f \notin L^1(\mathbb{R}^n)$ in general ($\hat{f}$ is *bounded* in $\mathbb{R}^n$), and consequently $\overline{\mathcal{F}}\hat{f} = \int_{\mathbb{R}^n}\hat{f}(\boldsymbol{\xi})e^{i2\pi\langle\mathbf{x},\boldsymbol{\xi}\rangle}d\boldsymbol{\xi}$ is not defined *in general*. But for functions of $S(\mathbb{R}^n)$, we have:

Theorem 7.7.1 (Fourier Inversion Theorem). $\forall f \in S(\mathbb{R}^n)$,

$$\overline{\mathcal{F}}\mathcal{F}f = \mathcal{F}\overline{\mathcal{F}}f = f. \tag{7.7.1}$$

In other words, $\overline{\mathcal{F}} = \mathcal{F}^{-1}$ *is the* inverse *to Fourier transform* $\mathcal{F}$*, and* $\mathcal{F} = \overline{\mathcal{F}}^{-1}$ *is the inverse to Fourier co-transform* $\overline{\mathcal{F}}$*:* $\forall f \in S(\mathbb{R}^n)$*,*

$$\mathcal{F} f = \hat{f} \implies \overline{\mathcal{F}} \hat{f} = f; \quad \overline{\mathcal{F}} f = g \implies \mathcal{F} g = f. \tag{7.7.2}$$

Proof. It is sufficient to show that $\forall \mathbf{a} \in \mathbb{R}^n$, $(\overline{\mathcal{F}} \mathcal{F} f)(\mathbf{a}) = f(\mathbf{a})\ \forall f \in S(\mathbb{R}^n)$, i.e.

$$\begin{aligned}(\overline{\mathcal{F}} \mathcal{F} f)(\mathbf{a}) &= [\overline{\mathcal{F}}(\mathcal{F} f)](\mathbf{a}) = \int_{\mathbb{R}^n} (\mathcal{F} f)(\boldsymbol{\xi}) e^{+i2\pi\langle \mathbf{a},\boldsymbol{\xi}\rangle} d\boldsymbol{\xi} \\ &= \int_{\mathbb{R}^n} \left(\int_{\mathbb{R}^n} f(\mathbf{x}) e^{-i2\pi\langle \mathbf{x},\boldsymbol{\xi}\rangle} d\mathbf{x} \right) e^{i2\pi\langle \mathbf{a},\boldsymbol{\xi}\rangle} d\boldsymbol{\xi} = f(\mathbf{a}),\end{aligned} \tag{7.7.3}$$

since following the steps of this proof and replacing 'i' with '$-i$', we will get $(\mathcal{F} \overline{\mathcal{F}} f)(\mathbf{a}) = f(\mathbf{a})\ \forall \mathbf{a} \in \mathbb{R}^n$, $\forall f \in S(\mathbb{R}^n)$.

But for the function $F(\mathbf{x}, \boldsymbol{\xi}) = f(\mathbf{x}) e^{-i2\pi\langle \mathbf{x},\boldsymbol{\xi}\rangle} e^{i2\pi\langle \mathbf{a},\boldsymbol{\xi}\rangle}$, $\int_{\mathbb{R}^n \times \mathbb{R}^n} |F(\mathbf{x}, \boldsymbol{\xi})| d\mathbf{x} d\boldsymbol{\xi} = \infty$, i.e. F is *not* summable (integrable) on $\mathbb{R}^n \times \mathbb{R}^n$.

Hence, we cannot apply Fubini's Theorem 7.1.2C and consequently cannot interchange the order of integration in (7.7.3). Therefore we will apply Riesz's Formula (Corollary 7.1.1): $\forall f(\mathbf{x}) \in L^1(\mathbb{R}^n)$, $\forall g(\boldsymbol{\xi}) \in L^1(\mathbb{R}^n)$,

$$\int_{\mathbb{R}^n} \hat{f}(\boldsymbol{\xi}) g(\boldsymbol{\xi}) e^{i2\pi\langle \mathbf{a},\boldsymbol{\xi}\rangle} d\boldsymbol{\xi} = \int_{\mathbb{R}^n} f(\mathbf{a} + \mathbf{x}) \hat{g}(\mathbf{x}) d\mathbf{x},$$

where

$$\hat{f}(\boldsymbol{\xi}) = (\mathcal{F} f)(\boldsymbol{\xi}) = \int_{\mathbb{R}^n} f(\mathbf{x}) e^{-i2\pi\langle \mathbf{x},\boldsymbol{\xi}\rangle} d\mathbf{x};$$

$$\hat{g}(\mathbf{x}) = (\mathcal{F} g)(\mathbf{x}) = \int_{\mathbb{R}^n} g(\boldsymbol{\xi}) e^{-i2\pi\langle \boldsymbol{\xi},\mathbf{x}\rangle} d\boldsymbol{\xi},$$

and Lebesgue's Dominated Convergence Theorem B.3.2.2 (Appendix B) for a sequence (g_k) of auxiliary functions such that $g_k \to 1$ as $k \to \infty$ in order to establish the result. Let $\phi \in S(\mathbb{R}^n)$. Define $g_k(\mathbf{x}) = \phi(\frac{\mathbf{x}}{k})\ \forall k \in \mathbb{N}$ such that $g_k(\mathbf{x}) \to \phi(\mathbf{0})$ as $k \to \infty$. ϕ will finally be chosen such that $\phi(\mathbf{0}) = 1$. Then, $\forall f, g_k \in S(\mathbb{R}^n) \subset L^1(\mathbb{R}^n)$, we can apply:

$$\int_{\mathbb{R}^n} \hat{f}(\boldsymbol{\xi}) g_k(\boldsymbol{\xi}) e^{i2\pi\langle \mathbf{a},\boldsymbol{\xi}\rangle} d\boldsymbol{\xi} = \int_{\mathbb{R}^n} f(\mathbf{a} + \mathbf{x}) \hat{g}_k(\mathbf{x}) d\mathbf{x},$$

where $g_k(\boldsymbol{\xi}) = \phi(\frac{\boldsymbol{\xi}}{k})$, $\hat{g}_k(\mathbf{x}) = \int_{\mathbb{R}^n} g_k(\boldsymbol{\xi}) e^{-i2\pi\langle \boldsymbol{\xi},\mathbf{x}\rangle} d\boldsymbol{\xi} = \int_{\mathbb{R}^n} \phi(\frac{\boldsymbol{\xi}}{k}) e^{-i2\pi\langle \boldsymbol{\xi},\mathbf{x}\rangle} d\boldsymbol{\xi}$.

Set $\mathbf{y} = \frac{\boldsymbol{\xi}}{k}$ with $k \in \mathbb{N}$. Then $\boldsymbol{\xi} = k\mathbf{y} \implies$ the Jacobian of this transformation $= k^n > 0\ \forall k \in \mathbb{N}$.

Hence,

$$\hat{g}_k(\mathbf{x}) = \int_{\mathbb{R}^n} \phi(\mathbf{y}) e^{-i2\pi\langle k\mathbf{y},\mathbf{x}\rangle} (k^n) d\mathbf{y} = k^n \int_{\mathbb{R}^n} \phi(\mathbf{y}) e^{-i2\pi\langle \mathbf{y},k\mathbf{x}\rangle} d\mathbf{y} = k^n \hat{\phi}(k\mathbf{x}),$$

and

$$\int_{\mathbb{R}^n} \hat{f}(\boldsymbol{\xi}) g_k(\boldsymbol{\xi}) e^{i2\pi\langle \mathbf{a},\boldsymbol{\xi}\rangle} d\boldsymbol{\xi} = k^n \int_{\mathbb{R}^n} f(\mathbf{a}+\mathbf{x})\hat{\phi}(k\mathbf{x})d\mathbf{x} \quad \forall k \in \mathbb{N}.$$

Set $\boldsymbol{\eta} = k\mathbf{x}$ with $k \in \mathbb{N}$. Then $\mathbf{x} = \frac{\boldsymbol{\eta}}{k} \Longrightarrow$ the Jacobian of this transformation is $\frac{1}{k^n} > 0\ \forall k \in \mathbb{N}$. Hence, $f \in S(\mathbb{R}^n), \phi \in S(\mathbb{R}^n) \Longrightarrow k^n \int_{\mathbb{R}^n} f(\mathbf{a}+\mathbf{x})\hat{\phi}(k\mathbf{x})d\mathbf{x} = k^n \int_{\mathbb{R}^n} f(\mathbf{a}+\frac{\boldsymbol{\eta}}{k})\hat{\phi}(\boldsymbol{\eta})\frac{1}{k^n}d\boldsymbol{\eta}$

$$\Longrightarrow \quad \int_{\mathbb{R}^n} \hat{f}(\boldsymbol{\xi}) g_k(\boldsymbol{\xi}) e^{i2\pi\langle \mathbf{a},\boldsymbol{\xi}\rangle} d\boldsymbol{\xi} = \int_{\mathbb{R}^n} f(a+\frac{\boldsymbol{\eta}}{k})\hat{\phi}(\boldsymbol{\eta})d\boldsymbol{\eta} \quad \forall k \in \mathbb{N}. \tag{7.7.4}$$

Now we let $k \to \infty$. Then the sequence $g_k(\boldsymbol{\xi}) = \phi(\frac{\boldsymbol{\xi}}{k}) \to \phi(\mathbf{0})$ and the sequence

$$\hat{f}(\boldsymbol{\xi}) g_k(\boldsymbol{\xi}) e^{i2\pi\langle \mathbf{a},\boldsymbol{\xi}\rangle} \to \hat{f}(\boldsymbol{\xi})\phi(\mathbf{0}) e^{i2\pi\langle \mathbf{a},\boldsymbol{\xi}\rangle} \quad \text{as } k \to \infty. \tag{7.7.5}$$

On the other hand, $\forall \boldsymbol{\xi} \in \mathbb{R}^n$, $\forall k \in \mathbb{N}$,

$$|\hat{f}(\boldsymbol{\xi}) g_k(\boldsymbol{\xi}) e^{i2\pi\langle \mathbf{a},\boldsymbol{\xi}\rangle}| = |\hat{f}(\boldsymbol{\xi})||g_k(\boldsymbol{\xi})| = |\hat{f}(\boldsymbol{\xi})| \cdot |\phi(\frac{\boldsymbol{\xi}}{k})| \le \|\phi\|_\infty |\hat{f}(\boldsymbol{\xi})|,$$

with $\|\phi\|_\infty = \sup_{\boldsymbol{\xi} \in \mathbb{R}^n} |\phi(\boldsymbol{\xi})|$. Moreover,

$$\int_{\mathbb{R}^n} \|\phi\|_\infty |\hat{f}(\boldsymbol{\xi})| d\boldsymbol{\xi} = \|\phi\|_\infty \int_{\mathbb{R}^n} |\hat{f}(\boldsymbol{\xi})| d\boldsymbol{\xi} < +\infty, \tag{7.7.6}$$

since $f \in S(\mathbb{R}^n) \subset L^1(\mathbb{R}^n) \Longrightarrow \hat{f} \in S(\mathbb{R}^n) \subset L^1(\mathbb{R}^n)$.

Hence, by virtue of (7.7.5) and (7.7.6), Lebesgue's Dominated Convergence Theorem B.3.2.2 (Appendix B) can be applied and we get:

$$\begin{aligned} \lim_{k\to\infty} \int_{\mathbb{R}^n} \hat{f}(\boldsymbol{\xi}) g_k(\boldsymbol{\xi}) e^{i2\pi\langle \mathbf{a},\boldsymbol{\xi}\rangle} d\boldsymbol{\xi} &= \int_{\mathbb{R}^n} \lim_{k\to\infty} (\hat{f}(\boldsymbol{\xi}) g_k(\boldsymbol{\xi}) e^{i2\pi\langle \mathbf{a},\boldsymbol{\xi}\rangle}) d\boldsymbol{\xi} \\ &= \int_{\mathbb{R}^n} \hat{f}(\boldsymbol{\xi})\phi(\mathbf{0}) e^{i2\pi\langle \mathbf{a},\boldsymbol{\xi}\rangle} d\boldsymbol{\xi} \end{aligned}$$

$$\Longrightarrow \quad \lim_{k\to\infty} \int_{\mathbb{R}^n} \hat{f}(\boldsymbol{\xi}) g_k(\boldsymbol{\xi}) e^{i2\pi\langle \mathbf{a},\boldsymbol{\xi}\rangle} d\boldsymbol{\xi} = \phi(\mathbf{0}) \int_{\mathbb{R}^n} \hat{f}(\boldsymbol{\xi}) e^{i2\pi\langle \mathbf{a},\boldsymbol{\xi}\rangle} d\boldsymbol{\xi}. \tag{7.7.7}$$

Again, for $f \in S(\mathbb{R}^n)$, the sequence $f(\mathbf{a}+\frac{\boldsymbol{\eta}}{k})\hat{\phi}(\boldsymbol{\eta}) \to f(\mathbf{a})\hat{\phi}(\boldsymbol{\eta})$ as $k \to \infty$ and $|f(\mathbf{a}+\frac{\boldsymbol{\eta}}{k})\hat{\phi}(\boldsymbol{\eta})| \le \|f\|_\infty |\hat{\phi}(\boldsymbol{\eta})|$ with $\|f\|_\infty = \sup_{\boldsymbol{\eta}\in\mathbb{R}^n, k\in\mathbb{N}} |f(\mathbf{a}+\frac{\boldsymbol{\eta}}{k})|$

$\Longrightarrow \int_{\mathbb{R}^n} \|f\|_\infty |\hat{\phi}(\boldsymbol{\eta})| d\boldsymbol{\eta} = \|f\|_\infty \int_{\mathbb{R}^n} |\hat{\phi}(\boldsymbol{\eta})| d\boldsymbol{\eta} < +\infty$, since $\hat{\phi} \in S(\mathbb{R}^n) \Longrightarrow \hat{\phi} \in L^1(\mathbb{R}^n) \Longrightarrow \int_{\mathbb{R}^n} |\hat{\phi}(\boldsymbol{\eta})| d\boldsymbol{\eta} < +\infty$.

Hence, again by Lebesgue's Dominated Convergence Theorem B.3.2.2 (Appendix B), we have, from (7.7.4):

$$\lim_{k\to\infty} \int_{\mathbb{R}^n} f\left(\mathbf{a}+\frac{\boldsymbol{\eta}}{k}\right)\hat{\phi}(\boldsymbol{\eta})d\boldsymbol{\eta} = \int_{\mathbb{R}^n} \lim_{k\to\infty} \left[f\left(\mathbf{a}+\frac{\boldsymbol{\eta}}{k}\right)\hat{\phi}(\boldsymbol{\eta})\right]d\boldsymbol{\eta} = \int_{\mathbb{R}^n} f(\mathbf{a})\hat{\phi}(\boldsymbol{\eta})d\boldsymbol{\eta}. \tag{7.7.8}$$

From (7.7.7) and (7.7.8),

$$\phi(\mathbf{0})\int_{\mathbb{R}^n}\hat{f}(\boldsymbol{\xi})e^{i2\pi\langle\mathbf{a},\boldsymbol{\xi}\rangle}d\boldsymbol{\xi} = f(\mathbf{a})\int_{\mathbb{R}^n}\hat{\phi}(\boldsymbol{\eta})d\boldsymbol{\eta} \quad \forall\phi\in S(\mathbb{R}^n). \tag{7.7.9}$$

In particular, for $\phi(\mathbf{x}) = e^{-\pi\|\mathbf{x}\|^2}$, $\hat{\phi}(\boldsymbol{\xi}) = e^{-\pi\|\boldsymbol{\xi}\|^2}$ and $\phi(\mathbf{0}) = 1$, $\int_{\mathbb{R}^n}\hat{\phi}(\boldsymbol{\eta})d\boldsymbol{\eta} = \int_{\mathbb{R}^n}e^{-\pi\|\boldsymbol{\eta}\|^2}d\boldsymbol{\eta} = 1$ (the Gauss integral) – see Example 7.1.5.

From (7.7.9), we get $\int_{\mathbb{R}^n}\hat{f}(\boldsymbol{\xi})e^{+i2\pi\langle\mathbf{a},\boldsymbol{\xi}\rangle}d\boldsymbol{\xi} = f(\mathbf{a}) \Longrightarrow (\overline{\mathcal{F}}\mathcal{F}f)(\mathbf{a}) = f(\mathbf{a})$ $\forall\mathbf{a}\in\mathbb{R}^n$ (by (7.7.3)) $\Longrightarrow \overline{\mathcal{F}}\mathcal{F}f = f$ in $S(\mathbb{R}^n)$. □

Remark 7.7.1.

- The proof shows that if $f \in L^1(\mathbb{R}^n)$ is *continuous* and *bounded* in $\mathbb{R}^n$ and $\hat{f}\in L^1(\mathbb{R}^n)$, then $\overline{\mathcal{F}}[\hat{f}] = f$ everywhere.
- Subsequently, when Fourier transforms of tempered distributions are introduced, it can be shown that the condition of the *boundedness* of f is superfluous.

Corollary 7.7.1. *Let $\mathcal{F} : S(\mathbb{R}^n) \longrightarrow S(\mathbb{R}^n)$. If $\mathcal{F}\phi = 0$ for $\phi\in S(\mathbb{R}^n)$, then $\phi = 0$ in $S(\mathbb{R}^n)$.*

Isomorphism of Fourier transform on $S(\mathbb{R}^n)$

Theorem 7.7.2. *Fourier transform $\mathcal{F} : S(\mathbb{R}^n) \longrightarrow S(\mathbb{R}^n)$ is a topological* isomorphism *from $S(\mathbb{R}^n)$ onto $S(\mathbb{R}^n)$.*

Proof. From Theorems 7.6.1 and 7.7.1, $\mathcal{F}$ is a linear bijective mapping from $S(\mathbb{R}^n)$ onto $S(\mathbb{R}^n)$, and hence an algebraic isomorphism from $S(\mathbb{R}^n)$ onto $S(\mathbb{R}^n)$. By Theorem 7.6.1, $\mathcal{F}$ and its inverse $\mathcal{F}^{-1} = \overline{\mathcal{F}} : S(\mathbb{R}^n) \longrightarrow S(\mathbb{R}^n)$ are continuous from $S(\mathbb{R}^n)$ onto $S(\mathbb{R}^n)$. Hence, $\mathcal{F}$ is a topological isomorphism from $S(\mathbb{R}^n)$ onto $S(\mathbb{R}^n)$. □

Application

We give here an alternative proof of Example 7.1.2 of the function $f(x) = e^{-\pi x^2}$, whose Fourier transform is the function itself, i.e. $f = \hat{f} = \mathcal{F}f$, if we neglect the role of variables x and ξ.

Let $f(x) = e^{-\pi x^2}$ be the function on $\mathbb{R}$ which satisfies the differential equation

$$f'(x) + 2\pi x f(x) = 0 \quad\Longrightarrow\quad f'(x) + i(-i2\pi x)f(x) = 0 \quad \forall x\in\mathbb{R}. \tag{7.7.10}$$

But $f(x) = e^{-\pi x^2} \in S(\mathbb{R}) \Longrightarrow xf(x) \in S(\mathbb{R}) \Longrightarrow \mathcal{F}[f(x)] = \hat{f}(\xi)$ and $\mathcal{F}[xf(x)] = \widehat{xf}(\xi)$ exist and belong to $S(\mathbb{R})$. But $\mathcal{F}[f'(x) + i(-i2\pi x)f(x)] = \mathcal{F}[0] = 0$, $\mathcal{F}[f'(x)] = i2\pi\xi\hat{f}(\xi)$ and $\mathcal{F}[(-i2\pi x)f(x)] = \frac{d\hat{f}}{d\xi}(\xi) \Longrightarrow i2\pi\xi\hat{f}(\xi) + i(\hat{f})'(\xi) = 0 \Longrightarrow (\hat{f})'(\xi) + 2\pi\xi\hat{f}(\xi) = 0$.

Thus, f and $\hat{f}$ satisfy the same first-order differential equation (7.7.10) in variables x and ξ respectively. Hence, a particular solution of the first-order equation in variable ξ is $e^{-\pi\xi^2}$, and its general solution is given by $\hat{f}(\xi) = Ce^{-\pi\xi^2}$, C being a constant. Now, we determine the constant C from the condition that $C = \hat{f}(0)$. In fact, $\hat{f}(\xi) = \int_{-\infty}^{\infty} f(x)e^{-i2\pi x\xi}dx$

$\implies \hat{f}(0) = \int_{-\infty}^{\infty} f(x)dx = \int_{-\infty}^{\infty} e^{-\pi x^2}d\mathbf{x} = 1$ (Gauss integral)

$\implies C = 1 \implies \hat{f}(\xi) = e^{-\pi\xi^2} \implies f = \hat{f}$.

Plancherel–Parseval theorem on isometry in $S(\mathbb{R}^n)$

Theorem 7.7.3. *Let $S(\mathbb{R}^n) \subset L^2(\mathbb{R}^n)$ be equipped with inner product $\langle\cdot,\cdot\rangle_{L^2(\mathbb{R}^n)}$ induced by $L^2(\mathbb{R}^n)$, i.e. $\forall\phi,\psi \in S(\mathbb{R}^n)$, $\langle\phi,\psi\rangle_{L^2(\mathbb{R}^n)} = \int_{\mathbb{R}^n}\phi(\mathbf{x})\overline{\psi(\mathbf{x})}d\mathbf{x}$, where $\overline{\psi(\mathbf{x})}$ is the complex conjugate of $\psi(\mathbf{x})$. Then, $\forall\phi,\psi \in S(\mathbb{R}^n)$ with $\hat{\phi} = \mathcal{F}\phi, \hat{\psi} = \mathcal{F}\psi \in S(\mathbb{R}^n)$,*

- $\langle\phi,\psi\rangle_{L^2(\mathbb{R}^n)} = \langle\hat{\phi},\hat{\psi}\rangle_{L^2(\mathbb{R}^n)}$, *i.e.*

$$\int_{\mathbb{R}^n}\phi(\mathbf{x})\overline{\psi(\mathbf{x})}d\mathbf{x} = \int_{\mathbb{R}^n}\hat{\phi}(\boldsymbol{\xi})\overline{\hat{\psi}(\boldsymbol{\xi})}d\boldsymbol{\xi}; \tag{7.7.11}$$

- $\|\phi\|_{L^2(\mathbb{R}^n)} = \|\hat{\phi}\|_{L^2(\mathbb{R}^n)}$, *i.e.*

$$\left(\int_{\mathbb{R}^n}|\phi(\mathbf{x})|^2d\mathbf{x}\right)^{\frac{1}{2}} = \left(\int_{\mathbb{R}^n}|\hat{\phi}(\boldsymbol{\xi})|^2d\boldsymbol{\xi}\right)^{\frac{1}{2}}. \tag{7.7.12}$$

Proof. Let $\phi,\psi \in S(\mathbb{R}^n)$. Then $\overline{\psi} \in S(\mathbb{R}^n)$ and $\exists\chi \in S(\mathbb{R}^n)$ such that $\overline{\psi} = \mathcal{F}\chi = \hat{\boldsymbol{\chi}}$, since $\mathcal{F}$ is an isomorphism on $S(\mathbb{R}^n)$. Hence, $\chi = \overline{\mathcal{F}}\mathcal{F}\chi = \overline{\mathcal{F}}\bar{\psi} = \overline{\mathcal{F}\psi}$. From Property 8 in (7.1.26), $\int_{\mathbb{R}^n}\hat{\phi}(\boldsymbol{\xi})\chi(\boldsymbol{\xi})d\boldsymbol{\xi} = \int_{\mathbb{R}^n}\phi(\mathbf{x})\hat{\chi}(\mathbf{x})d\mathbf{x} = \int_{\mathbb{R}^n}\phi(\mathbf{x})\overline{\psi(\mathbf{x})}d\mathbf{x}$.

But $\chi(\boldsymbol{\xi}) = \overline{\mathcal{F}\psi(\boldsymbol{\xi})} = \overline{\hat{\psi}(\boldsymbol{\xi})} \implies \int_{\mathbb{R}^n}\hat{\phi}(\boldsymbol{\xi})\overline{\hat{\psi}(\boldsymbol{\xi})}d\boldsymbol{\xi} = \int_{\mathbb{R}^n}\phi(\mathbf{x})\overline{\psi(\mathbf{x})}d\mathbf{x}$

$\implies \langle\phi,\psi\rangle_{L^2(\mathbb{R}^n)} = \langle\hat{\phi},\hat{\psi}\rangle_{L^2(\mathbb{R}^n)}$.

In particular, for $\psi = \phi$ with $\hat{\psi} = \hat{\phi}$, $\|\phi\|^2_{L^2(\mathbb{R}^n)} = \langle\phi,\phi\rangle_{L^2(\mathbb{R}^n)} = \langle\hat{\phi},\hat{\phi}\rangle_{L^2(\mathbb{R}^n)} = \|\hat{\phi}\|^2_{L^2(\mathbb{R}^n)}$

$\implies \|\phi\|_{L^2(\mathbb{R}^n)} = \|\hat{\phi}\|_{L^2(\mathbb{R}^n)}$. □

Chapter 8

Fourier transforms of distributions and Sobolev spaces of arbitrary order $H^S(\mathbb{R}^n)$

8.1 Motivation for a possible definition of the Fourier transform of a distribution

Let $f \in L^1(\mathbb{R}^n)$. Then $f \in L^1_{\text{loc}}(\mathbb{R}^n) \Longrightarrow f \in 423g\mathcal{D}'(\mathbb{R}^n)$ defines a regular distribution on $\mathbb{R}^n$: $\forall \phi \in \mathcal{D}(\mathbb{R}^n)$, $\langle f, \phi \rangle = \int_{\mathbb{R}^n} f(\mathbf{x})\phi(\mathbf{x})d\mathbf{x}$.

Since $f \in L^1(\mathbb{R}^n)$, its Fourier transform $\hat{f} = \mathcal{F} f$ is *bounded* and *continuous* in $\mathbb{R}^n$ and $\hat{f}(\boldsymbol{\xi}) \to 0$ as $\|\boldsymbol{\xi}\| \to \infty$ (see (7.1.6), (7.1.24), (7.1.36)), and consequently $\hat{f} \in L^1_{\text{loc}}(\mathbb{R}^n) \Longrightarrow \hat{f} \in \mathcal{D}'(\mathbb{R}^n)$ defines a distribution on $\mathbb{R}^n$, i.e. $\forall \phi \in \mathcal{D}(\mathbb{R}^n)$,

$$\begin{aligned}\langle \mathcal{F} f, \phi \rangle &= \int_{\mathbb{R}^n} (\mathcal{F} f)(\boldsymbol{\xi})\phi(\boldsymbol{\xi})d\boldsymbol{\xi} = \int_{\mathbb{R}^n} \left(\int_{\mathbb{R}^n} f(\mathbf{x})e^{-i2\pi\langle \mathbf{x},\boldsymbol{\xi}\rangle} d\mathbf{x} \right) \phi(\boldsymbol{\xi})d\boldsymbol{\xi} \\ &= \int_{\mathbb{R}^{2n}} f(\mathbf{x})\phi(\boldsymbol{\xi})e^{-i2\pi\langle \mathbf{x},\boldsymbol{\xi}\rangle} d\mathbf{x}d\boldsymbol{\xi},\end{aligned}$$

since the multiple integral over the product space $\mathbb{R}^{2n}$ exists.

In fact, $|e^{-i2\pi\langle \mathbf{x},\boldsymbol{\xi}\rangle} f(\mathbf{x})\phi(\boldsymbol{\xi})| = |f(\mathbf{x})\phi(\boldsymbol{\xi})|$ is integrable on $\mathbb{R}^{2n}$ as the product of an integrable (summable) function of $\mathbf{x}$ on $\mathbb{R}^n$ with another integrable (summable) function ϕ of $\boldsymbol{\xi}$ with *compact* support in $\mathbb{R}^n$. Consequently, we can apply Fubini's Theorem 7.1.2C to change the order of integration and write

$$\begin{aligned}\langle \mathcal{F} f, \phi \rangle &= \int_{\mathbb{R}^{2n}} f(\mathbf{x})\phi(\boldsymbol{\xi})e^{-i2\pi\langle \mathbf{x},\boldsymbol{\xi}\rangle} d\mathbf{x}d\boldsymbol{\xi} = \int_{\mathbb{R}^n} f(\mathbf{x}) \left(\int_{\mathbb{R}^n} \phi(\boldsymbol{\xi})e^{-i2\pi\langle \mathbf{x},\boldsymbol{\xi}\rangle} d\boldsymbol{\xi} \right) d\mathbf{x} \\ &= \int_{\mathbb{R}^n} f(\mathbf{x})(\mathcal{F}\phi)(\mathbf{x})d\mathbf{x} = \langle f, \mathcal{F}\phi \rangle \quad \forall \phi \in \mathcal{D}(\mathbb{R}^n). \qquad (8.1.1)\end{aligned}$$

(8.1.1) holds, even if $\phi \notin \mathcal{D}(\mathbb{R}^n)$, but ϕ must belong to $L^1(\mathbb{R}^n)$, since $\phi \in L^1(\mathbb{R}^n) \Longrightarrow \hat{\phi}(\mathbf{x}) = \mathcal{F}[\phi(\mathbf{x})]$ is bounded and continuous in $\mathbb{R}^n$ and $\hat{\phi}(\mathbf{x}) \to 0$ as $\|\mathbf{x}\| \to \infty$, and

$$\left| \int_{\mathbb{R}^n} f(\mathbf{x})(\mathcal{F}\phi)(\mathbf{x})d\mathbf{x} \right| \leq \sup_{\mathbf{x}\in\mathbb{R}^n} |(\mathcal{F}\phi)(\mathbf{x})| \int_{\mathbb{R}^n} |f(\mathbf{x})|d\mathbf{x} < +\infty. \qquad (8.1.2)$$

Hence, the two formulae:

$$\langle \mathcal{F} f, \phi \rangle = \int_{\mathbb{R}^n} \hat{f}(\boldsymbol{\xi})\phi(\boldsymbol{\xi})d\boldsymbol{\xi}, \tag{8.1.3}$$

$$\langle f, \mathcal{F}\phi \rangle = \int_{\mathbb{R}^n} f(\mathbf{x})\hat{\phi}(\mathbf{x})d\mathbf{x} \tag{8.1.4}$$

are *well defined* and *equal* for $\phi \in L^1(\mathbb{R}^n)$. Thus, (8.1.1) suggests that a *possible* definition of the Fourier transform $\mathcal{F}T$ for a distribution $T \in \mathcal{D}'(\mathbb{R}^n)$ should be

$$\langle \mathcal{F}T, \phi \rangle = \langle T, \mathcal{F}\phi \rangle \quad \forall \phi \in \mathcal{D}(\mathbb{R}^n) \text{ with } \phi = \phi(\boldsymbol{\xi}), \tag{8.1.5}$$

if the expressions on both sides of this equality (8.1.5) are meaningful. Unfortunately, the right-hand side of (8.1.5) has *no* meaning for arbitrary $\phi \in \mathcal{D}(\mathbb{R}^n)$, since, for $\phi \in \mathcal{D}(\mathbb{R}^n)$, $\mathcal{F}\phi \notin \mathcal{D}(\mathbb{R}^n)$ in general. In fact,we have the following result:

$$\text{If } \phi \in \mathcal{D}(\mathbb{R}^n), \phi \neq 0, \quad \text{then } \mathcal{F}\phi \notin \mathcal{D}(\mathbb{R}^n) \quad [7]. \tag{8.1.6}$$

Thus, the Fourier transform $\mathcal{F}T$ of arbitrary distribution $T \in \mathcal{D}'(\mathbb{R}^n)$ *does not exist.* (8.1.7)

But the right-hand side of (8.1.5) *becomes* meaningful for $\phi \in S(\mathbb{R}^n)$, since, by Theorem 7.6.1, $\forall \phi \in S(\mathbb{R}^n)$, $\mathcal{F}\phi \in S(\mathbb{R}^n)$ with $S(\mathbb{R}^n) \hookrightarrow L^1(\mathbb{R}^n)$.

Hence, we are going to identify a particular subclass of distributions containing $L^1(\mathbb{R}^n)$ as a *subspace*. This particular subclass of distributions, described by Laurent Schwartz, are called *tempered distributions*, or *distributions tempérés* in French.

8.2 Space $S'(\mathbb{R}^n)$ of tempered distributions

8.2.1 Tempered distributions

Definition 8.2.1. A distribution $T \in \mathcal{D}'(\mathbb{R}^n)$ is called a *tempered distribution* on $\mathbb{R}^n$ if and only if T can be extended to a *continuous linear functional* on $S(\mathbb{R}^n)$, and the extended continuous, linear functional will still be denoted by T.

Every tempered distribution T is a distribution in $\mathcal{D}'(\mathbb{R}^n)$.

Precisely speaking, if T is a tempered distribution, then its restriction to $\mathcal{D}(\mathbb{R}^n)$ denoted by $T{\downarrow}_{\mathcal{D}(\mathbb{R}^n)} \in \mathcal{D}'(\mathbb{R}^n)$, since $\mathcal{D}(\mathbb{R}^n) \hookrightarrow S(\mathbb{R}^n)$ and $\mathcal{D}(\mathbb{R}^n)$ is *dense* in $S(\mathbb{R}^n)$.

Characterization of tempered distributions

We state the results, which follow from Definition 8.2.1 and the notion of convergence in $S(\mathbb{R}^n)$ (see Definition 7.2.5), in the form of a proposition.

Proposition 8.2.1. *A distribution $T \in \mathcal{D}'(\mathbb{R}^n)$ is a tempered distribution if and only if*

I. *T is a* linear *functional on $S(\mathbb{R}^n)$: $T(\phi) \in \mathbb{C}\ \forall \phi \in S(\mathbb{R}^n)$,*

$$T(\alpha_1\phi_1 + \alpha_2\phi_2) = \alpha_1 T(\phi_1) + \alpha_2 T(\phi_2) \quad \forall \phi_i \in S(\mathbb{R}^n),\ \forall \alpha_i \in \mathbb{C}; \tag{8.2.1}$$

II. *T is* continuous *on $S(\mathbb{R}^n)$: $\exists$ a constant $C > 0$, independent of ϕ, and $\exists$ multi-index $\boldsymbol{\alpha}, \boldsymbol{\beta}$ such that*

$$|T(\phi)| \le C q_{\boldsymbol{\alpha},\boldsymbol{\beta}}(\phi) = C \sup_{\mathbf{x}\in\mathbb{R}^n} |\mathbf{x}^{\boldsymbol{\alpha}} \partial^{\boldsymbol{\beta}} \phi(\mathbf{x})| \quad \forall \phi \in S(\mathbb{R}^n), \tag{8.2.2}$$

or, equivalently, $\exists C > 0$, independent of ϕ, and $\exists l, m \in \mathbb{N}_0$ such that

$$|T(\phi)| \le C q^*_{l,m}(\phi) = C \sup_{|\boldsymbol{\beta}|\le m} \sup_{\mathbf{x}\in\mathbb{R}^n} |(1 + \|\mathbf{x}\|^2)^l \partial^{\boldsymbol{\beta}} \phi(\mathbf{x})| \quad \forall \phi \in S(\mathbb{R}^n). \tag{8.2.3}$$

Equivalent notations are:

$$T(\phi) = \langle T, \phi \rangle = [T, \phi] \quad \forall \phi \in S(\mathbb{R}^n). \tag{8.2.4}$$

Algebraic properties of tempered distributions

Sum of tempered distributions The sum $T_1 + T_2$ of tempered distributions T_1 and T_2 is a tempered distribution defined by:

$$(T_1 + T_2)(\phi) = T_1(\phi) + T_2(\phi) \quad \forall \phi \in S(\mathbb{R}^n), \tag{8.2.5}$$

equivalently written as $\langle T_1 + T_2, \phi \rangle = \langle T_1, \phi \rangle + \langle T_2, \phi \rangle \forall \phi \in S(\mathbb{R}^n)$.

Multiplication by a number The product αT of the multiplication of a tempered distribution T by a complex number $\alpha \in \mathbb{C}$ is a tempered distribution defined by:

$$(\alpha T)(\phi) = \alpha T(\phi) \quad \forall \phi \in S(\mathbb{R}^n), \tag{8.2.6}$$

equivalently written as $\langle \alpha T, \phi \rangle = \alpha \langle T, \phi \rangle\ \forall \phi \in S(\mathbb{R}^n)$.

Null tempered distribution A tempered distribution T is called a *null* tempered distribution, denoted by $0 \in S'(\mathbb{R}^n)$, if and only if $T(\phi) = 0\ \forall \phi \in S(\mathbb{R}^n)$, i.e.

$$T = 0 \text{ in } S'(\mathbb{R}^n) \quad \Longrightarrow \quad T(\phi) = 0\ \forall \phi \in S(\mathbb{R}^n). \tag{8.2.7}$$

Equality of tempered distributions Two tempered distributions T_1 and T_2 are called *equal* in $S'(\mathbb{R}^n)$ if and only if $T_1(\phi) = T_2(\phi)\ \forall \phi \in S(\mathbb{R}^n)$, i.e.

$$T_1 = T_2 \text{ in } S'(\mathbb{R}^n) \quad \Longleftrightarrow \quad T_1(\phi) = T_2(\phi)\ \forall \phi \in S(\mathbb{R}^n). \tag{8.2.8}$$

8.2.2 Space $S'(\mathbb{R}^n)$

Definition 8.2.2. Let the sum of tempered distributions, multiplication of a tempered distribution by a number and the null tempered distribution be defined by (8.2.5), (8.2.6) and (8.2.7) respectively. Then the set of all tempered distributions T on $\mathbb{R}^n$ form a linear space called the *space of tempered distributions on* $\mathbb{R}^n$, which is the *dual space* of $S(\mathbb{R}^n)$ and hence denoted by $S'(\mathbb{R}^n)$.

The celebrated French mathematician Laurent Schwartz introduced the space $S'(\mathbb{R}^n)$ of tempered distributions [7], [8].

Every tempered distribution $T \in S'(\mathbb{R}^n)$ is a distribution in $\mathcal{D}'(\mathbb{R}^n)$, i.e. $T \in S'(\mathbb{R}^n) \Longrightarrow T \in \mathcal{D}'(\mathbb{R}^n) \Longrightarrow$ the algebraic inclusion $S'(\mathbb{R}^n) \subset \mathcal{D}'(\mathbb{R}^n)$ (8.2.9) (see imbedding results later). But every distribution $T \in \mathcal{D}'(\mathbb{R}^n)$ is *not* a tempered distribution in $S'(\mathbb{R}^n)$, i.e. $\mathcal{D}'(\mathbb{R}^n) \not\subset S'(\mathbb{R}^n)$.

For example, for $n = 1$, $e^{x^2} \in \mathcal{D}'(]-\infty,\infty[)$, but $e^{x^2} \notin S'(]-\infty,\infty[)$. (8.2.10)

In fact, $e^{x^2} \in L^1_{\text{loc}}(]-\infty,\infty[) \Longrightarrow e^{x^2}$ defines a distribution $T_{e^{x^2}} \in \mathcal{D}'(]-\infty,\infty[)$ by $\langle T_{e^{x^2}}, \phi\rangle = \int_{-\infty}^{\infty} e^{x^2}\phi(x)dx \ \forall \phi \in \mathcal{D}(]-\infty,\infty[)$. But $\langle T_{e^{x^2}}, \phi\rangle$ is *not* defined $\forall \phi \in S(]-\infty,\infty[)$.

For instance, for $\phi(x) = e^{-x^2}$, $\langle T_{e^{x^2}}, e^{-x^2}\rangle = \int_{-\infty}^{\infty} e^{x^2}.e^{-x^2}dx = \infty \Longrightarrow T_{e^{x^2}}$ is *not* a tempered distribution on $]-\infty,\infty[$, i.e. $e^{x^2} \notin S'(]-\infty,\infty[)$. (8.2.11)

8.2.3 Examples of tempered distributions of $S'(\mathbb{R}^n)$

1. Function $f \in L^1(\mathbb{R}^n)$, integrable on $\mathbb{R}^n$, i.e. $\int_{\mathbb{R}^n} |f(\mathbf{x})|d\mathbf{x} < +\infty$, defines a tempered distribution $T_f \in S'(\mathbb{R}^n)$ by

$$T_f(\phi) = \langle f, \phi\rangle = \int_{\mathbb{R}^n} f(\mathbf{x})\phi(\mathbf{x})d\mathbf{x} \quad \forall \phi \in S(\mathbb{R}^n). \tag{8.2.12}$$

In fact, the integral in (8.2.12) exists $\forall \phi \in S(\mathbb{R}^n)$, since this integral exists for *bounded* ϕ and functions $\phi \in S(\mathbb{R}^n)$ are indeed bounded in $\mathbb{R}^n$: $\forall \mathbf{x} \in \mathbb{R}^n$,

$$|\phi(\mathbf{x})| \leq \sup_{\mathbf{x}\in\mathbb{R}^n} |\phi(\mathbf{x})| = q_{\mathbf{0},\mathbf{0}}(\phi) < +\infty$$

(see (7.2.12)) and

$$\left|\int_{\mathbb{R}^n} f(\mathbf{x})\phi(\mathbf{x})d\mathbf{x}\right| \leq q_{\mathbf{0},\mathbf{0}}(\phi)\int_{\mathbb{R}^n} |f(\mathbf{x})|d\mathbf{x} < +\infty \quad \forall f \in L^1(\mathbb{R}^n).$$

Moreover, we have shown later (see p. 433) that $L^1(\mathbb{R}^n) \hookrightarrow S'(\mathbb{R}^n)$, the imbedding operator $\hookrightarrow$ being a continuous one.

Hence, we identify $f \in L^1(\mathbb{R}^n)$ with $T_f \in S'(\mathbb{R}^n)$ defined by (8.2.12) and can write

$$f = T_f \in S'(\mathbb{R}^n), \quad \text{i.e. } L^1(\mathbb{R}^n) \subset S'(\mathbb{R}^n). \tag{8.2.13}$$

2. Every bounded function f in $\mathbb{R}^n$ defines a tempered distribution $T_f \in S'(\mathbb{R}^n)$ by the formula (8.2.12). In fact, f is bounded in $\mathbb{R}^n \implies \exists M > 0$ such that $|f(\mathbf{x})| \leq M \ \forall \mathbf{x} \in \mathbb{R}^n$, and $\phi \in S(\mathbb{R}^n) \implies \phi \in L^1(\mathbb{R}^n)$ (by Proposition 7.2.2), i.e. $\int_{\mathbb{R}^n} |\phi(\mathbf{x})| d\mathbf{x} < +\infty$.
Hence,
$$|T_f(\phi)| = \left| \int_{\mathbb{R}^n} f(\mathbf{x})\phi(\mathbf{x}) d\mathbf{x} \right| \leq \sup_{\mathbf{x}\in\mathbb{R}^n} \int_{\mathbb{R}^n} |\phi(\mathbf{x})| d\mathbf{x} < +\infty \quad \forall \phi \in S(\mathbb{R}^n). \tag{8.2.14}$$
The simplest examples of bounded functions on $\mathbb{R} =]-\infty, \infty[$ are the Heaviside function $H(x) = 1$ for $x > 0$ and $H(x) = 0$ for $x < 0$, constant function $f(x) = C \ \forall x \in \mathbb{R}, e^{-\|\mathbf{x}\|^2}$, trigonometric functions $\sin \alpha x$, $\cos \beta x \ \forall \alpha, \beta \in \mathbb{R}$, etc. All these functions define tempered distributions by (8.2.12).
3. Every locally integrable function $f \in L^1_{\text{loc}}(\mathbb{R}^n)$ with slow or polynomial growth at infinity defines a tempered distribution $T_f \in S'(\mathbb{R}^n)$ by the integral (8.2.12). f is locally integrable in $\mathbb{R}^n \implies \int_{\overset{\circ}{K}} |f(\mathbf{x})| d\mathbf{x} < +\infty \ \forall$ compact subsets K of $\mathbb{R}^n$.
f has slow/polynomial growth at infinity $\implies \exists$ an integer $k \in \mathbb{N}_0$ and a constant $C > 0$ such that
$$|f(\mathbf{x})| \leq C(1 + \|\mathbf{x}\|^2)^k \quad \text{for } \|\mathbf{x}\| \to \infty,$$
$$\implies \frac{|f(\mathbf{x})|}{(1+\|\mathbf{x}\|^2)^k} \leq C \text{ for } \|\mathbf{x}\| \to \infty. \tag{8.2.15}$$
In fact, for $\phi \in S(\mathbb{R}^n)$, $|(1 + \|\mathbf{x}\|^2)^{k+n}\phi(\mathbf{x})| \leq q^*_{k+n,0}(\phi) < +\infty$ (by (7.2.16)) $\implies |\phi(\mathbf{x})| \leq \frac{q^*_{k+n,0}(\phi)}{(1+\|\mathbf{x}\|^2)^{k+n}} \ \forall \mathbf{x} \in \mathbb{R}^n$ and $|f(\mathbf{x})\phi(\mathbf{x})| \leq C(1 + \|\mathbf{x}\|^2)^k |\phi(\mathbf{x})|$ (using (8.2.14))
$$\implies \quad |f(\mathbf{x})\phi(\mathbf{x})| \leq C((1 + \|\mathbf{x}\|^2)^k) \cdot \frac{q^*_{k+n,0}(\phi)}{(1 + \|\mathbf{x}\|^2)^{k+n}} = C \frac{q^*_{k+n,0}(\phi)}{(1 + \|\mathbf{x}\|^2)^n}.$$
Hence,
$$\begin{aligned} |T_f(\phi)| = |\langle f, \phi \rangle| &= \left| \int_{\mathbb{R}^n} f(\mathbf{x})\phi(\mathbf{x}) d\mathbf{x} \right| \\ &\leq C q^*_{k+n,0}(\phi) \int_{\mathbb{R}^n} \frac{d\mathbf{x}}{(1 + \|\mathbf{x}\|^2)^n} \leq C q^*_{k+n,0}(\phi)\pi^n < +\infty, \end{aligned}$$
since $\int_{\mathbb{R}^n} \frac{d\mathbf{x}}{(1+\|\mathbf{x}\|^2)^n} \leq \pi^n$ (see (7.2.26)).
Hence, every locally integrable function $f \in L^1_{\text{loc}}(\mathbb{R}^n)$ with slow growth at infinity defines a tempered distribution $T_f \in S'(\mathbb{R}^n)$, and we identify f with

T_f by writing $T_f = f \in S'(\mathbb{R}^n)$. The term *tempered* corresponds to this slow increase at infinity, and the corresponding distributions are thus called *tempered*.

Every locally summable function $f \in L^1_{\text{loc}}(\mathbb{R}^n)$ defines a regular distribution $T_f \in \mathcal{D}'(\mathbb{R}^n)$ by $T_f(\phi) = \int_{\mathbb{R}^n} f(\mathbf{x})\phi(\mathbf{x})d\mathbf{x}\ \forall\phi \in \mathcal{D}(\mathbb{R}^n)$, but $f \in L^1_{\text{loc}}(\mathbb{R}^n)$ does *not* define a tempered distribution of $S'(\mathbb{R}^n)$ in general.

For example, for $n = 1$, $f(x) = e^{x^2} \in L^1_{\text{loc}}(]-\infty,\infty[)$ defines a regular distribution such that $e^{x^2} \in \mathcal{D}'(]-\infty,\infty[)$, but e^{x^2} does *not* define a tempered distribution, i.e. $e^{x^2} \notin S'(]-\infty,\infty[)$ (see (8.2.10)). Similarly, $e^x \in \mathcal{D}'(]-\infty,\infty[)$, but $e^x \notin S'(]-\infty,\infty[)$.

Function $f(x) = e^x\cos(e^x)\ \forall x \in \mathbb{R}$ does *not* have slow/polynomial growth at infinity, since $\sup_{x\in\mathbb{R}}(|e^x\cos(e^x)|/(1+|x|^2)^k) = \infty\ \forall k \in \mathbb{N}$, although $e^x\cos(e^x) \in L^1_{\text{loc}}(]-\infty,\infty[)$, whereas $g(x) = \sin(e^x)\ \forall x \in \mathbb{R}$ has slow/polynomial growth at infinity, since

$$\sup_{x\in\mathbb{R}} \frac{|\sin(e^x)|}{(1+|x|^2)^k} \le 1\ \forall k \in \mathbb{N} \quad \text{and} \quad \sin(e^x) \in L^1_{\text{loc}}(]-\infty,\infty[). \tag{8.2.16}$$

Hence, $\sin(e^x)$ defines a tempered distribution, i.e. $\sin(e^x) \in S'(]-\infty,\infty[)$. Now we show that $e^x\cos(e^x)$ also defines a tempered distribution. In fact, integrating by parts, we have, $\forall\phi \in S(]-\infty,\infty[)$,

$$\begin{aligned}
|T_f(\phi)| &= \left|\int_{-\infty}^{\infty} e^x\cos(e^x)\phi(x)dx\right| = \left|-\int_{-\infty}^{\infty}\sin(e^x)\phi'(x)dx\right| \\
&\le \int_{-\infty}^{\infty}|\sin(e^x)|\,|\phi'(x)|dx \le \int_{-\infty}^{\infty}|\phi'(x)|dx \\
&\le C\sup_{x\in\mathbb{R}}((1+x^2)|\phi'(x)|) \le Cq^*_{1,1}(\phi)
\end{aligned}$$

with $C = \int_{-\infty}^{\infty}\frac{dx}{1+x^2}$,

$\Longrightarrow T_f \in S'(]-\infty,\infty[)$ by Proposition 8.2.1 $\Longrightarrow e^x\cos(e^x)$ is a tempered distribution in $S'(]-\infty,\infty[)$.

Moreover, we will show later that $\sin(e^x) \in S'(]-\infty,\infty[)$ will imply that $\frac{d}{dx}(\sin(e^x)) = e^x\cos(e^x) \in S'(]-\infty,\infty[)$ (see Section 8.2.5 later). Thus, $S'(\mathbb{R}^n)$ contains functions which may not have slow/polynomial growth at infinity.

4. Every distribution $T \in \mathcal{E}'(\mathbb{R}^n) \subset \mathcal{D}'(\mathbb{R}^n)$ with *compact support* in $\mathbb{R}^n$ is a tempered distribution of $S'(\mathbb{R}^n)$.

   In fact, $T \in \mathcal{D}'(\mathbb{R}^n)$ has compact support $\Longrightarrow T(\phi)$ is defined $\forall\phi \in C^\infty(\mathbb{R}^n)$. But $S(\mathbb{R}^n) \subset C^\infty(\mathbb{R}^n) \Longrightarrow T(\phi)$ is well defined $\forall\phi \in S(\mathbb{R}^n)$ (see (5.6.3)). For example, Dirac distribution $\delta_{\mathbf{a}} = \delta(\mathbf{x}-\mathbf{a}) \in \mathcal{D}'(\mathbb{R}^n)$ with mass/charge/

force etc. concentrated at $\mathbf{a} \in \mathbb{R}^n$ has compact support $= \{\mathbf{a}\}$. Hence, $\delta_{\mathbf{a}} = \delta(\mathbf{x}-\mathbf{a}) \in S'(\mathbb{R}^n)$ is defined by:

$$\langle \delta_{\mathbf{a}}, \phi \rangle = \langle \delta(\mathbf{x}-\mathbf{a}), \phi \rangle = \phi(\mathbf{a}) \quad \forall \phi \in S(\mathbb{R}^n). \tag{8.2.17}$$

5. Functions $\psi \in S(\mathbb{R}^n)$ define tempered distributions, i.e. $\psi \in S(\mathbb{R}^n)$

$$\Longrightarrow \quad T_\psi \in S'(\mathbb{R}^n) \quad \text{with } T_\psi(\phi) = \int_{\mathbb{R}^n} \psi(\mathbf{x})\phi(\mathbf{x})d\mathbf{x} \quad \forall \phi \in S(\mathbb{R}^n). \tag{8.2.18}$$

In fact, from (7.2.26), $S(\mathbb{R}^n) \hookrightarrow L^1(\mathbb{R}^n)$ with $\|\phi\|_{L^1(\mathbb{R}^n)} \le C q^*_{n,0}(\phi)$ $\forall \phi \in S(\mathbb{R}^n)$

$$\Longrightarrow \quad |T_\psi(\phi)| = \left| \int_{\mathbb{R}^n} \psi(\mathbf{x})\phi(\mathbf{x})d\mathbf{x} \right| \le \sup_{\mathbf{x}\in\mathbb{R}^n} |\psi(\mathbf{x})| \int_{\mathbb{R}^n} |\phi(\mathbf{x})|d\mathbf{x}$$
$$\le C q_{0,0}(\psi) q^*_{n,0}(\phi) < +\infty \quad \forall \phi \in S(\mathbb{R}^n)$$

$\Longrightarrow T_\psi$ is a tempered distribution.
The result also follows from (1) by virtue of the imbedding $S(\mathbb{R}^n) \hookrightarrow L^1(\mathbb{R}^n)$, and also from (2) by virtue of the *boundedness* of functions $\psi \in S(\mathbb{R}^n)$. Hence,

$$S(\mathbb{R}^n) \subset S'(\mathbb{R}^n). \tag{8.2.19}$$

For $n = 1$, $\psi(x) = x^m e^{-x^2}$, $\sin \alpha x \, e^{-x^2}$, $\cos \beta x e^{-x^2}$, e^{-x^2}, etc. belongs to $S(\mathbb{R})$, and consequently belongs to $S'(\mathbb{R})$. (8.2.20)
$\forall$ multi-index $\boldsymbol{\alpha}$,

$$\mathbf{x}^{\boldsymbol{\alpha}} e^{-\|\mathbf{x}\|^2} \in S(\mathbb{R}^n) \quad \Longrightarrow \quad \mathbf{x}^{\boldsymbol{\alpha}} e^{-\|\mathbf{x}\|^2} \in S'(\mathbb{R}^n). \tag{8.2.21}$$

8.2.4 Convergence of sequences in $S'(\mathbb{R}^n)$

Definition 8.2.3. A sequence $(T_k)_{k=1}^\infty$ of tempered distributions $T_k \in S'(\mathbb{R}^n)$ is said to *converge* to the tempered distribution $T \in S'(\mathbb{R}^n)$ if and only if

$$\langle T_k, \phi \rangle \to \langle T, \phi \rangle \text{ in } \mathbb{C} \ \forall \phi \in S(\mathbb{R}^n) \text{ as } k \to \infty. \tag{8.2.22}$$

Then we write $T_k \to T$ in $S'(\mathbb{R}^n)$ as $k \to \infty$, i.e.

$$T_k \to T \text{ in } S'(\mathbb{R}^n) \quad \Longrightarrow \quad \langle T_k, \phi \rangle \to \langle T, \phi \rangle \text{ in } \mathbb{C} \text{ as } k \to \infty \forall \phi \in S(\mathbb{R}^n). \tag{8.2.23}$$

Proposition 8.2.2. *Let $(T_k)_{k=1}^\infty$ be a sequence of tempered distributions $T_k \in S'(\mathbb{R}^n)$ $\forall k \in \mathbb{N}$ such that $\lim_{k\to\infty}\langle T_k, \phi \rangle$ exists in $\mathbb{C}$ $\forall \phi \in S(\mathbb{R}^n)$. Then the sequence $(T_k)_{k=1}^\infty$ has a limit in $S'(\mathbb{R}^n)$, i.e. $\exists$ a unique $T \in S'(\mathbb{R}^n)$ such that*

$$\langle T_k, \phi \rangle \to \langle T, \phi \rangle \textit{ in } \mathbb{C} \textit{ as } k \to \infty \ \forall \phi \in S(\mathbb{R}^n). \tag{8.2.24}$$

Convergence of series of tempered distributions in $S'(\mathbb{R}^n)$

Definition 8.2.4. Let $\sum_{k=1}^{\infty} T_k$ be a series of tempered distributions $T_k \in S'(\mathbb{R}^n)$ $\forall k \in \mathbb{N}$. Let $S_N = \sum_{k=1}^{N} T_k$ $\forall N \in \mathbb{N}$. If the sequence $(S_N)_{N=1}^{\infty}$ converges in $S'(\mathbb{R}^n)$, the series $\sum_{k=1}^{\infty} T_k$ is said to *converge* in $S'(\mathbb{R}^n)$, or equivalently $\sum_{k=1}^{\infty} T_k$ converges in $S'(\mathbb{R}^n)$, if and only if the series $\sum_{k=1}^{\infty}\langle T_k, \phi\rangle$ converges in $\mathbb{C}$ $\forall \phi \in S(\mathbb{R}^n)$.

Then the series $\sum_{k=1}^{\infty} T_k$ is called *summable* in $S'(\mathbb{R}^n)$.

$\sum_{k=1}^{\infty} T_k$ converges in $S'(\mathbb{R}^n)$ $\iff$ $\exists$ a unique $T \in S'(\mathbb{R}^n)$ such that $T = \sum_{k=1}^{\infty} T_k$ with

$$\langle T, \phi\rangle = \sum_{k=1}^{\infty}\langle T_k, \phi\rangle \quad \forall \phi \in S(\mathbb{R}^n). \tag{8.2.25}$$

$\sum_{i\in I} T_i$, I being a set of indices, converges in $S'(\mathbb{R}^n)$

$$\iff \quad \sum_{i\in I}\langle T_i, \phi\rangle \text{ converges in } \mathbb{C} \ \forall\ \phi \in S(\mathbb{R}^n). \tag{8.2.26}$$

Example 8.2.1. For real number $a > 0$, consider the following two series in $S'(\mathbb{R})$:

(A) $\sum_{n=0}^{\infty} a^n \delta_n$;

(B) $\sum_{n=-\infty}^{\infty} a^n \delta_n$,

where δ_n is the Dirac distribution with mass/charge/force etc. concentrated at $n = 0, \pm 1, \pm 2, \ldots$ defined by $\langle \delta_n, \phi\rangle = \phi(n)$ $\forall \phi \in S(\mathbb{R})$, $\forall n = 0, \pm 1, \pm 2, \ldots$ Show that

1. the series (A) converges in $S'(\mathbb{R})$ for $0 < a \le 1$ and diverges for $a > 1$;
2. the series (B) converges in $S'(\mathbb{R})$ only for $a = 1$.

Solution.

1. From Proposition 7.2.1, $\forall \phi \in S(\mathbb{R})$, $\exists M > 0$ such that $\forall k, l \in \mathbb{N}_0$, $|x^k \phi^{(l)}(x)| \le M$ $\forall x \in \mathbb{R}$. Hence, $\forall \phi \in S(\mathbb{R})$, $\exists M > 0$ such that $\forall k \in \mathbb{N}_0$, $|x^k \phi(x)| \le M$ $\forall x \in \mathbb{R} \implies |\phi(x)| \le \frac{M}{|x|^k}$ $\forall x \ne 0$ $\forall k \in \mathbb{N} \implies |\phi(n)| \le M n^{-k}$ $\forall n \in \mathbb{N}$, $\forall k \in \mathbb{N}$ and $|\phi(0)| \le M$. Hence, $\forall n \in \mathbb{N}$, $|\langle a^n \delta_n, \phi\rangle| = |a^n \phi(n)| \le M a^n n^{-k}$ $\forall k \in \mathbb{N}$ and, for $n = 0$, $|\langle \delta_0, \phi\rangle| = |\phi(0)| \le M$. Then

$$\begin{aligned}\left|\sum_{n=0}^{N}\langle a^n \delta_n, \phi\rangle\right| &\le \sum_{n=0}^{N} a^n |\langle \delta_n, \phi\rangle| = |\langle \delta_0, \phi\rangle| + \sum_{n=1}^{N} a^n |\langle \delta_n, \phi\rangle| \\ &\le M\left[1 + \sum_{n=1}^{N} a^n n^{-k}\right] \quad \forall k \in \mathbb{N},\ \forall N \in \mathbb{N},\ \forall \phi \in S(\mathbb{R}).\end{aligned}$$

In particular, this inequality holds $\forall k \geq 2$ and $\forall N \in \mathbb{N}$. But $\forall k \geq 2$, the sequence $(M(1 + \sum_{n=1}^{N} a^n n^{-k}))_{N=1}^{\infty}$ converges in $\mathbb{R}$ as $N \to \infty$ $\forall$ (strictly positive) $a \leq 1$. Then $\sum_{n=0}^{\infty} \langle a^n \delta_n, \phi \rangle$ converges (absolutely) in $\mathbb{R}$ $\forall a \leq 1$ and $\forall \phi \in S(\mathbb{R})$. Hence, by Definition 8.2.4, the series $\sum_{n=0}^{\infty} a^n \delta_n$ converges in $S'(\mathbb{R})$ for $0 < a \leq 1$.

Case $a > 1$: Let $b \in \mathbb{R}$ be any strictly positive real number such that $1 < b < a$, with $a/b > 1$. Then it is possible to construct a $\phi \in S(\mathbb{R})$ such that for $x > 0$, $\phi(x) = b^{-x}$ (which is a C^∞-function with rapid decay at infinity, since $x^k b^{-x} \to 0$ as $x \to \infty$ $\forall k \in \mathbb{N}$). Then $\langle a^n \delta_n, \phi \rangle = a^n \phi(n) = a^n b^{-n} = (\frac{a}{b})^n$ $\forall n \in \mathbb{N}$ with $a/b > 1$, and

$$\sum_{n=0}^{N} a^n \phi(n) = 1 + \sum_{n=1}^{N} \left(\frac{a}{b}\right)^n = \sum_{n=0}^{N} \left(\frac{a}{b}\right)^n \to \infty \quad \text{as } N \to \infty.$$

Hence, the series $\sum_{n=0}^{\infty} a^n \delta_n$ diverges in $S'(\mathbb{R})$ for $a > 1$. Combining the two cases, $\sum_{n=0}^{\infty} a^n \delta_n$ converges in $S'(\mathbb{R})$ for $0 < a \leq 1$ and diverges in $S'(\mathbb{R})$ for $a > 1$.

2. $\sum_{n=-\infty}^{\infty} a^n \delta_n$ will be convergent in $S'(\mathbb{R})$ if and only if both $\sum_{n=-\infty}^{-1} a^n \delta_n$ and $\sum_{n=0}^{\infty} a^n \delta_n$ converge in $S'(\mathbb{R})$. But $\sum_{n=0}^{\infty} a^n \delta_n$ is convergent in $S'(\mathbb{R})$ for $a \leq 1$ and divergent in $S'(\mathbb{R})$ for $a > 1$. Similarly, we can show that $\sum_{n=-\infty}^{-1} a^n \delta_n$ converges in $S'(\mathbb{R})$ for $a \geq 1$ and diverges in $S'(\mathbb{R})$ for $a < 1$. In fact, $\sum_{n=-\infty}^{-1} a^n \delta_n = \sum_{n=1}^{\infty} a^{-n} \delta_{-n}$ and $|\langle a^{-n} \delta_{-n}, \phi \rangle| = |a^{-n} \phi(-n)| \leq M a^{-n} n^{-k} = \frac{M}{a^n} n^{-k}$ $\forall \phi \in S(\mathbb{R})$, $\forall n \in \mathbb{N}$, $\forall k \in \mathbb{N}$. Following the steps of the proof of the convergence of series (A), we can show that $\sum_{n=1}^{\infty} \langle a^{-n} \delta_{-n}, \phi \rangle$ converges for $\frac{1}{a} \leq 1$, i.e. for $a \geq 1$. Hence, $\sum_{n=1}^{\infty} a^{-n} \delta_{-n} = \sum_{n=-\infty}^{-1} a^n \delta_n$ converges in $S'(\mathbb{R})$ for $a \geq 1$. Therefore, both the series $\sum_{n=-\infty}^{-1} a^n \delta_n$ and $\sum_{n=0}^{\infty} a^n \delta_n$ converge in $S'(\mathbb{R})$ *only* for $a = 1$, i.e. $\sum_{n=-\infty}^{\infty} a^n \delta_n$ converges in $S'(\mathbb{R})$ *only* for $a = 1$.

Example 8.2.2. Consider the series $\sum_{n=0}^{\infty} a_k \delta_k$, where $a_k \in \mathbb{R}$ $\forall k \in \mathbb{N}_0$, δ_k is the Dirac distribution with concentration at the point $k \in \mathbb{N}_0$. Show the conditions under which the series converges in $S'(\mathbb{R})$.

Proof. $\sum_{k=0}^{\infty} a_k \delta_k$ converges in $S'(\mathbb{R})$ if and only if $\sum_{k=0}^{\infty} a_k \langle \delta_k, \phi \rangle = \sum_{k=0}^{\infty} a_k \phi(k)$ converges $\forall \phi \in S(\mathbb{R})$. But $\forall$ fixed $\phi \in S(\mathbb{R})$, $\exists C > 0$ such that $|\phi(x)| \leq C|x|^{-l}$ $\forall l \in \mathbb{N}$, $\forall x \neq 0 \Longrightarrow |\phi(k)| \leq Ck^{-l}$ $\forall l \in \mathbb{N}$, $\forall k \in \mathbb{N}$.

Hence, $\forall$ fixed $\phi \in S(\mathbb{R})$, $\sum_{k=0}^{\infty} a_k \phi(k)$ will converge if $|a_k| \leq C_1 k^p$ $\forall k \in \mathbb{N}$ for some fixed $C_1 > 0$ and $p \in \mathbb{N}$, since the majoring series $M(1 + \sum_{k=1}^{\infty} k^{-(l-p)})$ will converge with $l - p \geq 2$, $M = \max\{|a_0||\phi(0)|, CC_1\}$. □

Remark 8.2.1. For $a_k = e^{k^2}$, the series $\sum_{k=0}^{\infty} a_k \delta_k$ will diverge, since for e^{k^2}, we can *not* write $e^{k^2} \leq C_1 k^p$ $\forall k \in \mathbb{N}$.

8.2.5 Derivatives of tempered distributions

Every tempered distribution $T \in S'(\mathbb{R}^n)$ is infinitely differentiable In fact, $T \in S'(\mathbb{R}^n) \Longrightarrow \forall$ multi-index $\boldsymbol{\alpha}$, $\partial^{\boldsymbol{\alpha}} T \in S'(\mathbb{R}^n)$ and the derivative $\partial^{\boldsymbol{\alpha}} T \in S'(\mathbb{R}^n)$ is defined by

$$\langle \partial^{\boldsymbol{\alpha}} T, \phi \rangle = (-1)^{|\boldsymbol{\alpha}|} \langle T, \partial^{\boldsymbol{\alpha}} \phi \rangle \quad \forall \phi \in S(\mathbb{R}^n). \tag{8.2.27}$$

Justification $\phi \in S(\mathbb{R}^n) \Longrightarrow \partial^{\boldsymbol{\alpha}} \phi \in S(\mathbb{R}^n)\ \forall |\boldsymbol{\alpha}| \in \mathbb{N} \Longrightarrow \langle T, \partial^{\boldsymbol{\alpha}} \phi \rangle$ is well defined $\forall T \in S'(\mathbb{R}^n) \Longrightarrow$ the left-hand side $\langle \partial^{\boldsymbol{\alpha}} T, \phi \rangle$ is also well defined $\forall \phi \in S(\mathbb{R}^n)$, and $\partial^{\boldsymbol{\alpha}} T$ is continuous on $S(\mathbb{R}^n) \Longrightarrow \partial^{\boldsymbol{\alpha}} T \in S'(\mathbb{R}^n)$ is a tempered distribution $\forall |\boldsymbol{\alpha}| \in \mathbb{N}$.

For $n = 1$, $f(x) = \sin(e^x)\ \forall x \in \mathbb{R} \Longrightarrow \sin(e^x)$ is a locally summable function of *slow/polynomial growth* at infinity (see (3)) $\Longrightarrow \sin(e^x) \in S'(]-\infty, \infty[)$ is a tempered distribution. Then, $\forall \phi \in S(]-\infty, \infty[)$,

$$\begin{aligned}
\left\langle \sin(e^x), \frac{d\phi}{dx} \right\rangle &= \int_{-\infty}^{\infty} \sin(e^x) \frac{d\phi}{dx}(x) dx \\
&= -\int_{-\infty}^{\infty} e^x \cos(e^x) \phi(x) dx \quad \forall \phi \in S(]-\infty, \infty[) \quad \text{(by integration by parts)} \\
&= -\langle e^x \cos(e^x), \phi \rangle = -\left\langle \frac{d}{dx}(\sin(e^x)), \phi \right\rangle
\end{aligned}$$

$\Longrightarrow \frac{d}{dx}(\sin(e^x)) = e^x \cos(e^x) \in S'(]-\infty, \infty[)$, since $\sin(e^x) \in S'(]-\infty, \infty[)$ implies that all its derivatives also belong to $S'(]-\infty, \infty[)$.

$\forall \boldsymbol{\alpha}$, $\partial^{\boldsymbol{\alpha}} : S'(\mathbb{R}^n) \longrightarrow S'(\mathbb{R}^n)$ is *linear* and *continuous*. (8.2.28)

Proof. Linearity:

$$\begin{aligned}
\langle \partial^{\boldsymbol{\alpha}}(\lambda_1 T_1 + \lambda_2 T_2), \phi \rangle &= (-1)^{|\boldsymbol{\alpha}|} \langle \lambda_1 T_1 + \lambda_2 T_2, \partial^{\boldsymbol{\alpha}} \phi \rangle \\
&= (-1)^{|\boldsymbol{\alpha}|} [\lambda_1 \langle T_1, \partial^{\boldsymbol{\alpha}} \phi \rangle + \lambda_2 \langle T_2, \partial^{\boldsymbol{\alpha}} \phi \rangle] \\
&= (-1)^{|\boldsymbol{\alpha}|} \lambda_1 \langle T_1, \partial^{\boldsymbol{\alpha}} \phi \rangle + (-1)^{|\boldsymbol{\alpha}|} \lambda_2 \langle T_2, \partial^{\boldsymbol{\alpha}} \phi \rangle \\
&= \lambda_1 (-1)^{|\boldsymbol{\alpha}|} \langle T_1, \partial^{\boldsymbol{\alpha}} \phi \rangle + \lambda_2 (-1)^{|\boldsymbol{\alpha}|} \langle T_2, \partial^{\boldsymbol{\alpha}} \phi \rangle \\
&= \lambda_1 \langle \partial^{\boldsymbol{\alpha}} T_1, \phi \rangle + \lambda_2 \langle \partial^{\boldsymbol{\alpha}} T_2, \phi \rangle \\
&= \langle \lambda_1 \partial^{\boldsymbol{\alpha}} T_1 + \lambda_2 \partial^{\boldsymbol{\alpha}} T_2, \phi \rangle \quad \forall \phi \in S(\mathbb{R}^n)
\end{aligned}$$

$\Longrightarrow \partial^{\boldsymbol{\alpha}}(\lambda_1 T_1 + \lambda_2 T_2) = \lambda_1 \partial^{\boldsymbol{\alpha}} T_1 + \lambda_2 \partial^{\boldsymbol{\alpha}} T_2\ \forall T_1, T_2 \in S'(\mathbb{R}^n), \forall \lambda_1, \lambda_2 \in \mathbb{C}$.

Continuity of operation of differentiation on $S'(\mathbb{R}^n)$: $\forall |\alpha| \in \mathbb{N}$, $\partial^{\boldsymbol{\alpha}} : \phi \in S(\mathbb{R}^n) \longmapsto \partial^{\boldsymbol{\alpha}} \phi \in S(\mathbb{R}^n)$ is linear and continuous by (7.3.5). We are to show that $T_k \to T$ in $S'(\mathbb{R}^n) \Longrightarrow \partial^{\boldsymbol{\alpha}} T_k \to \partial^{\boldsymbol{\alpha}} T$ in $S'(\mathbb{R}^n)$ as $k \to \infty$.

In fact, $\langle \partial^{\boldsymbol{\alpha}} T - \partial^{\boldsymbol{\alpha}} T_k, \phi \rangle = \langle \partial^{\boldsymbol{\alpha}}(T - T_k), \phi \rangle = (-1)^{|\boldsymbol{\alpha}|}\langle (T - T_k), \partial^{\boldsymbol{\alpha}} \phi \rangle$
$\Longrightarrow |\langle \partial^{\boldsymbol{\alpha}} T - \partial^{\boldsymbol{\alpha}} T_k, \phi \rangle| = |\langle T - T_k, \partial^{\boldsymbol{\alpha}} \phi \rangle| \to 0$ as $k \to \infty$, since $T_k \to T$ in $S'(\mathbb{R}^n) \Longrightarrow \partial^{\boldsymbol{\alpha}} T_k \to \partial^{\boldsymbol{\alpha}} T$ in $S'(\mathbb{R}^n)$ as $k \to \infty$ $\forall |\boldsymbol{\alpha}| \in \mathbb{N}$,
$\Longrightarrow \partial^{\boldsymbol{\alpha}} : S'(\mathbb{R}^n) \longrightarrow S'(\mathbb{R}^n)$ is continuous on $S'(\mathbb{R}^n)$. □

Multiplication of a tempered distribution by a polynomial

$T \in S'(\mathbb{R}^n)$, p a polynomial in n variables $x_1, x_2, \ldots, x_n \Longrightarrow pT \in S'(\mathbb{R}^n)$ defined by $\langle pT, \phi \rangle = \langle T, p\phi \rangle$ $\forall \phi \in S(\mathbb{R}^n)$. (8.2.29)

Indeed, $\phi \in S(\mathbb{R}^n) \Longrightarrow p\phi \in S(\mathbb{R}^n)$ $\forall$ polynomials p (by Proposition 7.2.1)
$\Longrightarrow \langle T, p\phi \rangle$ is well defined $\forall T \in S'(\mathbb{R}^n)$, $\forall \phi \in S(\mathbb{R}^n)$
$\Longrightarrow \langle pT, \phi \rangle$ is well defined $\forall \phi \in S(\mathbb{R}^n)$
$\Longrightarrow pT \in S'(\mathbb{R}^n)$.

Structure of the elements of $S'(\mathbb{R}^n)$

We agree to accept the following result without proof (see [8, p. 239]).

Theorem 8.2.1. *A distribution $T \in \mathcal{D}'(\mathbb{R}^n)$ is a tempered distribution of $S'(\mathbb{R}^n)$ if and only if there exists an integer $m \in \mathbb{N}_0$, a multi-index $\boldsymbol{\alpha}$ and a* bounded continuous *function f in $\mathbb{R}^n$ such that $T = \partial^{\boldsymbol{\alpha}}[(1 + \|\mathbf{x}\|^2)^m f]$ with*

$$\begin{aligned} \langle T, \phi \rangle &= \langle \partial^{\boldsymbol{\alpha}}[(1 + \|\mathbf{x}\|^2)^m f], \phi \rangle = (-1)^{|\boldsymbol{\alpha}|}\langle (1 + \|\mathbf{x}\|^2)^m f, \partial^{\boldsymbol{\alpha}} \phi \rangle \\ &= (-1)^{|\boldsymbol{\alpha}|}\langle f, (1 + \|\mathbf{x}\|^2)^m \partial^{\boldsymbol{\alpha}} \phi \rangle \quad \forall \phi \in S(\mathbb{R}^n). \end{aligned} \tag{8.2.30}$$

Imbedding results

1. $L^p(\mathbb{R}^n) \hookrightarrow S'(\mathbb{R}^n)$, $1 \le p \le \infty$, the imbedding operator $\hookrightarrow$ being a *continuous* one, i.e.

$$f_k \to f \text{ in } L^p(\mathbb{R}^n) \quad \Longrightarrow \quad T_{f_k} \to T_f \text{ in } S'(\mathbb{R}^n) \text{ as } k \to \infty. \tag{8.2.31}$$

Proof. Algebraic inclusion: $L^p(\mathbb{R}^n) \subset S'(\mathbb{R}^n) \forall p \in [1, \infty]$.
For $p = 1$, it has been shown in (8.2.12); now we prove it for general p. Let $f \in L^p(\mathbb{R}^n)$; we are to show that $\hookrightarrow f = T_f \in S'(\mathbb{R}^n)$. In fact, if, for $f \in L^p(\mathbb{R}^n)$, $\int_{\mathbb{R}^n} \frac{|f(\mathbf{x})|}{(1+\|\mathbf{x}\|^2)^n} d\mathbf{x} < +\infty$, then f will define a tempered distribution $T_f \in S'(\mathbb{R}^n)$ by $T_f(\phi) = \int_{\mathbb{R}^n} f(\mathbf{x})\phi(\mathbf{x}) d\mathbf{x}$ $\forall \phi \in S(\mathbb{R}^n)$. Applying Hölder's inequality and the proof of (7.4.1), we get

$$\begin{aligned} \int_{\mathbb{R}^n} \frac{|f(\mathbf{x})|}{(1 + \|\mathbf{x}\|^2)^n} d\mathbf{x} &\le \|f\|_{L^p(\mathbb{R}^n)} \left\| \frac{1}{(1 + \|\mathbf{x}\|^2)^n} \right\|_{L^q(\mathbb{R}^n)} \\ &\le \|f\|_{L^p(\mathbb{R}^n)} \pi^{n/q} < +\infty \end{aligned}$$

(with $\frac{1}{q} + \frac{1}{p} = 1$, $q = \infty$ for $p = 1$).

Hence, $f \in L^p(\mathbb{R}^n) \implies \hookrightarrow f = T_f \in S'(\mathbb{R}^n) \implies L^p(\mathbb{R}^n) \subset S'(\mathbb{R}^n)$ $\forall p \in [1, \infty]$.

$\hookrightarrow : L^p(\mathbb{R}^n) \longrightarrow S'(\mathbb{R}^n)$ *is linear*:

$$\begin{aligned}\hookrightarrow(\alpha_1 f_1 + \alpha_2 f_2) &= T_{\alpha_1 f_1 + \alpha_2 f_2} = \alpha_1 T_{f_1} + \alpha_2 T_{f_2} \\ &= \alpha_1 \hookrightarrow f_1 + \alpha_2 \hookrightarrow f_2 \quad \forall \alpha_1, \alpha_2 \in \mathbb{C},\ \forall f_1, f_2 \in L^p(\mathbb{R}^n).\end{aligned}$$

Continuity of $\hookrightarrow$: $L^p(\mathbb{R}^n) \longrightarrow S'(\mathbb{R}^n)$: For this we are to show that $f_k \longrightarrow f$ in $L^p(\mathbb{R}^n) \implies T_{f_k} \longrightarrow T_f$ in $S'(\mathbb{R}^n)$, where $T_{f_k} = \hookrightarrow f_k, T_f = \hookrightarrow f$. Set $T_k = T_{f_k}$ $\forall k \in \mathbb{N}$, $T = T_f$. Then, $\forall \phi \in S(\mathbb{R}^n)$,

$$\begin{aligned}|\langle T, \phi\rangle - \langle T_k, \phi\rangle| &= \left|\int_{\mathbb{R}^n} f\phi d\mathbf{x} - \int_{\mathbb{R}^n} f_k \phi d\mathbf{x}\right| = \left|\int_{\mathbb{R}^n} (f - f_k)\phi d\mathbf{x}\right| \\ &\le \int_{\mathbb{R}^n} \frac{|f(\mathbf{x}) - f_k(\mathbf{x})|}{(1 + \|\mathbf{x}\|^2)^n} |(1 + \|\mathbf{x}\|^2)^n \phi(\mathbf{x})| d\mathbf{x} \\ &\le \sup_{\mathbf{x}\in\mathbb{R}^n} |(1 + \|\mathbf{x}\|^2)^n \phi(\mathbf{x})| \int_{\mathbb{R}^n} \frac{|f(\mathbf{x}) - f_k(\mathbf{x})|}{(1 + \|\mathbf{x}\|^2)^n} d\mathbf{x} \\ &\le q^*_{n,0}(\phi)\pi^{n/q} \|f - f_k\|_{L^p(\mathbb{R}^n)} \to 0 \quad \text{as } k \to \infty,\end{aligned}$$

since $f_k \longrightarrow f$ in $L^p(\mathbb{R}^n)$. Hence, $T_k \longrightarrow T$ in $S'(\mathbb{R}^n) \implies T_{f_k} \longrightarrow T_f$ in $S'(\mathbb{R}^n)$.

Therefore, $\hookrightarrow : L^p(\mathbb{R}^n) \longrightarrow S'(\mathbb{R}^n)$ is continuous. □

2. $S'(\mathbb{R}^n) \hookrightarrow \mathcal{D}'(\mathbb{R}^n)$ with *continuous* imbedding operator $\hookrightarrow$. (8.2.32)

Proof. Since $\mathcal{D}(\mathbb{R}^n)$ is a *dense* subspace of $S(\mathbb{R}^n)$ (see (7.5.1)) and $\mathcal{D}(\mathbb{R}^n) \hookrightarrow S(\mathbb{R}^n)$, the imbedding being a continuous one (see (7.4.2)), we have $S'(\mathbb{R}^n) \hookrightarrow \mathcal{D}'(\mathbb{R}^n)$, the imbedding operator $\hookrightarrow$ being a continuous one (see also (4.2.7)). □

3. $\mathcal{D}(\mathbb{R}^n) \hookrightarrow S(\mathbb{R}^n) \hookrightarrow L^p(\mathbb{R}^n) \hookrightarrow S'(\mathbb{R}^n) \hookrightarrow \mathcal{D}'(\mathbb{R}^n)$, (8.2.33)

each imbedding operator $\hookrightarrow$ being *continuous* from the space on its left-hand side into the space on its right-hand side.

Proof. Combining the imbedding results (7.4.1), (7.4.2), (8.2.31) and (8.2.32), we get (8.2.33). □

8.3 Fourier transform of tempered distributions

Definition 8.3.1. Let $T \in S'(\mathbb{R}^n)$ be a tempered distribution. Then its *Fourier transform* $\mathcal{F}T \in S'(\mathbb{R}^n)$ and its *co-transform* $\bar{\mathcal{F}}T \in S'(\mathbb{R}^n)$ are tempered distributions defined by:

$$\langle \mathcal{F}T, \phi \rangle = \langle T, \mathcal{F}\phi \rangle \quad \forall \phi \in S(\mathbb{R}^n), \tag{8.3.1}$$

$$\langle \bar{\mathcal{F}}T, \phi \rangle = \langle T, \bar{\mathcal{F}}\phi \rangle \quad \forall \phi \in S(\mathbb{R}^n) \tag{8.3.2}$$

((8.3.1)–(8.3.2) were, in fact, suggested in (8.1.5)), with

$$(\mathcal{F}\phi)(\mathbf{x}) = \int_{\mathbb{R}^n} \phi(\boldsymbol{\xi}) e^{-i2\pi\langle \boldsymbol{\xi}, \mathbf{x}\rangle} d\boldsymbol{\xi}, \quad (\bar{\mathcal{F}}\phi)(\mathbf{x}) = \int_{\mathbb{R}^n} \phi(\boldsymbol{\xi}) e^{i2\pi\langle \boldsymbol{\xi}, \mathbf{x}\rangle} d\boldsymbol{\xi}. \tag{8.3.3}$$

Justification By Theorem 7.7.2, $\mathcal{F} : S(\mathbb{R}^n) \longrightarrow S(\mathbb{R}^n)$ is an *isomorphism* from $S(\mathbb{R}^n)$ onto $S(\mathbb{R}^n)$. Hence, $\forall \phi \in S(\mathbb{R}^n)$, $\mathcal{F}\phi \in S(\mathbb{R}^n)$ and consequently, $\forall T \in S'(\mathbb{R}^n)$, $\langle T, \mathcal{F}\phi \rangle$ is well defined $\forall \phi \in S(\mathbb{R}^n)$. Moreover, by Theorem 7.6.1, $\mathcal{F}\phi \in S(\mathbb{R}^n)$ and $\langle T, \mathcal{F}\phi \rangle$ represents a continuous, linear functional on $S(\mathbb{R}^n)$. Hence, the left-hand side $\langle \mathcal{F}T, \phi \rangle$ of (8.3.1) is well defined as a continuous, linear functional on $S(\mathbb{R}^n)$, and $\mathcal{F}T \in S'(\mathbb{R}^n)$ is also a tempered distribution.

Replacing $\mathcal{F}$ by $\bar{\mathcal{F}}$ and repeating the same arguments, we can justify that $\bar{\mathcal{F}}T \in S'(\mathbb{R}^n)$ and the formula (8.3.2) is well defined.

The Fourier transform defined on $S'(\mathbb{R}^n)$ by (8.3.1) extends that defined on $L^1(\mathbb{R}^n)$.

Proof. $\forall f \in L^1(\mathbb{R}^n) \subset S'(\mathbb{R}^n)$ and $\forall \phi \in S(\mathbb{R}^n) \subset L^1(\mathbb{R}^n)$, with $\mathcal{F}\phi \in S(\mathbb{R}^n)$ by Theorem 7.6.1,

$$\langle f, \mathcal{F}\phi \rangle = \int_{\mathbb{R}^n} f(\mathbf{x})(\mathcal{F}\phi)(\mathbf{x}) d\mathbf{x} = \int_{\mathbb{R}^n} (\mathcal{F}f)(\boldsymbol{\xi})\phi(\boldsymbol{\xi}) d\boldsymbol{\xi} = \langle \mathcal{F}f, \phi \rangle$$

(by (7.1.28)). □

Fourier transforms of functions of $L^2(\mathbb{R}^n)$

Since $L^2(\mathbb{R}^n) \hookrightarrow S'(\mathbb{R}^n)$ (see (8.2.31)), every function $f \in L^2(\mathbb{R}^n)$[1] defines a tempered distribution $T_f = f \in S'(\mathbb{R}^n)$, and hence its Fourier transform $\mathcal{F}f \in S'(\mathbb{R}^n)$ exists *in the sense of* (tempered) distribution. Long before the introduction of tempered distributions by Laurent Schwartz, Fourier transforms of functions $f \in L^2(\mathbb{R}^n)$ were defined by Plancherel–Riesz by a method of classical functional analysis. But their (Fourier transforms of L^2-functions) definition in the framework of tempered distributions is quite simple and we state it here.

[1] $L^2(\mathbb{R}^n) \not\subset L^1(\mathbb{R}^n)$. For example, $f(x) = \frac{1}{\sqrt{1+x^2}} \in L^2(\mathbb{R})$, but $f \notin L^1(\mathbb{R})$ (see (7.1.10)).

Theorem 8.3.1 (Plancherel–Riesz). *Fourier transform* $\mathcal{F} : S(\mathbb{R}^n) \longrightarrow S(\mathbb{R}^n)$ *(resp. co-transform* $\overline{\mathcal{F}} : S(\mathbb{R}^n) \longrightarrow S(\mathbb{R}^n)$*) can be extended to a unitary operator (see Definition* A.15.1.1 *and* (A.15.1.3)–(A.15.1.4) *in Appendix A) from* $L^2(\mathbb{R}^n)$ *onto* $L^2(\mathbb{R}^n)$ *(this extended operator will still be denoted by the same notation* $\mathcal{F}$ *(resp.* $\overline{\mathcal{F}}$*)) such that* $\mathcal{F} : f \in L^2(\mathbb{R}^n) \longmapsto \mathcal{F} f \in L^2(\mathbb{R}^n)$ *(resp.* $\overline{\mathcal{F}} : f \in L^2(\mathbb{R}^n) \longmapsto \overline{\mathcal{F}} f \in L^2(\mathbb{R}^n)$*) satisfies the following properties:*

I. $\langle \mathcal{F} f, \mathcal{F} g\rangle_{L^2(\mathbb{R}^n)} = \langle f, g\rangle_{L^2(\mathbb{R}^n)}\ \forall f, g \in L^2(\mathbb{R}^n)$ *(*$\langle \cdot,\cdot\rangle_{L^2(\mathbb{R}^n)}$ *being the inner product in* $L^2(\mathbb{R}^n)$*),* (8.3.4)

i.e. $\mathcal{F}$ *is a unitary operator from* $L^2(\mathbb{R}^n)$ *onto* $L^2(\mathbb{R}^n)$*;*

II. $\|\mathcal{F} f\|_{L^2(\mathbb{R}^n)} = \|f\|_{L^2(\mathbb{R}^n)}\ \forall f \in L^2(\mathbb{R}^n)$*;* (8.3.5)

III. $$\begin{aligned}\langle \mathcal{F} f, \phi\rangle_{S'(\mathbb{R}^n)\times S(\mathbb{R}^n)} &= \int_{\mathbb{R}^n} (\mathcal{F} f)(\boldsymbol{\xi})\phi(\boldsymbol{\xi})d\boldsymbol{\xi} = \int_{\mathbb{R}^n} f(\mathbf{x})(\mathcal{F}\phi)(\mathbf{x})d\mathbf{x}\\ &= \langle f, \mathcal{F}\phi\rangle_{S'(\mathbb{R}^n)\times S(\mathbb{R}^n)} \quad \forall f \in L^2(\mathbb{R}^n),\ \forall \phi \in S(\mathbb{R}^n),\end{aligned} \tag{8.3.6}$$

$\langle \cdot,\cdot\rangle_{S'(\mathbb{R}^n)\times S(\mathbb{R}^n)}$ *being the duality between* $S'(\mathbb{R}^n)$ *and* $S(\mathbb{R}^n)$*;*

Replacing $\mathcal{F}$ *by* $\overline{\mathcal{F}}$ *in the results above, the corresponding formulae for* $\overline{\mathcal{F}}$ *are obtained.*

IV. $\overline{\mathcal{F}}\mathcal{F} f = \mathcal{F}\overline{\mathcal{F}} f = f\ \forall f \in L^2(\mathbb{R}^n).$ (8.3.7)

Proof. By the Plancherel–Riesz Theorem 7.7.3 on $S(\mathbb{R}^n)$, $\mathcal{F} : S(\mathbb{R}^n) \longrightarrow S(\mathbb{R}^n)$ is a unitary operator from $S(\mathbb{R}^n)$ onto $S(\mathbb{R}^n)$ equipped with the norm of $L^2(\mathbb{R}^n)$, i.e. $\langle \mathcal{F}\phi, \mathcal{F}\psi\rangle_{L^2(\mathbb{R}^n)} = \langle \phi, \psi\rangle_{L^2(\mathbb{R}^n)}\ \forall \phi, \psi \in S(\mathbb{R}^n)$, and $\|\mathcal{F}\phi\|_{L^2(\mathbb{R}^n)} = \|\phi\|_{L^2(\mathbb{R}^n)} \forall \phi \in S(\mathbb{R}^n)$. Since $S(\mathbb{R}^n)$ is dense in $L^2(\mathbb{R}^n)$ in the norm of $L^2(\mathbb{R}^n)$ (see Proposition 7.5.1), by the principle of extension by density (see Theorem A.8.2.1 in Appendix A), $\mathcal{F} : S(\mathbb{R}^n) \longrightarrow S(\mathbb{R}^n)$ is extended to a unique continuous linear operator from $L^2(\mathbb{R}^n)$ onto $L^2(\mathbb{R}^n)$ such that this extended operator is still denoted by the same notation $\mathcal{F}$, i.e. $\mathcal{F} : f \in L^2(\mathbb{R}^n) \longmapsto \mathcal{F} f \in L^2(\mathbb{R}^n)$.

I. Again, by virtue of the density of $S(\mathbb{R}^n)$ in $L^2(\mathbb{R}^n)$, $\forall f, g \in L^2(\mathbb{R}^n)$, $\exists$ sequences (ϕ_k) and (ψ_k) in $S(\mathbb{R}^n)$ such that $\phi_k \to f$, $\psi_k \to g$ in $L^2(\mathbb{R}^n)$ as $k \to \infty \Longrightarrow \lim_{k\to\infty}\langle \phi_k, \psi_k\rangle_{L^2(\mathbb{R}^n)} = \langle f, g\rangle_{L^2(\mathbb{R}^n)}$. In fact, using the Schwarz inequality,

$$\begin{aligned}&|\langle f, g\rangle_{L^2(\mathbb{R}^n)} - \langle \phi_k, \psi_k\rangle_{L^2(\mathbb{R}^n)}| \le |\langle f, g - \psi_k\rangle_{L^2(\mathbb{R}^n)}| + |\langle f - \phi_k, \psi_k\rangle_{L^2(\mathbb{R}^n)}|\\ &\le (\|f\|_{L^2(\mathbb{R}^n)}\|g-\psi_k\|_{L^2(\mathbb{R}^n)} + \|f-\phi_k\|_{L^2(\mathbb{R}^n)}\|\psi_k\|_{L^2(\mathbb{R}^n)}) \to 0 \text{ as } k \to \infty,\end{aligned}$$

since $\psi_k \to g \Longrightarrow \|g - \psi_k\|_{L^2(\mathbb{R}^n)} \to 0$, $\phi_k \to f \Longrightarrow \|f - \phi_k\|_{L^2(\mathbb{R}^n)} \to 0$, $\|\psi_k\| \le \|\psi_k - g\| + \|g\| \to \|g\|_{L^2(\mathbb{R}^n)}$ as $k \to \infty$.

Hence, $\lim_{k\to\infty}\langle \phi_k, \psi_k\rangle_{L^2(\mathbb{R}^n)} = \langle f, g\rangle_{L^2(\mathbb{R}^n)}$. Since $\mathcal{F}$ is a continuous linear operator from $L^2(\mathbb{R}^n)$ onto $L^2(\mathbb{R}^n)$, $\phi_k \to f$, $\psi_k \to g$ in $L^2(\mathbb{R}^n) \Longrightarrow$

$\mathcal{F}\phi_k \to \mathcal{F}f, \mathcal{F}\psi_k \to \mathcal{F}g$ in $L^2(\mathbb{R}^n)$ as $k \to \infty$. Hence,

$$\lim_{k\to\infty} \langle \mathcal{F}\phi_k, \mathcal{F}\psi_k\rangle_{L^2(\mathbb{R}^n)} = \langle \mathcal{F}f, \mathcal{F}g\rangle_{L^2(\mathbb{R}^n)}. \tag{8.3.8}$$

But again by Plancherel–Riesz Theorem 8.3.1 on $S(\mathbb{R}^n)$, $\langle \mathcal{F}\phi_k, \mathcal{F}\psi_k\rangle_{L^2(\mathbb{R}^n)} = \langle \phi_k, \psi_k\rangle_{L^2(\mathbb{R}^n)}$ $\forall k \in \mathbb{N}$

$$\Longrightarrow \quad \lim_{k\to\infty} \langle \mathcal{F}\phi_k, \mathcal{F}\psi_k\rangle_{L^2(\mathbb{R}^n)} = \lim_{k\to\infty} \langle \phi_k, \psi_k\rangle_{L^2(\mathbb{R}^n)} = \langle f, g\rangle_{L^2(\mathbb{R}^n)}.$$

Thus, using (8.3.8), $\langle \mathcal{F}f, \mathcal{F}g\rangle_{L^2(\mathbb{R}^n)} = \langle f, g\rangle_{L^2(\mathbb{R}^n)}$ $\forall f, g \in L^2(\mathbb{R}^n)$
$\Longrightarrow \mathcal{F} : L^2(\mathbb{R}^n) \longrightarrow L^2(\mathbb{R}^n)$ is a unitary operator from $L^2(\mathbb{R}^n)$ onto $L^2(\mathbb{R}^n)$.

II. For $f = g$, $\langle \mathcal{F}f, \mathcal{F}f\rangle_{L^2(\mathbb{R}^n)} = \langle f, f\rangle_{L^2(\mathbb{R}^n)} \Longrightarrow \|\mathcal{F}f\|_{L^2(\mathbb{R}^n)} = \|f\|_{L^2(\mathbb{R}^n)}$ $\forall f \in L^2(\mathbb{R}^n)$.

III. For $f \in L^2(\mathbb{R}^n)$, let (ϕ_k) be a sequence in $S(\mathbb{R}^n)$ such that $\phi_k \to f$ in $L^2(\mathbb{R}^n)$ as $k \to \infty$. Then, $\forall \psi \in S(\mathbb{R}^n)$,

$$\begin{aligned}\langle f, \psi\rangle_{L^2(\mathbb{R}^n)} &= \lim_{k\to\infty} \langle \phi_k, \psi\rangle_{L^2(\mathbb{R}^n)} = \lim_{k\to\infty} \langle \mathcal{F}\phi_k, \mathcal{F}\psi\rangle_{L^2(\mathbb{R}^n)} \\ &= \langle \mathcal{F}f, \mathcal{F}\psi\rangle_{L^2(\mathbb{R}^n)} \forall f \in L^2(\mathbb{R}^n).\end{aligned} \tag{8.3.9}$$

Hence, from (8.3.9), $\int_{\mathbb{R}^n} f(\mathbf{x})\overline{\psi(\mathbf{x})}d\mathbf{x} = \int_{\mathbb{R}^n} (\mathcal{F}f)(\boldsymbol{\xi})\overline{(\mathcal{F}\psi)}(\boldsymbol{\xi})d\boldsymbol{\xi}$ $\forall \psi \in S(\mathbb{R}^n)$.
By Theorem 7.7.2 on isomorphism, $\exists \phi \in S(\mathbb{R}^n)$ such that $\psi(\mathbf{x}) = \overline{(\mathcal{F}\phi)} \Longrightarrow \mathcal{F}\psi = \mathcal{F}(\overline{\mathcal{F}}(\overline{\phi})) = \overline{\phi}$ (using (7.1.17) and the Inversion Theorem 7.7.1 on $S(\mathbb{R}^n)$) $\Longrightarrow \overline{\mathcal{F}\psi} = \phi \Longrightarrow \int_{\mathbb{R}^n} f(\mathbf{x})(\mathcal{F}\phi)(\mathbf{x})d\mathbf{x} = \int_{\mathbb{R}^n} (\mathcal{F}f)(\boldsymbol{\xi})\phi(\boldsymbol{\xi})d\boldsymbol{\xi}$

$$\begin{aligned}\Longrightarrow \quad &\langle \mathcal{F}f, \phi\rangle_{S'(\mathbb{R}^n)\times S(\mathbb{R}^n)} = \int_{\mathbb{R}^n} (\mathcal{F}f)(\boldsymbol{\xi})\phi(\boldsymbol{\xi})d\boldsymbol{\xi}) = \int_{\mathbb{R}^n} f(\mathbf{x})(\mathcal{F}\phi)(\mathbf{x})d\mathbf{x} \\ &= \langle f, \mathcal{F}\phi\rangle_{S'(\mathbb{R}^n)\times S(\mathbb{R}^n)} \quad \forall f \in L^2(\mathbb{R}^n), \ \forall \phi \in S(\mathbb{R}^n).\end{aligned}$$

IV. Since $S(\mathbb{R}^n)$ is dense in $L^2(\mathbb{R}^n)$, for (ϕ_k) in $S(\mathbb{R}^n)$ with $\phi_k \to f$ in $L^2(\mathbb{R}^n)$, by the Inversion Theorem 7.7.1 on $S(\mathbb{R}^n)$, we have $\overline{\mathcal{F}}\mathcal{F}\phi_k = \mathcal{F}\overline{\mathcal{F}}\phi_k = \phi_k$ $\forall k \in \mathbb{N}$. Hence, $\phi_k \to f$ in $L^2(\mathbb{R}^n) \Longrightarrow \mathcal{F}\phi_k \to \mathcal{F}f$ in $L^2(\mathbb{R}^n) \Longrightarrow \overline{\mathcal{F}}(\mathcal{F}\phi_k) \to \overline{\mathcal{F}}(\mathcal{F}f)$ in $L^2(\mathbb{R}^n)$ by virtue of the continuity of $\mathcal{F}$ on $L^2(\mathbb{R}^n)$. But $\overline{\mathcal{F}}(\mathcal{F}\phi_k) = \phi_k \to f$ in $L^2(\mathbb{R}^n)$. Hence, $\overline{\mathcal{F}}(\mathcal{F}f) = f$ $\forall f \in L^2(\mathbb{R}^n)$, since the limit is unique. Similarly, $\mathcal{F}\overline{\mathcal{F}}f = f$ $\forall f \in L^2(\mathbb{R}^n)$.

□

Remark 8.3.1. By an abuse of definitions, one often writes

$$(\mathcal{F}f)(\boldsymbol{\xi}) = \int_{\mathbb{R}^n} f(\mathbf{x})e^{-i2\pi\langle \mathbf{x},\boldsymbol{\xi}\rangle}d\mathbf{x} \quad \forall f \in L^2(\mathbb{R}^n), \tag{8.3.10}$$

which, in fact, must be understood in the following sense: for $f \in L^2(\mathbb{R}^n)$, $\mathcal{F} f \in L^2(\mathbb{R}^n)$ is the limit of the sequence $(f_k)_{k\in\mathbb{N}}$ defined by $f_k(\boldsymbol{\xi}) = \int_{B_k} f(\mathbf{x})e^{-i2\pi\langle \mathbf{x},\boldsymbol{\xi}\rangle} d\boldsymbol{\xi}$ in $L^2(\mathbb{R}^n)$, $(B_k)_{k\in\mathbb{N}}$ being a sequence of relatively compact subsets which tend to $\mathbb{R}^n$ as $k \to \infty$. Obviously, this formula (8.3.10) is meaningless in the usual sense *in general*, since the integral in (8.3.10) may *not* be defined for $f \in L^2(\mathbb{R}^n)$.

For example, for $f(x) = \frac{1}{\sqrt{1+x^2}} \in L^2(\mathbb{R})$, the integral in (8.3.10) does *not* exist, since $\frac{1}{\sqrt{1+x^2}} \notin L^1(\mathbb{R})$ (see (7.1.10)). But for $\frac{1}{\sqrt{1+x^2}} \in L^2(\mathbb{R})$, its Fourier transform $\mathcal{F}[\frac{1}{\sqrt{1+x^2}}]$ can be defined by (8.3.6): $\int_{\mathbb{R}} \mathcal{F}(\frac{1}{\sqrt{1+x^2}})\phi(\xi)d\xi = \int_{\mathbb{R}} \frac{1}{\sqrt{1+x^2}}(\mathcal{F}\phi)(x)dx$ $\forall\phi \in S(\mathbb{R})$, where $\frac{1}{\sqrt{1+x^2}}\mathcal{F}\phi \in L^1(\mathbb{R}) \implies$ the right-hand side integral is well defined $\forall\phi \in S(\mathbb{R})$.

Isometric isomorphism of Fourier transforms on $L^2(\mathbb{R}^n)$

Corollary 8.3.1. $\mathcal{F} : L^2(\mathbb{R}^n) \longrightarrow L^2(\mathbb{R}^n)$ *defined by* (8.3.6) *is an* isometric isomorphism *(see Definition* A.8.3.2 *in Appendix* A*) from* $L^2(\mathbb{R}^n)$ *onto* $L^2(\mathbb{R}^n)$.

Proof. The result follows from the *linearity* and *isometry* (8.3.5) of $\mathcal{F} : L^2(\mathbb{R}^n) \longrightarrow L^2(\mathbb{R}^n)$. □

8.3.1 Fourier transforms of Dirac distributions and their derivatives

Using the definitions of Fourier transform $\mathcal{F}$ and co-transform $\bar{\mathcal{F}}$ of tempered distributions of $S'(\mathbb{R}^n)$ in (8.3.1) and (8.3.2) respectively, we will find the Fourier transforms and co-transforms of Dirac distributions and their derivatives. For alternative methods, see Section 8.4 later.

1. $T = \delta = \delta(\mathbf{x}) \in \mathcal{E}'(\mathbb{R}^n) \subset \mathcal{D}'(\mathbb{R}^n)$ is the Dirac distribution with mass/charge/force etc. concentrated at $\mathbf{x} = \mathbf{0} \in \mathbb{R}^n$, i.e. with compact support $\{\mathbf{0}\}$. Hence, $\delta = \delta(\mathbf{x}) \in S'(\mathbb{R}^n) \subset \mathcal{D}'(\mathbb{R}^n)$, with $\langle\delta,\phi\rangle = \langle\delta(\mathbf{x}),\phi(\mathbf{x})\rangle = \phi(\mathbf{0})\forall\phi \in S(\mathbb{R}^n) \subset \mathcal{E}'(\mathbb{R}^n)$, is a tempered distribution and its Fourier transform $(\mathcal{F}\delta)(\boldsymbol{\xi})$ and co-transform $(\bar{\mathcal{F}}\delta)(\boldsymbol{\xi})$ are defined by (8.3.1) and (8.3.2) respectively.

$$\begin{aligned}\langle\mathcal{F}\delta,\phi\rangle &= \langle\delta,\mathcal{F}\phi\rangle = \langle\delta(\mathbf{x}),(\mathcal{F}\phi)(\mathbf{x})\rangle = (\mathcal{F}\phi)(\mathbf{0})\\ &= \int_{\mathbb{R}^n}\phi(\boldsymbol{\xi})e^{-i2\pi\langle\boldsymbol{\xi},\mathbf{0}\rangle}d\boldsymbol{\xi} = \int_{\mathbb{R}^n} 1\cdot\phi(\boldsymbol{\xi})d\boldsymbol{\xi} = \langle 1,\phi\rangle\ \forall\phi \in S(\mathbb{R}^n)\\ \implies\quad &\mathcal{F}\delta = \mathcal{F}[\delta(\mathbf{x})] = 1 \text{ in } S'(\mathbb{R}^n). \qquad (8.3.11)\end{aligned}$$

$$\begin{aligned}\langle\bar{\mathcal{F}}\delta,\phi\rangle &= \langle\delta,\bar{\mathcal{F}}\phi\rangle = \langle\delta(\mathbf{x}),(\bar{\mathcal{F}}\phi)(\mathbf{x})\rangle = \bar{\mathcal{F}}\phi(\mathbf{0})\\ &= \int_{\mathbb{R}^n}\phi(\boldsymbol{\xi})e^{i2\pi\langle\boldsymbol{\xi},\mathbf{0}\rangle}d\boldsymbol{\xi}\\ &= \int_{\mathbb{R}^n}\phi(\boldsymbol{\xi})1d\boldsymbol{\xi} = \langle 1,\phi\rangle\ \forall\phi \in S(\mathbb{R}^n)\end{aligned}$$

$$\Longrightarrow \quad \bar{\mathcal{F}}\delta = \mathcal{F}[\delta(\mathbf{x})] = 1 \text{ in } S'(\mathbb{R}^n). \tag{8.3.12}$$

$$\begin{aligned}
\langle \mathcal{F}[\partial_{\mathbf{x}}^{\boldsymbol{\alpha}}(\delta)], \phi\rangle &= \langle \partial_{\mathbf{x}}^{\boldsymbol{\alpha}}(\delta), \mathcal{F}\phi\rangle = (-1)^{|\boldsymbol{\alpha}|}\langle \delta(\mathbf{x}), \partial_{\mathbf{x}}^{\boldsymbol{\alpha}}(\mathcal{F}\phi)(\mathbf{x})\rangle \\
&\qquad \text{(using definition of derivative in (8.2.27))} \\
&= (-1)^{|\boldsymbol{\alpha}|}[\partial_{\mathbf{x}}^{\boldsymbol{\alpha}}(\mathcal{F}\phi)](\mathbf{0}) \quad \text{(applying the definition of } \delta = \delta(\mathbf{x})) \\
&= (-1)^{|\boldsymbol{\alpha}|}(\mathcal{F}[(-i2\pi\boldsymbol{\xi})^{\boldsymbol{\alpha}}\phi(\boldsymbol{\xi})])(\mathbf{0}) \\
&= (-1)^{|\boldsymbol{\alpha}|}\int_{\mathbb{R}^n}(-i2\pi\boldsymbol{\xi})^{\boldsymbol{\alpha}}\phi(\boldsymbol{\xi})e^{-i2\pi\langle\boldsymbol{\xi},\mathbf{0}\rangle}d\boldsymbol{\xi} \\
&= \int_{\mathbb{R}^n}(i2\pi\boldsymbol{\xi})^{\boldsymbol{\alpha}}\phi(\boldsymbol{\xi})d\boldsymbol{\xi} \\
&= \langle T_{(i2\pi\boldsymbol{\xi})^{\alpha}}, \phi\rangle = \langle (i2\pi\boldsymbol{\xi})^{\boldsymbol{\alpha}}, \phi\rangle \quad \forall \phi \in S(\mathbb{R}^n),
\end{aligned}$$

where $\partial_{\mathbf{x}}^{\boldsymbol{\alpha}} = \frac{\partial^{|\boldsymbol{\alpha}|}}{\partial x_1{}^{\alpha_1}\partial x_2{}^{\alpha_2}\cdots\partial x_n{}^{\alpha_n}}$; $(-i2\pi\boldsymbol{\xi})^{\boldsymbol{\alpha}} = (-i2\pi\xi_1)^{\alpha_1}\ldots(-i2\pi\xi_n)^{\alpha_n}$; $\partial_{\mathbf{x}}^{\boldsymbol{\alpha}}(\mathcal{F}\phi) = \mathcal{F}[(-i2\pi\boldsymbol{\xi}^{\boldsymbol{\alpha}}\phi(\boldsymbol{\xi}))]$ is given by Theorem 7.1.3,

$$\Longrightarrow \quad \mathcal{F}[\partial_{\mathbf{x}}^{\boldsymbol{\alpha}}(\delta)] = \mathcal{F}[\partial_{\mathbf{x}}^{\boldsymbol{\alpha}}(\delta(\mathbf{x}))] = (i2\pi\boldsymbol{\xi})^{\boldsymbol{\alpha}} \quad \text{in } S'(\mathbb{R}^n)\forall\boldsymbol{\alpha}. \tag{8.3.13}$$

$$\begin{aligned}
\langle \bar{\mathcal{F}}[\partial_{\mathbf{x}}^{\boldsymbol{\alpha}}(\delta)], \phi\rangle &= \langle \partial_{\mathbf{x}}^{\boldsymbol{\alpha}}(\delta), \bar{\mathcal{F}}\phi\rangle = (-1)^{|\boldsymbol{\alpha}|}\langle \delta(\mathbf{x}), \partial_{\mathbf{x}}^{\boldsymbol{\alpha}}(\bar{\mathcal{F}}\phi)(\mathbf{x})\rangle \\
&= (-1)^{|\boldsymbol{\alpha}|}[\partial_{\mathbf{x}}^{\boldsymbol{\alpha}}(\bar{\mathcal{F}}\phi)](\mathbf{0}) \\
&= (-1)^{|\boldsymbol{\alpha}|}(\bar{\mathcal{F}}[(i2\pi\boldsymbol{\xi})^{\boldsymbol{\alpha}}\phi(\boldsymbol{\xi})])(\mathbf{0}) \\
&= (-1)^{|\boldsymbol{\alpha}|}\int_{\mathbb{R}^n}(i2\pi\boldsymbol{\xi})^{\boldsymbol{\alpha}}\phi(\boldsymbol{\xi})e^{i2\pi\langle\boldsymbol{\xi},\mathbf{0}\rangle}d\boldsymbol{\xi} \\
&= \int_{\mathbb{R}^n}(-i2\pi\boldsymbol{\xi})^{\boldsymbol{\alpha}}\phi(\boldsymbol{\xi})d\boldsymbol{\xi} \\
&= \langle T_{(-i2\pi\boldsymbol{\xi})^{\alpha}}, \phi\rangle = \langle (-i2\pi\boldsymbol{\xi})^{\boldsymbol{\alpha}}, \phi\rangle \quad \forall \phi \in S(\mathbb{R}^n)
\end{aligned}$$

$$\Longrightarrow \quad \bar{\mathcal{F}}[\partial_{\mathbf{x}}^{\boldsymbol{\alpha}}(\delta)] = \bar{\mathcal{F}}[\partial_{\mathbf{x}}^{\boldsymbol{\alpha}}(\delta(\mathbf{x}))] = (-i2\pi\boldsymbol{\xi})^{\boldsymbol{\alpha}} \quad \text{in } S'(\mathbb{R}^n)\ \forall\boldsymbol{\alpha}. \tag{8.3.14}$$

2. $T = \delta_{\mathbf{a}} = \delta(\mathbf{x}-\mathbf{a}) \in \mathcal{D}'(\mathbb{R}^n)$ is the Dirac distribution with mass/charge/force etc. concentrated at $\mathbf{x} = \mathbf{a} \in \mathbb{R}^n$, i.e. with compact support $\{\mathbf{a}\}$.
   Hence, $\delta_{\mathbf{a}} = \delta(\mathbf{x}-\mathbf{a}) \in S'(\mathbb{R}^n) \subset \mathcal{E}'(\mathbb{R}^n) \subset \mathcal{D}'(\mathbb{R}^n)$ is a tempered distribution, and its Fourier transform $(\mathcal{F}\delta_{\mathbf{a}})(\boldsymbol{\xi}) = (\mathcal{F}[\delta(\mathbf{x}-\mathbf{a})(\boldsymbol{\xi})])$ and co-transform $(\bar{\mathcal{F}}\delta_{\mathbf{a}})(\boldsymbol{\xi}) = (\bar{\mathcal{F}}[\delta(\mathbf{x}-\mathbf{a})])(\boldsymbol{\xi})$ are defined by (8.3.1) and (8.3.2) respectively:

$$\begin{aligned}
\langle \mathcal{F}\delta_{\mathbf{a}}, \phi\rangle &= \langle \delta_{\mathbf{a}}, \mathcal{F}\phi\rangle = \langle \delta(\mathbf{x}-\mathbf{a}), (\mathcal{F}\phi)(\mathbf{x})\rangle = (\mathcal{F}\phi)(\mathbf{a}) \\
&= \int_{\mathbb{R}^n}\phi(\boldsymbol{\xi})e^{-i2\pi\langle\boldsymbol{\xi},\mathbf{a}\rangle}d\boldsymbol{\xi} = \langle e^{-i2\pi\langle\boldsymbol{\xi},\mathbf{a}\rangle}, \phi\rangle \quad \forall\phi \in S(\mathbb{R}^n)
\end{aligned}$$

$$\Longrightarrow \quad \mathcal{F}\delta_{\mathbf{a}} = \mathcal{F}[\delta(\mathbf{x}-\mathbf{a})] = e^{-i2\pi\langle\boldsymbol{\xi},\mathbf{a}\rangle} \text{ in } S'(\mathbb{R}^n), \tag{8.3.15}$$

   from which (8.3.11) is obtained for $\mathbf{a} = \mathbf{0}$.

$\langle \bar{\mathcal{F}}\delta_{\mathbf{a}}, \phi\rangle = \langle \delta_{\mathbf{a}}, \bar{\mathcal{F}}\phi\rangle = (\bar{\mathcal{F}}\phi)(\mathbf{a}) = \langle e^{i2\pi\langle \boldsymbol{\xi},\mathbf{a}\rangle}, \phi\rangle \ \forall \phi \in S(\mathbb{R}^n) \Longrightarrow \bar{\mathcal{F}}\delta_{\mathbf{a}} = \bar{\mathcal{F}}[\delta(\mathbf{x}-\mathbf{a})] = e^{i2\pi\langle \boldsymbol{\xi},\mathbf{a}\rangle}$ in $S'(\mathbb{R}^n)$, from which we get (8.3.12) for $\mathbf{a} = \mathbf{0}$.

$$\begin{aligned}
\langle \mathcal{F}[\partial_{\mathbf{x}}^{\boldsymbol{\alpha}}(\delta_{\mathbf{a}})], \phi\rangle &= \langle \partial_{\mathbf{x}}^{\boldsymbol{\alpha}}(\delta_{\mathbf{a}}), \mathcal{F}\phi\rangle = (-1)^{|\boldsymbol{\alpha}|}\langle \delta_{\mathbf{a}}, \partial_{\mathbf{x}}^{\boldsymbol{\alpha}}[\mathcal{F}\phi]\rangle \\
&= (-1)^{|\boldsymbol{\alpha}|}\langle \delta_{\mathbf{a}}, \partial_{\mathbf{x}}^{\boldsymbol{\alpha}}[\mathcal{F}\phi](\mathbf{x})\rangle \\
&= (-1)^{|\boldsymbol{\alpha}|}\langle \delta(\mathbf{x}-\mathbf{a}), (\mathcal{F}[(-i2\pi\boldsymbol{\xi})^{\boldsymbol{\alpha}}\phi(\boldsymbol{\xi})])(\mathbf{x})\rangle \\
&= (-1)^{|\boldsymbol{\alpha}|}(\mathcal{F}[(-i2\pi\boldsymbol{\xi})^{\boldsymbol{\alpha}}\phi(\boldsymbol{\xi})])(\mathbf{a}) \\
&= \int_{\mathbb{R}^n} (i2\pi\boldsymbol{\xi})^{\boldsymbol{\alpha}} e^{-i2\pi\langle \boldsymbol{\xi},\mathbf{a}\rangle}\phi(\boldsymbol{\xi})d\boldsymbol{\xi} \\
&= \langle T_{(i2\pi\boldsymbol{\xi})^{\boldsymbol{\alpha}} e^{-i2\pi\langle \boldsymbol{\xi},\mathbf{a}\rangle}}, \phi\rangle \\
&= \langle (i2\pi\boldsymbol{\xi})^{\boldsymbol{\alpha}} e^{-i2\pi\langle \boldsymbol{\xi},\mathbf{a}\rangle}, \phi\rangle \quad \forall \phi \in S(\mathbb{R}^n)
\end{aligned}$$

(see the proof of (8.3.13) for the intermediate steps)

$$\Longrightarrow \quad \mathcal{F}[\partial_{\mathbf{x}}^{\boldsymbol{\alpha}}(\delta_{\mathbf{a}})] = \mathcal{F}[\partial_{\mathbf{x}}^{\boldsymbol{\alpha}}(\delta(\mathbf{x}-\mathbf{a}))] = (i2\pi\boldsymbol{\xi})^{\boldsymbol{\alpha}} e^{-i2\pi\langle \boldsymbol{\xi},\mathbf{a}\rangle} \text{ in } S'(\mathbb{R}^n), \tag{8.3.16}$$

from which (8.3.13) is obtained for $\mathbf{a} = \mathbf{0}$.

$$\begin{aligned}
\langle \bar{\mathcal{F}}[\partial_{\mathbf{a}}^{\boldsymbol{\alpha}}(\delta_{\mathbf{a}})], \phi\rangle &= \langle \partial_{\mathbf{x}}^{\boldsymbol{\alpha}}(\delta_{\mathbf{a}}), \bar{\mathcal{F}}\phi\rangle = (-1)^{|\boldsymbol{\alpha}|}\langle \delta_{\mathbf{a}}, \partial_{\mathbf{x}}^{\boldsymbol{\alpha}}[\bar{\mathcal{F}}\phi]\rangle \\
&= \langle (-i2\pi\boldsymbol{\xi})^{\boldsymbol{\alpha}} e^{i2\pi\langle \boldsymbol{\xi},\mathbf{a}\rangle}, \phi\rangle \quad \forall \phi \in S(\mathbb{R}^n)
\end{aligned}$$

(see the proofs of (8.3.13) and (8.3.16))

$$\Longrightarrow \quad \bar{\mathcal{F}}[\partial_{\mathbf{x}}^{\boldsymbol{\alpha}}(\delta_{\mathbf{a}})] = (-i2\pi\boldsymbol{\xi})^{\boldsymbol{\alpha}} e^{i2\pi\langle \boldsymbol{\xi},\mathbf{a}\rangle} \text{ in } S'(\mathbb{R}^n), \tag{8.3.17}$$

from which (8.3.14) is obtained for $\mathbf{a} = \mathbf{0}$.

8.3.2 Inversion theorem for Fourier transforms on $S'(\mathbb{R}^n)$

Theorem 8.3.2 (Fourier Inversion Theorem). *Let* $\mathcal{F} : S'(\mathbb{R}^n) \longrightarrow S'(\mathbb{R}^n)$ *and* $\bar{\mathcal{F}} : S'(\mathbb{R}^n) \longrightarrow S'(\mathbb{R}^n)$ *be defined by* (8.3.1) *and* (8.3.2), *respectively:* $\forall \phi \in S(\mathbb{R}^n)$, $\forall T \in S'(\mathbb{R}^n)$, $\langle \mathcal{F}T, \phi\rangle = \langle T, \mathcal{F}\phi\rangle$, $\langle \bar{\mathcal{F}}T, \phi\rangle = \langle T, \bar{\mathcal{F}}\phi\rangle$.

Then $\bar{\mathcal{F}} = \mathcal{F}^{-1}$ *(resp.* $\mathcal{F} = \bar{\mathcal{F}}^{-1}$*) is the* inverse *to transform* $\mathcal{F}$ *(resp.* $\bar{\mathcal{F}}$*) and* $\mathcal{F}$ *is an isomorphism from* $S'(\mathbb{R}^n)$ *onto* $S'(\mathbb{R}^n)$,

$$\forall T \in S'(\mathbb{R}^n), \quad \bar{\mathcal{F}}\mathcal{F}T = \mathcal{F}\bar{\mathcal{F}}T = T. \tag{8.3.18}$$

Proof. From the Fourier Inversion Theorem 7.7.1 on $S(\mathbb{R}^n)$, $\bar{\mathcal{F}}\mathcal{F}\phi = \mathcal{F}\bar{\mathcal{F}}\phi \ \forall \phi \in S(\mathbb{R}^n)$. But $\forall T \in S'(\mathbb{R}^n)$, $\mathcal{F}T, \bar{\mathcal{F}}T \in S'(\mathbb{R}^n)$ and are defined by (8.3.1), (8.3.2), respectively. Hence, $\bar{\mathcal{F}}(\mathcal{F}T), \mathcal{F}(\bar{\mathcal{F}}T) \in S'(\mathbb{R}^n)$ and are given by:

$\langle \bar{\mathcal{F}}\mathcal{F}T, \phi\rangle = \langle \mathcal{F}T, \bar{\mathcal{F}}\phi\rangle = \langle T, \mathcal{F}\bar{\mathcal{F}}\phi\rangle = \langle T, \phi\rangle$ (since $\mathcal{F}\bar{\mathcal{F}}\phi = \phi$) $\forall \phi \in S(\mathbb{R}^n)$
$\Longrightarrow \bar{\mathcal{F}}\mathcal{F}T = T \; \forall T \in S'(\mathbb{R}^n)$;
$\langle \mathcal{F}\bar{\mathcal{F}}T, \phi\rangle = \langle \bar{\mathcal{F}}T, \mathcal{F}\phi\rangle = \langle T, \bar{\mathcal{F}}\mathcal{F}\phi\rangle = \langle T, \phi\rangle$ (since $\bar{\mathcal{F}}\mathcal{F}\phi = \phi$) $\forall \phi \in S(\mathbb{R}^n)$
$\Longrightarrow \mathcal{F}\bar{\mathcal{F}}T = T \; \forall T \in S'(\mathbb{R}^n)$. □

Continuity of $\mathcal{F}$ (resp. $\mathcal{F}^{-1}$): $T_k \to T$ in $S'(\mathbb{R}^n) \Longrightarrow \mathcal{F}T_k \to \mathcal{F}T$ in $S'(\mathbb{R}^n)$ (resp. $T_k \to T$ in $S'(\mathbb{R}^n) \Longrightarrow \mathcal{F}^{-1}T_k = \overline{\mathcal{F}}T_k \to \mathcal{F}^{-1}T = \overline{\mathcal{F}}^{-1}T$ in $S'(\mathbb{R}^n)$) as $k \to \infty$. Hence, $\mathcal{F}$ is an isomorphism from $S'(\mathbb{R}^n)$ onto itself.

Corollary 8.3.2.

I. $\mathcal{F}T = S \Longrightarrow \bar{\mathcal{F}}S = T;\; \bar{\mathcal{F}}S = T \Longrightarrow \mathcal{F}T = S.$ (8.3.19)

II. $\mathcal{F}T = 0 \Longrightarrow T = 0;\; \bar{\mathcal{F}}S = 0 \Longrightarrow S = 0.$ (8.3.20)

III. $(\mathcal{F}T)^\vee = \overline{\mathcal{F}}T;\; \overline{\mathcal{F}}\check{T} = \mathcal{F}T;\; \mathcal{F}\mathcal{F}T = \check{T},$ (8.3.21)

where $\check{T}$ *and* $(\mathcal{F}T)^\vee$ *are defined by:* $\forall \phi \in S(\mathbb{R}^n)$ *with* $\check{\phi}(\mathbf{x}) = \phi(-\mathbf{x})$, $\langle \check{T}, \phi\rangle = \langle T, \check{\phi}\rangle$, $\langle (\mathcal{F}T)^\vee, \phi\rangle = \langle \mathcal{F}T, \check{\phi}\rangle = \langle T, \mathcal{F}\check{\phi}\rangle$.

Proof.

I. $\mathcal{F}T = S \Longrightarrow \bar{\mathcal{F}}S = \bar{\mathcal{F}}\mathcal{F}T = T,\; \bar{\mathcal{F}}S = T \Longrightarrow \mathcal{F}T = \mathcal{F}\bar{\mathcal{F}}S = S.$

II. $\mathcal{F}T = S = 0 \Longrightarrow T = \bar{\mathcal{F}}S = \bar{\mathcal{F}}0 = 0;\; \bar{\mathcal{F}}S = T = 0 \Longrightarrow S = \mathcal{F}T = 0.$

III. $\langle (\mathcal{F}T)^\vee, \phi\rangle = \langle \mathcal{F}T, \check{\phi}\rangle = \langle T, \mathcal{F}\check{\phi}\rangle = \langle T, \bar{\mathcal{F}}\phi\rangle = \langle \bar{\mathcal{F}}T, \phi\rangle \Longrightarrow (\mathcal{F}T)^\vee = \bar{\mathcal{F}}T$;
$\langle \bar{\mathcal{F}}\check{T}, \phi\rangle = \langle \check{T}, \bar{\mathcal{F}}\phi\rangle = \langle \check{T}, (\mathcal{F}\phi)^\vee\rangle = \langle T, \mathcal{F}\phi\rangle = \langle \mathcal{F}T, \phi\rangle \Longrightarrow \bar{\mathcal{F}}\check{T} = \mathcal{F}T$;
$\mathcal{F}\mathcal{F}T = \mathcal{F}\bar{\mathcal{F}}\check{T} = \check{T}.$ □

Application From (8.3.11) and (8.3.12), $\mathcal{F}\delta = \bar{\mathcal{F}}\delta = 1$. Hence,

$$\mathcal{F}1 = \mathcal{F}(\bar{\mathcal{F}}\delta) = \delta; \quad \bar{\mathcal{F}}1 = \bar{\mathcal{F}}(\mathcal{F}\delta) = \delta. \tag{8.3.22}$$

8.3.3 Fourier transform of even and odd tempered distributions

Even and odd tempered distributions (see also Definition 1.7.1)

A tempered distribution $T \in S'(\mathbb{R})$ is called

- *even* if and only if $T(\check{\phi}) = T(\phi) \; \forall \phi \in S(\mathbb{R})$; (8.3.23)
- *odd* if and only if $T(\check{\phi}) = -T(\phi) \; \forall \phi \in S(\mathbb{R})$, (8.3.24)

where $\check{\phi}(x) = \phi(-x) \; \forall \phi \in S(\mathbb{R})$.

Examples of *even* tempered distributions on $\mathbb{R}$ are *even* functions $f \in S'(\mathbb{R})$, *even-order* derivatives $\delta^{(2m)}$ of the Dirac distribution δ, etc. (see (1.7.3)–(1.7.4)).

Examples of *odd* tempered distributions on $\mathbb{R}$ are *odd* functions $f \in S'(\mathbb{R})$, *odd-order* derivatives $\delta^{(2m+1)}$ of the Dirac distribution δ, etc. (see (1.7.6)–(1.7.7)).

$\forall T \in S'(\mathbb{R})$, $T = T_E + T_0$ such that $\forall \phi \in S(\mathbb{R})$, $T_E(\phi) = \frac{1}{2}[T(\phi) + T(\check{\phi})]$, $T_0(\phi) = \frac{1}{2}[T(\phi) - T(\check{\phi})]$ are *unique* even and odd tempered distributions respectively. The proof is exactly similar to that for (1.7.9) with $\phi \in \mathcal{D}(\mathbb{R})$ replaced by $\phi \in S(\mathbb{R})$.

Example 8.3.1. Show that c.p.v. $\frac{1}{x}$ is an odd tempered distribution.

Proof.

$$\begin{aligned}
\left\langle \text{c.p.v. } \frac{1}{x}, \check{\phi} \right\rangle &= \lim_{\varepsilon \to 0+} \int_{|x| \ge \varepsilon} \frac{\check{\phi}(x)}{x} dx = \lim_{\varepsilon \to 0+} \int_{|x| \ge \varepsilon} \frac{\phi(-x)}{x} dx \\
&= \lim_{\varepsilon \to 0+} \left[\int_{-\infty}^{-\varepsilon} \frac{\phi(-x)}{x} dx + \int_{\varepsilon}^{\infty} \frac{\phi(-x)}{x} dx \right] \\
&= \lim_{\varepsilon \to 0+} \left[\int_{\infty}^{\varepsilon} \frac{\phi(y)}{y} dy + \int_{-\varepsilon}^{-\infty} \frac{\phi(y) dy}{y} \right] \\
&= \lim_{\varepsilon \to 0+} \left[- \int_{-\infty}^{-\varepsilon} \frac{\phi(y)}{y} dy - \int_{\varepsilon}^{\infty} \frac{\phi(y)}{y} dy \right] \\
&= - \lim_{\varepsilon \to 0+} \int_{|x| \ge \varepsilon} \frac{\phi(x)}{x} dx = - \left\langle \text{c.p.v. } \frac{1}{x}, \phi \right\rangle \quad \forall \phi \in S(\mathbb{R})
\end{aligned}$$

$\Longrightarrow$ c.p.v. $\frac{1}{x}$ is an odd tempered distribution. □

Fourier transforms of even and odd distributions

$$T \in S'(\mathbb{R}) \text{ is } \textit{even} \text{ (resp. } \textit{odd}) \Longrightarrow \bar{\mathcal{F}} T = \mathcal{F} T \text{ (resp. } \bar{\mathcal{F}} T = -\mathcal{F} T) \text{ in } S'(\mathbb{R}). \tag{8.3.25}$$

Proof. $\langle \bar{\mathcal{F}} T, \phi \rangle = \langle T, \bar{\mathcal{F}} \phi \rangle$ $\forall \phi \in S(\mathbb{R})$ (by definition).

$$\begin{aligned}
(\bar{\mathcal{F}} \phi)(\xi) &= \int_{\mathbb{R}} \phi(x) e^{i2\pi x \xi} dx = \int_{\mathbb{R}} \phi(x) e^{-i2\pi x(-\xi)} dx \\
&= (\mathcal{F} \phi)(-\xi) = (\mathcal{F} \phi)^{\vee}(\xi) \quad \forall \xi \in \mathbb{R}^n
\end{aligned}$$

$\Longrightarrow \bar{\mathcal{F}} \phi = (\mathcal{F} \phi)^{\vee}$ $\forall \phi \in S'(\mathbb{R})$. Hence, $\langle \bar{\mathcal{F}} T, \phi \rangle = \langle T, (\mathcal{F} \phi)^{\vee} \rangle = \langle T, \mathcal{F} \phi \rangle$ (since T is even) $= \langle \mathcal{F} T, \phi \rangle$ $\forall \phi \in S(\mathbb{R}) \Longrightarrow \bar{\mathcal{F}} T = \mathcal{F} T$ in $S'(\mathbb{R})$.

Similarly, for *odd* distributions $T \in S'(\mathbb{R})$, $\langle \bar{\mathcal{F}} T, \phi \rangle = \langle T, (\mathcal{F} \phi)^{\vee} \rangle = -\langle T, \mathcal{F} \phi \rangle$ (since T odd) $= -\langle \mathcal{F} T, \phi \rangle \Longrightarrow \bar{\mathcal{F}} T = -\mathcal{F} T$ in $S'(\mathbb{R})$. □

$$\begin{aligned}
&T \in S'(\mathbb{R}) \text{ is } \textit{even} \Longrightarrow \mathcal{F} T \text{ is } \textit{even} \text{ (see also Property 4(a) in (7.1.19));} \\
&T \in S'(\mathbb{R}) \text{ is } \textit{odd} \Longrightarrow \mathcal{F} T \text{ is } \textit{odd}.
\end{aligned} \tag{8.3.26}$$

Proof. From (8.3.25), ∀ *even* $T \in S'(\mathbb{R})$, $\bar{\mathcal{F}} T = \mathcal{F} T$. But $\langle \mathcal{F} T, \check{\phi} \rangle = \langle T, \mathcal{F} \check{\phi} \rangle = \langle T, (\mathcal{F} \phi)^{\vee} \rangle$, since

$$(\mathcal{F} \check{\phi})(\xi) = \int_{\mathbb{R}} \check{\phi}(x) e^{-i2\pi x\xi} dx = \int_{-\infty}^{\infty} \phi(-x) e^{-i2\pi(-x)(-\xi)} dx$$
$$= \int_{-\infty}^{\infty} \phi(y) e^{-i2\pi y(-\xi)} dy = (\mathcal{F} \phi)(-\xi) = (\mathcal{F} \phi)^{\vee}(\xi)$$

$\Longrightarrow \mathcal{F} \check{\phi} = (\mathcal{F} \phi)^{\vee}$ $\forall \phi \in S(\mathbb{R})$. Hence, $\langle \mathcal{F} T, \check{\phi} \rangle = \langle T, (\mathcal{F} \phi)^{\vee} \rangle = \langle T, \mathcal{F} \phi \rangle$ (T is even) $= \langle \mathcal{F} T, \phi \rangle$ $\forall \phi \in S(\mathbb{R}) \Longrightarrow \mathcal{F} T$ is even.

For *odd* $T \in S'(\mathbb{R})$, $\langle \mathcal{F} T, \check{\phi} \rangle = \langle T, \mathcal{F} \check{\phi} \rangle = \langle T, (\mathcal{F} \phi)^{\vee} \rangle = -\langle T, \mathcal{F} \phi \rangle$ (since T is odd) $= -\langle \mathcal{F} T, \phi \rangle$ $\forall \phi \in S(\mathbb{R}) \Longrightarrow \langle \mathcal{F} T, \check{\phi} \rangle = -\langle \mathcal{F} T, \phi \rangle$ $\forall \phi \in S(\mathbb{R}) \Longrightarrow \mathcal{F} T$ is odd. □

Fourier transform of homogeneous tempered distributions

A tempered distribution $T \in S'(\mathbb{R}^n)$ is called *homogeneous of degree* $d \in \mathbb{R}$ if and only if, $\forall \lambda > 0$,

$$\langle T, \phi_\lambda \rangle = \lambda^{-(n+d)} \langle T, \phi \rangle \quad \forall \phi \in S(\mathbb{R}^n), \tag{8.3.27}$$

where $\phi_\lambda(\mathbf{x}) = \phi(\lambda \mathbf{x})$ $\forall \mathbf{x} \in \mathbb{R}^n$ (see (1.10.41) and (1.10.39) for the definition of homogeneous distributions of $\mathcal{D}'(\mathbb{R}^n)$).

Proposition 8.3.1. *Let $T \in S'(\mathbb{R}^n)$ be a homogeneous tempered distribution of degree $d \in \mathbb{R}$. Then its Fourier transform $\mathcal{F} T \in S'(\mathbb{R}^n)$ is a homogeneous tempered distribution of degree $-n-d$.*

Proof. $T \in S'(\mathbb{R}^n)$ is homogeneous of degree $d \in \mathbb{R} \Longrightarrow \langle T, \phi_\lambda \rangle = \lambda^{-(n+d)} \langle T, \phi \rangle$ $\forall \lambda > 0$, $\forall \phi \in S(\mathbb{R}^n)$, $\langle \mathcal{F} T, \phi_\lambda \rangle = \langle T, \mathcal{F} \phi_\lambda \rangle$ $\forall \phi \in S(\mathbb{R}^n)$, where

$$\mathcal{F} \phi_\lambda = \int_{\mathbb{R}^n} \phi_\lambda(\mathbf{x}) e^{-i2\pi \langle \mathbf{x}, \boldsymbol{\xi} \rangle} d\mathbf{x} = \int_{\mathbb{R}^n} \phi(\lambda \mathbf{x}) e^{-i2\pi \langle \mathbf{x}, \boldsymbol{\xi} \rangle} d\mathbf{x}$$
$$= \int_{\mathbb{R}^n} \phi(\mathbf{y}) e^{-i2\pi \langle \mathbf{y}, \frac{\boldsymbol{\xi}}{\lambda} \rangle} \frac{1}{\lambda^n} d\mathbf{y} \quad \text{(by change of variables: } \mathbf{y} = \lambda \mathbf{x} \text{ with } |J| = \lambda^{-n})$$
$$= \frac{1}{\lambda^n} (\mathcal{F} \phi)\left(\frac{\boldsymbol{\xi}}{\lambda}\right) = \frac{1}{\lambda^n} (\mathcal{F} \phi)_{1/\lambda}(\boldsymbol{\xi}).$$

Hence, $\langle \mathcal{F} T, \phi_\lambda \rangle = \langle T, \frac{1}{\lambda^n} (\mathcal{F} \phi)_{1/\lambda} \rangle = \lambda^{-n} \langle T, (\mathcal{F} \phi)_{1/\lambda} \rangle$.
But $\langle T, (\mathcal{F} \phi)_{1/\lambda} \rangle = (\frac{1}{\lambda})^{-(n+d)} \langle T, \mathcal{F} \phi \rangle$ (since T is homogeneous)
$\Longrightarrow \langle \mathcal{F} T, \phi_\lambda \rangle = \lambda^{-n} \cdot \lambda^{(n+d)} \langle T, \mathcal{F} \phi \rangle = \lambda^d \langle T, \mathcal{F} \phi \rangle$ $\forall \phi \in S(\mathbb{R}^n)$
$\Longrightarrow \mathcal{F} T$ is homogeneous of degree p with $-(n+p) = d \Longrightarrow p = -n-d$. □

Example 8.3.2. Show that $\frac{1}{z}$ with $z = x + iy$ defines a tempered distribution $T \in S'(\mathbb{R}^2)$ such that T and its Fourier transform $\mathcal{F}T \in S'(\mathbb{R}^2)$ are homogeneous of degree -1.

Proof. $\frac{1}{|z|} = \frac{1}{(x^2+y^2)^{1/2}}$ is integrable in the neighbourhood of the origin, since $1/r$ with $r = (x^2 + y^2)^{1/2}$ is integrable in the neighbourhood of the origin. Hence, $\frac{1}{|z|} = \frac{1}{r} \in L^1_{\text{loc}}(\mathbb{R}^2)$. Moreover, $\frac{1}{|z|}$ is bounded for $|z| \geq 1$. Hence, $T = \frac{1}{|z|} \in S'(\mathbb{R}^2)$.

$$\begin{aligned}\langle T, \phi_\lambda\rangle &= \int_{\mathbb{R}^2} \frac{1}{(x+iy)}\phi(\lambda x, \lambda y)dxdy = \int_{\mathbb{R}^2} \frac{\lambda}{\xi + i\eta}\phi(\xi,\eta)\frac{1}{\lambda^2}d\xi d\eta \\ &= \frac{1}{\lambda}\int_{\mathbb{R}^2} \frac{1}{\xi + i\eta}\phi(\xi,\eta)d\xi d\eta = \lambda^{-1}\langle T, \phi\rangle = \lambda^{-(2-1)}\langle T, \phi\rangle,\end{aligned}$$

which is obtained by the change of variables $\xi = \lambda x$, $\eta = \lambda y$ with $|J| = 1/\lambda^2$,

$\Longrightarrow T = \frac{1}{z} \in S'(\mathbb{R}^2)$ with $z = x + iy$ is homogeneous of degree $d = -1$ by definition in (8.3.27).

$\mathcal{F}T = \mathcal{F}(\frac{1}{z}) \in S'(\mathbb{R}^2)$ is also a homogeneous tempered distribution of order $p = -n - d = -2 - (-1) = -1$ by Proposition 8.3.1. □

Term-by-term differentiation of series of tempered distributions

For series of tempered distributions, we have a theorem similar to Theorem 2.11.1:

Theorem 8.3.3. *Let $\sum_{k=1}^{\infty} T_k$ be a series of tempered distributions $T_k \in S'(\mathbb{R}^n) \forall k \in \mathbb{N}$ with its sum $T = \sum_{k=1}^{\infty} T_k$ in $S'(\mathbb{R}^n)$. Then, $\forall$ multi-index $\boldsymbol{\alpha}$, $\partial^{\boldsymbol{\alpha}} T = \sum_{k=1}^{\infty} \partial^{\boldsymbol{\alpha}} T_k$ in $S'(\mathbb{R}^n)$.*

Proof. The proof is similar to that of Theorem 2.11.1, with '$\phi \in \mathcal{D}(\Omega)$' replaced by '$\phi \in S(\mathbb{R}^n)$' and '$\mathcal{D}'(\Omega)$' by '$S'(\mathbb{R}^n)$'. □

Fourier transform of a convergent series of tempered distributions

Theorem 8.3.4. *Let $\sum_{k=1}^{\infty} T_k$ be a series of tempered distributions $T_k \in S'(\mathbb{R}^n)$ $\forall k \in \mathbb{N}$ with $T = \sum_{k=1}^{\infty} T_k$ in $S'(\mathbb{R}^n)$. Then $\mathcal{F}T = \sum_{k=1}^{\infty} \mathcal{F}T_k$ in $S'(\mathbb{R}^n)$.*

Proof. Set $S_N = \sum_{k=1}^{N} T_k$ such that $S_N \to T$ in $S'(\mathbb{R}^n)$ as $N \to \infty$. But $\mathcal{F} : S'(\mathbb{R}^n) \longrightarrow S'(\mathbb{R}^n)$ is *continuous*, i.e. $S_N \to T$ in $S'(\mathbb{R}^n) \Longrightarrow \mathcal{F}S_N = \mathcal{F}[\sum_{k=1}^{N} T_k] = \sum_{k=1}^{N} \mathcal{F}T_k \to \mathcal{F}T$ in $S'(\mathbb{R}^n)$ as $N \to \infty$, from which the result follows. □

8.4 Fourier transform of distributions with compact support

Let $T \in \mathcal{E}'(\mathbb{R}^n) \subset S'(\mathbb{R}^n) \subset \mathcal{D}'(\mathbb{R}^n)$ be a distribution with compact support. Then, for $T = T_{\mathbf{x}} = T(\mathbf{x}) \in \mathcal{E}'(\mathbb{R}^n)$ and $\forall$ fixed $\boldsymbol{\xi} \in \mathbb{R}^n$, $e^{-i2\pi\langle \mathbf{x},\boldsymbol{\xi}\rangle} \in C^\infty(\mathbb{R}^n) = \mathcal{E}(\mathbb{R}^n)$ is a function of $\mathbf{x}$ and $\langle T_{\mathbf{x}}, e^{-i2\pi\langle \mathbf{x},\boldsymbol{\xi}\rangle}\rangle_{\mathcal{E}'(\mathbb{R}^n)\times\mathcal{E}(\mathbb{R}^n)}$ is well defined as a function of $\boldsymbol{\xi} \in \mathbb{R}^n$. Hence, we set

$$\widehat{T}(\boldsymbol{\xi}) = \langle T_{\mathbf{x}}, e^{-i2\pi\langle \mathbf{x},\boldsymbol{\xi}\rangle}\rangle_{\mathcal{E}'(\mathbb{R}^n)\times\mathcal{E}(\mathbb{R}^n)}. \tag{8.4.1}$$

But for fixed $\mathbf{x} \in \mathbb{R}^n$, $e^{-i2\pi\langle \mathbf{x},\boldsymbol{\xi}\rangle} \in C^\infty(\mathbb{R}^n)$ as a function of $\boldsymbol{\xi}$

$$\Longrightarrow \quad \widehat{T}(\boldsymbol{\xi}) \in C^\infty(\mathbb{R}^n) = \mathcal{E}(\mathbb{R}^n) \text{ as a function of } \boldsymbol{\xi} \tag{8.4.2}$$

$\Longrightarrow \forall$ multi-index $\boldsymbol{\alpha} = (\alpha_1, \alpha_2, \ldots, \alpha_n)$ with $\partial_{\boldsymbol{\xi}}^{\boldsymbol{\alpha}} = \frac{\partial^{|\boldsymbol{\alpha}|}}{\partial\xi_1^{\alpha_1}\partial\xi_2^{\alpha_2}\ldots\partial\xi_n^{\alpha_n}}$,

$$\begin{aligned}\partial_{\boldsymbol{\xi}}^{\boldsymbol{\alpha}}\widehat{T}(\boldsymbol{\xi}) &= \partial_{\boldsymbol{\xi}}^{\boldsymbol{\alpha}}\langle T_{\mathbf{x}}, e^{-i2\pi\langle \mathbf{x},\boldsymbol{\xi}\rangle}\rangle = \langle T_{\mathbf{x}}, \partial_{\boldsymbol{\xi}}^{\boldsymbol{\alpha}}(e^{-i2\pi\langle \mathbf{x},\boldsymbol{\xi}\rangle})\rangle = \langle T_{\mathbf{x}}, (-i2\pi\mathbf{x})^{\boldsymbol{\alpha}} e^{-i2\pi\langle \mathbf{x},\boldsymbol{\xi}\rangle}\rangle \\ &= (-i2\pi\mathbf{x})^{\boldsymbol{\alpha}}\langle T_{\mathbf{x}}, e^{-i2\pi\langle \mathbf{x},\boldsymbol{\xi}\rangle}\rangle = (-i2\pi\mathbf{x})^{\boldsymbol{\alpha}}\widehat{T}(\boldsymbol{\xi}),\end{aligned} \tag{8.4.3}$$

where $(-i2\pi\mathbf{x})^{\boldsymbol{\alpha}} = (-i2\pi x_1)^{\alpha_1}(-i2\pi x_2)^{\alpha_2}\ldots(-i2\pi x_n)^{\alpha_n}$, since $\frac{\partial^{|\boldsymbol{\alpha}|}}{\partial\xi_1^{\alpha_1}\ldots\partial\xi_n^{\alpha_n}}[e^{-i2\pi(x_1\xi_1+x_2\xi_2+\cdots+x_n\xi_n)}] = (-i2\pi x_1)^{\alpha_1}\ldots(-i2\pi x_n)^{\alpha_n}$.
$\widehat{T} = \mathcal{F}T$, i.e. the *Fourier transform* of $T \in \mathcal{E}'(\mathbb{R}^n)$:

Since $T \in \mathcal{E}'(\mathbb{R}^n) \Longrightarrow T \in S'(\mathbb{R}^n)$, $\mathcal{F}T \in S'(\mathbb{R}^n)$ is defined, $\forall \phi \in S(\mathbb{R}^n)$, by:

$$\begin{aligned}\langle \mathcal{F}T, \phi\rangle &= \langle T, \mathcal{F}\phi\rangle = \langle T_{\mathbf{x}}, (\mathcal{F}\phi)(\mathbf{x})\rangle = \left\langle T_{\mathbf{x}}, \int_{\mathbb{R}^n}\phi(\boldsymbol{\xi})e^{-i2\pi\langle\boldsymbol{\xi},\mathbf{x}\rangle}d\boldsymbol{\xi}\right\rangle \\ &= \left\langle T_{\mathbf{x}}, \int_{\mathbb{R}^n}\psi(\mathbf{x},\boldsymbol{\xi})d\boldsymbol{\xi}\right\rangle \quad \text{with } \psi(\mathbf{x},\boldsymbol{\xi}) = \phi(\boldsymbol{\xi})e^{-i2\pi\langle\boldsymbol{\xi},\mathbf{x}\rangle}.\end{aligned} \tag{8.4.4}$$

Since $T = T_{\mathbf{x}}$ is independent of $\boldsymbol{\xi}$, T can be written in the tensor product form: $T = T_{\mathbf{x}} \otimes 1_{\boldsymbol{\xi}}$ with $1_{\boldsymbol{\xi}}(\boldsymbol{\xi}) = 1 \forall \boldsymbol{\xi} \in \mathbb{R}^n$, and

$$\langle T_{\mathbf{x}} \otimes 1_{\boldsymbol{\xi}}, \psi(\mathbf{x},\boldsymbol{\xi})\rangle = \langle T_{\mathbf{x}}, \langle 1_{\boldsymbol{\xi}}, \psi(\mathbf{x},\boldsymbol{\xi})\rangle\rangle = \langle 1_{\boldsymbol{\xi}}, \langle T_{\mathbf{x}}, \psi(\mathbf{x},\boldsymbol{\xi})\rangle\rangle, \tag{8.4.5}$$

which follows from the definition of the tensor product $T_{\mathbf{x}} \otimes 1_{\boldsymbol{\xi}}$ of distributions in (6.1.19). But

$$\langle T_{\mathbf{x}}, \langle 1_{\boldsymbol{\xi}}, \psi(\mathbf{x},\boldsymbol{\xi})\rangle\rangle = \left\langle T_{\mathbf{x}}, \int_{\mathbb{R}^n} 1_{\boldsymbol{\xi}}(\boldsymbol{\xi})\psi(\mathbf{x},\boldsymbol{\xi})d\boldsymbol{\xi}\right\rangle = \left\langle T_{\mathbf{x}}, \int_{\mathbb{R}^n}\psi(\mathbf{x},\boldsymbol{\xi})d\boldsymbol{\xi}\right\rangle = \langle \mathcal{F}T, \phi\rangle,$$

which is obtained from (8.4.4). Again,

$$\begin{aligned}\langle 1_{\xi}, \langle T_{\mathbf{x}}, \psi(\mathbf{x}, \boldsymbol{\xi})\rangle\rangle &= \int_{\mathbb{R}^n} 1_{\xi}(\boldsymbol{\xi})\langle T_{\mathbf{x}}, \phi(\boldsymbol{\xi})e^{-i2\pi\langle\boldsymbol{\xi},\mathbf{x}\rangle}\rangle d\boldsymbol{\xi} \\ &= \int_{\mathbb{R}^n} \langle T_{\mathbf{x}}, e^{-i2\pi\langle\boldsymbol{\xi},\mathbf{x}\rangle}\rangle\phi(\boldsymbol{\xi})d\boldsymbol{\xi} = \int_{\mathbb{R}^n} \widehat{T}(\boldsymbol{\xi})\phi(\boldsymbol{\xi})d\boldsymbol{\xi} \\ &= \langle\widehat{T}, \phi\rangle \quad \text{(using (8.4.1)).} \end{aligned} \tag{8.4.6}$$

Then, from (8.4.4)–(8.4.6), we have $\langle\mathcal{F}T, \phi\rangle = \langle\widehat{T}, \phi\rangle \ \forall\phi \in S(\mathbb{R}^n) \Longrightarrow \mathcal{F}T = \widehat{T}$.

Now, we summarize the results:

Proposition 8.4.1. *For distributions $T \in \mathcal{E}'(\mathbb{R}^n)$ with compact support, the Fourier transform $\mathcal{F}T \in C^\infty(\mathbb{R}^n)$ and co-transform $\overline{\mathcal{F}}T \in C^\infty(\mathbb{R}^n)$ are defined by*

$$(\mathcal{F}T)(\boldsymbol{\xi}) = \widehat{T}(\boldsymbol{\xi}) = \langle T_{\mathbf{x}}, e^{-i2\pi\langle\mathbf{x},\boldsymbol{\xi}\rangle}\rangle; \tag{8.4.7}$$

$$(\overline{\mathcal{F}}T)(\boldsymbol{\xi}) = \langle T_{\mathbf{x}}, e^{i2\pi\langle\mathbf{x},\boldsymbol{\xi}\rangle}\rangle. \tag{8.4.8}$$

Fourier transforms of derivatives of distributions $T \in \mathcal{E}'(\mathbb{R}^n)$ with compact support and derivatives of their Fourier transforms

$$\left.\begin{aligned} \mathcal{F}[\partial_{\mathbf{x}}^{\boldsymbol{\alpha}}T] &= (i2\pi\boldsymbol{\xi})^{\boldsymbol{\alpha}}(\mathcal{F}T)(\boldsymbol{\xi}); \\ \overline{\mathcal{F}}[\partial_{\mathbf{x}}^{\boldsymbol{\alpha}}T] &= (-i2\pi\boldsymbol{\xi})^{\boldsymbol{\alpha}}(\overline{\mathcal{F}}T)(\boldsymbol{\xi}); \\ \partial_{\boldsymbol{\xi}}^{\boldsymbol{\alpha}}(\mathcal{F}T)(\boldsymbol{\xi}) &= \mathcal{F}[(-i2\pi\mathbf{x})^{\boldsymbol{\alpha}}T]; \\ \partial_{\boldsymbol{\xi}}^{\boldsymbol{\alpha}}(\overline{\mathcal{F}}T)(\boldsymbol{\xi}) &= \overline{\mathcal{F}}[(i2\pi\mathbf{x})^{\boldsymbol{\alpha}}T]. \end{aligned}\right\} \tag{8.4.9}$$

Proof.

$$\begin{aligned}\mathcal{F}[\partial_{\mathbf{x}}^{\boldsymbol{\alpha}}T](\boldsymbol{\xi}) &= \langle\partial_{\mathbf{x}}^{\boldsymbol{\alpha}}T, e^{-i2\pi\langle\mathbf{x},\boldsymbol{\xi}\rangle}\rangle = (-1)^{|\boldsymbol{\alpha}|}\langle T, \partial_{\mathbf{x}}^{\boldsymbol{\alpha}}(e^{-i2\pi\langle\mathbf{x},\boldsymbol{\xi}\rangle})\rangle \\ &= (-1)^{|\boldsymbol{\alpha}|}\langle T, (-i2\pi\boldsymbol{\xi})^{\boldsymbol{\alpha}}e^{-i2\pi\langle\mathbf{x},\boldsymbol{\xi}\rangle}\rangle = (-1)^{|\boldsymbol{\alpha}|}(-i2\pi\xi)^{\boldsymbol{\alpha}}\langle T, e^{-i2\pi\langle\mathbf{x},\boldsymbol{\xi}\rangle}\rangle \\ &= (i2\pi\boldsymbol{\xi})^{\boldsymbol{\alpha}}(\mathcal{F}T)(\boldsymbol{\xi}),\end{aligned}$$

since $\frac{\partial^{\alpha_i}}{\partial {x_i}^{\alpha_i}} e^{-i2\pi(x_1\xi_1+x_2\xi_2+\cdots+x_n\xi_n)} = (-i2\pi\xi_i)^{\alpha_i} e^{-i2\pi(x_1\xi_1+\cdots+x_n\xi_n)}$ and $(i2\pi\boldsymbol{\xi})^{\boldsymbol{\alpha}} = (i2\pi\xi_1)^{\alpha_1}(i2\pi\xi_2)^{\alpha_2}\dots(i2\pi\xi_n)^{\alpha_n}$.

$$\begin{aligned}\overline{\mathcal{F}}[\partial_{\mathbf{x}}^{\boldsymbol{\alpha}}T](\boldsymbol{\xi}) &= \langle\partial_{\mathbf{x}}^{\boldsymbol{\alpha}}T, e^{i2\pi\langle\mathbf{x},\boldsymbol{\xi}\rangle}\rangle = (-1)^{|\boldsymbol{\alpha}|}\langle T, \partial_{\mathbf{x}}^{\boldsymbol{\alpha}}(e^{i2\pi\langle\mathbf{x},\boldsymbol{\xi}\rangle})\rangle \\ &= (-1)^{|\boldsymbol{\alpha}|}\langle T, (i2\pi\boldsymbol{\xi})^{\boldsymbol{\alpha}}e^{i2\pi\langle\mathbf{x},\boldsymbol{\xi}\rangle}\rangle = (-i2\pi\boldsymbol{\xi})^{\boldsymbol{\alpha}}\overline{\mathcal{F}}T(\boldsymbol{\xi}). \\ \partial_{\boldsymbol{\xi}}^{\boldsymbol{\alpha}}(\mathcal{F}T)(\boldsymbol{\xi}) &= \partial_{\boldsymbol{\xi}}^{\boldsymbol{\alpha}}\langle T_{\mathbf{x}}, e^{-i2\pi\langle\mathbf{x},\boldsymbol{\xi}\rangle}\rangle = \langle T_{\mathbf{x}}, \partial_{\boldsymbol{\xi}}^{\boldsymbol{\alpha}}(e^{-i2\pi\langle\mathbf{x},\boldsymbol{\xi}\rangle})\rangle = \langle T_{\mathbf{x}}, (-i2\pi\mathbf{x})^{\boldsymbol{\alpha}}e^{-i2\pi\langle\mathbf{x},\boldsymbol{\xi}\rangle}\rangle \\ &= \langle(-i2\pi\mathbf{x})^{\boldsymbol{\alpha}}T_{\mathbf{x}}, e^{-i2\pi\langle\mathbf{x},\boldsymbol{\xi}\rangle}\rangle = \mathcal{F}[(-i2\pi\mathbf{x})^{\boldsymbol{\alpha}}T],\end{aligned}$$

since $T_{\mathbf{x}} = T \in \mathcal{E}'(\mathbb{R}^n)$, the product $(-i2\pi\mathbf{x})^{\boldsymbol{\alpha}} T \in \mathcal{E}'(\mathbb{R}^n)$ is a distribution with compact support $\forall$ multi-index $\boldsymbol{\alpha}$ and $(-i2\pi\mathbf{x})^{\alpha} \in C^{\infty}(\mathbb{R}^n)$.

$$\partial_{\boldsymbol{\xi}}^{\boldsymbol{\alpha}}(\overline{\mathcal{F}}T)(\boldsymbol{\xi}) = \partial_{\boldsymbol{\xi}}^{\boldsymbol{\alpha}}\langle T_{\mathbf{x}}, e^{i2\pi\langle\mathbf{x},\boldsymbol{\xi}\rangle}\rangle = \langle T_{\mathbf{x}}, \partial_{\boldsymbol{\xi}}^{\boldsymbol{\alpha}}(e^{i2\pi\langle\mathbf{x},\boldsymbol{\xi}\rangle})\rangle$$
$$= \langle (i2\pi\mathbf{x})^{\boldsymbol{\alpha}} T_{\mathbf{x}}, e^{i2\pi\langle\mathbf{x},\boldsymbol{\xi}\rangle}\rangle = \overline{\mathcal{F}}((i2\pi\mathbf{x})^{\boldsymbol{\alpha}} T). \qquad \square$$

Theorem 8.4.1 (Paley–Wiener–Schwartz). *Let $T \in \mathcal{E}'(\mathbb{R}^n)$ be a distribution with compact support in $\mathbb{R}^n$. Then its Fourier transform $\mathcal{F}T$ can be extended to an entire function F in $\mathbb{C}^n$ given by $F(\mathbf{z}) = \langle T_{\mathbf{x}}, e^{-i2\pi\langle\mathbf{x},\mathbf{z}\rangle}\rangle$, such that $F\downarrow_{\mathbb{R}^n}(\boldsymbol{\xi}) = \langle T_{\mathbf{x}}, e^{-i2\pi\langle\mathbf{x},\boldsymbol{\xi}\rangle}\rangle = (\mathcal{F}T)(\boldsymbol{\xi})\ \forall\boldsymbol{\xi} \in \mathbb{R}^n$. Moreover, $\exists$ constants $C, M > 0$ and an integer $n_0 \in \mathbb{N}$ such that, $\forall\mathbf{z} \in \mathbb{C}^n$, $|F(\mathbf{z})| \le C(1 + \|\mathbf{z}\|)^{n_0} e^{M\|\operatorname{Im}(\mathbf{z})\|}$. Conversely, $\forall$ entire functions F satisfying this inequality in $\mathbb{C}^n$, $\exists T \in \mathcal{E}'(\mathbb{R}^n)$ such that $(\mathcal{F}T)(\mathbf{z}) = F(\mathbf{z})\ \forall\mathbf{z} \in \mathbb{C}^n$.*

Example 8.4.1. Let $T \in \mathcal{E}'(\mathbb{R}^n)$ be a distribution with compact support in $\mathbb{R}^n$ such that $\langle T, \mathbf{x}^{\boldsymbol{\alpha}}\rangle = 0\ \forall$ multi-index $\boldsymbol{\alpha} = (\alpha_1, \alpha_2, \cdots, \alpha_n)$ with $\mathbf{x}^{\boldsymbol{\alpha}} = x_1{}^{\alpha_1} x_2{}^{\alpha_2} \cdots x_n{}^{\alpha_n}$, $|\boldsymbol{\alpha}| = \alpha_1 + \alpha_2 + \cdots + \alpha_n$. Then, using the Paley–Wiener–Schwartz Theorem 8.4.1, prove that $T = 0$ in $\mathcal{E}'(\mathbb{R}^n)$.

Proof. Since $T \in \mathcal{E}'(\mathbb{R}^n)$, by the Paley–Wiener–Schwartz Theorem 8.4.1, $\hat{T} = \mathcal{F}T$ can be extended to an entire function F on $\mathbb{C}^n$ defined by:

$$F(\mathbf{z}) = \langle T, e^{-i2\pi\langle\mathbf{x},\mathbf{z}\rangle}\rangle \quad \forall\mathbf{z} = z_1, z_2, \ldots, z_n \in \mathbb{C}^n.$$

By defining $\frac{\partial}{\partial z_k} = \frac{1}{2}(\frac{\partial}{\partial\xi_k} - i\frac{\partial}{\partial\eta_k})$ with $z_k = \xi_k + i\eta_k$, $1 \le k \le n$, the differentiation is reduced to that with respect to real variables ξ_k, η_k such that $\frac{\partial F}{\partial z_k}(\mathbf{z}) = \langle T, \frac{\partial}{\partial z_k} e^{-i2\pi\langle\mathbf{x},\mathbf{z}\rangle}\rangle$. Hence, $\forall$ multi-index $\boldsymbol{\alpha}$,

$$\partial_{\mathbf{z}}^{\boldsymbol{\alpha}} F(\mathbf{z}) = \langle T, \partial_{\mathbf{z}}^{\boldsymbol{\alpha}}(e^{-i2\pi\langle\mathbf{x},\mathbf{z}\rangle})\rangle = (-i2\pi)^{|\boldsymbol{\alpha}|}\langle T, \mathbf{x}^{\boldsymbol{\alpha}} e^{-i2\pi\langle\mathbf{x},\mathbf{z}\rangle}\rangle$$

$\Longrightarrow \forall\boldsymbol{\alpha}, \partial_{\mathbf{z}}^{\boldsymbol{\alpha}} F(\mathbf{0}) = (-i2\pi)^{|\boldsymbol{\alpha}|}\langle T, \mathbf{x}^{\boldsymbol{\alpha}} \cdot e^0\rangle = (i2\pi)^{|\boldsymbol{\alpha}|}\langle T, \mathbf{x}^{\boldsymbol{\alpha}}\rangle = 0$.

But F is an entire function $\Longrightarrow F(\mathbf{z}) = \sum_{\boldsymbol{\alpha}} \frac{\partial_{\mathbf{z}}^{\boldsymbol{\alpha}} F(\mathbf{0})}{\boldsymbol{\alpha}!} \cdot \mathbf{z}^{\boldsymbol{\alpha}}\ \forall\mathbf{z} \in \mathbb{C}^n$, where $\boldsymbol{\alpha}! = \alpha_1!\alpha_2! \cdot \alpha_n!$, $\mathbf{z}^{\boldsymbol{\alpha}} = z_1^{\alpha_1} z_2^{\alpha_2} \cdots z_n^{\alpha_n}$.

$\Longrightarrow F(\mathbf{z}) = 0\ \forall\mathbf{z} \in \mathbb{C}^n$

$\Longrightarrow F\downarrow_{\mathbb{R}^n} = \hat{T} = 0$ in $S'(\mathbb{R}^n) \Longrightarrow T = \overline{\mathcal{F}}\hat{T} = 0$ in $S'(\mathbb{R}^n)$. $\square$

Examples of Fourier transforms of Dirac distributions and their derivatives

Since the Dirac distribution (measure) $\delta = \delta_{\mathbf{x}} = \delta(\mathbf{x}) \in \mathcal{E}'(\mathbb{R}^n)$ with mass/charge/force etc. concentrated at $\mathbf{x} = \mathbf{0}$ has compact support $\{\mathbf{0}\}$, we can apply Proposition 8.4.1 to find its Fourier transform, and also the Fourier transform of its derivatives (see (8.3.11)–(8.3.17) for finding these directly from Definition 8.3.1).

1. $(\mathcal{F}\delta)(\boldsymbol{\xi}) = \widehat{\delta}(\boldsymbol{\xi}) = \langle \delta_{\mathbf{x}}, e^{-i2\pi\langle \mathbf{x},\boldsymbol{\xi}\rangle}\rangle = e^{-i2\pi\langle \mathbf{0},\boldsymbol{\xi}\rangle} = e^0 = 1\ \forall \boldsymbol{\xi} \in \mathbb{R}^n$
   $\Longrightarrow \mathcal{F}\delta = \mathcal{F}[\delta_{\mathbf{x}}] = \mathcal{F}[\delta(\mathbf{x})] = 1$ (see (8.3.11)), where $\delta_{\mathbf{x}}$ (resp. $\delta(\mathbf{x})$) denotes that the Dirac distribution δ with concentration at $\mathbf{x} = \mathbf{0}$ is associated with the variable $\mathbf{x}$.

$$(\overline{\mathcal{F}}\delta)(\boldsymbol{\xi}) = \langle \delta_{\mathbf{x}}, e^{i2\pi\langle \mathbf{x},\boldsymbol{\xi}\rangle}\rangle = e^{i2\pi\langle \mathbf{0},\boldsymbol{\xi}\rangle} = 1 \quad \forall \boldsymbol{\xi} \in \mathbb{R}^n$$

$\Longrightarrow \overline{\mathcal{F}}\delta = \overline{\mathcal{F}}[\delta_{\mathbf{x}}] = \overline{\mathcal{F}}[\delta(\mathbf{x})] = 1$ (see (8.3.12)).

$$\begin{aligned}\mathcal{F}[\partial^{\boldsymbol{\alpha}}_{\mathbf{x}}(\delta)](\boldsymbol{\xi}) &= \langle \partial^{\boldsymbol{\alpha}}_{\mathbf{x}}(\delta), e^{-i2\pi\langle \mathbf{x},\boldsymbol{\xi}\rangle}\rangle = (-1)^{|\boldsymbol{\alpha}|}\langle \delta_{\mathbf{x}}, \partial^{\boldsymbol{\alpha}}_{\mathbf{x}}(e^{-i2\pi\langle \mathbf{x},\boldsymbol{\xi}\rangle})\rangle\\ &= (-1)^{|\boldsymbol{\alpha}|}\langle \delta_{\mathbf{x}}, (-i2\pi\boldsymbol{\xi})^{\boldsymbol{\alpha}} e^{-i2\pi\langle \mathbf{x},\boldsymbol{\xi}\rangle}\rangle\\ &= (-1)^{|\boldsymbol{\alpha}|}(-i2\pi\boldsymbol{\xi})^{\boldsymbol{\alpha}}\langle \delta_{\mathbf{x}}, e^{-i2\pi\langle \mathbf{x},\boldsymbol{\xi}\rangle}\rangle\\ &= (i2\pi\boldsymbol{\xi})^{\boldsymbol{\alpha}}\cdot 1 = (i2\pi\boldsymbol{\xi})^{\boldsymbol{\alpha}} \quad \text{(see (8.3.13))}\end{aligned}$$

$\Longrightarrow \mathcal{F}[\partial^{\boldsymbol{\alpha}}_{\mathbf{x}}(\delta)] = \mathcal{F}[\partial^{\boldsymbol{\alpha}}_{\mathbf{x}}(\delta_{\mathbf{x}})] = \mathcal{F}[\partial^{\boldsymbol{\alpha}}_{\mathbf{x}}(\delta(\mathbf{x}))] = (i2\pi\boldsymbol{\xi})^{\boldsymbol{\alpha}}$.

$\overline{\mathcal{F}}[\partial^{\boldsymbol{\alpha}}_{\mathbf{x}}(\delta)] = \langle \partial^{\boldsymbol{\alpha}}_{\mathbf{x}}(\delta), e^{i2\pi\langle \mathbf{x},\boldsymbol{\xi}\rangle}\rangle = (-1)^{|\boldsymbol{\alpha}|}\langle \delta_{\mathbf{x}}, (i2\pi\boldsymbol{\xi})^{\boldsymbol{\alpha}} e^{i2\pi\langle \mathbf{x},\boldsymbol{\xi}\rangle}\rangle = (-i2\pi\boldsymbol{\xi})^{\boldsymbol{\alpha}}$

$\Longrightarrow \overline{\mathcal{F}}[\partial^{\boldsymbol{\alpha}}_{\mathbf{x}}(\delta)] = \overline{\mathcal{F}}[\partial^{\boldsymbol{\alpha}}_{\mathbf{x}}(\delta(\mathbf{x}))] = (-i2\pi\boldsymbol{\xi})^{\boldsymbol{\alpha}}$ (see (8.3.14)).

$$\begin{aligned}\partial^{\boldsymbol{\alpha}}_{\boldsymbol{\xi}}(\mathcal{F}\delta)(\boldsymbol{\xi}) &= \partial^{\boldsymbol{\alpha}}_{\boldsymbol{\xi}}\langle \delta_{\mathbf{x}}, e^{-i2\pi\langle \mathbf{x},\boldsymbol{\xi}\rangle}\rangle = \langle \delta_{\mathbf{x}}, \partial^{\boldsymbol{\alpha}}_{\boldsymbol{\xi}}(e^{-i2\pi\langle \mathbf{x},\boldsymbol{\xi}\rangle})\rangle\\ &= \langle \delta_{\mathbf{x}}, (-i2\pi\mathbf{x})^{\boldsymbol{\alpha}}\cdot e^{-i2\pi\langle \mathbf{x},\boldsymbol{\xi}\rangle}\rangle = -i2\pi 0\cdot e^{-i2\pi 0} = 0 \quad \forall \boldsymbol{\alpha} \neq \mathbf{0},\end{aligned}$$

which also follows from $\mathcal{F}\delta(\boldsymbol{\xi}) = 1 \Longrightarrow \partial^{\boldsymbol{\alpha}}_{\boldsymbol{\xi}} 1 = 0$. Here, $\delta_{\mathbf{x}}$ denotes that Dirac distribution δ is associated with $\mathbf{x}$, whereas $\mathcal{F}\delta$ is associated with $\boldsymbol{\xi}$.

$$\partial^{\boldsymbol{\alpha}}_{\boldsymbol{\xi}}(\overline{\mathcal{F}}\delta)(\boldsymbol{\xi}) = \partial^{\boldsymbol{\alpha}}_{\boldsymbol{\xi}}\cdot 1 = 0, \quad \text{since } \overline{\mathcal{F}}\delta = 1 \text{ (see (8.3.22))}.$$

2. For Dirac distribution (measure) $\delta_{\mathbf{a}} = \delta(\mathbf{x} - \mathbf{a})$ with mass/charge/force concentrated at $\mathbf{a}$, $\delta_{\mathbf{a}} \in \mathcal{E}'(\mathbb{R}^n)$ with compact support $\{\mathbf{a}\}$, we have

$$\begin{aligned}(\mathcal{F}\delta_{\mathbf{a}})(\boldsymbol{\xi}) = \widehat{\delta}_{\mathbf{a}}(\boldsymbol{\xi}) &= \langle \delta_{\mathbf{a}}, e^{-i2\pi\langle \mathbf{x},\boldsymbol{\xi}\rangle}\rangle = \langle \delta(\mathbf{x} - \mathbf{a}), e^{-i2\pi\langle \mathbf{x},\boldsymbol{\xi}\rangle}\rangle\\ &= e^{-i2\pi\langle \mathbf{a},\boldsymbol{\xi}\rangle} \quad \text{(see (8.3.15))},\end{aligned}$$

from which $(\mathcal{F}\delta)(\boldsymbol{\xi}) = 1$ is obtained for $\mathbf{a} = \mathbf{0}$.
Here, $\delta(\mathbf{x} - \mathbf{a})$ denotes Dirac distribution $\delta_{\mathbf{a}}$ with concentration at $\mathbf{a}$, which is associated with $\mathbf{x}$, whereas $\mathcal{F}\delta_{\mathbf{a}}$ is associated with $\boldsymbol{\xi}$.
$(\overline{\mathcal{F}}\delta_{\mathbf{a}})(\boldsymbol{\xi}) = \langle \delta_{\mathbf{a}}, e^{i2\pi\langle \mathbf{x},\boldsymbol{\xi}\rangle}\rangle = e^{i2\pi\langle \mathbf{a},\boldsymbol{\xi}\rangle}$, from which $(\overline{\mathcal{F}}\delta)(\boldsymbol{\xi}) = 1$ is obtained for $\mathbf{a} = \mathbf{0}$ (see (8.3.12)).

$$\begin{aligned}[\partial^{\boldsymbol{\alpha}}_{\mathbf{x}}(\delta_{\mathbf{a}})](\boldsymbol{\xi}) &= \langle \partial^{\boldsymbol{\alpha}}_{\mathbf{x}}(\delta_{\mathbf{a}}), e^{-i2\pi\langle \mathbf{x},\boldsymbol{\xi}\rangle}\rangle = (-1)^{|\boldsymbol{\alpha}|}\langle \delta_{\mathbf{a}}, \partial^{\boldsymbol{\alpha}}_{\mathbf{x}}(e^{-i2\pi\langle \mathbf{x},\boldsymbol{\xi}\rangle})\rangle\\ &= (-1)^{|\boldsymbol{\alpha}|}\langle \delta_{\mathbf{a}}, (-i2\pi\boldsymbol{\xi})^{\boldsymbol{\alpha}} e^{-i2\pi\langle \mathbf{x},\boldsymbol{\xi}\rangle}\rangle = (i2\pi\boldsymbol{\xi})^{\boldsymbol{\alpha}}\langle \delta_{\mathbf{a}}, e^{-i2\pi\langle \mathbf{x},\boldsymbol{\xi}\rangle}\rangle\\ &= (i2\pi\boldsymbol{\xi})^{\boldsymbol{\alpha}} e^{-i2\pi\langle \mathbf{a},\boldsymbol{\xi}\rangle} \quad \text{(see (8.3.14))},\end{aligned}$$

from which $\mathcal{F}[\partial_{\mathbf{x}}^{\boldsymbol{\alpha}}\delta] = (i2\pi\boldsymbol{\xi})^{\boldsymbol{\alpha}}$ is obtained for $\mathbf{a} = \mathbf{0}$.

$$\overline{\mathcal{F}}[\partial_{\mathbf{x}}^{\boldsymbol{\alpha}}(\delta_{\mathbf{a}})](\boldsymbol{\xi}) = \langle \partial_{\mathbf{x}}^{\boldsymbol{\alpha}}(\delta_{\mathbf{a}}), e^{i2\pi\langle \mathbf{x},\boldsymbol{\xi}\rangle}\rangle = (-1)^{|\boldsymbol{\alpha}|}(i2\pi\boldsymbol{\xi})^{\boldsymbol{\alpha}}\langle \delta_{\mathbf{a}}, e^{i2\pi\langle \mathbf{x},\boldsymbol{\xi}\rangle}\rangle$$
$$= (-i2\pi\boldsymbol{\xi})^{\boldsymbol{\alpha}} e^{i2\pi\langle \mathbf{a},\boldsymbol{\xi}\rangle},$$

from which $\overline{\mathcal{F}}[\partial_{\mathbf{x}}^{\boldsymbol{\alpha}}\delta] = (-i2\pi\boldsymbol{\xi})^{\boldsymbol{\alpha}}$ is obtained for $\mathbf{a} = \mathbf{0}$.

$$\partial_{\boldsymbol{\xi}}^{\boldsymbol{\alpha}}[\mathcal{F}\delta](\boldsymbol{\xi}) = \partial_{\boldsymbol{\xi}}^{\boldsymbol{\alpha}} 1 = 0; \quad \partial_{\boldsymbol{\xi}}^{\boldsymbol{\alpha}}[\overline{\mathcal{F}}\delta](\boldsymbol{\xi}) = \partial_{\boldsymbol{\xi}}^{\boldsymbol{\alpha}} 1 = 0.$$

3.
$$\mathcal{F}[e^{i2\pi\langle \mathbf{x},\mathbf{a}\rangle}](\boldsymbol{\xi}) = \mathcal{F}[(\overline{\mathcal{F}}\delta_{\mathbf{a}}(\boldsymbol{\xi}))(\mathbf{x})] = \mathcal{F}\overline{\mathcal{F}}\delta_{\mathbf{a}}(\boldsymbol{\xi}) = \delta_{\mathbf{a}}(\boldsymbol{\xi}). \tag{8.4.10}$$
$$\mathcal{F}[(-i2\pi\mathbf{x})^{\boldsymbol{\alpha}}](\boldsymbol{\xi}) = \mathcal{F}[(\bar{\mathcal{F}}(\partial_{\boldsymbol{\xi}}^{\boldsymbol{\alpha}}\delta(\boldsymbol{\xi})))(\mathbf{x})] = \mathcal{F}\overline{\mathcal{F}}(\partial_{\boldsymbol{\xi}}^{\boldsymbol{\alpha}}\delta(\boldsymbol{\xi})) = \partial_{\boldsymbol{\xi}}^{\boldsymbol{\alpha}}\delta(\boldsymbol{\xi}). \tag{8.4.11}$$

Fourier transform of $\delta_{S_R} \in S'(\mathbb{R}^3)$, the Dirac distribution of charges concentrated on a sphere $S_R \subset \mathbb{R}^3$

Example 8.4.2. Let δ_{S_R} denote a simple layer of charges on a sphere S_R of radius R with centre at the origin. Find the Fourier transform $\mathcal{F}[\delta_{S_R}]$.

Solution. δ_{S_R} is a Dirac distribution with compact support $S_R \subset\subset \mathbb{R}^3$. Hence, $\delta_{S_R} \in S'(\mathbb{R}^3)$ and its Fourier transform is given by $\mathcal{F}[\delta_{S_R}] = \langle \delta_{S_R}, e^{-i2\pi\langle \mathbf{x},\boldsymbol{\xi}\rangle}\rangle = \int_{S_R} e^{-i2\pi\langle \mathbf{x},\boldsymbol{\xi}\rangle} dS$. Choosing spherical coordinates R, θ, ϕ such that the $\boldsymbol{\xi}$ coincides with the axis of the cone $\theta =$ constant in order that $\forall \mathbf{x} \in S_R$, $\langle \mathbf{x}, \boldsymbol{\xi}\rangle = \|\mathbf{x}\|\|\boldsymbol{\xi}\|\cos\theta = R\|\boldsymbol{\xi}\|\cos\theta$, $\|\boldsymbol{\xi}\| = (\xi_1^2 + \xi_2^2 + \xi_3^2)^{1/2}$, we have

$$\begin{aligned}
\mathcal{F}[\delta_{S_R}] &= \int_0^{2\pi}\int_0^{\pi} e^{-i2\pi R\|\boldsymbol{\xi}\|\cos\theta} R^2 \sin\theta d\theta d\phi \\
&= R^2 \int_0^{2\pi} d\phi \int_0^{\pi} \frac{e^{-i2\pi R\|\boldsymbol{\xi}\|\cos\theta}}{i2\pi R\|\boldsymbol{\xi}\|} d(-i2\pi R\|\boldsymbol{\xi}\|\cos\theta) \\
&= 2\pi R^2 \cdot \frac{1}{i2\pi R\|\boldsymbol{\xi}\|} e^{-i2\pi R\|\boldsymbol{\xi}\|} \cos\theta\Big|_{\theta=0}^{\theta=\pi} = \frac{R}{i\|\boldsymbol{\xi}\|}(e^{+i\pi R\|\boldsymbol{\xi}\|} - e^{-i\pi R\|\boldsymbol{\xi}\|}) \\
&= \frac{R}{i\|\boldsymbol{\xi}\|} 2i\sin(2\pi R\|\boldsymbol{\xi}\|) = \frac{2R}{\|\boldsymbol{\xi}\|}\sin(2\pi R\|\boldsymbol{\xi}\|)
\end{aligned}$$

$\Longrightarrow \mathcal{F}[\delta_{S_R}] = \frac{2R}{\|\boldsymbol{\xi}\|}\sin(2\pi R\|\boldsymbol{\xi}\|)$.

Moreover, we have the important results used extensively in electrical engineering.

Example 8.4.3. Show that

1. $\delta(\xi) = 2\int_0^{\infty} \cos 2\pi x\xi dx$;
2. $\frac{\partial\delta}{\partial\xi} = -2\int_0^{\infty} 2\pi x \sin 2\pi x\xi dx.$ (8.4.12)

Solution.

1. $\delta(\xi) = \mathscr{F}[1(x)] = \int_{-\infty}^{\infty} e^{-i2\pi x\xi} dx = 2\int_0^{\infty} \cos 2\pi x\xi dx$.
2. $\frac{\partial\delta}{\partial\xi} = \mathscr{F}(-i2\pi x) = \int_{-\infty}^{\infty}(-i2\pi x)e^{-i2\pi x\xi} dx = -2\int_0^{\infty} 2\pi x \sin 2\pi x\xi dx$.

The integral in (1) (resp. (2)) is the limit in $S'(\mathbb{R})$ [8] of the corresponding integral over the finite interval $[-A, A]$ with $A > 0$, i.e. $\int_{-\infty}^{\infty}(\dots)dx = \lim_{A\to\infty}\int_{-A}^{A}(\dots)dx$ in $S'(\mathbb{R})$ (see also Examples 1.8.1–1.8.4 and Example 2.10.8).

(2) can be obtained directly from (1) by differentiating with respect to ξ under the integral sign $\int$.

8.5 Fourier transform of convolution of distributions

First of all, we summarize the convolution results (see Chapter 6 for details).

Convolution $\phi * \psi \in S(\mathbb{R}^n)$ of $\phi, \psi \in S(\mathbb{R}^n)$

$$\phi, \psi \in S(\mathbb{R}^n) \quad \Longrightarrow \quad (\phi * \psi)(\mathbf{x}) = \int_{\mathbb{R}^n} \phi(\mathbf{t})\psi(\mathbf{x}-\mathbf{t})d\mathbf{t}. \tag{8.5.1}$$

In fact, for $\phi \in S(\mathbb{R}^n)$, the mapping $\psi \in S(\mathbb{R}^n) \longmapsto \phi * \psi \in S(\mathbb{R}^n)$ is *continuous* from $S(\mathbb{R}^n)$ into $S(\mathbb{R}^n)$.

Convolution $\phi * T \in S(\mathbb{R}^n)$ of $\phi \in S(\mathbb{R}^n)$ and distribution $T \in \mathscr{E}'(\mathbb{R}^n)$ with compact support $\forall\phi \in C^{\infty}(\mathbb{R}^n) = \mathscr{E}(\mathbb{R}^n)$ and $\forall$ distributions $T \in \mathscr{E}'(\mathbb{R}^n)$ with compact support, $\phi * T = T * \phi \in C^{\infty}(\mathbb{R}^n)$ is defined by (6.4.1):

$$(\phi * T)(\mathbf{x}) = (T * \phi)(\mathbf{x}) = \langle T(\mathbf{t}), \phi(\mathbf{x}-\mathbf{t})\rangle = \langle \check{T}(\mathbf{t}), \phi(\mathbf{x}+\mathbf{t})\rangle \tag{8.5.2}$$

($\langle \check{T}(\mathbf{t}), \phi(\mathbf{t})\rangle = \langle T(\mathbf{t}), \check{\phi}(\mathbf{t})\rangle = \langle T(\mathbf{t}), \phi(-\mathbf{t})\rangle$, and for fixed $\mathbf{x} \in \mathbb{R}^n$, $\phi_{\mathbf{x}}(\mathbf{t}) = \phi(\mathbf{x}+\mathbf{t}) \Longrightarrow \check{\phi}_{\mathbf{x}}(\mathbf{t}) = \phi_{\mathbf{x}}(-\mathbf{t}) = \phi(\mathbf{x}-\mathbf{t})$ and $\langle \check{T}(\mathbf{t}), \phi(\mathbf{x}+\mathbf{t})\rangle = \langle \check{T}(\mathbf{t}), \phi_{\mathbf{x}}(\mathbf{t})\rangle = \langle T(\mathbf{t}), \check{\phi}_{\mathbf{x}}(\mathbf{t})\rangle = \langle T(\mathbf{t}), \phi(\mathbf{x}-\mathbf{t})\rangle$).

$\phi \in S(\mathbb{R}^n) \subset C^{\infty}(\mathbb{R}^n)$, $T \in \mathscr{E}'(\mathbb{R}^n) \Longrightarrow \phi * T \in S(\mathbb{R}^n)$ defined by:

$$(\phi * T)(\mathbf{x}) = \langle \check{T}(\mathbf{t}), \phi(\mathbf{x}+\mathbf{t})\rangle. \tag{8.5.3}$$

Convolution $\phi * T \in S'(\mathbb{R}^n)$ of $\phi \in S(\mathbb{R}^n)$ and $T \in S'(\mathbb{R}^n)$ $\phi \in S(\mathbb{R}^n)$, $T \in S'(\mathbb{R}^n) \Longrightarrow \phi * T \in S'(\mathbb{R}^n)$ is defined by:

$$\langle \phi * T, \psi\rangle = \langle T, \check{\phi} * \psi\rangle \quad \forall\psi \in S(\mathbb{R}^n). \tag{8.5.4}$$

In fact, $\phi \in S(\mathbb{R}^n) \Longrightarrow \check{\phi} \in S(\mathbb{R}^n) \Longrightarrow \check{\phi} * \psi \in S(\mathbb{R}^n)$ by (8.5.1) $\Longrightarrow$ the right-hand side of (8.5.4) is well defined $\forall T \in S'(\mathbb{R}^n) \Longrightarrow$ the left-hand side of (8.5.4) is well defined as a continuous linear functional on $S(\mathbb{R}^n) \Longrightarrow \phi * T \in S'(\mathbb{R}^n)$.

Convolution $S * T \in \mathcal{E}'(\mathbb{R}^n)$ of distribution $S, T \in \mathcal{E}'(\mathbb{R}^n)$ with compact supports $S, T \in \mathcal{E}'(\mathbb{R}^n)$ are distributions with compact supports $\Longrightarrow S * T \in \mathcal{D}'(\mathbb{R}^n)$ with compact support $\Longrightarrow S * T \in \mathcal{E}'(\mathbb{R}^n) \Longrightarrow S * T \in S'(\mathbb{R}^n)$ defined by:

$$\langle S * T, \phi \rangle = \langle S(\eta) \otimes T(\zeta), \phi(\eta + \zeta) \rangle \quad \forall \phi \in S(\mathbb{R}^n). \tag{8.5.5}$$

Convolution $S * T \in S'(\mathbb{R}^n)$ of tempered distribution $S \in S'(\mathbb{R}^n)$ and distribution $T \in \mathcal{E}'(\mathbb{R}^n)$ with compact support $S \in S'(\mathbb{R}^n)$, $T \in \mathcal{E}'(\mathbb{R}^n) \Longrightarrow S * T \in S'(\mathbb{R}^n)$ defined by

$$\langle T * S, \phi \rangle = \langle S, \check{T} * \phi \rangle \quad \forall \phi \in S(\mathbb{R}^n). \tag{8.5.6}$$

In fact, $T \in \mathcal{E}'(\mathbb{R}^n) \Longrightarrow \check{T} \in \mathcal{E}'(\mathbb{R}^n) \Longrightarrow \check{T} * \phi \in S(\mathbb{R}^n)$ by (8.5.3) and $\phi \in S(\mathbb{R}^n) \longmapsto \check{T} * \phi$ is continuous from $S(\mathbb{R}^n)$ into $S(\mathbb{R}^n) \Longrightarrow$ the right-hand side of (8.5.6) is well defined $\forall S \in S'(\mathbb{R}^n) \Longrightarrow$ the left-hand side of (8.5.6) is also well defined as a continuous linear functional on $S(\mathbb{R}^n)$ and $T * S \in S'(\mathbb{R}^n)$.

8.5.1 Fourier transforms of convolutions

Fourier transform of convolution $\phi * \psi \in S(\mathbb{R}^n)$ of $\phi, \psi \in S(\mathbb{R}^n)$

Theorem 8.5.1. $\forall \phi, \psi \in S(\mathbb{R}^n)$, *the following relations hold in* $S(\mathbb{R}^n)$*:*

I. $\bar{\mathcal{F}}(\phi\psi) = (\bar{\mathcal{F}}\phi) * (\bar{\mathcal{F}}\psi)$; $\mathcal{F}(\phi\psi) = \mathcal{F}\phi * \mathcal{F}\psi$; (8.5.7)

II. $\mathcal{F}(\phi * \psi) = (\mathcal{F}\phi) \cdot (\mathcal{F}\psi)$; $\bar{\mathcal{F}}(\phi * \psi) = (\bar{\mathcal{F}}\phi) \cdot (\bar{\mathcal{F}}\psi)$. (8.5.8)

(the Fourier transform of the convolution $\phi * \psi$ is the product of their Fourier transforms).

Proof.

I. From Riesz's formula in Corollary 7.1.1 we have, $\forall$ fixed $\mathbf{a} \in \mathbb{R}^n$, $\forall f, g \in S(\mathbb{R}^n) \subset L^1(\mathbb{R}^n)$ (changing variables: $\mathbf{y} = \mathbf{x} + \mathbf{a}$ with $|J| = 1$, J being the Jacobian),

$$\begin{aligned}
&\bar{\mathcal{F}}[(\mathcal{F} f)(\boldsymbol{\xi})g(\boldsymbol{\xi})](\mathbf{a}) = \int_{\mathbb{R}^n} (\mathcal{F} f)(\boldsymbol{\xi})g(\boldsymbol{\xi})e^{i2\pi\langle \mathbf{a},\boldsymbol{\xi}\rangle} d\boldsymbol{\xi} \\
&= \int_{\mathbb{R}^n} f(\mathbf{a}+\mathbf{x})(\mathcal{F} g)(\mathbf{x}) d\mathbf{x} \\
&= \int_{\mathbb{R}^n} f(\mathbf{y})(\mathcal{F} g)(\mathbf{y}-\mathbf{a}) d\mathbf{y} = \int_{\mathbb{R}^n} f(\mathbf{y})(\mathcal{F} g)^{\vee}(\mathbf{a}-\mathbf{y}) d\mathbf{y} \\
&= (f * (\mathcal{F} g)^{\vee})(\mathbf{a})) \quad (\text{since } (\mathcal{F} g)^{\vee}(\mathbf{a}-\mathbf{y}) = (\mathcal{F} g)(-(\mathbf{a}-\mathbf{y})) \\
\Longrightarrow \quad & \bar{\mathcal{F}}[(\mathcal{F} f)(\boldsymbol{\xi})g(\boldsymbol{\xi})](\mathbf{a}) = (f * (\mathcal{F} g)^{\vee})(\mathbf{a}) \quad \forall \text{fixed } \mathbf{a} \in \mathbb{R}^n \\
\Longrightarrow \quad & \bar{\mathcal{F}}[\mathcal{F} fg] = f * (\mathcal{F} g)^{\vee} = f * \bar{\mathcal{F}} g,
\end{aligned} \tag{8.5.9}$$

since

$$(\mathcal{F}g)^{\vee}(\mathbf{x}) = (\mathcal{F}g)(-\mathbf{x}) = \int_{\mathbb{R}^n} g(\boldsymbol{\xi})e^{-i2\pi\langle\boldsymbol{\xi},-\mathbf{x}\rangle}d\boldsymbol{\xi}$$
$$= \int_{\mathbb{R}^n} g(\boldsymbol{\xi})e^{i2\pi\langle\boldsymbol{\xi},\mathbf{x}\rangle}d\boldsymbol{\xi} = (\bar{\mathcal{F}}g)(\mathbf{x}) \quad \forall\mathbf{x}\in\mathbb{R}^n.$$

By Theorem 7.7.2, $\mathcal{F}$ is an isomorphism from $S(\mathbb{R}^n)$ onto $S(\mathbb{R}^n)$, and hence, for $\forall f \in S(\mathbb{R}^n)$, $\exists$ a unique $\phi \in S(\mathbb{R}^n)$ such that $\mathcal{F}f = \phi \Longrightarrow \bar{\mathcal{F}}\phi = f$. Set $\psi = g$. Then, from (8.5.9), $\bar{\mathcal{F}}[(\mathcal{F}f)g] = \bar{\mathcal{F}}[\phi\psi] = \bar{\mathcal{F}}\phi * \bar{\mathcal{F}}\psi \in S(\mathbb{R}^n)$.
$\mathcal{F}[\phi\psi] = \mathcal{F}\phi * \mathcal{F}\psi$: For $f, g \in S(\mathbb{R}^n)$, (8.3.2) has the form:

$$\int_{\mathbb{R}^n} (\overline{\mathcal{F}}f)(\boldsymbol{\xi})g(\boldsymbol{\xi})d\boldsymbol{\xi} = \int_{\mathbb{R}^n} f(\mathbf{x})\overline{\mathcal{F}}g(\mathbf{x})d\mathbf{x}.$$

For $\tau_{-\mathbf{a}}f$,

$$\int_{\mathbb{R}^n} \overline{\mathcal{F}}(\tau_{-\mathbf{a}}f)(\boldsymbol{\xi})g(\boldsymbol{\xi})d\boldsymbol{\xi} = \int_{\mathbb{R}^n} (\tau_{-\mathbf{a}}f)(\mathbf{x})(\overline{\mathcal{F}}g)(\mathbf{x})d\mathbf{x}$$
$$= \int_{\mathbb{R}^n} f(\mathbf{x}+\mathbf{a})(\overline{\mathcal{F}}g)(\mathbf{x})d\mathbf{x}$$
$$= \int_{\mathbb{R}^n} f(\mathbf{y})\overline{\mathcal{F}}g(-(\mathbf{a}-\mathbf{y}))d\mathbf{y} = f * (\overline{\mathcal{F}}g)^{\vee}$$

$\Longrightarrow \int_{\mathbb{R}^n} \overline{\mathcal{F}}(\tau_{-a}f)(\boldsymbol{\xi})g(\boldsymbol{\xi})d\boldsymbol{\xi} = f * \mathcal{F}g$, since $\overline{\mathcal{F}}g = (\mathcal{F}g)^{\vee} \Longrightarrow (\overline{\mathcal{F}}g)^{\vee} = (\mathcal{F}g)^{\vee} = \mathcal{F}g$.
But

$$\overline{\mathcal{F}}(\tau_{-\mathbf{a}}f)(\boldsymbol{\xi}) = \int_{\mathbb{R}^n} f(\mathbf{x}+\mathbf{a})e^{i2\pi\langle\mathbf{x},\boldsymbol{\xi}\rangle}d\mathbf{x} = \int_{\mathbb{R}^n} f(\mathbf{y})\cdot e^{i2\pi\langle\mathbf{y}-\mathbf{a},\boldsymbol{\xi}\rangle}d\mathbf{y}$$
$$= e^{-i2\pi\langle\boldsymbol{\xi},\mathbf{a}\rangle}\int_{\mathbb{R}^n} f(\mathbf{y})e^{i2\pi\langle\mathbf{y},\boldsymbol{\xi}\rangle}d\mathbf{y} = e^{-i2\pi\langle\boldsymbol{\xi},\mathbf{a}\rangle}(\overline{\mathcal{F}}f)(\boldsymbol{\xi}).$$

Hence, $\int_{\mathbb{R}^n} \overline{\mathcal{F}}(\tau_{-\mathbf{a}}f)(\boldsymbol{\xi})g(\boldsymbol{\xi})d\boldsymbol{\xi} = \int_{\mathbb{R}^n} (\overline{\mathcal{F}}f)(\boldsymbol{\xi})g(\boldsymbol{\xi})e^{-i2\pi\langle\boldsymbol{\xi},\mathbf{a}\rangle}d\boldsymbol{\xi} = f * \mathcal{F}g$ $\Longrightarrow \mathcal{F}[(\overline{\mathcal{F}}f)(\boldsymbol{\xi})g](\mathbf{a}) = (f * \mathcal{F}g)(\mathbf{a})\ \forall\mathbf{a} \Longrightarrow \mathcal{F}[(\overline{\mathcal{F}}f)(\boldsymbol{\xi})g] = f * \mathcal{F}g$.
Set $\psi = g$ and $\overline{\mathcal{F}}f = \phi$. Then $\mathcal{F}\phi = \mathcal{F}\overline{\mathcal{F}}f = f \Longrightarrow \mathcal{F}[\phi\psi] = \mathcal{F}\phi * \mathcal{F}\psi$.

II. $\phi, \psi \in S(\mathbb{R}^n) \Longrightarrow \phi\psi \in S(\mathbb{R}^n)$ and $\bar{\mathcal{F}}(\phi\psi) = \bar{\mathcal{F}}\phi * \bar{\mathcal{F}}\psi \in S(\mathbb{R}^n)$.
Hence, taking the Fourier transform of both sides, we get $\mathcal{F}\bar{\mathcal{F}}[\phi\psi] = \mathcal{F}[\bar{\mathcal{F}}\phi * \bar{\mathcal{F}}\psi]\ \forall\phi, \psi \in S(\mathbb{R}^n) \Longrightarrow \phi\psi = \mathcal{F}[\bar{\mathcal{F}}\phi * \bar{\mathcal{F}}\psi]$.
Set $\phi_1 = \bar{\mathcal{F}}\phi$ and $\psi_1 = \bar{\mathcal{F}}\psi$. Then $\phi = \mathcal{F}\phi_1$, $\psi = \mathcal{F}\psi_1 \Longrightarrow (\mathcal{F}\phi_1)(\mathcal{F}\psi_1) = \mathcal{F}(\phi_1 * \psi_1)$, ($\phi_1, \psi_1 \in S(\mathbb{R}^n)$ being arbitrary elements of $S(\mathbb{R}^n)$).
$\mathcal{F}(\phi_1\psi_1) = \mathcal{F}\phi_1 * \mathcal{F}\psi_1 \Longrightarrow \bar{\mathcal{F}}\mathcal{F}(\phi_1\psi_1) = \overline{\mathcal{F}}(\mathcal{F}\phi_1 * \mathcal{F}\psi_1)$. Set $\overline{\mathcal{F}}\phi = \phi_1$, $\overline{\mathcal{F}}\psi = \psi_1 \Longrightarrow \mathcal{F}\phi_1 = \phi$, $\mathcal{F}\psi_1 = \psi$. Then $(\overline{\mathcal{F}}\phi)\cdot(\overline{\mathcal{F}}\psi) = \overline{\mathcal{F}}(\phi * \psi)$, since $\overline{\mathcal{F}}\mathcal{F}\phi_1\psi_1 = \phi_1\psi_1$. □

Example 8.5.1. Let f_σ and f_τ be the normal probability distributions of Gauss defined by:

$$f_\sigma(x) = \frac{1}{\sigma\sqrt{2\pi}} e^{-\frac{x^2}{2\sigma^2}}, \quad f_\tau(x) = \frac{1}{\tau\sqrt{2\pi}} e^{-\frac{x^2}{2\tau^2}},$$

σ, τ being the standard deviations of the distributions f_σ and f_τ, respectively. Then $f_\sigma * f_\tau = f_{\sqrt{\sigma^2+\tau^2}}$, i.e.

$$\frac{1}{\sigma\sqrt{2\pi}} e^{-\frac{x^2}{2\sigma^2}} * \frac{1}{\tau\sqrt{2\pi}} e^{-\frac{x^2}{2\tau^2}} = \frac{1}{\sqrt{\sigma^2+\tau^2}\sqrt{2\pi}} e^{-\frac{x^2}{2(\sigma^2+\tau^2)}}.$$

Solution. From Example 7.1.2, $\mathcal{F}[e^{-\pi x^2}](\xi) = e^{-\pi\xi^2}$. Then, using (7.1.19), $\mathcal{F}[f(x)](\xi) = \hat{f}(\xi) \Longrightarrow \mathcal{F}[f(kx)] = \frac{1}{k}\hat{f}(\frac{\xi}{k})$ for real $k \neq 0$, we have:

$$\mathcal{F}[e^{-kx^2}](\xi) = \mathcal{F}[e^{-\pi(\sqrt{\frac{k}{\pi}}x)^2}](\xi) = \frac{1}{\sqrt{\frac{k}{\pi}}} e^{-\pi(\frac{\xi}{\sqrt{\frac{k}{\pi}}})^2} = \sqrt{\frac{\pi}{k}} e^{-\frac{\pi^2}{k}\xi^2}$$

$$\Longrightarrow \quad \mathcal{F}[e^{-\frac{x^2}{2\sigma^2}}](\xi) = \sqrt{\frac{\pi}{\frac{1}{2\sigma^2}}} e^{-\frac{\pi^2\xi^2}{\frac{1}{2\sigma^2}}} = \sigma\sqrt{2\pi} e^{-2\pi^2\sigma^2\xi^2} \quad (k = \tfrac{1}{2\sigma^2})$$

$$\Longrightarrow \quad \mathcal{F}[f_\sigma(x)](\xi) = \frac{1}{\sigma\sqrt{2\pi}} \mathcal{F}[e^{-\frac{x^2}{2\sigma^2}}](\xi) = e^{-2\pi^2\sigma^2\xi^2}.$$

Similarly, $\mathcal{F}[f_\tau(x)] = e^{-2\pi^2\tau^2\xi^2}$.

Hence

$$\begin{aligned}(e^{-2\pi^2\sigma^2\xi^2})(e^{-2\pi^2\tau^2\xi^2}) = e^{-2\pi^2(\sigma^2+\tau^2)\xi^2} &= \mathcal{F}[f_{\sqrt{\sigma^2+\tau^2}}(x)] \\ &= \mathcal{F}\left[\frac{1}{\sqrt{\sigma^2+\tau^2}\sqrt{2\pi}} e^{-\frac{x^2}{2(\sigma^2+\tau^2)}}\right].\end{aligned}$$

Then, since $f_\sigma \in S(\mathbb{R})$, $f_\tau \in S(\mathbb{R})$ and $f_\sigma * f_\tau \in S(\mathbb{R})$ by the Convolution Theorem 8.5.1 on $S(\mathbb{R})$:

$$\begin{aligned}&\mathcal{F}[f_\sigma] \cdot \mathcal{F}[f_\tau] = \mathcal{F}[f_\sigma * f_\tau] = \mathcal{F}[f_{\sqrt{\sigma^2+\tau^2}}] \\ \Longrightarrow \quad &\bar{\mathcal{F}}\mathcal{F}[f_\sigma * f_\tau] = \bar{\mathcal{F}}\mathcal{F}[f_{\sqrt{\sigma^2+\tau^2}}] \\ \Longrightarrow \quad &f_\sigma * f_\tau = f_{\sqrt{\sigma^2+\tau^2}} \quad \text{(by Theorem 8.3.2)}.\end{aligned}$$

Fourier Transform of Convolution $\phi * T \in S'(\mathbb{R}^n)$ with $\phi \in S(\mathbb{R}^n)$, $T \in S'(\mathbb{R}^n)$

Theorem 8.5.2. $\forall\phi \in S(\mathbb{R}^n)$, $\forall T \in S'(\mathbb{R}^n)$, *Fourier transform* $\mathcal{F}$ *and co-transform* $\bar{\mathcal{F}}$ *satisfy the following reciprocal relations in* $S'(\mathbb{R}^n)$*:*

I. $\mathcal{F}(\phi * T) = (\mathcal{F}\phi)(\mathcal{F}T)$; $\bar{\mathcal{F}}(\phi * T) = (\bar{\mathcal{F}}\phi)(\bar{\mathcal{F}}T)$; (8.5.10)

II. $\mathcal{F}(\phi T) = (\mathcal{F}\phi) * (\mathcal{F}T)$; $\bar{\mathcal{F}}(\phi T) = \bar{\mathcal{F}}\phi * \bar{\mathcal{F}}T$. (8.5.11)

Proof.

I. $\phi \in S(\mathbb{R}^n)$, $T \in S'(\mathbb{R}^n) \implies \phi * T \in S'(\mathbb{R}^n)$ and is defined by (8.5.4): $\langle \phi * T, \psi \rangle = \langle T, \check{\phi} * \psi \rangle \ \forall \psi \in S(\mathbb{R}^n)$. Hence, by the definition of Fourier transforms in $S'(\mathbb{R}^n)$, $\forall \psi \in S(\mathbb{R}^n)$,

$$\begin{aligned} \langle \mathcal{F}(\phi * T), \psi \rangle &= \langle \phi * T, \mathcal{F}\psi \rangle = \langle T, \check{\phi} * \mathcal{F}\psi \rangle \\ &= \langle \bar{\mathcal{F}}\mathcal{F}T, \check{\phi} * \mathcal{F}\psi \rangle \quad (\text{since } \bar{\mathcal{F}}\mathcal{F}T = T \text{ by Theorem 8.3.2}) \\ &= \langle \mathcal{F}T, \bar{\mathcal{F}}(\check{\phi} * \mathcal{F}\psi) \rangle \quad (\text{by definition of } \bar{\mathcal{F}} \text{ in (8.3.2)}). \end{aligned} \tag{8.5.11a}$$

But $\check{\phi}, \mathcal{F}\psi \in S(\mathbb{R}^n) \implies \check{\phi} * \mathcal{F}\psi \in S(\mathbb{R}^n)$ by (8.5.1)

$$\begin{aligned} \implies \quad \bar{\mathcal{F}}(\check{\phi} * \mathcal{F}\psi) &= (\bar{\mathcal{F}}\check{\phi})(\bar{\mathcal{F}}\mathcal{F}\psi) \quad (\text{by Theorem 8.5.1}) \\ &= (\bar{\mathcal{F}}\check{\phi})\psi = (\mathcal{F}\phi)\psi, \end{aligned}$$

since

$$\begin{aligned} (\bar{\mathcal{F}}\check{\phi})(\mathbf{x}) &= \int_{\mathbb{R}^n} \check{\phi}(\boldsymbol{\xi}) e^{i2\pi\langle \boldsymbol{\xi}, \mathbf{x} \rangle} d\boldsymbol{\xi} = \int_{\mathbb{R}^n} \phi(-\boldsymbol{\xi}) e^{i2\pi\langle \boldsymbol{\xi}, \mathbf{x} \rangle} d\boldsymbol{\xi} \\ &= \int_{\mathbb{R}^n} \phi(\boldsymbol{\zeta}) e^{-i2\pi\langle \boldsymbol{\zeta}, \mathbf{x} \rangle} d\boldsymbol{\zeta} = \mathcal{F}\phi \end{aligned}$$

(by change of variables, $\boldsymbol{\zeta} = -\boldsymbol{\xi}$).
Then, from (8.5.11a),

$$\begin{aligned} \langle \mathcal{F}(\phi * T), \psi \rangle &= \langle \mathcal{F}T, \bar{\mathcal{F}}(\check{\phi} * \mathcal{F}\psi) \rangle = \langle \mathcal{F}T, (\mathcal{F}\phi)\psi \rangle \\ &= \langle (\mathcal{F}\phi)(\mathcal{F}T), \psi \rangle \quad \forall \psi \in S(\mathbb{R}^n) \end{aligned}$$

$\implies \mathcal{F}(\phi * T) = (\mathcal{F}\phi)(\mathcal{F}T)$ in $S'(\mathbb{R}^n)$.
Replacing $\mathcal{F}$ by $\bar{\mathcal{F}}$ and vice versa, and 'i' by '$-i$' in this proof, we get the proof of $\bar{\mathcal{F}}(\phi * T) = (\bar{\mathcal{F}}\phi) \cdot (\bar{\mathcal{F}}T)$.

II. Set $T_1 = \mathcal{F}T$, $\phi_1 = \mathcal{F}\phi$. Then, applying the Fourier inversion Theorems 7.7.1 and 8.3.2 on $S(\mathbb{R}^n)$ and $S'(\mathbb{R}^n)$, respectively, we have $\bar{\mathcal{F}}T_1 = T$, $\bar{\mathcal{F}}\phi_1 = \phi$. Hence,

$$\begin{aligned} \mathcal{F}(\phi T) &= \mathcal{F}[(\bar{\mathcal{F}}\phi_1) \cdot (\bar{\mathcal{F}}T_1)] = \mathcal{F}[\bar{\mathcal{F}}(\phi_1 * T_1)] \\ &= \mathcal{F}\bar{\mathcal{F}}(\phi_1 * T_1) = \phi_1 * T_1 = \mathcal{F}\phi * \mathcal{F}T. \end{aligned}$$

Again, replacing $\mathcal{F}$ by $\bar{\mathcal{F}}$ and vice versa in this proof, we get $\bar{\mathcal{F}}(\phi T) = \bar{\mathcal{F}}\phi * \bar{\mathcal{F}}T$. □

Fourier transform of the convolution $S * T \in \mathcal{E}'(\mathbb{R}^n)$ of $S, T \in \mathcal{E}'(\mathbb{R}^n)$

$S, T \in \mathcal{E}'(\mathbb{R}^n)$ are distributions with compact support $\Longrightarrow S * T \in \mathcal{E}'(\mathbb{R}^n)$ (i.e. a distribution with compact support) $\Longrightarrow \mathcal{F}(S * T)$ is defined by (8.4.8):

$$\begin{aligned}\mathcal{F}(S * T)(\boldsymbol{\xi}) &= \langle S * T, e^{-i2\pi\langle \mathbf{x},\boldsymbol{\xi}\rangle}\rangle = \langle S(\mathbf{y}) \otimes T(\mathbf{z}), e^{-i2\pi\langle \mathbf{y}+\mathbf{z},\boldsymbol{\xi}\rangle}\rangle \\ &= \langle S(\mathbf{y}), \langle T(\mathbf{z}), e^{-i2\pi\langle \mathbf{z},\boldsymbol{\xi}\rangle} e^{-i2\pi\langle \mathbf{y},\boldsymbol{\xi}\rangle}\rangle\rangle = \langle S(\mathbf{y}), e^{-i2\pi\langle \mathbf{y},\boldsymbol{\xi}\rangle}\rangle \cdot \langle T(\mathbf{z}), e^{-i2\pi\langle \mathbf{z},\boldsymbol{\xi}\rangle}\rangle \\ &= \langle S(\mathbf{y}), e^{-i2\pi\langle \mathbf{y},\boldsymbol{\xi}\rangle}\rangle \cdot \langle T(\mathbf{z}), e^{-i2\pi\langle \mathbf{z},\boldsymbol{\xi}\rangle}\rangle = (\mathcal{F} S)(\mathcal{F} T)\end{aligned}$$

(since $S, T \in \mathcal{E}'(\mathbb{R}^n)$ and we can take out $\langle T(\mathbf{z}), e^{-i2\pi\langle \mathbf{z},\boldsymbol{\xi}\rangle}\rangle$, which is independent of $\mathbf{y}$)

$$\Longrightarrow \quad \mathcal{F}(S * T) = (\mathcal{F} S) \cdot (\mathcal{F} T) \quad \forall S, T \in \mathcal{E}'(\mathbb{R}^n). \tag{8.5.12}$$

Fourier transform of the convolution $S * T \in S'(\mathbb{R}^n)$ of $S \in S'(\mathbb{R}^n)$ and $T \in \mathcal{E}'(\mathbb{R}^n)$ with compact support

Theorem 8.5.3. *$\forall$ tempered distributions $S \in S'(\mathbb{R}^n)$ and $\forall$ distributions $T \in \mathcal{E}'(\mathbb{R}^n)$ with compact support, we have:*

$$\mathcal{F}(T * S) = (\mathcal{F} T)(\mathcal{F} S) \tag{8.5.13}$$

$$\bar{\mathcal{F}}(T * S) = (\bar{\mathcal{F}} T) \cdot (\bar{\mathcal{F}} S). \tag{8.5.14}$$

Remark 8.5.1. (8.5.12) follows from (8.5.13), since $S \in \mathcal{E}'(\mathbb{R}^n) \Longrightarrow S \in S'(\mathbb{R}^n)$.

Proof. For $T \in \mathcal{E}'(\mathbb{R}^n)$ and $S \in S'(\mathbb{R}^n)$, $T * S \in S'(\mathbb{R}^n)$ by (8.5.6). Then, by the definition of $\mathcal{F}$ in $S'(\mathbb{R}^n)$,

$$\begin{aligned}\langle \mathcal{F}(T * S), \phi\rangle &= \langle T * S, \mathcal{F}\phi\rangle = \langle S, \check{T} * \mathcal{F}\phi\rangle \quad \text{(by (8.4.6))} \\ &= \langle \bar{\mathcal{F}}\mathcal{F} S, \check{T} * \mathcal{F}\phi\rangle \quad \text{(by Theorem 8.3.2)} \\ &= \langle \mathcal{F} S, \bar{\mathcal{F}}(\check{T} * \mathcal{F}\phi)\rangle \quad \text{(by definition of } \bar{\mathcal{F}}) \\ &= \langle \mathcal{F} S, (\bar{\mathcal{F}}\check{T})(\bar{\mathcal{F}}\mathcal{F}\phi)\rangle \text{ (by Theorem 8.5.2)} \\ &= \langle \mathcal{F} S, (\mathcal{F} T)\phi\rangle = \langle (\mathcal{F} T)(\mathcal{F} S), \phi\rangle \quad \forall \phi \in S(\mathbb{R}^n),\end{aligned}$$

since $\bar{\mathcal{F}}\check{T} = \mathcal{F} T$ $(\langle \bar{\mathcal{F}}\check{T}, \phi\rangle = \langle \check{T}, \bar{\mathcal{F}}\phi\rangle = \langle T, (\bar{\mathcal{F}}\phi)^{\vee}\rangle = \langle T, \mathcal{F}\phi\rangle = \langle \mathcal{F} T, \phi\rangle$ $\forall \phi \in S(\mathbb{R}^n)$ and for $(\bar{\mathcal{F}}\phi)^{\vee} = \mathcal{F}\phi$ (see (7.1.17)) $) \Longrightarrow \mathcal{F}(T * S) = (\mathcal{F} T)(\mathcal{F} S)$.

Similarly, (8.5.14) can be proved. □

Example 8.5.2. Prove that for $a > 0$, $\alpha > 0$, $\beta > 0$ and the Heaviside function H,

$$H(x)e^{-ax}\frac{x^{\alpha-1}}{\Gamma(\alpha)} * H(x)e^{-ax}\frac{x^{\beta-1}}{\Gamma(\beta)} = H(x)e^{-ax}\frac{x^{\alpha+\beta-1}}{\Gamma(\alpha+\beta)}. \tag{8.5.14a}$$

Proof. For $a > 0$, $\alpha > 0$, $\beta > 0$, $H(x)e^{-ax}\frac{x^{\alpha-1}}{\Gamma(\alpha)}$, $H(x)e^{-ax}\frac{x^{\beta-1}}{\Gamma(\beta)} \in L^1(\mathbb{R})$ with *unbounded support* $\Longrightarrow$ (from (8.2.12)) $H(x)e^{|a|x}\frac{x^{\alpha-1}}{\Gamma(\alpha)}$, $H(x)e^{|a|x}\frac{x^{\beta-1}}{\Gamma(\beta)} \in S'(\mathbb{R})$ (see Remark 6.3.7). $(H(x)e^{|a|x}\frac{x^{\beta-1}}{\Gamma(\beta)} * H(x)e^{|a|x}\frac{x^{\beta-1}}{\Gamma(\beta)}) \in L^1(\mathbb{R}) \subset S'(\mathbb{R})$. Hence, Theorem 8.5.3 can be applied. In fact, from Example 7.1.6, $\mathcal{F}[H(x)e^{-ax}\frac{x^{\alpha-1}}{\Gamma(\alpha)}] = (\frac{1}{a+i2\pi\xi})^{\alpha}$ for $a > 0, \alpha > 0 \Longrightarrow$ for $a > 0, \alpha > 0, \beta > 0$,

$$\begin{aligned}\mathcal{F}\left[H(x)e^{-ax}\frac{x^{\alpha+\beta-1}}{\Gamma(\alpha+\beta)}\right] = \left(\frac{1}{a+i2\pi\xi}\right)^{\alpha+\beta} &= \left(\frac{1}{a+i2\pi\xi}\right)^{\alpha}\left(\frac{1}{a+i2\pi\xi}\right)^{\beta}\\ &= \mathcal{F}\left[H(x)e^{-ax}\frac{x^{\alpha-1}}{\Gamma(\alpha)}\right]\mathcal{F}\left[H(x)e^{-ax}\frac{x^{\beta-1}}{\Gamma(\beta)}\right]\\ &= \mathcal{F}\left[H(x)e^{-ax}\frac{x^{\alpha-1}}{\Gamma(\alpha)} * H(x)e^{-ax}\frac{x^{\beta-1}}{\Gamma(\beta)}\right]\end{aligned}$$

by the Convolution Theorem 8.5.3, from which the result follows:

$$\overline{\mathcal{F}}\mathcal{F}\left[H(x)e^{-ax}\frac{x^{\alpha-1}}{\Gamma(\alpha)} * H(x)e^{-ax}\frac{x^{\beta-1}}{\Gamma(\beta)}\right] = \overline{\mathcal{F}}\mathcal{F}\left[H(x)e^{-ax}\frac{x^{\alpha+\beta-1}}{\Gamma(\alpha+\beta)}\right]$$

$$\Longrightarrow \quad H(x)e^{-ax}\frac{x^{\alpha-1}}{\Gamma(\alpha)} * H(x)e^{-ax}\frac{x^{\beta-1}}{\Gamma(\beta)} = H(x)e^{-ax}\frac{x^{\alpha+\beta-1}}{\Gamma(\alpha+\beta)}. \qquad \square$$

Remark 8.5.2. This proof with the help of the Fourier transform holds *only* for $a > 0$, i.e. for $a < 0$ we cannot apply the Fourier transform since $H(x)e^{|a|x}\frac{x^{\alpha-1}}{\Gamma(\alpha)}$ is *not* a tempered distribution and, consequently, has *no* Fourier transform as a *tempered distribution*.

But the result (8.5.14a) has been proved for any complex a in (6.3.6).

Fourier transforms of *non-tempered* distributions have been studied in the framework of generalized functions by Gelfand and Schilov [1] (see also Schwartz [8, p. 233]).

Multiplier Set $\theta_M(\mathbb{R}^n)$ for $S(\mathbb{R}^n)$

Definition 8.5.1. The set $\theta_M(\mathbb{R}^n)$ or simply θ_M defined by

$$\begin{aligned}\theta_M(\mathbb{R}^n) = \{f : f \in C^{\infty}(\mathbb{R}^n),\ \forall\phi \in S(\mathbb{R}^n),\ \phi \longmapsto f\phi\\ \text{is a } \textit{continuous, linear} \text{ mapping from } S(\mathbb{R}^n) \text{ into } S(\mathbb{R}^n)\}\end{aligned} \quad (8.5.15)$$

is called the *multiplier set* for $S(\mathbb{R}^n)$. For example, $f(\mathbf{x}) = (-i2\pi\mathbf{x})^{\boldsymbol{\alpha}} \in \theta_M(\mathbb{R}^n)$.

Proposition 8.5.1. *Let $f \in \theta_M(\mathbb{R}^n)$. Then the mapping $T \longmapsto fT$ is a continuous, linear mapping from $S'(\mathbb{R}^n)$ into $S'(\mathbb{R}^n)$.*

- $f \in \theta_M(\mathbb{R}^n) \Longrightarrow \check{f}, \partial^{\boldsymbol{\alpha}} f, \mathbf{x}^{\boldsymbol{\alpha}} f \in \theta_M(\mathbb{R}^n)\ \forall \boldsymbol{\alpha}$, *and* $\forall \mathbf{a} \in \mathbb{R}^n$, $\tau_{\mathbf{a}} f \in \theta_M(\mathbb{R}^n)$. (8.5.16)
- $f \in \theta_M(\mathbb{R}^n) \Longleftrightarrow \partial^{\boldsymbol{\alpha}} f \in C^\infty(\mathbb{R}^n)$ *with* slow growth at infinity *(see also* (8.2.15)*), i.e.* $\exists k \in \mathbb{N}_0$ *such that* $\frac{\partial^{\alpha} f}{(1+\|\mathbf{x}\|^2)^k}$ *is* bounded *in* $\mathbb{R}^n$ $\forall |\boldsymbol{\alpha}| \in \mathbb{N}_0$.
- $f \in \theta_M(\mathbb{R}^n) \Longrightarrow f \in S'(\mathbb{R}^n)$.

Space $\theta'_C(\mathbb{R}^n)$ of distributions with rapid decay

Definition 8.5.2. The set $\theta'_C(\mathbb{R}^n)$ of tempered distributions,

$$\theta'_C(\mathbb{R}^n) = \{T : T \in S'(\mathbb{R}^n),\ T * \phi \in S(\mathbb{R}^n), \phi \in S(\mathbb{R}^n) \longmapsto \phi * T \in S(\mathbb{R}^n) \text{ is } \textit{continuous} \text{ from } S(\mathbb{R}^n) \text{ into } S(\mathbb{R}^n)\} \quad (8.5.17)$$

is called the *space of distributions with rapid decay*.

For $T \in \theta'_C(\mathbb{R}^n)$, $\check{T} \in \theta'_C(\mathbb{R}^n)$ and $\check{T} * \phi \in S(\mathbb{R}^n)\ \forall \phi \in S(\mathbb{R}^n)$ such that $\forall S \in S'(\mathbb{R}^n)$, $\langle T * S, \phi\rangle = \langle S, \check{T} * \phi\rangle\ \forall \phi \in S(\mathbb{R}^n)$. Hence, $\forall T \in \theta'_C(\mathbb{R}^n)$,

$$T * S \in S'(\mathbb{R}^n) \quad \forall S \in S'(\mathbb{R}^n). \quad (8.5.18)$$

Properties of $\theta'_C(\mathbb{R}^n)$

1. $\theta'_C(\mathbb{R}^n) \subset S'(\mathbb{R}^n)$ is a *subspace* of $S'(\mathbb{R}^n)$. (8.5.19)
2. $\mathcal{E}'(\mathbb{R}^n) \subset \theta'_C(\mathbb{R}^n)$ is a *subspace* of $\theta'_C(\mathbb{R}^n)$. (8.5.20)
3. $\theta'_C(\mathbb{R}^n)$ contains *locally integrable* functions with rapid decay.
4. $T \in \theta'_C(\mathbb{R}^n) \Longrightarrow \check{T}, \tau_{\mathbf{a}} T, \partial^{\boldsymbol{\alpha}} T, \mathbf{x}^{\boldsymbol{\alpha}} T \in \theta'_C(\mathbb{R}^n)\ \forall \mathbf{a} \in \mathbb{R}^n$, $\forall$ multi-index $\boldsymbol{\alpha}$. (8.5.21)

Theorem 8.5.4. *Let* $f \in \theta_M(\mathbb{R}^n)$, $T \in \theta'_C(\mathbb{R}^n)$ *and* $S \in S'(\mathbb{R}^n)$. *Then:*

I. $\mathcal{F} f \in \theta'_C(\mathbb{R}^n)$;

II. $\mathcal{F} T \in \theta_M(\mathbb{R}^n)$; (8.5.22)

III. $\mathcal{F}[fS] = (\mathcal{F} f) * (\mathcal{F} S)$ *and* $\mathcal{F}[S * T] = \mathcal{F}[S] \cdot \mathcal{F}[T]$. (8.5.23)

Proof.

I. $f \in \theta_M(\mathbb{R}^n) \Longrightarrow f\phi \in S(\mathbb{R}^n)\ \forall \phi \in S(\mathbb{R}^n)$ by definition. Set $T_1 = \mathcal{F} f$. Then, for $\psi \in S(\mathbb{R}^n)$, $\psi * T_1 = \mathcal{F}[\bar{\mathcal{F}}\psi] * \mathcal{F} f = \mathcal{F}[(\bar{\mathcal{F}}\psi) f] \in S(\mathbb{R}^n)$ (by (8.5.11a)), and the mapping $\psi \longmapsto \psi * T_1$ is continuous from $S(\mathbb{R}^n)$ into $S(\mathbb{R}^n)$ (since it is a composition of *three* continuous mappings: $\psi \longmapsto \overline{\mathcal{F}}\psi \longmapsto \overline{\mathcal{F}}\psi f \longmapsto \mathcal{F}[\overline{\mathcal{F}}\psi f]$) for $T_1 \in S'(\mathbb{R}^n)$. Hence $T_1 = \mathcal{F} f \in \theta'_C(\mathbb{R}^n)$ by definition (8.5.17).

II. $T \in \theta'_C(\mathbb{R}^n) \Longrightarrow \phi * T \in S(\mathbb{R}^n)$ and $\phi \longmapsto \phi * T$ is continuous from $S(\mathbb{R}^n)$ into $S(\mathbb{R}^n)$. Set $f_1 = \mathcal{F}T$. For $\psi \in S(\mathbb{R}^n)$, $\psi f_1 = \mathcal{F}\bar{\mathcal{F}}\psi f_1 = \mathcal{F}(\bar{\mathcal{F}}\psi) \cdot (\mathcal{F}T) = \mathcal{F}(\bar{\mathcal{F}}\psi * T) \in S(\mathbb{R}^n)$ and the mapping $\psi \in S(\mathbb{R}^n) \longmapsto \psi f_1 \in S(\mathbb{R}^n)$ is continuous from $S(\mathbb{R}^n)$ into $S(\mathbb{R}^n)$ (since it is a composition of *three* continuous mappings: $\psi \longmapsto \bar{\mathcal{F}}\psi \longmapsto \bar{\mathcal{F}}\psi * T \longmapsto \mathcal{F}(\bar{\mathcal{F}}\psi * T)$). Hence, $f_1 = \mathcal{F}T \in \theta_M(\mathbb{R}^n)$ by definition (8.5.15), since $\psi f_1 \in S(\mathbb{R}^n)$.

III. $\forall T \in \theta'_C(\mathbb{R}^n)$, $\forall S \in S'(\mathbb{R}^n)$, $T * S \in S'(\mathbb{R}^n)$ by (8.5.18). Following the steps of the proof of Theorem 8.5.2, we have, $\forall T \in \theta'_C(\mathbb{R}^n)$, $\forall S \in S'(\mathbb{R}^n)$,

$$\begin{aligned}\langle \mathcal{F}(T * S), \phi\rangle &= \langle T * S, \mathcal{F}\phi\rangle = \langle S, \check{T} * \mathcal{F}\phi\rangle = \langle \bar{\mathcal{F}}\mathcal{F}S, \check{T} * \mathcal{F}\phi\rangle \\ &= \langle \mathcal{F}S, \bar{\mathcal{F}}(\check{T} * \mathcal{F}\phi)\rangle = \langle \mathcal{F}S, (\bar{\mathcal{F}}\check{T}) \cdot (\bar{\mathcal{F}}\mathcal{F}\phi)\rangle = \langle \mathcal{F}S, (\mathcal{F}T)\phi\rangle \\ &= \langle (\mathcal{F}T)(\mathcal{F}S), \phi\rangle \quad \forall \phi \in S(\mathbb{R}^n)\end{aligned}$$

$\Longrightarrow \mathcal{F}(T * S) = (\mathcal{F}T) \cdot (\mathcal{F}S)$ (the second equality in (8.5.23)), since

$$\begin{aligned}\langle \overline{\mathcal{F}}\check{T}, \phi\rangle &= \langle \check{T}, \overline{\mathcal{F}}\phi\rangle = \langle T, (\overline{\mathcal{F}}\phi)^{\vee}\rangle = \langle T, \mathcal{F}\phi\rangle \quad \text{(by (7.1.17))} \\ &= \langle \mathcal{F}T, \phi\rangle \;\forall \phi \in S(\mathbb{R}^n)\end{aligned}$$

$\Longrightarrow \overline{\mathcal{F}}\check{T} = \mathcal{F}T$ in $S'(\mathbb{R}^n)$.

Set $R = \mathcal{F}S \in S'(\mathbb{R}^n)$. Then $T = \mathcal{F}f \Longrightarrow \bar{\mathcal{F}}T = f$ and $\bar{\mathcal{F}}R = S \Longrightarrow fS = (\bar{\mathcal{F}}T)(\bar{\mathcal{F}}R) = \bar{\mathcal{F}}(T * R) \Longrightarrow \mathcal{F}(fS) = \mathcal{F}\bar{\mathcal{F}}(T * R) = T * R = \mathcal{F}f * \mathcal{F}S$ (the first equality in (8.5.23)).

□

8.6 Derivatives of Fourier transforms and Fourier transforms of derivatives of tempered distributions

Theorem 8.6.1. *Let $T \in S'(\mathbb{R}^n)$ be a tempered distribution on $\mathbb{R}^n$. Then the following relations hold:*

I. $\forall$ *multi-index* $\boldsymbol{\alpha}$,

$$\mathcal{F}[\partial_{\mathbf{x}}^{\boldsymbol{\alpha}} T] = (i2\pi\boldsymbol{\xi})^{\boldsymbol{\alpha}}\mathcal{F}T; \quad \partial_{\boldsymbol{\xi}}^{\boldsymbol{\alpha}}(\mathcal{F}T)(\boldsymbol{\xi}) = \mathcal{F}[(-i2\pi\mathbf{x})^{\boldsymbol{\alpha}}T];$$

$$\mathcal{F}[\mathbf{x}^{\boldsymbol{\alpha}}T] = (-1)^{|\boldsymbol{\alpha}|}\frac{1}{(i2\pi)^{|\boldsymbol{\alpha}|}}\partial_{\boldsymbol{\xi}}^{\boldsymbol{\alpha}}(\mathcal{F}T); \tag{8.6.1}$$

II. $\forall \mathbf{a} \in \mathbb{R}^n$,

$$\mathcal{F}[\tau_{\mathbf{a}}T] = e^{-i2\pi\langle\boldsymbol{\xi},\mathbf{a}\rangle}\mathcal{F}T; \quad \tau_{\mathbf{a}}(\mathcal{F}T) = \mathcal{F}[e^{i2\pi\langle\mathbf{x},\mathbf{a}\rangle}T], \tag{8.6.2}$$

where $\partial_{\mathbf{x}}^{\boldsymbol{\alpha}} = \frac{\partial^{|\boldsymbol{\alpha}|}}{\partial x_1^{\alpha_1}\partial x_2^{\alpha_2}\ldots\partial x_n^{\alpha_n}}$, $\partial_{\boldsymbol{\xi}}^{\boldsymbol{\alpha}} = \frac{\partial^{|\boldsymbol{\alpha}|}}{\partial \xi_1^{\alpha_1}\partial \xi_2^{\alpha_2}\ldots\partial \xi_n^{\alpha_n}}$.

Proof.

I. $\forall T \in S'(\mathbb{R}^n) \subset \mathcal{D}'(\mathbb{R}^n)$, $\delta * T = T \Longrightarrow \partial_{\mathbf{x}}^{\boldsymbol{\alpha}} T = \partial_{\mathbf{x}}^{\boldsymbol{\alpha}}(\delta * T) = \partial_{\mathbf{x}}^{\boldsymbol{\alpha}}\delta * T$ (see (6.3.24), (6.7.4))

$$\begin{aligned}\Longrightarrow \quad \mathcal{F}[\partial_{\mathbf{x}}^{\boldsymbol{\alpha}} T] &= \mathcal{F}[\partial_{\mathbf{x}}^{\boldsymbol{\alpha}}\delta * T] = \mathcal{F}(\partial_{\mathbf{x}}^{\boldsymbol{\alpha}}\delta) \cdot \mathcal{F}T \quad \text{(by (8.5.13))}\\ &= (i2\pi\boldsymbol{\xi})^{\boldsymbol{\alpha}}\mathcal{F}T \quad \text{(by (8.3.13)).}\\ \mathcal{F}[(-i2\pi\mathbf{x})^{\boldsymbol{\alpha}} T] &= \mathcal{F}[(-i2\pi\mathbf{x})^{\boldsymbol{\alpha}}] * \mathcal{F}T \quad \text{(by (8.5.23))}\\ &= \partial_{\boldsymbol{\xi}}^{\boldsymbol{\alpha}}\delta * \mathcal{F}T \quad \text{(by (8.4.11))}\\ &= \partial_{\boldsymbol{\xi}}^{\boldsymbol{\alpha}}(\delta * \mathcal{F}T) = \partial_{\boldsymbol{\xi}}^{\boldsymbol{\alpha}}(\mathcal{F}T)(\boldsymbol{\xi}) \quad \text{(by (6.3.22), (6.7.4)).}\\ \mathcal{F}[(-i2\pi\mathbf{x})^{\boldsymbol{\alpha}} T] &= (-1)^{|\boldsymbol{\alpha}|}(i2\pi)^{|\boldsymbol{\alpha}|}\mathcal{F}[\mathbf{x}^{\boldsymbol{\alpha}} T] = \partial_{\boldsymbol{\xi}}^{\boldsymbol{\alpha}}(\mathcal{F}T)\end{aligned}$$

$$\Longrightarrow \mathcal{F}[\mathbf{x}^{\boldsymbol{\alpha}} T] = (-1)^{|\boldsymbol{\alpha}|}\frac{1}{(i2\pi)^{|\boldsymbol{\alpha}|}}\partial_{\boldsymbol{\xi}}^{\boldsymbol{\alpha}}(\mathcal{F}T).$$

II.
$$\begin{aligned}\mathcal{F}[\tau_{\mathbf{a}} T] &= \mathcal{F}(\delta_{\mathbf{a}} * T) \quad \text{(by (6.3.29))}\\ &= \mathcal{F}(\delta_{\mathbf{a}}) \cdot (\mathcal{F}T) = e^{-i2\pi\langle\boldsymbol{\xi},\mathbf{a}\rangle}(\mathcal{F}T) \quad \text{(by (8.3.15)).}\end{aligned}$$
$$\mathcal{F}(e^{i2\pi\langle\mathbf{x},\mathbf{a}\rangle}T) = \mathcal{F}(e^{i2\pi\langle\mathbf{x},\mathbf{a}\rangle}) * \mathcal{F}T = \delta_{\mathbf{a}} * \mathcal{F}T = \tau_{\mathbf{a}}(\mathcal{F}T) \quad \text{(by (6.3.29)),}$$

since $\mathcal{F}(e^{i2\pi\langle\mathbf{x},\mathbf{a}\rangle}) = \delta_{\mathbf{a}}$ by (8.4.10). □

Example 8.6.1. Let $u \in S'(\mathbb{R})$ such that $\frac{d^4u}{dx^4} + \lambda u \in L^2(\mathbb{R})$ $\forall$ constant $\lambda > 0$. Then show that $\frac{d^k u}{dx^k} \in L^2(\mathbb{R})$ for $0 \le k \le 4$.

Proof. From Corollary 8.3.1, $\mathcal{F} : L^2(\mathbb{R}) \longrightarrow L^2(\mathbb{R})$ is an *isometric isomorphism.*

Hence, $\frac{d^4u}{dx^4} + \lambda u \in L^2(\mathbb{R}) \Longrightarrow \mathcal{F}[\frac{d^4u}{dx^4} + \lambda u] \in L^2(\mathbb{R}) \subset S'(\mathbb{R})$. $u \in S'(\mathbb{R}) \Longrightarrow \frac{d^4u}{dx^4} \in S'(\mathbb{R}) \Longrightarrow \mathcal{F}[\frac{d^4u}{dx^4} + \lambda u] = (i2\pi\xi)^4\hat{u} + \lambda\hat{u} = [(2\pi)^4\xi^4 + \lambda]\hat{u}$ by Theorem 8.6.1. Hence, $\forall\lambda > 0$, $\lambda[\frac{(2\pi)^4}{\lambda}\xi^4 + 1]\hat{u} \in L^2(\mathbb{R}) \Longleftrightarrow (\xi^4 + 1)\hat{u} \in L^2(\mathbb{R})$, since, in particular, it must hold for $\lambda = (2\pi)^4 > 0$ and $\forall\xi \in \mathbb{R}$, $|\xi|^k \le 1 + \xi^4$ for $0 \le k \le 4$. In fact, $\xi \in \mathbb{R} \Longrightarrow |\xi| \le 1$ or $|\xi| > 1$. Then $|\xi| \le 1 \Longrightarrow |\xi|^k \le 1$ for $0 \le k \le 4 \Longrightarrow |\xi|^k \le 1 + \xi^4$ for $0 \le k \le 4$, and $|\xi| > 1 \Longrightarrow |\xi|^k \le \xi^4$ for $0 \le k \le 4 \Longrightarrow |\xi|^k \le 1 + \xi^4$ for $0 \le k \le 4$. Hence, $\int_{\mathbb{R}} |(i2\pi\xi)^k\hat{u}(\xi)|^2 d\xi \le \int_{\mathbb{R}} (2\pi)^{2k}|\xi|^{2k}|\hat{u}(\xi)|^2 d\xi \le (2\pi)^{2k}\int_{\mathbb{R}}(1+\xi^4)^2|\hat{u}(\xi)|^2 d\xi < +\infty$, since $(\xi^4+1)\hat{u} \in L^2(\mathbb{R}) \Longrightarrow (i2\pi\xi)^k\hat{u} \in L^2(\mathbb{R})$ for $0 \le k \le 4$. But $\mathcal{F}[\frac{d^k u}{dx^k}] = (i2\pi\xi)^k\hat{u} \in L^2(\mathbb{R})$ for $0 \le k \le 4 \Longrightarrow \frac{d^k u}{dx^4} \in L^2(\mathbb{R})$ for $0 \le k \le 4$ by Corollary 8.3.1. □

Fourier transform of polynomial $P(x)$

Example 8.6.2.

1. Show that every polynomial $P(x) = a_0 + a_1 x + \cdots + a_n x^n$ with real or complex coefficients a_k, $0 \le k \le n$, defines a tempered distribution.
2. Find the Fourier transform of $P(x)$.

Proof.

1. $\forall \phi \in S(\mathbb{R})$, $P(x)\phi(x) \in S(\mathbb{R})$ $\forall$ polynomials $P(x) \Longrightarrow P(x)\phi \in L^1(\mathbb{R}^n)$ $\Longrightarrow \int_{\mathbb{R}} P(x)\phi(x)dx$ is well defined $\forall \phi \in S(\mathbb{R})$ as a continuous linear functional on $S(\mathbb{R})$. Hence, $P(x)$ defines a tempered distribution T_P by:
$$\langle T_P, \phi\rangle = \int_{\mathbb{R}} P(x)\phi(x)dx \quad \forall \phi \in S(\mathbb{R}).$$
2. $$\begin{aligned}\mathcal{F}[x^k] = \mathcal{F}[x^k \cdot 1] &= (-1)^k \frac{1}{(i2\pi)^k}\partial_\xi^k(\mathcal{F}[1]) \\ &= (-1)^k \frac{1}{(i2\pi)^k}\partial_\xi^k \delta \quad \text{(by (8.3.22))} \\ &= (-1)^k \frac{\delta^{(k)}}{(i2\pi)^k}.\end{aligned}$$

By virtue of the linearity of $\mathcal{F}$,
$$\begin{aligned}\mathcal{F}[P(x)] &= a_0\mathcal{F}(1) + a_1\mathcal{F}[x] + \cdots + a_n\mathcal{F}[x^n] \\ &= a_0\delta + a_1\frac{(-1)}{i2\pi}\delta^{(1)} + a_2\frac{(-1)^2}{(i2\pi)^2}\delta^{(2)} + \cdots + a_n\frac{(-1)^n}{(i2\pi)^n}\delta^n \\ &= \sum_{k=0}^{n}(-1)^k\frac{a_k}{(i2\pi)^k}\delta^{(k)}\end{aligned}$$

with $\delta^{(0)} = \delta$,
$$\langle \delta^{(k)}, \phi\rangle = (-1)^k\langle \delta, \phi^{(k)}(\xi)\rangle = (-1)^k\phi^{(k)}(0) \quad \forall \phi \in S(\mathbb{R}).$$
□

Fourier transform of z^n with $z = x + iy$ in $S'(\mathbb{R}^2)$

Example 8.6.3.

1. Find the Fourier transform of the tempered distribution $T = z^n \in S'(\mathbb{R}^2)$ with $z = x + iy$ $\forall n \in \mathbb{N}_0$.
2. Using the identity $z.\frac{1}{z} = 1$, an elementary solution of the Cauchy–Riemann operator $\overline{\partial} = \frac{\partial}{\partial \bar{z}} = \frac{1}{2}(\frac{\partial}{\partial x} + i\frac{\partial}{\partial y})$ with $\bar{z} = x - iy$ and the homogeneity of $\frac{1}{z}$, prove that $\mathcal{F}(\frac{1}{z})(\zeta) = -\frac{i}{\zeta}$ with $\zeta = \xi + i\eta$.

Proof.

1. Since $\exists l \in \mathbb{N}$ such that $\sup_{(x,y)\in\mathbb{R}^2} \frac{|x+iy|^n}{[1+(x^2+y^2)]^l} < +\infty$ and $z^n = (x+iy)^n$ is a polynomial in two variables x, y with complex coefficients and with slow growth at infinity (8.2.15), z^n defines a tempered distribution in $\mathbb{R}^2$, i.e. $T = z^n \in S'(\mathbb{R}^2)$ and $\mathcal{F}(z^n) \in S'(\mathbb{R}^2)$ is well defined.

   But

$$\begin{aligned}
\mathcal{F}(z) &= \mathcal{F}[x+iy] = \mathcal{F}[x] + i\mathcal{F}[y] = \mathcal{F}[x.1] + i\mathcal{F}[y\cdot 1] \\
&= \frac{-1}{i2\pi}\frac{\partial}{\partial\xi}(\mathcal{F}1) + i\frac{-1}{i2\pi}\frac{\partial}{\partial\eta}(\mathcal{F}1) = -\frac{1}{i2\pi}\left(\frac{\partial}{\partial\xi} + i\frac{\partial}{\partial\eta}\right)\delta \\
\Longrightarrow \quad \mathcal{F}(z^n) &= \left[-\frac{1}{i2\pi}\left(\frac{\partial}{\partial\xi} + i\frac{\partial}{\partial\eta}\right)\delta\right] * \left[-\frac{1}{i2\pi}\left(\frac{\partial}{\partial\xi} + i\frac{\partial}{\partial\eta}\right)\delta\right] \\
&\quad * \cdots * \left[-\frac{1}{i2\pi}\left(\frac{\partial}{\partial\xi} + i\frac{\partial}{\partial\eta}\right)\delta\right] \\
&= \left(-\frac{1}{i2\pi}\right)^n \left(\frac{\partial}{\partial\xi} + i\frac{\partial}{\partial\eta}\right)^n [\delta * \delta * \cdots * \delta]
\end{aligned}$$

   (by repeated applications of (8.5.23)).

   Since $\delta * \delta = \delta \Longrightarrow \delta * \delta * \delta = \delta \Longrightarrow \ldots \Longrightarrow \delta * \delta * \cdots * \delta = \delta$, we have

$$\mathcal{F}(z^n) = \left(-\frac{1}{i2\pi}\right)^n \left(\frac{\partial}{\partial\xi} + i\frac{\partial}{\partial\eta}\right)^n \delta.$$

2. $T = \frac{1}{z} \in S'(\mathbb{R}^2)$, since $\frac{1}{z} \in L^1_{\text{loc}}(\mathbb{R}^2)$ and is bounded for $|z| \geq 1$. Then $z \cdot T = 1 \Longrightarrow \mathcal{F}(z\cdot T) = \mathcal{F}1 = \delta$ (by (8.3.22))

$$\Longrightarrow \quad \mathcal{F}[(x+iy)T] = \mathcal{F}[xT] + i\mathcal{F}[yT] = -\frac{1}{i2\pi}\left(\frac{\partial}{\partial\xi} + i\frac{\partial}{\partial\eta}\right)\mathcal{F}T = \delta.$$

   Now we will show that $\overline{\partial}(\frac{1}{z}) = \overline{\partial}(\frac{1}{x+iy}) = \pi\delta$, where $\overline{\partial} = \frac{1}{2}(\frac{\partial}{\partial x} + i\frac{\partial}{\partial y})$ is the Cauchy–Riemann operator [15]. In fact, for $x = r\cos\theta$, $y = r\sin\theta$, $J = r$,

$\phi(x, y) = \phi(r\cos\theta, r\sin\theta) = \tilde{\phi}(r, \theta), \overline{\partial}\phi = \frac{1}{2}(\frac{\partial\phi}{\partial x} + i\frac{\partial\phi}{\partial y})$,

$$\frac{\partial\phi}{\partial x} = \frac{\partial\tilde{\phi}}{\partial r}.\frac{\partial r}{\partial x} + \frac{\partial\tilde{\phi}}{\partial\theta}\cdot\frac{\partial\theta}{\partial x} = \frac{\partial\tilde{\phi}}{\partial r}\cos\theta - \frac{\partial\tilde{\phi}}{\partial\theta}\frac{\sin\theta}{r},$$

$$\frac{\partial\phi}{\partial y} = \frac{\partial\tilde{\phi}}{\partial r}\frac{\partial r}{\partial y} + \frac{\partial\tilde{\phi}}{\partial\theta}\cdot\frac{\partial\theta}{\partial y} = \frac{\partial\tilde{\phi}}{\partial r}\sin\theta + \frac{\partial\tilde{\phi}}{\partial\theta}\frac{\cos\theta}{r},$$

$$\left\langle\overline{\partial}\left(\frac{1}{z}\right), \phi\right\rangle = -\left\langle\frac{1}{z}, \overline{\partial}\phi\right\rangle$$

$$= -\frac{1}{2}\int_0^\infty\int_0^{2\pi}\frac{1}{r(\cos\theta + i\sin\theta)}\left[\left(\frac{\partial\tilde{\phi}}{\partial r}\cos\theta + \frac{\partial\tilde{\phi}}{\partial\theta}\left(-\frac{\sin\theta}{r}\right)\right)\right.$$

$$\left. + i\left(\frac{\partial\tilde{\phi}}{\partial r}\sin\theta + \frac{\partial\tilde{\phi}}{\partial\theta}\frac{\cos\theta}{r}\right)\right] r\,dr\,d\theta \quad \forall\phi\in\mathcal{D}(\mathbb{R}^2)$$

$$= -\frac{1}{2}\int_0^\infty\int_0^{2\pi}\frac{1}{\cos\theta + i\sin\theta}\left[\frac{\partial\tilde{\phi}}{\partial r}(\cos\theta + i\sin\theta)\right.$$

$$\left. + \frac{i}{r}\frac{\partial\tilde{\phi}}{\partial\theta}(\cos\theta + i\sin\theta)\right] dr\,d\theta$$

$$= -\frac{1}{2}\int_0^\infty\int_0^{2\pi}\left[\frac{\partial\tilde{\phi}}{\partial r} + \frac{i}{r}\frac{\partial\tilde{\phi}}{\partial\theta}\right] dr\,d\theta$$

$$= -\frac{1}{2}\int_0^{2\pi}\left(\int_0^\infty\frac{\partial\tilde{\phi}}{\partial r}dr\right)d\theta - \frac{i}{2}\int_0^\infty\frac{1}{r}\left(\int_0^{2\pi}\frac{\partial\tilde{\phi}}{\partial\theta}d\theta\right)dr.$$

But, $\forall\phi\in\mathcal{D}(\mathbb{R}^2)$, $\tilde{\phi}(0,\theta) = \phi(0,0)$, $\tilde{\phi}(r,\theta)$ is 2π-periodic and $\tilde{\phi}(r,\theta)\to 0$ as $r\to\infty$, $\Longrightarrow \int_0^\infty\frac{\partial\tilde{\phi}}{\partial r}dr = -\phi(0,0)$, $\int_0^{2\pi}\frac{\partial\tilde{\phi}}{\partial\theta}d\theta = 0$. Hence,

$$\left\langle\overline{\partial}\left(\frac{1}{z}\right), \phi\right\rangle = -\frac{1}{2}\int_0^{2\pi}[-\phi(0,0)]d\theta = \frac{1}{2}\phi(0,0)\cdot 2\pi$$

$$= \pi\phi(0,0) = \langle\pi\delta, \phi\rangle \quad \forall\phi\in\mathcal{D}(\mathbb{R}^2).$$

$\Longrightarrow \overline{\partial}(\frac{1}{z}) = \pi\delta$ in $\mathcal{D}'(\mathbb{R}^2) \Longrightarrow \frac{1}{\pi z}$ is an *elementary solution* of the Cauchy–Riemann operator $\overline{\partial} = \frac{1}{2}(\frac{\partial}{\partial\xi} + i\frac{\partial}{\partial\eta})$, since $\overline{\partial}(\frac{1}{\pi z}) = \delta$ in $\mathcal{D}'(\mathbb{R}^2)$. But $\pi\delta\in S'(\mathbb{R}^2) \Longrightarrow \overline{\partial}(\frac{1}{z}) = \pi\delta$ in $S'(\mathbb{R}^2)$. Then $\frac{1}{2\pi}[\frac{\partial}{\partial x}+i\frac{\partial}{\partial y}](\frac{1}{z}) = \delta \Longrightarrow -\frac{1}{i2\pi}[\frac{\partial}{\partial x}+ i\frac{\partial}{\partial y}](\frac{-i}{z}) = \delta$ in $S'(\mathbb{R}^2) \Longrightarrow -\frac{i}{z}\in S'(\mathbb{R}^2)$ is an elementary solution of $-\frac{1}{i2\pi}(\frac{\partial}{\partial x} + i\frac{\partial}{\partial y})$ in $S'(\mathbb{R}^2)$. Hence, for $\zeta = \xi + i\eta$, $E_0 = -\frac{i}{\zeta}$ is an elementary solution of $-\frac{1}{i2\pi}(\frac{\partial}{\partial\xi} + i\frac{\partial}{\partial\eta})(\mathcal{F}T) = \delta$ (8.6.2a). The general solution of the corresponding homogeneous equation $-\frac{1}{i2\pi}(\frac{\partial}{\partial\xi} + i\frac{\partial}{\partial\eta})(\mathcal{F}T) = 0$ is an analytic function $F = F(\zeta)$ with $\zeta = \xi + i\eta$. Hence, $\mathcal{F}T = \mathcal{F}(1/z) =$

$E_0 + F = -\frac{i}{\zeta} + F(\zeta)$. But $T = \frac{1}{z}$ is homogeneous of degree $-1 \Longrightarrow$ by Proposition 8.3.1, $\mathcal{F}T$ is also homogeneous of degree -1 (see Example 8.3.2). Since $-\frac{i}{\zeta}$ with $\zeta = \xi + i\eta$ is homogeneous of degree -1, analytic F is homogeneous of degree $-1 \Longrightarrow F \equiv 0$, i.e. $F(\zeta) = 0\ \forall \zeta \in \mathbb{C}$ with $\zeta = \xi + i\eta$. Hence,

$$\mathcal{F}\left(\frac{1}{z}\right) = -\frac{i}{\zeta} \quad \text{with } \zeta = \xi + i\eta.$$

In fact, if we suppose that $F \neq 0$, then $F(\lambda\zeta) = \frac{1}{\lambda}F(\zeta)\ \forall\zeta,\ \forall\lambda > 0$. In particular, for $\zeta = 1$, $F(\lambda) = \frac{1}{\lambda}F(1)\ \forall\lambda > 0$. But analytic F is a C^∞-function with $\lim_{\lambda\to 0^+,\lambda\in\mathbb{R}^+} F(\lambda) = \infty$, which is impossible. Hence, $F \equiv 0$ (See also [1] (page 168, pages 370–379) for more details). □

Fourier transforms of c.p.v. $\frac{1}{x}$, Heaviside function $H(x)$

Example 8.6.4. Find the Fourier transform of $T = \text{c.p.v.}\ \frac{1}{x}$. Then, using this result, find $\mathcal{F}H$ and $\overline{\mathcal{F}}H$, H being the Heaviside function with $H(x) = 1$ for $x > 0$ and $= 0$ for $x < 0$.

Solution. From Example 8.3.1, c.p.v. $\frac{1}{x}$ is an odd tempered distribution, and

$$\begin{aligned}\langle x \cdot \text{c.p.v.}\frac{1}{x}, \phi\rangle &= \langle \text{c.p.v.}\frac{1}{x}, x\phi\rangle = \lim_{\varepsilon\to 0^+}\int_{|x|\geq\varepsilon}\frac{x\phi}{x}dx \\ &= \lim_{\varepsilon\to 0}\int_{|x|\geq\varepsilon}\phi(x)dx = \int_{\mathbb{R}}\phi(x)dx = \langle 1, \phi\rangle \quad \forall\phi \in S(\mathbb{R})\end{aligned}$$

$\Longrightarrow x \cdot \text{c.p.v.}\ \frac{1}{x} = 1$ in $S'(\mathbb{R}) \Longrightarrow \mathcal{F}(xT) = \mathcal{F}1$ with $T = \text{c.p.v.}\ \frac{1}{x} \Longrightarrow (-1)\frac{1}{i2\pi}\frac{d}{d\xi}(\mathcal{F}T) = \delta$ (by Theorem 8.6.1 and (8.3.22)) $\Longrightarrow \frac{d}{d\xi}\widehat{T} = -i2\pi\delta \Longrightarrow \widehat{T}(\xi) = -i2\pi H(\xi) + C$ (see Example 2.7.4), where $H(\xi)$ is the Heaviside function in ξ and C is a constant to be determined, since $\widehat{T} = \mathcal{F}$ (c.p.v. $\frac{1}{x}$) is a unique tempered distribution. From Example 8.3.1, c.p.v. $\frac{1}{x}$ is an odd tempered distribution $\Longrightarrow \mathcal{F}(\text{c.p.v.}\ \frac{1}{x}) = \widehat{T}$ is also odd by (8.3.26) $\Longrightarrow \widehat{T}$ is an odd function of $\xi \Longrightarrow \widehat{T}(\xi) = -\widehat{T}(-\xi) \Longrightarrow -i2\pi H(\xi) + C = i2\pi H(-\xi) - C \Longrightarrow -i2\pi \cdot 1 + C = -C$ for $\xi > 0$, and $C = i2\pi - C$ for $\xi < 0 \Longrightarrow C = i\pi$.

Hence,

$$\widehat{T} = \mathcal{F}\left(\text{c.p.v.}\ \frac{1}{x}\right) = -i2\pi H(\xi) + i\pi = \begin{cases} -i\pi & \text{for } \xi > 0 \\ i\pi & \text{for } \xi < 0,\end{cases}$$

$$\mathcal{F}\widehat{T} = -i2\pi(\mathcal{F}H) + \mathcal{F}(i\pi).$$

But from (8.3.21), $\mathcal{F}\mathcal{F}T = \mathcal{F}\widehat{T} = \check{T}$ and $\mathcal{F}(i\pi) = i\pi\mathcal{F}[1] = i\pi\delta$ (by (8.3.22)). Hence,

$$(\mathcal{F}H)(x) = -\frac{1}{i2\pi}[\mathcal{F}\widehat{T} - \mathcal{F}(i\pi)] = -\frac{1}{i2\pi}[\check{T} - i\pi\delta] = -\frac{1}{i2\pi}\check{T} + \frac{1}{2}\delta.$$

$T = \text{c.p.v.}\,\frac{1}{x}$ is odd $\Longrightarrow \widehat{T} = \mathcal{F}T$ is odd $\Longrightarrow \mathcal{F}\widehat{T} = \check{T}$ is odd $\Longrightarrow \check{T} = -T$. In fact, $\langle \check{T}, \check{\phi}\rangle = -\langle \check{T}, \phi\rangle = -\langle T, \check{\phi}\rangle\ \forall \phi \in S(\mathbb{R}) \Longrightarrow \check{T} = -T \Longrightarrow (\mathcal{F}H)(x) = -\frac{1}{i2\pi}(-T) + \frac{1}{2}\delta = \frac{1}{2}\delta + \frac{1}{i2\pi}\,\text{c.p.v.}\,\frac{1}{x}$.

$\bar{\mathcal{F}}\mathcal{F}T = \bar{\mathcal{F}}(-i2\pi H + i\pi) = -i2\pi\bar{\mathcal{F}}H + i\pi\bar{\mathcal{F}}1 \Longrightarrow T = -i2\pi\bar{\mathcal{F}}H + i\pi\delta$ (by (8.3.22)) $\Longrightarrow \bar{\mathcal{F}}H = \frac{-1}{i2\pi}T + \frac{i\pi}{i2\pi}\delta$ with $T = \text{c.p.v.}\,\frac{1}{x} \Longrightarrow (\bar{\mathcal{F}}H)(x) = \frac{1}{2}\delta - \frac{1}{i2\pi}\,\text{c.p.v.}\,\frac{1}{x}$.

Thus, $(\mathcal{F}H)(x) = \frac{1}{2}\delta + \frac{1}{i2\pi}\,\text{c.p.v.}\,\frac{1}{x}$, $(\bar{\mathcal{F}}H)(x) = \frac{1}{2}\delta - \frac{1}{i2\pi}\,\text{c.p.v.}\,\frac{1}{x}$.

Example 8.6.5. Consider the function $H(x)e^{-ax}$, which is integrable on $\mathbb{R}$ for $a > 0$, $H(x)$ being the Heaviside function.

1. Find the Fourier transform of $H(x)e^{-ax}$.
2. Show the values of a for which $H(x)e^{-ax} \in S'(\mathbb{R})$.
3. Find the Fourier transform $\mathcal{F}H$ and co-transform $\bar{\mathcal{F}}H$ of the Heaviside function $H(x)$ using the Fourier transform of $H(x)e^{-ax}$ and the identity $H(x) + H(-x) = 1$.

Solution.

1. For $a > 0$, $H(x)e^{-ax} \in L^1(\mathbb{R}) \Longrightarrow \mathcal{F}[H(x)e^{-ax}]$ exists and is given by:

$$\mathcal{F}[H(x)e^{-ax}] = \int_{-\infty}^{\infty} H(x)e^{-ax}e^{-i2\pi x\xi}dx = \int_{0}^{\infty} e^{-(a+i2\pi\xi)x}dx$$
$$= \frac{e^{-(a+i2\pi\xi)x}}{-(a+i2\pi\xi)}\bigg|_{x=0}^{x=\infty} = \frac{1}{(a+i2\pi\xi)},$$

since $e^{-ax} \to 0$ as $x \to \infty$, $\Longrightarrow \mathcal{F}[H(x)e^{-ax}] = \frac{1}{a+i2\pi\xi}\ \forall a > 0$.

2. For $a \geq 0$, $H(x)e^{-ax} \in S'(\mathbb{R})$, since $H \in S'(\mathbb{R})$.
3. $\lim_{a\to 0} H(x)e^{-ax} = H(x) \notin L^1(\mathbb{R})$ and $\lim_{a\to 0}\frac{1}{a+i2\pi\xi} = \frac{1}{i2\pi\xi} \notin L^1(\mathbb{R})$. Since Fourier transform $\mathcal{F}\colon S'(\mathbb{R}) \longrightarrow S'(\mathbb{R})$ is continuous from $S'(\mathbb{R})$ onto $S'(\mathbb{R})$, we can find $\mathcal{F}[H(x)]$ by taking the limit of $\frac{1}{a+i2\pi\xi}$ in $S'(\mathbb{R})$ as $a \to 0^+$, i.e. $\mathcal{F}[H(x)] = \lim_{a\to 0^+}\frac{1}{a+i2\pi}$ in $S'(\mathbb{R})$.
   In fact, $H(x)e^{-ax} \in S'(\mathbb{R})\ \forall a \geq 0 \Longrightarrow H(x)e^{-ax} \to H(x)$ in $S'(\mathbb{R})$ as $a \to 0^+$
   $\Longrightarrow \mathcal{F}[H(x)e^{-ax}] \to \mathcal{F}[H(x)]$ in $S'(\mathbb{R})$ as $a \to 0^+$. Moreover, $\frac{d}{d\xi}$: $S'(\mathbb{R}) \longrightarrow S'(\mathbb{R})$ is also continuous on $S'(\mathbb{R})$.

But $\frac{d}{d\xi}\ln(a+i2\pi\xi) = \frac{i2\pi}{a+i2\pi\xi} \in S'(\mathbb{R})\ \forall a > 0 \Longrightarrow \frac{1}{a+i2\pi\xi} = \frac{1}{i2\pi}\frac{d}{d\xi}[\ln(a+i2\pi\xi)]\ \forall a > 0$.

Hence, by virtue of the continuity of $\frac{d}{d\xi}$ on $S'(\mathbb{R})$ (see (8.2.28)),

$$\begin{aligned}\lim_{a\to 0+}\frac{1}{a+i2\pi\xi} &= \frac{1}{i2\pi}\lim_{a\to 0+}\frac{d}{d\xi}\left[\ln(a^2+4\pi^2\xi^2)^{1/2} + i\arctan\frac{2\pi\xi}{a}\right]\\ &= \frac{1}{i2\pi}\frac{d}{d\xi}\left[\lim_{a\to 0+}\left\{\ln(a^2+4\pi^2\xi^2)^{1/2} + i\arctan\frac{2\pi\xi}{a}\right\}\right]\\ &= \frac{1}{i2\pi}\frac{d}{d\xi}\left[\ln 2\pi|\xi| + i\frac{\pi}{2}[H(\xi)-H(-\xi)]\right],\end{aligned}$$

since for $\xi > 0$, $a \to 0^+ \Longrightarrow \frac{2\pi\xi}{a} \to +\infty \Longrightarrow \arctan(\frac{2\pi\xi}{a}) \to \pi/2$, for $\xi < 0$, $a \to 0^+ \Longrightarrow \frac{2\pi\xi}{a} \to -\infty \Longrightarrow \arctan(\frac{2\pi\xi}{a}) \to -\pi/2$

$$\begin{aligned}\Longrightarrow\quad \lim_{a\to 0+}\arctan\left(2\pi\frac{\xi}{a}\right) &= \begin{cases}\pi/2 & \text{for } \xi > 0\\ -\pi/2 & \text{for } \xi < 0\end{cases}\\ &= (\pi/2)[H(\xi)-H(-\xi)]\quad \forall\xi\in\mathbb{R}\\ \Longrightarrow\quad \lim_{a\to 0+}\frac{1}{a+i2\pi\xi} &= \frac{1}{i2\pi}\frac{d}{d\xi}\left[\ln(2\pi|\xi|) + i\frac{\pi}{2}\Big(H(\xi)-H(-\xi)\Big)\right]\\ &= \frac{1}{i2\pi}\left[\text{c.p.v.}\,\frac{1}{\xi} + i\frac{\pi}{2}(\delta(\xi)+\delta(\xi))\right]\\ &= \frac{1}{i(2\pi)}\,\text{c.p.v.}\,\frac{1}{\xi} + \frac{1}{2}\delta,\end{aligned}$$

since $\frac{d}{d\xi}\ln(2\pi|\xi|) = \frac{d}{d\xi}[\ln 2\pi + \ln|\xi|] = 0 + \frac{d}{d\xi}ln|\xi| = \text{c.p.v.}\,\frac{1}{\xi}$ (see Example 2.3.6),

$$\left\langle\frac{dH(-x)}{dx},\phi\right\rangle = -\left\langle H(-x),\frac{d\phi}{dx}\right\rangle = -\int_{-\infty}^{0} 1\cdot\frac{d\phi}{dx}dx = -\phi(0) = -\langle\delta,\phi\rangle\quad\forall\phi$$

$\Longrightarrow \frac{dH(-x)}{dx} = -\delta \Longrightarrow \mathcal{F}[H(x)] = \frac{1}{i2\pi}\,\text{c.p.v.}\,\frac{1}{\xi} + \frac{1}{2}\delta$.

$H(x) + H(-x) = 1$ and $\mathcal{F}[H(x)] + \mathcal{F}[H(-x)] = \mathcal{F}[1] = \delta$.

But $H(-x) = \check{H}(x) \Longrightarrow \mathcal{F}[\check{H}(x)] = \overline{\mathcal{F}}[H(x)]$. Hence, $\mathcal{F}H + \bar{\mathcal{F}}H = \delta$ in $S'(\mathbb{R})$.

But

$$\langle \mathcal{F} H - \overline{\mathcal{F}} H, \phi\rangle = \int_{-\infty}^{\infty}\left(\int_{0}^{\infty} e^{-i2\pi x\xi} - e^{i2\pi x\xi} dx\right)\phi(\xi)d\xi$$
$$= -2i\int_{0}^{\infty}\int_{-\infty}^{\infty} \sin 2\pi x\xi\phi(\xi)d\xi dx \quad \forall\phi \in S(\mathbb{R})$$
$$= -2i \lim_{\substack{M\to\infty\\ \varepsilon\to 0^+}}\int_{0}^{M}\int_{|\xi|\geq\varepsilon} \sin 2\pi x\xi\phi(\xi)d\xi dx,$$

where the limit is to be understood in the sense of distribution. Applying Fubini's Theorem 7.1.2C, we can interchange the order of integration and get, by integration with respect to x:

$$\langle \mathcal{F} H - \overline{\mathcal{F}} H, \phi\rangle = i2 \lim_{\substack{M\to\infty\\ \varepsilon\to 0^+}}\int_{|\xi|\geq\varepsilon} \frac{\cos M2\pi\xi - 1}{2\pi\xi}\phi(\xi)d\xi$$
$$= \frac{-i}{\pi}\lim_{\varepsilon\to 0^+}\int_{|\xi|\geq\varepsilon}\frac{1}{\xi}\phi(\xi)d\xi = \frac{-i}{\pi}\left\langle \text{c.p.v.}\,\frac{1}{\xi}, \phi\right\rangle,$$

since $\lim_{M\to\infty}\cos M2\pi\xi = 0$ in the distribution sense (see (1.8.12)).
$\Longrightarrow \mathcal{F} H - \overline{\mathcal{F}} H = -\frac{i}{\pi}\,\text{c.p.v.}\,\frac{1}{\xi} = -\frac{i}{\pi}\,\text{c.p.v.}\,\frac{1}{\xi}$ in $S'(\mathbb{R})$ and $\mathcal{F} H + \overline{\mathcal{F}} H = \delta$ in $S'(\mathbb{R}) \Longrightarrow \quad (\mathcal{F} H)(\xi) = \frac{1}{2}\delta - \frac{i}{2\pi}\,\text{c.p.v.}\,\frac{1}{\xi}$, $(\bar{\mathcal{F}} H)(\xi) = \frac{1}{2}\delta + \frac{i}{2\pi}\,\text{c.p.v.}\,\frac{1}{\xi}$ (see also Example 8.6.4).

Fourier Transforms $\mathcal{F}[|x|]$ and $\mathcal{F}[\text{Pf}\,\frac{1}{\xi^2}]$

Example 8.6.6. Find $\mathcal{F}[|x|]$ and $\mathcal{F}[\text{Pf}\,\frac{1}{\xi^2}]$, where $\text{Pf}\,\frac{1}{\xi^2}$ is defined by (1.4.19).

Solution. Consider the identity $|x| = xH(x) - xH(-x)$, where H is the Heaviside function with $H(x) = 1$ for $x > 0$ and $= 0$ for $x < 0$, since the right-hand side equals x for $x > 0$ and $-x$ for $x < 0$. Hence,

$$\mathcal{F}[|x|] = \mathcal{F}[xH(x)] - \mathcal{F}[xH(-x)] = \frac{-1}{i2\pi}\frac{d}{d\xi}(\mathcal{F} H)(\xi) - \frac{-1}{i2\pi}\frac{d}{d\xi}[\mathcal{F} H(-x)].$$

But, from Example 8.6.5, $(\mathcal{F} H)(\xi) = \frac{1}{2}\delta + \frac{1}{i2\pi}\,\text{c.p.v.}\,\frac{1}{\xi}$. $\mathcal{F}[H(-x)] = \mathcal{F}\check{H} = \overline{\mathcal{F}} H$. In fact, $\langle \mathcal{F}\check{H}, \phi\rangle = \langle \check{H}, \mathcal{F}\phi\rangle = \langle H, (\mathcal{F}\phi)^{\vee}\rangle = \langle H, \overline{\mathcal{F}}\phi\rangle = \langle \overline{\mathcal{F}} H, \phi\rangle\ \forall\phi \in S(\mathbb{R})$, since $(\mathcal{F}\phi)^{\vee} = (\mathcal{F}\phi)(-\xi) = \int_{\mathbb{R}}\phi(x)e^{-i2\pi x(-\xi)}dx = \int_{\mathbb{R}}\phi(x)e^{i2\pi x\xi}dx = \overline{\mathcal{F}}\phi$.

From Example 8.6.5, $\overline{\mathcal{F}} H(\xi) = \frac{1}{2}\delta - \frac{1}{i2\pi}\,\text{c.p.v.}\,\frac{1}{\xi}$. Hence,

$$\mathcal{F}[|x|] = \frac{-1}{i2\pi}\frac{d}{d\xi}\left[\frac{1}{2}\delta + \frac{1}{i2\pi}\,\text{c.p.v.}\,\frac{1}{\xi}\right] + \frac{1}{i2\pi}\frac{d}{d\xi}\left[\frac{1}{2}\delta - \frac{1}{i2\pi}\,\text{c.p.v.}\,\frac{1}{\xi}\right]$$
$$= \frac{d}{d\xi}(\text{c.p.v.}\,\frac{1}{\xi})\left(\frac{2}{4\pi^2}\right) = \frac{1}{2\pi^2}\frac{d}{d\xi}(\text{c.p.v.}\,\frac{1}{\xi}) = \frac{1}{2\pi^2}\left(-\text{Pf}\,\frac{1}{\xi^2}\right) \quad \text{(see (2.3.32))}$$

$\Longrightarrow \mathcal{F}(|x|) = -\frac{1}{2\pi^2}\,\mathrm{Pf}(\frac{1}{\xi^2}) \Longrightarrow \mathcal{F}\mathcal{F}(|x|) = |\check{x}| = |x| = -\frac{1}{2\pi^2}\mathcal{F}[\mathrm{Pf}\,\frac{1}{\xi^2}]$ (see (8.3.21)) $\Longrightarrow \mathcal{F}[\mathrm{Pf}\,\frac{1}{\xi^2}] = -2\pi^2|x|$.

Fourier transform of e^{-x^2}, $e^{i\pi x^2}$

Instead of finding the Fourier transforms of e^{-x^2} and $e^{i2\pi x^2}$ directly from the definition, we will find them with the help of differential equations as shown in the following example.

Example 8.6.7.

1. Find the tempered distribution solutions of $\frac{du}{dx} + \lambda x u = 0$ in $S'(\mathbb{R})$ for complex $\lambda \neq 0$.
2. Let $u \in S'(\mathbb{R})$ be a tempered distribution solution of the equation in (1) and $\hat{u} = \mathcal{F}u$ be its Fourier transform. Find the differential equation satisfied by $\hat{u} \in S'(\mathbb{R})$. Then, show the tempered distribution solutions of this equation for $\hat{u}$.
3. Applying the results of (1) and (2), find the Fourier transforms $\mathcal{F}[e^{-x^2}]$ and $\mathcal{F}[e^{i\pi x^2}]$. Finally, using these results, find the Fourier transform of $[e^{-(x^2+y^2)+i2y}]$ in $S'(\mathbb{R}^2)$.

Solution.

1. From Theorem 2.7.2, the distribution solutions of $\frac{du}{dx} + \lambda x u = 0$ in $\mathcal{D}'(\mathbb{R})$ are the usual solutions: $u = Ce^{-\int \lambda x dx} = Ce^{-\frac{\lambda x^2}{2}}$, where C is a constant. For $\lambda \in \mathbb{C}, \mathrm{Re}(\lambda) \geq 0, e^{-\frac{\lambda x^2}{2}} \in L^1(\mathbb{R})$ is a tempered distribution in $S'(\mathbb{R})$. Hence, $\forall \lambda \in \mathbb{C}$ with $\mathrm{Re}(\lambda) \geq 0$, $u = Ce^{-\frac{\lambda x^2}{2}} \in S'(\mathbb{R})$ are tempered distribution solutions of the equation $u' + \lambda x u = 0$. For $\lambda \in \mathbb{C}$ with $\mathrm{Re}(\lambda) < 0$, classical solutions $u = Ce^{-\frac{\lambda x^2}{2}}$ are *not* tempered distributions on $\mathbb{R}$ and hence *not* tempered distribution solutions.
2. For $\mathrm{Re}(\lambda) \geq 0$, $\mathcal{F}u$ is well defined and given by $\mathcal{F}u = \int_{-\infty}^{\infty} u(x)e^{-i2\pi x\xi}dx$ and $\mathcal{F}[\frac{du}{dx}] = i2\pi\xi\mathcal{F}u$, $\mathcal{F}[xu] = -\frac{1}{i2\pi}\frac{d}{d\xi}(\mathcal{F}u)$. Set $\hat{u} = \mathcal{F}u$. Then, $\lambda\frac{d}{d\xi}\hat{u} = -i2\pi\lambda\mathcal{F}[xu] = -i2\pi\mathcal{F}[\lambda x u]$ and $i2\pi\xi(\mathcal{F}u) = \mathcal{F}[\frac{du}{dx}] \Longrightarrow -i2\pi(i2\pi\xi)\hat{u} = -i2\pi\mathcal{F}[\frac{du}{dx}] \Longrightarrow \lambda\frac{d\hat{u}}{d\xi} + 4\pi^2\xi\hat{u} = -i2\pi\mathcal{F}[\frac{du}{dx} + \lambda x u] = -i2\pi\mathcal{F}0 = 0$. Thus, $\hat{u}$ satisfies the differential equation $\lambda\frac{d\hat{u}}{d\xi} + 4\pi^2\xi\hat{u} = 0$.

   For $\lambda \neq 0$, $\frac{d\hat{u}}{d\xi} + \frac{4\pi^2}{\lambda}\xi\hat{u} = 0 \Longrightarrow \hat{u} = C_1e^{-\frac{4\pi^2}{\lambda}\int \xi d\xi} = C_1e^{\frac{-2\pi^2\xi^2}{\lambda}}$ is the general solution, C_1 being a constant. For $\mathrm{Re}(\lambda) \geq 0$, $e^{\frac{-2\pi^2\xi^2}{\lambda}} \in S'(\mathbb{R})$.
   For $\lambda = 0$, the differential equation reduces to $\xi\hat{u} = 0 \Longrightarrow \hat{u} = C_2\delta \in S'(\mathbb{R})$, C_2 being a constant.

3. Choose $\lambda = 2$, $C = 1$, i.e. $u = e^{-x^2}$. Then $\hat{u} = \mathcal{F}u = C_1 e^{-\frac{2\pi^2\xi^2}{2}} = C_1 e^{-\pi^2\xi^2}$, where C_1 is an unknown constant, which will be determined now.

$$C_1 = \hat{u}(0) = (\mathcal{F}u)(0) = \int_{-\infty}^{\infty} u(x)dx = \int_{-\infty}^{\infty} e^{-x^2}dx = \sqrt{\pi}.$$

Thus,

$$\mathcal{F}[e^{-x^2}] = \sqrt{\pi}e^{-\pi^2\xi^2}.$$

Now choose $\lambda = -i2\pi$ with $\mathrm{Re}(\lambda) = 0$, $C = 1$, i.e. $u(x) = e^{i\pi x^2}$.
Then $\hat{u} = \mathcal{F}u = C_1 e^{-\frac{2\pi^2}{-i2\pi}\xi^2} = C_1 e^{-i\pi\xi^2} = C_1\overline{u}(\xi)$, $\overline{u}(\xi)$ = complex conjugate of $u(\xi)$, $C_1 \in \mathbb{R}$.
$\overline{\mathcal{F}}\hat{u} = \overline{\mathcal{F}}\mathcal{F}u = u$ and $\overline{\mathcal{F}}\hat{u} = \overline{\mathcal{F}}[C_1\overline{u}] = C_1\overline{\mathcal{F}}[\overline{u}] = C_1\overline{\mathcal{F}u} = C_1C_1\overline{\overline{u}} = C_1^2u$
$\Longrightarrow C_1^2 u = u$
$\Longrightarrow C_1^2 = 1$ and $C_1 = 1$, since $C_1 > 0$.

$$\mathcal{F}[e^{i\pi x^2}] = e^{-i\pi\xi^2}.$$

$e^{-(x^2+y^2)+i2y}$ is integrable on $\mathbb{R}^2$, i.e. belongs to $L^1(\mathbb{R}^2)$

$$\Longrightarrow \quad \mathcal{F}[e^{-(x^2+y^2)+i2y}](\xi,\eta) = \iint_{\mathbb{R}^2} e^{-(x^2+y^2)+i2y}e^{-i2\pi(x\xi+y\eta)}dxdy$$
$$= \left(\int_{\mathbb{R}} e^{-x^2}e^{-i2\pi x\xi}d\xi\right)\cdot\int_{\mathbb{R}} e^{-y^2+i2y}e^{-i2\pi y\eta}dy$$
$$= \mathcal{F}[e^{-x^2}](\xi)\cdot\int_{\mathbb{R}} e^{-y^2}e^{-i2\pi(\eta-1/\pi)y}dy$$
$$= (\sqrt{\pi}e^{-\pi^2\xi^2})\cdot\mathcal{F}[e^{-y^2}](\eta-1/\pi)$$
$$= \sqrt{\pi}e^{-\pi^2\xi^2}\cdot\sqrt{\pi}e^{-\pi^2(\eta-1/\pi)^2}$$
$$= \pi e^{-\pi^2\xi^2}\cdot e^{-\pi^2(\eta^2+\frac{1}{\pi^2}-2\eta/\pi)}$$
$$= \pi e^{-\pi^2\xi^2}\cdot e^{-1}\cdot e^{-(\pi^2\eta^2-2\eta\pi)}.$$
$$\mathcal{F}[e^{-(x^2+y^2)+i2y}](\xi,\eta) = \pi e^{-\pi^2\xi^2}.e^{-1}\cdot e^{-(\pi^2\eta^2-2\eta\pi)}.$$

Fourier transform of $\sin x$, $\cos x$, $H(x)\sin x$

Example 8.6.8.

1. Find the Fourier transform of $\sin x$ and $\cos x$. Then, find $\mathcal{F}[H(x)\sin x]$, where H is the Heaviside function.

2. With the help of Fourier transforms, find the tempered distribution solution of the differential equation $\frac{d^2u}{dx^2} + \lambda u = H(x)\sin x$ for the admissible real values of λ.

Solution.

1. $\sin x = \frac{e^{ix}-e^{-ix}}{2i}$, $\cos x = \frac{e^{ix}+e^{-ix}}{2}$ are tempered distributions in $S'(\mathbb{R})$. In fact, for example,

$$|T(\phi)| = |\langle \sin x, \phi\rangle| = \left|\int_{-\infty}^{\infty} \sin x\phi(x)dx\right| \leq \int_{-\infty}^{\infty} |\phi(x)|dx < \infty \ \forall \phi \in S(\mathbb{R})$$

$\Longrightarrow T(\phi)$ is well defined $\forall \phi \in S(\mathbb{R}) \Longrightarrow \sin x$ is a tempered distribution on $\mathbb{R}$.

Similarly, $\cos x \in S'(\mathbb{R})$. Hence, their Fourier transforms $\mathcal{F}[\sin x]$ and $\mathcal{F}[\cos x]$ are well defined. In fact, from (8.4.10), we have $\mathcal{F}[e^{\pm ix}] = \mathcal{F}[e^{i2\pi x(\pm 1/(2\pi))}] = \delta_{\pm 1/2\pi}$, i.e. $\mathcal{F}[e^{ix}] = \delta_{1/2\pi}$, $\mathcal{F}[e^{-ix}] = \delta_{-1/2\pi}$, where $\delta_{\frac{1}{2\pi}}$ (resp. $\delta_{-\frac{1}{2\pi}}$) is the Dirac distribution with concentration at $a = \frac{1}{2\pi}$ (resp. $-\frac{1}{2\pi}$) defined by:

$$\langle \delta_{\frac{1}{2\pi}}, \phi\rangle = \phi\left(\frac{1}{2\pi}\right), \quad \langle \delta_{-\frac{1}{2\pi}}, \phi\rangle = \phi\left(-\frac{1}{2\pi}\right) \quad \forall \phi \in S(\mathbb{R}).$$

$$\mathcal{F}[\sin x] = \frac{1}{2i}(\mathcal{F}[e^{ix}] - \mathcal{F}[e^{-ix}]) = \frac{1}{2i}(\delta_{\frac{1}{2\pi}} - \delta_{-\frac{1}{2\pi}});$$

$$\mathcal{F}[\cos x] = \frac{1}{2}(\mathcal{F}[e^{ix}] + \mathcal{F}[e^{-ix}]) = \frac{1}{2}(\delta_{\frac{1}{2\pi}} + \delta_{-\frac{1}{2\pi}}).$$

$H(x)\sin x$ defines a tempered distribution on $\mathbb{R}$. Hence, $\mathcal{F}[H(x)\sin x]$ is well defined by (8.3.1). But $\mathcal{F}[\sin x] = \frac{1}{2i}(\delta_{1/2\pi} - \delta_{-\frac{1}{2\pi}}) \in \mathcal{E}'(\mathbb{R}) \subset S'(\mathbb{R})$ with *compact support* in $\mathbb{R} \Longrightarrow$ its convolution with $\mathcal{F}H = \frac{1}{2}\delta - \frac{i}{2\pi}$ c.p.v. $\frac{1}{\xi} \in S'(\mathbb{R})$ (see Example 8.6.4) is well defined. Hence, $\mathcal{F}[H(x)\sin x] = \mathcal{F}[H(x)] * \mathcal{F}[\sin x]$. In fact,

$$\bar{\mathcal{F}}(\mathcal{F}(H(x)\sin x) - \mathcal{F}[H(x)] * \mathcal{F}[\sin x]) = H(x)\sin x - \bar{\mathcal{F}}(\mathcal{F}[H(x)] * \mathcal{F}[\sin x])$$

$$= H(x)\sin x - \bar{\mathcal{F}}(\mathcal{F}[H(x)]) \cdot \bar{\mathcal{F}}(\mathcal{F}[\sin x]) = H(x)\sin x - H(x)\sin x = 0$$

$\Longrightarrow$ the equality (applying (8.5.14) and (8.3.18)). Hence,

$$\mathcal{F}[H(x)\sin x] = \mathcal{F}[H(x)] * \mathcal{F}[\sin x]$$

$$= \left(\frac{1}{2}\delta - \frac{i}{2\pi}\text{ c.p.v. }\frac{1}{\xi}\right) * \frac{1}{2i}(\delta_{\frac{1}{2\pi}} - \delta_{-\frac{1}{2\pi}})$$

$$= \frac{1}{2}\delta * \frac{1}{2i}(\delta_{\frac{1}{2\pi}} - \delta_{-\frac{1}{2\pi}}) - \frac{1}{4\pi}\text{ c.p.v. }\frac{1}{(\xi - \frac{1}{2\pi})} + \frac{1}{4\pi}\text{ c.p.v. }\frac{1}{(\xi + \frac{1}{2\pi})}.$$

$$\mathcal{F}[H(x)\sin x] = \frac{1}{4i}(\delta_{\frac{1}{2\pi}} - \delta_{-\frac{1}{2\pi}}) - \frac{1}{4\pi}\text{ c.p.v. }\frac{1}{(\xi - \frac{1}{2\pi})} + \frac{1}{4\pi}\text{ c.p.v. }\frac{1}{(\xi + \frac{1}{2\pi})},$$

where $\forall\phi \in S(\mathbb{R})$, $\langle \text{c.p.v.}(\frac{1}{\xi \mp \frac{1}{2\pi}}), \phi\rangle = \text{c.p.v.} \int_{-\infty}^{\infty} \frac{\phi(\xi)}{\xi \mp \frac{1}{2\pi}} d\xi$ (see (1.3.38)–(1.3.45)).

2. Let $u \in S'(\mathbb{R})$. Then $\mathcal{F} f = \hat{u} \in S'(\mathbb{R})$ and we have

$$\mathcal{F}\left[\frac{d^2u}{dx^2} + \lambda u\right] = \mathcal{F}[H(x)\sin x] \in S'(\mathbb{R})$$

$$\implies \mathcal{F}\left[\frac{d^2u}{dx^2}\right] + \lambda \mathcal{F}[u] = \mathcal{F}[H(x)\sin x].$$

But from (8.6.1), $\mathcal{F}[\frac{d^2u}{dx^2}] = (i2\pi\xi)^2 \mathcal{F} u = -4\pi^2\xi^2\hat{u} \implies (-4\pi^2\xi^2 + \lambda)\hat{u} = \mathcal{F}[H(x)\sin x]$, which has a unique tempered distribution solution, if the corresponding homogeneous equation $(-4\pi^2\xi^2 + \lambda)\hat{u} = 0$ has only the trivial solution $\hat{u} = 0$. But $(\lambda - 4\pi^2\xi^2)\hat{u} = 0$ has the trivial solution only if $\lambda - 4\pi^2\xi^2 \neq 0\ \forall\xi \in \mathbb{R}$, i.e. for $\lambda < 0$, $\lambda - 4\pi^2\xi^2 \neq 0\ \forall\xi \in \mathbb{R}$. Hence, for real $\lambda < 0$, the equation has a unique tempered distribution solution. Choose $\lambda = -a^2$ with $a > 0$. Then $a^2 + 4\pi^2\xi^2$ is a C^∞-function and $a^2 + 4\pi^2\xi^2 \neq 0$ $\forall\xi \in \mathbb{R}$. Hence, $\frac{1}{a^2+4\pi^2\xi^2}$ is also a C^∞-function on $\mathbb{R}$ and we can write $\hat{u} = \frac{-1}{a^2+4\pi^2\xi^2}\mathcal{F}[H(x)\sin x]$. Then $u = \overline{\mathcal{F}}\mathcal{F} u = \overline{\mathcal{F}}\hat{u}$

$$\implies u = \overline{\mathcal{F}}\left[-\frac{1}{a^2 + 4\pi^2\xi^2}\mathcal{F}[H(x)\sin x]\right]$$
$$= \overline{\mathcal{F}}\left[-\frac{1}{a^2 + 4\pi^2\xi^2}\right] * \overline{\mathcal{F}}\mathcal{F}[H(x)\sin x],$$

where it is *assumed* that the convolution is well defined, even though *neither* distribution has compact support. *Under this assumption*, we have, $\forall a > 0$,

$$u = \overline{\mathcal{F}}\left[-\frac{1}{a^2 + 4\pi^2\xi^2}\right] * H(x)\sin x.$$

We can further simplify using the result of Example 8.6.5.
In fact, $\frac{1}{a^2+4\pi^2\xi^2} = \frac{1}{2a}[\frac{1}{a+i2\pi\xi} + \frac{1}{a-i2\pi\xi}]$

$$\implies \overline{\mathcal{F}}\left[-\frac{1}{a^2 + 4\pi^2\xi^2}\right] = -\frac{1}{2a}\left(\overline{\mathcal{F}}\left[\frac{1}{a + i2\pi\xi}\right] + \overline{\mathcal{F}}\left[\frac{1}{a - i2\pi\xi}\right]\right)$$
$$= -\frac{1}{2a}[H(x)e^{-ax} + H(-x)e^{ax}],$$

since, from Example 8.6.5, $\mathcal{F}[H(x)e^{-ax}] = \frac{1}{a+i2\pi\xi}$, $\mathcal{F}[H(-x)e^{ax}] = \frac{1}{a-i2\pi\xi}$.
Hence, for $\lambda = -a^2 < 0$ with $a > 0$,

$$u = -\frac{1}{2a}[H(x)e^{-ax} + H(-x)e^{ax}] * H(x)\sin x.$$

Fourier transform of $\frac{1}{r^k}$

Let

$$r = r(\mathbf{x}) = \|\mathbf{x}\| = (x_1^2 + x_2^2 + \cdots + x_n^2)^{\frac{1}{2}} \quad \forall \mathbf{x} = (x_1, x_2, \ldots, x_n) \in \mathbb{R}^n \quad \text{and}$$

$$f(\mathbf{x}) = \frac{1}{r^k}. \tag{8.6.3}$$

For $0 < k < n$, $f \in S'(\mathbb{R}^n)$ and $\mathcal{F}[f] \in S'(\mathbb{R}^n)$. (8.6.4)

For $0 < k < n$, $\forall$ compact subsets $K \subset\subset \mathbb{R}^n$, $\int_K \frac{1}{r^k} d\mathbf{x} < +\infty \Longrightarrow f \in L^1_{\text{loc}}(\mathbb{R}^n)$
$\Longrightarrow f = \frac{1}{r^k} \in \mathcal{D}'(\mathbb{R}^n)$ is a distribution on $\mathbb{R}^n$. Let 1_B be the characteristic function of the unit ball $B = B(\mathbf{0}; 1) = \{\mathbf{x} : \|\mathbf{x}\| < 1\}$ in $\mathbb{R}^n$, $B^{\complement} = \{\mathbf{x} : \|\mathbf{x}\| \geq 1\}$ being the complement of B in $\mathbb{R}^n$, i.e.

$$1_B(\mathbf{x}) = \begin{cases} 1 & \text{for } \mathbf{x} \in B \text{ (i.e. for } \|\mathbf{x}\| < 1) \\ 0 & \text{for } \mathbf{x} \in B^{\complement} \text{ (i.e. for } \|\mathbf{x}\| \geq 1). \end{cases}$$

Then $f = 1_B f + (1 - 1_B) f = f_B + f_{B^{\complement}}$, where

$$f_B(\mathbf{x}) = 1_B(\mathbf{x}) f(\mathbf{x}) = \begin{cases} f(\mathbf{x}) & \text{for } \|\mathbf{x}\| < 1 \\ 0 & \text{for } \|\mathbf{x}\| \geq 1, \end{cases}$$

and $f_{B^{\complement}}(\mathbf{x}) = f(\mathbf{x}) - f_B(\mathbf{x})\ \forall \mathbf{x} \in \mathbb{R}^n$.

For $0 < k < n$, $\int_{\mathbb{R}^n} f_B(\mathbf{x}) d\mathbf{x} = \int_{\|\mathbf{x}\|<1} \frac{1}{r^k} d\mathbf{x} < +\infty \Longrightarrow f_B \in L^1(\mathbb{R}^n) \Longrightarrow f_B \in S'(\mathbb{R}^n)$. Hence, $f = f_B + f_{B^{\complement}} \in S'(\mathbb{R}^n) \Longrightarrow \mathcal{F}[f] = \mathcal{F}[\frac{1}{r^k}] \in S'(\mathbb{R}^n)$ for $0 < k < n$.

$f = f(r) = \frac{1}{r^k}$ is a radial (i.e. spherically symmetric) tempered distribution for $0 < k < n \Longrightarrow \mathcal{F}[f](\boldsymbol{\xi}) = \hat{f}(\rho)$, with $\rho = (\xi_1^2 + \xi_2^2 + \cdots + \xi_n^2)^{\frac{1}{2}}$, is also a radial (i.e. spherically symmetric) tempered distribution (see (7.1.20)), the proof for tempered distributions being exactly similar to that given for functions of $L^1(\mathbb{R}^n)$ in Property 4(c) (with necessary minor modifications).

For $2k > n$, i.e. $\frac{n}{2} < k < n$, $\mathcal{F}[f] = \hat{f}$ is a *function*: $f_{B^{\complement}}$ is a *square integrable function* on $\mathbb{R}^n$, i.e. $f_{B^{\complement}} \in L^2(\mathbb{R}^n)$, since

$$\int_{\mathbb{R}^n} |f_{B^{\complement}}(\mathbf{x})|^2 d\mathbf{x} = \int_{\|\mathbf{x}\|\geq 1} \frac{1}{r^{2k}} d\mathbf{x} = S_n \int_1^{\infty} \frac{1}{r^{2k}} r^{n-1} dr < +\infty$$

for $-2k + (n-1) + 1 < 0$, i.e. for $2k > n$, S_n being the surface area of the n-dimensional unit sphere, which is obtained as the result of transformation into spherical coordinates $\Longrightarrow \mathcal{F}[f_{B^{\complement}}] = \hat{f}_{B^{\complement}} \in L^2(\mathbb{R}^n)$ (by Theorem 8.3.1). $f_B \in L^1(\mathbb{R}^n) \Longrightarrow \mathcal{F}[f_B] = \hat{f}_B$ is a *continuous, bounded function* in $\mathbb{R}^n$ with $\|\hat{f}\|_\infty \to 0$

as $\|\boldsymbol{\xi}\| \to 0$ (see (7.1.24), (7.1.33), (7.1.36)). Hence, $\mathcal{F}[f] = \hat{f}_B + \hat{f}_{B^\complement} = \hat{f}$ is a *function*, which is *locally square integrable* on $\mathbb{R}^n$. For $\frac{n}{2} < k < n$, $\mathcal{F}[f] = \hat{f}$ is a radial (spherically symmetric) function in $S'(\mathbb{R}^n)$.

$f(r) = \frac{1}{r^k} \Longrightarrow f$ is a homogeneous radial function of degree $-k \in r$ $\Longrightarrow f(\lambda r) = \frac{1}{(\lambda r)^k} = \lambda^{-k}.\frac{1}{r^k} = \lambda^{-k} f(r)$ in $r \Longrightarrow \mathcal{F}[f] = \hat{f}$ is a homogeneous radial function of degree $-(n + (-k)) = k - n$ in ρ (see Proposition 8.3.1) $\Longrightarrow \hat{f} = \frac{C_{k,n}}{\rho^{n-k}}$, where $C_{k,n}$ is a constant to be determined now.

Calculation of $C_{k,n}$: For $\phi = e^{-\pi\rho^2}$, $\hat{\phi} = \mathcal{F}[\phi] = e^{-\pi r^2}$ (Example 7.1.2), $\langle \mathcal{F}[\frac{1}{r^k}], e^{-\pi\rho^2}\rangle = \langle \frac{1}{r^k}, e^{-\pi r^2}\rangle$ with $\mathcal{F}[\frac{1}{r^k}] = \frac{C_{k,n}}{\rho^{n-k}} \Longrightarrow \int_{\mathbb{R}^n} \frac{C_{k,n}}{\rho^{n-k}} e^{-\pi\rho^2} d\boldsymbol{\xi} = \int_{\mathbb{R}^n} \frac{e^{-\pi r^2}}{r^k} d\mathbf{x}$

$$\Longrightarrow \quad C_{k,n} = \frac{S_n \int_0^\infty \frac{e^{-\pi r^2}}{r^k} r^{n-1} dr}{S_n \int_0^\infty \frac{e^{-\pi\rho^2}}{\rho^{n-k}} \rho^{n-1} d\rho} = \frac{I_1}{I_2},$$

where

$$I_1 = \int_0^\infty e^{-\pi r^2} r^{n-k-1} dr = \frac{1}{2\pi^{\frac{n-k}{2}}} \Gamma\left(\frac{n-k}{2}\right) \quad \text{and}$$
$$I_2 = \int_0^\infty e^{-\pi\rho^2} \rho^{k-1} d\rho = \frac{1}{2\pi^{\frac{k}{2}}} \Gamma\left(\frac{k}{2}\right)$$

are obtained first by transformation into spherical coordinates and then again by introducing new variables. In fact, setting $t = \pi r^2$, $r = \sqrt{\frac{t}{\pi}}$, $dr = \frac{dt}{2\sqrt{\pi t}}$, we get

$$I_1 = \int_0^\infty e^{-t} \frac{(\sqrt{t})^{n-k-1}}{(\sqrt{\pi})^{n-k-1} 2\sqrt{\pi t}} dt = \frac{1}{2\pi^{\frac{n-k}{2}}} \int_0^\infty e^{-t} t^{\frac{n-k}{2}-1} dt = \frac{\Gamma(\frac{n-k}{2})}{2\pi^{\frac{n-k}{2}}},$$

and setting $s = \pi\rho^2$, $\rho = \sqrt{\frac{s}{\pi}}$, $ds = \frac{ds}{2\sqrt{\pi s}}$, we get

$$I_2 = \frac{1}{2\pi^{\frac{k}{2}}} \int_0^\infty e^{-s} s^{\frac{k}{2}-1} ds = \frac{\Gamma(\frac{k}{2})}{2\pi^{\frac{k}{2}}}.$$

Here,

$$C_{k,n} = \frac{2\pi^{\frac{k}{2}} \Gamma(\frac{n-k}{2})}{2\pi^{\frac{n-k}{2}} \Gamma(\frac{k}{2})} = \pi^{k-\frac{n}{2}} \frac{\Gamma(\frac{n-k}{2})}{\Gamma(\frac{k}{2})}.$$

Finally, for $\frac{n}{2} < k < n$,

$$\mathcal{F}\left[\frac{1}{r^k}\right] = \pi^{k-\frac{n}{2}} \frac{\Gamma(\frac{n-k}{2})}{\Gamma(\frac{k}{2})} . \frac{1}{\rho^{n-k}}. \tag{8.6.5}$$

Case $0 < 2k < n$, *i.e.* $0 < k < n/2$: Set $p = n - k$. Then, $k < \frac{n}{2} \Longrightarrow p = n - k > n - \frac{n}{2} = \frac{n}{2} \Longrightarrow \frac{n}{2} < p < n$, i.e. we are again in the earlier case for $\frac{1}{r^p} \Longrightarrow \mathcal{F}[\frac{1}{r^p}] = \frac{C_{p,n}}{\rho^{n-p}} \Longrightarrow \mathcal{F}[\frac{1}{r^{n-k}}] = C_{p,n}\frac{1}{\rho^k} \Longrightarrow \overline{\mathcal{F}}\mathcal{F}[\frac{1}{r^{n-k}}] = C_{p,n}\overline{\mathcal{F}}[\frac{1}{\rho^k}] = C_{p,n}\mathcal{F}[\frac{1}{\rho^k}]^\vee = C_{p,n}\mathcal{F}[\frac{1}{\rho^k}] \Longrightarrow \frac{1}{r^{n-k}} = C_{p,n}C_{k,n}\frac{1}{r^{n-k}} \Longrightarrow C_{p,n}C_{k,n} = 1 \Longrightarrow \frac{1}{C_{p,n}} = C_{k,n}$.

Hence, for $0 < 2k < n$, i.e. $0 < k < \frac{n}{2}$, $\mathcal{F}[\frac{1}{\rho^k}] = \frac{1}{C_{p,n}}\frac{1}{r^{n-k}} = \frac{C_{k,n}}{r^{n-k}}$. Interchanging the notation ρ with r and vice versa (Schwartz [8] uses the same notation r instead of ρ), we get, for $0 < 2k < n$, i.e. $0 < k < \frac{n}{2}$,

$$\mathcal{F}\left[\frac{1}{r^k}\right] = \frac{C_{k,n}}{\rho^{n-k}} = \frac{\pi^{k-\frac{n}{2}}\Gamma(\frac{n-k}{2})}{\Gamma(\frac{k}{2})}\frac{1}{\rho^{n-k}}.$$

Thus, for $0 < k < \frac{n}{2}$ and $\frac{n}{2} < k < n$, the formula (8.6.5) holds. Finally, for $k = \frac{n}{2}$,

$$\mathcal{F}\left[\frac{1}{r^{\frac{n}{2}}}\right] = \lim_{k\to\frac{n}{2}} \mathcal{F}\left[\frac{1}{r^k}\right] = \lim_{k\to\frac{n}{2}}\left[\frac{\pi^{k-\frac{n}{2}}\Gamma(\frac{n-k}{2})}{\Gamma(\frac{k}{2})}\cdot\frac{1}{\rho^{n-k}}\right]. \tag{8.6.6}$$

Combining (8.6.5) and (8.6.6), we get $\mathcal{F}[\frac{1}{r^k}]$ for $0 < k < n$:

$$\mathcal{F}\left[\frac{1}{r^k}\right] = \frac{\pi^{k-\frac{n}{2}}\Gamma(\frac{n-k}{2})}{\Gamma(\frac{k}{2})}\cdot\frac{1}{\rho^{n-k}}. \tag{8.6.7}$$

Singular values of k *in* $\frac{1}{r^k}$: Formula (8.6.5) with $0 < k < n$ does *not* hold for $\frac{n-k}{2} = -p$ and $k = -2p$ with integer $p \geq 0$, since $\Gamma(\frac{n-k}{2})$ (resp. $\Gamma(\frac{k}{2})$) has poles at $\frac{n-k}{2} = -p$ (resp. $k = -2p$). Hence, for these singular values of $k = n + 2p$ (resp. $k = -2p$) with $p \geq 0$, the formula for $\mathcal{F}[\frac{1}{r^k}]$ is to be modified by taking limits (starting with non-singular values of k as in (8.6.7)) and given by [8, p. 258]: for $k = -2p$ with $p \in \mathbb{N}_0$,

$$\mathcal{F}[r^{2p}] = \left(-\frac{\Delta_{\boldsymbol{\xi}}}{4\pi^2}\right)^p\delta, \tag{8.6.8}$$

where the notation 'Pf' is useless, since $2p \geq 0$; $\Delta_{\boldsymbol{\xi}} = \frac{\partial^2}{\partial\xi_1^2} + \cdots + \frac{\partial^2}{\partial\xi_n^2}$; $\langle\delta, \phi(\boldsymbol{\xi})\rangle = \phi(\mathbf{0})\ \forall\phi \in S(\mathbb{R}^n)$.

For $k = n + 2p$ with $p \in \mathbb{N}_0$, $n + 2p > n$, 'Pf' is to be used and we get, from [8, p. 258],

$$\mathcal{F}\left[\mathrm{Pf}\left(\frac{1}{r^{n+2p}}\right)\right] = \frac{\pi^{\frac{n}{2}+2p}}{\Gamma(\frac{n}{2}+p)}\cdot 2\frac{(-1)^p}{p!}\rho^{2p} \cdot\left[\ln\frac{1}{\pi\rho} + \frac{1}{2}\left(1 + \frac{1}{2} + \cdots + \frac{1}{p} - C\right) + \frac{1}{2}\frac{\Gamma'(\frac{n}{2}+p)}{\Gamma(\frac{n}{2}+p)}\right], \tag{8.6.9}$$

where C is Euler's constant, $\mathrm{Pf}(\frac{1}{r^{n+2p}})$ is defined by (3.3.42), the sum $(1+\frac{1}{2}+\cdots+\frac{1}{p})$ must be replaced by 0 for $p=0$.

For $k=n$ (i.e. $p=0$) we get, from (8.6.9),

$$\mathcal{F}\left[\mathrm{Pf}\left(\frac{1}{r^n}\right)\right]=\frac{2(\sqrt{\pi})^n}{\Gamma(\frac{n}{2})}\cdot\left[\ln\frac{1}{\rho}+A\right]\quad\text{with } A=\ln\frac{1}{\pi}-\frac{C}{2}+\frac{1}{2}\frac{\Gamma'(\frac{n}{2})}{\Gamma(\frac{n}{2})} \tag{8.6.10}$$

$$\Longrightarrow\quad \overline{\mathcal{F}}\mathcal{F}\left[\mathrm{Pf}\left(\frac{1}{r^n}\right)\right]=\frac{2(\sqrt{\pi})^n}{\Gamma(\frac{n}{2})}\cdot\left[\overline{\mathcal{F}}\left[\ln\frac{1}{\rho}\right]+A\overline{\mathcal{F}}[1]\right]$$

$$\Longrightarrow\quad \mathrm{Pf}\left(\frac{1}{r^n}\right)=\frac{2(\sqrt{\pi})^n}{\Gamma(\frac{n}{2})}\cdot\left[\mathcal{F}\left[\ln\frac{1}{\rho}\right]+A\delta\right],$$

since $\check{\rho}(\xi)=\rho(-\xi)=\rho(\xi)$, $\overline{\mathcal{F}}[\ln(\frac{1}{\rho})]=\mathcal{F}[\ln(\frac{1}{\rho})]$. Replacing ρ by r and vice versa, we can write

$$\mathcal{F}\left[\ln\left(\frac{1}{r}\right)\right]=\frac{\Gamma(\frac{n}{2})}{2(\sqrt{\pi})^n}\left[\mathrm{Pf}\left(\frac{1}{\rho^n}\right)\right]-A\delta \tag{8.6.11}$$

with A defined in (8.6.10).

For $n=1$,

$$\mathcal{F}[\ln|x|]=-\frac{1}{2}\,\mathrm{Pf}\left(\frac{1}{|\xi|}\right)-(C+\ln 2\pi)\delta,\quad [8] \tag{8.6.12}$$

C being Euler's constant.

Fourier transform of the tensor product $T(\mathbf{x})\otimes S(\mathbf{y})$ of tempered distributions $T(\mathbf{x})$ and $S(\mathbf{y})$

Let $T(\mathbf{x})$ and $S(\mathbf{y})$ ($T_{\mathbf{x}}$, $S_{\mathbf{y}}$ being alternative notations – see Chapter 6) be tempered distributions with Fourier transforms $\hat{T}(\boldsymbol{\xi})=\mathcal{F}T(\boldsymbol{\xi})$, $\hat{S}(\boldsymbol{\eta})=\mathcal{F}S(\boldsymbol{\eta})$, and $T(\mathbf{x})\otimes S(\mathbf{y})$ be their tensor product (see Chapter 6). Then the Fourier transform $\mathcal{F}[T(\mathbf{x})\otimes S(\mathbf{y})]$ of the tensor product $T(\mathbf{x})\otimes S(\mathbf{y})$ is equal to the tensor product of $\hat{T}(\boldsymbol{\xi})\otimes\hat{S}(\boldsymbol{\eta})$ of their Fourier transforms, i.e.

$$\mathcal{F}[T(\mathbf{x})\otimes S(\mathbf{y})]=\hat{T}(\boldsymbol{\xi})\otimes\hat{S}(\boldsymbol{\eta}). \tag{8.6.13}$$

For this, it is sufficient [1] to show that the equation (8.6.13) holds for functions $\phi(\boldsymbol{\xi},\boldsymbol{\eta})=\sum_i\phi_i(\boldsymbol{\xi})\psi_i(\boldsymbol{\eta})$ with $\phi_i(\boldsymbol{\xi}),\psi_i(\boldsymbol{\eta})\in\mathcal{D}(\mathbb{R}^n)$. In fact,

$$\begin{aligned}\left\langle\mathcal{F}[T(\mathbf{x})\otimes S(\mathbf{y})],\sum_i\phi_i(\boldsymbol{\xi})\psi_i(\boldsymbol{\eta})\right\rangle&=\left\langle T(\mathbf{x})\otimes S(\mathbf{y}),\mathcal{F}\left[\sum_i\phi_i(\boldsymbol{\xi})\psi_i(\boldsymbol{\eta})\right]\right\rangle\\&=\left\langle T(\mathbf{x})\otimes S(\mathbf{y}),\sum_i\hat{\phi}_i(\mathbf{x})\hat{\psi}_i(\mathbf{y})\right\rangle\end{aligned}$$

with $\hat{\phi}_i(\mathbf{x}) = (\mathcal{F}\phi_i)(\mathbf{x})$,

$$\hat{\psi}_i(\mathbf{y}) = (\mathcal{F}\psi_i)(\mathbf{y}) \quad \text{(see (7.1.15))}$$

$$= \left\langle T(\mathbf{x}), \left\langle S(\mathbf{y}), \sum_i \hat{\phi}_i(\mathbf{x})\hat{\psi}_i(\mathbf{y}) \right\rangle\right\rangle = \sum_i \langle T(\mathbf{x}), \hat{\phi}_i(\mathbf{x})\rangle \cdot \langle S(\mathbf{y}), \hat{\psi}_i(\mathbf{y})\rangle$$

$$= \sum_i \langle \mathcal{F}T(\boldsymbol{\xi}), \phi_i(\boldsymbol{\xi})\rangle\langle \mathcal{F}S(\boldsymbol{\eta}), \psi_i(\boldsymbol{\eta})\rangle = \left\langle \mathcal{F}T(\boldsymbol{\xi}) \otimes \mathcal{F}S(\boldsymbol{\eta}), \sum_i \phi_i(\boldsymbol{\xi})\psi_i(\boldsymbol{\eta})\right\rangle$$

$$\Longrightarrow \mathcal{F}[T(\mathbf{x}) \otimes S(\mathbf{y})] = \hat{T}(\boldsymbol{\xi}) \otimes \hat{S}(\boldsymbol{\eta}).$$

Example 8.6.9.

1. For

$$f(x, y) = H(x) \otimes H(y) = \begin{cases} 1 & \text{for } x > 0 \text{ and } y > 0 \\ 0 & \text{otherwise,} \end{cases}$$

$$\mathcal{F}[f(x, y)] = \mathcal{F}[H(x) \otimes H(y)] = \mathcal{F}H(\xi) \otimes \mathcal{F}H(\eta)$$
$$= \left(\frac{1}{2}\delta(\xi) + \frac{1}{i2\pi}\,\text{c.p.v.}\,\frac{1}{\xi}\right) \otimes \left(\frac{1}{2}\delta(\eta) + \frac{1}{i2\pi}\,\text{c.p.v.}\,\frac{1}{\eta}\right),$$

which is obtained from Example 8.6.4, $\delta(\xi)$ (resp. $\delta(\eta)$) being the Dirac distribution with force/charge/mass concentrated at $\xi = 0$ (resp. $\eta = 0$).

2. For $T = \delta(\mathbf{x})$, $\mathcal{F}[\delta(\mathbf{x}) \otimes S(\mathbf{y})] = \mathcal{F}\delta(\boldsymbol{\xi}) \cdot \mathcal{F}S(\boldsymbol{\eta}) = 1(\boldsymbol{\xi}) \otimes \hat{S}(\boldsymbol{\eta})$.

3. For $T = \delta(\mathbf{x})$, $S(\mathbf{y}) = 1(\mathbf{y})$, $\mathcal{F}[\delta(\mathbf{x}) \otimes 1(\mathbf{y})] = 1(\xi) \otimes \delta(\eta)$.

4. For $T = T(x_1, x_2, \dots, x_n) = 1(x_1, x_2, \dots, x_k) \otimes S(x_{k+1}, \dots, x_n)$, i.e. T is independent of the variables $x_1, x_2, \dots, x_k$, (see (6.1.18)),

$$(\mathcal{F}T)(\xi_1, \dots, \xi_n) = \mathcal{F}[1(x_1, x_2, \dots, x_k) \otimes S(x_{k+1}, \dots, x_n)]$$
$$= (\mathcal{F}1)(\xi_1, \dots, \xi_k) \otimes \mathcal{F}S(\xi_{k+1}, \dots, \xi_n)$$
$$= \delta(\xi_1, \dots, \xi_k) \otimes \mathcal{F}S(\xi_{k+1}, \dots, \xi_n),$$

i.e. Fourier transform $\mathcal{F}T = \hat{T}$ of a distribution T independent of the variables $x_1, x_2, \dots, x_k$ (see (6.1.18)) is the extension of the distribution $\hat{S}(\xi_{k+1}, \xi_{k+2}, \dots, \xi_n)$ defined on the subspace of the variables $\xi_{k+1}, \xi_{k+2}, \dots, \xi_n$ to $\delta(\xi_1, \dots, \xi_k) \otimes \hat{S}(\xi_{k+1}, \dots, \xi_n)$ defined on the whole space of variables $\xi_1, \xi_2, \cdots, \xi_n$.

8.7 Fourier transform methods for differential equations and elementary solutions in $S'(\mathbb{R}^n)$

Differential operators and Fourier transforms

For the sake of convenience and good notation, we will *differentiate* between the partial differential operators $\partial^{\boldsymbol{\alpha}}$ and $D^{\boldsymbol{\alpha}}$.

$$\partial^{\boldsymbol{\alpha}} = \partial^{(\alpha_1,\alpha_2,\dots,\alpha_n)} = \partial_1^{\alpha_1}\partial_2^{\alpha_2}\cdots\partial_n^{\alpha_n} = \frac{\partial^{|\boldsymbol{\alpha}|}}{\partial x_1^{\alpha_1}\partial x_2^{\alpha_2}\dots\partial x_n^{\alpha_n}}$$

with $\partial_k^{\alpha_k} = \frac{\partial^{\alpha_k}}{\partial x_k}$, $1 \le k \le n$;

$$\begin{aligned} D^{\boldsymbol{\alpha}} &= D^{(\alpha_1,\alpha_2,\dots,\alpha_n)} = D_1^{\alpha_1}D_2^{\alpha_2}\dots D_n^{\alpha_n} \\ &= \frac{1}{(i2\pi)^{\alpha_1}}\partial_1^{\alpha_1}\left(\frac{1}{(i2\pi)^{\alpha_2}}\partial_2^{\alpha_2}\left(\cdots\left(\frac{1}{(i2\pi)^{\alpha_n}}\partial_n^{\alpha_n}\right)\cdots\right)\right) \\ &= \frac{1}{(i2\pi)^{|\boldsymbol{\alpha}|}}\partial_1^{\alpha_1}\partial_2^{\alpha_2}\cdots\partial_n^{\alpha_n} = \frac{1}{(i2\pi)^{|\boldsymbol{\alpha}|}}\partial^{\boldsymbol{\alpha}} \quad \text{with} \qquad (8.7.1) \\ D^{\alpha_k} &= \frac{1}{(i2\pi)^{\alpha_k}}\partial_k^{\alpha_k} = \frac{1}{(i2\pi)^{\alpha_k}}\frac{\partial^{\alpha_k}}{\partial x_k^{\alpha_k}}, \qquad 1 \le k \le n. \end{aligned}$$

With each polynomial $P(\boldsymbol{\xi}) = \sum_{|\boldsymbol{\alpha}|\le m} a_{\boldsymbol{\alpha}}\boldsymbol{\xi}^{\boldsymbol{\alpha}}$ of degree $\le m$ in $\mathbb{R}^n$ with constant coefficients $a_{\boldsymbol{\alpha}} \in \mathbb{C}\ \forall|\boldsymbol{\alpha}| \le m$, we associate the partial differential operator $P(D)$ with constant coefficients, $P(D) = \sum_{|\boldsymbol{\alpha}|\le m} a_{\boldsymbol{\alpha}} D^{\boldsymbol{\alpha}}$, where $D^{\boldsymbol{\alpha}}$ is defined by (8.7.1).

Similarly, we can associate another partial differential operator $P(\partial)$ with the polynomial $P(\boldsymbol{\xi})$, defined by:

$$P(\partial) = \sum_{|\boldsymbol{\alpha}|\le m} a_{\boldsymbol{\alpha}}\partial^{\boldsymbol{\alpha}} = \sum_{|\boldsymbol{\alpha}|\le m} a_{\boldsymbol{\alpha}}\frac{\partial^{|\boldsymbol{\alpha}|}}{\partial x_1^{\alpha_1}\partial x_2^{\alpha_2}\dots\partial x_n^{\alpha_n}}.$$

Then $P(\partial) = \sum_{|\boldsymbol{\alpha}|\le m} a_{\boldsymbol{\alpha}}\partial^{\boldsymbol{\alpha}}$ is related to $P(D)$ by:

$$P(D) = \sum_{|\boldsymbol{\alpha}|\le m} a_{\boldsymbol{\alpha}} D^{\boldsymbol{\alpha}} = \sum_{|\boldsymbol{\alpha}|\le m} a_{\boldsymbol{\alpha}}\frac{1}{(i2\pi)^{|\boldsymbol{\alpha}|}}\partial^{\boldsymbol{\alpha}}. \qquad (8.7.2)$$

Definition 8.7.1A. A *linear partial differential operator* A with constant coefficients $a_{\boldsymbol{\alpha}} \in \mathbb{C}\ \forall|\boldsymbol{\alpha}| \le m$ defined by

$$Au = P(\partial)u = \sum_{|\boldsymbol{\alpha}|\le m} a_{\boldsymbol{\alpha}}\partial^{\boldsymbol{\alpha}}u = \sum_{|\boldsymbol{\alpha}|\le m} a_{\boldsymbol{\alpha}}\frac{\partial^{|\boldsymbol{\alpha}|}}{\partial x_1^{\alpha_1}\partial x_2^{\alpha_2}\cdots\partial x_n^{\alpha_n}} \qquad \forall u \in \mathcal{D}'(\mathbb{R}^n)$$

is called *elliptic* in $\mathbb{R}^n$ if and only if

$$P_m(\boldsymbol{\xi}) = \sum_{|\boldsymbol{\alpha}|=m} a_{\boldsymbol{\alpha}} \boldsymbol{\xi}^{\boldsymbol{\alpha}} \neq 0 \quad \forall \boldsymbol{\xi} \neq \mathbf{0} \text{ in } \mathbb{R}^n,$$

where $P_m(\boldsymbol{\xi})$ is a *homogeneous polynomial* of degree equal to m in the n variables $\xi_1, \xi_2, \ldots, \xi_n$.

Fourier transform of $P(D)u$ Let $u \in S'(\mathbb{R}^n)$. Then the Fourier transform of $P(D)u$ (resp. $P(\partial)u$) is defined by:

$$\begin{aligned}
\mathcal{F}[P(D)u] &= \sum_{|\boldsymbol{\alpha}|\le m} a_{\boldsymbol{\alpha}} \mathcal{F}[D^{\boldsymbol{\alpha}} u] = \sum_{|\boldsymbol{\alpha}|\le m} a_{\boldsymbol{\alpha}} \frac{1}{(i2\pi)^{|\boldsymbol{\alpha}|}} \mathcal{F}[\partial^{\boldsymbol{\alpha}} u] \\
&= \sum_{|\boldsymbol{\alpha}|\le m} a_{\boldsymbol{\alpha}} \frac{(i2\pi\boldsymbol{\xi})^{\boldsymbol{\alpha}}}{(i2\pi)^{|\boldsymbol{\alpha}|}} \mathcal{F}[u] = \sum_{|\boldsymbol{\alpha}|\le m} a_{\boldsymbol{\alpha}} \boldsymbol{\xi}^{\boldsymbol{\alpha}} \hat{u}(\boldsymbol{\xi}) = P(\boldsymbol{\xi})\hat{u}(\boldsymbol{\xi}), \qquad (8.7.3) \\
\mathcal{F}[P(\partial)u] &= \sum_{|\boldsymbol{\alpha}|\le m} a_{\boldsymbol{\alpha}} \mathcal{F}[\partial^{\boldsymbol{\alpha}} u] = \sum_{|\boldsymbol{\alpha}|\le m} a_{\boldsymbol{\alpha}} (i2\pi\boldsymbol{\xi})^{\boldsymbol{\alpha}} \mathcal{F}[u] = \sum_{|\boldsymbol{\alpha}|\le m} a_{\boldsymbol{\alpha}} (i2\pi)^{|\boldsymbol{\alpha}|} \boldsymbol{\xi}^{\boldsymbol{\alpha}} \hat{u}(\boldsymbol{\xi}), \\
& \qquad (8.7.4)
\end{aligned}$$

where $|\boldsymbol{\alpha}| = \alpha_1 + \alpha_2 + \cdots + \alpha_n$, $\boldsymbol{\xi}^{\boldsymbol{\alpha}} = \xi_1^{\alpha_1} \xi_2^{\alpha_2} \cdots \xi_n^{\alpha_n}$, $\hat{u} = \mathcal{F}[u]$.

Example 8.7.1. For $u \in S'(\mathbb{R}^2)$, find the Fourier transform of

1. Δu;
2. $\sum_{i=1}^{2} \sum_{j=1}^{2} \frac{\partial}{\partial x_j}(a_{ij} \frac{\partial u}{\partial x_i}) + a_0 u = a_{ij} \frac{\partial^2 u}{\partial x_i \partial x_j} + a_0 u$ with constant coefficients $a_0, a_{ij} \in \mathbb{R}$, $1 \le i, j \le 2$;[2]
3. $\Delta\Delta u$.

Solution.

1. $$\begin{aligned}
\mathcal{F}[\Delta u] &= \mathcal{F}\left[\frac{\partial^2 u}{\partial x_1^2} + \frac{\partial^2 u}{\partial x_2^2}\right] = \mathcal{F}[\partial^{(2,0)} u + \partial^{(0,2)} u] \\
&= (i2\pi)^2 \xi_1^2 \xi_2^0 \mathcal{F}[u] + (i2\pi)^2 \xi_1^0 \xi_2^2 \mathcal{F}[u] = -4\pi^2[\xi_1^2 + \xi_2^2]\hat{u} \\
\implies \quad & \mathcal{F}[\Delta u] = -4\pi^2 \rho^2 \hat{u} \quad \text{with } \rho^2 = \xi_1^2 + \xi_2^2. \qquad (8.7.5)
\end{aligned}$$

[2] Here we have followed on the right-hand side of the equality Einstein's summation convention with respect to twice-repeated indices to avoid the summation sign.

2. $$\mathcal{F}\left[a_{11}\frac{\partial^2 u}{\partial x_1^2} + (a_{12}+a_{21})\frac{\partial^2 u}{\partial x_1 \partial x_2} + a_{22}\frac{\partial^2 u}{\partial x_2^2} + a_0 u\right]$$
$$= a_{11}\mathcal{F}\left[\frac{\partial^2 u}{\partial x_1^2}\right] + (a_{12}+a_{21})\mathcal{F}\left[\frac{\partial^2 u}{\partial x_1 \partial x_2}\right] + a_{22}\mathcal{F}\left[\frac{\partial^2 u}{\partial x_2^2}\right] + a_0\mathcal{F}[u].$$
$$\frac{\partial^2 u}{\partial x_1^2} = \partial^{(2,0)}u \implies \mathcal{F}\left[\frac{\partial^2 u}{\partial x_1^2}\right] = (i2\pi)^2\xi_1^2\hat{u} = -4\pi^2\xi_1^2\hat{u},$$
$$\frac{\partial^2 u}{\partial x_1 \partial x_2} = \partial^{(1,1)}u \implies \mathcal{F}\left[\frac{\partial^2 u}{\partial x_1 \partial x_2}\right] = (i2\pi)^2\xi_1\xi_2\hat{u} = -4\pi^2\xi_1\xi_2\hat{u},$$
$$\frac{\partial^2 u}{\partial x_2^2} = \partial^{(0,2)}u \implies \mathcal{F}\left[\frac{\partial^2 u}{\partial x_2^2}\right] = (i2\pi)^2\xi_2^2\hat{u} = -4\pi^2\xi_2^2\hat{u}.$$

Hence,

$$\mathcal{F}\left[\sum_{i=1}^{2}\sum_{j=1}^{2} a_{ij}\frac{\partial^2 u}{\partial x_i \partial x_j} + a_0 u\right] = -4\pi^2[a_{11}\xi_1^2 + a_{12}\xi_1\xi_2$$
$$+ a_{21}\xi_2\xi_1 + a_{22}\xi_2^2]\hat{u} + a_0\hat{u}$$
$$= \left(-4\pi^2\sum_{i=1}^{2}\sum_{j=1}^{2} a_{ij}\xi_i\xi_j + a_0\right)\hat{u}. \quad (8.7.6)$$

3. $$\mathcal{F}\left[\frac{\partial^4 u}{\partial x_1^4} + 2\frac{\partial^4 u}{\partial x_1^2 \partial x_2^2} + \frac{\partial^4 u}{\partial x_2^4}\right] = \mathcal{F}[\partial^{(4,0)}u + 2\partial^{(2,2)}u + \partial^{(0,4)}u]$$
$$= (i2\pi)^{4+0}\xi_1^4\xi_2^0\hat{u} + 2(i2\pi)^{2+2}\xi_1^2\xi_2^2\hat{u} + (i2\pi)^{0+4}\xi_1^0\xi_2^4\hat{u}$$
$$= 16\pi^4[\xi_1^4 + 2\xi_1^2\xi_2^2 + \xi_2^4]\hat{u} = 16\pi^4\rho^4\hat{u} \quad \text{with } \rho^2 = \xi_1^2 + \xi_2^2. \quad (8.7.7)$$

Theorem 8.7.1. *Let $P(D) = \sum_{|\boldsymbol{\alpha}|\le m} a_{\boldsymbol{\alpha}} D^{\boldsymbol{\alpha}}$ with $D^{\boldsymbol{\alpha}}$ defined by* (8.7.1) *be a partial differential operator in $\mathbb{R}^n$ such that $P(\boldsymbol{\xi}) \ne 0\ \forall \boldsymbol{\xi} \ne \mathbf{0}$ in $\mathbb{R}^n$. Then show that the kernel of $P(D)$ in $S'(\mathbb{R}^n)$ consists of polynomials, i.e. every solution of the equation $P(D)u = 0$ in $S'(\mathbb{R}^n)$ is a polynomial.*

Proof. Let $u \in S'(\mathbb{R}^n)$. Then the kernel of $P(D)$ in $S'(\mathbb{R}^n) = \{u : u \in S'(\mathbb{R}^n), P(D)u = 0\}$. But $P(D)u = 0$ in $S'(\mathbb{R}^n) \implies \mathcal{F}[P(D)u] = 0 \implies P(\boldsymbol{\xi})\hat{u}(\boldsymbol{\xi}) = 0$ in $S'(\mathbb{R}^n)$ by (8.7.3). $P(\boldsymbol{\xi}) \ne 0$ for $\boldsymbol{\xi} \in \mathbb{R}^n \setminus \{\mathbf{0}\}$ and $\hat{u} \in S'(\mathbb{R}^n) \subset \mathcal{D}'(\mathbb{R}^n)$ with $\hat{u}\downarrow_{\mathbb{R}^n\setminus\{\mathbf{0}\}} = 0 \implies \text{supp}(\hat{u}) = \{\mathbf{0}\}$. Hence, by Theorem 5.7.1, $\hat{u} = \sum_{|\boldsymbol{\alpha}|\le m_0} C_{\boldsymbol{\alpha}}\partial_{\boldsymbol{\xi}}^{\boldsymbol{\alpha}}$ $\delta(\boldsymbol{\xi})$, m_0 being the order of $\hat{u}$, $C_{\boldsymbol{\alpha}} \in \mathbb{C}\ \forall\boldsymbol{\alpha}$, $\delta(\boldsymbol{\xi})$ being the Dirac distribution with concentration at $\boldsymbol{\xi} = \mathbf{0}$. But $\hat{u} \in S'(\mathbb{R}^n)$

$$\Longrightarrow \quad \overline{\mathcal{F}}\hat{u} = \sum_{|\boldsymbol{\alpha}|\le m_0} C_{\boldsymbol{\alpha}} \overline{\mathcal{F}}[\partial_{\boldsymbol{\xi}}^{\boldsymbol{\alpha}} \delta] \quad \text{(by (8.3.14))}$$

$$= \sum_{|\boldsymbol{\alpha}|\le m_0} C_{\boldsymbol{\alpha}} (-i2\pi\mathbf{x})^{\boldsymbol{\alpha}} = \sum_{|\boldsymbol{\alpha}|\le m_0} C_{\boldsymbol{\alpha}} (-i2\pi)^{|\boldsymbol{\alpha}|}\mathbf{x}^{\boldsymbol{\alpha}}.$$

Hence, $u = \overline{\mathcal{F}}\hat{u} = \sum_{|\boldsymbol{\alpha}|\le m_0} d_{\boldsymbol{\alpha}}\mathbf{x}^{\boldsymbol{\alpha}}$, where $d_{\boldsymbol{\alpha}} = C_{\boldsymbol{\alpha}}(-i2\pi)^{|\boldsymbol{\alpha}|}\ \forall|\boldsymbol{\alpha}| \le m_0$ is a polynomial of degree $\le m_0$. □

Example 8.7.2. Let $P(D) = \sum_{|\boldsymbol{\alpha}|\le m} a_{\boldsymbol{\alpha}} D^{\boldsymbol{\alpha}}$ with $D^{\boldsymbol{\alpha}}$ defined by (8.7.1) be a partial differential operator with constant coefficients $a_{\boldsymbol{\alpha}} \in \mathbb{C}\ \forall|\boldsymbol{\alpha}| \le m$ and $D^{\boldsymbol{\alpha}}$ defined by (8.7.1) such that the corresponding polynomial $P(\boldsymbol{\xi})$ is *not* identically zero in $\mathbb{R}^n$. If $u \in \mathcal{E}'(\mathbb{R}^n)$ (i.e. a distribution with compact support) such that $P(D)u = 0$, then $u = 0$.

Solution. Let $u \in \mathcal{E}'(\mathbb{R}^n)$. Then $u \in S'(\mathbb{R}^n) \Longrightarrow \mathcal{F}u = \hat{u} \in S'(\mathbb{R}^n)$ and $\hat{u} \in C^\infty(\mathbb{R}^n)$ by Proposition 8.4.1. Again, $P(D)u = 0 \Longrightarrow \mathcal{F}[P(D)u] = P(\boldsymbol{\xi})\hat{u}(\boldsymbol{\xi}) = 0$ (by (8.7.3)). Let Z denote the set of zeros of $P(\boldsymbol{\xi})$, i.e. $Z = \{\boldsymbol{\xi} : \boldsymbol{\xi} \in \mathbb{R}^n, P(\boldsymbol{\xi}) = 0\}$. Then $P(\boldsymbol{\xi}) \ne 0\ \forall\boldsymbol{\xi} \in \mathbb{R}^n \setminus Z \Longrightarrow \hat{u}(\boldsymbol{\xi}) = 0\ \forall\boldsymbol{\xi} \in \mathbb{R}^n \setminus Z$. But the set Z of zeros of a polynomial is a closed set with an empty interior. Hence, $\mathbb{R}^n \setminus Z$ is *dense* in $\mathbb{R}^n$, in which the continuous function $\hat{u} \in C^\infty(\mathbb{R}^n)$ vanishes $\Longrightarrow \hat{u}(\boldsymbol{\xi}) = 0$ $\forall\boldsymbol{\xi} \in \mathbb{R}^n \Longrightarrow \overline{\mathcal{F}}[\hat{u}] = 0 \Longrightarrow u = 0$ in $S'(\mathbb{R}^n)$.

Hypoelliptic operator *A*

Following Lions and Magenes [15], we define:

Definition 8.7.1B. A linear differential A with coefficients $a_{ij} \in C^\infty(\Omega)$ is called *hypoelliptic* if and only if $u \in \mathcal{D}'(\Omega)$ *and* $Au \in C^\infty(\Omega)$ implies that $u \in C^\infty(\Omega)$.

A is *elliptic* $\Longrightarrow$ A is *hypoelliptic*.

For more details, we refer to Hörmander [5].

Example 8.7.3. Let $P(D)$ be the operator defined by (8.7.2). If $P(D)$ is *elliptic* in $\mathbb{R}^n$, i.e.

$$P_m(\boldsymbol{\xi}) = \sum_{|\boldsymbol{\alpha}|=m} a_{\boldsymbol{\alpha}}\boldsymbol{\xi}^{\boldsymbol{\alpha}} \ne 0 \quad \forall\boldsymbol{\xi} \ne \mathbf{0} \text{ in } \mathbb{R}^n, \tag{8.7.8}$$

then show that

1. $\exists\alpha_0 > 0$ such that $|P_m(\boldsymbol{\xi})| \ge \alpha_0 \|\boldsymbol{\xi}\|_{\mathbb{R}^n}^m\ \forall\ \boldsymbol{\xi} \in \mathbb{R}^n$; (8.7.9)
2. $\exists\alpha_1 > 0$ and $R > 0$ such that $|P(\boldsymbol{\xi})| \ge \alpha_1 \|\boldsymbol{\xi}\|_{\mathbb{R}^n}^m\ \forall\|\boldsymbol{\xi}\|_{\mathbb{R}^n} > R$. (8.7.10)

Proof.

1. The homogeneous polynomial $P_m(\boldsymbol{\xi}) = \sum_{|\boldsymbol{\alpha}|=m} a_{\boldsymbol{\alpha}} \boldsymbol{\xi}^{\boldsymbol{\alpha}}$ is continuous on the unit sphere $S = S(\mathbf{0}; 1) = \{\hat{\boldsymbol{\xi}} : \hat{\boldsymbol{\xi}} \in \mathbb{R}^n, \|\hat{\boldsymbol{\xi}}\| = (\xi_1^2 + \xi_2^2 + \cdots + \xi_n^2)^{\frac{1}{2}} = 1\}$, which is a compact subset of $\mathbb{R}^n$. Hence, $\exists C > 0$ such that $|P_m(\hat{\boldsymbol{\xi}})| \geq C$ $\forall \hat{\boldsymbol{\xi}} \in S \subset \mathbb{R}^n$. $\forall \boldsymbol{\xi} \neq \mathbf{0}$, $\frac{\boldsymbol{\xi}}{\|\boldsymbol{\xi}\|} = \hat{\boldsymbol{\xi}} \in S \Longrightarrow P_m(\hat{\boldsymbol{\xi}}) = P_m(\frac{\boldsymbol{\xi}}{\|\boldsymbol{\xi}\|}) \geq C$ $\forall \boldsymbol{\xi} \neq \mathbf{0}$ in $\mathbb{R}^n$. But for $\boldsymbol{\xi} \neq \mathbf{0}$,

$$P_m\left(\frac{\boldsymbol{\xi}}{\|\boldsymbol{\xi}\|}\right) = \sum_{|\boldsymbol{\alpha}|=m} a_{\boldsymbol{\alpha}} \left(\frac{\boldsymbol{\xi}}{\|\boldsymbol{\xi}\|}\right)^{\boldsymbol{\alpha}} = \sum_{|\boldsymbol{\alpha}|=m} a_{\boldsymbol{\alpha}} \frac{1}{\|\boldsymbol{\xi}\|^{|\boldsymbol{\alpha}|}} \boldsymbol{\xi}^{\boldsymbol{\alpha}}$$
$$= \frac{1}{\|\boldsymbol{\xi}\|^m} \sum_{|\boldsymbol{\alpha}|=m} a_{\boldsymbol{\alpha}} \boldsymbol{\xi}^{\boldsymbol{\alpha}} = \frac{1}{\|\boldsymbol{\xi}\|^m} P_m(\boldsymbol{\xi})$$

$\Longrightarrow P_m(\hat{\boldsymbol{\xi}}) = \frac{1}{\|\boldsymbol{\xi}\|^m} |P_m(\boldsymbol{\xi})| \geq C > 0$
$\Longrightarrow \forall \boldsymbol{\xi} \in \mathbb{R}^n$, $|P_m(\boldsymbol{\xi})| \geq \alpha_0 \|\boldsymbol{\xi}\|^m$, with $\alpha_0 = C > 0$, since for $\boldsymbol{\xi} = \mathbf{0}$ the inequality becomes an equality with both sides equal to zero.

2. $P(\boldsymbol{\xi}) = \sum_{|\boldsymbol{\alpha}|\leq m} a_{\boldsymbol{\alpha}} \boldsymbol{\xi}^{\boldsymbol{\alpha}} = P_m(\boldsymbol{\xi}) + P_{m-1}(\boldsymbol{\xi}) + \cdots + P_0(\boldsymbol{\xi})$ with $P_k(\boldsymbol{\xi}) = \sum_{|\boldsymbol{\alpha}|=k} a_{\boldsymbol{\alpha}} \boldsymbol{\xi}^{\boldsymbol{\alpha}}$ $\forall k = 0, 1, 2, \ldots, m$. But

$$|P_k(\boldsymbol{\xi})| \leq \sum_{|\boldsymbol{\alpha}|=k} |a_{\boldsymbol{\alpha}}| \|\boldsymbol{\xi}\|^{|\boldsymbol{\alpha}|} = \left(\sum_{|\boldsymbol{\alpha}|=k} |a_{\boldsymbol{\alpha}}|\right) \|\boldsymbol{\xi}\|^k = C_k \, \|\boldsymbol{\xi}\|^k$$

with $C_k = \sum_{|\boldsymbol{\alpha}|=k} |a_{\boldsymbol{\alpha}}|$.

$$|P_m(\boldsymbol{\xi})| = |P(\boldsymbol{\xi}) - \sum_{k=0}^{m-1} P_k(\boldsymbol{\xi})| \leq |P(\boldsymbol{\xi})| + \sum_{k=0}^{m-1} |P_k(\boldsymbol{\xi})|$$

$\Longrightarrow |P(\boldsymbol{\xi})| \geq |P_m(\boldsymbol{\xi})| - \sum_{k=0}^{m-1} |P_k(\boldsymbol{\xi})| \geq C\|\boldsymbol{\xi}\|^m - \sum_{k=0}^{m-1} C^* \|\boldsymbol{\xi}\|^k$ with $C^* = \max_{0\leq k\leq m-1}\{C_k\}$. Hence, for $1 \leq R < \|\boldsymbol{\xi}\| \leq \|\boldsymbol{\xi}\|^2 \leq \cdots \leq \|\boldsymbol{\xi}\|^{m-1} \leq \|\boldsymbol{\xi}\|^{m-1} \cdot \frac{\|\boldsymbol{\xi}\|}{R} = \frac{\|\boldsymbol{\xi}\|^m}{R} \Longrightarrow |P(\boldsymbol{\xi})| \geq C\|\boldsymbol{\xi}\|^m - mC^* \frac{\|\boldsymbol{\xi}\|^m}{R} = (C - \frac{mC^*}{R})\|\boldsymbol{\xi}\|^m \geq \frac{C}{2}\|\boldsymbol{\xi}\|^m$, if $\frac{mC^*}{R} \leq \frac{C}{2}$ for sufficiently large R. Then the inequality (8.7.10) holds with $\alpha_1 = \frac{C}{2} > 0$ for sufficiently large $R > 0$ such that $\frac{mC^*}{R} \leq \alpha_1$.

□

Remark 8.7.1. Example 8.7.3 suggests to redefine the ellipticity:Let $A \equiv P(D)$ be the linear partial differential operator with constant coefficients defined by (8.7.2). Then A is called *elliptic* in $\mathbb{R}^n$ if and only if (8.7.8) holds, i.e. $P_m(\boldsymbol{\xi}) = \sum_{|\boldsymbol{\alpha}|=m} a_{\boldsymbol{\alpha}} \boldsymbol{\xi}^{\boldsymbol{\alpha}} \neq 0$ $\forall \boldsymbol{\xi} \neq \mathbf{0}$ in $\mathbb{R}^n$, or equivalently (8.7.9) holds, i.e. $\exists \alpha_0 > 0$ such that $|P_m(\boldsymbol{\xi})| \geq \alpha_0 \|\boldsymbol{\xi}\|_{\mathbb{R}^n}^m$ $\forall$ $\boldsymbol{\xi} \in \mathbb{R}^n$.

Laplacian $\Delta(\mathscr{F}\phi)$ and iterated Laplacian $\Delta^k(\mathscr{F}\phi)$ of Fourier transform $\mathscr{F}\phi$ with $\phi \in S(\mathbb{R}^n)$

Let $\Delta_{\mathbf{x}} = \frac{\partial^2}{\partial x_1^2} + \frac{\partial^2}{\partial x_2^2} + \cdots + \frac{\partial^2}{\partial x_n^2}$ and $\Delta_{\mathbf{x}}^k = (\frac{\partial^2}{\partial x_1^2} + \frac{\partial^2}{\partial x_2^2} + \cdots + \frac{\partial^2}{\partial x_n^2})^k$. Then we have

$$\begin{aligned}\Delta_{\mathbf{x}}(\mathscr{F}\phi)(\mathbf{x}) &= \Delta_{\mathbf{x}} \int_{\mathbb{R}^n} \phi(\boldsymbol{\xi}) e^{-i2\pi\langle\boldsymbol{\xi},\mathbf{x}\rangle} d\boldsymbol{\xi} = \int_{\mathbb{R}^n} \phi(\boldsymbol{\xi}) \Delta_{\mathbf{x}} e^{-i2\pi\langle\boldsymbol{\xi},\mathbf{x}\rangle} d\boldsymbol{\xi} \\ &= \int_{\mathbb{R}^n} \phi(\boldsymbol{\xi})[(-i2\pi\xi_1)^2 + (-i2\pi\xi_2)^2 + \cdots + (-i2\pi\xi_n)^2] e^{-i2\pi\langle\boldsymbol{\xi},\mathbf{x}\rangle}\, d\boldsymbol{\xi} \\ &= \int_{\mathbb{R}^n} (-4\pi^2)(\xi_1^2 + \xi_2^2 + \cdots + \xi_n^2)\phi(\boldsymbol{\xi}) e^{-i2\pi\langle\boldsymbol{\xi},\mathbf{x}\rangle} d\boldsymbol{\xi} \\ &= \int_{\mathbb{R}^n} (-4\pi^2\rho^2)\phi(\boldsymbol{\xi}) e^{-i2\pi\langle\boldsymbol{\xi},\mathbf{x}\rangle}\, d\boldsymbol{\xi} = \mathscr{F}[-4\pi^2\rho^2\phi(\boldsymbol{\xi})](\mathbf{x}) \quad \forall\phi \in S(\mathbb{R}^n).\end{aligned} \tag{8.7.11}$$

$$\Delta_{\mathbf{x}}(\mathscr{F}\phi)(\mathbf{0}) = \int_{\mathbb{R}^n} -4\pi^2\rho^2\phi(\boldsymbol{\xi}) d\boldsymbol{\xi} \quad \text{with } \rho^2 = \xi_1^2 + \xi_2^2 + \cdots + \xi_n^2. \tag{8.7.12}$$

$$\begin{aligned}\Delta_{\mathbf{x}}^k(\mathscr{F}\phi)(\mathbf{x}) &= \int_{\mathbb{R}^n} \phi(\boldsymbol{\xi})\ \Delta_{\mathbf{x}}^k(e^{-i2\pi\langle\boldsymbol{\xi},\mathbf{x}\rangle}) d\boldsymbol{\xi} \\ &= \int_{\mathbb{R}^n} [(-4\pi^2)(\xi_1^2 + \xi_2^2 + \cdots + \xi_n^2)]^k \phi(\boldsymbol{\xi})\ e^{-i2\pi\langle\boldsymbol{\xi},\mathbf{x}\rangle} d\boldsymbol{\xi} \\ &= \int_{\mathbb{R}^n} (-4\pi^2\rho^2)^k \phi(\boldsymbol{\xi}) e^{-i2\pi\langle\boldsymbol{\xi},\mathbf{x}\rangle}\, d\boldsymbol{\xi} = \mathscr{F}[(-4\pi^2\rho^2)^k\phi]\ \forall\phi \in S(\mathbb{R}^n).\end{aligned} \tag{8.7.13}$$

$$\Delta_{\mathbf{x}}^k(\mathscr{F}\phi)(\mathbf{0}) = \int_{\mathbb{R}^n} (-4\pi^2\rho^2)^k \phi(\boldsymbol{\xi}) d\boldsymbol{\xi} \quad \forall\phi \in S(\mathbb{R}^n). \tag{8.7.14}$$

Fourier transforms $\mathscr{F}[(\frac{\Delta}{4\pi^2})^k\delta]$, $\mathscr{F}[(1 - \frac{\Delta}{4\pi^2})^m\delta]$, $\mathscr{F}[(\Delta - \lambda)^m\delta]$, $\mathscr{F}[(\Delta + \lambda)^m\delta]$ with $\lambda > 0$

$$\begin{aligned}&\left\langle \mathscr{F}\left[\left(\frac{\Delta}{4\pi^2}\right)\delta\right], \phi\right\rangle = \left\langle\left(\frac{\Delta}{4\pi^2}\right)\delta, \mathscr{F}\phi\right\rangle = \frac{1}{4\pi^2}\langle\delta, \Delta(\mathscr{F}\phi)\rangle = \frac{1}{4\pi^2}\Delta_{\mathbf{x}}(\mathscr{F}\phi)(\mathbf{0}) \\ &\quad = \frac{1}{4\pi^2}\int_{\mathbb{R}^n} -4\pi^2\rho^2\phi(\boldsymbol{\xi})d\boldsymbol{\xi} = \langle -\rho^2, \phi\rangle \quad \forall\phi \in S(\mathbb{R}^n) \quad \text{(by (8.7.12))} \\ &\Longrightarrow \quad \mathscr{F}\left[\left(\frac{\Delta}{4\pi^2}\right)\delta\right] = -\rho^2 \quad \text{in } S'(\mathbb{R}^n)\end{aligned} \tag{8.7.15}$$

with $\rho^2 = \xi_1^2 + \xi_2^2 + \cdots + \xi_n^2$.

$$\left\langle \mathcal{F}\left[\left(\frac{\Delta}{4\pi^2}\right)^k \delta\right], \phi\right\rangle = \left\langle \left(\frac{\Delta}{4\pi^2}\right)^k \delta, \mathcal{F}\phi\right\rangle$$
$$= \frac{1}{(4\pi^2)^k}\langle \delta, \Delta^k(\mathcal{F}\phi)\rangle \quad (\text{since } (-1)^{2k} = 1)$$
$$= \frac{1}{(4\pi^2)^k}\Delta_{\mathbf{x}}^k(\mathcal{F}\phi)(\mathbf{0}) = \frac{1}{(4\pi^2)^k}\int_{\mathbb{R}^n}(-4\pi^2\rho^2)^k\phi(\boldsymbol{\xi})d\boldsymbol{\xi}(\text{by (8.7.14)})$$
$$= (-1)^k\int_{\mathbb{R}^n}\rho^{2k}\phi(\boldsymbol{\xi})d\boldsymbol{\xi}$$
$$\Longrightarrow \quad \langle \mathcal{F}\left[\left(\frac{\Delta}{4\pi^2}\right)^k \delta\right], \phi\rangle = \langle (-1)^k\rho^{2k}, \phi\rangle \forall \phi \in S(\mathbb{R}^n)$$
$$\Longrightarrow \quad \mathcal{F}\left[\left(\frac{\Delta}{4\pi^2}\right)^k \delta\right] = (-1)^k\rho^{2k} \quad \text{in } S'(\mathbb{R}^n) \tag{8.7.16}$$

with $\rho^2 = \xi_1^2 + \xi_2^2 + \cdots + \xi_n^2$.

$$\mathcal{F}\left[\left(-\frac{\Delta}{4\pi^2}\right)^k \delta\right] = \rho^{2k} \quad \text{in } S'(\mathbb{R}^n). \tag{8.7.17}$$

$$\mathcal{F}\left[\left(1-\frac{\Delta}{4\pi^2}\right)\delta\right] = \mathcal{F}\left[\delta - \frac{\Delta}{4\pi^2}\delta\right] = \mathcal{F}[\delta] - \mathcal{F}\left[\left(\frac{\Delta}{4\pi^2}\right)\delta\right] = 1 + \rho^2 \text{ in } S'(\mathbb{R}^n). \tag{8.7.18}$$

$$\mathcal{F}\left[\left(1-\frac{\Delta}{4\pi^2}\right)^m \delta\right] = \mathcal{F}\left[\left(1 + \sum_{k=1}^m (-1)^k \frac{m(m-1)\cdots(m-k+1)}{k!}\left(\frac{\Delta}{4\pi^2}\right)^k\right)\delta\right]$$
$$= \mathcal{F}[\delta] + \sum_{k=1}^m (-1)^k \frac{m(m-1)\cdots(m-k+1)}{k!}\mathcal{F}\left[\left(\frac{\Delta}{4\pi^2}\right)^k \delta\right]$$
$$= 1 + \sum_{k=1}^m (-1)^k \frac{m(m-1)\cdots(m-k+1)}{k!}(-1)^k\rho^{2k} \quad (\text{by (8.7.16)})$$
$$= 1 + \sum_{k=1}^m \frac{m(m-1)\cdots(m-k+1)}{k!}(\rho^2)^k$$
$$= (1+\rho^2)^m \quad \text{in } S'(\mathbb{R}^n). \tag{8.7.19}$$

Similarly, we have

$$\mathcal{F}[(\Delta - \lambda)^m\delta] = (-4\pi^2\rho^2 - \lambda)^m \quad \text{in } S'(\mathbb{R}^n); \tag{8.7.20}$$
$$\mathcal{F}[(\Delta + \lambda)^m\delta] = (-4\pi^2\rho^2 + \lambda)^m \quad \text{in } S'(\mathbb{R}^n). \tag{8.7.21}$$

Fourier transforms $\mathcal{F}[\Delta(\frac{1}{r^{n-2}})]$ for $n \neq 2$, $\mathcal{F}[\Delta(\ln\frac{1}{r})]$ for $n = 2$, $\mathcal{F}[\text{c.p.v.}\,\frac{1}{x}]$ for $n = 1$

For $n \neq 2$, $n-2 < C$, we are *not* in a singular situation, and hence the notation Pf in front of $\frac{1}{r^{n-2}}$ is *useless*. Consequently, using (8.6.6) and (8.7.5) with $\hat{u} = \mathcal{F}[\delta] = 1$, we have

$$\mathcal{F}\left[\Delta\left(\frac{1}{r^{n-2}}\right)\right] = \mathcal{F}\left[\Delta\delta * \frac{1}{r^{n-2}}\right] = \mathcal{F}[\Delta\delta]\cdot\mathcal{F}\left[\frac{1}{r^{n-2}}\right] = (-4\pi^2\rho^2)\cdot\mathcal{F}\left[\frac{1}{r^{n-2}}\right]$$
$$= -4\pi^2\rho^2\frac{\pi^{n-2-\frac{n}{2}}\Gamma(\frac{n-n+2}{2})}{\Gamma(\frac{n-2}{2})}\cdot\frac{1}{\rho^{(n-(n-2))}} = \frac{-4\pi^2\pi^{-2}\pi^{\frac{n}{2}}}{\frac{2\Gamma(\frac{n}{2})}{n-2}}$$
$$= -(n-2)2\cdot\frac{\pi^{\frac{n}{2}}}{\Gamma(\frac{n}{2})} = -(n-2)S_n, \left(S_n = \frac{2\pi^{\frac{n}{2}}}{\Gamma\left(\frac{n}{2}\right)}\right) \tag{8.7.22}$$

from which we can retrieve the formula (3.3.15), since

$$\Delta\left(\frac{1}{r^{n-2}}\right) = \overline{\mathcal{F}}\mathcal{F}\left[\Delta\left(\frac{1}{r^{n-2}}\right)\right] = -(n-2)S_n\overline{\mathcal{F}}[1] = -(n-2)S_n\delta, \quad \overline{\mathcal{F}}[1] = \delta.$$

For $n = 2$, using (8.7.5) and (8.6.11),

$$\mathcal{F}\left[\Delta\left(\ln\frac{1}{r}\right)\right] = (-4\pi^2\rho^2)\mathcal{F}\left[\ln\frac{1}{r}\right] = (-4\pi^2\rho^2)\left[\frac{\Gamma(1)}{2\pi}\,\text{Pf}\left(\frac{1}{\rho^2}\right) - A\delta\right]$$
$$= \frac{-4\pi^2}{2\pi}\rho^2\cdot\text{Pf}\left(\frac{1}{\rho^2}\right) + A4\pi^2\rho^2\delta$$
$$\Longrightarrow \quad \mathcal{F}\left[\Delta\left(\ln\frac{1}{r}\right)\right] = -2\pi, \tag{8.7.23}$$

from which we retrieve (3.3.14), since $\overline{\mathcal{F}}[1] = \delta$, $\rho^2\cdot\text{Pf}(\frac{1}{\rho^2}) = 1$ and $\langle\rho^2\delta,\phi(\xi)\rangle = \langle\delta,\rho^2\phi(\xi)\rangle = 0\ \forall\phi\in S(\mathbb{R}^2)$
$\Longrightarrow \rho^2\delta = 0 \Longrightarrow A4\pi^2\rho^2\delta = 0$ in $S'(\mathbb{R}^2)$.

For $n = 1$, (6.3.24), (8.3.13) and (8.6.12) give

$$\mathcal{F}\left[\frac{d}{dx}\ln|x|\right] = \mathcal{F}\left[\frac{d}{dx}\delta\right]\cdot\mathcal{F}[\ln|x|] = i2\pi\xi\mathcal{F}[\ln|x|]$$
$$= -i\pi\xi\left(\text{Pf}\left(\frac{1}{|\xi|}\right)\right) - [C + \ln 2\pi]i2\pi\xi\delta = -i\pi\xi\left(\text{Pf}\left(\frac{1}{|\xi|}\right)\right),$$

since $\xi\delta = 0$ ($\langle\xi\delta,\phi(\xi)\rangle = \langle\delta,\xi\phi(\xi)\rangle = 0\ \forall\phi\in S(\mathbb{R}) \Longrightarrow \xi\delta = 0$).

From Example 2.3.6, $\frac{d}{dx}\ln|x| = \text{c.p.v.}\ \frac{1}{x}$. Hence,

$$\mathcal{F}\left[\text{c.p.v.}\ \frac{1}{x}\right] = \mathcal{F}\left[\frac{d}{dx}\ln|x|\right] = -i\pi\xi\begin{cases}\text{Pf}(\frac{1}{\xi}) & \text{for } \xi > 0\\ \text{Pf}(\frac{1}{-\xi}) & \text{for } \xi < 0\end{cases}$$

$$\Longrightarrow \quad \mathcal{F}\left[\text{c.p.v.}\ \frac{1}{x}\right] = \begin{cases}-i\pi & \text{for } \xi > 0\\ +i\pi & \text{for } \xi < 0\end{cases} \quad \text{(see Example 8.6.4)}, \tag{8.7.24}$$

since $\xi\,\text{Pf}(\frac{1}{\pm\xi}) = \pm 1$.

Fourier transform $\mathcal{F}[(1+r^2)^{-\frac{m}{2}}]$

The Fourier transform $\mathcal{F}[(1+r^2)^{-\frac{m}{2}}]$ for $m > \frac{n}{2}$ is given by [8]: $\forall m \neq 0, -2, -4, \ldots,$

$$\begin{aligned}\mathcal{F}\left[\frac{1}{(1+r^2)^{\frac{m}{2}}}\right] &= \frac{2(\sqrt{\pi})^m}{\Gamma(\frac{m}{2})} r^{\frac{(m-n)}{2}} K_{\frac{(n-m)}{2}}(2\pi r)\\ &= \frac{2(\sqrt{\pi})^m}{\Gamma(\frac{m}{2})}\text{Pf}[r^{\frac{(m-n)}{2}} K_{\frac{(n-m)}{2}}(2\pi r)] = L_m,\end{aligned} \tag{8.7.25}$$

where L_m is already defined in (3.3.57), since $\frac{1}{(1+r^2)^{\frac{m}{2}}} \in S'(\mathbb{R}^n)\ \forall m \in \mathbb{N}$.

For $m = -2k$ with $k \in \mathbb{N}_0$, $(1+r^2)^{-\frac{m}{2}} = (1+r^2)^k$. Then, from (8.6.9),

$$\mathcal{F}[(1+r^2)^k] = \left(1 - \frac{\Delta}{4\pi^2}\right)^k \delta = L_{-2k} \quad \text{(see (3.3.58))}. \tag{8.7.26}$$

In fact, $(1+r^2)^k = 1 + \sum_{p=1}^k \frac{k(k-1)\cdots(k-p+1)}{p!}(r^2)^p$

$$\begin{aligned}\Longrightarrow \quad \mathcal{F}[(1+r^2)^k] &= \mathcal{F}\left[1 + \sum_{p=1}^k \frac{k(k-1)\cdots(k-p+1)}{p!} r^{2p}\right]\\ &= \mathcal{F}[1] + \sum_{p=1}^k \frac{k(k-1)\cdots(k-p+1)}{p!}\mathcal{F}[r^{2p}]\\ &= \delta + \sum_{p=1}^k \frac{k(k-1)\cdots(k-p+1)}{p!}\left(-\frac{\Delta_\xi}{4\pi^2}\right)^p \delta \quad \text{(by (8.6.9))}\\ &= \left[1 + \sum_{p=1}^k (-1)^p \frac{k(k-1)\cdots(k-p+1)}{p!}\left(\frac{\Delta_\xi}{4\pi^2}\right)^p\right]\delta\\ &= \left(1 - \frac{\Delta_\xi}{4\pi^2}\right)^k \delta.\end{aligned}$$

In particular,

$$\text{for } k = 0,\ \mathcal{F}[1] = \delta; \quad \text{for } k = 1,\ \mathcal{F}[(1 + r^2)] = \left(1 - \frac{\Delta}{4\pi^2}\right)\delta, \tag{8.7.27}$$

$\delta = \delta(\xi)$, which we already know.

Problems of division of tempered distribution by polynomials

Let $P(\boldsymbol{\xi})$ be a polynomial in $\boldsymbol{\xi} = (\xi_1, \xi_2, \ldots, \xi_n)$ and $S \in S'(\mathbb{R}^n)$ be a given tempered distribution. Then the question is: does there exist a tempered distribution $T \in S'(\mathbb{R}^n)$ such that $P(\boldsymbol{\xi})T = S$ in $S'(\mathbb{R}^n)$? In other words, can we write $T = \frac{S}{P(\boldsymbol{\xi})} \in S'(\mathbb{R}^n)$? In fact, we are asking whether S can be divided by a polynomial $P(\boldsymbol{\xi})$. If the answer to these equivalent questions is *affirmative*, then every partial differential equation with constant coefficients,

$$P(\partial)T = S \quad \text{in } S'(\mathbb{R}^n), \tag{8.7.28}$$

where

$$P(\partial) = \sum_{|\boldsymbol{\alpha}| \le m} a_{\boldsymbol{\alpha}} \partial^{\boldsymbol{\alpha}} \tag{8.7.29}$$

is a linear partial operator with constant coefficients $a_{\boldsymbol{\alpha}} \in \mathbb{R}\ \forall$ multi-index $\boldsymbol{\alpha}$ with $|\boldsymbol{\alpha}| \le m$, will have a tempered distribution solution $T \in S'(\mathbb{R}^n)$. In fact, taking the Fourier transform of both sides of equation (8.7.28), we have, from (8.7.4), $\mathcal{F}[P(\partial)T] = \mathcal{F}[S]$ in $S'(\mathbb{R}^n) \implies P^*(\boldsymbol{\xi})\mathcal{F}[T] = \hat{S}$ with $P^*(\boldsymbol{\xi}) = \sum_{|\boldsymbol{\alpha}| \le m} a_{\boldsymbol{\alpha}}$ $(i2\pi)^{|\boldsymbol{\alpha}|}\boldsymbol{\xi}^{\boldsymbol{\alpha}} \implies \mathcal{F}[T] = \hat{T} = \frac{\hat{S}}{P^*(\boldsymbol{\xi})} \in S'(\mathbb{R}^n)$ is well defined by the assumption that the division by polynomial $P^*(\boldsymbol{\xi})$ is well defined. Then

$$T = \overline{\mathcal{F}}[\hat{T}] = \overline{\mathcal{F}}\left[\frac{\hat{S}}{P^*(\boldsymbol{\xi})}\right] \in S'(\mathbb{R}^n) \tag{8.7.30}$$

is a tempered distribution solution of equation (8.7.28). This affirmative result for division of a tempered distribution by a polynomial was proved by Hörmander [31]. Thus, we have:

Theorem 8.7.2. $\forall S \in S'(\mathbb{R}^n)$, *the partial differential equation with constant coefficients in* (8.7.28) *has a tempered distribution solution* $T \in S'(\mathbb{R}^n)$ *defined by* (8.7.30).

Example 8.7.4. For given $f \in \mathcal{E}'(\mathbb{R}^3)$ (i.e. with compact support in $\mathbb{R}^3$), find $u \in S'(\mathbb{R}^3)$ such that $-\Delta u = f$ in $S'(\mathbb{R}^3)$.

Solution. Taking the Fourier transform of both sides of the equation, $\mathcal{F}[-\Delta u] = \mathcal{F}[f] \Longrightarrow 4\pi^2\rho^2\hat{u} = \hat{f} \Longrightarrow \hat{u} = \frac{\hat{f}}{4\pi^2\rho^2} \Longrightarrow u = \overline{\mathcal{F}}[\hat{u}] = \overline{\mathcal{F}}[\frac{\hat{f}}{4\pi^2\rho^2}]$.

In particular, for $f = \delta \in \mathcal{E}'(\mathbb{R}^3)$, i.e. $-\Delta u = \delta$, $\hat{f} = \mathcal{F}[\delta] = 1$ and

$$u = \overline{\mathcal{F}}\left[\frac{1}{4\pi^2\rho^2}\right] = \frac{1}{4\pi^2}\overline{\mathcal{F}}\left[\frac{1}{\rho^2}\right] = \frac{1}{4\pi^2}\cdot\frac{\pi^{2-\frac{3}{2}}\cdot\Gamma(\frac{1}{2})}{\Gamma(1)}\cdot\frac{1}{r} = \frac{1}{4\pi r} \in S'(\mathbb{R}^3)$$

(using (8.6.5)). Thus, we retrieve the elementary solution $u = -\frac{1}{4\pi r}$ of Δ given in Theorem 3.3.2 for $n = 3$.

Example 8.7.5. Show that the homogeneous elliptic partial differential equation $(1 - \frac{\Delta}{4\pi^2})^k T = 0$ in $S'(\mathbb{R}^n)$ has *no* solution in $S'(\mathbb{R}^n)$ other than the trivial solution $T = 0$ in $S'(\mathbb{R}^n)$.

Proof. $\mathcal{F}[(1 - \frac{\Delta}{4\pi^2})^k T] = \mathcal{F}[0] = 0$ in $S'(\mathbb{R}^n)$.

But $(1 - \frac{\Delta}{4\pi^2})^k T = (1 - \frac{\Delta}{4\pi^2})^k(\delta * T) = (1 - \frac{\Delta}{4\pi^2})^k\delta * T$.

Hence,

$$\mathcal{F}\left[\left(1 - \frac{\Delta}{4\pi^2}\right)^k T\right] = \mathcal{F}\left[\left(1 - \frac{\Delta}{4\pi^2}\right)^k \delta\right]\cdot\mathcal{F}[T] = (1+\rho^2)^k\mathcal{F}[T] = 0$$

using the Convolution Theorem 8.5.3 and (8.7.19). Then $\mathcal{F}[T] = 0$, since $(1 + \rho^2)$ does *not* vanish in $\mathbb{R}^n \Longrightarrow T = 0$ in $S'(\mathbb{R}^n)$.

In fact, $(1 - \frac{\Delta}{4\pi^2})^k T = 0$ has *only* the null tempered solution. But this equation has an *infinite number* of *non*-tempered solutions $T = e^{2\pi\langle\hat{\mathbf{h}},\mathbf{x}\rangle} \in \mathcal{D}'(\mathbb{R}^n)$ [8, p. 282] with $\hat{\mathbf{h}} = (h_1, h_2, \ldots, h_n)$, $\|\hat{\mathbf{h}}\|_{\mathbb{R}^n} = (h_1^2 + h_2^2 + \cdots + h_n^2)^{\frac{1}{2}} = 1$, since, for $T \in \mathcal{D}'(\mathbb{R}^n)$,

$$\begin{aligned}\Delta T &= \left(\frac{\partial^2}{\partial x_1^2} + \cdots + \frac{\partial^2}{\partial x_n^2}\right)e^{2\pi\langle\hat{\mathbf{h}},\mathbf{x}\rangle} = [(2\pi h_1)^2 + (2\pi h_2)^2 + \cdots + (2\pi h_n)^2]e^{2\pi\langle\hat{\mathbf{h}},\mathbf{x}\rangle} \\ &= 4\pi^2[(h_1)^2 + (h_2)^2 + \cdots + (h_n)^2]e^{2\pi\langle\hat{\mathbf{h}},\mathbf{x}\rangle} = 4\pi^2 e^{2\pi\langle\hat{\mathbf{h}},\mathbf{x}\rangle} \in \mathcal{D}'(\mathbb{R}^n)\end{aligned}$$

$(\|\hat{\mathbf{h}}\| = 1)$, and $(1 - \frac{\Delta}{4\pi^2})T = (1 - \frac{4\pi^2}{4\pi^2})e^{2\pi\langle\hat{\mathbf{h}},\mathbf{x}\rangle} = 0$ in $\mathcal{D}'(\mathbb{R}^n)$ $\forall\hat{\mathbf{h}}$ with $\|\hat{\mathbf{h}}\| = 1$ and $e^{2\pi\langle\hat{\mathbf{h}},\mathbf{x}\rangle} \notin S'(\mathbb{R}^n)$.

Hence, $(1 - \frac{\Delta}{4\pi^2})^k T = 0$ in $\mathcal{D}'(\mathbb{R}^n)$ with $T = e^{2\pi\langle\hat{\mathbf{h}},\mathbf{x}\rangle} \notin S'(\mathbb{R}^n)$ $\forall\hat{\mathbf{h}} \in \mathbb{R}^n$, $\|\hat{\mathbf{h}}\| = 1$. □

Convolution equations in $S'(\mathbb{R}^n)$

Consider the following convolution equation: for given $A, B \in S'(\mathbb{R}^n)$, find $T \in S'(\mathbb{R}^n)$ such that

$$A * T = B \quad \text{in } S'(\mathbb{R}^n), \tag{8.7.31}$$

where it is *assumed* that $A * T$ is well defined and $A * T \in S'(\mathbb{R}^n)$. Then, taking the Fourier transform of both sides of (8.7.31), we have

$$\mathcal{F}[A * T] = \mathcal{F}[B] \quad \Longrightarrow \quad \mathcal{F}[A] \cdot \mathcal{F}[T] = \mathcal{F}[B] = \hat{B}. \tag{8.7.32}$$

If $\mathcal{F}[A]$ is a *polynomial*, then $\exists$ a tempered distribution $T \in S'(\mathbb{R}^n)$ defined by

$$T = \overline{\mathcal{F}}[\hat{T}] = \overline{\mathcal{F}}\left[\frac{\hat{B}}{\mathcal{F}[A]}\right] \in S'(\mathbb{R}^n), \tag{8.7.33}$$

which follows from Hörmander's results [31].

Remark 8.7.2. Łojasiewicz (see [8, p. 126]) solved the problem of division of distributions by *analytic functions*.

Elementary (or fundamental) tempered distribution solution of a linear operator with constant coefficients

Definition 8.7.1C. A tempered distribution $E \in S'(\mathbb{R}^n)$ is called an *elementary* or *fundamental* solution of the linear operator $P(\partial)$ in (8.7.28)–(8.7.29) if and only if

$$P(\partial)E = \delta \quad \text{in } S'(\mathbb{R}^n). \tag{8.7.34}$$

Theorem 8.7.3. *Every partial differential operator $P(\partial)$ with constant coefficients as in* (8.7.29) *has* at least one *elementary tempered solution $E \in S'(\mathbb{R}^n)$.*

Proof. By Theorem 8.7.2, for $S = \delta \in S'(\mathbb{R}^n)$ in equation (8.7.28), elementary solution $T = E \in S'(\mathbb{R}^n)$ of $P(\partial)$ exists and is given by $E = \overline{\mathcal{F}}[\hat{E}] = \overline{\mathcal{F}}[\frac{1}{P^*(\boldsymbol{\xi})}] \in S'(\mathbb{R}^n)$, with $P^*(\boldsymbol{\xi}) = \sum_{|\boldsymbol{\alpha}| \le m} a_{\boldsymbol{\alpha}} (i2\pi)^{|\boldsymbol{\alpha}|} \boldsymbol{\xi}^{\boldsymbol{\alpha}}$. In fact, taking Fourier transforms of both sides of (8.7.34), we have $P^*(\boldsymbol{\xi})\mathcal{F}[E] = \mathcal{F}[\delta] = 1 \Longrightarrow \mathcal{F}[E] = \hat{E} = \frac{1}{P^*(\boldsymbol{\xi})} \in S'(\mathbb{R}^n)$, which is well defined by virtue of Hörmander's result [31] on division by the polynomial $P^*(\boldsymbol{\xi})$. Then an elementary solution is

$$E = \overline{\mathcal{F}}[\hat{E}] = \overline{\mathcal{F}}\left[\frac{1}{P^*(\boldsymbol{\xi})}\right] = \mathcal{F}\left[\frac{1}{P^*(\boldsymbol{\xi})}\right]^{\vee} \in S'(\mathbb{R}^n). \qquad \square$$

Elementary solution $E \in S'(\mathbb{R}^n)$ is *not* unique *in general*. If $\exists E_0 \in S'(\mathbb{R}^n)$ such that $P(\partial)E_0 = 0$ in $S'(\mathbb{R}^n)$, then $E + E_0 \in S'(\mathbb{R}^n)$ is also an elementary solution. In fact, $P(\partial)[E + E_0] = P(\partial)E + P(\partial)E_0 = \delta + 0 = \delta$ in $S'(\mathbb{R}^n)$.

Example 8.7.6. Find an elementary solution $E \in S'(\mathbb{R}^n)$ of the following operators with the help of Fourier transforms:

1. the iterated Laplace operator Δ^k for $k \geq 1$;
2. the elliptic operator $(1 - \frac{\Delta}{4\pi^2})^k$ with $k \geq 1$.

Solution.

1. $\Delta^k E_k = \delta$ in $S'(\mathbb{R}^n)$. But $\delta * E_k = E_k \Longrightarrow \Delta^k E_k = \Delta^k(\delta * E_k) = \Delta^k\delta * E_k = \delta \Longrightarrow \mathcal{F}[\Delta^k\delta * E_k] = \mathcal{F}[\delta] \Longrightarrow \mathcal{F}[\Delta^k\delta] \cdot \mathcal{F}[E_k] = 1$ (by Theorem 8.5.3). Then, using (8.7.16), $\mathcal{F}[\Delta^k\delta] = (-4\pi^2\rho^2)^k$ with $\rho^2 = \xi_1^2 + \xi_2^2 + \cdots + \xi_n^2 \Longrightarrow (-4\pi^2\rho^2)^k \hat{E}_k = 1$ with $\hat{E}_k = \mathcal{F}[E_k] \Longrightarrow \hat{E}_k = (-1)^k \frac{1}{2^{2k}\pi^{2k}} \operatorname{Pf}(\frac{1}{\rho^{2k}}) \Longrightarrow$ elementary solution $E = E_k = \overline{\mathcal{F}}[\hat{E}_k]$, since $\check{\rho}(\boldsymbol{\xi}) = \rho(-\boldsymbol{\xi}) = \rho(\boldsymbol{\xi}) \Longrightarrow \overline{\mathcal{F}}[\operatorname{Pf}(\frac{1}{\rho^{2k}})] = \mathcal{F}[\operatorname{Pf}(\frac{1}{\rho^{2k}})]$.
Hence,
$$E = E_k = (-1)^k \frac{1}{2^{2k}\,\pi^{2k}} \mathcal{F}\left[\operatorname{Pf}\left(\frac{1}{\rho^{2k}}\right)\right], \tag{8.7.35}$$
where $\mathcal{F}[\operatorname{Pf}(\frac{1}{\rho^{2k}})]$ is given by *two different* formulae (8.6.7) and (8.6.9) for *two different* cases:
Case I: If n is *odd* or if n is *even*, but $2k < n$, we are *not* in a singular case and the symbol Pf is *useless*, so can be dropped for $2k < n$. Then we get, from (8.6.7) and (8.7.35),
$$E = E_k = (-1)^k \frac{1}{2^{2k}\pi^{2k}} \mathcal{F}\left[\frac{1}{\rho^{2k}}\right] = (-1)^k \frac{1}{2^{2k}\pi^{2k}} \frac{\pi^{2k-\frac{n}{2}}\Gamma(\frac{n-2k}{2})}{\Gamma(\frac{2k}{2})} \cdot \frac{1}{r^{n-2k}}$$
$$= (-1)^k \frac{\Gamma(\frac{n}{2} - k)}{2^{2k}\pi^{\frac{n}{2}}(k-1)!} \cdot r^{2k-n} \quad \text{for } k \geq 1. \tag{8.7.36}$$
For $n = 3$, $k = 1$,
$$E = E_1 = (-1)\frac{\Gamma(\frac{3}{2} - 1)}{2^2\pi^{\frac{3}{2}}0!} \cdot \frac{1}{r} = \frac{-1 \cdot \sqrt{\pi}}{4\pi^{\frac{3}{2}} r} = -\frac{1}{4\pi r}, \tag{8.7.37}$$
which has been obtained in (3.3.22).
Case II: For *even n and* $2k \geq n$, we are in a singular case and get, from (8.6.9) and (8.7.35) with $2p = 2k - n \geq 0$:
$$E = E_k = (-1)^k \frac{1}{2^{2k}\,\pi^{2k}} \mathcal{F}\left[\operatorname{Pf}\left(\frac{1}{\rho^{n+(2k-n)}}\right)\right]$$
$$= (-1)^k \frac{1}{2^{2k}\pi^{2k}} \frac{\pi^{\frac{n}{2}+2k-n} \cdot 2 \cdot (-1)^{k-\frac{n}{2}}}{\Gamma(\frac{n}{2} + k - \frac{n}{2})(k - \frac{n}{2})!} \cdot r^{2k-n}\left[\ln\frac{1}{r} + B\right]$$
$$= \frac{(-1)^{\frac{n}{2}}}{2^{2k-1}\pi^{\frac{n}{2}}(k-1)!(k-\frac{n}{2})!} r^{2k-n} \ln\frac{1}{r} + Br^{2k-n}, \tag{8.7.38}$$

where

$$B = \left[\frac{(-1)^{\frac{n}{2}}}{2^{2k-1}\pi^{\frac{n}{2}}(k-1)!(k-\frac{n}{2})!} \right] \times \left[\ln\frac{1}{\pi} + \frac{1}{2}\left(1 + \frac{1}{2} + \cdots + \frac{1}{k-\frac{n}{2}} - C\right) + \frac{1}{2}\frac{\Gamma'(k)}{\Gamma(k)} \right] \tag{8.7.39}$$

is a constant, i.e., independent of $\mathbf{x}$. But this constant B is of *no importance* and *can be replaced by* 0, since $\Delta^k(Br^{2k-n}) = 0$, which leads to the addition of *polyharmonic polynomials* to the elementary solution. These polyharmonic polynomials can be neglected to consider the *simplest part* of (8.7.38):

$$E = E_k = \frac{(-1)^{\frac{n}{2}}}{2^{2k-1}\pi^{\frac{n}{2}}(k-1)!(k-\frac{n}{2})!} r^{2k-n} \ln\frac{1}{r}. \tag{8.7.40}$$

For $n = 2, k = 1$ with $2k = n$, we get from (8.7.40) that

$$E = E_1 = \frac{-1}{2\pi} \cdot \ln\frac{1}{r} = \frac{1}{2\pi}\ln r \tag{8.7.41}$$

is an elementary solution of the Laplace operator (see (3.3.23)), since $\ln\frac{1}{r} = -\ln r$.

For $n = 2, k = 2$ with $2k > n$, we get from (8.7.40) that

$$E = E_3 = \frac{-1}{2^3\pi 1!1!} \cdot r^{4-2}\ln\frac{1}{r} = -\frac{1}{8\pi}r^2\ln\frac{1}{r} = \frac{1}{8\pi}r^2\ln r \tag{8.7.42}$$

is an elementary solution of the *biharmonic operator* $\Delta\Delta$ in two variables (see also (3.3.65)).

For $n = 2, k = 3$ with $2k > n$,

$$E = E_3 = \frac{-1}{2^5\pi(3-1)!(3-1)!} \cdot r^4\ln\frac{1}{r} = -\frac{1}{128\pi}r^4\ln\frac{1}{r} = \frac{1}{128\pi}r^4\ln r \tag{8.7.43}$$

is an elementary solution of $\Delta\Delta\Delta$ in two variables (see also (3.3.67)).

For $n = 2, k = 4$ with $2k > n$, we get from (8.7.40) that

$$\begin{aligned} E = E_4 &= \frac{-1}{2^7\pi(4-1)!(4-1)!} \cdot r^6\ln\frac{1}{r} = -\frac{1}{128\cdot 6\cdot 6\pi}r^6\ln\frac{1}{r} \\ &= -\frac{1}{4608\pi}r^6\ln\frac{1}{r} = -\frac{1}{4608\pi}r^6\ln\frac{1}{r} \end{aligned} \tag{8.7.44}$$

is an elementary solution of $\Delta\Delta\Delta\Delta$ in two variables (see also (3.3.68)).

2. $(1 - \frac{\Delta}{4\pi^2})^k E_k = \delta$ in $S'(\mathbb{R}^n)$

$$\Longrightarrow \quad \mathcal{F}\left[\left(1-\frac{\Delta}{4\pi^2}\right)^k(\delta * E_k)\right] = \mathcal{F}\left[\left(1-\frac{\Delta}{4\pi^2}\right)^k \delta\right] \cdot \mathcal{F}[E_k] = \mathcal{F}[\delta] = 1.$$

From (8.7.19), $\mathcal{F}[(1-\frac{\Delta}{4\pi^2})^k\delta] = (1+\rho^2)^k \Longrightarrow (1+\rho^2)^k \hat{E}_k = 1 \Longrightarrow \hat{E}_k = \frac{1}{(1+\rho^2)^k} \Longrightarrow \overline{\mathcal{F}}[\hat{E}_k] = \mathcal{F}[\frac{1}{(1+\rho^2)^k}]$, since $\check{\rho}(\boldsymbol{\xi}) = \rho(\boldsymbol{\xi})$. Using (8.7.25) [8] with m replaced by $2k$,

$$\begin{aligned} E = E_k = \mathcal{F}\left[\frac{1}{(1+\rho^2)^k}\right] &= \frac{2\pi^{\frac{2k}{2}}}{\Gamma(\frac{2k}{2})} r^{\frac{2k-n}{2}} K_{\frac{n-2k}{2}}(2\pi r) \\ &= \frac{2\pi^k}{(k-1)!} r^{k-\frac{n}{2}} K_{\frac{n}{2}-k}(2\pi r), \end{aligned} \tag{8.7.45}$$

where $K_{\frac{n}{2}-k}$ from the theory of Bessel functions is defined by (3.3.60). This is the only tempered solution.

Elementary solution of $(\boldsymbol{\Delta} - \boldsymbol{\lambda})^{\boldsymbol{k}}$ with $\boldsymbol{\lambda} > \mathbf{0}$ $(\Delta - \lambda)^k E_k = \delta$ in $S'(\mathbb{R}^n)$. Using (8.7.20), we have

$$\overline{\mathcal{F}}[\hat{E}_k] = E_k = \mathcal{F}\left[\frac{1}{(-4\pi^2\rho^2 - \lambda)^k}\right] = \mathcal{F}\left[\frac{(-1)^k}{\lambda^k\left[1 + \frac{4\pi^2\rho^2}{\lambda}\right]^k}\right]$$

$$\Longrightarrow \quad E = E_k = \frac{(-1)^k}{\lambda^k}\mathcal{F}\left[\frac{1}{(1+\bar{\rho}^2)^k}\right] \quad \text{with } \bar{\rho} = \frac{2\pi\rho}{\sqrt{\lambda}}. \tag{8.7.46}$$

Elementary solution of elasticity operator and Stokes operator Without proof, we accept the elementary solutions of the *elasticity operator* and the *Stokes operator* given in [41].

The elasticity operator A is defined by

$$(A\mathbf{u})_i = \sum_{j=1}^{3} \frac{\partial}{\partial x_j}\sigma_{ij}(\mathbf{u}), \quad 1 \le i \le 3, \tag{8.7.47}$$

where $\mathbf{u} = (u_1, u_2, u_3)$ with $u_i(\mathbf{x})$ the *displacement vector* field; $\varepsilon_{ij}(\mathbf{u}) = \frac{1}{2}[u_{i,j} + u_{j,i}]$, $1 \le i \le 3$, is the *strain tensor* field; $\sigma_{ij}(\mathbf{u}) = \lambda\sum_{l=1}^{3}\varepsilon_{ll}(\mathbf{u}) + 2\mu\varepsilon_{ij}(\mathbf{u})$ is the *stress tensor* field, λ and μ being Lamé's parameters, satisfying the *equilibrium equations* $\sum_{j=1}^{3}\frac{\partial}{\partial x_j}(\sigma_{ij}(\mathbf{u}))(\mathbf{x}) = f_i(\mathbf{x})$ for $\mathbf{x} \in \mathbb{R}^3$, $1 \le i \le 3$, $\mathbf{f} = (f_1, f_2, f_3)$ being a force vector field.

An elementary solution of the elasticity operator A is a 3×3 symmetric matrix

$$E = (E_{ij}) \text{ with } E_{ij}(\mathbf{x}) = E_{ji}(\mathbf{x}) \text{ such that } AE = \delta I,\ 1 \leq i, j \leq 3, \tag{8.7.48}$$

where δ is the Dirac distribution with force concentrated at $\mathbf{0}$ and I is the identity matrix of order 3. Then an elementary solution $E = (E_{ij})$ of A is given by [41, pp. 68–69]:

$$E_{ij}(\mathbf{x}) = E_{ji}(\mathbf{x}) = \frac{1}{8\pi\mu(\lambda + 2\mu)\|\mathbf{x}\|}\left[(\lambda + 3\mu)\delta_{ij} + (\lambda + \mu)\frac{\partial(\|\mathbf{x}\|)}{\partial x_i} \cdot \frac{\partial(\|\mathbf{x}\|)}{\partial x_j}\right], \tag{8.7.49}$$

$$\delta_{ij} = \begin{cases} 1 & \text{for } i = j \\ 0 & \text{for } i \neq j, \end{cases} \quad 1 \leq i, j \leq 3, \quad \|\mathbf{x}\|^2 = x_1^2 + x_2^2 + x_3^2.$$

For further details, we refer to [41].

An elementary solution of the Stokes operator A defining the flow of incompressible viscous fluid in $\mathbb{R}^3$ with weak velocity (unknowns being the velocity field $\mathbf{u} = (u_1, u_2, u_3)$ and the pressure p at a point $\mathbf{x} \in \mathbb{R}^3$) is a 4×4 matrix

$$E = E_{i,j},\ 1 \leq i, j \leq 4 \text{ such that } AE = \delta I, \tag{8.7.50}$$

δ being the Dirac distribution with concentration at $\mathbf{0} \in \mathbb{R}^3$, I being the identity matrix of order 4. Then an elementary solution $E = (E_{i,j})$, $1 \leq i, j \leq 4$, of the Stokes operator A is given by [41, pp. 72–77]:

$$E_{ij}(\mathbf{x}) = \frac{1}{8\pi\nu}\left(\frac{\partial^2}{\partial x_i \partial x_j}(\|\mathbf{x}\|)\right) + 2\delta_{ij}\left(\frac{1}{\|\mathbf{x}\|} - \frac{\partial^2}{\partial x_i \partial x_j}(\|\mathbf{x}\|)\right), \quad 1 \leq i, j \leq 3, \tag{8.7.51}$$

$\delta_{ij} = 1$ for $i = j$ and $\delta_{ij} = 0$ for $i \neq j$, $\|\mathbf{x}\|^2 = x_1^2 + x_2^2 + x_3^2$;

$$E_{4i}(\mathbf{x}) = E_{i4}(\mathbf{x}) = P_i(\mathbf{x}) = \frac{1}{4\pi}\frac{\partial}{\partial x_i}\left(\frac{1}{\|\mathbf{x}\|}\right), \quad 1 \leq i \leq 3; \tag{8.7.52}$$

$$E_{44}(\mathbf{x}) = \delta, \quad \delta \text{ being the Dirac distribution.} \tag{8.7.53}$$

For further details, we refer to [41].

Remark 8.7.3. The author has *not* checked the correctness of the result in (8.7.49) (resp. (8.7.51)–(8.7.53)), which is left to the reader as an exercise.

8.8 Laplace transform of distributions on $\mathbb{R}$

8.8.1 Space $\mathcal{D}'^{+}$

Definition 8.8.1. $\mathcal{D}'^{+} \equiv \mathcal{D}'^{+}(\mathbb{R})$ is the space of distributions $T \in \mathcal{D}'(\mathbb{R})$ with $\operatorname{supp}(T) \subset \mathbb{R}_0^+ = \{x : x \in \mathbb{R}, x \geq 0\}$, i.e.

$$\mathcal{D}'^{+} = \mathcal{D}'^{+}(\mathbb{R}) = \{T : T \in \mathcal{D}'(\mathbb{R}) \text{ with } \operatorname{supp}(T) \subset \mathbb{R}_0^+\}. \tag{8.8.1}$$

Let $T \in \mathcal{D}'^{+}$ and $\xi_0 \in \mathbb{R}$ such that $e^{-\xi_0 x} T \in S'(\mathbb{R})$. Since $e^{-\xi_0 x}$ is a C^∞-function on $\mathbb{R}$, the product $e^{-\xi_0 x} T$ is a well defined distribution on $\mathbb{R}$, and is, in fact, a *tempered distribution* by assumption. Then,

$$T \in \mathcal{D}'^{+},\ \xi_0 \in \mathbb{R},\ e^{-\xi_0 x} T \in S'(\mathbb{R}) \quad \Longrightarrow \quad e^{-\xi x} T \in S'(\mathbb{R})\ \forall \xi > \xi_0. \tag{8.8.2}$$

In fact, $e^{-\xi x} T = e^{-(\xi-\xi_0)x} e^{-\xi x} T$ with $e^{-\xi_0 x} T \in S'(\mathbb{R})$, $\xi > \xi_0$. Although for $\xi > \xi_0$, $e^{-(\xi-\xi_0)x} \notin \theta_M$ (the multiplier set, see Definition 8.5.1), but $\alpha(x) e^{-(\xi-\xi_0)x} \in \theta_M$ such that $\alpha(x) e^{-(\xi-\xi_0)x} \phi \in S(\mathbb{R})$, where $\alpha(x)$ is defined by

$$\alpha \in C^\infty(\mathbb{R}), \quad \alpha(x) = 1 \quad \forall x \in U; \tag{8.8.3}$$

with $\operatorname{supp}(T) \subset U$, $\operatorname{supp}(\alpha)$ is bounded on the left.

Since $e^{-\xi_0 x} T \in S'(\mathbb{R})$ and $\alpha(x) e^{-(\xi-\xi_0)x} \phi \in S(\mathbb{R})$ with $\xi > \xi_0$, $\langle e^{-\xi_0 x} T, \alpha(x) e^{-(\xi-\xi_0)x} \phi \rangle$ is well defined ($\forall \phi \in S(\mathbb{R})$ with $\xi > \xi_0$) and independent of the choice of the function $\alpha(x)$.

For example, let β be another function such that $\beta \in C^\infty(\mathbb{R})$, $\beta(x) = 1\ \forall x \in U$ with $\operatorname{supp}(T) \subset U$ and $\operatorname{supp}(\beta)$ is bounded on the left. Then $\alpha(x) - \beta(x) = 0$ $\forall x \in \operatorname{supp}(T)$

$\Longrightarrow \langle e^{-\xi_0 x} T, (\alpha(x) - \beta(x)) e^{-(\xi-\xi_0)x} \phi \rangle = 0\ \forall \phi \in S(\mathbb{R})$ with $\xi > \xi_0$

$\Longrightarrow \langle e^{-\xi_0 x} T, \alpha(x) e^{-(\xi-\xi_0)x} \phi \rangle = \langle e^{-\xi_0 x} T, \beta(x) e^{-(\xi-\xi_0)x} \phi \rangle\ \forall \phi \in S(\mathbb{R})$.

Then, for $\xi > \xi_0$,

$$\langle e^{-\xi x} T, \phi \rangle = \langle e^{-\xi_0 x} T, e^{-(\xi-\xi_0)x} \phi \rangle = \langle e^{-\xi_0 x} T, \alpha(x) e^{-(\xi-\xi_0)x} \phi \rangle$$

$\forall \phi \in S(\mathbb{R})$ with $\alpha(x) e^{-(\xi-\xi_0)x} \phi \in S(\mathbb{R})$. Hence, $e^{-\xi x} T \in S'(\mathbb{R})$ is well defined $\forall \xi > \xi_0$.

Laplace transform $\mathcal{L}$

Definition 8.8.2. Let $T \in \mathcal{D}'^{+} \equiv \mathcal{D}'^{+}(\mathbb{R})$ be a distribution such that $e^{-\xi_0 x} T \in S'(\mathbb{R})$. Then the *Laplace transform* of T, denoted by $\mathcal{L}T$ or $\mathcal{L}(T)$, is defined, $\forall \xi > \xi_0$, by:

$$(\mathcal{L}T)(p) = \langle T, e^{-px} \rangle \quad \text{with } p = \xi + i\eta \in \mathbb{C},\ \xi > \xi_0. \tag{8.8.4}$$

$\mathcal{L}(T)(p) \equiv (\mathcal{L}T)(p)\ \forall p$ with $\operatorname{Re}(p) > \xi_0$.

Justification Although $e^{-px} \notin S(\mathbb{R})$ for $\operatorname{Re}(p) = \xi > \xi_0$, we can define a function $\alpha(x)$ by (8.8.3) such that $\alpha(x)e^{-(p-\xi_0)x} \in S(\mathbb{R})$. Then, (8.8.4) can be rewritten as

$$\langle T, e^{-px}\rangle = \langle e^{-\xi_0 x} T, \alpha(x)e^{-(p-\xi_0)x}\rangle, \tag{8.8.5}$$

which is well defined by virtue of the assumptions.

Example 8.8.1.

1. $\mathcal{L}[\delta] = 1$;
2. $\mathcal{L}[\delta_a] = e^{-ap}$ $(a \geq 0)$;
3. $\mathcal{L}[\delta^{(m)}] = p^m$;
4. $\mathcal{L}[H(x)] = 1/p$ for $\xi = \operatorname{Re}(p) > 0$;
5. $\mathcal{L}[f] = \int_0^\infty f(x)e^{-px}dx$ for $f \in L^1(\mathbb{R})$ with $\operatorname{supp}(f) \subset \mathbb{R}_0^+$ and $e^{-\xi_0 x} f \in S'(\mathbb{R})$;
6. $\mathcal{L}[H(x)\ln x] = \frac{-\ln p - C}{p}$, $C =$ Euler's constant $= -\int_0^\infty e^{-y}\ln y dy$.

Solution.

1. $\mathcal{L}[\delta](p) = \langle \delta, e^{-px}\rangle = e^{-p(x)}|_{x=0} = 1$;
2. $\mathcal{L}[\delta_a](p) = \langle \delta_a, e^{-px}\rangle = e^{-px}|_{x=a} = e^{-ap}$ for $a \geq 0$;
3. $\mathcal{L}[\delta^{(m)}](p) = \langle \delta^{(m)}, e^{-px}\rangle = (-1)^m\langle \delta, \frac{d^m}{dx^m}(e^{-px})\rangle = (-1)^m[(-1)^m p^m \cdot e^{-px}]_{x=0} = p^m$ $\forall m \in \mathbb{N}$;
4. $\mathcal{L}[H(x)](p) = \langle H, e^{-px}\rangle = \int_0^\infty 1.e^{-px}dx = -\frac{e^{-px}}{p}|_0^\infty = \frac{1}{p}$ for $\operatorname{Re}(p) > 0$;
5. $\mathcal{L}[f](p) = \langle T_f, e^{-px}\rangle = \int_0^\infty f(x)e^{-px}dx$ with $\operatorname{Re}(p) > \xi_0$, since $f(x) = 0$ for almost all $x < 0$;
6. $T = H(x)\ln x \Longrightarrow T \in S'(\mathbb{R})$ and $e^{-\xi_0 x}T \in S'(\mathbb{R})$ for $\xi_0 = 0$.
   Then, for $\operatorname{Re}(p) > 0$, $\mathcal{L}[H(x)\ln x](p) = \langle H(x)\ln x, e^{-px}\rangle = \int_0^\infty \ln x e^{-px} dx$. For real $p > 0$, set $y = px$. Then $x = \frac{y}{p}$, $dx = \frac{dy}{p}$

$$\Longrightarrow \quad \int_0^\infty \ln x e^{-px}dx = \int_0^\infty \ln\left(\frac{y}{p}\right)e^{-y}\frac{dy}{p} = \frac{1}{p}\int_0^\infty (\ln y - \ln p)e^{-y}dy$$
$$= \frac{1}{p}\left[-\ln p\int_0^\infty e^{-y}dy + \int_0^\infty \ln y \cdot e^{-y}dy\right] = \frac{-\ln p - C}{p},$$

   where $C = -\int_0^\infty \ln y \cdot e^{-y}dy = \lim_{n\to\infty}[1 + \frac{1}{2} + \cdots + \frac{1}{n} - \ln n] =$ Euler's constant.

$\mathcal{L}$ is linear $\mathcal{L}(T_1)(p), \mathcal{L}(T_2)(p)$ exist for $\mathrm{Re}(p) > \xi_0 \Longrightarrow \mathcal{L}[\alpha_1 T_1 + \alpha_2 T_2](p)$ exists for $\mathrm{Re}(p) > \xi_0$ and $\forall \alpha_1, \alpha_2 \in \mathbb{R}$;

$$\begin{aligned}\mathcal{L}[\alpha_1 T_1 + \alpha_2 T_2](p) &= \langle \alpha_1 T_1 + \alpha_2 T_2, e^{-px}\rangle = \alpha_1 \langle T_1, e^{-px}\rangle + \alpha_2 \langle T_2, e^{-px}\rangle \\ &= \alpha_1 \mathcal{L}(T_1)(p) + \alpha_2 \mathcal{L}(T_2)(p)\ \forall T_1, T_2 \in \mathcal{D}'^{+},\ \forall \alpha_1, \alpha_2 \in \mathbb{R}.\end{aligned} \tag{8.8.6}$$

Properties of the Laplace transform $\mathcal{L}$

Property 1: Relation with Fourier transform Instead of the definition of Fourier transform $\mathcal{F} f$ in (7.1.2) ($\int_{-\infty}^{\infty} f(x) e^{-i2\pi x\xi} dx$) followed up to now, we will define the Fourier transform $\mathcal{F} f$ here by (7.1.12):

$$g_3(\xi) = (\mathcal{F} f)(\xi) = \int_{-\infty}^{\infty} f(x) e^{-ix\xi} dx, \tag{8.8.7}$$

with the co-transform $\bar{\mathcal{F}}$ defined by:

$$f(x) = \bar{\mathcal{F}} g_3 = \frac{1}{2\pi} \int_{-\infty}^{\infty} g_3(\xi) e^{ix\xi} d\xi, \tag{8.8.8}$$

so that a simple relation between Fourier transform and Laplace transform can be obtained. Under the definition in (8.8.7), the Fourier transform $\mathcal{F} T$ of a tempered distribution $T \in S'(\mathbb{R})$ can be defined by the same formula (8.3.1):

$$\langle \mathcal{F} T, \phi \rangle = \langle T, \mathcal{F} \phi \rangle \quad \forall \phi \in S(\mathbb{R}), \tag{8.8.9}$$

where $\mathcal{F}\phi$ is defined by $(\mathcal{F}\phi)(x) = \int_{-\infty}^{\infty} \phi(\xi) e^{-ix\xi} d\xi$. Then we have:

Proposition 8.8.1. *Let $T \in \mathcal{D}'^{+} \equiv \mathcal{D}'^{+}(\mathbb{R})$ be a distribution with support in $\mathbb{R}_0^+$ such that $e^{-\xi_0 x} T \in S'(\mathbb{R})$ for some $\xi_0 \in \mathbb{R}$. Then the Laplace transform $\mathcal{L}$ is related to the Fourier transform $\mathcal{F}$ by: $\forall \xi > \xi_0$,*

$$(\mathcal{L} T)(p) = \mathcal{F}[e^{-\xi x} T](\eta) \quad \textit{with } p = \xi + i\eta \in \mathbb{C}. \tag{8.8.10}$$

Proof. Set $F(p) = \mathcal{L}(T)(p)$ with $p = \xi + i\eta$, $\mathrm{Re}(p) = \xi > \xi_0$. Let $\alpha \in C^\infty(\mathbb{R})$ be such that $\alpha(x) = 1\ \forall x \in U$, $\mathrm{supp}(T) \subset U$, $\mathrm{supp}(\alpha)$ is bounded on the left.

Then

$$\begin{aligned}F(p) &= \langle T, e^{-px}\rangle = \langle e^{-\xi_0 x} T(x), e^{-(\xi-\xi_0)x} e^{-i\eta x}\rangle \quad (\text{since } p = \xi + i\eta) \\ &= \langle e^{-\xi_0 x} T(x), \alpha(x) e^{-(\xi-\xi_0)x} e^{-i\eta x}\rangle.\end{aligned}$$

$\forall \phi \in S(\mathbb{R})$,

$$\begin{aligned}
\langle F, \phi \rangle &= \int_{\mathbb{R}} \langle e^{-\xi_0 x} T(x), \alpha(x) e^{-(\xi-\xi_0)x} e^{-i\eta x} \rangle \phi(\eta) d\eta \\
&= \langle e^{-\xi_0 x} T(x) \otimes \phi(\eta), \alpha(x) e^{-(\xi-\xi_0)x} e^{-i\eta x} \rangle \\
&= \langle e^{-\xi_0 x} T(x), \langle \phi(\eta), \alpha(x) e^{-(\xi-\xi_0)x} e^{-i\eta x} \rangle \rangle \\
&= \langle e^{-\xi_0 x} e^{-(\xi-\xi_0)x} T(x) \alpha(x) \langle \phi(\eta), e^{-i\eta x} \rangle \rangle = \langle e^{-\xi x} T(x) \langle \phi(\eta), e^{-i\eta x} \rangle \rangle \\
&= \langle e^{-\xi x} T(x), \mathcal{F}\phi(x) \rangle = \langle \mathcal{F}[e^{-\xi x} T](\eta), \phi(\eta) \rangle \quad \forall \phi \in S(\mathbb{R})
\end{aligned}$$

$\Longrightarrow F = \mathcal{F}[e^{-\xi x} T] \Longrightarrow \mathcal{L}(T)(\xi + i\eta) = \mathcal{F}[e^{-\xi x} T](\eta)$, since $F(p) = \mathcal{L}(T)(p)$. □

Property 2 Let $T \in \mathcal{D}'^{+} \equiv \mathcal{D}'^{+}(\mathbb{R})$ with $e^{-\xi_0 x} T \in S'(\mathbb{R})$ for some $\xi_0 \in \mathbb{R}$. Then, without proof we agree to accept that $F(p) = (\mathcal{L}T)(p)$ is an *analytic function* of $p = \xi + i\eta \in \mathbb{C}$ with $\mathrm{Re}(p) = \xi > \xi_0$.

For example, $T = \delta \in \mathcal{D}'^{+}(\mathbb{R})$ with $e^{-\xi_0 x}\delta \in S'(\mathbb{R})\ \forall \xi_0 \in \mathbb{R}$, and hence $F(p) = \mathcal{L}[\delta](p) = 1$ is an analytic function of $p = \xi + i\eta \in \mathbb{C}$.

Property 3

$$\mathcal{L}\left[\frac{d^m T}{dx^m}\right](p) = p^m \mathcal{L}(T)(p)\ \forall m \in \mathbb{N}, \tag{8.8.11}$$

where the derivative is in the *sense of distribution*. In fact,

$$\begin{aligned}
\mathcal{L}\left[\frac{d^m T}{dx^m}\right](p) &= \left\langle \frac{d^m T}{dx^m}, e^{-px} \right\rangle = (-1)^m \left\langle T, \frac{d^m}{dx^m}(e^{-px}) \right\rangle \\
&= (-1)^m \langle T, (-p)^m . e^{-px} \rangle = p^m \langle T, e^{-px} \rangle = p^m \mathcal{L}[T](p).
\end{aligned}$$

Property 4

$$\left[\frac{d^m}{dp^m}\mathcal{L}(T)\right](p) = \mathcal{L}[(-x)^m T](p). \tag{8.8.12}$$

Indeed,

$$\begin{aligned}
\left[\frac{d^m}{dp^m}\mathcal{L}(T)\right](p) &= \frac{d^m}{dp^m}\langle T, e^{-px} \rangle = \left\langle T, \frac{d^m}{dp^m}(e^{-px}) \right\rangle \\
&= \langle T, (-x)^m e^{-px} \rangle = \langle (-x)^m T, e^{-px} \rangle = \mathcal{L}[(-x)^m T](p).
\end{aligned}$$

Property 5: Convolution $S * T$ of distributions $S, T \in \mathcal{D}'^{+}$

Proposition 8.8.2. *Let $S, T \in \mathcal{D}'^{+} \equiv \mathcal{D}'^{+}(\mathbb{R})$ such that $e^{-\xi_1 x} S \in S'(\mathbb{R})$, $e^{-\xi_2 x} T \in S'(\mathbb{R})$ for some $\xi_1, \xi_2 \in \mathbb{R}$ and $\mathcal{L}(S)(p)$ and $\mathcal{L}(T)(p)$ exist for* $\mathrm{Re}(p) > \xi_1$ *and* $\mathrm{Re}(p) > \xi_2$ *respectively. Then their* convolution $S * T \in \mathcal{D}'^{+}$ *is defined and given,* $\forall \phi \in \mathcal{D}(\mathbb{R})$, *by:*

$$\langle S * T, \phi \rangle = \langle S_x \otimes T_y, \phi(x+y) \rangle = \langle S_x, \langle T_y, \phi(x+y) \rangle \rangle. \tag{8.8.13}$$

Proof. $\forall$ fixed $x \in \mathrm{supp}(S)$, $\langle T_y, \phi(x+y) \rangle$ is well defined, since $\tau_{-x}\phi(y) = \phi(x+y)$ has compact support in y. Again, the function $x \longmapsto \langle T_y, \phi(x+y) \rangle$ is a C^{∞}-function with compact support. Indeed, $x \in \mathrm{supp}(S) \Longrightarrow x \geq 0$, $y \in \mathrm{supp}(T) \Longrightarrow y \geq 0$ and $(x+y) \in \mathrm{supp}(\phi) \Longrightarrow \exists A > 0$ such that $|x+y| \leq A$. Hence, we have $0 \leq x \leq x+y \leq A$. So, the formula (8.8.13) is meaningful. Thus, $S * T \in \mathcal{D}'(\mathbb{R})$ is well defined, with

$$\mathrm{supp}(S * T) \subset \mathrm{supp}(S) + \mathrm{supp}(T) \subset [0, \infty[\quad \Longrightarrow \quad S * T \in \mathcal{D}'^{+}. \tag{8.8.14}$$

□

Since $\delta \in \mathcal{D}'^{+}$,

$$\delta * T = T \in \mathcal{D}'^{+} \quad \forall T \in \mathcal{D}'^{+}. \tag{8.8.15}$$

8.8.2 Distribution $T^{-1} \in \mathcal{D}'^{+}$ (see also convolution algebra $A = \mathcal{D}'^{+}$ (6.9.15b))

Definition 8.8.3. Let $T \in \mathcal{D}'^{+}$. Then the *unique* distribution $S \in \mathcal{D}'^{+}$ which satisfies the convolution equation $T * S = \delta$, δ being the Dirac distribution, is denoted by T^{-1}, i.e.

$$T * T^{-1} = \delta. \tag{8.8.16}$$

Example 8.8.2. Find

1. H^{-1};
2. $(\delta')^{-1}$;
3. $(\delta' - \lambda\delta)^{-1}$;

where $\delta' = \frac{d\delta}{dx}$.

Solution.

1. $H * S = \delta \Longrightarrow \frac{d}{dx}(H * S) = \frac{d\delta}{dx} \Longrightarrow S * \frac{dH}{dx} = \delta' \Longrightarrow S * \delta = \delta' \Longrightarrow S = \delta'$ by virtue of (8.8.15). Hence, $H^{-1} = \delta'$.
2. $S * \delta' = \delta$. But $S * \delta' = S' * \delta = \delta \Longrightarrow S' = \delta \Longrightarrow S = H \Longrightarrow (\delta')^{-1} = H$.

3. $S * (\delta' - \lambda\delta) = \delta \implies S * \delta' - \lambda S * \delta = \delta \implies \frac{dS}{dx} * \delta - \lambda S = \delta \implies (\frac{dS}{dx} - \lambda S) * \delta = \delta$.
   Set $S = e^{\lambda x} T$. Then $e^{\lambda x}[\lambda T + \frac{dT}{dx} - \lambda T] = \delta \implies e^{\lambda x}\frac{dT}{dx} = \delta \implies \frac{dT}{dx} = e^{-\lambda x}\delta = \delta$, since $\langle e^{-\lambda x}\delta, \phi\rangle = \langle \delta, e^{-\lambda x}\phi\rangle = (e^{-\lambda x}\phi(x))|_{x=0} = \phi(0) = \langle \delta, \phi\rangle \ \forall \phi \in \mathcal{D}(\mathbb{R}) \implies e^{-\lambda x}\delta = \delta$.
   But $\frac{dT}{dx} = \delta \implies T = H$, since $\frac{dH}{dx} = \delta$. Hence, $(\delta' - \lambda\delta)^{-1} = S = e^{\lambda x} H(x)$.

Theorem 8.8.1 (Convolution Theorem). *Let* $S, T \in \mathcal{D}'^{+} \equiv \mathcal{D}'^{+}(\mathbb{R})$ *such that* $e^{-\xi_1 x} S \in S'(\mathbb{R})$, $e^{-\xi_2 x} T \in S'(\mathbb{R})$ *for some* $\xi_1, \xi_2 \in \mathbb{R}$ *and* $\mathcal{L}(S)(p)$ *and* $\mathcal{L}(T)(p)$ *are defined for* $\mathrm{Re}(p) > \xi_1$ *and* $\mathrm{Re}(p) > \xi_2$ *respectively. Then, for* $\mathrm{Re}(p) > \max\{\xi_1, \xi_2\}$,

$$\mathcal{L}(S * T) = \mathcal{L}(S) \cdot \mathcal{L}(T). \tag{8.8.17}$$

Proof. For $\mathrm{Re}(p) > \max\{\xi_1, \xi_2\}$,

$$\begin{aligned}\mathcal{L}(S * T)(p) &= \langle S * T, e^{-px}\rangle = \langle S_x, \langle T_y, e^{-p(x+y)}\rangle\rangle = \langle S_x, e^{-px}\langle T_y, e^{-py}\rangle\rangle \\ &= \langle S_x, e^{-px}\rangle \cdot \langle T_y, e^{-py}\rangle = \mathcal{L}(S) \cdot \mathcal{L}(T),\end{aligned}$$

since $\mathrm{Re}(p) >$ both ξ_1 and ξ_2. □

Property 6

$$\mathcal{L}(T)(p) = 0 \text{ for } \mathrm{Re}(p) > \xi \quad \implies \quad T = 0. \tag{8.8.18}$$

8.8.3 Inverse $\mathcal{L}^{-1}$ of Laplace transform $\mathcal{L}$

Property 7 If $F(p) = \mathcal{L}(T)(p)$ for $p \in \mathbb{C}$ with $\mathrm{Re}(p) > \xi_0$,
$\mathcal{L}^{-1}(F(p)) = T(x)$. (8.8.19)

Example 8.8.3.

1. Find $\mathcal{L}(T)(p)$, when
   (a) $T = \mathrm{Pf}(\frac{H(x)}{x})$;
   (b) $T = x^k e^{\alpha x} H(x)$ for $k \in \mathbb{N}_0$, $\alpha \in \mathbb{C}$ with $\mathrm{Re}(p) > \mathrm{Re}(\alpha)$;
   (c) $T = H(x) \sin x$;
   (d) $T = H(x) e^{-\alpha x} \cos \beta x$ with $\mathrm{Re}(p) > -\alpha$;
   (e) $T = H(x) e^{-\alpha x} \sin \beta x$ with $\mathrm{Re}(p) > \mathrm{Re}(\alpha)$.

Then, using these results, find

(f) $\mathcal{L}^{-1}(\frac{p}{p+1})$ with $\mathrm{Re}(p) > -1$;

(g) $\mathcal{L}^{-1}(\frac{p^2+i}{p^2-3p+2})$.

2. Find $T \in \mathcal{D}'^{+}$ which satisfies $(xe^x H(x)) * T = H(x)\sin x$.

Solution.

1(a) From (1.4.26),

$$\langle T, \phi\rangle = \left\langle \mathrm{Pf}\left(\frac{H(x)}{x}\right), \phi\right\rangle = \lim_{\varepsilon\to 0^+}\left[\int_\varepsilon^\infty \frac{\phi(x)}{x}dx + \phi(0)\ln\varepsilon\right]. \quad (8.8.20)$$

Set $S = H(x)\ln x$. Then, $\forall\phi \in \mathcal{D}(\mathbb{R})$,

$$\begin{aligned}\left\langle \frac{dS}{dx}, \phi\right\rangle &= -\left\langle S, \frac{d\phi}{dx}\right\rangle = -\int_0^\infty \ln x\phi'(x)dx\\ &= -\lim_{\varepsilon\to 0^+}\left[\int_\varepsilon^\infty \ln x\phi'(x)dx\right] \quad \text{(by Lebesgue's Theorem)}\\ &= -\lim_{\varepsilon\to 0^+}\left[(\ln x\cdot\phi(x))|_\varepsilon^\infty - \int_\varepsilon^\infty \frac{\phi(x)}{x}dx\right]\\ &= \lim_{\varepsilon\to 0^+}\left[\int_\varepsilon^\infty \frac{\phi(x)}{x}dx + \ln\varepsilon\cdot\phi(\varepsilon)\right].\end{aligned}$$

But from (1.2.41) we get $\phi(\varepsilon) = \phi(0) + \varepsilon\psi(\varepsilon) \Longrightarrow \phi(\varepsilon)\ln(\varepsilon) = \phi(0)\ln\varepsilon + \varepsilon\ln\varepsilon\cdot\psi(\varepsilon) \Longrightarrow \lim_{\varepsilon\to 0^+}\ln\varepsilon\phi(\varepsilon) = \lim_{\varepsilon\to 0^+}\phi(0)\ln\varepsilon + \lim_{\varepsilon\to 0^+}\varepsilon\ln\varepsilon\cdot\psi(\varepsilon) = \lim_{\varepsilon\to 0^+}\phi(0)\ln\varepsilon$, where $\varepsilon\ln\varepsilon\psi(\varepsilon)\to 0$ as $\varepsilon\to 0^+$, since $\varepsilon\ln\varepsilon\to 0$ and $\psi(\varepsilon)\to\psi(0)$ as $\varepsilon\to 0^+$.

Hence,

$$\left\langle \frac{d}{dx}[H(x)\ln x], \phi\right\rangle = \lim_{\varepsilon\to 0^+}\left[\int_\varepsilon^\infty \frac{\phi(x)}{x}dx + \phi(0)\ln\varepsilon\right]. \quad (8.8.21)$$

From (8.8.20) and (8.8.21),

$$\begin{aligned}&\left\langle \frac{d}{dx}[H(x)\ln x], \phi\right\rangle = \left\langle \mathrm{Pf}\left(\frac{H(x)}{x}\right), \phi\right\rangle \quad \forall\phi\in\mathcal{D}(\mathbb{R})\\ \Longrightarrow\quad &\frac{d}{dx}[H(x)\ln x] = \mathrm{Pf}\left(\frac{H(x)}{x}\right) \text{ in } \mathcal{D}'(\mathbb{R}) \text{ with } \mathrm{supp}(\mathrm{Pf}(\tfrac{H(x)}{x})) \subset [0,\infty[\\ \Longrightarrow\quad &\mathcal{L}\left(\mathrm{Pf}\left(\frac{H(x)}{x}\right)\right) = \mathcal{L}\left[\frac{d}{dx}(H(x)\ln x)\right]\\ &= p\mathcal{L}(H(x)\ln x) \quad \text{(by (8.8.11))}\\ &= p\cdot\frac{-\ln p - C}{p} = -(\ln p + C) \quad \text{(see (6) in Example 8.8.1).}\end{aligned}$$

1(b)

$$\mathcal{L}[x^k e^{\alpha x} H(x)] = \langle x^k e^{\alpha x} H(x), e^{-px} \rangle = \underbrace{\int_0^\infty x^k e^{(\alpha-p)x} dx}_{I_k}, \quad \mathrm{Re}(p) > \mathrm{Re}(\alpha).$$

$$\begin{aligned}
I_k &= x^k \frac{e^{(\alpha-p)x}}{\alpha - p}\bigg|_{x=0}^{x=\infty} - \frac{k}{\alpha - p} \underbrace{\int_0^\infty x^{k-1} e^{(\alpha-p)x} dx}_{I_{k-1}} \\
&= \frac{k}{p-\alpha} I_{k-1} \quad (\text{since } e^{(\alpha-p)x} \to 0 \text{ as } x \to \infty \text{ for } \mathrm{Re}(p) > \mathrm{Re}(\alpha)) \\
&= \frac{k}{p-\alpha} \cdot \frac{k-1}{p-\alpha} \cdot I_{k-2} = \cdots = \frac{k!}{(p-\alpha)^k} I_0 = \frac{k!}{(p-\alpha)^k} \int_0^\infty e^{(\alpha-p)x} dx \\
&= \frac{k!}{(p-\alpha)^k} \cdot \frac{e^{(\alpha-p)x}}{\alpha - p}\bigg|_{x=0}^{x=\infty} = \frac{k!}{(p-\alpha)^k} \frac{1}{(p-\alpha)} = \frac{k!}{(p-\alpha)^{k+1}}.
\end{aligned}$$

Hence, for $\mathrm{Re}(p) > \mathrm{Re}(\alpha)$,

$$\mathcal{L}[x^k e^{\alpha x} H(x)](p) = \frac{k!}{(p-\alpha)^{k+1}}. \tag{8.8.22}$$

1(c)

$$\begin{aligned}
T = H(x) \sin x &= \frac{1}{2i}[(e^{ix} - e^{-ix}) H(x)] = \frac{1}{2i}[e^{ix} H(x) - e^{-ix} H(x)] \\
\mathcal{L}(H(x) \sin x) &= \frac{1}{2i}[\mathcal{L}(e^{ix} H(x)) - \mathcal{L}(e^{-ix} H(x)].
\end{aligned}$$

From (b) with $k = 0, \alpha = i$, $\mathcal{L}(e^{ix} H(x)) = \frac{1}{p-i}$, and with $k = 0, \alpha = -i$, $\mathrm{Re}(p) > 0$, $\mathcal{L}(e^{-ix} H(x)) = \frac{1}{p+i}$.
Hence,

$$\mathcal{L}(H(x) \sin x) = \frac{1}{2i}\left[\frac{1}{p-i} - \frac{1}{p+i}\right] = \frac{1}{p^2+1}. \tag{8.8.23}$$

1(d) For $\alpha > 0$,

$$\begin{aligned}
T &= H(x) e^{-\alpha x} \cos \beta x = H(x) e^{-\alpha x} \frac{e^{i\beta x} + e^{-i\beta x}}{2} \\
&= \frac{1}{2}[H(x) e^{-(\alpha - i\beta)x} + H(x) \cdot e^{-(\alpha + i\beta)x}]
\end{aligned}$$

$$\begin{aligned}
\implies \quad \mathcal{L}[H(x)e^{-\alpha x}\cos\beta x] &= \frac{1}{2}[\mathcal{L}(H(x)e^{-(\alpha-i\beta)x}) + \mathcal{L}(H(x)e^{-(\alpha+i\beta)})](p) \\
&= \frac{1}{2}\left[\frac{1}{p+(\alpha-i\beta)} + \frac{1}{p+(\alpha+i\beta)}\right] \\
&= \frac{1}{2}\cdot\frac{2(p+\alpha)}{(p+\alpha)^2+\beta^2} \\
&= \frac{p+\alpha}{(p+\alpha)^2+\beta^2} \quad \text{for } \operatorname{Re}(p) > -\alpha.
\end{aligned}$$

1(e) Similarly,

$$\mathcal{L}(H(x)e^{-\alpha x}\sin\beta x)(p) = \frac{1}{2i}[\mathcal{L}(H(x)\cdot e^{-(\alpha-i\beta)x}) - \mathcal{L}(H(x)e^{-(\alpha+i\beta)x})](p)$$

$$= \frac{1}{2i}\left[\frac{1}{p+(\alpha-i\beta)} - \frac{1}{p+(\alpha+i\beta)}\right]$$

$$= \frac{1}{2i}\cdot\frac{2i\beta}{(p+\alpha)^2+\beta^2} = \frac{\beta}{(p+\alpha)^2+\beta^2} \quad \text{for } \operatorname{Re}(p) > -\alpha.$$

1(f) From (b), we get, for $k = 0$, $\alpha = -1$, $\mathcal{L}(H(x)e^{-x})(p) = \frac{1}{p+1}$ with $\operatorname{Re}(p) > -1$. From (8.8.11),

$$\mathcal{L}\left[\frac{d}{dx}(H(x)e^{-x})\right](p) = p\mathcal{L}(H(x)e^{-x})(p) = \frac{p}{p+1} \quad \text{with } \operatorname{Re}(p) > -1$$

$$\begin{aligned}
\implies \quad \mathcal{L}^{-1}\left(\frac{p}{p+1}\right) &= \mathcal{L}^{-1}\left[\mathcal{L}\left(\frac{d}{dx}(H(x)e^{-x})\right)\right] = \frac{d}{dx}(H(x)e^{-x}) \\
&= e^{-x}\cdot\frac{dH}{dx} - H(x)e^{-x} = e^{-x}\cdot\delta - H(x)e^{-x},
\end{aligned}$$

$\delta = \delta(x)$ being the Dirac distribution with concentration at 0. But $\langle e^{-x}\delta, \phi\rangle = \langle\delta, e^{-x}\phi\rangle = (e^{-x}\phi(x))|_{x=0} = \phi(0) = \langle\delta,\phi\rangle \implies e^{-x}\delta = \delta$. Hence, $\mathcal{L}^{-1}(\frac{p}{p+1}) = \delta - H(x)e^{-x}$ for $\operatorname{Re}(p) > -1$.

1(g) $\frac{p^2+i}{p^2-3p+2} = \frac{p^2+i}{(p-2)(p-1)} = \frac{p^2}{p-2} + \frac{i}{p-2} - \frac{p^2}{p-1} - \frac{i}{p-1}$

$$\begin{aligned}
\implies \quad \mathcal{L}^{-1}\left(\frac{p^2+i}{p^2-3p+2}\right) &= \mathcal{L}^{-1}\left(\frac{p^2}{p-2}\right) + i\mathcal{L}^{-1}\left(\frac{1}{p-2}\right) \\
&\quad - \mathcal{L}^{-1}\left(\frac{p^2}{p-1}\right) - \mathcal{L}^{-1}\left(\frac{i}{p-1}\right). \qquad (8.8.24)
\end{aligned}$$

Then, from (8.8.22) with $k = 0$, $\alpha = 2$, $\mathcal{L}(e^{2x}H(x)) = \frac{1}{p-2}$, and $k = 0$, $\alpha = 1$, $\mathcal{L}(e^{x}H(x)) = \frac{1}{p-1}$.

Then

$$\mathcal{L}\left(\frac{d^2}{dx^2}[e^{2x}H(x)]\right) = p^2\mathcal{L}(e^{2x}H(x)) = \frac{p^2}{p-2} \quad \text{(by (8.8.11))}$$

and

$$\mathcal{L}\left(\frac{d^2}{dx^2}[e^{x}H(x)]\right) = p^2\mathcal{L}(e^{x}H(x)) = \frac{p^2}{p-1} \quad \text{(by (8.8.11))}.$$

Hence, from (8.8.24),

$$\mathcal{L}^{-1}\left(\frac{p^2+i}{p^2-3p+2}\right) = \frac{d^2}{dx^2}[e^{2x}H(x)] + ie^{2x}H(x) - \frac{d^2}{dx^2}[e^{x}H(x)] - i[e^{x}H(x)]. \tag{8.8.25}$$

But

$$\frac{d}{dx}[e^{2x}H(x)] = e^{2x}\delta + 2e^{2x}H(x) = \delta + 2H(x)e^{2x},$$
$$\frac{d}{dx}[e^{x}H(x)] = e^{x}\delta + e^{x}H(x) = \delta + H(x)e^{x},$$
$$\frac{d^2}{dx^2}[e^{2x}H(x)] = \delta' + 2\delta e^{2x} + 4H(x)e^{2x} = \delta' + 2\delta + 4H(x)e^{2x},$$
$$\frac{d^2}{dx^2}[e^{x}H(x)] = \delta' + \delta e^{x} + H(x)e^{x} = \delta' + \delta + H(x)e^{x},$$

where $\delta = \delta(x)$ is the Dirac distribution with concentration at 0, and $\delta' = \frac{d\delta}{dx}$.

$$\langle e^{kx}\delta, \phi\rangle = \langle \delta, e^{kx}\phi\rangle = (e^{kx}\phi(x))|_{x=0} = \phi(0) = \langle \delta, \phi\rangle \quad \forall \phi \in \mathcal{D}(\mathbb{R})$$

$\Longrightarrow e^{kx}\delta = \delta.$

So, from (8.8.25),

$$\begin{aligned}\mathcal{L}^{-1}\left(\frac{p^2+i}{p^2-3p+2}\right) &= (\delta' + 2\delta + 4e^{2x}H(x) + ie^{2x}H(x)) \\ &\quad + (-\delta' - \delta - e^{x}H(x)) - ie^{x}H(x) \\ &= \delta + (4+i)e^{2x}H(x) - (1+i)e^{x}H(x).\end{aligned}$$

2. $\mathcal{L}[xe^{x}H(x) * T] = \mathcal{L}[H(x)\sin x]$. From (8.8.17) and (8.8.22),

$$\mathcal{L}[xe^{x}H(x) * T](p) = \mathcal{L}[xe^{x}H(x)](p) \cdot \mathcal{L}[T](p) = \frac{1!}{(p-1)^2} \cdot \mathcal{L}(T)(p).$$

From (8.8.23), $\mathcal{L}(H(x)\sin x) = \frac{1}{p^2+1}$. Then $\frac{1}{(p-1)^2}\mathcal{L}(T)(p) = \frac{1}{p^2+1}$

$$\implies \quad \mathcal{L}(T)(p) = \frac{(p-1)^2}{p^2+1} = \frac{p^2-2p+1}{p^2+1} = 1 - 2\frac{p}{p^2+1}.$$

But

$$\mathcal{L}\left[\frac{d}{dx}(H(x)\sin x)\right](p) = \frac{p}{p^2+1}$$

$$\implies \quad T = \mathcal{L}^{-1}\left(1 - 2\frac{p}{p^2+1}\right) = \mathcal{L}^{-1}(1) - 2\mathcal{L}^{-1}\left(\frac{p}{p^2+1}\right)$$

$$\implies \quad T = \delta - 2\frac{d}{dx}[H(x)\sin x] = \delta - 2\sin x \cdot \delta - 2H(x)\cos x$$

$$= \delta - 0 - 2H(x)\cos x = \delta - 2H(x)\cos x$$

(since $\frac{d}{dx}H(x) = \delta$, and $\langle \sin x\delta, \phi\rangle = \langle \delta, \sin x\phi(x)\rangle = [\sin x\phi(x)]_{x=0} = 0$ $\forall \phi \in \mathcal{D}(\mathbb{R}) \implies \delta \sin x = 0$ in $\mathcal{D}'(\mathbb{R})$).

For more details, we refer to [6], [7], [8].

8.9 Applications

8.9.1 Sobolev spaces $H^s(\mathbb{R}^n)$

In Section 2.15 of Chapter 2, we defined Sobolev spaces $H^m(\Omega)$ *of integral order* $m \in \mathbb{N}$ for $\Omega \subset \mathbb{R}^n$. Now we are in a position to define Sobolev spaces $H^s(\mathbb{R}^n)$ for *arbitrary order* $s \in \mathbb{R}$ with the help of Fourier transforms. From this general definition we can retrieve the space $H^m(\mathbb{R}^n)$ for $\Omega = \mathbb{R}^n$ in Definition 2.15.1; this is proved in Proposition 8.9.2 of this section. Now we begin with the definition of Sobolev spaces $H^s(\mathbb{R}^n)$ of *arbitrary* order $s \in \mathbb{R}$.

Definition 8.9.1. $\forall s \in \mathbb{R}$, the set $H^s(\mathbb{R}^n)$ defined by

$$H^s(\mathbb{R}^n) = \{u : u \in S'(\mathbb{R}^n),\ (1 + \|\boldsymbol{\xi}\|^2)^{s/2}\hat{u} \in L^2(\mathbb{R}^n),$$
$$\hat{u}(\boldsymbol{\xi}) = (\mathcal{F}u)(\boldsymbol{\xi}),\ \|\boldsymbol{\xi}\|^2 = \xi_1^2 + \xi_1^2 + \cdots + \xi_n^2\} \tag{8.9.1}$$

is called the *Sobolev space* of arbitrary order $s \in \mathbb{R}$ on $\mathbb{R}^n$, which is equipped with inner product $\langle \cdot, \cdot \rangle_{s,\mathbb{R}^n}$ and norm $\|\cdot\|_{s,\mathbb{R}^n}$ given by: $\forall u, v \in H^s(\mathbb{R}^n)$ with $\hat{u} = \mathcal{F}u$, $\hat{v} = \mathcal{F}v$,

$$\langle u, v\rangle_{H^s(\mathbb{R}^n)} = \langle u, v\rangle_{s,\mathbb{R}^n} = \int_{\mathbb{R}^n} (1 + \|\boldsymbol{\xi}\|^2)^s \hat{u}(\boldsymbol{\xi})\bar{\hat{v}}(\boldsymbol{\xi})d\boldsymbol{\xi}, \tag{8.9.2}$$

where $\bar{\hat{v}}(\boldsymbol{\xi})$ is the complex conjugate of $\hat{v}(\boldsymbol{\xi})$;

$$\|u\|_{H^s(\mathbb{R}^n)} = \|u\|_{s,\mathbb{R}^n} = \langle u, u\rangle_{s,\mathbb{R}^n}^{1/2} = \left(\int_{\mathbb{R}^n} (1 + \|\boldsymbol{\xi}\|^2)^s |\hat{u}(\boldsymbol{\xi})|^2 d\boldsymbol{\xi}\right)^{\frac{1}{2}}. \quad (8.9.3)$$

Theorem 8.9.1. *$\forall s \in \mathbb{R}$, $H^s(\mathbb{R}^n)$ equipped with the inner product $\langle\cdot,\cdot\rangle_{s,\mathbb{R}^n}$ defined by* (8.9.2) *is a* Hilbert space.

Proof. For the proof we are to show that $H^s(\mathbb{R}^n)$ is complete, i.e. every Cauchy sequence in $H^s(\mathbb{R}^n)$ is convergent in $H^s(\mathbb{R}^n)$. Let (u_k) be any Cauchy sequence in $H^s(\mathbb{R}^n))$, i.e.

$$\begin{aligned}
\|u_k - u_m\|_{s,\mathbb{R}^n}^2 &= \int_{\mathbb{R}^n} (1 + \|\boldsymbol{\xi}\|^2)^s |\hat{u}_k - \hat{u}_m|^2 d\boldsymbol{\xi} \to 0 \quad \text{as } k, m \to \infty \\
\Longrightarrow \quad & \|(1 + \|\boldsymbol{\xi}\|^2)^{\frac{s}{2}} (\hat{u}_k - \hat{u}_m)\|_{L^2(\mathbb{R}^n)}^2 \\
&= \int_{\mathbb{R}^n} (1 + \|\boldsymbol{\xi}\|^2)^s |\hat{u}_k - \hat{u}_m|^2 d\boldsymbol{\xi} \to 0 \quad \text{as } k, m \to \infty
\end{aligned}$$

$\Longrightarrow ((1 + \|\boldsymbol{\xi}\|^2)^{s/2}\hat{u}_k)$ is a Cauchy sequence in $L^2(\mathbb{R}^n)$ which is a complete space $\Longrightarrow \exists w \in L^2(\mathbb{R}^n)$ such that $(1 + \|\boldsymbol{\xi}\|^2)^{\frac{s}{2}}\hat{u}_k \to w$ in $L^2(\mathbb{R}^n)$ as $k \to \infty$. But $L^2(\mathbb{R}^n) \hookrightarrow S'(\mathbb{R}^n)$ by (8.2.31) $\Longrightarrow w \in S'(\mathbb{R}^n) \Longrightarrow (1 + \|\boldsymbol{\xi}\|^2)^{\frac{-s}{2}} w \in S'(\mathbb{R}^n)$. By Theorem 8.3.2, $\mathcal{F} : S'(\mathbb{R}^n) \longrightarrow S'(\mathbb{R}^n)$ is an isomorphism $\Longrightarrow \exists u \in S'(\mathbb{R}^n)$ such that $\mathcal{F}u = \hat{u} = (1 + \|\boldsymbol{\xi}\|^2)^{\frac{-s}{2}} w \in S'(\mathbb{R}^n)$. $u \in S'(\mathbb{R}^n) \Longrightarrow (1 + \|\boldsymbol{\xi}\|^2)^{s/2}\hat{u} \in S'(\mathbb{R}^n)$. But $(1+\|\boldsymbol{\xi}\|^2)^{s/2}\hat{u} = w \in L^2(\mathbb{R}^n) \Longrightarrow u \in H^s(\mathbb{R}^n)$ and $(1+\|\boldsymbol{\xi}\|^2)^{\frac{s}{2}}\hat{u}_k \to w = (1 + \|\boldsymbol{\xi}\|^2)^{\frac{s}{2}}\hat{u}$ in $L^2(\mathbb{R}^n)$ as $k \to \infty$

$$\Longrightarrow \quad \|u - u_k\|_{s,\mathbb{R}^n}^2 = \int_{\mathbb{R}^n} (1 + \|\boldsymbol{\xi}\|^2)^s |\hat{u} - \hat{u}_k|^2 d\boldsymbol{\xi} \to 0 \quad \text{as } k \to \infty$$

$\Longrightarrow$ the Cauchy sequence (u_k) converges to $u \in H^s(\mathbb{R}^n)$. Hence, $H^s(\mathbb{R}^n)$ is complete, i.e. a Hilbert space. □

8.9.2 Imbedding result

Proposition 8.9.1. *For $s_1 \geq s_2$, $H^{s_1}(\mathbb{R}^n) \hookrightarrow H^{s_2}(\mathbb{R}^n)$ with*

$$\|u\|_{s_2,\mathbb{R}^n} \leq \|u\|_{s_1,\mathbb{R}^n} \quad \forall u \in H^{s_1}(\mathbb{R}^n). \quad (8.9.4)$$

Proof. For $s_1 \geq s_2$, $(1 + \|\boldsymbol{\xi}\|^2)^{s_2} \leq (1 + \|\boldsymbol{\xi}\|^2)^{s_1}$. Let $u \in H^{s_1}(\mathbb{R}^n)$. Then, $\int_{\mathbb{R}^n} (1 + \|\boldsymbol{\xi}\|^2)^{s_1} |\hat{u}(\boldsymbol{\xi})|^2 d\boldsymbol{\xi} < +\infty$. But $(1 + \|\boldsymbol{\xi}\|^2)^{s_2} |\hat{u}(\boldsymbol{\xi})|^2 \leq (1 + \|\boldsymbol{\xi}\|^2)^{s_1} |\hat{u}(\boldsymbol{\xi})|^2$ a.e. in $\mathbb{R}^n$

$$\Longrightarrow \quad \int_{\mathbb{R}^n} (1 + \|\boldsymbol{\xi}\|^2)^{s_2} |\hat{u}(\boldsymbol{\xi})|^2 d\boldsymbol{\xi} \leq \int_{\mathbb{R}^n} (1 + \|\boldsymbol{\xi}\|^2)^{s_1} |\hat{u}(\boldsymbol{\xi})|^2 d\boldsymbol{\xi} < +\infty.$$

Hence, $u \in H^{s_1}(\mathbb{R}^n) \subset S'(\mathbb{R}^n) \Longrightarrow (1 + \|\boldsymbol{\xi}\|^2)^{s_1/2}\hat{u} \in L^2(\mathbb{R}^n) \Longrightarrow (1 + \|\boldsymbol{\xi}\|^2)^{s_2/2}\hat{u} \in L^2(\mathbb{R}^n) \Longrightarrow u \in H^{s_2}(\mathbb{R}^n)$ with $\|u\|_{s_2,\mathbb{R}^n} \leq \|u\|_{s_1,\mathbb{R}^n} \Longrightarrow H^{s_1}(\mathbb{R}^n) \subset H^{s_2}(\mathbb{R}^n)$ with $\|u\|_{s_2,\mathbb{R}^n} \leq \|u\|_{s_1,\mathbb{R}^n}$
$\Longrightarrow H^{s_1}(\mathbb{R}^n) \hookrightarrow H^{s_2}(\mathbb{R}^n)$ with $\|u\|_{s_2,\mathbb{R}^n} \leq \|u\|_{s_1,\mathbb{R}^n}\ \forall u \in H^{s_1}(\mathbb{R}^n)$. □

Remark 8.9.1. For $s \geq 0$, $H^s(\mathbb{R}^n) \hookrightarrow L^2(\mathbb{R}^n) = H^0(\mathbb{R}^n)$. Hence, for $s \geq 0$, $H^s(\mathbb{R}^n)$ can equivalently be defined by:

Definition 8.9.2. For $s \geq 0$,

$$H^s(\mathbb{R}^n) = \{u : u \in L^2(\mathbb{R}^n), (1 + \|\boldsymbol{\xi}\|^2)^{s/2}\hat{u} \in L^2(\mathbb{R}^n)\} \tag{8.9.5}$$

with $\langle u, v\rangle_{s,\mathbb{R}^n}$ and $\|\cdot\|_{s,\mathbb{R}^n}$ defined by (8.9.2) and (8.9.3) respectively.

Example 8.9.1. Show that the partial differential operator $-\Delta + k^2 : H^{s+2}(\mathbb{R}^n) \longrightarrow H^s(\mathbb{R}^n)$ is an isomorphism from $H^{s+2}(\mathbb{R}^n)$ onto $H^s(\mathbb{R}^n)$ $\forall$ real $k \neq 0$, $\forall s \in \mathbb{R}$, $\Delta = \frac{\partial^2}{\partial x_1^2} + \cdots + \frac{\partial^2}{\partial x_n^2}$ being the n-dimensional Laplace operator.

Solution. Set $A = -\Delta + k^2$ defined by $\forall u \in H^{s+2}(\mathbb{R}^n)$, $Au \equiv -\Delta u + k^2 u \in H^s(\mathbb{R}^n)$ $\forall$ real $k \neq 0$, $\forall s \in \mathbb{R}$.
Continuity of $A : H^{s+2}(\mathbb{R}^n) \longrightarrow H^s(\mathbb{R}^n)$: Let $u \in H^{s+2}(\mathbb{R}^n) \subset S'(\mathbb{R}^n)$. Hence, $\mathcal{F}[-\Delta u + k^2 u] = (4\pi^2\|\boldsymbol{\xi}\|^2 + k^2)\hat{u} \in S'(\mathbb{R}^n)$ with $\hat{u} = \mathcal{F}u$

$$\Longrightarrow \quad (1 + \|\boldsymbol{\xi}\|^2)^{s/2}|\mathcal{F}[-\Delta u + k^2 u]| = (1 + \|\boldsymbol{\xi}\|^2)^{s/2}|(4\pi^2\|\boldsymbol{\xi}\|^2 + k^2)| \cdot |\hat{u}|.$$

But

$$\begin{aligned}(1 + \|\boldsymbol{\xi}\|^2)^{s/2}(4\pi^2\|\boldsymbol{\xi}\|^2 + k^2) &\leq \max\{4\pi^2, k^2\}(\|\boldsymbol{\xi}\|^2 + 1)(1 + \|\boldsymbol{\xi}\|^2)^{s/2} \\ &= C(1 + \|\boldsymbol{\xi}\|^2)^{s/2+1},\end{aligned}$$

with $C = \max\{4\pi^2, k^2\} > 0$.

Hence,

$$(1 + \|\boldsymbol{\xi}\|^2)^{s/2}|\mathcal{F}[-\Delta u + k^2 u]| \leq C(1 + \|\boldsymbol{\xi}\|^2)^{s/2+1}|\hat{u}|.$$

$u \in H^{s+2}(\mathbb{R}^n) \Longrightarrow$ from definition of $H^{s+2}(\mathbb{R}^n)$, $(1+\|\boldsymbol{\xi}\|^2)^{s/2+1}\hat{u} \in L^2(\mathbb{R}^n) \Longrightarrow (1 + \|\boldsymbol{\xi}\|^2)^{s/2}\mathcal{F}[-\Delta u + k^2 u] \in L^2(\mathbb{R}^n) \Longrightarrow (-\Delta u + k^2 u) \in H^s(\mathbb{R}^n)$.

Thus, $u \in H^{s+2}(\mathbb{R}^n) \Longrightarrow -\Delta u + k^2 u \in H^s(\mathbb{R}^n)$. Moreover,

$$\begin{aligned}\| - \Delta u + k^2 u\|^2_{s,\mathbb{R}^n} &= \|(1 + \|\boldsymbol{\xi}\|^2)^{s/2}|\mathcal{F}[-\Delta u + k^2 u]|\|^2_{L^2(\mathbb{R}^n)} \\ &\leq C^2\|(1 + \|\boldsymbol{\xi}\|)^{s/2+1}|\hat{u}|\|^2_{L^2(\mathbb{R}^n)}\end{aligned}$$

$\Longrightarrow \| - \Delta u + k^2 u\|_{s,\mathbb{R}^n} \leq C\|u\|_{s+2,\mathbb{R}^n}$ with $C > 0$, independent of u,

$\implies A = -\Delta u + k^2 : H^{s+2}(\mathbb{R}^n) \longrightarrow H^s(\mathbb{R}^n)$ is *continuous* from $H^{s+2}(\mathbb{R}^n)$ into $H^s(\mathbb{R}^n)$.

$A : H^{s+2}(\mathbb{R}^n) \longrightarrow H^s(\mathbb{R}^n)$ *is injective*: For $u \in H^{s+2}(\mathbb{R}^n)$, $Au = -\Delta u + k^2 u = 0$ in $H^s(\mathbb{R}^n) \implies (4\pi^2\|\boldsymbol{\xi}\|^2 + k^2)\hat{u} = 0$ in $S'(\mathbb{R}^n) \implies \hat{u} = 0$ in $S'(\mathbb{R}^n)$, since $(4\pi^2\|\boldsymbol{\xi}\|^2 + k^2) \neq 0\ \forall \boldsymbol{\xi} \in \mathbb{R}^n \implies \bar{\mathcal{F}}\hat{u} = 0 \implies u = 0$ in $S'(\mathbb{R}^n) \implies u = 0$ in $H^{s+2}(\mathbb{R}^n)$.

$A : H^{s+2}(\mathbb{R}^n) \longrightarrow H^s(\mathbb{R}^n)$ *is surjective*: Let $f \in H^s(\mathbb{R}^n)$. Then $f \in S'(\mathbb{R}^n) \implies \hat{f} \in S'(\mathbb{R}^n) \implies (4\pi^2\|\boldsymbol{\xi}\|^2 + k^2)^{-1}\hat{f} \in S'(\mathbb{R}^n)$.

Set $v = (4\pi^2\|\boldsymbol{\xi}\|^2 + k^2)^{-1}\hat{f}$. Then $v \in S'(\mathbb{R}^n) \implies \hat{f} = (4\pi^2\|\boldsymbol{\xi}\|^2 + k^2)v \in S'(\mathbb{R}^n)$

$\implies f = \bar{\mathcal{F}}\hat{f} = \bar{\mathcal{F}}[(4\pi^2\|\boldsymbol{\xi}\|^2 + k^2)v] = (-\Delta + k^2)\bar{\mathcal{F}}v.$

Define $u = \bar{\mathcal{F}}v \in S'(\mathbb{R}^n)$. Then $(-\Delta + k^2)u = f \in H^s(\mathbb{R}^n)$.

But $\mathcal{F}[(-\Delta + k^2)u] = (4\pi^2\|\boldsymbol{\xi}\|^2 + k^2)\hat{u} = \hat{f}$

$$\begin{aligned}\implies \quad (1 + \|\boldsymbol{\xi}\|^2)^{s/2+1}|\hat{u}| &= \frac{(1 + \|\boldsymbol{\xi}\|^2)^{s/2+1}}{(4\pi^2\|\boldsymbol{\xi}\|^2 + k^2)}|\hat{f}| \\ &\leq \frac{(1 + \|\boldsymbol{\xi}\|^2)}{(4\pi^2|\boldsymbol{\xi}\|^2 + k^2)} \cdot (1 + \|\boldsymbol{\xi}\|^2)^{s/2}|\hat{f}| \leq C_1(1 + \|\boldsymbol{\xi}\|^2)^{s/2}|\hat{f}|,\end{aligned}$$

since $(4\pi^2\|\boldsymbol{\xi}\|^2 + k^2) \geq \min\{4\pi^2, k^2\}(1 + \|\boldsymbol{\xi}\|^2)$

$\implies \frac{1}{4\pi^2\|\boldsymbol{\xi}\|^2+k^2} \leq C_1\frac{1}{(1+\|\boldsymbol{\xi}\|^2)}$ with $C_1 = \frac{1}{\min\{4\pi^2,k^2\}} > 0$.

Hence,

$$\int_{\mathbb{R}^n}(1 + \|\boldsymbol{\xi}\|^2)^{s+2}|\hat{u}|^2 d\boldsymbol{\xi} \leq C_1^2\int_{\mathbb{R}^n}(1 + \|\boldsymbol{\xi}\|^2)^s|\hat{f}|^2 d\boldsymbol{\xi} < +\infty$$

$\implies u \in H^{s+2}(\mathbb{R}^n)$. Thus, $\forall f \in H^s(\mathbb{R}^n)$, $\exists u \in H^{s+2}(\mathbb{R}^n)$ such that $Au = -\Delta u + k^2 u = f$ in $H^s(\mathbb{R}^n) \implies -\Delta + k^2$ is *surjective* from $H^{s+2}(\mathbb{R}^n)$ onto $H^s(\mathbb{R}^n)$.

Hence, $A = -\Delta + k^2$ is *bijective* from $H^{s+2}(\mathbb{R}^n)$ onto $H^s(\mathbb{R}^n)$. But $H^{s+2}(\mathbb{R}^n)$ and $H^s(\mathbb{R}^n)$ are Banach spaces, and $A = -\Delta + k^2$ is a *linear*, *bijective* mapping from Banach space $H^{s+2}(\mathbb{R}^n)$ onto Banach space $H^s(\mathbb{R}^n)$. Then, by Corollary A.8.1.1 of the Open Mapping Theorem A.8.1.3 in Appendix A, the continuity of the inverse follows. Hence, $-\Delta + k^2$ is an isomorphism from $H^{s+2}(\mathbb{R}^n)$ onto $H^s(\mathbb{R}^n)$.

Now we state two important lemmas which will be needed later.

Lemma 8.9.1 ([40]). *For $1 \leq |\boldsymbol{\alpha}| \leq m$, the following inequality holds: $\forall \boldsymbol{\xi} \in \mathbb{R}^n$, $\exists C = C(m) > 0$ such that*

$$\prod_{i=1}^{n}\xi_i^{2\alpha_i} \leq \left(1 + \sum_{i=1}^{n}\xi_i^2\right)^m \leq C\left[1 + \sum_{1\leq|\boldsymbol{\alpha}|\leq m}\left(\prod_{i=1}^{n}\xi_i^{2\alpha_i}\right)\right].$$

Proof. $0 \le \alpha_1 + \alpha_2 + \cdots + \alpha_n \le m$. Set $1 + \xi_1^2 + \cdots + \xi_n^2 = P \ge 1$. Hence, $P^m \ge P^{\alpha_1} P^{\alpha_2} \dots P^{\alpha_n}$. But $P \ge \xi_i^2 \; \forall i = 1, 2, \dots, n \Longrightarrow P^{\alpha_i} \ge \xi_i^{2\alpha_i} \; \forall i = 1, 2, \dots, n$

$$\Longrightarrow \quad \xi_1^{2\alpha_1} \xi_2^{2\alpha_2} \cdots \xi_n^{2\alpha_n} \le P^{\alpha_1} P^{\alpha_2} \cdots P^{\alpha_n} \le P^m = \left(1 + \sum_{i=1}^{n} \xi_i^2\right)^m$$

$$\Longrightarrow \prod_{i=1}^{n} \xi_i^{2\alpha_i} \le \left(1 + \sum_{i=1}^{n} \xi_i^2\right)^m.$$

For the proof of the second inequality, we apply binomial expansion. In fact,

$$(1 + \xi_1^2 + \cdots + \xi_n^2)^m$$
$$= \sum_{\beta=0}^{m} \sum_{\beta_1=0}^{\beta} \sum_{\beta_2=0}^{\beta-\beta_1} \cdots \sum_{\beta_{n-1}=0}^{\beta-\beta_1-\cdots-\beta_{n-2}} C_{\beta,\beta_1,\dots\beta_{n-1}} \xi_1^{2\beta_1} \cdots \xi_{n-1}^{2\beta_{n-1}} \xi_n^{2(\beta-\beta_1-\cdots-\beta_{n-1})},$$

where $C_{\beta,\beta_1,\dots,\beta_{n-1}}$ are constants with $C_{0,0,\dots,0} = 1$, $\beta_1 + \beta_2 + \cdots + \beta_{n-1} + (\beta - \beta_1 - \beta_2 - \cdots - \beta_{n-1}) = \beta \le m$

$$\Longrightarrow \quad (1 + \xi_1^2 + \cdots + \xi_n^2)^m \le C\left[1 + \sum_{|\boldsymbol{\alpha}|=1}^{m} \xi_1^{2\alpha_1} \xi_2^{2\alpha_2} \cdots \xi_n^{2\alpha_n}\right]$$

with $C \ge \max\{C_{\beta,\beta_1,\dots,\beta_{n-1}}\} > 0$. □

Lemma 8.9.2. *For $1 \le |\boldsymbol{\alpha}| \le m$, the following inequality holds: $\forall \boldsymbol{\xi} \in \mathbb{R}^n$, $\exists C = C(m) > 0$ such that*

$$\prod_{i=1}^{n} \xi^{2\alpha_i} \le \left(1 + \sum_{i=1}^{n} \xi_i^2\right)^m \le C\left[1 + \sum_{|\boldsymbol{\alpha}|=m} \left(\prod_{i=1}^{n} \xi_i^{2\alpha_i}\right)\right]. \tag{8.9.6}$$

Proof. For the proof of the first inequality, see the proof of Lemma 8.9.1. For the proof of the second inequality, set

$$f(\boldsymbol{\xi}) = \left(1 + \sum_{i=1}^{n} \xi_i^2\right)^m \Big/ \left(1 + \sum_{|\boldsymbol{\alpha}|=m} \left(\prod_{i=1}^{n} \xi_i^{2\alpha_i}\right)\right) > 0 \quad \forall \boldsymbol{\xi} \in \mathbb{R}^n.$$

Moreover, f is continuous in $\mathbb{R}^n$. Hence, in every compact subset of $\mathbb{R}^n$, f is bounded. Set $r^2 = \sum_{i=1}^{n} \xi_i^2$. Then,

$$f(\boldsymbol{\xi}) = \frac{(1 + r^2)^m}{1 + \sum_{|\boldsymbol{\alpha}|=m} (\prod_{i=1}^{n} \xi_i^{2\alpha_i})},$$

in which the denominator behaves as $1 + r^{2m} = 1 + (\sum_{i=1}^{n} |\xi_i|^2)^m$. Hence, $\exists r_0 > 0$ such that $\forall r \ge r_0$, $f(\boldsymbol{\xi}) \le C_1$ and for $0 \le r \le r_0$, $\exists C_2 > 0$ such that $f(\boldsymbol{\xi}) \le C_2$. Then $f(\boldsymbol{\xi}) \le C = \max\{C_1, C_2\} \Longrightarrow$ the result (8.9.6). □

8.9.3 Sobolev spaces $H^m(\mathbb{R}^n)$ of integral order m on $\mathbb{R}^n$

$H^m(\mathbb{R}^n)$ with $m \in \mathbb{N}$ can be defined in two *different* ways:

1. One based on Definition 2.15.1 in (2.15.1)–(2.15.3), which we agree to denote *temporarily* by a different notation $\mathcal{H}^m(\mathbb{R}^n)$ (instead of the natural one $H^m(\mathbb{R}^n)$ with $\Omega = \mathbb{R}^n$):

$$\mathcal{H}^m(\mathbb{R}^n) = \{u : u \in L^2(\mathbb{R}^n),\ \partial^{\boldsymbol{\alpha}} u \in L^2(\mathbb{R}^n)\ \forall |\boldsymbol{\alpha}| \le m\}. \tag{8.9.7}$$

2. The other one based on Fourier transforms given by Definition 8.9.1 with (8.9.1)–(8.9.3) or by (8.9.5), which we will denote by $H^m(\mathbb{R}^n)$ with $s = m$. Then we will show that $\mathcal{H}^m(\mathbb{R}^n) \equiv H^m(\mathbb{R}^n)$. In fact, we have:

Theorem 8.9.2. *For* $s = m \in \mathbb{N}$, $H^m(\mathbb{R}^n)$ *defined by* (8.9.1)/(8.9.5)*:*

$$H^m(\mathbb{R}^n) = \{u : u \in L^2(\mathbb{R}^n),\ (1 + \|\boldsymbol{\xi}\|^2)^{m/2}\hat{u} \in L^2(\mathbb{R}^n) \text{ with } \hat{u} = \mathcal{F}u\} \tag{8.9.8}$$

coincides *with the space* $\mathcal{H}^m(\mathbb{R}^n)$ *defined by* (8.9.7)*, i.e.* $H^m(\mathbb{R}^n) \equiv \mathcal{H}^m(\mathbb{R}^n)$, *where derivatives* $\partial^{\boldsymbol{\alpha}} u$ *are in the distributional sense:* $\forall \phi \in \mathcal{D}(\mathbb{R}^n)$,

$$\int_{\mathbb{R}^n} \partial^{\alpha} u(\mathbf{x})\phi(\mathbf{x})d\mathbf{x} = (-1)^{|\boldsymbol{\alpha}|} \int_{\mathbb{R}^n} u(\mathbf{x})\partial^{\boldsymbol{\alpha}}\phi(\mathbf{x})d\mathbf{x}.$$

Then

$$\|u\|_{\mathcal{H}(\mathbb{R}^n)} = \left(\sum_{0 \le |\boldsymbol{\alpha}| \le m} \int_{\mathbb{R}^n} |\partial^{\boldsymbol{\alpha}} u(\mathbf{x})|^2 d\mathbf{x} \right)^{1/2} \quad (\textit{see } (2.15.10)) \tag{8.9.9}$$

and

$$\|u\|_{m,\mathbb{R}^n} = \left(\int_{\mathbb{R}^n} (1 + \|\boldsymbol{\xi}\|^2)^m |\hat{u}(\boldsymbol{\xi})|^2 d\boldsymbol{\xi} \right)^{1/2} \quad (\textit{see } (8.9.3)) \tag{8.9.10}$$

are equivalent norms *in* $H^m(\mathbb{R}^n)$*, i.e.* $\exists C_1, C_2 > 0$ *such that*

$$C_1\|u\|_{\mathcal{H}^m(\mathbb{R}^n)} \le \|u\|_{m,\mathbb{R}^n} \le C_2\|u\|_{\mathcal{H}^m(\mathbb{R}^n)} \quad \forall u \in H^m(\mathbb{R}^n). \tag{8.9.11}$$

Proof. $H^m(\mathbb{R}^n) \subset \mathcal{H}^m(\mathbb{R}^n)$: Let $u \in H^m(\mathbb{R}^n)$; we will show that $u \in \mathcal{H}^m(\mathbb{R}^n)$. Then $u \in H^m(\mathbb{R}^n) \Longrightarrow (1 + \|\boldsymbol{\xi}\|^2)^{\frac{m}{2}}\hat{u} \in L^2(\mathbb{R}^n)$. Hence, $|\hat{u}(\boldsymbol{\xi})|^2 \le (1 + \|\boldsymbol{\xi}\|^2)^m \cdot |\hat{u}(\boldsymbol{\xi})|^2$ a.e. in $\mathbb{R}^n$

$$\Longrightarrow \quad \int_{\mathbb{R}^n} |\hat{u}(\boldsymbol{\xi}|^2 d\boldsymbol{\xi} \le \int_{\mathbb{R}^n} (1 + \|\boldsymbol{\xi}\|^2)^m |\hat{u}(\boldsymbol{\xi})|^2 d\boldsymbol{\xi} < +\infty$$

$\Longrightarrow \hat{u} \in L^2(\mathbb{R}^n)$.

Using Lemma 8.9.1, we have, $\forall$ multi-index $\boldsymbol{\alpha}$,

$$|\boldsymbol{\xi}^{\boldsymbol{\alpha}}\hat{u}(\boldsymbol{\xi})|^2 = \left(\prod_{i=1}^{n} \xi_i^{2\alpha_i}|\hat{u}(\boldsymbol{\xi})|\right)^2 \le (1+\|\boldsymbol{\xi}\|^2)^m|\hat{u}(\boldsymbol{\xi})|^2 \quad \text{a.e. in } \mathbb{R}^n$$

$$\Longrightarrow \quad \sum_{0\le|\boldsymbol{\alpha}|\le m} [\xi_1^{2\alpha_1}\xi_2^{2\alpha_2}\cdots\xi_n^{2\alpha_n}]|\hat{u}(\boldsymbol{\xi})|^2 \le \bar{C}(1+\|\boldsymbol{\xi}\|^2)^m|\hat{u}(\boldsymbol{\xi})|^2 \quad \text{a.e. in } \mathbb{R}^n$$

with $\bar{C} = \bar{C}(m) > 0$

$$\Longrightarrow \quad \sum_{0\le|\boldsymbol{\alpha}|\le m} \int_{\mathbb{R}^n} \xi_1^{2\alpha_1}\xi_2^{2\alpha_2}\cdots\xi_n^{2\alpha_n}|\hat{u}(\boldsymbol{\xi})|^2 d\boldsymbol{\xi} \le \bar{C}\int_{\mathbb{R}^n}(1+\|\boldsymbol{\xi}\|^2)^m|\hat{u}(\boldsymbol{\xi})|^2 d\boldsymbol{\xi}$$

$\Longrightarrow \boldsymbol{\xi}^{\boldsymbol{\alpha}}\hat{u} \in L^2(\mathbb{R}^n)$ $\forall$ multi-index $\boldsymbol{\alpha}$ with $|\boldsymbol{\alpha}| \le m \Longrightarrow \boldsymbol{\xi}^{\boldsymbol{\alpha}}\hat{u} \in S'(\mathbb{R}^n)$ $\forall |\boldsymbol{\alpha}| < m$.

But from Theorem 8.6.1, $\mathcal{F}(\partial_{\mathbf{x}}^{\boldsymbol{\alpha}} u) = (i2\pi\boldsymbol{\xi})^{\boldsymbol{\alpha}}\hat{u} = (i2\pi)^{|\boldsymbol{\alpha}|}\boldsymbol{\xi}^{\boldsymbol{\alpha}}\hat{u} \Longrightarrow \mathcal{F}(\partial_{\mathbf{x}}^{\boldsymbol{\alpha}} u) = (i2\pi)^{|\boldsymbol{\alpha}|}\boldsymbol{\xi}^{\boldsymbol{\alpha}}\hat{u} \in L^2(\mathbb{R}^n)$ $\forall$ multi-index $\boldsymbol{\alpha}$ with $|\boldsymbol{\alpha}| \le m \Longrightarrow$ by the Plancherel–Riesz Theorem 8.3.1, $\partial_{\mathbf{x}}^{\boldsymbol{\alpha}} u \in L^2(\mathbb{R}^n)$ with

$$\|\partial_{\mathbf{x}}^{\boldsymbol{\alpha}} u\|_{L^2(\mathbb{R}^n)}^2 = \|\mathcal{F}(\partial_{\mathbf{x}}^{\boldsymbol{\alpha}} u)\|_{L^2(\mathbb{R}^n)}^2 = \|(i2\pi)^{|\boldsymbol{\alpha}|}\boldsymbol{\xi}^{\boldsymbol{\alpha}}\hat{u}\|_{L^2(\mathbb{R}^n)}^2$$
$$= (2\pi)^{2|\boldsymbol{\alpha}|}\|\boldsymbol{\xi}^{\boldsymbol{\alpha}}\hat{u}\|_{L^2(\mathbb{R}^n)}^2 \quad \forall |\boldsymbol{\alpha}| \le m$$
$$\Longrightarrow \quad \|u\|_{\mathcal{H}^m(\mathbb{R}^n)}^2 = \sum_{0\le|\boldsymbol{\alpha}|\le m} \|\partial_{\mathbf{x}}^{\boldsymbol{\alpha}} u\|_{L^2(\mathbb{R}^n)}^2$$
$$= \sum_{0\le|\boldsymbol{\alpha}|\le m} (2\pi)^{2|\boldsymbol{\alpha}|}\int_{\mathbb{R}^n} \xi_1^{2\alpha_1}\xi_2^{2\alpha_2}\cdots\xi_n^{2\alpha_n}|\hat{u}(\boldsymbol{\xi})|^2 d\boldsymbol{\xi}$$
$$\le C_0\int_{\mathbb{R}^n}(1+\|\boldsymbol{\xi}\|^2)^m|\hat{u}(\boldsymbol{\xi})|^2 d\boldsymbol{\xi}$$
$$= C_0\|u\|_{m,\mathbb{R}^n}^2 < +\infty \quad \text{with } C_0 = (2\pi)^{2m}\bar{C} > 0$$

$\Longrightarrow u \in \mathcal{H}^m(\mathbb{R}^n)$ with $\|u\|_{\mathcal{H}^m(\mathbb{R}^n)}^2 \le C_0\|u\|_{m,\mathbb{R}^n}^2 \Longrightarrow u \in \mathcal{H}^m(\mathbb{R}^n)$ with $\|u\|_{\mathcal{H}(\mathbb{R}^n)} \le C_0^*\|u\|_{m,\mathbb{R}^n}$ with $C_0^* = \sqrt{C_0} > 0$. Hence, $u \in H^m(\mathbb{R}^n) \Longrightarrow u \in \mathcal{H}^m(\mathbb{R}^n)$

$$\Longrightarrow \quad H^m(\mathbb{R}^n) \subset \mathcal{H}^m(\mathbb{R}^n) \quad \text{with } C_1\|u\|_{\mathcal{H}^m(\mathbb{R}^n)} \le \|u\|_{m,\mathbb{R}^n} \text{ with } C_1 = \frac{1}{C_0^*} > 0. \tag{8.9.12}$$

$\mathcal{H}^m(\mathbb{R}^n) \subset H^m(\mathbb{R}^n)$: Let $u \in \mathcal{H}^m(\mathbb{R}^n)$; we will show that $u \in H^m(\mathbb{R}^n)$. Then $u \in \mathcal{H}^m(\mathbb{R}^n) \Longrightarrow u \in L^2(\mathbb{R}^n)$, $\partial^{\boldsymbol{\alpha}} u \in L^2(\mathbb{R}^n)$ $\forall$ multi-index $\boldsymbol{\alpha}$ with $|\boldsymbol{\alpha}| \le m$. Using Lemma 8.9.1 again, we get

$$(1+\|\boldsymbol{\xi}\|^2)^m|\hat{u}(\boldsymbol{\xi}|^2 \le C\left[|\hat{u}(\boldsymbol{\xi})|^2 + \sum_{1\le|\boldsymbol{\alpha}|\le m}|\boldsymbol{\xi}^{\boldsymbol{\alpha}}\hat{u}(\boldsymbol{\xi})|^2\right], \tag{8.9.13}$$

where $(\boldsymbol{\xi}^{\boldsymbol{\alpha}})^2 = \xi_1^{2\alpha_1}\xi_2^{2\alpha_2}\cdots\xi_n^{2\alpha_n} = \prod_{i=1}^n \xi_i^{2\alpha_i}$. But from Theorem 8.6.1, $\mathcal{F}(\partial_{\mathbf{x}}^{\boldsymbol{\alpha}} u) = (i2\pi\boldsymbol{\xi})^{\boldsymbol{\alpha}}\hat{u} \in L^2(\mathbb{R}^n)\ \forall|\boldsymbol{\alpha}| \le m$ and, by the Plancherel–Riesz Theorem 8.3.1,

$$\begin{aligned}
&\|\partial_{\mathbf{x}}^{\boldsymbol{\alpha}} u\|_{L^2(\mathbb{R}^n)} = \|\mathcal{F}(\partial_{\mathbf{x}}^{\boldsymbol{\alpha}} u)\|_{L^2(\mathbb{R}^n)} = (2\pi)^{|\boldsymbol{\alpha}|}\|\boldsymbol{\xi}^{\boldsymbol{\alpha}}\hat{u}\|_{L^2(\mathbb{R}^n)} \quad \forall|\boldsymbol{\alpha}| \le m \\
\Longrightarrow\quad &\|\boldsymbol{\xi}^{\boldsymbol{\alpha}}\hat{u}\|_{L^2(\mathbb{R}^n)} = \frac{1}{(2\pi)^{|\boldsymbol{\alpha}|}}\|\partial_{\mathbf{x}}^{\boldsymbol{\alpha}} u\|_{L^2(\mathbb{R}^n)} \quad \forall|\boldsymbol{\alpha}| \le m \\
\Longrightarrow\quad &\int_{\mathbb{R}^n} \boldsymbol{\xi}^{2\boldsymbol{\alpha}}|\hat{u}(\boldsymbol{\xi})|^2 d\boldsymbol{\xi} = \frac{1}{(2\pi)^{2|\boldsymbol{\alpha}|}}\int_{\mathbb{R}^n} |\partial_{\mathbf{x}}^{\boldsymbol{\alpha}} u(\mathbf{x})|^2 d\mathbf{x} \quad \forall|\boldsymbol{\alpha}| \le m. \qquad (8.9.14)
\end{aligned}$$

Hence, from (8.9.13) and (8.9.14),

$$\begin{aligned}
\int_{\mathbb{R}^n} (1 + \|\boldsymbol{\xi}\|^2)^m |\hat{u}(\boldsymbol{\xi})|^2 d\boldsymbol{\xi} &\le \tilde{C}\Big[\int_{\mathbb{R}^n} |u(\mathbf{x})|^2 d\mathbf{x} + \sum_{1\le|\boldsymbol{\alpha}|\le m}\int_{\mathbb{R}^n} |\partial^{\boldsymbol{\alpha}} u(\mathbf{x})|^2 d\mathbf{x}\Big] \\
&= \tilde{C}\|u\|^2_{\mathcal{H}^m(\mathbb{R}^n)} < +\infty, \quad \tilde{C} > 0
\end{aligned}$$

$$\begin{aligned}
&\Longrightarrow u \in H^m(\mathbb{R}^n) \text{ with } \|u\|^2_{m,\mathbb{R}^n} \le \tilde{C}\|u\|^2_{\mathcal{H}^m(\mathbb{R}^n)}. \text{ Then, } u \in \mathcal{H}^m(\mathbb{R}^n) \\
&\Longrightarrow u \in H^m(\mathbb{R}^n) \text{ with } \|u\|_{m,\mathbb{R}^n} \le C_2\|u\|_{\mathcal{H}^m(\mathbb{R}^n)} \text{ with } C_2 = \sqrt{\tilde{C}} > 0 \\
&\Longrightarrow \quad \mathcal{H}^m(\mathbb{R}^n) \subset H^m(\mathbb{R}^n) \text{ with } \|u\|_{m,\mathbb{R}^n} \le C_2\|u\|_{\mathcal{H}^m(\mathbb{R}^n)}. \qquad (8.9.15)
\end{aligned}$$

Combining (8.9.12) and (8.9.15), we have $H^m(\mathbb{R}^n) \subset \mathcal{H}^m(\mathbb{R}^n)$ and $\mathcal{H}^m(\mathbb{R}^n) \subset H^m(\mathbb{R}^n) \Longrightarrow H^m(\mathbb{R}^n) \equiv \mathcal{H}^m(\mathbb{R}^n)$ with $C_1\|u\|_{\mathcal{H}^m(\mathbb{R}^n)} \le \|u\|_{m,\mathbb{R}^n} \le C_2\|u\|_{\mathcal{H}^m(\mathbb{R}^n)}$. □

Theorem 8.9.3. *For non-integer $s > 0$ with $s = [s] + \sigma$, $[s] \in \mathbb{N}_0$ being the integral part of s, $\sigma \in\,]0,1[$ being the fractional part, $H^s(\mathbb{R}^n)$ defined by* (8.9.1) *coincides with the space defined by*

$$\{u : u \in H^{[s]}(\mathbb{R}^n),\ \partial^{\boldsymbol{\alpha}} u \in H^{\sigma}(\mathbb{R}^n)\ \forall|\boldsymbol{\alpha}| = [s]\} \text{ (see also (8.10.71))}. \qquad (8.9.16)$$

Then the mapping

$$u \in H^s(\mathbb{R}^n) \longmapsto \Big(\|u\|^2_{[s],\mathbb{R}^n} + \sum_{|\boldsymbol{\alpha}|=[s]} \|\partial^{\boldsymbol{\alpha}} u\|^2_{\sigma,\mathbb{R}^n}\Big)^{\frac{1}{2}} \qquad (8.9.17)$$

defines a norm *in $H^s(\mathbb{R}^n)$ equivalent to the original norm $\|\cdot\|_{s,\mathbb{R}^n}$, i.e. $\exists C_1, C_2 > 0$ such that, $\forall u \in H^s(\mathbb{R}^n)$,*

$$C_1\|u\|_{s,\mathbb{R}^n} \le \Big(\|u\|^2_{[s],\mathbb{R}^n} + \sum_{|\boldsymbol{\alpha}|=[s]} \|\partial^{\boldsymbol{\alpha}} u\|^2_{\sigma,\mathbb{R}^n}\Big)^{\frac{1}{2}} \le C_2\|u\|_{s,\mathbb{R}^n}, \qquad (8.9.18)$$

where $\|\cdot\|_{s,\mathbb{R}^n}$, $\|\cdot\|_{[s],\mathbb{R}^n}$ and $\|\cdot\|_{\sigma,\mathbb{R}^n}$ denote norms in $H^s(\mathbb{R}^n)$, $H^{[s]}(\mathbb{R}^n)$ and $H^{\sigma}(\mathbb{R}^n)$, respectively. $|\cdot\|_{[s],(\mathbb{R}^n)}$ and $\|\cdot\|_{\sigma,(\mathbb{R}^n)}$ are also defined by (8.9.3) *with 's' replaced by '[s]' and 'σ', respectively.*

Proof.
($\Longrightarrow$): Let $u \in H^s(\mathbb{R}^n)$. Since $0 \leq [s] < s$, $H^s(\mathbb{R}^n) \hookrightarrow H^{[s]}(\mathbb{R}^n)$ and $\|u\|_{[s],\mathbb{R}^n} \leq \|u\|_{s,\mathbb{R}^n}$ by Proposition 8.9.1. It remains to show that $\partial^{\boldsymbol{\alpha}} u \in H^\sigma(\mathbb{R}^n)$ $\forall|\boldsymbol{\alpha}| = [s]$. From (8.6.1), $\mathcal{F}(\partial^{\boldsymbol{\alpha}} u) = \widehat{\partial^{\boldsymbol{\alpha}} u} = (i2\pi\boldsymbol{\xi})^{\boldsymbol{\alpha}}\hat{u}$ $\forall|\boldsymbol{\alpha}| = [s]$. $H^s(\mathbb{R}^n) \hookrightarrow S'(\mathbb{R}^n) \Longrightarrow u \in S'(\mathbb{R}^n) \Longrightarrow \partial^{\boldsymbol{\alpha}} u \in S'(\mathbb{R}^n)$ and

$$\begin{aligned}(1 + \|\boldsymbol{\xi}\|^2)^\sigma |\widehat{\partial^{\boldsymbol{\alpha}} u}|^2 &= (1 + \|\boldsymbol{\xi}\|^2)^\sigma |(i2\pi\boldsymbol{\xi})^{\boldsymbol{\alpha}}|^2 |\hat{u}(\boldsymbol{\xi})|^2 \\ &\leq (2\pi)^{2|\boldsymbol{\alpha}|}(1 + \|\boldsymbol{\xi}\|^2)^{[s]+\sigma} |\hat{u}(\boldsymbol{\xi})|^2,\end{aligned}$$

since $\boldsymbol{\xi}^{2\boldsymbol{\alpha}} = \prod_{i=1}^n \xi_i^{2\alpha_i} \leq (1 + \|\boldsymbol{\xi}\|^2)^{|\boldsymbol{\alpha}|}$ with $|\boldsymbol{\alpha}| = [s]$ by Lemma 8.9.1. Hence, $\forall|\boldsymbol{\alpha}| = [s]$, $s = [s] + \sigma$,

$$\begin{aligned}&\int_{\mathbb{R}^n} (1 + \|\boldsymbol{\xi}\|^2)^\sigma |\widehat{\partial^{\boldsymbol{\alpha}} u}|^2 d\boldsymbol{\xi} \leq (2\pi)^{2[s]} \int_{\mathbb{R}^n} (1 + \|\boldsymbol{\xi}\|^2)^s |\hat{u}(\boldsymbol{\xi})|^2 d\boldsymbol{\xi}, \\ \Longrightarrow \quad & \|\partial^{\boldsymbol{\alpha}} u\|^2_{H^\sigma(\mathbb{R}^n)} \leq (2\pi)^{2[s]} \|u\|^2_{H^s(\mathbb{R}^n)} \\ \Longrightarrow \quad & \|\partial^{\boldsymbol{\alpha}} u\|_{\sigma,\mathbb{R}^n} \leq (2\pi)^{[s]} \|u\|_{s,\mathbb{R}^n} \quad \forall|\boldsymbol{\alpha}| = [s]. \end{aligned} \tag{8.9.19}$$

Hence, $\forall|\boldsymbol{\alpha}| = [s]$, $\partial^{\boldsymbol{\alpha}} u \in H^\sigma(\mathbb{R}^n)$. Thus, we have proved that $u \in H^s(\mathbb{R}^n) \Longrightarrow u \in H^{[s]}(\mathbb{R}^n)$, $\partial^{\boldsymbol{\alpha}} u \in H^\sigma(\mathbb{R}^n)$ $\forall|\boldsymbol{\alpha}| = [s]$ with

$$\begin{aligned}\left[\|u\|^2_{[s],\mathbb{R}^n} + \sum_{|\boldsymbol{\alpha}|=[s]} \|\partial^{\boldsymbol{\alpha}} u\|^2_{\sigma,\mathbb{R}^n}\right] &\leq \left(\|u\|^2_{s,\mathbb{R}^n} + \sum_{|\boldsymbol{\alpha}|=[s]} (2\pi)^{2|\boldsymbol{\alpha}|} \|u\|^2_{s,\mathbb{R}^n}\right) \\ &\leq \|u\|^2_{s,\mathbb{R}^n}\left(1 + \sum_{|\boldsymbol{\alpha}|=[s]} (2\pi)^{2|\boldsymbol{\alpha}|}\right) = C_2^2 \|u\|^2_{s,\mathbb{R}^n}\end{aligned}$$

$$\Longrightarrow \quad \left[\|u\|^2_{[s],\mathbb{R}^n} + \sum_{|\boldsymbol{\alpha}|=[s]} \|\partial^{\boldsymbol{\alpha}} u\|^2_{\sigma,\mathbb{R}^n}\right]^{\frac{1}{2}} \leq C_2 \|u\|_{s,\mathbb{R}^n}$$

with $C_2 = (1 + \sum_{|\boldsymbol{\alpha}|=[s]} (2\pi)^{2|\boldsymbol{\alpha}|})^{\frac{1}{2}} > 0$, i.e. the second inequality in (8.9.18) holds.
($\Longleftarrow$): Let $u \in H^{[s]}(\mathbb{R}^n)$ with $\partial^{\boldsymbol{\alpha}} u \in H^\sigma(\mathbb{R}^n)$ $\forall|\boldsymbol{\alpha}| = [s]$. We are to show that $u \in H^s(\mathbb{R}^n)$, satisfying the first inequality in (8.9.18). $u \in H^{[s]}(\mathbb{R}^n) \Longrightarrow u \in S'(\mathbb{R}^n)$ and $(1 + \|\boldsymbol{\xi}\|^2)^{[s]/2}\hat{u} \in L^2(\mathbb{R}^n)$. By Proposition 8.9.1, $H^{[s]}(\mathbb{R}^n) \hookrightarrow L^2(\mathbb{R}^n)$ $\forall[s] \geq 0$. Then $u \in H^{[s]}(\mathbb{R}^n) \Longrightarrow u \in L^2(\mathbb{R}^n) \Longrightarrow \hat{u} \in L^2(\mathbb{R}^n)$ by the Plancherel–Riesz Theorem 8.3.1. Besides, $(1 + \|\boldsymbol{\xi}\|^2)^s |\hat{u}|^2 = (1 + \|\boldsymbol{\xi}\|^2)^\sigma ((1 + \|\boldsymbol{\xi}\|^2)^{[s]}) |\hat{u}|^2$. Using the second inequality in (8.9.6):

$$(1 + \|\boldsymbol{\xi}\|^2)^{[s]} \leq C\left[1 + \sum_{|\boldsymbol{\alpha}|=[s]} \left(\prod_{i=1}^n |\xi_i|^{2\alpha_i}\right)\right].$$

But from (8.6.1), $(i2\pi\boldsymbol{\xi})^{\boldsymbol{\alpha}}\hat{u} = \widehat{\partial^{\boldsymbol{\alpha}} u}$ and, by assumption, $\partial^{\boldsymbol{\alpha}} u \in H^{\sigma}(\mathbb{R}^n)$ $\forall |\boldsymbol{\alpha}| = [s]$.

Hence,

$$(1 + \|\boldsymbol{\xi}\|^2)^{[s]}|\hat{u}|^2 \leq C\Bigg[1 + \sum_{|\boldsymbol{\alpha}|=[s]} \Bigg(\prod_{i=1}^{n} |\xi_i|^{2\alpha_i}\Bigg)\Bigg]|\hat{u}|^2$$
$$= C\Bigg[|\hat{u}|^2 + \sum_{|\boldsymbol{\alpha}|=[s]} |\boldsymbol{\xi}^{\boldsymbol{\alpha}}\hat{u}|^2\Bigg].$$

$|(2\pi)^{|\boldsymbol{\alpha}|}\boldsymbol{\xi}^{\boldsymbol{\alpha}}\hat{u}| = |\widehat{\partial^{\boldsymbol{\alpha}} u}| \implies |\boldsymbol{\xi}^{\boldsymbol{\alpha}}\hat{u}| = \frac{1}{(2\pi)^{|\boldsymbol{\alpha}|}}|\widehat{\partial^{\boldsymbol{\alpha}} u}| \implies |\boldsymbol{\xi}^{\boldsymbol{\alpha}}\hat{u}|^2 = \frac{|\widehat{\partial^{\boldsymbol{\alpha}} u}|^2}{(2\pi)^{2|\boldsymbol{\alpha}|}} \implies \sum_{|\boldsymbol{\alpha}|=[s]} |\boldsymbol{\xi}^{\boldsymbol{\alpha}}\hat{u}|^2 = \sum_{\boldsymbol{\alpha}|=[s]} \frac{1}{(2\pi)^{2|\boldsymbol{\alpha}|}}|\widehat{\partial^{\boldsymbol{\alpha}} u}|^2$

$$\implies \quad (1 + \|\boldsymbol{\xi}\|^2)^{s}|\hat{u}|^2 \leq (1 + \|\boldsymbol{\xi}\|^2)^{\sigma} C\Bigg[|\hat{u}|^2 + \sum_{|\boldsymbol{\alpha}|=[s]} \frac{|\widehat{\partial^{\boldsymbol{\alpha}} u}|^2}{(2\pi)^{2|\boldsymbol{\alpha}|}}\Bigg] \quad \text{a.e. in } \mathbb{R}^n$$
$$\leq C_0^2\Bigg[(1 + \|\boldsymbol{\xi}\|^2)^{\sigma}|\hat{u}|^2 + \sum_{|\boldsymbol{\alpha}|=[s]} (1 + \|\boldsymbol{\xi}\|^2)^{\sigma}|\widehat{\partial^{\boldsymbol{\alpha}} u}|^2\Bigg]$$

$$\implies \quad \int_{\mathbb{R}^n} (1 + \|\boldsymbol{\xi}\|^2)^{s}|\hat{u}|^2 d\boldsymbol{\xi}$$
$$\leq C_0^2\Bigg[\int_{\mathbb{R}^n} (1 + \|\boldsymbol{\xi}\|^2)^{\sigma}|\hat{u}|^2 d\boldsymbol{\xi} + \sum_{|\boldsymbol{\alpha}|=[s]} \int_{\mathbb{R}^n} (1 + \|\boldsymbol{\xi}\|^2)^{\sigma}|\widehat{\partial^{\boldsymbol{\alpha}} u}|^2 d\boldsymbol{\xi}\Bigg]$$

$$\implies \quad u \in H^s(\mathbb{R}^n) \quad \text{with } \|u\|_{s,\mathbb{R}^n}^2 \leq C_0^2\Bigg[\|u\|_{\sigma,\mathbb{R}^n}^2 + \sum_{|\boldsymbol{\alpha}|=[s]} \|\partial^{\boldsymbol{\alpha}} u\|_{\sigma,\mathbb{R}^n}^2\Bigg]$$

$$\implies \quad \|u\|_{s,\mathbb{R}^n} \leq C_0\Bigg[\|u\|_{\sigma,\mathbb{R}^n}^2 + \sum_{|\boldsymbol{\alpha}|=[s]} \|\partial^{\boldsymbol{\alpha}} u\|_{\sigma,\mathbb{R}^n}^2\Bigg]^{1/2}. \tag{8.9.20}$$

For $[s] = 0$, we get the *trivial* case, since $[s] = |\boldsymbol{\alpha}| = 0$ and $s = \sigma \in \,]0,1[$, and consequently $H^s(\mathbb{R}^n) \equiv H^{\sigma}(\mathbb{R}^n)$, and hence

$$u \in H^{\sigma}(\mathbb{R}^n) \hookrightarrow H^{[s]}(\mathbb{R}^n) \equiv L^2(\mathbb{R}^n). \tag{8.9.21}$$

For $[s] \geq 1$, $\sigma \in \,]0,1[$.

$$H^{[s]}(\mathbb{R}^n) \hookrightarrow H^{\sigma}(\mathbb{R}^n), \quad \text{with } \|u\|_{\sigma,\mathbb{R}^n} \leq \|u\|_{[s],\mathbb{R}^n} \quad \text{by (8.9.4).} \tag{8.9.22}$$

Hence, from (8.9.20) and (8.9.22), we get the first inequality in (8.9.18): for $[s] \geq 1$,

$$\|u\|_{s,\mathbb{R}^n} \leq C_0\Bigg[\|u\|_{[s],\mathbb{R}^n}^2 + |\sum_{|\boldsymbol{\alpha}|=[s]} \|\partial^{\boldsymbol{\alpha}} u\|_{\sigma,\mathbb{R}^n}^2\Bigg]^{\frac{1}{2}}. \tag{8.9.23}$$

Finally, from (8.9.19)–(8.9.23), the result (8.9.18) follows. □

8.9.4 Sobolev's Imbedding Theorem (see also imbedding results in Section 8.12)

Now we will identify conditions under which functions of $H^s(\mathbb{R}^n)$ will be continuous in $\mathbb{R}^n$, or will have continuous derivatives in the *usual pointwise sense in* $\mathbb{R}^n$. For this we will need:

Lemma 8.9.3. *Let* $r = r(\boldsymbol{\xi}) = \left(\sum_{i=1}^n \xi_i^2\right)^{1/2}$, $\Phi = \Phi(r) \geq 0\ \forall r \in]0, \infty[$ *and* $f = f(\boldsymbol{\xi}) = \Phi(r(\boldsymbol{\xi})) \geq 0 \forall \boldsymbol{\xi} \in \mathbb{R}^n$ *be integrable on* $\mathbb{R}^n$. *Then*

$$\int_{\mathbb{R}^n} f(\boldsymbol{\xi}) d\boldsymbol{\xi} = n V_n \int_0^\infty \Phi(r) r^{n-1} dr, \tag{8.9.24}$$

where V_n *is the volume of an n-dimensional unit sphere in* $\mathbb{R}^n$ [7, page 121].

Theorem 8.9.4 (Sobolev's Imbedding Theorem). *If* $s - \frac{n}{2} > k$ *with* $k \in \mathbb{N}_0$, *then*

$$H^s(\mathbb{R}^n) \hookrightarrow C^k(\mathbb{R}^n), \tag{8.9.25}$$

i.e. $u \in H^s(\mathbb{R}^n) \Longrightarrow u \in C^k(\mathbb{R}^n)$ *with* $\|u\|_{C^k(\mathbb{R}^n)} \leq C\|u\|_{H^s(\mathbb{R}^n)}$ *for some* $C > 0$ *and* $\lim_{\|\mathbf{x}\| \to \infty} |\partial^{\boldsymbol{\alpha}} u(\mathbf{x})| = 0\ \forall |\boldsymbol{\alpha}| \leq k$, *where* $C^k(\mathbb{R}^n)$ *is equipped with the notion of* uniform convergence *on every compact subset of* $\mathbb{R}^n$ *for all derivatives of order* $|\boldsymbol{\alpha}| \leq k$ *and* $\|u\|_{C^k(\mathbb{R}^n)} = \max_{0 \leq |\boldsymbol{\alpha}| \leq k} \sup_{\mathbf{x} \in \mathbb{R}^n} |\partial^{\boldsymbol{\alpha}} u(\mathbf{x})|$.

Proof. Let $u \in H^s(\mathbb{R}^n)$. Then, by the definition of $H^s(\mathbb{R}^n)$, $u \in S'(\mathbb{R}^n) \Longrightarrow \partial^{\boldsymbol{\alpha}} u \in S'(\mathbb{R}^n)$. We are to show that for $s - \frac{n}{2} > k$, $\partial^{\boldsymbol{\alpha}} u \in C^0(\mathbb{R}^n)\ \forall |\boldsymbol{\alpha}| \leq k$. For this, if $\mathcal{F}[\partial^{\boldsymbol{\alpha}} u] \in L^1(\mathbb{R}^n)\ \forall |\boldsymbol{\alpha}| \leq k$, we can apply (7.1.7) and Property 6 in (7.1.24) of Fourier transforms, which also hold for the Fourier co-transform $\bar{\mathcal{F}}$, of functions of $L^1(\mathbb{R}^n)$, i.e. $w_{\boldsymbol{\alpha}} = \widehat{\partial^{\boldsymbol{\alpha}} u} = \mathcal{F}[\partial^{\boldsymbol{\alpha}} u] \in L^1(\mathbb{R}^n) \subset S'(\mathbb{R}^n)$,

$$\Longrightarrow \quad \bar{\mathcal{F}} w_{\boldsymbol{\alpha}} = \bar{\mathcal{F}}\mathcal{F}[\partial^{\boldsymbol{\alpha}} u] = \partial^{\boldsymbol{\alpha}} u \in C^0(\mathbb{R}^n) \tag{8.9.26}$$

with $\|\partial^{\boldsymbol{\alpha}} u\|_{C^0(\mathbb{R}^n)} = \sup_{\mathbf{x} \in \mathbb{R}^n} |\partial^{\boldsymbol{\alpha}} u(\mathbf{x})| = \|\partial^{\boldsymbol{\alpha}} u\|_\infty \leq \|\widehat{\partial^{\boldsymbol{\alpha}} u}\|_{L^1(\mathbb{R}^n)}\ \forall |\boldsymbol{\alpha}| \leq k$ (since $\bar{\mathcal{F}}\mathcal{F}[\partial^{\boldsymbol{\alpha}} u] = \partial^{\boldsymbol{\alpha}} u \in S'(\mathbb{R}^n)$) by the Fourier Inversion Theorem 8.3.2 on $S'(\mathbb{R}^n)$) along with the Riemann–Lebesgue Property 11 in (7.1.36):

$$\lim_{\|\mathbf{x}\| \to \infty} \bar{\mathcal{F}} w_{\boldsymbol{\alpha}} = \lim_{\|\mathbf{x}\| \to \infty} \partial^{\boldsymbol{\alpha}} u(\mathbf{x}) = 0 \quad \forall |\boldsymbol{\alpha}| \leq k. \tag{8.9.27}$$

Hence, it is sufficient to show that $\widehat{\partial^{\boldsymbol{\alpha}} u} \in L^1(\mathbb{R}^n)\ \forall |\boldsymbol{\alpha}| \leq k$. Then, (8.9.26) and (8.9.27) will follow. In fact, by Theorem 8.6.1, $\widehat{\partial^{\boldsymbol{\alpha}} u} = \mathcal{F}[\partial^{\boldsymbol{\alpha}} u] = (i2\pi\boldsymbol{\xi})^{\boldsymbol{\alpha}} \hat{u} \in S'(\mathbb{R}^n)$, where

$$\begin{aligned}(i2\pi\boldsymbol{\xi})^{\boldsymbol{\alpha}} &= (i2\pi\xi_1)^{\alpha_1}(i2\pi\xi_2)^{\alpha_2} \cdots (i2\pi\xi_n)_n^{\alpha} \\ &= (i)^{|\boldsymbol{\alpha}|}(2\pi)^{|\boldsymbol{\alpha}|}\xi_1^{\alpha_1}\xi_2^{\alpha_2} \cdots \xi_n^{\alpha_n} = (i)^{|\boldsymbol{\alpha}|}(2\pi)^{|\boldsymbol{\alpha}|}\boldsymbol{\xi}^{\boldsymbol{\alpha}}.\end{aligned}$$

Hence, to prove that $\widehat{\partial^{\boldsymbol{\alpha}} u} \in L^1(\mathbb{R}^n)\ \forall |\boldsymbol{\alpha}| \leq k$, we are to show that

$$\int_{\mathbb{R}^n} |\widehat{\partial^{\boldsymbol{\alpha}} u(\boldsymbol{\xi})}| d\boldsymbol{\xi} = \int_{\mathbb{R}^n} |(i2\pi\boldsymbol{\xi})^{\boldsymbol{\alpha}} \hat{u}(\boldsymbol{\xi})| d\boldsymbol{\xi} = (2\pi)^{|\boldsymbol{\alpha}|} \int_{\mathbb{R}^n} |\boldsymbol{\xi}^{\boldsymbol{\alpha}} \hat{u}(\boldsymbol{\xi})| d\boldsymbol{\xi}$$
$$= (2\pi)^{|\boldsymbol{\alpha}|} \int_{\mathbb{R}^n} |\boldsymbol{\xi}^{\boldsymbol{\alpha}}| \cdot |\hat{u}(\boldsymbol{\xi})| d\boldsymbol{\xi} < +\infty. \tag{8.9.28}$$

But

$$|\boldsymbol{\xi}^{\boldsymbol{\alpha}}| = |\xi_1|^{\alpha_1} \cdot |\xi_2|^{\alpha_2} \cdots |\xi_n|^{\alpha_n} \leq \|\boldsymbol{\xi}\|^{\alpha_1} \|\boldsymbol{\xi}\|^{\alpha_2} \cdots \|\boldsymbol{\xi}\|^{\alpha_n} = \|\boldsymbol{\xi}\|^{|\boldsymbol{\alpha}|}$$

and

$$|\boldsymbol{\xi}^{\boldsymbol{\alpha}}\|\hat{u}(\boldsymbol{\xi}| \leq \underbrace{\|\boldsymbol{\xi}\|^{|\boldsymbol{\alpha}|}(1+\|\boldsymbol{\xi}\|^2)^{-\frac{s}{2}}}_{f(\boldsymbol{\xi})} \underbrace{(1+\|\boldsymbol{\xi}\|^2)^{\frac{s}{2}} |\hat{u}(\boldsymbol{\xi})|}_{g(\boldsymbol{\xi})}.$$

For $u \in H^s(\mathbb{R}^n)$, $g(\boldsymbol{\xi}) = (1+\|\boldsymbol{\xi}\|^2)^{s/2}\hat{u} \in L^2(\mathbb{R}^n)$. Now, if $f(\boldsymbol{\xi}) \in L^2(\mathbb{R}^n)$, then, by Hölder's inequality, their product $f(\boldsymbol{\xi})g(\boldsymbol{\xi}) \in L^1(\mathbb{R}^n)$, and consequently $\boldsymbol{\xi}^{\boldsymbol{\alpha}}\hat{u} \in L^1(\mathbb{R}^n)$ and we have, $\forall |\boldsymbol{\alpha}| \leq k$,

$$\int_{\mathbb{R}^n} |\boldsymbol{\xi}|^{\boldsymbol{\alpha}} |\hat{u}(\boldsymbol{\xi})| d\boldsymbol{\xi} \leq \int_{\mathbb{R}^n} \|\boldsymbol{\xi}\|^{|\boldsymbol{\alpha}|}(1+\|\boldsymbol{\xi}\|^2)^{-s/2}(1+\|\boldsymbol{\xi}\|^2)^{s/2}|\hat{u}(\boldsymbol{\xi})| d\boldsymbol{\xi}$$
$$\leq \left(\int_{\mathbb{R}^n} \|\boldsymbol{\xi}\|^{2|\boldsymbol{\alpha}|}(1+\|\boldsymbol{\xi}\|^2)^{-s} d\boldsymbol{\xi}\right)^{\frac{1}{2}} \left(\int_{\mathbb{R}^n} (1+\|\boldsymbol{\xi}\|^2)^{s} |\hat{u}(\boldsymbol{\xi})|^2 d\boldsymbol{\xi}\right)^{\frac{1}{2}} < +\infty, \tag{8.9.29}$$

and finally, from (8.9.28) and (8.9.29), $\widehat{\partial^{\boldsymbol{\alpha}} u} \in L^1(\mathbb{R}^n)$ with

$$\|\widehat{\partial^{\boldsymbol{\alpha}} u}\|_{L^1(\mathbb{R}^n)} \leq \left(\int_{\mathbb{R}^n} \|\boldsymbol{\xi}\|^{2|\boldsymbol{\alpha}|}(1+\|\boldsymbol{\xi}\|^2)^{-s} d\boldsymbol{\xi}\right)^{\frac{1}{2}} \|u\|_{s,\mathbb{R}^n}.$$

Now the whole proof reduces to the proof that $f(\boldsymbol{\xi}) = \|\boldsymbol{\xi}\|^{|\boldsymbol{\alpha}|}(1+\|\boldsymbol{\xi}\|^2)^{-s/2} \in L^2(\mathbb{R}^n)$. For this we apply Lemma 8.9.3 with $r = \|\boldsymbol{\xi}\|$ to show that

$$\int_{\mathbb{R}^n} \|\boldsymbol{\xi}\|^{2|\boldsymbol{\alpha}|}(1+\|\boldsymbol{\xi}\|^2)^{-s} d\boldsymbol{\xi} = nV_n \int_0^\infty (1+r^2)^{-s} r^{2|\boldsymbol{\alpha}|} r^{n-1} dr$$
$$= nV_n \Bigg[\underbrace{\int_0^1 (1+r^2)^{-s} r^{2|\boldsymbol{\alpha}|+n-1} dr}_{I_1} + \underbrace{\int_1^\infty (1+r^2)^{-s} r^{2|\boldsymbol{\alpha}|+n-1} dr}_{I_2}\Bigg] < +\infty,$$

i.e. $\|\boldsymbol{\xi}\|^{|\boldsymbol{\alpha}|}(1+\|\boldsymbol{\xi}\|^2)^{-s/2} \in L^2(\mathbb{R}^n)$ if both $I_1 < +\infty$, $I_2 < +\infty$.

For $0 < r < 1$, $(1+r^2)^{-s} \leq 1$, $I_1 \leq \int_0^1 r^{2|\boldsymbol{\alpha}|+n-1}\, dr < +\infty$ if $(2|\boldsymbol{\alpha}|+n-1)+1 > 0 \Longrightarrow 2|\boldsymbol{\alpha}| + n > 0$, which holds $\forall |\boldsymbol{\alpha}|$.

For $1 < r < \infty$, $(1 + r^2) \geq r^2 \Longrightarrow (1 + r^2)^{-s} \leq r^{-2s} \Longrightarrow I_2 \leq \int_1^\infty r^{-2s+2|\boldsymbol{\alpha}|+n-1}\, dr < +\infty$, if $(-2s + 2|\boldsymbol{\alpha}| + n - 1) + 1 < 0 \Longrightarrow -2s + 2|\boldsymbol{\alpha}| + n < 0 \Longrightarrow 2s - n > 2|\boldsymbol{\alpha}| \Longrightarrow s - \frac{n}{2} > |\boldsymbol{\alpha}|\ \forall |\boldsymbol{\alpha}| \leq k$.

Hence, $\|\boldsymbol{\xi}\|^{|\boldsymbol{\alpha}|}(1 + \|\boldsymbol{\xi}\|^2)^{-s/2} \in L^2(\mathbb{R}^n)$ if $s - \frac{n}{2} > |\boldsymbol{\alpha}|$ with $|\boldsymbol{\alpha}| \leq k \Longrightarrow \widehat{\partial^{\boldsymbol{\alpha}} u} \in L^1(\mathbb{R}^n)$ if $s - \frac{n}{2} > |\boldsymbol{\alpha}|$ with $|\boldsymbol{\alpha}| \leq k \Longrightarrow$ from (8.9.26), $\partial^{\boldsymbol{\alpha}} u \in C^0(\mathbb{R}^n)$ with $\|\partial^{\boldsymbol{\alpha}} u\|_{C^0(\mathbb{R}^n)} \leq \|\widehat{\partial^{\boldsymbol{\alpha}} u}\|_{L^1(\mathbb{R}^n)}\ \forall |\boldsymbol{\alpha}| \leq k$. Hence $u \in C^k(\mathbb{R}^n)$ and, using (8.9.28) and (8.9.29),

$$\|u\|_{C^k(\mathbb{R}^n)} = \max_{0\leq|\boldsymbol{\alpha}|\leq k} \|\partial^{\boldsymbol{\alpha}} u\|_\infty \leq \max_{0\leq|\boldsymbol{\alpha}|\leq k} \|\widehat{\partial^{\boldsymbol{\alpha}} u}\|_{L^1(\mathbb{R}^n)} \leq C\|u\|_{H^s(\mathbb{R}^n)},$$

and from (8.9.27)

$$\lim_{\|x\|\to\infty} \partial^{\boldsymbol{\alpha}} u(\mathbf{x}) = 0 \quad \forall |\boldsymbol{\alpha}| \leq k. \qquad \square$$

Remark 8.9.2. At this point some explanations on the results of the imbedding theorem are in order. Elements of $H^s(\mathbb{R}^n)$ are *equivalence classes* of functions defined almost everywhere in $\mathbb{R}^n$, i.e. functions of an equivalence class may differ on a set of points with n-dimensional Lebesgue measure equal to zero. Then, by virtue of the imbedding of type (8.9.25), we mean that every equivalence class $[u] \in H^s(\mathbb{R}^n)$ must contain a function u

(i) *belonging to* $C^k(\mathbb{R}^n)$, $C^k(\mathbb{R}^n)$ being the target space of imbedding, and

(ii) *bounded in* $C^k(\mathbb{R}^n)$ by $C\|u\|_{s,\mathbb{R}^n}$.

Hence, the imbedding $H^s(\mathbb{R}^n) \hookrightarrow C^k(\mathbb{R}^n)$ for $s - \frac{n}{2} > k \Longrightarrow$ each $u \in H^s(\mathbb{R}^n)$ can be considered as a function, which can be *redefined* on a set of points with n-dimensional Lebesgue measure equal to zero in $\mathbb{R}^n$ such that the *modified* function $\tilde{u}$ thus obtained has the properties: $\tilde{u} = u$ in $H^s(\mathbb{R}^n)$, $\tilde{u} \in C^k(\mathbb{R}^n)$, $\|\tilde{u}\|_{C^k(\mathbb{R}^n)} \leq C\|\tilde{u}\|_{H^s(\mathbb{R}^n)}$.

Example 8.9.2.

1. For $n = 1$, $s = 1$, $H^1(\mathbb{R})$, $s - \frac{n}{2} = 1 - \frac{1}{2} > 0$

$$\Longrightarrow H^1(\mathbb{R}) \hookrightarrow C^0(\mathbb{R}), \quad \text{i.e. functions of } H^1(\mathbb{R}) \text{ are } \textit{continuous} \text{ on } \mathbb{R}. \tag{8.9.30}$$

2. For $n = 2$, $s = 1$, $H^1(\mathbb{R}^2)$,

$$s - \frac{n}{2} = 1 - \frac{2}{2} = 0. \tag{8.9.31}$$

Hence, from Theorem 8.9.4, we cannot say anything. In fact, function u in Example 2.15.2 belongs to $H^1(\mathbb{R}^2)$, but there does *not* exist $\tilde{u} \in C^0(\mathbb{R}^2)$ such that $\tilde{u} = u$ in $H^1(\mathbb{R}^2)$ with $\|\tilde{u}\|_{C^0(\mathbb{R}^2)} \leq C\|\tilde{u}\|_{H^1(\mathbb{R}^2)}$ (see Remark 8.9.2). Hence, functions of $H^1(\mathbb{R}^2)$ are *not* continuous *in general*.

3. For $n = 2$, $s > 1$, $s - \frac{n}{2} = s - 1 > 0 \Longrightarrow H^s(\mathbb{R}^2) \hookrightarrow C^0(\mathbb{R}^2)$, i.e. for $s > 1$, functions of $H^s(\mathbb{R}^2)$ are *continuous* in $\mathbb{R}$. In particular, for $s = 2$,

$$H^2(\mathbb{R}^2) \hookrightarrow C^0(\mathbb{R}^2), \quad \text{i.e. functions of } H^2(\mathbb{R}^2) \text{ are continuous in } \mathbb{R}^2. \tag{8.9.32}$$

Translation property of imbedding results in Sobolev spaces

$s - \frac{n}{2} > k \in \mathbb{N}$

$$\Longrightarrow \quad H^s(\mathbb{R}^n) \hookrightarrow C^k(\mathbb{R}^n) \quad \Longrightarrow \quad H^{s+m}(\mathbb{R}^n) \hookrightarrow C^{k+m}(\mathbb{R}^n) \ \forall m \in \mathbb{N}. \tag{8.9.33}$$

In fact, $s - \frac{n}{2} > k \Longrightarrow (s+m) - \frac{n}{2} > k + m \ \forall m \in \mathbb{N}$. Hence, $H^s(\mathbb{R}^n) \hookrightarrow C^k(\mathbb{R}^n)$ $\Longrightarrow H^{s+m}(\mathbb{R}^n) \hookrightarrow C^{k+m}(\mathbb{R}^n)$.

Theorem 8.9.5. *If* $s - \frac{n}{2} = \lambda \in]0, 1[$, $H^s(\mathbb{R}^n) \hookrightarrow C^{0,\lambda}(\mathbb{R}^n)$, *i.e.*

$$u \in H^s(\mathbb{R}^n) \quad \Longrightarrow \quad u \in C^{0,\lambda}(\mathbb{R}^n), \tag{8.9.34}$$

u being a λ-Hölder continuous function in $\mathbb{R}^n$, $\forall \mathbf{x}, \mathbf{y} \in \mathbb{R}^n$, $\exists C > 0$ such that

$$|u(\mathbf{x} + \mathbf{y}) - u(\mathbf{x})| \leq C \|\mathbf{y}\|^\lambda, \tag{8.9.35}$$

and $\exists \tilde{C} > 0$ such that

$$\|u\|_{C^{0,\lambda}(\mathbb{R}^n)} = \|u\|_{C^0(\mathbb{R}^n)} + \sup_{\substack{\mathbf{x} \in \mathbb{R}^n \\ \mathbf{y} \neq \mathbf{0}}} \frac{|u(\mathbf{x} + \mathbf{y}) - u(\mathbf{x})|}{\|\mathbf{y}\|^\lambda} \leq \tilde{C} \|u\|_{H^s(\mathbb{R}^n)}. \tag{8.9.36}$$

Proof. We give the scheme of the proof.

$$\begin{aligned}
& u(\mathbf{x} + \mathbf{y}) = (\tau_{-\mathbf{y}} u)(\mathbf{x}) \\
\Longrightarrow \quad & \mathcal{F}[\tau_{-\mathbf{y}} u] = e^{i2\pi\langle \mathbf{y}, \boldsymbol{\xi} \rangle} \mathcal{F}[u] = e^{i2\pi\langle \mathbf{y}, \boldsymbol{\xi} \rangle} \hat{u} \quad (by (7.1.23)) \\
\Longrightarrow \quad & u(\mathbf{x} + \mathbf{y}) = \bar{\mathcal{F}}[e^{i2\pi\langle \mathbf{y}, \cdot \rangle} \hat{u}(\cdot)](\mathbf{x}) \\
\Longrightarrow \quad & u(\mathbf{x} + \mathbf{y}) - u(\mathbf{x}) = \bar{\mathcal{F}}[(e^{i2\pi\langle \mathbf{y}, \cdot \rangle} - 1)\hat{u}(\cdot)](\mathbf{x}) \\
\Longrightarrow \quad & |u(\mathbf{x} + \mathbf{y}) - u(\mathbf{x})| = \left| \int_{\mathbb{R}^n} \hat{u}(\boldsymbol{\xi})[e^{i2\pi\langle \mathbf{y}, \boldsymbol{\xi} \rangle} - 1] e^{i2\pi\langle \boldsymbol{\xi}, \mathbf{x} \rangle} d\boldsymbol{\xi} \right| \\
& \leq \int_{\mathbb{R}^n} \frac{|e^{i2\pi\langle \mathbf{y}, \boldsymbol{\xi} \rangle} - 1|}{\|\boldsymbol{\xi}\|^s} \|\boldsymbol{\xi}\|^s |\hat{u}(\boldsymbol{\xi})| d\boldsymbol{\xi}.
\end{aligned} \tag{8.9.37}$$

$\|\boldsymbol{\xi}\|^s \le (1+\|\boldsymbol{\xi}\|^2)^{s/2} \Longrightarrow \|\boldsymbol{\xi}\|^s|\hat{u}(\boldsymbol{\xi})| \in L^2(\mathbb{R}^n)$, since $u \in H^s(\mathbb{R}^n)$.

$$I = \int_{\mathbb{R}^n} \frac{|e^{i2\pi\langle \mathbf{y},\boldsymbol{\xi}\rangle} - 1|^2}{\|\boldsymbol{\xi}\|^{2s}} d\boldsymbol{\xi} \le C_1^2 \|\mathbf{y}\|^{2s-n} \quad \text{with } C_1 = C_1(s,n) > 0.$$

Avoiding the long proof of the estimate, we show again the scheme of the proof (see also [42]). The integral I is invariant under rotation, since a rotation preserves the inner product $\langle\cdot,\cdot\rangle$ and norm $\|\cdot\|$ in $\mathbb{R}^n$, the Jacobian J of such a transformation being 1. Moreover, for $\mathbf{y} \neq \mathbf{0}$, $\exists$ a rotation which transforms $\mathbf{y}$ into a vector collinear with $\hat{\mathbf{e}}_1 = (1,0,\dots,0)$. Hence, we choose $\mathbf{y} = (|h|,0,\dots,0)$ with $\|\mathbf{y}\| = |h|$ collinear with the first basis vector $\hat{\mathbf{e}}_1$, and set $\boldsymbol{\xi} = \frac{1}{|h|}\boldsymbol{\eta} \in \mathbb{R}^n$ with $\boldsymbol{\eta} = (\eta_1,\eta_2,\dots,\eta_n)$, $|J| = \frac{1}{|h|^n}$ and $h \neq 0$. Then $\langle \mathbf{y},\boldsymbol{\xi}\rangle = |h|\cdot\frac{\eta_1}{|h|} + 0 + \cdots + 0 = \eta_1$, and

$$I = \int_{\mathbb{R}^n} \frac{|e^{i2\pi\eta_1} - 1|^2}{\frac{1}{|h|^{2s}}\|\boldsymbol{\eta}\|^{2s}} \cdot \frac{1}{|h|^n} d\boldsymbol{\eta} \le C_1^2|h|^{2s-n} = C_1^2\|\mathbf{y}\|^{2s-n},$$

with $C_1^2 \ge \int_{\mathbb{R}^n} \frac{|e^{i2\pi\eta_1}-1|^2}{\|\boldsymbol{\eta}\|^{2s}} d\boldsymbol{\eta}$ (see Remark 8.9.3). Hence, $\frac{|e^{i2\pi\langle\mathbf{y},\boldsymbol{\xi}\rangle}-1|}{\|\boldsymbol{\xi}\|^s} \in L^2(\mathbb{R}^n)$ and $\|\boldsymbol{\xi}\|^s|\hat{u}(\boldsymbol{\xi})| \in L^2(\mathbb{R}^n)$, which was shown earlier.

$\Longrightarrow$ by Hölder's inequality, from (8.9.37), we have

$$\begin{aligned}
|u(\mathbf{x}+\mathbf{y}) - u(\mathbf{x})| &\le \int_{\mathbb{R}^n} \frac{|e^{i2\pi\langle\mathbf{y},\boldsymbol{\xi}\rangle} - 1|}{\|\boldsymbol{\xi}\|^s} \|\boldsymbol{\xi}\|^s |\hat{u}(\boldsymbol{\xi})| d\boldsymbol{\xi} \\
&\le \left(\int_{\mathbb{R}^n} \frac{|e^{i2\pi\langle\mathbf{y},\boldsymbol{\xi}\rangle} - 1|^2}{\|\boldsymbol{\xi}\|^{2s}} d\boldsymbol{\xi}\right)^{\frac{1}{2}} \left(\int_{\mathbb{R}^n} \|\boldsymbol{\xi}\|^{2s}|\hat{u}(\boldsymbol{\xi})|^2 d\boldsymbol{\xi}\right)^{\frac{1}{2}} \\
&\le C_1(\|\mathbf{y}\|^{2s-n})^{\frac{1}{2}}\|u\|_{s,\mathbb{R}^n} = C_1\|\mathbf{y}\|^{\lambda}\cdot\|u\|_{s,\mathbb{R}^n},
\end{aligned}$$

since $\int_{\mathbb{R}^n} \|\boldsymbol{\xi}\|^{2s}|\hat{u}(\boldsymbol{\xi})|^2 d\boldsymbol{\xi} \le \int_{\mathbb{R}^n}(1+\|\boldsymbol{\xi}\|^2)^s|\hat{u}(\boldsymbol{\xi})|^2 d\boldsymbol{\xi} = \|u\|^2_{s,\mathbb{R}^n}$. Then, from (8.9.37), for $s - \frac{n}{2} = \lambda \in]0,1[$ and $\mathbf{y} \neq \mathbf{0}$, $\exists C_1 > 0$, independent of u, such that $\frac{|u(\mathbf{x}+\mathbf{y})-u(\mathbf{x})|}{\|\mathbf{y}\|^\lambda} \le C_1\|u\|_{s,\mathbb{R}^n}$ $\forall \mathbf{x} \in \mathbb{R}^n$. Hence, $u \in C^{0,\lambda}(\mathbb{R}^n)$ and for the semi-norm $|\cdot|$ in $C^{0,\lambda}(\mathbb{R}^n)$,

$$|u|_{C^{0,\lambda}(\mathbb{R}^n)} = \sup_{\substack{\mathbf{x}\in\mathbb{R}^n \\ \mathbf{y}\neq\mathbf{0}}} \frac{|u(\mathbf{x}+\mathbf{y}) - u(\mathbf{x})|}{\|\mathbf{y}\|^\lambda} \le C_1\|u\|_{H^s(\mathbb{R}^n)}.$$

But $u \in H^s(\mathbb{R}^n)$ with $s - \frac{n}{2} = \lambda > 0 \Longrightarrow u \in C^0(\mathbb{R}^n)$ with $\|u\|_{C^0(\mathbb{R}^n)} \le C_2\|u\|_{H^s(\mathbb{R}^n)}$. Then

$$\|u\|_{C^{0,\lambda}(\mathbb{R}^n)} = \|u\|_{C^0(\mathbb{R}^n)} + |u|_{C^{0,\lambda}(\mathbb{R}^n)} \le C\|u\|_{H^s(\mathbb{R}^n)},$$

with $C = C_1 + C_2 > 0$. □

Remark 8.9.3. Although we have shown in the proof of Theorem 8.9.5 that

$$I = \int_{\mathbb{R}^n} \frac{|e^{-i2\pi\langle \mathbf{y},\boldsymbol{\xi}\rangle} - 1|^2}{\|\boldsymbol{\xi}\|^{2s}} d\boldsymbol{\xi} \le C_1^2 \|\mathbf{y}\|^{2s-n},$$

in fact, we have:

Lemma 8.9.4. *For* $(s - \frac{n}{2}) \in \,]0,1[$, *the following result holds:* $\exists C = C(s,n) > 0$ *such that*

$$I(\mathbf{y}, 2s) = \int_{\mathbb{R}^n} \frac{|e^{-i2\pi\langle \mathbf{y},\boldsymbol{\xi}\rangle} - 1|^2}{\|\boldsymbol{\xi}\|^{2s}} d\boldsymbol{\xi} = C\|\mathbf{y}\|^{2s-n}.$$

Then, for $s \in \,]0,1[$,

$$\iint_{\mathbb{R}^n \times \mathbb{R}^n} \frac{|u(\mathbf{x}+\mathbf{y}) - u(\mathbf{x})|^2}{\|\mathbf{y}\|^{n+2s}} d\mathbf{x} d\mathbf{y} = C \int_{\mathbb{R}^n} \|\boldsymbol{\xi}\|^{2s} |\hat{u}(\boldsymbol{\xi})|^2 d\boldsymbol{\xi}.$$

Proof. Assuming the first result, $I(\mathbf{y}, 2s) = C\|\mathbf{y}\|^{2s-n}$, we prove the second equality. From the proof of Theorem 8.9.5, $u(\mathbf{x}+\mathbf{y}) - u(\mathbf{x}) = \overline{\mathcal{F}}[(e^{i2\pi\langle \mathbf{y},\cdot\rangle} - 1)\hat{u}(\cdot)](\mathbf{x})$
$\Longrightarrow \mathcal{F}[u(\mathbf{x}+\mathbf{y}) - u(\mathbf{x})](\boldsymbol{\xi}) = (e^{i2\pi\langle \mathbf{y},\boldsymbol{\xi}\rangle} - 1)\hat{u}(\boldsymbol{\xi})$ a.e. in $\mathbb{R}^n$.

By the Plancherel–Riesz Theorem 8.3.1, $\forall \mathbf{y} \in \mathbb{R}^n$,

$$\begin{aligned} \int_{\mathbb{R}^n} |u(\mathbf{x}+\mathbf{y}) - u(\mathbf{x})|^2 d\mathbf{x} &= \int_{\mathbb{R}^n} |\mathcal{F}[u(\mathbf{x}+\mathbf{y}) - u(\mathbf{x})](\boldsymbol{\xi})|^2 d\boldsymbol{\xi} \\ &= \int_{\mathbb{R}^n} |e^{i2\pi\langle \mathbf{y},\boldsymbol{\xi}\rangle} - 1|^2 |\hat{u}(\boldsymbol{\xi})|^2 d\boldsymbol{\xi}. \end{aligned}$$

The functions $(\mathbf{x},\mathbf{y}) \in \mathbb{R}^n \times \mathbb{R}^n \longmapsto \frac{|u(\mathbf{x}+\mathbf{y}) - u(\mathbf{x})|^2}{\|\mathbf{y}\|^{n+2s}} \ge 0$ (a.e.), $(\boldsymbol{\xi},\boldsymbol{\eta}) \in \mathbb{R}^n \times \mathbb{R}^n \longmapsto \frac{|e^{i2\pi\langle \mathbf{y},\boldsymbol{\xi}\rangle} - 1|^2 |\hat{u}(\boldsymbol{\xi})|^2}{\|\mathbf{y}\|^{n+2s}} \ge 0$ (a.e.) are integrable on $\mathbb{R}^n \times \mathbb{R}^n$, and applying Fubini's Theorem 7.1.2C,

$$\begin{aligned} \iint_{\mathbb{R}^n \times \mathbb{R}^n} \frac{|u(\mathbf{x}+\mathbf{y}) - u(\mathbf{x})|^2}{\|\mathbf{y}\|^{n+2s}} d\mathbf{x} d\mathbf{y} &= \int_{\mathbb{R}^n} \frac{d\mathbf{y}}{\|\mathbf{y}\|^{n+2s}} \int_{\mathbb{R}^n} |u(\mathbf{x}+\mathbf{y}) - u(\mathbf{x})|^2 d\mathbf{x} \\ &= \int_{\mathbb{R}^n} \frac{d\mathbf{y}}{\|\mathbf{y}\|^{n+2s}} \int_{\mathbb{R}^n} |e^{i2\pi\langle \mathbf{y},\boldsymbol{\xi}\rangle} - 1|^2 |\hat{u}(\boldsymbol{\xi})|^2 d\boldsymbol{\xi} \\ &= \int_{\mathbb{R}^n} |\hat{u}(\boldsymbol{\xi})|^2 d\boldsymbol{\xi} \int_{\mathbb{R}^n} \frac{|e^{i2\pi\langle \mathbf{y},\boldsymbol{\xi}\rangle} - 1|^2}{\|\mathbf{y}\|^{n+2s}} d\mathbf{y}. \end{aligned}$$

Set $2s_1 = n + 2s > n$. Then $0 < 2s_1 - n = 2s < 2$, since $s \in \,]0,1[$. Hence, for $(s_1 - \frac{n}{2}) \in \,]0,1[$, from the first result of this lemma,

$$\int_{\mathbb{R}^n} \frac{|e^{i2\pi\langle \mathbf{y},\boldsymbol{\xi}\rangle} - 1|^2}{\|\mathbf{y}\|^{2s_1}} d\mathbf{y} = I(\boldsymbol{\xi}, 2s_1) = C\|\boldsymbol{\xi}\|^{2s_1-n} = C\|\boldsymbol{\xi}\|^{2s}$$

and

$$\iint_{\mathbb{R}^n\times\mathbb{R}^n} \frac{|u(\mathbf{x}+\mathbf{y})-u(\mathbf{x})|^2}{\|\mathbf{y}\|^{n+2s}}d\mathbf{x}d\mathbf{y} = C\int_{\mathbb{R}^n}\|\boldsymbol{\xi}\|^{2s}|\hat{u}(\boldsymbol{\xi})|^2d\boldsymbol{\xi}.$$

Now we prove the first result, $I(\mathbf{y},2s) = C\|\mathbf{y}\|^{2s-n}$. In fact, from the proof of Theorem 8.9.5, we have $\boldsymbol{xi} = \frac{\boldsymbol{\eta}}{|h|} \in \mathbb{R}^n$, $\|\mathbf{y}\| = |h|$, $\langle \mathbf{y},\boldsymbol{\xi}\rangle = \eta_1$, $|J| = \frac{1}{|h|^n}$ and

$$I(\mathbf{y},2s) = \int_{\mathbb{R}^n}\frac{|e^{i2\pi\langle \mathbf{y},\boldsymbol{\xi}\rangle}-1|^2}{\|\boldsymbol{\xi}\|^{2s}}d\boldsymbol{\xi} = |h|^{2s-n}\underbrace{\int_{\mathbb{R}^n}\frac{|e^{i2\pi\eta_1}-1|^2}{\|\boldsymbol{\eta}\|^{2s}}d\boldsymbol{\eta}}_{I(2s)} = |h|^{2s-n}I(2s).$$

Now we will show that for $s-\frac{n}{2} = \lambda \in \,]0,1[$, $I(2s) < +\infty$.
Again, we change the variables:

$$\eta_1 = t_1,\quad \eta_2 = t_1t_2,\quad \ldots,\quad \eta_n = t_1t_n,$$

such that $\boldsymbol{\eta} = (\eta_1,\eta_2,\ldots,\eta_n)$,

$$J(\boldsymbol{\eta},\mathbf{t}) = \begin{vmatrix} 1 & 0 & \cdots & \cdots & 0 \\ t_2 & t_1 & \ddots & & \vdots \\ t_3 & 0 & t_1 & \ddots & \vdots \\ \vdots & \vdots & \ddots & \ddots & 0 \\ t_n & 0 & \cdots & 0 & t_1 \end{vmatrix} = t_1^{n-1}$$

and

$$\begin{aligned} I(2s) &= \int_{\mathbb{R}^n}\frac{|e^{i2\pi t_1}-1|^2}{|t_1|^{2s}(1+\sum_{i=2}^n|t_i|^2)^s}|t_1|^{n-1}d\mathbf{t} \quad \text{with } d\mathbf{t} = dt_1\cdots dt_n \\ &= \underbrace{\int_{\mathbb{R}}\frac{|e^{i2\pi t_1}-1|^2}{|t_1|^{2s-n+1}}dt_1}_{I_1}\times\underbrace{\int_{\mathbb{R}^{n-1}}\frac{1}{(1+\sum_{i=2}^n|t_i|^2)^s}dt_2\ldots dt_n}_{I_2} = I_1\cdot I_2. \end{aligned}$$

Hence, $I(2s) < +\infty$ if $I_1 < +\infty$ and $I_2 < +\infty$.
Convergence of I_1:

- at *infinity*: if $-(2s-n+1)+1 < 0 \Longrightarrow 2s-n > 0 \Longrightarrow s-\frac{n}{2} > 0$ which holds, since $s-\frac{n}{2} = \lambda \in \,]0,1[$;
- at the *origin*: $|e^{i2\pi t_1}-1|^2 \approx t_1^2$ (since $\frac{e^{i2\pi t_1}-1}{t_1} \to$ finite limit) $\Longrightarrow -(2s-n+1-2)+1 > 0 \Longrightarrow -2s+n+2 > 0 \Longrightarrow 2 > 2s-n \Longrightarrow 1 > s-\frac{n}{2} = \lambda \in \,]0,1[$.

Hence, for $s-\frac{n}{2} = \lambda \in \,]0,1[$, $I_1 < +\infty$.

Convergence of I_2 with $n \geq 2$: $I_2 = \int_{\mathbb{R}^{n-1}} \frac{1}{(1+\rho^2)^s} dt_2 \dots dt_n$, for which we can apply Lemma 8.9.3. Then I_2 converges, if the corresponding integral $\int_0^\infty \frac{1}{(1+\rho^2)^s}$ $\rho^{n-2} d\rho$ converges at the origin and at infinity for $n \geq 2$:

- at the *origin*: For $0 < \rho < 1$, $1 + \rho^2 \geq 1 \Longrightarrow \frac{1}{(1+\rho^2)^s} \rho^{n-2} \leq \rho^{n-2} \Longrightarrow$ convergence for $n - 2 + 1 > 0$, i.e. for $n > 1$, which always holds;
- at *infinity*: For $1 \leq \rho < \infty$, $(1 + \rho^2)^s \geq \rho^{2s} \Longrightarrow$ the integral converges if $(-2s+n-2)+1 < 0 \Longrightarrow -2s+n-1 < 0 \Longrightarrow -2s+n < 1 \Longrightarrow s-\frac{n}{2} > -\frac{1}{2}$, which always holds, since $s - \frac{n}{2} = \lambda \in]0, 1[$.

Hence, for $s - \frac{n}{2} = \lambda \in]0, 1[$, $I_2 < +\infty$.

Thus, $I(\mathbf{y}, 2s) = C \|\mathbf{y}\|^{2s-n}$ with $C = C(s, n) = I(2s)$, since $\|\mathbf{y}\| = |h|$. □

Alternative definition of $H^s(\mathbb{R}^n)$ for $s \in]0, 1[$

Theorem 8.9.5 allows us to give an alternative definition of $H^s(\mathbb{R}^n)$ for $s \in]0, 1[$ (see also Definition 8.10.6) instead of the general definition of $H^s(\mathbb{R}^n)$ in (8.9.1) or (8.9.5).

Definition 8.9.3. For $0 < s < 1$, $H^s(\mathbb{R}^n)$ is defined by the space

$$H^s(\mathbb{R}^n) = \left\{ u : u \in L^2(\mathbb{R}^n), \iint_{\mathbb{R}^n \times \mathbb{R}^n} \frac{|u(\mathbf{x}+\mathbf{y}) - u(\mathbf{x})|^2}{\|\mathbf{y}\|^{n+2s}} d\mathbf{x} d\mathbf{y} < +\infty \right\} \tag{8.9.38}$$

equipped with the norm given by

$$\left(\|u\|^2_{L^2(\mathbb{R}^n)} + \int_{\mathbb{R}^n \times \mathbb{R}^n} \int \frac{|u(\mathbf{x}+\mathbf{y}) - u(\mathbf{x})|^2}{\|\mathbf{y}\|^{n+2s}} d\mathbf{x} d\mathbf{y} \right)^{\frac{1}{2}}, \tag{8.9.39}$$

which is equivalent to the original norm $\| \cdot \|_{s,\mathbb{R}^n}$ in (8.9.3), i.e. $\exists C_1, C_2 > 0$ such that

$$C_1 \|u\|_{s,\mathbb{R}^n} \leq \left(\|u\|^2_{L^2(\mathbb{R}^n)} + \iint_{\mathbb{R}^n \times \mathbb{R}^n} \frac{|u(\mathbf{x}+\mathbf{y}) - u(\mathbf{x})|^2}{\|\mathbf{y}\|^{n+2s}} d\mathbf{x} d\mathbf{y} \right)^{\frac{1}{2}} \leq C_2 \|u\|_{s,\mathbb{R}^n}. \tag{8.9.40}$$

Justification of Definition 8.9.3 For this we will need the following result:

Lemma 8.9.5. $\forall \rho \geq 0$, $\forall s \geq 0$, $\exists \alpha_1, \alpha_2 > 0$ *such that*

$$\alpha_1 \leq \frac{(1+\rho)^s}{1+\rho^s} \leq \alpha_2 \Longrightarrow \alpha_1(1+\rho^s) \leq (1+\rho)^s \leq \alpha_2(1+\rho^s).$$

Proof. The proof is similar to that of Lemma 8.9.2, with necessary modifications. □

Now we begin with the justification of Definition 8.9.3.

Set $\rho = \|\xi\|^2$ with $1+\rho^s = 1+\|\xi\|^{2s}$, $(1+\rho)^s = (1+\|\xi\|^2)^s$. Then, using Lemma 8.9.5, $\alpha_1(1+\|\xi\|^{2s}) \le (1+\|\xi\|^2)^s \le \alpha_2(1+\|\xi\|^{2s})$. But for $s \ge 0$, $u \in H^s(\mathbb{R}^n) \Longrightarrow u \in L^2(\mathbb{R}^n) \Longleftrightarrow \hat{u} \in L^2(\mathbb{R}^n)$ by the Plancherel–Riesz Theorem 8.3.1. Hence, we have

$$\alpha_1(1+\|\xi\|^{2s})|\hat{u}(\xi)|^2 \le (1+\|\xi\|^2)^s|\hat{u}(\xi)|^2 \le \alpha_2(1+\|\xi\|^{2s})|\hat{u}(\xi)|^2 \quad \text{a.e. in } \mathbb{R}^n$$

$$\Longrightarrow \quad \alpha_1\left[\int_{\mathbb{R}^n}|\hat{u}(\xi)|^2 d\xi + \int_{\mathbb{R}^n}\|\xi\|^{2s}|\hat{u}(\xi)|^2 d\xi\right] \le \int_{\mathbb{R}^n}(1+\|\xi\|^2)^s|\hat{u}(\xi)|^2 d\xi$$
$$\le \alpha_2\left[\int_{\mathbb{R}^n}|\hat{u}(\xi)|^2 d\xi + \int_{\mathbb{R}^n}\|\xi\|^{2s}|\hat{u}(\xi)|^2 d\xi\right].$$

But from Lemma 8.9.4, $\exists C > 0$ such that

$$\int_{\mathbb{R}^n}\|\xi\|^{2s}|\hat{u}(\xi)|^2 d\xi = \frac{1}{C}\iint_{\mathbb{R}^n\times\mathbb{R}^n}\frac{|u(\mathbf{x}+\mathbf{y})-u(\mathbf{x})|^2}{\|\mathbf{y}\|^{n+2s}}d\mathbf{x}d\mathbf{y}.$$

Then,

$$\alpha_1\left[\int_{\mathbb{R}^n}|\hat{u}(\xi)|^2 d\xi + \frac{1}{C}\int_{\mathbb{R}^n\times\mathbb{R}^n}\int\frac{|u(\mathbf{x}+\mathbf{y})-u(\mathbf{x})|^2}{\|\mathbf{y}\|^{n+2s}}d\mathbf{x}d\mathbf{y}\right] \le \|u\|^2_{H^s(\mathbb{R}^n)}$$
$$\le \alpha_2\left[\int_{\mathbb{R}^n}|\hat{u}(\xi)|^2 d\xi + \frac{1}{C}\int_{\mathbb{R}^n\times\mathbb{R}^n}\int\frac{|u(\mathbf{x}+\mathbf{y})-u(\mathbf{x})|^2}{\|\mathbf{y}\|^{n+2s}}d\mathbf{x}d\mathbf{y}\right]$$
$$\Longrightarrow \quad \beta_1\left[\|u\|^2_{L^2(\mathbb{R}^n)} + \int_{\mathbb{R}^n\times\mathbb{R}^n}\int\frac{|u(\mathbf{x}+\mathbf{y})-u(\mathbf{x})|^2}{\|\mathbf{y}\|^{n+2s}}d\mathbf{x}d\mathbf{y}\right]^{1/2} \le \|u\|_{H^s(\mathbb{R}^n)}$$
$$\le \beta_2\left[\|u\|^2_{L^2(\mathbb{R}^n)} + \int_{\mathbb{R}^n\times\mathbb{R}^n}\int\frac{|u(\mathbf{x}+\mathbf{y})-u(\mathbf{x})|^2}{\|\mathbf{y}\|^{n+2s}}d\mathbf{x}d\mathbf{y}\right]^{1/2},$$

where $\|u\|_{L^2(\mathbb{R}^n)} = \|\hat{u}\|_{L^2(\mathbb{R}^n)}$ by the Plancherel–Riesz Theorem 8.3.1, $\beta_1^2 = \min\{\alpha_1, \frac{\alpha_1}{C}\} > 0$, $\beta_2^2 = \max\{\alpha_2, \frac{\alpha_2}{C}\} > 0$. Hence,

$$C_1\|u\|_{H^s(\mathbb{R}^n)} \le \left[\|u\|^2_{L^2(\mathbb{R}^n)} + \int_{\mathbb{R}^n\times\mathbb{R}^n}\int\frac{|u(\mathbf{x}+\mathbf{y})-u(\mathbf{x})|^2}{\|\mathbf{y}\|^{n+2s}}d\mathbf{x}d\mathbf{y}\right]^{\frac{1}{2}}$$
$$\le C_2\|u\|_{H^s(\mathbb{R}^n)},$$

with $C_1 = \frac{1}{\beta_2}$, $C_2 = \frac{1}{\beta_1} > 0$.

Alternative definition of $H^s(\mathbb{R}^n)$ for $s > 0$ Instead of the definition of $H^s(\mathbb{R}^n)$ for non-integral $s = [s] + \sigma > 0$ given by Theorem 8.9.3, the following definition is a convenient one:

Definition 8.9.4. For $s > 0$ with $s = [s] + \sigma$, $[s] = k \in \mathbb{N}_0$ being the integral part of s, $\sigma \in]0, 1[$ the fractional part, $H^s(\mathbb{R}^n) \subset H^{[s]}(\mathbb{R}^n)$ defined by (8.9.16)–(8.9.18) can be defined alternatively by (see also Definition 8.10.7):

$$H^s(\mathbb{R}^n) = \{u : u \in H^{[s]}(\mathbb{R}^n),\ \partial^{\boldsymbol{\alpha}} u \in H^{\sigma}(\mathbb{R}^n)\ \forall |\boldsymbol{\alpha}| \le k,$$
$$H^{\sigma}(\mathbb{R}^n) \text{ is defined by (8.9.38)–(8.9.39)}\}.$$

Then the mapping

$$u \in H^s(\mathbb{R}^n) \longmapsto \left(\|u\|^2_{H^{[s]}(\mathbb{R}^n)} + \sum_{0 \le |\boldsymbol{\alpha}| \le k} \iint_{\mathbb{R}^n \times \mathbb{R}^n} \frac{|\partial^{\boldsymbol{\alpha}} u(\mathbf{x}+\mathbf{y}) - \partial^{\boldsymbol{\alpha}} u(\mathbf{x})|^2}{\|\mathbf{y}\|^{n+2s}} d\mathbf{x} d\mathbf{y} \right)^{1/2}$$

defines a norm equivalent to the original norm $\|\cdot\|_{s,\mathbb{R}^n}$ given by (8.9.3): $\exists C_1, C_2 > 0$ such that

$$C_1 \|u\|_{s,\mathbb{R}^n} \le \left[\|u\|^2_{H^{[s]}(\mathbb{R}^n)} + \sum_{0 \le |\boldsymbol{\alpha}| \le k} \iint_{\mathbb{R}^n \times \mathbb{R}^n} \frac{|\partial^{\boldsymbol{\alpha}} u(\mathbf{x}+\mathbf{y}) - \partial^{\boldsymbol{\alpha}} u(\mathbf{x})|^2}{\|\mathbf{y}\|^{n+2s}} d\mathbf{x} d\mathbf{y} \right]^{1/2}$$
$$\le C_2 \|u\|_{s,\mathbb{R}^n}. \tag{8.9.41}$$

8.9.5 Imbedding result: $S(\mathbb{R}^n) \hookrightarrow H^S(\mathbb{R}^n)$

Proposition 8.9.2. $\forall s \in \mathbb{R}$,

$$S(\mathbb{R}^n) \hookrightarrow H^s(\mathbb{R}^n), \tag{8.9.42}$$

the imbedding $\hookrightarrow$ being a continuous one from $S(\mathbb{R}^n)$ into $H^s(\mathbb{R}^n)$.

Proof.
$S(\mathbb{R}^n) \subset H^s(\mathbb{R}^n)$: Let $\phi \in S(\mathbb{R}^n) \subset S'(\mathbb{R}^n)$. Then, from Theorem 7.6.1, $\hat{\phi} = \mathcal{F}\phi \in S(\mathbb{R}^n)$. But $\hat{\phi} \in S(\mathbb{R}^n) \Longrightarrow (1 + \|\boldsymbol{\xi}\|^2)^{\frac{s}{2}} \hat{\phi} \in S(\mathbb{R}^n) \hookrightarrow L^2(\mathbb{R}^n)$ by Proposition 7.4.1 $\Longrightarrow (1 + \|\boldsymbol{\xi}\|^2)^{\frac{s}{2}} \hat{\phi} \in L^2(\mathbb{R}^n) \Longrightarrow \phi \in H^s(\mathbb{R}^n) \Longrightarrow S(\mathbb{R}^n) \subset H^s(\mathbb{R}^n)$.
Continuity of $\hookrightarrow$: Let (ϕ_m) be any sequence in $S(\mathbb{R}^n)$ such that $\phi_m \to 0$ in $S(\mathbb{R}^n)$ as $m \to \infty$. Then, by Theorem 7.6.1, $\hat{\phi}_m = \mathcal{F}\phi_m \to 0$ in $S(\mathbb{R}^n)$ as $m \to \infty \Longrightarrow (1 + \|\boldsymbol{\xi}\|^2)^{\frac{s}{2}} \hat{\phi}_m \to 0$ in $S(\mathbb{R}^n) \hookrightarrow L^2(\mathbb{R}^n)$ by Proposition 7.4.1 $\Longrightarrow (1 + \|\boldsymbol{\xi}\|^2)^{\frac{s}{2}} \hat{\phi}_m \to 0$ in $L^2(\mathbb{R}^n)$ as $m \to \infty \Longrightarrow \|\phi_m\|_{s,\mathbb{R}^n} = \|(1 + \|\boldsymbol{\xi}\|^2)^{\frac{s}{2}} \hat{\phi}_m\|_{L^2(\mathbb{R}^n)} \to 0$ as $m \to \infty \Longrightarrow \hookrightarrow \phi_m = \phi_m \in H^s(\mathbb{R}^n)$ and $\phi_m \to 0$ in $H^s(\mathbb{R}^n)$ as $m \to \infty \Longrightarrow \hookrightarrow : S(\mathbb{R}^n) \longrightarrow H^s(\mathbb{R}^n)$ is continuous. □

8.9.6 Density results $H^S(\mathbb{R}^n)$

Theorem 8.9.6.

I. $\forall s \in \mathbb{R}$, $S(\mathbb{R}^n)$ *is* dense *in* $H^s(\mathbb{R}^n)$; (8.9.43)

II. $\forall s \in \mathbb{R}$, $\mathcal{D}(\mathbb{R}^n)$ *is* dense *in* $H^s(\mathbb{R}^n)$. (8.9.44)

III. $\mathcal{D}(\mathbb{R}^n) \hookrightarrow S(\mathbb{R}^n) \hookrightarrow H^s(\mathbb{R}^n)$ *with* dense, continuous *imbeddings.* (8.9.45)

Proof.

I. Let $u \in H^s(\mathbb{R}^n)$. Then, from (8.9.1), $u \in S'(\mathbb{R}^n)$ and $(1 + \|\boldsymbol{\xi}\|^2)^{s/2}\hat{u} \in L^2(\mathbb{R}^n)$.
Since $\mathcal{D}(\mathbb{R}^n)$ is dense in $L^2(\mathbb{R}^n)$ by (1.2.26), $\exists$ a sequence (ϕ_k) in $\mathcal{D}(\mathbb{R}^n)$ such that $\phi_k \to (1 + \|\boldsymbol{\xi}\|^2)^{s/2}\hat{u}$ in $L^2(\mathbb{R}^n)$ as $k \to \infty$. Define ψ_k by $\hat{\psi}_k(\boldsymbol{\xi}) = \frac{\phi_k(\boldsymbol{\xi})}{(1+\|\boldsymbol{\xi}\|^2)^{\frac{s}{2}}}$ $\forall k \in \mathbb{N}$, $\forall \boldsymbol{\xi} \in \mathbb{R}^n$, which is well defined, since $\phi_k \in \mathcal{D}(\mathbb{R}^n) \Longrightarrow \frac{\phi_k}{(1+\|\boldsymbol{\xi}\|^2)^{\frac{s}{2}}} \in \mathcal{D}(\mathbb{R}^n) \subset S(\mathbb{R}^n)$ as $\frac{1}{(1+\|\boldsymbol{\xi}\|^2)^{\frac{s}{2}}} \in C^\infty(\mathbb{R}^n)$,

$$\Longrightarrow \quad \hat{\psi}_k \in S(\mathbb{R}^n) \quad \Longrightarrow \quad \bar{\mathcal{F}}[\hat{\psi}_k] = \psi_k \in S(\mathbb{R}^n) \quad \forall k \in \mathbb{N} \tag{8.9.46}$$

by Theorem 7.7.1.
But $\hat{\psi}_k \in S(\mathbb{R}^n) \Longrightarrow (1 + \|\boldsymbol{\xi}\|^2)^{s/2}\hat{\psi}_k \in S(\mathbb{R}^n) \Longrightarrow (1 + \|\boldsymbol{\xi}\|^2)^{\frac{s}{2}}\hat{\psi}_k \in L^2(\mathbb{R}^n)$, since $S(\mathbb{R}^n) \hookrightarrow L^2(\mathbb{R}^n)$ by (7.4.1). Hence, $\psi_k \in H^s(\mathbb{R}^n)$ $\forall k \in \mathbb{N}$. Again, by construction, $(1 + \|\boldsymbol{\xi}\|^2)^{\frac{s}{2}}\hat{\psi}_k = \phi_k \to (1 + \|\boldsymbol{\xi}\|^2)^{\frac{s}{2}}\hat{u}$ in $L^2(\mathbb{R}^n)$ as $k \to \infty \Longrightarrow \psi_k \to u$ in $H^s(\mathbb{R}^n)$ as $k \to \infty$ Hence, $\forall u \in H^s(\mathbb{R}^n)$, $\exists(\psi_k)$ in $S(\mathbb{R}^n)$ such that $\psi_k \to u$ in $H^s(\mathbb{R}^n)$, i.e. $S(\mathbb{R}^n)$ is *dense* in $H^s(\mathbb{R}^n)$. In other words, by virtue of (8.9.42),

$$S(\mathbb{R}^n) \hookrightarrow H^s(\mathbb{R}^n) \quad \text{with a } \textit{dense, continuous} \text{ imbedding.} \tag{8.9.47}$$

II. From Propositions 7.4.1 and 7.5.1, $\mathcal{D}(\mathbb{R}^n) \hookrightarrow S(\mathbb{R}^n)$ with *dense*, *continuous* imbeddings, which together with (8.9.42) gives the continuous imbeddings in $\mathcal{D}(\mathbb{R}^n) \hookrightarrow S(\mathbb{R}^n) \hookrightarrow H^s(\mathbb{R}^n)$ in (8.9.47). Since $S(\mathbb{R}^n)$ is dense in $H^s(\mathbb{R}^n)$ for $s \in \mathbb{R}$, for $u \in H^s(\mathbb{R}^n)$, $\forall \varepsilon > 0$, $\exists \psi \in S(\mathbb{R}^n)$ such that

$$\|u - \psi\|_{s,\mathbb{R}^n} < \frac{\varepsilon}{2}. \tag{8.9.48}$$

By Proposition 7.5.1, $\mathcal{D}(\mathbb{R}^n)$ is a dense subspace of $S(\mathbb{R}^n)$. Hence, for $\psi \in S(\mathbb{R}^n)$ satisfying (8.9.48), $\exists$ a sequence (ϕ_m) in $\mathcal{D}(\mathbb{R}^n)$ such that $\phi_m \to \psi$ in $S(\mathbb{R}^n)$ as $m \to \infty$, which in turn implies that $\phi_m \to \psi$ in $H^s(\mathbb{R}^n)$ by virtue of the continuous imbedding $S(\mathbb{R}^n) \hookrightarrow H^s(\mathbb{R}^n)$ (by Proposition 8.9.2), i.e. $\forall \varepsilon > 0$, $\exists m_0(\varepsilon) \in \mathbb{N}$ such that

$$\|\psi - \phi_m\|_{s,\mathbb{R}^n} < \frac{\varepsilon}{2} \quad \forall m \geq m_0. \tag{8.9.49}$$

Then, $\forall\varepsilon > 0$, $\exists m_0(\varepsilon) \in \mathbb{N}$ such that

$$\|u - \phi_m\|_{s,\mathbb{R}^n} \le \|u - \psi\|_{s,\mathbb{R}^n} + \|\psi - \phi_m\|_{s,\mathbb{R}^n} < \frac{\varepsilon}{2} + \frac{\varepsilon}{2} = \varepsilon \quad \forall m \ge m_0,$$

$$\text{i.e.}\quad \phi_m \to u \text{ in } H^s(\mathbb{R}^n) \text{ as } m \to \infty. \tag{8.9.50}$$

Hence, for any $u \in H^s(\mathbb{R}^n)$, $\exists(\phi_m)$ in $\mathcal{D}(\mathbb{R}^n)$ such that $\phi_m \to u$ in $H^s(\mathbb{R}^n)$ as $m \to \infty$, i.e. $\mathcal{D}(\mathbb{R}^n)$ is *dense* in $H^s(\mathbb{R}^n)$ $\forall s \in \mathbb{R}$.

III. We have already shown in the proof of II the *continuous* imbeddings in (8.9.45). Then (8.9.45) follows from I and II. □

8.9.7 Dual space $(H^s(\mathbb{R}^n))'$

Proposition 8.9.3. *The (topological) dual space $(H^s(\mathbb{R}^n))'$ is a* subspace of distributions, *i.e.* $(H^s(\mathbb{R}^n))' \hookrightarrow S'(\mathbb{R}^n) \hookrightarrow \mathcal{D}'(\mathbb{R}^n)$ *with continuous imbeddings.* (8.9.51)

Proof. From (8.9.45), $\mathcal{D}(\mathbb{R}^n) \hookrightarrow S(\mathbb{R}^n) \hookrightarrow H^s(\mathbb{R}^n)$ with *dense*, *continuous* imbeddings. Then, from (4.2.11), the result (8.9.51) follows. □

Theorem 8.9.7. *The (topological) dual space $(H^s(\mathbb{R}^n))'$ coincides algebraically and topologically with $H^{-s}(\mathbb{R}^n)$ in $\mathcal{D}'(\mathbb{R}^n)$.*

Proof. Since $H^s(\mathbb{R}^n)$ is a (complex) Hilbert space, by the Riesz Representation Theorem $\exists$ an *anti-linear, bijective, isometric* mapping $J : (H^s(\mathbb{R}^n))' \longrightarrow H^s(\mathbb{R}^n)$ defined, $\forall L \in (H^s(\mathbb{R}^n))'$, by:

$$JL = u_L \in H^s(\mathbb{R}^n) \tag{8.9.52}$$

such that $\forall v \in H^s(\mathbb{R}^n)$,

$$L(v) = \langle v, JL\rangle_{s,\mathbb{R}^n} = \langle v, u_L\rangle_{s,\mathbb{R}^n} = \int_{\mathbb{R}^n} (1 + \|\xi\|^2)^s\, \hat{v}(\xi)\overline{\hat{u}_L(\xi)}d\xi, \tag{8.9.53}$$

$$\left.\begin{aligned} &J(\alpha L) = \bar{\alpha} J(L)\ \forall\alpha \in \mathbb{C},\ \forall L \in (H^s(\mathbb{R}^n))' \\ &J(L_1 + L_2) = JL_1 + JL_2\ \forall L_i \in (H^s(\mathbb{R}^n))' \end{aligned}\right\} \text{ – \textit{anti-linearity};}$$

$$\left.\begin{aligned} &J(L) = 0 \text{ in } H^s(\mathbb{R}^n) \Longrightarrow L = 0 \text{ in } (H^s(\mathbb{R}^n))' \text{ – \textit{injectivity}} \\ &\forall u \in H^s(\mathbb{R}^n),\ \exists L \in (H^s(\mathbb{R}^n))' \text{ such that } JL = u \text{ – \textit{surjectivity}} \end{aligned}\right\} \text{ – \textit{bijectivity};}$$

$$\|JL\|_{H^s(\mathbb{R}^n)} = \|u_L\|_{H^s(\mathbb{R}^n)} = \|L\|_{(H^s(\mathbb{R}^n))'} \quad \text{ – \textit{isometry}.} \tag{8.9.54}$$

Obviously, (8.9.53) holds for $v = \phi \in \mathcal{D}(\mathbb{R}^n) \subset S(\mathbb{R}^n) \subset H^s(\mathbb{R}^n)$. For $L \in (H^s(\mathbb{R}^n))'$, $u_L \in H^s(\mathbb{R}^n)$. Hence, $(1 + \|\xi\|^2)^{\frac{s}{2}}\hat{u}_L \in L^2(\mathbb{R}^n) \hookrightarrow S'(\mathbb{R}^n) \Longrightarrow$ $(1 + \|\xi\|^2)^{\frac{s}{2}}\overline{(1 + \|\xi\|^2)^{\frac{s}{2}}\hat{u}_L} \in S'(\mathbb{R}^n)$

$$\Longrightarrow \mathcal{F}[(1 + \|\xi\|^2)^{\frac{s}{2}}\overline{(1 + \|\xi\|^2)^{\frac{s}{2}}\hat{u}_L}] = \mathcal{F}[(1 + \|\xi\|^2)^{\frac{s}{2}} w] \in S'(\mathbb{R}^n), \tag{8.9.55}$$

where $\overline{(\dots)}$ complex conjugate of $(\dots)$,

$$w = \overline{(1+\|\boldsymbol{\xi}\|^2)^{\frac{s}{2}}\hat{u}_L} \in L^2(\mathbb{R}^n) \hookrightarrow S'(\mathbb{R}^n). \tag{8.9.56}$$

We denote the tempered distribution in (8.9.55) by $T_{u_L} \in S'(\mathbb{R}^n)$, i.e. for $u_L \in H^S(\mathbb{R}^n)$

$$T_{u_L} = \mathcal{F}[(1+\|\boldsymbol{\xi}\|^2)^{\frac{s}{2}} w] \in S'(\mathbb{R}^n). \tag{8.9.57}$$

Now we will show that $T_{u_L} \in H^{-s}(\mathbb{R}^n)$, for which we are to show that $(1+\|\boldsymbol{\xi}\|^2)^{-\frac{s}{2}}\hat{T}_{u_L} \in L^2(\mathbb{R}^n)$ with $\hat{T}_{u_L} = \mathcal{F}[T_{u_L}]$. In fact,

$$\begin{aligned}(1+\|\boldsymbol{\xi}\|^2)^{-\frac{s}{2}}\hat{T}_{u_L} &= (1+\|\boldsymbol{\xi}\|^2)^{-\frac{s}{2}}\mathcal{F}[T_{u_L}] = (1+\|\boldsymbol{\xi}\|^2)^{-\frac{s}{2}}\mathcal{F}\mathcal{F}[(1+\|\boldsymbol{\xi}\|^2)^{\frac{s}{2}} w] \\ &= (1+\|\boldsymbol{\xi}\|^2)^{-\frac{s}{2}}(1+\|\boldsymbol{\xi}\|^2)^{\frac{s}{2}}\check{w} = \check{w} \in L^2(\mathbb{R}^n),\end{aligned}$$

since
$\mathcal{F}\mathcal{F}[(1+\|\boldsymbol{\xi}\|^2)^{\frac{s}{2}} w(\boldsymbol{\xi})] = [(1+\|-\boldsymbol{\xi}\|^2)^{\frac{s}{2}} w(-\boldsymbol{\xi})] = (1+\|\boldsymbol{\xi}\|^2)^{\frac{s}{2}}\check{w}(\boldsymbol{\xi})$ (see (8.3.21)).
Therefore,

$$(1+\|\boldsymbol{\xi}\|^2)^{-\frac{s}{2}}\hat{T}_{u_L} = \check{w} \in L^2(\mathbb{R}^n). \tag{8.9.58}$$

Hence, $T_{u_L} \in S'(\mathbb{R}^n)$, $(1+\|\boldsymbol{\xi}\|^2)^{-\frac{s}{2}}\hat{T}_{u_L} \in L^2(\mathbb{R}^n) \implies T_{u_L} \in H^{-s}(\mathbb{R}^n)$ by Definition 8.9.1. (8.9.59)

Now we are to show that $\|T_{u_L}\|_{H^{-s}(\mathbb{R}^n)} = \|u_L\|_{H^s(\mathbb{R}^n)}$.

In fact, from (8.9.58), (8.9.56) and (8.9.54), we have

$$\begin{aligned}\|(1+\|\boldsymbol{\xi}\|^2)^{-\frac{s}{2}}\hat{T}_{u_L}\|_{L^2(\mathbb{R}^n)} &= \|\check{w}\|_{L^2(\mathbb{R}^n)} = \|w\|_{L^2(\mathbb{R}^n)} = \|\overline{(1+\|\boldsymbol{\xi}\|^2)^{\frac{s}{2}}\hat{u}_L}\|_{L^2(\mathbb{R}^n)} \\ &= \|(1+\|\boldsymbol{\xi}\|^2)^{\frac{s}{2}}\hat{u}_L\|_{L^2(\mathbb{R}^n)} \\ &= \|u_L\|_{H^s(\mathbb{R}^n)} = \|L\|_{(H^s(\mathbb{R}^n))'}.\end{aligned}$$

Hence,

$$\|T_{u_L}\|_{H^{-s}(\mathbb{R}^n)} = \|(1+\|\boldsymbol{\xi}\|^2)^{-\frac{s}{2}}\hat{T}_{u_L}\|_{L^2(\mathbb{R}^n)} = \|u_L\|_{H^s(\mathbb{R}^n)} = \|L\|_{(H^s(\mathbb{R}^n))'}. \tag{8.9.60}$$

Now we are to show that $T_{u_L} = L$ in $\mathcal{D}'(\mathbb{R}^n)$ in order to show that $H^{-s}(\mathbb{R}^n)$ and $(H^s(\mathbb{R}^n))'$ coincide in $\mathcal{D}'(\mathbb{R}^n)$. In fact, $T_{u_L} \in S'(\mathbb{R}^n) \implies T_{u_L} \in \mathcal{D}'(\mathbb{R}^n)$. Then, $\forall \phi \in \mathcal{D}(\mathbb{R}^n) \hookrightarrow S(\mathbb{R}^n)$,

$$\begin{aligned}\langle T_{u_L}, \phi\rangle_{\mathcal{D}'(\mathbb{R}^n)\times\mathcal{D}(\mathbb{R}^n)} &= \langle T_{u_L}, \phi\rangle_{S'(\mathbb{R}^n)\times S(\mathbb{R}^n)} \\ &= \langle \overline{\mathcal{F}} T_{u_L}, \hat{\phi}\rangle_{S'(\mathbb{R}^n)\times S(\mathbb{R}^n)} = \langle \tilde{T}_{u_L}, \hat{\phi}\rangle_{S'(\mathbb{R}^n)\times S(\mathbb{R}^n)},\end{aligned}$$

with $\tilde{T}_{u_L} = \overline{\mathcal{F}} T_{u_L}$ and $\hat{\phi} = \mathcal{F}\phi$, since (by the definition of Fourier transforms and co-transforms in $S'(\mathbb{R}^n)$) $\langle \overline{\mathcal{F}} T_{u_L}, \hat{\phi}\rangle = \langle T_{u_L}, \overline{\mathcal{F}}\hat{\phi}\rangle = \langle T_{u_L}, \phi\rangle$ and $\mathcal{F}\tilde{T}_{u_L} = \mathcal{F}(\overline{\mathcal{F}} T_{u_L}) = T_{u_L}$. Hence,

$$\langle T_{u_L}, \phi\rangle_{\mathcal{D}'(\mathbb{R}^n)\times\mathcal{D}(\mathbb{R}^n)} = \langle (1+\|\xi\|^2)^{-\frac{s}{2}}\tilde{T}_{u_L}, (1+\|\xi\|^2)^{\frac{s}{2}}\hat{\phi}\rangle_{S'(\mathbb{R}^n)\times S(\mathbb{R}^n)}. \tag{8.9.61}$$

But $(1+\|\xi\|^2)^{-\frac{s}{2}}\tilde{T}_{u_L} \in L^2(\mathbb{R}^n)$. In fact, $T_{u_L} = \mathcal{F}[(1+\|\xi\|^2)^{\frac{s}{2}}w]$ (by (8.9.57)), and $T_{u_L} = \mathcal{F}[\tilde{T}_{u_L}] \Longrightarrow \tilde{T}_{u_L} = \overline{\mathcal{F}} T_{u_L} = (1+\|\xi\|^2)^{\frac{s}{2}}w$

$$\Longrightarrow \quad (1+\|\xi\|^2)^{-\frac{s}{2}}\tilde{T}_{u_L} = (1+\|\xi\|^2)^{-\frac{s}{2}}(1+\|\xi\|^2)^{\frac{s}{2}}w = w \in L^2(\mathbb{R}^n). \tag{8.9.62}$$

Then, from (8.9.61), (8.9.62) and (8.9.56),

$$\begin{aligned}\langle T_{u_L}, \phi\rangle_{\mathcal{D}'(\mathbb{R}^n)\times\mathcal{D}(\mathbb{R}^n)} &= \int_{\mathbb{R}^n} w(\xi)(1+\|\xi\|^2)^{\frac{s}{2}}\hat{\phi}(\xi)d\xi \\ &= \int_{\mathbb{R}^n} (1+\|\xi\|^2)^{\frac{s}{2}}\hat{\phi}(\xi)\overline{(1+\|\xi\|^2)^{\frac{s}{2}}\hat{u}_L(\xi)}d\xi \\ &= \int_{\mathbb{R}^n} (1+\|\xi\|^2)^{s}\hat{\phi}(\xi)\overline{\hat{u}_L(\xi)}d\xi. \end{aligned} \tag{8.9.63}$$

Hence, from (8.9.53) and (8.9.63), we have $\langle T_{u_L}, \phi\rangle_{\mathcal{D}'(\mathbb{R}^n)\times\mathcal{D}(\mathbb{R}^n)} = \langle L, \phi\rangle_{(H^s(\mathbb{R}^n))'\times H^s(\mathbb{R}^n)} \Longrightarrow \langle T_{u_L} - L, \phi\rangle_{\mathcal{D}'(\mathbb{R}^n)\times\mathcal{D}(\mathbb{R}^n)} = 0 \; \forall \phi \in \mathcal{D}(\mathbb{R}^n)$. But, by Theorem 8.9.6, $\mathcal{D}(\mathbb{R}^n)$ is dense in $H^s(\mathbb{R}^n)$ with $\mathcal{D}(\mathbb{R}^n) \hookrightarrow H^s(\mathbb{R}^n)$. Hence, $T_{u_L} - L = 0$ in $\mathcal{D}'(\mathbb{R}^n) \Longrightarrow T_{u_L} = L$ in $\mathcal{D}'(\mathbb{R}^n)$ with $T_{u_L} \in H^{-s}(\mathbb{R}^n)$ (by (8.9.59)), $L \in (H^s(\mathbb{R}^n))'$ (by (8.9.52)). Thus, $(H^s(\mathbb{R}^n))' \equiv H^{-s}(\mathbb{R}^n)$ in $\mathcal{D}'(\mathbb{R}^n)$ with $\|L\|_{(H^s(\mathbb{R}^n))'} = \|T_{u_L}\|_{H^{-s}(\mathbb{R}^n)}$ (by (8.9.60)). Consequently, the mapping $L \in (H^s(\mathbb{R}^n))' \longmapsto T_{u_L} \in H^{-s}(\mathbb{R}^n)$ is *linear* and *isometric*.

Conversely, let $T \in H^{-s}(\mathbb{R}^n)$. Then $T \in S'(\mathbb{R}^n)$ and $(1+\|\xi\|^2)^{-\frac{s}{2}}\hat{T} \in L^2(\mathbb{R}^n)$. Now we will show that $\exists \phi_T \in H^s(\mathbb{R}^n)$ and $T_{\phi_T} \in H^{-s}(\mathbb{R}^n)$ such that $T_{\phi_T} = T$ in $H^{-s}(\mathbb{R}^n)$ and

$$\|\phi_T\|_{H^s(\mathbb{R}^n)} = \|T_{\phi_T}\|_{H^{-s}(\mathbb{R}^n)} = \|T\|_{H^{-s}(\mathbb{R}^n)}. \tag{8.9.64}$$

For this, set $h = (1+\|\xi\|^2)^{\frac{-s}{2}}\hat{T} \in L^2(\mathbb{R}^n) \subset S'(\mathbb{R}^n)$ and define ϕ_T by $\phi_T = \mathcal{F}[(1+\|\xi\|^2)^{-\frac{s}{2}}h] \in S'(\mathbb{R}^n)$, since $h \in L^2(\mathbb{R}^n)$, $(1+\|\xi\|^2)^{-\frac{s}{2}}h \in L^2(\mathbb{R}^n) \hookrightarrow S'(\mathbb{R}^n)$ and its Fourier transform belongs to $S'(\mathbb{R}^n)$. Then $\hat{\phi}_T = \mathcal{F}[\phi_T] = \mathcal{F}\mathcal{F}[(1+\|\xi\|^2)^{\frac{-s}{2}}h] = (1+\|\xi\|^2)^{\frac{-s}{2}}\check{h} \Longrightarrow (1+\|\xi\|^2)^{\frac{s}{2}}\hat{\phi}_T = \check{h} \in L^2(\mathbb{R}^n) \Longrightarrow \phi_T \in H^s(\mathbb{R}^n)$ with

$$\begin{aligned}\|\phi_T\|_{H^s(\mathbb{R}^n)} &= \|(1+\|\xi\|^2)^{\frac{s}{2}}\hat{\phi}_T\|_{L^2(\mathbb{R}^n)} = \|\check{h}\|_{L^2(\mathbb{R}^n)} = \|h\|_{L^2(\mathbb{R}^n)} \\ &= \|(1+\|\xi\|^2)^{-\frac{s}{2}}\hat{T}\|_{L^2(\mathbb{R}^n)} = \|T\|_{H^{-s}(\mathbb{R}^n)}. \end{aligned} \tag{8.9.65}$$

Now, with $\phi_T \in H^s(\mathbb{R}^n)$ we associate T_{ϕ_T} defined by $T_{\phi_T} = \mathcal{F}[(1+\|\boldsymbol{\xi}\|^2)^{\frac{s}{2}}(1+\|\boldsymbol{\xi}\|^2)^{\frac{s}{2}}\hat{\phi}_T] = \mathcal{F}[(1+\|\boldsymbol{\xi}\|^2)^{\frac{s}{2}}\check{h}] \Longrightarrow \hat{T}_{\phi_T} = \mathcal{F}[T_{\phi_T}] = \mathcal{F}\mathcal{F}[(1+\|\boldsymbol{\xi}\|^2)^{\frac{s}{2}}\check{h}] = (1+\|\boldsymbol{\xi}\|^2)^{\frac{s}{2}}\check{\check{h}} = (1+\|\boldsymbol{\xi}\|^2)^{\frac{s}{2}}h$, since $\check{\check{h}} = h$.

Hence, $(1+\|\boldsymbol{\xi}\|^2)^{-\frac{s}{2}}\hat{T}_{\phi_T} = h \in L^2(\mathbb{R}^n) \Longrightarrow T_{\phi_T} \in H^{-s}(\mathbb{R}^n)$ with

$$\begin{aligned}\|T_{\phi_T}\|_{H^{-s}(\mathbb{R}^n)} &= \|(1+\|\boldsymbol{\xi}\|^2)^{-\frac{s}{2}}\hat{T}_{\phi_T}\|_{L^2(\mathbb{R}^n)} = \|h\|_{L^2(\mathbb{R}^n)} \\ &= \|(1+\|\boldsymbol{\xi}\|^2)^{-\frac{s}{2}}\hat{T}\|_{L^2(\mathbb{R}^n)} = \|T\|_{H^{-s}(\mathbb{R}^n)} \text{ (by (8.9.65))}. \qquad (8.9.66)\end{aligned}$$

But for T_{ϕ_T} and $T \in H^{-s}(\mathbb{R}^n)$ satisfying (8.9.66), $(1+\|\boldsymbol{\xi}\|^2)^{-\frac{s}{2}}\hat{T}_{\phi_T} = (1+\|\boldsymbol{\xi}\|^2)^{-\frac{s}{2}}\hat{T}$ in $L^2(\mathbb{R}^n) \hookrightarrow S'(\mathbb{R}^n)$. Hence, $\hat{T}_{\phi_T} = \hat{T}$ in $S'(\mathbb{R}^n)$ (since $(1+\|\boldsymbol{\xi}\|^2)^{-\frac{s}{2}} \neq 0\ \forall \boldsymbol{\xi} \in \mathbb{R}^n) \Longrightarrow T_{\phi_T} = T$ in $S'(\mathbb{R}^n) \subset \mathcal{D}'(\mathbb{R}^n)$, since $\mathcal{F}$ is an isomorphism on $S'(\mathbb{R}^n)$ by Theorem 8.3.2. Hence, $T_{\phi_T} = T$ in $H^{-s}(\mathbb{R}^n) \subset S'(\mathbb{R}^n) \subset \mathcal{D}'(\mathbb{R}^n)$ by Proposition 8.9.3. Finally, from (8.9.65) and (8.9.66), we get (8.9.64). □

L^p-properties of $H^s(\mathbb{R}^n)$

Theorem 8.9.8 (Sobolev–Peetre). *For* $0 < s < \frac{n}{2}$,

$$H^s(\mathbb{R}^n) \hookrightarrow L^p(\mathbb{R}^n) \quad \text{for } \tfrac{1}{p} = \tfrac{1}{2} - \tfrac{s}{n} > 0; \qquad (8.9.67)$$

$$H^1(\mathbb{R}^2) \hookrightarrow L^p(\mathbb{R}^2) \quad \forall p \in [2, \infty[. \qquad (8.9.68)$$

Proof. See [42], for example, and also (8.12.6c) in Theorem 8.12.1. □

8.9.8 Trace properties of elements of $H^s(\mathbb{R}^n)$

For $s \geq 0$, $u \in H^s(\mathbb{R}^n) \hookrightarrow L^2(\mathbb{R}^n) \Longrightarrow u(\mathbf{x})$ is defined a.e. in $\mathbb{R}^n$. Hence, $u(\mathbf{x})$ may *not* be defined on sets with an n-dimensional (Lebesgue) volume measure equal to zero. But the n-dimensional (Lebesgue) volume measure of $\mathbb{R}^{n-1}$ is zero. Thus, the following question arises:

Question. For $u \in H^s(\mathbb{R}^n)$ with $s \geq 0$, can we consider the restriction $u\!\downarrow_{\mathbb{R}^{n-1}}$ of u to $\mathbb{R}^{n-1}$?

For $0 \leq s \leq 1/2$, such a restriction is *not* defined. But the general answer for $s > \frac{1}{2}$ is *not* a straightforward one. It requires the introduction of a trace operator, which will be dealt with now.

Let $\mathbf{x} = (x_1, x_2, \ldots, x_n)$ denote a generic point in $\mathbb{R}^n$. For the sake of *notational convenience*, we write $\mathbf{x} = (\mathbf{y}, z)$ with $\mathbf{y} = (x_1, x_2, \ldots, x_{n-1}) \in \mathbb{R}^{n-1}$, $z = x_n \in \mathbb{R}$. Let Γ be the *hyperplane* in $\mathbb{R}^n$ defined by the equation $z = 0$ in $\mathbb{R}^n$, which is *identified* with $\mathbb{R}^{n-1} = \{\mathbf{y} : \mathbf{y} = (x_1, x_2, \ldots, x_{n-1})\}$, i.e. $\Gamma = \{\mathbf{x} : \mathbf{x} = (\mathbf{y}, z) \in \mathbb{R}^n,\ z = 0\}$ with $(\mathbf{y}, 0) \in \Gamma \Longleftrightarrow \mathbf{y} \in \mathbb{R}^{n-1}$. (8.9.69)

Trace operator

Definition 8.9.5. Let $\phi \in S(\mathbb{R}^n)$ with $\phi = \phi(\mathbf{y}, z), \mathbf{y} \in \mathbb{R}^{n-1}$, $z \in \mathbb{R}$ and, $\forall$ fixed $j \in \mathbb{N}_0$, $\gamma_j : S(\mathbb{R}^n) \longrightarrow S(\mathbb{R}^{n-1})$ be a mapping defined by $\gamma_j : \phi \in S(\mathbb{R}^n) \longmapsto (\gamma_j\phi) \in S(\mathbb{R}^{n-1})$ such that

$$(\gamma_j\phi)(\mathbf{y}) = \frac{\partial^j}{\partial z_j}\phi(\mathbf{y}, z)\bigg|_{z=0} = \frac{\partial^j}{\partial z_j}\phi(\mathbf{y}, 0) \quad \forall \mathbf{y} \in \mathbb{R}^{n-1}. \tag{8.9.70}$$

Then, $\forall$ fixed $j \in \mathbb{N}_0$, the function $\gamma_j\phi \in S(\Gamma) \equiv S(\mathbb{R}^{n-1})$ is called the *trace* of $\frac{\partial^j\phi}{\partial z^j} \in S(\mathbb{R}^n)$ on hyperplane Γ (which is *identified* with $\mathbb{R}^{n-1}$), γ_j being the *trace operator*. In particular, for $j = 0$, $\gamma_0\phi \in S(\mathbb{R}^{n-1})$ with

$$(\gamma_0\phi)(\mathbf{y}) = \phi(\mathbf{y}, z)\bigg|_{z=0} = \phi(\mathbf{y}, 0) \quad \forall\phi \in S(\mathbb{R}^n) \tag{8.9.71}$$

is called the trace of $\phi \in S(\mathbb{R}^n)$ on the hyperplane Γ.

Corresponding to variables $\boldsymbol{\xi} = (\xi_1, \xi_2, \dots, \xi_n)$ in $\mathbb{R}^n$, we define $\boldsymbol{\eta} = (\xi_1, \xi_2, \dots, \xi_{n-1}) \in \mathbb{R}^{n-1}$, $\zeta = \xi_n \in \mathbb{R}$ such that

$$\boldsymbol{\xi} = (\boldsymbol{\eta}, \zeta) \in \mathbb{R}^n. \tag{8.9.72}$$

Then, $\forall$ fixed $j \in \mathbb{N}_0$, we define another mapping $\mu_j : S(\mathbb{R}^n) \longrightarrow S(\mathbb{R}^{n-1})$ by taking the *moment* as follows [40]: $\forall\psi \in S(\mathbb{R}^n)$,

$$(\mu_j\psi)(\boldsymbol{\eta}) = \int_{\mathbb{R}} \zeta^j\psi(\boldsymbol{\eta}, \zeta)d\zeta \quad \forall\boldsymbol{\eta} \in \mathbb{R}^{n-1}. \tag{8.9.73}$$

Then $\mu_j\psi \in S(\mathbb{R}^{n-1})$ is well defined $\forall\psi \in S(\mathbb{R}^n)$.

In particular, for $j = 0$, $\mu_0\psi \in S(\mathbb{R}^{n-1})$ with

$$(\mu_0\psi)(\boldsymbol{\eta}) = \int_{\mathbb{R}} \psi(\boldsymbol{\eta}, \zeta)d\zeta = \chi(\boldsymbol{\eta}) \quad \forall\boldsymbol{\eta} \in \mathbb{R}^{n-1},\ \forall\psi \in S(\mathbb{R}^n). \tag{8.9.74}$$

Then, $\forall\phi \in S(\mathbb{R}^n)$, the trace $\gamma_j\phi \in S(\mathbb{R}^{n-1})$ and the moment $\mu_j\phi \in S(\mathbb{R}^{n-1})$ are related by:

Proposition 8.9.4. $\forall\phi \in S(\mathbb{R}^n)$,

$$(i2\pi)^j(\mu_j(\mathcal{F}\phi)) = \mathcal{F}[\gamma_j\phi], \tag{8.9.75}$$

where the first Fourier transform $\mathcal{F}\phi$ is in $\mathbb{R}^n$:

$$(\mathcal{F}\phi)(\boldsymbol{\xi}) = \int_{\mathbb{R}^n} \phi(\mathbf{x})e^{-i2\pi\langle\mathbf{x},\boldsymbol{\xi}\rangle}d\mathbf{x},$$

and the second one $\mathcal{F}[\gamma_j\phi]$ is in $\mathbb{R}^{n-1}$:

$$\mathcal{F}[\gamma_j\phi](\boldsymbol{\eta}) = \int_{\mathbb{R}^{n-1}} (\gamma_j\phi)(\mathbf{y})e^{-i2\pi\langle\mathbf{y},\boldsymbol{\eta}\rangle}d\mathbf{y}. \tag{8.9.76}$$

Proof. Let $\psi = \mathcal{F}\phi$ for $\phi \in S(\mathbb{R}^n)$. Then $\phi = \overline{\mathcal{F}}\psi$ with $\phi(\mathbf{x}) = \int_{\mathbb{R}^n} \psi(\boldsymbol{\xi})e^{i2\pi\langle\boldsymbol{\xi},\mathbf{x}\rangle} d\boldsymbol{\xi}$, $\langle\boldsymbol{\xi},\mathbf{x}\rangle = \langle\boldsymbol{\eta},\mathbf{y}\rangle + \zeta z$ for $\boldsymbol{\xi} = (\boldsymbol{\eta},\zeta) \in \mathbb{R}^n$ with $\boldsymbol{\eta} \in \mathbb{R}^{n-1}$, $\zeta \in \mathbb{R}$ and $\mathbf{x} = (\mathbf{y}, z) \in \mathbb{R}^n$ with $\mathbf{y} \in \mathbb{R}^{n-1}$, $z \in \mathbb{R}$.

Then

$$\begin{aligned}
\frac{\partial^j}{\partial z^j}\phi(\mathbf{y},z) &= \int_{\mathbb{R}^n} \psi(\boldsymbol{\xi})\frac{\partial^j}{\partial z^j}[e^{i2\pi[\langle\boldsymbol{\eta},\mathbf{y}\rangle+\zeta z]}]d\boldsymbol{\xi} \\
&= \int_{\mathbb{R}^n} (i2\pi\zeta)^j\psi(\boldsymbol{\eta},\zeta)e^{i2\pi[\langle\boldsymbol{\eta},\mathbf{y}\rangle+\zeta z]}d\boldsymbol{\eta}d\zeta \\
\Longrightarrow \quad (\gamma_j\phi)(\mathbf{y}) &= \frac{\partial^j}{\partial z^j}\phi(\mathbf{y},z)\bigg|_{z=0} \\
&= \int_{\mathbb{R}^n} (i2\pi\zeta)^j\psi(\boldsymbol{\eta},\zeta)e^{i2\pi\langle\boldsymbol{\eta},\mathbf{y}\rangle}d\boldsymbol{\eta}d\zeta \quad (\text{since } e^{i2\pi\zeta\cdot 0} = 1) \\
&= \int_{\mathbb{R}^{n-1}} e^{i2\pi\langle\boldsymbol{\eta},\mathbf{y}\rangle}\left(\int_{\mathbb{R}}(i2\pi\zeta)^j\psi(\boldsymbol{\eta},\zeta)d\zeta\right)d\boldsymbol{\eta} \\
&= (i2\pi)^j\int_{\mathbb{R}^{n-1}} e^{i2\pi\langle\boldsymbol{\eta},\mathbf{y}\rangle}(\mu_j\psi)(\boldsymbol{\eta})d\boldsymbol{\eta} \quad (\text{using (8.9.73)})
\end{aligned}$$

$\Longrightarrow (\gamma_j\phi)(\mathbf{y}) = (i2\pi)^j\overline{\mathcal{F}}[\mu_j\psi](\mathbf{y})$ (using (8.9.76))
$\Longrightarrow \mathcal{F}[\gamma_j\phi](\boldsymbol{\eta}) = (i2\pi)^j\mathcal{F}\overline{\mathcal{F}}[\mu_j\psi](\boldsymbol{\eta})$ (using (8.9.76))
$\Longrightarrow \mathcal{F}[\gamma_j\phi](\boldsymbol{\eta}) = (i2\pi)^j[\mu_j\psi](\boldsymbol{\eta}) = (i2\pi)^j\mu_j(\mathcal{F}\phi)(\boldsymbol{\eta})$, since $\psi = \mathcal{F}\phi$
$\Longrightarrow (i2\pi)^j\mu_j(\mathcal{F}\phi) = \mathcal{F}(\gamma_j\phi)\ \forall\phi \in S(\mathbb{R}^n)$. □

Theorem 8.9.9 (Trace Theorem A). *Let γ_0 be the trace operator from $S(\mathbb{R}^n) \subset H^s(\mathbb{R}^n)$ into $S(\Gamma) \equiv S(\mathbb{R}^{n-1}) \subset H^{s-\frac{1}{2}}(\mathbb{R}^{n-1})$ defined by* (8.9.71).

Then, for $s > \frac{1}{2}$, γ_0 is continuous from $S(\mathbb{R}^n)$ equipped with the norm $\|\cdot\|_{H^s(\mathbb{R}^n)}$ into $S(\Gamma) \equiv S(\mathbb{R}^{n-1})$ equipped with the norm $\|\cdot\|_{H^{s-\frac{1}{2}}(\mathbb{R}^{n-1})}$, i.e., for $s > \frac{1}{2}$, $\exists$ a constant $C > 0$ such that, $\forall\phi \in S(\mathbb{R}^n)$,

$$\|\gamma_0\phi\|_{H^{s-\frac{1}{2}}(\mathbb{R}^{n-1})} \le C\|\phi\|_{H^s(\mathbb{R}^n)}. \tag{8.9.77}$$

Hence, γ_0 can be extended to a unique, continuous, linear *operator from $H^s(\mathbb{R}^n)$ into $H^{s-\frac{1}{2}}(\mathbb{R}^{n-1})$, and this extended, continuous, linear operator will still be denoted by the same notation γ_0, i.e. $\gamma_0 : H^s(\mathbb{R}^n) \longrightarrow H^{s-\frac{1}{2}}(\mathbb{R}^{n-1})$ such that $\forall u \in H^s(\mathbb{R}^n)$, $\gamma_0 u \in H^{s-\frac{1}{2}}(\mathbb{R}^{n-1})$ and*

$$\|\gamma_0 u\|_{H^{s-\frac{1}{2}}(\mathbb{R}^{n-1})} \le C\ \|u\|_{H^s(\mathbb{R}^n)}. \tag{8.9.78}$$

Proof. From the density result (8.9.43), for $s > \frac{1}{2}$, $S(\mathbb{R}^n) \hookrightarrow H^s(\mathbb{R}^n)$ (resp. $S(\mathbb{R}^{n-1}) \hookrightarrow H^{s-\frac{1}{2}}(\mathbb{R}^{n-1})$) with a *dense*, continuous imbedding $\hookrightarrow$. Hence, for $\phi \in S(\mathbb{R}^n) \hookrightarrow H^s(\mathbb{R}^n)$, $\psi = \gamma_0\phi \in S(\mathbb{R}^{n-1}) \hookrightarrow H^{s-\frac{1}{2}}(\mathbb{R}^{n-1})$ for $s > \frac{1}{2}$.

Then, $(\mathcal{F}\phi)(\boldsymbol{\xi}) = \hat{\phi}(\boldsymbol{\xi}) = \hat{\phi}(\boldsymbol{\eta}, \zeta)$ and $\mathcal{F}[\gamma_0(\phi)](\boldsymbol{\eta}) = (\mathcal{F}\psi)(\boldsymbol{\eta}) = \hat{\psi}(\boldsymbol{\eta})$ with $\boldsymbol{\eta} = (\xi_1, \xi_2, \ldots, \xi_{n-1}) \in \mathbb{R}^{n-1}$, $\zeta = \xi_n \in \mathbb{R}$, where the second Fourier transform $\mathcal{F}[\gamma_0(\phi)](\boldsymbol{\eta})$ is defined by (8.9.76). From Proposition 8.9.4, $\mathcal{F}[\gamma_0(\phi)](\boldsymbol{\eta}) = \hat{\psi}(\boldsymbol{\eta}) = (\mu_0(\mathcal{F}\phi))(\boldsymbol{\eta}) = \int_{\mathbb{R}} \hat{\phi}(\boldsymbol{\eta}, \zeta) d\zeta$ by (8.9.73) for $j = 0$.

$$\implies \quad |\hat{\psi}(\boldsymbol{\eta})|^2 \le \left(\int_{\mathbb{R}} |\hat{\phi}(\boldsymbol{\eta}, \zeta)| d\zeta \right)^2 = \left(\int_{\mathbb{R}} \frac{(1 + \|\boldsymbol{\eta}\|^2 + \zeta^2)^{\frac{s}{2}} |\hat{\phi}(\boldsymbol{\eta}, \zeta)|}{(1 + \|\boldsymbol{\eta}\|^2 + \zeta^2)^{\frac{s}{2}}} d\zeta \right)^2,$$

$$\left| \left\langle (1 + \|\boldsymbol{\eta}\|^2 + \zeta^2)^{\frac{s}{2}} |\hat{\phi}(\boldsymbol{\eta}, \zeta)|, \frac{1}{(1 + \|\boldsymbol{\eta}\|^2 + \zeta^2)^{\frac{s}{2}}} \right\rangle_{L^2(\mathbb{R})} \right|^2$$

$$\le \|(1 + \|\boldsymbol{\eta}\|^2 + \zeta^2)^{\frac{s}{2}} |\hat{\phi}(\boldsymbol{\eta}, \zeta)|\|^2_{L^2(\mathbb{R})} \left\| \frac{1}{(1 + \|\boldsymbol{\eta}\|^2 + \zeta^2)^{\frac{s}{2}}} \right\|^2_{L^2(\mathbb{R})}$$

(by the Cauchy–Schwarz inequality)

$$\implies \quad |\hat{\psi}(\boldsymbol{\eta})|^2 \le \int_{\mathbb{R}} (1 + \|\boldsymbol{\eta}\|^2 + \zeta^2)^s |\hat{\phi}(\boldsymbol{\eta}, \zeta)|^2 d\zeta \int_{\mathbb{R}} \frac{d\zeta}{(1 + \|\boldsymbol{\eta}\|^2 + \zeta^2)^s}. \tag{8.9.79}$$

Changing the variable: $\zeta = t\sqrt{1 + \|\boldsymbol{\eta}\|^2}$ with $d\zeta = \sqrt{1 + \|\boldsymbol{\eta}\|^2} dt$ and $(1 + \|\boldsymbol{\eta}\|^2 + \zeta^2) = (1 + \|\boldsymbol{\eta}\|^2)(1 + t^2)$, we evaluate the second integral:

$$\int_{\mathbb{R}} \frac{d\zeta}{(1 + \|\boldsymbol{\eta}\|^2 + \zeta^2)^s} = \int_{\mathbb{R}} \frac{\sqrt{1 + \|\boldsymbol{\eta}\|^2} dt}{(1 + \|\boldsymbol{\eta}\|^2)^s (1 + t^2)^s}$$

$$= \frac{1}{(1 + \|\boldsymbol{\eta}\|^2)^{s - \frac{1}{2}}} \int_{\mathbb{R}} \frac{dt}{(1 + t^2)^s} \le \frac{C_1}{(1 + \|\boldsymbol{\eta}\|^2)^{s - \frac{1}{2}}}, \tag{8.9.80}$$

since $\exists C_1 > 0$ such that $\int_{\mathbb{R}} \frac{dt}{(1+t^2)^s} \le C_1$.

Then, from (8.9.79) and (8.9.80),

$$|\hat{\psi}(\boldsymbol{\eta})|^2 \le \frac{C_1}{(1 + \|\boldsymbol{\eta}\|^2)^{s - \frac{1}{2}}} \int_{\mathbb{R}} (1 + \|\boldsymbol{\eta}\|^2 + \zeta^2)^s |\hat{\phi}(\boldsymbol{\eta}, \zeta)|^2 d\zeta$$

$$\implies \quad (1 + \|\boldsymbol{\eta}\|^2)^{s - \frac{1}{2}} |\hat{\psi}(\boldsymbol{\eta})|^2 \le C_1 \int_{\mathbb{R}} (1 + \|\boldsymbol{\eta}\|^2 + \zeta^2)^s |\hat{\phi}(\boldsymbol{\eta}, \zeta)|^2 d\zeta$$

$$\implies \quad \int_{\mathbb{R}^{n-1}} (1 + \|\boldsymbol{\eta}\|^2)^{s - \frac{1}{2}} |\hat{\psi}(\boldsymbol{\eta})|^2 d\boldsymbol{\eta}$$

$$\le C_1 \int_{\mathbb{R}^{n-1}} \int_{\mathbb{R}} (1 + \|\boldsymbol{\eta}\|^2 + \zeta^2)^s |\hat{\phi}(\boldsymbol{\eta}, \zeta)|^2 d\boldsymbol{\eta} d\zeta$$

$$\implies \quad \|\psi\|^2_{H^{s-\frac{1}{2}}(\mathbb{R}^{n-1})} \le C_1 \int_{\mathbb{R}^n} (1 + \|\boldsymbol{\xi}\|^2)^s |\hat{\phi}(\boldsymbol{\xi})|^2 d\boldsymbol{\xi} = C_1 \|\phi\|^2_{H^s(\mathbb{R}^n)}$$

$$\implies \quad \|\psi\|_{H^{s-\frac{1}{2}}(\mathbb{R}^{n-1})} = \|\gamma_0 \phi\|_{H^{s-\frac{1}{2}}(\mathbb{R}^{n-1})} \le C \|\phi\|_{H^s(\mathbb{R}^n)}$$

with $C = \sqrt{C_1} > 0$.

Since $\gamma_0 : S(\mathbb{R}^n) \subset H^s(\mathbb{R}^n) \longrightarrow S(\mathbb{R}^{n-1}) \hookrightarrow H^{s-\frac{1}{2}}(\mathbb{R}^{n-1})$ is a *continuous, linear* operator from $S(\mathbb{R}^n)$ equipped with the norm $\|\cdot\|_{s,(\mathbb{R}^n)}$ into $H^{s-\frac{1}{2}}(\mathbb{R}^{n-1})$ and $S(\mathbb{R}^n)$ is a *dense* subspace of $H^s(\mathbb{R}^n)$ by (8.9.43), γ_0 can be extended to a *unique, continuous, linear* operator $\tilde{\gamma_0}$ from $H^s(\mathbb{R}^n)$ into $H^{s-\frac{1}{2}}(\mathbb{R}^{n-1})$ satisfying (8.9.78), i.e. $\tilde{\gamma_0}\phi = \gamma_0\phi \ \forall\phi \in S(\mathbb{R}^n)$ and $\|\tilde{\gamma_0}u\|_{H^{s-\frac{1}{2}}(\mathbb{R}^{n-1})} \le C\|u\|_{H^s(\mathbb{R}^n)}$. For the sake of notational simplicity, we agree to denote the *extended* continuous, linear operator $\tilde{\gamma_0}$ by the *same* notation γ_0 by setting $\tilde{\gamma_0} \equiv \gamma_0$ such that (8.9.78) holds. □

Theorem 8.9.10 (Trace Theorem B). *Let $j \in \mathbb{N}$ with $s - j - \frac{1}{2} > 0$. Let γ_j be the trace operator from $S(\mathbb{R}^n)$ into $S(\Gamma) \equiv S(\mathbb{R}^{n-1})$ defined by* (8.9.70)*:*

$$\gamma_j\phi(\mathbf{y}) = \frac{\partial^j}{\partial z^j}\phi(\mathbf{y}, z)\bigg|_{z=0} \qquad \forall\phi \in S(\mathbb{R}^n).$$

Then, $\forall$ fixed $j \in \mathbb{N}$ with $s - j - \frac{1}{2} > 0$, γ_j is a continuous, linear operator from $S(\mathbb{R}^n)$ equipped with the norm $\|\cdot\|_{H^s(\mathbb{R}^n)}$ into $S(\mathbb{R}^{n-1}) \subset H^{s-j-\frac{1}{2}}(\mathbb{R}^{n-1})$ equipped with the norm $\|\cdot\|_{H^{s-j-\frac{1}{2}}(\mathbb{R}^{n-1})}$, i.e. $\exists C > 0$ such that

$$\|\gamma_j\phi\|_{H^{s-j-\frac{1}{2}}(\mathbb{R}^{n-1})} \le C\|\phi\|_{H^s(\mathbb{R}^n)} \quad \forall\phi \in S(\mathbb{R}^n). \tag{8.9.81}$$

Moreover, γ_j can be extended to a unique, continuous, linear operator from $H^s(\mathbb{R}^n)$ into $H^{s-j-\frac{1}{2}}(\mathbb{R}^{n-1})$, and this extended unique, continuous, linear operator will still be denoted by the same notation γ_j, i.e. for $s - j - \frac{1}{2} > 0$, $\gamma_j : H^s(\mathbb{R}^n) \longmapsto H^{s-j-\frac{1}{2}}(\mathbb{R}^{n-1})$ such that $\forall u \in H^s(\mathbb{R}^n)$, $\gamma_j u \in H^{s-j-\frac{1}{2}}(\mathbb{R}^{n-1})$ and

$$\|\gamma_j u\|_{H^{s-j-\frac{1}{2}}(\mathbb{R}^{n-1})} \le C\|u\|_{H^s(\mathbb{R}^n)}. \tag{8.9.82}$$

Proof. Let $j \in \mathbb{N}$ such that $s - j - \frac{1}{2} > 0$. For $\phi \in S(\mathbb{R}^n) \hookrightarrow H^s(\mathbb{R}^n)$ with $s-j-\frac{1}{2} > 0$, set $\psi_j = \gamma_j\phi \in S(\mathbb{R}^{n-1}) \hookrightarrow H^{s-j-\frac{1}{2}}(\mathbb{R}^{n-1})$ with $\psi_j(\mathbf{y}) = \gamma_j\phi(\mathbf{y}) = \frac{\partial^j}{\partial z^j}\phi(\mathbf{y}, z)|_{z=0} \ \forall\mathbf{y} = (x_1, \ldots, x_{n-1}) \in \mathbb{R}^{n-1}$, $z = x_n \in \mathbb{R}$. Then $\mathcal{F}\phi \in S(\mathbb{R}^n)$ with $(\mathcal{F}\phi)(\boldsymbol{\xi}) = (\mathcal{F}\phi)(\boldsymbol{\eta}, \zeta)$, $\boldsymbol{\xi} = (\boldsymbol{\eta}, \zeta) \in (\mathbb{R}^n)$, $\boldsymbol{\eta} = (\xi_1, \xi_2, \ldots, \xi_{n-1}) \in \mathbb{R}^{n-1}$,

$\zeta = \xi_n \in \mathbb{R}$. Then, frcm Proposition 8.9.4 (see also Definition 8.9.6),

$$\begin{aligned}
&\hat{\psi}_j(\boldsymbol{\eta}) = (\mathcal{F}\psi_j)(\boldsymbol{\eta}) = \mathcal{F}[\gamma_j\phi](\boldsymbol{\eta}) = (i2\pi)^j\mu_j(\mathcal{F}\phi)\\
&\qquad = (i2\pi)^j\int_{\mathbb{R}}\zeta^j(\mathcal{F}\phi)(\boldsymbol{\eta},\zeta)d\zeta \quad \text{(by (8.9.73))}\\
&\qquad = (i2\pi)^j\int_{\mathbb{R}}\zeta^j\hat{\phi}(\boldsymbol{\eta},\zeta)d\zeta\\
\Longrightarrow\quad &|\hat{\psi}_j(\boldsymbol{\eta})| \le (2\pi)^j\int_{\mathbb{R}}|\zeta|^j|\hat{\phi}(\boldsymbol{\eta},\zeta)|d\zeta\\
\Longrightarrow\quad &|\hat{\psi}_j(\boldsymbol{\eta})|^2 \le (2\pi)^{2j}\left(\int_{\mathbb{R}}|\zeta|^j|\hat{\phi}(\boldsymbol{\eta},\zeta)|d\zeta\right)^2\\
&\qquad = (2\pi)^{2j}\underbrace{\left(\int_{\mathbb{R}}\frac{(1+\|\boldsymbol{\eta}\|^2+\zeta^2)^{\frac{s}{2}}|\zeta|^j|\hat{\phi}(\boldsymbol{\eta},\zeta)|}{(1+\|\boldsymbol{\eta}\|^2+\zeta^2)^{\frac{s}{2}}}d\zeta\right)^2}_{I_1} = (2\pi)^{2j}I_1
\end{aligned} \tag{8.9.83}$$

with

$$\|\boldsymbol{\eta}\|^2 = \xi_1^2+\xi_2^2+\cdots+\xi_{n-1}^2, \tag{8.9.84}$$

$$\begin{aligned}
I_1 &= \left|\left\langle(1+\|\boldsymbol{\eta}\|^2+\zeta^2)^{\frac{s}{2}}|\hat{\phi}(\boldsymbol{\eta},\zeta)|, \frac{|\zeta|^j}{(1+\|\boldsymbol{\eta}\|^2+\zeta^2)^{\frac{s}{2}}}\right\rangle_{L^2(\mathbb{R})}\right|^2\\
&\le \int_{\mathbb{R}}(1+\|\boldsymbol{\eta}\|^2+\zeta^2)^s|\hat{\phi}(\boldsymbol{\eta},\zeta)|^2d\zeta\cdot\underbrace{\int_{\mathbb{R}}\frac{|\zeta|^{2j}}{(1+\|\boldsymbol{\eta}\|^2+\zeta^2)^s}d\zeta}_{I_2}.
\end{aligned} \tag{8.9.85}$$

(by the Cauchy–Schwarz inequality).

Changing variable: $\zeta = t\sqrt{1+\|\boldsymbol{\eta}\|^2}$, $d\zeta = \sqrt{1+\|\boldsymbol{\eta}\|^2}dt$, we obtain

$$\begin{aligned}
I_2 &= \int_{\mathbb{R}}\frac{t^{2j}(1+\|\boldsymbol{\eta}\|^2)^j\sqrt{1+\|\boldsymbol{\eta}\|^2}dt}{(1+\|\boldsymbol{\eta}\|^2)^s(1+t^2)^s} = \frac{1}{(1+\|\boldsymbol{\eta}\|^2)^{s-j-\frac{1}{2}}}\int_{\mathbb{R}}\frac{t^{2j}}{(1+t^2)^s}dt\\
&\le C_0\frac{1}{(1+\|\boldsymbol{\eta}\|^2)^{s-j-\frac{1}{2}}}
\end{aligned} \tag{8.9.86}$$

(setting $\int_{\mathbb{R}}(\frac{t^2}{1+t^2})^j\frac{1}{(1+t^2)^{s-j}}dt = C_0 > 0$).

Then, from (8.9.83)–(8.9.85), we have, for $\boldsymbol{\xi} = (\boldsymbol{\eta}, \zeta)$ with $\boldsymbol{\eta} = (\xi_1, \xi_2, \cdots, \xi_{n-1}) \in \mathbb{R}^{n-1}$, $\zeta = \xi_n \in \mathbb{R}$ and $\|\boldsymbol{\eta}\|^2 = \xi_1^2 + \xi_2^2 + \cdots + \xi_{n-1}^2$,

$$|\hat{\psi}_j(\boldsymbol{\eta})|^2 \le C_0(2\pi)^{2j}\left(\int_{\mathbb{R}}(1 + \|\boldsymbol{\eta}\|^2 + \zeta^2)^s|\hat{\phi}(\boldsymbol{\eta}, \zeta)|^2 d\zeta\right)\frac{1}{(1 + \|\boldsymbol{\eta}\|^2)^{s-j-\frac{1}{2}}}$$

$$\Longrightarrow \quad \int_{\mathbb{R}^{n-1}}(1 + \|\boldsymbol{\eta}\|^2)^{s-j-\frac{1}{2}}|\hat{\psi}_j(\boldsymbol{\eta})|^2 d\boldsymbol{\eta}$$

$$\le C_0(2\pi)^{2j}\int_{\mathbb{R}^{n-1}}\int_{\mathbb{R}}(1 + \|\boldsymbol{\eta}\|^2 + \zeta^2)^s|\hat{\phi}(\boldsymbol{\eta}, \zeta)|^2 d\zeta d\boldsymbol{\eta}$$

$$\Longrightarrow \quad \|\psi_j\|^2_{H^{s-j-\frac{1}{2}}(\mathbb{R}^{n-1})} \le C_0(2\pi)^{2j}\ \|\phi\|^2_{H^s(\mathbb{R}^n)}$$

$$\Longrightarrow \quad \|\gamma_j\phi\|_{H^{s-j-\frac{1}{2}}(\mathbb{R}^{n-1})} \le C\|\phi\|_{H^s(\mathbb{R}^n)} \tag{8.9.87}$$

with $C = \sqrt{C_0}(2\pi)^{2j} > 0$. Hence, for fixed $j \in \mathbb{N}$ with $s - j - \frac{1}{2} > 0$, the trace operator $\gamma_j : \phi \in S(\mathbb{R}^n) \longmapsto \gamma_j\phi \in S(\mathbb{R}^{n-1})$ is *continuous* and *linear* from $S(\mathbb{R}^n)$ equipped with the norm $\|\cdot\|_{s,(\mathbb{R}^n)}$ into $S(\mathbb{R}^{n-1})$ equipped with the norm $\|\cdot\|_{s-j-\frac{1}{2},\mathbb{R}^{n-1}}$. Since $S(\mathbb{R}^n)$ is a *dense* subspace of $H^s(\mathbb{R}^n)$ and $\gamma_j : S(\mathbb{R}^n) \subset H^s(\mathbb{R}^n) \longrightarrow S(\mathbb{R}^{n-1}) \subset H^{s-j-\frac{1}{2}}(\mathbb{R}^{n-1})$ is a *continuous, linear* operator from $S(\mathbb{R}^n)$ equipped with the norm $\|\cdot\|_{H^s(\mathbb{R}^n)}$ into $H^{s-j-\frac{1}{2}}(\mathbb{R}^{n-1})$ for $j \in \mathbb{N}$ with $s - j - \frac{1}{2} > 0$, γ_j can be extended to a *unique, continuous, linear* operator $\tilde{\gamma}_j$ from $H^s(\mathbb{R}^n)$ into $H^{s-j-\frac{1}{2}}(\mathbb{R}^{n-1})$. □

For the sake of notational simplicity, we will denote this *extended, continuous, linear* trace operator $\tilde{\gamma}_j$ by the *same notation* γ_j by setting $\gamma_j = \tilde{\gamma}_j$, i.e. for $j \in \mathbb{N}$ with $s - j - \frac{1}{2} > 0$, $\gamma_j : H^s(\mathbb{R}^n) \longrightarrow H^{s-j-\frac{1}{2}}(\mathbb{R}^{n-1})$ such that $\forall u \in H^s(\mathbb{R}^n)$, $\gamma_j u \in H^{s-j-\frac{1}{2}}(\mathbb{R}^{n-1})$ with $\|\gamma_j u\|_{H^{s-j-\frac{1}{2}}(\mathbb{R}^{n-1})} \le C\|u\|_{H^s(\mathbb{R}^n)}$.

Surjectivity of the trace operator $\boldsymbol{\gamma_j : H^s(\mathbb{R}^n) \longrightarrow H^{s-j-\frac{1}{2}}(\mathbb{R}^{n-1})}$ For this we need the following results on *partial* or *truncated* Fourier transforms (see also Proposition 8.9.4).

Let $\phi \in S(\mathbb{R}^n)$. Then, $\forall$ fixed $\mathbf{y} = (x_1, \ldots, x_{n-1}) \in \mathbb{R}^{n-1}$ with $(\mathbf{y}, x_n) \in \mathbb{R}^n$, $\forall \phi \in S(\mathbb{R}^n)$,

- $\phi(\mathbf{y}, \cdot) : z = x_n \in \mathbb{R} \longmapsto \phi(\mathbf{y}, z)$ belongs to $S(\mathbb{R})$;
- $\forall$ fixed $z = x_n \in \mathbb{R}$ with $(\mathbf{y}, z) \in \mathbb{R}^n$, $\forall \phi \in S(\mathbb{R}^n)$, $\phi(\cdot, z) : \mathbf{y} = (x_1, \ldots, x_{n-1}) \in \mathbb{R}^{n-1} \longmapsto \phi(\mathbf{y}, z)$ belongs to $S(\mathbb{R}^{n-1})$.

Definition 8.9.6. $\forall \phi \in S(\mathbb{R}^n)$, $\mathcal{F}_1 : \phi(\mathbf{y}, \cdot) \in S(\mathbb{R}) \longmapsto \mathcal{F}_1[\phi(\mathbf{y}, \cdot)] \in S(\mathbb{R})$ defined by

$$(\mathcal{F}_1\phi)(\mathbf{y}, \zeta) = \int_{\mathbb{R}} e^{-i2\pi\zeta\cdot z}\phi(\mathbf{y}, z)dz = \hat{\phi}(\mathbf{y}, \zeta) \tag{8.9.88}$$

$\forall$ fixed $\mathbf{y} \in \mathbb{R}^{n-1}$, $\forall \zeta \in \mathbb{R}$, is the (one-dimensional) *partial* or *truncated* Fourier transform of $\phi(\mathbf{y}, \cdot) \in S(\mathbb{R})$ into $\hat{\phi}(\mathbf{y}, \cdot) \in S(\mathbb{R})$.

$\forall \phi \in S(\mathbb{R}^n)$, $\mathcal{F}_{n-1} : \phi(\cdot, z) \in S(\mathbb{R}^{n-1}) \longmapsto \mathcal{F}_{n-1}[\phi(\cdot, z)] \in S(\mathbb{R}^{n-1})$ defined by

$$(\mathcal{F}_{n-1}\phi)(\boldsymbol{\eta}, z) = \int_{\mathbb{R}^{n-1}} e^{-i2\pi\langle \boldsymbol{\eta}, \mathbf{y}\rangle} \phi(\mathbf{y}, z) d\mathbf{y} dz = \hat{\phi}(\boldsymbol{\eta}, z) \tag{8.9.89}$$

$\forall$ fixed $z \in \mathbb{R}$, $\forall \boldsymbol{\eta} = (\xi_1, \dots, \xi_{n-1}) \in \mathbb{R}^{n-1}$, $\langle \boldsymbol{\eta}, \mathbf{y}\rangle = \xi_1 x_1 + \xi_2 x_2 + \cdots + \xi_{n-1} x_{n-1}$, is the ($(n-1)$-dimensional) *partial* or *truncated* Fourier transform of $\phi(\cdot, z) \in S(\mathbb{R}^{n-1})$ into $\hat{\phi}(\cdot, z) \in S(\mathbb{R}^{n-1})$.

The Plancherel–Parseval Theorem 7.7.3 holds in each case:

$$\|\phi(\mathbf{y}, \cdot)\|_{L^2(\mathbb{R})} = \|\hat{\phi}(\mathbf{y}, \cdot)\|_{L^2(\mathbb{R})} \quad \forall \phi(\mathbf{y}, \cdot) \in S(\mathbb{R}), \tag{8.9.90}$$

$$\|\phi(\cdot, z)\|_{L^2(\mathbb{R}^{n-1})} = \|\hat{\phi}(\cdot, z)\|_{L^2(\mathbb{R}^{n-1})} \quad \forall \phi(\cdot, z) \in S(\mathbb{R}^{n-1}).$$

Then, $\forall \phi \in S(\mathbb{R}^n)$, $\mathcal{F} : \phi \in S(\mathbb{R}^n) \longmapsto \mathcal{F}\phi \in S(\mathbb{R}^n)$ defined by

$$(\mathcal{F}\phi)(\boldsymbol{\eta}, \zeta) = \int_{\mathbb{R}^n} e^{-i2\pi[\langle \boldsymbol{\eta}, \mathbf{y}\rangle + \zeta z]} \phi(\mathbf{y}, z) d\mathbf{y} dz = \hat{\phi}(\boldsymbol{\eta}, \zeta)$$

is related to $\mathcal{F}_1$ and $\mathcal{F}_{n-1}$. Since $S(\mathbb{R}^n) \subset L^1(\mathbb{R}^n)$, we can apply Fubini's Theorem 7.1.2C and we have

$$\begin{aligned} (\mathcal{F}\phi)(\boldsymbol{\eta}, \zeta) &= \int_{\mathbb{R}} e^{-i2\pi\zeta z} \left(\int_{\mathbb{R}^{n-1}} e^{-i2\pi\langle \boldsymbol{\eta}, \mathbf{y}\rangle} \phi(\mathbf{y}, z) d\mathbf{y} \right) dz \\ &= (\mathcal{F}_1 \mathcal{F}_{n-1}\phi)(\boldsymbol{\eta}, \zeta) = (\mathcal{F}_{n-1}\mathcal{F}\phi)(\boldsymbol{\eta}, \zeta) \\ \Longrightarrow \quad \mathcal{F}\phi &= \mathcal{F}_1\mathcal{F}_{n-1}\phi = \mathcal{F}_{n-1}\mathcal{F}_1\phi \quad \forall \phi \in S(\mathbb{R}^n). \end{aligned} \tag{8.9.91}$$

Using the density of $S(\mathbb{R})$ in $L^2(\mathbb{R})$ (resp. $S(\mathbb{R}^{n-1})$ in $L^2(\mathbb{R}^{n-1})$), we can extend the definition of $\mathcal{F}_1$ to $L^2(\mathbb{R})$ (resp. the definition of $\mathcal{F}_{n-1}$ to $L^2(\mathbb{R}^{n-1})$) such that the Plancherel–Riesz Theorem 8.3.1 holds in the corresponding cases: $\forall u \in L^2(\mathbb{R}^n)$ with $u(\mathbf{y}, \cdot) \in L^2(\mathbb{R})$ for *almost all* $\mathbf{y} \in \mathbb{R}^{n-1}$, $u(\cdot, z) \in L^2(\mathbb{R}^{n-1})$ for *almost all* $z \in \mathbb{R}$ such that

$$\mathcal{F}u = \mathcal{F}_1\mathcal{F}_{n-1}u = \mathcal{F}_{n-1}\mathcal{F}_1 u, \tag{8.9.92}$$

$\mathcal{F}_1 u(\mathbf{y}, \zeta) = \hat{u}(\mathbf{y}, \zeta)$ for *almost all* $\mathbf{y} \in \mathbb{R}^{n-1}$ with

$$\|\hat{u}(\mathbf{y}, \cdot)\|_{L^2(\mathbb{R})} = \|u(\mathbf{y}, \cdot)\|_{L^2(\mathbb{R})}; \tag{8.9.93}$$

$\mathcal{F}_{n-1} u(\boldsymbol{\eta}, z) = \hat{u}(\boldsymbol{\eta}, z)$ for *almost all* $z \in \mathbb{R}$ with

$$\|\hat{u}(\cdot, z)\|_{L^2(\mathbb{R}^{n-1})} = \|u(\cdot, z)\|_{L^2(\mathbb{R}^{n-1})}. \tag{8.9.94}$$

Moreover, we will use *spaces of vector-valued functions*, for which we refer to Sections 9.2–9.3, Chapter 9, for more details:

$$L^2(\mathbb{R}, L^2(\mathbb{R}^{n-1})) = \Big\{ v : v : z \in \mathbb{R} \longmapsto v(\cdot, z) \in L^2(\mathbb{R}^{n-1}) \quad \text{such that } \int_{\mathbb{R}} \|v(\cdot, z)\|^2_{L^2(\mathbb{R}^{n-1})} dz < +\infty \Big\}, \tag{8.9.95}$$

$$L^2(\mathbb{R}^{n-1}, L^2(\mathbb{R})) = \Big\{ v : v : \mathbf{y} \in \mathbb{R}^{n-1} \longmapsto v(\mathbf{y}, \cdot) \in L^2(\mathbb{R}) \quad \text{such that } \int_{\mathbb{R}^{n-1}} \|v(\mathbf{y}, \cdot)\|^2_{L^2(\mathbb{R})} d\mathbf{y} < +\infty \Big\} \tag{8.9.96}$$

with the property:

$$L^2(\mathbb{R}^n) \equiv L^2(\mathbb{R}, L^2(\mathbb{R}^{n-1})) \equiv L^2(\mathbb{R}^{n-1}, L^2(\mathbb{R})), \tag{8.9.97}$$

which follow from Fubini's Theorem 7.1.2C.

Moreover, we have:

Lemma 8.9.6. *For $s > 0$, the following* equivalent *properties hold:*

I. $u \in H^s(\mathbb{R}^n)$; (8.9.98)

II. (a) $u \in L^2(\mathbb{R}, H^s(\mathbb{R}^{n-1})) = \{v : z \in \mathbb{R} \longmapsto v(\cdot, z) \in H^s(\mathbb{R}^{n-1})$ *such that* $\int_{\mathbb{R}} \|v(\cdot, z)\|^2_{H^s(\mathbb{R}^{n-1})} dz < +\infty\}$;

(b) $\zeta \in \mathbb{R} \longmapsto |\zeta|^s \mathcal{F}_1 u(\cdot, \zeta) \in L^2(\mathbb{R}, L^2(\mathbb{R}^{n-1}))$; (8.9.99)

III. (a) $u \in L^2(\mathbb{R}, L^2(\mathbb{R}^{n-1}))$;

(b) $(1 + \|\boldsymbol{\eta}\|^2_{\mathbb{R}^{n-1}})^{s/2} \mathcal{F} u \in L^2(\mathbb{R}^n)$;

(c) $|\zeta|^s \mathcal{F} u \in L^2(\mathbb{R}^n)$, (8.9.100)

where $(\boldsymbol{\eta}, \zeta) \in \mathbb{R}^n$ *with* $\boldsymbol{\eta} = (\xi_1, \dots, \xi_{n-1}) \in \mathbb{R}^{n-1}$, $\zeta = \xi_n \in \mathbb{R}$, $\|\boldsymbol{\eta}\|^2 = \xi_1^2 + \xi_2^2 + \dots + \xi_{n-1}^2$.

Proof. We will *only* show that II $\Longrightarrow$ III $\Longrightarrow$ I, which we will need.

II $\Longrightarrow$ III: $u \in L^2(\mathbb{R}, H^s(\mathbb{R}^{n-1}))$ and $H^s(\mathbb{R}^{n-1}) \hookrightarrow L^2(\mathbb{R}^{n-1}) \Longrightarrow u \in L^2(\mathbb{R}, L^2(\mathbb{R}^{n-1})) \Longrightarrow$ III(a) holds. But $u \in L^2(\mathbb{R}, H^s(\mathbb{R}^{n-1})) \Longrightarrow \int_{\mathbb{R}} \|u(\cdot, z)\|^2_{H^s(\mathbb{R}^{n-1})} dz < +\infty$

$$\Longrightarrow \quad \int_{\mathbb{R}} dz \int_{\mathbb{R}^{n-1}} (1 + \|\boldsymbol{\eta}\|^2)^s |\mathcal{F}_{n-1} u(\boldsymbol{\eta}, z)|^2 d\boldsymbol{\eta} < +\infty. \tag{8.9.101}$$

$u \in L^2(\mathbb{R}^n) = L^2(\mathbb{R}, L^2(\mathbb{R}^{n-1}))$ (by (8.9.96)) $\Longrightarrow \mathcal{F}u = \hat{u} \in L^2(\mathbb{R}^n)$ by the Plancherel–Riesz Theorem 8.3.1. Hence, by Fubini's Theorem 7.1.2C,

$$\int_{\mathbb{R}} \|u(\cdot, z)\|^2_{H^s(\mathbb{R}^{n-1})} dz = \int_{\mathbb{R}} \int_{\mathbb{R}^{n-1}} (1 + \|\boldsymbol{\eta}\|^2)^s |\mathcal{F}_{n-1} u(\boldsymbol{\eta}, z)|^2 d\boldsymbol{\eta} dz$$
$$= \int_{\mathbb{R}^{n-1}} (1 + \|\boldsymbol{\eta}\|^2)^s \Big(\int_{\mathbb{R}} |\mathcal{F}_{n-1} u(\boldsymbol{\eta}, z)|^2 dz\Big) d\boldsymbol{\eta}. \tag{8.9.102}$$

But for *almost all* $\boldsymbol{\eta} \in \mathbb{R}^{n-1}$,

$$\int_{\mathbb{R}} |\mathcal{F}_{n-1} u(\boldsymbol{\eta}, z)|^2 dz = \int_{\mathbb{R}} |\mathcal{F}_1(\mathcal{F}_{n-1} u)(\boldsymbol{\eta}, \zeta)|^2 d\zeta \quad \text{(by (8.9.93))}$$
$$= \int_{\mathbb{R}} |\mathcal{F} u(\boldsymbol{\eta}, \zeta)|^2 d\zeta \quad \text{(by (8.9.69))}. \tag{8.9.103}$$

Hence, from (8.9.101)–(8.9.103),

$$\int_{\mathbb{R}} \|u(\cdot, z)\|^2_{H^s(\mathbb{R}^{n-1})} dz = \int_{\mathbb{R}^{n-1}} (1 + \|\boldsymbol{\eta}\|^2)^s d\boldsymbol{\eta} \int_{\mathbb{R}} |\mathcal{F} u(\boldsymbol{\eta}, \zeta)|^2 d\zeta$$
$$= \int_{\mathbb{R}^n} (1 + \|\boldsymbol{\eta}\|^2)^s |\mathcal{F} u(\boldsymbol{\eta}, \zeta)|^2 d\boldsymbol{\eta} d\zeta < +\infty$$

$\Longrightarrow (1 + \|\boldsymbol{\eta}\|^2)^{s/2} \mathcal{F} u \in L^2(\mathbb{R}^n) \Longrightarrow$ III(b) holds.

Again, from II(b), $\int_{\mathbb{R}^n} |\zeta|^{2s} \mathcal{F}_1 u(\mathbf{y}, \zeta)|^2 d\mathbf{y} d\zeta < +\infty$. Then

$$\int_{\mathbb{R}} |\zeta|^{2s} \Big(\int_{\mathbb{R}^{n-1}} |\mathcal{F}_1 u(\mathbf{y}, \zeta)|^2 d\mathbf{y}\Big) d\zeta \quad \text{(by Fubini's Theorem 7.1.2C)}$$
$$= \int_{\mathbb{R}} |\zeta|^{2s} d\zeta \int_{\mathbb{R}^{n-1}} |\mathcal{F}_{n-1}(\mathcal{F}_1 u)(\boldsymbol{\eta}, \zeta)|^2 d\boldsymbol{\eta} \quad \text{(by (8.9.93))}$$
$$= \int_{\mathbb{R}^n} |\zeta|^{2s} |\mathcal{F} u(\boldsymbol{\eta}, \zeta)|^2 d\boldsymbol{\eta} d\zeta < +\infty$$

$\Longrightarrow |\zeta|^s \mathcal{F} u \in L^2(\mathbb{R}^n) \Longrightarrow$ III(c) holds.

III $\Longrightarrow$ I: Let $u \in L^2(\mathbb{R}; L^2(\mathbb{R}^{n-1}))$. Then, $u \in L^2(\mathbb{R}^n)$ by (8.9.97).

Moreover, $\forall a, b > 0$, $\forall s \geq 1$, $(a + b)^s \leq 2^{s-1}(a^s + b^s)$;[3] $\forall s \in [0, 1]$, $(a + b)^s \leq a^s + b^s$.[4]

Hence,

$$(1 + \|\boldsymbol{\eta}\|^2 + \zeta^2)^s \leq 2^{s-1}[(1 + \|\boldsymbol{\eta}\|^2)^s + |\zeta|^{2s}] \quad \forall s \geq 1, \quad \text{or}$$
$$(1 + \|\boldsymbol{\eta}\|^2 + \zeta^2)^s \leq (1 + \|\boldsymbol{\eta}\|^2)^s + |\zeta|^{2s} \quad \text{for } s \in [0, 1].$$

[3]For $s \geq 1$, set $x = b/a \geq 0$ and $f(x) = (1 + x)^s/(1 + x^s)$, with $f(0) = 1 = \lim_{x\to\infty} f(x)$ and $f(x) > 1$ for $0 < x < \infty$. f has only one *critical point* on $]0, \infty[$ at $x = 1$, which is a *maximum*, i.e. $f(1) \geq f(x)$ $\forall x \in [0, \infty[$ with $f(1) = 2^{s-1} \Longrightarrow$ the inequality.

[4]For $0 < s < 1$, $s - 1 < 0 \Longrightarrow 1 - s > 0$ and $\frac{1}{2^{1-s}} \leq f(x) \leq 1$ $\forall x \in [0, \infty[\Longrightarrow$ inequality.

Then, from III(b) and (c), we have $(1 + \|\boldsymbol{\eta}\|^2)^{s/2}\mathcal{F}u \in L^2(\mathbb{R}^n)$ and $|\zeta|^s\mathcal{F}u \in L^2(\mathbb{R}^n)$

$$\Longrightarrow \quad \int_{\mathbb{R}^n} (1 + \|\boldsymbol{\eta}\|^2)^s |\mathcal{F}u(\boldsymbol{\eta},\zeta)|^2 d\boldsymbol{\eta} d\zeta < +\infty,$$

$$\int_{\mathbb{R}^n} |\zeta|^{2s} |\mathcal{F}u(\boldsymbol{\eta},\zeta)|^2 d\boldsymbol{\eta} d\zeta < +\infty$$

$$\Longrightarrow \quad \|u\|^2_{H^s(\mathbb{R}^n)} = \int_{\mathbb{R}^n} (1 + \|\boldsymbol{\eta}\|^2 + |\zeta|^2)^s |\mathcal{F}u(\boldsymbol{\eta},\zeta)|^2 d\boldsymbol{\eta} d\zeta < +\infty$$

$\Longrightarrow u \in H^s(\mathbb{R}^n) \Longrightarrow$ I holds. □

Lemma 8.9.7. *For fixed $j \in \mathbb{N}_0$ with $s - j - 1/2 > 0$, let $g_j \in H^{s-j-1/2}(\mathbb{R}^{n-1})$ and $\phi \in \mathcal{D}(\mathbb{R})$ such that $\phi^{(j)}(0) = 1$ and*

$$\hat{g}_j(\boldsymbol{\eta}) = \mathcal{F}_{n-1}[g_j(\mathbf{y})] = \int_{\mathbb{R}^{n-1}} e^{-i2\pi\langle\boldsymbol{\eta},\mathbf{y}\rangle} g_j(\mathbf{y}) d\mathbf{y},$$

with $\mathbf{y} = (x_1,\ldots,x_{n-1}) \in \mathbb{R}^{n-1}$, $\boldsymbol{\eta} = (\xi_1,\ldots,\xi_{n-1}) \in \mathbb{R}^{n-1}$, $(\mathbf{y},z) \in \mathbb{R}^n, z = x_n \in \mathbb{R}$. Let

$$\hat{v}(\boldsymbol{\eta},z) = (1 + \|\boldsymbol{\eta}\|^2)^{-j/2}\hat{g}_j(\boldsymbol{\eta})\phi(\sqrt{1 + \|\boldsymbol{\eta}\|^2}z) \quad \forall z \in \mathbb{R}, \tag{8.9.104}$$

$\boldsymbol{\eta} = (\xi_1,\ldots,\xi_{n-1}) \in \mathbb{R}^{n-1}$ with $\|\boldsymbol{\eta}\|^2 = \xi_1^2 + \cdots + \xi_{n-1}^2$,

$$\hat{g}_j(\boldsymbol{\eta}) = \mathcal{F}g_j(\boldsymbol{\eta}) = \int_{\mathbb{R}^{n-1}} e^{-i2\pi\langle\boldsymbol{\eta},\mathbf{y}\rangle} g_j(\mathbf{y}) d\mathbf{y}$$

($\hat{g}_j$ is not *a partial Fourier transform).*

Then:

I. *$\hat{v} \in L^2(\mathbb{R}^n)$ and $\exists v \in L^2(\mathbb{R}^n)$ such that*

$$\hat{v}(\boldsymbol{\eta},z) = \mathcal{F}_{n-1}[v(\mathbf{y},z)] = \int_{\mathbb{R}^{n-1}} e^{-i2\pi\langle\boldsymbol{\eta},\mathbf{y}\rangle} v(\mathbf{y},z) d\mathbf{y} \quad \forall z \in \mathbb{R};$$

II. *$v \in H^s(\mathbb{R}^n)$;*

III. *The mapping $g_j \in H^{s-j-1/2}(\mathbb{R}^{n-1}) \longmapsto v \in H^s(\mathbb{R}^n)$ defined by* (8.9.104) *is* continuous *and* linear *from $H^{s-j-1/2}(\mathbb{R}^{n-1})$ into $H^s(\mathbb{R}^n)$.*

Proof. We will show the scheme of the long, complicated proof, and the principal results.

I. Set $t = \sqrt{1 + \|\boldsymbol{\eta}\|^2} z$. Then

$$\int_{\mathbb{R}} |\phi(\sqrt{1 + \|\boldsymbol{\eta}\|^2} z)|^2 dz = \frac{1}{\sqrt{1 + \|\boldsymbol{\eta}\|^2}} \int_{\mathbb{R}} |\phi(t)|^2 dt < +\infty,$$

since $\phi \in \mathcal{D}(\mathbb{R})$.

Then

$$\begin{aligned}
&\int_{\mathbb{R}^n} (1 + \|\boldsymbol{\eta}\|^2)^s |\hat{v}(\boldsymbol{\eta}, z)|^2 d\boldsymbol{\eta} dz \\
&= \int_{\mathbb{R}^n} (1 + \|\boldsymbol{\eta}\|^2)^s \cdot (1 + \|\boldsymbol{\eta}\|^2)^{-j} |\hat{g}_j(\boldsymbol{\eta})|^2 |\phi(\sqrt{1 + \|\boldsymbol{\eta}\|^2} \cdot z)|^2 d\boldsymbol{\eta} dz \\
&= \int_{\mathbb{R}} |\phi(t)|^2 dt \int_{\mathbb{R}^{n-1}} (1 + \|\boldsymbol{\eta}\|^2)^{s-j-1/2} \cdot |\hat{g}_j(\boldsymbol{\eta})|^2 d\boldsymbol{\eta} \\
&= \int_{\mathbb{R}} |\phi(t)|^2 dt \|g_j\|^2_{H^{s-j-1/2}(\mathbb{R}^{n-1})} < +\infty \qquad (8.9.105)
\end{aligned}$$

(by Fubini's Theorem 7.1.2C).

Hence,

$$\int_{\mathbb{R}^n} |\hat{v}(\boldsymbol{\eta}, z)|^2 d\boldsymbol{\eta} dz \le \int_{\mathbb{R}^n} (1 + \|\boldsymbol{\eta}\|^2)^s |\hat{v}(\boldsymbol{\eta}, z)|^2 d\boldsymbol{\eta} dz < +\infty$$

$\Longrightarrow \hat{v} \in L^2(\mathbb{R}^n) \equiv L^2(\mathbb{R}, L^2(\mathbb{R}^{n-1}))$ (by (8.9.97)).

$\Longrightarrow \exists v \in L^2(\mathbb{R}^n) \equiv L^2(\mathbb{R}, L^2(\mathbb{R}^{n-1}))$ with $\hat{v}(\boldsymbol{\eta}, z) = \mathcal{F}_{n-1}[v(\mathbf{y}, z)]$ (by (8.9.94)).

II. Using Fubini's Theorem 7.1.2C repeatedly, we have

$$\int_{\mathbb{R}^n} (1 + \|\boldsymbol{\eta}\|^2)^s |\hat{v}(\boldsymbol{\eta}, z)|^2 d\boldsymbol{\eta} dz = \int_{\mathbb{R}} dz \int_{\mathbb{R}^{n-1}} (1 + \|\boldsymbol{\eta}\|^2)^s |\hat{v}(\boldsymbol{\eta}, z)|^2 d\boldsymbol{\eta} < +\infty$$

(by (8.9.105))

$$\Longrightarrow \quad \|v(\cdot, z)\|^2_{H^s(\mathbb{R}^{n-1})} = \int_{\mathbb{R}^{n-1}} (1 + \|\boldsymbol{\eta}\|^2)^s |\hat{v}(\boldsymbol{\eta}, z)|^2 d\boldsymbol{\eta} < +\infty$$

for *almost all* $z \in \mathbb{R}$ (by Fubini).

Hence, $\forall z \in \mathbb{R}$, $z \in \mathbb{R} \longmapsto \hat{v}(\cdot, z) \in L^2(\mathbb{R}^{n-1}) \subset S'(\mathbb{R}^{n-1})$ and $v(\cdot, z) \in H^s(\mathbb{R}^{n-1})$ $\forall z \in \mathbb{R}$ with

$$\int_{\mathbb{R}^n} (1 + \|\boldsymbol{\eta}\|^2)^s |\hat{v}(\boldsymbol{\eta}, z)| d\boldsymbol{\eta} dz = \int_{\mathbb{R}} \|v(\cdot, z)\|^2_{H^s(\mathbb{R}^{n-1})} dz < +\infty$$

$\Longrightarrow v \in L^2(\mathbb{R}, H^s(\mathbb{R}^{n-1}))$. Then, by II in Lemma 8.9.6, $v \in H^s(\mathbb{R}^n)$ if $\zeta \in \mathbb{R} \longmapsto |\zeta|^s \mathcal{F}_1 v(\cdot, \zeta) \in L^2(\mathbb{R}, L^2(\mathbb{R}^{n-1})) = L^2(\mathbb{R}^n)$ (by (8.9.97)). Hence, to prove this, we are to show that

$$\int_{\mathbb{R}^n} |\zeta|^{2s} |\mathcal{F}_1 v(\mathbf{y}, \zeta)|^2 d\mathbf{y} d\zeta < +\infty. \tag{8.9.106}$$

But

$$\begin{aligned}
\int_{\mathbb{R}^n} |\zeta|^{2s} |\mathcal{F}_1 v(\mathbf{y}, \zeta)|^2 d\mathbf{y} d\zeta &= \int_{\mathbb{R}} |\zeta|^{2s} d\zeta \int_{\mathbb{R}^{n-1}} |\mathcal{F}_1 v(\mathbf{y}, \zeta)|^2 d\mathbf{y} \quad \text{(by Fubini)} \\
&= \int_{\mathbb{R}} |\zeta|^{2s} d\zeta \int_{\mathbb{R}^{n-1}} |\mathcal{F}_{n-1}(\mathcal{F}_1 v)(\boldsymbol{\eta}, \zeta)|^2 d\boldsymbol{\eta} \quad \text{(by (8.9.94))} \\
&= \int_{\mathbb{R}} |\zeta|^{2s} d\zeta \int_{\mathbb{R}^{n-1}} |\mathcal{F}_1(\mathcal{F}_{n-1} v)(\boldsymbol{\eta}, \zeta)|^2 d\boldsymbol{\eta} \quad \text{(by (8.9.92))} \\
&= \int_{\mathbb{R}} |\zeta|^{2s} d\zeta \int_{\mathbb{R}^{n-1}} |(\mathcal{F}_1 \hat{v})(\boldsymbol{\eta}, \zeta)|^2 d\boldsymbol{\eta},
\end{aligned}$$

with $\hat{v}(\boldsymbol{\eta}, z) = \mathcal{F}_{n-1} v(\boldsymbol{\eta}, z)$ defined in (8.9.104), where

$$\begin{aligned}
\mathcal{F}_1 \hat{v} &= \int_{\mathbb{R}} e^{-i2\pi\zeta z} \hat{v}(\boldsymbol{\eta}, z) dz \\
&= \int_{\mathbb{R}} e^{-i2\pi\zeta z} (1 + \|\boldsymbol{\eta}\|^2)^{-j/2} \hat{g}_j(\boldsymbol{\eta}) \phi\left(\sqrt{1 + \|\boldsymbol{\eta}\|^2} z\right) dz \quad \text{(by (8.9.104))} \\
&= (1 + \|\boldsymbol{\eta}\|^2)^{-j/2} \hat{g}_j(\eta) \int_{\mathbb{R}} e^{-i2\pi\zeta z} \phi\left(\sqrt{1 + \|\boldsymbol{\eta}\|^2} z\right) dz.
\end{aligned} \tag{8.9.106a}$$

Hence, from (8.9.106a),

$$\begin{aligned}
&\int_{\mathbb{R}^n} |\zeta|^{2s} |\mathcal{F}_1 v(\mathbf{y}, \zeta)|^2 d\mathbf{y} d\zeta \\
&= \int_{\mathbb{R}} |\zeta|^{2s} d\zeta \\
&\quad \times \int_{\mathbb{R}^{n-1}} \left[(1 + \|\boldsymbol{\eta}\|^2)^{-j} |\hat{g}_j(\boldsymbol{\eta})|^2 | \underbrace{\int_{\mathbb{R}} e^{-i2\pi\zeta z} \phi\left(\sqrt{1 + \|\boldsymbol{\eta}\|^2} z\right) dz}_{I} |^2 \right] d\boldsymbol{\eta}.
\end{aligned} \tag{8.9.107}$$

$$\begin{aligned}
I &= \frac{1}{\sqrt{1 + \|\boldsymbol{\eta}\|^2}} \int_{\mathbb{R}} e^{-i2\pi\zeta t/\sqrt{1+\|\boldsymbol{\eta}\|^2}} \phi(t) dt \quad \left(\text{setting } t = \sqrt{1 + \|\boldsymbol{\eta}\|^2} z\right) \\
\Longrightarrow \quad |I|^2 &= \frac{1}{1 + \|\boldsymbol{\eta}\|^2} |\mathcal{F}\phi(\zeta/\sqrt{1 + \|\boldsymbol{\eta}\|^2})|^2.
\end{aligned}$$

Since $\phi \in \mathcal{D}(\mathbb{R}) \Longrightarrow \phi \in S(\mathbb{R}) \Longrightarrow \mathcal{F}\phi(\tau) = \int_{\mathbb{R}} e^{-i2\pi t\tau}\phi(t)dt$ with $\mathcal{F}\phi \in S(\mathbb{R})$ by Theorem 7.7.2 $\Longrightarrow (1+\tau^2)^{s/2}\mathcal{F}\phi \in S(\mathbb{R})$ by (7.2.17) $\Longrightarrow$ $(1+\tau^2)^{s/2}\mathcal{F}\phi \in L^2(\mathbb{R})$ by Theorem 7.7.3

$$\Longrightarrow \quad \int_{\mathbb{R}} (1+\tau^2)^s |\mathcal{F}\phi(\tau)|^2 d\tau < +\infty. \tag{8.9.108}$$

Set $\tau = \frac{\zeta}{\sqrt{1+\|\boldsymbol{\eta}\|^2}}$. Then $d\zeta = (\sqrt{1+\|\boldsymbol{\eta}\|^2})d\tau$, $|\zeta|^{2s} = (1+\|\boldsymbol{\eta}\|^2)^s \tau^{2s}$. $\tau^{2s} \le (1+\tau^2)^s \Longrightarrow \frac{\tau^{2s}}{(1+\tau^2)^s} \le 1$ and $|I|^2 = \frac{1}{1+\|\boldsymbol{\eta}\|^2}|\mathcal{F}\phi(\tau)|^2$. Hence, from (8.9.107),

$$\begin{aligned}
&\int_{\mathbb{R}^n} |\zeta|^{2s} |\mathcal{F}_1 v(\mathbf{y}, \zeta)|^2 d\mathbf{y} d\zeta \\
&= \int_{\mathbb{R}} (1+\|\boldsymbol{\eta}\|^2)^s \tau^{2s} \sqrt{1+\|\boldsymbol{\eta}\|^2} d\tau \\
&\times \int_{\mathbb{R}^{n-1}} (1+\|\boldsymbol{\eta}\|^2)^{-j} |\hat{g}_j(\boldsymbol{\eta})|^2 \frac{1}{1+\|\boldsymbol{\eta}\|^2} |\mathcal{F}\phi(\tau)|^2 d\boldsymbol{\eta} \\
&= \int_{\mathbb{R}^n} (1+\|\boldsymbol{\eta}\|^2)^{s-j-1/2} |\hat{g}_j(\boldsymbol{\eta})|^2 \\
&\quad \cdot \frac{\tau^{2s}}{(1+\tau^2)^s}(1+\tau^2)^s . |\mathcal{F}\phi(\tau)|^2 d\boldsymbol{\eta} d\tau \quad \text{(by Fubini)} \\
&\le \int_{\mathbb{R}^n} (1+\|\boldsymbol{\eta}\|^2)^{s-j-1/2} |\hat{g}_j(\boldsymbol{\eta})|^2 ((1+\tau^2)^s \cdot |\mathcal{F}\phi(\tau)|^2) d\boldsymbol{\eta} d\tau \\
&= \int_{\mathbb{R}} ((1+\tau^2)^s \cdot |\mathcal{F}\phi(\tau)|^2) d\tau \\
&\quad \times \int_{\mathbb{R}^{n-1}} (1+|\boldsymbol{\eta}\|^2)^{s-j-1/2} |\hat{g}_j(\eta)|^2 d\boldsymbol{\eta} \quad \text{(by Fubini)} \\
&= \int_{\mathbb{R}} ((1+\tau^2)^s \cdot |\mathcal{F}\phi(\tau)|^2) d\tau \cdot \|g_j\|^2_{H^{s-j-1/2}(\mathbb{R}^{n-1})} < +\infty \text{ (using (8.9.108))}.
\end{aligned} \tag{8.9.109}$$

Hence, from (8.9.106) and (8.9.109), we have $v \in H^s(\mathbb{R}^n)$, which has been constructed by (8.9.104).

III. Moreover,

$$\begin{aligned}
\|v\|^2_{H^s(\mathbb{R}^n)} &= \int_{\mathbb{R}^n} (1+\|\boldsymbol{\eta}\|^2+\zeta^2)^s |\mathcal{F}v(\boldsymbol{\eta},\zeta)|^2 d\boldsymbol{\eta} d\zeta \qquad (8.9.110) \\
&\le C_1 \Big[\int_{\mathbb{R}^n} (1+\|\boldsymbol{\eta}\|^2)^s |\mathcal{F}v(\boldsymbol{\eta},\zeta)|^2 d\boldsymbol{\eta} d\zeta + \int_{\mathbb{R}^n} |\zeta|^{2s} |\mathcal{F}v(\boldsymbol{\eta},\zeta)|^2 d\boldsymbol{\eta} d\zeta \Big]
\end{aligned}$$

with $C_1 = \max\{1, 2^{s-1}\} > 0$.

Repeatedly applying Fubini's Theorem 7.1.2C, we have

$$\int_{\mathbb{R}^n} (1+\|\boldsymbol{\eta}\|^2)^s |\mathcal{F}v(\boldsymbol{\eta},\zeta)|^2 d\boldsymbol{\eta} d\zeta = \int_{\mathbb{R}^{n-1}} (1+\|\boldsymbol{\eta}\|^2)^s d\boldsymbol{\eta} \int_{\mathbb{R}} |\mathcal{F}v(\boldsymbol{\eta},\zeta)|^2 d\zeta$$
$$= \int_{\mathbb{R}^{n-1}} (1+\|\boldsymbol{\eta}\|^2)^s d\boldsymbol{\eta} \int_{\mathbb{R}} |\mathcal{F}_1[(\mathcal{F}_{n-1}v)(\boldsymbol{\eta},z)]|^2 d\zeta \quad \text{(by (8.9.92))}$$
$$= \int_{\mathbb{R}^{n-1}} (1+\|\boldsymbol{\eta}\|^2)^s d\boldsymbol{\eta} \int_{\mathbb{R}} |\mathcal{F}_{n-1}v(\boldsymbol{\eta},z)]|^2 d\zeta$$
$$\text{(by the Plancherel–Riesz Theorem 8.3.1 for } \mathcal{F}_1)$$
$$= \int_{\mathbb{R}^{n-1}} (1+\|\boldsymbol{\eta}\|^2)^s d\boldsymbol{\eta} \int_{\mathbb{R}} |\hat{v}(\boldsymbol{\eta},z)|^2 dz$$
$$= \int_{\mathbb{R}} |\phi(t)|^2 dt \|g_j\|^2_{H^{s-j-1/2}(\mathbb{R}^{n-1})} \quad \text{(by (8.9.105))} \tag{8.9.111}$$
$$\int_{\mathbb{R}^n} |\zeta|^{2s} |\mathcal{F}v(\boldsymbol{\eta},\zeta)|^2 d\boldsymbol{\eta} d\zeta = \int_{\mathbb{R}^n} |\zeta|^{2s} |\mathcal{F}_1[\mathcal{F}_{n-1}v](\boldsymbol{\eta},\zeta)|^2 d\boldsymbol{\eta} d\zeta$$
$$= \int_{\mathbb{R}} |\zeta|^{2s} d\zeta \int_{\mathbb{R}^{n-1}} |(\mathcal{F}_1 v)(\mathbf{y},\zeta)|^2 d\mathbf{y}$$
$$\text{(by Fubini and then by the Plancherel–Riesz Theorem 8.3.1 for } \mathcal{F}_{n-1})$$
$$\leq \int_{\mathbb{R}} (1+\tau^2)^s |\mathcal{F}\phi(\tau)|^2 d\tau \|g_j\|^2_{H^{s-j-1/2}(\mathbb{R}^{n-1})} \quad \text{(by (8.9.109)).} \tag{8.9.112}$$

Combining (8.9.111) and (8.9.112), we get, from (8.9.110),

$$\|v\|^2_{H^s(\mathbb{R}^n)} \leq C_1 \Big[\int_{\mathbb{R}} |\phi(t)|^2 dt + \int_{\mathbb{R}} (1+\tau^2)^s |\mathcal{F}\phi(\tau)|^2 d\tau\Big] \cdot \|g_j\|^2_{H^{s-j-1/2}(\mathbb{R}^{n-1})} \tag{8.9.113}$$

$\Longrightarrow \|v\|_{H^s(\mathbb{R}^n)} \leq C\|g_j\|_{H^{s-j-1/2}(\mathbb{R}^{n-1})}$ with $C > 0$ and independent of g_j and v.

$\Longrightarrow$ The mapping $g_j \in H^{s-j-1/2}(\mathbb{R}^{n-1}) \longmapsto v \in H^s(\mathbb{R}^n)$ defined by (8.9.104) is *continuous* and *linear* from $H^{s-j-1/2}(\mathbb{R}^{n-1})$ into $H^s(\mathbb{R}^n)$.

□

Proposition 8.9.5. *For fixed $j \in \mathbb{N}_0$ with $s-j-1/2 > 0$, let $g_j \in H^{s-j-1/2}(\mathbb{R}^{n-1})$ and $\gamma_j : H^s(\mathbb{R}^n) \longrightarrow H^{s-j-1/2}(\mathbb{R}^{n-1})$ be the trace operator γ_j defined in Theorem* 8.9.10. *Then,*

I. *$\exists v \in H^s(\mathbb{R}^n)$ defined by* (8.9.104) *such that $\gamma_j v = g_j$ in $H^{s-j-1/2}(\mathbb{R}^{n-1})$, i.e. the trace operator γ_j is a continuous, linear* surjection *from $H^s(\mathbb{R}^n)$ onto $H^{s-j-1/2}(\mathbb{R}^{n-1})$;*

II. *$\exists$ a continuous, linear operator $\rho_j : H^{s-j-1/2}(\mathbb{R}^{n-1}) \longrightarrow H^s(\mathbb{R}^n)$ defined by: $\forall g_j \in H^{s-j-1/2}(\mathbb{R}^{n-1})$, $\rho_j g_j \in H^s(\mathbb{R}^n)$ such that $\gamma_k(\rho_j g_j) = \delta_{kj} g_j$, where the Kronecker $\delta_{kj} = 1$ for $k = j$ and $\delta_{kj} = 0$ for $k \neq j$, $0 \le k \le m$, $0 \le m < s - \frac{1}{2}$.*

ρ_j is called the right-hand inverse *of γ_j;* (8.9.114)

III. *The trace operator $\boldsymbol{\gamma} = (\gamma_0, \gamma_1, \ldots, \gamma_m) : H^s(\mathbb{R}^n) \longrightarrow \prod_{j=0}^m H^{s-j-1/2}(\mathbb{R}^{n-1})$ is* surjective, *and the right-hand inverse $\boldsymbol{\rho}$ of $\boldsymbol{\gamma}$ defined by $\rho : \prod_{j=0}^m H^{s-j-1/2}(\mathbb{R}^{n-1}) \longrightarrow H^s(\mathbb{R}^n)$ is* continuous *for $s - m - 1/2 > 0$.*

Proof.

I. By Lemma 8.9.7, for $g_j \in H^{s-j-1/2}(\mathbb{R}^{n-1})$, $\exists v \in H^s(\mathbb{R}^n)$ defined by (8.9.104). Then, for *almost all* $\boldsymbol{\eta} \in \mathbb{R}^{n-1}$, $\forall z \in \mathbb{R}$,

$$\frac{\partial^j \hat{v}}{\partial z^j}(\boldsymbol{\eta}, z)\downarrow_{z=0} = \frac{\partial^j \hat{v}}{\partial z^j}(\boldsymbol{\eta}, 0) = (1 + \|\boldsymbol{\eta}\|^2)^{-j/2}\hat{g}_j(\boldsymbol{\eta})(1 + \|\boldsymbol{\eta}\|)^{j/2}\phi^{(j)}(0) = \hat{g}_j(\boldsymbol{\eta}), \tag{8.9.115}$$

since $\phi^{(j)}(0) = 1$ in (8.9.104). The proof will be complete if we can show that

$$\frac{\partial^j \hat{v}}{\partial z^j}(\cdot, 0) = \widehat{\frac{\partial^j v}{\partial z^j}}, \tag{8.9.116}$$

where $\widehat{\frac{\partial^j v}{\partial z^j}}(\boldsymbol{\eta}, z) = \mathscr{F}_{n-1}(\frac{\partial^j v}{\partial z^j})(\boldsymbol{\eta}, z)$ and $\widehat{\frac{\partial^j v}{\partial z^j}}(\cdot, 0) = \mathscr{F}_{n-1}[\frac{\partial^j v}{\partial z^j}](\cdot, 0)$, since $\widehat{\frac{\partial^j v}{\partial z^j}}(\cdot, 0) = \frac{\partial^j \hat{v}}{\partial z^j}(\cdot, 0) = \hat{g}_j \in L^2(\mathbb{R}^{n-1}) \Longrightarrow \frac{\partial^j v}{\partial z^j}(\cdot, 0) = g_j$

$$\Longrightarrow \quad \gamma_j v = \frac{\partial^j v}{\partial z^j}(\cdot, 0) = g_j \in H^{s-j-1/2}(\mathbb{R}^{n-1}) \hookrightarrow L^2(\mathbb{R}^{n-1}), \tag{8.9.117}$$

$\Longrightarrow \widehat{\gamma_j v} = \hat{g}_j$ is well defined in $L^2(\mathbb{R}^{n-1})$.

Step 1: $\forall \phi \in S(\mathbb{R}^n)$, $\frac{\partial^j \hat{\phi}}{\partial z^j}(\cdot, 0) = \widehat{\frac{\partial^j \phi}{\partial z^j}}(\cdot, 0)$. In fact, $\forall \phi \in S(\mathbb{R}^n)$,

$$\mathscr{F}_1\left[\frac{\partial^j \phi}{\partial z^j}\right](\mathbf{y}, \zeta) = \int_{\mathbb{R}} e^{-i2\pi\zeta z}\frac{\partial^j \phi}{\partial z^j}(\mathbf{y}, z)dz = (i2\pi\zeta)^j(\mathscr{F}_1\phi)(\mathbf{y}, \zeta)$$

(by integrating by parts). Since $\overline{\mathscr{F}_1}\mathscr{F}_1[\frac{\partial^j \phi}{\partial z^j}](\mathbf{y}, z) = \frac{\partial^j \phi}{\partial z^j}(\mathbf{y}, z)$, $\overline{\mathscr{F}_1}$ being *the one-dimensional partial Fourier co-transform* defined by (7.1.3), (8.3.2),

$$\frac{\partial^j \phi}{\partial z^j}(\mathbf{y}, z) = \overline{\mathscr{F}_1}[(i2\pi\zeta)^j(\mathscr{F}_1\phi)(\mathbf{y}, \zeta)] = \int_{\mathbb{R}} e^{i2\pi z\zeta}(i2\pi\zeta)^j(\mathscr{F}_1\phi)(\mathbf{y}, \zeta)d\zeta.$$

Then

$$\mathcal{F}_{n-1}\left[\frac{\partial^j \phi}{\partial z^j}\right](\boldsymbol{\eta}, z) = \int_{\mathbb{R}^{n-1}} e^{-i2\pi\langle \boldsymbol{\eta},\mathbf{y}\rangle} \frac{\partial^j \phi}{\partial z^j}(\mathbf{y}, z)d\mathbf{y}$$

$$= \int_{\mathbb{R}^{n-1}} e^{-i2\pi\langle \boldsymbol{\eta},\mathbf{y}\rangle}\left(\int_{\mathbb{R}} (i2\pi\zeta)^j e^{i2\pi z\zeta}(\mathcal{F}_1\phi)(\mathbf{y}, \zeta)d\zeta\right)d\mathbf{y}$$

$$= (i2\pi)^j \int_{\mathbb{R}} e^{i2\pi z\zeta}\zeta^j d\zeta \int_{\mathbb{R}^{n-1}} e^{-i2\pi\langle \boldsymbol{\eta},\mathbf{y}\rangle}(\mathcal{F}_1\phi)(\mathbf{y}, \zeta)d\mathbf{y}$$

(by Fubini, since $S(\mathbb{R}^n) \subset L^1(\mathbb{R}^n)$)

$$= (i2\pi)^j \int_{\mathbb{R}} e^{i2\pi z\zeta}\zeta^j (\mathcal{F}_{n-1}\mathcal{F}_1\phi)(\boldsymbol{\eta}, \zeta)d\zeta = \int_{\mathbb{R}} e^{i2\pi z\zeta}(i2\pi\zeta)^j(\mathcal{F}\phi)(\boldsymbol{\eta}, \zeta)d\zeta \tag{8.9.118}$$

$$\implies \quad \widehat{\frac{\partial^j \phi}{\partial z^j}}(\boldsymbol{\eta}, 0) = \mathcal{F}_{n-1}\left[\frac{\partial^j \phi}{\partial z^j}\right](\boldsymbol{\eta}, 0) = \int_{\mathbb{R}} (i2\pi\zeta)^j \mathcal{F}\phi(\boldsymbol{\eta}, \zeta)d\zeta \ \forall \eta \in \mathbb{R}^{n-1}. \tag{8.9.119}$$

For $j = 0$,

$$\hat{\phi}(\boldsymbol{\eta}, z) = (\mathcal{F}_{n-1}\phi)(\boldsymbol{\eta}, z) = \int_{\mathbb{R}} e^{i2\pi z\zeta}(\mathcal{F}\phi)(\boldsymbol{\eta}, \zeta)d\zeta \quad \text{(by (8.9.117))}$$

$$\implies \quad \frac{\partial^j \hat{\phi}}{\partial z^j}(\boldsymbol{\eta}, z) = \int_{\mathbb{R}} \frac{\partial^j}{\partial z^j}[e^{i2\pi z\zeta}(\mathcal{F}\phi)(\boldsymbol{\eta}, \zeta)]d\zeta$$

$$= \int_{\mathbb{R}} (i2\pi\zeta)^j e^{i2\pi z\zeta}(\mathcal{F}\phi)(\boldsymbol{\eta}, \zeta)d\zeta$$

$$\implies \quad \frac{\partial^j \hat{\phi}}{\partial z^j}(\boldsymbol{\eta}, 0) = \frac{\partial^j \hat{\phi}}{\partial z^j}(\boldsymbol{\eta}, z)\downarrow_{z=0} = \int_{\mathbb{R}} (i2\pi\zeta)^j \mathcal{F}\phi(\boldsymbol{\eta}, \zeta)d\zeta \quad \forall \boldsymbol{\eta} \in \mathbb{R}^{n-1}. \tag{8.9.120}$$

Then, from (8.9.118) and (8.9.119), $\widehat{\frac{\partial^j \phi}{\partial z^j}}(\cdot, 0) = \frac{\partial^j \hat{\phi}}{\partial z^j}(\cdot, 0)\ \forall \phi \in S(\mathbb{R}^n)$.

Step 2: $\gamma_j v = g_j \in H^{s-j-1/2}(\mathbb{R}^{n-1})$. By Theorem 8.9.6, $S(\mathbb{R}^{n-1})$ is *dense* in $H^{s-j-1/2}(\mathbb{R}^{n-1})$. Hence, for $g_j \in H^{s-j-1/2}(\mathbb{R}^{n-1})$, $\exists$ a sequence (ψ_k) with $\psi_k = \psi_{jk}\ \forall k \in \mathbb{N}$ in $S(\mathbb{R}^{n-1})$ such that $\psi_k \to g_j$ in $H^{s-j-1/2}(\mathbb{R}^{n-1}) \hookrightarrow L^2(\mathbb{R}^{n-1})$ and $\mathcal{F}\psi_k = \hat{\psi}_k \to \hat{g}_j$ in $L^2(\mathbb{R}^{n-1})$ as $k \to \infty$. (8.9.121)

Since $S(\mathbb{R}^{n-1}) \subset H^{s-j-1/2}(\mathbb{R}^{n-1})$, by Lemma 8.9.7, $\forall k \in \mathbb{N}, \exists \chi_k \in H^s(\mathbb{R}^n)$ defined by (8.9.104) in terms $\hat{\psi}_k \in S(\mathbb{R}^{n-1})$:

$$\hat{\chi}_k(\boldsymbol{\eta}, z) = (1 + \|\boldsymbol{\eta}\|^2)^{-j/2}\hat{\psi}_k(\boldsymbol{\eta})\phi(1\sqrt{1 + \|\boldsymbol{\eta}\|^2}z) \tag{8.9.122}$$

with $\phi \in \mathcal{D}(\mathbb{R})$, $\phi^{(j)}(0) = 1$ and

$$\frac{\partial^j \hat{\chi}_k}{\partial z^j}(\boldsymbol{\eta}, 0) = \hat{\psi}_k(\boldsymbol{\eta}). \tag{8.9.123}$$

Then, by III of Lemma 8.9.7, $\psi_k \to g_j$ in $H^{s-j-1/2}(\mathbb{R}^{n-1})$ as $k \to \infty \Longrightarrow \chi_k \to v$ in $H^s(\mathbb{R}^n)$. Then $\gamma_j \chi_k \to \gamma_j v$ in $H^{s-j-1/2}(\mathbb{R}^{n-1}) \hookrightarrow L^2(\mathbb{R}^{n-1})$ by Theorem 8.9.10 as $k \to \infty$. Hence,

$$\widehat{\gamma_j v} = \lim_{k\to\infty} \widehat{\gamma_j \chi_k} \quad \text{in } L^2(\mathbb{R}^{n-1})$$

$$\Longrightarrow \quad \widehat{\gamma_j v} = \widehat{\frac{\partial^j v}{\partial z^j}}(\cdot, 0) = \lim_{k\to\infty} \widehat{\frac{\partial^j \chi_k}{\partial z^j}}(\cdot, 0) \quad \text{in } L^2(\mathbb{R}^{n-1}).$$

But $\frac{\partial^j \hat{\chi}_k}{\partial z^j}(\boldsymbol{\eta}, 0) = \hat{\psi}_k(\boldsymbol{\eta})$ (by (8.9.123) and (8.9.115)) $\Longrightarrow \frac{\partial^j \hat{\chi}_k}{\partial z^j}(\cdot, 0) = \hat{\psi}_k(\cdot)$ $\forall k \in \mathbb{N}$. From (8.9.120),

$$\begin{aligned}
\hat{g}_j &= \lim_{k\to\infty} \hat{\psi}_k = \lim_{k\to\infty} \frac{\partial^j \hat{\chi}_k}{\partial z^j}(\cdot, 0) = \lim_{k\to\infty} \widehat{\frac{\partial^j \chi_k}{\partial z^j}}(\cdot, 0) \quad \text{(from Step 1)} \\
&= \widehat{\frac{\partial^j v}{\partial z^j}}(\cdot, 0) \quad \text{(by (8.9.117))} \\
&= \widehat{\gamma_j v} \quad \Longrightarrow \quad \gamma_j v = g_j.
\end{aligned}$$

II. For $g_j \in H^{s-j-1/2}(\mathbb{R}^{n-1})$ with $0 \le j \le m$, $s - m - 1/2 > 0$, and the corresponding $v \in H^s(\mathbb{R}^n)$ defined by (8.9.104),

$$\hat{v}(\boldsymbol{\eta}, z) = (\mathcal{F}_{n-1} v)(\boldsymbol{\eta}, z) = (1 + \|\boldsymbol{\eta}\|^2)^{-j/2} \hat{g}_j(\boldsymbol{\eta}) \phi(\sqrt{1 + \|\boldsymbol{\eta}\|^2} z) \quad \forall z \in \mathbb{R}$$

with $\phi^{(j)}(0) = 1$, for fixed j with $0 \le j \le m$, define

$$\rho_j g_j(\mathbf{y}, z) = \sum_{i=1}^{m+1} \lambda_i v(\mathbf{y}, iz) \; \forall z \in \mathbb{R}$$

and for *almost all* $\mathbf{y} = (x_1, \dots, x_{n-1}) \in \mathbb{R}^{n-1}$, $\forall z = x_n \in \mathbb{R}$, $(\lambda_i)_{i=1}^{m+1}$ being real numbers to be made precise later.

Then, by III of Lemma 8.9.7, $\rho_j g_j \in H^s(\mathbb{R}^n)$ and consequently, $0 \le k \le m$, $\gamma_k(\rho_j g_j) \in H^{s-j-1/2}(\mathbb{R}^{n-1})$ by Theorem 8.9.10, and

$$\begin{aligned}
\gamma_k(\rho_j g_j)(\mathbf{y}) &= \frac{\partial^k}{\partial z^k}(\rho_j g_j)(\mathbf{y}, 0) = \sum_{i=1}^{m+1} \lambda_i \frac{\partial^k v}{\partial z^k}(\mathbf{y}, iz).i^k|_{z=0} \\
&= \sum_{i=1}^{m+1} \lambda_i i^k \frac{\partial^k v}{\partial z^k}(\mathbf{y}, 0) = \Big(\sum_{i=1}^{m+1} \lambda_i i^k\Big) \gamma_k v(\mathbf{y})
\end{aligned}$$

for *almost all* $\mathbf{y} \in \mathbb{R}^{n-1}$

$$\Longrightarrow \quad \gamma_k(\rho_j g_j) = \Big(\sum_{i=1}^{m+1} \lambda_i i^k\Big)\gamma_k v \quad \forall 0 \le k \le m. \tag{8.9.124}$$

We can choose *uniquely* $(\lambda_i)_{i=1}^{m+1}$ such that $\sum_{i=1}^{m+1} i^k\lambda_i = \delta_{kj}$, $0 \le k \le m$, $0 \le j \le m$, since for *fixed* j with $0 \le j \le m$, the system of $(m+1)$ linear equations for $(\lambda_i)_{i=1}^{m+1}$: $1^k\lambda_1 + 2^k\lambda_2 + \cdots + j^k\lambda_j + \cdots + (m+1)^k\lambda_{m+1} = \delta_{kj}$ for $0 \le k \le m$, or in matrix form, for *fixed* j with $0 \le j \le m$,

$$\begin{pmatrix} 1^0 & 2^0 & \cdots & (m+1)^0 \\ 1^1 & 2^1 & \cdots & (m+1)^1 \\ \vdots & \vdots & & \vdots \\ 1^j & 2^j & \cdots & (m+1)^j \\ \vdots & \vdots & & \vdots \\ 1^m & 2^m & \cdots & (m+1)^m \end{pmatrix} \begin{pmatrix} \lambda_1 \\ \lambda_2 \\ \vdots \\ \lambda_j \\ \vdots \\ \lambda_{m+1} \end{pmatrix} = \begin{pmatrix} \delta_{0j} \\ \delta_{1j} \\ \vdots \\ \delta_{jj} \\ \vdots \\ \delta_{mj} \end{pmatrix} = \begin{pmatrix} 0 \\ 0 \\ \vdots \\ 1 \\ \vdots \\ 0 \end{pmatrix}$$

with $\delta_{kj} = 0$ for $k \ne j$, $\delta_{jj} = 1$, is a Vandermonde system of equations, which has a unique solution $(\lambda_i)_{i=1}^{m+1}$. For this choice of solution $(\lambda_i)_{i=1}^{m+1}$, from (8.9.124), we have: for *fixed* j with $0 \le j \le m$, $\gamma_k(\rho_j g_j) = \delta_{kj}\gamma_k v = \delta_{kj} g_k \Longrightarrow \gamma_j\rho_j g_j = g_j \Longrightarrow \rho_j : H^{s-j-1/2}(\mathbb{R}^{n-1}) \longrightarrow H^s(\mathbb{R}^n)$ is the *right-hand inverse* of $\gamma_j : H^s(\mathbb{R}^n) \longrightarrow H^{s-j-1/2}(\mathbb{R}^{n-1})$.

The *continuity* of ρ_j follows from III in Lemma 8.9.7.

III. Let $\mathbf{g} = (g_k)_{k=0}^m = (g_0, g_1, \ldots, g_m) \in \prod_{j=0}^m H^{s-j-1/2}(\mathbb{R}^{n-1})$.

Let $\rho_j : H^{s-j-1/2}(\mathbb{R}^{n-1}) \longrightarrow H^s(\mathbb{R}^n)$ be the right-hand inverse of $\gamma_j : H^s(\mathbb{R}^n) \longrightarrow H^{s-j-1/2}(\mathbb{R}^{n-1})$ defined in II. Set $\boldsymbol{\rho}\mathbf{g} = \sum_{j=0}^m \rho_j g_j \in H^s(\mathbb{R}^n)$ with $\rho_j g_j \in H^s(\mathbb{R}^n)$ $\forall j = 0, \ldots, m$. Then $\boldsymbol{\rho}\mathbf{g} \in H^s(\mathbb{R}^n)$. Hence, for fixed k with $0 \le k \le m$,

$$\gamma_k(\boldsymbol{\rho}\mathbf{g}) = \gamma_k \sum_{j=0}^m \rho_j g_j = \sum_{j=0}^m \gamma_k\rho_j g_j = \sum_{j=0}^m \delta_{kj} g_j = g_k$$

$$\Longrightarrow \quad \boldsymbol{\gamma}(\boldsymbol{\rho}\mathbf{g}) = (\gamma_k(\boldsymbol{\rho}\mathbf{g}))_{k=0}^m = (g_k)_{k=0}^m = \mathbf{g} \in \prod_{j=0}^m H^{s-j-1/2}(\mathbb{R}^{n-1}). \ \square$$

Finally, we have:

Theorem 8.9.11 (Trace Theorem C). *Let $j \in \mathbb{N}_0$ with $0 \le j \le m$, $s - m - \frac{1}{2} > 0$. Then,*

I. *the mapping $\gamma_j : H^s(\mathbb{R}^n) \longrightarrow H^{s-j-\frac{1}{2}}(\mathbb{R}^{n-1})$ is a* surjection *from $H^s(\mathbb{R}^n)$ onto $H^{s-j-1/2}(\mathbb{R}^{n-1})$;* (8.9.125)

II. $\exists$ *a continuous, linear operator* ρ_j *(i.e. the* right-hand inverse *of* γ_j*):* $\rho_j : H^{s-j-\frac{1}{2}}(\mathbb{R}^{n-1}) \longrightarrow H^s(\mathbb{R}^n)$ *such that* $\forall g_j \in H^{s-j-\frac{1}{2}}(\mathbb{R}^{n-1})$, $\gamma_j(\rho_j g_j) = g_j$ *and* $\gamma_k(\rho_j g_j) = \delta_{kj} g_j$, $k = 0, 1, \ldots, m$; (8.9.126)

III. $\boldsymbol{\gamma} = (\gamma_0, \gamma_1, \cdots, \gamma_m) : H^s(\mathbb{R}^n) \longrightarrow \prod_{j=0}^m H^{s-j-\frac{1}{2}}(\mathbb{R}^{n-1})$ *is a continuous, linear, surjection from* $H^s(\mathbb{R}^n)$ *onto* $\prod_{j=0}^m H^{s-j-\frac{1}{2}}(\mathbb{R}^{n-1})$ *with* $s - m - \frac{1}{2} > 0$. (8.9.127)

Moreover, we have the following basic results which will be used later in Section 8.14 in the study of traces for $\Omega \subsetneq \mathbb{R}^n$.

Theorem 8.9.12. $\forall s > 0$, *let* $u \in H^s(\mathbb{R}^n)$. *Then we have:*

I. *For almost all* $\mathbf{y} = (x_1, x_2, \ldots, x_{n-1}) \in \mathbb{R}^{n-1}$ *with* $(\mathbf{y}, z) \in \mathbb{R}^n$, $z = x_n \in \mathbb{R}$, $u(\mathbf{y}, \cdot) \in H^s(\mathbb{R})$ *and for almost all* $z = x_n \in \mathbb{R}$, $u(\cdot, z) \in H^s(\mathbb{R}^{n-1})$; (8.9.128)

II. *For* $z = x_n \in \mathbb{R}$, $\forall j = 0, 1, \ldots, m$ *with* $s - m - 1/2 > 0$*:*
- $\frac{\partial^j u}{\partial z^j}(\cdot, z) \in H^{s-j-1/2}(\mathbb{R}^{n-1})$;
- $\exists C > 0$, *independent of* z *and* u, *such that* $\|\frac{\partial^j u}{\partial z^j}(\cdot, z)\|_{H^{s-j-1/2}(\mathbb{R}^{n-1})} \le C\|u\|_{H^s(\mathbb{R}^n)}$; (8.9.129)

III. *The mapping* $z = x_n \in \mathbb{R} \longmapsto \frac{\partial^j u}{\partial z^j}(\cdot, z) \in H^{s-j-1/2}(\mathbb{R}^{n-1})$ *is* continuous *from* $\mathbb{R}$ *into* $H^{s-j-1/2}(\mathbb{R}^{n-1})$. (8.9.130)
In particular, by virtue of this continuity, $z \to 0$ *in* $\mathbb{R}$ *with* $z > 0$

$$\Longrightarrow \quad \lim_{z \to 0+} \frac{\partial^j u}{\partial z^j}(\cdot, z) = \frac{\partial^j u}{\partial z^j}(\cdot, 0) = \gamma_j u \in H^{s-j-1/2}(\mathbb{R}^{n-1}). \quad (8.9.131)$$

See also [12, 15] *for more details.*

Example 8.9.3. For $n = 2$, let $u : \mathbb{R}^2 \longrightarrow \mathbb{R}$ be defined by:

$$u(x_1, x_2) = \begin{cases} \ln|\ln(x_1^2 + x_2^2)^{1/2}| & \text{for } \|\mathbf{x}\| = (x_1^2 + x_2^2)^{1/2} < 1/e \\ 0 & \text{for } \|\mathbf{x}\| \ge 1/e. \end{cases}$$

Then the trace $\gamma_0 u = u\downarrow_{\mathbb{R}} = u_1 \in H^{1/2}(\mathbb{R})$ with

$$u_1(x_1) = \begin{cases} \ln|\ln|x_1|| & \text{for } |x_1| < 1/e \\ 0 & \text{for } |x_1| \ge 1/e. \end{cases}$$

Solution. From Example 2.15.2 $u \in H^1(\mathbb{R}^2)$ (see also (2.3.24)). Then, for $n = 2$, applying Trace Theorem A (Theorem 8.9.9), we have $u_1 = u\downarrow_{\mathbb{R}} = \gamma_0 u \in H^{1-1/2}(\mathbb{R})$, i.e. $u_1 \in H^{1/2}(\mathbb{R})$, since for $j = 0$, $\gamma_0 : H^1(\mathbb{R}^2) \longrightarrow H^{1-1/2}(\mathbb{R}) = H^{1/2}(\mathbb{R})$.

8.10 Sobolev spaces on $\Omega \neq \mathbb{R}^n$ revisited

We have already studied some elementary properties of Sobolev spaces $H^m(\Omega)$ of integral order $m \in \mathbb{N}$ on *arbitrary* domains $\Omega \subset \mathbb{R}^n$ in Section 2.15 of Chapter 2, and in the previous section of this chapter we have defined Sobolev spaces $H^s(\mathbb{R}^n)$ of arbitrary real order $s \in \mathbb{R}$ by means of Fourier transforms. Then, in Theorem 8.9.2, we have shown that for $s = m \in \mathbb{N}$, these two different definitions give the *same* Sobolev space $H^m(\mathbb{R}^n)$ with *norm equivalence*. Now we study the case of $\Omega \neq \mathbb{R}^n$. We again have *two* different definitions of Sobolev spaces on $\Omega \neq \mathbb{R}^n$ for $s = m \in \mathbb{N}$:

- Sobolev space $H^m(\overline{\Omega})$ as the space of restrictions to Ω of the (*equivalent classes* of) elements of $H^m(\mathbb{R}^n)$;
- Sobolev space $H^m(\Omega)$ defined in Section 2.15 for $\Omega \subset \mathbb{R}^n$, i.e.

$$H^m(\Omega) = \{u : \partial^{\boldsymbol{\alpha}} u \in L^2(\Omega) \quad \forall |\boldsymbol{\alpha}| \leq m\}$$

with

$$\|u\|_{H^m(\Omega)} = \Big(\sum_{0 \leq |\boldsymbol{\alpha}| \leq m} \|\partial^{\boldsymbol{\alpha}} u\|^2_{L^2(\Omega)} \Big)^{1/2}.$$

Then the natural question arises:

Is $H^m(\overline{\Omega}) \equiv H^m(\Omega)$ for $\Omega \neq \mathbb{R}^n$, $m \in \mathbb{N}$? The answer is *not* an affirmative one, i.e. *in general*, $H^m(\overline{\Omega}) \not\equiv H^m(\Omega)$. But for some particular cases (for example, when $\Omega = \mathbb{R}^n_+$, Ω is an angular sector, Ω is a polygonal domain, or Ω is a C^m-regular domain, Ω has a Lipschitz continuous boundary, etc.), $H^m(\overline{\Omega}) \equiv H^m(\Omega)$ with norm equivalence.

For this we first define $H^s(\overline{\Omega})$.

8.10.1 Space $H^s(\overline{\Omega})$ with $s \in \mathbb{R}$, $\Omega \subsetneq \mathbb{R}^n$

Definition 8.10.1. Let $\Omega \subsetneq \mathbb{R}^n$ be an open subset of $\mathbb{R}^n$ with boundary Γ. Then, by means of $H^s(\mathbb{R}^n)$ defined in (8.9.1)–(8.9.3), we define $H^s(\overline{\Omega})$: $\forall s \in \mathbb{R}$,

$$H^s(\overline{\Omega}) = \{u : u \in \mathcal{D}'(\Omega) \text{ and } \exists w \in H^s(\mathbb{R}^n) \text{ such that } u = w{\downarrow_\Omega}\} \tag{8.10.1}$$

equipped with the norm $\|\cdot\|_{H^s(\overline{\Omega})}$

$$\|u\|_{H^s(\overline{\Omega})} = \inf_{\substack{w \in H^s(\mathbb{R}^n) \\ w\downarrow_\Omega = u}} \|w\|_{H^s(\mathbb{R}^n)}, \tag{8.10.2}$$

where

$$\|w\|_{H^s(\mathbb{R}^n)} = \|(1 + \|\boldsymbol{\xi}\|^2)^{s/2} \hat{w}\|_{L^2(\mathbb{R}^n)} \tag{8.10.3}$$

with $\hat{w} = \mathcal{F} w$ = Fourier transform of $w \in H^s(\mathbb{R}^n)$, $w = w(\mathbf{x})$, $\hat{w} = \hat{w}(\boldsymbol{\xi})$.

If we let $\Omega \equiv \mathbb{R}^n$, trivially

$$H^s(\overline{\mathbb{R}^n}) \equiv H^s(\mathbb{R}^n). \tag{8.10.4}$$

Important properties of $H^s(\overline{\Omega})$

Property 1

Proposition 8.10.1.

I. $\forall s \in \mathbb{R}$, $H^s(\overline{\Omega})$ *defined by* (8.10.1) *and* (8.10.2) *is a* Hilbert space. (8.10.5)

II. *For* $s \geq 0$, $H^s(\overline{\Omega}) \hookrightarrow L^2(\Omega)$. (8.10.6)

Proof.

I. $H^s(\overline{\Omega})$ is, obviously, a linear space. Define a subspace $\mathcal{M}$ of $H^s(\mathbb{R}^n)$ by:

$$\mathcal{M} = \{w : w \in H^s(\mathbb{R}^n) \text{ such that } w{\downarrow_\Omega} = 0\}. \tag{8.10.7}$$

$\mathcal{M}$ is a closed subspace of $H^s(\mathbb{R}^n)$: Let $(w_k)_{k=1}^\infty$ be a sequence in $\mathcal{M}$ such that $w_k \to w \in H^s(\mathbb{R}^n)$ as $k \to \infty$. We are to show that $w \in \mathcal{M}$. Since $H^s(\mathbb{R}^n) \hookrightarrow \mathcal{S}'(\mathbb{R}^n) \hookrightarrow \mathcal{D}'(\mathbb{R}^n)$, $w_k \to w$ in $H^s(\mathbb{R}^n) \Longrightarrow w_k \to w$ in $\mathcal{D}'(\mathbb{R}^n) \Longrightarrow \langle w_k, \phi\rangle_{\mathcal{D}'(\mathbb{R}^n)\times\mathcal{D}(\mathbb{R}^n)} \to \langle w, \phi\rangle_{\mathcal{D}'(\mathbb{R}^n)\times\mathcal{D}(\mathbb{R}^n)}$ as $k \to \infty$ $\forall \phi \in \mathcal{D}(\mathbb{R}^n)$. But $w_k \in \mathcal{M}$ $\forall k \in \mathbb{N} \Longrightarrow w_k{\downarrow_\Omega} = 0 \Longrightarrow \langle w_k{\downarrow_\Omega}, \phi\rangle = 0$ $\forall \phi \in \mathcal{D}(\Omega)$ $\forall k \in \mathbb{N}$ i.e. $\forall \phi \in \mathcal{D}(\mathbb{R}^n)$ with $\operatorname{supp}(\phi) \subset\subset \Omega$ (see Definition 5.3.1) $\Longrightarrow \lim_{k\to\infty}\langle w_k{\downarrow_\Omega}, \phi\rangle = 0 = \langle w{\downarrow_\Omega}, \phi\rangle$ $\forall \phi \in \mathcal{D}(\Omega)$, i.e. $\forall \phi \in \mathcal{D}(\mathbb{R}^n)$ with $\operatorname{supp}(\phi) \subset\subset \Omega \Longrightarrow w{\downarrow_\Omega} = 0$ in $\mathcal{D}'(\Omega)$ with $w \in H^s(\mathbb{R}^n) \Longrightarrow w \in H^s(\mathbb{R}^n)$ with $w{\downarrow_\Omega} = 0 \Longrightarrow w \in \mathcal{M} \Longrightarrow \mathcal{M}$ is a closed subspace of $H^s(\mathbb{R}^n)$.

Hence, $\mathcal{M}$ is a Hilbert space. Then $H^s(\overline{\Omega})$ can be identified with the *quotient space* $H^s(\mathbb{R}^n)/\mathcal{M}$ equipped with the norm

$$\|u\|_{H^s(\overline{\Omega})} = \inf_{\substack{w\in H^s(\mathbb{R}^n)\\ w\downarrow_\Omega = u}} \|w\|_{H^s(\mathbb{R}^n)} = \inf_{v\in\mathcal{M}} \|w_0 + v\|_{H^s(\mathbb{R}^n)} \tag{8.10.8}$$

for any fixed $w_0 \in H^s(\mathbb{R}^n)$ with $w_0{\downarrow_\Omega} = u$. Hence, $H^s(\overline{\Omega})$ is a Hilbert space.

II. From Remark 8.9.1, $H^s(\mathbb{R}^n) \hookrightarrow L^2(\mathbb{R}^n)$ $\forall s \geq 0$, we get the continuous imbedding $H^s(\overline{\Omega}) \hookrightarrow L^2(\Omega)$. In fact, $u \in H^s(\overline{\Omega}) \Longrightarrow \exists w \in H^s(\mathbb{R}^n)$ with $w{\downarrow_\Omega} = u \Longrightarrow w \in L^2(\mathbb{R}^n)$ with $w{\downarrow_\Omega} = u \in L^2(\Omega)$, since $\int_\Omega |u(\mathbf{x})|^2 d\mathbf{x} \leq \int_{\mathbb{R}^n} |w(\mathbf{x})|^2 d\mathbf{x} < +\infty \Longrightarrow \|u\|^2_{L^2(\Omega)} \leq \|w\|^2_{L^2(\mathbb{R}^n)} \Longrightarrow \|u\|_{L^2(\Omega)} \leq \|w\|_{L^2(\mathbb{R}^n)} \leq \|w\|_{H^s(\mathbb{R}^n)}$

$$\Longrightarrow \quad \|u\|_{L^2(\Omega)} \leq \inf_{\substack{w\in H^s(\mathbb{R}^n)\\ w\downarrow_\Omega = u}} \|w\|_{H^s(\mathbb{R}^n)} = \|u\|_{H^s(\overline{\Omega})}.$$

□

Property 2 For $s = m$, $\Omega = \mathbb{R}^n$,

$$H^m(\overline{\mathbb{R}^n}) \equiv H^m(\mathbb{R}^n), \tag{8.10.9}$$

where $H^m(\mathbb{R}^n) = \{u : u \in L^2(\mathbb{R}^n), \partial^{\boldsymbol{\alpha}} u \in L^2(\mathbb{R}^n)\ \forall |\boldsymbol{\alpha}| \leq m\}$ ((2.15.1)–(2.15.3)) with

$$\|u\|_{m,\mathbb{R}^n} = \Big(\sum_{0 \leq |\boldsymbol{\alpha}| \leq m} \int_{\mathbb{R}^n} |\partial^{\boldsymbol{\alpha}} u(\mathbf{x})|^2 d\mathbf{x} \Big)^{1/2} \quad \forall u \in H^m(\mathbb{R}^n). \tag{8.10.10}$$

The result follows from (8.10.4) with $s = m$ and Theorem 8.9.2, the norms (8.10.3) and (8.10.10) being *equivalent* ones.

Property 3 For $m \in \mathbb{N}$, $\Omega \subsetneq \mathbb{R}^n$, we define $H^m(\overline{\Omega})$ and $H^m(\Omega)$ by

$$H^m(\overline{\Omega}) = \{u : u \in \mathcal{D}'(\Omega) \text{ and } \exists w \in H^m(\mathbb{R}^n) \text{ (8.10.10) such that } u = w\!\downarrow_{\Omega}\}, \tag{8.10.11}$$

with

$$\|u\|_{H^m(\overline{\Omega})} = \inf_{\substack{w \in H^m(\mathbb{R}^n) \\ w\downarrow_{\Omega} = u}} \|w\|_{H^m(\mathbb{R}^n)}; \tag{8.10.12}$$

$H^m(\Omega) = \{u : u \in L^2(\Omega), \partial^{\boldsymbol{\alpha}} u \in L^2(\Omega)\ \forall |\boldsymbol{\alpha}| \leq m\}$, with

$$\|u\|_{m,\Omega} = \Big(\sum_{0 \leq |\boldsymbol{\alpha}| \leq m} \int_{\Omega} |\partial^{\boldsymbol{\alpha}} u(\mathbf{x})|^2 d\mathbf{x} \Big)^{1/2} \quad \forall u \in H^m(\Omega). \tag{8.10.13}$$

Then we have:

Proposition 8.10.2. *$H^m(\overline{\Omega}) \hookrightarrow H^m(\Omega)$, $\hookrightarrow$ being a continuous imbedding.*

Proof. Let $u \in H^m(\overline{\Omega})$ with $m \geq 1$. Then, by the definition of $H^m(\overline{\Omega})$, $\exists w \in H^m(\mathbb{R}^n)$ such that $w\!\downarrow_{\Omega} = u \in L^2(\Omega)$ (by (8.10.6)). But $w \in H^m(\mathbb{R}^n) \Longrightarrow \forall |\boldsymbol{\alpha}| \leq m, \partial^{\boldsymbol{\alpha}} w \in L^2(\mathbb{R}^n)$

$$\Longrightarrow \quad (\partial^{\boldsymbol{\alpha}} w)\!\downarrow_{\Omega} = \partial^{\boldsymbol{\alpha}}(w\!\downarrow_{\Omega}) \in L^2(\Omega). \tag{8.10.14}$$

In fact, $\forall \phi \in \mathcal{D}(\mathbb{R}^n)$ with $\operatorname{supp}(\phi) \subset\subset \Omega$, i.e. $\phi \in \mathcal{D}(\Omega)$,

$$\langle \partial^{\boldsymbol{\alpha}} w, \phi \rangle_{\mathcal{D}'(\mathbb{R}^n)\times\mathcal{D}(\mathbb{R}^n)} = \int_{\mathbb{R}^n} (\partial^{\boldsymbol{\alpha}} w)\phi d\mathbf{x} = \int_{\Omega} (\partial^{\boldsymbol{\alpha}} w)\phi d\mathbf{x} = (-1)^{|\boldsymbol{\alpha}|} \int_{\Omega} w \partial^{\boldsymbol{\alpha}} \phi d\mathbf{x}$$
$$= (-1)^{|\boldsymbol{\alpha}|} \langle w\!\downarrow_{\Omega}, \partial^{\boldsymbol{\alpha}} \phi \rangle = \langle \partial^{\boldsymbol{\alpha}}(w\!\downarrow_{\Omega}), \partial^{\boldsymbol{\alpha}} \phi \rangle_{\mathcal{D}'(\Omega)\times\mathcal{D}(\Omega)} \quad \forall \phi \in \mathcal{D}(\Omega). \tag{8.10.15}$$

But $\langle(\partial^{\boldsymbol{\alpha}} w)\downarrow_\Omega, \phi\rangle = \langle\partial^{\boldsymbol{\alpha}} w, \phi\rangle_{\mathcal{D}'(\mathbb{R}^n)\times\mathcal{D}(\mathbb{R}^n)}$ $\forall\phi \in \mathcal{D}(\mathbb{R}^n)$ with $\text{supp}(\phi) \subset\subset \Omega$
$\Longrightarrow \langle(\partial^{\boldsymbol{\alpha}} w)\downarrow_\Omega, \phi\rangle_{\mathcal{D}'(\Omega)\times\mathcal{D}(\Omega)} = \langle\partial^{\boldsymbol{\alpha}}(w\downarrow_\Omega), \phi\rangle_{\mathcal{D}'(\Omega)\times\mathcal{D}(\Omega)}$ $\forall\phi \in \mathcal{D}(\Omega)$ (using (8.10.15))
$\Longrightarrow \partial^{\boldsymbol{\alpha}}(w\downarrow_\Omega) = (\partial^{\boldsymbol{\alpha}} w)\downarrow_\Omega \in L^2(\Omega)$ $\forall|\boldsymbol{\alpha}| \leq m$ (by (8.10.14))
$\Longrightarrow \partial^{\boldsymbol{\alpha}} u \in L^2(\Omega) \forall|\boldsymbol{\alpha}| \leq m \Longrightarrow u \in H^m(\Omega) \Longrightarrow H^m(\overline{\Omega}) \subset H^m(\Omega)$.

Continuity of $\hookrightarrow$: Let $u \in H^m(\overline{\Omega})$ with $\|u\|_{H^m(\overline{\Omega})}$ given by (8.10.12). Then $u \in H^m(\Omega)$ and

$$\|u\|^2_{m,\Omega} = \sum_{0\leq|\boldsymbol{\alpha}|\leq m} \int_\Omega |\partial^{\boldsymbol{\alpha}} u|^2 d\mathbf{x} = \sum_{0\leq|\boldsymbol{\alpha}|\leq m} |\partial^{\boldsymbol{\alpha}}(w\downarrow_\Omega)|^2 d\mathbf{x}$$

$$= \sum_{0\leq|\boldsymbol{\alpha}|\leq m} \int_\Omega ((\partial^{\boldsymbol{\alpha}} w)\downarrow_\Omega)^2 d\mathbf{x} \leq \sum_{0\leq|\boldsymbol{\alpha}|\leq m} \int_{\mathbb{R}^n} |\partial^{\boldsymbol{\alpha}} w|^2 d\mathbf{x} = \|w\|^2_{m,\mathbb{R}^n}$$

$$\Longrightarrow \quad \|u\|_{m,\Omega} \leq \|w\|_{m,\mathbb{R}^n} \forall w \in H^m(\mathbb{R}^n) \quad \text{with } w\downarrow_\Omega = u$$

$$\Longrightarrow \quad \|u\|_{m,\Omega} \leq \inf_{\substack{w\in H^m(\mathbb{R}^n)\\ w\downarrow_\Omega = u}} \|w\|_{m,\mathbb{R}^n} = \|u\|_{H^m(\overline{\Omega})} \quad \text{(by (8.10.10))} \tag{8.10.16}$$

$$\Longrightarrow \quad \hookrightarrow \text{ is continuous from } H^m(\overline{\Omega}) \text{ into } H^m(\Omega). \tag{8.10.17}$$

□

Remark 8.10.1. If we use the original norm $\|w\|_{m,\mathbb{R}^n}$ defined in (8.10.3) instead of that in (8.10.10), then (8.10.16) will be replaced by: $\|u\|_{m,\Omega} \leq C\|u\|_{H^m(\overline{\Omega})}$ for some $C > 0$ by virtue of Theorem 8.9.2.

Property 4: Density result

Space $\mathcal{D}(\overline{\Omega})$

Definition 8.10.2. Let $\Omega \subsetneq \mathbb{R}^n$ be an open subset of $\mathbb{R}^n$. Then $\mathcal{D}(\overline{\Omega})$ is the linear space of restrictions to Ω of elements of $\mathcal{D}(\mathbb{R}^n)$, i.e.

$$\mathcal{D}(\overline{\Omega}) = \{\psi : \exists\phi \in \mathcal{D}(\mathbb{R}^n) \text{ such that } \phi\downarrow_\Omega = \psi\}. \tag{8.10.18}$$

Theorem 8.10.1. $\forall s \in \mathbb{R}$, $\mathcal{D}(\overline{\Omega})$ *is* dense *in* $H^s(\overline{\Omega})$.

Proof. Let $u \in H^s(\overline{\Omega})$. Then $\exists v \in H^s(\mathbb{R}^n)$ such that $v\downarrow_\Omega = u$. But $\mathcal{D}(\mathbb{R}^n)$ is dense in $H^s(\mathbb{R}^n)$ by Theorem 8.9.6. Hence, $\exists$ a sequence $(\phi_m)_{m=1}^\infty$ in $\mathcal{D}(\mathbb{R}^n)$ such that $\|v - \phi_m\|_{H^s(\mathbb{R}^n)} \to 0$ as $m \to \infty$. Define a sequence $(\psi_m)_{m=1}^\infty$ in $\mathcal{D}(\overline{\Omega})$ by $\psi_m = \phi_m\downarrow_\Omega$ $\forall\, m \in \mathbb{N}$. Then $\psi_m \in H^s(\overline{\Omega})$ $\forall\, m \in \mathbb{N}$ and

$$\|u - \psi_m\|_{H^s(\overline{\Omega})} \leq \|v - \phi_m\|_{H^s(\mathbb{R}^n)} \to 0 \quad \text{as } m \to \infty. \tag{8.10.19}$$

Hence, for any $u \in H^s(\overline{\Omega})$, $\exists$ a sequence $(\psi_m)_{m=1}^\infty$ in $\mathcal{D}(\overline{\Omega})$ such that $\|u-\psi_m\|_{H^s(\overline{\Omega})} \to 0$ as $m \to \infty$, i.e. $\mathcal{D}(\overline{\Omega})$ is *dense* in $H^s(\overline{\Omega})$. □

Property 5: Imbedding results

Proposition 8.10.3.

I. *For* $s \in \mathbb{R}$, $s - \frac{n}{2} > k$, $k \in \mathbb{N}_0$, $H^s(\overline{\Omega}) \hookrightarrow C^k(\overline{\Omega})$. (8.10.20)

II. *For* $s \in \mathbb{R}$, $s - \frac{n}{2} = \lambda \in]0,1[$, $H^s(\overline{\Omega}) \hookrightarrow C^{0,\lambda}(\overline{\Omega})$. (8.10.21)

Proof.

I. $\forall s \in \mathbb{R}$ with $s - \frac{n}{2} > k$, from Sobolev's Imbedding Theorem 8.9.4, $u \in H^s(\mathbb{R}^n) \Longrightarrow u \in C^k(\mathbb{R}^n)$ with $\|u\|_{C^k(\mathbb{R}^n)} \le C\|u\|_{H^s(\mathbb{R}^n)}$, i.e. $H^s(\mathbb{R}^n) \hookrightarrow C^k(\mathbb{R}^n)$ equipped with *uniform* convergence on *every compact* subset of $\mathbb{R}^n$ for all derivatives of order $\le k$, $H^s(\mathbb{R}^n)$ being the space defined in (8.9.1)–(8.9.3). Hence, by virtue of Definition 8.10.1, $H^s(\overline{\Omega}) \hookrightarrow C^k(\overline{\Omega})$, since $C^k(\overline{\Omega})$ is the space of restrictions to Ω of elements of $C^k(\mathbb{R}^n)$ equipped with *uniform* convergence on *every compact* subset of $\mathbb{R}^n$ for all derivatives of order $\le k$.

II. Similarly, from Theorem 8.9.5, the result (8.10.21) follows.

□

Property 6 For $m \in \mathbb{N}$ and $\Omega \subsetneq \mathbb{R}^n$, $H^m(\overline{\Omega}) \ne H^m(\Omega)$ in general, $H^m(\Omega)$ being the space defined in (8.10.13). We repeat again Property 2: $H^m(\overline{\mathbb{R}^n}) \equiv H^m(\mathbb{R}^n)$.

Now, we will identify the additional property, called the *m-extension property*, which $\Omega \subsetneq \mathbb{R}^n$ must possess so that $H^m(\overline{\Omega}) \equiv H^m(\Omega)$.

8.10.2 *m*-extension property of Ω

Definition 8.10.3. Let $\Omega \subsetneq \mathbb{R}^n$ be an open subset of $\mathbb{R}^n$. Then Ω is said to possess the *m-extension property* if and only if there exists a *continuous, linear extension operator* $P : H^m(\Omega) \longrightarrow H^m(\mathbb{R}^n)$ such that $(Pu)(\mathbf{x}) = u(\mathbf{x})$ a.e. on Ω $\forall u \in H^m(\Omega)$. (8.10.22)

Continuity of P: $\exists C > 0$ such that

$$\|Pu\|_{m,\mathbb{R}^n} \le C\|u\|_{m,\Omega} \quad \forall u \in H^m(\Omega)). \tag{8.10.23}$$

m-extension operator P *for* Ω: The continuous, linear extension operator $P : H^m(\Omega) \longrightarrow H^m(\mathbb{R}^n)$ satisfying (8.10.22) and (8.10.23) is called an *m-extension operator* for Ω.

Theorem 8.10.2. *If* Ω *possesses the m-extension property satisfying* (8.10.22) *and* (8.10.23), *then* $H^m(\overline{\Omega}) \equiv H^m(\Omega)$ *with norm equivalence:* (8.10.24)

$\exists C_1, C_2 > 0$ *such that*

$$C_1\|u\|_{H^m(\overline{\Omega})} \le \|u\|_{H^m(\Omega)} \le C_2\|u\|_{H^m(\overline{\Omega})}, \tag{8.10.25}$$

where $H^m(\overline{\Omega}) = \{u : \exists v \in H^m(\mathbb{R}^n)$ *such that* $u = v{\downarrow_\Omega}\}$ *(see* (8.10.1)*) with*

$$\|u\|_{H^m(\overline{\Omega})} = \inf_{\substack{v \in H^m(\mathbb{R}^n) \\ v\downarrow_\Omega = u}} \|v\|_{H^m(\mathbb{R}^n)} \quad \forall u \in H^m(\overline{\Omega}).$$

Proof.
$H^m(\overline{\Omega}) = H^m(\Omega)$: It has already been shown in Proposition 8.10.2 that $H^m(\overline{\Omega}) \hookrightarrow H^m(\Omega)$. Hence, for the equality $H^m(\overline{\Omega}) \equiv H^m(\Omega)$, we are to show only that $H^m(\Omega) \subset H^m(\overline{\Omega})$. Let $u \in H^m(\Omega)$. Since Ω has the m-extension property, there exists a continuous, linear operator $P : H^m(\Omega) \longrightarrow H^m(\mathbb{R}^n)$ such that $\forall u \in H^m(\Omega)$, $Pu = w \in H^m(\mathbb{R}^n)$ with $w{\downarrow_\Omega} = (Pu){\downarrow_\Omega} = u$. Hence, $u \in H^m(\overline{\Omega})$ by Definition 8.10.1 and $H^m(\Omega) \subset H^m(\overline{\Omega})$. Thus, we have proved that $H^m(\overline{\Omega}) \equiv H^m(\Omega)$.
Norm equivalence (8.10.25): From the definition of $H^m(\overline{\Omega})$,

$$\|u\|_{H^m(\overline{\Omega})} = \inf_{\substack{w \in H^m(\mathbb{R}^n) \\ w\downarrow_\Omega = u}} \|w\|_{H^m(\mathbb{R}^n)} \leq \|Pu\|_{H^m(\mathbb{R}^n)} \leq C_0 \|u\|_{H^m(\Omega)} \tag{8.10.26}$$

for some $C_0 > 0$ by the definition of the m-extension operator P for Ω satisfying (8.10.22) and (8.10.23),
$\Longrightarrow \frac{1}{C_0}\|u\|_{H^m(\overline{\Omega})} \leq \|u\|_{H^m(\Omega)} \leq C\|u\|_{H^m(\overline{\Omega})}$ (by Proposition 8.10.2 and Remark 8.10.1), from which the norm equivalence (8.10.25) follows with $C_1 = \frac{1}{C_0} > 0$, $C_2 = C > 0$. □

Remark 8.10.2.

- Any arbitrary domain $\Omega \subsetneq \mathbb{R}^n$ does *not* possess the *m-extension property* (see also [14]). For example, $\Omega = \{(x_1, x_2) : 0 < |x_1| < 1, 0 < x_2 < 1\} \subset \mathbb{R}^2$ with a *slit/cut* along the boundary segment $\Gamma_0 : 0 < x_2 < 1$ on the x_2-axis does *not* possess the 1-extension property for $m = 1$, since it is locally a *two-sided* domain along the slit/cut Γ_0 (see Appendix D).The function u defined by $u(x_1, x_2) = 0$ for $(x_1, x_2) \in \Omega_1 =]0, 1[\times]0, 1[$ and $u(x_1, x_2) = 1$ for $(x_1, x_2) \in \Omega_2 =]-1, 0[\times]0, 1[$ belongs to $H^1(\Omega)$, but u can *not* be extended to $H^1(\mathbb{R}^2)$, i.e. $\nexists$ (there does *not* exist) any $w \in H^1(\mathbb{R}^2)$ such that $w{\downarrow_\Omega} = u \in H^1(\Omega)$. See Figure 8.1 for a diagram of this scenario.

- $u \in H^1(\Omega)$: $\int_\Omega |u(x_1, x_2)|^2 dx_1 dx_2 = \int_{\Omega_2} 1^2 dx_1 dx_2 = \text{area meas}(\Omega_2) = 1 \Longrightarrow u \in L^2(\Omega)$.

- The distributional derivative $\frac{\partial u}{\partial x_1} = 0 \in L^2(\Omega)$. In fact, $\forall \phi \in \mathcal{D}(\Omega)$, ϕ and $\partial^\alpha \phi$

vanish along the boundary Γ with $\Gamma_0 \subset \Gamma\ \forall|\boldsymbol{\alpha}| \in \mathbb{N}$, i.e. $\phi\downarrow_{\Gamma_0} = 0$, $\partial^{\boldsymbol{\alpha}}\phi\downarrow_{\Gamma_0} = 0$,

$$\begin{aligned}\left\langle \frac{\partial u}{\partial x_1}, \phi \right\rangle_{\mathcal{D}'(\Omega)\times\mathcal{D}(\Omega)} &= -\left\langle u, \frac{\partial \phi}{\partial x_1} \right\rangle_{\mathcal{D}'(\Omega)\times\mathcal{D}(\Omega)} = -\int_\Omega u \frac{\partial \phi}{\partial x_1} dx_1 dx_2 \\ &= -\int_{\Omega_1} 0 \cdot \frac{\partial \phi}{dx_1} dx_1 dx_2 - \int_{\Omega_2} 1 \cdot \frac{\partial \phi}{\partial x_1} dx_1 dx_2 \\ &= -\int_0^1 dx_2 \int_{-1}^0 \frac{\partial \phi}{\partial x_1} dx_1 = -\int_0^1 [\phi(0, x_2) - \phi(-1, x_2)] dx_2 \\ &= -\int_0^1 0 dx_2 = 0 \quad \forall \phi \in \mathcal{D}(\Omega)\end{aligned}$$

$\Longrightarrow \frac{\partial u}{\partial x_1} = 0 \in \mathcal{D}'(\Omega) \Longrightarrow \frac{\partial u}{\partial x_1} = 0 \in L^2(\Omega) \hookrightarrow \mathcal{D}'(\Omega)$.

Similarly, $\langle \frac{\partial u}{\partial x_2}, \phi \rangle = 0\ \forall \phi \in \mathcal{D}(\Omega) \Longrightarrow \frac{\partial u}{\partial x_2} = 0 \in L^2(\Omega)$. Hence, $u \in H^1(\Omega)$.

$\nexists w \in H^1(\mathbb{R}^2)$ with $w\downarrow_\Omega = u$: Suppose that the contrary holds, i.e. $\exists w \in H^1(\mathbb{R}^2)$ with $w\downarrow_\Omega = u \in H^1(\Omega)$. Then $w = u$ a.e. on Ω. But $w \in H^1(\mathbb{R}^2) \Longrightarrow w, \frac{\partial w}{\partial x_i} \in L^2(\mathbb{R}^2)$ for $i = 1, 2$. Hence, for $\Omega_0 = \Omega \cup \Gamma_0 =]-1, 1[\times]0, 1[$, $w\downarrow_{\Omega_0} = w_0 \in L^2(\Omega_0)$, $\frac{\partial w}{\partial x_i}\downarrow_{\Omega_0} \in L^2(\Omega_0)$ for $i = 1, 2$. (8.10.26a)

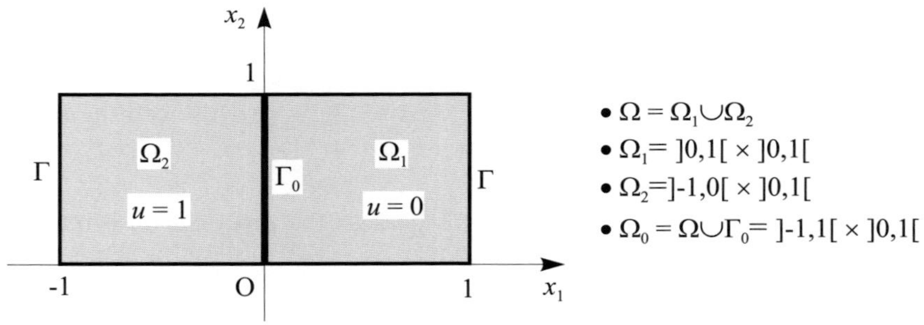

Figure 8.1 $\Omega = \{(x_1, x_2) : 0 < |x_1| < 1, 0 < x_2 < 1\} \subset \mathbb{R}^2$ with a slit/cut along the boundary segment $\Gamma_0 : 0 < x_2 < 1$

Since $\frac{\partial w}{\partial x_1}\downarrow_{\Omega_0} \in \mathcal{D}'(\Omega_0)$, we have, $\forall \phi \in \mathcal{D}(\Omega_0)$,

$$\left\langle \frac{\partial w}{\partial x_1}\downarrow_{\Omega_0}, \phi \right\rangle_{\mathcal{D}'(\Omega_0)\times\mathcal{D}(\Omega_0)} = \int_{\Omega_0} \frac{\partial w}{\partial x_1} \phi dx_1 dx_2 = -\int_{\Omega_0 = \Omega\cup\Gamma_0} w_0 \frac{\partial \phi}{\partial x_1} dx_1 dx_2$$

$$= -\int_\Omega w \frac{\partial \phi}{\partial x_1} dx_1 dx_2 \quad \text{(since area meas}(\Gamma_0) = 0)$$

$$= -\int_{\Omega_2 =]-1,0[\times]0,1[} u \frac{\partial \phi}{\partial x_1} dx_1 dx_2$$

$$\text{(since } w = u = 0 \text{ in } \Omega_1 =]0,1[\times]0,1[)$$

$$= -\int_0^1 dx_2 \int_{-1}^0 \frac{\partial \phi}{\partial x_1} dx_1 = -\int_0^1 [\phi(0,x_2) - \phi(-1,x_2)] dx_2$$

$$= -\int_0^1 \phi(0,x_2) dx_2 \quad \text{(since } \phi(-1,x_2) = 0 \ \forall x_2 \in]0,1[)$$

$$= -\int_{\Gamma_0 : x_1 = 0, 0 < x_2 < 1} \phi(x_1,x_2) dx_2 = \langle -\delta_{\Gamma_0}, \phi \rangle$$

(see Example 1.3.4 and Section 1.11 for details)
$\Longrightarrow \frac{\partial w}{\partial x_1}\downarrow_{\Omega_0} = -\delta_{\Gamma_0} \in \mathcal{D}'(\Omega_0)$, δ_{Γ_0} being the Dirac distribution with concentration on Γ_0, i.e. $\frac{\partial w}{\partial x_1}\downarrow_{\Omega_0} = -\delta_{\Gamma_0}$ is *not* a (locally summable) function by Proposition 1.3.2, and consequently $\frac{\partial w}{\partial x_1}\downarrow_{\Omega_0} \notin L^2(\Omega_0)$, which contradicts our hypothesis that $\frac{\partial w}{\partial x_1}\downarrow_{\Omega_0} \in L^2(\Omega_C)$ (by (8.10.26a)). Hence, our assumption is wrong, i.e. there does *not* exist any $w \in H^1(\mathbb{R}^2)$ with $w\downarrow_\Omega = u \in H^1(\Omega)$.

The m-extension results are powerful tools for extending several important results valid for $\mathbb{R}^n$ to similar ones on domains $\Omega \subsetneq \mathbb{R}^n$. Hence, we will give a good number of examples of Ω with the m-extension property.

m-extension property of $\Omega = \mathbb{R}^n_+$

For the proof of the m-extension property of $\Omega = \mathbb{R}^n_+$, we will need the following density result:

Density of $\mathcal{D}(\overline{\mathbb{R}^n_+})$ in $H^m(\mathbb{R}^n_+)$ Define $\mathbb{R}^n_+, \mathbb{R}^n_-, \mathbb{R}^n_0, \mathbb{R}^n_\alpha$ by:

$$\begin{aligned} &\mathbb{R}^n_+ = \{\mathbf{x} : \mathbf{x} = (x_1, x_2, \ldots, x_{n-1}, x_n) \in \mathbb{R}^n \text{ with } x_n > 0\}, \\ &\mathbb{R}^n_- = \{\mathbf{x} : \mathbf{x} \in \mathbb{R}^n \text{ with } x_n < 0\}, \quad \mathbb{R}^n_0 = \{\mathbf{x} : \mathbf{x} \in \mathbb{R}^n \text{ with } x_n = 0\}, \\ &\mathbb{R}^n_\alpha = \{\mathbf{x} : \mathbf{x} \in \mathbb{R}^n \text{ with } x_n > \alpha\} \quad \forall \alpha > 0, \quad \text{and} \end{aligned} \tag{8.10.27}$$

$$\overline{\mathbb{R}^n_+} = \mathbb{R}^n_+ \cup \mathbb{R}^n_0, \quad \overline{\mathbb{R}^n_-} = \mathbb{R}^n_- \cup \mathbb{R}^n_0, \quad \overline{\mathbb{R}^n_\alpha} = \{\mathbf{x} : \mathbf{x} \in \mathbb{R}^n \text{ with } x_n \geq \alpha\}. \tag{8.10.28}$$

For $\Omega = \mathbb{R}^n_+$, $\mathcal{D}(\overline{\mathbb{R}^n_+})$ is *dense* in $H^m(\overline{\mathbb{R}^n_+})$ defined by (8.10.1)–(8.10.3) with $\overline{\Omega} = \overline{\mathbb{R}^n_+}$ by Theorem 8.10.1. Now we show the density of $\mathcal{D}(\overline{\mathbb{R}^n_+})$ in $H^m(\mathbb{R}^n_+)$ defined by (2.15.1)–(2.15.3) with $\Omega = \mathbb{R}^n_+$:

Theorem 8.10.3. *$\mathcal{D}(\overline{\mathbb{R}^n_+})$ is* dense *in $H^m(\mathbb{R}^n_+)$, where $H^m(\mathbb{R}^n_+)$ is defined by (2.15.1)–(2.15.3) with $\Omega = \mathbb{R}^n_+$, i.e. $H^m(\mathbb{R}^n_+) = \{u : \partial^{\boldsymbol{\alpha}} u \in L^2(\mathbb{R}^n_+)\ \forall \boldsymbol{\alpha}$ with $0 \leq |\boldsymbol{\alpha}| \leq m\}$ with*

$$\|u\|_{H^m(\mathbb{R}^n_+)} = \|u\|_{m,\mathbb{R}^n_+} = \left(\sum_{0\leq|\boldsymbol{\alpha}|\leq m} \|\partial^{\boldsymbol{\alpha}} u\|^2_{L^2(\mathbb{R}^n_+)} \right)^{1/2}. \tag{8.10.29}$$

Proof. Let $\tilde{u}$ be the null extension to $\mathbb{R}^n$ of $u \in H^m(\mathbb{R}^n_+)$, i.e.

$$\tilde{u}(\mathbf{x}) = u(\mathbf{x}) \text{ for } \mathbf{x} \in \mathbb{R}^n_+ \quad \text{and} \quad \tilde{u}(\mathbf{x}) = 0 \text{ for } \mathbf{x} \in \mathbb{R}^n \setminus \mathbb{R}^n_+. \tag{8.10.30}$$

Hence,

$$\int_{\mathbb{R}^n} |\tilde{u}(\mathbf{x})|^2 d\mathbf{x} = \int_{\mathbb{R}^n_+} |u(\mathbf{x})|^2 d\mathbf{x} < +\infty \quad \Longrightarrow \quad \tilde{u} \in L^2(\mathbb{R}^n). \tag{8.10.31}$$

Let $\varrho \in \mathcal{D}(\mathbb{R}^n)$ such that $\forall \mathbf{x} \in \mathbb{R}^n$, $0 \leq \varrho(\mathbf{x}) \leq 1$, $\int_{\mathbb{R}^n} \varrho(\mathbf{x}) d\mathbf{x} = 1$, $\text{supp}(\varrho) \subset \mathbb{R}^n_-$, i.e. $\text{supp}(\varrho) \cap \overline{\mathbb{R}^n_+} = \emptyset$ – see Figure 8.2. (8.10.32)

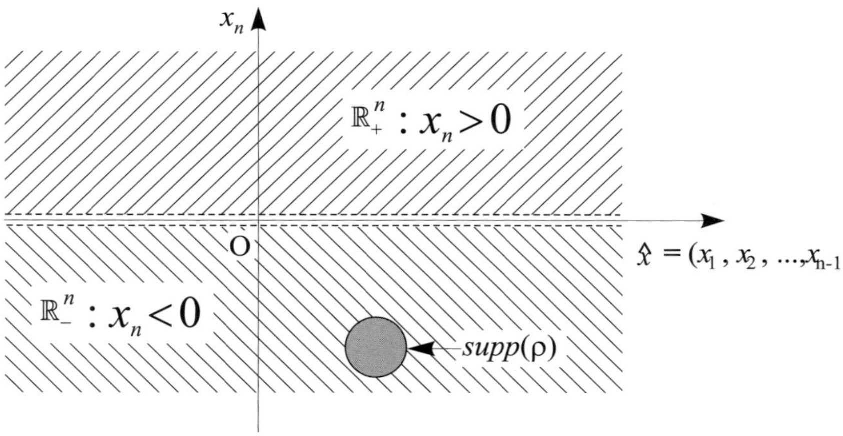

Figure 8.2 supp(ϱ) as defined in (8.10.32)

Define a family $(\varrho_\varepsilon)_{\varepsilon>0}$ of regularizing functions with the help of ϱ, i.e. $\forall \varepsilon > 0$, $\varrho_\varepsilon(\mathbf{x}) = \frac{1}{\varepsilon^n} \varrho(\frac{\mathbf{x}}{\varepsilon})$ with $\varrho_\varepsilon \in \mathcal{D}(\mathbb{R}^n)$, $\varrho_\varepsilon(\mathbf{x}) \geq 0\ \forall \mathbf{x} \in \mathbb{R}^n$, $\int_{\mathbb{R}^n} \varrho_\varepsilon(\mathbf{x}) d\mathbf{x} = 1$, volume measure $\mu(\text{supp}(\varrho_\varepsilon)) \to 0$ as $\varepsilon \to 0^+$. Then $\tilde{u} \in L^2(\mathbb{R}^n) \Longrightarrow \varrho_\varepsilon * \tilde{u} \in \mathcal{D}(\mathbb{R}^n)$, since $\varrho_\varepsilon \in \mathcal{D}(\mathbb{R}^n)\ \forall \varepsilon > 0$ (by Corollary 6.2.1) $\Longrightarrow (\varrho_\varepsilon * \tilde{u})\downarrow_{\mathbb{R}^n_+} \in \mathcal{D}(\overline{\mathbb{R}^n_+})$ $\forall \varepsilon > 0$. Thus, $\forall u \in H^m(\mathbb{R}^n_+)$ we have constructed a family $\{(\varrho_\varepsilon * \tilde{u})\downarrow_{\mathbb{R}^n_+}\}_{\varepsilon>0}$ in $\mathcal{D}(\overline{\mathbb{R}^n_+})$. (8.10.33)

Hence, the proof will be complete if we can show that $(\varrho_\varepsilon * \tilde{u})\downarrow_{\mathbb{R}^n_+} \to u$ in $H^m(\mathbb{R}^n_+)$ as $\varepsilon \to 0^+$.

From (8.10.30), $\tilde{u} \equiv L^2(\mathbb{R}^n) \subset \mathcal{D}'(\mathbb{R}^n) \Longrightarrow \partial^{\boldsymbol{\alpha}}\tilde{u} \in \mathcal{D}'(\mathbb{R}^n)\ \forall|\boldsymbol{\alpha}| \in \mathbb{N}$, and $u \in H^m(\mathbb{R}^n_+) \Longrightarrow \partial^{\boldsymbol{\alpha}} u \in L^2(\mathbb{R}^n_+)\ \forall|\boldsymbol{\alpha}| \le m \Longrightarrow$ its null extension $\widetilde{\partial^{\boldsymbol{\alpha}} u} \in L^2(\mathbb{R}^n) \subset \mathcal{D}'(\mathbb{R}^n)\ \forall|\boldsymbol{\alpha}| \le m$, i.e. $\widetilde{\partial^{\boldsymbol{\alpha}} u} = \partial^{\boldsymbol{\alpha}} u$ in $\mathbb{R}^n_+$ and $= 0$ in $\mathbb{R}^n \setminus \mathbb{R}^n_+$. Hence, their difference $\partial^{\boldsymbol{\alpha}}\tilde{u} - \widetilde{\partial^{\boldsymbol{\alpha}} u}$ will be a distribution $T_{\boldsymbol{\alpha}}$ $\forall|\boldsymbol{\alpha}| \le m$, i.e.

$$T_{\boldsymbol{\alpha}} = \partial^{\boldsymbol{\alpha}}\tilde{u} - \widetilde{\partial^{\boldsymbol{\alpha}} u} \in \mathcal{D}'(\mathbb{R}^n). \tag{8.10.34}$$

$\mathrm{supp}(T_{\boldsymbol{\alpha}}) \subset \mathbb{R}^n_0$: For this, it suffices to show that $T\!\downarrow_{(\mathbb{R}^n_0)^{\complement}} = 0$, ($(\mathbb{R}^n_0)^{\complement}$ = the complement of $\mathbb{R}^n_0$ in $\mathbb{R}^n$). In fact, $\forall\phi \in \mathcal{D}((\mathbb{R}^n_0)^{\complement})$,

$$\begin{aligned}
\langle T_{\boldsymbol{\alpha}}, \phi\rangle &= \langle \partial^{\boldsymbol{\alpha}}\tilde{u} - \widetilde{\partial^{\boldsymbol{\alpha}} u}, \phi\rangle_{\mathcal{D}'((\mathbb{R}^n_0)^{\complement})\times\mathcal{D}((\mathbb{R}^n_0)^{\complement})} \\
&= \langle \partial^{\boldsymbol{\alpha}}\tilde{u}, \phi\rangle_{\mathcal{D}'((\mathbb{R}^n_0)^{\complement})\times\mathcal{D}((\mathbb{R}^n_0)^{\complement})} - \langle \widetilde{\partial^{\boldsymbol{\alpha}} u}, \phi\rangle & (8.10.35)\\
&= (-1)^{|\boldsymbol{\alpha}|}\langle \tilde{u}, \partial^{\boldsymbol{\alpha}}\phi\rangle_{\mathcal{D}'((\mathbb{R}^n_0)^{\complement})\times\mathcal{D}((\mathbb{R}^n_0)^{\complement})} - \int_{\mathbb{R}^n_+} \partial^{\boldsymbol{\alpha}} u\phi d\mathbf{x} & (8.10.36)\\
&= (-1)^{|\boldsymbol{\alpha}|}\int_{\mathbb{R}^n_+} u\partial^{\boldsymbol{\alpha}}\phi d\mathbf{x} - \int_{\mathbb{R}^n_+} \partial^{\boldsymbol{\alpha}} u\phi d\mathbf{x} = 0 \quad \forall\phi \in \mathcal{D}((\mathbb{R}^n_0)^{\complement}). & (8.10.37)
\end{aligned}$$

Since $u \in H^m(\mathbb{R}^n_+) \Longrightarrow \forall|\boldsymbol{\alpha}| \le m$, $\int_{\mathbb{R}^n_+} \partial^{\boldsymbol{\alpha}} u\phi d\mathbf{x} = (-1)^{|\boldsymbol{\alpha}|}\int_{\mathbb{R}^n_+} u\partial^{\boldsymbol{\alpha}}\phi d\mathbf{x}\ \forall\phi \in \mathcal{D}(\mathbb{R}^n_+)$, $\forall|\boldsymbol{\alpha}| \le m$, $(-1)^{|\boldsymbol{\alpha}|}\int_{\mathbb{R}^n_+} u\partial^{\boldsymbol{\alpha}}\phi d\mathbf{x} - \int_{\mathbb{R}^n_+} \partial^{\boldsymbol{\alpha}} u\phi d\mathbf{x} = 0\ \forall\phi \in \mathcal{D}((\mathbb{R}^n_0)^{\complement})$ with $\mathrm{supp}(\phi) \subset \mathbb{R}^n_+$ and $(-1)^{|\boldsymbol{\alpha}|}\int_{\mathbb{R}^n_-} \tilde{u}\partial^{\boldsymbol{\alpha}}\phi d\mathbf{x} - \int_{\mathbb{R}^n_-} \widetilde{\partial^{\boldsymbol{\alpha}} u}\phi d\mathbf{x} = 0\ \forall\phi \in \mathcal{D}((\mathbb{R}^n_0)^{\complement})$ with $\mathrm{supp}(\phi) \subset \mathbb{R}^n_-$. (8.10.38)

Hence, from (8.10.37) and (8.10.38), $\langle T_{\boldsymbol{\alpha}}, \phi\rangle = 0\ \forall\phi \in \mathcal{D}((\mathbb{R}^n_0)^{\complement})$ with support in $\mathbb{R}^n_+$ or $\mathbb{R}^n_-$ $\forall|\boldsymbol{\alpha}| \le m$, i.e. $\forall\phi \in \mathcal{D}(\mathbb{R}^n_+ \cup \mathbb{R}^n_-) \Longrightarrow T_{\boldsymbol{\alpha}} = 0$ in $\mathcal{D}'((\mathbb{R}^n_0)^{\complement})$ by (5.1.1) (see Figure 8.3)

$$\Longrightarrow \quad T_{\boldsymbol{\alpha}}\!\downarrow_{(\mathbb{R}^n_0)^{\complement}} = 0 \quad \Longrightarrow \quad \mathrm{supp}(T_{\boldsymbol{\alpha}}) \subset ((\mathbb{R}^n_0)^{\complement})^{\complement} \equiv \mathbb{R}^n_0\ \forall|\boldsymbol{\alpha}| \le m. \tag{8.10.39}$$

$\forall\varepsilon > 0$, $\varrho_\varepsilon \in \mathcal{E}'(\mathbb{R}^n)$ is a distribution with *compact* support in $\mathbb{R}^n \Longrightarrow \forall\varepsilon > 0$, $T_{\boldsymbol{\alpha}} * \varrho_\varepsilon$ is well defined and $T_{\boldsymbol{\alpha}} * \varrho_\varepsilon \in C^\infty(\mathbb{R}^n) \subset \mathcal{D}'(\mathbb{R}^n)\ \forall|\boldsymbol{\alpha}| \le m$ by Theorem 6.4.1.

Moreover, $\forall$ sufficiently small $\varepsilon > 0$,

$$(T_{\boldsymbol{\alpha}} * \varrho_\varepsilon)\!\downarrow_{\mathbb{R}^n_+} = 0 \quad \text{in } \mathcal{D}'(\mathbb{R}^n_+). \tag{8.10.40}$$

Since $\mathrm{supp}(\varrho) \subset \mathbb{R}^n_-$ by (8.10.31), ϱ can be chosen such that $\mathrm{supp}(\varrho_\varepsilon) \subset \mathbb{R}^n_-$ for all sufficiently small $\varepsilon > 0$. Then, by Theorem 6.2.2, $\forall|\boldsymbol{\alpha}| \le m$,

$$\mathrm{supp}(T_{\boldsymbol{\alpha}} * \varrho_\varepsilon) \subset \mathrm{supp}(T_{\boldsymbol{\alpha}}) + \mathrm{supp}(\varrho_\varepsilon) \subset \mathbb{R}^n_0 + \mathbb{R}^n_- \subset \mathbb{R}^n_-$$

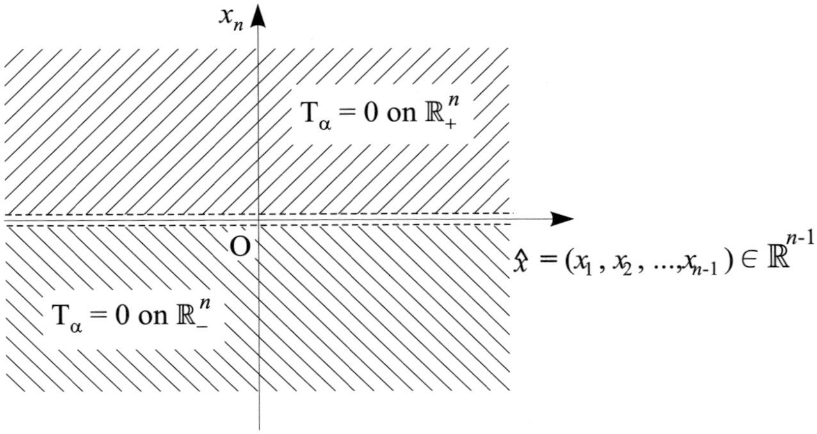

Figure 8.3 $T_{\boldsymbol{\alpha}} = 0$ in $\mathcal{D}'((\mathbb{R}^n_0)^{\complement})$

for all sufficiently small $\varepsilon > 0$, since $\operatorname{supp}(T_{\boldsymbol{\alpha}}) \subset \mathbb{R}^n_0$ and $\operatorname{supp}(\varrho_\varepsilon) \subset \mathbb{R}^n_-$ for all sufficiently small $\varepsilon > 0$. Hence, for all sufficiently small $\varepsilon > 0$, $T_{\boldsymbol{\alpha}} * \varrho_\varepsilon = 0$ in $(\overline{\mathbb{R}^n_-})^{\complement} \equiv \mathbb{R}^n_+ \Longrightarrow (T_{\boldsymbol{\alpha}} * \varrho_\varepsilon)\downarrow_{\mathbb{R}^n_+} = 0$ in $\mathcal{D}'(\mathbb{R}^n_+)$.

An *alternative proof* of (8.10.40) is as follows:

In fact, $\forall \phi \in \mathcal{D}(\mathbb{R}^n)$ with $\operatorname{supp}(\phi) \subset \mathbb{R}^n_+$, by the definition of convolution of distributions in (6.3.14), we have, $\forall \varepsilon > 0$, $\forall |\boldsymbol{\alpha}| \le m$,

$$\begin{aligned}\langle T_{\boldsymbol{\alpha}} * \varrho_\varepsilon, \phi\rangle_{\mathcal{D}'(\mathbb{R}^n)\times\mathcal{D}(\mathbb{R}^n)} &= \langle T_{\boldsymbol{\alpha}}(\mathbf{x}), \langle \varrho_\varepsilon(\mathbf{y}), \phi(\mathbf{x}+\mathbf{y})\rangle\rangle \\ &= \langle T_{\boldsymbol{\alpha}}(\mathbf{x}), \psi_\varepsilon(\mathbf{x})\rangle_{\mathcal{D}'(\mathbb{R}^n)\times\mathcal{D}(\mathbb{R}^n)},\end{aligned}$$

where

$$\begin{aligned}\psi_\varepsilon(\mathbf{x}) = \langle \varrho_\varepsilon(\mathbf{y}), \phi(\mathbf{x}+\mathbf{y})\rangle_{\mathcal{D}'(\mathbb{R}^n)\times\mathcal{D}(\mathbb{R}^n)} &= \int_{\mathbb{R}^n} \varrho_\varepsilon(\mathbf{y})\phi(\mathbf{x}+\mathbf{y})d\mathbf{y} \\ &= \int_{\operatorname{supp}(\varrho_\varepsilon)\cap\operatorname{supp}(\phi(\mathbf{x}+\mathbf{y}))} \varrho_\varepsilon(\mathbf{y})\phi(\mathbf{x}+\mathbf{y})d\mathbf{y}.\end{aligned}$$

Since $\operatorname{supp}(\varrho) \subset \mathbb{R}^n_-$ by the definition of ϱ in (8.10.32) and $\operatorname{supp}(\phi) \subset \mathbb{R}^n_+$, $\forall \mathbf{y} \in \operatorname{supp}(\varrho_\varepsilon)$, $\phi(\mathbf{x}+\mathbf{y}) = 0$ for $\mathbf{x}+\mathbf{y} = (x_1+y_1, \ldots, x_n+y_n) \notin \operatorname{supp}(\phi)$, i.e. for $\mathbf{x}+\mathbf{y}$ with $x_n + y_n \le 0$. Then $\exists \delta > 0$, dependent on the choice of the support of ϱ satisfying (8.10.32) and also on the choice of ϕ, such that $\forall \varepsilon > 0$,

$$\psi_\varepsilon(\mathbf{x}) = \int_{\operatorname{supp}(\varrho_\varepsilon)\cap\operatorname{supp}(\phi(\mathbf{x}+\mathbf{y}))} \varrho_\varepsilon(\mathbf{y})\phi(\mathbf{x}+\mathbf{y})d\mathbf{y} = 0 \quad \forall \mathbf{x} \in \mathbb{R}^n \text{ with } |x_n| \le \delta. \tag{8.10.41}$$

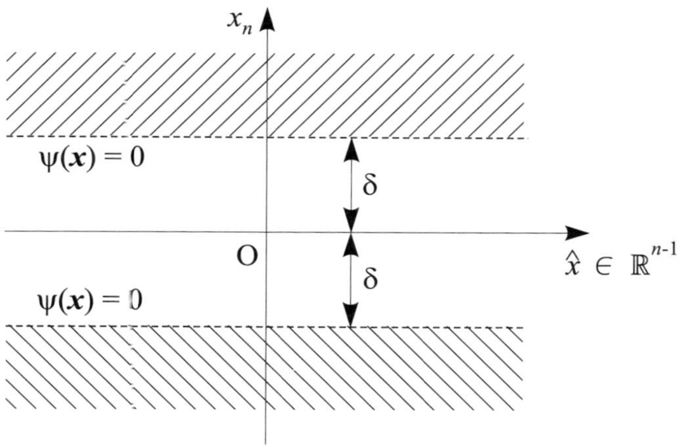

Figure 8.4 $\text{supp}(\psi_\varepsilon) \subset$ the complement of $\{\mathbf{x} : \mathbf{x} \in \mathbb{R}^n$ with $|x_n| \leq \delta\}$

Hence, $\forall \varepsilon > 0$, $\psi_\varepsilon \in \mathcal{D}(\mathbb{R}^n)$ with $\text{supp}(\psi_\varepsilon) \subset$ complement of $\{\mathbf{x} : \mathbf{x} \in \mathbb{R}^n$ with $|x_n| \leq \delta\}$ (see Figure 8.4), $\langle T_{\boldsymbol{\alpha}}, \psi_\varepsilon \rangle_{\mathcal{D}'(\mathbb{R}^n)\times\mathcal{D}(\mathbb{R}^n)} = 0$, since $\text{supp}(T_{\boldsymbol{\alpha}}) \subset \mathbb{R}^n_0$ (by (8.10.39)) and $\psi_\varepsilon(\mathbf{x}) = 0 \; \forall \mathbf{x} \in \mathbb{R}^n_0$ (by (8.10.41)) $\Longrightarrow \forall \varepsilon > 0$, $\langle T_{\boldsymbol{\alpha}} * \varrho_\varepsilon, \phi \rangle_{\mathcal{D}'(\mathbb{R}^n)\times\mathcal{D}(\mathbb{R}^n)} = \langle T_{\boldsymbol{\alpha}}, \psi_\varepsilon \rangle_{\mathcal{D}'(\mathbb{R}^n)\times\mathcal{D}(\mathbb{R}^n)} = 0 \; \forall \phi \in \mathcal{D}(\mathbb{R}^n)$ with $\text{supp}(\phi) \subset \mathbb{R}^n_+ \Longrightarrow \forall \varepsilon > 0$, $\forall |\boldsymbol{\alpha}| \leq m$, $\langle (T_{\boldsymbol{\alpha}} * \varrho_\varepsilon)\downarrow_{\mathbb{R}^n_+}, \phi \rangle_{\mathcal{D}'(\mathbb{R}^n)\times\mathcal{D}(\mathbb{R}^n)} = 0$ $\forall \phi \in \mathcal{D}(\mathbb{R}^n_+)$ (see (5.1.1)), $\Longrightarrow \forall$ sufficiently small $\varepsilon > 0$, $\forall |\boldsymbol{\alpha}| \leq m$, $(T_{\boldsymbol{\alpha}} * \varrho_\varepsilon)\downarrow_{\mathbb{R}^n_+} = 0$ in $\mathcal{D}'(\mathbb{R}^n_+)$.

$(\partial^{\boldsymbol{\alpha}}_{\mathbf{x}}(\varrho_\varepsilon * \tilde{u}))\downarrow_{\mathbb{R}^n_+} = (\varrho_\varepsilon * \widetilde{\partial^{\boldsymbol{\alpha}}_{\mathbf{y}} u})\downarrow_{\mathbb{R}^n_+}$ *with* $\partial^{\boldsymbol{\alpha}}_{\mathbf{x}} = \dfrac{\partial^{|\boldsymbol{\alpha}|}}{\partial x_1{}^{\alpha_1} \ldots \partial x_n{}^{\alpha_n}}$, $\partial^{\boldsymbol{\alpha}}_{\mathbf{y}} = \frac{\partial^{|\boldsymbol{\alpha}|}}{\partial y_1{}^{\alpha_1} \ldots \partial y_n{}^{\alpha_n}}$:
In fact, we have

$$\begin{aligned}
\partial^{\boldsymbol{\alpha}}_{\mathbf{x}}(\varrho_\varepsilon * \tilde{u})(\mathbf{x}) &= \partial^{\boldsymbol{\alpha}}_{\mathbf{x}} \int_{\mathbb{R}^n} \varrho_\varepsilon(\mathbf{x}-\mathbf{y})\tilde{u}(\mathbf{y})d\mathbf{y} = \int_{\mathbb{R}^n} \partial^{\boldsymbol{\alpha}}_{\mathbf{x}}[\varrho_\varepsilon(\mathbf{x}-\mathbf{y})]\tilde{u}(\mathbf{y})d\mathbf{y} \\
&= (-1)^{|\boldsymbol{\alpha}|} \int_{\mathbb{R}^n} \partial^{\boldsymbol{\alpha}}_{\mathbf{y}}[\varrho_\varepsilon(\mathbf{x}-\mathbf{y})]\tilde{u}(\mathbf{y})d\mathbf{y} \\
&= (-1)^{2|\boldsymbol{\alpha}|} \int_{\mathbb{R}^n} \varrho_\varepsilon(\mathbf{x}-\mathbf{y})\partial^{\boldsymbol{\alpha}}\tilde{u}(\mathbf{y})d\mathbf{y} = (\varrho_\varepsilon * \partial^{\boldsymbol{\alpha}}_{\mathbf{y}}\tilde{u})(\mathbf{x}),
\end{aligned}$$

since $\varrho_\varepsilon \in \mathcal{D}(\mathbb{R}^n) \; \forall \varepsilon > 0$.

Hence,

$$\begin{aligned}
\partial^{\boldsymbol{\alpha}}_{\mathbf{x}}(\varrho_\varepsilon * \tilde{u})\downarrow_{\mathbb{R}^n_+} &= (\varrho_\varepsilon * \partial^{\boldsymbol{\alpha}}_{\mathbf{y}}\tilde{u})\downarrow_{\mathbb{R}^n_+} = (\varrho_\varepsilon * (T_{\boldsymbol{\alpha}} + \widetilde{\partial^{\boldsymbol{\alpha}}_{\mathbf{y}} u}))\downarrow_{\mathbb{R}^n_+} \quad \text{(by (8.10.34))} \\
&= (\varrho_\varepsilon * T_{\boldsymbol{\alpha}} + \varrho_\varepsilon * \widetilde{\partial^{\boldsymbol{\alpha}}_{\mathbf{y}} u})\downarrow_{\mathbb{R}^n_+}
\end{aligned}$$

$\Longrightarrow \partial_{\mathbf{x}}^{\boldsymbol{\alpha}}(\varrho_\varepsilon * \tilde{u})\downarrow_{\mathbb{R}^n_+} = (T_{\boldsymbol{\alpha}} * \varrho_\varepsilon)\downarrow_{\mathbb{R}^n_+} + (\varrho_\varepsilon * \widetilde{\partial_{\mathbf{y}}^{\boldsymbol{\alpha}} u})\downarrow_{\mathbb{R}^n_+} = (\varrho_\varepsilon * \widetilde{\partial_{\mathbf{y}}^{\boldsymbol{\alpha}} u})\downarrow_{\mathbb{R}^n_+}$, since $(T_{\boldsymbol{\alpha}} * \varrho_\varepsilon)\downarrow_{\mathbb{R}^n_+} = 0 \ \forall$ sufficiently small $\varepsilon > 0$ by (8.10.40). Then, for $u \in H^m(\mathbb{R}^n_+)$, $\partial^{\boldsymbol{\alpha}} u \in L^2(\mathbb{R}^n_+) \ \forall |\boldsymbol{\alpha}| \leq m \Longrightarrow \widetilde{\partial^{\boldsymbol{\alpha}} u} \in L^2(\mathbb{R}^n) \Longrightarrow (\varrho_\varepsilon * \widetilde{\partial^{\boldsymbol{\alpha}} u})\downarrow_{\mathbb{R}^n_+} \to \widetilde{\partial^{\boldsymbol{\alpha}} u}\downarrow_{\mathbb{R}^n_+} = \partial^{\boldsymbol{\alpha}} u \in L^2(\mathbb{R}^n_+)$ as $\varepsilon \to 0^+$

$\Longrightarrow \forall |\boldsymbol{\alpha}| \leq m,\ \partial^{\boldsymbol{\alpha}}(\varrho_\varepsilon * \tilde{u})\downarrow_{\mathbb{R}^n_+} = (\varrho_\varepsilon * \widetilde{\partial^{\boldsymbol{\alpha}} u})\downarrow_{\mathbb{R}^n_+} \to \partial^{\boldsymbol{\alpha}} u \in L^2(\mathbb{R}^n_+)$ as $\varepsilon \to 0^+$ (by Theorem 6.8.2, Property II) $\Longrightarrow (\varrho_\varepsilon * \tilde{u})\downarrow_{\mathbb{R}^n_+} \to u$ in $H^m(\mathbb{R}^n_+)$ as $\varepsilon \to 0^+$.

Hence, $u \in H^m(\mathbb{R}^n_+)$ and, from (8.10.33), $\exists (\varrho_\varepsilon * \tilde{u})\downarrow_{\mathbb{R}^n_+} \in \mathcal{D}(\overline{\mathbb{R}^n_+}) \ \forall \varepsilon > 0$ such that $(\varrho_\varepsilon * \tilde{u})\downarrow_{\mathbb{R}^n_+} \to u$ in $H^m(\mathbb{R}^n_+)$ as $\varepsilon \to 0^+$. Consequently, $\mathcal{D}(\overline{\mathbb{R}^n_+})$ is dense in $H^m(\mathbb{R}^n_+)$. For an alternative proof of this theorem using cut-off functions (but not convolutions), see Lions [14]. □

8.10.3 *m*-extension property of $\mathbb{R}^n_+$

Theorem 8.10.4. $\mathbb{R}^n_+$ *has the* m-extension property *for* $m \geq 0$.

Proof.
Case $m = 0$: For $u \in H^0(\mathbb{R}^n_+) \equiv L^2(R^n_+)$, let $\tilde{u}$ be the null extension to $\mathbb{R}^n$ of u. Then

$$\int_{\mathbb{R}^n} |\tilde{u}(\mathbf{x})|^2 d\mathbf{x} = \int_{\mathbb{R}^n_+} |u(\mathbf{x})|^2 d\mathbf{x} + \int_{\mathbb{R}^n \setminus \mathbb{R}^n_+} 0 d\mathbf{x} = \int_{\mathbb{R}^n_+} |u(\mathbf{x})|^2 d\mathbf{x} < +\infty$$

$\Longrightarrow \tilde{u} \in L^2(\mathbb{R}^n)$ with $\|\tilde{u}\|_{L^2(\mathbb{R}^n)} = \|u\|_{L^2(\mathbb{R}^n_+)}$. Hence, define the 0-extension operator $P : L^2(\mathbb{R}^n_+) \longrightarrow L^2(\mathbb{R}^n)$ by $Pu = \tilde{u}$ and $(Pu)\downarrow_{\mathbb{R}^n_+} = u$ with

$$\|Pu\|_{L^2(\mathbb{R}^n)} = \|u\|_{L^2(\mathbb{R}^n_+)} \quad \forall u \in L^2(\mathbb{R}^n_+). \tag{8.10.42}$$

Case $m \geq 1$: First of all, we will show that $\exists P : \mathcal{D}(\overline{\mathbb{R}^n_+}) \longrightarrow H^m(\mathbb{R}^n)$ such that $\forall u \in \mathcal{D}(\overline{\mathbb{R}^n_+})$, $Pu \in H^m(\mathbb{R}^n)$ and P is a *continuous*, *linear* operator from $\mathcal{D}(\overline{\mathbb{R}^n_+})$ equipped with the norm $\|\cdot\|_{H^m(\mathbb{R}^n_+)}$ defined in (8.10.29) into $H^m(\mathbb{R}^n)$. Then the result will follow from the density of $\mathcal{D}(\overline{\mathbb{R}^n_+})$ in $H^m(\mathbb{R}^n_+)$ by Theorem 8.10.3.

Let $u \in \mathcal{D}(\overline{\mathbb{R}^n_+})$. We set $\mathbf{x} = (x_1, \ldots, x_{n-1}, x_n) = (\mathbf{y}, z) \in \mathbb{R}^n$ with $\mathbf{y} = (x_1, \ldots, x_{n-1}) \in \mathbb{R}^{n-1}$ and $z = x_n \in \mathbb{R}$. Define Pu by

$$(Pu)(\mathbf{y}, z) = \begin{cases} u(\mathbf{y}, z) & \text{for } z \geq 0, \\ v(\mathbf{y}, z) & \text{for } z < 0, \end{cases} \tag{8.10.43}$$

where $v(\mathbf{y}, z)$ is defined in terms of the values $\{u(\mathbf{y}, -kz)\}_{k=1}^m$ with $z < 0$, which are well defined for $u \in \mathcal{D}(\overline{\mathbb{R}^n_+})$ as follows: $\forall \mathbf{y} = (x_1, \dots, x_{n-1}) \in \mathbb{R}^{n-1}$, $z = x_n < 0$,

$$v(\mathbf{y}, z) = \lambda_1 u(\mathbf{y}, -z) + \lambda_2 u(\mathbf{y}, -2z) + \cdots + \lambda_m u(\mathbf{y}, -mz) = \sum_{k=1}^m \lambda_k u(\mathbf{y}, -kz), \tag{8.10.44}$$

λ_k being constants, which are to be determined from the condition that $Pu \in H^m(\mathbb{R}^n)$. □

Remark 8.10.3. Other choices of $v(\mathbf{y}, z)$ are also possible. For example,

$$v(\mathbf{y}, z) = \sum_{k=1}^m \lambda_k u\left(\mathbf{y}, -\frac{1}{k} z\right). \tag{8.10.45}$$

In order that $Pu \in H^m(\mathbb{R}^n)$, the *usual* partial derivatives $[\frac{\partial^j (Pu)}{\partial z_j}(\mathbf{y}, z)]$ of order $j \leq m-1$ with respect to z in the *pointwise sense* must be *continuous* across the boundary $\Gamma \equiv \mathbb{R}^n_0$, i.e. for $z = 0$, since otherwise the *distributional derivative* $\frac{\partial^m (Pu)}{\partial z^m}$ of mth order with respect to z will involve the Dirac distribution δ_Γ concentrated on Γ (i.e. on $\mathbb{R}^n_0$), and then $\frac{\partial^m (Pu)}{\partial z^m}$ will *not* belong to $L^2(\mathbb{R}^n)$ and, consequently, Pu will *not* belong to $H^m(\mathbb{R}^n)$ (see Chapter 3 for more details). Thus, for $u \in \mathcal{D}(\overline{\mathbb{R}^n_+})$, we must have, for $0 \leq j \leq m-1$,

$$\left[\frac{\partial^j (Pu)}{\partial z^j}(\mathbf{y}, z)\right]_{z=0-} = \left[\frac{\partial^j u}{\partial z^j}(\mathbf{y}, z)\right]_{z=0+} = \left[\frac{\partial^j u}{\partial z^j}(\mathbf{y}, z)\right]_{z=0}$$
$$\Longrightarrow \quad \left[\frac{\partial^j v}{\partial z^j}(\mathbf{y}, z)\right]_{z=0-} = \left[\frac{\partial^j u}{\partial z^j}(\mathbf{y}, 0)\right] \quad \text{for } 0 \leq j \leq m-1. \tag{8.10.46}$$

Then, using the definition of $v(\mathbf{y}, z)$ in (8.10.44), we have:

$$j = 0 \quad \Longrightarrow \quad v(\mathbf{y}, z)\downarrow_{z=0-} = \lim_{z \to 0^-} v(\mathbf{y}, z) = \sum_{k=1}^m \lambda_k u(\mathbf{y}, 0);$$

$$j = 1 \quad \Longrightarrow \quad \left[\frac{\partial v}{\partial z}(\mathbf{y}, z)\right]_{z=0-} = \sum_{k=1}^m (-k)\lambda_k \left[\frac{\partial u}{\partial z}(\mathbf{y}, 0)\right];$$

$$\vdots$$

$$j = m-1 \quad \Longrightarrow \quad \left[\frac{\partial^{m-1} v}{\partial z^{m-1}}(\mathbf{y}, z)\right]_{z=0-} = \sum_{k=1}^m (-k)^{m-1}\lambda_k \left[\frac{\partial^{m-1} u}{\partial z^{m-1}}(\mathbf{y}, 0)\right]. \tag{8.10.47}$$

From (8.10.46) and (8.10.47), we have

$$\sum_{k=1}^{m} \lambda_k u(\mathbf{y},0) = u(\mathbf{y},0) \qquad \Longrightarrow \qquad \sum_{k=1}^{m} \lambda_k = 1;$$

$$\sum_{k=1}^{m} (-k)\lambda_k[\frac{\partial u}{\partial z}(\mathbf{y},0)] = \left[\frac{\partial u}{\partial z}(\mathbf{y},0)\right] \qquad \Longrightarrow \qquad \sum_{k=1}^{m} (-k)\lambda_k = 1;$$

$$\vdots$$

$$\sum_{k=1}^{m} (-k)^{m-1}\lambda_k\left[\frac{\partial^{m-1} u}{\partial z^{m-1}}(\mathbf{y},0)\right] = \left[\frac{\partial^{m-1} u}{\partial z^{m-1}}(\mathbf{y},0)\right] \Longrightarrow \sum_{k=1}^{m} (-k)^{m-1}\lambda_k = 1,$$

which we write in matrix form:

$$\begin{bmatrix} 1 & 1 & \cdots & 1 \\ -1 & -2 & \cdots & -m \\ (-1)^2 & (-2)^2 & \cdots & (-m)^2 \\ \vdots & \vdots & & \vdots \\ (-1)^{m-1} & (-2)^{m-1} & \cdots & (-m)^{m-1} \end{bmatrix} \begin{bmatrix} \lambda_1 \\ \lambda_2 \\ \lambda_3 \\ \vdots \\ \lambda_m \end{bmatrix} = \begin{bmatrix} 1 \\ 1 \\ 1 \\ \vdots \\ 1 \end{bmatrix}. \tag{8.10.48}$$

This is the well-known Vandermonde system of equations with $x_0 = -1$, $x_1 = -2$, $\ldots$, $x_{m-1} = -m$, having a unique solution which completely determines $(\lambda_k)_{k=1}^m$.

Since $u \in \mathcal{D}(\overline{\mathbb{R}^n_+})$,

$$u \in C^\infty(\overline{\mathbb{R}^n_+}) \quad \text{and} \quad v \in C^\infty(\mathbb{R}^n_-) \tag{8.10.49}$$

by definition (8.10.44). From the condition (8.10.46), the partial derivative $[\frac{\partial^j (Pu)}{\partial z^j}(\mathbf{y},z)]$ of order $j \le m-1$ in the *usual pointwise sense* is continuous in $\overline{R^n_+} \cup \mathbb{R}^n_- \equiv \mathbb{R}^n$, i.e. $Pu \in C_0^{m-1}(\mathbb{R}^n)$ with *compact* support in $\mathbb{R}^n$, since $u \in \mathcal{D}(\overline{\mathbb{R}^n_+})$ has compact support in $\mathbb{R}^n$. Hence, the distributional derivative

$$\partial^{\boldsymbol{\alpha}}(Pu) \in L^2(\mathbb{R}^n) \quad \forall |\boldsymbol{\alpha}| \le m-1. \tag{8.10.50}$$

Moreover, the *usual* partial derivatives $[\partial^{\boldsymbol{\alpha}} Pu(\mathbf{y},z)] \in L^2(\mathbb{R}^n)$ $\forall |\boldsymbol{\alpha}| \le m$ (although defined in $\mathbb{R}^n \setminus \mathbb{R}^n_0$, the n-dimensional Lebesgue volume measure of $\mathbb{R}^n_0$ is zero).

It remains to show that the mth-order distributional derivative $\frac{\partial^m (Pu)}{\partial z^m}$ coincides with the mth-order usual partial derivative $[\frac{\partial^m (Pu)}{\partial z^m}(\mathbf{y},z)]$ in the pointwise sense, $[\frac{\partial^m (Pu)}{\partial z^m}(\mathbf{y},z)]$ being an element of $L^2(\mathbb{R}^n)$. Then, combining with (8.10.50), we shall have $Pu \in H^m(\mathbb{R}^n)$. Hence, we show that

$$\frac{\partial^m (Pu)}{\partial z^m} = \left[\frac{\partial^m (Pu)}{\partial z^m}(\mathbf{y},z)\right] = \begin{cases} [\frac{\partial^m u}{\partial z^m}(\mathbf{y},z)] & \text{for } z > 0 \\ [\frac{\partial^m v}{\partial z^m}(\mathbf{y},z)] & \text{for } z < 0 \end{cases} \quad \text{in } \mathcal{D}'(\mathbb{R}^n). \tag{8.10.51}$$

In fact, $\forall \phi \in \mathcal{D}(\mathbb{R}^n)$,

$$\begin{aligned}
&\left\langle Pu, \frac{\partial^m \phi}{\partial z^m} \right\rangle_{\mathcal{D}'(\mathbb{R}^n)\times\mathcal{D}(\mathbb{R}^n)} = (-1)^{m-1}\left\langle \frac{\partial^{m-1}(Pu)}{\partial z^{m-1}}, \frac{\partial \phi}{\partial z} \right\rangle_{\mathcal{D}'(\mathbb{R}^n)\times\mathcal{D}(\mathbb{R}^n)} \\
&= (-1)^{m-1}\left[\int_{\mathbb{R}^n_+} \left[\frac{\partial^{m-1} u}{\partial z^{m-1}}(\mathbf{y}, z) \right] \frac{\partial \phi}{\partial z} d\mathbf{y} dz + \int_{\mathbb{R}^n_-} \left[\frac{\partial^{m-1} v}{\partial z^{m-1}}(\mathbf{y}, z) \right] \frac{\partial \phi}{\partial z} d\mathbf{y} dz \right] \\
&= (-1)^{m-1}\left[\int_{\mathbb{R}^{n-1}} d\mathbf{y} \left(\int_{0+}^{\infty} \left[\frac{\partial}{\partial z}(u^{(m-1)}(\mathbf{y}, z)\phi(\mathbf{y}, z)) \right] dz \right.\right. \\
&\quad \left. - \int_{0+}^{\infty} \left[\frac{\partial^m u}{\partial z^m}(\mathbf{y}, z) \right] \phi(\mathbf{y}, z) dz \right) + \int_{\mathbb{R}^{n-1}} d\mathbf{y} \left(\int_{-\infty}^{0^-} \left[\frac{\partial}{\partial z}(v^{(m-1)}(\mathbf{y}, z)\phi(\mathbf{y}, z)) \right] dz \right. \\
&\quad \left.\left. - \int_{-\infty}^{0^-} \left[\frac{\partial^m v}{\partial z^m}(\mathbf{y}, z) \right] \phi(\mathbf{y}, z) dz \right) \right] \\
&= (-1)^m \left[\int_{\mathbb{R}^{n-1}} \left(\int_{0+}^{\infty} \left[\frac{\partial^m u}{\partial z^m}(\mathbf{y}, z) \right] \phi(\mathbf{y}, z) dz - \int_{-\infty}^{0^-} \left[\frac{\partial^m v}{\partial z^m}(\mathbf{y}, z) \right] \cdot \phi(\mathbf{y}, z) dz \right) d\mathbf{y} \right] \\
&\quad + (-1)^{m-1} \int_{\mathbb{R}^{n-1}} [(u^{(m-1)}(\mathbf{y}, z)\phi(\mathbf{y}, z))|_{z=0+}^{z=\infty} + (v^{(m-1)}(\mathbf{y}, z)\phi(\mathbf{y}, z))|_{z=-\infty}^{z=0^-}] d\mathbf{y} \\
&= (-1)^m \int_{\mathbb{R}^n} \left[\frac{\partial^m (Pu)}{\partial z^m}(\mathbf{y}, z) \right] \phi(\mathbf{y}, z) d\mathbf{y} dz
\end{aligned}$$

(using (8.10.51)), since $\phi(\mathbf{y}, z)\downarrow_{z=\pm\infty} = 0$, $\phi(\mathbf{y}, 0^+) = \phi(\mathbf{y}, 0^-) = \phi(\mathbf{y}, 0)$, $u^{(m-1)}$ $(\mathbf{y}, 0) = v^{(m-1)}(\mathbf{y}, 0^-)$ by (8.10.46),
$\Longrightarrow \langle Pu, \frac{\partial^m \phi}{\partial z^m} \rangle = (-1)^m \langle [\frac{\partial^m (Pu)}{\partial z^m}(\mathbf{y}, z)], \phi \rangle$ $\forall \phi \in \mathcal{D}(\mathbb{R}^n)$
$\Longleftrightarrow$ $\frac{\partial^m (Pu)}{\partial z^m} = [\frac{\partial^m (Pu)}{\partial z^m}(\mathbf{y}, z)] \in \mathcal{D}'(\mathbb{R}^n)$ (by the definition of distributional derivatives) with

$$\frac{\partial^m (Pu)}{\partial z^m} \in L^2(\mathbb{R}^n). \tag{8.10.52}$$

Hence, $Pu \in H^m(\mathbb{R}^n)$ $\forall u \in \mathcal{D}(\overline{\mathbb{R}^n_+})$. From the definition (8.10.43)–(8.10.44) of Pu, it is obvious that $P : \mathcal{D}(\overline{\mathbb{R}^n_+}) \longrightarrow H^m(\mathbb{R}^n)$ is linear. To complete the proof, we are to show that $\|Pu\|_{m,\mathbb{R}^n} \leq C_0 \|u\|_{m,\mathbb{R}^n_+}$ $\forall u \in \mathcal{D}(\overline{\mathbb{R}^n_+})$ for some $C_0 > 0$. In fact, $\forall |\boldsymbol{\alpha}| \leq m$,

$$\int_{\mathbb{R}^n} |\partial^{\boldsymbol{\alpha}}(Pu)|^2 d\mathbf{x} = \int_{\mathbb{R}^n_+} |\partial^{\boldsymbol{\alpha}} u|^2 d\mathbf{y} dz + \int_{\mathbb{R}^n_-} |\partial^{\boldsymbol{\alpha}} v|^2 d\mathbf{y} dz, \tag{8.10.53}$$

where the distributional derivatives and the *usual* derivatives in the pointwise sense are identical in $L^2(\mathbb{R}^n) \forall |\boldsymbol{\alpha}| \leq m$.

Set $\mathbf{y} = \mathbf{y}$, $z = -k\xi$ with $1 \le k \le m$, $\xi < 0$, Jacobian $J = -k$ given by:

$$J = \begin{vmatrix} 1 & 0 & \cdots & 0 \\ 0 & \ddots & \ddots & \vdots \\ \vdots & \ddots & 1 & 0 \\ 0 & \cdots & 0 & -k \end{vmatrix} = -k.$$

Define $u(\mathbf{y}, z) \equiv u(\mathbf{y}, -k\xi) \equiv v_k(\mathbf{y}, \xi)$ with $\xi = -\frac{1}{k}z < 0$, $\mathbf{y} = (x_1, x_2, \ldots, x_{n-1}) \in \mathbb{R}^{n-1}$.

Hence, $\frac{\partial u}{\partial z}(\mathbf{y}, z) = \frac{\partial v_k}{\partial \xi}(\mathbf{y}, \xi(z))\frac{\partial \xi}{\partial z} = -\frac{1}{k}\frac{\partial v_k}{\partial \xi}(\mathbf{y}, \xi)$

$\Longrightarrow \frac{\partial^{\alpha_n} u}{\partial z^{\alpha_n}}(\mathbf{y}, z) = (-\frac{1}{k})^{\alpha_n}\frac{\partial^{\alpha_n} v_k}{\partial \xi^{\alpha_n}}(\mathbf{y}, \xi) \Longrightarrow \forall \boldsymbol{\alpha} = (\alpha_1, \alpha_2, \ldots, \alpha_{n-1}, \alpha_n)$ with $|\boldsymbol{\alpha}| \le m$,

$$\frac{\partial^{\alpha_1+\cdots+\alpha_{n-1}}}{\partial x_1^{\alpha_1} \ldots \partial x_{n-1}^{\alpha_{n-1}}}\left(\frac{\partial^{\alpha_n} u}{\partial z^{\alpha_n}}\right)(\mathbf{y}, z) = \left(-\frac{1}{k}\right)^{\alpha_n} \frac{\partial^{\alpha_1+\cdots+\alpha_{n-1}}}{\partial x_1^{\alpha_1} \ldots \partial x_{n-1}^{\alpha_{n-1}}}\left(\frac{\partial^{\alpha_n} v_k}{\partial \xi^{\alpha_n}}\right)(\mathbf{y}, \xi)$$

$\Longrightarrow |\partial^{\boldsymbol{\alpha}} u(\mathbf{y}, z)|^2 = \frac{1}{k^{2\alpha_n}}|\partial^{\boldsymbol{\alpha}} v_k(\mathbf{y}, \xi)|^2$ with $\xi < 0$, $\mathbf{y} = (x_1, \ldots, x_{n-1})$, $|\boldsymbol{\alpha}| \le m$.

$\Longrightarrow \int_{\mathbb{R}^n_+} |\partial^{\boldsymbol{\alpha}} u(\mathbf{y}, z)|^2 d\mathbf{y}dz = \int_{\mathbb{R}^n_-} \frac{1}{k^{2\alpha_n}}|\partial^{\boldsymbol{\alpha}} v_k(\mathbf{y}, \xi)|^2 |J| d\mathbf{y}d\xi.$

Replacing ξ by z and putting $|J| = k$, we get

$$\int_{\mathbb{R}^n_-} |\partial^{\boldsymbol{\alpha}} v_k(\mathbf{y}, z)|^2 d\mathbf{y}dz = k^{2\alpha_n - 1}\int_{\mathbb{R}^n_+} |\partial^{\boldsymbol{\alpha}} u(\mathbf{y}, z)|^2 d\mathbf{y}dz. \tag{8.10.54}$$

Then, from (8.10.44), (8.10.53) and (8.10.54), $\forall |\boldsymbol{\alpha}| \le m$, $|v(\mathbf{y}, z)| \le \sum_{k=1}^m |\lambda_k| \cdot |v_k(\mathbf{y}, z)|$ for $z = x_n < 0$, and

$$\int_{\mathbb{R}^n} |\partial^{\boldsymbol{\alpha}}(Pu)|^2 d\mathbf{y}dz \le \int_{\mathbb{R}^n_+} |\partial^{\boldsymbol{\alpha}} u|^2 d\mathbf{y}dz + 2^{m-1}(\max_{1\le k\le m}\{|\lambda_k|\})^2$$

$$\sum_{k=1}^{m} k^{2\alpha_n - 1}\int_{\mathbb{R}^n_+} |\partial^{\boldsymbol{\alpha}} u|^2 d\mathbf{y}dz \le C\int_{\mathbb{R}^n_+} |\partial^{\boldsymbol{\alpha}} u|^2 d\mathbf{y}dz$$

with $C = C(\lambda_1, \lambda_2, \ldots, \lambda_m, \alpha_n, m) > 0$

$$\Longrightarrow \quad \sum_{|\boldsymbol{\alpha}|\le m}\int_{\mathbb{R}^n} |\partial^{\boldsymbol{\alpha}}(Pu)|^2 d\mathbf{y}dz \le C_1 \sum_{|\boldsymbol{\alpha}|\le m}\int_{\mathbb{R}^n_+} |\partial^{\boldsymbol{\alpha}} u|^2 d\mathbf{y}dz$$

$\Longrightarrow \|Pu\|^2_{m,\mathbb{R}^n} \le C_1\|u\|^2_{m,\mathbb{R}^n_+}$ $\forall u \in \mathcal{D}(\overline{\mathbb{R}^n_+}) \Longrightarrow \exists C_0 > 0$ with $C_0 = \sqrt{C_1}$ such that $\|Pu\|_{H^m(\mathbb{R}^n)} \le C_0\|u\|_{H^m(\mathbb{R}^n_+)}$.

Thus, the linear operator $P : \mathcal{D}(\overline{\mathbb{R}^n_+}) \to H^m(\mathbb{R}^n)$ is *continuous* from $\mathcal{D}(\overline{\mathbb{R}^n_+})$ equipped with the norm $\|\cdot\|_{H^m(\mathbb{R}^n_+)}$ into $H^m(\mathbb{R}^n)$. But $\mathcal{D}(\overline{\mathbb{R}^n_+})$ is dense in $H^m(\mathbb{R}^n_+)$ by Theorem 8.10.3. Hence, P has a *unique*, *continuous*, *linear* extension $\tilde{P}$ (which will be ultimately denoted by the earlier original notation P itself) from $H^m(\mathbb{R}^n_+)$ into $H^m(\mathbb{R}^n)$. Thus, $\mathbb{R}^n_+$ has the m-extension property.

Babitch–Nikolski extension The extension of $u \in \mathcal{D}(\overline{\mathbb{R}^n_+})$ to $Pu \in H^m(\mathbb{R}^n)$ defined by (8.10.43)–(8.10.44) will be called the *Babitch–Nikolski extension* ([16], [43]) here, although some authors, for example Grisvard [18], call this the *Nikolski technique of extension by reflection*.

m-extension property of $\mathbb{R}^{n+}$

Define

$$\begin{aligned}
\mathbb{R}^{n+} &= \{\mathbf{x} : \mathbf{x} \in \mathbb{R}^n,\ x_i > 0,\ 1 \le i \le n\};\\
\overline{\mathbb{R}^{n+}} &= \{\mathbf{x} : \mathbf{x} \in \mathbb{R}^n,\ x_i \ge 0,\ 1 \le i \le n\};\\
\mathbb{R}^{n-} &= \{\mathbf{x} : \mathbf{x} \in \mathbb{R}^n,\ x_i < 0,\ 1 \le i \le n\};\\
\overline{\mathbb{R}^{n-}} &= \{\mathbf{x} : \mathbf{x} \in \mathbb{R}^n,\ x_i \le 0,\ 1 \le i \le n\};\\
\mathbb{R}^{n0} &= \{\mathbf{x} : \mathbf{x} \in \mathbb{R}^n,\ x_i = 0,\ 1 \le i \le n\};\\
\overline{\mathbb{R}^{n+}} &= \mathbb{R}^{n+} \cup \mathbb{R}^{n0}; \quad \overline{\mathbb{R}^{n-}} = \mathbb{R}^{n-} \cup \mathbb{R}^{n0}.
\end{aligned} \tag{8.10.55}$$

For $n = 2$, $\mathbb{R}^{2+}$ is the *first quadrant*, $\mathbb{R}^{2-}$ the *third quadrant* in $\mathbb{R}^2$, $\mathbb{R}^2_+$ being the *upper half-plane* in $\mathbb{R}^2$. Then $\mathbb{R}^2_{+,-} = \{\mathbf{x} : \mathbf{x} \in \mathbb{R}^2_+, x_1 < 0\}$ is the *second quadrant* in $\mathbb{R}^2$ (this last notation is *not* standard).

Proposition 8.10.4. $\mathcal{D}(\overline{\mathbb{R}^{n+}})$ *is a* dense *subspace of* $H^m(\mathbb{R}^{n+})$ *with*

$$\|u\|_{H^m(\mathbb{R}^{n+})} = \Bigg(\sum_{0 \le |\boldsymbol{\alpha}| \le m} \|\partial^{\boldsymbol{\alpha}} u\|^2_{L^2(\mathbb{R}^{n+})} \Bigg)^{1/2}.$$

Proof. The proof is exactly similar to that of Theorem 8.10.3 with *minor modifications*:

- Choose $\varrho \in \mathcal{D}(\mathbb{R}^n)$ with $\operatorname{supp}(\varrho) \subset \mathbb{R}^{n-}$ and define $\varrho_\varepsilon = \frac{1}{\varepsilon^n}\varrho(\frac{\mathbf{x}}{\varepsilon})$ such that $\operatorname{supp}(\varrho_\varepsilon) \subset \mathbb{R}^{n-}$ for *all* sufficiently small $\varepsilon > 0$;
- $\forall |\boldsymbol{\alpha}| \le m$, $T_{\boldsymbol{\alpha}} * \varrho_\varepsilon \in C^\infty(\mathbb{R}^n)$ with $\operatorname{supp}(T_{\boldsymbol{\alpha}} * \varrho_\varepsilon) \subset \operatorname{supp}(T_{\boldsymbol{\alpha}}) + \operatorname{supp}(\varrho_\varepsilon) \subset \mathbb{R}^{n-} \implies T_{\boldsymbol{\alpha}} * \varrho_\varepsilon\downarrow_{\mathbb{R}^{n+}} = 0\ \forall |\boldsymbol{\alpha}| \le m$ and for *all* sufficiently small $\varepsilon > 0$, the other steps being almost the same, with $\mathbb{R}^n_+$ replaced by $\mathbb{R}^{n+}$. □

Proposition 8.10.5. $\mathbb{R}^{n+}$ *has the* m-extension property.

Proof. For the sake of simplicity, consider the case $n = 2$, i.e. $\mathbb{R}^{2+}$ is the *first quadrant* in $\mathbb{R}^2$. Let $u \in \mathcal{D}(\overline{\mathbb{R}^{2+}})$. Following the steps of the Babitch–Nikolski extension procedure in (8.10.44)–(8.10.52), we define

$$(P_1u)(x_1,x_2) = \begin{cases} u(x_1,x_2) & \text{for } (x_1,x_2) \in \mathbb{R}^{2+} \\ v(x_1,x_2) = \sum_{k=1}^m \lambda_k v_k(x_1,x_2) & \\ \qquad = \sum_{k=1}^m \lambda_k u(-kx_1,x_2) & \text{for } x_1 < 0, x_2 > 0, \end{cases} \tag{8.10.56}$$

where $\{\lambda_k\}_{k=1}^m$ are uniquely determined by solving $\sum_{k=1}^m (-k)^j \lambda_k = 1$ for $j = 0, 1, \ldots, m-1$ (see (8.10.48)) and $\partial^{\boldsymbol{\alpha}}(P_1u) \in L^2(\mathbb{R}^2_+)$ $\forall |\boldsymbol{\alpha}| \le m$ with

$$\begin{aligned} \int_{\mathbb{R}^2_+:x_2>0} |\partial^{\boldsymbol{\alpha}} P_1u|^2 dx_1dx_2 &= \int_{\mathbb{R}^{2+}:x_1>0,x_2>0} |\partial^{\boldsymbol{\alpha}} u|^2 dx_1dx_2 \\ &\quad + \int_{\mathbb{R}^2_{+,-}:x_1<0,x_2>0} |\partial^{\boldsymbol{\alpha}} v|^2 dx_1dx_2 \\ &\le C \int_{\mathbb{R}^{2+}} |\partial^{\boldsymbol{\alpha}} u|^2 dx_1dx_2, \end{aligned}$$

since from (8.10.56) and (8.10.54),

$$\begin{aligned} &\int_{\mathbb{R}^2_{+,-}:x_1<0,x_2>0} |\partial^{\boldsymbol{\alpha}} u(-kx_1,x_2)|^2 dx_1dx_2 \\ &= k^{2\alpha_1-1} \int_{\mathbb{R}^{2+}:x_1>0,x_2>0} |\partial^{\boldsymbol{\alpha}} u(x_1,x_2)|^2 dx_1dx_2 \quad (v_k(x_1,x_2) = u(-kx_1,x_2)) \end{aligned}$$

$\forall \boldsymbol{\alpha} = (\alpha_1,\alpha_2)$ with $|\boldsymbol{\alpha}| = \alpha_1 + \alpha_2 \le m$ (see (8.10.54)).

Hence, $\forall u \in \mathcal{D}(\overline{\mathbb{R}^{2+}})$, $\|P_1u\|_{H^m(\mathbb{R}^2_+)} \le C_0\|u\|_{H^m(\mathbb{R}^{2+})}$. Then, by virtue of the density of $\mathcal{D}(\overline{\mathbb{R}^{2+}})$ in $H^m(\mathbb{R}^{2+})$, $\exists$ a *unique, continuous, linear extension* of P_1 to $H^m(\mathbb{R}^{2+})$, which will still be denoted by $P_1 \in \mathcal{L}(H^m(\mathbb{R}^{2+}); H^m(\mathbb{R}^2_+))$ such that $\forall u \in H^m(\mathbb{R}^{2+})$, $P_1u \in H^m(\mathbb{R}^2_+)$ with $(P_1u)\downarrow_{\mathbb{R}^{2+}} = u$ and $\|P_1u\|_{H^m(\mathbb{R}^2_+)} \le C_0\|u\|_{H^m(\mathbb{R}^{2+})}$. Again, by Theorem 8.10.4, $\exists$ an m-extension operator $P_2 : P_1u \in H^m(\mathbb{R}^2_+) \longmapsto P_2(P_1u) \in H^m(\mathbb{R}^2)$ with $(P_2(P_1,u))\downarrow_{\mathbb{R}^2_+} = P_1u$ and $\|P_2(P_1u)\|_{H^m(\mathbb{R}^2)} \le C_1\|P_1u\|_{H^m(\mathbb{R}^2_+)} \le C_1C_0\|u\|_{H^m\mathbb{R}^{2+}}$ $\forall u \in H^m(\mathbb{R}^{2+})$. Define $P = P_2 \circ P_1$. Then $P : H^m(\mathbb{R}^{2+}) \longrightarrow H^m(\mathbb{R}^2)$ with $Pu = P_2 \circ P_1u = P_2(P_1u) \in H^m(\mathbb{R}^2)$, $Pu\downarrow_{\mathbb{R}^{2+}} = (P_2(P_1u))\downarrow_{\mathbb{R}^{2+}} = (P_1u)\downarrow_{\mathbb{R}^{2+}} = u$, and $\|Pu\|_{H^m(\mathbb{R}^2)} \le C_2\|u\|_{H^m(\mathbb{R}^{2+})}$ with $C_2 = C_1C_0 > 0$. Hence $P \in \mathcal{L}(H^m(\mathbb{R}^{2+}); H^m(\mathbb{R}^2))$ is the required m-extension operator. See Figure 8.5 for an illustration of this proof. □

Remark 8.10.4. In many situations, repeated applications of the Babitch–Nikolski extension technique, possibly including a change of variables, will give the required

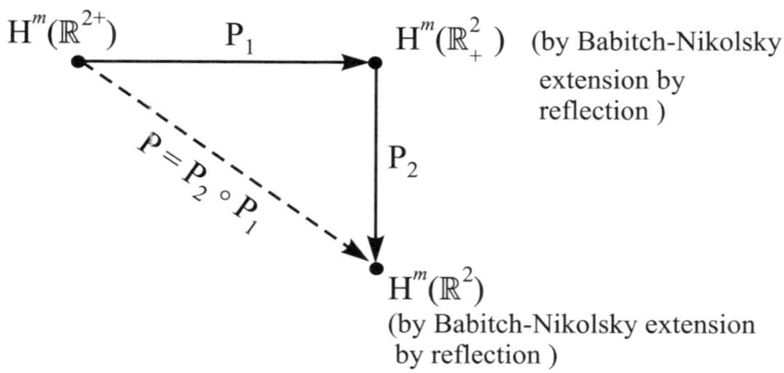

Figure 8.5 The m-extension operator P of $\mathbb{R}^{2+}$ constructed by Babitch–Nikolsky extension

m-extension operator. For example, for $\Omega = \{\mathbf{x} : \mathbf{x} \in \mathbb{R}^n, x_{n-1} > 0, x_n > 0\}$, we can define the m-extension operators $P_1 : H^m(\Omega) \longrightarrow H^m(\mathbb{R}^n_+)$ with $\mathbb{R}^n_+ = \{\mathbf{x} : \mathbf{x} \in \mathbb{R}^n, x_n > 0\}$ and $P_2 : H^m(\mathbb{R}^n_+) \longrightarrow H^m(\mathbb{R}^n)$ such that $P = P_2 \circ P_1$ is the required m-extension operator from $H^m(\Omega)$ into $H^m(\mathbb{R}^n)$. Similarly, we have:

Proposition 8.10.6. *Let Ω be an infinite angular sector in $\mathbb{R}^2$ with angular measure $0 < \omega < \pi/2$. Then Ω has the m-*extension property.

Proof. Without loss of generality, we consider the angular sector Ω with vertex at the origin and bounded by the x_1-axis, $x_2 = 0$, $x_1 > 0$, and the half-line $x_2 = mx_1$, $x_1 > 0$, $m > 0$, i.e. $\Omega = \{(x_1, x_2) : x_1 > 0, 0 < x_2 < mx_1, m > 0\}$.

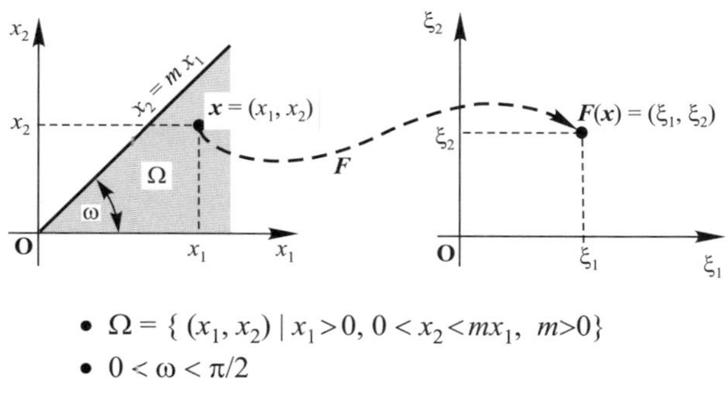

Figure 8.6 Defining an m-extension operator through a change of variables

Let $\mathbf{F} : \mathbb{R}^2 \longrightarrow \mathbb{R}^2$ be an *invertible* linear mapping (see Figure 8.6) defined, $\forall \mathbf{x} = (x_1, x_2) \in \mathbb{R}^2$, by:

$$\mathbf{F}(\mathbf{x}) = \begin{pmatrix} 1 & -1/m \\ 0 & 1 \end{pmatrix} \begin{pmatrix} x_1 \\ x_2 \end{pmatrix} = \begin{pmatrix} \xi_1 \\ \xi_2 \end{pmatrix} \in \mathbb{R}^2 \qquad (8.10.57)$$

such that $\forall \mathbf{x} \in \overline{\Omega}$, $\mathbf{F}(\mathbf{x}) = \xi \in \overline{\mathbb{R}^{2+}}$; $\forall \boldsymbol{\xi} = (\xi_1, \xi_2) \in \overline{\mathbb{R}^{2+}}$, $\mathbf{F}^{-1}(\boldsymbol{\xi}) = \mathbf{x} \in \overline{\Omega}$, where, $\forall \boldsymbol{\xi} \in \overline{\mathbb{R}^{2+}}$,

$$\mathbf{F}^{-1}(\boldsymbol{\xi}) = \begin{pmatrix} 1 & 1/m \\ 0 & 1 \end{pmatrix} \begin{pmatrix} \xi_1 \\ \xi_2 \end{pmatrix} = \begin{pmatrix} x_1 \\ x_2 \end{pmatrix} \in \overline{\Omega}. \tag{8.10.58}$$

Then the linear mappings $\mathbf{F}\downarrow_{\Omega} = \mathbf{F} : \Omega \longrightarrow \mathbb{R}^{2+}$, $\mathbf{F}^{-1} : \mathbb{R}^{2+} \longrightarrow \Omega$ are *continuously, infinitely differentiable*, i.e. $\mathbf{F}$ and $\mathbf{F}^{-1}$ are C^{∞}-diffeomorphisms.

Let $u \in \mathcal{D}(\overline{\Omega})$. $\forall (x_1, x_2) \in \Omega$, $u(x_1, x_2) = u(\mathbf{F}^{-1}(\xi_1, \xi_2)) = (u \circ \mathbf{F}^{-1})(\xi_1, \xi_2)$. Define $v = u \circ \mathbf{F}^{-1}$ and $u = v \circ \mathbf{F}$ with $u(x_1, x_2) = v(\xi_1, \xi_2)$. Then, for $u \in H^m(\Omega)$, $v \in H^m(\mathbb{R}^{2+})$ with

$$\begin{aligned} \|u\|_{H^m(\Omega)} &= \|v \circ \mathbf{F}\|_{H^m(\mathbb{R}^{2+})} \leq \gamma \|v\|_{H^m(\mathbb{R}^{2+})}, \\ \|v\|_{H^m(\mathbb{R}^{2+})} &= \|u \circ \mathbf{F}^{-1}\|_{H^m(\Omega)} \leq \gamma \|u\|_{H^m(\Omega)}, \end{aligned}$$

where $\gamma = \gamma(m, \mathbf{F}, \mathbf{F}^{-1}) > 0$.

By Proposition 8.10.5, $\exists$ an extension operator $P_1 \in \mathcal{L}(H^m(\mathbb{R}^{2+}); H^m(\mathbb{R}^2))$ such that $(P_1 v)\downarrow_{\mathbb{R}^{2+}}(\xi_1, \xi_2) = v(\xi_1, \xi_2) = u(x_1, x_2)$ with $(x_1, x_2) \in \Omega$ and $\|P_1 v\|_{H^m(\mathbb{R}^2)} \leq C \|v\|_{H^m(\mathbb{R}^{2+})}$.

Define $Pu = P_1(u \circ \mathbf{F}^{-1}) \in H^m(\mathbb{R}^2)$ such that

$$\begin{aligned} (Pu)\downarrow_{\Omega}(x_1, x_2) &= (P_1(u \circ \mathbf{F}^{-1}))\downarrow_{\mathbb{R}^{2+}}(\xi_1, \xi_2) \\ &= (P_1 v)\downarrow_{\mathbb{R}^{2+}}(\xi_1, \xi_2) = v(\xi_1, \xi_2) = u(x_1, x_2) \end{aligned}$$

for almost all $\mathbf{x} \in \Omega$, i.e. $Pu\downarrow_{\Omega} = u$. Moreover, $\|Pu\|_{H^m(\mathbb{R}^2)} = \|P_1 v\|_{H^m(\mathbb{R}^2)} \leq C \|v\|_{H^m(\mathbb{R}^{2+})} \leq C\gamma \|u\|_{H^m(\Omega)} \Longrightarrow \|Pu\|_{H^m(\mathbb{R}^2)} \leq C_0 \|u\|_{H^m(\Omega)}$. Hence, $P : H^m(\Omega) \longrightarrow H^m(\mathbb{R}^2)$ is the required m-extension operator. □

m-extension property of a non-convex planar sector with $\omega = 3\pi/2$

Now we will consider the example of a non-convex planar sector with angular measure $\omega = 3\pi/2$.

Proposition 8.10.7. *Let Ω be a planar sector with angular measure $\omega = \frac{3\pi}{2}$. Then Ω has the m-*extension property.

Proof. We orient the coordinate axes in such a way that

$$\Omega = \{(x_1, x_2) : x_1 > 0 \textit{ or } x_2 > 0\} \quad \text{(see Figure 8.7)}, \tag{8.10.59}$$

$$\Omega_1 = \{\mathbf{x} : x_1 > 0\}, \quad \Omega_2 = \{\mathbf{x} : x_2 > 0\} = \mathbb{R}^2_+, \quad \Omega_3 = \{\mathbf{x} : x_1 > 0, x_2 < 0\}.$$

Let $u \in H^m(\Omega)$ such that $v = u\downarrow_{\Omega_2} \in H^m(\Omega_2)$. Since $\Omega_2 = \mathbb{R}^2_+$, by Theorem 8.10.4 $\exists$ an m-extension operator $P_1 \in \mathcal{L}(H^m(\Omega_2); H^m(\mathbb{R}^2))$ such that $P_1 v = \underset{\sim}{v} \in H^m(\mathbb{R}^2)$ with $\underset{\sim}{v}\downarrow_{\Omega_2} = v$. Then $(u - \underset{\sim}{v}\downarrow_{\Omega})\downarrow_{\Omega_2} = u\downarrow_{\Omega_2} - \underset{\sim}{v}\downarrow_{\Omega_2} = v - v = 0$.

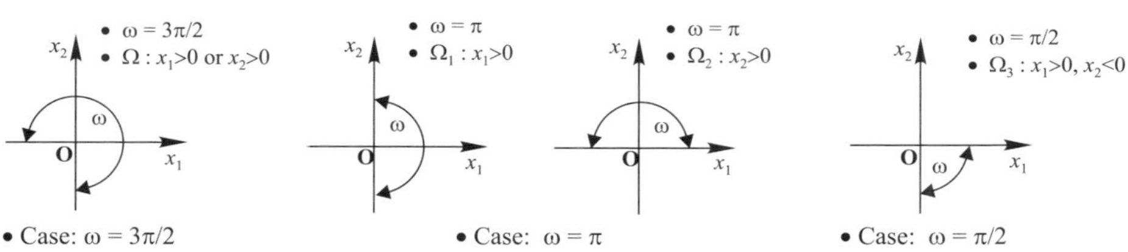

Figure 8.7 Planar sectors with angular measures ω with $\omega = \pi/2$, π and $3\pi/2$

Define $w = (u - \underline{v}\downarrow_{\Omega})\downarrow_{\Omega_1} \in H^m(\Omega_1)$ with $w = 0$ in Ω_2, i.e. for $x_2 > 0$. Then, again by Theorem 8.10.5, $\exists$ an m-extension operator $P_2 \in \mathcal{L}(H^m(\Omega_1);\ H^m(\mathbb{R}^2))$ such that $P_2 w = \underline{w} \in H^m(\mathbb{R}^2)$ with $\underline{w}\downarrow_{\Omega_2} = w = 0$, i.e. for $x_2 > 0$. Finally, define $Pu = \underline{u}$ by $\underline{u} = \underline{v} + \underline{w} \in H^m(\mathbb{R}^2)$. Then $P : H^m(\Omega) \longrightarrow H^m(\mathbb{R}^2)$ is a continuous, linear operator. Moreover, for $\Omega_3 : x_1 > 0$, $x_2 < 0$ and $\Omega_2 : x_2 > 0$, $\underline{u}\downarrow_{\Omega_3} = \underline{v}\downarrow_{\Omega_3} + \underline{w}\downarrow_{\Omega_3} = \underline{v}\downarrow_{\Omega_3} + w\downarrow_{\Omega_3 \subset \Omega_1} = \underline{v}\downarrow_{\Omega_3} + (u\downarrow_{\Omega_3} - \underline{v}\downarrow_{\Omega_3}) = u\downarrow_{\Omega_3}$ and $\underline{u}\downarrow_{\Omega_2} = \underline{v}\downarrow_{\Omega_2} + \underline{w}\downarrow_{\Omega_2} = v + 0 = v = u\downarrow_{\Omega_2}$, i.e. $Pu = u$ a.e. on Ω (since area meas$(\Omega \setminus (\Omega_2 \cup \Omega_3)) = 0$) and $Pu \in \mathcal{L}(H^m(\Omega); H^m(\mathbb{R}^2))$ is the required m-extension operator of $\Omega : x_1 > 0$ or $x_2 > 0$. □

m-extension property of polygonal domains in $\mathbb{R}^2$

Since the extension results for planar sectors, quadrants, half-planes, etc. can be locally 'patched' or 'glued' together, the results given in the earlier Propositions and Theorems can be used to construct the required extension for any polygonal domain (see also Grisvard [18]).

Proposition 8.10.8. *Every bounded polygonal domain $\Omega \subset \mathbb{R}^2$ has the m-*extension property.

Proof. We will give the scheme of the proof using a partition of unity (see Appendix C). Since $\overline{\Omega}$ is compact in $\mathbb{R}^2$, $\exists$ a finite *open cover* $\{\Omega_i\}_{i=0}^N$ of $\overline{\Omega} \subset \bigcup_{i=0}^N \Omega_i$ such that $\Gamma \subset \bigcup_{i=1}^N \Omega_i$, $\Omega_i \cap \overline{\Omega} = \Omega_i \cap \overline{S}_i$, $0 \le i \le N$, where $\overline{\Omega}_0 \subset \Omega \subset \mathbb{R}^2$, $S_0 = \mathbb{R}^2$ and S_i, $1 \le i \le N$, is an infinite angular *sector*, possibly also a *quadrant* or a *half-plane*, etc., for which we have already shown the extension results – see Figure 8.8 for an illustration.

With this finite open cover $\{\Omega_i\}_{i=0}^N$ of $\overline{\Omega}$ we associate a partition $\{\phi_i\}_{i=0}^N$ of unity with the properties: $\sum_{i=0}^N \phi_i = 1$ in $\overline{\Omega}$ (i.e. $\sum_{i=0}^N \phi_i(x_1, x_2) = 1\ \forall (x_1, x_2) \in \overline{\Omega}$); $\phi_i \in \mathcal{D}(\mathbb{R}^n)$ with supp$(\phi_i) \subset\subset \Omega_i$ (i.e. $\phi_i \in \mathcal{D}(\Omega_i)$, $0 \le i \le N$) (see Appendix C).

Let $u \in H^m(\Omega)$ and $\tilde{u}$ be its *null extension* to $\mathbb{R}^2$. Then, for $1 \le i \le N$, denoting the restriction $\phi_i\tilde{u}\downarrow_{S_i}$ by $\phi_i\tilde{u}$, we have $\phi_i\tilde{u}\downarrow_{S_i} = \phi_i\tilde{u} \in H^m(S_i)$ with supp$(\phi_i\tilde{u}) \subset\subset S_i$, $\{S_i\}_{i=1}^N$ being infinite angular sectors. For $i = 0$, $S_0 = \mathbb{R}^2$ and let $\phi_0\tilde{u} = w_0$. Then $w_0 \in H^m(\mathbb{R}^2)$, since supp$(\phi_0\tilde{u}) = $ supp$(\phi_0) \cap$ supp$(\tilde{u}) \subset$

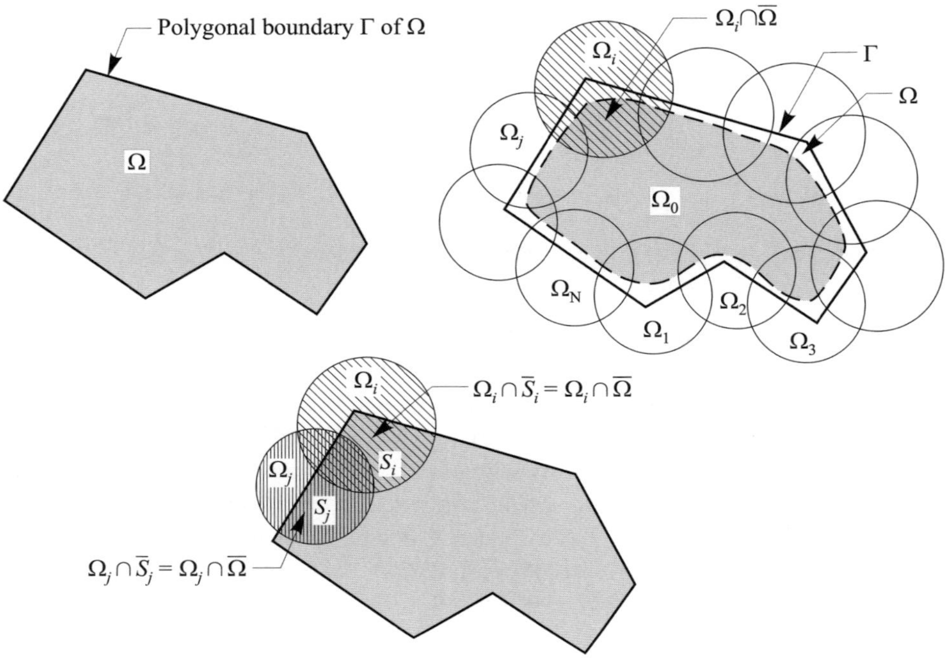

Figure 8.8 Finite open cover $\{\Omega_i\}_{i=0}^N$ with $\Omega_i \cap \overline{\Omega} = \Omega_i \cap \overline{S}_i$, $1 \le i \le N$, $\{S_i\}_{i=1}^N$ being infinite angular sectors

$\mathrm{supp}(\phi_0) \subset\subset \Omega_0 \subset \Omega \subset \mathbb{R}^2$, the result follows from Theorem 2.15.5. For $i = 0$, the extension operator $P_0 \in \mathcal{L}(H^m(\Omega); H^m(\mathbb{R}^2))$ is defined by $P_0 u = w_0 \in H^m(\mathbb{R}^2)$ with $w_0 = \phi_0 \tilde{u}$ for $u \in H^m(\Omega)$. Hence, for $1 \le i \le N$, $\exists$ an *extension* operator $P_i \in \mathcal{L}(H^m(S_i); H^m(\mathbb{R}^2))$ and $\exists w_i \in H^m(\mathbb{R}^2)$ such that $P_i(\phi_i \tilde{u}) = w_i \in H^m(\mathbb{R}^2)$ with $P_i(\phi_i \tilde{u})\downarrow_{S_i} = w_i \downarrow_{S_i} = \phi_i \tilde{u}$, $\forall u \in H^m(\Omega)$. Then the required *m-extension operator* $P \in \mathcal{L}(H^m(\Omega); H^m(\mathbb{R}^2))$ is defined by $Pu = w_0 + \sum_{i=1}^N w_i = \sum_{i=0}^N w_i = w \in H^m(\mathbb{R}^2)$. In fact, $(x_1, x_2) \in \Omega \Longrightarrow \exists i$ with $0 \le i \le N$ such that $(x_1, x_2) \in S_i$ and $(P_i(\phi_i \tilde{u}))(x_1, x_2) = w_i(x_1, x_2) = \phi_i \tilde{u}(x_1, x_2)$. Then

$$
\begin{aligned}
(Pu)(x_1, x_2) = w(x_1, x_2) &= \sum_{i=0}^N w_i(x_1, x_2) = \sum_{i=0}^N (\phi_i \tilde{u})(x_1, x_2) \\
&= \left(\sum_{i=0}^N \phi_i(x_1, x_2) \right) \tilde{u}(x_1, x_2) = 1 \cdot \tilde{u}(x_1, x_2) = u(x_1, x_2)
\end{aligned}
$$

for almost all $(x_1, x_2) \in \Omega \Longrightarrow (Pu)\downarrow_\Omega = u$ a.e. on Ω. □

8.10.4 m-extension property of C^m-regular domains Ω

C^m-regular domains Ω

Definition 8.10.4. An open subset $\Omega \subset \mathbb{R}^n$ with boundary Γ is called *C^m-regular* with $m \in \mathbb{N}$ if and only if

1. Ω is bounded;
2. Ω is *locally* on one side of its boundary Γ (i.e. Ω is a *one-sided* domain);
3. Γ is an $(n-1)$-dimensional *manifold* belonging to *C^m-class* (see Appendix D).

Some explanations on Properties 2 and 3 are in order.

Property 2 excludes the possibility of Ω having a *slit/cut/crack*. For example, $\Omega = \{(x_1, x_2) : 0 < |x_1| < 1, 0 < x_2 < 1\} \subset \mathbb{R}^2$ is a bounded domain in $\mathbb{R}^2$ with a slit/cut along the boundary segment $\Gamma_0 : x_1 = 0, 0 < x_2 < 1$ (see Remark 8.10.2). Hence, it is locally on *two* sides of the boundary segment Γ_0, i.e. it is a *two-sided* domain *locally* along Γ_0 and Property 2 does *not* hold. Similarly, $\Omega = \{(x_1, x_2) : x_1^2 + x_2^2 < 1,$ $x_2 \neq 0$ for $0 < x_1 < 1\} \subset \mathbb{R}^2$ has a cut/crack along $\Gamma_0 : x_2 = 0, 0 < x_1 < 1$ and is locally on both sides of Γ_0. Hence, Property 2 does not hold in this case, either. For more examples, see Dauge [44].

Property 3: we define an $(n-1)$-dimensional boundary Γ of C^m-class in a manner suitable for Theorem 8.10.5 as follows (for an alternative definition and other details, see Appendix D).

Boundary Γ of C^m-class

Definition 8.10.5. Let $\{\Omega_i\}_{i=0}^N$ be a finite *open cover* of the compact set $\overline{\Omega} = \Omega \cup \Gamma \subset \mathbb{R}^n$ such that $\overline{\Omega}_0 \subset \Omega$, and $\{\Omega_i\}_{i=1}^N$ be an *open cover* of the boundary Γ of Ω (see Appendix C). Then Γ *belongs to C^m-class with $m \in \mathbb{N}$* if and only if. $\forall i = 1, 2, \dots, N$, $\exists$ an *m-times continuously differentiable* and *bounded* (i.e. the derivatives of *all* orders $\leq m$ are *continuous* and *bounded*), *invertible* mapping $\boldsymbol{\Phi}_i$:

$$\boldsymbol{\Phi}_i : \mathbf{x} \in \Omega_i \longmapsto \boldsymbol{\Phi}_i(\mathbf{x}) = \boldsymbol{\xi} \in \widehat{Q} \subset \mathbb{R}^n, \tag{8.10.60}$$

where $\widehat{Q}$ is a *hypercube* in $\mathbb{R}^n : \widehat{Q} = \{\boldsymbol{\xi} : \boldsymbol{\xi} = (\xi_1, \dots, \xi_n) \in \mathbb{R}^n, |\xi_j| < 1$ $\forall j = 1, 2, \dots, n\}$. $\forall i$, the same hypercube $\widehat{Q}$ can be chosen (and will be chosen) such that its inverse

$$\boldsymbol{\Phi}_i^{-1} : \boldsymbol{\xi} \in \widehat{Q} \longmapsto \boldsymbol{\Phi}_i^{-1}(\boldsymbol{\xi}) = \mathbf{x} \in \Omega_i \tag{8.10.61}$$

is also an *m-times continuously differentiable* and *bounded* mapping from $\widehat{Q}$ onto Ω_i and the following additional conditions hold: $\boldsymbol{\Phi}_i(\Omega_i \cap \Omega) = \widehat{Q} \cap \mathbb{R}^n_+ \equiv \widehat{Q}_+$ with $\mathbb{R}^n_+ = \{\boldsymbol{\xi} : \boldsymbol{\xi} = (\widehat{\boldsymbol{\xi}}, \xi_n) \in \mathbb{R}^n$ with $\widehat{\boldsymbol{\xi}} = (\xi_1, \dots, \xi_{n-1}) \in \mathbb{R}^{n-1}, \xi_n > 0\}$,

$$\boldsymbol{\Phi}_i(\Omega_i \cap \Gamma) = \widehat{Q} \cap \{\boldsymbol{\xi} : \boldsymbol{\xi} = (\widehat{\boldsymbol{\xi}}, \xi_n) \text{ with } \widehat{\boldsymbol{\xi}} \in \mathbb{R}^{n-1}, \xi_n = 0\} \equiv \widehat{Q}_0. \tag{8.10.62}$$

m-extension property of C^m-regular bounded domain Ω

Theorem 8.10.5. *Let $\Omega \subset \mathbb{R}^n$ be a bounded open subset of $\mathbb{R}^n$ with a C^m-regular boundary Γ for $m \geq 1$. Then Ω has the m-extension property.*

Proof. Since $\overline{\Omega} = \Omega \cup \Gamma \subset \mathbb{R}^n$ is a compact subset of $\mathbb{R}^n$, $\exists$ a finite open cover $\{\Omega_i\}_{i=0}^N$ of $\overline{\Omega}$ with open subsets Ω_i, $0 \leq i \leq N$, such that $\overline{\Omega}_0 \subset \Omega$, $\overline{\Omega} \subset \bigcup_{i=0}^N \Omega_i$ and $\Gamma \subset \bigcup_{i=1}^N \Omega_i$; see Figure 8.9.

Then, by Theorem C.1.1.1 (Appendix C), $\exists$ a C_0^∞-partition $\{\phi_i\}_{i=0}^N$ of unity 1 (i.e. $\sum_{i=0}^N \phi_i(\mathbf{x}) = 1 \ \forall \mathbf{x} \in \Omega$) subordinate to this finite open cover $\{\Omega_i\}_{i=0}^N$ with $\phi_i \in \mathcal{D}(\Omega_i)$ and $\operatorname{supp}(\phi_i) = K_i \subset\subset \Omega_i, 0 \leq i \leq N$. Let $u \in H^m(\Omega)$. Considering the null extension $\tilde{\phi_i}$ to $\mathbb{R}^n$ of ϕ_i and then taking the restriction $\tilde{\phi_i}\downarrow_\Omega \in \mathcal{D}(\overline{\Omega})$, $0 \leq i \leq N$, we have $u_i = \tilde{\phi_i}\downarrow_\Omega \cdot u \in H^m(\Omega)$, which we agree to denote by $u_i = \phi_i u \in H^m(\Omega)$ with $\phi_i \in \mathcal{D}(\Omega_i), u \in H^m(\Omega)$,

$$\operatorname{supp}(u_i) = \operatorname{supp}(\phi_i) \cap \operatorname{supp}(u) \subset \operatorname{supp}(\phi_i) = K_i \subset\subset \Omega_i, 0 \leq i \leq N. \quad (8.10.63)$$

Then

$$\sum_{i=0}^N u_i = \Big(\sum_{i=0}^N \phi_i\Big) u = 1 \cdot u = u \quad \text{in } \Omega. \quad (8.10.64)$$

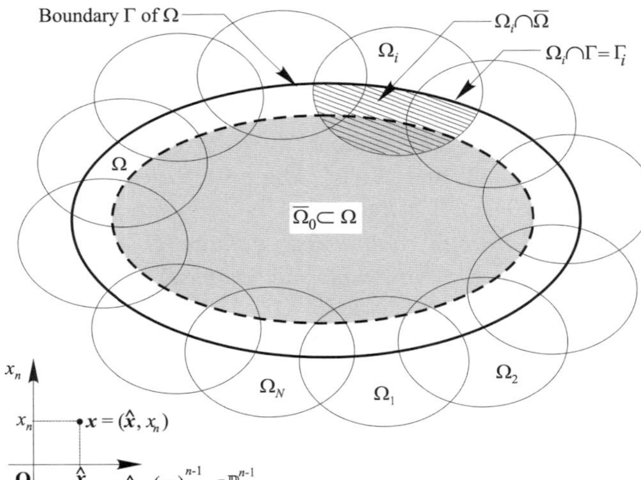

Figure 8.9 Finite open cover $\{\Omega_i\}_{i=0}^N$ with $\overline{\Omega}_0 \subset \Omega$, $\Omega \subset \bigcup_{i=0}^N \Omega_i$, $\Gamma \subset \bigcup_{i=1}^N \Omega_i$, $\Gamma \subset \bigcup_{i=1}^N \Gamma_i$

For $i = 0$, $u_0 \in H^m(\Omega)$ with $\operatorname{supp}(u_0) = K_0 \subset\subset \Omega_0 \subset \Omega \Longrightarrow$ its null extension $\tilde{u}_0 = \widetilde{\phi_0 u} \in H^m(\mathbb{R}^n)$ with $\tilde{u}_0(\mathbf{x}) = u_0(\mathbf{x})$ for $\mathbf{x} \in \Omega_0$ and $\tilde{u}_0(\mathbf{x}) = 0$ for $\mathbf{x} \in \mathbb{R}^n \setminus \Omega_0$.

Define $P_0 : H^m(\Omega) \longrightarrow H^m(\mathbb{R}^n)$ by

$$P_0 u = \tilde{u}_0 \quad \forall u \in H^m(\Omega), \tag{8.10.65}$$

since $u \in H^m(\Omega) \longmapsto \phi_0 u = u_0 \in H^m(\Omega)$ with $\operatorname{supp}(u_0) = K_0 \subset\subset \Omega_0 \subset \Omega$, and consequently $\tilde{u}_0 = \widetilde{\phi_0 u} \in H^m(\mathbb{R}^n)$ (see Theorem 2.15.5). Then P_0 is a *continuous*, *linear* operator from $H^m(\Omega)$ into $H^m(\mathbb{R}^n)$ with $P_0 u{\downarrow_\Omega} = \tilde{u}_0{\downarrow_\Omega} = u_0$ a.e. on Ω.

$\forall i = 1, 2, \dots, N$, let $\mathbf{\Phi}_i : \Omega_i \longrightarrow \widehat{Q}$ (such a choice of $\widehat{Q}$ $\forall$ fixed i can be made) such that $\mathbf{\Phi}_i(K_i) = L_i$ with $K_i = \operatorname{supp}(\phi_i) \subset\subset \Omega_i$, $L_i \subset \widehat{Q}$, K_i and L_i being *compact* subsets of Ω_i and $\widehat{Q}$, respectively, where $\mathbf{\Phi}_i$ is a C^m-*diffeomorphism*[5] from Ω_i onto $\widehat{Q}$ such that the derivatives of all orders $\leq m$ of the components ϕ_k^i and ψ_k^i $(1 \leq k \leq n)$ of $\mathbf{\Phi}_i = (\phi_k^i)_{k=1}^n$ and $\mathbf{\Psi}_i = \mathbf{\Phi}_i^{-1} = (\psi_k^i)_{k=1}^n$ are *bounded* in Ω_i and $\widehat{Q}$, respectively. Then, $\forall i = 1, 2, \dots, N$, $u_i \circ \mathbf{\Phi}_i^{-1} \in H^m(\widehat{Q})$ with $\operatorname{supp}(u_i \circ \mathbf{\Phi}_i^{-1}) \subset L_i \subset \widehat{Q}$; define $v_i = (u_i \circ \mathbf{\Phi}_i^{-1}){\downarrow_{\widehat{Q}_+}}$ $\forall i = 1, 2, \dots, N$ with $\widehat{Q}_+ = \widehat{Q} \cap \mathbb{R}^n_+$. Then $v_i \in H^m(\widehat{Q}_+)$ with $\operatorname{supp}(v_i) \subset \overline{L_i^+} = L_i \cap \overline{\widehat{Q}_+} \subset\subset \widehat{Q}$, since $u_i \in H^m(\Omega)$ and $\mathbf{\Phi}_i$ is a C^m-diffeomorphism from Ω_i onto $\widehat{Q}$ and the derivatives of all orders $\leq m$ of the components of $\mathbf{\Phi}_i$ and of $\mathbf{\Phi}_i^{-1}$ are *bounded* in Ω_i and $\widehat{Q}$ respectively, $\operatorname{supp}(u_i) \subset K_i = \mathbf{\Phi}_i^{-1}(L_i)$ and, consequently, $\operatorname{supp}(u_i \circ \mathbf{\Phi}_i^{-1}) \subset L_i \subset \widehat{Q}$. Then, $\forall i = 1, 2, \dots, N$, the null extension $\widetilde{u_i \circ \Phi_i^{-1}}$ to $\mathbb{R}^n$ of $u_i \circ \Phi_i^{-1}$ is defined by $\widetilde{u_i \circ \Phi_i^{-1}}(\boldsymbol{\xi}) = u_i \circ \Phi_i^{-1}(\boldsymbol{\xi})$ for $\boldsymbol{\xi} \in \widehat{Q}$ and $\widetilde{u_i \circ \Phi_i^{-1}}(\boldsymbol{\xi}) = 0$ for $\boldsymbol{\xi} \in \mathbb{R}^n \setminus \widehat{Q}$. Since $u_i \circ \Phi_i^{-1} \in H^m(\widehat{Q})$ and $\operatorname{supp}(u_i \circ \Phi_i^{-1}) = \operatorname{supp}(\widetilde{u_i \circ \Phi_i^{-1}}) \subset L_i \subset\subset \widehat{Q}$, $\widetilde{u_i \circ \Phi_i^{-1}} \in H^m(\mathbb{R}^n)$ (by Theorem 2.15.5), whose restriction $\widetilde{u_i \circ \Phi_i^{-1}}{\downarrow_{\mathbb{R}^n_+}} \in H^m(\mathbb{R}^n_+)$. Then, for $v_i = (u_i \circ \Phi_i^{-1}){\downarrow_{\widehat{Q}_+}}$, define $\tilde{v}_i = \widetilde{u_i \circ \Phi_i^{-1}}{\downarrow_{\mathbb{R}^n_+}}$ such that $\tilde{v}_i \in H^m(\mathbb{R}^n_+)$ with $\operatorname{supp}(\tilde{v}_i) = \operatorname{supp}(v_i) \subset \overline{L_i^+}$, $L_i^+ = L_i \cap \widehat{Q}_+$, $\overline{L_i^+} = L_i \cap \overline{\widehat{Q}_+}$; see Figure 8.10.

Hence, by Theorem 8.10.4 on the m-extension property of $\mathbb{R}^n_+$, $\exists$ a *unique* Babitch–Nikolski extension $P\tilde{v}_i \in H^m(\mathbb{R}^n)$ of $\tilde{v}_i \in H^m(\mathbb{R}^n_+)$ with

$$\operatorname{supp}(P\tilde{v}_i) = L_i^* \subset \widehat{Q} \quad \forall \text{ fixed } i, \tag{8.10.66}$$

$\Longrightarrow (P\tilde{v}_i){\downarrow_{\widehat{Q}}} \in H^m(\widehat{Q})$ with $\operatorname{supp}(P\tilde{v}_i) = L_i^* \subset \widehat{Q}$.

But $\mathbf{\Phi}_i : \Omega_i \longrightarrow \widehat{Q}$ is a C^m-diffeomorphism with the derivatives of the components ϕ_k^i, $1 \leq k \leq n$, of $\mathbf{\Phi}_i$ *bounded* in Ω_i. Hence, $\forall i = 1, 2, \dots, N$, $((P\tilde{v}_i) \circ \mathbf{\Phi}_i) \in$

[5] C^m-*diffeomorphism*: $\forall$ fixed i, $1 \leq i \leq N$, $\mathbf{\Phi}_i : \Omega_i \longrightarrow \widehat{Q}$ is called a C^m-*diffeomorphism* from Ω_i onto $\widehat{Q}$ if and only if

1. $\mathbf{\Phi}_i = (\phi_1^i, \phi_2^i, \dots, \phi_n^i)$ is a *bijection* and of C^m-class (i.e. the derivatives of *all* orders $\leq m$ of the components ϕ_k^i, $1 \leq k \leq n$, of $\mathbf{\Phi}_i$ are *continuous* from Ω_i onto $\widehat{Q}$);
2. its inverse $\mathbf{\Phi}_i^{-1} = \mathbf{\Psi}_i = (\psi_1^i, \psi_2^i, \dots, \psi_n^i) : \widehat{Q} \longrightarrow \Omega_i$ is also of C^m-class (i.e. the derivatives of *all* orders $\leq m$ of the components ψ_k^i, $1 \leq k \leq n$, of $\mathbf{\Psi}_i$ are *continuous* from $\widehat{Q}$ onto Ω_i).

$H^m(\Omega_i)$ with $\operatorname{supp}(P\tilde{v}_i \circ \mathbf{\Phi}_i) \subset\subset \Omega_i$. Let $\widetilde{(P\tilde{v}_i \circ \mathbf{\Phi}_i)}$ be the null extension to $\mathbb{R}^n$ of $P\tilde{v}_i \circ \mathbf{\Phi}_i$, i.e. $\widetilde{(P\tilde{v}_i \circ \mathbf{\Phi}_i)}(\mathbf{x}) = (P\tilde{v}_i \circ \mathbf{\Phi}_i)(\mathbf{x})$ for $\mathbf{x} \in \Omega_i$ and $\widetilde{(P\tilde{v}_i \circ \mathbf{\Phi}_i)}(\mathbf{x}) = 0$ for $\mathbf{x} \in \mathbb{R}^n \setminus \Omega_i$. Then $\widetilde{(P\tilde{v}_i \circ \mathbf{\Phi}_i)} \in H^m(\mathbb{R}^n)$ (see the proof of Theorem 2.15.5).

$\forall u \in H^m(\Omega)$, define

$$P_i u = \widetilde{P\tilde{v}_i \circ \mathbf{\Phi}_i} \in H^m(\mathbb{R}^n), \tag{8.10.67}$$

with $v_i = \widetilde{(u_i \circ \mathbf{\Phi}_i^{-1})}\downarrow_{\widehat{Q}_+}$, $\tilde{v}_i = \widetilde{(u_i \circ \mathbf{\Phi}_i^{-1})}\downarrow_{\mathbb{R}^n_+}$ $\forall i = 1, 2, \ldots, N$.

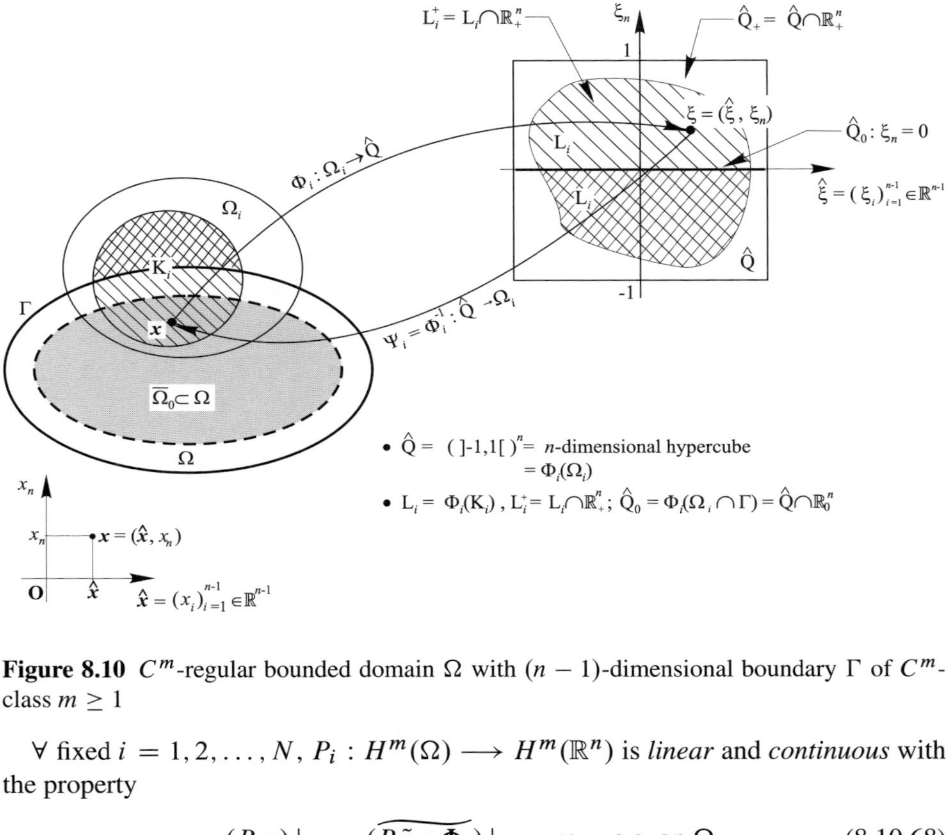

Figure 8.10 C^m-regular bounded domain Ω with $(n-1)$-dimensional boundary Γ of C^m-class $m \geq 1$

$\forall$ fixed $i = 1, 2, \ldots, N$, $P_i : H^m(\Omega) \longrightarrow H^m(\mathbb{R}^n)$ is *linear* and *continuous* with the property

$$(P_i u)\downarrow_\Omega = \widetilde{(P\tilde{v} \circ \mathbf{\Phi}_i)}\downarrow_\Omega = u_i \quad \text{a.e. on } \Omega, \tag{8.10.68}$$

$u_i \in H^m(\Omega)$ with $\operatorname{supp}(u_i) \subset\subset \Omega_i$ (by (8.10.63)).

Now we define $P_m : H^m(\Omega) \longrightarrow H^m(\mathbb{R}^n)$ by

$$P_m u = P_0 u + \sum_{i=1}^{N} P_i u \quad \forall u \in H^m(\Omega), \tag{8.10.69}$$

where P_0 and P_i, $1 \leq i \leq N$, are defined by (8.10.65) and (8.10.68) respectively.

Then, $\forall u \in H^m(\Omega)$, $(P_m u)\downarrow_\Omega = (P_0 u)\downarrow_\Omega + \sum_{i=1}^N (P_i u)\downarrow_\Omega = u_0 + \sum_{i=1}^N u_i = u$ a.e. on Ω by (8.10.64), and $P_m : H^m(\Omega) \longrightarrow H^m(\mathbb{R}^n)$ is a *linear, continuous* operator from $H^m(\Omega)$ into $H^m(\mathbb{R}^n)$ by virtue of the *linearity* and *continuity* of P_0 and $\{P_i\}_{i=1}^N$ with $(P_m u)\downarrow_\Omega = u$ a.e. on Ω. Hence, P_m is an m-extension operator of Ω. Thus, any C^m-regular domain with $m \geq 1$ has the m-extension property.

Justification of (8.10.66), supp$(P\tilde{v}_i) = L_i^* \subset\subset \widehat{Q}$ For *fixed* i, $1 \leq i \leq N$, $\tilde{v}_i \in H^m(\mathbb{R}^n_+)$ with $\mathrm{supp}(\tilde{v}_i) = \mathrm{supp}(v) \subset \overline{L_i^+} \subset\subset \widehat{Q}$, $L_i^+ = L_i \cap \widehat{Q}_+$, $\overline{L_i^+} = L_i \cap \overline{\widehat{Q}_+}$, and let $P\tilde{v}_i \in H^m(\mathbb{R}^n)$ be its Babitch–Nikolski extension to $\mathbb{R}^n$. Then $\mathrm{supp}(P\tilde{v}_i) = L_i^* \subset\subset \widehat{Q}$. In fact, by virtue of the density of $\mathcal{D}(\overline{\mathbb{R}^n_+})$ in $H^m(\mathbb{R}^n_+)$, $\exists$ a sequence $(\psi_k)_{k=1}^\infty$ in $\mathcal{D}(\overline{\mathbb{R}^n_+})$ such that $\psi_k \to \tilde{v}_i$ in $H^m(\mathbb{R}^n_+)$ as $k \to \infty$. But $\forall \psi_k \in \mathcal{D}(\overline{\mathbb{R}^n_+})$, its natural extension (denoted by ψ_k itself) $\psi_k \in \mathcal{D}(\mathbb{R}^n)$ may *not* have the support in $\widehat{Q}$. Hence, we are to construct a new sequence $(w_k)_{k=1}^\infty$ in $\mathcal{D}(\mathbb{R}^n)$ ($w_k\downarrow_{\mathbb{R}^n_+} \in \mathcal{D}(\overline{\mathbb{R}^n_+})$ also denoted by w_k itself) with the desired additional properties as follows. Let $\phi \in \mathcal{D}(\mathbb{R}^n)$ with $\phi = 1$ in $\overline{L_i^+}$, $0 \leq \phi(\boldsymbol{\xi}) \leq 1$ $\forall \boldsymbol{\xi} \in \mathbb{R}^n$, $\mathrm{supp}(\phi) \subset E_0 \subset \overline{E}_0 \subset\subset \widehat{Q}$ (see Theorem 6.2.4 for such a ϕ). Then $\phi\tilde{v}_i = \tilde{v}_i \in H^m(\mathbb{R}^n_+)$ $\forall$ *fixed* i. Set $w_k = \phi\psi_k \in \mathcal{D}(\overline{\mathbb{R}^n_+})$ with $\mathrm{supp}(w_k) \subset \mathrm{supp}(\phi) \subset E_0 \subset \widehat{Q}$ $\forall k \in \mathbb{N}$ and $w_k \to \tilde{v}_i$ in $H^m(\mathbb{R}^n_+)$ as $k \to \infty$, since the mapping $\psi_k \longmapsto \phi\psi_k = w_k$ is continuous on $\mathcal{D}(\overline{\mathbb{R}^n_+})$ in the norm of $H^m(\mathbb{R}^n_+)$, $\psi_k \to \tilde{v}_i \Longrightarrow \phi\psi_k \to \phi\tilde{v}_i = \tilde{v}_i \in H^m(\mathbb{R}^n_+)$ as $k \to \infty$. But for *fixed* i, $1 \leq i \leq N$, by the construction of *Babitch–Nikolski extension* (8.10.43), for $w_k = \phi\psi_k \in \mathcal{D}(\overline{\mathbb{R}^n_+})$:

- $Pw_k \in H^m(\mathbb{R}^n)$ with $\mathrm{supp}(Pw_k) \subset E_0 \subset \widehat{Q}$ $\forall k \in \mathbb{N}$;
- $Pw_k \to P\tilde{v}_i$ in $H^m(\mathbb{R}^n)$ (since $P \in \mathcal{L}(H^m(\mathbb{R}^n_+); H^m(\mathbb{R}^n))$ and $w_k \to \tilde{v}_i$ in $H^m(\mathbb{R}^n_+)) \Longrightarrow Pw_k \to P\tilde{v}_i$ in $L^2(\mathbb{R}^n)$ as $k \to \infty \Longrightarrow \mathrm{supp}(P\tilde{v}_i) = L_i^* \subset E_0 \subset \widehat{Q}$ $\forall$ fixed i.

In fact, for fixed i,

$$\|P\tilde{v}_i - Pw_k\|^2_{L^2(\mathbb{R}^n)} = \lim_{k\to\infty}\left[\int_{E_0} |P\tilde{v}_i - Pw_k|^2 d\boldsymbol{\xi} + \int_{\mathbb{R}^n\setminus E_0} |P\tilde{v}_i - 0|^2 d\boldsymbol{\xi}\right] = 0.$$

Hence, $\int_{\mathbb{R}^n\setminus E_0} |P\tilde{v}_i|^2 d\boldsymbol{\xi} = 0 \Longrightarrow P\tilde{v}_i = 0$ a.e. on $\mathbb{R}^n \setminus E_0 \Longrightarrow L_i^* = \mathrm{supp}(P\tilde{v}_i) \subset E_0 \subset \widehat{Q}$. □

8.10.5 Space $H^s(\Omega)$ with $s \in \mathbb{R}^+$, $\Omega \subset \mathbb{R}^n$

For an arbitrary open set $\Omega \subset \mathbb{R}^n$ and $s = [s] + \sigma$ with $[s] =$ the integral part of s $= m \in \mathbb{N}_0$ and $\sigma =$ the *fractional* part of s, $0 < \sigma < 1$, an alternative definition of $H^s(\mathbb{R}^n)$ was given in Definition 8.9.4 as the set $\{u : u \in H^{[s]}(\mathbb{R}^n), \partial^{\boldsymbol{\alpha}} u \in H^\sigma(\mathbb{R}^n)$

$\forall|\boldsymbol{\alpha}| = [s] = m\}$, where $H^\sigma(\mathbb{R}^n)$ is defined by (8.9.38)–(8.9.39) for $0 < \sigma < 1$:

$$H^\sigma(\mathbb{R}^n) = \left\{u : u \in L^2(\mathbb{R}^n), \iint_{\substack{\mathbb{R}^n\times\mathbb{R}^n\\ \mathbf{x}\neq\mathbf{y}}} \frac{|u(\mathbf{x}) - u(\mathbf{y})|^2}{\|\mathbf{x}-\mathbf{y}\|^{n+2\sigma}} d\mathbf{x}d\mathbf{y} < +\infty\right\}.$$

We will now extend this definition to *arbitrary* open sets $\Omega \subset \mathbb{R}^n$ by replacing $\mathbb{R}^n$ by Ω (see Grisvard [17], [18], [19], Nečas [16], etc.).

Case $0 < s < 1$ with $s = \sigma$

Definition 8.10.6. For an arbitrary open set $\Omega \subset \mathbb{R}^n$ and $0 < s = \sigma < 1$, $H^\sigma(\Omega)$ is defined by

$$H^\sigma(\Omega) = \left\{u : u \in L^2(\Omega), \iint_{\substack{\Omega\times\Omega\\ \mathbf{x}\neq\mathbf{y}}} \frac{|u(\mathbf{x}) - u(\mathbf{y})|^2}{\|\mathbf{x}-\mathbf{y}\|^{n+2\sigma}} d\mathbf{x}d\mathbf{y} < +\infty\right\},$$

which is equipped with the *inner product* $\langle\cdot,\cdot\rangle_{\sigma,\Omega}$, *norm* $\|\cdot\|_{\sigma,\Omega}$ and *semi-norm* $|\cdot|_{\sigma,\Omega}$, for $u, v \in H^\sigma(\Omega)$, given by:

$$\langle u, v\rangle_{\sigma,\Omega} = \langle u, v\rangle_{L^2(\Omega)} + \iint_{\substack{\Omega\times\Omega\\ \mathbf{x}\neq\mathbf{y}}} \frac{(u(\mathbf{x}) - u(\mathbf{y}))\cdot(v(\mathbf{x}) - v(\mathbf{y}))}{\|\mathbf{x}-\mathbf{y}\|^{n+2\sigma}} d\mathbf{x}d\mathbf{y}; \tag{8.10.70a}$$

$$\|u\|^2_{\sigma,\Omega} = \langle u, u\rangle_{\sigma,\Omega} = \|u\|^2_{L^2(\Omega)} + \iint_{\substack{\Omega\times\Omega\\ \mathbf{x}\neq\mathbf{y}}} \frac{|u(\mathbf{x}) - u(\mathbf{y})|^2}{\|\mathbf{x}-\mathbf{y}\|^{n+2\sigma}} d\mathbf{x}d\mathbf{y}; \tag{8.10.70b}$$

$$|u|^2_{\sigma,\Omega} = \iint_{\substack{\Omega\times\Omega\\ \mathbf{x}\neq\mathbf{y}}} \frac{|u(\mathbf{x}) - u(\mathbf{y})|^2}{\|\mathbf{x}-\mathbf{y}\|^{n+2\sigma}} d\mathbf{x}d\mathbf{y}. \tag{8.10.70c}$$

Case $s > 0$

Definition 8.10.7. For an *arbitrary* open set $\Omega \subset \mathbb{R}^n$ and for $s = [s] + \sigma$ with $[s] = m \in \mathbb{N}_0$, $0 < \sigma < 1$, $H^s(\Omega)$ is defined by (Grisvard [17], [18], [19], Nečas [16]):

$$H^s(\Omega) = \{u : u \in H^m(\Omega),\ \partial^{\boldsymbol{\alpha}} u \in H^\sigma(\Omega)\ \forall|\boldsymbol{\alpha}| = m = [s]\}, \tag{8.10.71}$$

which is equipped with the *inner product* $\langle\cdot,\cdot\rangle_{s,\Omega}$, *norm* $\|\cdot\|_{s,\Omega}$ and *semi-norm* $|\cdot|_{s,\Omega}$:

$$\begin{aligned}\langle u,v\rangle_{s,\Omega} &= \langle u,v\rangle_{H^m(\Omega)} \\ &\quad + \sum_{|\boldsymbol{\alpha}|=m} \iint_{\substack{\Omega\times\Omega \\ \mathbf{x}\neq\mathbf{y}}} \frac{(\partial^{\boldsymbol{\alpha}}u(\mathbf{x})-\partial^{\boldsymbol{\alpha}}u(\mathbf{y}))\cdot(\partial^{\boldsymbol{\alpha}}v(\mathbf{x})-\partial^{\boldsymbol{\alpha}}v(\mathbf{y}))}{\|\mathbf{x}-\mathbf{y}\|^{n+2\sigma}} d\mathbf{x}d\mathbf{y} \\ &= \sum_{0\le|\boldsymbol{\alpha}|\le m} \langle\partial^{\boldsymbol{\alpha}}u,\partial^{\boldsymbol{\alpha}}v\rangle_{L^2(\Omega} \\ &\quad + \sum_{|\boldsymbol{\alpha}|=m} \iint_{\substack{\Omega\times\Omega \\ \mathbf{x}\neq\mathbf{y}}} \frac{(\partial^{\boldsymbol{\alpha}}u(\mathbf{x})-\partial^{\boldsymbol{\alpha}}u(\mathbf{y}))\cdot(\partial^{\boldsymbol{\alpha}}v(\mathbf{x})-\partial^{\boldsymbol{\alpha}}v(\mathbf{y}))}{\|\mathbf{x}-\mathbf{y}\|^{n+2\sigma}} d\mathbf{x}d\mathbf{y};\end{aligned} \tag{8.10.72}$$

$$\|u\|^2_{s,\Omega} = \sum_{0\le|\boldsymbol{\alpha}|\le m} \|\partial^{\boldsymbol{\alpha}}u\|^2_{L^2(\Omega)} + \sum_{|\boldsymbol{\alpha}|=m} \iint_{\substack{\Omega\times\Omega \\ \mathbf{x}\neq\mathbf{y}}} \frac{|\partial^{\boldsymbol{\alpha}}u(\mathbf{x})-\partial^{\boldsymbol{\alpha}}u(\mathbf{y})|^2}{\|\mathbf{x}-\mathbf{y}\|^{n+2\sigma}} d\mathbf{x}d\mathbf{y}; \tag{8.10.73}$$

$$|u|^2_{s,\Omega} = \sum_{|\boldsymbol{\alpha}|=m} \iint_{\substack{\Omega\times\Omega \\ \mathbf{x}\neq\mathbf{y}}} \frac{|\partial^{\boldsymbol{\alpha}}u(\mathbf{x})-\partial^{\boldsymbol{\alpha}}u(\mathbf{y})|^2}{\|\mathbf{x}-\mathbf{y}\|^{n+2\sigma}} d\mathbf{x}d\mathbf{y}. \tag{8.10.74}$$

Theorem 8.10.6. *For $s > 0$, $H^s(\Omega)$ equipped with the* inner product $\langle\cdot,\cdot\rangle_{s,\Omega}$ *is a* Hilbert space.

Proof. Let $s = [s] + \sigma$ with the integral part $[s] = m \in \mathbb{N}_0$ and fractional part σ, $0 < \sigma < 1$. Let $(u_n)_{n=1}^{\infty}$ be a Cauchy sequence in $H^s(\Omega)$. We are to show that (u_n) converges to $u \in H^s(\Omega)$ as $n \to \infty$. Since (u_n) is a Cauchy sequence in $H^s(\Omega)$, from the definition of $H^s(\Omega)$, (u_n) is a Cauchy sequence in $H^m(\Omega)$ with $m = [s] \in \mathbb{N}_0$ and, $\forall|\boldsymbol{\alpha}| = m$, $I_\sigma[\partial^\alpha u_l - \partial^\alpha u_n] \to 0$ as $l, n \to \infty$, i.e.

$$I_\sigma[\partial^\alpha u_l - \partial^\alpha u_n] = \iint_{\substack{\Omega\times\Omega \\ \mathbf{x}\neq\mathbf{y}}} \frac{|\partial^\alpha(u_l-u_n)(\mathbf{x}) - \partial^\alpha(u_l-u_n)(\mathbf{y})|^2}{\|\mathbf{x}-\mathbf{y}\|^{n+2\sigma}} d\mathbf{x}d\mathbf{y} \to 0 \tag{8.10.75}$$

as $l, n \to \infty$, $\forall|\boldsymbol{\alpha}| = m$, $0 < \sigma < 1$.

Since $H^m(\Omega)$ is a Hilbert space by Theorem 2.15.1, $\exists u \in H^m(\Omega)$ such that $u_n \to u \in L^2(\Omega)$ and $\partial^\alpha u_n \to \partial^\alpha u$ in $L^2(\Omega)$ $\forall|\boldsymbol{\alpha}| \le m$. Then, we can apply the following result.

Lemma 8.10.1 ([20, Theorem 2, p. 305], [27, Theorem 3.12, p. 68]). *If $(u_n)_{n=1}^{\infty}$ converges to $u \in L^p(\Omega)$ such that $\|u-u_n\|_{L^p(\Omega)} \to 0$ as $n \to \infty$, $1 \le p \le \infty$, then* $\exists$ *a* subsequence $(u_{n_k})_{k=1}^{\infty}$ *of the sequence (u_n) which converges* pointwise almost everywhere *to u, i.e. $\lim_{k\to\infty} u_{n_k}(\mathbf{x}) = u(\mathbf{x})$ a.e. on Ω.*

Hence, by Lemma 8.10.1, $\exists$ a subsequence $(\partial^\alpha u_{n_k})$ of $(\partial^\alpha u_n)_{n=1}^\infty$ in $L^2(\Omega)$ such that

$$\lim_{k\to\infty} \partial^\alpha u_{n_k}(\mathbf{x}) = \partial^\alpha u(\mathbf{x}) \quad \text{a.e. on } \Omega \ \forall|\boldsymbol{\alpha}| = m. \tag{8.10.76}$$

Then, for $\mathbf{x} \neq \mathbf{y}$, $\forall$ fixed $\boldsymbol{\alpha}$ with $|\boldsymbol{\alpha}| = m$, the sequence

$$\left(\frac{|\partial^\alpha u_{n_k}(\mathbf{x}) - \partial^\alpha u_{n_k}(\mathbf{y})|^2}{\|\mathbf{x}-\mathbf{y}\|^{n+2\sigma}}\right) \to \left(\frac{|\partial^\alpha u(\mathbf{x}) - \partial^\alpha u(\mathbf{y})|^2}{\|\mathbf{x}-\mathbf{y}\|^{n+2\sigma}}\right) \tag{8.10.77}$$

a.e. on $\Omega \times \Omega$ as $k \to \infty$.

In fact,

$$\begin{aligned} &|(\partial^\alpha u(\mathbf{x}) - \partial^\alpha u(\mathbf{y})) - (\partial^\alpha u_{n_k}(\mathbf{x}) - \partial^\alpha u_{n_k}(\mathbf{y}))| \\ &\leq |\partial^\alpha u(\mathbf{x}) - \partial^\alpha u_{n_k}(\mathbf{x})| + |\partial^\alpha u(\mathbf{y}) - \partial^\alpha u_{n_k}(\mathbf{y})| \to 0 \end{aligned}$$

a.e. in $\Omega \times \Omega$ as $k \to \infty$ $\forall|\boldsymbol{\alpha}| = m$ (by virtue of (8.10.76))
$\Longrightarrow \forall|\boldsymbol{\alpha}| = m$,

$$\lim_{k\to\infty} |\partial^\alpha u_{n_k}(\mathbf{x}) - \partial^\alpha u_{n_k}(\mathbf{y})| = |\partial^\alpha u(\mathbf{x}) - \partial^\alpha u(\mathbf{y})|$$

a.e. in $\Omega \times \Omega$
$\Longrightarrow \forall|\boldsymbol{\alpha}| = m$,

$$\lim_{k\to\infty} \frac{|\partial^\alpha u_{n_k}(\mathbf{x}) - \partial^\alpha u_{n_k}(\mathbf{y})|^2}{\|\mathbf{x}-\mathbf{y}\|^{n+2\sigma}} = \frac{|\partial^\alpha u(\mathbf{x}) - \partial^\alpha u(\mathbf{y})|^2}{\|\mathbf{x}-\mathbf{y}\|^{n+2\sigma}}$$

a.e. in $\Omega \times \Omega$ for $\mathbf{x} \neq \mathbf{y}$
$\Longrightarrow \forall|\boldsymbol{\alpha}| = m$,

$$\liminf_{k\to\infty} \frac{|\partial^\alpha u_{n_k}(\mathbf{x}) - \partial^\alpha u_{n_k}(\mathbf{y})|^2}{\|\mathbf{x}-\mathbf{y}\|^{n+2\sigma}} = \frac{|\partial^\alpha u(\mathbf{x}) - \partial^\alpha u(\mathbf{y})|^2}{\|\mathbf{x}-\mathbf{y}\|^{n+2\sigma}} \tag{8.10.78}$$

a.e. in $\Omega \times \Omega$ for $\mathbf{x} \neq \mathbf{y}$.

Since (u_n) is a Cauchy sequence in $H^s(\Omega)$ satisfying (8.10.75), i.e. $\forall \varepsilon > 0$, $\exists n_0 = n_0(\varepsilon) \in \mathbb{N}$ such that $\forall|\boldsymbol{\alpha}| = m$, $I_\sigma[\partial^\alpha u_l - \partial^\alpha u_n] < \varepsilon \ \forall\, l, n > n_0(\varepsilon) \in \mathbb{N}$ (8.10.78a)
$\Longrightarrow I_\sigma[\partial^\alpha u_l - \partial^\alpha u_{n_k}] < \varepsilon \ \forall l > n_0$, $\forall k > k_0$ with $n_k > n_{k_0} > n_0$, $\forall|\boldsymbol{\alpha}| = m$
$\Longrightarrow I_\sigma[\partial^\alpha u_{n_k}]$ is *bounded* $\forall k \in \mathbb{N}$, i.e. $\forall k \in \mathbb{N}$, $\exists M > 0$ such that, $\forall|\boldsymbol{\alpha}| = m$,

$$I_\sigma[\partial^\alpha u_{n_k}] = \iint_{\Omega\times\Omega} \frac{|\partial^\alpha u_{n_k}(\mathbf{x}) - \partial^\alpha u_{n_k}(\mathbf{y})|^2}{\|\mathbf{x}-\mathbf{y}\|^{n+2\sigma}} d\mathbf{x}d\mathbf{y} \leq M < +\infty \tag{8.10.79}$$

$$\begin{aligned} \Longrightarrow \quad \liminf_{k\to\infty} I_\sigma[\partial^\alpha u_{n_k}] &= \liminf_{k\to\infty} \iint_{\Omega\times\Omega} \frac{|\partial^\alpha u_{n_k}(\mathbf{x}) - \partial^\alpha u_{n_k}(\mathbf{y})|^2}{\|\mathbf{x}-\mathbf{y}\|^{n+2\sigma}} d\mathbf{x}d\mathbf{y} \\ &\leq M < +\infty \quad \forall|\boldsymbol{\alpha}| = m. \end{aligned} \tag{8.10.80}$$

Hence, for the sequence $\left(\frac{|\partial^\alpha u_{n_k}(\mathbf{x})-\partial^\alpha u_{n_k}(\mathbf{y})|^2}{\|\mathbf{x}-\mathbf{y}\|^{n+2\sigma}}\right)_{k=1}^{\infty}$ of *non-negative* functions integrable on Ω satisfying (8.10.78) and (8.10.79), we can apply Fatou's Lemma:

Lemma 8.10.2. *Let* (f_n) *be a sequence of* non-negative, integrable *functions on* Ω *(i.e.* $f_n \geq 0$, $\int_\Omega f_n(\mathbf{x})d\mathbf{x} \leq M < \infty\ \forall n \in \mathbb{N}$*) such that*

$$\liminf_{n\to\infty} \int_\Omega f_n(\mathbf{x})d\mathbf{x} < \infty, \quad \liminf_{n\to\infty} f_n(\mathbf{x}) = f(\mathbf{x}) \quad \textit{a.e. on } \Omega.$$

Then

I. f *is integrable on* Ω, *i.e.* $\int_\Omega f(\mathbf{x})d\mathbf{x} < \infty$;

II. $\int_\Omega f(\mathbf{x})d\mathbf{x} \leq \liminf_{n\to\infty} \int_\Omega f_n(\mathbf{x})d\mathbf{x}$.

In other words, $\int_\Omega \liminf_{n\to\infty} f_n(\mathbf{x})d\mathbf{x} \leq \liminf_{n\to\infty} \int_\Omega f_n(\mathbf{x})d\mathbf{x}$.

Remark 8.10.5. If $\lim_{n\to\infty} f_n(\mathbf{x}) = f(\mathbf{x})$ a.e. on Ω, then $\liminf_{n\to\infty} f_n(\mathbf{x}) = \lim_{n\to\infty} f_n(\mathbf{x}) = f(\mathbf{x})$ a.e. on Ω by the property of the *limit inferior* (see Definitions A.0.8.1 and A.0.8.2 in Appendix A).

Consequently, by virtue of (8.10.78) and (8.10.79), I and II of Fatou's Lemma 8.10.2 follow, i.e. $\frac{|\partial^\alpha u(\mathbf{x})-\partial^\alpha u(\mathbf{y})|^2}{\|\mathbf{x}-\mathbf{y}\|^{n+2\sigma}}$ is integrable on $\Omega \times \Omega\ \forall |\boldsymbol{\alpha}| = m$ and

$$I_\sigma[\partial^\alpha u] = \iint_{\substack{\Omega\times\Omega \\ \mathbf{x}\neq\mathbf{y}}} \frac{|\partial^\alpha u(\mathbf{x}) - \partial^\alpha u(\mathbf{y})|^2}{\|\mathbf{x}-\mathbf{y}\|^{n+2\sigma}} d\mathbf{x}d\mathbf{y} \leq \liminf_{k\to\infty} I_\sigma[\partial^\alpha u_{n_k}] < \infty, \quad \forall |\boldsymbol{\alpha}| = m.$$

Hence, $u \in H^m(\Omega)$ with $m = [s]$ and $I_\sigma[\partial^\alpha u] < \infty\ \forall |\boldsymbol{\alpha}| = m$ and for $0 < \sigma < 1 \Longrightarrow u \in H^s(\Omega)$.

Now we are to show that for $u \in H^s(\Omega)$,

$$I_\sigma[\partial^\alpha u - \partial^\alpha u_n] = \iint_{\Omega\times\Omega \mathbf{x}\neq\mathbf{y}} \frac{|\partial^\alpha(u-u_n)(\mathbf{x}) - \partial^\alpha(u-u_n)(\mathbf{y})|^2}{\|\mathbf{x}-\mathbf{y}\|^{n+2\sigma}} d\mathbf{x}d\mathbf{y} \to 0$$

as $n \to \infty$, $\forall |\boldsymbol{\alpha}| = m$.

In fact, from (8.10.78a), we also have $I_\sigma[\partial^\alpha u_{l_k} - \partial^\alpha u_n] < \varepsilon\ \forall n > n_0$, $\forall k > k_0$, with $l_k > l_{k_0} > n_0\ \forall |\boldsymbol{\alpha}| = m$. We *fix* $n > n_0$. Then, $\forall$ *fixed* $n > n_0$, $I_\sigma[\partial^\alpha u_{l_k} - \partial^\alpha u_n] < \varepsilon\ \forall k > k_0$ with $l_k > l_{k_0} > n_0\ \forall |\boldsymbol{\alpha}| = m$

$$\Longrightarrow \quad \liminf_{k\to\infty} I_\sigma[\partial^\alpha u_{l_k} - \partial^\alpha u_n] < \varepsilon \quad \forall \textit{ fixed } n > n_0,\ \forall |\boldsymbol{\alpha}| = m. \tag{8.10.81}$$

Applying (8.10.76) with $l_k = n_k\ \forall k \in \mathbb{N}$, $\forall$ *fixed* $n > n_0$, $\liminf_{k\to\infty} \partial^\alpha(u_{l_k} - u_n)(\mathbf{x}) = \partial^\alpha(u-u_n)(\mathbf{x})$ a.e. on $\Omega\ \forall |\boldsymbol{\alpha}| = m$.

Then,

$$\liminf_{k\to\infty} \frac{|\partial^\alpha(u_{l_k}-u_n)(\mathbf{x}) - \partial^\alpha(u_{l_k}-u_n)(\mathbf{y})|^2}{\|\mathbf{x}-\mathbf{y}\|^{n+2\sigma}} = \frac{|\partial^\alpha(u-u_n)(\mathbf{x}) - \partial^\alpha(u-u_n)(\mathbf{y})|^2}{\|\mathbf{x}-\mathbf{y}\|^{n+2\sigma}} \tag{8.10.82}$$

a.e. on $\Omega \times \Omega\ \forall$ *fixed* $n > n_0$, $\forall |\boldsymbol{\alpha}| = m$ (see (8.10.78)).

Again, for the sequence $\left(\frac{|\partial^\alpha(u_{l_k}-u_n)(\mathbf{x})-\partial^\alpha(u_{l_k}-u_n)(\mathbf{y})|^2}{\|\mathbf{x}-\mathbf{y}\|^{n+2\sigma}}\right)_{k=1}^\infty$ $\forall$ *fixed* $n > n_0$ satisfying (8.10.81) and (8.10.82), we can apply Fatou's Lemma 8.10.2 once more and get, $\forall \varepsilon > 0$, $\exists n_0 = n_0(\varepsilon) \in \mathbb{N}$ such that, $\forall |\boldsymbol{\alpha}| = m$,

$$\begin{aligned}I_\sigma[\partial^\alpha u - \partial^\alpha u_n] &= \iint_{\substack{\Omega\times\Omega\\ \mathbf{x}\neq\mathbf{y}}} \frac{|\partial^\alpha(u-u_n)(\mathbf{x}) - \partial^\alpha(u-u_n)(\mathbf{y})|^2}{\|\mathbf{x}-\mathbf{y}\|^{n+2\sigma}} d\mathbf{x}d\mathbf{y}\\ &\le \liminf_{k\to\infty} I_\sigma[\partial^\alpha u_{l_k} - \partial^\alpha u_n] < \varepsilon \quad \forall n > n_0\end{aligned}$$

$\Longrightarrow \lim_{n\to\infty} I_\sigma[\partial^\alpha u - \partial^\alpha u_n] = 0 \ \forall |\boldsymbol{\alpha}| = m.$

Thus we have proved that for $u \in H^s(\Omega)$, $\|u - u_n\|_{H^m(\Omega)} \to 0$ with $m = [s]$ and $I_\sigma[\partial^\alpha u - \partial^\alpha u_n] \to 0$ as $n \to \infty$ $\forall |\boldsymbol{\alpha}| = m \Longrightarrow u_n \to u$ in $H^s(\Omega)$ as $n \to \infty$. Hence, $H^s(\Omega)$ is a Hilbert space. □

Equivalent norm in Hilbert space $H^s(\Omega)$ with $s > 0$

In some situations it may be convenient to use the norm in $H^s(\Omega)$ with $s = [s] + \sigma$, $[s] = m \in \mathbb{N}_0$, $0 \le \sigma < 1$, defined by (see also Definition 8.9.4 and (8.9.41)):

$$\begin{aligned}\|u\|^2_{H^s(\Omega)} &= \|u\|^2_{H^m(\Omega)} + \sum_{1\le|\boldsymbol{\alpha}|\le m} \|\partial^{\boldsymbol{\alpha}} u\|^2_{H^\sigma(\Omega)}\\ &= \sum_{0\le|\boldsymbol{\alpha}|\le m} \|\partial^{\boldsymbol{\alpha}} u\|^2_{L^2(\Omega)} + \sum_{1\le|\boldsymbol{\alpha}|\le m} \iint_{\substack{\Omega\times\Omega\\ \mathbf{x}\neq\mathbf{y}}} \frac{|\partial^\alpha u(\mathbf{x}) - \partial^\alpha u(\mathbf{y})|^2}{\|\mathbf{x}-\mathbf{y}\|^{n+2\sigma}} d\mathbf{x}d\mathbf{y}\end{aligned} \tag{8.10.83}$$

which is *equivalent* to the original norm:

$$\|u\|^2_{H^s(\Omega)} = \sum_{0\le|\boldsymbol{\alpha}|\le m} \|\partial^{\boldsymbol{\alpha}} u\|^2_{L^2(\Omega)} + \sum_{|\boldsymbol{\alpha}|=m} \iint_{\substack{\Omega\times\Omega\\ \mathbf{x}\neq\mathbf{y}}} \frac{|\partial^\alpha u(\mathbf{x}) - \partial^\alpha u(\mathbf{y})|^2}{\|\mathbf{x}-\mathbf{y}\|^{n+2\sigma}} d\mathbf{x}d\mathbf{y}. \tag{8.10.84}$$

Hence, in the following we will use either of these two *equivalent* norms interchangeably, according to our convenience.

8.10.6 Density results in $H^s(\Omega)$

Although $\mathcal{D}(\mathbb{R}^n)$ is dense in $H^s(\mathbb{R}^n)$ $\forall s > 0$ (see Theorem 8.9.6) (in fact, this density result holds $\forall s \in \mathbb{R}$), $\mathcal{D}(\Omega)$ is *not* dense in $H^s(\Omega)$ for $\Omega \neq \mathbb{R}^n$ if $s > 1/2$ (see also (2.15.26d)). But we have:

Theorem 8.10.7. *Let* $\Omega \subset \mathbb{R}^n$ *be an open set of* $\mathbb{R}^n$ *with* Lipschitz continuous boundary Γ *(see Definition* D.2.3.1, *Appendix* D*). Then* $\mathcal{D}(\bar{\Omega})$ *is* dense *in* $H^s(\Omega)$ $\forall s > 0$, *where* $\mathcal{D}(\bar{\Omega}) = \{\phi : \exists\psi \in \mathcal{D}(\mathbb{R}^n)$ *such that* $\phi = \psi\downarrow_\Omega\}$ *(see Definition* 8.10.2*).*

Proof. For *Lipschitz continuous* boundary Γ of Ω, the result follows from Theorem 8.10.1 and Theorem 8.10.9 given later, since $H^s(\Omega) \equiv H^s(\overline{\Omega})$ for $s > 0$.

For arbitrary Ω,

$$\mathcal{D}(\overline{\Omega}) \cap H^s(\Omega) \text{ is } \textit{dense} \text{ in } H^s(\Omega) \tag{8.10.85}$$

for $s > 0$ [12]. □

Remark 8.10.6. Theorem 8.10.7 also holds for a continuous boundary [18]. Domains with slits do not have a continuous boundary (see Appendix D). For example, $\Omega = \{(x_1, x_2) : 0 < |x_1| < 1, 0 < |x_2| < 1\}$ in Remark 8.10.2 has a slit along $\Gamma_0 : x_1 = 0, 0 < x_2 < 1$ and consequently does *not* possess a continuous boundary Γ, and $\mathcal{D}(\overline{\Omega})$ is *not* dense in $H^s(\Omega)$ for $s > 0$ [12].

Proposition 8.10.9 ([17, p. 5], [18]). *Let $\Omega \subset \mathbb{R}^n$ be a* bounded *domain with* Lipschitz continuous *boundary* Γ. *Then $\mathcal{D}(\Omega)$ is* dense *in $H^s(\Omega)$ for $0 \leq s \leq 1/2$.*

In particular,

$$\mathcal{D}(\Omega) \textit{ is } \text{dense } \textit{in } H^{1/2}(\Omega). \tag{8.10.86}$$

8.10.7 Dual space $H^{-s}(\Omega)$

For $\Omega \neq \mathbb{R}^n$ *and* $s > 1/2$, the dual space $(H^s(\Omega))'$ of $H^s(\Omega)$ cannot be identified with a *space of distributions* (see Remark 4.2.1, Chapter 4), although $(H^s(\mathbb{R}^n))' \equiv H^{-s}(\mathbb{R}^n)$ is a *space of distributions* $\forall s > 0$ (see Theorem 8.9.7 and Theorem 8.9.6). Consequently, we need:

8.10.8 Space $H_0^s(\Omega)$ with $s > 0$

Definition 8.10.8. For $s > 0$, $H_0^s(\Omega) \equiv \overline{\mathcal{D}(\Omega)}$ in the *norm* of $H^s(\Omega)$, i.e. $H_0^s(\Omega)$ is the *closure* of $\mathcal{D}(\Omega)$ in $H^s(\Omega)$.

Theorem 8.10.8. *For $s > 0$, $H_0^s(\Omega)$ is a* Hilbert space *equipped with the inner product $\langle \cdot, \cdot \rangle_{s,\Omega}$ induced by $H^s(\Omega)$.*

Proof. The result follows from Definition 8.10.8, since $H_0^s(\Omega)$ is a closed subspace of Hilbert space $H^s(\Omega)$. □

Corollary 8.10.1. *Let $\Omega \subset \mathbb{R}^n$ be a bounded open subset with* Lipschitz continuous *boundary* Γ. *Then, for $0 < s \leq 1/2$, $H_0^s(\Omega) = H^s(\Omega)$.*

Proof. The result follows from Proposition 8.10.9 and Definition 8.10.8. □

Now we are in a position to define $H^{-s}(\Omega)$ for $s > 0$.

8.10.9 Space $H^{-s}(\Omega)$ with $s > 0$

Definition 8.10.9. For $s > 0$, $H^{-s}(\Omega)$ is defined as the (topological) *dual space* of $H_0^s(\Omega)$, i.e. $(H_0^s(\Omega))' \equiv H^{-s}(\Omega) \hookrightarrow \mathcal{D}'(\Omega)$, i.e. for $s > 0$, $H^{-s}(\Omega)$ is a *space of distributions* ($\mathcal{D}(\Omega) \hookrightarrow H_0^s(\Omega)$, the imbedding $\hookrightarrow$ being a *dense*, *continuous* one).

For Ω with *Lipschitz continuous* boundary Γ and for $0 < s \leq 1/2$, $\mathcal{D}(\Omega)$ is *dense* in $H^s(\Omega) \equiv H_0^s(\Omega)$ and

$$H^{-s}(\Omega) = (H_0^s(\Omega))' = (H^s(\Omega))' \hookrightarrow \mathcal{D}'(\Omega). \tag{8.10.87}$$

Proposition 8.10.10 ([18]). *For* arbitrary $\Omega \subset \mathbb{R}^n$, $\mathcal{D}(\bar{\Omega})$ *is* dense *in* $H^{-s}(\Omega)$ $\forall s > 0$.

Imbedding result For $0 < s_1 < s_2$,

$$H^{s_2}(\Omega) \hookrightarrow H^{s_1}(\Omega) \text{ with } \textit{dense, continuous} \text{ imbedding } \hookrightarrow. \tag{8.10.88}$$

Relation between $H^s(\overline{\Omega})$ and $H^s(\Omega)$ for $s > 0$

Theorem 8.10.9 ([18]). *Let* $\Omega \subset \mathbb{R}^n$ *be a* bounded *domain with* Lipschitz continuous *boundary* Γ. *Then,* $\forall s > 0$, $\exists$ *an* s-extension operator $P \in \mathcal{L}(H^s(\Omega), H^s(\mathbb{R}^n))$ *such that* $(Pu)\downarrow_\Omega = u$, $\|Pu\|_{H^s(\mathbb{R}^n)} \leq C\|u\|_{H^s(\Omega)}$ $\forall u \in H^s(\Omega)$ *and for some* $C > 0$. *Consequently,*

$$H^s(\Omega) \equiv H^s(\bar{\Omega}) \textit{ with norm equivalence}, \tag{8.10.89}$$

where $H^s(\bar{\Omega})$ *is defined by* (8.10.1).

For *bounded polygonal domains* $\Omega \subset \mathbb{R}^2$, the results of Theorem 8.10.9 hold, since bounded polygons are *Lipschitz continuous*. For $s = m$, see the *independent proof* of Proposition 8.10.8.

Remark 8.10.7. For integral $m \in \mathbb{N}_0$ and $1 \leq p \leq \infty$, $W^{m,p}(\Omega)$ has been defined in (2.15.29)–(2.15.34) with $W^{m,2}(\Omega) \equiv H^m(\Omega)$, i.e. by giving a *natural* extension to the definition of $H^m(\Omega)$. Similarly, Definition 8.10.6 of $H^s(\Omega)$ for $s > 0$ can be given a *natural* extension to define the space $W^{s,p}(\Omega)$ for real $s > 0$ and $1 \leq p < \infty$ such that $W^{s,2}(\Omega) \equiv H^s(\Omega)$.

8.10.10 Space $W^{s,p}(\Omega)$ for real $s > 0$ and $1 \leq p < \infty$

Definition 8.10.10. For real $s = [s] + \sigma$ with $[s] = m \in \mathbb{N}_0$, $0 < \sigma < 1$, and $1 \leq p < \infty$, $W^{s,p}(\Omega)$ is defined as a subspace of $W^{m,p}(\Omega)$ (see (2.15.28)) by:

$$W^{s,p}(\Omega) = \left\{u : u \in W^{m,p}(\Omega), \iint_{\substack{\Omega\times\Omega \\ \mathbf{x}\neq\mathbf{y}}} \frac{|\partial^{\boldsymbol{\alpha}} u(\mathbf{x}) - \partial^{\boldsymbol{\alpha}} u(\mathbf{y})|^p}{\|\mathbf{x}-\mathbf{y}\|^{n+\sigma p}} d\mathbf{x}d\mathbf{y} < +\infty \ \forall |\boldsymbol{\alpha}| = m\right\}$$

equipped with the *Slobodetskii norm* $\|\cdot\|_{s,p,\Omega}$ [45] and *semi-norm* $|\cdot|_{s,p,\Omega}$: $\forall u \in W^{s,p}(\Omega)$,

$$\|u\|_{s,p,\Omega} = \left[\|u\|^p_{m,p,\Omega} + \sum_{|\boldsymbol{\alpha}|=m} \iint_{\substack{\Omega\times\Omega \\ \mathbf{x}\neq\mathbf{y}}} \frac{|\partial^{\boldsymbol{\alpha}} u(\mathbf{x}) - \partial^{\boldsymbol{\alpha}} u(\mathbf{y})|^p}{\|\mathbf{x}-\mathbf{y}\|^{n+\sigma p}} d\mathbf{x} d\mathbf{y} \right]^{1/p}; \tag{8.10.90a}$$

$$\|u\|^p_{m,p,\Omega} = \|u\|^p_{W^{m,p}(\Omega)} = \sum_{0\le|\boldsymbol{\alpha}|\le m} \|\partial^{\boldsymbol{\alpha}} u\|^p_{L^p(\Omega)} \quad \text{(see (2.15.29))}; \tag{8.10.90b}$$

$$|u|_{s,p,\Omega} = \left[\sum_{|\boldsymbol{\alpha}|=m} \iint_{\substack{\Omega\times\Omega \\ \mathbf{x}\neq\mathbf{y}}} \frac{|\partial^{\boldsymbol{\alpha}} u(\mathbf{x}) - \partial^{\boldsymbol{\alpha}} u(\mathbf{y})|^p}{\|\mathbf{x}-\mathbf{y}\|^{n+\sigma p}} d\mathbf{x} d\mathbf{y} \right]^{1/p}. \tag{8.10.90c}$$

Remark 8.10.8. For $p = \infty$, $s = m + \sigma$ with $m = [s]$, $0 < \sigma < 1$, *modifying the values* of the functions on a set of points with *measure zero*, we can identify

$$\begin{aligned} W^{s,\infty}(\Omega) &\equiv C^{m,\sigma}(\Omega) \\ &\equiv \{u : u \in C^m(\Omega),\ \forall \text{ open } \Omega_1 \text{ with } \overline{\Omega}_1 \subset\subset \Omega,\ u{\downarrow}_{\Omega_1} \in C^{m,\sigma}(\overline{\Omega}_1)\}. \end{aligned} \tag{8.10.91}$$

Elements of $C^{m,\sigma}(\Omega)$ are of C^m-class such that all the mth-order derivatives are *locally σ-Hölder continuous* functions in Ω (see (8.9.35)–(8.9.36) and also Section A.5.1, Appendix A).

Remark 8.10.9. For $s = 0$, we write $W^{0,p}(\Omega) \equiv L^p(\Omega)$, $1 \le p \le \infty$ (see (2.15.37)). Hence, we may omit the case $s = 0$, even when the result holds for $s = 0$.

Important properties of $W^{s,p}(\Omega)$ with $s > 0$

Here we collect the important properties of $W^{s,p}(\Omega)$.

Property 1 Since, for $p = \infty$, by modifying the values of the functions on a set of points with measure zero, we can identify $W^{s,\infty}(\Omega)$ with $C^{m,\sigma}(\Omega)$ and write $W^{s,\infty}(\Omega) \equiv C^{m,\sigma}(\Omega)$ with $s = m + \sigma$, $m = [s]$ (see Remark 8.10.8), for $s > 0$, the case $1 \le p < \infty$ is of interest. Hence, for $s > 0$, we will discuss $W^{s,p}(\Omega)$ for $1 \le p < \infty$ (in fact, for $1 < p < \infty$ later).

Theorem 8.10.10. $\forall$ *real* $s > 0$ *and* $1 \le p < \infty$, $W^{s,p}(\Omega)$ *equipped with the* Slobodetskii norm $\|\cdot\|_{s,p,\Omega}$ *is a* Banach space *and* separable.

Proof. See the proof of Theorem 8.10.6 for $H^s(\Omega) \equiv W^{s,2}(\Omega)$, i.e. $p = 2$, since the proof for $1 \le p < \infty$ is similar to the one given for $p = 2$. □

For $1 < p < \infty$, $W^{s,p}(\Omega)$ is *reflexive* for $s > 0$, but

$$\text{for } p = 1,\ s > 0,\ W^{s,p}(\Omega) \text{ is } \textit{not} \text{ reflexive.} \tag{8.10.92}$$

Hence, in the following we will consider only $1 < p < \infty$, unless otherwise stated.

For $p = 2$, $W^{s,2}(\Omega) = H^s(\Omega)$ is a *Hilbert* space.

Equivalent norm in $W^{s,p}(\Omega)$ with $s > 0, 1 \le p < \infty$ $\forall u \in W^{s,p}(\Omega)$,

$$\|u\|^p_{s,p,\Omega} = \|u\|^p_{m,p,\Omega} + \sum_{1 \le |\alpha| \le m} \iint_{\substack{\Omega\times\Omega \\ \mathbf{x}\neq\mathbf{y}}} \frac{|\partial^\alpha u(\mathbf{x}) - \partial^\alpha u(\mathbf{y})|^p}{\|\mathbf{x}-\mathbf{y}\|^{n+\sigma p}} d\mathbf{x}d\mathbf{y} \tag{8.10.93}$$

defines a *norm* in $W^{s,p}(\Omega)$ with $s = [s] + \sigma$, $[s] = m \in \mathbb{N}_0$, $0 \le \sigma < 1$, *equivalent* to the original *Slobodetskii norm* in (8.10.90a). Hence, (8.10.90a) and (8.10.93) will be used interchangeably according to our convenience.

Property 2 For $s = m \in \mathbb{N}_0$, $\|\cdot\|_{m,p,\Omega}$ is given by (8.10.90b).

Property 3: Space $W_0^{s,p}(\Omega)$ with $s > 0, 1 < p < \infty$ Since $\mathcal{D}(\Omega)$ is *not* dense in $W^{s,p}(\Omega)$ for $\Omega \ne \mathbb{R}^n$ *or* for $s > 1/p$, the dual space $(W^{s,p}(\Omega))'$ of $W^{s,p}(\Omega)$ will *not* be a space of distributions (see (4.2.8)). Hence, we need:

Definition 8.10.11. For $s > 0$ and $1 < p < \infty$, $W_0^{s,p}(\Omega) \equiv \overline{\mathcal{D}(\Omega)}$ in the norm of $W^{s,p}(\Omega)$, i.e. $\mathcal{D}(\Omega)$ is *dense* in $W_0^{s,p}(\Omega)$ $\forall s > 0$, $1 < p < \infty$.

Theorem 8.10.11. *$\forall s > 0$ and for $1 < p < \infty$, $W_0^{s,p}(\Omega)$ is a* Banach space *equipped with the* Slobodetskii norm (8.10.90a) *or the equivalent norm* (8.10.93) *induced by $W^{s,p}(\Omega)$.*

In general, $W_0^{s,p}(\Omega) \subsetneq W^{s,p}(\Omega)$.

$$\begin{aligned} \Omega = \mathbb{R}^n \quad &\Longrightarrow \quad W_0^{s,p}(\mathbb{R}^n) \equiv W^{s,p}(\mathbb{R}^n)\ \forall s > 0,\ 1 < p < \infty, \\ \Omega \subset \mathbb{R}^n,\ &0 \le s \le 1/2,\ p = 2,\ H_0^s(\Omega) \equiv H^s(\Omega) \end{aligned} \tag{8.10.94}$$

(see Proposition 8.10.9).

For bounded Ω with *Lipschitz continuous* boundary Γ, $H_0^{\frac{1}{2}}(\Omega) \equiv H^{\frac{1}{2}}(\Omega)$ (see (8.10.86) and also Property 6 below).

Equivalent norm in $W_0^{s,p}(\Omega)$ with $s > 0, 1 < p < \infty$

$$\|u\|_{W_0^{s,p}(\Omega)} = \left[\sum_{|\alpha|=m} \|\partial^\alpha u\|^p_{L^p(\Omega)} + \sum_{|\alpha|=m} \frac{|\partial^\alpha u(\mathbf{x}) - \partial^\alpha u(\mathbf{y})|^p}{\|\mathbf{x}-\mathbf{y}\|^{n+\sigma p}} d\mathbf{x} d\mathbf{y} \right]^{1/p} \tag{8.10.95}$$

defines a norm in $W_0^{s,p}(\Omega)$ with $s = [s] + \sigma$, $[s] = m \in \mathbb{N}_0$, $0 < \sigma < 1$, *equivalent* to the original norm induced by $W^{s,p}(\Omega)$. Hence, (8.10.90a), (8.10.93) and (8.10.95) will be used interchangeably according to our convenience, i.e. the notations $\|\cdot\|_{s,p,\Omega}$ and $\|\cdot\|_{W_0^{s,p}(\Omega)}$ will mean any one of the *equivalent norms* in $W_0^{s,p}(\Omega)$.

Property 4: Dual space $W^{-s,q}(\Omega)$ with $s > 0$, $\frac{1}{p} + \frac{1}{q} = 1$, $1 < p < \infty$, $1 < q < \infty$

Definition 8.10.12. For $s > 0$, $1 < p < \infty$, $1 < q < \infty$ with $\frac{1}{p} + \frac{1}{q} = 1$, $W^{-s,q}(\Omega) \equiv (W_0^{s,p}(\Omega))'$, i.e. for $s > 0$, $W^{-s,q}(\Omega)$ is identified with the *dual space* $(W_0^{s,p}(\Omega))'$ of $W_0^{s,p}(\Omega)$ and is a *space of distributions*. Thus, for $s > 0$, $u \in W^{-s,q}(\Omega) \implies \langle u, v\rangle_{W^{-s,q}(\Omega)\times W_0^{s,p}(\Omega)}$ is well defined $\forall v \in W_0^{s,p}(\Omega)$, $\langle\cdot,\cdot\rangle$ being the duality pairing between $W^{-s,q}(\Omega)$ and $W_0^{s,p}(\Omega)$.

For $p = q = 2$ and $\Omega = \mathbb{R}^n$, see Theorem 8.9.3.
For $s > 0$, $\frac{1}{p} + \frac{1}{q} = 1$, $1 < p < \infty$,

$$\mathcal{D}(\Omega) \hookrightarrow W_0^{s,p}(\Omega) \hookrightarrow W^{-s,q}(\Omega) \hookrightarrow \mathcal{D}'(\Omega). \tag{8.10.96}$$

Property 5: Imbedding results For *fixed* $p \in [1, \infty[$ and $0 < s_1 < s_2$, $W^{s_2,p}(\Omega) \hookrightarrow W^{s_1,p}(\Omega)$, $W_0^{s_2,p}(\Omega) \hookrightarrow W_0^{s_1,p}(\Omega)$ with *dense*, *continuous* imbedding $\hookrightarrow$. (8.10.97)

Property 6: Density results

- $\mathcal{D}(\Omega)$ is *dense* in $W_0^{s,p}(\Omega)$ for $s > 0$, $1 < p < \infty$, by Definition 8.10.11. (8.10.98a)
- $\mathcal{D}(\mathbb{R}^n)$ is *dense* in $W^{s,p}(\mathbb{R}^n)$ and consequently $W_0^{s,p}(\mathbb{R}^n) \equiv W^{s,p}(\mathbb{R}^n)$ for $s > 0$, $1 < p < \infty$. (8.10.98b)
- For $\Omega \subset \mathbb{R}^n$ with *Lipschitz continuous* boundary Γ, $\mathcal{D}(\Omega)$ is *dense* in $W^{s,p}(\Omega)$, and consequently $W_0^{s,p}(\Omega) \equiv W^{s,p}(\Omega)$ for $0 < s \leq 1/p$, $1 < p < \infty$. (8.10.98c)
- For $\Omega \subset \mathbb{R}^n$ with a *continuous* boundary Γ, $\mathcal{D}(\overline{\Omega}) = \{\psi : \exists \phi \in \mathcal{D}(\mathbb{R}^n)$ such that $\psi = \phi{\downarrow_\Omega}\}$ is *dense* in $W^{s,p}(\Omega)$ for $s > 0$, $1 < p < \infty$. (8.10.98d)

Remark 8.10.10. Domains Ω with slits do *not* have a continuous boundary. For example, $\Omega = \{(x_1, x_2) : 0 < |x_1| < 1, 0 < |x_2| < 1\}$ in Remark 8.10.2 does *not* have a continuous boundary owing to the slit along $\Gamma_0 : x_1 = 0, 0 < x_2 < 1$, and (8.10.98d) does **not** hold.

For *arbitrary* Ω,

$$\mathcal{D}(\bar{\Omega}) \cap W^{s,p}(\Omega) \text{ is } \textit{dense} \text{ in } W^{s,p}(\Omega) \tag{8.10.99}$$

$\forall s > 0$ [17], [18].

Property 7: Extension results

Theorem 8.10.12. *Let* $\Omega \subset \mathbb{R}^n$ *be a bounded open set with* Lipschitz continuous *boundary* Γ. *Then,* $\forall s > 0$ *and for* $1 \leq p < \infty$, $\exists$ *a* continuous, linear, extension operator $P_s : W^{s,p}(\Omega) \longrightarrow W^{s,p}(\mathbb{R}^n)$ *such that* $(P_s u)\downarrow_\Omega = u$ *and* $\|P_s u\|_{s,p,\mathbb{R}^n} \leq C\|u\|_{s,p,\Omega}$ $\forall u \in W^{s,p}(\Omega)$ *and for some* $C > 0$.

Property 8: Space $W^{s,p}(\bar{\Omega})$, $s > 0$, $1 \leq p < \infty$

Definition 8.10.13. $\forall s > 0$, $1 \leq p < \infty$, let $\Omega \subsetneq \mathbb{R}^n$ be an open subset and $W^{s,p}(\bar{\Omega})$ be defined by (see also (8.10.1)):

$$W^{s,p}(\bar{\Omega}) = \{u : u \in \mathcal{D}'(\Omega),\ \exists w \in W^{s,p}(\mathbb{R}^n) \text{ such that } w\downarrow_\Omega = u\}. \tag{8.10.100}$$

Then $W^{s,p}(\bar{\Omega})$ equipped with the norm $\|\cdot\|_{s,p,\overline{\Omega}}$ defined by (see also (8.10.1)):

$$\|u\|_{s,p,\overline{\Omega}} = \inf_{\substack{w \in W^{s,p}(\mathbb{R}^n) \\ w\downarrow_\Omega = u}} \|w\|_{s,p,\mathbb{R}^n} \quad \forall u \in W^{s,p}(\bar{\Omega}) \tag{8.10.101}$$

is a *Banach space*.

Theorem 8.10.13. *Let* $\Omega \subset \mathbb{R}^n$ *be a* bounded *open set with* Lipschitz continuous *boundary* Γ. *Then,* $\forall s > 0$ *and for* $1 \leq p < \infty$, $W^{s,p}(\Omega) = W^{s,p}(\bar{\Omega})$ *with* $\|u\|_{s,p,\overline{\Omega}} \leq C\|u\|_{s,p,\Omega}$ *for some* $C > 0$, $\forall u \in W^{s,p}(\Omega)$.

Proof.
$W^{s,p}(\overline{\Omega}) \subset W^{s,p}(\Omega)$: Let $u \in W^{s,p}(\overline{\Omega})$ for $s > 0$, $1 \leq p < \infty$. Then, from (8.10.100), $\exists w \in W^{s,p}(\mathbb{R}^n)$ such that $w\downarrow_\Omega = u$. Hence, $\|u\|_{s,p,\Omega} = \|w\downarrow_\Omega\|_{s,p,\Omega} \leq \|w\|_{s,p,\mathbb{R}^n} < \infty \Longrightarrow u \in W^{s,p}(\Omega)$, i.e. $W^{s,p}(\overline{\Omega}) \subset W^{s,p}(\Omega)$.
$W^{s,p}(\Omega) \subset W^{s,p}(\overline{\Omega})$: Let $u \in W^{s,p}(\Omega)$ for $s > 0$, $1 \leq p < \infty$. Then, by Theorem 8.10.13, $\exists P_s u \in W^{s,p}(\mathbb{R}^n)$ with $\|P_s u\|_{s,p,\mathbb{R}^n} \leq C\|u\|_{s,p,\Omega}$ for some $C > 0$. Hence, from (8.10.101), $\|u\|_{s,p,\overline{\Omega}} \leq \|P_s u\|_{s,p,\mathbb{R}^n} \leq C\|u\|_{s,p,\Omega} < \infty \Longrightarrow u \in W^{s,p}(\overline{\Omega})$, i.e. $W^{s,p}(\Omega) \subset W^{s,p}(\overline{\Omega})$.

Hence, combining these results, we get $W^{s,p}(\overline{\Omega}) \equiv W^{s,p}(\Omega)$ with $\|u\|_{s,p,\overline{\Omega}} \leq C\|u\|_{s,p,\Omega}$. □

8.10.11 Space $H^s_{00}(\Omega)$ with $s > 0$

In some problems, for example in dealing with traces of functions in polygonal domains or domains with corners, we will require a subspace $H^s_{00}(\Omega)$ (also denoted by $\widetilde{H}^s(\Omega)$ in Grisvard [17], [18]) of $H^s(\Omega)$ for $s > 0$ defined in (8.10.71)–(8.10.72). We have accepted the notation $H^s_{00}(\Omega)$ of Lions–Magenes [15].

Definition 8.10.14. Let $\Omega \subsetneq \mathbb{R}^n$ be an open subset of $\mathbb{R}^n$. Then, for real $s > 0$, the space $H^s_{00}(\Omega)$ is defined by the set

$$H^s_{00}(\Omega) = \{u : u \in H^s(\Omega) \text{ such that its null extension } \tilde{u} \in H^s(\mathbb{R}^n)\}, \quad (8.10.102)$$

which is equipped with the inner product $\langle \cdot, \cdot \rangle_{00,s,\Omega}$ and norm $\| \cdot \|_{00,s,\Omega}$ given by: $\forall u \in H^s(\Omega)$ with

$$\tilde{u}(\mathbf{x}) = u(\mathbf{x}) \ \forall \mathbf{x} \in \Omega \quad \text{and} \quad \tilde{u}(\mathbf{x}) = 0 \forall \ \mathbf{x} \in \mathbb{R}^n \setminus \Omega, \quad (8.10.103)$$

$$\langle u, v \rangle_{00,s,\Omega} = \langle \tilde{u}, \tilde{v} \rangle_{s,\mathbb{R}^n}; \quad \|u\|_{00,s,\Omega} = \langle u, u \rangle^{1/2}_{00,s,\Omega} = \|\tilde{u}\|_{s,\mathbb{R}^n}. \quad (8.10.104)$$

Theorem 8.10.14. *For $s > 0$, let $H^s_{00}(\Omega)$ be defined by* (8.10.102)–(8.10.104). *Then $H^s_{00}(\Omega)$ is a* Hilbert space.

Important properties of $H^s_{00}(\Omega)$ with $s > 0$

Property 1 $H^s_{00}(\Omega) \hookrightarrow H^s(\overline{\Omega}) \hookrightarrow H^s(\Omega)$ $\forall s > 0$, where $H^s(\overline{\Omega})$ and $H^s(\Omega)$ are defined by (8.10.1) and (8.10.71), respectively, with

$$\|u\|_{s,\Omega} \le \|u\|_{s,\overline{\Omega}} \le \|u\|_{00,s,\Omega} \quad \forall u \in H^s_{00}(\Omega). \quad (8.10.105)$$

Property 2 For $s = m \in \mathbb{N}$, $H^m_0(\Omega) \subset H^m_{00}(\Omega)$ (see (8.10.113) for their equality for Ω with *continuous boundary*), since $u \in H^m_0(\Omega) \Longrightarrow \tilde{u} \in H^m(\mathbb{R}^n)$ by Theorem 2.15.5 $\Longrightarrow u \in H^m_{00}(\Omega) \Longrightarrow H^m_0(\Omega) \subset H^m_{00}(\Omega)$.

Property 3 The norm (8.10.104) is induced by $H^s(\Omega)$ only when $s = m \in \mathbb{N}$. In fact, we have the following result:

Lemma 8.10.3 (Grisvard [18]).

I. *For $s = m \in \mathbb{N}$, $\|u\|_{H^m_{00}(\Omega)} = \|u\|_{m,\Omega}$.* (8.10.106)

II. *For $s = m + \sigma$, $0 < \sigma < 1$, the norm $\| \cdot \|_{00,s,\Omega}$ in* (8.10.104) *is* equivalent *to mapping*

$$u \in H^s_{00}(\Omega) \longmapsto \|u\|_{H^s_{00}(\Omega)}$$

$$= \left[\|u\|^2_{s,\Omega} + \sum_{|\boldsymbol{\alpha}|=m} \int_\Omega w_\sigma(\mathbf{x}) |\partial^{\boldsymbol{\alpha}} u(\mathbf{x})|^2 d\mathbf{x} \right]^{1/2} \ge \|u\|_{s,\Omega}, \quad (8.10.107)$$

where $w_\sigma(\mathbf{x})$ *is a suitable weight, i.e.*

$$\exists\, c_1, c_2 > 0 \text{ such that } c_1\|u\|_{00,s,\Omega} \leq \|u\|_{H^s_{00}(\Omega)} \leq c_2\|u\|_{00,s,\Omega}. \tag{8.10.108}$$

In order to distinguish the formulae for equivalent norms we have used $\|\cdot\|_{00,s,\Omega}$ *and* $\|\cdot\|_{H^s_{00}(\Omega)}$ *defined by* (8.10.104) *and* (8.10.107), *respectively.*

III. *For* bounded $\Omega \subset \mathbb{R}^n$ *with* Lipschitz continuous *boundary* Γ, *the weight function* w_σ *in* (8.10.107) *satisfies the following equivalent condition:* $\exists \tilde{c}_1, \tilde{c}_2 > 0$ *with* $\tilde{c}_1 \leq \tilde{c}_2$ *such that*

$$\tilde{c}_1 \frac{1}{\rho^{2\sigma}_{(\mathbf{x})}} \leq w_\sigma(\mathbf{x}) \leq \tilde{c}_2 \frac{1}{\rho^{2\sigma}_{(\mathbf{x})}} \quad \text{for } \mathbf{x} \in \Omega, \tag{8.10.109}$$

where $\rho(\mathbf{x}) = d(\mathbf{x}, \Gamma) = \inf_{\mathbf{y}\in\Gamma} d(\mathbf{x}, \mathbf{y}) = \inf_{\mathbf{y}\in\Gamma} \|\mathbf{x} - \mathbf{y}\|_{\mathbb{R}^n}\ \forall \mathbf{x} \in \Omega$ *denotes the distance from* $\mathbf{x} \in \Omega$ *to the boundary* Γ *of* Ω *and we have,* $\forall u \in H^s_{00}(\Omega)$,

$$\begin{aligned} \tilde{c}_1\left(\|u\|^2_{s,\Omega} + \sum_{|\boldsymbol{\alpha}|=m} \int_\Omega \frac{|\partial^{\boldsymbol{\alpha}} u|^2}{\rho^{2\delta}} d\mathbf{x}\right)^{1/2} &\leq \|u\|_{H^s_{00}(\Omega)} \\ &\leq \tilde{c}_2\left(\|u\|^2_{s,\Omega} + \sum_{|\boldsymbol{\alpha}|=m} \int_\Omega \frac{|\partial^{\boldsymbol{\alpha}} u|^2}{\rho^{2\sigma}} d\mathbf{x}\right), \end{aligned} \tag{8.10.110}$$

where $\|u\|_{H^s_{00}}(\Omega)$ *is defined by* (8.10.107).

Consequently, $(\|u\|^2_{s,\Omega} + \sum_{|\boldsymbol{\alpha}|=m} \int_\Omega \frac{|\partial^{\boldsymbol{\alpha}} u|^2}{\rho^{2\sigma}} d\mathbf{x})^{1/2}$, $\|u\|_{H^s_{00}(\Omega)}$ *and* $\|u\|_{00,s,\Omega}$ *are* equivalent *norms in* $H^s_{00}(\Omega)$ *for* $s = m + \sigma$, $0 < \sigma < 1$. *Finally,*

$$\begin{aligned} c_1^*\left(\|u\|^2_{s,\Omega} + \sum_{|\boldsymbol{\alpha}|=m} \int_\Omega \frac{|\partial^{\boldsymbol{\alpha}} u|^2}{\rho^{2\sigma}} d\mathbf{x}\right)^{1/2} &\leq \|u\|_{00,s,\Omega} \\ &\leq c_2^*\left(\|u\|^2_{s,\Omega} + \sum_{|\boldsymbol{\alpha}|=m} \int_\Omega \frac{|\partial^{\boldsymbol{\alpha}} u|^2}{\rho^{2\sigma}} d\mathbf{x}\right)^{1/2} \end{aligned} \tag{8.10.111}$$

with $c_1^*, c_2^* > 0$.

Property 4: Density result

Theorem 8.10.15 (Grisvard [18]). *Let* $\Omega \subset \mathbb{R}^n$ *be an open subset of* $\mathbb{R}^n$ *with* continuous *boundary* Γ. *Then,* $\forall s > 0$, $\mathcal{D}(\Omega)$ *is* dense *in* $H^s_{00}(\Omega) = \overline{\mathcal{D}(\Omega)}$ *in the norm* $\|\cdot\|_{00,s,\Omega}$. (8.10.112)

For $s = m \in \mathbb{N}$,

$$H^m_{00}(\Omega) = H^m_0(\Omega). \tag{8.10.113}$$

In fact, $H^m_{00}(\Omega) = \overline{\mathcal{D}(\Omega)}$ *in the norm* $\|\cdot\|_{00,m,\Omega}$, *and* $H^m_0(\Omega) = \overline{\mathcal{D}(\Omega)}$ *in the norm* $\|\cdot\|_{m,\Omega}$, *but* $\|\cdot\|_{m,\Omega} = \|\cdot\|_{00,m,\Omega}$ *(by Lemma* 8.10.3*). Hence,* $\overline{\mathcal{D}(\Omega)} = H^m_{00}(\Omega) = H^m_0(\Omega)$.

Property 5: Alternative characterization of $H^s_{00}(\Omega)$ for bounded domains with Lipschitz continuous boundary Γ For $s > 0$ with $s = m + \sigma$, $m \in \mathbb{N}$, $0 < \sigma < 1$, we define the space $X^s(\Omega)$ by $X^s(\Omega) = \{u : u \in H^s_0(\Omega), \frac{\partial^{\alpha} u}{\rho^{\sigma}} \in L^2(\Omega)\ \forall |\alpha| = m\}$ equipped with the norm $\|\cdot\|_{X^s(\Omega)}$:

$$\|u\|_{X^s(\Omega)} = \left[\|u\|^2_{s,\Omega} + \sum_{|\alpha|=m} \int_{\Omega} \frac{|\partial^{\alpha} u(\mathbf{x})|^2}{\rho^{2\sigma}} d\mathbf{x}\right]^{1/2}, \tag{8.10.114}$$

where $\rho = \rho(\mathbf{x})$ denotes the distance of Γ from $\mathbf{x} \in \Omega$, Γ being the *Lipschitz continuous* boundary of bounded $\Omega \subset \mathbb{R}^n$.

Case $s - \frac{1}{2} \neq$ *an integer:* Using Hardy's inequality (Grisvard [18], Tartar [42]), we have:

Theorem 8.10.16 (Grisvard [18]). *For* bounded $\Omega \subset \mathbb{R}^n$ *with* Lipschitz continuous *boundary* Γ, *and for* $s > 0$, $s - \frac{1}{2} \neq$ *an integer and* $u \in H^s_0(\Omega)$,

$$\frac{\partial^{\alpha} u}{\rho^{s-|\alpha|}} \in L^2(\Omega) \quad \forall |\alpha| \leq s. \tag{8.10.115}$$

In particular, for $0 < s < \frac{1}{2}$ *and*

$$u \in H^s(\Omega) = H^s_0(\Omega) \tag{8.10.116}$$

(see Corollary 8.10.1*),* $\frac{u}{\rho^s} \in L^2(\Omega)$, *and for* $\frac{1}{2} < s \leq 1$ *and* $u \in H^s_0(\Omega)$,

$$\frac{u}{\rho^s} \in L^2(\Omega). \tag{8.10.117}$$

Corollary 8.10.2. *Let* $\Omega \subset \mathbb{R}^n$ *be a* bounded *open subset of* $\mathbb{R}^n$ *with* Lipschitz continuous *boundary* Γ.

Then, for $s > 0$ *with* $s - \frac{1}{2} \neq$ *an integer,*

$$X^s(\Omega) \equiv H^s_0(\Omega) \equiv H^s_{00}(\Omega) \tag{8.10.118}$$

with norm equivalence.

In particular, for $0 < s < \frac{1}{2}$,

$$H^s_{00}(\Omega) \equiv H^s_0(\Omega) \equiv H^s(\Omega). \tag{8.10.119}$$

Proof. Since $X^s(\Omega) \subset H_0^s(\Omega)$ by definition, for $X^s(\Omega) \equiv H_0^s(\Omega)$ it is sufficient to show that $H_0^s(\Omega) \subset X^s(\Omega)$. In fact, let $u \in H_0^s(\Omega)$. Then, for $s - \frac{1}{2} \neq$ *an integer*, by Theorem 8.10.16, $\frac{\partial^\alpha u}{\rho^\sigma} \in L^2(\Omega)\ \forall |\boldsymbol{\alpha}| = m$, $\sigma = s - |\boldsymbol{\alpha}| \Longrightarrow u \in X^s(\Omega) \Longrightarrow H_0^s(\Omega) \subset X^s(\Omega)$. Hence, $X^s(\Omega) \equiv H_0^s(\Omega)$ with norm equivalence, i.e. $\|\cdot\|_{X^s(\Omega)}$ and $\|\cdot\|_{s,\Omega}$ are equivalent.

For $s - \frac{1}{2} \neq$ an integer, the norms $\|\cdot\|_{X^s(\Omega)}$ and $\|\cdot\|_{00,s,\Omega}$ are *equivalent* on $\mathcal{D}(\Omega)$ by Lemma 8.10.3, since $\mathcal{D}(\Omega) \subset H_{00}^s(\Omega)$, $\mathcal{D}(\Omega) \subset H_0^s(\Omega)$ and $\|\cdot\|_{s,\Omega}$ and $\|\cdot\|_{X^s(\Omega)}$ are equivalent on $H_0^s(\Omega)$. But $\mathcal{D}(\Omega)$ is *dense* in $H_{00}^s(\Omega)$ in the norm $\|\cdot\|_{00,s,\Omega}$ (by Theorem 8.10.15) and *dense* in $H_0^s(\Omega)$ in the norm $\|\cdot\|_{X^s(\Omega)}$ (equivalent to $\|\cdot\|_{s,\Omega}$). Hence, the completions $\overline{\mathcal{D}(\Omega)}$ of $\mathcal{D}(\Omega)$ in two equivalent norms $\|\cdot\|_{00,s,\Omega}$ and $\|\cdot\|_{X^s(\Omega)}$ (by Lemma 8.10.3) represent the same space, i.e. $\overline{\mathcal{D}(\Omega)} \equiv H_{00}^s(\Omega) \equiv H_0^s(\Omega) \equiv X^s(\Omega)$ with norm equivalence.

For $0 < s < \frac{1}{2}$, $s - \frac{1}{2} \neq$ an integer $\Longrightarrow H_{00}^s(\Omega) \equiv H_0^s(\Omega)$ and $H_0^s(\Omega) = H^s(\Omega)$ by Corollary 8.10.1, and the result follows. □

Remark 8.10.11. For $s - \frac{1}{2} \neq$ an integer, $H_{00}^s(\Omega)$ loses its importance by virtue of (8.10.118).

For *unbounded* $\Omega = \mathbb{R}_+^n = \{\mathbf{x} : \mathbf{x} = (x_1, \ldots, x_{n-1}, x_n) \in \mathbb{R}^n,\ x_n > 0\}$, we have:

Theorem 8.10.17. *Let* $\Omega = \mathbb{R}_+^n$. *Then* $H_{00}^m(\mathbb{R}_+^n) = H_0^m(\mathbb{R}_+^n)$.

Proof. From the following facts:

1. $H_0^m(\mathbb{R}_+^n)$ is a *closed* subspace of $H_{00}^m(\mathbb{R}_+^n)$;
2. $X_{00}^m(\mathbb{R}_+^n) = \{u : u \in H_{00}^m(\mathbb{R}_+^n),\ \text{supp}(u) \subset\subset \overline{\mathbb{R}_+^n}\}$ is a *dense* subspace of $H_{00}^m(\mathbb{R}_+^n)$, which follows from Theorem 6.8.6 and Remark 6.8.1 with necessary modifications;
3. $X_{00}^m(\mathbb{R}_+^n) \subset H_0^m(\mathbb{R}_+^n)$, which is proved with the help of convolutions with suitable regularizing functions, Theorem 6.2.2 and Remark 6.2.2;

and from $X_{00}^m(\mathbb{R}_+^n) \subset H_0^m(\mathbb{R}_+^n) \subset H_{00}^m(\mathbb{R}_+^n)$, the closed subspace $H_0^m(\mathbb{R}_+^n)$ is *dense* in $H_{00}^m(\mathbb{R}_+^n) \Longrightarrow H_0^m(\mathbb{R}_+^n) = H_{00}^m(\mathbb{R}_+^n)$. □

Case $s - \frac{1}{2} =$ *an integer:* This exceptional case arises in trace theorems on bounded polygonal domains with corners and, hence, is a very important one. First of all, we have

$$s - \tfrac{1}{2} = \text{an integer} \Longrightarrow \text{Corollary 8.10.2 does } \textit{not} \text{ hold}$$

and $X^s(\Omega) \neq H_0^s(\Omega)$. But we have the important result:

Theorem 8.10.18. *Let* $\Omega \subset \mathbb{R}^n$ *be a* bounded *open subset of* $\mathbb{R}^n$ *with* Lipschitz continuous *boundary* Γ. *Then, for* $s = m + \frac{1}{2}$ *with* $\sigma = \frac{1}{2}$, $s - \frac{1}{2}$ = an integer $m \in \mathbb{N}_0$, $H^s_{00}(\Omega) \equiv X^s(\Omega)$ *with norm equivalence (see Lemma* 8.10.3*), where*

$$X^s(\Omega) = \{u : u \in H^s_0(\Omega), \frac{\partial^{\boldsymbol{\alpha}} u}{\sqrt{\rho}} \in L^2(\Omega)\ \forall |\boldsymbol{\alpha}| = m\} \subsetneq H^s_0(\Omega) \qquad (8.10.120)$$

with $\|u\|_{X^s(\Omega)} = [\|u\|^2_{s,\Omega} + \sum_{|\boldsymbol{\alpha}|=m} \int_\Omega \frac{|\partial^{\boldsymbol{\alpha}} u(\mathbf{x})|^2}{\rho} d\mathbf{x}]^{\frac{1}{2}}$.

Proof.
$H^s_{00}(\Omega) \subset X^s(\Omega)$: Let $u \in H^s_{00}(\Omega)$ with $s - \frac{1}{2} = m \in \mathbb{N}_0$. Then, by virtue of the density of $\mathcal{D}(\Omega)$ in $H^s_{00}(\Omega)$ (Theorem 8.10.15) $\exists$ a sequence $(\phi_k)_{k=1}^\infty$ in $\mathcal{D}(\Omega)$ such that $\|u - \phi_k\|_{0C,s,\Omega} \to 0$ as $k \to \infty$. By Lemma 8.10.3, $[\|u - \phi_k\|^2_{s,\Omega} + \sum_{|\boldsymbol{\alpha}|=m} \int_\Omega \frac{|\partial^{\boldsymbol{\alpha}} u - \partial^{\boldsymbol{\alpha}} \phi_k|^2}{\rho} d\mathbf{x}] \to 0$ as $k \to \infty \Longrightarrow \phi_k \to u$ in $H^s(\Omega)$, and $\frac{\partial^{\boldsymbol{\alpha}} \phi_k}{\sqrt{\rho}} \to \frac{\partial^{\boldsymbol{\alpha}} u}{\sqrt{\rho}}$ in $L^2(\Omega)\ \forall |\boldsymbol{\alpha}| = m$ as $k \to \infty$. But (ϕ_k) is a sequence in $\mathcal{D}(\Omega)$, which is dense in $H^s_0(\Omega) \Longrightarrow u \in H^s_0(\Omega)$ and $\frac{\partial^{\boldsymbol{\alpha}} u}{\sqrt{\rho}} \in L^2(\Omega)\ \forall |\boldsymbol{\alpha}| = m \Longrightarrow u \in X^s(\Omega) \Longrightarrow H^s_{00}(\Omega) \subset X^s(\Omega)$.
$X^s(\Omega) \subset H^s_{00}(\Omega)$: Let $u \in X^s(\Omega)$ with $\|u\|_{X^s(\Omega)}$ defined by (8.10.120). Then $u \in H^s_0(\Omega)$ and $\frac{\partial^{\boldsymbol{\alpha}} u}{\sqrt{\rho}} \in L^2(\Omega)\ \forall |\boldsymbol{\alpha}| = m$. But $u \in H^s_0(\Omega) \Longrightarrow u \in H^m_0(\Omega)$, since $H^s_0(\Omega) \hookrightarrow H^m_0(\Omega) \Longrightarrow \partial^{\boldsymbol{\alpha}} \tilde{u} = \widetilde{\partial^{\boldsymbol{\alpha}} u}\ \forall |\boldsymbol{\alpha}| \le m$ (see the proof of Theorem 2.15.5). But $\partial^{\boldsymbol{\alpha}} u \in L^2(\Omega)\ \forall |\boldsymbol{\alpha}| \le m \Longrightarrow \widetilde{\partial^{\boldsymbol{\alpha}} u} = \partial^{\boldsymbol{\alpha}} \tilde{u} \in L^2(\mathbb{R}^n)\ \forall |\boldsymbol{\alpha}| \le m \Longrightarrow \tilde{u} \in H^m(\mathbb{R}^n)$. To show that $u \in H^s_{00}(\Omega)$, we are to prove that $\tilde{u} \in H^s(\mathbb{R}^n)$ with $s = m + \frac{1}{2}$. Hence, $\tilde{u} \in H^m(\mathbb{R}^n)$ will belong to $H^s(\mathbb{R}^n)$ if $\partial^{\boldsymbol{\alpha}} \tilde{u} \in H^{\frac{1}{2}}(\mathbb{R}^n)\ \forall |\boldsymbol{\alpha}| = m$, which implies that $\partial^{\boldsymbol{\alpha}} \tilde{u} = \widetilde{\partial^{\boldsymbol{\alpha}} u} \in H^{\frac{1}{2}}(\mathbb{R}^n)\ \forall |\boldsymbol{\alpha}| = m$ (since we have already shown that $\partial^{\boldsymbol{\alpha}} \tilde{u} = \widetilde{\partial^{\boldsymbol{\alpha}} u} \in L^2(\mathbb{R}^n)\ \forall |\boldsymbol{\alpha}| = m$), i.e. $\partial^{\boldsymbol{\alpha}} u \in H^{\frac{1}{2}}_{00}(\Omega)\ \forall |\boldsymbol{\alpha}| = m$. For this, we are to prove that $\forall |\boldsymbol{\alpha}| = m$, $\|\partial^{\boldsymbol{\alpha}} u\|_{H^{\frac{1}{2}}_{00}(\Omega)} < +\infty$. From Lemma 8.10.3, $\forall$ fixed $|\boldsymbol{\alpha}| = m$,

$$\begin{aligned} \|\partial^{\boldsymbol{\alpha}} u\|^2_{H^{\frac{1}{2}}_{00}(\Omega)} &= \|\partial^{\boldsymbol{\alpha}} u\|^2_{H^{\frac{1}{2}}(\Omega)} + \|\frac{\partial^{\boldsymbol{\alpha}} u}{\sqrt{\rho}}\|_{L^2(\Omega)} \\ &\le \|u\|^2_{s,\Omega} + \sum_{|\boldsymbol{\alpha}|=m} \|\frac{\partial^{\boldsymbol{\alpha}} u}{\sqrt{\rho}}\|_{L^2(\Omega)} = \|u\|^2_{X^s(\Omega)} < +\infty \quad \text{(by (8.10.120))}. \end{aligned}$$

Hence, $\forall |\boldsymbol{\alpha}| = m$, $\partial^{\boldsymbol{\alpha}} u \in H^{\frac{1}{2}}_{00}(\Omega) \Longrightarrow u \in H^s_{00}(\Omega)$ with $s - \frac{1}{2} = m \in \mathbb{N}_0$. □

An alternative definition of $H^s_{00}(\Omega)$ for $s > 0$

As a consequence of Theorem 8.10.18, Theorem 8.10.16 and Corollary 8.10.2, we have:

Definition 8.10.15. For $s > 0$ with $s = m + \sigma$, $m \in \mathbb{N}_0$, $0 \le \sigma < 1$ ($\sigma = 0 \Longrightarrow s = m \in \mathbb{N}$) and $\Omega \subset \mathbb{R}^n$ with *Lipschitz continuous* boundary Γ, $H^s_{00}(\Omega)$ is defined by:

$$H^s_{00}(\Omega) = X^s(\Omega) = \left\{ u : u \in H^s_0(\Omega),\ \frac{\partial^{\boldsymbol{\alpha}} u}{\rho^\sigma} \in L^2(\Omega)\ \forall\ |\boldsymbol{\alpha}| = m \right\} \tag{8.10.121}$$

with *equivalent* norms $\|\cdot\|_{00,s,\Omega}$, $\|\cdot\|_{H^s_{00}(\Omega)}$ and $\|\cdot\|_{X^s(\Omega)}$ defined by (8.10.104), (8.10.107) and (8.10.120), respectively (see (8.10.111) also).

Then the following hold:

- for $s = m$, with $\sigma = 0$, $H^m_{00}(\Omega) = X^m(\Omega) = H^m_0(\Omega)$ with $\|u\|_{00,m,\Omega} = \|u\|_{m,\Omega}$ (by (8.10.106));
- for $s - \frac{1}{2} \neq$ *an integer*, $H_{00}(\Omega) = X^s(\Omega) = H^s_0(\Omega)$ (by (8.10.118)); (8.10.122)
- for $s - \frac{1}{2} =$ *an integer*, i.e. $\sigma = \frac{1}{2}$, $H^s_{00}(\Omega) = X^s(\Omega) \subsetneq H^s_0(\Omega)$ with

$$\|u\|_{H^s_{00}(\Omega)} = \|u\|_{X^s(\Omega)} = \left(\|u\|^2_{s,\Omega} + \sum_{|\boldsymbol{\alpha}|=m} \int_\Omega \frac{|\partial^{\boldsymbol{\alpha}} u|^2}{\rho} \right)^{\frac{1}{2}}, \tag{8.10.123}$$

the norms $\|\cdot\|_{H^s_{00}(\Omega)}$ and $\|\cdot\|_{00,s,\Omega}$ being equivalent ones (see also Lemma 8.10.3).

Let $\Omega \subset \mathbb{R}^n$ be a *bounded* open subset of $\mathbb{R}^n$ with *Lipschitz continuous* boundary Γ. Then, $s = \frac{1}{2}, \frac{3}{2}$ and so on $\Longrightarrow s - \frac{1}{2} =$ an *integer* and Theorem 8.10.18 is applicable, and

- $H^{\frac{1}{2}}_{00}(\Omega) = \{u : u \in H^{\frac{1}{2}}_0(\Omega) \equiv H^{\frac{1}{2}}(\Omega),\ \frac{u}{\sqrt{\rho}} \in L^2(\Omega)\} \subsetneq H^{\frac{1}{2}}_0(\Omega)$ with

$$\|u\|_{00,\frac{1}{2},\Omega} = \left(\|u\|^2_{\frac{1}{2},\Omega} + \int_\Omega \frac{|u(\mathbf{x})|^2}{\rho(\mathbf{x})} d\mathbf{x} \right)^{\frac{1}{2}}; \tag{8.10.124}$$

- $H^{\frac{3}{2}}_{00}(\Omega) = \{u : u \in H^{\frac{3}{2}}_0(\Omega),\ \frac{\partial^\alpha u}{\sqrt{\rho}} \in L^2(\Omega)\ \forall |\boldsymbol{\alpha}| = 1\} \subsetneq H^{\frac{3}{2}}_0(\Omega)$ with

$$\|u\|_{00,\frac{3}{2},\Omega} = \left(\|u\|^2_{\frac{3}{2},\Omega} + \sum_{|\boldsymbol{\alpha}|=1} \frac{|\partial^{\boldsymbol{\alpha}} u(\mathbf{x})|^2}{\rho(\mathbf{x})} d\mathbf{x} \right)^{\frac{1}{2}}; \tag{8.10.125}$$

and so on.

8.10.12 Dual space $(H^s_{00}(\Omega))'$ for $s > 0$

Since $\mathcal{D}(\Omega)$ is *dense* in $H^s_{00}(\Omega)$ for $s > 0$ by Theorem 8.10.15, the dual space $(H^s_{00}(\Omega))'$ of $H^s_{00}(\Omega)$ is a *space of distributions* in $\mathcal{D}'(\Omega)$. For bounded $\Omega \subset \mathbb{R}^n$ with *Lipschitz continuous* boundary Γ, we have, for $s > 0$:

- $s - \frac{1}{2} \neq$ *an integer,*

$$H^s_{00}(\Omega) = H^s_0(\Omega) \quad \Longrightarrow \quad (H^s_{00}(\Omega))' = (H^s_0(\Omega))' = H^{-s}(\Omega); \tag{8.10.126a}$$

- $s - \frac{1}{2} =$ *an integer,*

$$H^s_{00}(\Omega) \subsetneq H^s_0(\Omega) \quad \Longrightarrow \quad H^{-s}(\Omega) \subset (H^s_{00}(\Omega))'; \tag{8.10.126b}$$

- $\mathcal{D}(\Omega) \hookrightarrow H^s_{00}(\Omega) \hookrightarrow H^s_0(\Omega) \hookrightarrow H^{-s}(\Omega) \hookrightarrow (H^s_{00}(\Omega))' \hookrightarrow \mathcal{D}'(\Omega)$ with continuous imbedding $\hookrightarrow$. (8.10.127)
- For $(H^s_{00}(\Omega))'$ with $s > 0$, $\langle u, v\rangle_{(H^s_{00}(\Omega))'\times H^s_{00}(\Omega)}$ is well defined $\forall u \in (H^s_{00}(\Omega))'$ and $\forall v \in H^s_{00}(\Omega)$, $\langle \cdot, \cdot\rangle_{(H^s_{00}(\Omega))'\times H^s_{00}(\Omega)}$ being the *duality* between $(H^s_{00}(\Omega))'$ and $H^s_{00}(\Omega)$. (8.10.128)

Remark 8.10.12. We can define $W^{s,p}_{00}(\Omega)$ for $s > 0$, $1 < p < \infty$ by extending the definition and properties of $H^s_{00}(\Omega)$ for $p \neq 2$ as follows.

8.10.13 Space $W^{s,p}_{00}(\Omega)$ for $s > 0, 1 < p < \infty$

Definition 8.10.16. For $s > 0$, $1 < p < \infty$, $W^{s,p}_{00}(\Omega)$ is defined by:

$$W^{s,p}_{00}(\Omega) = \{u : u \in W^{s,p}(\Omega) \text{ such that its } \textit{null extension } \tilde{u} \in W^{s,p}(\mathbb{R}^n)\}, \tag{8.10.129}$$

with $\|u\|_{00,s,p,\Omega} = \|\tilde{u}|_{s,p,\mathbb{R}^n}$. For $p = 2$, $W^{s,2}_{00}(\Omega) \equiv H^s_{00}(\Omega)$.

For $s > 0$, $1 < p < \infty$, $W^{s,p}_{00}(\Omega) \subset W^{s,p}(\overline{\Omega})$ (see (8.10.100)–(8.10.101)), $W^{m,p}_0(\Omega) \subset W^{m,p}_{00}(\Omega)$ for $1 < p < \infty$ (see (8.10.134) for their equality). (8.10.130)

The properties of $W^{s,p}_{00}(\Omega)$ are similar to those of $H^s_{00}(\Omega)$.

Theorem 8.10.19. *For $s > 0$, $1 < p < \infty$, $W^{s,p}_{00}(\Omega)$ is a* Banach space *which is* separable *and* reflexive.

Lemma 8.10.4. *For $s = m \in \mathbb{N}$ and $u \in W^{m,p}_{00}(\Omega)$, $1 < p < \infty$,*

$$\|u\|_{00,m,p,\Omega} = \|u\|_{m,p,\Omega}. \tag{8.10.131}$$

For $s = m + \sigma$, $m \in \mathbb{N}_0$, $0 < \sigma < 1$, *and* $\Omega \subset \mathbb{R}^n$ *with* Lipschitz continuous *boundary* Γ, $\exists C_1, C_2 > 0$ *with* $C_1 \le C_2$ *such that*

$$C_1\left(\|u\|^p_{s,p,\Omega} + \sum_{|\boldsymbol{\alpha}|=m} \int_\Omega \frac{|\partial^{\boldsymbol{\alpha}} u(\mathbf{x})|^p}{\rho^{\sigma p}} d\mathbf{x}\right)^{\frac{1}{p}}$$

$$\le \|u\|_{00,s,p,\Omega} \le C_2\left(\|u\|^p_{s,p,\Omega} + \sum_{|\boldsymbol{\alpha}|=m} \int_\Omega \frac{|\partial^{\boldsymbol{\alpha}} u(\mathbf{x})|^p}{\rho^{\sigma p}} d\mathbf{x}\right)^{\frac{1}{p}}, \qquad (8.10.132)$$

where $\rho = \rho(\mathbf{x}) = d(\mathbf{x}, \Gamma)$ *is defined by* (8.10.107).

Density result

For $s > 0$, $1 < p < \infty$, and $\Omega \subset \mathbb{R}^n$ with a *continuous* boundary Γ,

$$\mathcal{D}(\Omega) \text{ is } \textit{dense} \text{ in } W^{s,p}_{00}(\Omega). \qquad (8.10.133)$$

As a consequence of Lemma 8.10.4 and (8.10.130), for $s = m$, $1 < p < \infty$,

$$W^{m,p}_{00}(\Omega) = W^{m,p}_0(\Omega). \qquad (8.10.134)$$

Space $X^s_p(\Omega)$

For $s = m + \sigma$, $m \in \mathbb{N}_0$, $0 \le \sigma < 1$, $1 < p < \infty$, and $\Omega \subset \mathbb{R}^n$ with *Lipschitz continuous* boundary Γ, define $X^s_p(\Omega) = \{u : u \in W^{s,p}_0(\Omega),\ \frac{\partial^{\alpha} u}{\rho^{\sigma}} \in L^p(\Omega)\ \forall |\boldsymbol{\alpha}| = m\}$ with

$$\|u\|_{X^s_p(\Omega)} = \left(\|u\|^p_{s,p,\Omega} + \sum_{|\boldsymbol{\alpha}|=m} \int_\Omega \frac{|\partial^{\boldsymbol{\alpha}} u(\mathbf{x})|^p}{\rho^{\sigma p}} d\mathbf{x}\right)^{\frac{1}{p}}. \qquad (8.10.135)$$

For $p = 2$, $X^s_2(\Omega) = X^s(\Omega)$ defined by (8.10.120).
From Lemma 8.10.4 and (8.10.135),

$$C_1\|u\|_{X^s_p(\Omega)} \le \|u\|_{00,s,p,\Omega} \le C_2\|u\|_{X^s_p(\Omega)} \quad \forall u \in W^{s,p}_{00}(\Omega). \qquad (8.10.136)$$

Theorem 8.10.20. *Let* $s > 0$, $1 < p < \infty$ *and* $\Omega \subset \mathbb{R}^n$ *be an open subset with* Lipschitz continuous *boundary* Γ. *Then:*

I. *For* $s - \frac{1}{p} \ne$ an integer *and* $u \in W^{s,p}_0(\Omega)$, $\frac{\partial^{\alpha} u}{\rho^{s-|\alpha|}} \in L^p(\Omega)\ \forall |\boldsymbol{\alpha}| \le s$ *with* $\|\frac{\partial^{\alpha} u}{\rho^{s-|\alpha|}}\|_{L^p(\Omega)} \le C\|u\|_{s,p,\Omega}$ *for some* $C > 0$. (8.10.137)

II. *For* $s - \frac{1}{p} \ne$ *an integer,* $W^{s,p}_{00}(\Omega) = W^{s,p}_0(\Omega) = X^s_p(\Omega)$ *with norm equivalence.* (8.10.138)

III. *For* $0 < s < \frac{1}{p}$, $W_{00}^{s,p}(\Omega) = W_0^{s,p}(\Omega) = W^{s,p}(\Omega)$. (8.10.139)

IV. *For* $s - \frac{1}{p}$ = an integer, $W_{00}^{s,p}(\Omega) = X_p^s(\Omega) \subsetneq W_0^{s,p}(\Omega)$ *with* equivalent *norms and* $W_{00}^{s,p}(\Omega) \neq W_0^{s,p}(\Omega)$. (8.10.140)

Proof. For I, we refer to Grisvard [18]. Proofs for II and III are exactly similar to those given in the proof of Corollary 8.10.2 with $p = 2$ replaced by general $p \in \,]1, \infty[$. For IV, see the proof of Theorem 8.10.18 with L^2-spaces replaced by L^p-spaces and $H^m(\Omega)$ by $W^{m,p}(\Omega)$, etc. □

An alternative definition of $W_{00}^{s,p}(\Omega)$ is given by:

Definition 8.10.17. For $s = m + \sigma, m \in \mathbb{N}_0, 0 \leq \sigma < 1, 1 < p < \infty$, and $\Omega \subset \mathbb{R}^n$ with *Lipschitz continuous* boundary Γ,

$$W_{00}^{s,p}(\Omega) = X_p^s(\Omega) = \left\{ u : u \in W_0^{s,p}(\Omega), \ \frac{\partial^{\boldsymbol{\alpha}} u}{\rho^\sigma} \in L^p(\Omega) \ \forall |\boldsymbol{\alpha}| = m \right\} \quad (8.10.141)$$

with $\|u\|_{W_{00}^{s,p}(\Omega)} = \|u\|_{X_p^s(\Omega)}$, the norms $\|\cdot\|_{X_p^s(\Omega)}$ and $\|\cdot\|_{00,s,p,\Omega}$ in (8.10.135) and (8.10.129), respectively, being equivalent ones by Lemma 8.10.4. Finally, for $p = 2$, $W_{00}^{s,2}(\Omega) = H_{00}^s(\Omega)$ is a Hilbert space.

Dual space $(W_{00}^{s,p}(\Omega))'$ for $s > 0$, $1 < p < \infty$, is a *space of distributions* in $\mathcal{D}'(\Omega)$, since $\mathcal{D}(\Omega)$ is *dense* in $W_{00}^{s,p}(\Omega)$ by (8.10.133).

For $s > 0$, $1 < p < \infty$, $s - \frac{1}{p}$ $\neq$ *an integer*, $(W_{00}^{s,p}(\Omega))' = W^{-s,q}(\Omega)$, since $W_{00}^{s,p}(\Omega) = W_0^{s,p}(\Omega)$ by (8.10.138) and

$$W^{-s,q}(\Omega) = (W_0^{s,p}(\Omega))', \quad \frac{1}{p} + \frac{1}{q} = 1. \quad (8.10.142)$$

For $s > 0$, $s - \frac{1}{p}$ = *an integer*, $1 < p < \infty$ and $\Omega \subset \mathbb{R}^n$ with *Lipschitz continuous* boundary,

$$\mathcal{D}(\Omega) \subset W_{00}^{s,p}(\Omega) \subset W_0^{s,p}(\Omega) \subset W^{-s,q}(\Omega) \subset (W_{00}^{s,p}(\Omega))' \subset \mathcal{D}'(\Omega), \quad (8.10.143)$$

$\frac{1}{p} + \frac{1}{q} = 1$.

8.10.14 Restrictions of distributions in Sobolev spaces

Restrictions of distributions play an important role in the study of the differentiation of functions in $H^s(\overline{\Omega})$ (resp. $W^{s,p}(\overline{\Omega})$), and in the study of traces of functions on a part of the boundary.

Restrictions of distributions in Sobolev spaces defined on $\mathbb{R}^n$

Theorem 8.10.21. *Let* $\Omega \subset \mathbb{R}^n$ *be a* bounded *open subset of* $\mathbb{R}^n$ *with* Lipschitz continuous *boundary* Γ. *For* $s > 0$, *let* $T \in H^{-s}(\mathbb{R}^n)$ *be a distribution defined on* $\mathbb{R}^n$ *and* $T{\downarrow_\Omega}$ *be its restriction to* Ω. *Then*

$$T{\downarrow_\Omega} \in (H^s_{00}(\Omega))', \tag{8.10.144}$$

$(H^s_{00}(\Omega))'$ *being the* dual space *of* $H^s_{00}(\Omega)$ *defined by* (8.10.121).

For $s > 0$, $1 < p < \infty$, *and for* $T \in W^{-s,p}(\mathbb{R}^n)$, $T{\downarrow_\Omega}$ *belongs to* $(W^{s,q}_{00}(\Omega))'$, $\frac{1}{p} + \frac{1}{q} = 1$. (8.10.145)

Proof. For $s > 0$, $T \in H^{-s}(\mathbb{R}^n) \Longrightarrow |\langle T, v\rangle| \le \|T\|_{H^{-s}(\mathbb{R}^n)}\|v\|_{H^s(\mathbb{R}^n)}\ \forall v \in H^s(\mathbb{R}^n)$. Since $\mathcal{D}(\mathbb{R}^n)$ is *dense* in $H^s(\mathbb{R}^n)$ with $s > 0$, $H^{-s}(\mathbb{R}^n)$ is a *space of distributions* in $\mathcal{D}'(\mathbb{R}^n)$ (see Section 4.2, Chapter 4) with $\mathcal{D}(\mathbb{R}^n) \hookrightarrow H^s(\mathbb{R}^n) \hookrightarrow H^{-s}(\mathbb{R}^n) \hookrightarrow \mathcal{D}'(\mathbb{R}^n)$. Hence, $\forall \phi \in \mathcal{D}(\Omega)$ with its null extension $\tilde{\phi} \in \mathcal{D}(\mathbb{R}^n)$, $\langle T{\downarrow_\Omega}, \phi\rangle_{\mathcal{D}'(\Omega)\times\mathcal{D}(\Omega)} = \langle T, \tilde{\phi}\rangle_{\mathcal{D}'(\mathbb{R}^n)\times\mathcal{D}(\mathbb{R}^n)}$ and $|\langle T{\downarrow_\Omega}, \phi\rangle| = |\langle T, \tilde{\phi}\rangle| \le \|T\|_{H^{-s}(\mathbb{R}^n)}\|\tilde{\phi}\|_{H^s(\mathbb{R}^n)} = \|T\|_{H^{-s}(\mathbb{R}^n)}\|\phi\|_{00,s,\Omega}\ \forall \phi \in \mathcal{D}(\Omega) \Longrightarrow T{\downarrow_\Omega}$ is a *continuous, linear* functional on $\mathcal{D}(\Omega)$ in the norm $\|\cdot\|_{00,s,\Omega}$ of $H^s_{00}(\Omega)$. But since Lipschitz continuity of the boundary Γ of $\Omega \Longrightarrow$ its continuity, $\mathcal{D}(\Omega)$ is *dense* in $H^s_{00}(\Omega)$ by Theorem 8.10.15. Hence, $T{\downarrow_\Omega}$ can be given a *unique, continuous, linear extension* to $H^s_{00}(\Omega)$, and this unique *extended, continuous, linear* functional on $H^s_{00}(\Omega)$ will still be denoted by the same notation $T{\downarrow_\Omega}$, i.e. for $s > 0$, $T{\downarrow_\Omega} \in (H^s_{00}(\Omega))'$ with $\langle T{\downarrow_\Omega}, v\rangle_{(H^s_{00}(\Omega))'\times H^s_{00}(\Omega)} = \langle T, \tilde{v}\rangle_{H^{-s}(\mathbb{R}^n)\times H^s(\mathbb{R}^n)}$ and $|\langle T{\downarrow_\Omega}, v\rangle| \le \|T\|_{H^{-s}(\mathbb{R}^n)}\|\tilde{v}\|_{H^s(\mathbb{R}^n)} = \|T\|_{H^{-s}(\mathbb{R}^n)}\|v\|_{H^s_{00}(\Omega)}\ \forall v \in H^s_{00}(\Omega)$ with $\|T{\downarrow_\Omega}\|_{(H^s_{00}(\Omega))'} \le \|T\|_{H^{-s}(\mathbb{R}^n)}$.

For (8.10.145), we see that for $s > 0$, $T \in W^{-s,p}(\mathbb{R}^n) = (W^{s,q}(\mathbb{R}^n))'$, $\frac{1}{p}+\frac{1}{q} = 1$. But, $\forall \phi \in \mathcal{D}(\Omega)$,

$$|\langle T{\downarrow_\Omega}, \phi\rangle| \le \|T\|_{W^{-s,p}(\mathbb{R}^n)}\|\tilde{\phi}\|_{W^{s,q}(\mathbb{R}^n)} = \|T\|_{W^{-s,p}(\mathbb{R}^n)}\|\phi\|_{W^{s,q}_{00}(\mathbb{R}^n)}$$

(by (8.10.129)) $\Longrightarrow |\langle T{\downarrow_\Omega}, v\rangle_{(W^{s,q}_{00}(\Omega))'\times W^{s,q}_{00}(\Omega)} = \langle T, \tilde{v}\rangle_{W^{-s,p}(\mathbb{R}^n)\times W^{s,q}(\mathbb{R}^n)}\ \forall v \in W^{s,q}_{00}(\Omega) \equiv \overline{\mathcal{D}(\Omega)}$ (by (8.10.133)). □

Restriction of distributions defined on open subsets of $\mathbb{R}^n$

So far, we have considered restrictions of distributions defined on $\mathbb{R}^n$. Now we will consider restrictions of distributions defined on *open subsets* of $\mathbb{R}^n$.

Theorem 8.10.22. *Let* Ω_1 *and* Ω_2 *be open subsets of* $\mathbb{R}^n$ *with* $\Omega_1 \subset \Omega_2$. *Then the* restriction *mapping* $u \longmapsto u{\downarrow_{\Omega_1}}$ *is* continuous *from:*

I. $W^{s,p}(\Omega_2)$ *into* $W^{s,p}(\Omega_1)$ *if* $s > 0$, $1 \le p \le \infty$; (8.10.146a)

II. $W^{-k,q}(\Omega_2)$ *into* $W^{-k,q}(\Omega_1)$ *if* $s = k \in \mathbb{N}$, $1 < q < \infty$. (8.10.146b)

In particular, for $p = 2$, $W^{s,2}(\Omega_i) = H^s(\Omega_i)$, $i = 1, 2$, *and the restriction mapping* $u \longmapsto u{\downarrow}_{\Omega_1}$ *is* continuous *from*

III. $H^s(\Omega_2)$ *into* $H^s(\Omega_1)$ *if* $s > 0$; (8.10.146c)

IV. $H^{-k}(\Omega_2)$ *into* $H^{-k}(\Omega_1)$ *if* $s = k \in \mathbb{N}$. (8.10.146d)

Proof.

I. The result follows from Definition 8.10.9, since $\|u{\downarrow}_{\Omega_1}\|_{s,p,\Omega_1} \le \|u\|_{s,p,\Omega_2}$ $\forall u \in W^{s,p}(\Omega_1)$ with $s > 0, 1 \le p \le \infty$.

II. For $s = k \in \mathbb{N}$, $1 < p, q < \infty$, $\frac{1}{p} + \frac{1}{q} = 1$, $u \in W^{-k,q}(\Omega_2) \equiv (W_0^{k,p}(\Omega_2))'$

$$\Longrightarrow \quad |\langle u, v\rangle_{W^{-k,q}(\Omega_2)\times W_0^{k,p}(\Omega_2)} \le \|u\|_{W^{-k,q}(\Omega_2)} \cdot \|v\|_{W_0^{k,p}(\Omega_2)} \quad \forall v \in W_0^{k,p}(\Omega_2). \tag{8.10.147}$$

But $\mathcal{D}(\Omega_i)$ is *dense* in $W_0^{k,p}(\Omega_i)$ $\forall k \in \mathbb{N}$, $1 < p < \infty \Longrightarrow W^{-k,q}(\Omega_i) \hookrightarrow \mathcal{D}'(\Omega_i)$ is a space of distributions on Ω_i, $i = 1, 2$, $\frac{1}{p} + \frac{1}{q} = 1$, $1 < p, q < \infty$. Hence, $u \in W^{-k,q}(\Omega_2)$ is a distribution on Ω_2 and its restriction to Ω_1 is defined by (5.3.1): $\langle u{\downarrow}_{\Omega_1}, \phi\rangle_{\mathcal{D}'(\Omega_1)\times\mathcal{D}(\Omega_1)} = \langle u, \widetilde{\phi}\rangle_{\mathcal{D}'(\Omega_2)\times\mathcal{D}(\Omega_2)}$ $\forall \phi \in \mathcal{D}(\Omega_1)$, with null extension $\tilde{\phi}$ to Ω_2 belonging to $\mathcal{D}(\Omega_2)$ and

$$\begin{aligned} |\langle u{\downarrow}_{\Omega_1}, \phi\rangle| &= |\langle u, \widetilde{\phi}\rangle| \le \|u\|_{-k,q,\Omega_2} \|\widetilde{\phi}\|_{W_0^{k,p}(\Omega_2)} \quad \text{(by (8.10.147))} \\ &= \|u\|_{-k,q,\Omega_2} \|\phi\|_{W_0^{k,p}(\Omega_1)} \quad \forall \phi \in \mathcal{D}(\Omega_1) \text{ with } \tilde{\phi} \in \mathcal{D}(\Omega_2) \end{aligned}$$

(see the proof of Theorem 2.15.5). Hence, $u{\downarrow}_{\Omega_1}$ is a *continuous*, linear functional on $\mathcal{D}(\Omega_1)$ in the norm $\|\cdot\|_{W_0^{k,p}(\Omega_1)}$, $\mathcal{D}(\Omega_1)$ being a *dense* subspace of $W_0^{k,p}(\Omega_1)$, $1 < p < \infty$. Therefore, $u{\downarrow}_{\Omega_1}$ can be given a *unique*, *continuous* linear extension to $W_0^{k,p}(\Omega_1)$ and will still be denoted by the same notation $u{\downarrow}_{\Omega_1} \in W^{-k,q}(\Omega_1) \equiv (W_0^{k,p}(\Omega_1))'$, i.e. $\langle u{\downarrow}_{\Omega_1}, v\rangle_{W^{-k,q}(\Omega_1)\times W_0^{k,p}(\Omega_1)} = \langle u, \tilde{v}\rangle_{W^{-k,q}(\Omega_2)\times W_0^{k,p}(\Omega_2)}$ $\forall u \in W^{-k,q}(\Omega_2)$ and $\forall v \in W_0^{k,p}(\Omega_1)$ with the *continuous* null extension $\tilde{v} \in W_0^{k,p}(\Omega_2)$ of v, since $\|\tilde{v}\|_{W_0^{k,p}(\Omega_2)} = \|v\|_{W_0^{k,p}(\Omega_1)}$ (see the proof of Theorem 2.15.5).

□

Theorem 8.10.23. *Let* Ω_1 *and* Ω_2 *be open subsets of* $\mathbb{R}^n$ *such that* $\Omega_1 \subset \Omega_2$ *and the boundaries* Γ_1 *and* Γ_2 *of* Ω_1 *and* Ω_2, *respectively, are Lipschitz continuous. Then the* restriction *mapping* $u \longmapsto u{\downarrow}_{\Omega_1}$ *is* continuous *from:*

I. $(W_{00}^{s,p}(\Omega_2))'$ *into* $(W_{00}^{s,p}(\Omega_1))'$ *if* $s > 0$, $s - \frac{1}{p} =$ *an integer*, $1 < p < \infty$; (8.10.148a)

II. $W^{-s,q}(\Omega_2)$ *into* $(W_{00}^{s,p}(\Omega_1))'$ *if* $s > 0$, $s - \frac{1}{p}$ = *an integer*, $1 < p < \infty$; (8.10.148b)

III. $W^{-s,q}(\Omega_2)$ *into* $W^{-s,q}(\Omega_1)$ *if* $s > 0$, $s - \frac{1}{p} \neq$ *an integer*, $1 < p, q < \infty$, $\frac{1}{p} + \frac{1}{q} = 1$. (8.10.148c)

In particular, for $p = 2$, $W^{s,2}(\Omega_i) = H^s(\Omega_i)$, $i = 1, 2$, *and the restriction mapping* $u \longmapsto u\downarrow_{\Omega_1}$ *is* continuous *from:*

IV. $(H_{00}^s(\Omega_2))'$ *into* $(H_{00}^s(\Omega_1))'$ *if* $s > 0$, $s - \frac{1}{2}$ = *an integer;* (8.10.148d)

V. $H^{-s}(\Omega_2)$ *into* $(H_{00}^s(\Omega_1))'$ *if* $s > 0$, $s - \frac{1}{2}$ = *an integer;* (8.10.148e)

VI. $H^{-s}(\Omega_2)$ *into* $H^{-s}(\Omega_1)$ *if* $s > 0$, $s - \frac{1}{2} \neq$ *an integer.* (8.10.148f)

Proof.

I. For $s > 0$, $s - \frac{1}{p}$ = an integer, $1 < p < \infty$, $W_{00}^{s,p}(\Omega_2) \subsetneq W_0^{s,p}(\Omega_2)$ by (8.10.140). $u \in (W_{00}^{s,p}(\Omega_2))'$

$$\begin{aligned} \Longrightarrow \quad & |\langle u, v\rangle|_{(W_{00}^{s,p}(\Omega_2))' \times (W_{00}^{s,p}(\Omega_2))} \\ & \leq \|u\|_{(W_{00}^{s,p}(\Omega_2))'} \|v\|_{(W_{00}^{s,p}(\Omega_2))} \\ & = \|u\|_{(W_{00}^{s,p}(\Omega_2))'} \|\tilde{v}\|_{W^{s,p}(\mathbb{R}^n)} \quad \text{(by (8.10.129))}, \end{aligned} \tag{8.10.149}$$

where $\tilde{v}$ is the null extension to $\mathbb{R}^n$ of $v \in W_{00}^{s,p}(\Omega_2)$.

But, for $s > 0$, $1 < p < \infty$ and (Lipschitz) continuous boundary Γ_i, $\mathcal{D}(\Omega_i)$ is *dense* in $W_{00}^{s,p}(\Omega_i)$ by (8.10.133). Hence, $(W_{00}^{s,p}(\Omega_i))' \hookrightarrow \mathcal{D}'(\Omega_i)$ is a *space of distributions* on Ω_i ($i = 1, 2$) (see Section 4.3, Chapter 4). Hence, $u \in (W_{00}^{s,p}(\Omega_2))'$ is a distribution on Ω_2 and its restriction to Ω_1 is defined by (5.3.1): $\forall \phi \in \mathcal{D}(\Omega_2)$ with $\operatorname{supp}(\phi) \subset\subset \Omega_1$,

$$\langle u\downarrow_{\Omega_1}, \phi\rangle_{\mathcal{D}'(\Omega_1)\times\mathcal{D}(\Omega_1)} = \langle u, \phi\rangle_{\mathcal{D}'(\Omega_2)\times\mathcal{D}(\Omega_2)}. \tag{8.10.150}$$

$\forall \phi \in \mathcal{D}(\Omega_2)$ with $\operatorname{supp}(\phi) \subset\subset \Omega_1, \tilde{\phi} \in \mathcal{D}(\mathbb{R}^n)$ is the null extension to $\mathbb{R}^n$ of ϕ. Since $\operatorname{supp}(\phi) \subset\subset \Omega_1$, considering $\phi = \phi\downarrow_{\Omega_1} \in \mathcal{D}(\Omega_1)$, $\tilde{\phi} \in \mathcal{D}(\mathbb{R}^n)$ is also the null extension of $\phi \in \mathcal{D}(\Omega_1)$ to $\mathbb{R}^n$ with $\tilde{\phi}\downarrow_{\Omega_2} \in \mathcal{D}(\Omega_2)$. Then, using (8.10.149) and (8.10.150), $\forall \phi \in \mathcal{D}(\Omega_1)$,

$$\begin{aligned} |\langle u\downarrow_{\Omega_1}, \phi\rangle| = |\langle u, \tilde{\phi}\rangle| &\leq \|u\|_{(W_{00}^{s,p}(\Omega_2))'} \|\tilde{\phi}\|_{W^{s,p}(\mathbb{R}^n)} \\ &= \|u\|_{(W_{00}^{s,p}(\Omega_2))'} \|\phi\|_{W_{00}^{s,p}(\Omega_1)}. \end{aligned} \tag{8.10.151}$$

Hence, $u\downarrow_{\Omega_1}$ is a *continuous*, linear functional on $\mathcal{D}(\Omega_1)$ in the norm $\|\cdot\|_{W_{00}^{s,p}(\Omega_1)}$, $\mathcal{D}(\Omega_1)$ being a *dense* subspace of $W_{00}^{s,p}(\Omega_1)$ by (8.10.133). Therefore, $u\downarrow_{\Omega_1}$ can be given a *unique, continuous*, linear extension to $W_{00}^{s,p}$

(Ω_1), and this *extended*, *continuous*, *linear* functional on $W_{00}^{s,p}(\Omega_1)$ will be denoted by the same notation $u\downarrow_{\Omega_1} \in (W_{00}^{s,p}(\Omega_1))'$ with

$$\langle u\downarrow_{\Omega_1}, v\rangle_{(W_{00}^{s,p}(\Omega_1))'\times W_{00}^{s,p}(\Omega_1)} = \langle u, \tilde{v}\rangle_{(W_{00}^{s,p}(\Omega_2))'\times W_{00}^{s,p}(\Omega_2)} \quad \forall u \in (W_{00}^{s,p}(\Omega_2))',$$

$\forall v \in W_{00}^{s,p}(\Omega_1)$ with *continuous* null extension $\tilde{v} \in W_{00}^{s,p}(\Omega_2)$ ($\|\tilde{v}\|_{W_{00}^{s,p}(\Omega_2)} = \|\tilde{\tilde{v}}\|_{W^{s,p}(\mathbb{R}^n)} = \|\tilde{v}\|_{W^{s,p}(\mathbb{R}^n)} = \|v\|_{W_{00}^{s,p}(\Omega_1)}$, where $\tilde{\tilde{v}}$ is the null extension to $\mathbb{R}^n$ of $\tilde{v}$, which is, in fact, the null extension to $\mathbb{R}^n$ of v).

II. First of all, $\mathcal{D}(\Omega_i)$ is *dense* in $W_0^{s,p}(\Omega_i)$, and for Lipschitz continuous Γ_i, $s - \frac{1}{p}$ = an integer, $1 < p < \infty$, $W_{00}^{s,p}(\Omega_i) \subsetneq W_0^{s,p}(\Omega_i)$ by (8.10.140) $\Longrightarrow W^{-s,q}(\Omega_i) \hookrightarrow (W_{00}^{s,p}(\Omega_i))' \hookrightarrow \mathcal{D}'(\Omega_i)$, $\hookrightarrow$ being a continuous one. Hence, $W^{-s,q}(\Omega_2) \hookrightarrow (W_{00}^{s,p}(\Omega_2))'$, $\hookrightarrow$ being a continuous one for $s > 0$, $s - \frac{1}{p}$ = an integer, $1 < p, q < \infty$, $\frac{1}{p} + \frac{1}{q} = 1$. The composition of the *continuous* mappings

$$u \in W^{-s,q}(\Omega_2) \xmapsto{\text{continuous}} u \in (W_{00}^{s,p}(\Omega_2))' \xmapsto{\text{continuous}} u \in (W_{00}^{s,p}(\Omega_1))'$$

(the continuity of the last mapping following from I) is *continuous* from $W^{-s,q}(\Omega_2)$ into $(W_{00}^{s,p}(\Omega_1))'$.

III. For $s > 0$, $s - \frac{1}{p} \neq$ an integer, $1 < p < \infty$ and Lipschitz continuous Γ_i, $W_{00}^{s,p}(\Omega_i) = W_0^{s,p}(\Omega_i)$ with norm equivalence (by (8.10.138) and (8.10.136)), i.e. $\exists C_1, C_2 > 0$ such that

$$C_1\|v\|_{W_0^{s,p}(\Omega_i)} \le \|v\|_{W_{00}^{s,p}(\Omega_i)} \le C_2\|v\|_{W_0^{s,p}(\Omega_i)} \quad \forall v \in W_{00}^{s,p}(\Omega_i). \tag{8.10.152}$$

Then $u \in W^{-s,q}(\Omega_2)$

$$\begin{aligned} \Longrightarrow \quad |\langle u, v\rangle_{W^{-s,q}(\Omega_2)\times W_0^{s,p}(\Omega_2)}| &\le \|u\|_{W^{-s,q}(\Omega_2)}\|v\|_{W_0^{s,p}(\Omega_2)} \\ &\le (C_1^*\|u\|_{W^{-s,q}(\Omega_2)})\|v\|_{W_{00}^{s,p}(\Omega_2)} \\ &= (C_1^*\|u\|_{W^{-s,q}(\Omega_2)})\|\tilde{v}\|_{W^{s,p}(\mathbb{R}^n)} \quad \forall v \in W_0^{s,p}(\Omega_2). \end{aligned}$$

Since $\mathcal{D}(\Omega_i)$ is *dense* in $W_0^{s,p}(\Omega_i)$, $W^{-s,q}(\Omega_i) \equiv (W_0^{s,p}(\Omega_i))' \hookrightarrow \mathcal{D}'(\Omega_i)$ is a *space of distributions* on Ω_i. Hence, $\forall\phi \in \mathcal{D}(\Omega_1)$ with null extension on $\tilde{\phi} \in \mathcal{D}(\mathbb{R}^n)$, i.e. $\tilde{\phi}\downarrow_{\Omega_2} \in \mathcal{D}(\Omega_2)$,

$$\begin{aligned} &\langle u\downarrow_{\Omega_1}, \phi\rangle_{\mathcal{D}'(\Omega_1)\times\mathcal{D}(\Omega_1)} = \langle u, \tilde{\phi}\rangle_{\mathcal{D}'(\Omega_2)\times\mathcal{D}(\Omega_2)} \\ \Longrightarrow \quad &|\langle u\downarrow_{\Omega_1}, \phi\rangle| = |\langle u, \tilde{\phi}\rangle| \le C_1^*\|u\|_{W^{-s,q}(\Omega_2)} \cdot \|\tilde{\phi}\|_{W^{s,p}(\mathbb{R}^n)} \\ \Longrightarrow \quad &|\langle u\downarrow_{\Omega_1}, \phi\rangle| \le (C_1^*\|u\|_{W^{-s,q}(\Omega_2)})\|\phi\|_{W_{00}^{s,p}(\Omega_1)} \\ &\qquad\qquad \le (C_2^*\|u\|_{W^{-s,q}(\Omega_2)})\|\phi\|_{W_0^{s,p}(\Omega_1)} \quad \forall\phi \in \mathcal{D}(\Omega_1) \end{aligned}$$

(by (8.10.151) and (8.10.152))
$\Longrightarrow u\downarrow_{\Omega_1}$ is a *continuous*, linear functional on $\mathcal{D}(\Omega_1)$ in the norm $\|\cdot\|_{W_0^{s,p}(\Omega_1)}$. Hence, by virtue of the density of $\mathcal{D}(\Omega_1)$ in $W_0^{s,p}(\Omega_1)$, $u\downarrow_{\Omega_1}$ can be given a *unique*, *continuous*, linear extension to $W_0^{s,p}(\Omega_1)$, and the extended, continuous, linear functional will still be denoted by $u\downarrow_{\Omega_1}$, i.e. $u\downarrow_{\Omega_1} \in W^{-s,q}(\Omega_1)$ with $\langle u\downarrow_{\Omega_1}, v\rangle_{W^{-s,q}(\Omega_1)\times W_0^{s,p}(\Omega_1)} = \langle u, \tilde{v}\rangle_{W^{-s,q}(\Omega_2)\times W_0^{s,p}(\Omega_2)}$ $\forall u \in W^{-s,q}(\Omega_2)$, $\forall v \in W_0^{s,p}(\Omega_1)$, with $\|\tilde{v}\|_{W_0^{s,p}(\Omega_2)} = \|\tilde{\tilde{v}}\|_{W^{s,p}(\mathbb{R}^n)} = \|\tilde{v}\|_{W^{s,p}(\mathbb{R}^n)} = \|v\|_{W_{00}^{s,p}(\Omega_1)} \le C_2\|v\|_{W_0^{s,p}(\Omega_1)}$ (by (8.10.152)). □

Warning Under the assumptions of Theorem 8.10.23, the restriction mapping $u \longmapsto u\downarrow_{\Omega_1}$ is *not* continuous from:

- $H^{-s}(\Omega_2)$ into $H^{-s}(\Omega_1)$ if $s > 0$, $s - \frac{1}{2} =$ *an integer*; (8.10.153)
- $W^{-s,q}(\Omega_2)$ into $W^{-s,q}(\Omega_1)$ if $s > 0$, $s - \frac{1}{p} =$ *an integer*, $1 < p, q < \infty$, $\frac{1}{p} + \frac{1}{q} = 1$.

8.10.15 Differentiation of distributions in $H^s(\Omega)$ with $s \in \mathbb{R}$

First of all, we refer to the results stated in (4.2.13) (Section 4.2, Chapter 4).

Case $\Omega = \mathbb{R}^n$

Theorem 8.10.24. $\forall s \in \mathbb{R}$, $\forall$ *multi-index* $\boldsymbol{\alpha}$ *with* $|\boldsymbol{\alpha}| = m \in \mathbb{N}_0$, *the linear operator*

$$\partial^{\boldsymbol{\alpha}} : H^s(\mathbb{R}^n) \longrightarrow H^{s-|\boldsymbol{\alpha}|}(\mathbb{R}^n) \textit{ is } \text{continuous} \textit{ from } H^s(\mathbb{R}^n) \textit{ into } H^{s-|\boldsymbol{\alpha}|}(\mathbb{R}^n). \tag{8.10.154}$$

Proof. Let $u \in H^s(\mathbb{R}^n)$ for $s \in \mathbb{R}$. Then $\int_{\mathbb{R}^n}(1 + \|\boldsymbol{\xi}\|^2)^s|\hat{u}(\boldsymbol{\xi})|^2 d\boldsymbol{\xi} < +\infty$ with $\hat{u} = \mathcal{F}[u]$. From Theorem 8.6.1, $|\widehat{\partial^{\boldsymbol{\alpha}} u}(\boldsymbol{\xi})| = \mathcal{F}[\partial_{\mathbf{x}}^{\boldsymbol{\alpha}} u] = (i2\pi\boldsymbol{\xi})^{\boldsymbol{\alpha}}\mathcal{F}[u] = (i2\pi\boldsymbol{\xi})^{\boldsymbol{\alpha}}\hat{u}$. We are to show that $\partial^{\boldsymbol{\alpha}} u \in H^{s-|\boldsymbol{\alpha}|}(\mathbb{R}^n)$ $\forall$ multi-index $\boldsymbol{\alpha}$, $\forall s \in \mathbb{R}$, i.e. $\int_{\mathbb{R}^n}(1 + \|\boldsymbol{\xi}\|^2)^{s-|\boldsymbol{\alpha}|}|\widehat{\partial^{\boldsymbol{\alpha}} u}(\boldsymbol{\xi})|^2 d\boldsymbol{\xi} < +\infty$. In fact,

$$\int_{\mathbb{R}^n}(1 + \|\boldsymbol{\xi}\|^2)^{s-|\boldsymbol{\alpha}|}|\widehat{\partial^{\boldsymbol{\alpha}} u}(\boldsymbol{\xi})|^2 d\boldsymbol{\xi} = \int_{\mathbb{R}^n}(1 + \|\boldsymbol{\xi}\|^2)^{s-|\boldsymbol{\alpha}|}(|(i2\pi\boldsymbol{\xi})^{\boldsymbol{\alpha}}|^2|)\hat{u}(\boldsymbol{\xi})|^2 d\boldsymbol{\xi}$$

$$\le \int_{\mathbb{R}^n}(1 + \|\boldsymbol{\xi}\|^2)^{s-|\boldsymbol{\alpha}|}(2\pi)^{2|\boldsymbol{\alpha}|}(1 + \|\boldsymbol{\xi}\|^2)^{|\boldsymbol{\alpha}|}|\hat{u}(\boldsymbol{\xi})|^2 d\boldsymbol{\xi}$$

$$\left(\text{since } \boldsymbol{\xi}^{2\boldsymbol{\alpha}} = \prod_{i=1}^{n} \xi_i^{2\alpha_i} \le (1 + \|\boldsymbol{\xi}\|^2)^{|\boldsymbol{\alpha}|} \text{ by Lemma 8.9.1}\right)$$

$$\le C \int_{\mathbb{R}^n}(1 + \|\boldsymbol{\xi}\|^2)^s|\hat{u}(\boldsymbol{\xi})|^2 d\boldsymbol{\xi} < +\infty$$

with $C > (2\pi)^{2|\boldsymbol{\alpha}|} > 0$. Hence, $\forall u \in H^s(\mathbb{R}^n)$, $\partial^{\boldsymbol{\alpha}} u \in H^{s-|\boldsymbol{\alpha}|}(\mathbb{R}^n)$ $\forall \boldsymbol{\alpha}$, $\forall s \in \mathbb{R}$, and $\|\partial^{\boldsymbol{\alpha}} u\|_{H^{s-|\boldsymbol{\alpha}|}(\mathbb{R}^n)} \leq C_1 \|u\|_{H^s(\mathbb{R}^n)}$ with $C_1 = \sqrt{C} > 0$. Thus, $\partial^{\boldsymbol{\alpha}} : H^s(\mathbb{R}^n) \longrightarrow H^{s-|\boldsymbol{\alpha}|}(\mathbb{R}^n)$ is *continuous* from $H^s(\mathbb{R}^n)$ into $H^{s-|\boldsymbol{\alpha}|}(\mathbb{R}^n)$. □

Case $\Omega \subsetneq \mathbb{R}^n$ The situation is *different* in this case. In fact, an analogous result holds if both *s and* $s - |\boldsymbol{\alpha}|$ have the *same sign*. In other words, for positive $s \geq 0$, if $s - |\boldsymbol{\alpha}| < 0$ then the result does *not* hold, in general, but for $s \leq 0$ *and* $s - |\boldsymbol{\alpha}| < 0$, the result *does hold*.

Theorem 8.10.25. *For* $s - |\boldsymbol{\alpha}| \geq 0$*, the following results hold:*

I. $\partial^{\boldsymbol{\alpha}} : H^s(\Omega) \longrightarrow H^{s-|\boldsymbol{\alpha}|}(\Omega)$ *is a* continuous, linear *operator from* $H^s(\Omega)$ *into* $H^{s-|\boldsymbol{\alpha}|}(\Omega)$. (8.10.155)

II. $\partial^{\boldsymbol{\alpha}} : H_0^s(\Omega) \longrightarrow H_0^{s-|\boldsymbol{\alpha}|}(\Omega)$ *is a* continuous, linear *operator from* $H_0^s(\Omega)$ *into* $H_0^{s-|\boldsymbol{\alpha}|}(\Omega)$. (8.10.156)

III. *For* $s \geq 0$, $\forall$ *multi-index* $\boldsymbol{\alpha}$*, the linear operator* $\partial^{\boldsymbol{\alpha}} : H^{-s}(\Omega) \longrightarrow H^{-s-|\boldsymbol{\alpha}|}(\Omega)$ *is* continuous *from* $H^{-s}(\Omega)$ *into* $H^{-s-|\boldsymbol{\alpha}|}(\Omega)$. (8.10.157)

For bounded Ω *with Lipschitz continuous boundary* Γ*, the linear operator*

IV. $\partial^{\boldsymbol{\alpha}} : H_{00}^s(\Omega) \longrightarrow H_{00}^{s-|\boldsymbol{\alpha}|}(\Omega)$ *is* continuous *from* $H_{00}^s(\Omega)$ *into* $H_{00}^{s-|\boldsymbol{\alpha}|}(\Omega)$*;* (8.10.158)

V. $\partial^{\boldsymbol{\alpha}} : (H_{00}^s(\Omega))' \longrightarrow (H_{00}^{s+|\boldsymbol{\alpha}|}(\Omega))'$ *is* continuous *from* $(H_{00}^s(\Omega))'$ *into* $(H_{00}^{s+|\boldsymbol{\alpha}|}(\Omega))'$ $\forall s > 0$, $\forall |\boldsymbol{\alpha}| \in \mathbb{N}$. (8.10.159)

Proof.

I. For $s - |\boldsymbol{\alpha}| \geq 0$, there are two cases: $s = m \in \mathbb{N}_0$ and $s = m + \sigma$ with $m = [s]$, $0 < \sigma < 1$.

Case $s = m \in \mathbb{N}_0$: $\forall |\boldsymbol{\alpha}| \leq m$ with $m - |\boldsymbol{\alpha}| \geq 0$, $u \in H^m(\Omega) \Longrightarrow \partial^{\boldsymbol{\alpha}} u = u_{\boldsymbol{\alpha}} \in H^{m-|\boldsymbol{\alpha}|}(\Omega)$, since $u_{\boldsymbol{\alpha}} = \partial^{\boldsymbol{\alpha}} u \in L^2(\Omega)$ $\forall |\boldsymbol{\alpha}| \leq m \Longrightarrow \partial^{\boldsymbol{\beta}} u_{\boldsymbol{\alpha}} \in L^2(\Omega)$ $\forall |\boldsymbol{\beta} + \boldsymbol{\alpha}| = |\boldsymbol{\beta}| + |\boldsymbol{\alpha}| \leq m$, i.e. $\partial^{\boldsymbol{\beta}} u_{\boldsymbol{\alpha}} = \partial^{\boldsymbol{\beta}}(\partial^{\boldsymbol{\alpha}} u) \in L^2(\Omega)$ $\forall |\boldsymbol{\beta}| \leq m - |\boldsymbol{\alpha}| \Longrightarrow \partial^{\boldsymbol{\alpha}} u \in H^{m-|\boldsymbol{\alpha}|}(\Omega)$ $\forall |\boldsymbol{\alpha}| \leq m$.

Case $s = m + \sigma$ *with* $m = [s] \in \mathbb{N}_0$, $0 < \sigma < 1$: From Definition 8.10.6 $u \in H^s(\Omega) \Longrightarrow u \in H^m(\Omega)$ *and* $\partial^{\boldsymbol{\beta}} u \in H^{\sigma}(\Omega)$ $\forall |\boldsymbol{\beta}| = m \Longrightarrow \partial^{\boldsymbol{\alpha}} u \in H^{m-|\boldsymbol{\alpha}|}(\Omega)$ *and* $\partial^{\boldsymbol{\beta}}(\partial^{\boldsymbol{\alpha}} u) \in H^{\sigma}(\Omega)$ $\forall |\boldsymbol{\beta} + \boldsymbol{\alpha}| = m$, i.e. $\forall |\boldsymbol{\beta}| = m - |\boldsymbol{\alpha}| \Longrightarrow \partial^{\boldsymbol{\alpha}} u \in H^{m-|\boldsymbol{\alpha}|+\sigma}(\Omega) = H^{m+\sigma-|\boldsymbol{\alpha}|}(\Omega) = H^{s-|\boldsymbol{\alpha}|}(\Omega)$ with $\|\partial^{\boldsymbol{\alpha}} u\|_{H^{s-|\boldsymbol{\alpha}|}(\Omega)} \leq \|u\|_{H^s(\Omega)}$. Hence, $\partial^{\boldsymbol{\alpha}} : H^s(\Omega) \longrightarrow H^{s-|\boldsymbol{\alpha}|}(\Omega)$ is a *continuous*, *linear* operator from $H^s(\Omega)$ into $H^{s-|\boldsymbol{\alpha}|}(\Omega)$.

II. For $s - |\boldsymbol{\alpha}| \geq 0$, $\mathcal{D}(\Omega)$ is *dense* in $H_0^s(\Omega)$ (resp. $H_0^{s-|\boldsymbol{\alpha}|}(\Omega)$) in the norm $\|\cdot\|_{H^s(\Omega)}$ (resp. $\|\cdot\|_{H^{s-|\boldsymbol{\alpha}|}(\Omega)}$). Hence, $\forall u \in H_0^s(\Omega)$, $\exists (\phi_n)$ in $\mathcal{D}(\Omega)$ such that $\phi_n \to u$ in $H^s(\Omega)$. But $\partial^{\boldsymbol{\alpha}} : H^s(\Omega) \longrightarrow H^{s-|\boldsymbol{\alpha}|}(\Omega)$ is *continuous* for

$s - |\boldsymbol{\alpha}| \geq 0$. $\partial^{\boldsymbol{\alpha}} : \phi_n \in \mathcal{D}(\Omega) \subset H^s(\Omega) \longmapsto \partial^{\boldsymbol{\alpha}}\phi_n \in \mathcal{D}(\Omega) \subset H^{s-|\boldsymbol{\alpha}|}(\Omega)$. But $\phi_n \longrightarrow u$ in $H^s(\Omega) \Longrightarrow \partial^{\boldsymbol{\alpha}}\phi_n \longrightarrow \partial^{\boldsymbol{\alpha}}u$ in $H^{s-|\boldsymbol{\alpha}|}(\Omega)$ as $n \to \infty$ by virtue of the continuity of $\partial^{\boldsymbol{\alpha}}$ from $H^s(\Omega)$ into $H^{s-|\boldsymbol{\alpha}|}(\Omega)$ for $s - |\boldsymbol{\alpha}| \geq 0 \Longrightarrow \partial^{\boldsymbol{\alpha}}u \in H_0^{s-|\boldsymbol{\alpha}|}(\Omega)$ for $s - |\boldsymbol{\alpha}| \geq 0$. Hence, $\partial^{\boldsymbol{\alpha}} : u \in H_0^s(\Omega) \longmapsto \partial^{\boldsymbol{\alpha}}u \in H_0^{s-|\boldsymbol{\alpha}|}(\Omega)$ $\forall \boldsymbol{\alpha}$ with $s - |\boldsymbol{\alpha}| \geq 0$.

III. For $A \in \mathcal{L}(V_1; V_2)$, V_1 and V_2 being Hilbert spaces, its unique *transpose* (also called *dual*) $A^* \in \mathcal{L}(V_2'; V_1')$, V_1' and V_2' being the *topological* and *algebraic* duals of V_1 and V_2, respectively, is defined by: $\langle A^*u, v\rangle_{V_1' \times V_1} = \langle u, Av\rangle_{V_2' \times V_2}$ $\forall u \in V_2'$, $\forall v \in V_1$ ([46]).

Let $V_1 = H_0^{s+|\boldsymbol{\alpha}|}(\Omega)$, $V_2 = H_0^s(\Omega)$, in which $\mathcal{D}(\Omega)$ is *dense* in the norm $\|\cdot\|_{H_0^{s+|\boldsymbol{\alpha}|}(\Omega)}$ and $\|\cdot\|_{H_0^s(\Omega)}$, respectively, such that $V_1' = (H_0^{s+|\boldsymbol{\alpha}|}(\Omega))' = H^{-s-|\boldsymbol{\alpha}|}(\Omega) \hookrightarrow \mathcal{D}'(\Omega)$, $V_2' = H^{-s}(\Omega) \hookrightarrow \mathcal{D}'(\Omega)$ are *spaces of distributions* for $s > 0$. Then $\partial^{\boldsymbol{\alpha}} : H_0^{s+|\boldsymbol{\alpha}|}(\Omega) \longrightarrow H_0^s(\Omega)$ is *continuous* by II. For any $v \in H_0^{s+|\boldsymbol{\alpha}|}(\Omega)$, $\exists(\phi_n)$ in $\mathcal{D}(\Omega)$ such that $\phi_n \to v$ in $H_0^{s+|\boldsymbol{\alpha}|}(\Omega)$ as $n \to \infty$, which implies that $\partial^{\boldsymbol{\alpha}}\phi_n \to \partial^{\boldsymbol{\alpha}}v$ in $H_0^s(\Omega)$ by II. Set $A = (-1)^{|\boldsymbol{\alpha}|}\partial^{\boldsymbol{\alpha}} \in \mathcal{L}(H_0^{s+|\boldsymbol{\alpha}|}(\Omega); H_0^s(\Omega))$. Then its transpose $((-1)^{|\boldsymbol{\alpha}|}\partial^{\boldsymbol{\alpha}})^* \in \mathcal{L}(H^{-s}(\Omega); H^{-s-|\boldsymbol{\alpha}|}(\Omega))$ for $s > 0$ is given by:

$$
\begin{aligned}
&\langle ((-1)^{|\boldsymbol{\alpha}|}\partial^{\boldsymbol{\alpha}})^* u, v\rangle_{H^{-s-|\boldsymbol{\alpha}|}(\Omega) \times H_0^{s+|\boldsymbol{\alpha}|}(\Omega)} = \langle u, (-1)^{|\boldsymbol{\alpha}|}\partial^{\boldsymbol{\alpha}}v\rangle_{H^{-s}(\Omega) \times H_0^s(\Omega)} \\
&= \lim_{n\to\infty} \langle u, (-1)^{|\boldsymbol{\alpha}|}\partial^{\boldsymbol{\alpha}}\phi_n\rangle_{H^{-s}(\Omega) \times H_0^s(\Omega)} = \lim_{n\to\infty} \langle \partial^{\boldsymbol{\alpha}}u, \phi_n\rangle_{\mathcal{D}'(\Omega) \times \mathcal{D}(\Omega)} \\
&= \langle \partial^{\boldsymbol{\alpha}}u, v\rangle_{H^{-s-|\boldsymbol{\alpha}|}(\Omega) \times H_0^{s+|\boldsymbol{\alpha}|}(\Omega)} \qquad \forall v \in H_0^{s+|\boldsymbol{\alpha}|}(\Omega),
\end{aligned}
$$

since $\phi_n \to v$ in $H_0^{s+|\boldsymbol{\alpha}|}(\Omega) \Longrightarrow \partial^{\boldsymbol{\alpha}}\phi_n \to \partial^{\boldsymbol{\alpha}}v$ in $H_0^s(\Omega)$ by II $\Longrightarrow \partial^{\boldsymbol{\alpha}}\phi_n \to \partial^{\boldsymbol{\alpha}}v$ *weakly* in $H_0^s(\Omega)$ and a fortiori $\phi_n \to v$ *weakly* in $H_0^{s+|\boldsymbol{\alpha}|}(\Omega)$, and $H^{-s-|\boldsymbol{\alpha}|}(\Omega) \hookrightarrow \mathcal{D}'(\Omega)$, $H^{-s}(\Omega) \hookrightarrow \mathcal{D}'(\Omega)$ $\forall s > 0$ as stated earlier. Hence, $\partial^{\boldsymbol{\alpha}}u = ((-1)^{|\boldsymbol{\alpha}|}\partial^{\boldsymbol{\alpha}})^*u$ in $H^{-s-|\boldsymbol{\alpha}|}(\Omega)$ $\forall u \in H^{-s}(\Omega) \Longrightarrow \partial^{\boldsymbol{\alpha}} = ((-1)^{|\boldsymbol{\alpha}|}\partial^{\boldsymbol{\alpha}})^* \in \mathcal{L}(H^{-s}(\Omega); H^{-s-|\boldsymbol{\alpha}|}(\Omega))$, i.e. $\partial^{\boldsymbol{\alpha}} : H^{-s}(\Omega) \longrightarrow H^{-s-|\boldsymbol{\alpha}|}(\Omega)$ is a *continuous, linear* operator from $H^{-s}(\Omega)$ into $H^{-s-|\boldsymbol{\alpha}|}(\Omega)$.

IV. Let $u \in H_{00}^s(\Omega)$ for $s - |\boldsymbol{\alpha}| \geq 0$. Then its null extension $\tilde{u} \in H^s(\mathbb{R}^n)$ and $\partial^{\boldsymbol{\alpha}}\tilde{u} \in H^{s-|\boldsymbol{\alpha}|}(\mathbb{R}^n)$ for $s - |\boldsymbol{\alpha}| \geq 0$ (in fact, $\forall s \in \mathbb{R}$, $\forall |\boldsymbol{\alpha}| \in \mathbb{N}$) by (8.10.154). But for $s - |\boldsymbol{\alpha}| \geq 0$, $\partial^{\boldsymbol{\alpha}}\tilde{u} = \widetilde{\partial^{\boldsymbol{\alpha}}u} \in H^{s-|\boldsymbol{\alpha}|}(\mathbb{R}^n)$, $\widetilde{\partial^{\boldsymbol{\alpha}}u}$ being the null extension to $\mathbb{R}^n$ of $\partial^{\boldsymbol{\alpha}}u$, since $\forall \phi \in \mathcal{D}(\Omega)$ with $\tilde{\phi} \in \mathcal{D}(\mathbb{R}^n)$, $\partial^{\boldsymbol{\alpha}}\tilde{\phi} = \widetilde{\partial^{\boldsymbol{\alpha}}\phi}$ $\forall |\boldsymbol{\alpha}| \in \mathbb{N}$ and for $u \in H^s_{00}(\Omega) \hookrightarrow \mathcal{D}'(\Omega)$, $\mathcal{D}(\Omega)$ being a *dense* subspace in $H^s_{00}(\Omega)$,

$$\begin{aligned}\langle \partial^{\boldsymbol{\alpha}} \tilde{u}, \tilde{\phi}\rangle &= (-1)^{|\boldsymbol{\alpha}|}\langle \tilde{u}, \partial^{\boldsymbol{\alpha}} \tilde{\phi}\rangle_{\mathcal{D}'(\mathbb{R}^n)\times\mathcal{D}(\mathbb{R}^n)} = (-1)^{|\boldsymbol{\alpha}|}\langle \tilde{u}, \widetilde{\partial^{\boldsymbol{\alpha}}\phi}\rangle_{\mathcal{D}'(\mathbb{R}^n)\times\mathcal{D}(\mathbb{R}^n)} \\ &= (-1)^{|\boldsymbol{\alpha}|}\int_\Omega u\partial^{\boldsymbol{\alpha}}\phi d\mathbf{x} \quad (\text{since } u \in H^s_{00}(\Omega)) \\ &= \int_\Omega (\partial^{\boldsymbol{\alpha}} u)\phi d\mathbf{x} = \langle \widetilde{\partial^{\boldsymbol{\alpha}} u}, \tilde{\phi}\rangle_{\mathcal{D}'(\mathbb{R}^n)\times\mathcal{D}(\mathbb{R}^n)}\end{aligned}$$

$\forall \phi \in \mathcal{D}(\Omega)$ with $\tilde{\phi} \in \mathcal{D}(\mathbb{R}^n) \iff \partial^{\boldsymbol{\alpha}}\tilde{u} = \widetilde{\partial^{\boldsymbol{\alpha}} u}$ in $\mathcal{D}'(\mathbb{R}^n)$ with $\partial^{\boldsymbol{\alpha}}\tilde{u} \in H^{s-|\boldsymbol{\alpha}|}(\mathbb{R}^n) \Longrightarrow \widetilde{\partial^{\boldsymbol{\alpha}} u} \in H^{s-|\boldsymbol{\alpha}|}(\mathbb{R}^n)$ with $s - |\boldsymbol{\alpha}| \geq 0 \Longrightarrow \partial^{\boldsymbol{\alpha}} u \in H^{s-|\boldsymbol{\alpha}|}_{00}(\Omega)$ for $s - |\boldsymbol{\alpha}| \geq 0$.

Thus, $u \in H^s_{00}(\Omega) \Longrightarrow \partial^{\boldsymbol{\alpha}} u \in H^{s-|\boldsymbol{\alpha}|}_{00}(\Omega)$ for $s - |\boldsymbol{\alpha}| \geq 0$ and $\exists C > 0$ such that

$$\begin{aligned}\|\partial^{\boldsymbol{\alpha}} u\|_{H^{s-|\boldsymbol{\alpha}|}_{00}(\Omega)} &= \|\widetilde{\partial^{\boldsymbol{\alpha}} u}\|_{H^{s-|\boldsymbol{\alpha}|}(\mathbb{R}^n)} = \|\partial^{\boldsymbol{\alpha}}\tilde{u}\|_{H^{s-|\boldsymbol{\alpha}|}(\mathbb{R}^n)} \\ &\leq C\|\tilde{u}\|_{H^s(\mathbb{R}^n)} = C\|u\|_{H^s_{00}(\Omega)},\end{aligned}$$

since $\partial^{\boldsymbol{\alpha}} : H^s(\mathbb{R}^n) \longrightarrow H^{s-|\boldsymbol{\alpha}|}(\mathbb{R}^n)$ is *continuous* by (8.10.154). Hence, $\exists$ a constant $C > 0$ such that $\forall u \in H^s_{00}(\Omega)$, $\|\partial^{\boldsymbol{\alpha}} u\|_{H^{s-|\boldsymbol{\alpha}|}_{00}(\Omega)} \leq C\|u\|_{H^s_{00}(\Omega)}$ for $s - |\boldsymbol{\alpha}| \geq 0$, i.e. $\partial^{\boldsymbol{\alpha}} : H^s_{00}(\Omega) \longrightarrow H^{s-|\boldsymbol{\alpha}|}_{00}(\Omega)$ is *continuous* for $s - |\boldsymbol{\alpha}| \geq 0$.

V. The proof is almost identical to that of III. Replacing $H^s_0(\Omega)$ by $H^s_{00}(\Omega)$, $H^{s+|\boldsymbol{\alpha}|}_0(\Omega)$ by $H^{s+|\boldsymbol{\alpha}|}_{00}(\Omega)$, $H^{-s}(\Omega)$ by $(H^s_{00}(\Omega))'$, $H^{-s-|\boldsymbol{\alpha}|}(\Omega)$ by $(H^{s+|\boldsymbol{\alpha}|}_{00}(\Omega))'$ in the proof of III, using the *density* of $\mathcal{D}(\Omega)$ in $H^{s+|\boldsymbol{\alpha}|}_{00}(\Omega)$ (resp. $H^s_{00}(\Omega)$) $\forall s > 0$ and the continuity of $\partial^{\boldsymbol{\alpha}} : H^{s+|\boldsymbol{\alpha}|}_{00}(\Omega) \longrightarrow H^s_{00}(\Omega)$ for $s > 0$ and the imbeddings $(H^{s+|\boldsymbol{\alpha}|}_{00}(\Omega))' \hookrightarrow \mathcal{D}'(\Omega)$, $(H^s_{00}(\Omega))' \hookrightarrow \mathcal{D}'(\Omega)$, the result is obtained.

□

8.10.16 Differentiation of distributions $u \in H^s(\overline{\Omega})$ with $s > 0$

The differentiation of distributions $u \in H^s(\overline{\Omega})$ with $s > 0$ (defined in (8.10.1)) is closely related to the properties of *restrictions* (see Theorem 8.10.21) and *extension* (see Theorem 8.10.9). In fact, for the distribution $u \in H^s(\overline{\Omega})$, $\exists$ (extension) $w \in H^s(\mathbb{R}^n)$ such that $w{\downarrow_\Omega} = u$. But $\forall s \in \mathbb{R}$, $\partial^{\boldsymbol{\alpha}} : \Omega \in H^s(\mathbb{R}^n) \longmapsto \partial^{\boldsymbol{\alpha}}\Omega \in H^{s-|\boldsymbol{\alpha}|}(\mathbb{R}^n)$ is *continuous* $\forall$ multi-index $\boldsymbol{\alpha}$.

Hence, for $u \in H^s(\overline{\Omega})$, $\partial^{\boldsymbol{\alpha}} u = (\partial^{\boldsymbol{\alpha}} w){\downarrow_\Omega}$ is the *restriction* of a distribution $\partial^{\boldsymbol{\alpha}} w \in H^{s-|\boldsymbol{\alpha}|}(\mathbb{R}^n)$, which will be governed by Theorem 8.10.21 for Ω with a *Lipschitz continuous* boundary Γ. Indeed, for $u \in H^s(\overline{\Omega})$, $\exists$ a distribution $w \in H^s(\mathbb{R}^n)$ such that $u = w\downarrow_\Omega$, which is defined, $\forall\phi \in \mathcal{D}(\Omega)$ with $\tilde{\phi} \in \mathcal{D}(\mathbb{R}^n)$, by $\langle w\downarrow_\Omega, \phi\rangle = \langle w, \tilde{\phi}\rangle$ (see (5.3.1)). Thus,

$$\begin{aligned}\langle \partial^{\boldsymbol\alpha} u, \phi\rangle &= (-1)^{|\boldsymbol\alpha|}\langle u, \partial^{\boldsymbol\alpha}\phi\rangle = (-1)^{|\boldsymbol\alpha|}\langle w\downarrow_\Omega, \partial^{\boldsymbol\alpha}\phi\rangle = (-1)^{|\boldsymbol\alpha|}\langle w, \widetilde{\partial^{\boldsymbol\alpha}\phi}\rangle \\ &= (-1)^{|\boldsymbol\alpha|}\langle w, \partial^{\boldsymbol\alpha}\tilde{\phi}\rangle \quad (\text{since } \widetilde{\partial^{\boldsymbol\alpha}\phi} = \partial^{\boldsymbol\alpha}\tilde{\phi} \text{ for } \phi \in \mathcal{D}(\Omega) \text{ with } \tilde{\phi} \in \mathcal{D}(\mathbb{R}^n)) \\ &= \langle \partial^{\boldsymbol\alpha} w, \tilde{\phi}\rangle = \langle (\partial^{\boldsymbol\alpha} w)\downarrow_\Omega, \phi\rangle \quad \forall\phi \in \mathcal{D}(\Omega),\ \forall\boldsymbol\alpha\end{aligned}$$

$\Longrightarrow \partial^{\boldsymbol\alpha} u = (\partial^{\boldsymbol\alpha} w)\downarrow_\Omega$. But by Theorem 8.10.9, for *bounded* domains Ω with *Lipschitz continuous* boundary Γ, $H^s(\overline{\Omega}) = H^s(\Omega)\ \forall s > 0$. Hence, for $s > 0$ with $s - |\boldsymbol\alpha| < 0$ the final results on *differentiation* on *bounded* $\Omega \subset \mathbb{R}^n$ with *Lipschitz continuous* boundary Γ will follow from the restriction results of Theorem 8.10.21 (for $s - |\boldsymbol\alpha| \geq 0$, see Theorem 8.10.25).

Theorem 8.10.26. *For bounded* $\Omega \subset \mathbb{R}^n$ *with* Lipschitz continuous *boundary* Γ *and for* $s > 0$ *with* $s - |\boldsymbol\alpha| < 0$, *the linear operator* $\partial^{\boldsymbol\alpha} : H^s(\Omega) \longrightarrow (H_{00}^{|\boldsymbol\alpha|-s}(\Omega))'$ *is* continuous *from* $H^s(\Omega)$ *into* $(H_{00}^{|\boldsymbol\alpha|-s}(\Omega))'$.

In particular,

- *for* $s - \frac{1}{2} \neq$ *an integer,* $\partial^{\boldsymbol\alpha} : H^s(\Omega) \longrightarrow H^{s-|\boldsymbol\alpha|}(\Omega)$ *for* $s - |\boldsymbol\alpha| < 0$;
- *for* $s - \frac{1}{2} =$ *an integer,* $\partial^{\boldsymbol\alpha} : H^s(\Omega) \longrightarrow (H_{00}^{|\boldsymbol\alpha|-s}(\Omega))'$ *for* $s - |\boldsymbol\alpha| < 0$.

Proof. For $s > 0$, for Ω with a *Lipschitz continuous* boundary, $H^s(\Omega) = H^s(\overline{\Omega})$ by Theorem 8.10.9. Hence, for $s > 0$, $u \in H^s(\Omega) \Longrightarrow u \in H^s(\overline{\Omega}) \Longrightarrow \exists w \in H^s(\mathbb{R}^n)$ such that $u = w\downarrow_\Omega$ and $\partial^{\boldsymbol\alpha} u = \partial^{\boldsymbol\alpha}(w\downarrow_\Omega) = (\partial^{\boldsymbol\alpha} w)\downarrow_\Omega)$ (see the proof given earlier) with $\partial^{\boldsymbol\alpha} w \in H^{s-|\boldsymbol\alpha|}(\mathbb{R}^n)$ by Theorem 8.10.24. Since $s - |\boldsymbol\alpha| < 0$, i.e. $-(|\boldsymbol\alpha| - s) < 0$ with $|\boldsymbol\alpha| - s > 0$, by Theorem 8.10.21, $\partial^{\boldsymbol\alpha} u = \partial^{\boldsymbol\alpha} w\downarrow_\Omega) \in (H_{00}^{|\boldsymbol\alpha|-s}(\Omega))'$.

For $s - \frac{1}{2} \neq$ an integer $\Longrightarrow |\boldsymbol\alpha| - s - \frac{1}{2} \neq$ an integer $\Longrightarrow$ by Corollary 8.10.2, $H_{00}^{|\boldsymbol\alpha|-s}(\Omega) = H_0^{|\boldsymbol\alpha|-s}(\Omega) \Longrightarrow (H_{00}^{|\boldsymbol\alpha|-s}(\Omega))' = H^{s-|\boldsymbol\alpha|}(\Omega)$ (by (8.10.126a)).

For $s - \frac{1}{2} =$ an integer, $|\boldsymbol\alpha| - s - \frac{1}{2} =$ an integer $\Longrightarrow$ by (8.10.123), $H_{00}^{|\boldsymbol\alpha|-s}(\Omega) \subsetneq H_0^{|\boldsymbol\alpha|-s}(\Omega) \Longrightarrow H^{s-|\boldsymbol\alpha|}(\Omega) \subset (H_{00}^{|\boldsymbol\alpha|-s}(\Omega))'$ (by (8.10.126b)).

By Theorem 8.10.25, for $s \geq 1$ *or* $s \leq 0$, $\partial_j : H^s(\Omega) \longrightarrow H^{s-1}(\Omega)$ is *continuous*. Hence, it remains to prove the result for $0 < s < 1$ with $s \neq \frac{1}{2}$, and also the exceptional case $s = \frac{1}{2}$. □

Theorem 8.10.27. *For bounded* $\Omega \subset \mathbb{R}^n$ *with* Lipschitz continuous *boundary* Γ *and* $0 < s < 1$,

- *with* $s \neq \frac{1}{2}$, $\partial_j = \frac{\partial}{\partial x_j}$ *is a* continuous, linear *operator from* $H^s(\Omega)$ *into* $H^{s-1}(\Omega)$;

- *with* $s = \frac{1}{2}$, $\partial_j : H^{\frac{1}{2}}(\Omega) \longrightarrow (H_{00}^{\frac{1}{2}}(\Omega))'$ *is a* continuous, linear *operator from* $H^{\frac{1}{2}}(\Omega)$ *into* $(H_{00}^{\frac{1}{2}}(\Omega))'$ $((H_{00}^{\frac{1}{2}}(\Omega))'$ cannot *be identified with* $H^{-\frac{1}{2}}(\Omega) = (H_0^{\frac{1}{2}}(\Omega))'$, *although* $H^{\frac{1}{2}}(\Omega) = H_0^{\frac{1}{2}}(\Omega)$ *by Corollary* 8.10.1*)*.

Proof. First of all, by Theorem 8.10.9, for $0 < s < 1$, $H^s(\overline{\Omega}) = H^s(\Omega)$. Consequently, for $H^s(\Omega) = H^s(\overline{\Omega})$, for $u \in H^s(\Omega)$, $\exists w \in H^s(\mathbb{R}^n)$ with $\partial_j u = (\partial_j w)\downarrow_\Omega$, with $\partial_j w \in H^{s-1}(\mathbb{R}^n)$. Hence, $\forall \phi \in \mathcal{D}(\Omega)$ with its null extension $\tilde{\phi} \in \mathcal{D}(\mathbb{R}^n)$, $\langle \partial_j u, \phi \rangle = \langle \partial_j w\downarrow_\Omega, \phi \rangle = \langle \partial_j w, \tilde{\phi} \rangle \Longrightarrow |\langle \partial_j u, \phi \rangle| \leq \|\partial_j w\|_{H^{s-1}(\mathbb{R}^n)} \|\tilde{\phi}\|_{H^{-(s-1)}(\mathbb{R}^n)} = \|\partial_j w\|_{H^{s-1}(\mathbb{R}^n)} \|\phi\|_{00,1-s,\Omega}$ $\forall \phi \in \mathcal{D}(\Omega) \Longrightarrow \partial_j u$ is a *continuous, linear* functional on $\mathcal{D}(\Omega)$ in the norm $\|\cdot\|_{00,1-s,\Omega}$ of $H_{00}^{1-s}(\Omega)$ with $1-s > 0$. But $\mathcal{D}(\Omega)$ is *dense* in $H_{00}^{1-s}(\Omega)$ in the norm $\|\cdot\|_{00,1-s,\Omega}$. Hence, $\partial_j u$ has a *unique, continuous, linear* extension, which will still be denoted by $\partial_j u$, to $H_{00}^{1-s}(\Omega)$, i.e. for $0 < s < 1$, $\partial_j : u \in H^s(\Omega) \longmapsto \partial_j u \in (H_{00}^{1-s}(\Omega))'$.

Now we consider the two cases $s \neq \frac{1}{2}$, $0 < s < 1$, and $s = \frac{1}{2}$.

$s \neq \frac{1}{2}$, $0 < s < 1 \Longrightarrow 1 - s > 0$ and $H_{00}^{1-s}(\Omega) = H_0^{1-s}(\Omega)$ by Corollary 8.10.2 and $(H_{00}^{1-s}(\Omega))' = (H_0^{1-s}(\Omega))' = H^{s-1}(\Omega)$, i.e. for $0 < s < 1$ with $s \neq \frac{1}{2}$, $\partial_j : u \in H^s(\Omega) \longmapsto \partial_j u \in H^{s-1}(\Omega)$.

$s = \frac{1}{2} \Longrightarrow H_{00}^{\frac{1}{2}}(\Omega) \subsetneq H_0^{\frac{1}{2}}(\Omega) = H^{\frac{1}{2}}(\Omega)$ by Corollary 8.10.1 $\Longrightarrow (H_0^{\frac{1}{2}}(\Omega))' = H^{-\frac{1}{2}}(\Omega) \subsetneq (H_{00}^{\frac{1}{2}}(\Omega))'$ (see (8.10.126b)), for $s = \frac{1}{2}$,

$$\partial_j : u \in H^{\frac{1}{2}}(\Omega) \longmapsto (H_{00}^{\frac{1}{2}}(\Omega))'. \tag{8.10.160}$$

□

Remark 8.10.13. For $\Omega =]0,1[$, $\frac{d}{dx} : H^{\frac{1}{2}}(]0,1[) \longrightarrow (H_{00}^{\frac{1}{2}}(]0,1[))'$, i.e. for $u \in H^{\frac{1}{2}}(]0,1[)$, $\frac{du}{dx}$ is *not* necessarily in $H^{-\frac{1}{2}}(]0,1[) = (H_0^{\frac{1}{2}}(]0,1[))' = (H^{\frac{1}{2}}(]0,1[))'$ in general.

Example 8.10.1. Consider the function $u : \mathbb{R}^2 \longrightarrow \mathbb{R}$ in Example 8.9.1 defined by $u(x_1, x_2) = \ln|\ln|\mathbf{x}||$ for $\|\mathbf{x}\| < 1/e$, and 0 for $\|\mathbf{x}\| \geq 1/e$.

Let $u_1 = u\downarrow_{\mathbb{R}}$ be defined by $u_1(x) = \ln|\ln|x||$ for $|x| < 1/e$, and 0 for $|x| \geq 1/e$ where $x = x_1$. Then show that

1. $\frac{du_1}{dx} = u_1' \in H^{-1/2}(\mathbb{R})$;
2. $u_1'\downarrow_{]0,1/e[} \in (H_{00}^{1/2}(]0,1/e[))'$ and $u_1'\downarrow_{]0,1/e[} \notin H^{-1/2}(]0,1/e[)$.

Solution.

1. From Example 2.3.9, $u \in H^1(\mathbb{R}^2)$. Then, by Trace Theorem C 8.9.11, for $j = 0$, $\gamma_0 u = u\downarrow_{\mathbb{R}} = u_1 \in H^{1/2}(\mathbb{R})$ (see Example 8.9.1 also) with $u_1(x) = \ln|\ln|x||$ for $|x| < 1/e$ and $= 0$ for $|x| \geq 1/e$. But $u_1 \in H^{1/2}(\mathbb{R}) \Longrightarrow$

its distributional derivative $\frac{du_1}{dx} \in H^{1/2-1}(\mathbb{R})$ by Theorem 8.10.24 for $s = \frac{1}{2}$, $\alpha = 1$, i.e. $u_1' \in H^{-1/2}(\mathbb{R})$.

2. For $s = \frac{1}{2} > 0$ and $u_1' \in H^{-1/2}(\mathbb{R})$ and $\Omega =]0, 1/e[\subset \mathbb{R}$ (Lipschitz continuity of the boundary consisting of *two* points $\{0, 1/e\}$ *vacuously* holds), by Theorem 8.10.21, $u_1'\downarrow_{]0,1/e[} \in (H_{00}^{1/2}(]0, 1/e[))'$ and $u_1'\downarrow_{]0,1/e[} \notin H^{-1/2}(]0, 1/e[)$.

Remark 8.10.14. Similar results (8.10.154)–(8.10.157) hold for $W^{s,p}(\Omega)$, $1 < p < \infty$, and we have: the mappings

1. $\partial^{\boldsymbol{\alpha}} : W^{s,p}(\mathbb{R}^n) \longrightarrow W^{s-|\boldsymbol{\alpha}|,p}(\mathbb{R}^n)\ \forall s \in \mathbb{R},\ 1 < p < \infty$;

2. $\partial^{\boldsymbol{\alpha}} : W^{s,p}(\Omega) \longrightarrow W^{s-|\boldsymbol{\alpha}|,p}(\Omega)$ for $s - |\boldsymbol{\alpha}| \geq 0$, $1 < p < \infty$;

3. $\partial^{\boldsymbol{\alpha}} : W_0^{s,p}(\Omega) \longrightarrow W_0^{s-|\boldsymbol{\alpha}|,p}(\Omega)$ for $s - |\boldsymbol{\alpha}| \geq 0$, $1 < p < \infty$;

4. $\partial^{\boldsymbol{\alpha}} : W^{-s,q}(\Omega) \longrightarrow W^{-s-|\boldsymbol{\alpha}|,q}(\Omega)$ for $s \geq 0$, $1 < p, q < \infty$, $\frac{1}{p} + \frac{1}{q} = 1$

are all *continuous*, *linear* operators from the left-hand side space into the right-hand space in all cases, for example from $W_0^{s,p}(\Omega)$ into $W_0^{s-|\boldsymbol{\alpha}|,p}(\Omega)$ in (3).

For bounded Ω with *Lipschitz continuous* boundary Γ and for $s > 0$ with $s - |\boldsymbol{\alpha}| < 0$, the linear operator

5. $\partial^{\boldsymbol{\alpha}} : W^{s,p}(\Omega) \longrightarrow (W_{00}^{|\boldsymbol{\alpha}|-s,q}(\Omega))'$

is *continuous* for $1 < p < \infty$, $\frac{1}{p} + \frac{1}{q} = 1$, which follows from the restriction results (8.10.145).

For $s \geq 1$ *or* $s \leq 0$, the results of Theorem 8.10.25 hold for $W^{s,p}(\Omega)$ with $1 < p < \infty$. Hence, it suffices to show it for $0 < s < 1$, $1 < p < \infty$, for which $W^{s,p}(\overline{\Omega}) = W^{s,p}(\Omega)$ for a bounded domain Ω with *Lipschitz continuous* boundary Γ, by Theorem 8.10.13.

Theorem 8.10.28. *For bounded* $\Omega \in \mathbb{R}^n$ *with* Lipschitz continuous *boundary* Γ *and for* $0 < s < 1$ *with* $s \neq \frac{1}{p}$, $1 < p < \infty$, $\partial_j = \frac{\partial}{\partial x_j}$ *is a* continuous, linear *operator from* $W^{s,p}(\Omega)$ *into* $W^{s-1,p}(\Omega)$, *and for* $0 < s < 1$ *with* $s = \frac{1}{p}$, $1 < p, q < \infty$, *with* $\frac{1}{p} + \frac{1}{q} = 1$, $\partial_j : W^{\frac{1}{p},p}(\Omega) \longrightarrow (W_{00}^{\frac{1}{q},q}(\Omega)')$ *is a* continuous, linear *operator from* $W^{\frac{1}{p},p}(\Omega)$ *into* $(W_{00}^{\frac{1}{q},q}(\Omega))'$.

Proof. Let $u \in W^{s,p}(\Omega)$ with $0 < s < 1$, $1 < p < \infty$. $\exists w \in W^{s,p}(\mathbb{R}^n)$ such that $u = w\downarrow_\Omega$ and $\partial_j u = (\partial_j w)\downarrow_\Omega$ with $\partial_j w \in W^{s-1,p}(\mathbb{R}^n)$ (see the proof of Theorem 8.10.26). Then, for $s \neq \frac{1}{p}$ with $\frac{1}{p} + \frac{1}{q} = 1$, $1 < p < \infty, 0 < s < 1$,

$$|\langle \partial_j u, \phi \rangle| \leq \|\partial_j w\|_{W^{s-1,p}(\mathbb{R}^n)} \|\tilde{\phi}\|_{W^{1-s,q}(\mathbb{R}^n)} = \|\partial_j w\|_{W^{s-1,p}(\mathbb{R}^n)} \|\phi\|_{W_{00}^{1-s,q}(\Omega)} \tag{8.10.161}$$

$\forall \phi \in \mathcal{D}(\Omega)$ with $\tilde{\phi} \in \mathcal{D}(\mathbb{R}^n)$, and $\mathcal{D}(\Omega)$ is *dense* in $W_{00}^{1-s,q}(\Omega)$ with $0 < s < 1$, $1 < q < \infty$, $s \neq \frac{1}{p}$. Hence, for $s \neq \frac{1}{p}$, $0 < s < 1$, $1 < p < \infty$, $u \in W^{s,p}(\Omega) \Longrightarrow \partial_j u \in (W_{00}^{1-s,q}(\Omega))' = (W_0^{1-s,q}(\Omega))' = W^{s-1,p}(\Omega)$ by (8.10.142) and (8.10.138).

Finally, for $s = \frac{1}{p}$, $u \in W^{\frac{1}{p},p}(\Omega) \Longrightarrow \partial_j u \in (W_{00}^{1-s,q}(\Omega))'$ with $1-s = 1-\frac{1}{p} = \frac{1}{q}$. □

Multiplication by a function

In Theorem 6.8.7, we have proved that $u \in H^m(\Omega) \longmapsto \phi u \in H^m(\Omega)$ for $\phi \in C^m(\Omega)$ with *bounded usual* derivatives $\partial^{\boldsymbol{\alpha}}\phi$ in Ω $\forall |\boldsymbol{\alpha}| \leq m$. Moreover, we have shown in Section 4.2, Chapter 4, that for Banach spaces V, W with $\mathcal{D}(\Omega)$ *dense* in V, V *dense* in W, if the mapping $u \in V \longmapsto \phi u \in W$ is continuous and linear, then the mapping is *continuous* and *linear* from W' into V'. Obviously, a complete answer will be possible with the help of the imbedding results. Meanwhile, we state the following results (see, for example, Grisvard [18]). For $k \in \mathbb{N}_0$, $\lambda \in [0,1]$, define $C_0^{k,\lambda}(\overline{\Omega})$ by $C_0^{k,\lambda}(\overline{\Omega}) = \{\phi : \exists \psi \in C^{k,\lambda}(\mathbb{R}^n)$ such that $\psi\downarrow_{\overline{\Omega}} = \phi$ and $\operatorname{supp}(\psi) \subset\subset \mathbb{R}^n$ is compact$\}$.

Theorem 8.10.29 (Grisvard [18]). *Let $\phi \in C_0^{k,\lambda}(\overline{\Omega})$ with $k + \lambda \geq |s|$, when $s \in \mathbb{Z}$ (resp. $k + \lambda > |s|$, when $s \notin \mathbb{Z}$). Then the following* continuous, linear *mapping $u \longmapsto \phi u$ holds with:*

I. *$\phi u \in W^{s,p}(\Omega)$ $\forall u \in W^{s,p}(\Omega)$ with $\|\phi u\|_{s,p,\Omega} \leq c_1 \|u\|_{s,p,\Omega}$;*

(II. *$\phi u \in W_0^{s,p}(\Omega)$ $\forall u \in W_0^{s,p}(\Omega)$, $s \geq 0$, $s - \frac{1}{p} \notin \mathbb{Z}$, $1 < p < \infty$, with $\|\phi u\|_{s,p,\Omega} \leq c_2 |u\|_{s,p,\Omega}$;*

III. *$\phi u \in W^{s,p}(\overline{\Omega})$ $\forall u \in W^{s,p}(\overline{\Omega})$, $s \geq 0$, $1 < p < \infty$, with $\|\phi u\|_{s,p,\overline{\Omega}} \leq c_3 \|u\|_{s,p,\overline{\Omega}}$;*

IV. *$\phi u \in W_{00}^{s,p}(\Omega)$ $\forall u \in W_{00}^{s,p}(\Omega)$, $s \geq 0$, $s - \frac{1}{p} \in \mathbb{Z}$, $1 < p < \infty$, with $\|\phi u\|_{00,s,p,\Omega} \leq c_4 \|u\|_{00,s,p,\Omega}$.*

8.11 Compactness results in Sobolev spaces

The compact[6] imbedding results are of extreme importance in many applications, especially in the spectral theory of elliptic boundary value problems. From the classical theory of separable Hilbert spaces, which are always reflexive, we know the following result (see the Eberlein–Schmulyan Theorem A.11.1.2, Appendix A):

[6] Let X and Y be Banach spaces and $A : X \longrightarrow Y$ be a linear operator from X into Y. Then, A is called *compact* from X into Y if and only if $x_n \rightharpoonup x$ *weakly* in X *implies* $Ax_n \rightarrow Ax$ *strongly* in Y as $n \rightarrow \infty$. For $X \hookrightarrow\hookrightarrow Y$, the imbedding operator $\hookrightarrow\hookrightarrow : X \rightarrow \hookrightarrow\hookrightarrow X \subset Y$ is called *compact* from X into Y if and only if $x_n \rightharpoonup x$ *weakly* in X *implies* $x_n \rightarrow x$ in Y as $n \rightarrow \infty$, since $x_n \in X \longmapsto \hookrightarrow\hookrightarrow x_n = x_n \in Y$, $x \in X \longmapsto \hookrightarrow\hookrightarrow x = x \in Y$.

Every *bounded* set of a Hilbert space H is *weakly relatively compact* in H.

In other words, for any given *bounded* sequence $(u_k)_{k=1}^{\infty}$ in H, $\exists$ an element $u \in H$ and a *subsequence* $(u_{k_m})_{m=1}^{\infty}$ of this bounded sequence $(u_k)_{k=1}^{\infty}$ such that $u_{k_m} \rightharpoonup u$ *weakly* in H as $m \to \infty$; i.e. $\forall f \in H$,

$$\langle u_{k_m}, f\rangle_H \to \langle u, f\rangle_H \quad \text{as } m \to \infty. \tag{8.11.1}$$

Now we prepare some elementary auxiliary results which will be used later.

1. Let $(u_{k_m})_{m=1}^{\infty}$ be a subsequence of the *bounded* sequence $(u_k)_{k=1}^{\infty}$ such that $u_{k_m} \rightharpoonup u$ *weakly in* H *as* $m \to \infty$. Then we can define a new *bounded* sequence $(v_m)_{m=1}^{\infty}$ by

$$v_m = u - u_{k_m} \quad \forall m \in \mathbb{N}, \tag{8.11.2}$$

such that $u_{k_m} \rightharpoonup u$ weakly in $H \iff v_m \rightharpoonup 0$ *weakly* in H as $m \to \infty$, since $\langle v_m, f\rangle_H = [\langle u, f\rangle_H - \langle u_{k_m}, f\rangle_H] \to 0 \; \forall f \in H$ as $m \to \infty$, and

$$\|v_m\|_H \le \|u\|_H + \|u_{k_m}\|_H \le M \quad \forall m \in \mathbb{N} \tag{8.11.3}$$

(i.e. $(v_m)_{m=1}^{\infty}$ is *bounded* in H).

2. If $u_m \rightharpoonup u$ *weakly* in $L^2(\mathbb{R}^n)$, $\operatorname{supp}(u_m) \subset K \; \forall \, m \in \mathbb{N}$, $K \subset\subset \mathbb{R}^n$ being a compact subset of $\mathbb{R}^n$, then

$$u_m \rightharpoonup u \textit{ weakly} \text{ in } L^2(\mathring{K}) \text{ with } \operatorname{supp}(u) \subset K, \; \mathring{K} = \operatorname{int}(K). \tag{8.11.4}$$

In fact, from (8.11.1) with $H = L^2(\mathbb{R}^n)$, $\operatorname{supp}(u_m) \subset K \; \forall m \in \mathbb{N}$, $\forall f \in L^2(\mathbb{R}^n)$,

$$\begin{aligned} &\lim_{m\to\infty}\left[\int_{\mathbb{R}^n} uf d\mathbf{x} - \int_{\mathring{K}} u_m f d\mathbf{x}\right] \\ &= \lim_{m\to\infty}\left[\int_{\mathbb{R}^n\setminus\mathring{K}} uf d\mathbf{x} + \int_{\mathring{K}} uf d\mathbf{x} - \int_{\mathring{K}} u_m f d\mathbf{x}\right] = 0, \end{aligned} \tag{8.11.5}$$

since $\operatorname{supp}(u_m) \subset K \; \forall m \in \mathbb{N} \Longrightarrow u_m(\mathbf{x}) = 0$ a.e. in $\mathbb{R}^n \setminus K \; \forall m \in \mathbb{N}$.

But $\forall f \in L^2(\mathring{K})$, its *null extension* (i.e. $\tilde{f}(\mathbf{x}) = 0$ for $\mathbf{x} \in \mathbb{R}^n \setminus \mathring{K}$ and $\tilde{f}(\mathbf{x}) = f(\mathbf{x})$ a.e. in $\mathring{K}$), $\tilde{f} \in L^2(\mathbb{R}^n)$. Then, from (8.11.5), $u_m \rightharpoonup u$ *weakly* in $L^2(\mathbb{R}^n)$ with $\operatorname{supp}(u_m) \subset K \; \forall m \in \mathbb{N} \Longrightarrow \lim_{m\to\infty}[\int_{\mathring{K}} uf d\mathbf{x} - \int_{\mathring{K}} u_m f d\mathbf{x}] = 0$ $\forall f \in L^2(\mathring{K})$ with $\tilde{f} \in L^2(\mathbb{R}^n) \Longrightarrow \langle u_m, f\rangle_{L^2(\mathring{K})} \to \langle u, f\rangle_{L^2(\mathring{K})} \; \forall f \in L^2(\mathring{K})$

$$\Longrightarrow \quad u_m \rightharpoonup u \textit{ weakly} \text{ in } L^2(\mathring{K}) \text{ as } m \to \infty. \tag{8.11.6}$$

It remains to show that $\text{supp}(u) \subset K$. $\forall f \in L^2(\mathbb{R}^n)$, (8.11.5) and (8.11.6) must hold $\Longrightarrow \lim_{m\to\infty} \int_{\mathbb{R}^n \setminus \mathring{K}} u_m f d\mathbf{x} = \int_{\mathbb{R}^n \setminus \mathring{K}} u f d\mathbf{x} = 0 \ \forall f \in L^2(\mathbb{R}^n)$. In particular, for $f = u \in L^2(\mathbb{R}^n)$,

$$\int_{\mathbb{R}^n \setminus \mathring{K}} |u(\mathbf{x})|^2 d\mathbf{x} = 0 \quad \Longrightarrow \quad u = 0 \text{ a.e. in } \mathbb{R}^n \setminus \mathring{K} \Longrightarrow \text{supp}(u) \subset K. \tag{8.11.7}$$

3. For $v_m = u - u_{k_m}$ with $\text{supp}(u_{k_m}) \subset K \ \forall m \in \mathbb{N}$, $\text{supp}(u) \subset K$ by (8.11.7), $u(\mathbf{x}) = 0$ in $K^{\complement}$, $u_{K_m}(\mathbf{x}) = 0$ in $K^{\complement} \ \forall m \in \mathbb{N}$, $v_m(\mathbf{x}) = 0$ in $K^{\complement}$

$$\Longrightarrow \quad \text{supp}(v_m) = \text{supp}(u - u_{k_m}) \subset K \quad \forall\, m \in \mathbb{N}. \tag{8.11.8}$$

Lemma 8.11.1. *Let V and W be any two Hilbert spaces and $A \in \mathcal{L}(V;W)$ be a continuous linear operator from V into W. Then if $u_m \rightharpoonup u$ weakly in V as $m \to \infty$, $Au_m \rightharpoonup Au$ weakly in W as $m \to \infty$. In particular, for $V \hookrightarrow W$, i.e. the imbedding operator $\hookrightarrow \in \mathcal{L}(V;W)$, $u_m \rightharpoonup u$ weakly in V as $m \to \infty \Longrightarrow u_m \rightharpoonup u$ weakly in W as $m \to \infty$.*

Proof. Since $A \in \mathcal{L}(V;W)$ is a bounded, linear operator from V into W, we can define its adjoint $A^* \in \mathcal{L}(W;V)$ by $\langle Av, w\rangle_W = \langle v, A^*w\rangle_V \ \forall v \in V$ with $Av \in W$, $\forall w \in W$ with $A^*w \in V$, $\langle\cdot,\cdot\rangle_V$ and $\langle\cdot,\cdot\rangle_W$ being inner products in V and W, respectively. Let $u_m \rightharpoonup u$ *weakly* in V as $m \to \infty$. Then, $\langle u_m, f\rangle_V \to \langle u, f\rangle_V \ \forall f \in V$. Hence, $\langle u_m, A^*w\rangle_V \to \langle u, A^*w\rangle_V$ as $m \to \infty \ \forall w \in W$. But $\langle u_m, A^*w\rangle_V = \langle Au_m, w\rangle_W \ \forall w \in W$, $\forall m \in \mathbb{N}$, and $\langle u, A^*w\rangle_V = \langle Au, w\rangle_W \Longrightarrow \langle Au_m, w\rangle_W \to \langle Au, w\rangle_W$ as $m \to \infty \ \forall w \in W \Longrightarrow Au_m \rightharpoonup Au$ *weakly* in W. For $A \equiv \hookrightarrow \in \mathcal{L}(V;W)$, $u_m \rightharpoonup u$ *weakly* in $V \Longrightarrow \hookrightarrow u_m = u_m \rightharpoonup \hookrightarrow u = u$ *weakly* in W as $m \to \infty$. □

Lemma 8.11.2. *Let V_1, V_2 and V_3 be Hilbert spaces. Let $E \in \mathcal{L}(V_1;V_2)$ and $F \in \mathcal{L}(V_2;V_3)$ be continuous, linear operators from V_1 into V_2 and from V_2 into V_3, respectively. If* either *of them, i.e. E or F, is* compact, *then $F \circ E \in \mathcal{L}(V_1;V_3)$ is* compact.

Proof. Let $E \in \mathcal{L}(V_1;V_2)$ be continuous and $F \in \mathcal{L}(V_2;V_3)$ be a *compact* operator. Let $(u_k)_{k=1}^{\infty}$ be a bounded sequence in Hilbert space V_1, which is reflexive. Hence, $\exists$ a subsequence $(u_{k_m})_{m=1}^{\infty}$ of this sequence $(u_k)_{k=1}^{\infty}$ such that $u_{k_m} \rightharpoonup u$ weakly in V_1. But $E \in \mathcal{L}(V_1;V_2)$ and $u_{k_m} \rightharpoonup u$ *weakly* in $V_1 \Longrightarrow Eu_{k_m} \rightharpoonup Eu$ *weakly* in V_2 by Lemma 8.11.1. But $F \in \mathcal{L}(V_2;V_3)$ is *compact*. Hence, $F(Eu_{k_m}) \to F(Eu)$ *strongly* in $V_3 \Longrightarrow (F \circ E)u_{k_m} \to (F \circ E)u$ *strongly* in V_3. Thus, $u_{k_m} \rightharpoonup u$ *weakly* in $V_1 \Longrightarrow (F \circ E)u_{k_m} \to (F \circ E)u$ *strongly* in $V_3 \Longrightarrow F \circ E \in \mathcal{L}(V_1;V_3)$ is *compact*.

Alternatively, let $E \in \mathcal{L}(V_1; V_2)$ be *compact* and $F \in \mathcal{L}(V_2; V_3)$ be continuous from V_2 into V_3. Let $(u_k)_{k=1}^{\infty}$ be any sequence *weakly* convergent to u in V_1; i.e. $u_k \rightharpoonup u$ *weakly* in V_1 as $k \to \infty$. But $E \in \mathcal{L}(V_1; V_2)$ is compact $\Longrightarrow Eu_k \longrightarrow Eu$ *strongly* in V_2 as $k \to \infty \Longrightarrow F(Eu_k) \longrightarrow F(Eu)$ *strongly* in V_3 as $k \to \infty$, since $F \in \mathcal{L}(V_2; V_3)$ is continuous from V_2 into $V_3 \Longrightarrow (F \circ E)u_k \longrightarrow (F \circ E)u$ *strongly* in V_3 as $k \to \infty$. Thus, $u_k \rightharpoonup u$ *weakly* in $V_1 \Longrightarrow (F \circ E)u_k \longrightarrow (F \circ E)u$ *strongly* in V_3 as $k \to \infty \Longrightarrow F \circ E \in \mathcal{L}(V_1; V_3)$ is *compact*. □

Now we state the compactness result for $H^1(\mathbb{R}^n)$:

Theorem 8.11.1 (Compactness Theorem I). *Let $(u_k)_{k=1}^{\infty}$ be a bounded sequence in $H^1(\mathbb{R}^n)$ such that* $\operatorname{supp}(u_k) \subset K\ \forall k \in \mathbb{N}$, $K \subset\subset \mathbb{R}^n$ *being a* fixed compact *subset of* $\mathbb{R}^n$.

Then a subsequence $(u_{k_m})_{m=1}^{\infty}$ can be extracted from this bounded sequence $(u_k)_{k=1}^{\infty}$ such that $(u_{k_m})_{m=1}^{\infty}$ converges strongly *in $L^2(\mathbb{R}^n)$, i.e. $\exists u \in L^2(\mathbb{R}^n)$ such that $\|u - u_{k_m}\|_{L^2(\mathbb{R}^n)} \to 0$ as $m \to \infty$.*

Proof. Let $(u_k)_{k=1}^{\infty}$ be a *bounded* sequence in Hilbert space $H^1(\mathbb{R}^n)$, with $\operatorname{supp}(u_k) \subset K \subset\subset \mathbb{R}^n$, i.e. $\exists C_0 > 0$ such that $\|u_k\|_{1,\mathbb{R}^n} \le C_0\ \forall\ k \in \mathbb{N}$. Hence, by (8.11.1) $\exists$ a subsequence $(u_{k_m})_{m=1}^{\infty}$ of $(u_k)_{k=1}^{\infty}$, which converges *weakly* to $u \in H^1(\mathbb{R}^n)$ as $m \to \infty$, i.e. $u_{k_m} \rightharpoonup u$ *weakly* in $H^1(\mathbb{R}^n)$ as $m \to \infty$. Define a new *bounded* sequence by (8.11.2): $v_m = u - u_{k_m}$ in $H^1(\mathbb{R}^n)$ with $\|v_m\|_{1,\mathbb{R}^n} \le C_1$ with $\operatorname{supp}(v_m) \subset K\ \forall\ m \in \mathbb{N}$. Then $v_m \rightharpoonup 0$ *weakly* in $H^1(\mathbb{R}^n)$ as $m \to \infty$. But $H^1(\mathbb{R}^n) \hookrightarrow L^2(\mathbb{R}^n)$ with the imbedding operator $\hookrightarrow\ \in \mathcal{L}(H^1(\mathbb{R}^n); L^2(\mathbb{R}^n))$ and $v_m \rightharpoonup 0$ *weakly* in $H^1(\mathbb{R}^n)$ by (8.11.2)–(8.11.8). Hence, by Lemma 8.11.1, $v_m \rightharpoonup 0$ *weakly* in $L^2(\mathbb{R}^n)$ as $m \to \infty$ and the proof will be complete if we can show that $v_m \to 0$ *strongly* in $L^2(\mathbb{R}^n)$ as $m \to \infty$, i.e.

$$\|v_m\|_{L^2(\mathbb{R}^n)} \to 0 \quad \text{as } m \to \infty. \tag{8.11.9}$$

Set $\hat{v}_m(\boldsymbol{\xi}) = \mathcal{F} v_m(\boldsymbol{\xi})$ (the Fourier transform $\hat{v}_m$ of v_m) $\forall m \in \mathbb{N}$. Then, $v_m \in L^2(\mathbb{R}^n)$, $\operatorname{supp}(v_m) \subset K \subset\subset \mathbb{R}^n$ (by (8.11.8)) $\forall m \in \mathbb{N} \Longrightarrow v_m \in L^1(\mathbb{R}^n)\ \forall m \in \mathbb{N}$. Hence, from (7.1.2),

$$\hat{v}_m(\boldsymbol{\xi}) = \int_{\mathbb{R}^n} v_m(\mathbf{x}) e^{-i2\pi\langle \mathbf{x}, \boldsymbol{\xi}\rangle} d\mathbf{x} \quad \forall m \in \mathbb{N}. \tag{8.11.10}$$

Then, by the Plancherel–Riesz Theorem 8.3.1,

$$\begin{aligned}
\|v_m\|^2_{L^2(\mathbb{R}^n)} &= \|\hat{v}_m\|^2_{L^2(\mathbb{R}^n)} = \int_{\mathbb{R}^n} |\hat{v}_m(\boldsymbol{\xi})|^2 d\boldsymbol{\xi} \\
&= \int_{\|\boldsymbol{\xi}\| \le r} |\hat{v}_m(\boldsymbol{\xi})|^2 d\boldsymbol{\xi} + \int_{\|\boldsymbol{\xi}\| > r} |\hat{v}_m(\boldsymbol{\xi})|^2 d\boldsymbol{\xi} \\
&\le \int_{\|\boldsymbol{\xi}\| \le r} |\hat{v}_m(\boldsymbol{\xi})|^2 d\boldsymbol{\xi} + \frac{1}{1+r^2} \int_{\|\boldsymbol{\xi}\| > r} (1 + \|\boldsymbol{\xi}\|^2) |\hat{v}_m(\boldsymbol{\xi})|^2 d\boldsymbol{\xi}\ \forall m \in \mathbb{N},
\end{aligned}$$

since $\|\boldsymbol{\xi}\| > r \Longrightarrow 1 + \|\boldsymbol{\xi}\|^2 > 1 + r^2 \Longrightarrow \frac{1+\|\boldsymbol{\xi}\|^2}{1+r^2} > 1 \;\forall\; \|\boldsymbol{\xi}\| > r$

$$\Longrightarrow \quad \|v_m\|^2_{L^2(\mathbb{R}^n)} \leq \underbrace{\int_{\|\boldsymbol{\xi}\|\leq r} |\hat{v_m}(\boldsymbol{\xi})|^2 d\boldsymbol{\xi}}_{I_m(r)} + \underbrace{\frac{1}{1+r^2}\|v_m\|^2_{1,\mathbb{R}^n}}_{J_m(r)} \quad \text{(by (8.9.3))}. \tag{8.11.11}$$

Estimate for $J_m(r)$: Since $\|v_m\|_{1,\mathbb{R}^n} \leq C_1 \;\forall\; m \in \mathbb{N}$, $J_m(r) \leq \frac{C_1^2}{1+r^2} \to 0$ as $r \to \infty$ $\forall m \in \mathbb{N}$. Hence, $\forall$ given $\varepsilon > 0$, $\exists$ sufficiently large $R > 0$ such that

$$J_m(r) \leq \frac{C_1^2}{1+r^2} \leq \frac{\varepsilon}{2} \quad \forall r > R,\ \forall m \in \mathbb{N}. \tag{8.11.12}$$

Estimate for $I_m(r)$: Since $\operatorname{supp}(v_m) \subset K$, $\forall\; m \in \mathbb{N}$,

$$\hat{v}_m(\boldsymbol{\xi}) = \int_{\mathbb{R}^n} v_m(\mathbf{x}) e^{-i2\pi\langle\mathbf{x},\boldsymbol{\xi}\rangle} d\mathbf{x} = \int_{\mathbb{R}^n} v_m(\mathbf{x})\chi_K(\mathbf{x}) e^{-i2\pi\langle\mathbf{x},\boldsymbol{\xi}\rangle} d\mathbf{x}, \tag{8.11.13}$$

where χ_K is the characteristic function of the compact set K with $\chi_K(\mathbf{x}) = 1$ for $\mathbf{x} \in K$ and $\chi_K(\mathbf{x}) = 0$ for $\mathbf{x} \in \mathbb{R}^n \setminus K$, i.e. $\forall m \in \mathbb{N}$, $\chi_K(\mathbf{x})v_m(\mathbf{x}) = v_m(\mathbf{x})$ for $\mathbf{x} \in K$ and $\chi_K(\mathbf{x})v_m(\mathbf{x}) = 0$ for $\mathbf{x} \in \mathbb{R}^n \setminus K$. Although $e^{-i2\pi\langle\cdot,\boldsymbol{\xi}\rangle}$ does *not* belong to $L^2(\mathbb{R}^n)$, the function $\mathbf{x} \in \mathbb{R}^n \longmapsto \chi_K(\mathbf{x})e^{-i2\pi\langle\mathbf{x},\boldsymbol{\xi}\rangle}$ belongs to $L^2(\mathbb{R}^n)$, since

$$\int_{\mathbb{R}^n} |\chi_K(\mathbf{x})e^{-i2\pi\langle\mathbf{x},\boldsymbol{\xi}\rangle}|^2 d\mathbf{x} = \int_{\mathbb{R}^n} |\chi_K(\mathbf{x})|^2 d\mathbf{x} = \int_K 1^2 d\mathbf{x} = \operatorname{meas}(K) < +\infty.$$

From (8.11.13),

$$\hat{v}_m(\boldsymbol{\xi}) = \langle v_m(\cdot), \chi_K(\cdot)e^{i2\pi\langle\cdot,\boldsymbol{\xi}\rangle}\rangle_{L^2(\mathbb{R}^n)} \to 0 \quad \text{as } m \to \infty, \tag{8.11.14}$$

since $v_m \rightharpoonup 0$ *weakly* in $L^2(\mathbb{R}^n)$ and $\chi_K(\cdot)e^{i2\pi\langle\cdot,\boldsymbol{\xi}\rangle} \in L^2(\mathbb{R}^n)$. Moreover, for almost all $\boldsymbol{\xi} \in \mathbb{R}^n$,

$$\begin{aligned} |\hat{v}_m(\boldsymbol{\xi})| &= \left| \int_{\mathbb{R}^n} v_m(\mathbf{x}) e^{-i2\pi\langle\mathbf{x},\boldsymbol{\xi}\rangle} d\mathbf{x} \right| \leq \int_{\mathbb{R}^n} |v_m(\mathbf{x})| d\mathbf{x} = \int_K |v_m(\mathbf{x})| d\mathbf{x} \\ &\leq \|v_m\|_{L^2(\mathring{K})} \|1\|_{L^2(\mathring{K})} \leq \|v_m\|_{L^2(\mathbb{R}^n)} \cdot (\operatorname{meas}(K))^{1/2} \quad \forall m \in \mathbb{N} \end{aligned}$$

$\Longrightarrow$ for almost all $\boldsymbol{\xi} \in \mathbb{R}^n$,

$$|\hat{v}_m(\boldsymbol{\xi})|^2 \leq \|v_m\|^2_{L^2(\mathbb{R}^n)} \cdot \operatorname{meas}(K) \leq C \quad \forall\; m \in \mathbb{N} \text{ with } \int_{\|\boldsymbol{\xi}\|\leq r} C d\boldsymbol{\xi} < +\infty. \tag{8.11.15}$$

Thus, by virtue of (8.11.14) and (8.11.15), we can apply Lebesgue's Dominated Convergence Theorem B.3.2.2 in Appendix B and get

$$\lim_{m\to\infty}\int_{\|\boldsymbol{\xi}\|\le r}|\hat{v}_m(\boldsymbol{\xi})|^2 d\boldsymbol{\xi} = \int_{\|\boldsymbol{\xi}\|\le r}\lim_{m\to\infty}|\hat{v}_m(\boldsymbol{\xi})|^2 d\boldsymbol{\xi} = 0 \quad \forall \text{ fixed } r > 0.$$

Hence, $\forall$ given $\varepsilon > 0$, $\exists m_0 \in \mathbb{N}$ such that

$$I_m(r) = \int_{\|\boldsymbol{\xi}\|\le r}|\hat{v}_m(\boldsymbol{\xi})|^2 d\boldsymbol{\xi} \le \varepsilon/2 \quad \forall m \ge m_0 \text{ and } \forall \text{ fixed } r > 0. \tag{8.11.16}$$

Then, from (8.11.11), (8.11.12) and (8.11.16), we have: $\forall \varepsilon > 0$, $\exists R > 0$ and $m_0 \in \mathbb{N}$ such that $\|v_m\|^2_{L^2(\mathbb{R}^n)} \le I_m(r) + J_m(r) \le \frac{\varepsilon}{2} + \frac{\varepsilon}{2} = \varepsilon\ \forall\ m \ge m_0$, $\forall$ fixed $r \ge R$.

Thus, $\lim_{m\to\infty}\|v_m\|^2_{L^2(\mathbb{R}^n)} = 0 \Longrightarrow \|u - u_{k_m}\|_{L^2(\mathbb{R}^n)} = \|v_m\|_{L^2(\mathbb{R}^n)} \to 0$ as $m \to \infty$, i.e. $u_{k_m} \to u$ *strongly* in $L^2(\mathbb{R}^n)$ as $m \to \infty$. □

Remark 8.11.1. Theorem 8.11.1 does *not* hold without the condition that $\text{supp}(u_m)$ is contained in a *fixed compact* set $K \subset\subset \mathbb{R}^n$ $\forall m \in \mathbb{N}$. Consider the interesting example given by Tartar [42]. Let $\phi \in \mathcal{D}(\mathbb{R}^n) \subset H^1(\mathbb{R}^n)$ with $\phi \ne 0$. Define, $\forall\ m \in \mathbb{N}$, $\phi_m(\mathbf{x}) = \phi_m(x_1, x_2, \ldots, x_n) = \phi(x_1 + m, x_2, \ldots, x_n)$. Then, $\forall m \in \mathbb{N}$, $\text{supp}(\phi_m)$ is *not* contained in a *fixed* compact set $K \subset\subset \mathbb{R}^n$. But $\forall\ m \in \mathbb{N}$,

$$\begin{aligned}\|\phi_m\|^2_{L^2(\mathbb{R}^n)} = \int_{\mathbb{R}^n}|\phi_m(\mathbf{x})|^2 d\mathbf{x} &= \int_{\mathbb{R}^n}|\phi(\xi_1, \xi_2, \ldots, \xi_n)|^2 d\xi_1 d\xi_2 \ldots d\xi_n \\ &= \|\phi\|^2_{L^2(\mathbb{R}^n)},\end{aligned} \tag{8.11.17}$$

which is obtained by changing variables: $\xi_1 = x_1 + m$, $\xi_i = x_i$, $2 \le i \le n$, with $|J| = 1$, $d\boldsymbol{\xi} = d\xi_1 d\xi_2 \ldots d\xi_n \Longrightarrow \|\phi_m\|_{L^2(\mathbb{R}^n)} = \|\phi\|_{L^2(\mathbb{R}^n)}$ $\forall m \in \mathbb{N}$. But $\forall f \in L^2(\mathbb{R}^n)$,

$$\lim_{m\to\infty}\langle\phi_m, f\rangle_{L^2(\mathbb{R}^n)} = \lim_{m\to\infty}\int_{\mathbb{R}^n}\phi_m(\mathbf{x})f(\mathbf{x})d\mathbf{x} = 0.$$

Since $\phi \in \mathcal{D}(\mathbb{R}^n) \Longrightarrow \text{supp}(\phi) \subset\subset \mathbb{R}^n$ is compact in $\mathbb{R}^n$, $\phi(x_1, x_2, \ldots, x_n) = 0$ $\forall (x_1, x_2, \ldots, x_n) \notin \text{supp}(\phi) \Longrightarrow \forall \mathbf{x} \in \mathbb{R}^n$, $\exists m_0 \in \mathbb{N}$ such that $(x_1 + m, x_2, \ldots, x_n) \notin \text{supp}(\phi)$ $\forall m \ge m_0 \Longrightarrow \forall \mathbf{x} \in \mathbb{R}^n$, $\exists m_0 \in \mathbb{N}$ such that $|\phi(x_1 + m, x_2, \ldots, x_n)| = 0$ $\forall m \ge m_0 \Longrightarrow \lim_{m\to\infty}\phi(x_1 + m, x_2, \ldots, x_n) = 0$. Hence,

$$\lim_{m\to\infty}[\phi_m(\mathbf{x})f(\mathbf{x})] = 0 \quad \text{a.e in } \mathbb{R}^n. \tag{8.11.18}$$

$\forall$ fixed $m \in \mathbb{N}$,

$$|\phi_m(\mathbf{x})f(\mathbf{x})| \le |g(\mathbf{x})| \quad \text{a.e. in } \mathbb{R}^n, \tag{8.11.19}$$

where $g(\mathbf{x}) = \phi_m(\mathbf{x}) f(\mathbf{x})$ for $(x_1 + m, x_2, \ldots, x_n) = \mathbf{x}_m \in K$, $g(\mathbf{x}) = 0$ otherwise, and

$$\int_{\mathbb{R}^n} |g(\mathbf{x})| d(\mathbf{x}) = \int_{\mathbf{x}_m \in K} |\phi_m(\mathbf{x}) f(\mathbf{x})| d\mathbf{x} \le \max_{\mathbf{x} \in \mathbb{R}^n} |\phi(\mathbf{x})| \int_{\mathbf{x}_m \in K} |f(\mathbf{x})| d\mathbf{x} < +\infty.$$

Hence, by virtue of (8.11.18) and (8.11.19), we can apply Lebesgue's Dominated Convergence Theorem B.3.2.2 (Appendix B) and get

$$\begin{aligned} \lim_{m \to \infty} \langle \phi_m, f \rangle_{L^2(\mathbb{R}^n)} &= \lim_{m \to \infty} \int_{\mathbb{R}^n} \phi_m(\mathbf{x}) f(\mathbf{x}) d\mathbf{x} \\ &= \int_{\mathbb{R}^n} \lim_{m \to \infty} \phi_m(\mathbf{x}) f(\mathbf{x}) d\mathbf{x} = 0 \quad \forall f \in L^2(\mathbb{R}^n). \end{aligned}$$

Hence, $\phi_m \rightharpoonup 0$ *weakly* in $L^2(\mathbb{R}^n)$ as $m \to \infty$. Thus, $(\phi_m)_{m=1}^{\infty}$ is a sequence in $H^1(\mathbb{R}^n)$, for which there does *not* exist any fixed *compact* set K with $\operatorname{supp}(\phi_m) \subset K$ $\forall m \in \mathbb{N}$, and $(\phi_m)_{m=1}^{\infty}$ converges *weakly* to 0 in $L^2(\mathbb{R}^n)$. But $(\phi_m)_{m=1}^{\infty}$ does *not* converge to 0 *strongly* in $L^2(\mathbb{R}^n)$. In fact, from (8.11.17), $\|\phi_m\|_{L^2(\mathbb{R}^n)} = \|\phi\|_{L^2(\mathbb{R}^n)}$ $\neq 0$ $\forall$ $m \in \mathbb{N} \Longrightarrow \lim_{m \to \infty} \|\phi_m\|_{L^2(\mathbb{R}^n)} \neq 0$.

Theorem 8.11.2 (Compactness Theorem II). *For an* arbitrary bounded domain $\Omega \subset \mathbb{R}^n$, *the imbedding* $H_0^1(\Omega) \hookrightarrow L^2(\Omega)$ *is* compact.

Proof. Let $(u_k)_{k=1}^{\infty}$ be a *bounded* sequence in Hilbert space $H_0^1(\Omega)$. Then, by (8.11.1), $\exists u \in H_0^1(\Omega)$ and a subsequence $(u_{k_m})_{m=1}^{\infty}$ of $(u_k)_{k=1}^{\infty}$ such that $u_{k_m} \rightharpoonup u$ *weakly* in $H_0^1(\Omega)$ as $m \to \infty$. Define a new bounded sequence $(v_m)_{m=1}^{\infty}$ by $v_m = u - u_{k_m}$ $\forall m \in \mathbb{N}$. By (8.11.2), $v_m \rightharpoonup 0$ *weakly* in $H_0^1(\Omega)$ as $m \to \infty$. Now we define a *continuous linear* operator $P \in \mathcal{L}(H_0^1(\Omega); H^1(\mathbb{R}^n))$ by $Pv = \tilde{v} \in H^1(\mathbb{R}^n)$ $\forall v \in H_0^1(\Omega)$, $\tilde{v}$ being the *null extension* of v to $\mathbb{R}^n$, i.e. $\tilde{v}(\mathbf{x}) = v(\mathbf{x})$ for $\mathbf{x} \in \Omega$ and $\tilde{v}(\mathbf{x}) = 0$ for $\mathbf{x} \in \mathbb{R}^n \setminus \Omega$, with $\|Pv\|_{H^1(\mathbb{R}^n)} = \|\tilde{v}\|_{H^1(\mathbb{R}^n)} = \|v\|_{H^1(\Omega)}$ $\forall v \in H_0^1(\Omega)$, which follows from Theorem 2.15.5. Hence, by Lemma 8.11.1, $v_m \rightharpoonup 0$ *weakly* in $H_0^1(\Omega) \Longrightarrow Pv_m = \tilde{v}_m \rightharpoonup P0 = 0$ *weakly* in $H^1(\mathbb{R}^n)$. Since Ω is a *bounded* domain in $\mathbb{R}^n$ and $\tilde{v}_m(\mathbf{x}) = 0$ outside Ω, $\exists$ a *fixed compact* set K such that $\operatorname{supp}(\tilde{v}_m) \subset K$ $\forall m \in \mathbb{N}$. Thus, $\tilde{v}_m \rightharpoonup 0$ *weakly* in $H^1(\mathbb{R}^n)$ as $m \to \infty$ and $\operatorname{supp}(\tilde{v}_m) \subset K \subset\subset \mathbb{R}^n$ $\forall m \in \mathbb{N}$. Hence, by Theorem 8.11.1, $\tilde{v}_m \to 0$ *strongly* in $L^2(\mathbb{R}^n)$, i.e. $\|\tilde{v}_m\|_{L^2(\mathbb{R}^n)} \to 0$ as $m \to \infty$. Then, $\|v_m\|_{L^2(\Omega)}^2 = \int_\Omega |v_m(\mathbf{x})|^2 d\mathbf{x} = \int_{\mathbb{R}^n} |\tilde{v}_m(\mathbf{x})|^2 d\mathbf{x} = \|\tilde{v}_m\|_{L^2(\mathbb{R}^n)}^2 \to 0 \Longrightarrow \|v_m - 0\|_{L^2(\Omega)} = \|v_m\|_{L^2(\Omega)} \to 0$ as $m \to \infty \Longrightarrow v_m \to 0$ *strongly* in $L^2(\Omega) \Longrightarrow u_{k_m} \to u$ *strongly* in $L^2(\Omega)$ as $m \to \infty$. Thus, $u_{k_m} \rightharpoonup u$ *weakly* in $H_0^1(\Omega) \Longrightarrow u_{k_m} \to u$ *strongly* in $L^2(\Omega) \Longrightarrow$ the imbedding operator $\hookrightarrow$: $H_0^1(\Omega) \longrightarrow L^2(\Omega)$ is *compact* from $H_0^1(\Omega)$ into $L^2(\Omega)$. □

Theorem 8.11.3. *For an* arbitrary bounded *domain* $\Omega \subset \mathbb{R}^n$, *the imbedding* $H_0^{m+1}(\Omega) \hookrightarrow H_0^m(\Omega)$ *is* compact $\forall\ m \in \mathbb{N}_0$, *where* $H_0^k(\Omega) = \overline{\mathcal{D}(\Omega)}$ *in the norm* $\|\cdot\|_{H^k(\Omega)}$ (2.15.10) *(resp. semi-norm* $|\cdot|_{H^k(\Omega)}$ (2.15.11)*)*, $k = m, m+1$.

Proof. For $m = 0$, $H_0^0(\Omega) \equiv L^2(\Omega)$, the result is proved in Theorem 8.11.2.

Now we consider the case $m \in \mathbb{N}$. Let $(u_k)_{k=1}^\infty$ be a *bounded* sequence in Hilbert space $H_0^{m+1}(\Omega)$ with $\|u_k\|_{H_0^{m+1}(\Omega)} \leq C_0\ \forall\ k \in \mathbb{N}$. Hence, by (8.11.1), $\exists u \in H_0^{m+1}(\Omega)$ and a subsequence $(u_{k_\mu})_{\mu=1}^\infty$ of $(u_k)_{k=1}^\infty$ such that $u_{k_\mu} \rightharpoonup u$ *weakly* in $H_0^{m+1}(\Omega)$ as $\mu \to \infty$. Define a new *bounded* sequence $(v_\mu)_{\mu=1}^\infty$ by

$$v_\mu = u - u_{k_\mu} \quad \forall \mu \in \mathbb{N} \text{ in } H_0^{m+1}(\Omega) \text{ with } \|v_\mu\|_{H^{m+1}(\Omega)} \leq C_1\ \forall\ \mu \in \mathbb{N}. \tag{8.11.20}$$

Then $v_\mu \rightharpoonup 0$ *weakly* in $H_0^{m+1}(\Omega)$ as $\mu \to \infty$. Let $\tilde{v}_\mu = v_\mu$ in Ω and $\tilde{v}_\mu = 0$ in $\mathbb{R}^n \setminus \Omega$ be the null extension of $v_\mu \in H_0^{m+1}(\Omega)$ to $\mathbb{R}^n$ $\forall \mu \in \mathbb{N}$. Then, by Theorem 2.15.5, $\tilde{v}_\mu \in H^{m+1}(\mathbb{R}^n)$ with $\operatorname{supp}(\tilde{v}_\mu) = \operatorname{supp}(v_\mu) = K\ \forall \mu \in \mathbb{N}$. Hence $\tilde{v}_\mu \rightharpoonup 0$ *weakly* in $H^{m+1}(\mathbb{R}^n)$. But $H^{m+1}(\mathbb{R}^n) \hookrightarrow H^m(\mathbb{R}^n)$ with the imbedding operator $\hookrightarrow\ \in \mathcal{L}(H^{m+1}(\mathbb{R}^n); H^m(\mathbb{R}^n))$ and $\tilde{v}_\mu \rightharpoonup 0$ *weakly* in $H^{m+1}(\mathbb{R}^n) \Longrightarrow \tilde{v}_\mu \rightharpoonup 0$ *weakly* in $H^m(\mathbb{R}^n)$ by Lemma 8.11.1.

Similarly, $\hookrightarrow\ \in \mathcal{L}(H^{m+1}(\mathbb{R}^n); L^2(\mathbb{R}^n))$ and $\tilde{v}_\mu \rightharpoonup 0$ weakly in $H^{m+1}(\mathbb{R}^n)$

$$\Longrightarrow \quad \tilde{v}_\mu \rightharpoonup 0 \textit{ weakly} \text{ in } L^2(\mathbb{R}^n) \text{ as } \mu \to \infty \tag{8.11.21}$$

by Lemma 8.11.1.

Then, using first Theorem 2.15.5 on null extension and then Theorem 8.9.3 on *equivalent* norms in $H^s(\mathbb{R}^n)$ with $s \in \mathbb{R}$, we have

$$\|v_\mu\|_{H_0^m(\Omega)}^2 = \|\tilde{v}_\mu\|_{H^m(\mathbb{R}^n)}^2 = \int_{\mathbb{R}^n} (1 + \|\boldsymbol{\xi}\|^2)^m |\hat{\tilde{v}}_\mu(\boldsymbol{\xi})|^2 d\boldsymbol{\xi}$$

with

$$\hat{\tilde{v}}_\mu(\boldsymbol{\xi}) = \mathcal{F}[\tilde{v}_\mu](\boldsymbol{\xi}) \quad \forall \mu \in \mathbb{N}. \text{ Then,}$$

$$\|v_\mu\|_{H_0^m(\Omega)}^2 = \underbrace{\int_{\|\boldsymbol{\xi}\| \leq r} (1 + \|\boldsymbol{\xi}\|^2)^m |\hat{\tilde{v}}_\mu(\boldsymbol{\xi})|^2 d\boldsymbol{\xi}}_{I_\mu(r)} + \underbrace{\int_{\|\boldsymbol{\xi}\| > r} (1 + \|\boldsymbol{\xi}\|^2)^m |\hat{\tilde{v}}_\mu(\boldsymbol{\xi})|^2 d\boldsymbol{\xi}}_{J_\mu(r)}. \tag{8.11.22}$$

Estimate for $J_\mu(r)$:

$$\begin{aligned}
J_\mu(r) &= \int_{\|\boldsymbol{\xi}\|>r} \frac{(1+\|\boldsymbol{\xi}\|^2)^{m+1}}{(1+\|\boldsymbol{\xi}\|^2)} |\hat{\tilde{v}}_\mu(\boldsymbol{\xi})|^2 d\boldsymbol{\xi} \\
&\le \frac{1}{1+r^2} \int_{\|\boldsymbol{\xi}\|>r} (1+\|\boldsymbol{\xi}\|^2)^{m+1} |\hat{\tilde{v}}_\mu(\boldsymbol{\xi})|^2 d\boldsymbol{\xi} \\
&\le \frac{1}{1+r^2} \int_{\mathbb{R}^n} (1+\|\boldsymbol{\xi}\|^2)^{m+1} |\hat{\tilde{v}}_\mu(\boldsymbol{\xi})|^2 d\boldsymbol{\xi} \\
&= \frac{1}{1+r^2} \|\tilde{v}_\mu\|^2_{H^{m+1}(\mathbb{R}^n)} \to 0 \quad \text{as } r \to \infty
\end{aligned}$$

$\forall$ fixed $\mu \in \mathbb{N}$, since $\|\boldsymbol{\xi}\| > r \Longrightarrow 1 + \|\boldsymbol{\xi}\|^2 > 1 + r^2 \Longrightarrow \frac{1}{1+\|\boldsymbol{\xi}\|^2} < \frac{1}{1+r^2} \to 0$ as $r \to \infty$.

Hence, $\forall$ fixed $\mu \in \mathbb{N}$, $\forall \varepsilon > 0$, $\exists$ a sufficiently large $R > 0$ such that

$$J_\mu(r) \le \frac{1}{1+r^2} \|\tilde{v}_\mu\|^2_{H^{m+1}(\mathbb{R}^n)} \le \frac{\varepsilon}{2} \quad \forall r > R. \tag{8.11.23}$$

Estimate for $I_\mu(r)$: Since Ω is a *bounded* domain in $\mathbb{R}^n$, $L^2(\Omega) \hookrightarrow L^1(\Omega)$ with continuous imbedding $\hookrightarrow$ and $\forall \mu \in \mathbb{N}$, $\tilde{v}_\mu \in H^{m+1}(\mathbb{R}^n) \Longrightarrow \tilde{v}_\mu \in L^2(\mathbb{R}^n) \Longrightarrow v_\mu \in L^2(\Omega) \Longrightarrow v_\mu \in L^1(\Omega) \Longrightarrow \tilde{v}_\mu \in L^1(\mathbb{R}^n)$ with $\operatorname{supp}(\tilde{v}_\mu) = \operatorname{supp}(v_\mu) \subset K$, $K \subset \Omega$ with $K \subset\subset \mathbb{R}^n$. Hence, $\forall \mu \in \mathbb{N}$,

$$\hat{\tilde{v}}_\mu(\boldsymbol{\xi}) = \int_{\mathbb{R}^n} \tilde{v}_\mu(\mathbf{x}) e^{-i2\pi\langle \mathbf{x}, \boldsymbol{\xi}\rangle} d\mathbf{x} = \int_{\mathbb{R}^n} \tilde{v}_\mu(\mathbf{x}) \chi_K(\mathbf{x}) e^{-i2\pi\langle \mathbf{x}, \boldsymbol{\xi}\rangle} d\mathbf{x} \quad \text{for } \boldsymbol{\xi} \in \mathbb{R}^n, \tag{8.11.24}$$

where χ_K is the characteristic function of the *compact set* K with $K \subset \Omega$ with $K \subset\subset \Omega \subset \mathbb{R}^n$ such that, $\forall \mu \in \mathbb{N}$,

$$\chi_K(\mathbf{x}) \tilde{v}_\mu(\mathbf{x}) = \begin{cases} \tilde{v}_\mu(\mathbf{x}) & \text{for } \mathbf{x} \in K \\ 0 & \text{for } \mathbf{x} \in \mathbb{R}^n \setminus K, \end{cases}$$

since $\chi_K(\mathbf{x}) = 1$ $\forall \mathbf{x} \in K$ and $\chi_K(\mathbf{x}) = 0$ otherwise. Although $e^{-i2\pi\langle \cdot, \boldsymbol{\xi}\rangle}$ does *not* belong to $L^2(\mathbb{R}^n)$, the function $\mathbf{x} \in \mathbb{R}^n \longmapsto \chi_K(\mathbf{x}) e^{-i2\pi\langle \mathbf{x}, \boldsymbol{\xi}\rangle}$ belongs to $L^2(\mathbb{R}^n)$. Hence, (8.11.24) can be rewritten as follows: for almost all $\boldsymbol{\xi} \in \mathbb{R}^n$, $\hat{\tilde{v}}_\mu(\boldsymbol{\xi}) = \langle \tilde{v}_\mu(\cdot), \chi_K(\cdot) e^{i2\pi\langle \cdot, \boldsymbol{\xi}\rangle} \rangle_{L^2(\mathbb{R}^n)} \to 0$ as $\mu \to \infty$, since $\tilde{v}_\mu \rightharpoonup 0$ *weakly* in $L^2(\mathbb{R}^n)$ by (8.11.21),

$$|\hat{\tilde{v}}_\mu(\boldsymbol{\xi})|^2 = |\langle \tilde{v}_\mu, \chi_K e^{i2\pi\langle \cdot, \boldsymbol{\xi}\rangle} \rangle_{L^2(\mathbb{R}^n)}|^2 \to 0 \quad \text{as } \mu \to \infty. \tag{8.11.25}$$

Moreover, *for almost all* $\boldsymbol{\xi} \in \mathbb{R}^n$,

$$|\hat{\tilde{v}}_\mu(\boldsymbol{\xi})| = \left| \int_{\mathbb{R}^n} \tilde{v}_\mu(\mathbf{x}) e^{-i2\pi\langle \mathbf{x},\boldsymbol{\xi}\rangle} d\mathbf{x} \right| \leq \int_{\mathbb{R}^n} |\tilde{v}_\mu(\mathbf{x})| d\mathbf{x} = \int_\Omega |v_\mu(\mathbf{x})| d\mathbf{x}$$
$$\leq \|v_\mu\|_{L^2(\Omega)} \|1\|_{L^2(\Omega)} \quad \text{(by the Cauchy–Schwarz inequality)}$$
$$\leq \|v_\mu\|_{H_0^{m+1}(\Omega)} \cdot (\text{meas}(\Omega))^{1/2} \leq C_1 (\text{meas}(\Omega))^{1/2} \quad \forall \mu \in \mathbb{N}$$

(by the boundedness of (v_μ) in $H_0^{m+1}(\Omega) \hookrightarrow L^2(\Omega)$) by (8.11.21) $\Longrightarrow$ for *almost all* $\boldsymbol{\xi} \in \mathbb{R}^n$,

$$|\hat{\tilde{v}}_\mu(\boldsymbol{\xi})|^2 \leq C_1^2 \cdot \text{meas}(\Omega) \leq C, \tag{8.11.26}$$

with $C > 0$ and $\int_{\|\boldsymbol{\xi}\| \leq r} C d\boldsymbol{\xi} < +\infty$ $\forall \mu \in \mathbb{N}$, $\forall$ fixed $r \in]0, \infty[$. Thus, by virtue of (8.11.25) and (8.11.26), we can apply Lebesgue's Dominated Convergence Theorem B.3.2.2 (see Appendix B) and get, $\forall$ fixed $r \in]0, \infty[$,

$$\lim_{\mu \to \infty} \int_{\|\boldsymbol{\xi}\| \leq r} |\hat{\tilde{v}}_\mu(\boldsymbol{\xi})|^2 d\boldsymbol{\xi} = \int_{\|\boldsymbol{\xi}\| \leq r} \lim_{\mu \to \infty} |\hat{\tilde{v}}_\mu(\boldsymbol{\xi})|^2 d\boldsymbol{\xi} = 0. \tag{8.11.27}$$

Hence,

$$I_\mu(r) = \int_{\|\boldsymbol{\xi}\| \leq r} (1 + \|\boldsymbol{\xi}\|^2)^m |\hat{\tilde{v}}_\mu(\boldsymbol{\xi})|^2 d\boldsymbol{\xi} \leq (1 + r^2)^m \int_{\|\boldsymbol{\xi}\| \leq r} |\hat{\tilde{v}}_\mu(\boldsymbol{\xi})|^2 d\boldsymbol{\xi} \to 0$$

as $\mu \to \infty$ (by (8.11.27)). Then, $\forall \varepsilon > 0$, $\exists$ a sufficiently large $m_0 \in \mathbb{N}$ such that

$$I_\mu(r) \leq \frac{\varepsilon}{2} \quad \forall \mu \geq m_0. \tag{8.11.28}$$

Then, combining (8.11.22), (8.11.23) and (8.11.28), we get, $\forall \varepsilon > 0$, $\exists R > 0$ and $m_0 \in \mathbb{N}$ such that $\|v_\mu\|^2_{H_0^m(\Omega)} \leq I_\mu(r) + J_\mu(r) \leq \frac{\varepsilon}{2} + \frac{\varepsilon}{2} = \varepsilon$ $\forall \mu \geq m_0$, $\forall r \geq R$.

Hence, $\lim_{\mu \to \infty} \|v_\mu\|_{H_0^m(\Omega)} = 0 \Longrightarrow \|u - u_{k_\mu}\|_{H_0^m(\Omega)} = \|v_\mu\|_{H_0^m(\Omega)} \to 0$ as $\mu \to \infty$, i.e. $u_{k_\mu} \to u$ *strongly* in $H_0^m(\Omega)$ and the imbedding $\hookrightarrow : H_0^{m+1}(\Omega) \hookrightarrow H_0^m(\Omega)$ is *compact.* □

Theorem 8.11.4 (Rellich–Kondraschov [47]). *Let* $\Omega \subset \mathbb{R}^n$ *be a* bounded *domain with* Lipschitz continuous *boundary* Γ. *Then the imbedding* $H^{m+1}(\Omega) \hookrightarrow H^m(\Omega)$ *is* compact.

In general, for $0 \leq k \leq m$, *the imbedding* $H^{m+1}(\Omega) \hookrightarrow H^k(\Omega)$ *is* compact.

Proof. By Theorem 8.10.9, a *bounded* $\Omega \subset \mathbb{R}^n$ with Lipschitz continuous boundary Γ has the $(m+1)$-extension property, i.e. $\exists$ an $(m+1)$-extension operator $P \in \mathcal{L}(H^{m+1}(\Omega); H^{m+1}(\mathbb{R}^n))$ such that, $\forall u \in H^{m+1}(\Omega)$,

$$Pu \in H^{m+1}(\mathbb{R}^n) \text{ with } Pu{\downarrow_\Omega} = u \quad \text{and} \quad \|Pu\|_{H^{m+1}(\mathbb{R}^n)} \leq C_1 \|u\|_{H^{m+1}(\Omega)}. \tag{8.11.29}$$

Then $H^{m+1}(\overline{\Omega}) = H^{m+1}(\Omega)$ (resp. $H^m(\overline{\Omega}) = H^m(\Omega)$) with norm equivalence.

Let $\Omega_1 = B(\mathbf{0}; r)$ with centre at $\mathbf{0}$ and radius $r > 0$ such that $\overline{\Omega} \subset \Omega_1$, $\overline{\Omega}_1 = \bar{B}(\mathbf{0}; r)$ is *compact* in $\mathbb{R}^n$. Let $\phi \in \mathcal{D}(\mathbb{R}^n)$ such that $\phi = 1$ in Ω and $\operatorname{supp}(\phi) \subset \Omega_1$ (see Theorem 6.2.4). Then $\phi\downarrow_{\Omega_1} \in \mathcal{D}(\Omega_1)$. For $u \in H^{m+1}(\Omega)$, $Pu \in H^{m+1}(\mathbb{R}^n)$ satisfying (8.11.29) such that $\phi \cdot (Pu) \in H^{m+1}(\mathbb{R}^n)$ and

$$(\phi \cdot (Pu))\downarrow_{\Omega} = \phi\downarrow_{\Omega} \cdot (Pu)\downarrow_{\Omega} = 1 \cdot u = u \quad \text{a.e. on } \Omega \tag{8.11.30}$$

and $\phi \cdot (Pu) = 0$ outside Ω_1, since $\operatorname{supp}(\phi) \subset \Omega_1$. Hence, $\phi(Pu)\downarrow_{\Omega_1} \in H^{m+1}(\overline{\Omega}_1)$ (by Definition 8.10.1) and its null extension $\widetilde{\phi(Pu)\downarrow_{\Omega_1}} \in H^{m+1}(\mathbb{R}^n) \Longrightarrow \phi(Pu)\downarrow_{\Omega_1} \in H_{00}^{m+1}(\Omega_1)$ by Definition 8.10.14. Since $m+1-\frac{1}{2} \neq$ an integer and Γ is Lipschitz continuous, $H_{00}^{m+1}(\Omega_1) = H_0^{m+1}(\Omega_1)$ by (8.10.113). Hence, $\forall u \in H^{m+1}(\Omega)$, $\phi(Pu)\downarrow_{\Omega_1} \in H_0^{m+1}(\Omega_1)$ with $\|\phi(Pu)\downarrow_{\Omega_1}\|_{H_{00}^{m+1}(\Omega_1)} = \|\phi(Pu)\downarrow_{\Omega_1}\|_{H^{m+1}(\Omega_1)}$ (by (8.10.106) in Lemma 8.10.3), $\|\cdot\|_{H^{m+1}(\Omega_1)}$ being the norm in $H_0^{m+1}(\Omega_1)$ induced by $H^{m+1}(\Omega_1)$. Hence, we can define a *linear* extension operator $E : H^{m+1}(\Omega) \longrightarrow H_0^{m+1}(\Omega_1)$ by $Eu = (\phi Pu)\downarrow_{\Omega_1} \in H_0^{m+1}(\Omega_1)\ \forall u \in H^{m+1}(\Omega)$, where P is defined in (8.11.29). Using Leibniz's formula ($\partial^{\boldsymbol{\alpha}}[\phi(Pu)] = \sum_{\boldsymbol{\beta} \leq \boldsymbol{\alpha}} \binom{\boldsymbol{\alpha}}{\boldsymbol{\beta}} \partial^{\boldsymbol{\beta}}(Pu)\partial^{\boldsymbol{\alpha}-\boldsymbol{\beta}}\phi\ \forall |\boldsymbol{\alpha}| \leq m$), the *boundedness* of $\partial^{\boldsymbol{\gamma}}\phi\ \forall |\boldsymbol{\gamma}| = |\boldsymbol{\alpha} - \boldsymbol{\beta}| \leq m+1$, the fact that $\partial^{\boldsymbol{\beta}}(Pu) \in L^2(\mathbb{R}^n)\ \forall \boldsymbol{\beta} \leq m+1$ and the inequality in (8.11.29), we have $\|Eu\|_{H^{m+1}(\Omega_1)} = \|\phi(Pu)\|_{H^{m+1}(\Omega_1)} \leq C_2\|Pu\|_{H^{m+1}(\mathbb{R}^n)} \leq C\|u\|_{H^{m+1}(\Omega)}$, $C = C_2 \cdot C_1$.

Hence,

$$E \text{ is } \textit{continuous} \text{ from } H^{m+1}(\Omega) \text{ into } H_0^{m+1}(\Omega_1). \tag{8.11.31}$$

But $\forall u \in H_0^m(\Omega_1)$, $\partial^{\boldsymbol{\alpha}}(u\downarrow_{\Omega}) = (\partial^{\boldsymbol{\alpha}}u)\downarrow_{\Omega} \in L^2(\Omega)\ \forall |\boldsymbol{\alpha}| \leq m$. In fact,

$$\begin{aligned}
\langle(\partial^{\boldsymbol{\alpha}}u)\downarrow_{\Omega}, \psi\rangle_{\mathcal{D}'(\Omega)\times\mathcal{D}(\Omega)} &= \int_{\Omega} (\partial^{\boldsymbol{\alpha}}u)\psi d\mathbf{x} = (-1)^{|\boldsymbol{\alpha}|} \int_{\Omega} u\partial^{\boldsymbol{\alpha}}\psi d\mathbf{x} \\
&= (-1)^{|\boldsymbol{\alpha}|}\langle u\downarrow_{\Omega}, \partial^{\boldsymbol{\alpha}}\psi\rangle_{\mathcal{D}'(\Omega)\times\mathcal{D}(\Omega)} \\
&= \langle\partial^{\boldsymbol{\alpha}}(u\downarrow_{\Omega}), \psi\rangle_{\mathcal{D}'(\Omega)\times\mathcal{D}(\Omega)} \quad \forall \psi \in \mathcal{D}(\Omega)
\end{aligned}$$

$\Longrightarrow \partial^{\boldsymbol{\alpha}}(u\downarrow_{\Omega}) = (\partial^{\boldsymbol{\alpha}}u)\downarrow_{\Omega}$ in $\mathcal{D}'(\Omega)$, but $(\partial^{\boldsymbol{\alpha}}u)\downarrow_{\Omega} \in L^2(\Omega) \hookrightarrow \mathcal{D}'(\Omega) \Longrightarrow \partial^{\boldsymbol{\alpha}}(u\downarrow_{\Omega}) = (\partial^{\boldsymbol{\alpha}}u)\downarrow_{\Omega} \in L^2(\Omega)\ \forall |\boldsymbol{\alpha}| \leq m$. Hence, we can define a *linear*, *continuous* restriction operator $\rho : u \in H_0^m(\Omega_1) \longmapsto \rho u = u\downarrow_{\Omega} \in H^m(\Omega)$ with

$$\|\rho u\|_{H^m(\Omega)} = \|u\|_{H^m(\Omega)} \leq \|u\|_{H^m(\Omega_1)} \quad \forall u \in H_0^m(\Omega_1). \tag{8.11.32}$$

By Theorem 8.11.3, the imbedding $\hookrightarrow : H_0^{m+1}(\Omega_1) \longrightarrow H_0^m(\Omega_1)$ is *compact*, since Ω_1 is bounded. Then, by Lemma 8.11.2, the composition $\hookrightarrow \circ E : H^{m+1}(\Omega) \longrightarrow H_0^m(\Omega_1)$ is *compact* from $H^{m+1}(\Omega)$ into $H_0^m(\Omega_1)$, since $E : H^{m+1}(\Omega) \longrightarrow H_0^{m+1}(\Omega_1)$ is continuous. Once again we apply Lemma 8.11.2 to prove that the new composition mapping $\rho \circ (\hookrightarrow \circ E) : H^{m+1}(\Omega) \longrightarrow H^m(\Omega)$ is *compact* from $H^{m+1}(\Omega)$ into $H^m(\Omega)$, since $\hookrightarrow \circ E$ is *compact* from $H^{m+1}(\Omega)$ into $H_0^m(\Omega_1)$ and the restriction mapping $\rho : H_0^m(\Omega_1) \longrightarrow H^m(\Omega)$ is *continuous*. Then, $\forall u \in H^{m+1}(\Omega)$, $(\rho \circ \hookrightarrow \circ E)u = \rho \circ \hookrightarrow (Eu) = \rho(Eu) = \rho((\phi P u)\downarrow_\Omega) = \rho u$ (using (8.11.30)) $= u \in H^m(\Omega)$, i.e. $\rho \circ \hookrightarrow \circ E : u \in H^{m+1}(\Omega) \longmapsto u \in H^m(\Omega)$. Hence, $\rho \circ \hookrightarrow \circ E$ defines the imbedding $H^{m+1}(\Omega) \hookrightarrow H^m(\Omega)$, which is, therefore, *compact*. □

8.11.1 Compact imbedding results in $H^s(\Omega)$, $H_0^s(\Omega)$ and $H_{00}^s(\Omega)$

Now we collect the following general compactness results for any real $s \geq 0$, the proofs for which are *almost* similar to those given in Theorems 8.11.1–8.11.4 for integral order $s = m$.

Theorem 8.11.5. *For* bounded $\Omega \subset \mathbb{R}^n$ *with Lipschitz continuous boundary* Γ *and* $s \geq 0$, *let* $H^s(\Omega)$, $H_0^s(\Omega)$, $H_{00}^s(\Omega)$ *be defined by Definitions 8.10.7, 8.10.8 and 8.10.14, respectively. Let* $0 \leq s_1 < s_2$ *and positive integers* m_1, m_2 *with* $0 \leq m_1 < m_2$. *Then the following* dense,compact imbeddings $\hookrightarrow\hookrightarrow$ *hold:*

I. $H^{s_2}(\Omega) \hookrightarrow\hookrightarrow H^{s_1}(\Omega)$; (8.11.32a)

II. $H_0^{s_2}(\Omega) \hookrightarrow\hookrightarrow H_0^{s_1}(\Omega)$; (8.11.32b)

III. $H_{00}^{m_2+\frac{1}{2}}(\Omega) \hookrightarrow\hookrightarrow H_{00}^{m_1+\frac{1}{2}}(\Omega)$; (8.11.32c)

IV. $H_{00}^{m_2+\frac{1}{2}}(\Omega) \hookrightarrow\hookrightarrow H_0^{s_1}(\Omega)(s_1 < m_2 + \frac{1}{2})$; (8.11.32d)

V. $H_0^{s_2}(\Omega) \hookrightarrow\hookrightarrow H_{00}^{m_1+\frac{1}{2}}(\Omega)\ (m_1 + \frac{1}{2} < s_2)$, (8.11.32e)

the left-hand side space *being* dense *in the* right-hand side space *in all cases.*

Remark 8.11.2. For bounded Ω with *Lipschitz continuous boundary* Γ, there exists an *s-extension operator* from $H^s(\Omega)$ into $H^s(\mathbb{R}^n)$ by Theorem 8.10.9. Then, by (8.10.89), $H^s(\overline{\Omega}) \equiv H^s(\Omega)$ with *norm equivalence* for $s \geq 0$.

Consequently, from I, for $0 \leq s_1 < s_2$,

$$H^{s_2}(\overline{\Omega}) \hookrightarrow\hookrightarrow H^{s_1}(\overline{\Omega}). \tag{8.11.33}$$

For $s_i - \frac{1}{2} \neq$ an integer, $H_{00}^{s_i}(\Omega) \equiv H_0^{s_i}(\Omega)$ by (8.10.118). Hence, for $0 \leq s_1 < s_2$ with $s_i - \frac{1}{2} \neq$ an integer, from II,

$$H_{00}^{s_2}(\Omega) \hookrightarrow\hookrightarrow H_{00}^{s_1}(\Omega). \tag{8.11.34}$$

For $s_i - \frac{1}{2} = m_i \in \mathbb{N}_0$, by Theorem 8.10.18,

$$H_{00}^{m_i+\frac{1}{2}} \subsetneq H_0^{m_i+\frac{1}{2}}(\Omega). \tag{8.11.35}$$

Compactness results in $W^{s,p}$-spaces

We collect compact imbedding results in $W^{s,p}-$ spaces in the following theorem.

Theorem 8.11.6. *For* bounded $\Omega \subset \mathbb{R}^n$ *with Lipschitz continuous boundary* Γ, $s \geq 0$ *and* $1 < p < \infty$, *let* $W^{s,p}(\Omega)$, $W_0^{s,p}(\Omega)$, $W_{00}^{s,p}(\Omega)$ *be defined by Definitions* 8.10.9, 8.10.10 *and* 8.10.15/8.10.16, *respectively. Let* $0 \leq s_1 < s_2$ *and positive integers* m_1, m_2 *with* $0 \leq m_1 < m_2$. *Then, for* fixed p *with* $1 < p < \infty$, $s_1 < m_2 + \frac{1}{p}$ *and* $s_2 > m_1 + \frac{1}{p}$, *the following* compact *imbeddings* $\hookrightarrow\hookrightarrow$ *hold:*

I. $W^{s_2,p}(\Omega) \hookrightarrow\hookrightarrow W^{s_1,p}(\Omega)$;

II. $W_0^{s_2,p} \hookrightarrow\hookrightarrow W_0^{s_1,p}(\Omega)$;

III. $W_{00}^{m_2+\frac{1}{p},p}(\Omega) \hookrightarrow\hookrightarrow W_{00}^{m_1+\frac{1}{p},p}(\Omega)$;

IV. $W_{00}^{m_2+\frac{1}{p},p}(\Omega) \hookrightarrow\hookrightarrow W_0^{s_1,p}(\Omega)$;

V. $W_0^{s_2,p}(\Omega) \hookrightarrow\hookrightarrow W_{00}^{m_1+\frac{1}{p},p}(\Omega)$.

Remark 8.11.3. For $\Omega \subset \mathbb{R}^n$ with *Lipschitz continuous boundary* Γ, the different cases $s_i - \frac{1}{p} \neq$ an integer and $s_i - \frac{1}{p} =$ an integer, $1 < p < \infty$, are to be considered.

For $s_i - \frac{1}{p} \neq$ an integer, $W_{00}^{s_i,p}(\Omega) \equiv W_0^{s_i,p}(\Omega)$ with norm equivalence by Theorem 8.10.20, and consequently

$$W_{00}^{s_2,p}(\Omega) \hookrightarrow\hookrightarrow W_{00}^{s_1,p}(\Omega). \tag{8.11.37}$$

For $s_i = m_i + \frac{1}{p}$, $s_i - \frac{1}{p} = m_i =$ an integer,

$$W_{00}^{m_i+\frac{1}{p},p}(\Omega) \subsetneq W_0^{m_i+\frac{1}{p},p}(\Omega) \tag{8.11.38}$$

by Theorem 8.10.20.

8.12 Sobolev's imbedding results

Imbedding results were given in Section 2.15 and in Section 4.3 (see (4.3.14)–(4.3.19)). Then we gave some of the most important (*continuous*) imbedding results with proofs for the case $\Omega = \mathbb{R}^n$ in Theorems 8.9.4 and 8.9.5, and for $\Omega \subsetneq \mathbb{R}^n$ in Proposition 8.10.3, and *compact* imbedding results in Theorems 8.11.1–8.11.4 (with proofs) and in Theorems 8.11.5–8.11.6 (without proof).

Now we collect here all the *continuous* and *compact* imbedding results in a more general setting, since these results are of fundamental importance in applications. For this we begin with:

Lemma 8.12.1.

I. *Let* $(f_i)_{i=1}^k$ *be k functions on* $\Omega \subset \mathbb{R}^n$ *such that* $f_i \in L^{p_i}(\Omega)$, $1 \le i \le k$, *with* $\frac{1}{p} = \frac{1}{p_1} + \frac{1}{p_2} + \cdots + \frac{1}{p_k} \le 1$, $k \ge 2$. *Then* $f = f_1 \cdot f_2 \cdots f_k$ *belongs to* $L^p(\Omega)$, *with*

$$\|f\|_{L^p(\Omega)} \le \|f_1\|_{L^{p_1}(\Omega)} \cdot \|f_2\|_{L^{p_2}(\Omega)} \cdots \|f_k\|_{L^{p_k}(\Omega)}. \tag{8.12.1}$$

II. *Let* $f \in L^p(\Omega) \cap L^q(\Omega)$ *with* $1 \le p \le q \le \infty$ *and* $\alpha \in [0,1]$ *such that* $p \le r \le q$ *with* $\frac{1}{r} = \frac{\alpha}{p} + \frac{1-\alpha}{q}$ *(i.e.* $\alpha = 0 \Longrightarrow r = q$, $\alpha = 1 \Longrightarrow r = p$ *and* $0 < \alpha < 1 \Longrightarrow p < r < q$*). Then* $f \in L^r(\Omega)$, *with*

$$\|f\|_{L^r(\Omega)} \le \|f\|_{L^p(\Omega)}^{\alpha} \|f\|_{L^q(\Omega)}^{1-\alpha}, \tag{8.12.2}$$

which is called the interpolation inequality.

Proof.

I. For $k = 2$, set $f = f_1 \cdot f_2$ with $f_i \in L^{p_i}(\Omega)$, $\frac{1}{p} = \frac{1}{p_1} + \frac{1}{p_2} \le 1$. Then $|f_i|^p \in L^{p_i/p}(\Omega)$ with $p_i > p$ for $i = 1, 2$, and $|f|^p = |f_1|^p |f_2|^p \in L^1(\Omega)$, since $\frac{p}{p_1} + \frac{p}{p_2} = 1$, i.e. $\frac{p_1}{p}$ and $\frac{p_2}{p}$ are *conjugate indices*. By Hölder's inequality,

$$\begin{aligned}
\|f\|_{L^p(\Omega)}^p &= \int_\Omega |f|^p d\mathbf{x} \\
&= \int_\Omega |f_1|^p \cdot |f_2|^p d\mathbf{x} \le \||f_1|^p\|_{L^{p_1/p}(\Omega)} \||f_2|^p\|_{L^{p_2/p}(\Omega)} \\
&= \left(\int_\Omega (|f_1|^p)^{p_1/p} d\mathbf{x}\right)^{p/p_1} \cdot \left(\int_\Omega (|f_2|^p)^{p_2/p} d\mathbf{x}\right)^{p/p_2} \\
&= \left(\left(\int_\Omega |f_1|^{p_1} d\mathbf{x}\right)^{1/p_1}\right)^p \cdot \left(\left(\int_\Omega |f_2|^{p_2} d\mathbf{x}\right)^{1/p_2}\right)^p \\
&= \|f_1\|_{L^{p_1}(\Omega)}^p \|f_2\|_{L^{p_2}(\Omega)}^p
\end{aligned}$$

$$\Longrightarrow \quad \|f\|_{L^p(\Omega)}^p \le (\|f_1\|_{L^{p_1}(\Omega)} \|f_2\|_{L^{p_2}(\Omega)})^p$$

$$\Longrightarrow \quad \|f\|_{L^p(\Omega)} \le \|f_1\|_{L^{p_1}(\Omega)} \|f_2\|_{L^{p_2}(\Omega)}. \tag{8.12.3}$$

By the method of induction, we assume that the result holds for k, i.e. (8.12.1) holds for k. Then we are to show that it holds for $k+1$, i.e. for $f_i \in L^{p_i}$ with $1 \le i \le k+1$,

$$\frac{1}{r} = \underbrace{\frac{1}{p_1} + \cdots + \frac{1}{p_k}}_{1/p} + \frac{1}{p_{k+1}} \le 1 \text{ and } f = \underbrace{f_1 \cdot f_2 \cdots f_k}_{F_k} \cdot f_{k+1} = F_k \cdot f_{k+1}$$

with $\frac{1}{r} = \frac{1}{p} + \frac{1}{p_{k+1}}$, $F_k \in L^p(\Omega)$,

$$\|F_k\|_{L^p(\Omega)} \le \|f_1\|_{L^{p_1}(\Omega)} \|f_2\|_{L^{p_2}(\Omega)} \cdots \|f_k\|_{L^{p_k}(\Omega)} \tag{8.12.4}$$

(since the result holds for k by assumption) and $f_{k+1} \in L^{p_{k+1}}(\Omega)$. Then, by (8.12.3), $f = F_k \cdot f_{k+1} \in L^r(\Omega)$ with

$$\begin{aligned} \|f\|_{L^r(\Omega)} &\le \|F_k\|_{L^p(\Omega)} \|f_{k+1}\|_{L^{p_{k+1}}(\Omega)} \\ &\le \|f_1\|_{L^{p_1}(\Omega)} \cdots \|f_k\|_{L^{p_k}(\Omega)} \cdot \|f_{k+1}\|_{L^{p_{k+1}}(\Omega)} \end{aligned}$$

(by (8.12.4)). Hence, the result also holds for $k+1$, and I is proved.

II. $\alpha = 0$ and $\alpha = 1$ are trivial cases. Hence, we prove for $0 < \alpha < 1$. $f \in L^p(\Omega) \cap L^q(\Omega) \Longrightarrow |f|^\alpha \in L^{p/\alpha}(\Omega)$, $|f|^{1-\alpha} \in L^{q/(1-\alpha)}(\Omega)$, $p/\alpha > 1$ and $q/(1-\alpha) > 1$. Define $f_1 = |f|^\alpha$, $f_2 = |f|^{1-\alpha}$, $p_1 = p/\alpha$ and $p_2 = q/(1-\alpha)$. Then $f_1 \in L^{p_1}(\Omega)$, $f_2 \in L^{p_2}(\Omega)$ with $\frac{1}{p_1} + \frac{1}{p_2} = \frac{\alpha}{p} + \frac{1-\alpha}{q} = \frac{1}{r}$ (by hypothesis). Hence, from (8.12.1), $f = f_1 \cdot f_2 \in L^r(\Omega)$, and

$$\|f\|_{L^r(\Omega)} \le \||f|^\alpha\|_{L^{p/\alpha}(\Omega)} \cdot \||f|^{1-\alpha}\|_{L^{q/(1-\alpha)}(\Omega)} = \|f\|^\alpha_{L^p(\Omega)} \|f\|^{1-\alpha}_{L^q(\Omega)},$$

since, for example,

$$\begin{aligned} \||f|^\alpha\|_{L^{p/\alpha}(\Omega)} &= \left(\int_\Omega (|f|^\alpha)^{p/\alpha} d\mathbf{x}\right)^{\alpha/p} \\ &= \left(\left(\int_\Omega |f|^p d\mathbf{x}\right)^{1/p}\right)^\alpha = \|f\|^\alpha_{L^p(\Omega)}. \end{aligned}$$ □

Lemma 8.12.2. *For $n \ge 2$, let $\hat{\mathbf{x}}_i = (x_1, \ldots, x_{i-1}, x_{i+1}, \ldots, x_n) \in \mathbb{R}^{n-1}$ $\forall \mathbf{x} = (x_1, \ldots, x_i, \ldots, x_n) \in \mathbb{R}^n$, $f_i : \hat{\mathbf{x}}_i \in \mathbb{R}^{n-1} \longmapsto f_i(\hat{\mathbf{x}}_i) \in \mathbb{R}$ belong to $L^{n-1}(\mathbb{R}^{n-1})$ for $1 \le i \le n$. Then the function f defined by $f(\mathbf{x}) = f_1(\hat{\mathbf{x}}_1) f_2(\hat{\mathbf{x}}_2) \cdots f_n(\hat{\mathbf{x}}_n)$ $\forall \mathbf{x} \in \mathbb{R}^n$ belongs to $L^1(\mathbb{R}^n)$, with*

$$\|f\|_{L^1(\mathbb{R}^n)} \le \|f_1\|_{L^{n-1}(\mathbb{R}^{n-1})} \cdots \|f_n\|_{L^{n-1}(\mathbb{R}^{n-1})} = \prod_{i=1}^{n-1} \|f_i\|_{L^{n-1}(\mathbb{R}^{n-1})}. \tag{8.12.5}$$

Proof. For $n = 2$, $|f(x_1, x_2)| = |f_1(x_2)| \cdot |f_2(x_1)|$ with $f_i \in L^1(\mathbb{R})$. Hence, applying Fubini's Theorem, we get the result. In fact,

$$\begin{aligned} \|f\|_{L^1(\mathbb{R}^2)} &= \int_{\mathbb{R}^2} |f| dx_1 dx_2 = \int_{\mathbb{R}^2} |f_1(x_2)| \cdot |f_2(x_1)| dx_1 dx_2 \\ &= \int_{\mathbb{R}} |f_1(x_2)| dx_2 \int_{\mathbb{R}} |f_2(x_1)| dx_1 = \|f_1\|_{L^1(\mathbb{R})} \|f_2\|_{L^1(\mathbb{R})}. \end{aligned}$$

For $n = 3$, $|f(x_1, x_2, x_3)| = |f_1(x_2, x_3)| \cdot |f_2(x_1, x_3)| \cdot |f_3(x_1, x_2)|$ with $f_i \in L^2(\mathbb{R}^2)$. Then

$$\int_{\mathbb{R}} |f(x_1, x_2, x_3)| dx_3 = |f_3(x_1, x_2)| \int_{\mathbb{R}} |f_1(x_2, x_3)| \cdot |f_2(x_1, x_3)| dx_3$$
$$\leq |f_3(x_1, x_2)| \cdot \|f_1(x_2, \cdot)\|_{L^2(\mathbb{R})} \cdot \|f_2(x_1, \cdot)\|_{L^2(\mathbb{R})}$$

(by the Cauchy–Schwarz inequality in $L^2(\mathbb{R})$)

$$\implies \int_{\mathbb{R}^3} |f(\mathbf{x})| dx_1 dx_2 dx_3$$
$$\leq \int_{\mathbb{R}^2} \|f_1(x_2, \cdot)\|_{L^2(\mathbb{R})} \cdot \|f_2(x_1, \cdot)\|_{L^2(\mathbb{R})} \cdot |f_3(x_1, x_2)| dx_1 dx_2$$
$$\leq \|f_1\|_{L^2(\mathbb{R}^2)} \cdot \|f_2\|_{L^2(\mathbb{R}^2)} \cdot \|f_3\|_{L^2(\mathbb{R}^2)}$$

(again by the Cauchy–Schwarz inequality in $L^2(\mathbb{R}^2)$), since

$$\left(\int_{\mathbb{R}^2} \|f_1(x_2, \cdot)\|_{L^2(\mathbb{R})} \cdot \|f_2(x_1, \cdot)\|_{L^2(\mathbb{R})} dx_1 dx_2 \right)^{1/2}$$
$$= \left(\int_{\mathbb{R}^2} \left[\left(\int_{\mathbb{R}} |f_1(x_2, x_3)|^2 dx_3 \right)^{1/2} \cdot \left(\int_{\mathbb{R}} |f_2(x_1, x_3)|^2 dx_3 \right)^{1/2} \right]^2 dx_1 dx_2 \right)^{1/2}$$
$$= \left(\int_{\mathbb{R}} \int_{\mathbb{R}} |f_1(x_2, x_3)|^2 dx_2 dx_3 \cdot \int_{\mathbb{R}} \int_{\mathbb{R}} |f_2(x_1, x_3)|^2 dx_1 dx_3 \right)^{1/2}$$
$$= (\|f_1\|^2_{L^2(\mathbb{R}^2)} \cdot \|f_2\|^2_{L^2(\mathbb{R}^2)})^{1/2} = \|f_1\|_{L^2(\mathbb{R}^2)} \cdot \|f_2\|_{L^2(\mathbb{R}^2)}$$

$\implies \|f\|_{L^1(\mathbb{R}^3)} \leq \|f_1\|_{L^2(\mathbb{R}^2)} \cdot \|f_2\|_{L^2(\mathbb{R}^2)}$.

The general result for arbitrary $n \geq 4$ can be proved by the method of induction. □

Now we begin with the principal result for integral $m \in \mathbb{N}$, for which we will follow Brezis [26]:

Theorem 8.12.1.

I. *For* $1 \leq p < n$, $W^{1,p}(\mathbb{R}^n) \hookrightarrow L^{p^*}(\mathbb{R}^n)$ *with* $\frac{1}{p^*} = \frac{1}{p} - \frac{1}{n}$ *such that* $\exists C = C(p, n) > 0$ *such that* $\|u\|_{L^{p^*}(\mathbb{R}^n)} \leq C\|\nabla u\|_{(L^p(\mathbb{R}^n))^n}$ *(see* (8.12.6f) *and* (8.12.6g)*)* $\forall u \in W^{1,p}(\mathbb{R}^n)$. (8.12.6a)

II. *For* $1 \leq p < n$, $W^{1,p}(\mathbb{R}^n) \hookrightarrow L^r(\mathbb{R}^n)$ $\forall r \in [p, p^*]$ *with* p^* *defined in* I,

$$\|u\|_{L^r(\mathbb{R}^n)} \leq C\|u\|_{1,p,\mathbb{R}^n} \quad \forall u \in W^{1,p}(\mathbb{R}^n). \tag{8.12.6b}$$

III. Limit case $p = n$: *For* $p = n$, $W^{1,n}(\mathbb{R}^n) \hookrightarrow L^r(\mathbb{R}^n)$ $\forall r$ *with* $n \leq r < \infty$,

$$\|u\|_{L^r(\mathbb{R}^n)} \leq C\|u\|_{1,n,\mathbb{R}^n} \ \forall u \in W^{1,n}(\mathbb{R}^n). \tag{8.12.6c}$$

IV. *For $p > n$, $W^{1,p}(\mathbb{R}^n) \hookrightarrow L^\infty(\mathbb{R}^n)$ with*

$$\|u\|_{L^\infty(\mathbb{R}^n)} \le C\|u\|_{1,p,\mathbb{R}^n} \quad \forall u \in W^{1,p}(\mathbb{R}^n). \tag{8.12.6d}$$

Moreover, for $\sigma = 1 - \frac{n}{p}$ with $\sigma \in]0,1[$ and $\forall u \in W^{1,p}(\mathbb{R}^n)$, we have

$$|u(\mathbf{x}) - u(\mathbf{y})| \le C\|\mathbf{x}-\mathbf{y}\|^{\sigma}_{\mathbb{R}^n}\|\nabla u\|_{(L^p(\mathbb{R}^n))^n} \tag{8.12.6e}$$

for almost all $\mathbf{x}, \mathbf{y} \in \mathbb{R}^n$.

Remark 8.12.1. The same constant $C = C(p,n) > 0$ has been used in all the inequalities (8.12.6a)–(8.12.6e), and will be used in all the following inequalities for the sake of convenience to denote *different* values of the constant C in *different* inequalities in *different* steps of a proof;

$$\|\nabla u\|_{(L^p(\mathbb{R}^n))^n} = \sum_{i=1}^{n}\left\|\frac{\partial u}{\partial x_i}\right\|_{L^p(\mathbb{R}^n)} \le \|u\|_{1,p,\mathbb{R}^n} = \|u\|_{W^{1,p}(\mathbb{R}^n)}$$

with

$$\|u\|_{1,p,\mathbb{R}^n} = \|u\|_{L^p(\mathbb{R}^n)} + \|\nabla u\|_{(L^p(\mathbb{R}^n))^n}, \tag{8.12.6f}$$

or, *equivalently*,

$$\|\nabla u\|_{(L^p(\mathbb{R}^n))^n} = \left(\sum_{i=1}^{n}\left\|\frac{\partial u}{\partial x_i}\right\|^p_{L^p(\mathbb{R}^n)}\right)^{1/p} \le \|u\|_{1,p,\mathbb{R}^n} = \|u\|_{W^{1,p}(\mathbb{R}^n)}$$

with

$$\|u\|_{1,p,\mathbb{R}^n} = (\|u\|^p_{L^p(\mathbb{R}^n)} + \|\nabla u\|^p_{L^p(\mathbb{R}^n)})^{1/p}. \tag{8.12.6g}$$

Hence, we will use interchangeably either of the two *equivalent* norms in (8.12.6f) and (8.12.6g), and denote these two by the *same* notation $\|u\|_{1,p,\mathbb{R}^n}$. Similarly, we will use $\|\nabla u\|_{(L^p(\mathbb{R}^n))^n}$ for *either* of the two *equivalent* formulae in (8.12.6f) and (8.12.6g).

Remark 8.12.2 (on inequality (8.12.6e)). Let $A \subset \mathbb{R}^n$ denote a set of points with measure $\mu(A) = 0$ where the inequality (8.12.6e) does *not* hold, and it holds for *all* $\mathbf{x}, \mathbf{y} \in \mathbb{R}^n \setminus A$. Hence, the restriction $u{\downarrow_{\mathbb{R}^n\setminus A}}$ of u to $\mathbb{R}^n \setminus A$ is defined $\forall \mathbf{x}, \mathbf{y} \in \mathbb{R}^n \setminus A$, and $\mathbb{R}^n \setminus A$ is *dense* in $\mathbb{R}^n$. Moreover, $\forall \mathbf{x}, \mathbf{y} \in \mathbb{R}^n \setminus A$, $\forall \varepsilon > 0$, $\exists \delta = \delta(\varepsilon) = (\frac{\varepsilon}{C\|\nabla u\|_{(L^p(\mathbb{R}^n))^n}})^{1/\sigma} > 0$ such that $\|\mathbf{x}-\mathbf{y}\| < \delta \Longrightarrow |u(\mathbf{x}) - u(\mathbf{y})| < \varepsilon \Longrightarrow u{\downarrow_{\mathbb{R}^n\setminus A}}$ is *continuous* on $\mathbb{R}^n \setminus A$. By virtue of the density of $\mathbb{R}^n \setminus A$ in $\mathbb{R}^n$, $u{\downarrow_{\mathbb{R}^n\setminus A}}$ has a *unique, continuous* extension $\tilde{u}$ to $\mathbb{R}^n$ such that $u(\mathbf{x}) = \tilde{u}{\downarrow_{\mathbb{R}^n\setminus A}}(\mathbf{x})$ $\forall \mathbf{x} \in \mathbb{R}^n \setminus A$, and $u = \tilde{u}$ a.e. in $\mathbb{R}^n$. Thus, in every *equivalence* class $[u]$ of functions $u \in W^{1,p}(\mathbb{R}^n)$

with $p > n$, $\exists$ a *continuous* function $\tilde{u} \in C^0(\mathbb{R}^n)$ such that $u = \tilde{u}$ a.e. in $\mathbb{R}^n$. Hence, we agree to write

$$W^{1,p}(\mathbb{R}^n) \hookrightarrow C^0(\mathbb{R}^n) \quad \text{for } p > n, \tag{8.12.6h}$$

where $C^0(\mathbb{R}^n)$ is equipped with

$$\|u\|_{C^0(\mathbb{R}^n)} = \|u\|_{L^\infty(\mathbb{R}^n)} = \sup_{\mathbf{x}\in\mathbb{R}^n} |u(\mathbf{x})| \quad \forall u \in C^0(\mathbb{R}^n). \tag{8.12.6i}$$

Moreover, $u \in C^0(\mathbb{R}^n)$ satisfying (8.12.6e) with a possible change of values on a set of points with measure zero will belong to $C^{0,\sigma}(\mathbb{R}^n)$ with $\sigma \in]0,1[$. Hence, for $\sigma = 1 - \frac{n}{p} > 0$ with $p > n$, we agree to write

$$W^{1,p}(\mathbb{R}^n) \hookrightarrow C^{0,\sigma}(\mathbb{R}^n) \hookrightarrow C^0(\mathbb{R}^n)$$

with

$$\|u\|_{C^0(\mathbb{R}^n)} \le \|u\|_{C^{0,\sigma}(\mathbb{R}^n)} \le C\|u\|_{W^{1,p}(\mathbb{R}^n)}. \tag{8.12.6j}$$

Proof of Theorem 8.12.1.

I. We prove the inequality (8.12.6a) for the *dense* subspace $\mathcal{D}(\mathbb{R}^n)$ of $W^{1,p}(\mathbb{R}^n)$ (see Theorem 6.8.9), $1 \le p < \infty$.

Case $p = 1$: Let $u \in \mathcal{D}(\mathbb{R}^n) \subset W^{1,1}(\mathbb{R}^n)$. Then, $\forall i = 1, 2, \dots, n$,

$$\begin{aligned}
&|u(x_1,\dots,x_{i-1},x_i,x_{i+1},\dots,x_n)| \\
&= \left| \int_{-\infty}^{x_i} \frac{\partial u}{\partial x_i}(x_1,\dots,x_{i-1},\tau,x_{i+1},\dots,x_n)d\tau \right| \\
&\le \int_{-\infty}^{\infty} \left| \frac{\partial u}{\partial x_i}(x_1,\dots,x_{i-1},\tau,x_{i+1},\dots,x_n)\right| d\tau = f_i(\hat{\boldsymbol{x}}_i),
\end{aligned} \tag{8.12.7}$$

say, with $\hat{\boldsymbol{x}}_i = (x_1,\dots,x_{i-1},x_{i+1},\dots,x_n)$. Then, for $u \in \mathcal{D}(\mathbb{R}^n)$, $|u(\mathbf{x})|^n \le \prod_{i=1}^n f_i(\hat{\boldsymbol{x}}_i) \Longrightarrow |u(\mathbf{x})|^{n/(n-1)} \le \prod_{i=1}^n |f_i|^{1/(n-1)}$ with $|f_i|^{1/(n-1)} \in L^{n-1}(\mathbb{R}^{n-1})$. By Lemma 8.12.2, we have

$$\begin{aligned}
&\| |u|^{n/(n-1)} \|_{L^1(\mathbb{R}^n)} \le \prod_{i=1}^n \| |f_i|^{1/(n-1)} \|_{L^{n-1}(\mathbb{R}^{n-1})} \\
&= \prod_{i=1}^n \Big(\int_{\mathbb{R}^{n-1}} |f_i| d\hat{\boldsymbol{x}}_i\Big)^{1/(n-1)} \\
&= \prod_{i=1}^n \left(\int_{\mathbb{R}^{n-1}} \left(\int_{-\infty}^{\infty} \left| \frac{\partial u}{\partial x_i}(x_1,\dots,x_{i-1},\tau,x_{i+1},\dots,x_n) \right| d\tau \right) d\hat{\boldsymbol{x}}_i \right)^{1/(n-1)}
\end{aligned}$$

$$= \prod_{i=1}^{n} \left(\int_{\mathbb{R}^n} \left| \frac{\partial u}{\partial x_i} \right| d\mathbf{x} \right)^{1/(n-1)} \quad \text{(by Fubini's Theorem 7.1.2C)}$$

$$= \prod_{i=1}^{n} \left\| \frac{\partial u}{\partial x_i} \right\|_{L^1(\mathbb{R}^n)}^{1/(n-1)}$$

$$\implies \quad \|u\|_{L^{n/(n-1)}(\mathbb{R}^n)}^{n/(n-1)} \leq \left(\prod_{i=1}^{n} \left\| \frac{\partial u}{\partial x_i} \right\|_{L^1(\mathbb{R}^n)} \right)^{1/(n-1)}$$

$$\implies \quad \|u\|_{L^{p^*}(\mathbb{R}^n)} \leq \prod_{i=1}^{n} \left\| \frac{\partial u}{\partial x_i} \right\|_{L^1(\mathbb{R}^n)}^{1/n} \quad \forall u \in \mathcal{D}(\mathbb{R}^n), \tag{8.12.8}$$

since $p = 1 \implies \frac{1}{p^*} = 1 - \frac{1}{n} \implies p^* = \frac{n}{n-1}$.

Case $p \geq 1$: Let $t \geq 1$. $\forall u \in \mathcal{D}(\mathbb{R}^n)$ choose $u_t = |u|^{t-1} \cdot u$, for which the inequality (8.12.8) holds (the precise value of such a $t \geq 1$ will be chosen soon), i.e. $\|u_t\|_{L^{n/(n-1)}(\mathbb{R}^n)} \leq \prod_{i=1}^{n} \|\frac{\partial u_t}{\partial x_i}\|_{L^1(\mathbb{R}^n)}^{1/n}$ for $u \in \mathcal{D}(\mathbb{R}^n)$.

$$\||u_t|\|_{L^{n/(n-1)}(\mathbb{R}^n)} = \left(\int_{\mathbb{R}^n} \left(|u|^t\right)^{n/(n-1)} d\mathbf{x} \right)^{\frac{(n-1)}{n}}$$

$$= \left(\int_{\mathbb{R}^n} |u|^{tn/(n-1)} d\mathbf{x} \right)^{\frac{(n-1)}{tn} \cdot t} = \|u\|_{L^{tn/(n-1)}}^{t}. \tag{8.12.9}$$

The absolute value of the distributional derivative $|\frac{\partial u_t}{\partial x_i}| = t|u|^{t-1}|\frac{\partial u}{\partial x_i}| \in C^0(\mathbb{R}^n)$ for $u \in \mathcal{D}(\mathbb{R}^n)$, $1 \leq i \leq n$. Hence,

$$\left\| \frac{\partial u_t}{\partial x_i} \right\|_{L^1(\mathbb{R}^n)}^{1/n} = \left(\int_{\mathbb{R}^n} t|u|^{t-1} \left| \frac{\partial u}{\partial x_i} \right| d\mathbf{x} \right)^{1/n} = t^{1/n} \left\| |u|^{t-1} \cdot |\frac{\partial u}{\partial x_i}| \right\|_{L^1(\mathbb{R}^n)}^{1/n}$$

$$\implies \quad \prod_{i=1}^{n} \left\| \frac{\partial u_t}{\partial x_i} \right\|_{L^1(\mathbb{R}^n)}^{1/n} = t \prod_{i=1}^{n} \left\| |u|^{t-1} \cdot \left| \frac{\partial u}{\partial x_i} \right| \right\|_{L^1(\mathbb{R}^n)}^{1/n}$$

$$= t \prod_{i=1}^{n} \left(\int_{\mathbb{R}^n} \underbrace{|u|^{t-1}}_{L^q(\mathbb{R}^n)} \cdot \underbrace{\left| \frac{\partial u}{\partial x_i} \right|}_{L^p(\mathbb{R}^n)} \right)^{1/n}$$

$$\leq t \prod_{i=1}^{n} \left(\||u|^{t-1}\|_{L^q(\mathbb{R}^n)} \cdot \left\| \frac{\partial u}{\partial x_i} \right\|_{L^p(\mathbb{R}^n)} \right)^{1/n}$$

$$\text{(by Hölder's inequality with } \frac{1}{p} + \frac{1}{q} = 1)$$

$$= t \prod_{i=1}^{n} \left(\|u\|_{L^{q(t-1)}(\mathbb{R}^n)}^{(t-1)/n} \cdot \left\| \frac{\partial u}{\partial x_i} \right\|_{L^p(\mathbb{R}^n)}^{1/n} \right)$$

$$= t\|u\|^{t-1}_{L^{q(t-1)}(\mathbb{R}^n)} \prod_{i=1}^{n} \left\| \frac{\partial u}{\partial x_i} \right\|^{1/n}_{L^p(\mathbb{R}^n)} \tag{8.12.10}$$

$$(\text{since } \underbrace{\|u\|^{(t-1)/n} \cdot \|u\|^{(t-1)/n} \cdots \|u\|^{(t-1)/n}}_{n \text{ times}} = \|u\|^{t-1}).$$

Then, from (8.12.9) and (8.12.10), we have

$$\|u\|^t_{L^{tn/(n-1)}(\mathbb{R}^n)} \leq t\|u\|^{t-1}_{L^{q(t-1)}(\mathbb{R}^n)} \prod_{i=1}^{n} \left\| \frac{\partial u}{\partial x_i} \right\|^{1/n}_{L^p(\mathbb{R}^n)}. \tag{8.12.11}$$

In order to cancel $\|u\|^{t-1}_{L^{q(t-1)}(\mathbb{R}^n)}$, we *choose* $\frac{tn}{n-1} = q(t-1)$ with $q = \frac{p}{p-1}$, from which $t[\frac{n}{n-1} - \frac{p}{p-1}] = \frac{-p}{p-1} \Longrightarrow t = \frac{n-1}{n} p^* \geq p \geq 1$.
In fact, with $\frac{tn}{n-1} = p^*$, $\frac{1}{p^*} = \frac{1}{p} - \frac{1}{n} \Longrightarrow p^* = \frac{np}{n-p} \Longrightarrow \frac{(n-p)}{n} p^* = p \Longrightarrow t = \frac{n-1}{n} p^* \geq \frac{n-p}{n} \cdot p^* = p \geq 1$. Then we have:

$$\|u\|_{L^{p^*}(\mathbb{R}^n)} \leq t \prod_{i=1}^{n} \left\| \frac{\partial u}{\partial x_i} \right\|^{1/n}_{L^p(\mathbb{R}^n)} \leq C \prod_{i=1}^{n} \left\| \frac{\partial u}{\partial x_i} \right\|^{1/n}_{L^p(\mathbb{R}^n)}$$

$$\leq C \underbrace{\left(\sum_{i=1}^{n} \left\| \frac{\partial u}{\partial x_i} \right\|_{L^p(\mathbb{R}^n)} \right)^{1/n} \cdots \left(\sum_{i=1}^{n} \left\| \frac{\partial u}{\partial x_i} \right\|_{L^p(\mathbb{R}^n)} \right)^{1/n}}_{n \text{ times}}$$

$$\Longrightarrow \quad \|u\|_{L^{p^*}(\mathbb{R}^n)} \leq C \sum_{i=1}^{n} \left\| \frac{\partial u}{\partial x_i} \right\|_{L^p(\mathbb{R}^n)} = C\|\nabla u\|_{(L^p(\mathbb{R}^n))^n} \quad \forall u \in \mathcal{D}(\mathbb{R}^n) \tag{8.12.12}$$

(by (8.12.6f)).
Let $u \in W^{1,p}(\mathbb{R}^n)$. Then, $\exists$ a sequence $(u_k)_{k\in\mathbb{N}}$ in $\mathcal{D}(\mathbb{R}^n)$ such that $u_k \to u$ in $W^{1,p}(\mathbb{R}^n)$ and we can also suppose (by extracting a subsequence, if necessary [26]) that $u_k \to u$ pointwise a.e. in $\mathbb{R}^n$.
Hence, $\lim_{k\to\infty} \|u - u_k\|_{L^p(\mathbb{R}^n)} = 0$, i.e. $\lim_{k\to\infty} \|u_k\|_{L^p(\mathbb{R}^n)} = \|u\|_{L^p(\mathbb{R}^n)}$ and $\lim_{k\to\infty} \|\nabla u - \nabla u_k\|_{(L^p(\mathbb{R}^n))^n} = 0$

$$\Longrightarrow \quad \lim_{k\to\infty} \|\nabla u_k\|_{(L^p(\mathbb{R}^n))^n} = \|\nabla u\|_{(L^p(\mathbb{R}^n))^n}. \tag{8.12.13}$$

But $\forall k \in \mathbb{N}$, $\|u_k\|_{L^{p^*}(\mathbb{R}^n)} \leq C\|\nabla u_k\|_{(L^p(\mathbb{R}^n))^n}$ by (8.12.12). Hence, $\|u_k - u_l\|_{L^{p^*}(\mathbb{R}^n)} \leq C\|\nabla u_k - \nabla u_l\|_{(L^p(\mathbb{R}^n))^n} \to 0$ as $k, l \to \infty \Longrightarrow (u_k)_{k\in\mathbb{N}}$ is a Cauchy sequence in $L^{p^*}(\mathbb{R}^n)$, which is a Banach space. Thus, $\exists v \in L^{p^*}(\mathbb{R}^n)$ such that $\|v - u_k\|_{L^{p^*}(\mathbb{R}^n)} \to 0$ as $k \to \infty$. Hence, $\exists$ a subsequence $(u_{k_l})_{l=1}^{\infty}$ of $(u_k)_{k\in\mathbb{N}}$ such that $u_{k_l} \to v$ pointwise a.e. in $\mathbb{R}^n$. But $u_k \to u$ pointwise a.e.

in $\mathbb{R}^n$ as $k \to \infty \Longrightarrow u_{k_l} \to u$ pointwise a.e. in $\mathbb{R}^n$ as $l \to \infty \Longrightarrow v = u$ a.e. in $\mathbb{R}^n$ by virtue of the uniqueness of the limit. Hence, $u_k \to u$ in $L^{p^*}(\mathbb{R}^n)$

$$\begin{aligned} \Longrightarrow \quad \|u\|_{L^{p^*}(\mathbb{R}^n)} &= \lim_{k\to\infty} \|u_k\|_{L^{p^*}(\mathbb{R}^n)} \\ &\le C \lim_{k\to\infty} \|\nabla u_k\|_{(L^p(\mathbb{R}^n))^n} \\ &= C\|\nabla u\|_{(L^p(\mathbb{R}^n))^n} \quad \text{(by (8.12.13))}. \end{aligned} \tag{8.12.14}$$

II. For $1 \le p < n$, let $p \le r \le p^*$ with $\frac{1}{r} = \frac{\alpha}{p} + \frac{1-\alpha}{p^*}$ $(0 \le \alpha \le 1)$, $\frac{1}{p^*} = \frac{1}{p} - \frac{1}{n}$. Then, using (8.12.2), we have

$$\begin{aligned} \|u\|_{L^r(\mathbb{R}^n)} &\le \|u\|^{\alpha}_{L^p(\mathbb{R}^n)} \|u\|^{1-\alpha}_{L^{p^*}(\mathbb{R}^n)} \\ &\le \|u\|_{L^p(\mathbb{R}^n)} + \|u\|_{L^{p^*}(\mathbb{R}^n)} \quad \text{(by Young's inequality)} \\ &\le \|u\|_{L^p(\mathbb{R}^n)} + C\|\nabla u\|_{(L^p(\mathbb{R}^n))^n} \quad \text{(by (8.12.6a))} \\ &\le C\|u\|_{W^{1,p}(\mathbb{R}^n)}. \end{aligned}$$

(Young's inequality: $ab \le \frac{1}{p}a^p + \frac{1}{q}b^q$ with $\frac{1}{p} + \frac{1}{q} = 1$ is used with $a = \|u\|^{\alpha}_{L^p(\mathbb{R}^n)}$, $b = |u\|^{1-\alpha}_{L^{p^*}(\mathbb{R}^n)}$, $\frac{1}{p} = \alpha$, $\frac{1}{q} = 1-\alpha$ such that $\frac{1}{p} + \frac{1}{q} = \alpha + 1 - \alpha = 1$. Then

$$\begin{aligned} \|u\|^{\alpha}_{L^p(\mathbb{R}^n)} \|u\|^{1-\alpha}_{L^{p^*}(\mathbb{R}^n)} &\le \alpha(\|u\|^{\alpha}_{L^p(\mathbb{R}^n)})^{1/\alpha} + (1-\alpha)(\|u\|^{1-\alpha}_{L^{p^*}(\mathbb{R}^n)})^{\frac{1}{1-\alpha}} \\ &= \alpha\|u\|_{L^p(\mathbb{R}^n)} + (1-\alpha)\|u\|_{L^{p^*}(\mathbb{R}^n)} \\ &\le \|u\|_{L^p(\mathbb{R}^n)} + \|u\|_{L^{p^*}(\mathbb{R}^n)}.) \end{aligned}$$

III. Let $u \in \mathcal{D}(\mathbb{R}^n)$. Then, for $p = n$, $\frac{1}{q} = 1 - \frac{1}{n} = \frac{n-1}{n} \Longrightarrow q(t-1) = \frac{(t-1)n}{n-1}$ with $t \ge 1$ and, from (8.12.11), we have, for $t \ge 1$, $\|u\|^t_{L^{tn/(n-1)}(\mathbb{R}^n)} \le t\|u\|^{t-1}_{L^{(t-1)n/(n-1)}(\mathbb{R}^n)} \|\nabla u\|_{(L^n(\mathbb{R}^n))^n}$, since

$$\prod_{i=1}^{n} \Big\|\frac{\partial u}{\partial x_i}\Big\|^{1/n}_{L^n(\mathbb{R}^n)} \le \sum_{i=1}^{n} \Big\|\frac{\partial u}{\partial x_i}\Big\|_{L^n(\mathbb{R}^n)} \quad \text{(see (8.12.12))}$$

$\Longrightarrow \|u\|_{L^{tn/(n-1)}(\mathbb{R}^n)} \le t^{1/t} \|u\|^{(t-1)/t}_{L^{(t-1)n/(n-1)}(\mathbb{R}^n)} \|\nabla u\|^{1/t}_{(L^n(\mathbb{R}^n))^n}$ $\forall u \in \mathcal{D}(\mathbb{R}^n)$, $\forall t \ge 1$. Applying Young's inequality with $a = \|u\|^{(t-1)/t}_{L^{(t-1)n/(n-1)}(\mathbb{R}^n)}$, $b = \|\nabla u\|^{1/t}_{(L^n(\mathbb{R}^n))^n}$, $p_1 = \frac{t}{t-1}$, $q_1 = t$ such that $\frac{1}{p_1} + \frac{1}{q_1} = 1$,

$$\|u\|_{L^{tn/(n-1)}(\mathbb{R}^n)} \le t^{1/t}\left[\frac{t-1}{t}\|u\|_{L^{(t-1)n/(n-1)}(\mathbb{R}^n)} + \frac{1}{t}\|\nabla u\|_{(L^n(\mathbb{R}^n))^n}\right]$$

$$\Longrightarrow \quad \|u\|_{L^{tn/(n-1)}(\mathbb{R}^n)} \le C[\|u\|_{L^{(t-1)n/(n-1)}(\mathbb{R}^n)} + \|\nabla u\|_{(L^n(\mathbb{R}^n))^n}].$$

Then, for $t = n$,

$$\|u\|_{L^{n^2/(n-1)}(\mathbb{R}^n)} \le C[\|u\|_{L^n(\mathbb{R}^n)} + \|\nabla u\|_{(L^n(\mathbb{R}^n))^n}] \\ \le C\|u\|_{W^{1,n}(\mathbb{R}^n)} \quad \forall u \in \mathcal{D}(\mathbb{R}^n),$$

and for $n \le r \le \frac{n^2}{n-1}$, $\|u\|_{L^r(\mathbb{R}^n)} \le \|u\|^{\alpha}_{L^n(\mathbb{R}^n)}\|u\|^{1-\alpha}_{L^{n^2/(n-1)}(\mathbb{R}^n)}$ for $0 \le \alpha \le 1$ by (8.12.2). Hence, $\|u\|_{L^r(\mathbb{R}^n)} \le \|u\|_{L^{n2(n-1)}(\mathbb{R}^n)} \le C\|u\|_{W^{1,n}(\mathbb{R}^n)}$ for $\alpha = 0$, $\forall u \in \mathcal{D}(\mathbb{R}^n)$. Setting $t = n+1, n+2, \ldots$ and repeating the steps done for $t = n$, we have $\|u\|_{L^r(\mathbb{R}^n)} \le C\|u\|_{W^{1,n}(\mathbb{R}^n)}$ $\forall u \in \mathcal{D}(\mathbb{R}^n)$, $\forall r$ with $r \in [n, \infty[$. Then the result (8.12.6c) follows from the density of $\mathcal{D}(\mathbb{R}^n)$ in $W^{1,n}(\mathbb{R}^n)$ (by Theorem 6.8.9).

IV. Let $\hat{Q}$ be an open cube with the origin $\mathbf{0} \in \hat{Q}$ such that its edges are parallel to the coordinate axes and have length $r > 0$. Let $u \in \mathcal{D}(\mathbb{R}^n)$. Then, for $\mathbf{x} \in \hat{Q}$, we have $\int_0^1 \frac{du}{dt}(t\mathbf{x})dt = u(t\mathbf{x})|_{t=0}^{t=1} = u(\mathbf{x}) - u(\mathbf{0})$. But $\frac{d}{dt}u(tx_1, tx_2, \ldots, tx_n) = \sum_{i=1}^n x_i \frac{\partial}{\partial x_i}u(t\mathbf{x})$. Hence,

$$|u(\mathbf{x}) - u(\mathbf{0})| = \left|\int_0^1 \frac{du}{dt}(t\mathbf{x})dt\right| \le \int_0^1 \sum_{i=1}^n |x_i|\left|\frac{\partial}{\partial x_i}u(t\mathbf{x})\right|dt \\ \le r\sum_{i=1}^n \int_0^1 \left|\frac{\partial}{\partial x_i}u(t\mathbf{x})\right|dt. \tag{8.12.15}$$

Let $\bar{u} \in \mathbb{R}$ denote the *mean* of u on $\hat{Q}$ defined by $\bar{u} = \frac{1}{\text{meas}\,\mu(\hat{Q})}\int_{\hat{Q}} u(\mathbf{x})d\mathbf{x}$ with $\mu(\hat{Q}) = r^n$. Then,

$$|\bar{u} - u(\mathbf{0})| = \left|\frac{1}{\mu(\hat{Q})}\int_{\hat{Q}} u(\mathbf{x})d\mathbf{x} - \frac{1}{\mu(\hat{Q})}\int_{\hat{Q}} u(\mathbf{0})d\mathbf{x}\right| \\ \le \frac{1}{r^n}\int_{\hat{Q}}\left[r\sum_{i=1}^n\left(\int_0^1\left|\frac{\partial u}{\partial x_i}(t\mathbf{x})\right|dt\right)\right]d\mathbf{x} \quad \text{(using (8.12.15))} \\ = \frac{1}{r^{n-1}}\int_0^1 dt\int_{\hat{Q}}\sum_{i=1}^n\left|\frac{\partial u}{\partial x_i}(t\mathbf{x})\right|d\mathbf{x} \\ = \frac{1}{r^{n-1}}\int_0^1 dt\int_{t\hat{Q}}\sum_{i=1}^n\left|\frac{\partial u}{\partial x_i}(\boldsymbol{\xi})\right|\frac{d\boldsymbol{\xi}}{t^n}$$

(by change of variables: $\mathbf{x} = \frac{\boldsymbol{\xi}}{t}$ with $|J| = \frac{1}{t^n}$, $\boldsymbol{\xi} = t\mathbf{x}$ and $t\hat{Q} \subset \hat{Q}$, $\mu(t\hat{Q}) \le \mu(\hat{Q})$ with $0 < t < 1$). Hence,

$$|\bar{u} - u(\mathbf{0})| \le \frac{1}{r^{n-1}}\int_0^1 dt\int_{t\hat{Q}}\sum_{i=1}^n\left|\frac{\partial u}{\partial x_i}(\boldsymbol{\xi})\right|\frac{d\boldsymbol{\xi}}{t^n}. \tag{8.12.16a}$$

For $u \in \mathcal{D}(\mathbb{R}^n)$, $\frac{\partial u}{\partial x_i} \in L^p(\mathbb{R}^n)$ with $p > n$, and by Hölder's inequality, we have

$$\int_{t\hat{Q}} \left|\frac{\partial u}{\partial x_i}(\boldsymbol{\xi})\right| d\boldsymbol{\xi} \le \left\|\frac{\partial u}{\partial x_i}\right\|_{L^p(t\hat{Q})} \cdot \|1\|_{L^q(t\hat{Q})}$$

$$= \left(\int_{t\hat{Q}} \left|\frac{\partial u}{\partial x_i}\right|^p d\boldsymbol{\xi}\right)^{1/p} \cdot \left(\int_{t\hat{Q}} 1 \cdot d\boldsymbol{\xi}\right)^{1/q}$$

$$\Longrightarrow \quad \int_{t\hat{Q}} \left|\frac{\partial u}{\partial x_i}(\boldsymbol{\xi})\right| d\boldsymbol{\xi} \le \left(\int_{\hat{Q}} \left|\frac{\partial u}{\partial x_i}\right|^p d\mathbf{x}\right)^{1/p} \cdot \left(\mu(t\hat{Q})\right)^{1/q} \tag{8.12.16b}$$

$$= \left\|\frac{\partial u}{\partial x_i}\right\|_{L^p(\hat{Q})} \cdot t^{n/q} r^{n/q} \tag{8.12.16c}$$

(since $(\mu(t\hat{Q}))^{1/q} = [(tr)^n]^{1/q} = t^{n/q} r^{n/q}$). For $\frac{n}{q} - n = -\frac{n}{p}$ with $p > n$, we have from (8.12.6f), (8.12.16a) and (8.12.16c):

$$|\bar{u} - u(\mathbf{0})| \le \frac{r^{n/q}}{r^{n-1}} \sum_{i=1}^{n} \left\|\frac{\partial u}{\partial x_i}\right\|_{L^p(\hat{Q})} \int_0^1 \frac{t^{n/q}}{t^n} dt$$

$$\le r^{n/q-n+1} \left(\frac{t^{n/q-n+1}}{n/q - n + 1}\right)\Bigg|_{t=0}^{t=1} \|\nabla u\|_{(L^p(\hat{Q}))^n}$$

$$= \frac{r^{1-n/p}}{1 - n/p} \|\nabla u\|_{(L^p(\hat{Q}))^n} \quad \forall u \in \mathcal{D}(\mathbb{R}^n). \tag{8.12.16d}$$

But (8.12.16d) remains valid for all cubes Q obtained by the 'rigid-body translation' of $\hat{Q}$ such that the edges of Q are always parallel to the coordinate axes and are of length r. Then, for $p > n$, $\forall \mathbf{x} \in Q$,

$$|\bar{u} - u(\mathbf{x})| \le \frac{r^{1-n/p}}{1 - n/p} \|\nabla u\|_{(L^p(Q))^n} \quad \forall u \in \mathcal{D}(\mathbb{R}^n). \tag{8.12.17}$$

Hence, $\forall \mathbf{x}, \mathbf{y} \in Q$, $\forall u \in \mathcal{D}(\mathbb{R}^n)$,

$$|u(\mathbf{x}) - u(\mathbf{y})| \le |u(\mathbf{x}) - \bar{u}| + |\bar{u} - u(\mathbf{y})| \le \frac{2r^{1-n/p}}{1 - n/p} \|\nabla u\|_{(L^p(Q))^n}. \tag{8.12.18}$$

But for *any two* points $\mathbf{x}, \mathbf{y} \in \mathbb{R}^n$, $\exists$ a cube Q with the length of its edges $= r = 2\|\mathbf{x} - \mathbf{y}\|$ such that $\mathbf{x}, \mathbf{y} \in Q$ and the inequality (8.12.18) holds $\forall u \in \mathcal{D}(\mathbb{R}^n)$. But $\mathcal{D}(\mathbb{R}^n)$ is *dense* in $W^{1,p}(\mathbb{R}^n)$ by Theorem 6.8.9 for $1 \le p < \infty$. Hence, $\exists$ a sequence (u_k) in $\mathcal{D}(\mathbb{R}^n)$ such that $u_k \to u$ in $W^{1,p}(\mathbb{R}^n)$ and $u_k \to u$ (if necessary, a subsequence can be chosen) pointwise a.e. in $\mathbb{R}^n$. Then

$|u_k(\mathbf{x}) - u_k(\mathbf{y})| \leq C\|\mathbf{x}-\mathbf{y}\|^{1-n/p}\|\nabla u_k\|_{(L^p(\mathbb{R}^n))^n}$ $\forall k \in \mathbb{N}$ for *almost all* $\mathbf{x}, \mathbf{y} \in \mathbb{R}^n$. But $u_k \to u$ in $W^{1,p}(\mathbb{R}^n) \Longrightarrow \nabla u_k \to \nabla u$ in $(L^p(\mathbb{R}^n))^n \Longrightarrow \|\nabla u_k\|_{(L^p(\mathbb{R}^n))^n} \to \|\nabla u\|_{(L^p(\mathbb{R}^n))^n}$ as $k \to \infty$. For $u \in W^{1,p}(\mathbb{R}^n)$ with $p > n$,

$$\begin{aligned}|u(\mathbf{x})-u(\mathbf{y})| &= \lim_{k\to\infty} |u_k(\mathbf{x}) - u_k(\mathbf{y})| \\ &\leq C\|\mathbf{x}-\mathbf{y}\|^\sigma \lim_{k\to\infty} \|\nabla u_k\|_{(L^p(\mathbb{R}^n))^n} \leq C\|\mathbf{x}-\mathbf{y}\|^\sigma \|\nabla u\|_{(L^p(\mathbb{R}^n))^n}\end{aligned}$$

for *almost all* $\mathbf{x}, \mathbf{y} \in \mathbb{R}^n$.

Proof of $W^{1,p}(\mathbb{R}^n) \hookrightarrow L^\infty(\mathbb{R}^n)$ *for* $p > n$: For $u \in \mathcal{D}(\mathbb{R}^n)$ and $\mathbf{x} \in Q$, we get, from (8.12.18),

$$\begin{aligned}|u(\mathbf{x})| &\leq |u(\mathbf{x}) - \bar{u}| + |\bar{u}| \leq |\bar{u}| + C\|\nabla u\|_{(L^p(Q))^n} \leq C\|u\|_{W^{1,p}(Q)} \\ &\leq C\|u\|_{W^{1,p}(\mathbb{R}^n)}.\end{aligned}$$

Hence, $\forall u \in \mathcal{D}(\mathbb{R}^n)$, $\|u\|_{L^\infty(\mathbb{R}^n)} = \operatorname{ess\,sup}_{\mathbf{x}\in\mathbb{R}^n} |u(\mathbf{x})| = \max_{\mathbf{x}\in\mathbb{R}^n} |u(\mathbf{x})| \leq C\|u\|_{W^{1,p}(\mathbb{R}^n)}$. Then, by virtue of the density of $\mathcal{D}(\mathbb{R}^n)$ in $W^{1,p}(\mathbb{R}^n)$ with $p > n$ by Theorem 6.8.9, for $u \in W^{1,p}(\mathbb{R}^n)$, $\exists$ a sequence $(u_k)_{k\in\mathbb{N}}$ in $\mathcal{D}(\mathbb{R}^n)$ such that $u_k \to u$ in $W^{1,p}(\mathbb{R}^n)$ and $u_k \to u$ (if necessary, a subsequence can be chosen) pointwise a.e. in $\mathbb{R}^n$, from which the result follows: $\|u\|_{L^\infty(\mathbb{R}^n)} \leq C\|u\|_{W^{1,p}(\mathbb{R}^n)}$.

□

Consequence of the imbedding result (8.12.6d) For $u \in W^{1,p}(\mathbb{R}^n)$, $\exists$ a sequence (u_k) in $\mathcal{D}(\mathbb{R}^n)$ such that $u_k \longrightarrow u$ in $W^{1,p}(\mathbb{R}^n)$. But by (8.12.6d), $\|u-u_k\|_{L^\infty(\mathbb{R}^n)} \leq C\|u-u_k\|_{1,p,\mathbb{R}^n} \to 0$ as $k \to \infty \Longrightarrow (u_k)_{k=1}^\infty$ converges *uniformly* to u as $k \to \infty$, and $u_k(\mathbf{x}) \to 0$ as $\|\mathbf{x}\| \to \infty$ $\forall k \in \mathbb{N}$, since $\operatorname{supp}(u_k) \subset\subset \mathbb{R}^n$ $\forall k \in \mathbb{N}$. Hence,

$$\lim_{\|\mathbf{x}\|\to\infty} u(\mathbf{x}) = 0. \tag{8.12.19}$$

Now we extend the imbedding ($\hookrightarrow$) results of Theorem 8.12.1 to $W^{m,p}(\mathbb{R}^n)$ as follows:

Theorem 8.12.2. *For* $m \in \mathbb{N}$ *and* $1 \leq p < \infty$, *the following* continuous *imbedding* ($\hookrightarrow$) *results hold:*

I. $\frac{1}{p} - \frac{m}{n} > 0$ *(i.e.* $mp < n$*)* $\Longrightarrow W^{m,p}(\mathbb{R}^n) \hookrightarrow L^q(\mathbb{R}^n)$ *with* $\frac{1}{q} = \frac{1}{p} - \frac{m}{n}$,

$$\|u\|_{L^q(\mathbb{R}^n)} \leq C\|u\|_{W^{m,p}(\mathbb{R}^n)} \quad \forall u \in W^{m,p}(\mathbb{R}^n); \tag{8.12.20a}$$

II. $\frac{1}{p} - \frac{m}{n} = 0$ *(i.e.* $mp = n$*)* $\Longrightarrow W^{m,p}(\mathbb{R}^n) \hookrightarrow L^q(\mathbb{R}^n)$ $\forall q \in [p, \infty[$,

$$\|u\|_{L^q(\mathbb{R}^n)} \leq C\|u\|_{W^{m,p}(\mathbb{R}^n)} \quad \forall u \in W^{m,p}(\mathbb{R}^n); \tag{8.12.20b}$$

III. $\frac{1}{p} - \frac{m}{n} < 0$ *(i.e. $mp > n$)* $\Longrightarrow W^{m,p}(\mathbb{R}^n) \hookrightarrow L^\infty(\mathbb{R}^n)$ *with*

$$\|u\|_{L^\infty(\mathbb{R}^n)} \le C\|u\|_{W^{m,p}(\mathbb{R}^n)} \quad \forall u \in W^{m,p}(\mathbb{R}^n). \tag{8.12.20c}$$

Moreover, for $m - \frac{n}{p} = k + \sigma$ with $k = [m - \frac{n}{p}] =$ integral part of $(m - \frac{n}{p})$, $\sigma =$ fractional part of $(m - \frac{n}{p})$, $0 < \sigma < 1$, $k \in \mathbb{N}_0$, we have, $\forall u \in W^{m,p}(\mathbb{R}^n)$:

IV.
$$\|\partial^{\boldsymbol{\alpha}} u\|_{L^\infty(\mathbb{R}^n)} \le C\|u\|_{W^{m,p}(\mathbb{R}^n)} \quad \forall |\boldsymbol{\alpha}| \le k, \tag{8.12.20d}$$

with

$$|\partial^{\boldsymbol{\alpha}} u(\mathbf{x}) - \partial^{\boldsymbol{\alpha}} u(\mathbf{y})| \le C\|\mathbf{x}-\mathbf{y}\|_{\mathbb{R}^n}\|u\|_{W^{m,p}(\mathbb{R}^n)} \quad \forall \mathbf{x},\mathbf{y} \in \mathbb{R}^n,\ \forall |\boldsymbol{\alpha}| < k; \tag{8.12.20e}$$

$$|\partial^{\boldsymbol{\alpha}} u(\mathbf{x}) - \partial^{\boldsymbol{\alpha}} u(\mathbf{y})| \le C\|\mathbf{x}-\mathbf{y}\|^{\sigma}_{\mathbb{R}^n}\|u\|_{W^{m,p}(\mathbb{R}^n)} \tag{8.12.20f}$$

for almost all $\mathbf{x}, \mathbf{y} \in \mathbb{R}^n$, $\forall |\boldsymbol{\alpha}| = k$.

V. *In particular, for $m - \frac{n}{p} = k + \sigma$, $W^{m,p}(\mathbb{R}^n) \hookrightarrow C^{k,\sigma}(\mathbb{R}^n) \hookrightarrow C^k(\mathbb{R}^n)$ with*

$$\|u\|_{C^k(\mathbb{R}^n)} \le \|u\|_{C^{k,\sigma}(\mathbb{R}^n)} \le C\|u\|_{W^{m,p}(\mathbb{R}^n)}. \tag{8.12.20g}$$

The same constant C has been used everywhere to denote different *values (see Remark* 8.12.1*).*

Proof. Repeated applications of (8.12.6a), (8.12.6c) and (8.12.6d) of Theorem 8.12.1 will give the results (8.12.20a), (8.12.20b) and (8.12.20c) respectively. Similarly, (8.12.20d)–(8.12.20f) are proved using (8.12.6e) of Theorem 8.12.1 and Remark 8.12.2. As an example, we show the proof of (8.12.20a) using (8.12.6a). Let $u \in W^{m,p}(\mathbb{R}^n)$ with $m \in \mathbb{N}$, $1 \le p < n$, $\frac{1}{p} - \frac{m}{n} > 0$ (or $mp < n$). Then, $\forall |\boldsymbol{\alpha}| = m-1$, $\partial^{\boldsymbol{\alpha}} u \in W^{1,p}(\mathbb{R}^n) \hookrightarrow L^{p_1^*}(\mathbb{R}^n)$ with $\frac{1}{p_1^*} = \frac{1}{p} - \frac{1}{n}$ by (8.12.6a), i.e. $\forall |\boldsymbol{\alpha}| = m-1$, $\partial^{\boldsymbol{\alpha}} u \in L^{p_1^*}(\mathbb{R}^n) \Longrightarrow u \in W^{m-1,p_1^*}(\mathbb{R}^n)$ (by definition). Hence, $u \in W^{m,p}(\mathbb{R}^n) \Longrightarrow u \in W^{m-1,p_1^*}(\mathbb{R}^n)$ and $W^{m,p}(\mathbb{R}^n) \hookrightarrow W^{m-1,p_1^*}(\mathbb{R}^n)$. Again, $u \in W^{m-1,p_1^*}(\mathbb{R}^n) \Longrightarrow \forall |\boldsymbol{\alpha}| = m-2$, $\partial^{\boldsymbol{\alpha}} u \in W^{1,p_1^*}(\mathbb{R}^n) \hookrightarrow L^{p_2^*}(\mathbb{R}^n)$ with $\frac{1}{p_2^*} = \frac{1}{p_1^*} - \frac{1}{n} = (\frac{1}{p} - \frac{1}{n}) - \frac{1}{n} = \frac{1}{p} - \frac{2}{n}$ by (8.12.6a) and $\forall |\boldsymbol{\alpha}| = m-2$, $\partial^{\boldsymbol{\alpha}} u \in L^{p_2^*}(\mathbb{R}^n) \Longrightarrow u \in W^{m-2,p_2^*}(\mathbb{R}^n)$. Thus, $u \in W^{m,p}(\mathbb{R}^n) \Longrightarrow u \in W^{m-2,p_2^*}(\mathbb{R}^n)$ and $W^{m,p}(\mathbb{R}^n) \hookrightarrow W^{m-2,p_2^*}(\mathbb{R}^n)$. Repeating this procedure $(m-1)$ times, we get $u \in W^{m,p}(\mathbb{R}^n) \Longrightarrow u \in W^{1,p_{m-1}^*}(\mathbb{R}^n)$ and $W^{m,p}(\mathbb{R}^n) \hookrightarrow W^{1,p_{m-1}^*}(\mathbb{R}^n)$ with $\frac{1}{p_{m-1}^*} = \frac{1}{p} - \frac{m-1}{n}$.

Finally, again applying (8.12.6a), we get $W^{1,p_{m-1}^*}(\mathbb{R}^n) \hookrightarrow L^{p_m^*}(\mathbb{R}^n)$ with $\frac{1}{p_m^*} = \frac{1}{p_{m-1}^*} - \frac{1}{n} = (\frac{1}{p} - \frac{m-1}{n}) - \frac{1}{n} = \frac{1}{p} - \frac{m}{n}$ and $u \in W^{m,p}(\mathbb{R}^n) \Longrightarrow u \in L^{p_m^*}(\mathbb{R}^n)$ and $W^{m,p}(\mathbb{R}^n) \hookrightarrow L^{p_m^*}(\mathbb{R}^n)$, and the result (8.12.20a) follows with $q = p_m^*$ and $\frac{1}{q} = \frac{1}{p} - \frac{m}{n} > 0$. □

Remark 8.12.3. For $p = 1$, $m = n$, $W^{n,1}(\mathbb{R}^n) \hookrightarrow L^\infty(\mathbb{R}^n)$, although (8.12.20b) holds for $q \in [n, \infty[$. In fact, $\forall u \in \mathcal{D}(\mathbb{R}^n)$,

$$u(x_1, \dots, x_n) = \int_{-\infty}^{x_1} \int_{-\infty}^{x_2} \cdots \int_{-\infty}^{x_n} \frac{\partial^n u}{\partial x_1 \cdots \partial x_n}(t_1, t_2, \dots t_n) dt_1 dt_2 \dots dt_n$$

$$\Longrightarrow \quad \|u\|_{L^\infty(\mathbb{R}^n)} = \operatorname*{ess\,sup}_{\mathbf{x} \in \mathbb{R}^n} |u(\mathbf{x})| = \sup_{\mathbf{x} \in \mathbb{R}^n} |u(\mathbf{x})|$$

$$\leq \left\| \frac{\partial^n u}{\partial x_1 \dots \partial x_n} \right\|_{L^1(\mathbb{R}^n)} \leq \|u\|_{W^{n,1}(\mathbb{R}^n)} \quad \forall u \in \mathcal{D}(\mathbb{R}^n).$$

Then, by the property of the density of $\mathcal{D}(\mathbb{R}^n)$ in $W^{n,1}(\mathbb{R}^n)$ (by Theorem 6.8.9), the result follows, i.e.

$$\|u\|_{L^\infty(\mathbb{R}^n)} \leq \|u\|_{W^{n,1}(\mathbb{R}^n)} \ \forall u \in W^{n,1}(\mathbb{R}^n). \tag{8.12.21}$$

Theorem 8.12.3. *Let $\Omega \subsetneq \mathbb{R}^n$ be a domain with the m-*extension property *with $m = 1$ (see Definition* 8.10.3 *for $p = 2$), i.e. $\exists$ a continuous, linear operator $P \in \mathcal{L}(W^{1,p}(\Omega); W^{1,p}(\mathbb{R}^n))$ such that $Pu\downarrow_\Omega = u$ and*

$$\|Pu\|_{W^{1,p}(\mathbb{R}^n)} \leq C\|u\|_{W^{1,p}(\Omega)} \quad \forall u \in W^{1,p}(\Omega),\ 1 \leq p \leq \infty \tag{8.12.22a}$$

(see Theorem 8.10.12*).*

Then, for $1 \leq p \leq \infty$, the following continuous *imbedding ($\hookrightarrow$) results hold:*

I. *for $1 \leq p < n$, $W^{1,p}(\Omega) \hookrightarrow L^{p^*}(\Omega)$ with $\frac{1}{p^*} = \frac{1}{p} - \frac{1}{n}$,*

$$\|u\|_{L^{p^*}(\Omega)} \leq C\|u\|_{W^{1,p}(\Omega)} \quad \forall u \in W^{1,p}(\Omega); \tag{8.12.22b}$$

II. *for $p = n$, $W^{1,p}(\Omega) \hookrightarrow L^q(\Omega)$ with $q \in [p, \infty[$,*

$$\|u\|_{L^q(\Omega)} \leq C\|u\|_{W^{1,p}(\Omega)} \quad \forall u \in W^{1,p}(\Omega); \tag{8.12.22c}$$

III. *for $p > n$, $W^{1,p}(\Omega) \hookrightarrow L^\infty(\Omega)$ with*

$$\|u\|_{L^\infty(\Omega)} \leq C\|u\|_{W^{1,p}(\Omega)} \quad \forall u \in W^{1,p}(\Omega). \tag{8.12.22d}$$

Moreover, for $u \in W^{1,p}(\Omega)$ with $p > n$ and $\sigma = 1 - \frac{n}{p} \in]0, 1[$:

IV. $|u(\mathbf{x}) - u(\mathbf{y})| \leq C\|\mathbf{x} - \mathbf{y}\|_{\mathbb{R}^n}^\sigma \|u\|_{1,p,\Omega}$ for almost all $\mathbf{x}, \mathbf{y} \in \Omega$. (8.12.22e)
In particular,

$$W^{1,p}(\Omega) \hookrightarrow C^{0,\sigma}(\overline{\Omega}) \hookrightarrow C^0(\overline{\Omega}) \tag{8.12.22f}$$

with $\|u\|_{C^0(\overline{\Omega})} \leq \|u\|_{C^{0,\sigma}(\overline{\Omega})} \leq C\|u\|_{W^{1,p}(\Omega)}$ $\forall u \in W^{1,p}(\Omega)$.

The same constant C has been used to denote different *values (see Remark* 8.12.1*).*

Proof. Using (8.12.22a), we have, $\forall u \in W^{1,p}(\Omega)$, $Pu \in W^{1,p}(\mathbb{R}^n)$, and we apply Theorem 8.12.1 for $Pu \in W^{1,p}(\mathbb{R}^n)$ and finally, again applying the properties of P in (8.12.22a), all the *continuous* imbedding results (8.12.22b)–(8.12.22f) are obtained. As an example, we prove (8.12.22d) in III. For $p > n$, let $u \in W^{1,p}(\Omega)$. Then $Pu \in W^{1,p}(\mathbb{R}^n)$ with (8.12.22a). Then, from (8.12.6d), $\|Pu\|_{L^\infty(\mathbb{R}^n)} \le C\|Pu\|_{W^{1,p}(\mathbb{R}^n)} \le C\|u\|_{W^{1,p}(\Omega)}$ (by (8.12.22a)). Hence,

$$\begin{aligned}\|u\|_{L^\infty(\Omega)} &= \|(Pu)\downarrow_\Omega\|_{L^\infty(\Omega)} \quad \text{(by (8.12.22a))}\\ &\le \|Pu\|_{L^\infty(\mathbb{R}^n)} \le C\|u\|_{W^{1,p}(\Omega)}\end{aligned}$$

$\Longrightarrow$ for $p > n$, $W^{1,p}(\Omega) \hookrightarrow L^\infty(\Omega)$. □

Remark 8.12.4. $\Omega = \mathbb{R}^n_+, \mathbb{R}^{n+}$, angular sectors and polygonal domains, C^m-regular bounded domains and bounded domains with Lipschitz continuous boundary Γ all possess the m-extension property (see Section 8.10 for details).

Now we extend the imbedding ($\hookrightarrow$) results of Theorem 8.12.3 to $W^{m,p}(\Omega)$ as follows:

Theorem 8.12.4. *Let* $\Omega \subsetneq \mathbb{R}^n$ *be a domain with the* m-extension property *(see* (8.12.22a)*):* $\exists P \in \mathcal{L}(W^{m,p}(\Omega); W^{m,p}(\mathbb{R}^n))$ *with* $Pu\downarrow_\Omega = u$,

$$\|Pu\|_{W^{m,p}(\mathbb{R}^n)} \le C\|u\|_{W^{m,p}(\Omega)} \tag{8.12.23a}$$

$\forall u \in W^{m,p}(\Omega)$, $1 \le p \le \infty$, $\forall m \in \mathbb{N}$ *(see Theorem* 8.10.2 *for* $W^{m,p}(\Omega)$*, and for* $p = 2$*, see Theorems* 8.10.2, 8.10.5, 8.10.9*, etc.).*

Then the following continuous *imbedding* ($\hookrightarrow$) *results hold:*

I. *for* $\frac{1}{p} - \frac{m}{n} > 0$ *(i.e.* $mp < n$*),* $W^{m,p}(\Omega) \hookrightarrow L^q(\Omega)$ *with* $\frac{1}{q} = \frac{1}{p} - \frac{m}{n}$,

$$\|u\|_{L^q(\Omega)} \le C\|u\|_{W^{m,p}(\Omega)} \quad \forall u \in W^{m,p}(\Omega); \tag{8.12.23b}$$

II. *for* $\frac{1}{p} - \frac{m}{n} = 0$ *(i.e.* $mp = n$*),* $W^{m,p}(\Omega) \hookrightarrow L^q(\Omega)$ *with* $q \in [p, \infty[$,

$$\|u\|_{L^q(\Omega)} \le C\|u\|_{W^{m,p}(\Omega)} \quad \forall u \in W^{m,p}(\Omega); \tag{8.12.23c}$$

III. *for* $\frac{1}{p} - \frac{m}{n} < 0$ *(i.e.* $mp > n$*),* $W^{m,p}(\Omega) \hookrightarrow L^\infty(\Omega)$ *with*

$$\|u\|_{L^\infty(\Omega)} \le C\|u\|_{W^{m,p}(\Omega)} \quad \forall u \in W^{m,p}(\Omega). \tag{8.12.23d}$$

Moreover, for $m - \frac{n}{p} = k + \sigma$ *with* $k = [m - \frac{n}{p}] \in \mathbb{N}_0$, $0 < \sigma < 1$*:*

IV. $\forall u \in W^{m,p}(\Omega)$,

$$\|\partial^{\boldsymbol{\alpha}} u\|_{L^\infty(\Omega)} \le C\|u\|_{W^{m,p}(\Omega)} \ \forall |\boldsymbol{\alpha}| \le k, \tag{8.12.23e}$$

$$|\partial^{\boldsymbol{\alpha}} u(\mathbf{x}) - \partial^{\boldsymbol{\alpha}} u(\mathbf{y})| \le C\|\mathbf{x} - \mathbf{y}\|^\sigma \|u\|_{m,p,\sigma} \tag{8.12.23f}$$

for almost all $\mathbf{x}, \mathbf{y} \in \Omega$, *and in particular,*

$$W^{m,p}(\Omega) \hookrightarrow C^{k,\sigma}(\overline{\Omega}) \hookrightarrow C^k(\overline{\Omega}) \tag{8.12.23g}$$

with $\|u\|_{C^k(\overline{\Omega})} \le \|u\|_{C^{k,\sigma}(\overline{\Omega})} \le C\|u\|_{W^{m,p}(\Omega)} \ \forall u \in W^{m,p}(\Omega)$.

The same constant C has been used everywhere to denote different *values (see Remark* 8.12.1*).*

Proof. Using the m-extension operator P in (8.12.23a), then applying Theorem 8.12.2 for $Pu \in W^{m,p}(\mathbb{R}^n)$, and finally applying (8.12.23a) again, the imbedding results (8.12.23b)–(8.12.23g) are obtained (see the proof of Theorem 8.12.3, where an example has been given). □

8.12.1 Compact imbedding results

See Section 8.11, where *compact* imbedding results in Sobolev spaces for fixed p with $p = 2$ (i.e. Hilbert spaces) or $1 \le p < \infty$ are given. Now we state the general results without proof (For $\Omega \subset \mathbb{R}^n$ and $p = 2$, see the proof of Theorem 8.11.4).

Theorem 8.12.5 (Rellich–Kondraschov). *Let* $\Omega \subset \mathbb{R}^n$ *be a* bounded *domain with* Lipschitz continuous boundary Γ *(see Appendix D). Then the following* compact imbedding $\hookrightarrow\hookrightarrow$ *results hold:*

I. *for* $p < n$, $W^{1,p}(\Omega) \hookrightarrow\hookrightarrow L^q(\Omega)\ \forall q \in [1, p^*[$ *with*

$$\frac{1}{p^*} = \frac{1}{p} - \frac{1}{n}; \tag{8.12.24}$$

II. *for* $p = n$,

$$W^{1,p}(\Omega) \hookrightarrow\hookrightarrow L^q(\Omega) \quad \forall q \in [1, \infty[; \tag{8.12.25}$$

III. *for* $p > n$,

$$W^{1,p}(\Omega) \hookrightarrow\hookrightarrow C^0(\overline{\Omega}). \tag{8.12.26}$$

We have previously discussed the continuous imbedding $\hookrightarrow$ and compact imbedding $\hookrightarrow\hookrightarrow$ results for $W^{m,p}$-spaces of integral order $m \in \mathbb{N}$. Now we give the results for $W^{s,p}$-spaces for *arbitrary* $s \ge 0$ without proof.

Theorem 8.12.6. *Let* $0 \le r \le s$, $k \in \mathbb{N}$ *and let* $\Omega \subset \mathbb{R}^n$ *possess the* s-extension property *(see Theorem* 8.10.12*). Then, for* $1 < p \le q < \infty$ and $0 < \sigma < 1$, *we have,* $\forall u \in W^{s,p}(\Omega)$*:*

I. *for* $r - \frac{n}{q} \le s - \frac{n}{p}$, $W^{s,p}(\Omega) \hookrightarrow W^{r,q}(\Omega)$ *with*

$$\|u\|_{W^{r,q}(\Omega)} \le C \|u\|_{W^{s,p}(\Omega)}; \tag{8.12.27}$$

II. *for* $k + \sigma \le s - \frac{n}{p}$, $W^{s,p}(\Omega) \hookrightarrow C^{k,\sigma}(\overline{\Omega})$ *with*

$$\|u\|_{C^{k,\sigma}(\overline{\Omega})} \le C \|u\|_{W^{s,p}(\Omega)}; \tag{8.12.28}$$

III. Limit cases: *If* $p = 1$ *and* $q = \infty$, *then* (8.12.27) *holds: for* $r \le s - n$,

$$W^{s,1}(\Omega) \hookrightarrow W^{r,\infty}(\Omega); \tag{8.12.29}$$

For $p = 1$ *with* $q < \infty$, (8.12.27) *holds if either*

(a) both r *and* s *are integers, or*

(b) s *is* not *an integer, or*

(c) $s - r \ge 1$. (8.12.30)

IV. Compact imbeddings $\hookrightarrow\hookrightarrow$: *For* $1 \le p \le q \le \infty$, $0 < \sigma < 1$ *and* bounded Ω *with* Lipschitz continuous boundary Γ, *the following* compact *imbedding* $\hookrightarrow\hookrightarrow$ *results hold:*

$$r - \frac{n}{q} < s - \frac{n}{p} \quad \Longrightarrow \quad W^{s,p}(\Omega) \hookrightarrow\hookrightarrow W^{r,q}(\Omega); \tag{8.12.31}$$

$$k + \sigma < s - \frac{n}{p} \quad \Longrightarrow \quad W^{s,p}(\Omega) \hookrightarrow\hookrightarrow C^{k,\sigma}(\overline{\Omega}). \tag{8.12.32}$$

Translation property of imbedding results (see (8.9.33))

$$W^{1,p}(\mathbb{R}^n) \hookrightarrow L^q(\mathbb{R}^n) \quad \Longrightarrow \quad W^{m+1,p}(\mathbb{R}^n) \hookrightarrow W^{m,q}(\mathbb{R}^n); \tag{8.12.33}$$

$$W^{1,p}(\mathbb{R}^n) \hookrightarrow C^0(\mathbb{R}^n) \quad \Longrightarrow \quad W^{m+1,p}(\mathbb{R}^n) \hookrightarrow C^m(\mathbb{R}^n); \tag{8.12.34}$$

$$W^{m,p}(\Omega) \hookrightarrow L^q(\Omega) \quad \Longrightarrow \quad W^{m+j,p}(\Omega) \hookrightarrow W^{j,q}(\Omega) \quad \forall j \in \mathbb{N}; \tag{8.12.35}$$

$$W^{m,p}(\Omega) \hookrightarrow C^0(\overline{\Omega}) \quad \Longrightarrow \quad W^{m+j,p}(\Omega) \hookrightarrow C^j(\overline{\Omega}) \quad \forall j \in \mathbb{N}; \tag{8.12.36}$$

$$W^{m,p}(\Omega) \hookrightarrow C^{0,\sigma}(\overline{\Omega}) \quad \Longrightarrow \quad W^{m+j,p}(\Omega) \hookrightarrow C^{j,\sigma}(\overline{\Omega}) \quad \forall j \in \mathbb{N}; \tag{8.12.37}$$

and so on.

All the above imbedding results (8.12.27)–(8.12.37) will hold for $W_0^{s,p}(\Omega)$ with $s > 0$, $1 \le p < \infty$. (8.12.38)

8.13 Sobolev spaces $H^s(\Gamma)$, $W^{s,p}(\Gamma)$ on a manifold boundary Γ

In the study of boundary value problems of partial differential equations on $\overline{\Omega} = \Omega \cup \Gamma$, Γ being the boundary of $\Omega \subsetneq \mathbb{R}^n$, boundary conditions are prescribed on Γ which involve a prescription of functions on Γ. Hence, in order to place these functions in suitable Sobolev spaces on Γ, we will need trace theorems involving trace operators; these will be introduced in the following section, Section 8.14.

The definition and basic properties of Sobolev spaces on Γ can be obtained from those given for $\Omega \subset \mathbb{R}^n$ in earlier sections and chapters by replacing 'Ω' by 'Γ', keeping in mind the *definition*, *smoothness* and *regularity* properties of Γ, which will play a vital, decisive role. Hence, for all important details of these properties of Γ, we refer to Appendix D, which will be our constant source of reference. So we begin with:

8.13.1 Surface integrals on boundary Γ of bounded $\Omega \subset \mathbb{R}^n$

Let $\Omega \subset \mathbb{R}^n$ be a bounded, open subset of $\mathbb{R}^n$ with a *Lipschitz continuous* boundary Γ.

Let $(x_i)_{i=1}^n$ and $(\xi_i^r)_{i=1}^n$ be *fixed* and *local* coordinate systems related by $\mathbf{x} = A_r^{-1}\boldsymbol{\xi}^r + \tilde{\boldsymbol{b}}_r$, A_r being an orthogonal matrix, $1 \le r \le N$ (see (D.2.2.6) and Figure D.8, Appendix D), and $(\Gamma_r, \phi_r)_{r=1}^N$ be the *atlas* or *local charts* defined by (D.2.3.10) such that

$$\Gamma_r = \{(\hat{\boldsymbol{\xi}}^r, \phi_r(\hat{\boldsymbol{\xi}}^r)) : \forall \hat{\boldsymbol{\xi}}^r = (\xi_1^r, \dots, \xi_{n-1}^r) \in \hat{Q}_r \subset \mathbb{R}^{n-1}, \quad \xi_n^r = \phi_r(\hat{\boldsymbol{\xi}}^r)\}, \quad 1 \le r \le N, \tag{8.13.1}$$

$$\Gamma = \bigcup_{r=1}^N \Gamma_r \quad \text{and} \quad \Gamma \in C^{0,1}\text{-class (Lipschitz continuous)}, \tag{8.13.2}$$

i.e. $\forall r = 1, \dots, N$, $\phi_r \in C^{0,1}(\hat{Q}_r)$ (see (D.2.3.6)).

Then we have:

Theorem 8.13.1. *Let* Γ *be* Lipschitz continuous *(i.e.* $\phi_r \in C^{0,1}(\hat{Q}_r)$ $\forall r = 1, \dots, N$*) satisfying* (8.13.1) *and* (8.13.2)*. Then:*

I. $\forall r = 1, \dots, N$, ϕ_r *has a total differential* $d\phi_r$ *almost everywhere in* $\hat{Q}_r$*, with* $\frac{\partial \phi_r}{\partial \xi_i^r} \in L^\infty(\hat{Q}_r)$*,* $1 \le i \le n$*;*

II. *The unit normal* $\hat{\mathbf{n}}$ *to* Γ *(exterior to* Ω*) exists almost everywhere on* Γ *and is defined,* $\forall \hat{\boldsymbol{\xi}}^r \in \hat{Q}_r$, $1 \le r \le N$, *by:*

$$\hat{\mathbf{n}}(\hat{\boldsymbol{\xi}}^r, \phi_r(\hat{\boldsymbol{\xi}}^r)) = \frac{(-\frac{\partial \phi_r}{\partial \xi_1^r}(\hat{\boldsymbol{\xi}}^r), \ldots, -\frac{\partial \phi_r}{\partial \xi_{n-1}^r}(\hat{\boldsymbol{\xi}}^r), 1)}{\sqrt{1 + \sum_{i=1}^{n-1}(\frac{\partial \phi_r}{\partial \xi_i^r}(\hat{\boldsymbol{\xi}}^r))^2}}; \tag{8.13.3}$$

III. *In particular, for* Γ *belonging to the* $C^{m,1}$*-class (see* (D.2.3.8)*), the normal vector field* $\hat{\mathbf{n}}$ *is only of* $C^{m-1,1}$*-class.* (8.13.4)

Space $L^p(\Gamma)$, $1 \le p < \infty$

For $\mathbf{x} \in \Gamma_r$, $\mathbf{x} = A_r^{-1}\boldsymbol{\xi}^r + \tilde{\boldsymbol{b}}_r$ (see (D.2.2.6)) with $\boldsymbol{\xi}^r = (\hat{\boldsymbol{\xi}}^r, \xi_n^r)$, $\xi_n^r = \phi_r(\hat{\boldsymbol{\xi}}^r)$ $\forall \hat{\boldsymbol{\xi}}^r \in \hat{Q}_r \subset \mathbb{R}^{n-1}$ (see (D.2.3.1)),

$$u(\mathbf{x}) = u(A_r^{-1}\boldsymbol{\xi}^r + \tilde{\boldsymbol{b}}_r) = u_r(\boldsymbol{\xi}^r) = u_r[(\hat{\boldsymbol{\xi}}^r, \phi_r(\hat{\boldsymbol{\xi}}^r))]. \tag{8.13.5}$$

Definition 8.13.1. Let $\Gamma \in C^{0,1}$-class (i.e. $\phi_r \in C^{0,1}(\hat{Q}_r)$ $\forall r = 1, \ldots, N$, Lipschitz continuous), and $u : \Gamma \to \mathbb{R}$ be a function defined a.e. on Γ satisfying (8.13.5). Then $u \in L^p(\Gamma)$ with $1 \le p < \infty$ if and only if $u_r \in L^p(\hat{Q}_r)$ $\forall r = 1, \ldots, N$, i.e.

$$\begin{aligned} u \in L^p(\Gamma) \quad &\Longleftrightarrow \quad u_r(\cdot, \phi_r(\cdot)) \in L^p(\hat{Q}_r) \\ &\Longleftrightarrow \quad \int_{\hat{Q}_r} |u_r(\hat{\boldsymbol{\xi}}^r, \phi_r(\hat{\boldsymbol{\xi}}^r))|^p d\hat{\boldsymbol{\xi}}^r < +\infty \quad \forall r = 1, \ldots, N. \end{aligned} \tag{8.13.6}$$

In particular, for $p = 2$,

$$u \in L^2(\Gamma) \quad \Longleftrightarrow \quad \int_{\hat{Q}_r} |u_r(\hat{\boldsymbol{\xi}}^r, \phi_r(\hat{\boldsymbol{\xi}}^r))|^2 d\hat{\boldsymbol{\xi}}^r < +\infty \ \forall r = 1, \ldots, N. \tag{8.13.7}$$

Let $\Gamma \in C^{0,1}$-class (i.e. $\phi_r \in C^{0,1}(\hat{Q}_r)$ $\forall r = 1, \ldots, N$, Lipschitz continuous), and $u \in L^1(\Gamma)$ (i.e. $p = 1$). Let $(\beta_r)_{r=1}^N$ be a C_0^∞-partition of 1 with respect to the open cover $(\Gamma_r)_{r=1}^N$ of Γ such that, for $\mathbf{x} \in \Gamma$,

$$\begin{aligned} u(\mathbf{x}) &= \sum_{r=1}^N \beta_r(\mathbf{x})u(\mathbf{x}) = \sum_{r=1}^N \beta_r(A_r^{-1}\boldsymbol{\xi}^r + \tilde{\boldsymbol{b}}_r)u(A_r^{-1}\boldsymbol{\xi}^r + \tilde{\boldsymbol{b}}_r) \\ &= \sum_{r=1}^N \beta_r(\hat{\boldsymbol{\xi}}^r, \phi_r(\hat{\boldsymbol{\xi}}^r))u_r(\hat{\boldsymbol{\xi}}^r, \phi_r(\hat{\boldsymbol{\xi}}^r)) \quad \text{(by (8.13.5))}. \end{aligned} \tag{8.13.8}$$

Let dS be the surface area measure on Γ induced by the $(n-1)$-dimensional (Lebesgue) volume measure $d\mathbf{x} = dx_1 dx_2 \ldots dx_{n-1}$. (For $n = 2$, $dS = ds$

is called the *arc length measure* on Γ induced by the length measure dx.) Then $dS = [1 + \sum_{i=1}^{n-1} (\frac{\partial \phi_r}{\partial \xi_i^r})^2]^{1/2} d\hat{\boldsymbol{\xi}}^r$ with $d\hat{\boldsymbol{\xi}}^r = d\xi_1^r \dots d\xi_{n-1}^r$, $1 \le r \le N$.

Then we have:

Definition 8.13.2. For $\Gamma \in C^{0,1}$-class and $u \in L^1(\Gamma)$ $(p = 1)$, dS the *surface area measure* defined by $dS = [1 + \sum_{i=1}^{n-1} (\frac{\partial \phi_r}{\partial \xi_i^r}(\hat{\boldsymbol{\xi}}^r))^2]^{1/2} d\hat{\boldsymbol{\xi}}_r$ with $d\hat{\boldsymbol{\xi}}^r = d\hat{\xi}_1^r d\hat{\xi}_2^r \dots d\hat{\xi}_{n-1}^r$.

Then, the *surface integral* $\int_\Gamma u(\mathbf{x}) dS$ is defined by

$$\int_\Gamma u(\mathbf{x}) dS = \sum_{r=1}^N \int_{\Gamma_r} (\beta_r u)(\mathbf{x}) dS$$
$$= \sum_{r=1}^N \int_{\hat{Q}_r} \beta_r(\hat{\boldsymbol{\xi}}^r, \phi_r(\hat{\boldsymbol{\xi}}^r)) u_r(\hat{\boldsymbol{\xi}}^r, \phi_r(\hat{\boldsymbol{\xi}}^r)) \left[1 + \sum_{i=1}^{n-1} \left(\frac{\partial \phi_r}{\partial \xi_i^r}\right)^2\right]^{1/2} d\hat{\boldsymbol{\xi}}^r. \tag{8.13.9}$$

Theorem 8.13.2. *Let* $\Gamma \in C^{0,1}$-class *and* $L^p(\Gamma)$, $1 \le p < \infty$, *be defined by Definition* 8.13.1. *Then* $L^p(\Gamma)$ *equipped with the norm* $\|\cdot\|_{L^p(\Gamma)}$:

$$\|u\|_{L^p(\Gamma)} = \left(\int_\Gamma |u(\mathbf{x})|^p dS\right)^{1/p}$$
$$= \left(\sum_{r=1}^N \int_{\hat{Q}_r} |\beta_r(\hat{\boldsymbol{\xi}}^r, \phi_r(\hat{\boldsymbol{\xi}}^r)) u_r(\hat{\boldsymbol{\xi}}^r, \phi_r(\hat{\boldsymbol{\xi}}^r))|^p \left[1 + \sum_{i=1}^{n-1} \left(\frac{\partial \phi_r}{\partial \xi_i^r}\right)^2\right]^{1/2} d\hat{\boldsymbol{\xi}}\right)^{1/p} \tag{8.13.10}$$

is a separable Banach space *for* $1 \le p < \infty$, *and*

$$\|u\|_{0,p,\Gamma} = \left(\sum_{r=1}^N \int_{\hat{Q}_r} |u_r(\hat{\boldsymbol{\xi}}^r, \phi_r(\hat{\boldsymbol{\xi}}^r))|^p d\hat{\boldsymbol{\xi}}^r\right)^{1/p} \tag{8.13.11}$$

defines a norm on $L^p(\Gamma)$ *equivalent to the original norm* $\|\cdot\|_{L^p(\Gamma)}$ *in* (8.13.10) [16, Lemma 1.2, Chapter 3], *i.e.* $\exists C_1, C_2 > 0$ *such that* $\forall u \in L^p(\Gamma)$, $C_1 \|u\|_{L^p(\Gamma)} \le \|u\|_{0,p,\Gamma} \le C_2 \|u\|_{L^p(\Gamma)}$.

Space $W^{s,p}(\Gamma)$, $1 \le p < \infty$, $s \ge 0$

Let $\Omega \subset \mathbb{R}^n$ be a bounded, open subset of $\mathbb{R}^n$ with boundary Γ of $C^{m,1}$-class, i.e. all *mth-order* derivatives of ϕ_r are *Lipschitz continuous*, $(\Gamma_r, \phi_r)_{r=1}^N$ being the *atlas* or *local charts* of Γ(see (D.2.3.8) and Definition D.2.3.1, Appendix D). Then, by (8.13.5), for $u \in L^p(\Gamma)$, for almost all $\mathbf{x} \in \Gamma_r$, $u(\mathbf{x}) = u_r(\hat{\boldsymbol{\xi}}^r, \phi_r(\hat{\boldsymbol{\xi}}^r))$ with $\hat{\boldsymbol{\xi}}^r \in \hat{Q}_r \subset \mathbb{R}^{n-1}$.

Definition 8.13.3. For $\Gamma \in C^{m,1}$-class and $0 \le s \le m+1$, $W^{s,p}(\Gamma)$ with $1 \le p < \infty$ is the subspace of $L^p(\Gamma)$:

$$W^{s,p}(\Gamma) = \{u : u \in L^p(\Gamma),\ u_r(\cdot, \phi_r(\cdot)) \in W^{s,p}(\hat{Q}_r),\ 1 \le r \le N\}$$

equipped with the norm $\|\cdot\|_{W^{s,p}(\Gamma)}$ defined, for $1 \le p < \infty$, by:

$$\|u\|_{W^{s,p}(\Gamma)} = \left(\sum_{r=1}^{N} \|u_r(\cdot, \phi_r(\cdot))\|^p_{W^{s,p}(\hat{Q}_r)} \right)^{1/p}, \tag{8.13.12}$$

where for $s = k \in \mathbb{N}$ with $k \le m+1$, $\|\cdot\|_{W^{k,p}(\hat{Q}_r)}$ is obtained from (2.15.29) with $\Omega = \hat{Q}_r \subset \mathbb{R}^{n-1}$, and for $s = k + \sigma$ with $[s] = k \in \mathbb{N}_0$, $0 < \sigma < 1$, $\|\cdot\|_{W^{s,p}(\hat{Q}_r)}$ is given by (8.10.90a):

$$\begin{aligned}
&\|u_r(\cdot, \phi_r(\cdot))\|_{W^{s,p}(\hat{Q}_r)} \\
&= \Bigg[\|u(\cdot, \phi_r(\cdot))\|^p_{k,p,\hat{Q}_r} \\
&\quad + \sum_{|\boldsymbol{\alpha}|=k} \int_{\hat{Q}_r} \int_{\hat{Q}_r} \frac{|\partial^{\boldsymbol{\alpha}} u(\hat{\boldsymbol{\xi}}, \phi_r(\hat{\boldsymbol{\xi}})) - \partial^{\boldsymbol{\alpha}} u(\hat{\boldsymbol{\zeta}}, \phi_r(\hat{\boldsymbol{\zeta}}))|^p}{\|\hat{\boldsymbol{\xi}} - \hat{\boldsymbol{\zeta}}\|^{(n-1)+\sigma p}_{\mathbb{R}^{n-1}}} dS(\hat{\boldsymbol{\xi}}) dS(\hat{\boldsymbol{\zeta}}) \Bigg]^{1/p}.
\end{aligned} \tag{8.13.13}$$

Theorem 8.13.3. $W^{s,p}(\Gamma)$ *defined by Definition* 8.13.3 *is a* separable Banach space. *In particular, for* $p=2$, $W^{s,2}(\Gamma) = H^s(\Gamma)$ *is a* separable Hilbert space. (8.13.14)

8.13.2 Alternative definition of $H^s(\Gamma)$ with $\Gamma \in C^m$-class (resp. C^∞-class)

Let $\Omega \subsetneq \mathbb{R}^n$ be a C^m-regular (resp. C^∞-regular, i.e. C^m-regular $\forall m \in \mathbb{N}$) domain in $\mathbb{R}^n$ with boundary Γ of C^m-class (resp. C^∞-class) with $m \in \mathbb{N}$ such that Ω is locally on one side of Γ. (8.13.15)

Following Definition D.4.1.1, Appendix D, let $\{\Omega_i\}_{i=1}^N$ be an *open cover* of Γ such that $(\Gamma_i, \Phi_i)_{i=1}^N$ with $\Gamma_i = \Omega_i \cap \Gamma$ is an *atlas* (or *local charts*) (see (D.5.1.1)) defining Γ, with $\Phi_i : \Omega_i \longrightarrow \hat{Q}$ and its *inverse* $\Phi_i^{-1} : \hat{Q} \longrightarrow \Omega_i$ belonging to C^m-diffeomorphisms (see also Definition 8.10.5 and the proof of Theorem 8.10.5) (resp. C^∞-*class*) (see (D.4.1.2)). (8.13.16)

Then, $\forall \mathbf{x} \in \Gamma_i = \Omega_i \cap \Gamma$, $1 \le i \le N$, $\Phi_i(\mathbf{x}) = (\phi_1^i(\mathbf{x}), \ldots, \phi_{n-1}^i(\mathbf{x}), \phi_n^i(\mathbf{x})) = (\xi_1, \ldots, \xi_{n-1}, 0) \in \hat{Q} \cap \mathbb{R}^n_0$ which we identify with $\hat{Q}_0 \subset \mathbb{R}^{n-1}$, i.e. $\phi_n^i(\mathbf{x}) = 0$ is the *equation to the boundary* Γ *locally*. (8.13.17)

Let $(\theta_i)_{i=1}^N$ be a C_0^∞-partition of unity with respect to the open cover $\{\Omega_i\}_{i=1}^N$ of Γ (see Figures 8.9 and 8.10) such that $\theta_i \in \mathcal{D}(\Omega_i)$ with $\beta_i = \theta_i \downarrow_{\Gamma_i} \in \mathcal{D}(\Gamma_i)$, $\operatorname{supp}(\beta_i) \subset\subset \Gamma_i$, $\sum_{i=1}^N \beta_i(\mathbf{x}) = 1\ \forall \mathbf{x} \in \Gamma$. (8.13.18)

Hence, $(\beta_i)_{i=1}^N$ with (8.13.18) is a C_0^∞-partition of unity on Γ such that for a function $u : \Gamma \longrightarrow \mathbb{R}$, $u = \sum_{i=1}^N \beta_i u = \sum_{i=1}^N u_i$ with $\operatorname{supp}(\beta_i u) \subset\subset \Gamma_i$. (8.13.19)

Then, $\forall i = 1, \dots, N$, $\beta_i u \circ \Phi_i^{-1}(\hat{\boldsymbol{\xi}}, 0) = \beta_i u(\mathbf{x})$ for $\mathbf{x} \in \Gamma_i$, $\hat{\boldsymbol{\xi}} = (\xi_1, \dots, \xi_{n-1}) \in \hat{Q}_0 \subset \mathbb{R}^{n-1}$.

Considered as a function of $\hat{\boldsymbol{\xi}} \in \hat{Q}_0 \subset \mathbb{R}^{n-1}$, $\beta_i u \circ \Phi_i^{-1}$ vanishes outside $\operatorname{supp}(\beta_i u \circ \Phi_i^{-1}) \subset\subset \hat{Q}_0 \subset \mathbb{R}^{n-1}$ and is given a *null extension* to $\mathbb{R}^{n-1}$ (i.e. extension by 0 in $\mathbb{R}^{n-1} \setminus \hat{Q}_0$), which is denoted by

$$\widetilde{(\beta_i u \circ \Phi_i^{-1})} \quad \forall i = 1, \dots, N. \tag{8.13.20}$$

Space $L^2(\Gamma)$ with Γ in C^m-class (resp. C^∞-class)

We redefine $L^2(\Gamma)$ by:

Definition 8.13.4. For $\Gamma \in C^m$-class (resp. C^∞-class) satisfying (8.13.16)–(8.13.20),

$$L^2(\Gamma) = \{u : u : \Gamma \longrightarrow \mathbb{R} \text{ such that } \widetilde{(\beta_i u \circ \Phi_i^{-1})} \in L^2(\mathbb{R}^{n-1}) \ \forall i = 1, \dots, N\} \tag{8.13.21}$$

is equipped with a new norm $\|\cdot\|_{0,\Gamma}$ defined by

$$\|u\|_{0,\Gamma} = \left(\sum_{i=1}^N \|\widetilde{(\beta_i u \circ \Phi_i^{-1})}\|_{L^2(\mathbb{R}^{n-1})}^2 \right)^{\frac{1}{2}} \quad \forall u \in L^2(\Gamma), \tag{8.13.22}$$

which is *equivalent* to the original norm $\|\cdot\|_{L^2(\Gamma)}$ obtained from (8.13.10) for $p = 2$.

8.13.3 Space $H^s(\Gamma)$ ($s > 0$) with Γ in C^m-class (resp. C^∞-class)

Now we extend this new definition of $L^2(\Gamma)$ to $H^s(\Gamma)$ by:

Definition 8.13.5. For $\Gamma \in C^m$-class (resp. C^∞-class) and for $0 < s \leq m$ (resp. $\forall s > 0$),

$$H^s(\Gamma) = \{u : u \in L^2(\Gamma), \widetilde{(\beta_i u \circ \Phi_i^{-1})} \in H^s(\mathbb{R}^{n-1}) \ \forall i = 1, \dots, N\} \tag{8.13.23}$$

is equipped with the norm $\|\cdot\|_{H^s(\Gamma)}$ or $\|\cdot\|_{s,\Gamma}$ defined by:

$$\|u\|_{H^s(\Gamma)} = \|u\|_{s,\Gamma} = \left(\sum_{i=1}^N \|\widetilde{(\beta_i u \circ \Phi_i^{-1})}\|_{H^s(\mathbb{R}^{n-1})}^2 \right)^{\frac{1}{2}} \quad \forall u \in H^s(\Gamma), \tag{8.13.24}$$

where Φ_i^{-1} is of C^m-class (resp. C^∞-class) by (8.13.16).

For $\Gamma \in C^m$-class (resp. C^∞-class) with $m \in \mathbb{N}$, we let

$$H^0(\Gamma) = L^2(\Gamma). \tag{8.13.25}$$

Remark 8.13.1. Norm $\|\cdot\|_{s,\Gamma}$ in (8.13.24) with $s \geq 0$ depends on the choice of C_0^∞-partition of unity on Γ and open cover of Γ. But *different* norms $\|\cdot\|_{s,\Gamma}$ for *different* open covers of Γ and partitions of unity on Γ are *equivalent.*

Remark 8.13.2. In Definition 8.13.5, we identify the space $H^s(\Gamma)$ with a *closed subspace* of $(H^s(\mathbb{R}^{n-1}))^N$ for $0 \leq s \leq m$ (resp. $\forall s \geq 0$) such that to every function u defined on the boundary Γ of Ω, there corresponds an N-tuple of functions $(\widetilde{\beta_i u \circ \Phi_i^{-1}})$ defined on $\mathbb{R}^{n-1}$.

Important properties of $H^s(\Gamma)$

Property 1

Theorem 8.13.4. *For $\Gamma \in C^m$-class (resp. C^∞-class) with $m \geq 1$ and $0 \leq s \leq m$ (resp. $\forall s \geq 0$), $H^s(\Gamma)$ equipped with $\|\cdot\|_{s,\Gamma}$ defined in* (8.13.24) *is a* separable Hilbert space.

Property 2: Imbedding results For $\Gamma \in C^m$-class (resp. C^∞-class) with $m \geq 1$, $0 \leq s_2 < s_1 \leq m$ (resp. $0 \leq s_2 < s_1 < \infty$), $H^{s_1}(\Gamma) \hookrightarrow H^{s_2}(\Gamma)$ with *continuous* imbedding $\hookrightarrow$. (8.13.26)

Theorem 8.13.5. *For $\Gamma \in C^m$-class (resp. C^∞-class) and for $k \in \mathbb{N}_0$ with $0 \leq k < s - \frac{n-1}{2}$, $0 < s \leq m$ (resp. $\forall s > 0$), $H^s(\Gamma) \hookrightarrow C^k(\Gamma)$.*

Proof. $u \in H^s(\Gamma) \implies \forall i = 1, \ldots, N$, $(\widetilde{\beta_i u \circ \Phi_i^{-1}}) \in H^s(\mathbb{R}^{n-1})$. Then, by Sobolev's Imbedding Theorem 8.9.4, for $0 \leq k < s - \frac{n-1}{2}$ with $s \leq m$ (resp. $s > 0$), $(\widetilde{\beta_i u \circ \Phi_i^{-1}}) \in C^k(\mathbb{R}^{n-1})$, the imbedding $\hookrightarrow$ being a continuous one. Hence, $\beta_i u \circ \Phi_i^{-1} \in C^k(\hat{Q}_0)$ $\forall i = 1, \ldots, N$ (by (8.13.20)) $\implies \beta_i u\downarrow_{\Gamma_i} \circ \Phi_i^{-1} \in C^k(\hat{Q}_0) \implies u = \sum_{i=1}^N \beta_i u \in C^k(\Gamma)$ with continuous $\hookrightarrow$. □

Property 3: Density result for $\Gamma \in C^m$-class

Theorem 8.13.6. *For Γ of C^m-class, $\forall s \in [0, m]$, $C^m(\Gamma)$ is* dense *in $H^s(\Gamma)$.*

In particular, for $\Gamma \in C^\infty$-class, $C^m(\Gamma)$ is dense *in $H^s(\Gamma)$ $\forall m \in \mathbb{N}$, $\forall s \geq 0$ (see also Theorem* 8.13.7*).* (8.13.27)

Spaces $\mathcal{D}(\Gamma)$ and $\mathcal{D}'(\Gamma)$ for $\Gamma \in C^\infty$-class

For $\Gamma \in C^\infty$-class, following Definitions 1.10.2 and 1.10.3 for circle Γ, we have:

Definition 8.13.6. For $\Gamma \in C^\infty$-class, $\mathcal{D}(\Gamma)$ defined by $\mathcal{D}(\Gamma) = \{\phi : \phi : \Gamma \longrightarrow \mathbb{R}, \phi \in C^\infty(\Gamma)\}$ is the linear space of *test functions* $\phi \in C^\infty(\Gamma)$. (8.13.28)

$\mathcal{D}(\Gamma)$ is equipped with the notions of *convergence* and *continuity* given in Definition 1.10.2.

Then $\mathcal{D}'(\Gamma)$ is the *dual space* of $\mathcal{D}(\Gamma)$ and is a *space of distributions* on Γ. (8.13.29)

In general, $\phi \in \mathcal{D}(\Gamma)$ is *not* periodic.

Remark 8.13.3. For $\Gamma \in C^m$-class (but $\Gamma \notin C^\infty$-class), $\mathcal{D}(\Gamma)$ can *not* be defined.

Property 4: Density result for $\Gamma \in C^\infty$-class

Theorem 8.13.7. *For* $\Gamma \in C^\infty$*-class,* $\mathcal{D}(\Gamma)$ *is* dense *in* $H^s(\Gamma)$ $\forall s \geq 0$.

In particular, $\mathcal{D}(\Gamma)$ *is* dense *in* $H^{s-j-1/2}(\Gamma)$ *for* $s - j - 1/2 > 0$, j *being an integer* ≥ 0. (8.13.30)

Proof. The result basically follows from the density of $\mathcal{D}(\mathbb{R}^{n-1})$ in $H^s(\mathbb{R}^{n-1})$ $\forall s \geq 0$. Let $u \in H^s(\Gamma)$. Then, $\forall i = 1, \dots, N$, $\tilde{\beta}_i u \in H^s(\Gamma)$ since, by (8.13.18), the null extension $\tilde{\beta}_i \in \mathcal{D}(\Gamma)$ with $\text{supp}(\tilde{\beta}_i) = \text{supp}(\beta_i) \subset\subset \Gamma_i$. By the definition of $H^s(\Gamma)$, in (8.13.23) $\widetilde{(\beta_i u \circ \Phi_i^{-1})} \in H^s(\mathbb{R}^{n-1})$ $\forall i = 1, \dots, N$. By virtue of the density of $\mathcal{D}(\mathbb{R}^{n-1})$ in $H^s(\mathbb{R}^{n-1})$ $\forall s \geq 0$ by Theorem 8.9.6, $\forall i = 1, \dots, N$, $\exists$ a sequence $(\psi_{im})_{m=1}^\infty$ in $\mathcal{D}(\mathbb{R}^{n-1})$ with $\text{supp}(\psi_{im}) \subset\subset \hat{Q}_0$ by virtue of (8.13.20) such that $\psi_{im} \longrightarrow \widetilde{(\beta_i u \circ \Phi_i^{-1})}$ in $H^s(\mathbb{R}^{n-1})$ as $m \to \infty$. (8.13.31)

Since $\Phi_i : \Omega_i \cap \Gamma \longrightarrow \hat{Q}_0$ is *infinitely* differentiable from $\Omega_i \cap \Gamma$ onto $\hat{Q}_0 \subset \mathbb{R}^{n-1}$ $\forall i = 1, 2, \dots, N$, $(\psi_{im} \circ \Phi_i) \in \mathcal{D}(\Gamma)$ $\forall m \in \mathbb{N}$ with $\text{supp}(\psi_{im} \circ \Phi_i) \subset\subset \Gamma_i$. Hence, using (8.13.31), $\forall i = 1, \dots, N$, $\widetilde{(\psi_{im} \circ \Phi_i)} \in \mathcal{D}(\Gamma) \to \widetilde{(\beta_i u \circ \Phi_i^{-1}) \circ \Phi_i} = \tilde{\beta}_i u \in H^s(\Gamma)$ as $m \to \infty$ and $\text{supp}(\tilde{\beta}_i u) \subset\subset \Gamma_i$.

Thus, $\forall u \in H^s(\Gamma)$ with $u = \sum_{i=1}^N \beta_i u$, $\exists$ a sequence $\widetilde{(\psi_{im} \circ \Phi_i)}$ in $\mathcal{D}(\Gamma)$ with $\lim_{m\to\infty} \widetilde{\psi_{im} \circ \Phi_i} = \tilde{\beta}_i u$ in $H^s(\Gamma)$ and $\sum_{i=1}^N \lim_{m\to\infty} \widetilde{(\psi_{im} \circ \Phi_i)} = \sum_{i=1}^N \tilde{\beta}_i u = u \in H^s(\Gamma)$. □

Property 5 For $\Gamma \in C^\infty$-class, since $\mathcal{D}(\Gamma)$ is *dense* in $H^s(\Gamma)$ $\forall s > 0$ (by Theorem 8.13.7),

$$H^s(\Gamma) \equiv H_0^s(\Gamma) = \overline{\mathcal{D}(\Gamma)} \quad \text{in } H^s(\Gamma). \tag{8.13.32}$$

Property 6: Space $H^{-s}(\Gamma)$ with $\Gamma \in C^\infty$-class, $s > 0$

Definition 8.13.7. For $\Gamma \in C^\infty$-class, $s > 0$, $(H^s(\Gamma))' = H^{-s}(\Gamma)$ is the *dual* of $H^s(\Gamma)$. Then $H^{-s}(\Gamma) \hookrightarrow \mathcal{D}'(\Gamma)$, i.e. $H^{-s}(\Gamma)$ is a *space of distributions* on Γ (see Section 5.1, Chapter 5).

Remark 8.13.4. For $\Gamma \in C^m$-class (but $\Gamma \notin C^\infty$-class), by virtue of Remark 8.13.3, for $0 < s \le m$, $(H^s(\Gamma))' =$ *dual* of $H^s(\Gamma)$ is *not* a *space of distributions*, which are, in fact, *not* defined on $\Gamma \in C^m$-class.

8.13.4 Sobolev spaces on boundary curves Γ in $\mathbb{R}^2$

Based on Definitions D.2.3.1 and D.4.1.1, Appendix D, $W^{s,p}(\Gamma)$ and $H^s(\Gamma)$ were defined in earlier sections. These definitions give the *global* properties of the boundary Γ of a domain $\Omega \subset \mathbb{R}^2$. For example, polygonal boundary Γ is *globally* Lipschitz continuous (i.e. of $C^{0,1}$-class *globally*) and can never be of $C^{1,1}$-class globally. But polygons are *piecewise of C^∞-class*. In many applications, boundaries Γ will be *piecewise sufficiently smooth*. For these situations, *parametric representation* of Γ is very convenient to utilize this *local* regularity of Γ, which suggests a third *alternative* definition of $W^{s,p}(\Gamma)$, $s \ge 0$, $1 \le p < \infty$, based on parametrization of Γ by its arc length parameter t. We consider first a *smooth* boundary Γ and then polygonal Γ in $\mathbb{R}^2$.

Case I: Smooth boundary Γ of $\Omega \subset \mathbb{R}^2$

Let $\Gamma \subset \mathbb{R}^2$ be a *smooth* (i.e. of C^∞-class globally), *simple*, *closed* curve of length L, and $\mathbf{x}_0 \in \Gamma$ be *any fixed* point. Then the equation of Γ in *parametric form* is given, $\forall \mathbf{x} \in \Gamma$, by:

$$\mathbf{x} = \phi_{\mathbf{x}_0}(t), \quad 0 \le t < L, \tag{8.13.33}$$

t being arc length measured along Γ from $\mathbf{x}_0 \in \Gamma$ such that $\phi_{\mathbf{x}_0} : [0, L[\longrightarrow \Gamma \subset \mathbb{R}^2$ and its *inverse* $\psi_{\mathbf{x}_0} : \Gamma \longrightarrow [0, L[$ with $\phi_{\mathbf{x}_0}(0) = \mathbf{x}_0$ and $\psi_{\mathbf{x}_0}(\mathbf{x}_0) = 0$. (8.13.34)

Smooth open arc $\Gamma_0 \subset \Gamma$ Let $\Gamma_0 \subset \Gamma$ be an *open* arc of Γ such that $L_0 =$ length measure of Γ_0, $P_0 = \mathbf{x}_0$ and $P_1 = \mathbf{x}_1$ are the end points of Γ_0. Then $\overline{\Gamma}_0 = \Gamma_0 \cup \{P_0, P_1\}$ is defined by (8.13.33) and (8.13.34) with L replaced by L_0, i.e.

$$\phi_{\mathbf{x}_0}(0) = \mathbf{x}_0 = P_0,\ \phi_{\mathbf{x}_0}(L_0) = P_1 \quad \text{and} \quad \psi_{\mathbf{x}_0}(\mathbf{x}_0) = 0,\ \psi_{\mathbf{x}_0}(\mathbf{x}_1) = L_0. \tag{8.13.35}$$

Remark 8.13.4a. $\phi_{\mathbf{x}_0}$ in (8.13.33)–(8.13.35) is not shown in bold face and also later.

Example 8.13.1. Consider the unit circle Γ in $\mathbb{R}^2$, which has the parametric representation: $x_1 = \cos(\theta_0 + t)$, $x_2 = \sin(\theta_0 + t)$, i.e. $\mathbf{x} = \phi_{\mathbf{x}_0}(t) = (\cos(\theta_0 + t), \sin(\theta_0 + t))$ for $0 \le t < 2\pi$, if $\mathbf{x}_0 = (\cos\theta_0, \sin\theta_0)$. Choosing $\mathbf{x}_0 = (1, 0)$, we have $\phi_0(t) = (\cos t, \sin t) = (x_1, x_2)$ with $x_1 = \cos t$, $x_2 = \sin t$, $0 \le t < 2\pi$. Then Γ is *smooth*, i.e. of C^∞-class *globally*.

Spaces $W^{s,p}(\Gamma)$, $W^{s,p}(\Gamma_0)$, $1 \le p < \infty$, $s \ge 0$ with $\Gamma, \Gamma_0 \in C^\infty$-class

Definition 8.13.8. Let Γ be a *smooth* (i.e. C^∞-class) boundary of $\Omega \subset \mathbb{R}^2$ defined by (8.13.33)–(8.13.34). Then, for $s \ge 0$, $1 \le p < \infty$, $W^{s,p}(\Gamma)$ is defined by

$$W^{s,p}(\Gamma) = \{u : \forall \mathbf{x}_0 \in \Gamma,\ u \circ \phi_{\mathbf{x}_0} \in W^{s,p}(]0, L[)\}, \tag{8.13.36}$$

with $\|u\|_{W^{s,p}(\Gamma)} = \|u \circ \phi_{\mathbf{x}_0}\|_{W^{s,p}(]0,L[)}$ and $\|\cdot\|_{W^{s,p}(]0,L[)}$ given by (8.10.90a). (8.13.37)

Let $\Gamma_0 \subset \Gamma$ be a *smooth* arc of Γ defined in (8.13.33)–(8.13.35). Then, for $s \ge 0$, $1 \le p < \infty$,

$$W^{s,p}(\Gamma_0) = \{u : u \circ \phi_{\mathbf{x}_0} \in W^{s,p}(]0, L_0[)\}, \tag{8.13.38}$$

with

$$\|u\|_{W^{s,p}(\Gamma_0)} = \|u \circ \phi_{\mathbf{x}_0}\|_{W^{s,p}(]0,L_0[)}. \tag{8.13.39}$$

Then $W^{s,p}(\Gamma)$ (resp. $W^{s,p}(\Gamma_0)$) is a *Banach space*. (8.13.40)

For $p = 2$, $H^s(\Gamma) = W^{s,2}(\Gamma)$, $H^s(\Gamma_0) = W^{s,2}(\Gamma_0)$ with $\|u\|_{H^s(\Gamma)} = \|u \circ \phi_{\mathbf{x}_0}\|_{H^s(]0,L[)} = \|u \circ \phi_{\mathbf{x}_0}\|_{W^{s,2}(]0,L[)}$, $\|u\|_{H^s(\Gamma_0)} = \|u \circ \phi_{\mathbf{x}_0}\|_{W^{s,2}(]0,L_0[)}$ and $\|u\|_{H^s(]0,L_0[)}$ given by (8.10.73) are *Hilbert spaces*. (8.13.41)

Proposition 8.13.1. *Let Γ be a smooth boundary of $\Omega \subset \mathbb{R}^2$. Then, $\forall k \in \mathbb{N}_0$, the mapping $u \in C^k(\overline{\Omega}) \longmapsto u\!\downarrow_\Gamma \in W^{s,p}(\Gamma)$ is a* continuous, *linear mapping from $C^k(\overline{\Omega})$ into $W^{s,p}(\Gamma)$ for $0 \le s \le k$, $1 \le p < \infty$.*

In particular, for $u \in C^\infty(\overline{\Omega})$, $u\!\downarrow_\Gamma \in W^{s,p}(\Gamma)$ $\forall s \ge 0$, $1 \le p < \infty$. (8.13.42)

Spaces $W_0^{s,p}(\Gamma_0)$ and $W_{00}^{s,p}(\Gamma_0)$

Definition 8.13.9. Let $\Gamma_0 \subset \Gamma$ be a *smooth* arc (i.e. of C^∞-class) of the boundary Γ of a bounded domain $\Omega \subset \mathbb{R}^2$, which is parametrized by (8.13.33)–(8.13.35). Then:

1. for $s > 0$ and $1 \le p < \infty$, $W_0^{s,p}(\Gamma_0)$ is the *closure* of the set $\{v : v = u\!\downarrow_{\Gamma_0}$ with $u \in \mathcal{D}(\Gamma)$, $u(\mathbf{x}) = 0$ for $\mathbf{x}$ belonging to *some* neighbourhood of $\Gamma \setminus \Gamma_0\}$ in $W^{s,p}(\Gamma_0)$; (8.13.43)
2. for $s > 0$ and $1 \le p < \infty$ such that $s - 1/p =$ an integer, we define $W_{00}^{s,p}(\Gamma_0)$ by: $W_{00}^{s,p}(\Gamma_0) = \{u : u \in W^{s,p}(\Gamma_0)$, its null extension $\tilde{u}$ to Γ belongs to $W^{s,p}(\Gamma)\}$ with $\|u\|_{W^{s,p}(\Gamma_0)} = \|\tilde{u}\|_{W^{s,p}(\Gamma)} = \|\tilde{u}\|_{W_{00}^{s,p}(]0,L[)}$ (see also (8.10.129)). (8.13.44)

Example 8.13.2. For $\Gamma \in C^\infty$- class in $\mathbb{R}^2$ and $\Gamma_0 \subset \Gamma$ with $L_0 < L$ as in (8.13.44), we define, for $p = 2$, $H_0^{1/2}(\Gamma_0) = W_0^{s,2}(\Gamma_0)$, $H_{00}^{1/2}(\Gamma_0) = W_{00}^{s,2}(\Gamma_0)$ by:

- $H_0^{1/2}(\Gamma_0) = \overline{\text{Closure}\{v : v = u\downarrow_{\Gamma_0} \text{ with } u \in \mathcal{D}(\Gamma), u \text{ vanishes in } \textit{some}}$ $\overline{\text{neighbourhood of } \Gamma \setminus \Gamma_0\}}$ in $H^{1/2}(\Gamma_0)$; (8.13.45)
- $H_{00}^{1/2}(\Gamma_0) = \{u : u \in H^{1/2}(\Gamma_0) \text{ such that } \tilde{u} \in H^{1/2}(\Gamma)\}$ with

$$\langle u, v\rangle_{H_{00}^{1/2}(\Gamma_0)} = \langle \tilde{u}, \tilde{v}\rangle_{1/2,\Gamma}; \quad \|u\|_{H_{00}^{1/2}(\Gamma_0)} = \|\tilde{u}\|_{1/2,\Gamma} \quad \forall u, v \in H_{00}^{1/2}(\Gamma_0); \tag{8.13.46}$$

- $H_0^{3/2}(\Gamma_0) = \overline{\text{Closure}\{v : v = u\downarrow_{\Gamma_0} \text{ with } u \in \mathcal{D}(\Gamma), u \text{ vanishes in } \textit{some}}$ $\overline{\text{neighbourhood of } \Gamma \setminus \Gamma_0\}}$ in $H^{3/2}(\Gamma_0)$; (8.13.47)
- $H_{00}^{3/2}(\Gamma_0) = \{u : u \in H^{3/2}(\Gamma_0) \text{ such that } \tilde{u} \in H^{3/2}(\Gamma)\}$ with

$$\langle u, v\rangle_{H_{00}^{3/2}(\Gamma_0)} = \langle \tilde{u}, \tilde{v}\rangle_{3/2,\Gamma}; \quad \|u\|_{H_{00}^{3/2}(\Gamma_0)} = \|\tilde{u}\|_{3/2,\Gamma} \quad \forall u, v \in H_{00}^{3/2}(\Gamma_0); \tag{8.13.48}$$

Space $W^{-s,q}(\Gamma_0)$ with $s > 0$, $1 < p < \infty$, $\frac{1}{p} + \frac{1}{q} = 1$

Definition 8.13.10. For the smooth open arc $\Gamma_0 \subset \Gamma$ with $L_0 < L$ in (8.13.33)–(8.13.35), $s > 0$ and $1 < p < \infty$, $\frac{1}{p} + \frac{1}{q} = 1$, we define $W^{-s,q}(\Gamma_0)$ by:

$$W^{-s,q}(\Gamma_0) = \text{dual space of } W_0^{s,p}(\Gamma_0) = (W_0^{s,p}(\Gamma_0))'. \tag{8.13.49}$$

Remark 8.13.5. For smooth Γ (i.e. of C^∞-class), $\mathcal{D}(\Gamma)$ is well defined by (8.13.28) and *dense* in $W_0^{s,p}(\Gamma) = \overline{\mathcal{D}(\Gamma)} = W^{s,p}(\Gamma)$, and

$$(W^{s,p}(\Gamma))' = (W_0^{s,p}(\Gamma))' = W^{-s,q}(\Gamma) \hookrightarrow \mathcal{D}'(\Omega) \tag{8.13.50}$$

with $\frac{1}{p} + \frac{1}{q} = 1$.

In particular, for $p=2$, $H^{-s}(\Gamma) = (H^s(\Gamma))'$ for *smooth* Γ (see (Definition 8.13.7)).

Imbedding results for $\Gamma \in C^\infty$-class with $\Gamma \subset \mathbb{R}^2$ (resp. $\Gamma_0 \subset \Gamma$ with (8.13.35))

For easy reference, we have:

Theorem 8.13.8. *Let $\Gamma \in C^\infty$-class be a boundary curve in $\mathbb{R}^2$ (resp. $\Gamma_0 \subset \Gamma \subset \mathbb{R}^2$, $\Gamma_0 \in C^\infty$-class satisfying* (8.13.35)*) and $k \in \mathbb{N}_0$, $s_i \geq 0$, $i = 1, 2$. Then:*

- *The following imbeddings* I–IV *are* dense *and* continuous*:*

  I. $s_1 < s_2 \Longrightarrow H^{s_2}(\Gamma) \hookrightarrow H^{s_1}(\Gamma)$; $H^{s_2}(\Gamma_0) \hookrightarrow H^{s_1}(\Gamma_0)$;
  $H_0^{s_2}(\Gamma_0) \hookrightarrow H_0^{s_1}(\Gamma_0)$; (8.13.51a)

II. $k_1 < k_2 \Longrightarrow H_{00}^{k_2+\frac{1}{2}}(\Gamma_0) \hookrightarrow H_{00}^{k_1+\frac{1}{2}}(\Gamma_0)$.

In particular, $H_{00}^{3/2}(\Gamma_0) \hookrightarrow H_{00}^{1/2}(\Gamma_0)$; (8.13.51b)

III. $s_1 < k_2 + 2 \Longrightarrow H_{00}^{k_2+1/2}(\Gamma_0) \hookrightarrow H_0^{s_1}(\Gamma_0)$.

In particular, $s_1 = 0, k_2 = 0 \Longrightarrow H_{00}^{1/2}(\Gamma_0) \hookrightarrow L^2(\Gamma_0); s = 1, k_2 = 1 \Longrightarrow H_{00}^{3/2}(\Gamma_0) \hookrightarrow H_0^1(\Gamma_0)$; (8.13.51c)

IV. $k_1 + \frac{1}{2} < s_2 \Longrightarrow H^{s_2}(\Gamma_0) \hookrightarrow H_{00}^{k_1+1/2}(\Gamma_0)$.

In particular, $k_1 = 0$, $s_2 = 1 \Longrightarrow H_0^1(\Gamma_0) \hookrightarrow H_{00}^{1/2}(\Gamma_0)$. (8.13.51d)

- *The following imbeddings* V–VIII *are* continuous*:*

V. $s_1 < s_2 \Longrightarrow H^{-s_1}(\Gamma) \hookrightarrow H^{-s_2}(\Gamma)$*;* $H^{-s_1}(\Gamma_0) \hookrightarrow H^{s_2}(\Gamma_0)$*;* (8.13.51e)

VI. $k_1 < k_2 \Longrightarrow (H_{00}^{k_1+1/2}(\Gamma_0))' \hookrightarrow (H_{00}^{k_2+1/2}(\Gamma_0))'$.

In particular, $(H_{00}^{1/2}(\Gamma_0))' \hookrightarrow (H_{00}^{3/2}(\Gamma_0))'$; (8.13.51f)

VII. $s_1 < k_2 + 1/2 \Longrightarrow H^{-s_1}(\Gamma_0) \hookrightarrow (H_{00}^{k_2+1/2}(\Gamma_0))'$; (8.13.51g)

VIII. $k_1 + 1/2 < s_2 \Longrightarrow (H_{00}^{k_1+1/2}(\Gamma_0))' \hookrightarrow H^{-s_2}(\Gamma_0)$. (8.13.51h)

Restriction results for functions in $H^s(\Gamma)$ for $\Gamma \in C^\infty$-class ($\Gamma \subset \mathbb{R}^2$)

Let $\Gamma \in C^\infty$-class be the smooth boundary of a bounded $\Omega \subset \mathbb{R}^2$, $\Gamma_0 \subsetneq \Gamma$ satisfying (8.13.35), $u \in H^s(\Gamma)$ and v be its restriction to Γ_0, i.e. $v = u{\downarrow}_{\Gamma_0}$. Then we have:

1. $\forall s \geq 0$, the *restriction mapping* $u \longmapsto u{\downarrow}_{\Gamma_0}$ is continuous from $H^s(\Gamma)$ into $H^s(\Gamma_0)$ (see Theorem 8.10.23). (8.13.52)

   But for $u \in H^{-1/2}(\Gamma)$ (resp. $H^{-3/2}(\Gamma)$) the restriction mapping is *not* continuous from $H^{-1/2}(\Gamma)$ into $H^{-1/2}(\Gamma_0)$ (resp. *not* continuous from $H^{-3/2}(\Gamma)$ into $H^{-3/2}(\Gamma_0)$) (see (8.10.153)).

2. The restriction mapping $u \longmapsto u{\downarrow}_{\Gamma_0} = v$ is *continuous* from $H^{-1/2}(\Gamma)$ into $(H_{00}^{1/2}(\Gamma_0))'$ (resp. from $H^{-3/2}(\Gamma)$ into $(H_{00}^{3/2}(\Gamma_0))'$) (see Theorem 8.10.23). (8.13.53)

Case II: Polygonal and curvilinear polygonal boundary Γ

Let $\Omega \subset \mathbb{R}^2$ be a bounded, open subset of $\mathbb{R}^2$ and Γ be the *polygonal* (resp. *curvilinear polygonal*) boundary of Ω with *vertices* or *corner points* $\{P_i\}_{i=1}^N$ and straight line segments $\overline{\Gamma}_i$ (resp. curves $\overline{\Gamma}_i$) joining the vertices P_i and P_{i+1}, which are the *end points* of the straight (resp. curved) sides $\overline{\Gamma}_i$ with $P_{i+1} = P_1 \pmod N$ such that $\Gamma_i = \mathrm{int}(\overline{\Gamma}_i)$, i.e. $\overline{\Gamma}_i$ with end points P_i and P_{i+1} *excluded*, and $\omega_i \in]0, 2\pi[$, $\omega_i \neq \pi$ is the measure of the *interior* angle at P_{i+1} between $\overline{\Gamma}_{i+1}$ and $\overline{\Gamma}_i(\overline{\Gamma}_{i+1} = \overline{\Gamma}_1$ for $i = N)$ in the trigonometric sense. The length measure of $\overline{\Gamma}_i = L_i$, $1 \leq i \leq N$, and

$L = \bigcup_{i=1}^{N} L_i$ is the length measure of Γ. For *polygonal* (resp. *curvilinear polygonal*) boundary Γ, the parametric representation of Γ in (8.13.33)–(8.13.34) will hold *in principle*.

In general, $\mathbf{x}_0$ in $\phi_{\mathbf{x}_C}(s) = \mathbf{x} \in \Gamma$ is chosen as the first vertex P_1 of Γ with $\phi_{P_1}(s) = \mathbf{x} \in \Gamma$, $0 \leq s < L$, such that $\phi_{P_1}(0) = P_1$, $\phi_{P_1}(s) = \mathbf{x} \in \overline{\Gamma}_1$ for $0 \leq s \leq L_1, \ldots,$ $\phi_{P_1}(s) = \mathbf{x} \in \overline{\Gamma}_i$ for $\sum_{j=1}^{i-1} L_j \leq s \leq \sum_{j=1}^{i} L_j, \cdots,$ $\phi_{P_1}(s) = \mathbf{x} \in \overline{\Gamma}_N$ for $\sum_{i=1}^{N-1} L_i \leq s \leq L.$ (8.13.54)

But a *local* definition of $\overline{\Gamma}_i$ is obviously more convenient, if we use $\phi_{P_i}(s) = \mathbf{x} \in \overline{\Gamma}_i$ for $0 \leq s \leq L_i$ with $\phi_{P_i}(0) = P_i \in \overline{\Gamma}_i$, $\phi_{P_i}(L_i) = P_{i+1} \in \overline{\Gamma}_i$, $1 \leq i \leq N$, $i+1 = 1$ for $i = N$. (8.13.55)

In fact, Γ can be described by parametric equations in *various* ways. For example, for polygonal Γ with vertices $P_i = (x_1^i, x_2^i)$, $1 \leq i \leq N$,

$$\phi_i(s) = \phi_{P_i}(s) = P_i + \frac{P_{i+1} - P_i}{L_i} s = \left(x_1^i + \frac{x_1^{i+1} - x_1^i}{L_i} s,\, x_2^i + \frac{x_2^{i+1} - x_2^i}{L_i} s \right) \in \Gamma_i \tag{8.13.56}$$

for $0 \leq s \leq L_i$, $1 \leq i \leq N$, $i+1 = 1 \pmod N$.

Case II(a): $\Gamma_0 \subsetneq \Gamma$ (see (8.13.35) also) For polygonal (resp. curvilinear polygonal) Γ, Γ_0 is an *open* subset of Γ with $\overline{\Gamma}_0 \subsetneq \Gamma$, L_0 = length measure of $\overline{\Gamma}_0$ such that $L_0 < L$, for $i = 1, \ldots, N_0$,

$$\Gamma_i \subset \Gamma_0, \tag{8.13.57}$$

P_1 and P_N being the end points of Γ_0 with $1 < N_0 < N$ (after a *possible* renumbering of end-points and vertices) and $\bigcup_{i=1}^{N_0} \overline{\Gamma}_i = \overline{\Gamma}_0$.

Properties of function $\phi_{P_1} : [0, L[\longrightarrow \Gamma$ in (8.13.54) For polygonal boundary Γ (resp. $\Gamma_0 \subset \Gamma$), ϕ_{P_1} is *globally Lipschitz continuous* on $[0, L[$, i.e. of $C^{0,1}$-class, but *never* of $C^{1,1}$-class (globally). ϕ_{P_1} is globally *piecewise* smooth (i.e. *piecewise* of C^∞-class), since each side Γ_i is of C^∞-class. (8.13.58)

For curvilinear polygons Γ (resp. $\Gamma_0 \subset \Gamma$), each curve $\overline{\Gamma}_i$ may have different *orders of regularity*. We make the following assumption:

Each curve defining the sides Γ_i are of $C^{k,1}$-class with $k \in \mathbb{N}_0$, i.e. kth-order derivatives are *Lipschitz continuous* (see (D.2.3.8), Appendix D).

Then, Γ will be called *piecewise* of $C^{k,1}$-class. (8.13.59)

Spaces $W^{s,p}(\Gamma)$ and $W^{s,p}(\Gamma_0)$ for polygonal Γ and $\Gamma_0 \subset \Gamma$

Since polygons Γ are *piecewise* of C^∞-class, for smooth functions $\phi \in \mathcal{D}(\overline{\Omega})$ the restriction $\phi{\downarrow}_{\Gamma_i} \in \mathcal{D}(\overline{\Gamma}_i)$ $\forall$ sides Γ_i, $1 \leq i \leq N$. Moreover, each open side Γ_i of Γ can be viewed *locally* as an open segment of the real line $\mathbb{R}$. (8.13.60)

Then the results of Sobolev spaces $W^{s,p}(\Omega)$ with $\Omega = \mathbb{R}$ or $\Omega \subset \mathbb{R}$ given in earlier sections can be applied to $\Gamma_i \subset \mathbb{R}$. In particular, for two neighbouring sides $\overline{\Gamma}_i$ and $\overline{\Gamma}_{i+1}$ with a common vertex P_{i+1} (mod N), we may define $\overline{\Gamma}_i$ by

$$\phi_{P_{i+1}}(s) = \mathbf{x} \in \overline{\Gamma}_i \quad \text{for } s \in [-L_i, 0],$$

and $\overline{\Gamma}_{i+1}$ by

$$\phi_{P_{i+1}}(s) = \mathbf{x} \in \overline{\Gamma}_{i+1} \quad \text{for } s \in [0, L_{i+1}], \tag{8.13.61}$$

where $\phi_{P_{i+1}}$ is a *smooth* function of s on $]-L_i, 0[$ (resp. $]0, L_{i+1}[$) and $\phi_{P_{i+1}}(0)$ is defined and $\phi_{P_{i+1}}$ is *continuous* at $s = 0$, but $\phi_{P_{i+1}}$ will *not* be differentiable at $s = 0$ which corresponds to the origin $0 \in \mathbb{R}$. Moreover, separately on each open side Γ_i, *normal derivatives of smooth functions can be prescribed*. Then, instead of the *global* approach to defining a function on the *whole* polygon, it will be more prudent and useful to define *locally* functions $v_i \in W^{s,p}(\Gamma_i)$, $1 \le i \le N, s > 0$ and then find the property of the *globally* defined function u on Γ with $u\downarrow_{\Gamma_i} = v_i \in W^{s,p}(\Gamma_i)$, $1 \le i \le N$. The first question that arises is: Will u also belong to $W^{s,p}(\Gamma)$ for any $s > 0$? The answer is *negative* in general. The reason is that for this, the functions v_i defined *separately* on *each* side Γ_i must satisfy *additional* conditions, i.e. 'compatibility conditions', near the corner points or vertices. We explain with the help of a simple example.

Example 8.13.3. Consider the unit square with $N = 4$ vertices: $P_1 = (0,0)$, $P_2 = (1,0)$, $P_3 = (1,1)$, $P_4 = (0,1)$, and N sides: $\overline{\Gamma}_1 = [P_1, P_2]$, $\overline{\Gamma}_2 = [P_2, P_3]$, $\overline{\Gamma}_3 = [P_3, P_4]$, $\overline{\Gamma}_4 = [P_4, P_1]$ following the orientation of the boundary Γ enclosing $\Omega =]0,1[\times]0,1[\subset \mathbb{R}^2$. We consider the adjacent sides $\overline{\Gamma}_1$ and $\overline{\Gamma}_4$ with the common vertex $\overline{\Gamma}_1 \cap \overline{\Gamma}_4 = \{P_1 = (0,0)\}$, the parametrized equation for which is given by (8.13.61): for $\mathbf{x} \in \overline{\Gamma}_1, \mathbf{x} = \phi_{P_1}(t) = (t, 0)$ for $0 \le t \le 1$; for $\mathbf{x} \in \overline{\Gamma}_4$, $\mathbf{x} = \phi_{P_1}(t) = (0, t)$ for $-1 \le t \le 0$, $\phi_{P_1}(0) = P_1 = (0,0)$. Let $u(\mathbf{x}) = x_1 \ \forall (x_1, x_2) \in \overline{\Omega} = [0,1] \times [0,1]$, which is of C^∞-class.

Then $u \circ \phi_{P_1}(t) = t$ for $0 \le t \le 1$, $u \circ \phi_{P_1}(t) = 0$ for $-1 \le t \le 0$.

Set $v(t) = u \circ \phi_{P_1}(t)$ for $-1 \le t \le 1$. Then v has the following properties: $v(0) = v(0^+) = v(0^-) = 0$, i.e. $v \in C^0([-1,1])$; $v'(t) = (u \circ \phi_{P_1})'(t) = 1$ for $0 < t < 1$ and $v'(t) = 0$ for $-1 < t < 0$ with $v'(0^-) = 0$, $v'(0^+) = 1$, i.e. $v'(t) = (u \circ \phi_{P_1})'(t)$ has a *non-zero* jump at $t = 0$. Hence $v' = (u \circ \phi_{P_1})'$ is *not* continuous on $]-1,1[$, i.e. $v' \notin C^0([-1,1])$, but $v' \in L^p(]-1,1[) = W^{0,p}(]-1,1[)$ $\forall p$ with $1 \le p \le \infty$. Then $v' = (u \circ \phi_{P_1})' \notin W^{s,p}(]-1,1[)$ $\forall s > 0, 1 \le p \le \infty$ (for $n = 1,]-1,1[\subset \mathbb{R}$, $W^{s,p}(]-1,1[)$ is defined by Definition 8.10.10 for arbitrary $s > 0$, $1 \le p < \infty$). In fact, v' will belong to $W^{s,p}(]-1,1[)$ for those $s > 0$, $1 \le p \le \infty$, for which $W^{s,p}(]-1,1[) \not\subset C^0([-1,1])$ with $n = 1$, i.e. for $s > 0$ with $s - \frac{1}{p} < 0$, $W^{s,p}(]-1,1[) \not\subset C^0([-1,1])$ (see imbedding theorem). Hence,

$v' = (u \circ \phi_{P_1})' \in W^{s,p}(]-1,1[)$ iff $0 \le s < 1/p$, $1 \le p \le \infty$,

$$\Longrightarrow \quad v = (u \circ \phi_{P_1}) \in W^{s,p}(]-1,1[) \text{ iff } 0 \le s < 1 + \frac{1}{p},\ 1 \le p \le \infty, \quad (8.13.62)$$

even when $u \in C^\infty$-class. Hence, for each corner point P_i, $1 \le i \le 4$, $v \circ \phi_{P_i} \in W^{s,p}(]-1,1[)$ iff $0 \le s < 1 + \frac{1}{p}$, $1 \le p \le \infty$. Then, for Γ with

$$\begin{aligned} &\overline{\Gamma}_1\colon\ \mathbf{x} = \phi_{P_1}(t) = (t,0) && \text{for } 0 \le t \le 1;\\ &\overline{\Gamma}_2\colon\ \mathbf{x} = \phi_{P_1}(t) = (1,t-1) && \text{for } 1 \le t \le 2;\\ &\overline{\Gamma}_3\colon\ \mathbf{x} = \phi_{P_1}(t) = (3-t,1) && \text{for } 2 \le t \le 3;\\ &\overline{\Gamma}_4\colon\ \mathbf{x} = \phi_{P_1}(t) = (0,4-t) && \text{for } 3 \le t \le 4, \end{aligned} \quad (8.13.63)$$

$u \circ \phi_{P_1} \in C^0([0,4])$, $u \circ \phi_{P_1} \in L^p(]0,4[) = W^{0,p}(]0,4[)$ $\forall p$ with $1 \le p \le \infty$, but $u \circ \phi_{P_1} \in W^{s,p}(]0,4[$ $\forall s$ with $0 \le s < 1 + \frac{1}{p}$, $1 \le p \le \infty$. (8.13.64)

Thus, for higher values of $s > 0$, with $0 < s < 1 + \frac{1}{p}$, the definition of $W^{s,p}(]0,4[)$ will imply *only* the *continuity* at the corners, while *jumps* of tangential derivatives will be allowed.

Now we state the interesting result which will permit us to combine (or glue) the *piecewise* locally defined functions belonging to $W^{s,p}$-spaces for $0 \le s < 1 + 1/p$ into a *globally* defined function having the *same* properties possibly under some additional compatibility conditions at the corner points.

Theorem 8.13.9. *Let $1 \le p \le \infty$, $0 \le s < 1 + 1/p$ and $u = u(t)$ for $-1 < t < 1$ such that $u \in L^p(]-1,1[)$, $v_1 = u\downarrow_{]-1,0[} \in W^{s,p}(]-1,0[)$, $v_2 = u\downarrow_{]0,1[} \in W^{s,p}(]0,1[)$, where $W^{s,p}$-spaces are defined by Definition* 8.10.10. *Then $u \in W^{s,p}(]-1,1[)$*

I. *for $s < 1/p$ (i.e. no additional conditions necessary);* (8.13.65)

II. *for $s > 1/p$ iff $u(0^-) = u(0^+)$;* (8.13.66)

III. *for $s = 1/p$ iff $\int_{-\delta}^{\delta} \frac{|u(t)-u(-t)|^p}{|t|} dt < +\infty$ or, equivalently, iff $\int_0^{\delta} \frac{|u(t)-u(-t)|^p}{t} dt < +\infty$ with $0 < \delta < \delta_0$, $0 < \delta_0 < 1$.* (8.13.67)

In particular, for $p = 2$ and $0 \le s < 3/2$, and $u = u(t)$ for $-1 < t < 1$ such that $u \in L^2(]-1,1[)$,

$$\begin{aligned} v_1 &= u\downarrow_{]-1,0[} \in H^s(]-1,0[) = W^{s,2}(]-1,0[);\\ v_2 &= u\downarrow_{]0,1[} \in H^s(]0,1[) = W^{s,2}(]0,1[) \end{aligned}$$

$\Longrightarrow u \in H^s(]-1,1[)$ *for*

IV. $0 \le s < 1/2$ *(no additional conditions);* (8.13.68)

V. $1/2 < s < 3/2$ *if and only if* $u(0^-) = u(0^+)$; (8.13.69)

VI. $s = 1/2$ *iff* $\int_{-\delta}^{\delta} \frac{|u(t)-u(-t)|^2}{|t|} dt < +\infty$ *or equivalently, iff* $\int_0^{\delta} \frac{|u(t)-u(-t)|^2}{t} dt < +\infty$ *with* $0 < \delta < \delta_0$, $0 < \delta_0 < 1$. (8.13.70)

Conditions (8.13.65)–(8.13.67) (resp. (8.13.68)–(8.13.70)) are the corresponding compatibility conditions associated with the point $t = 0$, which must hold for $t = 0$.

Remark 8.13.6. The proof and justification of these compatibility conditions are intimately linked with trace results to be discussed in Section 8.14, but some explanations without trace results are in order. We give them for $p = 2$. For $s = 1/2$, $H^{1/2}(]-1, 1[) \not\subset C^0([-1, 1])$, i.e. functions of $H^{1/2}(]-1, 1[)$ are *not* continuous in general. Condition (8.13.69) is *not* meaningful for $s = 1/2$ in general, but the compatibility condition (8.13.70) or III will guarantee that $u \in H^{1/2}(]-1, 1[)$, which follows from its Definition 8.10.6. Again, for $s > 1/2$ with $s - 1/2 > 0 \Longrightarrow H^s(]-1, 1[) \hookrightarrow C^0([-1, 1])$ by (8.12.28), i.e. functions of $H^s(]-1, 1[)$ are *continuous*, and hence (8.13.69) holds at $t = 0$ and condition (8.13.69) is meaningful.

If we translate the results of Theorem 8.13.9 to the case of polygon Γ in Example 8.13.3, $t = 0$ corresponds to a vertex of Γ.

Based on the results of Theorem 8.13.9 and Example 8.13.3, we can now define:

Spaces $W^{s,p}(\Gamma)$, $W^{s,p}(\Gamma_0)$

Definition 8.13.11. Let Γ be a *polygonal boundary* of a bounded, open subset $\Omega \subset \mathbb{R}^2$ and $\Gamma_0 \subset \Gamma$ be defined by (8.13.33)–(8.13.35). Then $W^{s,p}(\Gamma)$ and $W^{s,p}(\Gamma_0)$ are defined as follows:

1. For $0 < s < 1/p$ with $1 \le p < \infty$ (or $s = 0$ *and* $p = \infty$),

$$W^{s,p}(\Gamma) = \{u : u \circ \phi_{P_1} \in W^{s,p}(]0, L[)\}$$

with

$$\|u\|_{W^{s,p}(\Gamma)} = \|u\|_{s,p,\Gamma} = \|u \circ \phi_{P_1}\|_{s,p,]0,L[} \tag{8.13.71}$$

(see (8.13.37));

$$W^{s,p}(\Gamma_0) = \{u : u \circ \phi_{P_1} \in W^{s,p}(]0, L_0[)\}$$

with

$$\|u\|_{W^{s,p}(\Gamma_0)} = \|u\|_{s,p,\Gamma_0} = \|u \circ \phi_{P_1}\|_{s,p,]0,L_0[}, \tag{8.13.72}$$

where $\Gamma : \mathbf{x} = \phi_{P_1}(t)$ for $0 \le t < L =$ length measure of Γ; for $0 < s < 1/p$, $\Gamma_0 : \mathbf{x} = \phi_{P_1}(t)$ for $0 \le t \le L_0 =$ length measure of Γ_0; for $0 < s < 1/p$,

$$\|u \circ \phi_{P_1}\|_{s,p,]0,L[} = \Bigg(\|u \circ \phi_{P_1}\|^p_{0,p,]0,L[} + \int_0^L \int_0^L_{t \neq \tau} \frac{|u(\phi_{P_1}(t)) - u(\phi_{P_1}(\tau))|^p}{|t-\tau|^{1+sp}} dt\, d\tau \Bigg)^{1/p}; \quad (8.13.73)$$

replacing L by L_0 in (8.13.73), we get $\|\cdot\|_{s,p,]0,L_0[}$.

2. For $1/p < s$ with $1 < p \le \infty$, $L_j =$ length measure of $\overline{\Gamma}_j = [P_j, P_{j+1}]$,

$$W^{s,p}(\Gamma) = \{u : u \in L^p(\Gamma_j),$$

$u \circ \phi_{P_j} \in W^{s,p}(]0, L_j[)$ and $u \circ \phi_{P_j}(L_j^-) = u \circ \phi_{P_j}(L_j^+)$, $1 \le j \le N\}$ with

$$\|u\|_{s,p,\Gamma} = \Bigg(\sum_{j=1}^{N} \|u \circ \phi_{P_j}\|^p_{s,p,]0,L_j[} \Bigg)^{1/p}; \quad (8.13.74)$$

$$W^{s,p}(\Gamma_0) = \{u : u \in L^p(\Gamma_j), \quad (8.13.75)$$

$u \circ \phi_{P_j} \in W^{s,p}(]0, L_j[)$ and $u \circ \phi_{P_j}(L_j^-) = u \circ \phi_{P_j}(L_j^+)$, $1 \le j \le N_0 - 1\}$ with

$$\|u\|_{s,p,\Gamma_0} = \Bigg(\sum_{j=1}^{N_0} \|u \circ \phi_{P_j}\|^p_{s,p,]0,L_j[} \Bigg)^{1/p},$$

with

$$s = [s] + \sigma < 1 + 1/p, \quad [s] \in \mathbb{N}_0, \quad 0 < \sigma < 1, \quad (8.13.76)$$

where for $1 \le j \le N_0$, $\overline{\Gamma}_j : \mathbf{x} = \phi_{P_j}(t)$ for $0 \le t \le L_j =$ length measure of $\overline{\Gamma}_j$, $\Gamma_0 = \bigcup_{j=1}^{N_0} \overline{\Gamma}_j$; for $s = [s] + \sigma$ with $[s] = k \in \mathbb{N}_0$ with $0 < \sigma < 1$,

$$\|u \circ \phi_{P_j}\|_{s,p,]0,L_j[} = \Bigg(\|u \circ \phi_{P_j}\|^p_{k,p,]0,L_j[} + \int_0^{L_j} \int_0^{L_j}_{t \neq \tau} \frac{|(u(\phi_{P_j}))^{(k)}(t) - (u(\phi_{P_j}))^{(k)}(\tau)|^p}{|t-\tau|^{1+\sigma p}} dt\, d\tau \Bigg)^{1/p}. \quad (8.13.77)$$

3. For $1 < p < \infty$, $s = 1/p$,

$$W^{1/p,p}(\Gamma) = \Big\{u : u \in L^p(\Gamma_j),\ u \circ \phi_{P_j} \in W^{1/p,p}(]0, L_j[) \text{ and } \int_{-\delta}^{\delta} \frac{|u(\phi_{P_j}(L_j + t)) - u(\phi_{P_j}(L_j - t))|^p}{|t|} dt < +\infty,$$

$1 \le j \le N, \delta = \min_{1 \le j \le N} L_j\}$ with

$$\|u\|_{1/p,p,\Gamma} = \Bigg[\sum_{j=1}^{N} \Bigg(\|u \circ \phi_{P_j}\|^p_{1/p,p,]0,L_j[} + \int_{-\delta}^{\delta} \frac{|u(\phi_{P_j}(L_j + t)) - u(\phi_{P_j}(L_j - t))|^p}{|t|} dt\Bigg)\Bigg]^{1/p};$$

$$W^{1/p,p}(\Gamma_0) = \Big\{u : u \in L^p(\Gamma_j),\ u \circ \phi_{P_j} \in W^{1/p,p}(]0, L_j[) \text{ and } \int_{-\delta}^{\delta} \frac{|u(\phi_{P_j}(L_j + t)) - u(\phi_{P_j}(L_j - t))|^p}{|t|} dt < \infty,\ 1 \le j \le N_0 - 1, \delta = \min_{1 \le j \le N_0} L_j\Big\}$$

with

$$\|u\|_{1/p,p,\Gamma_0} = \Bigg[\sum_{j=1}^{N_0} \Bigg(\|u \circ \phi_{P_j}\|^p_{1/p,p,]0,L_j[} + \int_{-\delta}^{\delta} \frac{|u(\phi_{P_j}(L_j + t)) - u(\phi_{P_j}(L_j - t))|^p}{|t|} dt\Bigg)\Bigg]^{1/p};$$

$$\|u \circ \phi_{P_j}\|_{1/p,p,]0,L_j[} = \Bigg(\|u \circ \phi_{P_j}\|^p_{0,p,]0,L_j[} + \int_0^{L_j} \int_{\substack{0 \\ t \ne \tau}}^{L_j} \frac{|u(\phi_{P_j}(t)) - u(\phi_{P_j}(\tau))|^p}{|t - \tau|^2} dt d\tau\Bigg)^{1/p} \tag{8.13.78}$$

for $1 \le j \le N$. See subsection 8.14.4 and [17, 18, 19] for more details.

In particular, for $p = 2$ in (1)–(3), we get $H^s(\Gamma) = W^{s,2}(\Gamma)$ with

$$\|\cdot\|_{s,\Gamma} = \|\cdot\|_{s,2,\Gamma} \quad \text{and} \quad \textit{inner product } \langle \cdot, . \rangle_{H^s(\Gamma)} = \langle \cdot, \cdot \rangle_{s,\Gamma}, \tag{8.13.79}$$

and $H^s(\Gamma_0) = W^{s,2}(\Gamma_0)$, with $\langle \cdot, \cdot \rangle_{s,\Gamma_0}$, $\|\cdot\|_{s,\Gamma_0} = \|\cdot\|_{s,2,\Gamma_0}$. (8.13.80)

8.13.5 Spaces $H_0^s(\Gamma_i)$, $H_{00}^s(\Gamma_i)$ for polygonal sides $\Gamma_i \in C^\infty$-class, $1 \le i \le N$

For a bounded, polygonal boundary $\Gamma \subset \mathbb{R}^2$ with $\Gamma = \bigcup_{i=1}^N \overline{\Gamma}_i$, *open* sides $\Gamma_i \in C^\infty$-class, $\mathcal{D}(\Gamma_i)$ is well defined, $1 \le i \le N$. Then Γ_i can be viewed as an *open segment* of $\mathbb{R}$ and we can extend all earlier results for Sobolev spaces on $\Omega \subset \mathbb{R}$. Hence, considering $\Gamma_i = \Gamma_0$ and $\Gamma_i \subset \mathbb{R}$, we can extend the results of Theorems 8.10.15 and 8.13.7, and also (8.10.127) and (8.13.44) and get the following results:

1. $\mathcal{D}(\Gamma_i)$ is *dense* in $H_0^s(\Gamma_i)$, $1 \le i \le N$, for $s - 1/2 \neq$ an integer;
2. $\mathcal{D}(\Gamma_i)$ is *dense* in $H_{00}^s(\Gamma_i)$, $1 \le i \le N$, for $s - 1/2 =$ an integer (for example, for $s = 1/2, 3/2$); (8.13.81)
3. $H^{-s}(\Gamma_i) = (H_0^s(\Gamma_i))'$ is the *dual space* of $(H_0^s(\Gamma_i))$ for $s - \frac{1}{2} \neq$ an integer and is a *space of distributions* on Γ_i, $1 \le i \le N$; (8.13.82)
4. For $s - \frac{1}{2} =$ an integer, (for example, $s = 1/2, 3/2$), $(H_{00}^s(\Gamma_i))'$ is the *dual space* of $(H_{00}^s(\Gamma_i))$ and is a *space of distributions* on Γ_i, with

$$\begin{aligned}\mathcal{D}(\Gamma_i) &\subset (H_{00}^s(\Gamma_i)) \subset H_0^s(\Gamma_i) \\ &\subset H^{-s}(\Gamma_i) \subset (H_{00}^s(\Gamma_i))' \subset \mathcal{D}'(\Gamma_i), \quad 1 \le i \le N. \end{aligned} \tag{8.13.83}$$

Remark 8.13.7. Imbeddings (8.13.51a)–(8.13.51h) hold for each side $\Gamma_i \in C^\infty$-class of a bounded, polygonal boundary $\Gamma \subset \mathbb{R}^2$, if we put $\Gamma_i = \Gamma_0$.

8.14 Trace results in Sobolev spaces on $\Omega \subsetneq \mathbb{R}^n$

In Definition 8.9.5, we have already defined *traces* $u\!\downarrow_\Gamma$ of functions u on $\Gamma \equiv \mathbb{R}^{n-1}$, which is, in fact, a *hyperplane* in $\mathbb{R}^n$ defined by $x_n = 0$, and also *trace operators* γ_j from $S(\mathbb{R}^n)$ into $S(\mathbb{R}^{n-1})$, and then applying the principle of extension by density we stated the final *trace results* in Theorems 8.9.9–8.9.11 for functions $u \in H^s(\mathbb{R}^n)$ for $s - m - 1/2 > 0$, $0 \le j \le m - 1$. Now we will give similar *optimal* trace results, for example in $H^m(\Omega)$ $(s = m)$ with $\Omega \subsetneq \mathbb{R}^n$, which is motivated by the following reasons. For *sufficiently* smooth Γ, $u \in C^0(\overline{\Omega})$ with $\overline{\Omega} = \Omega \cup \Gamma$ implies that its restriction $u\!\downarrow_\Gamma =$ trace of u on Γ is well defined and $u\!\downarrow_\Gamma \in C^0(\Gamma) \subset L^2(\Gamma)$. Such simple, obvious trace results do *not* exist for functions u in Sobolev spaces in general. First of all, functions $u \in H^1(\Omega)$ with $\Omega \subsetneq \mathbb{R}^n$, $n \ge 2$, may be highly *pathological* ones (see Examples 2.3.9 and 2.15.2) and need *not* be defined on a line segment $L \subset \Omega$ with Lebesgue measure $\mu(L) = 0$. Sobolev's imbedding results (see Section 8.11) are extremely useful tools and provide very important results in this direction. But for $\Omega \subsetneq \mathbb{R}^2$ with *sufficiently* smooth boundary Γ, $H^1(\Omega) \not\subset C^0(\overline{\Omega})$ (resp. $H^2(\Omega) \not\subset C^1(\overline{\Omega})$) $\Longrightarrow u\!\downarrow_\Gamma$ (resp. $\frac{\partial u}{\partial n}\!\downarrow_\Gamma$) *cannot* be defined in $H^1(\Omega)$ (resp. $H^2(\Omega)$) by Sobolev's imbedding results. Hence, for $u \in H^1(\Omega)$ (resp. $H^2(\Omega)$), the trace $u\downarrow_\Gamma$ (resp. $\frac{\partial u}{\partial n}\downarrow_\Gamma$) of u (resp. $\frac{\partial u}{\partial n}$) on Γ is *not* obvious and needs both interpretation and justification.

Moreover, consider the inverse problem: For a given function g on the boundary Γ of $\Omega \subsetneq \mathbb{R}^2$,

I. does there exist a function $u \in H^1(\Omega)$ such that $u\downarrow_\Gamma = g$?

II. Can we identify the class of functions g on Γ for which the answer to I is *affirmative*?

Then, in the case of affirmative answers to both I and II, we can use the result in the weak or *variational* formulation of second-order elliptic boundary value problems, which have solutions in $H^1(\Omega)$. The answers to questions I and II are given with the help of *trace operators* and *trace theorems* which will be dealt with in this section. We begin with:

8.14.1 Trace results in $H^m(\mathbb{R}^n_+)$

Let

$$\Omega = \mathbb{R}^n_+ = \{\mathbf{x} : \mathbf{x} = (\hat{\mathbf{x}}, x_n) \in \mathbb{R}^n,\ \hat{\mathbf{x}} \in \mathbb{R}^{n-1},\ x_n > 0\} \tag{8.14.1}$$

with boundary $\Gamma = \mathbb{R}^{n-1}\times\{0\}$, which is identified with $\mathbb{R}^{n-1}$, i.e. Γ is the hyperplane in $\mathbb{R}^n$ defined by $x_n = 0$. Then

$$(\hat{\mathbf{x}}, 0) \in \Gamma \iff \hat{\mathbf{x}} \in \mathbb{R}^{n-1}. \tag{8.14.2}$$

The following important results associated with $\mathbb{R}^n_+$ have been proved earlier:

- $\mathcal{D}(\overline{\mathbb{R}^n_+})$ is *dense* in $H^m(\mathbb{R}^n_+)$ by Theorem 8.10.3; (8.14.3)
- $\mathbb{R}^n_+$ has the *m-extension* property by Theorem 8.10.4 $\Longrightarrow \exists$ an m-extension continuous, linear operator $P \in \mathcal{L}(H^m(\mathbb{R}^n_+); H^m(\mathbb{R}^n))$ such that $\exists C > 0$, independent of u, with

$$\|Pu\|_{H^m(\mathbb{R}^n)} \le C\|u\|_{H^m(\mathbb{R}^n_+)} \quad \forall u \in H^m(\mathbb{R}^n_+) \quad \text{and}$$
$$Pu\downarrow_{\mathbb{R}^n_+} = u \quad \text{in } H^m(\mathbb{R}^n_+). \tag{8.14.4}$$

By Trace Theorems 8.9.10 and 8.9.11, for $0 \le j \le m-1$, $\exists$ a *continuous, linear trace operator* $\gamma_j : H^m(\mathbb{R}^n) \longrightarrow H^{m-j-1/2}(\Gamma) = H^{m-j-1/2}(\mathbb{R}^{n-1})$ defined by

$$u \in H^m(\mathbb{R}^n) \longmapsto \gamma_j u = \frac{\partial^j u}{\partial n^j}\downarrow_\Gamma \in H^{m-j-1/2}(\mathbb{R}^{n-1}), \tag{8.14.5}$$

such that

$$\|\gamma_j u\|_{H^{m-j-1/2}(\mathbb{R}^{n-1})} \le C\|u\|_{H^m(\mathbb{R}^n)}\ \forall u \in H^m(\mathbb{R}^n) \tag{8.14.6}$$

and $\boldsymbol{\gamma} = (\gamma_0, \gamma_1, \dots, \gamma_{m-1}) : H^m(\mathbb{R}^n) \longrightarrow \prod_{j=0}^{m-1} H^{m-j-1/2}(\mathbb{R}^{n-1})$ is defined by: $u \in H^m(\mathbb{R}^n) \longmapsto \boldsymbol{\gamma} u = (\gamma_0 u, \gamma_1 u, \dots, \gamma_{m-1} u) \in \prod_{j=0}^{m-1} H^{m-j-1/2}(\mathbb{R}^{n-1})$ with

$$\gamma_0 u(\hat{\mathbf{x}}) = u\downarrow_\Gamma(\hat{\mathbf{x}}, 0); \quad \gamma_1 u(\hat{\mathbf{x}}) = \frac{\partial u}{\partial x_n}\downarrow_\Gamma(\hat{\mathbf{x}}, 0); \quad \dots;$$

$$\gamma_{m-1} u(\hat{\mathbf{x}}) = \frac{\partial^{m-1} u}{\partial x_n^{m-1}}\downarrow_\Gamma(\hat{\mathbf{x}}, 0), \tag{8.14.7}$$

where $\frac{\partial^j u}{\partial x_n^j}\downarrow_\Gamma(\hat{\mathbf{x}}, 0)$ is the jth-order normal derivative of u at $(\hat{\mathbf{x}}, 0) \in \Gamma$ in the direction of $x_n > 0$.

$$\boldsymbol{\gamma} \text{ is a } \textit{continuous, linear surjection} \text{ from } H^m(\mathbb{R}^n) \textit{ onto } \prod_{j=1}^{m-1} H^{m-j-1/2}(\mathbb{R}^{n-1}), \tag{8.14.8}$$

which is *right invertible* from $\prod_{j=0}^{m-1} H^{m-j-1/2}(\mathbb{R}^{n-1})$ into $H^m(\mathbb{R}^n)$. (8.14.9)

Note that the same notation C has been used and will continue to be used to denote *different* values of the constant at *different* steps or inequalities.

Let $\phi \in \mathcal{D}(\overline{\mathbb{R}^n_+})$. Then, by Definition 8.10.2 of $\mathcal{D}(\overline{\mathbb{R}^n_+})$ with $\overline{\Omega} = \overline{\mathbb{R}^n_+}$ and by (8.14.4), $\exists \psi = P\phi \in \mathcal{D}(\mathbb{R}^n) \subset S(\mathbb{R}^n) \subset H^m(\mathbb{R}^n)$. By Trace Theorem 8.9.10, $\forall j = 0, 1, \dots, m-1, \exists$ a continuous, linear trace operator $\gamma_j : (P\phi) \in \mathcal{D}(\mathbb{R}^n) \longmapsto \gamma_j(P\phi) \in \mathcal{D}(\mathbb{R}^{n-1}) \subset S(\mathbb{R}^{n-1}) \subset H^{m-j-1/2}(\mathbb{R}^{n-1})$ defined, $\forall \phi \in \mathcal{D}(\overline{\mathbb{R}^n_+})$, by $\gamma_j(P\phi)(\hat{\mathbf{x}}) = \frac{\partial^j(P\phi)}{\partial x_n^j}(\hat{\mathbf{x}}, x_n)\downarrow_{x_n=0} = \frac{\partial^j(P\phi)}{\partial x_n^j}(\hat{\mathbf{x}}, 0)$ with

$$\|\gamma_j(P\phi)\|_{H^{m-j-1/2}(\mathbb{R}^{n-1})} \le C\|P\phi\|_{H^m(\mathbb{R}^n)} \le C\|\phi\|_{H^m(\mathbb{R}^n_+)} \tag{8.14.10}$$

(by (8.14.4)).

Definition 8.14.1. For $0 \le j \le m-1$, we define the *trace operator* by: $\gamma_j^* : \mathcal{D}(\overline{\mathbb{R}^n_+}) \subset H^m(\mathbb{R}^n_+) \longrightarrow \mathcal{D}(\mathbb{R}^{n-1}) \subset H^{m-j-1/2}(\mathbb{R}^{n-1})$ such that, $\forall \phi \in \mathcal{D}(\overline{\mathbb{R}^n_+})$,

$$\gamma_j^* \phi = \gamma_j(P\phi) = \frac{\partial^j(P\phi)}{\partial x_n^j}\downarrow_\Gamma \in H^{m-j-1/2}(\mathbb{R}^{n-1}) \tag{8.14.11}$$

with $\|\gamma_j^*\phi\|_{H^{m-j-1/2}(\mathbb{R}^{n-1})} = \|\gamma_j(P\phi)\|_{H^{m-j-1/2}(\mathbb{R}^{n-1})} \le C\|\phi\|_{H^m(\mathbb{R}^n_+)}$, where

$$\gamma_j^*\phi(\hat{\mathbf{x}}) = \frac{\partial^j \phi}{\partial x_n^j}(\hat{\mathbf{x}}, 0) = \gamma_j(P\phi)(\hat{\mathbf{x}}) = \frac{\partial^j(P\phi)}{\partial x_n^j}(\hat{\mathbf{x}}, 0) \quad \forall \hat{\mathbf{x}} \in \mathbb{R}^{n-1}. \tag{8.14.12}$$

Since $\mathcal{D}(\overline{\mathbb{R}^n_+})$ is *dense* in $H^m(\mathbb{R}^n_+)$ and $\forall j = 0, 1, \dots, m-1$, γ_j^* defined by (8.14.11) and (8.14.12) is a *continuous, linear* operator from $\mathcal{D}(\overline{\mathbb{R}^n_+})$ equipped with the norm $\|\cdot\|_{H^m(\mathbb{R}^n_+)}$ into $\mathcal{D}(\mathbb{R}^{n-1})$ equipped with the norm $\|\cdot\|_{H^{m-j-1/2}(\mathbb{R}^{n-1})}$, γ_j^* can be extended to a *unique, continuous, linear* operator $\widetilde{\gamma}_j^*$ from $H^m(\mathbb{R}^n_+)$ into $H^{m-j-1/2}(\mathbb{R}^{n-1})$ for $j = 0, 1, \ldots, m-1$. For the sake of *notational simplicity*, we will denote this extended operator $\widetilde{\gamma}_j^*$ by the same notation γ_j^*, i.e. $\widetilde{\gamma}_j^* = \gamma_j^* : H^m(\mathbb{R}^n_+) \longrightarrow H^{m-j-1/2}(\mathbb{R}^{n-1})$ for $0 \le j \le m-1$. Finally, we have:

Theorem 8.14.1 (Trace Theorem in $H^m(\mathbb{R}^n_+)$). *For $0 \le j \le m-1$, the mapping*

I. $\gamma_j^* : H^m(\mathbb{R}^n_+) \longrightarrow H^{m-j-1/2}(\mathbb{R}^{n-1})$ *is a* continuous, linear surjection*;* (8.14.13)

II. $\boldsymbol{\gamma}^* = (\gamma_0^*, \gamma_1^*, \ldots, \gamma_{m-1}^*) : H^m(\mathbb{R}^n_+) \longrightarrow \prod_{j=0}^{m-1} H^{m-j-1/2}(\mathbb{R}^{n-1})$ *is a* continuous, linear surjection *from* $H^m(\mathbb{R}^n_+)$ onto $\prod_{j=0}^{m-1} H^{m-j-1/2}(\mathbb{R}^{n-1})$ *such that* $\forall u \in H^m(\mathbb{R}^n_+)$, $\gamma_j^* u(\cdot) = \gamma_j(Pu)(\cdot) = \frac{\partial^j Pu}{\partial x_n^j}(\cdot, 0) = \frac{\partial^j u}{\partial x_n^j}(\cdot, 0)$, $0 \le j \le m-1$; $\boldsymbol{\gamma}^* u = (\gamma_0^* u, \gamma_1^* u, \ldots, \gamma_{m-1}^* u) \in \prod_{j=0}^{m-1} H^{m-j-1/2}(\mathbb{R}^{n-1})$ *with* $\|\boldsymbol{\gamma}^* u\|_{\prod_{j=0}^{m-1} H^{m-j-1/2}(\mathbb{R}^{n-1})} \le C\|u\|_{H^m(\mathbb{R}^n_+)}$ *for some* $C > 0$*;* (8.14.14)

III. $\boldsymbol{\gamma}^*$ *is* right invertible*, i.e.* $\exists$ *a* continuous, linear *operator* $\boldsymbol{\rho} : \prod_{j=0}^{m-1} H^{m-j-1/2}(\mathbb{R}^{n-1}) \longrightarrow H^m(\mathbb{R}^n_+)$ *such that* $\boldsymbol{\gamma}^* \boldsymbol{\rho} \mathbf{g} = \mathbf{g} = (g_0, g_1, \ldots, g_{m-1}) \in \prod_{j=0}^{m-1} H^{m-j-1/2}(\mathbb{R}^{n-1})$. (8.14.15)

Proof. For I–III, it remains to show the surjectivity of γ_j^*, $0 \le j \le m-1$, which, in fact, also follows from Theorem 8.9.11. Let $\mathbf{g} = (g_0, g_1, \ldots, g_{m-1}) \in \prod_{j=0}^{m-1} H^{m-j-1/2}(\mathbb{R}^{n-1})$. By Theorem 8.9.11, $\exists w \in H^m(\mathbb{R}^n)$ such that $\boldsymbol{\gamma} w = \mathbf{g}$. Since $\mathbb{R}^n_+$ has the m-extension property by Theorem 8.10.4, $H^m(\mathbb{R}^n_+) \equiv H^m(\overline{\mathbb{R}^n_+})$ by Theorem 8.10.2, $H^m(\overline{\mathbb{R}^n_+})$ being the space defined in (8.10.1). Hence, $w = Pu \in H^m(\mathbb{R}^n)$ with $u = Pu\downarrow_{\mathbb{R}^n_+} \in H^m(\mathbb{R}^n_+)$ and $\boldsymbol{\gamma} w = \mathbf{g} = \boldsymbol{\gamma}^* u$ with $\frac{\partial^j w}{\partial x_n^j}(\cdot, 0) = \frac{\partial^j u}{\partial x_n^j}(\cdot, 0) = g_j \in H^{m-j-1/2}(\mathbb{R}^{n-1})$ (using Theorem 8.9.12). □

8.14.2 Trace results in $H^m(\Omega)$ with bounded domain $\Omega \subsetneq \mathbb{R}^n$

First of all we consider an interesting *simple* trace result.

An elementary first trace result in $H^1(\Omega)$

Proposition 8.14.1. *Let Ω be the unit circular disc: $\Omega = B(\mathbf{0}; 1) = \{\mathbf{x} : \mathbf{x} \in \mathbb{R}^2, \|\mathbf{x}\| = (x_1^2 + x_2^2)^{1/2} < 1\}$ with boundary Γ, Γ being the unit circle, and $u \in H^1(\Omega)$. Then the* trace *of u on the boundary Γ, i.e. the restriction $u\downarrow_\Gamma$, may be considered to* *be a function in* $L^2(\Gamma)$ *satisfying the following condition:* $\exists C > 0$ *such that*

$$\|u\|_{L^2(\Gamma)} \le C\|u\|_{H^1(\Omega)}. \tag{8.14.16}$$

For the proof of Proposition 8.14.1, we need the following result:

Lemma 8.14.1. *For* Ω *with boundary* Γ *defined in Proposition* 8.14.1 *and* $u \in \mathcal{D}(\overline{\Omega})$, $u\downarrow_\Gamma \in L^2(\Gamma)$ *with* $\|u\|_{L^2(\Gamma)} \le \sqrt[4]{8}\|u\|_{0,\Omega}^{1/2}\|u\|_{H^1(\Omega)}^{1/2}$, *i.e.* $\exists C > 0$, *independent of* u, *such that* $\forall u \in \mathcal{D}(\overline{\Omega})$,

$$\|u\|_{L^2(\Gamma)} \le C\|u\|_{H^1(\Omega)}. \tag{8.14.17}$$

Proof. Introducing polar coordinates, Γ is defined by: $x_1 = 1 \cdot \cos\theta$, $x_2 = 1 \cdot \sin\theta$ $\forall(x_1, x_2) \in \Gamma$, $\forall\theta \in [0, 2\pi[$. We are to show that

$$\int_\Gamma u^2 ds = \int_0^{2\pi} [u(1,\theta)]^2 d\theta \le \sqrt{8}\|u\|_{0,\Omega}\|u\|_{1,\Omega}.$$

For $0 < r \le 1$, $0 \le \theta < 2\pi$,

$$\int_0^1 \frac{\partial}{\partial r}[(ru(r,\theta))^2]dr = (ru(r,\theta))^2|_{r=0}^{r=1} = [u(1,\theta)]^2 - 0 = [u(1,\theta)]^2.$$

Then, for fixed $\theta \in [0, 2\pi[$,

$$\begin{aligned}[u(1,\theta)]^2 &= \int_0^1 \frac{\partial}{\partial r}[(ru(r,\theta))^2]dr = \int_0^1 2ru(r,\theta))[u(r,\theta) + r\frac{\partial u}{\partial r}(r,\theta)]dr \\ &= 2\int_0^1 [r^2 u\frac{\partial u}{\partial r} + ru^2](r,\theta)dr.\end{aligned}$$

For $\hat{\mathbf{x}} = \frac{\mathbf{x}}{r}$ with $r = \|\mathbf{x}\|$ and $\|\hat{\mathbf{x}}\| = 1$,

$$\begin{aligned}\left|\frac{\partial u}{\partial r}\right| &= |\nabla u \cdot \hat{\mathbf{x}}| \le \|\nabla u\| \cdot \|\hat{\mathbf{x}}\| = \|\nabla u\| \\ &= \left[\left(\frac{\partial u}{\partial x_1}\right)^2 + \left(\frac{\partial u}{\partial x_2}\right)^2\right]^{1/2} \text{ (by the Cauchy–Schwarz inequality).}\end{aligned} \tag{8.14.18}$$

Hence,

$$[u(1,\theta)]^2 \le 2\int_0^1 \left(r^2|u|\left|\frac{\partial u}{\partial r}\right| + ru^2\right)(r,\theta)dr \le 2\int_0^1 (r|u|\|\nabla u\| + u^2)rdr$$
$$\le 2\int_0^1 (|u|\|\nabla u\| + |u|^2)rdr \quad (\text{since } r \le 1)$$
$$\Longrightarrow \quad \int_0^{2\pi} [u(1,\theta)]^2 d\theta \le 2\int_0^{2\pi}\int_0^1 (|u|\|\nabla u\| + u^2)rdrd\theta)$$
$$= 2\int_\Omega [|u|\|\nabla u\| + u^2](x_1,x_2)dx_1dx_2$$
$$\Longrightarrow \quad \|u\|^2_{L^2(\Gamma)} = \int_\Gamma u^2 ds = \int_0^{2\pi} [u(1,\theta)]^2 d\theta \le 2[\langle |u|, \|\nabla u\|\rangle_{L^2(\Omega)} + \|u\|^2_{L^2(\Omega)}]$$
$$\le 2\left[\|u\|_{L^2(\Omega)} \cdot \left\| \left[\left(\frac{\partial u}{\partial x_1}\right)^2 + \left(\frac{\partial u}{\partial x_2}\right)^2\right]^{1/2} \right\|_{L^2(\Omega)} + \|u\|^2_{L^2(\Omega)}\right]$$

(by the Cauchy–Schwarz inequality in $L^2(\Omega)$ and (8.14.18)), where

$$\left\| \left[\left(\frac{\partial u}{\partial x_1}\right)^2 + \left(\frac{\partial u}{\partial x_2}\right)^2\right]^{1/2} \right\|_{L^2(\Omega)}$$
$$= \left(\int_\Omega \left[\left(\frac{\partial u}{\partial x_1}\right)^2 + \left(\frac{\partial u}{\partial x_2}\right)^2\right] dx_1dx_2\right)^{1/2} = |u|_{H^1(\Omega)}$$

(semi-norm $|\cdot|_{1,\Omega} = |\cdot|_{H^1(\Omega)}$ (2.15.18) in $H^1(\Omega)$)

$$\Longrightarrow \quad \|u\|^2_{L^2(\Gamma)} \le 2(\|u\|_{L^2(\Omega)}|u|_{H^1(\Omega)} + \|u\|^2_{L^2(\Omega)})$$
$$= 2\|u\|_{L^2(\Omega)}(|u|_{1,\Omega} + \|u\|_{L^2(\Omega)})$$
$$\le 2\|u\|_{L^2(\Omega)}\sqrt{2}(|u|^2_{1,\Omega} + \|u\|^2_{L^2(\Omega)})^{1/2} = \sqrt{8}\|u\|_{L^2(\Omega)}\|u\|_{H^1(\Omega)}$$
$$\Longrightarrow \quad \|u\|_{L^2(\Gamma)} \le \sqrt[4]{8}\|u\|^{1/2}_{L^2(\Omega)}\|u\|^{1/2}_{H^1(\Omega)} \le \sqrt[4]{8}\|u\|^{1/2}_{H^1(\Omega)}\|u\|^{1/2}_{H^1(\Omega)}$$
$$= C\|u\|_{H^1(\Omega)} \quad \forall u \in \mathcal{D}(\overline{\Omega})$$

with $C = \sqrt[4]{8} > 0$. □

Remark 8.14.1.

1. The proof remains valid for $u \in C^1(\overline{\Omega})$, which is also *dense* in $H^1(\Omega)$, since for $u \in C^1(\overline{\Omega})$, $u{\downarrow_\Gamma} \in C^1(\Gamma) \subset L^2(\Gamma)$, the circle Γ being a C^∞-manifold.
2. The trace $u{\downarrow_\Gamma}$ = the restriction $u{\downarrow_\Gamma}$ makes sense as a function in $L^2(\Gamma)$.

3. Using the mapping $\theta \in [0, 2\pi[\longmapsto (\cos\theta, \sin\theta) \in \Gamma$, $L^2(\Gamma)$ has been identified with $L^2(]0, 2\pi[)$ (see also Definition 8.13.8 with $p = 2$, $s = 0$ and $L = 2\pi$). (8.14.19)

Proof of Proposition 8.14.1. Since Ω is a C^∞-regular domain with boundary Γ of C^∞-class (i.e. C^m-regular domain with Γ of C^m-class $\forall m \in \mathbb{N}$), $H^1(\overline{\Omega}) = H^1(\Omega)$ (see Definition 8.10.1 for $H^1(\overline{\Omega})$) and $\mathcal{D}(\overline{\Omega})$ is *dense* in $H^1(\Omega)$ by Theorem 8.10.1. Let $u \in H^1(\Omega)$. Then $\exists$ a sequence $(u_m)_{m=1}^\infty$ in $\mathcal{D}(\overline{\Omega})$ such that $\|u - u_m\|_{1,\Omega} \to 0$ as $m \to \infty$. Hence, $(u_m)_{m=1}^\infty$ is a Cauchy sequence in $H^1(\Omega)$ such that $\|u_m - u_n\|_{1,\Omega} \to 0$ as $m, n \to \infty$. But $u_m, u_n \in \mathcal{D}(\overline{\Omega})$ $\forall m, n \in \mathbb{N}$ and, by Lemma 8.14.1, $\|(u_m - u_n)\downarrow_\Gamma\|_{L^2(\Gamma)} = \|u_m - u_n\|_{L^2(\Gamma)} \le C\|u_m - u_n\|_{1,\Omega} \to 0$ as $m, n \to \infty \implies (u_n\downarrow_\Gamma)_{n=1}^\infty$ is a Cauchy sequence in $L^2(\Gamma)$, which is a Hilbert space. Hence, $\exists u_0 \in L^2(\Gamma)$ such that $\|u_0 - u_n\|_{L^2(\Gamma)} \to 0$ as $n \to \infty$. Define $u\downarrow_\Gamma = u_0 \in L^2(\Gamma)$. This definition is independent of the choice of the sequence $(u_m)_{m=1}^\infty$, since for any other choice of $(\widetilde{u}_m)_{m=1}^\infty$ with $\lim_{m\to\infty} \widetilde{u}_m = \widetilde{u}_0$ in $H^1(\Omega)$ and $\widetilde{u}_0 \in L^2(\Omega)$, $u_0 - \widetilde{u}_0 = \lim_{m\to\infty}(u_m - \widetilde{u}_m)$ in $L^2(\Gamma)$,

$$\begin{aligned}\|u_m - \widetilde{u}_m\|_{L^2(\Gamma)} &\le C\|u_m - \widetilde{u}_m\|_{H^1(\Omega)}\\ &\le C(\|u_m - u\|_{H^1(\Omega)} + \|u - \widetilde{u}_m\|_{H^1(\Omega)}) \to 0\end{aligned}$$

as $m \to \infty \implies u_0 - \widetilde{u}_0 = 0$ in $L^2(\Gamma) \implies u_0 = \widetilde{u}_0$ a.e. on Γ. Hence, $u\downarrow_\Gamma = u_0$ is well defined in $L^2(\Gamma)$. (8.14.20)

Using Lemma 8.14.1 and (8.14.20), we have

$$\begin{aligned}\|u\downarrow_\Gamma\|_{L^2(\Gamma)} = \|u_0\|_{L^2(\Gamma)} &= \lim_{m\to\infty} \|u_m\|_{L^2(\Gamma)} \le \lim_{m\to\infty}(C\|u_m\|_{H^1(\Omega)})\\ &\le C \lim_{m\to\infty} \|u_m\| = C\|u\|_{H^1(\Omega)} \quad \forall u \in H^1(\Omega).\end{aligned}$$

□

Remark 8.14.2.

1. Although for functions $u \in \mathcal{D}(\overline{\Omega})$, the trace defined here is the *same* as the ordinary pointwise restrictions to the boundary Γ. In fact, for circle Γ (which is of C^∞-class – see Definitions D.2.3.1 and D.4.1.1, Appendix DD), $u \in \mathcal{D}(\overline{\Omega}) \implies u\downarrow_\Gamma \in \mathcal{D}(\Gamma)$.

2. But Proposition 8.14.1 does *not* assure us that *pointwise values* $u(\mathbf{x})$ of u are defined on Γ. It shows only that $u\downarrow_\Gamma = u_0$ is a *square integrable* function on Γ, which allows u_0 to achieve even values $\pm\infty$ on a set of points on Γ with arc length measure zero.

3. The trace result $u \in H^1(\Omega) \longmapsto \gamma_0 u = u\downarrow_\Gamma \in L^2(\Gamma)$ is *not* a *surjection* and hence is *not optimal*, and we are now going to use more sophisticated methods to obtain *optimal* trace results.

Trace results for C^m-regular domains $\Omega \subsetneq \mathbb{R}^n$

Let $\Omega \subsetneq \mathbb{R}^n$ be a C^m-regular domain with boundary Γ of C^m-class (see Definitions 8.10.4 and 8.10.5, respectively). We will follow in principle the notations used in Definition 8.10.5 and in the proof of Theorem 8.10.5 (see also Figures 8.9 and 8.10).

Result 1 Let $(\theta_i)_{i=0}^N$ be a C_0^∞-partition of 1 (see Appendices C and D) with respect to the open cover $\{\Omega_i\}_{i=0}^N$ of $\overline{\Omega}$ with $\overline{\Omega}_0 \subset \Omega$ such that $\theta_i \in \mathcal{D}(\Omega_i)$, and $(\beta_i)_{i=1}^N$ with $\beta_i = \theta_i\downarrow_{\Gamma_i}$, $\Gamma_i = \Omega_i \cap \Gamma$, $1 \le i \le N$, is a partition of 1 with respect to the open cover $\{\Gamma_i\}_{i=1}^N$ of Γ (see also Section 8.13.2).

Let $\phi \in \mathcal{D}(\overline{\Omega})$ such that $\forall \mathbf{x} \in \overline{\Omega}$, $\phi(\mathbf{x}) = \sum_{i=0}^N \theta_i(\mathbf{x})\phi(\mathbf{x}) = \sum_{i=0}^N \psi_i(\mathbf{x})$ with

$$\psi_i = \theta_i \phi \tag{8.14.21}$$

and $\forall \mathbf{x} \in \Gamma$,

$$\phi(\mathbf{x}) = \sum_{i=1}^N \theta_i(\mathbf{x})\phi(\mathbf{x}) = \sum_{i=1}^N \psi_i(\mathbf{x}), \tag{8.14.22}$$

where

$$\sum_{i=0}^N \theta_i(\mathbf{x}) = 1 \ \forall \mathbf{x} \in \overline{\Omega}, \quad \sum_{i=1}^N \theta_i(\mathbf{x}) = \sum_{i=1}^N \beta_i(\mathbf{x}) = 1 \ \forall \mathbf{x} \in \Gamma, \tag{8.14.23}$$

($\beta_i = \theta_i\downarrow_{\Gamma_i} \in C^m(\Gamma_i)$, $1 \le i \le N$, used in Section 8.13).

Result 2 $\forall i = 0, 1, \ldots, N$, let Φ_i and its inverse Φ_i^{-1} be defined by (8.10.60)–(8.10.62) (see also Figures 8.9 and 8.10): $\Phi_i : \mathbf{x} \in \Omega_i \longmapsto \Phi_i(\mathbf{x}) = \boldsymbol{\xi} = (\hat{\boldsymbol{\xi}}, \xi_n) \in \hat{Q} \subset \mathbb{R}^n$ with

$$\hat{Q} = \{(\hat{\boldsymbol{\xi}}, \xi_n) : \hat{\boldsymbol{\xi}} = (\xi_1, \ldots, \xi_{n-1}) \in \mathbb{R}^{n-1}, |\xi_j| < 1 \text{ for } 1 \le j \le n\} \tag{8.14.24}$$

a *hypercube* in $\mathbb{R}^n$;

$$\Phi_i^{-1} : \boldsymbol{\xi} = (\hat{\boldsymbol{\xi}}, \xi_n) \in \hat{Q} \longmapsto \Phi_i^{-1}(\boldsymbol{\xi}) = \mathbf{x} \in \Omega_i, \tag{8.14.25}$$

such that Φ_i and Φ_i^{-1} are C^m-diffeomorphisms from Ω_i *onto* $\hat{Q}$ and from $\hat{Q}$ onto Ω_i, respectively (see the proof of Theorem 8.10.5 and the footnote there), and $(\Omega_i, \Phi_i)_{i=1}^N$ is a system of *local charts* or *atlas* of Γ (see (D.5.1.1), Appendix D). More precisely, we choose the local charts in a suitable way such that Lemma 8.14.1 (given later) will hold as follows: $\forall i = 1, 2, \ldots, N$, $\forall \mathbf{x} \in \Omega_i$,

$$\Phi_i(\mathbf{x}) = (\phi_i^1(\mathbf{x}), \cdots, \phi_i^j(\mathbf{x}), \ldots, \phi_i^n(\mathbf{x})) = \boldsymbol{\xi} = (\hat{\boldsymbol{\xi}}, \xi_n) \in \hat{Q},$$

$$\Phi_i^{-1}(\boldsymbol{\xi}) = (\psi_i^1(\boldsymbol{\xi}), \ldots, \psi_i^j(\boldsymbol{\xi}), \ldots, \psi_i^n(\boldsymbol{\xi})) = \mathbf{x} \in \Omega_i$$

such that $\forall j = 1, 2, \ldots, n, \phi_i^j$ (resp. $\psi_i^j(\boldsymbol{\xi})$) is *m times continuously differentiable* and *bounded* on Ω_i (resp. $\hat{Q}$). Define, $\forall i = 1, 2, \ldots, N$,

$$\xi_j = \phi_i^j(\mathbf{x}) \quad \forall \mathbf{x} \in \Omega_i,\ 1 \le j \le n-1; \tag{8.14.26}$$

$$\xi_n = \phi_i^n(\mathbf{x}) \quad \forall \mathbf{x} \in \Omega_i, \tag{8.14.27}$$

where $\xi_n = \phi_i^n(\mathbf{x})$, $d(\mathbf{x}, \Gamma) \ge 0$ for $\mathbf{x} \in \overline{\Omega} \cap \Omega_i$, i.e. for $\mathbf{x} \in \Omega_i \cap \Gamma$, $\xi_n = 0$ for $d(\mathbf{x}, \Gamma) = 0$ and (8.10.62) holds; $(\phi_i^j)_{j=1}^{n-1}$ are *independent* functions satisfying the equations: $\frac{\partial \phi_i^j}{\partial n} = \sum_{k=1}^n \cos(\hat{\mathbf{n}}, \hat{\mathbf{i}}_k) \frac{\partial \phi_i^j}{\partial x_k} = 0$ on Γ_i, $1 \le j \le n-1$

$$\Longrightarrow \frac{\partial \phi_i^j}{\partial n} = \sum_{k=1}^n \alpha_k \frac{\partial \phi_i^j}{\partial x_k} = 0 \text{ on } \Gamma_i,\ 1 \le j \le n-1. \tag{8.14.28}$$

The rectangular matrix $(\frac{\partial \phi_i^j(\mathbf{x})}{\partial x_k})_{1 \le j \le n-1, 1 \le k \le n}$ has rank $(n-1)$ at each $\mathbf{x} \in \Gamma_i$, $1 \le i \le N$. (8.14.29)

$\alpha_k = \cos(\hat{\mathbf{n}}, \hat{\mathbf{i}}_k) = \alpha_k(\mathbf{x})$ $\forall \mathbf{x} \in \Gamma$, $1 \le i \le N$, $1 \le k \le n$, $\hat{\mathbf{n}}$ being the unit normal to Γ and $\hat{\mathbf{i}}_k$ being the unit vector in the direction of x_k-axis.

$\frac{\partial}{\partial n} = \sum_{k=1}^n \alpha_k \frac{\partial}{\partial x_k}$ is the *normal derivative* operator on Γ. (8.14.30)

Result 3 For $\psi_i = \theta_i \phi$ with $\phi \in \mathcal{D}(\overline{\Omega})$, $1 \le i \le N$, defined in (8.14.22), let $\frac{\partial \psi_i}{\partial n}$ be the normal derivative of ψ_i at $\mathbf{x} \in \Gamma_i$, $1 \le i \le N$. Then, by (8.14.25), $\forall i = 1, 2, \ldots, N$, for $\xi_n = 0$, $\Phi_i^{-1}(\hat{\boldsymbol{\xi}}, 0) = \mathbf{x} \in \Gamma_i$ (see (8.10.62)). Hence, we have, $\forall \mathbf{x} \in \Gamma_i$,

$$\frac{\partial \psi_i}{\partial n}(\mathbf{x}) = \frac{\partial \psi_i}{\partial n}(\Phi_i^{-1}(\hat{\boldsymbol{\xi}}, 0)) = \frac{\partial \psi_i}{\partial n} \circ \Phi_i^{-1}(\hat{\boldsymbol{\xi}}, 0). \tag{8.14.31}$$

Again, $\forall \mathbf{x} \in \Omega_i$, $\psi_i(\mathbf{x}) = \psi_i(\Phi_i^{-1}(\hat{\boldsymbol{\xi}}, \xi_n)) = \psi_i \circ \Phi^{-1}(\hat{\boldsymbol{\xi}}, \xi_n)$ with $(\hat{\boldsymbol{\xi}}, \xi_n) \in \hat{Q}$, $1 \le i \le N$

$$\Longrightarrow \quad \forall \mathbf{x} \in \Gamma_i,\ \psi_i(\mathbf{x}) = \psi_i \circ \Phi_i^{-1}(\hat{\xi}, 0),\ 1 \le i \le N$$

$$\Longrightarrow \quad \theta_i \phi\downarrow_{\Gamma_i} = \psi_i \downarrow_{\Gamma_i} = (\psi_i \circ \Phi_i^{-1})\downarrow_{\xi_n = 0} = (\theta_i \phi \circ \Phi_i^{-1})\downarrow_{\xi_n = 0}. \tag{8.14.32}$$

Then, from Results 1–3 above, we have:

Lemma 8.14.2. $\forall i = 1, 2, \ldots, N$, *for* $\xi_n = 0$,

$$\frac{\partial \psi_i}{\partial n} \circ \Phi_i^{-1}(\hat{\boldsymbol{\xi}}, 0) = \frac{\partial}{\partial \xi_n}(\psi_i \circ \Phi_i^{-1})(\hat{\boldsymbol{\xi}}, \xi_n)\downarrow_{\xi_n = 0} = \frac{\partial(\psi_i \circ \Phi_i^{-1})}{\partial \xi_n}(\hat{\boldsymbol{\xi}}, 0). \tag{8.14.33}$$

In general, for $1 \le i \le N$, $1 \le j \le m-1$ *and* $\xi_n = 0$,

$$\frac{\partial^j \psi_i}{\partial n^j} \circ \Phi_i^{-1}(\hat{\boldsymbol{\xi}}, 0) = \frac{\partial^j}{\partial \xi_n^j}(\psi_i \circ \Phi_i^{-1})(\hat{\boldsymbol{\xi}}, 0),$$

i.e.

$$\frac{\partial^j \psi_i}{\partial n_j}\downarrow_{\Gamma_i} = (\frac{\partial^j \psi_i}{\partial n^j} \circ \Phi_i^{-1})\downarrow_{\xi_n=0} = \frac{\partial^j}{\partial \xi_n^j}(\psi_i \circ \Phi_i^{-1})\downarrow_{\xi_n=0}. \tag{8.14.34}$$

Proof. We will prove the result for $j = 1$; it can be proved similarly for $j = 2, \ldots, m-1$. We will also use the notation: $\mathbf{x} = \mathbf{x}(\hat{\boldsymbol{\xi}}, \xi_n) = \Phi_i^{-1}(\hat{\boldsymbol{\xi}}, \xi_n)\ \forall(\hat{\boldsymbol{\xi}}, \xi_n) \in \hat{Q}$, i.e. $x_k = x_k(\hat{\boldsymbol{\xi}}, \xi_n)\ \forall k = 1, 2, \ldots, n$. Using (8.14.30),

$$\begin{aligned}\frac{\partial \psi_i}{\partial n} \circ \Phi_i^{-1}(\hat{\xi}, \xi_n) &= \left(\sum_{k=1}^n \alpha_k \frac{\partial \psi_i}{\partial x_k}\right) \circ \Phi_i^{-1}(\hat{\xi}, \xi_n)\\ &= \sum_{k=1}^n \alpha_k(\Phi_i^{-1}(\hat{\xi}, \xi_n))\frac{\partial \psi_i}{\partial x_k}(\Phi_i^{-1}(\hat{\xi}, \xi_n)); \end{aligned}\tag{8.14.35}$$

$$\begin{aligned}\frac{\partial}{\partial \xi_n}[\psi_i \circ \Phi_i^{-1}](\hat{\xi}, \xi_n) &= \frac{\partial \psi_i}{\partial \xi_n}(x_1(\hat{\xi}, \xi_n), \cdots, x_n(\hat{\xi}, \xi_n))\\ &= \sum_{k=1}^n \frac{\partial \psi_i}{\partial x_k}(\mathbf{x}(\hat{\xi}, \xi_n))\frac{\partial x_k}{\partial \xi_n}(\hat{\xi}, \xi_n)\\ &= \sum_{k=1}^n \frac{\partial \psi_i}{\partial x_k}(\Phi_i^{-1}(\hat{\xi}, \xi_n))\frac{\partial x_k}{\partial \xi_n}(\hat{\xi}, \xi_n). \end{aligned}\tag{8.14.36}$$

In order to prove (8.14.33), it is sufficient to show that (8.14.35) and (8.14.36) are equal for $\xi_n = 0$, i.e. we are to show that for $\xi_n = 0$,

$$\alpha_k = \frac{\partial x_k}{\partial \xi_n} \quad \forall 1 \le k \le n. \tag{8.14.37}$$

For this, consider the system of $(n-1)$ equations (8.14.26): $\phi_i^j(\Phi_i^{-1}(\hat{\xi}, \xi_n)) = \xi_j$, $1 \le j \le n-1$

$$\begin{aligned}&\Longrightarrow \quad \frac{\partial \phi_i^j}{\partial \xi_n}(\mathbf{x}(\hat{\xi}, \xi_n)) = 0,\ 1 \le j \le n-1\\ &\Longrightarrow \quad \sum_{k=1}^n \frac{\partial \phi_i^j}{\partial x_k} \cdot \frac{\partial x_k}{\partial \xi_n} = 0,\ 1 \le j \le n-1. \end{aligned}\tag{8.14.38}$$

Set $\zeta_k = \frac{\partial x_k}{\partial \xi_n}$ for $1 \le k \le n$, $a_{jk} = \frac{\partial \phi_i^j}{\partial x_k}$ for $1 \le j \le n-1$, $1 \le k \le n$. (8.14.39)
Then equations (8.14.38) and (8.14.28), which must be satisfied by $(n-1)$ *independent* functions $(\phi_i^j)_{j=1}^{n-1}$, can be rewritten in matrix form:

$$[A][\zeta] = 0, \quad [A][\alpha] = 0,$$

where the rectangular matrix $[A] = (a_{jk})_{1\le j\le n-1, 1\le k\le n}$ with a_{jk} defined by (8.14.39) has rank $(n-1)$ at $\mathbf{x}$ by (8.14.29); $[\zeta] = (\zeta_k)_{k=1}^n$ with $\zeta_k = \frac{\partial x_k}{\partial \xi_n}$ by (8.14.39); $[\alpha] = (\alpha_k)_{k=1}^n$ with $\alpha_k = \cos(\hat{\mathbf{n}}, \hat{\mathbf{i}}_k)$.

Since $[A]_{(n-1)\times n}$ has rank $(n-1)$ $\forall \mathbf{x} \in \Gamma_i$, $[\zeta]$ and $[\alpha]$ will be *dependent* at every $\mathbf{x} \in \Gamma_i$, i.e. $\exists \lambda = \lambda(\mathbf{x})$ such that

$$[\zeta] = \lambda[\alpha] \quad \Longrightarrow \quad \zeta_k = \lambda\alpha_k \quad \Longrightarrow \quad \frac{\partial x_k}{\partial \xi_n} = \lambda\alpha_k \quad \forall k = 1, \dots, n. \tag{8.14.40}$$

Now we find the value of $\lambda = \lambda(\mathbf{x}(\hat{\xi}, \xi_n))$ for $\xi_n = 0$. From (8.14.27), we have $\phi_i^n(\mathbf{x}) = \phi_i^n[\mathbf{x}(\hat{\xi}, \xi_n)] = \xi_n$. Differentiating both sides of the equality with respect to ξ_n, we have

$$\sum_{k=1}^n \frac{\partial \phi_i^n}{\partial x_k}\frac{\partial x_k}{\partial \xi_n} = 1 \quad \Longrightarrow \quad \sum_{k=1}^n \frac{\partial \phi_i^n}{\partial x_k} \cdot (\lambda\alpha_k) = 1 \quad \text{(by (8.14.40))}$$

$$\Longrightarrow \quad \left(\sum_{k=1}^n \frac{\partial \phi_i^n}{\partial x_k}\alpha_k\right)\lambda = 1. \tag{8.14.41}$$

Now we evaluate $(\sum_{k=1}^n \frac{\partial \phi_i^n}{\partial x_k}\alpha_k)$ at $\mathbf{x} \in \Omega_i \cap \overline{\Omega}$. Let $\mathbf{x}_0 \in \Gamma$ be the projection of $\mathbf{x} \in \Omega_i \cap \overline{\Omega}$ onto Γ_i such that $\mathbf{x} = \mathbf{x}_0 + \hat{\mathbf{n}} \cdot t$ with

$$t = d(\mathbf{x}, \Gamma) = \phi_i^n(\mathbf{x}) = \phi_i^n(\mathbf{x}_0 + \hat{\mathbf{n}}t). \tag{8.14.42}$$

Again, differentiating (8.14.42) with respect to t, we have $1 = \frac{\partial}{\partial t}[\phi_i^n(\mathbf{x}_0 + \hat{\mathbf{n}}t)] = \sum_{k=1}^n \frac{\partial \phi_i^n}{\partial x_k} \cdot \frac{\partial x_k}{\partial t}$ with $x_k = x_0 + \alpha_k t \Longrightarrow 1 = \sum_{k=1}^n \frac{\partial \phi_i^n}{\partial x_k}\alpha_k \Longrightarrow$ from (8.14.41), for $\xi_n = 0$, we get $\lambda = 1$. Then, from (8.14.40), $\frac{\partial x_k}{\partial \xi_n} = \zeta_k = \alpha_k$ $\forall k = 1, \dots, n$ for $\xi_n = 0$ and the result (8.14.33) follows from (8.14.37). □

Lemma 8.14.3. *Let Results* 1–3 *in* (8.14.21)–(8.14.32) *hold. Then* $\exists C > 0$ *such that* $\forall \phi \in \mathcal{D}(\overline{\Omega})$,

$$\sum_{i=1}^N \|\theta_i\phi\|^2_{H^m(\Omega\cap\Omega_i)} \le C^2\|\phi\|^2_{H^m(\Omega)}. \tag{8.14.43}$$

Proof. Since $\theta_i \in \mathcal{D}(\Omega_i)$, $\phi \in \mathcal{D}(\overline{\Omega})$, $1 \le i \le N$, $\forall|\boldsymbol{\alpha}| \le m$,

$$\partial^{\boldsymbol{\alpha}}(\theta_i\phi) = \sum_{|\boldsymbol{\beta}|\le|\boldsymbol{\alpha}|} \partial^{\boldsymbol{\beta}}\phi\,\partial^{\boldsymbol{\alpha}-\boldsymbol{\beta}}\theta_i = \sum_{|\boldsymbol{\beta}|\le|\boldsymbol{\alpha}|} \partial^{\boldsymbol{\beta}}\phi\, p_{\boldsymbol{\beta}}(\theta_i), \tag{8.14.44}$$

where $p_{\boldsymbol{\beta}}(\theta_i)$ is a polynomial in partial derivatives $\partial^{\boldsymbol{\gamma}}\theta_i$ ($\boldsymbol{\gamma} = \boldsymbol{\alpha} - \boldsymbol{\beta}$) with real constant coefficients. Hence, $\forall|\boldsymbol{\beta}| \le |\boldsymbol{\alpha}|$, $p_{\boldsymbol{\beta}}(\theta_i)$ is *bounded* on Ω_i, i.e. $\exists$ a constant

$C(\boldsymbol{\alpha}, \theta_i, \Omega) > 0$ such that $\max_{|\boldsymbol{\beta}|\le m} \max_{\mathbf{x}\in\Omega_i} |p_{\boldsymbol{\beta}}(\theta_i)(\mathbf{x})| \le C$ and $\forall |\boldsymbol{\alpha}| \le m$, $|\partial^{\boldsymbol{\alpha}}(\theta_i\phi)(\mathbf{x})| \le C \sum_{|\boldsymbol{\beta}|\le|\boldsymbol{\alpha}|} |\partial^{\boldsymbol{\beta}}\phi(\mathbf{x})|$ in $\Omega \cap \Omega_i \Longrightarrow \forall |\boldsymbol{\alpha}| \le m$,

$$\int_{\Omega\cap\Omega_i} |\partial^{\boldsymbol{\alpha}}(\theta_i\phi)(\mathbf{x})|^2 d\mathbf{x} \le C^2 \int_{\Omega\cap\Omega_i} \sum_{|\boldsymbol{\beta}|\le|\boldsymbol{\alpha}|} |\partial^{\boldsymbol{\beta}}\phi(\mathbf{x})|^2 d\mathbf{x}.$$

We will follow the usual convention of using the same notation C to denote *different* values of C at *different* steps or inequalities.

$$\begin{aligned}
\Longrightarrow \quad \sum_{|\boldsymbol{\alpha}|\le m} \int_{\Omega\cap\Omega_i} |\partial^{\boldsymbol{\alpha}}(\theta_i\phi)(\mathbf{x})|^2 d\mathbf{x} &\le C^2 \sum_{|\boldsymbol{\alpha}|\le m} \sum_{|\boldsymbol{\beta}|\le|\boldsymbol{\alpha}|} \int_{\Omega\cap\Omega_i} |\partial^{\boldsymbol{\beta}}\phi(\mathbf{x})|^2 d\mathbf{x} \\
&\le C^2 \sum_{|\boldsymbol{\beta}|\le m} \int_{\Omega\cap\Omega_i} |\partial^{\boldsymbol{\beta}}\phi(\mathbf{x})|^2 d\mathbf{x} \\
&\le C^2 \sum_{|\boldsymbol{\beta}|\le m} \int_{\Omega} |\partial^{\boldsymbol{\beta}}\phi(\mathbf{x})|^2 d\mathbf{x} = C^2 \|\phi\|^2_{H^m(\Omega)}
\end{aligned}$$

$\Longrightarrow \forall \phi \in \mathcal{D}(\overline{\Omega}),\ \theta_i \in \mathcal{D}(\Omega_i),\ \|\theta_i\phi\|^2_{H^m(\Omega\cap\Omega_i)} \le C^2\|\phi\|^2_{H^m(\Omega)}$
$\Longrightarrow \sum_{i=1}^N \|\theta_i\phi\|^2_{H^m(\Omega\cap\Omega_i)} \le C^2 \cdot N\|\phi\|^2_{H^m(\Omega)} \le C^2\|\phi\|^2_{H^m(\Omega)}$. □

Trace results in $\mathcal{D}(\overline{\Omega})$ for C^m-regular Ω

Let $\phi \in \mathcal{D}(\overline{\Omega})$. Using (8.14.21)–(8.14.23) and following the proof of Theorem 8.10.5, define $v_i = \theta_i\phi \circ \Phi_i^{-1}$ for $i = 1, 2, \ldots, N$. Since $\phi \in \mathcal{D}(\overline{\Omega})$, $\theta_i \in \mathcal{D}(\Omega_i)$ with its *null extension* (denoted by θ_i itself) $\theta_i \in \mathcal{D}(\mathbb{R}^n)$, $\Phi_i^{-1} \in C^m(\overline{\hat{Q}})\ \forall i = 1, 2, \ldots, N$,

$$v_i = \theta_i\phi \circ \Phi_i^{-1} \in C^m(\overline{\hat{Q}_+}) \subset H^m(\hat{Q}_+), \tag{8.14.45}$$

and v_i vanishes in *some neighbourhood* of the boundary of $\hat{Q}_+$ excluding

$$\hat{Q}_0 = \{(\hat{\boldsymbol{\xi}}, 0) : \hat{\boldsymbol{\xi}} \in \mathbb{R}^{n-1},\ |\xi_i| < 1,\ 1 \le i \le n-1\} \tag{8.14.45a}$$

(see Figures 8.9 and 8.10).

Hence, $\forall i = 1, \ldots, N$, v_i can be extended by 0 outside $\hat{Q}_+$. Let $\widetilde{v}_i = \widetilde{(\theta_i\phi \circ \Phi_i^{-1})}$ be the *null extension* of v_i to $\mathbb{R}^n_+$, i.e. $\widetilde{v}_i(\boldsymbol{\xi}) = 0$ for $\boldsymbol{\xi} \in \mathbb{R}^n_+ \setminus \hat{Q}_+$ and $\widetilde{v}_i(\boldsymbol{\xi}) = v_i(\boldsymbol{\xi})$ for $\boldsymbol{\xi} \in \hat{Q}_+$ for $1 \le i \le N$. Then, following the steps of the proof of Theorem 2.15.5, we have, $\forall 1 \le i \le N$,

$$\widetilde{v}_i \in H^m(\mathbb{R}^n_+) \quad \text{with } \|v_i\|_{H^m(\hat{Q}_+)} = \|\widetilde{v}\|_{H^m(\mathbb{R}^n_+)}. \tag{8.14.46}$$

Since Φ_i and Φ_i^{-1} are C^m-diffeomorphisms from $\Omega \cap \Omega_i$ onto $\hat{Q}_+$ and from $\hat{Q}_+$ onto $\Omega \cap \Omega_i$, respectively, and $\Phi_i^{-1}(\hat{Q}_+) = \Omega \cap \Omega_i$, $1 \le i \le N$, by virtue of (8.14.45) $\|v_i\|_{H^m(\hat{Q}_+)} = \|\theta_i \phi \circ \Phi_i^{-1}\|_{H^m(\hat{Q}_+)} \le C\|\theta_i \phi\|_{H^m(\Omega\cap\Omega_i)}$

$$\implies \quad \|\widetilde{v}_i\|_{H^m(\mathbb{R}^n_-)} = \|v_i\|_{H^m(\hat{Q}_+)} \le C\|\theta_i\phi\|_{H^m(\Omega\cap\Omega_i)} \quad \text{(by (8.14.46))}. \tag{8.14.47}$$

Now we can apply Trace Theorem 8.14.1 for $\widetilde{v}_i \in H^m(\mathbb{R}^n_+)$. In fact, for $m - j - 1/2 > 0$ with $j = 0, 1, \ldots, m-1$, $\exists$ a *continuous, linear trace operator* $\gamma_j^* : H^m(\mathbb{R}^n_+) \longrightarrow H^{m-j-1/2}(\mathbb{R}^{n-1})$ defined, $\forall \widetilde{v} \in H^m(\mathbb{R}^n_+)$, by:

$$\gamma_j^*\widetilde{v}_i = \frac{\partial^j \widetilde{v}_i}{\partial \xi_n^j}\downarrow_{\xi_n=0} \in H^{m-j-1/2}(\mathbb{R}^{n-1}), \quad 1 \le i \le N, \tag{8.14.48}$$

and $\exists C > 0$ such that

$$\|\gamma_j^*\widetilde{v}_i\|_{H^{m-j-1/2}(\mathbb{R}^{n-1})} = \|\frac{\partial^j \widetilde{v}_i}{\partial \xi_n^j}\downarrow_{\xi_n=0}\|_{H^{m-j-1/2}(\mathbb{R}^{n-1})} \le C\|\widetilde{v}_i\|_{H^m(\mathbb{R}^n_+)}, \tag{8.14.49}$$

where $\frac{\partial^j \widetilde{v}_i}{\partial \xi_n^j}(\hat{\xi}, \xi_n)\downarrow_{\xi_n=0} = \frac{\partial^j \widetilde{v}_i}{\partial \xi_n^j}(\hat{\boldsymbol{\xi}}, 0) \implies \frac{\partial^j \widetilde{v}_i}{\partial \xi_n^j}(\cdot, 0)$ is considered to be a function of $\hat{\boldsymbol{\xi}} \in \mathbb{R}^{n-1}$ only. (8.14.50)

The same constant C has been used and will continue to be used to denote *different* values of C at *different* steps or inequalities.

Then, from (8.14.49) and (8.14.47),

$$\|\gamma_j^*\widetilde{v}_i\|_{H^{m-j-1/2}(\mathbb{R}^{n-1})} \le C\|v_i\|_{H^m(\hat{Q}_+)} \le C\|\theta_i\phi\|_{H^m(\Omega\cap\Omega_i)} \tag{8.14.51}$$

for $0 \le j \le m-1$, $1 \le i \le N$.

For $0 \le j \le m-1$, $\frac{\partial^j \widetilde{v}_i}{\partial \xi_n^j}\downarrow_{\xi_n=0}(\hat{\boldsymbol{\xi}}) = \frac{\partial^j v_i}{\partial \xi_n^j}\downarrow_{\xi_n=0}(\hat{\boldsymbol{\xi}})$ with $\hat{\boldsymbol{\xi}} \in \mathbb{R}^{n-1}$, $(\hat{\boldsymbol{\xi}}, 0) \in \hat{Q}_0 \subset \mathbb{R}^{n-1} \times \{0\}$, $1 \le i \le N$

$$\implies \quad \frac{\partial^j \widetilde{v}}{\partial \xi_n^j}\downarrow_{\xi_n=0}(\hat{\boldsymbol{\xi}}) = \widetilde{\frac{\partial^j v_i}{\partial \xi_n^j}}\downarrow_{\xi_n=0}(\hat{\boldsymbol{\xi}}) \quad \forall \hat{\boldsymbol{\xi}} \in \mathbb{R}^{n-1},\ 0 \le j \le m-1, 1 \le i \le N, \tag{8.14.52}$$

where

$$\widetilde{\frac{\partial^j v_i}{\partial \xi_n^j}}\downarrow_{\xi_n=0}(\hat{\boldsymbol{\xi}}) = \begin{cases} \frac{\partial^j v_i}{\partial \xi_n^j}\downarrow_{\xi_n=0}(\hat{\boldsymbol{\xi}}) & \text{for } (\hat{\boldsymbol{\xi}}, 0) \in \hat{Q}_0; \\ 0 & \forall \hat{\boldsymbol{\xi}} \in \mathbb{R}^{n-1} \text{ with } (\hat{\boldsymbol{\xi}}, 0) \notin \hat{Q}_0. \end{cases} \tag{8.14.53}$$

Hence, from (8.14.48), (8.14.50), (8.14.52) and (8.14.53), we have $\widetilde{\frac{\partial^j v_i}{\partial \xi_n^j}} \downarrow_{\xi_n=0} = \frac{\partial^j \tilde{v}}{\partial \xi_n^j} \downarrow_{\xi_n=0} \in H^{m-j-1/2}(\mathbb{R}^{n-1})$ for $0 \le j \le m-1$, $1 \le i \le N$, i.e. $v_i = \theta_i \phi \circ \Phi_i^{-1}$ with $\widetilde{\frac{\partial^j v_i}{\partial \xi_n^j}} \downarrow_{\xi_n=0} \in H^{m-j-1/2}(\mathbb{R}^{n-1})$, $0 \le j \le m-1$. (8.14.54)

But by Lemma 8.14.1, we have

$$\frac{\partial^j(\theta_i\phi)}{\partial n^j} \circ \Phi_i^{-1}(\hat{\boldsymbol{\xi}}, \xi_n)\downarrow_{\xi_n=0} = \frac{\partial^j(\theta_i\phi)}{\partial n^j} \circ \Phi_i^{-1}(\hat{\boldsymbol{\xi}}, 0) = \frac{\partial^j(\theta_i\phi \circ \Phi_i^{-1})}{\partial \xi_n^j}(\hat{\boldsymbol{\xi}}, 0)$$
$$= \frac{\partial^j v_i}{\partial \xi_n^j}(\hat{\boldsymbol{\xi}}, 0) = \frac{\partial^j v_i}{\partial \xi_n^j}(\hat{\boldsymbol{\xi}}, \xi_n)\downarrow_{\xi_n=0} \qquad (8.14.54a)$$

with $(\hat{\boldsymbol{\xi}}, 0) \in \hat{Q}_0 \subset \mathbb{R}^{n-1} \times \{0\}$.

Let $(\frac{\partial^j(\theta_i\phi)}{\partial n^j} \circ \Phi_i^{-1}\downarrow_{\xi_n=0})^\sim$ denote the *null extension*, i.e. extension by 0 to $\mathbb{R}^{n-1} \times \{0\} \setminus \hat{Q}_0$ of $(\frac{\partial^j(\theta_i\phi)}{\partial n^j} \circ \Phi_i^{-1}\downarrow_{\xi_n=0})$ in (8.14.54a). Hence, by virtue of (8.14.50), (8.14.53) and (8.14.54a), $\frac{\partial^j(\theta_i\phi)}{\partial n^j} \circ \Phi_i^{-1}\downarrow_{\xi_n=0}$ is considered to be a function of $\hat{\boldsymbol{\xi}}$ *only* for $(\hat{\boldsymbol{\xi}}, 0) \in \hat{Q}_0 \subset \mathbb{R}^{n-1} \times \{0\}$, and its *null extension* $(\frac{\partial^j(\theta_i\phi)}{\partial n^j} \circ \Phi_i^{-1}\downarrow_{\xi_n=0})^\sim$ is also a function of $\hat{\boldsymbol{\xi}}$ only $\forall \hat{\boldsymbol{\xi}} \in \mathbb{R}^{n-1}$ (i.e. $\forall (\hat{\boldsymbol{\xi}}, 0) \in \mathbb{R}^{n-1} \times \{0\}$). Then

$$\left(\frac{\partial^j(\theta_i\phi)}{\partial n^j} \circ \Phi_i^{-1}\downarrow_{\xi_n=0}\right)^\sim = \widetilde{\left(\frac{\partial^j v_i}{\partial \xi_n^j}\downarrow_{\xi_n=0}\right)} \in H^{m-j-1/2}(\mathbb{R}^{n-1}) \qquad (8.14.55)$$

by (8.14.54) for $0 \le j \le m-1$, $\forall 1 \le i \le N$.

But $\phi \in \mathcal{D}(\overline{\Omega})$ and Γ is of C^m-class (Definition 8.10.5) $\Longrightarrow \frac{\partial^j \phi}{\partial n^j}\downarrow_\Gamma \in C^m(\Gamma) \subset L^2(\Gamma)$ $\forall j = 0, 1, \ldots, m-1$, and $\frac{\partial^j \phi}{\partial n^j}\downarrow_\Gamma = \sum_{i=1}^N \theta_i\phi\downarrow_\Gamma \in L^2(\Gamma)$ with $\theta_i\phi\downarrow_\Gamma \in L^2(\Gamma)$ (since $\theta_i \in \mathcal{D}(\Omega_i)$ can always be given a null extension such that $\theta_i \in \mathcal{D}(\mathbb{R}^n)$ and $(\frac{\partial^j(\theta_i\phi)}{\partial n^j} \circ \Phi_i^{-1})^\sim \in H^{m-j-1/2}(\mathbb{R}^{n-1})$ $\forall i = 1, 2, \ldots, N$, by (8.14.55)), $0 \le j \le m-1$. Hence, for $j = 0, 1, \ldots, m-1$, by Definition 8.13.7 of $H^s(\Gamma)$ with

$s = m - j - 1/2 > 0$, $\frac{\partial^j \phi}{\partial n^j}\downarrow_\Gamma \in H^{m-j-1/2}(\Gamma)$ with the norm given by (8.13.24):

$$\begin{aligned}\left\|\frac{\partial^j \phi}{\partial n^j}\right\|^2_{H^{m-j-1/2}(\Gamma)} &= \sum_{i=1}^{N} \left\| \left(\frac{\partial^j(\theta_i \phi)}{\partial n^j} \circ \Phi_i^{-1} \downarrow_{\xi_n=0} \right)^{\sim} \right\|^2_{H^{m-j-1/2}(\mathbb{R}^{n-1})} \\ &= \sum_{i=1}^{N} \left\| \widetilde{\frac{\partial^j v_i}{\partial \xi_n^j}} \downarrow_{\xi_n=0} \right\|^2_{H^{m-j-1/2}(\mathbb{R}^{n-1})} \\ &= \sum_{i=1}^{N} \left\| \frac{\partial^j \widetilde{v}_i}{\partial \xi_n^j} \downarrow_{\xi_n=0} \right\|^2_{H^{m-j-1/2}(\mathbb{R}^{n-1})} \qquad \text{(by (8.14.52))} \\ &\le \sum_{i=1}^{N} C^2 \|\widetilde{v}_i\|^2_{H^m(\mathbb{R}^n_+)} \qquad \text{(by (8.14.49))} \\ &\le C^2 \sum_{i=1}^{N} \|\theta_i \phi\|^2_{H^m(\Omega \cap \Omega_i)} \qquad \text{(by (8.14.47))} \\ &\le C^2 \|\phi\|^2_{H^m(\Omega)} \qquad \text{(by Lemma 8.14.2)} \end{aligned}$$

$$\Longrightarrow \quad \left\|\frac{\partial^j \phi}{\partial n^j}\right\|_{H^{m-j-1/2}(\Gamma)} \le C\|\phi\|_{H^m(\Omega)} \quad \forall j = 0, 1, \dots, m-1. \tag{8.14.56}$$

Hence, for $0 \le j \le m-1$, $\forall \phi \in \mathcal{D}(\overline{\Omega})$, $\frac{\partial^j \phi}{\partial n^j}\downarrow_\Gamma \in H^{m-j-1/2}(\Gamma)$ with (8.14.56) and we have:

Definition 8.14.2. Let $\Omega \subsetneq \mathbb{R}^n$ be a C^m-regular domain with boundary Γ of C^m-class (see Definitions 8.10.4 and 8.10.5). Then, $\forall j = 0, 1, \dots m-1$, we define the *trace operator* γ_j^* by:

$$\gamma_j^* : \mathcal{D}(\overline{\Omega}) \longrightarrow C^m(\Gamma) \subset H^{m-j-1/2}(\Gamma) \tag{8.14.57}$$

and

$$\phi \in \mathcal{D}(\overline{\Omega}) \longmapsto \gamma_j^* \phi = \frac{\partial^j \phi}{\partial n^j}\downarrow_\Gamma \in H^{m-j-1/2}(\Gamma) \tag{8.14.58}$$

such that γ_j^* is a *continuous*, *linear operator* from $\mathcal{D}(\overline{\Omega})$ equipped with the norm $\|\cdot\|_{H^m(\Omega)}$ into $C^m(\Gamma)$ equipped with the norm $\|\cdot\|_{H^{m-j-1/2}(\Gamma)}$, i.e. $\exists C > 0$, independent of ϕ, such that

$$\|\gamma_j^* \phi\|_{H^{m-j-1/2}(\Gamma)} \le C\|\phi\|_{H^m(\Omega)} \quad \forall \phi \in \mathcal{D}(\overline{\Omega}), \tag{8.14.59}$$

where

$$\begin{aligned}\gamma_0^*\phi &= \phi\downarrow_\Gamma \in H^{m-1/2}(\Gamma),\\ \gamma_1^* &= \frac{\partial\phi}{\partial n}\downarrow_\Gamma \in H^{m-3/2}(\Gamma),\\ &\vdots\\ \gamma_{m-1}^*\phi &= \frac{\partial^{m-1}\phi}{\partial n^{m-1}}\downarrow_\Gamma \in H^{1/2}(\Gamma).\end{aligned} \tag{8.14.60}$$

But for C^m-regular domains, $\mathcal{D}(\overline{\Omega})$ is *dense* in $H^m(\Omega)$ and hence, $\forall j = 0, 1, \ldots, m-1$, the *continuous, linear* operator $\gamma_j^* : \mathcal{D}(\overline{\Omega}) \longrightarrow H^{m-j-1/2}(\Gamma)$ can be extended by density to a *unique, continuous, linear* operator $\widetilde{\gamma}_j^* : H^m(\Omega) \longrightarrow H^{m-j-1/2}(\Gamma)$ defined, $\forall u \in H^m(\Omega)$, by:

$$\widetilde{\gamma}_j^* u = \frac{\partial^j u}{\partial n^j}\downarrow_\Gamma \in H^{m-j-1/2}(\Gamma) \quad \text{with } \|\widetilde{\gamma}_j^* u\|_{H^{m-j-1/2}(\Gamma)} \le C\|u\|_{H^m(\Omega)}. \tag{8.14.61}$$

$\widetilde{\gamma}_j^*$ will be denoted by *the same notation* γ_j^*, i.e. $\gamma_j^* = \widetilde{\gamma}_j^*$ in (8.14.61) $\forall j = 0, 1, \ldots, m-1$. Then we have:

Theorem 8.14.2 (Trace theorem for C^m-regular domains). *Let $\Omega \subsetneq \mathbb{R}^n$ be a C^m-regular domain with boundary Γ of C^m-class. Then, for $0 \le j \le m-1$, the mapping:*

I. $\gamma_j^* : H^m(\Omega) \longrightarrow H^{m-j-1/2}(\Gamma)$ *is a* surjection *from* $H^m(\Omega)$ onto $H^{m-j-1/2}(\Gamma)$; (8.14.62)

II. $\boldsymbol{\gamma}^* = (\gamma_0^*, \gamma_1^*, \ldots, \gamma_{m-1}^*) : H^m(\Omega) \longrightarrow \prod_{j=0}^{m-1} H^{m-j-1/2}(\Gamma)$ *is a* continuous, linear, surjection *from* $H^m(\Omega)$ onto $\prod_{j=0}^{m-1} H^{m-j-1/2}(\Gamma)$; (8.14.63)

III. $\exists$ *a* continuous, linear operator

$$\boldsymbol{\rho} : \prod_{j=0}^{m-1} H^{m-j-1/2}(\Gamma) \longrightarrow H^m(\Omega) \tag{8.14.64}$$

such that $\boldsymbol{\gamma}^*\boldsymbol{\rho} = I_{\prod_{j=0}^{m-1} H^{m-j-1/2}(\Gamma)}$, *which is the identity operator I on* $\prod_{j=0}^{m-1} H^{m-j-1/2}(\Gamma)$, *i.e.* $\forall \mathbf{g} = (g_0, g_1, \ldots, g_{m-1}) \in \prod_{j=0}^{m-1} H^{m-j-1/2}(\Gamma)$, $\boldsymbol{\gamma}^*\boldsymbol{\rho}\mathbf{g} = \mathbf{g}$. (8.14.65)

Proof. It remains to prove the surjection in I and II and the right invertibility of $\boldsymbol{\gamma}^*$ in III. For defining the trace operator $\boldsymbol{\gamma}_j^*$ in (8.14.62), we have used the trace results of Theorem 8.14.1(see (8.14.48)–(8.14.50)). Obviously, the *surjectivity* will follow from Theorem 8.14.1 (see also Theorem 8.9.11 and Proposition 8.9.5 on the surjectivity of the trace operator $\boldsymbol{\gamma}^*$ and for more details on the *explicit construction* of $\boldsymbol{\rho}\mathbf{g}$ such that $\boldsymbol{\gamma}^*\boldsymbol{\rho}\mathbf{g} = \mathbf{g}$). □

Characterization of the kernel of trace operator $\boldsymbol{\gamma^*} : \boldsymbol{H^m(\Omega) \to \prod_{j=0}^{m-1} H^{m-j-1/2}(\Gamma)}$

Theorem 8.14.3. *Let $\Omega \subset \mathbb{R}^n$ be a C^m-regular domain with boundary $\Gamma \in C^m$-class and $\boldsymbol{\gamma}^* = (\gamma_0^*, \dots, \gamma_{m-1}^*) : H^m(\Omega) \longrightarrow \prod_{j=0}^{m-1} H^{m-j-1/2}(\Gamma)$ be the trace operator defined by Theorem* 8.14.2. *Let* Ker$(\boldsymbol{\gamma}^*)$ *denote the* kernel *of $\boldsymbol{\gamma}^*$ such that*

$$\operatorname{Ker}(\boldsymbol{\gamma}^*) = \{u : u \in H^m(\Omega),\ \gamma_0^* u = \gamma_1^* u, \dots, \gamma_{m-1}^* u = 0\}. \tag{8.14.66}$$

Then

$$H_0^m(\Omega) = \operatorname{Ker}(\boldsymbol{\gamma}^*). \tag{8.14.67}$$

In particular, for $m = 1$,

$$H_0^1(\Omega) = \operatorname{Ker}(\gamma_0^*) = \{u : u \in H^1(\Omega),\ \gamma_0^* u = u{\downarrow_\Gamma} = 0\};$$

for $m = 2$,

$$H_0^2(\Omega) = \operatorname{Ker}(\boldsymbol{\gamma}^*) = \{u : u \in H^2(\Omega),\ \gamma_0^* u = u{\downarrow_\Gamma} = 0,\ \gamma_1^* u = \frac{\partial u}{\partial n}{\downarrow_\Gamma} = 0\}. \tag{8.14.68}$$

Proof. For the sake of simplicity, we give the proof for $m = 1$. First of all, we show that $H_0^1(\Omega) \subset \operatorname{Ker}(\gamma_0^*) = \{u : u \in H^1(\Omega), \gamma_0^* u = u{\downarrow_\Gamma} = 0 \in H^{1/2}(\Gamma)$. Let $u \in H_0^1(\Omega)$. Since $\mathcal{D}(\Omega)$ is dense in $H_0^1(\Omega)$, $\exists$ a sequence $(\phi_k)_{k=1}^\infty$ in $\mathcal{D}(\Omega)$ such that $\phi_k \longrightarrow u$ in $H_0^1(\Omega)$ as $k \to \infty$. But $\phi_k \in \mathcal{D}(\Omega) \Longrightarrow \operatorname{supp}(\phi_k) \subset\subset \Omega$ is a *compact* subset of $\Omega \Longrightarrow \gamma_0^*(\phi_k) = \phi_k{\downarrow_\Gamma} = 0\ \forall k \in \mathbb{N}$. Since the trace operator $\gamma_0^* : H^1(\Omega) \longrightarrow H^{1/2}(\Gamma) \hookrightarrow L^2(\Gamma)$ is *continuous* from $H^1(\Omega)$ onto $H^{1/2}(\Gamma)$ and is *continuously* imbedded in $L^2(\Gamma)$ by (8.13.26), i.e. γ_0^* is *continuous* from $H^1(\Omega)$ into $L^2(\Gamma)$. Hence, $\gamma_0^* u = \gamma_0^*(\lim_{k\to\infty} \phi_k) = \lim_{k\to\infty} \gamma_0^* \phi_k = \lim_{k\to\infty} 0 = 0$ in $L^2(\Gamma)$. Thus, $u \in H_0^1(\Omega) \Longrightarrow u \in H^1(\Omega)$ and $\gamma_0^* u = u{\downarrow_\Gamma} = 0 \Longrightarrow u \in \operatorname{Ker}(\gamma_0^*) \Longrightarrow H_0^1(\Omega) \subset \operatorname{Ker}(\gamma_0^*)$.

Now we prove the *converse* part, i.e. $u \in H^1(\Omega)$ and $\gamma_0^* u = u{\downarrow_\Gamma} = 0 \Longrightarrow u \in H_0^1(\Omega)$, the proof of which is quite involved. For this we introduce the local charts $(\Gamma_i, \Phi_i)_{i=1}^N$ and partition $(\theta_i)_{i=1}^N$ of 1 with respect to the open cover $\{\Omega_i\}_{i=1}^N$ of Γ (see (8.14.21)–(8.14.26) for all details) and apply the *density* of $\mathcal{D}(\overline{\Omega})$ in $H^1(\Omega)$ (see (8.14.57)–(8.14.62) with $j = 0$) to deduce the final results. Without repeating the intermediate steps, we state the final results and the scheme of the proof: $u \in H^1(\Omega)$, $\gamma_0^* u = u{\downarrow_\Gamma} = 0$ will imply (of course, using the density of $\mathcal{D}(\overline{\Omega})$ in $H^1(\Omega)$) that new functions $w_i = w_i(\boldsymbol{\xi})$ with $\boldsymbol{\xi} = (\hat{\boldsymbol{\xi}}, \xi_n)$, $\hat{\boldsymbol{\xi}} = (\xi_1, \dots, \xi_{n-1}) \in \mathbb{R}^{n-1}$ will have the properties: $w_i = \widetilde{v_i} \in H^1(\mathbb{R}_+^n)$ (see (8.14.46)) with *compact support* $\operatorname{supp}(w_i) \subset\subset \overline{\mathbb{R}_+^n}$ (see (8.14.46)) satisfying the additional condition:

$$\gamma_0^* w_i = w_i(\cdot, 0) = w_i(\cdot, \xi_n){\downarrow_{\xi_n=0}} = 0, \tag{8.14.69}$$

since $\gamma_0^* u = u\downarrow_{\Gamma} = 0$ will imply that $w_i \downarrow_{\hat{Q}_0}(\cdot, 0) = w_i(\cdot, 0) = 0\ \forall i = 1, \dots, N$ (see (8.14.45a)). If we can prove that w_i satisfying the properties shown in (8.14.69) belongs to $H_0^1(\mathbb{R}_+^n)\ \forall i = 1, \dots, N$, then $u \in H_0^1(\Omega)$. Thus, the proof will be complete if we can show that $w_i \in H_0^1(\mathbb{R}_+^n)\ \forall i = 1, \dots, N$. Let $\widetilde{w}_i$ be the *null extension* to $\mathbb{R}^n$ of $w_i \in H^1(\mathbb{R}_+^n)$ satisfying (8.14.69), i.e. $\widetilde{w}_i = 0$ in $\mathbb{R}^n \setminus \mathbb{R}_+^n$. But $\widetilde{w}_i$ will belong to $H^1(\mathbb{R}^n)$ if we can show that

$$\frac{\partial \widetilde{w}_i}{\partial x_j} = \widetilde{\frac{\partial w_i}{\partial x_j}} \quad \forall j = 1, \dots, n,\ \forall i = 1, \dots, N, \tag{8.14.70}$$

in the *sense of distributions* on $\mathbb{R}^n$, i.e. in $\mathcal{D}'(\mathbb{R}^n)$. For this, we need:

Lemma 8.14.4. *For $v, w \in H^1(\mathbb{R}_+^n)$, we have*

$$\int_{\mathbb{R}_+^n} w \frac{\partial v}{\partial x_j} d\mathbf{x} = -\int_{\mathbb{R}_+^n} \frac{\partial w}{\partial x_j} \cdot v\, d\mathbf{x}, \quad 1 \le j \le n-1; \tag{8.14.71}$$

$$\begin{aligned} \int_{\mathbb{R}_+^n} w \frac{\partial v}{\partial x_n} d\mathbf{x} &= -\int_{\mathbb{R}_+^n} \frac{\partial w}{\partial x_n} v\, d\mathbf{x} - \int_{\mathbb{R}^{n-1}} w(\hat{\mathbf{x}}, 0) v(\hat{\mathbf{x}}, 0)\, d\hat{\mathbf{x}} \\ &= -\int_{\mathbb{R}^{n-1}} (\gamma_0^* w)(\hat{\mathbf{x}})(\gamma_0^* v)(\hat{\mathbf{x}})\, d\hat{\mathbf{x}} - \int_{\mathbb{R}_+^n} \frac{\partial w}{\partial x_n} v\, d\mathbf{x}, \end{aligned} \tag{8.14.72}$$

where $d\mathbf{x} = dx_1 \dots dx_n$, $d\hat{\mathbf{x}} = dx_1, \dots dx_{n-1}$.

Proof. $\forall \phi, \psi \in \mathcal{D}(\overline{\mathbb{R}_+^n})$, we have, for $1 \le j \le n$, $\phi \frac{\partial \psi}{\partial x_j} = \frac{\partial}{\partial x_j}(\phi\psi) - \psi \frac{\partial \phi}{\partial x_j}$. Hence, for $1 \le j \le n$, $\int_{\mathbb{R}_+^n} \phi \frac{\partial \psi}{\partial x_j} d\mathbf{x} = \int_{\mathbb{R}_+^n} \frac{\partial}{\partial x_j}(\phi\psi) d\mathbf{x} - \int_{\mathbb{R}_+^n} \psi \frac{\partial \phi}{\partial x_j} d\mathbf{x}$.

But $(\phi\psi) \in \mathcal{D}(\overline{\mathbb{R}_+^n}) \implies \phi\psi \in \mathcal{D}(\mathbb{R}^n)$ with $\operatorname{supp}(\phi\psi) \subset\subset \mathbb{R}^n \implies (\phi\psi)(x_1, \dots, x_j, \cdots, x_n) = 0$ for $x_j \to \pm\infty, 1 \le j \le n$. (8.14.73)

Hence, for $j = n$,

$$\begin{aligned} \int_{\mathbb{R}_+^n = \mathbb{R}^{n-1} \times \mathbb{R}_+} \frac{\partial}{\partial x_n}(\phi\psi)\, d\hat{\mathbf{x}}\, dx_n &= \int_{\mathbb{R}^{n-1}} d\hat{\mathbf{x}} \int_0^\infty \frac{\partial}{\partial x_n}(\phi\psi)\, dx_n \\ = \int_{\mathbb{R}^{n-1}} (\phi\psi)(\hat{\mathbf{x}}, x_n)\big|_{x_n=0}^{x_n=\infty} d\hat{\mathbf{x}} &= -\int_{\mathbb{R}^{n-1}} \phi(\hat{\mathbf{x}}, 0)\psi(\hat{\mathbf{x}}, 0)\, d\hat{\mathbf{x}}; \end{aligned} \tag{8.14.74}$$

and for $1 \le j \le n-1$,

$$\begin{aligned} \int_{\mathbb{R}_+^n} \frac{\partial}{\partial x_j}(\phi\psi)\, d\check{\mathbf{x}}\, dx_j &= \int_{\mathbb{R}^{n-2} \times \mathbb{R}_+} d\check{\mathbf{x}} \int_{-\infty}^{\infty} \frac{\partial}{\partial x_j}(\phi\psi)\, dx_j \\ &= \int_{\mathbb{R}^{n-2} \times \mathbb{R}_+} (\phi\psi)(\dots, x_j, \dots)\big|_{x_j=-\infty}^{x_j=+\infty} d\check{\mathbf{x}} \\ &= 0 \quad \text{(by (8.14.73))}, \end{aligned} \tag{8.14.75}$$

where $\check{\mathbf{x}} = (x_1, \ldots, x_{j-1}, x_{j+1}, \ldots, x_n)$ with x_j omitted; $d\check{\mathbf{x}} = dx_1 \ldots dx_{j-1} dx_{j+1} \ldots dx_n$.

Then, from (8.14.67) and (8.14.75) (resp. (8.14.67) and (8.14.74)), we get, $\forall \phi, \psi \in \mathcal{D}(\overline{\mathbb{R}^n_+})$,

$$\int_{\mathbb{R}^n_+} \phi \frac{\partial \psi}{\partial x_j} d\mathbf{x} = -\int_{\mathbb{R}^n_+} \frac{\partial \phi}{\partial x_j} \psi d\mathbf{x} \quad \text{for } 1 \le j \le n-1;$$

$$\int_{\mathbb{R}^n_+} \phi \frac{\partial \psi}{\partial x_n} d\mathbf{x} = -\int_{\mathbb{R}^n_+} \frac{\partial \phi}{\partial x_n} \psi d\mathbf{x} - \int_{\mathbb{R}^{n-1}} (\gamma_0^* \phi)(\gamma_0^* \psi) d\hat{\mathbf{x}}.$$

Since $\mathcal{D}(\overline{\mathbb{R}^n_+})$ is *dense* in $H^1(\mathbb{R}^n_+)$ and γ_0^* is continuous from $H^1(\mathbb{R}^n_+)$ onto $H^{1/2}(\mathbb{R}^{n-1}) \hookrightarrow L^2(\mathbb{R}^{n-1})$, the results follow from the *extension* by density. □

Proof of Theorem 8.14.3 *(continued).* We now prove (8.14.70). $\forall i = 1, 2, \ldots, N$, $\forall \phi \in \mathcal{D}(\mathbb{R}^n)$,

$$\begin{aligned}\left\langle \frac{\partial \widetilde{w}_i}{\partial \xi_j}, \phi \right\rangle &= -\left\langle \widetilde{w}_i, \frac{\partial \phi}{\partial \xi_j} \right\rangle = -\int_{\mathbb{R}^n} \widetilde{w}_i \frac{\partial \phi}{\partial \xi_j} d\boldsymbol{\xi} = -\int_{\mathbb{R}^n_+} w_i \frac{\partial \phi}{\partial \xi_j} d\boldsymbol{\xi} \\ &= \int_{\mathbb{R}^n_+} \phi \frac{\partial w_i}{\partial \xi_j} d\boldsymbol{\xi} \quad \text{(by (8.14.71))} \\ &= \int_{\mathbb{R}^n} \widetilde{\frac{\partial w_i}{\partial \xi_j}} \phi d\boldsymbol{\xi}\end{aligned}$$

$\Longrightarrow \frac{\partial \widetilde{w}_i}{\partial \xi_j} = \widetilde{\frac{\partial w_i}{\partial \xi_j}}$ in $\mathcal{D}'(\mathbb{R}^n)$, $1 \le j \le n-1$;

$$\begin{aligned}\left\langle \frac{\partial \widetilde{w}_i}{\partial \xi_n}, \phi \right\rangle &= -\int_{\mathbb{R}^n_+} w_i \frac{\partial \phi}{\partial \xi_n} d\boldsymbol{\xi} \\ &= \int_{\mathbb{R}^n_+} \phi \frac{\partial w_i}{\partial \xi_n} d\boldsymbol{\xi} + \int_{\mathbb{R}^{n-1}} \widetilde{w}_i(\hat{\boldsymbol{\xi}}, 0) \phi(\hat{\boldsymbol{\xi}}, 0) d\hat{\boldsymbol{\xi}} \quad \text{(by (8.14.72))} \\ &= \int_{\mathbb{R}^n} \widetilde{\frac{\partial w_i}{\partial \xi_n}} \phi d\boldsymbol{\xi} \quad \forall \phi \in \mathcal{D}(\mathbb{R}^n),\end{aligned}$$

since $\widetilde{w}_i(\hat{\boldsymbol{\xi}}, 0) = w_i(\hat{\boldsymbol{\xi}}, 0) = 0$ (by (8.14.69)). Hence, $\frac{\partial \widetilde{w}_i}{\partial \xi_n} = \widetilde{\frac{\partial w_i}{\partial \xi_n}}$ in $\mathcal{D}'(\mathbb{R}^n)$, $1 \le i \le N$. But, $\forall i = 1, 2, \ldots, N$, $w_i \in H^1(\mathbb{R}^n_+) \Longrightarrow \forall i = 1, 2, \ldots, N$, $w_i \in L^2(\mathbb{R}^n_+)$ with $\frac{\partial w_i}{\partial \xi_j} \in L^2(\mathbb{R}^n_+)$ $\forall j = 1, 2, \ldots, n \Longrightarrow \forall i = 1, 2, \ldots, N$, $\widetilde{w}_i \in L^2(\mathbb{R}^n)$ and $\widetilde{\frac{\partial w_i}{\partial \xi_j}} \in L^2(\mathbb{R}^n)$ and consequently, $\frac{\partial \widetilde{w}_i}{\partial \xi_j} = \widetilde{\frac{\partial w_i}{\partial \xi_j}} \in L^2(\mathbb{R}^n) \hookrightarrow \mathcal{D}'(\mathbb{R}^n)$ $\forall j = 1, 2, \ldots, n$. Hence, $\forall i = 1, 2, \ldots, N$, $\widetilde{w}_i \in L^2(\mathbb{R}^n)$ with $\frac{\partial \widetilde{w}_i}{\partial \xi_j} \in L^2(\mathbb{R}^n)$ $\forall j = 1, 2, \ldots, n \Longrightarrow \forall i = 1, 2, \ldots, N$, $\widetilde{w}_i \in H^1(\mathbb{R}^n)$. (8.14.76)

Then $w_i \in H^1_{00}(\mathbb{R}^n_+)$ by Definition 8.10.14. But for $\mathbb{R}^n_+$, $H^1_{00}(\mathbb{R}^n_+) = H^1_0(\mathbb{R}^n_+)$ by Theorem 8.10.17 (see also (8.10.106)). Thus, $\forall i = 1, \dots, N$, $w_i \in H^1_0(\mathbb{R}^n_+)$ and consequently, $u \in H^1_0(\Omega)$ and the proof is complete. □

Space $H^m_0(\Omega)$ defined with the help of trace operator γ^*

In Definition 2.15.2, we defined $H^m_0(\Omega)$ as the *closure* of $\mathcal{D}(\Omega)$ in $H^m(\Omega)$, where $\Omega \subset \mathbb{R}^n$ is *any open subset* of $\mathbb{R}^n$. As stated in the alternative characterization of $H^m_0(\Omega)$ in (2.15.23), for a C^m-regular domain $\Omega \subset \mathbb{R}^n$ with boundary $\Gamma \in C^m$-class (see Definitions 8.10.4 and 8.10.5), Theorem 8.14.3 allows us to give an alternative characterization or definition of $H^m_0(\Omega)$ by:

Definition 8.14.3. Let $\Omega \subset \mathbb{R}^n$ be a C^m-regular domain with $\Gamma \in C^m$-class. Then H^m_0 is *equivalently* defined as the *kernel* of the trace operator $\boldsymbol{\gamma}^* = (\gamma^*_0, \dots, \gamma^*_{m-1}) : H^m(\Omega) \longrightarrow \prod_{j=0}^{m-1} H^{m-j-1}(\Gamma)$, i.e.

$$H^m_0(\Omega) = \mathrm{Ker}(\boldsymbol{\gamma}^*) = \{u : u \in H^m(\Omega),\ \gamma^*_0 u = \cdots = \gamma^*_j u = \dots \gamma^*_{m-1} u = 0\}. \tag{8.14.77}$$

8.14.3 Trace results in $W^{s,p}$-spaces

Without repeating almost similar proofs with necessary modifications, we will give the *final* trace results for $W^{s,p}$-spaces. We begin with trace results in $W^{s,p}(\mathbb{R}^n)$, which are similar to those for $H^s(\mathbb{R}^n)$ in Theorem 8.9.11, since $\mathcal{D}(\mathbb{R}^n)$ is *dense* in $W^{s,p}(\mathbb{R}^n)$ for $s > 0$, $1 \le p < \infty$.

Let $s - \frac{1}{p} = k + \sigma$ with $k \in \mathbb{N}_0$, $0 < \sigma < 1$. Then, for $0 \le j \le k$, the linear trace operator $\gamma_j : \mathcal{D}(\mathbb{R}^n) \longrightarrow \mathcal{D}(\mathbb{R}^{n-1})$ defined, $\forall u \in \mathcal{D}(\mathbb{R}^n)$, by

$$\gamma_j u(\cdot) = \frac{\partial^j u}{\partial n^j}(\cdot, x_n)\downarrow_{x_n=0} = \frac{\partial^j u}{\partial n^j}(\cdot, 0) \in W^{s-j-1/p,p}(\mathbb{R}^{n-1}),$$

with $\gamma_0 u(\cdot) = u(\cdot, 0) \in W^{s-1/p,p}(\mathbb{R}^{n-1})$, is *continuous* from $\mathcal{D}(\mathbb{R}^n)$ equipped with the norm $\|\cdot\|_{s,p,\mathbb{R}^n}$ into $\mathcal{D}(\mathbb{R}^{n-1})$ equipped with the norm $\|\cdot\|_{s-1/p,p,\mathbb{R}^{n-1}}$ and can be given a *unique, continuous, linear extension* to $W^{s,p}(\mathbb{R}^n)$. This unique extended operator will still be denoted by $\gamma_j : W^{s,p}(\mathbb{R}^n) \longrightarrow W^{s-1/p,p}(\mathbb{R}^{n-1})$, and we have (see also Theorems 8.9.10 and 8.9.11):

Theorem 8.14.4. *Let $s - 1/p \neq$ an integer such that $s - 1/p = k + \sigma$ with $k \in \mathbb{N}_0, 0 < \sigma < 1$. Then:*

I. *For $0 \le j \le k$, the* trace operator $\gamma_j : W^{s,p}(\mathbb{R}^n) \longrightarrow W^{s-p-1/p,p}(\mathbb{R}^{n-1})$ *is a* continuous, linear surjection *from* $W^{s,p}(\mathbb{R}^n)$ onto $W^{s-j-1/p,p}(\mathbb{R}^n)$*;* (8.14.78)

II. $\boldsymbol{\gamma} = (\gamma_0, \ldots, \gamma_k) : W^{s,p}(\mathbb{R}^n) \longrightarrow \prod_{j=0}^{k} W^{s-j-1/p,p}(\mathbb{R}^{n-1})$ (8.14.79)
is a continuous, linear surjection *from* $W^{s,p}(\mathbb{R}^n)$ onto $\prod_{j=0}^{k} W^{s-j-1/p,p}(\mathbb{R}^{n-1})$, *which has a continuous, linear right inverse* $\boldsymbol{\rho}$ *from* $\prod_{j=0}^{k} W^{s-j-1/p,p}(\mathbb{R}^{n-1})$ *into* $W^{s,p}(\mathbb{R}^n)$.

Following the notation $\gamma_j^* = \gamma_j P$ in Definition 8.14.1, P being the *m-extension operator*, we will use asterisk $*$ to denote that an *extension operator* has been used to define the trace operator γ_j^*. Then we have the *global* trace result (i.e. on the whole of the boundary Γ):

Theorem 8.14.5. *Let $\Omega \subset \mathbb{R}^n$ be a bounded, open subset of $\mathbb{R}^n$ with boundary $\Gamma \in C^{k,1}$-class with $k \geq 0$ (see* (D.2.3.8), *Definition* D.2.3.1, *Appendix* D*).*

Let $s - 1/p \neq$ an integer such that $s \leq k + 1$ and $s - 1/p = l + \sigma$ with $l \in \mathbb{N}_0$, $0 < \sigma < 1$. Then, using the density of $C^{k,1}(\overline{\Omega})$ in $W^{s,p}(\Omega)$, we have:

I. $\forall u \in W^{s,p}(\Omega)$, $\gamma_j^* u = \frac{\partial^j u}{\partial n^j}\downarrow_\Gamma \in W^{s-j-1/p,p}(\Gamma), 0 \leq j \leq l$ *such that* $\boldsymbol{\gamma}^* = (\gamma_0^*, \gamma_1^*, \cdots, \gamma_l^*) : W^{s,p}(\Omega) \longrightarrow \prod_{j=0}^{l} W^{s-j-1/p,p}(\Gamma)$ *is a* continuous, linear surjection *from* $W^{s,p}(\Omega)$ onto $\prod_{j=0}^{l} W^{s-j-1/p,p}(\Gamma)$ (8.14.80)
and $\boldsymbol{\gamma}^* = (\gamma_0^*, \ldots, \gamma_l^*)$ *has a continuous, linear right inverse*
$\boldsymbol{\rho} : \prod_{j=0}^{l} W^{s-j-1/p,p}(\Gamma) \longrightarrow W^{s,p}(\Omega)$ *with*
$\boldsymbol{\gamma}^* \boldsymbol{\rho} \mathbf{g} = \mathbf{g} = (g_0, g_1, \ldots, g_l) \in \prod_{j=0}^{l} W^{s-j-1/p,p}(\Gamma)$. (8.14.81)

II. *The* kernel $\operatorname{Ker}(\boldsymbol{\gamma}^*)$ *of the trace operator* $\boldsymbol{\gamma}^*$ *in* (8.14.80) *is characterized by:*

$$\begin{aligned}\operatorname{Ker}(\boldsymbol{\gamma}^*) &= W_0^{s,p}(\Omega) = W_{00}^{s,p}(\Omega) \\ &= \{u : u \in W^{s,p}(\Omega),\ \gamma_0^* u = \gamma_1^* u = \cdots = \gamma_l^* u = 0\}. \end{aligned} \quad (8.14.82)$$

III. *In particular, for Lipschitz continuous boundary $\Gamma \in C^{0,1}$-class, we have, for $s = 1$, $k = 0$, $p > 1$ and $\Gamma \in C^{0,1}$-class (Γ is Lipschitz continuous by* (D.2.3.6), *Appendix* D*), the mapping $u \longmapsto \gamma_0^* u = u\downarrow_\Gamma$, which is defined for $C^{0,1}(\overline{\Omega})$, has a unique, continuous, linear extension γ_0^* (denoted by the same notation) to $W^{1,p}(\Omega)$ with $1 - 1/p = \sigma \in]0, 1[$, i.e.*

$$\gamma_0^* : W^{1,p}(\Omega) \longrightarrow W^{1-1/p,p}(\Gamma) \quad (8.14.83)$$

with a continuous, linear, right inverse $\rho : W^{1-1/p,p}(\Gamma) \longrightarrow W^{1,p}(\Omega)$ with
$\gamma_0^* \rho g = g \in W^{1-1/p,p}(\Gamma)$. (8.14.84)

IV. $\operatorname{Ker}(\gamma_0^*) = W_0^{1,p}(\Omega) = \{u : u \in W^{1,p}(\Omega),\ \gamma_0^* u = u\downarrow_\Gamma = 0\}$. (8.14.85)

8.14.4 Trace results for polygonal domains $\Omega \subset \mathbb{R}^2$

For all notations, geometric details, etc. for polygonal (resp. curvilinear polygonal) boundary $\Gamma = \bigcup_{j=1}^{N} \overline{\Gamma}_j \subset \mathbb{R}^2$, we refer to Section 8.13.4, Case II.

Each open side Γ_j, $1 \le j \le N$, can be viewed as an *open segment* of $\mathbb{R}$, and the results of Sobolev spaces can be applied with $\Gamma_j = \Omega \subset \mathbb{R}$, $1 \le j \le N$, as shown in Example 8.13.3, for example.

We will now follow Grisvard [17], [18] to present some basic trace results for bounded polygonal domains $\Omega \subset \mathbb{R}^2$.

Traces of smooth functions $u \in \mathcal{D}(\bar{\Omega})$ for bounded polygonal Γ are defined as follows: $\forall$ open $\Gamma_j, 1 \le j \le N$, traces $\gamma_{0j}u$ and $\gamma_{kj}u$ are defined by the *restrictions* to Γ_j of u and $\frac{\partial^k u}{\partial n_j^k}$, $1 \le k \le l$, i.e. $\forall u \in \mathcal{D}(\overline{\Omega})$,

$$\gamma_{0j}u = u\!\downarrow_{\Gamma_j} \in \mathcal{D}(\overline{\Gamma}_j), \quad \gamma_{kj}u = \frac{\partial^k u}{\partial n_j^k}\!\downarrow_{\Gamma_j} \in \mathcal{D}(\overline{\Gamma}_j), \tag{8.14.86}$$

$1 \le k \le l$, $1 \le j \le N$, where $\mathcal{D}(\overline{\Omega}) \subset H^s(\Omega)$ with $s - l - 1/2 > 0$, $\mathcal{D}(\overline{\Gamma}_j) \subset H^{s-k-1/2}(\Gamma_j)$ with $0 \le k \le l$ (see also (8.13.60)), $\hat{\mathbf{n}}_j$ being the unit exterior normal to Γ_j, $1 \le j \le N$.

Since bounded polygons Γ are *Lipschitz continuous*, i.e. $\Gamma \in C^{0,1}$-class, but *never* of $C^{1,1}$-class owing to the presence of corners, for the *whole* polygonal Γ, Trace Theorem 8.14.5 holds *only* for the particular case of $k = 0$ in (8.14.82). But for *each open side* Γ_j, $1 \le j \le N$, we have:

Theorem 8.14.6. *Let $\Gamma \subset \mathbb{R}^2$ be the polygonal boundary of the bounded $\Omega \subset \mathbb{R}^2$. Then, $\forall \Gamma_j$, $1 \le j \le N$, for $s - l - 1/2 > 0$, mapping $u \in \mathcal{D}(\overline{\Omega}) \longmapsto (\gamma_{0j}u, \gamma_{1j}u, \ldots, \gamma_{lj}u) \in \prod_{k=0}^{l} H^{s-k-1/2}(\Gamma_j)$ with $\gamma_{0j}u = u\!\downarrow_{\Gamma_j} \in H^{s-1/2}(\Gamma_j)$, $\ldots$, $\gamma_{lj}u = \frac{\partial^l u}{\partial n_j^l}\!\downarrow_{\Gamma_j} \in H^{s-l-1/2}(\Gamma_j)$ defined in* (8.14.86) *has a unique, continuous, linear extension as a trace operator from $H^s(\Omega)$ onto $\prod_{k=0}^{l} H^{s-k-1/2}(\Gamma_j)$. The extended operator will be denoted by the same notation, i.e.*

$$\begin{aligned} u \in H^s(\Omega) \longmapsto \gamma_{0j}u &= u\!\downarrow_{\Gamma_j} \in H^{s-1/2}(\Gamma), \\ &\vdots \\ \gamma_{lj}u &= \frac{\partial^l u}{\partial n_j^l}\!\downarrow_{\Gamma_j} \in H^{s-l-1/2}(\Gamma_j). \end{aligned} \tag{8.14.87}$$

Remark 8.14.3. Some explanations are in order. For a C^2-regular domain $\Omega \subset \mathbb{R}^2$, by Theorem 8.14.2, we have for $m = 1$, $u \in H^1(\Omega)$, $\gamma_0 u = u\!\downarrow_\Gamma \in H^{1/2}(\Gamma)$; for $m = 2$, $u \in H^2(\Omega)$, $\gamma_1 u = \frac{\partial u}{\partial n}\!\downarrow_\Gamma \in H^{1/2}(\Gamma)$, i.e. $\gamma_0(H^1(\Omega)) = \gamma_1(H^2(\Omega)) = H^{1/2}(\Gamma)$, where the $*$ in γ_k^* has been dropped.

Such results do *not* hold if Ω is a *polygonal domain*, i.e. for polygonal Ω, the images $\gamma_0(H^1(\Omega))$ and $\gamma_1(H^2(\Omega))$ will no longer be the same. Hence, it will be prudent to define traces of functions on each open side Γ_j *separately* and identify the compatibility conditions which must be satisfied near the corner points $P_j = \overline{\Gamma}_j \cap \overline{\Gamma}_{j-1}$ of two adjacent sides $\overline{\Gamma}_j$ and $\overline{\Gamma}_{j-1}$ of Γ. These compatibility conditions were explained in Example 8.13.3, Theorem 8.13.9 and Definition 8.13.11.

As shown in the proof of Proposition 8.10.8 (see also Figure 8.2), we use a finite open cover $\{\Omega_i\}_{i=0}^N$ of $\overline{\Omega} \subset \mathbb{R}^2$ such that polygon $\Gamma \subset \bigcup_{i=1}^N \Omega_i$, $\Omega_i \cap \overline{\Omega} = \Omega_i \cap \overline{S}_i$, $\{S_i\}_{i=1}^N$ being infinite angular sectors/quadrants/half-planes etc. Then, introducing a partition of unity (see the proof of Proposition 8.10.8), the whole problem of trace results for polygonal Ω reduces to the study of one corner/vertex, for example, of an angular sector. Again, by change of variables by affine transformation, the corresponding vertex of the polygonal domain/angular sector can be replaced by the origin or zero in new variables, angle ω_j at P_j between $\overline{\Gamma}_{j-1}$ and $\overline{\Gamma}_j$ by $\pi/2$ or $3\pi/2$ (see Figure 8.1) and the sides $\overline{\Gamma}_{j-1}$ and $\overline{\Gamma}_j$ by the axes of coordinates $x_2 = 0$ and $x_1 = 0$ (or $x_1 = 0$ and $x_2 = 0$) respectively. Thus, for each vertex $P_j = \overline{\Gamma}_{j-1} \cap \overline{\Gamma}_j$, the problem is reduced to the trace problem for the first quadrant $\mathbb{R}_+ \times \mathbb{R}_+ = \{(x_1, x_2) : x_1 > 0, x_2 > 0\}$ or the *complement* of $\overline{\mathbb{R}_+ \times \mathbb{R}_+}$ with angular measure $\omega = 3\pi/2 : \complement(\overline{\mathbb{R}_+ \times \mathbb{R}_+}) \equiv \mathbb{R}^2 \setminus (\overline{\mathbb{R}_+ \times \mathbb{R}_+})$.

Since the results for the first quadrant $\mathbb{R}_+ \times \mathbb{R}_+$ with $\omega = \pi/2$ will also hold for the complement of $\overline{\mathbb{R}_+ \times \mathbb{R}_+}$ with angular measure $3\pi/2$ by virtue of its m-extension property (see Proposition 8.10.7), the problem of identification of *compatibility conditions* at the vertices of a polygonal Ω is reduced to that for the first quadrant with a *corner* at the origin. Hence, for the sake of clarity in understanding the compatibility conditions, it is sufficient to study the trace results for the first quadrant with a corner point at $(0,0)$. For this we will follow the treatment of the problem given in Grisvard [17].

Trace theorem for the first quadrant with corner at the origin $(0,0)$

Let $\Omega = \mathbb{R}_+ \times \mathbb{R}_+ = \{(x_1, x_2) : x_1 > 0, x_2 > 0\} \subsetneq \mathbb{R}^2_+$ be the *first quadrant* in $\mathbb{R}^2$, $\mathbb{R}^2_+$ being the *upper half plane* $x_2 > 0$, i.e. $\mathbb{R}^2_+ = \mathbb{R} \times \mathbb{R}_+$. For smooth $u \in \mathcal{D}(\overline{\mathbb{R}_+ \times \mathbb{R}_+})$, the traces of u and its derivatives $\frac{\partial^k u}{\partial x_2^k} = \partial_2^k u$ on $\mathbb{R}_+ : x_2 = 0$, $x_1 > 0$ and $\frac{\partial^l u}{\partial x_1^l} = \partial_1^l u$ on $\mathbb{R}_+ : x_1 = 0$, $x_2 > 0$ are defined by:

$$f_k(\cdot) = \frac{\partial^k u}{\partial x_2^k}(\cdot, x_2)\downarrow_{x_2=0} = \partial_2^k u(\cdot, 0) \quad \text{for } 0 \le k \le m-1; \tag{8.14.88}$$

$$g_l(\cdot) = \frac{\partial^l u}{\partial x_1^l}(x_1, \cdot)\downarrow_{x_1=0} = \partial_1^l u(0, \cdot), \quad 0 \le l \le m-1.$$

Auxiliary subspace M of $H^m(\mathbb{R}_+ \times \mathbb{R}_+)$ Define the *subspace* M of $H^m(\mathbb{R}_+ \times \mathbb{R}_+)$ by:

$$M = \{u : u \in H^m(\mathbb{R}_+ \times \mathbb{R}_+),\ g_l = 0 \text{ for } 0 \le l \le m-1\}, \tag{8.14.89}$$

where $g_l = \partial_1^l u\downarrow_{x_!=0} \in H^{m-l-1/2}(\mathbb{R}_+)$ for $0 \le l \le m-1$.

In fact, $\Omega = \mathbb{R}_+ \times \mathbb{R}_+$ is an m-extension domain (Proposition 8.10.5). Hence, $\forall u \in H^m(\mathbb{R}_+ \times \mathbb{R}_+)$, $\exists U \in H^m(\mathbb{R}^2)$ such that $U\downarrow_{\mathbb{R}_+\times\mathbb{R}_+} = u$.

Then, $\forall U \in H^m(\mathbb{R}^2)$, $\gamma_l U = \frac{\partial^l U}{\partial x_1^l}\downarrow_{x_1=0} \in H^{m-l-1/2}(\mathbb{R})$ by Trace Theorem 8.9.11 $\Longrightarrow g_l = \gamma_l U\downarrow_{\mathbb{R}_+:x_2>0} \in H^{m-l-1/2}(\mathbb{R}_+)$ for $0 \le l \le m-1$.

Let $u \in M$ and $\tilde{u}$ be its null extension to $\mathbb{R}^2_+$, i.e. $\tilde{u}(\mathbf{x}) = 0$ for $\mathbf{x} \in \mathbb{R}^2_+ \setminus (\mathbb{R}_+ \times \mathbb{R}_+)$ and $\tilde{u}(\mathbf{x}) = u(\mathbf{x})$ for $\mathbf{x} \in \mathbb{R}_+ \times \mathbb{R}_+$. Then $\tilde{u} \in H^m(\mathbb{R}^2_+)$ (apply Theorem 2.15.5 with necessary modifications).

Let $\tilde{f}_k = \gamma_k^* \tilde{u} = \frac{\partial^k \tilde{u}}{\partial x_2^k}\downarrow_{\Gamma\equiv\mathbb{R}}$ be the traces of $\partial_2^k \tilde{u}$ on the boundary Γ with $\Gamma = \{(x_1, x_2) : x_1 \in \mathbb{R},\ x_2 = 0\} \equiv \mathbb{R}$ (i.e. $x_2 = 0$).

Then, by Theorem 8.14.1, $\tilde{f}_k \in H^{m-k-1/2}(\mathbb{R})$ with $\tilde{f}_k\downarrow_{\mathbb{R}_+:x_1>0} = f_k$ for $0 \le k \le m-1$. (8.14.90)

Now we state two lemmas which will be essential to prove the Trace Theorem 8.14.7 for the first quadrant $\mathbb{R}_+ \times \mathbb{R}_+$, which is stated later.

Lemma 8.14.5. *Let $u \in M$ with M defined by* (8.14.89) *and $(f_k)_{k=0}^{m-1}$ be defined by* (8.14.90). *Then the mapping $u \in M \longmapsto (f_k)_{k=0}^{m-1}$ has a unique, continuous extension from M onto the trace subspace T_0 characterized by $T_0 \subset \prod_{k=0}^{m-1} H^{m-k-1/2}(\mathbb{R}_+)$ defined by:*

I. $\partial_1^l f_k(0) = 0$ *for* $l + k < m-1$; (8.14.91a)

II. $\int_0^\infty \frac{|\partial_1^l f_k(t)|^2}{t} dt < +\infty$ *for* $l + k = m-1$. (8.14.91b)

Proof.
Necessity: From (8.14.90), $\tilde{f}_k \in H^{m-k-1/2}(\mathbb{R})$ with $m-k-1/2 > 0$ for $0 \le k \le m-1 \Longrightarrow f_k = \tilde{f}_k\downarrow_{\mathbb{R}_+} \in H^{m-k-1/2}(\mathbb{R}_+)$ for $0 \le k \le m-1 \Longrightarrow (f_k)_{k=0}^{m-1} \in \prod_{k=0}^{m-1} H^{m-k-1/2}(\mathbb{R}_+)$.

Moreover, by Sobolev's Imbedding Theorem 8.9.4, $H^{m-k-1/2}(\mathbb{R}) \hookrightarrow C^p(\mathbb{R})$ with $p \ge 0$ for $(m-k-1/2)-1/2 > p$, i.e. $m-k-1 > p$ with $p = m-k-2 \in \mathbb{N}_0$. Thus, functions $\tilde{f}_k \in H^{m-k-1/2}(\mathbb{R})$ are $(m-k-2)$ times *continuously differentiable*. Hence, $\partial_1^l f_k(0) = \partial_1^l \tilde{f}_k(0) = 0$ for $l \le m-k-2 < m-k-1$, i.e. $\partial_1^l f_k(0) = 0$ for $l + k < m-1$ and the identity I holds.

Now we prove II for $k + l = m-1$. Since $\tilde{f}_k \in H^{m-k-1/2}(\mathbb{R})$ for $0 \le k \le m-1$,

$$\partial_1^l \tilde{f}_k \in H^{m-k-1/2-l}(\mathbb{R}) = H^{m-(k+l)-1/2}(\mathbb{R}) = H^{m-(m-1)-1/2}(\mathbb{R}) = H^{1/2}(\mathbb{R}).$$

Hence, by Definition 8.10.6 of $H^{1/2}(\mathbb{R})$, we have

$$\iint_{\substack{\mathbb{R}\times\mathbb{R}\\ t\neq\tau}} \frac{|\partial_1^l \tilde{f}_k(t) - \partial_1^l \tilde{f}_k(\tau)|^2}{|t-\tau|^{1+2\cdot 1/2}}\,dt\,d\tau = \iint_{\substack{\mathbb{R}\times\mathbb{R}\\ t\neq\tau}} \frac{|\partial_1^l \tilde{f}(t) - \partial_1^l \tilde{f}(\tau)|^2}{|t-\tau|^{2}}\,dt\,d\tau < \infty.$$

But

$$\begin{aligned}\iint_{\mathbb{R}\times\mathbb{R}} (\dots)\,dt\,d\tau &= \int_{-\infty}^{0}\int_{-\infty}^{0} (\dots)\,dt\,d\tau + \int_{-\infty}^{0}\int_{0}^{\infty} (\dots)\,dt\,d\tau \\ &\quad + \int_{0}^{\infty}\int_{-\infty}^{0} (\dots)\,dt\,d\tau + \int_{0}^{\infty}\int_{0}^{\infty} (\dots)\,dt\,d\tau < +\infty,\end{aligned} \tag{8.14.92}$$

where the integrand $(\dots) \geq 0$ in each case. Hence, for $]0,\infty[\times]-\infty,0[$ in particular, we have $\partial_1^l \tilde{f}_k(\tau) = 0$ for $-\infty < \tau < 0$

$$\begin{aligned}&\int_0^\infty \int_{-\infty}^{0}{}_{(t\neq\tau)} \frac{|\partial_1^l f_k(t) - 0|^2}{|t-\tau|^2}\,dt\,d\tau < +\infty \\ \Longrightarrow\quad &\int_0^\infty |\partial_1^l f_k(t)|^2 \left(\int_{-\infty}^{0} \frac{d\tau}{(t-\tau)^2}\right) dt < +\infty \\ \Longrightarrow\quad &\int_0^\infty \frac{|\partial_1^l f_k(t)|^2}{t}\,dt < +\infty,\end{aligned}$$

since $\int_{-\infty}^{0} \frac{d\tau}{(t-\tau)^2} = \frac{1}{t-\tau}\Big|_{\tau=-\infty}^{\tau=0} = \frac{1}{t}$.

Hence, for $k + l = m - 1$, the condition II must hold. Thus, the necessity part for the characterisation of T_0 is proved.

Sufficiency: Let $(f_k)_{k=0}^{m-1} \in T_0 \subset \prod_{k=0}^{m-1} H^{m-k-1/2}(\mathbb{R}_+)$ and both (8.14.91a) and (8.14.91b) hold. Then $\partial_1^l f_k \in H^{1/2}(\mathbb{R}_+)$ for $l + k = m - 1$ and, by definition of $H^{1/2}(\mathbb{R}_+)$,

$$\int_0^\infty \int_0^\infty{}_{(t\neq\tau)} \frac{|\partial_1^l f_k(t) - \partial_1^l f_k(\tau)|^2}{|t-\tau|^2}\,dt\,d\tau < +\infty. \tag{8.14.93}$$

Since $\partial_1^l f_k(0) = 0$ for $l+k < m-1$ by I, $f_k \in H^{m-k-1/2}(\mathbb{R}_+)$ for $0 \leq k \leq m-1$, its null extension $\tilde{f}_k \in H^{m-k-1/2}(\mathbb{R})$ for $0 \leq k \leq m-1$ if $\partial_1^l \tilde{f}_k \in H^{1/2}(\mathbb{R})$. For this we are to show that $\iint_{\mathbb{R}\times\mathbb{R}\,(t\neq\tau)} \frac{|\partial_1^l \tilde{f}_k(t)-\partial_1^l \tilde{f}_k(\tau)|^2}{|t-\tau|^2} dt\,d\tau < +\infty$. In fact, using (8.14.92),

$$\begin{aligned}\iint_{\mathbb{R}\times\mathbb{R}} (\dots) dt\,d\tau &= \int_{-\infty}^{0}\int_{-\infty}^{0} 0\,dt\,d\tau + \int_{-\infty}^{0}\int_{0}^{\infty} \frac{|\partial_1^l f_k(t)|^2}{|t-\tau|^2} dt\,d\tau \\ &\quad + \int_{0}^{\infty}\int_{-\infty}^{0} \frac{|\partial_1^l f_k(t)|^2}{|t-\tau|^2} dt\,d\tau + \int_{0}^{\infty}\int_{0}^{\infty} \frac{|\partial_1^l f_k(t)-\partial_1^l f_k(\tau)|^2}{|t-\tau|^2} dt\,d\tau \\ &= 2\int_{0}^{\infty}\int_{-\infty}^{0} \frac{|\partial_1^l f_k(t)|^2}{|t-\tau|^2} dt\,d\tau + \int_{0}^{\infty}\int_{0}^{\infty} \frac{|\partial_1^l f_k(t)-\partial_1^l f_k(\tau)|^2}{|t-\tau|^2} dt\,d\tau \\ &= 2\int_{0}^{\infty} \frac{|\partial_1^l f_k(t)|^2}{t} dt + \int_{0}^{\infty}\int_{0}^{\infty} \frac{|\partial_1^l f_k(t)-\partial_1^l f_k(\tau)|^2}{|t-\tau|^2} dt\,d\tau < +\infty\end{aligned}$$

(by condition II and (8.14.93)).

Hence, $\partial_1^l \tilde{f}_k \in H^{1/2}(\mathbb{R})$ and $\tilde{f}_k \in H^{m-k-1/2}(\mathbb{R})$ for $0 \le k \le m-1$. Then, by the surjectivity of the trace operator in Theorem 8.14.1, $\exists v \in H^m(\mathbb{R}^2_+)$ such that $\gamma_k^* v = \tilde{f}_k = \frac{\partial^k v}{\partial x_2^k}\downarrow_{x_2=0} \in H^{m-k-1/2}(\mathbb{R})$ for $0 \le k \le m-1$. So the proof will be complete, if we can show that:

(i) $v\downarrow_{\mathbb{R}_+\times\mathbb{R}_+} = u \in M$ with M defined by (8.14.89), and

(ii) $\partial_2^k u\downarrow_{x_2=0} = f_k$ for $0 \le k \le m-1$.

Although $v \in H^m(\mathbb{R}^2_+) \implies v\downarrow_{\mathbb{R}_+\times\mathbb{R}_+} \in H^m(\mathbb{R}_+\times\mathbb{R}_+)$ by (8.10.146c), $v \in H^m(\mathbb{R}^2_+)$ does *not* imply that $\partial_1^l u\downarrow_{x_1=0} = 0$ for $0 \le l \le m-1$, i.e. $v\downarrow_{\mathbb{R}_+\times\mathbb{R}_+} = u$ may *not* belong to M. But we can construct a suitable restriction $v\downarrow_{\mathbb{R}_+\times\mathbb{R}_+} \in M$ by a method similar to that of the *Babitch–Nikolski extension* technique (see (8.10.43)–(8.10.45) and thereafter for all details) as follows.

Define u by: for $x_1 > 0$, $x_2 > 0$, $u(x_1,x_2) = v(x_1,x_2) - \sum_{k=1}^m \lambda_k v(-kx_1,x_2)$ where the unknown numbers $(\lambda_k)_{k=1}^m$ are *uniquely* determined by the m conditions: for $0 \le l \le m-1$, $\partial_1^l u(0,x_2) = \partial_1^l v(0,x_2) - \sum_{k=1}^m (-k)^l \lambda_k \partial_1^l v(0,x_2) = 0$, i.e. by solving $\sum_{k=1}^m (-k)^l \lambda_k = 1$ for $0 \le l \le m-1$ (see (8.10.48)). Then $u \in H^m(\mathbb{R}_+\times\mathbb{R}_+)$ and $\partial_1^l u\downarrow_{x_1=0} = 0$ for $0 \le l \le m-1$, i.e. $u \in M$. But for $x_1 > 0$,

$$\begin{aligned}\partial_2^k u(x_1,0) &= \partial_2^k v(x_1,0) - \sum_{j=1}^m \lambda_j \partial_2^k v(-jx_1,0) \\ &= \tilde{f}_k(x_1) - \sum_{j=1}^m \lambda_j \tilde{f}_k(-jx_1) = f_k(x_1) - 0 = f_k(x_1)\end{aligned}$$

for $0 \le k \le m-1$, since for $x_1 > 0$, $\tilde{f}_k(-jx_1) = 0$. Hence, $\partial_2^k u\downarrow_{x_2=0} = f_k$ for $0 \le k \le m-1$ and the proof is complete. □

Lemma 8.14.6. *For $m \geq 1$, let $u \in H^m(\mathbb{R}_+ \times \mathbb{R}_+)$ and f_{lk} and g_{kl} be defined by: $f_{lk}(\cdot) = \partial_1^l f_k(\cdot)$ and $g_{kl}(\cdot) = \partial_2^k g_l$ for $k + l \leq m - 1$, where $f_k(\cdot) = \partial_2^k u(\cdot, x_2)\downarrow_{x_2=0} = \partial_2^k u(\cdot, 0)$ with $0 \leq k \leq m-1$ and $g_l(\cdot) = \partial_1^l u(x_1, \cdot)\downarrow_{x_1=0} = \partial_1^l u(0, \cdot)$ for $0 \leq l \leq m-1$. Then the following hold:*

I. $f_{lk}(0) = g_{kl}(0)$ *for* $l + k < m - 1$; (8.14.94a)

II. $\int_0^\infty \frac{|f_{lk}(t)-g_{kl}(t)|^2}{t} dt < +\infty$ *for* $l + k = m - 1$, (8.14.94b)

which are the compatibility conditions *at the origin.*

In particular, for $k = l = 0$, $f_0 = f_{00} = u\downarrow_{x_2=0}$, $g_0 = g_{00} = u\downarrow_{x_1=0}$, we have:

I′ $f_0(0) = g_0(0)$ *for* $m > 1$ *(since $0 < m - 1 \Longrightarrow m > 1$);* (8.14.95a)

II′ $\int_0^\infty \frac{|f_0(t)-g_0(t)|^2}{t} dt < +\infty$ *for* $m = 1$ *(since $0 = m - 1 \Longrightarrow m = 1$).* (8.14.95b)

Obviously, for $m < 1$, i.e. $m = 0$, Trace Theorem 8.14.6 *does* not *hold.*

Proof.

I. Since $f_{lk}(0) = \partial_1^l \partial_2^k u(0,0)$ and $g_{kl}(0) = \partial_2^k \partial_1^l u(0,0)$, $f_{lk}(0) = g_{kl}(0) = \partial_1^l \partial_2^k u(0,0)$ will hold only if $\partial_1^l \partial_2^k u = \partial_2^k \partial_1^l u \in C^0(\overline{\mathbb{R}_+ \times \mathbb{R}_+})$, i.e. only if $H^{m-(k+l)}(\mathbb{R}_+ \times \mathbb{R}_+) \hookrightarrow C^0(\overline{\mathbb{R}_+ \times \mathbb{R}_+})$, which holds only for $m - (k+l) - 2/2 > 0$ by Sobolev's imbedding results, i.e. $f_{lk}(0) = g_{kl}(0)$ for $k + l < m - 1$.

II. For the sake of notational simplicity, we give the proof of II′, i.e. for $l = k = 0$, $\int_0^\infty \frac{|f_0(t)-g_0(t)|^2}{t} dt < +\infty$ for $m = 1$.

Let $u \in \mathcal{D}(\overline{\mathbb{R}_+ \times \mathbb{R}_+})$. Then, $\forall$ *fixed* $t > 0$,

$$
\begin{aligned}
f_0(t) - g_0(t) &= u(t,0) - u(0,t) = u(t,0) - u(t,t) + u(t,t) - u(0,t) \\
&= \int_0^t \frac{\partial u}{\partial x_1}(s,t)ds - \int_0^t \frac{\partial u}{\partial x_2}(t,s)ds \\
\Longrightarrow \quad |f_0(t) - g_0(t)| &\leq \left| \int_0^t \frac{\partial u}{\partial x_1}(s,t)ds \right| + \left| \int_0^t \frac{\partial u}{\partial x_2}(t,s)ds \right| \\
&= \left| \left\langle \frac{\partial u}{\partial x_1}(\cdot,t), 1 \right\rangle_{L^2(]0,t[)} \right| + \left| \left\langle \frac{\partial u}{\partial x_2}(t,\cdot), 1 \right\rangle_{L^2(]0,t[)} \right| \\
&\leq \left(\left\| \frac{\partial u}{\partial x_1}(\cdot,t) \right\|_{L^2(]0,t[)} + \left\| \frac{\partial u}{\partial x_2}(t,\cdot) \right\|_{L^2(]0,t[)} \right) \cdot \|1\|_{L^2(]0,t[)} \\
&= \sqrt{t} \left[\left(\int_0^t \left| \frac{\partial u}{\partial x_1}(s,t) \right|^2 ds \right)^{1/2} + \left(\int_0^t \left| \frac{\partial u}{\partial x_2}(t,s) \right|^2 ds \right)^{1/2} \right]
\end{aligned}
$$

$$\Longrightarrow \quad \frac{|f_0(t)-g_0(t)|^2}{t} \leq 2\Big[\int_0^t \Big(\Big|\frac{\partial u}{\partial x_1}(s,t)\Big|^2 + \Big|\frac{\partial u}{\partial x_2}(t,s)\Big|^2\Big)ds\Big]$$

$$\leq 2\int_0^\infty \Big(\Big(\Big|\frac{\partial u}{\partial x_1}(s,t)\Big|^2 + \Big|\frac{\partial u}{\partial x_2}(t,s)\Big|^2\Big)\Big)ds$$

(since $u \in H^m(\mathbb{R}_+ \times \mathbb{R}_+)$)

$$\Longrightarrow \quad \int_0^\infty \frac{|f_0(t)-g_0(t)|^2}{t}dt$$

$$\leq 2\int_0^\infty \int_0^\infty \Big(\Big|\frac{\partial u}{\partial x_1}(s,t)\Big|^2 \Big|\frac{\partial u}{\partial x_2}(t,s)\Big|^2\Big)dsdt \tag{8.14.96}$$

$$\leq C\|u\|^2_{H^1(\mathbb{R}_+\times\mathbb{R}_+)} \quad \forall u \in \mathcal{D}(\overline{\mathbb{R}_+ \times \mathbb{R}_+}) \text{ with } C > 0.$$

By virtue of the *density* of $\mathcal{D}(\overline{\mathbb{R}_+ \times \mathbb{R}_+})$ in $H^1(\mathbb{R}_+ \times \mathbb{R}_+)$, the inequality (8.14.96) remains valid also $\forall u \in H^1(\mathbb{R}_+ \times \mathbb{R}_+)$, i.e. $\forall u \in H^1(\mathbb{R}_+ \times \mathbb{R}_+)$, $\int_0^\infty \frac{|f_0(t)-g_0(t)|^2}{t}dt \leq C\|u\|^2_{H^1(\mathbb{R}_+\times\mathbb{R}_+)} < +\infty$, and the lemma is proved.

□

Trace results for the first quadrant $\mathbb{R}_+ \times \mathbb{R}_+$

Theorem 8.14.7. *Let $\Omega = \mathbb{R}_+ \times \mathbb{R}_+ = \{(x_1,x_2) : x_1 > 0, x_2 > 0\}$ be the first quadrant in $\mathbb{R}^2$. The mapping*

$$u \in \mathcal{D}(\overline{\mathbb{R}_+ \times \mathbb{R}_+}) \longmapsto (\{f_k\}_{k=0}^{m-1}, \{g_l\}_{l=0}^{m-1}) \in \prod_{k=0}^{m-1} H^{m-k-1/2}(\mathbb{R}_+) \times \prod_{l=0}^{m-1} H^{m-l-1/2}(\mathbb{R}_+) \tag{8.14.97}$$

with $f_k = \partial_2^k u\downarrow_{x_2=0}$, $g_l = \partial_1^l u\downarrow_{x_1=0}$ has a unique, continuous linear extension as an operator from $H^m(\mathbb{R}_+ \times \mathbb{R}_+)$ onto the trace subspace T_1 *of the space*

$$T = \prod_{k=0}^{m-1} H^{m-k-1/2}(\mathbb{R}_+) \times \prod_{l=0}^{m-1} H^{m-l-1/2}(\mathbb{R}_+), \tag{8.14.98}$$

with T_1 defined by the compatibility conditions*:*

I. $\partial_1^l f_k(0) = \partial_2^k g_l(0)$ *for* $k + l < m - 1$; (8.14.99a)

II. $\int_0^\infty \frac{|\partial_1^l f_k(t)-\partial_2^k g_l(t)|^2}{t}dt < +\infty$ *for* $k + l = m + 1$. (8.14.99b)

Proof.
Necessity: For $u \in \mathcal{D}(\overline{\mathbb{R}_+ \times \mathbb{R}_+})$, $\partial_2^k u\downarrow_{x_2=0} = f_k \in \mathcal{D}(\overline{\mathbb{R}_+}) \subset H^{m-k-1/2}(\mathbb{R}_+)$ for $0 \leq k \leq m-1$, $\partial_1^l u\downarrow_{x_1=0} = g_l \in \mathcal{D}(\overline{\mathbb{R}_+}) \subset H^{m-l-1/2}(\mathbb{R}_+)$ for $0 \leq l \leq m-1$,

which justifies (8.14.97), $\mathcal{D}(\overline{\mathbb{R}_+ \times \mathbb{R}_+})$ being a dense subspace of $H^m(\mathbb{R}_+ \times \mathbb{R}_+)$. Hence, we start with $u \in H^m(\mathbb{R}_+ \times \mathbb{R}_+)$. Since $\mathbb{R}_+ \times \mathbb{R}_+$ is an m-extension domain, $\exists$ an m-extension operator $P : H^m(\mathbb{R}_+ \times \mathbb{R}_+) \longrightarrow H^m(\mathbb{R}^2)$ such that $\exists U = Pu \in H^m(\mathbb{R}^2)$ with $U\downarrow_{\mathbb{R}_+\times\mathbb{R}_+} = u \in H^m(\mathbb{R}_+ \times \mathbb{R}_+)$, by Proposition 8.10.5. Then, by the Trace Theorem 8.9.11, for $U \in H^m(\mathbb{R}^2)$, $\gamma_k^2 U = \partial_2^k U\downarrow_{x_2=0} = F_k \in H^{m-k-1/2}(\mathbb{R})$ for $0 \leq k \leq m-1$ and $\gamma_l^1 U = \partial_1^l U\downarrow_{x_1=0} = G_l \in H^{m-l-1/2}(\mathbb{R})$ for $0 \leq l \leq m-1$. Hence, by (8.10.146c), $F_k\downarrow_{\mathbb{R}_+:x_1>0} = f_k \in H^{m-k-1/2}(\mathbb{R}_+)$ for $0 \leq k \leq m-1$ and $G_l\downarrow_{\mathbb{R}_+:x_2>0} = g_l \in H^{m-l-1/2}(\mathbb{R}_+)$ for $0 \leq l \leq m-1$.

Hence,

$$(\{f_k\}_{k=0}^{m-1}, \{g_l\}_{l=0}^{m-1}) \in T = \prod_{k=0}^{m-1} H^{m-k-1/2}(\mathbb{R}_+) \times \prod_{l=0}^{m-1} H^{m-l-1/2}(\mathbb{R}_+).$$

$\partial_1^l f_k(x_1) = \frac{\partial^{l+k} u(x_1,0)}{\partial x_1^l \partial x_2^k}$; $\partial_2^k g_l(x_2) = \frac{\partial^{l+k} u(0,x_2)}{\partial x_2^k \partial x_1^l}$ for $0 \leq k+l \leq m-1$. For $u \in H^m(\mathbb{R}_+ \times \mathbb{R}_+)$, $\partial_2^k \partial_1^l u = \partial_1^l \partial_2^k u \in H^{m-k-l}(\mathbb{R}_+ \times \mathbb{R}_+)$ with $m-k-l \geq 1$. Then, $\partial_1^l f_k(0) = \partial_2^k g_l(0) = \partial_2^k \partial_1^l u(0,0)$ will hold only if $\partial_1^l \partial_2^k u = \partial_2^k \partial_1^l u \in C^0(\overline{\mathbb{R}_+ \times \mathbb{R}_+})$, i.e. for $H^{m-k-l}(\mathbb{R}_+ \times \mathbb{R}_+) \hookrightarrow C^0(\overline{\mathbb{R}_+ \times \mathbb{R}_+})$, which holds only for $(m-k-l) - 2/2 > 0 \Longrightarrow m-(k+l)-1 > 0 \Longrightarrow k+l < m-1 \Longrightarrow$ Condition I in (8.14.99a).

By Lemma 8.14.6, which we apply to the operator $\partial_1^l \partial_2^k$ with $\partial_1^l(\partial_2^k u\downarrow_{x_2=0}) = \partial_1^l f_k$, $\partial_2^k(\partial_1^l u\downarrow_{x_1=0}) = \partial_2^k g_l$ for $k+l \leq m-1$, we get $\partial_1^l f_k(0) = \partial_2^k g_l(0)$ for $k+l < m-1$, which is already proved above, and $\int_0^\infty \frac{|\partial_1^l f_k(t) - \partial_2^k g_l(t)|^2}{t} dt < +\infty$ for $l+k = m-1 \Longrightarrow$ Condition II in (8.14.94b).

Sufficiency: Let

$$(\{f_k\}^{m-1} k = 0, \{g_l\}_{l=0}^{m-1}) \in T = \prod_{k=0}^{m-1} H^{m-k-1/2}(\mathbb{R}_+) \times \prod_{l=0}^{m-1} H^{m-l-1/2}(\mathbb{R}_+)$$

such that Conditions I and II hold. We are to show that $\exists u \in H^m(\mathbb{R}_+ \times \mathbb{R}_+)$ such that $\partial_2^k u\downarrow_{x_2=0} = f_k$, $\partial_1^l u\downarrow_{x_1=0} = g_l$ for $0 \leq k, l \leq m-1$. For this we will find $u \in H^m(\mathbb{R}_+ \times \mathbb{R}_+)$ as the sum of *two* functions: $u = w + v$ on $\mathbb{R}_+ \times \mathbb{R}_+$ with $w \in M$ (M being the space in (8.14.89)) and $v \in H^m(\mathbb{R}_+^2)$. Since $\mathbb{R}_+$ is an m-extension domain by Theorem 8.10.4, $\exists$ a continuous $P_1 : H^{m-l-1/2}(\mathbb{R}_+) \longrightarrow H^{m-l-1/2}(\mathbb{R})$ with $G_l^* = P_1 g_l \in H^{m-l-1/2}(\mathbb{R})$ such that $G_l^*\downarrow_{\mathbb{R}_+} = g_l \in H^{m-l-1/2}(\mathbb{R}_+)$ for $0 \leq l \leq m-1$. Then, by the surjectivity of the Trace Theorem 8.14.1, $\exists v \in H^m(\mathbb{R}_+^2)$ with $\mathbb{R}_+^2 : x_1 > 0$ (the *right half-plane*) such that $\partial_1^l v\downarrow_{\mathbb{R}:x_1=0} = G_l^*$ and $\partial_1^l(v\downarrow_{\mathbb{R}_+\times\mathbb{R}_+})\downarrow_{x_1=0} = G_l^*\downarrow_{\mathbb{R}_+} = g_l$ for $0 \leq l \leq m-1$.

It remains to prove the existence of $w \in M$ with M defined in (8.14.89) such that $w = u - v\downarrow_{\mathbb{R}_+\times\mathbb{R}_+}$ and $\partial_2^k w\downarrow_{x_2=0} = \partial_2^k u\downarrow_{x_2=0} - \partial_2^k(v\downarrow_{\mathbb{R}_+\times\mathbb{R}_+})\downarrow_{x_2=0} = f_k - \phi_k$ with $\phi_k = \partial_2^k(v\downarrow_{\mathbb{R}_+\times\mathbb{R}_+})\downarrow_{x_2=0}$ for $0 \leq k \leq m-1$. Hence, for $v\downarrow_{\mathbb{R}_+\times\mathbb{R}_+} \in$

$H^m(\mathbb{R}_+ \times \mathbb{R}_+)$ with $\phi_k = \partial_2^k(v\downarrow_{\mathbb{R}_+\times\mathbb{R}_+})\downarrow_{x_2=0}$ for $0 \le k \le m-1$ and $g_l = \partial_1^l(v\downarrow_{\mathbb{R}_+\times\mathbb{R}_+})\downarrow_{x_1=0}$ for $0 \le l \le m-1$, we get from the proof in the necessity part of this theorem: $\partial_1^l \phi_k(0) = \partial_2^k g_l(0)$ for $k+l < m-1$ and

$$\int_0^\infty \frac{|\partial_1^l \phi_k(t) - \partial_2^k g_l(t)|^2}{t} dt < +\infty \quad \text{for } k+l = m-1. \tag{8.14.100}$$

Now, define $\psi_k = \partial_2^k w\downarrow_{x_2=0} = f_k - \phi_k$ for $0 \le k \le m-1$. Then:

i. $\partial_1^l \psi_k(0) = \partial_1^l f_k(0) - \partial_1^l \phi_k(0) = \partial_1^l f_k(0) - \partial_2^k g_k(0) = 0$ for $k+l < m-1$ by the hypothesis that I holds;

ii. $\int_0^\infty \frac{|\partial_1^l \psi_k(t)|^2}{t} dt < +\infty$, since

$$\begin{aligned} |\partial_1^l \psi_k(t)|^2 &= |\partial_1^l f_k(t) - \partial_1^l \phi_k(t)|^2 \\ &\le C(|\partial_1^l f_k(t) - \partial_2^k g_l(t)|^2 + |\partial_2^k g_l(t) - \partial_1^l \phi_k(t)|^2) \\ \implies \quad & \int_0^\infty \frac{|\partial_1^l \psi_k(t)|^2}{t} dt \\ &\le C\left[\int_0^\infty \frac{|\partial_1^l f_k(t) - \partial_2^k g_l(t)|^2}{t} dt + \int_0^\infty \frac{|\partial_1^l \phi_k(t) - \partial_2^k g_l(t)|^2}{t} dt\right] \\ &< +\infty \end{aligned}$$

(by hypothesis that II holds and by (8.14.100), respectively),

i and ii being the conditions I and II of Lemma 8.14.5, according to which $\exists w \in M$, i.e. $w \in H^m(\mathbb{R}_+ \times \mathbb{R}_+)$ with $\partial_1^l w\downarrow_{x_1=0} = 0$ for $0 \le l \le m-1$, and by definition $\partial_2^k w\downarrow_{x_2=0} = \psi_k = f_k - \phi_k$ for $0 \le k \le m-1$. Then $u = w + v\downarrow_{\mathbb{R}_+\times\mathbb{R}_+}$ is the required $u \in H^m(\mathbb{R}_+ \times \mathbb{R}_+)$, with $\partial_2^k u\downarrow_{x_2=0} = \partial_2^k w\downarrow_{x_2=0} + \partial_2^k(v\downarrow_{\mathbb{R}_+\times\mathbb{R}_+})\downarrow_{x_2=0} = (f_k - \phi_k) + \phi_k = f_k$ for $0 \le k \le m-1$ and $\partial_1^l u\downarrow_{x_1=0} = \partial_1^l w\downarrow_{x_1=0} + \partial_1^l(v\downarrow_{\mathbb{R}_+\times\mathbb{R}_+})\downarrow_{x_1=0} = 0 + g_l = g_l$ for $0 \le l \le m-1$, and the proof is complete. □

Trace theorem for a general bounded polygonal domain

Let $\overline{\Gamma}_j = [P_j, P_{j+1}]$ and $\overline{\Gamma}_{j-1} = [P_{j-1}, P_j]$ $\forall 1 \le j \le N$ such that P_j is the common end point of $\overline{\Gamma}_{j-1}$ and $\overline{\Gamma}_j$. Let t be the distance measured from P_j along Γ and $\phi_j(t) = \phi_{P_j}(t)$ denote the point on Γ whose distance to P_j is t. Then, for sufficiently small $|t|$ (i.e. $\exists \delta_j > 0$ with $|t| \le \delta_j$), we have $\phi_j(t) \in \Gamma_j$ for $t > 0$ and $\phi_j(t) \in \Gamma_{j-1}$ for $t < 0$. As a consequence of Theorem 8.14.7, following Grisvard [17] we have the following definition.

Two functions f_{j-1} and f_j defined on Γ_{j-1} and Γ_j respectively are called *equivalent* at the vertex P_j if and only if $\int_0^{\delta_j} \frac{|f_{j-1}(\phi_j(-t)) - f_j(\phi_j(t))|^2}{t} dt < +\infty$.

Then we write

$$f_{j-1} \equiv f_j \quad \text{at } P_j. \tag{8.14.101}$$

Consequence For Hölder continuous functions f_{j-1} and f_j in a neighbourhood of P_j, the equivalence (8.14.101) $\Longrightarrow f_{j-1}(P_j) = f_j(P_j)$.

Let L be a differential operator of order $d \le m-1$ with constant coefficients and $Q_{j,l}$ be the differential operators involving tangential derivatives $\frac{\partial}{\partial \tau_j}$ along Γ_j such that $L = \sum Q_{j,l} \frac{\partial^j}{\partial n_j^l}$ (see [17, p. 117]). (8.14.102)

Now we can state the trace theorem for general polygonals without proof.

Theorem 8.14.8 ([17]). *Let $\Gamma \subset \mathbb{R}^2$ be the polygonal boundary of the bounded $\Omega \subset \mathbb{R}^2$. For $u \in H^m(\Omega)$, let $\{\gamma_{kj}\}_{k=0}^{m-1}$ be the continuous, linear trace operators from $H^m(\Omega)$ onto $\prod_{k=0}^{m-1} H^{m-k-1/2}(\Gamma_j)$ $\forall$ sides Γ_j defined by Theorem* 8.14.6: $\gamma_{kj}u = \frac{\partial^k u}{\partial n_j^k}\downarrow_{\Gamma_j} = f_{j,k} \in H^{m-k-1/2}(\Gamma_j)$, $0 \le k \le m-1$, $1 \le j \le N$. *Then the mapping $u \in H^m(\Omega) \longmapsto \{f_{j,k}\}_{k=0}^{m-1}$, $1 \le j \le N$, is a* continuous, linear *mapping from $H^m(\Omega)$ onto the trace subspace T_0 of $T = \prod_{j=1}^{N} \prod_{k=0}^{m-1} H^{m-k-1/2}(\Gamma_j)$ with $T_0 \subset T$ defined, for* $1 \le j \le N$, *by:*

I. $\sum_l (Q_{j,l} f_{j,l})(P_j) = \sum_l (Q_{j-1,l} f_{j-1,l})(P_j)$ *for* $d \le m-2$; (8.14.103a)

II. $\sum_l (Q_{j,l} f_{j,l})(P_j) \equiv \sum_l (Q_{j-1,l} f_{j-1,l})(P_j)$ *for* $d = m-1$, (8.14.103b)

where $\sum_l Q_{j,l} f_{j,l}$ (resp. $\sum_l Q_{j-1,l} f_{j-1,l}$) is of order d and defined in (8.14.102); *'$\equiv$' in* II *denotes the equivalence defined in* (8.14.101). □

We do not like to repeat the outline of the proof explained and given in Remark 8.14.3 with sufficient explanations in the justification for Theorem 8.14.7 for the first quadrant, in which we dealt with the operator of the form $\partial_1^l \partial_2^k$ instead of L. For more details, we refer to [17].

Since the compatibility conditions I and II of Theorem 8.14.8 at the common end point P_j are not as clear and obvious as those for the operator $\partial_1^l \partial_2^k$ in Theorem 8.14.7 at the origin, we give an example to illustrate the results of Theorem 8.14.8.

Example 8.14.1. Let Γ_j be any *fixed* open side of the polygonal boundary $\Gamma \subset \mathbb{R}^2$. With this fixed side Γ_j, we associate the space $U_j \subset H^2(\Omega)$ defined by:

$$U_j = \Big\{ u : u \in H^2(\Omega),\ \gamma_{0k} u = u\downarrow_{\Gamma_k} = f_{k,0} = 0, \\ \gamma_{1k} u = \frac{\partial u}{\partial n_k}\downarrow_{\Gamma_k} = f_{k,1} = 0\ \forall k \ne j,\ 1 \le k \le N \Big\}. \tag{8.14.104}$$

Let $f_{j,0} = \gamma_{0j} u = u\downarrow_{\Gamma_j}$ and $f_{j,1} = \gamma_{1j} u = \frac{\partial u}{\partial n_j}\downarrow_{\Gamma_j}$.

Then the mapping $u \in U_j \longmapsto (f_{j,0}; f_{j,1}) \in T_0$, T_0 being the *subspace* of $T = H^{3/2}(\Gamma_j) \times H^{1/2}(\Gamma_j)$ with T_0 defined by the compatibility conditions I and II of Theorem 8.14.8, which will be explicitly shown in this case. Since $\overline{\Gamma}_j = [P_j, P_{j+1}]$, we are to consider the end points P_j and P_{j+1} only, but $P_j = \overline{\Gamma}_k \cap \overline{\Gamma}_j$ with $k = j-1$ and $P_{j+1} = \overline{\Gamma}_j \cap \overline{\Gamma}_k$ with $k = j+1$. For the traces $f_{j,0}$ and $f_{j,1}$ on Γ_j, we are to consider also traces $f_{k,0}$ and $f_{k,1}$ on Γ_k with $k = j-1,\ j+1$. But the traces $f_{k,0} = 0$, $f_{k,1} = 0$ for $k = j-1$ or $k = j+1$.

Since $U_j \subset H^2(\Omega)$, $u \in U_j \Longrightarrow f_{j,0} = u\downarrow_{\Gamma_j} \in H^{3/2}(\Gamma_j)$ and $f_{j,1} = \frac{\partial u}{\partial n_j}\downarrow_{\Gamma_j} \in H^{1/2}(\Gamma_j)$ with the additional compatibility conditions I and II in (8.14.103a) and (8.14.103b) defining T_0 as the subspace of $H^{3/2}(\Gamma_j) \times H^{1/2}(\Gamma_j)$.

Compatibility conditions I and II for $f_{j,0} = u\downarrow_{\Gamma_j} \in H^{3/2}(\Gamma_j)$ at P_j and P_{j+1} : $m = 2$

1. for $d \le m-2$, i.e. $d = 0$ and $l = 0$, I gives
   - $f_{j,0}(P_j) = f_{k,0}(P_j) = 0$ with $k = j-1 \Longrightarrow f_{j,0}(P_j) = 0$;
   - $f_{j,0}(P_{j+1}) = f_{k,0}(P_{j+1}) = 0$ with $k = j+1 \Longrightarrow f_{j,0}(P_{j+1}) = 0$, i.e. $f_{j,0}(P_j) = f_{j,0}(P_{j+1}) = 0$;
2. for $d = m-1$, i.e. $d = 1$, $l = 0$, $Q_{j,0} = \frac{\partial}{\partial \tau_j}$, II gives
   - $\frac{\partial f_{j,0}}{\partial \tau_j} \equiv \frac{\partial f_{k,0}}{\partial \tau_j} = 0$ at P_j with $k = j-1 \Longrightarrow \frac{\partial f_{j,0}}{\partial \tau_j} \equiv 0$ at P_j
   $$\Longrightarrow \int_0^{\delta_j} \frac{|\frac{\partial}{\partial \tau_j} f_{j,0}(\phi_{P_j}(t))|^2}{t} dt < +\infty \text{ (by (8.14.101))}; \tag{8.14.105}$$
   - $\frac{\partial f_{j,0}}{\partial \tau_j} \equiv \frac{\partial f_{k,0}}{\partial \tau_j} = 0$ at P_{j+1} with $k = j+1 \Longrightarrow \frac{\partial f_{j,0}}{\partial \tau_j} \equiv 0$ at P_{j+1}
   $$\Longrightarrow \int_0^{\delta_{j+1}} \frac{|\frac{\partial}{\partial \tau_j} f_{j,0}(\phi_{P_{j+1}}(-t))|^2}{t} dt < +\infty \text{ (by (8.14.101))}. \tag{8.14.106}$$

Hence, $f_{j,0} \in H^{3/2}(\Gamma_j)$ with $f_{j,0}(P_j) = f_{j,0}(P_{j+1}) = 0 \Longrightarrow f_{j,0} \in H_0^{3/2}(\Gamma_j)$ and $\frac{\frac{\partial f_{j,0}}{\partial \tau_j}}{\sqrt{t}} \in L^2(\Gamma_j)$ by virtue of (8.14.105) and (8.14.106) $\Longrightarrow f_{j,0} \in H_{00}^{3/2}(\Gamma_j)$ by (8.10.125).

Compatibility conditions for $f_{j,1} = \frac{\partial u}{\partial n_j}\downarrow_{\Gamma_j} \in H^{1/2}(\Gamma_j)$ at P_j and P_{j+1} are given by II:

- $f_{j,1} \equiv f_{k,1} = 0$ at P_j with $k = j-1 \Longrightarrow f_{j,1} \equiv 0$ at $P_j \Longrightarrow$
  $$\int_0^{\delta_j} \frac{|f_{j,1}(\phi_{P_j}(t))|^2}{t} dt < +\infty \text{ (by (8.14.101))}; \tag{8.14.107}$$
- $f_{j,1} \equiv f_{k,1} = 0$ at P_{j+1} with $k = j+1 \Longrightarrow f_{j,1} \equiv 0$ at $P_{j+1} \Longrightarrow$
  $$\int_0^{\delta_{j+1}} \frac{|f_{j,1}(\phi_{P_{j+1}}(-t))|^2}{t} dt < +\infty \text{ (by (8.14.101))}. \tag{8.14.108}$$

Hence, $f_{j,1} \in H^{1/2}(\Gamma_j) \equiv H_0^{1/2}(\Gamma_j)$ (by Corollary 8.10.1) and $\frac{f_{j,1}}{\sqrt{t}} \in L^2(\Gamma_j)$ by virtue of (8.14.107) and (8.14.108) $\Longrightarrow f_{j,1} \in H_{00}^{1/2}(\Gamma_j)$ by (8.10.124). Thus, $(f_{j,0}; f_{j,1}) \in T_0 = H_{00}^{3/2}(\Gamma_j) \times H_{00}^{1/2}(\Gamma_j) \subset H^{3/2}(\Gamma_j) \times H^{1/2}(\Gamma_j)$.

Characterization of the kernel of the trace operator $\{\gamma_{kj}\}_{k=0,1\le j\le N}^{m-1}$ for polygonal Γ

Proposition 8.14.2. *Let $\Omega \subset \mathbb{R}^2$ be a bounded polygonal domain with boundary Γ and the mapping $u \in H^m(\Omega) \longmapsto \{\gamma_{kj}u\}_{k=0,j=1}^{m-1,N} \in T_0 \subset \prod_{j=1}^N \prod_{k=0}^{m-1} H^{m-k-1/2}(\Gamma_j)$ be defined by Theorem 8.14.8. Then $H_0^m(\Omega)$ is the* kernel *of this mapping from $H^m(\Omega)$ onto T_0, i.e.*

$$H_0^m(\Omega) = \Big\{u : u \in H^m(\Omega),\ \gamma_{0j}u = u\downarrow_{\Gamma_j} = 0,\ \gamma_{1j}u = \frac{\partial u}{\partial n_j}\downarrow_{\Gamma_j} = 0, \ldots,$$
$$\gamma_{m-1,j}u = \frac{\partial^{m-1}u}{\partial n_j^{m-1}}\downarrow_{\Gamma_j} = 0\ \forall \text{ sides } \Gamma_j,\ 1 \le j \le N\Big\}. \tag{8.14.109}$$

Proof. For the sake of simplicity, we give the proof for $m = 1$, i.e.

$$H_0^1(\Omega) = \{u : u \in H^1(\Omega),\ \gamma_{0j}u = u\downarrow_{\Gamma_j} = 0\ \forall \Gamma_j, 1 \le j \le N\}.$$

Let $u \in H_0^1(\Omega)$. We are to show that $\gamma_{0j}u = u\downarrow_{\Gamma_j} = 0\ \forall j = 1,\ldots,N$. Since $\mathcal{D}(\Omega)$ is *dense* in $H_0^1(\Omega)$, $\exists$ a sequence (ϕ_m) in $\mathcal{D}(\Omega)$ such that $\phi_m \longrightarrow u$ in $H^1(\Omega)$ as $m \to \infty$. But $\phi_m \in \mathcal{D}(\Omega)\ \forall m \in N \Longrightarrow \phi_m\downarrow_\Gamma = 0\ \forall m \in \mathbb{N} \Longrightarrow \gamma_{0j}\phi_m = \phi_m\downarrow_{\Gamma_j} = 0\ \forall j = 1,\ldots,N\ \forall m \in \mathbb{N}$. By virtue of the continuity of the trace operator γ_{0j}, $\phi_m \to u$ in $H^1(\Omega) \Longrightarrow \gamma_{0j}\phi_m \to \gamma_{0j}u$ in $H^{1/2}(\Gamma_j)\ \forall j = 1,2,\ldots,N$ as $m \to \infty \Longrightarrow \|\gamma_{0j}u - \gamma_{0j}\phi_m\|_{H^{1/2}(\Gamma_j)} = \|\gamma_{0j}u - 0\|_{H^{1/2}(\Gamma_j)} = \|\gamma_{0j}u\|_{H^{1/2}(\Gamma_j)} = 0$ as $m \to \infty\ \forall 1 \le j \le N$. Hence, $\gamma_{0j}u = 0\ \forall 1 \le j \le N$. Thus, $u \in H_0^1(\Omega) \Longrightarrow u \in H^1(\Omega)$ with $\gamma_{0j}u = 0\ \forall j = 1,2,\cdots,N \Longrightarrow H_0^1(\Omega) \subset \{u : u \in H^1(\Omega),\ \gamma_{0j}u = 0\ \forall j = 1,\ldots,N\}$.

Conversely, let $u \in H^1(\Omega)$ with $\gamma_{0j}u = u\downarrow_{\Gamma_j} = 0\ \forall j = 1,\ldots,N$.

We are to show that $u \in H_0^1(\Omega)$. Let $\tilde{u}$ be its *null extension* to $\mathbb{R}^2$. We will show that $\tilde{u} \in H^1(\mathbb{R}^2)$. In fact, $u \in H^1(\Omega) \Longrightarrow u \in L^2(\Omega)$ and $\frac{\partial u}{\partial x_i} \in L^2(\Omega)$ for $i = 1,2 \Longrightarrow \tilde{u} \in L^2(\mathbb{R}^2)$, $\widetilde{\frac{\partial u}{\partial x_i}} \in L^2(\mathbb{R}^2)$, $\widetilde{\frac{\partial u}{\partial x_i}}$ being the null extension of $\frac{\partial u}{\partial x_i} \in L^2(\Omega)$ for $i = 1,2$. Hence, we are to show that $\frac{\partial \tilde{u}}{\partial x_i} = \widetilde{\frac{\partial u}{\partial x_i}}$ in $L^2(\mathbb{R}^2)$ for $i = 1,2$ in the distributional sense on $\mathbb{R}^2$.

Let $u \in \mathcal{D}(\overline{\Omega})$ and $\tilde{u}$ be its null extension to $\mathbb{R}^2$. Then, $\forall \phi \in \mathcal{D}(\mathbb{R}^2)$,

$$\begin{aligned}\left\langle \frac{\partial \tilde{u}}{\partial x_i}, \phi \right\rangle_{\mathcal{D}'(\mathbb{R}^2)\times\mathcal{D}(\mathbb{R}^2)} &= -\left\langle \tilde{u}, \frac{\partial \phi}{\partial x_i} \right\rangle_{\mathcal{D}'(\mathbb{R}^2)\times\mathcal{D}(\mathbb{R}^2)} = -\int_{\mathbb{R}^2} \tilde{u}\frac{\partial \phi}{\partial x_i} dx_1 dx_2 \\ &= -\int_{\Omega} u \frac{\partial \phi}{\partial x_i} dx_1 dx_2 \\ &= -\int_{\Omega} \frac{\partial}{\partial x_i}(u\phi) dx_1 dx_2 + \int_{\Omega} \frac{\partial u}{\partial x_i}\phi dx_1 dx_2 \\ &= \int_{\mathbb{R}^2} \widetilde{\frac{\partial u}{\partial x_i}} \phi dx_1 dx_2 - \sum_{j=1}^{N} \int_{\Gamma_j} (\gamma_{0j} u)\phi n_{ij} ds,\end{aligned}$$

where $\widetilde{\frac{\partial u}{\partial x_i}}$ is the null extension of $\frac{\partial u}{\partial x_i}$ to $\mathbb{R}^2$; n_{ij} is the ith component of unit vector $\hat{\mathbf{n}}_j = (n_{1j}, n_{2j})$ normal to Γ_j, $1 \le j \le N$.

Hence, $\forall u \in \mathcal{D}(\overline{\Omega})$, $\forall \phi \in \mathcal{D}(\mathbb{R}^2)$, for $1 \le i \le 2$,

$$\left\langle \frac{\partial \tilde{u}}{\partial x_i}, \phi \right\rangle_{\mathcal{D}'(\mathbb{R}^2)\times\mathcal{D}(\mathbb{R}^2)} = \left\langle \widetilde{\frac{\partial u}{\partial x_i}}, \phi \right\rangle_{\mathcal{D}'(\mathbb{R}^2)\times\mathcal{D}(\mathbb{R}^2)} - \sum_{j=1}^{N} \langle T_{\gamma_{0j} u}, \phi \rangle_{\mathcal{D}'(\mathbb{R}^2)\times\mathcal{D}(\mathbb{R}^2)},$$

where $T_{\gamma_j u}$ is a distribution on $\mathbb{R}^2$ defined by

$$\langle T_{\gamma_{0j} u}, \phi \rangle_{\mathcal{D}'(\mathbb{R}^2)\times\mathcal{D}(\mathbb{R}^2)} = \int_{\Gamma_j} (\gamma_{0j} u\, n_{ij})\phi ds \quad \forall j = 1, \dots, N,$$

since, by virtue of the continuity of the trace operator γ_{0j} from $H^1(\Omega)$ onto $H^{1/2}(\Gamma_j)$ $\forall j = 1, \dots, N$,

$$\begin{aligned}\left| \int_{\Gamma_j} (\gamma_{0j} u) n_{ij} \phi ds \right| &\le C \max_{\mathbf{x}\in \mathrm{supp}(\phi)} |\phi(\mathbf{x})| \int_{\Gamma_j} |\gamma_{0j} u| ds \\ &\le C \max_{\mathbf{x}\in K} |\phi(\mathbf{x})| \cdot \|u\|_{1,\Omega} \le (C\|u\|_{1,\Omega}) p_{K,0}(\phi) \quad \forall \phi \in \mathcal{D}(\mathbb{R}^2)\end{aligned}$$

with $\mathrm{supp}(\phi) = K \Longrightarrow$ the mapping $\phi \in \mathcal{D}(\mathbb{R}^2) \longmapsto \langle T_{\gamma_{0j} u}, \phi \rangle = \int_{\Gamma_j} (\gamma_{0j} u\, n_{ij}) \phi ds$ is *linear* and *continuous* on $\mathcal{D}(\mathbb{R}^2)$ $\forall j = 1, \dots, N$ by Proposition 1.3.1, and $T_{\gamma_{0j} u} \in \mathcal{D}'(\mathbb{R}^2)$ $\forall j = 1, \dots, N$. $\forall u \in \mathcal{D}(\bar{\Omega})$ with $\gamma_{0j} u \in H^{1/2}(\Gamma_j) \hookrightarrow L^2(\Gamma_j)$, for $1 \le i \le 2$,

$$\left\langle \frac{\partial \tilde{u}}{\partial x_i} - \widetilde{\frac{\partial u}{\partial x_i}} - \sum_{j=1}^{N} T_{\gamma_{0j} u}, \phi \right\rangle_{\mathcal{D}'(\mathbb{R}^2)\times\mathcal{D}(\mathbb{R}^2)} = 0 \quad \forall \phi \in \mathcal{D}(\mathbb{R}^2). \tag{8.14.110}$$

Since the mapping $u \in \mathcal{D}(\bar{\Omega}) \subset H^1(\Omega) \longmapsto [\frac{\partial \tilde{u}}{\partial x_i} - \widetilde{\frac{\partial u}{\partial x_i}} - \sum_{j=1}^{N} T_{\gamma_{0j} u}] \in \mathcal{D}'(\mathbb{R}^2)$ is *linear* and *continuous* from $\mathcal{D}(\bar{\Omega})$ equipped with the norm of $\|\cdot\|_{H^1(\Omega)}$

into $\mathcal{D}'(\mathbb{R}^2)$ and $\mathcal{D}(\bar{\Omega})$ is *dense* in $H^1(\Omega)$, the equality (8.14.110) holds $\forall u \in H^1(\Omega)$ with $\gamma_{0j}u = u \downarrow_{\Gamma_j} \in H^{1/2}(\Gamma_j)$, $1 \le j \le N$, i.e. $\forall u \in H^1(\Omega)$, $\langle \frac{\partial \tilde{u}}{\partial x_i} - \widetilde{\frac{\partial u}{\partial x_i}} - \sum_{j=1}^N T_{\gamma_{0j}u}, \phi\rangle = 0\ \forall \phi \in \mathcal{D}(\mathbb{R}^2) \iff \forall u \in H^1(\Omega)$, $\frac{\partial \tilde{u}}{\partial x_i} - \widetilde{\frac{\partial u}{\partial x_i}} - \sum_{j=1}^N T_{\gamma_{0j}u} = 0$ in $\mathcal{D}'(\mathbb{R}^2)$.

But by hypothesis, $\gamma_{0j}u = u \downarrow_{\Gamma_j} = 0\ \forall j = 1,\dots,N$ for $u \in H^1(\Omega) \implies \langle T_{\gamma_{0j}u}, \phi\rangle = \int_{\Gamma_j}(\gamma_{0j}u)\phi n_{ij}\,ds = 0\ \forall \phi \in \mathcal{D}(\mathbb{R}^2)\ \forall 1 \le j \le N \implies T_{\gamma_{0j}u} = 0$ in $\mathcal{D}'(\mathbb{R}^2)\ \forall j = 1,\dots,N$.

Hence, for $u \in H^1(\Omega)$ with $\gamma_{0j}u = u \downarrow_{\Gamma_j} = 0\ \forall j = 1,\dots,N$, $\frac{\partial \tilde{u}}{\partial x_i} = \widetilde{\frac{\partial u}{\partial x_i}}$ in $\mathcal{D}'(\mathbb{R}^2)$. But $\widetilde{\frac{\partial u}{\partial x_i}} \in L^2(\mathbb{R}^2)$ for $i = 1, 2$. Hence, $\frac{\partial \tilde{u}}{\partial x_i} \in L^2(\mathbb{R}^2)$ for $i = 1, 2$. Thus, $\tilde{u} \in L^2(\mathbb{R}^2)$, $\frac{\partial \tilde{u}}{\partial x_i} \in L^2(\mathbb{R}^2) \implies \tilde{u} \in H^1(\mathbb{R}^2) \implies u \in H^1_{00}(\Omega)$ by Definition 8.10.14, but $H^1_{00}(\Omega) = H^1_0(\Omega)$ by Theorem 8.10.15. Hence, $u \in H^1_0(\Omega)$, i.e. $\{u : u \in H^1(\Omega), \gamma_{0j}u = u \downarrow_{\Gamma_j} = 0, 1 \le j \le N\} \subset H^1_0(\Omega)$, and the result (8.14.109) is proved for $m = 1$. □

8.14.5 Trace results for bounded domains with curvilinear polygonal boundary Γ in $\mathbb{R}^2$

For all notations and geometric details, we refer to Section 8.13.4, Case II. Without repeating proofs, we state the following trace theorems for easy reference. For more details, we refer to [18].

Theorem 8.14.9. *Let $\Omega \subset \mathbb{R}^2$ be a bounded domain with curvilinear polygonal boundary Γ such that each boundary curved side $\Gamma_j \in C^{k,1}$-class, $1 \le j \le N$. Then, $\forall \Gamma_j$, $1 \le j \le N$, the mapping $u \in \mathcal{D}(\bar{\Omega}) \longmapsto (\gamma_{0j}u, \gamma_{1j}u, \dots, \gamma_{lj}u) \in \Pi_{i=0}^l W^{m-i-1/p,p}(\Gamma_j)$ with $\gamma_{0j}u = u \downarrow_{\Gamma_j}, \gamma_{1j}u = \frac{\partial u}{\partial n} \downarrow_{\Gamma_j}, \dots, \gamma_{lj}u = \frac{\partial^l u}{\partial n_j^l} \downarrow_{\Gamma_j}$ with $l \le m-1 \le k$ has a unique, continuous, linear extension as an operator from $W^{m,p}(\Omega)$ onto $\Pi_{i=0}^l W^{m-i-1/p,p}(\Gamma_j)$, $1 \le j \le N$.*

Theorem 8.14.10 ([18]). *Under the assumptions of Theorem 8.14.9 with $k = 1$, i.e. $\Gamma_j \in C^{1,1}$-class (C^1-class is also permissible), the mapping $u \longmapsto \{f_j\}_{j=1}^N \in T_0$ with $f_j = \gamma_{0j}u = u \downarrow_{\Gamma_j}$ is a linear continuous operator from $W^{1,p}(\Omega)$, $1 < p < \infty$,* onto *the trace subspace T_0 with $T_0 \subset \Pi_{j=1}^N W^{1-1/p,p}(\Gamma_j)$ defined by the compatibility conditions:*

I. *No additional condition is necessary for $1 < p < 2$;*

II. *$f_j(P_j) = f_{j-1}(P_j)$, $1 \le j \le N$, for $p > 2$;*

III. *$\int_0^{\delta_j} \frac{|f_j(\phi_{P_j}(t)) - f_{j-1}(\phi_{P_j}(-t))|^2}{t}dt < +\infty$, $1 \le j \le N$, for $p = 2$, where $\phi_{P_j}(t) \in \Gamma_j$ for $t > 0$ and $\phi_{P_j}(t) \in \Gamma_{j-1}$ for $t < 0$ with $|t| < \delta_j$ (see also Theorems 8.13.9 and 8.14.7).*

Remark 8.14.4.

1. For $p > 2$, $W^{1,p}(\Omega) \hookrightarrow C^0(\overline{\Omega})$ by Sobolev's imbedding results and II is meaningful, since f_j and f_{j-1} are *continuous* at P_j.
2. For $p = 2$, II is *not always* meaningful, since functions in $H^{1/2}(\Gamma_j)$ are *not* always continuous. But if for some $u \in H^1(\Omega)$, f_j and f_{j-1} happen to be *continuous* near P_j, then condition III reduces to condition II. For more details, see [18].

8.14.6 Traces of normal components in $L^p(\mathrm{div};\Omega)$

In Theorem 8.14.5, we studied traces of functions in $W^{s,p}(\Omega)$ with $s - \frac{1}{p} > 0$. Now we will study traces of functions belonging to spaces for which $s - \frac{1}{p} > 0$ does *not* hold, i.e. $s - \frac{1}{p} < 0$, but some *additional smoothness* conditions hold.

Space $L^p(\mathrm{div};\Omega)$

Definition 8.14.4. Let $\Omega \subset \mathbb{R}^n$ be a C^∞-regular domain with boundary $\Gamma \in C^\infty$-class. Then, for $1 < p < \infty$, $L^p(\mathrm{div};\Omega)$ is the subspace of

$$(L^p(\Omega))^n = \underbrace{L^p(\Omega) \times \cdots \times L^p(\Omega)}_{n \text{ times}}$$

consisting of functions $\mathbf{v} = (v_1, \ldots, v_n) \in (L^p(\Omega))^n$ such that $\operatorname{div} \mathbf{v} = \sum_{i=1}^n \frac{\partial v_i}{\partial x_i} \in L^p(\Omega)$, the derivative being in the sense of distributions, i.e.

$$L^p(\mathrm{div};\Omega) = \{\mathbf{v} : \mathbf{v} = (v_1, \ldots, v_n) \in (L^p(\Omega))^n \text{ such that } \operatorname{div} \mathbf{v} \in L^p(\Omega)\}. \tag{8.14.111}$$

Then,

$$(W^{1,p}(\Omega))^n \subset L^p(\mathrm{div};\Omega) \subset (L^p(\Omega))^n. \tag{8.14.112}$$

In fact, $\mathbf{v} \in (W^{1,p}(\Omega))^n \Longrightarrow v_i \in W^{1,p}(\Omega)$ for $1 \le i \le n$, $v_i \in L^p(\Omega)$ and $\frac{\partial v_i}{\partial x_j} \in L^p(\Omega)$ for $1 \le i, j \le n \Longrightarrow \mathbf{v} = (v_1, \ldots, v_n) \in (L^p(\Omega))^n$ and $\operatorname{div} \mathbf{v} = \sum_{i=1}^n \frac{\partial v_i}{\partial x_i} \in L^p(\Omega) \Longrightarrow \mathbf{v} \in L^p(\mathrm{div};\Omega)$. Hence, $(W^{1,p}(\Omega))^n \subset L^p(\mathrm{div};\Omega)$.

Proposition 8.14.3. *Let $\Omega \subset \mathbb{R}^n$ be a C^∞-regular domain with boundary $\Gamma \in C^\infty$-class. Then, for $1 < p < \infty$, $L^p(\mathrm{div};\Omega)$ equipped with the graph norm $\|\cdot\|_{L^p(\mathrm{div};\Omega)}$:*

$$\begin{aligned}
\|\mathbf{v}\|_{L^p(\mathrm{div};\Omega)} &= (\|\mathbf{v}\|^p_{(L^p(\Omega))^n} + \|\operatorname{div} \mathbf{v}\|^p_{L^p(\Omega)})^{1/p} \\
&= ([\|v_1\|^p_{L^p(\Omega)} + \cdots + \|v_n\|^p_{L^p(\Omega)}] + \left\| \sum_{i=1}^n \frac{\partial v_i}{\partial x_i} \right\|^p_{L^p(\Omega)})^{1/p}
\end{aligned} \tag{8.14.113}$$

is a Banach space.

In particular, for $p = 2$, $H(\mathrm{div};\Omega) \equiv L^2(\mathrm{div};\Omega)$ *is a* Hilbert space, *i.e.* $H(\mathrm{div};\Omega) = \{\mathbf{v} : \mathbf{v} = (v_1,\dots,v_n) \in (L^2(\Omega))^n$ *with* $\mathrm{div}\,\mathbf{v} \in L^2(\Omega)\}$ *equipped with inner product* $\langle\cdot,\cdot\rangle_{H(\mathrm{div};\Omega)}$:

$$\langle \mathbf{u},\mathbf{v}\rangle_{H(\mathrm{div};\Omega)} = \langle \mathbf{u},\mathbf{v}\rangle_{(L^2(\Omega))^n} + \langle \mathrm{div}\,\mathbf{u}, \mathrm{div}\,\mathbf{v}\rangle_{L^2(\Omega)}$$
$$= \sum_{i=1}^{n} \langle u_i, v_i\rangle_{L^2(\Omega)} + \langle \mathrm{div}\,\mathbf{u}, \mathrm{div}\,\mathbf{v}\rangle_{L^2(\Omega)};$$
$$\|\mathbf{v}\|_{H(\mathrm{div};\Omega)} = \langle \mathbf{v},\mathbf{v}\rangle^{1/2}_{H(\mathrm{div};\Omega)} = (\|\mathbf{v}\|^2_{(L^2(\Omega))^n} + \|\,\mathrm{div}\,\mathbf{v}\|^2_{L^2(\Omega)})^{1/2} \quad (8.14.114)$$

is a Hilbert space.

Density result For a C^∞-regular domain Ω and $1 < p < \infty$, $(\mathcal{D}(\overline{\Omega}))^n$ is *dense* in $L^p(\mathrm{div};\Omega)$. (8.14.115)

For $\mathbf{u} \in (W^{1,p}(\Omega))^n \subset L^p(\mathrm{div};\Omega)$, $1 < p < \infty$, each component $u_i \in W^{1,p}(\Omega)$ with $1 - 1/p > 0$, and consequently, by Theorem 8.14.5, each $u_i \in W^{1,p}(\Omega)$ has trace on Γ: $\gamma_0^* u_i = u_i\!\downarrow_\Gamma \in W^{1-1/p,p}(\Gamma)$, since $\Gamma \in C^\infty$-class $\Longrightarrow \Gamma \in C^{0,1}$-class and $1 - 1/p \neq$ an integer. Then, $\forall i = 1,\dots,n$, each component $(u_i n_i)\!\downarrow_\Gamma$ is well defined as an element of $W^{1-1/p,p}(\Gamma)$, n_i being the ith component of the exterior unit normal $\hat{\mathbf{n}}$ to $\Gamma \in C^\infty$-class. But for $\mathbf{u} \in L^p(\mathrm{div};\Omega)$, these trace results on Γ do *not* hold in general, since for $\mathbf{u} \in L^p(\mathrm{div};\Omega)$, $u_i \in L^p(\Omega)$ (and $\notin W^{1,p}(\Omega)$ in general) with $s - 1/p = 0 - 1/p < 0$, and Theorem 8.14.5 is *not* applicable and $(u_i n_i)\!\downarrow_\Gamma$ is *not* well defined $\forall i = 1,\cdots,n$ in general. For $\mathbf{u} \in L^p(\mathrm{div};\Omega)$, the additional property of *smoothness* that $\mathrm{div}\,\mathbf{u} \in L^p(\Omega)$ must hold, which, in fact, relates to an *elliptic operator* (to be shown later), as a result of which the trace of the normal component $(\mathbf{u}\cdot\hat{\mathbf{n}})\!\downarrow_\Gamma$ is defined for $\mathbf{u} \in L^p(\mathrm{div};\Omega)$, and we have:

Theorem 8.14.11 (Trace of normal component). *Let* $\Omega \subset \mathbb{R}^n$ *be a* C^∞*-regular domain with boundary* $\Gamma \in C^\infty$*-class. Then the mapping* $\mathbf{v} \longmapsto \gamma\mathbf{v} = \mathbf{v}\cdot\hat{\mathbf{n}}\!\downarrow_\Gamma$ *from* $(\mathcal{D}(\overline{\Omega})^n$ *has a* unique, continuous, linear *extension as an operator from* $L^p(\mathrm{div};\Omega)$ onto $W^{-1/p,p}(\Gamma)$, $1 < p < \infty$, *i.e.* $\gamma : \mathbf{v} \in L^p(\mathrm{div};\Omega) \longmapsto \gamma\mathbf{v} \in W^{-1/p,p}(\Gamma)$ *such that*

$$\gamma\mathbf{v} = \mathbf{v}\cdot\hat{\mathbf{n}}\!\downarrow_\Gamma \in W^{-1/p,p}(\Gamma). \quad (8.14.116)$$

Moreover, γ *has a continuous, linear right inverse* $\rho : W^{-1/p,p}(\Gamma) \longrightarrow L^p(\mathrm{div};\Omega)$ *such that*

$$\gamma\rho g = g \quad \forall g \in W^{-1/p,p}(\Gamma). \quad (8.14.117)$$

In particular, for $p = 2$*, the trace operator* $\gamma : \mathbf{v} \in H(\mathrm{div};\Omega) \longmapsto \gamma\mathbf{v} = \mathbf{v}\cdot\hat{\mathbf{n}}\!\downarrow_\Gamma \in H^{-1/2}(\Gamma)$ *is a* continuous, linear surjection *from* $H(\mathrm{div};\Omega)$ *onto* $H^{-1/2}(\Gamma)$. (8.14.118)

Proof. For $p = 2$, we refer to [46, p. 288]. □

Remark 8.14.5.

1. As stated earlier, $\mathbf{u} \cdot \hat{\mathbf{n}}\downarrow_\Gamma$ is related to an elliptic operator and $(\mathcal{D}(\overline{\Omega}))^n$ is *dense* in $H(\text{div}; \Omega)$ by (8.14.115), the proof of Theorem 8.14.11 for $p = 2$ in [46] is based on the use of Green's formula, which is as follows: for $1 < p, q < \infty$ with $1/p + 1/q = 1$, $\forall \mathbf{w} \in L^p(\text{div}; \Omega)$, $\forall v \in W^{1,q}(\Omega)$ with $\nabla v = (\frac{\partial v}{\partial x_1}, \frac{\partial v}{\partial x_2}, \dots, \frac{\partial v}{\partial x_n}) \in (L^q(\Omega))^n$,
$$\int_\Omega \mathbf{w} \cdot \nabla v d\mathbf{x} + \int_\Omega v \operatorname{div} \mathbf{w} d\mathbf{x} = \langle (\mathbf{w} \cdot \hat{\mathbf{n}})\downarrow_\Gamma, v\downarrow_\Gamma \rangle_{W^{-1/p,p}(\Gamma)\times W^{1/p,q}(\Gamma)}, \tag{8.14.119}$$
where $\gamma_0 v = v\downarrow_\Gamma \in W^{1-1/q,q}(\Gamma) = W^{1/p,q}(\Gamma)$ by Theorem 8.14.5; $\mathbf{w} \cdot \hat{\mathbf{n}}\downarrow_\Gamma \in (W^{1/p,q}(\Gamma))' = W^{-1/p,p}(\Gamma)$.
2. Setting $v = 1$ in (8.14.119), we get the Generalized Divergence Theorem:
$$\forall \mathbf{w} \in H(\text{div}; \Omega), \qquad \int_\Omega \operatorname{div} \mathbf{w} \, d\mathbf{x} = \langle (\mathbf{w} \cdot \hat{\mathbf{n}})\downarrow_\Gamma, 1 \rangle_{H^{-1/2}(\Gamma)\times H^{1/2}(\Gamma)}. \tag{8.14.120}$$

Physical interpretation (8.14.120) states that *the flow of a vector field across the boundary* Γ *is equal to the integral of the divergence of the vector field on* Ω.

Now we show in more detail the relation between $\mathbf{w}.\hat{\mathbf{n}}\downarrow_\Gamma$ and the *conormal derivative* with respect to an elliptic operator in a general situation.

Let A be an *elliptic operator* of the following type:
$$Au = -\sum_{i=1}^n \sum_{j=1}^n \frac{\partial}{\partial x_j}(a_{ij} \frac{\partial u}{\partial x_i}) + \sum_{i=1}^n a_i \frac{\partial u}{\partial x_i} + a_0 u, \tag{8.14.121}$$
with $a_{ij} \in C^{0,1}(\overline{\Omega})$, $a_i \in C^{0,1}(\overline{\Omega})$ and $a_0 \in L^\infty(\Omega)$ such that $\exists \alpha > 0$ satisfying the inequality, $\forall \mathbf{x} \in \Omega$,,
$$\sum_{i=1}^n \sum_{j=1}^n a_{ij}(\mathbf{x})\xi_i \xi_j \geq \alpha \|\boldsymbol{\xi}\|^2_{\mathbb{R}^n} = \alpha(\xi_1^2 + \dots + \xi_n^2) \quad \forall \boldsymbol{\xi} = (\xi_1, \cdots, \xi_n) \in \mathbb{R}^n.$$

Then the adjoint operator A^* is defined by:
$$A^* u = -\sum_{i=1}^n \sum_{i=1}^n \frac{\partial}{\partial x_j}\left(a_{ji} \frac{\partial u}{\partial x_i}\right) - \sum_{i=1}^n \frac{\partial}{\partial x_i}(a_i u) + a_0 u. \tag{8.14.122}$$

See also Example 2.14.2, in which the transpose or formal adjoint A' is defined for C^∞-coefficients a_{ij}, a_i and a_0. But since the condition that $a_{ij}, a_1 \in C^{0,1}(\overline{\Omega})$ and $a_0 \in L^\infty(\Omega)$ are sufficient for our purpose, in the final definition of A' we replace them by $a_{ij}, a_i \in C^{0,1}(\overline{\Omega})$ and $a_0 \in L^\infty(\Omega)$ and the resultant operator is denoted by A^*.

Let $a(\cdot,\cdot)$ denote the bilinear form defined by:

$$a(u,v) = \sum_{i=1}^n \sum_{j=1}^n \int_\Omega a_{ij} \frac{\partial u}{\partial x_i} \frac{\partial v}{\partial x_j} d\mathbf{x} + \int_\Omega a_0 u v d\mathbf{x},$$

which is well defined for $u \in W^{1,p}(\Omega)$, $v \in W^{1,q}(\Omega)$ with $1/p + 1/q = 1$, $1 < p,q < \infty$, and $a_{ij} \in C^{0,1}(\overline{\Omega})$, $a_0 \in L^\infty(\Omega)$. (8.14.123)

Conormal derivatives $\frac{\partial u}{\partial n_A}$, $\frac{\partial u}{\partial n_{A^*}}$ and conormal directions with respect to A and A^* and their traces $\frac{\partial u}{\partial n_A}\downarrow_\Gamma$, $\frac{\partial u}{\partial n_{A^*}}\downarrow_\Gamma$

Let $u \in W^{1,p}(\Omega)$ with $Au \in L^p(\Omega)$, A being the elliptic operator defined by (8.14.121). Define vector $\mathbf{w} = (w_1, \cdots, w_n)$ with components $w_j = \sum_{i=1}^n a_{ij} \frac{\partial u}{\partial x_i} \in L^p(\Omega)$, $1 \le j \le n$. Hence, $\mathbf{w} \in (L^p(\Omega))^n$ and

$$\begin{aligned}\operatorname{div} \mathbf{w} &= \sum_{j=1}^n \frac{\partial w_j}{\partial x_j} = \sum_{j=1}^n \frac{\partial}{\partial x_j}\left(\sum_{i=1}^n a_{ij} \frac{\partial u}{\partial x_i}\right) \\ &= \sum_{i=1}^n \sum_{j=1}^n \frac{\partial}{\partial x_j}\left(a_{ij} \frac{\partial u}{\partial x_i}\right) = -Au + \sum_{j=1}^n a_j \frac{\partial u}{\partial x_j} + a_0 u \in L^p(\Omega),\end{aligned}$$

since $Au \in L^p(\Omega)$, $a_j \frac{\partial u}{\partial x_j} \in L^p(\Omega)$, $a_0 u \in L^p(\Omega)$.

Hence, $\mathbf{w} \in (L^p(\Omega))^n$ with $\operatorname{div} \mathbf{w} \in L^p(\Omega)$, i.e. $\mathbf{w} \in L^p(\operatorname{div};\Omega)$, and by Theorem 8.14.11, $\gamma \mathbf{w} = \mathbf{w} \cdot \hat{\mathbf{n}}\downarrow_\Gamma$ is well defined as an element of $W^{-1/p,p}(\Gamma)$, i.e.

$$\gamma \mathbf{w} = \mathbf{w} \cdot \hat{\mathbf{n}}\downarrow_\Gamma = \sum_{j=1}^n w_j n_j \downarrow_\Gamma \in W^{-1/p,p}(\Gamma). \tag{8.14.124}$$

For smooth $u \in \mathcal{D}(\overline{\Omega})$, $\mathbf{w} \cdot \hat{\mathbf{n}}\downarrow_\Gamma = \sum_{j=1}^n w_j n_j \downarrow_\Gamma = (\sum_{j=1}^n \sum_{i=1}^n a_{ij} \frac{\partial u}{\partial x_i} n_j)\downarrow_\Gamma$, which is, in fact, a *first-order boundary operator* denoted by $\frac{\partial u}{\partial n_A}\downarrow_\Gamma$ and called the *conormal derivative of u with respect to the elliptic operator A*, i.e. the *conormal derivative* $\frac{\partial u}{\partial n_A}\downarrow_\Gamma$ is given by:

$$\frac{\partial u}{\partial n_A}\downarrow_\Gamma = \sum_{j=1}^n \sum_{i=1}^n a_{ij} \frac{\partial u}{\partial x_i} n_j \downarrow_\Gamma \in W^{-1/p,p}(\Gamma), \tag{8.14.125}$$

which is, in fact, the derivative of u in the direction of the vector $(\sum_{j=1}^{n} a_{ij}n_j)_{i=1}^{n}$ with components $\sum_{j=1}^{n} a_{1j}n_j, \sum_{j=1}^{n} a_{2j}n_j, \ldots, \sum_{j=1}^{n} a_{nj}n_j$, and the direction of this vector $\left(\sum_{j=1}^{n} a_{ij}n_j\right)_{i=1}^{n}$ is called the *conormal direction* associated with the operator A. (8.14.126)

Thus, for $u \in W^{1,p}(\Omega)$ with $Au \in L^p(\Omega)$, $1 < p < \infty$, the conormal derivative $\frac{\partial u}{\partial n_A}$ has a *trace on* Γ:

$$\frac{\partial u}{\partial n_A}\downarrow_\Gamma \in W^{-1/p,p}(\Gamma), \tag{8.14.127}$$

but *first-order* derivatives $\frac{\partial u}{\partial x_i}$ of u might have *no* traces on Γ.

Remark 8.14.6. The *conormal direction*, and consequently the *conormal derivative* of u in (8.14.125) is *not unique*, i.e. for the *same* A, we can have different representations with different choices of a_{ij} in (8.14.125). Hence, for the *same* A in (8.14.121), we have *several* conormal directions and *several* conormal derivatives with different choices of a_{ij}. For example, for $n = 2$,

$$-\Delta u = -\frac{\partial^2 u}{\partial x_1^2} - \frac{\partial^2 u}{\partial x_2^2} = -\sum_{i=1}^{2}\sum_{j=1}^{2} \frac{\partial}{\partial x_j}(\delta_{ij})\frac{\partial u}{\partial x_i},$$

the corresponding conormal derivative of u is

$$\frac{\partial u}{\partial n_\Delta}\downarrow_\Gamma = \sum_{j=1}^{n}\sum_{i=1}^{n} \delta_{ij}\frac{\partial u}{\partial x_i}n_j\downarrow_\Gamma = \sum_{i=1}^{n}\frac{\partial u}{\partial x_i}n_i\downarrow_\Gamma = \frac{\partial u}{\partial n}\downarrow_\Gamma,$$

i.e. the conormal derivative is the *usual normal derivative* on Γ.

But for $a_{11} = +1, a_{12} = -1, a_{21} = 1, a_{22} = +1$, we have $-(\frac{\partial^2 u}{\partial x_1^2} - \frac{\partial^2 u}{\partial x_1 \partial x_2} + \frac{\partial^2 u}{\partial x_2 \partial x_1} + \frac{\partial^2 u}{\partial x_2^2}) = -\Delta u$, and the corresponding *conormal derivative*

$$\begin{aligned}
\frac{\partial u}{\partial n_\Delta}\downarrow_\Gamma &= \left(a_{11}n_1\frac{\partial u}{\partial x_1} + a_{12}n_2\frac{\partial u}{\partial x_1} + a_{21}n_1\frac{\partial u}{\partial x_2} + a_{22}n_2\frac{\partial u}{\partial x_2}\right)\downarrow_\Gamma \\
&= \left(n_1\frac{\partial u}{\partial x_1} - n_2\frac{\partial u}{\partial x_1} + n_1\frac{\partial u}{\partial x_2} + n_2\frac{\partial u}{\partial x_2}\right)\downarrow_\Gamma \\
&= \left(\frac{\partial u}{\partial x_1}n_1 + \frac{\partial u}{\partial x_2}n_2\right)\downarrow_\Gamma + \left(\frac{\partial u}{\partial x_1}(-n_2) - \frac{\partial u}{\partial x_2}n_1\right)\downarrow_\Gamma \\
&= \frac{\partial u}{\partial n}\downarrow_\Gamma + \nabla u \cdot (-n_2, n_1)\downarrow_\Gamma = \frac{\partial u}{\partial n}\downarrow_\Gamma + \frac{\partial u}{\partial \tau}\downarrow_\Gamma
\end{aligned}$$

with $\hat{\tau} = (-n_2, n_1), \hat{\mathbf{n}} \cdot \hat{\boldsymbol{\tau}} = 0$.

Hence, $\frac{\partial u}{\partial n_\Delta}\downarrow_\Gamma = (\frac{\partial u}{\partial n} + \frac{\partial u}{\partial \tau})\downarrow_\Gamma$ is *another* conormal derivative of $-\Delta$.

Similarly, for $u \in W^{1,p}(\Omega)$ with $Au \in L^p(\Omega)$, $1 < p < \infty$, the conormal derivative $\frac{\partial u}{\partial n_{A^*}}$ associated with the adjoint A^* is defined by $\frac{\partial u}{\partial n_{A^*}} = \sum_{j=1}^n \sum_{i=1}^n a_{ji} \frac{\partial u}{\partial x_i} n_j$, which has a trace on Γ, i.e.

$$\frac{\partial u}{\partial n_{A^*}}\downarrow_\Gamma = \left(\sum_{j=1}^n \sum_{i=1}^n a_{ji} \frac{\partial u}{\partial x_i} n_j\right)\downarrow_\Gamma \in W^{-1/p,p}(\Gamma). \tag{8.14.128}$$

Now we will state general trace results based on Green's formula.

8.14.7 Trace theorems based on Green's formula

Let $\Omega \subset \mathbb{R}^n$ be a bounded domain with sufficiently smooth boundary Γ. Previously, except in Theorem 8.14.11, we have given trace results under the crucial assumption $s - 1/p > 0$. But if u is a solution of an *elliptic* equation, i.e. $Au \in L^p(\Omega)$, A being an elliptic operator (see (8.14.121)), then $u \in W^{s,p}(\Omega)$ has traces on a *sufficiently smooth* boundary *even when* $s - 1/p < 0$. The method of proof of this important result is based on the use of Green's formula, so we state Green's formula for a bounded domain with *Lipschitz continuous boundary* Γ.

Proposition 8.14.4. *Let $\Omega \subset \mathbb{R}^n$ be a bounded domain with Lipschitz continuous boundary $\Gamma \in C^{0,1}$-class (see* (D.2.3.6), *Appendix* D*). Then, for $1 < p, q < \infty$ with $1/p + 1/q = 1$, $\forall u \in W^{1,p}(\Omega)$, $\forall v \in W^{1,q}(\Omega)$,*

$$\int_\Omega \frac{\partial u}{\partial x_i} v \, d\mathbf{x} = -\int_\Omega u \frac{\partial v}{\partial x_i} d\mathbf{x} + \int_\Gamma (\gamma_0 u)(\gamma_0 v) n_i \, dS, \tag{8.14.129}$$

where $\gamma_0 u = u\downarrow_\Gamma \in W^{1-1/p,p}(\Gamma)$, $\gamma_0 v = v\downarrow_\Gamma \in W^{1-1/q,q}(\Gamma)$ by Theorem 8.14.5*; n_i is the ith component of the exterior unit normal $\hat{\mathbf{n}}$, which is defined a.e. on Lipschitz continuous Γ (see* (8.13.3)*).*

Proof. We refer to Nečas [16, p. 121] for the details of the proof. □

Applying (8.14.129), we have Green's formula given by:

Proposition 8.14.5. *Let $\Omega \subset \mathbb{R}^n$ be a bounded domain with Lipschitz continuous boundary $\Gamma \in C^{0,1}$-class (see* (D.2.3.6), *Appendix* D*). Then, $\forall u \in W^{2,p}(\Omega)$, $\forall v \in W^{2,q}(\Omega)$ with $1 < p, q < \infty$, $\frac{1}{p} + \frac{1}{q} = 1$, we have Green's formula:*

$$\int_\Omega vAu \, d\mathbf{x} - \int_\Omega uA^* v \, d\mathbf{x}$$
$$= \int_\Gamma \gamma_0 u \frac{\partial v}{\partial n_{A^*}} dS - \int_\Gamma \frac{\partial u}{\partial n_A} \cdot \gamma_0 v \, dS - \int_\Gamma \left(\sum_{i=1}^n a_i n_i\right) \gamma_0 u \gamma_0 v \, dS, \tag{8.14.130}$$

where A and A^* are defined by (8.14.121) *and* (8.14.122), *respectively; conormal derivatives* $\frac{\partial u}{\partial n_A}$ *and* $\frac{\partial v}{\partial n_{A^*}}$ *are defined by* (8.14.125) *and* (8.14.128), *respectively;* $\gamma_0 u = u\downarrow_\Gamma$*;* $\gamma_0 v = v\downarrow_\Gamma$*;* $\mathbf{a}\cdot\hat{\mathbf{n}} = \sum_{i=1}^n a_i n_i$*,* a_i *being the coefficients of* A *(resp.* A^**).*

In particular, for $A = -\Delta$*,* $u \in H^2(\Omega)$*,* $v \in H^2(\Omega)$*, we have*

$$\int_\Omega (-\Delta u) v d\mathbf{x} = \int_\Omega \nabla u \cdot \nabla v d\mathbf{x} - \int_\Gamma (\gamma_1 u)(\gamma_0 v) dS, \tag{8.14.131}$$

where, $\forall u \in H^2(\Omega)$*,* $\gamma_1 u = \frac{\partial u}{\partial n}\downarrow_\Gamma \in H^{2-1-1/2}(\Gamma) \equiv H^{1/2}(\Gamma) \hookrightarrow L^2(\Gamma)$*;* $\forall v \in H^2(\Omega)$*,* $\gamma_0 v = v\downarrow_\Gamma \in H^{3/2}(\Gamma) \hookrightarrow C^0(\Gamma)$*;* $\nabla u \cdot \nabla v = \sum_{i=1}^n \frac{\partial u}{\partial x_i}\frac{\partial v}{\partial x_i}$.

Proof. For $u \in W^{2,p}(\Omega)$, $\frac{\partial u}{\partial x_i} \in W^{1,p}(\Omega)$ and $a_{ij}\frac{\partial u}{\partial x_i} \in W^{1,p}(\Omega) \Longrightarrow \frac{\partial}{\partial x_j}(a_{ij}\frac{\partial u}{\partial x_i}) \in W^{0,p}(\Omega) \equiv L^p(\Omega) \Longrightarrow Au \in L^p(\Omega)$. Similarly, $v \in W^{2,q}(\Omega) \Longrightarrow A^* v \in L^q(\Omega)$.

Then $vAu \in L^1(\Omega)$, $uA^*v \in L^1(\Omega) \Longrightarrow \int_\Omega vAu\, d\mathbf{x}$, $\int_\Omega uA^*v d\mathbf{x}$ are well defined. Hence, we will show only the outline of the proof. By Proposition 8.14.4, we have

$$\begin{aligned}
\int_\Omega vAud\mathbf{x} &= -\sum_{i=1}^n\sum_{j=1}^n \int_\Omega \frac{\partial}{\partial x_j}(a_{ij}\frac{\partial u}{\partial x_i})vd\mathbf{x} + \sum_{i=1}^n \int_\Omega a_i \frac{\partial u}{\partial x_i} vd\mathbf{x} + \int_\Omega a_0 uvd\mathbf{x} \\
&= a(u,v) + \sum_{i=1}^n \int_\Omega \frac{\partial}{\partial x_i}(a_i uv)d\mathbf{x} - \sum_{i=1}^n \int_\Omega \frac{\partial}{\partial x_i}(a_i v)\cdot ud\mathbf{x} \\
&\quad - \sum_{i=1}^n\sum_{j=1}^n \int_\Gamma a_{ij}\frac{\partial u}{\partial x_i} n_j vdS \\
&= a(u,v) - \sum_{i=1}^n \int_\Omega \frac{\partial}{\partial x_i}(a_i v)\cdot ud\mathbf{x} + \int_\Gamma \left(\sum_{i=1}^n a_i n_i\right) uvdS - \int_\Gamma \frac{\partial u}{\partial n_A} vdS \\
&= a(u,v) - \sum_{i=1}^n \int_\Omega \frac{\partial}{\partial x_i}(a_i v)\cdot ud\mathbf{x} - \int_\Gamma \frac{\partial u}{\partial n_A}\cdot vdS - \int_\Gamma \left(\sum_{i=1}^n a_i n_i\right) uvdS;
\end{aligned} \tag{8.14.132}$$

$$\int_\Omega uA^*vd\mathbf{x} = a(u,v) - \sum_{i=1}^n \int_\Omega \frac{\partial}{\partial x_i}(a_i v) ud\mathbf{x} - \int_\Gamma \frac{\partial v}{\partial n_{A^*}}\cdot udS,$$

where $a(u,v) = \sum_{i=1}^n\sum_{j=1}^n \int_\Omega a_{ij}\frac{\partial u}{\partial x_i}\frac{\partial v}{\partial x_j} d\mathbf{x} + \int_\Omega a_0 uvd\mathbf{x}$ is defined in (8.14.123) $\Longrightarrow$ the result (8.14.130).

Again, by writing

$$\int_\Omega (-\Delta u) vd\mathbf{x} = -\sum_{i=1}^n\sum_{j=1}^n \int_\Omega \frac{\partial}{\partial x_j}\left(\delta_{ij}\frac{\partial u}{\partial x_i}\right) vd\mathbf{x}$$

with $A = -\Delta u$, $a_i = 0$, $a_0 = 0$, we get, from (8.14.132),

$$-\int_\Omega (\Delta u) v d\mathbf{x} = \sum_{i=1}^{n}\sum_{j=1}^{n}\int_\Omega \delta_{ij}\frac{\partial u}{\partial x_i}\frac{\partial v}{\partial x_j} d\mathbf{x} - \int_\Gamma \frac{\partial u}{\partial n}\cdot v dS. \qquad \square$$

8.14.7(a) Trace results in $H(\Delta;\Omega)$ based on Green's formula

Space $H(\Delta;\Omega)$

Definition 8.14.5. $H(\Delta;\Omega)$ is the linear space of functions $u \in L^2(\Omega)$ such that $\Delta u = \frac{\partial^2 u}{\partial x_1^2} + \cdots + \frac{\partial^2 u}{\partial x_n^2} \in L^2(\Omega)$:

$$H(\Delta;\Omega) = \{u : u \in L^2(\Omega),\ \Delta u \in L^2(\Omega)\}. \tag{8.14.133}$$

$H(\Delta;\Omega)$ equipped with the graph *norm* $\|\cdot\|_{H(\Delta;\Omega)} = \|\cdot\|_{0,\Delta,\Omega}$ and *inner product* $\langle\cdot,\cdot\rangle_{0,\Delta,\Omega}$:

$$\|u\|_{0,\Delta,\Omega} = (\|u\|_{0,\Omega}^2 + \|\Delta u\|_{0,\Omega}^2)^{1/2}; \tag{8.14.134}$$

$$\langle u, v\rangle_{0,\Delta,\Omega} = \langle u, v\rangle_{0,\Omega} + \langle \Delta u, \Delta v\rangle_{0,\Omega}\ \forall u, v \in H(\Delta;\Omega) \tag{8.14.135}$$

is a *Hilbert space* with the inclusion $H^2(\Omega) \subset H(\Delta;\Omega)$.

Question. For $u \in H(\Delta;\Omega)$, is it possible to define the traces $u{\downarrow_\Gamma}, \frac{\partial u}{\partial n}{\downarrow_\Gamma}$ on $\Gamma \in C^\infty$-class?

The answer is an *affirmative* one, although for $u \in L^2(\Omega)$ neither $\frac{\partial u}{\partial n}{\downarrow_\Gamma}$ nor $u{\downarrow_\Gamma}$ is defined. But it requires a proof, for which we first prove two lemmas, following the outline of the proof given in [13].

Lemma 8.14.7. *For a C^∞-regular domain, $\mathcal{D}(\overline{\Omega})$ is* dense *in $H(\Delta;\Omega)$.*

Proof. Let l be a continuous, linear functional on $H(\Delta;\Omega)$, which is a Hilbert space. Then

$$l(u) = \langle f, u\rangle_{0,\Omega} + \langle g, \Delta u\rangle_{0,\Omega} \quad \forall u \in H(\Delta;\Omega) \text{ and } f, g \in L^2(\Omega). \tag{8.14.136}$$

Suppose that $l(\phi) = \langle f,\phi\rangle + \langle g, \Delta\phi\rangle = 0\ \forall \phi \in \mathcal{D}(\overline{\Omega})$. As a consequence of the Hahn–Banach Theorem (Appendix A), the proof will be complete if we can show that $l(u) = 0\ \forall u \in H(\Delta;\Omega)$. Let $\tilde{f}, \tilde{g}$ be the null extensions of f and g to $\mathbb{R}^n$, respectively. Then $\tilde{f}, \tilde{g} \in L^2(\mathbb{R}^n)$. For $\phi \in \mathcal{D}(\overline{\Omega})$, $\exists \psi \in \mathcal{D}(\mathbb{R}^n)$ such that $\psi{\downarrow_\Omega} = \phi$. Hence, $l(\phi) = \langle \tilde{f}, \psi\rangle_{0,\mathbb{R}^n} + \langle \tilde{g}, \Delta\psi\rangle_{0,\mathbb{R}^n}$ with $\psi, \Delta\psi \in \mathcal{D}(\mathbb{R}^n) \subset$

$L^2(\mathbb{R}^n) \Longrightarrow \forall \psi \in \mathcal{D}(\mathbb{R}^n)$ with $\psi\downarrow_\Omega = \phi \in \mathcal{D}(\bar{\Omega})$,

$$\begin{aligned} l(\phi) &= \langle \tilde{f}, \psi\rangle_{\mathcal{D}'(\mathbb{R}^n)\times\mathcal{D}(\mathbb{R}^n)} + \langle \Delta(\tilde{g}), \psi\rangle_{\mathcal{D}'(\mathbb{R}^n)\times\mathcal{D}(\mathbb{R}^n)} \\ &= \langle \tilde{f} + \Delta(\tilde{g}), \psi\rangle_{\mathcal{D}'(\mathbb{R}^n)\times\mathcal{D}(\mathbb{R}^n)} = 0 \\ \Longrightarrow &\quad \langle \tilde{f} + \Delta(\tilde{g}), \psi\rangle = 0 \quad \forall \psi \in \mathcal{D}(\mathbb{R}^n) \\ \Longrightarrow &\quad \tilde{f} + \Delta(\tilde{g}) = 0 \quad \text{in } \mathcal{D}'(\mathbb{R}^n). \end{aligned} \tag{8.14.137}$$

But $\tilde{f} \in L^2(\mathbb{R}^n) \Longrightarrow \Delta(\tilde{g}) = -\tilde{f} \in L^2(\mathbb{R}^n)$. Hence, $\tilde{g} \in L^2(\mathbb{R}^n)$ and $\Delta(\tilde{g}) \in L^2(\mathbb{R}^n) \Longrightarrow \tilde{g} \in H^2(\mathbb{R}^n)$, since $-\Delta(\tilde{g}) + \tilde{g} \in L^2(\mathbb{R}^n) \Longrightarrow$ Fourier transform $\mathcal{F}[-\Delta(\tilde{g}) + \tilde{g}] \in L^2(\mathbb{R}^n)$ by the Plancherel–Riesz Theorem 8.3.1 $\Longrightarrow 4\pi^2\|\boldsymbol{\xi}\|^2\hat{\tilde{g}} + \hat{\tilde{g}} = (4\pi\|\boldsymbol{\xi}\|^2 + 1)\hat{\tilde{g}} \in L^2(\mathbb{R}^n)$ with $\hat{\tilde{g}} = \mathcal{F}[\tilde{g}] \Longrightarrow (1 + \|\boldsymbol{\xi}\|^2)\hat{\tilde{g}} \in L^2(\mathbb{R}^n) \Longrightarrow \tilde{g} \in H^2(\mathbb{R}^n)$ by Definition 8.9.1. Thus, $g \in L^2(\Omega)$ with $\tilde{g} \in H^2(\mathbb{R}^n)$. Then $\partial^{\boldsymbol{\alpha}}\tilde{g} = \widetilde{\partial^{\boldsymbol{\alpha}} g} \in L^2(\mathbb{R}^n)\ \forall|\boldsymbol{\alpha}| \le 2$. Hence, $g \in L^2(\Omega), \partial^{\boldsymbol{\alpha}} g = \widetilde{\partial^{\boldsymbol{\alpha}} g}\downarrow_\Omega \in L^2(\Omega)\ \forall|\boldsymbol{\alpha}| \le 2$. Thus, $g \in H^2(\Omega)$ with $\tilde{g} \in H^2(\mathbb{R}^n) \Longrightarrow g \in H^2_{00}(\Omega)$ by Definition 8.10.14, and $H^2_{00}(\Omega) \equiv H^2_0(\Omega)$ by Theorem 8.10.15. So $g \in H^2_0(\Omega)$. Since $\mathcal{D}(\Omega)$ is dense in $H^2_0(\Omega)$, for $g \in H^2_0(\Omega)$ $\exists$ a sequence $(\phi_k)_{k=1}^\infty$ in $\mathcal{D}(\Omega)$ such that $\phi_k \to g$ in $H^2(\Omega)$ as $k \to \infty$. Hence, $\langle \Delta u, \phi_k\rangle = \langle u, \Delta\phi_k\rangle\ \forall k \in \mathbb{N}$. Then, by taking limits, we have $\lim_{k\to\infty}\langle \phi_k, \Delta u\rangle = \langle g, \Delta u\rangle = \langle \Delta g, u\rangle$, since $g \in H^2_0(\Omega)$. Hence, from (8.14.136), $l(u) = \langle f, u\rangle + \langle \Delta g, u\rangle = \langle f + \Delta g, u\rangle\ \forall u \in H(\Delta;\Omega)$. But from (8.14.137), $\tilde{f} + \Delta(\tilde{g}) = 0$ in $L^2(\mathbb{R}^n) \subset \mathcal{D}'(\mathbb{R}^n) \Longrightarrow (\tilde{f} + \widetilde{\Delta g})\downarrow_\Omega = 0$ a.e. on Ω. Hence $l(u) = 0\ \forall u \in H(\Delta;\Omega)$, i.e. $\mathcal{D}(\overline{\Omega})$ is *dense* in $H(\Delta;\Omega)$. □

Trace results in $\mathcal{D}(\overline{\Omega})$ equipped with the norm $\|\cdot\|_{H(\Delta;\Omega)}$

Let $\phi_0 \in H^{3/2}(\Gamma)$, $\phi_1 \in H^{1/2}(\Gamma)$ be any two elements in $H^{3/2}(\Gamma)$ and $H^{1/2}(\Gamma)$, respectively. Then, by Theorem 8.14.2, $\exists$ a *continuous*, *linear* operator $\rho : H^{3/2}(\Gamma)\times H^{1/2}(\Gamma) \longrightarrow H^2(\Omega)$ such that $\phi = (\phi_0, \phi_1) \longmapsto \rho\phi \in H^2(\Omega)$ with $\boldsymbol{\gamma}\rho\phi = (\gamma_0\rho, \gamma_1\rho)\phi = (\gamma_0\rho(\phi), \gamma_1\rho(\phi)) = (\phi_0, \phi_1)$, i.e. $\gamma_0\rho(\phi) = \phi_0$, $\gamma_1\rho(\phi) = \phi_1$.

For fixed $u \in H(\Delta;\Omega)$, define T_u, $\forall \phi = (\phi_0, \phi_1) \in H^{3/2}(\Gamma) \times H^{1/2}(\Gamma)$, by:

$$T_u(\phi) = \langle \Delta u, \rho(\phi)\rangle_{L^2(\Omega)} - \langle u, \Delta\rho(\phi)\rangle_{L^2(\Omega)}. \tag{8.14.138}$$

$T_u(\phi)$ does *not* depend on the choice of $\rho(\phi) \in H^2(\Omega)$:

Let $v_1, v_2 \in H^2(\Omega)$ such that $v_i = \rho(\phi)$, $\gamma_0 v_1 = \gamma_0 v_2 = \phi_0$ and $\gamma_1 v_1 = \gamma_1 v_2 = \phi_1$. Then $\gamma_0(v_1 - v_2) = 0$, $\gamma_1(v_1 - v_2) = 0$. Hence, $(v_1 - v_2) \in H^2_0(\Omega)$ by Theorem 8.14.3. We are to show that $T_u(\phi) = \langle \Delta u, v_1\rangle_{0,\Omega} - \langle u, \Delta v_1\rangle_{0,\Omega} = \langle \Delta u, v_2\rangle - \langle u, \Delta v_2\rangle_{0,\Omega}$, or equivalently to show that $\langle \Delta u, v_1 - v_2\rangle = \langle u, \Delta(v_1 - v_2)\rangle$, since it holds for $(v_1 - v_2) \in \mathcal{D}(\Omega)$, which is *dense* in $H^2_0(\Omega)$. Hence, $T_u(\phi)$ is well defined by (8.14.138) $\forall \phi \in H^{3/2}(\Gamma) \times H^{1/2}(\Gamma)$.

T_u is *linear* and *continuous* on $H^{3/2}(\Omega) \times H^{1/2}(\Gamma)$:

Linearity: $T_u(\alpha_1\phi + \alpha_2\psi) = \alpha_1 T_u(\phi) + \alpha_2 T_u(\psi)\ \forall \phi, \psi \in H^{3/2}(\Gamma) \times H^{1/2}(\Gamma)$.

Continuity:

$|T_u(\phi)| \le |\langle \Delta u, \rho(\phi)\rangle_{0,\Omega}| + |\langle u, \Delta(\rho(\phi))\rangle_{0,\Omega}| \le C\|u\|_{0,\Delta,\Omega} |||\phi|||_{H^{3/2}(\Gamma)\times H^{1/2}(\Gamma)}$
with

$$|||\phi|||_{H^{3/2}(\Gamma)\times H^{1/2}(\Gamma)} = (\|\phi_0\|^2_{H^{3/2}(\Gamma)} + \|\phi_1\|^2_{H^{1/2}(\Gamma)})^{1/2}, \ C > 0$$

$\Longrightarrow$ T_u is a *continuous*, linear functional on $H^{3/2}(\Gamma) \times H^{1/2}(\Gamma)$ with $\|T_u\|_{H^{-3/2}(\Gamma)\times H^{-1/2}(\Gamma)} \le C\|u\|_{0,\Delta,\Omega}$.

Hence, we can rewrite (8.14.138) as follows: $\forall \phi = (\phi_0, \phi_1) \in H^{3/2}(\Gamma) \times H^{1/2}(\Gamma)$,

$$T_u(\phi) = \langle h, \phi_0\rangle_{H^{-3/2}(\Gamma)\times H^{3/2}(\Gamma)} - \langle g, \phi_1\rangle_{H^{-1/2}(\Gamma)\times H^{1/2}(\Gamma)}, \tag{8.14.139}$$

with $h \in H^{-3/2}(\Gamma)$, $g \in H^{-1/2}(\Gamma)$.

Lemma 8.14.8. *For* $u \in \mathcal{D}(\overline{\Omega})$, $u\downarrow_\Gamma = g \in H^{-1/2}(\Gamma)$, $\frac{\partial u}{\partial n}\downarrow_\Gamma = h \in H^{-3/2}(\Gamma)$.

Proof. For $u \in \mathcal{D}(\overline{\Omega}) \subset H(\Delta;\Omega)$, $\rho(\phi) \in H^2(\Omega)$, we have the usual Green's formula:

$$\langle \Delta u, \rho(\phi)\rangle_{0,\Omega} - \langle u, \Delta\rho(\phi)\rangle_{0,\Omega} = \int_\Gamma \frac{\partial u}{\partial n}\gamma_0(\rho\phi)ds - \int_\Gamma u\gamma_1(\rho\phi)ds$$

$$= \int_\Gamma \frac{\partial u}{\partial n}\phi_0 ds - \int_\Gamma u\phi_1 ds = \left\langle \frac{\partial u}{\partial n}\downarrow_\Gamma, \phi_0\right\rangle - \langle u\downarrow_\Gamma, \phi_1\rangle.$$

Hence, for $u \in \mathcal{D}(\overline{\Omega}) \subset H(\Delta;\Omega)$, $\forall \phi = (\phi_0, \phi_1) \in H^{3/2}(\Gamma) \times H^{1/2}(\Gamma)$,

$$T_u(\phi) = \left\langle \frac{\partial u}{\partial n}\downarrow_\Gamma, \phi_0\right\rangle - \langle u\downarrow_\Gamma, \phi_1\rangle$$

$$= \langle h, \phi_0\rangle_{H^{-3/2}(\Gamma)\times H^{3/2}(\Gamma)} - \langle g, \phi_1\rangle_{H^{-1/2}(\Gamma)\times H^{1/2}(\Gamma)}$$

$\Longrightarrow \frac{\partial u}{\partial n}\downarrow_\Gamma = h \in H^{-3/2}(\Gamma)$; $u\downarrow_\Gamma = g \in H^{-1/2}(\Gamma)$ for $u \in \mathcal{D}(\overline{\Omega})$. □

As a consequence of Lemma 8.14.8, we can define trace operator $\gamma : u \in \mathcal{D}(\overline{\Omega}) \longrightarrow \gamma u = (\gamma_0 u, \gamma_1 u) \in H^{-1/2}(\Gamma) \times H^{-3/2}(\Gamma)$ by:

$$\gamma_0 u = u\downarrow_\Gamma \in H^{-1/2}(\Gamma), \quad \gamma_1 u = \frac{\partial u}{\partial n}\downarrow_\Gamma \in H^{-3/2}(\Gamma), \tag{8.14.140}$$

with $\gamma u = (\gamma_0 u, \gamma_1 u) = (u\downarrow_\Gamma, \frac{\partial u}{\partial n}\downarrow_\Gamma) \in H^{-1/2}(\Gamma) \times H^{-3/2}(\Gamma)$, and

$$\langle \gamma_0 u, \phi_1\rangle_{H^{-1/2}(\Gamma)\times H^{1/2}(\Gamma)} = T_u(0, -\phi_1) \quad \forall \phi_1 \in H^{1/2}(\Gamma);$$

$$\langle \gamma_1 u, \phi_0\rangle_{H^{-3/2}(\Gamma)\times H^{3/2}(\Gamma)} = T_u(\phi_0, 0) \quad \forall \phi_0 \in H^{3/2}. \tag{8.14.141}$$

Lemma 8.14.9. *The trace operator* γ *defined by* (8.14.140)–(8.14.141) *is* linear *and* continuous *from* $\mathcal{D}(\overline{\Omega})$ *equipped with* $\|\cdot\|_{H(\Delta;\Omega)}$ *into* $H^{-1/2}(\Gamma) \times H^{-3/2}(\Gamma)$.

Proof. The linearity of γ_0 and γ_1 and consequently of γ is obvious from their definitions; their continuity is shown as follows:

$$\begin{aligned}|\langle \gamma_0 u, \phi_1\rangle_{H^{-1/2}(\Gamma)\times H^{1/2}(\Gamma)}| &= |T_u(0,-\phi_1)|\\ &\le C\|u\|_{H(\Delta;\Omega)}\|(0,-\phi_1)\|_{H^{3/2}(\Gamma)\times H^{1/2}(\Gamma)}\\ &\le C\|u\|_{H(\Delta;\Omega)}\|\phi_1\|_{H^{1/2}(\Gamma)} \quad \forall \phi_1 \in H^{1/2}(\Gamma),\ \forall u \in H(\Delta;\Omega)\end{aligned}$$

$\Longrightarrow \|\gamma_0 u\|_{H^{-1/2}(\Gamma)} \le C\|u\|_{H(\Delta;\Omega)}$ $\forall u \in H(\Delta;\Omega)$.

Hence, $\gamma_0 : \mathcal{D}(\overline{\Omega}) \longrightarrow H^{-1/2}(\Gamma)$ is *continuous* from $\mathcal{D}(\overline{\Omega})$ into $H^{-1/2}(\Gamma)$.

Similarly, $\|\gamma_1 u\|_{H^{-3/2}(\Gamma)} \le C\|u\|_{H(\Delta;\Omega)}$ $\forall u \in H(\Delta;\Omega) \Longrightarrow$ the continuity of $\gamma_1 : \mathcal{D}(\overline{\Omega}) \longrightarrow H^{-3/2}(\Gamma)$.

Consequently,

$$\begin{aligned}\|\gamma u\|_{H^{-1/2}(\Gamma)\times H^{-3/2}(\Gamma)} &= \|(\gamma_0 u, \gamma_1 u)\|_{H^{-1/2}(\Gamma)\times H^{-3/2}(\Gamma)}\\ &\le C\|u\|_{H(\Delta;\Omega)} \quad \forall u \in H(\Delta;\Omega)\end{aligned}$$

$\Longrightarrow \gamma : \mathcal{D}(\overline{\Omega}) \longrightarrow H^{-1/2}(\Gamma) \times H^{-3/2}(\Gamma)$ is also *continuous* from $\mathcal{D}(\overline{\Omega})$ equipped with $\|\cdot\|_{H(\Delta;\Omega)}$ into $H^{-1/2}(\Gamma) \times H^{-3/2}(\Gamma)$. □

Trace results in $H(\Delta;\Omega)$

Theorem 8.14.12. *Let $\Omega \subset \mathbb{R}^n$ be a C^∞-regular domain with the boundary $\Gamma \in C^\infty$-class (see* (D.2.3.8)*, Appendix* D*, and Definition* 8.10.4*). Then the linear, continuous trace operator $\gamma : u \in \mathcal{D}(\overline{\Omega}) \longmapsto \gamma u = (\gamma_0 u, \gamma_1 u) \in H^{-1/2}(\Gamma) \times H^{-3/2}(\Gamma)$ defined by* (8.14.140)–(8.14.141) *has a* unique, continuous, linear extension *as an operator from $H(\Delta;\Omega)$ into $H^{-1/2}(\Gamma) \times H^{-3/2}(\Gamma)$, and this extended operator will still be denoted by the same notation γ, i.e.*

$$\gamma : u \in H(\Delta;\Omega) \longmapsto \gamma u = (\gamma_0 u, \gamma_1 u) \in H^{-1/2}(\Gamma) \times H^{-3/2}(\Gamma)$$

with $\gamma_0 u = u{\downarrow_\Gamma} \in H^{-1/2}(\Gamma)$, $\gamma_1 u = \frac{\partial u}{\partial n}{\downarrow} \in H^{-3/2}(\Gamma)$ and $\gamma u = (\gamma_0 u, \gamma_1 u) = (u{\downarrow_\Gamma}, \frac{\partial u}{\partial n}{\downarrow_\Gamma}) \in H^{-1/2}(\Gamma) \times H^{-3/2}(\Gamma)$.

Moreover, the following Green's formula holds: $\forall u \in H(\Delta;\Omega)$, $\forall v \in H^2(\Omega)$,

$$\begin{aligned}&\int_\Omega (\Delta u)v\,d\mathbf{x} - \int_\Omega u(\Delta v)\,d\mathbf{x}\\ &= \left\langle \frac{\partial u}{\partial n}{\downarrow_\Gamma}, v{\downarrow_\Gamma}\right\rangle_{H^{-3/2}(\Gamma)\times H^{3/2}(\Gamma)} - \left\langle u{\downarrow_\Gamma}, \frac{\partial v}{\partial n}{\downarrow_\Gamma}\right\rangle_{H^{-1/2}(\Gamma)\times H^{1/2}(\Gamma)}\\ &= \langle \gamma_1 u, \gamma_0^* v\rangle_{H^{-3/2}(\Gamma)\times H^{3/2}(\Gamma)} - \langle \gamma_0 u, \gamma_1^* v\rangle_{H^{-1/2}(\Gamma)\times H^{1/2}(\Gamma)},\end{aligned} \tag{8.14.142}$$

where $\gamma_0^* v = v{\downarrow_\Gamma} \in H^{3/2}(\Gamma)$, $\gamma_1^* v = \frac{\partial v}{\partial n}{\downarrow_\Gamma} \in H^{1/2}(\Gamma)$ $\forall v \in H^2(\Omega)$ by Theorem 8.14.2*, and $\gamma_0 u = u{\downarrow_\Gamma} \in H^{-1/2}$, $\gamma_1 u = \frac{\partial u}{\partial n}{\downarrow_\Gamma} \in H^{-3/2}(\Gamma)$ $\forall u \in H(\Delta;\Omega)$.*

Proof. By Lemma 8.14.7, $\mathcal{D}(\overline{\Omega})$ is dense in $H(\Delta;\Omega)$. Then the result follows from Lemma 8.14.9 by virtue of the density of $\mathcal{D}(\overline{\Omega})$ in $H(\Delta;\Omega)$, and Green's formula follows from (8.14.130) with $A = A^* = \Delta$. □

8.14.7(b) Trace results based on Green's formula for bounded polygonal domains $\Omega \subset \mathbb{R}^2$ [18, 17]

For bounded polygonal domains $\Omega \in \mathbb{R}^2$, Green's formula in Proposition 8.14.4 holds, since bounded polygonal $\Gamma \in C^{0,1}$-class and we can rewrite it as follows: $\forall u \in H^1(\Omega)$ with $\gamma_{0j}u = u \downarrow_{\Gamma_j} \in H^{1/2}(\Gamma_j)$, $\forall v \in H^2(\Omega)$ with $\gamma_{1j}v = \frac{\partial v}{\partial n_j} \downarrow_{\Gamma_j} \in H^{1/2}(\Gamma_j)$, $1 \le j \le N$,

$$\int_\Omega u\Delta v dx_1 dx_2 = -\int_\Omega \nabla u \cdot \nabla v dx_1 dx_2 + \sum_{j=1}^{N} \int_{\Gamma_j} (\gamma_{0j}u)(\gamma_{1j}v)ds, \quad (8.14.143)$$

where $\nabla u \cdot \nabla v = \frac{\partial u}{\partial x_1}\frac{\partial v}{\partial x_1} + \frac{\partial u}{\partial x_2}\frac{\partial v}{\partial x_2}$.

$\forall u \in H^2(\Omega)$ with $\gamma_{0j}u = u \downarrow_{\Gamma_j} \in H^{3/2}(\Gamma_j)$, $\forall v \in H^2(\Omega)$ with $\gamma_{0j}v = v \downarrow_{\Gamma_j} \in H^{3/2}(\Gamma_j)$, $\gamma_{1j}v = \frac{\partial v}{\partial n_j} \downarrow_{\Gamma_j} \in H^{1/2}(\Gamma_j)$, $1 \le j \le N$,

$$\int_\Omega u\Delta v dx_1 dx_2 - \int_\Omega v\Delta dx_1 dx_2 = \sum_{j=1}^{N} \left[\int_\Gamma (\gamma_{0j}u)(\gamma_{1j}v)ds - \int_\Gamma (\gamma_{0j}v)(\gamma_{1j}u)ds\right]. \quad (8.14.144)$$

Trace results in $H(\Delta;\Omega)$ for bounded polygonal Ω [18, 17]

For bounded polygonal $\Omega \subset \mathbb{R}^2$, the space $H(\Delta;\Omega)$ is defined by (8.14.133)–(8.14.135) is a *Hilbert space* with norm $\|v\|_{0,\Delta,\Omega} = (\|v\|^2_{0,\Omega} + \|\Delta v\|^2_{0,\Omega})^{1/2}$. But the trace results for $v \in H(\Delta;\Omega)$ given by Theorem 8.14.12 for C^∞-regular domain Ω based on Green's formula will *no longer hold* for polygonal Ω. In fact, for a *fixed* side Γ_j, we are to find traces $v \downarrow_{\Gamma_j}$ and $\frac{\partial v}{\partial n_j} \downarrow_{\Gamma_j}$ for $v \in H(\Delta;\Omega)$ using Green's formula (8.14.144). For this, we will follow [18] and first prove the density of $H^2(\Omega)$ in $H(\Delta;\Omega)$.

Lemma 8.14.10. *$H^2(\Omega)$ is a* dense *subspace of* $H(\Delta;\Omega)$.

Proof. $v \in H^2(\Omega) \implies v, \partial^\alpha v \in L^2(\Omega)\ \forall|\alpha| \le 2 \implies v; v_{,11}; v_{,22} \in L^2(\Omega) \implies v \in L^2(\Omega)$, $\Delta v \in L^2(\Omega) \implies v \in H(\Delta;\Omega)$. Hence, $H^2(\Omega) \subset H(\Delta;\Omega)$. Let l be a bounded, linear functional on $H(\Delta;\Omega)$. Then we show that $l(u) = 0\ \forall u \in H^2(\Omega) \implies l = 0$ on $H(\Delta;\Omega)$. This result follows from the density of $\mathcal{D}(\overline{\Omega})$ in $H^2(\Omega)$ (since polygonal Ω is a 2-extension domain) and from the density of $\mathcal{D}(\overline{\Omega})$ in $H(\Delta;\Omega)$ proved in Lemma 8.14.7.

With any *fixed* side Γ_j of Γ, we associate the space $U_j \subset H^2(\Omega)$ defined by:

$$U_j = \Big\{u : u : H^2(\Omega),\ \gamma_{0k} u = u \downarrow_{\Gamma_k} = f_{k,0} = 0,$$
$$\gamma_{1k} u = \frac{\partial u}{\partial n_k} \downarrow_{\Gamma_k} = f_{k,1} = 0\ \forall k \neq j,\ 1 \le k \le \mathbb{N}\Big\}.$$

Hence, $u \in U_j \implies u \in H^2(\Omega)$ with $u \downarrow_{\Gamma \setminus \overline{\Gamma}_j} = 0$, $\frac{\partial u}{\partial n_\Gamma} \downarrow_{\Gamma \setminus \overline{\Gamma}_j} = 0$ a.e. on $\Gamma \setminus \overline{\Gamma}_j$. Then, from Theorem 8.14.8 and Example 8.14.1, the mapping $u \in U_j \mapsto (f_{j,0}; f_{j,1}) \in T_0$ is a *linear, continuous surjection* from U_j *onto* the trace subspace $T_0 = H_{00}^{3/2}(\Gamma_j) \times H_{00}^{1/2}(\Gamma_j)$, $T_0 \subset H^{3/2}(\Gamma_j) \times H^{1/2}(\Gamma_j)$ such that $f_{j,0} = u \downarrow_{\Gamma_j} = \gamma_{0,j} u \in H_{00}^{3/2}(\Gamma_j)$; $f_{j,1} = \frac{\partial u}{\partial n_j} \downarrow_{\Gamma_j} = \gamma_{1j} u \in H_{00}^{1/2}(\Gamma_j)$. Let $\phi_0 \in H_{00}^{3/2}(\Gamma_j)$, $\phi_1 \in H_{00}^{1/2}(\Gamma_j)$ be any two elements. Then $\exists u \in U_j$ such that $u \downarrow_{\Gamma_j} = \phi_0 \in H_{00}^{3/2}(\Gamma_j)$, $\frac{\partial u}{\partial n_j} \downarrow_{\Gamma_j} = \phi_1 \in H_{00}^{1/2}(\Gamma_j)$ $\forall \phi = (\phi_0, \phi_1) \in H_{00}^{3/2}(\Gamma_j) \times H_{00}^{1/2}(\Gamma_j)$ with $\rho_j \phi = u \in U_j$ such that $\gamma_j(\rho_j \phi) = (\gamma_{0j}(\gamma_j \phi), \gamma_{1j}(\gamma_j \phi)) = (\phi_0, \phi_1)$ and $\Delta(\rho_j \phi) \in L^2(\Omega)$ (ϕ not shown in bold here and also later). Hence, $\langle (\rho_j \phi), \Delta v \rangle_{0,\Omega}$ and $\langle v, \Delta(\rho_j \phi) \rangle_{0,\Omega}$ are well defined $\forall v \in H(\Delta; \Omega)$. Define T_v by:

$$T_v(\phi) = \langle \rho_j \phi, \Delta v \rangle_{0,\Omega} - \langle v, \Delta(\rho_j \phi) \rangle_{0,\Omega}, \tag{8.14.145}$$

which does *not* depend on the choice of $\rho_j \phi \in U_j$. Then T_v is *linear* and *continuous* on $H_{00}^{3/2}(\Gamma_j) \times H_{00}^{1/2}(\Gamma_j)$, since

$$\begin{aligned}
|T_v(\phi)| &\le |\langle \rho_j \phi, \Delta v \rangle_{0,\Omega}| + |\langle v, \Delta(\rho_j \phi) \rangle| \\
&\le \|\rho_j \phi\|_{0,\Omega} \|\Delta v\|_{0,\Omega} + \|v\|_{0,\Omega} \|\Delta(\rho_j \phi)\|_{0,\Omega} \\
&\le \langle (\|\Delta v\|_{0,\Omega}, \|v\|_{0,\Omega}), (\|\rho_j \phi\|_{0,\Omega}, \|\Delta(\rho_j \phi)\|_{0,\Omega}) \rangle_{\mathbb{R}^2} \\
&\le (\|v\|_{0,\Omega}^2 + \|\Delta v\|_{0,\Omega}^2)^{1/2} (\|\rho_j \phi\|_{0,\Omega}^2 + \|\Delta(\rho_j \phi)\|_{0,\Omega}^2)^{1/2} \\
&\le C \|v\|_{0,\Delta,\Omega} \cdot \|\rho_j \phi\|_{2,\Omega} \le (C \|v\|_{0,\Delta,\Omega}) |||\phi|||,
\end{aligned}$$

where $|||\phi|||_{H_{00}^{3/2}(\Gamma_j) \times H_{00}^{1/2}(\Gamma_j)} = (\|\phi_0\|_{H_{00}^{3/2}(\Gamma_j)}^2 + \|\phi_1\|_{H_{00}^{1/2}(\Gamma_j)}^2)^{1/2}$ with $\|T_v\| \le C \|v\|_{0,\Delta,\Omega}$, since $H_{00}^{3/2}(\Gamma_j) \times H_{00}^{1/2}(\Gamma_j) \hookrightarrow H^{3/2}(\Gamma_j) \times H^{1/2}(\Gamma_j)$ with $|||\phi|||_{H^{3/2}(\Gamma_j) \times H^{1/2}(\Gamma_j)} \le C |||\phi|||_{H_{00}^{3/2}(\Gamma_j) \times H_{00}^{1/2}(\Gamma_j)}$.

Hence, T_v is a continuous, linear functional on $H_{00}^{3/2}(\Gamma_j) \times H_{00}^{1/2}(\Gamma_j)$ and we can rewrite (8.14.145) as follows: $\forall \phi = (\phi_0, \phi_1) \in H_{00}^{3/2}(\Gamma_j) \times H_{00}^{1/2}(\Gamma_j)$,

$$T_v(\phi) = \langle h, \phi_0 \rangle_{(H_{00}^{3/2}(\Gamma_j))' \times H_{00}^{3/2}(\Gamma_j)} - \langle g, \phi_1 \rangle_{(H_{00}^{1/2}(\Gamma_j))' \times H_{00}^{1/2}(\Gamma_j)}$$

with $h \in (H_{00}^{3/2}(\Gamma_j))'$, $g = (H_{00}^{3/2}(\Gamma_j))'$. □

Lemma 8.14.11. *For* $v \in H^2(\Omega)$,

$$\gamma_{0j} v = v \downarrow_{\Gamma_j} = g \in (H_{00}^{1/2}(\Gamma_j))',$$
$$\gamma_{1j} v = \frac{\partial v}{\partial n_j} \downarrow_{\Gamma_j} = h \in (H_{00}^{3/2}(\Gamma_j))',$$

Γ_j *being the fixed open side of* Γ.

Proof. For $v \in H^2(\Omega)$ and $u \in U_j$ with $u \downarrow_{\Gamma_j} = \phi_0 \in H_{00}^{3/2}(\Gamma_j)$, $\frac{\partial u}{\partial n_j} \downarrow_{\Gamma_j} = \phi_1 \in H_{00}^{1/2}(\Gamma_j)$, and $u \downarrow_{\Gamma_k} = 0$, $\frac{\partial u}{\partial n_k} \downarrow_{\Gamma_k} = 0 \ \forall k \neq j$, Green's formula (8.14.144) becomes

$$\langle \Delta v, u \rangle_{0,\Omega} - \langle v, \Delta u \rangle_{0,\Omega} = \int_{\Gamma_j} \phi_0 \frac{\partial v}{\partial n} ds - \int_{\Gamma_j} \phi_1 v ds$$
$$= \left\langle \frac{\partial v}{\partial n} \downarrow_{\Gamma_j}, \phi_0 \right\rangle - \langle v \downarrow_{\Gamma_j}, \phi_1 \rangle = \langle h, \phi_0 \rangle - \langle g, \phi_1 \rangle$$
$$\iff \quad \frac{\partial v}{\partial n} \downarrow_{\Gamma_j} = h \in (H_{00}^{3/2}(\Gamma_j))', v \downarrow_{\Gamma_j} = g \in (H_{00}^{1/2}(\Gamma_j))'. \qquad \square$$

As a consequence of Lemma 8.14.11, applying the density of $H^2(\Omega)$ in $H(\Delta; \Omega)$, we have:

Theorem 8.14.13. *Let* $\Gamma \subset \mathbb{R}^2$ *be the bounded polygonal boundary of* $\Omega \subset \mathbb{R}^2$. *Then, for any fixed side* Γ_j *of* Γ, *the mapping* $v \in H^2(\Omega) \longmapsto (\gamma_{0j} v, \gamma_{1j} v) \in (H_{00}^{1/2}(\Gamma_j))' \times (H_{00}^{3/2}(\Gamma_j))'$ *with* $\gamma_{0j} v = v \downarrow_{\Gamma_j} \in (H_{00}^{1/2}(\Gamma_j))'$; $\gamma_{1j} v = \frac{\partial v}{\partial n_j} \downarrow_{\Gamma_j} \in (H_{00}^{3/2}(\Gamma_j))'$, *has a* unique, continuous, linear *extension as an operator from* $H(\Delta; \Omega)$ *into* $(H_{00}^{1/2}(\Gamma_j))' \times (H_{00}^{3/2}(\Gamma_j))'$, *and the extended operator is denoted by the same notation, i.e.*

$$\gamma_j : v \in H(\Delta; \Omega) \longmapsto \gamma_j v = (\gamma_{0j} v, \gamma_{1j} v) \in (H_{00}^{1/2}(\Gamma_j))' \times (H_{00}^{3/2}(\Gamma_j))'. \tag{8.14.146}$$

8.14.7(c) Trace results based on Green's formula for fourth-order elliptic equations of aniso-/ortho-/isotropic plate bending problems

Let $\Omega \in \mathbb{R}^2$ be a C^∞-regular domain with the boundary $\Gamma \in C^\infty$-class, whose parametric representation is given by: $\phi_{\mathbf{x}_0}$: $s \in [0, L[\subset \mathbb{R} \longmapsto \phi_{\mathbf{x}_0}(s) = \mathbf{x} \in \Gamma \subset \mathbb{R}^2$ with $\phi_{\mathbf{x}_0}(0) = \mathbf{x}_0 = \phi_{\mathbf{x}_0}(L^-)$, s being the arc length of Γ measured from $\mathbf{x}_0 \in \Gamma$, L being the length measure of Γ. (8.14.147)

Let Λ be the aniso-/ortho-/isotropic elastic thin plate bending operator for the small displacement field $u = u(x_1, x_2)$ with $\mathbf{x} = (x_1, x_2) \in \overline{\Omega}$ by (see [48] for more details). For $u \in H^2(\Omega)$,

$$(\Lambda u)(\mathbf{x}) \equiv \sum_{i=1}^{2}\sum_{j=1}^{2}\sum_{k=1}^{2}\sum_{l=1}^{2} \frac{\partial^2}{\partial x_k \partial x_l}\left(a_{ijkl}(\mathbf{x})\frac{\partial^2 u}{\partial x_i \partial x_j}\right)(\mathbf{x})$$
$$\equiv (a_{ijkl}u_{,ij})_{,kl}(\mathbf{x}) \text{ in } \Omega.^7 \qquad (8.14.148)$$

The variable coefficients a_{ijkl} representing the aniso-/ortho-/isotropic elastic properties and the variable thickness $h = h(x_1, x_2)$ of the thin plate satisfy the following conditions: $\forall i, j, k, l = 1, 2$,

(A1) $a_{ijkl} \in C^\infty(\overline{\Omega})$; $a_{ijkl} \geq 0$; (8.14.149a)

(A2) $a_{ijkl}(\mathbf{x}) = a_{klij}(\mathbf{x}) = a_{lkij}(\mathbf{x}) = a_{lkji}(\mathbf{x})$ $\forall \mathbf{x} \in \overline{\Omega}$; (8.14.149b)

(A3) $\exists \alpha_0 > 0$ such that $\forall \boldsymbol{\xi} = (\xi_{11}, \xi_{12}, \xi_{21}, \xi_{22}) \in \mathbb{R}^4$ with $\xi_{12} = \xi_{21}$,
$a_{ijkl}(\mathbf{x})\xi_{ij}\xi_{kl} \geq \alpha_0 \|\boldsymbol{\xi}\|^2_{\mathbb{R}^4}$ $\forall \mathbf{x} \in \overline{\Omega}$. (8.14.149c)

$(\frac{\partial(\cdot)}{\partial x_i} \equiv (\cdot)_{,i}$; $\frac{\partial^2(\cdot)}{\partial x_i \partial x_j} = (\cdot)_{,ij}$; $\frac{\partial^2(\cdot)}{\partial x_k \partial x_l} \equiv (\cdot)_{,kl}.)$

Remark 8.14.7. The standard ellipticity condition has been replaced by a more convenient condition (A3), which will be sufficient for all cases of plate bending problems.

Conditions (A2) could have been replaced by the less restrictive condition $a_{ijkl} = a_{klij}$ only (see [48] for more details).

For $a_{ijkl} = \delta_{ij}\delta_{kl}$ and also for several *other* choices of a_{ijkl}, $\Lambda = \Delta\Delta$, the well-known *biharmonic operator*. (8.14.150)

Tensor-valued functions $\boldsymbol{\Phi}$

$\mathbf{x} \in \overline{\Omega} \longmapsto \Phi(\mathbf{x}) = (\phi_{ij}(\mathbf{x}))_{1 \leq i,j \leq 2}$ with $\phi_{12}(\mathbf{x}) = \phi_{21}(\mathbf{x})$, where $\Phi(\mathbf{x})$ defines a *symmetric tensor* of rank 2 with 4 components $(\phi_{ij}(\mathbf{x}))_{1 \leq i,j \leq 2}$.

To every admissible displacement field $v = v(\mathbf{x})$ of the bent thin plate at $\mathbf{x} = (x_1, x_2) \in \overline{\Omega}$, we associate a *symmetric, tensor-valued* function $(\phi_{ij}(\mathbf{x}))_{1 \leq i,j \leq 2} = \Phi$ defined, $\forall v \in H^2(\Omega)$, by:

$$\phi_{ij} = a_{ijkl}v_{,kl} = a_{jikl}v_{,kl} = \phi_{ji} \quad \text{such that } \phi_{ij} = \phi_{ji} \in L^2(\Omega). \qquad (8.14.151)$$

In plate bending theory $\Phi = (\phi_{ij})$ with ϕ_{ij} defined by (8.14.151) is called the *bending moment tensor field*, an *alternative* notation being $M = (M_{ij})$ with $M_{12} = M_{21} = H =$ twisting moment.

Normal moment M_n, twisting moment M_{nt}, vertical shear Q_n and *Kirchhoff force K_n* in a bent plate are defined below.

[7] In (8.14.148) and what follows, Einstein's summation convention with respect to twice-repeated indices $i, j, k, l = 1, 2$ has been followed.

With each $\Phi = (\phi_{ij})$ in (8.14.151) and unit normal field $\hat{\mathbf{n}}$ and unit tangent vector $\hat{\mathbf{t}}$, we associate $M_n(\Phi)$, $M_{nt}(\Phi)$, $Q_n(\Phi)$ and $K_n(\Phi)$ defined, for $\hat{\mathbf{n}} = (n_1, n_2)$, $\hat{\mathbf{t}} = (t_1, t_2)$, by:

$$M_n(\Phi) = \phi_{ij} n_i n_j = a_{ijkl} v_{,kl} n_i n_j; \tag{8.14.152a}$$

$$M_{nt}(\Phi) = \phi_{ij} n_i t_j = a_{ijkl} v_{,kl} n_i t_j; \tag{8.14.152b}$$

$$\begin{aligned} K_n(\Phi) &= \frac{\partial}{\partial t}[M_{nt}(\Phi)] + Q_n(\Phi) = \frac{\partial}{\partial t}[\phi_{ij} n_i t_j] + (\phi_{ij,i}) n_j \\ &= \frac{\partial}{\partial t}[a_{ijkl} v_{,kl} n_i t_j] + (a_{ijkl} v_{,kl})_{,i} n_j; \end{aligned} \tag{8.14.152c}$$

$$Q_n(\Phi) = (\phi_{ij,i}) n_j = (a_{ijkl} v_{,kl})_{,i} n_j; \tag{8.14.152d}$$

where $\frac{\partial(\cdot)}{\partial t}$ denotes the derivative of $(\cdot)$ in the direction of unit tangent $\hat{\mathbf{t}}$ and summations with respect to twice-repeated indices $i, j, k, l = 1, 2$ are always assumed.

Traces $v\downarrow_\Gamma$, $\frac{\partial v}{\partial n}\downarrow_\Gamma$ of admissible displacement $v \in H^2(\Omega)$ on Γ are defined by:

$$\gamma_0 v = v\downarrow_\Gamma \in H^{3/2}(\Gamma); \quad \gamma_1 v = \frac{\partial v}{\partial n}\downarrow_\Gamma \in H^{1/2}(\Gamma) \tag{8.14.153}$$

by Theorem 8.14.2.

Green's formula First of all, we define the space $\mathbb{D}(\overline{\Omega})$ of *smooth symmetric tensor-valued* functions $\Phi = (\phi_{ij})$ by:

$$\mathbb{D}(\bar{\Omega}) = \{\Phi : \Phi = (\phi_{ij})_{1 \le i,j \le 2} \text{ with } \phi_{ij} = \phi_{ji} \in \mathcal{D}(\overline{\Omega})\}. \tag{8.14.154}$$

Proposition 8.14.6. *For C^∞-regular domain $\Omega \subset \mathbb{R}^2$, $\forall \Psi = (\psi_{ij}) \in \mathbb{D}(\bar{\Omega})$, $\forall v \in H^2(\Omega)$, the following Green's formula holds:*

$$\int_\Omega \psi_{ij,ij} v \, d\mathbf{x} - \int_\Omega \psi_{ij}(v_{,ij}) d\mathbf{x} = \int_\Gamma K_n(\Psi)\gamma_0 v \, ds - \int_\Gamma M_n(\Psi)\gamma_1 v \, ds, \tag{8.14.155}$$

where $K_n(\Psi)$, $M_n(\Psi)$ are defined by (8.14.152c) *and* (8.14.152a), *respectively;*

$$\gamma_0 v = v\downarrow_\Gamma \in H^{3/2}(\Gamma) \subset C^0(\Gamma); \quad \gamma_1 v = \frac{\partial v}{\partial n}\downarrow_\Gamma \in H^{1/2}(\Gamma) \subset L^2(\Gamma). \tag{8.14.156}$$

Proof. $\forall \Psi = (\psi_{ij}) \in \mathbb{D}(\overline{\Omega})$, $\forall v \in H^2(\Omega)$, the *usual* Green's formula holds:

$$\int_\Omega \psi_{ij,ij} v \, d\mathbf{x} - \int_\Omega \psi_{ij}(v_{,ij}) d\mathbf{x} = \int_\Gamma (\psi_{ij,i} n_j) v \, ds - \int_\Gamma (\psi_{ij} . v_{,i} n_j) ds. \tag{8.14.157}$$

Since Γ is smooth, $v_{,i} = \frac{\partial v}{\partial x_i} = \frac{\partial v}{\partial n} n_i + \frac{\partial v}{\partial t} t_i$ $(i = 1, 2)$, $\frac{\partial(\cdot)}{\partial n}$ and $\frac{\partial(\cdot)}{\partial t}$ being the derivatives in the directions of $\hat{\mathbf{n}}$ and $\hat{\mathbf{t}}$, respectively.

Then

$$\begin{aligned}\int_\Gamma (\psi_{ij} v_{,i} n_j) ds &= \int_\Gamma \psi_{ij} \left(\frac{\partial v}{\partial n} n_i + \frac{\partial v}{\partial t} t_i \right) n_j ds \\ &= \int_\Gamma (\psi_{ij} n_i n_j) \frac{\partial v}{\partial n} ds + \int_\Gamma (\psi_{ij} t_i n_j) \frac{\partial v}{\partial t} ds \\ &= \int_\Gamma M_n(\Psi) \frac{\partial v}{\partial n} ds + \int_\Gamma M_{nt}(\Psi) \frac{\partial v}{\partial t} ds \\ &= \int_\Gamma M_n(\Psi) \frac{\partial v}{\partial n} ds + \int_\Gamma \frac{\partial}{\partial t}[M_{nt}(\Psi) \cdot v] ds - \int_\Gamma \frac{\partial}{\partial t}[M_{nt}(\Psi)] \cdot v ds.\end{aligned}$$

Then the right-hand side of (8.14.157)

$$\begin{aligned}&\int_\Gamma (\psi_{ij,i} n_j) v ds - \int_\Gamma (\psi_{ij} v_{,i} n_j) ds \\ &= \int_\Gamma Q_n(\Psi) v ds + \int_\Gamma \frac{\partial}{\partial t}[M_{nt}(\Psi)] \cdot v ds - \int_\Gamma M_n(\Psi) \frac{\partial v}{\partial n} ds - \int_\Gamma \frac{\partial}{\partial t}[M_{nt}(\Psi) v] ds \\ &= \int_\Gamma [\frac{\partial}{\partial t}[M_n(\Psi)] + Q_n(\Psi)] v ds - \int_\Gamma M_n(\Psi) \frac{\partial v}{\partial n} ds \\ &= \int_\Gamma K_n(\Psi) v ds - \int_\Gamma M_n(\Psi) \frac{\partial v}{\partial n} dS,\end{aligned}$$

since $\int_\Gamma \frac{\partial}{\partial t}[M_n(\Psi) v] ds = 0$. In fact, using the parametric representation of Γ in (8.14.147), we have

$$\int_\Gamma \frac{\partial}{\partial t}[M_{nt}(\Psi) v] ds = [M_{nt}(\Psi) v](\phi_{\mathbf{x}_0}(s))|_{s=0}^{s=L^-} = 0.$$

□

Remark 8.14.8. For $u \in H^2(\Omega)$, $\psi_{ij} = a_{ijkl} u_{,kl} \in L^2(\Omega)$, and Theorem 8.14.2 does *not* hold for $\psi_{ij} \in L^2(\Omega)$, $\psi_{ij}|_\Gamma$ is *not* defined and $M_n(\Psi)\downarrow_\Gamma$, $K_n(\Psi)\downarrow_\Gamma$ are *not* defined in general and Green's formula (8.14.155) *loses its meaning*.

This suggests we consider the natural subspace $H^2(\Lambda, \Omega)$ of $H^2(\Omega)$ associated with the fourth-order anisotropic plate bending operator Λ defined in (8.14.148):

$$H^2(\Lambda, \Omega) = \{v : v \in H^2(\Omega),\ \Lambda v = (a_{ijkl},\ v_{,ij})_{,kl} \in L^2(\Omega)\}. \qquad (8.14.158)$$

$H^2(\Lambda, \Omega)$ equipped with the graph norm $\|\cdot\|_{2,\Lambda,\Omega}$ and corresponding inner product $\langle \cdot, \cdot \rangle_{2,\Lambda,\Omega}$:

$$\begin{aligned}\|v\|_{2,\Lambda,\Omega} &= (\|v\|^2_{2,\Omega} + \|\Lambda v\|^2_{0,\Omega})^{1/2}; \\ \langle u, v \rangle_{2,\Lambda,\Omega} &= \langle u, v \rangle_{2,\Omega} + \langle \Lambda u, \Lambda v \rangle_{0,\Omega}; \qquad (8.14.159)\end{aligned}$$

is a *Hilbert space* with the continuous imbedding $H^2(\Lambda, \Omega) \hookrightarrow H^2(\Omega)$. We will extend Green's formula (8.14.155) to functions $u \in H^2(\Lambda, \Omega)$, i.e. for functions $u \in H^2(\Lambda, \Omega)$, the traces $M_n(\Psi)\downarrow_\Gamma$ and $K_n(\Gamma)\downarrow_\Gamma$ of $M_n(\Psi)$ and $K_n(\Psi)$ on Γ will be well defined. For this, a complete proof is essential and given below. Although the proof given here is completely different from that of Theorem 8.14.12, the general outline of the proof has a lot of similarity with that of Theorem 8.14.12 for functions of $H(\Delta; \Omega)$. First of all, we define:

Space $\mathbb{L}^2(\Omega)$ We define $\mathbb{L}^2(\Omega)$ as follows:

$$\mathbb{L}^2(\Omega) = \{\Phi : \Phi = (\phi_{ij})_{1\le i,j\le 2} \text{ with } \phi_{ij} = \phi_{ji} \in L^2(\Omega); \phi_{ij,ij} \in L^2(\Omega)\} \tag{8.14.160}$$

is equipped the graph norm $|||\cdot|||_{0,\Omega}$ and inner product $[\cdot,\cdot]_{0,\Omega}$:

$$\begin{aligned}|||\Phi|||_{0,\Omega}^2 &= \|\Phi\|_{0,\Omega}^2 + \|\phi_{ij,ij}\|_{0,\Omega}^2 \\ &= \int_\Omega (\phi_{11}^2 + 2\phi_{12}^2 + \phi_{22}^2)d\mathbf{x} + \int_\Omega (\phi_{11,11} + 2\phi_{12,12} + \phi_{22,22})^2 d\mathbf{x};\end{aligned} \tag{8.14.161}$$

$$[\Psi, \Phi]_{0,\Omega} = \langle\langle \Psi, \Phi\rangle\rangle_{0,\Omega} + \langle \psi_{ij,ij}, \phi_{ij,ij}\rangle_{0,\Omega},$$

with

$$\langle\langle \Psi, \Phi\rangle\rangle_{0,\Omega} = \langle\langle(\psi_{ij}), (\phi_{ij})\rangle\rangle_{0,\Omega} = \int_\Omega (\psi_{11}\phi_{11} + 2\psi_{12}\phi_{12} + \psi_{22}\phi_{22})d\mathbf{x}; \tag{8.14.162a}$$

$$\begin{aligned}\langle \psi_{ij,ij}, \phi_{ij,ij}\rangle_{0,\Omega} = \int_\Omega &[(\psi_{11,11} + 2\psi_{12,12} + \psi_{22,22}) \\ &\times (\phi_{11,11} + 2\phi_{12,12} + \phi_{22,22})]d\mathbf{x}.\end{aligned} \tag{8.14.162b}$$

Then $\mathbb{L}^2(\Omega)$ equipped with the inner product (8.14.162) and norm (8.14.161) is a *Hilbert space*.

$\mathbb{L}^2(\Omega)$ and $H^2(\Lambda, \Omega)$ are related by: $u \in H^2(\Lambda, \Omega) \implies \psi_{ij} = a_{ijkl}u_{,kl} \in L^2(\Omega)$ and $\psi_{ij,ij} = (a_{ijkl}u_{,kl})_{,ij} = \Lambda u \in L^2(\Omega) \implies \Psi = (\psi_{ij}) \in \mathbb{L}^2(\Omega)$ with $\psi_{ij} = a_{ijkl}u_{,kl}\ \forall i, j = 1, 2.$ (8.14.163)

Hence, our first aim is to give trace results $M_n(\Psi)\downarrow_\Gamma$ and $K_n(\Psi)\downarrow_\Gamma$ for $\Psi \in \mathbb{L}^2(\Omega)$ and extend Green's formula (8.14.155) to $\Psi \in \mathbb{L}^2(\Omega)$ and $v \in H^2(\Omega)$ using the *density* of $\mathbb{D}(\overline{\Omega})$ in $\mathbb{L}^2(\Omega)$, which will be proved first.

Lemma 8.14.12. $\mathbb{D}(\overline{\Omega})$ *is* dense *in* $\mathbb{L}^2(\Omega)$.

Proof. Let l be a *continuous*, *linear* functional on Hilbert space $\mathbb{L}^2(\Omega)$ defined by: $\forall \Phi \in \mathbb{L}^2(\Omega), l(\Phi) = \langle\langle \Psi, \Phi \rangle\rangle_{0,\Omega} + \langle q, \phi_{ij,ij} \rangle_{0,\Omega}$ with $q \in L^2(\Omega)$, $\Psi = (\psi_{ij})_{1 \le i,j \le 2}$ with $\psi_{ij} = \psi_{ji} \in L^2(\Omega)$. (8.14.164)

As a consequence of the Hahn–Banach Theorem, the density result will follow if we can prove that $l(\Phi) = 0 \ \forall \Phi \in \mathbb{D}(\overline{\Omega}) \Longrightarrow l = 0$ on $\mathbb{L}^2(\Omega)$. Assume that

$$l(\Phi) = \langle\langle (\psi_{ij}), (\phi_{ij}) \rangle\rangle_{0,\Omega} + \langle q, \phi_{ij,ij} \rangle_{0,\Omega} = 0 \quad \forall \Phi \in \mathbb{D}(\overline{\Omega}). \tag{8.14.164a}$$

Then $l(\Phi) = 0 \ \forall \Phi \in \mathbb{D}(\Omega) = \{\Phi : \Phi = (\phi_{ij}) \text{ with } \phi_{ij} = \phi_{ji} \in \mathcal{D}(\Omega)\} \subset \mathbb{D}(\overline{\Omega}) \Longrightarrow \sum_{i,j=1}^{2} \{[\psi_{ij}, \phi_{ij}]_{\mathcal{D}'(\Omega) \times \mathcal{D}(\Omega)} + [q_{,ij}, \phi_{ij}]_{\mathcal{D}'(\Omega) \times \mathcal{D}(\Omega)}\} = 0 \ \forall \Phi = (\phi_{ij}) \in \mathbb{D}(\Omega)$ with $\phi_{ij} \in \mathcal{D}(\Omega) \Longrightarrow \psi_{ij} + q_{,ij} = 0$ in $\mathcal{D}'(\Omega), 1 \le i, j \le 2 \Longrightarrow$
$q_{,ij} = -\psi_{ij} \in L^2(\Omega) \ \forall i, j = 1, 2,$ (8.14.164b)
$\Longrightarrow q \in H^2(\Omega)$ (for an alternative proof, see [48]). (8.14.165)

Now we will show that $q \in H_0^2(\Omega)$. For this we are to show that $\widetilde{q} \in H^2(\mathbb{R}^n)$. $\forall \phi_{ij} \in \mathcal{D}(\overline{\Omega})$, $\exists \chi_{ij} \in \mathcal{D}(\mathbb{R}^2)$ such that $\chi_{ij}\downarrow_{\Omega} = \phi_{ij}$, $1 \le i, j \le 2$. Let $\widetilde{\Psi} = (\widetilde{\psi_{ij}})_{1 \le i,j \le 2}$ with $\widetilde{\psi}_{ij} = \widetilde{\psi}_{ji}$, $\widetilde{\psi}_{ij}$ being the null-extension of ψ_{ij} to $\mathbb{R}^2, 1 \le i, j \le 2$, and $\widetilde{q}$ being the null extension of q to $\mathbb{R}^2$.

Then, from (8.14.164a), we have, $\forall \chi_{ij} \in \mathcal{D}(\mathbb{R}^2)$ with $\chi_{ij}\downarrow_{\Omega} = \phi_{ij}, 1 \le i, j \le 2$, $l(\Phi) = \langle\langle (\widetilde{\psi}_{ij}), (\chi_{ij}) \rangle\rangle_{0,\mathbb{R}^2} + \langle \widetilde{q}, \chi_{ij,ij} \rangle_{0,\mathbb{R}^2} = 0$

$\Longrightarrow \langle\langle (\widetilde{\psi}_{ij}), (\chi_{ij}) \rangle\rangle_{0,\mathbb{R}^2} + \langle \widetilde{q}_{,ij}, \chi_{ij} \rangle_{0,\mathbb{R}^2} = 0 \ \forall \chi_{ij} \in \mathcal{D}(\mathbb{R}^2)$ with $\chi_{ij}\downarrow_{\Omega} = \phi_{ij}$

$\Longrightarrow \int_{\mathbb{R}^2} \widetilde{\psi}_{ij} \chi_{ij} d\mathbf{x} + \int_{\mathbb{R}^2} \widetilde{q}_{,ij} \chi_{ij} d\mathbf{x} = 0 \Longrightarrow \langle\langle \widetilde{\psi}_{ij} + \widetilde{q}_{,ij}, \chi_{ij} \rangle\rangle_{0,\mathbb{R}^2} = 0$

$\Longrightarrow \widetilde{\psi}_{ij} + \widetilde{q}_{,ij} = 0$ in $L^2(\mathbb{R}^2)$, $1 \le i, j \le 2 \Longrightarrow \widetilde{q}_{,ij} = -\widetilde{\psi}_{ij} \in L^2(\mathbb{R}^2)$ $\forall i, j = 1, 2$.

Thus, $\widetilde{q} \in L^2(\mathbb{R}^2)$ and $\widetilde{q}_{,ij} \in L^2(\mathbb{R}^2) \ \forall i, j = 1, 2 \Longrightarrow \widetilde{q} \in H^2(\mathbb{R}^2)$. From (8.14.165), $q \in H^2(\Omega)$ with $\widetilde{q} \in H^2(\mathbb{R}^2) \Longrightarrow q \in H_{00}^2(\Omega)$ by Definition 8.10.14. But by Theorem 8.10.15, $H_0^2(\Omega) \equiv H_{00}^2(\Omega)$. Hence, $q \in H_0^2(\Omega)$ (see an alternative proof in [48]).

Since $\mathcal{D}(\Omega)$ is *dense* in $H_0^2(\Omega)$, $\exists$ a sequence (q_n) in $\mathcal{D}(\Omega)$ such that $q_n \longrightarrow q$ in $H^2(\Omega)$ as $n \to \infty$. Then, $\forall \Phi \in \mathbb{L}^2(\Omega)$,

$$\begin{aligned} &\lim_{n\to\infty} [-\langle\langle (q_{n,ij}), (\phi_{ij}) \rangle\rangle_{0,\Omega} + \langle q_n, \phi_{ij,ij} \rangle_{0,\Omega}] \\ &= -\langle\langle (q_{,ij}), (\phi_{ij}) \rangle\rangle_{0,\Omega} + \langle q, \phi_{ij,ij} \rangle_{0,\Omega} = l(\Phi) \end{aligned}$$

by (8.14.164), since $-(q_{,ij}) = (\psi_{ij}) \ \forall i, j = 1, 2$ by (8.14.164b). But $\forall \Phi = (\phi_{ij}) \in \mathbb{L}^2(\Omega)$ and $\forall n \in \mathbb{N}$, $\langle q_n, \phi_{ij,ij} \rangle_{0,\Omega} = \langle\langle (q_{n,ij}), (\phi_{ij}) \rangle\rangle_{0,\Omega}$
$\Longrightarrow -\langle\langle (q_{n,ij}), (\phi_{ij}) \rangle\rangle_{0,\Omega} + \langle q_n, \phi_{ij,ij} \rangle_{0,\Omega} = 0 \ \forall n \in \mathbb{N}, \forall \Phi \in \mathbb{L}^2(\Omega)$
$\Longrightarrow \lim_{n\to\infty} [-\langle\langle (q_{n,ij}), (\phi_{ij}) \rangle\rangle_{0,\Omega} + \langle q_n, \phi_{ij,ij} \rangle_{0,\Omega}] = 0 \ \forall \Phi \in \mathbb{L}^2(\Omega)$
$\Longrightarrow l(\Phi) = 0 \ \forall \Phi \in \mathbb{L}^2(\Omega)$ and the density of $\mathbb{D}(\overline{\Omega})$ in $\mathbb{L}^2(\Omega)$ follows. □

Trace results in $\mathbb{L}^2(\Omega)$

For fixed $\Psi = (\psi_{ij}) \in \mathbb{L}^2(\Omega)$, define $T_\Psi : H^{3/2}(\Gamma) \times H^{1/2}(\Gamma) \longrightarrow \mathbb{R}$ by: $\forall \chi = (\chi_0, \chi_1) \in H^{3/2}(\Gamma) \times H^{1/2}(\Gamma)$,

$$T_\Psi(\chi) = \int_\Omega \psi_{ij,ij}\rho(\chi)d\mathbf{x} - \int_\Omega \psi_{ij}(\rho(\chi))_{,ij}d\mathbf{x}, \tag{8.14.166}$$

where $\rho : H^{3/2}(\Gamma) \times H^{1/2}(\Gamma) \longrightarrow H^2(\Omega)$ is a linear, continuous operator defined (see Theorem 8.14.2) by $\rho : \chi = (\chi_0, \chi_1) \in H^{3/2}(\Gamma) \times H^{1/2}(\Gamma) \longmapsto \rho(\chi) \in H^2(\Omega)$ with

$$\gamma_0(\rho(\chi)) = (\rho(\chi)) \downarrow_\Gamma = \chi_0 \in H^{3/2}(\Gamma);\ \gamma_1(\rho(\chi)) = \frac{\partial(\rho(\chi))}{\partial n} \downarrow_\Gamma = \chi_1 \in H^{1/2}(\Gamma), \tag{8.14.167}$$

$$\|\rho(\chi)\|_{2,\Omega} \leq C\|\chi\|_{H^{3/2}(\Gamma)\times H^{1/2}(\Gamma)} = C(\|\chi_0\|^2_{H^{3/2}(\Gamma)} + \|\chi_1\|^2_{H^{1/2}(\Gamma)})^{1/2}. \tag{8.14.168}$$

In (8.14.166), $\rho(\chi) \in H^2(\Omega)$ satisfying (8.14.167) and (8.14.168) is *not* unique.

Lemma 8.14.13.

I. $T_\Psi(\chi)$ *in* (8.14.166) *does* not *depend on the choice of* $\rho(\chi) \in H^2(\Omega)$ *satisfying* (8.14.167) *and* (8.14.168)*;*

II. T_Ψ *is a* linear, continuous functional *on* $H^{3/2}(\Gamma) \times H^{1/2}(\Gamma)$ *with*

$$\|T_\Psi\|_{H^{-3/2}(\Gamma)\times H^{-1/2}(\Gamma)} \leq C\|\!|\Psi|\!\|_{\mathbb{L}^2(\Omega)} (C > 0), \tag{8.14.169}$$

and consequently can be written as follows: $\forall \chi = (\chi_0, \chi_1) \in H^{3/2}(\Gamma) \times H^{1/2}(\Gamma)$,

$$T_\Psi(\chi) = [h, \chi_0]_{H^{-3/2}(\Gamma)\times H^{3/2}(\Gamma)} - [g, \chi_1]_{H^{-1/2}(\Gamma)\times H^{1/2}(\Gamma)} \tag{8.14.170}$$

with $g \in H^{-1/2}(\Gamma)$, $h \in H^{-3/2}(\Gamma)$*;*

III. $\forall \Psi = (\psi_{ij}) \in \mathbb{D}(\overline{\Omega})$, $K_n(\Psi) \downarrow_\Gamma = h \in H^{-3/2}(\Gamma)$*;* $M_n(\Psi) \downarrow_\Gamma = g \in H^{-1/2}(\Gamma)$, (8.14.171)

where $K_n(\Psi)$ *and* $M_n(\Psi)$ *are defined by* (8.14.152c) *and* (8.14.152a)*, respectively.*

Proof.

I. Suppose that $\exists u_1, u_2 \in H^2(\Omega)$ with $\rho(\chi) = u_i$ and $\gamma_0 u_i = \chi_0$, $\gamma_1 u_i = \chi_1$ $(i = 1, 2)$. Set $w^1_{ij} = u_{1,ij} \in L^2(\Omega)$ and $w^2_{ij} = (u_{2,ij}) \in L^2(\Omega)$ $\forall i, j = 1, 2$.

Then we are to show that $T_\Psi(\boldsymbol{\chi}) = \langle \psi_{ij,ij}, u^i\rangle_{0,\Omega} - \langle\!\langle \Psi, w^i_{ij}\rangle\!\rangle_{0,\Omega}$ $(i = 1, 2)$ or, equivalently, that

$$\langle \psi_{ij,ij}, u_1 - u_2\rangle_{0,\Omega} = \langle\!\langle (\psi_{ij}), (w^1_{ij}) - (w^2_{ij})\rangle\!\rangle_{0,\Omega} = \langle\!\langle (\psi_{ij}), (w^1_{ij} - w^2_{ij})\rangle\!\rangle_{0,\Omega}.$$

But $(u_1 - u_2) \in H^2(\Omega)$ with $\gamma_0(u_1 - u_2) = 0$, $\gamma_1(u_1 - u_2) = 0 \implies (u_1 - u_2) \in H^2_0(\Omega) \implies \langle \psi_{ij,ij}, u_1 - u_2\rangle_{0,\Omega} = \langle\!\langle (\psi_{ij}), (u_{1,ij} - u_{2,ij})\rangle\!\rangle_{0,\Omega} = \langle\!\langle (\psi_{ij}), (w^1_{ij} - w^2_{ij})\rangle\!\rangle_{0,\Omega}$ for $(u_1 - u_2) \in H^2_0(\Omega)$.

II. The linearity of T_Ψ is obvious from (8.14.166), and its *continuity* follows from: $\forall \boldsymbol{\chi} = (\chi_0, \chi_1) \in H^{3/2}(\Gamma) \times H^{1/2}(\Gamma)$ satisfying (8.14.167) and (8.14.168),

$$\begin{aligned} |T_\Psi(\boldsymbol{\chi})| &\le \|\psi_{ij,ij}\|_{0,\Omega}\|\rho(\chi)\|_{0,\Omega} + \|\Psi\|_{0,\Omega}\|(\rho(\chi))_{,ij}\|_{0,\Omega} \\ &\le (\|\Psi\|^2_{0,\Omega} + \|\psi_{ij,ij}\|^2_{0,\Omega})^{1/2}(|\rho(\chi)\|^2_{0,\Omega} + \|\rho(\chi)_{,ij}\|^2_{0,\Omega})^{1/2} \\ &\le |\!|\!|\Psi|\!|\!|_{0,\Omega}\|\rho(\chi)\|_{2,\Omega} \le (C|\!|\!|\Psi|\!|\!|_{0,\Omega})\|\chi\|_{H^{3/2}(\Gamma)\times H^{1/2}(\Gamma)} \quad \text{(by (8.14.168))} \end{aligned}$$

$\implies T_\Psi \in H^{-3/2}(\Gamma)\times H^{-1/2}(\Gamma)$ with $\|T_\Psi\|_{H^{-3/2}(\Gamma)\times H^{-1/2}(\Gamma)} \le C|\!|\!|\Psi|\!|\!|_{0,\Omega}$ $\forall$ fixed $\Psi \in \mathbb{L}^2(\Omega)$.

III. Let $\Psi = (\psi_{ij}) \in \mathbb{D}(\overline{\Omega}) \subset \mathbb{L}^2(\Omega)$. Then, $\forall \boldsymbol{\chi} = (\chi_0, \chi_1) \in H^{3/2}(\Gamma) \times H^{1/2}(\Gamma)$ with (8.14.167) and (8.14.168), from (8.14.166) and Green's formula (8.14.155),

$$\begin{aligned} T_\Psi(\boldsymbol{\chi}) &= \int_\Omega \psi_{ij,ij}\rho(\chi)d\mathbf{x} - \int_\Omega \psi_{ij}\rho(\chi)_{,ij}d\mathbf{x} \\ &= \int_\Gamma K_n(\Psi)\gamma_0(\rho(\chi))ds - \int_\Gamma M_n(\Psi)\gamma_1(\rho(\chi))ds \\ &= \int_\Gamma K_n(\Psi)\chi_0 ds - \int_\Gamma M_n(\Psi)\chi_1 ds, \end{aligned} \tag{8.14.172}$$

where $\chi_0 \in H^{3/2}(\Gamma)$, $\chi_1 \in H^{1/2}(\Gamma)$.

Then the result (8.14.171) follows from a comparison of (8.14.172) and (8.14.171).

□

Trace results in $\mathbb{D}(\overline{\Omega})$

By virtue of the results in (8.14.171), we can now define a *continuous*, *linear trace operator* $\boldsymbol{\delta} = (\delta_0, \delta_1)$ for *tensor-valued smooth functions* in $\mathbb{D}(\overline{\Omega})$ as follows:

The mapping $\Psi \in \mathbb{D}(\bar{\Omega}) \longrightarrow (\delta_0\Psi, \delta_1\Psi) \in H^{-3/2}(\Gamma) \times H^{-1/2}(\Gamma)$ is defined, $\forall \Psi \in \mathbb{D}(\bar{\Omega})$, by:

$$\delta_0\Psi = K_n(\Psi)\downarrow_\Gamma = \left(\frac{\partial}{\partial t}(\psi_{ij}n_i t_j) + \psi_{ij,i}n_j\right)\downarrow_\Gamma \in H^{-3/2}(\Gamma); \tag{8.14.173}$$

$$\delta_1\Psi = M_n(\Psi)\downarrow_\Gamma = (\psi_{ij}n_i n_j)\downarrow_\Gamma \in H^{-1/2}(\Gamma) \tag{8.14.174}$$

such that $\forall \boldsymbol{\chi} = (\chi_0, \chi_1) \in H^{3/2}(\Gamma) \times H^{1/2}(\Gamma)$,

$$T_\Psi(\chi) = [\delta_0 \Psi, \chi_0]_{H^{-3}(\Gamma)\times H^{-3/2}(\Gamma)} - [\delta_1 \Psi, \chi_1]_{H^{-1/2}(\Gamma)\times H^{1/2}(\Gamma)}.$$

Then,

$$T_\Psi((\chi_0, 0)) = [\delta_0 \Psi, \chi_0]_{H^{3/2}(\Gamma)\times H^{1/2}(\Gamma)};$$
$$T_\Psi((0, -\chi_1)) = [\delta_1 \Psi, \chi_1]_{H^{-1/2}(\Gamma)\times H^{1/2}(\Gamma)} \quad \forall \chi_0 \in H^{3/2}(\Gamma),\ \forall \chi_1 \in H^{1/2}(\Gamma).$$

Hence, $\delta_0 : \mathbb{D}(\bar{\Omega}) \longrightarrow H^{-3/2}(\Gamma)$, $\delta_1 : \mathbb{D}(\bar{\Omega}) \longrightarrow H^{-1/2}(\Gamma)$ defined by (8.14.173) and (8.14.174) are *linear* and *continuous* from $\mathbb{D}(\bar{\Omega})$ equipped with the norm $|||\cdot|||_{0,\Omega}$ of $\mathbb{L}^2(\Omega)$. Their linearity is obvious from the definitions (8.14.173) and (8.14.174), and their *continuity* follows from that of T_Ψ in (8.14.169). In fact, $\forall \chi_0 \in H^{3/2}(\Gamma)$,

$$\begin{aligned}|[\delta_0 \Psi, \chi_0]| &= |T_\Psi((\chi_0, 0))| \le (C|||\Psi|||_{0,\Omega})|||(\chi_0, 0)|||_{H^{3/2}(\Gamma)\times H^{1/2}(\Gamma)} \\ &= (C|||\Psi|||_{0,\Omega})\|\chi_0\|_{H^{3/2}(\Gamma)}\end{aligned}$$

$\Longrightarrow$ $\|\delta_0 \Psi\|_{H^{-3/2}(\Gamma)} \le C|||\Psi|||_{0,\Omega}$ $\forall \Psi \in \mathbb{D}(\bar{\Omega})$ for some $C > 0$

$\Longrightarrow$ δ_0 is continuous from $\mathbb{D}(\bar{\Omega})$ into $H^{-3/2}(\Gamma)$.

Similarly, we can show that $\|\delta_1 \Psi\|_{H^{-1/2}(\Gamma)} \le C|||\Psi|||_{0,\Omega} \forall \Psi \in \mathbb{D}(\bar{\Omega})$ for some $C > 0$

$\Longrightarrow$ δ_1 is *continuous* from $\mathbb{D}(\bar{\Omega})$ into $H^{-1/2}(\Gamma)$.

Define

$$\boldsymbol{\delta} : \Psi \in \mathbb{D}(\bar{\Omega}) \longmapsto \boldsymbol{\delta}\Psi = (\delta_0 \Psi, \delta_1 \Psi) \in H^{-3/2}(\Gamma) \times H^{-1/2}(\Gamma). \tag{8.14.175}$$

Then $\boldsymbol{\delta}$ is *linear* and *continuous* from $\mathbb{D}(\bar{\Omega})$ into $H^{-3/2}(\Gamma) \times H^{-1/2}(\Gamma)$, since

$$\begin{aligned}\|\boldsymbol{\delta}\Psi\|_{H^{-3/2}(\Gamma)\times H^{-1/2}(\Gamma)} &= \|(\delta_0 \Psi, \delta_1 \Psi)\|_{H^{-3/2}(\Gamma)\times H^{-1/2}(\Gamma)} \\ &\le C|||\Psi|||_{0,\Omega} \quad \forall \Psi \in \mathbb{D}(\bar{\Omega}) \subset \mathbb{L}^2(\Omega).\end{aligned} \tag{8.14.176}$$

Note that both here and later, the same constant $C > 0$ has been used to denote different values at different inequalities.

Moreover, Green's formula (8.14.155) can be rewritten as follows: $\forall \Psi \in \mathbb{D}(\bar{\Omega})$, $\forall v \in H^2(\Omega)$,

$$\begin{aligned}&\langle \psi_{ij,ij}, v\rangle_{0,\Omega} - \langle\langle(\psi_{ij}), v_{,ij}\rangle\rangle \\ &= [K_n(\Psi), \gamma_0 v]_{H^{-3/2}(\Gamma)\times H^{3/2}(\Gamma)} - [M_n(\psi), \gamma_1 v]_{H^{-1/2}(\Gamma)\times H^{1/2}(\Gamma)} \\ &= [\delta_0 \Psi, \gamma_0 v]_{H^{-3/2}(\Gamma)\times H^{3/2}(\Gamma)} - [\delta_1 \Psi, \gamma_1 v]_{H^{-1/2}(\Gamma)\times H^{1/2}(\Gamma)}.\end{aligned} \tag{8.14.177}$$

Trace results and Green's formula in $\mathbb{L}^2(\Omega)$

Since $\mathbb{D}(\bar{\Omega})$ is dense in $\mathbb{L}^2(\Omega)$, the *continuous linear* trace operator $\boldsymbol{\delta} : \Psi \in \mathcal{D}(\bar{\Omega}) \longmapsto \boldsymbol{\delta}\psi = (\delta_0\Psi, \delta_1\Psi) \in H^{-3/2}(\Gamma) \times H^{-1/2}(\Gamma)$ defined by (8.14.175):

$$\delta_0\Psi = K_n(\Psi) \downarrow_\Gamma = \left[\frac{\partial(\psi_{ij}n_i t_j)}{\partial t} + \psi_{ij,i}n_j\right] \downarrow_\Gamma \in H^{-3/2}(\Gamma);$$

$$\delta_1\Psi = M_n(\Psi) \downarrow_\Gamma = (\psi_{ij}n_i nj) \downarrow_\Gamma \in H^{-1/2}(\Gamma)$$

satisfying (8.14.175)–(8.14.176) can be given a *unique*, *continuous*, *linear* extension to an operator from $\mathbb{L}^2(\Omega)$ into $H^{-3/2}(\Gamma) \times H^{-1/2}(\Gamma)$, which will still be denoted by the same notation, i.e. $\boldsymbol{\delta} = (\delta_0, \delta_1) : \Psi \in \mathbb{L}^2(\Omega) \longrightarrow \boldsymbol{\delta}\Psi = (\delta_0\Psi, \delta_1\Psi) \in H^{-3/2}(\Gamma) \times H^{-1/2}(\Gamma)$ with

$$\delta_0\psi = K_n(\psi) \downarrow_\Gamma \in H^{-3/2}(\Gamma), \quad \delta_1(\psi) = M_n(\Psi) \downarrow_\Gamma \in H^{-1/2}(\Gamma) \tag{8.14.178}$$

such that $\forall$ fixed $\Psi = \mathbb{L}^2(\Omega), \forall \boldsymbol{\chi} = (\chi_0, \chi_1) \in H^{3/2}(\Gamma) \times H^{1/2}(\Gamma)$,

$$T_\Psi(\boldsymbol{\chi}) = [\delta_0\psi, \chi_0]_{H^{-3/2}(\Gamma)\times H^{3/2}(\Gamma)} - [\delta_1\Psi, \chi_1]_{H^{-1/2}(\Gamma)\times H^{1/2}(\Gamma)}. \tag{8.14.179}$$

Then Green's formula (8.14.177) has the form: $\forall\Psi \in \mathbb{L}^2(\Omega)$, $\forall v \in H^2(\Omega)$ with $\gamma_0 v = v \downarrow_\Gamma \in H^{3/2}(\Gamma)$, $\gamma_1 v = \frac{\partial v}{\partial n} \downarrow_\Gamma \in H^{1/2}(\Gamma)$,

$$\begin{aligned}&\langle\psi_{ij,ij}, v\rangle_{0,\Omega} - \langle\!\langle(\psi_{ij}), (v_{,ij})\rangle\!\rangle_{0,\Omega}\\ &= [\delta_0\Psi, \gamma_0 v]_{H^{-3/2}(\Gamma)\times H^{3/2}(\Gamma)} - [\delta_1\Psi, \gamma_1 v]_{H^{-1/2}(\Gamma)\times H^{1/2}(\Gamma)},\end{aligned} \tag{8.14.180}$$

where $\delta_0\Psi = K_n(\Psi) \downarrow_\Gamma \in H^{-3/2}(\Gamma)$, $\delta_1\Psi = M_n(\Psi) \downarrow_\Gamma \in H^{-1/2}(\Gamma)$ with

$$\|\delta_0\Psi\|_{H^{-3/2}(\Gamma)} \le C\,|\!|\!|\Psi|\!|\!|_{0,\Omega}; \quad \|\delta_1\Psi\|_{H^{-1/2}(\Gamma)} \le C\,|\!|\!|\Psi|\!|\!|_{0,\Omega} \quad \forall\Psi \in \mathbb{L}^2(\Omega); \tag{8.14.181}$$

and

$$\begin{aligned}&\|\boldsymbol{\delta}\Psi\|_{H^{-3/2}(\Gamma)\times H^{-1/2}(\Gamma)}\\ &= \|(\delta_0\Psi, \delta_1\Psi)\|_{H^{-3/2}(\Gamma)\times H^{-1/2}(\Gamma)} \le C\,|\!|\!|\Psi|\!|\!|_{0,\Omega} \quad \forall\Psi \in \mathbb{L}^2(\Omega).\end{aligned} \tag{8.14.182}$$

Trace results in the space $H^2(\Lambda, \Omega)$

Let $u \in H^2(\Lambda, \Omega)$. Then $\psi_{ij} = a_{ijkl}u_{,kl} \in L^2(\Omega)$ with $\psi_{ij} = \psi_{ji}\ \forall i, j = 1, 2$ and $\psi_{ij,ij} = (a_{ijkl}u_{,kl})_{,ij} = \Lambda u \in L^2(\Omega) \Longrightarrow \Psi = (\psi_{ij}) \in \mathbb{L}^2(\Omega)$.

Hence, we can now define the *trace operator* $\boldsymbol{\nu} : H^2(\Lambda, \Omega) \longrightarrow H^{-3/2}(\Gamma) \times H^{-1/2}(\Gamma)$ by: $\forall u \in H^2(\Lambda, \Omega)$ with $\psi_{ij} = a_{ijkl}u_{,kl} \in L^2(\Omega)$,

$$\boldsymbol{\nu}u = (\nu_0 u, \nu_1 u) = (\delta_0\Psi, \delta_1\Psi) \in H^{-3/2}(\Gamma) \times H^{-1/2}(\Gamma) \tag{8.14.183}$$

with

$$\nu_0 u = \delta_0 \Psi = K_n(\Psi) \downarrow_\Gamma$$
$$= \left(\frac{\partial}{\partial t}[(a_{ijkl}u_{,kl})n_i t_j] + [((a_{ijkl}u_{,kl})_{,i})n_j]\right) \downarrow_\Gamma \in H^{-3/2}(\Gamma); \quad (8.14.184)$$

$$\nu_1 u = \delta_1 \Psi = M_n(\Psi) \downarrow_\Gamma = ((a_{ijkl}u_{,kl})n_i n_j) \downarrow_\Gamma \in H^{-1/2}(\Gamma). \quad (8.14.185)$$

Now we can state the final trace results:

Theorem 8.14.14 (Trace theorem in $H^2(\Lambda, \Omega)$). *Let Ω be a C^∞-regular domain with the boundary $\Gamma \in C^\infty$-class and Λ be the aniso-/ortho-/isotropic elastic thin plate bending operator defined by* (8.14.148) *with coefficients satisfying* (A1)–(A3). *Then:*

I. *The trace operator $\boldsymbol{\nu} : H^2(\Lambda, \Omega) \longrightarrow H^{-3/2}(\Gamma) \times H^{-1/2}(\Gamma)$ defined by* (8.14.183)–(8.14.185) *is* linear *and* continuous *from $H^2(\Lambda, \Omega)$ into $H^{-3/2}(\Gamma) \times H^{-1/2}(\Gamma)$.*

II. *Consequently, Green's formula* (8.14.180) *has the following form: $\forall u \in H(\Lambda, \Omega)$ with $\psi_{ij} = a_{ijkl}u_{,kl}$ $\forall i, j = 1, 2$, and $\forall v \in H^2(\Omega)$,*

$$\int_\Omega (\Lambda u) v \, d\mathbf{x} - \int_\Omega \psi_{ij} v_{,ij} \, d\mathbf{x}$$
$$= [\nu_0 u, \gamma_0 v]_{H^{-3/2}(\Gamma) \times H^{3/2}(\Gamma)} - [\nu_1 u, \gamma_1 v]_{H^{-1/2}(\Gamma) \times H^{1/2}(\Gamma)},$$

where

$$\nu_0 u = K_n(\Psi) \downarrow_\Gamma$$
$$= \left(\frac{\partial}{\partial t}[(a_{ijkl}u_{,kl})n_i t_j] + [((a_{ijkl}u_{,kl})_{,i})n_j]\right) \downarrow_\Gamma \in H^{-3/2}(\Gamma);$$
$$\nu_1 u = M_n(\Psi) \downarrow_\Gamma = (a_{ijkl}u_{,kl})n_i n_j) \downarrow_\Gamma \in H^{-1/2}(\Gamma);$$

$K_n(\Psi)$ and $M_n(\Psi)$ are defined by (8.14.152c) *and* (8.14.152a), *respectively.* $\gamma_0 v = v \downarrow_\Gamma \in H^{3/2}(\Gamma)$, $\gamma_1 v = \frac{\partial v}{\partial n} \downarrow_\Gamma \in H^{1/2}(\Gamma)$.

Proof. The linearity of $\boldsymbol{\nu} = (\nu_0, \nu_1)$ follows immediately from the definition in (8.14.183)–(8.14.185). For its *continuity*, we have: $\forall u \in H^2(\Lambda, \Omega)$ with $\psi_{ij} = a_{ijkl}u_{,kl}$ and $\Psi = (\psi_{ij}) \in \mathbb{L}^2(\Omega)$, from (8.14.181)–(8.14.182), we have

$$\|\nu_0 u\|_{H^{-3/2}(\Gamma)} = \|\delta_0 \Psi\|_{H^{-3/2}(\Gamma)} \le C \, |||\Psi|||_{0,\Omega};$$
$$\|\nu_1 u\|_{H^{-1/2}(\Gamma)} = \|\delta_1 \Psi\|_{H^{-1/2}(\Gamma)} \le C \, |||\Psi|||_{0,\Omega};$$

and

$$\|\boldsymbol{\nu} u\|_{H^{-3/2}(\Gamma) \times H^{-1/2}(\Gamma)} = \|(\nu_0 u, \nu_1 u)\|_{H^{-3/2}(\Gamma) \times H^{-1/2}(\Gamma)} \le C \, |||\Psi|||_{0,\Omega}. \quad (8.14.186)$$

Using the properties (A1) and (A2) of the coefficients a_{ijkl}, we have

$$\psi_{ij}^2 = (a_{ijkl}u_{,kl})^2 = (a_{ij11}u_{,11} + 2a_{ij12}u_{,12} + a_{ij22}u_{,22})^2$$

$\Longrightarrow\ \psi_{ij}^2 \le C_0[u_{,11}^2 + (u_{,12})^2 + (u_{,22})^2]$ for *some* $C_0 > 0$, $\forall i, j = 1, 2$ and $\psi_{ij,ij} = (a_{ijkl}u_{,kl})_{,ij} = \Lambda u$ with Λ defined by (8.14.148).

Then

$$|||\Psi|||_{0,\Omega}^2 = \|\Psi\|_{0,\Omega}^2 + \|\psi_{ij,ij}\|_{0,\Omega}^2 = \int_\Omega \sum_{i,j=1}^{2} \psi_{ij}^2 d\mathbf{x} + \|\Lambda u\|_{0,\Omega}^2$$

$$\le C_1(\|u\|_{2,\Omega}^2 + \|\Lambda u\|_{0,\Omega}^2) = C_1\|u\|_{2,\Lambda,\Omega}^2 \quad \text{with some } C_1 > 1.$$

Hence, from (8.14.186), $\forall u \in H^2(\Lambda, \Omega)$, $\|\nu_0 u\|_{H^{-3/2}(\Gamma)} \le C\|u\|_{2,\Lambda,\Omega}$; $\|\nu_1 u\|_{H^{-1/2}(\Gamma)} \le C\|u\|_{2,\Lambda,\Omega}$ and

$$\|\boldsymbol{\nu} u\|_{H^{-3/2}(\Gamma)\times H^{-1/2}(\Gamma)} = \|(\nu_0 u, \nu_1 u)\|_{H^{-3/2}(\Gamma)\times H^{-1/2}(\Gamma)} \le C\|u\|_{2,\Lambda,\Omega}.$$

(The same $C > 0$ is used to denote *different* values of C at *different* inequalities.)

$\Longrightarrow$ the trace operator $\boldsymbol{\nu}$ is *continuous* from $H^2(\Lambda, \Omega)$ into $H^{-3/2}(\Gamma)\times H^{-1/2}(\Gamma)$.

□

8.14.8 Traces on $\Gamma_0 \subset \Gamma$

Let $\Gamma \in C^\infty$-class be the smooth boundary of a bounded domain $\Omega \subset \mathbb{R}^n$, and $\Gamma_0 \subsetneq \Gamma$ be an *open* subset of Γ. Then we may apply the *restriction* results of Theorem 8.10.23 (see also Theorem 8.13.8 and (8.13.52)–(8.13.53), (8.13.81)–(8.13.82)) to define the trace results on Γ_0 by:

$$u\!\downarrow_{\Gamma_0} = (u\!\downarrow_\Gamma)\!\downarrow_{\Gamma_0}, \quad \frac{\partial u}{\partial n}\!\downarrow_{\Gamma_0} = \left(\frac{\partial u}{\partial n}\!\downarrow_\Gamma\right)\!\downarrow_{\Gamma_0}, \quad \frac{\partial u}{\partial n_A}\!\downarrow_{\Gamma_0} = \left(\frac{\partial u}{\partial n_A}\!\downarrow_\Gamma\right)\!\downarrow_{\Gamma_0},$$

where the restrictions $(\cdot)\!\downarrow_\Gamma$ are to be understood as the *traces* of $(\cdot)$ on Γ, which are given by the *trace theorems* and *trace results* of this section, and the restrictions $(\cdot)\!\downarrow_{\Gamma_0}$ to Γ_0 of the distributions $(\cdot)\!\downarrow_\Gamma$ on Γ are obtained by applying Theorem 8.10.23. For example, we have:

1. for $s - \frac{1}{p} \neq$ an integer, $s = l + \sigma$ with $l \in \mathbb{N}_0$, $0 < \sigma < 1$ and $s - l - \frac{1}{p} > 0$, and for $u \in W^{s,p}(\Omega)$, $u\!\downarrow_\Gamma = \gamma_0^* u \in W^{s-1/p,p}(\Gamma)$ and $\frac{\partial^j u}{\partial n^j}\!\downarrow_\Gamma = \gamma_j^* u \in W^{s-j-1/p,p}(\Gamma)$, $1 \le j \le l$ by Trace Theorem 8.14.5 (since $\Gamma \in C^\infty$-class $\Longrightarrow \Gamma \in C^{k,1}$-class $\forall k \in \mathbb{N}$) $\Longrightarrow u\!\downarrow_{\Gamma_0} = (u\!\downarrow_\Gamma)\!\downarrow_{\Gamma_0} \in W^{s-1/p,p}(\Gamma_0)$ and $\frac{\partial^j u}{\partial n^j}\!\downarrow_{\Gamma_0} = (\frac{\partial^j u}{\partial n^j}\!\downarrow_\Gamma)\!\downarrow_{\Gamma_0} \in W^{s-j-1/p,p}(\Gamma_0)$ by Theorem 8.10.23; (8.14.187)

2. for $\mathbf{w} \in L^p(\text{div}; \Omega)$, $\mathbf{w}\cdot\hat{\mathbf{n}}\!\downarrow_\Gamma \in W^{-1/p,p}(\Gamma)$ by Trace Theorem 8.14.11 $\Longrightarrow$ for $p \neq 2$, $\mathbf{w}\cdot\hat{\mathbf{n}}\!\downarrow_{\Gamma_0} = (\mathbf{w}\cdot\hat{\mathbf{n}}\!\downarrow_\Gamma)\!\downarrow_{\Gamma_0} \in W^{-1/p,p}(\Gamma_0)$, and for $p = 2$, $\mathbf{w}\cdot\hat{\mathbf{n}}\!\downarrow_{\Gamma_0} = (\mathbf{w}\cdot\hat{\mathbf{n}}\!\downarrow_\Gamma)\!\downarrow_{\Gamma_0} \in (H_{00}^{1/2}(\Gamma_0))'$ by Theorem 8.10.23; (8.14.188)

3. for $u \in W^{1,p}(\Omega)$ with $Au \in L^p(\Omega)$, $1 < p < \infty$, A being the *elliptic* operator in (8.14.121), the *conormal derivative* $\frac{\partial u}{\partial n_A}$ defined by (8.14.125) has a *trace* on Γ : $\frac{\partial u}{\partial n_A}\downarrow_\Gamma \in W^{-1/p,p}(\Gamma)$ by (8.14.127) $\Longrightarrow$ $\frac{\partial u}{\partial n_A}\downarrow_{\Gamma_0} \in W^{-1/p,p}(\Gamma_0)$ for $p \neq 2$ and $\frac{\partial u}{\partial n_A}\downarrow_{\Gamma_0} \in (H_{00}^{1/2}(\Gamma_0))'$ for $p = 2$ by Theorem 8.10.23 (see also (8.14.188)); (8.14.189)
4. for $u \in H^2(\Lambda, \Omega)$ with $H^2(\Lambda, \Omega)$ defined by (8.14.158)–(8.14.159) and symmetric *tensor-valued function* $\Psi = (\psi_{ij})_{\leq i,j \leq 2}$ with $\psi_{ij} = \psi_{ji} = a_{ijkl}u_{,kl}$, $1 \leq i, j, k, l \leq 2$, defined in (8.14.151), $\nu_1 u = M_n(\Psi)\downarrow_\Gamma \in H^{-1/2}(\Gamma)$, $\nu_0 u = K_n(\Psi)\downarrow_\Gamma \in H^{-3/2}(\Gamma)$ by Trace Theorem 8.14.14 $\Longrightarrow$ $M_n(\Psi)\downarrow_{\Gamma_0} \in (H_{00}^{1/2}(\Gamma_0))'$, $K_n(\Psi)\downarrow_{\Gamma_0} \in (H_{00}^{3/2}(\Gamma_0))'$ by Theorem 8.10.23, (8.14.190)
   where $M_n(\Psi)$ and $K_n(\Psi)$ are defined by (8.14.152a) and (8.14.152c) respectively; $(H_{00}^{1/2}(\Gamma_0))'$ and $(H_{00}^{3/2}(\Gamma_0))'$ are defined by (8.13.83).

Chapter 9
Vector-valued distributions

9.1 Motivation

Up to now we have studied distributions $T \in \mathcal{D}'(\Omega)$ having scalar values, i.e. $T(\phi) \in \mathbb{R}$ (resp. $\mathbb{C}$) $\forall \phi \in \mathcal{D}(\Omega)$. But in the study of evolution problems (i.e. time-dependent problems) of parabolic and hyperbolic equations of mathematical physics, vector-valued functions and vector-valued distributions arise. Hence, our approach to the presentation of vector-valued distributions will be motivated by the specific requirement that they will be used in the study of evolution (time-dependent) equations and we will completely refrain from a general treatment, which has some additional new difficulties, for which we refer to the original source as usual, Laurent Schwartz [8], [29] and also [12].

9.2 Vector-valued functions

We begin with the simplest example of vector-valued functions.

Consider the scalar-valued functions $v_i : t \in \mathbb{R} \longmapsto v_i(t) \in \mathbb{R}$, $1 \leq i \leq n$, the value $v_i(t)$ of the function v_i being a real number, i.e. v_i is a scalar-valued function. Let $\mathbf{v}(t)$ denote the vector $(v_1(t), v_2(t), \ldots, v_n(t)) \in \mathbb{R}^n$, i.e.

$$\mathbf{v}(t) = (v_1(t), v_2(t) \ldots v_n(t)) \quad \forall t. \tag{9.2.1}$$

Then the mapping $t \in \mathbb{R} \longmapsto \mathbf{v}(t) \in \mathbb{R}^n$ is a *vector-valued function*, since the value $\mathbf{v}(t)$ of the function $t \longmapsto \mathbf{v}(t)$ is an element of $\mathbb{R}^n$ and hence a *vector* (*not* a number). Now we can generalize this definition by replacing $\mathbb{R}^n$ by a *Banach space* V, whose elements are vectors. Then a mapping $t \in]0,T[\longmapsto w(t) \in V$ defines a *vector-valued function* from $]0,T[$ into V with $0 < T \leq \infty$, since the value $w(t)$ of the function is an element of V, i.e. a vector in V for $t \in]0,T[$ ($w(t)$ is *not* a number). To fix our ideas, we give some examples:

1. Let $V = L^2(\Omega)$, where $\Omega \subset \mathbb{R}^n$ is an open subset of $\mathbb{R}^n$.
   Let $u : (\mathbf{x}, t) \in \Omega \times]0,T[\longmapsto u(\mathbf{x}, t) \in \mathbb{R}$ be a scalar-valued function from $\Omega \times]0,T[$ into $\mathbb{R}$, i.e. a function of space variable $\mathbf{x} \in \Omega$ and time variable $t \in]0,T[\subset \mathbb{R}$, such that:
   - *for almost all* $t \in]0,T[$, the mapping $t \in]0,T[\longmapsto w(t) = u(\cdot, t) \in L^2(\Omega)$ is a vector-valued function with its value $w(t) = u(\cdot, t) \in L^2(\Omega)$ for almost all $t \in]0,T[$; and (9.2.2)

- $$\|w(t)\|^2_{L^2(\Omega)} = \|u(\cdot\,,t)\|^2_{L^2(\Omega)} = \int_\Omega |u(\mathbf{x},t)|^2 d\mathbf{x} \tag{9.2.3}$$

  for almost all $t \in \,]0,T[$;
- *for almost all* $t \in \,]0,T[$,

$$w(t) = u(\cdot\,,t), \quad z(t) = v(\cdot\,,t) \in L^2(\Omega),$$

$$\langle w(t), z(t)\rangle_{L^2(\Omega)} = \langle u(\cdot\,,t), v(\cdot\,,t)\rangle_{L^2(\Omega)} = \int_\Omega u(\mathbf{x},t)v(\mathbf{x},t)d\mathbf{x}. \tag{9.2.4}$$

Here, we have considered the mapping $t \in \,]0,T[\, \longmapsto w(t)$ whose values $w(t)$ are *vectors* of $L^2(\Omega)$ (and *not* numbers).

2. Let $V = C^0(\overline{\Omega})$, where Ω is an open, bounded subset of $\mathbb{R}^n$ and $\overline{\Omega}$ is its *closure*. Let $u : (\mathbf{x},t) \in \overline{\Omega} \times [0,T] \longmapsto u(\mathbf{x},t) \in \mathbb{R}$ be a scalar-valued function from $\overline{\Omega} \times [0,T]$ into $\mathbb{R}$, i.e. a function of space variable $\mathbf{x} \in \overline{\Omega}$ and time variable $t \in [0,T] \subset \mathbb{R}$. Then:

   - $\forall t \in [0,T]$, the mapping $t \in [0,T] \longmapsto w(t) = u(\cdot\,,t) \in C^0(\overline{\Omega})$ is a vector-valued function with its value $w(t) = u(\cdot\,,t) \in C^0(\overline{\Omega})$; (9.2.5)
   - $$\|w(t)\|_{C^0(\overline{\Omega})} = \|u(\cdot\,,t)\|_{C^0(\overline{\Omega})} = \sup_{\mathbf{x}\in\overline{\Omega}} |u(\mathbf{x},t)| \quad \forall t \in [0,T]. \tag{9.2.6}$$

   Here again, we have considered the mapping $t \in [0,T,] \longmapsto w(t)$ whose values $w(t)$ are *vectors* of $C^0(\overline{\Omega})$ (and *not* numbers).

Now we will study relations between scalar-valued functions in $C^0(\overline{\Omega} \times [0,T])$ and $L^2(\Omega \times]0,T[)$ which are Banach spaces with appropriate norms, and the corresponding vector-valued functions in (9.2.5)–(9.2.6) and (9.2.2)–(9.2.4), respectively.

$C^0(\overline{\Omega} \times [0,T])$ is a Banach space of scalar-valued functions $u : (\mathbf{x};t) \in \Omega \times \,]0,T[\, \longmapsto u(\mathbf{x};t) \in \mathbb{R}$ which are bounded and uniformly continuous in $\Omega \times \,]0,T[$ and have unique continuous extension to $\overline{\Omega} \times [0,T]$ (the extended continuous function is denoted by the same notation u) with norm

$$\|u\|_{C^0(\overline{\Omega}\times[0,T])} = \sup_{(\mathbf{x};t)\in\Omega\times]0,T[} |u(\mathbf{x};t)|. \tag{9.2.7}$$

Then $u \in C^0(\overline{\Omega} \times [0,T])$ if and only if

- $\forall t \in [0,T]$, function $t \in [0,T] \longmapsto w(t) = u(\cdot\,,t) \in C^0(\overline{\Omega})$; (9.2.8)
- $\forall t$ and $\forall$ sequences (t_n) in $[0,T]$, $t_n \to t$ in $[0,T] \Longrightarrow w(t_n) \longrightarrow w(t)$ in $C^0(\overline{\Omega}) \Longrightarrow \|w(t) \to w(t_n)\|_{C^0(\overline{\Omega})} \to 0 \Longrightarrow \|u(\cdot\,,t) - u(\cdot\,,t_n)\|_{C^0(\overline{\Omega})} \to 0.$ (9.2.9)

Here, instead of the original scalar-valued function $u : (\mathbf{x}; t) \longmapsto u(\mathbf{x}; t) \in \mathbb{R}$, we are considering the vector-valued function $t \in [0, T] \longmapsto w(t) \in C^0(\overline{\Omega})$, $w(t)$ being a vector in $C^0(\overline{\Omega})$. (9.2.8)–(9.2.9) are the properties of this vector-valued function $t \longrightarrow w(t)$ with (9.2.9) defining the continuity of the mapping $t \longmapsto w(t)$ from $[0, T]$ into $C^0(\overline{\Omega})$.

$L^2(\Omega \times]0, T[)$ is a Banach (in fact, Hilbert) space of (equivalence classes of) scalar-valued functions $u : (\mathbf{x}; t) \in \Omega \times]0, T[\longmapsto u(\mathbf{x}; t) \in \mathbb{R}$, which are defined a.e. and measurable[1] in $\Omega \times]0, T[$ and

$$\iint_{\Omega \times]0,T[} |u|^2 d\mathbf{x} dt = \iint_{\Omega \times]0,T[} |u(\mathbf{x}; t)|^2 d\mathbf{x} dt < +\infty. \tag{9.2.10}$$

Then, using Fubini's Theorem 7.1.2C, and under the necessary measurability conditions, we have: $u \in L^2(\Omega \times]0, T[)$ if and only if

- *for almost all* $t \in]0, T[$, $t \in]0, T[\longmapsto u(\mathbf{x}; t)$ is measurable and

$$\int_\Omega |u(\mathbf{x}; t)|^2 d\mathbf{x} < +\infty; \tag{9.2.11}$$

- $$\int_0^T \left(\int_\Omega |u(\mathbf{x}; t)|^2 d\mathbf{x} \right) dt = \iint_{\Omega \times]0,T[} |u|^2 d\mathbf{x} dt < +\infty, \tag{9.2.12}$$

which can be replaced by the following conditions, i.e. $u \in L^2(\Omega \times]0, T[)$ if and only if

- *for almost all* $t \in]0, T[$, $t \longmapsto w(t) = u(\cdot, t) \in L^2(\Omega)$ with

$$\|w(t)\|^2_{L^2(\Omega)} = \int_\Omega |w(t)|^2 d\mathbf{x} = \|u(\cdot, t)\|^2_{L^2(\Omega)} = \int_\Omega |u(\mathbf{x}; t)|^2 d\mathbf{x} < +\infty; \tag{9.2.13}$$

- $$\int_0^T \|w(t)\|^2_{L^2(\Omega)} dt = \int_0^T \|u(\cdot, t)\|^2_{L^2(\Omega)} dt < +\infty. \tag{9.2.14}$$

Here again, instead of the original scalar valued function $(\mathbf{x}; t) \longmapsto u(\mathbf{x}; t)$, we are considering the vector-valued function $t \longmapsto w(t) = u(\cdot, t) \in L^2(\Omega)$ for almost all $t \in]0, T[$ in (9.2.13)–(9.2.14), the property (9.2.14) being the square integrability of the function $t \longmapsto w(t) \in L^2(\Omega)$ from $]0, T[$ into $L^2(\Omega)$ for almost all $t \in]0, T[$.

[1] The reader who is not acquainted with measure theory may *safely* ignore the standard statements on measurability throughout this chapter, considering the fact that all functions in applications are measurable, and non-measurable functions can *not* be written in explicit form and their existence is based on the axiom of choice. Otherwise, see Appendix B, which contains elementary results on Lebesgue measures and integration. See also Rudin [27], Halmos [49].

Continuity of vector-valued functions

Definition 9.2.1. A vector-valued function $w :]0,T[\longrightarrow V$, V being a Banach space, is called *continuous* at $t \in]0,T[$ if and only if $t_n \to t$ in $]0,T[\Longrightarrow w(t_n) \to w(t)$ in V, i.e.

$$\|w(t) - w(t_n)\|_V \to 0 \quad \text{as } n \to \infty. \tag{9.2.15}$$

A vector-valued function $w :]0,T[\longrightarrow V$ is called *continuous* on $]0,T[$ iff it is continuous at each $t \in]0,T[$.

Derivative of vector-valued functions (in the strong sense)

Definition 9.2.2. The usual derivative $w'(t)$ (in the *strong sense*), if it exists, of $w :]0,T[\longrightarrow V$ is defined by:

$$w'(t) = \lim_{\Delta t \to 0} \frac{w(t+\Delta t) - w(t)}{\Delta t} \quad \text{in } V. \tag{9.2.16}$$

Then higher-order derivatives are defined by:

$$w'' = (w')',\ w''' = (w'')',\ \ldots \quad \text{with } w^{(0)} = w. \tag{9.2.17}$$

9.3 Spaces of vector-valued functions

Let V be a Banach space and V' be its (topological) dual space (i.e. also a Banach space) with $\langle \cdot,\cdot \rangle_{V' \times V}$ denoting the duality pairing between V' and V.

- $C^0(]0,T[;V)$ is the linear space of *vector-valued functions* $w : t \in]0,T[\longmapsto w(t) \in V$ which are continuous from $]0,T[$ into V in the sense of (9.2.15), i.e. $w \in C^0(]0,T[;V)$

$$\Longrightarrow \quad w(t) \in V\ \forall t \in]0,T[\text{ and } t_n \to t \text{ in }]0,T[\Longrightarrow w(t_n) \to w(t) \text{ in } V; \tag{9.3.1}$$

- $C^0([0,T];V)$ is the linear space of vector-valued functions $w \in C^0(]0,T[;V)$ which have continuous extension to $[0,T]$ (the extended vector-valued function will be denoted by the same notation w); (9.3.2)
- $C^k([0,T];V)$ is the linear space of vector-valued functions $w \in C^0([0,T];V)$ whose jth derivative (in the sense of (9.2.16)–(9.2.17)) $w^{(j)} \in C^0([0,T];V)$, $0 \le j \le k$; (9.3.3)
- $C^k(]0,T[;V)$ is the linear space of vector-valued functions $w \in C^0(]0,T[;V)$ whose jth derivative (in the sense of (9.2.16)–(9.2.17)) $w^{(j)} \in C^0(]0,T[;V)$, $0 \le j \le k$; (9.3.4)

- $C^\infty(]0,T[;V)$ is the linear space of vector-valued functions $w \in C^k(]0,T[;V)$ $\forall k \in \mathbb{N}$; (9.3.5)
- $\mathcal{D}(]0,T[;V) = C_0^\infty(]0,T[;V)$ is the linear space of vector-valued functions $\phi \in C^\infty(]0,T[;V)$ which have *compact* support in $]0,T[$. (9.3.6)

Space $L^p(0,T;V) = L^p(]0,T[;V)$, $1 \le p \le \infty$

Measurability A vector-valued function $w :]0,T[\longrightarrow V$ defined a.e. on $]0,T[$ is *measurable* if and only if $\exists$ a sequence (w_n) with $w_n \in C^0(]0,T[;V)$ $\forall n \in \mathbb{N}$ such that $w_n(t) \to w(t)$ *strongly* in V (i.e. $\|w(t) - w_n(t)\|_V \to 0$) *for almost all* $t \in]0,T[$. (9.3.7)

Consequence

- $t \longmapsto w(t)$ is measurable $\Longrightarrow t \longmapsto \|w(t)\|_V$ is measurable; (9.3.8)
- $\int_0^T \|w(t)\|_V^p dt$ is defined (possibly $= \infty$) $\forall$ measurable w and for $1 \le p < \infty$. (9.3.9)

Definition 9.3.1. For $1 \le p \le \infty$, $L^p(0,T;V) \equiv L^p(]0,T[;V)$ is the linear space of all (*equivalence classes* of) vector-valued functions $w :]0,T[\longmapsto w(t) \in V$ which are measurable such that $\|w(t)\|_V \in L^p(]0,T[)$, i.e.

- for $1 \le p < \infty$,

$$L^p(0,T;V) = \left\{ w : w :]0,T[\longmapsto V \text{ is measurable, } \int_0^T \|w(t)\|_V^p dt < +\infty \right\} \tag{9.3.10}$$

equipped with its natural norm: $\forall w \in L^p(0,T;V)$,

$$\|w\|_{L^p(0,T;V)} = \left(\int_0^T \|w(t)\|_V^p dt \right)^{\frac{1}{p}}; \tag{9.3.11}$$

- for $p = \infty$,

$$L^\infty(0,T;V) = \{ w : w :]0,T[\longmapsto V \text{ is measurable, } \|w(t)\|_V \text{ is } \textit{essentially} \\ \textit{bounded} \text{ on }]0,T[, \text{ i.e. } \|w(t)\|_V \in L^\infty(]0,T[) \} \tag{9.3.12}$$

is equipped with the norm: $\forall w \in L^\infty(0,T;V)$,

$$\|w\|_{L^\infty(0,T;V)} = \operatorname*{ess\,sup}_{0<t<T} \|w(t)\|_V. \tag{9.3.13}$$

Proposition 9.3.1.

I. *For* $1 \leq p \leq \infty$, $L^p(0, T; V)$ *defined in* (9.3.10)–(9.3.13) *is a* Banach space.

II. *If* V *is a Hilbert space with inner product* $\langle \cdot, \cdot \rangle_V$ and $p = 2$, *then* $L^p(0, T; V) = L^2(0, T; V)$ *is a* Hilbert space *equipped with inner product* $\langle \cdot, \cdot \rangle_{L^2(0,T;V)}$ *defined,* $\forall u, v \in L^2(0, T; V)$, *by:*

$$\langle u, v \rangle_{L^2(0,T;V)} = \int_0^T \langle u(t), v(t) \rangle_V dt. \tag{9.3.14}$$

III. *For a* separable *Banach space* V *and for* $1 \leq p < \infty$, $(L^p(0, T; V))' = L^q(0, T; V')$ *with* $\frac{1}{p} + \frac{1}{q} = 1$, $1 < q \leq \infty$ *such that*

$$\langle w, z \rangle_{L^q(0,T;V') \times L^p(0,T;V)} = \int_0^T \langle w(t), z(t) \rangle_{V' \times V} dt. \tag{9.3.15}$$

IV. *If* V *is* reflexive $(V'' = V)$ *and* $1 < p < \infty$, *then* $L^p(0, T; V)$ *is* reflexive. (9.3.16)

V. *For* $p = 1$, *the following properties hold in* $L^1(0, T; V)$: (9.3.17)

- $w \in L^1(0, T; V)$

$$\Longrightarrow \int_0^T w(t)dt \in V \text{ is defined and belongs to } V. \tag{9.3.18}$$

- $$\left\| \int_0^T w(t)dt \right\|_V \leq \int_0^T \|w(t)\|_V dt.$$

- $\forall L \in V'$,

$$\left\langle L, \int_0^T w(t)dt \right\rangle_{V' \times V} = \int_0^T \langle L, w(t) \rangle_{V' \times V} dt. \tag{9.3.19}$$

Remark 9.3.1.

- For $V = V(\Omega)$, i.e. a Banach space of functions (of space variable $\mathbf{x}$) defined (a.e.) on Ω, $u \in L^1(0, T; V(\Omega)) \Longrightarrow w(t) = u(\cdot, t) \in V(\Omega)$ and

$$\int_0^T w(t)dt = \int_0^T u(\cdot, t)dt \in V(\Omega) \tag{9.3.20}$$

has the usual meaning: as a result of integration with respect to t, a function $v(\cdot) = \int_0^T u(\cdot, t)dt$ of the space variable $\mathbf{x} \in \Omega$ is obtained, which belongs to $V(\Omega)$. (9.3.21)

- For *bounded* $]0, T[\subset \mathbb{R}$, $L^p(]0, T[) \hookrightarrow L^1(]0, T[)$ for $1 \leq p \leq \infty$

$$\Longrightarrow \quad L^p(0, T; V) \hookrightarrow L^1(0, T; V), 1 \leq p \leq \infty \tag{9.3.22}$$

$$\Longrightarrow \quad \langle L, \int_0^T w(t)dt \rangle_{V' \times V} = \int_0^T \langle L, w(t) \rangle_{V' \times V} dt \quad \forall L \in V', \tag{9.3.23}$$

$\forall w \in L^p(0, T; V) \hookrightarrow L^1(0, T; V)$, $1 \leq p \leq \infty$.

9.4 Vector-valued distributions

Let

$$\mathcal{D}(]0,T[) = C_0^\infty(]0,T[) = \{\phi : \phi \in C^\infty(]0,T[), \mathrm{supp}(\phi) \subset\subset]0,T[\}, \quad (9.4.1)$$

a compact subset of $]0,T[$.

Definition 9.4.1. A *continuous, linear mapping* $S : \mathcal{D}(]0,T[) \longmapsto V$ from $\mathcal{D}(]0,T[)$ into Banach space V is called a *vector-valued distribution on* $]0,T[$ with its value in a Banach space V, i.e.

- $S : \mathcal{D}(]0,T[) \longmapsto V$ is a *linear* mapping with its value $S(\phi) \in V \ \forall \phi \in \mathcal{D}(]0,T[)$;
- *continuity* of S:

$$\begin{aligned} \phi_n \to \phi \text{ on } \mathcal{D}(]0,T[) \quad &\Longrightarrow \quad S(\phi_n) \to S(\phi) \text{ in } V \\ &\Longrightarrow \quad \|S(\phi) - S(\phi_n)\|_V \to 0. \end{aligned} \quad (9.4.2)$$

Space $\mathcal{D}'(]0,T[;V)$ of vector-valued distributions

Definition 9.4.2. The set of all *continuous, linear mappings* S from $\mathcal{D}(]0,T[)$ into V form a *vector space of vector-valued distributions on* $(]0,T[)$ with their values in V, which is denoted by $\mathcal{D}'(]0,T[;V)$. Then, by definition,

$$\mathcal{D}'(]0,T[;V) = \mathcal{L}(\mathcal{D}(]0,T[);V), \quad (9.4.3)$$

$\mathcal{L}(\mathcal{D}(]0,T[);V)$ being the linear space of continuous linear mappings from $(\mathcal{D}(]0,T[)$ into V. $\forall S \in \mathcal{D}'(]0,T[;V)$, $\forall \phi \in \mathcal{D}(]0,T[)$,

$$S(\phi) = S \cdot \phi = \langle S, \phi \rangle \in V, \quad (9.4.4)$$

$S \cdot \phi$ and $\langle S, \phi \rangle$ being alternative, *equivalent* notations for $S(\phi)$, which will be used interchangeably.

Equality of vector-valued distributions in $\mathcal{D}'(]0,T[;V)$

$S_1, S_2 \in \mathcal{D}'(]0,T[;V)$,

$$S_1 = S_2 \quad \Longleftrightarrow \quad S_1 \cdot \phi = S_2 \cdot \phi \text{ in } V \ \forall \phi \in \mathcal{D}(]0,T[). \quad (9.4.5)$$

Characterization of vector-valued distributions

Proposition 9.4.1. *Let* $S : \mathcal{D}(]0,T[) \longmapsto V$ *be a linear mapping from* $\mathcal{D}(]0,T[)$ *into Banach space* V. *Then the following properties are equivalent:*

I. S *is a vector-valued distribution on* $]0,T[$ *with its value in* V, *i.e.* $S \in \mathcal{D}'(]0,T[; V) = \mathcal{L}(\mathcal{D}(]0,T[); V)$.

II. $\forall$ *compact intervals* $K = [a,b] \subset\subset]0,T[$, *the restriction* $S_K = S\downarrow_{\mathcal{D}_K(]0,T[)}$ *of* S *to* $\mathcal{D}_K(]0,T[) = \{\phi : \phi \in \mathcal{D}(]0,T[), \operatorname{supp}(\phi) \subset K\}$ *is* continuous, *i.e.* $\forall K \subset\subset]0,T[$,

$$S_K \cdot \phi = S \cdot \phi \quad \forall \phi \in \mathcal{D}_K(]0,T[) \subset \mathcal{D}(]0,T[),$$
$$\|S_K \cdot \phi\|_V \leq C_K \sup_{t \in K} |\phi(t)| \quad \forall \phi \in \mathcal{D}_K(]0,T[). \tag{9.4.6}$$

($\Longrightarrow$ *[*$\forall K \subset\subset]0,T[$, $\phi_n \to \phi$ *in* $\mathcal{D}_K(]0,T[) \Longrightarrow \|S_K \cdot \phi - S_K \cdot \phi_n\|_V \leq C_K \sup_{t\in K} |\phi(t) - \phi_n(t)| \to 0$ *in* $n \to \infty$*]* $\Longrightarrow S_K : \mathcal{D}_K(]0,T[) \longrightarrow V$ *is continuous* $\forall K \subset\subset]0,T[\Longrightarrow S : \mathcal{D}(]0,T[) \longmapsto V$ *is continuous.)*

Convergence in $\mathcal{D}'(]0,T[; V)$

Definition 9.4.3. A sequence (S_n) in $\mathcal{D}'(]0,T[;V)$ is said to converge to $S \in \mathcal{D}'(]0,T[; V)$ if and only if

$$S_n \cdot \phi \to S \cdot \phi \text{ in } V \quad \text{as } n \to \infty \quad \forall \phi \in \mathcal{D}(]0,T[), \tag{9.4.7}$$

i.e.

$$\|S \cdot \phi - S_n \cdot \phi\|_V \to 0 \quad \text{as } n \to \infty \quad \forall \phi \in \mathcal{D}(]0,T[). \tag{9.4.8}$$

Proposition 9.4.2. *Let* V, W *be any two Banach spaces such that* $V \hookrightarrow W$, *the imbedding* $\hookrightarrow$ *being a continuous one from* V *into* W. *Then*

$$\mathcal{D}'(]0,T[; V) \hookrightarrow \mathcal{D}'(]0,T[; W), \tag{9.4.9}$$

i.e.

I. $S \in \mathcal{D}'(]0,T[; V) \Longrightarrow S \in \mathcal{D}'(]0,T[; W)$; (9.4.10)

II. $S_n \to S$ *in* $\mathcal{D}'(]0,T[; V) \Longrightarrow S_n \to S$ *in* $\mathcal{D}'(]0,T[; W) \Longrightarrow S_n \cdot \phi \to S \cdot \phi$ *in* W

$$\Longrightarrow \quad \|S \cdot \phi - S_n \cdot \phi\|_W \to 0 \quad \forall \phi \in \mathcal{D}(]0,T[). \tag{9.4.11}$$

Proof.

I. Let $S \in \mathcal{D}'(]0,T[;V)$. Then, $\forall \phi \in \mathcal{D}(]0,T[)$, $S\cdot\phi \in V \Longrightarrow S\cdot\phi \in W$, since $V \hookrightarrow W$. Hence, $S : \mathcal{D}(]0,T[) \longrightarrow W$ is linear. But $V \hookrightarrow W \Longrightarrow \exists C > 0$ such that $\|S\cdot\phi\|_W \le C\|S\cdot\phi\|_V$ $\forall\phi \in \mathcal{D}(]0,T[)$. $S \in \mathcal{D}'(]0,T[;V) \Longrightarrow S \in \mathcal{L}(\mathcal{D}'(]0,T[;V) \Longrightarrow [\phi_n \to \phi$ in $\mathcal{D}(]0,T[) \Longrightarrow S\cdot\phi_n \to S\cdot\phi$ in $V \Longrightarrow \|S\cdot\phi - S\cdot\phi_n\|_V \to 0$ as $n\to\infty]$.

Hence, $\phi_n \to \phi$ in $\mathcal{D}(]0,T[) \Longrightarrow \|S\cdot\phi - S\cdot\phi_n\|_W \le C\|S\cdot\phi - S\cdot\phi_n\|_V \to 0 \Longrightarrow S \in \mathcal{L}(\mathcal{D}(]0,T[;W) \Longrightarrow S \in \mathcal{D}'(]0,T[;W)$.

II. $S_n \to S$ in $\mathcal{D}'(]0,T[;V) \Longrightarrow S_n\cdot\phi \to S\cdot\phi$ in V $\forall\phi \in \mathcal{D}(]0,T[) \Longrightarrow \|S\cdot\phi - S_n\cdot\phi\|_V \to 0 \Longrightarrow \|S\cdot\phi - S_n\cdot\phi\|_W \le C\|S\cdot\phi - S_n\cdot\phi\|_V \to 0$ $\forall\phi \in \mathcal{D}(]0,T[) \Longrightarrow S_n \to S$ in $\mathcal{D}'(]0,T[;W)$.

Hence, $\mathcal{D}'(]0,T[;V) \hookrightarrow \mathcal{D}'(]0,T[;W)$. □

Proposition 9.4.3. *Let $L^1_{\text{loc}}(0,T;V) = \{u : \forall$ compact intervals $K = [a,b] \subset\subset]0,T[, \int_K \|u(t)\|_V dt < +\infty\}$. Then*

$$L^1_{\text{loc}}(0,T;V) \subset \mathcal{D}'(]0,T[;V), \tag{9.4.12}$$

i.e. $u \in L^1_{\text{loc}}(0,T;V) \Longrightarrow S_u \in \mathcal{D}'(]0,T[;V)$ is a vector-valued distribution on $]0,T[$ with values in V and defined by:

$$S_u\cdot\phi = \langle S_u,\phi\rangle = \int_0^T u(t)\phi(t)dt \in V \quad \forall\phi \in \mathcal{D}(]0,T[). \tag{9.4.13}$$

Proof. Let $u \in L^1_{\text{loc}}(0,T;V)$. Then, $\forall$ compact intervals $K = [a,b] \subset\subset]0,T[$, $\int_K \|u(t)\|_V dt = \int_a^b \|u(t)\|_V dt < +\infty$. But $\forall\phi \in \mathcal{D}(]0,T[)$, $u \in L^1_{\text{loc}}(0,T;V) \Longrightarrow \phi u \in L^1(0,T;V)$, since

$$\int_0^T \|\underbrace{\phi(t)}_{\in\mathbb{R}}\ \underbrace{u(t)}_{\in V}\|_V dt = \int_0^T |\phi(t)|\|u(t)\|_V dt$$

$$\le \sup_{t\in\text{supp}(\phi)} |\phi(t)| \int_{\text{supp}(\phi)} \|u(t)\|_V dt < +\infty.$$

Then, $\forall\phi \in \mathcal{D}(]0,T[)$, $\phi u \in L^1(0,T;V) \Longrightarrow \int_0^T u(t)\phi(t)dt$ is well defined as an element (vector) of V (by (9.3.18)). Hence, $u \in L^1_{\text{loc}}(0,T;V) \Longrightarrow S_u.\phi = \int_0^T u(t)\phi(t)dt$. $S_u : \mathcal{D}(]0,T[) \to V$ defined by $S_u\cdot\phi = \int_0^T u(t)\phi(t)dt \in V$ is a linear mapping from $\mathcal{D}(]0,T[)$ into V. Moreover, $\forall$ compact intervals $K = [a,b] \subset\subset]0,T[$, $\forall\phi \in \mathcal{D}_K(]0,T[)$,

$$\|S_u\cdot\phi\|_V = \left\|\int_0^T u(t)\phi(t)dt\right\| \le \int_0^T \|u(t)\phi(t)\|_V dt$$

$$\le \sup_{t\in K}|\phi(t)| \int_{K=[a,b]} \|u(t)\|_V dt = C_K \sup_{t\in K}|\phi(t)|,$$

with $C_K = \int_K \|u(t)\|_V dt$.

Hence, $\forall K \subset\subset]0,T[$, $\phi_n \to \phi$ in $\mathcal{D}_K(]0,T[) \implies \|S_u\phi - S_u\phi_n\|_V \le C_K \sup_{t\in K} |\phi(t) - \phi_n(t)| \to 0 \implies \forall K \subset\subset]0,T[$, $S_u : \mathcal{D}_K(]0,T[) \to V$ is continuous from $\mathcal{D}_K(]0,T[)$ into $V \implies S_u : \mathcal{D}(]0,T[) \longrightarrow V$ is continuous from $\mathcal{D}(]0,T[)$ into V (by (9.4.6)), i.e. $S_u \in \mathcal{D}'(]0,T[;V)$ by Proposition 9.4.1. Hence, $u \in L^1_{\text{loc}}(0,T;V)$ defines a vector-valued distribution $S_u \in \mathcal{D}'(]0,T[;V)$. □

Theorem 9.4.1. *Let $u, v \in L^1_{\text{loc}}(0,T;V)$. Then u and v will define the same vector-valued distribution in $\mathcal{D}'(]0,T[;V)$ if and only if*

$$\langle w', u(t)\rangle_{V'\times V} = \langle w', v(t)\rangle_{V'\times V} \quad \forall w' \in V' \tag{9.4.14}$$

and for almost all $t \in]0,T[$, i.e.

$$S_u = S_v \tag{9.4.15}$$

in $\mathcal{D}'(]0,T[;V)$ if and only if (9.4.14) *holds.*

Proof. Let $u, v \in L^1_{\text{loc}}(0,T;V)$. Then $S_u, S_v \in \mathcal{D}'(]0,T[;V)$ are defined by:

$$S_u \cdot \phi = \int_0^T \underbrace{u(t)}_{\in V} \underbrace{\phi(t)}_{\in \mathbb{R}} dt \in V;$$

$$S_v \cdot \phi = \int_0^T \underbrace{v(t)}_{\in V} \underbrace{\phi(t)}_{\in \mathbb{R}} dt \in V \quad \forall \phi \in \mathcal{D}(]0,T[).$$

We show that $S_u = S_v \iff \langle w', u(t) - v(t)\rangle_{V'\times V} = 0\ \forall w' \in V'$ and for almost all $t \in]0,T[$.

But $S_u = S_v$ in $\mathcal{D}'(]0,T[;V) \iff S_u \cdot \phi = S_v \cdot \phi$ in V $\forall \phi \in \mathcal{D}(]0,T[)$

$$\iff \int_0^T u(t)\phi(t)dt = \int_0^T v(t)\phi(t)dt \quad \forall \phi \in \mathcal{D}(]0,T[)$$

$$\iff \left\langle w', \int_0^T u(t)\phi(t)dt \right\rangle_{V'\times V} = \left\langle w', \int_0^T v(t)\phi(t)dt \right\rangle_{V'\times V}$$

$\forall w' \in V'$, $\forall \phi \in \mathcal{D}(]0,T[)$

$$\iff \int_0^T \langle w', u(t) - v(t)\rangle_{V'\times V}\phi(t)dt = 0 \quad \forall \phi \in \mathcal{D}(]0,T[) \quad \text{(by (9.3.20))}$$

$\iff \langle w', u(t) - v(t)\rangle_{V'\times V} = 0\ \forall w' \in V'$ and for almost all $t \in]0,T[$, since $t \in]0,T[\longmapsto \langle w', u(t) - v(t)\rangle_{V'\times V} = f(t)$ is a nice function of t belonging to $L^1_{\text{loc}}(]0,T[)$ and $\int_0^T f(t)\phi(t)dt = 0\ \forall \phi \in \mathcal{D}(]0,T[) \implies f(t) = 0$ a.e. on $]0,T[$ (by Theorem 1.2.3). □

Corollary 9.4.1. *Under the assumptions of Theorem* 9.4.1, *if V is a* separable *Banach space, then $u = v$ a.e. on* $]0, T[$.

For a *separable* Banach space V we agree to denote the vector-valued distribution $S_u \in \mathcal{D}'(]0,T[;V)$ (defined by $u \in L^1_{\text{loc}}(0,T;V)$) by the same notation u and set $S_u = u \in \mathcal{D}'(]0,T[;V)$. (9.4.16)

$$\begin{aligned} & L^p(0,T;V) \subset L^1_{\text{loc}}(0,T;V) \quad \text{for } (1 \le p \le \infty) \\ \Longrightarrow \quad & L^p(0,T;V) \subset \mathcal{D}'(]0,T[;V) \quad \text{for } (1 \le p \le \infty). \end{aligned} \tag{9.4.17}$$

Proposition 9.4.4. *Let V be a separable Banach space and $L^2(0,T;V)$ be the space of vector-valued functions defined by* (9.3.10)–(9.3.11) *for $p = 2$. Then*

$$L^2(0,T;V) \hookrightarrow \mathcal{D}'(]0,T[;V), \tag{9.4.18}$$

i.e.

I. $u \in L^2(0,T;V) \Longrightarrow S_u \in \mathcal{D}'(]0,T[;V)$ *with*

$$S_u \cdot \phi = \int_0^T u(t)\phi(t)dt \quad \forall \phi \in \mathcal{D}(]0,T[); \tag{9.4.19}$$

II. *for* $u, v \in L^2(0,T;V)$, $S_u = S_v$ *in* $\mathcal{D}'(]0,T[;V)$

$$\Longrightarrow \quad u = v \text{ a.e. in }]0,T[; \tag{9.4.20}$$

III. $u_n \to u$ *in* $L^2(0,T;V) \Longrightarrow S_{u_n} \to S_u$ *in* $\mathcal{D}'(]0,T[;V)$

$$\Longrightarrow \quad S_{u_n} \cdot \phi \to S_u \cdot \phi \text{ in } V \quad \forall \phi \in \mathcal{D}(]0,T[). \tag{9.4.21}$$

Proof. $L^2(0,T;V) \subset L^1_{\text{loc}}(0,T;V) \Longrightarrow$ I and II by Theorem 9.4.1 and Corollary 9.4.1, since V is separable. It remains to prove III. From Definition 9.3.1, $u_n \to u$ in $L^2(0,T;V)$

$$\Longrightarrow \quad \int_0^T \|u(t) - u_n(t)\|_V^2 dt \to 0. \tag{9.4.22}$$

We are to show that $S_{u_n} \cdot \phi \to S_u \cdot \phi$ in V $\forall \phi \in \mathcal{D}(]0,T[)$. In fact,

$$\begin{aligned} \|S_u \cdot \phi - S_{u_n} \cdot \phi\|_V &= \left\| \int_0^T [u(t)\phi(t) - u_n(t)\phi(t)]dt \right\|_V \\ &= \left\| \int_0^T [u(t) - u_n(t)]\phi(t)dt \right\|_V \le \int_0^T \|u(t) - u_n(t)\|_V |\phi(t)| dt \\ &= \langle \|u(t) - u_n(t)\|_V, |\phi| \rangle_{L^2(]0,T[)} \\ &\le \left(\int_0^T \|u(t) - u_n(t)\|_V^2 dt \right)^{\frac{1}{2}} \|\phi\|_{L^2(]0,T[)} \\ &\qquad \text{(by the Schwarz inequality),} \end{aligned}$$

since $t \longmapsto \|u(t) - u_n(t)\|_V$ is a function in $L^2(]0,T[)$ and $\phi \in \mathcal{D}(]0,T[) \subset L^2(]0,T[)$, and hence the Schwarz inequality in $L^2(]0,T[)$ is applicable. Consequently,

$$\|S_u \cdot \phi - S_{u_n} \cdot \phi\|_V \leq \left(\int_0^T \|u(t) - u_n(t)\|_V^2 dt \right)^{\frac{1}{2}} \|\phi\|_{L^2(]0,T[)} \to 0 \quad \text{as } n \to \infty,$$

since $u_n \to u$ in $L^2(]0,T[;V) \Longrightarrow$ (9.4.22) holds.

Therefore, $u_n \to u$ in $L^2(]0,T[;V) \Longrightarrow S_{u_n}\cdot\phi \to S_u\cdot\phi$ in V $\forall \phi \in \mathcal{D}(]0,T[) \Longrightarrow S_{u_n} \to S_u$ in $\mathcal{D}'(]0,T[;V)$. □

9.5 Derivatives of vector-valued distributions

Let V be a Banach space. The derivatives of vector-valued distributions $S \in \mathcal{D}'(]0,T[;V)$ will be introduced in the same way as was done for scalar-valued distributions $T \in \mathcal{D}'(\Omega)$ in Chapter 2.

Definition 9.5.1. Let $S \in \mathcal{D}'(]0,T[;V)$ be a vector-valued distribution with its values in V. Then the mapping $\phi \in \mathcal{D}(]0,T[) \longmapsto (-1)^m\langle S, \frac{d^m\phi}{dt^m}\rangle \in V$ is a vector-valued distribution with its value in V, which we will denote by $\frac{d^m S}{dt^m} \in \mathcal{D}'(]0,T[;V)$ and call the *mth-order derivative of S*. Then the mth-order derivative $\frac{d^m S}{dt^m} \in \mathcal{D}'(]0,T[;V)$ is defined, $\forall m \in \mathbb{N}$, by:

$$\left\langle \frac{d^m S}{dt^m}, \phi \right\rangle = (-1)^m \left\langle S, \frac{d^m \phi}{dt^m} \right\rangle \quad \text{in } V \ \forall \phi \in \mathcal{D}(]0,T[), \tag{9.5.1}$$

i.e.*vector-valued distributions of* $\mathcal{D}'(]0,T[;V)$ *can be differentiated indefinitely.* (9.5.2)

Proposition 9.5.1. $\forall m \in \mathbb{N}$, $\frac{d^m}{dt^m} : \mathcal{D}'(]0,T[;V) \longmapsto \mathcal{D}'(]0,T[;V)$ *is* continuous *from* $\mathcal{D}'(]0,T[;V)$ *into* $\mathcal{D}'(]0,T[;V)$, *i.e.*

$$S_n \to S \text{ in } \mathcal{D}'(]0,T[;V) \quad \Longrightarrow \quad \frac{d^m S_n}{dt^m} \to \frac{d^m S}{dt^m} \text{ in } \mathcal{D}'(]0,T[;V). \tag{9.5.3}$$

Proof. The proof is similar to that for the (scalar-valued) distributions of $\mathcal{D}'(\Omega)$ given earlier in Chapter 2. □

Distributional derivatives of vector-valued functions of $L^2(0,T;V)$

Let V be a *separable* Banach space. Let $u \in L^2(0,T;V)$. Then $u = S_u \in \mathcal{D}'(]0,T[;V) \Longrightarrow \frac{du}{dt} \in \mathcal{D}'(]0,T[;V)$ is defined, $\forall \phi \in \mathcal{D}(]0,T[)$, by:

$$\left\langle \frac{du}{dt}, \phi \right\rangle = -\left\langle u, \frac{d\phi}{dt} \right\rangle \quad \text{in } V. \tag{9.5.4}$$

$u \in L^2(0,T;V)$ does *not* imply $\frac{du}{dt} \in L^2(0,T;V)$, since $\frac{du}{dt} \in \mathcal{D}'(]0,T[;V)$ is a vector-valued *distribution* and *not* a vector-valued function *in general*. But we have:

Definition 9.5.2. Let V, W be *separable* Banach spaces with $V \hookrightarrow W$, the imbedding $\hookrightarrow$ being a *continuous* one from V into W, and $u \in L^2(0,T;V)$. Then, if $\exists v \in L^2(0,T;W)$ such that

$$\langle v, \phi \rangle = -\left\langle u, \frac{d\phi}{dt} \right\rangle \quad \text{in } W \ \forall \phi \in \mathcal{D}(]0,T[), \tag{9.5.5}$$

i.e.

$$\int_0^T \underbrace{v(t)}_{\in W} \underbrace{\phi(t)}_{\in \mathbb{R}}\, dt = -\int_0^T \underbrace{u(t)}_{\in V \hookrightarrow W} \underbrace{\frac{d\phi}{dt}(t)}_{\in \mathbb{R}}\, dt \quad \text{in } W, \tag{9.5.6}$$

then we call

$$v = \frac{du}{dt} = u' \in L^2(0,T;W) \hookrightarrow \mathcal{D}'(]0,T[;W). \tag{9.5.7}$$

Justification By (9.4.18), we have

$$L^2(0,T;V) \hookrightarrow \mathcal{D}'(]0,T[;V), \tag{9.5.8}$$

$$L^2(0,T;W) \hookrightarrow \mathcal{D}'(]0,T[;W), \tag{9.5.9}$$

and

$$V \hookrightarrow W \implies \mathcal{D}'(]0,T[;V) \hookrightarrow \mathcal{D}'(]0,T[;W) \quad \text{(by (9.4.9))} \tag{9.5.10}$$

$$\implies \left[\frac{du}{dt} \in \mathcal{D}'(]0,T[;V) \implies \frac{du}{dt} \in \mathcal{D}'(]0,T[;W)\right]. \tag{9.5.11}$$

But $\frac{du}{dt} = v \in L^2(0,T;W) \hookrightarrow \mathcal{D}'(]0,T[;W)$.

In particular,

$$W = V \implies \frac{du}{dt} \in L^2(0,T;V). \tag{9.5.12}$$

9.6 Applications

Some important spaces of vector-valued distributions

We now define some important spaces of vector-valued distributions which will be useful in the study of evolution problems in general. In the study of Navier–Stokes equations of fluid mechanics, one meets with some of these spaces [36], [42].

9.6.1 Space $\mathbb{E}(0, T; V, W)$

Definition 9.6.1. Let V and W be *separable* Banach spaces such that $V \hookrightarrow W$, the imbedding $\hookrightarrow$ being a *dense* and *continuous* one from V into W. Then the space $\mathbb{E}(0, T; V, W)$ is defined, for $1 < p, q < \infty$ with $\frac{1}{p} + \frac{1}{q} = 1$, by:

$$\mathbb{E}(0, T; V, W) = \left\{ u : u \in L^p(0, T; V), \ \frac{du}{dt} \in L^q(0, T; W) \right\}, \tag{9.6.1}$$

which is equipped with the norm defined by:

$$\|u\|_{\mathbb{E}(0,T;V,W)} = \|u\|_{L^p(0,T;V)} + \left\| \frac{du}{dt} \right\|_{L^q(0,T;W)}, \tag{9.6.2}$$

where $\frac{du}{dt} \in L^q(0, T; W)$ is to be understood in the following sense: for $u \in L^p(0, T; V)$, if $\exists v \in L^q(0, T; W)$ such that

$$\int_0^T \underbrace{v(t)\phi(t)}_{\in W} \, dt = -\int_0^T \underbrace{u \frac{d\phi}{dt}}_{\in V \hookrightarrow W} \, dt \quad \forall \phi \in \mathcal{D}(]0, T[), \tag{9.6.3}$$

then $v = \frac{du}{dt}$ in $L^q(0, T; W)$.

For $1 < p, q < \infty$ with $\frac{1}{p} + \frac{1}{q} = 1$ and $V \hookrightarrow W$ with *dense*, *continuous* imbedding, $\mathbb{E}(0, T; V, W)$ is a Banach space. (9.6.4)

9.6.2 Hilbert space $\mathbb{W}_1(0, T; V)$

Let H and V be any two (separable) *Hilbert spaces* such that $V \hookrightarrow H$, the imbedding $\hookrightarrow$ being a *dense*, *continuous* one, i.e.

1. $V \subset H$;
2. V is *dense* in H;
3. $u_n \to u$ in $V \Longrightarrow u_n \to u$ in H. (9.6.5)

Now, identifying H with its dual H', i.e. $H \equiv H'$, we have $V \hookrightarrow H = H' \hookrightarrow V'$ with the corresponding continuous imbeddings of V in H and H in V', V being a dense subspace of H. (9.6.6)

Definition 9.6.2. For $0 < T < \infty$, $\mathbb{W}_1(0, T; V)$ is the linear space

$$\mathbb{W}_1(0, T; V) = \left\{ u : u \in L^2(0, T; V), \ \frac{du}{dt} \in L^2(0, T; V') \right\}, \tag{9.6.7}$$

equipped with norm $\|\cdot\|_{\mathbb{W}_1(0,T;V)}$ defined by:

$$\|u\|_{\mathbb{W}_1(0,T;V)} = \left(\|u\|^2_{L^2(0,T;V)} + \left\|\frac{du}{dt}\right\|^2_{L^2(0,T;V')}\right)^{\frac{1}{2}}, \tag{9.6.8}$$

where $\frac{du}{dt} \in L^2(0,T;V')$ is defined by (9.6.3) with $V' = W$ and $p = q = 2$, i.e. for $v \in L^2(0,T;V')$,

$$\int_0^T \underbrace{v(t)\phi(t)}_{\in V'}\, dt = -\left\langle u, \frac{d\phi}{dt}\right\rangle = -\int_0^T \underbrace{u\frac{d\phi}{dt}}_{\in V \hookrightarrow V'}\, dt \quad \forall \phi \in \mathcal{D}(]0,T[); \tag{9.6.9}$$

$$L^2(0,T;V) \hookrightarrow \mathcal{D}'(]0,T[;V) \hookrightarrow \mathcal{D}'(]0,T[;V'); \tag{9.6.10}$$

$$L^2(0,T;V') \hookrightarrow \mathcal{D}'(]0,T[,V'). \tag{9.6.11}$$

Theorem 9.6.1. $\mathbb{W}_1(0,T;V)$ *defined by* (9.6.7)–(9.6.8) *is a* Hilbert space.

Proof. It is sufficient to show that every Cauchy sequence in $\mathbb{W}_1(0,T;V)$ is a convergent one in $\mathbb{W}_1(0,T;V)$.

Let(u_n) be any Cauchy sequence in $\mathbb{W}_1(0,T;V)$. Then

- (u_n) is a Cauchy sequence in $L^2(0,T;V)$ which is a Hilbert space (see Proposition 9.3.1), since V is a Hilbert space;
- (u_n') is a Cauchy sequence in $L^2(0,T;V')$ which is also a Hilbert space (see Proposition 9.3.1), since V' is a Hilbert space.

Hence, $\exists u \in L^2(0,T;V)$ and $v_1 \in L^2(0,T;V')$ such that $u_n \to u$ in $L^2(0,T;V)$, $u_n' \to v_1$ in $L^2(0,T;V')$.

The proof will be complete if we can show that $v_1 = \frac{du}{dt} = u'$. In fact, $u_n \to u$ in $L^2(0,T;V) \hookrightarrow \mathcal{D}'(]0,T[;V) \implies u_n \to u$ in $\mathcal{D}'(]0,T[;V) \implies \frac{du_n}{dt} = u_n' \to \frac{du}{dt}$ in $\mathcal{D}'(]0,T[,V)$, by Proposition 9.5.1. Since $V \hookrightarrow H = H' \hookrightarrow V'$, with *continuous*, *dense* imbedding of V in H and the continuous imbedding of H' in V', $\mathcal{D}'(]0,T[,V) \hookrightarrow \mathcal{D}'(]0,T[;V')$ by Proposition 9.4.2,

$$\implies \quad u_n' \to \frac{du}{dt} = u' \quad \text{in } \mathcal{D}'(]0,T[,V') \quad \text{(by (9.4.10) and (9.4.11))}. \tag{9.6.12}$$

But $u_n' \to v_1$ in $L^2(0,T;V') \hookrightarrow \mathcal{D}'(]0,T[,V')$

$$\implies \quad u_n' \to v_1 \quad \text{in } \mathcal{D}'(]0,T[,V'). \tag{9.6.13}$$

Hence, from (9.6.12) and (9.6.13), $u_n' \to \frac{du}{dt}$ in $\mathcal{D}'(]0,T[;V')$ and $u_n' \to v_1$ in $\mathcal{D}'(]0,T[,V') \iff v_1 = \frac{du}{dt} = u'$ in $\mathcal{D}'(]0,T[;V')$, since the limit is unique. Hence, $u_n \to u$ in $L^2(0,T;V)$ and $u_n' \to \frac{du}{dt} \in L^2(0,T;V') \implies u_n \to u$ in $\mathbb{W}_1(0,T;V)$. Therefore, $\mathbb{W}_1(0,T;V)$ is a Hilbert space. □

Remark 9.6.1. Spaces $\mathbb{E}(0, T; V, W)$ and $\mathbb{W}_1(0, T; V)$ are used in evolution problems of parabolic equations.

Now we state an important result used in the study of parabolic evolution equations.

Proposition 9.6.1. *Let H, V be Hilbert spaces with $V \hookrightarrow H = H' \hookrightarrow V'$, the imbedding $V \hookrightarrow H$ being a dense and continuous one and $H' \hookrightarrow V'$ being a continuous one. For $u \in \mathbb{W}_1(0, T; V)$ defined by* (9.6.7)–(9.6.8), *$v \in V$, $0 < T < \infty$,*

$$\langle u'(\cdot), v\rangle_{V' \times V} = \frac{d}{dt}\langle u(\cdot), v\rangle_H \quad \text{in } \mathcal{D}'(]0, T[), \tag{9.6.14}$$

where $\langle \cdot, \cdot \rangle_{V' \times V}$ is the duality pairing between V' and V; $\langle \cdot, \cdot \rangle_H$ is the inner product in H.

Proof. Since $u' \in L^2(0, T; V')$, $t \longmapsto \langle u'(t), v\rangle_{V' \times V}$ is a nice function of $L^2(]0, T[)$, and we have

$$\begin{aligned}\int_0^T \langle u'(t), v\rangle_{V' \times V}\phi(t)dt &= \int_0^T \langle u'(t)\phi(t), v\rangle_{V' \times V} dt = \left\langle \int_0^T u'(t)\phi(t)dt, v\right\rangle_{V' \times V} \\ &= \left\langle -\int_0^T u(t)\phi'(t)dt, v\right\rangle_{V' \times V} \qquad (9.6.15) \\ &= -\int_0^T \langle u(t)\phi'(t), v\rangle_{V' \times V} dt \quad \forall \phi \in \mathcal{D}(]0, T[).\end{aligned}$$

(9.6.16)

But $u(t) \in V \hookrightarrow H = H' \Longrightarrow u(t) \in H \equiv H'$ for almost all $t \in]0, T[$, and $v \in V \Longrightarrow v \in H \Longrightarrow \langle u(t), v\rangle_{V' \times V} = \langle u(t), v\rangle_H$, $\langle \cdot, \cdot \rangle_H$ being an inner product in $H = H'$ (since $V \hookrightarrow H = H' \hookrightarrow V'$ with continuous imbeddings, V being dense in H). (9.6.17)

Hence, from (9.6.15) and (9.6.16), $\forall \phi \in \mathcal{D}(]0, T[)$,

$$\begin{aligned}&\int_0^T \langle u'(t), v\rangle_{V' \times V}\phi(t)dt = -\int_0^T \langle u(t), v\rangle_H \phi'(t)dt = \int_0^T \frac{d}{dt}\langle u(t), v\rangle_H \phi(t)dt \\ &\Longrightarrow \quad \int_0^T \left[\langle u'(t), v\rangle_{V' \times V} - \frac{d}{dt}(\langle u(t), v\rangle_H)\right]\phi(t)dt = 0 \quad \forall \phi \in \mathcal{D}(]0, T[)\end{aligned}$$

$\Longrightarrow \langle \langle u'(\cdot), v\rangle_{V' \times V} - \frac{d}{dt}\langle u(\cdot), v\rangle_H, \phi\rangle_{\mathcal{D}'(]0,T[) \times \mathcal{D}(]0,T[)} = 0 \ \forall \phi \in \mathcal{D}(]0, T[)$
$\Longrightarrow \langle u'(\cdot), v\rangle_{V' \times V} - \frac{d}{dt}\langle u(\cdot), v\rangle_H = 0$ in $\mathcal{D}'(]0, T[)$
$\Longrightarrow \langle u'(\cdot), v\rangle_{V' \times V} = \frac{d}{dt}\langle u(\cdot), v\rangle_H$ in $\mathcal{D}'(]0, T[)$. □

9.6.3 Hilbert space $\mathbb{W}_2(0, T; V)$

Definition 9.6.3. Let V be a separable Hilbert space such that $V \hookrightarrow H = H' \hookrightarrow V'$ with *dense, continuous* imbeddings (see (9.6.5) and (9.6.6)).

Then, for $0 < T < \infty$, $\mathbb{W}_2(0, T; V)$ is the linear space

$$\mathbb{W}_2(0, T; V) = \left\{ u : u \in L^2(0, T; V), \frac{d^2u}{dt^2} \in L^2(0, T; V') \right\} \tag{9.6.18}$$

equipped with the norm $\| \cdot \|_{\mathbb{W}_2(0,T;V)}$ defined by

$$\|u\|_{\mathbb{W}_2(0,T;V)} = \left(\|u\|^2_{L^2(0,T;V)} + \left\| \frac{d^2u}{dt^2} \right\|^2_{L^2(0,T;V')} \right)^{\frac{1}{2}}, \tag{9.6.19}$$

where

$$\int_0^T \underbrace{\frac{d^2u}{dt^2}(t)\phi(t)\, dt}_{\in V'} = \int_0^T \underbrace{u(t)\frac{d^2\phi}{dt^2}\, dt}_{\in V \hookrightarrow V'} \quad \forall \phi \in \mathcal{D}(]0, T[).$$

Theorem 9.6.2. $\mathbb{W}_2(0, T; V)$ *defined by* (9.6.18)–(9.6.19) *is a* Hilbert space.

Proof. The proof is exactly similar to that for $\mathbb{W}_1(0, T; V)$ and is obtained by replacing 'u_n'' by 'u_n''' and 'v_1' by 'v_2', '$\frac{du}{dt}$' by '$\frac{d^2u}{dt^2}$', etc. in the steps of the proof. □

Remark 9.6.2. $\mathbb{W}_2(0, T; V)$ is met with in the study of evolution equations of hyperbolic type.

Density and imbedding results

We collect here the important results on density and imbedding, without proof.

Density results

Theorem 9.6.3 (Density). *Let* V, W *be separable Banach spaces with* $V \hookrightarrow W$*, the imbedding* $\hookrightarrow$ *being a* dense, continuous *one from* V *into* W*. Let the space* $\mathbb{E}(0, T; V, W)$ *be defined by* (9.6.1)–(9.6.4).

Let

$$\mathcal{D}([0, T]; V) \equiv C^\infty([0, T]; V) = \{\phi : \phi \in C^k([0, T]; V)\ \forall k \in \mathbb{N}\ (\text{see } (9.3.3))\} \tag{9.6.20}$$

or, alternatively,

$$\mathcal{D}([0, T]; V) = \{\phi : \exists \psi \in \mathcal{D}(]0, \infty[\,; V) \text{ with compact support in } \mathbb{R},\ \phi = \psi\downarrow_{[0,T]}\}. \tag{9.6.21}$$

Then $\mathcal{D}([0, T]; V)$ *is* dense *in* $\mathbb{E}(0, T; V, W)$.

Theorem 9.6.4 (Density). *$\mathcal{D}([0,T];V)$ defined by* (9.6.20) *is* dense *in* $\mathbb{W}_1(0,T;V)$ *(resp.* $\mathbb{W}_2(0,T;V)$*) defined by* (9.6.7)–(9.6.8) *(resp.* (9.6.18)–(9.6.19)*).*

Imbedding results

Theorem 9.6.5. *Let H, V be separable Hilbert spaces with $V \hookrightarrow H = H' \hookrightarrow V'$ (see* (9.6.6)*). Then $\mathbb{W}_1(0,T;V) \hookrightarrow C^0([0,T];H)$, i.e. for $u \in \mathbb{W}_1(0,T;V)$, $\exists v \in C^0([0,T];H)$ such that $u = v$ a.e on $]0,T[$ and $\exists C > 0$ such that*

$$\|u\|_{C^0([0,T];H)} \leq C\|u\|_{\mathbb{W}_1(0,T;V)} \quad \forall u \in \mathbb{W}_1(0,T;V). \tag{9.6.22}$$

Consequence of the imbedding result For $u \in \mathbb{W}_1(0,T;V) \hookrightarrow C^0([0,T];H)$ with $[0,T] \subset [0,\infty[$, $u(0) \in H$, $u(T) \in H$ can be called *traces* of u. (9.6.23)

We agree to accept the following result without proof:

The mapping $u \in \mathbb{W}_1(0,T;V) \longmapsto u(0) \in H$ is a *surjection* from $\mathbb{W}_1(0,T;V)$ onto H. (9.6.24)

9.6.4 Green's formula

Theorem 9.6.6. *Let V, H be two separable Hilbert spaces such that V is a* dense *subspace of H and $V \hookrightarrow H = H' \hookrightarrow V'$, the corresponding imbeddings being continuous ones. For $u, v \in \mathbb{W}_1(0,T;V)$ with $[0,T] \subset [0,\infty[$, the following formula called Green's formula holds:*

$$\int_0^T \langle u'(t), v(t)\rangle_{V'\times V} dt - \int_0^T \langle u(t), v'(t)\rangle_{V\times V'} dt = \langle u(T), v(T)\rangle_H - \langle u(0), v(0)\rangle_H, \tag{9.6.25}$$

where $\langle\cdot,\cdot\rangle_{V'\times V}$ (resp. $\langle\cdot,\cdot\rangle_{V\times V'}$) denotes the duality pairing between V' and V (resp. V and V'); $\langle\cdot,\cdot\rangle_H$ denotes the inner product in Hilbert space H.

Proof. The mapping

$$\begin{aligned} f_1 &: t \in]0,T[\longmapsto \langle u'(t), v(t)\rangle_{V'\times V} = f_1(t) \text{ belongs to } L^2(]0,T[), \\ f_2 &: t \in]0,T[\longmapsto \langle u(t), v'(t)\rangle_{V\times V'} = f_2(t) \text{ belongs to } L^2(]0,T[) \end{aligned} \tag{9.6.26}$$

$\forall u, v \in \mathbb{W}_1(0,T;V)$.

From (9.6.23), $u, v \in \mathbb{W}_1(0,T;V) \Longrightarrow u(0), v(0) \in H$ and $u(T), v(T) \in H$.

From Definition 9.6.2, $u, v \in \mathbb{W}_1(0,T;V) \Longrightarrow u, v \in L^2(0,T;V)$ *and* $u', v' \in L^2(0,T;V')$

$$\Longrightarrow \quad \int_0^T \langle u'(t), v(t)\rangle_{V'\times V} dt \in \mathbb{R}, \quad \int_0^T \langle u(t), v'(t)\rangle_{V\times V'} dt \in \mathbb{R}. \tag{9.6.27}$$

Again, from (9.6.20)/(9.6.21), $u, v \in \mathcal{D}([0,T];V) \Longrightarrow u^{(k)}(t), v^{(k)}(t) \in V \hookrightarrow H \ \forall k \in \mathbb{N}, \forall t \in [0,T] \Longrightarrow u^{(k)}(t), v^{(k)}(t) \in H \ \forall k \in \mathbb{N}, \forall t \in [0,T]$. Hence, $\forall u, v \in \mathcal{D}([0,T];V)$,

$$\langle u'(t), v(t)\rangle_{V'\times V} = \langle u'(t), v(t)\rangle_H, \qquad (9.6.28)$$
$$\langle u(t), v'(t)\rangle_{V\times V'} = \langle u(t), v'(t)\rangle_H,$$

and

$$\int_0^T \langle u'(t), v(t)\rangle_H dt + \int_0^T \langle u(t), v'(t)\rangle_H dt = \int_0^T [\langle u'(t), v(t)\rangle_H + \langle u(t), v'(t)\rangle_H] dt$$
$$= \int_0^T \frac{d}{dt}[\langle u(t), v(t)\rangle_H] dt = \langle u(t), v(t)\rangle_H \big|_0^T = \langle u(T), v(T)\rangle_H - \langle u(0), v(0)\rangle_H. \qquad (9.6.29)$$

But $\mathcal{D}([0,T];V)$ is *dense* in $\mathbb{W}_1(0,T;V)$ by Theorem 9.6.4, and $\mathbb{W}_1(0,T;V) \hookrightarrow C^0([0,T];H)$ by Theorem 9.6.5. Hence, $\forall u, v \in \mathbb{W}_1(0,T;V), u(T), v(T), u(0), v(0) \in H$ and the continuous bilinear functional vanishing on $(\mathcal{D}([0,T];V))^2$, i.e.

$$(u,v) \in (\mathcal{D}([0,T];V))^2 \longmapsto \Big[\int_0^T \langle u'(t), v(t)\rangle_H dt + \int_0^T \langle u(t), v'(t)\rangle_H$$
$$- \langle u(T), v(T)\rangle_H + \langle u(0), v(0)\rangle_H \Big] = 0, \quad (9.6.30)$$

can be extended by *density* to a continuous, bilinear functional vanishing on $\mathbb{W}_1(0,T;V) \times \mathbb{W}_1(0,T;V)$, i.e. $\forall u, v \in \mathbb{W}_1(0,T;V)$,

$$\int_0^T \langle u'(t), v(t)\rangle_{V'\times V} dt + \int_0^T \langle u(t), v'(t)\rangle_{V\times V'} dt$$
$$- \langle u(T), v(T)\rangle_H + \langle u(0), v(0)\rangle_H = 0, \quad (9.6.31)$$

since the *extension* of the inner product $\langle u'(t), v(t)\rangle_H$ (resp. $\langle u(t), v'(t)\rangle_H$) is identified with $\langle u'(t), v(t)\rangle_{V'\times V}$ (resp. $\langle u(t), v'(t)\rangle_{V\times V'}$) (9.6.32)
by virtue of the *continuous* imbeddings $V \hookrightarrow H = H' \hookrightarrow V'$ and *density* of V in H.

Then the result (9.6.25) follows from (9.6.31) $\forall u, v \in \mathbb{W}_1(0,T;V)$. □

Appendix A
Functional analysis (basic results)

A.0 Preliminary results

A.0.1 An important result on logical implication ($\Longrightarrow$) and non-implication ($\not\Longrightarrow$)

We deal with theorems, lemmas and propositions which have the form $(\mathcal{P}) \Longrightarrow (\mathcal{Q})$, or 'if the statement $(\mathcal{P})$ holds, then the statement $(\mathcal{Q})$ holds'. The assertion 'if $(\mathcal{P})$, then $(\mathcal{Q})$' is the same as the assertion '$(\mathcal{P})$ holds *only if* $(\mathcal{Q})$ holds' or, equivalently speaking, '*if not* $(\mathcal{Q})$ then *not* $(\mathcal{P})$', i.e. '*not* $(\mathcal{Q})$ implies *not* $(\mathcal{P})$'. Hence, the *positive* assertion that '$(\mathcal{P})$ holds *only if* $(\mathcal{Q})$ holds' and the *contrapositive* assertion 'if *not* $(\mathcal{Q})$ then *not* $(\mathcal{P})$' are *equivalent*, i.e.

$$[(\mathcal{P}) \Longrightarrow (\mathcal{Q})] \iff [\text{not } (\mathcal{Q}) \Longrightarrow \text{not } (\mathcal{P})]. \tag{A.0.1.1}$$

Theorem A.0.1.1. *The proposition $(\mathcal{P}) \Longrightarrow (\mathcal{Q})$ is true $\iff$ the proposition $(\sim\mathcal{Q}) \Longrightarrow (\sim\mathcal{P})$, where the negation of a statement $\mathcal{P}$ is denoted by $\sim\mathcal{P}$ or by $\not\mathcal{P}$ (resp. $\sim\mathcal{Q}$ or by $\not\mathcal{Q}$).*

If $(\mathcal{P})$ is true then its negation $(\sim\mathcal{P})$ (or $(\not\mathcal{P})$ is not *true, i.e.* false. *Conversely, if $(\mathcal{P})$ is* false, *then $(\sim\mathcal{P})$ or $(\not\mathcal{P}\,)$ is* true.

Now we explain *the rule of construction of negation of a statement*. We will write $\mathcal{P}$ to denote $(\mathcal{P})$. Let $\mathcal{P}$ be the statement '$\forall x$ (for all x) satisfying a condition $\mathcal{C}$, $\mathcal{Q}(x)$ is *true*'. Then $\sim\mathcal{P} \iff \sim(\forall x$ (for all x) satisfying a condition $\mathcal{C}$, $\mathcal{Q}(x)$ is *true*) $\iff$ the statement '$\forall x$ satisfying a condition $\mathcal{C}$, $\mathcal{Q}(x)$ is true' is *false* $\iff$ ($\exists x$ (there exists x or for *some* x) satisfying a condition $\mathcal{C}$ such that $\mathcal{Q}(x)$ does *not* hold).

Conversely, let $\mathcal{P}$ be the statement '$\exists x$ satisfying a condition $\mathcal{C}$ such that $\mathcal{Q}(x)$ is *false*'. Then its negation $\sim\mathcal{P}$ is given by: $\sim(\exists x$ satisfying a condition $\mathcal{C}$ such that $\mathcal{Q}(x)$ is *false*) $\iff \forall x$ satisfying the condition $\mathcal{C}$, $\mathcal{Q}(x)$ is *true*.

Hence, the negation of a statement $\mathcal{P}$ is obtained by replacing '$\forall$' by '$\exists$', '$\exists$' by '$\forall$' and '$\mathcal{Q}$' *by* '$\sim\mathcal{Q}$', but the *condition* $\mathcal{C}$ to be satisfied remains *unchanged*.

Instead of a statement defined by $\forall x,\ Q(x)$ (holds), let us assume that *the contrary*, i.e. the negation, holds. Then we have: $\sim(\forall x, Q(x)) \iff \exists x$ such that $(\sim Q(x)) \iff \exists x$ such that $Q(x)$ does *not* hold.

Example A.0.1.1. Statement $\mathscr{P}$: f is *continuous* at $a \in \mathbb{R} \iff \forall \varepsilon > 0$, $\exists \delta = \delta(\varepsilon) > 0$ such that $\forall x \in \mathbb{R}$ with $|x - a| < \delta$,

$$\underbrace{|f(x) - f(a)| < \varepsilon}_{\mathscr{Q}(x)}.$$

Then $\sim\mathscr{P}$: f is *discontinuous* at $a \in \mathbb{R} \iff \exists \varepsilon > 0$, $\forall \delta > 0$ such that $\exists x \in \mathbb{R}$ with $|x - a| < \delta$,

$$\underbrace{|f(x) - f(a)| \geq \varepsilon}_{\sim\mathscr{Q}(x)}.$$

Here, the negation $\sim\mathscr{P}$ is obtained by replacing '$\forall \varepsilon > 0$' by '$\exists \varepsilon > 0$', '$\exists \delta > 0$' by '$\forall \delta > 0$', '$\forall x \in \mathbb{R}$ with $|x - a| < \delta$' by '$\exists x \in \mathbb{R}$ with $|x - a| < \delta$' and '$\mathscr{Q}$' by '$\sim\mathscr{Q}$', all the *conditions*, such as '> 0', '$\in \mathbb{R}$', 'with $|x - a| < \delta$', remain *unaltered*.

A.0.2 Supremum (l.u.b.) and infimum (g.l.b.)

Definition A.0.2.1. Let $X \subset \mathbb{R}$ be a set of real numbers *bounded from above* (resp. *bounded from below*), i.e. $\exists a \in \mathbb{R}$ (resp. $b \in \mathbb{R}$) such that $x \leq a$ $\forall x \in X$ (resp. $x \geq b$ $\forall x \in X$). Then the *unique smallest* (resp. *largest*) real number $\alpha \in \mathbb{R}$ (resp. $\beta \in \mathbb{R}$) is called the *least upper bound* (*l.u.b.*) or, equivalently, the *supremum* of X (resp. the *greatest lower bound* (*g.l.b.*) or, equivalently, the *infimum* of X), if and only if

1. $\alpha \geq x$ $\forall x \in X$;
2. $\forall n \in \mathbb{N}$, $\exists x \in X$ such that $\alpha - \frac{1}{n} < x \leq \alpha$. (A.0.2.1)

(resp.

1. $\beta \leq x$ $\forall x \in X$;
2. $\forall n \in \mathbb{N}$, $\exists x \in X$ such that $\beta \leq x < \beta + \frac{1}{n}$). (A.0.2.2)

Then we write $\alpha = \sup X$ and $\beta = \inf X$.

Definition A.0.2.2. Let $a = \sup X$ (resp. $b = \inf X$) such that $a \in X$ (resp. $b \in X$). Then, $a = \sup X = \max X$ (resp. $b = \inf X = \min X$). (A.0.2.3)

A.0.3 Metric spaces and important results therein

Definition A.0.3.1. Let X be any *nonempty* set. Then, $\forall$ pairs of elements $x, y \in X$, we define the *distance* (function) $d(x, y)$ between them satisfying the following *properties of distance* or *metric* d: $\forall x, y, z \in X$,

1. $d(x, y) \geq 0$; $d(x, y) = 0 \iff x = y$ (*positive definiteness*);
2. $d(x, y) = d(y, x)$ (*symmetry*);

3. $d(x, z) \leq d(x, y) + d(y, z)$ (*triangle inequality*). (A.0.3.1)

Then the pair (X, d), or briefly X, is called a *metric space* equipped with the *metric* d.

Example A.0.3.1.

1. $A \subset \mathbb{R}$ (any *nonempty* subset of $\mathbb{R}$) equipped with the *usual* metric $d(x, y) = |x - y|$ is a *metric space*. (A.0.3.2)

2. Let $\Omega \subset \mathbb{R}^n$ be any *nonempty* subset of $\mathbb{R}^n$ equipped with any one of the metrics $d_1(\cdot,\cdot)$, $d_2(\cdot,\cdot)$, $d_\infty(\cdot,\cdot)$ defined, $\forall \mathbf{x} = (x_1, \dots, x_n) \in \mathbb{R}^n$, $\mathbf{y} = (y_1, \dots, y_n) \in \mathbb{R}^n$, by:

$$d_1(\mathbf{x}, \mathbf{y}) = \sum_{i=1}^{n} |x_i - y_i|; \quad d_2(\mathbf{x}, \mathbf{y}) = \left(\sum_{i=1}^{n} |x_i - y_i|^2\right)^{1/2};$$

$$d_\infty(\mathbf{x}, \mathbf{y}) = \max_{1 \leq i \leq n} |x_i - y_i|.$$

This, in fact, defines three *different metric spaces*: (Ω, d_1), (Ω, d_2), (Ω, d_∞). (A.0.3.3)

3. Let $C^0([0, 1]) = \{f : f \text{ is } \textit{continuous} \text{ on } [0, 1]\} \equiv X$ equipped with the metric $d(f, g) = \max_{0 \leq x \leq 1} |f(x) - g(x)|$. Then X is a *metric space*. (A.0.3.4)

Remark A.0.3.1. *All* metrics $d(\cdot,\cdot)$ on $\mathbb{R}^n$ are *equivalent*, i.e. for *any two* metrics $d_i(\cdot,\cdot)$ and $d_j(\cdot,\cdot)$ on $\mathbb{R}^n$, $i \neq j$, $\exists C_1, C_2 > 0$ such that $C_1 d_i(\mathbf{x}, \mathbf{y}) \leq d_j(\mathbf{x}, \mathbf{y}) \leq C_2 d_i(\mathbf{x}, \mathbf{y})$ $\forall \mathbf{x}, \mathbf{y} \in \mathbb{R}^n$.

Sequences and subsequences in X

Let $(X, d) \equiv X$ be a metric space and A, $B \subset X$ be *nonempty* subsets of X. Then A, B are also metric spaces with metric d inherited from X.

Definition A.0.3.2. A mapping $m \in \mathbb{N} \longmapsto \mathbf{x}_m \in X$ (from $\mathbb{N}$ into X) is called a *sequence* in X and denoted by $(x_m)_{m=1}^{\infty}$ or (x_m) or simply x_m.

A *subsequence* or *extracted sequence* $(x_{m_k})_{k=1}^{\infty}$ of $(x_m)_{m=1}^{\infty}$ is obtained by means of a *composite mapping*: $k \in \mathbb{N} \longmapsto m_k \in \mathbb{N} \longmapsto x_{m_k} \in X$, with $k_1 < k_2 \Longrightarrow m_{k_1} < m_{k_2}$ in $\mathbb{N}$. (A.0.3.5)

For the sequence $(x_m)_{m=1}^{\infty} = 1, \frac{1}{2}, 2, \frac{1}{4}, 3, \dots$ in $\mathbb{R}$, a subsequence is $(x_{m_k})_{k=1}^{\infty} = \frac{1}{2}, \frac{1}{4}, \frac{1}{8}, \dots$, with $x_{m_1} = \frac{1}{2}$, $x_{m_2} = \frac{1}{4}, \dots$ for $m_k = 2, 4, 6, \dots$ with $k = 1, 2, \dots$, respectively. (A.0.3.6)

Convergence of a sequence in X

Definition A.0.3.3. A sequence $(x_m)_{m=1}^{\infty}$ in X is said to *converge* to $x_0 \in X$ iff $\forall \varepsilon > 0$, $\exists n_0 = n_0(\varepsilon) \in \mathbb{N}$ such that $d(x_0, x_m) < \varepsilon \ \forall m > n_0$. Then we write $\lim_{m\to\infty} x_m = x_0$ in X, where $x_0 \in X$ is called the *limit* of (x_m) in X as $m \to \infty$, or is equivalently denoted by: $x_m \to x_0$ in X as $m \to \infty$. The limit x_0 is *unique*. (A.0.3.7)

Definition A.0.3.4. A sequence $(x_m)_{m=1}^{\infty}$ in X is called a *Cauchy sequence* in X iff $\forall \varepsilon > 0$, $\exists n_0 = n_0(\varepsilon) \in \mathbb{N}$ such that $d(x_m, x_n) < \varepsilon \ \forall m, n > n_0$.

A Cauchy sequence $(x_m)_{m=1}^{\infty}$ *may not* be a *convergent* sequence in X, but every *convergent* sequence is a Cauchy sequence in X. (A.0.3.8)

Definition A.0.3.5. A metric space X is called *complete* iff every Cauchy sequence (x_m) in X converges to an element $x \in X$ as $m \to \infty$.

Example A.0.3.2. *Each* of the metric spaces in $\mathbb{R}$ in (A.0.3.2), $\mathbb{R}^n$ in (A.0.3.3) and $C^0([0,1])$ in (A.0.3.4) is *complete*. (A.0.3.9)

More examples of complete metric spaces will be given later.

Distance $d(A, B)$ between nonempty sets $A, B \subset X$

$$d(A, B) = \inf_{x \in A, y \in B} d(x, y). \tag{A.0.3.10}$$

For $A = \{x\}$,

$$d(x, B) = d(\{x\}, B) = \inf_{y \in B} d(x, y) \tag{A.0.3.11}$$

is the *distance from $x \in X$ to B*, and $d(A, B) = \inf_{x\in A} d(x, B)$.

Diameter $\mathrm{dia}(A)$ of a nonempty subset $A \subset X$

The *diameter* $\mathrm{dia}(A)$ of a nonempty subset $A \subset X$ is given by:

$$\mathrm{dia}(A) = \sup_{x, y \in A} d(x, y), \tag{A.0.3.12}$$

with values in $[0, \infty]$.

$A \subset X$ is *bounded* $\Longrightarrow \mathrm{dia}(A) < +\infty$. (A.0.3.13)

A.0.4 Important subsets of a metric space $X \equiv (X, d)$

The *open ball* $B(x_0; r)$, *closed ball* $\overline{B}(x_0; r)$ and *sphere* $S(x_0; r)$ in X with *centre* x_0 and *radius* $r > 0$ are defined by:

$$B(x_0; r) = \{x : x \in X,\ d(x, x_0) < r\}; \tag{A.0.4.1}$$

$$\overline{B}(x_0; r) = \{x : x \in X,\ d(x, x_0) \le r\}; \tag{A.0.4.2}$$

$$S(x_0; r) = \{x : x \in X,\ d(x, x_0) = r\} = \overline{B}(x_0; r) - B(x_0; r). \tag{A.0.4.3}$$

Definition A.0.4.1. A nonempty subset $A \subset X$ is called *open* in X iff $\forall x \in A$, $\exists$ an *open* ball $B(x; r)$ contained in A, i.e. $B(x; r) \subset A$.

- An *arbitrary* union $\bigcup_{i \in I} A_i$ of open sets $\{A_i\}_{i \in I}$ is *open*. (A.0.4.4)
- Any *finite* intersection $\bigcap_{i=1}^{N} A_i$ of open sets $\{A_i\}_{i=1}^{N}$ is *open*. (A.0.4.5)
- The *empty set* $\emptyset$ and the whole space X are *open* (also *closed* (to be shown later)). (A.0.4.6)

Definition A.0.4.2. A *neighbourhood* (resp. *open neighbourhood*) of a nonempty set $A \subset X$ is *any* set B with $A \subset B$ (resp. any *open* set $\mathcal{N}(A) \subset X$ with $A \subset \mathcal{N}(A)$). In particular, an *open neighbourhood* of $x_0 \in X$ is an open set $\mathcal{N}(x_0) = \mathcal{N}(\{x_0\}) \subset X$ with $x_0 \in \mathcal{N}(x_0)$. (A.0.4.7)

- $\forall$ nonempty sets $A \subset X$ and $r > 0$, the set $\mathcal{N}_r(A) = \{x : x \in X, d(x, A) < r\}$ is an *open neighbourhood* of A in X. (A.0.4.8)
- For $A = \{x_0\}$, $\mathcal{N}_r(x_0) = \mathcal{N}_r(\{x_0\}) \equiv B(x_0; r)$. (A.0.4.9)
- $\mathcal{N}(x_0; \delta) = \{x : x \in X, 0 < d(x, x_0) < \delta\} = B(x_0; \delta) - \{x_0\}$ is called the *deleted open neighbourhood* of $x_0 \in X$. (A.0.4.10)

Definition A.0.4.3. A point $x_0 \in A \subset X$ is called an *interior* point of A iff $\exists$ an *open* ball $B(x_0; r)$ contained in A. Then the set of all interior points $x \in A$ is called the *interior* of A and denoted by $\overset{\circ}{A}$ or $\operatorname{int}(A)$. (A.0.4.11)

Closed sets, cluster points and closure of a set in X

Definition A.0.4.4. A nonempty subset $A \subset X$ is called *closed* in X iff $A^{\complement}$ is *open* in X.

For example, a *closed* interval $[a, b]$ on $X = \mathbb{R}$ and a *closed* ball $\overline{B}(x_0; r)$ in X are *closed subsets* of X. The sphere $S(x_0, r) \subset X$ is a *closed* set, since $[S(x_0, r)]^{\complement} = B(x_0; r) \cup (\overline{B}(x_0; r))^{\complement}$ is *open* by (A.0.4.4). (A.0.4.12)

Definition A.0.4.5. $A \subset X$ is *closed* in X

$$\iff \quad [\forall (x_n) \text{ in } A \text{ with } \lim_{n\to\infty} x_n = x \text{ in } X \textit{ and } \text{the limit } x \in A]. \tag{A.0.4.13}$$

Proposition A.0.4.1. *Let X be a* complete *metric space. Then $A \subset X$ is* closed *in $X \iff A$ is also a* complete *metric space.*

- *The arbitrary intersection $\bigcap_{i \in I} A_i$ of* closed *sets $\{A_i\}_{i \in I}$ is* closed. (A.0.4.14)
- *Any finite union $\bigcup_{i=1}^{N} A_i$ of closed sets $\{A_i\}_{i=1}^{N}$ is* closed. (A.0.4.15)
- *The empty set $\emptyset$ and the whole space X are* closed *(see also* (A.0.4.6)*).* (A.0.4.16)

Definition A.0.4.6. A point $x \in X$ is called a *cluster point* (or *limit point* or *accumulation point*) of a set $A \subset X$ iff *every* open neighbourhood $\mathcal{N}(x)$ of x contains points of A, i.e. $\mathcal{N}(x) \cap A \neq \emptyset$ or, equivalently, $\exists$ a sequence (x_n) in A such that $\lim_{n \to \infty} x_n = x$ in X. (A.0.4.17)

Definition A.0.4.7. The *closure* of $A \subset X$ is the set of all *cluster points* of A in X and is denoted by $\overline{A}$.

- $A \subset X$ is *closed* in $X \iff A = \overline{A}$. (A.0.4.18)
- $x \notin \overline{A} \Longrightarrow x \in \operatorname{int}(X \setminus A)$. (A.0.4.19)

Definition A.0.4.8. A point $x \in X$ is called the *cluster point* of a sequence (x_n) in $A \subset X$ iff $\exists$ a *subsequence* $(x_{n_k})_{k=1}^{\infty}$ of (x_n) such that $\lim_{k \to \infty} x_{n_k} = x$ in X. (A.0.4.20)

- $A \subset B \Longrightarrow \overline{A} \subset \overline{B}$;
- $\overline{A \cup B} = \overline{A} \cup \overline{B}$;
- $\overline{A \cap B} \subset \overline{A} \cap \overline{B}$. (A.0.4.21)

Dense subset of X and separability of X

Definition A.0.4.9. A nonempty subset $A \subset X$ is called *dense* in X iff $\forall x \in X$, $\exists$ a sequence (x_n) in A such that $x_n \to x$ in X as $n \to \infty$, i.e. $\lim_{n \to \infty} x_n = x$ in X or, equivalently, $\forall x \in X$, $\forall \varepsilon > 0$, $\exists y_\varepsilon \in A$ such that $d(x, y_\varepsilon) < \varepsilon$. (A.0.4.22)

A is *dense* in $X \iff \overline{A} = X$, i.e. the *closure* $\overline{A}$ is X itself $\iff$ every $x \in X$ is a *cluster point* of $A \iff X \subset \overline{A}$. (A.0.4.23)

Definition A.0.4.10. A metric space X is *separable* iff $\exists$ a *countable dense* set in X.

For example, $X \equiv \mathbb{R}$ in (A.0.3.2) is *separable*, since the set $\mathbb{Q}$ of rationals is *dense* in $\mathbb{R}$ and countable. Similarly, $X \equiv \mathbb{R}^n$ in (A.0.3.3) is *separable*.

A.0.5 Compact sets in $\mathbb{R}^n$ with the *usual* metric d_2

Definition A.0.5.1. A subset $K \subset \mathbb{R}^n$ is called *compact* if and only if K is *closed* and *bounded* in $\mathbb{R}^n$, and the *compact inclusion* is denoted by $K \subset\subset \mathbb{R}^n$ (this is a consequence of the Heine–Borel Theorem of real analysis).

- $K \subset\subset \mathbb{R}^n$
  $\iff$ *every* sequence $(x_n)_{n=1}^{\infty}$ in K has a *convergent* subsequence $(x_{n_k})_{k=1}^{\infty}$ in K, i.e. has a *cluster point* $x \in K$; (A.0.5.1)
  $\iff$ every *infinite* open cover $\{\Omega_\lambda\}_{\lambda\in\Lambda}$ of K has a *finite* open *subcover* $\{\Omega_{\lambda_k}\}_{k=1}^{N}$ of K with $\lambda_k \in \Lambda$. (A.0.5.2)
- $A \subset K$ is *closed*, $K \subset\subset \mathbb{R}^n \implies A \subset\subset \mathbb{R}^n$ (A.0.5.3)
- $A \subset B \subset \mathbb{R}^n$, $\overline{B}$ is *compact* $\implies \overline{A}$ is *compact*; (A.0.5.4)
- A is *closed* in $\mathbb{R}^n$, $K \subset\subset \mathbb{R}^n \implies$ both $A + K$ and $A \cap K$ are *compact*. (A.0.5.5)
- $A, B \subset\subset \mathbb{R}^n \implies A \cup B$ is *compact* in $\mathbb{R}^n$. (A.0.5.6)

Theorem A.0.5.1 ([27]). *Let $\Omega \subset \mathbb{R}^n$ be an open subset of $\mathbb{R}^n$, $K \subset\subset \Omega$ be a compact subset of Ω and $\mathbf{x} \in K^{\mathsf{C}}$. Then $\exists$* disjoint, open *subsets U and V in Ω such that $K \subset V$, $\mathbf{x} \in U$ with $U \cap V = \emptyset$.*

Theorem A.0.5.2 ([27]). *Let $\{K_\lambda\}_{\lambda\in\Lambda}$ be an infinite system of* compact *subsets of $\mathbb{R}^n$ such that $\bigcap_{\lambda\in\Lambda} K_\lambda = \emptyset$. Then $\exists$ a* finite *subset of $\{K_\lambda\}_{\lambda\in\Lambda}$ with an* empty intersection.

Theorem A.0.5.3. *Let $K \subset\subset \Omega \subset \mathbb{R}^n$ be a* compact *subset of an* open *set Ω in $\mathbb{R}^n$. Then $\exists$ an open set U with* compact closure $\overline{U}$ *such that $K \subset U \subset \overline{U} \subset\subset \Omega$.*

Proof.
Case $\Omega = \mathbb{R}^n$: Let $\{\mathbf{x}_i\}_{i=1}^{N}$ be N *distinct* points in K whose open neighbourhoods (open sets containing $\mathbf{x}_i$) $\{\Omega_{\mathbf{x}_i}\}_{i=1}^{N}$ have *compact* closures $\{\overline{\Omega}_{\mathbf{x}_i}\}_{i=1}^{N}$ (since $\mathbb{R}^n$ is *locally compact*) such that $\{\Omega_{\mathbf{x}_i}\}_{i=1}^{N}$ is an *open cover* of K, i.e. $K \subset \bigcup_{i=1}^{N} \Omega_{\mathbf{x}_i}$. Define the *open* set $W = \bigcup_{i=1}^{N} \Omega_{\mathbf{x}_i}$ with *compact* closure $\overline{W}$. Then, choosing $U = W$, we have $K \subset U \subset \overline{U} \subset\subset \Omega \equiv \mathbb{R}^n$.
Case $\Omega \subsetneq \mathbb{R}^n$: Let $W = \bigcup_{i=1}^{N} \Omega_{\mathbf{x}_i}$ with *compact* closure $\overline{W}$, both of which we have already constructed in the previous case. Although W is open and $K \subset W$, W *may not* be in Ω. Let $\Omega^{\mathsf{C}} = \mathbb{R}^n \setminus \Omega$. Then, by Theorem A.0.5.1, $\forall \mathbf{y} \in \Omega^{\mathsf{C}}$ $\exists$ an *open* set $V_{\mathbf{y}} \subset \Omega$ with $\overline{V}_{\mathbf{y}} \subset \Omega$ such that $K \subset V_{\mathbf{y}}$ and $\mathbf{y} \notin \overline{V}_{\mathbf{y}} \subset \mathbb{R}^n$. But $\overline{W}$ is *compact* in $\mathbb{R}^n$ and $\overline{V}_{\mathbf{y}}$ is *closed* $\forall \mathbf{y} \in \Omega^{\mathsf{C}}$. Hence, by (A.0.5.5), $\forall \mathbf{y} \in \Omega^{\mathsf{C}}$, Ω^{C} being a closed set, $\Omega^{\mathsf{C}} \cap \overline{W} \cap \overline{V}_{\mathbf{y}}$ is *compact* and $\{\Omega^{\mathsf{C}} \cap \overline{W} \cap \overline{V}_{\mathbf{y}}\}_{\mathbf{y}\in\Omega^{\mathsf{C}}}$ is an *infinite* collection of compact sets with *empty* intersection, i.e. $\Omega^{\mathsf{C}} \cap \overline{W} \bigcap_{\mathbf{y}\in\Omega^{\mathsf{C}}} \overline{V}_{\mathbf{y}} = \emptyset$. Then, by Theorem A.0.5.2,

$\exists$ a *finite* subcollection $\{\Omega^{\complement} \cap \overline{W} \cap \overline{V}_{\mathbf{y}_k}\}_{k=1}^{N}$ with $\mathbf{y}_k \in \Omega^{\complement}$, $K \subset V_{\mathbf{y}_k} \subset \Omega$ and $\mathbf{y}_k \notin \overline{V}_{\mathbf{y}_k}$ such that $\Omega^{\complement} \cap \overline{W} \cap \overline{V}_{\mathbf{y}_1} \cap \overline{V}_{\mathbf{y}_2} \cap \cdots \cap \overline{V}_{\mathbf{y}_N} = \emptyset$. But by construction, $K \subset W$, $K \subset V_{\mathbf{y}_k}$, W and $V_{\mathbf{y}_k}$ being *open* sets such that $\overline{W}$ is *compact* and $\mathbf{y}_k \notin \overline{V}_{\mathbf{y}_k}$, $\mathbf{y}_k \in \Omega^{\complement}$, $1 \le k \le N$. Now we define $U = W \cap V_{\mathbf{y}_1} \cap V_{\mathbf{y}_2} \cap \cdots \cap V_{\mathbf{y}_N}$. Then the closure $\overline{U} \subset \overline{W} \cap \overline{V}_{\mathbf{y}_1} \cap \overline{V}_{\mathbf{y}_2} \cap \cdots \cap \overline{V}_{\mathbf{y}_N}$ by (A.0.4.21) such that U is *open* and $\overline{U}$ is *compact*, since $\overline{W}$ is *compact* and $\bigcap_{k=1}^{N} \overline{V}_{\mathbf{y}_k}$ is *closed* by (A.0.4.14) $\Longrightarrow$ $\overline{W} \cap (\bigcap_{k=1}^{N} \overline{V}_{\mathbf{y}_k})$ is *compact* by (A.0.5.5) $\Longrightarrow$ the closure $\overline{U}$ is *compact* by (A.0.5.3). Thus we have $K \subset U \subset \overline{U} \subset\subset \Omega \subsetneq \mathbb{R}^n$. □

A.0.6 Elementary properties of functions of real variables

Let $\Omega \subset \mathbb{R}^n$ be an *open* subset of $\mathbb{R}^n$ equipped with the usual metric $d_2(\mathbf{x}, \mathbf{y}) = (\sum_{i=1}^{n} |x_i - y_i|^2)^{1/2}$ in (A.0.3.3) $\forall \mathbf{x}, \mathbf{y} \in \Omega$. Define the set $\mathcal{F}(\Omega)$:

$$\mathcal{F}(\Omega) = \{f : f : \mathbf{x} \in \Omega \longmapsto f(\mathbf{x}) \in \mathbb{R} \text{ is a } \textit{real-valued} \text{ function on } \Omega\}.$$

Definition A.0.6.1. A function $f \in \mathcal{F}(\Omega)$ is *bounded from above* (resp. *from below*) iff $\exists M \in \mathbb{R}$ (resp. $\exists m \in \mathbb{R}$) such that $f(\mathbf{x}) \le M \ \forall \mathbf{x} \in \Omega$ (resp. $m \le f(\mathbf{x}) \ \forall \mathbf{x} \in \Omega$). Then f is called *bounded* on Ω iff f is bounded from above *and* from below, i.e. $\exists M > 0$ such that $|f(\mathbf{x})| \le M \ \forall \mathbf{x} \in \Omega$. (A.0.6.1)

Definition A.0.6.2.

- $\alpha = \sup_{\mathbf{x}\in\Omega} f(\mathbf{x})$ if and only if
  1. $\alpha \ge f(\mathbf{x}) \ \forall \mathbf{x} \in \Omega$;
  2. $\forall n \in N$, $\exists \mathbf{x}_n \in \Omega$ such that $f(\mathbf{x}_n) > \alpha - \frac{1}{n}$; (A.0.6.2)
- $\beta = \inf_{\mathbf{x}\in\Omega} f(\mathbf{x})$ iff
  1. $\beta \le f(\mathbf{x}) \ \forall \mathbf{x} \in \Omega$;
  2. $\forall n \in N$, $\exists \mathbf{y}_n \in \Omega$ such that $f(\mathbf{y}_n) < \beta + \frac{1}{n}$; (A.0.6.3)
- $f(\mathbf{x}) \le g(\mathbf{x}) \ \forall \mathbf{x} \in \Omega \Longrightarrow \sup_{\mathbf{x}\in\Omega} f(\mathbf{x}) \le \sup_{\mathbf{x}\in\Omega} g(\mathbf{x})$; $\inf_{\mathbf{x}\in\Omega} f(\mathbf{x}) \le \inf_{\mathbf{x}\in\Omega} g(\mathbf{x})$; (A.0.6.4)
- $\alpha = \sup_{\mathbf{x}\in\Omega} f(\mathbf{x}) = f(\mathbf{x}_0)$ for $\mathbf{x}_0 \in \Omega \Longleftrightarrow \sup_{\mathbf{x}\in\Omega} f(\mathbf{x}) = \max_{\mathbf{x}\in\Omega} f(\mathbf{x}) = f(\mathbf{x}_0)$;
- $\sup_{\mathbf{x}\in\Omega} f(\mathbf{x}) = -\inf_{\mathbf{x}\in\Omega}(-f(\mathbf{x}))$; $\inf_{\mathbf{x}\in\Omega} f(\mathbf{x}) = -\sup_{\mathbf{x}\in\Omega}(-f(\mathbf{x}))$. (A.0.6.5)

A.0.7 Limit of a function at a cluster point $\mathbf{x}_0 \in \mathbb{R}^n$

Let $f \in \mathcal{F}(\Omega)$, $\mathbf{x}_0 \in \mathbb{R}^n$ be a *cluster point* (see Definition A.0.4.6) and $\mathcal{N}_0(\mathbf{x}_0; \delta) = B(\mathbf{x}_0; \delta) - \{\mathbf{x}_0\}$ be a *deleted neighbourhood* of $\mathbf{x}_0$ (see (A.0.4.10)) in Ω.

Definition A.0.7.1. The *limit* of $f \in \mathcal{F}(\Omega)$ at $\mathbf{x}_0$ is the *real number* $L \in R$ iff $\forall \varepsilon > 0$, $\exists \delta > 0$ such that $\forall \mathbf{x} \in \mathcal{N}_0(\mathbf{x}_0; \delta)$, $|f(\mathbf{x}) - L| < \varepsilon$. Then we write $\lim_{\mathbf{x}\to\mathbf{x}_0} f(\mathbf{x}) = L \in \mathbb{R}$, the limit L being a *unique* number.

Cauchy criterion for the existence of limit $\lim_{\mathbf{x}\to\mathbf{x}_0} f(\mathbf{x})$ exists iff for $\varepsilon > 0$, $\exists\delta > 0$ such that

$$|f(\mathbf{x}) - f(\mathbf{y})| < \varepsilon \quad \forall\mathbf{x}, \mathbf{y} \in \mathcal{N}_0(\mathbf{x}_0; \delta) \cap \Omega. \tag{A.0.7.1}$$

$\lim_{\mathbf{x}\to\mathbf{x}_0} f(\mathbf{x}) = L \in \mathbb{R}$ with $L \neq 0 \Longrightarrow \exists$ some $\mathcal{N}(\mathbf{x}_0; \delta)$ of $\mathbf{x}_0$ such that:

1. $f(\mathbf{x})$ is *bounded* in $\mathcal{N}(\mathbf{x}_0; \delta)$;
2. $|f(\mathbf{x})| > \frac{|L|}{2}$ $\forall\mathbf{x} \in \mathcal{N}(\mathbf{x}_0; \delta)$. (A.0.7.2)

A.0.8 Limit superior and limit inferior of a sequence in $\mathbb{R}$

Let $\overline{\mathbb{R}} = \mathbb{R} \cup \{-\infty, \infty\}$ denote the *extended* real number system.

Definition A.0.8.1. Let $(x_n)_{n=1}^{\infty}$ be a sequence on $\mathbb{R}$. $\forall k \in \mathbb{N}$, set

$$\alpha_k = \sup_{n>k}\{x_n\} = \sup\{x_{k+1}, x_{k+2}, \ldots\};$$

$$\beta_k = \inf_{n>k}\{x_n\} = \inf\{x_{k+1}, x_{k+2}, \ldots\},$$

with $\alpha_1 \geq \alpha_2 \geq \cdots \geq \alpha_k \geq \cdots$ such that $\alpha = \inf_{k\in\mathbb{N}} \alpha_k \in \overline{\mathbb{R}}$ and $\beta_1 \leq \beta_2 \leq \cdots \leq \beta_k \leq \cdots$ such that $\beta = \sup_{k\in\mathbb{N}} \beta_k \in \overline{\mathbb{R}}$. (A.0.8.1)

Then α is called the *limit superior* or *upper limit* of (x_n) and denoted by:

$$\alpha = \limsup_{n\to\infty} x_n = \overline{\lim_{n\to\infty}}\, x_n. \tag{A.0.8.2}$$

β is called the *limit inferior* or *lower limit* of (x_n) and denoted by:

$$\beta = \liminf_{n\to\infty} x_n = \underline{\lim_{n\to\infty}}\, x_n. \tag{A.0.8.3}$$

Theorem A.0.8.1.

I. *For* every *sequence* $(x_n)_{n=1}^{\infty}$ *on* $\mathbb{R}$, $\limsup_{n\to\infty} x_n$ *and* $\liminf_{n\to\infty} x_n$ always exist, *but may possibly be* infinite.

II. $\limsup_{n\to\infty} x_n = \liminf_{n\to\infty} x_n \Longrightarrow \lim_{n\to\infty} x_n$ *exists in the* usual sense *on* $\mathbb{R}$. *Then, all the three coincide, i.e.*

$$\limsup_{n\to\infty} x_n = \liminf_{n\to\infty} x_n = \lim_{n\to\infty} x_n. \tag{A.0.8.4}$$

Definition A.0.8.2. Let $f \in \mathcal{F}(\Omega)$ and $\alpha_k = \sup_{n>k} f(\mathbf{x}_n)$, $\beta_k = \inf_{n>k} f(\mathbf{x}_n)$ with $\alpha_1 \geq \alpha_2 \geq \cdots \geq \alpha_k \geq \cdots$ and $\beta_1 \leq \beta_2 \leq \cdots \beta_k \leq \cdots$ such that $\alpha = \inf_{k\in\mathbb{N}} \alpha_k$ and $\beta = \sup_{k\in\mathbb{N}} \beta_k$. Then $\alpha = \limsup_{n\to\infty} f(\mathbf{x}_n) = \overline{\lim}_{n\to\infty} f(\mathbf{x}_n)$, and $\beta = \liminf_{n\to\infty} f(\mathbf{x}_n) = \underline{\lim}_{n\to\infty} f(\mathbf{x}_n)$, which will *always* exist and may possibly be *infinite*. (A.0.8.5)

Moreover, $\overline{\lim}_{n\to\infty} f(\mathbf{x}_n) = \underline{\lim}_{n\to\infty} f(\mathbf{x}_n) \Longrightarrow \lim_{n\to\infty} f(\mathbf{x}_n)$ exists in $\mathbb{R}$ and the *three coincide*. (A.0.8.6)

A.0.9 Pointwise and uniform convergence of sequences of functions

Definition A.0.9.1. A sequence $(f_n)_{n=1}^{\infty}$ of functions $f_n \in \mathcal{F}(\Omega)$ *converges pointwise* on Ω to a function $f \in \mathcal{F}(\Omega)$ if and only if, $\forall \mathbf{x} \in \Omega$, the sequence $(f_n(\mathbf{x}))_{n=1}^{\infty}$ of *real numbers* converges to the number $f(\mathbf{x})$ in $\mathbb{R}$ as $n \to \infty$, i.e. $\forall \mathbf{x} \in \Omega$, $\forall \varepsilon > 0$, $\exists n_0 = n_0(\varepsilon; \mathbf{x}) \in \mathbb{N}$ (i.e. n_0 depends on *both* ε and $\mathbf{x}$) such that $|f(\mathbf{x}) - f_n(\mathbf{x})| < \varepsilon$ $\forall n > n_0$.

Moreover, we say that the sequence $(f_n)_{n=1}^{\infty}$ converges *uniformly* on Ω to $f \in \mathcal{F}(\Omega)$ if and only if

$$\sup_{\mathbf{x}\in\Omega} |f(\mathbf{x}) - f_n(\mathbf{x})| < \varepsilon \quad \forall n \geq n_0, \tag{A.0.9.1}$$

i.e. $\forall \mathbf{x} \in \Omega$, $\forall \varepsilon > 0$, $\exists n_0 = n_0(\varepsilon) \in \mathbb{N}$ (i.e. n_0 depends *only* on ε) such that

$$|f(\mathbf{x}) - f_n(\mathbf{x})| \leq \sup_{\mathbf{x}\in\Omega} |f(\mathbf{x}) - f_n(\mathbf{x})| = \|f - f_n\|_{\infty} < \varepsilon \quad \forall n > n_0. \tag{A.0.9.2}$$

Remark A.0.9.1. The notation $\|f - f_n\|_{\infty}$ is also used to denote $\sup_{\mathbf{x}\in\Omega} |f(\mathbf{x}) - f_n(\mathbf{x})|$. (A.0.9.3)

Theorem A.0.9.1. *A sequence (f_n) of continuous functions $f_n \in \mathcal{F}(\Omega)$ converges uniformly to $f \Longrightarrow f$ is continuous on Ω.*

A.0.10 Continuity and uniform continuity of $f \in \mathcal{F}(\Omega)$

Definition A.0.10.1. A function $f \in \mathcal{F}(\Omega)$ is said to be *continuous* at a point $\mathbf{x}_0 \in \Omega$ if and only if, $\forall \varepsilon > 0$, $\exists \delta = \delta(\varepsilon; \mathbf{x}_0) > 0$ (i.e. δ *depends* on both ε and $\mathbf{x}_0$) such that $d(\mathbf{x}, \mathbf{x}_0) < \delta$ with $\mathbf{x} \in \Omega \Longrightarrow |f(\mathbf{x}) - f(\mathbf{x}_0)| < \varepsilon$, or equivalently iff $\forall (\mathbf{x}_n)$ in Ω with $\mathbf{x}_n \to \mathbf{x}_0$ in $\Omega \Longrightarrow f(\mathbf{x}_n) \to f(\mathbf{x}_0)$ in $\mathbb{R}$ as $n \to \infty$. (A.0.10.1)

Then f is *continuous* on Ω iff f is continuous at each $\mathbf{x} \in \Omega$.

Moreover, f is said to be *uniformly continuous* on Ω iff $\forall \varepsilon > 0$, $\exists \delta = \delta(\varepsilon) > 0$ (i.e. δ depends *only* on ε) such that $d(\mathbf{x}, \mathbf{y}) < \delta$ with $\mathbf{y} \in \Omega \Longrightarrow |f(\mathbf{x}) - f(\mathbf{y})| < \varepsilon$.

Then f is *uniformly continuous* on $\Omega \Longrightarrow f$ is *continuous* on Ω, and the converse is *false*.

For example, f defined by $f(x) = x^2$ $\forall x \in \mathbb{R}^2$ is *continuous* on $\mathbb{R}$, but *not uniformly continuous* on $\mathbb{R}$.

A.1 Important properties of continuous functions

A.1.1 Some remarkable properties on compact sets in $\mathbb{R}^n$

Theorem A.1.1.1. *Let* $f \in \mathcal{F}(\Omega)$ *be* continuous *on* $\Omega \subset \mathbb{R}^n$ *and* $K \subset\subset \Omega$ *be* compact *in* Ω. *Then* $f(K)$ *is* compact *in* $\mathbb{R}$*;*

I. f *is* bounded *on* K*;*

II. $\exists \mathbf{x}_0 \in K$ *(resp.* $\exists \mathbf{y}_0 \in K$*) such that* $f(\mathbf{x}_0) = \max_{\mathbf{x}\in K} f(\mathbf{x})$ *(resp.* $f(\mathbf{y}_0) = \min_{\mathbf{x}\in K} f(\mathbf{x})$*);*

III. f *is* uniformly continuous *on* K.

IV. *Moreover, let* $(f_n)_{n=1}^{\infty}$ *be a sequence of continuous functions on* Ω *such that* (f_n) *converges uniformly to* f *on* K. *Then* f *is* bounded *and* uniformly continuous *on* K.

A.1.2 $C_0^\infty(\Omega)$-partition of unity on compact set $K \subset\subset \Omega \subset \mathbb{R}^n$

For details of the $C_0^\infty(\Omega)$-partition of unity on a compact set $K \subset\subset \Omega \subset \mathbb{R}^n$, see Appendix C.

A.1.3 Continuous extension theorems

Theorem A.1.3.1 (Tietze–Urysohn Extension Theorem, Dieudonné [28]). *Let* X *be a metric space,* $A \subset X$ *be a* closed *subset of* X *and* $f : A \to \mathbb{R}$ *be a* bounded, continuous *real-valued function on* A. *Then there is a* continuous, *real-valued function* $g : X \longrightarrow \mathbb{R}$ *such that* $g(x) = f(x)\ \forall x \in A$ *with*

$$\sup_{x\in X} g(x) = \sup_{y\in A} f(y) \quad \textit{and} \quad \inf_{x\in X} g(x) = \inf_{y\in A} f(y). \tag{A.1.3.1}$$

Proposition A.1.3.1. *Let* A *and* B *be two* nonempty, disjoint, closed *subsets of a metric space* X, *i.e.* $A \cap B = \emptyset$. *Then* $\exists$ *a continuous function* g *defined on* X *with values* $g(x) \in [0,1]$ *such that* $g(x) = 0$ *for* $x \in A$, $g(x) = 1$ *for* $x \in B$.

Proof. $A, B \subset X$ are *non-empty, disjoint* and *closed* $\Longrightarrow \overline{A \cup B} = \overline{A} \cup \overline{B} = A \cup B \Longrightarrow A \cup B$ is closed. Define function f on $A \cup B$ such that $f(x) = 0$ for $x \in A$, $f(x) = 1$ for $x \in B$. Then f is continuous in A and in $B \Longrightarrow f$ is continuous in $A \cup B$ with values $f(x) = 0$ for $x \in A$ and $f(x) = 1$ for $x \in B$, $|f(x)| = f(x) \le 1$ for $x \in A \cup B$. Hence, f is a *bounded, continuous* function on the closed set $A \cup B \subset X$. Then, by the Tietze–Urysohn Extension Theorem A.1.3.1, $\exists$ a *continuous* function $g : X \longrightarrow \mathbb{R}$ with $g(x) = f(x) \in \{0,1\} \subset [0,1] \subset \mathbb{R}\ \forall x \in A \cup B$, $\inf_{x\in X} g(x) = \inf_{x\in A\cup B} f(x) = 0$, $\sup_{x\in X} g(x) = \sup_{x\in A\cup B} f(x) = 1$, i.e. $0 \le g(x) \le 1\ \forall x \in X$, since g is *continuous* on X. Thus, $g : X \longrightarrow [0,1]$ is *continuous* from X into $[0,1]$. (For an alternative, interesting proof, see [4].) □

Proposition A.1.3.2. *Let* $\Omega \subset \mathbb{R}^n$ *be a* bounded, open *subset of* $\mathbb{R}^n$ *and* $\psi \in C_0(\Omega)$ *be a continuous function in* Ω *with compact support* $K = \text{supp}(\psi) \subset\subset \Omega$. *Let* $A, B \subset \Omega$ *be defined, for fixed* $\varepsilon > 0$, *by:*

$$A = \{\mathbf{x} : \mathbf{x} \in \Omega,\ \psi(\mathbf{x}) \geq \varepsilon\}, \quad B = \{\mathbf{x} : \mathbf{x} \in \Omega,\ \psi(\mathbf{x}) \leq -\varepsilon\}. \tag{A.1.3.2}$$

Then $\exists$ *a continuous function* $\phi_0 \in C_0(\Omega)$ *with compact support in* Ω *such that* $\phi_0(\mathbf{x}) = +1\ \forall \mathbf{x} \in A$, $\phi_0(\mathbf{x}) = -1\ \forall \mathbf{x} \in B$, *and* $-1 \leq \phi_0(\mathbf{x}) \leq 1\ \forall \mathbf{x} \in \Omega$.

Proof. First we show that A (resp. B) is closed in Ω. Let $(\mathbf{x}_n)_{n=1}^{\infty}$ be any sequence in A such that $\mathbf{x}_n \to \mathbf{x}$ in Ω as $n \to \infty$. For the proof it is sufficient to show that $\mathbf{x} \in A$. In fact, for fixed $\varepsilon > 0$, $\psi(\mathbf{x}_n) \geq \varepsilon\ \forall n \in \mathbb{N}$. Then $\lim_{n\to\infty} \psi(\mathbf{x}_n) \geq \varepsilon$. But $\lim_{n\to\infty} \psi(\mathbf{x}_n) = \psi(\mathbf{x})$, since ψ is continuous in Ω and $\mathbf{x}_n \to \mathbf{x}$ in Ω. Hence, $\psi(\mathbf{x}) \geq \varepsilon \Longrightarrow \mathbf{x} \in A$. Similarly, B is closed in Ω. Moreover, $\mathbf{x} \in A \Longrightarrow \mathbf{x} \notin B$, and $\mathbf{x} \in B \Longrightarrow \mathbf{x} \notin A$. Hence, A, B are *disjoint, bounded, closed*, proper subsets of K, i.e. $A, B \subsetneq K \subset\subset \Omega$. Then, following the steps of the proof of Proposition A.1.3.1, define a *bounded, continuous* function $f : A \cup B \longrightarrow \mathbb{R}$ such that $f(\mathbf{x}) = +1$ $\forall \mathbf{x} \in A$, $f(\mathbf{x}) = -1\ \forall \mathbf{x} \in B$, $|f(\mathbf{x})| \leq 1\ \forall \mathbf{x} \in A \cup B$, $A \cup B \subset K$ being a bounded, closed subset of $K \subset\subset \Omega$. By Theorem A.1.3.1, $\exists$ a *bounded, continuous* function g on Ω which extends f to Ω, i.e. $g(\mathbf{x}) = f(\mathbf{x})\ \forall \mathbf{x} \in A \cup B$, with values $g(\mathbf{x}) \in [-1, 1] \subset \mathbb{R}$, such that $g(\mathbf{x}) = f(\mathbf{x}) = +1\ \forall \mathbf{x} \in A$, $g(\mathbf{x}) = f(\mathbf{x}) = -1$ $\forall \mathbf{x} \in B$ and $-1 \leq g(\mathbf{x}) \leq 1\ \forall \mathbf{x} \in \Omega$. By Theorem 6.2.4 of Chapter 6, $\exists \phi \in C_0(\Omega)$ with support contained in some neighbourhood K_0 of K such that $\phi(\mathbf{x}) = 1\ \forall \mathbf{x} \in K$ with $K \subset K_0$ and $0 \leq \phi(\mathbf{x}) \leq 1\ \forall \mathbf{x} \in \Omega$. Hence, $\phi(\mathbf{x}) = 1$ for $\mathbf{x} \in A \cup B \subset K$ with $K \subset K_0$. Define $\phi_0 = \phi g$. Then, $\phi_0 \in C_0(\Omega)$ with

$$\phi_0(\mathbf{x}) = \phi(\mathbf{x})g(\mathbf{x}) = g(\mathbf{x}) = \begin{cases} +1 & \text{for } \mathbf{x} \in A \\ -1 & \text{for } \mathbf{x} \in B, \end{cases}$$

$-1 \leq \phi_0(\mathbf{x}) \leq 1\ \forall \mathbf{x} \in \Omega$, and $\text{supp}(\phi_0) \subset (\text{supp}(\phi) \cap \text{supp}(g)) \subset\subset \Omega$. □

Theorem A.1.3.2 (Continuous extension by density). *Let* $\Omega \subset \mathbb{R}^n$ *be an* open *subset of* $\mathbb{R}^n$, $\overline{\Omega}$ *be its* closure *in* $\mathbb{R}^n$ *and* f *be a* uniformly continuous *function on* Ω. *Then* f *has a* unique, continuous extension $\tilde{f}$ *to* $\overline{\Omega}$, *i.e.* $\tilde{f}(\mathbf{x}) = f(\mathbf{x})\ \forall \mathbf{x} \in \Omega$ *and* $\tilde{f}$ *is* continuous *on* $\overline{\Omega}$. *(In fact,* $\tilde{f}$ *is* uniformly continuous *on* $\overline{\Omega}$*).*

Theorem A.1.3.3 (Extension of bounded and uniformly continuous functions). *Let* $\overline{\Omega}$ *be the* closure *of an* open *subset* $\Omega \subset \mathbb{R}^n$ *of* $\mathbb{R}^n$. *Then every* bounded, uniformly continuous *function* f *on* Ω *has a* unique *extension* $\tilde{f}$ *to* $\overline{\Omega}$ *which is (*uniformly*)* continuous *and* bounded *on* $\overline{\Omega}$.

Proof. By Theorem A.1.3.2, f has a *unique, continuous* extension $\tilde{f}$ to $\overline{\Omega}$. $\forall \mathbf{x} \in \overline{\Omega}$, $\exists$ a sequence $(\mathbf{x}_n)$ in Ω such that $\mathbf{x}_n \to \mathbf{x}$ as $n \to \infty \Longrightarrow \lim_{n\to\infty} \tilde{f}(\mathbf{x}_n) = \tilde{f}(\mathbf{x})$ by virtue of the continuity of $\tilde{f}$. But $\mathbf{x}_n \in \Omega \ \forall n \in \mathbb{N} \Longrightarrow \tilde{f}(\mathbf{x}_n) = f(\mathbf{x}_n) \ \forall n \in \mathbb{N}$. Since f is bounded on Ω, $\exists M > 0$ such that $|f(\mathbf{x}_n)| \leq M \ \forall n \in \mathbb{N}$. Hence, $\lim_{n\to\infty} |\tilde{f}(\mathbf{x}_n)| = \lim_{n\to\infty} |f(\mathbf{x}_n)| \leq M \Longrightarrow \| \tilde{f}(\mathbf{x})| - |\tilde{f}(\mathbf{x}_n)\| \leq |\tilde{f}(\mathbf{x}) - \tilde{f}(\mathbf{x}_n)| \to 0$ as $n \to \infty \Longrightarrow |\tilde{f}(\mathbf{x})| = \lim_{n\to\infty} |\tilde{f}(\mathbf{x}_n)| \leq M \ \forall \mathbf{x} \in \overline{\Omega} \Longrightarrow \tilde{f}$ is bounded on $\overline{\Omega}$.

By virtue of the uniqueness of the continuous extension $\tilde{f}$ to $\overline{\Omega}$ of the *uniform, continuous* function f on Ω, the *extended function* will still be denoted by the *same notation* f (instead of $\tilde{f}$) in order to simplify notations. □

A.2 Finite and infinite dimensional linear spaces

A.2.1 Linear spaces

Definition A.2.1.1. Let X be a *nonempty* set and $\mathbb{F}$ be a number field, i.e. $\mathbb{F} = \mathbb{R}$ or $\mathbb{C}$, such that X is *closed* under the following *binary* operations:

$$x, y \in X \Longrightarrow x + y = z \in X; \quad \alpha \in \mathbb{F}, x \in X \Longrightarrow \alpha x = w \in X; \quad \text{(A.2.1.1)}$$

and $\exists$ a *null element* $0 \in X$ such that

$$0 + x = x \quad \forall x \in X. \quad \text{(A.2.1.2)}$$

Then, if the following properties hold $\forall x, y, z \in X$, $\forall \alpha, \beta, \gamma \in \mathbb{F}$,

1. $x + y = y + x$;
2. $(x + y) + z = x + (y + z)$;
3. $\forall x \in X, \exists -x \in X$ such that $x + (-x) = 0 \in X$; (A.2.1.3)
4. $1 \cdot x = x$;
5. $\alpha(x + y) = \alpha x + \alpha y$;
6. $(\alpha + \beta)x = \alpha x + \beta x$;
7. $\alpha(\beta x) = (\alpha\beta)x$, (A.2.1.4)

the pair $(X, \mathbb{F})$ or $X(\mathbb{F})$ or simply X is called a *vector space* or *linear space* or *linear set* over $\mathbb{F}$; these names will be used interchangeably as *synonyms*. Elements of a vector space X are called *vectors* or *points* and will be denoted, in general, in bold $\mathbf{x} \in X, \mathbf{0} \in X$, etc.

- A vector space will contain either one element, i.e. $X = \{\mathbf{0}\}$ is a vector space, or an *infinite* number of elements including $\mathbf{0}$. (A.2.1.5)
- $\mathbb{F} = \mathbb{R} \Longrightarrow (X, \mathbb{R})$ or X is a *real vector space*. (A.2.1.6)
- $\mathbb{F} = \mathbb{C} \Longrightarrow (X, \mathbb{C})$ or X is a *complex vector space*. (A.2.1.7)

We will be primarily concerned with real vector spaces.

Remark A.2.1.1. In all situations which we have dealt with in this book, if the conditions in (A.2.1.1)–(A.2.1.2) hold in X, the other conditions 1–7 will automatically hold, i.e. we will deal with sets for which (A.2.1.1) and (A.2.1.2) will determine whether or not X is a vector space.

Subspace of a vector space X Let $M \subset X$ be a nonempty subset of X. If M is a *vector space* itself over the same field $\mathbb{F} = \mathbb{R}$ or $\mathbb{C}$, then M is called a *subspace* of X.

$\{\mathbf{0}\}$ and X are *trivial* subspaces of X. (A.2.1.8)

Linear independence and dependence, dimension, basis of a vector space

Definition A.2.1.2. Let $(\mathbf{x}_i)_{i=1}^m$ be a system of vectors $\mathbf{x}_i$ in X. Then $(\mathbf{x}_i)_{i=1}^m$ are called *linearly independent* in X if the equation $\sum_{i=1}^m \alpha_i \mathbf{x}_i = \mathbf{0}$ for $\alpha_i \in \mathbb{F}$, $1 \le i \le m$, has *only* the *trivial* solution $\alpha_1 = \alpha_2 = \cdots = \alpha_m = 0$. Otherwise, $(\mathbf{x}_i)_{i=1}^m$ are called *linearly dependent* in X.

Definition A.2.1.3. Let $(\mathbf{x}_i)_{i=1}^m$ be *linearly independent* in X, but $(\mathbf{x}_1, \mathbf{x}_2, \dots, \mathbf{x}_m; \mathbf{y})$ be *linearly dependent* $\forall \mathbf{y} \in X$. Then:

1. $\exists$ *at most m* linearly independent vectors in X, and X is a *finite-dimensional vector space* with the *dimension* of
$$X = \dim(X) = m; \tag{A.2.1.9}$$
2. $(\mathbf{x}_i)_{i=1}^m$ is a *basis* in X and $\forall \mathbf{y} \in X$,
$$\mathbf{y} = \sum_{i=1}^m \alpha_i \mathbf{x}_i, \tag{A.2.1.10}$$
the representation being a *unique* one.

Definition A.2.1.4. Let $(\mathbf{x}_i)_{i=1}^m$ be linearly *independent* in X with $\dim(X) = m$. Then, $\forall k \le m$, $M_k = \{\mathbf{x} : \mathbf{x} = \sum_{i=1}^k \alpha_i \mathbf{x}_i,\ \alpha_i \in \mathbb{F}\}$ is called a *k-dimensional subspace* of X *spanned* by or generated by $(\mathbf{x}_i)_{i=1}^k$, and we write $M_k = \operatorname{span}\{\mathbf{x}_i\}_{i=1}^k$ with $k \le m$.

Definition A.2.1.5. If $\forall m \in \mathbb{N}$, $\exists$ *m linearly independent* vectors $(\mathbf{x}_i)_{i=1}^m$ in X, then X is called an *infinite-dimensional vector space* and we write $\dim(X) = \infty$.

$M \subset X$ is a *subspace* of $X \Longrightarrow \dim(M) \le \dim(X)$. (A.2.1.11)

Example A.2.1.1.

1. $$\mathbb{R}^n = \underbrace{\mathbb{R} \times \cdots \times \mathbb{R}}_{n \text{ times}} = \{\mathbf{x} : \mathbf{x} = (x_1, \ldots, x_n) \text{ with } x_i \in \mathbb{R},\ 1 \le i \le n\}$$

   is an *n-dimensional linear space* with $\dim(\mathbb{R}^n) = n$, $(\hat{\mathbf{e}}_i)_{n=1}^n$ being the *canonical* or *standard* basis:

   $$\begin{aligned} \hat{\mathbf{e}}_1 &= (1, 0, \ldots, 0, \ldots), \\ \hat{\mathbf{e}}_i &= (0, \ldots, 0, 1, 0, \ldots 0) \in \mathbb{R}^n, \quad 1 \le i \le n. \end{aligned} \tag{A.2.1.12}$$

2. $P_n([a,b]) = \{p : p$ is a polynomial of *degree* $\le n$ in variable x defined on $[a,b] \subset \mathbb{R}\}$ is an $(n+1)$*-dimensional vector space* with a *basis* $\{p_k\}_{k=0}^n$ with

   $$p_k(x) = x^k \quad \forall k = 0, 1, \ldots, n,\ \forall x \in [a,b]; \tag{A.2.1.13}$$

3. $\mathcal{P}([a,b]) = \{p : p$ is a polynomial of *arbitrary* degree in variable x defined on $[a,b]\}$ is an *infinite-dimensional vector space* with a *linearly independent system* $\{p_k\}_{k=0}^\infty = \{p_0, p_1, p_2, \ldots\}$, i.e. $\{p_k\}_{k=0}^\infty$ is linearly independent $\forall k \in \mathbb{N}_0$,

   $$p_k(x) = x^k \quad \forall x \in [a,b],\ \forall k = 0, 1, 2, \ldots, \tag{A.2.1.14}$$

   i.e. $\dim(\mathcal{P}([a,b])) = \infty$; (the notation used to denote that $\mathcal{P}([a,b])$ is an infinite dimensional vector space) (A.2.1.14a)

4. $C^0([a,b]) = \{f : f \in \mathcal{F}(\Omega),\ f$ is a *continuous* function on $[a,b] \subset\subset \mathbb{R}\}$ is an *infinite-dimensional vector space* with a *linearly independent system* $\{p_k\}_{k=0}^\infty = \{p_0, p_1, \ldots\}$, $p_k \in C^0([a,b])$ with

   $$p_k(x) = x^k \quad \forall k \in \mathbb{N},\ \forall x \in [a,b], \tag{A.2.1.15}$$

   i.e. $\dim(C^0([a,b])) = \infty$.

Elements (i.e. *polynomials* in $\mathcal{P}([a,b])$ or *functions* in $C^0([a,b])$) are *vectors*, i.e. abstract, mathematical vectors.

Remark A.2.1.2. All the spaces $C^\infty(\Omega)$, $C_0^\infty(\Omega)$, $\mathcal{D}(\Omega)$, $C^k(\overline{\Omega})$, $L^p(\Omega)$, $1 \le p \le \infty$, $H^m(\Omega)$, $W^{m,p}(\Omega)$, $\mathcal{D}'(\Omega)$, $S(\mathbb{R}^n)$, $S'(\mathbb{R}^n)$, etc. which we deal with in this book are *infinite-dimensional vector spaces.*

Mapping on vector spaces There are two kinds of mappings on vector spaces V:

- *vector-valued* mappings called *operators*;
- *scalar-valued* mappings called *functionals*, which will be dealt with first.

A.2.2 Linear functionals

Let V be a *real* (resp. *complex*) *vector space*.

Definition A.2.2.1. A mapping $J : V \longrightarrow \mathbb{R}$ (resp. $\mathbb{C}$), which assigns to each $\mathbf{v} \in V$ a *unique real* (resp. *complex*) number $\alpha \in \mathbb{R}$ (resp. $\mathbb{C}$) with $J(\mathbf{v}) = \alpha$, is called a *functional* from V into $\mathbb{R}$ (resp. $\mathbb{C}$).

Definition A.2.2.2. A functional $L : V \longrightarrow \mathbb{R}$ (resp. $\mathbb{C}$) is called *linear* on V iff $\forall \alpha_i \in \mathbb{R}$ (resp. $\mathbb{C}$), $\forall \mathbf{x}, \mathbf{y} \in V$,

$$L(\alpha_1 \mathbf{x} + \alpha_2 \mathbf{y}) = \alpha_1 L(\mathbf{x}) + \alpha_2 L(\mathbf{y}). \tag{A.2.2.1}$$

Otherwise, L is called *nonlinear* on V.
A linear functional on V is also called a *linear form*. (A.2.2.2)

Except in Fourier analysis, we will only deal with *real-valued* linear functionals. Hence, from now on, L is a *real-valued linear functional* unless stated otherwise.

Algebraic dual space V^* of V Let V^* be the set of all *linear functionals* L defined on V, i.e.

$$V^* = \{L : L : V \longrightarrow \mathbb{R} \text{ is a } \textit{linear functional} \text{ on } V\} \tag{A.2.2.3}$$

such that $\forall L_1, L_2 \in V^*$, $\forall \mathbf{x} \in V$, $\forall \alpha \in \mathbb{R}$,

$$(L_1 + L_2)(\mathbf{x}) = L_1(\mathbf{x}) + L_2(\mathbf{x}); \quad (\alpha L)(\mathbf{x}) = \alpha . L(\mathbf{x}); \tag{A.2.2.4}$$

and $\exists$ a *null functional* $L = 0 \in V^*$ defined by

$$0(\mathbf{x}) = 0 \in \mathbb{R} \quad \forall \mathbf{x} \in X. \tag{A.2.2.5}$$

Definition A.2.2.3. V^* with (A.2.2.4) and (A.2.2.5) is a *linear space* called the algebraic *dual space* of V.

Theorem A.2.2.1. *Let V be a finite-dimensional vector space with* $\dim(V) = n$ *and basis* $(\mathbf{x}_i)_{i=1}^n$, *and V^* be the algebraic dual space of V. Then:*

I. *for arbitrary real numbers $\alpha_1, \alpha_2, \ldots, \alpha_n$, $\exists$ a unique $l \in V^*$ such that*

$$l(\mathbf{x}_i) = \alpha_i, \quad 1 \le i \le n; \tag{A.2.2.6}$$

II. *$\exists$ a unique basis $(l_i)_{i=1}^n$ in V^*, called the* dual basis *of $(\mathbf{x}_i)_{i=1}^n$, such that*

$$l_i(\mathbf{x}_j) = \delta_{ij}, \quad 1 \le i, j \le n; \tag{A.2.2.7}$$

III. *for $V^{**} = (V^*)^*$, V can always be identified with V^{**} by writing $V \equiv V^{**}$, i.e. V is* reflexive. (A.2.2.8)

For interesting results on linear functionals on *infinite-dimensional* spaces, see later in Section A.7.

A.2.3 Linear operators

Let V and W be real vector spaces.

Definition A.2.3.1. The mapping $A : V \longrightarrow W$ from V into W, which assigns to each $\mathbf{v} \in V$ a *unique* $\mathbf{w} \in W$ with $A\mathbf{v} = \mathbf{w}$, is called an *operator* with its *domain of definition* $D(A) = V$ and range

$$\mathcal{R}(A) = \{\mathbf{w} : \mathbf{w} \in W, \exists \mathbf{v} \in V \text{ such that } A\mathbf{v} = \mathbf{w}\} \subset W. \tag{A.2.3.1}$$

Then $A : V \longrightarrow W$ is called a *linear operator* from V into W iff

$$A(\alpha_1 \mathbf{v}_1 + \alpha_2 \mathbf{v}_2) = \alpha_1 A\mathbf{v}_1 + \alpha_2 A\mathbf{v}_2 \quad \forall \alpha_i \in \mathbb{R},\ \forall \mathbf{v}_i \in V,\ i = 1, 2. \tag{A.2.3.2}$$

Otherwise, A is called nonlinear.

$A, B : V \longrightarrow W$ are *equal* $\Longleftrightarrow$ $A\mathbf{v} = B\mathbf{v}$ in W $\forall \mathbf{v} \in V$. (A.2.3.3)

The *null operator* is 0 such that $0\mathbf{v} = \mathbf{0} \in W$ $\forall \mathbf{v} \in V$. (A.2.3.4)

For $W = V$, $I : V \longrightarrow V$ is an *identity operator* defined by $I\mathbf{v} = \mathbf{v}$ $\forall \mathbf{v} \in V$. (A.2.3.5)

The *null space* $\mathcal{N}(A)$ or *kernel* $\operatorname{Ker}(A)$ of a *linear* operator $A : V \longrightarrow W$ is defined by:

$$\mathcal{N}(A) = \operatorname{Ker}(A) = \{\mathbf{v} : \mathbf{v} \in V, A\mathbf{v} = \mathbf{0} \in W\} \subset V. \tag{A.2.3.6}$$

Proposition A.2.3.1. *$\mathcal{N}(A)$ and $\mathcal{R}(A)$ are subspaces of V and W, respectively.*

A is called

- *surjective* or *onto* $\Longleftrightarrow$ $\mathcal{R}(A) = W$; (A.2.3.7)
- *injective* or *one-to-one* $\Longleftrightarrow$ $\operatorname{Ker}(A) = \{\mathbf{0}\}$; (A.2.3.8)
- *bijective* or *one-to-one* and *onto* $\Longleftrightarrow$ $\operatorname{Ker}(A) = \{\mathbf{0}\}$ *and* $\mathcal{R}(A) = W$. (A.2.3.9)

$M \subset V \Longrightarrow A(M) \subset W$ is called the *image* of M under A. (A.2.3.10)

$A(M) = \{\mathbf{w} : \mathbf{w} = A\mathbf{v} \text{ with } \mathbf{v} \in M\}$ is a *subspace* of W when M is a subspace of V.

$\mathcal{N} \subset W \Longrightarrow A^{-1}(\mathcal{N})$ is called the *pre-image* of $\mathcal{N}$ under A (possibly the *empty set* $\emptyset$, even when A^{-1} does *not* exist).

$A^{-1}(\mathcal{N}) = \{\mathbf{x} : \mathbf{x} \in V \text{ such that } A\mathbf{x} \in \mathcal{N}\} \subset V$ is a *subspace* of V when $\mathcal{N}$ is a subspace of W. (A.2.3.11)

$\mathcal{R}(A) = A(V) \Longrightarrow A : V \longrightarrow A(V) \subset W$ is *surjective* from V onto $A(V)$. (A.2.3.12)

A is a *surjection* $\Longrightarrow$ $A\mathbf{x} = \mathbf{y} \in W$ has *at least one* solution $\mathbf{x} \in V$. (A.2.3.13)

A is an *injection* $\Longrightarrow$ $A\mathbf{x} = \mathbf{y} \in W$ has *at most one* solution $\mathbf{x} \in V$. (A.2.3.14)

A is a *bijection* $\Longrightarrow$ $A\mathbf{x} = \mathbf{y} \in W$ has *exactly one* solution $\mathbf{x} \in V$. (A.2.3.15)

A is *bijective* from V onto W $\Longrightarrow$ $A^{-1} : W \longrightarrow V$ exists and is *linear* such that $AA^{-1} = I_W$, $A^{-1}A = I_V$. (A.2.3.16)

Linear operator $A : V \longrightarrow W$ is *onto* (surjective) $\Longrightarrow \exists$ a *right-hand inverse* $B : W \longrightarrow V$ defined by $AB\mathbf{w} = \mathbf{w}\ \forall \mathbf{w} \in W$ (left-hand inverse may not exist). (A.2.3.17)

Algebraic isomorphism and algebraically isomorphic vector spaces

Definition A.2.3.2. Let V and W be vector spaces. Then a mapping $T : V \longrightarrow W$ is called an *algebraic isomorphism* from vector space V *onto* vector space W iff

1. T is *linear*;
2. T is *bijective* from V onto W (i.e. $T^{-1} : W \longrightarrow V$ exists and is *linear* from W *onto* V), and the two vector spaces V and W are called *algebraically isomorphic*.

Theorem A.2.3.1. *Every* n-dimensional vector space V *is* algebraically isomorphic *to* $\mathbb{R}^n$.

Additional results on linear operators will be given later, in Section A.8.

A.3 Normed linear spaces

A.3.1 Semi-norm and norm

Let V be a real vector space.

Definition A.3.1.1. A *semi-norm* p in V is a mapping $\mathbf{x} \in V \longmapsto p(\mathbf{x}) \in \mathbb{R}_0^+$ from V into $\mathbb{R}_0^+ = [0, \infty[$ satisfying:

1. $p(\mathbf{x}+\mathbf{y}) \le p(\mathbf{x}) + p(\mathbf{y})\ \forall \mathbf{x}, \mathbf{y} \in V$;
2. $p(\alpha\mathbf{x}) = |\alpha| p(\mathbf{x})\ \forall \alpha \in \mathbb{R}, \forall \mathbf{x} \in V$. (A.3.1.1)

Definition A.3.1.2. A *norm* $\|\cdot\|_V$ in V is a *semi-norm* p in V with the *additional property* $p(\mathbf{x}) = 0 \Longrightarrow \mathbf{x} = \mathbf{0}$ in V, i.e.

1. $\|\mathbf{x}\|_V = p(\mathbf{x}) \ge 0\ \forall \mathbf{x} \in V$; $\|\mathbf{x}\|_V = p(\mathbf{x}) = 0 \Longrightarrow \mathbf{x} = 0$ in V;
2. $\|\alpha\mathbf{x}\|_V = p(\alpha\mathbf{x}) = |\alpha|\|\mathbf{x}\|\ \forall \alpha \in \mathbb{R}, \forall \mathbf{x} \in V$;
3. $\|\mathbf{x}+\mathbf{y}\|_V = p(\mathbf{x}+\mathbf{y}) \le \|\mathbf{x}\| + \|\mathbf{y}\|\ \forall \mathbf{x}, \mathbf{y} \in V$.

(1)–(3) are the *properties of norm* $\|\cdot\|_V$. (A.3.1.2)

Then the linear space V equipped with the norm $\|\mathbf{x}\|_V = \|\mathbf{x}\|\ \forall \mathbf{x} \in V$ is called the *normed linear space* V. Hence, the pair $(V, \|\cdot\|)$ is the normed linear space V.

Let $\|\cdot\|_1$ and $\|\cdot\|_2$ be any two different norms on the same linear space V. Then the two norms $\|\cdot\|_1, \|\cdot\|_2$ are called *equivalent* on V iff $\exists C_1, C_2 > 0$ such that

$$C_1\|\mathbf{x}\|_1 \le \|\mathbf{x}\|_2 \le C_2\|\mathbf{x}\|_1 \quad \forall \mathbf{x} \in V. \tag{A.3.1.3}$$

Hence, $(V, \|\cdot\|_1)$, $(V, \|\cdot\|_2)$ are two *different* normed linear spaces. (A.3.1.4)

All norms on a finite-dimensional vector space are *equivalent.* (A.3.1.5)

Every normed linear space V is *metrizable* and becomes a *metric space* if we define the *metric* d on V by:

$$d(\mathbf{x}, \mathbf{y}) = \|\mathbf{x} - \mathbf{y}\|_V \quad \forall \mathbf{x}, \mathbf{y} \in V, \tag{A.3.1.6}$$

and the pair (V, d) is the corresponding metric space V. (A.3.1.7)

All the results in Sections A.0.3 and A.0.4 can be extended to normed linear space V by setting $d(\mathbf{x}, \mathbf{y}) = \|\mathbf{x} - \mathbf{y}\|_V$ in the corresponding formulae. (A.3.1.8)

Banach spaces

Definition A.3.1.3. A normed linear space V is called *complete* iff *every* Cauchy sequence in V is a *convergent* sequence in V. Then V is called a *Banach space.*

Convergence in a Banach space V A sequence $(\mathbf{u}_n)_{n=1}^{\infty}$ in a *Banach space* V converges iff $(\mathbf{u}_n)_{n=1}^{\infty}$ is a *Cauchy sequence* in V, i.e. iff

$$d(\mathbf{u}_m, \mathbf{u}_n) = \|\mathbf{u}_m - \mathbf{u}_n\|_V \to 0 \quad \text{as } m, n \to \infty. \tag{A.3.1.9}$$

Example A.3.1.1.

1. $\mathbb{R}^n$ equipped with any one of the norms $\|\cdot\|_1$, $\|\cdot\|_2$, $\|\cdot\|_\infty$ defined, $\forall \mathbf{x} = (x_1, x_2, \ldots, x_n) \in \mathbb{R}^n$, by $\|\mathbf{x}\|_1 = \sum_{i=1}^n |x_i|$; $\|\mathbf{x}\|_2 = (\sum_{i=1}^n |x_i|^2)^{1/2}$; $\|\mathbf{x}\|_\infty = \max_{1\le i\le n}\{|x_i|\}$, is a *Banach space*, i.e. $(\mathbb{R}^n, \|\cdot\|_1)$, $(\mathbb{R}^n, \|\cdot\|_2)$, $(\mathbb{R}^n, \|\cdot\|_\infty)$ are *Banach spaces.* (A.3.1.10)
2. $C^0(\overline{\Omega}) = \{f : f \in C^0(\Omega),\ f$ is *bounded* and *uniformly continuous* on $\Omega\}$ equipped with the norm $\|\cdot\|_\infty$: $\|f\|_{C^0(\overline{\Omega})} = \|f\|_\infty = \sup_{\mathbf{x}\in\Omega} \|f(\mathbf{x})\|$ is a *Banach space* (see Theorem A.4.1.1). (A.3.1.11)
3. For $1 \le p < \infty$, $L^p(\Omega) = \{u : u$ is an *equivalence class* of Lebesgue measurable functions on Ω such that $\int_\Omega |u(\mathbf{x})|^p d\mathbf{x} < +\infty\}$ equipped with the norm $\|\cdot\|_{L^p(\Omega)}$: (A.3.1.12)

$$1 \le p < \infty, \quad \|u\|_{L^p(\Omega)} = \left(\int_\Omega |u(\mathbf{x})|^p d\mathbf{x}\right)^{1/p}, \tag{A.3.1.13}$$

and, for $p = \infty$, $L^\infty(\Omega) = \{u : u$ is an *equivalence* class of Lebesgue measurable functions on Ω such that $\operatorname{ess\,sup}_{\mathbf{x}\in\Omega} |u(\mathbf{x})| < +\infty\}$ equipped with the norm $\|\cdot\|_{L^\infty(\Omega)}$: (A.3.1.14)

$$\|u\|_{L^\infty(\Omega)} = \operatorname*{ess\,sup}_{\mathbf{x}\in\Omega} |u(\mathbf{x})|, \tag{A.3.1.15}$$

are *Banach spaces* (see Theorem B.4.1.2).

A.3.2 Closed subspace, dense subspace, Banach space and its separability

Let V be a *Banach space*.

Definition A.3.2.1. Let $M \subset V$ be a *normed linear space* with $\|\mathbf{x}\|_M = \|\mathbf{x}\|_V$ $\forall \mathbf{x} \in M$. If $(\mathbf{v}_n)$ is any sequence in M such that $\mathbf{v}_n \to \mathbf{v}$ in V as $n \to \infty$ implies that $\mathbf{v} \in M$, then M is a *closed subspace* of V. Hence, *every closed subspace* of V is also a *Banach space* for the same norm.

All finite-dimensional subspaces of a Banach space V are *closed* subspaces of V. (A.3.2.1)

Definition A.3.2.2. A subspace $M \subset V$ is called *dense* in Banach space V iff, $\forall \mathbf{u} \in V$, $\exists$ a sequence $(\mathbf{u}_n)$ in M such that $\mathbf{u}_n \to \mathbf{u}$ in V as $n \to \infty$ or, equivalently, iff $\overline{M} \equiv V$, $\overline{M}$ being the *closure* of M in V, or iff $\forall \mathbf{u} \in V$, $\forall \varepsilon > 0$, $\exists \mathbf{v}_\varepsilon \in M$ such that

$$d(\mathbf{u}, \mathbf{v}_\varepsilon) = \|\mathbf{u} - \mathbf{v}_\varepsilon\|_V < \varepsilon. \tag{A.3.2.2}$$

For example, $\forall$ compact $[a, b] \subset\subset \mathbb{R}$, $\mathcal{P}([a, b])$ defined in (A.2.1.14) is *dense* in the Banach space $C^0([a, b])$ equipped with the norm

$$\|f\|_{C^0([a,b])} = \sup_{x\in]a,b[} |f(x)| = \max_{x\in[a,b]} |f(x)| \tag{A.3.2.3}$$

by Weierstrass's Approximation Theorem A.4.1.2, and $C^0([a, b])$ is *dense* in $L^2(]a, b[) \implies \mathcal{P}([a, b])$ is *dense* in $L^2(]a, b[)$, which can be easily shown. (A.3.2.4)

Definition A.3.2.3. V is called *separable* iff V contains a *countable dense* subset in V.

Every *finite-dimensional* Banach space is *separable*. (A.3.2.5)

V_i is a *separable* Banach space, $i = 1, \ldots, n \implies V = V_1 \times \cdots \times V_n$ is also a *separable* Banach space with the *product norm*. (A.3.2.6)

$W \subset V$ is a *subspace* of a separable Banach space $\implies W$ is *separable*. (A.3.2.7)

For compact $\overline{\Omega} \subset\subset \mathbb{R}^n$, $C^0(\overline{\Omega})$, $L^p(\Omega)$, $1 \le p < \infty$, are *separable*. (A.3.2.8)

A.4 Banach spaces of continuous functions

A.4.1 Banach spaces $C^0(\overline{\Omega})$, $C^k(\overline{\Omega})$

Definition A.4.1.1. Let $\Omega \subset \mathbb{R}^n$ be an *open* subset of $\mathbb{R}^n$. Then $C^0(\overline{\Omega}) \equiv C(\overline{\Omega})$ is the linear space of functions f which are *bounded* and *uniformly continuous* on Ω, equipped with the norm:

$$\|f\|_{C^0(\overline{\Omega})} = \|f\|_\infty = \sup_{\mathbf{x}\in\Omega} |f(\mathbf{x})|. \tag{A.4.1.1}$$

$f \in C^0(\overline{\Omega}) \Longrightarrow$ by Theorem A.1.3.3, f has a *unique*, *bounded* and *continuous* extension to the closure $\overline{\Omega}$ of Ω.

Theorem A.4.1.1. *The normed linear space* $C^0(\overline{\Omega})$ *equipped with the norm* (A.4.1.1) *is a* Banach space.

Proof. Let $(f_n)_{n=1}^{\infty}$ be a *Cauchy sequence* in $C^0(\overline{\Omega})$, i.e. $\forall \varepsilon > 0$, $\exists n_0 = n_0(\varepsilon) \in \mathbb{N}$ such that $\|f_m - f_n\|_{C^0(\overline{\Omega})} = \sup_{\mathbf{x}\in\boldsymbol{\Omega}} |f_m(\mathbf{x}) - f_n(\mathbf{x})| < \varepsilon \; \forall m, n > n_0$

$$\Longrightarrow \quad |f_m(\mathbf{x}) - f_n(\mathbf{x})| \leq \sup_{\mathbf{x}\in\boldsymbol{\Omega}} |f_m(\mathbf{x}) - f_n(\mathbf{x})| < \varepsilon \quad \forall m, n > n_0 \qquad \text{(A.4.1.1a)}$$

$\Longrightarrow \forall$ *fixed* $\mathbf{x} \in \boldsymbol{\Omega}$, $(f_n(\mathbf{x}))_{n=1}^{\infty}$ is a Cauchy sequence on $\mathbb{R}$, which is *complete*, i.e. every Cauchy sequence converges on $\mathbb{R} \Longrightarrow \forall \mathbf{x} \in \boldsymbol{\Omega}$, $(f_n(\mathbf{x}))$ converges to a *unique* real number *dependent* on $\mathbf{x} \in \boldsymbol{\Omega}$, which defines a mapping

$$f : \mathbf{x} \in \boldsymbol{\Omega} \longmapsto f(\mathbf{x}) = \lim_{n\to\infty} f_n(\mathbf{x}) \quad \text{in } \mathbb{R}. \qquad \text{(A.4.1.1b)}$$

We are to show that f is *bounded* and *uniformly continuous* on Ω, i.e. $f \in C^0(\overline{\Omega})$. For this, fix $\varepsilon > 0$, $m, n > n_0$ and $\mathbf{x} \in \boldsymbol{\Omega}$. Then, using (A.4.1.1a), we get $|f(\mathbf{x}) - f_n(\mathbf{x})| \leq |f(\mathbf{x}) - f_m(\mathbf{x})| + |f_m(\mathbf{x}) - f_n(\mathbf{x})| < \varepsilon + |f(\mathbf{x}) - f_m(\mathbf{x})| \; \forall$ *fixed* $\mathbf{x} \in \boldsymbol{\Omega}$ and $m, n > n_0$. Now, letting $m \to \infty$ with fixed $n > n_0$ and $\mathbf{x} \in \boldsymbol{\Omega}$ and using (A.4.1.1b), we have $|f(\mathbf{x}) - f_n(\mathbf{x})| \leq \varepsilon + \lim_{m\to\infty} |f(\mathbf{x}) - f_m(\mathbf{x})| = \varepsilon + 0 = \varepsilon$
$\Longrightarrow \forall$ *fixed* $\mathbf{x} \in \boldsymbol{\Omega}$, $|f(\mathbf{x}) - f_n(\mathbf{x})| \leq \varepsilon \; \forall n > n_0 \Longrightarrow \forall \varepsilon > 0$, $\exists n_0 = n_0(\varepsilon) \in \mathbb{N}$ such that

$$\sup_{\mathbf{x}\in\boldsymbol{\Omega}} |f(\mathbf{x}) - f_n(\mathbf{x})| \leq \varepsilon \quad \forall n > n_0 \qquad \text{(A.4.1.1c)}$$

$\Longrightarrow (f_n)$ converges *uniformly* to f on Ω. Hence, (f_n) is a sequence of *bounded* and *uniformly continuous* functions, which converges *uniformly* to f on Ω. Then, by Theorem A.0.9.1, f is *continuous* on Ω. For fixed $\varepsilon > 0$ with $n > n_0$, $|f(\mathbf{x}) - f_n(\mathbf{x})| \leq \varepsilon \Longrightarrow |f(\mathbf{x})| \leq |f(\mathbf{x}) - f_n(\mathbf{x})| + |f_n(\mathbf{x})| \leq \varepsilon + M \; \forall \mathbf{x} \in \Omega \Longrightarrow$ the continuous f is *bounded* on Ω. But f_n is *uniformly continuous* on $\Omega \; \forall n \in \mathbb{N} \Longrightarrow \forall \varepsilon > 0$, $\exists \delta = \delta(\varepsilon) > 0$ such that for $\|\mathbf{x} - \mathbf{y}\| < \delta$ with $\mathbf{x}, \mathbf{y} \in \Omega$,

$$|f_n(\mathbf{x}) - f_n(\mathbf{y})| < \varepsilon/3 \quad \forall n \in N. \qquad \text{(A.4.1.1d)}$$

Then, using (A.4.1.1d) and (A.4.1.1c) with $n = n_1 > n_0$, $\forall \varepsilon > 0$, $\exists \delta = \delta(\varepsilon) > 0$ such that for $\|\mathbf{x} - \mathbf{y}\| < \delta$ with $\mathbf{x}, \mathbf{y} \in \Omega$,

$$\begin{aligned} |f(\mathbf{x}) - f(\mathbf{y})| &\leq |f(\mathbf{x}) - f_{n_1}(\mathbf{x})| + |f_{n_1}(\mathbf{x}) - f_{n_1}(\mathbf{y})| + |f_{n_1}(\mathbf{y}) - f(\mathbf{y})| \\ &< \varepsilon/3 + \varepsilon/3 + \varepsilon/3 = \varepsilon \end{aligned}$$

$\Longrightarrow \forall \varepsilon > 0, \exists \delta = \delta(\varepsilon) > 0$ such that for $\|\mathbf{x}-\mathbf{y}\| < \delta$ with $\mathbf{x}, \mathbf{y} \in \Omega$, $|f(\mathbf{x}) - f(\mathbf{y})| < \varepsilon \Longrightarrow f$ is *uniformly continuous* on Ω. Thus, f is bounded *and* uniformly continuous on Ω, i.e. $f \in C^0(\overline{\Omega})$ and $\|f - f_n\|_{C^0(\overline{\Omega})} = \sup_{\mathbf{x}} \in \mathbf{\Omega} |f(\mathbf{x}) - f_n(\mathbf{x})| \to 0$ as $n \to \infty$. Hence, (arbitrary) Cauchy sequence (f_n) in $C^0(\overline{\Omega})$ converges to $f \in C^0(\overline{\Omega})$. So, $C^0(\overline{\Omega})$ is a *complete normed linear space*, i.e. a *Banach space*. □

Remark A.4.1.1. For *bounded* Ω, $\overline{\Omega} \subset\subset \mathbb{R}^n$ is *compact*, and the *continuity* of f on $\overline{\Omega} \Longrightarrow$

1. its *boundedness*, and
2. *uniform continuity* on $\overline{\Omega}$, and
3. $\sup_{\mathbf{x}\in\Omega} |f(\mathbf{x})| = \max_{\mathbf{x}\in\overline{\Omega}} |f(\mathbf{x})|$ by Theorem A.1.1.1.

Hence, we have:

Definition A.4.1.2. For *bounded* $\Omega \subset \mathbb{R}^n$, $C^0(\overline{\Omega})$ is the *linear space* of *restrictions* $f\downarrow_\Omega$ to Ω of all functions *continuous* on $\overline{\Omega}$ and equipped with the norm

$$\|f\|_{C^0(\overline{\Omega})} = \|f\|_\infty = \max_{\mathbf{x}\in\overline{\Omega}} |f(\mathbf{x})|. \tag{A.4.1.2}$$

Theorem A.4.1.2 (Weierstrass's Approximation Theorem). *For* compact $\overline{\Omega} \subset\subset \mathbb{R}^n$, $\mathcal{P}(\overline{\Omega}) = \{p : p$ *is a polynomial of arbitrary degree on* $\overline{\Omega}$ *in variables* $x_1, \ldots, x_n\}$ *is* dense *in* $C^0(\overline{\Omega})$.

Moreover, for compact $\overline{\Omega}$, $C^0(\overline{\Omega})$ *is* separable. (A.4.1.3)

Definition A.4.1.3. The space

$$\begin{aligned} C^k(\overline{\Omega}) = \{f : f \in C^0(\overline{\Omega}), &\text{ partial derivatives } \partial^{\boldsymbol{\alpha}} f \in C^0(\overline{\Omega}) \\ &\text{of all orders } |\boldsymbol{\alpha}| = \alpha_1 + \cdots + \alpha_n \le k\} \end{aligned} \tag{A.4.1.4}$$

equipped with the norm $\|\cdot\|_{C^k(\overline{\Omega})}$ (resp. $\|\cdot\|_{C^k(\overline{\Omega})}$) defined by:

$$\|f\|_{C^k(\overline{\Omega})} = \sum_{|\boldsymbol{\alpha}|=0}^{k} \|\partial^{\boldsymbol{\alpha}} f\|_{C^0(\overline{\Omega})} = \sum_{|\boldsymbol{\alpha}|=0}^{k} \sup_{\mathbf{x}\in\Omega} |\partial^{\boldsymbol{\alpha}} f(\mathbf{x})|; \tag{A.4.1.5}$$

(resp.

$$|||f|||_{C^k(\overline{\Omega})} = \max_{0\le|\boldsymbol{\alpha}|\le k} |||\partial^{\boldsymbol{\alpha}} f|||_{C^0(\overline{\Omega})} = \max_{0\le|\boldsymbol{\alpha}|\le k} \sup_{\mathbf{x}\in\Omega} |\partial^{\boldsymbol{\alpha}} f(\mathbf{x})|) \tag{A.4.1.6}$$

is a *Banach space*, the two norms $\|\cdot\|_{C^k(\overline{\Omega})}$ and $|||\cdot|||_{C^k(\overline{\Omega})}$ being *equivalent* ones, i.e. $\exists C_1, C_2 > 0$ such that

$$C_1\|f\|_{C^k(\overline{\Omega})} \le |||f|||_{C^k(\overline{\Omega})} \le C_2|||f|||_{C^k(\overline{\Omega})} \quad \forall f \in C^k(\overline{\Omega}). \tag{A.4.1.7}$$

For *bounded* Ω, we define partial derivatives $\partial^{\boldsymbol{\alpha}} f$ with $1 \leq |\boldsymbol{\alpha}| \leq k$ on $\overline{\Omega}$ in the *generalized sense*, i.e. as the *unique*, *bounded*, (*uniformly*) continuous extension to the *closure* $\overline{\Omega}$ of $\partial^{\boldsymbol{\alpha}} f$ defined, *bounded* and *uniformly continuous* on Ω, $1 \leq |\boldsymbol{\alpha}| \leq k$. Then, we can define:

Definition A.4.1.4. For *bounded* Ω, $C^k(\overline{\Omega}) = \{f\downarrow_\Omega : f\downarrow_\Omega$ is the *restriction* to Ω of the function f, whose partial derivaties $\partial^{\boldsymbol{\alpha}} f$ are *continuous* on $\overline{\Omega}$ $\forall |\boldsymbol{\alpha}| \leq k\}$ equipped with the *equivalent* norms $\|\cdot\|_{C^k(\overline{\Omega})}$ (resp. $\|\cdot\|_{C^k(\overline{\Omega})}$):

$$\|f\|_{C^k(\overline{\Omega})} = \sum_{0 \leq |\boldsymbol{\alpha}| \leq k} \max_{\mathbf{x} \in \overline{\Omega}} \left| \frac{\partial^{|\boldsymbol{\alpha}|} f(\mathbf{x})}{\partial x_1^{\alpha_1} \cdot \partial x_2^{\alpha_2} \cdots \partial x_n^{\alpha_n}} \right|; \tag{A.4.1.8}$$

$$\|f\|_{C^k(\overline{\Omega})} = \max_{0 \leq |\boldsymbol{\alpha}| \leq k} \left\{ \max_{\mathbf{x} \in \overline{\Omega}} \left| \frac{\partial^{|\boldsymbol{\alpha}|} f(\mathbf{x})}{\partial x_1^{\alpha_1} \cdot \partial x_2^{\alpha_2} \cdots \partial x_n^{\alpha_n}} \right| \right\}. \tag{A.4.1.9}$$

Semi-norm $|\cdot|_{C^k(\overline{\Omega})}$ Corresponding to $\|\cdot\|_{C^k(\overline{\Omega})}$, a *semi-norm* p: $C^k(\overline{\Omega}) \longrightarrow \mathbb{R}$ is defined by:

$$p(f) = |f|_{C^k(\overline{\Omega})} = \sum_{|\boldsymbol{\alpha}|=k} \|\partial^{\boldsymbol{\alpha}} f\|_{C^0(\overline{\Omega})} = \sum_{|\boldsymbol{\alpha}|=k} \sup_{\mathbf{x} \in \Omega} |\partial^{\boldsymbol{\alpha}} f(\mathbf{x})|. \tag{A.4.1.10}$$

Corresponding to the *norm* $\|\cdot\|_{C^k(\overline{\Omega})}$, a *semi-norm* $p_0 : C^k(\overline{\Omega}) \rightarrow \mathbb{R}$ is defined by:

$$p_0(f) = |f|_{C^k(\overline{\Omega})} = \max_{|\boldsymbol{\alpha}|=k} \{\sup_{\mathbf{x} \in \Omega} |\partial^{\boldsymbol{\alpha}} f(\mathbf{x})|\}. \tag{A.4.1.11}$$

A.5 Banach spaces $C^{0,\lambda}(\overline{\Omega})$, $0 < \lambda < 1$, of Hölder continuous functions

A.5.1 Hölder continuity and Lipschitz continuity

Definition A.5.1.1. For $0 < \lambda < 1$, a function $f \in \mathcal{F}(\Omega)$ is called *Hölder continuous of order* λ (or λ-*Hölder continuous*) on $\Omega \subset \mathbb{R}^n$ if and only if $\exists M > 0$ such that

$$|f(\mathbf{x}) - f(\mathbf{y})| \leq M \|\mathbf{x} - \mathbf{y}\|_{\mathbb{R}^n}^{\lambda} \quad \forall \mathbf{x}, \mathbf{y} \in \Omega \tag{A.5.1.1}$$

or, equivalently, the function $\frac{|f(\mathbf{x}) - f(\mathbf{y})|}{\|\mathbf{x}-\mathbf{y}\|_{\mathbb{R}^n}^{\lambda}}$ is *bounded* on $\{(\mathbf{x}, \mathbf{y}) : (\mathbf{x}, \mathbf{y}) \in \Omega \times \Omega$ with $\mathbf{x} \neq \mathbf{y}\}$, where $\|\cdot\|_{\mathbb{R}^n} = (\sum_{i=1}^n x_i^2)^{1/2}$.

Lipschitz continuity

Definition A.5.1.2. A function $f \in \mathcal{F}(\Omega)$ is called *Lipschitz continuous* on $\Omega \subset \mathbb{R}^n$ if and only if $\exists L > 0$ such that

$$|f(\mathbf{x}) - f(\mathbf{y})| \leq L\|\mathbf{x} - \mathbf{y}\|_{\mathbb{R}^n} \quad \forall \mathbf{x}, \mathbf{y} \in \Omega. \tag{A.5.1.2}$$

Then we say that f satisfies the *Lipschitz condition* (A.5.1.2) with *Lipschitz constant* $L > 0$.

Moreover, if (A.5.1.2) holds with $0 < L < 1$, then f is called a *contraction mapping*. (A.5.1.3)

f is *Lipschitz continuous* on $\Omega \Longrightarrow f$ is *Hölder continuous* of order $\lambda = 1$ on Ω.

Every function that is Hölder continuous with $0 < \lambda < 1$ (resp. *Lipschitz continuous*) on Ω is *uniformly continuous* and, hence, *continuous* on Ω. (A.5.1.4)

A.5.2 Hölder space $C^{0,\lambda}(\overline{\Omega})$

Definition A.5.2.1. Let $\Omega \subset \mathbb{R}^n$ be an *open* subset of $\mathbb{R}^n$. For $0 < \lambda \leq 1$, $C^{0,\lambda}(\overline{\Omega})$ is the subspace of $C^0(\bar{\Omega})$ consisting of those functions which are *Hölder continuous* of order λ on Ω, i.e. for $0 < \lambda \leq 1$,

$$C^{0,\lambda}(\overline{\Omega}) = \{f : f \in C^k(\bar{\Omega}) \text{ such that } \exists M > 0, \text{ for which } |f(\mathbf{x}) - f(\mathbf{y})| \leq M\|\mathbf{x} - \mathbf{y}\|^{\lambda}_{\mathbb{R}^n} \ \forall \mathbf{x}, \mathbf{y} \in \Omega\}. \tag{A.5.2.1}$$

Semi-norm $|\cdot|_{C^{0,\lambda}(\overline{\Omega})}$ and *norm* $\|\cdot\|_{C^{0,\lambda}(\overline{\Omega})}$ in $C^{0,\lambda}(\overline{\Omega})$ are defined by:

$$|f|_{C^{0,\lambda}(\overline{\Omega})} = \sup_{\mathbf{x},\mathbf{y}\in\Omega,\mathbf{x}\neq\mathbf{y}} \frac{|f(\mathbf{x}) - f(\mathbf{y})|}{\|\mathbf{x} - \mathbf{y}\|^{\lambda}_{\mathbb{R}^n}}; \tag{A.5.2.2}$$

$$\|f\|_{C^{0,\lambda}(\overline{\Omega})} = \|f\|_{C^0(\bar{\Omega})} + |f|_{C^{0,\lambda}(\overline{\Omega})}, \tag{A.5.2.3}$$

with $\|f\|_{C^0(\bar{\Omega})}$ defined by (A.4.1.1) (resp. (A.4.1.2) for *bounded* domains).

Theorem A.5.2.1. $C^{0,\lambda}(\overline{\Omega})$, $0 < \lambda \leq 1$, *equipped with the norm* $\|\cdot\|_{C^{0,\lambda}(\overline{\Omega})}$, *is a* Banach space.

A.5.3 Space $C^{k,\lambda}(\overline{\Omega})$, $0 < \lambda \leq 1$

Definition A.5.3.1. For $k \in \mathbb{N}$ *and* $0 < \lambda \leq 1$, $C^{k,\lambda}(\overline{\Omega})$ is defined by:

$$C^{k,\lambda}(\overline{\Omega}) = \{f : f \in C^k(\bar{\Omega}),\ \partial^{\boldsymbol{\alpha}} f \in C^{0,\lambda}(\overline{\Omega}) \text{ for } |\boldsymbol{\alpha}| \leq k\}, \tag{A.5.3.1}$$

which is equipped with the *semi-norm* $|\cdot|_{C^{k,\lambda}(\overline{\Omega})}$ (resp. $\mathbf{|}\cdot\mathbf{|}_{C^{k,\lambda}(\overline{\Omega})}$) and *norm* $\|\cdot\|_{C^{k,\lambda}(\overline{\Omega})}$ (resp. $|||\cdot|||_{C^{k,\lambda}(\overline{\Omega})}$):

$$|f|_{C^{k,\lambda}(\overline{\Omega})} = \sum_{|\boldsymbol{\alpha}|=k} |\partial^{\boldsymbol{\alpha}} f|_{C^{0,\lambda}(\overline{\Omega})} = \sum_{|\boldsymbol{\alpha}|=k} \sup_{\mathbf{x},\mathbf{y}\in\Omega,\mathbf{x}\neq\mathbf{y}} \frac{|\partial^{\boldsymbol{\alpha}} f(\mathbf{x}) - \partial^{\boldsymbol{\alpha}} f(\mathbf{y})|}{\|\mathbf{x}-\mathbf{y}\|^{\lambda}_{\mathbb{R}^n}}; \tag{A.5.3.2}$$

$$\|f\|_{C^{k,\lambda}(\overline{\Omega})} = \sum_{0\le|\boldsymbol{\alpha}|\le k} \|\partial^{\boldsymbol{\alpha}} f\|_{C^{0,\lambda}(\overline{\Omega})}; \tag{A.5.3.3}$$

(resp.

$$\mathbf{|}f\mathbf{|}_{C^{k,\lambda}(\overline{\Omega})} = \max_{|\boldsymbol{\alpha}|=k} \left\{ \sup_{\mathbf{x},\mathbf{y}\in\Omega,\mathbf{x}\neq\mathbf{y}} \frac{|\partial^{\boldsymbol{\alpha}} f(\mathbf{x}) - \partial^{\boldsymbol{\alpha}} f(\mathbf{y})|}{\|\mathbf{x}-\mathbf{y}\|^{\lambda}_{\mathbb{R}^n}} \right\}; \tag{A.5.3.4}$$

$$|||f|||_{C^{k,\lambda}(\overline{\Omega})} = \|f\|_{C^{k}(\bar{\Omega})} + \mathbf{|}f\mathbf{|}_{C^{k,\lambda}(\overline{\Omega})}). \tag{A.5.3.5}$$

Theorem A.5.3.1. $\forall k \in \mathbb{N}$, $0 < \lambda \le 1$, $C^{k,\lambda}(\overline{\Omega})$ *defined by* (A.5.3.1) *and* (A.5.3.3) *(resp.* (A.5.3.1) *and* (A.5.3.5)*) is a* Banach space.

$C^{k,\lambda}(\overline{\Omega})$, $0 < \lambda < 1$, *is* not *reflexive. For* bounded Ω, $C^{k,\lambda}(\overline{\Omega})$, $0 < \lambda < 1$ *is* separable.

Proposition A.5.3.1. *The norms* $\|\cdot\|_{C^{k,\lambda}(\overline{\Omega})}$ *and* $|||\cdot|||_{C^{k,\lambda}(\overline{\Omega})}$ *defined by* (A.5.3.3) *and* (A.5.3.5) *respectively are* equivalent *on* $C^{k,\lambda}(\overline{\Omega})$, $0 < \lambda \le 1$.

Imbedding results $\hookrightarrow$ See (A.8.1.6) for details of the imbedding operator.

For $k \in \mathbb{N}_0$ and $0 < \mu < \lambda \le 1$, we have:

1. (a) $C^{k+1}(\overline{\Omega}) \hookrightarrow C^{k}(\bar{\Omega})$;

   (b) $C^{k,\lambda}(\overline{\Omega}) \hookrightarrow C^{k}(\bar{\Omega})$;

   (c) $C^{k,\lambda}(\overline{\Omega}) \hookrightarrow C^{k,\mu}(\overline{\Omega})$. (A.5.3.6)

2. For *bounded* Ω, the imbeddings (b) and (c) are *compact*:

   (d) $C^{k,\lambda}(\overline{\Omega}) \hookrightarrow\hookrightarrow C^{k}(\bar{\Omega})$;

   (e) $C^{k,\lambda}(\overline{\Omega}) \hookrightarrow\hookrightarrow C^{k,\mu}(\overline{\Omega})$. (A.5.3.7)

A.6 Quotient space V/M

Let V be a *Banach space* and M be a *closed* subspace of V.

Definition A.6.1.1. The *quotient space* V/M (i.e. the *quotient* of V *by* M) is the *normed linear space* of *equivalence classes* $[u]$ of elements $u \in V$ satisfying the properties:

1. $u, v \in [u] \Longrightarrow u - v \in M \Longrightarrow u = v \pmod{M}$, i.e. $[u] = u + M$, $u + M$ being the *coset* of u relative to M; (A.6.1.1)
2. $\forall [u], [v] \in V/M$, $[u] + [v] = [u + v] \in V/M$; $\forall \alpha \in \mathbb{R}$, $\alpha[u] = [\alpha u] \in V/M$; (A.6.1.2)

   $\exists$*null-element* $[0] = 0 + M \in V/M$; (A.6.1.3)
3. $\forall [u] \in V/M$, the *quotient norm* $\|[u]\|_{V/M}$ is defined by: $\|[u]\|_{V/M} = \inf_{u \in [u]} \|u\|_V = \inf_{w \in M} \|u + w\|_V$, which satisfies the *properties of norm.* (A.6.1.4)

Theorem A.6.1.1. *Let V be a Banach space and M be a closed subspace of V. Then the* quotient normed linear space *V/M is a* Banach space.

A.7 Continuous linear functionals on normed linear spaces

A.7.1 Space V'

Let V be a *normed linear space*.

Definition A.7.1.1. A linear functional $L : V \longrightarrow \mathbb{R}$ is called *continuous* on V iff $\forall$ sequences $(\mathbf{u}_n)_{n=1}^{\infty}$ with $\mathbf{u}_n \to \mathbf{u}$ in $V \Longrightarrow L(\mathbf{u}_n) \to L(\mathbf{u})$ in $\mathbb{R}$ as $n \to \infty$.

$L : V \longrightarrow \mathbb{R}$ is called *bounded* on $V \Longleftrightarrow \exists M > 0$ such that

$$|L(\mathbf{u})| \leq M \|\mathbf{u}\|_V \quad \forall \mathbf{u} \in V. \tag{A.7.1.1}$$

Theorem A.7.1.1. *A linear functional $L : V \longrightarrow \mathbb{R}$ is* continuous *on $V \Longleftrightarrow L$ is* bounded *on V.*

Let V' be the set of all *continuous* (equivalently *bounded*) linear functionals on a *normed linear space* V, i.e. $V' = \{L : L : V \longrightarrow \mathbb{R}$ is *linear* and *continuous* on $V\}$. (A.7.1.2)

Then, V' becomes a *linear space* if we define $\forall L_1, L_2 \in V'$, $\mathbf{u} \in V$, $\forall \alpha \in \mathbb{R}$,

$$(L_1 + L_2)(\mathbf{u}) = L_1(\mathbf{u}) + L_2(\mathbf{u}); \quad (\alpha L)(\mathbf{u}) = \alpha \cdot L(\mathbf{u})$$

and the null element in V' by $L = 0$ in $V' \Longrightarrow L(\mathbf{u}) = 0\ \forall \mathbf{u} \in V$, the norm $\|\cdot\|_{V'} = \|\cdot\|'$ by:

$$\|L\|_{V'} = \sup_{\mathbf{u} \in V - \{\mathbf{0}\}} \frac{|L(\mathbf{u})|}{\|\mathbf{u}\|_V} = \sup_{\|\hat{\mathbf{u}}\|_V = 1} |L(\mathbf{u})| = \sup_{\|\mathbf{u}\| \leq 1} |L(\mathbf{u})|$$

and

$$|L(\mathbf{u})| \leq \|L\|_{V'} \|\mathbf{u}\|_V. \tag{A.7.1.3}$$

The *normed linear space* V' is also denoted by $\mathcal{L}(V; \mathbb{R})$, i.e.

$$V' \equiv \mathcal{L}(V; \mathbb{R}) \quad \text{with } \|\cdot\|_{V'} = \|\cdot\|_{\mathcal{L}(V;\mathbb{R})}. \tag{A.7.1.4}$$

Since $\mathbb{R}$ is complete, we have:

Theorem A.7.1.2. *$V' \equiv \mathcal{L}(V; \mathbb{R})$ with the norm $\|\cdot\|_{V'} = \|\cdot\|_{\mathcal{L}(V;\mathbb{R})}$ is a* Banach space *(even when normed linear space V is* not *a Banach space).*

Definition A.7.1.2. For a *normed linear space* V, the *Banach space* $V' \equiv \mathcal{L}(V; \mathbb{R})$ is called the (*strong topological*) *dual space* of V (also called the *conjugate space*).

Remark A.7.1.1. For a linear space V (possibly *not* a *normed* linear space), we have used V^* to denote the *algebraic dual* space of V (Definition A.2.2.3). Hence, we have used V' to denote the *algebraic* and *topological* dual space of V, i.e. linear functionals in V' are endowed with the *additional* property of their *continuity* (*boundedness*) on V. Hence, for $L \in V'$, $\mathbf{v} \in V$, the value $L(\mathbf{v})$ is also denoted by $\langle L, \mathbf{v}\rangle$ or $\langle \mathbf{v}, L\rangle$, i.e.

$$L(\mathbf{v}) = \langle L, \mathbf{v}\rangle_{V' \times V} = \langle \mathbf{v}, L\rangle_{V \times V'}, \tag{A.7.1.5}$$

where $\langle \cdot, . \rangle_{V' \times V}$ (resp. $\langle \cdot, \cdot \rangle_{V \times V'}$) is called the *duality* pairing between V' and V (resp. V and V'). Other equivalent notations are $[L, \mathbf{v}]_{V' \times V}$ or $[\mathbf{v}, L]_{V \times V'}$. (A.7.1.6)

A.7.2 Hahn–Banach extension of linear functionals in analytic form

One of the most important theorems of functional analysis is the celebrated Hahn–Banach theorem. We will present here the most general form of the theorem in analytic form.

Let V be a real vector space (which may *not* be normable).

Definition A.7.2.1. A *functional* $p : V \longrightarrow \mathbb{R}$ on a real vector space V is called *sublinear* iff:

1. $p(\lambda \mathbf{x}) = \lambda p(\mathbf{x})\ \forall \mathbf{x} \in V, \forall \lambda \in [0, \infty[$;
2. $p(\mathbf{x} + \mathbf{y}) \leq p(\mathbf{x}) + p(\mathbf{y})\ \forall \mathbf{x}, \mathbf{y} \in V$ (*subadditive*).

Example A.7.2.1. A *semi-norm* p on V defined by $p(\mathbf{x}) = |\mathbf{x}|\ \forall \mathbf{x} \in V$ is a *sublinear* functional on V since $|\lambda \mathbf{x}| = \lambda|\mathbf{x}|\ \forall \mathbf{x} \in V, \forall \lambda \in [0, \infty[, |\mathbf{x}+\mathbf{y}| \leq |\mathbf{x}|+|\mathbf{y}|\ \forall \mathbf{x}, \mathbf{y} \in V$.

Example A.7.2.2. Let the *semi-norm* p be a *norm* $\|\cdot\|_V$ in V. Then $p(\mathbf{x}) = \|\mathbf{x}\|_V$ $\forall \mathbf{x} \in V$ is also a *sublinear* functional on V, which follows from the properties of norm.

Definition A.7.2.2. A *linear functional* $L : V \longrightarrow \mathbb{R}$ is called *bounded from above* or *majorized* by a *sublinear* functional p on V if and only if $L(\mathbf{x}) \leq p(\mathbf{x})\ \forall \mathbf{x} \in V$.

Theorem A.7.2.1 (Hahn–Banach). *Let $M \subset V$ be a subspace of V. Let $p : V \longrightarrow \mathbb{R}$ be a sublinear functional on V and $l : M \longrightarrow \mathbb{R}$ be a linear functional on $M \subset V$ such that l is bounded above by p on M, i.e. $l(\mathbf{x}) \leq p(\mathbf{x})\ \forall \mathbf{x} \in M$. Then $\exists$ a linear functional $L : V \longrightarrow \mathbb{R}$ on V such that L is bounded above by p on V and L extends l, i.e. $L(\mathbf{x}) \leq p(\mathbf{x})\ \forall \mathbf{x} \in V$ and $L(\mathbf{x}) = l(\mathbf{x})\ \forall \mathbf{x} \in M$.*

A.7.3 Consequences of the Hahn–Banach theorem in normed linear spaces

Let V be a *normed linear space* and $V' \equiv \mathcal{L}(V;\mathbb{R})$ be the *dual space* of all *continuous* (equivalently *bounded*), *linear* functionals l defined on V and equipped with the norm $\|\cdot\|_{V'}$ defined by

$$\|l\|_{V'} = \sup_{\mathbf{x}\in V-\{\mathbf{0}\}} \frac{|l(\mathbf{x})|}{\|\mathbf{x}\|_V} = \sup_{\|\mathbf{x}\|\leq 1} |l(\mathbf{x})|.$$

Then we have the following corollaries.

Corollary A.7.3.1. *Let M be a subspace of V with $\|\mathbf{x}\|_M = \|\mathbf{x}\|_V\ \forall \mathbf{x} \in M$. Let $l \in M'$ be a continuous linear functional on M. Then $\exists$ a* norm-preserving, continuous, linear extension *$L \in V'$ such that $L(\mathbf{x}) = l(\mathbf{x})\ \forall \mathbf{x} \in M$ (i.e. L extends l) and $\|L\|_{V'} = \|l\|_{M'}$.*

Corollary A.7.3.2. *For fixed $\mathbf{x} \neq \mathbf{0}$ in V, $\exists L \in V'$ such that $L(\mathbf{x}) = \|\mathbf{x}\|_V$ with $\|L\|_{V'} = 1$.*

Corollary A.7.3.3. *$\forall$ fixed $\mathbf{x}_0 \in V$, $\exists L_0 \in V'$ with $\|L_0\|_{V'} = \|\mathbf{x}_0\|_V$ and $L(\mathbf{x}_0) = \|\mathbf{x}_0\|_V^2$.*

Corollary A.7.3.4. *$L(\mathbf{x}) = 0\ \forall L \in V' \Longrightarrow \mathbf{x} = 0$ in V.*

Corollary A.7.3.5. *Let M_0 be a* closed *subspace of V. Let $\mathbf{x}_0 \in V$ with $\mathbf{x}_0 \notin M_0$. Then $\exists L \in V'$ such that $L(\mathbf{x}) = 0\ \forall \mathbf{x} \in M_0$* and *$L(\mathbf{x}_0) \neq 0$.*

Corollary A.7.3.6. *Let M be a subspace of a* Banach space *V. Then M is* dense *in V if and only if $L(\mathbf{x}) = 0 \ \forall \mathbf{x} \in M \implies L = 0$ in V' (i.e. $L(\mathbf{x}) = 0 \ \forall \mathbf{x} \in V$).*

Proof. Let M be *dense* in V. Then, by Definition A.3.2.2, $\forall \mathbf{x} \in V$, $\exists$ a sequence $(\mathbf{x}_n)$ in M such that $\mathbf{x}_n \to \mathbf{x}$ in V as $n \to \infty$. But $L \in V' \implies L$ is *continuous* on V. Hence, $\mathbf{x}_n \to \mathbf{x}$ in $V \implies L(\mathbf{x}_n) \to L(\mathbf{x})$ as $n \to \infty \implies L(\mathbf{x}) = \lim_{n\to\infty} L(\mathbf{x}_n) = \lim_{n\to\infty} 0 = 0$ (since $\mathbf{x}_n \in M \ \forall n \in \mathbb{N}$ and $L(\mathbf{x}_n) = 0 \ \forall n \in \mathbb{N} \implies L(\mathbf{x}) = 0 \ \forall \mathbf{x} \in V \implies L = 0$ in V').

Conversely, assume that $L(\mathbf{x}) = 0 \ \forall \mathbf{x} \in M \implies L = 0$ in V', but the contrary holds, i.e. M is *not* dense in $V \implies \exists \mathbf{x}_0 \in V$ with $\mathbf{x}_0 \notin M$, $d(\mathbf{x}_0, M) = \inf_{\mathbf{y}\in M} d(\mathbf{x}_0, \mathbf{y}) = d > 0$. Then, following the proof (*not given*) of Corollary A.7.3.5, we can show that $\exists L \in V'$, $L \neq 0$ such that $L(\mathbf{x}) = 0 \ \forall \mathbf{x} \in M$ and $L(\mathbf{x}_0) \neq 0$ for $\mathbf{x}_0 \in V$ with $\mathbf{x}_0 \notin M$, i.e. $L \neq 0$ in V', which contradicts the hypothesis that $L(\mathbf{x}) = 0 \ \forall \mathbf{x} \in M \implies L(\mathbf{x}) = 0 \ \forall \mathbf{x} \in V$. Hence, our assumption is wrong and the result follows: M is *dense* in V. □

Corollary A.7.3.7 (Extension of continuous linear functional by density). *Let V be a Banach space and X be a* dense *subspace of V. Let $l : X \longrightarrow \mathbb{R}$ be a* continuous, linear *functional on X. Then $\exists$ a* unique, continuous, linear *functional $\tilde{l} : V \longrightarrow \mathbb{R}$ extending l such that $\tilde{l}(\mathbf{x}) = l(\mathbf{x}) \ \forall \mathbf{x} \in X$ and $\|\tilde{l}\|_{V'} = \|l\|_{X'}$.*

Alternative version of this corollary on extension by density for spaces X which are *not* normable:

Corollary A.7.3.8 (Extension by density). *Let $(p_i)_{i\in I}$ be a family of semi-norms defining a topology on the vector space V such that X is* dense *in V in this topology. Let l be a linear functional* continuous *on X in this topology, i.e. $\exists (p_{i_k})_{k=1}^N$ and $C_k > 0$ such that $|l(\mathbf{x})| \leq C_k p_{i_k}(\mathbf{x}) \ \forall \mathbf{x} \in X$, $i_k \in I \ \forall k$, $1 \leq k \leq N$. Then $\exists$ a* unique, continuous, linear *functional $\tilde{l}$ on V, which extends l, such that $\tilde{l}(\mathbf{x}) = l(\mathbf{x})$, $\mathbf{x} \in X$, $|\tilde{l}(\mathbf{x})| \leq C_k p_{i_k}(\mathbf{x}) \ \forall \mathbf{x} \in V$, $i_k \in I \ \forall k$, $1 \leq k \leq N$.*

Proof. By Corollary A.7.3.1, $\exists$ a *continuous linear* extension $\tilde{l} : V \longrightarrow \mathbb{R}$ of $l \in X'$ such that $\tilde{l}(\mathbf{x}) = l(\mathbf{x}) \ \forall \mathbf{x} \in X$ and $\|\tilde{l}\|_{V'} = \|l\|_{X'}$.
Uniqueness of $\tilde{l}$: Suppose that $\exists$ another *continuous, linear* functional $l^* \in V'$, i.e. $l^* : V \longrightarrow \mathbb{R}$ with $l^*(\mathbf{x}) = \tilde{l}(\mathbf{x}) = l(\mathbf{x}) \ \forall \mathbf{x} \in X$. Then $(l^* - \tilde{l})(\mathbf{x}) = 0 \ \forall \mathbf{x} \in X$, X being a *dense* subspace of $V \implies$ by Corollary A.7.3.6, $l^* - \tilde{l} = 0$ in $V' \implies l^* = \tilde{l}$ in V'. □

Remark A.7.3.1. This unique, continuous, linear extension $\tilde{l}$ of l is usually denoted by the same notation l itself, i.e. we write $\tilde{l} = l$.

Corollary A.7.3.9. *Let V be a normed linear space. Then, $\forall \mathbf{x} \in V$,*

$$\|\mathbf{x}\|_V = \sup_{L\in V', \|L\|_{V'}\leq 1} |L(\mathbf{x})| = \max_{L\in V', \|L\|_{V'}\leq 1} |L(\mathbf{x})|.$$

A.8 Continuous linear operators on normed linear spaces

A.8.1 Space $\mathcal{L}(V;W)$

Let V and W be *normed linear spaces.*

Theorem A.8.1.1. *A* linear *operator* $A : V \longrightarrow W$ *is* bounded *from* V *into* $W \Longleftrightarrow A$ *is* continuous *from* V *into* $W \Longleftrightarrow \exists M > 0$ *such that*

$$\|A\mathbf{x}\|_W \leq M\|\mathbf{x}\|_V \quad \forall \mathbf{x} \in V. \tag{A.8.1.1}$$

Definition A.8.1.1. Let $\mathcal{L}(V;W)$ denote the set of all *continuous* (equivalently *bounded*), *linear* operators from the *normed linear space* V into the *normed linear space* W, i.e.

$$\mathcal{L}(V;W) = \{A : A : V \longrightarrow W \text{ is linear, } A \text{ is } \textit{continuous} \text{ from } V \text{ into } W\}, \tag{A.8.1.2}$$

is a *linear* space. The *norm* of a *bounded* linear operator $A \in \mathcal{L}(V;W)$ is defined by:

$$\|A\|_{\mathcal{L}(V;W)} = \sup_{\mathbf{x}\in V-\{0\}} \frac{\|A\mathbf{x}\|_W}{\|\mathbf{x}\|_V} \quad \Longrightarrow \quad \|A\mathbf{x}\|_W \leq \|A\|_{\mathcal{L}(V;W)}\|\mathbf{x}\|_V \quad \forall \mathbf{x} \in V \tag{A.8.1.3}$$

and

$$\|A\mathbf{x}\|_W \leq M\|\mathbf{x}\|_V \ \forall \mathbf{x} \in V \quad \Longrightarrow \quad \|A\|_{\mathcal{L}(V;W)} \leq M.$$

Theorem A.8.1.2.

I. $\mathcal{L}(V;W)$ *equipped with the norm* $\|\cdot\|_{\mathcal{L}(V;W)}$ *is a* normed linear space.

II. *If* W *is a* Banach space, $\mathcal{L}(V;W)$ *is a* Banach *space, even when* V *is a normed linear space but* not *a Banach space. For* $W = V$,

$$\mathcal{L}(V;V) = \mathcal{L}(V). \tag{A.8.1.4}$$

Theorem A.8.1.3 (Open Mapping Theorem of Banach). *Let* $A \in \mathcal{L}(V;W)$ *such that* A *is a* surjection *from Banach space* V onto *Banach space* W. *Then* $\exists r_0 > 0$ *such that* $B_W(\mathbf{0}; r_0) \subset A(B_V(\mathbf{0}; 1))$, *where* $A(B_V(\mathbf{0}; 1)) \subset W$ *is the image of the* unit *open ball* $B_V(\mathbf{0}; 1)$ *in* V *under* A; $B_W(\mathbf{0}; r_0)$ *is the* open *ball in* W *with centre* $\mathbf{0}$ *and radius* r_0.

Corollary A.8.1.1. *Let* V *and* W *be Banach spaces and* $A \in \mathcal{L}(V;W)$ *be* bijective *from* V onto W. *Then* A^{-1} *is also* continuous *from* W *onto* V.

Continuous, linear right-hand inverse of $A \in \mathcal{L}(V;W)$ Let V and W be Banach spaces and $A \in \mathcal{L}(V;W)$ be a *surjection* from V *onto* W. Then the *continuous, linear* A has the *continuous, linear, right-hand inverse* $B : W \longrightarrow V$ defined by $A(Bw) = w \ \forall w \in W$ (A may *not* have the *left-hand* inverse). (A.8.1.5)

Imbedding operator $\hookrightarrow$ or canonical injection For $V \subset W$, the *canonical injection* or, equivalently, the *imbedding operator* $\hookrightarrow$ is defined by:

$$\hookrightarrow : V \longrightarrow (\hookrightarrow V) = V \subset W \quad \text{with} \hookrightarrow : \mathbf{v} \in V \longmapsto (\hookrightarrow \mathbf{v}) = \mathbf{v} \in V \subset W. \tag{A.8.1.6}$$

Other equivalent notations are I, J_V, etc.

In fact, $\hookrightarrow \ \equiv \ \subset + \longrightarrow$, where $\subset$ means *algebraic inclusion* ($V \subset W$) and $\longrightarrow$ implies a (*continuous*) mapping from V into W with $V \subset W$ (with topological properties such as the *continuity* of the mapping). (A.8.1.7)

A.8.2 Continuous extension of continuous linear operators by density

Theorem A.8.2.1 (Extension of a continuous linear operator by density). *Let V and W be Banach spaces and X be a* dense *subspace of V (i.e. $\overline{X} \equiv V$). Let $A \in \mathcal{L}(X;W)$ be a* continuous, linear *operator from X into W. Then $\exists$ a* unique, *continuous, linear operator $\tilde{A} \in \mathcal{L}(V;W)$ such that $\tilde{A}$ is the unique continuous, linear extension of A: $\tilde{A}\mathbf{x} = A\mathbf{x} \ \forall \mathbf{x} \in X$ and $\|\tilde{A}\|_{\mathcal{L}(V;W)} = \|A\|_{\mathcal{L}(X;W)}$.*

Proof. $\forall \mathbf{x} \in V$, $\exists$ a sequence $(\mathbf{x}_n)$ in X such that $\mathbf{x}_n \to \mathbf{x}$ in V. Hence, $(\mathbf{x}_n)$ is a *Cauchy sequence* in $V \implies (A\mathbf{x}_n)$ is a *Cauchy sequence* in Banach space W, since $A \in \mathcal{L}(X;W) \implies \exists \mathbf{y} \in W$ such that $\lim_{n\to\infty} A\mathbf{x}_n = \mathbf{y}$ in W, which does *not* depend on the choice of $(\mathbf{x}_n)$ in X. Thus, to *each* $\mathbf{x} \in V$ with $\mathbf{x}_n \to \mathbf{x}$ in V, we associate a *unique* $\mathbf{y} \in W$ by $\mathbf{y} = \lim_{n\to\infty} A\mathbf{x}_n$, which, in fact, defines a *mapping* from V into W, which we denote by $\widetilde{A}$, i.e. $\widetilde{A}\mathbf{x} = \mathbf{y} = \lim_{n\to\infty} A\mathbf{x}_n$ in W and $\widetilde{A}\mathbf{x} = A\mathbf{x} \ \forall \mathbf{x} \in X$.
$\widetilde{A}$ is linear: $\widetilde{A}(\alpha_1\mathbf{x}_1 + \alpha_2\mathbf{x}_2) = \lim_{n\to\infty} A(\alpha_1\mathbf{x}_1^n + \alpha_2\mathbf{x}_n^n) = \alpha_1\widetilde{A}\mathbf{x}_1 + \alpha_2\widetilde{A}\mathbf{x}_2$;
$\widetilde{A}$ is continuous from V into W: $\mathbf{x}_n \to \mathbf{x}$ in $V \implies \lim_{n\to\infty} A\mathbf{x}_n = \widetilde{A}\mathbf{x}$ in W

$$\begin{aligned}\implies \quad \|\widetilde{A}\mathbf{x}\|_W &= \lim_{n\to\infty} \|A\mathbf{x}_n\|_W \le \lim_{n\to\infty} (\|A\|_{\mathcal{L}(X;W)} \|\mathbf{x}_n\|_V) \\ &= \|A\|_{\mathcal{L}(X;W)} \|\mathbf{x}\|_V \quad \forall \mathbf{x} \in V\end{aligned}$$

$\implies \widetilde{A}$ is *continuous* from V into W with $\|\widetilde{A}\|_{\mathcal{L}(V;W)} \le \|A\|_{\mathcal{L}(X;W)} \le \|\widetilde{A}\|_{\mathcal{L}(V;W)}$, i.e. with $\|\widetilde{A}\|_{\mathcal{L}(V;W)} = \|A\|_{\mathcal{L}(X;W)}$.

The *uniqueness* of $\widetilde{A}$ obviously follows from the definition, and the proof is complete. □

A.8.3 Isomorphisms and isometric isomorphisms

Definition A.8.3.1. Let V and W be any two *normed linear spaces*. A map $T : V \longrightarrow W$ is called an *algebraic* and *topological isomorphism* from V *onto* W if and only if:

1. T is an *algebraic isomorphism*, i.e. T is *linear* and *bijective* from V *onto* W;
2. T and T^{-1} are *continuous* from V *onto* W and from W *onto* V, respectively.

Two *normed linear spaces* V and W are called (both *algebraically* and *topologically*) *isomorphic* iff there exists an (*algebraic* and *topological*) *isomorphism* T from V *onto* W.

Theorem A.8.3.1. *Let T be an algebraic and topological* isomorphism *from V onto W. Then T maps*

I. open, closed, bounded, dense *subsets of V* onto *open, closed, bounded and dense subsets of W;*

II. Cauchy *sequences,* convergent *sequences etc. in V onto Cauchy sequences, convergent sequences, etc. in W.*

Definition A.8.3.2. Let $T \in \mathcal{L}(V; W)$ be an *isomorphism*. Then T is called an *isometric isomorphism* iff $\|T\mathbf{x} - T\mathbf{y}\|_W = \|\mathbf{x} - \mathbf{y}\|_V \ \forall \mathbf{x}, \mathbf{y} \in V$.

The equivalence of two norms $\|\cdot\|_1$ and $\|\cdot\|_2$ on V If $I : (V, \|\cdot\|_1) \longrightarrow (V, \|\cdot\|_2)$ is *continuous*, the *identity operator I* is an *isomorphism* from $(V, \|\cdot\|_1)$ *onto* $(V, \|\cdot\|_2)$ under the *equivalence* of two norms: $\exists C_1, C_2 > 0$ such that

$$C_1\|\mathbf{x}\|_1 \le \|\mathbf{x}\|_2 \le C_2\|\mathbf{x}\|_1 \quad \forall \mathbf{x} \in V \quad \text{(see also (A.3.1.3))}. \tag{A.8.3.1}$$

A.8.4 Graph of an operator $A \in \mathcal{L}(V; W)$ and graph norm

Let $A : V \longrightarrow W$ be a *linear* operator from Banach space V into Banach space W equipped with the norm $\|\cdot\|_V$ and $\|\cdot\|_W$, respectively.

Definition A.8.4.1. The *graph* of the operator A, denoted by $G(A)$, is defined by the *subspace* $G(A) = \bigcup_{\mathbf{x}\in V}\{(\mathbf{x}, A\mathbf{x})\} = \{(\mathbf{x}, \mathbf{y}) : \mathbf{x} \in V, \mathbf{y} = A\mathbf{x} \in W\}$ of the *product Banach space* $V \times W$, the *graph* $G(A)$ is called *closed* in $V \times W$ iff

$$\mathbf{x}_n \to \mathbf{x} \text{ in } V \textit{ and } A\mathbf{x}_n \to \mathbf{y} \text{ in } W \quad \text{implies} \quad \mathbf{y} = A\mathbf{x} \text{ in } W, \tag{A.8.4.1}$$

and the *graph norm* $\|\cdot\|_p$ on V is defined, $\forall \mathbf{x} \in V$, by:

$$\|\mathbf{x}\|_p = (\|\mathbf{x}\|_V^p + \|A\mathbf{x}\|_W^p)^{1/p} \quad \forall p \in [1, \infty]. \tag{A.8.4.2}$$

$$A \in \mathcal{L}(V; W) \Longrightarrow \text{the graph } G(A) \text{ of } A \text{ is } \textit{closed} \text{ in } V \times W. \tag{A.8.4.3}$$

A.9 Reflexivity of Banach spaces

Definition A.9.1.1. Let V be a *normed linear space*. Then the dual space $(V')'$ of V' ($V' = \mathcal{L}(V;\mathbb{R})$, a *Banach space* by Theorem A.7.1.2) is called the *second dual space* of V and denoted by $V'' \equiv (V')'$. V'' is also a *Banach space* with $\|L\|_{V''} = \sup_{l \in V' - \{0\}} \frac{|L(l)|}{\|l\|_{V'}}$.

To each $\mathbf{v} \in V$, we can associate a linear functional $L_{\mathbf{v}} : V' \longrightarrow \mathbb{R}$ on V' defined by $L_{\mathbf{v}}(l) = l(\mathbf{v})\ \forall l \in V'$. Then $L_{\mathbf{v}}$ belongs to $(V')' \equiv V''$. (A.9.1.1)

Define $J : V \longrightarrow V''$ by $J\mathbf{v} = L_{\mathbf{v}} \in V''$ with $J\mathbf{v}(l) = L_{\mathbf{v}}(l) = l(\mathbf{v})\ \forall l \in V'$. (A.9.1.2)

Then J is *linear*, *continuous*, *injective* and *isometric* from V *onto* $J(V) \subset V''$, i.e. an *isometric*, *isomorphism* from V *onto* $J(V) \subset V''$.

Definition A.9.1.2. A *normed linear space* V is called *reflexive* iff $J : V \longrightarrow V''$ defined by (A.9.1.2) is *onto* V'', i.e. $V'' \equiv JV$. Then, $\forall L \in V''$, $\exists \mathbf{v} \in V$ such that $L(l) = l(\mathbf{v})\ \forall l \in V'$.

Proposition A.9.1.1. *Every reflexive normed linear space is a Banach space and* isometrically isomorphic *to its second dual space V''. Hence, for reflexive V, V and V'' can be* identified, *i.e. $V \equiv V''$ under J.*

I. *Every* closed *subspace of a reflexive V is* reflexive. (A.9.1.3)

II. *V is reflexive* $\Longrightarrow$ *V' is* reflexive. (A.9.1.4)

III. *Every* uniformly convex *(i.e. $\forall \varepsilon \in]0, 2]$, $\exists \delta = \delta(\varepsilon) > 0$ such that* $[\mathbf{u}, \mathbf{v} \in V$, $\|\mathbf{u}\| = \|\mathbf{v}\| = 1$ and $\|\mathbf{u} - \mathbf{v}\|_V \geq \varepsilon] \Longrightarrow \|\frac{\mathbf{u}+\mathbf{v}}{2}\|_V \leq 1 - \delta$*) Banach space V is* reflexive. (A.9.1.5)

For example, Banach spaces $L^p(\Omega)$, $1 < p < \infty$, are *uniformly convex*, and hence *reflexive*, but $L^1(\Omega)$, $L^\infty(\Omega)$, $C^0(\overline{\Omega})$, $C^k(\overline{\Omega})$, etc. are *not* reflexive. (A.9.1.6)

A.10 Strong, weak and weak-* convergence in Banach space V

A.10.1 Strong convergence $\rightarrow$

$\rightarrow$ implies the usual *convergence* in the norm $\|\cdot\|_V$: $\mathbf{x}_n \rightarrow \mathbf{x}$ *strongly* in $V \Longleftrightarrow \|\mathbf{x} - \mathbf{x}_n\|_V \rightarrow 0$ as $n \rightarrow \infty$. Then $\mathbf{x}$ is called the (*strong*) *limit* of the sequence $(\mathbf{x}_n)$ in V. (A.10.1.1)

A.10.2 Weak convergence $\rightharpoonup$

A sequence $(\mathbf{x}_n)$ in Banach space V is said to *converge weakly* to $\mathbf{x} \in V$ (or converge to $\mathbf{x} \in V$ in the *weak topology*) iff $L(\mathbf{x}_n) \to L(\mathbf{x})$ in $\mathbb{R}$ $\forall L \in V'$ as $n \to \infty$, i.e. $\mathbf{x}_n \overset{\text{weakly}}{\rightharpoonup} \mathbf{x}$ in $V \iff \langle L, \mathbf{x}_n \rangle_{V' \times V} \to \langle L, \mathbf{x} \rangle_{V' \times V}$ in $\mathbb{R}$ $\forall L \in V'$, as $n \to \infty$. Then $\mathbf{x} \in V$ is called the *unique weak limit* of $(\mathbf{x}_n)$ in V. (A.10.2.1)

Weak convergence of a sequence in V does *not* imply its *strong convergence* in V. (A.10.2.2)

Strong convergence of $(\mathbf{x}_n)$ in $V \Longrightarrow$ its *weak convergence* in V. (A.10.2.3)

$\mathbf{x}_n \rightharpoonup \mathbf{x} \in V$ *weakly and* $l_n \to l \in V'$ *strongly* $\Longrightarrow \langle l_n, \mathbf{x}_n \rangle_{V' \times V} \to \langle l, \mathbf{x} \rangle_{V' \times V}$ as $n \to \infty$. (A.10.2.4)

$\mathbf{x}_n \rightharpoonup \mathbf{x} \in V$ *weakly* $\Longrightarrow (\|\mathbf{x}_n\|_V)_{n=1}^{\infty}$ is *bounded* in $\mathbb{R}$. (A.10.2.5)

In a finite-dimensional Banach space V, *weak* and *strong* convergences *coincide*. (A.10.2.6)

A.10.3 Weak-* convergence $\rightharpoonup^*$ in Banach space V'

Definition A.10.3.1. A sequence $(l_n)_{n=1}^{\infty}$ in Banach space V' is said to converge to $l \in V'$ in the *weak-* sense* (or in the *weak-* topology* of V') iff $l_n(\mathbf{x}) \to l(\mathbf{x})$ in $\mathbb{R}$ $\forall \mathbf{x} \in V$ as $n \to \infty$, i.e. $l_n \rightharpoonup^* l$ in $V' \iff \langle l_n, \mathbf{x} \rangle \to \langle l, \mathbf{x} \rangle$ $\forall \mathbf{x} \in V$ as $n \to \infty$. Then $l \in V'$ is the *unique weak-* limit* of (l_n). (A.10.3.1)

Weak convergence in $V' \Longrightarrow$ *weak-* convergence* in V', i.e. $l_n \rightharpoonup l$ in $V' \Longrightarrow l_n \rightharpoonup^*$ in V' as $n \to \infty$. (A.10.3.2)

In a *reflexive* Banach space V, weak-* and weak convergences *coincide*. (A.10.3.3)

$l_n \rightharpoonup^* l$ in $V' \Longrightarrow (\|l_n\|_{V'})_{n=1}^{\infty}$ is *bounded* in $\mathbb{R}$. (A.10.3.4)

$l_n \rightharpoonup^* l$ in V' *and* $\mathbf{x}_n \to \mathbf{x}$ in V *strongly* $\Longrightarrow \langle l_n, \mathbf{x}_n \rangle_{V' \times V} \to \langle l, \mathbf{x} \rangle_{V' \times V}$. (A.10.3.5)

A.11 Compact linear operators in Banach spaces

Let V, W and Z be Banach spaces.

Definition A.11.1.1. A *linear* operator $A : V \longrightarrow W$ is called *compact* iff, $\forall$ *bounded* $U \subset V$, $A(U)$ is *relatively compact* in W, i.e. $\overline{A(U)}$ in W is *compact*, or, equivalently, iff $\forall$ *bounded* sequences $(\mathbf{x}_n)_{n=1}^{\infty}$ in V, $\exists$ a subsequence $(\mathbf{x}_{n_k})_{k=1}^{\infty}$ such that $(A\mathbf{x}_{n_k})$ converges in W as $k \to \infty$.

Theorem A.11.1.1.

I. *A, B are* compact *from V into W* $\Longrightarrow A + B$, αA *with* $\alpha \in \mathbb{R}$ *are* compact.

II. $A \in \mathcal{L}(V;W)$, $B \in \mathcal{L}(W;Z) \Longrightarrow B \circ A = BA : V \longrightarrow Z$ *is* compact *if either* B or A *is compact, i.e.* composition *of a* continuous *operator and a* compact *operator is* compact.

Definition A.11.1.2. A subset $A \subset V$ is called *weakly sequentially compact* in a *Banach space* V, if and only if every sequence $(\mathbf{x}_n)_{n=1}^{\infty}$ in A has a subsequence $(\mathbf{x}_{n_k})_{n=1}^{\infty}$ such that $\mathbf{x}_{n_k} \overset{\text{weakly}}{\rightharpoonup} \mathbf{x} \in A$ in V (i.e. $\Longleftrightarrow l(\mathbf{x}_{n_k}) \to l(\mathbf{x})$ in $\mathbb{R}$ as $k \to \infty\ \forall l \in V'$).

Theorem A.11.1.2 (Eberlein–Schmulyan). *Let V be a* reflexive *Banach space. Then, $\forall$ bounded sequences $(\mathbf{v}_n)$ in V (i.e. $\exists C > 0$ such that $\|\mathbf{v}_n\|_V \le C\ \forall n \in \mathbb{N}$), $\exists$ a subsequence $(\mathbf{v}_{n_k})$ of $(\mathbf{v}_n)$ such that $\mathbf{v}_{n_k} \overset{\text{weakly}}{\rightharpoonup} \mathbf{v} \in V$ as $k \to \infty$, i.e. every bounded set of a reflexive Banach space is weakly (sequentially) compact.*

A.12 Hilbert space V

Definition A.12.1.1. A *real* (resp. *complex*) vector space V equipped with an *inner product* $\langle \cdot, \cdot \rangle_V : (\mathbf{u}, \mathbf{v}) \in V \times V \longmapsto \langle \mathbf{u}, \mathbf{v} \rangle_V \in \mathbb{R}$ (resp. $\mathbb{C}$) satisfying the following *properties of inner product*: $\forall \mathbf{u}, \mathbf{v}, \mathbf{w} \in V$, $\forall \alpha \in \mathbb{R}$ (resp. $\mathbb{C}$),

- $\langle \mathbf{u}, \mathbf{u} \rangle_V \ge 0$; $\langle \mathbf{u}, \mathbf{u} \rangle_V = 0 \Longleftrightarrow \mathbf{u} = \mathbf{0}$ in V (*positive-definiteness*);
- $\langle \mathbf{u}, \mathbf{v} \rangle_V = \langle \mathbf{v}, \mathbf{u} \rangle_V$ (*symmetry*) (resp. $\langle \mathbf{u}, \mathbf{v} \rangle_V = \overline{\langle \mathbf{v}, \mathbf{u} \rangle_V}$ (*Hermitian symmetry*));
- $\langle \mathbf{u}, \alpha\mathbf{v} \rangle_V = \alpha \langle \mathbf{u}, \mathbf{v} \rangle_V\ \forall \alpha \in \mathbb{R}$ (*linearity*) (resp. $\langle \mathbf{u}, \alpha\mathbf{v} \rangle_V = \overline{\alpha} \langle \mathbf{u}, \mathbf{v} \rangle_V\ \forall \alpha \in \mathbb{C}$ (*antilinearity*));
  $\langle \alpha\mathbf{u}, \mathbf{v} \rangle_V = \alpha \langle \mathbf{u}, \mathbf{v} \rangle_V\ \forall \alpha \in \mathbb{R}$ (resp. $\alpha \in \mathbb{C}$) (*linearity*);
- $\langle \mathbf{u}, \mathbf{v} + \mathbf{w} \rangle_V = \langle \mathbf{u}, \mathbf{v} \rangle_V + \langle \mathbf{u}, \mathbf{w} \rangle_V$; $\langle \mathbf{u} + \mathbf{v}, \mathbf{w} \rangle_V = \langle \mathbf{u}, \mathbf{w} \rangle_V + \langle \mathbf{v}, \mathbf{w} \rangle_V$ (*additive and distributive*),

is called an *inner product space* or *Euclidean space* or *pre-Hilbert space*.

Proposition A.12.1.1. *Every (real or complex) inner product space V can be equipped with a* norm $\|\cdot\|_V$ *and* metric $d(\cdot,\cdot)$ *defined, $\forall \mathbf{u} \in V$, by:*

$$\|\mathbf{u}\|_V = \langle \mathbf{u}, \mathbf{u} \rangle_V^{1/2}; \tag{A.12.1.1}$$

$$d(\mathbf{u}, \mathbf{v}) = \|\mathbf{u} - \mathbf{v}\|_V \quad \forall \mathbf{u}, \mathbf{v} \in V, \tag{A.12.1.2}$$

such that V equipped with the norm $\|\cdot\|_V$ becomes a normed linear space, *and V equipped with the metric $d(\cdot,\cdot)$ becomes a* metric space.

$$|\langle \mathbf{u}, \mathbf{v} \rangle_V| \le \langle \mathbf{u}, \mathbf{u} \rangle_V^{1/2} \langle \mathbf{v}, \mathbf{v} \rangle_V^{1/2} \text{ (the *Cauchy–Schwarz inequality*).} \tag{A.12.1.3}$$

The *parallelogram law* holds:

$$\|\mathbf{u} + \mathbf{v}\|_V^2 + \|\mathbf{u} - \mathbf{v}\|_V^2 = 2(\|\mathbf{u}\|_V^2 + \|\mathbf{v}\|_V^2) \quad \forall \mathbf{u}, \mathbf{v} \in V. \tag{A.12.1.4}$$

For non-null vectors $\mathbf{u}, \mathbf{v} \in V$, $\mathbf{u}$ is *orthogonal* to $\mathbf{v}$ in $V \iff \langle \mathbf{u}, \mathbf{v} \rangle_V = 0$. (A.12.1.5)

$$\langle \mathbf{u}, \mathbf{v} \rangle_V = 0 \ \forall \mathbf{v} \in V \iff \mathbf{u} = \mathbf{0} \text{ in } V. \tag{A.12.1.6}$$

Definition A.12.1.2. A system $(\hat{\mathbf{e}}_i)_i$ of vectors in V is called *orthonormal* in V if and only if $\langle \hat{\mathbf{e}}_i, \hat{\mathbf{e}}_j \rangle_V = \delta_{ij} = 0$ for $i \neq j$ and $\delta_{ij} = 1$ for $i = j$. (A.12.1.7)

Definition A.12.1.3. Let $M \subset V$ be a nonempty subset of an inner product space V. A non-null vector $\mathbf{x} \in V$ is *orthogonal* to M iff $\langle \mathbf{x}, \mathbf{y} \rangle_V = 0 \ \forall \mathbf{y} \in M$. Then we write $\mathbf{x} \perp M$.

Definition A.12.1.4. An inner product space V is called *complete* iff every *Cauchy sequence* in V *converges* in V in the norm $\| \cdot \|_V$ defined by (A.12.1.1), and a *complete* inner product space V is called a *Hilbert space*.

Example A.12.1.1. For *Hilbert space* $V \equiv L^2(\Omega)$ (see $L^p(\Omega)$ with $p = 2$ in Section B.4 and also Table B.3 in Appendix B), the *inner product* $\langle \cdot, . \rangle_{L^2(\Omega)}$, *norm* $\| \cdot \|_{L^2(\Omega)}$ and *metric* $d(\cdot, \cdot)$ in $L^2(\Omega)$ are defined, $\forall u, v \in L^2(\Omega)$, by:

$$\langle u, v \rangle_{L^2(\Omega)} = \int_\Omega u(\mathbf{x}) v(\mathbf{x}) d\mathbf{x} \quad (\textit{real case}), \tag{A.12.1.7a}$$

(resp.

$$\langle u, v \rangle_{L^2(\Omega)} = \int_\Omega u(\mathbf{x}) \overline{v(\mathbf{x})} d\mathbf{x} \quad (\textit{complex case}))$$

$$\|u\|_{L^2(\Omega)} = \langle u, u \rangle_{L^2(\Omega)}^{1/2} = \left(\int_\Omega |u(\mathbf{x})|^2 d\mathbf{x} \right)^{1/2};$$

$$d(u, v) = \langle u - v, u - v \rangle_{L^2(\Omega)}^{1/2} = \left(\int_\Omega |u(\mathbf{x}) - v(\mathbf{x})|^2 d\mathbf{x} \right)^{1/2}. \tag{A.12.1.7b}$$

- $(u_n)_{n=1}^\infty$ converges in $L^2(\Omega) \iff (u_n)_{n=1}^\infty$ is a Cauchy sequence in $L^2(\Omega) \iff \lim_{m \to \infty} \|u_m - u_n\|_{L^2(\Omega)} = 0$; (A.12.1.7c)
- $u_n \to u$ in $L^2(\Omega) \Longrightarrow \lim_{n \to \infty} \int_\Omega |u(\mathbf{x}) - u_n(\mathbf{x})|^2 d\mathbf{x} = 0$. (A.12.1.7d)

Proposition A.12.1.2. *Every* Hilbert space V *is a* Banach space *with the norm* $\| \cdot \|_V$ *defined by* $\|\mathbf{v}\|_V = \langle \mathbf{v}, \mathbf{v} \rangle_V^{1/2} \ \forall \mathbf{v} \in V$. *Conversely, a Banach space V satisfying the parallelogram law* (A.12.1.4) *and equipped with the inner product* $\langle \cdot, \cdot \rangle_V$ *defined,* $\forall \mathbf{u}, \mathbf{v} \in V$*, by:*

(real case):

$$\langle \mathbf{u}, \mathbf{v} \rangle_V = \frac{1}{4} [\|\mathbf{u} + \mathbf{v}\|_V^2 - \|\mathbf{u} - \mathbf{v}\|_V^2]$$

(resp. (complex case):

$$\langle \mathbf{u}, \mathbf{v} \rangle_V = \frac{1}{4}[(\|\mathbf{u}+\mathbf{v}\|_V^2 - \|\mathbf{u}-\mathbf{v}\|_V^2) + i(\|\mathbf{u}+i\mathbf{v}\|_V^2 - \|\mathbf{u}-i\mathbf{v}\|_V^2)]) \quad \text{(A.12.1.8)}$$

is called a Hilbert space.

But $C^0(\overline{\Omega})$ with $\|\cdot\|_\infty$ is a Banach space in which the parallelogram law (A.12.1.4) does *not* hold, and hence *cannot* be equipped with an inner product corresponding to $\|\cdot\|_\infty$.

All the results for Banach spaces given earlier will also hold for Hilbert spaces. (A.12.1.9)

Proposition A.12.1.3. *Every* finite-dimensional *inner product space is a* Hilbert space.

Theorem A.12.1.1. *Every* closed *subspace of a Hilbert space V is a* Hilbert space.

Definition A.12.1.5. Let M be a *closed subspace* of a Hilbert space V. Then

$$M^\perp = \{\mathbf{w} : \mathbf{w} \in V,\ \langle \mathbf{w}, \mathbf{v} \rangle_V = 0\ \forall \mathbf{v} \in M\} \quad \text{(A.12.1.10)}$$

is called the *orthogonal complement* of M in V, and we write $V = M \oplus M^\perp$, i.e. $V = M + M^\perp$ *and* $M \cap M^\perp = \{\mathbf{0}\}$, or, equivalently, $\forall \mathbf{u} \in V$, $\exists$ a *unique* $\mathbf{v} \in M$ and a *unique* $\mathbf{w} \in M^\perp$ such that $\mathbf{u} = \mathbf{v} + \mathbf{w}$. Then M and $M^\perp$ denote a *decomposition* of Hilbert space V into *mutually orthogonal* closed subspaces of V.

Definition A.12.1.6. For $V = M \oplus M^\perp$, the correspondence $\mathbf{u} \in V \longmapsto \mathbf{v} \in M$ (resp. $\mathbf{u} \in V \longmapsto \mathbf{w} \in M^\perp$) defines a mapping called the *orthogonal projection operator* or *projection operator* from V *onto* M (resp. $M^\perp$) denoted by P_M (resp. $P_{M^\perp}$), which is characterized by the equation

$$\langle \mathbf{u} - P_M \mathbf{u}, \mathbf{z} \rangle_V = 0 \quad \forall \mathbf{z} \in M, \quad \text{(A.12.1.11)}$$

i.e. $\exists$ a *unique* $\mathbf{v} \in V$ such that $\langle \mathbf{u} - \mathbf{v}, \mathbf{z} \rangle_V = 0\ \forall \mathbf{z} \in M \Longrightarrow \mathbf{v} = P_M \mathbf{u}$ and vice versa (resp. $\langle \mathbf{u} - P_{M^\perp}\mathbf{u}, \mathbf{w} \rangle_V = 0\ \forall \mathbf{w} \in M^\perp$). Hence, $\forall \mathbf{u} \in V$, $\mathbf{u} = P_M \mathbf{u} + P_{M^\perp}\mathbf{u}$, the representation being a unique one. $P_M + P_{M^\perp} = I_V$, I_V being the *identity operator* on V; P_M is a *bounded* (hence, *continuous*), *linear* operator from V onto M with $\mathcal{N}(P_M) =$ the *null space* of $P_M = M^\perp$ by definition. (A.12.1.12)

Definition A.12.1.7. A *subspace* $M \subset V$ of a *Hilbert space* V is called *dense* in V iff

$$\langle \mathbf{u}, \mathbf{v} \rangle_V = 0 \quad \forall \mathbf{v} \in M \Longrightarrow \mathbf{u} = 0 \text{ in } V. \quad \text{(A.12.1.13)}$$

Definition A.12.1.8. A Hilbert space V is called *separable* iff $\exists$ a *complete* (resp. *closed*) *orthonormal system* in V.

An *orthonormal system* $(\hat{\mathbf{e}_n})_{n=1}^{\infty}$ is *complete* (resp. *closed*) in V iff $\langle \mathbf{f}, \hat{\mathbf{e}}_n \rangle_V = 0$ $\forall n \in \mathbb{N} \Longrightarrow \mathbf{f} = 0$ in V (resp. $\|\mathbf{f}\|_V^2 = \sum_{k=1}^{\infty} |\langle \mathbf{f}, \hat{\mathbf{e}}_k \rangle_V|^2$ $\forall \mathbf{f} \in V$).

Remark A.12.1.1. In applications, all Hilbert spaces used are separable. Henceforth, by Hilbert space, we mean the separable one.

A.13 Dual space V' of a Hilbert space V, reflexivity of V

Let V be a *real* Hilbert space. Then V is a *Banach space* with the norm $\|\mathbf{v}\|_V = \langle \mathbf{v}, \mathbf{v} \rangle_V^{1/2}$ $\forall \mathbf{v} \in V$, and

$$V' = \mathcal{L}(V; \mathbb{R}) = \{l : l : V \longrightarrow \mathbb{R} \text{ is a } \textit{continuous} \text{ linear functional on } V\} \tag{A.13.1.1}$$

is the *dual space* of V with

$$\|l\|_{V'} = \sup_{\mathbf{v} \in V - \{\mathbf{0}\}} \frac{|l(\mathbf{v})|}{\|\mathbf{v}\|_V} \quad \forall l \in V'. \tag{A.13.1.2}$$

Theorem A.13.1.1 (Riesz Representation Theorem). *Let V be a* real *Hilbert space and $l \in V'$ be a* continuous, linear *functional on V. Then $\exists$ a unique $\mathbf{u}_l \in V$ such that $l(\mathbf{v}) = \langle \mathbf{v}, \mathbf{u}_l \rangle_V$ $\forall \mathbf{v} \in V$ with $\|l\|_{V'} = \|\mathbf{u}_l\|_V$, $\langle \cdot, . \rangle_V$ being the* inner product *in V.*

Conversely, every $\mathbf{u} \in V$ defines a continuous, linear *functional $l_{\mathbf{u}} \in V'$ such that $l_{\mathbf{u}}(\mathbf{v}) = \langle \mathbf{v}, \mathbf{u} \rangle_V$ $\forall \mathbf{v} \in V$ and $\|l_{\mathbf{u}}\|_{V'} = \|\mathbf{u}\|_V$.*

For a *real* Hilbert space V, Theorem A.13.1.1 allows us to define a mapping $J : V \longrightarrow V'$ such that $\mathbf{u} \in V \longmapsto J\mathbf{u} = l \in V'$ with

$$J\mathbf{u}(\mathbf{v}) = \langle \mathbf{v}, \mathbf{u} \rangle_V \quad \forall \mathbf{u}, \mathbf{v} \in V. \tag{A.13.1.3}$$

Proposition A.13.1.1. *$J : V \longrightarrow V'$ is an* isometric isomorphism *from V* onto *V'.*

Definition A.13.1.1. For the isometric isomorphism $J : V \longrightarrow V'$, its *inverse* $J^{-1} : V' \longrightarrow V$ is called the *Riesz map* from V' *onto* V such that $J^{-1}l$ is a *Riesz representer* of $l \in V'$ satisfying

$$l(\mathbf{v}) = \langle \mathbf{v}, J^{-1}l \rangle_V \quad \forall \mathbf{v} \in V. \tag{A.13.1.4}$$

Proposition A.13.1.2. *The dual space V' of V is also a real* Hilbert space *equipped with the inner product $\langle l_1, l_2 \rangle_{V'} = \langle J^{-1}l_1, J^{-1}l_2 \rangle_V$ $\forall l_1, l_2 \in V'$, where $J^{-1}l_1$, $J^{-1}l_2 \in V$ are Riesz representers of l_1 and $l_2 \in V'$, respectively.*

Remark A.13.1.1. As a result of this *isometric isomorphism*, *real* Hilbert space V can be identified with its dual space V' in *some* situations (but *not* always) such that $V = V'$. Then V is called a *pivot* space.

Remark A.13.1.2. For *complex* Hilbert spaces V, Theorem A.13.1.1 holds. But the mapping $J : V \longrightarrow V'$ is *anti-linear*, since $\forall \alpha \in \mathbb{C}$, $J(\alpha \mathbf{u}) = \overline{\alpha} J\mathbf{u}$ and J is *bijective* from the set V onto the set V'. Hence, J is *not* an *isomorphism* and V *cannot* be identified with $V' = \mathcal{L}(V;\mathbb{C})$.

Every Hilbert space V (*real* or *complex*) is *uniformly convex* and hence *reflexive*. (A.13.1.5)

A.14 Strong, weak and weak-* convergences in a Hilbert space

See also Section A.10.

By virtue of Theorem A.13.1.1, *weak* convergence $\rightharpoonup$ in V can be defined by means of the *inner product* $\langle \cdot, . \rangle_V$, i.e. $\mathbf{x}_n \rightharpoonup \mathbf{x}$ in $V \iff \langle \mathbf{x}_n, \mathbf{u} \rangle_V \to \langle \mathbf{x}, \mathbf{u} \rangle_V \ \forall \mathbf{u} \in V$ as $n \to \infty$, and $\mathbf{x} \in V$ is called the unique *weak limit* of $(\mathbf{x}_n)$ in V. (A.14.1.1)

*Weak-** and *weak* convergences *coincide* in a Hilbert space. (A.14.1.2)

A.15 Self-adjoint and unitary operators in Hilbert space V

Definition A.15.1.1. Let V be a *complex* Hilbert space and $A \in \mathcal{L}(V) \equiv \mathcal{L}(V;V)$. Then $\exists$ a *unique*, *bounded*, *linear* operator $B \in \mathcal{L}(V)$ called the *adjoint* of the *bounded*, *linear* operator A such that $\langle A\mathbf{u}, \mathbf{v} \rangle_V = \langle \mathbf{u}, B\mathbf{v} \rangle_V \ \forall \mathbf{u}, \mathbf{v} \in V$. B is usually denoted by $A^* \in \mathcal{L}(V)$ such that

$$\langle A\mathbf{u}, \mathbf{v} \rangle_V = \langle \mathbf{u}, A^*\mathbf{v} \rangle_V \quad \forall \mathbf{u}, \mathbf{v} \in V. \tag{A.15.1.1}$$

If $A = A^*$, i.e. $\langle A\mathbf{u}, \mathbf{v} \rangle_V = \langle \mathbf{u}, A\mathbf{v} \rangle_V \ \forall \mathbf{u}, \mathbf{v} \in V$, then A is called *self-adjoint* on V. (A.15.1.2)

If $AA^* = A^*A = I$, then A is called a *unitary* operator. For a unitary operator U on V with $UU^* = U^*U = I$ on V, we have:

1. $\langle U\mathbf{x}, U\mathbf{y} \rangle_V = \langle \mathbf{x}, U^*U\mathbf{y} \rangle_V = \langle \mathbf{x}, I\mathbf{y} \rangle_V = \langle \mathbf{x}, \mathbf{y} \rangle_V \ \forall \mathbf{x}, \mathbf{y} \in V$; (A.15.1.3)
2. $\|U\mathbf{x}\|_V = \|\mathbf{x}\|_V \ \forall \mathbf{x} \in V$ (i.e. U is *isometric* on V). (A.15.1.4)

A.16 Compact linear operators in Hilbert spaces

The *most frequently used* definition in applications is the following equivalent one in Hilbert spaces:

Definition A.16.1.1. A linear operator $A : V \longrightarrow W$ is called *compact* iff $\mathbf{x}_n \overset{\text{weakly}}{\rightharpoonup} \mathbf{x}$ in $V \Longrightarrow A\mathbf{x}_n \overset{\text{strongly}}{\rightarrow} A\mathbf{x}$ in W, V and W being Hilbert spaces.

For example, the identity operator $I \in \mathcal{L}(V)$ is *continuous* on V, but *not* compact, since $\mathbf{x}_n \overset{\text{weakly}}{\rightharpoonup} \mathbf{x}$ in V does *not* imply

$$I\mathbf{x}_n = \mathbf{x}_n \rightarrow \mathbf{x} = I\mathbf{x} \quad \text{in } V. \tag{A.16.1.1}$$

Appendix B
L^p-spaces

B.1 Lebesgue measure μ on $\mathbb{R}^n$

B.1.1 Lebesgue-measurable sets in $\mathbb{R}^n$

Definition B.1.1.1. A family τ of *subsets* of $\mathbb{R}^n$ satisfying the properties:

1. $\mathbb{R}^n \in \tau$;
2. $A \in \tau \Longrightarrow A^{\complement} \in \tau$;
3. $A_k \in \tau \ \forall k \in \mathbb{N} \Longrightarrow \bigcup_{k=1}^{\infty} A_k \in \tau$, (B.1.1.1)

is called a σ-*algebra* of subsets of $\mathbb{R}^n$.

Definition B.1.1.2. Let μ be the *positive measure* on τ defined by:

1. $\mu : A \in \tau \longmapsto \mu(A) \in [0, \infty] = \overline{\mathbb{R}}_+$ such that μ is *countably additive*, i.e. for $\{A_i\}_{i=1}^{\infty}$ with $A_i \in \tau$, $A_i \cap A_j = \emptyset \ \forall i \neq j$ (all the A_is are *mutually disjoint*),

$$\mu\left(\bigcup_{i=1}^{\infty} A_i\right) = \sum_{i=1}^{\infty} \mu(A_i), \tag{B.1.1.2}$$

where at least one $A_i \in \tau$ has $\mu(A_i) < +\infty$, with the additional properties:

2. (a) $\mu(\emptyset) = 0$;
   (b) $\mu(\bigcup_{i=1}^{m} A_i) = \sum_{i=1}^{m} \mu(A_i)$ for $A_i \in \tau$ with $A_i \cap A_j = \emptyset \ \forall i \neq j$;
   (c) $\forall A, B \in \tau$ with $A \subset B$, $\mu(A) \leq \mu(B)$;
   (d) $A, B \in \tau$, $A \subset B$ with $\mu(B) = 0 \Longrightarrow \mu(A) = 0$; (B.1.1.3)
   (e) for $A_1 \subset A_2 \subset \cdots \subset A_k \subset \cdots$ such that $\bigcup_{k=1}^{\infty} A_k = A$ is *bounded*, $\lim_{k\to\infty} \mu(A_k) = \mu(A)$; (B.1.1.4)
   (f) for $A_1 \supset A_2 \supset \cdots \supset A_k \supset \cdots$ with $\mu(A_1) < +\infty$ such that $A = \bigcap_{k=1}^{\infty} A_k$, $\lim_{k\to\infty} \mu(A_k) = \mu(A)$; (B.1.1.5)
3. for $A =$ the n-dimensional *open rectangular parallelepiped* R:

$$A = R = \{\mathbf{x} : \mathbf{x} = (x_1, \ldots, x_n) \in \mathbb{R}^n,\ a_i < x_i < b_i \text{ for } 1 \leq i \leq n\} \subset \mathbb{R}^n,$$

$$\mu(A) = \mu(R) = (b_1 - a_1)(b_2 - a_2) \cdots (b_n - a_n) = \prod_{i=1}^{n} (b_i - a_i) > 0 \tag{B.1.1.6}$$

is the *n-dimensional volume measure* $\mu(A) = \mu(R)$ of the open parallelepiped R.

4. If $\{R_k\}_{k=1}^m$ with $R_k = \prod_{i=1}^n]a_i^k, b_i^k[$ (the superscript k in a_i^k, b_i^k corresponds to the subscript k in R_k, i.e. k is *not* an exponent) is a family of *disjoint* ($R_i \cap R_j = \emptyset\ \forall i \neq j$) *open* rectangular parallelepipeds R_k, then the *n-dimensional volume measure* of $A = \bigcup_{k=1}^m R_k$ is given by:
$$\mu(A) = \mu\left(\bigcup_{k=1}^m R_k\right) = \sum_{k=1}^m \mu(R_k) = \sum_{k=1}^m \left[\prod_{i=1}^n (b_i^k - a_i^k)\right] > 0; \quad \text{(B.1.1.7)}$$
5. $\forall \mathbf{x} \in \mathbb{R}^n, \forall A \in \tau, \mu(\mathbf{x} + A) = \mu(A)$, i.e. the positive measure μ is *translation invariant*. (B.1.1.8)

Then the *positive measure* $\mu(A)$ of $A \in \tau$ satisfying properties 1–5 is called the *Lebesgue measure* of $A \in \tau$ or the *measure* of A, other equivalent notations used being $m(A)$, mes(A), meas(A), etc. Hence, the *Lebesgue measure* is the natural generalization of the length measure

$$\mu([a, b]) = \mu(]a, b[) = b - a > 0 \quad \text{(B.1.1.9)}$$

of the *interval* $[a, b]$ or $]a, b[$, respectively, on $\mathbb{R}$; the *area measure* $\mu([a, b] \times [c, d])$ (resp. $\mu(]a, b[\times]c, d[)$) of the *closed rectangle* $\overline{R} = [a, b] \times [c, d]$ (resp. the *open* rectangle $R =]a, b[\times]c, d[$) given by

$$\mu(\overline{R}) = \mu(R) = (b - a) \times (d - c) > 0 \quad \text{in } \mathbb{R}^2; \quad \text{(B.1.1.10)}$$

and the *volume measure* $\mu(\overline{R})$(resp. $\mu(R)$) of the *closed* rectangular parallelepiped $\overline{R} = \prod_{i=1}^3 [a_i, b_i]$ (resp. the *open* rectangular parallelepiped $R = \prod_{i=1}^3]a_i, b_i[$) such that

$$\mu(\overline{R}) = \mu(R) = \prod_{i=1}^3 (b_i - a_i) > 0 \quad \text{in } \mathbb{R}^3. \quad \text{(B.1.1.11)}$$

Proposition B.1.1.1. *For an* arbitrary *family* $\{A_i\}_{i=1}^\infty$ *of subsets* $A_i \subset \mathbb{R}^n$ *with* $\mu(A_i) < +\infty\ \forall i \in \mathbb{N}$, $\mu(\bigcup_{i=1}^\infty A_i) \leq \sum_{i=1}^\infty \mu(A_i)$. *In particular,*

$$\mu\left(\bigcup_{i=1}^m A_i\right) \leq \sum_{i=1}^m \mu(A_i). \quad \text{(B.1.1.12)}$$

B.1.2 Sets with zero (Lebesgue) measure in $\mathbb{R}^n$

The empty set $\emptyset$ has measure 0, i.e. $\mu(\emptyset) = 0$ (see property 2(a) in Definition B.1.1.2). (B.1.2.1)

Definition B.1.2.1 (Case $n = 1$). A set $A \in \tau$ with $A \subset \mathbb{R}$ is said to have (*Lebesgue length*) *measure zero*, $\mu(A) = 0$, if and only if $\forall \varepsilon > 0, \exists$ an *open* set B with $A \subset B$ such that B has measure $\mu(B) < \varepsilon$.

Example B.1.2.1. Let $A = \{x\} \subset \mathbb{R}$ be a *singleton* set. Then $\mu(\{x\}) = 0$, since $\forall \varepsilon > 0, \exists$ an open set $B =]x - \frac{\varepsilon}{4}, x + \frac{\varepsilon}{4}[$ with $\{x\} \subset]x - \frac{\varepsilon}{4}, x + \frac{\varepsilon}{4}[$ such that

$$\mu(B) = \mu\left(\left]x - \frac{\varepsilon}{4}, x + \frac{\varepsilon}{4}\right[\right) = x + \frac{\varepsilon}{4} - x + \frac{\varepsilon}{4} = \frac{\varepsilon}{2} < \varepsilon$$

(by (B.1.1.6) or (B.1.1.9)). Hence, by 2(c) in Definition B.1.1.2, $\mu(\{x\}) \leq \mu(B) < \varepsilon$ and the result follows from Definition B.1.2.1.

Definition B.1.2.2. A set $A \in \tau$ with $A \subset \mathbb{R}$ is said to have (*Lebesgue length*) *measure zero*, i.e. $\mu(A) = 0$, if and only if $\forall \varepsilon > 0, \exists$ a *finite system* $(I_k)_{k=1}^n$ or a *countably infinite system* $(I_k)_{k=1}^\infty$ of *open* intervals $I_k \subset \mathbb{R}$ covering A (i.e. $A \subset \bigcup_{k=1}^n I_k$ or $A \subset \bigcup_{k=1}^\infty I_k$) such that the sum of (*Lebesgue length*) *measures* of the open intervals I_k is less than ε, i.e. $\sum_{k=1}^n \mu(I_k) < \varepsilon$ or $\sum_{k=1}^\infty \mu(I_k) < \varepsilon$.

Example B.1.2.2. Let $A = \{x_1, \ldots, x_n, \ldots\} \subset \mathbb{R}$, with $x_i \neq x_j$ for $i \neq j$, be a set containing a *countably infinite number* of elements $x_k \in \mathbb{R}$. Then the *length measure*

$$\mu(\{x_1, \ldots, x_n, \ldots\}) = 0. \qquad \text{(B.1.2.2)}$$

In fact, $\forall \varepsilon > 0$, $\forall k \in \mathbb{N}$, $\{x_k\}_{k=1}^\infty$ can be covered by an *open* interval $]x_k - \frac{\varepsilon}{2^{k+2}}, x_k + \frac{\varepsilon}{2^{k+2}}[$ with *length measure* $\mu(]x_k - \frac{\varepsilon}{2^{k+2}}, x_k + \frac{\varepsilon}{2^{k+2}}[) = \frac{\varepsilon}{2^{k+1}}$ by (B.1.1.9)/(B.1.1.6). Then the *countably infinite* system $\{]x_k - \frac{\varepsilon}{2^{k+2}}, x_k + \frac{\varepsilon}{2^{k+2}}[\}_{k=1}^\infty$ of *open* intervals is a *cover* of A $\forall \varepsilon > 0$, i.e. $\forall \varepsilon > 0$, $\{x_1, \ldots, x_k, \ldots\} \subset \bigcup_{k=1}^\infty (]x_k - \frac{\varepsilon}{2^{k+2}}, x_k + \frac{\varepsilon}{2^{k+2}}[)$ with

$$\begin{aligned}
\mu\left(\bigcup_{k=1}^{\infty}\left(\left]x_k - \frac{\varepsilon}{2^{k+2}}, x_k + \frac{\varepsilon}{2^{k+2}}\right[\right)\right) &\leq \sum_{k=1}^{\infty} \mu\left(\left]x_k - \frac{\varepsilon}{2^{k+2}}, x_k + \frac{\varepsilon}{2^{k+2}}\right[\right) \\
&= \sum_{k=1}^{\infty} \frac{\varepsilon}{2^{k+1}} \\
&= \frac{\varepsilon}{4}\left(1 + \frac{1}{2} + \ldots\right) = \frac{\varepsilon}{4} \cdot 2 = \frac{\varepsilon}{2} < \varepsilon.
\end{aligned}$$

Hence, the result follows from Definition B.1.2.2.

Example B.1.2.3. The set $\mathbb{Q}$ of all rational numbers, which is *countably infinite*, has (Lebesgue) length measure $\mu(\mathbb{Q}) = 0$. (B.1.2.3)

Remark B.1.2.1. There are subsets of $\mathbb{R}$ containing *uncountably infinite* numbers of elements whose (Lebesgue) *length measure* is zero. For example, the *Cantor set*[1] $E \subset [0,1]$ is an *uncountably infinite* subset of $[0,1]$ with (Lebesgue) length measure $\mu(E) = 0$.

Proposition B.1.2.1. *Let* $A \subset \mathbb{R}$ *be the* union *of a* finite *system* $(I_k)_{k=1}^n$ *(resp.* countably infinite *system* $(I_k)_{k=1}^\infty$*) of intervals* $I_k \subset \mathbb{R}$ *with (Lebesgue)* length measure $\mu(I_k) = 0\ \forall k \in \mathbb{N}$. *Then the (Lebesgue)* length measure $\mu(A) = 0$.

Proof. $A = \bigcup_k I_k \Longrightarrow \mu(A) = \mu(\bigcup_k I_k) \le \sum_k \mu(I_k) = 0 \Longrightarrow \mu(A) = 0$, since $\mu(A) \ge 0$. □

Proposition B.1.2.1 does *not* hold in general if A is the union of an *uncountably infinite number* of sets I_λ with $\mu(I_\lambda) = 0\ \forall\lambda$. For example, for $A =]0,1[= \bigcup_{0<x<1}\{x\}$ with $\mu(\{x\}) = 0$, $\mu(A) = 1 - 0 \ne 0$.

Now we generalize the results on sets with *zero measure* in $\mathbb{R}^n$.

Definition B.1.2.3 (Case n). Let $R_k \subset \mathbb{R}^n$ be an n-dimensional *open* rectangular parallelepiped $\forall k \in \mathbb{N}$ defined in (B.1.1.7) with (Lebesgue) *n-dimensional volume measure* $\mu(R_k) = \prod_{i=1}^n (b_i^k - a_i^k) > 0\ \forall k \in \mathbb{N}$. Then a measurable set $A \subset \mathbb{R}^n$ is said to have *(Lebesgue) measure* zero if and only if, $\forall \varepsilon > 0$, $\exists$ a *finite* system $\{R_k\}_{k=1}^m$ or a *countably infinite* system $\{R_k\}_{k=1}^\infty$ such that:

1. $A \subset \bigcup_{k=1}^n R_k$ (resp. $A \subset \bigcup_{k=1}^\infty R_k$);
2. $\sum_{k=1}^n \mu(R_k) < \varepsilon$ (resp. $\sum_{k=1}^\infty \mu(R_k) < \varepsilon$). (B.1.2.4)

Proposition B.1.2.2. *Let* $A \subset \mathbb{R}^n$ *be a measurable set with* $A = \bigcup_{k=1}^m E_k$ *(resp.* $A = \bigcup_{k=1}^\infty E_k$*),* $E_k \subset \mathbb{R}^n$ *with* $\mu(E_k) = 0\ \forall k$. *Then* $\mu(A) = 0$.

Proposition B.1.2.2 does *not* hold when A is the union of an *uncountably infinite system* $(E_\lambda)_{\lambda\in\Lambda}$ of sets E_λ with $\mu(E_\lambda) = 0\ \forall\lambda \in \Lambda$. (B.1.2.5)

Some examples of sets in $\mathbb{R}^2$ with zero Lebesgue area measure μ_2

Example B.1.2.4. $A \subset \mathbb{R} \Longrightarrow$ (Lebesgue) *area measure*

$$\mu_2(A) = \mu(A) = 0. \tag{B.1.2.6}$$

[1] Set $E_0 = [0,1]$. Delete the 'middle third open' interval $]\frac{1}{3}, \frac{2}{3}[$ and obtain closed interval $E_1 = [0, \frac{1}{3}] \cup [\frac{2}{3}, 1]$. Now delete the 'open middle third' of $[0, \frac{1}{3}]$ and $[\frac{2}{3}, 1]$, i.e. delete $]\frac{1}{9}, \frac{2}{9}[$ from $[0, \frac{1}{3}]$ and $]\frac{7}{9}, \frac{8}{9}[$ from $[\frac{2}{3}, 1]$ and set $E_2 = [0, \frac{1}{9}] \cup [\frac{2}{9}, \frac{1}{3}] \cup [\frac{2}{3}, \frac{7}{9}] \cup [\frac{8}{9}, 1]$, and so on. Then a sequence of *closed* sets $E_0 \supset E_1 \supset \cdots \supset E_n \supset \cdots$ is obtained and *Cantor set* $E = \bigcap_{k=0}^\infty E_k$ is a *closed* subset of $[0,1]$ which contains, in addition to an *uncountably infinite number of points*, the *countably infinite* number of points $0, 1, \frac{1}{3}, \frac{2}{3}, \frac{1}{9}, \frac{2}{9}, \frac{7}{9}, \frac{8}{9}, \ldots,$ which are the interval end points.

Example B.1.2.5. Let $A = \Gamma \subset \mathbb{R}^2$ denote any *line segment* or any *rectifiable plane* curve with *finite or infinite length* in $\mathbb{R}^2$. Then the *Lebesgue (area) measure*

$$\mu_2(\Gamma) = \mu(\Gamma) = 0.$$

Example B.1.2.6. Let $A = \{\mathbf{x}_1, \cdots, \mathbf{x}_n, \cdots\} \subset \mathbb{R}^2$. Then the *Lebesgue area measure* $\mu_2(A) = \mu(A) = 0$.

Remark B.1.2.2. The *Peano curve* $\Gamma \subset [0,1]^2$, which passes through *all* points of the *unit square* $[0,1]^2$ in $\mathbb{R}^2$ has *Lebesgue area measure* $\mu_2(\Gamma) = \mu([0,1]^2) = (1-0)^2 = 1 > 0$.

Some examples of sets in $\mathbb{R}^3$ with zero Lebesgue volume measure μ_3

Example B.1.2.7. A set A consisting of a countable set of points $\{\mathbf{x}_k\}_{k=1}^{\infty}$, or consisting of a *countable* set of *orientable* piecewise smooth curves $\{\Gamma_k\}_{k=1}^{\infty}$, or consisting of a *countable* system of *orientable smooth* surfaces $\{S_k\}_{k=1}^{\infty}$, is an example of a set with *Lebesgue volume measure zero*, i.e.

$$\begin{aligned}
\mu_3(\{\mathbf{x}_k\}_{k=1}^{\infty}) &= \mu(\{\mathbf{x}_k\}_{k=1}^{\infty}) = 0;\\
\mu_3(\{\Gamma_k\}_{k=1}^{\infty}) &= \mu(\{\Gamma_k\}_{k=1}^{\infty}) = 0;\\
\mu_3(\{S_k\}_{k=1}^{\infty}) &= \mu(\{S_k\}_{k=1}^{\infty}) = 0
\end{aligned}$$

by Proposition B.1.2.2, since

$$\mu_3(\{\mathbf{x}_k\}) = 0, \quad \mu_3(\Gamma_k) = 0, \quad \mu_3(S_k) = 0 \quad \forall k \in \mathbb{N}. \tag{B.1.2.7}$$

B.1.3 Property P holds almost everywhere (a.e.) on Ω

Definition B.1.3.1. A property P relating to *points* $\mathbf{x} \in \Omega \subset \mathbb{R}^n$ is said to hold *almost everywhere* (a.e.), or *for almost all* $\mathbf{x}$ in Ω, if and only if the set of points $\mathbf{x} \in \Omega$, at which the property P does *not* hold has *Lebesgue n-dimensional measure zero*, i.e.

$$P \text{ holds } a.e. \text{ on } \Omega \iff \mu(\{\mathbf{x} : \mathbf{x} \in \Omega,\ P \text{ does } not \text{ hold at } \mathbf{x}\}) = 0. \tag{B.1.3.1}$$

By property $P = P(\mathbf{x})$ relating to points $\mathbf{x} \in \Omega$ we mean, for example, 'equality of two functions f and g', 'function f is defined', 'function f is bounded', 'function f vanishes', 'sequence $(f_n(\mathbf{x}))_{n=1}^{\infty}$ converges' relating to points $\mathbf{x} \in \Omega \subset \mathbb{R}^n$, etc., i.e.

- f is *defined* a.e. on $\Omega \iff \mu(\{\mathbf{x} : \mathbf{x} \in \Omega, f(\mathbf{x}) \text{ is } \textit{not defined}\}) = 0$;
- $f = g$ a.e. on $\Omega \iff \mu(\{\mathbf{x} : \mathbf{x} \in \Omega, f(\mathbf{x}) \neq g(\mathbf{x})\}) = 0$;

- f is *bounded* a.e. on $\Omega \iff \mu(\{\mathbf{x} : \mathbf{x} \in \Omega, f(\mathbf{x})$ is *not bounded* at $\mathbf{x}\}) = 0$;
- f is *continuous* a.e. on $\Omega \iff \mu(\{\mathbf{x} : \mathbf{x} \in \Omega, f(\mathbf{x})$ is *not continuous* at $\mathbf{x}\}) = 0$;
- f *vanishes* a.e. on $\Omega \iff \mu(\{\mathbf{x} : \mathbf{x} \in \Omega, f(\mathbf{x}) \neq 0\}) = 0$;
- $(f_n(\mathbf{x}))_{n=1}^{\infty}$ *converges* to $f(\mathbf{x})$ a.e. on Ω as $n \to \infty \iff \mu(\{\mathbf{x} : \mathbf{x} \in \Omega, \lim_{n\to\infty} f_n(\mathbf{x}) \neq f(\mathbf{x})\}) = 0$.

Example B.1.3.1.

1. $f(x) = 1 \ \forall$ *rational* $x \in [0, 1]$ and $f(x) = 0$ for all *irrational* $x \in [0, 1] \Longrightarrow f = 0$ a.e. on $[0, 1]$ or f *vanishes* a.e. on $[0, 1]$, since

$$\mu(\{x : x \in [0,1], f(x) \neq 0\}) = \mu(\mathbb{Q} \cap [0,1]) = \mu(\mathbb{Q}_{[0,1]}) = 0. \quad \text{(B.1.3.2)}$$

2. $f(x_1, x_2) = x_1 + x_2 \ \forall (x_1, x_2) \in \Omega =]0,1[\times]0,1[$,

$$g(x_1,x_2) = \begin{cases} 0 & \forall (x_1, 0) \text{ with } 0 < x_1 < 1 \\ 1 & \forall (0, x_2) \text{ with } 0 < x_2 < 1 \\ x_1 + x_2 & \forall (x_1,x_2) \in \Omega - \{(x_1,x_2) : 0 < x_i < 1 \\ & \textit{and } x_j = 0 \text{ for } 1 \le i \neq j \le 2\} \end{cases}$$

$\Longrightarrow f = g$ a.e. on Ω, since $f(x_1, x_2) \neq g(x_1, x_2)$ on *two* line segments, $x_2 = 0$ with $0 < x_1 < 1$, $x_1 = 0$ with $0 < x_2 < 1$, each of which has *Lebesgue area measure* zero, i.e.

$$\mu(\{(x_1,x_2) : (x_1,x_2) \in \Omega,\ f(x_1,x_2) \neq g(x_1,x_2)\}) = 0.$$

B.2 Space $\mathcal{M}(\Omega)$ of Lebesgue-measurable functions on Ω

B.2.1 Measurable functions and space $\mathcal{M}(\Omega)$

Definition B.2.1.1. A function $f : \Omega \longrightarrow \overline{\mathbb{R}}$ is called *measurable* if and only if $\exists$ a sequence $(f_n)_{n=1}^{\infty}$ of *continuous* functions f such that $f(\mathbf{x}) = \lim_{n\to\infty} f_n(\mathbf{x})$ a.e. on Ω.

In particular, a function is *measurable* on Ω if it is *continuous* except at a set of points $\mathbf{x} \in \Omega$ with measure zero.

At the time of writing, not a *single* non-measurable function is explicitly known to us.

Nonmeasurable functions *cannot* be explicitly written. Their existence is established mathematically by the axiom of choice. (B.2.1.1)

Hence, one may safely assume that *all* functions in applied problems are measurable and, without any practical loss, the measurability condition may be ignored and the reader may skip the following explanations. But for mathematically alert readers, we will present here the mathematical concepts associated with measurable functions.

Let $\mathcal{M}(\Omega)$ be the set of all Lebesgue-measurable functions on Ω with the following notion of equality in $\mathcal{M}(\Omega)$:

$$f = g \text{ in } \mathcal{M}(\Omega) \quad \Longleftrightarrow \quad f = g \text{ a.e. on } \Omega, \tag{B.2.1.2}$$

which defines an *equivalence relation* on $\mathcal{M}(\Omega)$, and partitions $\mathcal{M}(\Omega)$ into *equivalence classes* $[f]$ of measurable functions f on Ω, i.e.

- $[f] = \{g : g \in \mathcal{M}(\Omega),\, g = f \text{ a.e. on } \Omega\}$. (B.2.1.3)
- $[f], [g]$ with $[f] \neq [g] \Longrightarrow [f] \cap [g] = \emptyset$, i.e. *distinct equivalent classes* $[f], [g]$ are *disjoint*, and $[f] \neq [g] \Longleftrightarrow f \neq g$ in $\mathcal{M}(\Omega)$. (B.2.1.4)
- Every $f \in \mathcal{M}(\Omega)$ will belong to a *unique* equivalence class $[f]$. (B.2.1.5)
- The set $\mathcal{E}([\cdot])$ of all *equivalence classes* of measurable functions of $\mathcal{M}(\Omega)$ is a partition of $\mathcal{M}(\Omega)$, i.e. $\mathcal{M}(\Omega) = \bigcup_{[f]\in\mathcal{E}([\cdot])}[f]$ and $[f], [g] \in \mathcal{E}$ with $[f] \neq [g] \Longrightarrow [f] \cap [g] = \emptyset$. (B.2.1.6)

Notational convention For the sake of notational simplicity, we will use $f \in \mathcal{M}(\Omega)$ to denote the *unique equivalence class* $[f] \in \mathcal{E}$ which f belongs to, keeping in mind the *equivalence relation*.

Warning The usual statements: 'f is defined *everywhere* on Ω', '$f = g \Longleftrightarrow f(\mathbf{x}) = g(\mathbf{x})\ \forall x \in \Omega$', '$f$ is defined at a point $\mathbf{x} \in \Omega$ or on a $\Gamma \subset \Omega \subset \mathbb{R}^n$ $(n \geq 2)$', '$f_k(\mathbf{x}) \to f(x)$ for all $\mathbf{x} \in \Omega$ as $n \to \infty$', '$\sup_{x\in\Omega} f(\mathbf{x})$', '$\inf_{\mathbf{x}\in\Omega} f(\mathbf{x})$', etc. do *not* make sense for functions $f, g, (f_k)_{k=1}^{\infty}$ in $\mathcal{M}(\Omega)$. (B.2.1.7)

In fact, $f = g$ in $\mathcal{M}(\Omega)$ ($\Longleftrightarrow [f] = [g]$ in $\mathcal{E}[\cdot]$) does *not* imply $\sup_{\mathbf{x}\in\Omega} f(\mathbf{x}) = \sup_{\mathbf{x}\in\Omega} g(\mathbf{x})$ (resp. $\inf_{\mathbf{x}\in\Omega} f(\mathbf{x}) = \inf_{\mathbf{x}\in\Omega} g(\mathbf{x})$). Hence, such statements will involve a *change of values of $f(\mathbf{x})$ on a set of points with measure zero.* (B.2.1.8)

Example B.2.1.1. Let $f(x) = x^2\ \forall x \in [0,1]$ and let $g(x) = x^2$ for *irrational* $x \in [0,1]$ and $= 1 + x^2$ for *rational* $x \in [0,1]$. Then $f = g$ a.e. on $[0,1]$, since $\mu(\{x : x \in [0,1], x \text{ is } \textit{rational}\}) = 0$. But $\sup_{x\in[0,1]} f(x) = 1 \neq \sup_{x\in[0,1]} g(x) = 2$.

Definition B.2.1.2. The *essential supremum*, i.e. ess sup (resp. *essential infimum*, i.e. ess inf) of a function $f \in \mathcal{M}(\Omega)$ is defined by:

$$\operatorname*{ess\,sup}_{\mathbf{x}\in\Omega} f(\mathbf{x}) = \inf\{M : f(\mathbf{x}) \leq M \text{ a.e. on } \Omega,\ -\infty < M \leq \infty\}; \tag{B.2.1.9}$$

$$\operatorname*{ess\,inf}_{\mathbf{x}\in\Omega} f(\mathbf{x}) = \sup\{M : f(\mathbf{x}) \geq M \text{ a.e. on } \Omega,\ -\infty \leq M < \infty\}. \tag{B.2.1.10}$$

Definition B.2.1.3. A function $f \in \mathcal{M}(\Omega)$ is called *essentially bounded from above* (resp. *from below*) in Ω if and only if $\text{ess sup}_{\mathbf{x}\in\Omega} f(\mathbf{x}) < +\infty$ (resp. $\text{ess sup}_{\mathbf{x}\in\Omega} f(\mathbf{x}) > -\infty$).

Then $f \in \mathcal{M}(\Omega)$ is called *essential bounded* in Ω if and only if both $\text{ess sup}_{\mathbf{x}\in\Omega} f(\mathbf{x})$ and $\text{ess inf}_{\mathbf{x}\in\Omega} f(\mathbf{x})$ are *finite*, or, equivalently, if and only if $\exists M_1 > 0$ such that $|f(\mathbf{x})| \le M_1$ a.e. on Ω, i.e.

$$\mu(\{\mathbf{x} : \mathbf{x} \in \Omega, |f(\mathbf{x})| > M_1\}) = 0. \tag{B.2.1.11}$$

A function f is *bounded* on $\Omega \Longrightarrow f$ is *essentially bounded* on Ω, but the converse is *false*. (B.2.1.12)

Example B.2.1.2. For $f(x) = x \ \forall x = n \in \mathbb{Z}$ and $f(x) = e^{-x^2}$ otherwise, $\sup_{x\in\mathbb{R}} f(x) = \infty$ and $\inf_{x\in\mathbb{R}} f(x) = -\infty$. Hence, f is *unbounded* on $\mathbb{R}$, but f is *essentially bounded* on $\mathbb{R}$, since $\mu(\{x : x = n \in \mathbb{Z}\}) = 0$, and consequently $|f(x)| = e^{-x^2} \le 1$ a.e. in $\mathbb{R}$ with $M_1 = 1$.

B.2.2 Pointwise convergence a.e. on Ω

Definition B.2.2.1. A sequence $(f_n)_{n=1}^{\infty}$ in $\mathcal{M}(\Omega)$ is said to *converge to f pointwise a.e. on* Ω if and only if $\lim_{n\to\infty} f_n(\mathbf{x}) = f(\mathbf{x})$ *for almost all* $\mathbf{x} \in \Omega$, i.e.

$$\mu(\{\mathbf{x} : \mathbf{x} \in \Omega, \lim_{n\to\infty} f_n(\mathbf{x}) \neq f(\mathbf{x})\}) = 0.$$

Example B.2.2.1. The sequence (f_n) defined by $f_n(x) = x^n \ \forall x \in [0,1]$ *converges a.e.* on $[0,1]$ to $f = 0$, since it does *not* converge to $f = 0$ *only at* $x = 1$ and $\mu(\{1\}) = 0$.

Definition B.2.2.2. A sequence $(f_n)_{n=1}^{\infty}$ of measurable functions f_n on a measurable set $E \subset \mathbb{R}^n$ is said to converge to a measurable f *in measure* on E if and only if $\forall \delta > 0$, the *measure* of the set $E_k = \{\mathbf{x} : \mathbf{x} \in E, |f(\mathbf{x}) - f_k(\mathbf{x})| > \delta\}$ tends to 0 as $k \to \infty$, i.e. $\mu(E_k) \to 0$ as $k \to \infty$. (B.2.2.1)

Pointwise a.e. convergence $\Longrightarrow$ convergence *in measure*. (B.2.2.2)

B.3 Lebesgue integrals and their important properties

B.3.1 Lebesgue integral of a bounded function on bounded domain Ω

Let $\Omega \subset \mathbb{R}^n$ be a *bounded*, *open* subset of $\mathbb{R}^n$ and $f : \Omega \longrightarrow \mathbb{R}$ be a Lebesgue-measurable function on Ω with the *range* $R(f) \subset]c, d[$, $-\infty < c < d < +\infty$, i.e. f is a *bounded* function on Ω and $-\infty < c < f(\mathbf{x}) < d < +\infty$ and $|f(\mathbf{x})| \le \max\{|c|, |d|\} \ \forall \mathbf{x} \in \Omega$. (B.3.1.1)

Subdivide the *closure* $[c, d]$ with $R(f) \subset]c, d[\subset [c, d]$ (in the case of Riemann integrals, the closure of the *domain* of *definition* $\mathcal{D}(f) = \overline{\Omega}$ is *subdivided*) into subintervals with the help of real numbers $\{a_i\}_{i=0}^n$ such that

$$\Pi([c, d]) : c = a_0 < a_1 < \cdots < a_{i-1} < a_i < \cdots < a_n = d, \tag{B.3.1.2}$$

which define a *partition* Π of $[c, d]$ containing $R(f)$ into closed subintervals $I_k = [a_{k-1}, a_k]$ with (length) measure $\mu_1(I_k) = a_k - a_{k-1} > 0$, $1 \le k \le n$.

For each partition $\Pi([c, d])$ we define a *strictly* positive number $\Delta(\Pi) > 0$ by:

$$\Delta(\Pi) = \max_{1 \le k \le n} \{\mu(I_k)\} = \max_{1 \le k \le n} \{(a_k - a_{k-1})\} > 0. \tag{B.3.1.3}$$

Then we can refine the partition Π in (B.3.1.2) by making $n \to \infty$ such that $\Delta(\Pi) \to 0^+$, i.e. the length of *each* subinterval tends to 0^+. Define measurable, *disjoint* subsets E_k of Ω by:

$$E_k = \{\mathbf{x} : \mathbf{x} \in \Omega,\ a_{k-1} \le f(\mathbf{x}) < a_k\} \quad \text{with } \mu(E_k) \ge 0\ \forall k = 1, 2, \ldots, n, \tag{B.3.1.4}$$

some of which (i.e. E_k) may possibly be *empty sets*.

Define $\underline{S}_L(\Pi; f) = \sum_{k=1}^n a_{k-1}\mu(E_k)$, $\overline{S}_L(\Pi; f) = \sum_{k=1}^n a_k\mu(E_k)$, which are called *Lebesgue lower* and *upper integral sums* of f corresponding to the partition $\Pi([c, d])$, respectively, with the property $\underline{S}_L(\Pi; f) \le \overline{S}_L(\Pi; f)$ $\forall \Pi$. Now, refine the partition Π of $[c, d]$ by letting $n \to \infty$ such that $\Delta(\Pi) \to 0^+$.

Definition B.3.1.1. $\forall$ partitions Π, if $\lim_{\Delta(\Pi)\to 0+} \underline{S}_L(\Pi; f) = \lim_{\Delta(\Pi)\to 0+} \overline{S}_L(\Pi; f)$ in $\mathbb{R}$, then the limit is called the *Lebesgue integral* of f *on* Ω and denoted by $\int_\Omega f(\mathbf{x})d\mu$ or, equivalently, by $\int_\Omega f(\mathbf{x})d\mathbf{x}$, i.e.

$$\int_\Omega f(\mathbf{x})d\mathbf{x} = \int_\Omega f(\mathbf{x})d\mu = \lim_{\Delta(\Pi)\to 0+} \underline{S}_L(\Pi; f) = \lim_{\Delta(\Pi)\to 0+} \overline{S}_L(\Pi; f), \tag{B.3.1.5}$$

where $d\mathbf{x} = d\mu$ is the Lebesgue volume measure in $\mathbb{R}^n$ and f is called *Lebesgue integrable* on Ω.

Theorem B.3.1.1 (Kolmogorov and Fomin [20]). *Let* $\Omega \subset \mathbb{R}^n$ *be a* bounded, measurable *subset of* $\mathbb{R}^n$ *(i.e.* $\mu(\Omega) < +\infty$*) and* f *be a* bounded, measurable *function on* Ω*. Then*

$$\int_\Omega f(\mathbf{x})d\mathbf{x} = \int_\Omega f(\mathbf{x})d\mu = \lim_{\Delta(\Pi)\to 0+} \underline{S}_L(\Pi; f) = \lim_{\Delta(\Pi)\to 0+} \overline{S}_L(\Pi; f) \in \mathbb{R}, \tag{B.3.1.6}$$

i.e. the Lebesgue integral $\int_\Omega f(\mathbf{x})d\mu = \int_\Omega f(\mathbf{x})d\mathbf{x}$ *exists. Hence,* f *is* Lebesgue integrable *on* Ω.

Example B.3.1.1. Let f be the *Dirichlet function* on [0, 1] defined by $f(x) = 1$ for *rational* $x \in [0, 1]$ and $f(x) = 0$ for *irrational* $x \in [0, 1]$. Then f is *Lebesgue integrable* on [0, 1], but f is *not Riemann integrable* on [0, 1].

Theorem B.3.1.2. *If a* bounded *function* $f \in \mathcal{F}(\Omega)$ *on a* bounded, closed $\overline{\Omega}$ *(i.e.* $\mu(\overline{\Omega}) < +\infty$*) is* Riemann integrable *on* $\overline{\Omega}$*, then* f *is also* Lebesgue integrable *on* $\overline{\Omega}$*, but the converse is* false.

The *Lebesgue integrability* of a (measurable) *bounded* function f on a *bounded* $\overline{\Omega}$ does *not* imply its *Riemann integrability* on $\overline{\Omega}$. The class of Lebesgue-integrable bounded functions f on bounded domains Ω (i.e. $\mu(\Omega) < +\infty$) is *much wider* than that of Riemann-integrable bounded functions on bounded Ω. (B.3.1.7)

Lebesgue integrals on unbounded domains Ω Let $\Omega \subset \mathbb{R}^n$ be an *unbounded* domain, with $\mu(\Omega) = +\infty$, such that $\exists$ a sequence $(\Omega_m)_{m=1}^{\infty}$ of *measurable, bounded* subdomains Ω_m with $\mu(\Omega_m) < +\infty$ $\forall m \in \mathbb{N}$ with the following properties:

1. $\Omega = \bigcup_{m=1}^{\infty} \Omega_m$;
2. $\Omega_1 \subset \Omega_2 \subset \cdots \subset \Omega_m \subset \cdots$;
3. the Lebesgue integral $\int_{\Omega_m} f(\mathbf{x})d\mathbf{x}$ exists $\forall m \in \mathbb{N}$. (B.3.1.8)

Then, if $\lim_{m\to\infty} \int_{\Omega_m} f(\mathbf{x})d\mathbf{x}$ exists, the *limit* is called the *Lebesgue integral* of f on Ω, denoted by $\int_\Omega f(\mathbf{x})d\mu$ or, equivalently, by $\int_\Omega f(\mathbf{x})d\mathbf{x}$ with $d\mathbf{x} = d\mu$, and f is called *Lebesgue integrable* on Ω, i.e.

$$\int_\Omega f(\mathbf{x})d\mu = \int_\Omega f(\mathbf{x})d\mathbf{x} = \lim_{m\to\infty} \int_{\Omega_m} f(\mathbf{x})d\mathbf{x}. \tag{B.3.1.9}$$

$L^1(\Omega)$ is the set of all *Lebesgue-integrable functions* on $\Omega \subset \mathbb{R}^n$. The exponent 1 in $L^1(\Omega)$ is used to denote that it is the space $L^p(\Omega)$ with $p = 1$, to be dealt with later. (B.3.1.10)

B.3.2 Important properties of Lebesgue integrals (Kolmogorov and Fomin [20])

1. If $f = C \in \mathbb{R}$ a.e. on Ω with $\mu(\Omega) < +\infty$, $\int_\Omega C d\mathbf{x} = C \int_\Omega d\mathbf{x} = C.\mu(\Omega)$.
2. $f \in L^1(\Omega)$ and $f = g$ a.e. on $\Omega \Longrightarrow \int_\Omega f(\mathbf{x})d\mathbf{x} = \int_\Omega g(\mathbf{x})d\mathbf{x}$.
3. If f is *bounded* on Ω with $\mu(\Omega) < +\infty$, $f \in L^1(\Omega)$. (B.3.2.1)
4. $f \in L^1(\Omega) \Longleftrightarrow |f| \in L^1(\Omega)$. (B.3.2.2)
5. $f, g \in L^1(\Omega)$, $\alpha_1, \alpha_2 \in \mathbb{R} \Longrightarrow \int_\Omega (\alpha_1 f + \alpha_2 g)(\mathbf{x})d\mathbf{x} = \alpha_1 \int_\Omega f(\mathbf{x})d\mathbf{x} + \alpha_2 \int_\Omega g(\mathbf{x})d\mathbf{x}$.

6. $\Omega = A \cup B$ with $A \cap B = \emptyset$, $f \in L^1(A)$, $f \in L^1(B) \Longrightarrow f \in L^1(\Omega)$ and $\int_{A\cup B} f(\mathbf{x})d\mathbf{x} = \int_A f(\mathbf{x})d\mathbf{x} + \int_B f(\mathbf{x})d\mathbf{x}$. Conversely, $f \in L^1(\Omega)$, $A \subset \Omega \Longrightarrow f \in L^1(A)$.
7. $f \in L^1(\Omega)$, $f(\mathbf{x}) \geq 0$ a.e. on $\Omega \Longrightarrow \int_\Omega f(\mathbf{x})d\mathbf{x} \geq 0$.
8. $f(\mathbf{x}) \geq g(\mathbf{x})$ a.e. on Ω and $f, g \in L^1(\Omega) \Longrightarrow \int_\Omega f(\mathbf{x})d\mathbf{x} \geq \int_\Omega g(\mathbf{x})d\mathbf{x}$.
9. $f \in L^1(\Omega) \Longrightarrow |\int_\Omega f(\mathbf{x})d\mathbf{x}| \leq \int_\Omega |f(\mathbf{x})|d\mathbf{x}$.
10. $m \leq f(\mathbf{x}) \leq M$ a.e. on Ω with $\mu(\Omega) < +\infty \Longrightarrow m\mu(\Omega) \leq \int_\Omega f(\mathbf{x})d\mathbf{x} \leq M\mu(\Omega)$.
11. *Mean value result for integrals*: Let $f \in \mathcal{M}(\Omega)$ and $g \in L^1(\Omega)$ and $\alpha, \beta \in \mathbb{R}$ such that $\alpha \leq f(\mathbf{x}) \leq \beta$ a.e. on Ω. Then $\exists \gamma \in \mathbb{R}$ with $\alpha \leq \gamma \leq \beta$ such that $\int_\Omega f(\mathbf{x})|g(\mathbf{x})|d\mathbf{x} = \gamma \int_\Omega |g(\mathbf{x})|d\mathbf{x}$.
12. Let $f(\mathbf{x}) \geq 0$ a.e. on Ω and $\int_\Omega f(\mathbf{x})d\mathbf{x} = 0$. Then $f = 0$ a.e. on Ω.
13. Let $f \in L^1(\Omega)$ and $A \subset \Omega$ with $\mu(A) = 0$. Then $\int_A f(\mathbf{x})d\mathbf{x} = 0$. (B.3.2.3)
14. Let $f(\mathbf{x}) > 0$ a.e. on $A \subset \Omega$ and $\int_A f(\mathbf{x})d\mathbf{x} = 0$. Then $\mu(A) = 0$.
15. $f \in L^1(\Omega)$, $\int_A f(\mathbf{x})d\mathbf{x} = 0$ $\forall$ measurable $A \subset \Omega \Longrightarrow f = 0$ a.e. on Ω.
16. *Improper Riemann integrals on bounded Ω:*
    - *$f \geq 0$ in Ω with $\mu(\Omega) < +\infty$:*
      For the sake of simplicity, consider $n = 1$ and $\Omega = [a,b] \subset \mathbb{R}$ with $\mu([a,b]) = b - a > 0$. Suppose that $\lim_{x\to a} f(x) = +\infty$ and the (definite) Riemann integral $\int_{a+\varepsilon}^b f(x) = I(\varepsilon)$ exists $\forall \varepsilon > 0$ and $\lim_{\varepsilon\to 0+} = \lim_{\varepsilon\to 0+} I(\varepsilon) = I \in \mathbb{R}$. Then, for $f \geq 0$, the *improper Riemann integral* $\int_a^b f(x)dx = \lim_{\varepsilon\to 0+} \int_{a+\varepsilon}^b f(x)dx$ exists $\Longleftrightarrow$ the *non-negative* f is *Lebesgue-integrable* on $[a,b]$ and the *Lebesgue integral* $\int_{[a,b]} f(x)dx =$ the *improper Riemann integral* $\int_a^b f(x)dx$. (B.3.2.4)
    - For *non-negative* $f \geq 0$, the *improper Riemann integral* $\int_a^b f(x)dx$ does *not* exist $\Longrightarrow$ the *Lebesgue integral* $\int_{[a,b]} f(x)dx$ does *not* exist. (B.3.2.5)
    - For $f(\mathbf{x})$ *with variable sign*, i.e. $f(\mathbf{x}) \in [-\infty, \infty]$ and $\mu(\Omega) < +\infty$ or $\mu(\Omega) = +\infty$, *analogous results for improper integrals hold.*

In fact, we have the general result for *improper integrals*.

Theorem B.3.2.1. *Let f be Riemann integrable on $\Omega \subset \mathbb{R}^n$ in the* improper sense. *Then, f will be* Lebesgue integrable *on Ω if and only if $|f|$ is also Riemann integrable on Ω in the* improper sense. *When both Riemann and Lebesgue integrals exist, their values will coincide.*

Example B.3.2.1. For $f(x) = \frac{1}{\sqrt{x}} > 0$ on $[0, 1]$, the *improper Riemann integral*

$$\int_0^1 \frac{1}{\sqrt{x}} dx = \lim_{\varepsilon \to 0+} \int_\varepsilon^1 \frac{1}{\sqrt{x}} dx = \lim_{\varepsilon \to 0+} 2(1 - \sqrt{\varepsilon}) = 2 < +\infty.$$

Hence, $\frac{1}{\sqrt{x}}$ is also *Lebesgue integrable* on $[0, 1]$, and their values coincide.

Example B.3.2.2. The improper Riemann integral

$$\int_0^1 \frac{1}{x} \sin \frac{1}{x} dx = \lim_{\varepsilon \to 0+} \int_\varepsilon^1 \frac{1}{x} \sin \frac{1}{x} dx$$

exists, but $\lim_{\varepsilon \to 0+} \int_\varepsilon^1 |\frac{1}{x} \sin \frac{1}{x}| dx$ does *not* exist. Hence, both $\frac{1}{x} \sin \frac{1}{x}$ and $|\frac{1}{x} \sin \frac{1}{x}|$ are *not* Lebesgue integrable on $[a, b]$.

Example B.3.2.3. The *improper* Riemann integral $\int_{-\infty}^{\infty} \frac{\sin x}{x} dx$ exists and $\int_{-\infty}^{\infty} \frac{\sin x}{x} dx = \pi$, but the *improper* Riemann integral $\int_{-\infty}^{\infty} |\frac{\sin x}{x}| dx = +\infty$, i.e. does *not* exist. Hence, $|\frac{\sin x}{x}|$ is *not Lebesgue integrable* on $]-\infty, \infty[$ and, consequently, $\frac{\sin x}{x}$ is *not Lebesgue integrable* on $]-\infty, \infty[$ by Property 4 in (B.3.2.2).

Theorem B.3.2.2 (Lebesgue's Dominated Convergence Theorem). *Let $(f_n)_{n=1}^{\infty}$ be a sequence of Lebesgue-integrable functions f_n on Ω (i.e. $f_n \in L^1(\Omega)\ \forall n \in \mathbb{N}$) such that $f_n(\mathbf{x}) \to f(\mathbf{x})$ pointwise a.e. on Ω and $|f_n(\mathbf{x})| \leq g(\mathbf{x})$ a.e. on $\Omega\ \forall n \in \mathbb{N}$, where $g \in L^1(\Omega)$ is a non-negative Lebesgue-integrable function on Ω. Then:*

I. *$f \in L^1(\Omega)$, i.e. f is Lebesgue integrable on Ω;* (B.3.2.6)

II. $\lim_{n\to\infty} \int_\Omega f_n(\mathbf{x}) d\mathbf{x} = \int_\Omega \lim_{n\to\infty} f_n(\mathbf{x}) d\mathbf{x} = \int_\Omega f(\mathbf{x}) d\mathbf{x}$.

Proof.

I. The proof depends on the result (B.2.2.2), according to which the pointwise convergence of a sequence a.e. on Ω implies its *convergence in measure* in Ω. Let $\Omega_0 \subset \Omega$ be the largest measurable subset of Ω in which $f_n(\mathbf{x})$ is *finite* $\forall n \in \mathbb{N}$, $g(\mathbf{x})$ is *finite* and $|f_n(\mathbf{x})| \leq g(\mathbf{x}) \leq M\ \forall n \in \mathbb{N}$ and $f_n(\mathbf{x}) \to f(\mathbf{x})$ as $n \to \infty$. Hence, $f_n(\mathbf{x})$ is *infinite* in $\Omega - \Omega_0\ \forall n \in \mathbb{N}$ and f_n will belong to $L^1(\Omega)\ \forall n \in \mathbb{N}$ if and only if

$$\int_\Omega |f_n(\mathbf{x})| d\mathbf{x} = \int_{\Omega_0} |f_n(\mathbf{x})| d\mathbf{x} + \int_{\Omega - \Omega_0} |f_n(\mathbf{x})| d\mathbf{x} < +\infty$$
$$\iff \int_{\Omega - \Omega_0} |f_n(\mathbf{x})| d\mathbf{x} = 0 \iff \mu(\Omega - \Omega_0) = 0.$$

Therefore, $\mu(\Omega_0) + \mu(\Omega - \Omega_0) = \mu(\Omega) \Longrightarrow \mu(\Omega_0) = \mu(\Omega)$.

f is measurable on $\Omega_0 \subset \Omega$. $|f_n(\mathbf{x})| \leq g(\mathbf{x})\ \forall n \in \mathbb{N}$

$$\Longrightarrow \quad \lim_{n\to\infty} |f_n(\mathbf{x})| = |f(\mathbf{x})| \leq g(\mathbf{x}) \text{for } \mathbf{x} \in \Omega_0 \tag{B.3.2.7}$$

$\Longrightarrow \int_{\Omega_0} |f(\mathbf{x})| d\mathbf{x} \leq \int_{\Omega_0} g(\mathbf{x}) d\mathbf{x} < +\infty \Longrightarrow f \in L^1(\Omega_0) \Longrightarrow f \in L^1(\Omega)$, since $\mu(\Omega - \Omega_0) = 0$ and $\int_{\Omega-\Omega_0} |f(\mathbf{x})| d\mathbf{x} = 0$, which establishes I.

II. Set $\Omega_0 = \Omega'{}_n \cup \Omega''_n$, where for a *chosen* $\delta > 0$ (to be made precise later),

$$\begin{aligned} \Omega'_n &= \{\mathbf{x} : \mathbf{x} \in \Omega_0,\ |f(\mathbf{x}) - f_n(\mathbf{x})| > \delta\}; \\ \Omega''_n &= \{\mathbf{x} : \mathbf{x} \in \Omega_0,\ |f(\mathbf{x}) - f_n(\mathbf{x})| \leq \delta\} \end{aligned} \tag{B.3.2.8}$$

with $\Omega'_n \cap \Omega''_n = \emptyset$.

Since $f_n(\mathbf{x}) \longrightarrow f(\mathbf{x})$ a.e. on Ω, (f_n) converges to f *in measure* in Ω by (B.2.2.2). Hence, by Definition B.2.2.2, (f_n) converges *in measure* to f in $\Omega'_n \subset \Omega$, i.e. $\mu(\Omega'_n) = \mu(\{\mathbf{x} : \mathbf{x} \in \Omega_0, |f(\mathbf{x}) - f_n(\mathbf{x})| > \delta\}) \to 0$ as $n \to \infty$. Hence,

$$\begin{aligned} &\left| \int_\Omega f(\mathbf{x}) d\mathbf{x} - \int_\Omega f_n(\mathbf{x}) d\mathbf{x} \right| = \left| \int_\Omega [f(\mathbf{x}) - f_n(\mathbf{x})] d\mathbf{x} \right| \\ &= \left| \int_{\Omega_0} [f(\mathbf{x}) - f_n(\mathbf{x})] d\mathbf{x} \right| \quad (\text{since } \mu(\Omega - \Omega_0) = 0) \\ &\leq \int_{\Omega'_n} (|f(\mathbf{x})| + |f_n(\mathbf{x})|) d\mathbf{x} + \int_{\Omega''_n} |f(\mathbf{x}) - f_n(\mathbf{x})| d\mathbf{x} \\ &\leq \int_{\Omega'_n} g(\mathbf{x}) d\mathbf{x} + \int_{\Omega'_n} g(\mathbf{x}) d\mathbf{x} + \delta \int_{\Omega''_n} d\mathbf{x} \\ &= 2 \int_{\Omega'_n} g(\mathbf{x}) d\mathbf{x} + \delta \mu(\Omega''_n) \quad (\text{since } |f_n(\mathbf{x})| \leq g(\mathbf{x}) \forall n \text{ a.e.}, |f(\mathbf{x})| \leq g(\mathbf{x}) \text{ a.e.}). \end{aligned} \tag{B.3.2.9}$$

Now we will choose $\delta > 0$ for (B.3.2.8) and (B.3.2.9) such that $\delta \cdot \mu(\Omega''_n) \leq \delta[\mu(\Omega)] < \frac{\varepsilon}{2}$, i.e. $\delta < \frac{\varepsilon}{2\mu(\Omega)}$ with $\varepsilon > 0$ being an arbitrary positive number. Since $\mu(\Omega'_n) \to 0$ as $n \to \infty$, $\int_{\Omega'_n} g(\mathbf{x}) d\mathbf{x} \to 0$ as $n \to \infty$. Hence, $\forall \varepsilon > 0$, $\exists n_0 = n_0(\varepsilon) \in \mathbb{N}$ such that $\int_{\Omega'_n} g(\mathbf{x}) d\mathbf{x} < \frac{\varepsilon}{4}\ \forall n > n_0$. Then, from (B.3.2.9), $\forall \varepsilon > 0$, $\exists n_0 = n_0(\varepsilon) \in \mathbb{N}$ such that $\forall n > n_0$,

$$\left| \int_\Omega f(\mathbf{x}) d\mathbf{x} - \int_\Omega f_n(\mathbf{x}) d\mathbf{x} \right| < 2 \cdot \frac{\varepsilon}{4} + \frac{\varepsilon}{2\mu(\Omega)} \cdot \mu(\Omega) = \varepsilon,$$

i.e.

$$\lim_{n\to\infty} \int_\Omega f_n(\mathbf{x}) d\mathbf{x} = \int_\Omega f(\mathbf{x}) d\mathbf{x}. \qquad \square$$

Theorem B.3.2.3 (Lebesgue's Bounded Convergence Theorem). *Let* $\Omega \subset \mathbb{R}^n$ *be a* bounded *domain, i.e.* $\mu(\Omega) < +\infty$. *Let* $(f_n)_{n=1}^{\infty}$ *be a measurable sequence* bounded *a.e. on* Ω, *i.e.* $\exists M > 0$ *such that* $|f_n(\mathbf{x})| \leq M$ *a.e. on* Ω $\forall n \in \mathbb{N}$ *such that* $\lim_{n\to\infty} f_n(\mathbf{x}) = f(\mathbf{x})$ *a.e. on* Ω. *Then:*

I. $f \in L^1(\Omega)$, *i.e.* $\int_\Omega |f(\mathbf{x})| d\mathbf{x} < +\infty$;

II. $\lim_{n\to\infty} \int_\Omega f_n(\mathbf{x}) d\mathbf{x} = \int_\Omega f(\mathbf{x}) d\mathbf{x}$. (B.3.2.10)

Proof. Define $g(\mathbf{x}) = M$ on Ω. Then $\int_\Omega g(\mathbf{x}) d\mathbf{x} = M \int_\Omega d\mathbf{x} = M\mu(\Omega) < +\infty$, i.e. $g \in L^1(\Omega)$. $|f_n(\mathbf{x})| \leq g(\mathbf{x}) = M$ a.e. on Ω $\forall n \in \mathbb{N}$ and $\lim_{n\to\infty} f_n(\mathbf{x}) = f(\mathbf{x})$ a.e. on Ω. Hence, by Lebesgue's Dominated Convergence Theorem B.3.2.2, $f \in L^1(\Omega)$ and $\lim_{n\to\infty} \int_\Omega f_n(\mathbf{x}) d\mathbf{x} = \int_\Omega \lim_{n\to\infty} f_n(\mathbf{x}) d\mathbf{x} = \int_\Omega f(\mathbf{x}) d\mathbf{x}$. □

Theorem B.3.2.4 (Monotone Convergence Theorem of Beppo Levi). *Let* $(f_n)_{n=1}^{\infty}$ *be a non-decreasing sequence of non-negative functions, i.e.* $0 \leq f_1(\mathbf{x}) \leq f_2(\mathbf{x}) \leq \dots \leq f_n(\mathbf{x}) \leq \dots$ *on* $\Omega \subset \mathbb{R}^n$ *such that* f_n *is* integrable *on* Ω $\forall n \in \mathbb{N}$, *i.e.* $\int_\Omega f_n(\mathbf{x}) d\mathbf{x} \leq M < +\infty$ $\forall n \in \mathbb{N}$. *Then:*

I. $\lim_{n\to\infty} f_n(\mathbf{x}) = f(\mathbf{x})$ *a.e. on* Ω;

II. $f \in L^1(\Omega)$, *i.e.* f *is integrable on* Ω *and* $\int_\Omega f(\mathbf{x}) d\mathbf{x} = \lim_{n\to\infty} \int_\Omega f_n(\mathbf{x}) d\mathbf{x}$. *In other words,* $\int_\Omega (\lim_{n\to\infty} f_n(\mathbf{x})) d\mathbf{x} = \lim_{n\to\infty} \int_\Omega f_n(\mathbf{x}) d\mathbf{x}$.

B.3.3 Some important approximation and density results in $L^1(\Omega)$

Simple functions and their approximation properties

Definition B.3.3.1. A *real-valued function* $s : \Omega \subset \mathbb{R}^n \longrightarrow \{\alpha_1, \alpha_2, \dots, \alpha_m\} \subset \mathbb{R}_0^+ = [0, \infty[$ is called a *non-negative, simple function* if and only if its range is a *finite* set $\{\alpha_1, \alpha_2, \dots, \alpha_m\}$ of *non-negative* numbers $\alpha_i \geq 0$, $1 \leq i \leq m$, i.e. $\forall \mathbf{x} \in \Omega \subset \mathbb{R}^n$, $s(\mathbf{x}) \in \{\alpha_1, \alpha_2, \dots, \alpha_m\} \subset \mathbb{R}_0^+$

$$\Longrightarrow \quad s(\mathbf{x}) = \sum_{i=1}^{m} \alpha_i \chi_{A_i}(\mathbf{x}), \tag{B.3.3.1}$$

where $A_i = \{\mathbf{x} : \mathbf{x} \in \Omega,\ s(\mathbf{x}) = \alpha_i\}$ is a Lebesgue-measurable subset of Ω for $1 \leq i \leq m$; χ_{A_i} is the *characteristic function* of set A_i with $\chi_{A_i}(\mathbf{x}) = 1$ for $\mathbf{x} \in A_i$ and $= 0$ for $\mathbf{x} \in A_i^{\complement}$ i.e. $s \geq 0$ if and only if $s(\mathbf{x}) = \alpha_i \geq 0$ for $\mathbf{x} \in \mathbb{R}^n$. (B.3.3.2)

Theorem B.3.3.1 (Approximation of a measurable function by simple functions). *Let* $f : \Omega \subset \mathbb{R}^n \longrightarrow [0, \infty]$ *be a* measurable, non-negative *function on* Ω. *Then* $\exists$ *a sequence* $(s_n)_{n=1}^{\infty}$ *of (measurable)* non-negative *simple functions monotonically* increasing *at each point* $\mathbf{x} \in \Omega$ *which converges* pointwise *to* f *on* Ω, *i.e.* (possibly with a change of values on a set of points with measure zero *(see* (B.2.1.8)*)*

I. $0 \le s_1 \le s_2 \le \cdots \le s_n \le \cdots \le f$;

II. $\lim_{n\to\infty} s_n(\mathbf{x}) = f(\mathbf{x})\ \forall \mathbf{x} \in \Omega$.

Proof. First of all, we will construct a sequence $(\phi_n)_{n=1}^{\infty}$ of *non-negative* functions on $[0,\infty]$ defined by $\phi_n : t \in [0,\infty] \longmapsto \phi_n(t) \in [0,\infty[\ \forall n \in \mathbb{N}$ such that $0 \le \phi_1(t) \le \phi_2(t) \le \cdots \le \phi_n(t) \cdots \le t\ \forall t \in [0,\infty]$, $\lim_{n\to\infty} \phi_n(t) = t\ \forall t \in [0,\infty]$. Then the sequence (s_n) of simple functions s_n defined by $s_n = \phi_n \circ f : \Omega \longrightarrow [0,\infty[$ will have all the required properties: $\forall n \in \mathbb{N}$, $s_n(\mathbf{x}) = \phi_n(f(\mathbf{x})) = \phi_n(t) \in [0,\infty[$ with $f(\mathbf{x}) = t\ \forall \mathbf{x} \in \Omega$, $t \in [0,\infty[$, $0 \le \phi_1(t) = s_1(\mathbf{x}) \le \cdots \le \phi_n(t) = s_n(\mathbf{x}) \le \cdots \le t = f(\mathbf{x})$, i.e. $0 \le s_1(\mathbf{x}) \le \cdots \le s_n(\mathbf{x}) \le \cdots \le f(\mathbf{x})\ \forall \mathbf{x} \in \Omega$, and $\lim_{n\to\infty} \phi_n(t) = t \Longrightarrow \lim_{n\to\infty} s_n(\mathbf{x}) = f(\mathbf{x})\ \forall \mathbf{x} \in \Omega$. (B.3.3.3)

Hence, it is sufficient to construct $(\phi_n)_{n=1}^{\infty}$ with the above properties such that $\lim_{n\to\infty} \phi_n(t) = t\ \forall t \in [0,\infty]$ as follows. With each $n \in \mathbb{N}$ and every real number $t \in [0,\infty[$ we associate a *unique, non-negative integer* $k_n = k_n(t) \in \mathbb{N}_0$ such that $\frac{k_n}{2^n} \le t < \frac{k_n+1}{2^n}$.

For example, for $n = 1$, $k_1 = 0$ for $t \in [0, \frac{1}{2}[$, $k_1 = 1$ for $t \in [\frac{1}{2}, 1[$, $k_1 = 2$ for $t \in [1, \frac{3}{2}[, \ldots, k_1 = m$ for $t \in [\frac{m}{2}, \frac{m+1}{2}[, \ldots$; for $n = 2$, $k_2 = 0$ for $t \in [0, \frac{1}{2^2}[$, $k_2 = 1$ for $t \in [\frac{1}{2^2}, \frac{2}{2^2}[$, $k_2 = 2$ for $t \in [\frac{2}{2^2}, \frac{3}{2^2}[, \ldots, k_2 = m$ for $t \in [\frac{m}{2^2}, \frac{m+1}{2^2}[, \ldots$, and so on.

Now, $\forall n \in \mathbb{N}$, we define $\phi_n(t) = \frac{k_n(t)}{2}$ for $0 \le t < n$ and $\phi_n(t) = n$ for $n \le t \le \infty$. Then $\phi_n : t \in [0,\infty] \longmapsto \phi_n(t) \in [0,n] \subset [0,\infty[$. ϕ_n is also a *non-negative simple* function on $[0,\infty]$. But $\frac{k_n(t)}{2^n} \le t < \frac{k_n(t)}{2^n} + \frac{1}{2^n}\ \forall n \in \mathbb{N} \Longrightarrow t - \frac{1}{2^n} < \frac{k_n(t)}{2^n}$ $\forall n \in \mathbb{N}$, $\forall t \in [0,\infty[$. Then $t - \frac{1}{2^n} < \frac{k_n(t)}{2^n} = \phi_n(t)$ for $0 \le t < n$ and $\phi_n(t) = n$ for $n \le t \le \infty$.

Combining these, we have $t - \frac{1}{2^n} < \phi_n(t) = \frac{k_n(t)}{2^n} \le t\ \forall t \in [0,\infty]$. Hence, $0 \le \phi_1(t) \le \cdots \le \phi_n(t) \le \cdots \le t\ \forall t \in [0,\infty]$. But $\lim_{n\to\infty}(t - \frac{1}{2^n}) \le \lim_{n\to\infty} \phi_n(t) \le t\ \forall t \in [0,\infty] \Longrightarrow t \le \lim_{n\to\infty} \phi_n(t) \le t\ \forall t \in [0,\infty] \Longleftrightarrow \lim_{n\to\infty} \phi_n(t) = t$ $\forall t \in [0,\infty]$, which completes the proof of (B.3.3.3). □

Interesting relation between measurable functions and continuous functions is given by

Theorem B.3.3.2 (Lusin's Theorem [27]). *Let $\Omega \subset \mathbb{R}^n$ be an open subset of $\mathbb{R}^n$ and $A \subset \Omega$ be a* bounded *subset of Ω with $\mu(A) < +\infty$, $\mu(A)$ being the n-dimensional Lebesgue volume measure of A. Let f be a measurable function on Ω such that $f(\mathbf{x}) = 0\ \forall \mathbf{x} \in A^{\complement}$, and $\varepsilon > 0$ be any given number. Then $\exists g \in C_0(\Omega)$ such that*

$$\mu(\{\mathbf{x} : \mathbf{x} \in \Omega, f(\mathbf{x}) \ne g(\mathbf{x})\}) < \varepsilon,$$

$$\sup_{\mathbf{x}\in\Omega} |g(\mathbf{x})| \le sup_{\mathbf{x}\in\Omega} |f(\mathbf{x})| = \|f\|_\infty,$$

$$|g(\mathbf{x})| \le \|f\|_\infty \quad \forall \mathbf{x} \in \Omega$$

(possibly after changing values of $f(\mathbf{x})$ *on a set of points with measure* 0 *(see* (B.2.1.8)*)).*

Space $S(\Omega)$ of non-negative simple functions with bounded support

Definition B.3.3.2. $S(\Omega)$ is the space of measurable *non-negative*, *simple* functions s on $\Omega \subset \mathbb{R}^n$ with *bounded support*, i.e. Lebesgue measure $\mu(\{\mathbf{x} : \mathbf{x} \in \Omega, s(\mathbf{x}) \neq 0\}) < +\infty$.

Proposition B.3.3.1. $S(\Omega) \subset L^1(\Omega)$.

Proof. Let s in $S(\Omega)$. Then $s(\mathbf{x}) = \sum_{i=1}^m \beta_i \chi_{B_i}(\mathbf{x})$ for $\mathbf{x} \in \Omega$, where $B_i = \{\mathbf{x} : \mathbf{x} \in \Omega, s(\mathbf{x}) = \beta_i\}$, $1 \leq i \leq m$, and

$$\int_\Omega s(\mathbf{x}) d\mathbf{x} = \int_\Omega \sum_{i=1}^m \beta_i \chi_{B_i}(\mathbf{x}) d\mathbf{x} \leq \sum_{i=1}^m \beta_i \mu(B_i) = \sum_{k=1}^l \beta_{i_k} \mu(B_{i_k}) < +\infty,$$

where $\beta_{i_k} \neq 0$, $\mu(B_{i_k}) < +\infty \ \forall k = 1, 2, \ldots, l \Longrightarrow s \in L^1(\Omega) \Longrightarrow S(\Omega) \subset L^1(\Omega)$. □

Theorem B.3.3.3 (Approximation Theorem in $L^1(\Omega)$). *Let $f \geq 0$ be a* non-negative *real-valued function on $\Omega \subset \mathbb{R}^n$ such that $f \in L^1(\Omega)$. Then $\exists$ a sequence $(s_n)_{n=1}^\infty$ of* non-negative, simple *functions $s_n \geq 0$ in $S(\Omega)$ such that $s_n \longrightarrow f$ in $L^1(\Omega)$ as $n \to \infty$.*

Proof. Let $f \geq 0$ a.e. on Ω such that $f \in L^1(\Omega)$. Then, by Theorem B.3.3.1, $\exists$ a sequence $(s_n)_{n=1}^\infty$ of *non-negative, simple* functions $s_n \geq 0$ such that

I. $0 \leq s_1(\mathbf{x}) \leq s_2(\mathbf{x}) \leq \cdots \leq s_n(\mathbf{x}) \leq \cdots \leq f(\mathbf{x})$ for $\mathbf{x} \in \Omega$ and
II. $\lim_{n\to\infty} s_n(\mathbf{x}) = f(\mathbf{x})$ for $\mathbf{x} \in \Omega$.

Hence, $\forall n \in \mathbb{N}$, $\int_\Omega s_n(\mathbf{x}) d\mathbf{x} \leq \int_\Omega f(\mathbf{x}) d\mathbf{x} < +\infty \Longrightarrow s_n \in L^1(\Omega) \ \forall n \in \mathbb{N}$. Then $|f(\mathbf{x}) - s_n(\mathbf{x})| \leq f(\mathbf{x}) \ \forall n \in \mathbb{N}$ a.e. on Ω, $f \in L^1(\Omega)$ and $\lim_{n\to\infty} |f(\mathbf{x}) - s_n(\mathbf{x})| = 0$ a.e. on Ω. Hence, by Lebesgue's Dominated Convergence Theorem B.3.2.2,

$$\lim_{n\to\infty} \int_\Omega |f(\mathbf{x}) - s_n(\mathbf{x})| d\mathbf{x} = \int_\Omega \lim_{n\to\infty} |f(\mathbf{x}) - s_n(\mathbf{x})| d\mathbf{x} = \int_\Omega 0 \cdot d\mathbf{x} = 0.$$

Hence, $s_n \longrightarrow f$ in $L^1(\Omega)$. □

Corollary B.3.3.1. *Let $S(\Omega)$ be the space of all measurable,* complex-valued *simple functions in Ω such that $\mu(\{\mathbf{x} : \mathbf{x} \in \Omega, s(\mathbf{x}) \neq 0\}) < +\infty$. Then, $\forall$* complex-valued *$f \in L^1(\Omega)$, $\exists$ a sequence (s_n) of* complex-valued *simple functions in $S(\Omega)$ such that $s_n \to f$ in $L^1(\Omega)$ as $n \to \infty$.*

Proof. $\forall$ complex-valued f in $L^1(\Omega)$ we can write

$$f(\mathbf{x}) = u(\mathbf{x}) + iv(\mathbf{x}) = (u^+ - u^-)(\mathbf{x}) + i(v^+ - v^-)(\mathbf{x}),$$

where

$$u^+(\mathbf{x}) = \max\{u(\mathbf{x}), 0\}, \qquad v^+(\mathbf{x}) = \max\{v(\mathbf{x}), 0\},$$
$$u^-(\mathbf{x}) = -\min\{0, u(\mathbf{x})\}, \quad v^-(\mathbf{x}) = -\min\{0, v(\mathbf{x})\}$$

are *all non-negative* real-valued functions. Hence, we can apply Theorem B.3.3.1 for u^+, v^+, u^-, v^- to get a corresponding sequence of *non-negative* simple functions $(s_n^+)_{n=1}^\infty$, $(s_n^-)_{n=1}^\infty$, $(t_n^+)_{n=1}^\infty$ and $(t_n^-)_{n=1}^\infty$, respectively, in $S(\Omega)$ such that $s_n^+ \to u^+$, $s_n^- \to u^-$, $t_n^+ \to v^+$, $t_n^- \to v^-$ in $L^1(\Omega)$ as $n \to \infty$. Then $(s_n^+ - s_n^-) + i(t_n^+ - t_n^-) \to (u^+ - u^-) + i(v^+ - v^-) = f$ in $L^1(\Omega) \Longrightarrow s_n \to f$ in $L^1(\Omega)$. □

Density of $C_0(\Omega)$ in $L^1(\Omega)$

Theorem B.3.3.4. $C_0(\Omega) = \{\phi : \phi$ *is continuous with* compact *support in* $\Omega\}$ *is* dense *in* $L^1(\Omega)$.

Proof. Let $u \in L^1(\Omega)$. We are to show that $\exists$ a sequence $(\phi_n)_{n=1}^\infty$ in $C_0(\Omega) \subset L^1(\Omega)$ such that $\phi_n \to u \in L^1(\Omega)$ as $n \to \infty \Longrightarrow \forall \varepsilon > 0$, $\exists n_0 = n_0(\varepsilon) \in \mathbb{N}$ such that $\|u - \phi_n\|_{L^1(\Omega)} < \varepsilon \ \forall n \geq n_0$. Without loss of generality, we assume that u is a *non-negative*, real-valued function, since for the complex-valued case, $u = (u^+ - u^-) + i(v^+ - v^-)$, where u^+, u^-, v^+, v^- are all *non-negative*, real-valued functions, the proof given will hold for *all* the four functions, and the final result can be obtained with minor modifications (see also the proof of Corollary B.3.3.1). Then, by Theorem B.3.3.3 on approximation by *non-negative* simple functions with *bounded support* in Ω, $\exists$ a sequence $(s_n)_{n=1}^\infty$ of *non-negative*, measurable simple functions $s_n \in S(\Omega) = \{s : s$ is a *non-negative*, measurable simple function with $\mu(\{\mathbf{x} : \mathbf{x} \in \Omega, s_n(\mathbf{x}) \neq 0\}) < +\infty\}$ such that $\|u - s_n\|_{L^1(\Omega)} \to 0$ as $n \to \infty$, i.e. $\forall \varepsilon > 0$, $\exists n_0 = n_0(\varepsilon) \in \mathbb{N}$ such that

$$\|u - s_n\|_{L^1(\Omega)} < \frac{\varepsilon}{2} \quad \forall n \geq n_0. \tag{B.3.3.4}$$

Hence, $\forall$ *fixed* $n \geq n_0$ with (B.3.3.4), $s_n \in S(\Omega)$ with $\mu(\{\mathbf{x} : \mathbf{x} \in \Omega, s_n(\mathbf{x}) \neq 0\}) < +\infty$. Then, by Theorem B.3.3.2 of Lusin, $\forall$ *fixed* $n \geq n_0$ with (B.3.3.4), $\exists \phi_n \in C_0(\Omega)$ such that $|\varphi_n(\mathbf{x})| \leq \|s_n\|_\infty \ \forall \mathbf{x} \in \Omega$, $\|s_n\|_\infty = \sup_{\mathbf{x}\in\Omega} |s_n(\mathbf{x})|$, and $\mu(\{\mathbf{x} : \mathbf{x} \in \Omega, s_n(\mathbf{x}) \neq \phi_n(\mathbf{x})\}) < \frac{\varepsilon}{4\|s_n\|_\infty}$. Since $|s_n(\mathbf{x}) - \phi_n(\mathbf{x})| \leq |s_n(\mathbf{x})| +$

$|\phi_n(\mathbf{x})| \le \|s_n\|_\infty + \|s_n\|_\infty = 2\|s_n\|_\infty$, for $\mathbf{x} \in \Omega$, $\forall$ *fixed* $n \ge n_0$ with (B.3.3.4),

$$\begin{aligned}\|s_n - \phi_n\|_{L^1(\Omega)} &= \int_\Omega |s_n(\mathbf{x}) - \phi_n(\mathbf{x})| d\mathbf{x} = \int_{\{\mathbf{x}:\mathbf{x}\in\Omega,\ s_n(\mathbf{x})\ne\phi_n(\mathbf{x})\}} |s_n(\mathbf{x}) - \phi_n(\mathbf{x})| d\mathbf{x} \\ &\le 2\|s_n\|_\infty \int_{\{\mathbf{x}:\mathbf{x}\in\Omega,\ s_n(\mathbf{x})\ne\phi_n(\mathbf{x})\}} d\mathbf{x} \\ &\le 2\|s_n\|_\infty \mu\left(\{\mathbf{x} : \mathbf{x} \in \Omega,\ s_n(\mathbf{x}) \ne \phi_n(\mathbf{x})\}\right) \\ &< 2\|s_n\|_\infty \cdot \frac{\varepsilon}{4\|s_n\|_\infty} = \frac{\varepsilon}{2}\end{aligned}$$

$\Longrightarrow \|s_n - \phi_n\|_{L^1(\Omega)} < \frac{\varepsilon}{2}$ $\forall$ *fixed* $n \ge n_0$. Hence, $\exists$ a sequence (ϕ_n) in $C_0(\Omega)$ with the property $\forall \varepsilon > 0$, $\exists n_0 = n_0(\varepsilon) \in \mathbb{N}$ such that

$$\begin{aligned}\|u - \phi_n\|_{L^1(\Omega)} &= \|u - s_n + s_n - \phi_n\|_{L^1(\Omega)} \le \|u - s_n\|_{L^1(\Omega)} + \|s_n - \phi_n\|_{L^1(\Omega)} \\ &< \frac{\varepsilon}{2} + \frac{\varepsilon}{2} = \varepsilon \quad \forall n \ge n_0\end{aligned}$$

$\Longrightarrow \phi_n \to u$ in $L^1(\Omega)$ as $n \to \infty$. Hence, $C_0(\Omega)$ is *dense* in $L^1(\Omega)$. □

B.4 Spaces $L^p(\Omega)$, $1 \le p \le \infty$

B.4.1 Basic properties

Definition B.4.1.1. Let $\Omega \subset \mathbb{R}^n$ be an *open* subset of $\mathbb{R}^n$. Then:

1. for $1 \le p < \infty$,

$$L^p(\Omega) = \left\{u : u \in \mathcal{M}(\Omega),\ \int_\Omega |u(\mathbf{x})|^p d\mathbf{x} < +\infty\right\}, \tag{B.4.1.1}$$

   where the (multiple) integral of $|u(\mathbf{x})|^p$ is in the *Lebesgue sense*;

2. for $p = \infty$,

$$L^\infty(\Omega) = \{u : u \in \mathcal{M}(\Omega), \operatorname*{ess\,sup}_{\mathbf{x}\in\Omega} |u(\mathbf{x})| < +\infty\}. \tag{B.4.1.2}$$

Precisely speaking, elements of $L^p(\Omega)$ are *equivalence classes* $[u]$ of functions u, but we have written functions u to mean $u \in L^p(\Omega) \Longleftrightarrow [u] \in L^p(\Omega)$ (see the discussion of notational conventions in B.2.1 above);

$$u \ne v \text{ in } L^p(\Omega) \Longleftrightarrow [u] \ne [v]. \tag{B.4.1.3}$$

We will consider $L^p(\Omega)$, $1 \le p \le \infty$, consisting of *real-valued functions* unless stated otherwise.

Properties of functions in $L^p(\Omega)$, $1 \le p \le \infty$

1. For $u \in L^p(\Omega)$, $1 \le p \le \infty$, u is defined *a.e.* on Ω.
2. $u, v \in L^p(\Omega)$, $u = v$ in $L^p(\Omega) \iff u(\mathbf{x}) = v(\mathbf{x})$ *a.e.* on Ω.
   $u \neq v$ in $L^p(\Omega) \iff \exists A \subset \Omega$ with measure $\mu(A) > 0$ such that $u(\mathbf{x}) \neq v(\mathbf{x})$ for $\mathbf{x} \in A$.
3. The *null function* 0 in $L^p(\Omega)$ is defined by $u = 0$ in $L^p(\Omega) \iff u(\mathbf{x}) = 0 \in \mathbb{R}$ *a.e.* on Ω. (B.4.1.4)
4. $u, v \in L^p(\Omega) \implies (u+v)(\mathbf{x}) = u(\mathbf{x}) + v(\mathbf{x})$ *a.e.* on Ω; $(\alpha u)(\mathbf{x}) = \alpha u(\mathbf{x})$ *a.e.* on Ω $\forall \alpha \in \mathbb{R}$.
5. $L^p(\Omega)$, $1 \le p \le \infty$, is a *linear space*.

Definition B.4.1.2. The linear space $L^p(\Omega)$, $1 \le p \le \infty$, equipped with the *norm* $\|\cdot\|_{L^p(\Omega)}$ defined by:

- for $1 \le p < \infty$,

$$\|u\|_{L^p(\Omega)} = \left(\int_\Omega |u(\mathbf{x})|^p d\mathbf{x}\right)^{\frac{1}{p}}; \tag{B.4.1.5}$$

- for $p = \infty$,

$$\|u\|_{L^\infty(\Omega)} = \operatorname*{ess\,sup}_{\mathbf{x}} \in \mathbf{\Omega}\, |u(\mathbf{x})| \quad \forall u \in L^\infty(\Omega) \tag{B.4.1.6}$$

is a *normed linear space*.

Theorem B.4.1.1 (Hölder inequality). *Let $p, q \in [1, \infty]$ be* conjugate indices, *i.e.* $\frac{1}{p} + \frac{1}{q} = 1$. *Then,* $\forall u \in L^p(\Omega)$, $\forall v \in L^q(\Omega)$, $uv \in L^1(\Omega)$,

$$\int_\Omega |u(\mathbf{x})v(\mathbf{x})| d\mathbf{x} \le \|u\|_{L^p(\Omega)} \|v\|_{L^q(\Omega)}, \tag{B.4.1.7}$$

where

$$\|u\|_{L^p(\Omega)} = \left(\int_\Omega |u(\mathbf{x})|^p d\mathbf{x}\right)^{1/p}, \quad \|v\|_{L^q(\Omega)} = \left(\int_\Omega |v(\mathbf{x})|^q d\mathbf{x}\right)^{1/q}; \tag{B.4.1.8}$$

$$\|u\|_{L^\infty(\Omega)} = \operatorname*{ess\,sup}_{\mathbf{x}\in\Omega} |u(\mathbf{x})|, \qquad \|v\|_{L^\infty(\Omega)} = \operatorname*{ess\,sup}_{\mathbf{x}\in\Omega} |v(\mathbf{x})|. \tag{B.4.1.9}$$

Proof. For $u = 0$ or $v = 0$, the result follows *trivially*. Hence, for $u \neq 0$, $v \neq 0$, $u \in L^p(\Omega)$, $v \in L^q(\Omega)$ with $1 < p, q < \infty$ and $\frac{1}{p} + \frac{1}{q} = 1$, setting $a = \frac{|u(\mathbf{x})|}{\|u\|_{L^p(\Omega)}} >$

0, $b = \frac{|v(\mathbf{x})|}{\|v\|_{L^q(\Omega)}} > 0$ a.e. on Ω and applying *Young's inequality* (for $a > 0$, $b > 0$, $ab \le \frac{1}{p}a^p + \frac{1}{q}b^q$), we get the result. In fact,

$$\frac{\int_\Omega |u(\mathbf{x})v(\mathbf{x})|d\mathbf{x}}{\|u\|_{L^p(\Omega)}\|v\|_{L^q(\Omega)}} \le \frac{1}{p}\cdot\frac{\int_\Omega |u(\mathbf{x})|^p d\mathbf{x}}{\|u\|^p_{L^p(\Omega)}} + \frac{1}{q}\cdot\frac{\int_\Omega |v(\mathbf{x})|^q d\mathbf{x}}{\|v\|^q_{L^q(\Omega)}} = \frac{1}{p} + \frac{1}{q} = 1$$

$\Longrightarrow$ the result.

For $p = 1, q = \infty$, $|u(\mathbf{x})v(\mathbf{x})| \le |u(\mathbf{x})|\|v\|_{L^\infty(\Omega)}$ a.e. on Ω

$$\Longrightarrow \quad \int_\Omega |u(\mathbf{x})v(\mathbf{x})|d\mathbf{x} \le \|v\|_{L^\infty(\Omega)} \int_\Omega |u(\mathbf{x})|d\mathbf{x} = \|u\|_{L^1(\Omega)}\|v\|_{L^\infty(\Omega)} < +\infty$$

$\Longrightarrow$ the result.

Similarly, for $p = \infty$, $q = 1$ the result follows. □

Minkowski inequality

$$\|u + v\|_{L^p(\Omega)} \le \|u\|_{L^p(\Omega)} + \|v\|_{L^p(\Omega)} \quad \forall u, v \in L^p(\Omega),\ 1 \le p \le \infty. \tag{B.4.1.10}$$

Theorem B.4.1.2. *The normed linear space $L^p(\Omega)$, $1 \le p \le \infty$, equipped with the norm $\|\cdot\|_{L^p(\Omega)}$ is a* Banach space.

Proof.
Case $p = \infty$: Let $(u_n)_{n=1}^\infty$ be a *Cauchy sequence* in $L^\infty(\Omega)$, i.e. $\forall k \in \mathbb{N}$, $\exists n_k \in \mathbb{N}$ such that $\|u_m - u_n\|_{L^\infty(\Omega)} < \frac{1}{k}$ $\forall m, n > n_k$. Hence, $\forall k \in \mathbb{N}$, $\exists E_k \subset \Omega$ with $\mu(E_k) = 0$ (possibly $E_k = \emptyset$ for some k) such that $|u_m(\mathbf{x}) - u_n(\mathbf{x})| \le \|u_m - u_n\|_{L^\infty(\Omega)} < \frac{1}{k}$ $\forall \mathbf{x} \in \Omega \setminus E_k$, $\forall m, n > n_k$. Set $E = \bigcup_{k=1}^\infty E_k$. Then, by Proposition B.1.1.1, $\mu(E) \le \sum_{k=1}^\infty \mu(E_k) = 0 \Longrightarrow \mu(E) = 0$ (by (B.1.1.3)). Then, $\forall k \in \mathbb{N}$, $\forall \mathbf{x} \in \Omega \setminus E$ with $\Omega \setminus E \subset \Omega \setminus E_k$, $\exists n_k \in \mathbb{N}$ such that $|u_m(\mathbf{x}) - u_n(\mathbf{x})| < \frac{1}{k}$ $\forall m, n > n_k$. Hence, $\forall \mathbf{x} \in \Omega \setminus E$, $(u_n(\mathbf{x}))_{k=1}^\infty$ is a Cauchy sequence in $\mathbb{R}$, which is *complete*. Therefore, $\forall$ fixed $\mathbf{x} \in \Omega \setminus E$, $\exists$ a *unique real* number $= \lim_{n\to\infty} u_n(\mathbf{x})$, and the correspondence $\mathbf{x} \in \Omega \setminus E \longmapsto \lim_{n\to\infty} u_n(\mathbf{x})$ defines a mapping denoted by $u : \mathbf{x} \in \Omega \setminus E \longmapsto u(\mathbf{x}) = \lim_{n\to\infty} u_n(\mathbf{x})$. Then, $\forall \mathbf{x} \in \Omega \setminus E$, $\forall k \in \mathbb{N}$, $\exists \bar{n}_k \in \mathbb{N}$ such that

$$|u(\mathbf{x}) - u_n(\mathbf{x})| < \frac{1}{k} \quad \forall n > \bar{n}_k. \tag{B.4.1.10a}$$

Set $u(\mathbf{x}) = 0$ $\forall \mathbf{x} \in E$. Fix $n > \bar{n}_k$ such that (B.4.1.10a) holds and we have, $\forall \mathbf{x} \in \Omega \setminus E$,

$$|u(\mathbf{x})| \le |u(\mathbf{x}) - u_n(\mathbf{x})| + |u_n(\mathbf{x})| < \frac{1}{k} + \|u_n\|_{L^\infty(\Omega)}$$

$\Longrightarrow u$ is *essentially bounded* on Ω with $\operatorname{ess\,sup}_{\mathbf{x}\,\in\,\boldsymbol{\Omega}} |u(\mathbf{x})| \le \frac{1}{k} + \|u_n\|_{L^\infty(\Omega)} < +\infty\ \forall n > \bar{n}_k \Longrightarrow u \in L^\infty(\Omega)$. Moreover, from (B.4.1.10a), $\operatorname{ess\,sup}_{\mathbf{x}\,\in\,\boldsymbol{\Omega}} |u(\mathbf{x}) - u_n(\mathbf{x})| = \|u - u_n\|_{L^\infty(\Omega)} < \frac{1}{k}\ \forall n > \bar{n}_k \Longrightarrow \lim_{n\to\infty} u_n = u$ in $L^\infty(\Omega)$. Hence, $L^\infty(\Omega)$ is a *complete normed linear space*, i.e. a *Banach space*.

Case $1 \le p < \infty$: Let $(u_n)_{n=1}^\infty$ be a *Cauchy sequence* in $L^p(\Omega)$, $1 \le p < \infty$, i.e. $\forall \varepsilon > 0, \exists n_0 \in \mathbb{N}$ such that

$$\|u_m - u_n\|_{L^p(\Omega)} < \varepsilon \quad \forall m, n > n_0. \tag{B.4.1.10b}$$

Hence, for $\varepsilon = \frac{1}{2^j}$ with $j \in \mathbb{N}$, $\exists n_j \in \mathbb{N}$ such that $\|u_m - u_n\|_{L^p(\Omega)} < \frac{1}{2^j}$ $\forall m, n > n_j$. We extract a subsequence $(u_{n_k})_{k=1}^\infty$ of the sequence $(u_n)_{n=1}^\infty$ such that

$$\|u_{n_{k+1}} - u_{n_k}\|_{L^p(\Omega)} < \frac{1}{2^k} \quad \forall k \in \mathbb{N}. \tag{B.4.1.10c}$$

Set

$$v_m(\mathbf{x}) = \sum_{k=1}^{m} |u_{n_{k+1}}(\mathbf{x}) - u_{n_k}(\mathbf{x})| \quad \text{a.e. on } \Omega. \tag{B.4.1.10d}$$

Then, $\forall m \in \mathbb{N}$,

$$\|v_m\|_{L^p(\Omega)} \le \sum_{k=1}^{m} \|u_{n_{k+1}} - u_{n_k}\|_{L^p(\Omega)} < \frac{1}{2} + \frac{1}{2^2} + \cdots + \frac{1}{2^m} < 1$$

$\Longrightarrow \forall m \in \mathbb{N}$, $v_m \in L^p(\Omega)$ with $\|v_m\|_{L^p(\Omega)} < 1 \Longrightarrow v_m^p \in L^1(\Omega)$, $0 \le (v_1(\mathbf{x}))^p \le \cdots \le (v_m(\mathbf{x}))^p \le \cdots$ a.e. on Ω and $\sup_{m\in\mathbb{N}} \int_\Omega v_m^p(\mathbf{x})d\mathbf{x} \le 1$. Then, by the Monotone Convergence Theorem B.3.2.4, $(v_m^p(\mathbf{x}))$ converges to a finite limit denoted by $(v(\mathbf{x}))^p$ a.e. on Ω, i.e.

$$\lim_{m\to\infty} (v_m(\mathbf{x}))^p = (v(\mathbf{x}))^p \ge 0 \quad \text{a.e. on } \Omega \text{ with } v^p \in L^1(\Omega). \tag{B.4.1.10e}$$

Then, for $m > k \ge 2$,

$$\begin{aligned}
|u_{n_m}(\mathbf{x}) - u_{n_k}(\mathbf{x})| &\le |u_{n_m}(\mathbf{x}) - u_{n_{m-1}}(\mathbf{x})| + |u_{n_{m-1}}(\mathbf{x}) - u_{n_{m-2}}(\mathbf{x})| \\
&\quad + \cdots + |u_{n_{k+1}}(\mathbf{x}) - u_{n_k}(\mathbf{x})| \\
&= \sum_{l=k}^{m-1} |u_{n_{l+1}}(\mathbf{x}) - u_{n_l}(\mathbf{x})| \\
&= \sum_{l=1}^{m-1} |u_{n_{l+1}}(\mathbf{x}) - u_{n_l}(\mathbf{x})| - \sum_{l=1}^{k-1} |u_{n_{l+1}}(\mathbf{x}) - u_{n_l}(\mathbf{x})|
\end{aligned}$$

$\Longrightarrow$ for $m > k \geq 2$, $|u_{n_m}(\mathbf{x}) - u_{n_k}| \leq v_{m-1}(\mathbf{x}) - v_{k-1}(\mathbf{x}) \leq v(\mathbf{x}) - v_{k-1}(\mathbf{x})$ a.e. on Ω (by (B.4.1.10d)). But by (B.4.1.10e), $[v(\mathbf{x}) - v_{k-1}(\mathbf{x})] \to 0$ a.e. on Ω as $k \to \infty$. Hence, *for almost all* $\mathbf{x} \in \Omega$, $|u_{n_m}(\mathbf{x}) - u_{n_k}(\mathbf{x})| \to 0$ as $m, k \to \infty \Longrightarrow (u_{n_k}(\mathbf{x}))$ is a *Cauchy sequence* in $\mathbb{R}$ *for almost all* $\mathbf{x} \in \Omega$, but $\mathbb{R}$ is complete. Hence, we can define a function u a.e. on Ω by:

$$u(\mathbf{x}) = \lim_{k\to\infty} u_{n_k}(\mathbf{x}) \quad \textit{for almost all } \mathbf{x} \in \Omega. \tag{B.4.1.10f}$$

Then, $\forall m > k \geq 2$, *for almost all* $\mathbf{x} \in \boldsymbol{\Omega}$,

$$\begin{aligned}|u(\mathbf{x}) - u_{n_k}(\mathbf{x})| &\leq |u(\mathbf{x}) - u_{n_m}(\mathbf{x})| + |u_{n_m}(\mathbf{x}) - u_{n_k}(\mathbf{x})| \\ &\leq |u(\mathbf{x}) - u_{n_m}(\mathbf{x})| + v(\mathbf{x}) - v_{k-1}(\mathbf{x}) \leq |u(\mathbf{x}) - u_{n_m}(\mathbf{x})| + v(\mathbf{x}),\end{aligned}$$

since $v(\mathbf{x}) \geq v_{k-1}(\mathbf{x}) \geq 0$ a.e. on Ω (by (B.4.1.10e)). Letting $m \to \infty$ and using (B.4.1.10f), we have, for *almost all* $\mathbf{x} \in \boldsymbol{\Omega}$,

$$|u(\mathbf{x}) - u_{n_k}(\mathbf{x})| \leq \lim_{m\to\infty} |u(\mathbf{x}) - u_{n_m}(\mathbf{x})| + v(\mathbf{x}) = 0 + v(\mathbf{x}) = v(\mathbf{x}).$$

Set $f_k(\mathbf{x}) = |u(\mathbf{x}) - u_{n_k}(\mathbf{x})|^p \leq (v(\mathbf{x}))^p$ a.e. on Ω with $f_k = |u - u_{n_k}|^p \ \forall k \geq 2$. Then $\lim_{k\to\infty} f_k(\mathbf{x}) = 0$ a.e. on Ω by (B.4.1.10f) and $f_k(\mathbf{x}) \leq (v(\mathbf{x}))^p$ a.e. on Ω, $\forall k \geq 2$ and $v^p \in L^1(\Omega)$ by (B.4.1.10e). Hence, by Lebesgue's Dominated Convergence Theorem B.3.2.2, $\lim_{k\to\infty} \int_\Omega f_k(\mathbf{x})d\mathbf{x} = \int_\Omega \lim_{k\to\infty} f_k(\mathbf{x})d\mathbf{x} = 0 \Longrightarrow$ $\lim_{k\to\infty} f_k = 0$ in $L^1(\Omega)$

$$\Longrightarrow \quad \lim_{k\to\infty} \int_\Omega |u - u_{n_k}|^p d\mathbf{x} = \lim_{k\to\infty} \|u - u_{n_k}\|^p_{L^p(\Omega)} = 0. \tag{B.4.1.10g}$$

Then, $\forall$ fixed $\varepsilon_0 > 0$, $\exists k_0 \in \mathbb{N}$ such that $\|u - u_{n_k}\|^p_{L^p(\Omega)} < \varepsilon_0 \ \forall k > k_0$. Thus, for *fixed* $k > k_0$,

$$\begin{aligned}\|u\|^p_{L^p(\Omega)} &\leq (\|u - u_{n_k}\|_{L^p(\Omega)} + \|u_{n_k}\|_{L^p(\Omega)})^p \\ &\leq 2^{p-1}[\|u - u_{n_k}\|^p_{L^p(\Omega)} + \|u_{n_k}\|^p_{L^p(\Omega)}] \\ &\leq 2^{p-1}(\varepsilon_0 + \|u_{n_k}\|^p_{L^p(\Omega)}) < +\infty\end{aligned}$$

$\Longrightarrow u \in L^p(\Omega)$ and $\|u - u_{n_k}\|_{L^p(\Omega)} \to 0$ as $k \to \infty$ (by (B.4.1.10g)), i.e. $u_{n_k} \to u$ in $L^p(\Omega)$ as $k \to \infty$. Hence, $\forall \varepsilon > 0$, $\exists k^* \in \mathbb{N}$ such that $\|u - u_{n_k}\|_{L^p(\Omega)} < \frac{\varepsilon}{2}$ $\forall k > k^*$. From (B.4.1.10b), $\forall \varepsilon > 0$, $\exists n_0 \in \mathbb{N}$ such that $\|u_{n_k} - u_n\| < \frac{\varepsilon}{2} \ \forall n, n_k >$ n_0. Then, $\forall$ *fixed* $k > \bar{k}_0 = \max\{k_0, k^*\}$ with $n_k > \bar{n}_0 = \max\{n_0, n_{\bar{k}_0}\}$, $\forall \varepsilon > 0$, $\|u - u_n\|_{L^p(\Omega)} \leq \|u - u_{n_k}\|_{L^p(\Omega)} + \|u_{n_k} - u_n\|_{L^p(\Omega)} < \frac{\varepsilon}{2} + \frac{\varepsilon}{2} = \varepsilon \ \forall n > \bar{n}_0$, i.e. $\lim_{n\to\infty} u_n = u$ in $L^p(\Omega)$. Hence, $L^p(\Omega)$ is a *complete normed linear space*, i.e. a *Banach space* for $1 \leq p < \infty$. Therefore the proof of the theorem is complete. □

Definition B.4.1.3. A sequence $(u_n)_{n=1}^{\infty}$ in $L^p(\Omega)$ *converges* (*strongly*) to $u \in L^p(\Omega)$ if and only if $\lim_{n\to\infty} \|u - u_n\|_{L^p(\Omega)} = 0$, $1 \le p \le \infty$.

Theorem B.4.1.3. *Let* $u_n \to u$ *in* $L^p(\Omega)$, $1 \le p \le \infty$, strongly *as* $n \to \infty$. *Then* $\exists$ *a subsequence* $(u_{n_k})_{k=1}^{\infty}$ *such that* $u_{n_k}(\mathbf{x}) \to u(\mathbf{x})$ pointwise a.e. *on* Ω *as* $k \to \infty$.

Imbedding results for bounded domains $\Omega \subset \mathbb{R}^n$ Let $\Omega \subset \mathbb{R}^n$ be a *bounded*, *open* subset with $\mu(\Omega) < +\infty$. Then the following results hold:

1. For $1 \le p \le q \le \infty$, $L^q(\Omega) \hookrightarrow L^p(\Omega)$, i.e. $L^q(\Omega) \subset L^p(\Omega)$ (*algebraic inclusion*) *and* $\exists C = C(\Omega) > 0$ such that

$$\|u\|_{L^p(\Omega)} \le C\|u\|_{L^q(\Omega)} \quad \forall u \in L^q(\Omega). \tag{B.4.1.11}$$

   In other words, the *imbedding* operator $\hookrightarrow$ is *continuous* from $L^q(\Omega)$ into $L^p(\Omega)$.

   In general, for $1 \le p \le q \le \infty$,

$$L^\infty(\Omega) \hookrightarrow L^q(\Omega) \hookrightarrow L^p(\Omega) \hookrightarrow L^1(\Omega). \tag{B.4.1.12}$$

2. If $u \in L^p(\Omega)$, $1 \le p < \infty$, and $\exists$ a constant $M > 0$ such that $\|u\|_{L^p(\Omega)} \le M$ $\forall p \in [1, \infty[$, then $u \in L^\infty(\Omega)$ and $\|u\|_{L^\infty(\Omega)} \le M$. (B.4.1.13)

Example B.4.1.1. Let $\Omega =]1, \infty[\subset \mathbb{R}$, $\mu(]1, \infty[) = +\infty$ and $f(x) = x^{-\alpha}$ $\forall x \in]1, \infty[$, $\alpha \in \mathbb{R}$. Then:

$$\text{Lebesgue integral} \int_{]1,\infty[} x^{-\alpha}\,dx = \text{improper Riemann integral} \int_1^\infty x^{-\alpha}\,dx$$
$$= \lim_{R\to\infty} \int_1^R x^{-\alpha}\,dx = \lim_{R\to\infty} \left(\frac{x^{-\alpha+1}}{-\alpha+1}\bigg|_{x=1}^{x=R} \right)$$
$$\begin{cases} < +\infty & \text{for } \alpha > 1 \\ = +\infty & \text{for } \alpha \le 1. \end{cases}$$

Hence, $x^{-\alpha} \in L^1(]1, \infty[)$ for $\alpha > 1$ and $x^{-\alpha} \notin L^1(]1, \infty[)$ for $\alpha \le 1$. (B.4.1.14)
Now we will find out for which values of α $x^{-\alpha} \in L^2(]1, \infty[)$. Again,

$$\text{Lebesgue integral} \int_{]1,\infty[} (x^{-\alpha})^2\,dx = \text{improper Riemann integral} \int_1^\infty x^{-2\alpha}\,dx$$
$$= \lim_{R\to\infty} \left(\frac{x^{-2\alpha+1}}{-2\alpha+1}\bigg|_{x=1}^{x=R} \right)$$
$$\begin{cases} < +\infty & \text{for } \alpha > 1/2; \\ = +\infty & \text{for } \alpha \le 1/2. \end{cases}$$

Hence, $x^{-\alpha} \in L^2(]1,\infty[)$ for $\alpha > 1/2$ and $x^{-\alpha} \notin L^2(]1,\infty[)$ for $\alpha \leq 1/2$. (B.4.1.15)

Combining, (B.4.1.14) and (B.4.1.15), we have

- for $\alpha \leq \frac{1}{2}$, $x^{-\alpha} \notin L^2(]1,\infty[)$ and $x^{-\alpha} \notin L^1(]1,\infty[)$;
- for $\frac{1}{2} < \alpha \leq 1$, $x^{-\alpha} \in L^2(]1,\infty[)$, but $x^{-\alpha} \notin L^1(]1,\infty[)$;
- for $\alpha > 1$, $x^{-\alpha} \in L^2(]1,\infty[)$ and $x^{-\alpha} \in L^1(]1,\infty[)$.

Thus, for *unbounded* $\Omega =]1,\infty[$, $L^1(]1,\infty[) \not\subset L^2(]1,\infty[)$, $L^2(]1,\infty[) \not\subset L^1(]1,\infty[)$, i.e. *no* inclusion results hold. (B.4.1.16)

B.4.2 Dual space $(L^p(\Omega))'$ of $L^p(\Omega)$ for $1 \leq p \leq \infty$

Let $1 \leq p, q \leq \infty$ with $\frac{1}{p} + \frac{1}{q} = 1$. Then, $\forall$ *fixed* $v \in L^q(\Omega)$, the mapping $l_v : u \in L^p(\Omega) \longmapsto l_v(u) \in \mathbb{R}$ defined by

$$l_v(u) = \int_\Omega u(\mathbf{x})v(\mathbf{x})d\mathbf{x} \tag{B.4.2.1}$$

is a *linear functional* on $L^p(\Omega)$, since by Hölder's Theorem B.4.1.1, $uv \in L^1(\Omega) \Longrightarrow$ *linear* functional $l_v : L^p(\Omega) \longrightarrow \mathbb{R}$ is *bounded* and, hence, *continuous* on $L^p(\Omega)$.

Let $v_1, v_2 \in L^q(\Omega)$. Then, define *continuous, linear* functionals l_1, l_2 corresponding to v_1 and v_2, respectively, by:

$$l_1(u) = \int_\Omega u(\mathbf{x})v_1(\mathbf{x})d\mathbf{x}, \quad l_2(u) = \int_\Omega u(\mathbf{x})v_2(\mathbf{x})d\mathbf{x} \quad \forall u \in L^p(\Omega). \tag{B.4.2.2}$$

Let $(L^p(\Omega))'$, $1 \leq p \leq \infty$, denote the *linear space* of all *continuous, linear functionals* l_v defined on the Banach space $L^p(\Omega)$ by (B.4.2.1) for all functions $v \in L^q(\Omega)$.

Definition B.4.2.1. For $1 \leq p \leq \infty$, the linear space $(L^p(\Omega))'$ of *continuous, linear* functionals on $L^p(\Omega)$ is called the *dual space* of $L^p(\Omega)$. Then $(L^p(\Omega))'$ is a *normed linear space* equipped with the norm $\|\cdot\|'_{(L^p(\Omega))'}$, or simply $\|\cdot\|'$, defined by:

$$\|l\|' = \|l\|_{(L^p(\Omega))'} = \sup_{\substack{u \in L^p(\Omega) \\ u \neq 0}} \frac{|l(u)|}{\|u\|_{L^p(\Omega)}} \geq \frac{|l(u)|}{\|u\|_{L^p(\Omega)}}$$

$$\Longrightarrow \quad |l(u)| \leq \|l\|'\|u\|_{L^p(\Omega)} \quad \forall u \in L^p(\Omega). \tag{B.4.2.3}$$

From (B.4.2.1) and (B.4.2.3), we find that

$$\|l_v\|' \leq \|v\|_{L^q(\Omega)} \quad \forall \text{ fixed } v \in L^q(\Omega), \tag{B.4.2.4}$$

since $\|l_v\|' = \inf\{M : |l_v(u)| \le M\|u\|_{L^p(\Omega)}\ \forall u \in L^p(\Omega)\}$. In fact, a strict *equality* holds in (B.4.2.4).

Define a mapping $J : L^q(\Omega) \to (L^p(\Omega))'$ by $J : v \in L^q(\Omega) \longmapsto Jv = l_v \in (L^p(\Omega))'$ with $l_v(u) = \int_\Omega uv d\mathbf{x}$. Then J is *linear, one-to-one, onto* $J(L^q(\Omega)) \subset (L^p(\Omega))'$ and J is *isometric* from $L^q(\Omega)$ onto $J(L^q(\Omega)) : \|Jv\|_{(L^p(\Omega))'} = \|l_v\|' = \|v\|_{L^q(\Omega)}$. Hence, J is an *isometric isomorphism* from $L^q(\Omega)$ onto the *subspace* $J(L^q(\Omega))$ of $(L^p(\Omega))'$. (B.4.2.5)

Theorem B.4.2.1 (Riesz Representation Theorem). *For* $1 \le p < \infty$, $\forall l \in (L^p(\Omega))'$, $\exists$ *a* unique $v_l \in L^q(\Omega)$ *with* $1 < q \le \infty$, $\frac{1}{p} + \frac{1}{q} = 1$, *such that*

$$l(u) = \int_\Omega u(\mathbf{x})v_l(\mathbf{x})d\mathbf{x}\ \forall u \in L^p(\Omega) \quad \textit{and} \quad \|l\|' = \|v_l\|_{L^q(\Omega)}. \tag{B.4.2.6}$$

Conversely, $\forall v \in L^q(\Omega)$, $\exists$ *a* unique $l_v \in (L^p(\Omega))'$ *such that*

$$l_v(u) = \int_\Omega u(\mathbf{x})v(\mathbf{x})d\mathbf{x}\ \forall u \in L^p(\Omega) \quad \textit{and} \quad \|l_v\|' = \|v\|_{L^q(\Omega)}. \tag{B.4.2.7}$$

Identification of $(L^p(\Omega))'$ with $L^q(\Omega)$

For $p \ne \infty$, by virtue of the Riesz Representation Theorem B.4.2.1, the mapping $J : u \in L^q(\Omega) \longmapsto l_v \in (L^p(\Omega))'$ is *linear, one-to-one, onto* and *isometric*, i.e. J is an *isometric isomorphism* from $L^q(\Omega)$ onto $(L^p(\Omega))' = J(L^q(\Omega))$. Hence, for $p \ne \infty$, $1 \le p < \infty$, $(L^p(\Omega))'$ is *identified with* $L^q(\Omega)$ by writing:

$(L^p(\Omega))' \equiv L^q(\Omega)$, $1 \le p < \infty$, $1 < q \le \infty$, $\frac{1}{p} + \frac{1}{q} = 1$ and $l_v = v$ $\forall v \in L^q(\Omega)$ such that

$$l_v(u) = \langle v, u\rangle_{L^q(\Omega)\times L^p(\Omega)} = \int_\Omega u(\mathbf{x})v(\mathbf{x})d\mathbf{x} \quad \forall u \in L^p(\Omega). \tag{B.4.2.8}$$

For $p = \infty$, the mapping $J : L^1(\Omega) \longrightarrow (L^\infty(\Omega))'$ is *not onto*, i.e. $J(L^1(\Omega)) \subsetneq (L^\infty(\Omega))'$. (B.4.2.9)

Hence, $(L^\infty(\Omega))'$ *cannot* be identified with $L^1(\Omega)$, i.e. $(L^\infty(\Omega))' \ne L^1(\Omega)$. (B.4.2.10)

In fact, $(L^\infty(\Omega))'$ is *larger* than $L^1(\Omega)$ (see [4]).

Reflexivity and separability of L^p-spaces

Definition B.4.2.2. $L^p(\Omega)$ is called *reflexive* if and only if $L^p(\Omega) \equiv (L^p(\Omega))'' = ((L^p(\Omega))')'$, i.e. $L^p(\Omega)$ can be *identified* with its *second dual* $(L^p(\Omega))''$.

Definition B.4.2.3. $L^p(\Omega)$ is called *separable* if and only if $\exists$ a *countable*, *dense* subset in $L^p(\Omega)$.

For the *reflexivity* and *separability* of L^p-spaces, we consider separately three different cases, for which we state the final results (see Table B.1):

1. $1 < p < \infty$, $(L^p(\Omega))' \equiv L^q(\Omega)$; $L^p(\Omega)$ is *reflexive* and *separable*. (B.4.2.11)
2. $p = 1$, $(L^1(\Omega))' \equiv L^\infty(\Omega)$; $L^1(\Omega)$ is *not* reflexive, but *separable*. (B.4.2.12)
3. $p = \infty$, $(L^\infty(\Omega))' \neq L^1(\Omega)$; $L^\infty(\Omega)$ is *neither* reflexive *nor* separable. (B.4.2.13)

Strong, weak, weak-* convergences in L^p-spaces

See Table B.2 for details of convergence in L^p-spaces.

Strong convergence in $L^p(\Omega)$, $1 \leq p \leq \infty$ $u_n \overset{\text{strongly}}{\to} u$ in $L^p(\Omega)$, $1 \leq p \leq \infty$

$$\Longrightarrow \quad \|u - u_n\|_{L^p(\Omega)} \to 0 \quad \text{as } n \to \infty. \tag{B.4.2.14}$$

Weak convergence in L^p-spaces Let $(u_k)_{k=1}^\infty$ be a sequence in $L^p(\Omega)$, $1 \leq p < \infty$.

Definition B.4.2.4. $u_k \overset{\text{weakly}}{\rightharpoonup} u$ in $L^p(\Omega) \Longleftrightarrow \int_\Omega u_k(\mathbf{x})v(\mathbf{x})d\mathbf{x} \to \int_\Omega u(\mathbf{x})v(\mathbf{x})d\mathbf{x}$ $\forall v \in L^q(\Omega)$ in $\mathbb{R}$ as $k \to \infty$, $1 < q \leq \infty$, $\frac{1}{p} + \frac{1}{q} = 1$.

In particular, $u_k \overset{\text{weakly}}{\rightharpoonup}$ in $L^1(\Omega)$

$$\Longleftrightarrow \quad \int_\Omega u_k(\mathbf{x})v(\mathbf{x})d\mathbf{x} \to \int_\Omega u(\mathbf{x})v(\mathbf{x})d\mathbf{x} \quad \forall v \in L^\infty(\Omega) \equiv (L^1(\Omega))' \tag{B.4.2.15}$$

in $\mathbb{R}$ as $k \to \infty$.

Strong convergence $\Longrightarrow$ weak convergence, *if defined*, but the converse is *false*. (B.4.2.16)

Weak convergence is *not* defined in $L^\infty(\Omega)$, since $(L^\infty(\Omega))' \neq L^1(\Omega)$. (B.4.2.17)

Weak−* convergence in $L^\infty(\Omega)$ Since $(L^p(\Omega))' = L^q(\Omega)$ for $1 \leq p < \infty$ with $1 < q \leq \infty$, $\frac{1}{p} + \frac{1}{q} = 1$, we can discuss *weak-* convergence* in $L^q(\Omega)$, i.e. $u_k \rightharpoonup^* u$ in $L^q(\Omega)$ (in the *weak-* sense*), $1 < q \leq \infty \Longleftrightarrow \langle u_k, v\rangle_{L^q(\Omega)\times L^p(\Omega)} \to \langle u, v\rangle_{L^q(\Omega)\times L^p(\Omega)}$ $\forall v \in L^p(\Omega)$, $1 \leq p < \infty$, as $k \to \infty$, where $\langle\cdot,\cdot\rangle_{L^q(\Omega)\times L^p(\Omega)}$ is the *duality pairing* between $L^q(\Omega) \times L^p(\Omega)$, $1 \leq p < \infty$, $1 < q \leq \infty$, $\frac{1}{p} + \frac{1}{q} = 1$. (B.4.2.18)

Since for $1 < q < \infty$, $L^q(\Omega)$ is *reflexive* by (B.4.2.11), weak-* convergence and weak convergence *coincide* for $1 < q < \infty$.

But for $q = \infty$, weak convergence is *not* defined in $L^\infty(\Omega)$, whereas weak-* convergence in $L^\infty(\Omega)$ *is* meaningful and defined by (B.4.2.18), i.e.

$$u_k \rightharpoonup^* u \text{ in } L^\infty(\Omega) \iff \int_\Omega u_k(\mathbf{x})v(\mathbf{x})d\mathbf{x} \to \int_\Omega u(\mathbf{x})v(\mathbf{x}) \quad \forall v \in L^1(\Omega) \tag{B.4.2.19}$$

in $\mathbb{R}$ as $k \to \infty$, where

$$\langle u_k, v\rangle_{L^\infty(\Omega)\times L^1(\Omega)} = \int_\Omega u_k(\mathbf{x})v(\mathbf{x})d\mathbf{x};$$
$$\langle u, v\rangle_{L^\infty(\Omega)\times L^1(\Omega)} = \int_\Omega u(\mathbf{x})v(\mathbf{x})d\mathbf{x} \quad \forall v \in L^1(\Omega).$$

See Tables B.1 and B.2 for the properties of L^p-spaces.

B.4.3 Space $L^2(\Omega)$

Owing to the Hilbert space structure of $L^2(\Omega)$ by virtue of the inner product $\langle\cdot,\cdot\rangle_{L^2(\Omega)}$:

$$\langle u, v\rangle_{L^2(\Omega)} = \int_\Omega u(\mathbf{x})v(\mathbf{x})d\mathbf{x}$$
$$(\text{resp. } \langle u, v\rangle_{L^2(\Omega)} = \int_\Omega u(\mathbf{x})\overline{v(\mathbf{x})}d\mathbf{x}) \quad \forall u, v \in L^2(\Omega), \tag{B.4.3.1a}$$

there are very important additional properties in $L^2(\Omega)$ which are listed below (see Table B.3 for details).

Property 1: Cauchy–Schwarz inequality $|\langle u, v\rangle_{L^2(\Omega)}| \le \|u\|_{L^2(\Omega)}\|v\|_{L^2(\Omega)}$ $\forall u, v \in L^2(\Omega)$, i.e.

$$\left|\int_\Omega u(\mathbf{x})v(\mathbf{x})d\mathbf{x}\right| \le \left(\int_\Omega |u(\mathbf{x})|^2dx\right)^{1/2}\left(\int_\Omega |v(\mathbf{x})|^2dx\right)^{1/2}$$
$$(\text{resp. } \left|\int_\Omega u(\mathbf{x})\overline{v(\mathbf{x})}d\mathbf{x}\right| \le \left(\int_\Omega |u(\mathbf{x})|^2dx\right)^{1/2}\left(\int_\Omega |v(\mathbf{x})|^2dx\right)^{1/2}). \tag{B.4.3.1}$$

Property 2

Theorem B.4.3.1 (Riesz Representation Theorem). *Let l be a* continuous *linear functional on $L^2(\Omega)$. Then $\exists$ a unique $u_l \in L^2(\Omega)$ such that*

$$l(v) = \langle v, u_l\rangle_{L^2(\Omega)} \; \forall v \in L^2(\Omega) \quad \textit{with } \|l\|' = \|u_l\|_{L^2(\Omega)}. \tag{B.4.3.2}$$

Conversely, $\forall v \in L^2(\Omega)$, $\exists$ *a unique, continuous, linear functional* l_v *on* $L^2(\Omega)$ *such that*

$$l_v(u) = \langle u, v\rangle_{L^2(\Omega)} \ \forall u \in L^2(\Omega) \quad \textit{with } \|l_v\|' = \|v\|_{L^2(\Omega)}. \tag{B.4.3.3}$$

Consequently, $(L^2(\Omega)) = (L^2(\Omega))'$ *for* real *Hilbert space* $L^2(\Omega)$ *(see Remark* A.13.1.1*).* (B.4.3.4)

Property 3 $L^2(\Omega)$ is *reflexive* and *separable*. (B.4.3.5)

Property 4: Weak convergence in $L^2(\Omega)$ $u_k \overset{\text{weakly}}{\rightharpoonup} u$ in $L^2(\Omega) \Longleftrightarrow \langle u_k, v\rangle_{L^2(\Omega)} \to \langle u, v\rangle_{L^2(\Omega)}$ $\forall v \in L^2(\Omega)$ in $\mathbb{R}$ as $k \to \infty$, i.e.

$$\left|\int_\Omega u(\mathbf{x})v(\mathbf{x})d\mathbf{x} - \int_\Omega u_k(\mathbf{x})v(\mathbf{x})d\mathbf{x}\right| = \left|\int_\Omega (u(\mathbf{x}) - u_k(\mathbf{x}))\cdot v(\mathbf{x})d\mathbf{x}\right| \to 0 \tag{B.4.3.6}$$

as $k \to \infty$ $\forall v \in L^2(\Omega)$. See also Table B.2 for weak convergence.

Property 5: Orthogonality in $L^2(\Omega)$ $u, v \in L^2(\Omega)$ with $u \neq 0, v \neq 0$ are *orthogonal* if and only if

$$\langle u, v\rangle_{L^2(\Omega)} = \int_\Omega u(\mathbf{x})v(\mathbf{x})d\mathbf{x} = 0 \tag{B.4.3.7}$$

(resp. $\langle u, v\rangle_{L^2(\Omega)} = \int_\Omega u(\mathbf{x})\overline{v(\mathbf{x})}d\mathbf{x} = 0$ in a *complex* Hilbert space $L^2(\Omega)$).

Interesting geometric properties of $L^2(\Omega)$ See Table B.3.

B.4.4 Some negative properties of $L^\infty(\Omega)$

See also Tables B.1 and B.2.

1. $(L^\infty(\Omega))'$ *cannot* be identified with $L^1(\Omega)$, i.e. $L^1(\Omega) \subsetneq (L^\infty(\Omega))'$. (B.4.4.1)
2. $L^\infty(\Omega)$ is *neither* reflexive *nor* separable. (B.4.4.2)
3. $C_0(\Omega), C_0^\infty(\Omega)$ are *not* dense in $L^\infty(\Omega)$. (B.4.4.3)
4. $C^0(\overline{\Omega})$ is a *closed* (but *not* dense) subspace of $L^\infty(\Omega)$ with $C^0(\overline{\Omega}) \subsetneq L^\infty(\Omega)$. (B.4.4.4)

B.4.5 Some nice properties of $L^\infty(\Omega)$

See also Tables B.1 and B.2.

1. The *closed unit ball* $\overline{B}(0;1)$ in $L^\infty(\Omega)$ is *compact* in *weak-* topology*. (B.4.5.1)
2. Every *bounded* sequence $(v_n)_{n=1}^\infty$ in $L^\infty(\Omega)$ contains a subsequence $(v_{n_k})_{k=1}^\infty$ which converges to an element $v \in L^\infty(\Omega)$ in the *weak-* sense*, i.e. $v_{n_k} \to^* v$ in $L^\infty(\Omega)$ in the weak-* sense as $k \to \infty$

$$\iff \quad \lim_{k\to\infty} \int_\Omega v_{n_k}(\mathbf{x})u(\mathbf{x})d\mathbf{x} = \int_\Omega v(\mathbf{x})u(\mathbf{x})d\mathbf{x} \quad \forall u \in L^1(\Omega). \quad \text{(B.4.5.2)}$$

B.4.6 Space $L^p_{\text{loc}}(\Omega)$ inclusion results

Definition B.4.6.1. For $1 \le p \le \infty$,

$$L^p_{\text{loc}}(\Omega) = \{u : \forall\ \textit{compact}\ K \subset\subset \Omega,\ u\downarrow_{\mathring{K}} \in L^p(\mathring{K}),\ \mathring{K} = \text{int}(K)\}. \quad \text{(B.4.6.1)}$$

$\forall$ open sets Ω, for $1 \le p \le q \le \infty$, the following inclusions hold by virtue of imbedding results (B.4.1.11) and (B.4.1.12) (see also Table B.1 for imbedding results).

1. $L^p(\Omega) \subset L^p_{\text{loc}}(\Omega)$; (B.4.6.2)
2. $C^0(\overline{\Omega}) \subset L^\infty_{loc}(\Omega) \subset L^q_{\text{loc}}(\Omega) \subset L^p_{\text{loc}}(\Omega) \subset L^1_{\text{loc}}(\Omega)$.

Table B.1 Important properties of $L^p(\Omega)$-space, $1 \le p \le \infty$

p	$1 < p < \infty, \quad L^p(\Omega)$	$p = 1, \quad L^1(\Omega)$	$p = \infty, \quad L^\infty(\Omega)$
Banach space	Banach space	Banach space	Banach space
Hilbert space	$L^2(\Omega)$ is a Hilbert space *only for* $p = 2$	*Not* a Hilbert space	*Not* a Hilbert space
Hölder's inequality	• For arbitrary $\Omega \subset \mathbb{R}^n$ and $1 \le p, q \le \infty$ with $\frac{1}{p} + \frac{1}{q} = 1$, $u \in L^p(\Omega)$, $v \in L^q(\Omega)$ $\Longrightarrow uv \in L^1(\Omega)$ and $\int_\Omega \lvert uv \rvert \, d\mathbf{x} \le \lVert u \rVert_{L^p(\Omega)} \lVert v \rVert_{L^q(\Omega)}$; • For $p = q = 2$, Hölder's inequality becomes Cauchy–Schwarz inequality: $\lvert \langle u, v \rangle_{L^2(\Omega)} \rvert \le \lVert u \rVert_{L^2(\Omega)} \lVert v \rVert_{L^2(\Omega)}$ $\forall u, v \in L^2(\Omega)$.		
Imbedding results	• For *bounded* $\Omega \subset \mathbb{R}^n$, for $1 \le p \le q \le \infty$, $L^\infty(\Omega) \hookrightarrow L^q(\Omega) \hookrightarrow L^p(\Omega) \hookrightarrow L^1(\Omega)$ with continuous imbedding $\hookrightarrow$: $\lVert u \rVert_{L^q(\Omega)} \le C \lVert u \rVert_{L^p(\Omega)}$ $\forall u \in L^q(\Omega)$for *some* $C > 0$; • For *unbounded* $\Omega \subset \mathbb{R}^n$ (i.e. $\mathrm{vol}(\Omega) = \infty$), $L^2(\Omega) \not\subset L^1(\Omega)$, $L^1(\Omega) \not\subset L^2(\Omega)$.		
Dual space $(L^p(\Omega))'$ of $L^p(\Omega)$	$(L^p(\Omega))' \equiv L^q(\Omega)$, $\frac{1}{p} + \frac{1}{q} = 1$	$(L^1(\Omega))' \equiv L^\infty(\Omega)$	$(L^\infty(\Omega))' \not\equiv L^1(\Omega)$, $L^1(\Omega) \subsetneq (L^\infty(\Omega))'$,
Riesz Representation Theorem	$\forall u \in L^p(\Omega)$, $\forall v \in L^q(\Omega)$, $1 < p, q < \infty$, $\frac{1}{p} + \frac{1}{q} = 1$, $L_u(v) = \langle u, v \rangle_{L^p(\Omega) \times L^q(\Omega)} = \int_\Omega uv \, d\mathbf{x}$	$\forall u \in L^1(\Omega)$, $\forall v \in L^\infty(\Omega)$, $L_u(v) = \langle u, v \rangle_{L^1(\Omega) \times L^\infty(\Omega)} = \int_\Omega uv \, d\mathbf{x}$	Does *not* hold
Density	• $\mathbb{C}_0^0(\Omega), C_0^k(\Omega), \mathbb{C}_0^\infty(\Omega)$ are *dense* in $L^p(\Omega)$ $\forall$ open subsets $\Omega \subset \mathbb{R}^n$; • $\mathbb{C}^0(\bar{\Omega}), C^k(\overline{\Omega}), \mathbb{C}^\infty(\bar{\Omega})$ are *dense* in $L^p(\Omega)$ $\forall$ *compact* subsets $\overline{\Omega} \subset\subset \mathbb{R}^n$.	• $\mathbb{C}_0^0(\Omega), C_0^k(\Omega), \mathbb{C}_0^\infty(\Omega)$ are *dense* in $L^1(\Omega)$ $\forall$ open subsets $\Omega \subset \mathbb{R}^n$; • $\mathbb{C}^0(\bar{\Omega}), C^k(\overline{\Omega}), \mathbb{C}^\infty(\bar{\Omega})$ are *dense* in $L^1(\Omega)$ $\forall$ *compact* subsets $\overline{\Omega} \subset\subset \mathbb{R}^n$.	• $\mathbb{C}_0^0(\Omega), \mathbb{C}_0^\infty(\Omega)$ *not* dense in $L^\infty(\Omega)$; • $\mathbb{C}^0(\bar{\Omega}), \mathbb{C}^\infty(\bar{\Omega})$ *not* dense in $L^\infty(\Omega)$; • $\mathbb{C}^0(\bar{\Omega})$ is a *closed* subspace of $L^\infty(\Omega)$.
Reflexivity	*Reflexive* $(L^p(\Omega))'' = (L^q(\Omega))' = L^p(\Omega)$	*Not reflexive* $(L^1(\Omega))'' = (L^\infty(\Omega))' \not\equiv L^1(\Omega)$	*Not reflexive* $(L^\infty(\Omega))'' \not\equiv (L^\infty(\Omega))'$
Separability	*Separable* $\exists$ a countable dense subset in $L^p(\Omega)$ $\forall$ compact subsets $\overline{\Omega} \subset\subset \mathbb{R}^n$.	*Separable* $\exists$ a countable dense subset in $L^1(\Omega)$ $\forall$ compact subsets $\overline{\Omega} \subset\subset \mathbb{R}^n$.	*Not separable* $\nexists$ a countable dense subset in $L^\infty(\Omega)$

Table B.2 Different notions of convergence in $L^p(\Omega)$, $1 \le p \le \infty$

p	$1 < p < \infty, \quad L^p(\Omega)$	$p = 1, \quad L^1(\Omega)$	$p = \infty, \quad L^\infty(\Omega)$
Strong convergence ($\to$) $u_n \to u$ as $n \to \infty$	$u_n \to u$ in $L^p(\Omega)$ as $n \to \infty$ $\iff \|u - u_n\|^p_{L^p(\Omega)} = \int_\Omega \lvert u - u_n\rvert^p \, d\mathbf{x} \to 0$	$u_n \to u$ in $L^1(\Omega)$ as $n \to \infty$ $\iff \int_\Omega \lvert u - u_n\rvert \, d\mathbf{x} \to 0$	$u_n \to u$ in $L^\infty(\Omega)$ as $n \to \infty$ $\iff \operatorname{supp} \lvert u(\mathbf{x}) - u_n(\mathbf{x})\rvert \to 0$
Weak convergence ($\rightharpoonup$) $u_n \rightharpoonup u$ as $n \to \infty$	$u_n \rightharpoonup u$ in $L^p(\Omega)$ as $n \to \infty$ $\iff \int_\Omega u_n v \, d\mathbf{x} \to \int_\Omega uv \, d\mathbf{x}$ in $\mathbb{R}$ (resp. $\mathbb{C}$) $\forall v \in L^q(\Omega), \quad \frac{1}{p} + \frac{1}{q} = 1$	$u_n \rightharpoonup u$ in $L^1(\Omega)$ as $n \to \infty$ $\iff \int_\Omega u_n v \, d\mathbf{x} \to \int_\Omega uv \, d\mathbf{x}$ in $\mathbb{R}$ (resp. $\mathbb{C}$) $\forall v \in L^\infty(\Omega)$	*No weak convergence,* since $(L^\infty(\Omega))' \not\equiv L^1(\Omega)$
Weak-* convergence ($\rightharpoonup^*$) $u_n \rightharpoonup^* u$ as $n \to \infty$	$u_n \rightharpoonup^* u$ in $L^p(\Omega)$ as $n \to \infty$ $\iff u_n \rightharpoonup u$ in $L^p(\Omega)$ as $n \to \infty$, since $L^p(\Omega)$ is *reflexive* $\iff \int_\Omega u_n v \, d\mathbf{x} \to \int_\Omega uv \, d\mathbf{x}$ in $\mathbb{R}$ (resp. $\mathbb{C}$) $\forall v \in L^q(\Omega), \quad \frac{1}{p} + \frac{1}{q} = 1$	*No weak-* convergence,* since $L^1(\Omega) \not\equiv (L^\infty(\Omega))'$	$u_n \rightharpoonup^* u$ in $L^\infty(\Omega)$ as $n \to \infty$ $\iff \int_\Omega u_n v \, d\mathbf{x} \to \int_\Omega uv \, d\mathbf{x}$ in $\mathbb{R}$ (resp. $\mathbb{C}$) $\forall v \in L^1(\Omega)$

Table B.3 Important properties of Hilbert space $L^2(\Omega)$

Inner product $\langle\cdot,\cdot\rangle_{L^2(\Omega)}$	• $\langle f,g\rangle_{L^2(\Omega)} = \begin{cases} \int_\Omega f(\mathbf{x})\overline{g(\mathbf{x})}\,d\mathbf{x} & \text{for complex-valued functions } f \text{ and } g; \\ \int_\Omega f(\mathbf{x})g(\mathbf{x})\,d\mathbf{x} & \text{for real-valued functions } f \text{ and } g. \end{cases}$
Norm $\lVert\cdot\rVert_{L^2(\Omega)}$	• $\lVert f\rVert_{L^2(\Omega)} = \langle f,f\rangle^{\frac{1}{2}}_{L^2(\Omega)} = (\int_\Omega \lvert f(\mathbf{x})\rvert^2\,d\mathbf{x})^{\frac{1}{2}}\ \forall f \in L^2(\Omega)$; • $\hat{f} = \frac{f}{\lVert f\rVert_{L^2(\Omega)}}$ with $\lVert\hat{f}\rVert = 1$.
Function $\hat{f}$ **with unit norm**	• $\lVert\hat{f}\rVert_{L^2(\Omega)} = 1,\ \hat{f} = \frac{f}{\lVert f\rVert_{L^2(\Omega)}}$.
Cauchy–Schwarz inequality	• $\lvert\langle f,g\rangle_{L^2(\Omega)}\rvert \le \lVert f\rVert_{L^2(\Omega)}\lVert g\rVert_{L^2(\Omega)}\ \forall f,g \in L^2(\Omega)$
Orthogonality, orthogonal system $(x_i)_{i=1}^n$	• $f\perp g$ in $L^2(\Omega) \Longleftrightarrow f$ *orthogonal* to g in $L^2(\Omega) \Longleftrightarrow \langle f,g\rangle_{L^2(\Omega)} = 0$ • $(x_i)_{i=1}^n$ is an *orthogonal system* in $L^2(\Omega)$ $\Longleftrightarrow \langle x_i,x_j\rangle_{L^2(\Omega)} = 0\ \forall i \ne j,\ 1 \le i,j \le n$
Orthonormal system $(\hat{x}_i)_{i=1}^n$	• $(\hat{x}_i)_{i=1}^n$ is an *orthonormal system* (o.n.s.) in $L^2(\Omega)$ $\Longleftrightarrow \langle\hat{x}_i,\hat{x}_j\rangle = \delta_{ij} = \begin{cases} 1 & \text{for } i=j \\ 0 & \text{for } i \ne j \end{cases}\ 1 \le i,j \le n$; • $(\hat{x}_i)_{i=1}^\infty$ is a *complete* o.n.s. in $L^2(\Omega)$ $\Longleftrightarrow \langle f,\hat{x}_i\rangle = 0\ \forall i \in \mathbb{N} \Longrightarrow f = 0$ in $L^2(\Omega)$; • $(\hat{x}_i)_{i=1}^\infty$ is a *closed* o.n.s. in $L^2(\Omega)$ $\Longleftrightarrow \lVert f\rVert^2 = \sum_{i=1}^\infty \lvert\langle f,\hat{x}_i\rangle\rvert^2\ \forall f \in L^2(\Omega)$
$M \subset L^2(\Omega)$ is a **closed** subspace, $\mathbf{x}\perp M$	$\mathbf{x}\perp \mathrm{M} \Longleftrightarrow \langle\mathbf{x},\mathbf{y}\rangle_{L^2(\Omega)} = 0\quad \forall\mathbf{y} \in M$
Orthogonal complement $M^\perp$ of a *closed* subspace $M \subset L^2(\Omega)$	• $M^\perp = \{\mathbf{y} : \mathbf{y}\perp M,\ M$ is a closed subspace of $L^2(\Omega)\}$; • $M \cap M^\perp = \{\mathbf{0}\}$; • $L^2(\Omega) = M \oplus M^\perp$ such that $\forall\mathbf{x} \in L^2(\Omega),\ \mathbf{x} = \mathbf{y} + \mathbf{z}$ with $\mathbf{y} \in M,\ \mathbf{z} \in M^\perp$.
Orthogonal projection operator $P_M : L^2(\Omega) \to M$	• M is a *closed* subspace of $L^2(\Omega)$; • $P_M : L^2(\Omega) \to M$ with $P_M(L^2(\Omega)) = M$ is a *linear, continuous* operator called a *projection operator* from $L^2(\Omega)$ *onto* M, *range* $\mathcal{R}(P_M)$ of $P_M = M$; • $\mathrm{Ker}(P_M) = M^\perp,\ \mathcal{R}(P_M) = M,\ \lVert P_M\rVert < 1,\ P_M^2 = P_M P_M = P_M$
Isometric isomorphism	• $L^2(\Omega)$ is *isometrically isomorphic* to $l^2 : (\hat{x}_k)_{k=1}^\infty$ is a complete o.n.s in $L^2(\Omega)$; • $f \in L^2(\Omega) \Longleftrightarrow (c_1,c_2,\ldots,c_k,\ldots) = (c_k)_{k=1}^\infty \in l^2$ with $c_k \in \mathbb{C}$, $f = \sum\langle f,\hat{x}_k\rangle_{L^2(\Omega)}\hat{x}_k;\ \lVert f\rVert_{L^2(\Omega)} = \lVert(c_k)_{k=1}^\infty\rVert_{l^2} = (\sum_{k=1}^\infty \lvert c_k\rvert^2)^{\frac{1}{2}} < +\infty$; • $g \in L^2(\Omega) \Longleftrightarrow (d_1,d_2,\ldots,d_k,\ldots) = (d_k)_{k=1}^\infty \in l^2$ with $d_k \in \mathbb{C}$; $g = \sum\langle g,\hat{x}_k\rangle_{L^2(\Omega)}\hat{x}_k;\ \lVert g\rVert_{L^2(\Omega)} = \lVert(d_k)_{k=1}^\infty\rVert_{l^2} = (\sum_{k=1}^\infty \lvert d_k\rvert^2)^{\frac{1}{2}} < +\infty$; • $\langle f,g\rangle_{L^2(\Omega)} = \langle(c_k)_{k=1}^\infty,(d_k)_{k=1}^\infty\rangle_{l^2} = \sum_{k=1}^\infty c_k\bar{d}_k$.
Riesz Representation Theorem	• $\forall$ continuous linear functionals $l : L^2(\Omega) \longrightarrow \mathbb{R},\ \exists$ a unique $u_l \in L^2(\Omega)$ such that $l(v) = \langle v,u_l\rangle_{L^2(\Omega)}\ \forall v \in L^2(\Omega)$; with $\lVert l\rVert_{(L^2(\Omega))'} = \lVert u_l\rVert_{L^2(\Omega)}$; • $\forall u \in L^2(\Omega),\ \exists$ a unique continuous linear functional $l_u : L^2(\Omega) \longrightarrow \mathbb{R}$ such that $l_u(v) = \langle u,v\rangle_{L^2(\Omega)}\ \forall v \in L^2(\Omega)$; with $\lVert l_u\rVert_{(L^2(\Omega))'} = \lVert u\rVert_{L^2(\Omega)}$.

Appendix C
Open cover and partition of unity

C.1 $C_0^\infty(\Omega)$-partition of unity theorem for compact sets

In many proofs, it is necessary to write test function $\phi \in C_0^\infty(\Omega)$ as the *sum* of test functions $\{\phi_i\}$ with *smaller compact supports*. For this, we need a partition of *unity* (i.e. 1) over a *compact* set. For basic results on the topological properties of compact sets K in $\mathbb{R}^n$, see Section A.0.5 in Appendix A.

Open cover or covering of a subset of $\mathbb{R}^n$

Definition C.1.1.1. Let $\Omega \subset \mathbb{R}^n$ be an open subset of $\mathbb{R}^n$ and $\{\Omega_i\}_{i\in I}$ be a collection of *open* subsets $\Omega_i \subset \mathbb{R}^n$ $\forall i \in I$, I being a set of indices.

Then $\{\Omega_i\}_{i\in I}$ is called an *open cover* or *covering* of Ω iff

$$\Omega \subset \bigcup_{i\in I} \Omega_i. \tag{C.1.1.1}$$

Finite open cover of a subset $\Omega \subset \mathbb{R}^n$ If I contains a finite number of indices $i_1, i_2, \ldots, i_m$ for which (C.1.1.1) holds, i.e. $I = \{i_1, i_2, \ldots, i_m\}$ *and* $\Omega \subset \bigcup_{k=1}^m \Omega_{i_k}$, then $\{\Omega_{i_k}\}_{k=1}^m$ is called a *finite open cover* or *covering* of Ω. Otherwise, $\{\Omega_i\}_{i\in I}$ is called an *infinite open cover* or *covering* of Ω. (C.1.1.2)

K is compact in $\mathbb{R}^n$ $\Longrightarrow$ from any *infinite* open cover $\{\Omega_i\}_{i\in I}$ of K, a *finite* open cover $\{\Omega_{i_k}\}_{k=1}^m$ of K can be extracted, i.e. $K \subset \bigcup_{k=1}^m \Omega_{i_k}$. (C.1.1.3)

$\mathbb{R}^n$ is *not* compact, but $\mathbb{R}^n$ is *locally* compact. Hence, from every infinite open cover of $\mathbb{R}^n$, we can extract a *locally finite cover* of $\mathbb{R}^n$.

Definition C.1.1.2. A collection $\{\Omega_i\}_{i\in I}$ of open subsets $\Omega_i \subset \mathbb{R}^n$ is called a *locally finite open cover* iff, $\forall \mathbf{x} \in \mathbb{R}^n$, $\exists$ a compact neighbourhood $\mathcal{N}(\mathbf{x})$ with $\mathcal{N}(\mathbf{x}) \subset \bigcup_{k=1}^m \Omega_{i_k}$, $i_k \in I$ for $1 \le k \le m$.

Let $\{\Omega_i\}_{i\in I}$ be an open cover of a set $A \subset \mathbb{R}^n$. Then the open cover $\{\Omega_i\}_{i\in I}$ is called *locally finite* if and only if any *compact* set $K \subset\subset A$ can intersect *at most* finitely many elements of $\{\Omega_i\}_{i\in I}$.

If $A \subset \mathbb{R}^n$ is a *closed* subset of $\mathbb{R}^n$, then any *open cover* of A possesses a *locally finite subcover*. (C.1.1.3a)

In general, we will require the partition of unity for *compact sets* in many proofs. Hence, we will just state the result for *arbitrary* sets without proof, which can be found, for example, in [11], [12].

Theorem C.1.1.1 (Partition of unity for an arbitrary set). *Let $\Omega \subset \mathbb{R}^n$ be an arbitrary subset of $\mathbb{R}^n$ and $\{\Omega_i\}_{i\in I}$ be a collection of open subsets of $\mathbb{R}^n$ which cover Ω, I being a set of indices, i.e. $\Omega \subset \bigcup_{i\in I} \Omega_i$. Then $\exists$ a collection $\{\phi_i\}_{i\in I}$ of functions ϕ_i with the following properties:*

I. $\forall i \in I$, $0 \leq \phi_i(\mathbf{x}) \leq 1$ $\forall \mathbf{x} \in \mathbb{R}^n$; $\phi_i \in C_0^\infty(\mathbb{R}^n)$;

II. $\text{supp}(\phi_i) \subset \Omega_i$ $\forall i \in I$ *(I and II* $\Longrightarrow \phi_i \in \mathcal{D}(\Omega_i)$ $\forall i \in I$*);* (C.1.1.4)

III. $\forall \mathbf{x} \in \Omega$, $\sum_{i\in I} \phi_i(\mathbf{x}) \equiv 1$.

The collection $\{\phi_i\}_{i\in I}$ is called a C_0^∞-partition of unity for Ω subordinate to $\{\Omega_i\}_{i\in I}$.

Theorem C.1.1.2 (C_0^∞-partition of unity for a compact set). *Let $K \subset\subset \mathbb{R}^n$ be a compact subset of $\mathbb{R}^n$ and $\{\Omega_i\}_{i=1}^N$ be a finite open cover of K, i.e. each Ω_i is an open subset of $\mathbb{R}^n$ and $K \subset \bigcup_{i=1}^N \Omega_i$. Then $\exists \phi_i \in C_0^\infty(\Omega_i)$, $1 \leq i \leq N$, such that:*

I. $0 \leq \phi_i(\mathbf{x}) \leq 1$ $\forall \mathbf{x} \in \Omega_i$ *and* $\sum_{i=1}^N \phi_i(\mathbf{x}) \equiv 1$ $\forall \mathbf{x} \in K$. (C.1.1.5)

$\{\phi_i\}_{i=1}^N$ *is called a* $\mathbf{C_0^\infty}$-partition of unity for compact set K subordinate to the open cover $\{\Omega_i\}_{i=1}^N$ of **K**. (C.1.1.6)

II. *Moreover, for $\phi \in C_0^\infty(\Omega)$ with* $\text{supp}(\phi) = K \subset\subset \Omega$,

$\exists\{\chi_i\}_{i=1}^N$ *with $\chi_i \in C_0^\infty(\Omega_i)$, $1 \leq i \leq N$, such that $\phi = \sum_{i=1}^N \chi_i$.* (C.1.1.7)

III. *Finally, for $\phi \in C_0^\infty(\Omega)$ with $\phi \geq 0$ and* $\text{supp}(\phi) = K \subset\subset \Omega$, $\exists \chi_i \in C_0^\infty(\Omega_i)$ *with $\chi_i \geq 0$, $1 \leq i \leq N$, such that $\phi = \sum_{i=1}^N \chi_i$, i.e. $\{\chi_i\}_{i=1}^N$ in* (C.1.1.7) *is a C_0^∞-partition of $\phi \in C_0^\infty(\Omega)$ subordinate to the open cover $\{\Omega_i\}_{i=1}^N$ of the compact support K of ϕ.* (C.1.1.8)

Proof. We will follow the proof in [39] with minor modifications (see also [5]). For compact set $K \subset\subset \mathbb{R}^n$, $\exists$ a finite *open cover* $\{\Omega_i\}_{i=1}^N$ of open subsets $\Omega_i \subset \mathbb{R}^n$, i.e. $K \subset \bigcup_{i=1}^N \Omega_i$. From this open cover $\{\Omega\}_{i=1}^N$ of K, we will construct a *compact* cover $\{K_i\}_{i=1}^N$ of K, $K_i \subset\subset \Omega_i$ being a compact subset of Ω_i, $1 \leq i \leq N$.

Construction of compact cover of K: Let $\mathbf{x} \in K$. Then $\exists i$, $1 \leq i \leq N$, such that $\mathbf{x} \in \Omega_i$, $\partial\Omega_i$ being the boundary of Ω_i. Define the distance function $d(\mathbf{x}) = d(\mathbf{x}, \partial\Omega_i) = \inf_{\mathbf{y}\in\partial\Omega_i} d(\mathbf{x}, \mathbf{y}) > 0$. Let $B(\mathbf{x}; r(\mathbf{x}))$ be an open ball with centre at $\mathbf{x}$ and radius $r = r(\mathbf{x})$ such that $0 < r(\mathbf{x}) < d(\mathbf{x})$ and $B(\mathbf{x}; r(\mathbf{x})) \subset \Omega_i$.

Then $K \subset \bigcup_{\mathbf{x}\in K} B(\mathbf{x}; r(\mathbf{x}))$. But $K \subset\subset \mathbb{R}^n$ is a compact set. Hence, we can find a finite number of balls $B(\mathbf{x}_j; r(\mathbf{x}_j))$ with centres at $\mathbf{x}_j$ and radii $r(\mathbf{x}_j)$, $1 \leq j \leq m$, such that $K \subset \bigcup_{j=1}^m B(\mathbf{x}_j; r(\mathbf{x}_j))$. Let $\overline{B}(\mathbf{x}_j; r(\mathbf{x}_j))$ be the closed ball corresponding to $B(\mathbf{x}_j; r(\mathbf{x}_j))$, $1 \leq j \leq m$.

Define compact sets $K_i = \bigcup_{\mathbf{x}_j \in \Omega_i, 1 \le j \le m} \overline{B}(\mathbf{x}_j; r(\mathbf{x}_j)) \subset \Omega_i$, $1 \le i \le N$. Then $K \subset \bigcup_{i=1}^{N} K_i$ with $K_i \subset \Omega_i$, $1 \le i \le N$. (C.1.1.9)

Construction of two auxiliary compact sets K_i^* and $\tilde{K}_i$ $\forall$ fixed i: With every $i = 1, 2, \dots, N$, we associate two auxiliary compact sets K_i^*, $\tilde{K}_i \subset\subset \mathbb{R}^n$ such that

$$K_i \subset \mathring{K}_i^* \subset K_i^* \subset \mathring{\tilde{K}}_i \subset \tilde{K}_i \subset \Omega_i, \tag{C.1.1.10}$$

since we can always construct such compact sets, and $\mathring{K}_i = \operatorname{int}(K_i)$, $\mathring{K}_i^* = \operatorname{int}(K_i^*)$, $\mathring{\tilde{K}} = \operatorname{int}(\tilde{K}_i)$, $K_i = \mathring{K}_i \cup \partial K_i$, $K_i^* = \mathring{K}_i^* \cup \partial K_i^*$, $\tilde{K}_i = \mathring{\tilde{K}}_i \cup \partial \tilde{K}_i$, ∂K_i, ∂K_i^* and $\partial \tilde{K}_i$ being the boundaries of K_i, K_i^* and $\tilde{K}_i$, respectively, $\forall i = 1, 2, \dots, N$.

Set

$$d_i = d(\partial K_i, \partial K_i^*) = \inf_{\substack{\mathbf{x} \in \partial K_i \\ \mathbf{y} \in \partial K_i^*}} d(\mathbf{x}, \mathbf{y}) > 0,$$

$$d_i^* = d(\partial K_i^*, \partial \tilde{K}_i) = \inf_{\substack{\mathbf{x} \in \partial K_i^* \\ \mathbf{y} \in \partial \tilde{K}_i}} d(\mathbf{x}, \mathbf{y}) > 0,$$

$$d^* = \min_{1 \le i \le N} \{d_i\} > 0, \quad \tilde{d} = \min_{1 \le i \le N} \{d_i^*\} > 0.$$

Define

$$d = \min\{d^*, \tilde{d}\} > 0. \tag{C.1.1.11}$$

Let χ_i^* be the *characteristic function* of K_i^* defined by $\chi_i^*(\mathbf{x}) = 1$ for $\mathbf{x} \in K_i^*$ and $\chi_i^*(\mathbf{x}) = 0$ for $\mathbf{x} \in \mathbb{R}^n \setminus K_i^* (1 \le i \le N)$.

Then each χ_i^* is discontinuous in $\mathbb{R}^n$, but belongs to $L^2(\mathbb{R}^n)$. Hence, we will regularize them with the help of regularizing functions $\rho_\varepsilon \in C_0^\infty(\mathbb{R}^n)$ with $\varepsilon > 0$ (6.2.18)–(6.2.19). For this, we choose $\varepsilon > 0$ such that $\operatorname{supp}(\rho_\varepsilon) \subset B(\mathbf{0}; d)$ with $d > 0$ defined by (C.1.1.11). Then, $\forall i = 1, 2, \dots, N$, $\chi_i^* \in L^2(\mathbb{R}^n) \Longrightarrow$ for such an $\varepsilon > 0$, $\rho_\varepsilon * \chi_i^* \in C^\infty(\mathbb{R}^n)$ by Theorem 6.2.3. Moreover, $\operatorname{supp}(\chi_i^*) = K_i^* \subset\subset \mathbb{R}^n \Longrightarrow \operatorname{supp}(\rho_\varepsilon * \chi_i^*) \subset\subset \mathbb{R}^n$ for this $\varepsilon > 0 \Longrightarrow \rho_\varepsilon * \chi_i^* \in C_0^\infty(\mathbb{R}^n) \equiv \mathcal{D}(\mathbb{R}^n)$ by Theorem 6.2.3. Again, for this $\varepsilon > 0$, $\forall \mathbf{x} \in K_i$, $\forall \mathbf{y} \in \operatorname{supp}(\rho_\varepsilon) \subset B(\mathbf{0}; d)$,

$$\mathbf{x} - \mathbf{y} \in K_i^* \quad \Longrightarrow \quad \chi_i^*(\mathbf{x} - \mathbf{y}) = 1. \tag{C.1.1.12}$$

Set $\psi_i = \rho_\varepsilon * \chi_i^* \in C_0^\infty(\mathbb{R}^n)$ $\forall i = 1, 2, \dots, N$. Then $0 \le \psi_i(\mathbf{x}) \le 1$ $\forall \mathbf{x} \in \mathbb{R}^n$, since $\rho_\varepsilon \ge 0$, $\chi_i^* \ge 0$, $\forall \mathbf{x} \in K_i$, and

$$\begin{aligned}
\psi_i(\mathbf{x}) &= \int_{\mathbb{R}^n} \rho_\varepsilon(\mathbf{y}) \chi_i^*(\mathbf{x} - \mathbf{y}) d\mathbf{y} = \int_{\operatorname{supp}(\rho_\varepsilon) \subset B(\mathbf{0}; d)} \rho_\varepsilon(\mathbf{y}) \chi_i^*(\mathbf{x} - \mathbf{y}) d\mathbf{y} \\
&= \int_{\operatorname{supp}(\rho_\varepsilon)} \rho_\varepsilon(\mathbf{y}) d\mathbf{y} \quad \text{(using (C.1.1.12))} \\
&= 1 \quad \text{(by definition of } \rho_\varepsilon \text{ (6.2.18))}.
\end{aligned}$$

$\forall \mathbf{x} \notin \tilde{K}_i$, $\psi_i(\mathbf{x}) = \int_{\mathbb{R}^n} \rho_\varepsilon(\mathbf{y})\chi_i^*(\mathbf{x}-\mathbf{y})d\mathbf{y} = 0$, since $\mathbf{x} \notin \tilde{K}_i \Longrightarrow \mathbf{x}-\mathbf{y} \notin K_i^*$ $\forall \mathbf{y} \in \operatorname{supp}(\rho_\varepsilon)$ with $\operatorname{supp}(\rho_\varepsilon) \subset B(\mathbf{0}; d) \Longrightarrow \chi_i^*(\mathbf{x}-\mathbf{y}) = 0 \ \forall \mathbf{y} \in \operatorname{supp}(\rho_\varepsilon)$, $\forall \mathbf{x} \in \Omega_i, 0 \le \psi_i(\mathbf{x}) \le 1$.

Hence, $\operatorname{supp}(\psi_i) \subset\subset \Omega_i \ \forall i = 1, 2, \dots, N$, since $\tilde{K}_i \subset \Omega_i \ \forall i = 1, 2, \dots, N$ by construction. Thus, we have constructed $\psi_i \in C_0^\infty(\mathbb{R}^n)$ with $\operatorname{supp}(\psi_i) \subset\subset \Omega_i \Longrightarrow$ $\psi_i \in C_0^\infty(\Omega_i)$ with $\psi_i(\mathbf{x}) = 1 \ \forall \mathbf{x} \in K_i$, and $0 \le \psi_i(\mathbf{x}) \le 1 \ \forall \mathbf{x} \in \Omega_i, 1 \le i \le N$. Set

$$\begin{aligned} &\phi_1 = \psi_1, \quad \phi_2 = (1-\psi_1)\psi_2, \quad \dots, \\ &\phi_i = (1-\psi_1)(1-\psi_2)\cdots(1-\psi_{i-1})\psi_i, \quad 1 \le i \le N, \end{aligned} \tag{C.1.1.13}$$

with $0 \le \phi_i(\mathbf{x}) \le 1 \ \forall \mathbf{x} \in \Omega_i, 1 \le i \le N$.

Then $\operatorname{supp}(\phi_i) \subset \Omega_i$ and $\phi_i \in C_0^\infty(\Omega_i)$ with the property

- $\forall \mathbf{x} \in K_1, \phi_1(\mathbf{x}) = \psi_1(\mathbf{x}) = 1, \phi_2(\mathbf{x}) = 0, \dots, \phi_i(\mathbf{x}) = 0$ for $i = 3, 4, \dots, N$;
- $\forall \mathbf{x} \in K_2$, $\psi_2(\mathbf{x}) = 1 \Longrightarrow \phi_2(\mathbf{x}) = (1-\psi_1(\mathbf{x})) \cdot 1 = 1 - \phi_1(\mathbf{x}) \Longrightarrow$ $\phi_1(\mathbf{x}) + \phi_2(\mathbf{x}) = 1$;
- $\forall \mathbf{x} \in K_3, \phi_3(\mathbf{x}) = (1-\psi_1(\mathbf{x}))(1-\psi_2(\mathbf{x}))\psi_3(\mathbf{x}) = (1-\psi_1(\mathbf{x}))(1-\psi_2(\mathbf{x})) =$ $(1-\psi_1(\mathbf{x})) - (1-\psi_1(\mathbf{x}))\psi_2(\mathbf{x}) = 1 - \phi_1(\mathbf{x}) - \phi_2(\mathbf{x})$
$\Longrightarrow \phi_3(\mathbf{x}) + \phi_2(\mathbf{x}) + \phi_1(\mathbf{x}) = 1 \ \forall \mathbf{x} \in K_3$, since $\psi_3(\mathbf{x}) = 1 \Longrightarrow 1 - \psi_3(\mathbf{x}) = 0$
$\Longrightarrow \phi_j(\mathbf{x}) = 0 \ \forall \mathbf{x} \in K_3, \forall j \ge 4$, $(1-\psi_3(\mathbf{x}))$ being a factor of all $\phi_j(\mathbf{x})$ with $j \ge 4$.

Hence, $\forall \mathbf{x} \in K_i$, $3 \le i \le N$, $\psi_i(\mathbf{x}) = 1 \Longrightarrow 1 - \psi_i(\mathbf{x}) = 0 \Longrightarrow \phi_j(\mathbf{x}) = 0$ $\forall j \ge i+1$, $(1-\psi_i)$ being a factor of all ϕ_j with $j \ge i+1$ and

$$\begin{aligned} \phi_i(\mathbf{x}) &= [(1-\psi_1(\mathbf{x}))\cdots(1-\psi_{i-2}(\mathbf{x}))](1-\psi_{i-1}(\mathbf{x}))\psi_i(\mathbf{x}) \\ &= [(1-\psi_1(\mathbf{x}))\cdots(1-\psi_{i-2}(\mathbf{x}))]1 - \phi_{i-1}(\mathbf{x}) \\ &= [(1-\psi_1(\mathbf{x}))\cdots(1-\psi_{i-3}(\mathbf{x}))]1 - \phi_{i-2}(\mathbf{x}) - \phi_{i-1}(\mathbf{x}) \\ &\ \ \vdots \\ &= 1 - \phi_1(\mathbf{x}) - \phi_2(\mathbf{x}) - \cdots - \phi_{i-2}(\mathbf{x}) - \phi_{i-1}(\mathbf{x}) \end{aligned}$$

$\Longrightarrow \sum_{i=1}^N \phi_i(\mathbf{x}) = \sum_{i=1}^i \phi_i(\mathbf{x}) = 1 \ \forall \mathbf{x} \in K_i$. Hence, $\{\phi_i\}_{i=1}^N$ with $\phi_i \in C_0^\infty(\Omega_i)$ and $\sum_{i=1}^N \phi_i(\mathbf{x}) = 1 \ \forall \mathbf{x} \in K$ is the *required C_0^∞-partition of unity for the compact set K subordinate to the open cover $\{\Omega_i\}_{i=1}^N$* of K, with $K \subset \bigcup_{i=1}^N K_i$, $K_i \subset \Omega_i$ $\forall i = 1, 2, \dots, N$, since $\forall \mathbf{x} \in K$, $\exists i$ with $1 \le i \le N$ such that $\mathbf{x} \in K_i$, and $\sum_{j=1}^N \phi_j(\mathbf{x}) = \sum_{j=1}^i \phi_j(\mathbf{x}) = 1, \phi_j(\mathbf{x}) = 0 \ \forall j \ge i+1$.

Moreover, for $\phi \in C_0^\infty(\Omega)$, define χ_i, $1 \le i \le N$, by: $\chi_1 = \phi \cdot \phi_1$, $\chi_i = \phi \cdot \phi_i$, $2 \le i \le N$, where $\{\phi_i\}_{i=1}^N$ is the C_0^∞-partition of unity for the compact set K subordinate to the open cover $\{\Omega_i\}_{i=1}^N$ and is given by (C.1.1.13). Then $\chi_i \in C_0^\infty(\Omega_i)$, $1 \le i \le$

N, such that $\phi = \sum_{i=1}^{N} \chi_i$. In fact, $\sum_{i=1}^{N} \chi_i = \sum_{i=1}^{N} \phi\phi_i = \phi \sum_{i=1}^{N} \phi_i = \phi \cdot 1 = \phi$ in K, since $\sum_{i=1}^{N} \phi_i(\mathbf{x}) = 1 \ \forall \mathbf{x} \in K \subset \bigcup_{i=1}^{N} \Omega_i$.

Finally, since $\phi_i \geq 0$, $\exists \chi_i \geq 0$ for $\phi \geq 0$, $\forall i = 1, 2, \ldots, N$, such that $\phi = \sum_{i=1}^{N} \chi_i$. □

Appendix D
Boundary geometry

D.1 Boundary geometry

Since not only the m-extension property (see Definition 8.10.3) but also many other properties of Sobolev spaces depend on the geometric properties and regularity of the boundary Γ of a domain, we include here some of these results for immediate use in the study of the m-extension property of bounded $\Omega \subset \mathbb{R}^n$ and also in other problems of Sobolev spaces.

D.1.1 Locally one-sided and two-sided bounded domains Ω

If one moves along the oriented boundary Γ of Ω such that Ω always remains on the left-hand side, then Ω is called a *one-sided domain* (see Figure D.1(a)).

If the domain Ω has a *slit* or *cut* along a part $\Gamma_0 \subset \Gamma$, then Ω lies *locally* on *both sides* of Γ_0. Such a domain is called *locally two-sided* along $\Gamma_0 \subset \Gamma$. For example (see Figure D.1(b)), $\Omega = \{(x_1, x_2) : 0 < |x_1|, 0 < x_2 < 1\}$ is a *two-sided* domain with a slit along $\Gamma_0 \subset \Gamma$:

$$\Gamma_0 = \{(0, x_2) : 0 < x_2 < 1\}. \qquad \text{(D.1.1.1)}$$

Locally *two-sided* domains are pathological cases which will *not* be considered by us. For an alternative *analytic* definition, see Definition D.3.1.1. For more examples, we refer to Dauge [44].

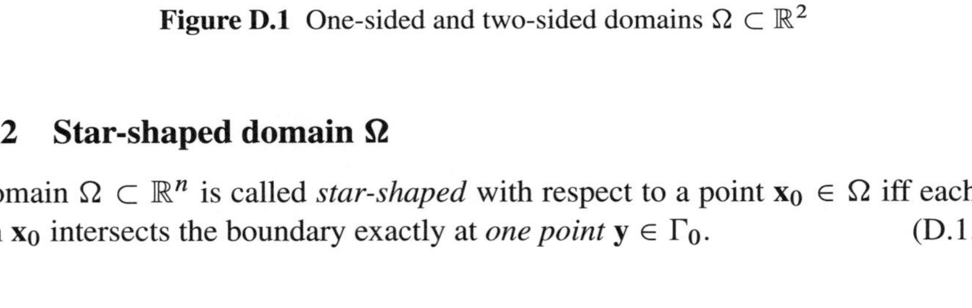

Figure D.1 One-sided and two-sided domains $\Omega \subset \mathbb{R}^2$

D.1.2 Star-shaped domain Ω

A domain $\Omega \subset \mathbb{R}^n$ is called *star-shaped* with respect to a point $\mathbf{x}_0 \in \Omega$ iff each ray from $\mathbf{x}_0$ intersects the boundary exactly at *one point* $\mathbf{y} \in \Gamma_0$. (D.1.2.1)

A domain $\Omega \subset \mathbb{R}^n$ is called *star-shaped* with respect to an open ball $B(\mathbf{x}_0; r) \subset \Omega$ contained in Ω if and only if it is *star-shaped* with respect to every $\mathbf{x} \in B(\mathbf{x}_0; r)$. (D.1.2.2)

Every *convex* domain Ω is *star-shaped*, but a *two-sided* domain is *not* star-shaped. (D.1.2.3)

D.1.3 Cone property and uniform cone property

Cones in R^n Let $\mathbf{x}_0 \in \mathbb{R}^n$ be a *fixed* point. Then, roughly speaking, an *open*, *infinite cone* $C_{\mathbf{x}_0}$ with the vertex $\mathbf{x}_0$ (see Figure D.2) is an *open* set with the property:

$$\mathbf{y} \in C_{\mathbf{x}_0} \quad \Longrightarrow \quad \text{the element } \mathbf{x}_0 + \lambda(\mathbf{y} - \mathbf{x}_0) \in C_{\mathbf{x}_0} \ \forall \lambda > 0 \qquad \text{(D.1.3.1)}$$

($\mathbf{x}_0 \notin C_{\mathbf{x}_0}$, since $\lambda > 0$).

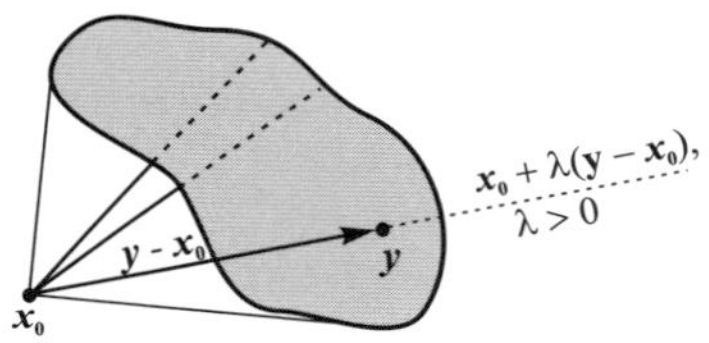

Figure D.2 Open infinite cone $C_{\mathbf{x}_0}$

Finite open cone $C_{\mathbf{x}_0} = C_{\mathbf{x}_0}(r; \Sigma)$ For fixed $\mathbf{x}_0 \in \mathbb{R}^n$, let $B(\mathbf{x}_0; r)$ be an open ball and $S(\mathbf{x}_0; r)$ be the sphere (the boundary of $B(\mathbf{x}_0; r)$), respectively:

$$B(\mathbf{x}_0; r) = \{\mathbf{x} : \mathbf{x} \in \mathbb{R}^n, \ \|\mathbf{x} - \mathbf{x}_0\| < r\}, \quad S(\mathbf{x}_0; r) = \{\mathbf{x} : \mathbf{x} \in \mathbb{R}^n, \ \|\mathbf{x} - \mathbf{x}_0\| = r\}.$$

Let $\Sigma \subset S(\mathbf{x}_0; r)$ be any *nonempty* subset of $S(\mathbf{x}_0; r)$. Then by a *finite open cone* in $\mathbb{R}^n$ with vertex at $\mathbf{x}_0$, we mean the *open* set $C_{\mathbf{x}_0}(r; \Sigma)$ defined by:

$$C_{\mathbf{x}_0}(r; \Sigma) = B(\mathbf{x}_0; r) \cap \{\mathbf{x} : \mathbf{x} = \mathbf{x}_0 + \lambda(\mathbf{y} - \mathbf{x}_0) \text{ with } \mathbf{y} \in \Sigma, \ \lambda > 0\}. \quad \text{(D.1.3.2)}$$

Hence, $\mathbf{x}_0 \notin C_{\mathbf{x}_0}(r; \Sigma)$. See Figure D.3 for examples in $\mathbb{R}^2$ and $\mathbb{R}^3$.

Standard or reference open cone C_0 For $\mathbf{x}_0 = \mathbf{0} \in \mathbb{R}^n$, $C_0(r; \Sigma) \equiv C_0$ is called the *standard* or *reference open cone* in $\mathbb{R}^n$ (where $r > 0$ and $\Sigma \subset S(\mathbf{x}_0; r)$ may be chosen conveniently). (D.1.3.3)

Then the set $C_{\mathbf{x}} = C_{\mathbf{x}}(r; \Sigma) = \mathbf{x} + C_0 = \{\mathbf{z} : \mathbf{z} = \mathbf{x} + \mathbf{y} \text{ with } \mathbf{y} \in C_0\}$ denotes a *finite cone with vertex at* $\mathbf{x}$ which is obtained by parallel translation of the standard or reference cone C_0 through the vector $\mathbf{x}$. For example (see Figure D.4), for $n = 2$, $C_{\mathbf{x}} = C_{\mathbf{x}}(r; \Sigma) \cong C_0$ (i.e. *congruent to* C_0) $\Longleftrightarrow$ $C_{\mathbf{x}}$ can be obtained from $C_0 = C_0(r; \Sigma)$ by 'rigid body motion' (i.e. parallel translation and rotation). (D.1.3.4)

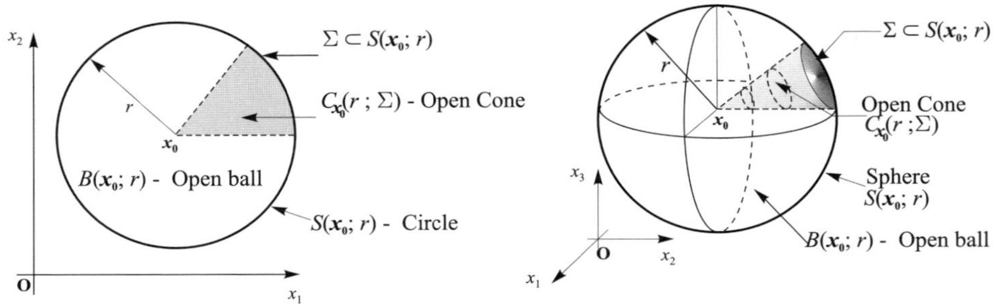

Figure D.3 Finite open cones in $\mathbb{R}^2$ and $\mathbb{R}^3$

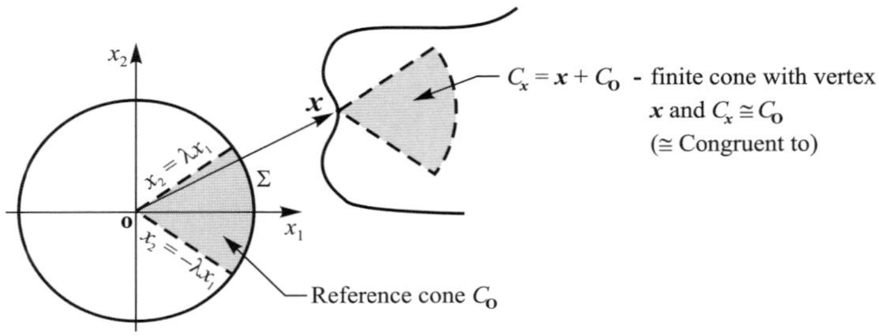

Figure D.4 Finite cone $C_{\mathbf{x}}$ and reference cone C_0 in $\mathbb{R}^2$

Definition D.1.3.1. A bounded domain $\Omega \subset \mathbb{R}^n$ with boundary Γ is said to have the *cone property* (or to satisfy the *cone condition*) if and only if $\forall \mathbf{x} \in \Gamma$, $\exists$ an open cone $C_{\mathbf{x}}(r; \Sigma) \subset \Omega$ which is *congruent* to a given *fixed* standard cone C_0, i.e. $C_{\mathbf{x}} \cong C_0$. See Figure D.5 for examples in $\mathbb{R}^2$.

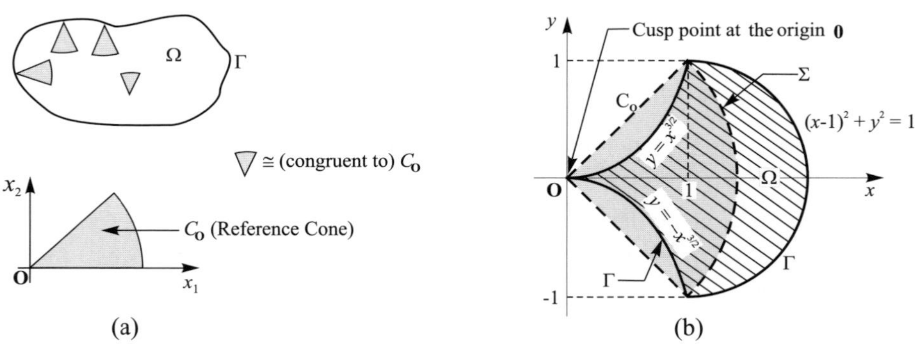

Figure D.5 (a) Ω satisfying the cone property; (b) Ω does *not* satisfy the cone condition at **0**, since **0** is a *cusp point* and $\nexists$ any $C_0 \subset \Omega$

Definition D.1.3.2. A bounded domain $\Omega \subset \mathbb{R}^n$ with boundary Γ is said to have the *uniform* or *restricted cone property* if and only if, $\forall \mathbf{x} \in \Gamma$, $\exists$ a neighbourhood $N_{\mathbf{x}}$ of $\mathbf{x}$ in $\mathbb{R}^n$ containing an open cone $C_{\mathbf{x}}$ such that:

1. $C_{\mathbf{x}}$ is *congruent* to a given, *fixed standard* or *reference* cone $C_{\mathbf{0}}$,
2. $\forall \mathbf{y} \in \overline{\Omega} \cap N_{\mathbf{x}}$, the subset $\mathbf{y} + C_{\mathbf{x}} \subset \Omega$.

Ω has the uniform/restricted cone property $\Longrightarrow$ Ω has the cone property. (D.1.3.5)
See Figure D.6 for examples of the uniform cone property in $\mathbb{R}^2$.

Remark D.1.3.1. While laying the foundation of Sobolev spaces and imbedding results in them, Sobolev himself used the (uniform) cone property of domains Ω (see, for example, Adams [12], Nečas [16], Grisvard [18]). This trend was followed by other authors too during the early stages of the theory of Sobolev spaces.

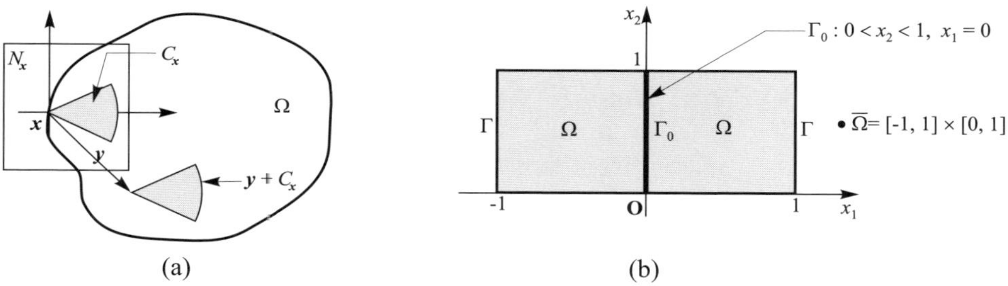

Figure D.6 (a) Uniform cone property of Ω; (b) Locally two-sided domain Ω *satisfies* the cone property along Γ_0, but does *not* satisfy the uniform cone property, since $\forall N_{\mathbf{x}}$ of $\mathbf{x} \in \Gamma_0$, $\exists \mathbf{y} \in \overline{\Omega} \cap N_{\mathbf{x}}$ such that $\mathbf{y} + C_{\mathbf{x}} \not\subset \Omega$

D.1.4 Segment property

A bounded domain $\Omega \subset \mathbb{R}^n$ with boundary Γ is said to have the *segment property* if and only if, $\forall \mathbf{x} \in \Gamma$, $\exists$ a linear segment L with origin at $\mathbf{x}$ such that $L \setminus \{\mathbf{x}\}$ is contained in Ω.

Definition D.1.4.1. A bounded domain $\Omega \subset \mathbb{R}^n$ with boundary Γ is said to have the *restricted* or *uniform segment property* if and only if, $\forall \mathbf{x} \in \Gamma$, $\exists$ a *neighbourhood* $N_{\mathbf{x}}$ of $\mathbf{x}$ in $\mathbb{R}^n$ and $\exists$ a *non-null* vector $\mathbf{y}_{\mathbf{x}} \neq \mathbf{0}$ such that $\forall \mathbf{z} \in \overline{\Omega} \cap N_{\mathbf{x}}$, the point $\mathbf{z} + \lambda \mathbf{y}_{\mathbf{x}} \in \Omega$ $\forall \lambda \in]0, \lambda_0[$ with $0 < \lambda_0 < 1$.

The *restricted segment property* of Ω implies its *segment* property. (D.1.4.1)

Locally *two-sided* domains satisfy *neither* the *restricted segment property nor* the *uniform cone property*, despite satisfying the cone property (see the example in (D.1.1.1)). (D.1.4.2)

See Figure D.7 for examples of the uniform segment property in $\mathbb{R}^2$.

Remark D.1.4.1. Agmon [39] used the uniform segment property of a domain Ω to prove many basic results.

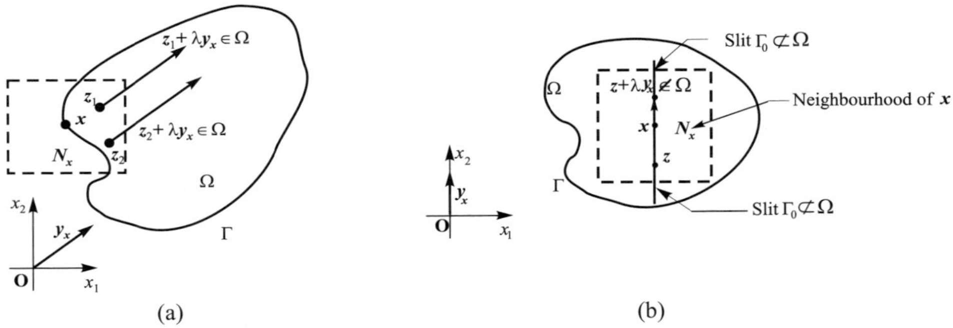

Figure D.7 (a) Uniform segment property of Ω; (b) Locally two-sided domain Ω with a slit along Γ_0 does *not* satisfy the *uniform* segment property, since $\exists \mathbf{z} \in N_{\mathbf{x}} \cap \overline{\Omega}$ such that $\mathbf{z} + \lambda \mathbf{y}_{\mathbf{x}} \in \Gamma_0 \not\subset \Omega \ \forall \lambda \in]0, \lambda_0[$ with $0 < \lambda_0 < 1$

D.2 Continuity and differential properties of a boundary

D.2.1 Continuity and differential properties

Γ as a manifold in $\mathbb{R}^n$

An elementary definition of a manifold Let E and F be vector spaces and $\phi : U \subset E \longrightarrow F$ be a mapping from U into F. Then its *graph* is the set $\mathscr{G}$ of ordered pairs $(\mathbf{x}, \phi(\mathbf{x})) \in E \times F \ \forall \mathbf{x} \in U$, i.e. $\mathscr{G} = \{(\mathbf{x}, \mathbf{y}) : \mathbf{x} \in U \subset E, \mathbf{y} = \phi(\mathbf{x}) \in F\}$ is called a *manifold* in $E \times F$ defined by the equation $\mathbf{y} = \phi(\mathbf{x}) \ \forall \mathbf{x} \in U \subset E$, the *dimension* of the manifold $\mathscr{G}$ being that of E. (D.2.1.1)

Example D.2.1.1.

1. For $E = \mathbb{R}^2, F = \mathbb{R}^1 = \mathbb{R}$, the manifold $\mathscr{G} = S$ is a *two-dimensional* one and a *surface* in $E \times F = \mathbb{R}^3$ defined by $z = \phi(x, y)$ with $(x, y) \in U \subset \mathbb{R}^2, z \in \mathbb{R}$. In particular, $z = \sqrt{1 - x^2 - y^2}$ defines the *upper half* S^+ of the *unit sphere* in $\mathbb{R}^3$ for $z > 0$.

2. For $E = \mathbb{R}, F = \mathbb{R}^2$, the manifold $\mathscr{G}$ is of *dimension* 1 and a *curve* defined by $y = \phi_1(x), z = \phi_2(x) \ \forall x \in I \subset \mathbb{R}$, i.e. $\Phi = (\phi_1, \phi_2)$ and $\Phi(\mathbf{x}) = (\phi_1(\mathbf{x}), \phi_2(\mathbf{x})) \ \forall x \in I \subset \mathbb{R}$.

3. For $E = \mathbb{R}^{n-1}, F = \mathbb{R}$, the manifold $\mathscr{G}$ is of *dimension* $n - 1$ and defined by $x_n = \phi(x_1, x_2, \ldots, x_{n-1}) \ \forall (x_1, x_2, \ldots, x_{n-1}) \in U \subset \mathbb{R}^{n-1}$. (D.2.1.2)

Definition D.2.1.1. The manifold $\mathcal{G}$ defined by $y = \phi(x) \in F \ \forall x \in U \subset E$ in (D.2.1.1) is called *continuous/Lipschitz continuous/k times continuously differentiable* ($k \geq 1$) and so on if and only if ϕ is *continuous/Lipschitz continuous/k times continuously differentiable* ($k \geq 1$) and so on, respectively.

For $k = 0$, i.e. ϕ is *continuous*, the manifold $\mathcal{G}$ is usually called a *topological manifold.* (D.2.1.3)

Remark D.2.1.1. Unfortunately, this *elementary* definition of a *differential* manifold can *not* be applied even to the unit sphere S_1, since there does *not* exist a *single* equation in the explicit form $x_n = \sqrt{1 - x_1^2 - x_2^2 - \cdots - x_{n-1}^2}$ for the *entire* (whole) *unit* sphere S_1 as an $(n-1)$*-dimensional differential manifold of* C^∞-class (i.e. $k = \infty$). Introducing an *open cover* $\{\Gamma_r\}_{r=1}^N$ for the boundary Γ, *local coordinate systems* $\{\xi^r\}$ in each Γ_r, $1 \leq r \leq N$, and local mappings ϕ_r on each Γ_r, we can overcome these difficulties as follows.

D.2.2 Open cover $\{\Gamma_r\}_{r=1}^N$ of Γ, local coordinate systems $\{\xi_i^r\}_{i=1}^n$ and mappings $\{\phi_r\}_{r=1}^N$

In the fixed/global orthogonal coordinate system $x_1, x_2, \ldots, x_n$ used to define points $\mathbf{x} \in \overline{\Omega} = \Omega \cup \Gamma \subset \mathbb{R}^n$, the representation of the boundary Γ is *not* a convenient one for the study of its differential properties, for example, the differentiability of Γ (see Remark D.2.1.1). This suggests that we should introduce new, local, orthogonal systems of coordinates. For this, we first introduce an open cover $\{\Gamma_r\}_{r=1}^N$ of Γ with $\Gamma_r \subset \Gamma$, $\Gamma = \bigcup_{r=1}^N \Gamma_r$, and then introduce a *new, local, orthogonal system of coordinates* $(\xi_i^r)_{i=1}^n$ with origin at $\mathbf{0}_r$ to define Γ_r, $1 \leq r \leq N$, as follows: $\{\Gamma_r\}_{r=1}^N$ is an *open cover* of Γ with $\Gamma_r \subset \Gamma$, $\Gamma = \bigcup_{r=1}^N \Gamma_r$, such that

1. for $1 \leq r \neq s \leq N$, $\Gamma_r \cap \Gamma_s$ = a common part of Γ_r and Γ_s in Γ, or $\emptyset$, i.e. open sets Γ_r and Γ_s may be *overlapping* ($\Gamma_r \cap \Gamma_s \neq \emptyset$) or *disjoint* ($\Gamma_r \cap \Gamma_s = \emptyset$); (D.2.2.1)
2. $\forall \mathbf{x} \in \Gamma$, $\exists$ at least one Γ_r with $\mathbf{x} \in \Gamma_r, 1 \leq r \leq N$; (D.2.2.2)
3. $\Gamma_r = \{(\hat{\boldsymbol{\xi}}^r, \xi_n^r) : \hat{\boldsymbol{\xi}}^r = (\xi_1^r, \xi_2^r, \ldots, \xi_{n-1}^r) \in \mathbb{R}^{n-1}, \xi_n^r = \phi_r(\hat{\boldsymbol{\xi}}^r)\}$, $1 \leq r \leq N$, (D.2.2.3)

   $\boldsymbol{\xi}^r = (\hat{\boldsymbol{\xi}}^r, \xi_n^r) = (\xi_1^r, \xi_2^r, \ldots, \xi_n^r)$ being the *local coordinate system* with origin at $\mathbf{0}_r$ to define Γ_r, $\phi_r : \mathbb{R}^{n-1} \longrightarrow \mathbb{R}$ defined by $\hat{\boldsymbol{\xi}}^r \in \mathbb{R}^{n-1} \longmapsto \phi_r(\hat{\boldsymbol{\xi}}^r) \in \mathbb{R}$ (D.2.2.4)

   being a function of the $(n-1)$ *local* variables $\xi_1^r, \xi_2^r, \ldots, \xi_{n-1}^r$ used to define Γ_r. The properties of ϕ_r will be explained below.

First of all, we show in Figure D.8 *three* local coordinate systems $(\xi_i^r)_{i=1}^n$, $(\xi_i^s)_{i=1}^n$ and $(\xi_i^t)_{i=1}^n$, with origins at $\mathbf{0}_r$, $\mathbf{0}_s$ and $\mathbf{0}_t$, respectively, $1 \leq r \neq s \neq t \leq N$, such that *locally* $\Gamma_r, \Gamma_s, \Gamma_t \subset \Gamma$ are defined by:

$$\Gamma_r\colon \xi_n^r = \phi_r(\hat{\boldsymbol{\xi}^r}); \quad \Gamma_s\colon \xi_n^s = \phi_s(\hat{\boldsymbol{\xi}^s}); \quad \Gamma_t\colon \xi_n^t = \phi_t(\hat{\boldsymbol{\xi}^t}) \tag{D.2.2.5}$$

(for the sake of *clarity*, non-overlapping Γ_r and Γ_s have been shown in the figure) with $\hat{\boldsymbol{\xi}^k} = (\xi_1^k, \xi_2^k, \ldots, \xi_{n-1}^k) \in \mathbb{R}^{n-1}$, functions $\phi_r, \phi_s, \phi_t : \mathbb{R}^{n-1} \longrightarrow \mathbb{R}$ $(k = r, s, t)$ having the properties described later.

For $1 \le r \le N$, the *local* and *fixed* (global) orthogonal coordinate systems are related by: $\boldsymbol{\xi}^r = A_r\mathbf{x} + \mathbf{b}_r$ and $\mathbf{x} = A_r^t\boldsymbol{\xi}^r + \tilde{\mathbf{b}}_r$, $1 \le r \le N$, where A_r is an *orthogonal matrix* of order n with $A_r^t = A_r^{-1}$, $\mathbf{b}_r \in \mathbb{R}^n$, $\tilde{\mathbf{b}}_r = -A_r^t\mathbf{b}_r$, and the *origin* $\mathbf{0}$ of the *fixed* system $(x_i)_{i=1}^n$ is translated through the vector $\mathbf{b}_r$ to obtain the origin $\mathbf{0}_r$ of the local system $(\xi_i^r)_{i=1}^n$. (D.2.2.6)

D.2.3 Properties of the mappings $\phi_r : \mathbb{R}^{n-1} \longrightarrow \mathbb{R}, 1 \le r \le N$

Now, suppose that $\exists$ two positive numbers $\alpha > 0$, $\beta > 0$ and $\exists$ an $(n-1)$-dimensional hypercube $\widehat{Q}_r \subset \mathbb{R}^{n-1}$ in the new local coordinates $(\xi_i^r)_{i=1}^n$ for $1 \le r \le N$,

$$\widehat{Q}_r = \{\hat{\boldsymbol{\xi}}^r : \hat{\boldsymbol{\xi}}^r = (\xi_1^r, \xi_2^r, \ldots, \xi_{n-1}^r) \in \mathbb{R}^{n-1},\ |\xi_i^r| < \alpha,\ 1 \le i \le n-1\}, \tag{D.2.3.1}$$

such that function $\phi_r : \widehat{Q}_r \subset \mathbb{R}^{n-1} \longrightarrow \mathbb{R}$ of $(n-1)$-variables $\xi_1^r, \xi_2^r, \ldots, \xi_{n-1}^r$ defined on $\widehat{Q}_r$ satisfies the following conditions:

1. $\phi_r \in C^0(\widehat{Q}_r)$, i.e. ϕ_r is *continuous* on $\widehat{Q}_r$ as a function of $(n-1)$ local variables $(\xi_i^r)_{i=1}^{n-1}$; (D.2.3.1a)
2. Γ_r is *locally* defined by $\xi_n^r = \phi_r(\xi_1^r, \xi_2^r, \ldots, \xi_{n-1}^r) = \phi_r(\hat{\boldsymbol{\xi}}^r)$, i.e. Γ_r is an $(n-1)$-dimensional manifold defined by $\xi_n^r = \phi_r(\hat{\boldsymbol{\xi}}^r)$ with $\Gamma_r = \{(\hat{\boldsymbol{\xi}}^r, \xi_n^r) : \xi_n^r = \phi_r(\hat{\boldsymbol{\xi}}^r)\ \forall \hat{\boldsymbol{\xi}}^r \in \widehat{Q}_r\}$; (D.2.3.1b)
3. If the part Γ_r of Γ is moved by a *parallel translation* in the positive direction by $\beta > 0$, then the translated part Γ_r still remains inside Ω, i.e. for $1 \le r \le N$,

$$\{(\hat{\boldsymbol{\xi}}^r, \xi_n^r) : \hat{\boldsymbol{\xi}}^r = (\xi_1^r, \xi_2^r, \ldots, \xi_{n-1}^r),\ \phi_r(\hat{\boldsymbol{\xi}}^r) < \xi_n^r < \phi_r(\hat{\boldsymbol{\xi}}^r) + \beta\ \forall \hat{\boldsymbol{\xi}}^r \in \hat{Q}_r\} \subset \Omega. \tag{D.2.3.2}$$

4. On the contrary, if the part Γ_r of Γ is moved by a *parallel translation* (in the *reverse* direction) by $-\beta$, then the translated part Γ_r immediately goes *outside* $\overline{\Omega}$ or falls inside $\overline{\Omega}^C \equiv$ complement of $\overline{\Omega}$, i.e. for $1 \le r \le N$,

$$\{(\hat{\boldsymbol{\xi}}^r, \xi_n^r) : \hat{\boldsymbol{\xi}}^r = (\xi_1^r, \xi_2^r, \ldots, \xi_{n-1}^r),\ \phi_r(\hat{\boldsymbol{\xi}}^r) - \beta < \xi_n^r < \phi_r(\hat{\boldsymbol{\xi}}^r)\ \forall \hat{\boldsymbol{\xi}}^r \in \hat{Q}_r\} \subset \overline{\Omega}^C. \tag{D.2.3.3}$$

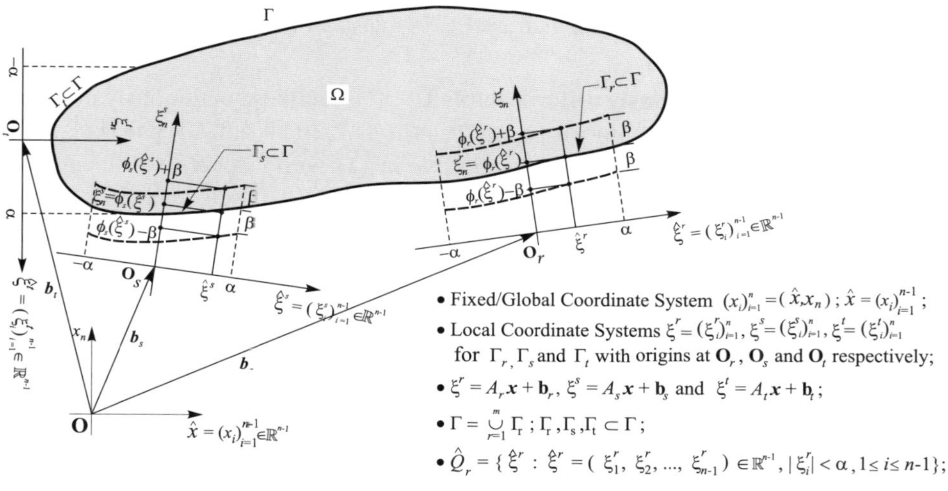

Figure D.8 Local coordinates at $\mathbf{0}_r$, $\mathbf{0}_s$ and $\mathbf{0}_t$ for $\Gamma_r, \Gamma_s, \Gamma_t \subset \Gamma$ respectively

Definition D.2.3.1. Let $\Omega \subset \mathbb{R}^n$ be an open subset with boundary Γ and $\{\Gamma_r\}_{r=1}^N$ be an open cover of $\Gamma \subset \bigcup_{r=1}^N \Gamma_r$ and $(\xi_i^r)_{i=1}^n$ be *local*, *orthogonal* coordinates for each Γ_r with origin at $\mathbf{0}_r$, $\hat{Q}_r \subset \mathbb{R}^{n-1}$ be an $(n-1)$-dimensional hypercube defined by (D.2.3.1)–(D.2.3.3) and $\phi_r : \widehat{Q}_r \longrightarrow \mathbb{R}$ be a function of $(n-1)$ local variables $\xi_1^r, \xi_2^r, \ldots, \xi_{n-1}^r$, $1 \le r \le N$. Then the following important cases arise:

Case 1: Continuous Γ Γ is called *continuous*, or Γ is of C^0*-class*, if and only if the properties 1–4 (in (D.2.2.1)–(D.2.2.4) and (D.2.3.1)–(D.2.3.3)) of the function ϕ_r defining Γ_r hold $\forall$ fixed $r = 1, 2, \ldots, N$. This fact is *also* denoted by:

$$\Omega \in C^0\text{-class}. \tag{D.2.3.4}$$

Case 2: Uniformly Lipschitz continuous Γ Γ is called *uniformly Lipschitz continuous*, or briefly *Lipschitz continuous*, if and only if

1. Γ is *continuous* (i.e. Properties 1–4 (in (D.2.2.1)–(D.2.2.4) and (D.2.3.1)–(D.2.3.3)) of ϕ_r hold $\forall$ fixed $r = 1, 2, \ldots, N$);
2. ϕ_r satisfies the *uniformly Lipschitz condition* in $\hat{Q}_r$ as a function of local coordinates $(\xi_i^r)_{i=1}^{n-1}$: $\forall$ fixed $r = 1, 2, \ldots, N$, $\forall \hat{\boldsymbol{\xi}}^r = (\xi_1^r, \ldots, \xi_{n-1}^r)$, $\hat{\boldsymbol{\eta}}^r = (\eta_1^r, \ldots, \eta_{n-1}^r) \in \widehat{Q}_r$, $\exists L > 0$, independent of $\hat{\boldsymbol{\xi}}^r$, $\hat{\boldsymbol{\eta}}^r$ and r, such that

$$|\phi_r(\hat{\boldsymbol{\xi}}^r) - \phi_r(\hat{\boldsymbol{\eta}}^r)| \le L \|\hat{\boldsymbol{\xi}}^r - \hat{\boldsymbol{\eta}}^r\|_{\mathbb{R}^{n-1}}, \tag{D.2.3.5}$$

where $L > 0$ is the *Lipschitz constant*, i.e. $\forall$ fixed $r = 1, 2, \ldots, N$, ϕ_r is *uniformly Lipschitz continuous in* $\hat{\mathbf{Q}}_\mathbf{r}$.

We also denote this result by writing: Γ is of $C^{0,1}$-class, or $\Omega \in C^{0,1}$-class. (D.2.3.6)

Case 3: k-times continuously differentiable Γ Γ is k-times continuously differentiable if and only if ϕ_r satisfying properties 1–4 (in (D.2.2.1)–(D.2.2.4) and (D.2.3.1)–(D.2.3.3)) is *k-times continuously differentiable* in $\hat{Q}_r$ with respect to all local variables $(\xi_i^r)_{i=1}^{n-1}$ in $\hat{Q}_r$ $\forall r = 1, 2, \ldots, N$, i.e. $\phi_r \in C^k(\hat{Q}_r)$. Then we write: Γ is of C^k-class or $\Omega \in C^k$-class. (D.2.3.7)

Case 4: Γ is of class $C^{k,1}$ Γ is of $C^{k,1}$-class (or $\Omega \in C^{k,1}$-class) if and only if Γ is of C^k-class (D.2.3.7) and all *kth-order derivatives* $\partial^{\boldsymbol{\alpha}}\phi_r$ with $|\boldsymbol{\alpha}| = k$ are *uniformly Lipschitz continuous* in $\hat{Q}_r$ in local variables defined (D.2.3.5)–(D.2.3.6): $\forall |\boldsymbol{\alpha}| = k$, $1 \leq r \leq N$,

$$|\partial^{\boldsymbol{\alpha}}\phi_r(\hat{\boldsymbol{\xi}}^r) - \partial^{\boldsymbol{\alpha}}\phi_r(\hat{\boldsymbol{\eta}}^r)| \leq L_1 \|\hat{\boldsymbol{\xi}}^r - \hat{\boldsymbol{\eta}}^r\|_{\mathbb{R}^{n-1}} \quad \forall \hat{\boldsymbol{\xi}}^r, \hat{\boldsymbol{\eta}}^r \in \hat{Q}_r, \tag{D.2.3.8}$$

where $L_1 > 0$ is the *Lipschitz constant*, i.e. $\phi_r \in C^{k,1}(\widehat{Q}_r)$ $\forall r = 1, 2, \ldots, m$.

Case 5: Γ is smooth/continuously infinitely differentiable, i.e. Γ is of C^∞-class or $\Omega \in C^\infty$-class Γ is an *infinitely differentiable $n-1$-dimensional manifold* iff Γ is of C^k-class $\forall k \in \mathbb{N}$. (D.2.3.9)

Atlas or local charts of Γ The system $(\Gamma_r, \phi_r)_{r=1}^N$ in Definition D.2.3.1 is called an *atlas* or *local charts* defining the boundary Γ. (D.2.3.10)

D.3 Alternative definition of locally one-sided domain

Instead of the definition given at the beginning of this appendix, we can now provide this alternative.

Definition D.3.1.1. A bounded domain $\Omega \subset \mathbb{R}^n$ with boundary $\Gamma \subset \bigcup_{r=1}^N \Gamma_r$, $\{\Gamma_r\}_{r=1}^N$ being an open cover of Γ, lies *locally on one* side $\Gamma_r \subset \Gamma$ if and only if the properties (D.2.3.2) and (D.2.3.3) of the function $\phi_r = \phi_r(\xi_1^r, \xi_2^r, \ldots, \xi_{n-1}^r)$ defining Γ_r, $1 \leq r \leq N$, in (D.2.3.1) hold.

For example, for $n = 2$ and for $\Omega = \{(x_1, x_2) : 0 < |x_1| < 1, 0 < x_2 < 1\}$, choose $\Gamma_r = \Gamma_0 = \{(0, x_2) : 0 < x_2 < 1\} \subset \Gamma$. Then, we define Γ_0 in local coordinates (ξ_1, ξ_2) by: $\xi_2 = \phi_0(\xi_1) = C_0$ for $|\xi_1| < \alpha$ with $\alpha = \frac{1}{2}$ with $\hat{Q}_0 =]-\frac{1}{2}, \frac{1}{2}[$ (see Figure D.9). Then for $\beta \in]0, 1[$, $\{(\xi_1, \xi_2) : \phi_0(\xi_1) < \xi_2 < \phi_0(\xi_1) + \beta \ \forall \xi_1\} \subset \Omega$ (i.e. property (D.2.3.2) holds), but $\{(\xi_1, \xi_2) : \phi_0(\xi_1) - \beta < \xi_2 < \phi_0(\xi_1) \ \forall \xi_1 \in \hat{Q}_0\} \subset \overline{\Omega}$ (i.e. property (D.2.3.3) does *not* hold). Hence, Ω is *not* one-sided along Γ_0, i.e. is a *two-sided* domain along Γ_0.

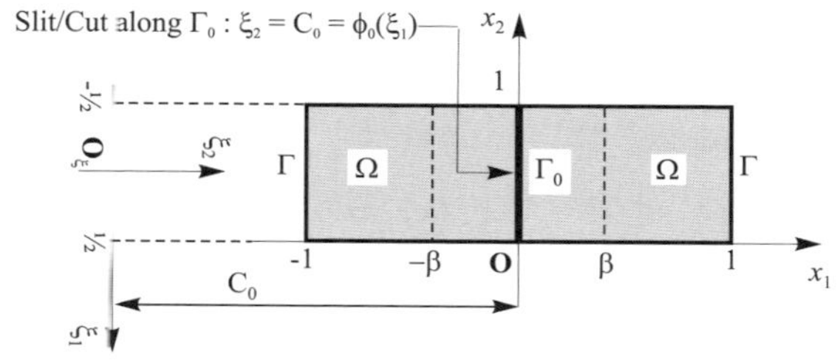

Figure D.9 Local coordinates ξ_1, ξ_2 with origin at $\mathbf{0}_\xi$ to define Γ_0 by $\xi_2 = C_0 = \phi_0(\xi_1)$, $\forall \xi_1 \in]-\frac{1}{2}, \frac{1}{2}[$

D.4 Alternative definition of continuity and differential properties of $\overline{\Omega}$ as a manifold in $\mathbb{R}^n$

Considering $\overline{\Omega} = \Omega \cup \Gamma$ as an n-dimensional manifold with boundary Γ embedded in $\mathbb{R}^n$, along with some regularity assumptions, we can make the following definition (see [15], [18]).

Definition D.4.1.1. Let $\overline{\Omega}$ be the closure of a *bounded*, open subset Ω, Γ being its boundary and $\{\Omega_i\}_{i=0}^N$ be a *finite open cover* of $\overline{\Omega}$ such that $\overline{\Omega}_0 \subset \Omega$, $\Omega_i \subset \mathbb{R}^n$ $\forall i = 1, 2, \ldots, N$, $\overline{\Omega} \subset \bigcup_{i=0}^N \Omega_i$ and $\{\Omega_i\}_{i=1}^N$ is an *open cover* of Γ (i.e. $\Gamma \subset \bigcup_{i=1}^N \Omega_i$). Then the n-dimensional manifold $\overline{\Omega}$ is called *continuous* (resp. *uniformly Lipschitz continuous/m-times continuously differentiable*, or $\overline{\Omega}$ is of *C^m-class*/is of *$C^{m,1}$-class*/ is of *C^∞-class* or *smooth*) if and only if, $\forall i = 1, 2, \ldots, N$, $\exists$ an *invertible* mapping $\Phi_i : \Omega_i \longrightarrow \hat{Q} \subset \mathbb{R}^n$, $\hat{Q}$ being a *fixed* n-dimensional hypercube in variables $(\xi_i)_{i=1}^n$:

$$\hat{Q} = \{(\hat{\boldsymbol{\xi}}, \xi_n) : \hat{\boldsymbol{\xi}} = (\xi_1, \xi_2, \ldots, \xi_{n-1}) \in \mathbb{R}^{n-1},\ |\xi_i| < 1,\ 1 \leq i \leq n\} \subset \mathbb{R}^n, \tag{D.4.1.1}$$

such that, $\forall i = 1, 2, \ldots, N$,

1. Φ_i together with its inverse $\Phi_i^{-1} : \hat{Q} \longrightarrow \Omega_i$ is *continuous* (resp. *uniformly Lipschitz continuous/m-times continuously differentiable* or is of *C^m-class*/is of *$C^{m,1}$-class* (i.e. Φ_i and Φ_i^{-1} are of C^m-class and all mth-order derivatives are *uniformly Lipschitz continuous in* Ω_i)/is of *C^∞-class* or *smooth*); (D.4.1.2)
2. $\Phi_i(\Omega_i \cap \Omega) = \hat{Q} \cap \{(\hat{\boldsymbol{\xi}}, \xi_n) : \hat{\boldsymbol{\xi}} = (\xi_1, \xi_2, \ldots, \xi_{n-1}) \in \mathbb{R}^{n-1}, \xi_n > 0\} = \hat{Q} \cap \mathbb{R}^n_+ = \hat{Q}_+$;
3. $\Phi_i(\Omega_i \cap \Gamma) = \hat{Q} \cap \{(\hat{\boldsymbol{\xi}}, \xi_n) : \hat{\boldsymbol{\xi}} = (\xi_1, \xi_2, \ldots, \xi_{n-1}) \in \mathbb{R}^{n-1}, \xi_n = 0\} = \hat{Q} \cap \mathbb{R}^n_0$, which we identify with $\hat{Q}_0 \subset \mathbb{R}^{n-1}$ and write $\boldsymbol{\Phi_i}(\Omega_i \cap \Gamma) = \hat{Q}_0 \subset \mathbb{R}^{n-1}$. (D.4.1.3)

Remark D.4.1.1. Since $\forall \mathbf{x} \in \Omega_i \cap \Gamma$, $1 \leq i \leq N$,

$$\Phi_i(\mathbf{x}) = (\phi_1^i(\mathbf{x}), \phi_2^i(\mathbf{x}), \ldots, \phi_n^i(\mathbf{x})) = (\xi_1, \xi_2, \ldots, \xi_{n-1}, 0) \in \hat{Q}_0 \cap \mathbb{R}^n, \quad \text{(D.4.1.4)}$$

which we identify with $\hat{Q}_0 \subset \mathbb{R}^{n-1}$, $\phi_n^i(\mathbf{x}) = 0$ is the equation to the boundary Γ *locally*, i.e. for the part $\Gamma_i = \Omega_i \cap \Gamma$ of Γ which lies in Ω_i, $1 \leq i \leq N$.

D.5 Atlas/local charts of Γ

The system $(\Omega_i, \Phi_i)_{i=1}^N$ with $\Gamma_i = \Omega_i \cap \Gamma$ is called a *system of local charts* or an *atlas* defining the boundary Γ. (D.5.1.1)

Bibliography

[1] Gelfand IM, Schilov GE. Generalized Functions. vol. 1. New York: Academic Press; 1964.

[2] Courant R, Hilbert D. Methods of Mathematical Physics. vol. 2. New Delhi: Wiley Eastern Private Ltd.; 1975.

[3] Rudin W. Functional Analysis. New Delhi: Tata-McGraw Hill Co. Ltd.; 1975.

[4] Yoshida K. Functional Analysis. Springer Verlag; 1965.

[5] Hörmander L. Linear Partial Differential Equations. Springer Verlag; 1963.

[6] Vladimirov VS. Generalized Functions in Mathematical Physics. Moscow: Mir Publishers; 1979.

[7] Schwartz L. Mathematics for Physical Sciences. Reading: Addison-Wesley; 1966.

[8] Schwartz L. Théorie des distributions. Paris: Hermann; 1966.

[9] Horvath J. Topological Vector Spaces and Distributions. Reading: Addison-Wesley; 1966.

[10] Trèves F, Pisier G, Yor M. Laurent Schwartz (1915-2002). Notices of the AMS. 2002;50(9).

[11] Friedman A. Generalized Functions and Partial Differential Equations. New Jersey: Prentice Hall; 1963.

[12] Adams RA. Sobolev Spaces. New York: Academic Press; 1975.

[13] Lions JL. Problémes aux limites dans les équations aux dérivés partielles. Canada: Univ. de Montréal Presse; 1965.

[14] Lions JL. Lectures on Elliptic Partial Differential Equations. Bombay: TIFR; 1957.

[15] Lions JL, Magenes E. Nonhomogeneous Boundary Value Problems. vol. 1. Berlin: Springer Verlag; 1970.

[16] Nečas J. Les Méthods directes en théorie des équations elliptiques. Paris: Masson; 1967.

[17] Grisvard P. Singularities in Boundary Value Problems. Berlin: Springer Verlag; 1992.

[18] Grisvard P. Boundary Value Problems in Non-smooth Domains. Part I, Lecture Notes No. 19. College Park: Univ. of Maryland; 1980.

[19] Grisvard P. Elliptic Problems in Non-smooth Domains. Monographs and studies in Mathematics, 24. London: Pitman; 1985.

[20] Kolmogorov AN, Fomin SV. Introductory Real Analysis. Translated by R.A. Silverman. New York: Dover Publications; 1970.

[21] Dirac PAM. The Principles of Quantum Mechanics. 4th ed. London: Oxford University Press; 1958.

[22] Dirac PAM. Quantized singularities in the electro-magnetic field. Proc Roy Soc, London, Ser A. 1931;p. 60–72.

[23] Dirac PAM. The physical interpretation of the quantum dynamics. Proc Roy Soc, London, Ser A. 1927;113:621–641.

[24] Horvath J. An introduction to distributions. Notices of the AMS. 1970;p. 227–240.

[25] Trèves F. Topological Vector Spaces, Distributions and Kernels. New York: Academic Press; 1967.

[26] Brezis H. Analyse fonctionnelle : Théorie et applications. Paris: Masson; 1983.

[27] Rudin W. Real and Complex Analysis. McGraw-Hill Company; 1986.

[28] Dieudonné J. Foundations of Modern Analysis. New York: Academic Press Inc.; 1960.

[29] Schwartz L. Distributions à valeurs vectorielles. Ann de l'Inst Fourier, t7, 1957 et t8. 1959;.

[30] Nikolski SM. A Course of Mathematical Analysis. vol. 1. Moscow: Mir Publishers; 1985.

[31] Hörmander L. On the division of distributions by polynomials. Ark Math. 1958;3:558–568.

[32] Widder DV. Advanced Calculus. 2nd ed. New Delhi: Prentice-Hall of India Ltd.; 1974.

[33] Apostol Tom M. Mathematical Analysis. 2nd ed. New Delhi: Addison-Wesley Publishing Company/Narosa Publishing House; 1993.

[34] Bernadou M. Méthodes d'approximations pour les problèmes elliptiques. Contributions à l'Encyclopédie: Méthodes des Mathématiques Appliquées. C.E.A, Paris: J.L.Lions et R.Dautray (Ed.); 1981.

[35] Ciarlet PG. The Finite Element Method for Elliptic Problems. Amsterdam: North-Holland; 1978.

[36] Pironneau O. Méthodes des éléments finis pour les fluides. Paris: Masson; 1988.

[37] Jaswon MA, Symm GT. Integral Equation Methods in Potential Theory and Elastostatics. London: Academic Press; 1977.

[38] Malgrange B. Équations aux dérivés partielles à coefficients constants. CR Ac Sc, Paris, t237. 1953;.

[39] Agmon S. Lectures on Elliptic Boundary Value Problems. Van Norstrand-Reinhold; 1965.

[40] Khoan Vk. Distributions, analyse de Fourier, opérateurs aux dérivés partielles. Paris: Librairie Vuibert; 1972.

[41] Nedlec JC. Notions sur les équations intégrales de la physique: Théorie et Approximation. Palaiseau, France: Centre de Math. Appl., École Polytechnique; 1978.

[42] Tartar L. Topics in Nonlinear Analysis. Publications Math. d'Orsay, Dépt. de Math.. Orsay, France: Univ. de Paris-Sud; 1978.

[43] Babitch VM. On the Extension of Functions(in Russian). Uspekhi Mat Nauk. 1953 August;p. 111–113.

[44] Dauge M. Elliptic Boundary Value Problems on Corner Domains. Lecture Notes in Mathematics, 1341. Springer Verlag; 1988.

[45] Slobodetski LN. Sobolev spaces of fractional order and their applications to the boundary value problems of partial differential equations (in Russian). Dokl Akad Nauk, USSR. 1958;118:243–246.

[46] Aubin JP. Applied Functional Analysis. New York: John Wiley and Sons; 1979.

[47] Kondraschov VI. On the Properties of Functions of L^p-spaces (in Russian). Dokl Akad Nauk, USSR. 1960;130:536–566.

[48] Bhattacharyya PK, Gopalsamy S. On Existence and Uniqueness of Solutions of Boundary Value Problems of Fourth Order Elliptic Partial Differential Equations with Variable Coefficients. Journal of Mathematical Analysis and Applications. 1988 December;136:589–608.

[49] Halmos P. Measure Theory. Berlin: Springer Verlag; 1974.

Index